ADVANCED
GENERAL ORGANIC CHEMISTRY

A Modern Approach

PART I

For Undergraduate students majoring in Chemistry and for Postgraduate students

ADVANCED
GENERAL ORGANIC CHEMISTRY

A Modern Approach

PART I

Sachin Kumar Ghosh

Formerly Head of the Department of Chemistry, New Alipore College
Affiliated to University of Calcutta

New Central Book Agency (P) Ltd

LONDON

HYDERABAD ERNAKULAM BHUBANESWAR

NEW DELHI KOLKATA PUNE GUWAHATI

NCBA

REGD OFFICE
8/1 Chintamoni Das Lane, Kolkata 700 009, India
email: ncbapvtltd@eth.net

OVERSEAS
NCBA (UK) Ltd, 149 Park Avenue North, Northampton, NN3 2HY, UK
email: ncbauk@yahoo.com

EXPORT
NCBA Exports Pvt. Ltd, 212 Shahpur Jat, New Delhi 110 049
email: ncbaexports@ncbapvtltd.com

BRANCHES

2/27 Ansari Road, 1st Floor
Daryaganj, **New Delhi** 110 002
email: ncbadel@ncbapvtltd.com

Shop Nos. 3 & 4, Vinayak Towers
681/B Budhwar Peth, Appa Balwant Chowk
Pune 411 002
email: ncbapune@ncbapvtltd.com

Shop No. 15, Bharati Towers, Ground Floor
Forest Park
Bhubaneswar 751 001
email: ncbabub@ncbapvtltd.com

House No. 3-1-315, 1st Floor
Nimboliadda, Kachiguda, **Hyderabad** 500 027
email: ncbahydb@ncbapvtltd.com

GSS Shopping Complex, 1st Floor
Opposite College Post Office, Convent Road
Ernakulam 682 035
0484-4021054 / 53

Radhamayee Complex [Second Floor]
Opp. Axis Bank, Panbazar, Jaswanta Road
Guwahati 781 001
email: ncbaguwahati@ncbapvtltd.com

ADVANCED GENERAL ORGANIC CHEMISTRY [Part I] • Sachin Kumar Ghosh

First Published: 1986
Reprinted: 1992, 1994
Second Edition: 1998
Third Edition [thoroughly revised & updated]: 2009
Revised Reprint: 2010
Reprinted: 2011
Revised Reprint: 2013
Reprinted: 2016, 2017, **2019**

PUBLISHER
New Central Book Agency (P) Ltd
8/1 Chintamoni Das Lane, Kolkata 700 009

TYPESETTER
Anin, Kolkata

PRINTER
New Central Book Agency (P) Ltd
Web-Offset Division, Dhulagarh, Sankrail, Howrah

COVER DESIGNER
Soumya Jyoti Ghosh

TECHNICAL EDITOR
Pradip Kr. Chowdhury and Dipan Roy

PROJECT TEAM
Joydeep Ghosh and Prabhat Jas

ISBN: 978 81 7381 441 9

Price of Part I & Part II togetl ₹1500.

DEDICATION

To my parents and brother
To my maternal grandfather,
To others, and my wife
Who helped me build my life,
To my sons and daughters
Who bring me joy and cheers,
To my teachers and the taught,
To authors of home and abroad,
To my friends and foes,
To whomsoever the book owes,
To the people who toil,
To make fruitful the soil.
For all times and now
I bow, bow and bow.

THIRD EDITION

To the hands of my grandchildren

Akash, Hindole and Kopai

THIRD EDITION

To the hands of my grandchildren

CONTENTS

Foreword ix
Preface to the Third Edition xi
Preface to the First Edition xiii

PART I

Structure, Isomerism, Spectroscopy and Mechanism

Chapter 1 Introduction ... 3
Chapter 2 Structure of Organic Molecules 9
Chapter 3 Classification and Nomenclature of Organic Compounds 65
Chapter 4 Factors Influencing Electron-availability 147
Chapter 5 Elementary Thermodynamics 197
Chapter 6 Kinetics of a Reaction 207
Chapter 7 Intermolecular Forces and Related Properties 223
Chapter 8 Acids and Bases 245
Chapter 9 Isomerism 269
Chapter 10 UV-Visible, IR and ^{1}HNMR Spectroscopy 413
Chapter 11 Fundamentals of Organic Reaction Mechanism 499
Chapter 12 Nucleophilic Substitution Reactions 561
Chapter 13 Addition to C—C Multiple Bond 613
Chapter 14 Rearrangement Reactions 697
Chapter 15 1, 2-Elimination Reactions 783
Chapter 16 Reactions Involving >C=O 823
Chapter 17 Aromatic Electrophilic Substitutions 931

Foreword

The first edition of the present text, published in 1986, has been quite popular amongst the advanced undergraduate students and teachers of organic chemistry throughout the country, especially in West Bengal. To my knowledge, this has been the only comprehensive text by an Indian author aimed at advanced undergraduate-level, having a wide coverage of almost all major aspects of organic chemistry, viz., modern structural theories, stereoisomerism, reaction mechanism and general functional group chemistry. A large number of problems, solved and unsolved, in most of the chapters have made the relevant topics more intelligible and interesting, arousing scientific curiosity.

The author deserves congratulations for the untiring endeavour he has undertaken to write this text and now to make it updated and more comprehensive in its present second edition. Objectives given at the beginning of each chapter is a characteristic feature of this text, wherein the topics and concepts to be known and understood have been outlined. Many chapters are revised versions of the original ones of the first edition, of which Chapter 4 'Factors Influencing Electron Availability', Chapter 9 'Isomerism', Chapter 11 'Fundamentals of Organic Reaction Mechanism', and Chapter 13 'Addition to C—C Multiple Bond' deserve special mention. The concepts regarding 'The cause of optical activity' and of topicity, prochirality, chirotopicity and stereogenecity introduced in Chapter 9 would give further insight to stereoisomerism. Inclusion of separate chapters on 'Elementary Thermodynamics' (Chapter 5) and 'Kinetics of a Reaction' (Chapter 6) (taking these topics out of Chapter 11 of the first edition) has been appropriate, as only this knowledge can lead to the basic concepts of organic reaction mechanism and the position of equilibrium. Emphasis is usually given on these topics in any good organic chemistry text.

Chapter 21 'Planning a Synthesis' (64 pages) is a noteworthy and very useful addition. This chapter deals with disconnection of different types of simple organic molecules and rational ways of reconnection illustrating chemoselective, regioselective and stereoselective routes. Protection and deprotection of functional groups involved in synthons and synthetic equivalents and common routes to synthesise mono and polyfunctional molecules—aliphatic, aromatic and heterocyclic—have been discussed, incorporating a number of tables in order to enable the readers to assimilate these matters at a glance.

Two more new chapters on 'Chemistry of Amino Acids and Peptides' (Chapter 22) (32 pages) and 'An Introduction to Carbohydrates' (Chapter 23) (55 pages) have been incorporated in the present edition, giving all necessary salient features. These subject-matters are indispensable parts of any modern comprehensive text on organic chemistry. The inclusion of Appendices A, B and C at the end, and the Subject Index and Compound Index, prepared with utmost care, would be useful to the readers for quickly finding out the desired information. Chemistry of a large number of organic compounds of different classes of everyday use (Appendix C), a new addition, will make the subject-matter lively and interesting.

However, the inclusion in an elementary way of some basic mechanistic concepts, as of Curtin-Hammett principle in Chapter 11, of some important topics of stereochemistry like Sharpless epoxidation, Prelog's rule, Octant rule, Axial haloketone rule etc. and also of the principles and applications of ^{1}H NMR and mass spectra for the structural elucidation of simple organic molecules would have enhanced the standard and utility of this text at the expense of increasing the volume to some extent. I feel that this should be done in a future edition, in which several chapters could be significantly condensed without losing the essence and understandability of the subject-matter. The book is already big enough (~1450 pages) to be split into two volumes for convenience of handling.

The author has made commendable efforts to present the subject-matter in each chapter in lucid language with sufficient numbers of illustrations and problems.

In conclusion, I recommend this textbook to the undergraduate students majoring in chemistry, beginning postgraduate students and also to the teachers of organic chemistry, who, I hope, will find this book very thorough and useful.

July 1998

SUNIL K. TALAPATRA
UGC Professor of Chemistry
Centre of Advanced Studies on Natural Products
University of Calcutta

Preface to the Third Edition

I am to write the preface to this edition with a deep heart. Meanwhile I have lost three respectable personalities. My elder brother, Late Rathin Ghosh, who taught me and bore the whole financial burden of my education in Kolkata. My wife, Late Jyotsna Ghosh, to whom I could not give my company for long eight years during the preparation of the book. She was a pillar of support to me. And Late Jogendra Nath Sen, the then Director of New Central Book Agency (P) Ltd, who brought me in the limelight. My heart bleeds for them. I have had many predicaments but, inspired by New Central Book Agency (P) Ltd, I have completed the book at last.

I have tried to update the book as far as possible and the following topics have been included in this edition:

1. Homoaromatics [Chapter 4]
2. Phase Transfer Catalyst (PTC) and Crown ethers [Chapter 7]
3. General and Specific acid-base catalysis [Chapter 8]
4. Prelog's Rule [Chapter 9]
5. ^{1}HNMR (PMR) Spectroscopy [Chapter 10] and Mass Spectra [Appendix E]
6. Curtain-Hammett Principle [Chapter 11]
7. Designations of addition creaction to C$=$C [Chapter 13]
8. Sigmatropic [i, j] hydrogen shift [Chapter 14]
9. 1, 3-Dithianes [Chapter 16]
10. Protein and nucleic acids [Chapter 22]
11. Green Chemistry [Appendix D]

Chapters on Organometallics and Active methylenes have not been written separately since organometallics have been discussed in several chapters (2, 16, 18, 20, 21, etc.). Active methylenes have also been discussed in Chapters 9, 16, 20 and 21—as required.

However, the Hammett equation has been dropped from Chapter 17 for personal reasons. Large volume of the book is a problem to handle. Now, the book and its contents have been separated into two parts—I and II. Part II contains all the indexes. The continuity of page numbers is maintained throughout the volumes. Readers are requested to buy both the parts together for fruitful use of the book, since the two parts together complete the whole text. I would request the readers to offer their valuable suggestions. Every suggestion and helpful criticism will be gratefully acknowledged.

In preparing the indexes my elder son, Soumya Jyoti, and his wife, Sayani, have done a lot. Soumya Jyoti has also designed the cover of the book. I am indebted to both of them. I am also indebted to Mr Ranjit Chaudhuri, the father-in-law of my elder son, who presented me a high-class magnifying glass without the help of which the proof reading would not have been possible for me. My younger son, Subhojyoti, has warned me several times not to do so much hard work with ill-health, yet I have completed the book having the full support, physical and mental, from my daughter-in-law and elder son. My grandson, Akash, always keeps me cheerful.

I am thankful to Principal Dilip Kumar Sarkar, formerly of Darjeeling Govt. College for correcting some printing mistakes and language of the revised third edition.

Mr Amitabha Sen, the Director of New Central Book Agency (P) Ltd, Mr Soumen Paul and Mr Prabhat Jas of New Central Book Agency (P) Ltd have taken a keen interest in the publication of the revised third edition. Without their hard labour it was impossible to publish this edition. The second edition is the foundation of the third edition. Mr Arunabha Sen published the second edition. I, most humbly, convey my gratitude to Prof. Sunil Kr. Talapatra for writing the 'Foreward' to the second edition of the book. I am also grateful to him. Dr P. K. Maity and his workers of 'Anin' have done their job neatly. I am really grateful to all of them and convey my heartfelt thanks to them. Good wishes to all.

AUTHOR

76B Rishi Arabinda Sarani
Haridevpur, Kolkata 700 082

Preface to the First Edition

The book is meant for advanced degree students. The syllabi prescribed by the universities for them in general organic chemistry include (i) the modern structural theories of organic chemistry; (ii) the stereoisomerism, (iii) the reaction mechanism; and (iv) the general functional group of chemistry.

Unfortunately, no single book covers all the aspects; naturally, students have to plod through several specific books of eminent authors within a short span of time and, as a consequence, they very often fail to assimilate the subject matter. It is this difficulty of the students that led me to undertake this concise and systematic work on general organic chemistry—in the light of the modern structural theories.

All the four aspects of the syllabi and the theoretical basis of the practical organic chemistry have been dealt with in detail in the book to minimise the labour and save the time of the students. Hundreds of related problems have also been solved to make the subject matter lucid and, at the same time, to develop a mind of scientific inquisitiveness. Thus the book is expected to be a dependable guide to the advanced degree students of organic chemistry.

The field of organic chemistry being inconceivably vast, 'What to Study' and 'How to Study' are the two common questions asked by the students. In chapter one, titled "INTRODUCTION", effort has been made to cope up with such questions. Since organic compounds mostly occur in nature and organic chemistry is an experimental science, isolation and purification of organic compounds, and their analyses—qualitative, quantitative and instrumental—have been discussed in the second, third and tenth chapters, respectively. Analytical portions related to the functional group chemistry have been discussed in their right places in chapter twenty as well as in the Appendix. Chapters four to seventeen deal with the structure, nomenclature, physical properties, electronic condition, acidity-basicity, intermolecular forces, stereo-chemistry and reaction mechanisms of the organic compounds.

Chapters one to seventeen constitute the first part of the book and they deal with the modern basic concepts. The chemistry of a few hydrocarbons, heterocycles with one hetero atom and functional groups, and uses of organic compounds have been included in the second part of the book which have three chapters, namely, chapters eighteen, nineteen and twenty. Notes on a few modern terms and IR spectroscopy and properties of a few organic compounds have been incorporated in the appendix. Thus it is a comprehensive textbook covering all the different aspects of the syllabi. Nevertheless, the students are advised to consult the books on specific aspects written by eminent chemists and established authors.

It is my duty to seek apology of the readers for the printing mistakes in the book which could not be avoided; these are, however, easily detectable and will not create any confusion to the readers. I also hope that all these errors will be rectified in the next edition of the book. In this connection I request the readers to offer their valuable suggestions for making the book more useful and up-to-date. Any such suggestion and helpful criticism will be gratefully acknowledged.

It is a great plesure to thank Dr Ramaprasad Sarkar who excited my feeling to undertake the present work. My thanks are also due to Dr Sankar Chandra Ghosh, Prof. R. N. Sinha Roy, Dr Swapan Kumar Ghosh, Dr Dhananjay Chatterjee, Dr G. P. Chitnis, Dr Niren Chandra Gayen, Dr Gour Gopal Basak and Prof. Bibek Ranjan Sarkar who occasionally read the manuscript and gave valuable suggestions for the betterment of the book.

I also owe a debt of gratitude to my students Arabinda, Manas, Santanu, Falguni, Dhiman, and Manabendra, who supplied me with books and papers on several aspects for preparing the manuscript; and to Sri Jogendranath Sen (Partner, New Central Book Agency Pvt. Ltd), to Sri Biren Das (Artist), to the workers of Books & Allied (P) Ltd and to the staff of New Central Book Agency (P) Ltd who made possible the publication.

I owe the index to Mrs. Suchitra Ganguly (Librarian, New Alipore College), and to my students, Manabendra, Goutam, Sanhita, Nibedita, Arijit, Abhijit, Bhanu, Jyoti and Arup.

I pay my respect to Dr. Atul Chandra Chatterjee who took much pain in editing the language of a few chapters (chapter 9 to chapter 12).

My wife, Smt Jyotsna Ghosh, and my sons, Bubun and Pat, gave me enough time for completing the book, I am indebted to them.

New Alipore College
'L' Block, New Alipore
Calcutta 700 053

AUTHOR

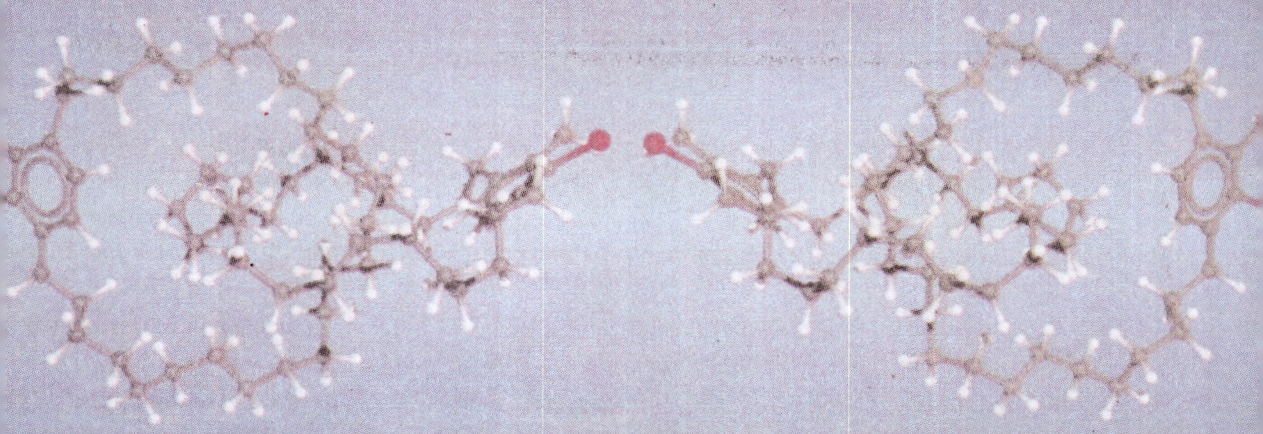

Part I

ructure, Isomerism, Mechanism and Spectroscop

Chapter 1

Introduction

Objectives

- To understand the property of catenation [1.2] p. 5
- To understand the role of organic compounds [1.4] p. 6
- To understand the points of difference between organic and inorganic compounds [1.5] p. 6
- To understand the importance of structure [1.7] p. 7
- To follow the stepwise approach to study the science of organic chemistry [1.8] p. 7

Chapter 1

Introduction

Objectives

- To understand the property of catenation [1.2]
- To understand the role of organic compounds [1.4]
- To understand the points of difference between organic and inorganic compounds [1.5]
- To understand the importance of structure [1.7]
- To follow the stepwise approach to study the science of organic chemistry [1.8]

1.1 Organic Compounds

The aim of organic chemistry is to study the science of organic compounds. Most of the covalent compounds of carbon may be called **organic compounds**. Of course, some of the authors define organic compounds simply as carbon compounds. According to many others, **hydrocarbons** (binary compounds of C and H) and their derivatives are organic compounds. Each of the definitions has some exception. However, more than three millions of organic compounds are known and almost ninety thousand new compounds are discovered every year. These include a fantastic variety of substances ranging from simple hydrocarbons used as fuels to complex nucleic acids of the genetic material. Their properties markedly differ from those of ionic inorganic compounds. They are a part and parcel of the life processes (nutrition, growth, respiration, reproduction, etc.). Organic compounds make up living bodies, food-stuff, clothes, drugs, paper, films, dyes, and almost everything of daily use.

1.2 A Unique Property of Carbon

$$^{12}_{6}\text{C}\,1s^2 2s^2 2p^2$$

The electron configuration, atomic volume, atomic mass of C atoms and oxygen of the earth's atmosphere have developed in carbon almost a unique property of catenation. By **catenation** we mean the formation of chains of identical atoms. C atoms are able to combine among themselves indefinitely to produce wide and varied types of carbon chains, short and long, straight and branched, open and closed (Fig. 1.1). Sulphur and silicon also have a tendency to catenation.

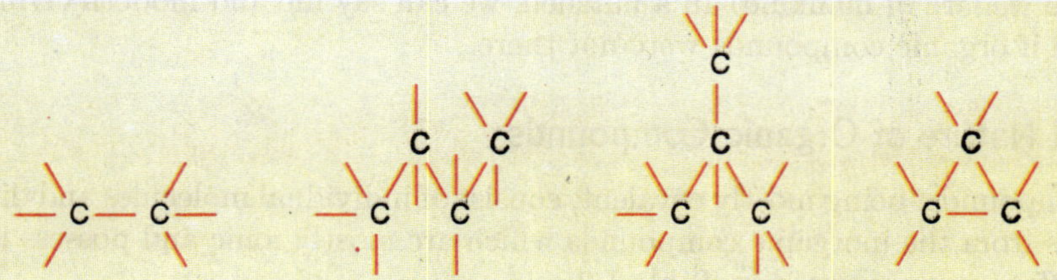

Fig. 1.1. Catenation property of carbon

But the catenated carbon compounds are unusually stable in comparison to their sulphur and silicon analogues. This is because of the high intrinsic energy of C—C bond and its relative stability towards oxidation compared to those of Si—Si and S—S bonds. C—C and C—O bonds have comparable stability but Si—O and S—O bonds are considerably stronger than Si—Si and S—S bonds respectively. So, compounds with Si—Si linkage get converted into compounds with Si—O linkage easily and exothermically. Same is the case with catenated compounds of sulphur. These can be understood from the bond energy values as given in Table 1.1.

Table 1.1 Bond energy

Bond	Energy, kJ/mole	Bond	Energy, kJ/mole	Bond	Energy, kJ/mole
C—C	348.6	Si—Si	228.4	S—S	224.2
C—O	344.4	Si—O	369.6	S—O	336.0

Therefore, it may be possible for Si and S to form catenated compounds in the absence of oxygen atmosphere outside the earth.

1.3 Huge Number of Organic Compounds

Besides catenation, carbon atoms form stable bonds with many other chemical elements, the most important being hydrogen, oxygen, nitrogen, sulphur and the halogens. This ability and catenation of C atoms explain the existence of about 3 millions of organic compounds, simple and complex. Being devoid of these properties, the rest of the elements form only 55,000 compounds.

1.4 Role of Organic Compounds

Plant and animal bodies are composed of organic compounds. Intensive knowledge about them, therefore, is very helpful to understand the life processes. Living cells, their constituents, amino acids, fats, hormones, carbohydrates, proteins, enzymes, vitamins, etc., are synthesised within the body from organic compounds. Hence study of organic compounds helps the study of biochemistry, the chemistry of living matter. Green plants synthesise carbohydrate from carbon dioxide and water with the help of sunlight and chlorophyll. They convert carbohydrates into fats and proteins and animals take in them. We eat meat, egg, fish and pulse for protein; rice, fruits, wheat, sugar for carbohydrate. Cotton, woollen and synthetic fibre like polyester, nylon, vinylon, etc., are made up of organic compounds. Sulphur drugs, antimalarials, antibiotics, antiseptics are nothing but organic compounds. Petrol, diesel, natural gas, i.e., fuels are organic substances. To make our life comfortable and easy-going soaps, cosmetics, dyes, polythene, photographic film, paper, etc., are playing their valuable roles. Explosives like nitroglycerine, T.N.T. are used in war and for the welfare of mankind. In a nutshell, we can say life and modern civilization would not be possible if organic compounds were not there.

1.5 General Nature of Organic Compounds

Organic compounds, being mostly covalent, consist of individual molecules and differ largely in their properties from the inorganic compounds which are mostly ionic and possess ionic crystals. Some of the differences are noted in Table 1.2.

Table 1.2 Organic vs Inorganic compound

Organic compounds (covalent)	Inorganic compounds (ionic)
1. Volatile and distil steadily	1. Non-volatile
2. Low melting	2. High melting
3. Soluble in non-aqueous liquids	3. Insoluble in non-aqueous liquids
4. Insoluble in water	4. Soluble in water
5. Non-electrolytes	5. Electrolytes
6. Less stable to heat and decompose above 700°C	6. More stable to heat
7. More susceptible to oxidation	7. Less susceptible to oxidation
8. Most reactions take place slowly and need temperature, catalyst, pressure, etc.	8. Reactions are rapid and spontaneous
9. Reaction products are often mixtures of several compounds	9. Reaction products contain less number of compounds
10. Isomeric compounds are often found	10. Isomeric compounds are rare

1.6 Organic Chemistry

The huge number of organic compounds, their application in every sphere of life and wide differences in properties with inorganic compounds have compelled the chemists to study the organic compounds as a separate branch of chemistry, called **organic chemistry**. **Organic chemistry**, therefore, is the study of covalent compounds of carbon.

1.7 Organic Chemistry is a Systematic Subject

Because of the large number of organic compounds, their systematic study seems almost impossible. But in reality, similar structural features in a comparatively small number of compounds help classify them into a group of families. Most of the members of each family or class have similar methods of preparation, called **general methods of preparation** and similar chemical properties, called **general properties**. Again, physical properties like m.p., b.p., solubility, etc., change from member to member in a regular order within a class. On the other hand, the classes differ remarkably in their methods of preparation and in chemical properties owing to the differences in structures. Structural theory, therefore, is the basis of the science of organic chemistry which makes the nomenclature, classification and the study of these compounds relatively simple and easy.

1.8 Method of Studying Organic Chemistry

The sources of organic compounds are plant and animal bodies, coal, natural gas and petroleum. Therefore, the isolation of a compound from the natural source is very important. A study of an impure compound gives faulty results. Therefore, the purification of an organic compound is another important part of study. We cannot name a compound, classify it and study its reactions unless we know its structure. In determining the structure of a compound elemental analysis, molecular weight determination and the knowledge of the structural theory are essential. The elucidation of a structure further requires a study of the properties and synthesis of the compound. **Synthesis** means preparation of compounds from their constituent elements or from simple compounds of known structure so that the structure of the target molecule and that of each intermediate will be known to us. Sometimes an isolation of a compound from its natural source is costly, troublesome and time consuming. Again, it may not be available in the market. In that case, the compound is to be prepared in the laboratory. So the methods of their preparation deserve mention. The course of a reaction of a compound gives valuable information about the reaction products. Hence mechanism of organic reactions is another subject of study. Organic compounds being widely used in bulk in industry, we need to know the manufacturing processes. The **manufacturing** or **industrial process** means large-scale production of a compound with an optimum of time and cost. An extensive use of the organic compounds in medicine and surgery directs us to study their physiological activity.

In order to systematise the realm of study, therefore, the subject needs the following heads:

(i) Isolation and purification (See Appendix A)

(ii) Elemental analysis and molecular formula

(iii) Structural theory, nature of bond, physical properties and chemical reactivity

(iv) Structural formula, classification and naming of compounds

(v) Classification of organic reactions and their mechanism

(vi) General methods of preparation with exceptions and laboratory method of preparation

(vii) Structure and general properties with exceptions

(viii) Synthesis and manufacture

(ix) Applications

The readers are requested to consult any standard practical book for the second head of the above index and the last four heads will be discussed together under the chemistry of individual class in this book. The rest of the index will be discussed in appropriate places.

Exercise

1. Define : organic compound, organic chemistry, catenation, synthesis, manufacture, general preparations, general properties.
2. Why are there so many carbon compounds in nature?
3. Why does silicon form catenated compounds of very low stability?
4. What are the main differences between organic and inorganic compounds?
5. What makes the organic chemistry a systematic subject?
6. How can the study of organic compounds be made systematic?
7. What role do the organic compounds play in our life?

Chapter 2

Structure of Organic Molecules

Objectives

- To write electron configuration of atoms [2.2] p. 11
- To draw pure and hybrid atomic orbitals [2.3 and 2.4] p. 15 & 19
  - Lewis theory of covalent bond formation [2.6.3] p. 26
- To draw electron-dot and condensed structural formulae of molecules [2.7 and 2.17] p. 28 & 60
- To understand:
  - V.B. and M.O. theories [2.8] p. 31
  - Bond length, bond strength, polarity of bonds and bond angles [2.9–2.14] p. 40
  - Bond dissociation energy and bond energy [2.10] p. 41
  - Electronegativity and group electronegativity [2.11] p. 44
  - Bond moments and group moments [2.12] p. 46
  - Shape of the molecules and directional nature of covalent bond [2.15] p. 49
  - Nature of bonds in carbon compounds [2.16] p. 52

2.1 Introduction

The mode of arrangement of the constituent atoms in a molecule regardless of their orientation in space gives its molecular constitution. When this is projected on the plane of a paper, we get the constitutional or gross structural formula for the molecule. The gross structural formula can explain the properties of a compound because they depend upon its gross structure. Hence, the molecular formula has nothing to do with the chemistry of the molecule unless its gross structural formula is known. The aim of this chapter is, however, to study the structures of organic molecules.

The structure of a chemical entity (an ion or a molecule) is a description of the sequential bond connectivities of the constituent atoms of the entity including their orientation in space.

To understand the meaning of structure of a molecule we need to go through the structural theory. This deals with

I. the forces that hold the atoms together in a molecule
II. the shapes and sizes of the molecules
III. the distribution of electrons over the molecule
IV. the intermolecular forces

Obviously, the electronic structure of atoms has a key position in the above study because the chemical combination between the atoms takes place only by the rearrangement of one or more of the electrons which the combining atoms possess. Hence this chapter should begin with the concise discussion on the electronic structure of atoms.

2.2 The Electronic Structure of Atoms

Positively charged protons, chargeless neutrons, and negatively charged electrons(e) are the three fundamental particles, amongst others, that constitute atoms. The atomic number of an element represents the number of protons and that of electrons in its atom. The amount of charge in a proton being equal to that in an electron, an atom is an electrically neutral particle.

Protons and neutrons together occupy a very small space within an atom, called nucleus. The positively charged nucleus is surrounded by negatively charged electrons arranged in shells or energy levels. The shells are designated as K, L, M, N, O, P and Q. The K shell being the first one, it is at the least distance from the nucleus in comparison to others. The maximum number of electrons that a shell can accommodate is obtained by the expression $2n^2$; where, n indicates the ordinal number (shell number) of the shell. The ordinal numbers of the shells are: K, 1; L, 2; M, 3; N, 4 and so on. Hence, two electrons can reside in the K shell. 8, 18, 32 are, therefore, the maximum number of electrons that L, M and N shells respectively can contain.

There are subshells within a shell. They are designated as s, p, d and f ; 2, 6, 10 and 14 respectively are the maximum number of electrons which these subshells can accommodate. So, K shell has the s subshell only; L shell contains s and p subshells; s, p, d are the three subshells present in M shell. The subshells differ amongst themselves in energy they contain.

Subshells are composed of orbitals. An orbital is a three-dimensional space within an atom in which an electron is most likely to be found. It may contain one or two electrons, but it cannot have more than two electrons; an orbital may remain empty also. Therefore, s subshell has one orbital, p, d and f subshells have 3, 5 and 7 orbitals respectively. The orbitals in a subshell do not differ in their energy content. The lone orbital of s subshell is designated by the same letter s. The three p orbitals are represented as, p_x, p_y and p_z. The number of orbitals that a shell can have is given by the expression n^2; where, n is the ordinal number of the shell. Thus, K shell has one orbital 1^2; L, M and N shells contain 4, 9 and 16 orbitals respectively. An orbital of a particular shell is represented by placing the ordinal number of the shell before the designation of the orbital. Thus, s orbital of L shell is written as $2s$.

The discussion made above on the electronic structure of atom may be summarised in Table 2.1.

Table 2.1 Subshells and orbitals of K, L and M shells

Designation of shell with ordinal number, n	K–1	L–2	M–3
Maximum no. of e, $2n^2$	2	8	18
Subshells with designation	$1s$	$2s, 2p$	$3s, 3p, 3d$
Maximum no. of e per subshell	s–2	s–2, p–6	s–2, p–6, d–10
No. of orbital per shell, n^2	1	4	9
No. of orbital per subshell	s–1	s–1, p–3	s–1, p–3, d–5
Designation of orbitals	$1s^2$	$2s^2, 2p_x^2, 2p_y^2, 2p_z^2$	$3s^2, 3p_x^2, 3p_y^2, 3p_z^2$ $3d_{xy}^2, 3d_{xz}^2, 3d_{yz}^2$ $3d_{x^2-y^2}^2, 3d_{z^2}^2$

When two electrons reside in an orbital, they are called **paired**. An electron has spinning motion about its axis. It may spin either clockwise or anti-clockwise. Since a moving charged particle is always accompanied by a magnetic field, a spinning electron behaves as a tiny bar magnet and consequently possesses magnetic moment. When two oppositely spinning electrons come very close to each other, a strong attractive force generates instead of usual repulsive force between them. This arises from the cancellation of their magnetic moments which are equal in magnitude but opposite in direction. Owing to this attractive force, two electrons with anti-parallel spins can pair up in an orbital. On the other hand, two electrons with parallel spins cannot cancel their respective magnetic moments being equal in magnitude and in direction; hence, they cannot form a pair. If a dash (—) indicates an orbital and a vertical arrow, downward (↓) or upward (↑), represents an electron with its spin, then $\underline{\downarrow\uparrow}$ is the symbol of paired electrons in an orbital. **Incomplete orbital**, i.e., an orbital containing a single electron is written as $\underline{\uparrow}$ and an empty or **unoccupied orbital** is denoted by a dash only.

The distribution of the total number of electrons of an atom is done following three principles. These principles taken together are called Aufbau or Building-up principle.

1. Ascending order of energy levels of orbitals

Orbitals are filled up by electrons in an order of increasing energy.

The increasing order of energy levels of orbitals in atoms is: $1s, 2s, 2p, 3s, 3p, 4s, 3d, 4p, 5s, 4d, 5p, 6s, 4f, 5d, 6p, \ldots$ (Fig. 2.1).

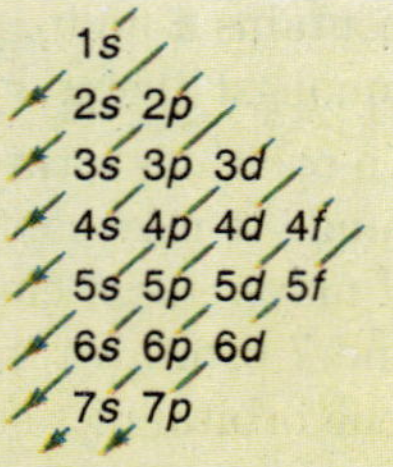

Fig. 2.1. Order of energy levels of orbitals

The atomic number of hydrogen is 1. So, an atom of hydrogen contains one proton and one electron. Normally, this electron will remain in the lowest energy level, i.e., the 1s level, and the hydrogen atom is said to be in the **ground state**, i.e., the **lowest energy state**.

Every material body, macro or micro, subatomic or terrestrial, is always associated with energy. Matter without energy is inconceivable. At the same time the form in which a particular body exists should have a minimum of energy. Minimum energy content of a body means maximum stability of the body and **vice versa**. Similarly, the arrangement of electrons in an atom in its ground state is that which makes its energy minimum, i.e., its stability maximum. Perhaps, for this reason the pairing of electrons in an atom starts gradually from the lowest energy level to the higher. Thus the ground state of any atom is the most stable state of all else.

2. Pauli's exclusion principle

No orbital can have more than two electrons and then they should be of opposite spins.

Its application involves the conception of electron spin. Two electrons of parallel spins cannot cancel the magnetic moment of each other. Hence, a repulsive force between the two negatively charged electrons develops and consequently two electrons of parallel spins cannot reside in an orbital.

3. Hund's rule of maximum multiplicity

Two electrons do not occupy a particular orbital in a subshell until each of the orbitals of the subshell has at least one electron. In other words, one electron is placed in each equal-energy orbital before pairing.

Thus, if three electrons need to occupy p orbitals in any particular shell, one electron will go into each of the three available p orbitals. Similarly, if four electrons are to go to the three p orbitals of the same shell, then the fourth electron will pair up with any one of the three electrons each occupying a single orbital, i.e., the electron of one orbital will be paired.

The interpretation of this empirical rule is that: Owing to the repulsive force between two electrons, they want to keep themselves as far away as possible. So, each of the electrons occupies a separate equal-energy orbital.

Let us take the example of C atom to show the distribution of electrons in its atomic orbitals, i.e., its electronic structure.

The atomic number of C is 6. One C atom has, therefore, six electrons.

According to the 1st principle of **Aufbau**, two electrons of anti-parallel spins should go to the 1s orbital. Pauli's exclusion principle tells us that the third electron should not go to 1s orbital, instead it will occupy the next higher energy orbital, the 2s orbital. The fourth electron will pair with the third electron in the 2s orbital. The fifth electron will occupy any one of the three 2p orbitals. The sixth one will now go to the second 2p orbital. According to Hund's principle, it should not pair with the fifth electron. The third 2p orbital will remain vacant. This description of electron distribution of C atom may be written as follows:

Electron	Energy level downward increasing order	
1st e	1s	$\uparrow$
2nd e	1s	$\uparrow\downarrow$
3rd e	2s	$\uparrow$
4th e	2s	$\uparrow\downarrow$
5th e	$2p_x$	$\uparrow$
6th e	$2p_y$	$\uparrow$

Hence, the electron configuration of C atom is

$$E \uparrow \Big| \; \underset{1s}{\underline{\uparrow\downarrow}} \;\; \underset{2s}{\underline{\uparrow\downarrow}} \;\; \underset{2p_x}{\underline{\uparrow}} \;\; \underset{2p_y}{\underline{\uparrow}} \;\; \underset{2p_z}{\underline{}}$$

Although p_x, p_y and p_z orbitals are of same energy, it is customary to introduce an electron in p_x orbital before the others.

The electronic structure of carbon as represented above is called the **detailed electron configuration** of C atom. Two abbreviated forms of this are also used. They are:

1. In which separate subshell orbitals and spin designations are omitted, i.e., only subshell electrons are shown. Thus, the electron configuration of C is $1s^2\ 2s^2\ 2p^2$; where, coefficients of subshells represent the shell numbers and their superscripts indicate the number of electrons each subshell contains.
2. In which a part of the electronic configuration of the element is replaced by the symbol of the preceding inert gas. The rest is written in the same way as stated above. Thus, the electronic configuration of C is He, $2s^2\ 2p^2$ and that of a sodium atom is Ne, $3s^1$. He stands for $1s^2$ and Ne stands for $1s^2, 2s^2, 2p^6$.

NOTE

The electronic structure of atoms as discussed in this section is a mere approximation. Actual picture is obtained from a branch of higher mathematics, known as **quantum mechanics**. This uses four **quantum numbers** to describe the actual state of an electron in an atom. n, the ordinal number of a shell is called the principal quantum number. The other three are:

1. Azimuthal quantum number, l,
2. Magnetic quantum number, m and
3. Spin quantum number, s.

The four quantum numbers of an electron may be called the four degrees of freedom, i.e., the coordinates of the electron. They have the following significance:

I. n indicates the principal energy level of the orbital of the electron.

II. l indicates the shape of the orbital.

III. m indicates the orientation of the angular momentum of the orbital in an applied magnetic field.

IV. s indicates the spin momentum of the electron.

The principal quantum number, n, of an electron can have any one of the values $1, 2, 3, 4, \ldots$. For a particular value of n, values of l are $0, 1, 2, \ldots, (n-1)$. Thus for $n = 1$, l is 0; when $n = 2$, $l = 0$ and 1; and so on. Again, for a definite value of l, m values are $-l \ldots 0 \ldots +l$. This means when $n = 2$, $l = 0$; for $l = 0$, $m = 0$ and for $l = 1$, $m = -1, 0, +1$. For any m value, s values are $+\frac{1}{2}$ and $-\frac{1}{2}$. The value of $n = 2$ means the electron belongs to the 2nd shell; the value of $l = 0$ means an s orbital holds the electron and the value of $l = 1$ means three p orbitals. Therefore, each electron possesses one value of n, one value of l, one value of m, and one value of s.

The Pauli's exclusion principle, based on experimental observation, may be stated in the quantum number terminology and this is: no two electrons can have the four quantum numbers alike. This means: when all the quantum numbers are assigned for each electron involved in an atom, each electron can be distinguished from the rest at least by one quantum number. Let us take the example of He atom. It contains two electrons; n value for them is 1. Therefore, the value of $l = 0$, m value is also 0 and their s values are $+\frac{1}{2}$ and $-\frac{1}{2}$. Hence, for the 1st electron: $n = 1$, $l = 0$, $m = 0$ and $s = +\frac{1}{2}$; for the 2nd electron: $n = 1$, $l = 0$, $m = 0$ and $s = -\frac{1}{2}$. So, we see that the two electrons of $1s$ orbital have three identical quantum numbers (n, l, m), but they differ in the spin quantum number value: for one it is $+\frac{1}{2}$ and for the other it is $-\frac{1}{2}$.

Problem 2.1. Write down the detailed electronic configuration of **(a)** nitrogen atom and **(b)** oxygen atom. What are their abbreviated electron configurations?

Solution:

(a) Atomic number of N is 7 electron (e)	Energy level downward increasing order	(b) Atomic number of O is 8 electron (e)	Energy level downward increasing order
1st *e*	$1s$	1st *e*	$1s$
2nd *e*	$1s$	2nd *e*	$1s$
3rd *e*	$2s$	3rd *e*	$2s$
4th *e*	$2s$	4th *e*	$2s$
5th *e*	$2p_x$	5th *e*	$2p_x$
6th *e*	$2p_y$	6th *e*	$2p_y$
7th *e*	$2p_z$	7th *e*	$2p_z$
		8th *e*	$2p_x$

(a) So, the detailed electron configuration of N is:

$$\underset{1s}{\uparrow\downarrow}\quad \underset{2s}{\uparrow\downarrow}\quad \underset{2p_x}{\uparrow}\quad \underset{2p_y}{\uparrow}\quad \underset{2p_z}{\uparrow}$$

Its abbreviated forms are:

(1) $1s^2, 2s^2, 2p^3$; (2) He,$2s^2, 2p^3$

(b) So, the detailed electron configuration of O is:

$$\underset{1s}{\uparrow\downarrow}\quad \underset{2s}{\uparrow\downarrow}\quad \underset{2p_x}{\uparrow\downarrow}\quad \underset{2p_y}{\uparrow}\quad \underset{2p_z}{\uparrow}$$

Its abbreviated forms are

(1) $1s^2, 2s^2, 2p^4$; (2) He,$2s^2, 2p^4$

2.3 Shape, Size and Orientation of Atomic Orbitals

The conception of atomic orbitals has originated from wave mechanics, the developed form of quantum mechanics. The chief contributors of the theory are L. de Broglie, E. Schrödinger and W. Heisenberg. This tells that a moving particle is always associated with wave properties, i.e., it has a dual character of matter, particle and wave. According to the nature of an experiment, it may behave as a particle or as a wave; but it cannot behave as both at the same time.

This idea is true for all particles, but the wavelike aspect is significant for small particles only as they have large wavelength. The wavelength of a large particle being very small, its wave properties are not measurable and hence insignificant.

Being a small particle in motion, an electron has particle-wave duality. Hence, wave mechanics is applicable to it. Applying wave mechanics, Schrödinger worked out mathematical expressions, called wave equations, to describe the motion of an electron in an atom in terms of energy. The Schrödinger wave equation which was proposed to describe the behaviour of a subatomic particle is as follows:

$$\frac{\partial^2\psi}{\partial x^2} + \frac{\partial^2\psi}{\partial y^2} + \frac{\partial^2\psi}{\partial z^2} + \frac{8\pi^2 m}{h^2}(E - V)\psi = 0$$

It is a second order differential equation in three dimensions; where,

$$\begin{aligned} \psi &= \text{wave function, the solution of the wave equation} \\ x, y, z &= \text{co-ordinates of the space} \\ m &= \text{mass of the subatomic particle} \\ h &= \text{planck's constant} \\ E &= \text{total energy} \\ V &= \text{potential energy.} \end{aligned}$$

This wave equation describes the state of the wave, i.e., its energy, intensity, etc. For waves representing electromagnetic radiation (wave motion of photon) ψ corresponds to the amplitude of the wave; the intensity of the wave at any point is given by the square of the amplitude at the point. So, the greater the value of the (amplitudes)^2, the greater will be the intensity and the greater will be the probability of a photon to be found there. Similarly, the wave equation for the wave of an electron or any other subatomic particle, ψ is analogous to amplitude and the probability of finding the electron (or any subatomic particle) is proportional to the square of its wave function (ψ), i.e., ψ^2. However, as ψ may be a complex number, the probability is expressed by considering its conjugate complex ψ^* and the probability is then given by $\psi\psi^*$.

However, if we solve the Schrödinger wave equation for an electron in atom, we may get a number of values for ψ; but all the values of ψ will not be acceptable to us. Only those values will be accepted which will obey the specific conditions characteristic to the system and those acceptable values of ψ will be called eigenvalues (eigen = characteristic) and the functions will be called eigenfunctions. The following limitations are imposed on the wave function to get the acceptable values:

(a) **ψ must be single-valued.** Since the probability of finding an electron in space is proportional to ψ^2, ψ must be single-valued; if it were not so, there would be two or more probabilities of finding the electron at the same point in space—which should not be.

(b) **ψ must be finite.** If it were infinite, it would mean that the electron is fixed at a point which is against the wave motion concept.

(c) **ψ must be continuous.** Real waves are continuous; to obtain physically meaningful solutions, this limitation has also been imposed for the subatomic particles.

(d) **ψ must be normalised.** By normalisation we mean the total probability of finding the electron over all space must be unity, i.e., hundred per cent.

Each eigenvalue of the eigenfunction, ψ, obtained by solving the Schrödinger wave equation for an electron in an atom, represents an orbital. Orbitals are three-dimensional regions or contours within an atom in which the probability of finding the electron is greatest. If an electron is considered as a charge cloud, then also an orbital represents a space in an atom within which the charge density of the electron cloud is maximum. It may also be looked upon as a space in which an electron remains most of the time.

The shape, size and orientation of different energy orbitals are different. On the other hand, equal energy orbitals have the same size and shape. Organic chemistry concerns with the electronic structure of C atom mainly. It contains, $1s$, $2s$ and three $2p$ orbitals. So, we shall discuss here the shape, size and orientation of these orbitals only. Since there is a probability factor, an orbital should not have any boundary. However, as the probability of finding an electron decreases very rapidly beyond a certain distance, we can consider the boundary. From this consideration the question of shape and size have arisen.

2.3.1 1*s* Orbital

This orbital is spherical in shape. The centre of the sphere is at the nucleus of the atom. The orbital being spherical, its charge density is symmetrically oriented about the nucleus, but decreases with increasing distance from the nucleus (Fig. 2.2a). Spherical symmetry of the orbital gives rise to its two important properties:

(1) It is a non-directional orbital.

(2) Electrons of this orbital are very strongly bound to the nucleus.

For simplicity, 1*s* orbital may be represented by Fig. 2.2b, with solid lines as its boundary surface. The + sign comes from the wave equation and it is not associated with the electrical or ionic charge. It indicates the symmetry of the orbital. As there is no electronic charge density at the centre, where the nucleus of the atom is present, it is called a nodal point, the point of no-charge density.

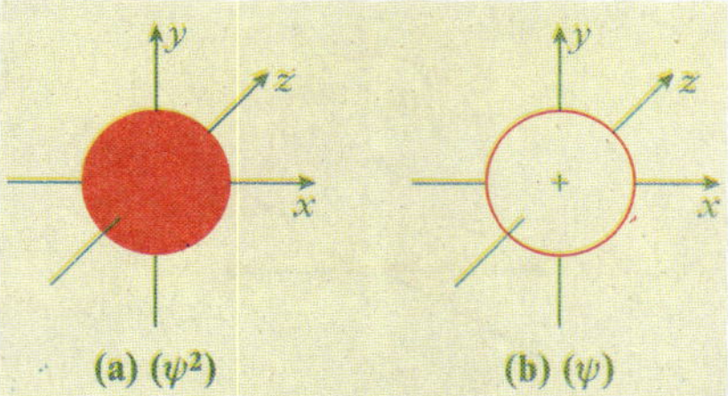

Fig. 2.2. 1*s* orbital

The size of 1*s* orbital of different elements is different. The decreasing order of 1*s* orbital size is H > Li > Be > B > C > N > O > F.

2.3.2 2*s* Orbital

This is also a non-directional orbital and spherically symmetrical about the nucleus as the 1*s* orbital. It differs from the 1*s* orbital in size and in energy content. 2*s* orbital being of 2nd (L) shell, it is bigger than the 1*s* orbital in size and the former lies at a greater distance from the nucleus than the latter (Fig. 2.3).

1*s* orbital is surrounded by the 2*s* orbital. There is a region between the two concentric spheres where the probability of finding an electron is zero. This region is called spherical nodal surface (Fig. 2.4).

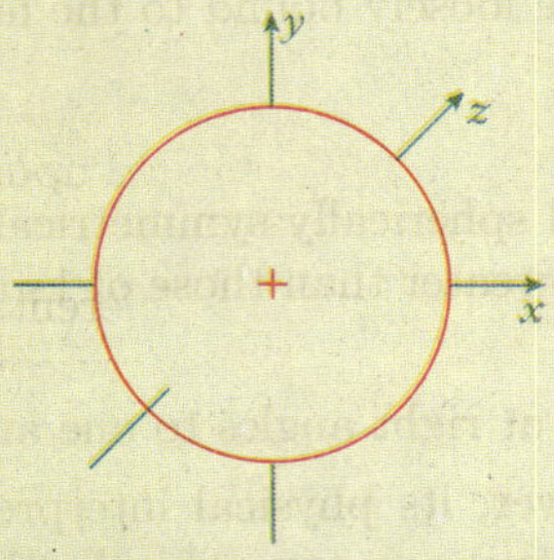

Fig. 2.3. 2*s* orbital (ψ)

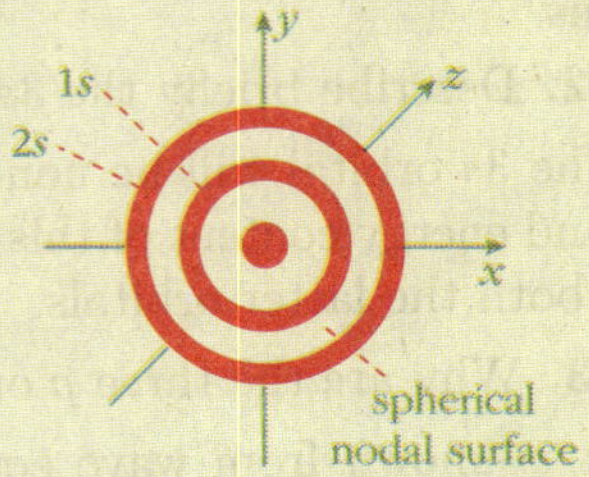

Fig. 2.4. 1*s* and 2*s* orbitals together

2.3.3 2p Orbitals

The shape of a 2*p* orbital resembles a dumb-bell (Fig. 2.5).

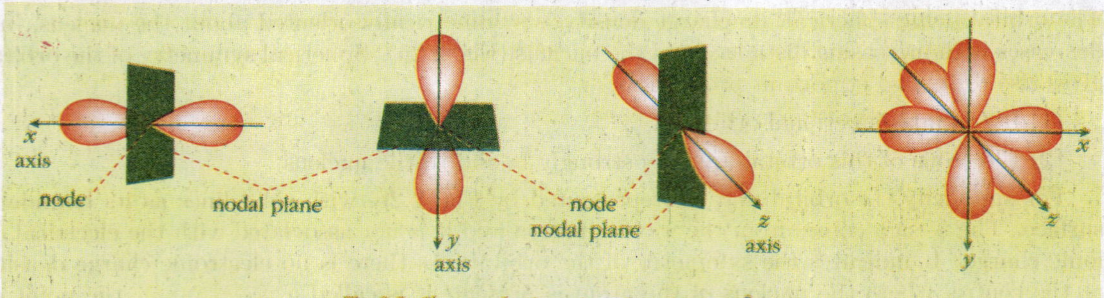

Fig. 2.5. Shape and orientation of *p* orbitals (ψ^2)

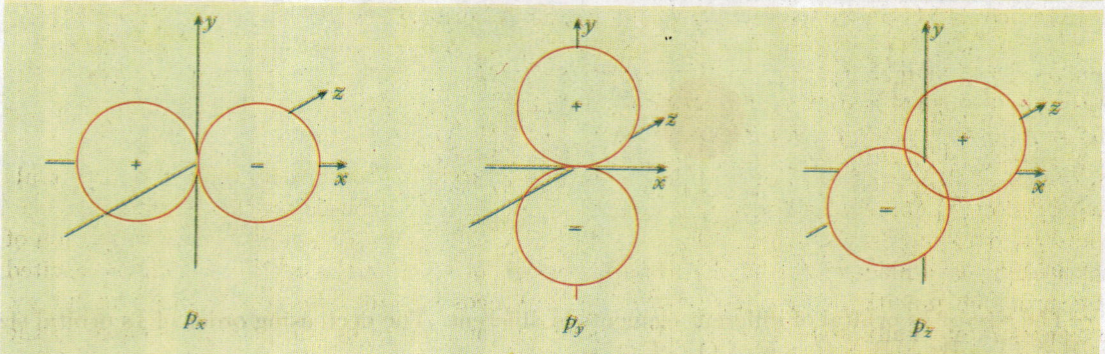

Fig. 2.6. *p* orbitals (ψ)

The three 2*p* orbitals being of equal energy (degenerated), they are identical in shape and size; but they differ from each other in their spatial orientation about the nucleus. These three orbitals are oriented mutually perpendicular along x, y and z axes. Hence, they have directions and are denoted as p_x, p_y and p_z respectively. In such an orbital, there is no chance of finding an electron at the nucleus. The nucleus, therefore, is called the nodal point. These orbitals also possess nodal planes, in which the probability of finding the electron is zero. The nodal planes pass through the nucleus and are at right angles to x, y, z axes respectively. Regions of such an orbital separated by the nucleus are assigned by + and − signs as per wave equations; the signs indicate the symmetry factor (Fig. 2.6).

A *p* orbital being non-spherical, its electrons are more loosely bound to the nucleus than the *s* orbital electrons.

Problem 2.2. Describe briefly the 3*s* orbital.

Solution: The 3*s* orbital will be non-directional and spherically symmetrical about the nucleus. The size and energy content of this orbital will be greater than those of 1*s* and 2*s* orbitals. It will surround both the latter orbitals.

Problem 2.3. Why are the three *p* orbitals oriented at right angles to one another?

Solution: It is known from wave equations. However, its physical interpretation may be as follows: In order to minimise the repulsive force between any two orbitals, they try to keep themselves as far away as possible. This is possible if they are directed in space at right angles to each other.

2.4 Hybridised Atomic Orbitals

The s and p orbitals discussed so far are pure atomic orbitals (A.O.s). To explain the molecular structure of C compounds, we need to consider mixed orbitals, called hybrid orbitals, instead of pure A.O.s. It should be remembered that all the A.O.s pure or mixed, are the results of wave equations.

The A.O.s of a ground-state atom do not mix up. This usually occurs when an atom absorbs sufficient energy for unpairing its paired electrons and promoting it to the available higher energy orbital. Not all such promotions of electrons are permitted. There exists a definite rule of selection; s to s and d to s transitions are not permitted, but $s \to p$, $p \to s$ or d are allowed.

When such a transition of one or more electrons occurs, the atom is said to be in the excited state[1]. An atom in the excited state contains more energy than that in its ground state and hence it becomes less stable. The ground and excited states of a C atom are shown below.

Here promotion of one $2s$ electron to the vacant $2p_z$ orbital occurs.

↑↓	↑↓	↑	↑			↑↓	↑	↑	↑	↑
$1s^2$	$2s^2$	$2p_x$	$2p_y$	$2p_z$		$1s^2$	$2s$	$2p_x$	$2p_y$	$2p_z$
Ground state						Excited state				

Vacant, half-filled or filled orbitals of the excited atom may then mix up to form new orbitals having different charge densities from the pure A.O.s. The new orbitals are then known as hybrid orbitals and the process by which they are obtained is called hybridisation. The hybridisation of orbitals may take place either by the reshuffling of all or some of the orbitals which the excited atom provides. In either case the hybrid orbitals formed have the same shape, size and energy, i.e., they are mutually equivalent to each other. The number of hybrid orbitals is equal to the number of pure A.O.s reshuffled; but they differ from each of the pure A.O.s in their shape, size, orientation and energy content. The hybrid orbitals, though equivalent, have different orientations about the nucleus of the atom.

When s and p orbitals undergo the process of hybridisation, it is called s-p hybridisation; d-s-p hybridisation is the name given to the process if it does occur among d, s and p orbitals. However, we are much concerned with the s-p type as carbon atoms are supposed to do so. Let us discuss this in relation to a C atom.

An excited C atom, as shown above, possesses four incomplete orbitals, one $2s$ and three $2p$ orbitals. These four A.O.s may mix up in three ways to form sp^3, sp^2 and sp hybridised orbitals as discussed below.

2.4.1 sp^3 Hybrid Atomic Orbitals

When the redistribution of electron density occurs among one s and three p orbitals, four hybrid orbitals are formed. These are designated as sp^3, indicating their origin. Each of the four sp^3 hybrid A.O.s possesses one-fourth s character and three-fourths p character.

The shape of an individual sp^3 orbital is given in Fig. 2.7. + and − signs have the usual meanings.

It has two lobes with a common axis which passes through the nucleus; one lobe is larger than the other. To minimise the repulsive force amongst themselves sp^3 A.O.s remain as far apart as possible. The angle between any two sp^3 hybrid orbitals is thus 109°28′. If the centre of a regular

[1]Excitation of an atom may also occur without promotion of electron (see Problem 2.5).

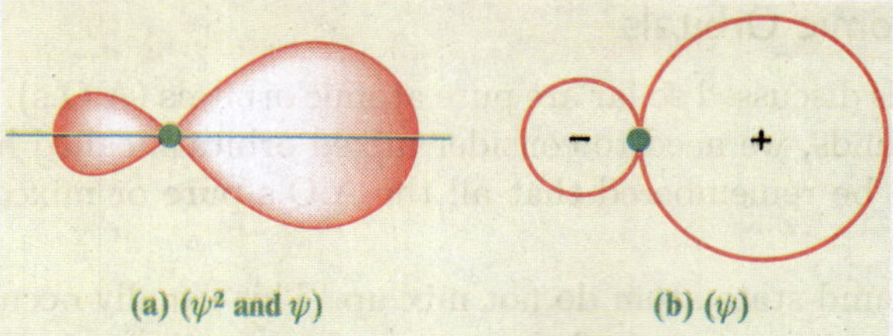

Fig. 2.7. Shape of an sp^3 A.O.

tetrahedron (a four-based body) is joined to its four corners by drawing imaginary lines, then the angle between any two lines is found to be 109°28′. Hence, it may be supposed that the four sp^3 hybrid orbitals are directed to the four corners of a regular tetrahedron, the centre of which coincides with the nucleus of the atom. This supposition leads to call the orientation of these orbitals as tetrahedral and the angle made by two such orbitals as tetrahedral angle. The overall picture of these orbitals is shown in Fig. 2.8 with small lobes omitted.

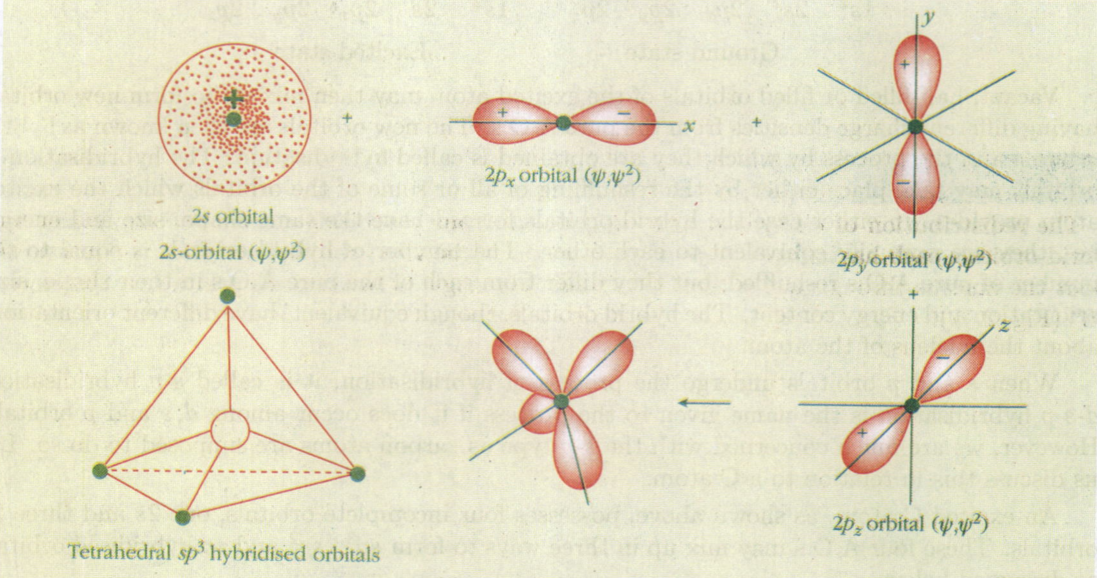

Fig. 2.8. Orientation of four sp^3 A.O.s

2.4.2 sp^2 Hybrid Orbitals

These are formed out of one *s* and two *p* A.O.s. By this type of hybridisation, we get three equivalent hybrid orbitals, the designation (sp^2) of which is obvious from the origin. Each of them has one-third *s* character and two-thirds *p* character.

The shape of sp^2 hybrid atomic orbitals is given in Fig. 2.9.

It has two lobes, one of them is larger than the other; they have a common axis passing through the nucleus. The repulsive force amongst the electrons makes these three orbitals to orient in a plane trigonally. Therefore, the angle between any two of these sp^2 orbitals is 120° (Fig. 2.9).

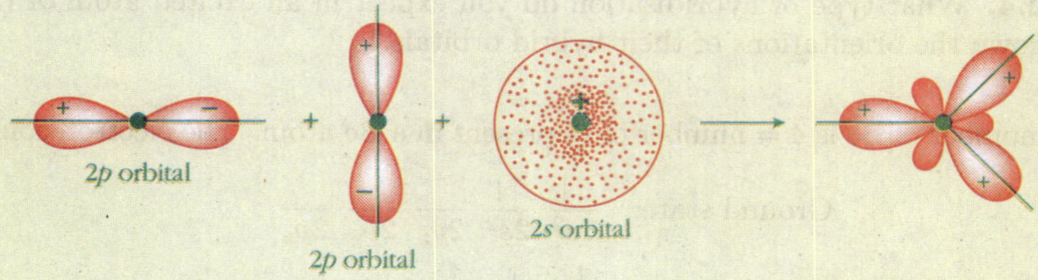

Fig. 2.9. Shape and orientation of sp^2 A.O.s

If in an excited C atom sp^2 hybridisation occurs, one *p* orbital remains perpendicular to the plane of three hybrid A.O.s and hence it is at right angle to each of the three hybrid sp^2 A.O.s (Fig. 2.10).

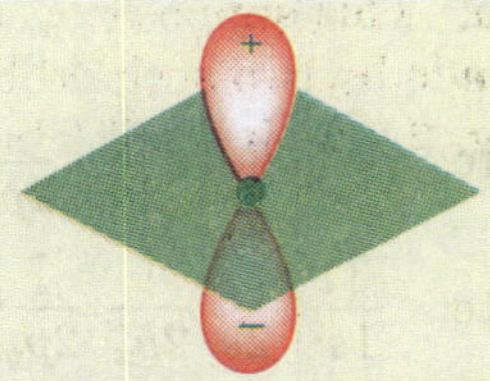

Fig. 2.10. Orientation of the p_z orbital of an sp^2-hybridised C atom

2.4.3 *sp* Hybrid Orbitals

The redistribution of electron density of one *s* and one *p* orbitals creates two equivalent *sp* hybrid orbitals each having two lobes of unequal size. The two orbitals have linear orientations about the nucleus through which their axes pass. The angle between these orbitals is, therefore, 180° (Fig. 2.11).

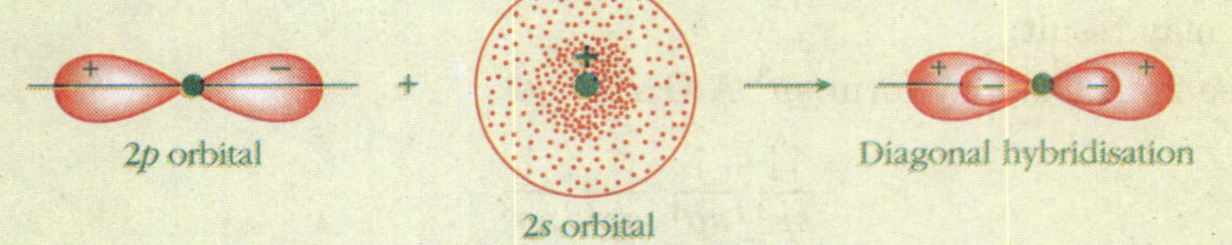

Fig. 2.11. Shape and Orientation of *sp* A.O.s

An *sp* orbital has one-half *s* character and one-half *p* character.

The combination of one *s* and one *p* orbitals left two *p* orbitals, p_y and p_z, of the C atom undisturbed. Two pure A.O.s are mutually perpendicular to each other and each of them is at right angle to each of the two *sp* hybrid orbitals (Fig. 2.12).

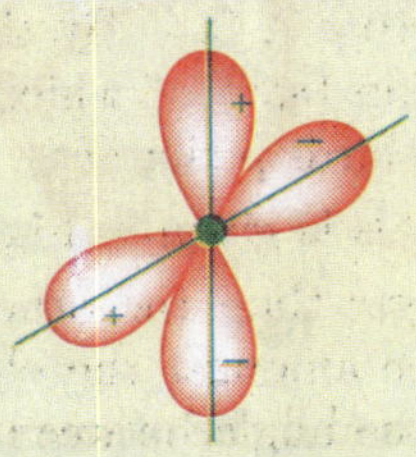

Fig. 2.12. Orientation of p_y and p_z orbitals of an *sp*-hybridised C atom

Problem 2.4. What type of hybridisation do you expect in an excited atom of **(i)** Be and **(ii)** B? Write down the orientations of their hybrid orbitals.

Solution:

(i) Atomic number of Be is 4 = number of *e*s present in a Be atom. The electron configuration of Be atom is:

Ground state: $\underset{1s^2}{\uparrow\downarrow}\ \underset{2s^2}{\uparrow\downarrow}\ \underset{2p_x}{\_\_}\ \underset{2p_y}{\_\_}\ \underset{2p_z}{\_\_}$

Excited state: $\underset{1s^2}{\uparrow\downarrow}\ \underset{2s}{\uparrow}\ \underset{2p_x}{\uparrow}\ \underset{2p_y}{\_\_}\ \underset{2p_z}{\_\_}$

Hence, two equivalent *sp* orbitals will be formed by the process of hybridisation and their orientation about the nucleus will be linear making 180° angle between them.

(ii) Atomic number of B is 5 = number of electrons that an atom of B contains. The electronic configuration of a B atom is:

Ground state: $\underset{1s^2}{\uparrow\downarrow}\ \underset{2s^2}{\uparrow\downarrow}\ \underset{2p_x}{\uparrow}\ \underset{2p_y}{\_\_}\ \underset{2p_z}{\_\_}$

Excited state: $\underset{1s^2}{\uparrow\downarrow}\ \underset{2s}{\uparrow}\ \underset{2p_x}{\uparrow}\ \underset{2p_y}{\uparrow}\ \underset{2p_z}{\_\_}$

Therefore, three sp^2 orbitals will be the result of hybridisation and their orientation will be planar trigonal; each orbital angle will be 120°.

Problem 2.5. Discuss and draw different types of hybrid A.O.s of N atom.

Solution: Ground state electron configuration of N:

$\underset{1s^2}{\uparrow\downarrow}\ \underset{2s^2}{\uparrow\downarrow}\ \underset{2p_x}{\uparrow}\ \underset{2p_y}{\uparrow}\ \underset{2p_z}{\uparrow}$

In the excited state the outer shell orbitals may mix up in several ways and thus several types of hybrid orbitals may result:

(i) 2*s* and three 2*p* A.O.s may form sp^3 A.O.s

$\underset{sp^3}{\uparrow\downarrow}\ \underset{sp^3}{\uparrow}\ \underset{sp^3}{\uparrow}\ \underset{sp^3}{\uparrow}$

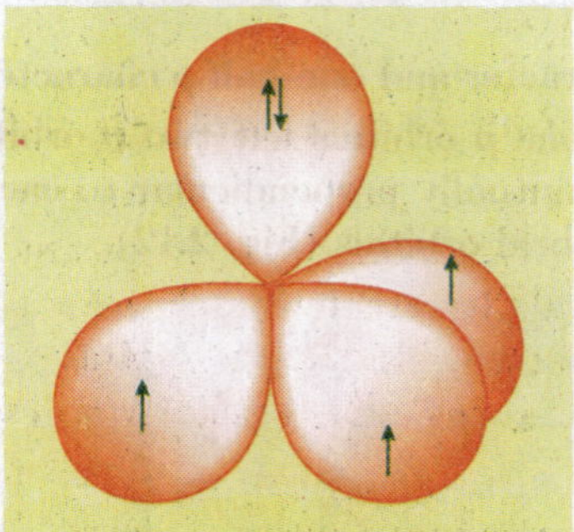

Fig. a

one sp^3 A.O. contains a pair of electrons and the others one each. Therefore, the hybrid A.O.s will be tetrahedrally oriented around the nucleus and Fig. a represents them.

(ii) In the excited state the electron configuration of N atom may be

$$\underset{1s^2}{\underline{\uparrow\downarrow}}\ \underset{2s}{\underline{\uparrow}}\ \underset{2p_x}{\underline{\uparrow}}\ \underset{2p_y}{\underline{\uparrow}}\ \underset{2p_z}{\underline{\uparrow\downarrow}}$$

Now $2s$, $2p_x$ and $2p_y$ may mix up to form three sp^2 A.O.s and a pair of electrons may remain in the p_z A.O. which is perpendicular to each of the three sp^2 A.O.s. The sp^2 A.O.s have planar trigonal orientations as usual (Fig. b).

$$\underset{sp^2}{\underline{\uparrow}}\ \underset{sp^2}{\underline{\uparrow}}\ \underset{sp^2}{\underline{\uparrow}}\ \underset{p_z}{\underline{\uparrow\downarrow}}$$

(iii) The sp^2 A.O.s may also form in a manner in which one sp^2 A.O. contains a pair of electron and the p_z A.O. contains one electron only instead of two as above. The orientation of A.O.s, of course, remains the same (Fig. c).

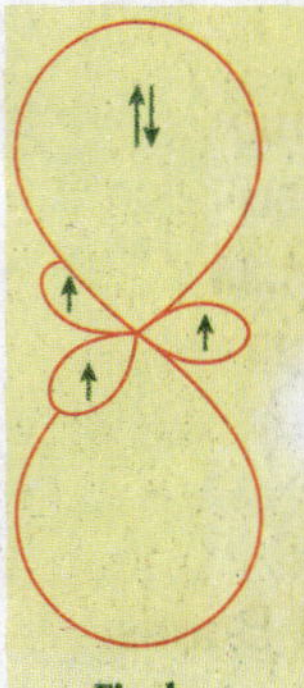

Fig. b

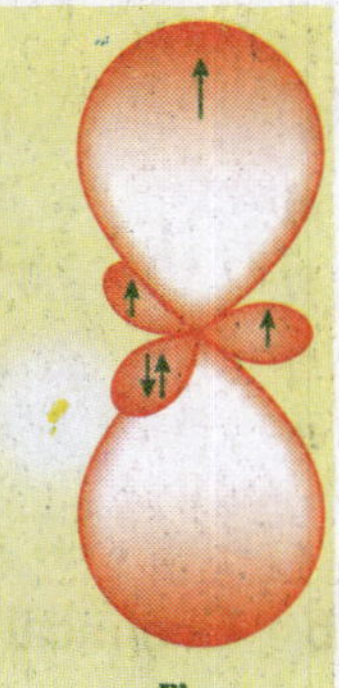

Fig. c

$$\underset{sp^2}{\underline{\uparrow\downarrow}}\ \underset{sp^2}{\underline{\uparrow}}\ \underset{sp^2}{\underline{\uparrow}}\ \underset{p_z}{\underline{\uparrow}}$$

(iv) sp hybridisation is also possible in an N atom. In this case one sp A.O. will contain a pair of electrons and one sp will have one electron. p_y and p_z A.O.s too will accommodate one electron each. Two sp A.O.s will be oriented linearly.

The p_y and p_z A.O.s will remain perpendicular to each other and with each of the sp A.O.s (Fig. d).

$$\underset{sp}{\underline{\uparrow\downarrow}}\ \underset{sp}{\underline{\uparrow}}\ \underset{p_y}{\underline{\uparrow}}\ \underset{p_z}{\underline{\uparrow}}$$

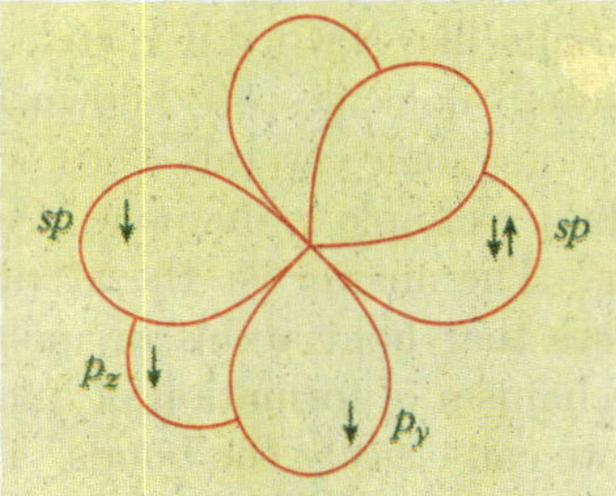

Fig. d

2.5 The Relative Energy Content of s, p and s-p Hybridised Orbitals

An s orbital has lower energy than a p orbital if both of them belong to one and the same shell. Again, s-p hybrid orbitals (sp^3, sp^2 and sp) being produced by the combination of s and p orbitals, their energy as well as stability are dependent on the percentage of s and p characters they possess. The more the s character of the hybrid orbital, the lower is the energy content and the higher is the stability. The increasing order of energy and consequently the decreasing order of stability of the orbitals are:

$$\xrightarrow{\text{increasing order of energy,}}$$
$$s,\ sp,\ sp^2,\ sp^3,\ p$$
$$\xrightarrow[\text{decreasing order of stability.}]{}$$

Table 2.2 Comparative study of the s, p and s-p hybrid orbitals

Orbital type	% of s character	Number of orbitals	Orientation	Orbital angle	Number of p orbitals left (in the case of s-p hybridisation)
s	100	1	Spherically symmetrical	×	×
sp	50	2	Linear	180°	2
sp^2	33.33	3	Planar trigonal	120°	1
sp^3	25	4	Tetrahedral	109°28′	0
p	0	3	Mutually perpendicular to each other	90°	×

2.6 Chemical Bonds

Chemical substances, elements and compounds, are assemblies of either molecules or atoms or ions. Molecules, in their turn, are composed of atoms. Gases consist of molecules—mono-, di- or polyatomic. In a gas they move at random motion as a very minimum of cohesive force exists amongst them. Solid and liquid states, on the other hand, contain atoms or ions or molecules. Therefore, atoms, ions and molecules are the basic chemical units.

The chemical units of a substance in the liquid state have a tendency to form loosely ordered array amongst themselves. When molecules are involved in such an arrangement, the liquid is called a molecular liquid. In the solid state of substances these chemical units remain in rigidly ordered three-dimensional arrays having definite shapes, called crystals. A crystal consisting of ions only is known as ionic crystal. Salts like NaCl, LiCl, etc., have ionic crystals. Crystalline solids that are formed by the individual molecules are known as molecular crystals. Organic compounds belong to this type. There are some crystals whose constituents are atoms only and they are called atomic crystals. Its best example is diamond. Now the question is: how do the basic chemical units remain together? The answer is: the attractive force amongst themselves holds these units together. The forces that hold atoms or molecules or ions together may be called chemical bonds. From the mechanism of formation of bonds and properties possessed by several types of chemical substances we can classify the bonds. The important bonds, i.e., forces are: (i) electrovalent bond, (ii) covalent and dative bonds, (iii) metallic bond, (iv) hydrogen bond and (v) van der Waals forces.

In this section we shall discuss electrovalent, covalent and dative bonds. Hydrogen bond and van der Waals' force will be discussed under intermolecular forces along with other types of bonds.

2.6.1 Electrovalent, Covalent and Dative Bonds

Bonds in molecules, ionic crystals and atomic crystals are produced by the regrouping and rearrangement of outer electrons, called valence electrons, of the constituent atoms. The rules of such bonding are guided by the following basic principles in most of the cases:

1. Atoms do so as to attain stability, i.e., the combined atoms in a molecule or in a crystal have lower energy than what they had before combination.
2. An atom stabilises if it takes up the electronic configuration of the nearest inert gas.
3. Like charges repel and unlike charges attract each other

It is found that inert gas atoms do not normally react with other atoms and their molecules too are mono-atomic. It is assumed, therefore, that atoms of inert gases are extremely stable and their energy being so favourable that by compound formation they cannot be taken into still lower energy state. Now what makes these atoms so stable? As outer electrons of an atom are involved in bond formation, it is said that the outermost shell of an inert gas atom is saturated with electrons and for this reason it has no tendency to react. This saturation of outermost shell with electrons brings stability to the inert gases.

Helium has two electrons in its only shell. The outermost shells of other inert gases have eight electrons. So, two electrons in the first shell and eight electrons in the other outer shells make them saturated. Atoms of elements other than inert gases have less than eight electrons in their outermost shells. They want eight electrons in those shells to attain energetically more stable states by assuming electron configurations of their nearest inert gases in the periodic table. These atoms can do so by combining with other atom or atoms through bond formation. This may take place in two main ways depending upon the electron configuration of atoms involved. These are :

I. By the transfer of electron or electrons from one atom to the other; this gives rise to an electrovalent bond.

II. By the mutual sharing of electrons between atoms; in this case a covalent bond is the result.

2.6.2 Electrovalent or Ionic Bond

An electropositive element, i.e., an element placed closely after an inert gas in the periodic table could achieve the electron configuration of the inert gas by giving up electron or electrons; this leads to the formation of a positively charged ion, called cation.

On the other hand, an electronegative element placed closely before an inert gas in the periodic table, might attain the electron configuration of the inert gas by taking in electron or electrons; this gives rise to a negatively charged ion, known as anion.

When an atom of an electropositive element reacts with that of an electronegative element, the former gives up electrons and the latter takes them in and thus transfer of electrons takes place with the formation of cation and anion. These ions, being oppositely charged, attract each other and form an ion pair; this ion pair may or may not have lower energy than the individual atoms. Ultimately, these oppositely charged ions get set in an ionic lattice with the liberation of energy, called lattice energy. In this case we see that ions remain together in an ionic lattice by the electrostatic force of attraction. The attractive force that holds oppositely charged ions together is called ionic bond, or electrovalent bond or more simply electrovalence.

The withdrawal of electrons from an atom or an ion requires energy, called ionisation energy (I.E.). The energy needed to remove the most loosely bound electron from the atom in the gaseous state to the infinity against the nuclear charge is known as first ionisation energy. If a second electron is removed from the atom in the same way as above, the energy required is called the second ionisation energy. Because of the greater nuclear charge and smaller volume of the ion than those of the neutral atom, the second I.E. value is much greater than the first I.E. value. The increasing order of I.E. values is:

$$\text{1st I.E.} < \text{2nd I.E.} << \text{3rd I.E.} <<< \cdots$$

I.E. values are expressed either in kcal/mole or in kJ/mole or in electron volt (eV). The I.E. value may be called the heat of reaction of the change.

$$A + \text{energy (1st I.E.)} \rightarrow A^{+} + e;\ A^{+} + \text{energy (2nd I.E.)} \rightarrow A^{+2} + e.$$

On the other hand, the process of taking in electron by an atom takes place with the liberation of energy. The amount of energy given out when an isolated atom or ion takes in an electron is known as **electron affinity (E.A.)**. The E.A. value is also expressed in kJ/mole, kcal/mole and in **electron-volt (eV)**. Because of electron-electron repulsion, the E.A. value of an atom is much greater than that of its uninegative ion.

An atom can only undergo ionic bond formation with another atom, if the I.E. value of one atom is compensated by the sum of (i) E.A. value of the other atom, (ii) lattice energy liberated, and (iii) **solvation energy**—it is the energy liberated when reactants interact with the chemical units of the solvent (used as medium of the reaction).

Let us now consider the well-known example of sodium chloride crystal formation from the sodium atom and chlorine atom in the gaseous state.

The electron configuration of Na is $1s^2, 2s^2, 2p^6, 3s^1$. Therefore, if it gives up one e, it will get the electron configuration of the nearest inert gas, Ne.

Similarly, if Cl atom $(1s^2, 2s^2, 2p^6, 3s^2, 3p^5)$ takes in this electron, it will attain the electron configuration of argon. This transfer of electron will lead to the formation of Na^+ and Cl^- ions which ultimately get set in an ionic crystal through ionic bond.

$$\underset{1s^2, 2s^2, 2p^6, 3s^1}{Na} + 493.7 \text{ kJ mole}^{-1} \text{ (I.E. of Na)} \longrightarrow \underset{1s^2, 2s^2, 3p^6}{Na^+} + e$$

$$\underset{1s^2, 2s^2, 2p^6, 3s^2, 3p^5}{Cl} + e \longrightarrow \underset{1s^2, 2s^2, 2p^6, 3s^2, 2p^6}{Cl^-} + 362.1 \text{ kJ mole}^{-1} \text{ (E.A. of Cl)}$$

$$\underset{\text{gas}}{Na} + \underset{\text{gas}}{Cl} \longrightarrow Na^+ + Cl^- \longrightarrow \underset{\text{crystal}}{Na^+Cl^-} + \underset{\text{lattice energy}}{753.7 \text{ kJ mole}^{-1}}$$

So, the energy released in the formation of ionic crystal of NaCl from gaseous atom is:

$$362.1 \text{ kJ mole}^{-1} + 753.7 \text{ kJ mole}^{-1} - 493.7 \text{ kJ mole}^{-1} = 622.1 \text{ kJ mole}^{-1}$$

The crystal being of lower energy than that of the combining atoms in the gaseous state, it is stable and very rigid. As the attractive force of an ion is directed in all possible directions, ionic bonds are comparatively weak and non-directional in nature.

Generally, atoms of comparatively low I.E. value form ionic bonds with atoms of comparatively high E.A. value. IA and IIA elements of the periodic table have low ionisation energy. VIB and VIIB elements have high electron affinity. So, IA and IIA elements form ionic bonds with VIB and VIIB elements. Compounds containing ionic bonds are called **ionic compounds**. This type of compounds do not have individual molecules. They have ionic lattice. In an NaCl ionic lattice each of the Cl^- ion is surrounded by six Na^+ ion and **vice versa**.

2.6.3 Covalent Bond and Dative Bonds (before 1926)

There are a large number of substances which cannot be formulated with ionic bonds because of the reasons given below:

(a) If one of the bonding atoms fails to attain the stable electron configuration of the nearest inert gas by transfer of electrons, an ionic bond cannot form. For example, in the formation of H_2 molecule if one H atom transfers one electron to the other H atom, the latter H atom gets the electron configuration of He $(1s^2)$ but the former one becomes a bare proton $(H^{\oplus})$, a very unstable species.

(b) The transfer of electron cannot take place and consequently ionic bond cannot form if the difference between I.E. value of one of the combining atoms and E.A. value of the other is too high to be compensated by the lattice energy and/or salvation energy. The formation of ionic bond between H and F atoms is not energetically favourable, because the difference between the I.E. value of H (1306 kJ $mole^{-1}$) and E.A. value of F (410.3 kJ $mole^{-1}$) is very high.

In 1916, Lewis suggested that atoms could attain inert gas structure without complete transference of electron but by sharing of electrons. The sharing of electron generates a bond (between the atoms involved) which is known as non-ionic or covalent bond. So, covalent bond is a force that holds the atoms together by sharing of electrons. The shared electrons remain in between the two nuclei of the bonded atom. Thus, both the positively charged nuclei attract the negatively charged electrons strongly. So, the binding force in a covalent bond is again electrostatic force of attraction. This force outweighs nucleus-nucleus repulsive force as well as electron-electron repulsive force.

Let us now take the example of fluorine molecule (F_2) to understand the process. It contains a covalent bond, i.e., a shared pair of electrons, one being provided by each atom. The shared pair is occupied equally by both the atoms and hence both of them attain the electronic configuration of neon ($1s^2, 2s^2, 2p^6$). If F represents the fluorine nucleus with inner shell electrons and seven dots (or crosses) represent seven outer electrons, then the process of bond formation may be written as

Thus, a shared electron pair may be represented by two dots or by a dash. This type of formula is known as Lewis electron-dot formula.

Generally, atoms containing one to four electrons less than the nearest inert gas form covalent bonds. Covalent bonds are found in individual molecules and in atomic crystals which are also called covalent crystals. Compounds containing covalent bonds are known as covalent compounds and they remain in the solid state as molecular crystals.

The shared pair of electrons of a covalent bond may also come from a single atom; in this case that atom is called a donor and the atom which also shares the pair of electrons is called an acceptor. The bond generated by the donation of a pair of electrons from a single atom but shared equally by the two atoms is called a dative bond or coordinate bond or coordinate covalent bond. A dative bond is represented by an arrow ($\longrightarrow$), the tail of which starts from the donor and the head ends to the acceptor. Partial positive and negative charges generate on the donor and acceptor atoms respectively. For this reason a dative bond is also known as a semipolar bond. An ammonia molecule with a lone pair of electrons forms a dative bond with boron trifluoride molecule in which born atom has two electrons less than the nearest inert gas, neon. Through coordination the boron atom shares two more electrons from the N atom of ammonia and gets stabilised by attaining the electron configuration of Ne.

The coordinated compound formed by this process may be written as

$$H_3N : + BF_3 \longrightarrow H_3N \rightarrow BF_3.$$

This molecule may also be represented as

$$H_3\overset{\oplus}{N}\text{---}\overset{-}{B}F_3.$$

The + and − charges placed on the N and B atoms respectively are called formal charges.

2.7 Lewis Electron-dot Formula

Following the method of Lewis, we can write down the gross structural formula for molecules and ions. Of course, covalent linkages, normally called bonds, amongst the constituent atoms in the chemical unit should be known to us.

In this method the following considerations are to be made for writing the gross structural formula of chemical entities:

1. An atomic symbol represents the atomic kernel, i.e., the nucleus of the atom with its inner shell electrons.

A nitrogen atom has two electrons in its inner shell ($1s^2$); seven protons and seven neutrons are there in its nucleus. Hence the symbol of nitrogen, N, indicates the nucleus of nitrogen atom with two $1s$ electrons.

2. Outer shell electrons of an atom are indicated by dots; a pair of dots represents paired electrons and a single dot stands for an unpaired electron.

There are five electrons in the outer shell (2nd shell) of an N atom, two of which are paired ($2s^2$) and the rest three are unpaired ($2p_x^1, 2p_y^1, 2p_z^1$). So, Lewis' electron-dot method of representation of nitrogen atom is $\cdot\dot{\underset{\cdot}{N}}\!:$.

3a. A covalent bond between two atoms is represented either by a pair of dots or by a dash between the atoms involved.

In a molecule of hydrogen, there is a single bond in between two H atoms; so a hydrogen molecule may be written as H··H or H—H.

3b. If two atoms are linked to each other by two covalent bonds, called a double bond, two pairs of dots or two parallel dashes (=) are to be placed between the symbols of the atoms.

Thus, a double bond between two oxygen atoms in an oxygen molecule is written as O::O or O=O.

3c. Similarly, for three covalent linkages between two atoms, i.e., for a triple bond, we are to write three pairs of dots or three parallel dashes between them.

A molecule of nitrogen possesses a triple bond between two N atoms. This triple bond is shown as N⋮⋮N or N≡N.

4. The total number of shared electrons (i.e., bonded electrons) and unshared electrons should be equal to the sum of the outer electrons of the constituent atoms of the molecule in its gross structural formula.

For example, there are two chlorine atoms in a molecule of chlorine. These are linked to each other by a single covalent bond. A chlorine atom contains seven outer electrons ($3s^2, 3p^5$). So in the structural formula of Cl_2, fourteen electrons (7+7) are to be shown. One covalent bond contains two electrons; hence twelve unshared electrons must be represented by dots. From the outer electron configuration, it is obvious that each of the Cl atoms should bear three pairs of dots for three unshared pairs of electrons. Thus, Cl_2 is:

$$:\ddot{\underset{..}{Cl}}\cdot\ \cdot\ddot{\underset{..}{Cl}}: \quad \text{or} \quad :\ddot{\underset{..}{Cl}}—\ddot{\underset{..}{Cl}}:$$

5a. The total number of electrons, bonded and non-bonded, in a polyatomic anion should be equal to the sum of the outer electrons of the constituent atoms plus the number of the negative charges on the anion, e.g.,

A hydroxide ion has one O atom covalently linked to one H atom by a single bond and has one negative charge. Therefore, the number of total electrons to be shown in the gross structural formula of hydroxide ion are six for O atom, one for H atom and one for a negative charge, i.e., eight electrons. Hence, the gross structural formula for a hydroxide ion is :Ö—H.

5b. Similarly, the total number of electrons, shared and unshared, of a polyatomic cation should be equal to the sum of the outer electrons of the involved atoms **minus** the number of positive charges on the cation.

An ammonium ion has four H atoms linked to the N atom by four covalent bonds and the species bears a positive charge. In its gross structure, we shall have to show eight electrons (5e for N + 4e for 4 H atoms $-1e$ for a +ve charge). Hence, ammonium ion is written as

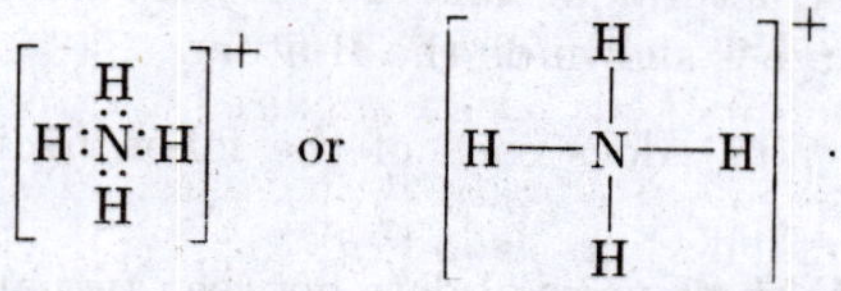

6. Atoms possessing second quantum shell (L), i.e., Li to Ne in the periodic table should not have more than eight outer electrons shared, or unshared, in any gross structural formula. On the other hand, atoms possessing third or higher quantum levels may have more than eight such electrons in their gross structural formula.

Following these six rules about Lewis electron-dot formula, we are in a position to write down the structural formula of molecules. Let us take the example of nitrogen molecule. In a molecule of nitrogen there are ten outer electrons (five on each N atom). Out of these a triple bond is formed between two N atoms. Hence, Lewis' electron-dot formula for N_2 is

:N⋮⋮N: or :N≡N:

Each of the N atoms uses three electrons for the formation of the triple bond and thereby each of them gets the electronic configuration of Ne. Now out of five outer electrons two are then left on each N and these are shown by a pair of dots.

7. In the case of a polyatomic ion though we assign the charge on the whole of the ion, yet it actually appears on an atom or atoms that constitute the ion. These charges are called **formal charges** and may be defined as the charges the atoms would have in a chemical entity if all were of the same electronegativity (Section 2.11).

In the calculation of the formal charge on an atom, it is assumed that all bonded electrons are shared equally by the atoms involved and that each atom possesses one-half of the bonded electrons as its own. Therefore, the formal charge on an atom in a chemical species is

$$\text{Formal charge} = e_n - \left(\frac{1}{2}e_b + e_{nb}\right), \tag{2.1}$$

where, e_n = number of outer electrons of the isolated atom,

e_b = number of bonded, i.e., shared electrons,

e_{nb} = number of unshared (non-bonded) electrons.

In other words, the **formal charge** on an atom equals the number of outer electrons **minus** the sum of unshared electrons and one-half of the shared electrons.

A hydroxide ion is represented as $:\ddot{\underset{\cdot\cdot}{\mathrm{O}}}$—H. Why is a negative charge placed on the O atom? In the isolated state, an O atom and an H atom have six and one outer electrons respectively. From the gross structural formula of the hydroxide ion the formal charge on O and H atoms can be calculated using the Equation 2.1.

For H atom: $e_n = 1,\ e_b = 2$ and $e_{nb} = 0$.

$\therefore$ Formal charge on the H atom: $1 - \left(\dfrac{1}{2} \times 2 + 0\right) = 0.$

For O atom: $e_n = 6,\ e_b = 2$ and $e_{nb} = 2 \times 3$.

$\therefore$ Formal charge on the O atom: $6 - \left(\dfrac{1}{2} \times 2 + 2 \times 3\right) = -1.$

Hence in the gross structural formula of hydroxide ion one negative charge is placed on the O atom and no charge is shown on the H atom.

Problem 2.6. (a) What does each of the following indicate in electron-dot formula of chemical entities?

(i) F, **(ii)** $:\ddot{\mathrm{F}}\cdot$ and **(iii)** three pairs of dots between two atoms.

(b) Write down the electron-dot formula for hydrazine in which two N atoms are linked by a single bond and each of the N atoms holds two H atoms by covalent bonds.

(c) Calculate the formal charge on each of the constituent atoms of 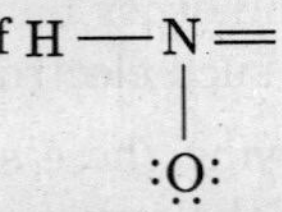 and hence write the complete electron-dot formula for the species.

Solution : (a) (i) It indicates the fluorine nucleus with two K-shell electrons.

(ii) It indicates the fluorine nucleus with nine electrons, of which seven are outer electrons represented by seven dots. Amongst these seven electrons, six are paired (as three pairs of dots are written) and one is unpaired.

(iii) Three pairs of dots between two atoms indicate a triple bond ($\equiv$) between the atoms.

(b) The molecular formula of hydrazine, as per our problem, is N_2H_4. Therefore, in its electron-dot formula we are to show 14*es* (2×5 for 2 N atoms and 4×1 for 4 H atoms). As there are five covalent bonds in the molecule (one N—N and four N—H), 10*es* are shared electrons. Hence 4*es* should remain unshared. Now each N atom has three covalent bonds, one with another N atom and two with two H atoms. Therefore, each N atom has six shared electrons of which three belong to each N atom. So each N atom has $(5-3)$ or 2 electrons left unshared. Therefore, the electron-dot formula of hydrazine is:

$$H_2\ddot{N}—\ddot{N}H_2$$

(c) Formal charge on H : $1 - \left(\dfrac{1}{2} \times 2 + 0\right) = 0$

Formal charge on double-bonded O : $6 - \left(\dfrac{1}{2} \times 4 + 4\right) = 0$

Formal charge on single-bonded O : $6 - \left(\dfrac{1}{2} \times 2 + 6\right) = -1$

Formal charge on N atom : $5 - \left(\dfrac{1}{2} \times 8 + 0\right) = +1.$

Hence, the complete electron-dot formula of the given species is:

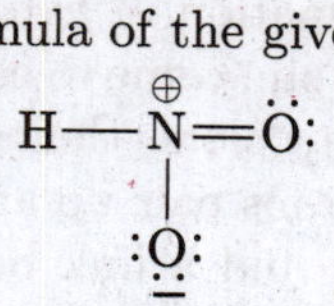

2.8 Modern Concept of the Covalent Bond

Lewis theory of covalent bond formation does not provide any explanation for the following:

(i) Why pairing and sharing of electrons bring stability to the system, i.e., lower the energy of the system.

(ii) The molecular geometry, i.e., the shape of the molecules containing the covalent bonds.

After the advent of quantum mechanics two theories about the formation of covalent bond have been advanced. These theories are capable of explaining the stability brought to the system by the covalent bond formation as well as the geometry of the simple molecules. As these theories are highly mathematical and only rigidly applicable to simple molecules, we shall discuss here qualitatively the assumptions made therein and outcome of these assumptions. These theories are known as

I. valence bond theory: Proposed by Heitler and London and extended by Pauling and Slater.

II. molecular orbital theory: Proposed by Hund and Mullikan.

2.8.1 Valence Bond Theory (V.B. theory)

V.B. theory is the logical quantum mechanical extension of Lewis' theory of electron pair formation. Naturally, this theory considers particle-wave duality of electrons. As particles the electrons possess spin motion and as waves they have orbital diagrams. Both of these properties are utilised in the V.B. theory to describe a covalent bond.

The following are the assumptions and salient points of the V.B. theory :

(a) First assumption, its discussion and salient points

A covalent bond between two atoms results from the maximum overlap of two pure or hybrid A.O.s, each containing one electron of antiparallel spins; thus a larger common orbital of two atoms, called bond orbital, is formed having a pair of electrons of opposed spins. The probability of finding the electron pair is maximum between the two nuclei of the bonded atoms.

When the two bonding atoms come to each other at an optimum distance, maximum overlapping of A.O.s, i.e., pairing of electrons can occur; thus neutralisation of the magnetic moments of the two oppositely spinning electrons and generation of attractive force result (Pauli). The probability of finding the electron-pair being maximum between the two nuclei, it is attracted by both the nuclei and therefore, just two atoms are bonded. Hence the force, i.e., the covalent bond that holds the two atoms together is electrostatic in nature. If the two atoms have electrons of the same spin or if the two atoms come to each other at a distance shorter or longer than the optimum distance, instead of attraction, electron-electron repulsion and proton-proton repulsion become predominating factors; thus bonding does not occur. Now, the bond orbital being larger in volume than those of the overlapping A.O.s, the electron-pair has greater space to move which lowers the energy of the system and thus brings stability to the system. So, a covalent bond formation is always accompanied with release of energy.

Let us take the example of the formation of hydrogen molecule. When two H atoms, each containing one electron of anti-parallel spin, come to each other at an optimum distance (0.074 nm for H atoms), the two *s* orbitals overlap along their radii. Thus, maximum overlap of the two orbitals occurs; consequently, two electrons pair up and a bond orbital results. The electron pair now is attracted by both the H nuclei and a link between them is formed, known as covalent bond. This liberates 432.2 kJ/mole (103 kcal/mole) of energy.

From the discussion made above, we get :

(i) To form a covalent bond an atom must have a stable orbital with unpaired electron.

(ii) Two overlapping orbitals should have one electron each of opposed spins.

(iii) The number of covalent bonds that an atom can form is determined by the number of incomplete orbitals it has or it can provide for.

(iv) A covalent bond links just two atoms.

(v) The covalent bond is a force which is electrostatic in nature.

(vi) The bond orbital has lower energy than those of the overlapping A.O.s.

(b) Second assumption, its discussion and salient points therein

Each electron of the shared pair has an equal probability of being found in its own atom as well as in the other atom and each nucleus may be associated equally with both the electrons, i.e., exchange of electron between bonding atoms is a condition to form a covalent bond.

To understand the meaning of the second assumption of the V.B. theory, let us consider a diatomic molecule, AB, containing only one covalent bond between A and B atoms. Let us further consider that the electron of A which pairs with the electron of B to form the covalent bond is not identical but different; the electrons being different, let us represent the electron of A by an asterisk (*) and that of B by a dot (.). Now, the second assumption directs us to conceive that :

(i) Each electron may remain in its own orbital with its own atom; thus participating A.O.s do not lose their identity by the process of overlap, and as no transfer of electrons takes place, a non-ionic linkage between the atoms is established. Hence, the bond between A* and .B is A*.B or A*——.B.

(ii) The two electrons may be exchanged between the two atoms; thus, each of the atoms contains the same number of electrons as before hence, here too a non-ionic linkage is established between the bonding atoms. But, if the participating electrons are supposed to be different, then the non-ionic linkage formed by the exchanged electrons is a bit different from that in (i) and may be represented as

A.*B or A.——*B

(iii) The electron pair may remain with one atom only and the other atom remains without bonding electrons; thus, the transfer of electrons is conceived and the atom which now has both the electrons becomes an anion and the other a cation. This process gives rise to two ionic bond structures—the cation and the anion in one such structure are the anion and the cation respectively in the other structure. The two ionic structures are represented as:

$\overset{+}{A}\ {}^{*}_{.}\overset{-}{B}$ (c) and $\overset{-}{A}{}^{*}_{.}\ \overset{+}{B}$ (d)

The foregoing discussion tells us to assume four structures of two nuclei system joined by a single covalent bond; two of them are non-ionic and two others are ionic. The actual structure of the system will, therefore, be a combination of these four structures. Of course, owing to the electron-electron repulsion and large difference between the ionisation energy value and electron affinity value, there is a small probability of finding both the electrons in one and the same atom. However, the actual structure of the system being a combination of four structures, it is not possible to represent it by a single Lewis' electron-dot structure and no other method has yet been developed to do so. Chemists project such a structure by writing down all the four structures, covalent and ionic, and joining them by double-headed arrows (↔). The actual structure is then called the resonance hybrid of the four hypothetical structures, each of them being known as resonating or canonical structure. The so-called resonance hybrid should have a lower energy than any of the canonical structures. In the formation of resonance hybrid with maximum stability and hence lowered energy state, it is essential that energy be released in the process. As resonating structures are considered from the idea of exchange of electrons between the bonding atoms, the energy released by the covalent bond formation is called exchange energy or resonance energy which brings stability to the system.

Now following the second assumption of the V.B. theory, we can write down four canonical structures for the molecule AB. These are

$$\text{(a) } A^{*} \text{—} .B,\ \text{(b) } A^{\cdot} \text{—} {}_{*}B,\ \text{(c) } \overset{-}{A}{\dot{}}{}_{*}\ \overset{+}{B},\ \text{(d) } \overset{+}{A}\ {}^{\cdot}_{*}\overset{-}{B}$$

The actual structure of the molecule AB is a combination of these four structures and called resonance hybrid which is represented as

$$A^{*} \text{—} .B \longleftrightarrow A^{\cdot} \text{—} {}_{*}B \longleftrightarrow \overset{-}{A}{}^{*}_{\cdot}\ \overset{+}{B} \longleftrightarrow \overset{+}{A}\ {}^{*}_{\cdot}\overset{-}{B}$$

Here, ionic structures (c and d) are unimportant for electron-electron repulsion. As the formation of covalent bond releases energy, structures (a) and (b) will be the best approximations of the molecule AB, i.e., of the resonance hybrid. Again, hypothetically the presence of each electron to its own atom is considered, but it is impossible to distinguish them; hence, one covalent structure is enough to describe the AB molecule, i.e., A—B.

The salient points that are obtained from the second assumption are:

(i) The overlapping A.O.s preserve their character though they form a bond orbital.

(ii) The actual structure of a two nuclei covalent system is a combination of some hypothetical, covalent and ionic, structures none of which is capable of describing the structure completely. The combination is called the resonance hybrid and each hypothetical structure is known as a resonating or canonical structure.

(iii) The process of combination of resonating structures releases energy which brings stability to the system. The released energy is called resonance energy, an exchange energy.

(iv) The structure that possesses the largest number of covalent bonds will be the most stable.

2.8.2 Shape and Orientation of Bond Orbitals

Bond orbitals may be of different types, viz., sigma (σ), pi (π), delta (δ), banana, etc. Organic chemistry concerns mostly with the sigma and pi bond orbitals and to some extent with the banana or bent orbitals; so these are discussed below.

Sigma bond orbital

A sigma bond orbital is formed when two A.O.s, pure or hybrid, overlap around their axes, i.e., head-to-head. An axis is the imaginary line which joins the two nuclei and is known as bond axis. Sigma bond orbital is bigger in size than that of the overlapping A.O.s. It has cylindrical charge density around the bond axis and hence symmetrical in nature. The charge density is maximum between the nuclei (Fig. 2.13). The bond formed by head-to-head overlap is called sigma (σ) bond.

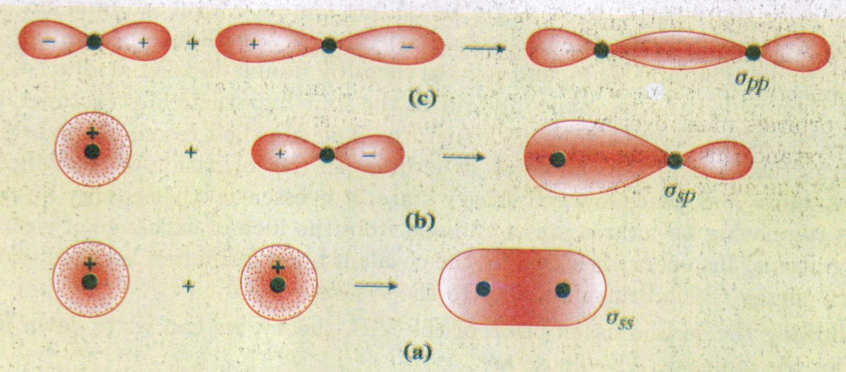

Fig. 2.13. σ-**bond orbitals**

Single bonds are sigma bonds. These may result by the overlap of several types of A.O.s. Some of them are:

s-s, s-p, p_x-p_x, s-sp, p-sp, s-sp^2, p-sp^2, s-sp^3, sp^2-sp^2, sp-sp^3, sp-sp^2, sp-sp, sp^3-sp^3, sp^2-sp^3, etc.

Actually, a head-to-head overlap of lobes of like sign (+ or −) gives rise to sigma bond orbital. A sigma bond has no nodal point or plane or surface between the two nuclei and maximum overlap of lobes occurs along the orbital axes; furthermore, a sigma bond orbital has a symmetrical cylindrical charge density around the bond axis—for these reasons a sigma bond is strong.

Pi bond orbital

The bond orbital formed by the side-to-side overlap of two parallel A.O.s is known as pi bond orbital. It has a maximum charge density in its cross-sectional plane. The plane perpendicular to it has no π charge density and is called a nodal plane. The line which joins the two nuclei is called bond axis and it lies in the nodal plane, i.e., in a plane of no charge density. The bond formed by side-to-side overlap is known as pi bond. A double bond is considered to consist of one sigma and one pi bonds, triple bonds may be supposed to have one sigma and two pi bonds. Pi bonds are formed by p_y-p_y or p_z-p_z overlap (Fig. 2.14).

A side-to-side overlap of lobes of the same phase of two orbitals does not involve maximum overlap and a π bond orbital has an angular momentum of one unit ($h/2k$), and furthermore the π bond axis lies in the plane of no charge density, i.e., π bond orbital has a node; for these reasons π bonds are weak and easy to break.

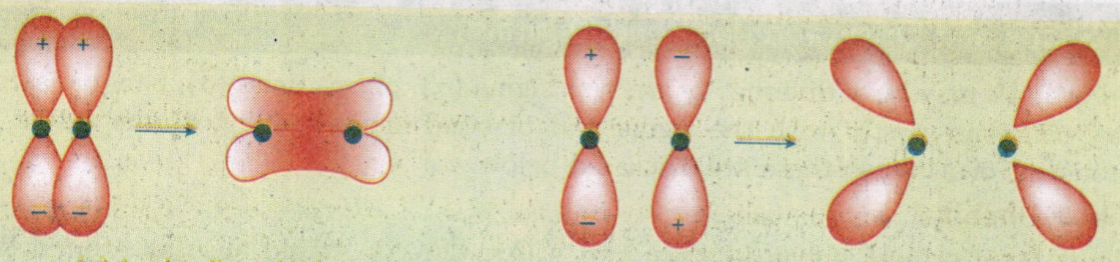

(a) A π-bonding orbital **(b) A π-antibonding orbital**

Fig. 2.14

Banana bond orbital

An oblique overlap of two A.O.s produces bent bond or banana bond. Obviously, this type of overlap is neither head-to-head nor side-to-side and may be taken as an intermediate between the two types. Naturally, the charge density along the bond axis is very minimum and hence the bond is weak. sp^3-sp^3 and sp^2-sp^2 overlap can form such type of bond orbitals.

In this connection it is important to note that to form a bonding orbital lobes of the same phase of two orbitals must overlap. Now, the question is: what do we mean by the phase of a lobe? To understand this, let us represent the wave motion developed when a stone is thrown in a calm pond by the curve as shown below (Fig. 2.15).

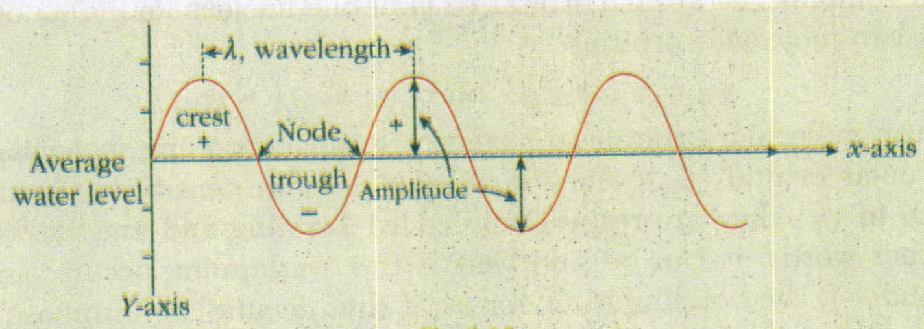

Fig. 2.15

If we consider $y = \sin x$ as the equation for the wave motion graphically respresented by the curve, then y is the wave function. However, waves formed in the pond have crests and troughs; crests and troughs are regions where the waves rise above and below the average level of water in the calm pond respectively. In crests the wave function y will be taken as (+) and in troughs this will be (−) and places just at the average level of water in the calm pond y will be zero (node, a point of no disturbance); + and − signs are used to indicate the conditions of the wave above and below the average level. So far as intensity, energy, etc., are concerned, a crest and a trough are same but the former differs from the latter in phase; if the phase of a crest is indicated by (+) sign, a trough is indicated by (−) sign. Then the question arises: why do we differentiate a crest and a trough by (+) phase and (−) phase? If two waves of the same phase approach and overlap, a wave larger than the overlapping waves develops; on the other hand, two waves of opposite phases form a smaller wave, smaller than the approaching waves, through their combination as shown in Fig. 2.16.

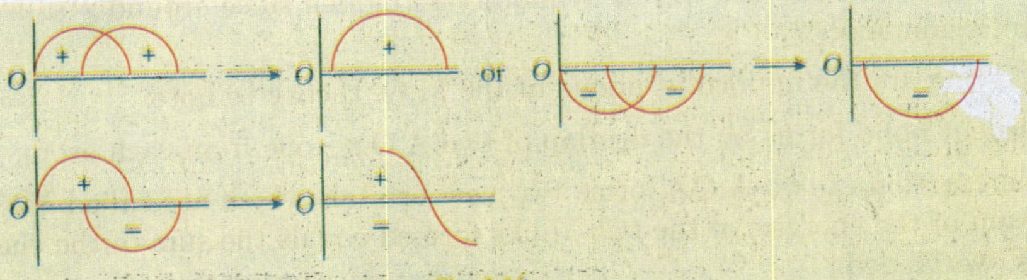

Fig. 2.16

Similarly, when A.O.s with the same phased lobe overlap, a bonding orbital forms; if it is otherwise, an antibonding orbital results.

2.8.3 Molecular Orbital Theory (M.O. Theory)

M.O. theory starts with the consideration that after the covalent bond formation the concerned electrons do not belong to their individual atoms, instead they associate themselves with the molecule as a whole embracing all the nuclei; the nuclei, in their turn, remain in equilibrium positions. However, the problem of an electron moving in the multi-nuclei field cannot be solved exactly. For this reason we shall have to begin with an approximate form for one-electron molecular orbitals. Linear Combination of Atomic Orbitals (LCAO) method is a qualitatively useful but a simple approximation.

The basis of the LCAO approximation is: In the case of one-electron two-nuclei system, the electron at one time may come very close to one nucleus and far from the other and that when it is close to a particular nucleus, it will be more or less controlled by that nucleus and naturally its behaviour will be more or less as if it were in an atomic orbital belonging to that nucleus only. Therefore, we can combine the atomic orbital (ψ_A) of one nucleus and (ψ_B) of the other nucleus linearly to obtain two molecular orbitals:

$$\psi_b = \psi_A + \psi_B \quad \text{and} \quad \psi_a = \psi_A - \psi_B.$$

The one-electron molecular orbitals so formed are called bonding molecular orbital (ψ_b) and antibonding molecular orbital (ψ_a); since in ψ_b the electron density between the two nuclei is much greater than in ψ_a, they are respectively called **bonding** and **antibonding** M.O.s between two A.O.s. In other words, it can be said that if the overlapping occurs "in the same phase" (+ and + or, − and −), the bonding M.O. forms; if that occurs "in antiphase", the antibonding M.O. forms.

There are certain conditions for effective combination of A.O.s as given below:

(a) To form M.O.s the combining A.O.s should have comparable energies. For homonuclear molecules 1*s* orbital should combine with 1*s* orbital and not with any other A.O. However for heteronuclear system such combinations are accepted.

(b) The charge cloud of the combining A.O.s must overlap one another as much as possible and the principle is called the **Principle of Maximum Overlap**.

(c) The A.O.s must have the same symmetry about the bond forming axis and this may be called the **Symmetry Condition** for the combination of A.O.s. For example, if an 1*s* orbital of an atom combines with a *p* A.O. of another atom and if the bonding axis is the x-axis, the s-p_x combination will be allowed since p_x A.O. is symmetric with respect to x-axis (if we rotate the p_x A.O. around the x-axis; no phase change will take place and the *s* A.O. will also have the same symmetry property with respect to the x-axis). But s-p_y and s-p_z combinations in that case will be symmetry-forbidden since around x-axis p_y and p_z A.O.s are antisymmetric.

The following are the important points of the M.O. theory to note:

(a) A covalent bond forms by the overlap of two A.O.s—one from each atom.

(b) The interaction of two A.O.s forms two new orbitals which are called **Molecular Orbitals**. The sum of the energies of the two M.O.s formed equals the sum of the energies of the two A.O.s overlapped.

(c) One of these M.O.s has lower energy and hence greater stability than the individual A.O.s and is known as **bonding M.O.**; the other M.O. possesses greater energy, i.e., lower stability than the participating A.O.s and is called **antibonding M.O.** The bonding M.O. is the result of the overlap of A.O.s with lobes of like signs; antibonding M.O.s are formed when A.O.s with lobes of unlike signs overlap. In a bonding M.O. the probability of finding maximum charge density is the space between the two nuclei, whereas in the antibonding M.O. the region between the two nuclei has no charge density.

(d) There are different types of M.O.s, viz., sigma M.O., pi M.O., delta M.O. etc. Sigma M.O.s are formed by head-on overlap of A.O.s and pi M.O.s are the result of side-on overlap of A.O.s. In fact, bond orbitals in the V.B. theory are the bonding molecular orbitals in the M.O. theory. The shape and orientation of sigma and pi M.O.s are identical with those of the sigma and pi bond orbitals in the V.B. theory.

(e) Outer A.O.s of similar energy only overlap, inner A.O.s are supposed to remain undisturbed.

(f) Outer A.O.s of each atom may mix up before forming M.O.s. This is not actually equal to the hybridisation of A.O.s but for convenience it may be taken as such.

(g) The energy order of M.O.s in a molecule is as follows:

$\sigma 1s < \sigma^* 1s < \sigma 2s < \sigma^* 2s < \sigma 2p_x < \pi_y 2p = \pi_z 2p < \pi_y^* 2p = \pi_z^* 2p < \sigma^* 2p_x \cdots$

σ denotes the M.O. forming σ-bond.

π denotes the M.O. forming π-bond.

$*$ denotes an antibonding M.O.

$\sigma 1s$, $\sigma^* 1s$ denote the M.O.s formed from $1s$ A.O.s.

$\sigma 2s$, $\sigma^* 2s$ denote the M.O.s. formed from $2s$ A.O.s.

$\sigma 2p_x$, $\sigma^* 2p_x$ denote the M.O.s formed from $2p_x$ A.O.s.

$\pi_y 2p$, $\pi_y^* 2p$ denote the M.O.s formed from $2p_y$ A.O.s.

$\pi_z 2p$, $\pi_z^* 2p$ denote the M.O.s formed from $2p_z$ A.O.s.

However, spectroscopic experiments show that up to nitrogen molecule the energy order of M.O.s is as follows:

$$\sigma 1s < \sigma^* 1s < \sigma 2s < \sigma^* 2s < \pi_y 2p = \pi_z 2p < \sigma 2p_x < \pi_y^* 2p = \pi_z^* 2p < \sigma^* 2p_x$$

Orbital notations may be changed according as the A.O.s taking part in the formation of M.O.s. Thus, if $1s$ orbital overlaps with one $2p$ A.O., the two M.O.s formed may be designated as σsp and $\sigma^* sp$.

(h) The process of filling in of M.O.s is identical to that of A.O.s and governed by the three same principles:

- **(i)** Pouring on electrons should start from the lowest energy M.O. and gradually to the higher energy M.O.s.
- **(ii)** Not more than two electrons can remain in an M.O. If two electrons occupy the same M.O., then they should have antiparallel spins.
- **(iii)** Electrons of equal energy orbitals cannot pair up until all such M.O.s have one electron each.

However, from the M.O. theory we get:

(i) When two atoms form a molecule, the number of bonding electrons should be greater than the number of antibonding electrons and then only a release of energy will occur; the release of energy will bring stability to the system.

(ii) The number of covalent bonds in a molecule is one-half of the difference between the number of bonding and antibonding electrons and this is known as bond order.

$$\text{Bond order} = \frac{1}{2}\,(\text{No. of bonding } e - \text{No. of antibonding } e) \qquad (2.2)$$

Two $1s$ orbitals of the two H atoms, one from each, when overlap, two M.O.s will be formed; one is bonding ($\sigma 1s$) and the other is antibonding ($\sigma^* 1s$). $\sigma 1s$ being of lower energy than the$\sigma^* 1s$, the first electron will go to $\sigma 1s$ M.O. The second electron too will go to this orbital as two oppositely spinning electrons can remain in an M.O. (Pauli). The $\sigma^* 1s$ orbital will thus remain vacant. $\sigma 1s$ orbital has less energy than individual $1s$ A.O.s. Therefore, the formation of a covalent bond will occur with the release of energy and hence hydrogen molecule will be stable. The number of bonds between two H atoms will be according to Equation (2.2).

Bond order $= \frac{1}{2}(2-0) = 1.$

$\therefore$ H_2 molecule may be represented as H—H.

The whole thing is represented pictorially in Fig. 2.17 and 2.18.

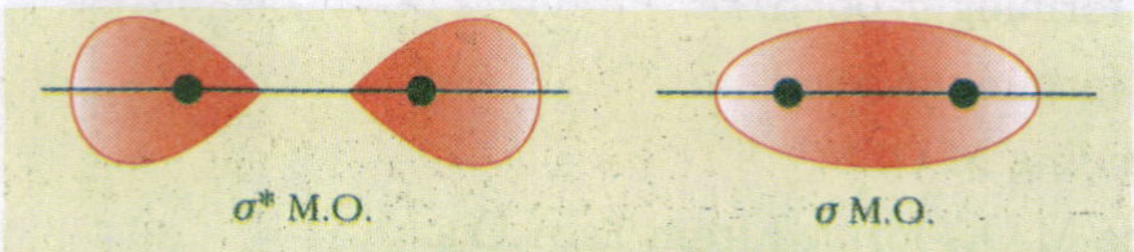

Fig. 2.17. σ*1s M.O. and σ1s M.O.

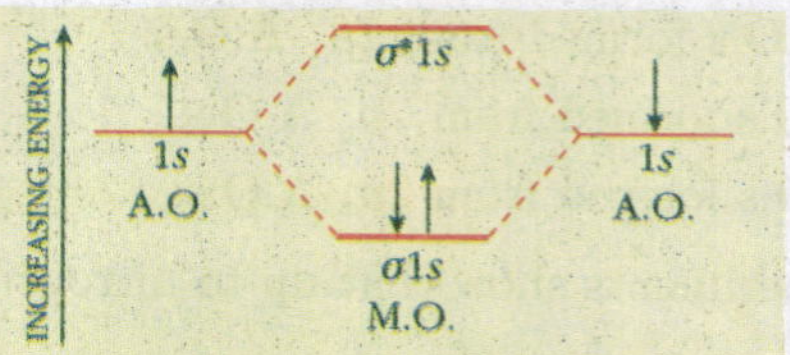

Fig. 2.18. M.O. energy diagram for H_2

Let us now consider the fate of a diatomic He_2 molecule if at all it forms. The electronic configuration of He is $1s^2$. Hence two $1s$ orbitals, each from one He atom, will overlap to form two M.O.s, $\sigma 1s$ and $\sigma^*(1s)$. As two He atoms has 4 electrons, these are to be placed in the two M.O.s so formed. According to the Aufbau principle the first and second electrons will go to the $\sigma 1s$ M.O. Naturally, the third and fourth electrons will occupy the $\sigma^* 1s$ M.O. Therefore, both the M.O.s are filled. Again, we know that the sum of the energies of the overlapping A.O.s, equals the sum of the energies of the M.O.s formed. In fact, σ^* is more antibonding than σ is bonding.

Hence, no energy will be released by the formation of He_2 and therefore it will be unstable. Actually, He_2 does not exist. We can also arrive at the conclusion from Equation (2.2). Thus the bond order in $He_2 = \frac{1}{2}(2-2) = 0$.

2.8.4 Rotation Around σ-Bonds

A σ-bond being symmetrical in so far as its charge is concerned, one of the bonded atoms can rotate around the bond axis keeping the other atom fixed. This rotation does not destroy the charge symmetry of the σ-bond orbital and the bond remains intact. On the other hand, a double bond, consisting of a σ-bond and a π-bond, is very rigid and rotation of an atom about the double bond is not normally possible. This is because rotation of one of the atoms will destroy the parallelism of the A.O.s and will break the π-bond orbital which is formed by the side-on overlap of orbitals. Of course, if energy is supplied, π-bond will break and an electron will go to the π^* orbital. When it will come down to the ground state again π-bond formation will occur through spin inversion. This bond-breaking requires input of energy.

Rotation around bonds gives rise to certain important phenomena like conformational isomerism and diastereomerism (see Chapter 9) and many properties of the molecule. These will be discussed in detail later on.

Problem 2.7. Describe the bonding in diatomic nitrogen molecule by **(i)** VB theory and **(ii)** that of oxygen molecule by M.O. theory.

Solution: **(i)** Ground state electron configuration of the N atom is

$$1s^2, 2s^2, 2p_x^1, 2p_y^1, 2p_z^1.$$

In the excited state, each of the N atoms forms two linear *sp* hybridised orbitals; p_y and p_z orbitals are oriented mutually perpendicular to each other and to each of the *sp* hybrid A.O.s. The picture of this is given below (Fig. e).

$$\underset{sp}{\underline{\uparrow\downarrow}}\ \underset{sp}{\underline{\uparrow}}\ \underset{p_y}{\underline{\uparrow}}\ \underset{p_z}{\underline{\uparrow}}$$

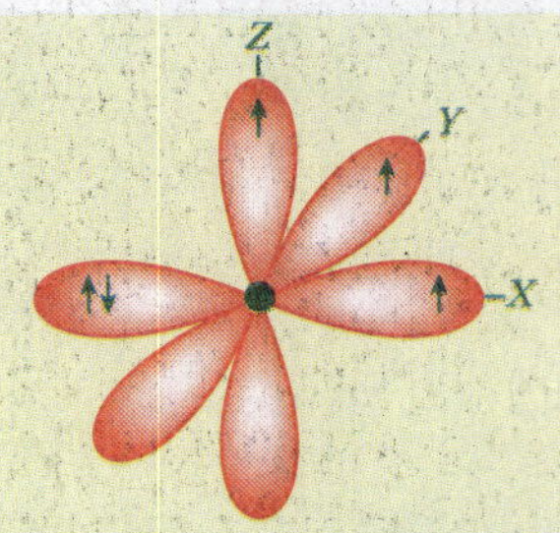

Fig. e. *sp*, p_y and p_z A.O.s of N atom

The *sp* hybrid A.O.s, one from each N atom, overlap head-to-head to form a bond orbital, i.e., σ bond. The other *sp* A.O. of each N atom contains a lone pair of electrons. Now, p_y orbital of each of the N atom being parallel to each other, they overlap side-to-side to form a π-bond orbital. Similarly, p_z orbitals of two N atoms overlap side-to-side to form another π-bond orbital, i.e., p_y-p_y and p_z-p_z overlaps produce two π-bonds. Owing to the formation of two π-bonds and greater *s* character of the σ-bond (for involvement of *sp* A.O.s), the internuclear distance between two N atoms is very small and they are, therefore, held together very strongly by the Coulombic force of attraction between nuclei and electrons; hence, the N_2 molecule will be inert in nature. The formation of three covalent bonds release much of energy and the molecule is, therefore, exceptionally stable. The bonding between two N atoms are given in Fig. f.

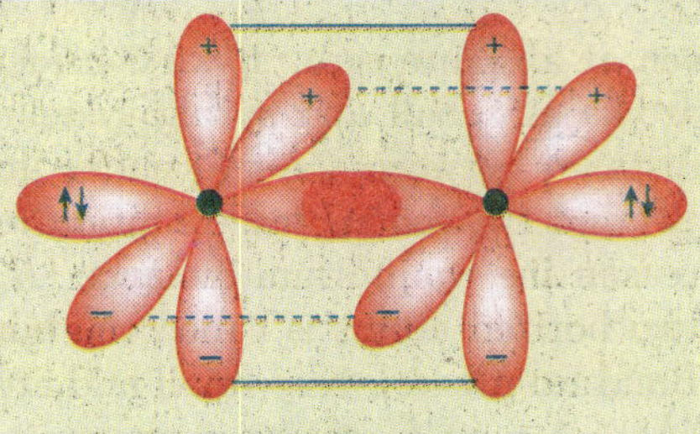

Fig. f. Bonding in N_2

(ii) Each O atom contains six outer electrons in four orbitals. In a diatomic oxygen molecule, therefore, 8 M.O.s (4 A.O.s from each O atom) will develop; and twelve electrons will be accommodated to these M.O.s. Now first two electrons, one from the 2*s* A.O. of each O atom, will go to the $\sigma 2s$ M.O. The second two electrons from the same A.O.s will occupy σ^*2s M.O. The third pair of electrons, one from each $2p_x$ orbital, will go to $\sigma 2p_x$ M.O. The fourth pair of electrons, one from each of the $2p_y$ A.O.s, will occupy $\pi_y 2p$ and $\pi_z 2p$ M.O.s (being of same energy). The fifth pair of electrons from the $2p_z$ A.O.s will pair up with the electrons of $\pi_y 2p$ and $\pi_z 2p$ M.O.s. The last two electrons will go to $\pi_y^* 2p$ and $\pi_z^* 2p$, one each, since these two M.O.s are degenerated. The energy diagram is given below (Fig. g).

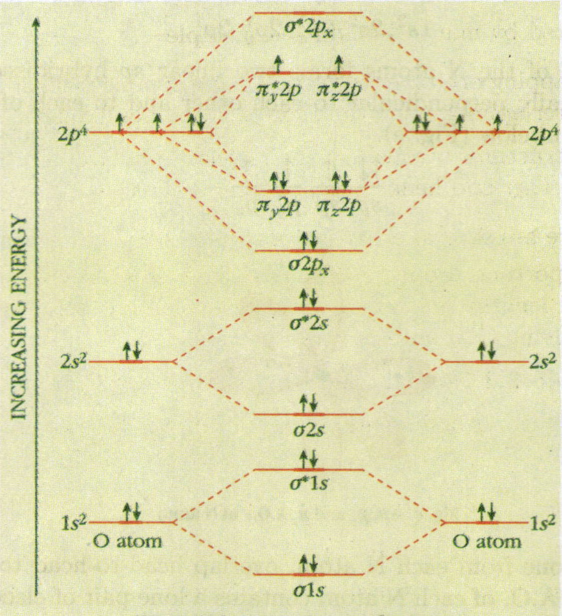

Fig. g. M.O. energy diagram for O_2

Following Equation (2.2) we can now determine the bond order between the two O atoms. Here number of bonding electrons is equal to eight (2 from $\sigma 2s + 2$ from $\sigma 2p_x + 2$ from $\pi_y 2p + 2$ from $\pi_z 2p$). The number of antibonding electrons are only four (two in σ^*2s and one each in π_y^*2p and in π_z^*2p). Therefore, the bond order $= \frac{1}{2}(8-4) = 2$. Hence the O_2 molecules have two covalent bonds, between two O atoms and is called a double bond. Thus, O_2 is O═O.

2.9 Bond Length

The bonded atoms vibrate even at absolute zero. If we consider the two covalently bonded atoms as a ball and spring system as shown in Fig. 2.19, the distance between the two nuclei under the stretched condition (b) and the contracted condition (c) will be different. For this reason the equilibrium distance (R_e) between the nuclei of the two bonded atoms is called the **bond distance** or **bond length.** Bond lengths are very small and are measured in angstrom unit (Å; 1 Å$= 10^{-8}$ cm) or in nanometre (1 nm $= 10^{-7}$ cm) or in picometre (1 pm $= 10^{-10}$ cm).

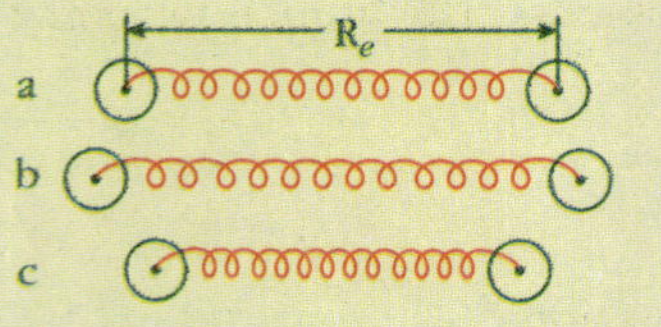

Fig. 2.19. Vibration of two bonded atoms

Each covalently bonded atom has been assigned a characteristic covalent radius. The sum of the two covalent radii of the two bonded atoms gives the bond length since the covalent radius of a particular atom is reasonably constant over a wide range of structures.

The single bond lengths between two particular atoms A and B, are found to be almost constant in their different types of compounds, e.g., C—C bond length in diamond as well as in propane is found to be the same and it is 0.154 nm. This is also true for double and triple bonds.

If two atoms are capable of forming single, double and triple bonds between them, then it is found that:

Single bond length > Double bond length > Triple bond length.

Bond lengths are affected by many a factor; for example—

(i) Nature of the overlapping orbitals—*s* orbitals being symmetrical in nature, the greater the *s* character in an overlapping A.O., the shorter is the bond length.

(ii) Resonance—ionic structures involved in the resonance hybrid of a molecule make the ions come close to each other and hence bond distance decreases.

Besides these two, there are several other factors which will be dealt with in other sections.

Bond length is an important property of a molecule; this provides information about its structure and hence bond lengths are very helpful in explaining the properties of compounds. Normal bond lengths involving C atoms are given in Table 2.3.

Table 2.3 Normal bond lengths involving C atoms

Bond	Bond length (nm)	Bond	Bond length (nm)
sp^3C—H	0.109	C—N	0.147
sp^2C—H	0.108	C=N	0.130
spC—H	0.106	C≡N	0.116
C—C(sp^3-sp^3)	0.154	C=O	0.122
C—C(sp^2-sp^2)	0.148	C—S	0.182
C—C(sp-sp)	0.138	C—F	0.132
C=C(sp^2-sp^2)	0.133	C—Cl	0.177
C=C(sp^2-sp)	0.131	C—Br	0.194
C≡C(sp-sp)	0.120	C—I	0.214
C—O	0.142	N—H	0.100

Though Si and C possess the same outer electron configuration, yet Si—Si bonds are weaker than C—C bonds. Owing to the large size of Si atom, Si—Si bond is long and the electrostatic force holding the atoms together is therefore weak. Thus, C atom possesses the unique property of catenation but Si does so to a minimum extent (Section 1.2).

2.10 Bond Dissociation Energy and Bond Energy

The vibrational energy of a diatomic molecule is given by the equation

$$\epsilon_v = h\nu\left(v + \frac{1}{2}\right) \tag{2.3}$$

where ϵ_v is the energy, h = Planck's constant, ν = frequency of vibration and v = vibrational quantum number $(0, 1, 2, 3, \ldots)$.

The lowest possible vibrational level is $v = 0$. Putting this value of v in Equation (2.3), we get the lowest possible vibrational energy, called the zero-point vibrational energy.

$$\epsilon_0 = \frac{1}{2}h\nu$$

The next higher vibrational energy ϵ_1 is

$$\epsilon_1 = h\nu\left(1 + \frac{1}{2}\right) = \frac{3}{2}h\nu.$$

Now, if we go on supplying energy to a bond, its vibrational energy will go on increasing and at one time it will reach its highest vibrational energy state and beyond this level the bond will not exist as shown below (the Morse curve; Fig. 2.20).

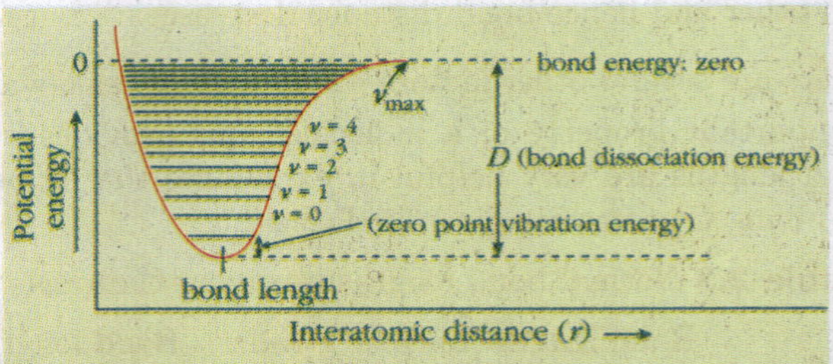

Fig. 2.20. The Morse curve

So, the vertical distance between the lowest vibrational level of a bond and the energy zero on the Morse potential energy curve (Fig. 2.20) may be called the dissociation energy of the bond. However, the homolytic bond dissociation energy is defined as the energy required to break a bond of a molecule (in the gaseous state) into free radicals in the gaseous state, i.e., the energy required for the reaction A—B(g) → A·(g) + B·(g). It is a measure of bond strength. The greater the bond dissociation energy, the greater is the stability of the bond. It is expressed in kcal/mole or kJ/mole. When 432.2 kJ of energy is given to 1 mole of H_2, 2 moles of H atoms are formed. Therefore, the bond dissociation energy of H—H is 432.2 kJ/mole. Again, we know that two moles of H atoms combine together to form 1 mole of H_2 and liberate 432.2 kJ/mole of energy. Hence, the bond dissociation energy is nothing but the energy evolved during its formation.

In methane there are four C—H bonds, 431.2 kJ/mole of energy is required to break one of these bonds. But 1653.4 kJ/mole of energy is needed to break all the bonds. This value indicates that the average C—H bond dissociation energy is 1653.4/4 kJ/mole or 413.3 kJ/mole. The average bond dissociation energy is called bond energy.

Bond energy values are very important to determine the structure and hence to explain the chemical reactions of compounds. The greater the bond dissociation energy, the greater is the stability of the bond and the less is its probability of taking part in chemical reactions.

The bond dissociation energy of N≡N is very high and it is 892 kJ/mole. So, it is expected that nitrogen will be inert in nature and actually, it is so. On the other hand, the bond dissociation energy of F—F is 154.9 kJ/mole. Therefore, fluorine should be highly reactive and we know that fluorine reacts with almost all the elements. In fact, the high bond energy of C—C bond is responsible for the catenation property of carbon (Section 1.2).

Bond dissociation energy and bond energy of some bonds involving C atoms is given in Table 2.4 (see the next page).

If *s* character of the involved A.O.s in the bond formation increases, the bond strength, i.e., bond energy increases. For example, the bond energy of a C—H in methane is 413.3 kJ/mole, whereas that in H—C=C—H is 506.4 kJ/mole. C—H in methane is formed by the overlap of sp^3-*s* orbitals and C—H in ethylene is formed by the sp^2-*s* overlap (Section 2.14).

Normally, the greater the overlap of orbitals to form the bond, the greater is its stability and the higher is its bond dissociation energy.

Table 2.4 Bond dissociation energy and bond energy at 0 K

	Bond	Bond dissociation energy kJ/mole		Bond energy kJ/mole	
1	C—H			1	413.3
2	H_3C—H	2	431.2		
3	C=C—H			3	506.4
4	H_3C—CH_3	4	334.94		
5	C—C			5	347.5
6	C=C			6	598.7
7	C≡C			7	812.3
8	CH_3O—H	8	418.68		
9	H_3C—OH	9	372.7		
10	H_2C=O	10	686.8		
11	R_2C=O	11	728.5		
12	CH_3NH—H	12	439.6		
13	C—N			13	293.09
14	C—F			14	439.6
15	C—Cl			15	330.75
16	H_3C—Cl	16	334.94		
17	C_6H_5—Cl	17	360.06		
18	C—Br			18	276.3
19	H_3C—Br	19	280.5		
20	C—I			20	238.6

In the case of a ball-spring system the more the spring is stretched, the more it resists further extension. A restoring force is exerted by the spring in order to return to its equilibrium length R_e; such is the case with two covalently bonded atoms. If the bond is not stretched too far, the magnitude of the restoring force (F) is proportional to the displacement.

$$F = -fx \tag{2.4}$$

where $x = R - R_e$, and f, the proportionality constant, is called the force constant. The relation shows that as the value of the force constant increases, the restoring force also increases, so the larger the force constant, the stiffer is the bond. The force constant of a bond is calculated by Equation (2.5)

$$\nu = \frac{1}{2\pi}\sqrt{\frac{f}{m_1m_2/m_1+m_2}} \tag{2.5}$$

or, $$f = 4\pi^2\nu^2(m_1m_2/m_1+m_2) \tag{2.6}$$

where ν = vibrational frequency, f = force constant of the bond m_1 and m_2 are the masses of the bonded atoms and m_1m_2/m_1+m_2 is called the reduced mass.

Equation (2.5) shows that

(i) The larger the force constant, the higher is the vibrational frequency.

(ii) The higher the reduced mass, the lower is the vibrational frequency. This tells us that a C—H bond having lower reduced mass than a C—D bond will vibrate at a higher frequency than the latter bond.

On the other hand, Equation (2.6) shows that the higher the reduced mass, the larger is the force constant and hence, the higher is the bond dissociation energy. So, a C—D bond has a higher bond dissociation energy than a C—H bond.

However, a bond may break heterolytically giving rise to two ions, a cation and an anion. The heterolytic bond dissociation energies have also been determined.

Table 2.5 Heterolytic bond dissociation energy in kJ mol^{-1}

1.	CH_3—Cl	949.2	5.	CH_3—I	886.5	9.	CH_3—OH	1145.7
2.	C_2H_5—Cl	798.6	6.	C_2H_5—I	735.9	10.	CH_3CH_2—OH	1011.94
3.	CH_3—Br	915.7	7.	C_6H_5—Cl	915.7	11.	CH_2=CH—CH_2—Cl	865.6
4.	C_2H_5—Br	769.4	8.	C_6H_5Br	878.1	12.	CH_2=CH·CH_2Br	690.0

Problem 2.8. Distinguish between sigma and pi bonds.

Solution:

Sigma bond	Pi bond
1. Formed by head-on overlap of A.O.s.	1. Formed by side-on overlap of p_y-p_y and p_z-p_z (and d) orbitals.
2. Possesses cylindrical charge density around the bond axis.	2. Maximum charge density lies on the cross-sectional plane of the orbitals.
3. Only one σ bond can exist between two particular atoms	3. One or two π bonds can exist between two atoms.
4. Free rotation around a sigma bond can occur.	4. Free rotation about a π bond cannot occur.
5. σ bonds are strong and difficult to break.	5. π bonds are weak and easy to break.
6. It has no node.	6. It has a node.

2.11 Electronegativity of Elements and Group Electronegativity

The maximum probability of finding the electron pair of a covalent bond is the region that lies between the two nuclei of the bonded atoms. When the two atoms are of the same element, the electron pair is equidistant from the two nuclei; thus, H—H bond may be represented as H : H. On the other hand, the electron pair of a covalent bond between two different atoms should be drawn close to the atom which possesses greater ionisation energy and electron affinity values in comparison to the other. For example, in H—F molecule the I.E. and E.A. values of F being higher than those of H, the bonded pair of electrons remain closer to F than H, and may be represented as H :F. Obviously, this develops a partial negative charge on the atom which holds the electrons nearer to it and a partial positive charge on the other atom. Hence, the covalent bond gets partial ionic character and is said to be polarised. The polarity so developed is indicated by the symbol $H \longleftrightarrow F$ or $\overset{\delta+}{H}—\overset{\delta-}{F}$ or $\overset{\delta+}{H} \rightleftarrows \overset{\delta-}{F}$. The property by virtue of which a covalently bonded atom attracts bonded electrons towards itself is known as electronegativity (E.N.) of the element. The E.N. values of: F — 4 eV, O — 3.5 eV, N and Cl — 3 eV, C — 2.5 eV and H — 2.2 eV. Therefore, the decreasing order of attraction on bonded electrons is F, O, N = Cl, Br, I = C, H, Mg, Na.

This property of electronegativity is responsible for the development of two poles within a molecule consisting of atoms of different E.N. and such a molecule is said to be a dipole. As there is a difference in electrical charge separated by a distance (bond distance), the molecule has an electrical dipole moment. Similarly, the property by virtue of which a bonded group (Me, Et, CN, etc.) can attract the bonded *e*s towards itself may be called group electronegativity. Obviously, it is a relative property.

Kharasch was one of the earliest scientists who reported a series of relative group electronegativities. His sequence of group electronegativities of some groups is as follows:

$$\underset{\text{i}}{p\text{-MeO—}C_6H_4\text{—}} > \underset{\text{ii}}{p\text{-Me—}C_6H_4\text{—}} > \text{Ph—} > \underset{\text{iii}}{p\text{-Cl—}C_6H_4\text{—}} > \text{Me—} > \underset{\text{iv}}{\text{Et—}}$$

$$> \underset{\text{v}}{\text{Ph—}CH_2\text{—}} > \underset{\text{vi}}{Me_3C\text{—}} > \underset{\text{vii}}{Ph_3C\text{—}}$$

He gave the above order by measuring the relative amounts of R′H and RH produced by the cleavage of R—Hg and Hg—R′ bonds of R—Hg—R′ in the aqueous hydrochloric acid solution.

$$\text{R—Hg—R}' \rightleftharpoons \text{R—Hg}^{\oplus} + \text{R}'^{-} \xrightarrow{\text{HCl}} \text{RHgCl} + \text{R}'\text{H}$$

$$\text{R—Hg—R}' \rightleftharpoons \text{Hg}^{\oplus}\text{—R}' + \text{R}^{-} \xrightarrow{\text{HCl}} \text{RH} + \text{R}'\text{HgCl}$$

The same series has also been obtained by Brown by arranging groups in the order of C—Cl bond moments in the chloro derivatives. He extended the series and got the following order of group electronegativities:

$$\text{PhC}\equiv\text{C—} > \text{PhCH}=\text{CH—} > \text{MeCH}=\text{CH—} > CH_2=\text{CH—}$$

$$> p\text{-MeO—}C_6H_4\text{—} > p\text{-Me—}C_6H_4\text{—} > \text{Ph—} > p\text{-Cl—}C_6H_4\text{—}$$

$$> \text{Me—} > \text{Et—} > i\text{-Pr} > \text{cyclohexyl} > \text{alkyl} > \text{benzyl} > Me_3C\text{—} > Ph_3C\text{—}$$

Thus from a chemical as well as a physical method the same order of relative group electronegativities has been obtained.

However, the group electronegativity values in eV of a few groups are given below:

—Me = 2.3 —CN = 3.3 —CCl_3 = 3.0 —Ph = 3.0

—CF_3 = 3.35 —NO_2 = 3.4 —COOH = 2.85

The relative group electronegativities have profound influence on the bond moment and hence on the dipole moment of a compound. Its effect being a permanent effect, physical and chemical properties of a compound depend on the group electronegativities of the constituent groups.

However, electronegativity of an element increases with the increasing *s* character of the hybridised state of the atom. A cationic atom has greater electronegativity than its neutral state. An anionic atom has less electronegativity than its neutral state.

2.12 Bond Moment and Group Moment

When two atoms of different electronegativities, A and B, are bonded to each other by a covalent bond, two electrical poles, positive and negative, develop at the two ends of the bond; the more electronegative atom becomes the negative pole while the less electronegative atom becomes the positive pole and the bond is said to be polar. It is needless to mention that the amounts of charges on two poles are equal in magnitude but opposite in sign. Since the bond is dipolar, it has dipole moment which is called bond moment; the product of the negative charge (e of the negative pole which is equal to the positive charge of the positive pole) and the distance between the two poles (d in Angstrom unit, 10^{-8} cm) gives the bond moment. The charge of an electron being of the order 10^{-10} e.s.u., the dipole moment (μ) of the bond is of the order $10^{-10} \times 10^{-8}$ or 10^{-18} e.s.u. cm which is called a Debye (D), the unit of dipole moment.

A few bond moments are given in Table 2.6.

Table 2.6 Some bond moments

Bond	Bond moment μ in D	Bond	Bond moment μ in D	Bond	Bond moment μ in D
$\overset{\delta+}{A}—\overset{\delta-}{B}$		$\overset{\delta+}{A}—\overset{\delta-}{B}$		$\overset{\delta+}{A}—\overset{\delta-}{B}$	
H—C(sp^3)	0.31	C—Br	1.48	C—F	1.51
H—C(sp^2)	0.63	C—I	1.29	C—Cl	1.56
H—C(sp)	1.05	C—N	0.40	C(sp^3)—C(sp^2)	0.68
H—N	1.31	C=N	0.90	C(sp^3)—C(sp)	1.48
H—O	1.53	C≡N	3.60	C(sp^2)—C(sp)	1.15
H—F	1.98	C—O	0.86	N^+—O^-	3.20
H—Cl	1.03	C=O	2.40	N=O	2.00
H—Br	0.78	C—S	2.95		
H—I	0.38	C=S	2.80		
H—S	0.68	N—O	0.30		

Table 2.6 clearly shows that the dipole moment of a bond depends on the hybridised states of the bonded atom. An example is given below:

$$\mu_{\text{H—C}(sp^3)} = 0.31 \text{ D whereas } \mu_{\text{H—C}(sp^2)} = 0.63 \text{ D}.$$

However, the dipole moment of a diatomic molecule is the bond moment of the bond between the two atoms; while that of a polyatomic molecule is the resultant of the bond moments of the bonds in the molecule. Since dipole moments are considered as vectors, the resultant vector can be calculated by using the formula of vectorial addition for two bond moments as given below:

$$\mu_r = \sqrt{\mu_1^2 + \mu_2^2 + 2\mu_1\mu_2 \cos\alpha} \tag{2.7}$$

where μ_1 = moment of one bond, μ_2 = moment of the other bond, μ_r = the resultant moment and α = angle between the two vectors.

If there are four bonds in a molecule, the resultant of the resultant vectors of two pairs of bond moments is to be calculated which is a time-consuming process. To avoid this the moment of individual group has been determined by summing up the bond moments of the bonds present in the group which is called the group moment. The organic chemists being very much interested in hydrocarbon groups, functional groups, etc., to them group moments have greater significance than the bond moments have, particularly in the cases of aromatic compounds. However, in Table

2.7 group moments of a few groups have been listed. It is important to note that the groups have different moments in different situations but for general purposes we have considered two sets of group moments, one for alkyl C-bond groups and the other for aryl C-bond groups.

Table 2.7 Group moments of some groups

Group	μ in D		Group	μ in D	
	Alkyl	Aryl		Alkyl	Aryl
—OMe	1.22	1.35	—COOH	1.68	1.73
—OH	1.53	1.40	—CHO	2.73	2.76
$—NH_2$	1.20	1.48	$—COCH_3$	2.78	4.21
$—NO_2$	3.68	4.21	—C≡N	4.00	4.39
—Br	2.01	1.73	—Cl	2.05	1.70

Perhaps participation of the above groups in resonance involving the aryl group makes their group moments different from the values when they are present on an alkyl or cycloalkyl chain.

However, group moments can be used to calculate molecular moments. Let us calculate the dipole moment of CH_3OH molecule in which C—O—H bond angle is taken as 108°. The O—H bond moment is 1.53 D, C—O bond moment is 0.86 D and C—H bond moment is 0.3 D; the group moment of CH_3 group is 0.3 D. Therefore, O—CH_3 bond moment is the sum of 0.3 D and 0.86 D, i.e., 1.16 D. So from the figure given below we can write:

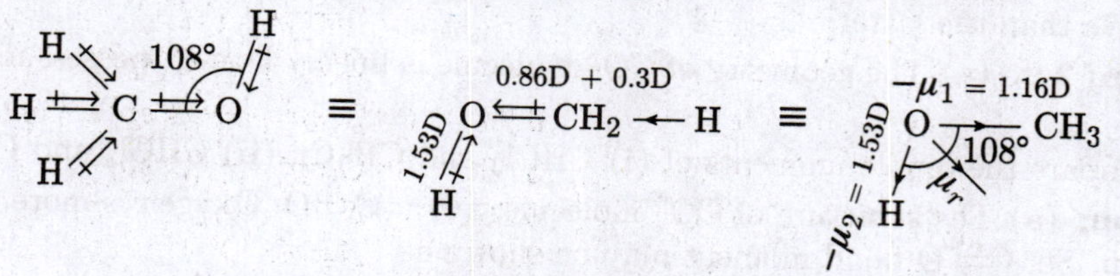

$$\mu_{\text{resultant}} = \sqrt{(-\mu_1)^2 + (-\mu_3)^2 + 2(-\mu_1)(-\mu_2)\cos 108°}$$
$$= \sqrt{(1.16\text{ D})^2 + (1.53\text{ D})^2 + (2 \times 1.16\text{ D} \times 1.53\text{ D})\cos 108°}$$
$$= 1.67\text{ D.}$$

Bond moments and group moments are thus used to determine the dipole moment of a molecule. The molecular dipole in its turn helps us to determine the structure (cis-trans) of the molecule, its conformation, etc. However, to calculate molecular moment non-bonded orbital moments are also to be considered (See Sec. 4.8).

Bond dipoles are intimately associated with physical and chemical properties of a molecule and these determine the kind of reaction that can take place at a bond. Bond polarity develops polarity in a molecule. The dipole moment of H—F molecule is 1.75 D. A molecule may have zero dipole moment, even if its bonds are polar. Thus in carbon tetrachloride a C—Cl bond has polarity but the molecule as a whole has no dipole moment.

Table 2.8 Dipole moments of some simple molecules

Formula	Dipole moment (D)	Formula	Dipole moment (D)	Formula	Dipole moment (D)
H—F	1.75	CH_4	0	CCl_4	0
H—Cl	1.08	CH_3Cl	1.87	NF_3	0.23
H—Br	0.80	CH_2Cl_2	1.55	NH_3	1.47
H—I	0.42	$CHCl_3$	1.12	H_2O	1.85

2.13 Bond Polarity and Polarisability of a Bond

Polarity of a bond is a state of having two opposite electrical poles in the bond. The difference between electronegativities of the bonded atoms or groups is responsible for the development of polarity. It is a permanent property of the bond and therefore it determines the physical and chemical properties of the molecule holding the bond. The bond dipole moment is a measure of the bond polarity.

The polarisability of a bond is the ease of polarisation of the bond, i.e., the ease with which the charge cloud of the bond orbital gets distored by the external electrical field effect or any ion in the solvent medium. The polarisability of a bond depends on the volume of the bonded atoms or groups and on the strength of the external field. The larger the volume of the bonded atoms, the greater is the polarisability of the bond. Similarly, the greater the field strength, the greater is the polarisability. For example, the order of polarisability of C—X bond is:

$$C—I > C—Br > C—Cl$$

though the bond-polarity order is exactly the reverse.

Polarisability of a bond is a temporary property of a molecule. So it has nothing to do with most of the physical properties but it determines chemical properties of a molecule. For example, MeI is more reactive than MeCl though the C—Cl bond is more polar than the C—I bond. The greater polarisability of the C—I bond than that of the C—Cl bond makes the former more reactive than the latter.

Problem 2.9. (a) The geometry of CO_2 molecule is linear. Does it possess molecular dipole moment?

(b) Compare the dipole moments of (i) $CHCl_3$ and CH_3Cl; (ii) $CHCl_3$ and CH_2Cl_2.

Solution: (a) The structure of CO_2 molecule is O=C=O. Oxygen is more electronegative than carbon. So, C=O bond moment may be shown as

$$O \overset{\leftarrow\!\!+}{=} C \overset{+\!\!\rightarrow}{=} O.$$

The individual C=O bond moment being equal and oppositely directed will cancel each other. Thus, the molecular dipole moment will be zero Debye.

(b) (i) μ_{CH_4} and μ_{CCl_4} are zero. Therefore, $\mu_{CH_3} = \mu_{C—H}$ (in magnitude) and $\mu_{CCl_3} = \mu_{C—Cl}$ (in magnitude). But actually μ_{CCl_3} is less than $\mu_{C—Cl}$. This is because of the induced moment due to one C—Cl bond on two other decreases $\mu_{C—Cl}$ of each C—Cl bond and hence μ_{CCl_3} is less than 1.56 D, $\mu_{C—Cl}$.

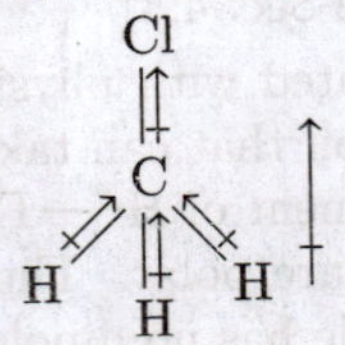

$$\mu_{C—Cl} + \mu_{CH_3} = \mu_{CH_3Cl} \text{ or } \mu_{C—Cl} + \mu_{C—H} = \mu_{CH_3Cl},$$

$$\mu_{CH_3Cl} = 1.56\text{ D} + 0.31\text{ D} = 1.87\text{ D}.$$

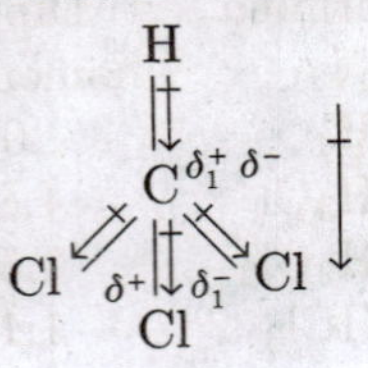

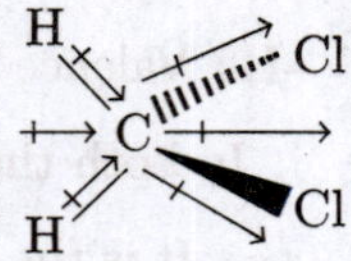

$\delta_1^- > \delta^+$ so, $\mu_{CCl_3} < \mu_{C—Cl}$; here δ_1^- and δ_1^+ are charges for the C—Cl bond polarity but δ^- and δ^+ are the charges owing to the induction effect of other C—Cl bonds.

$\therefore$ $\mu_{CHCl_3} = \mu_{CCl_3} + \mu_{C—H}$ = a value less than $\mu_{C—Cl} + 0.31$ D = (0.71 + 0.31) D = 1.02 D. ($\mu_{CCl_3} = 0.71$ D).

So, $\mu_{CHCl_3} < \mu_{CH_3Cl}$.

(ii) $\mu_{CH_2Cl_2} = \mu_{CH_2} + \mu_{CCl_2}$

$\mu_{CH_2} > \mu_{C—H}$ and $\mu_{CCl_2} > \mu_{CCl_3}$. [$\because$ induced moment on CCl_2 < on CCl_3]

$\therefore$ $\mu_{CH_2Cl_2} > \mu_{CHCl_3}$.

Problem 2.10. The dipole moment of water molecule is 1.85 D. What is its shape?

Solution: In a molecule of water 2 H atoms are linked to one O atom. If the two O—H bonds are at 180° angle, then the molecular dipole moment of H_2O would have been zero. The molecular dipole moment value of water indicates that the molecule must be of bent shape.

2.14 Bond Angle

A polyatomic ion or molecule generally contains at least one atom with two or more covalent bonds and that atom in the chemical entity is said to be the central atom. For example, nitrogen is the central atom in NH_3 molecule and in $NH_4^{\oplus}$ ion. The angle between any two bond axes of a central atom in a molecule is called its bond angle; the bonds may be single, double or triple. If there are more than one polycovalent atoms in a molecule, then for each of the atoms bond angle may arise. In berillium chloride, Cl—Be—Cl bond angle is 180°; H—C=O bond angle in H_2C=O is 121.8°, its H—C—H bond angle is 116.5°.

The values of different bond angles in a molecule depend primarily on the shape of the molecule, i.e., on the orientation of the constituent atoms of the molecule in space. The following three factors have strong influence on bond angles :

(i) Electron-electron repulsion involving bond-pair and lone-pair electrons,

(ii) Electronegativity of the bonded atoms,

(iii) Steric effect.

Still if we survey the bond angles in a large number of covalent molecules, we shall see that they have more or less some constant values like, 109.5°, 120°, 180°, 90°, etc. In the next section we shall discuss bond angles of some compounds along with their molecular geometry.

2.15 Shape of the Molecules and Directional Nature of the Covalent Bonds

It is found that covalent molecules and ions have characteristic shapes, e.g., $BeCl_2$, BCl_3 and $NH_4^{\oplus}$ ion have linear, planar trigonal and tetrahedral shapes respectively. Pyramidal, square planar, bi-pyramidal, octahedral and angular shapes are also very common amongst covalent molecules. Owing to the directional nature of the covalent bonds the molecules have special shapes.

As most of the covalent bonds are formed by the overlap of directional A.O.s (except *s* orbitals all other orbitals, pure and hybrid, are directional), they must also have directional character. The directional nature of covalent bonds and the shape of the molecules are explained by:

I. Hybridisation of orbitals and

II. Valence Shell Electron Pair Repulsion (VSEPR) theory.

In both the cases the guiding principles are:

(a) It is the outer electronic arrangement of the central atom in a molecule that determines its shape.

(b) Owing to the electron-electron repulsion, the bonded and lone-pair electrons want to take up positions in space as far apart from each other as possible.

The shapes of the molecules will be explained here in the light of hybridisation of orbitals.

$BeCl_2$ molecule is linear in shape and Cl—Be—Cl bond angle is 180°; these two are the experimental facts. How this is explained by hybridisation of orbitals of Be atom is given below:

The electron configuration of Be is $1s^2 2s^2$. Lewis theory tells us that as there is no unpaired electron in Be atom, it should not form covalent bond and it should be inert in nature. But the extended V.B. theory, which incorporates the idea of hybridisation of A.O.s, explains the shape of $BeCl_2$ molecule.

An excited Be atom possesses two unpaired electrons; one in $2s$ orbital and the other in $2p$ orbital. These two orbitals hybridise to form two equivalent *sp* A.O.s making 180° angle between their orbital axes (see Problem 2.4i). When each of the *sp* orbitals overlaps with one *p* orbital of Cl atom, a linear molecule results. Obviously, Cl—Be—Cl bond angle will be 180°. The energy required to excite the Be atom is certainly compensated by the excess of energy released in the formation of two Be—Cl single bonds (Fig. 2.21).

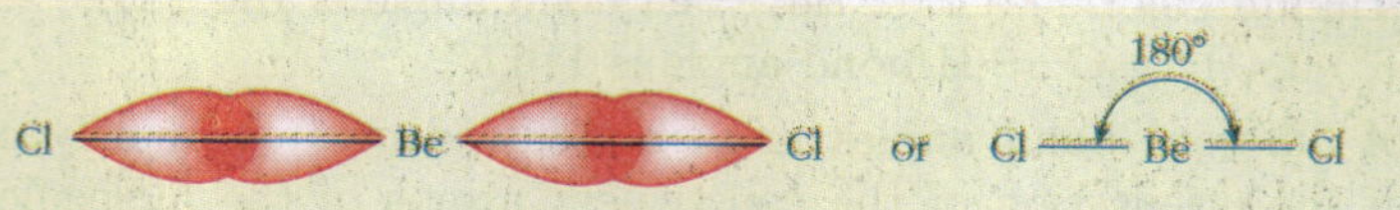

Fig. 2.21. Shape of the $BeCl_2$ molecule

Similarly, the planar trigonal shape of BCl_3 molecule with Cl—B—Cl bond angle 120° is explained as follows:

B atom in excited state forms three equivalent sp^2-hybridised orbitals having 120° angle between any two such orbital axes. When the Cl atoms with their *p* orbitals overlap with the three sp^2 hybrid A.O.s, a planar trigonal molecule, BCl_3, is formed. Naturally, the Cl—B—Cl bond angles are 120° (Fig. 2.22).

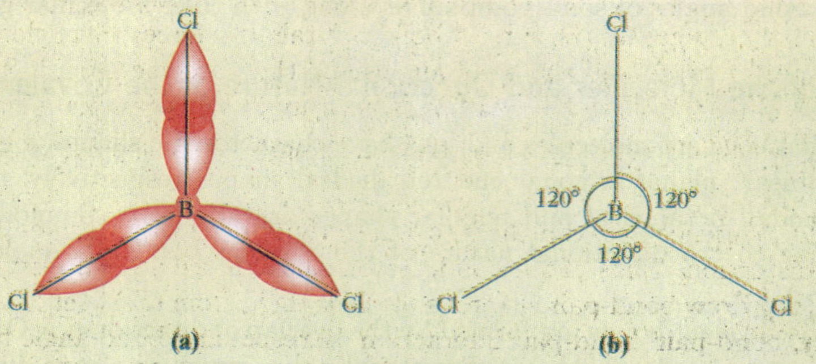

Fig. 2.22. Shape of the BCl_3 molecule

Again, the tetrahedral shape of $NH_4^{\oplus}$ ion may be explained by considering hybrid A.O.s of N atom in its excited state; nitrogen atom forms four equivalent sp^3 hybrid A.O.s—one of which contains a lone pair of electrons and other three contain one electron each (see Problem 2.5). These three sp^3 A.O.s form three covalent bonds with H atoms by sp^3-s overlap. The lone pair of electrons of the fourth sp^3 orbital coordinates with a proton ($H^{\oplus}$) to produce $NH_4^{\oplus}$ ion. sp^3 hybrid A.O.s being tetrahedrally arranged in space, the $NH_4^{\oplus}$ ion is tetrahedral in shape with H—N—H bond angle 109.5° roughly (Fig. 2.23).

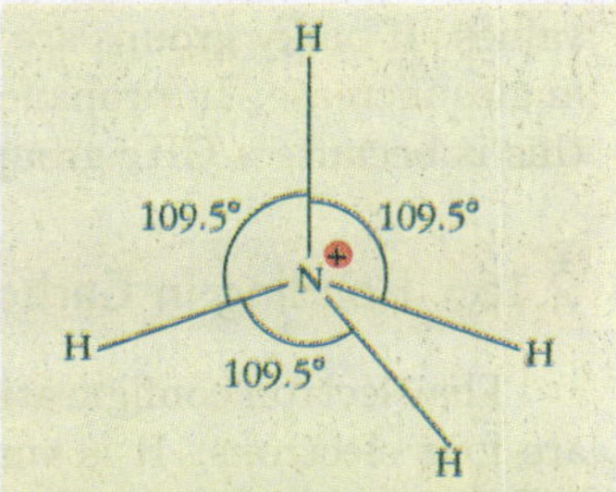

Fig. 2.23. Shape of the ammonium ion

Thus in the light of hybridisation of orbitals the shapes of molecules and bond angles can be explained. Shapes of molecules and corresponding bond angles at the central atom help us suppose the nature of hybridisation of the central atom. A list of molecular geometry of some molecules, bond angles and expected state of hybridisation of the central atom are given in Table 2.9.

Table 2.9 Some hybrid A.O.s, molecular geometry and bond angle

Molecular geometry	Bond angle	Example	Hybrid orbital type
Tetrahedral	109°28′	Methane (CH_4)	sp^3
Planar trigonal	120°	Boron trichloride (BCl_3)	sp^2
Linear	180°	Berillium hydride (BeH_2)	sp
Octahedral	90°	Sulphur hexafluoride (SF_6)	d^2sp^3

The ideal bond angles as expected from hybrid types are seldom found in compounds, because some of the hybrid A.O.s are occupied by lone-pair electrons. The lone-pair electrons require greater space than bonding pairs and force the bonds closer together. In fact, electron-electron repulsion decreases in the order:

lone-pair – lone-pair > lone-pair – bonded pair > bonded pair – bonded pair.

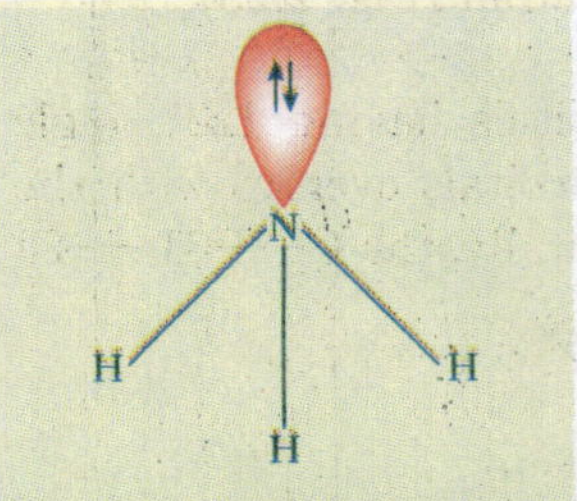

Fig. 2.24. Shape of NH_3 molecule

For example, in an ammonia molecule H—N—H bond angle is 107.3°. How can this be explained? In NH_3, the central N atom has $4sp^3$ hybrid A.O.s; one of which contains a lone-pair of electrons and the rest one each (see Problem 2.5i). Each of the three sp^3 hybrid A.O.s overlaps with 1s A.O. of hydrogen to form NH_3 molecule. Hence we should expect a tetrahedral shape of NH_3 molecule with H—N—H bond angle 109°28′. But the sp^3 A.O. containing a lone-pair of electrons repulses the three σ bond orbitals and forces them closer together. This decreases the bond angles and they become 107.3°. The orientation of the three N—H bonds appears to be pyramidal (Fig. 2.24).

Another factor which distorts the expected bond angles is electronegativity of the bonded atoms.

The greater the difference between the electronegativitives of the bonded atoms, the greater is the bond angle distortion. Thus, if the bonded atoms have much greater electronegativity than the central atom, they draw bond-pair electrons towards them from the valence shell of the central atom. Accordingly, bond-pair–bond-pair interaction decreases and bond angle too decreases. For example, in NF_3, F—N—F bond angle is 102.1°, whereas H—N—H bond angle in NH_3 is 107.3°.

Steric factor is also important in explaining the deviation of bond angle values from the ideal values. If bulky groups are bonded to the central atom, they require greater space and thus bond angles increase. In propane ($CH_3{\cdot}CH_2{\cdot}CH_3$), C—C—C bond angle is 112° instead of 109.5°, this is because a CH_3 group requires more space than an H atom.

2.16 Bonding in Carbon Compounds

The electron configuration of C atom is $1s^2\ 2s^2\ 2p_x^1\ 2p_y^1\ 2p_z^0$. Therefore in its outer shell, there are four electrons. It is very difficult to remove these four electrons as ionisation energy values will be very high. On the other hand, C atom cannot take in four electrons in order to attain next inert gas electronic configuration; this is because of very high electron-electron repulsion. Hence normally, C atom will not form ionic compounds.

So, we shall expect covalent bonding in carbon compounds. Again, as there are only two unpaired electrons ($2p_x^1$ and $2p_y^1$) in the outer shell of a C atom, it will form only two covalent bonds (according to Lewis theory). But in most of the carbon compounds, it shows four covalent bonds. In the excited state of a C atom one of the two $2s$ electrons is promoted to $2p_z$ orbital and thus it can provide four unpaired electrons for bonding. In this case, of course, four orbitals will not be equivalent and four bond orbitals formed by the overlap of orbitals from other atoms will not also be equivalent. The tetracovalency of C atom and the equivalency of bond orbitals can be explained by hybridisation. In this connection it is noteworthy that the constancy of four bonds for carbon was first predicted by Couper in Scotland and Kekulé in Germany in 1859, i.e., long before the advent of quantum mechanics. However, the nature of bonds in organic molecules and their disposition in space, i.e., their geometries are discussed below.

2.16.1 Carbon-Hydrogen Bonds

C—H bond is a single bond, hence it is a sigma bond. Let us discuss this bond in methane.

Methane is a tetrahedral molecule having four H atoms at the four corners of the tetrahedron and the centre of the tetrahedron is occupied by the nucleus of the C atom (Fig. 2.25).

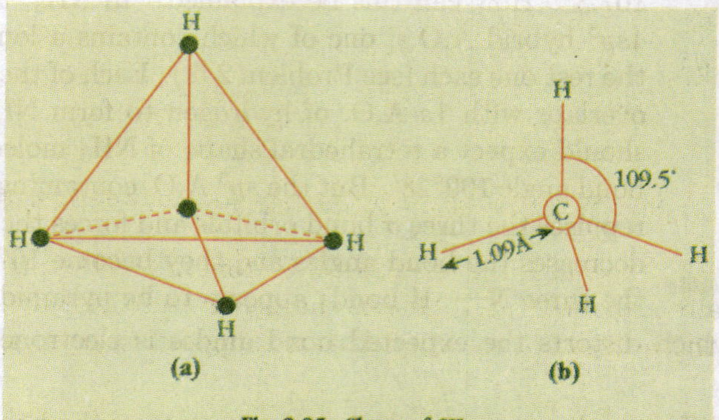

Fig. 2.25. Shape of CH_4

sp^3 hybridisation of C atom in methane explains its very shape. The end-on overlap of four sp^3 hybrid A.O.s with four $1s$ orbitals of H atoms gives rise to the four equivalent C—H sigma bonds. Therefore, C—H sigma bond length and bond dissociation energy will be the same for each of them. C—H bond length is 0.109 nm and bond dissociation energy is 431.2 kJ/mole.

A C—H bond may also be formed by sp^2-s and sp-s overlap. As we know that an increase in s character of an overlapping orbital decreases the bond distance and increases the bond dissociation energy, the strength of the C—H single bond will be of the order:

$$\xrightarrow{\text{Decreasing energy and strength}}$$
$$sp\text{-}s \quad sp^2\text{-}s \quad sp^3\text{-}s$$
$$\xrightarrow[\text{Increasing bond length}]{}$$

C—H bond length in $H_2C{=}CH_2$ is 0.108 nm and the bond energy of C=C—H bond is 506.4 kJ/mole, whereas that of H_3C—H is 413.3 kJ/mole. This indicates the greater stability of the bond formed by sp^2-s overlap than that formed by sp^3-s overlap.

The electronegativity of sp^3 C is 2.5 eV and that of H is 2.2 eV. Therefore, electronegativity difference between them is very small. This makes the C—H bond almost non-polar. The non-polarity of the C—H bond in alkanes and cycloalkanes and its sigma bond character make the bond very strong; it does not react easily and drastic conditions are required for the same.

However, HC≡C—H bond is polar and its length is 0.106 nm, because C atom is in sp hybridised state. An sp-hybridised C atom has 3.0 eV electronegativity. So it is reactive.

Note

It is very difficult to represent a tetrahedron in a drawing. The general practice is to form models with ball and stick; where balls are atoms and sticks are bonds (Fig. 2.26a). Then projection formula for the model is drawn on the plane of the paper; this shows two dimensions only, the other dimension is to be imagined. Let us project the ball and stick model of methane by the wedge projection as shown below (Fig. 2.26).

(i) In a Wedge projection formula, the tetrahedral C atom lies on the plane of the paper; dotted lines represent bonds below it and the wedges (◁) indicate those bonds above the plane of the paper (2.26c).

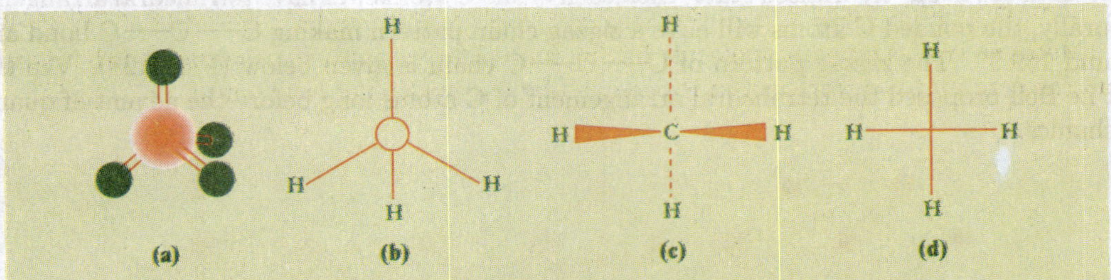

Fig. 2.26. Ball and stick model and Wedge projection formula for CH_4

(ii) In a Fisher projection formula, the tetrahedral C atom lies on the plane of the paper; bonds drawn horizontally are above the plane of the paper and those drawn vertical are below it (Fig. 2.26d); the position of the C nucleus is supposed to be at the crossing of the two lines.

C—C single bonds are formed by the head-on overlap of two hybrid A.O.s, one from each C atom, and thus form a strong bond between them. The hybrid A.O.s may be either sp^3 or sp^2 or sp. In ethane (H_3C—CH_3), C—C σ bond is formed by the overlap of two sp^3 A.O.s, one from each of the C atoms. The C—C bond length is 0.154 nm. Each of the C atoms in ethane is in the tetrahedral state; two such tetrahedra are joined at their apexes (Fig. 2.27). C—H bonds are formed by sp^3-s overlap. Therefore, there are seven σ bonds—six C—H and one C—C.

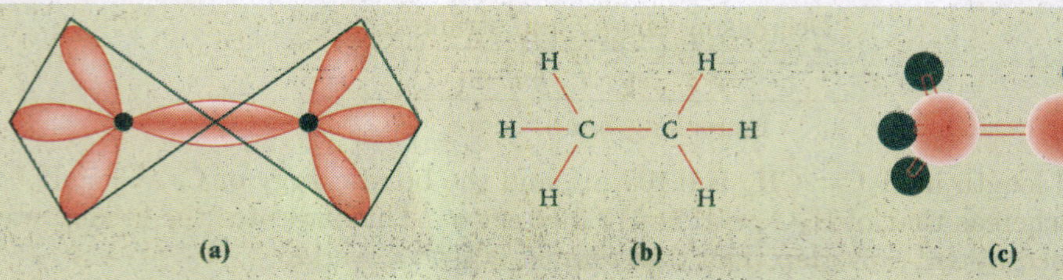

Fig. 2.27. Ethane molecule

If a C—C single bond is formed by sp^2-sp^2 overlap, the bond length decreases and the bond dissociation energy increases. The C—C bond length is, in this case, 0.148 nm. Such a bond between two sp-hybridised carbons is still shorter and its value is 0.138 nm. In fact, C—C single bond length and bond dissociation energy are of the order (see Table 2.3 and Table 2.4).

$$\xrightarrow{\text{Decreasing bond length}}$$
$$sp^3\text{-}sp^3,\ sp^2\text{-}sp^2,\ sp\text{-}sp$$
$$\xrightarrow[\text{Increasing bond dissociation energy}]{}$$

C—C bonds being sigma, they are strong and do not react easily; drastic conditions are required to perform reactions at a C—C single bond. Bond dissociation energy values also support this view. If other effects are disregarded, C—C bond will be non-polar as the bonded atoms are of the same electronegativity. If three or more C atoms are joined by single bonds and if all of them are in sp^3-hybridised state, then each of the C atoms will have tetrahedral arrangement; naturally, the bonded C atoms will have a zigzag chain pattern making C—C—C bond angles around 109.5°. The zigzag pattern of C—C—C chain is given below (Fig. 2.28). Van't Hoff and Le Bell proposed the tetrahedral arrangement of C atoms long before the advent of quantum mechanics.

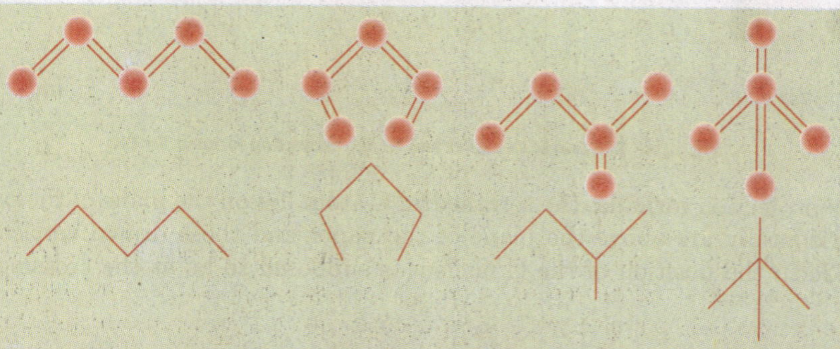

Fig. 2.28. Zigzag pattern of open carbon chain

If we ignore the three-dimensional description of the molecular structures, these can be represented in a straight way (Fig. 2.29). This sort of representation was first used by Kekulé and is known as graphic representation.

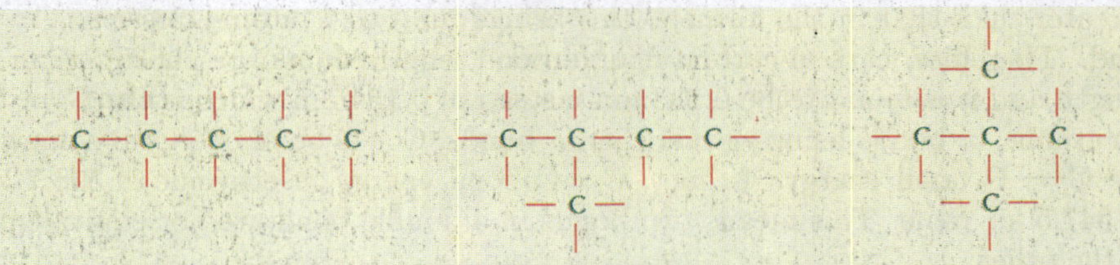

Fig. 2.29. Graphic representation of open carbon chains

2.16.3 Carbon-Carbon Double Bond

According to the V.B. theory, two sp^2-hybridised C atoms when come to each other at an optimum distance, two sp^2-hybridised A.O.s, one from each C atom, overlap head-to-head to form a sigma bond orbital; the remaining two p_z orbitals, one of each C atom, overlap side-to-side to produce a π bond orbital. Thus a C=C bond consists of a σ bond and a π bond.

M.O. description of the C=C bond also starts with sp^2 hybridisation of C atoms; two sp^2 A.O.s, one from each C atom, overlap end-to-end to form one σ M.O. and the two p_z A.O.s, one from each C atoms, overlap sideways to produce two M.O.s, π and π^*. The two electrons occupy π M.O., because it has lower energy than π^* M.O. Thus according to the M.O. theory, a C=C possesses one σ bond and one π bond.

If we consider the case of ethylene ($CH_2{=}CH_2$), two of the three sp^2 A.O.s of each C atom form two C—H sigma bonds. Thus the ethylene molecule contains 5 σ bonds; four of them are C—H sigma bonds and one of them is C—C sigma bond. Over and above these 5 σ bonds, there is a π bond. As each of the C atoms is in the sp^2-hybridised state, each CH_2 part should have planar trigonal shape, i.e., all the six atoms should lie in a single plane. In fact, an ethylene molecule is planar. The coplanarity of the atoms can also be explained by the fact that the maximum sidewise overlap of two p orbitals can only occur when all of the bonded atoms are planar (Fig. 2.30). This figure does not show actual overlap; See Fig. 2.14a for π bonding.

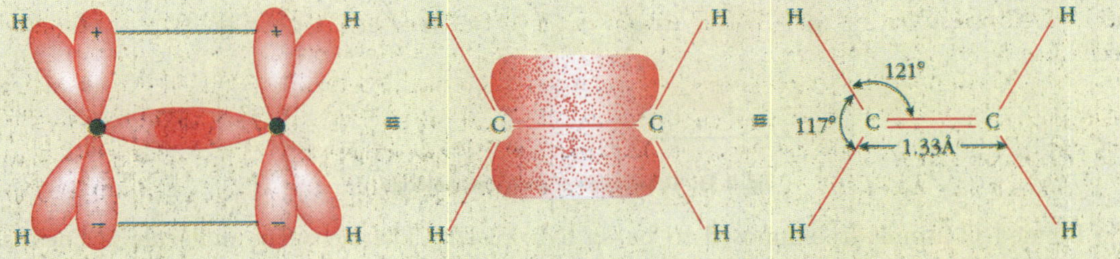

Fig. 2.30. Description of ethylene molecule

Each of the C atoms in ethylene being in a planar trigonal state, the H—C—H and H—C—C bond angles are expected to be one and the same, and the value should be 120°. But as the π bond electron pair exercises strong repulsions towards its neighbouring bonded electron pairs, the two C—H single bonds on each atom come closer to each other than expected; thus H—C—H bond angle decreases and H—C—C bond angle increases—the respective bond angle values are 116.7° and 121.6°.

In a single bond an electron pair holds the two nuclei by coulombic force of attraction. But in a double bond two electron pairs hold the two nuclei together; hence the force of attraction in the latter is greater than that in the former—this brings the two C atoms closer in C═C than in C—C bond. Therefore, carbon-carbon double bond length should be shorter than C—C single bond length and it is found to be 0.133 nm instead of 0.154 nm which is the C—C single bond length in ethane. π bonds being weaker than σ bonds, C═C bond energy is somewhat less than twice the C—C bond energy. In fact, C═C bond energy in ethylene is 598.7 kJ/mole instead of 2×347.5 kJ/mole. This means two bonds in a double bond are not equivalent; one is stronger than the other.

M.O. theory tells us that π bond electron cloud is distributed above and below the plane of the molecule. This means a region of negative charge is effectively waiting there and this invites positively charged species to react. Oxidising agents accept electrons and hence they react with a carbon-carbon π bond but not with a C—C sigma bond.

As a π bond is a sheet of negative charge, a cation attacks first and then attack by an anion occurs.

2.16.4 Carbon-Carbon Triple Bonds

A carbon-carbon triple bond consists of a C—C σ bond and two π bonds. The *sp* hybridisation of two bonded C atoms explains this. Two *sp* hybrid A.O.s, one from each C atom, ovelap to form one C—C σ bond. Each of the two C atoms possesses one $2p_y$ and one $2p_z$ orbitals. By the p_y-p_y overlap and p_z-p_z overlap two π bonds are formed; they are mutually perpendicular to each other and also perpendicular to the C—C σ bond.

Acetylene molecule, HC≡C—H, possesses three σ bonds: one C—C σ bond and two C—H σ bonds (formed by *sp*-*s* overlap). Lateral, i.e., side-on overlap of p_y and p_z orbitals of one C atom with p_y and p_z orbitals of the other C atom respectively produce two π bonds (Fig. 2.31). Here again overlap of *p* A.O.s are not shown properly (See. Fig. 2.14a).

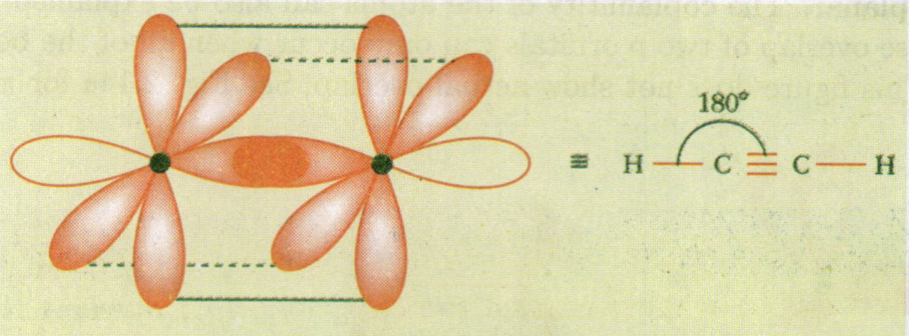

Fig. 2.31. Description of acetylene molecule

So bonded C atoms are supposed to be *sp* hybridised. The acetylene molecule is linear in shape and H—C—C bond angle is 180°. As three pairs of electrons are present between the two C nuclei, the bond distance is still shorter than the double bond distance and it is found to be 0.120 nm. π bonds being weaker than σ bonds, C≡C triple bond energy is less than thrice the C—C single bond energy but it is greater than carbon-carbon single bond energy and carbon-carbon double bond energy; C≡C bond energy is 812.3 kJ/mole.

The charge density of a triple bond is cylindrically symmetrical. The two π bonds together give an electron cloud which has a tubular cross-section at right angle to the σ bond. On the carbon-carbon bond axis π electron density is zero as usual; the σ electron density is very high and the total electron density of π and σ together looks like that of σ 1*s* bond orbital. For the π clouds, C≡C triple bonds are attacked by cations and addition reactions occur at the triple bond.

The H atoms of acetylene are more acidic than H atoms of ethylene and ethane. In fact, when acetylene is heated with sodamide, a strong base, sodium acetylide is formed. The explanation of the acidity is given below:

If one proton is taken out from each of ethane, ethylene and acetylene, then ethide, ethylide and acetylide ions respectively will be produced. Ethide ion contains a lone pair of electrons in an sp^3 A.O., the lone pair of electrons of ethylide ion occupies an sp^2 A.O. and that of acetylide ion remains in an sp A.O. As the order of stability of electrons is: $s > sp > sp^2 > sp^3$, the base character decreases in the order:

$$\text{ethide} > \text{ethylide} > \text{acetylide.}$$

We know that, if a conjugate base is weaker, the acid is stronger. Hence, acidity increases in the order: ethane, ethylene, acetylene.

Acid		Conjugate base
$CH_3—CH_3$	$\xrightarrow{-H}$	$CH_3—\bar{C}H_2$ (Ethide ion)
$CH_2═CH_2$	$\xrightarrow{-H}$	$CH_2═\bar{C}H$ (Ethylide ion)
$CH≡CH$	$\xrightarrow{-H}$	$CH≡C^-$ (Acetylide ion)

This can also be explained as follows:

The electronegativity order of the C atoms in ethane, ethylene, acetylene is:

$$C(sp) > C(sp^2) > C(sp^3).$$

So the bond polarity order is:

$$C(sp)—H > C(sp^2)—H > C(sp^3)—H.$$

The greater the bond polarity, the higher is the acidity. Hence, the order of acidity is as written above.

2.16.5 Carbon-Halogen Bonds

A p or s-p-hybridised orbital of a halogen atom ($ns^2, np_x^2, np_y^2, np_z^1$) overlaps head-to-head with any one of the sp^3 or sp^2 or sp hybrid A.O.s of a C atom and thus a σ bond is formed. As F, Cl and Br are more electronegative than C, C—F, C—Cl, C—Br bonds are essentially polar; polarity decreases in the order of decreasing electronegativity.

$$C—F > C—Cl > C—Br > C—I.$$

Iodine and C are of the same electronegativity but C—I bond has polarity owing to the presence of lone pair of *es* on I.

The difference in electronegativity between carbon and halogen atoms has two effects on reactivity:

(i) Halogen being the negative pole of the dipole, it is easy for the halogen atom to leave as a halide ion.

(ii) Partial positive charge developed on the C atom tends to pull the electrons away from the adjacent C atom, if any, which in its turn tends to give up a proton if it has an H atom.

As the volume of halogen atoms decreases in the order I > Br > Cl > F, bond length decreases in the order:

$$C—I > C—Br > C—Cl > C—F.$$

Naturally, bond energy decreases in the order:

$$C—F > C—Cl > C—Br > C—I.$$

Hence the bond polarisability and reactivity of the C—X bond are in the order:

$$C—I > C—Br > C—Cl > C—F.$$

2.16.6 Carbon-Oxygen Bonds

In organic molecules, oxygen forms two types of bonds with carbon:

I. Carbon-oxygen single bond,

II. Carbon-oxygen double bond.

The electron configuration of oxygen is $1s^2 2s^2 2p_x^2 2p_y^1 2p_z^1$. The relative overlapping powers of A.O.s are found to be: $sp^3 > sp^2 > sp > p > s$. So, oxygen is supposed to utilise hybrid A.O.s to form strongest possible bonds.

I. Carbon-oxygen single bond is formed by the head-on overlap of one sp^3 A.O. of oxygen with one sp^3 or sp^2 or sp A.O. of a C atom. Thus a sigma bond is formed. As there are two lone pairs of electrons on oxygen atom, C—O—C portion of an organic molecule is angular in shape. In dimethyl ether, C—O—C bond angle is 110°, C—O bond length is 0.142 nm and bond energy is 360.06 kJ/mole.

Because of the greater electronegativity of oxygen than that of carbon, the C—O bond is polar and easy to break. On the other hand, if an H atom is attached to oxygen, it draws electron from O—H bond and tends to give up a proton and thus behaves as an acid. The lone pairs of electrons may also be donated by the oxygen atom and therefore it may show base character also.

II. In the formation of carbon-oxygen double bond, C=O, oxygen and carbon use *sp* and sp^2 hybridisations. It consists of one $\sigma(sp^2\text{-}sp)$ bond and one $\pi(p_z\text{-}p_z)$ bond; the two pairs of non-bonded *es* reside in one *sp* and in p_y A.O.s. $H_2C{=}O$ (formaldehyde) molecule is planar trigonal in shape. C=O bond length is 0.122 nm, the bond energy is 686.8 kJ/mole and H—C—O bond angle is about 121.8°.

It is interesting to note that C=O bond energy is slightly greater than twice that of C—O. This may be attributed to (i) the greater electronegativity of O in *sp*-hybridised state than that in sp^3-hybridised state which brings stability to the system and to (ii) the resonance stabilisation of the C=O group as shown below:

$$C=O \longleftrightarrow \overset{\oplus}{C}—\overset{-}{O}.$$

As the electronegativity of O atom is greater than that of C, C=O bond has polar character; the carbon atom is the positive pole and oxygen atom is the negative pole. Since there is a π bond in C=O, addition reactions occur at the double bond. C being positive pole, anions attack that C.

2.16.7 Carbon-Nitrogen Bonds

Nitrogen forms three types of bonds with carbon: (i) C—N single bond, (ii) C=N double bond and (iii) C≡N triple bond.

Nitrogen utilises hybrid A.O.s in forming bonds with carbon (see Problem 2.5). The electron configuration of nitrogen is $1s^2, 2s^2, 2p_x^1, 2p_y^1, 2p_z^1$. One 2*s* and three 2*p* A.O.s may mix up to produce $4sp^3$ A.O.s; one sp^3 hybrid A.O. contains a lone pair of electrons and the rest three A.O.s contain one electron each (sp^3 ↑↓, sp^3 ↑, sp^3 ↑, sp^3 ↑). The 2s A.O. may mix up with two 2*p* A.O.s to give rise to three sp^2 A.O.s; one of the sp^2 A.O.s may contain one electron pair (sp^2 ↑↓, sp^2 ↑, sp^2 ↑, p_z ↑). On the other hand, one of the two 2*s* electrons may get promoted to p_z A.O. and form a pair and each of the sp^2 A.O.s may contain one electron (sp^2 ↑, sp^2 ↑, sp^2 ↑, p_z ↑↓). The 2*s* A.O. and one of the 2*p* A.O.s may also form two *sp* hybrid A.O.s, of which one *sp* hybrid A.O. contains one lone pair of *e*s and the other one electron only (*sp* ↑↓, *sp* ↑, p_y, ↑, p_z ↑).

(i) Carbon-nitrogen single bond is a sigma bond. It is usually formed by the head-to-head overlap of one sp^3 hybrid A.O. of N (containing a single electron) with one hybrid A.O. of C atom. Because of the greater electronegativity of N atom than C atom, the C—N bond is polar. Its average bond length is 0.147 nm and bond energy is 293.09 kJ/mole.

(ii) A carbon-nitrogen double bond consists of a sigma bond and a pi bond. Sigma bond is formed by the head-to-head overlap of two sp^2 A.O.s, one from C atom and the other from N atom. The remaining p_z orbital of C atom overlaps with the p_z orbital of N atom side-to-side. Bond energy and average bond length of C=N bond are 615.45 kJ/mole and 0.130 nm, respectively.

(iii) Carbon-nitrogen triple bond contains one sigma and two pi bonds. *sp*-hybridised A.O.s of carbon and nitrogen overlap head-to-head to form the sigma bond; side-on p_y-p_y and p_z-p_z overlaps develop two π bonds. C≡N bond length is about 0.116 nm and its bond energy is 891.79 kJ/mole.

In C—N multiple bond addition reactions occur. Cations are added to N and anions to C atom.

2.16.8 Carbon-Metal Bonds

Li, Na, K, Mg, etc., metals form sigma bonds with carbon atoms. As the electronegativity of carbon is greater than those of the metals, the carbon-metal bond is highly polar; carbon is the negative pole and metal atom is the positive pole in this bond. In an isolated molecule, this bond is stable, but in a reaction the bond breaks with the liberation of metal cation. Actually, in the resonance hybrid of the C-metal bond, ionic structure has a large contribution; thus lithium methide is a resonance hybrid of the structures I and II as written below and the contribution of structure II is very large. Therefore, chemical reaction occurs very easily at carbon-metal bonds.

$$\underset{\text{I}}{CH_3\text{—}Li} \longleftrightarrow \underset{\text{II}}{\bar{C}H_3Li^{\oplus}}$$

From the foregoing discussion it is clear that a molecule consisting of several atoms has different shapes at different parts. Thus, sp^3 hybrid atomic part is tetrahedral, sp^2 hybrid part is planar trigonal and *sp* hybrid part is linear (see Problem 2.12).

2.17 Conventions for Representing the Gross Structural Formulae for Molecules

If the structure of an organic molecule is known, i.e., the linkages amongst the constituent atoms in the molecule along with their spatial arrangement are known, then its gross structural formula can be represented by Lewis' electron-dot formula. This formula may be converted into dash and dot formula; where a dash represents a covalent bond and a pair of dots represent a lone pair of electrons. When lone pairs of electrons are not shown, the gross structural formula is reduced to a graphic formula. The dash and dot formula and graphic formula may further be written by omitting either all the dashes or most of them and such representations are called condensed formulae. Thus, the gross structural formula for methyl alcohol and methyl cyanide are written below in various ways as stated above :

	Methyl alcohol	Methyl cyanide
Molecular formula	CH_4O	C_2H_3N
Electron-dot formula	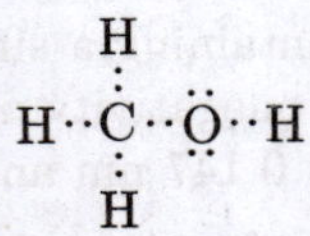	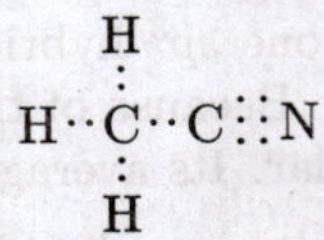
Dash-dot formula	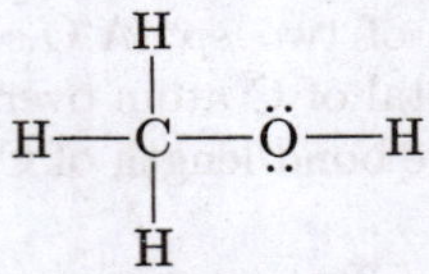	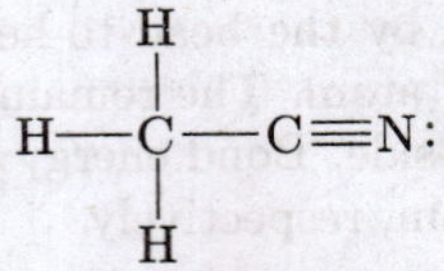
Graphic formula	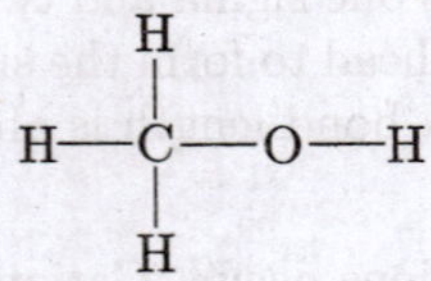	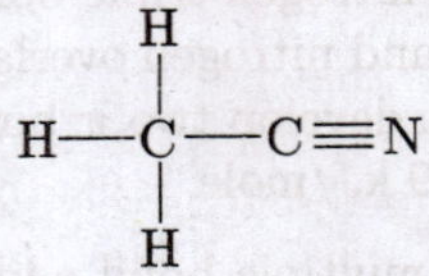
Condensed form of dash and dot formula	$H_3C—\ddot{O}—H$	$H_3C—C{\equiv}N:$
Condensed formula	CH_3OH	CH_3CN

Condensed formulae are to be used throughout this book in most of the cases. These sorts of representation do not indicate anything about the shape of the molecules.

Problem 2.11. Write down the gross structural formula for CH_2O.

Solution: (a) Hydrogen, carbon and oxygen atoms possess unicovalency, tetracovalency and bicovalency respectively. Hence, in CH_2O molecule the C atom should have four covalent bonds and oxygen should have two covalent bonds. This is possible if the C atom forms two single bonds with two H atoms and one double bond with the O atom. So, structural formula for CH_2O may be written as :

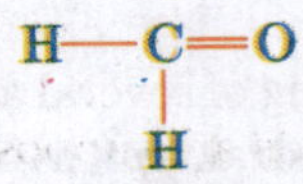

Graphic formula

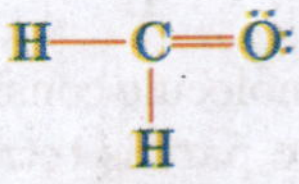

Dash-dot formula

Electron-dot formula

H_2CO

Condensed formula

Problem 2.12. What types of spatial arrangements of atoms do you expect in $CH_3CO{\cdot}C{\equiv}CH$?

Solution: In the said molecule, the C atom of CH_3— is supposed to be sp^3-hybridised, the C atom of C=O part is considered as sp^2-hybridised and that of C≡CH part as sp-hybridised. Thus, —CH_3 part is tetrahedral and H—C—H bond angle is about 109.5°; C=O part is planar trigonal having O—C—C bond angle 120° and C≡CH part has linear arrangement with H—C—C bond angle 180°. The molecule may thus be represented as:

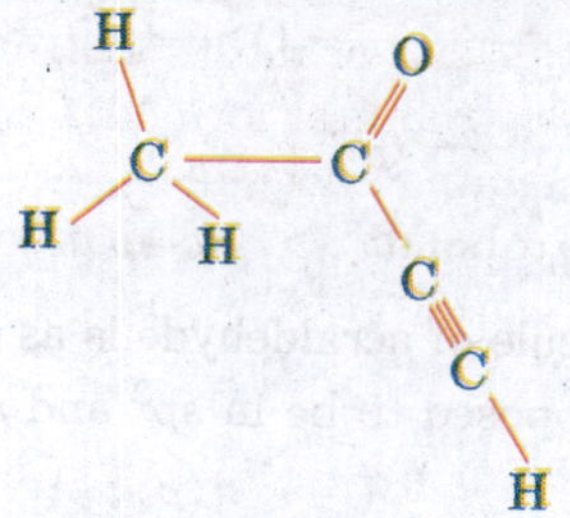

2.18 The Distribution of Electrons over the Molecules

The discussion made on molecular structure of organic compounds tells us minimum about the electronic distribution over the molecules. This requires detailed study and Chapter 4 is meant for this.

2.19 The Intermolecular Forces

The forces that hold molecules together may be called **intermolecular forces**. To understand the physical properties (b.p., m.p. and solubility) of organic compounds, intermolecular forces are to be studied extensively along with the molecular structure. Chapter 7 is intended for this purpose.

Problem 2.13. Draw the orbital pictures for the following compounds:

(a) $CH_2{=}C{=}CH_2$; (b) $CH_2{=}C{=}O$; (c) $CH_2{=}CH{-}CHO$.

Solution : (a) Experiments show that allene is a linear molecule and H—C—H bond angles are approximately 120°. To account for the above-said geometry, the middle C is supposed to be in sp-hybridised state and the terminal Cs are in sp^2-hybridised state.

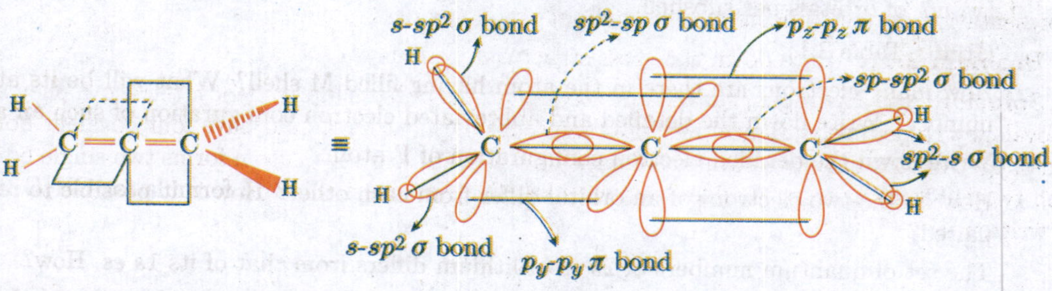

(b) Ketene is a linear molecule so, C—C and C—O σ bonds are supposed to be sp^2-sp and sp-sp. sp and a p A.O. hold non-bonding electron pairs. The middle C atom is considered to be in sp-hybridised state, terminal C and O atoms are in sp^2 and sp-hybridised states respectively.

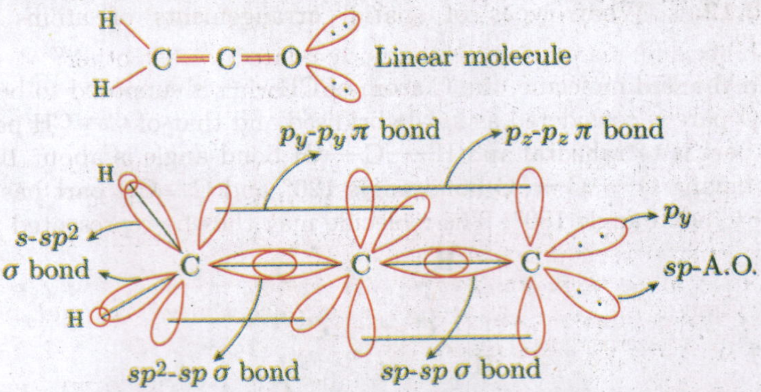

(c) The geometry of the molecule of acraldehyde is as shown in the picture.

All the C and O atoms are supposed to be in sp^2 and sp-hybridised state respectively.

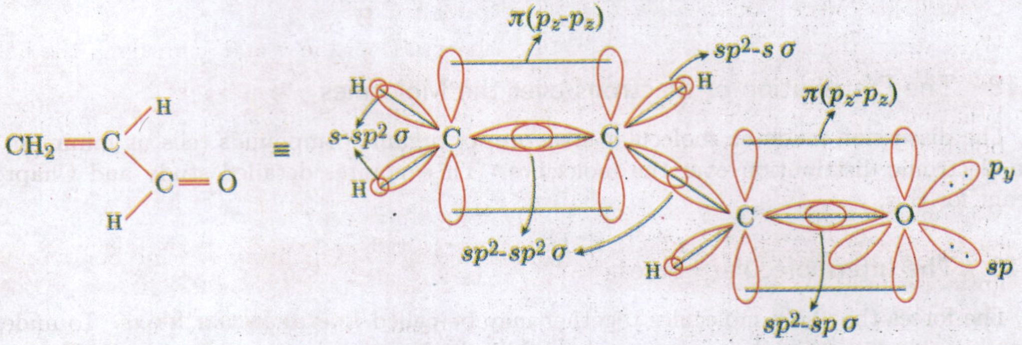

Two C—C σ bonds : sp^2-sp^2. One C—O σ bond : sp^2-sp.

Four C—H σ bonds : sp^2-s. C—C π and C—O π bonds : p-p; lone pairs of es are in sp and p_y A.O.s.

Exercise

1. Make a table on distribution of electrons in filled N shell showing : (i) maximum no. of es, (ii) subshells, (iii) maximum no. of es per subshell, (iv) no. of orbitals in the shell and (v) no. of orbitals per subshell.

   (**Hint** : Table 2.1)
2. How many electrons are there in the atom having filled M shell? What will be its atomic number? Write down the detailed and abbreviated electron configuration of such an atom.
3. Write down the detailed electron configuration of F atom.
4. How do the two electrons of an orbital differ from each other? How is it possible to remain paired?
5. The set of quantum numbers of 2s es of lithium differs from that of its 1s es. How?
6. What do the + and − signs indicate when they are placed in an orbital, pure or hybrid?
7. Write down the designations of three p and s-p hybridised orbitals. Draw these orbitals. What are their orientations in space? Why are they oriented in such ways?

8. How do $3p$ orbitals differ from $2p$ orbitals?
9. How are stability and energy content of a body related to each other? Write down the order of stability and energy content of s, p and s-p hybridised orbitals. Give reasons in favour of your answer.
10. Define : Ionic crystal, molecular crystal, atomic crystal. What do you mean by chemical bonds? How many types of bonds are known to you? Name them.
11. Define : Covalent, electrovalent and dative bonds. Discuss the nature of the forces that are present in these bonds.
12. Define: Ionisation energy, electron affinity, electronegativity. Write down the conditions for ionic and covalent bond formation.
13. Define formal charge. How can you calculate this? Write down the electron-dot formula, dash-dot formula, graphic formula and condensed formula for each of the following chemical units:

    Cyanide ion, nitric acid, acetaldehyde, acetamide, ethoxide ion.
14. Write down the assumptions made in V.B. theory. Describe σ and π bonds in the light of the V.B. theory.
15. What do you mean by hybrid A.O.s? How is it possible for a C atom to form four covalent bonds though it contains only two unpaired electrons in its outer shell? How many types of hybridisation may occur in an O atom?
16. Why are the four sp^3 A.O.s tetrahedrally oriented in space?
17. Discuss the formation of N_2 molecule in the light of M.O. theory. Define bonding M.O., antibonding M.O., sigma M.O., pi M.O.
18. Write short notes on: Bond length, polarity of bond, bond angle, bond energy and dipole moment.
19. Predict the shapes and bond angles of the following molecules:

    $$Cl—C{\equiv}C—Cl,\ CHCl{=}CHCl,\ CCl_4.$$
20. Describe the nature of the following bonds:

    $$C{=}O,\ C—I,\ C—Na,\ —C{=}C—,\ C—S.$$

8. How do sp orbitals differ from $2p$ orbitals?
9. How are stability and energy content of a body related to each other? Write down the order of stability and energy content of s, p and sp hybridised orbitals. Give reasons in favour of your answer.
10. Define : Ionic crystal, molecular crystal, atomic crystal. What do you mean by chemical bonds? How many types of bonds are known to you? Name them.
11. Define : Covalent, electrovalent and dative bonds. Discuss the nature of the forces that are present in these bonds.
12. Define : Ionisation energy, electron affinity, electronegativity. Write down the conditions for ionic and covalent bond formation.
13. Define formal charge. How can you calculate it? Write down the electron-dot formula, dash-dot formula, graphic formula and condensed formula for each of the following chemical units :

Cyanide ion, nitric acid, acetaldehyde, acetamide, chloride ion.

14. Write down the assumptions made in V.B. theory. Describe σ and π bonds in the light of the V.B. theory.
15. What do you mean by hybrid A.O.s. How is it possible for a C atom to form four covalent bonds though it contains only two unpaired electrons in its outer shell? How many types of hybridisation may occur in an C atom?
16. Why are the four sp^3 A.O.s tetrahedrally oriented in space?
17. Discuss the formation of N_2 molecule in the light of M.O. theory. Define bonding M.O., antibonding M.O., sigma M.O., pi M.O.
18. Write short notes on: Bond length, polarity of bond, bond angle, bond energy and dipole moment.
19. Predict the shapes and bond angles of the following molecules:

$$Cl—C \equiv C—Cl,\ CHCl=CHCl,\ CCl_4$$

20. Describe the nature of the following bonds:

groups C=O, C—I, C—N, —C=C—, C—S.

o classify

Chapter 3

Classification and Nomenclature of Organic Compounds

Objectives

- To understand:
  - Elementary classification of organic compounds [3.2] p. 67
  - What is meant by functional groups and homologous series; and the method of writing structural formulae of the functional groups [3.3] p. 68
- To classify:
  - Aliphatic organic compounds on the basis of characteristic functional group [Table 3.1] p. 68
  - Cs and Hs of an organic compound [3.4] p. 70
  - Hydrocarbons and hydrocarbon groups [3.5 and 3.6] p. 71 & 76
- To memorise:
  - The common names of the hydrocarbons and hydrocarbon groups [3.5 and 3.6] p. 71 & 76

* The order of precedence of functional groups [Table 3.11] p. 98

- To use:

  * The common names of the carboxylic acids, aromatic compounds and heterocyclic compounds [3.7] p. 80

    The IUPAC nomenclature rules for naming organic compounds [3.8] p. 93

- To draw:

  * Structural formulae of the compounds from their names [3.9] p. 140

3.1 Introduction

The number of organic compounds discovered to date has been estimated to be more than three millions. It is impossible for a person to remember the name and chemistry of so many organic compounds. But scientific classification of such a large number of compounds into a comparatively small number of families on the basis of structural theory has made these relatively simple and easy (Sec. 1.7). This is because compounds within a class have general methods of preparation and general chemical properties; the physical properties of the compounds within a class change from member to member but in a regular manner. So we need not study the chemistry of individual members of a class; instead, we can learn the chemistry of the class as a whole.

For the reasons noted above we shall discuss in this chapter the classification and nomenclature of organic compounds.

The name of an organic compound should be such as can represent its structural feature and hence can provide information on the properties of the compound. This we find in IUPAC system, the modern international system of nomenclature. Therefore, IUPAC names of the compounds are to be learnt with utmost care.

In the early days an organic compound was named either according to the name of its natural source or according to the name of its characteristic property. These names are called trivial or common names. For example, HCOOH is obtained from ants. The Latin synonym of ant is formicum and HCOOH behaves as an acid; therefore, HCOOH is named as formic acid.

As the number of organic compounds increased, it became a burden to remember individual names of so many compounds. Then came the era of substituted or derived system of naming. In this system, the simplest member of a class was given the trivial name and other compounds of the class were named as substituted derivatives of the simplest compound.

Thus, $CH_2{=}CH_2$ is ethylene but $CH_3{\cdot}CH{=}CH_2$ is called methylethylene.

As the complexity and number of organic compounds increased, the derived system of nomenclature too was found inadequate. In 1892, in Geneva the International Chemical Congress accepted a systematic nomenclature based on structural theory. This is known as Geneva System. The chemists and chemical traders refused to accept the Geneva System because short trivial names were replaced by big Geneva names. The rule makers retreated. International Union of Chemists met in 1930 at Liege. They revised Geneva names and approved certain new rules; the new rules came to be called IUC system. That system too was not accepted by the chemical world. Then International Union of Pure and Applied Chemistry in 1947 appointed a new commission for further work on nomenclature of organic compounds. Since then the modern international nomenclature has been called IUPAC system. It is needless to mention that the original version of the IUPAC rules was formulated in Geneva in 1892. Of course, it may be called a substituted system of nomenclature. However, IUPAC systems are revised and extended time to time as per necessity.

3.2 Elementary Classification of Organic Compounds

Organic compounds are classified into three main groups depending on the nature of carbon chains they possess:

1. Organic compounds possessing open carbon chains are called open chain or acyclic compounds or more commonly aliphatic compounds (aleiphatos is a Greek word meaning fat;

fats have open chain structures and so open chain compounds are called aliphatic compounds).

2. A large number of organic compounds possess rings consisting of carbon atoms only; these are called carbocyclic compounds. These can be divided under three subheads:
   (a) Some of the carbocyclic compounds behave like aliphatic compounds and these are known as alicyclic compounds.
   (b) There are other carbocyclic compounds having benzene ring and these are said to be benzenoid aromatic compounds as they possess characteristic smell (**aroma** is a Latin word meaning odour).
   (c) There is also a group of aromatic compounds which do not have benzenoid rings; these are called non-benzenoid aromatic compounds.
3. Organic compounds containing rings composed of carbon and any other atom O, N, S, etc., are termed as heterocyclic compounds.

3.3 Functional Groups and Classification of Organic Compounds

The main constituents of organic compounds are C and H atoms. C—C and C—H sigma bonds, formed by the head-on overlap of two atomic orbitals, are stable and do not cleave easily. On the other hand, a sigma bond between a C atom and any electronegative atom like F, Cl, Br, I, O, N, S, etc., being polar, is a cite for chemical reaction. Moreover C=C, C≡N, C=O, etc., multiple bonds are composed of sigma and pi bonds. π bonds have π electrons and these are less firmly held between the two bonding nuclei. So π bonds take part in a chemical reaction readily. Therefore, the part of a molecule, which contains an electronegative element or an electropositive element, i.e., metal atom or C=C or any other multiple bond, is the reactive part of the molecule. The reactive part of a molecule largely determines its properties. On the basis of the structural feature of this part, organic compounds can be grouped into several families. For this reason, the reactive part of a molecule—either an atom or group of atoms—that determines the properties of the molecule and defines the structure of a particular family of organic compounds is called the functional group or the function (the group that functions). Thus each elementary class of compounds as grouped in Sec. 3.2 can be further classified according to their functions.

The names of some of the functions and the corresponding classes of compounds in aliphatic series are given in Table 3.1; G and I are used in parentheses to indicate respectively the common and IUPAC names of the classes. The dashes written in the graphic formulae of the functions indicate the number of C or H atoms that can be bonded to the group.

Table 3.1 Names and formulae of functions and classification of acyclic compounds with examples

Name of the function	Graphic formula	Class name of the acyclic compounds	Example	
×	$-\overset{\vert}{\underset{\vert}{C}}-\overset{\vert}{\underset{\vert}{C}}-$	Paraffin (G) Alkane (I)	$CH_3—CH_3$	: Ethane Ethane
Carbon-carbon double bond	$-\overset{\vert}{C}=\overset{\vert}{C}-$	Olefin (G) Alkene (I)	$CH_2=CH_2$	: Ethylene Ethene
Carbon-carbon triple bond	$-C\equiv C-$	Acetylene (G) Alkyne (I)	$HC\equiv CH$	: Acetylene Ethyne

Contd.

Name of the function	Graphic formula	Class name of the acyclic compounds		Example
Halo	—X(F, Cl, Br, I)	Alkyl halide (G) Haloalkane (I)	CH_3Cl	: Methyl chloride Chloromethane
Hydroxyl	—OH	Alcohol (G) Alkanol (I)	CH_3OH	: Methyl alcohol Methanol
Ether linkage	—O—	Ether (G) Alkoxyalkane (I)	$(CH_3)_2O$	: Methyl ether Methoxymethane
Aldehyde, oxo or formyl	$-\underset{\underset{H}{\vert}}{C}=O$	Aldehyde (G) Alkanal (I)	HCHO	: Formaldehyde Methanal
Keto or oxo	$>C=O$	Ketone (G) Alkanone (I)	$(CH_3)_2CO$	: Acetone Propanone
Carboxyl	$-\underset{\underset{OH}{\vert}}{C}=O$	Carboxylic acid (G) Alkanoic acid (I)	HCOOH	: Formic acid Methanoic acid
Amido or carbamoyl	$-\underset{\underset{NH_2}{\vert}}{C}=O$	Acid amide (G) Alkanamide (I)	$HCONH_2$	: Formamide Methanamide
Ester or carbalkoxy	$-\underset{\underset{OR}{\vert}}{C}=O$	Ester (G) Alkyl alkanoate (I)	$HCOOCH_3$	: Methyl formate Methyl methanoate
Acid halide group, or haloformyl group, or carbonyl halide group	$-\underset{\underset{X}{\vert}}{C}=O$	Acid halide (G) Alkanoyl halide (I)	CH_3COCl	: Acetyl chloride Ethanoyl chloride
Acid anhydride	$-\underset{\underset{O}{\Vert}}{C}-O-\underset{\underset{O}{\Vert}}{C}-$	Acid anhydride (G) Alkanoic anhydride (I)	$(CH_3CO)_2O$	: Acetic anhydride Ethanoic anhydride
Cyano	$-C\equiv N$	Alkyl cyanide (G) Alkanenitrile (I)	CH_3CN	: Methyl cyanide Ethanenitrile
Isocyano	$-\overset{\oplus}{N}\equiv\overset{-}{C}:$	Alkyl isocyanide (G) Alkylcarbylamine (I)	CH_3NC	: Methyl isocyanide Methylcarbylamine
Nitro	$-\underset{\underset{O^-}{\vert}}{\overset{\oplus}{N}}=O$	Nitroalkane (G&I)	CH_3NO_2	: Nitromethane
Amino	$-\underset{\vert}{N}-$	Alkylamine (G) Alkanamine (I)	CH_3NH_2	: Methylamine Methanamine
Nitroso	$-N=O$	Nitrosoalkane (G&I)	CH_3NO	: Nitrosomethane
Sulphonic acid	$-\underset{\underset{OH}{\vert}}{SO_2}$	Alkanesulphonic acid (G&I)	CH_3SO_3H	: Methanesulphonic acid
Mercapto	—SH	Thioalcohol (G) Alkanethiols (1)	CH_3SH	: Methyl mercaptan Methanethiol

Starting from the simplest member, if the molecular formulae of the compounds of a particular class are arranged in an increasing number of C atoms, a series is obtained in which one member differs from the next member by a methylene group (—CH_2— group); such a series is called a homologous series and each member of the series is known as homolog (Gr; meaning similarly related). As the members of a homologous series contain one and the same function, they have general chemical properties. The formula weight of methylene group being 14.026 amu, the molecular weight of the homologs in the series increases by 14.026 amu. Hence, the physical properties like m.p., b.p., solubility, etc., differ from homolog to homolog in a regular manner, because, with the increasing molecular weight, London force (Ch. 7) increases. An example of homologous series in given in Table 3.2.

3.4 Classification of C and H Atoms of an Organic Compound

C and H atoms of an organic compound may be classified in two ways :

1. **Depending on the mode of arrangement of C atoms in a chain:** According to this classification we get four types of C atoms and three types of H atoms. These are:

(a) Primary or 1° C and H atoms; (b) Secondary or 2° C and H atoms;

(c) Tertiary or 3° C and H atoms; (d) Quarternary or 4° C atoms.

If a C atom in a compound is bonded to one other C atom only, the former C atom is said to be a primary or 1° C atom. On the other hand, if in a carbon chain a C atom is bonded to two other C atoms, then it is known as a secondary C atom or 2° C atom. Similarly, tertiary or 3° C and quarternary or 4° C atoms have bonds with three other and four other C atoms, respectively. The H atoms attached to primary, secondary and tertiary C atoms are said to be primary or 1°, secondary or 2° and tertiary or 3° H atoms, respectively. As the valency of a quarternary C atom is satisfied by four C atoms, it has no H atom attached to it. Thus, the C and H atoms are marked according to the above classification in the example given below (structure 1):

1. 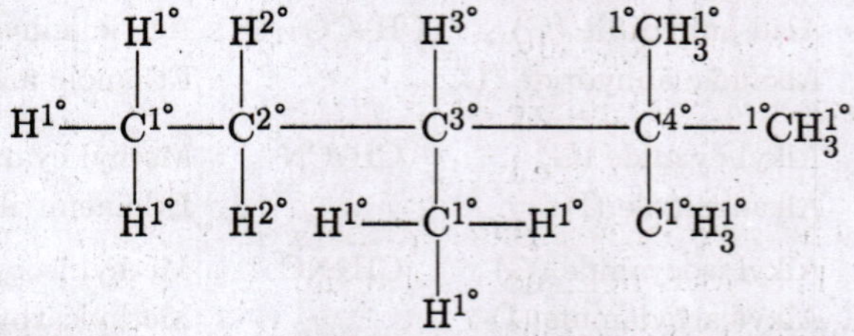

2. **Depending on the functional group:** In this method the relative positions of different C atoms in a molecule are designated with respect to the functional group it contains. These designations are done by the small letters of Greek alphabet. The C atom directly attached to the functional group is designated as αC atom; the C atom next to it is indicated as βC atom and that next to βC is γC atom and so on. Each of the H atoms is given the similar α, β, γ or δ designation as the C atom to which it is attached. Problem 3.1 will clarify this method of classification of C and H atoms.

Problem 3.1. Designate the C and H atoms of the hydrocarbon part of the compound $(CH_3)_3C{\cdot}CHO$.

Solution: The C atoms of the three CH_3 groups are 1° or primary as each of them is bonded to one C atom only; their H atoms are therefore 1° H atoms. The C atom attached to the functional group (—CHO) is 4° C because it is attached to four C atoms.

With respect to the —CHO group the 4° C atom is α; this is because it is the adjacent C atom to the functional group. Each of the rest C atoms of the compound is placed next to the αC atom; hence all of them are βC atoms. Therefore, nine H atoms of the hydrocarbon part of the said compound are, βH atoms. The designations are shown below in structure 2:

2. 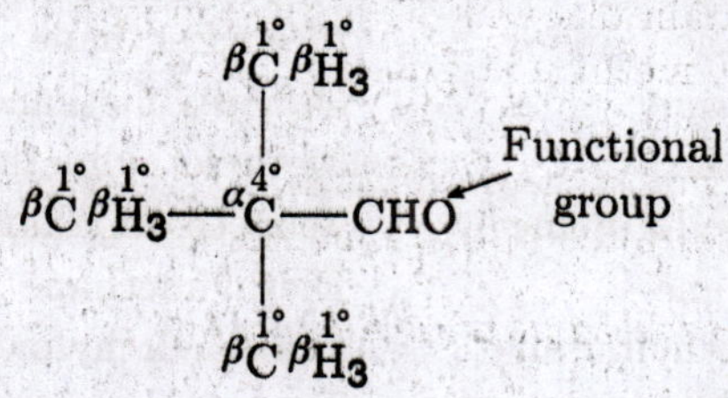

Note

The functions present in a molecule may similarly be designated as it is done in the case of H atoms. Thus $Cl{\cdot}CH_2{\cdot}CH_2OH$ is a primary alcohol as the OH group is bonded to 1° C atom and the position of the Cl atom may be designated as β, because it is bonded to βC atom with respect to the OH group. The Cl atom is bonded to a CH_2 group which is bonded to another C, so it is 1° chloride also.

3.5 Hydrocarbons

It has already been said that hydrocarbons, the binary compounds of C and H are the parent compounds of the organic domain. These may be classified into three main groups according to the elementary classification of organic compounds; the groups are:

1. Aliphatic hydrocarbons; 2. Alicyclic hydrocarbons; 3. Aromatic benzenoid hydrocarbons.

Aliphatic and alicyclic hydrocarbons may further be subdivided into three classes:

(a) Hydrocarbons containing C—C single bonds only.

(b) Hydrocarbons containing at least one C=C double bond.

(c) Hydrocarbons containing at least one C≡C triple bond.

Aromatic benzenoid hydrocarbons consist of benzene rings; a benzene ring contains alternate C—C single and double bonds only. Therefore, benzenoid hydrocarbons are of one type. Of course, a large number of these compounds possess aliphatic or alicyclic parts attached to the benzene ring; in those parts, one may find C—C single, double or triple bonds. However, in this section we are going to discuss the hydrocarbons in a nutshell because these are the parent compounds.

3.5.1 Paraffins or Alkanes

If all the C—C bonds in an acyclic hydrocarbon are single bonds, the hydrocarbon is said to be an alkane (IUPAC name of the class) or a paraffin (trivial name; param affinis is a Latin term meaning little affinity, i.e., inactivity). The general formula of alkanes is $C_nH_{2n+2} (n = 1, 2, 3,$ etc.). These are represented by RH; where R represents C_nH_{2n+1}. The simplest member of the alkanes is methane, $CH_4 (n = 1, R{=}CH_3)$. The molecular formulae, condensed structural formulae and names of some of the alkanes are written in Table 3.2.

Table 3.2 Some simple alkanes or paraffins

Molecular formula	Condensed structural formula	Common name	IUPAC name	Substituted name or derived name
CH_4	CH_4	Methane	Methane	Methane
C_2H_6	$CH_3{\cdot}CH_3$	Ethane	Ethane	Methylmethane
C_3H_8	$CH_3{\cdot}CH_2{\cdot}CH_3$	Propane	Propane	Dimethylmethane
C_4H_{10}	$CH_3{\cdot}CH_2{\cdot}CH_2{\cdot}CH_3$	*n*-Butane	Butane	Methylethylmethane
C_4H_{10}	$(CH_3)_2CH{\cdot}CH_3$	Isobutane	Methylpropane	Trimethylmethane
C_5H_{12}	$CH_3{\cdot}CH_2{\cdot}CH_2{\cdot}CH_2{\cdot}CH_3$	*n*-Pentane	Pentane	Diethylmethane
C_5H_{12}	$(CH_3)_2CH{\cdot}CH_2{\cdot}CH_3$	Isopentane	2-Methylbutane	Dimethylethylmethane
C_5H_{12}	$(CH_3)_4C$	Neopentane	2, 2-Dimethylpropane	Tetramethylmethane

3.5.2 Olefins or Alkenes

If an acyclic hydrocarbon contains one C═C double bond only, the hydrocarbon is called an alkene (IUPAC name of the class) or an olefin (**olefin** is a common name, it is a Latin term meaning oil producing). The general formula of olefins is C_nH_{2n}; as they contain two H atoms less than the alkanes, alkenes are called unsaturated hydrocarbons. The molecular formulae, structural formulae and names of some alkenes are noted in Table 3.3.

Table 3.3 Some simple alkenes or olefins

Molecular formula	Condensed structural formula	Common name	IUPAC name	Substituted name or derived name
C_2H_4	$CH_2{=}CH_2$	Ethylene	Ethene	Ethylene
C_3H_6	$CH_3{\cdot}CH{=}CH_2$	Propylene	Propene	Methylethylene
C_4H_8	$CH_2{=}CH{\cdot}CH_2{\cdot}CH_3$	α-Butylene	But-1-ene	Ethylethylene
C_4H_8	$CH_3CH{=}CH{\cdot}CH_3$	β-Butylene	But-2-ene	Sym* Dimethylethylene
C_4H_8	$(CH_3)_2C{=}CH_2$	Isobutylene	2-Methylpropene	Unsym** Dimethylethylene
C_5H_{10}	$CH_2{=}CH{\cdot}CH_2{\cdot}CH_2{\cdot}CH_3$	α-Pentylene or α-amylene	Pent-1-ene	*n*-Propylethylene
C_5H_{10}	$CH_3{\cdot}CH{=}CH{\cdot}CH_2{\cdot}CH_3$	β-Pentylene or β-amylene	Pent-2-ene	Methylethylethylene
C_5H_{10}	$(CH_3)_2C{=}CH{\cdot}CH_3$	Isopentylene Isoamylene	2-Methyl-but-2-ene	Trimethylethylene

* Sym = Symmetrical. **Unsym = Unsymmetrical.

3.5.3 Acetylenes or Alkynes

If there is a C≡C triple bond in an acyclic hydrocarbon, it is known as an acetylene (common name) or alkyne (IUPAC name). The general formula of acetylenes is C_nH_{2n-2}. These are also unsaturated hydrocarbons. Some alkynes are listed in Table 3.4.

Table 3.4 Some alkynes or acetylenes

Molecular formula	Condensed structural formula	Common name	IUPAC name	Substituted name or derived name
C_2H_2	$HC{\equiv}CH$	Acetylene	Ethyne	Acetylene
C_3H_4	$CH_3{\cdot}C{\equiv}CH$	×	Propyne	Methylacetylene
C_4H_6	$CH_3{\cdot}CH_2{\cdot}C{\equiv}CH$	×	But-1-yne	Ethylacetylene
C_4H_6	$CH_3{\cdot}C{\equiv}C{\cdot}CH_3$	×	But-2-yne	Dimethylacetylene
C_5H_8	$CH_3{\cdot}CH_2{\cdot}CH_2{\cdot}C{\equiv}CH$	×	Pent-1-yne	*n*-Propylacetylene
C_5H_8	$CH_3{\cdot}CH_2{\cdot}C{\equiv}CCH_3$	×	Pent-2-yne	Methylethylacetylene
C_5H_8	$(CH_3)_2CH{\cdot}C{\equiv}CH$	×	3-Methylbut-1-yne	Isopropylacetylene

3.5.4 Alicyclic Hydrocarbons

These may further be classified as:

I. Cycloalkanes or cycloparaffins, II. Cycloalkenes, and III. Cycloalkynes.

Cycloalkanes are the simplest type of alicyclic compounds. These are also called polymethylenes as their structure possesses more than two methylene ($—CH_2—$) groups. The general formula of cycloalkanes is C_nH_{2n}; the first member of the series is cyclopropane, C_3H_6. Such hydrocarbons bear the names of their nonacyclic analogs (having the same number of C atoms) to which the prefix 'cyclo' is added.

The general formula of the cycloalkenes is C_nH_{2n-2}. The cycloalkynes have the general formula C_nH_{2n-4}. The names of cycloalkenes and cycloalkynes are derived in the same way as done for cycloalkanes. Cycloalkynes of less than 8 C atoms are very unstable because of high strain due to the presence of $—C{\equiv}C—$, a linear part, in the structure. So we need not consider them here.

An alicyclic hydrocarbon may have an acyclic hydrocarbon part; the acyclic part is then called a side chain. The structures and names of some alicyclic hydrocarbons are listed in Table 3.5.

Table 3.5 Some simple alicyclic hydrocarbons

Mol. formula	Structural formula	IUPAC name	Mol. formula	Structural formula	IUPAC name
C_3H_6	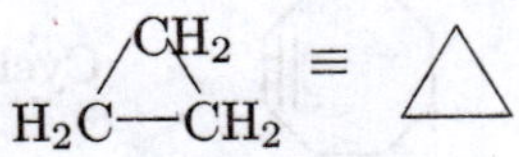	Cyclopropane	C_5H_8	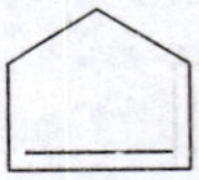	Cyclopentene
C_4H_8	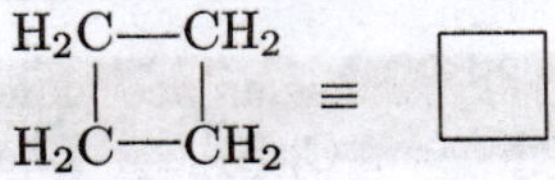	Cyclobutane	C_5H_8	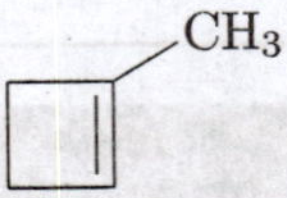	1-Methyl-cyclobutene
C_4H_8	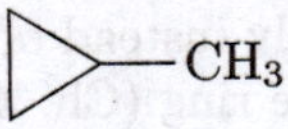	Methylcyclo-propane	C_5H_8	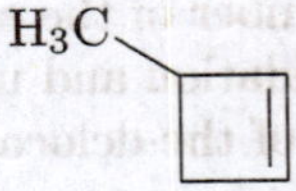	3-Methyl-cyclobutene

Contd.

Mol. formula	Structural formula	IUPAC name	Mol. formula	Structural formula	IUPAC name
C_5H_{10}	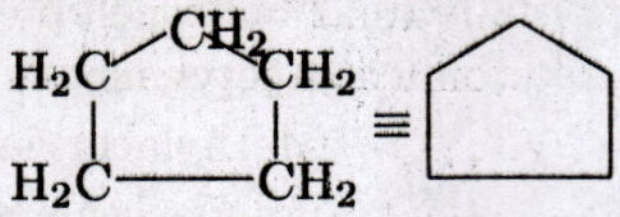	Cyclopentane	C_6H_{10}	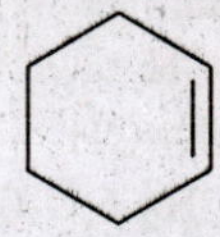	Cyclohexene
C_5H_{10}	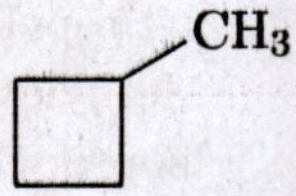	Methylcyclobutane	C_6H_{10}	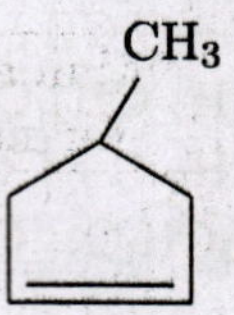	4-Methylcyclopentene
C_5H_{10}	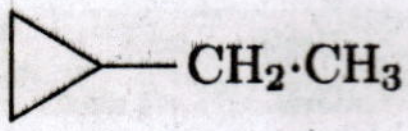	Ethylcyclopropane	C_6H_{10}	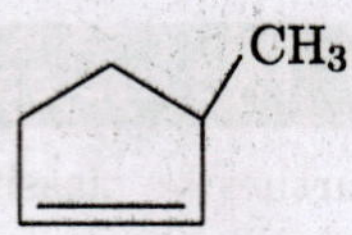	3-Methylcyclopentene
C_5H_{10}		1,2-Dimethylcyclopropane	C_6H_{10}	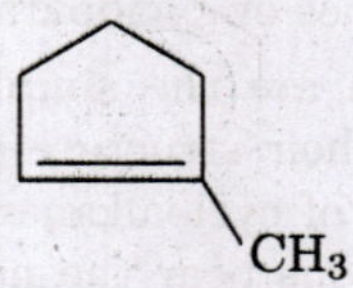	1-Methylcyclopentene
C_5H_{10}	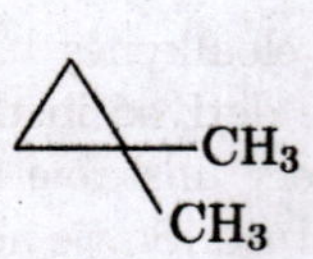	1,1-Dimethylcyclopropane	C_6H_{10}	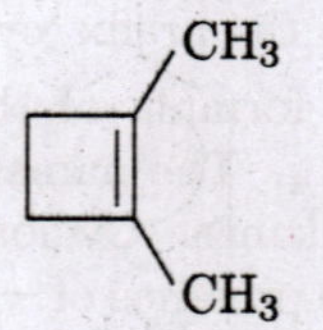	1,2-Dimethylcyclobutene
C_6H_{12}		Cyclohexane			
C_3H_4		Cyclopropene	C_6H_{10}	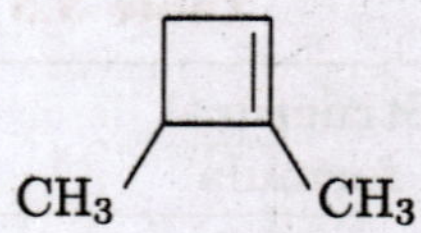	2,3-Dimethylcyclobutene
C_4H_6		Cyclobutene	C_8H_{12}		Cyclooctyne

3.5.5 Aromatic Benzenoid Hydrocarbons

The simplest member of the aromatic benzenoid hydrocarbons is benzene, C_6H_6. It is highly resistant towards oxidation and undergoes substitution reaction mostly instead of addition reaction; this is because of the delocalisation of π electrons of the benzene ring (Ch. 4).

If one of the H atoms of benzene is replaced by an aliphatic or alicyclic hydrocarbon part, it is called a side chain.

An aromatic hydrocarbon may have more than one benzene ring, this is known as a polynuclear aromatic hydrocarbon.

Polynuclear aromatic hydrocarbons may consist of either individual benzene rings or fused rings having two C atoms common to two rings. Here are some aromatic hydrocarbons with their names and structures in Table 3.6.

Table 3.6 Some common benzenoid aromatic hydrocarbons (Arenes)

Molecular formula	Structural formula	Common name	IUPAC name
C_6H_6	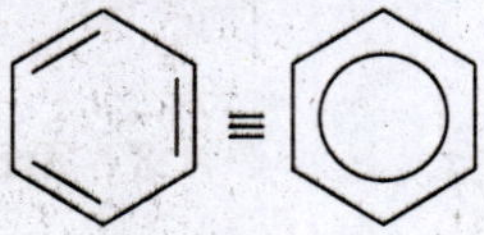	Benzol	Benzene
C_7H_8		Toluene	Methylbenzene, Toluene
C_8H_{10}	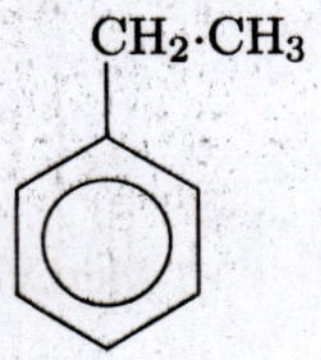	×	Ethylbenzene
C_8H_{10}	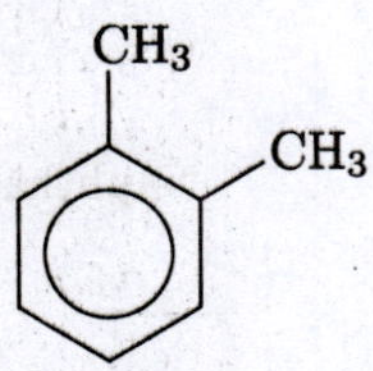	*o*-xylene(ortho-xylene)[1]	1,2-Dimethylbenzene*
C_8H_{10}	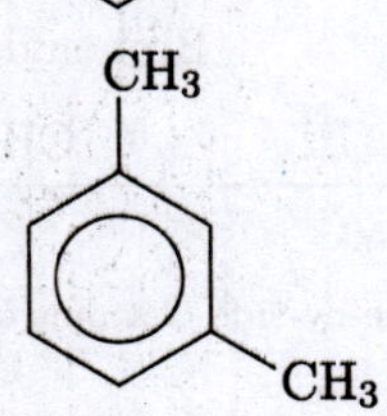	*m*-xylene(meta-xylene)[2]	1,3-Dimethylbenzene*
C_8H_{10}	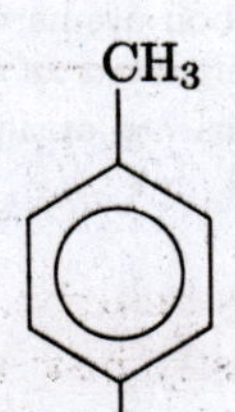	*p*-xylene(para-xylene)[3]	1,4-Dimethylbenzene*

Contd.

Molecular formula	Structural formula	Common name	IUPAC name
C_8H_8	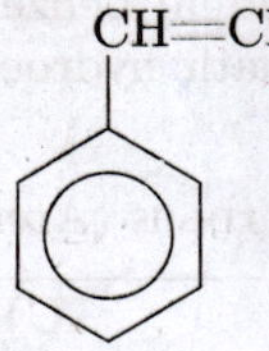	Styrene	Phenylethene
C_9H_{12}	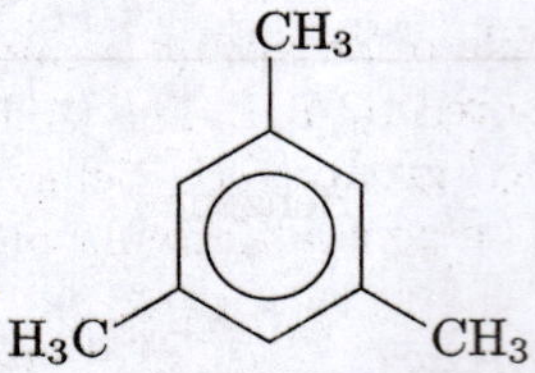	Mesitylene	1,3,5-Trimethylbenzene
C_9H_{12}	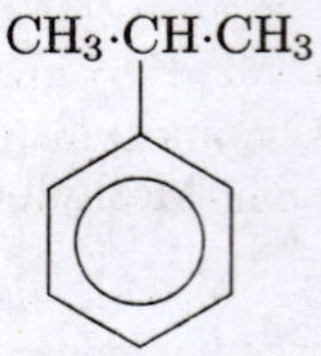	Cumene	Isopropylbenzene
C_9H_{12}	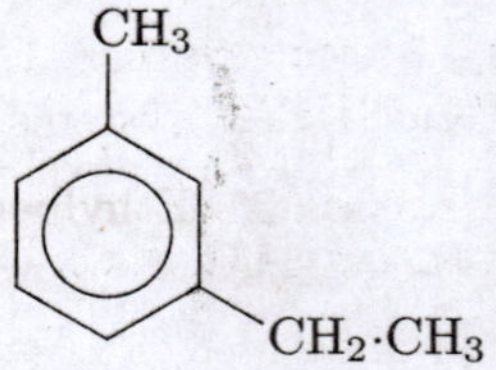	*m*-methylethylbenzene	3-Methylethylbenzene
$C_{10}H_8$	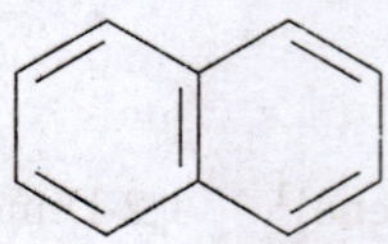	Naphthalene	Naphthalene
$C_{12}H_{10}$	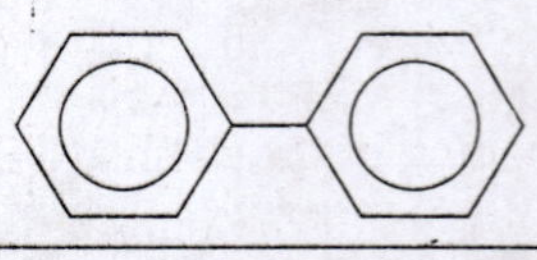	Biphenyl	Biphenyl

1*Common names are also used as their IUPAC names.

1. When two groups of a derivative of benzene are present on two side-by-side C atoms of a benzene nucleus, the compound is denoted as ortho-, i.e., *o*-compound.
2. If the two substituents present in a benezene ring are attached to two carbon atoms which are themselves separated by one C atom, the relative positions of the substituents are designated as meta, i.e., *m*-.
3. Disubstituted derivatives of benzene in which substituted groups or atoms are attached to two opposite carbons atoms of the benzene ring are said to be para or *p*-compounds.

3.6 Hydrocarbon Groups

If one or more H atoms are taken out from a hydrocarbon, the group left is known as a hydrocarbon group. As hydrocarbons are of three major types, hydrocarbon groups too belong. to three main classes; these are:

I. Acyclic hydrocarbon groups; II. Alicyclic hydrocarbon groups, and III. Aromatic benzenoid hydrocarbon groups.

3.6.1 Acyclic Hydrocarbon Groups

Depending on the classification of acyclic hydrocarbons, these are subdivided into three heads:

(a) Alkyl groups, (b) Alkenyl groups, and (c) Alkynyl groups.

The group obtained by removing one H atom from the formula of an alkane is said to be an alkyl group. These are represented by R— as alkanes are denoted by RH. The suffix 'ane' of the parent alkane is replaced by the suffix 'yl' in naming an alkyl group. CH_4 being named as methane, CH_3— is to be called methyl (meth $+yl$) group. The general formula of the alkyl group is C_nH_{2n-1}. Alkyl groups are of five types:

1. **Normal alkyl group:** This is formed by the removal of one 1° H atom (H atom from 1° C) from the formula of a straight chain hydrocarbon. A normal alkyl group is written as n-alkyl group in the common naming system; and in its IUPAC nomenclature, the prefix n- is dropped. Thus, $CH_3{\cdot}CH_2{\cdot}CH_2$— is n-propyl group in the common naming system; but it is propyl group in the IUPAC system.

2. **Secondary alkyl group:** It has one 2° H atom less than the parent hydrocarbon and it is denoted by sec-alkyl or s-alkyl group in both the systems of nomenclature. $CH_3{\cdot}CH_2{\cdot}\underset{|}{C}H{\cdot}CH_3$ is thus s-butyl group in the common naming as well as in IUPAC systems.

3. **Tertiary alkyl group:** We get such a group by removing one 3° H atom, from the corresponding alkane; it is denoted by tert- or, t-alkyl group. Thus, the common name and IUPAC name of $(CH_3)_3C$— is tert- or t-Butyl group.

4. **Isoalkyl group:** An alkyl group containing one terminal CH_2— group and $(CH_3)_2CH$— group on the other end with no other branching is said to be an isoalkyl or i-alkyl group. $(CH_3)_2CH{\cdot}CH_2$— is known as isobutyl group in both the systems, but $(CH_3)_2CH{\cdot}\underset{|}{C}HCH_3$ is not isopentyl group as it does not have terminal CH_2— group.

5. **Neoalkyl group:** A neoalkyl group contains one CH_2— group on one end and one $(CH_3)_3C$— group on the other end with no other branching in the chain. $(CH_3)_3C{\cdot}CH_2$— is a neopentyl group (common name).

IUPAC and common names of sec-, tert-, iso- and neo-alkyl groups up to C_5 chain are the same.

Hydrocarbon groups containing C—C double and triple bonds may be called alkenyl and alkynyl groups, respectively. Their common names are accepted in the IUPAC system in most of the cases.

Hydrocarbon groups obtained by removing two H atoms either from a single C atom (alkylidene group) or from two side-by-side C atoms or from two separated C atoms (alkylene groups) are also well known; e.g., $CH_3{\cdot}CH{=}$, $CH_3{\cdot}\underset{|}{C}H{\cdot}CH_2$—, —$CH_2{\cdot}CH_2{\cdot}CH_2$—, etc.

Some of the acyclic hydrocarbon groups are listed in Table 3.7.

Table 3.7 Acyclic hydrocarbon groups

Condensed structural formula	Common name with its abbreviated form	IUPAC name with its abbreviated form
CH_3-	Methyl (Me)	Methyl (Me)
$CH_3\cdot CH_2-$	Ethyl (Et)	Ethyl (Et)
$CH_3\cdot CH_2\cdot CH_2-$	*n*-Propyl (*n*-Pr)	Propyl (Pr)
$CH_3\underset{\vert}{C}H\cdot CH_3$	Isopropyl (*i*-Pr)	Isopropyl (*i*-Pr)
$CH_3\cdot CH_2\cdot CH_2\cdot CH_2-$	*n*-Butyl (*n*-Bu)	Butyl (Bu)
$CH_3\cdot \underset{\vert}{C}H\cdot CH_2\cdot CH_3$	sec-Butyl (*s*-Bu)	sec-Butyl (*s*-Bu)
$(CH_3)_2\cdot CH\cdot CH_2-$	Isobutyl (*i*-Bu)	Isobutyl (*i*-Bu)
$(CH_3)_3C-$	tert-Butyl (*t*-Bu)	tert-Butyl (*t*-Bu)
$CH_3\cdot CH_2\cdot CH_2\cdot CH_2\cdot CH_2-$	*n*-Pentyl	Pentyl
$CH_3\cdot CH_2\cdot CH_2\underset{\vert}{C}H\cdot CH_3$	sec-Pentyl	sec-Pentyl
$(CH_3)_2CH\cdot CH_2CH_2-$	Isoamyl	Isopentyl
$(CH_3)_2\underset{\vert}{C}\cdot CH_2\cdot CH_2$	tert-pentyl	tert-pentyl
$(CH_3)_3C\cdot CH_2-$	Neopentyl	Neopentyl
$-CH_2-$	Methylene	Methylene
$-CH_2\cdot CH_2-$	Ethylene	Ethylene
$CH_2{=}CH-$	Vinyl	Vinyl or Ethenyl
$CH_3-CH{=}$	Ethylidene	Ethylidene
$CH_2{=}CH-CH_2-$	Allyl	Allyl
$CH_3CH{=}CH-$	... ...	Propenyl
$CH_3-\underset{\vert}{C}{=}CH_2$	... ...	Isopropenyl
$CH_3\cdot CH_2\cdot CH{=}$	Propylidene	Propylidene
$(CH_3)_2C{=}$	Isopropylidene	Isopropylidene
$-CH_2\cdot CH_2\cdot CH_2-$	Propylene	Propylene
$HC\equiv$	Methyne	Methyne
$CH\equiv C-$	Ethynyl	Ethynyl
$HC\equiv C-CH_2-$	Propargyl	Propargyl
$CH_3\cdot C\equiv C-$	Propynyl	Propynyl

3.6.2 Alicyclic Hydrocarbon Groups

These are obtained when one or more H atoms are removed from the formulae of alicyclic hydrocarbons and these may be classified as:

(a) Cycloalkyl groups, (b) Cycloalkenyl groups, and (c) Cycloalkynyl groups.

IUPAC names of some alicyclic hydrocarbon groups are written below:

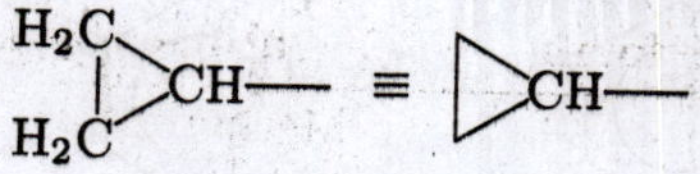

Cyclopropyl

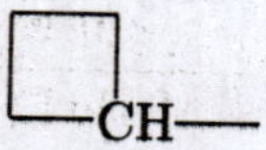

Cyclobutyl

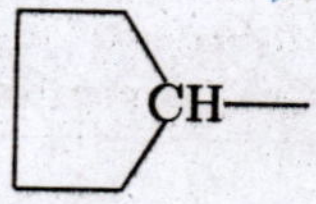

Cyclopentyl

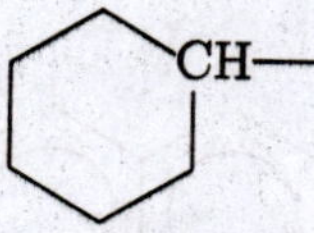

Cyclohexyl

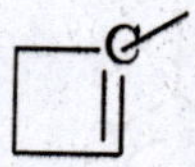

1-Cyclobutenyl

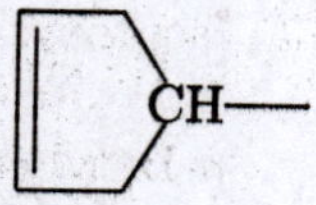

3-Cyclopentenyl

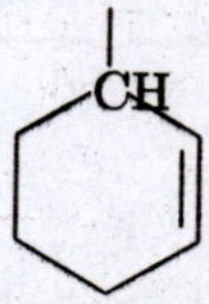

2-Cyclohexenyl

3.6.3 Aromatic Benzenoid Hydrocarbon Groups

Aromatic hydrocarbon groups have one or more H atoms less than the parent hydrocarbons. These are, in general, denoted by Ar- and called aryl groups. The simplest aryl group is phenyl group (—C_6H_5). This is obtained by removing one H atom from the benzene molecule and denoted by Ph- or ϕ. Some of the aryl groups are noted below:

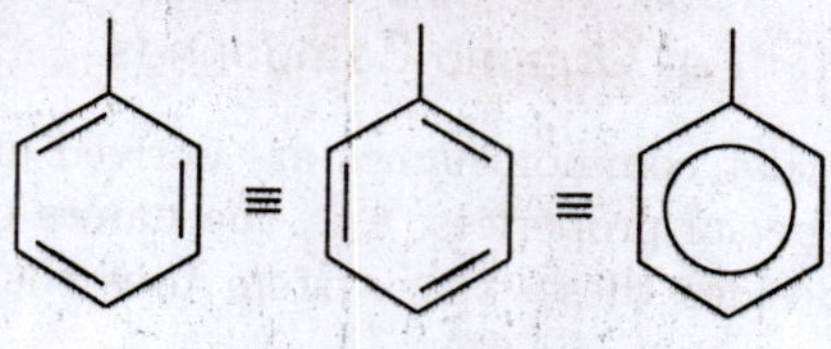

C_6H_5 or Ph or ϕ

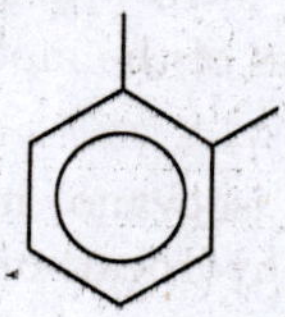

o-Phenylene

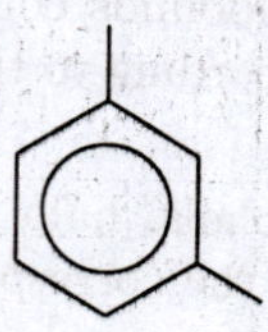

m-Phenylene

p-Phenylene

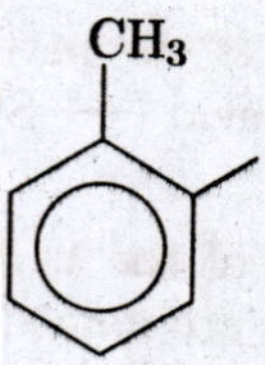

o-Tolyl

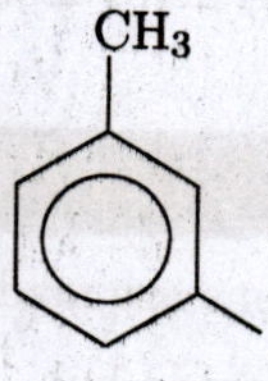

m-Tolyl

p-Tolyl

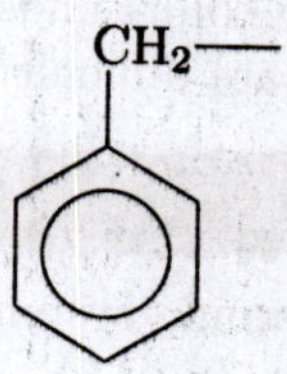

Benzyl

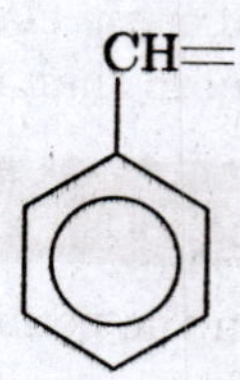

Benzal(G)
Benzylidene (I)

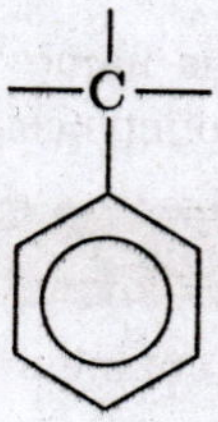

Benzo

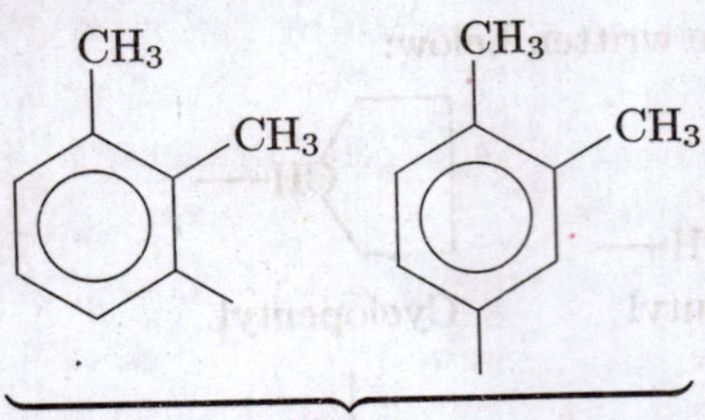

Xylyl

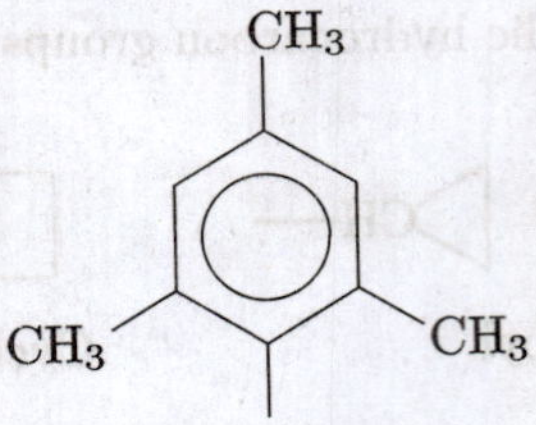

Mesityl

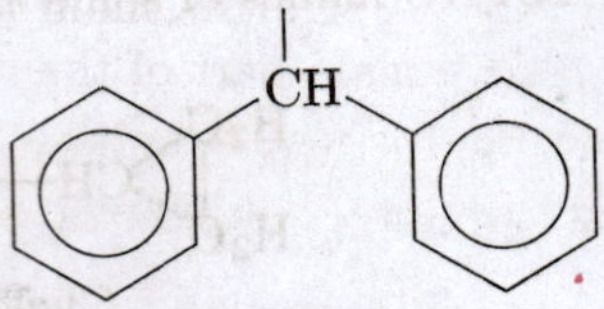

Benzhydryl

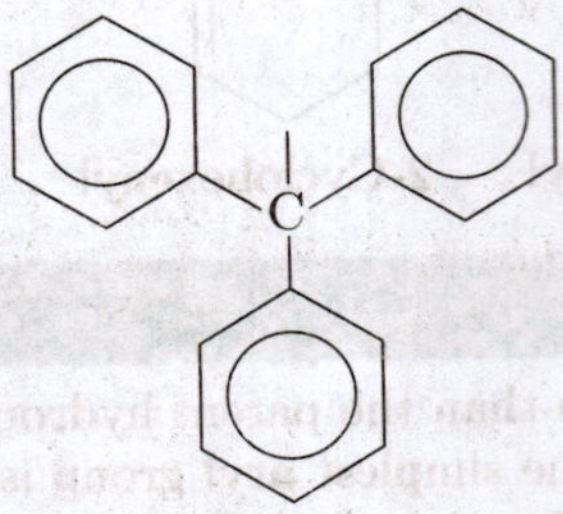

Trityl

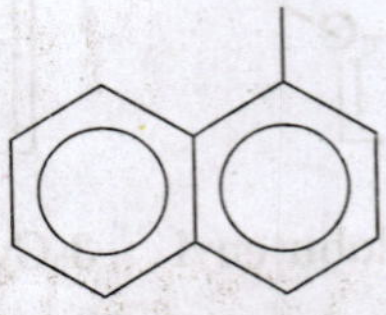

α-Naphthyl (G)
1-Naphthyl (I)

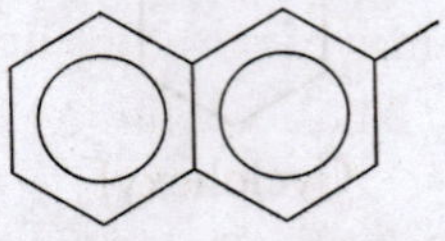

β-Naphthyl (G)
2-Naphthyl (I)

3.7 Common Naming System of Organic Compounds

It has already been stated that common names are derived either from the names of their natural sources or from their special properties, e.g., the names of carboxylic acids are mostly derived from their sources. Besides these, this system uses the following in naming organic compounds:

(a) Names of the parent hydrocarbons, e.g., compounds containing nitro ($—NO_2$) and sulphonic acid ($—SO_3H$) groups are named according to the names of the mother hydrocarbons.

(b) Names of the hydrocarbon groups, e.g., alcohols, halo, cyano, isocyano, marcapto compounds are named according to the names of the corresponding hydrocarbon groups.

(c) Names of the carboxylic acids, e.g., aldehydes, ketones, nitrites, amides, acid halides, acid anhydrides and esters have names derived from the names of the parent carboxylic acids.

The names of the different classes of compounds are discussed below. In general aliphatic compounds containing more than six Cs are difficult to name in this system.

3.7.1 Paraffins and Cycloparaffins

To name a paraffin in the common naming system the following considerations are made:

I. The total number of C atoms in the compound is counted and this is indicated by:

meth- for one C atom;	but- for four C atoms;	hept- for seven C atoms;
eth- for two C atoms;	pent- for five C atoms;	oct- for eight C atoms;
prop- for three C atoms;	hex- for six C atoms;	non- for nine C atoms.

II. To that the syllable **ane** is added as the name ending of the paraffins; this constitutes the main part of the name.

III. A straight chain paraffin containing more than three Cs is named by prefixing '*n*-' to the main part of the name.

IV. If there is an isopropyl fork, $(CH_3)_2CH$—, on one end with no other branching, the prefix 'iso' is used instead of '*n*-' for the paraffin to be named.

V. The prefix becomes 'neo' if the paraffin contains a tert-butyl group, $(CH_3)_3C$—, on one end and no other side chain.

All the paraffins consisting of one to five Cs and some of the paraffins containing six Cs have common names. Higher paraffins with complex structures do not have common names. Common names of paraffins up to five Cs are listed in Table 3.2.

Problem 3.2. Write down the common names of the following compounds:

(a) $CH_3{\cdot}CH_2{\cdot}CH_2{\cdot}CH_2{\cdot}CH_2{\cdot}CH_3$; **(b)** $(CH_3)_2CH{\cdot}CH_2{\cdot}CH_2{\cdot}CH_3$;

(c) $(CH_3)_3C{\cdot}CH_2{\cdot}CH_3$; **(d)** $(CH_3)_2CH{\cdot}CH(CH_3)_2$.

Solution: (a) It contains six Cs; so the main part of its name is hexane (hex + ane). As it is a straight chain compound, its full name is *n*-Hexane (vide rule III).

(b) Since it has six Cs, the main part of its name is hexane. It contains $(CH_3)_2CH$— group on one end and there is no other side chain; hence its full name is Isohexane.

(c) As there are six Cs and $(CH_3)_3C$— group on one end with no other branching, its full name is Neohexane.

(d) This compound has no common name.

Cycloparaffins are known as polymethylenes in common naming system. Number of Cs present in the ring are indicated by the name of the Greek numerals, and it is written before the word—methylene. Common names of cycloparaffins are not used today.

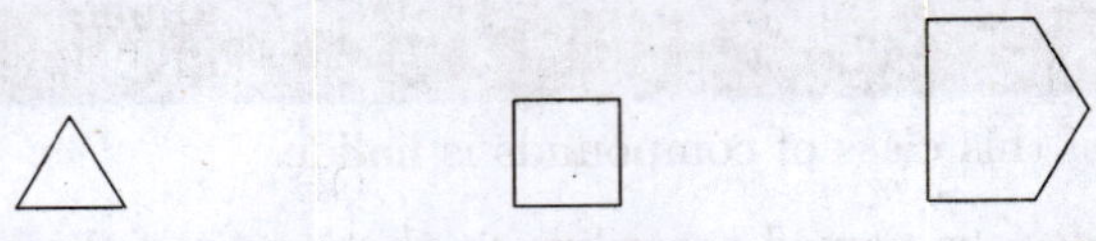

Trimethylene Tetramethylene Pentamethylene

3.7.2 Olefins

In the common naming system an olefin is named according to the following rules:

(a) The total number of C atoms in the olefin is counted and the name of the corresponding alkane is determined.

(b) The suffix 'ane' is then dropped from the name of the paraffin and suffix **'ylene'** is added to the prefix.

(c) The position of the double bond is indicated by Greek letters, $\alpha, \beta, \gamma, \delta, \ldots$; the end C atom nearest to the double bond is denoted by α, next to it is β and so on. These letters are then known as locants.

(d) The locant of the double-bonded C atom is then placed before the name of the olefin as obtained from (a) and (b). A hyphen is written in between the locant and the name. The locants are used only to name olefins containing more than three carbon atoms.

Olefins of low molecular weights only have common names.

Problem 3.3. (i) Write down the common names of the following compounds:

(a) $CH_3{\cdot}CH{=}CH{\cdot}CH_3$; **(b)** $CH_3{\cdot}CH_2{\cdot}CH_2{\cdot}CH{=}CH_2$.

(ii) Write down the structures of the following compounds:

(a) Isobutylene; **(b)** β-pentylene (β-amylene).

Solution: (i) (a) It has four Cs; so the name of the parent paraffin is butane. Therefore, the class name of the compound is butylene (but + ylene). Now the locant of the double-bonded C atom nearest to one end is as shown below:

$$\overset{\alpha}{C}H_3{\cdot}\overset{\beta}{C}H{=}CH{\cdot}CH_3.$$

Hence the full name of the compound is: β-Butylene.

(b) The name of the corresponding paraffin: Pentane.

The class name of the compound: Pentylene or Amylene

$$CH_3CH_2{\cdot}CH_2\overset{\beta}{C}H{=}\overset{\alpha}{C}H_2.$$

The full name of the compound: α-Pentylene or α-Amylene

(ii) (a) $CH_3{-}\underset{\displaystyle CH_3}{\underset{|}{C}}{=}CH_2$; **(b)** $\overset{\alpha}{C}H_3{-}\overset{\beta}{C}H{=}CH{\cdot}CH_2CH_3$.

3.7.3 Acetylenes

The simplest member, $CH{\equiv}CH$, has its common name acetylene.

3.7.4 Alkyl Halides

The name ending of this class of compounds is **halide**.

(i) Alkyl monohalides are named according to the names of the alkyl groups they contain; a space is left in between the name of the alkyl group and the word 'halide'.

$(CH_3)_2CH{\cdot}CH_2Cl$ should be named as isobutyl chloride, because it contains isobutyl group. On the other hand $CH_3CH_2{\cdot}CHI{\cdot}CH_3$ is sec-Butyl iodide.

(ii) The dihalide of a paraffin with both of the halogen atoms on a single C atom is known as a geminal (abbreviated form is **gem**; it is a Latin word meaning paired) dihalide. In common naming system a geminal dihalide is named as alkylidene halide, i.e., the suffix 'yl' of the parent alkyl group is replaced by **'ylidene'** in naming a gem-dihalide; e.g., $CH_3{\cdot}CBr_2{\cdot}CH_3$ may be supposed to be a derivative of isopropylidene group and accordingly the compound is named as isopropylidene bromide.

(iii) If the dihalide of a paraffin contains one halogen atom on each of the two side-by-side C atoms, the dihalide is said to be a vicinal dihalide (vicinal is an adjective of the Latin word **vicinous** meaning neighbour; its abbreviated form is vic). A vicinal dihalide is named as a derivative of the corresponding olefin and hence the word 'halide' is placed after the name of the olefin leaving a space between them. Thus, $CH_3{\cdot}CH{\cdot}Cl{\cdot}CH{\cdot}Cl{\cdot}CH_3$ is β-butylene chloride; here β-butylene, $CH_3CH{=}CH{\cdot}CH_3$, is supposed to be the mother olefin from which the dihalide is derived.

(iv) When each of the two terminal C atoms of a dihalide contains a halogen atom, it is called a α-ω dihalide or polymethylene halide, e.g., $CH_2I{\cdot}CH_2{\cdot}CH_2{\cdot}CH_2{\cdot}CH_2I$ is Pentamethylene iodide.

(v) Only the trihalo derivatives of methane have common names; these are: $CHCl_3$, chloroform; $CHBr_3$, bromoform; CHI_3, iodoform.

Problem 3.4. Write down the common names of the following halides:
(a) CCl_4, **(b)** $(CH_3)_2CH{\cdot}CHF_2$, **(c)** CH_3CHICH_2I.

Solution: **(a)** It is named according to the inorganic system of nomenclature and called carbon tetrachloride.

(b) It's a gem-difluoride and the parent paraffin is isobutane as it has $(CH_3)_2CH$— fork; hence its name is isobutylidene fluoride.

(c) It is a vic-dihalide of propylene, so its name is propylene iodide.

3.7.5 Nitroparaffins

These are obviously nitro derivatives of paraffins possessing nitro (—NO_2) group.

The name of a nitroparaffin is expressed as a single word in which 'nitro' is used as a prefix to the name of the parent paraffin; nitroparaffins of low molecular weights having a terminal nitro group are named by this system. Thus, $CH_3{\cdot}CH_2NO_2$ is nitroethane.

3.7.6 Ethers

Compounds containing two hydrocarbon groups joined to an O atom are known as ethers. They are of two types:

1. Simple ethers—having two identical hydrocarbon groups; R—O—R is a simple ether.
2. Mixed ethers—having two different hydrocarbon groups; R—O—R′ is a mixed ether.

Simple alkyl ethers are named as dialkyl ethers: a space is left between the name of the alkyl group and the word ether. Thus, $CH_3{\cdot}O{\cdot}CH_3$ is dimethyl ether. Similarly, $CH_3CH_2{\cdot}O{\cdot}CH_2CH_3$ is diethyl ether.

Mixed ethers are named by writing the names of the two hydrocarbon groups separately in an increasing order of complexity, a space is left and then the word ether is written.

$CH_3{\cdot}O{\cdot}CH(CH_3)_2$ is thus methyl isopropyl ether.

Problem 3.5. Write down the names of the following ethers:

(a) $CH_2{=}CH{\cdot}O{\cdot}CH{=}CH_2$;

(b) 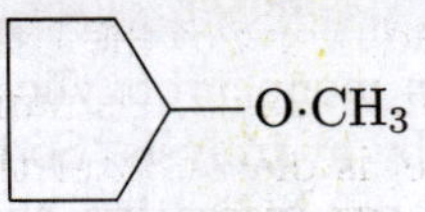

; **(c)** 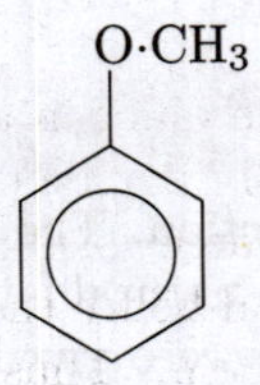

; **(d)** 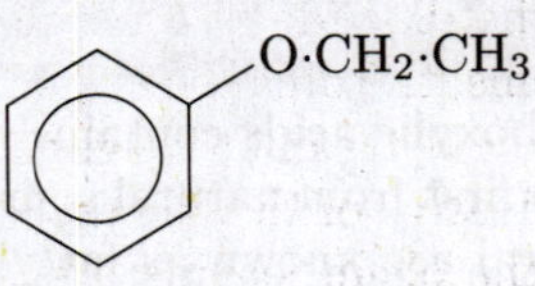

Solution: **(a)** This is a simple ether and it contains vinyl group; hence its name is divinyl ether.

(b) This is a mixed ether. It has a methyl group and a cyclopentyl group; the latter group being more complex than the former one, the given compound is to be named as methyl cyclopentyl ether.

(c) Its trivial name is anisole. It is also a mixed ether; its name is methyl phenyl ether.

(d) The trivial name of the compound is phenetole. It is also called ethyl phenyl ether.

3.7.7 Alcohols

A monohydric alcohol contains one —OH group. To name a monohydric alcohol in common naming system, one writes the name of the hydrocarbon group and the word alcohol separately, e.g., CH_3OH is said to be methyl alcohol (wood spirit is its trivial name as it is obtained from wood). Similarly, the trivial name of CH_3CH_2OH is grain spirit (obtained from food grains); its common name is ethyl alcohol.

Problem 3.6. Write down the names of the following alcohols:

(a) $(CH_3)_2CH{\cdot}CH_2OH$; **(b)** $CH_3{\cdot}CHOH{\cdot}CH_2{\cdot}CH_3$; **(c)** $(CH_3)_3C{\cdot}OH$;

(d) 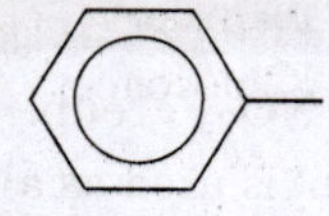 $CH_2{\cdot}OH$ **(e)** $CH_2{=}CH{\cdot}CH_2OH$.

Solution: **(a)** It contains isobutyl group; so its name is isobutyl alcohol.

(b) The hydroxyl group is attached to the 2°C atom and hence the hydrocarbon group is sec-butyl group; the name of the given compound is, therefore, sec-butyl alcohol.

(c) It is tert-butyl alcohol as its hydroxyl group is joined to the tert-butyl group.

(d) $C_6H_5CH_2$— is benzyl group, hence the name of the compound is benzyl alcohol.

(e) It is allyl alcohol, because $CH_2{=}CH{-}CH_2{-}$ is allyl group.

Note

Dihydric alcohols are called glycols. A glycol may be α or β or γ or δ, etc., according to the relative positions of two —OH groups. If two —OH groups are present on two -side-by-side C atoms, it is an α-glycol. On the other hand, if these are present on alternate C atoms, it is termed as a β-glycol. Similarly, if two —OH groups are on two Cs separated by two other C atoms, the compound is said to be a γ-glycol.

The name of α-glycols are derived from the names of corresponding olefins. Thus, CH_2OHCH_2OH is ethylene glycol; $CH_3{\cdot}CHOH{\cdot}CH_2OH$ is propylene glycol.

α-ω glycols are named as polymethylene glycols, e.g., $OH{\cdot}CH_2{\cdot}CH_2{\cdot}CH_2{\cdot}CH_2OH$ is tetramethylene glycol.

The trihydric alcohol which deserves mention is $CH_2OH{\cdot}CHOH{\cdot}CH_2OH$. Its common name is glycerol.

3.7.8 Carboxylic Acids

Carboxylic acids contain —COOH function. The straight-chain monocarboxylic acids were isolated first from natural sources; e.g., from natural fats and waxes by hydrolysis. So these acids in general are known as fatty acids. These were given common names indicating their source. To name a fatty acid, the last syllable from the Latin (L.) or Greek (Gr.) name of its source is dropped and to that 'ic acid' is added. Table 3.8 contains condensed structural formulae of

C_1—C_5 carboxylic acids, their sources and names. Compounds containing isopropyl fork are named by adding the prefix 'iso' to the common names.

The names of dicarboxylic acids and some of the substituted monocarboxylic acids are also derived from their sources. Other substituted acids are named according to the name of their parent acids.

Table 3.8 Common names of aliphatic carboxylic acids

	Condensed structural formula	English name of the source	Latin/Greek synonym of the source	Syllable dropped	Name of the carboxylic acid
1.	HCOOH	Ant	L. Formica	ica	Formic acid
2.	CH_3COOH	Vinegar	L. Acetum	um	Acetic acid
3.	CH_3CH_2COOH	First fat	Gr. Proto pion	to	Propionic acid I
4.	$CH_3CH_2CH_2COOH$	Butter	L. Butyrum	um	Butyric acid
5.	$(CH_2)_2CH{\cdot}COOH$	Butter	L. Butyrum	um	Isobutyric acid
6.	$CH_3(CH_2)_3COOH$	Valerian root	Gr. Valerian	ian	Valeric acid
7.	$(CH_3)_2CHCH_2COOH$	Valerian root	Gr. Valerian	ian	Isovaleric acid
8.	$CH_2Cl{\cdot}COOH$	... ...	... ...	...	Chloroacetic acid
9.	$CH_3CHOH{\cdot}COOH$	Milk	L. Lactis	is	Lactic acid
10.	$CH_3CHCl{\cdot}COOH$	... ...	... ...	...	α-Chloropropionic acid
11.	$COOH{\cdot}COOH$	Wood sorrel	Gr. Oxalis	is	Oxalic acid
12.	$COOH{\cdot}CH_2{\cdot}COOH$	... ...	... ...	...	Malonic acid II
13.	$COOH(CH_2)_2COOH$	Amber	L. Succinum	um	Succinic acid
14.	$COOH{\cdot}CHOH{\cdot}CH_2COOH$	Apple	Malum	um	Malic acid
15.	$COOH{\cdot}(CH_2)_3COOH$	... ...	... ...	...	Glutaric acid III
16.	$COOH(CH_2)_2CH(NH_2){\cdot}COOH$	Glue	Gluten	en	Glutamic acid IV
17.	$CH_2{=}CH{\cdot}COOH$	... ...	... ...	...	Acrylic acid V

I. 'to' from 'proto' is dropped and 'pion' is then joined to 'pro'.

II. It is derived from Malic acid.

III. It is obtained from Glutamic acid.

IV. To indicate the presence of aminogroup, the name becomes glutamic acid.

V. It is a sharp acid. The Latin synonym of sharp is acrid.

3.7.9 Aldehydes

The aldehydes contain —CHO function. The common name of an aldehyde is derived from the common name of the corresponding acid it would yield on oxidation. The '**ic acid**' ending of the name of the acid is dropped and the word '**aldehyde**' is added to the remainder; thus the name becomes one word.

For example, one gets acetic acid when CH_3CHO is oxidised. So, to name CH_3CHO:

(a) The name ending, 'ic acid' is dropped from 'acetic acid'—the remainder is thus '**acet**'

(b) The word 'aldehyde' is then added to the remainder 'acet'; thus the common name of CH_3CHO becomes acetaldehyde.

Problem 3.7. What are the common names of **(a)** $(CH_3)_2CH{\cdot}CHO$, **(b)** $CH_3CH{=}CH{\cdot}CHO$.

Solution : (a) The structural formula of the corresponding acid is $(CH_3)_2CH{\cdot}COOH$; its common name is isobutyric acid. Hence the common name of the given compound is isobutyraldehyde ('ic acid' is replaced by 'aldehyde').

(b) This is crotonaldehyde, because the name of the corresponding acid ($CH_3CH{=}CH{\cdot}COOH$) is crotonic acid.

3.7.10 Ketones

The ketones contain >C=O function. A ketone possessing two identical hydrocarbon groups attached to the ketonic C atom is called a simple ketone and that having two different hydrocarbon groups is known as a mixed ketone.

A simple ketone may be prepared by the pyrolysis of a suitable carboxylic acid and accordingly the name of the ketone is derived from the name of that acid; the name ending 'ic acid' of the acid is replaced by the suffix 'one'. CH_3COCH_3 is prepared from acetic acid by pyrolysis; hence if one drops 'ic acid' ending from acetic acid and adds the suffix 'one' to 'acet', then the common name of the ketone, CH_3COCH_3, becomes acetone (acet + one).

Common names of the mixed ketones have 'ketone' as the name ending. According to this system of nomenclature, the names of the different hydrocarbon groups are written separately in an increasing order of complexity before the word ketone. Thus, $CH_3COCH(CH_3)_2$ is methyl isopropyl ketone.

Simple ketones may also be named in the same way as discussed for the mixed ketones; in this case the name of the hydrocarbon group should have 'di' as prefix. Thus, $CH_3CH_2{\cdot}CO{\cdot}CH_2{\cdot}CH_3$ is diethyl ketone as the compound contains two ethyl groups ($CH_3{\cdot}CH_2$).

Problem 3.8. Write down the common names of

(a) 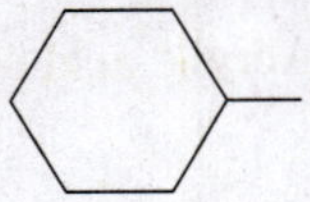 $CO{\cdot}CH_3$; **(b)** $CH_2{=}CH{\cdot}CO{\cdot}CH{=}CH_2$.

Solution: (a) It is a mixed ketone and the names of the hydrocarbon groups, in an increasing order of complexity, are methyl and cyclohexyl; hence the name of the compound is methyl cyclohexyl ketone.

(b) It is a simple ketone, the hydrocarbon group present is vinyl; therefore, the name of the compound is divinyl ketone.

3.7.11 Esters

Organic esters have —COOR as function (R being any hydrocarbon group). These are formed by the reaction between an organic carboxylic acid and an alcohol with the loss of water.

In the common naming system the hydrocarbon group attached to the O atom is named first, a space is left and then the common name of the acid (from which it is obtained) is written with the 'ic acid' ending changed to 'ate'; thus $CH_3CH_2COOCH_3$ is methyl propionate. Similarly, $CH_3COOCH{=}CH_2$ is vinyl acetate.

Ring-substituted alicyclic carboxylic esters are named as derivatives of formic acid and have 'formate' as name ending. Thus, 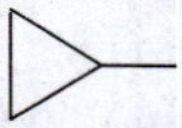 $CO{\cdot}O{\cdot}CH_2{\cdot}CH_3$ is ethyl cyclopropylformate.

Inorganic esters (formed by the reaction between an inorganic acid and an alcohol) are named as a salt. Hydrocarbon groups are named first with the prefixes, di, tri, etc. (according to its number), followed by the name of the anion of the inorganic acid involved. Thus, $(CH_3CH_2)_2SO_4$ is called diethyl sulphate. Sometimes the prefixes are omitted.

Problem 3.9. Write down the names of the following esters:

(a) $CH_2{=}CH{-}COOCH_3$, **(b)** $(CH_3)_2CH{\cdot}COO{-}$  ;

(c) $(CH_3)_2CH{\cdot}CH_2{\cdot}ONO_2$; **(d)** 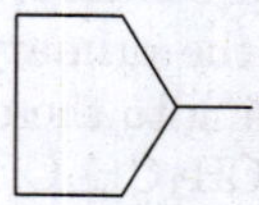 $-COO{\cdot}CH_2{\cdot}CH{=}CH_2$.

Solution: (a) O-hydrocarbon group is methyl; the parent acid is acrylic acid. So its name is methyl acrylate.

(b) O-hydrocarbon group is cyclohexyl; the common name of the acid is isobutyric acid. Hence it is cyclohexyl isobutyrate.

(c) Hydrocarbon group is isobutyl and the anion is nitrate; so it is named as isobutyl nitrate.

(d) O-hydrocarbon group is allyl ($CH_2{=}CH{-}CH_2{-}$); the common name of the acid is cyclopentylformic acid. Therefore, the name of the ester is allyl cyclopentylformate.

3.7.12 Acid halides

Compounds belonging to this class contain $-COX$, haloformyl, function (where X=Cl, Br, I). The common name of an acid halide is derived from the common name of the corresponding acid as it is obtained from the acid by replacing its $-OH$ group by the halogen atom; the 'ic acid' ending is replaced by the '**yl halide**' to name the acid halide. Thus, $CH_3{\cdot}CH_2{\cdot}COCl$ is **propionyl** chloride. Similarly, $(CH_3)_2CH{\cdot}CH_2COBr$ is **isovaleryl** bromide.

Problem 3.10. Write down the names of the following acid halides:

(a) $CH_3{\cdot}CH{=}CH{\cdot}COI$; **(b)** 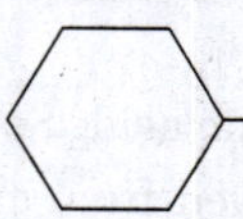 $-COCl$; **(c)** $(CH_3)_2CH{\cdot}COBr$; **(d)** CH_3COCl.

Solution: (a) The corresponding acid is crotonic acid and the halogen is iodine; hence its name is crotonyl iodide.

(b) The parent acid is cyclohexylformic acid. So, the name of the acid halide is cyclohexylformyl chloride.

(c) Isobutyryl bromide.

(d) Acetyl chloride.

3.7.13 Acid amides

This class of compounds possess $-CONH_2$ function ($-OH$ of the $-COOH$ is replaced by $-NH_2$). The common name of an amide is given according to the name of the parent acid; the 'ic acid' ending is replaced by 'amide' to form one-word name. Thus, $CH_3\cdot CONH_2$ is acetamide. $CH_2{=}CH\cdot CONH_2$ is acrylamide.

Alicyclic amides are named as derivatives of formamides. Thus, 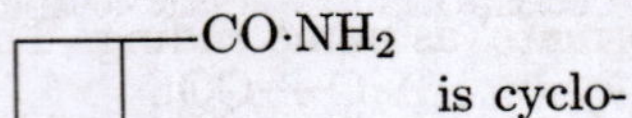 is cyclobutylformamide.

When H atoms of the amino group are replaced by the hydrocarbon groups, these are indicated by attaching the prefix N to the name of the hydrocarbon group; the full name of the amide is given by prefixing this to the name of the amide as derived from the parent acid. Thus, $H\cdot CON(CH_3)_2$ is N, N-dimethylformamide; because the parent acid is HCOOH (formic acid) and two H.s of the —$CONH_2$ group are replaced by two —CH_3 (Methyl) groups.

Secondary and tertiary amides are supposed to be derived from ammonia. When its two or three H atoms are replaced by acyl groups, secondary and tertiary amides respectively are formed. To name these compounds the name of the primary amide is first determined and then to this name prefix 'di' or 'tri' is attached according to the number of acyl groups the compound contains. Thus, $(CH_3\cdot CO)_2NH$ is diacetamide; $(CH_3CH_2CO)_3N$ is tripropionamide.

Cyclic secondary amides are called imides. These are named according to the name of the corresponding dicarboxylic acid; the 'ic acid' ending on the name is replaced by imide. For example:

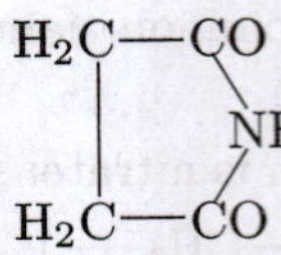

is succinimide.

Problem 3.11. Write down the names of the following compounds:

(a) $(CH_3)_2CH\cdot CONH_2$; **(b)** $HCONH\cdot CH_3$; **(c)** 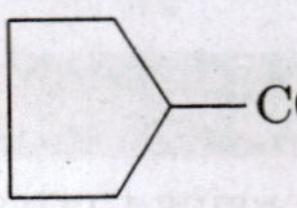 $CO\cdot NEt_2$;

(d) $CH_3\cdot CH_2\cdot CO\cdot N$(Me)(Et); **(e)** $(CH_3\cdot CH_2\cdot CO)_2NH$; **(f)** 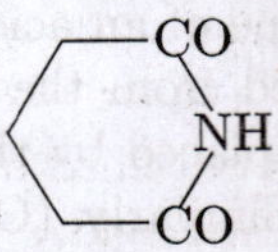

Solution: **(a)** It is a simple primary amide derived from isobutyric acid; hence its name is isobutyramide.

(b) It is an N-substituted amide derived from formic acid; So its name is N-methylformadide.

(c) It is an N,N-disubstituted alicyclic amide derived from cyclopentylformic acid; therefore, its name is N,N-diethylcyclopentylformamide.

(d) N-methyl,N-ethylpropionamide.

(e) It is a secondary amide consisting of two propionyl groups; so, it is dipropionamide.

(f) It is a cyclic secondary amide, i.e., imide; the parent acid is glutaric acid; hence, its common name is glutarimide.

3.7.14 Acid anhydrides

Acid anhydrides contain —CO—O—CO— function. These compounds are formed by the dehydration of two molecules of a monocarboxylic or one of a dicarboxylic acid; for this reason their names are derived from the parent acids; the word 'acid' is replaced by the word '**anhydride**'. Thus, $CH_3{\cdot}CO{\cdot}O{\cdot}CO{\cdot}CH_3$ is obtained by the dehydration of acetic acid; so the name of the compound is acetic anhydride.

Similarly, 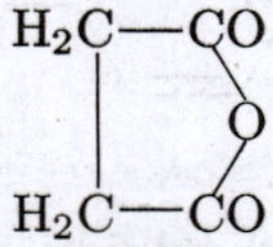 is obtained by the dehydration of succinic acid; so it is called succinic anhydride.

Problem 3.12. Compound A is called phthalic acid in the common naming system. What will be the name of compound B and C?

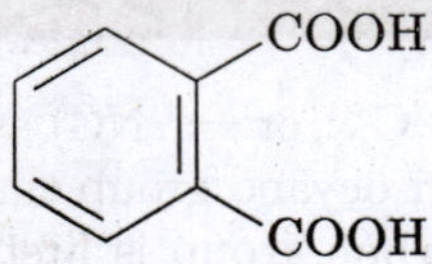
A

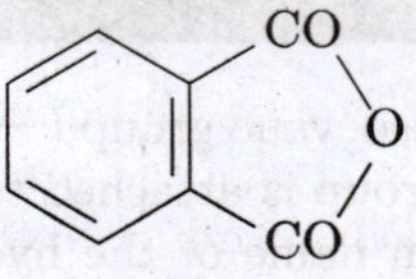
B

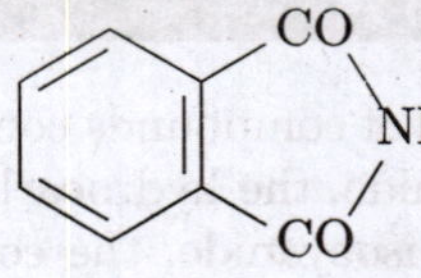
C

Solution: Compound B is the anhydride of phthalic acid obtained from the latter compound; hence its name will be phthalic anhydride.

Compound C is an imide. The parent compound from which it is prepared is phthalic acid. Hence to name this compound, '**ic acid** ending from phthalic acid is to be dropped and the suffix '**imide**' is to be added to the remainder 'phthal'. Thus, the name of the compound C is phthalimide.

3.7.15 Carbonitriles

This class of compounds contain —CN(—C≡N:, cyano) function. Carbonitriles when hydrolysed give carboxylic acids; so their common names are derived from the common names of the corresponding carboxylic acids. The '**ic acid**' ending on the common name of the acid is replaced by '**onitrile**' to name a carbonitrile.

Thus, acetic acid is obtained from CH_3CN on hydrolysis; hence the common name of CH_3CN will be acetonitrile (acetic acid—'ic acid' + 'onitrile').

Problem 3.13. Write down the names of the following compounds:

(a) $(CH_3)_2CH{\cdot}CN$; **(b)** $CH_2{=}CH{\cdot}CN$; **(c)** 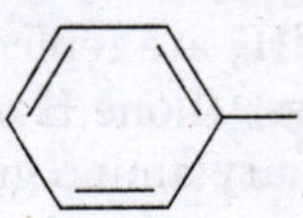CN.

Compound C on hydrolysis gives benzoic acid.

Solution: **(a)** This on hydrolysis gives $(CH_3)_2CH{\cdot}COOH$, isobutyric acid; therefore, the name of the compound is isobutyronitrile.

(b) This produces $CH_2{=}CH{\cdot}COOH$, acrylic acid, on hydrolysis; hence its name will be acrylonitrile.

(c) As benzoic acid is the product of hydrolysis of the given compound, its common name should be benzonitrile.

Carbonitriles may also be named according to the name of the hydrocarbon group attached to the cyano group. Thus, acetonitrile, CH_3CN, may be called methyl cyanide, the hydrocarbon group attached to —CN being methyl group.

Problem 3.14. Give the common names to the following compounds:

(a) 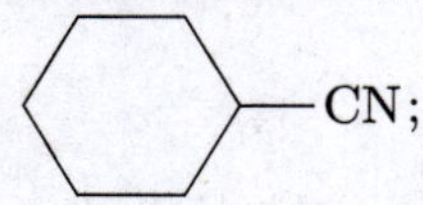

; **(b)** $(CH_3)_3C{\cdot}CN$; **(c)** $CH_2{=}CH{\cdot}CN$; **(d)** 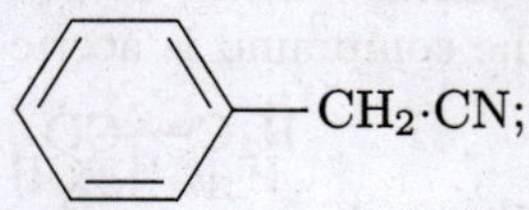

;

(e) $CH_3{\cdot}CH_2{\cdot}CH_2{\cdot}CN$.

Solution: **(a)** Cyclohexyl cyanide. **(b)** tert-Butyl cyanide. **(c)** Vinyl cyanide. **(d)** Benzyl cyanide. **(e)** *n*-Propyl cyanide.

3.7.16 Isocyanides

This class of compounds contain isocyano group ($-\overset{\oplus}{N}{\equiv}C^{-}$: or $-NC$) as functional group. In an isocyanide, the hydrocarbon group is attached to the isocyano group through the N atom. To name an isocyanide, the common name of the hydrocarbon group is first written and then the word isocyanide leaving a space in between the two. Thus, CH_3NC is methyl isocyanide and $(CH_3)_2CH{\cdot}NC$ is isopropyl isocyanide. Methyl isocyanide may also be named as methyl isonitrile.

Problem 3.15. What are the common names of the following compounds?

(a) $CH_2{=}CH{\cdot}NC$; **(b)** 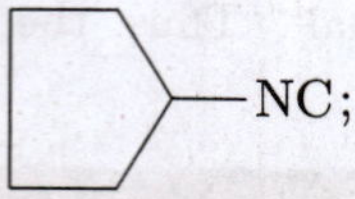

; **(c)** 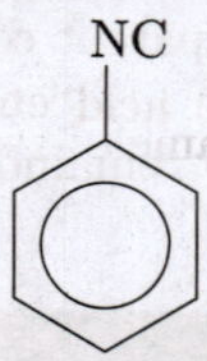

; **(d)** $CH_3{\cdot}CH_2{\cdot}CH_2{\cdot}CH_2{\cdot}NC$.

Solution: **(a)** The name of the hydrocarbon group being vinyl, its common name is vinyl isocyanide.

(b) Cyclopentyl isocyanide.

(c) Phenyl isocyanide.

(d) *n*-Butyl isocyanide.

3.7.17 Amines

When one or more Hs of NH_3 are replaced by a hydrocarbon group or groups, the compounds so formed are known as amines. If one H of NH_3 is replaced, it is called a primary (1°) amine; it contains $-NH_2$ group (primary amino group). Replacing 2 or 3 Hs of NH_3, one gets secondary (2°) or tertiary (3°) amines respectively; they have $-NH-$ (secondary) and $-\underset{|}{N}-$ (tertiary) amino functions respectively.

Amines have one-word common names which are obtained by combining the name of each hydrocarbon group on the N atom with the suffix 'amine'.

$CH_3{\cdot}NH_2$

Methylamine

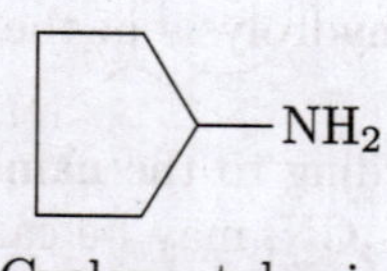

Cyclopentylamine

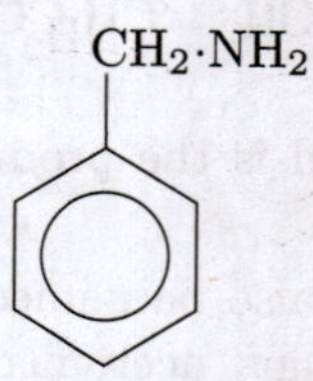

Benzylamine

$$\begin{array}{l} CH_3\cdot CH_2\cdot CH\cdot NH_2 \\ \qquad\qquad\quad | \\ \qquad\qquad\; CH_3 \end{array}$$

sec-Butylamine

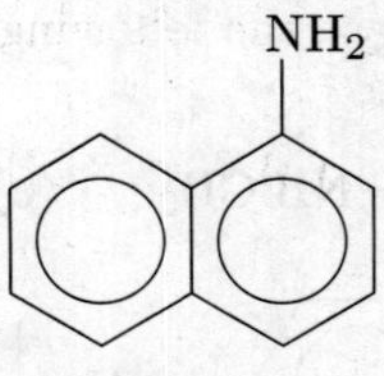

α-Naphthylamine

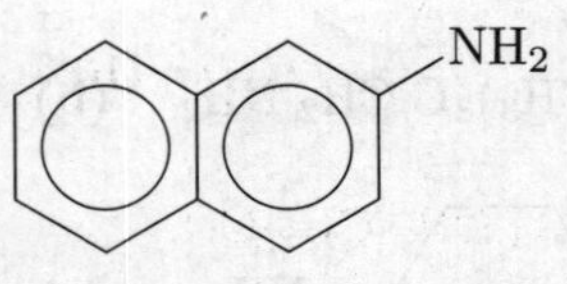

β-Naphthylamine

Some of the aromatic amines have trivial names

Aniline

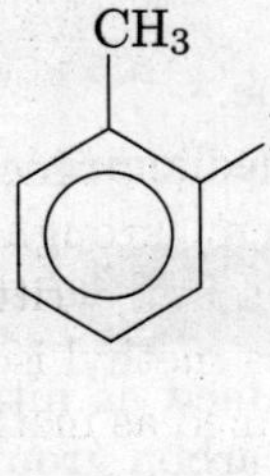

o-Toluidine

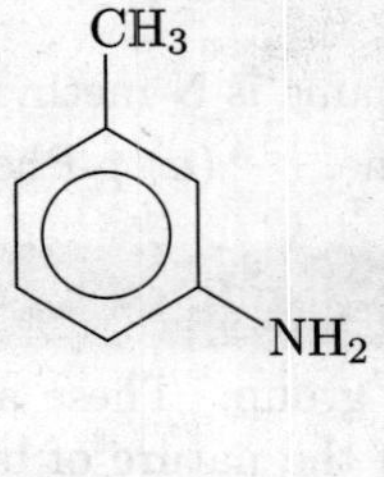

m-Toluidine

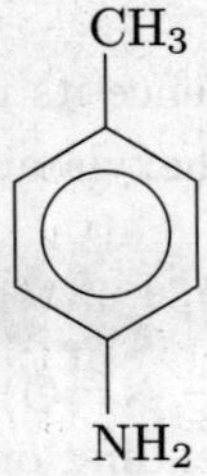

p-Toluidine

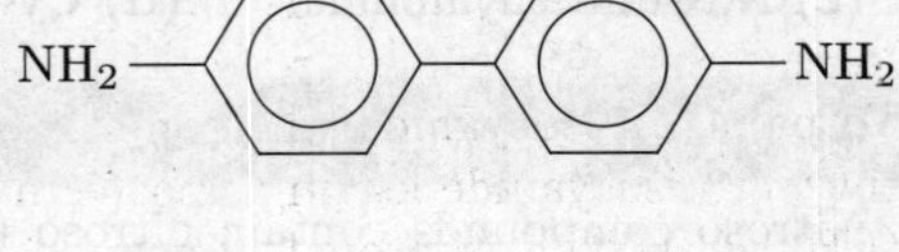

Benzidine

If a hydrocarbon group happens to occur more than once in an amine, a prefix 'di' or 'tri' is to be used before the one-word name of the amine; thus,

$CH_2\cdot NH\cdot CH_3$ — Dimethylamine

$(CH_3\cdot CH_2)_3N$ — Triethylamine

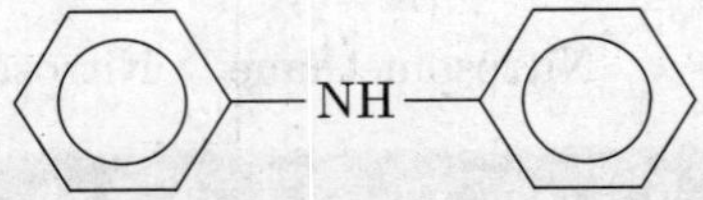

Diphenylamine

When different groups are attached to the N atom, the name of the least complex group is written first and that of the most complex group comes just before the name ending 'amine' in the name of a particular amine; for example,

$CH_3—NH—CH_2CH_3$ — Methylethylamine

$(CH_3)_2CH\cdot NH\cdot CH_2CH_3$ — Ethylisopropylamine

Aliphatic saturated diamines are named as polymethylene derivatives; thus,

$H_2N\cdot CH_2\cdot CH_2\cdot CH_2NH_2$ — Trimethylenediamine

$H_2N(CH_2)_4NH_2$ — Tetramethylenediamine

Aromatic nuclear substituted diamines are named according to the name of the hydrocarbon groups; e.g.,

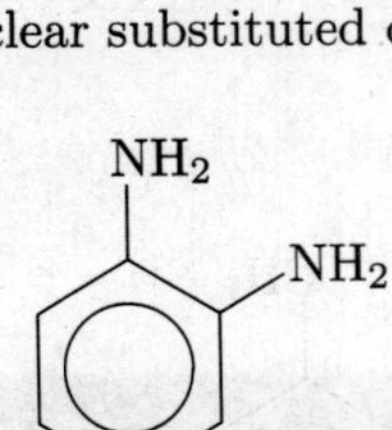

o-Phenylenediamine

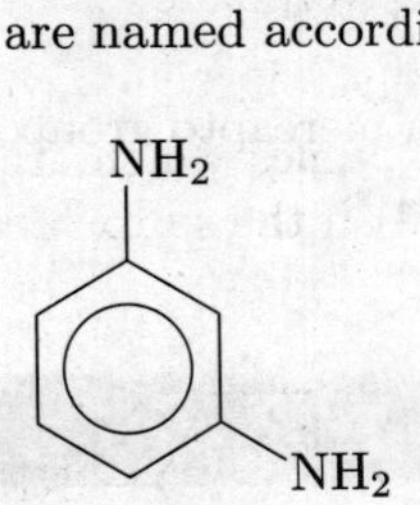

m-Phenylenediamine

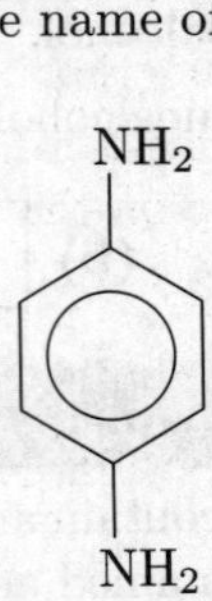

p-Phenylenediamine

Problem 3.16. Name the following amines:

(a) $(CH_3)_3C{\cdot}CH_2{\cdot}NH_2$ **(b)** 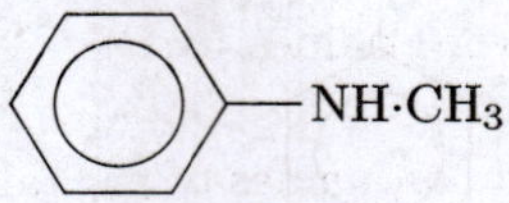 **(c)** 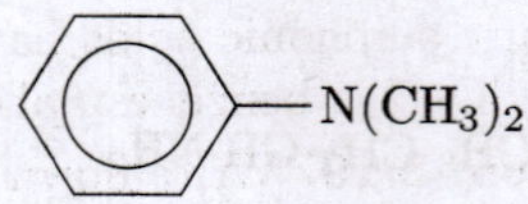

(d) 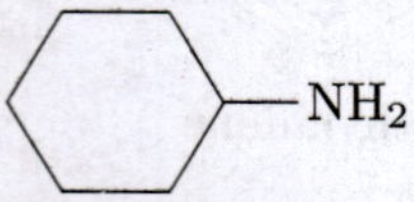 **(e)** 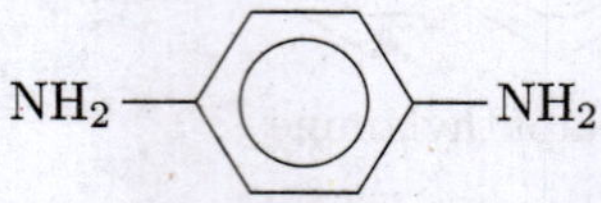

Solution: (a) $(CH_3)_3C{\cdot}CH_2$— is neopentyl group, so the name of the compound is neopentylamine.

(b) The mother compound is aniline; hence its name is N-methylaniline.

(c) N,N-dimethylaniline. **(d)** Cyclohexylamine. **(e)** *p*-Phenylenediamine.

3.7.18 Nitroso Compounds

Nitroso compounds contain nitroso (—N=O) group. These are named as nitrosoalkane, nitrosoarene or as nitrosocycloalkane (depending on the nature of hydrocarbon group the compounds possess); thus,

CH_3—N=O 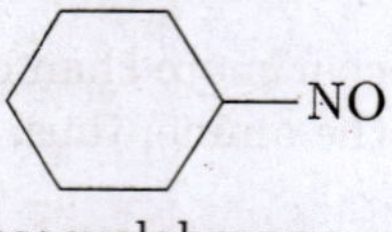 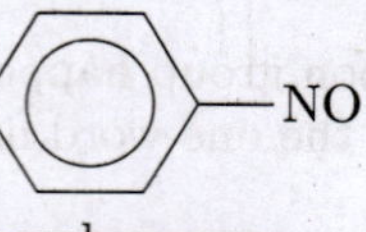

Nitrosomethane Nitrosocyclohexane Nitrosobenzene

3.7.19 Mercaptans (Thioalcohols)

These compounds contain mercapto function (—SH). Thioalcohols, i.e., mercaptans are named according to the name of the hydrocarbon group attached to the mercapto group. Thus, $(CH_3)_2CH{\cdot}CH_2SH$ is isobutyl mercaptan or isobutyl thioalcohol.

Problem 3.17. Write down the common names of the following compounds:

(a) $(CH_3)_3C{\cdot}SH$; **(b)** 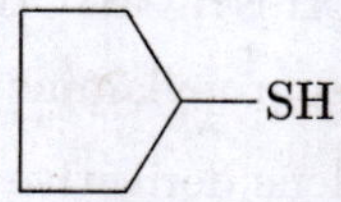 ; **(c)** 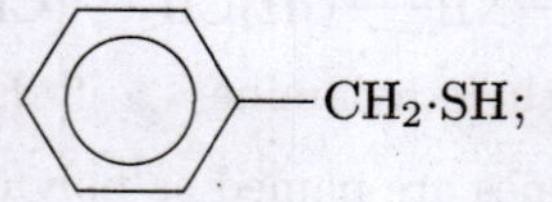 ; **(d)** $CH_2{=}CH$—$CH_2{\cdot}SH$.

Solution: (a) tert-Butyl mercaptan, $(CH_3)_3C$— being a ter-butyl group. **(b)** Cyclopentyl mercaptan. **(c)** Benzyl mercaptan. **(d)** Allyl mercaptan.

Problem 3.18. Write down the structure of the following compounds: **(a)** Ethyl thioalcohol; **(b)** Cyclobutyl thioalcohol.

Solution: As thioalcohols contain mercapto group (—SH),

(a) $CH_3{\cdot}CH_2SH$; **(b)** 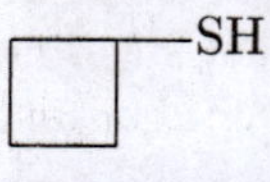

3.7.20 Sulphonic Acids

Sulphonic acids contain an —SO_3H group which is known as sulphonic acid group. Aliphatic sulphonic acids are named according to the name of the hydrocarbon group joined to the SO_3H group; these have one-word names; thus CH_3SO_3H is methylsulphonic acid.

Aromatic sulphonic acids have one-word names derived from the name of the parent arene; e.g., $C_6H_5SO_3H$ is benzenesulphonic acid.

Problem 3.19. Write down the common names of the following compounds:

(a) $(CH_3)_2CH{\cdot}SO_3H$; **(b)** 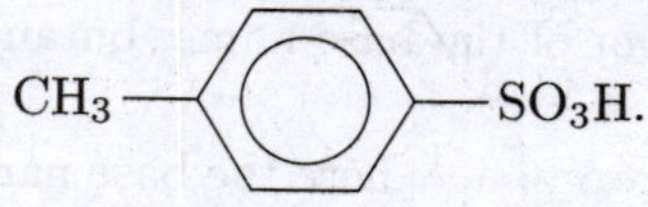

Solution: **(a)** Isopropylsulphonic acid. **(b)** *p*-Toluenesulphonic acid.

3.7.21 Phenols

Phenols are aromatic compounds having nuclear substituted hydroxyl (—OH) group. Hydroxybenzene is the simplest compound of this class and called phenol (C_6H_5OH). Common names of some of the phenols are given below. They bear 'ol' as name ending.

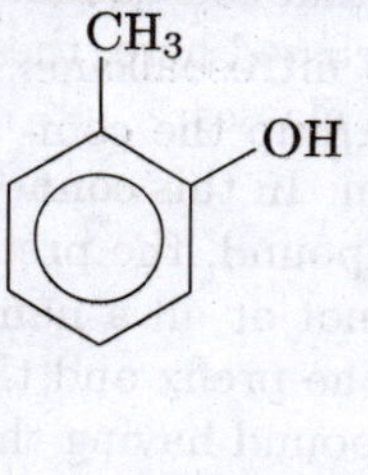
o-Cresol

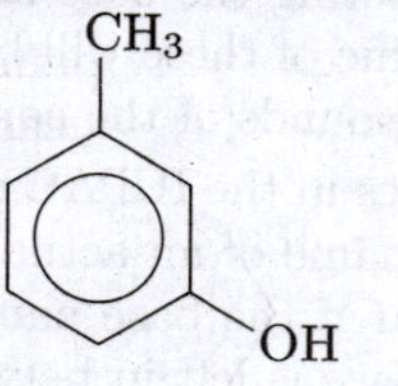
m-Cresol

p-Cresol

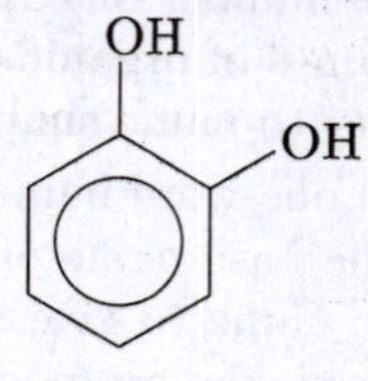
Catechol

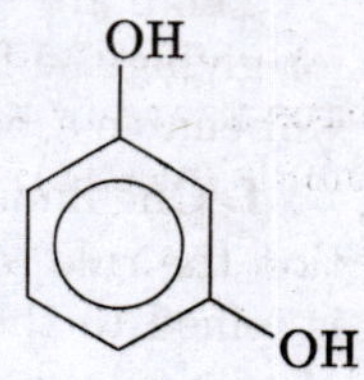
Resorcinol

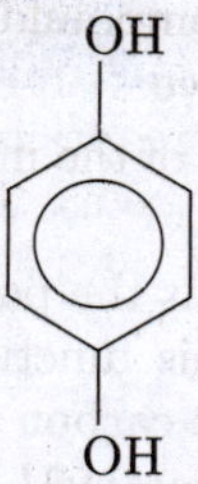
Hydroquinone

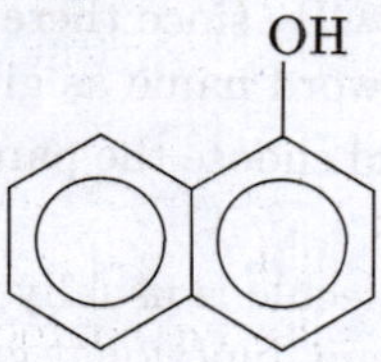
α-Naphthol

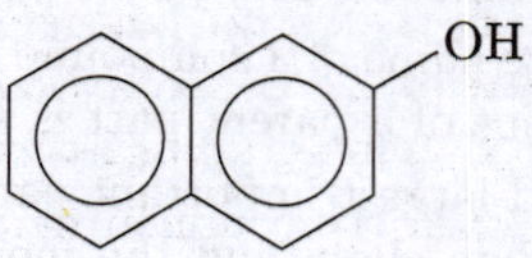
β-Naphthol

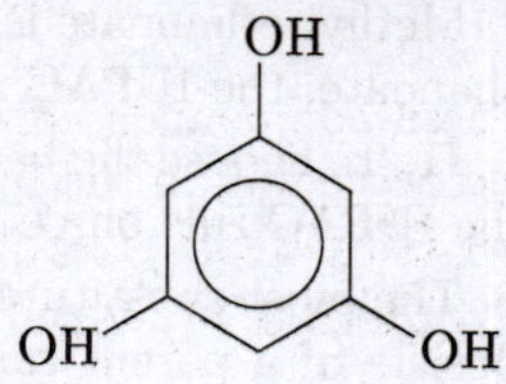
Phloroglucinol

3.8 IUPAC System of Nomenclature

This derived system of nomenclature is highly systematic but fails to cover all the organic compounds. However, for a given IUPAC name only one structural formula can be written; this helps one translate a name into a structural formula and vice versa.

The IUPAC name of a compound consists of a base name with or without a prefix; such a prefix may be called a prefix to the base name. Thus, the IUPAC name of 1 is 2, 3, 3-Trimethyl-1-butanol or, 2, 3, 3-Trimethylbutan-1-ol. Here the base name is butanol and the rest is a prefix to this base name.

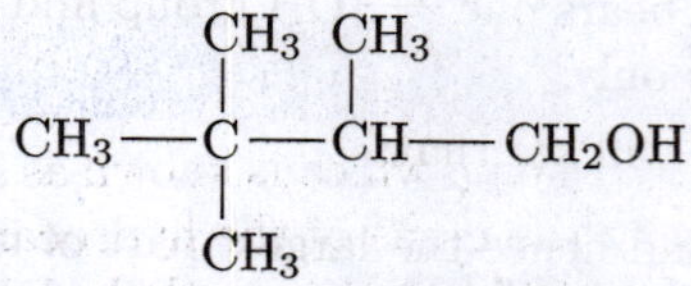

2, 3, 3-Trimethyl-l-butanol (USA) or 2, 3, 3-Trimethylbutan-l-ol (IUPAC).

1

The base name of a compound may be its full name, i.e., it may not have any prefix to its base name; e.g., CH_3CH_2OH is ethanol—it is its base name.

The base name, in its turn, consists of (i) a root or stem and (ii) a prefix or a suffix. In the IUPAC name of **1**, butan is the root of the base name, butanol, and ol is the suffix to the root, butan.

Compound **2** is 2-Methyl-1-nitropropane; here the base name is nitropropane—propane is the root of the base name and nitro is a prefix to the root, propane.

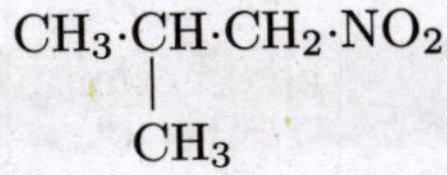

2-Methyl-1-nitropropane.
2

There are a large number of IUPAC rules for constructing the base names and consequently for giving the full names of organic compounds. Only some of these will be discussed here, these will enable a beginner to name simple and complex compounds of the common classes:

I. One finds both one-word names and two-word names in the IUPAC system. In this connection the rule is: if the base name of a compound is the name of an actual compound, the prefix is joined to the base name to give a one-word name; but if the base name is not at all a name of an actual compound, the prefix stands alone and a space is left in between the prefix and the base name. For example, in **2** the base name is nitropropane; as there is a compound having this name, **2** is given a one-word name. Salts and esters do not have one-word names.

Methyl ethanoate is the name for the ester CH_3COOCH_3 since there is no compound named ethanoate, the IUPAC name for CH_3COOCH_3 is a two-word name as given above.

II. To choose the base name of a compound one should choose the parent part of the molecule. The IUPAC rule on choice of a parent part is as follows:

The most substituted largest important part of a molecule is usually chosen as the parent. It consists of a parent carbon chain and the most important functional group; this function may also be considered as a part of the parent carbon chain, the base chain, when the carbon atom of the function is directly linked to a carbon atom of the chain. Thus in 2, 3, 3-Trimethyl-l-butanol we have chosen $CH_3{\cdot}C{\cdot}CH{\cdot}CH_2OH$ as the parent part of the molecule. Let us discuss whether the choice is correct or incorrect as per the rule stated above.

(a) $CH_3{\cdot}C{\cdot}CH{\cdot}CH_2$—may be chosen as the parent part; but this choice is wrong because this does not include the —OH group, the most important part of the molecule, which determines most of the properties of the compound.

(b) One may choose $CH_3{\cdot}C{\cdot}CH{\cdot}CH_3$ as the parent part, but this too will be wrong for the same reason as given in (a).

(c) If one chooses $CH_3{\cdot}CH{\cdot}CH_2OH$ as the parent part, it will also be wrong; because this is neither the largest part nor the most substituted part of the molecule though it contains the hydroxyl group. This part bears one —OH group and one tert-butyl group, $(CH_3)_3C$— and it is composed of 3 Cs only.

(d) In $CH_3{\cdot}C{\cdot}CH{\cdot}CH_2OH$ one will see that:

  (i) It consists of 4 Cs and hence the largest part of the molecule.

  (ii) It contains three methyl branches and the functional group, —OH; thus it is the most substituted important part.

III. The nature of the base chain and the number of C atoms present in it provide the root for the base name. The following are the general principles used in deciding the root for the base name:

A. Open chain compounds are named as derivatives of straight chain saturated hydrocarbons, called alkanes: Therefore, in the case of an open chain compound the base chain is a straight carbon chain. Naturally the root for the base name is derived from the name of the alkane corresponding to this chain (Table 3.2). In deriving the root, the name ending of the alkane, i.e., ane is changed in some cases and in some other cases it is kept as such. Whether 'ane' will be changed or not depends on the nature of the function present on the parent part. Thus, the base chain $CH_3{\cdot}C{\cdot}CH{\cdot}CH_2$— of the molecule $(CH_3)_3C{\cdot}CH(CH_3){\cdot}CH_2OH$ contains 4 Cs; hence the root for its base name is derived from the corresponding alkane, butane, and it is here butan.

B. Alicyclic compounds are usually named as derivatives of cycloalkanes, the cycloalkane ring being considered as the base chain. Hence the root is derived from the name of the cycloalkane that constitutes the compound (Table 3.5). Thus, **3, 4, 5** possess a cyclopentane ring; hence the roots for their base names are derived from the word cyclopentane.

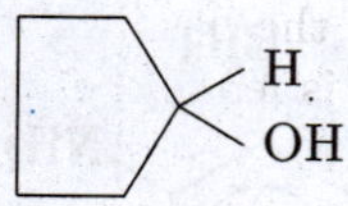

3

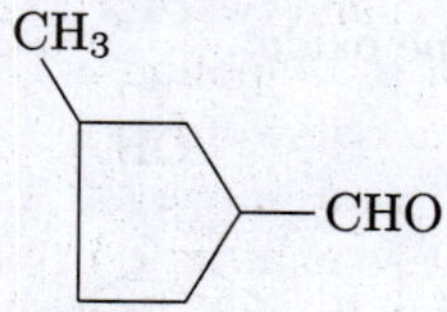

4

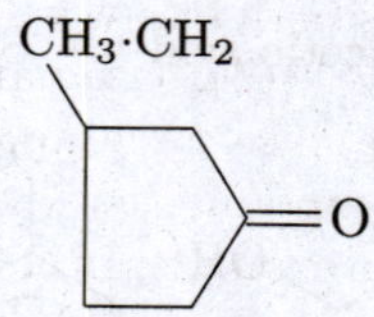

5

C. The base chains of aromatic benzenoid compounds may contain a benzene ring, a naphthalene ring, an antharacene ring, biphenyl system, etc. Naturally, the root for the base name of such a compound is derived from the name of the ring it has. Thus **6** is a derivative of naphthalene; hence the root for its base name is derived from the word naphthalene.

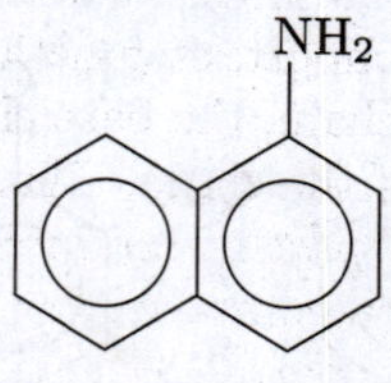 is 1-Naphthylamine.

6

The common names of some of the aromatic compounds are accepted as their IUPAC names (Table 3.9). The derivatives of these compounds have their roots derived from the names of these compounds.

Table 3.9 Common as well as IUPAC names of some aromatic Compounds

	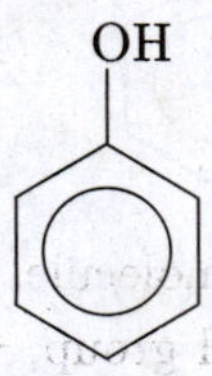	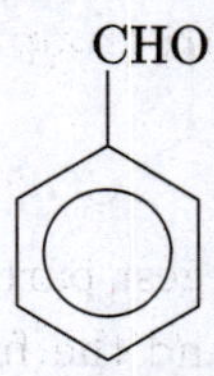	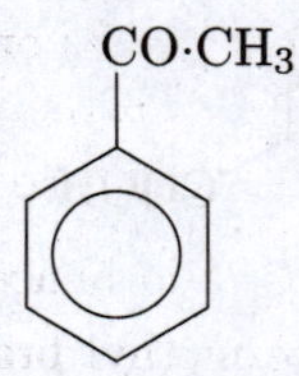	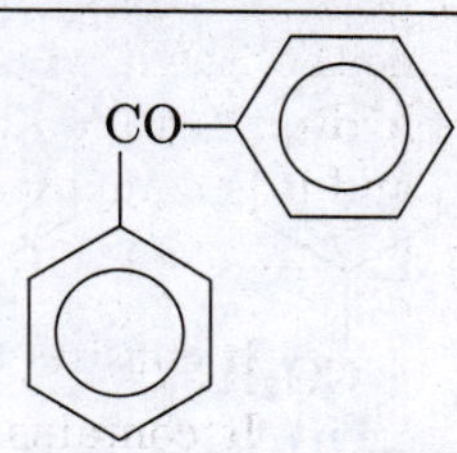
Toluene	Phenol	Benzaldehyde	Acetophenone	Benzophenone

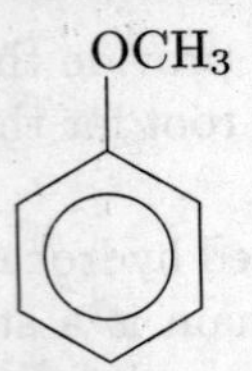

Anisole

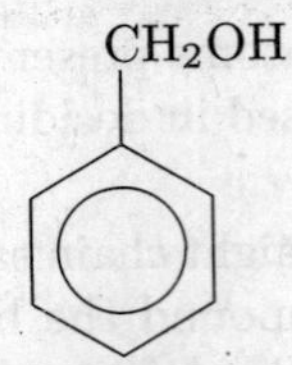

Benzyl alcohol

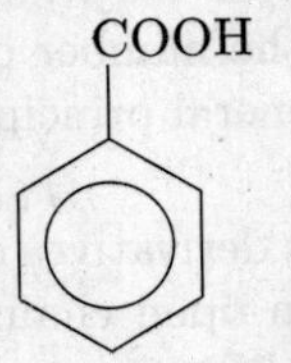

Benzoic acid

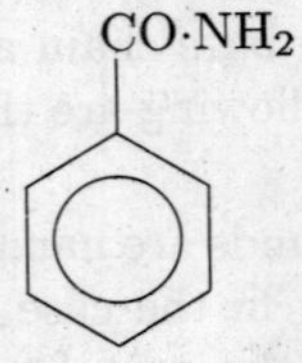

Benzamide

Aniline

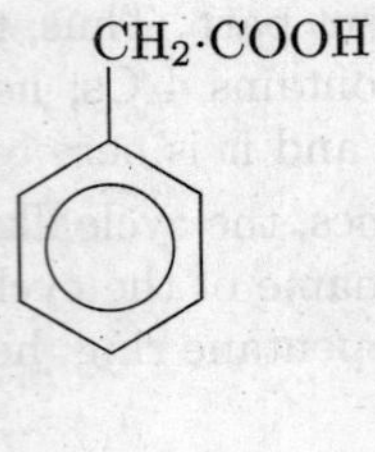

Phenylacetic acid

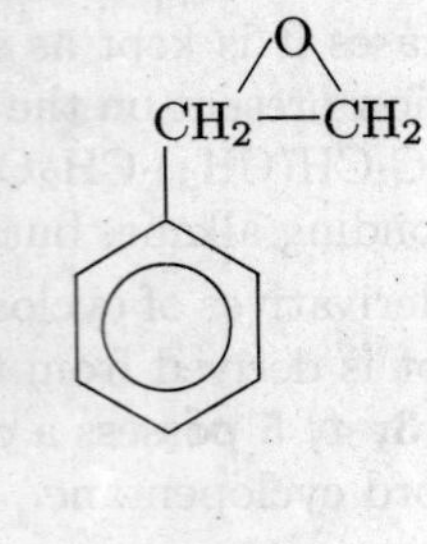

Styrene oxide

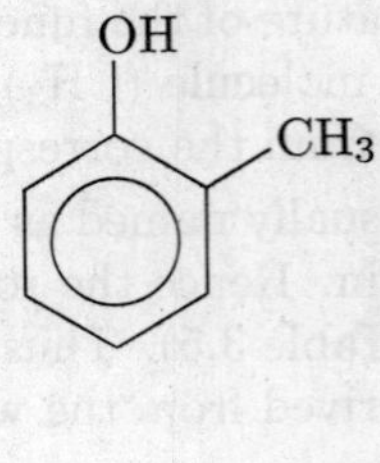

o-Cresol

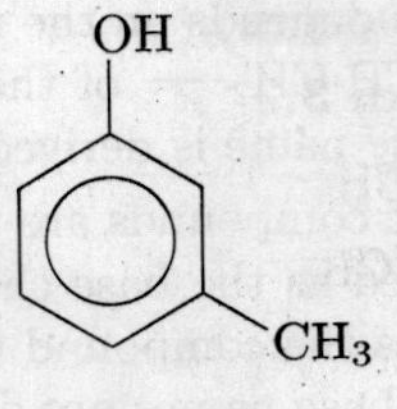

m-Cresol

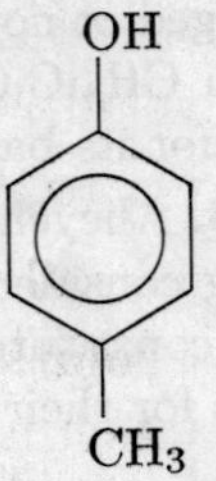

p-Cresol

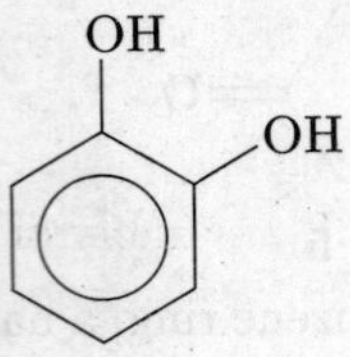

Catechol

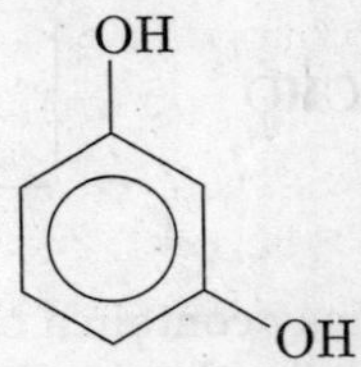

Resorcinol

Hydroquinone

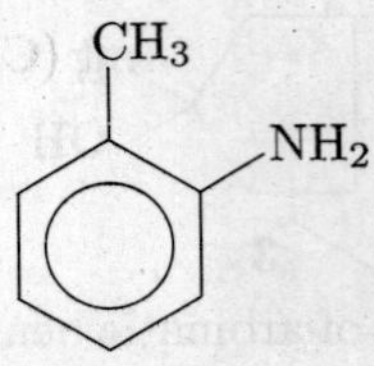

o-Toluidine

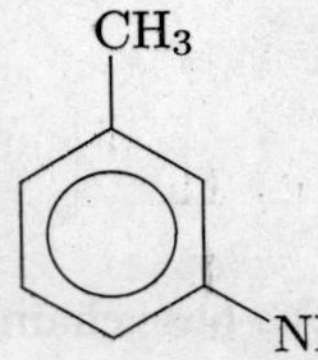

m-Toluidine

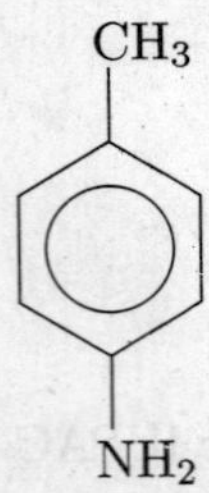

p-Toluidine

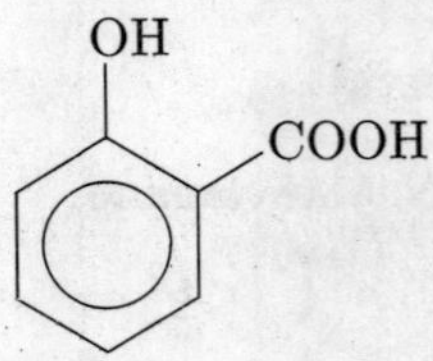

Salicylic acid

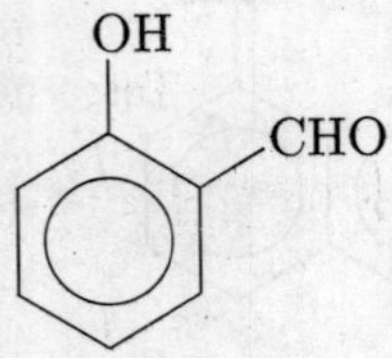

Salicylaldehyde

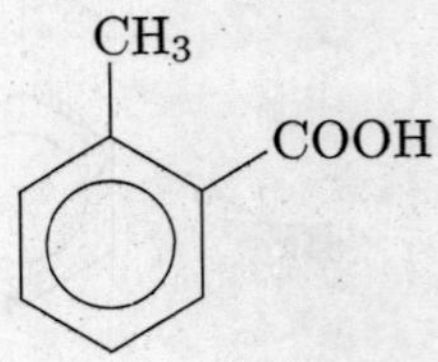

o-Toluic acid

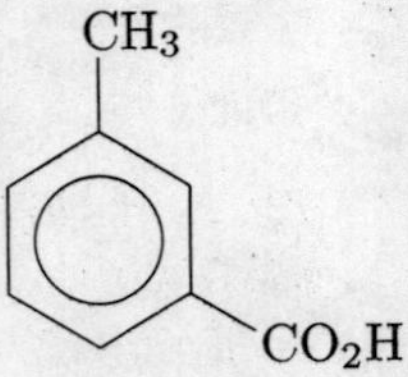

m-Toluic acid

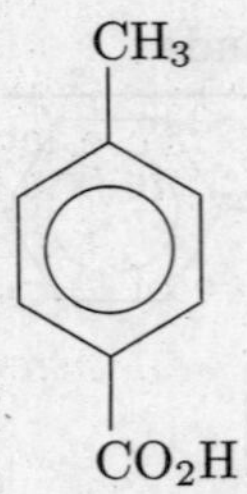

p-Toluic acid

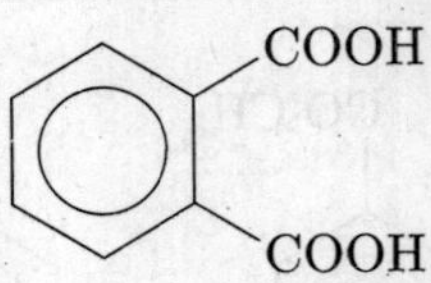

Phthalic acid

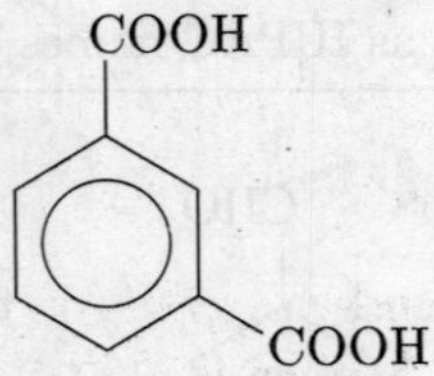

Isophthalic acid

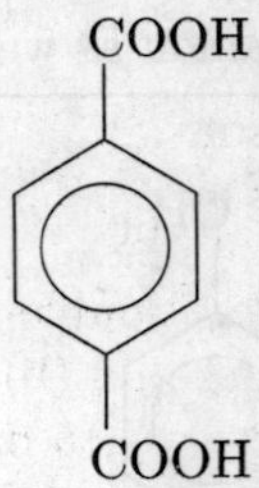

Terephthalic acid

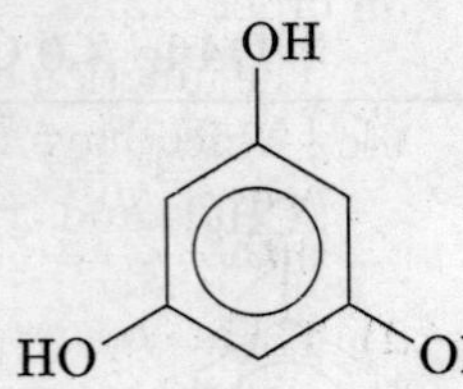

Phloroglucinol

Thus, **7** is a derivative of aniline and hence the root for its base is aniline.

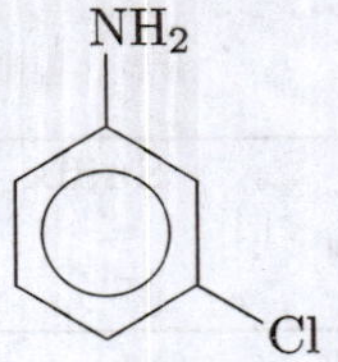

is *m*-chloroaniline or 3-chloroaniline.

7

D. The roots for heterocyclic compounds are also derived from their parent rings. The IUPAC sanctioned names of some heterocyclic compounds are given in Table 3.10.

Table 3.10 IUPAC names of some heterocyclic compounds

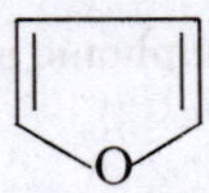
Furan

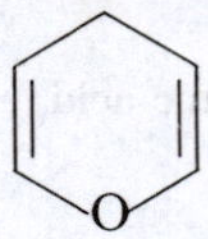
Pyran

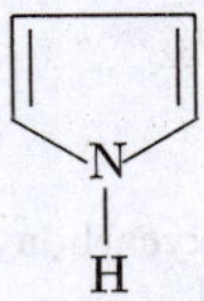
Pyrrole

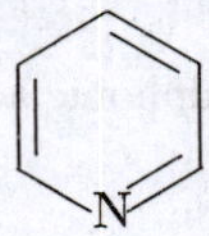
Pyridine

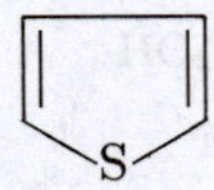
Thiophen(e)

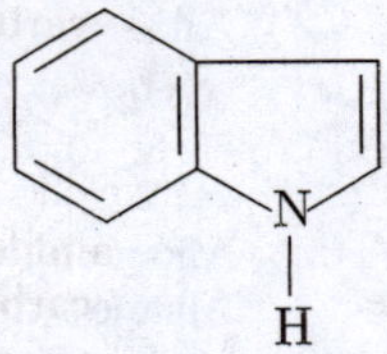
Indole

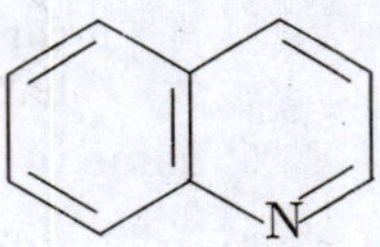
Quinoline

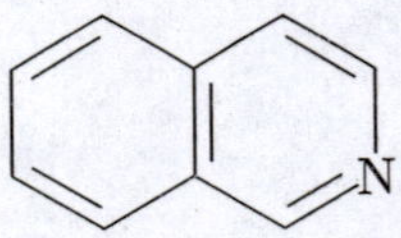
Isoquinoline

Thus, **8** is a derivative of pyrrole and its name is 5-bromopyrrole-2-carboxylic acid; here the base name is pyrrolecarboxylic acid—pyrrole is the root and carboxylic acid is the suffix.

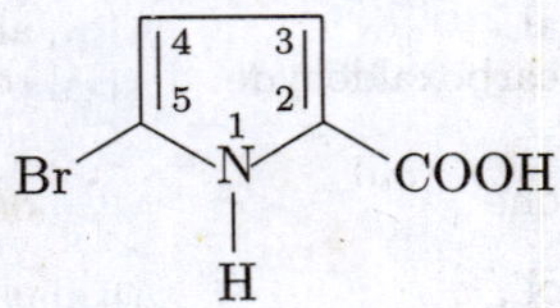

is 5-Bromopyrrole-2-carboxylic acid.

8

IV. The most important function, i.e., the primary function of a molecule determines the suffix or prefix to the root for constructing the base name. The primary function is decided by an order of precedence which is shown in Table 3.11.

The function of a monofunctional compound is usually denoted by suffix except for halo, nitro, etc., which have no suffixes.

Double and triple bonds are always indicated by suffixes only as there are no prefixes for them (Table 3.11).

Table 3.11 List of some functions with prefixes and suffixes in order of precedence (the order is followed up to $-\underset{|}{\overset{\bullet}{N}}-$ group)

Graphic formula of the function	Name of the function	Prefix	Suffix	The process of forming* base names of acyclic compounds
$\overset{\oplus}{N}R_3$, $\overset{\oplus}{S}R_2$, etc.	Ammonium Sulphonium		-onium (halide, sulphide)	Trialkyl ammonium
OOH		hydroperoxyl	Hydroperoxide	
$-\underset{\underset{OH}{\vert}}{C}=O$	Carboxyl	Carboxy	oic acid or carboxylic acid	Alkanoic acid Alkanecarboxylic acid
$-SO_2OH$	Sulphonic acid	Sulpho	sulphonic acid	Alkanesulphonic acid
$-\overset{\overset{OR}{\vert}}{C}=O$	Ester	Alkoxycarbonyl* Alkyl	carbalkoxylate . . .oate	Alkanecarbalkoxylate Alkyl alkanoate
$-\underset{\underset{X}{\vert}}{C}=O$	Carbonyl halide	Haloformyl	oyl halide carbonyl halide	Alkanoyl halide Alkanecarbonyl halide
$-\overset{\overset{NH_2}{\vert}}{C}=O$	Amido	Carbamoyl	amide carboxamide	Alkanamide Alkanecarboxamide
$-C\equiv N$	Carbonitrile	Cyano	nitrile	Alkanenitrile
$-\overset{+}{N}\equiv\overset{-}{C}$	Isonitrile	Isocyanide	carbylamine	Alkylcarbylamine
$-\overset{\overset{H}{\vert}}{C}=O$	Formyl	Formyl, oxo, aldo	al carboxaldehyde	Alkanal Alkanecarboxaldehyde
$C-\overset{\overset{O}{\Vert}}{C}-C$	Keto	Oxo or Keto	one	Alkanone
$-OH$	Hydroxyl	Hydroxy	ol	Alkanol
$-SH$	Mercapto	Mercapto	thiol	Alkanethiol
$-\overset{\vert}{N}-$	Amino	Amino	amine	Alkanamine Alkylamine
RO—	Ether linkage	Alkoxy	alkane	Alkoxyalkane Alkyl ether
RSO_2-	Alkyl Sulphonyl	Alkyl Sulphonyl	sulphone	Alkanesulphone
C=C	double bond	. . .	ene	Alkene
C≡C	Triple bond	. . .	yne	Alkyne
—Cl	Chloro	Chloro	. . .	Chloroalkane

Contd.

Graphic formula of the function	Name of the function	Prefix	Suffix	The process of forming* base names of acyclic compounds
—Br	Bromo	Bromo	...	Bromoalkane
—I	Iodo	Iodo	...	Iodoalkane
—N=N—	Azo	Azo	...	Azoalkane
$—NO_2$	Nitro	Nitro	...	Nitroalkane
—NO	Nitroso	Nitroso	...	Nitrosoalkane

IUPAC sanctioned common names of aromatic and heterocyclic compounds are used as the base names for naming their derivatives.

The suffix or prefix so determined is to be placed after or before the root respectively to construct the base name.

Thus, 2, 3, 3-trimethylbutan-1-ol is a monofunctional compound; the function —OH has -ol as its suffix. Butan is the root derived from butane as the base chain contains 4 Cs; the root, butan, when joined to the suffix ol, it becomes butanol, the base name of the compound.

V. Each of the carbon atoms in the base carbon chain is numbered sequentially by Arabic numerals; the positions of functions and hydrocarbon groups in the chain are indicated by these numbers. The assigned number to a C atom which bears a substituent is called the location number or locant of the substituent. Thus, in 1 the —OH group is attached to number 1 C atom and the three methyl groups are attached to number 2 and 3 C atoms in the base chain; therefore, in 1 the location number of the —OH group is 1 and those of the three methyl groups are 2, 3 and 3.

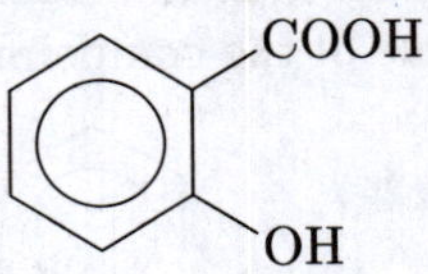

9

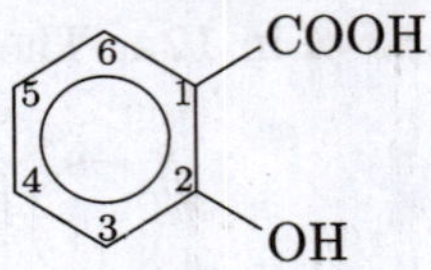

9a

Let us take another example: in **9a** the location number of —COOH group is 1 and that of —OH group is 2.

VI. The method of numbering base chains is different for different types of chains:

A. In the case of open chain compounds the numbering on Cs of the base chain is done from the end C that will give the lowest number to the primary function. If the primary function is equidistant from both the ends, the numbering starts from that end C which gives the lowest sum of the numbers for those Cs which carry the substituents. This may be called the lowest sum rule. For example, if numbering in **10** is done from the right end C, the locant of the —OH group will be 3 and that of the $—CH_3$ group will be 4 as shown in **10a**.

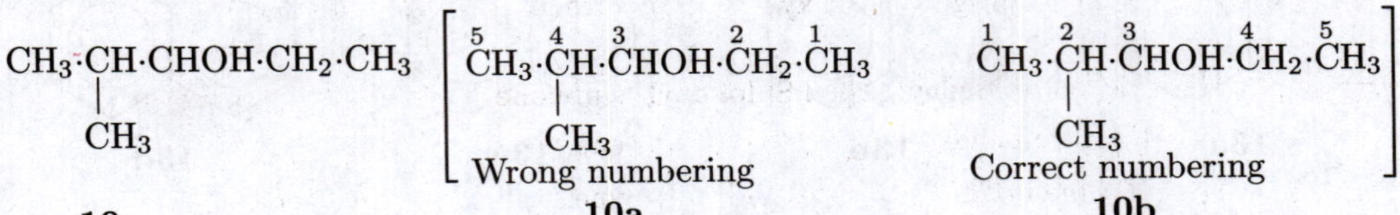

10 **10a** **10b**

On the other hand, if one numbers the Cs in the chain from the left end, the locant of —OH group and that of $—CH_3$ group will be 3 and 2 respectively as shown in **10b**. Thus in both the cases the locant of the —OH group is 3, i.e., it is equidistant from both of the ends; but in **10b**, it is found that the sum of the locants of —OH group and —CH_3 group is (3 + 2) or 5 and that in **10a** is (3 + 4) or 7. Therefore, according to the lowest sum rule, the numbering of Cs in **10** should start from the left end C.

B. In the case of alicyclic compounds consisting of one ring and also for benzene derivatives, the numbering of Cs starts from that C which carries the primary substituent and to that direction which gives the lowest sum of the locants.

Thus in **11** the numbering of Cs in the benzene ring starts from that C which carries the —COOH group, the primary function, and in the clockwise direction as shown in **11a**. If the numbering is done in anticlockwise direction, it will not follow the lowest sum rule as shown in **11b**.

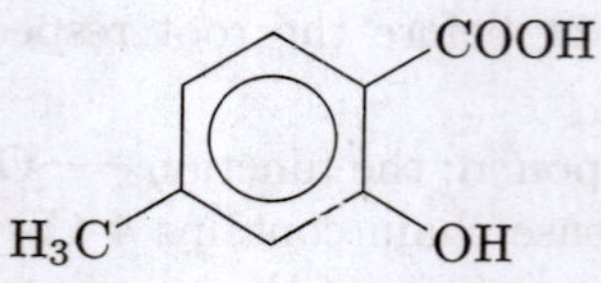

11

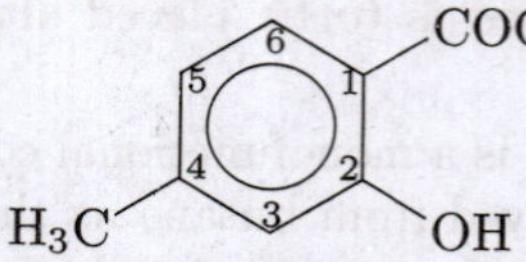

Right numbering

11a

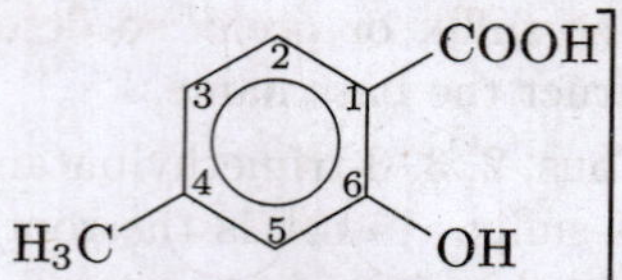

Wrong numbering

11b

There is another important point in this connection which tells us that like groups should be treated similarly and numbering on ring Cs is to be done accordingly, e.g., **12** cannot be named as a derivative of aniline because the two —NH_2 groups then are not treated similarly; in this case the unlike group, the nitro group, is to be considered as principal function and consequently the numbering should start from the NO_2— substituted C atom; the lowest sum rule should also be followed as shown in **12a**. The base name of the compound will be then nitrobenzene.

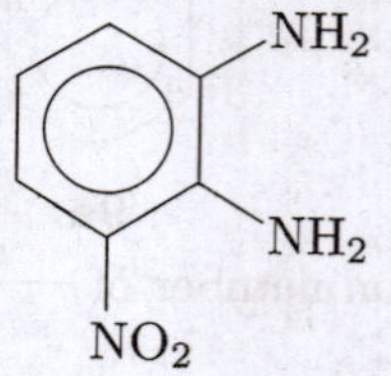

12

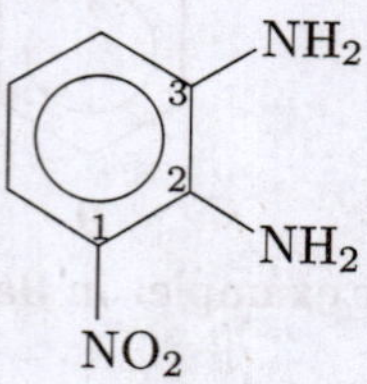

12a

C. In the case of polynuclear aromatic and alicyclic compounds special numbering system are adopted; these will be discussed in detail in the proper chapters. Here we shall deal with naphthalene only as an example. Numbering on Cs of naphthalene ring is done as shown in **13a, 13b, 13c, 13d,** four 1 positions being actually one and the same in these structures of naphthalene.

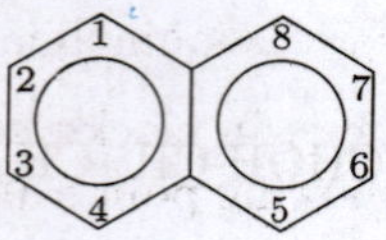

13a

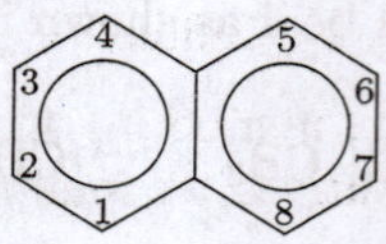

13b

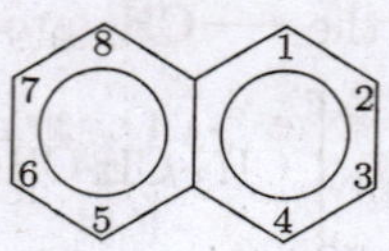

13c

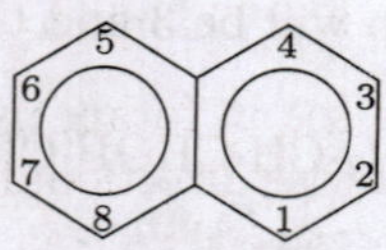

13d

To number naphthalene derivative we shall have to consider the following points:

(a) The numbering will start from that 1 position which will give the smallest locant to the highest important group. This is also followed in numbering C atoms of anthracene derivatives, where numbering on Cs of anthracene being done as follows in **14a, 14b**.

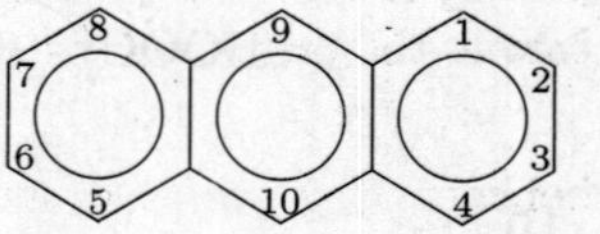

14a

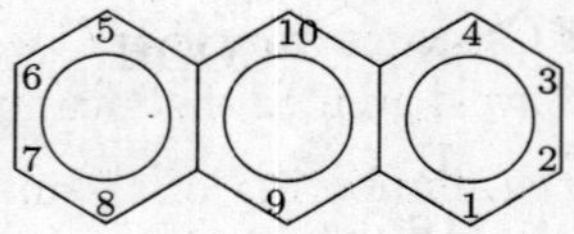

14b

For example, the Cs on compound **15, 16** and **17** are numbered as shown below :

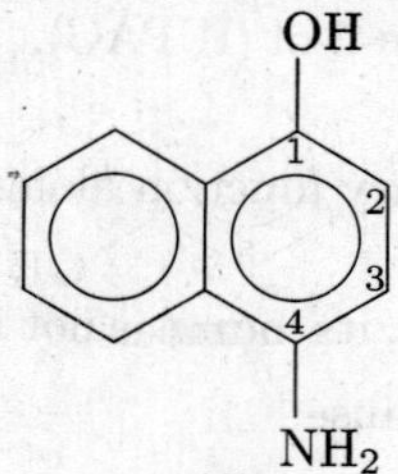

15

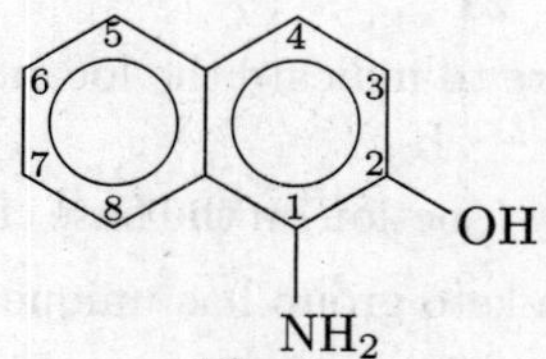

16

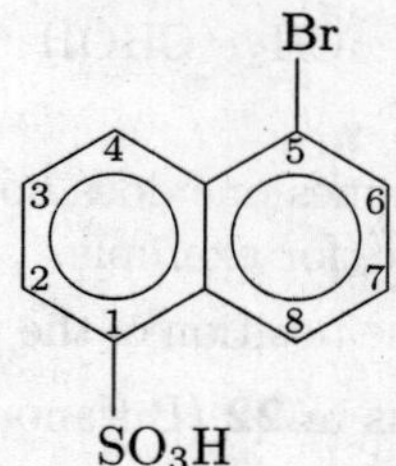

17

(b) Lowest sum of the locants should also be followed.

D. The constituent atoms of a heterocyclic compound consisting of one hetero atom are usually numbered from that atom, i.e., the hetero atom is given number 1 and others are numbered sequentially:

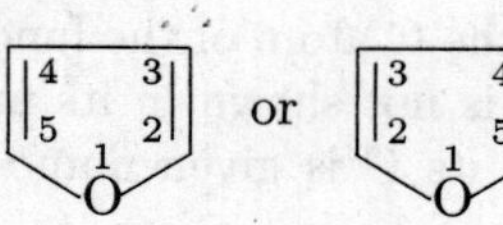

Furan

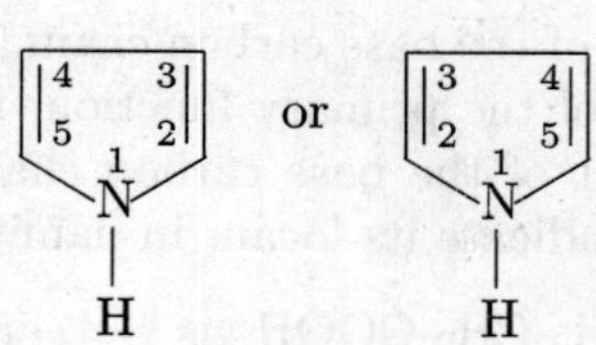

Pyrrole

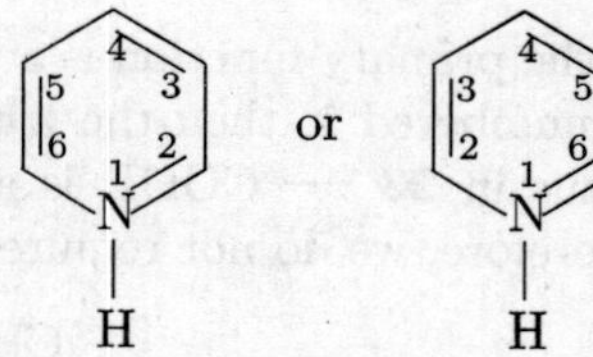

Pyridine

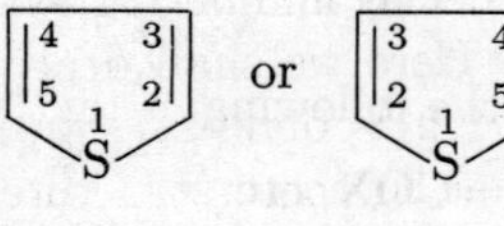

Thiophen

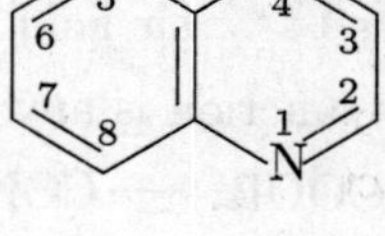

Quinoline

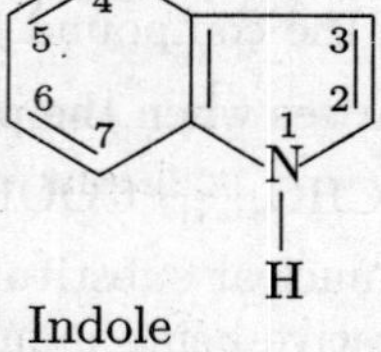

Indole

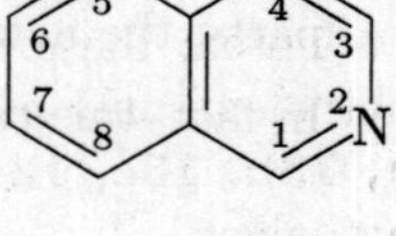

Isoquinoline

Thus we see in furan, pyrrole, thiophen, pyridine, etc., numbering may be done in clockwise or anticlockwise direction; for numbering Cs of their derivatives, we follow the following points along with the general rule as stated above :

a. The lowest sum rule and b. more important function should get the lower number.

Thus **18** is numbered as shown below; this is done so because carboxyl group has priority over Br. For the same reason **19** and **20** are numbered as shown below.

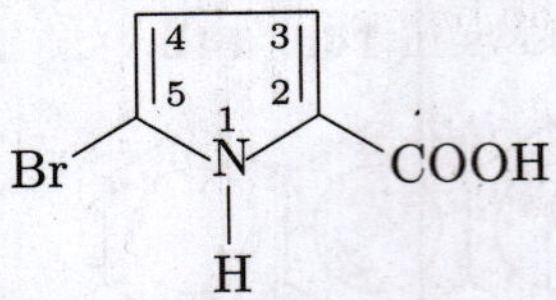

18

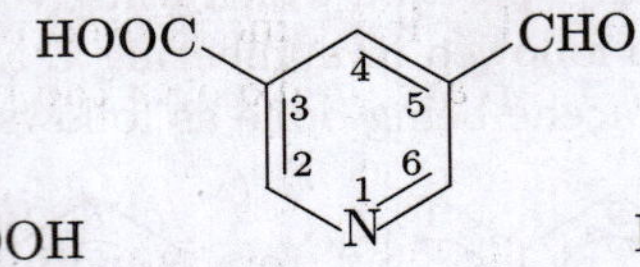

19

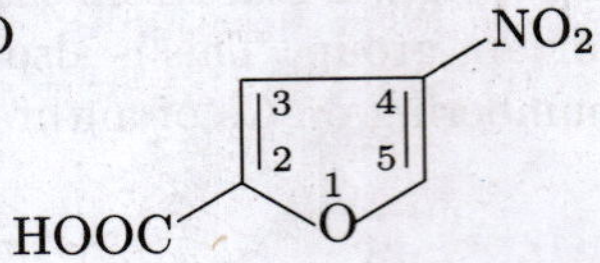

20

VII. Knowing the locant of the primary function, it is joined to the base name by a hyphen. Thus in **21** the locant of —OH group is 2; so the base name is written as 2-butanol or butan-2-ol.

$^1CH_3 \cdot {}^2CHOH \cdot {}^3CH_2 \cdot {}^4CH_3$ is 2-butanol (U.S.A) or; Butan-2-ol (IUPAC).

21

Sometimes one does not require to indicate the locant of the primary function along with the base name, for example,

(a) If the position of the primary function on the base chain is unique, its locant is not indicated.

Thus in **22** (Butanone), the keto group has unique position because:

(i) The end C cannot be a keto group as its valency is two.

(ii) In whatever way one can write the structure of 22, the second C atom will be the ketonic C atom as it should be given the lowest locant.

$^1CH_3 \cdot {}^2CO \cdot {}^3CH_2 \cdot {}^4CH_3$ $\quad$ $^4CH_3 \cdot {}^3CH_2 \cdot {}^2CO \cdot {}^1CH_3$

22 $\quad$ **22**

(b) If the primary function is a part of the base carbon chain and if the C atom of the function is numbered 1, then the locant of the primary function, i.e., 1, is not shown in its name. Thus in **23** —COOH is a part of the base carbon chain and its C is given number 1; therefore, we do not require to indicate its locant in naming **23**.

$CH_3 \cdot CH_2 \cdot CH_2 \cdot COOH$ is Butanoic acid.

23

Here butan is the root derived from butane and oic acid is the suffix; uniting these two parts, the name of the compound becomes butanoic acid.

In fact, this case arises when the primary function is any one of the following:

—CHO, —COOH, —COOR, —$CONH_2$, —COX, etc.

(c) The locant of the nuclear substituted primary function of an alicyclic compound and that of a benzene derivative being number 1, it is understood in many a case. Thus, in **24** the primary function is OH and its base name is cyclopentanol; here the locant of the —OH group is not shown

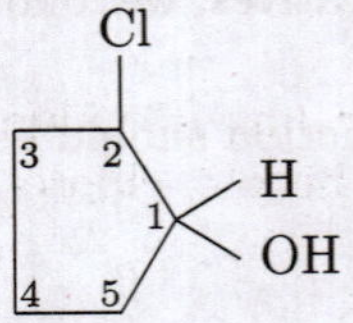

is 2-Chlorocyclopentanol.

24

VIII. The names of the functions other than primary one and those of branches, i.e., hydrocarbon groups of the base carbon chain are denoted by prefixes to the base name of a compound. Thus in **25** the primary function is —CHO, its name is used to construct the base name. The names of —OH group and —CH_3 group are prefixed to the base name.

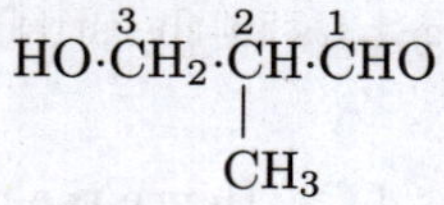

is 3-Hydroxy-2-methylpropanal.

25

This rule has exceptions:

(i) Double bond, (ii) triple bond and (iii) two or more identical functions are denoted by suffixes only.

IX. A. Branches are always indicated by prefixes. To indicate a branch or a function as a prefix to the base name, its locant is prefixed to its name by a hyphen and the combined part is joined to the base name, e.g., in **26** the —CH_3 branch has locant 2; when the locant of —CH_3 and its name is combined, it becomes 2-Methyl. The base name of the compound is butanal; therefore, the full name of **26** is 2-Methylbutanal.

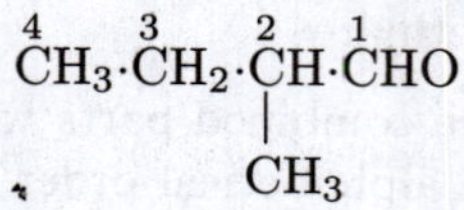

is 2-Methylbutanal.

26

B. When the base name does not bear any locant, the prefix is joined to the base name without using hyphen; but, if the base name bears the locant of the primary function, the prefix is joined to the base name by a hyphen. Thus in **27** the prefix 3-methyl is joined to the base name, 1-butanol, by a hyphen

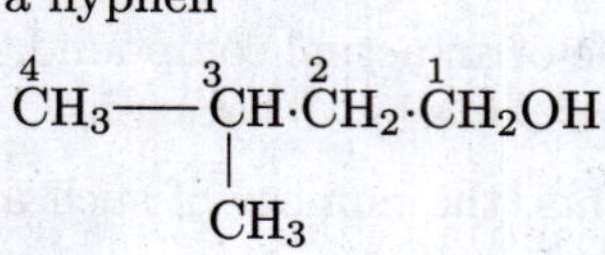

is 3-Methyl-1-butanol.
or 3-Methylbutan-1-ol.

27

X. If there are two or more substituents, branches or functions, each of them is to be combined with its locant and then the combined parts are arranged either alphabetically or in order of increasing complexity and joined to each other by hyphens. The last combined part is joined to the base name without hyphen if the base name does not have any locant; but if there is a locant before the base name, the last combined part is joined to the latter by a hyphen—this will give the full IUPAC name of the compound.

IUPAC accepted names of the branches, i.e., hydrocarbon groups are given in Secs. 3.6.1, 3.6.2 and 3.6.3. The first letter of tert-alkyl group and that of sec-alkyl group, i.e., t- and s- are not considered in determining alphabetical order; but I of isoalkyl group and N of neoalkyl group are considered.

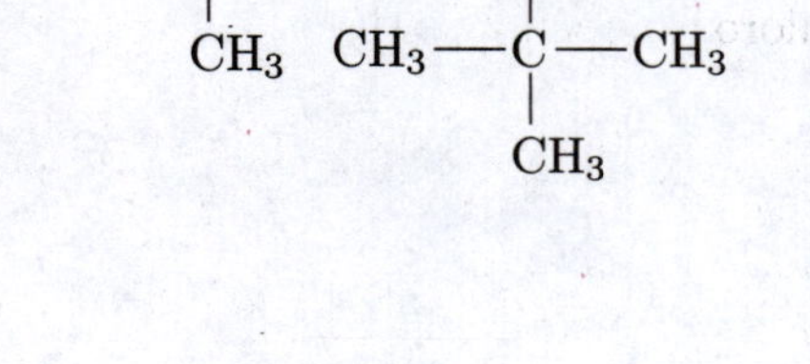

is 2-tert-butyl-3-chloro-3-methyl-1-butanol.
or, 2-t-butyl-3-chloro-3-methylbutan-1-ol.

28

In **28**

(i) The parent part is $CH_3C{\cdot}CH{\cdot}CH_2OH$ because

(a) it contains the primary function, —OH

(b) it is the most substituted largest chain containing 4 Cs in the straight chain and 4 substituents ($Cl, CH_3, (CH_3)_3C$— and —OH).

(ii) The parent chain is thus $CH_3{\cdot}C{\cdot}CH{\cdot}CH_2$—; it contains 4 Cs. Hence the root for the base name is **butan**.

(iii) The primary function is —OH; its name as suffix is **ol**. Therefore, the base name is **butanol**.

(iv) To indicate the —OH group by a locant, numbering is done from the right end C. Therefore, the base name with the locant of —OH group is 1-**butanol** or, butan-1-ol.

(v) The locant of tert-butyl branch is 2 and this is denoted by 2-tert-butyl.

(vi) The locant of Cl is 3; this is to be denoted by 3-chloro.

(vii) The locant of —CH_3 is 3; it is denoted by 3-methyl.

(viii) 2-tert-butyl, 3-chloro and 3-methyl are thus three combined parts which are to be joined and prefixed to the base name, 1-butanol, in an alphabetical order placing hyphen after each combined part.

As 't' of tert-butyl is not considered to determine alphabetical order, **b** of **b**utyl comes before **c** of **c**hloro, and **m** of **m**ethyl comes after **c** of **c**hloro; therefore, the prefix to the base name when arranged alphabetically is 2-tert-butyl-3-chloro-3-methyl.

(ix) As the base name of **28**, i.e., **butanol** is a name of an actual compound, **28** should be given one-word name; thus it is 2-tert-**butyl**-3-**chloro**-3-**methyl**-1-**butanol**.

XI. If there are two or more identical substituents, the number of such a group is indicated by a prefix to its name as shown below:

di = two, tri = three, tetra = four, penta = five, hexa = six, etc.

The location numbers of such a substituent are all retained and written in an increasing order; the numbers are separated by commas amongst themselves and the last number is separated from the name of the substituent by a hyphen.

The first letters of di, tri, tetra, penta, etc., do not get any importance in considering the alphabetical order for assembling prefixes to the base name.

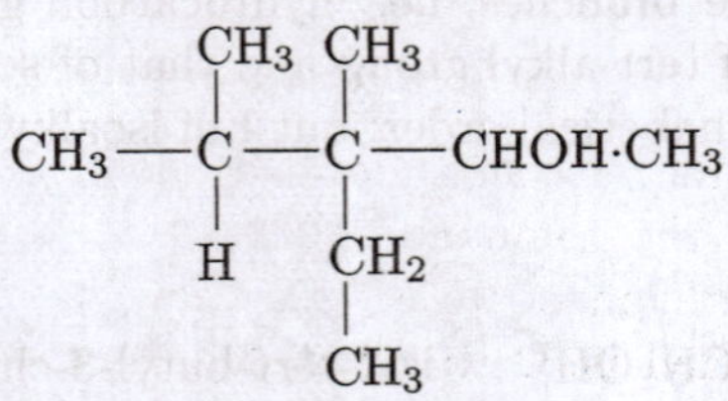

29

In **29**

(i) The parent part: $CH_3CHC{\cdot}CHOH{\cdot}CH_3$.

(ii) The base chain: $CH_3 \cdot CH \cdot C \cdot CH \cdot CH_3$; thus root is pentan

(iii) The primary function: —OH; its name as suffix is ol.

(iv) The base name: pentanol.

(v) The base name with the locant of the primary function: 2-pentanol or pentan-2-ol.

(vi) There are two $—CH_3$ branches; its name is methyl and its locants are 3 and 4 as Cs are numbered from the right end with respect to the reader.

Therefore, these two branches are to be represented as 3,4-dimethyl.

(vii) There is one $—CH_2CH_3$ branch; its name is ethyl and its locant is 3; hence it is to be represented as 3-ethyl.

(viii) Two methyl branches and one ethyl branch are to be used as prefixes to the base name, pentanol; for this they are to be arranged alphabetically. As di of dimethyl does not get importance in deciding alphabetical order and as *e* of ethyl comes before *m* of methyl in the alphabet, the prefixes are to be combined as 3-ethyl-3,4-dimethyl.

(ix) The base name, 2-pentanol, being a name of an actual compound, **29** should have a one-word IUPAC name; thus it is 3-ethyl-3,4-dimethylpentan-2-ol.

XII. If a substituent on a base chain contains substituted groups, the substituent is called a compound substituent. Thus, **31** is the compound substituent in **30**.

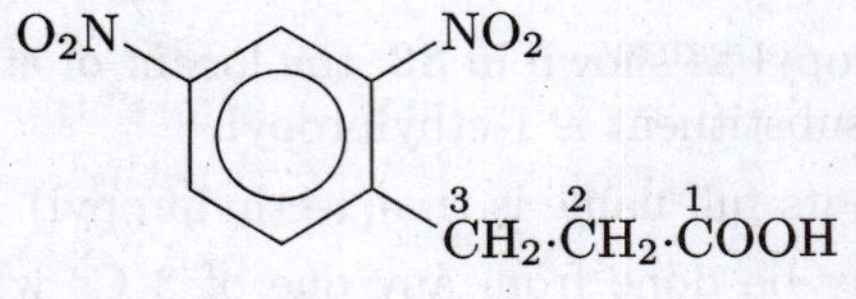

30

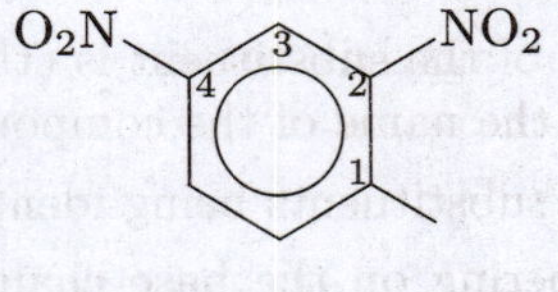

31

Compound substituents are indicated as given below:

(a) The name of a compound substituent is given as one gives the name of a compound. Thus the base chain in **31** is phenyl ring; it contains two nitro groups; hence its name is dinitrophenyl.

(b) The substituted groups of a compound substituent are to be located; to do this, numbering within the compound substituent is done and it starts from the point of attachment of the substituent to the chain, i.e., the C atom of the substituent which is attached to the base chain is given number 1 and the lowest sum rule is followed. Thus, in **31** the locants of two nitro groups are 2 and 4 as number 1 is the C attached to the main chain. Hence the full name of the compound substituent is 2,4-dinitrophenyl.

(c) To avoid ambiguity, the name of the compound substituent is generally enclosed in the parentheses. The locant of the compound substituent with respect to the base chain is joined before its name by a hyphen. In **30** the base chain is $CH_2 \cdot CH_2 \cdot COOH$; therefore, the locant of 2,4-dinitrophenyl is 3. Hence this is represented as 3-(2, 4-dinitrophenyl) propanoic acid.

(d) The first letters of di, tri, etc., get importance in alphabetisation purpose in considering substituent groups in a compound substituent.

(e) If a compound possesses two or more identical compound substituents, the prefixes di, tri, tetra, penta, etc., are replaced by 'bis', 'tris', 'tetrakis', 'pentakis', etc., respectively.

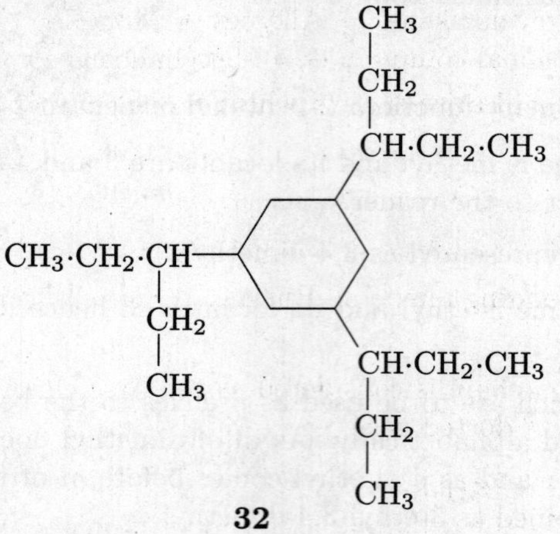

32

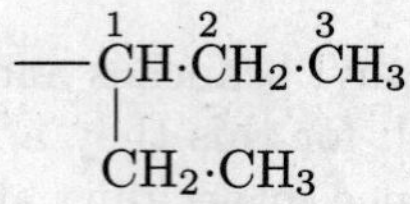

33

In **32**

(i) The most substituted part is cyclohexane ring and, therefore, it is the base chain. Hence the base name is cyclohexane.

(ii) There are three identical substituents, each of which bears substituted groups, i.e., these are compound substituents.

(iii) The name of the substituent is ethylpropyl as shown in **33**; the locant of ethyl group is 1. Therefore the name of the compound substituent is 1-ethylpropyl.

(iv) The three substituents being identical, its full name is tris-(1-ethylpropyl).

(v) The numbering on the base chain may be done from any one of 3 Cs which bear substituents; therefore, the locants of the three compound groups are 1, 3 and 5. Thus the full name of **32** is 1,3,5-tris-(1-ethylpropyl)cyclohexane.

We are now in a position to draw a sequential procedure to name an organic compound in the IUPAC system on the basis of the rules discussed so far:

(a) Select the parent part of the molecule (Rule-II).

(b) Select the base carbon chain and determine the root for the base name (Rule III).

(c) Determine the primary function and thus decide the suffix or prefix to the root for constructing the base name (Rule IV and Table 3.11).

(d) Number the Cs of the base carbon chain. (Rules V and VI).

(e) Represent the base name with the locant of the primary function if necessary (Rule VII).

(f) Decide whether the IUPAC name of the given compound will be a one-word name or not (Rule I).

(g) Decide the names of the prefixes to the base name (Rule VIII and Table 3.11).

(h) Combine each of the prefixes with its locant (Rule IX).

(i) Assemble the prefixes with locants in the alphabetical order (Rules X, XI and XII).

(j) Join the combined prefix to the base name and this will give the full IUPAC name of the given compound (Rule X).

Following this general procedure we shall now discuss the nomenclature of some organic compounds under the heads given below:

• Alkanes and cycloalkanes, • Alkenes and cycloalkenes, • Alkynes, • Monofunctional compounds, • Bifunctional compounds, • Polyfunctional compounds, • Miscellaneous compounds.

In this connection a few more IUPAC rules will arise each of which is specially applicable to a particular class of compounds.

3.8.1 Alkanes and Cycloalkanes

(1a) Alkanes and cycloalkanes have the name ending ane, e.g., Butane, Pentane, Cyclopropane, Cyclohexane.

(1b) For alkanes, the longest continuous carbon chain is considered as the base chain. Thus, in **34** the longest continuous chain is indicated by a dotted cage.

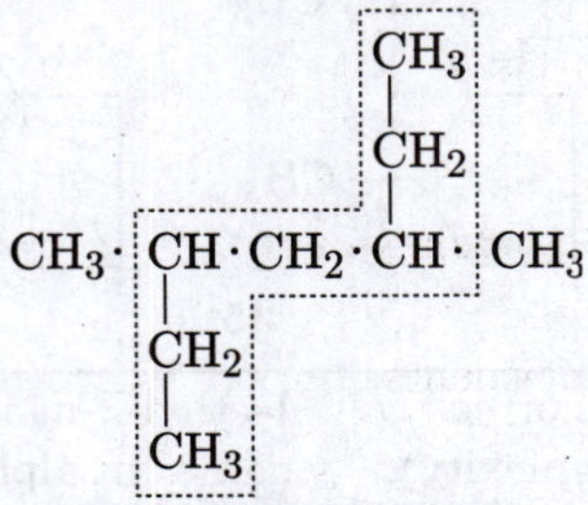

34

The root for the name of the alkane is determined by the number of Cs present in the base chain (Sec. 3.7.1, Rule I). Thus, in **34** the number of Cs in the base chain is 7; hence the root is hept. The base name is, therefore, heptane.

(1c) For cycloalkane, the alicyclic ring is the base chain. Its base name is that of the alkane containing the same number of Cs as the alicyclic ring with prefix cyclo. In **35** the base chain is a cyclopentane ring; hence its base name is a cyclopentane.

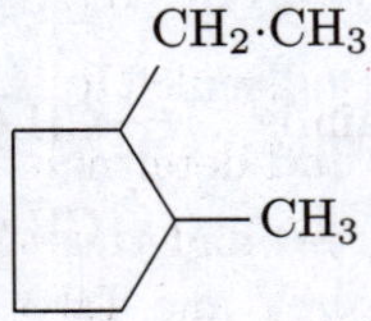

35

(1d) The lowest sum rule is followed in numbering the Cs of the base chain (Rule VI). Thus, the Cs of the base chain of **34** and **35** are numbered as given below:

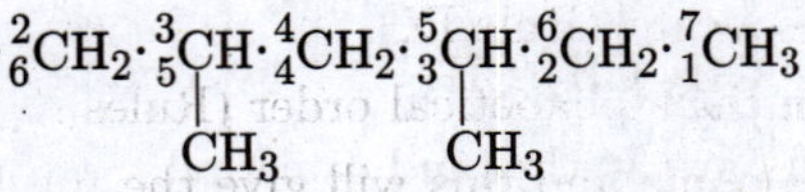

34 (base chain)

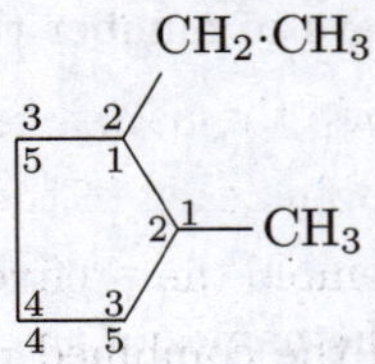

35 (base chain)

(1e) Branch group names along with their locants are prefixed either alphabetically or in an order of increasing complexity (Rules IX to XII). Thus, in **34** the two branch groups are 3- methyl and 5-methyl (Rule XI). Therefore, the full name of **34** is 3,5-dimethylheptane (Rule X). In **35**, the two branch groups are either 1-methyl and 2-ethyl (when arranged in an order of increasing complexity) or 1-ethyl and 2-methyl (when arranged alphabetically). Thus, the name of **35** is either 1-ethyl-2-methylcyclopentane or 1-methyl-2-ethylcyclopentane.

(1f) When the alkyl groups are in equivalent positions, the one written first in the name should be given lower locant. Thus, if in the name of **35**, the prefixes are arranged alphabetically, then the C atom of cyclopentane ring bearing the ethyl group should be given number 1; on the other hand if it is numbered 2 and the C atom bearing the methyl group is numbered 1, the nomenclature of **35** will be wrong. Similarly, when the branches are arranged in the order of increasing complexity, the last complex group should bear the lowest locant; thus 2-methyl-1-ethylcyclopentane and 2-ethyl-1-methylcyclopentane cannot be the names of **35**.

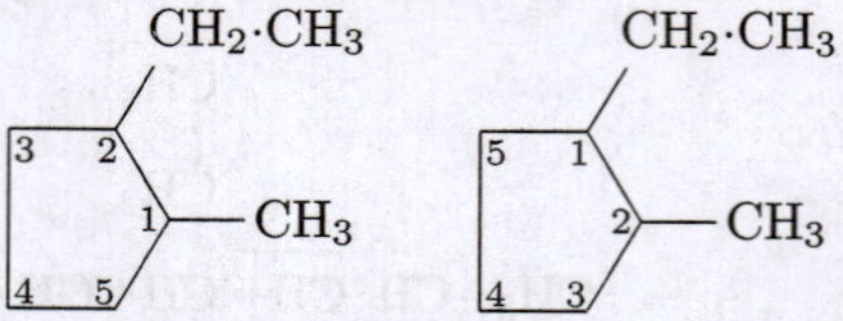

35

1-methyl-2-ethylcyclopentane is the correct name (in the order of increasing complexity).

2-ethyl-1-methyleyclopentane is a wrong name (in alphabetical order).

1-ethyl-2-methylcyclopentane is the correct name (in alphabetical order).

2-methyl-1-ethylcyclopentane is a wrong name (in the order of increasing complexity).

Problem 3.20. What are the IUPAC names of a and b?

a. 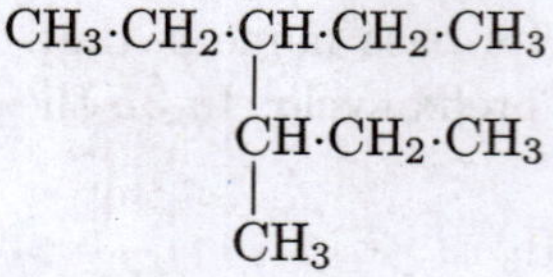

b. 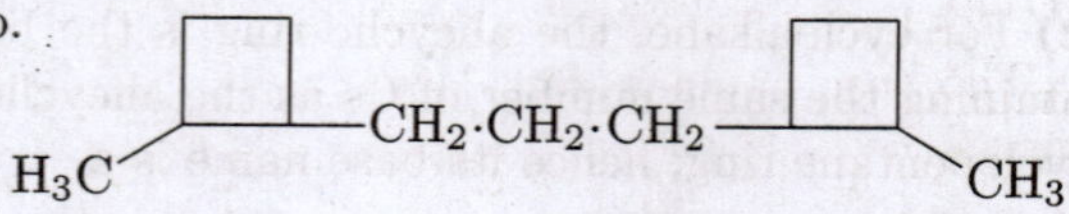

Solution: (a) (i) The base chain is 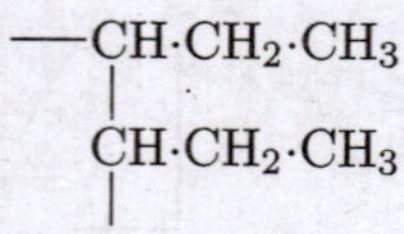

(ii) It consists of 6 Cs. The base name is, therefore, hexane.

(iii) The numbering of Cs in the base chain may be done from either end as the branches have equivalent positions. Now, if we want to arrange the names of the branches alphabetically, we shall have to number the Cs from right end C (see **1f**).

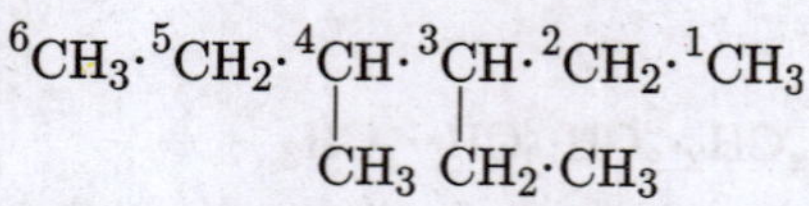

(iv) The names of the branches as prefix are 3-ethyl and 4-methyl.

(v) Hence the full name of (a) is 3-ethyl-4-methylhexane.

(b)(i) As like groups should get similar treatment (Rule VI), this compound is to be considered as a derivative of propane.

(ii) There are two identical substituents; it is represented by **36**. It is a compound substituent. Therefore, we shall have to number the Cs within this also as shown in **36** and the name is 2-methylcyclobutyl (Rule XII a, b).

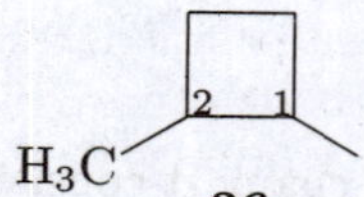

36

(iii) The base chain is —$CH_2 \cdot CH_2 \cdot CH_2$—. The base name of (b) is thus propane. The Cs of it are numbered as in **37**

$$—{}^{1}CH_2 \cdot {}^{2}CH_2 \cdot {}^{3}CH_2—$$

37

(iv) Name of the compound substituent will be 1-(2-methylcyclobutyl) and 3-(2-methylcyclobutyl) [Rule XII(c)].

(v) As two compound substituents are identical, it cannot be indicated by di, we should denote this by bis [Rule XII(e)]. Thus, the combined prefix to the base name is 1, 3-bis-(2-methylcyclobutyl). Hence the full name of the compound b is, 1, 3-bis-(2-methylcyclobutyl)propane.

3.8.2 Alkenes and Cycloalkenes

(2a) The base chain of an alkene is the most substituted longest C chain including two double-bonded C atoms.

(2b) The base chain of a cycloalkene is the alicyclic ring.

(2c) The base name is determined from the name of the alkane having the same number of C atoms as the base chain. The name ending ane is only changed to ene.

(2d) To number the Cs of the base chain in an alkene, the double-bonded C atom nearest to the end should have the lowest possible number, the lowest sum rule is also followed.

(2e) In the case of cycloalkene, both of the double-bonded Cs must be given small numbers; but one does not need to indicate the numbers of the double-bonded Cs.

Problem 3.21. What are the names of the following compounds in the IUPAC system?

a. 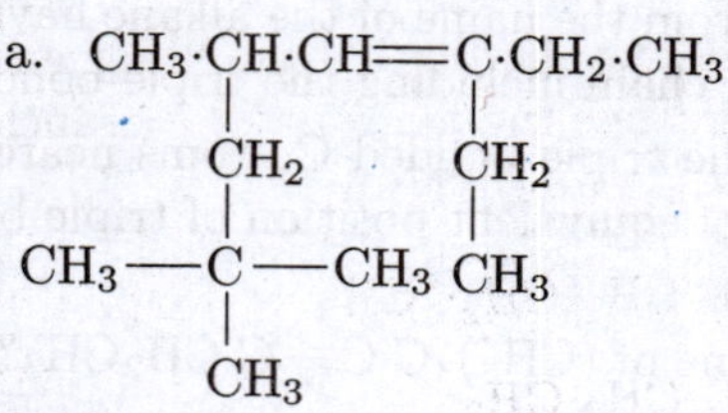

b. 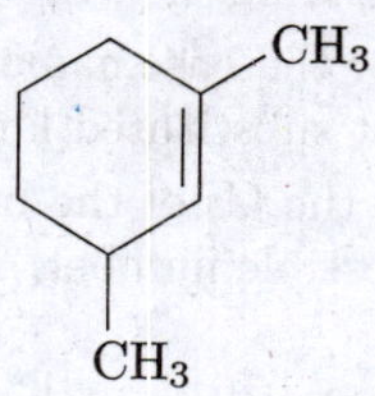

Solution: (a)(i) The parent part:

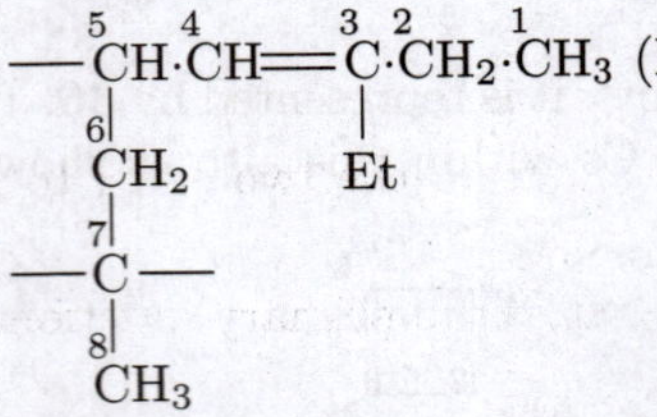 (Rule II)

(ii) The parent part is here the base chain; it consists of 8 Cs. The base name, therefore, is to be derived from octane. The name ending of alkenes being ene, the base name of the compound is octene; the root is thus oct (Rules III and IV).

(iii) To give the lowest number to doubly bonded Cs the numbering is done from the right end C. The first double-bonded C from this side is thus numbered 3. Hence the base name of the compound with the locant is 3-octene or, oct-3-ene. (Rules IV and VII).

(iv) There are four branches: One ethyl branch on C atom number 3, three methyl branches, one on 5th and 2 on 7th C atoms. These are to be indicated as prefixes to the base name. Their names with their locants are: 3-ethyl, 5-methyl, 7-methyl and 7-methyl (Rule IX).

(v) These prefixes when combined, becomes 3-ethyl-5,7,7-trimethyl (Rule XI).

(vi) Hence the full name is 3-ethyl-5,7,7-trimethyl-3-octene, or, 3-ethyl-5,7,7-trimethyloct-3-ene.

(b)(i) The base chain: 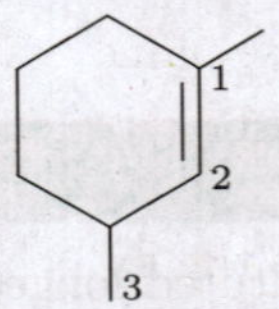

It consists of six Cs; the base name is then to be derived from cyclohexane. Its root is, therefore, cyclohex. Name ending is ene. Hence the base name cyclohexene.

(ii) The numbering on Cs of the base chain is done as shown above; this is so done because both of the double-bonded Cs should be given small numbers [See **(2e)**].

(iii) There are two methyl branches and their locants are 1 and 3; therefore, these are to be prefixed as 1,3-dimethyl.

(iv) Hence the full name of b is 1,3-dimethylcyclohexene.

3.8.3 Alkynes

(3a) The name ending, i.e., the base name suffix is yne.

(3b) The root for the base name is derived from the name of the alkane having the same number of Cs as the most substituted largest carbon chain including the triple-bonded Cs of the alkyne.

(3c) To number the Cs of the base chain, the triple bonded C atoms nearest to the end should get the lowest possible number. In the case of equivalent position of triple bond, lowest sum rule is followed.

Problem 3.22. What is the IUPAC name of $(CH_3)_3C{\cdot}C{\equiv}C{\cdot}CH_2CH_3$?

Solution: (i) The base chain:

(ii) The root for the base name: Hex.

(iii) The base name suffix: yne.

(iv) Hence the base name: Hexyne.

(v) The triple-bonded Cs are equidistant from the two ends; to follow the lowest sum rule, the base chain is numbered from the left end C.

(vi) Hence the base name with the locant of the primary function is denoted by: 3-hexyne or Hex-3-yne.

(vii) Branches are: methyl groups; their names with locants: 2,2-dimethyl.

(viii) Therefore, the full name of the compound is 2,2-dimethyl-3-hexyne or 2,2-dimethylhex-3-yne.

3.8.4 Carboxylic Acids

A. For fatty acids:

4(a) When the C atom of the —COOH group is considered as a part of the base chain in naming a fatty acid, the following points are important:

4(a)(i) The root for the base name is derived from the name of the base C chain (Rule IIIA) by dropping the end e; e.g., if the base chain consists of 5 Cs, the root will be derived from pentane and it will be pentan.

4(a)(ii) The suffix to the root is oic acid; the base name is thus root +oic acid.

4(a)(iii) The numbering of Cs starts from the carboxyl C, i.e., it is given number 1, but this is not indicated in the name.

4(a)(iv) Rules IX to XII are followed to indicate the branches.

Problem 3.23. What is the IUPAC name of $(CH_3)_2CH{\cdot}CH(CH_2CH_3){\cdot}CH_2COOH$?

Solution: 1. The base chain:

$$_5CH_3{\cdot}\,_4\underset{|}{CH}{\cdot}\,_3\overset{|}{C}H{\cdot}\,_2CH_2{\cdot}\,_1COOH$$

It consists of 5 Cs; hence the root is pentan. [4(a)(i)]

2. The suffix to the root: oic acid

The base name is, therefore, pentanoic acid. [4(a)(ii)]

3. The numbering of Cs is done as shown above. [4(a)(iii)]

4. The branches are: 3-ethyl and 4-methyl. [Rule IX]

5. The full name: 3-ethyl-4-methylpentanoic acid. [Rule X]

4(b) When the —COOH group is considered as a substituent, it does not belong to the base chain; in this case:

4(b)(i) The root is the name of the base chain, e.g., if there are 4 Cs in the base C chain, the root is butane.

4(b)(ii) The suffix to the root is carboxylic acid. The base name is thus, root +carboxylic acid.

4(b)(iii) The numbering of Cs starts from the C which bears the —COOH group, but the locant of the —COOH group, i.e., 1, is not shown in the name.

4(b)(iv) Rules IX to XII are followed to indicate the branches.

Problem 3.24. Considering the —COOH group as a substituent, write down the name of

$$(CH_3)_2CH\ \underset{\underset{Et}{|}}{CH}\ CH_2COOH$$

Solution: 1. The base chain: $^{4}CH_3\cdot{}^{3}\underset{|}{C}H\cdot{}^{2}\underset{|}{C}H\cdot{}^{1}CH_2$— ; it consists of 4 Cs and hence the root for the base name is butane. [4(b)(i)]

2. The suffix to the root is carboxylic acid and the base name is butanecarboxylic acid. [4(b)(ii)]

3. The numbering of Cs is done as shown above in 1. [4(b)(iii)]

4. The branches are 2-ethyl and 3-methyl. [Rule IX]

5. The full name is 2-ethyl-3-methylbutanecarboxylic acid. [Rule X]

B. For carboxyl derivatives of cycloalkanes:

4(c)(i) The root is the name of the cycloalkane ring.

4(c)(ii) The suffix to the root is carboxylic acid.

4(c)(iii) The numbering of Cs starts from that C of the cycloalkane ring which carries the —COOH group and the lowest sum rule is followed. The locant of the carboxyl group is not indicated in the name.

4(c)(iv) The branches are indicated as usual (Rules IX to XII).

Problem 3.25. Write down the IUPAC name of 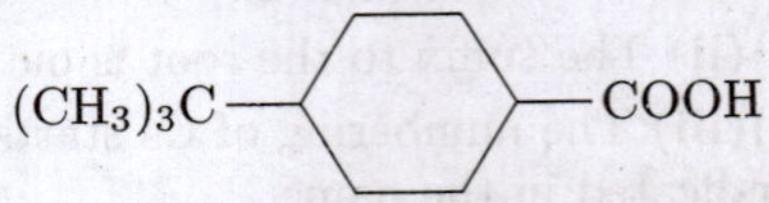

Solution:

1. The base chain: 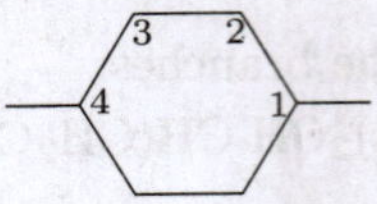 ; so the root is cyclohexane. [4(c)(i)]

2. The suffix to the root is carboxylic acid; the base name is thus, cyclohexanecarboxylic acid. [4(c)(ii)]

3. The numbering is shown above.

4. The branch is 4-tert-butyl.

5. The full name: 4-tert-Butylcyclohexanecarboxylic acid.

C. For carboxyl derivatives of benzene:

4(d)(i) The base name is benzoic acid.

4(d)(ii) The numbering of Cs start from the C atom of the ring bearing the —COOH group and the lowest sum rule is followed.

4(d)(iii) Rules IX to XII are followed to indicate the side chains.

Problem 3.26. What is the IUPAC name of 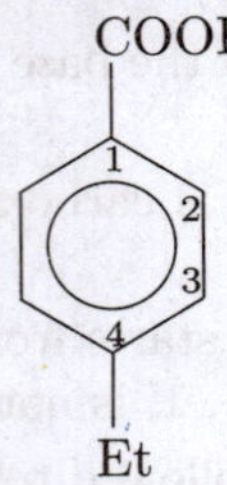 ?

Solution: 1. The base name: Benzoic acid.

2. Numbering of Cs is done as shown:

3. The branch is 4-ethyl or *p*-ethyl (Rule IX).

4. The full name: 4-ethylbenzoic acid.

D. For carboxyl derivatives of naphthalene and anthracene:

4(e)(i) The roots are napth and anthr, respectively.

4(e)(ii) The suffix to the root is oic acid.

Therefore, the base names are naphthoic acid and anthroic acid.

4(e)(iii) Rule VI C is followed to number the Cs; the locant of the carboxyl group is indicated in the base name.

4(e)(iv) Rules IX to XII are followed to indicate the side chains.

Problem 3.27. What is the IUPAC name of

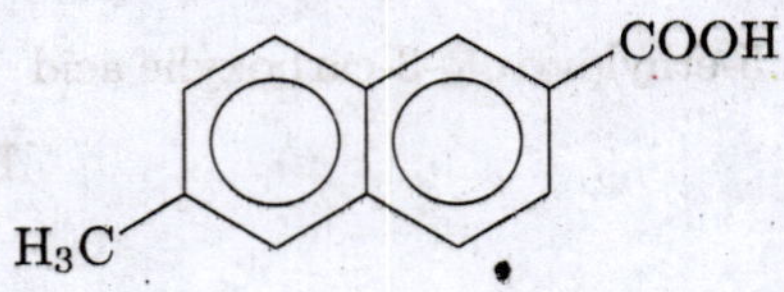

Solution: 1. The base C chain is

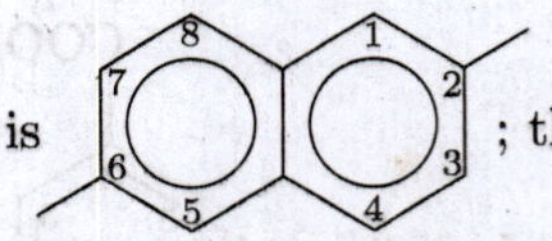

; the root is, therefore, naphth. [4(e)(i)]

2. The suffix being oic acid; the base name is naphthoic acid. [4(e)(ii)]

3. The base name with the locant is 2-naphthoic acid. [4(e)(iii)]

4. The branch is 6-methyl. [Rule ix]

5. The full name is 6-methyl-2-naphthoic acid.

E. For carboxyl derivatives of heterocycles:

4(f)(i) The root is the name of the heterocyclic ring present in the compound.

4(f)(ii) The suffix to the root is carboxylic acid.

4(f)(iii) The numbering of Cs is done according to the Rule VID. The locant of the carboxyl group is indicated in the name.

4(f)(iv) Rules IX to XII are followed to indicate the side chains.

Problem 3.28. Write down the IUPAC name of the following compounds:

(a) 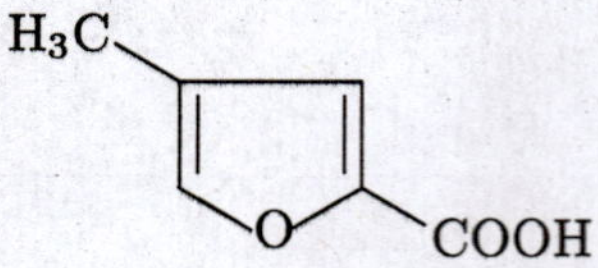 **(b)** 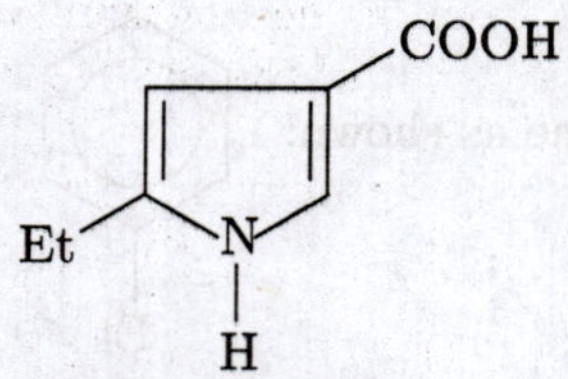 **(c)** 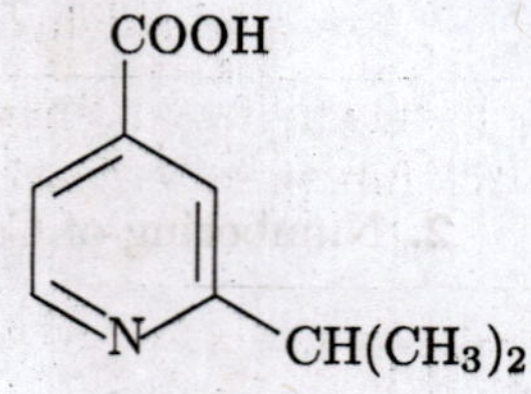

Solution: (a) The base chain is furan; the root is, therefore, furan. The suffix to the root is carboxylic acid; the base name is thus furancarboxylic acid. The numbering is done as shown:

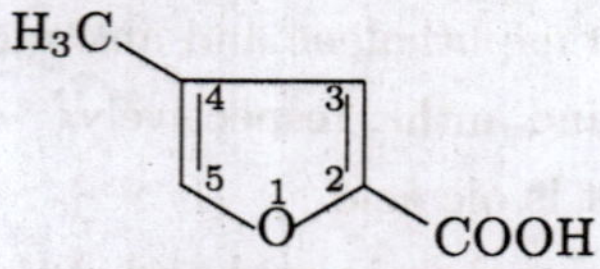

the base name with its locant is Furan-2-carboxylic acid. The side chain is 4-methyl. Hence, the full name: 4-methylfuran-2-carboxylic acid. It is also named as 4-methyl-2-furoic acid.

(b) It is 5-ethylpyrrole-3-carboxylic acid

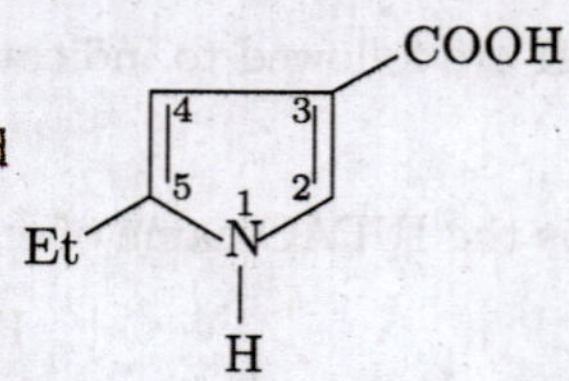

(c) It is 2-Isopropylpyridine-4-carboxylic acid

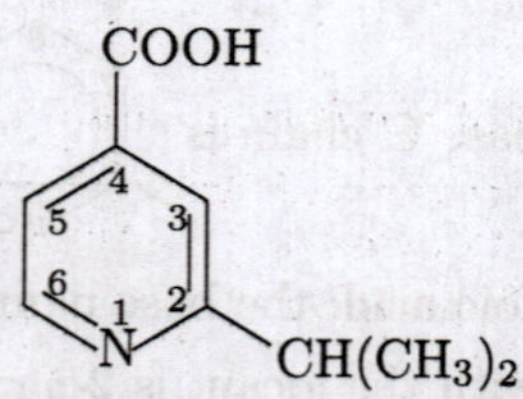

3.8.5 Acid Halides, Amides, Acid Anhydrides, Esters, Nitriles, Aldehydes and Acid Salts

The general rules and procedures adopted to name carboxylic acids of aliphatic, alicyclic, aromatic and heterocyclic classes are also applicable to name these compounds; only difference is that each class of compounds has its own suffix for constructing the base name.

The suffixes are noted in Table 3.12.

Table 3.12 The suffixes for constructing the base names of acid halides, amides, esters, nitriles and aldehydes

Formula of the function	When the C atom of the function is considered as a part of the base C chain, the suffix is:	When the function is considered as a substituent to the base C chain, the suffix is:
—C(=O)—X	-oyl halide	carbonyl halide
—C(=O)—NH_2	-amide	-carboxamide
—C(=O)—O—C(=O)—	oic anhydride	×
—C(=O)—OR	Name of the R -oate	Name of the R carboxylate
—C≡N	-nitrile	-carbonitrile
—C(=O)—H	-al	carboxaldehyde
—COO^-M^+	Name of the cation oate	Name of the cation carboxylate

Problem 3.29. Write down the IUPAC names of the following compounds:

(a) $(CH_3)_3C{\cdot}COCl$; **(b)** $(CH_3)_2CH{\cdot}CH_2CH(Et){\cdot}CONH_2$; **(c)** $CH_3{\cdot}CH_2{\cdot}CH(CH_3){\cdot}COOEt$

(d) CH_3CH_2CN; **(e)** $(CH_3)_2CH{\cdot}CH_2CHO$.

Solution:

(a) 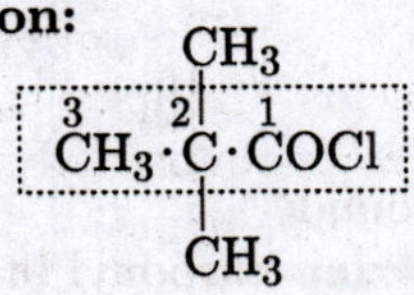

1. Root: Propan
2. Suffix: oyl chloride
3. Base name: Propanoyl chloride
4. Branches: 2,2-dimethyl
5. Full name: 2,2-Dimethylpropanoyl chloride

(b) 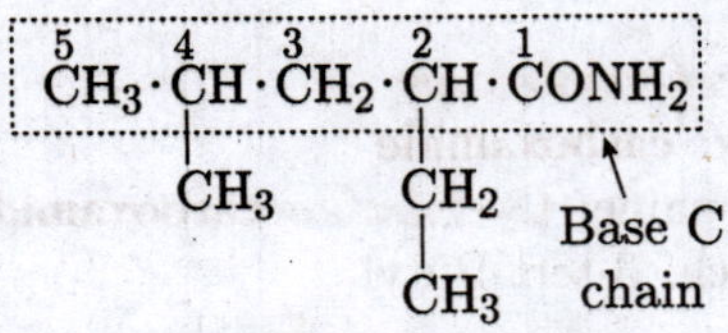

1. Root: Pentan
2. Suffix: amide
3. Base name: Pentanamide
4. Branches: 2-Ethyl,4-methyl

Name: 2-Ethyl-4-methylpentanamide

(c) $\overset{4}{C}H_3\cdot\overset{3}{C}H_2\cdot\overset{2}{C}H(CH_3)\cdot\overset{1}{C}OOEt$

1. Root: Butan
2. Suffix: oate
3. Base name: Ethyl butanoate
4. Branch: 2-methyl

Ethyl 2-methylbutanoate

(d) $\overset{3}{C}H_3\cdot\overset{2}{C}H_2\cdot\overset{1}{C}N$

1. Root: Propane
2. Suffix: nitrile
3. Base name: Propanenitrile

Propanenitrile

(e) $\overset{4}{C}H_3\cdot\overset{3}{C}H(CH_3)\cdot\overset{2}{C}H_2\cdot\overset{1}{C}HO$

1. Root: Butan
2. Suffix: al
3. Base name: Butanal
4. Branch: 3-Methyl

3-Methylbutanal

Problem 3.30. Write down the IUPAC names of the following alicyclic compounds:

(a) 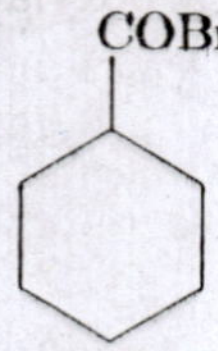

(b) 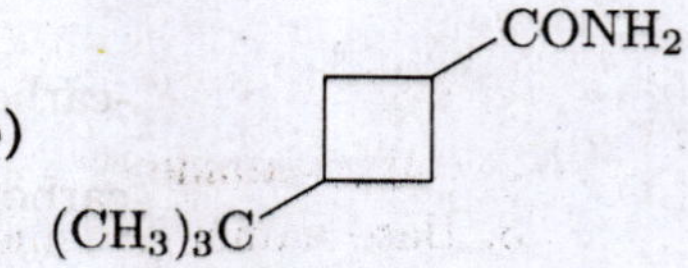

(c) 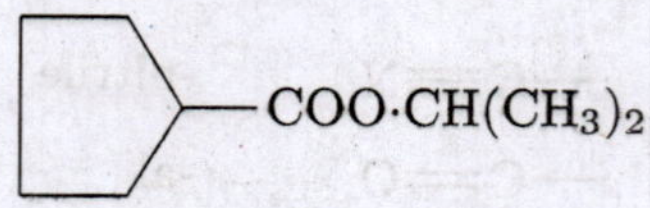

(d) 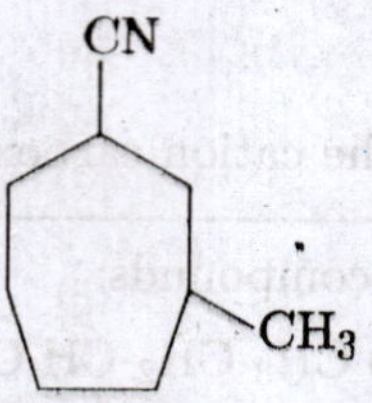

(e) 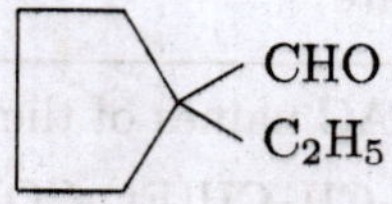

(f) 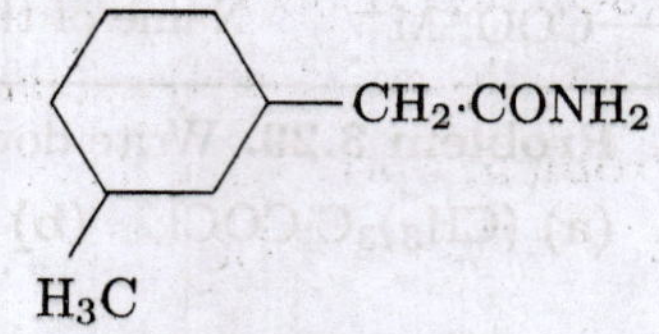

Solution: In the following figures the base chain is shown by dotted cage.

(a) 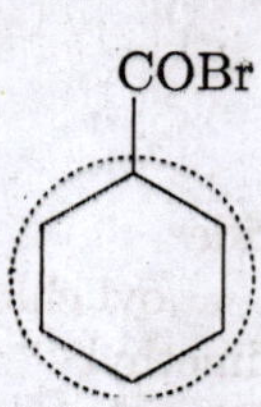

1. Root: Cyclohexane
2. Suffix: carbonyl bromide
3. Base name: Cyclohexanecarbonyl bromide

Cyclohexanecarbonyl bromide

(b) 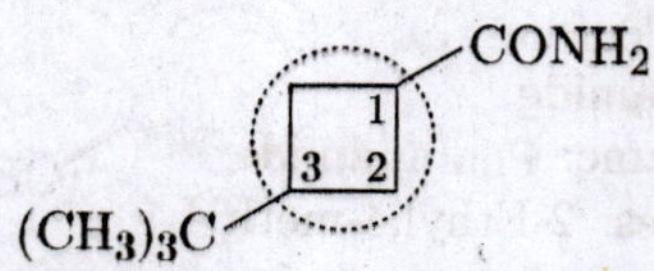

1. Root: Cyclobutane
2. Suffix: carboxamide
3. Base name: Cyclobutanecarboxamide
4. Branch: 3-tert-Butyl

3-tert-Butylcyclobutanecarboxamide

(c)

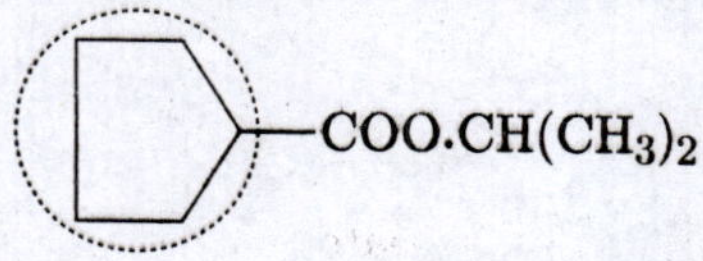

1. Root: Cyclopentane
2. Suffix: carboxylate
3. Base name: Isopropyl cyclopentanecarboxylate

Isopropyl Cyclopentanecarboxylate

(d)

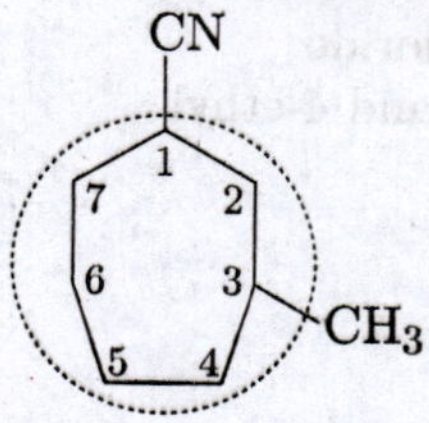

1. Root: Cycloheptane
2. Suffix: carbonitrile
3. Base name: Cycloheptanecarbonitrile
4. Branch: 3-Methyl

3-Methylcycloheptanecarbonitrile

(e)

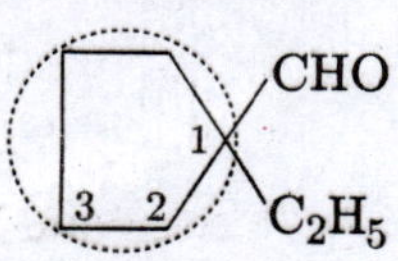

1. Root: Cyclopentane
2. Suffix: carboxaldehyde
3. Base name: Cyclopentanecarboxaldehyde
4. Branch: 1-ethyl

1-Ethylcyclopentanecarboxaldehyde

(f)

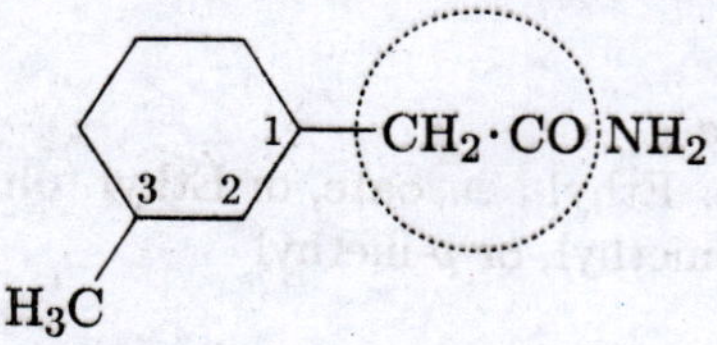

1. Root: Ethan
2. Suffix: amide
3. Base name: Ethanamide
4. Branch: (3-methylcyclohexyl)

(3-Methylcyclohexyl)ethanamide

(3-Methylcyclohexyl)acetamide

Problem 3.31. Write down the IUPAC names of the following compounds:

(a)

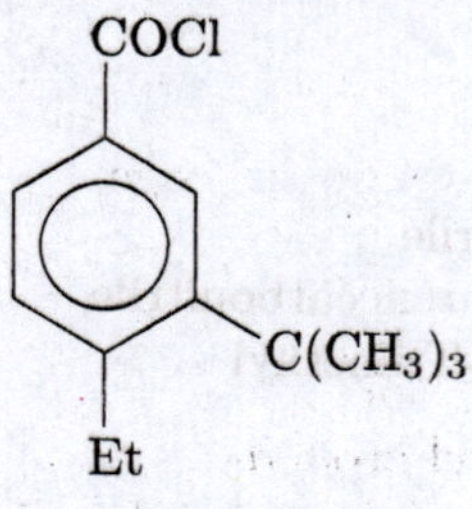

(b)

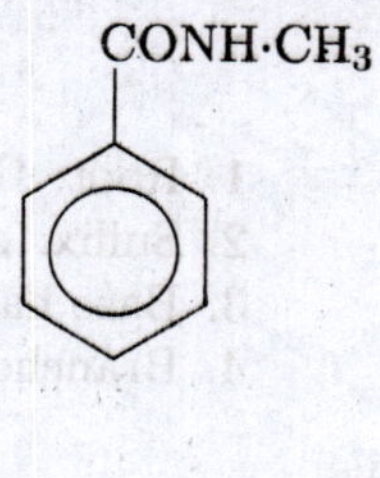

(c)

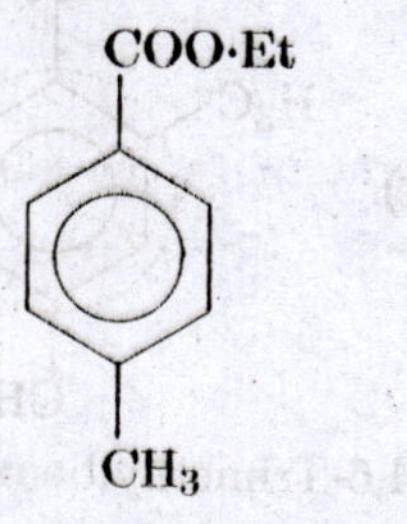

(d)

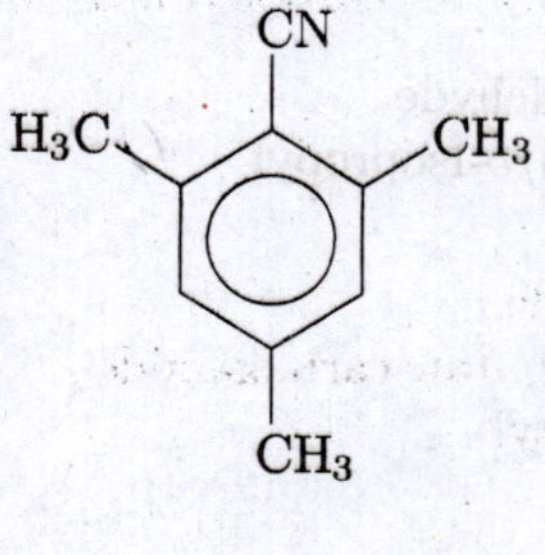

(e)

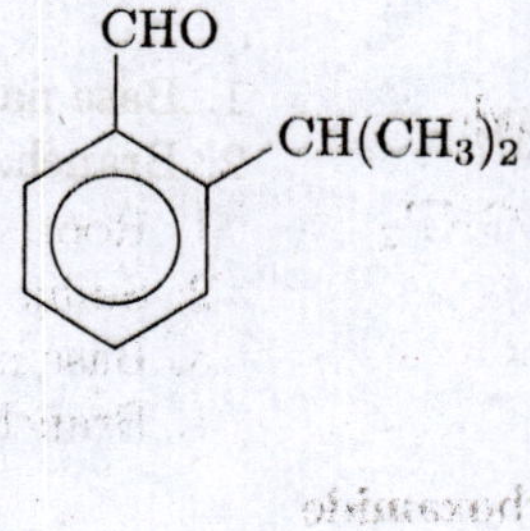

(f)

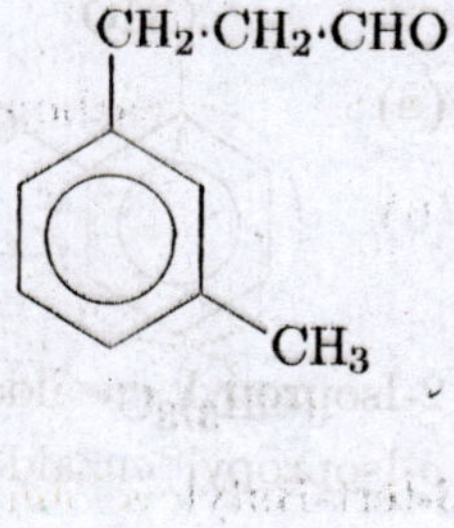

Solution:

(a)

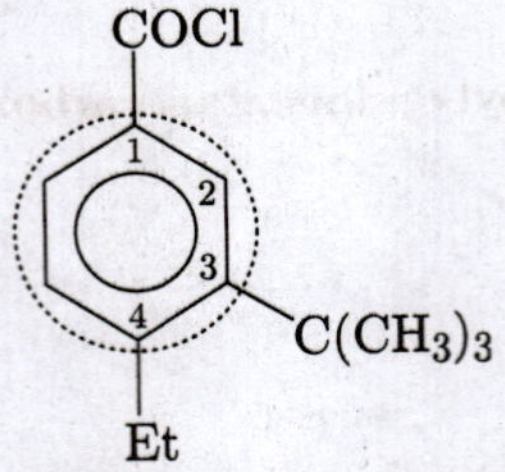

3-tert-Butyl-4-ethylbenzoyl chloride

1. Root: Benz
2. Suffix: oyl chloride
3. Base name: Benzoyl chloride
4. Branches: 3-tert-Butyl and 4-ethyl

(b)

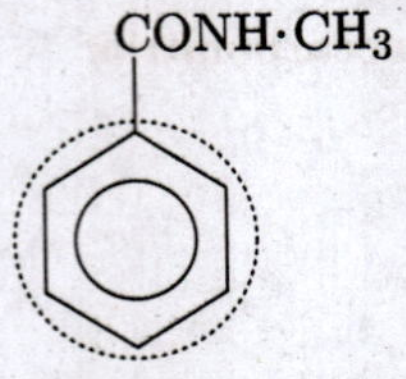

N-methylbenzamide

1. Root: Benz
2. Suffix: amide
3. Base name: benzamide
4. Alkyl group: N-methyl

(c)

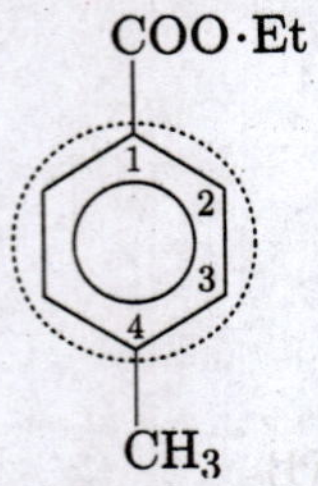

Ethyl *p*-methylbenzoate
Ethyl 4-methylbenzoate
Ethyl *p*-toluate

1. Root: Benz
2. Suffix: oate
3. Base name: Ethyl benzoate, or Ethyl toluate
4. Branch: 4-methyl, or *p*-methyl

(d)

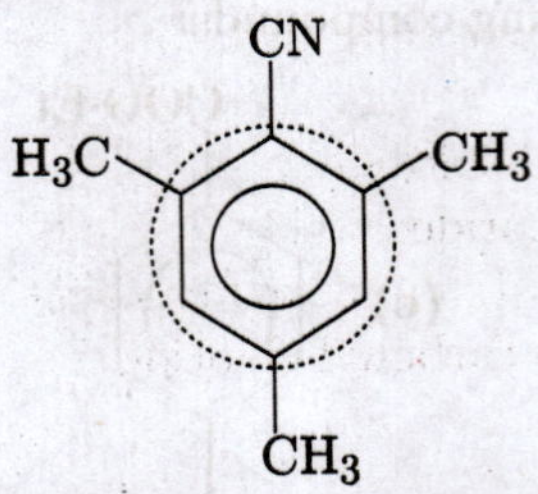

2,4,6-Trimethylbenzenecarbonitrile

1. Root: Benzene
2. Suffix: carbonitrile
3. Base name: Benzenecarbonitrile
4. Branches: 2,4,6-Trimethyl

(e)

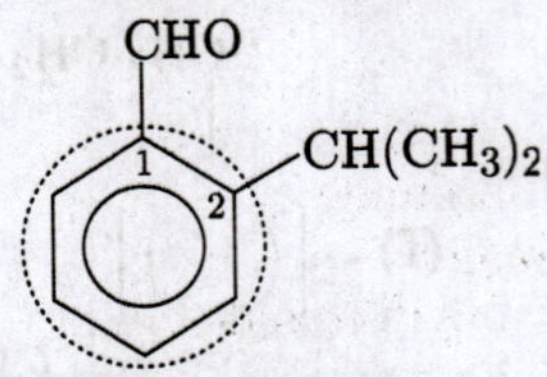

2-Isopropylbenzaldehyde
o-Isopropylbenzaldehyde

1. Base name: Benzaldehyde
2. Branch: 2-Isopropyl/*o*-Isopropyl

(f)

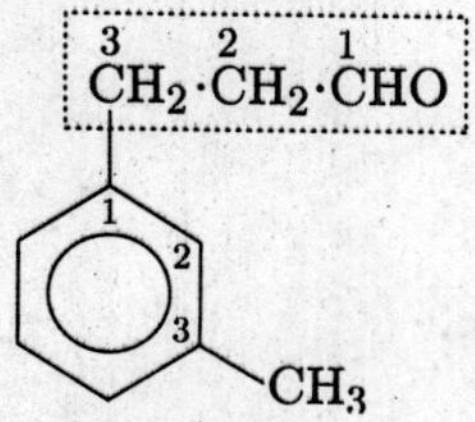

1. Root: Propan
2. Suffix: al
3. Base name: Propanal.
4. Branch: 3-(*m*-Tolyl)

3-(*m*-Tolyl)propanal

Problem 3.32. Write down the names of the following compounds:

(a)

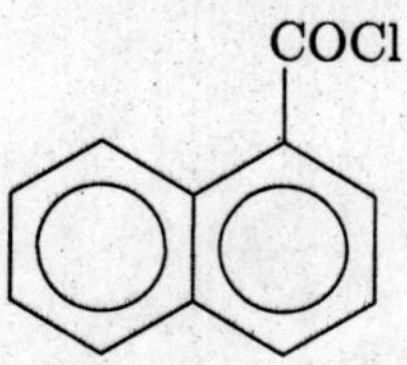

(b)

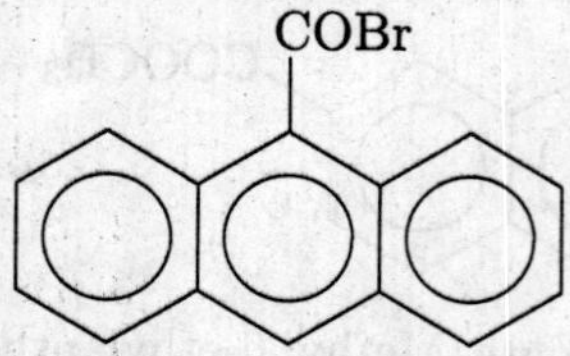

(c)

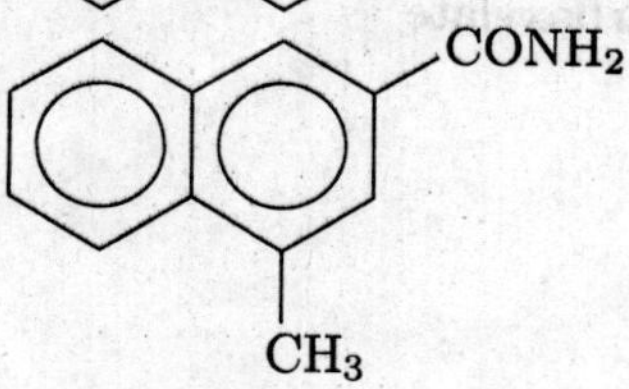

(d)

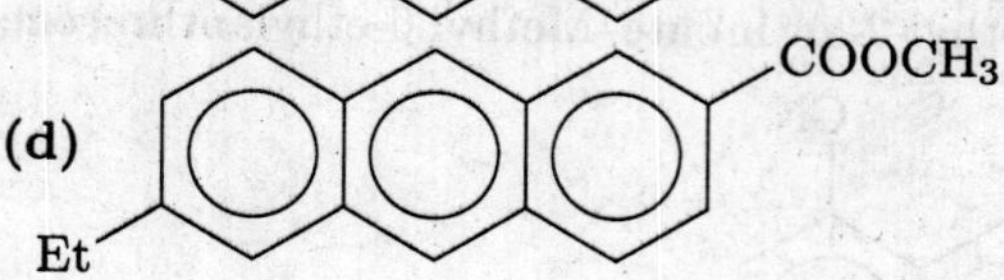

(e)

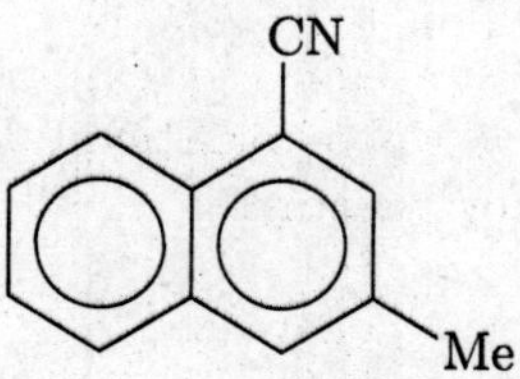

(f)

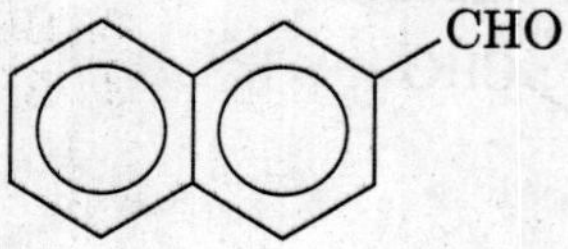

Solution:

(a)

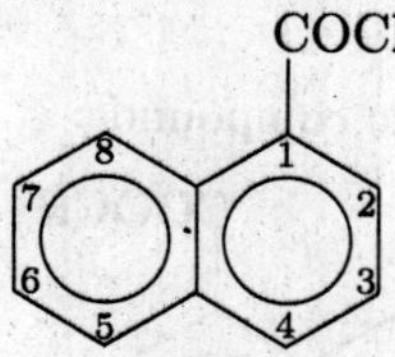

1. Root: (i) Naphth
(ii) Naphthalene
2. Suffix: (i) oyl chloride or
(ii) carbonyl chloride
3. Base name: (i) Naphthoyl chloride
(ii) Naphthalenecarbonyl chloride

(i) 1-Naphthoyl chloride
or
(ii) 1-Naphthalenecarbonyl chloride

(a)

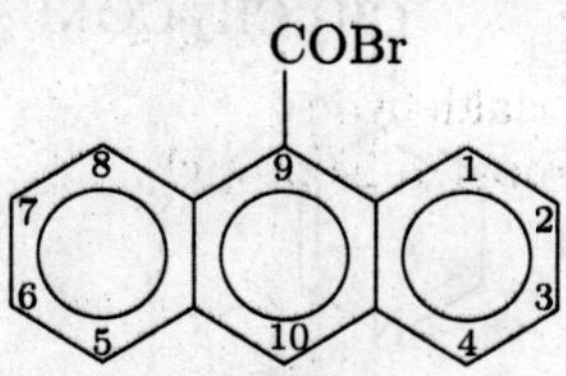

1. Root: (i) Anthr (ii) Anthracene
2. Suffix: (i) oyl bromide
(ii) carbonyl bromide
3. Base: (i) Anthroyl bromide

(i) 9-Anthroyl Bromide

(ii) Anthracene-9-carbonyl bromide

Similarly, the names of the other compounds are:

(c) 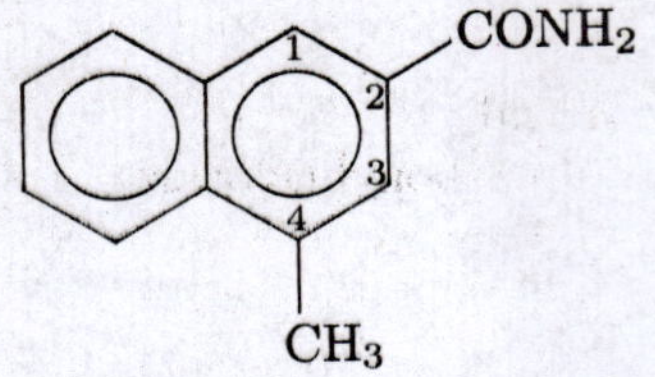

4-Methyl-2-naphthamide
4-Methylnaphthalene-2-carboxamide

(d) 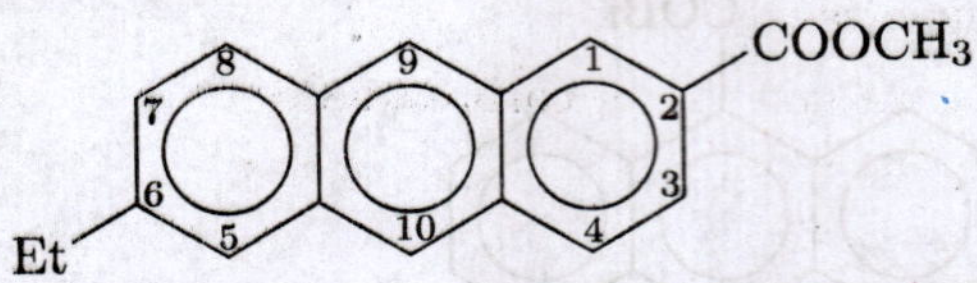

Methyl 6-ethyl-2-anthroate/Methyl 6-ethylanthracene-2-carboxylate

(e) 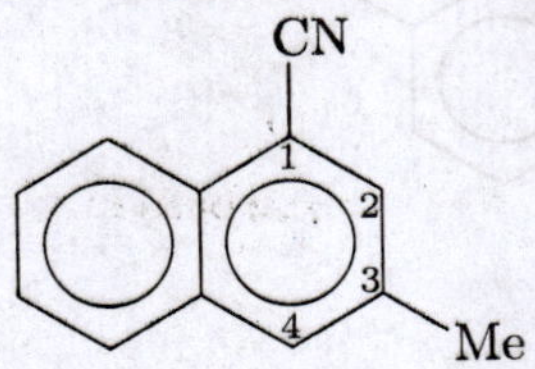

3-Methylnaphthalene-1-carbonitrile

(f) 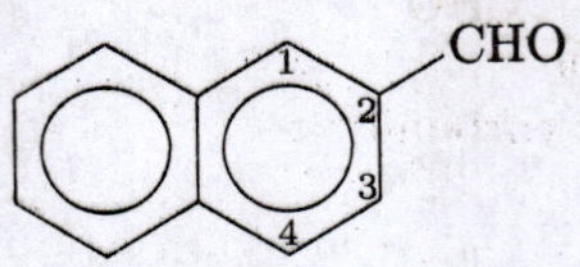

Naphthalene-2-carboxaldehyde

Problem 3.33. Write down the names of the following heterocyclic compounds:

(a) 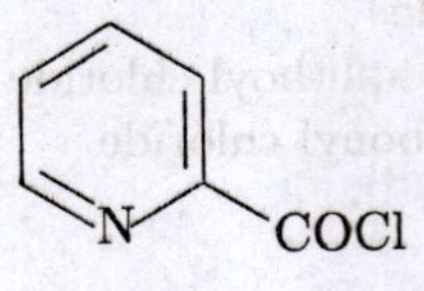

(b) 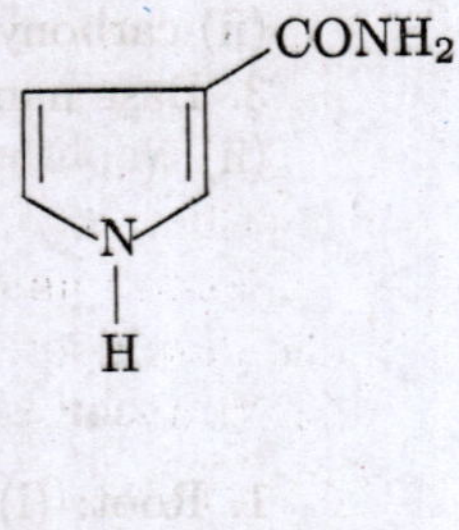

(c) 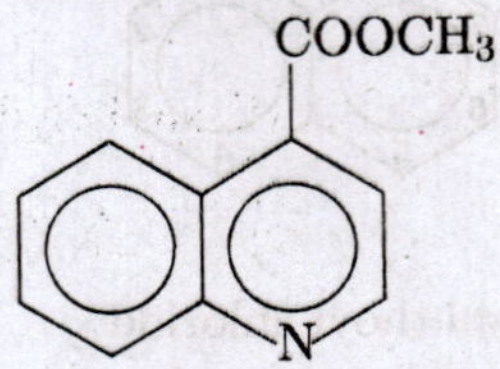

(d) 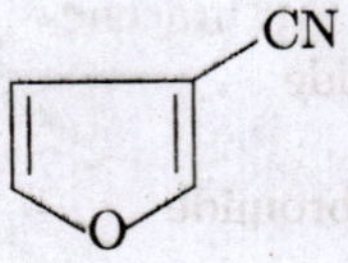

(e) 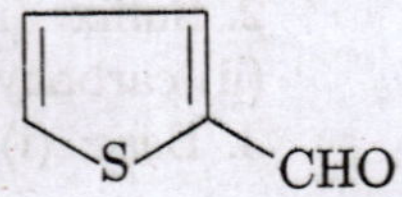

(f) 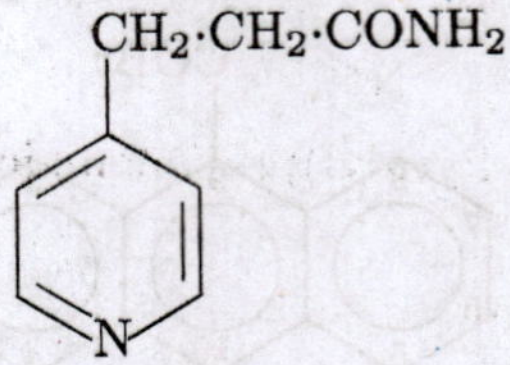

Solution:

(a) 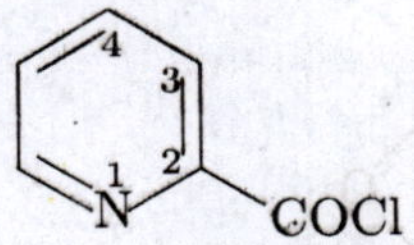

1. Root: Pyridine
2. Suffix: **carbonyl chloride**
3. Base name: Pyridine**carbonyl chloride**

Pyridine-2-**carbonyl chloride**

(b) 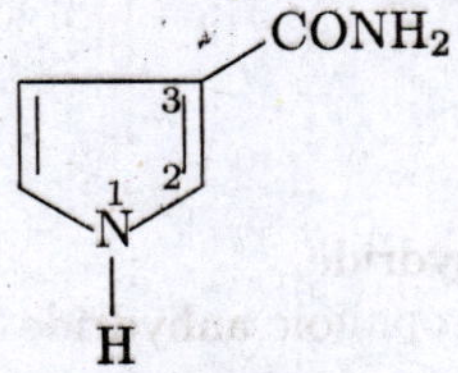

1. Root: Pyrrole
2. Suffix: **carboxamide**
3. **Base name:** Pyrrole**carboxamide**

Pyrrole-3-**carboxamide**

(c) 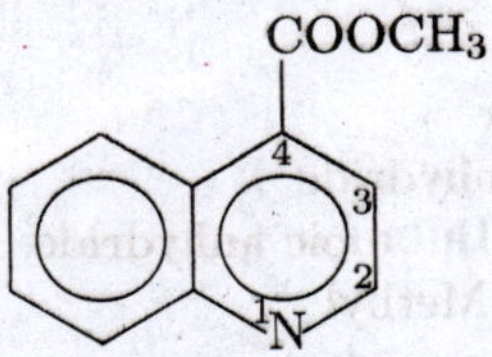

1. Root: Quinoline
2. Suffix: **carboxylate**
3. Alkyl group: Methyl
4. Base name: Methyl quinoline**carboxylate**

Methyl quinoline-4-**carboxylate**

(d) 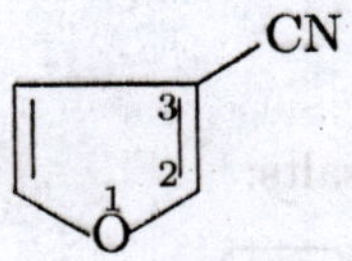

1. Root: Furan
2. Suffix: **carbonitrile**
3. Base name: Furan**carbonitrile**

Furan-3-**carbonitrile**

(e) 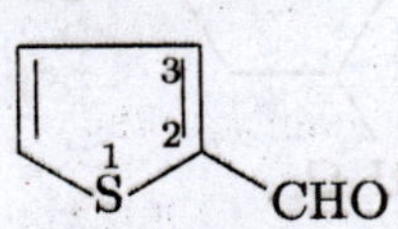

1. Root: Thiophen
2. Suffix: **carboxaldehyde**
3. Base name: Thiophen**carboxaldehyde**

Thiophen-2-**carboxaldehyde**

(f) 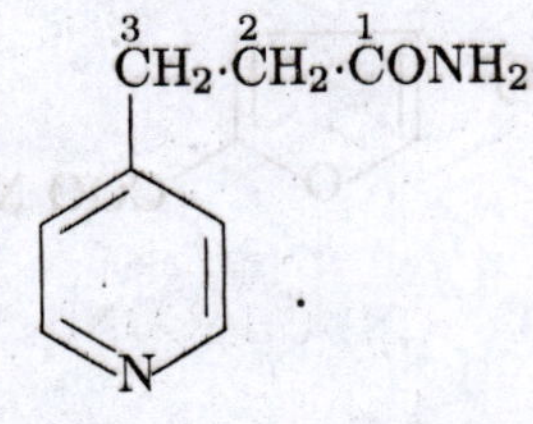

1. Root: Propan
2. Suffix: **amide**
3. Base name: Propan**amide**
4. Branch: 3-(4-Pyridyl)

3-(4-Pyridyl)propan**amide**

Problem 3.34. Write down the IUPAC names of the following anhydrides:

(a) 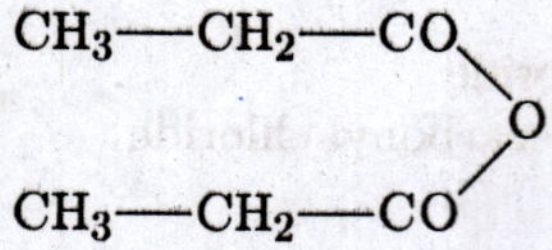

(b) 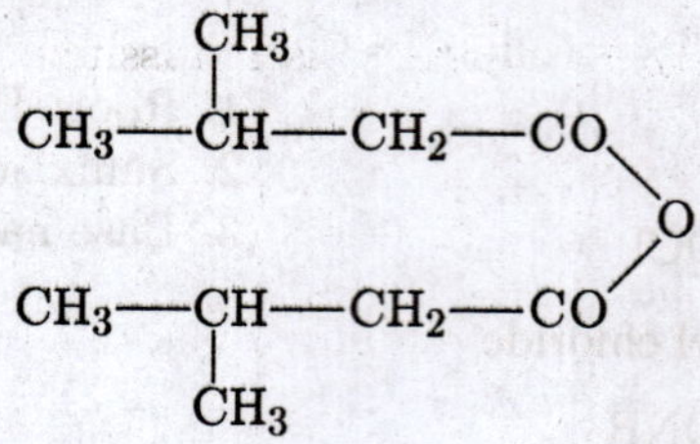

Solution:

(a) 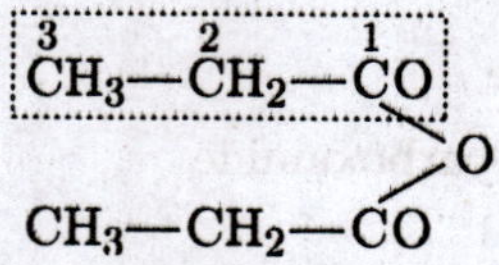

1. Root: Propan
2. Suffix: oic anhydride
3. Base name: Propanoic anhydride

Propanoic anhydride

(b) 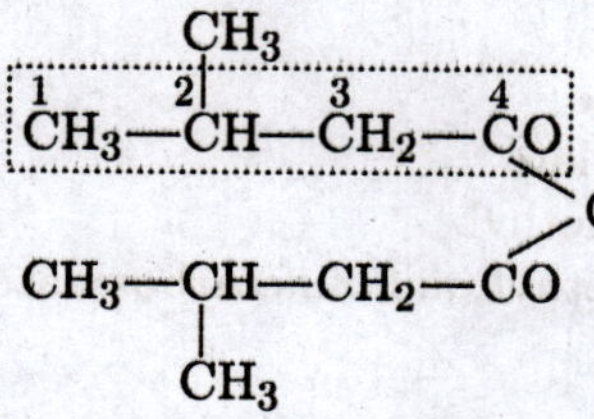

1. Root: Butan
2. Suffix: oic anhydride
3. Base name: Butanoic anhydride
4. Branches: 3-Methyl

(Wrong numbering)

3-Methylbutanoic anhydride (Correct name)

Problem 3.35. Write down the IUPAC names of the following acid salts:

(a) $CH_3CO\bar{O}NH_4^{\oplus}$ (b) $CH_3{\cdot}CH(CH_3){\cdot}CO\bar{O}Na^{\oplus}$ (c) 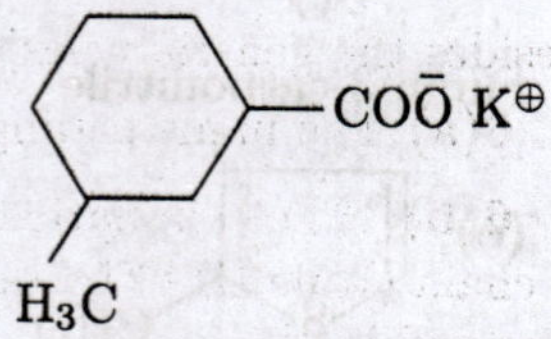

(d) 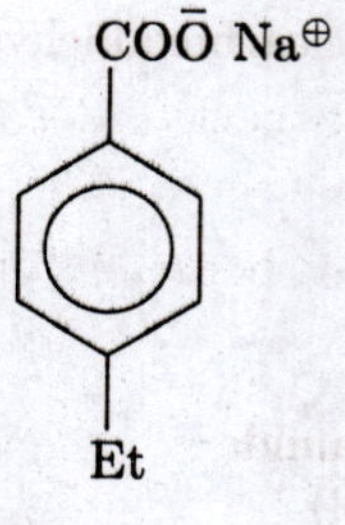 (e) 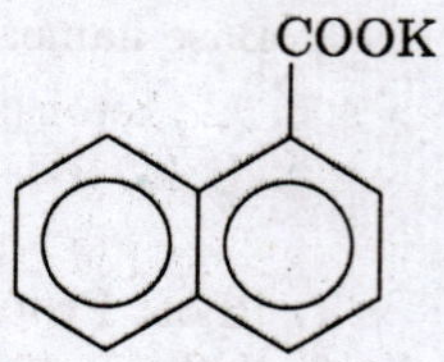 (f) 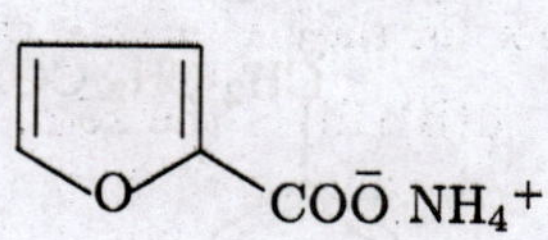

Solution:

(a) $CH_3CO\bar{O}\,NH_4^{\oplus}$ is Ammonium ethanoate

(b) $\overset{3}{C}H_3{\cdot}\overset{2}{C}H(CH_3){\cdot}\overset{1}{C}O\bar{O}\,Na^{\oplus}$ is Sodium 2-methylpropanoate

(c) 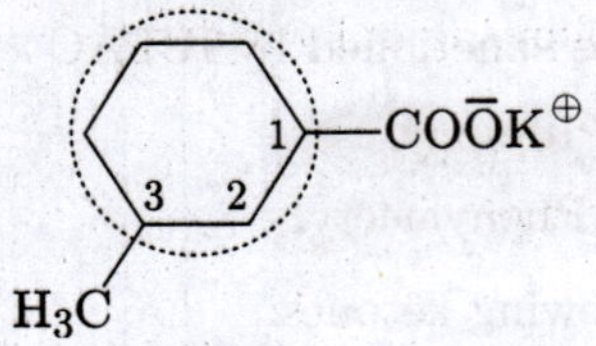

is Potassium 3-methylcyclohexanecarboxylate.

(d) 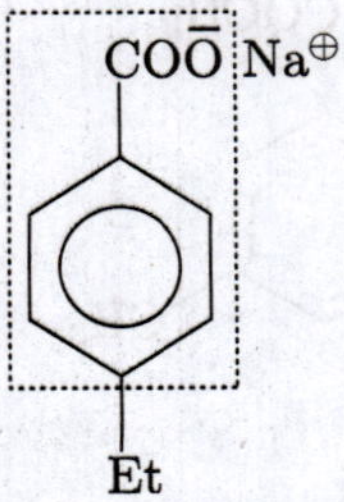

is Sodium *p*-ethylbenzoate.

(e) 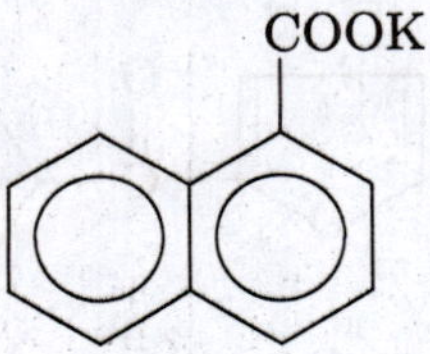

is Potassium 1-naphthoate or

Potassium naphthalene-1-carboxylate.

(f) 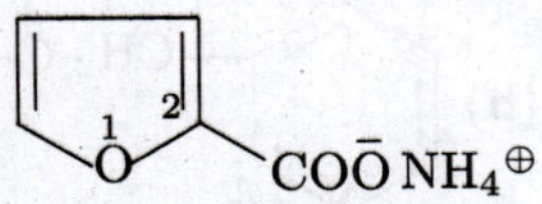

is Ammonium 2-furoate or
Ammonium furan-2-carboxylate

3.8.6 Ketones

Besides the general rules, following are to be noted in giving IUPAC name to a ketone.

6(a) The name ending, i.e., suffix to the root is one.

6(b)(i) For aliphatic ketones, the root for the base name is derived from the name of the base C chain (Rule IIIA) by dropping the end e; for example, if the base C chain consists of 6 Cs, the root will be hexan.

6(b)(ii) If the ketonic C atom of an alicyclic ketone is a constituent of the alicyclic ring, the root for the base name will be cycloalkan (Rule III-B).

6(b)(iii) If the ketonic C atom of a compound is attached to a benzene ring, the compound may be named as a phenone, e.g.,

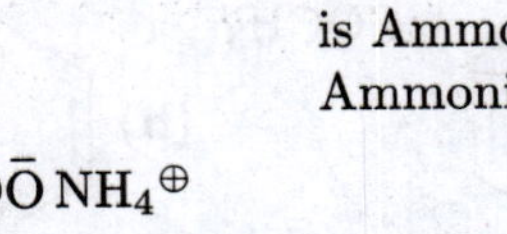

Acetophenone

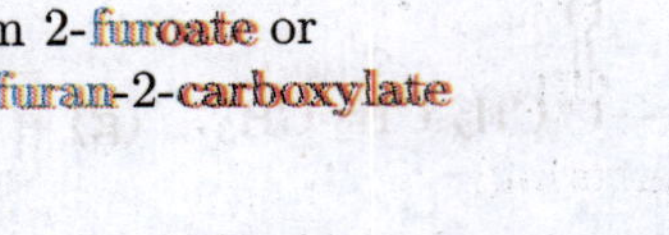

Benzophenone

Acetophenone and benzophenone are used as base names for their derivatives.

6(b)(iv) Ketones containing cycloalkyl, aryl and heterocyclyl groups may be named as derivatives of alkanones or as their derivatives. The IUPAC sanctioned names of some such groups are listed in Table 3.13.

Table 3.13 Names of some substituted keto groups sanctioned by IUPAC

$CH_3 \cdot CO$— Acetyl	CH_3CH_2CO— Propionyl
C_6H_5CO— Benzoyl	$C_6H_5CH_2CO$— Phenylacetyl

Problem 3.36. Write down the IUPAC names of the following ketones:

(a) 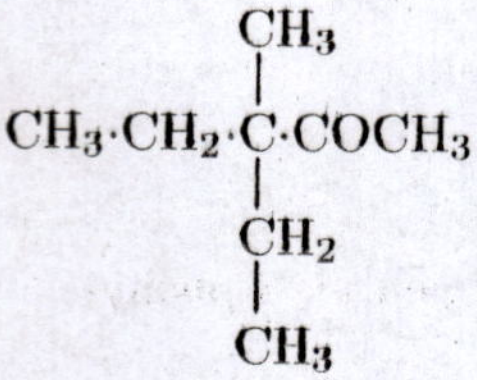

(b) 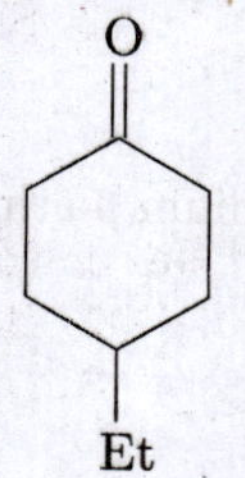

(c) 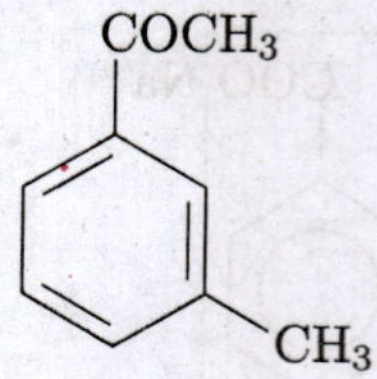

(d) 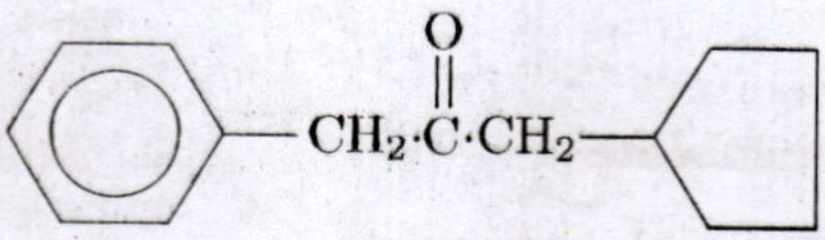

(e) 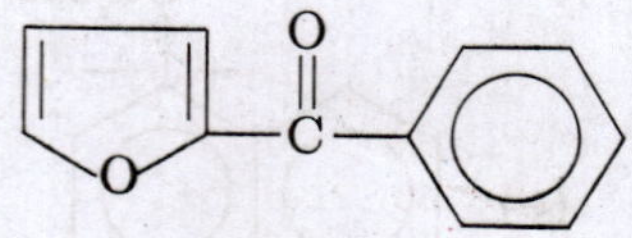

(f) 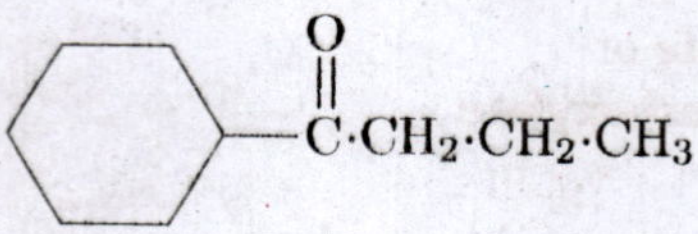

(g) 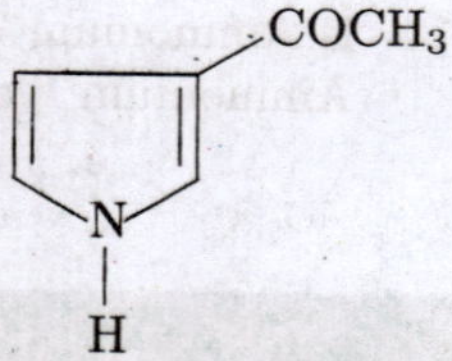

(h) 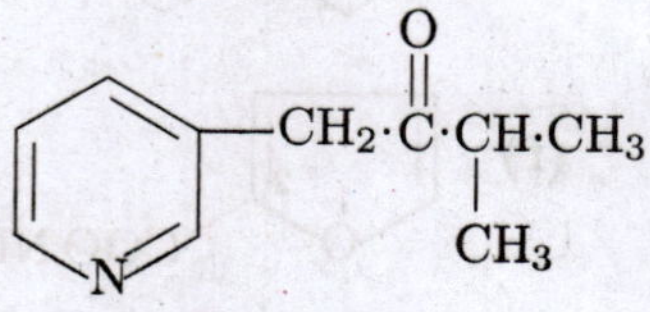

Solution:

(a) 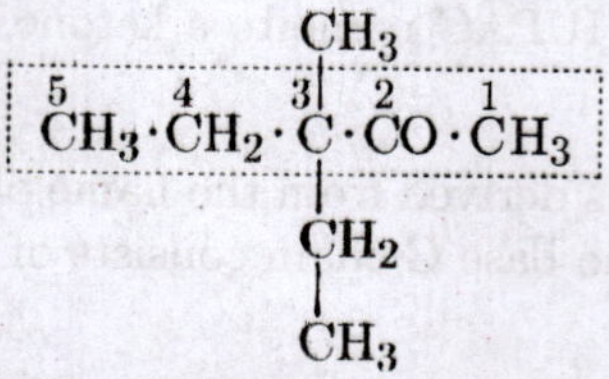

1. Root: Pentan
2. Suffix: one
3. Base name: Pentanone
4. Branches: 3-Ethyl and 3-Methyl

3-Ethyl-3-methyl-2-pentanone or,
3-Ethyl-3-methylpentan-2-one

(b) 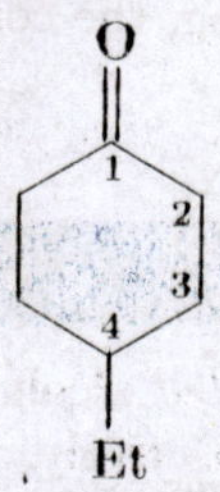

1. Root: Cyclohexan
2. Suffix: one
3. Base name: Cyclohexanone
4. Side chain: 4-Ethyl

4-Ethylcyclohexanone

(c) 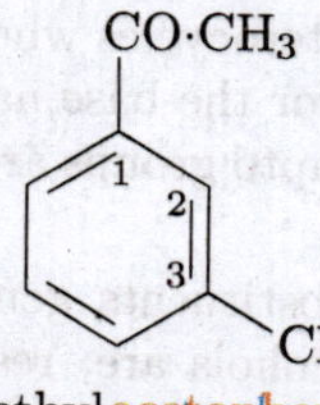

3-Methylacetophenone

1. Base name: Acetophenone
2. Side chain: 3-Methyl or *m*-Methyl

(d) 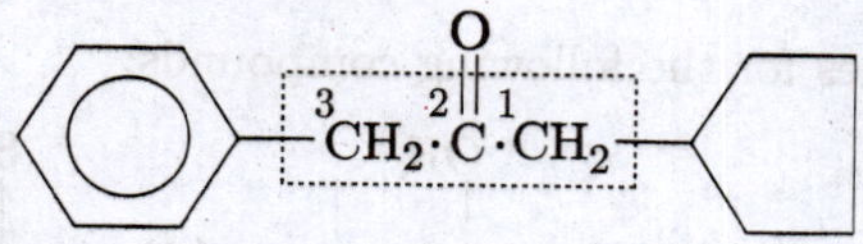

1-Cyclopentyl-3-phenylpropanone

1. Root : Propan
2. Base name: Propanone
3. Branches: 1-Cyclopentyl, 3-phenyl

(e) 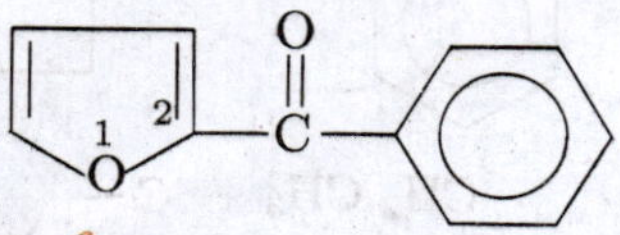

2-Benzofuran

1. Base name: Furan
2. Branch: Benzoyl group named as 2-benzo

(f) 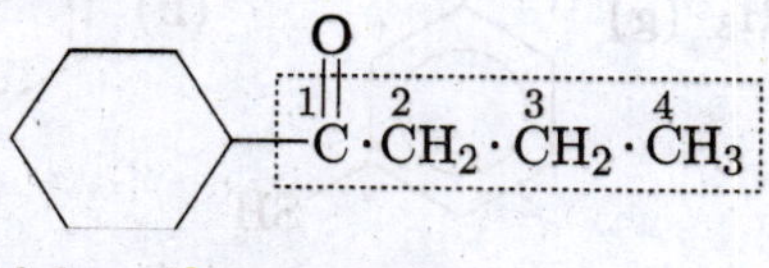

1-Cyclohexylbutan-1-one

1. Root: Butan
2. Suffix: one
3. Base name: Butanone
4. Branch: 1-Cyclohexyl

(g) 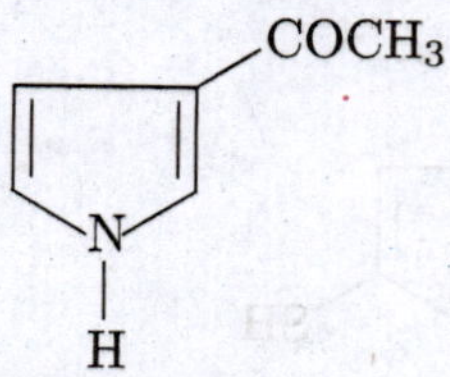

3-Acetylpyrrole

1. Base name: Pyrrole
2. Side chain: 3-Acetyl

(h) 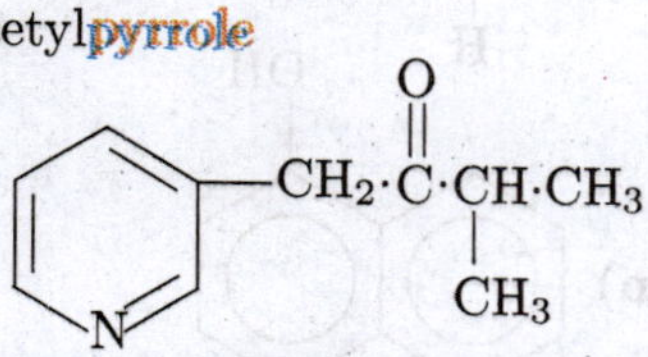

3-Methyl-l-(3-pyridyl)-2-butanone or,
3-Methyl-l-(3-pyridyl)-butan-2-one

1. Base name: Butanone
2. Branches: 3-Methyl and 1-(3-Pyridyl)

3.8.7 Hydroxy and Mercapto Derivatives

7(a)(i) Hydroxy and mercapto derivatives of acyclic and alicyclic series are called alcohols and thioalcohols, respectively. The suffixes of these compounds are **ol** and **thiol**, respectively.

7(a)(ii) Nuclear substituted hydroxy derivatives of benzene and its homologues are called phenols, and their mercapto derivatives are known as thiophenols. **Phenol** and **thiophenol** are used as base names in naming their derivatives. Side chain substituted hydroxy and mercapto derivatives of benzene are termed alcohols and thioalcohols and naturally suffixes used for the functions are **ol** and **thiol**, respectively.

7(a)(iii) Nuclear substituted hydroxy and mercapto derivatives of heterocycles when named, the prefix name of the functions are used as prefix to the root; the root for the base name is the name of the heterocyclic ring. The prefix name of the hydroxyl and mercapto groups are hydroxy and mercapto, respectively.

7(b) In determining the names of the roots and locants of the substituents general rules and procedures are to be adopted. The roots for alcohols and thioalcohols are, respectively, Alkan/Cycloalkan and Alkane/Cycloalkane.

Problem 3.37. Write down the IUPAC names for the following compounds:

(a) 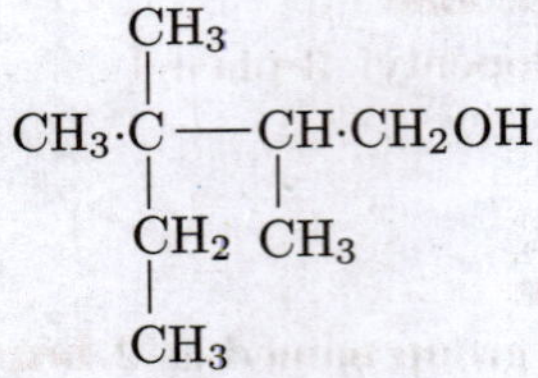 (b) 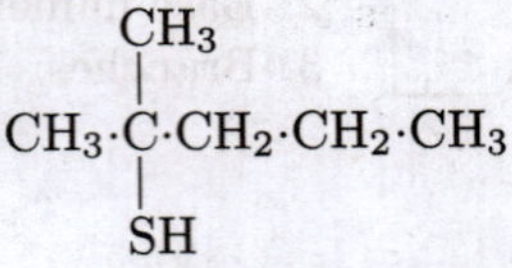 (c) 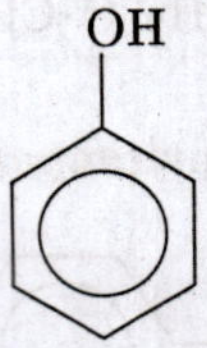 (d) 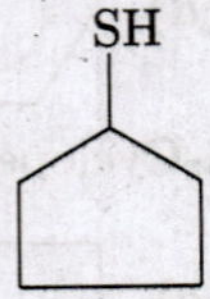

(e) 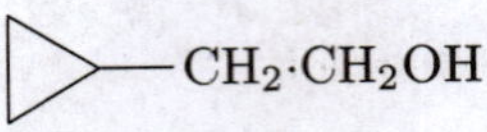 (f) 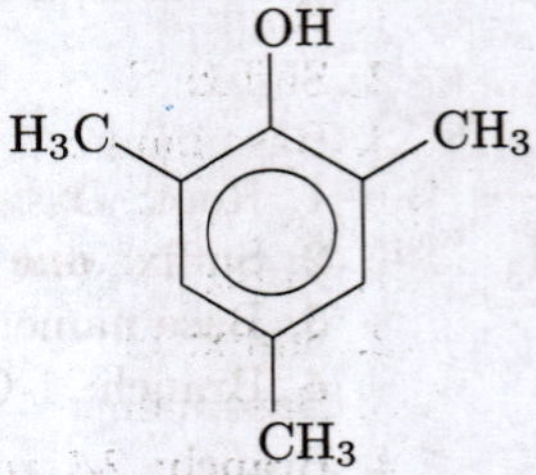 (g) 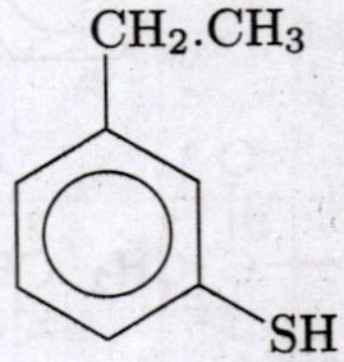 (h) 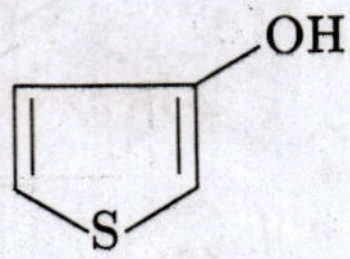

(i) 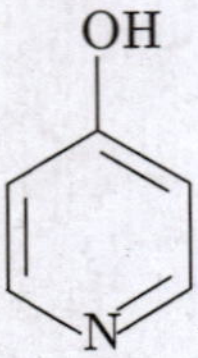 (j) 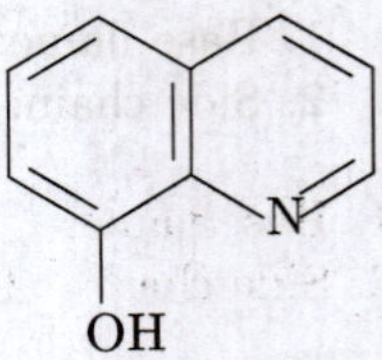 (k) 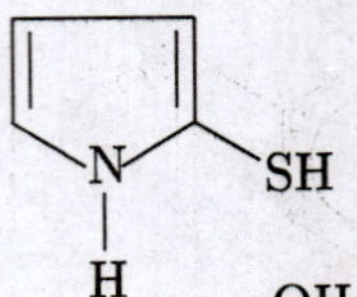

(l) 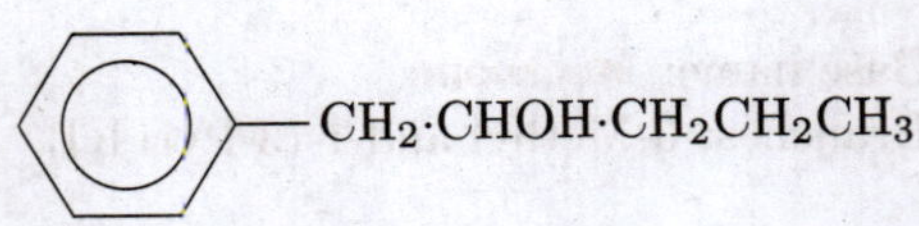 (m) 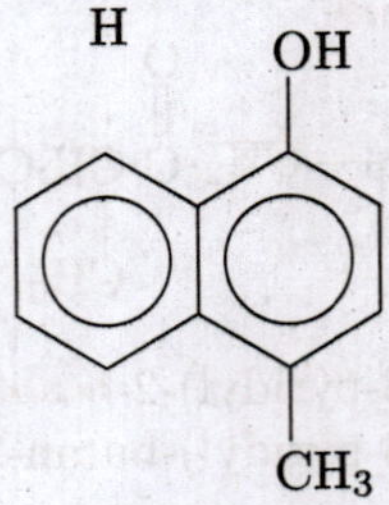

Solution:

(a) 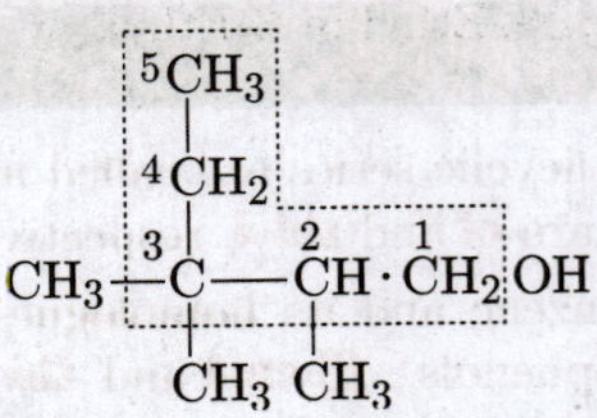

1. Root: Pentan
2. Suffix: ol
3. Base name: Pentanol
4. Branches: 2,3,3-Trimethyl

2,3,3-Trimethyl-1-pentanol
(2,3,3-Trimethylpentan-1-ol)

(b) 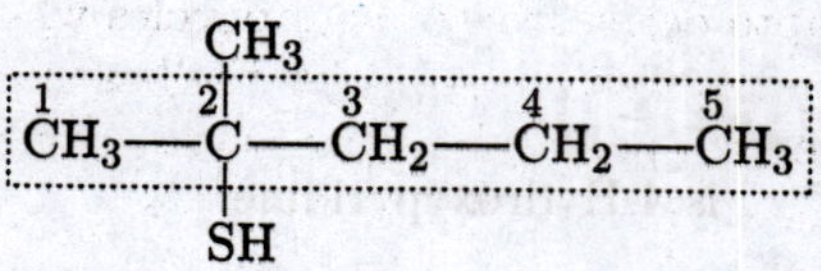

1. Root: Pentane
2. Suffix: thiol
3. Base name: Pentanethiol
4. Branch: 2-Methyl

2-Methyl-2-pentanethiol
2-Methylpentane-2-thiol

(c) 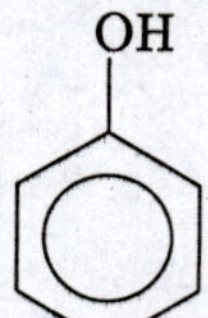

is Phenol.

(d) 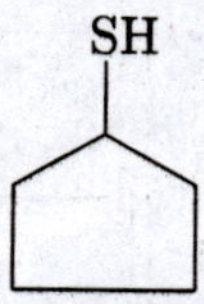

1. Root: Cyclopentane
2. Suffix: thiol
3. Base name: Cyclopentanethiol

Cyclopentanethiol

(e) 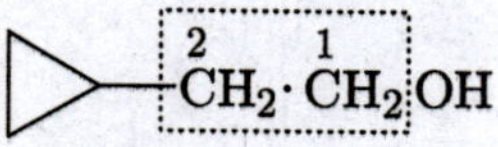

1. Root: Ethan
2. Suffix: ol
3. Base name: Ethanol
4. Branch: 2-Cyclopropyl

2-Cyclopropylethanol

(f) 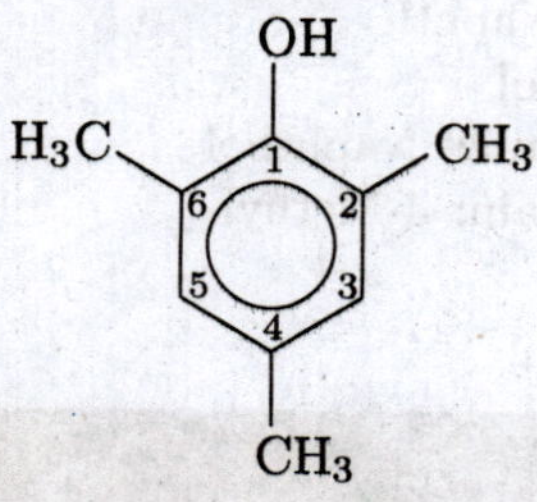

1. Base name: Phenol
2. Side chains: 2,4,6-Trimethyl

2,4,6-Trimethylphenol

(g) 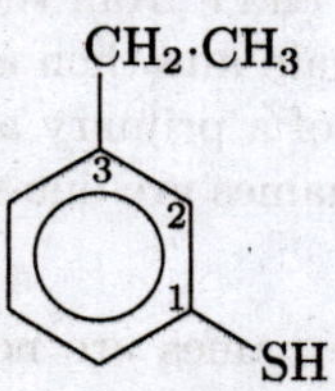

1. Base name: Thiophenol
2. Side chain: *m*-Ethyl or 3-Ethyl

m-Ethylthiophenol
3-Ethylthiophenol

(h) 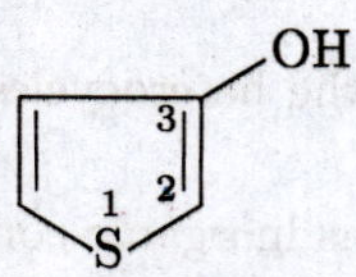

1. Base name: Thiophen
2. Prefix: 3-Hydroxy

3-Hydroxythiophen

(i) 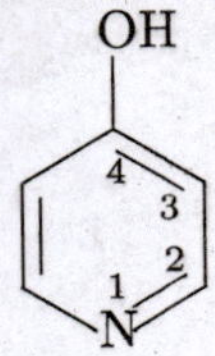

is 4-Hydroxypyridine

(j) 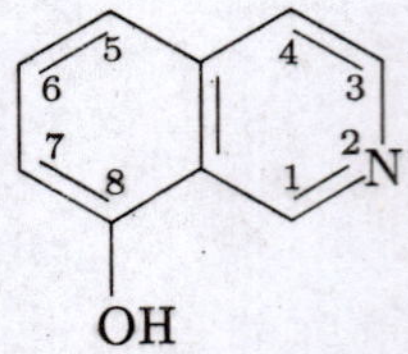

is 8-Hydroxyquinoline

(k) 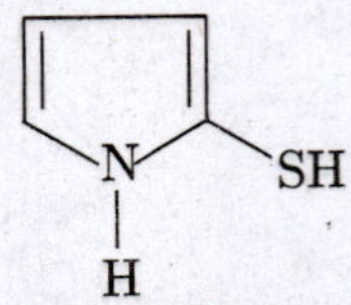

is 2-Mercaptopyrrole

(l) 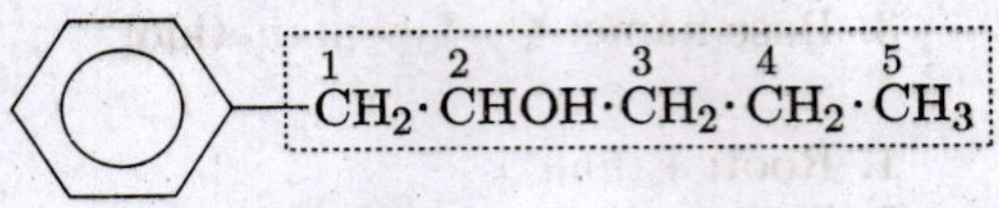

is 1-Phenylpentan-2-ol

(m) 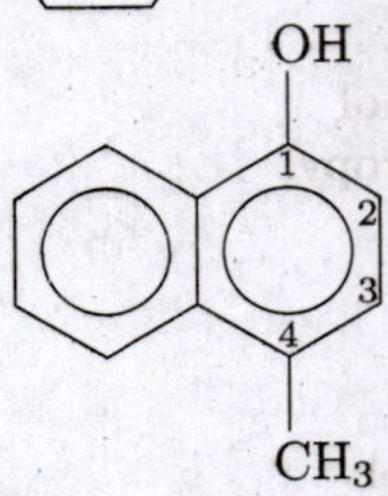

1. Root: Naphth
2. Suffix: ol
3. Base name: Naphthol
4. Side chain: 4-Methyl

4-Methyl-l-naphthol

3.8.8 Amines

8(a) In the IUPAC system, primary amines are named by dropping the end e from the name of the alkane or cycloalkane containing same number of Cs as the base C chain and then adding the suffix amine. 2° and 3° amines are named as N-substituted derivatives of a primary amine; the most complex alkyl group is chosen as the parent primary amine. The names are one-word.

Note

Some amines are named as aminoalkanes, alkylaminoalkanes; these names are not IUPAC recommended.

8(b) Aromatic amines have common names. In naming their derivatives, those names are used as base names.

8(c) The base names of the heterocyclic amines are the names of the heterocycles and the name of the amino group is used as a prefix.

8(d) Quarternary ammonium salts and diazonium salts are named as inorganic compounds.

8(e) The locants are determined by the general rules.

Problem 3.38. Write down the names of the following compounds:

(a) 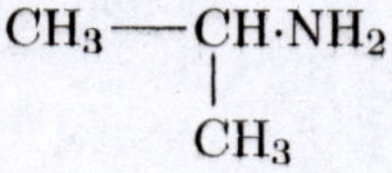 (b) 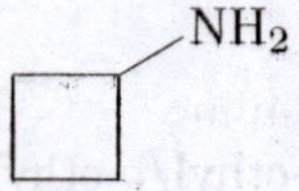 (c) 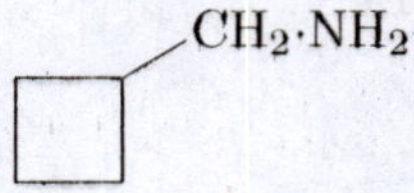 (d) 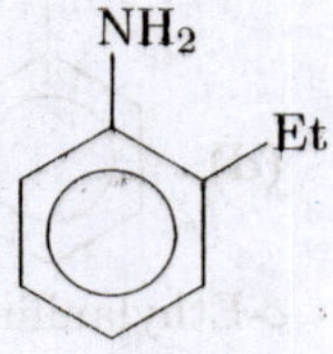

(e) 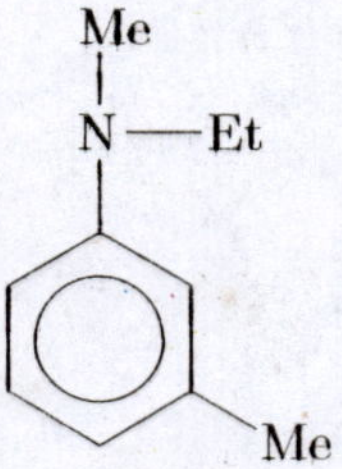 (f) 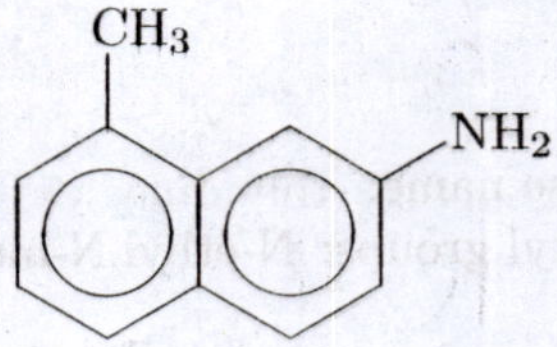 (g) 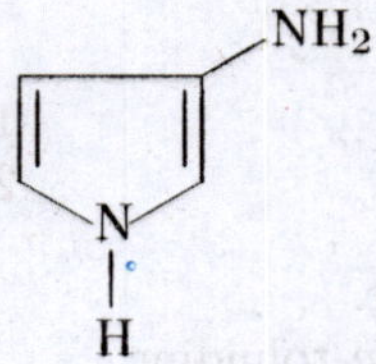 (h) $CH_3\cdot NH\cdot CH_2\cdot CH_3$

(i) 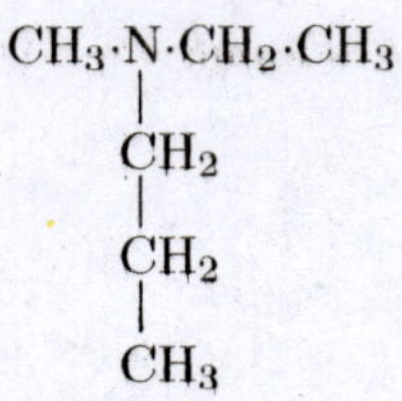 (j) $CH_3\cdot\overset{\oplus}{N}H_3Cl^-$ (k) 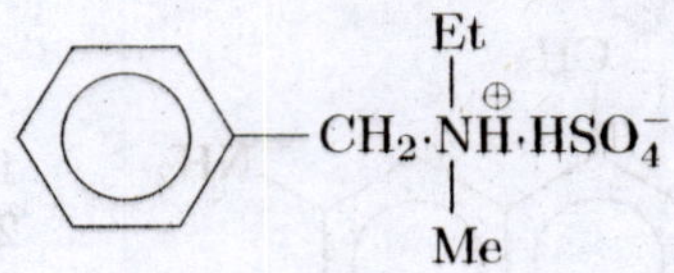

(l) 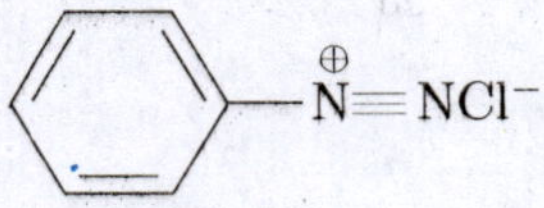

Solution:

(a) 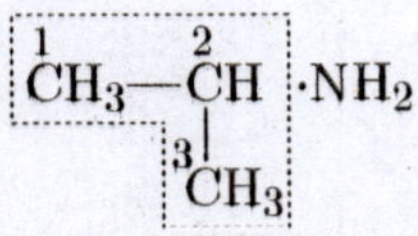

1. Root: Propan
2. Suffix: amine
3. Base name: Propanamine

2-Propanamine
(2-Aminopropane)

(b) 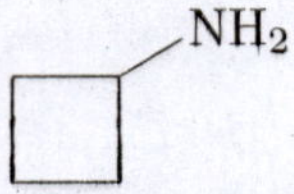

1. Root: Cyclobutan
2. Suffix: amine
3. Base name: Cyclobutanamine

Cyclobutanamine

(c) 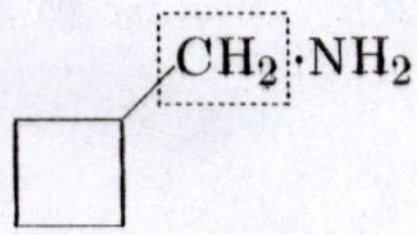

1. Root: Methan
2. Suffix: amine
3. Base name: Methanamine
4. Side chain: Cyclobutyl

Cyclobutylmethanamine (Cyclobutylmethylamine)

(d) 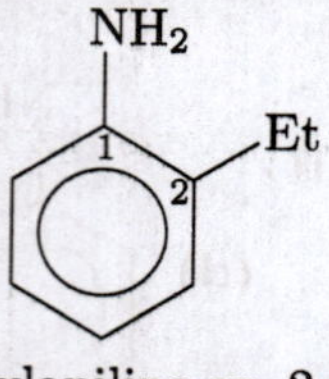

1. Base name: aniline
2. Side chain: 2-ethyl/*o*-ethyl

o-Ethylaniline or, 2-Ethylaniline

(e) 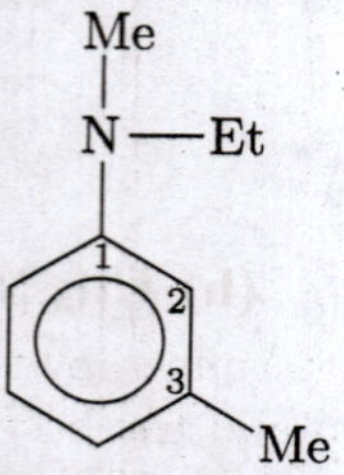

1. Base name: Toluidine
2. Alkyl groups: N-ethyl,N-methyl

N-Ethyl-N-methyl-*m*-toluidine

(f) 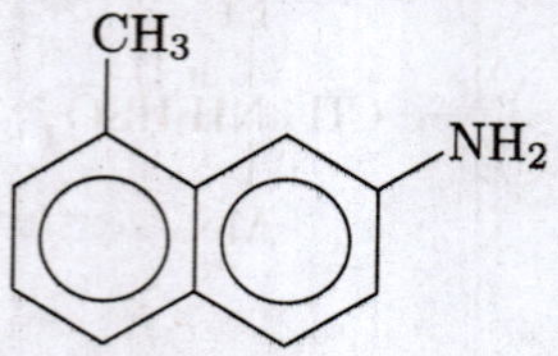

1. Root: Naphthyl
2. Suffix: amine
3. Base name: Naphthylamine
4. Side chain: 8-Methyl

8-Methyl-2-naphthylamine

(g) 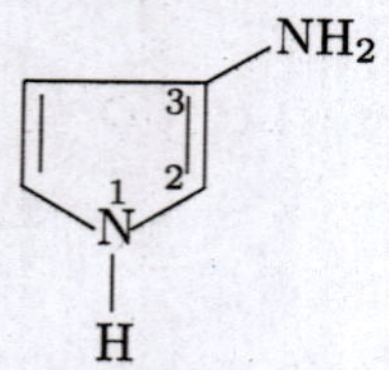

1. Root: Pyrrole
2. Suffix: amino
3. Base name: Aminopyrrole

3-Aminopyrrole

(h) $CH_3{\cdot}NH{\cdot}CH_2{\cdot}CH_3$

1. Root: Ethyl or ethan
2. Suffix: amine
3. Base name: Ethylamine, ethanamine
4. Substituent: N-Methyl

N-Methylethylamine
N-Methylethanamine

(i) 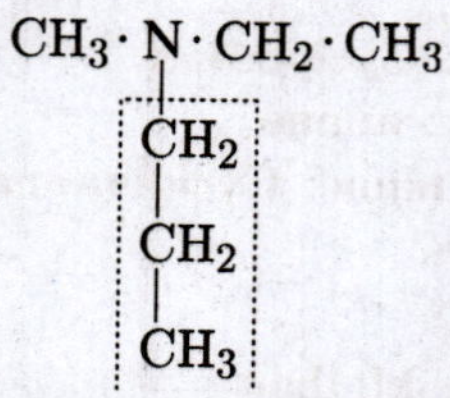

1. Root: Propan
2. Suffix: amine
3. Base name: Propanamine
4. Substituent: N-Ethyl,N-Methyl

N-Ethyl-N-methylpropanamine

(j) $CH_3{\cdot}\overset{\oplus}{N}H_3Cl^-$

Methylammonium chloride

(k) 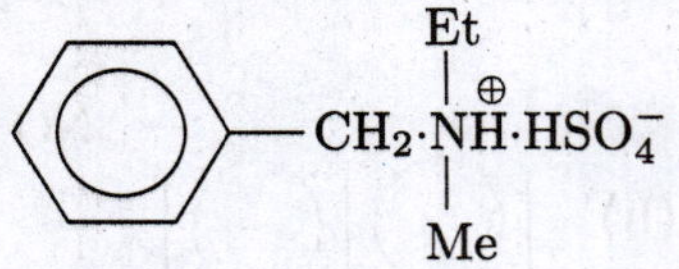 Benzylethylmethylammonium hydrogensulphate

(l) 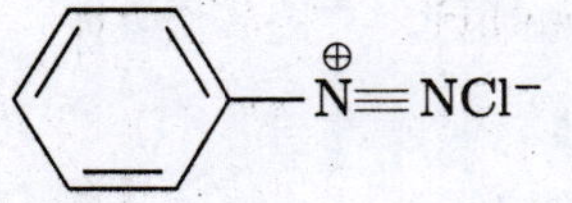 Benzenediazonium chloride

3.8.9 Ethers, Halogen Derivatives, Nitro Derivatives, Nitroso Derivatives

In these compounds the names of the functions are prefixed to the names of the base C chains. Prefix names are given in Table 3.11. Locants and branched groups are determined by the general rules. Now-a-days total numbers of atoms of the main chain are counted and the base name for the hydrocarbon is fixed; the position of oxygen is then indicated by its locant followed by the word 'oxa'. For example,

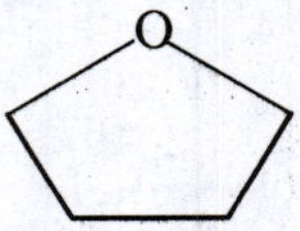

Oxacyclopentane

$\overset{1}{C}H_3\overset{2}{O}\overset{3}{C}H_2\overset{4}{C}H_2\overset{5}{C}H_3$

2-oxapentane

Names of some hydrocarbon groups linked to oxygen are given below; these are helpful in naming ethers:

CH_3O	$CH_3{\cdot}CH_2O$	$CH_3—CH(CH_3)—O$	$CH_3{\cdot}CH_2{\cdot}CH_2{\cdot}CH_2{\cdot}O$
Methoxy	Ethoxy	Isopropoxy	Butoxy

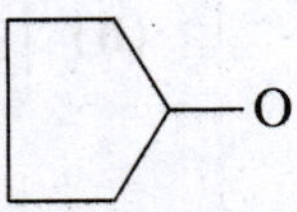

Cyclopentoxy

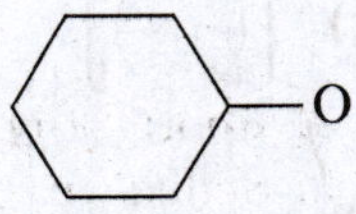

Cyclohexoxy

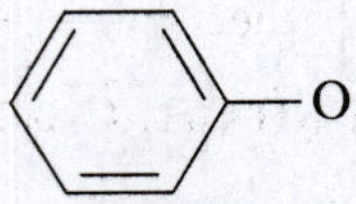

Phenoxy

Problem 3.39. Write down the IUPAC names of the following compounds:

(a) $CH_3{\cdot}O{\cdot}CH_2{\cdot}CH_2{\cdot}CH_3$

(b) 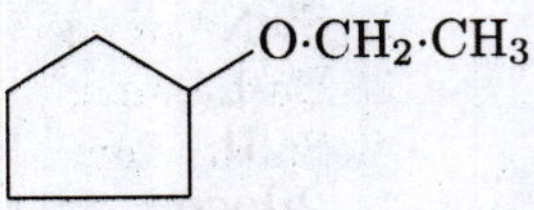

(c) 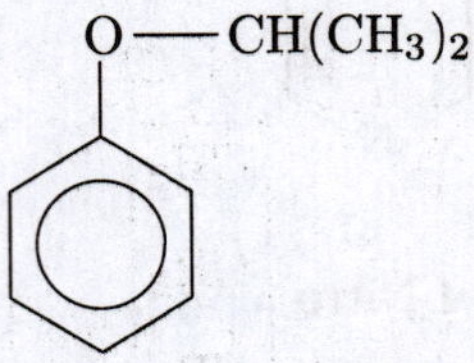

(d) 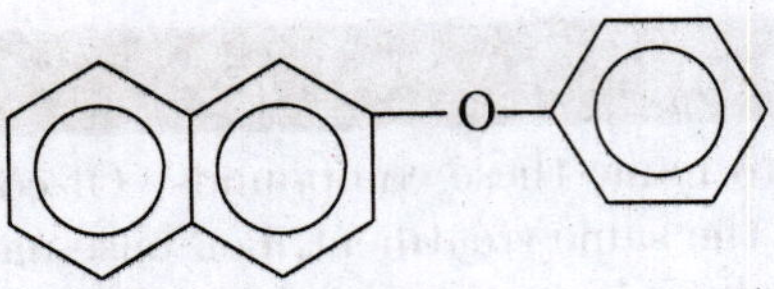

(e) $CH_3{\cdot}CH(Cl){\cdot}CH_2{\cdot}CH_3$

(f) 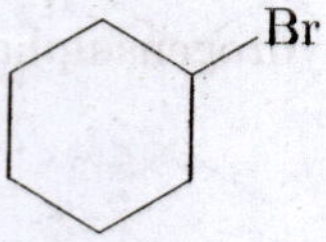

(g) 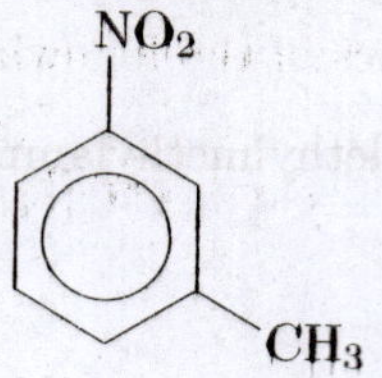

(h) 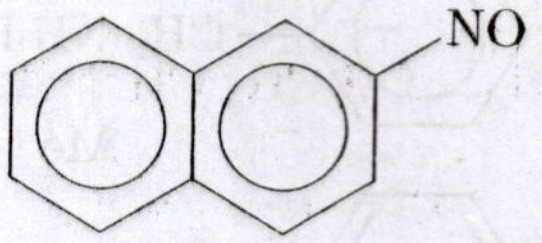

(i) 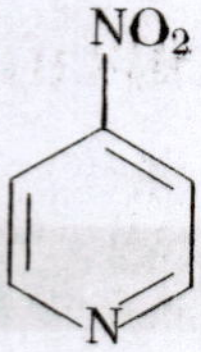

(j) 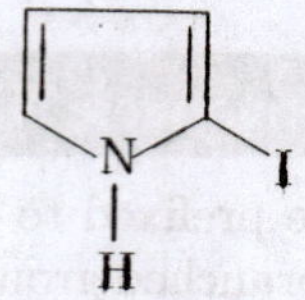

Solution:

(a) $CH_3 \cdot O \cdot \overset{1}{C}H_2 \cdot \overset{2}{C}H_2 \cdot \overset{3}{C}H_3$

1-Methoxypropane

(b) 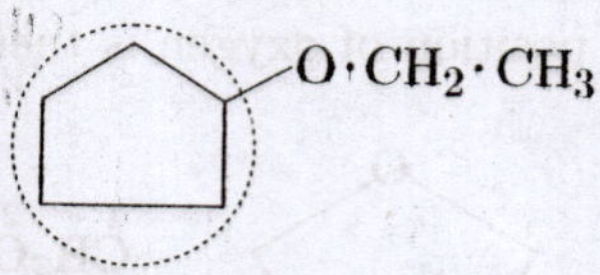

Ethoxycyclopentane

(c) 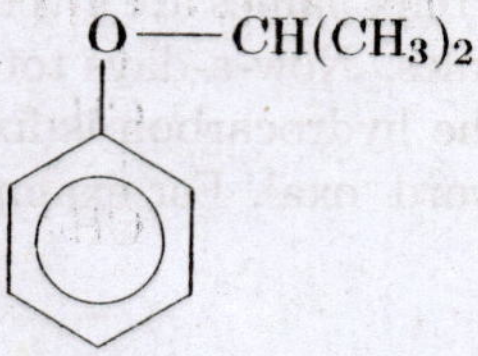

Isopropoxybenzene

(d) 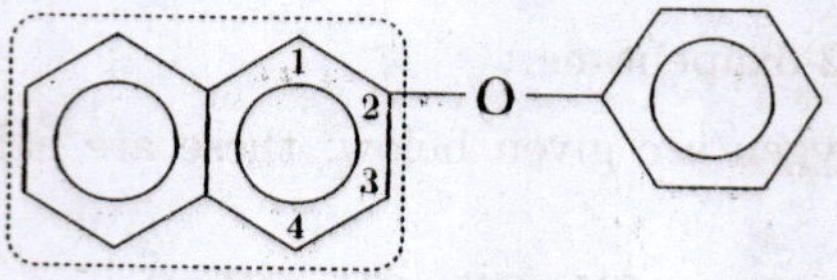

2-Phenoxynaphthalene

(e) $\overset{1}{C}H_3 \cdot \overset{2}{C}H(Cl) \cdot \overset{3}{C}H_2 \cdot \overset{4}{C}H_3$

2-Chlorobutane

(f) 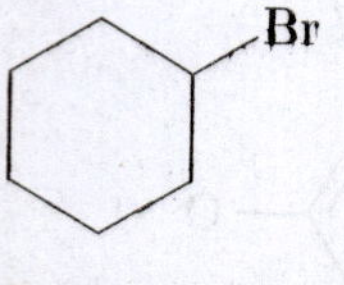

Bromocyclohexane

(g) 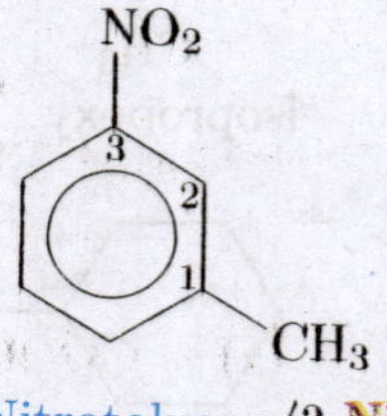

m-Nitrotoluene/3-Nitrotoluene

(h) 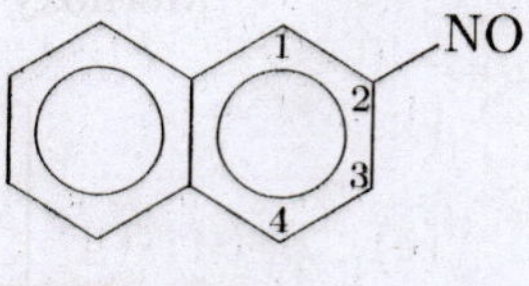

2-Nitrosonaphthalene

(i) 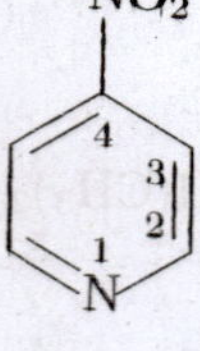

4-Nitropyridine

(j) 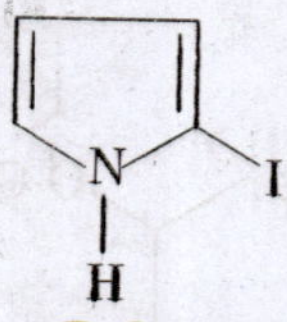

2-Iodopyrrole

3.8.10 Compounds Containing Two Like Functions

The general rules and procedures are also applicable to name these compounds. Of course, one thing deserves mention. The two functions should get the same treatment, i.e., base name suffix must indicate both the functions; this is done by prefixing 'di' to the function suffix.

Problem 3.40. Write down the IUPAC names of the following compounds:

(a) $CH_2{=}CH{\cdot}CH_2{\cdot}CH{=}CH_2$ **(b)** $CH_3{\cdot}C{\equiv}C{\cdot}C{\equiv}C{\cdot}CH_3$ **(c)** 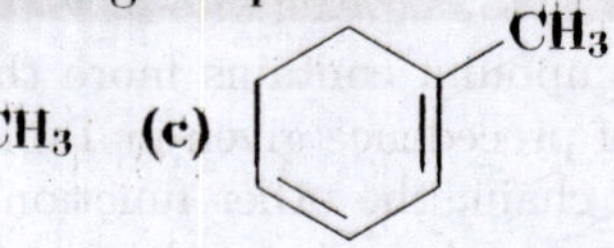

(d) 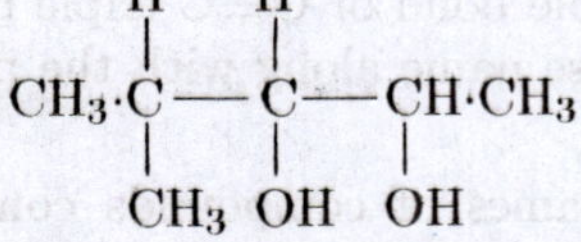 **(e)** 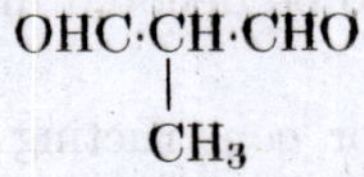 **(f)** $COOH{\cdot}CH_2{\cdot}CH_2{\cdot}COOH$

(g) 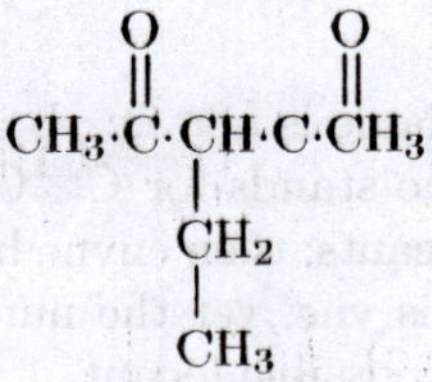 **(h)** 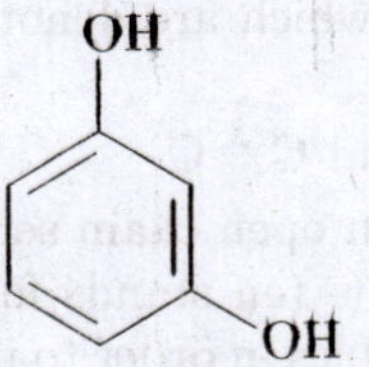 **(i)** $H_2N{\cdot}CH_2{\cdot}CH_2{\cdot}NH_2$

(j) 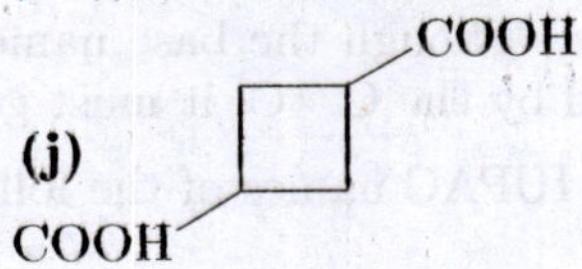

Solution:

(a) $^{1}CH_2{=}^{2}CH{\cdot}^{3}CH_2{\cdot}^{4}CH{=}^{5}CH_2$

Penta-1,4-diene

(b) $^{6}CH_3{\cdot}^{5}C{\equiv}^{4}C{\cdot}^{3}C{\equiv}^{2}C{\cdot}^{1}CH_3$

Hexa-2,4-diyne

(c) 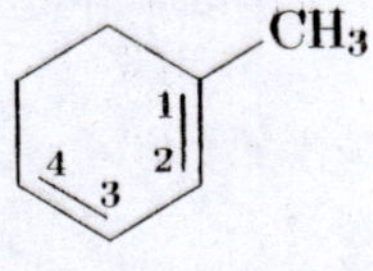

1-Methylcyclohexa-1,3-diene
(1-Methyl-1,3-cyclohexadiene)

(d) 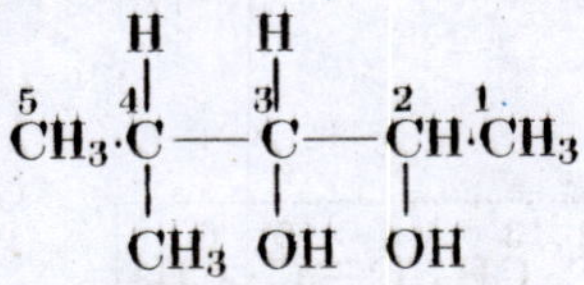

4-Methyl-2,3-pentanediol
(4-Methylpentane-2,3-diol)

(e) 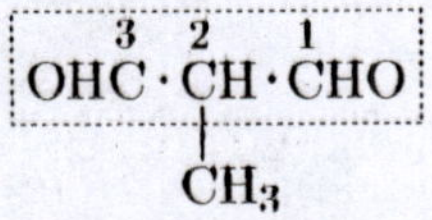

2-Methylpropanedial

(f) $COOH{\cdot}CH_2{\cdot}CH_2{\cdot}COOH$

Butanedioic acid

(g) 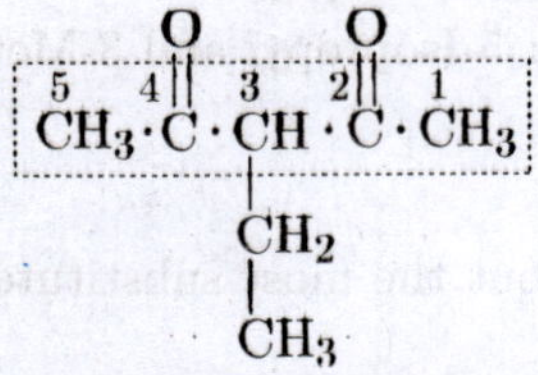

3-Ethylpentane-2,4-dione

(h) 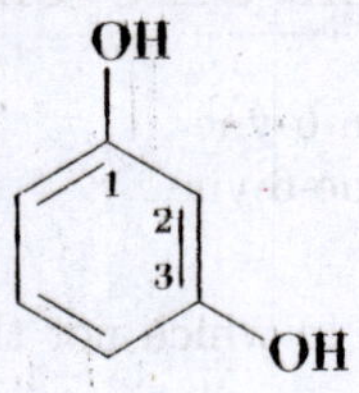

1,3-dihydroxybenzene or Resorcinol

(i) $H_2N{\cdot}CH_2{\cdot}CH_2{\cdot}NH_2$

1,2-Ethanediamine

(j) 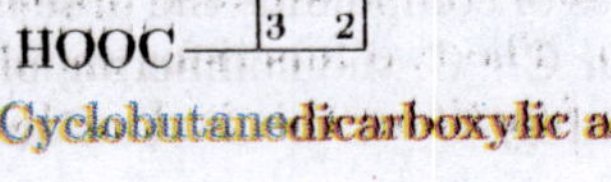

1,3-Cyclobutanedicarboxylic acid

3.8.11 Compounds Containing two Different Functions

When a compound contains more than one function, the primary function (as determined by the order of precedence given in Table 3.11) controls the base name and the numbering of Cs in the base C chain; the other function is indicated as a prefix to the base name. The case is of course different when one of the functions is either a C═C double bond or C≡C triple bond as they have no prefix name; they must also be denoted in the base name along with the primary function.

Naturally, this involves new rules for constructing base names of compounds containing C═C or C≡C and another function which are denoted by suffix names. These are discussed below:

A. Compounds containing C═C and C≡C

This type of compounds are found in open chain series. The root name is alk, i.e., but, pent, hex, etc. The suffix to the root is enyne (en stands for C═C and yne stands for C≡C). This name ending is written as separate syllables in order to indicate their locants; then enyne becomes -ene or -en-(locant of the C≡C)-yne. Though the base name ending is yne, yet the numbering of Cs of the base C chain is governed by the C═C, it must get lowest possible locant.

Problem 3.41. Write down the IUPAC names of the following compounds:

(a) 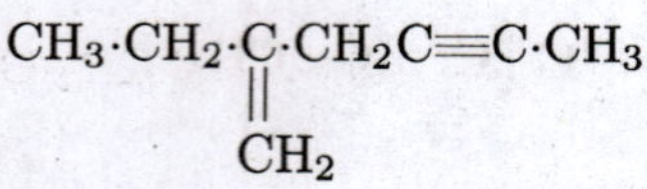

(b) 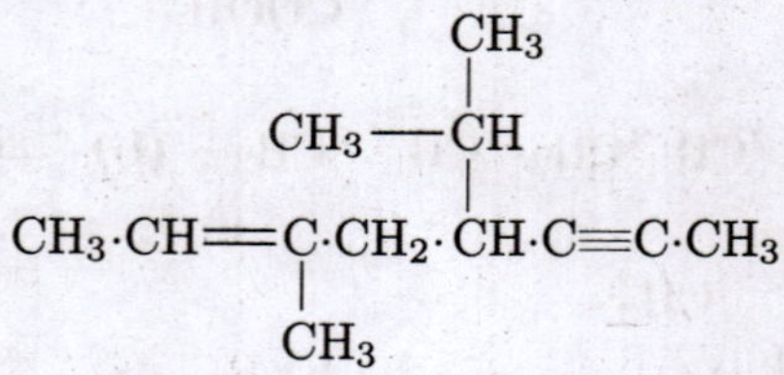

Solution:

(a) 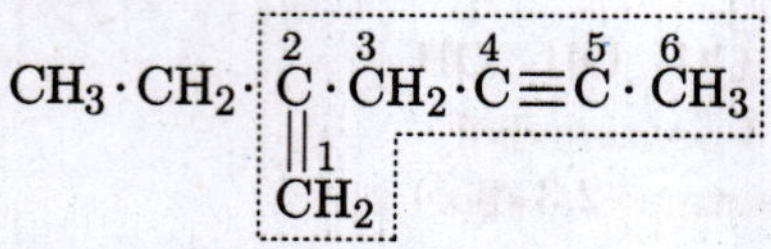

1. Root: Hex
2. Suffix: enyne
3. Base name: Hexenyne
4. Branch: 2-Ethyl

2-Ethyl-1-hexen-4-yne
2-Ethylhex-1-ene-4-yne

(b) 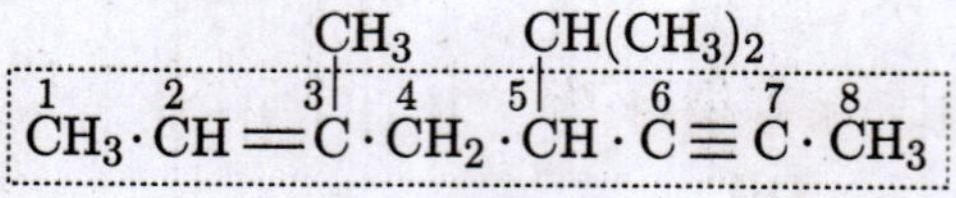

1. Root: Oct
2. Suffix: enyne
3. Base name: Octenyne
4. Branches: 5-Isopropyl and 3-Methyl

5-Isopropyl-3-methyl-2-octen-6-yne
5-Isopropyl-3-methyloct-2-ene-6-yne

Note

Compound (a) is an example in which not the longest chain but the most substituted chain is chosen as the base chain.

B. Compounds containing C═C or C≡C and any one of the following functions: —COOH, —COOR, —$CONH_2$, —CN, —COX, —CHO, >C═O, —OH.

In naming these types of compounds the base name ending is the suffix name of these functions and not that of C═C or C≡C, the numbering of Cs of the base C chain is also governed by the primary function and not by C═C or C≡C. Compounds containing C═C or C≡C and —OH or >C=O when named, the locant of C=C or C≡C and that of —OH or >C=O are shown; but in other cases the locant of the primary function remain understood as it is 1 then. In Table 3.14 the suffixes used to construct the base names are given.

Table 3.14 Suffix for the compounds containing C=C or C≡C and any one of —COOH, —COOR, —$CONH_2$, —COX, —CHO, —CN, >C=O, —OH groups

Function	Suffix	Function	Suffix
C=C & —COOH	—enoic acid	C≡C & —COOH	—ynoic acid
C=C & —COOR	—enoate	C≡C & —COOR	ynoate
C=C & —$CONH_2$	—enamide	C≡C & —$CONH_2$	ynamide
C=C & —COX	—enoyl	C≡C & —COX	ynoyl
C=C & —CHO	—enal	C≡C & —CHO	—ynal
C=C & —CN	ennitrile	C≡C & —CN	ynnitrile
C=C & =C=O	enone, ene, -one	C≡C & =C=O	ynone yne, -one
C=C & —OH	enol, ene, -ol	C≡C & —OH	ynol yne, -ol

Problem 3.42. Write down the IUPAC names of the following compounds:

(a) 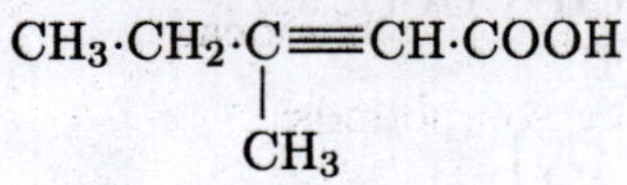

(b) 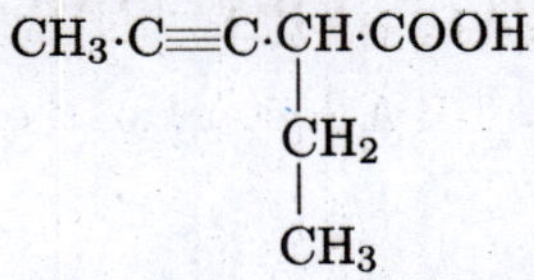

(c) $CH_2{=}CH{\cdot}CH_2{\cdot}COOCH_3$

(d) 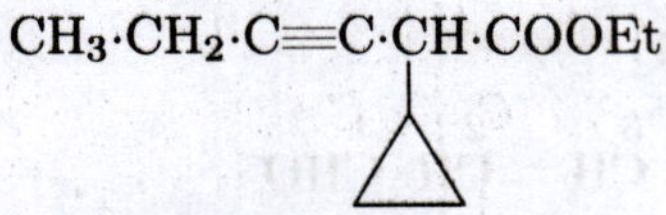

(e) 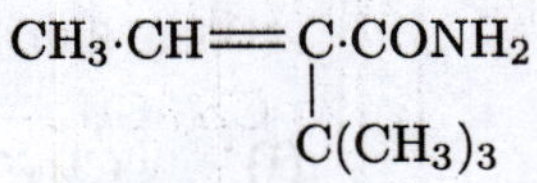

(f) $CH_3{\cdot}C{\equiv}C{\cdot}CH_2{\cdot}CONH_2$

(g) 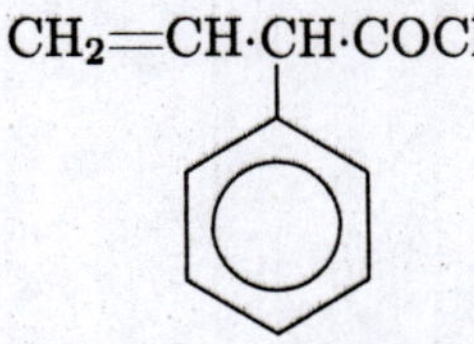

(h) 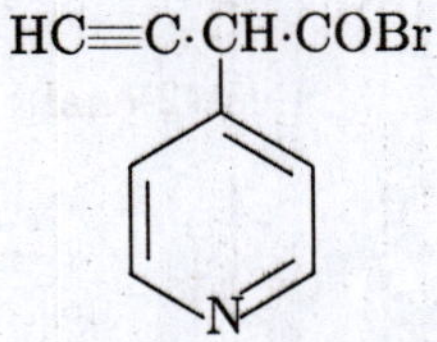

(i) $CH_3{\cdot}CH_2{\cdot}CH{=}CH{\cdot}CHO$

(j) $CH_3{\cdot}CH_2{\cdot}C{\equiv}C{\cdot}CHO$

(k) $CH_3{\cdot}CH{=}CH{\cdot}CN$

(l) 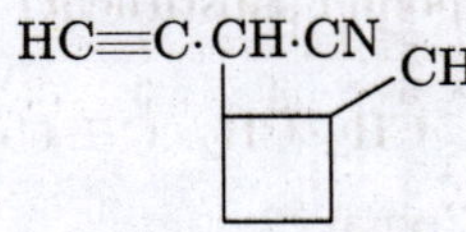

(m) $CH_3{\cdot}CH{=}CH{\cdot}COCH_3$

(n) $CH_3{\cdot}C{\equiv}C{\cdot}CO{\cdot}CH_2{\cdot}CH_3$

(o) 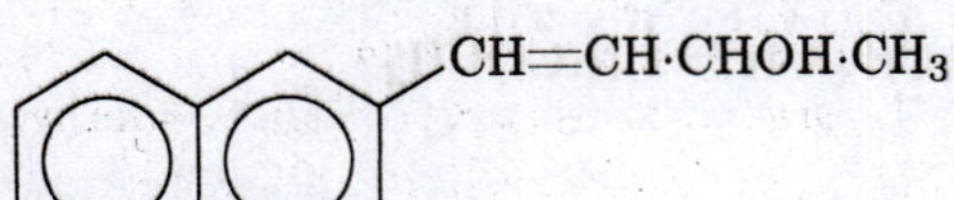

(p) 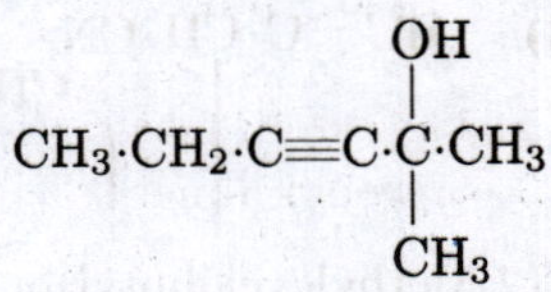

Solution:

(a) 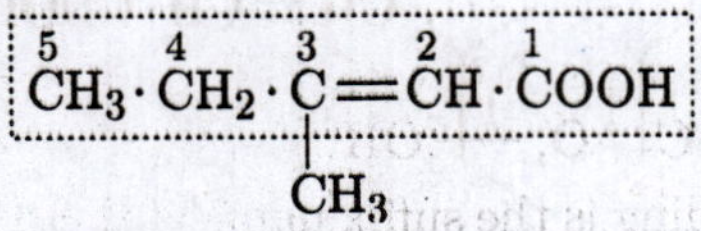

1. Root: Pent
2. Suffix: enoic acid
3. Base name: Pentenoic acid
4. Branch: 3-Methyl

3-Methyl-2-pentenoic acid or
3-Methylpent-2-enoic acid

Similarly,

(b) 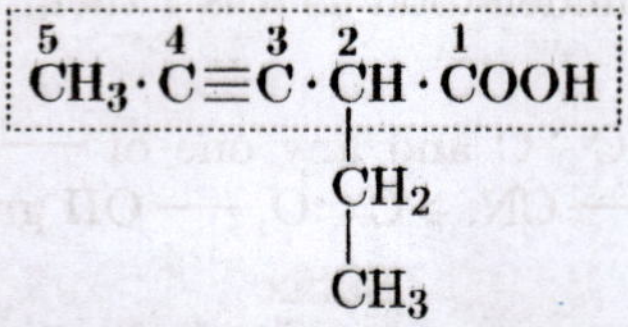

2-Ethyl-3-pentynoic acid

(c) $CH_2=CH\cdot CH_2\cdot COOCH_3$

Methyl 3-butenoate

(d) 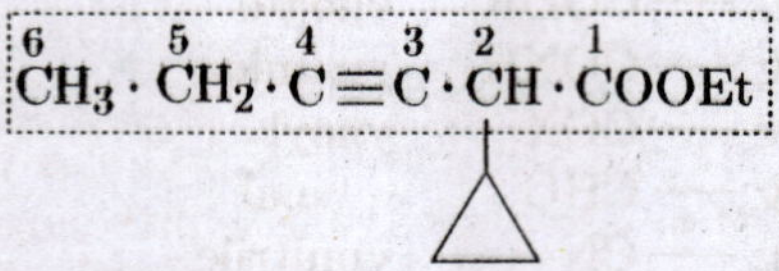

Ethyl 2-cyclopropyl-3-hexynoate
(hex-3-ynoate)

(e) 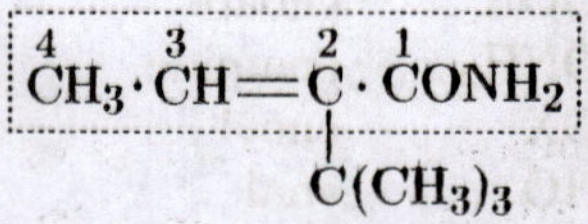

2-tert-Butylbut-2-enamide

(f) $CH_3\cdot C\equiv C\cdot CH_2\cdot CONH_2$

Pent-3-ynamide

(g) 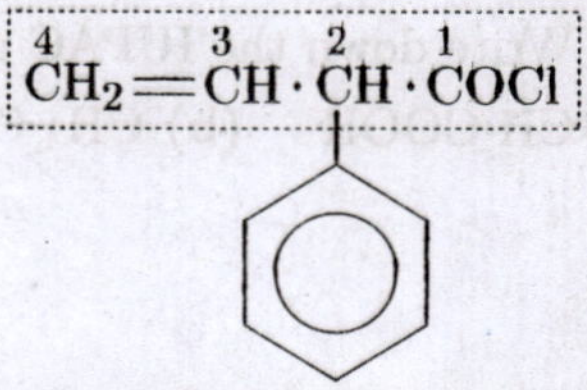

2-Phenylbut-3-enoyl chloride

(h) 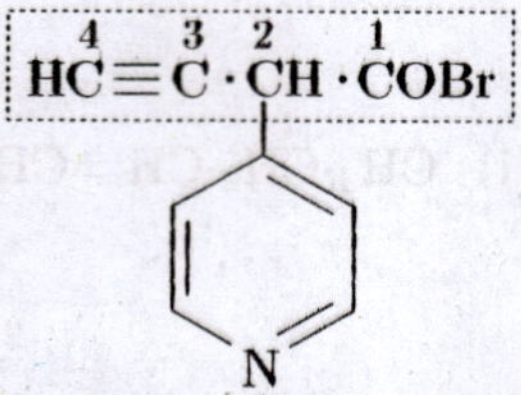

2-(4-Pyridyl)but-3-ynoyl bromide
(Compound substituent)

(i) $CH_3\cdot CH_2\cdot CH=CH\cdot CHO$

Pent-2-enal

(j) $CH_3\cdot CH_2\cdot C\equiv C\cdot CHO$

Pent-2-ynal

(k) $CH_3\cdot CH=CH\cdot CN$

But-2-ennitrile

(l) 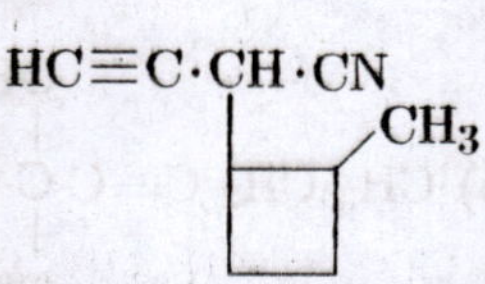

2-(2-Methylcyclobutyl)but-3-ynnitrile

(m) $CH_3\cdot CH=CH\cdot COCH_3$

3-Penten-2-one/pent-3-ene-2-one

(n) $CH_3\cdot C\equiv C\cdot CO\cdot CH_2\cdot CH_3$

4-Hexyn-3-one
Hex-4-yne-3-one

(o) 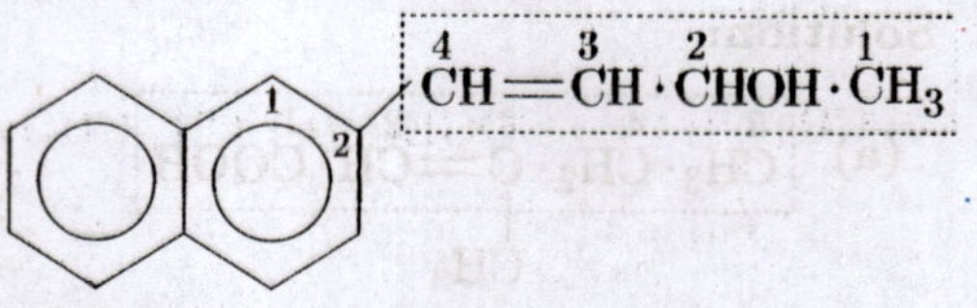

4-(2-Naphthyl)-3-buten-2-ol
4-(2-Naphthyl)-but-3-ene-2-ol

(p) $\overset{6}{CH_3}\cdot\overset{5}{CH_2}\cdot\overset{4}{C}\equiv\overset{3}{C}\cdot\underset{\underset{CH_3}{|}}{\overset{\overset{OH}{|}}{\overset{2}{C}}}\cdot\overset{1}{CH_3}$

2-Methyl-3-hexyn-2-ol

2-Methylhex-3-yne-2-ol

C. Compounds containing C=C or C≡C bond and any of —NH₂, —OR, —OX, —NO, etc., groups

In naming these compounds the suffixes to the roots are **ene** and **yne** for C=C and C≡C, respectively; the other groups are indicated as prefixes to the base names. The numbering of Cs is controlled by C=C and C≡C in all cases except in the case of amines.

Problem 3.43. Write down the IUPAC names of the following compounds:

(a) $CH_2{=}CH{\cdot}CH_2{\cdot}NH_2$ **(b)** $CH_2{\cdot}C{\equiv}C{\cdot}CH_2N\begin{matrix}\diagup CH_3\\ \diagdown CH_3\end{matrix}$ **(c)** $CH_3{\cdot}CH{=}CH{\cdot}CH_2NO_2$

(d) $CH{\equiv}C{\cdot}CH_2Cl.$ (e) $CH_3O{\cdot}CH{=}CH{\cdot}CH_3$

Solution:

(a) $\overset{3}{CH_2}=\overset{2}{CH}\cdot\overset{1}{CH_2}\cdot NH_2$

1-Amino-2-propene

(b) $\overset{4}{CH_3}\cdot\overset{3}{C}\equiv\overset{2}{C}\cdot\overset{1}{CH_2}\cdot N\begin{matrix}\diagup CH_3\\ \diagdown CH_3\end{matrix}$

1-(N,N-dimethylamino)-2-butyne

(c) $\overset{4}{CH_3}\cdot\overset{3}{CH}=\overset{2}{CH}\cdot\overset{1}{CH_2}NO_2$

1-Nitro-2-butene

(d) $\overset{1}{CH}\equiv\overset{2}{C}\cdot\overset{3}{CH_2}Cl$

3-Chloropropyne

(e) $CH_3O\cdot\overset{1}{CH}=\overset{2}{CH}\cdot\overset{3}{CH_3}$

1-Methoxypropene

D. Compounds containing two functions other than C=C and C≡C

The general rules and procedure are sufficient to name these types of compounds; Table 3.11 should be consulted in determining the prefixed names of the functions.

Problem 3.44. Write down the names of the following compounds:

(a) $OHC{\cdot}\underset{\underset{CH_3}{|}}{CH}{\cdot}CH_2{\cdot}COOH$ **(b)** $CH_3{\cdot}\overset{\overset{O}{\|}}{C}{\cdot}CH_2{\cdot}COOH$ **(c)** $CONH_2{\cdot}CH_2{\cdot}CH_2{\cdot}SO_3H$

(d) $CN{\cdot}CH_2{\cdot}\underset{\underset{\underset{\underset{CH_3}{|}}{CH_2}}{|}}{CH}{\cdot}CH_2{\cdot}COOEt$ **(e)** $CH_3{\cdot}CHOH{\cdot}CH_2{\cdot}\underset{\underset{CH_3}{|}}{CH}{\cdot}CHO$ **(f)** $SH{\cdot}CH_2{\cdot}CH_2{\cdot}CH_2OH$

(g) 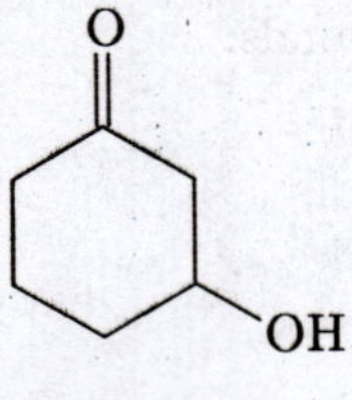

(h) 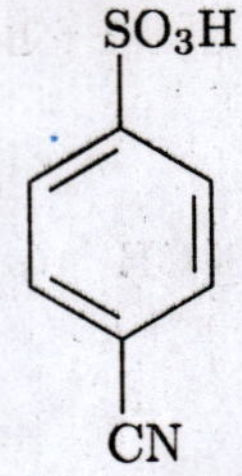

(i) 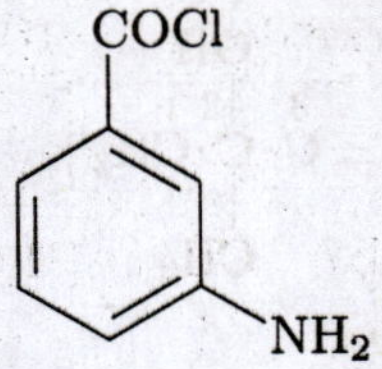

Solution:

(a) $\overset{4}{OHC}\cdot\overset{3}{CH}(CH_3)\cdot\overset{2}{CH_2}\cdot\overset{1}{COOH}$

1. Root: butan (—CHO is a substituent, Table 3.11).
2. Suffix: oic acid
3. Base name: Butanoic acid
4. Substituent: 3-Methyl, 4-oxo

3-Methyl-4-oxobutanoic acid

Similarly,

(b) $\overset{4}{CH_3}\cdot\overset{3}{C}(=O)\cdot\overset{2}{CH_2}\cdot\overset{1}{COOH}$

3-Oxobutanoic acid

(c) $CONH_2\cdot\overset{2}{CH_2}\cdot\overset{1}{CH_2}\cdot SO_3H$

2-Carbamoylethanesulphonic acid

(d) $CN\cdot\overset{4}{CH_2}\cdot\overset{3}{CH}(CH_2CH_3)\cdot\overset{2}{CH_2}\cdot\overset{1}{COOEt}$

Ethyl 4-cyano-3-ethylbutanoate

(e) $\overset{5}{CH_3}\cdot\overset{4}{CHOH}\cdot\overset{3}{CH_2}\cdot\overset{2}{CH}(CH_3)\cdot\overset{1}{CHO}$

4-Hydroxy-2-methylpentanal

(f) $SH\cdot\overset{3}{CH_2}\cdot\overset{2}{CH_2}\cdot\overset{1}{CH_2}\cdot OH$

3-Mercapto-1-propanol

(g) 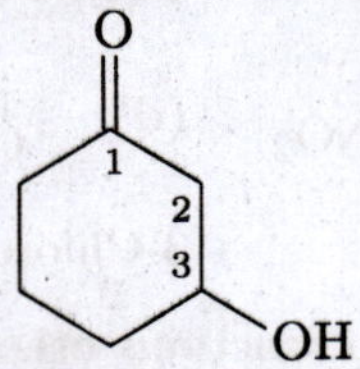

3-Hydroxycyclohexanone

(h) 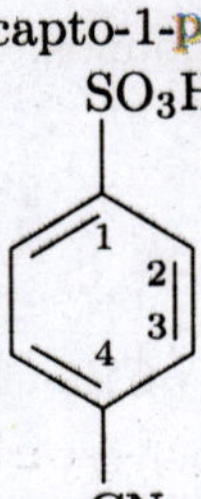

4-Cyanobenzenesulphonic acid

(i) 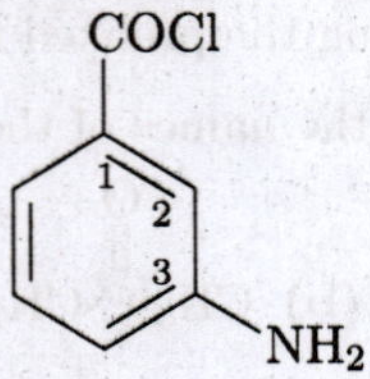

3-Aminobenzoyl chloride

3.8.12 Polyfunctional Compounds

Table 3.11, general rules and all the discussions made above are helpful in giving IUPAC name to a polyfunctional compound.

Problem 3.45. Write down the IUPAC names of the following compounds:

(a) 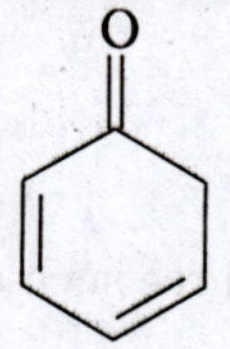

(b) $CH_2{=}CH{\cdot}CH{=}CH{\cdot}CH_2OH$

(c) 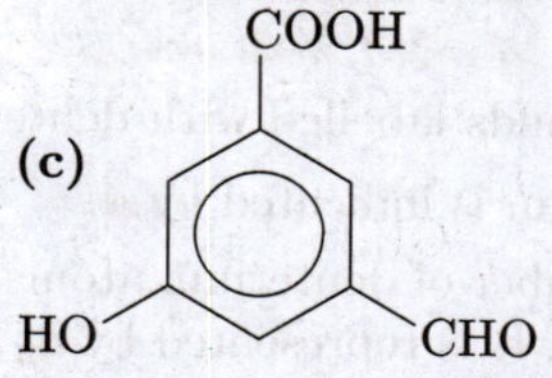

(d) $Br{\cdot}CH_2{\cdot}CH{=}CH{\cdot}CH(C(CH_3)_3){\cdot}COOH$

(e) 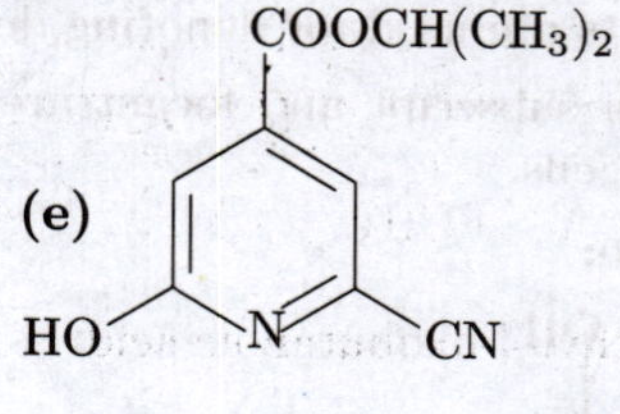

(f) 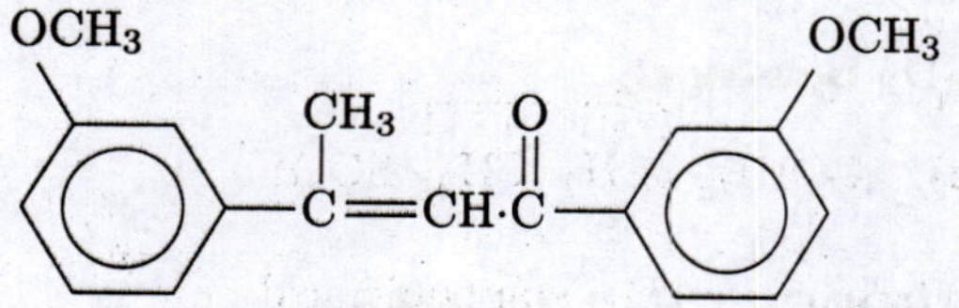

Solution:

(a) 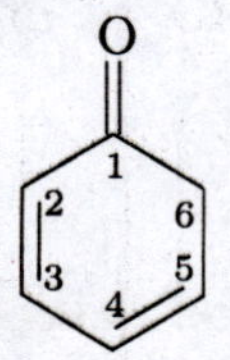

cyclohexa-2,4-dienone

(b) $\overset{5}{C}H_2{=}\overset{4}{C}H{\cdot}\overset{3}{C}H{=}\overset{2}{C}H{\cdot}\overset{1}{C}H_2OH$

Penta-2,4-dien-1-ol

(c) 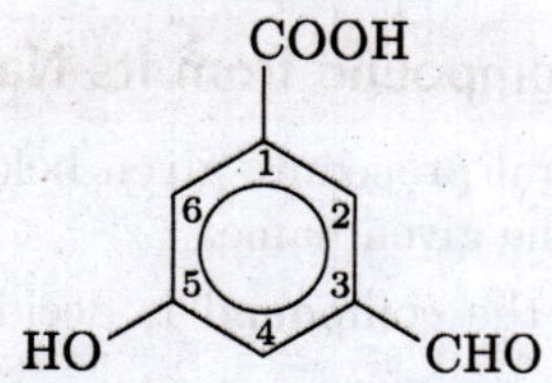

3-Formyl-5-hydroxybenzoic acid

(d) $Br{\cdot}\overset{5}{C}H_2{\cdot}\overset{4}{C}H{=}\overset{3}{C}H{\cdot}\overset{2}{C}H(C(CH_3)_3){\cdot}\overset{1}{C}OOH$

5-Bromo-2-tert-butyl-3-pentenoic acid

(e) 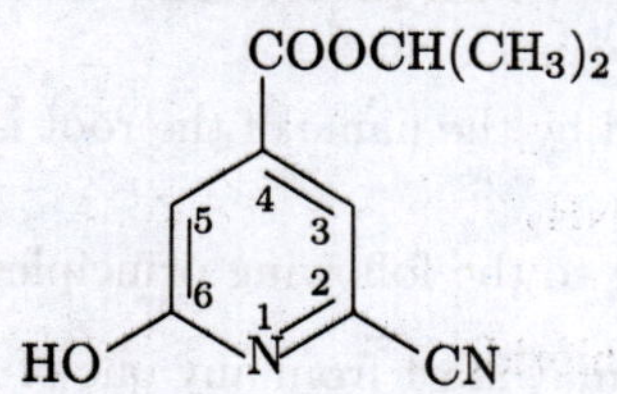

Isopropyl 2-cyano-6-hydroxypyridine-4-carboxylate

(f) 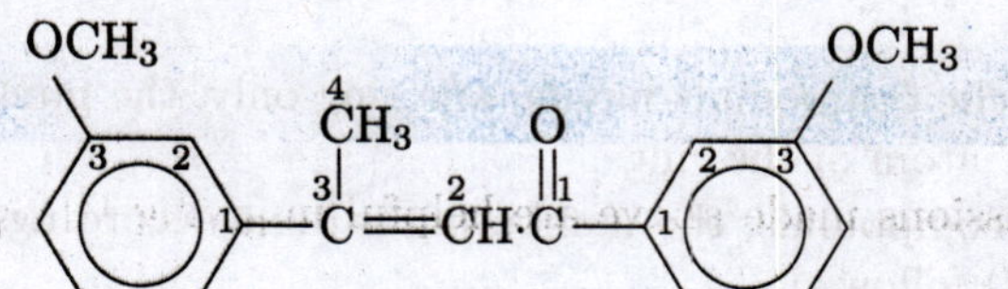

1,3-Bis(3-anisyl)-2-butenone

3.8.13 Miscellaneous Compounds

A. Compounds labelled with isotopes

(a) Compounds labelled with deuterium

(i) Deuterium is indicated by **d**.

(ii) The number of deuterium atoms used are denoted by a subscript; thus, if three such atoms are used, it is represented by $\mathbf{d_3}$.

(iii) Locants are used for denoting deuterium atoms if necessary. Locants are prefixed to **d**.

(iv) **d** with subscript and locant are placed after the part of the name designating replaced hydrogens.

Example:

1. 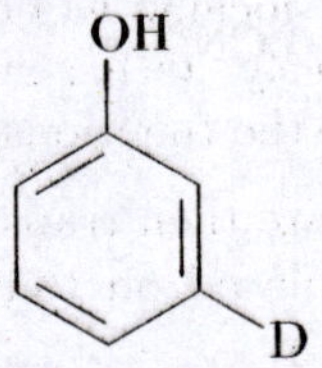

phen-*m*-**d-ol** 2. $C_6H_4D_2$ **Benzene-d_2**

3. $CH_3{\cdot}CH_2OD$ **Ethanol-d** 4. $CH_3CHODCH_2CH_3$ 2-**Butanol-d**

(b) Isotopic C, N, O, etc., are also denoted by the same procedure as used in the deuterium labelled compounds.

$CH_3{\cdot}CO{\cdot}CH_2{\cdot}{}^{14}CH_3$ is **Butanone**-4-^{14}C.

B. Nomenclature of stereoisomers

This will be discussed in Chapter 9.

C. Nomenclature of carbenium ions (carbocations), carbanions and carbon radicals: These are discussed in Chapter 11.

3.9 Drawing of The Structural Formula of an Organic Compound from Its Name

There are no general rules in this connection; however, the general procedure given below may be followed to draw the structural formula of a compound from the given name.

1. Whether it is the IUPAC name or the common name of the compound is decided and accordingly the name is split up into the following parts:

   (a) Root of the base name, **(b)** suffix or suffixes to the root with their locants and **(c)** the prefixed names of the different substituents with their locants.

2. The carbon skeleton of the base carbon chain as indicated by the name of the root is drawn on the plane of the paper.

3. The Cs of the chain are numbered sequentially according to the following principles:

   (a) For an open chain compound, the numbering of Cs may start from any one of the two ends of the base chain.

   (b) For benzene derivatives and alicyclic compounds having one ring only, the numbering may start from any constituent C atom of the rings.

   (c) For heterocyclic and polynuclear compounds, the numbering is done according to the conventions (Sec. 3.8 VI C and D) followed.

4(a) From the names of the suffixes to the root, the structural formula of the primary function is drawn; Table 3.11 indicates the primary function and its structure.

4(b) If the primary function is a part of the base carbon chain, the C atom bearing the locant of the primary function as its number is converted into the function.

4(c) On the other hand, if the primary function is not a part of the base C chain, i.e., it is a mere substituent, then its structural formula is written on the C atom that bears the locant of the primary function as its number.

In this connection one must remember that if the primary function does not carry any locant with its name, then either the position of the function is unique in the chain or its locant is 1; in both the cases one should not mention the locant.

5. To indicate suffix names of C═C and C≡C, the Cs bearing their locants as number are converted into those groups inserting one and two bonds, respectively. Other suffixes are also indicated in the structure by converting corresponding Cs into the functions.

6. The prefixed names of the different substituents with their locants are then translated into the corresponding structural formulae and then each of them is placed on that C atom whose number is the locant of the substituent it carries.

7. To adjust tetracovalency of the Cs of the base C chain, an adequate number of H atom is bonded to each C atom of the chain.

8. The structural formula so drawn is then given a name in the same system of nomenclature as used in the question. If these two names are identical, the structural formula so written is correct; if otherwise, it is not.

If one wants to follow this general procedure in drawing the structural formula for a given name, one must learn all the rules followed in naming a compound in all the systems with utmost care.

Example: What is the structure of 3-hepten-5-yn-2-one(hept-3-en-5-yn-2-one)?

(i) The name is written in the IUPAC system.

(a) Root: Hept, (b) Suffix: 3-en, 5-yn and 2-one (Rule 1), (c) Substituents: Nil.

(ii) The root being hept the C skeleton of the base chain is C—C—C—C—C—C—C

A (Rule 2)

(iii) The C skeleton is numbered as: $^{7}C—^{6}C—^{5}C—^{4}C—^{3}C—^{2}C—^{1}C$

B (Rule 3a)

(iv) According to Table 3.11, the primary function is a ketonic carbonyl function and its structure is >C═O (Rule 4a).

(v) It is a part of the base carbon chain and its locant being 2, the C skeleton becomes

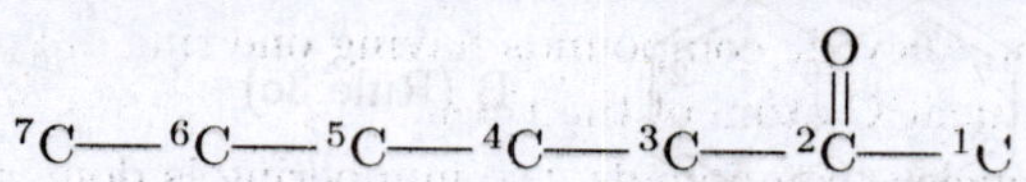

C (Rule 4b)

(vi) The other suffixes are 3-en, i.e., C═C on C_3 atom and 5-yn, i.e., C≡C on C_5 atom. Hence, C skeleton becomes

$$^{7}C—^{6}C\equiv{}^{5}C—^{4}C={}^{3}C—^{2}\overset{\overset{O}{\|}}{C}—^{1}C$$

D (Rule 5)

(vii) As each of the Cs should have tetracovalency, the C skeleton becomes

$$H_3C—C\equiv C—CH=CH\cdot\overset{\overset{O}{\|}}{C}\cdot CH_3$$

E (Rule 6)

(viii) The IUPAC name of this structure is

$CH_3\cdot C\equiv C\cdot CH=CH\cdot COCH_3$

(a) Root : hept
(b) Suffix: -en, -yn, -one
(c) Substituents: nil
(d) Primary function: >C═O

3-Hepten-5-yn-2-one

(ix) Hence the structure of the given compound is $CH_3\cdot C\equiv C—CH=CH\cdot COCH_3$.

Problem 3.46. Write down the structures of the following compounds:

(a) 5-vinyl-2-naphthalenesulphonic acid.

(b) *m*-Nitrobenzamide.

(c) 2-Formyl-3-tert-butylpyrrole.

(d) 2-(2,4-Diaminophenyl)butanedioic acid.

(e) Vinyl allyl ketone.

(f) Ethyl 3,5-dimethylcyclohexanecarboxylate.

(g) 2,4,6-Tris(3-cyclopentenyl)phenol.

(h) Diethyl malonate.

Solution: (a)(i) IUPAC name—Root: Naphthalene; Suffix: Sulphonic acid, its locant is 2. Substituent: 5-Vinyl (Rule 1).

(ii) C skeleton:

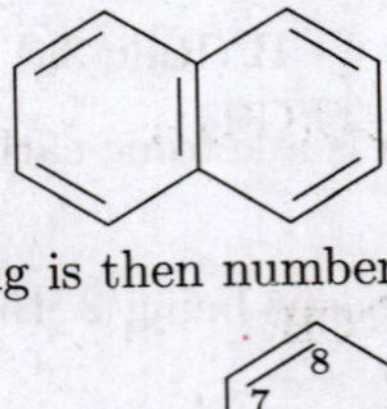

A (Rule 2)

(iii) Naphthalene ring is then numbered.

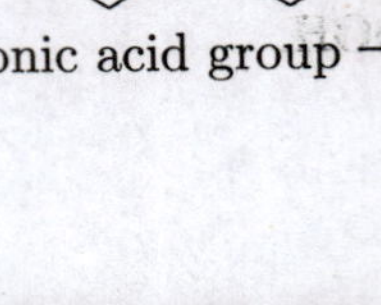

B (Rule 3c)

(iv) Primary function: sulphonic acid group —SO_3H (Rule 4a).

(v) When —SO_3H function is placed on the C_2 atom of the naphthalene ring it becomes

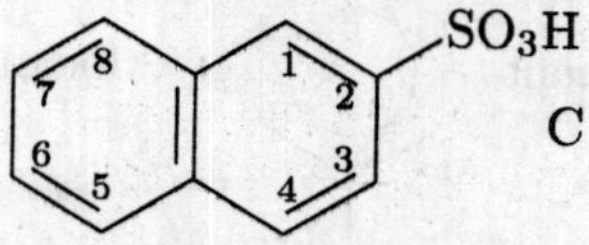

C (Rule 4c)

(vi) The structure of the substituent is: —CH=CH_2 its locant being 5, the structure C becomes

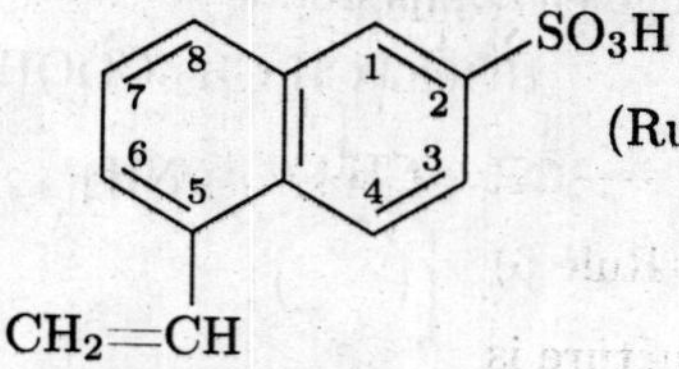

(Rule 6)

Hence, the structure of the given compound is

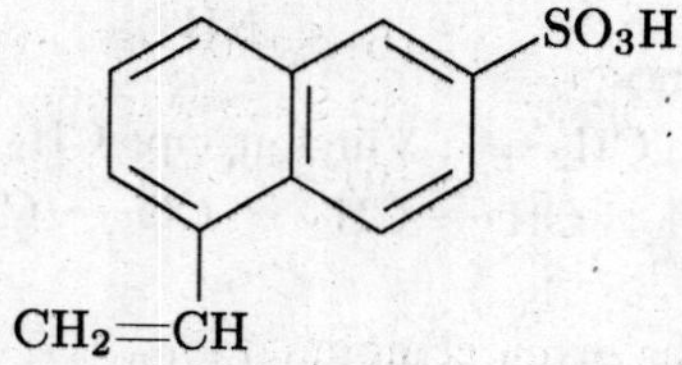

(b) (i)  ; **(ii)** 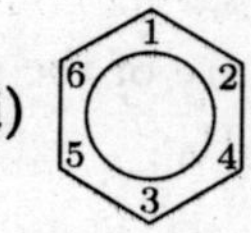 ; **(iii)** —$CONH_2$; **(iv)** 

Benz (Rule 2) (Rule 3b) (Rule 4a)

(v) *m*-Nitro = 3-nitro; Nitro group —NO_2.

The structure of the compound is then

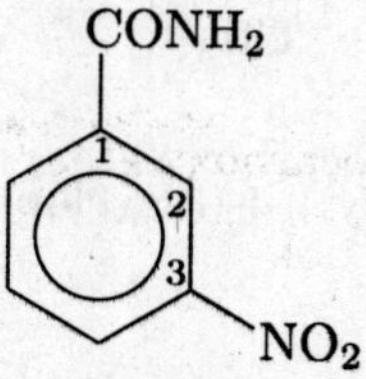

(c) (i) 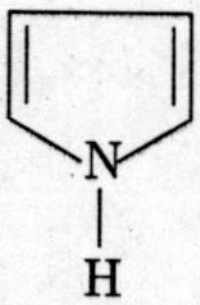 ; **(ii)**  ; **(iii)** Formyl —CHO; t-Butyl—$C(CH_3)_3$.

The structure of the compound is then

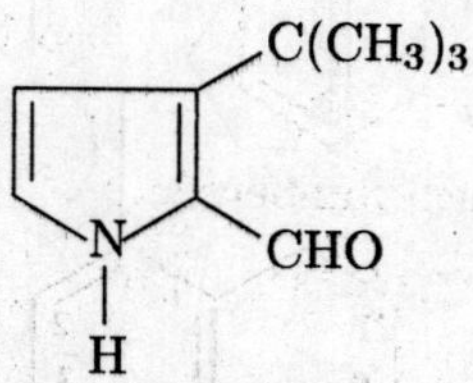

(d) Root; Butane: C—C—C—C; Primary functions: —COOH and —COOH. The C skeleton becomes HOOC—C—C—COOH.

Prefix: It is a compound substituent 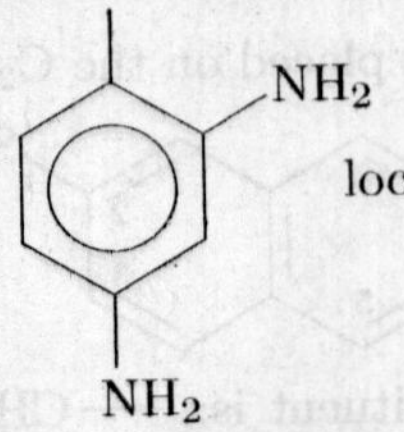 locant: 2

Therefore, the structure of the given compound is

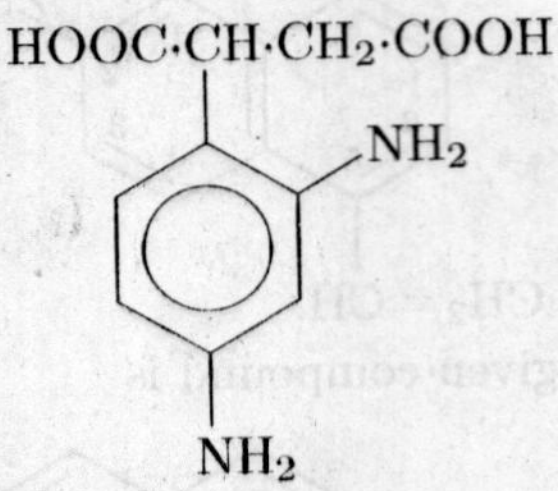

(e) It is a substituted name.

Allyl group: $CH_2{=}CH{-}CH_2{-}$; Vinyl group: $CH_2{=}CH{-}$; Function: >C=O.

Hence the Structure is $CH_2{=}CH{-}CH_2{-}CO{-}CH{=}CH_2$.

(f) Root: 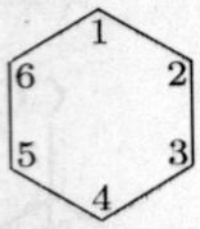 Suffix: Ethyl carboxylate, i.e., $-COOC_2H_5$.

Substituents: 3-methyl and 5-methyl; so, the structure of the given compound is

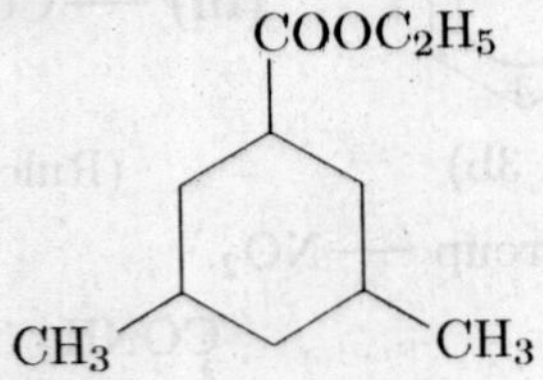

(g) Base name: Phenol,

Substituents: 2-(3-cyclopentenyl), 4-(3-cyclopentenyl) and 6-(3-cyclopentenyl); 3-cyclopentenyl group is

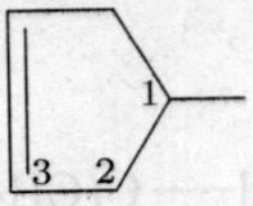

Hence, the structure of the given compound is

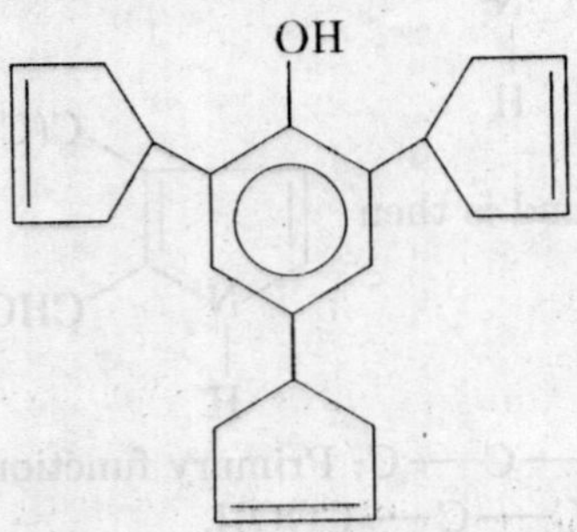

(h) It is a common name of the ethyl ester of malonic acid. The name ending 'ate' indicates that it is either an ester or a salt. As there is no indication of cation in the name, it must be an ester. The structure of malonic acid is $COOH{\cdot}CH_2{\cdot}COOH$. Therefore, the structure of the given compound is $C_2H_5OOC{\cdot}CH_2{\cdot}COOC_2H_5$.

Exercise

1. Write notes on: Functional group and Homologous series.
2. Write down the names of the following functions: >C=C< , —C=C—OH, —NH_2, —NO, —CN, —NC, —CO_2H, —$CONH_2$, —CHO, —COCl.
3. What are the prefixed and suffixed names of the following functions in the IUPAC system: —SO_3H, —CHO, —COO R, —SH, —OH?
4. Arrange the following functions in an increasing order of preference:

   —OOH, —NO_2, —COOH, OH, —CN, >C=O, —SH.
5. Write down the structure of the following groups: Vinyl, 2-pyridyl, Benzoyl, O-naphthyl, 2-cyclopentenyl.
6. Write down the class names of the following compounds in IUPAC system: RCN, RNH_2, $RCONH_2$, RCO·O·CO·R, RCOOR′.
7. Write down the sequential steps for translating a name into a structure.
8. Write down the structures of the following compounds: (a) 1-Methyl-2-propyl-cyclobutanol, (b) *m*-Bromobenzamide, (c) 2-Furoic acid, (d) Aniline-^{15}N, (e) Phen-*p*-d-ol.

Chapter **4**

Factors Influencing Electron-availability

Objectives

- To understand:
  - Inductive effect [4.2] p. 149
  - Inductiomeric effect [4.2.3] p. 152
  - Electromeric effect [4.3] p. 152
- To know:
  - The method of writing resonance hybrid [4.4.1] p. 154
  - Which resonating structure has major contribution in a resonance hybrid [4.4.2] p. 155
  - Resonance interaction [Problem 4.21c] p. 178 [4.8.5] p. 189
- To know the method of calculating resonance energy [4.4.3] p. 161
- To understand:
  - Mesomeric or conjugative effect [4.4.5] p. 165
  - Hyperconjugative effect [4.5] p. 171

* Steric inhibition of resonance [4.4.4] p. 161 and steric effect [4.6] p. 176
* Aromaticity [4.7] p. 179
* Molecular moments [4.8] p. 185

◈ To apply:

* Inductive effect [4.2.2] p. 150
* Electromeric effect [4.3.1] p. 153
* Resonance effect [4.4.7] p. 167
* Hyperconjugative effect [Problems 4.17-4.20] p. 173-176

4.1 Introduction

The discussion made on bonding in carbon compounds (Sec. 2.16) tells us that electron-availability in a particular bond or at a particular atom in a molecule is a dominant factor which determines the reactivity of the compound. Electron-availability depends upon five factors. These are:

I. Inductive effect

II. Electromeric effect

III. Resonance, mesomeric effect or conjugative effect and its M.O. equivalent

IV. Hyperconjugative effect

V. Steric effect

One or more of these factors may operate in a molecule, in opposition or in support of each other. To understand the reactivity and physical properties of a molecule, it is, therefore, essential to determine the direction and magnitude of these five effects qualitatively and to study aromaticity.

4.2 Inductive Effect

The permanent dipole induced in a bond by another directly along a chain of atoms within a molecule is called inductive effect.

Inductive effects may also operate through space, by solvent molecules or ions and are often called field effects.

The inductive effect is a permanent effect in the ground state of a molecule and is supposed to operate through single bonds. So the inductive effect is called permanent polarisation of single bonds.

The C_1—Li bond in the chain C_4—C_3—C_2—C_1—Li is a permanent dipole because of the greater electronegativity of C than that of Li. This dipole induces a permanent dipole in C_2—C_1 bond and C_2 becomes slightly negatively charged, the magnitude of which is smaller than that on C_1, and may be represented by the symbol $\delta\delta-$, where $\delta\delta- < \delta-$. In this way the induction of negative charge may transmit down the chain ; but the effect decreases with the increasing distance from the starting dipole. Thus the example cited above may be represented as follows:

$$\overset{\delta\delta\delta\delta-}{C_4}\leftarrow\overset{\delta\delta\delta-}{C_3}\leftarrow\overset{\delta\delta-}{C_2}\leftarrow\overset{\delta-}{C_1}\leftarrow\overset{\delta+}{Li}$$

where the arrows indicate the electron displacements in the bonds and

$$\delta- > \delta\delta- > \delta\delta\delta- > \delta\delta\delta\delta-$$

In fact inductive effect is important up to second atom (here C_3) and dies down beyond the fourth atom.

Let us take another example to understand the inductive effect.

In the chain C_4—C_3—C_2—C_1—F, F is the negative pole and C_1 is the positive pole as the electronegativity of F is greater than that of C. Therefore, induction of positive charge will occur in the carbon chain as shown below :

$$\overset{\delta\delta\delta\delta+}{C_4}\rightarrow\overset{\delta\delta\delta+}{C_3}\rightarrow\overset{\delta\delta+}{C_2}\rightarrow\overset{\delta+}{C_1}\rightarrow\overset{\delta-}{F}$$

where $\delta+ > \delta\delta+ > \delta\delta\delta+ > \delta\delta\delta\delta+$

Most of the atoms or groups of atoms like F, Cl, Br, —OH, etc., draw electrons towards themselves, i.e., away from the carbon chain; these are known as electron-withdrawing groups and the effect is called electron-withdrawing effect which is symbolically represented as −I effect.

Metals, silicon and alkyl groups repel electrons and are known as electron-releasing or electron-repelling groups; the effect is called electron-releasing effect and its symbolic representation is +I.

The relative strength and direction of inductive effects of different atoms and groups of atoms are given below (Ingold's scale):

Decreasing order of +I effect:

COO^-, R_3C, R_2CH, RCH_2, CH_3, D, H.

Increasing order of −I effect:

H, $R_2C{=}CH$, Ar, $RC{\equiv}C$, OH, SR, SH, OR, COOR, OAr, I, Br, Cl, F, COOH, CN, NO_2, $\overset{\oplus}{N}H_3$, $\overset{\oplus}{N}R_3$.

4.2.1 Scale of measuring inductive effects

Ingold proposed a scale for measurement of relative inductive effects of atoms and groups of atoms in which H atom is taken as reference in the molecule R_3C—H.

Let the H atom be replaced by a group, G, i.e., R_3C—G is formed. Now, if electron density in the R_3C part of R_3C—G is less than that in R_3C—H, G is said to be electron-withdrawing group and its electron-withdrawing effect is termed as −I effect. On the other hand, if the electron density in the R_3C part of R_3C—G is greater than that in R_3C—H, the G is said to have electron-releasing effect and this is denoted by +I.

Robinson's terminology is exactly the opposite to that of Ingold. Ingold's terminology is used mostly; we shall follow this too.

4.2.2 Application

(a) Dipole moment: Inductive effect is intimately associated with dipole moments of the molecules. Generally, dipole moment increases with the increase in inductive effect provided this effect due to more than one group do not act in opposition to each other. Thus the dipole moment of cis-2-butene (1) is greater than that of trans-2-butene (2) because in the latter compound +I effects of —CH_3 groups are cancelled as they act in opposition to each other.

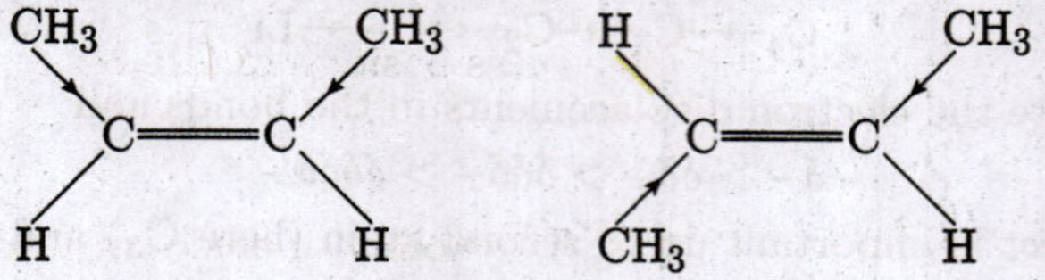

1 **2**

(b) Chemical reactivity: Inductive effect has a dominant role on the reactivity of a compound. Thus meta-dinitrobenzene (3) when heated with fuming HNO_3 and fuming H_2SO_4 gives s-trinitrobenzene (4) and the reaction takes five days to complete; but toluene (5) is very easily nitrated with conc. H_2SO_4 and conc. HNO_3 to produce ortho- and para-nitrotoluenes (6 and 7). The −I effect of —NO_2 groups in meta-dinitrobenzene makes the molecule much less reactive than the toluene containing a methyl group which has +I effect. For this reason nitration on *m*-dinitrobenzene requires drastic condition but that on toluene does not. Not only that, the deactivating effect of nitro groups invites the incoming nitro group to the meta position; but the activating effect of methyl group directs the incoming nitro group into ortho and para positions in toluene.

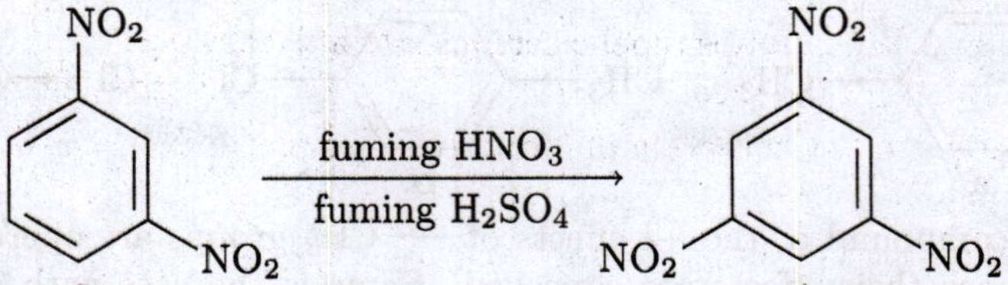

3 4

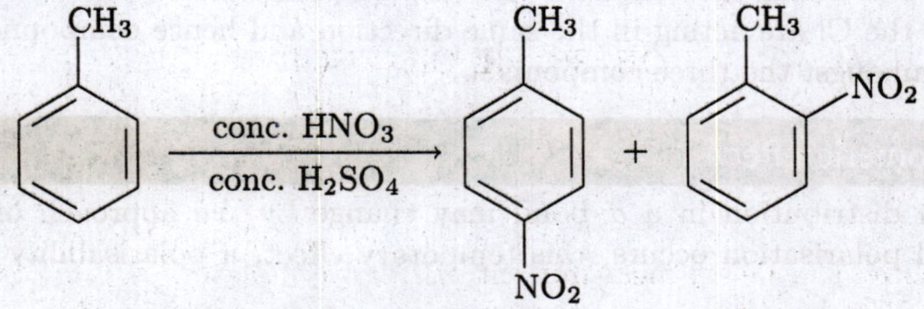

5 6 7

(c) In comparing basicity of bases: According to Lewis, a base is a chemical entity which is capable of donating electron. Organic species containing N or O atom with donatable lone-pair of electrons may be considered as organic bases. If any electron-releasing group is attached to such an N atom or O atom in a base, it will increase the electron density on the atom and hence the basicity of the base will increase. Thus, owing to the +I effect of —CH_3 group, methylamine is more basic than ammonia. Similarly, for the same reason methoxide ion is more basic than hydroxide ion.

Basicity of $CH_3 \rightarrow NH_2$ > basicity of NH_3

Basicity of $CH_3 \rightarrow \bar{O}$ > basicity of $\bar{O}—H$

On the other hand, any electron-withdrawing group attached to N or O atom in an organic base decreases the electron density on the atom and thereby decreases its basicity. Thus, trifluoromethylamine (8) has a very low basicity with respect to methylamine; this is because of the –I effect of the three fluorine atoms.

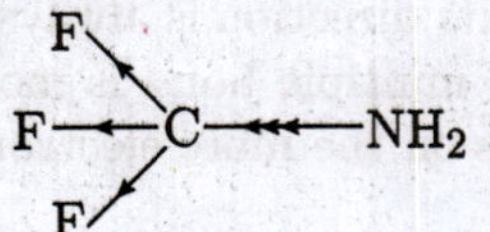 is less basic than CH_3NH_2

8

(d) In comparing acidity of organic acids: The ability of donating proton of a chemical entity may be called acidity of the entity and the entity is said to be an acid. If we compare the acidity of formic acid, acetic acid and chloroacetic acid, we shall see that the acidity of chloroacetic acid is the highest of the three and that of acetic acid is the lowest amongst them. This is because of the fact that any electron-releasing group attached to the carboxyl C atom decreases its ability to donate proton. Methyl group in acetic acid is an electron-releasing group, it lowers the acidity of acetic acid than that of formic acid. But electron-withdrawing effect of Cl atom, increases this ability of chloroacetic acid largely. For this reason the acidity of

$$ClCH_2COOH > HCOOH > CH_3COOH.$$

(e) Stability of carbenium ion: See Problem 4.20.

Problem 4.1. Which one of the following will have the highest dipole moment and why?

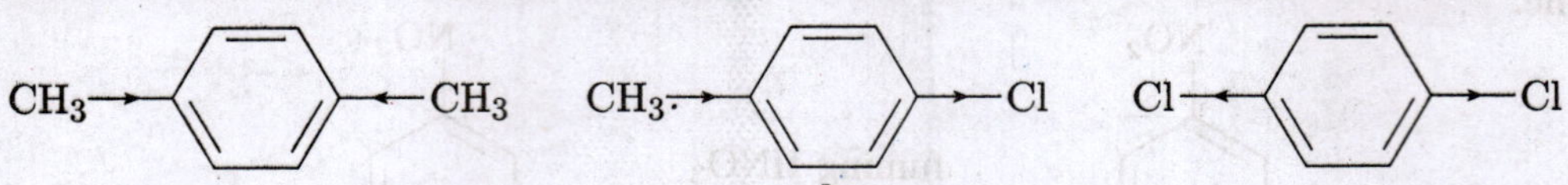

a b c

Solution: In compound **a**, the +I effects of —CH_3 groups are operating in opposition to each other and hence their effects are cancelled. Same is the case with compound **c**; here −I effects of the two C atoms are cancelled. But in compound **b**, +I effect of the —CH_3 group and −I effect of the Cl are acting in the same direction and hence compound b has the highest dipole moment amongst the three compounds.

4.2.3 Inductomeric effect

The electron distribution in a σ bond may change by the approach of a charged species and thus partial polarisation occurs; this temporary effect, a polarisability effect, is known as inductomeric effect.

4.3 Electromeric Effect

The temporary or time-variable effect involving the complete transfer of a shared pair of π electrons to one of the atoms joined by a multiple bond, double or triple, at the requirement of an attacking reagent is known as electromeric effect. As soon as the attacking reagent is removed, the transferred π electron pair again form the bond and the molecule reverts to its ground state electronic condition. For this reason, the electromeric effect may be called polarisability of multiple bond. There is, of course, permanent polarisation of the sigma bond in the multiple bond, but this inductive effect is negligible with respect to the electromeric effect.

To understand this time-variable effect let us now consider a proton ($H^{\oplus}$) which comes close to a carbon-carbon double bond. To provide electron to the proton, π bond will break and electron-pair will be transferred completely to one of the C atoms as shown below:

$$>C{=}C< \;\rightleftharpoons\; >\overset{\oplus}{C}{-}\overset{-}{\ddot{C}}<$$

The electromeric effect has no specific direction. Its direction is always that which favours the reaction. However, if one of the atoms joined to the multiple bond is more electronegative than the other, the transfer of electron-pair usually occurs on the more electronegative atom; e.g., the electromeric effect on >C=O bond is as follows:

$$>C{=}O \;\rightleftharpoons\; >\overset{\oplus}{C}{-}\overset{-}{O}$$

This is because of the greater electronegativity of O than that of C.

Inductive effect does also play an important role in determining the direction of the electromeric effect. If a group with +I or −I effect is attached to a multiple bonded C atom of the system >C=C<, the direction of the electromeric effect is the same as that of the inductive effect. Thus in propene (9) the transfer of electron-pair will occur on the first C atom (10).

$$\underset{\mathbf{9}}{CH_3{-}CH{=}CH_2} \;\overset{H^{\oplus}}{\rightleftharpoons}\; \underset{\mathbf{10}}{CH_3{-}\overset{\oplus}{C}H{-}\overset{-}{\ddot{C}}H_2}$$

Similarly, in allyl alcohol (11), the –I effect of —OH reverses the electromeric effect as shown in the structure 12. Here too the direction of the electromeric effect is the same as that of the inductive effect.

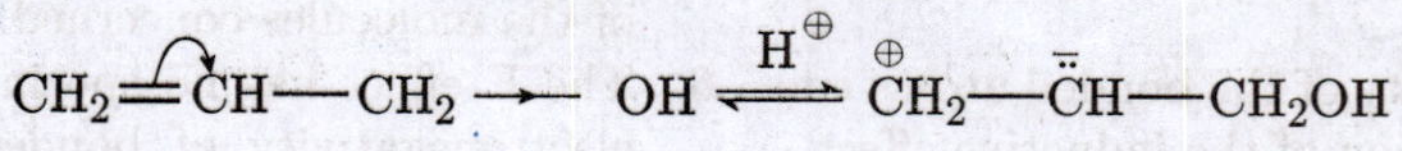

11 **12**

The electromeric effect being temporary, it is not reflected in the physical properties of the concerned molecules.

The electromeric effect is denoted by the symbol E. It is said to be –E or +E as the transfer of π electron-pair is away from or towards the carbon chain respectively.

4.3.1 Application

The electromeric effect, i.e., the close approach of a reagent to a multiple bond enhances the reactivity of the reactant molecule and explains its addition reaction product. Thus the addition of ammonia to acetaldehyde (13) may be cited. The greater electronegativity of O atom and +I effect of the methyl group in acetaldehyde help transfer the π electrons on O atom at the requirement of the ammonia molecule which coordinates with the double-bonded C atom.

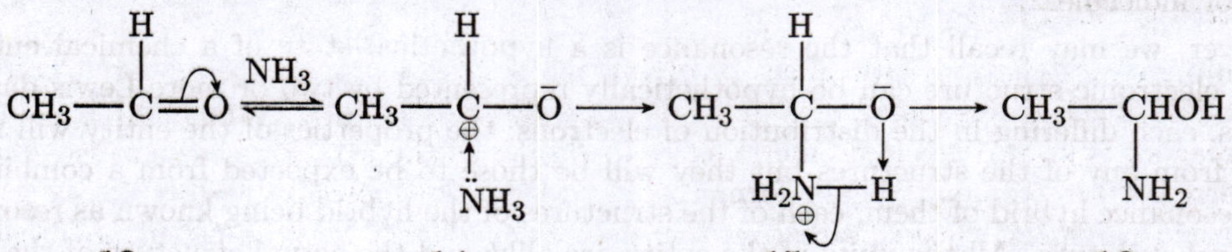

13 **14** **15** **16**

Problem 4.2. Predict the direction of the electromeric effect in the compound COOH.CH=CHOH (I effects are to be considered only).

Solution: Both the —OH and —COOH group have –I effect. So, each of them will try to draw the π electrons to its side. As the –I effect of —COOH group is greater than that of the OH group, the transfer of π electrons will take place towards —COOH group.

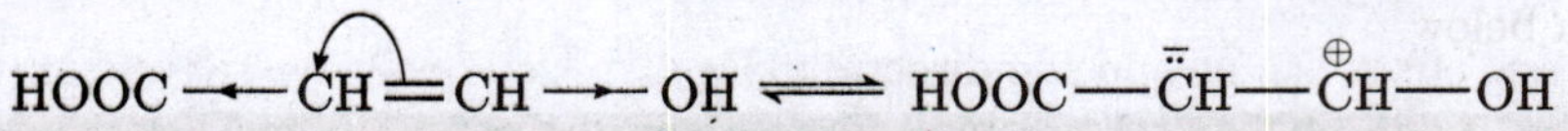

Problem 4.3. Distinguish between inductive effect and electromeric effect.

Solution:

Inductive effect	Electromeric effect
1. Permanent effect.	1. Temporary effect.
2. It is the permanent polarisation of a single bond.	2. It is the polarisability of a multiple bond.
3. σ electrons are involved in this case.	3. π electrons are involved in this case.
4. Only electron displacement takes place in the inductive effect. Partial charge separation occurs.	4. Complete transfer of π electrons takes place in the electromeric effect.

Contd.

Inductive effect	Electromeric effect
5. It determines both the physical properties and chemical reactivity of the molecules concerned.	5. It has nothing to do with the physical property but it enhances the chemical reactivity of the molecules concerned.
6. Electronegativity of the bonded atoms controls the direction of the inductive effect.	6. The E effect has no specific direction; the electronegativity of bonded atoms and I effects of the attached groups may determine the direction of the electromeric effect. In fact, its direction is that which favours the reaction.

4.4 Resonance, Mesomeric Effect or Conjugative Effect and M.O. Equivalent of Resonance

In the discussion of the second assumption of the V.B. theory we have come to know something about the resonance of covalent bonds involving σ electrons. There we have found that ionic resonating structures are unimportant but resonance involving π electrons are very useful in the prediction of molecular geometry, bond parameters and chemical reactivity of chemical entities, i.e., ions or molecules.

However, we may recall that the resonance is a hypothetical state of a chemical entity in which its electronic structure can be hypothetically represented by two or more Lewis dash-dot structures, each differing in the distribution of electrons; the properties of the entity will not be expected from any of the structures but they will be those to be expected from a combination called a resonance hybrid of them, each of the structures of the hybrid being known as resonating or canonical structure. All the units of the entity are alike and the actual structure of the entity neither does shift from one canonical form to other, nor does it consist of a mixture of them. The entity has a single structure of its own which cannot be written down by the conventional Lewis method. So we take the help of other possible, though non-existent, structures which could be represented by the Lewis method. Generally, the resonance hybrid is projected on the plane of the paper by writing down all the resonating structures and placing a double-headed arrow between each pair. Resonating structures, in turn, are to be written down following certain conditions which are given below.

4.4.1 Method of writing resonance hybrid

The following conditions are to be followed for writing resonating structures:

(a) All the canonical structures must be written according to the Lewis method showing bonds, lone-pair of electrons and formal charges where required. (See Section 2.7.)

(b) The positions of the atomic nuclei of the chemical entity must remain unchanged in all the resonating structures. For example, in an ethylene molecule, each of the two C atoms have two H atoms; the condition stated above tells us that we cannot write any structure for ethylene which shows one H atom linked to one C and three H atoms linked to the other C. Thus, structure II cannot be a resonating structure of ethylene.

$$\underset{\textbf{I}}{H_2C{=}CH_2} \longleftrightarrow\!\!\!\!\!\!\!\times\;\; \underset{\textbf{II}}{H_3C{-}\ddot{C}H}$$

(c) To write resonating structures, only the π bonds can be converted to π bonds and to lone-pairs of electrons and vice versa. Thus, the resonance hybrid of ethylene may be written as (though not proper):

$$\underset{\textbf{I}}{H_2C{=}CH_2} \longleftrightarrow \underset{\textbf{III}}{H_2\overset{\oplus}{C}-\overset{=}{C}H_2} \longleftrightarrow \underset{\textbf{IV}}{H_2\overset{=}{C}-\overset{\oplus}{C}H_2}$$

(d) The number of unpaired electrons in each of the structures must be the same. Thus, structure V cannot be a resonating form in the resonance hybrid of ethylene as it contains two unpaired electrons which no other structure possesses.

$$\underset{\textbf{I}}{CH_2{=}CH_2} \longleftrightarrow \underset{\textbf{III}}{\overset{\oplus}{C}H_2-\overset{=}{C}H_2} \longleftrightarrow \underset{\textbf{IV}}{\overset{=}{C}H_2-\overset{\oplus}{C}H_2} \longleftarrow\times\longrightarrow \underset{\textbf{V}}{\dot{C}H_2-\dot{C}H_2}$$

(e) The several resonating structures must be of comparable energies; thus, for ethylene, structure III and IV have such a high energy in comparison to I that these should not be accepted as resonating structures. This means that structures III and IV do not explain the properties of ethylene but structure I explains.

(f) All atoms involved in the resonance must lie on the same plane or nearly on the same plane; any change in structure which prevents planarity will diminish resonance simply because maximal overlap of p orbitals cannot be possible then to form the π bond orbital. Thus, in allene the two $=CH_2$ units are lying in two mutually perpendicular planes and hence no interaction between the two π bonds can take place in allene (VI).

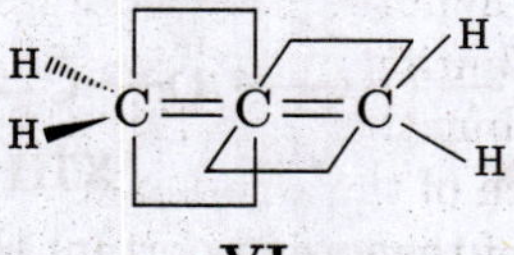

VI

4.4.2 Contributing resonating structures

Following the conditions said above one may write down a resonance hybrid of an entity. But it is of importance to tick out those canonical structures which contribute significantly in the hybrid. The contribution of each form is proportional to its stability; the most stable structure contributes the most and the least stable structure contributes the least. Here are some empirical rules which are helpful in deciding the relative stability and contribution of structures.

1. Resonating structures with more covalent bonds are normally more contributing than those with a fewer number of covalent bonds since more the number of covalent bonds, the less is the energy of the species and more is the stability. Thus, the structure VII of CO_2 has major contribution in its resonance hybrid and the structures VIII and IX have minor contributions.

$$\underset{\textbf{VII}}{:\ddot{O}{=}C{=}\ddot{O}:} \longleftrightarrow \underset{\textbf{VIII}}{:\overset{=}{\underset{..}{O}}-\overset{\oplus}{C}{=}\ddot{O}:} \longleftrightarrow \underset{\textbf{IX}}{:\ddot{O}{=}\overset{\oplus}{C}-\overset{=}{\underset{..}{O}}:}$$

2. If the different resonating structures have the same number of covalent bonds, then uncharged structures have major contribution in the resonance hybrid. Thus, structure VII has much greater contribution than structures X and XI of CO_2 in its resonance hybrid.

$$\underset{\textbf{VII}}{:\ddot{O}{=}C{=}\ddot{O}:} \longleftrightarrow \underset{\textbf{X}}{:\overset{=}{\underset{..}{O}}-C{\equiv}\overset{\oplus}{O}:} \longleftrightarrow \underset{\textbf{XI}}{:\overset{\oplus}{O}{\equiv}C-\overset{=}{\underset{..}{O}}:}$$

3. Structures with isolated unlike charges have less contribution than those in which they are close to each other. Thus, structures X and XI of CO_2 have less contribution than structures VIII and IX in its resonance hybrid; contribution to this by structure VII being the most (as per rule I). Structures X and XI have good contribution as per octet rule.

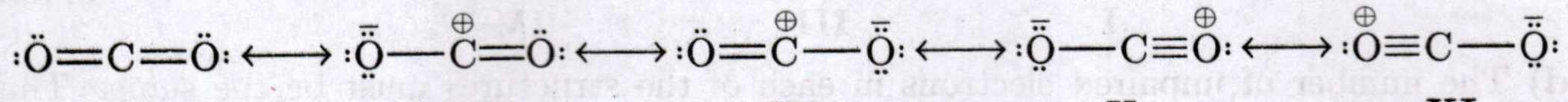

VII VIII IX X XI

4. Structures with like formal charges crowded together have very low stability and hence these are less contributing than those in which like charges remain separated. Structure XII has no contribution to the resonance hybrid of CO_2 molecule.

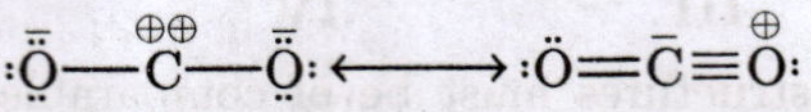

XII XIIc

5. Any structure in which a second row element of the periodic table (Li to Ne) has more than eight electrons in its outermost shell contributes nothing. Structure XIIc of CO_2 possesses a C atom with ten electrons and hence this structure contributes nothing to its resonance hybrid.

6. Structures with a negative formal charge on the most electronegative atom and a positive formal charge on the least electronegative atom have significant contribution; structures with opposite charge distributions are less contributing. Thus, structures VIII and IX of CO_2 are more contributing than structures XIII and XIV.

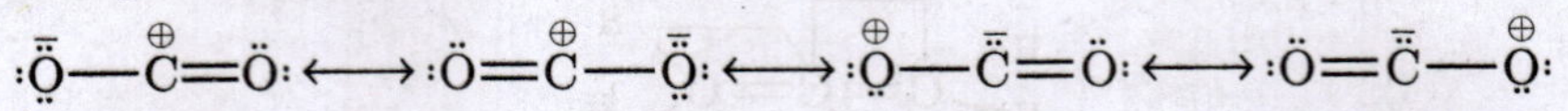

VIII IX XIII XIV

7. Equivalent resonating structures are more important than non-equivalent resonating structures, because in the former case the equivalent structures have the same amount of energy but in the latter case non-equivalent structures do not. Thus, resonance in the carbonate ion is more important than that in the carbonic acid. In the structures XV, XVI and XVII of CO_3^{--} ion we see that each of them possesses two negative formal charges over two O atoms, which are single-bonded to C atom, and one O atom is double-bonded to C atom. Hence these structures are equivalent structures. On the other hand structure XVIII, XIX and XX of carbonic acid are non-equivalent.

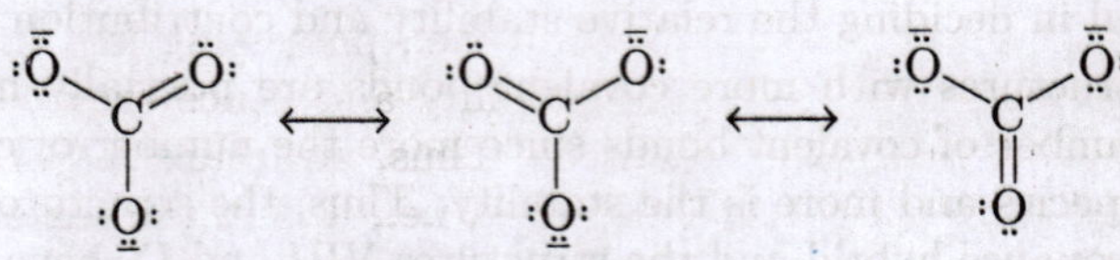

XV XVI XVII

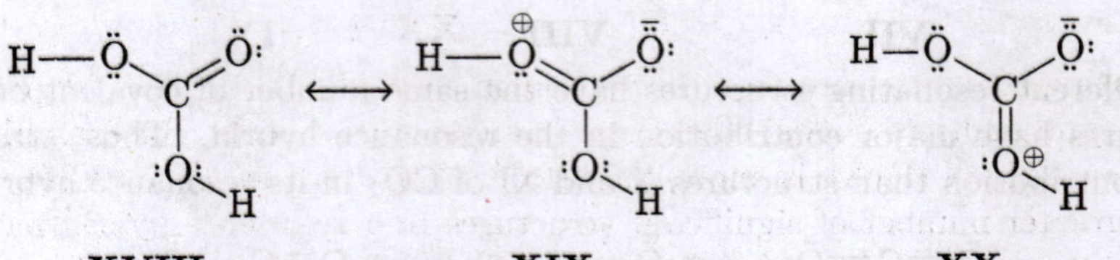

XVIII XIX XX

Note

To show equivalent resonating structures, the resonance hybrid of CO_3^{--} may be represented as:

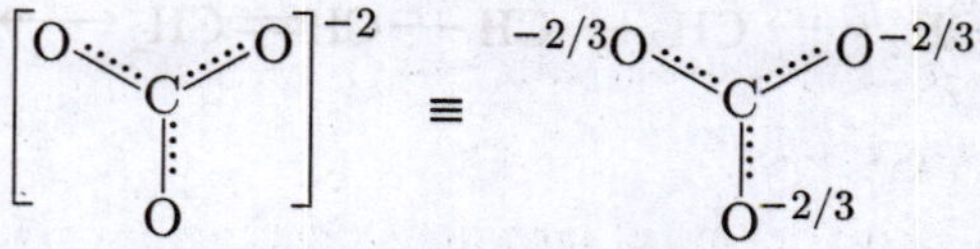

This means each of the C—O bonds in CO_3'' is neither double nor single but an intermediate between the two; the two negative charges may be shown by distributing the charge equally on each of the three O atoms and thus each of them will have 2/3 unit of negative charge.

8. Isovalent resonating structures, i.e., structures having the same number of covalent bonds are more significant than **heterovalent resonating structures** which contain different numbers of covalent bonds. Thus, resonating structures in the resonance hybrid of benzene are very important.

9. Structures with **distorted** bond lengths and bond angles are unstable and less contributing. Benzene has a planar regular hexagonal structure in which all the C—C bonds are of the same length; structures XXIV, XXV and XXVI have central bonds; its length is 0.23 nm, i.e., almost twice the length of a C—C bond (0.139 nm). It is too long for significant bonding. Therefore, these three structures have minor contributions to the resonance hybrid of benzene.

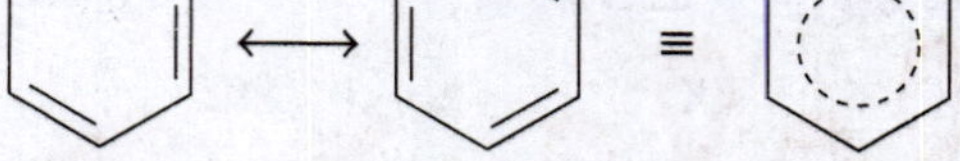

XXI **XXII** **XXIII**

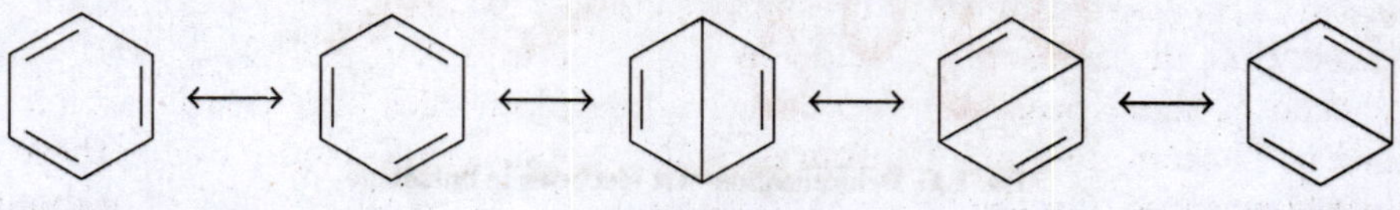

XXI **XXII** **XXIV** **XXV** **XXVI**

10. Structures with electron deficient atom, i.e., possessing less than eight electrons in its outermost shell, have high energy, i.e., low stability and hence they contribute less than those structures in which **octet rule** is not violated. Thus, structure XXVII contributes more than structure XXVIII (the terminal N has 6 electrons, i.e., it is in the sextet state) in the resonance hybrid of diazomethane.

$$CH_2=\overset{\oplus}{N}=\bar{N}: \longleftrightarrow CH_2=\bar{N}-\ddot{N}:$$

XXVII **XXVIII**

11. Resonating structures are more important in a **conjugated system** (structures possessing alternate single and multiple bonds) than those in an isolated system. For example, the resonance in $CH_2=CH-CH=CH_2$(A) is more important than that in $CH_2=CH-CH_2-CH=CH_2$(B). The greater number of significant structures in a resonance hybrid means the greater importance of the resonance in comparison to another hybrid possessing a fewer number of resonating structures. The resonance hybrid of compound A contains a greater number of resonating structures than the hybrid of compound B as shown below:

$$\textbf{A.}\ CH_2{=}CH{-}CH{=}CH_2 \longleftrightarrow \overset{-}{\ddot{C}}H_2{-}CH{=}CH{-}\overset{\oplus}{C}H_2 \longleftrightarrow \overset{\oplus}{C}H_2{-}CH{=}CH{-}\overset{-}{\ddot{C}}H_2$$

$$\updownarrow$$

$$CH_2{=}CH{-}\overset{\oplus}{C}H{-}\overset{-}{\ddot{C}}H_2 \longleftrightarrow \overset{-}{\ddot{C}}H_2{-}\overset{\oplus}{C}H{-}CH{=}CH_2 \longleftrightarrow \overset{\oplus}{C}H_2{-}\overset{-}{\ddot{C}}H{-}CH{=}CH_2$$

$$\updownarrow$$

etc.

$$\textbf{B.}\ CH_2{=}CH{-}CH_2{-}CH{=}CH_2 \longleftrightarrow \overset{\oplus}{C}H_2{-}\overset{-}{\ddot{C}}H{-}CH_2{-}CH{=}CH_2$$

$$\updownarrow$$

$$CH_2{=}CH{-}CH_2{-}\overset{\oplus}{C}H{-}\overset{-}{\ddot{C}}H_2 \longleftrightarrow \text{etc.}$$

Note

Thus, we see that in A electrons are distributed over whole of the molecule embracing all the C atoms and do not remain localised; this phenomenon is called delocalisation of electrons. It diminishes the reactivity of π electrons and lowers the energy of the system, i.e., brings stability to the system. In an isolated system delocalisation also occurs between two adjacent atoms only, but such a wide range of delocalisation of electrons involving large number of atoms cannot take place as in a conjugated system. So we can conclude that the more the delocalisation of electrons or positive charge, the more is the stability of the entity and the lower is its energy. The delocalisation of electrons in butadiene may be shown as in Fig. 4.1.

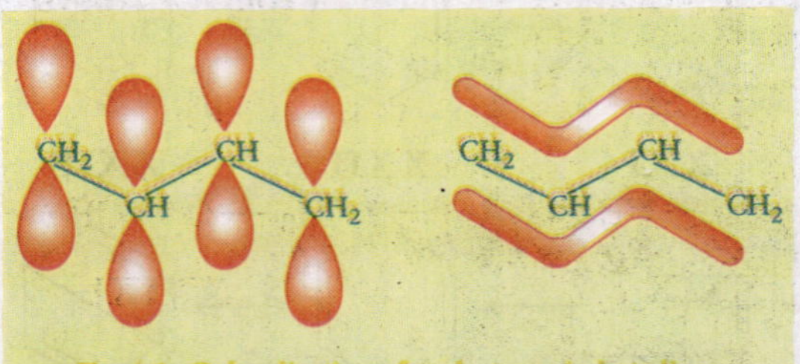

Fig. 4.1. Delocalisation of π electrons in butadiene

The delocalisation of electrons in benzene makes the molecule stable. Each of the six C atoms is sp^2 hybridised and each of them possesses one p_z orbital with an unpaired electron. Three sp^2 A.O.s are utilised to form one C—H σ bond and two C—C σ bonds. In this way six C atoms form a planar hexagonal ring structure and six H atoms also lie on this plane. These constitute the σ skeleton of benzene unit.

Therefore six p_z orbitals of six C atoms are now parallel to each other. By side-on overlapping of two p_z orbitals of two adjacent C atoms a π bond is formed. In this way three π bonds are formed in five different ways in a molecule of benzene. The whole picture may be obtained from Figure 4.2.

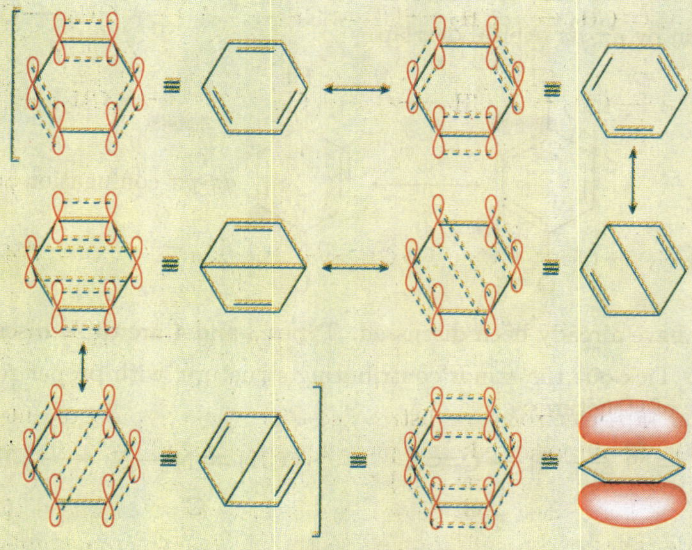

Fig. 4.2. Delocalisation of π electrons in Benzene

The delocalisation of electrons does not only take place in alternate single and multiple bonds involving C atoms, but it may also occur in a conjugated system involving C and other atoms, e.g.,

$$CH_3-\overset{\overset{O}{\|}}{C}-CH=CH_2 \longleftrightarrow CH_3-\overset{\overset{:\bar{\ddot{O}}:}{|}}{C}=CH-\overset{\oplus}{C}H_2$$

XXIXa **XXIXb**

$$CH_2=CH-\ddot{\underset{\cdot\cdot}{Cl}}: \longleftrightarrow \bar{C}H_2-CH=\overset{\oplus}{\underset{\cdot\cdot}{Cl}}:$$

XXXa **XXXb**

$$CH_2=CH-\overset{\oplus}{C}H_2 \longleftrightarrow \overset{\oplus}{C}H_2-CH=CH_2$$

XXXIa **XXXIb**

There are also examples in which π orbital and p orbital (vacant or filled) overlap and the delocalisation of charge occurs. Thus, in the structure XXXa, the filled p orbital of Cl atom is involved in the process of delocalisation with the electrons of C═C bond.

Similarly in the structure XXXIa, the vacant p orbital of C atom in $\overset{\oplus}{C}H_2$ part overlaps the π orbital of C═C bond.

Thus, delocalisation of charge or electrons are of several types and a few of them are:

1. Delocalisation by π-π orbital overlap
2. Delocalisation by π-p orbital overlap
3. Delocalisation by σ-π orbital overlap

4. Delocalisation by σ-p orbital overlap

5. Delocalisation by $d\pi$-$p\pi$ orbital overlap

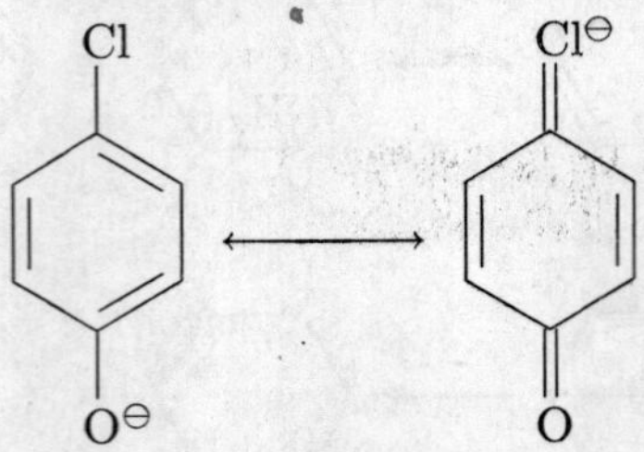

$d\pi$-$p\pi$ conjugation

Types 1 and 2 have already been discussed. Types 3 and 4 are to be discussed later on.

Problem 4.4. Tick out the minor contributing structure, with proper reasoning, in each of the following pairs of structures:

(a) $R-\overset{\oplus}{C}=\ddot{N}: \longleftrightarrow R-\overset{\ominus}{C}=\overset{\oplus}{N}:$ (A, B)

(b) $CH_2=\ddot{N}-\ddot{N}: \longleftrightarrow CH_2=\overset{\oplus}{N}=\overset{\ominus}{N}:$ (C, D)

(c) 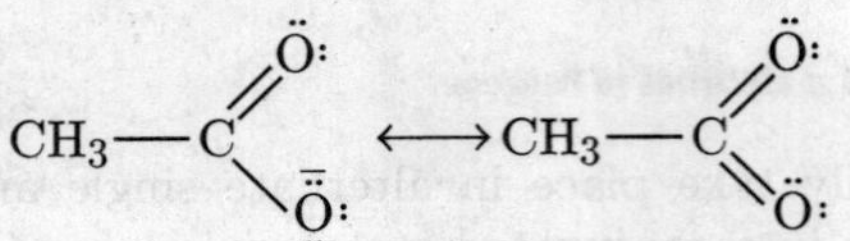

E F

(d) $:\ddot{O}=\ddot{N}-\ddot{O}: \longleftrightarrow :\overset{\oplus}{O}-\overset{\ominus}{N}-\ddot{O}:$ (G, H)

(e) 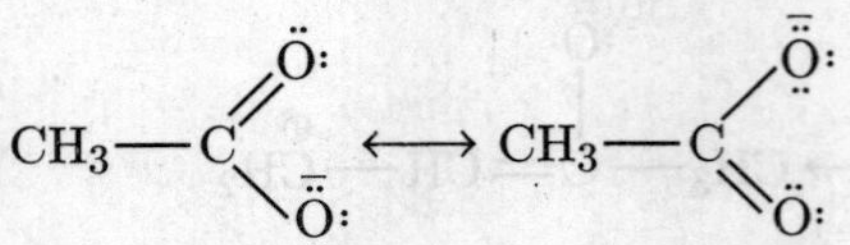

I J

Solution: (a) Structure B; as more electronegative element N has positive charge and less electronegative carbon has negative charge.

(b) Structure C; because it has a fewer number of covalent bonds and electron deficient N atom which is in the sextet state.

(c) Structure F; since oxygen-bonded C atom is pentacovalent, i.e., possesses 10 electrons in its outermost shell.

(d) Structure H; here it contains a fewer number of covalent bonds as well as like charges (−ve) on adjacent atoms.

(e) None of the two structures is less contributing than the other as they are equivalent structures.

Problem 4.5. Write down at least one resonating structure for each of the following structures:

A. 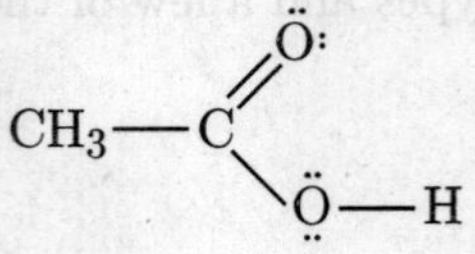

B. 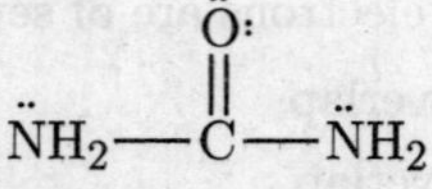

C. $CH_2=CH-\overset{\ominus}{C}H_2$

Solution:

A. 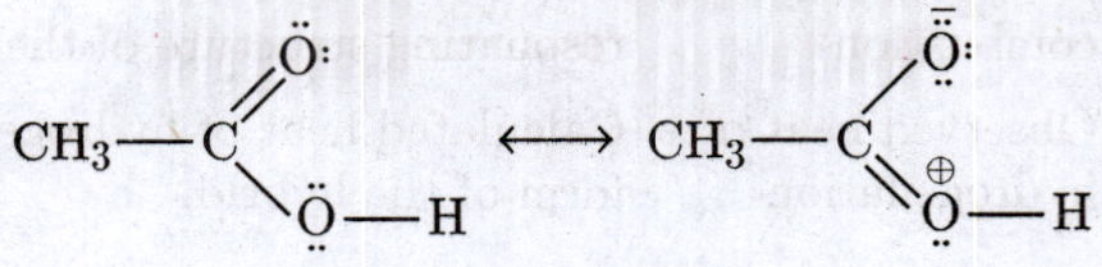

B. 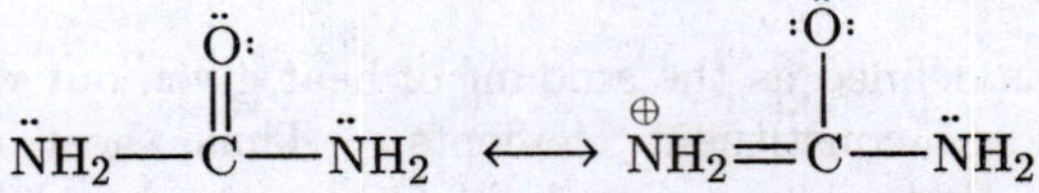

C. 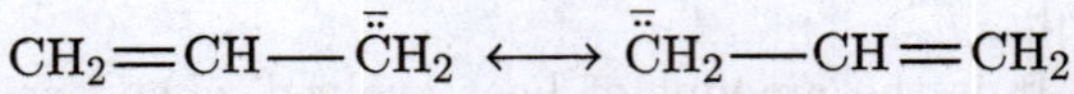

Problem 4.6. Represent the resonance hybrids of the following structures:

(a) 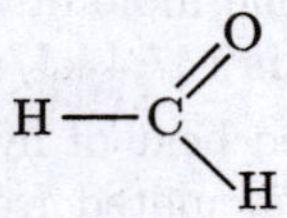 (b) 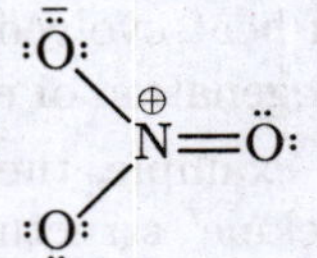 (c) $CH_2=CH-CH=CH_2$

Solution:

(a) 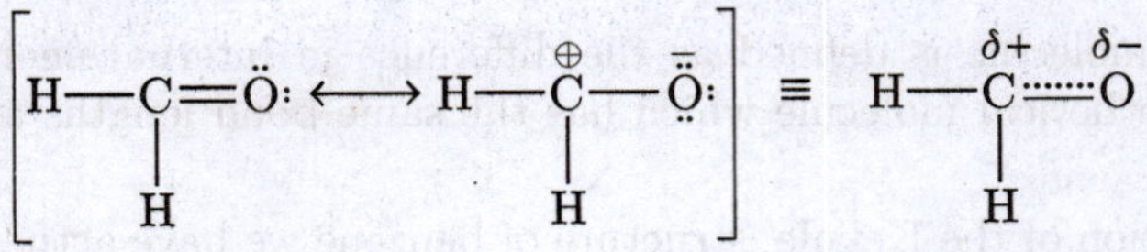

(b) 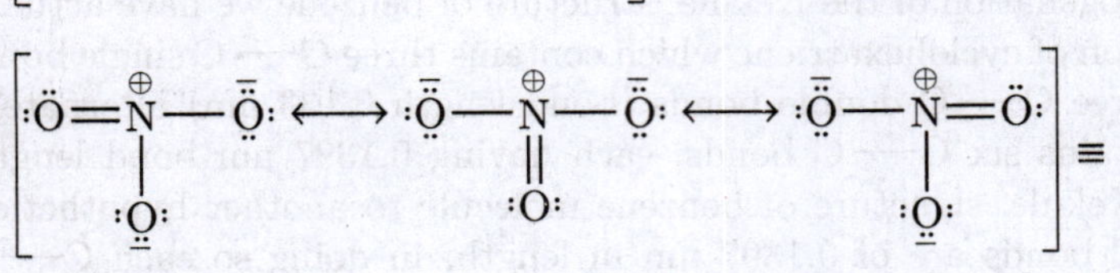

(c) 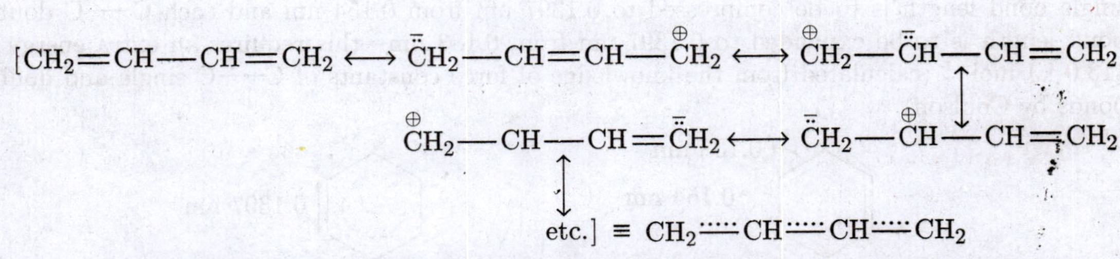

4.4.3 Resonance energy

The resonance hybrid of an entity is a more stable structure than any one of the resonating structures contributing to it. The **resonance energy** is a measure of the extra stability of the resonance hybrid. It is defined as the difference in energy between the actual structure of the entity and the most stable structure of the hypothetical structures. The resonance energy of a molecule may be calculated by either of the following methods :

1. Resonance energy = Observed heat of formation − Calculated heat of formation of the stablest resonating structure of the hybrid (4.1)

2. Resonance energy = Observed heat of combustion − Calculated heat of combustion of the stablest resonating structure of the hybrid (4.2)

3. Resonance energy = Observed heat of hydrogenation − Calculated heat of hydrogenation of the stablest form of the hybrid. (4.3)

Note

[Heat of formation: It is defined as the amount of heat given out when one mole of the substance is produced from its constituent elements. Thus, heat of formation of CO_2 is 394.9 kJ/mole; this means that when one mole of CO_2 is produced from C and O_2, 394.9 kJ amount of heat is released.

Heat of combustion: It is the amount of heat evolved when one mole of the substance is completely burnt down in oxygen. Thus, heat of combustion of benzene is 330.5 kJ/mole,

Heat of hydrogenation: It is the amount of heat evolved when one mole of an unsaturated compound is hydrogenated. Thus, heat of hydrogenation of ethylene is 137.4 kJ/mole.]

So resonance energy is a relative term. For example, the observed heat of hydrogenation of benzene is 208.5 kJ mol^{-1} and that of the Kekule′ structure (as calculated from the heat of hydrogenation of cyclohexene; see p. 165 Problem 4.8) is 359.1 kJ mol^{-1}; hence the resonance energy of benzene is (208.5 – 359.1) or 150.6 kJ mol^{-1}.

The **vertical resonance energy** of a molecule is defined as the difference in internal energy between the actual molecule and a hypothetical molecule which has the same bond lengths and angles as those of the actual molecule.

In calculating the heat of hydrogenation of the Kekule′ structure of benzene we have actually calculated the heat of hydrogenation of cyclohexatriene which contains three C—C single bonds (bond length : 0.154 nm) and three C—C double bonds (bond length 0.133 nm) alternately; but actual benzene molecule contains six C—C bonds, each having 0.1397 nm bond length. So we shall have to convert the Kekule′ structure of benzene molecule to another hypothetical structure in which all the C—C bonds are of 0.1397 nm in length; in doing so each C—C single bond length is to be compressed to 0.1397 nm from 0.154 nm and each C—C double bond length is to be expanded to 0.1397 nm from 0.133 nm—this requires an extra energy of 113.0 kJ mol^{-1} (calculated from the knowledge of force constants of C—C single and double bonds by Coulson).

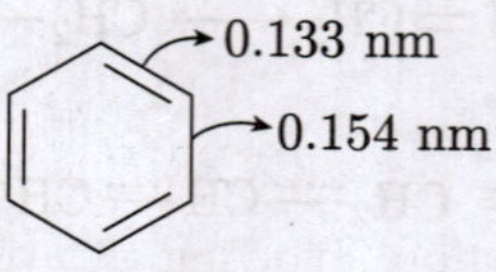

Cyclohexatriene
Kekule′ benzene

I

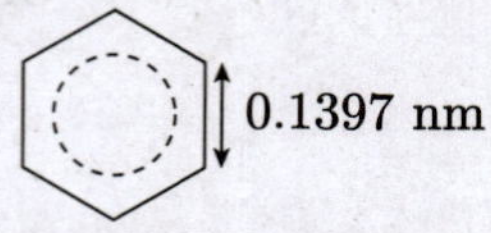

Hypothetical molecule having the same bond lengths as those of the actual benzene molecule

II

Therefore, the vertical resonance energy of benzene is [208.5 – (359.1 + 113.0)] or 264.6 kJ mol^{-1} where the reference molecule is II instead of I.

So, the vertical resonance energy is also a relative energy. However, this is always different from the resonance energy of the entity.

4.4.4 Steric inhibition of resonance

The most important condition for resonance to occur is that the involved atoms in resonating structures must be coplanar or nearly coplanar for maximum resonance energy. If this condition does not fulfil, the involved orbitals cannot be parallel to each other and as a consequence delocalisation of electrons or positive charge cannot occur. There are many examples in which resonance is inhibited or reduced because the involved atoms and orbitals are sterically forced out of planarity and the phenomenon is called steric inhibition of resonance. For example, in N,N-dimethyl-2,4,6-trinitroaniline —NMe_2 group is out of the plane of the ring owing to the presence of the three closely-placed bulky groups and consequently the lone pair of electrons on the N atom of the —NMe_2 group cannot get delocalised through p-π conjugation.

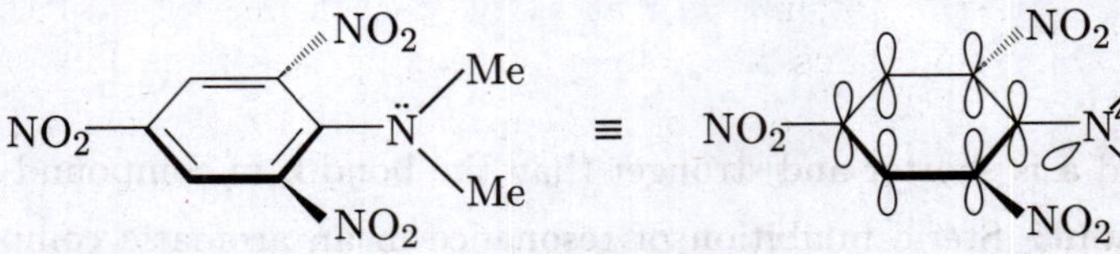

The steric inhibition of resonance has profound effects on:

(a) physical properties, (b) acidity and basicity, (c) the reactivity of organic compounds.

These are discussed below in brief :

(a) Physical properties and steric inhibition of resonance:

(i) Bond length and bond strength: Owing to the involvement of resonance, a single bond assumes a partial double bond character while a double assumes a partial single bond character. If steric effect inhibits resonance, the shortening or lengthening of such a bond does not occur. Since the shortening of a bond increases its strength and the lengthening of a bond decreases the same, steric inhibition of resonance changes the strength and length of a particular bond. Thus, in the examples given below C—N bond **a** is shorter than the C—N bond **b** and hence the former is stronger than the latter.

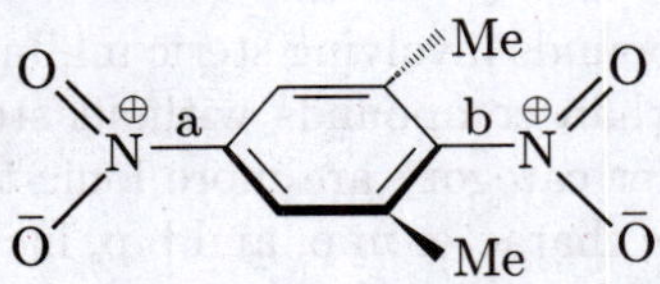

XXXII

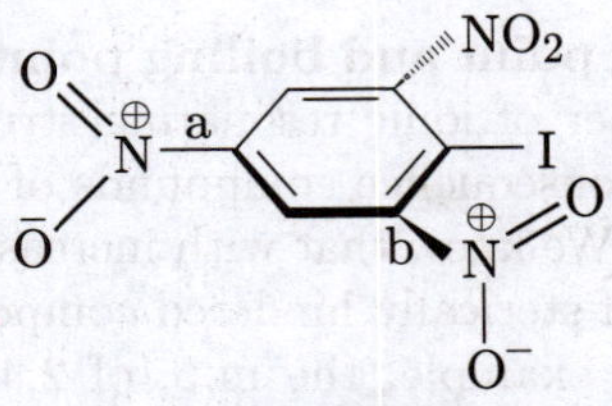

XXXIII

In the compound XXXII owing to the steric congestion involving two methyl and one nitro groups, the O atoms of the nitro group and the vacant p A.O. of its N atom are sterically forced out of the plane of the ring. Thus, the −R effect of the nitro group cannot play its role and the C—N bond, **b**, fails to assume a partial double bond character. On the other hand, the vacant p. A.O. of the p-nitro group and its O atoms are on their usual plane because of the absence of steric congestion.

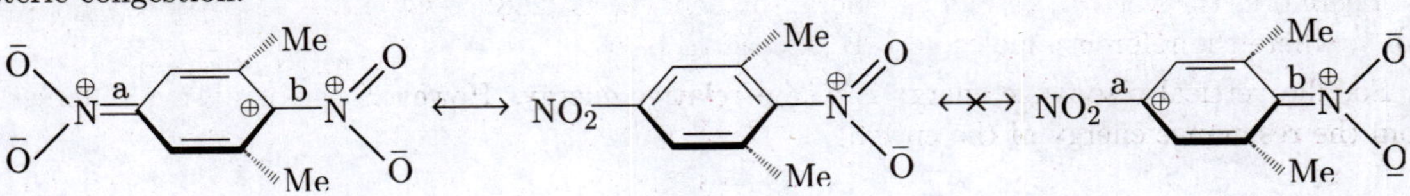

So, the C——N bond, **a**, has a partial double bond character because of the −R effect of the nitro group. Therefore, the bond **a** is shorter than the bond **b**, i.e., the former is stronger than the latter.

Similarly, in the compound XXXIII the bond b does not have a partial double bond character owing to steric inhibition of resonance; whereas the bond a in this compound has a partial double bond character because of resonance.

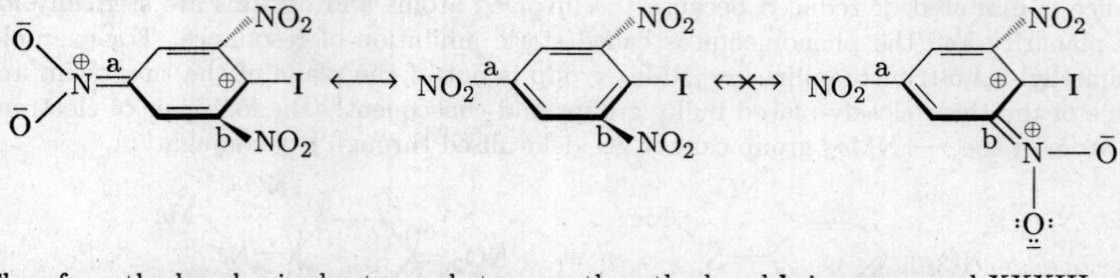

Therefore, the bond **a** is shorter and stronger than the bond **b** in compound XXXIII.

(ii) Dipole moment: Steric inhibition of resonance in an aromatic compound lowers its dipole moment since the distance between the two poles of such a compound is shorter than that in a similar compound without steric inhibition of resonance. For example, the dipole moment of N,N-dimethylaniline is 1.61D whereas that of *o,o′*, N,N-tetramethylaniline is only 0.94 D.

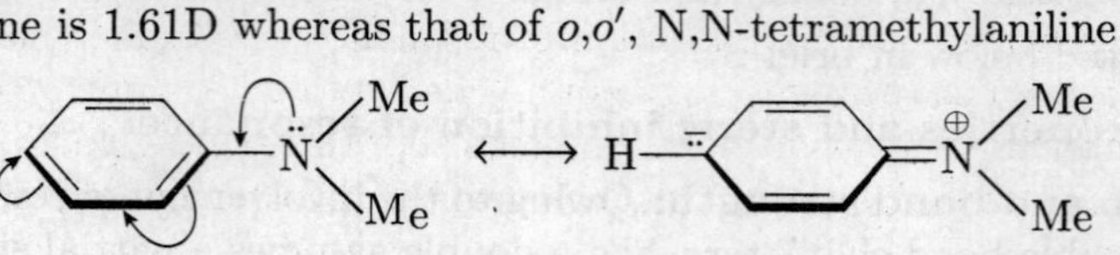

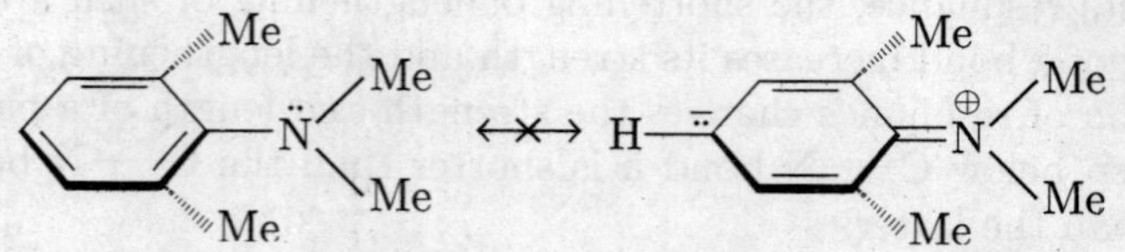

(iii) Melting point and boiling point: Compounds involving steric inhibition of resonance have lesser number of ionic resonating structures than compounds without steric inhibition of resonance; as a consequence compounds of the latter category are more ionic than those of the former category. We know that with increasing ionic character m.p. and b.p. increase. Therefore, m.p.s and b.p.s of sterically hindered compounds are less than those of the sterically unhindered compounds. For example, the m.p. of 2,4,6-trichloroaniline is 78.5°C where as that of 2,4,6-trichloro-N-methylaniline is only 32°C and 2,4,6-trichloro-N,N-dimethylaniline is a liquid under ordinary conditions.

(b) Acidity and basicity: Since steric hindrance reduces or prevents delocalisation of positive charge or electrons through resonance, steric inhibition of resonance has appreciable effect on the acidity and basicity of compounds.

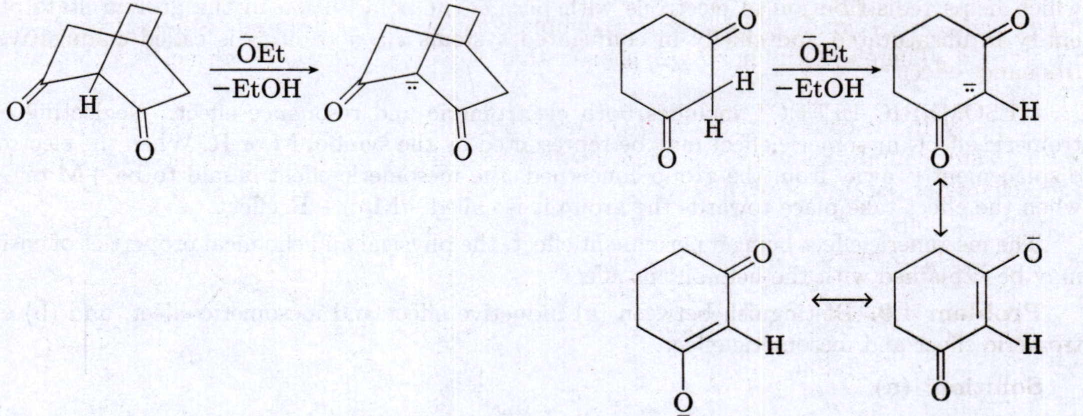

Cyclohexane-1,3-dione is much more acidic than bicyclo [2.2.2]-octane-2,6-dione simply because the enolate ion of the former gets stability through resonance while that of the latter does not; the filled p orbital of the enolate ion of the latter compound being out of plane of the C=O groups, it cannot overlap with the π electrons of the carbonyl groups assuming planar trigonal geometry.

(c) Reactivity: Owing to steric inhibition of resonance, electronic charge density decreases at a particular C atom of an aromatic compound, i.e., its nucleophilicity decreases. Therefore reactivity of the compound as a nucleophile decreases. For this reason N,N,*o*,*o*′-tetramethylaniline does not couple with benzenediazonium ion (see problem 17.23).

Problem 4.7. The observed heat of combustion of 1,3-butadiene is −1250 kcal/mole. Theoretical heat of combustion values in kcal/mole for C—H bond: −99.5; C—C: −81.0 and C=C; −146.1. Find out the resonance energy of the compound.

Solution: 1,3-Butadiene is CH_2=CH—CH=CH_2. It has six C—H σ bonds. The heat of combustion of six C—H bonds = 99.5 × 6 or 597.3 kcal mol^{-1}. For 1 C—C σ bond the heat of combustion is 81.0 kcal mol^{-1} and that for 2 C=C part is 146.1 × 2 or, 292.2 kcal mol^{-1}. The sum of the heat of combustion values is 970.5 kcal mol^{-1}. Resonance energy = (1250 − 970.5) or, 279.5 kcal mol^{-1}.

Problem 4.8. Heat of hydrogenation of cyclohexene to cyclohexane is −28.6 kcal/mole. The observed heat of hydrogenation of benzene to cyclohexane is −49.8 kcal/mole. Find out the resonance energy of benzene.

Solution: Cyclohexene contains one C=C bond but benzene contains three C=C bonds.

∴ Calculated heat of hydrogenation of benzene is 3 × −28.6 = −85.8 kcal/mole.

Hence resonance energy of benzene

$$= [-49.8 - (-85.8)] \text{ kcal/mole (See Eqn. 4.3)}$$
$$= 36 \text{ kcal/mole}$$

4.4.5 Mesomeric or conjugative effect or resonance effect

The discussion on resonance involving π electrons tells us that it decreases the electron density at one position and increases the same elsewhere in a chemical entity; this redistribution of electrons takes place with permanent polarisation in the ground state of the entity. The effect which helps redistribution of electrons with permanent polarisation in the ground state of an entity in unsaturated and chiefly in conjugated systems via π orbitals is called conjugative or resonance effect.

MESOMERIC EFFECT includes both electromeric and resonance effect. Neglecting electromeric effect, mesomeric effect may be represented by the symbol **M** or **R**. When the electronic displacement is away from the group concerned, the mesomeric effect is said to be +**M** or +**R**; when the effect take place towards the group it is called −**M** or −**R** effect.

The mesomeric effect being a permanent effect, the physical and chemical properties of entities may be explained with the help of this effect.

Problem 4.9. Distinguish between (a) inductive effect and mesomeric effect, and (b) electromeric effect and mesomeric effect.

Solution: (a)

Mesomeric (M or R) effect	Inductive (I) effect
1. This is a permanent polarisation effect.	1. This is also a permanent polarisation effect.
2. This involves π electrons.	2. This involves σ electrons.
3. This occurs in the unsaturated and especially in the conjugated systems.	3. This occurs essentially in a saturated system.
4. This effect is transmitted over a wide range, provided conjugation is present.	4. It is transmitted through a short distance in a saturated chain.

(b)

Electromeric effect (E)	Mesomeric effect (M or R)
1. Temporary polarisability effect.	1. Permanent polarisation effect.
2. This involves π electrons.	2. This also involves π electrons.
3. This occurs involving two atoms only at the requirement of an attacking reagent.	3. This may occur over a wide range, provided conjugation is there.
4. Being a temporary effect, it has nothing to do with the physical properties of an entity.	4. It can explain both the physical and chemical properties of entities.

4.4.6 Resonance versus tautomerism

Constitutional isomers existing in rapid equilibrium are known as tautomers and the equilibrium reaction is called tautomerism. For example, acetaldehyde (XXXIV) and vinyl alcohol (XXXIV) are tautomers that exist in solution. We detect the presence of both the compounds in

$$\underset{\text{XXXIV}}{CH_3-\underset{\displaystyle H}{\underset{|}{C}}=O} \rightleftharpoons \underset{\text{XXXIV}}{CH_2=CH-OH}$$

solution by chemical tests. Therefore, tautomers have real existence and they are isolable too. On the other hand, resonating structures are essentially hypothetical and non-existent. Still we may draw some main points of differences between the two as stated in the next page :

Resonance	Tautomerism
1. It is a hypothetical state of chemical entity.	1. It is a phenomenon which occurs really.
2. Resonating structures are hypothetical and non-isolable.	2. Tautomers have existence and they are isolable.
3. In all resonating structures of an entity the positions of the atomic nuclei remain unchanged.	3. The positions of the atomic nuclei are not the same in tautomers.
4. Only redistribution of electron is involved.	4. Redistribution of electrons also occurs.
5. Actual structure of the entity is one and the same though different resonating structures are written.	5. Tautomers have different structures.
6. There is no equilibrium between resonating structures.	6. There is a rapid equilibrium between tautomers.
7. Resonance lowers the energy of the entity.	7. As tautomers are different compounds, the question of lowering of energy does not arise.

Problem 4.10. Tick out the pair of compounds as tautomers or resonating structures:

a. 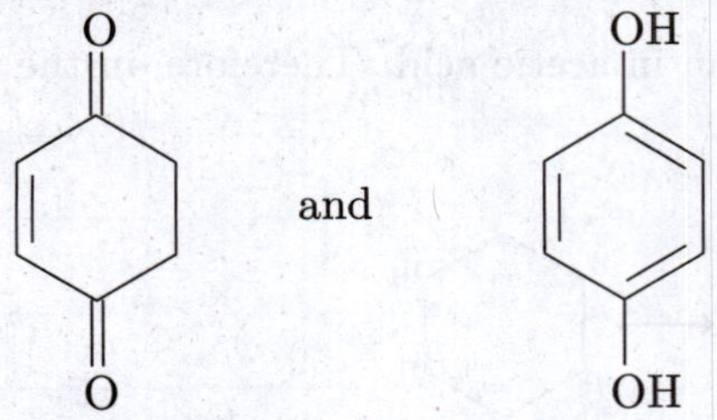

b. 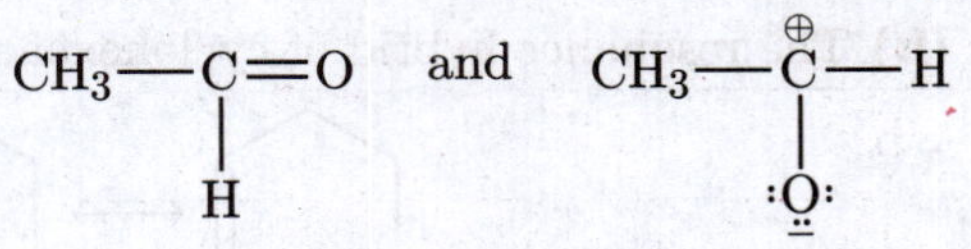

c. 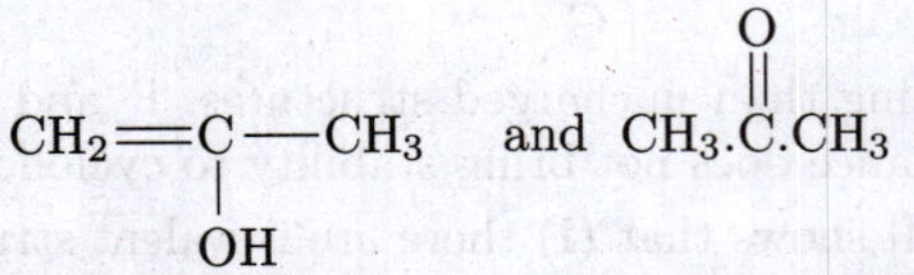

d. 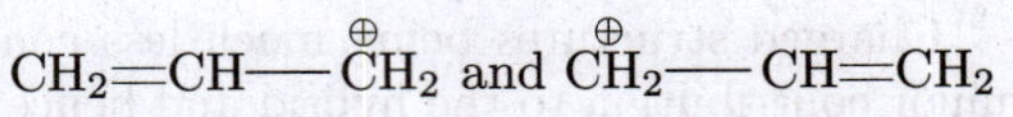

Solution: **(a)** Tautomers; as there are changes in positions of two H atoms. **(b)** Resonating forms; because the positions of the atomic nuclei remain the same and only electron redistribution has occurred. **(c)** Tautomers. **(d)** Resonating forms.

4.4.7 Application

A. Resonance is used to explain the stability of an entity :

(i) A resonance hybrid possessing a greater number of resonating structures is more stable than the resonance hybrid consisting of a fewer number of such structures.

(ii) Electron redistribution over a wider range brings greater stability to a system than that over a shorter range.

(iii) The stability of the resonance hybrid increases with the increasing number of significant structures in it.

Problem 4.11. Which one is more stable—

(a) Acetic acid or acetate ion in the aqueous solution? **(b)** Benzene or cyclohexene?

Solution: (a) The resonance hybrid of acetic acid is:

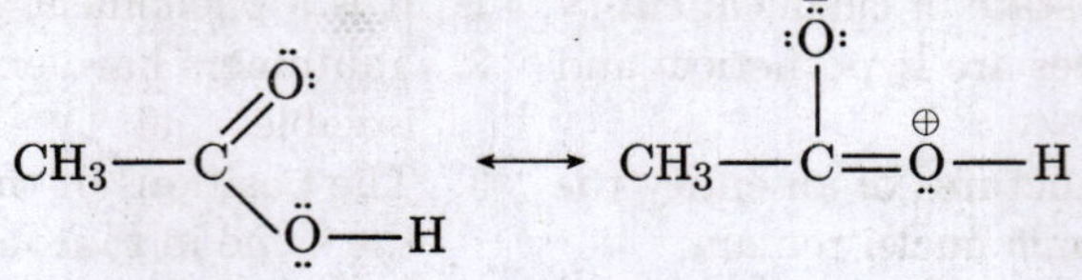

1　　2

Here structure 2 has minor contribution because the latter structure possesses a positive charge on the electronegative O atom.

The resonance hybrid of acetate ion is:

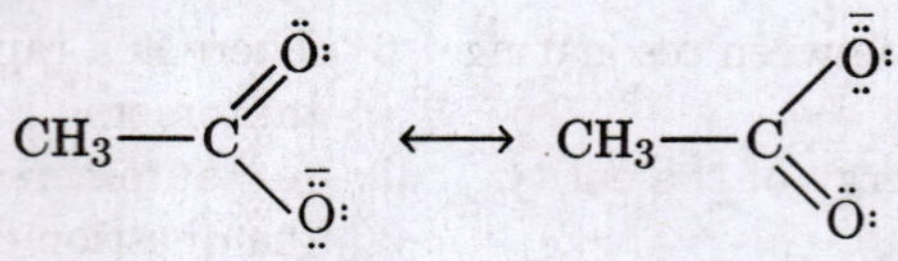

3　　4

Here the two structures are equivalent and, therefore, they have equal contribution to the hybrid.

So resonance is more important in acetate ion than in acetic acid. Therefore, in the aqueous solution acetate ion is more stable than acetic acid.

(b) The resonance hybrid of cyclohexene is:

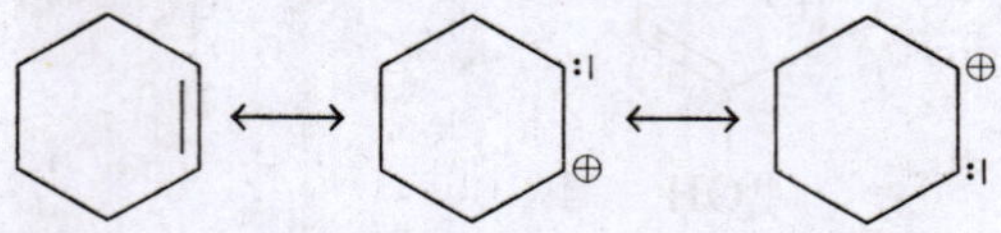

E　　F　　G

Charged structures being much less contributing than uncharged structures, F and G have minor contribution to the hybrid and hence resonance does not bring stability to cyclohexene.

The resonance hybrid of benzene [XXI—XXVI] shows that **(i)** there are isovalent structures, **(ii)** delocalisation occurs over a wide range and **(iii)** the number of resonating structures are also more than those in the hybrid of cyclohexene. Hence benzene is more stable than the latter.

B. Resonance can be used to explain bond length values:

(i) If resonance occurs between single and double bonds then the single bond length decreases but double bond length increases.

(ii) If double and triple bonds are involved in resonance, this decreases the double bond length and increases the triple bond length.

Problem 4.12. C—N bond length in NH_2CONH_2 is 0.137 nm instead of 0.147 nm. How can you explain this shortening of bond length?

Solution: Structure of NH_2CONH_2 may be represented by the resonance hybrid as written below:

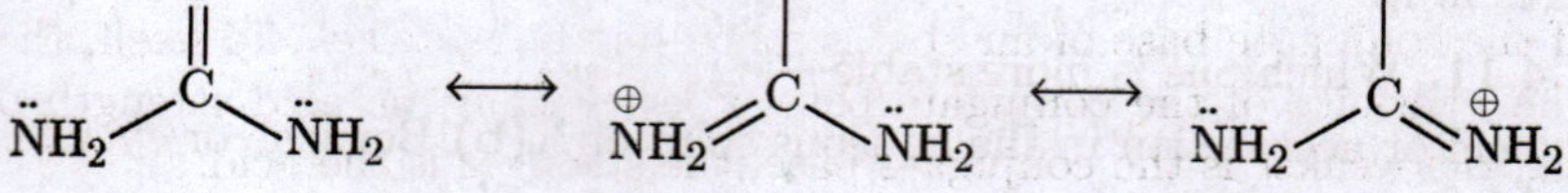

Thus, we see that each of the two C—N bonds in the compound is single in two structures and double in one structure. So, it should be shorter than carbon-nitrogen single bond but longer than carbon-nitrogen double bond. Therefore, the given value of the bond length, 0.137 nm, is expected.

C. Resonance can be used to explain the dipole moment of an entity:

The dipole moment (μ) is directly proportional to the distance between the two poles (r). This means as the distance between the two poles increases, the dipole moment increases. If resonance happens to occur over a wide range, the high dipole moment will be the result; on the other hand, resonance over a short distance gives rise to a low dipole moment.

Problem 4.13. The dipole moments of nitromethane and nitrobenzene are 3.60D and 3.95D respectively—explain.

Solution: The resonance hybrid of nitromethane is:

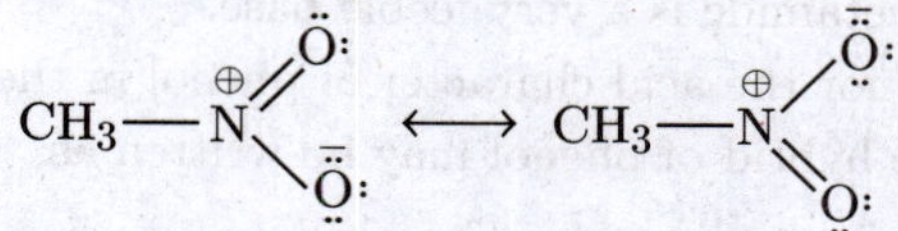

The resonance hybrid of nitrobenzene is:

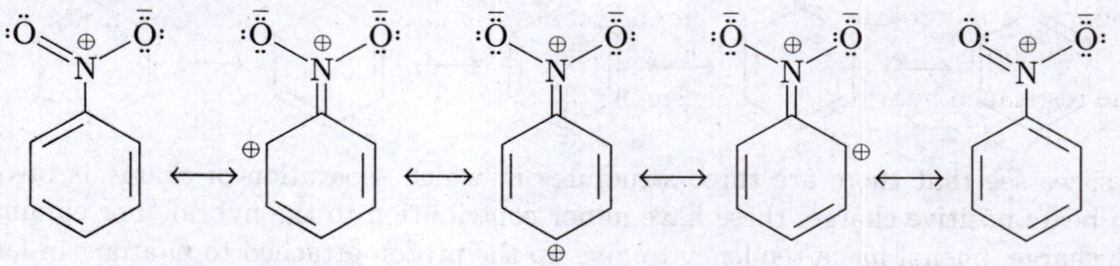

Thus, the distance between positive and negative charges in nitrobenzene is greater than that in nitromethane. Hence, the dipole moment of nitrobenzene should be greater than that of nitromethane.

D. Resonance can be used to explain the symmetry of an entity:

If all the resonating structures are equivalent in the resonance hybrid of an entity, the entity will be symmetrical.

Problem 4.14. Explain the symmetrical nature of carbonate ion.

Solution: The carbonate ion may be represented by a resonance hybrid possessing three equivalent resonating structures as written below. For this reason carbonate ion is a symmetrical entity.

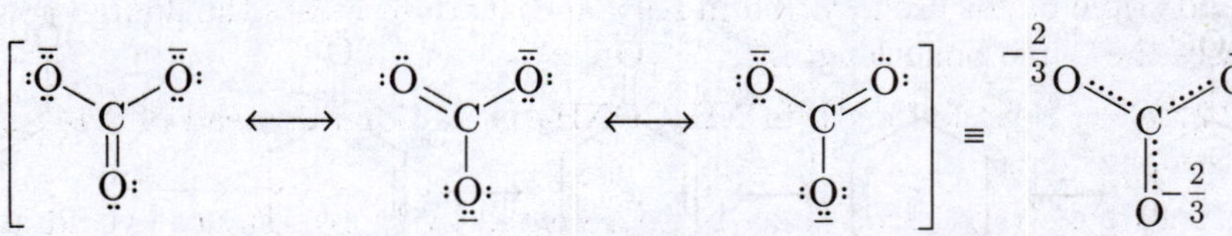

E. Resonance can be used to explain the relative acidity and basicity of acids and bases respectively:

(i) Acidity: If the conjugate base of an acid is more stable than the acid itself, then the base strength, i.e., basicity of the conjugate base is lower than the acid strength of the acid. This means, the weaker is the conjugate base, the stronger is the acid.

(ii) Basicity: If the lone-pair of electrons of a base gets delocalised, it is not readily available for donation and hence the entity will be of low basicity. As the delocalisation of lone-pair of electrons increases, the basicity decreases and **vice versa**.

Problem 4.15. Acetamide is feebly basic, though the N atom possesses a lone-pair of electrons—explain.

Solution: The resonance hybrid of acetamide is:

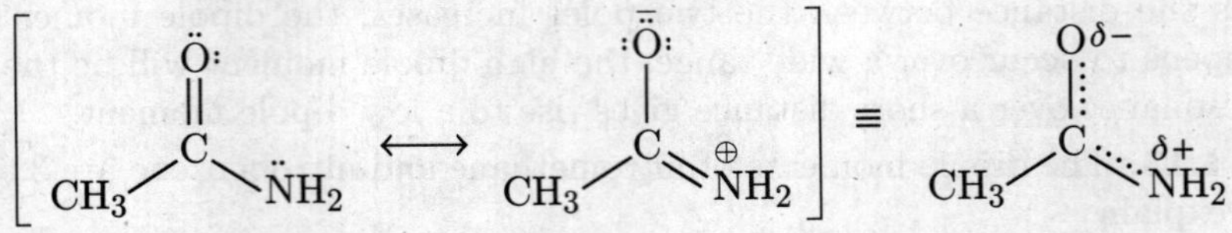

Thus, the delocalisation of lone-pair of electrons of N atom occurs and hence it is not easily available for donation. So, acetamide is a very feeble base.

Problem 4.16. Account for the acid character of phenol in the aqueous medium.

Solution: The resonance hybrid of phenol may be written as:

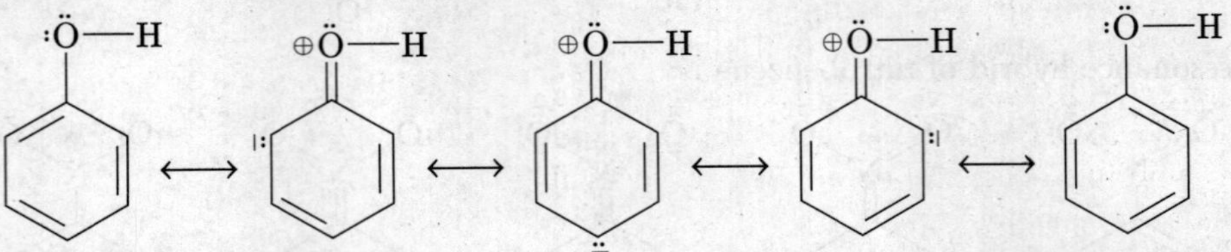

Thus, we see that there are three structures in which separation of charge is involved and O atom bears positive charge; these have minor contribution to the hybrid. For eliminating the positive charge, phenol has a tendency to give up the proton attached to O atom; in fact, in the aqueous medium it does so and produces phenolate ion, the conjugate base.

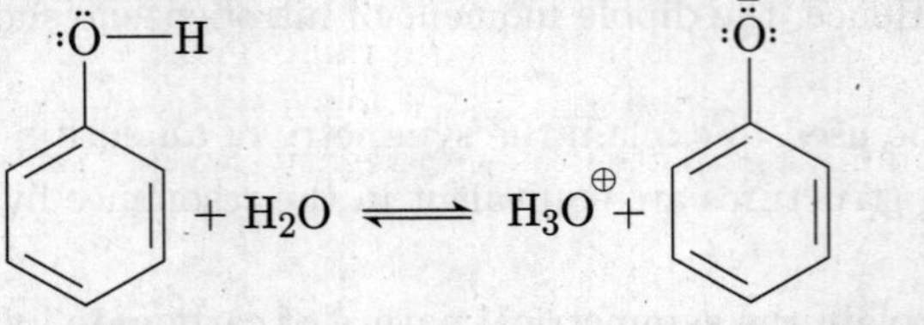

Phenol (acid) Phenolate ion (conjugate base)

The resonance hybrid of the phenolate ion is:

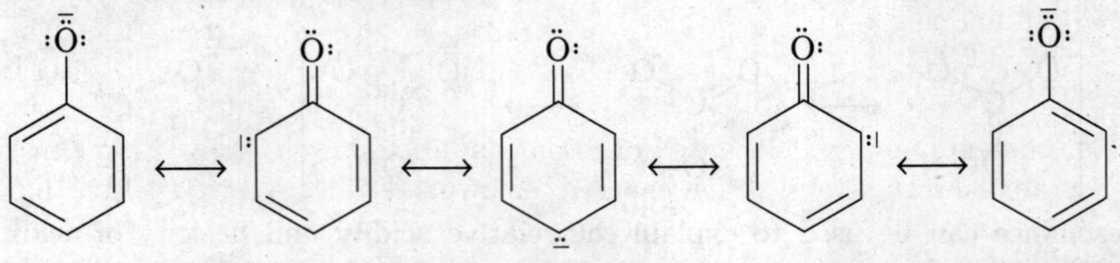

All the structures in the resonance hybrid are isovalent and no structure bears a positive charge on O atom; the lone-pair of electrons on O atom gets delocalised over a wide range. Hence the ion is stable and does not want to donate electrons readily. So it is a weak base. This means the acid is stronger in relation to the conjugate base.

4.4.8 M.O. equivalent of resonance

To explain the delocalisation of electrons in unsaturated and conjugated systems the V.B. theory takes the help of resonance. On the other hand, the M.O. theory tells us that when two or more atoms combine to form molecules, the bonding electrons do not belong to individual atoms but these form molecular orbitals which embrace all the atoms of the molecule. Thus, delocalisation of electrons is an obvious phenomenon in the M.O. theory. Let us now take the example of benzene to understand the meaning of delocalisation by the M.O. theory.

From experiment it is known to us that benzene has a planar regular hexagonal structure with bond angle of 120°; sp^2 hybridisation of C atoms is considered to describe such a framework. Six p orbitals, one from each C atom, remain perpendicular to the plane of the carbon ring and parallel to each other. All the six p A.O.s could overlap to form six delocalised π M.O.s—three bonding and three antibonding. The three bonding orbitals will accommodate the six electrons. Fig. 4.3 shows the energy scheme for the π electrons of benzene.

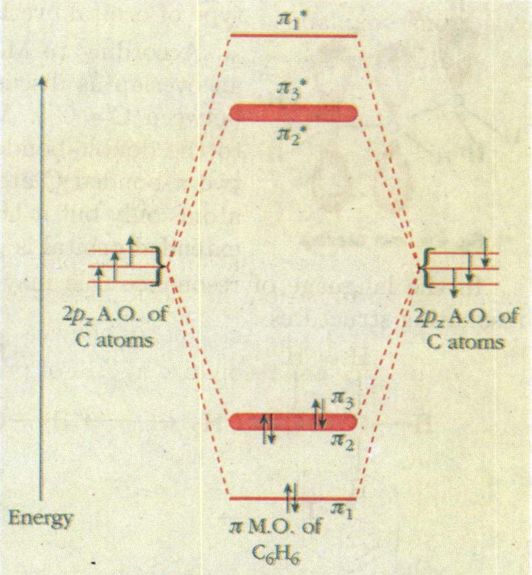

Fig. 4.3. M.O. energy diagram of benzene molecule

The bonding M.O. of lowest energy (π_1) is cyclic and embraces all the six C atoms. π_1 M.O. has a nodal plane, the plane of the carbon ring, and two lobes—one below the nodal plane and the other above it (Fig. 4.4); π_2 and π_3 are degenerated M.O.s; they too encompass all the six C atoms. Besides the nodal plane said above, each of the π_2 and π_3 M.O.s possesses another nodal plane which is perpendicular to the plane of the ring. Thus, each of the degenerated M.O.s has four lobes. The net result of combination of ten lobes (two lobes of π_1 M.O. + 2 × 4 lobes of π_2 and π_3 M.O.s) is an annular electron cloud, i.e., negative charge sheet, below and above the plane of the ring (Fig. 4.5).

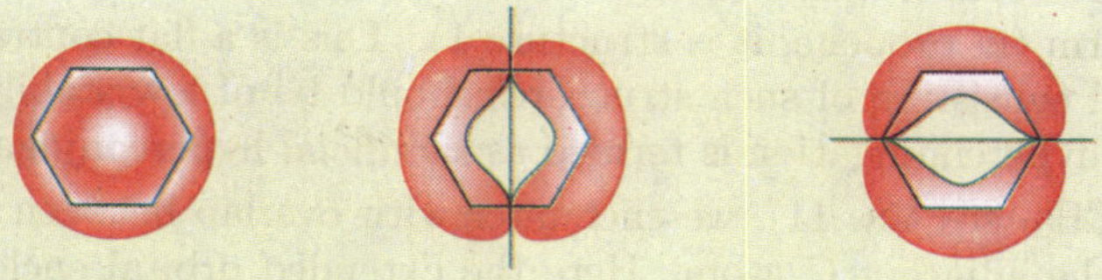

π_1 π_2 π_3

Fig. 4.4

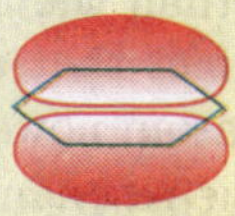

Side view

Fig. 4.5

The electrons are now in delocalised orbitals containing lower energy than expected. Hence, benzene is stable and the stabilisation energy is known as delocalisation energy, the M.O. equivalent of resonance energy.

4.5 Hyperconjugative Effect

All of the examples of delocalisation, i.e., conjugation cited so far, involve π electrons. There are a few examples on conjugation involving σ electrons. This type of conjugation may be called extended conjugation. The prefix, hyper-, of the title means excessive. The excessive, i.e., extended conjugation involving σ orbitals, generally of C—H or C—X (X=F, Cl) or C—C bond, is called **hyperconjugation**. We know that π–π or π–p (vacant or filled) type of orbital overlap gives rise to conjugation. But σ–π or σ–p (incomplete or vacant) type of orbital overlap develops hyperconjugation.

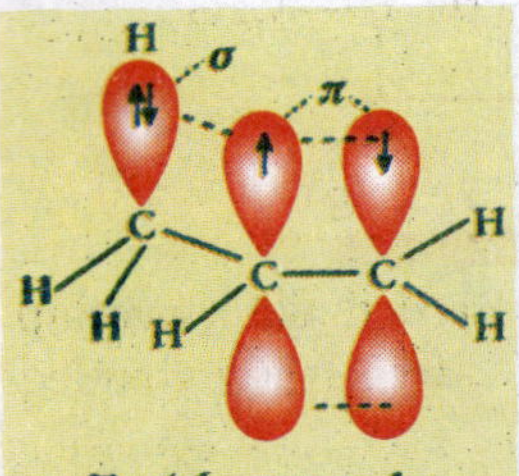

Fig. 4.6. σ-π **overlap**

According to M.O. theory, an extended orbital is the result of such an overlap as discussed above. Thus, in propene, delocalisation occurs between C=C π M.O. and C—H σ M.O.s of the C atom adjacent to the double-bonded C; here the extended orbital does not encompass two π-bonded C atoms or two σ-bonded C atoms or σ-bonded C and H atoms only but it holds all of the four atoms together. The picture of the extended orbital is given in Fig. 4.6.

In the language of resonance this may be pictured as a resonance hybrid of the following resonating structures :

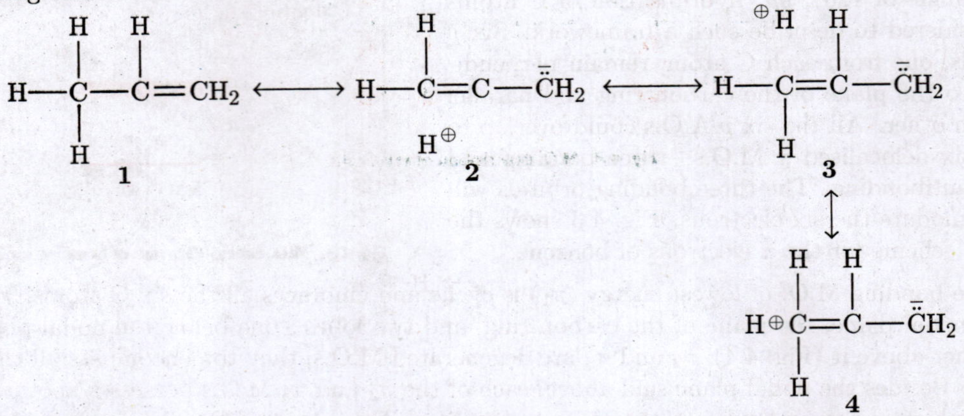

As the structures 2, 3 and 4 do not have any formal bond between $H^{\oplus}$ ion and C atom, the state of the molecule is known as **no-bond resonance**. The $H^{\oplus}$ ion does not leave the species. Each of these structures is hypothetical and non-existent. This sort of resonance too stabilises the entity, though no-bond resonating structures bear isolated charges and have a fewer number of covalent bonds than the main uncharged form (in propene, it is structure 1). This is a flat contradiction of the fact that a resonance hybrid consisting of such structures should be of low stability. For this reason, nowadays this type of hyperconjugation is termed as **sacrificial hyperconjugation**.

In ethyl free radical ($H_2\dot{C}$—CH_3, see Ch. 11), we encounter with overlap between C—H orbital and incomplete orbital of the adjacent C atom. Here the extended orbital encloses the two C and one H atoms as shown in Fig. 4.7.

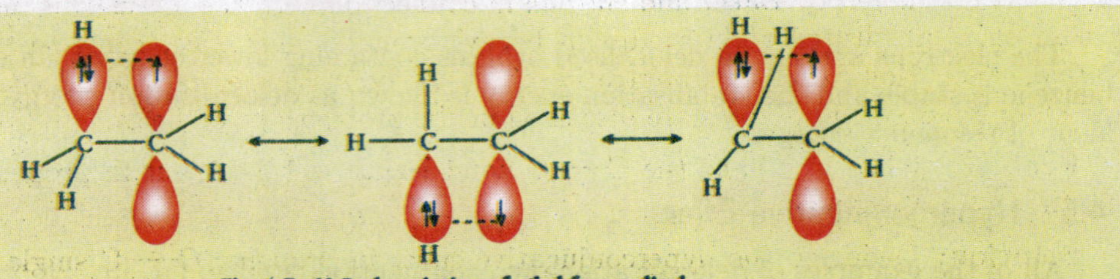

Fig. 4.7. M.O. description of ethyl free radical

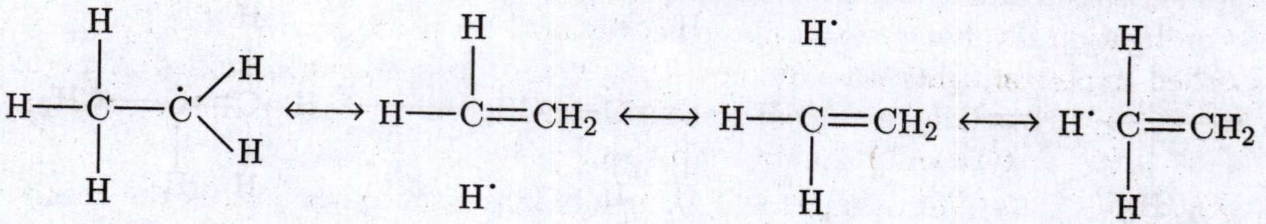

5 **6** **7** **8**

Similarly, in ethyl cation ($CH_3 \cdot \overset{\oplus}{C}H_2$, see Ch. 11) hyperconjugation is regarded as an overlap of the C—H orbital and the vacant *p* orbital of the positively charged C atom; the extended orbital formed holds the two C atoms and the H atom together (Fig. 4.8).

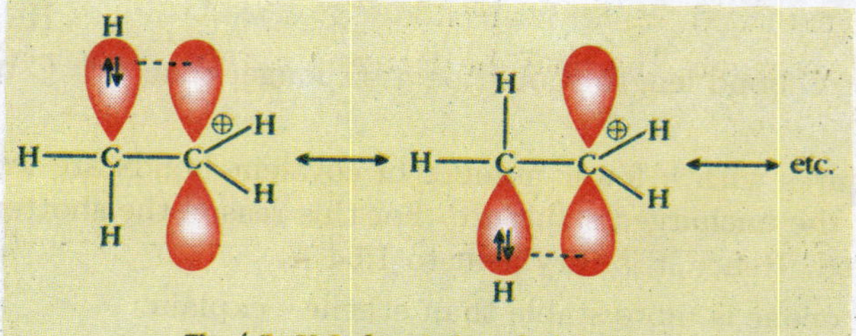

Fig. 4.8. M.O. description of ethyl cation

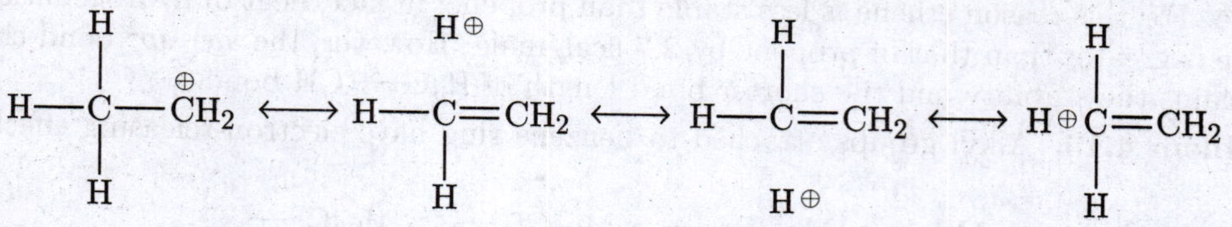

9 **10** **11** **12**

Thus, in free radicals and in carbenium ions with adjacent C—H σ bonds, the resonating structures display no more odd electron or charge separation that do the main structures (here 5 and 9 respectively) and all of the structures possess the same number of covalent bonds. Hence this type of hyperconjugation is termed as **isovalent hyperconjugation**.

The effect of hyperconjugation is that of electron release from the C atom adjacent to the π-bonded C atom or electron-deficient C atom (i.e., a C atom with odd electron or positive charge). When it happens to occur in the ground state of a molecule, some of the physical and chemical properties of the molecule can be explained with its help. On the other hand, when hyperconjugation takes place in the excited state of the entity, it can then only be helpful in explaining chemical reactivity of the entity. It is here worth-mentioning that the hyperconjugative effect was first noted by **Baker and Nathan** in connection with the anomalous inductive electron release by *p*-substituted alkyl groups in benzyl bromide; this effect came to be called **Baker-Nathan effect**. However, hyperconjugative effect can be utilised to explain the chemical reactivity and stability of entities. These are discussed in the problems given below:

Problem 4.17. In ethane the C—C bond length is 1.54 Å but in propene C—C bond length is 1.50 Å—explain.

Solution: Owing to the hyperconjugative effect in propene, C—C single bond has some double bond character as seen from the following resonating structures:

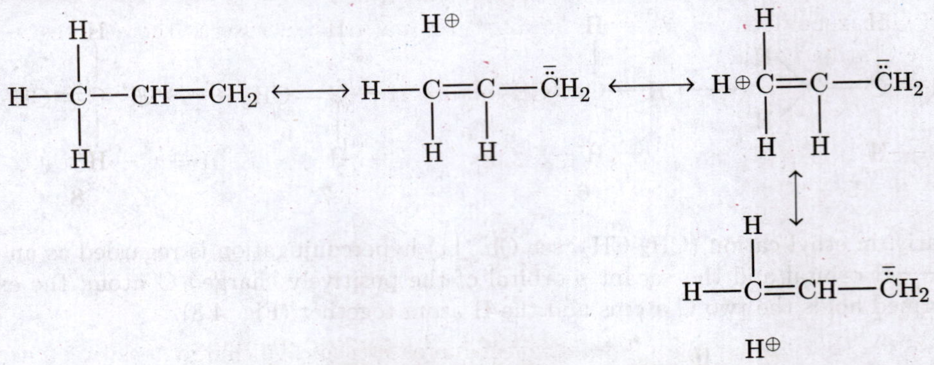

Therefore, the C—C bond length should be a bit longer than C=C bond and shorter than C—C single bond.

But charged structures with a fewer number of covalent bonds are less contributing to the resonance hybrid than the uncharged structure. For this reason the shortening of bond length is not so much as expected. Hence it is very near to 1.54 Å.

Problem 4.18. Propene is more stable than ethene—explain.

Solution: Hyperconjugative effect plays its role in bringing stability to propene; this is absent in ethene. For this reason ethene is less stable than propene; in fact, heat of hydrogenation value of ethene is greater than that of propene by 2.7 kcal/mole. However, the sp^3–sp^2 bond character also explains the stability and the shorter bond length of H_3C—CH bond.

Problem 4.19. Alkyl groups attached to benzene ring have electron-releasing effect in the order:

$$\text{Me—} > \text{MeCH}_2\text{—} > \text{Me}_2\text{CH—} > \text{Me}_3\text{C—}$$

This is quite opposite to that expected from inductive electron release. Explain this anomaly.

Solution: Hyperconjugative effect alone can explain this anomaly.

When an alkyl group is attached to a benzene ring, the C—H σ bond of the C atom, which is directly joined to the benzene ring, overlaps with the π orbitals of the ring. For each C—H σ bond three charged resonating structures are found. As a methyl group has three C—H σ bonds, we get nine charged resonating structures and two isovalent uncharged resonating structures in the resonance hybrid.

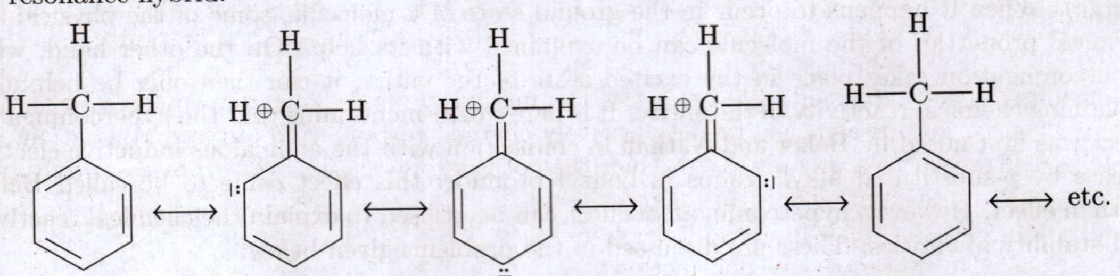

In an ethyl group there are two such C—H σ bonds. We should have eight resonating structures only for ethyl benzene, six of them are charged structures and the rest two are uncharged structures. In the resonance hybrid of isopropylbenzene, we find only five resonating structures, two uncharged and three charged structures. This is because of the presence of only one C—H σ bond α to the benzene ring.

tert-Butylbenzene does not have any no-bond resonating structure as there is no C—H σ bond on the C atom bonded to the benzene ring.

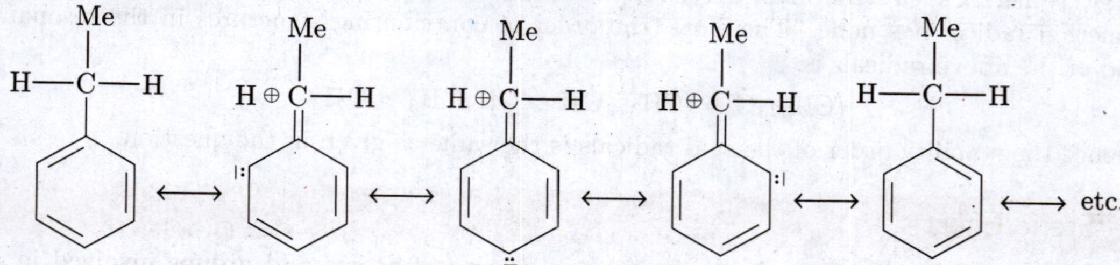

Thus, we see that the hyperconjugative effect increases negative charge density, i.e., electron density in the benzene ring. Therefore, it is expected that the greater the number of negatively charged resonating structures in a hybrid, the greater will be the electron density in the benzene ring.

The decreasing order of the number of charged resonating structures of the alkylbenzenes discussed above is :

Methylbenzene : 9, ethylbenzene : 6, isopropylbenzene : 3, tert-butylbenzene : 0.

Hence, the electron-releasing effects of the said alkyl groups are in the order of:

methyl > ethyl > isopropyl > tert-butyl.

Problem 4.20. The order of stability of

a. $(CH_3)_3\overset{\oplus}{C} > (CH_3)_2\overset{\oplus}{C}H > CH_3\overset{\oplus}{C}H_2 > \overset{\oplus}{C}H_3$

b. $(CH_3)_3\dot{C} > (CH_3)_2\dot{C}H > CH_3\dot{C}H_2 > \dot{C}H_3$

Explain the stability order in both the cases.

Solution: (a) Two effects are to be considered for explaining the given order:

(i) Hyperconjugative effect

We know that the more the number of contributing structures of comparable energy in a resonance hybrid, the greater is its stability. In tert-butyl cation, there are nine α-C—H sigma bonds and hence its resonance hybrid consists of ten resonating forms, nine of which are uncharged structures; we shall have six similar resonating forms in isopropyl cation, three such in ethyl cation and none in methyl cation. Therefore, the order of stability of the carbenium ions is as given in the question.

(ii) Inductive effect

This can also be explained by +I effect of the methyl groups. Thus, due to the +I effect of the three methyl groups attached to the positively charged carbon atom of tert-butyl cation, its charge is neutralised to a greater extent than that of isopropyl cation which possesses only two methyl groups for such an effect; this charge neutralisation effect is still less in the case of ethyl cation as it contains only one methyl group and it is least in methyl cation because it does not have any electron-releasing group attached to the positively charged C atom. This is represented pictorially in Fig. 4.9.

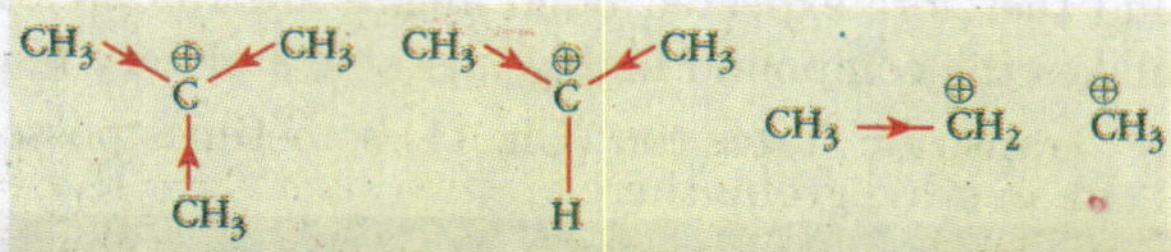

Fig. 4.9. +I effect on alkyl cations

(b) The given stability order can be explained by hyperconjugative effect.

$(CH_3)_3\dot{C}$ has nine resonating forms without odd electron in its resonance hybrid, $(CH_3)_2\dot{C}H$ has six such structures, $CH_3\dot{C}H_2$ possesses only three structures without odd electron, and methyl radical has none. Therefore, the order of contributing structures in the resonance hybrid of the above radicals is:

$$(CH_3)_3\dot{C} > (CH_3)_2\dot{C}H > CH_3\dot{C}H_3 > \dot{C}H_3$$

Hence the stability order of the said radicals is the same as given in the question.

4.6 Steric Effect

The factor which considers the sheer bulk or volume of atoms and groups involved in an active part of a chemical species and their spatial arrangement in connection with its stability, acidity, basicity, the rate and stereochemistry (Ch. 9) of its reaction as well as the yield of the reaction product may be called the steric factor. It plays an important role in understanding reaction mechanism (Ch. 11). The volume of atoms or groups on the reacting part of an organic species has varied types of effect on the stability and reactivity of the species and those effects may be called **steric effects**. Four types of steric effects may be considered here: (i) Steric strain, (ii) Steric acceleration, (iii) Steric retardation, (iv) Electron-availability.

4.6.1 Steric strain

If the constituent atoms and groups of a chemical species require more space than what is available for them, owing to their bulky nature, then mechanical interference amongst the groups and atoms takes place and the species is said to be under steric strain. Consequently, distortion of normal bond angle occurs, which in its turn, brings instability to it. Owing to the great bulk of the tert-butyl group and small available space for it, 1,2,3-tri-tert-butylbenzene is under strain and very difficult to prepare. However, the strain will be discussed in the Isomerism chapter (Chapter 9) in detail.

Note

Even small groups may cause steric strain owing to the overlap of their van der Waals radii known as van der Waals strain. This strain develops repulsion between two such non-bonded atoms or groups which is called **steric repulsion**. For example, anti-conformation of 2,3-dibromobutane is the most stable conformation since amongst other factors least amount of steric repulsion is found in this conformation in comparison to that in fully eclipsed and gauche conformations.

Sometimes not the bulk of the groups and atoms but their very spatial arrangement in a species create distortion of bond angles and hence develop steric strain in it which is known as Internal strain or I-strain. This is found in small rings. Thus, cyclopropane has planar triangular ring structure having C—C—C bond angle 60°, though all the Cs are in sp^3-hybridised state and therefore expected bond angles are 109°28′. Hence cyclopropane ring has I-strain and this makes the compound highly unstable and very reactive.

Transition states too (Ch. 11) sometimes possess I-strain.

4.6.2 Steric acceleration

Strained chemical species try to avoid the steric strain and thus react readily to produce less strained species. When a steric strain speeds up a reaction, it is said to be steric acceleration and

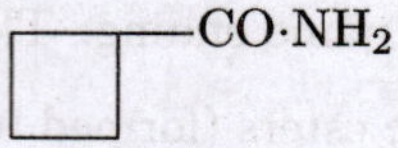 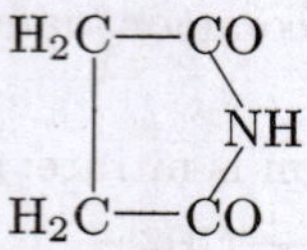 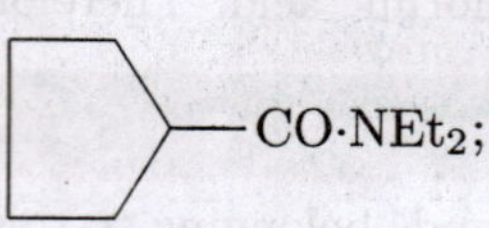

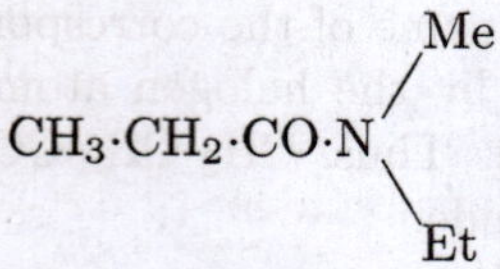 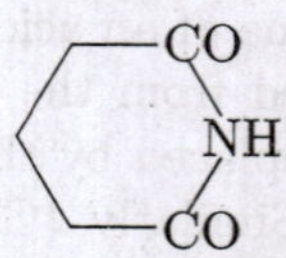

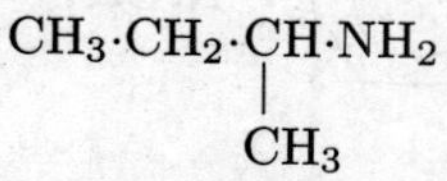 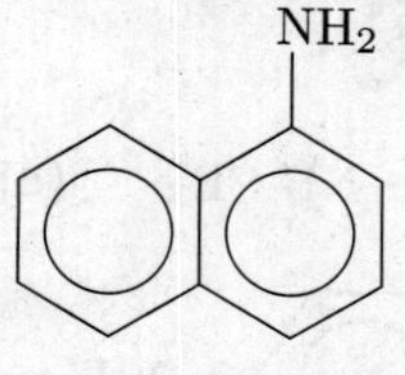 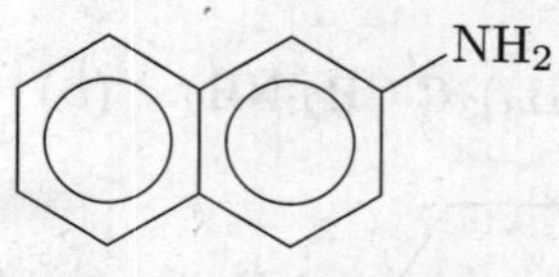

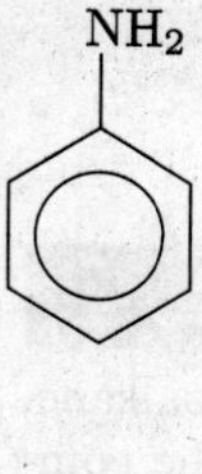 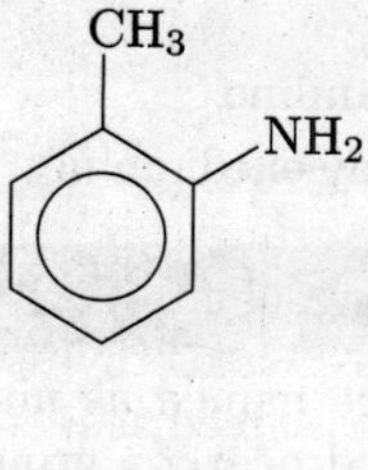 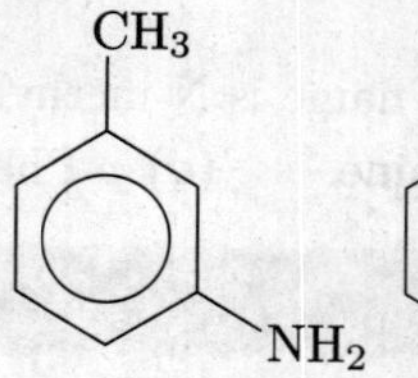 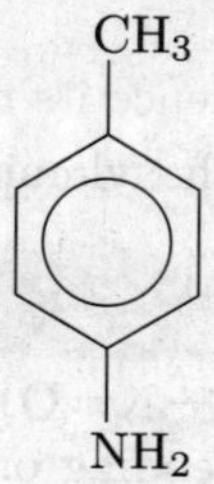 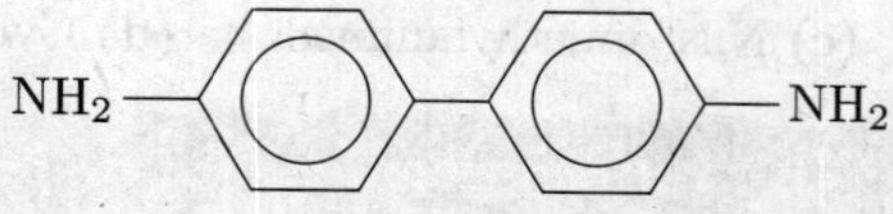

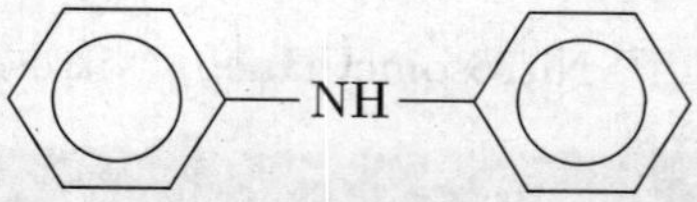

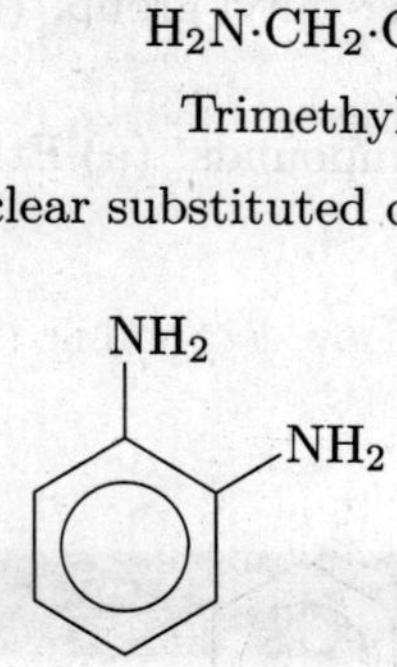 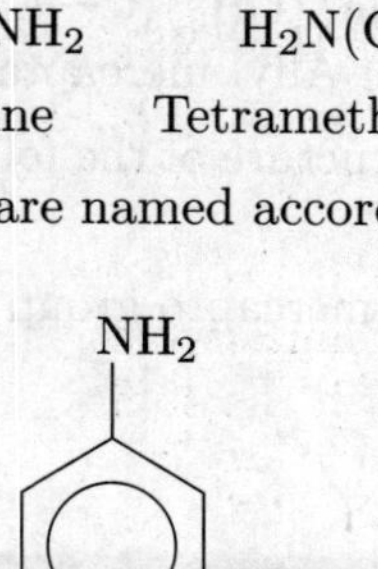 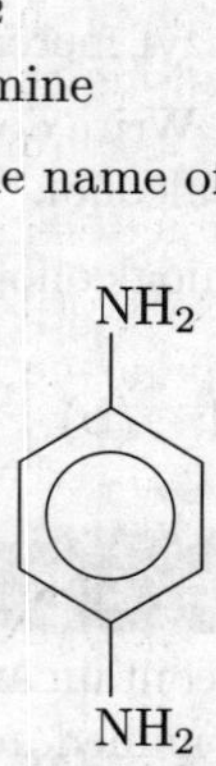

Problem 3.16. Name the following amines:

(a) $(CH_3)_3C{\cdot}CH_2{\cdot}NH_2$ **(b)** 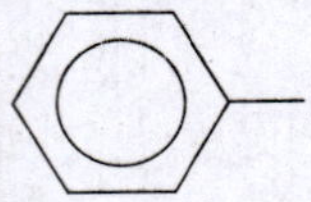 $-NH{\cdot}CH_3$ **(c)** 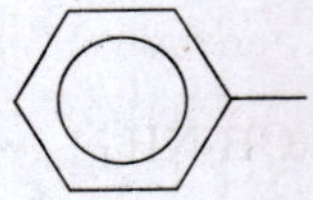 $-N(CH_3)_2$

(d) 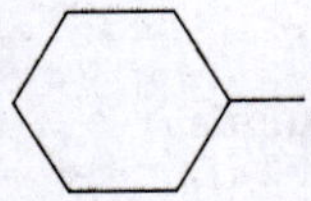 $-NH_2$ **(e)** NH_2- 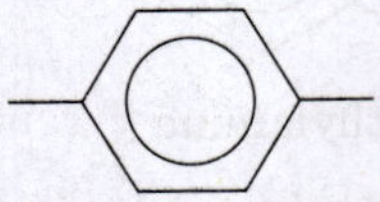 $-NH_2$

Solution: (a) $(CH_3)_3C{\cdot}CH_2-$ is neopentyl group, so the name of the compound is neopentylamine.

(b) The mother compound is aniline; hence its name is N-methylaniline.

(c) N,N-dimethylaniline. **(d)** Cyclohexylamine. **(e)** *p*-Phenylenediamine.

3.7.18 Nitroso Compounds

Nitroso compounds contain nitroso (—N=O) group. These are named as nitrosoalkane, nitrosoarene or as nitrosocycloalkane (depending on the nature of hydrocarbon group the compounds possess); thus,

$CH_3-N{=}O$  $-NO$ 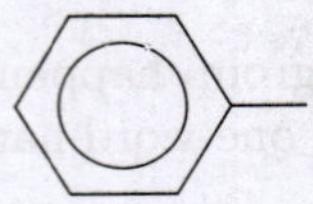 $-NO$

Nitrosomethane Nitrosocyclohexane Nitrosobenzene

3.7.19 Mercaptans (Thioalcohols)

These compounds contain mercapto function (—SH). Thioalcohols, i.e., mercaptans are named according to the name of the hydrocarbon group attached to the mercapto group. Thus, $(CH_3)_2CH{\cdot}CH_2SH$ is isobutyl mercaptan or isobutyl thioalcohol.

Problem 3.17. Write down the common names of the following compounds:

(a) $(CH_3)_3C{\cdot}SH$; **(b)** 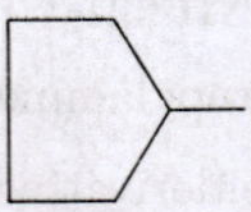 $-SH$; **(c)** 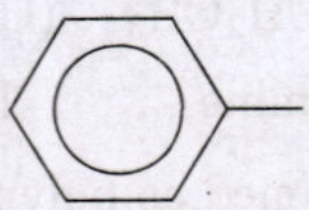 $-CH_2{\cdot}SH$; **(d)** $CH_2{=}CH-CH_2{\cdot}SH$.

Solution: (a) tert-Butyl mercaptan, $(CH_3)_3C-$ being a ter-butyl group. **(b)** Cyclopentyl mercaptan. **(c)** Benzyl mercaptan. **(d)** Allyl mercaptan.

Problem 3.18. Write down the structure of the following compounds: **(a)** Ethyl thioalcohol; **(b)** Cyclobutyl thioalcohol.

Solution: As thioalcohols contain mercapto group (—SH),

(a) $CH_3{\cdot}CH_2SH$; **(b)** 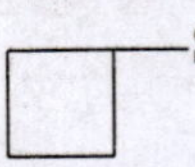 $-SH$

3.7.20 Sulphonic Acids

Sulphonic acids contain an $-SO_3H$ group which is known as sulphonic acid group. Aliphatic sulphonic acids are named according to the name of the hydrocarbon group joined to the SO_3H group; these have one-word names; thus CH_3SO_3H is methylsulphonic acid.

The base name of a compound may be its full name, i.e., it may not have any prefix to its base name; e.g., CH_3CH_2OH is ethanol—it is its base name.

The base name, in its turn, consists of (i) a root or stem and (ii) a prefix or a suffix. In the IUPAC name of **1**, butan is the root of the base name, butanol, and ol is the suffix to the root, butan.

Compound **2** is 2-Methyl-1-nitropropane; here the base name is nitropropane—propane is the root of the base name and nitro is a prefix to the root, propane.

$$\begin{array}{l} CH_3\cdot\underset{\displaystyle |}{C}H\cdot CH_2\cdot NO_2 \\ \quad\;\; CH_3 \end{array}$$

2-Methyl-1-nitropropane.

2

There are a large number of IUPAC rules for constructing the base names and consequently for giving the full names of organic compounds. Only some of these will be discussed here, these will enable a beginner to name simple and complex compounds of the common classes:

I. One finds both one-word names and two-word names in the IUPAC system. In this connection the rule is: if the base name of a compound is the name of an actual compound, the prefix is joined to the base name to give a one-word name; but if the base name is not at all a name of an actual compound, the prefix stands alone and a space is left in between the prefix and the base name. For example, in **2** the base name is nitropropane; as there is a compound having this name, **2** is given a one-word name. Salts and esters do not have one-word names.

Methyl ethanoate is the name for the ester CH_3COOCH_3 since there is no compound named ethanoate, the IUPAC name for CH_3COOCH_3 is a two-word name as given above.

II. To choose the base name of a compound one should choose the parent part of the molecule. The IUPAC rule on choice of a parent part is as follows:

The most substituted largest important part of a molecule is usually chosen as the parent. It consists of a parent carbon chain and the most important functional group; this function may also be considered as a part of the parent carbon chain, the base chain, when the carbon atom of the function is directly linked to a carbon atom of the chain. Thus in 2, 3, 3-Trimethyl-l-butanol we have chosen $CH_3\cdot C\cdot CH\cdot CH_2OH$ as the parent part of the molecule. Let us discuss whether the choice is correct or incorrect as per the rule stated above.

(a) $CH_3\cdot C\cdot CH\cdot CH_2$—may be chosen as the parent part; but this choice is wrong because this does not include the —OH group, the most important part of the molecule, which determines most of the properties of the compound.

(b) One may choose $CH_3\cdot C\cdot CH\cdot CH_3$ as the parent part, but this too will be wrong for the same reason as given in (a).

(c) If one chooses $CH_3\cdot CH\cdot CH_2OH$ as the parent part, it will also be wrong; because this is neither the largest part nor the most substituted part of the molecule though it contains the hydroxyl group. This part bears one —OH group and one tert-butyl group, $(CH_3)_3C$— and it is composed of 3 Cs only.

(d) In $CH_3\cdot C\cdot CH\cdot CH_2OH$ one will see that:

(i) It consists of 4 Cs and hence the largest part of the molecule.

(ii) It contains three methyl branches and the functional group, —OH; thus it is the most substituted important part.

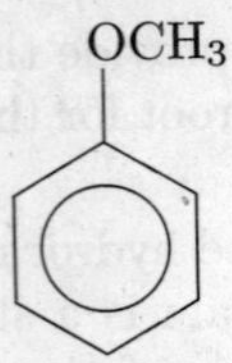
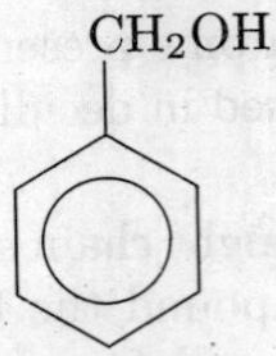
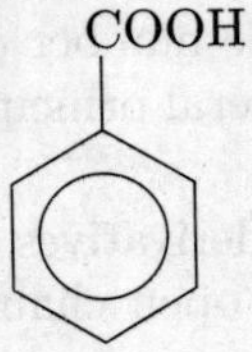
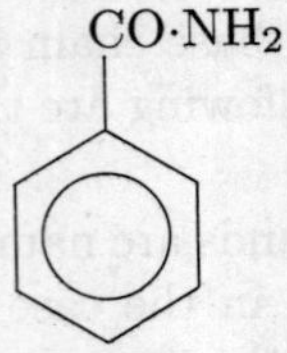
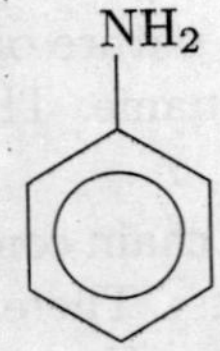
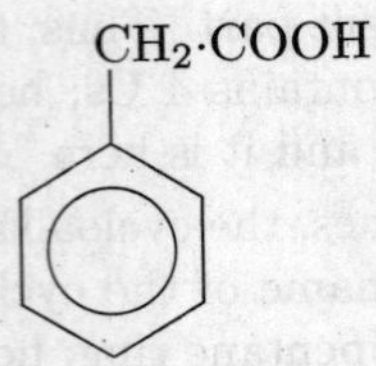
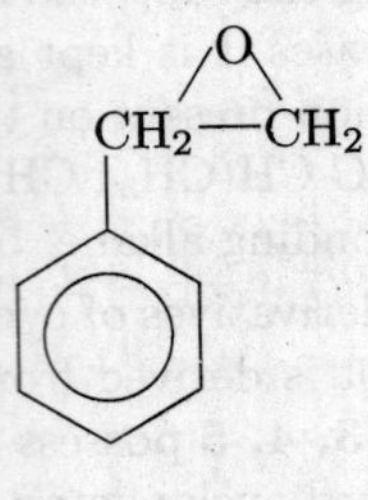
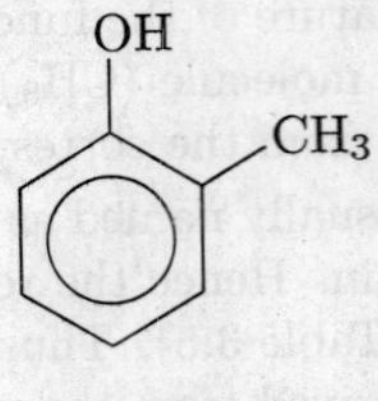
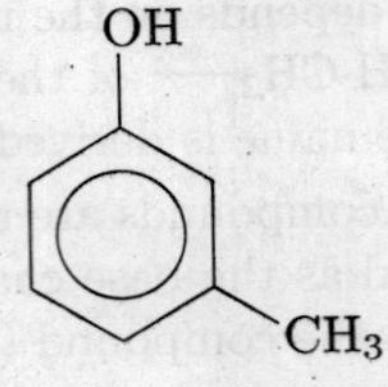
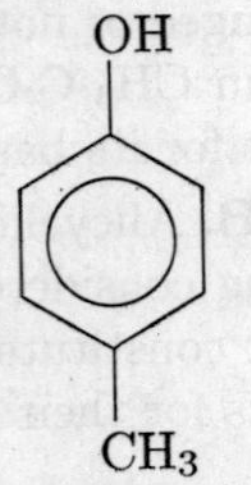
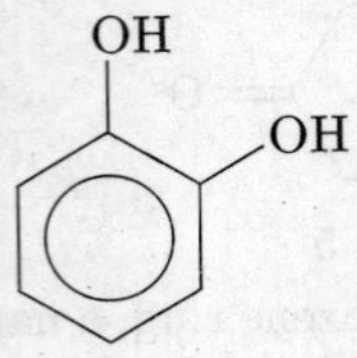
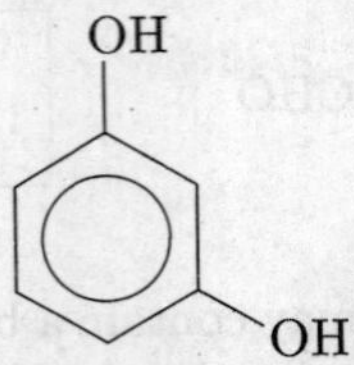
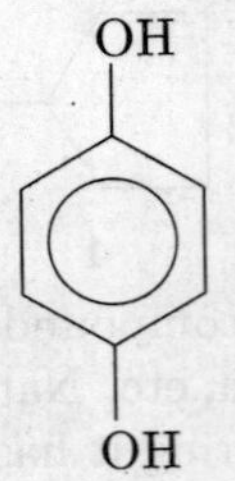
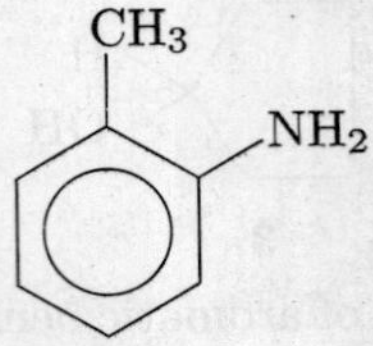
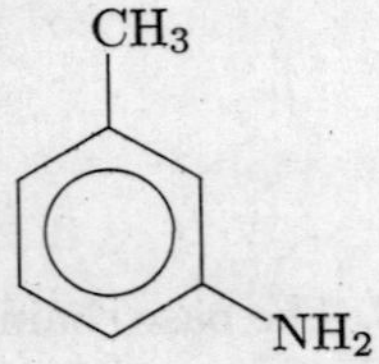
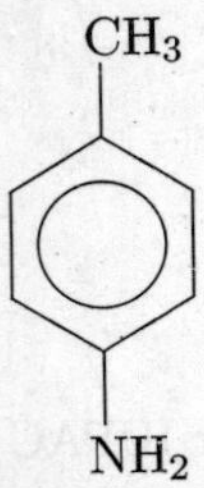
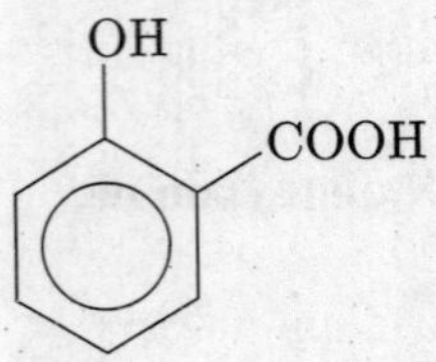
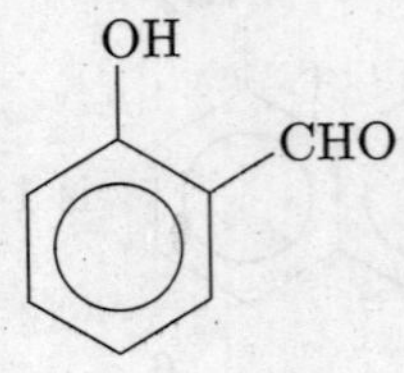
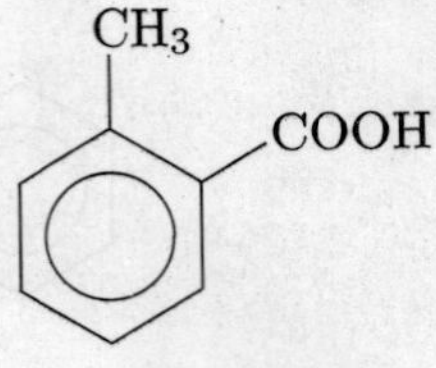
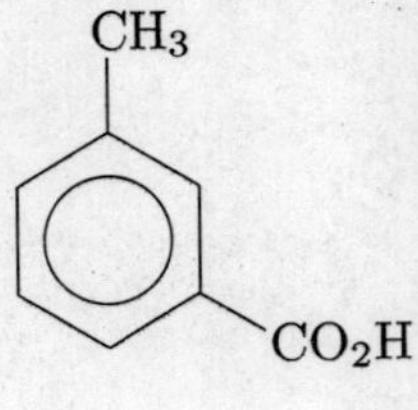
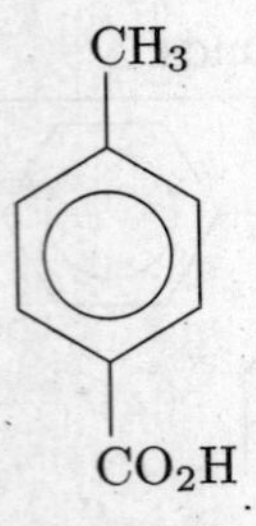
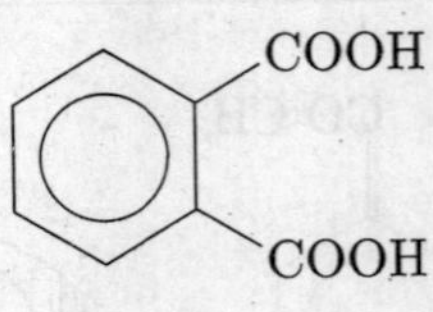
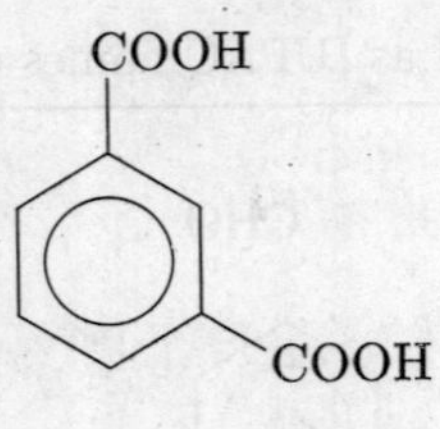
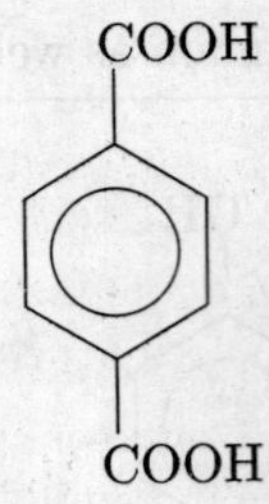
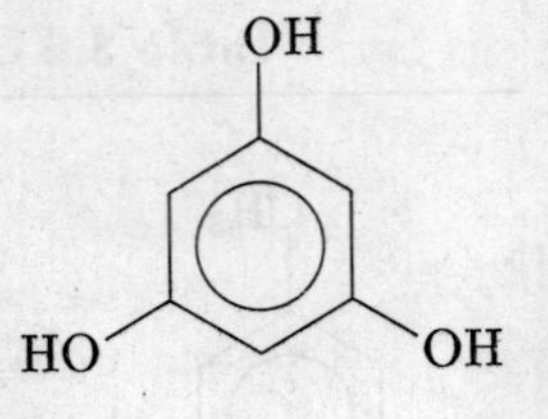
$$\mu_{m-XY} = \sqrt{(-\mu_X)^2 + (\mu_Y)^2 + 2(-\mu_X)\mu_Y \cos 120^\circ} = \sqrt{\mu_X^2 + \mu_Y^2 + 2(-\mu_X)\mu_Y \times -\frac{1}{2}}$$

$$= \sqrt{\mu_X^2 + \mu_Y^2 + \mu_X\mu_Y} = \sqrt{(0.4 \text{ D})^2 + (3.95 \text{ D})^2 + 0.4 \text{ D} \times 3.95 \text{ D}^2}$$

$$= \sqrt{0.16 \text{ D}^2 + 15.6025 \text{ D}^2 + 1.58 \text{ D}^2} = \sqrt{17.3425 \text{ D}^2}$$

$$= 4.16 \text{ D}$$

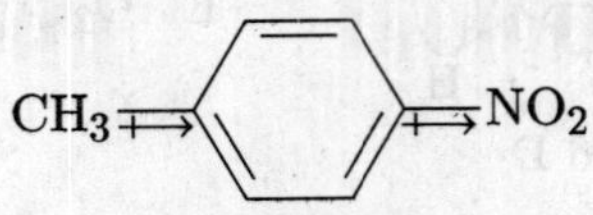

$$\mu_{P-XY} = \sqrt{(-\mu_X)^2 + (\mu_Y)^2 + 2(-\mu_X)\mu_Y \cos 180^\circ} = \sqrt{\mu_X^2 + \mu_Y^2 + 2\mu_X\mu_Y}$$

$$= \sqrt{(\mu_X + \mu_Y)^2} = 0.4 \text{ D} + 3.95 \text{ D} = 4.35 \text{ D}$$

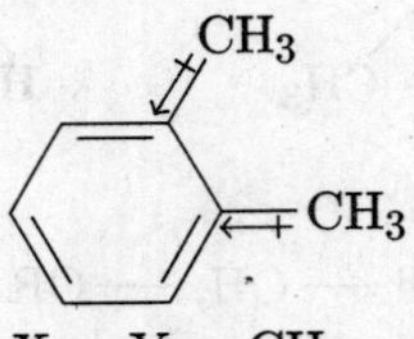

$X = Y = CH_3$

$\mu_X = \mu_Y = 0.4$ D

(iii) Example 3.

$$\mu_{o-XY} = \sqrt{\mu_X^2 + \mu_X^2 + 2\mu_X\mu_X \cos 60^\circ} = \sqrt{\mu_X^2 + \mu_X^2 + \mu_X^2}$$

$$= \mu_X\sqrt{3} = 0.4 \times 1.73 \text{ D} = 0.692 \text{ D}.$$

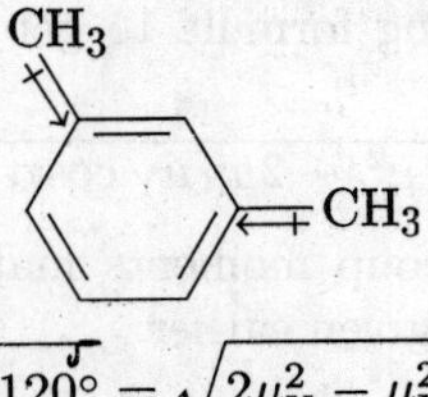

$$\mu_{m-XY} = \sqrt{\mu_X^2 + \mu_X^2 + 2\mu_X\mu_X \cos 120^\circ} = \sqrt{2\mu_X^2 - \mu_X^2}$$

$$= \mu_X = 0.4 \text{ D}$$

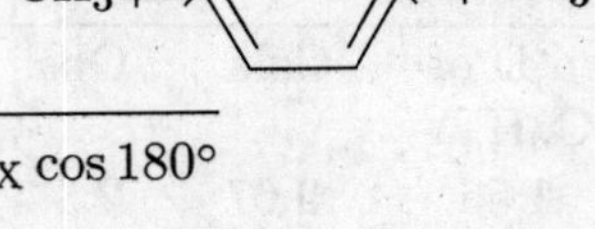

$$\mu_{P-XY} = \sqrt{\mu_X^2 + \mu_X^2 + 2\mu_X \cdot \mu_X \cos 180^\circ}$$

$$= \sqrt{2\mu_X^2 - 2\mu_X^2} = 0.$$

The calculations made above tell us that when two same groups of the above types are oriented para to each other, the resultant dipole moment of the para-disubstituted benzenes will be zero, as is the case with para-xylene or para-dinitrobenzene.

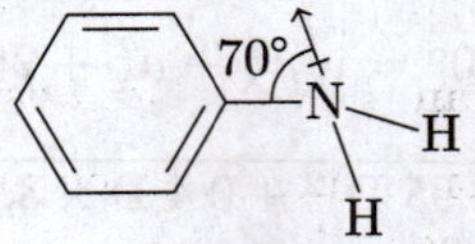

$\mu = 1.53$ D

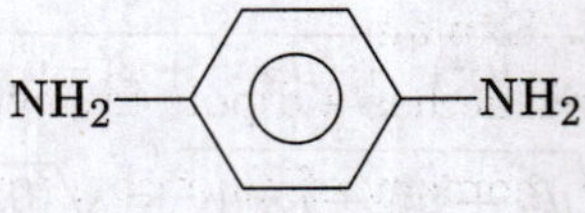

$\mu = 2.3$ D

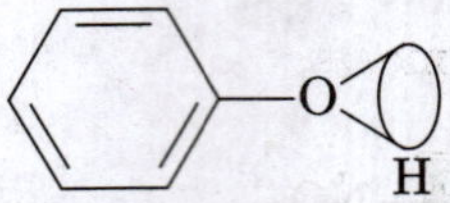

$\mu = 1.6$ D

$\mu = 1.73$ D

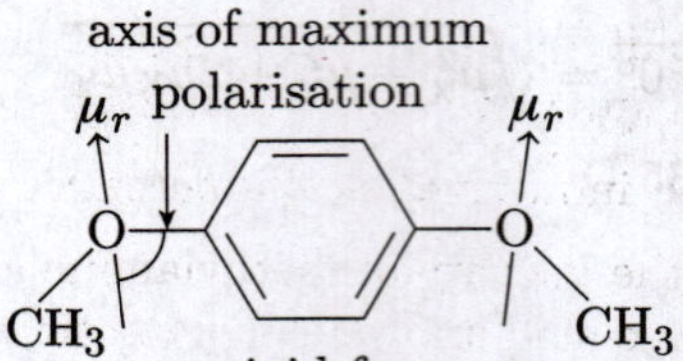

rigid form
$\mu = 1.73$ D

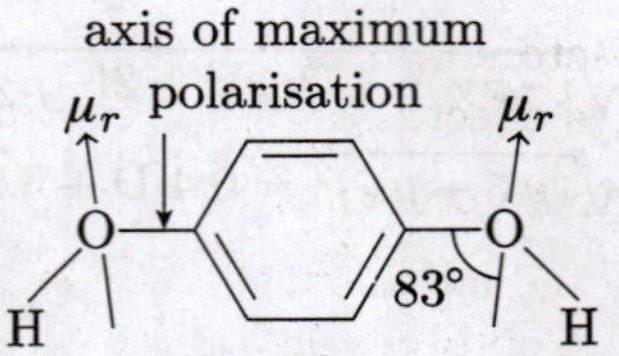

rigid form
$\mu > 0$ D

However, when the substituents are —OH, —OR, —NH_2, —NHR, —N R_2, —CO_2R, etc., the direction of the resultant group moments does not lie on the plane of the ring and makes an angle with its plane in which lies the axis of maximum polarisation (the line joining nuclear C and O or N or C atoms); for example the resultant group moment of —OH makes an angle of 83° with the axis of maximum polarisation which lies on the plane of the nucleus, and —NH_2 makes 70° angle. For substituted benzenes having these groups at para-positions have appreciable dipole moments.

Here the —OH group rotates around the O—C bond axis.

We shall have to use the following formula to get the resultant moment for these types of compounds.

$$\mu_{XY} = \sqrt{\mu_X^2 + \mu_Y^2 + 2\mu_X\mu_Y \cos\alpha \cdot \cos\phi_1 \cdot \cos\phi_2} \qquad (4.4)$$

where ϕ_1, ϕ_2 = angles which the group moments make with their axes of rotation and other symbols bear the same meanings as given earlier.

A list of calculated and observed moments are given in Table 4.2.

Table 4.2 List of molecular moments of benzene derivatives

Monosubstituted benzene				Moment of C_6H_4XY in debye					
				ortho		meta		para	
X	μ D of C_6H_5X	Y	μ D of C_6H_5Y	Calc.	Obs.	Calc.	Obs.	Calc.	Obs.
Cl	1.55	Cl	1.55	2.67	2.3	1.55	1.48	0	0
CH_3	0.4	CH_3	0.4	0.7	0.5	0.4	0.4	0	0
CH_3	0.4	NO_2	3.95	3.76	3.66	4.16	4.17	4.35	4.4
Cl	1.55	NO_2	3.95	4.91	4.3	3.44	3.4	2.4	2.5
CH_3	0.4	Cl	1.55	1.39	1.3	1.79	1.78	1.95	1.90
Br	1.52	Cl	1.55	2.67	2.2	1.54	1.5	0.03	0

Problem 4.25. Explain:

(i) *o*-Dinitrobenzene has a dipole moment lower than the calculated value.

(ii) The dipole moment of chlorobenzene and that of nitrobenzene are 1.55 D and 3.95 D respectively. The dipole moment of *p*-chloronitrobenzene is 2.4 D.

Solution: (i) Induced dipole moment is held responsible.

(ii) Cl and —NO_2 have opposite directions of moments.

Exercise

1. Name the factors that influence electron-availability within organic molecules. Write notes on: Inductive effect, Conjugative effect, Hyperconjugative effect, Steric effect, Aromaticity.
2. What is inductomric effect? Distinguish between inductive and inductomric effects.
3. Considering inductive effects only **(a)** compare the basicity of methylamine and cyclohexylamine. **(b)** Compare the acidity of trifluoroacetic acid and trimethylacetic acid.
4. Write down the conditions for writing the resonating structures of a chemical species. How can one judge which structures will be contributing in the resonance hybrid of an entity? How are resonance hybrids represented?
5. Why is butadiene more stable than ethylene? Explain this with M.O. and V.B. theories.
6. All the C—C bond lengths in benzene have equal value—explain with proper reasoning.
7. NO_3' is a symmetrical species—explain.
8. Write down the resonating structures of the following compounds:

$H_2C{=}O$ CH_3CHO

CH_3CH_2CHO $CH_3{-}CH{=}CH{-}CHO$

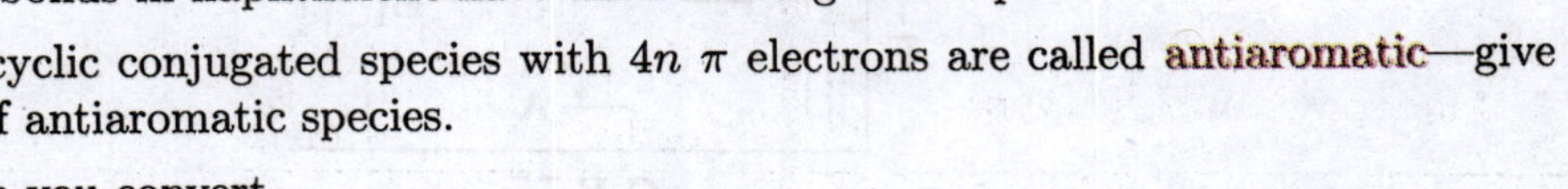

$CH_2{=}CH{-}Cl$ $CH{\equiv}C{-}Cl$

9. Write notes on resonance energy and tautomerism.
10. C—C bonds in naphthalene have different lengths—explain.
11. Planar cyclic conjugated species with $4n$ π electrons are called antiaromatic—give an example of antiaromatic species.
12. How can you convert—
    - (i) Tropolidene to tropolium bromide,
    - (ii) Tropolone into hydroxytropolium chloride,
    - (iii) Cyclopentadiene to sodium cyclopentadienide,
    - (iv) Diphenylcyclopropene to diphenylcyclopropenyl cation? Why are these conversions possible?
    - (v) Cyclooctatetraene to cyclooctatetraenyl dication.

Ans. (i)

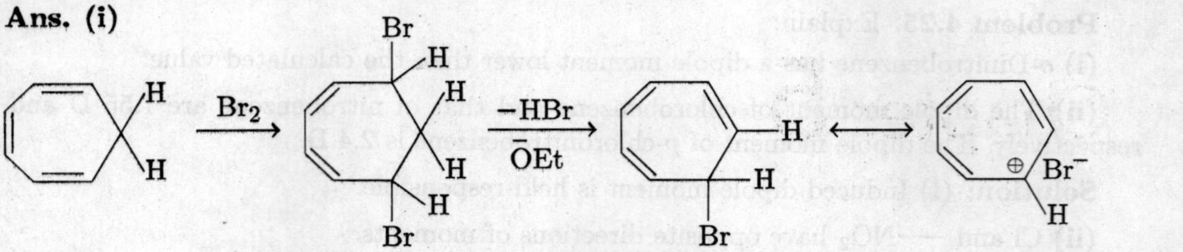

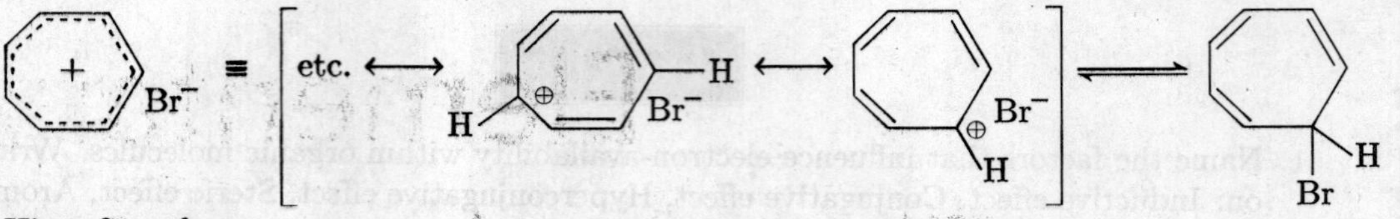

Hints for others

The substrates of the above conversions are non-aromatic compounds but all the products are aromatic compounds. Aromatic compounds being more stable than their non-aromatic counterparts, the substrates have a tendency to get converted into the products.

(v)

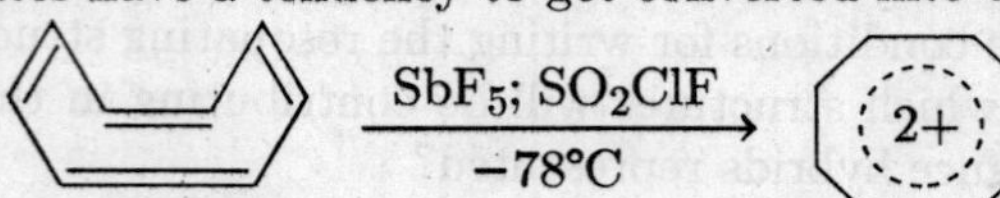

13. **Cyclooctatetraene forms salt with metallic potassium without evolving hydrogen gas—give an explanation for the reaction.**

Solution: **Cyclooctatetraene is an acyclic, 'tub'-shaped, non-planar, non-aromatic compound having 8 ($4n$; $n = 2$)π electrons. When it reacts with metallic potassium, one molecule of cyclooctatetraene takes up two electrons from two potassium atoms and produces a salt consisting of dianion of cyclooctatetraene ($C_8H_8^{--}$) and $2K^{\oplus}$. $C_8H_8^{--}$ is a cyclic planar ion having 10 delocalised π electrons, i.e., ($4n+2$) delocalised π electrons when $n = 2$. So it becomes an aromatic species and hence cyclooctatetraene gets stability through salt formation. The energy diagram for the π electrons is given below considering the electronic configuration of the dianion as $(\psi_A)^2(\psi_{E_1})^4(\psi_{E_2})^4$, but that of planar cyclooctatetraene is $(\psi_A)^2(\psi_{E_1})^4(\psi_{E_2})^2$.**

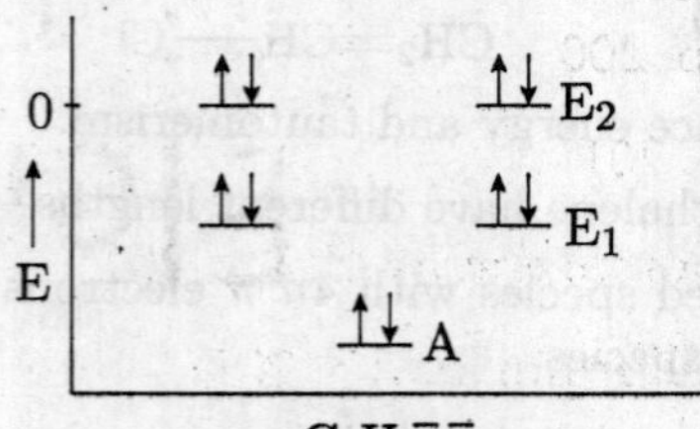

$C_8H_8^{--}$

To gain stability, thus, cyclooctatetraene reacts with K without evolving hydrogen gas as shown below:

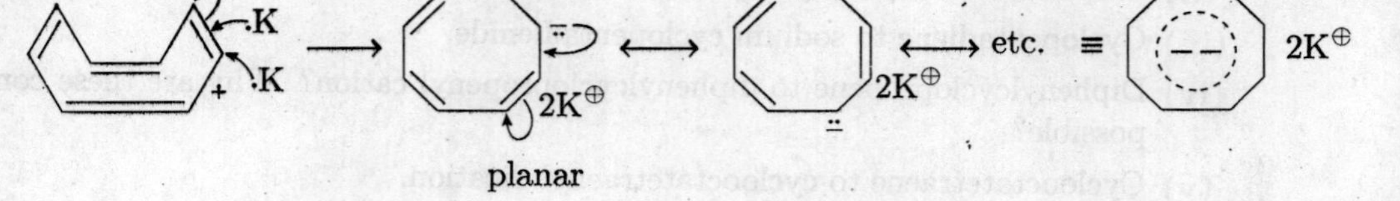

planar

Chapter 5

Elementary Thermodynamics

Objectives

◈ To understand:

* Internal energy [5.2] p. 199
* Enthalpy [5.3] p. 200
* Entropy [5.4] p. 201
* Gibbs energy or free energy [5.5] p. 202
* Position of equilibrium [5.6] p. 203
* The effect of temperature on equilibrium [5.7] p. 204

5.1 Introduction

Thermodynamics deals with the laws governing the energy interchanges in macroscopic systems, i.e., matters in bulk. There are three such laws. On the basis of these laws we can understand the energy relationships for physical and chemical processes. These laws also help predict:

(i) Whether a chemical reaction will occur or not when chemical substances are mixed together.

(ii) The amount of energy theoretically required by or released from the reaction.

(iii) The extent to which a chemical reaction will go over towards the products, i.e., the ratio of products to reactants when it goes to the equilibrium. In fact chemical thermodynamics concerns mainly with the position of the equilibrium of a reaction though it fails to predict the speed at which a chemical reaction proceeds.

Thus the understanding of the energy relation between reactants and products, the estimation of the energy requirements for a reaction to occur and the determination of the position of equilibrium may be called the thermodynamics of a reaction. To understand this, the following four important thermodynamic functions are to be discussed :

(a) Internal energy of a system (U), (b) Enthalpy of a system (H), (c) Entropy of a system (S) and (d) Gibbs energy or free energy (G) of a system.

In a chemical reaction, reactants are converted into products and consequently these four thermodynamic functions get changed. We are interested in those changes and so these functions are to be considered for reactants as well as for products. From this consideration the relation amongst the functions can be drawn which helps to understand the thermodynamics of a reaction.

In a chemical reaction, the system consists of some chemical substances; the reactants are the initial state of the system and the products are its final state.

5.2 Internal Energy (U)

The total energy associated with a system is called its internal energy. For a molecule it is a sum of several terms such as, nuclear energies (potential), electronic energies (potential and kinetic) and molecular energies (kinetic).

When heat is absorbed by or evolved from a system, it is utilised to change the internal energy of the system and to perform work. Thus,

$$q = \Delta U + W \quad \text{or,} \quad \Delta U = q - W. \tag{5.1}$$

where, q is the quantity of heat involved, W is the amount of work and ΔU is the change in the internal energy of the system, i.e.,

$$\Delta U = U_{\text{Final}} - U_{\text{Initial}} \tag{5.2}$$

In a chemical reaction U_{Final} is $U_{\text{Product}}(U_{\text{Prodt}})$ and U_{Initial} is $U_{\text{Reactant}}(U_{\text{React}})$ and the work is either electrical involving electron-transfer or pressure-volume type involving volume change. When the reaction takes place at a constant volume and without electrical work, W is zero and heat involved in the reaction at constant volume (q_v),

$$q_v = \Delta U = U_{\text{Prodt}} - U_{\text{React}} \tag{5.3}$$

Again, we know that the greater the energy content of a molecule the less is its stability, and vice versa. Thus for a reaction occurring at constant volume:

(i) If ΔU has a large positive value, the internal energy of the products must be greater than that of the reactants; in that case the product will be less stable than the reactant and consequently the amount of product molecules will be very small and for practical purposes it can be said that the reaction will not proceed.

(ii) If ΔU is zero, the reactant and the product have the same stability, and the ratio of the product to the reactant will naturally be 1.

(iii) If ΔU has a high negative value, the reaction will proceed practically irreversibly because the product is more stable than the reactant.

But only some simple reactions involving liquids and solids happen to occur at a constant or nearly constant volume. For these reactions only U values can be applied. On the other hand, most of the chemical reactions are carried out under a constant (atmospheric) pressure in open vessels rather than at a constant volume. For such a reaction another thermodynamic function, enthalpy, deserves consideration. This is not only applicable to reactions occurring at a constant pressure but also to reactions occurring at a constant volume.

5.3 Enthalpy (H)

The heat content of a system is called its **enthalpy**; it is, in fact, a modified internal energy function which is applicable to a system at a constant pressure. The internal energy, U, and enthalpy, H, of a system are related to each other by the equation $H = U + PV$, where P is the pressure and V is the volume of the system.

If a reaction is carried out at a constant pressure, then—

Enthalpy of the reactant, $H_{\text{React}} = U_{\text{React}} + PV_{\text{React}}$ and Enthalpy of the product, $H_{\text{Prodt}} = U_{\text{Prodt}} + PV_{\text{Prodt}}$. The difference in the enthalpies,

$$\begin{aligned}\Delta H &= H_{\text{Prodt}} - H_{\text{React}}\\ &= (U_{\text{Prodt}} + PV_{\text{Prodt}}) - (U_{\text{React}} + PV_{\text{React}})\\ &= (U_{\text{Prodt}} - U_{\text{React}}) + P(V_{\text{Prodt}} - V_{\text{React}})\\ &= \Delta U + P\Delta V \end{aligned} \tag{5.4}$$

If the reaction happens to occur at a constant volume also, then ΔV, the volume change $= 0$, i.e., $\Delta H = \Delta U$.

The difference in enthalpy of the product and that of the reactant is nothing but the heat energy involved in the system at a constant pressure, q_p.

Therefore, $q_p = \Delta H = H_{\text{Prodt}} - H_{\text{React}}$ 5.5

The condition for a chemical change to occur at a constant pressure is $\Delta H < 0$, i.e, ΔH must have a negative value; this implies that a chemical reaction has a tendency to occur by decrease in enthalpy. This is possible only when $H_{\text{Prodt}} < H_{\text{React}}$. The heat content of the reactant being greater than that of the product, heat is evolved in such a reaction and it is said to be an exothermic reaction.

The enthalpy of a molecule may be considered to reside in its bond energies. If so, the sum of the bond energies (enthalpies) of the product molecule minus the sum of the bond energies (enthalpies) of the reactant molecule gives the value of ΔH (with sign changed), the enthalpy change (bond energy values are in kcal/mole or kJ/mole; each such value should be multiplied by number of moles of the substance involved). If the bonds in the product are more stable than those in the reactant, heat is evolved in the reaction and ΔH becomes negative; on the other hand, if the reverse is true, heat is absorbed and ΔH becomes positive, i.e., the reaction is endothermic. From the enthalpy consideration it is clear that endothermic reactions cannot proceed spontaneously as ΔH values of the reactions require to be negative for spontaneity.

Practical experience tells us that there are some exothermic reactions in which product/reactant ratio is very low whereas in some endothermic reactions it is very large which is contrary to the enthalpy consideration. Thus decrease in enthalpy is not the only driving force for a chemical change; there is definitely certain other factor which too can control the spontaneity of the reaction, i.e., the possibility of a chemical reaction and this is entropy of the system.

5.4 Entropy (S)

Entropy, a term introduced by the second law of thermodynamics, is a measure of the disorder or randomness of a system. The more the disorder, the greater is the entropy; the greater the order, the less is the entropy. If a system tends to attain the most stable state, it must tend towards the maximum entropy, i.e., the maximum disorder. It may also be defined as a measure of the probability for the existence of a system in a chosen state; all other things remaining equal, if the existence of the products is more probable than the reactants in a chemical change, then the products must have a greater entropy, i.e., a greater disorder than the reactants. For such a reaction the change in entropy, S, must be a positive quantity and it is represented as

$$\Delta S = S_{\text{Prodt}} - S_{\text{React}} > 0 \tag{5.6}$$

In a chemical reaction which of the two, the reactant or the product, has the greater entropy, i.e., disorder may be predicted from the following considerations :

(i) If in a chemical change the number of moles of the product is greater than that of the reactants, the product molecules will be in a more disordered state than the reactant molecules as there has been an increase in the degrees of freedom of the product molecules; hence the entropy of the product must be greater than that of the reactant.

(ii) If the reactant is in the solid state and the product is in the liquid state, the product will have greater disorder and hence greater entropy than the reactant. This is because in a crystal lattice the molecules and ions have no freedom to move (except vibrational motion), whereas in a liquid state, they have every freedom to move.

(iii) For the same reason as stated in (ii) gaseous products must have higher disorder, i.e., entropy than the liquid reactants.

(iv) An open chain product must have greater disorder than a cyclic reactant as the former molecule can attend a greater number of conformations than the latter owing to rotation about one or more single bonds.

(v) In a solvated condition the system is less disordered than when a chemical species is removed from solvation; this is because many of the solvent molecules are then set free.

Thus we see that if a system wants to attain its most stable state, it must attain a state of minimum enthalpy (H) but maximum entropy (S). Relating these two opposing thermodynamic functions, a fourth such function, known as Gibbs or free energy function, needs consideration to determine the direction of the particular chemical reaction.

5.5 Gibbs Energy or Free Energy (G)

The energy available for a useful work from a system at a constant temperature and pressure is known as Gibbs energy or free energy of the system. At a constant temperature and pressure the enthalpy (H) of a system has two components : available energy for a useful work, i.e., Gibbs energy and unavailable energy for a useful work. Absolute temperature (T) of the system multiplied by its entropy (S) gives the energy unavailable for useful work. Thus

$$H = G + TS \quad \text{or,} \quad G = H - TS \tag{5.7}$$

For a chemical reaction :

$$G_{\text{React}} = H_{\text{React}} - TS_{\text{React}}$$
$$G_{\text{Prodt}} = H_{\text{Prodt}} - TS_{\text{Prodt}}$$

The change in free energy of the system for a chemical change is thus,

$$\begin{aligned}\Delta G &= (H_{\text{Prodt}} - TS_{\text{Prodt}}) - (H_{\text{React}} - TS_{\text{React}}) \\ &= (H_{\text{Prodt}} - H_{\text{React}}) - T(S_{\text{Prodt}} - S_{\text{React}}) \\ &= \Delta H - T\Delta S\end{aligned} \tag{5.8}$$

During a chemical change at a constant pressure, the reactants and products remain at the same temperature. So the free energy change can be applied for chemical processes. In fact, it has more practical use than entropy alone though the latter is more fundamental than the free energy.

The important uses of the free energy change are:

1. to obtain spontaneity criterion of a chemical change and
2. to determine the position of the equilibrium of the reaction.

Spontaneity of a reaction

For a chemical process carried out at constant temperature and pressure, the condition for the spontaneous change is given by $\Delta G < 0$, i.e., the value of ΔG must be negative. When thermodynamic equilibrium is reached in the system, $\Delta G = 0$. This means a chemical reaction takes place spontaneously by decrease in free energy, i.e., the system wants to attain a state of minimum G. Thus for an isothermal chemical change:

(i) If ΔG is negative and very large, the reaction will be practically irreversible.

(ii) If ΔG is negative but small, the reaction will proceed spontaneously.

(iii) If ΔG has a large positive value, the reaction will not at all take place under the condition applied for.

(iv) If ΔG has a very small positive value, the reaction will not be spontaneous. For such a reaction to occur energy is to be supplied.

As ΔG determines the direction and the fate of a chemical reaction, it is sometimes called chemical potential.

However, ΔG is composed of two terms : ΔH and $T\Delta S$. Therefore, the sign of ΔG depends on the sign of ΔH and $T\Delta S$. Thus:

(i) For an exothermic reaction, ΔH is negative and it alone determines the sign of ΔG; it is then negative and the reaction proceeds.

(ii) For an endothermic reaction, ΔH is positive and then the sign of ΔG depends on that of $T\Delta S$. If $T\Delta S$ is positive, ΔG can be made negative by carrying out the reaction at a high temperature as $T\Delta S$ increases with increasing T.

(iii) If, for a reaction, ΔH is positive and ΔS is negative then the reaction will not at all take place under the given condition; but if one couples this reaction with another having so large negative ΔG value that the sum of the two ΔGs becomes negative, then the reaction can be made to proceed.

5.6 Position of Equilibrium

In most of the cases the chemical reactions are reversible. The extent to which a chemical reaction proceeds depends on the equilibrium constant K, for the reaction.

For a reaction as written below:

$$xA + yB \rightleftharpoons wC + zD$$

$$K_C = \frac{[C]^w \cdot [D]^z}{[A]^x \cdot [B]^y} \quad \text{and} \quad K_P = \frac{p_C^w \cdot p_D^z}{p_A^x \cdot p_B^y} \tag{5.9}$$

where the symbols within crochets ([ ]) represent the concentrations of the products and those of the reactants in mole per litre when equilibrium is reached; of course, here activity coefficients are neglected. p^n terms are partial pressures of the involved substances. Conventionally the equilibrium concentration and equilibrium partial pressure terms of the products of a reaction are written in the numerator of the expression and those of the reactants in the denominator as shown above. However, the expression shows that a large value of K indicates that the reaction $xA+yB \rightleftharpoons wC+zD$ goes to completion. In practice, if the final equilibrium mixture contains less than about 0.1% of the reactants, the reaction may be considered to go to completion. Since in most of the organic reactions pressure and temperature of the products at the end of the reaction are the same as those of the reactants at the start, the equilibrium constant, K, can be related to ΔG, the free energy change, by the equation

$$\Delta G = \Delta G^\circ + RT \ln K. \tag{5.10}$$

Equation 5.10, referred to as the reaction isotherm, gives the free energy change of a reaction in terms of the free energy in the standard state and the reactants and product at any constant temperature T.

Here R is the universal gas constant. T is temperature in Kelvin scale while ΔG° is the standard free energy change of the reaction with the components in their standard states. The standard state for a gas is generally the corresponding ideal gas at 1 atm pressure. The standard state for a solution (l) is normally taken as the ideal 1 M solution. The temperature for the standard state is usually 298 K. The standard free energy change (ΔG°) of a system is defined as the free energy change (ΔG) of an ideal solution in which the concentration of each reactant and product is 1 M at 298 K. Under this condition $RT \ln K$ is zero and ΔG becomes equal to ΔG°. When the reaction reaches equilibrium, the free energy change of the system, $\Delta G = 0$. So, the equation 5.10 becomes

$$0 = \Delta G^\circ + RT \ln K \quad \text{or,} \quad \Delta G^\circ = -RT \ln K. \tag{5.11}$$

Now, knowing the standard free energy values of the components we can easily calculate the equilibrium constant of a reaction and thereby we can predict whether the given set of reactants will likely to yield desired set of products or not. Thus:

(i) If $\Delta G^\circ = 0$, $K = 1$; then the reaction proceeds in both the directions equally in the absence of an outside force.

(ii) If $\Delta G^\circ < 0$, i.e., negative, $K > 1$; then at equilibrium concentrations of the products will be greater than those of the reactants, and the equilibrium lies over to the right.

(iii) If $\Delta G^\circ > 0$, i.e., positive, then $K < 1$; in this case the equilibrium concentration of the reactants will be greater than those of the products and the equilibrium lies over to the left.

Table 5.1 Relation between ΔG° in kJmol^{-1} at 298 K for reactions of the type $A \rightleftharpoons B$

ΔG°	−1.71	−2.72	−3.42	−3.97	−5.85	−7.52	−11.29	−17.14	−22.99
K	2	3	4	5	10	20	100	1,000	10,000
% of B	67	75	80	83	91	95	99	99.9	99.99

Table 5.1 shows that if in a reaction $\Delta G^\circ \geq -17.14$ kJmol^{-1}, the product yield will be 99.9% or more and we shall conclude that the reaction practically goes to completion.

However,

$$\Delta G^\circ = -RT \ln K \quad \text{or,} \quad \ln K = -\frac{\Delta G^\circ}{RT}, \quad \text{i.e.,} \quad K = e^{-\Delta G^\circ/RT}$$

$$\text{or,} \quad 2.303 \log K = -\frac{\Delta G^\circ}{RT} \quad \text{or,} \quad \log K = -\frac{\Delta G^\circ}{2.303 \times RT}$$

$$\therefore \quad K = 10^{-\Delta G^\circ/2.303RT} \qquad (5.12)$$

5.7 The Effect of Temperature on Equilibrium

We know,

$$\Delta G^\circ = -RT \ln K$$

$$\text{or,} \quad RT \ln K = -\Delta G^\circ = -(\Delta H^\circ - T\Delta S^\circ) = -\Delta H^\circ + T\Delta S^\circ$$

$$\text{or,} \quad \ln K = -\frac{\Delta H^\circ}{RT} + \frac{T\Delta S^\circ}{RT} = -\frac{\Delta H^\circ}{RT} + \frac{\Delta S^\circ}{R}$$

$$\text{or,} \quad \frac{d}{dT}(\ln K) = \frac{\Delta H^\circ}{RT^2} \quad \left[\because \frac{\Delta S^\circ}{R} = \text{const.}\right] \qquad (5.13)$$

Equation 5.13 is known as Van't Hoff isochore. It shows how the equilibrium constant varies with temperature : in an exothermic reaction ($\Delta H^\circ = -$ve) the equilibrium constant decreases with the increase in temperature and that in an endothermic reaction (with positive ΔH°) increases. This is actually the mathematical expression of the application of Le Chatelier's principle to temperature changes. However, using the equation and determining equilibrium constants at a series of temperatures, we can determine the value of ΔH° of a reaction.

Problem 5.1. Calculate ΔH° of the reaction and predict whether the reaction will take place or not.

$$CH_4(g) + Br_2(g) \rightarrow CH_3Br(g) + HBr(g)$$

Given, $D_{Br—Br} = 192.35$ kJ mol^{-1}, $D_{C—H} = 414$ kJ mol^{-1}

$D_{C—Br} = 276$ kJ mol^{-1} and $D_{H—Br} = 363.8$ kJ mol^{-1}

Solution: To break one mole of C—H bond and one mole of Br—Br bond (192.35 + 414) or 606.35 kJ of energy is to be supplied.

To form one mole of C—Br bond and one mole of H—Br, bond (276 + 363.8) or 639.8 kJ energy will be liberated.

Thus, the heat change = 639.8 kJ −606.35 kJ = 33.45 kJ

$\therefore\ \Delta H^\circ = -33.45$ kJ

Since ΔH° is negative, the reaction will occur.

Problem 5.2. Calculate ΔG° and K for the reaction given in the Problem 5.1. In which direction the equilibrium will lie?

Solution: We know, $\Delta G^\circ = \Delta H^\circ - T\Delta S^\circ$.

Since the reactant side and the product side involve the same number of moles having similar molecular geometry and state of aggregation (gaseous), the ΔS° is negligible, so we can ignore $T\Delta S$ term. Therefore,

$$\Delta G^\circ = \Delta H^\circ = -33.45 \text{ kJ}$$

Eqn. 5.12 shows $K = 10^{-\Delta G^\circ/2.303RT}$

$$\therefore\quad K = 10^{-\Delta H^\circ/2.303RT} = 10^{33.45\times1000/2.303\times8.314\times298}$$

$$= 10^{5.862} = 727779.80.$$

The large value of K and $\Delta G^\circ > -17.14$ kJ indicate that the equilibrium will lie far over to the right and the reaction will go to completion.

Exercise

1. Calculate ΔH° for the reaction

   $CH_4 + Cl_2 \rightarrow CH_3—Cl + HCl$; given that

   $D_{C—H} = 104$ kcal mol^{-1} $\quad D_{Cl—Cl} = 58$ kcal mol^{-1}

   $D_{C—Cl} = 83.5$ kcal mol^{-1} and $D_{H—Cl} = 103$ kcal mol^{-1}

2. State with reason whether the following reactions have positive or negative ΔS°:

   (i) $Br_2 + CH_2 = CH_2 \rightarrow CH_2Br—CH_2Br$

   (ii) 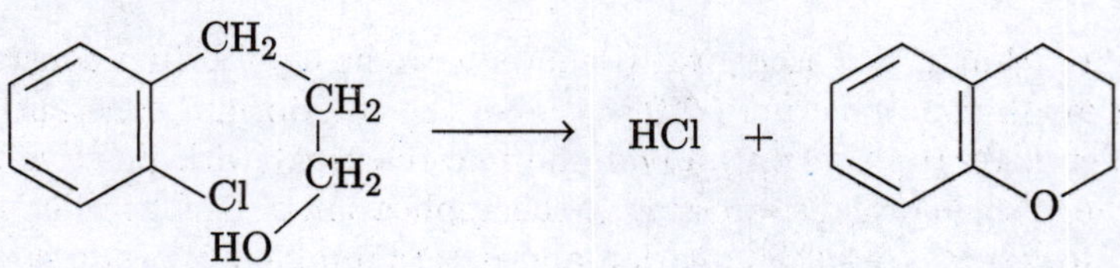

3. Predict whether the following reaction has a positive or negative ΔG° :

$$HI + CH_3I \rightarrow I_2 + CH_4.$$

   Given that bond dissociation energies in kcal/mole H—I = 71, C—I = 57, I—I = 36 and C—H = 99. In which direction the equilibrium will lie?

Chapter 6

Kinetics of a Reaction

Objectives

- To understand:
    - The rate, the rate constant and the rate equation of a reaction [6.2] p. 209
    - Kinetics of a reaction in the presence of a catalyst [6.3] p. 211
    - Collision theory for the reaction mechanism [6.4.1] p. 212
    - Transition state theory for the reaction mechanism [6.4.2] p. 214
- To determine the activation energy of a reaction [6.5] p. 216
- To learn the drawing of energy profiles for reactions [6.6] p. 217
- To correlate nature of products and energetics of a reaction [6.7] p. 220

6.1 Introduction

It has already been discussed that negative ΔG value is a necessary condition for a reaction to occur under certain set of conditions. It has also been stated that the value of K, the equilibrium constant of a reaction, determines how far the reaction will proceed. These are not all for a reaction to occur. The ΔG value of the reaction between cellulose and oxygen to carbon dioxide and water is negative, still we handle papers, cotton cloths in air and do not find them burning spontaneously. This is because, these two quantities tell nothing about how fast or how slow the reactants are converted into the products. In fact, the speed of a reaction is another important factor for a reaction to occur, which, in its turn, depends on the temperature of the reaction amongst other criteria. Ionic reactions like acid-base neutralisation are very fast, while reactions of covalent molecules are slow. Moreover, the starting materials may react in several ways and may give a number of different products. In fact, under a given set of conditions the fastest reaction gives rise to the predominating product, provided it is also thermodynamically stable. Of course, if one adjusts the reaction variables, i.e., the concentration of the reactants, the temperature, the solvent, the catalyst, etc. the desired product can be obtained preferentially. Study of the speed of a reaction and all these which enable us to understand the mechanism of the reaction belong to a special branch of physical chemistry known as chemical kinetics. This part is meant for the study of the kinetics of a reaction.

6.2 The Rate of a Reaction

The velocity of a reaction is called its rate. It is the amount of chemical change taking place per unit time. Chemical change means conversion of the reactant into the product. The rate of a reaction is dependent on the following factors as obtained from experiments:

(a) Concentration of the reactant: As the time proceeds, the concentrations of the reactants decrease and those of the products increase till the equilibrium of the reaction is reached. Hence the rate is generally expressed either as the decrease in concentration of the reactants or as the increase in concentration of the products. If C is the concentration of a reactant and x is that of the product at a constant temperature at any time t, the rate of the reaction is either $-dC/dt$ or dx/dt; the negative sign of the term dC/dt indicates the decrease in concentration of the reactant (dC/dt is negative and $-dC/dt$ is positive). The concentration is expressed in g-mole/litre and time in second; so the unit of the rate is mol/litre/second or mole litre^{-1} second^{-1}.

The dependence of the rate of a reaction on the concentrations of the reactants is quantitatively represented by a new term, the order of the reaction. The order of a reaction is a number obtained by summing up the powers of the concentration terms involved in the reaction rate equation. If the rate of a reaction is dependent on the single formula unit of the reactant A, then the rate of the reaction is proportional to C_A, the concentration of A, at a given temperature. Mathematically,

$$\text{the rate of the reaction} \propto C_A \text{ or, the rate} = k \cdot C_A.$$

Here the power of the concentration term is unity and the reaction is said to be 1st order, i.e., its order is one with respect to A. On the other hand, if the rate of the reaction is found to be proportional to C_A^2, the order of the reaction will be 2 with respect to A. Similarly, if the rate is proportional to the third power of the concentration of A, it is a third order reaction with respect to A.

If there be a reaction involving A, B and D as reactants and if the rate of the reaction is found to be as follows:

$$\text{the rate} = kC_A^{n_1} \cdot C_B^{n_2} \cdot C_D^{n_3} \tag{6.1}$$

Eqn. (6.1) is said to be the rate equation where k, the proportionality constant, is called the rate constant or specific reaction rate. Here the order of the reaction is $(n_1 + n_2 + n_3)$ and the reaction is said to be $(n_1 + n_2 + n_3)$th order.

Equation (6.1) shows that how the rate constant, k, relates the rate of the reaction to the actual concentrations of the reactants. k, infact, is a numerical measure of how fast a reaction occurs when the reactants are brought together. The greater the value of k, the faster is the reaction. The order of a particular reaction determines its unit. For a first order reaction the unit is $\sec^{-1}$,

Since $k = \dfrac{2.303}{t} \log \dfrac{C}{C-x}$, where C = initial concentration and $C - x$ is the concentration of the reactant at a time t.

For a second order reaction the rate constant, $k = \dfrac{1}{t} \dfrac{x}{C(C-x)}$; hence the unit of k is litre/mole/second.

However, for a reversible reaction as shown,

$$\mathrm{A} \underset{k_2 \text{ or } k_{-1}}{\overset{k_1}{\rightleftharpoons}} \mathrm{B},$$

where k_1 and k_2 (or k_{-1}) are the rate constants for the forward and reverse reaction respectively.

The rate equation becomes :

Either, $-dC_A/dt = k_1C_A - k_{-1}C_B$

or, $-dC_A/dt = k_1C_A - k_{-1}(C_{Ao} - C_A)$, (6.2)

where C_{Ao}, is the initial concentration. At the beginning, $C_{Ao} = C_A$; then Eqn. (6.2) becomes:

$-dC_A/dt = k_1C_A$ and the reaction becomes 1st order with respect to A. As the reaction proceeds, concentration of A decreases and the term $C_{Ao} - C_A$ becomes more and more important until $k_1C_A = k_{-1}(C_{Ao} - C_A)$. At this time, $-dC_A/dt = 0$.

This means the concentrations of A and B do not change thereafter although these two species are in dynamic equilibrium. The order of the reaction will no longer be 1st order and the extent of deviation depends upon the numerical values of k_1 and k_{-1}

(b) Activation energy: Every reaction generally involves an energy barrier that must be overcome in going from reactant to products. Chemical units (molecules, atoms or ions) usually repel each other. To bring them together for breaking and remaking of bonds, force must be applied. This results to an energy barrier which is called the activation energy or the energy of activation. We shall discuss this later on.

(c) Particle size: With decreasing particle size, i.e., increasing surface area of solids the reaction rate increases.

(d) Temperature: The empirical rule is that the value of the rate constant k becomes double to triple for every 10 °C rise of temperature.

Arrhenius showed that the rate constant, k, of a reaction increases exponentially with temperature. He gave the relation as:

$$k = Ae^{-E/RT} \quad \text{or,} \quad \frac{d\ln k}{dT} = \frac{E}{RT^2} \tag{6.3}$$

where A is a frequency factor, E is activation energy.

(e) Catalyst: A catalyst increases the rate but an inhibitor decreases the same though they have nothing to do with the position of the equilibrium.

The order of a reaction is an experimental property and hence it must be determined experimentally. Let us suppose that we want to determine the order of the reaction involving A and B as reactants. At first all the variables, the temperature, the solvent, the catalyst and the concentration of B are kept constant and the rates of the reactions for different concentrations of A are measured. It is found that the rate doubles when the concentration of A is doubled and it becomes half when the concentration of A is half. This indicates that the rate of the reaction is directly proportional to the concentration of A. Mathematically,

$$\text{the rate} \propto C_A.$$

Now keeping the concentration of A and other variables constant and varying the concentration of B, it is found that the rate becomes four times when the concentration of B is doubled and that becomes one-fourth when the same is made half. Thus we can say that the rate of the reaction is proportional to C_B^2.

Hence the overall rate of the reaction is proportional to $C_A \cdot C_B^2$

or, the overall rate $= kC_A \cdot C_B^2$.

Therefore, the order of the reaction is $(1+2)$ or 3. On the other hand, if one finds no change in the rate with the change of the concentration of B, the reaction is said to be a zero order reaction with respect to B; then

$$\text{the rate of the reaction} = kC_A \cdot C_B^0 = kC_A.$$

Consequently the order of the reaction will be $(1+0)$ or 1, i.e., the reaction follows the first order kinetics.

In a multi-step reaction, the speeds of the reaction on all steps are not the same and cannot be measured. One of the steps will be the fastest and other one will be the slowest. The slowest step of a multi-step reaction is said to be the rate-determining step (R.D. step). The number of molecules involved in the rate-determining step may be called molecularity of a reaction. If one molecule is involved, it is called unimolecular reaction. When two or three molecules are involved in the rate-determining step, the reactions are termed as bimolecular or termolecular respectively.

Thus molecularity of a reaction is a number just like the order of a reaction; but the former is a theoretical prediction considered to satisfy the experimental fact; whereas the latter is obtained from the experimental result. The molecularity of a reaction is always an integer, while the order of a reaction may be an integer, zero or a fraction. For some simple reactions the order and molecularity are one and the same, but they are not so for most of the reactions.

6.3 Kinetics of a Reaction in the Presence of a Catalyst

Let us now describe the kinetics of a reaction in the presence of a catalyst D. Let us assume the reaction $A \rightarrow B$ will be faster in the presence of D through the formation of an intermediate product AD. A very readily combines with D but the intermediate AD slowly converts to the final product $B+D$.

$$\mathrm{A + D} \underset{k_{-1}}{\overset{k_1\ (\text{fast})}{\rightleftharpoons}} \mathrm{AD} \underset{\text{slow}}{\overset{k_2}{\longrightarrow}} \mathrm{B + D} \qquad \text{(Expression a)}$$

The above expression tells us that

$$-\frac{dC_A}{dt} = k_2 C_{AD}. \tag{6.4}$$

Since AD changes to B and D slowly but AD forms very fast, the concentration of the intermediate AD very negligibly changes in the course of the reaction and practically we can say that the concentration of AD is constant and it is in the steady state. So from the steady state postulates we can write,

$$\frac{dC_{AD}}{dt} = 0. \tag{6.5}$$

Using the expression and the law of mass action we can write:

$$\frac{dC_{AD}}{dt} = k_1 C_A C_D - (k_{-1} + k_2) C_{AD}, \tag{6.6}$$

where $k_1 C_A C_D$: the rate at which AD is formed and $(k_{-1} + k_2)C_{AD}$: the rate at which AD breaks down. According to Eqn. (6.5), the left side of Eqn. (6.6) is zero. Therefore,

$$k_1 C_A C_D - (k_{-l} + k_2) C_{AD} = 0$$

or, $$C_{AD} = \frac{k_1}{(k_{-1} + k_2)} C_A C_D.$$

Putting this value in Eqn. (6.4)

$$-\frac{dC_A}{dt} = \frac{k_1 k_2}{k_{-1} + k_2} C_A C_D. \tag{6.7}$$

Equation (6.7) clearly shows that the rate is dependent on the concentration of the catalyst. However, usually the formation of AD is the slow step while decomposition of AD is the fast one.

6.4 Theory of Reaction Mechanism

The actual changes undergone by the reacting chemical units in order to reach the product state gradually may be called the reaction pathway or reaction coordinate or reaction mechanism. This involves breaking and remaking of bonds.

However, there are two general theories to explain the mechanism of reactions, their rates, order, etc.; they are: (i) Collision theory and (ii) Transition State theory.

6.4.1 Collision theory

This theory on reaction mechanism tells us that:

1. Collision amongst the reacting species (atoms, ions, molecules) is a necessary condition for a chemical reaction to occur.
2. The number of collisions per unit time, i.e., the collision frequency is a factor for the rate of a reaction. The higher the collision frequency, the higher is the rate of the reaction. The collision frequency, in its turn, depends on the following factors:
   (a) The concentration or pressure of the reactants: With increasing concentration the collision frequency increases. The greater the concentration or pressure, the faster will be the reaction.
   (b) Size of the species: The greater the size of the chemical units (atoms, ions or molecules), the greater is the collision frequency.

(c) Temperature of the reacting species: The higher the temperature of the colliding particles, the greater will be their speed. The faster the speed of the species, the greater is the collision frequency. The speed, in its turn, depends on the temperature and on the formula weight of the species.

3. Energy of activation: Not all such collisions give rise to a chemical change; only a fraction of them is effective. A chemical change occurs through those collisions which take place between the reactants containing certain critical amount of energy above the average value. This critical amount of energy which is necessary for a mole of reactant to undergo a chemical change is called energy of activation and the species impregnated with the energy of activation is said to be the activated species (complex). Therefore, the activated species only undergoes chemical change on collision and hence the rate of a reaction is dependent on the energy of activation of the reactants. The relationship between the energy of activation (E_{act}), and fraction of collisions with energy greater than that is:

$$e^{-E_{act}/RT} = \text{Fraction of collisions with energy greater than } E_{act},$$

where E_{act} = Energy of activation

e = Base of natural logarithms

R = Gas constant

T = Temperature in Kelvin scale.

The greater the energy of activation, the more difficult is the reaction to occur.

4. Orientation of activated units: Collisions involving activated species give rise to a chemical reaction, only if the orientations of the activated species are proper. For example, AB molecules decompose to A_2 and B_2, when activated AB molecules collide with each other having orientations as given below:

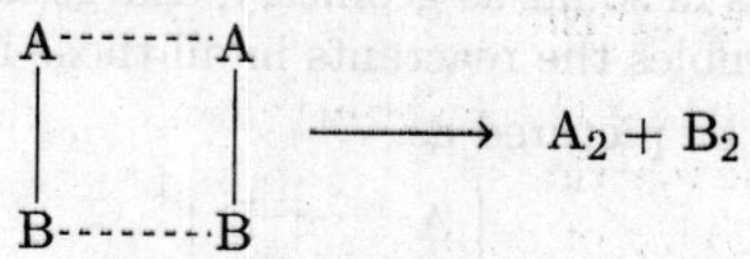

If, on the other hand, the orientations of two colliding AB molecules are as shown below, decomposition will not take place at all.

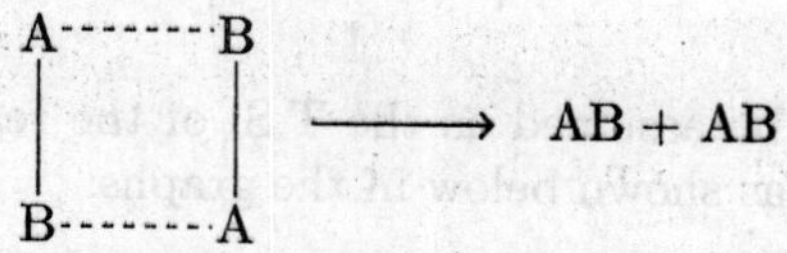

Thus proper orientation of colliding species is a factor for calculating the rate of a reaction. This factor may be called probability factor or orientation factor.

Taking all the considerations as stated above the rate of a chemical reaction may be represented by the equation :

$$\text{the rate} = \text{collision frequency} \times \text{energy factor} \times \text{orientation factor}.$$

Amongst these, the collision frequency and the energy factor are very important. The greater the energy factor, the slower will be the reaction. By varying reaction variables one can change the collision frequency and energy factor and thus one is able to get a particular product preferentially amongst others from the given reactants.

6.4.2 Transition state theory

A chemical reaction may be considered as a gradual change from reactants to products via a very unstable state called a transition state. Let us consider the following reaction in which all the bond breaking and bond making processes occur simultaneously:

$$A—B + C—D = A—D + B—C.$$

To obtain AD and BC from AB and CD we require to break A—B and C—D bonds. The bond breaking process requires absorption of energy from the surroundings and energy will be given off to the surroundings when A—D and B—C bonds will form. By absorbing energy some of the A—B and C—D molecules will be activated being impregnated with certain amount of energy much greater than the average energy of the reactants. The energy of the activated species (activated complex which will undergo the reaction) relative to the average energy of the reactant molecules is called the energy of activation. In reality, the bond dissociation energy is always greater than the activation energy. Therefore, the reaction proceeds so that the energy of cleavage of the bonds (A—B and C—D) is partly compensated by the energy liberated during the formation of new bonds (A—D and B—C). Pictorially it may be shown as follows: As atoms B and C approach each other partial bonds A······D and B······C are forming and A—B and C—D bonds are breaking partially. During this process energy transfers partly from one section to the other. As the reacting atoms approach, sooner or later, all bonds will equally become loose. This state is called the Transition State. The transition state is not at all a discrete molecular species but it possesses a definite geometry and charge distribution. It is highly unstable and transient and has both partial and complete bonds (in the case of polyatomic species as reactants); some bonds may also be forming and some bonds may be breaking in this stage. T.S.s are of very high energy and for a single step reaction the T.S. possesses the highest energy amongst reactants, products and the T.S. According to Hammond, in an endothermic reaction, the T.S. resembles the products in so far as geometry, charge distribution and energy content are concerned, while the T.S. resembles the reactants in all those in an exothermic reaction.

However, the reaction may be pictured as.

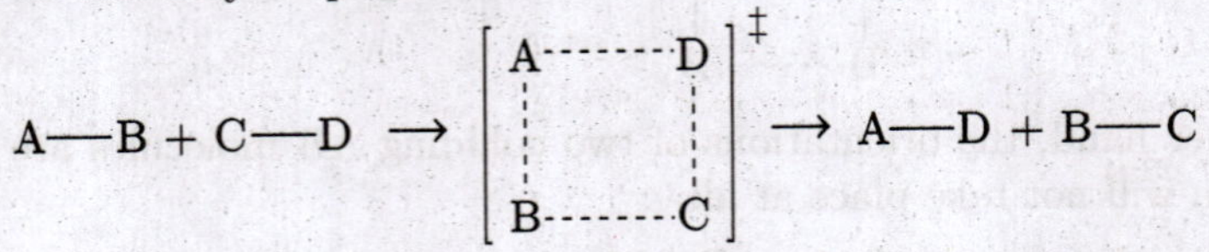

I

where ‡ refers to T.S. I may be assumed as the T.S. of the reaction. Its energy content is the summit of the energy barrier as shown below in the graphs:

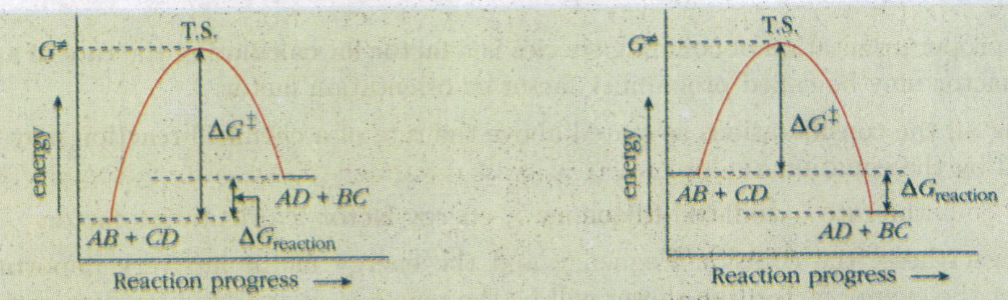

(a) Energy profile for a one-step endothermic reaction (b) Energy profile for a one-step exothermic reaction

Fig. 6.1

The basic idea of the transition state theory is as follows :

(a) The reacting species must be forced to come very close to each other when each of the species loses its individual identity and assumes the most unstable state containing high energy, known as transition state, which equilibrates the reacting species and decomposes into the product.

(b) The energy required for the creation of the transition state from the reactants is called the energy of activation. The transition state (T.S.) is practically the activated complex in the collision theory. Only those chemical units may form the T.S. which acquire the energy of activation.

The reactants that do not acquire the energy of activation fall back to compensate their store of energy. The ascent to the summit is the most difficult task, but as soon as the chemical units reach the peak, they irreversibly roll down to the product side. Thus the theory assumes that all the T.S.s are converted to the product at the same rate and thus the rate constant of the reaction depends on the position of the equilibrium between reactants and the T.S. only.

$$K^{\ddagger} = \frac{\left[\begin{array}{ccc} A & \cdots & D \\ \vdots & & \vdots \\ B & \cdots & C \end{array}\right]^{\ddagger}}{[A - B][C - D]} \quad \text{where } \ddagger \text{ refers to T.S.}$$

(c) The higher the number of chemical units that reach the summit, the higher is the rate of the reaction. This, in its turn, tells us that the rate constant of the reaction may be represented as the product of the two values, $a^{\ddagger}$ and $K^{\ddagger}$, i.e.,

$$k = a^{\ddagger} K^{\ddagger} \tag{6.8}$$

$a^{\ddagger}$ is a constant of a monomolecular conversion of an activated complex (transition state) into reaction products; it is measured in frequency units and at 300 K the value of $a^{\ddagger}$ is 6.3×10^{12}. sec^{-1}.

$K^{\ddagger}$ is related to $\Delta G^{\ddagger}$, the free energy of activation (the free energy required by a reacting system to pass into the T.S.), by the usual relation.

$$\Delta G^{\ddagger} = -RT \ln K^{\ddagger}$$

or, $$K^{\ddagger} = e^{-\Delta G^{\ddagger}/RT}. \tag{6.9}$$

Putting the value of $K^{\ddagger}$ to equation (6.8) we get,

$$k = a^{\ddagger} \cdot e^{-\Delta G^{\ddagger}/RT}. \tag{6.10}$$

Equation (6.10) shows that as the free energy of activation increases, the rate constant decreases. This means the lower the energy barrier, the higher is the rate constant of the reaction. However,

$$\Delta G^{\ddagger} = \Delta H^{\ddagger} - T\Delta S^{\ddagger}, \tag{6.11}$$

where $\Delta G^{\ddagger}$ and $\Delta S^{\ddagger}$ are enthalpy of activation and entropy of activation respectively.

Hence equation (6.10) may be written as:

$$\begin{aligned} k &= a^{\ddagger} e^{(-\Delta H^{\ddagger} + T\Delta S^{\ddagger})/RT} \\ &= a^{\ddagger} e^{(T\Delta S^{\ddagger} - \Delta H^{\ddagger})/RT} \\ &= a^{\ddagger} e^{\Delta S^{\ddagger}/R} \cdot e^{-\Delta H^{\ddagger}/RT}. \end{aligned} \tag{6.12}$$

For chemical reactions at room temperature the activation energy usually varies from 42 to 125 kJ mol^{-1}.

(d) The number of chemical entities involved in the T.S. of the R.D. (rate-determining) step is called the molecularity of the reaction.

Unimolecular reactions involve one molecule or ion in the said T.S. Similarly bimolecular and termolecular reactions involve two and three chemical entities respectively in the T.S. of the R.D. step.

The merit of the Transition State Theory is that it rationalises the orientation factor of the Collision Theory. A chemical reaction in which the T.S. is highly ordered will have a large negative entropy of activation ($\Delta S^{\ddagger} \ll 0$, i.e., $-$ve) corresponding to a small orientation factor.

6.5 Exothermic and Endothermic Reactions and the Determination of Activation Energy

Let us now recall Arrhenius equation (6.3). For a reversible single-step reaction ($A \rightleftharpoons B$) having $E_f^{\ddagger}$ and $E_r^{\ddagger}$ as the energies of activation for the forward and the reverse reactions respectively and k_1, and k_{-1} are the corresponding rate constants at a temperature T, then,

$$\frac{d}{dT}(\ln k_1) = \frac{E_f^{\ddagger}}{RT^2} \quad \text{and} \quad \frac{d}{dT}(\ln k_{-1}) = \frac{E_r^{\ddagger}}{RT^2}.$$

Integrating both the equations we get,

$$\ln k_1 = \frac{-E_f^{\ddagger}}{RT} + Z_1 \text{ (const.)} \quad \text{and} \quad \ln k_{-1} = \frac{-E_r^{\ddagger}}{RT} + Z_2 \text{ (const.)}$$

By subtraction,

$$\ln \frac{k_1}{k_{-1}} = \frac{E_r^{\ddagger} - E_f^{\ddagger}}{RT} + Z; \ (Z_1 - Z_2 = Z, \text{ a const.}) \qquad (6.13)$$

Again at any time t,

$$-\frac{dC_A}{dt} = k_1 C_A \quad \text{and} \quad -\frac{dC_B}{dt} = k_{-1} C_B.$$

If the equilibrium of the reaction is established at the time t, then the rate of the forward reaction = the rate of the reverse reaction.

$$\therefore \quad k_1 C_A = k_{-1} C_B \quad \text{or,} \quad \frac{k_1}{k_{-1}} = \frac{C_B}{C_A} = K \text{ (the equilibrium const.)}$$

Replacing $\frac{k_1}{k_{-1}}$ by K in equation (6.13), we get,

$$\ln K = \frac{E_r^{\ddagger} - E_f^{\ddagger}}{RT} + Z.$$

$$\text{Hence,} \quad \frac{d}{dT} \ln K \quad \text{or} \quad \frac{d \ln K}{dT} = -\frac{(E_r^{\ddagger} - E_f^{\ddagger})}{RT^2}$$

$$= \frac{(E_f^{\ddagger} - E_r^{\ddagger})}{RT^2}. \qquad (6.14)$$

But from Van't Hoff isochore (Eqn. 5.13) we know that, $\frac{d}{dT} \ln K = \frac{\Delta H}{RT^2}$, where ΔH = enthalpy change of the reaction.

Comparing eqn. (6.14) and Van't Hoff isochore,

$$\Delta H = E_f^\ddagger - E_r^\ddagger. \tag{6.15}$$

So, the enthalpy change of a chemical reaction is the difference in the activation energies of two opposing processes.

Now, if $E_f^\ddagger > E_r\ddagger$, the forward reaction will be endothermic and if $E_f^\ddagger < E_r^\ddagger$, the same reaction will be exothermic.

Again, if we replace $E_f^\ddagger$ and $E_r^\ddagger$ by $G_f^\ddagger$ and $G_r^\ddagger$ (free energy of activation of the forward and the reverse reactions) respectively, then

if $G_f^\ddagger > G_r^\ddagger$, the forward reaction is endothermic while the reverse reaction is exothermic, and

if $G_f^\ddagger < G_r^\ddagger$, the forward reaction is exothermic and the reverse reaction will be endothermic.

However, the activation energy for a reaction can directly be calculated from rate constant values measured at two different temperatures. We know that,

$$k = Ae^{-E^\ddagger/RT} \quad \text{or,} \quad \ln k = \ln A - E^\ddagger/RT; \text{ where } E^\ddagger = \text{Activation energy.}$$

At temperature, T_1, $\ln k_1 = \ln A - E^\ddagger/RT_1$. (6.16)

At temperature T_2, $\ln k_2 = \ln A - E^\ddagger/RT_2$. (6.17)

Now, subtracting eqn. (6.17) from eqn. (6.16) we get,

$$\ln \frac{k_1}{k_2} = \frac{E^\ddagger}{R}\left(\frac{1}{T_2} - \frac{1}{T_1}\right)$$

$$\text{or,} \quad E^\ddagger = R\left(\frac{T_2T_1}{T_1 - T_2}\right)\ln\frac{k_1}{k_2}$$

$$= \frac{2.303RT_1T_2}{T_1 - T_2}\log\frac{k_1}{k_2}. \tag{6.18}$$

Since T_1, T_2 are known and k_1 and k_2 are measurable quantities, $E^\ddagger$, the activation energy, can be calculated by using equation (6.18).

6.6 Energy Profile for a Reaction

Two energy profiles, 6.la and 6.lb have already been drawn in the previous section; those are for single-step reactions.

There are reactions involving reaction intermediates and intermediate molecular species. The formation of an intermediate from the reactant or conversion of the intermediate into a product is said to be an elementary process or a step of the overall reaction, the total reaction being called multi-step reaction. In such a reaction each step has its own T.S. Energy profile of this type of reaction has (i) an energy 'well' and (ii) an intermediate between two successive steps. Each T.S. has higher energy than the intermediate. The deeper the energy well, the more stable is the intermediate. The energy profile of a two-step reaction with an intermediate is drawn in Fig. 6.2.

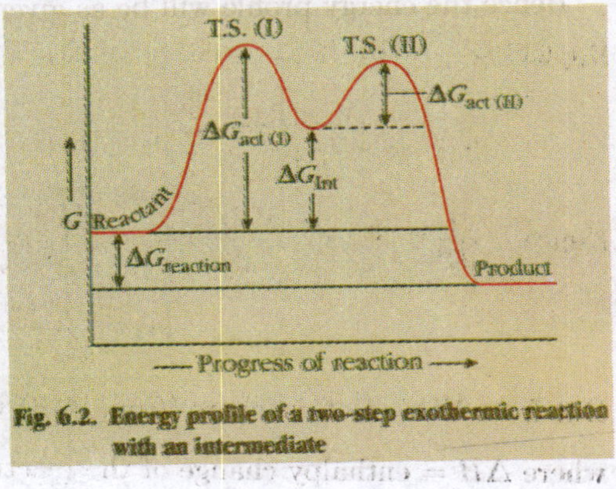

Fig. 6.2. Energy profile of a two-step exothermic reaction with an intermediate

From the three figures, 6.1a, 6.1b and 6.2, it is clear that the progress of the reaction is indicated along the abscissa and free energy (or potential energy) is represented along the ordinate.

The abscissa axis is also indicated as "reaction coordinate" which shows that how involved particles are located relative to one another.

However, to draw an energy profile for a reaction we require the relative energies of the reactants, transition states, intermediates and products. Instead of relative energies, corresponding rate constants can help us draw energy profiles since rate constants are related to energy barriers—the higher the energy barrier, the lower is the rate constant.

Let us consider some examples:

(i) A single step conversion of A to B with $\Delta G_{\text{reaction}} = -\text{ve}$

• $\Delta G_{\text{reaction}}$ is negative; this means the reaction is exothermic, a single-step reaction means that there is only one transition state. According to Hammond the T.S. is more alike to the reactant than the product. Hence the energy profile will be alike to Fig. 6.1b.

(ii) A single-step conversion of A to B; the reaction is endothermic.

• Since the reaction involves one step, there is only one T.S. The reaction being endothermic, G_{react} is less than G_{prodt}. The T.S. will be alike to the product. Hence $\Delta G_{\text{reaction}}$ is positive. So in the energy profile the order of free energies is:

$$G^{\ddagger} \gg G_{\text{product}} > G_{\text{reactant}}.$$

Hence Fig. 6.1a will represent the conversion.

(iii) Conversion of A to C via a reaction intermediate, B, the reaction is endothermic and the second step is the rate-determining step, i.e.,

$$A \xrightarrow[\text{fast}]{k_1} B \xrightarrow[\text{slow}]{k_2} C \text{ (endothermic reaction)}$$

• Since the reaction is endothermic $G_A < G_C$. B being a reaction intermediate, $G_B > G_C > G_A$. Now step I, i.e., A to B is fast and step II is slow. A slow step indicates, a relatively large energy barrier and a fast step indicates a relatively small energy barrier. Hence $\Delta G_1^{\ddagger}$ (free energy change for the conversion of A to the T.S. of Step I) for the conversion of A to B is smaller than $\Delta G_2^{\ddagger}$ (free energy change for the conversion of B to the T.S. of Step II) for the conversion of B to C. So the order of free energies involved in the forward reaction is

$$G_2^{\ddagger} > G_1^{\ddagger} > G_B > G_C > G_A \quad \text{and} \quad \Delta G_2^{\ddagger} > \Delta G_1^{\ddagger}.$$

Hence the energy profile will be as given below:

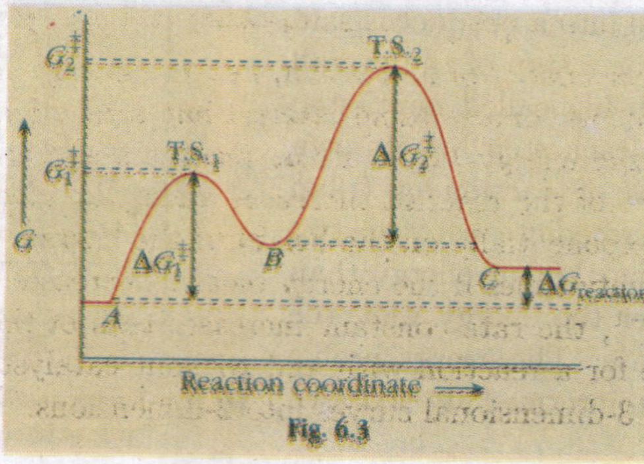

Fig. 6.3

(iv) $A \underset{k_{-1}}{\overset{k_1}{\rightleftharpoons}} B$ (Intermediate) $\underset{k_{-2}}{\overset{k_2}{\rightleftharpoons}} C$; $k_{-1} > k_1 > k_2 > k_{-2}$ the reaction is exothermic.

- The reaction involves two T.S.s; A to B has T.S.$_1$ and B to C has T.S.$_2$.

Since $k_{-1} > k_1 > k_2 > k_{-2}$ and since $k = a^{\ddagger}e^{-\Delta G^{\ddagger}/RT}$, B to T.S.$_1$ is the shortest path. C to T.S.$_2$ is the longest path, k_{-2} being of the lowest value; k_1 has the second highest value; so A to T.S.$_1$ is the shorter path while B to T.S.$_2$ is the longer one (k_2 has the lower value). The reaction being exothermic, G_A is greater than G_C. Hence the energy profile for the reaction is (Fig. 6.4):

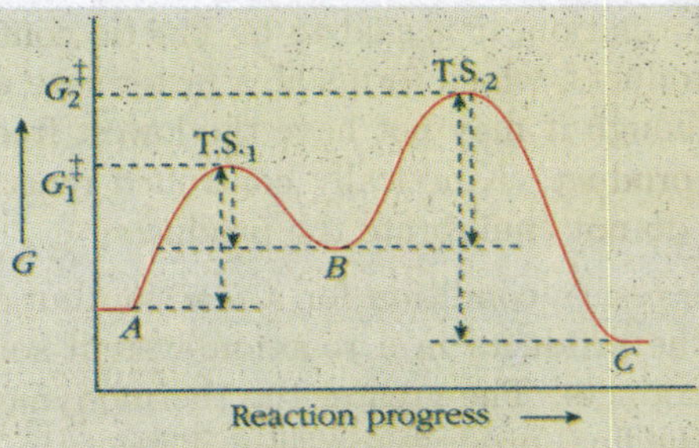

Fig. 6.4

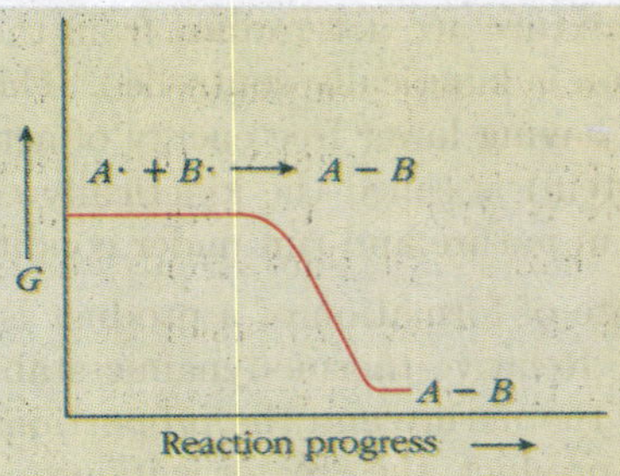

Fig. 6.5

There are reactions whose T.S.s have the energy of the reactants. Fig. 6.5 represents the energy profile for such a reaction.

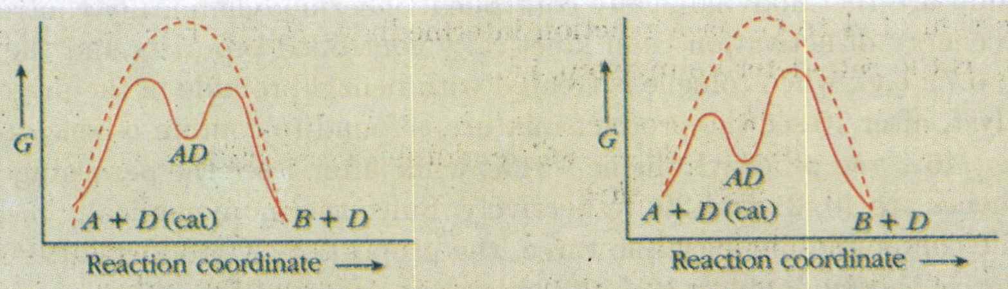

(a) 1st step (R.D.) **(b) 2nd step (R.D.)**

Fig. 6.6. Energy profile for an uncatalysed reaction (broken line) and that for the catalysed reaction (solid line)

A significant property of the energy curve is that the positions of reactants and products in the curve do not alter if certain conditions like pressure, temperature, etc. are maintained and at the same time their positions do not also depend on the height of the energy barrier. On the other hand, the height of the energy barrier can be altered: this means the reaction may be carried out in more than one pathway having different energies of activation including that at which energy barrier is zero (Fig. 6.5 : $k = a^{\ddagger}e^{-\Delta G^{\ddagger}/RT}$; if $\Delta G^{\ddagger} = 0$, $k = a^{\ddagger}e^0$ or, $k = a^{\ddagger}$. Therefore, even when the free energy of activation, i.e., the energy barrier is zero, the reaction occurs at a definite rate). The use of a suitable catalyst into a reaction mixture lowers the energy barrier without disturbing the energy levels of the reactants and the products; as the energy barrier comes down, the rate of the reaction increases under the given set of conditions. Since, the rate constant depends exponentially on the height of the barrier (Eqn. 6.10), the rate of the reaction increases ten to twenty times if the energy barrier decreases by 4 kJ mol^{-1}; if the same decreases by 20–25 kJ mol^{-1}, the rate constant increases tens of thousands of times. Fig. 6.6 describes the energy profile for a reaction with and without catalyst (see Sec. 6.3). The above diagrams are projections of 3-dimensional curves into 2-dimensions.

6.7 The Nature of Products and Energetics of a Reaction

When a chemical reaction gives rise to alternative sets of products under a given set of conditions, the nature of the products and their proportion depend on the energetics of the reaction, i.e., on the kinetics, and thermodynamics of the reaction.

The proportions of different products of a reaction are very often determined by their relative rates of formation. Usually the faster a product is formed, the greater is its proportion in the product mixture. When the nature of the product or the proportions of the products in the final product mixture are determined from the rate of formation, it is said to be **kinetic control**, i.e., the reaction is kinetically controlled. The set of products, which forms at a faster rate than the other set, having lower free energy of activation (though it may not have the lowest free energy of the system) is called the **kinetically controlled product**. Kinetically controlled reactions are reversible in nature and run under conditions that do not equilibrate the products.

The rate of formation of a product is thus a necessary condition for a reaction but it is not sufficient. Relative thermodynamic stabilities of the products in a reaction system sometimes determine the nature and the proportions of the products. The greater the thermodynamic stability of a product, the greater is its proportion in the final product mixture. Thus if the amount of product is determined by their relative thermodynamic stabilities, it is known as **thermodynamic control** or **equilibrium control**. The products which have the lowest free energy in the system are called **thermodynamically controlled products**. Reactions run under conditions that achieve equilibrium are thermodynamically controlled. Thermodynamically controlled products have higher free energy of activation than kinetically controlled products and the reactions are irreversible (Fig. 6.7); e.g., when toluene is treated with benzyl bromide in the presence of galium bromide as catalyst, after 10 sec the product mixture is found to contain *o*-benzyltoluene : 23%, *m*-benzyltoluene : 46% and *p*-benzyltoluene : 31%; while after 1 sec the percentage of the *o*-, *m*- and *p*-benzyltoluenes are 40, 21 and 39, respectively. Since thermodynamically *m*-benzyltoluene is the most stable compound amongst the three, the proportion of products after 10 sec is definitely due to relative thermodynamic stability though the rates of formation of *o*- and *p*-isomers are almost equal and that of *m*-isomer lags behind; while control of the latter reaction being kinetic, the percentages of *o*-, *m*- and *p*-benzyltoluene are found to be 40, 21 and 39, respectively.

The nature of control, kinetic or thermodynamic, is sometimes influenced by the condition of the reaction; the formation of napthalenesulphonic acids at 80 °C give 1-napthalenesulphonic acid as the main product; while at 160 °C the 2-napthalenesulphonic acid is the predominating product. Here the former reaction is kinetically controlled whereas the latter reaction is thermodynamically controlled. The possible explanation for their relative stabilities is as follows: in 1-naphthalenesulphonic acid a steric repulsion between the H atom in 8-position and —SO_3H group in 1-position makes the compound unstable at high temperature. The repulsion is absent in 2-naphthalenesulphonic acid and thus the former compound is less stable than the latter at 160 °C.

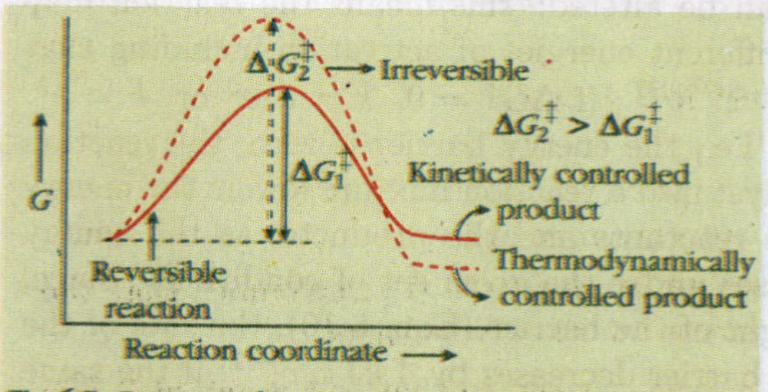

Fig. 6.7. Energy profile : Kinetic control vs thermodynamic control

The broken line is the energy profile for the thermodynamically controlled reaction, while the solid line is that of the kinetically controlled reaction.

In a given reaction the favourable electronic, stereoelectronic and steric factors give rise to kinetically controlled products usually at a lower temperature and much before the equilibrium is reached. If the above said factors are less favourable but the products are more stable, we get the thermodynamically controlled, i.e., stable products at a higher temperature when the equilibrium is established. So, if we want to get a kinetically controlled product, we shall have to carry out the reaction at a lower temperature and the product must be separated as early as possible after the completion of the reaction, whereas a thermodynamically controlled product can be obtained by carrying out the reaction at a higher temperature and separating the product after a prolonged standing after the completion of the reaction. However, sometimes the kinetically controlled products are also thermodynamically controlled.

Exercise

1. What do you mean by the term rate constant of a reaction? How is the rate constant of a reaction related to the concentrations of the reactant?

2. What is the relation between

(i) the rate constant and the absolute temperature of a reaction?

(ii) the rate constant and the energy of activation of a reaction?

3. Deduce the rate constant for the reaction,

$$A + B \underset{k_{-1}}{\overset{k_1(\text{fast})}{\rightleftharpoons}} AB \xrightarrow[\text{slow}]{k_2} C.$$

4. Which step is said to be rate-determining or rate-limiting in a multi-step reaction?

5. Distinguish between the molecularity and the order of a reaction.

6. What do you mean by the transition state of a reaction? Write down the postulates of Hammond regarding transition states. Why is the Transition State Theory superior to the Collision Theory?

7. Define $\Delta E^{\ddagger}$, $\Delta G^{\ddagger}$, $\Delta H^{\ddagger}$, $\Delta S^{\ddagger}$.

8. **(a)** Draw the energy profiles for the following reactions:

(i) $R \underset{k_{-1}}{\overset{k_1}{\rightleftharpoons}} I \underset{k_{-2}}{\overset{k_2}{\rightleftharpoons}} P$; $G_I > G_R > G_P$ and $k_2 > k_{-1} > k_1 > k_{-2}$

(ii) $R \underset{k_2}{\overset{k_1}{\rightleftharpoons}} I$; $I \xrightarrow{k_3} P$; stability order: $R > P > I$

k_1 small; k_2 and k_3 large, $k_2 > k_3$.

(b) Which steps are the R.D. steps of the reactions given in (a)?

(c) Write down the order of $\Delta G^{\ddagger}$s of the reactions given in (a) (i); remember that the forward reaction have two $\Delta G^{\ddagger}$s and the reverse reaction two others.

9. What will be the minimal free energy of the T.S. with respect to that of a reactant ? [see Fig. 6.5]

10. Explain with illustration :

(i) Kinetically controlled reactions,

(ii) Thermodynamically controlled reactions.

11. See the graphs given below and answer the following questions:

(i) Which one is the R.D. step?

(ii) Is the forward reaction exothermic?

(iii) Write down the order of the rate constants for the forward and reverse reactions.

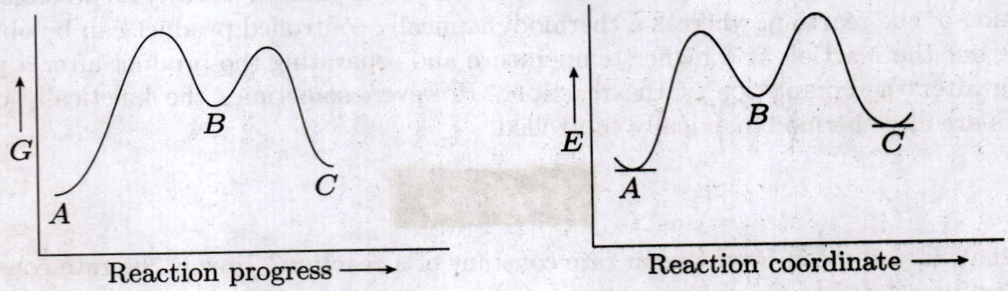

12. Explain : $\Delta G^{\ddagger}$ of a reaction is at least equal to the ΔH of an endothermic reaction.

Chapter 7

Intermolecular Forces and Related Properties

Objectives

- To understand:
  - Dipole-dipole interaction [7.2] p. 225
  - Induced-dipole interaction [7.3] p. 226
  - Instantaneous dipole-induced dipole interaction : London force [7.4] p. 226
  - H-bonding force [7.5] p. 227
  - Free energy aspects of a solution [7.7.3] p. 236
  - The effect of solvation on equilibria [7.7.4] p. 236
  - The effect of solvation on rates of reactions [7.7.5] p. 237
- To compare:
  - Melting and boiling points of organic compounds [7.7.1] p. 231
  - Solubilities of organic compounds [7.7.2] p. 233
- To know about:
  - Phase transfer catalyst [7.8] p. 237
  - Crown ethers [7.9] p. 239

7.1 Introduction

The forces that hold molecules together may be called **intermolecular forces**. Here too all such forces are electrostatic in nature as they invariably involve dipoles, i.e., these forces arise from the attraction between negative and positive charges. Hence **intermolecular forces** may be defined as **dipole-interactions**.

Again, van der Waals proposed that these intermolecular forces are responsible for the deviation of the real gases from the ideal behaviour and accordingly he modified the gas equation,

$$PV = RT \text{ to } \left(P + \frac{a}{V^2}\right)(V - b) = RT \text{ for a mole of gas;}$$

here the constant 'a' has arisen for the intermolecular forces and the constant 'b' determines the total volume of the molecules. In his honour the intermolecular forces are also called **van der Waals** forces which are considerably important in liquid and solid states. However, for the purpose of better understanding of the physical properties of organic molecules, we are classifying these forces as follows:

1. Dipole-dipole interaction
2. Induced-dipole interaction
3. Instantaneous dipole-induced dipole interaction, i.e., London force
4. The strongest dipole-dipole interaction, i.e., hydrogen bonding force.

It is needless to mention that these forces are much weaker in comparison to those between ions in ionic compounds and atoms in covalent molecules or crystals. Let us now discuss these forces in brief.

7.2 Dipole-dipole Interaction

We have discussed bond dipoles and molecular dipoles in Sec. 4.8. Here we shall discuss the action of a dipole on another dipole. The consequence of such an action is the generation of an attractive force between the molecules involved.

The attraction of the positive end of a molecular dipole for the negative end of another dipole may be called **dipole-dipole interaction**. Hence these interactions are also, in one sense, directional. It is inversely proportional to the third power of the distance between the two dipoles (r). Hence it is effective over a short intermolecular distance though it is moderately strong. Dipole-dipole bond dissociation energy is of the order 8.37 kJ/mole. The force $\propto \frac{1}{r^3}$.

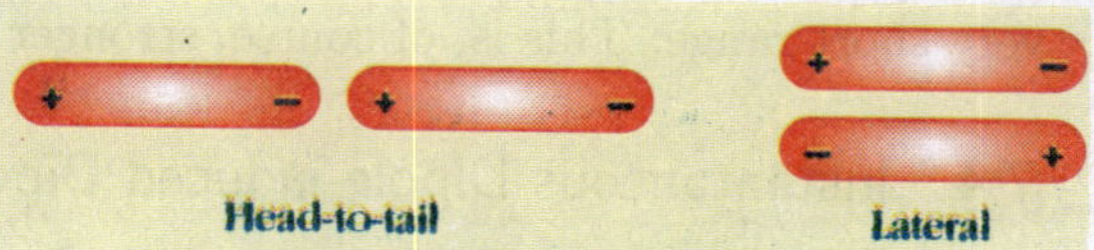

Fig. 7.1. Dipole-dipole interaction

Dipole-dipole interaction may occur either head-to-tail or laterally as shown in Figure 7.1.

Dipole-dipole interactions are weaker than ion-dipole interactions (see note) but stronger than those present in non-polar molecules of comparable molecular weights. The strength of this kind of bond is reflected in the physical properties of the compounds concerned. Hydrogen chloride is a polar molecule. There are dipole-dipole interactions amongst hydrogen chloride molecules. Argon is a non-polar entity and of the same molecular weight as that of HCl. Owing to the dipole-dipole interaction, HCl boils at a much higher temperature than argon. The difference in their b.p., is 101°C.

Polar compounds dissolve in polar solvents by dipole-dipole interactions.

Note

Ions have strong but short-range effect on dipoles. These are known as ion-dipole interactions.

When an ion approaches a dipole, the latter orients itself in such a way that its attractive end (the end with charge opposite to that of the ion) is directed toward the ion and the other is directed away. Thus an electrostatic force of attraction is established between the ion and the dipole; this force obeys Coulomb's law of electric force. It is found to be inversely proportional to the square of the distance between the ion and the dipole (r). Thus the force acts over a short range though it is strong. Considering orientation of the dipole, this force may be called directional. Here the force $\propto \frac{1}{r^2}$.

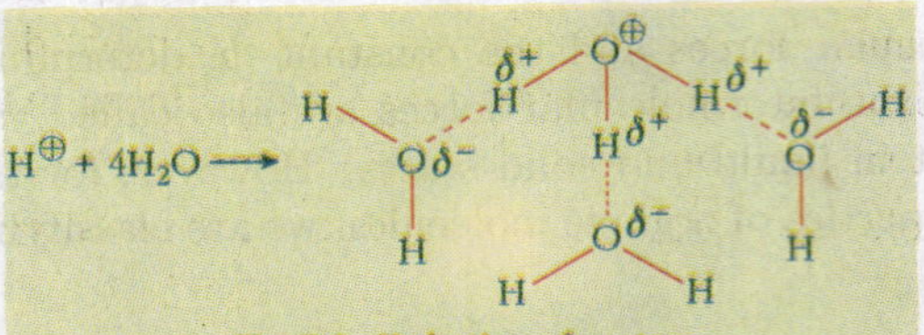

Fig. 7.2. Hydration of proton

Hydration of proton to form hydronium ion, the solvation of ions and dissolution of ionic lattices in polar solvents are the examples of ion-dipole interactions. Stabilisation of ions by the attraction of solvent molecules is called solvation and hydration is a special case of solvation where water is the solvent (Fig. 7.2).

7.3 Induced-dipole Interaction

If polar and non-polar molecules are present together, then polar molecules, i.e., dipoles can induce dipoles in non-polar molecules. The net result is the bonding between the polar and non-polar molecules, i.e., bonding between the dipoles and the induced dipoles. This is called dipole-induced dipole bond. This force varies inversely as the sixth power of the distance (r) between the dipole and the induced dipole. It is effective over an extremely short range and very weak in nature.

Non-polar molecules dissolve in polar solvents by dipole-induced dipole interactions.

Note

A charged particle, i.e., an ion can induce dipole in a non-polar molecule when they come close to each other. This occurs because the ion distorts the electron-cloud of the molecule. This may be called an ion-induced dipole and the result is the bonding between the ion and the induced-dipole which is known as ion-induced dipole bond. This type of force varies as the fourth power of the distance between the ion and the induced-dipole. It is a weak force and acts over a very short range. This is, of course, stronger than dipole-induced dipole force.

7.4 Instantaneous Dipole-induced Dipole Interaction : London Force

A non-polar molecule is electrically neutral. Of course, owing to the motion of the electrons within such a molecule, the centre of negative charge density due to electrons does not always coincide with the centre of positive charge density of the protons. This disturbs momentarily the electrical symmetry of the molecule and an instantaneous dipole is formed. This instantaneous dipole, in its turn, induces dipole in another adjacent molecule. Thus the instantaneous dipole and the induced dipole remain together by an attractive force or bonding, known as instantaneous dipole-induced dipole interaction. It is true that the momentary dipoles and induced dipoles are changing constantly to the neutral state, still the attractive force so generated is able to bind the molecules together and is known as London force. Some of the authors mean van der Waals forces as London forces. However, when non-polar molecules come very close to each other, the London forces act between the surface of the concerned molecules. Therefore, London force is directly proportional to the molecular volume. As the force originates from the motion of the electrons within molecules, the strength of the force also varies directly as the number of polarisable electrons in the molecule. Again, as the molecular volume and the number of electrons in a molecule increase, the molecular weight increases. Thus London force increases rapidly with the increasing molecular weight. It is found that London force is inversely proportional to the seventh power of the distance (r) between the two dipoles so formed. So, London force is extremely short ranged in action. But if the molecules happen to come to each other below an optimum distance, a repulsive force develops between the outer electrons of the molecules concerned. On the other hand, London force is found to be maximum when the molecules are separated from each other by the optimum distance. In fact, half of the distance between the nuclei of two non-bonded atoms of the same element over which van der Waals force (i.e., London force) is maximum may be called van der Waals radius. This is always greater than the atomic radius of the element.

London force is the weakest of all kinds of intermolecular forces. The bond dissociation energy is only of the order 4 kJ/mole or something like that.

7.5 The Strongest Dipole-dipole Interaction : Hydrogen Bonding Force

The electron configuration of H atom is $1s^1$. It precedes the nearest inert gas, He ($1s^2$). Therefore, it normally shows monovalency, either electrovalency or covalency.

There are many well-known examples in which a hydrogen atom forms two bonds with two highly electronegative atoms from the group F, O, N, etc.—one of the bonds is covalent and the other is thought to be mainly electrostatic; these two atoms may be identical or different.

Thus in a molecule of hydrogen fluoride, the hydrogen atom forms a covalent bond with fluorine atom. This H atom also links another F atom of the adjacent HF molecule by a comparatively weak electrostatic force and is represented by a dotted line as shown in Fig. 7.3.

Fig. 7.3. H-bonding

From Figure 7.3, we can say that the H atom serves as a bridge between the two electronegative atoms. For this reason the weak electrostatic bond between the H atom of one molecule and the F atom of the other molecule here is called the hydrogen bond.

Thus **H-bond** may be defined as a weak electrostatic chemical bond which is found to act between one covalently bonded H atom of a species and a highly electronegative atom like F, O, N, Cl, S of another species.

When a hydrogen atom is covalently linked to a strong electronegative element, the molecule formed is a dipole. Thus H—F is a dipole. In an aggregate of H—F molecules, dipole-dipole interactions take place when they come very close to each other. The obvious result is the orientation of poles and a weak bond is established between two unlike poles ($H^{\delta+}$ and $F^{\delta-}$ as usual. So, H-bond is a result of dipole-dipole interactions. This is represented in figure 7.4.

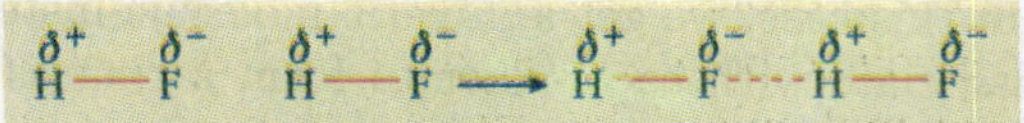

Fig. 7.4. Formation of H-bond by dipole-dipole interaction

Now the question : Why do not we get such strong dipole-dipole interactions with more electropositive atoms than hydrogen? An H atom, covalently bonded to a strong electronegative atom, has relatively high positive charge density but minimal coordination number (only two) because it has a very small volume which makes H-bonding possible; while other electropositive atoms have neither high positive charge density nor minimal coordination number because of their large volume.

The H-bond dissociation energy varies from 21 to 42 kJ/mole. Hence it is stronger than usual dipole-dipole interactions. The strength of the bond depends upon the electronegativity of the atom bonded to the H atom; it decreases with the decreasing electronegativity within the series F, O, N, Cl, S.

The H-bond length is found to be within the limit 0.12 nm to 0.34 nm; this is always less than the sum of the van der Waals radii of the atoms involved, i.e., H atom and another electronegative atom. The H-bond length is the longest in H··· S bond and it is the shortest in H··· F bond.

H-bonds are of two types:

1. Intermolecular H-bond,
2. Intramolecular or internal H-bond.

When H-bonding happens to occur between two or more molecules, it is known as **intermolecular H-bond**. It gives rise to association of molecules and species, e.g.,

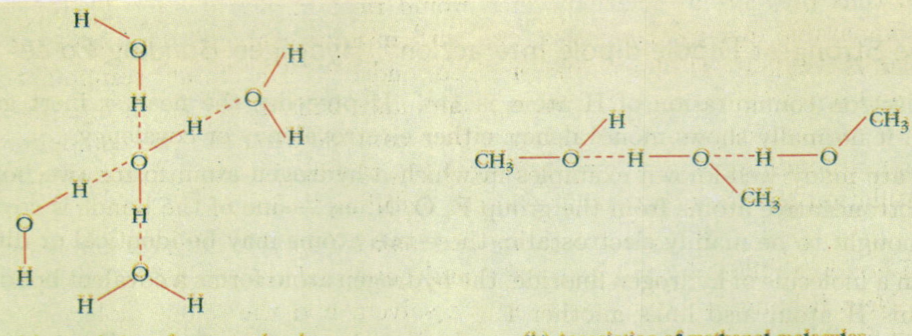

(a) Association of water molecules (b) Association of methanol molecules

Fig. 7.5

When H-bond is formed between two atoms within a molecule, it is called an internal or intramolecular H-bond. The result is the formation of a ring known as chelation. We find intramolecular H-bonding in salicylaldehyde (1), ortho-nitro-phenol (2), etc. (Fig. 7.6).

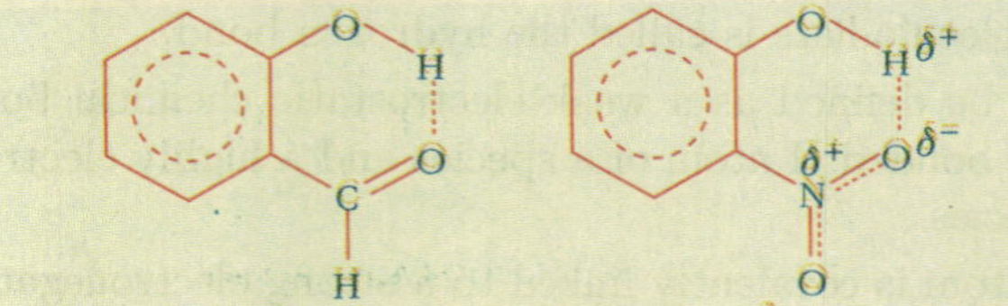

1 2

Fig. 7.6. Intramolecular H-bond

Intermolecular H-bonding can also form ring. In this case, we get a dimeric molecule instead of a polymer, e.g., two molecules of acetic acid associate with each other to form a ring as in Figure 7.7.

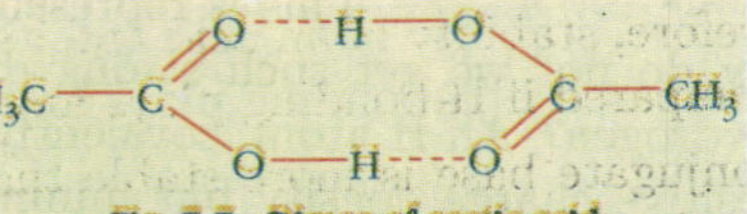

Fig. 7.7. Dimer of acetic acid

7.5.1 The effect of H-bonding force

(a) Effect on m.p. and b.p.: Owing to the association of identical molecules by intermolecular H-bond, the intermolecular forces amongst the molecules increase to a relatively greater extent. This has a marked effect on m.p. and b.p. of the compounds. An extra amount of heat energy is required to break the H-bond. Hence, compounds with intermolecular H-bond have greater m.p. and b.p. than those of compounds without intermolecular H-bonds, provided they have comparable molecular weights.

CH_4, HF and H_2O molecules have comparable molecular weights. CH_4 cannot form H-bond, because C atom has low electronegativity (C − 2.5 and H − 2.2 eV); HF can utilise its lone H atom to form H-bond but a water molecule forms H-bond with both the H atoms. Consequently, the m.p. and b.p. orders of the said compounds are as follows:

m.p. order:	H_2O(0°C)	HF(−110°C)	CH_4(−190°C)
b.p. order:	H_2O(100°C)	HF(19.5°C)	CH_4(−161.5°C)

On the other hand, intramolecular H-bonding in a compound prevents intermolecular H-bonding and thus prevents association which would raise m.p. and b.p. Therefore, two compounds of comparable molecular weights, one with intermolecular H-bond and the other with intramolecular H-bond, should have different m.p. and b.p. The former compound will possess higher m.p. and b.p. than those of the latter compound. This difference in b.p. often helps separation of two compounds from their mixture by steam distillation. The compound possessing intramolecular H-bonding being of lower boiling point, this distils out with steam when subjected to steam distillation but the other remains in the distilling flask.

(b) Effect on solubility: H-bond in a compound makes it soluble in water and in other solvents possessing H-bonding. The H-bond in solvent molecules and that in solute molecules break on dissolution. New H-bonding occurs between the solvent and the solute molecules because the energy gained by this new bond formation is more than that lost by the bond breaking processes.

(c) Effect on stability of molecules: Intramolecular H-bonding brings stability to a compound. Acetylacetone molecules in liquid state remain in two tautomeric forms keto and enol (3 and 4 respectively).

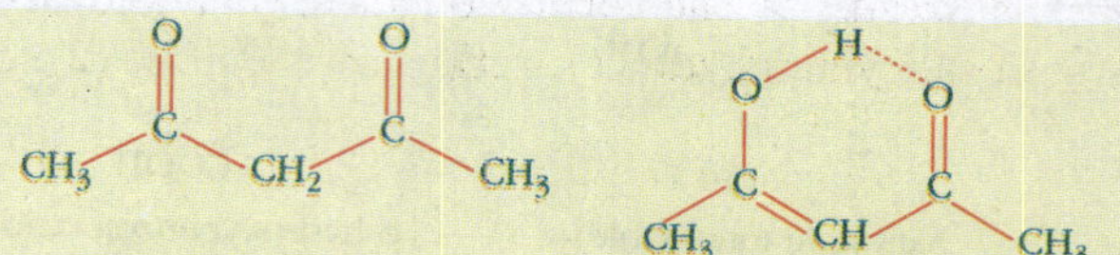

Keto form **3** Enol form **4**

Fig. 7.8. Tautomeric forms of Acetylacetone

Owing to the intramolecular H-bonding and conjugation in the enol form, it is more stable than the keto form in the liquid acetylacetone and in an n-hexane solution and their percentage is 80% and 20% respectively. In the aqueous medium, however, enol form is 16% and keto form is 84%.

(d) Effect on acidity and basicity: Hydrogen bonding not only brings stability to molecules but also it does so to ions. Therefore, stability of an acid and its conjugate base or that of a base and its conjugate acid can be compared if H-bonding plays any role in the species concerned.

Again, we know that if a conjugate base is more stable than the parent acid, the base will be relatively weaker than the acid. If the order of their relative stability is a reverse one, then the order of their relative strength will change accordingly. The relation between stability and strength of a base and its conjugate acid follows the same principle as stated above.

Thus dimethylamine is a stronger base than trimethylamine in aqueous medium as the conjugate acid of the former base is more stable than that of the latter compound. The stability order is so because of the greater hydration effect on dimethylammonium ion than that on trimethylammonium ion. This is revealed from the structures of hydrated dimethylammonium ion and that of trimethylammonium ion (Fig. 7.9).

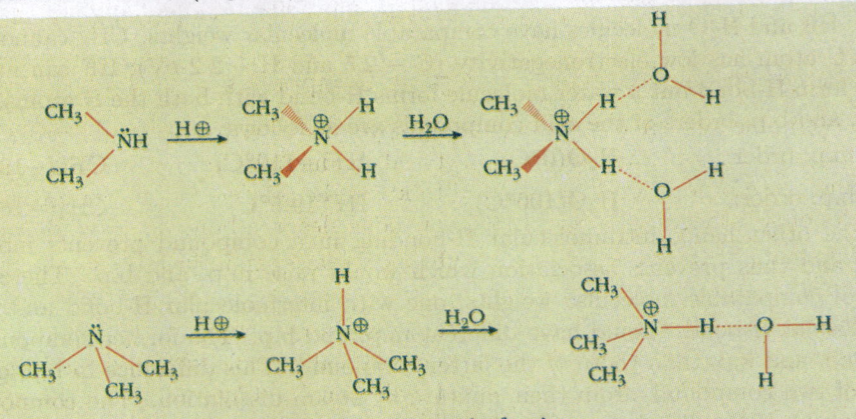

Fig. 7.9. Hydration of conjugatacids of amines

Similarly, the conjugate base of salicylic acid gets stabilised by the intraionic H-bonding, but *p*-hydroxybenzoate ion, the conjugate base of the *p*-hydroxybenzoic acid, does not stabilise by such a hydrogen bonding. Hence, we can easily conclude that salicylate ion is a weaker base than *p*-hydroxybenzoate ion. This means: salicylic acid is a stronger acid than *p*-hydroxybenzoic acid.

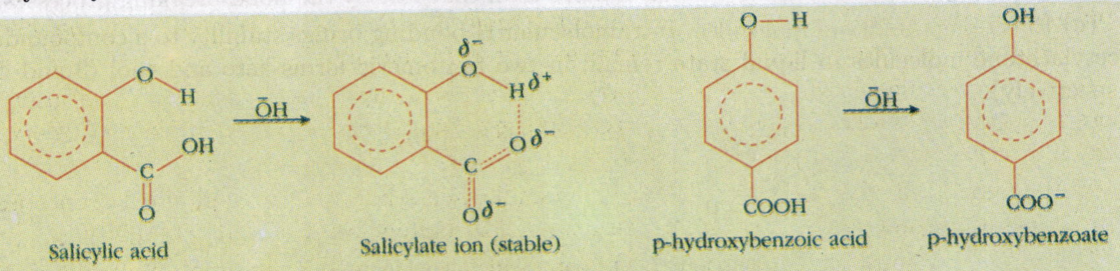

Fig. 7.10. Intramolecular H-bonding in salicylate ion

(e) Effect on biological system: Both inter- and intra-molecular H-bonding play important role in biological system. The long helical structure of protein molecules and double helical structure of DNA molecules are stabilised by the existence of H-bonding, e.g., a protein molecule contains a large number of >C=O and >N—H groups, and H-bonding occurs between these two groups (Fig. 7.11).

Fig. 7.11. H-bonding in protein

A single protein molecule contains a large number of these H-bonds. Although each of them is very weak (the dissociation energy around 21 kJ/mole), yet the overall effect of all the H-bonds is highly significant and brings stability to the molecule of protein.

The H-bonding force in protein and cellulose makes them able to absorb dyes.

7.6 The Relative Strength of Intermolecular Forces

From the discussion made above on intermolecular forces, it is obvious that H-bonding force is the strongest and London force is the weakest amongst them. The strength of dipole-dipole interaction lies between the strengths of the other two mentioned above. Thus the order of strength of the three forces is:

H-bonding force >	Dipole-dipole interaction >	London force
~ 21-42 kJ/mole	~8 kJ/mole	~4 kJ/mole

7.7 Some Physical Properties of Organic Compounds

A considerable amount of substance is employed in order to study its physical properties like melting point, boiling point, solubility, electrical conductivity refractive index, etc. Therefore, to understand these properties, it is essential to know the relative amount of forces which hold the basic units of substances., i.e., atoms, ions or molecules in a bulk.

An aggregate of the basic units of a substance may remain in the gaseous state or in liquid state or in solid state.

The basic units of organic compounds are mostly molecules. In the *solid state*, the organic molecules are held in highly ordered arrangements, called molecular crystals (Sec 2.6). In such a crystal structure, London, dipole-dipole and hydrogen bonding forces hold the molecules in fixed positions within the lattice. The molecules, in their turn, possess vibrational motion as they have very little of thermal energy.

In the liquid state, the organic molecules still have ordered arrangements but it is less regular than those in the corresponding solid. In fact, we find disrupted lattice structures, which may be called ordered clusters of molecules, within the bulk of the liquid. Besides vibrational motion, here the molecules have certain amount of rotational and translational motions, although London, dipole-dipole and hydrogen bonding forces hold the molecules together as they do so in the solid state. Thermal energy of the molecules here is greater than that in the solid state.

Organic molecules in the gaseous state are in all sorts of rapid random motion. Naturally, there is negligible amount of attractive forces amongst the molecules but their thermal energy content is very high.

In the light of the discussion made above, we shall discuss here the three physical properties of practical importance; these are: (a) melting point, (b) boiling point, and (c) solubility.

Melting and boiling points are very useful for identifying organic compounds. Solubilities of organic compounds are also very important for the fact that the study of chemical properties, isolation and separation of compounds are generally done in solutions.

When heat is gradually added to a solid, its temperature as well as the vibrational motion of the molecules within the lattice gradually increase. Increasing vibrational motion of the molecules makes the lattice somewhat loose. At the melting point of the solid, the added heat is utilised mostly in disrupting the lattice, allowing the molecules to attain the rotational and translational motions which are characteristics of the liquid state. For this reason, the heat absorbed by the solid during the process of melting does not increase the temperature of the substance. Melting is the change from the highly ordered arrangement of particles in the crystal lattice to the less regular arrangement that characterises a liquid, and the temperature at which this change occurs is called melting point. Disruption of the lattice occurs by the dislocation of a few particles from their position in the lattice.

As the intermolecular forces in the molecular crystals are very weak, the disruption of the lattice occurs at a low temperature. Non-polar compounds of low molecular weights which have only London force, generally remain in the gaseous state under ordinary conditions of temperature and pressure. But as molecular weight increases within a class of compounds, the London force also increases and hence melting point too increases.

A compound with intermolecular dipole-dipole interactions has higher melting point than a compound possessing London forces only, if they have comparable molecular weights.

If hydrogen bonding force is present within a molecular crystal, it certainly requires comparatively high heat energy to disrupt the lattice. Hence such a compound should have higher melting point than those possessing London force or dipole-dipole interactions.

Further addition of heat to the melt increases the temperature of the liquid and then serves to increase all sorts of motion of the organic molecules within the liquid. As liquids are heated, the regularity of ordered molecular clusters decreases. At the boiling point, the added heat energy is absorbed but temperature of the liquid remains constant. The supplied heat acts to break down the remaining molecular clusters and helps the molecules attaining rapid random motions. Thus, the molecules from the liquid state enter into the gas phase. Therefore, boiling is the rapid change from a more or less ordered arrangements of particles in a liquid state to a state of complete disorder that characterises a gas phase, and the temperature at which the change occurs is called boiling point.

As stated earlier, in the liquid state the organic compounds possess London, dipole-dipole and hydrogen bonding forces. These forces being weak, breaking down of the molecular clusters and setting free of the molecules require small amount of energy. Hence, organic compounds have low boiling points as compared with ionic compounds.

Amongst these three forces hydrogen bonding force is the strongest one and London force is the weakest, dipole-dipole force has an intermediate position in strength. So a compound involving intermolecular hydrogen bonding has higher boiling point than a compound possessing dipole-dipole interactions as intermolecular forces; of course they should have identical or nearly identical molecular weights. If London force is involved in a compound, it should have very low boiling point. As London force increases with the increasing molecular weight, the boiling point also increases with the increasing molecular weight. Since the order of surface area amongst 1°, 2° and 3° isomeric compounds is $1° > 2° > 3°$, the order of London force amongst these compounds is $3° < 2° < 1°$ and consequently the b.p. order is $1° > 2° > 3°$.

Problem 7.1. Account for the following progression in b.p.,

$$CH_4 : -161.5°C, CH_3Cl : -24°C, CH_3OH : 65°C$$

Solution: The increasing order of molecular weights of the given compounds is

$$CH_4 < CH_3OH < CH_3Cl$$

From the consideration of molecular weight, i.e., London forces, CH_3Cl should have the highest b.p. and that of CH_4 should be the lowest. On the other hand, there is intermolecular hydrogen bonding in CH_3OH; there is no such bonding in CH_3Cl, but it is a polar molecule. CH_4 is definitely a non-polar molecule and London force is only acting in an aggregate of methane molecules. Therefore, the increasing order of intermolecular force in the given set of compounds is:

London < dipole-dipole < H-bond : hence, the b.P., of CH_3OH will be the highest, then comes the position of CH_3Cl, and CH_4 should have the lowest b.p.

7.7.2 Solvents, Solutes and Solubility

If a homogeneous mixture is obtained on mixing two or more substances together, the mixture is called a solution. If the physical state of aggregation of one of the components of the solution does not change but that of others get changed, the former is called the solvent and the others are called solutes of the solution. In an aqueous solution of sugar, water is the solvent and sugar is the solute. If the physical state of aggregations of the components of a solution do not change, then the component which is present in excess is called the solvent and the others are solute.

However, solvents are classified into two main types : **Protic solvents and aprotic solvents.** Those solvents which have H atom on a highly electronegative element (O, N, etc.) are capable of forming H-bonds and are called protic solvents; ammonia, water and alcohols are protic solvents. Water and alcohols are also called **hydroxylic solvents** since they have hydroxyl groups. Those solvents that have no H atom on an electronegative element (O, N, etc.) are called aprotic solvents; dimethylsulphoxide (DMSO), dimethylformamide (DMF), hexamethyl phosphoric triamide (HMPT) are the important aprotic solvents. Again a solvent may be polar or non-polar. A solvent of relativity high dipole moment and **dielectric constant** is called a polar solvent; water, alcohol, DMF, DMSO, HMPT are polar solvents but carbon tetrachloride, benzene, n-hexane are non-polar solvents. Dielectric means a non-conductor of electricity, an insulator. The dielectric constant, ϵ, of a substance x is given by

$$\epsilon = \frac{C_x}{C_{\text{air}}};\ \text{where } C_x = \text{capacitance of the substance and}$$
$$C_{\text{air}} = \text{capacitance of air.}$$

Capacitance is the property of a substance that allows it to store an electrical charge. In other words, the dielectric constant of a medium is equal to the attractive force between opposite charges in the medium relative to that in a vacuum or air. So it is a factor by which an electrostatic interaction in a vacuum or air is reduced by the medium. If two opposite charges Z_1 and Z_2 are separated by a distance r in a medium of dielectric constant ϵ, electrostatic interaction energy (E) will be related to ϵ by the following equation:

$$\epsilon = \frac{Z_1 Z_2}{Er}. \tag{7.1}$$

This equation shows that as the dielectric constant decreases, the electrostatic interaction energy between opposite ions increases and the possibility of their combination to form ion-pairs and insoluble crystals increase, while the greater the value of dielectric constant of a medium, the greater is the ability of the solvent medium to insulate both ions and developing ions. The insulating ability facilitates separation of positive and negative ions from each other in the solution. Solvent polarity order and dielectric constants are given in Table 7.1.

Table 7.1 Polarity of solvents

Solvent	b.p.(°C)	μ(D)(g)	ϵ(at 25°C)	
n-Hexane	68.7	~ 0	2	Increasing order of solvent polarity ↓
Carbon tetrachloride	76.7	0	2	
Benzene	80	0	2	
Acetic acid	118	1.76	6	

Contd.

Solvent	b.p.(°C)	μ(D)(g)	ϵ(at 25°C)	
Acetone	56.25	2.88	20.7	Increasing order of solvent polarity ↓
Ethyl alcohol	78.5	1.67	24.3	
HMPT[$(CH_3)_2N]_3PO$	232	4.50	30.0	
Methyl alcohol	64.5	1.70	32.6	
Nitrobenzene	210.85	3.95	35	
DMF ($Me_2NCH{=}O$)	153	3.82	36.7	
DMSO ($Me_2S{=}O$)	189	3.96	49.0	
Formic acid	101	1.35	59	
Water	100	1.85	78.5	
Hydrogen cyanide	25.6	2.95	118	

A solvent capable of exerting strong electrostatic interactions between ionic or dipolar solutes and solvent dipoles may be called polar solvent. The degree to which a solvent is capable of solvating and stabilising ionic and dipolar solutes is called its polarity. The ability to solvate depends on steric factors (shape, size) of solute and solvent molecules amongst other factors. The greater the dielectric constant, the greater is the polarity of the solvent. **Dipole moment is not a measure of the solvent polarity**.

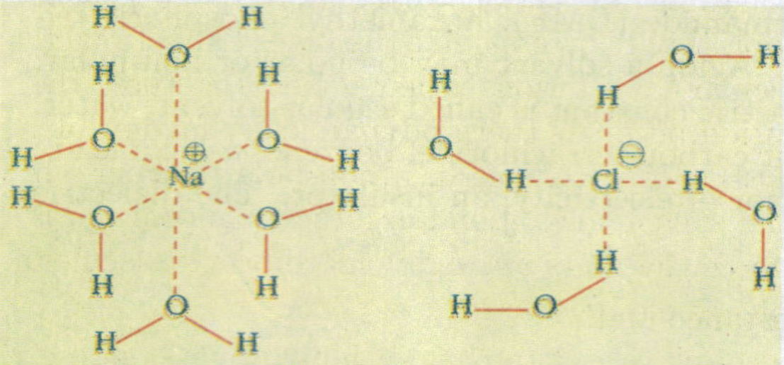

Hydration of ions of an ionic compound

Fig. 7.12. Hydration of ions of an ionic compound

However, in a solution, solvent molecules combine with solute molecules or ions with a release of energy and the process is called **solvation**. When the solvent is water, it is called hydration. In the cases of polar hydroxylic solvents, the cation and the anion of an ionic compound get solvated through weak forces as shown with NaCl (Fig. 7.12):

The smaller the ionic size, the more concentrated is the charge on the ion and hence more strongly it is solvated than a larger ion. For this reason the solvation order amongst halide ions in water : $F^- > Cl^- > Br^- > I^-$. This means in the aqueous medium I^- is the most free and F^- is the least free ions to move.

Polar aprotic solvents solvate cations very well but owing to their inability to form H-bond, they do not solvate anions appreciably. So anions are poorly stabilised and free to move and therefore, highly reactive both as bases and nucleophiles; while cations being strongly stabilised, their reactivity is remarkably low. Solvation of cations in an aprotic polar solvent is shown in Fig. 7.13.

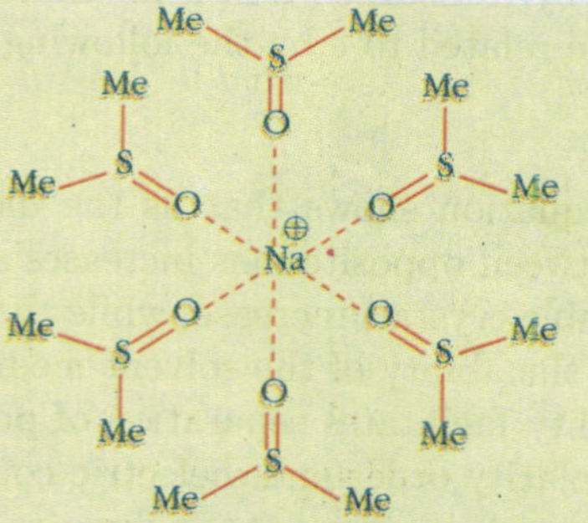

Fig. 7.13. Solvation of cations by an aprotic polar solvent, DMSO

Anions being poorly solvated in an aprotic polar solvent, their reactivity is not disturbed by solvation. Thus in such a solvent the reactivity order amongst halides is $F^- > Cl^- > Br^- > I^-$ while in the hydroxylic (protic) solvent the order is $I^- > Br^- > Cl^- > F^-$. This is because the smaller the volume of the anion, the higher is the charge density and more stabilised is the solvated anion.

Aprotic non-polar solvents do not dissolve ionic compounds because of their inability to solvate the ions. They dissolve non-polar compounds through London forces.

Now if we want to predict whether two compounds can form a solution or not, it is essential to consider the following three interactions:

1. Solute-solute interaction
2. Solvent-solvent interaction
3. Solvent-solute interaction.

The necessary condition for dissolution of a solute in a solvent is:

Solvent-solute interactions should **equal** or **exceed** the combined solute-solute and solvent-solvent interactions.

As stated earlier, in organic compounds solute-solute interactions, in general, are:

(a) London force
(b) Dipole-dipole interactions
(c) Hydrogen bond.

It is found that a solute possessing any of the above three interactions gets dissolved in a solvent having the same intermolecular forces and this is popularly known as 'like dissolves like'. In fact, two compounds having identical type of intermolecular interactions when mixed together, the entropy of the system increases and thus the dissolution becomes thermodynamically favourable. Carbon tetrachloride is a non-polar molecule; so it should be soluble in a non-polar solvent—it dissolves in benzene. Here both of the solute and the solvent have London forces. Chloroform is a polar compound; dipole-dipole interaction is the force that acts amongst chloroform molecules. It is soluble in acetone, a polar solvent having dipole-dipole interactions.

There is intermolecular hydrogen bond in ethyl alcohol. Water too has such hydrogen bonding. We know that ethyl alcohol and water can mix homogeneously in all proportions.

Ionic compounds dissolve in solvents where ion-dipole interactions occur. Thus, in an ionic crystal, ions in the surface are loosely bound to the lattice and strongly attracted by the polar solvent molecules when the compound is placed in a polar solvent. The solvent-solute interaction is much greater than the combined ion-ion interaction in the crystal and dipole-dipole interaction in the solvent molecules. Water is a good solvent for ionic compounds because it is a polar solvent.

In most of the cases with increasing molecular weight within a class of compounds, the solubility decreases. This is very important in the case of organic compounds. As molecular weight increases, London force also increases; consequently solute-solute interactions increase with the increasing molecular weight. Thus, the solvent-solute interaction cannot outweigh the combined solute-solute and solvent-solvent forces.

Sometimes dissolution occurs through chemical reaction; organic acids can ionise and by ion-ion interaction they get dissolved in alkali with the formation of salt. Such is the case with bases; they are soluble in acids and form salts.

Problem 7.2 Cyclohexane dissolves in carbon tetrachloride but not in water—explain.

Solution: Water is a polar compound in which its molecules are held together by strong hydrogen bonding.

Cyclohexane and carbon tetrachloride molecules are nonpolar and each of them possesses weak London forces.

We know that 'like dissolves like'; because entropy increases in such a dissolution. Therefore, cyclohexane is miscible with carbon tetrachloride.

On the other hand, cyclohexane cannot form H-bonding with water. If it at all dissolves in water, there might generate dipole-induced dipole attraction. This is much weaker than H-bonding forces. Hence, solvent-solute forces cannot exceed or equal to the sum of the solute-solute London forces and solvent- solvent hydrogen bonding forces. Hence cyclohexane does not dissolve in water.

7.7.3 Solution and free energy

From the thermodynamical standpoint, the necessary condition for the spontaneous dissolution of a solute to a solvent is that the free energy change (ΔG) must be negative, i.e., the free energy of the solution must be less than the sum of the free energies of the solute and the solvent. It has already been stated that when solute-solvent interaction energy is greater than the sum of the solute-solute and solvent-solvent interaction energies, the solute dissolves in a solvent.

Again we know, $\Delta G = \Delta H - T\Delta S$.

The three factors, solute-solute interaction, solvent-solvent interaction and solute-solvent interaction, contribute to ΔH and hence they also contribute to ΔG; the first two factors have positive ΔH and the last one has negative contribution. Besides enthalpy factor there is entropy factor. Usually when a solute gets dissolved in a solvent spontaneously, the system goes from the less ordered system to a more ordered system and hence the entropy change (ΔS) is negative.

If, however, the dissolution involves dissociation of the solute and formation of solvated units, the entropy change may be positive or negative depending upon the amount of solvation of the dissociated and undissociated systems. If the undissociated system solvates more than the dissociated one, the former will be more stable than the latter and, therefore, the undissociated system will have less entropy than the dissociated system and naturally the ΔS will be positive; whereas if the dissociated system gets more solvated than the undissociated one, the entropy change will be negative. For an example, when acetic acid is dissolved in water, acetic acid gets solvated less than the acetate ion and hydroxium ion system; and, therefore, in this reaction the entropy change is negative while enthalpy change is also negative (-92 JK^{-2} mol^{-1} and 0.46 kJ mol^{-2} respectively).

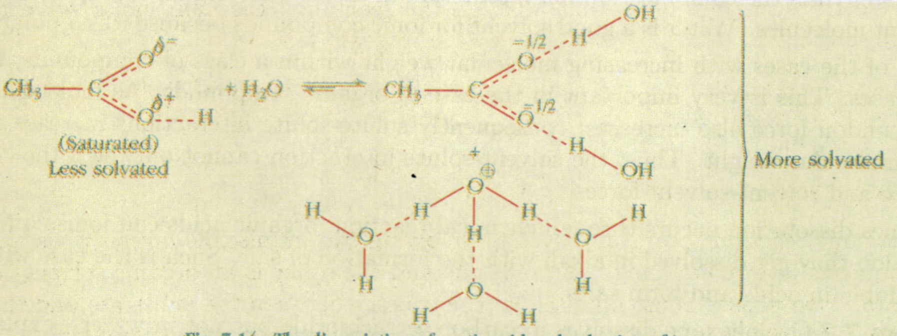

Fig. 7.14. The dissociation of acetic acid and solvation of ions

7.7.4 The effect of solvation on equilibria

In general, an equilibrium will be shifted to that side by a change in solvent which will stabilise the side more than the other. An example has already been given in connection with H-bonding force. Here is another example.

Water forms H-bond with the carbonyl groups and brings stability to the 1, 3-diketone system while hexane does not form H-bond and does not bring stability to the keto form, as a result of which when hexane is replaced by water the equilibrium of the keto-enol tautomerism shifts from the enol side to the keto side.

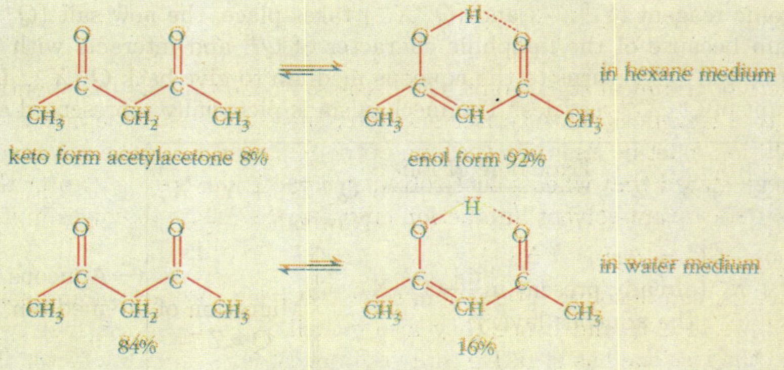

Fig. 7.15. The effect of hydration on the keto-enol system

7.7.5 The effect of solvation on rates of reactions

In the sixth chapter it has been stated that the lower the energy barrier, i.e., the free energy of activation, the higher is the rate. Therefore, if solvation stabilises the T.S. more than the reactants, the $\Delta G^{\ddagger}$ will come down and there will be an increase in rate. Now the T.S. of a reaction may have:

(i) more electrical charges on it than that on the reactants: the more polar solvent will bring more stability to the T.S. than to the reactants and the rate will increase since a more polar solvent stabilises more a more charged species than a less charged species.

(ii) less electrical charges on it than that on the reactants, the more polar solvent will less stabilise the T.S. than the reactant and the rate will fall.

We shall find examples when we shall discuss the substitution and the elimination reactions in detail (Ch 12 & 15).

7.8 Phase Transfer Catalysts (PTC)

Many organic reactions occur in a system involving two immiscible layers (phases). Usually one is an aqueous layer containing ionic reagents and the other is an organic solvent containing the substrate. Two phases being immiscible, the contact between the substrate and the reagent occurs at the interface only, as a result of which the reaction occurs but very slowly. Vigorous stirring increases the rate but to a small extent. In fact, the rate of the reaction is dependent on the rate of stirring.

There are certain catalysts which transfer the requisite part of the ionic reagent from the aqueous phase to the organic phase and thus help the reagent interact with the substrate in the organic phase and thereby the rate of the reaction increases; these catalysts, since transfer one component from one phase to another, are known as phase transfer catalysts.

7.8.1 Mechanism involving PTC

Quaternary ammonium, phosphonium, arsonium, etc. salts ($Q^{\oplus}X^{-}$) are used as PTC. The hydrocarbon part ($Q^{\oplus}$) of $Q^{\oplus}X^{-}$ being lipophilic (lipid seeking), it is attracted by the organic solvent whereas $Q^{\oplus}x^{-}$ is soluble in water being ionic. In the aqueous medium exchange of ions between the ionic reagent ($Y^{\oplus}Z^{-}$) and ($Q^{\oplus}X^{-}$) takes place, the new salt ($Q^{\oplus}Z^{-}$) goes to the organic medium because of the lipophilic character of $Q^{\oplus}$ and interacts with the substrate to give the product and $Q^{\oplus}$ returns to the aqueous medium to give back $Q^{\oplus}X^{-}$. In this reaction a catalytic amount of $Q^{\oplus}X^{-}$ recycles. The mechanism is pictorially represented as given below:

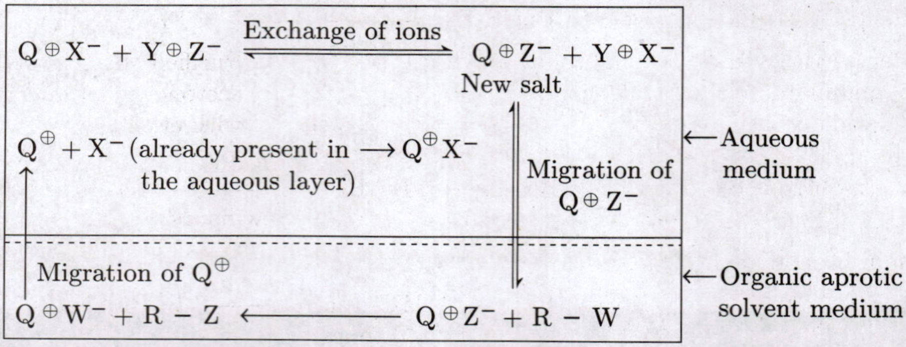

7.8.2 Examples of phase transfer catalytic reactions

1. Substitution reaction :

(i) $\underset{(l)}{CH_3CH_2Br} \xrightarrow[NaNO_2(aq.)]{(n\text{-}Bu)_4N^{\oplus}Cl^{-}(aq.)} CH_3CH_2NO_2$

(ii) $\underset{(l)}{R{-}X} + KF \xrightarrow{PTC} RF + KX$

(iii) $\underset{(l)}{R{-}Cl} + NaCl^{36} \xrightarrow{PTC} R{-}Cl^{36} + NaCl$

2. N—C bond formation: $RNH_2 + CHCl_3 \xrightarrow{PhCH_2N^{\oplus}Et_3Cl^{-}(aq.)} R{-}N{\equiv}C$

3. C—N bond formation: $NH_2NH_2 + CHCl_3 + NaOH \xrightarrow[CH_2Cl_2\ (solvent)]{(n\text{-}Bu)_4N^{\oplus}Cl^{-}(aq.)} CH_2N_2$

4. Addition reaction:

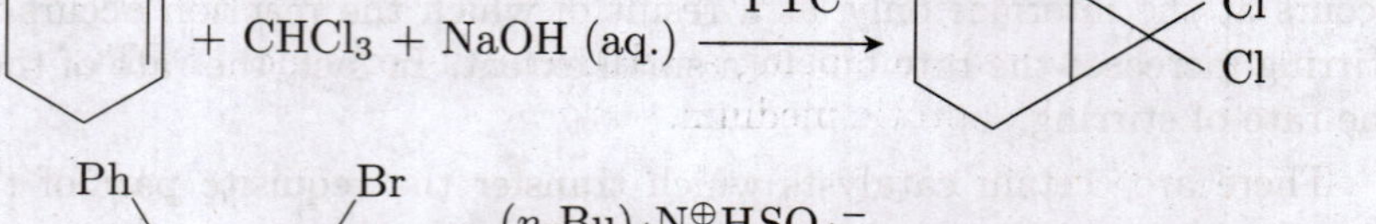

5. Elimination reaction:

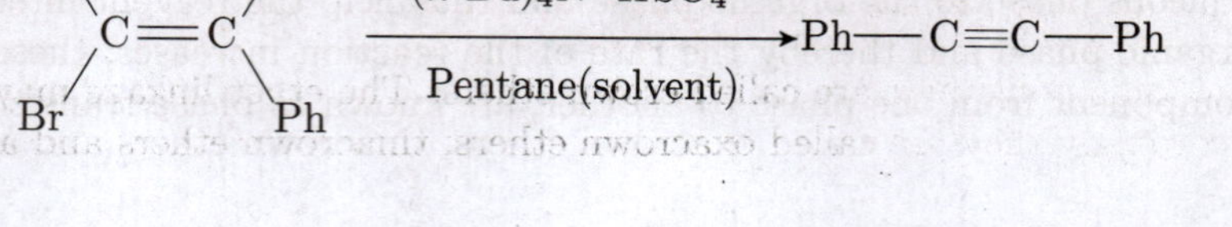

6. C—C bond formation:

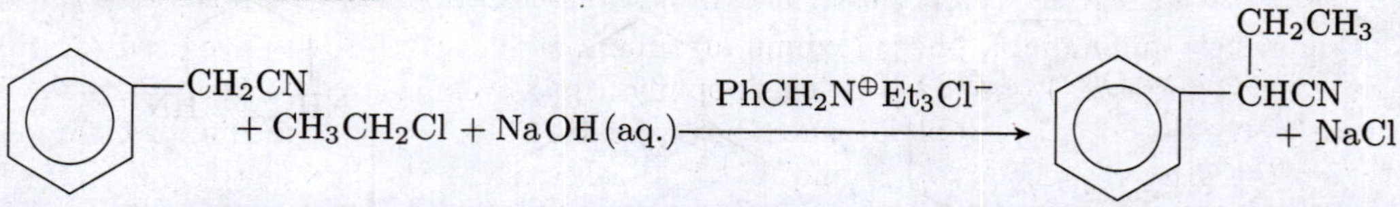

7.8.3 Preparation of PTC

(i) $PhCH_2N^{\oplus}Et_3Cl^-$: $Et_3N + PhCH_2Cl \xrightarrow[\text{64 hours}]{\text{refluxed}} PhCH_2N^{\oplus}Et_3Cl^-$

Benzyltriethylammonium chloride; in absolute ethyl alcohol; Purified by dissolving in acetone and reprecipitated with ether

(ii) $(n\text{-Bu})_4N^{\oplus}Br^-$: $(n\text{-Bu})_3N: + n\text{-Bu-Br} \xrightarrow[\text{a few hours}]{50°C\text{-}100°C} (n\text{-Bu})_4N^{\oplus}Br^-$

Tetra *n*-butylammonium bromide

(iii) $(n\text{-Bu})_4N^{\oplus}Cl^-$: $(n\text{-Bu})_4N^{\oplus}Br^-$ in CH_2Cl_2 + NaCl (aq.) 40 times $\longrightarrow (n\text{-Bu})_4N^{\oplus}Cl^-$

(iv) $(PhCH_2)_3P^{\oplus}EtI^-$: $(PhCH_2)_3P + EtI \xrightarrow[\text{MeCN (sol.)}]{\text{refluxed}} (PhCH_2)_3P^{\oplus}EtI^-$

Ethyltribenzylphosphonium iodide

(v) $(n\text{-Bu})_4N^{\oplus}HSO_4^-$:

$$(n\text{-Bu})_4N^{\oplus}Br^- \text{ in Ph—Cl} \xrightarrow[80°C\text{–}85°C]{Me_2SO_4} \text{mixture} \xrightarrow[\text{refluxed for 48 hours}]{\text{conc. } H_2SO_4 \text{ in} H_2O} (n\text{-Bu})_4N^{\oplus}HSO_4^-$$

7.8.4 Advantages of PTC

(a) Slow reactions can be made fast in the presence of PTC.
(b) Its use requires mild condition.
(c) Expensive aprotic solvents are not required.
(d) The reactions can be carried out in the presence of water.
(e) Weaker bases like $\bar{O}H$ can replace stronger bases like $:\bar{\ddot{N}}H_2$, $\bar{O}R$, H^-, etc.
(f) Anions of the reagents being available in the presence of PTC, weaker nucleophiles become stronger in the aprotic medium.

7.9 Crown Ethers

Cyclic polyethers are called crown ethers. The ether linkage may have O or S or N atom and accordingly they are called oxacrown ethers, thiacrown ethers and azacrown ethers, respectively.

Oxacrown ether | Thiacrown ether | Azacrown ether

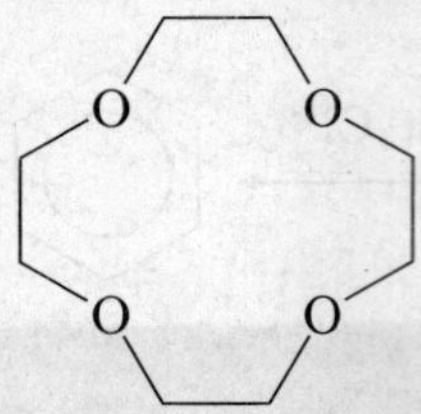

[12]-Crown-4

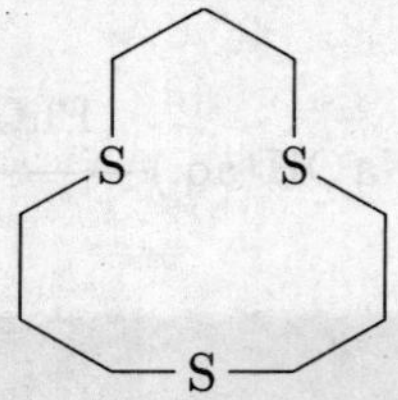

[12]-Crown-3

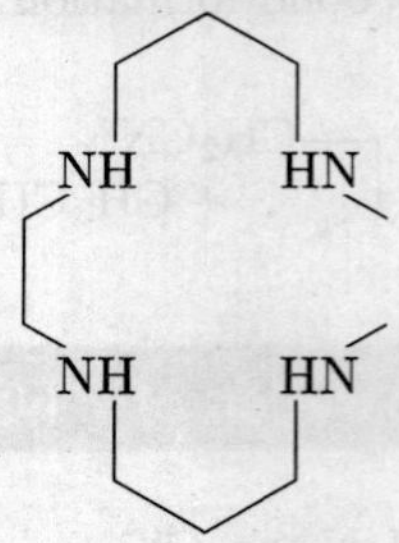

[14]-Crown-4

In the above names, the first number describes the number of constituent atoms of the ring (crown) and the second one designates the number of heteroatoms that constitute the ether linkage in the ring. Some more examples are as follows:

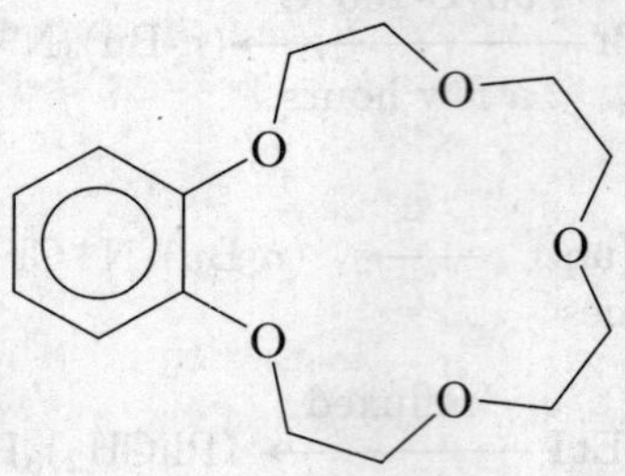

Benzo-[15]-Crown-5
B[15]C-5

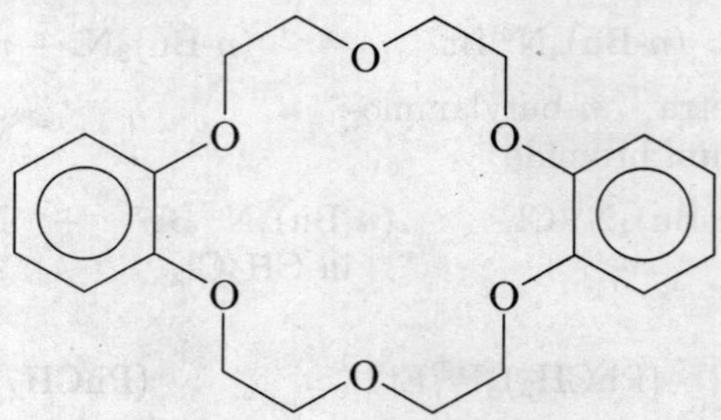

Dibenzo-[18]-Crown-6
DB[18]C-6

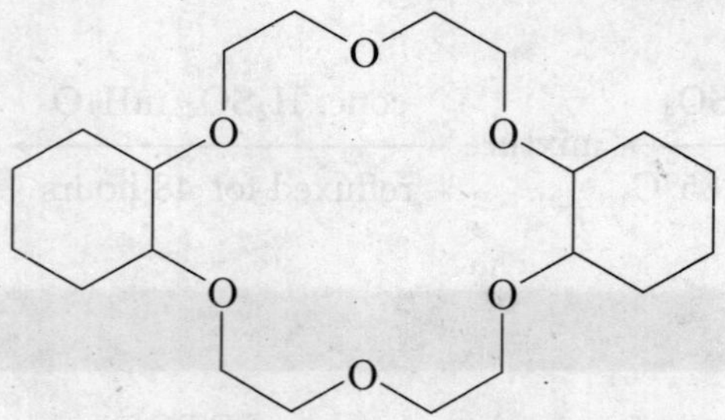

Dicyclohexano-[18]-Crown-6
DCH[18]C-6

Every crown ether has a cavity within it. The dimension of the cavity depends on the size of the ring. For oxacrown ethers the cavity diameters are given below:

Crown ether	Cavity diameter in pm
[12]-Crown-4	120-125
[15]-Crown-5	170-220
[18]-Crown-6	260-320

The interior of the Crown ethers contains hetero atoms (O, N, S) with lone-pair of electrons. While the exterior of the cavity is built up by the C atom. Alkali metal and alkaline earth metal cations have fine accommodation in these cavities as given below:

Metal ion	Diameter of the ion in pm	Oxacrown ether
$Li^{\oplus}$	120	[12]-Crown-4
$Na^{\oplus}$	180	[15]-Crown-5
$K^{\oplus}$	265	[18]-Crown-6

The exterior of the cavity being hydrocarbon-like (lipophilic), it is attracted by organic solvents; while interior of the cavity attracts cation through their l.p. *es*. For this reason water soluble compounds like NaOH, $KMnO_4$ etc. after dissolution in water when treated with appropriate crown ether, the cation gets its accommodation in the cavity to form a complex which owing to its lipophilic exterior goes to the organic medium. If in a reaction whose substrate remains in the organic solvent but the reagent remains in the aqueous medium, the reaction can readily be carried out by adding a catalytic amount of crown ether which will aptly take the reagent to the organic medium. For example, the reaction between an alkyl halide in an aprotic solvent and aqueous potassium cyanide could safely be converted to alkyl cyanide by treating the system with a catalytic amount of [18]-Crown-6. Thus crown ethers also act as PTC. They transfer reagents from the aqueous phase to the organic solvent phase where substrates are present and make them react with the substrate. Some of the examples are given below:

(i) 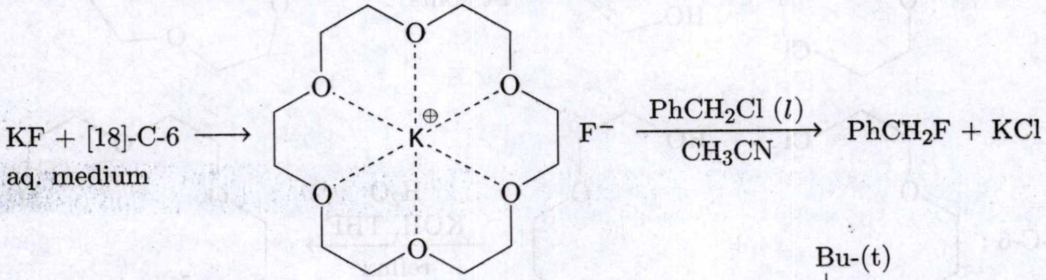

(ii) 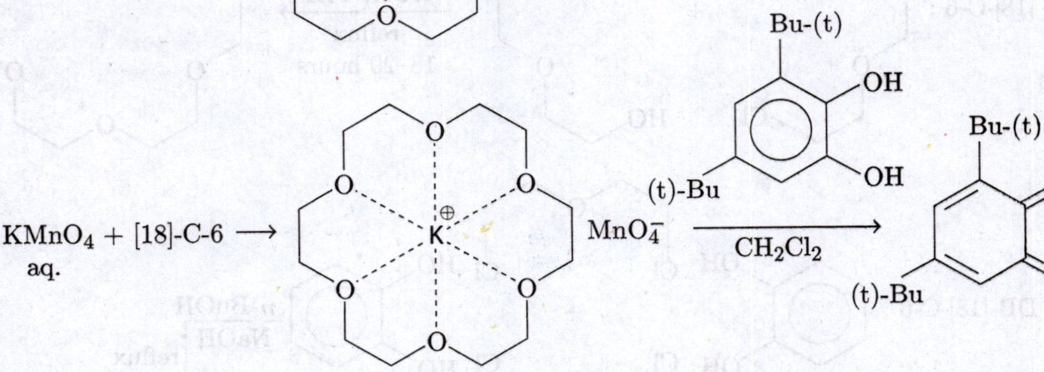

(iii) 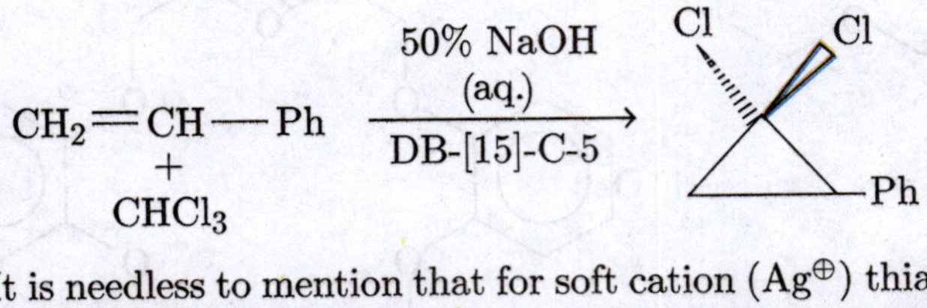

It is needless to mention that for soft cation ($Ag^{\oplus}$) thiacrowns are to be used since S atoms are soft donors; while for $Na^{\oplus}$, $K^{\oplus}$ oxacrowns are the right choice because both cations and oxygen atoms are hard centers. Thus the cavity size, the size of the cation and the hard or soft nature of the interacting ions and atoms are to be considered in selecting the right crown ethers.

7.9.1 Synthesis of crown ethers

(a) [12]-C-4 :

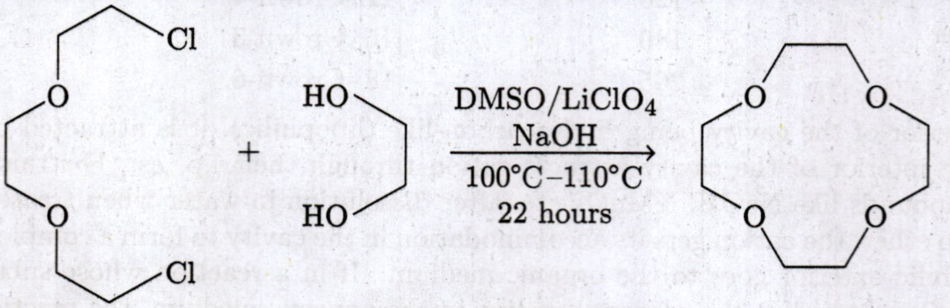

(b) [15]-C-5 :

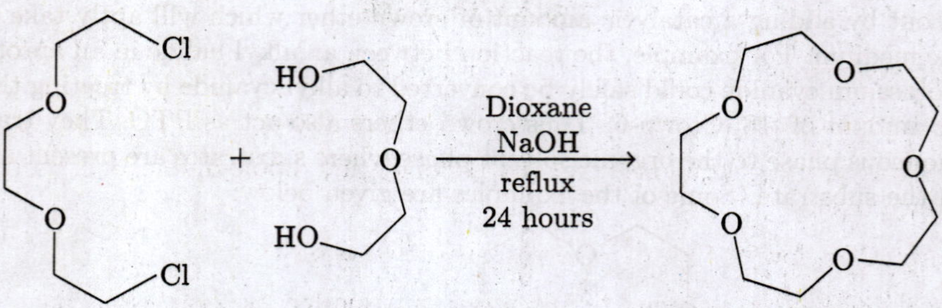

(c) [18]-C-6 :

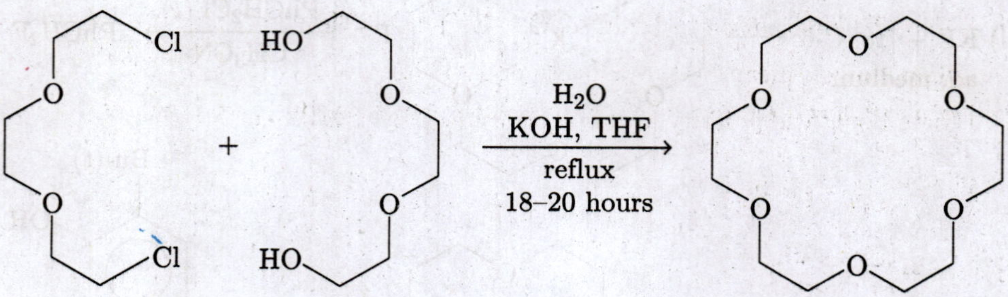

(d) DB-[18]-C-6 :

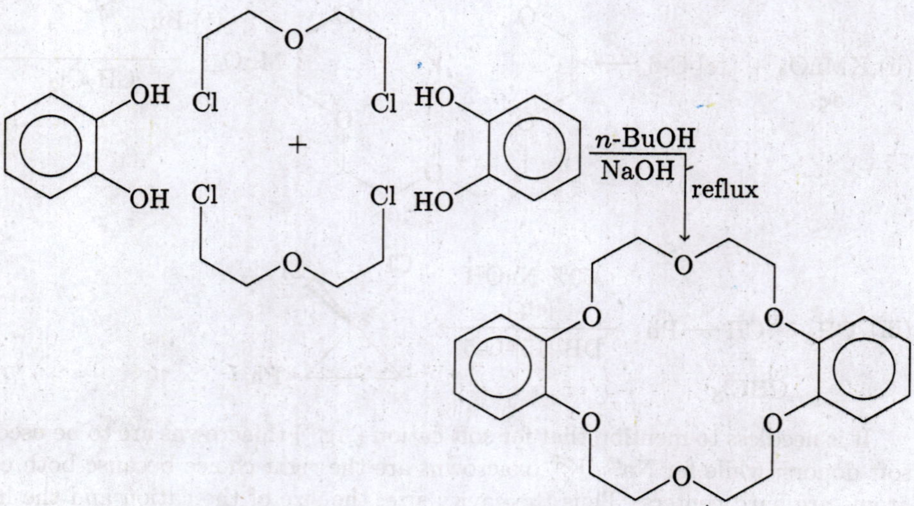

(e) DCH-[18]-C-6 : DB-[18]-C-6 $\xrightarrow[\text{Ru on } Al_2O_3 \text{ (cat)}]{H_2;\ 5\%}$ 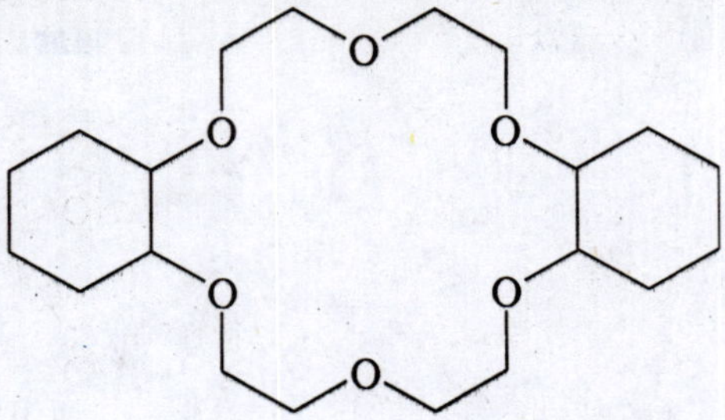

Exercise

1. Why do ionic compounds dissolve in polar solvents?
2. Non-polar molecules dissolve in non-polar solvents—explain.
3. Write notes on : (i) London force, (ii) H-bond, (iii) dipole-dipole interaction, (iv) PTC, (v) Crown ethers.
4. Why are ethers low boiling than corresponding alcohols?
5. Ethyl alcohol boils at a higher temperature than methyl alcohol—explain.
6. Write down solvation order of MeO^-, EtO^- and Me_3CO^-.
7. What do you mean by the polarity of a solvent ?
8. Write down the polarity order of DMSO, DMF, HMPT.

(c) DB-[18]-C-6 $\xrightarrow[\text{Ru on } Al_2O_3(\text{cat})]{H_2,\ 5\%}$ DCH-[18]-C-6

EXERCISES

1. Why do ionic compounds dissolve in polar solvents?
2. Non-polar molecules dissolve in non-polar solvents : explain.
3. Write notes on : (i) London force; (ii) H-bond, (iii) dipole-dipole interaction, (iv) PTC, (v) Crown ethers.
4. Why are ethers low boiling than corresponding alcohols?
5. Ethyl alcohol boils at a higher temperature than methyl alcohol : explain.
6. Write down solvation order of MeO^-, EtO^- and Me_3CO^-.
7. What do you mean by the polarity of a solvent ?
8. Write down the polarity order of DMSO, DMF, HMPT.

Chapter 8

Acids and Bases

Objectives

◈ To understand:

* What are organic acids and bases [8.1] p. 247
* The measure of acid-base strength [8.2] p. 247
* Energy involved in acid-base equilibria [8.3] p. 249
* The scale of acidity [8.4] p. 252
* Relative strength of acids and bases from the consideration of structural effects [8.5] p. 254
* Acid and base catalysis [8.6] p. 262

8.1 Introduction

According to Bronsted and Lowry, acids are proton donors and bases are proton acceptors. Lewis defined an acid as an electron-pair acceptor and a base as an electron-pair donor. Organic chemists are much concerned with the Bronsted's definition of acids and bases. In general, organic acids consist of —COOH (carboxyl) group and bases contain $-\ddot{N}-$ or $-\ddot{O}-$ group. But, if we can take out a proton ($H^{\oplus}$) from any organic species, we can call it an acid. Similarly, if ($H^{\oplus}$) can be added to a species, the latter may be called a base. For example, if a proton is taken out from acetylene (1), it forms acetylide ion (2); acetylene is then termed as an acid and the acetylide ion is named as a base because the latter species can accept a proton to form acetylene back.

$$\underset{\mathbf{1}}{HC\equiv C-H} \underset{+H^{\oplus}}{\overset{-H^{\oplus}}{\rightleftharpoons}} \underset{\mathbf{2}}{HC\equiv C:^{-}}$$

Here, of course, for the forward reaction acetylide ion is called the conjugate base of the acid acetylene; but for the reverse reaction, acetylene is the conjugate acid of the base acetylide. Acetylene and acetylide ion may thus be termed as conjugate acid-base pair.

Protons may be donated and accepted by C, N and O, etc. atoms of organic substances and accordingly we designate the organic acids and bases as carbon acids and bases, nitrogen acids and bases and oxygen, etc. acids and bases respectively. Acetylene is a carbon acid and acetylide ion is a carbon base.

Problem 8.1. Show the conjugate acids and bases in the reaction of methyl cyanide with hydrochloric acid.

Solution: $CH_3C\ddot{N} + HCl \rightleftharpoons CH_3C\equiv\overset{\oplus}{N}H + Cl^{-}$

There are two conjugate acid-base pairs:

1. For the forward reactions:
   - **(a)** CH_3CN is a base as it accepts proton and $CH_3C\overset{\oplus}{N}H$ is its conjugate acid.
   - **(b)** HCl is an acid because it donates proton, and Cl^{-} is its conjugate base.
2. For the backward reaction:
   - **(a)** $CH_3C\overset{\oplus}{N}H$ is an acid and CH_3CN: is its conjugate base.
   - **(b)** Cl^{-} is a base and HCl is its conjugate acid.

8.2 Measures of Acid-Base Strength

The strength of an acid, i.e., acidity is measured by the extent to which it can lose proton in a solvent medium which acts as a base. Water is the common solvent used for this purpose. Other non-aqueous solvents can also be used if required. Thus we get the following equilibrium for the acid-solvent system, because organic acids are weak acids and do not dissociate completely :

$$HA + \text{solvent} \rightleftharpoons \text{solvent}.H^{\oplus} + A^{-}.$$

Here, HA is the acid and A^{-} is its conjugate base. Now the equilibrium constant,

$$K_a = \frac{[\text{solvent}.H^{\oplus}][A^{-}]}{[HA]}. \tag{8.1}$$

Here, (1) concentration terms are used instead of activity and (2) the concentration of solvent is taken as unity, because in the system, the solvent is present in large excess and a very little of it takes part in the reaction.

The K_a is called the dissociation constant or ionisation constant of the acid and is a measure of its acidity. The higher the value of K_a, the greater is the acidity. Thus, in the aqueous medium, the K_a value of benzoic acid is 6.1×10^{-5} and that of acetic acid is 1.75×10^{-5}; hence, acetic acid is a weaker acid than benzoic acid.

Although the dissociation constant of an acid expresses the acidity, it is inconvenient to express it in words because of its exceedingly small numerical value. To avoid this the acidity is expressed as the negative logarithm of K_a with respect to the base ten, i.e., $-\log_{10} K_a$ and this is called **pK_a**.

In water medium at 25°C, K_a of acetic acid is 1.75×10^{-5}. Therefore, $pK_a = \log_{10}(1.75 \times 10^{-5})$ of 4.76.

The pK_a of benzoic acid under the identical condition is 4.20. Thus the smaller the value of pK_a, the stronger is the acid and vice versa. In fact, if two acids differ in their pK_a values by one unit, the acidity of one acid will be ten times that of the other acid. Therefore, benzoic acid is $(4.76 - 4.20) \times 10$ or 5.6 times stronger than acetic acid.

Thus K_a and pK_a values can be used to compare the acidity of weak acids but these cannot be used to compare the same for strong acids. Organic acids being weak, pK_a values are used to compare their relative strengths.

Basicity, i.e., the strength of a base is measured by the extent to which it can gain proton in a solvent that acts as an acid. Water being amphoteric in nature, it is also the common solvent for the purpose of measuring basicity. If necessary, other solvents too can be used. Thus the base-solvent equilibrium may be represented as:

$$B: + HA \text{ (solvent)} \rightleftharpoons \overset{\oplus}{HB} + A^-.$$

The ionisation constant or equilibrium constant of the base, B : is

$$K_b = \frac{[\overset{\oplus}{HB}][A^-]}{[B:]} \qquad (8.2)$$

Here too the activity terms are replaced by the concentration terms, and the concentration of HA, the solvent, is taken as unity.

K_b values of bases may be used to express and compare basicity of a series of bases. As the **K_b** increases, the basicity increases. Thus, **K_b** of pyridine is 2.3×10^{-9} and that of ammonia is 1.8×10^{-5}; this means ammonia is a stronger base than pyridine.

For the same reason as with K_a, the basicity is expressed by **pK_b**; naturally the greater the numerical value of **pK_b**, the weaker is the base and vice versa. Thus, **pK_b** value of CH_3NH_2 and $(CH_3)_2NH$ are 3.36 and 3.23 respectively. Therefore, dimethylamine is a stronger base than methylamine; the former base is $(3.36 - 3.23) \times 10$ or 1.3 times stronger than the latter base in the aqueous medium.

Problem 8.2. Write down the formulae for calculating the dissociation constants of **(a)** acetic acid and **(b)** methylamine in the aqueous medium.

Solution: (a) $CH_3COOH + H_2O \rightleftharpoons CH_3COO^- + H_3O^{\oplus}$

$$K_a = \frac{[CH_3COO^-][H_3\overset{\oplus}{O}]}{[CH_3COOH]}$$

(b) $CH_3NH_2 + H_2O \rightleftharpoons CH_3\overset{\oplus}{N}H_3 + \overset{-}{O}H$

$$K_b = \frac{[CH_3\overset{\oplus}{N}H_3][\overset{-}{O}H^-]}{[CH_3NH_2]}.$$

Problem 8.3. Calculate the pK_a and pK_b for formic acid and ethylamine respectively in the aqueous medium; given that K_a for formic acid is 2.0×10^{-4} and K_b of ethylamine is 4.6×10^{-4}.

Solution:

$$\begin{aligned} pK_a \text{ of formic acid} &= -\log_{10} K_a \\ &= -\log_{10}(2.0 \times 10^{-4}) \\ &= 3.75. \end{aligned}$$

$$\begin{aligned} pK_b \text{ of ethylamine} &= -\log_{10} K_b \\ &= -\log_{10}(4.6 \times 10^{-4}) \\ &= 3.34. \end{aligned}$$

8.3 Acidity Constants and Standard Free Energy

i. In solution:

The relation between the equilibrium constant and a standard free energy change can safely be used in the case of dissociation of acids. The relation is,

$$\Delta G^\circ = -RT \ln K_a. \tag{8.3}$$

Where ΔG° is the standard free energy change; K_a, the equilibrium constant, i.e., the dissociation constant of the acid at 298 K(T).

Equation (8.3) may be written as:

$$\begin{aligned} \Delta G^\circ &= -8.314 \times 298 \times 2.303 \times \log K_a \text{ J mol}^{-1}; \text{ R} = 8.314 \text{ J degree}^{-1} \text{ mol}^{-1} \\ &= -5.706 \text{ kJ} \log K_a \text{ mol}^{-1} \text{ [dividing the product by 1000]} \\ &= 5.706 \text{ kJ} \times pK_a \text{ mol}^{-1} \quad [\because -\log_{10} K_a = pK_a] \end{aligned} \tag{8.4}$$

$$\therefore \quad pK_a = \frac{\Delta G^\circ}{5.706 \text{ kJ mol}^{-1}}. \tag{8.5}$$

The unit of ΔG° is kJ mol^{-1}, so both the denominator and the numerator have the same unit; pK_a has, therefore, no unit.

For example, ΔG° of ionization of benzoic acid is 24 kJ mol^{-1}.

$$\therefore \quad pK_a \text{ (benzoic acid)} = \frac{24}{5.706} \quad \text{or,} \quad 4.206.$$

Thus systems having different ΔG° values must have different pK_a values and hence different K_a values.

Again, we know that, $\Delta G^\circ = \Delta H^\circ - T\Delta S^\circ$,

where ΔH° is the standard enthalpy change for O—H or N—H bond breaking of the acids and new bond formation with the solvent molecule (H_2O in the aqueous medium) in the cases of oxygen and nitrogen acids, and ΔS° is the standard entropy change which mostly depends on the solvation factor. When the acid-solvent system does not involve ions but the dissociated system involves and the solvent is polar, the acid- solvent system is less stable and hence possesses higher entropy than the dissociated system. Of course, in the liquid state, the solvent is in a well-ordered state and, therefore, the entropy change, though negative, is not of high value. On the other hand, if the conjugate base developed gets less solvated than the undissociated acid or if the acid-solvent system involves ions but the dissociated system does not, then the standard entropy of the acid-solvent system will be less than the standard entropy of the dissociated system and there will be a positive standard entropy change. Actually standard entropy change is the determining factor of the acid constants in many a case.

Let us consider the acetic acid-water and trichloroacetic acid-water systems:

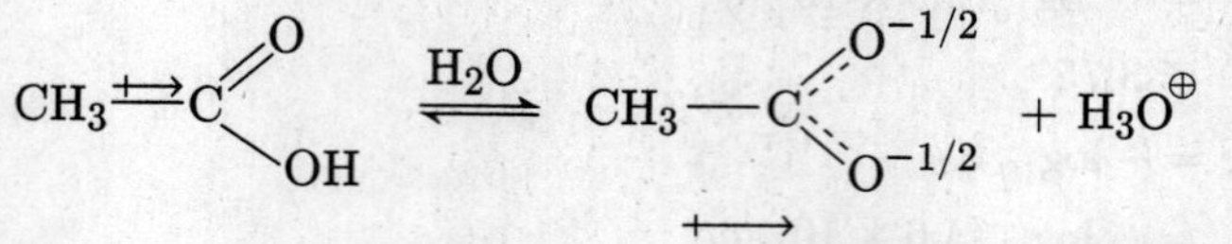

μ increases in solvation, more solvated

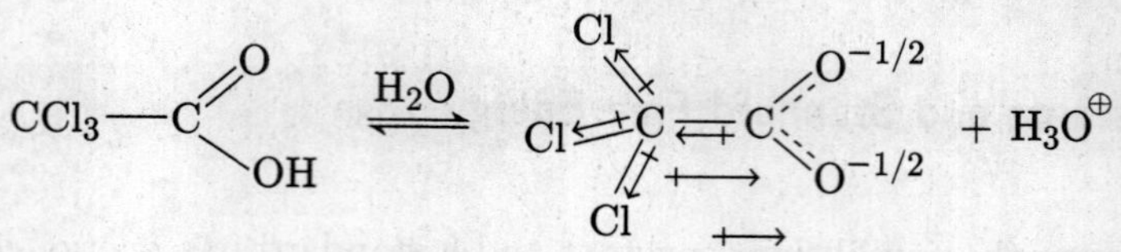

small change of μ in solvation, less solvated

In the case of acetic acid-water system, the products being more solvated than the reactants, $\Delta S°$ is negative (-0.092 kJ deg^{-1} mol^{-1}); while in the case of trichloroacetic acid the products being less solvated than the reactants, $\Delta S°$ is positive (+0.0084 unit). In the latter case dipolar interactions between CCl_3 and CO_2^- in trichloroacetate ion are in opposite directions and almost equal in magnitude and, therefore, the negative charge disperses over the whole of the species making it less prone to ion-solvent interactions than trichloroacetic acid itself. In the case of acetic acid the negative charge in the acetate ion is less dispersed and hence it undergoes more ion-solvent interaction than trichloroacetate ion while hydroxium ($H_3O^{\oplus}$) ion is solvated to the same extent in both the cases. In so far as $\Delta H°$ is concerned, the former has a small negative value (-0.46 kJ mol^{-1}) and the latter has a positive value (6.14 kJ mol^{-1}). Now computing these values of $\Delta H°$ and $\Delta S°$ we get,

$$\Delta G°_{CH_3COOH/H_2O} = [-0.46 - (-0.092)298] \text{ kJ mol}^{-1}$$

$$= [298 \times 0.092 - 0.46] \text{ kJ mol}^{-1}$$

$$= 26.95 \text{ kJ mol}^{-1}$$

$$\therefore \quad pK_{a_{CH_3CO_2H}} = \frac{26.95}{5.706} \text{ [using Equation 8.5]} = 4.72$$

$$\Delta G°_{CCl_3CO_2H/H_2O} = [6.14 - (0.0084) \times 298] \text{ kJ mol}^{-1} = 3.64 \text{ kJ mol}^{-1}$$

$$\therefore \quad pK_{a_{CCl_3CO_2H}} = \frac{3.64}{5.706} = 0.64.$$

Though the calculated pK_a values do not tally with the experimental pK_a values, yet it can be concluded that trichloroacetic acid is a stronger acid than acetic acid.

Table 8.1 A few pK_a and the corresponding $\Delta G°$, $\Delta H°$, $\Delta S°$, $T\Delta S°$ values of some organic acids.

Acid	$\Delta G°$ in kJ mol^{-1}	$\Delta H°$ in kJ mol^{-1}	$\Delta S°$ in kJ deg^{-1} mol^{-1}	$T\Delta S°$ in kJ mol^{-1}	pK_a
Formic	+21.5	− 0.17	−0.073	−21.75	3.75
Acetic	+27.14	− 0.46	−0.092	−27.4	4.75
Propionic	+27.8	− 0.96	−0.096	−28.6	4.87
Trimethylacetic	+28.8	− 3.02	−0.107	−31.9	5.03
Trichloroacetic	+ 3.64	+ 6.14	+0.0084	+ 2.50	0.64
Ammonium	+52.7	−51.80	+0.0029	− 0.86	9.245
Methylammonium	+60.5	+54.70	+0.0196	− 5.8	10.62
Dimethylammonium	+61.5	+49.60	−0.040	−11.9	10.77
Trimethylammonium	+55.8	+36.90	−0.063	−18.7	9.80
Phenylammonium	+26.23	+30.90	+0.0155	+ 4.6	4.60
Benzoic	+24.00	+ 0.37	−0.080	−23.8	4.213
p-Hydroxybenzoic	+26.10	+ 2.25	−0.080	−23.8	4.58
p-Toluic	+24.70	+ 1.25	−0.079	−23.5	4.334
m-Toluic	+24.10	+ 0.29	−0.080	−23.8	4.243
p-Bromobenzoic	+22.8	+ 0.46	−0.075	−22.3	4.002
p-Chlorobenzoic	+22.7	+ 0.83	−0.073	−21.7	3.986
m-Chlorobenzoic	+21.8	− 0.084	−0.073	−21.7	3.827
p-Cyanobenzoic	+20.3	+ 0.12	−0.068	−20.3	3.551

Putting the standard thermodynamic values (as given in Table 8.1) one can at ease calculate the pK_a value for an acid. For example, putting the $\Delta G°$ values from the table and using Eqn. 8.5, we get,

$$pK_{a_{CH_3COOH}} = \frac{27.14}{5.706} = 4.75 \quad \text{and} \quad pK_{a_{CCl_3CO_2H}} = \frac{3.64}{5.706} = 0.6379$$

ii. In the gas-phase :

To avoid solvent-acid interactions, nowadays gas-phase measurements of ΔH has been done as per equations given below:

$$HA(g) \xrightarrow[\text{energy}]{\text{bond dissociation}} H\cdot(g) + A\cdot(g) \xrightarrow{\text{electron affinity}} H^{\oplus}(g) + A^{\overline{\cdot}}(g)$$

$$B(g) + H^{\oplus}(g) \xrightarrow{\text{proton affinity}} BH^{\oplus}(g)$$

If we consider ionisation energy of hydrogen and entropy change, the values given in Table 8.2 will change a lot.

Table 8.2 ΔH values for calculating gas-phase acidities and basicities

A. Acid	ΔH in kJ mol^{-1}	B. Base	ΔH in kJ mol^{-1}
Formic acid	+ 132	NH_3	+ 828
Acetic acid	+ 146	$MeNH_2$	+ 878
Trifluoroacetic acid	+ 38	$EtNH_2$	+ 891
Benzoic acid	+ 112	Me_2NH	+ 906
Phenol	+ 151	Me_3N	+ 923

Gas-phase acidities increase with the increasing bulk of the alkyl group; this is because of the greater polarisability of the O—H bond with increasing bulk of the alkyl group bonded to the carboxyl group. For example:

$$\text{Gas-phase}: \quad \underset{\text{increasing acidity}}{\xrightarrow{CH_3COOH < CH_3CH_2COOH < CH_3CH_2CH_2CO_2H}}$$

$$\text{Aqueous medium}: \quad \underset{\text{increasing acidity}}{\xrightarrow{CH_3CH_2CH_2CO_2H < CH_3CH_2CO_2H < CH_3CO_2H}}$$

However, the basicity order of the following is the same in the gas-phase as well as in chlorobenzene (a non-hydroxylic non-polar solvent): $NH_3 < MeNH_2 < Me_2NH < Me_3N$.

In the cases of carbon acids, the rate of C—H bond-breaking can be measured by determining the rate of hydrogen isotope exchange with appropriate deuteron $D^{\oplus}$ donors like D_2O, EtOD and the measured acidity is called **kinetic acidity** which does not correlate with the **thermodynamic acidity** as discussed above.

Table 8.3 Common acids and bases used in organic reactions.

	Acids	pK_a	Bases	pK_a	
Decreasing pK_a ↓	CH_3COOH	4.75	Pyridine	5.2	Increasing pK_a ↓
	CF_3COOH	0.0	CH_3CO_2Na	9.25	
	HCl	−7	Et_3N	~ 10	
	$C_6H_5SO_3H$		NaOH	15.7	
	BF_3		$NaOCH_3$	16	
	H_2SO_4	−9	$KOC(CH_3)_3$	19	
	HBF_4		ϕ_3CNa	35	
	FSO_3H		$NaNH_2$	36	
	$FSO_3H.SbF_5$	< −20	$LiNEt_2$		
			n-BuLi		
			t-BuLi		
			NaH		

8.4 Scale of Acidity

$pK_a^{\oplus} + pK_b = 14$; when the medium is water ($\because$ $pK_w = 14$) and where $pK_a^{\oplus}$ and pK_b are the constants for the conjugate acid-base pair ($BH^{\oplus}$ and B). Nowadays the $pK_a^{\oplus}$ values are written as pK_a as shown in Table 8.4. pK_a is the measure of the acid strength of the conjugate acid $BH^{\oplus}$. pK_a of aniline is 4.59; this means pK_a^{+} of $Ph\overset{\oplus}{N}H_3$ is 4.59 in water but

$$pK_a^{\oplus} + pK_b^{\oplus} = 14 \quad \text{or,} \quad 4.59 + pK_b = 14, \quad \therefore \ pK_b \text{ (aniline)} = 9.41.$$

Table 8.4 pK_a, values of some compounds (correlated to the aqueous medium)

Acid($BH^{\oplus}$ or, BH)	pK_a	Conjugate Base (B)	Acid ($BH^{\oplus}$ or, BH)	pK_a	Conjugate Base (B)
Cyclohexane	52	Cyclohexyl anion	p-Nitrophenol	7.2	p-nitrophenoxide
Et—H	50	$Et:^-$	H_2S	7.0	HS^-
Me—H	49	$Me:^-$	H_2CO_3	6.4	HCO_3^-
$CH_2{=}CH_2$	44	$CH_2{=}\bar{\ddot{C}}H$	$(MeCO)_3C—H$	5.9	$(MeCO)_3C:^-$
Ph—H	43	$Ph:^-$	Pyridinium ion	5.2	Pyridine
$PhCH_2—H$	41	$Ph\bar{\ddot{C}}H_2$	$Ph\overset{\oplus}{N}HMe_2$	5.06	$PhNMe_2$
$(Me_2CH)_2N—H$	38	$(Me_2CH)_2\bar{\ddot{N}}$	$CH_3COO—H$	4.75	CH_3COO^-
$:NH_3$	36	$:\bar{\ddot{N}}H_2$	2, 4-Dinitrophenol	4.0	2, 4-Dinitrophenoxide
$Et\ddot{N}H_2$	35	$Et\bar{\ddot{N}}H$	$HCOO—H$	3.7	$HCOO^-$
$H—CH_2CH{=}CH_2$	35	$:\bar{C}H_2—CH{=}CH_2$	$CH_2(NO_2)_2$	3.6	$:\bar{C}H(NO_2)_2$
$Ph_3C—H$	35	$Ph_3C:^-$	$ClCH_2CO_2H$	2.8	$ClCH_2COO^-$
$Ph\ddot{N}H_2$	27	$Ph\bar{\ddot{N}}H$	$RCH(\overset{\oplus}{N}H_3)CO_2H$	2.4	$RCH(NH_2)CO_2H$
$H—CH_2CN$	25	$:\bar{C}H_2CN$	$Cl_2CHCOOH$	1.3	Cl_2CHCOO^-
$H—CH_2CO_2Me$	25	$:\bar{C}H_2CO_2Me$	$Ph_2\overset{\oplus}{N}H_2$	0.8	Ph_2NH
$HC{\equiv}C—H$	25	$H—C{\equiv}\bar{C}:$	CCl_3COOH	0.7	CCl_3COO^-
$CH_3COCH_2—H$	20	$CH_3CO\bar{\ddot{C}}H_2$	$PhSO_3H$	0.7	$PhSO_3^-$
$Me_3CO—H$	18	Me_3CO^-	2, 4, 6-Trinitrophenol	0.4	2, 4, 6-Trinitrophenoxide
MeOH	16	MeO^-	CF_3CO_2H	0.2	CF_3COO^-
$MeCH_2OH$	16	$MeCH_2O^-$	$CH_3CO\overset{\oplus}{N}H_3$	0.3	CH_3CONH_2
H_2O	15.7	HO^-	HNO_3	−1.4	NO_3^-
Cyclopetadiene	15	Cyclopentadienyl anion	$H_3O^{\oplus}$	−1.7	H_2O
$CH_2(CO_2R)_2$	13.5	$:\bar{C}H(CO_2R)_2$	$PhCO\overset{\oplus}{N}H_3$	−2	$PhCONH_2$
$(H_2N)_2C{=}\overset{\oplus}{N}H_2$	13.4	$(H_2N)_2C{=}NH$	$Me\overset{\oplus}{O}H_2$	−2.2	CH_3OH
$(CN)_2CH_2$	11.2	$(CN)_2\bar{\ddot{C}}H$	$Et\overset{\oplus}{O}H_2$	−2.4	EtOH
$CH_3COCH_2CO_2R$	11	$CH_3CO\bar{\ddot{C}}HCO_2R$	$Me_3C\overset{\oplus}{O}H_2$	−3.8	Me_3COH
$EtS—H$	10.6	EtS^-	$Me_2\overset{\oplus}{O}H$	−3.8	Me_2O
$H—CH_2NO_2$	10.2	$:\bar{C}H_2NO_2$	$Me_2\overset{\oplus}{S}H$	−5.2	Me_2S
HCO_3^-	10.2	CO_3^{--}	$CH_3CO\overset{\oplus}{O}H_2$	−6.2	CH_3COOH
$R\overset{\oplus}{N}H_3, R_2\overset{\oplus}{N}H_2$	~10	RNH_2, R_2NH	$Ph\overset{\oplus}{O}H_2$	−6.7	PhOH
$R_3\overset{\oplus}{N}H$	~10	R_3N	$PhCH\overset{\oplus}{O}H$	−7.1	PhCHO
PhOH	9.98	PhO^-	$R—C{\equiv}\overset{\oplus}{N}—H$	~−10.0	$R—C{\equiv}N$
$\overset{\oplus}{N}H_4$	9.2	NH_3	$R—\overset{\oplus}{N}O_2H$	~11.0	RNO_2
$H—CN$	9.1	$:\bar{C}N$	H_2SO_4	—	HSO_4^-
$R—COCH_2CO—R$	8.8	$RCO\bar{\ddot{C}}HCO\,R$	$HClO_4$	—	ClO_4^-
			$SbF_5.FSO_3H$	< −20	$SbF_5FSO_3^-$

Table 8.4 tells us that an acid listed in the table will react with a base above it. Since acids are arranged in decreasing pK_a values, the lower the position of the acid, the higher is its acidity. A base (actually its conjugate acid) above it has a pK_a value more than the acid. This means the conjugate acid of the base is a weaker acid than the acid under consideration. Therefore, if an acid reacts with a base above it, the base will generate a weaker conjugate acid and this, in its turn, will dissociate less than the acid under consideration. Thus cyclopentadiene ($pK_a \sim 15$) will react with methoxide ion ($pK_a \sim 16$; its conjugate acid is CH_3OH) to produce cyclopentadienyl anion and methyl alcohol and the equilibrium will be over to the right.

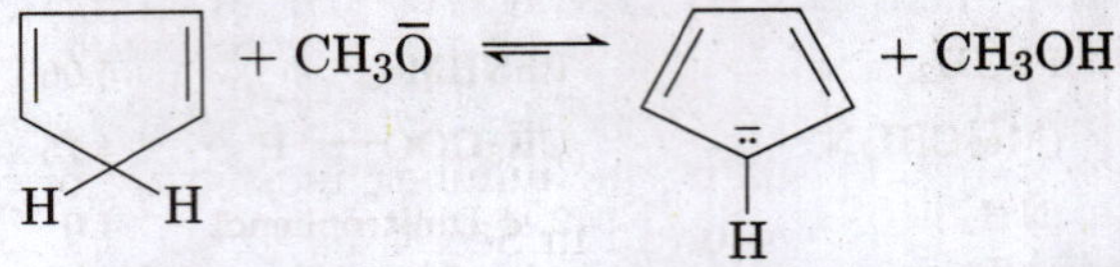

However, if we want to shift the equilibrium almost totally to the right a base of much higher pK_a is to be used. In the reaction of cyclopentadiene as an acid if we use sodium triphenylmethide ($pK_a \sim 35$ of the conjugate acid, Ph_3CH), the reaction may be considered almost irreversible.

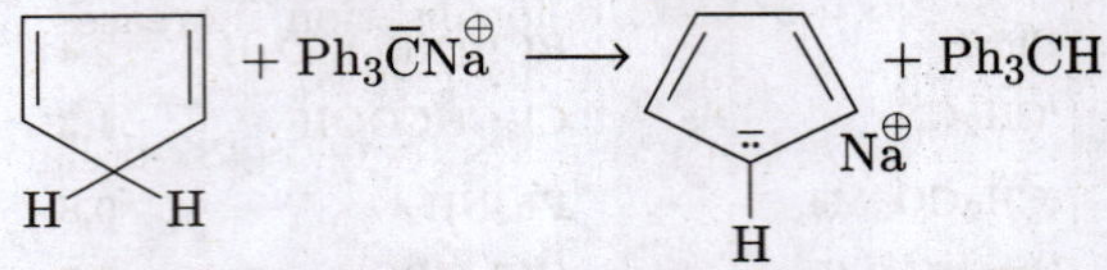

8.5 Relative Strengths of Acids and Bases and Structural Effects

Relative strengths of acids and that of bases in a given series may be assessed with the help of the following considerations:

1. Conjugate acid-base pairs of the given species are to be ascertained.
2. In the case of acids, relative stabilities of the conjugate bases and for bases those of conjugate acids are to be determined in general.

The most stable conjugate base in the series of the conjugate bases has the least tendency to accept a proton, i.e., donate electrons. Hence, it will be the weakest base of the series. Again, if a conjugate base is weak, the corresponding acid will be strong and vice versa. Therefore, the acid which corresponds to the weakest conjugate base will be the strongest acid in the given series. Thus, the stability of conjugate base varies directly with the acidity of the parent acid.

Similarly, the stablest of all the conjugate acids in a series is the most reluctant species to give up proton. This means that the stablest conjugate acid is the weakest one amongst them in acidity. Therefore, the corresponding base will be the strongest in the given series of the bases. So the stability of conjugate acid varies directly with the basicity of the parent base.

Thus, the basicity increases in a series of bases as the acidity of the corresponding conjugate acids decreases and vice versa. In the same way we can say that as the acidity increases in a series of acids, the basicity of the corresponding conjugate bases decreases and vice versa. So acidity and basicity of the conjugate acid-base pair vary inversely with each other.

Relative stabilities of species can be assessed by considering the following effects :

(a) Resonance effect
(b) Inductive effect
(c) Internal H-bonding effect

(d) Effect due to the orbital nature of electron-pair
(e) Steric effect
(f) Solvation effect
(g) Bond length
(h) Electronegativity of the bonded atoms.

More than one of these effects may act simultaneously on an acid-base conjugate pair, but generally one is found to be predominant.

8.5.1 Resonance effect

The stability of a species is determined by the number and nature of contributing structures in a resonance hybrid of the species as discussed in Sec. 4.4.7A.

The more the number of contributing structures in a resonance hybrid of a neutral or negatively charged species, the more is the stability of the hybrid and the greater is the delocalisation of electrons. The greater the delocalisation, the lower is the probability of donating electron-pair. Therefore, as the resonance stabilisation, i.e., delocalisation increases, basicity decreases and hence relatively acidity increases. A few examples are given below :

(i) $Ph\bar{O}$ is a weaker base than $Et\bar{O}$. In these cases the electron-donating atom is oxygen. In $Ph\bar{O}$ +R effect operates as shown below:

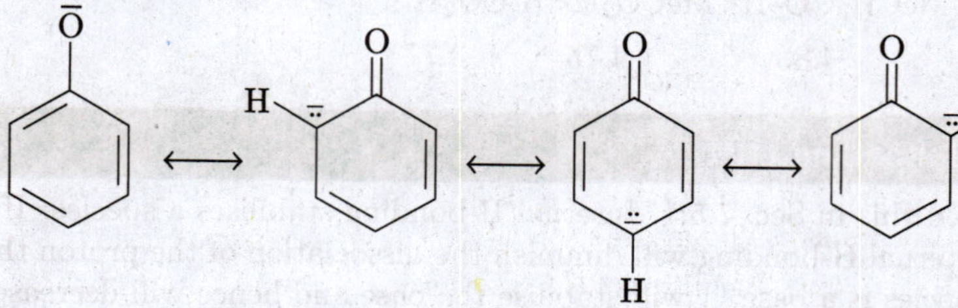

So the electron-pair is less available in the phenoxide ion. In the ethoxide ion, the delocalisation of charge does not occur and at the same time +I effect of Me— increases the base character.

For the same reason aniline is a stronger base than *p*-nitroaniline; N atoms of $—NH_2$ are electron donor in these compounds. In other words, we can say *p*-nitroaniline is a stronger acid than aniline because of resonance interaction involving $—NH_2$ and $—NO_2$ groups.

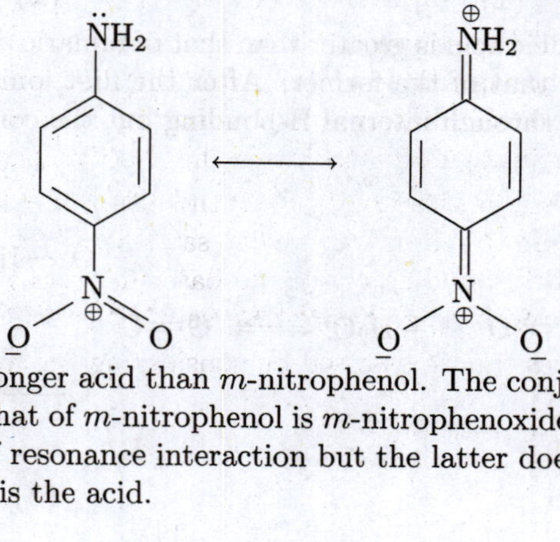

(ii) *p*-nitrophenol is a stronger acid than *m*-nitrophenol. The conjugate base of *p*-nitrophenol is *p*-nitrophenoxide ion and that of *m*-nitrophenol is *m*-nitrophenoxide ion. The former conjugate base gets stability because of resonance interaction but the latter does not. The more stable the conjugate base, the stronger is the acid.

8.5.2 Inductive effect

This has already been discussed in Sec. 4.2(c) & (d). In a nutshell, the following are the two principles used for comparing acidity and basicity of organic species :

(i) +I effect increases the availability of electrons on a particular atom and thus increases the basicity but decreases the acidity of a species. A greater +I effect means a greater basicity and a lower acidity.

(ii) −I effect decreases the electron density on a particular atom of the species and hence decreases the basicity but increases the acidity of the species. The higher the −I effect, the higher is the acidity and the lower is the basicity.

The following pK_a values support the above statements:

(i) $CH_3COOH < ICH_2CO_2H < BrCH_2CO_2H \sim Cl\text{-}CH_2CO_2H < FCH_2CO_2H$

pK_a : 4.75 3.12 2.90 2.86 2.66

(ii) $CCl_3COOH > Cl_2CHCO_2H > ClCH_2CO_2H$

pK_a : 0.65 1.30 2.86

(iii) $ClCH_2CH_2CH_2CO_2H < ClCH_2CH_2CO_2H < ClCH_2CO_2H$

pK_a : 4.52 4.05 2.86

(iv) $Me_3CCO_2H < MeCH_2CO_2H < MeCO_2H < H\text{-}CO_2H$

pK_a : 5.05 4.88 4.75 3.77

8.5.3 Internal H-bonding effect

We have discussed this in Sec. 7.5.1. Internal H-bonding stabilises a species. If the species is an acid, certainly internal H-bonding will diminish the dissociation of the proton therein. On the other hand if the species is a base, it will stabilise the base and hence will decrease its basicity.

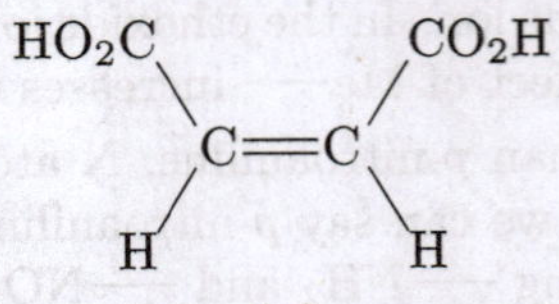

(1)

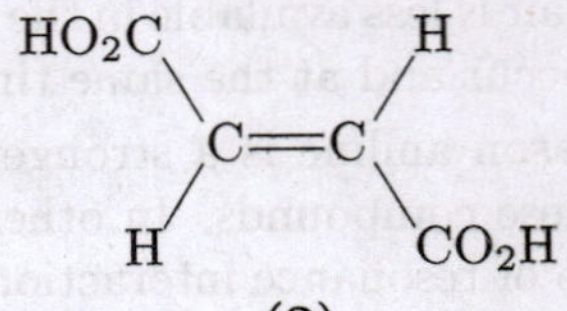

(2)

The first ionisation of maleic acid is greater than that of fumaric acid but the second ionisation of the latter is greater than that of the former. After the first ionisation the conjugate base of the former (3) gets stability through internal H-bonding but the conjugate base of the latter (4) does not.

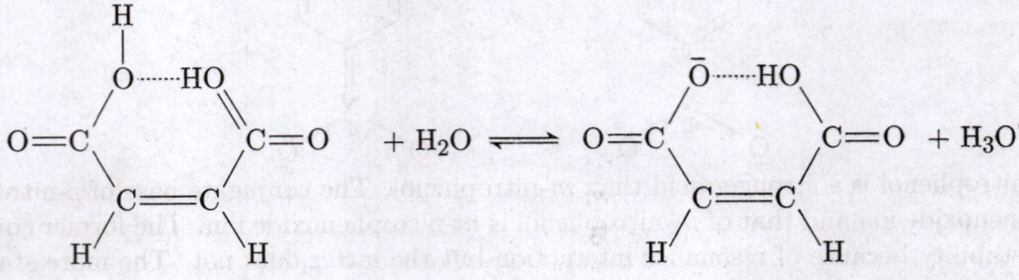

(1) (3)

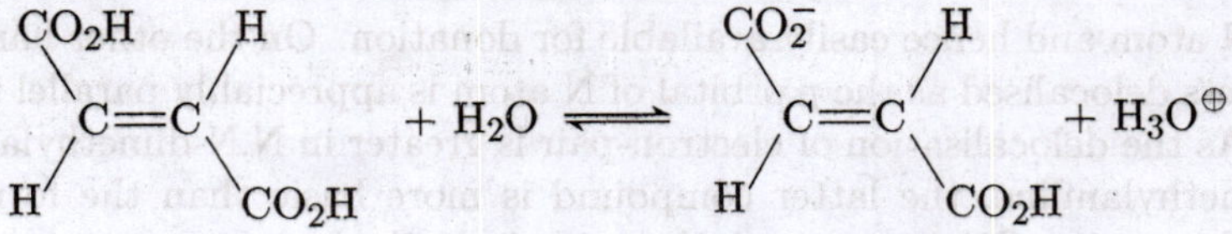

(2) (4)

Hence K_{a_1} of (1) is greater than that of (2); for the same reason K_{a_2} of (2) is greater than that of (1).

8.5.4 Effect due to the orbital nature of electron-pair

As the *s* character of an orbital increases, the attractive force between the electron-pair and the nucleus increases; this means the availability of electron-pair for donation decreases, i.e., basicity decreases and acidity increases.

The order of availability of electron-pair in orbitals is as follows:

$$p > sp^3 > sp^2 > sp > s.$$

Thus methyl cyanide is definitely a very weak base though it contains a lone pair of electrons on N atom. Here the N atom has a triple bond with a C atom and hence N is in the *sp* hybridised state. The lone pair of electrons reside in one *sp* hybrid A.O. Therefore, the lone pair of electrons are strongly bound to the nucleus of N atom and hence it is not readily available for donation. Obviously the basicity of methyl cyanide will be very low.

8.5.5 Steric effect

Steric hindrance sometimes inhibits delocalisation of electrons and thus electrons are then readily available for donation (See Sec. 4.6.4). For example :

Each of the N,N-dimethylaniline and N,N,2,6-tetramethylaniline contains a lone pair of electrons on N atom but the former is six times weaker in basicity than the latter compound. The explanation is as follows:

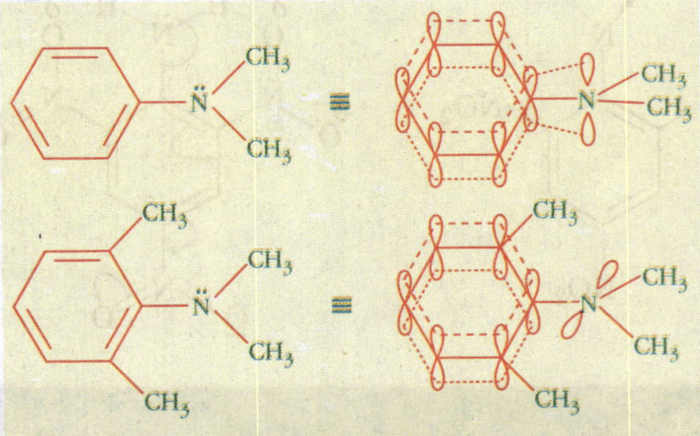

Fig. 8.1. Steric effect on delocalisation of electrons

In N,N,2,6-tetramethylaniline, the interaction amongst N-methyl groups and C-methyl groups rotates the —$N(CH_3)_2$ group around C—$N(CH_3)_2$ bond in such a way that the *p* orbital possessing the lone pair of electrons is no longer parallel to the *p* orbitals of benzene ring. This is because of the steric hindrance developed by the interaction of the neighbouring methyl groups. Thus, delocalisation of electron-pair on N atom is inhibited, i.e., the electron-pair remains more or less localised on N atom and hence easily available for donation. On the other hand, that in N,N-dimethylaniline gets delocalised as the *p* orbital of N atom is appreciably parallel to the *p* orbitals of benzene ring. As the delocalisation of electron-pair is greater in N,N-dimethylaniline than that in N,N,2,6-tetramethylaniline, the latter compound is more basic than the former though the inductive effects due to —CH_3 groups are the same in both the cases.

However, *o*-nitrobenzoic acid is more acidic than *p*-nitrobenzoic acid. Here steric inhibition of resonance does not do any harm since the two groups do not undergo resonance interaction. Here –I effect of the —NO_2 group in *o*-nitrobenzoic acid is more than that in *p*-nitrobenzoic acid.

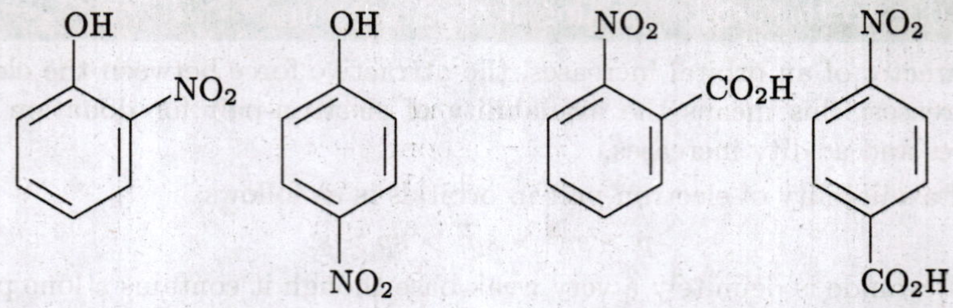

pK_a 7.23 7.15 2.17 3.43

In the cases of *o*-and *p*-nitrophenols, the conjugate bases, the corresponding phenoxides undergo resonance interaction between O^- and —NO_2 and thereby brings stability. However, the resonance interaction gets hindered in *o*-nitrophenoxide ion. So *o*-nitrophenol is less acidic than *p*-nitrophenol.

On the other hand 2,4,6-trinitro-N,N-dimethylaniline is 40 000 times more basic than 2,4,6-trinitroaniline. In the former case strong steric inhibition of resonance attributes to the ready availability of *n es* of the NMe_2 group; while in the latter case, strong internal H-bonding with nitro groups at 2 and 6 positions makes the molecule planar and thus help delocalise the lone pair of *es* on NH_2 group as shown below:

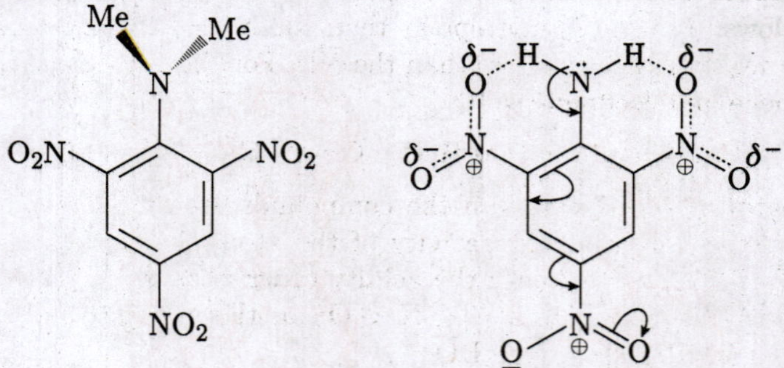

8.5.6 Solvation effect

Solvation effect helps to stabilise a species. A greater stability means a lower reactivity. Thus when a base gets stabilised by solvation, the availability of electrons decreases and hence its basicity will be lower than expectation. Similarly, if an acid is stabilised by solvation, its acidity also decreases. For example, hydration effect stabilises dimethylammonium ion to a greater extent than trimethylammonium ion. Thus the latter is more acidic than the former in the aqueous medium though +I effect of three methyl groups stabilises the trimethylammonium ion to a greater extent than the dimethylammonium ion. Therefore, we should expect a greater basicity of dimethylamine than trimethylamine in the aqueous solution.

On the other hand, in chlorobenzene medium there is no remarkable solvation effect on each of dimethylammonium and trimethylammonium ions; but +I effect of $-CH_3$ groups in the latter species being greater than that in the former, trimethylammonium ion gets more stabilised than dimethylammonium ion. This makes the latter species more acidic than the former. Therefore, trimethylamine, the conjugate base of trimethylammonium ion, is more basic than dimethylamine in chlorobenzene medium.

8.5.7 Effect of bond length

The longer the bond, the weaker is the bond and it breaks readily. Between R—S—H (thiols) and R—O—H (alcohols), the S—H bond is longer than the O—H bond. So S—H bond breaks readily than O—H bond. For this reason R—S—H is a stronger acid than R—O— H. However, this can be explained in another way : the corresponding conjugate bases are RS^- and RO^-; RS^- accommodates the negative charge better than RO^- because of the larger volume of S atom than that of O atom. So RS^- is a weaker base than RO^-; hence the conjugate acid RSH is stronger than the conjugate acid ROH.

8.5.8 Electronegativity of the bonded atoms

The greater the electronegativity of the atom bonded to H, the higher is the acidity; and the greater the electronegativity of atom holding the donatable *e*s, the lower is the basicity.

$$R-\overset{\overset{O}{\|}}{C}-NH_2 \text{ is feebly acidic with respect to } R-\overset{\overset{O}{\|}}{C}-OH$$

Here H atom is bonded to N and O atoms respectively. O atom being more electronegative than an N atom, the O—H bond is more polar than the N—H bond and at the same time $\overline{O}$ accommodates the negative charge better than the $\overline{N}H$. For the same reason the acidity order amongst ethane, ethylene and acetylene is:

$$H-C\equiv C-H > H_2C=CH_2 > CH_3-CH_3$$

The hybridised states of the C atoms in the compounds are *sp*, sp^2 and sp^3 respectively. As the *s* character changes, the electronegativity of the atom changes. The electronegativity order is C(*sp*) > C(sp^2) > C(sp^3). Hence the acidity order is as given. The electronegativity order of the three second row elements is C < N < O; for this reason, the basicity order is (i) $\overline{C}H_3 > \overline{N}H_2 > \overline{O}H$ and (ii) $R_3\overline{C} > R_2\overline{N} > R\overline{O}$.

Problem 8.4. Why does acetic acid behave as an acid in the aqueous medium?

Solution: $CH_3COOH + H_2O \rightleftharpoons CH_3COO^- + H_3O^{\oplus}$

Acetic acid is a resonance hybrid of the following canonical forms (1 & 2).

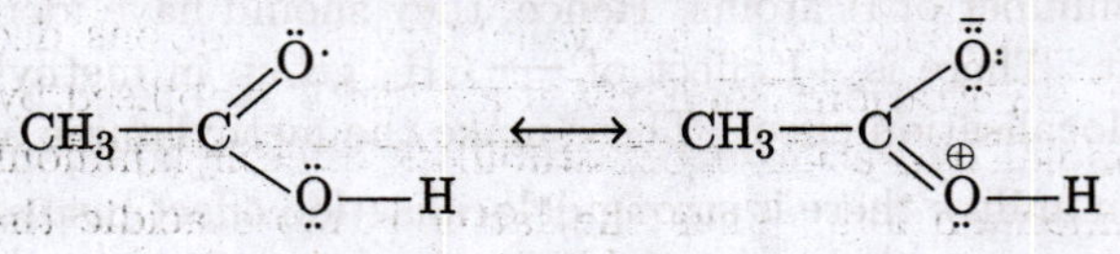

1 **2**

The conjugate base of the acetic acid, the acetate ion, is also a resonance hybrid of the two resonating forms (3 & 4).

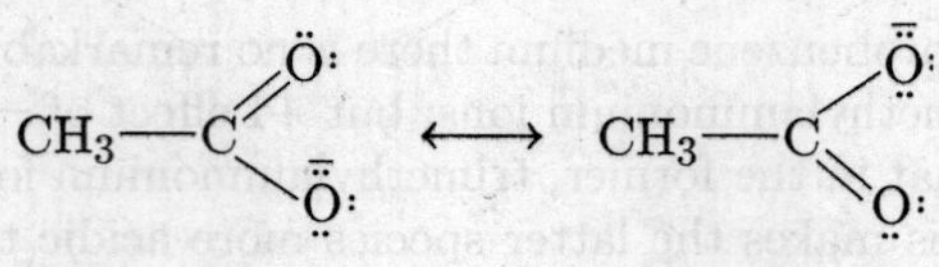

3 **4**

Thus, the hybrid of acetic acid contains one charged structure in which the more electronegative O atom bears a positive charge; this makes the hybrid of acetic acid somewhat unstable. On the other hand, the hybrid of acetate ion possesses two equivalent structures and hence it is stable. In order to attain stability, acetic acid ionises to proton and acetate ion; hence it behaves as an acid.

Problem 8.5. Why is phenol less acidic than acetic acid?

Solution: The comparative study of the resonance stabilisation of acetate ion and phenoxide ion, the conjugate bases of acetic acid and phenol respectively, shows that the latter is less stable than the former. This is because the hybrid of the phenoxide ion contains structures possessing negative charge on the less electronegative element, C. Hence those resonating structures are less contributing in the hybrid. Therefore, the resonance hybrid of phenoxide ion possesses low resonance energy.

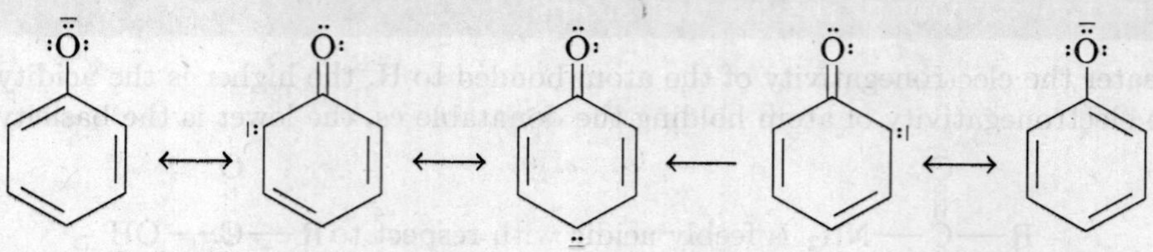

On the other hand, because of the equivalent resonating structures in the resonance hybrid of the acetate ion, it has high resonance energy (see Problem 8.4—structure 3 and 4).

Since basicity increases with decreasing stability, phenoxide ion is more basic than acetate ion. Therefore, phenol is less acidic than acetic acid.

Problem 8.6. Aniline is less basic than methylamine in the aqueous medium.—Explain.

Solution:

$CH_3—\ddot{N}H_2$

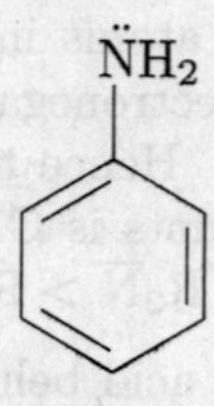

Methylamine **Aniline**

The N atom of methylamine is sp^3 hybridised and in aniline this is sp^2 hybridised, but both of them hold the same number of H atoms. Hence, they should have identical basicity from the point of hydration effect. There is +I effect of —CH_3 group in methylamine but there is no resonance effect, i.e., delocalisation effect. These make the methylamine a strong base.

On the other hand, in aniline there is strong delocalisation effect instead of +I effect as shown below; these make the electron-pair less available for donation. Hence it is a weak base.

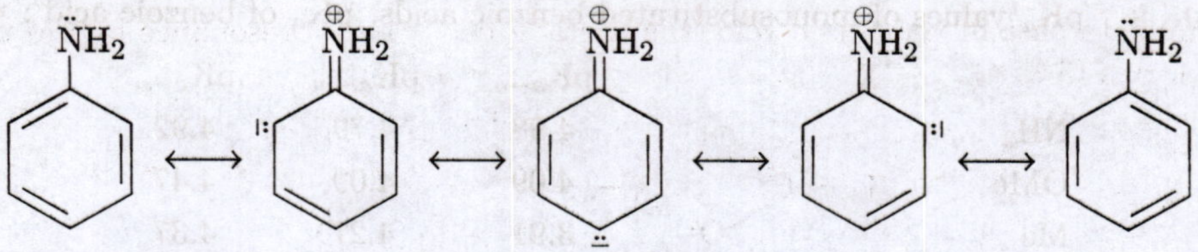

Therefore, methylamine is a stronger base than aniline.

Note:

Acidity and basicity are relative terms. A species may act as an acid in one medium, it may behave as a base in another medium. The acid or base character depends on the medium used. If the medium is more basic than the species concerned, the latter will act as an acid and vice versa.

Thus sulphuric acid behaves as an acid in the aqueous as well as in nitric acid medium, but it acts as a base in perchloric acid medium because perchloric acid is stronger than sulphuric acid in so far as acidity is concerned.

A few tables for pK_a values of certain classes of compounds are given below:

Table 8.5. pK_{ax} values of saturated dicarboxylic acids

		pK_{a_1}	pK_{a_2}
↓	Oxalic	1.27	4.26
	Malonic	2.86	5.70
	Succinic	4.21	5.64
	Glutaric	4.34	5.27
	Adipic	4.41	5.28
	Maleic	1.53	6.07
decreasing strength	Fumaric	3.03	4.44

Table 8.6 pK_a values of monosubstituted monohydric phenols.

pK_a of phenol = 9.98

	X	pK_{a-o}	pK_{a-m}	pK_{a-p}
Increasing strength	Me:	10.28	10.08	10.14
↓	MeO:	9.98	9.65	10.21
	NH_2:	9.71	9.87	10.30
	OH:	9.48	9.44	9.96
	F:	8.81	9.28	9.95
	Cl:	8.48	9.03	9.38
	NO_2:	7.23	8.40	7.15
	Picric acid: 0.71			

Table 8.7 pK_a values of monosubstituted benzoic acids. pK_a of benzoic acid : 4.17

X	pK_{a-o}	pK_{a-m}	pK_{a-p}
NH_2	4.98	4.79	4.92
OMe	4.09	4.09	4.47
Me	3.91	4.27	4.37
F	3.27	3.87	4.14
Cl	2.94	3.83	3.98
OH	2.98	4.08	4.56
NO_2	2.17	3.49	3.43
Trinitrobenzoic acid: 0.65			

Table 8.8 pK_a values of monosubstituted anilines pK_a of aniline : 4.58

(i) for N-substituted anilines.

	pK_a		pK_a
PhNHMe	4.85	$PhNEt_2$	6.56
$PhNMe_2$	5.06	Ph_2NH	0.9
PhNHEt	5.11	PhNHAc	0.4

(ii) for nuclear substituted anilines.

X	pK_{a-0}	pK_{a-m}	pK_{a-p}
MeO	4.49	4.20	5.29
NH_2	4.47	4.20	6.08
Me	4.39	4.69	5.12
Cl	2.64	3.34	3.98
Br	2.60	3.51	3.91
NO_2	0.29	2.50	1.02

Note: Try to explain the above pK_a values in Tables given above.

8.6 Acid and Base Catalysis

Catalysts play a fundamental role in determining mechanism of reactions. The most important amongst catalysts are acids and bases. Acid-catalysed hydration of alkenes, acid- and base-catalysed hydrolysis of esters, amides and cyanides, enolisation of carbonyls, formation of esters, formation and hydrolysis of acetals are some of the examples of acid and base catalysis.

(i) Acid-catalysed hydration of alkenes:

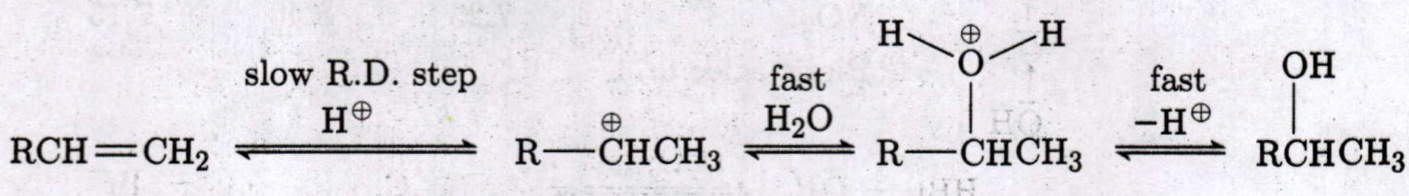

The backward reaction is acid catalysed dehydration of an alcohol.

(ii) Acid-catalysed formation and hydrolysis of acetals:

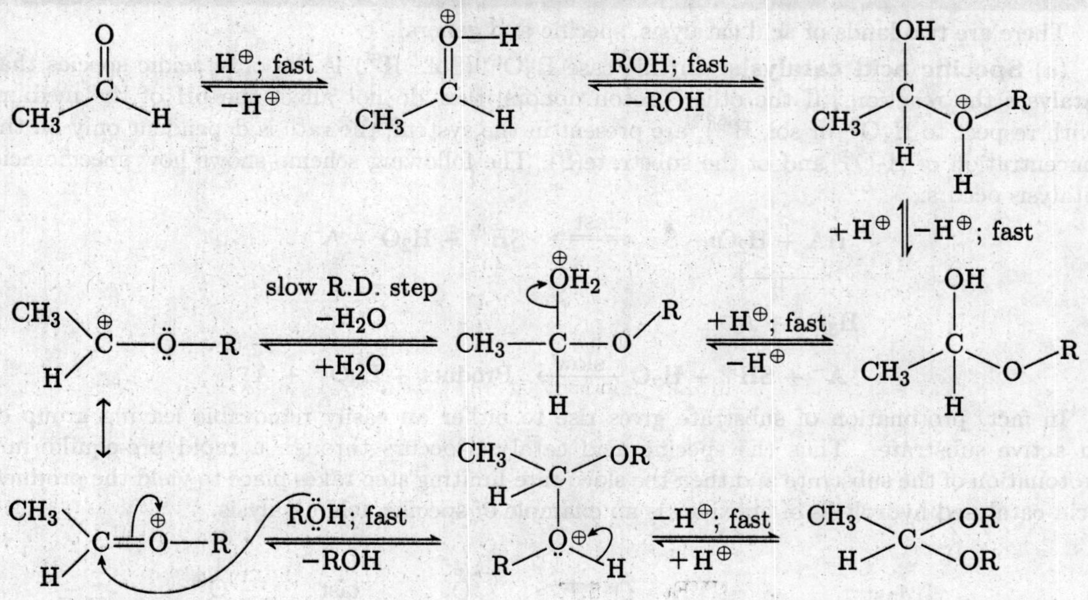

(iii) Base-catalysed reverse aldol reaction:

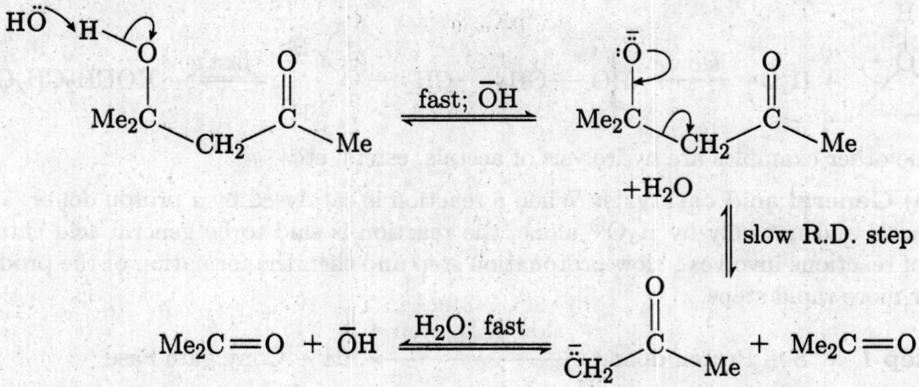

(iv) Base-catalysed α-bromination of carbonyl compounds:

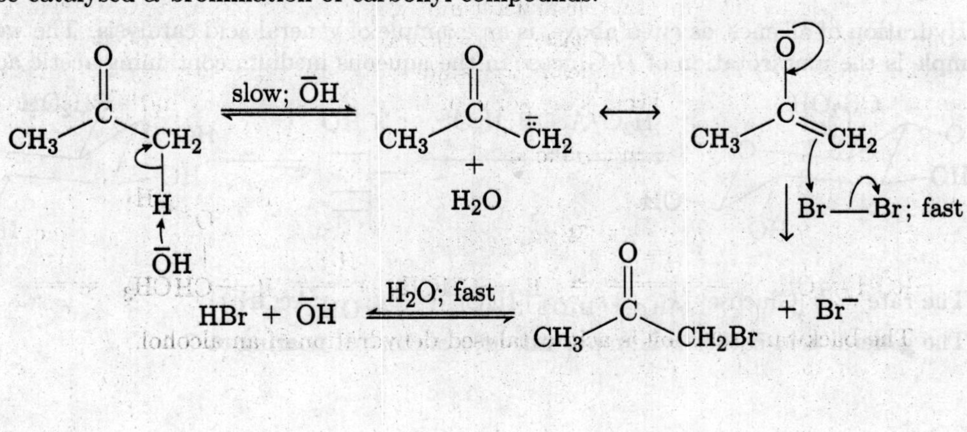

8.6.1 Acid Catalysis

There are two kinds of acid catalysis, specific and general.

(a) **Specific acid catalysis:** In this case $H_3O^{\oplus}$ (Sol. $H^{\oplus}$) is the only acidic species that catalyses the reaction. If the other proton donors, that do not affect the pH of the medium (with respect to $H_3O^{\oplus}$ or sol. $H^{\oplus}$), are present in the system, the rate is dependent only on the concentration of $H_3O^{\oplus}$ and of the substrate(S). The following scheme shows how specific acid catalysis occurs:

$$\underbrace{HA + H_2O}_{H_3O^{\oplus} + A^-} + S \xrightleftharpoons{\text{fast}} SH^{\oplus} + H_2O + A^-$$

$$A^- + SH^{\oplus} + H_2O \xrightarrow{\text{slow}} \text{Product} + H_3O^{\oplus} + A^-$$

In fact, protonation of substrate gives rise to either an easily removable leaving group or an active substrate. Thus the specific acid catalysis occurs through a rapid pre-equilibrium protonation of the substrate and then the slow, rate-limiting step takes place to yield the product. Acid-catalysed hydrolysis of epoxides is an example of specific acid catalysis.

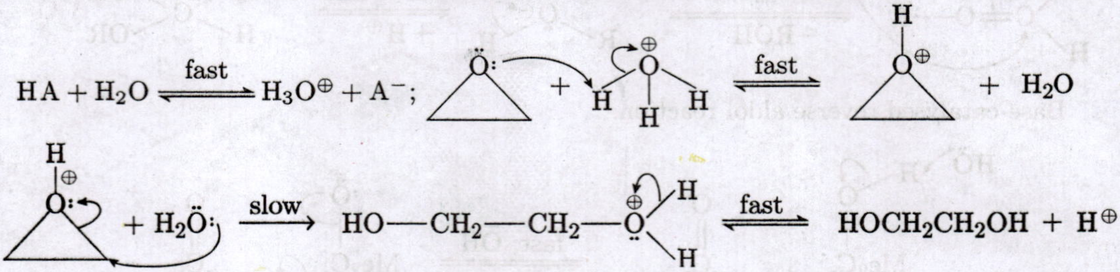

The other examples are hydrolysis of acetals, esters, etc.

(b) **General acid catalysis:** When a reaction is catalysed by a proton donor, a protic acid, in general and not only by $H_3O^{\oplus}$ alone, the reaction is said to be general acid catalysed. This type of reactions involves : slow protonation step and then the formation of the product through one or more rapid steps,

Step I $\quad S + \text{Proton donor} \xrightarrow{\text{slow R.D. step}} \overset{\oplus}{S}H + \text{Conjugate base}$

Step II $\quad \overset{\oplus}{S}H \rightleftharpoons \text{Product} + H^{\oplus}$

Hydration of alkenes, as cited above, is an example of general acid catalysis. The well-studied example is the mutarotation of *D*-Glucose in the aqueous medium containing acetic acid.

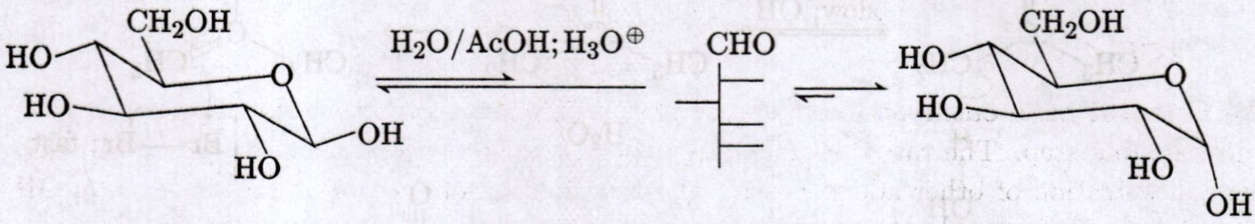

The rate = k [Glucose] $(k_{H_2O}+k_{H_3O^{\oplus}})[H_3O^{\oplus}] + k_{AcOH}[AcOH])$

The general acid catalysis is important at a higher pH nearing to 7.

8.6.2 Base catalysis

Base-Catalysed reactions are also of two kinds, specific and general.

(a) Specific base-catalysed reaction: The reaction is specific catalysed in the sense that the rate of reaction depends only on the concentration of $\bar{O}H$ even if other bases are added in such a way that the pH of the medium remains more or less constant. In this type of reaction the proton abstraction step is fast. The R.D. step may be any one of the other steps. The base catalysed reverse aldol condensation reaction is an example of specific base-catalysed reactions as shown below:

Aldol condensation:

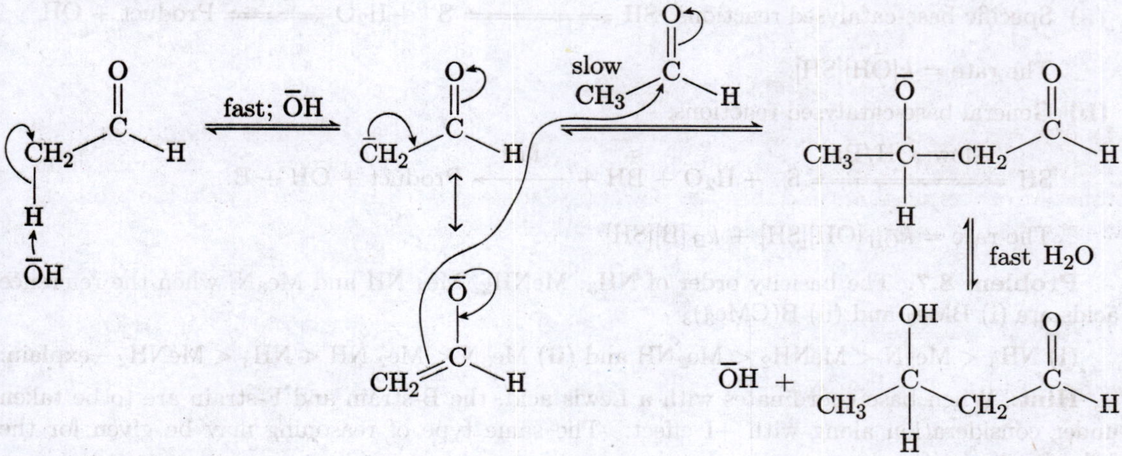

Reverse aldol:

The backward reaction of aldol condensation is called retro-aldol or reverse aldol condensation

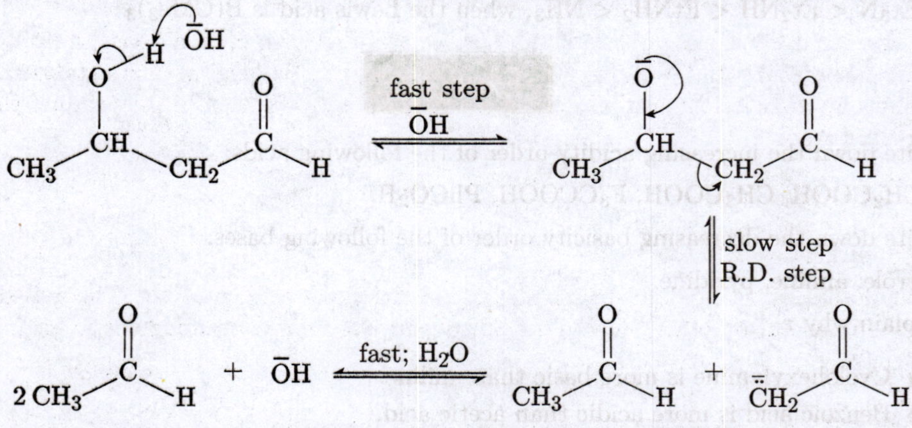

(b) General base-catalysed reactions: In this case the abstraction of proton is the slow, rate determining step. The rate does not only depend on the concentration of $\bar{O}H$ ion but also on the concentration of other added bases. The bromination of acetone in the presence of $\bar{O}H$ and AcO^- ion is an example of general base catalysed reactions.

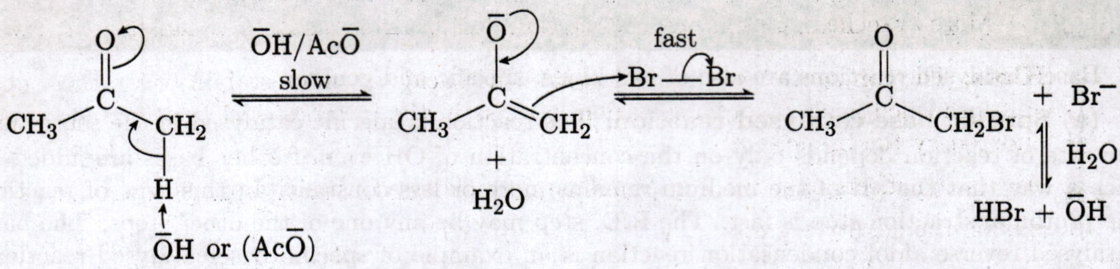

The schemes for:

(a) Specific base-catalysed reactions: $SH \underset{}{\overset{\text{fast; } \bar{O}H}{\rightleftharpoons}} S^- + H_2O \underset{}{\overset{\text{slow}}{\rightleftharpoons}} \text{Product} + \bar{O}H$

The rate $= k[\bar{O}H][SH]$

(b) General base-catalysed reactions:

$$SH \overset{\text{slow; } \bar{O}H/B:}{\rightleftharpoons} S^- + H_2O + \overset{\oplus}{B}H + \overset{\text{fast}}{\longrightarrow} \text{Product} + \bar{O}H + B:$$

The rate $= k_{\bar{O}H}[\bar{O}H][SH] + k_B[B][SH]$

Problem 8.7. The basicity order of NH_3, $MeNH_2$, Me_2NH and Me_3N, when the reference acids are (i) BMe_3 and (ii) $B(CMe_3)_3$

(i) $NH_3 < Me_3N < MeNH_2 < Me_2NH$ and **(ii)** $Me_3N < Me_2NH < NH_3 < MeNH_2$—explain.

Hint: When base coordinates with a Lewis acid, the B-strain and F-strain are to be taken under consideration along with +I effect. The same type of reasoning may be given for the following basicity order.

(iii) $Et_3N < NH_3 < Et_2NH < EtNH_2$, when the Lewis acid is BMe_3

(iv) $Et_3N < Et_2NH < EtNH_2 < NH_3$, when the Lewis acid is $B(CMe_3)_3$

Exercise

1. Write down the increasing acidity order of the following acids:
   $ClCH_2COOH$, CH_3COOH, F_3CCOOH, $PhCO_2H$.
2. Write down the decreasing basicity order of the following bases:
   Pyrrole, aniline, pyridine.
3. Explain why :
   (a) Cyclohexylamine is more basic than aniline.
   (b) Benzoic acid is more acidic than acetic acid.
   (c) Urea is a monoacidic base.
   (d) Acetamide is a weak base.
   (e) 2, 6-Dimethyl-4-nitrophenol is a stronger acid than 3, 5-Dimethyl-4-nitrophenol.
   (f) Picric acid is more acidic than phenol.
   (g) Acetylene is more acidic than ethylene.

(h) More branched acids are less acidic than less branched acids.

[**Hints:** The conjugate bases of the less branched acids get more stability than those of the more branched acids through solvation because in the latter species effective solvation is hindered by the branches.]

4. Discuss the effects of structural factors on the acidity and basicity of species.

5. Explain the acidity order:

(i)

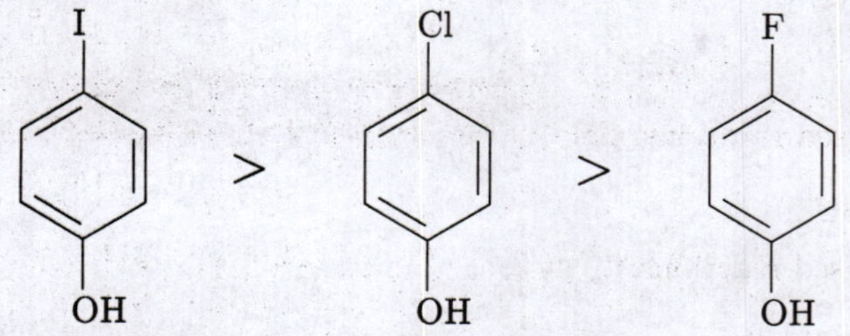

(ii) $CHCl_3 > CHF_3$

(iii) $\overset{\oplus}{N}H_4 > NH_3$

(iv) $CH_3COCH_2—H, (CH_3CO)_2CH—H, (CH_3CO)_3C—H$

pK$_a$ 20 9 6

(v) $CH_3COCH_2COCH_3 > CH_3COCH_2CO_2Et > CH_2(CO_2Et)_2$

6. (a) Guanidine is a very strong base.—Explain.

(b) Cyclopentadine behaves as an acid.—Explain.

(c) *m*-Chlorobenzoic acid is a stronger acid than *m*-methoxybenzoic acid but *p*-chlorobenzoic acid is a weaker acid than *m*-chlorobenzoic acid.

7. Explain the basicity order:

(i) $Me_3CO^- > CH_3O^- > PhO^- > CH_3CO_2^-$;

(ii) $CH_2{=}CH—NH_2 < CH_3NH_2$

8. $\overset{\oplus}{H}$ reacts faster with $\bar{O}H$ than with $CH_2{=}CH_2$ — why?

9. The acidity order of *t*-BuOH and MeOH is opposite in the gas-phase than what it is in the aqueous medium.—Explain.

[**Hint:** In the aqueous medium $Me_3C\bar{O}$ gets less stability through hydration than $Me\bar{O}$ because of its bulk; in the gas-phase the greater +I effect of *t*-Bu group helps more to polarise O—H bond.]

10. With respect to $B(Et)_3$ the basicity order of the following amines is : $MeNH_2 > Me_2NH > Me_3N$.—Explain.

[**Hint:** In $B(Et)_3$ bulky Et groups are present; when bulky amines come to coordinate with $B(Et)_3$, F-strain develops and the coordinated complexes experience B-strain.]

[Consult strain factors in organic compounds (Chapter 9).]

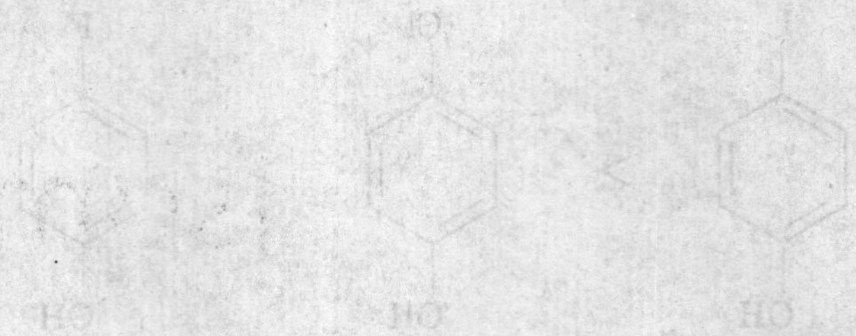

(ii) $CHCl_3 > CHF_3$

(iii) $CH_4 > NH_3$

(iv) CH_3COCH_3 — H, $(CH_3CO)_2CH$ — H, $(CH_3CO)_3C$ — H

pKa 20 9 6

[Consult steric effects in organic compounds (Chapter 6) for details.]

Chapter 9

Isomerism

Objectives

- To understand:
  - * The meaning of the terms isomerism and isomers [9.1] p. 271
  - * The basis on which the isomers are classified [9.1, 9.2 & 9.3] p. 271-286
  - * Tautomerism in detail [9.2.2] p. 274
  - * The meaning of the terms conformations, conformers and conformational analysis [9.4] p. 286
  - * Chiral molecules, enantiomers and diastereomers [9.6] p. 300
  - * The cause of optical activity [9.11.3] p. 342
- To draw:
  - * The constitutional formulae of chain, position and functional group isomers [9.2] p. 272
  - * The Newman, Sawhorse and flying wedge projections of anti, staggered, fully eclipsed and eclipsed conformations [9.4] p. 286

- To understand the terms configuration and configurational isomerism [9.5] p. 297
- To memorise the priority sequence rule and to designate the configurational isomers as cis, trans, Z, E and R, S, D, L [9.10] p. 324
- To understand the terms:

  Polarised light, plane of polarisation, optical activity, dextro and levo-rotations, enantiomers, diastereomers, specific rotation, molecular rotation [9.11, 9.12, 9.13, 9.14] p. 324-353
- To find out:
  - The symmetry elements (plane of symmetry, centre of symmetry, C_n and S_n axes), chiral elements and chirality of molecules [9.7, 9.8] and point groups [9.9] p. 301-324
  - The isomer number of chiral molecules [9.12] p. 344
- To make oneself equipped with the following processes :
  - Racemisation [9.15], * Resolution [9.16], * Asymmetric synthesis [9.17], * Substitution reactions at a chiral C [9.18] p. 353-368
- To understand diasteromerism involving double bond and ring [9.19] p. 369
- To understand the stereoisomerism of cyclohexane derivatives and anchoring substituents [9.20] p. 375
- To know:
  - Homotopicity, heterotopicity and prochiral centre [9.21] p. 399
  - Stereoselective and stereospecific reactions [9.22] p. 403
  - Strains in organic molecules [9.23] p. 405
  - Stereoelectronic and steric factors of a reaction [9.24] p. 407

9.1 Introduction

Isomers are nowadays defined as chemical entities which have the same molecular formula (in the case of covalent molecules) or formula unit (in the case of ionic compounds) but differ in structure and are separated by energy barriers. The phenomenon of existence of isomers is called isomerism. We may recall here that the structure of a chemical entity is a description of the sequential bond connectivities of the constituent atoms of the entity including their orientation in space; whereas the constitution of an entity is a description of the sequential bond connectivities of the constituent atoms regardless of their direction in space. So, the structural difference may arise either owing to the difference in constitution or owing to the difference in spacial arrangements of constituent atoms and groups, the constitution i.e., bond connectivities remaning the same.

Properties of a chemical compound depend on its structure. If two chemical entities differ in structure, either in constitution or in the way their atoms and groups are directed in space, they will differ at least in one property, physical or chemical. The difference in property will help us identify them as two different chemical entities provided they are separable by energy barriers.

Let us consider some compounds for the shake of clarity :

(i) $CH_3—C{\equiv}C—CH_3$ and $CH_3CH_2C{\equiv}CH$

a — But-2-yne

b — But-1-yne

These two compounds have the same molecular formula, C_4H_6, but they have different bond connectivities i.e., *a* and *b* differ in constitution and hence in structure. These two compounds can be separated from their mixture. Therefore, by definition, they constitute a pair of isomers; *a* and *b* have different properties e.g., *b* gives white precipitate with ammonical silver nitrate solution whereas *a* does not.

(ii)

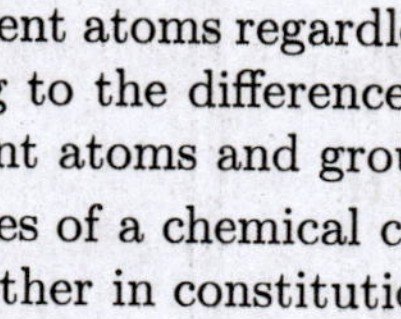

c

cis-1, 2-Dichloroethene

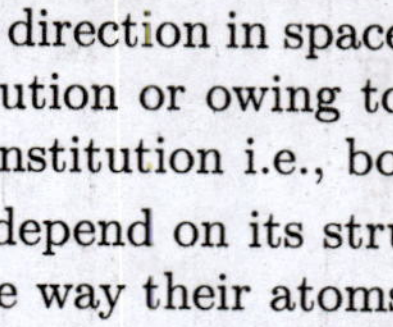

d

trans-1, 2-Dichlorethene

These two compounds have the same molecular formula and constitution but they differ in the direction of the H and Cl atoms in space. So, they are structurally different though they have the same constitution. They are also separable from their mixture. Hence *c* and *d* are two isomeric compounds. They have different properties; *c* has appreciable dipole moment but *d* has zero dipole moment.

(iii)

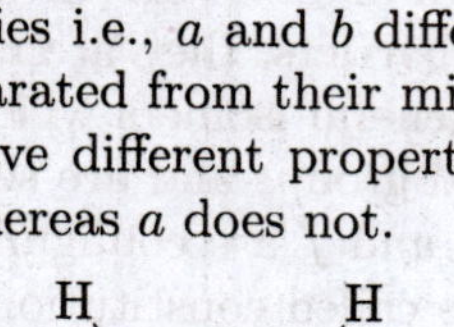

e

(S)-Methylethylisopropylisobutylammonium chloride

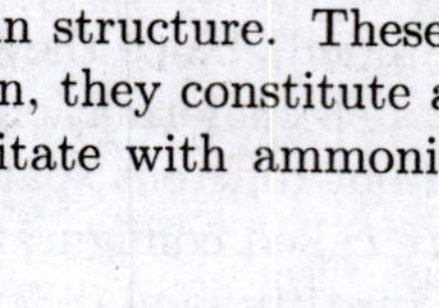

f

(R)-Methylethylisopropylisobutylammonium chloride

We find no difference in these two compounds insofar as formula unit and bond connectivities are concerned but *i*-pr and Et groups are directed in space in different ways. Hence they have difference in structure. The two compounds are isolable from their mixture. So, they are isomers. One of them rotates the plane of the plane polarised light to the right and the other to the left; thus they differ at least in one property.

(iv)

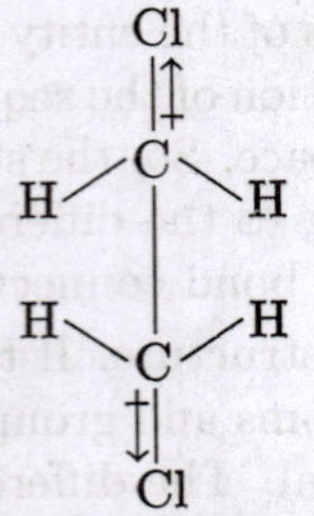

g (Staggered form)
1,2-dichloroethane

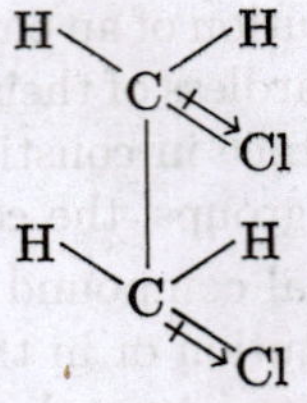

h (Eclipsed form)
1,2-dichloroethane

Here, *g* and *h* have the same molecular formula and constitution but different spatial arrangements of the H and Cl atoms. Therefore, *g* and *h* differ in structure. The dipole moment of *g* is zero but that of *h* is of appreciable amount. So one may expect that these two are isomers, but they are not. We cannot separate the two forms from their mixture i.e., they are non-isolable and one form gets converted to the other owing to the rotation of one —CH_2Cl unit around the C—C single bond. Since they are not separable by energy barriers, they are not isomers. We designate them as conformations of 1, 2-dichloroethane. In general isomers which have the same constitution but different spatial arrangements of atoms and groups and are separated by energy barriers are called configurational isomers; *c* and *d*, and *e* and *f* are configurational isomers. There are constitutional differences in *a* and *b* and they are called constitutional isomers. Since conformations and configurational isomers regard the directions of their constituent atoms in space, they constitute a separate branch of isomerism, called stereoisomerism. Isomers which have difference in constitution are called constitutional isomers. Thus isomerism may broadly be classified under two heads:

(i) Constitutional isomerism and (ii) Stereoisomerism.

Modern stereochemistry deals with the studies of conformations (dynamic stereoisomers) and configurations (static stereoisomers). 'Stereos' is a Greek word which is used with reference to three dimensionality. So, by stereochemistry we mean the study of chemistry with reference to three dimensionality of the chemical units.

9.2 Constitutional Isomerism

The phenomenon of existence of two or more compounds with the same molecular formula but different in constitution (and are separable by energy barriers) is known as constitutional isomerism and the compounds are called constitutional isomers. Propanone (CH_3COCH_3) and propanal (CH_3CH_2CHO) are constitutional isomers because they have different constitutions; propanone contains a keto group bonded to two methyl groups, whereas propanal contains a formyl group bonded to an ethyl group.

Depending on the nature of constitutional difference in isomeric compounds, constitutional isomerism is further classified as given on the next page.

1. Chain isomerism, 2. Functional group isomerism and 3. Position isomerism.

1. Chain isomerism: Isomeric substances that differ only in the arrangement of atoms forming the base chain are known as **chain isomers** and the phenomenon of their existence is said to be **chain isomerism**. Butane (C_4H_{10}) and methylpropane (C_4H_{10}) are the examples of chain isomers.

$CH_3.CH_2.CH_2CH_3$

C—C—C—C

Butane

(a straight chain compound)

1

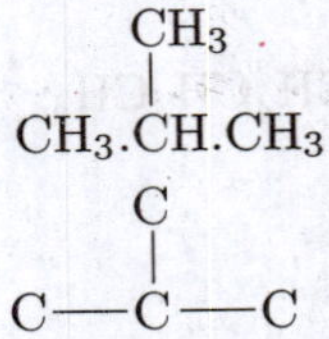

Methylpropane

(a branched chain compound)

2

2. Functional group isomerism: Compounds with the same molecular formula but differing in the type of functional group they possess are classed as **functional group isomers** and the phenomenon of their existence is called **functional group isomerism**. Ethyl alcohol and dimethyl ether exhibit functional group isomerism.

$CH_3.CH_2OH$(C_2H_6O ; it contains hydroxyl function, —OH)

$CH_3.O.CH_3$ (C_2H_6O; it possesses ether linkage, —O—)

3. Position isomerism: If compounds with the same molecular formula differ in the position of the same function on the identical base chain, the compounds are called position isomers and the phenomenon of their existence is known as position isomerism. Thus 1-propanol and 2-propanol have the same molecular formula, base chain and functional group, but the position of OH group is different in the two compounds; it is on the terminal C in the former compound, whereas in the latter compound, the group is attached to the middle C.

$CH_3.CH_2.CH_2OH$

1-Propanol

5

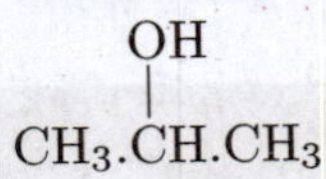

2-Propanol

6

Sometimes many isomeric compounds are designated as metamers and some others are called tautomers; these are nothing but constitutional isomers:

9.2.1 Metamerism

Metamerism is a type of constitutional isomerism exhibited by the organic compounds of the same class caused by the attachment of different hydrocarbon groups to a central atom and the constitutional isomers are then called **metamers**. Diethyl ether and methyl propyl ether are the examples of metamers; here the central atom is oxygen; the different hydrocarbon groups are two ethyl groups in one and in the other there are methyl and *n*-propyl groups; these are nothing but constitutional isomers

$CH_3.CH_2.O.CH_2.CH_3$ — Diethyl ether($C_4H_{10}O$) — 7

$CH_3.O.CH_2.CH_2.CH_3$ — Methyl propyl ether($C_4H_{10}O$) — 8

Problem 9.1. Write down the types of isomerism the following pairs of compounds exhibit.

a(i). $CH_2{=}CH.CH_2.CH{=}CH_2$ **a(ii).** $CH_3.CH_2C{\equiv}C.CH_3$

b(i). $Br.CH_2.CH_2.CH_2.Br$ **b(ii).** $CH_3.CH_2.CHBr_2$

c(i). $(CH_3)_4C$ **c(ii).** $(CH_3)_2.CH.CH_2CH_3$

d(i). $CH_3.CO.CH_2CH_2CH_3$ **d(ii).** $CH_3.CH_2.CO.CH_2.CH_3$

e(i). $CH_3.\underset{\underset{O}{\|}}{C}.CH_3$ **e(ii).** $CH_2{=}\overset{\overset{OH}{|}}{C}.CH_3$

Solution: (a) Functional group isomerism ; C=C and C≡C are the different functions.

(b) Position isomerism; in (b)(i) the positions of the Br atoms are 1 and 3 but in (b)(ii) these are 1 and 1.

(c) Chain isomerism; the chain in (i) is 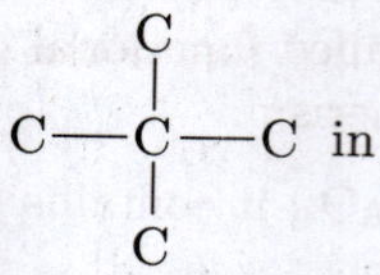 in (ii) is 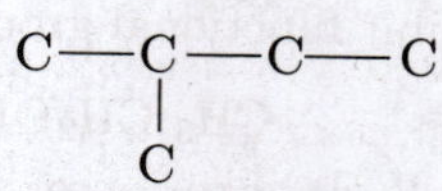

(d) Position isomerism and metamerism also; in (i), C of >C=O is the C_2 atom and this is C_3 in (ii). These are metamers because there are different hydrocarbon groups attached to the central carbonyl (C) atom [methyl and *n*-propyl in (i) but two ethyl groups in (ii)].

(e) Functional group isomerism; (i) contains keto function, but (ii) contains C=C and hydroxyl function, i.e., enol (C=C—OH) function. These are also tautomers as (ii) is formed from (i) by the migration of one H from C_1 to the O atom; besides this, these two remain in rapid equilibrium in solution.

9.2.2 Tautomerism

Tautomers are constitutional isomers existing in rapid equilibrium in a solution or in the liquid state and the equilibrium reaction is known as **tautomerism**; the constitutional change occurs owing to the migration of an atom or a group from one position of the skeleton to the other with necessary change in the bond order. Thus nitroethane and aci-nitroethane are the two constitutional isomers which remain in rapid equilibrium in solution and so these are tautomers.

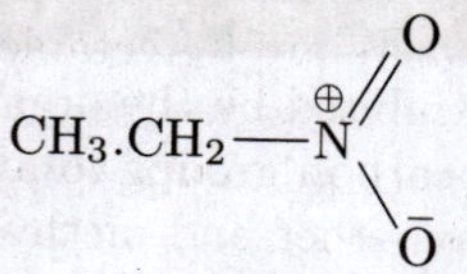

Nitroethane

9

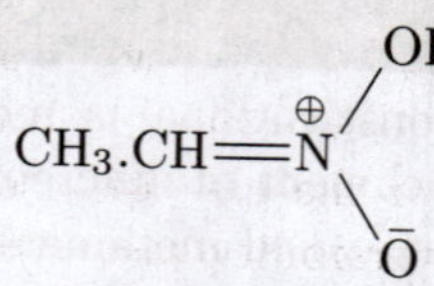

Aci-nitroethane

10

Another example is a keto-enol tautomerism; in this case one isomer is a carbonyl compound and the other is an enol and the equilibrium reaction is said to be keto-enol tautomerism. The classical example of the keto-enol tautomerism is the equilibrium reaction of acetoacetic ester which remains in solution in two structures, A and B.

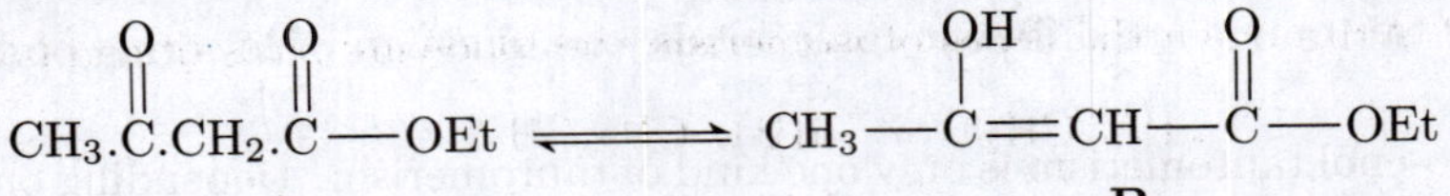

A **B**

Compound A is a β-keto ester while compound B is an enolic ester. The former contains a keto group in addition to the carbethoxy group and the latter contains an enol moiety (ene means C═C and ol means a hydroxyl group) along with the carbethoxy group. When an H atom from the active methylene (—CH_2—) group of the β-keto ester migrates to the O atom of the carbonyl group, the enolic ester forms; whereas when the H atom from the hydroxyl group of the enolic ester migrates to the ═CH— moiety, the β-keto ester develops.

However, tautomers being constitutional isomers, they are detectable as well as isolable. When acetoacetic ester is treated with:

(i) sodium, hydrogen gas evolves;

(ii) ethanolic solution of bromine, the colour of bromine immediately gets discharged;

(iii) ferric chloride, a reddish-violet colour develops.

The first reaction indicates the presence of a hydroxyl group, the second reaction is a positive test for olefinic double bonds and the third reaction is a test for —C(OH)═C— (enolic OH) moiety. Thus the solution of acetoacetic ester gives the tests for enols.

Again the ester forms:

(i) a bisulphite compound when it is treated with sodium bisulphite;

(ii) a cyanohydrin with hydrogen cyanide.

These two tests indicate the presence of a carbonyl group. So, acetoacetic ester is a keto ester. Thus tautomers are detectable.

Knorr in 1911 succeeded in isolating both the forms. He isolated the keto form exclusively as crystals by cooling the solution of acetoacetic ester in light petrol at −78 °C; the crystals did not give any test of enols. On the other hand, when Knorr treated the suspension of sodium derivative of acetoacetic ester in light petrol at −78 °C with just enough of hydrogen chloride, he obtained a glassy solid, instead of a crystalline product, which did not give any test of a keto group but gave the tests of enols. So knorr isolated the enol form of the acetoacetic ester under the conditions as stated above. Thus tautomers are isolable.

Here it is important to note that resonance and tautomerism are two quite different subjects. Resonating structures are hypothetical but tautomers have real existence. Resonating structures being hypothetical, they are not isolable; whereas tautomers are detectable as well as isolable. Tautomers remain in dynamic equilibrium in the solution or in the liquid state; the question of equilibrium does not arise in the case of resonating structures. For this reason tautomers are written, with a sign of reversibility ($\rightleftharpoons$) between two tautomers, while the resonating structures of a resonance hybrid are written by placing a double-headed arrow between every two such structures. However, in all resonating structures the positions of the atomic nuclei remain the same but in tautomers at least an atom or a group of atoms changes its position. Resonance involves redistribution of bonded and non-bonded electrons which also occurs in tautomerism. In resonance the actual structure of the entity, which is one and the same, is expressed by writing several hypothetical Lewis electron-dot structures. In tautomerism, the tautomers differ widely from each other in their constitutions. The concept of resonance involves lowering of energy of of the entity. As tautomers are different compounds, the question of lowering of energy does not arise.

However, keto-enol tautomerism is only one kind of tautomerism. Depending on the important function of the tautomers tautomerism may be classified as:

1. Nitro-acinitro tautomerism: 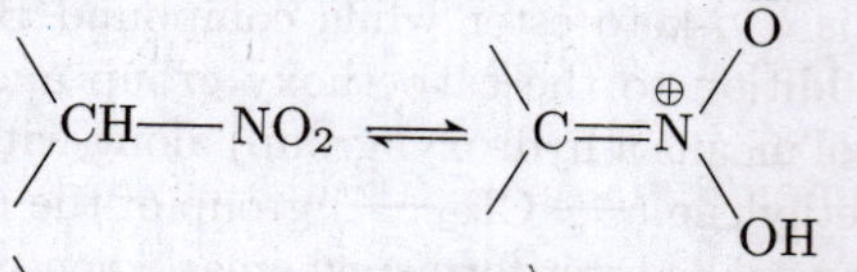

2. Nitroso-oximino tautomerism: 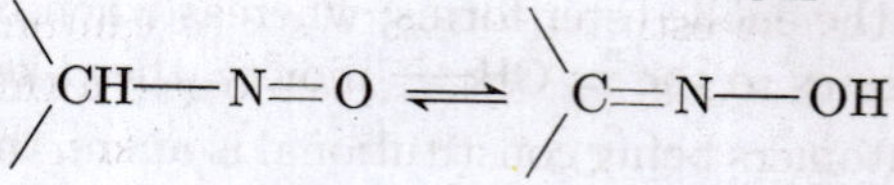

3. Amidine tautomerism: 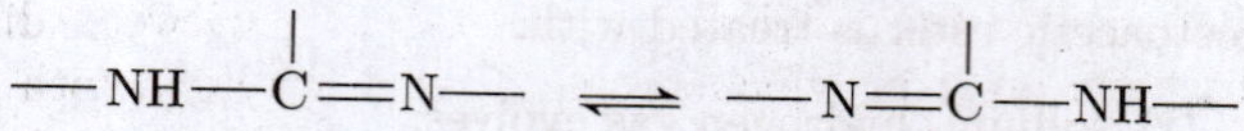

4. Amido-imidol tautomerism: 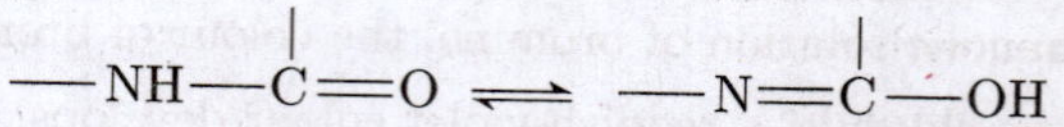

5. Azo-hydrazo tautomerism: 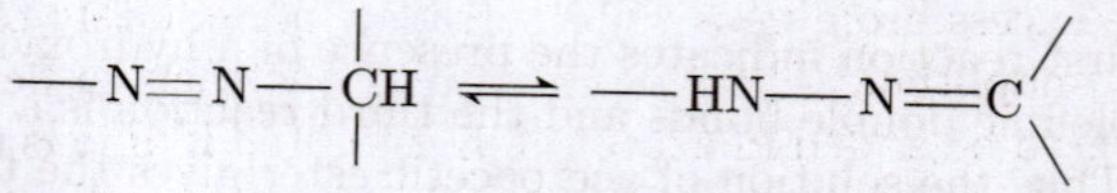

6. Diazo-amino tautomerism: 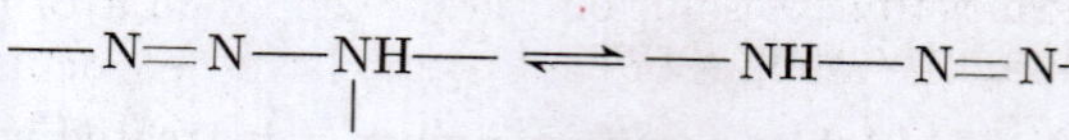

7. Diazo-nitrosoamine tautomerism: Ar—N=NOH ⇌ Ar—NH—N=O

8. Three carbon tautomerism: 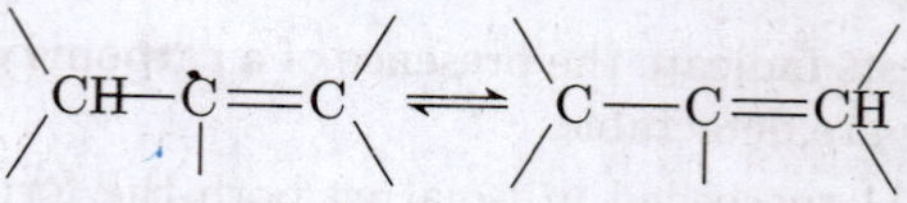

In all the above examples the migrating atom migrates to the alternate atom and thus migration origin and migration terminus may be numbered 1 and 3 and as a consequence these are called triad systems.

Besides open chain tautomerism, there is ring-chain tautomerism which we shall discuss in this section later on.

Nowadays tautomerism is classified on the basis of the nature of migrating group. When a cation migrates the phenomenon is called cationotropy; whereas when an anion migrates, it is said to be anionotropy. The keto-enol tautomerism is an example of cationotropy—here a proton ($H^{\oplus}$) migrates and so it is also called prototropy. The equilibrium mixture of 3-chloro-1-butene and 1-chloro-2-butene is an example of anionotropy (Cl^- ion being the migrating group). However, when an —OH group migrates, the system will be an example of anionotropy and may be called oxotropy.

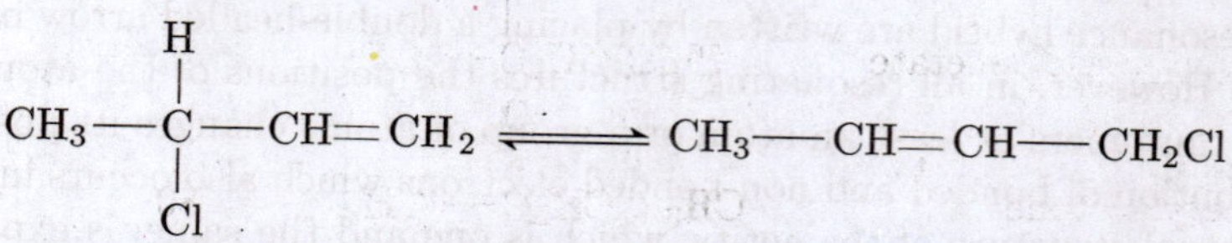

The amounts of tautomers in an equilibrium mixture may quantitatively be determined both by physical and chemical methods.

(i) By measuring refractive index of the equilibrium mixture experimentally and then calculating the same from the table of atomic refractions, one can calculate the amounts of all tautomers in the mixture.

(ii) In the cases of electrolytic tautomers, the conductivity measurement experiment helps to calculate the relative amounts in the mixture.

(iii) Optical rotation experiments and proton-magnetic resonance also helps to determine the amounts of tautomers in an equilibrium mixture.

Chemical methods of estimating amounts of tautomers utilise those chemical reactions in which one form reacts much faster than the other form and than the conversion of the latter form to the former. For example, the amounts of keto and enol forms of acetoacetic ester may be determined by carrying out the reaction between dilute ethanolic solution of bromine and acetoacetic ester in which bromine reacts rapidly with the enol form while keto form acts so slowly that the reaction may be ignored; at the same time the rate of conversion of keto to enol is much slower with respect to the rate of the reaction between bromine and the enol form.

A weighed sample of acetoacetic ester is to be rapidly treated with a slight excess of bromine at 0°C and the excess bromine is to be removed immediately by treating the solution with ethanolic solution of β-naphthol. The reaction mixture is then to be acidified with hydrochloric acid and KI solution is to be added in excess. The liberated iodine is then to be titrated against a standard sodium thiosulphate solution. From the amount of iodine liberated the amount of bromo-ester formed i.e., the amount of enol present can be calculated.

$$CH_3{-}\overset{OH}{\overset{|}{C}}{=}CH{-}\overset{O}{\overset{\|}{C}}OEt + Br_2 \longrightarrow CH_3{-}\overset{O}{\overset{\|}{C}}{-}\overset{Br}{\overset{|}{C}}HCO_2Et + HBr$$

$$CH_3.CO.CHBr\ CO_2Et + KI \longrightarrow CH_3CO\ CHI\ CO_2Et + KBr$$

$$CH_3COCHICO_2Et + HI\ (KI + HCl) \longrightarrow CH_3COCH_2CO_2Et + I_2$$

$$2Na_2S_2O_3 + I_2 = Na_2S_4O_6 + 2NaI.$$

1 ml of N $Na_2S_2O_3$ solution = 0.065 g of ethyl aectoacetate.

However, the enol contents of several keto-enol systems have been determined in pure liquid which are tabulated in Table 9.1.

Table 9.1 % Enol in a few keto-enol systems in pure liquid

Name	Formula	% Enol
Acetone	CH_3COCH_3	2.5×10^{-4}
Ethylacetoacetate	$CH_3.\overset{O}{\overset{\|}{C}}.CH_2\overset{O}{\overset{\|}{C}}{-}OEt$	8
Ethyl benzoylacetate	$Ph.\overset{O}{\overset{\|}{C}}.CH_2\overset{O}{\overset{\|}{C}}{-}OEt$	21
Acetylacetone	$CH_3.\overset{O}{\overset{\|}{C}}CH_2\overset{O}{\overset{\|}{C}}{-}CH_3$	80
Acetoacetaldehyde	$CH_3.\overset{O}{\overset{\|}{C}}.CH_2.\overset{O}{\overset{\|}{C}}{-}H$	98

Now the question is: why does the enol content varies so largely? There is no clear-cut answer to this question. Relative stabilities of the enols, however, may be taken under consideration.

Stability of an enol depends on the following factors:

(a) Conjugative effect, (b) Inductive effect, (c) Steric effect, (d) Solvation effect, (e) Internal hydrogen bonding effect, etc.

If we consider these effects to determine the stabilities of the resonance hybrids of the enols (Table 9.1), we shall find that the resonance hybrid of the enol form of acetoacetaldehyde is the most stable and that of the acetoacetic ester is the least stable. These effects explain why the enol content of acetoacetaldehyde is the topmost and that of acetoacetic ester is the least.

(i) 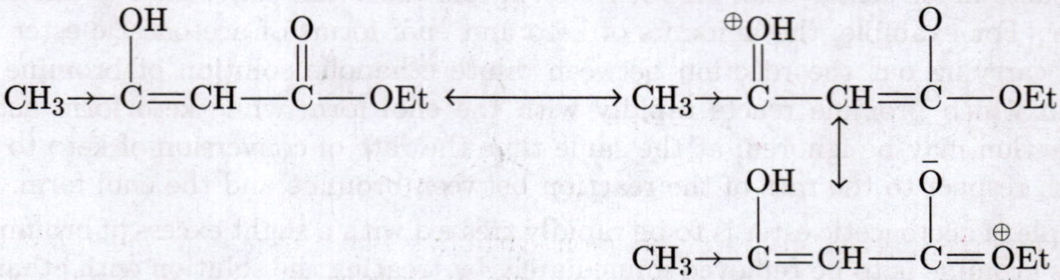

(ii) 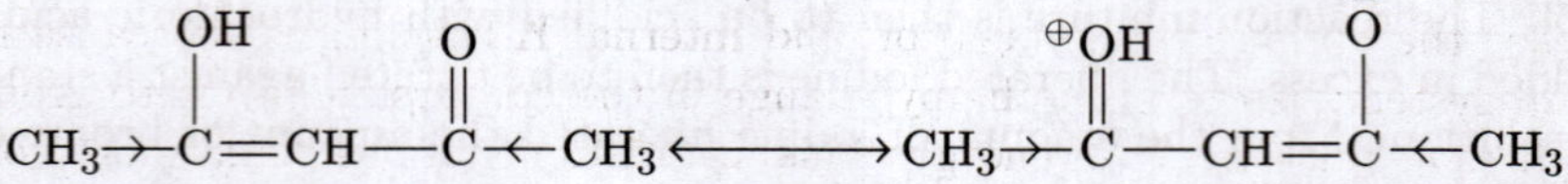

(iii) 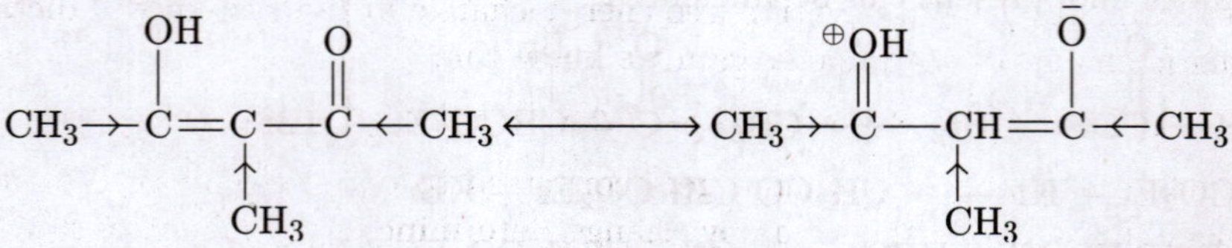

(iv) 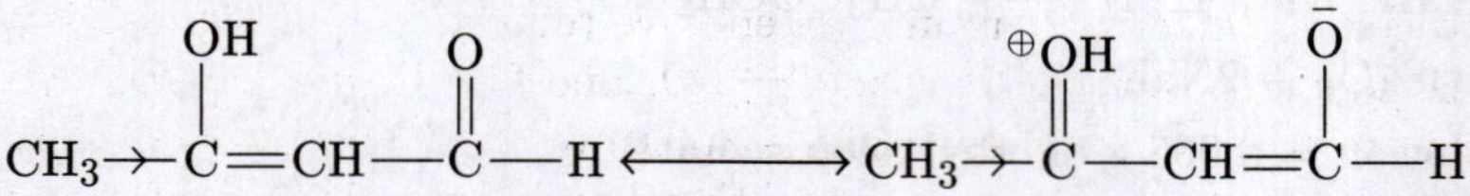

In the compound (i), the enol of acetoacetic ester, the +R effect of the OEt group inhibits the delocalisation of the π es of the —C(OH)=CH— to an appreciable extent and thus brings instability; whereas there is no +R effect to inhibit resonance in other enols. +I and hyperconjugative effects of the methyl group is found minimum in the enol of acetoacetaldehyde (iv) while this effect is maximum in the case of the enol of α-methylacetylacetone (iii). The +I effect and hyperconjugative effect of α-methyl group being acting against conjugation, the delocalisation of π es of —C(OH)=CH moiety is the least in compound (iii) and the highest in compound (iv) among compounds (iv), (iii) and (ii). So the enol content of compound (iv) is 98%, that of compound (ii) is 80% and that of compound (iii) is 40%. The enol content of ethyl acetoacetate is only 8%—this is owing to the +R effect of the —OEt group as stated above. In compound (iii) there is also steric strain owing to the presence of 3 methyl groups on 3 adjacent C atoms in the chelated form.

(i) 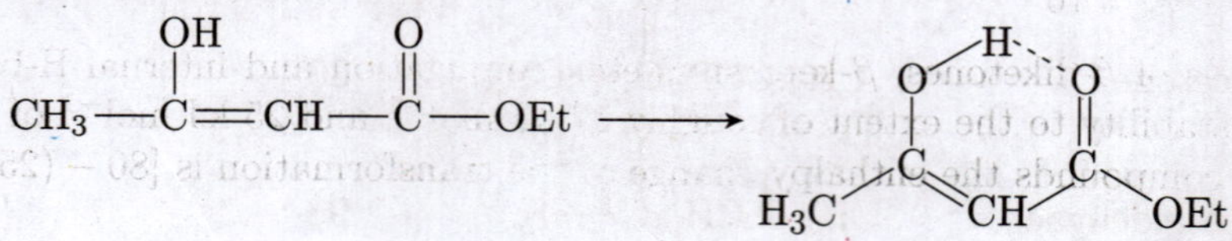

(ii) 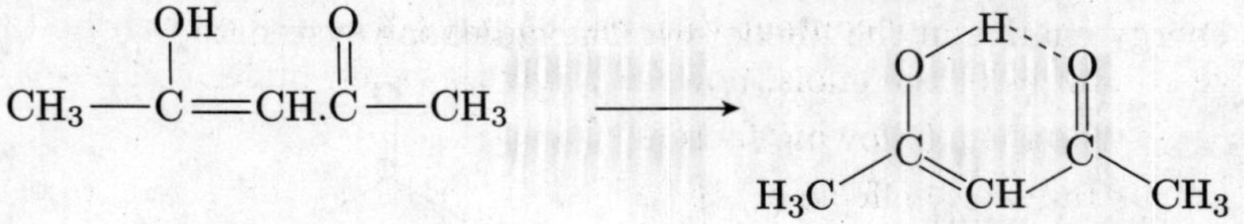

(iii) 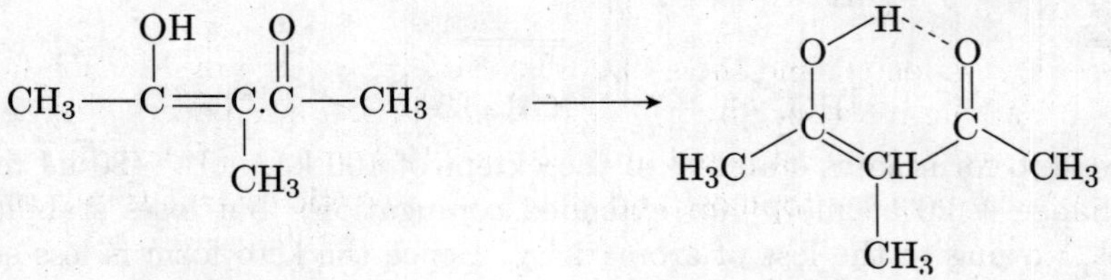

(iv) 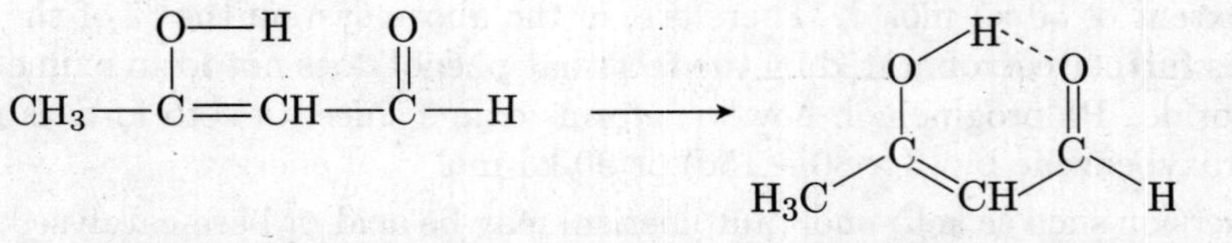

In this context it is important to note that monocarbonyls like acetone has much less of enol form (10^{-4}%) in the liquid state. This means the enol of acetone is much less stable than its keto form; in this case the absence of conjugation and internal H-bonding is the cause for the instability of the enol system. However, enthalpy change in the conversion of a keto group into an enol roughly requires +80 kJ mol^{-1} of energy. This means the enol has 80 kJ mol^{-1} of more energy than the keto form. As usual a negative free energy change in the system will dictate the spontaneous tautomeric change of a system. Again we know that

$$\Delta G^\circ = \Delta H^\circ - T\Delta S^\circ$$

Since one molecule of a carbonyl compound gets converted to one molecule of the enol, the entropy change is negligible. Hence the enthalpy change determines the amount of free energy change of the system. To convert a keto group into an enol we require to replace one C=O, one C—C, and one C—H bond respectively by one C—O, one C=C and one O—H bond. The breaking of (i) C=O to C—O requires (686.8 – 360.0) or 326.8 kJ mol^{-1} of energy, (ii) C—H bond requires 413 kJ mol^{-1} of energy; on the other hand the formation of (i) a C—C π bond gives up (598.7 – 347.5) or 251.2 kJ mol^{-1} and (ii) one O—H bond liberates 419 kJ mol^{-1}). So the enthalpy change = [(251.2 + 419) – (326.8 + 413)] kJ mol^{-1}

$$= (670.2 - 739.8)\ \text{kJ mol}^{-1}$$

$$= -69.6\ \text{kJ mol}^{-1}$$

$\therefore\ \Delta H^\circ = +69.6$ kJ mol^{-1}; hence $\Delta G^\circ = +69.6$ kJ mol^{-1} Since $\Delta S^\circ = 0$

Again,

$$K = 10^{\frac{-\Delta G^\circ \times 1000}{2.303 \times 8.31 \times 2.98}} \quad [R = 8.31\ \text{J deg}^{-1}\ \text{mol}^{-1} = 0.0083\text{lkJ deg}^{-1}\ \text{mol}^{-1}]$$

Here,

$$K = 10^{\frac{-69.6 \times 1000}{2.303 \times 8.31 \times 298}} = 10^{-12.2}$$

However, in the cases of β-diketones, β-ketoesters etc., conjugation and internal H-bonding in the enol form bring stability to the extent of roughly 17 kJ mol^{-1} and 25 kJ mol^{-1} of energy respectively. For these compounds the enthalpy change of the transformation is [80 – (25 + 17)] or +38 kJ mol^{-1}.

Let us apply the free energy change in the phenol-cyclohexadienone system :

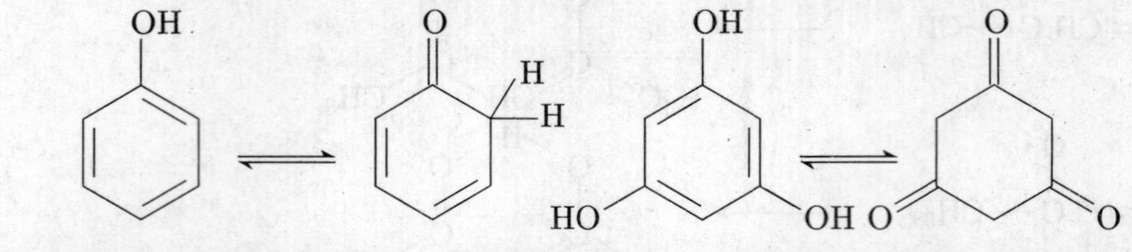

In this conversion the keto form gains stability to the extent of 100 kJ mol^{-1} (80 kJ mol^{-1} for enol-keto enthalpy change + 20 kJ mol^{-1} for extended conjugation); but loses stability to the extent of 150 kJ mol^{-1} owing to the loss of aromaticity. Hence the keto form is less stable than the phenol to the extent of 50 kJ mol^{-1}. Therefore, in the above system the % of the keto form is almost nil. This is further corroborated by the fact that phenol does not form oxime with hydroxylamine hydrochloride. Phloroglucinol, however, forms oxime since its keto form is more stable than the *s*-trihydroxybenzene by (3 × 80 – 150) or 90 kJ mol^{-1} of energy.

Prototropic interconversion such as keto-enol tautomerism may be acid-or base-catalysed, The base- or acid-catalysed reaction may occur in two discrete steps or in a single concerted step.

Base-catalysed conversion of a keto form to the enol form and vice versa:

Step I: Abstraction of proton from the more active α-position

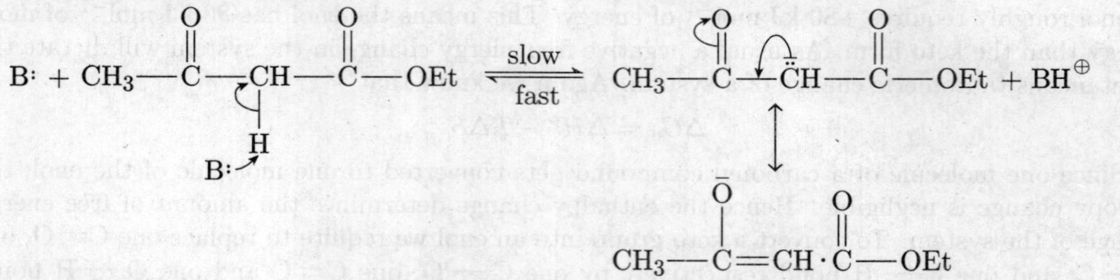

Step II: Protonation of the anionic O by the conjugate acid, BH$^{\oplus}$, of the base.

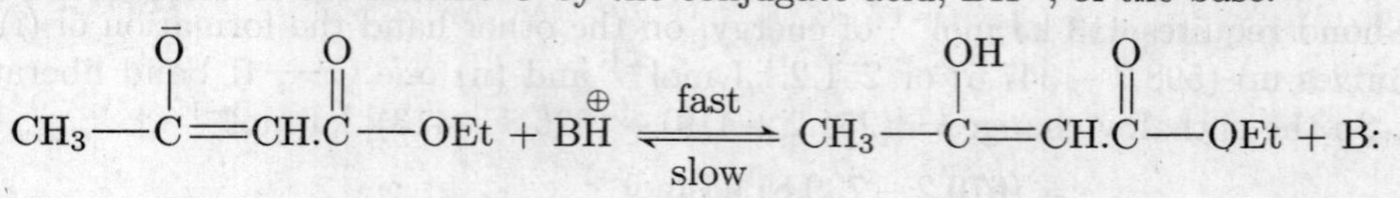

The first step is the slow step and the second step is the fast step.

The conversion of the enol to the keto form follows the reverse pathway which is in accordance with the Principle of Microscopic Reversibility.

When the reaction is carried out with $CH_3—\overset{O}{\overset{\|}{C}}—CD_2\overset{O}{\overset{\|}{C}}—OEt$, it shows primary kinetic isotope effect which is consistent with a mechanism in which a proton is lost in the rate-determining step.

Acid-catalysed conversion of a keto form into the enol form and vice versa:

Step I: Protonation of the carbonyl O atom.

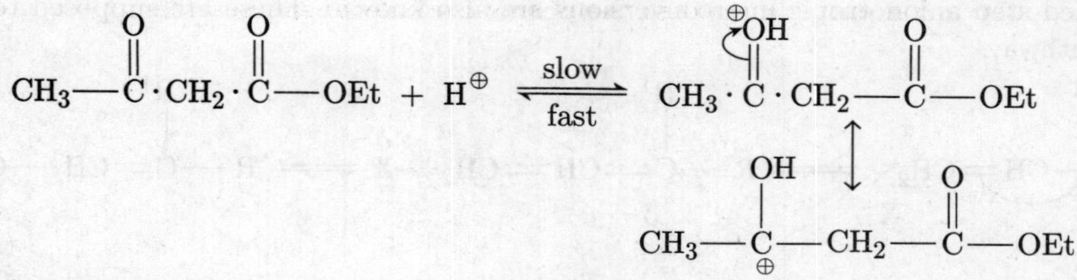

Step II: Elimination of a proton from the more active α-position.

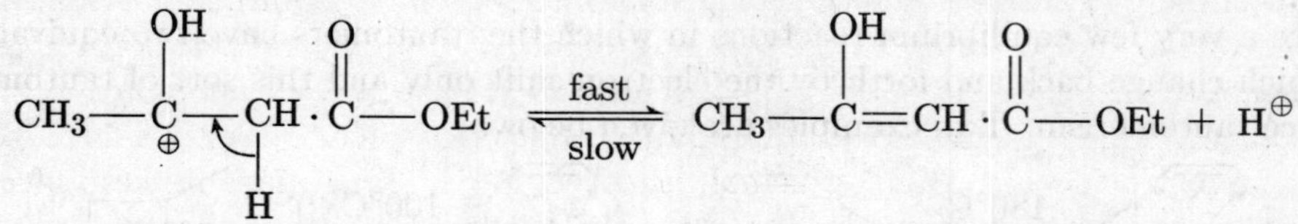

In this case also the conversion of the enol to the keto form follows the reverse pathway.

Concerted step mechanism

The concerted step mechanism is supposed to involve both acid and base. The acid pushes a proton to the electron-rich centre while the base pulls a proton from the active position.

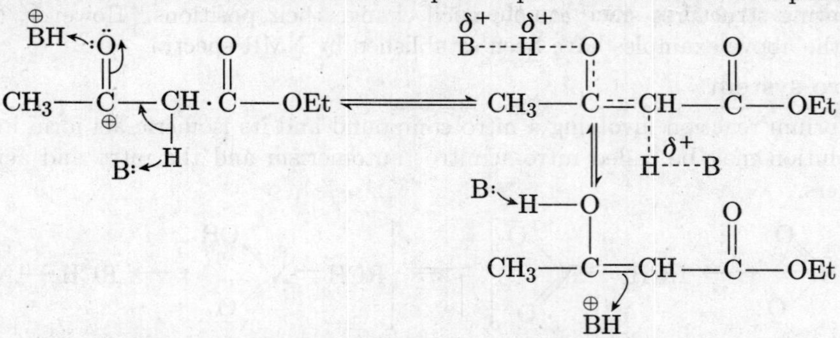

Anionotropic interconversions usually follow the S_N1' pathway involving resonace-stabilised carbenium ion.

Step I:

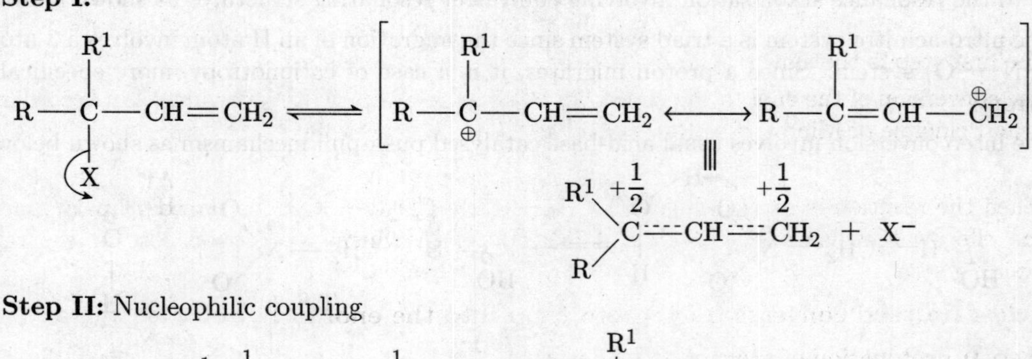

Step II: Nucleophilic coupling

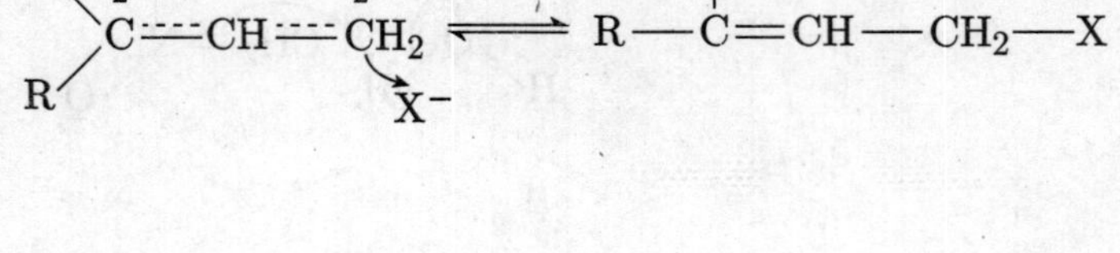

Concerted step anionotropic interconversions are also known. These are supposed to follow the S_N2' pathway.

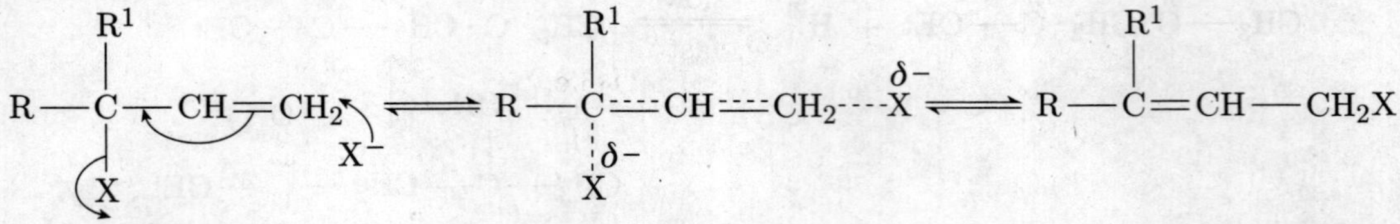

Valence tautomerism

There are a very few equilibrium reactions in which the, tautomers have two equivalent constitution which change back and forth by the electron shift only and this sort of tautomerism is called valence tautomerism. Two examples are given below:

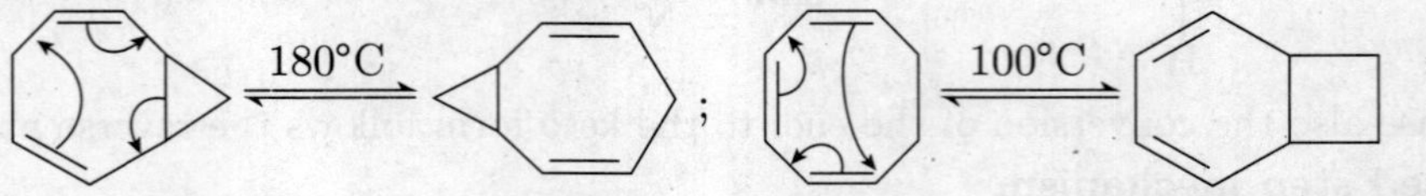

85% 15%

Though the interchange between two valence tautomers involve electron shit only, yet they are not resonating structures since atomic nucli change their positions. However, the valence tautomers of the above examples have been established by NMR spectra.

Nitro-acinitro system

The equilibrium reaction involving a nitro compound and its isomeric aci form in the liquid state or in solution may be called nitro-acinitro tautomerism and the nitro and aci forms are called tautomers.

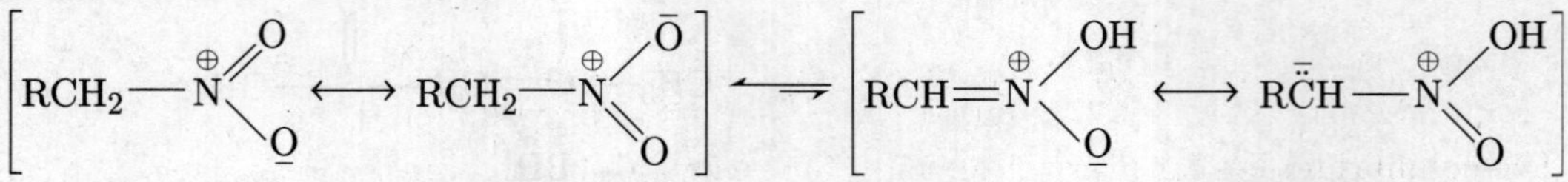

Nitro form Aci form

The equilibrium lies far over to the left because the nitro form is more stable than the aci form owing to the resonance stabilisation involving equivalent resonating structures as shown above.

The nitro-acinitro system is a triad system since the migration of an H atom involves a 3-atom (C—N—O) system. Since a proton migrates, it is a case of cationotropy, more specifically prototropy.

The interconversion involves usual acid-base catalysed push-pull mechanism as shown below :

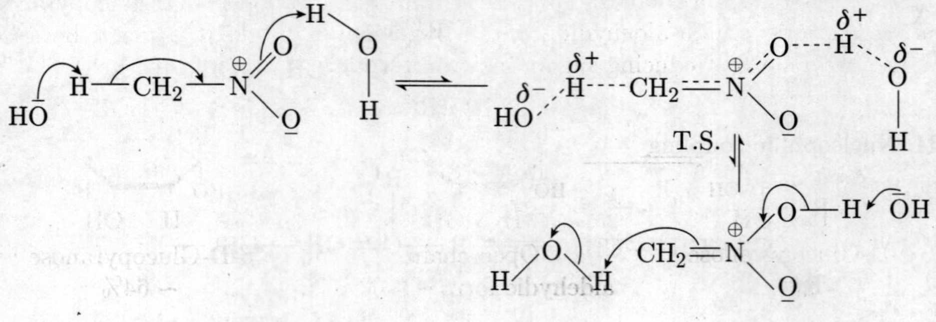

The presence of aci form in the system is supposed in the reactions of 1° and 2° nitroalkanes with nitrous acid.

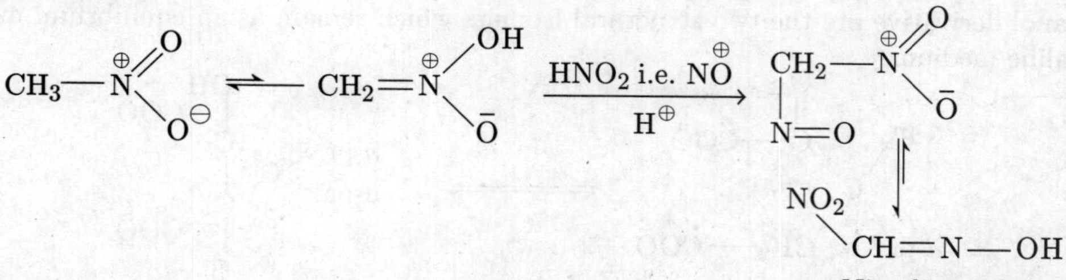

Nitrolic acid

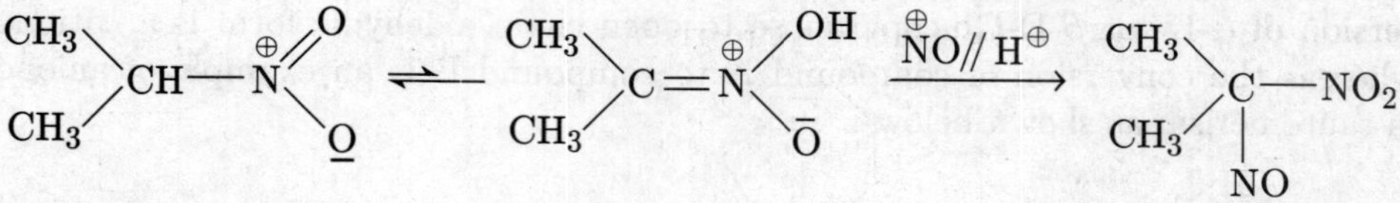

Pseudonitrol

When the sodium salt of a 1° or 2° nitroalkane is acidified at a low temperature, there is not always an immediate separation of oil drops of the nitro compounds; it has been argued that the protonation occurs to the aci form which then changes to the nitro form and hence an extra-time is required for the separation of oil.

Oximino-nitroso tautomerism

This is also a prototropic triad system as shown below:

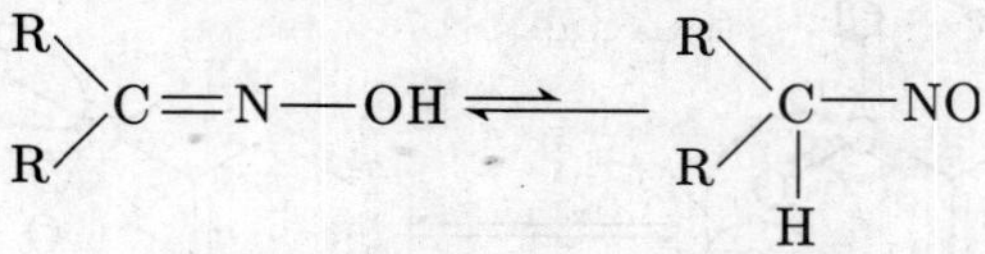

oximino nitroso

The equilibrium lies far over to the left. The reaction is almost inaccessible since tautomerisation of an oxime to a nitroso compound has not yet been known. Only 3° nitroso compounds exist. As a rule nitroso compounds exist when there is no α-H.

Ring-chain tautomerism

When a closed chain compound and an open chain compound are constitutional isomers and exist in rapid equilibrium in liquid or in a solution, the constitutional isomers are called ring-chain tautomers and the equilibrium reaction is said to be ring-chain tautomerism.

Ring-chain tautomerism is exhibited by all reducing sugars. When α-D- or β-D-glucose is dissolved in water and kept for half an hour, an equilibrium mixture of α-D-glucopyranose, β-D-glucopyranose and open chain aldehydic form of D-glucose is obtained. In fact, because of the ring-chain tautomerism the reducing sugars show mutarotation.

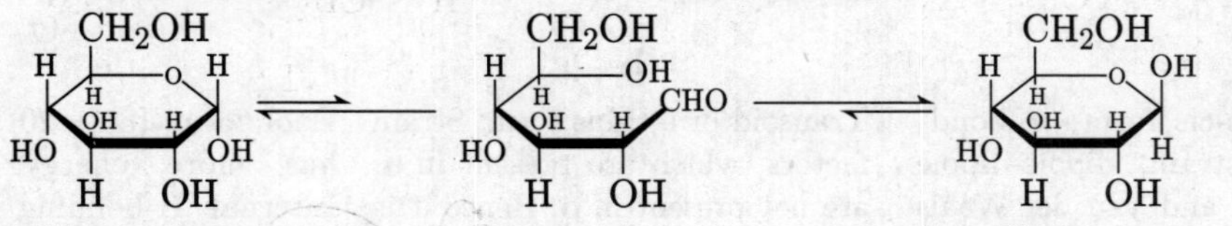

α-D-Glucopyranose ~36% Open chain aldehydic form ~ 0.02% β-D-Glucopyranose ~ 64%

However, this sort of tautomerism is usually exhibited by compounds containing at least two functions. For example β, β-di-n-propyl derivative of α-keto glutaric acid and the corresponding cyclopropanol derivative are the two structural isomers which remain as an equilibrium mixture in the alkaline medium.

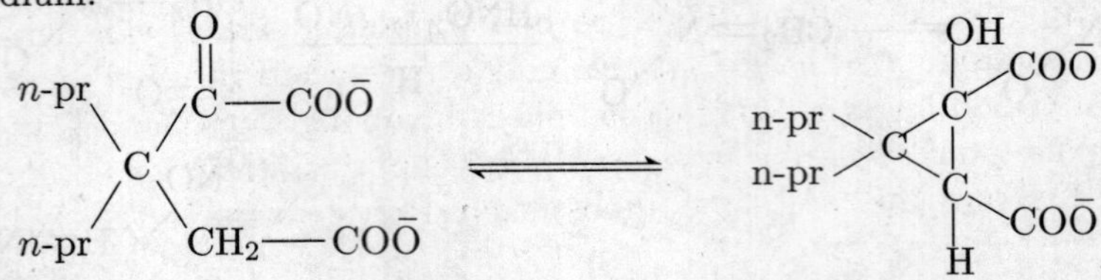

The interconversion of ring-chain tautomers may either be cationotropic or aninotropic. The interconversion of α-D- or β-D-Glucopyranose to open chain aldehydic form is a cationotropic process whereas the conversion of compound A to compound B is an example of anionotropic ring-chain tautomerism as shown below :

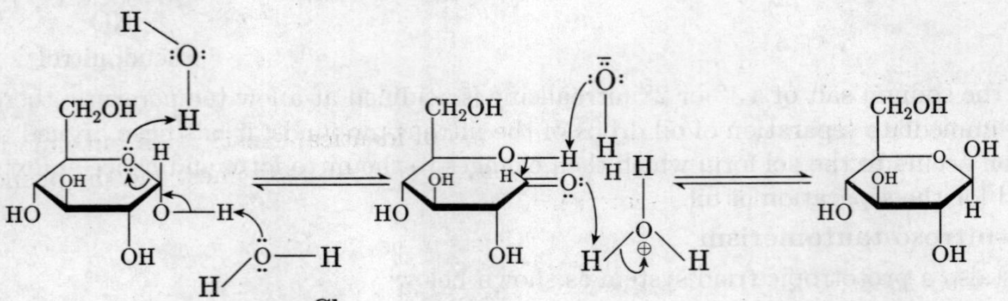

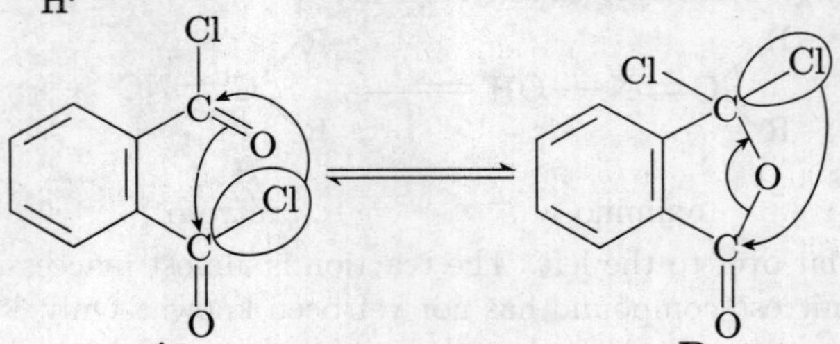

A B

Problem 9.2. Explain the following facts:

(i) Butane-2, 3-dione forms $5.6 \times 10^{-3}\%$ of enol i.e., keto form is almost cent per cent.

(ii) Cyclopantane-1, 2-dione exists almost cent per cent in the enol form.

Solution:

(i)

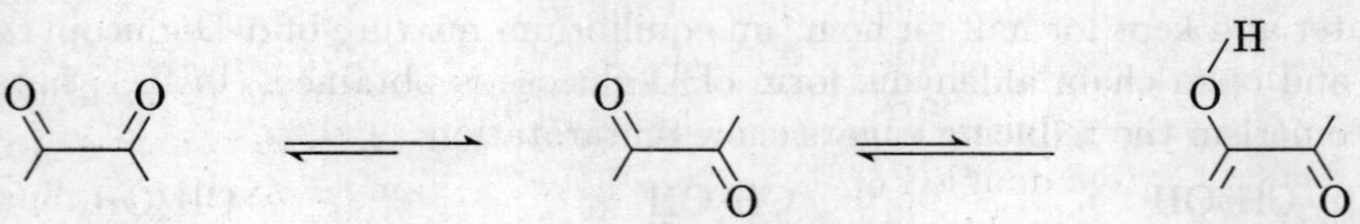

a b c

a: Cisoid or s-cis form; π bond-opposing strain, dipole-dipole interaction and van der Waals strain are present. Highly unstable.

b: Transoid or s-trans form. Strain factors, which are present in **a**, are not present in **b**. Hence it is stable.

c: Enol form; (80 − 20) kJ mol^{-1} has more energy than b. Internal H-bonding does not occur. So it is not stable.

For the reasons noted below each structure, butane -2, 3-dione remains mainly in the keto form (transoid) [consult conformational analysis : 9.4.5].

(ii) Cyclopentane-1, 2-dione has a rigid cyclic structure in the cisoid form; it cannot assume transoid form. In the cisoid form there are bond-opposing strain, van der Waals strain and dipole-dipole interaction. When the molecule gets converted to the enol form, it gains stability through internal H-bonding and conjugative effect involving C═C and C═O bonds; besides these, dipole-dipole interaction dies down and bond-opposing strain gets reduced. So **e** is much less stable than **f**. Hence the compound almost exclusively remains in the enol form.

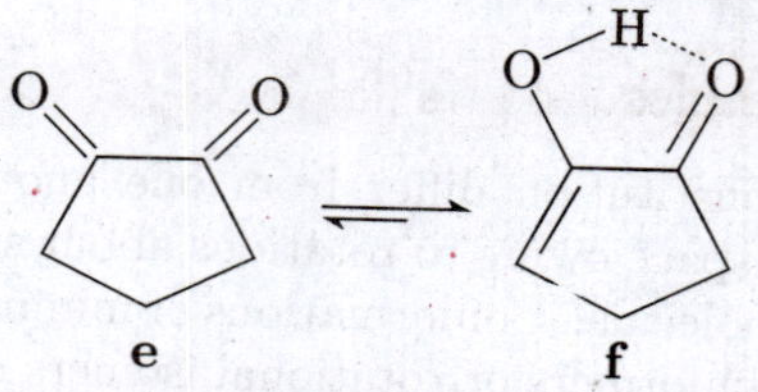

9.3 Stereoisomerism

Compounds with the same molecular formula and of identical constitution but differing only in the spatial arrangement of the constituent atoms are called stereoisomers and the phenomenon of their existence is known as **stereoisomerism**.

Butenedioic acid (COOH.CH═CH.COOH) exists in nature in two forms:

(i) In which the two H atoms are on the same side of the C═C double bond and two —COOH groups are on the other side of the C═C double bond; this form is named as maleic acid.

(ii) In which the two H atoms lie on the opposite sides of the C═C double bond and two —COOH groups also lie similarly; this form is named as fumaric acid.

(a) Maleic acid **(b) Fumaric acid**

Fig. 9.1

Maleic and fumaric acids have different properties. For example, maleic acid is a highly water soluble solid melting at 130 °C but fumaric acid is a slightly water soluble solid melting at 287 °C. Maleic acid forms anhydride when it is heated at 140 °C, but fumaric acid forms maleic anhydride at 275 °C.

It is needless to mention that maleic and fumaric acids have the same molecular formula ($C_4H_4O_4$); their bond connectivities are also identical as each of them contains two carboxyl groups and one carbon-carbon double bond. As they have the same molecular formula, they are isomers. Their isomerism is caused by the different spatial arrangements of the two H atoms and that of the two —COOH groups about the C═C double bond though their constitutions are identical. Hence maleic and fumaric acids are stereoisomers.

The causes of different spatial arrangements in stereoisomers are :

1. Restricted rotation of a part of the molecule about a C—C single bond; this gives rise to the rotational isomerism, or conformational isomerism.

2. Restricted rotation of a part of the molecule about a double bond or a ring and chirality, i.e., dissymmetry or asymmetry of a molecule. These causes give rise to configurational isomerism.

So, stereoisomerism may be subdivided under two heads :

1. Conformational (or dynamic stereo-) isomerism
2. Configurational (or static stereo-) isomerism.

Usually configurational isomers are known as stereoisomers.

9.4 Conformation of Molecules

If molecules of identical constitution differ from one another only in arrangements of its constituent groups or atoms in space owing to rotations about single bonds, then they are said to be conformations of the same molecule. Conformations of minimum energy are sometimes termed as conformational isomers or conformers or rotational isomers or rotamers and the phenomenon of their existence may be called conformational or rotational isomerism.

As rotation about a single bond is possible because of its symmetrical charge density, one of the C atoms in a C—C bond may rotate with its substituents around the single bond when the other C atom remains static. This develops innumerable spatial arrangements of the groups attached to the rotating C atom in relation to the position of the groups on the static C atom. Each of these spatial arrangements is known as a conformation of the molecule. The two extreme cases amongst these are called eclipsed and staggered forms. Any intermediate arrangement between these two conformations is known as a skewed (partially eclipsed) conformation.

If the ball-stick model (Fig. 2.27c) of each of the extreme conformations is viewed along the C—C bond axis, then in one of the conformations the atoms and groups attached to the nearer C atom appear to be as far apart as possible from the atoms and groups attached to the farther C atom and the conformation is known as **staggered conformation**.

On the other hand, when the atoms and groups of the nearer C lie exactly opposite with respect to the atoms and groups of the farther C atom, the conformation is known as **eclipsed conformation**.

9.4.1 Projection of conformations

The conformations are generally projected on the plane of the paper in three ways:

(i) Newman projection formula,
(ii) Flying-wedge projection formula,
(iii) Sawhorse projection formula.

To understand the projection formula of a particular compound one should handle its ball-stick model. In this model balls of different colours are considered as atoms of different elements and sticks as bonds.

(i) In Newman projection formulae, the C atom nearer to the observer is represented by the centre of the circle and its 3 σ bonds are shown by drawing solid lines from points outside the circumference to the centre of the circle; the three bonds are at 120° to one another. The circle represents the farther C atom and the three sigma bonds of this C atom are shown by solid lines which are drawn from the circumference of the circle outside it. Thus the two extreme conformations of ethane molecule and one partially eclipsed form may be represented by the Newman projection formula as shown in Fig. 9.2.

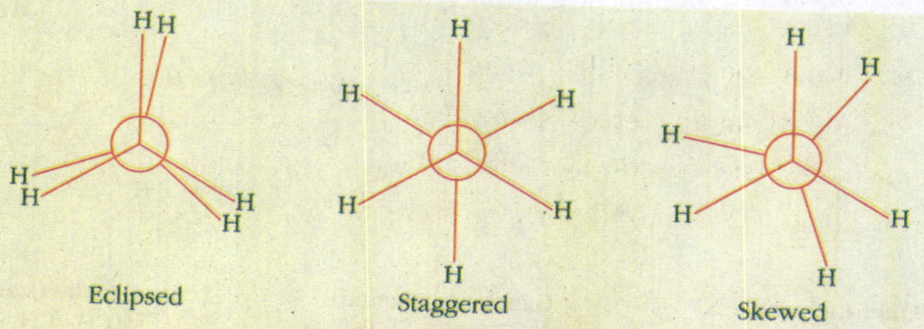

Fig. 9.2. Newman projection formulae for ethane

In Newman projection formula for ethane, the angle between an H on the nearer C atom and an H on the farther C atom in the plane of the paper is called **dihedral angle** or **angle of torsion** (as this originates from the rotation, i.e., torsion about a single bond) and denoted by the symbol ϕ. 'Di' means two and 'hedron' means base or plane. So dihedral angle means an angle between two planes of two tetrahedra.

If we project an organic molecule (or a part of it) having two-carbon system with four atoms or groups of each carbon such as C_1ABD—$C_2A'B'D'$ on the plane of the paper, then the angle between one plane defined by a designated C_1—A (or C_1—B or C_1—D) bond and C_1—C_2 bond and the other plane defined by C_1—C_2 bond and a designated C_2—A' (or C_2—B' or C_2—D') bond is called **dihedral angle** ϕ, Fig. 9.3. As rotation about a single bond may vary from 0° to 360°, this angle also varies from 0° to 360°. If this angle is 0° or 120° or 240°, eclipsed arrangements result; on the other hand when the dihedral angle is 60° or 180° or 300°, staggered conformations are obtained. For ethane the three staggered conformations are the same; its three eclipsed conformations are also identical. Klyne-Prelog terminology for dihedral angle (actually designated by θ, here written as ϕ) is the angle of torsion (ϕ) with a '+' or '−' sign; '+' for clockwise measurement of the angle and '−' for otherwise. Though dihedral angle is synonymous with torsion angle yet there is a difference. Torsion angle has a direction. When it is measured clockwise the sign of the angle '+'; '−' sign indicates the measurement of the angle is made anticlockwise. Its range is, therefore, 0° to ±180°; while that of dihedral angle is 0° to 360°.

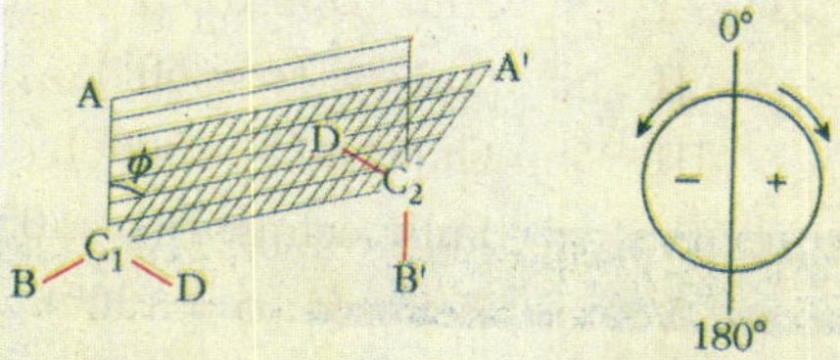

Fig. 9.3. Dihedral angle and torsion angle

If each of the single-bonded C atoms contains substituents other than H, then the molecule will have three different staggered conformations ($\phi = 60°, 180°, 300°$) and three different eclipsed conformations ($\phi = 0°, 120°, 240°$). When the dihedral angle between the two largest substituents, one on each C, is 180°, one staggered conformation will result and this is known as **anti** or **anti-periplanar conformation**. In two other staggered forms the dihedral angles between the two largest groups are 60° and 300°; these two are said to be **gauche** (pronounce as 'goss') or **syn-clinal** or **skew conformations**. Similarly if the value of ϕ is 0° with respect to the two largest substituents, one on each C, then the resulting eclipsed conformation is called **fully eclipsed** or **syn-periplanar conformation**. The other two eclipsed forms have 120° and 240° as the value of dihedral angle; these are simply called **eclipsed or anti-clinal** forms. Thus for n-butane there are one anti form (CH_3/CH_3, ϕ = 180°), two gauche forms (CH_3/CH_3, ϕ = 60°, 300°), one fully eclipsed form

(CH_3/CH_3, $\phi = 0°$) and two other eclipsed forms (CH_3/CH_3, $\phi = 120°, 240°$); all these are shown in Fig. 9.4.

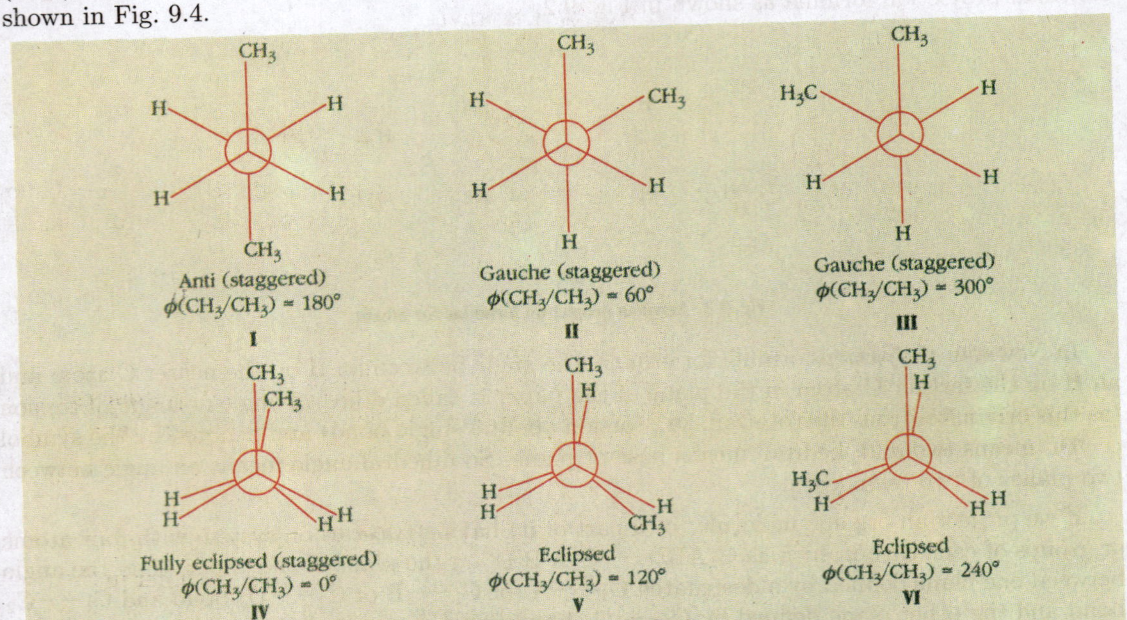

Fig. 9.4. Conformations of *n*-butane in Newman projection

According to the **Klyne-Prelog terminology** the conformations with their angle of torsion are designated as follows:

Conformation (Fig 9.4) with dihedral angle (θ) though written as ϕ		Corresponding Klyne-Prelog terminology with angle of torsion (ϕ)
I. :	Anti ($\phi = 180°$)	$\pm$ anti-periplanar ($\pm$ap, $\phi = 180° \pm 30°$)
II. :	Gauche ($\phi = 60°$)	+ syn-clinal (+sc, $\phi = +60° \pm 30°$)
III. :	Gauche ($\phi = 300°$)	− syn-clinal (−sc, $\phi = -60° \pm 30°$)
IV. :	Fully eclipsed ($\phi = 0°$)	$\pm$ syn-periplanar ($\pm$sp, $\phi = 0° \pm 30°$)
V. :	Eclipsed ($\phi = 120°$)	+ anti-clinal (+ac, $\phi = +120° \pm 30°$)
VI. :	Eclipsed ($\phi = 240°$)	− anti-clinal (−ac, $\phi = -120° \pm 30°$)

'+' for clockwise measurement of angle of torsion.

'−' for anticlockwise measurement of angle of torsion.

Since conformations are seldom ideal, ϕ = ideal angle of torsion $\pm 30°$,

anti-periplanar	:	around planar and on opposite sides
syn-periplanar	:	around planar and on the same sides
anti-clinal	:	with a slope and on opposite sides
syn-clinal	:	with a slope and on the same sides

(ii) In the flying-wedge projection, the molecule is viewed sideways. So, C—C bond is drawn horizontally. The convention is that the solid lines represent bonds on the plane of the paper; the bonds directed behind the plane of the paper are projected by broken wedges and dark wedges represent bonds above the plane of the paper.

In Fig. 9.5 the staggered conformation of ethane is shown.

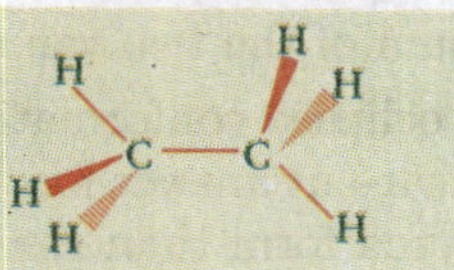

Fig. 9.5. Flying-wedge projection of ethane (staggered)

(iii) In the sawhorse projection, the molecule is viewed from an angle. So, the C—C bond is drawn backward obliquely (Fig. 9.6). In this drawing the left hand carbon is nearer to and the right hand carbon is farther from the viewer.

In this method, the eclipsed groups are drawn parallel to each other on the same side of the oblique line; the staggered groups are also drawn parallel to each other, but on the opposite sides of the oblique line. An eclipsed form gets converted into the staggered form when a C atom is rotated by a 180° angle around C—C bond. A partially eclipsed from is generally drawn form a staggered or an eclipsed form by rotating one of the Cs in C—C bond around it through any angle other than 60°, 120°, 180°, 240°, 300° and 360°.

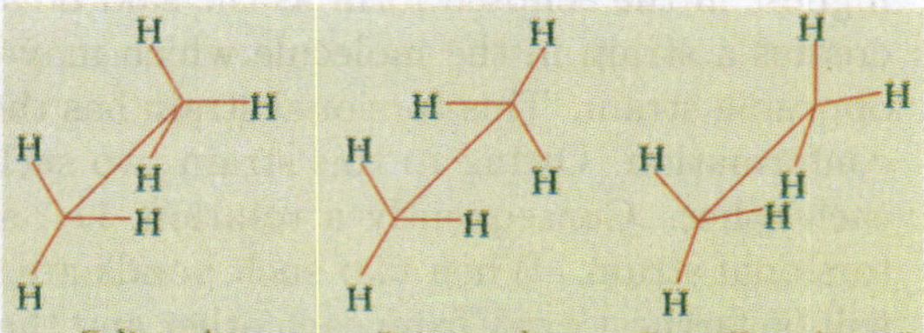

Fig. 9.6. Sawhorse projections of ethane

Problem 9.3. Represent the anti and fully eclipsed conformations of 2, 3-butanediol by Newman and sawhorse projection formulae.

Solution:

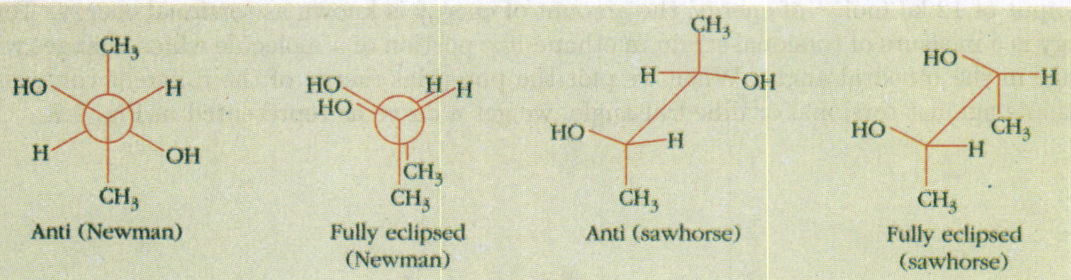

Fig. 9.7

If rotation about a C—C bond is free, one of the methyl groups in ethane could rotate with respect to the other without any change in the energy of the molecule. Assuming free rotation about C—C bond, E. Teller and B. Topyely (1935 A.D.) calculated the heat capacity of ethane theoretically and found that it was significantly greater than the experimental value. They, therefore, suggested that the rotation about a single bond is not entirely free but is hindered by an energy barrier, i.e., the rotation is somewhat restricted. If this be correct, then one of the conformations of ethane must have the lowest energy, i.e., the greatest stability. Calculations on energy content of the different conformations of the ethane show that the staggered conformation is the most stable one and the eclipsed conformation is the least stable. The staggered form contains less energy than the eclipsed form by 12.5 kJ mol^{-1}. The distance between an H atom on C_1 from its nearest neighbour on C_2 is 230 pm (2.3Å) in the eclipsed form and 250 pm (2.5Å) in the staggered form whereas the sum of the two van der waals radii is 240 pm.(2.4Å).

Now the question is: Why is the eclipsed conformation of ethane less stable than its staggered conformation? Up to this date no clear-cut answer has been given. There is a minor contribution to the greater stability of the staggered form than the eclipsed one by the van der Waals strain (See Sec. 9.4.4) since the distance between 2 Hs, one on each C, is 250 pm and 230 pm respectively whereas there sum of van der Waals radii is 240 pm. This means in the eclipsed conformation of ethane there is van der Waals strain. However, the electronic charge cloud of any two C—H bonds, one on each C atom of a C—C bond, must repel each other. The repulsion will be the highest in the eclipsed form as the said bonds are then closest to each other; this repulsive force creates a strain in the molecule which may be called eclipsing strain or torsional strain or bond opposing strain. This torsional strain has the major contribution to the stability of the staggered conformation. Owing to this strain two such bonds tend to remain as far away as possible form each other. Consequently a rotation, i.e., a torsion about a C—C bond takes place to avoid torsional strain. When two such bonds are in anti position in the staggered conformation, they will be farthest away from each other and the torsional strain will be the least. This means : so far as torsional strain is concerned, any staggered conformation is always a more stable conformation than its any eclipsed or partially eclipsed form. A partially eclipsed conformation has a stability in between the eclipsed and the staggered conformations. For this reason, the staggered form of ethane is more stable than its eclipsed form.

Modern theories tell us that the rotational energy barrier may be ascribed to the π-bonding M.O.s of ethane which are more bonding in the staggered form than in the eclipsed form.

As the staggered conformation of ethane can be converted into the eclipsed conformation by the input of 12 kJ mol^{-1} of energy, the amount of energy is known as torsional energy. Torsional energy is a measure of torsional strain in ethane-like portion of a molecule which changes with the change in the dihedral angle. When we plot the potential energy of the different conformations of ethane against torsional or dihedral angle, we get a curve as represented in Fig. 9.8.

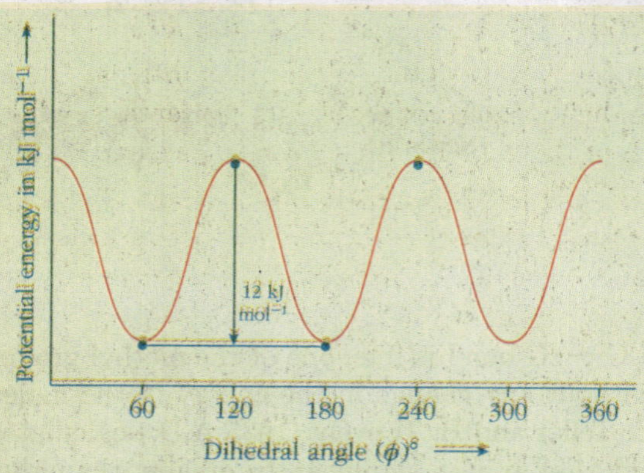

Fig. 9.8. Potential energy diagram of conformations of ethane as a function of dihedral angle in degree

Calculation on difference in energy between staggered and eclipsed conformations for propane shows that the energy barrier is 13.53 kJ mol^{-1} (3.3 k cal/mole), i.e., staggered form contains 13.53 kJ mol^{-1} (3.3 k cal/mole) of energy less than the eclipsed form. Such is the case with n-butane, here the energy barrier is 25 kJ (6 k cal) mol^{-1}. There are two stable gauche conformations for n-butane which have identical energy and this is greater than that of the anti form by 3.36 kJ (0.8

k cal) mol^{-1}. The interaction between two CH_3 groups in the gauche forms ($\phi = 60°, 300°$) makes them less stable than the anti form; this intraction is said to be n-butane gauche interaction. The potential energy of the different conformations of n-butane are plotted against dihedral angle in Fig. 9.9. This indicates the relative stability of the different forms. From the curve it is seen that the two gauche conformations of n-butane have much greater stability than the eclipsed conformations though formers are less stable than the anti form.

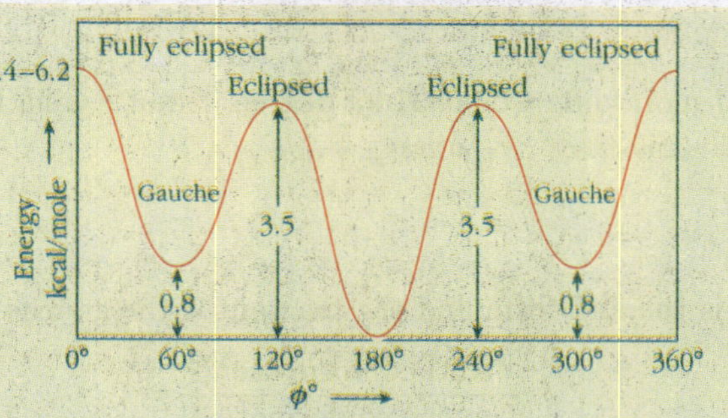

Fig. 9.9. Potential energy of n-butane as a function of dihedral angle. At the point of reference the potential energy is that of anti form : dihedral angle is measured in degree

9.4.3 Conformational isomerism

Stable conformations, i.e., conformations of minimum energy are called conformational isomers or rotational isomers or conformers or rotamers. Two gauche forms and the anti form of n-butane are thus its three conformers or rotamers. If we closely look at the two gauche forms, we shall see that they are non-super-imposable mirror images of each other and conformations which bear such a relation are said to be conformational enantiomers; the anti form does not bear any mirror-image relation to any of the gauche conformations and the anti form is said to be diastereomeric (conformational) with each of the two gauche forms, i.e., a gauche conformer and the anti conformer of n-butane form a pair of conformational diastereomers (Fig. 9.10).

As these three conformers are stable, they will remain as an equilibrium mixture in a bulk of n-butane. Anti-form being the most stable, its percentage will be the highest in the mixture. The percentage of each of the gauche forms will be the same because of their identical stability.

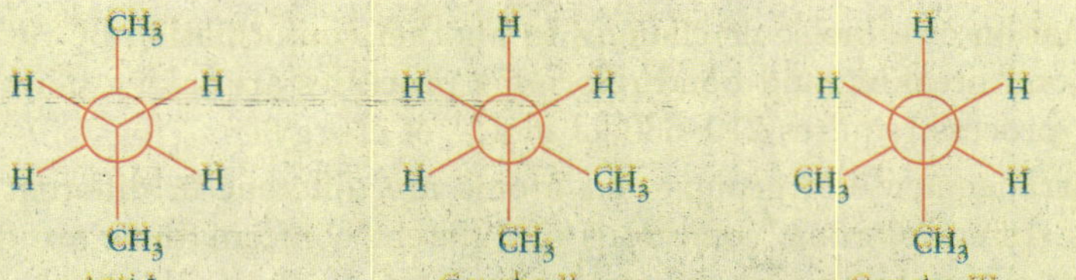

Fig. 9.10. Anti and gauche conformations of n-butane. I, II, and I, III are diastereomers (conformational). II and III are enantiomers (conformational)

Although the anti conformation of n-butane is more stable than the gauche conformation to the extent of 3.3 kJ (0.8 k cal) $mole^{-1}$, yet the statistical probability of existence of gauche conformations is twice that of the anti conformation. For this reason the $T\Delta S°$ term has an appreciable amount of contribution to the $\Delta G°$ of the equation.

$$\Delta G° = \Delta H° - T\Delta S° \tag{9.1}$$

Where $\Delta G°$, $\Delta H°$ and $\Delta S°$ are the conformational free energy change, enthalpy change and entropy change respectively when the gauche conformations get converted to the anti conformation under standard conditions.

The calculated value of the stable free energy change is $\Delta G° = -1.7$ kJ mol^{-1} only because of the statistical probability factor mentioned above.

Again we know that,

$$K = 10^{-\Delta G°/RT2.303}$$

Here $-\Delta G° = 1.7$ kJ mol^{-1}, R = 0.00831 kJ degree^{-1} mol^{-1} and T = 298 Kelvin

$$K = 10^{1.7/0.00831 \times 298 \times 2.303} = 10^{0.298} = 1.986 \simeq 2$$

But, $K = \dfrac{[\text{product}]}{[\text{reactant}]}$

Here, the product is the anti form and the reactant is the gauche form. So

$$\frac{[\text{Anti conformation}]}{[\text{Gauche conformation}]} = 2$$

Thus at room temperature the ratio of anti and gauche conformations is 2:1.

Conformations of simple molecules have low energy barrier; but that in complicated molecules, specially in alicyclic compounds, is comparatively high and it may go upto 46 kJ mol^{-1} (11 k cal/mole).

Of course molecules in a bulk can easily overcome this energy barrier : every molecule in a bulk is associated with kinetic energy owing to its motion and the energy increases with the increasing temperature; when molecules collide with each other amongst themselves, their energies also change—even at room temperature 63-84 kJ mol^{-1} of energy change may occur. Thus at room temperature conformations get interconverted. So we may say that in a bulk of *n*-butane, its molecules spend maximum time in anti conformation and minimum time in fully eclipsed conformation.

Isolation of stereoisomers require more than 84 kJ of energy per mole. As the difference in energy between the least stable and the most stable conformations is only 4.2 to 46 kJ mol^{-1}, which they can obtain from their molecular motion, they get interconverted into one another and hence they are non-isolable.

One conformation of a molecule changes to another conformation by rotation about a single bond only; no bond breaking and bond reforming processes are taking place during interconversion. The latter process requires 210–630 kJ mol^{-1} of energy.

As spatial arrangements of groups and atoms are different in different conformations, the steric as well as stereoelectronic conditions of different conformations are different. These two factors determine the stability and chemical reactivity of a molecule and hence these factors are discussed in Sec. 9.4.4 and 9.24.

9.4.4 Conformational analysis

Physical properties and chemical reactivity of a compound are largely dependent on steric and stereoelectronic conditions of its molecules; again, different conformations of a molecule have different steric and stereoelectronic conditions. Hence physical and chemical properties of a compound depend on the conformations; as the percentage of the most stable conformation in a bulk is the highest, the properties of the compound will largely depend on this conformation. Therefore, determination of relative stabilities of the conformations of a compound and interpretation of its properties in terms of conformations is an important field of study and is referred to as conformational analysis.

The stability of a conformation depends on four main factors. These are:

1. Torsional strain, 2. Angle strain, 3. van der Waals strain and 4. Dipole-Dipole interaction.

Hence to determine relative stabilities of conformations the four factors said above are to be considered.

1. Torsional strain: Strain developed in a molecule by the bond-pair bond-pair repulsion, one on each C atom of a C—C single bond, may be called torsional strain. The greater the torsional strain, the lower is the stability of a conformation and vice versa.

2. Angle strain: Every C atom in an organic molecule tends to have bond angles as indicated by the nature of its bonding orbitals (the actual geometry of molecules as experimentally determined helps us to suppose the nature of the bonding orbital); e.g., an sp^3 hybridised C atom tends to have 109°28′ bond angle, sp^2 and sp hybridised Cs have a tendency to make 120° and 180° bond angles respectively. Any deviation from the usual bond angle (as determined by the nature of bonding orbitals of the concerned atom) will create a strain in the molecule which is known an angle strain. The greater the deviation of the bond angle from the usual value, the greater is the strain and consequently the less is the stability. For example, in cyclopropane the C—C—C bond angles are 60° though each of the Cs is supposed to be in sp^3 hybridised state; this orbital nature indicates that each C atom tends to have 109°28′ as the (C—C—C) bond angle value. Hence the molecule is under great strain and it has low stability.

3. van der Waals strain or steric strain: van der Waals radius of a group or atom is defined as one-half of the equilibrium distance between two equivalent groups or atoms, not bonded to each other, at the point of energy-minimum in their molecular crystals; the van der Waals radius of a group actually is a measure of its effective size. The van der Waals radius of a few atoms and groups are given below:

Table 9.2 van der Waals radii in pm

H	N	O	F	P	S	Cl	Br	I	CH_2	CH_3
120	150	140	135	190	185	180	195	215	200	200

If the distance between two such substituents is less than the sum of their van der Waals radii, then a repulsive force generates and this creates a strain in the conformation which is known as van der Waals strain or steric strain. To avoid van der Waals strain, a molecule takes up that conformation which has the least van der Waals strain and usually this is the anti conformation. The greater the size of the non-bonded substituents, the greater is the van der Waals strain in a conformation and the less is its stability. Fully eclipsed conformations have the highest van der Waals strain. However, in the gauche or skew conformation sometimes **van der Waals attraction** becomes more important than repulsion; when the distance between two non-bonded atoms is equal to or greater than the sum of their van der Waals radii, the atoms experience attraction instead of repulsion. The attractive force falls down with increasing distance. To understand this we shall discuss the conformation of a few haloalkanes in this section later on.

4. Dipole-dipole interaction and the H-bonding force: Non-bonded substituents in a C—C bond may undergo dipole-dipole interactions, the greater form of which is hydrogen bonding force. If the non-bonded atoms have opposite partial charge densities, + and −, then an attractive force develops between them and consequently this brings stability to the conformation; if the interaction leads to hydrogen bonding, stabilisation of the conformation will be of higher order. On the other hand, if the non-bonded atoms have like charges + and + or – and –, then repulsive force plays its role and the result is the destabilisation of the conformation.

Thus all these four factors may work together to stabilise a particular conformation or to destabilise it by opposing each other.

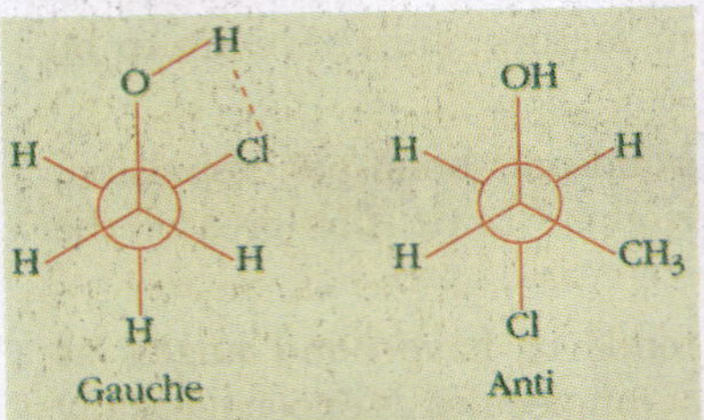

Fig. 9.11. Conformations of 2-chloroethane

Usually molecules tend to get a relief of van der Waals strain and to do so they are ready to accept other strains if it is required. But as dipole-dipole attractive force, specially H-bonding force, stabilises a conformation appreciably, a conformation can receive even a part of van der Waals strain if it facilitates dipole-dipole attraction and H-bonding. For example in the case of 2-chloroethanol, the gauche form is the most stable conformation since this involves H-bonding between the Cl and H—O groups as shown in Fig. 9.11.

The gauche conformation of this compound is more stable than the anti form by 4 kJ mol^{-1}.

Let us now apply conformational analysis in determining dipole moment of 1, 2-dibromoethane. The anti, fully eclipsed and gauche conformations of 1, 2-dibromoethane are projected by Newman projection formulae as shown in Fig. 9.12.

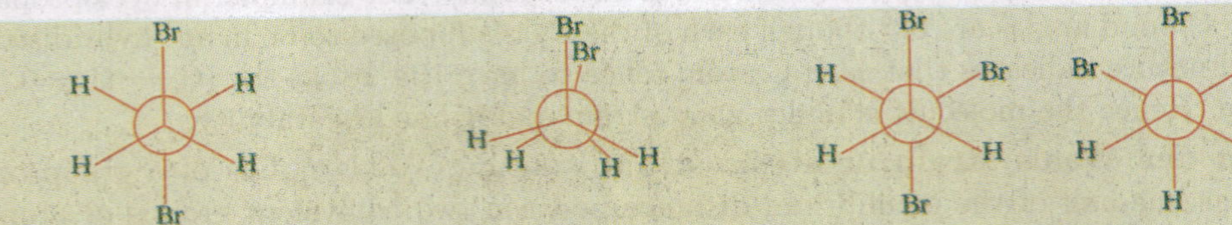

Anti 1,2-dibromoethane **I** Fully eclipsed-1,2-dibromoethane **II** Gouche-1,2-dibromoethane **III** **IV**

Fig. 9.12. Newman projection formulae of 1,2-dibromoethane

1, 2-dibromoethane contains two sp^3-hybridised C atoms; therefore expected bond angles, C—C—C and Br—C—C, should be nearly 109.5°. These are almost so in the compound and angle strain is practically nil in its conformations. Therefore, to determine relative stabilities of different conformations, we are to consider torsional strain, van der Waals strain and dipole-dipole interaction.

(a) Torsional strain is the least in the anti form and hence this form is the most stable insofar as torsional strain is concerned.

(b) van der Waals strain will also be the least in the anti form as two bulky Br groups are farthest apart from each other in this form.

(c) B is the negative pole of the $C^{\delta+}$—$Br^{\delta-}$ bond dipole. Hence $Br^{\delta-}$ and $Br^{\delta-}$ will repel each other and this repulsive force will be the least in the anti form. Therefore the anti form will be the most stable form from the consideration of dipole-dipole interaction.

Thus all the three factors said above indicate that the anti form is the most stable form. Naturally its percentage in the equilibrium will be the highest at any moment at room temperature. As the two C—Br bond dipole moments are equal in magnitude but opposite in direction in the anti form, they will cancel their dipole moments. The dipole moment of the compound will then be shown by its eclipsed and gauche forms. These being present in minimum of amount at any instant at room temperature, the dipole moment of the compound will be very small. Observed dipole moment of 1, 2-dibromethane is really very small. However, with increase in temperature, the population of eclipsed and gauche forms increases and consequently dipole moment of 1,2-dibromethane increases.

The anti conformation of 1, 2-dichloroethane in the vapour phase has been found to be more stable than its gauche conformation by 5.04 kJ mol^{-1} but in the liquid phase the energy barrier is zero. To account for this observation two opposing forces are to be considered; the dipole-dipole repulsion and van der Waals attraction between two Cl atoms. In the vapour phase the dipole-dipole repulsion is stronger than the van der Waals attraction. In the liquid phase the latter becomes equal to the former. Owing to the presence of many other 1, 2-dichloroethane molecules in the neighbourhood of a particular molecule, the intermolecular dipole-dipole interactions reduce the intramolecular dipole-dipole interaction to such an extent that the van der Waals attraction just cancels the intramolecular dipolar interaction in the gauche forms and as a consequence the energy barrier between the anti and gauche conformations is zero (Fig. 9.13).

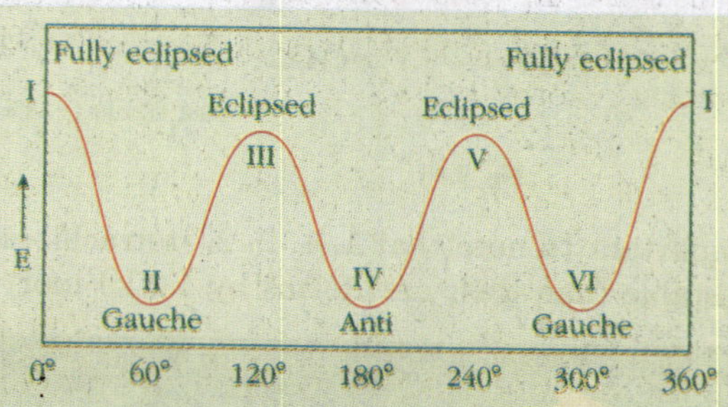

Fig. 9.13. Potential energy of 1,2-dichloroethane (*l*) as a function of ϕ. I : Fully eclipsed, II & VI : Gauche, III & V : Eclipsed, IV : Anti in the liquid state

In the case of n-propyl chloride, the anti form exists exclusively in the solid crystalline state; n-propyl bromide and n-butyl bromide also exist in the anti form in the crystalline state. However, in the liquid state all these compounds exist in anti and skew forms and the latter forms are more stable than the anti form in the liquid as well as in the gaseous states. Perhaps dipolar attraction and van der Waals attraction between C—Cl and C—CH_3 units are the causes for the greater stability of gauche forms.

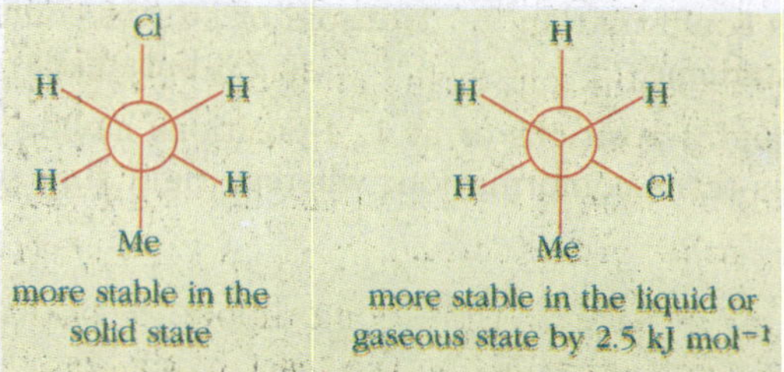

Fig. 9.14

For the same reason the following observations have been found in the case of n-butyl chloride.

Et—CH_2—CH_2—Cl or Me—CH_2—CH_2—CH_2—Cl

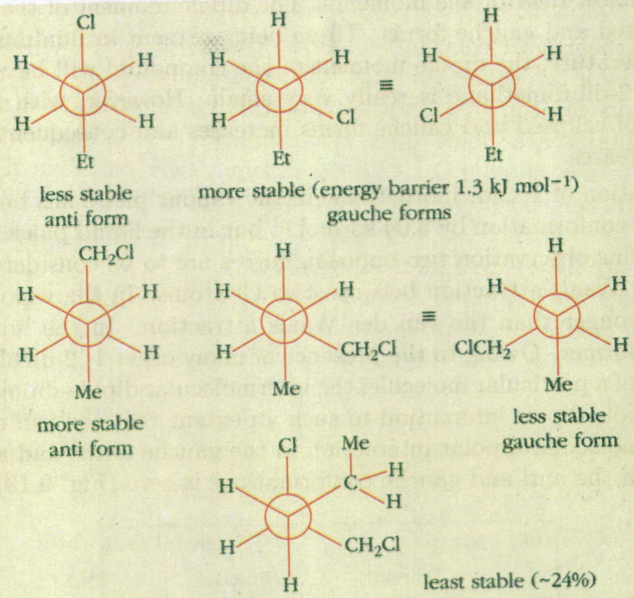

Fig. 9.15

In this connection it is important to note that 1, 1, 2, 2, -tetrachloro and tetrabromoethanes have the gauche forms more stable than their anti forms by 4.2 kJ mol^{-1},

9.4.5 Cisoid and Transoid conformations

When there are two groups on two carbons of a C—C single bond, they can remain on the same side of the C—C single bond or on opposite sides; the former conformation is said to be cisoid (cis-like) or *s*-cis and the latter transoid (trans-like) or *s*-trans. In the case of 1, 3-butadine *s*-trans is more stable than *s*-cis by 10 kJ/mol^{-1} but their interchange requires 25 kJ mol^{-1} of energy.

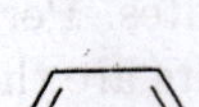

cisoid or s-cis conformation
1,3-butadiene

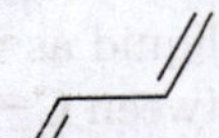

transoid or s-trans conformation
1,3-butadiene

The bulky groups on 2- and 3-positions or at 1, 4-positions having unfavourable orientation make a molecule to assume s-trans-conformation; whereas rigid ring structures are forced into the cisoid form.

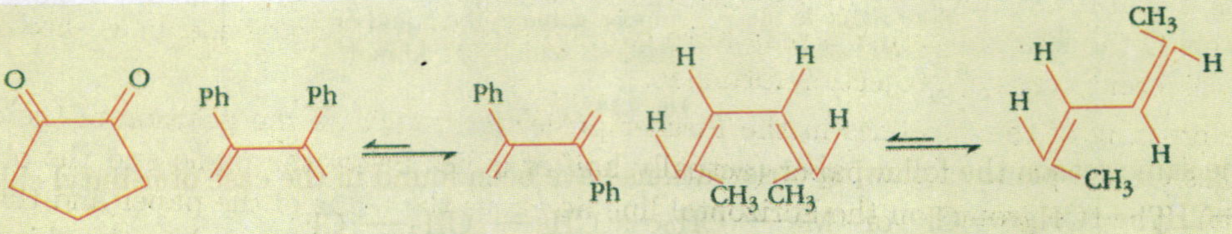

Fig. 9.16. Cisoid and transoid conformations of some compounds

9.5 Configuration of Molecules

Configuration is a theoretical concept related to three dimensional structure of a chemical species and expressed by a projection formula. Conventionally configuration of a species means the spatial arrangements of atoms or groups around its rigid or dissymmetric part. The part of a species which contains a double bond or a ring structure is said to be the rigid one as rotation about them is not at all free; while a dissymmetric part usually possesses a chiral (pronounced as kiral) atom—an atom carrying different single-bonded atoms and groups with a non-planar disposition. A chiral C has four different atoms or groups attached to it in a tetrahedral arrangement; the C atom then is supposed to be in sp^3-hybridised state. A chiral atom is said to be stereogen or chiral centre since its environment is dissymmetric.

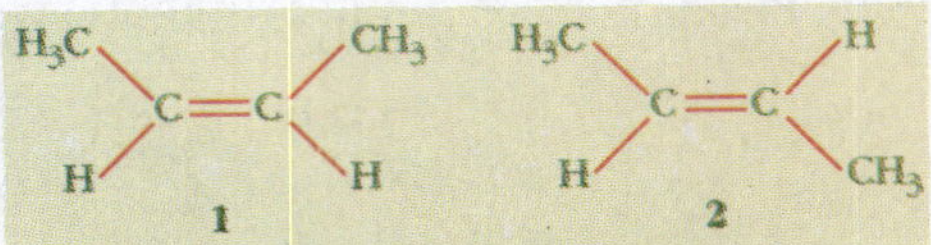

Fig. 9.17. Configurations of 2-butene

2-butene, in which 4 Cs and 2 Hs are planar, have two different spatial arrangements around the C=C bond as shown in Fig. 9.17.

1 and 2 in Fig. 9.17 may also be projected as shown below:

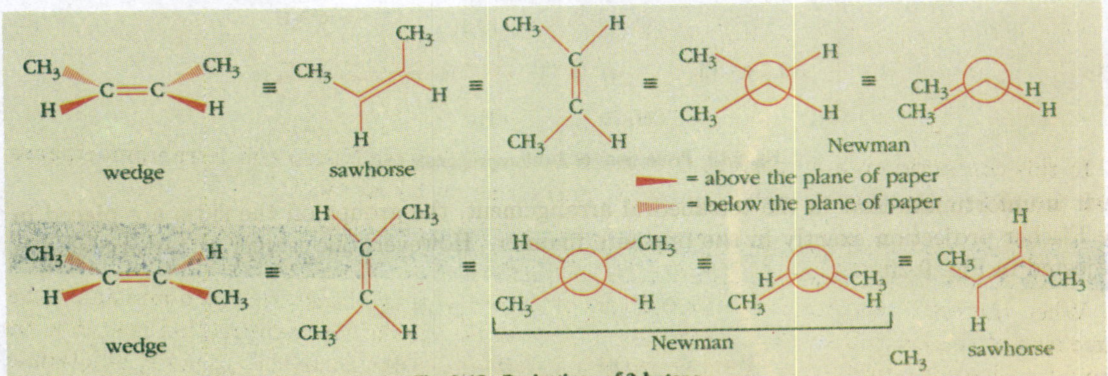

Fig. 9.18. Projections of 2-butene

In 1 two CH_3 groups are on the same side of the double bond; but in 2, those groups are on the opposite sides of the C=C bond. Thus 2-butene exists in two different configurations. These two configurations in fact represent two different compounds of identical constitution but differing in many a physical property like dipole-moment, b.p., m.p., etc., and hence they represent a pair of configurational isomers. It is important to note that configuration 1 cannot be converted into configuration 2 without breaking of any bond and reforming it and both of the configurational isomers are isolable.

A molecule having a chiral atom also can exist in two different spatial arrangements about the chiral atom, the dissymmetric part. For example, $CH_3\overset{*}{C}HCl\ COOH$ (2-chloropropanoic acid) contains a chiral C (indicated by asterisk) as it has four different groups (CH_3,H,Cl,COOH); about this C the four substituents may have two different spatial arrangements as shown in Fig. 9.19 by different types of projection formulae.

The crossing of the two lines in the Fischer projection indicates the position of the chiral C which lies on the plane of the paper (actually half of it lies below the plane and the other half above it). The two groups on the horizontal line lie above the plane of the paper and the groups on the vertical line are below it. Usually the carbon atom directly bonded to the chiral C and bearing the lowest locant in the IUPAC system of nomenclature is placed at the top of the vertical line while that having the lower locant is placed at the bottom. Thus when a ball and stick model of a molecule with a chiral C is projected on the paper following the Fischer method, the model is to be held by the viewer as directed above and the groups on the left and right sides of the viewer are to be placed on the left and right sides respectively of the horizontal line. When two chiral Cs are present in a molecule, both should lie on the plane of the paper as shown below:

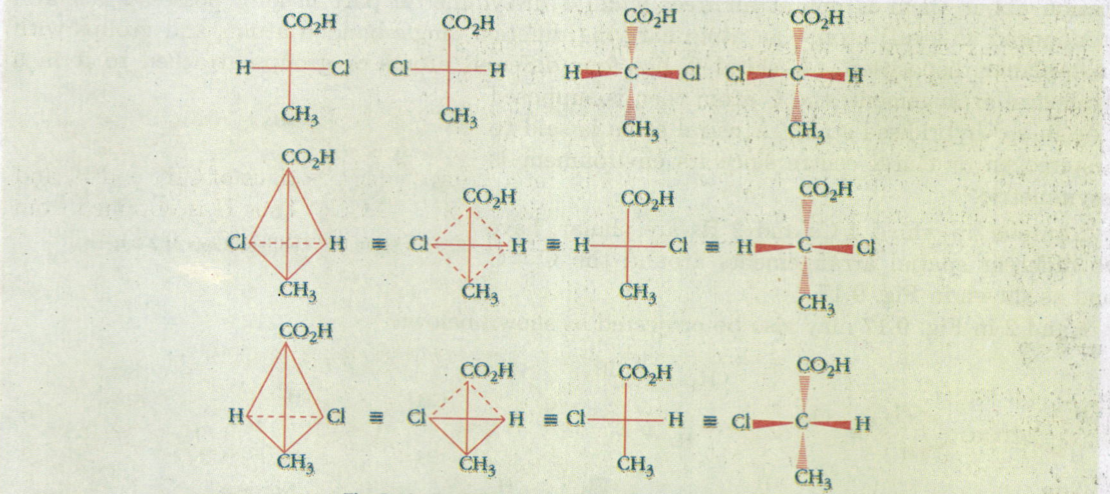

Fig. 9.19. Projections of 2-chloropropanoic acid

It is unfortunate that in the tetrahedral arrangement, the groups on the sides are placed in the Fischer projection exactly in the opposite manner. However, one isomer of tartaric acid is projected in Fig. 9.20.

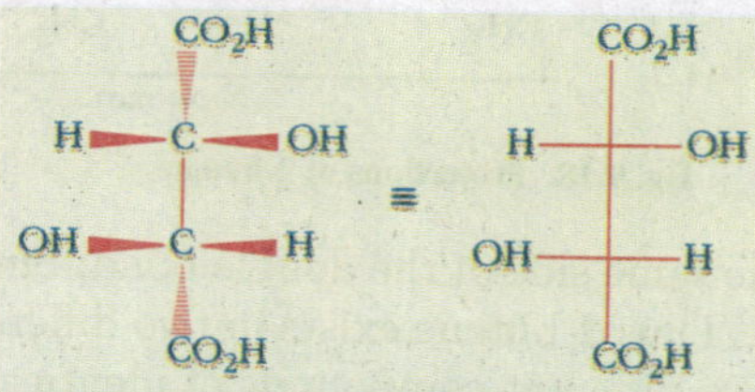

Fig. 9.20. Projection of tartaric acid

2-Chloropropanoic acid may also be projected as shown in Fig. 9.21.

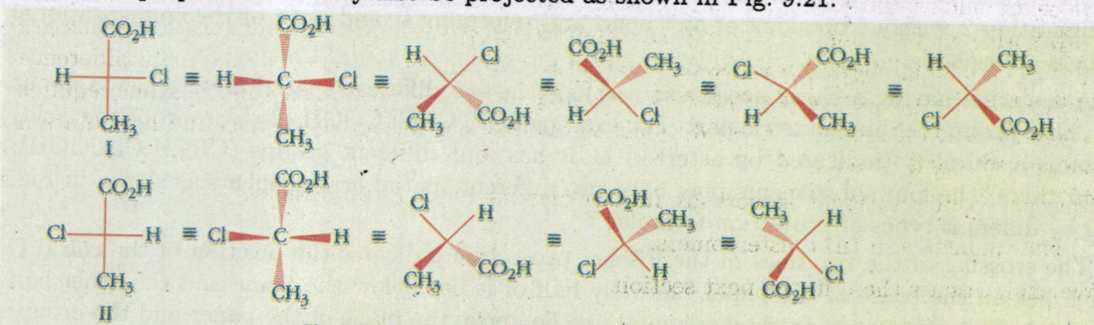

Fig. 9.21. Projections of molecules wtih one chiral carbon

Here the groups joined to the chiral C by the solid lines are on the plane of the paper.

One can draw a number of so-called other configurations for 2-chloropropanoic acid, but each of them will be a repetition of either I or II. If one draws a configuration apperently different from either I or II, then the groups on the chiral C of the new configuration are to be interchanged. If one gets I by an odd number of interchanges of position, then the new configuration will not be I but it will be II, the mirror-image of I. On the other hand, if I is obtained from the new configuration by an even number of interchanges of positions of groups on the chiral C atom, then the new configuration is I. Thus 2-chloropropanoic acid may also be projected as III and IV as given in Figure 9.22.

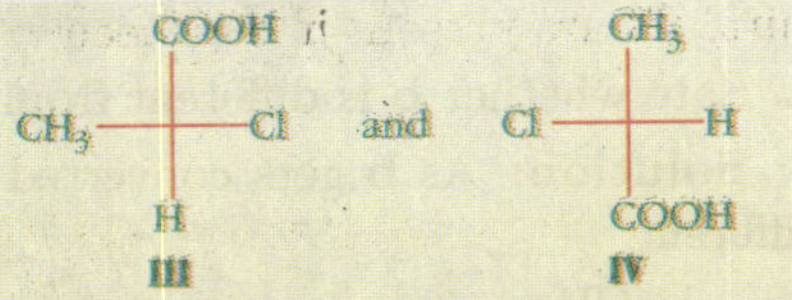

Fig. 9.22. Projections of 2-chloropropanoic acid

Form III can be converted into form I only by interchanging the positions of CH_3 and H, and I can be converted into II by interchanging the positions of H and Cl. Thus II is obtained from III by two (an even number) interchanges and hence III and II are identical (Fig. 9.23).

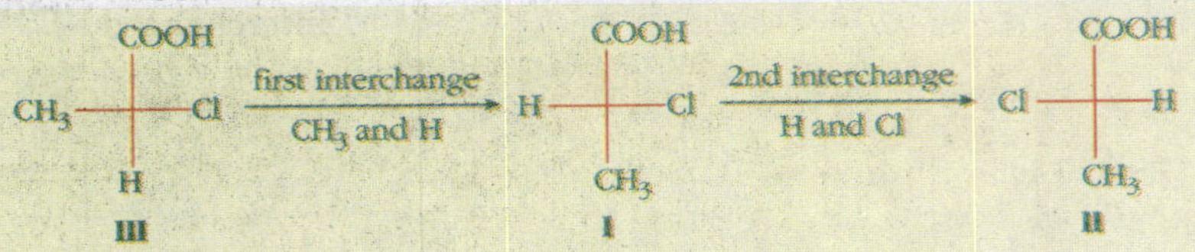

Fig. 9.23. Projection formulae

This means form III is not the third configuration of the compound. Similarly, it can be shown that IV and I are identical and they represent one and the same configuration (Fig. 9.24).

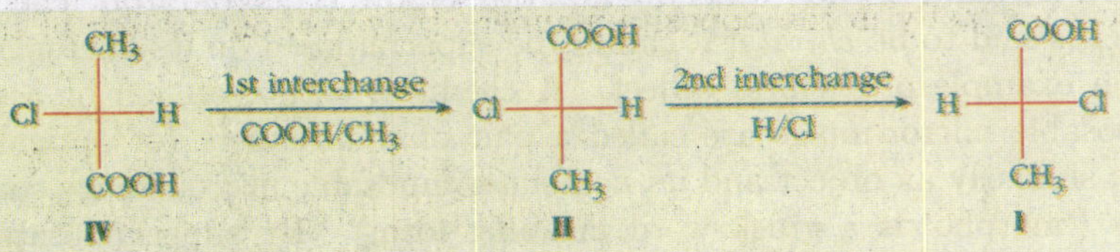

Fig. 9.24. Interconversion of Projection formulae

Therefore, we see that only two different configurations of 2-chloropropanoic acid is possible about the chiral C. These two represent two different compounds though they are of identical constitutional formula. These are configurational isomers, known as enantiomers (because one form is a non-superimposable mirror-image of the other). One cannot convert one enantiomer into the other without breaking a bond and then reforming another. Hence configurational changes are always associated with bond breaking and bond making processes. Thus the main difference between conformation and configuration is: to change one configuration into another, one requires the breaking and reforming of a bond, while to change one conformation into another rotation about a single bond is required.

However, configurational isomers are further classified under two heads:

(i) enantiomers and (ii) diastereomers.

We shall discuss these in the next section.

Problem 9.4. Two Fischer projection formulae for CH_3CHDCl are drawn below :

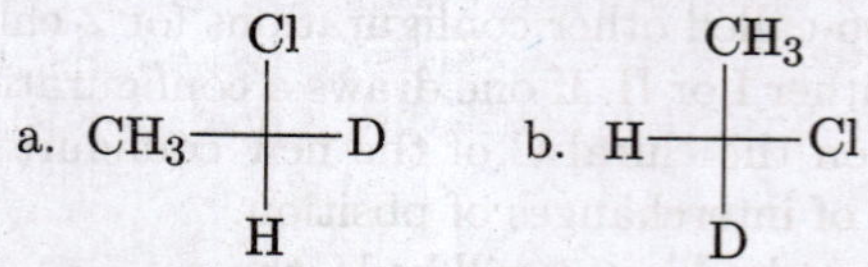

Sate whether **b** is different from or identical with **a**.

Solution: As **b** gets converted into a by three (an ood number) interchanges, **a** and **b** are different.

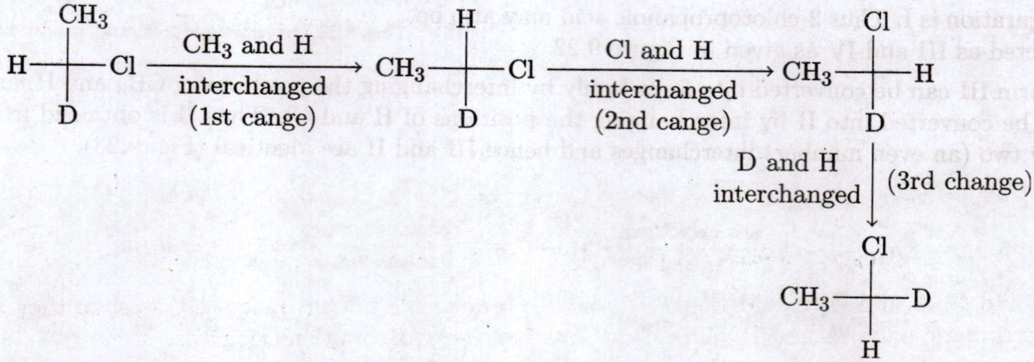

9.6 Chiral Molecules, Enantiomers And Diastereomers

The Greek word for hand is 'cheir' (kier). The word 'chiral' is derived from 'cheir'. A chemical entity or an object is said to be *chiral* if it cannot be superimposed on its mirror image. A hand is a very common example of a chiral object. A chemical entity, an ion or a molecule, and its non-superimposable mirror-image are called a pair of enantiomers (or optical antipodes or optical antimers); similarly an object and its non-superimposable mirror-image constitute a pair of enantiomorphs ('morpho' is a greek word meaning form). By superimposability we mean when the molecule is placed over its mirror-image, either directly or indirectly after rotating the molecule through a certain angle around an axis which passes through the molecule or after rotating a part of the molecule around one or more single bonds, atoms and bonds of the object molecule will lie exactly on identical atoms and bonds of the image molecule and the molecule is called achiral. Ethyne and 1-chloroethyne are achiral molecules whereas 1, 3-dichloropropadiene is a chiral molecule; let us verify the statements.

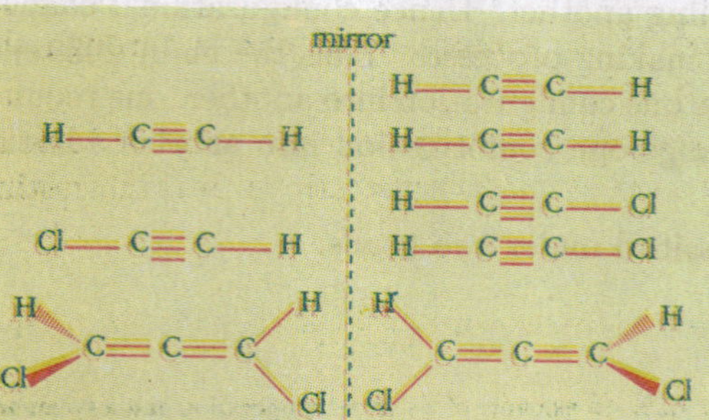

The object is transposed directly

The object is transposed and rotated 180° through an axis which is perpendicular to the plane of the paper and passes through the mid-point of C≡C

Non-superimposable; no matter in whatever way we may rotate the molecule, it will not be superimposable on the image

Fig. 9.25

If two stereoisomers do not bear any mirror-image relation between them, they are called a pair of diastereomers. 1, 2-dichloroethene exists in two forms, I and II; they do not have a mirror-image relation and, therefore, they constitute a pair of diastereomer.

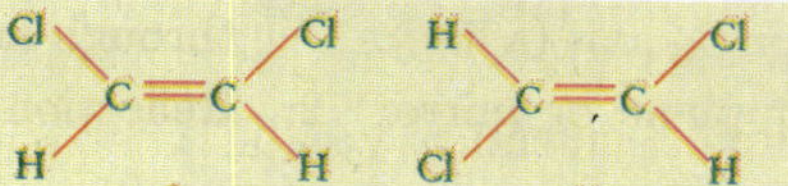

I II

Fig. 9.26. Projection of 1,2-dichloroethane

Similarly, 3-chlorobutan-2-ol exists in four stereoisomeric forms, A, B, C, D.

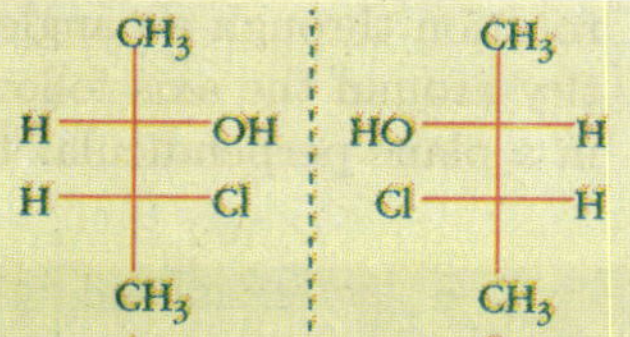

A B

Fig. 9.27. A pair of enantiomers

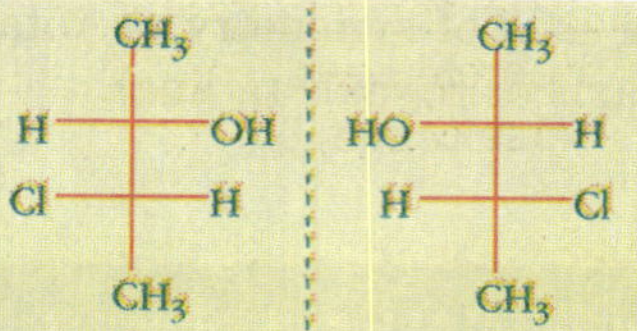

C D

Fig. 9.28. A pair of enantiomers

A and C, A and D, B and C, and B and D do not have mirror-image relations; so they are diastereomeric pairs; A and B, and C and D are two sets of enantiomeric pairs.

When a molecule (or an object) which is superimposable on its mirror-image is a symmetric molecule (or object) and which is non-superimposable on its mirror-image is a dissymmetric or an asymmetric molecule. A chemical unit (molecule or ion) having a superposable mirror-image relation with another chemical unit, the two chemical units constitute a pair of homomers (homo means similar, mer means unit).

9.7 Symmetry Operations, Symmetry Elements and Molecular Symmetry

Actually four kinds of symmetry elements and symmetry operations are required to specify molecular symmetry (Table 9.3). In this section we shall devote our attention to this. Let us first know the meaning of 'symmetry operation' and 'symmetry element'.

Symmetry operations: A symmetry operation is a movement of a body (having a specific position and orientation) in such a way that, after the completion of the movement, every point of the body coincides with an equivalent point of the body in its previous position and orientation i.e. after the symmetry operation we shall get a body which is indistinguishable from the original insofar as its position and orientation are concerned. The 180° rotation of the ethyne molecule around the axis which is perpendicular to the plane of the molecule and passes through the mid-point of the C≡C bond is a symmetry operation.

Symmetry elements: A symmetry element is a geometrical entity such as a point or a line or a plane with respect to which one can carry out one or more symmetry operations. It is interesting to note that the symmetry operations can be defined only with respect to the symmetry elements and appropriate symmetry operations help us note the existence of the symmetry elements. In Fig. 9.29, the axis around which the rotation operation is carried out is a symmetry element.

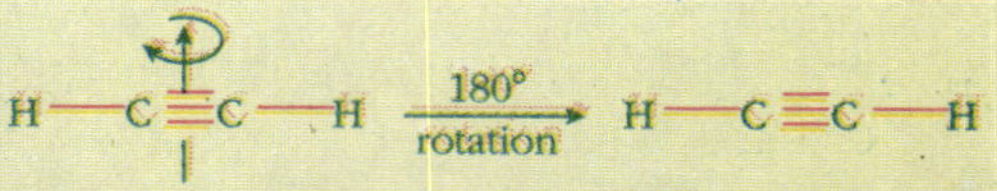

Fig. 9.29. An example of a symmetry operation and a symmetry element

Table 9.3 Symmetry elements and operations

Symmetry elements	Symmetry operations (and notations)
1. Proper rotation axis of order n or n-fold simple axis of symmetry or C_n axis (a line).	1. One or several rotations about the axis through an angle $2\pi/n$ (C_n).
2. Plane of symmetry or σ plane or mirror plane (a plane).	2. Reflection in the plane (σ).
3. Centre of symmetry or centre of inversion (i, a point).	3. Inversion of all atoms through the centre (i).
4. Improper rotation axis of order n or n-fold alternating axis of symmetry or rotation-reflection axis or S_n axis (a line).	4. One or several repetition of the sequence: rotation through an angle $2\pi/n$ of the entity around the axis followed by reflection in a plane perpendicular to the axis (S_n).

9.7.1 Proper rotation axis of order n or n-fold simple axis of symmetry or C_n axis

If a chemical entity rotating about an axis through an angle of 360°(2π) coincides with itself n times, the axis is referred to as the proper rotation axis of order n or n-fold simple axis of symmetry. It is denoted by C_n; the letter 'C' has been derived from the Latin word 'circulate'. In other words an entity possesses a proper rotation axis of order n if the rotation of the entity about the axis of $2\pi/n$ or $\left(\frac{360}{n}\right)^\circ$ produces a structure indistinguishable from the original. The symmetry operation is denoted by C_n or $\hat{C}_n$.

If we carry out k times successive rotations of the entity through an angle $2\pi/n$ i.e., if we rotate, the entity $\frac{2\pi k}{n}$, the operation is denoted by C_n^k. If the value of $k = n$, the entity will come back into the initial position and orientation and we shall get an identical structure and C_n^n is said to be the identity operation.

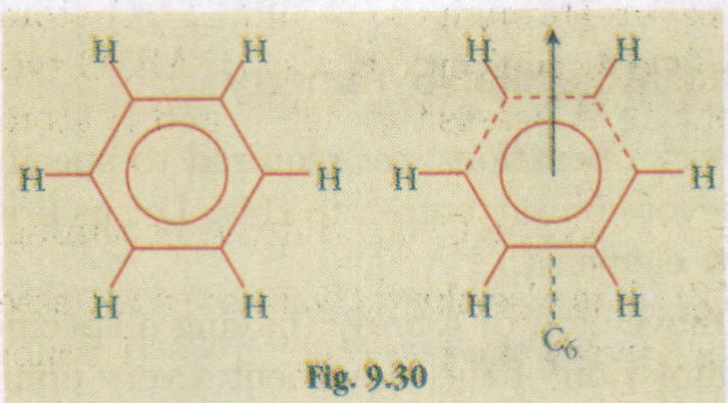

Fig. 9.30

$C_n^n \equiv E$ (identity operation); E is also the symbol for the identity element.

Every entity has an infinite number of trivial C_1 axes since 360° rotation around any axis restores the original structure.

If an entity has more than one symmetry axis, the one with the highest value of n and the one which passes through the largest number of constituent atoms is called the **principal axis**. Let us consider the case of benzene molecule; it has a planar hexagonal (regular) structure as shown below:

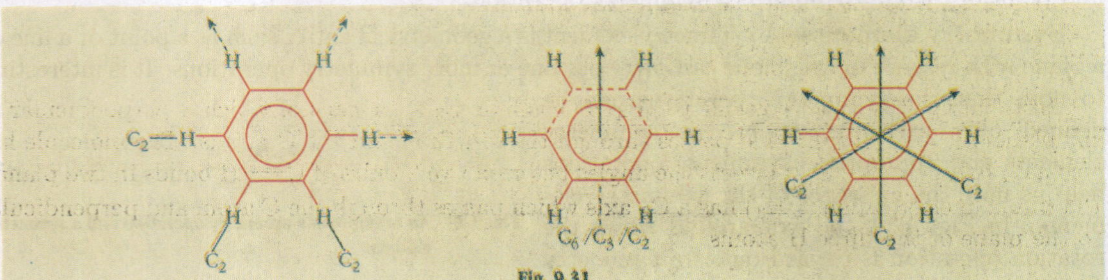

Fig. 9.31

If we now consider an axis, which is perpendicular to the plane of the molecule and passes through the centre of the regular hexagon, and rotate the molecule by 60° or $\left(\frac{360}{6}\right)^\circ$, we shall get an equivalent structure indistinguishable from the original. So benzene has a C_6 axis and $C_6^6 \equiv E$. Again if we consider an axis which passes diagonally through the centre and lies on the molecular plane and 180° rotation of the molecule around this axis will produce an equivalent structure; so benzene molecule has a C_2 axis—infact it has seven C_2 axes. The molecule has a C_3 axis also; C_6, C_2 and C_3 are one and the same axis. Thus in the case of benzene C_6 is the principal axis.

Since all atoms of a linear molecule like acetylene, nitrogen, hydrogen cyanide, carbon dioxide etc., lie on the plane of molecular axis, rotation of the molecule around this axis by any angle imaginable i.e., by infinite number of angles, develops a configuration indistinguishable from the original. Hence linear molecules have ∝-fold axis of proper rotation i.e., they have a C_∞ axis.

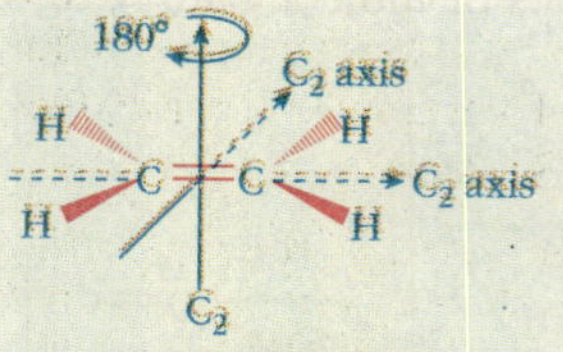

Fig. 9.32

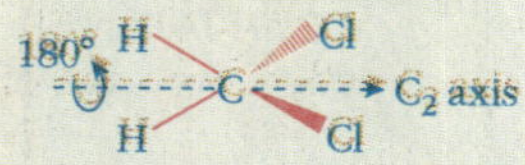

Fig. 9.33

Ethylene molecule has three C_2 axes. One is collinear with the C—C bond axis; the second one is perpendicular to the molecular plane and bisects the C—C line, and the third C_2 axis is perpendicular to the former two and intersects both at the mid-point of the C—C line.

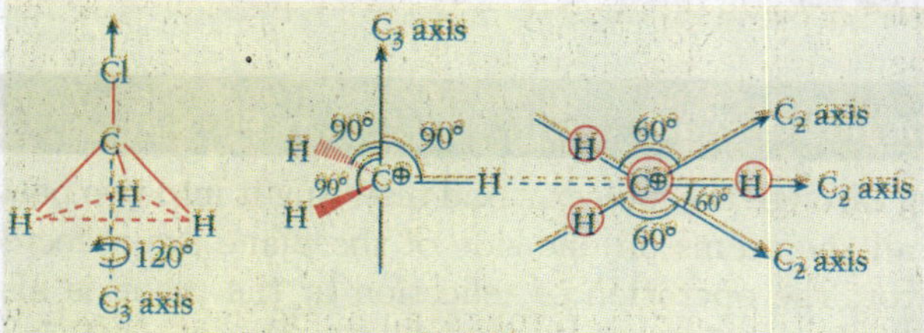

Fig. 9.34

CH_2Cl_2 (tetrahedral; AB_2C_2 type) molecule has a single C_2 axis; it bisects Cl—C—Cl and H—C—H angles.

Tetrahedral CH_3Cl (AB_3C type) has one C_3 axis; this passes through the C and Cl (A) atoms and perpendicular to the plane of the three H(B) atoms. $CHCl_3$ molecule is also of AB_3C type and tetrahedral in structure. It has, therefore, only one C_3 axis; it passes through C and H atoms and perpendicular to the plane of the three Cl atoms.

Planar methyl cation (AB_3 type; A = carbon and B = H) has one C_3 axis; it passes through the C atom and perpendicular to the plane of the entity. $\overset{\oplus}{C}H_3$ has also three C_2 axes; each passes through the C atom and one H atom and bisects the angle between the two other C—H bonds.

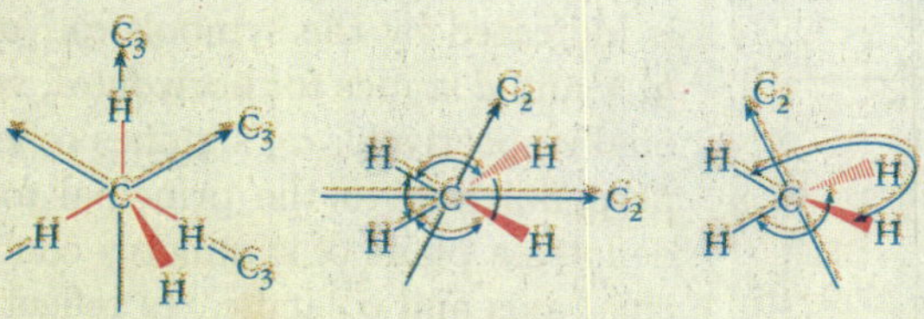

Fig. 9.35

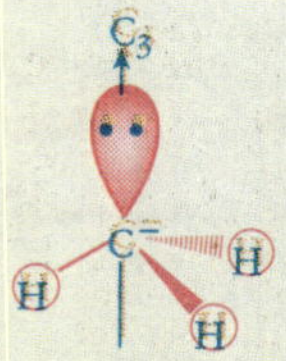

Fig. 9.36

Tetrahedral AB_4 type molecule is CH_4. It has four C_3 axes, each of which is perpendicular to the plane of three H atoms and passes through the fourth H and the C atoms. The molecule has also 3 C_2 axes; each C_2 axis bisects the angles between two 2 pairs of C—H bonds in two planes. Pyramidal methyl anion ($\bar{C}H_3$) has a C_3 axis which passes through the C atom and perpendicular to the plane of the three H atoms.

There are chemical entities which do not have C_n (except trivial C_1) axis. Molecules with one chiral C do not have C_n axis other than C_1. Two examples are given below :

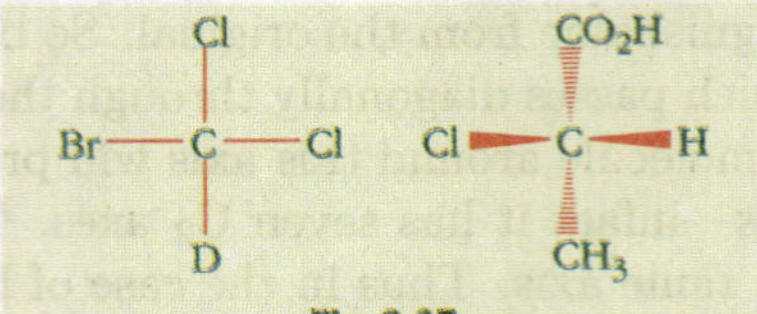

Fig. 9.37

A few more examples of molecules having C_2 axis are given below:

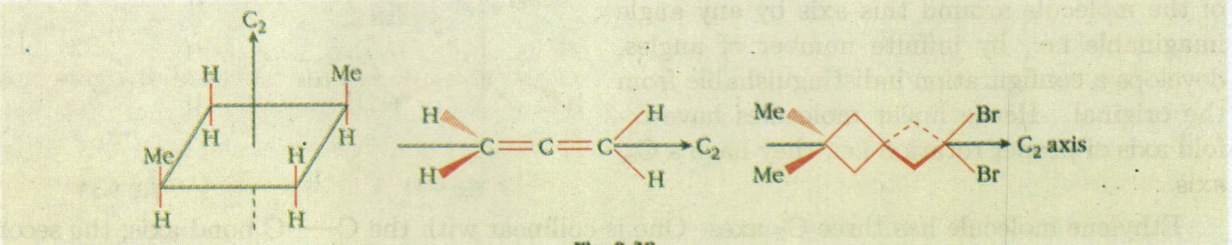

Fig. 9.38

9.7.2 Plane of symmetry or σ plane or mirror plane

A plane of symmetry (σ) is a plane passing through the molecule and bisecting it into two parts which are mirror-images of each other i.e., all the atoms on one side of the plane are reflected through the plane as though it were a mirror; the operation of reflection in the plane is also designated as σ or $\hat{\sigma}$. The σ operation produces a configuration equivalent to the original. It is obvious that two successive reflections in the same plane produce a configuration identical with the original.

$\sigma^2 \equiv$ E; where E is called the identity operation.

In fact, any combination of operations which takes the molecule to a configuration identical with the original configuration is an identity operation, E.

Thus, $\sigma^n =$ E; when n is even and $\sigma^n = \sigma$ when n is odd.

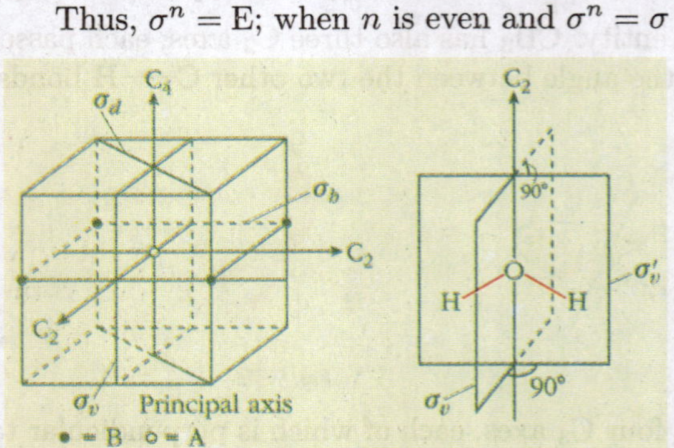

Fig. 9.39

The position of the plane of symmetry (σ) with respect to the principal axis of a chemical entity is indicated by the symbols σ_h, σ_v and σ_d where h, v and d stands for horizontal, vertical and diagonal respectively. σ_h signifies reflection in a plane perpendicular to the principal axis; similarly σ_v denotes a plane of symmetry containing the principal axis and σ_d indicates reflection in the plane which contains the principal axis and bisects the angle between two C_2 axes. Let us consider the case of a planar AB_4 molecule.

Molecules with cylindrical symmetry, such as acetylene, carbon dioxide, oxygen, nitrogen, etc. have an infinite number of planes of symmetry which intersect at the C_∞ axis. Besides these planes, there also has a σ plane which, is perpendicular to C_∞; this plane contains an infinite number of C_2 axes.

AB_2C_2 type tetrahedral molecules, for example CH_2Cl_2, have two orthogonal σ planes, one contains AB_2 (CH_2) and the other contains AC_2 (CCl_2).

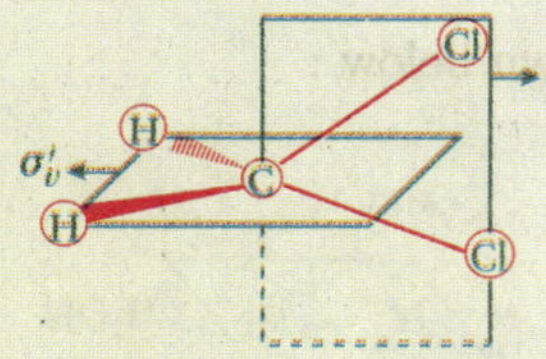

Fig. 9.40

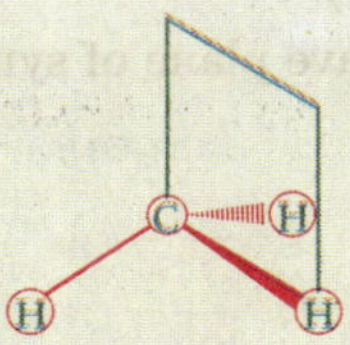

Fig. 9.41

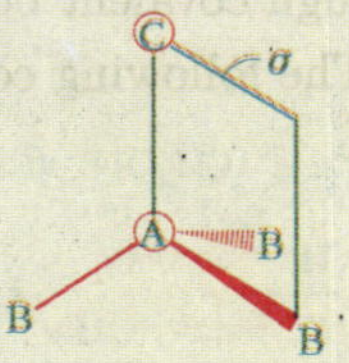

Fig. 9.42

A pyramidal entity of the type AB_3 ($\bar{C}H_3$; methide ion) has three planes of symmetry. Every plane includes the A(C) atom and one of the three B(H) atoms and bisects the angle between the two other A—B (C—H) bonds. A tetrahedral AB_3C type molecule, such as $CHCl_3$, has also three σ planes. Each plane includes A and C and one of the three Bs and the plane bisects the angle between remaining two A—B bonds.

AB_4 type regular tetrahedral molecules (CH_4, CCl_4) have six σ planes. For clarity, let us designate four Bs as B_1, B_2, B_3 and B_4; the six planes will contain the following groups of atoms:

1. AB_1B_2 : the plane will bisect the angle between B_3 and B_4 so that B_3 will be the mirror-image of B_4.
2. AB_1B_3 : the plane will bisect the angle between B_2 and B_4 etc.
3. AB_1B_4 : σ plane will bisect $\angle B_2AB_3$ etc.
4. AB_2B_3 : σ plane will be the bisector of the $\angle B_1AB_4$ etc.
5. AB_2B_4 : the plane will bisect the angle between B_1 and B_3 etc.
6. AB_3B_4 : σ plane will bisect the angle between B_1 and B_2.

A planar AB_3 type entity ($\overset{\oplus}{C}H_3$; carbocation) has four σ planes; one σ plane containing all the four atoms. Three σ planes each of which contains A and one B and is perpendicular to the plane of the entity and bisects the angle between other two Bs.

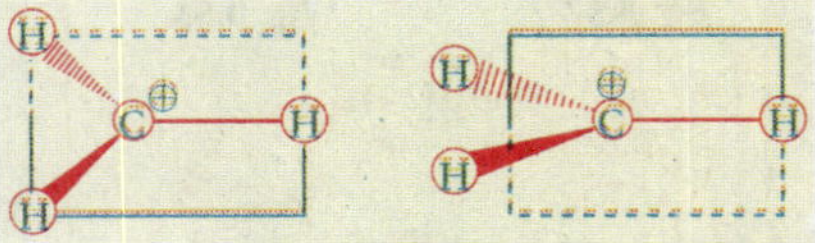

Fig. 9.43

AB_4 type planar molecules, like $PtCl_4$, have five σ planes; one σ plane containing all the five atoms. Two mutually perpendicular σ planes containing A and two diagonal Bs which are also perpendicular to molecular plane. The other two mutually perpendicular σ planes which are also perpendicular to the molecular plane and they pass through A and bisect BAB angles.

Allenes with like groups or atoms on the same carbon have plane of symmetry. The plane passing through three carbons and A and B, and bisecting the angle ACA is the σ plane.

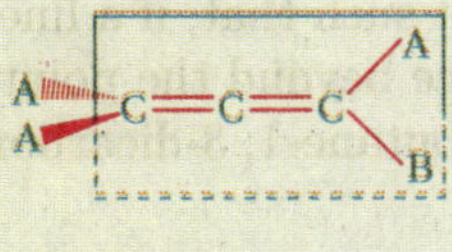

Fig. 9.44

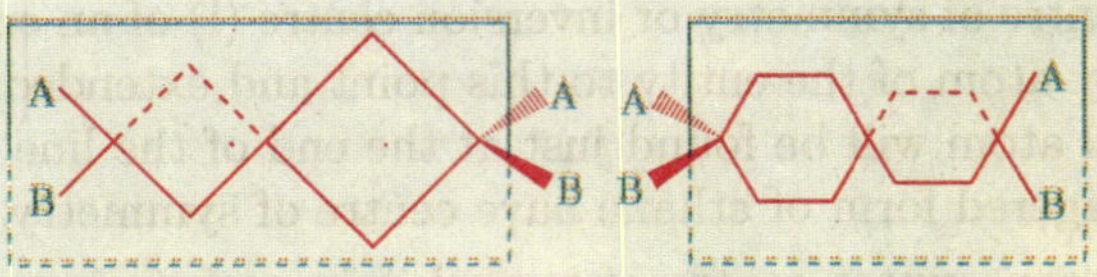

Fig. 9.45

The same is the case with spirans (derivative of spiranes) having even number of ring constituent C atoms and the same ligands (groups and atoms surrounding the central atom directly through covalent bonds or indirectly).

The following compounds also have plane of symmetry as shown below :

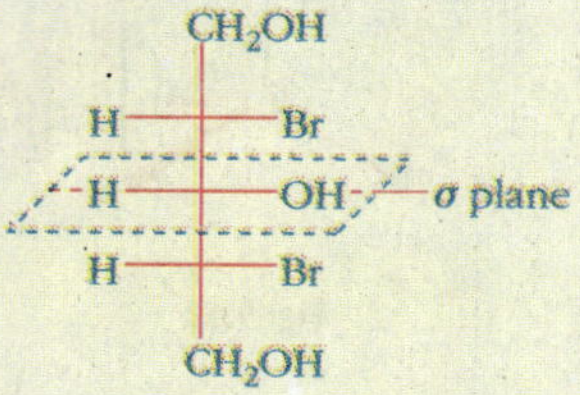

Fig. 9.46

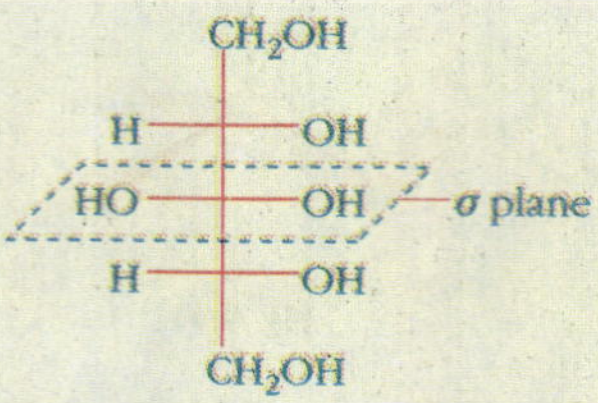

Fig. 9.47

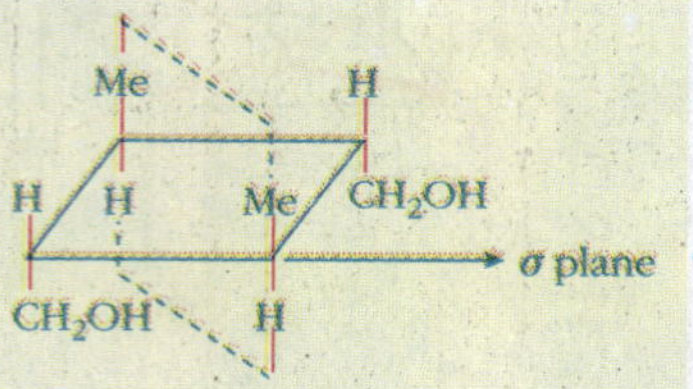

Fig. 9.48

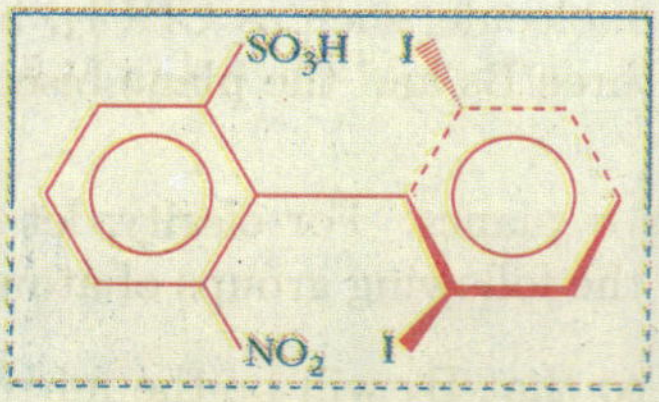

Fig. 9.49

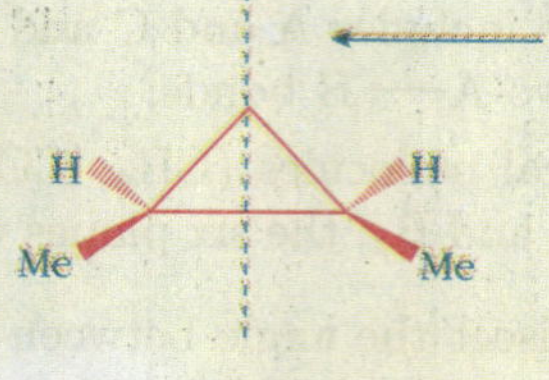

Fig. 9.50

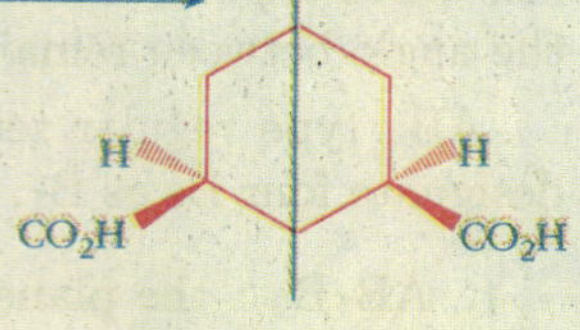

Fig. 9.51

However, there are molecules which do not have σ planes as given below:

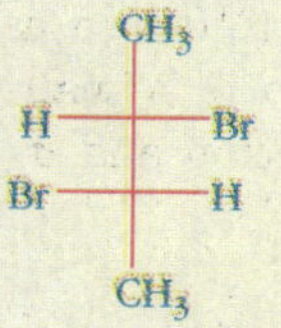

Fig. 9.52

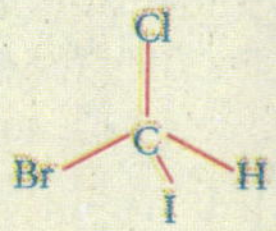

Fig. 9.53

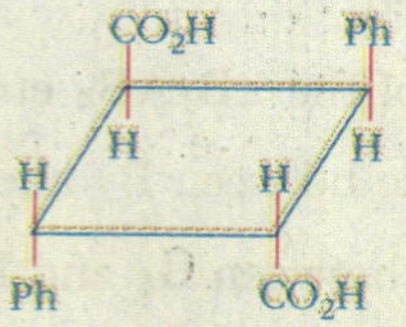

Fig. 9.54

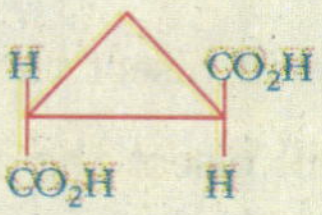

Fig. 9.55

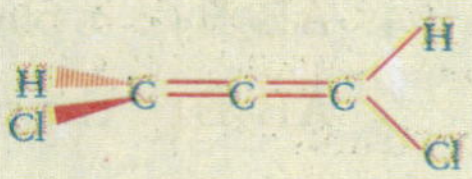

Fig. 9.56

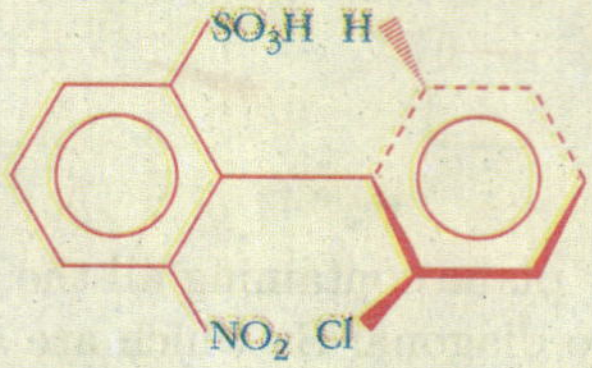

Fig. 9.57

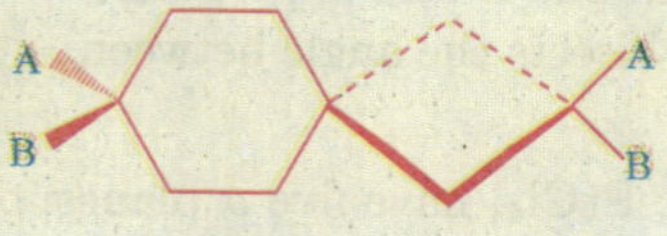

Fig. 9.58

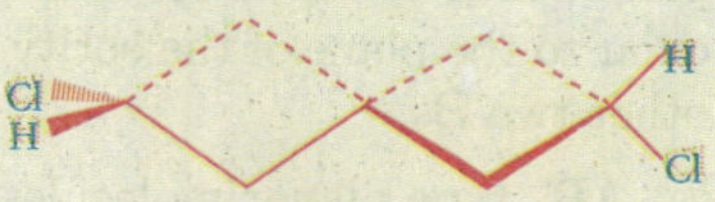

Fig. 9.59

9.7.3 Centre of symmetry or inversion centre (i)

A centre of symmetry or inversion centre (i) of an entity is a point such that, if a line is drawn from any atom of the entity to this point and extended equal distance beyond the point, another identical atom will be found just at the end of the line. Trans-cyclobutane-1, 3-dicarboxylic acid and staggered form of ethane have centre of symmetry.

One can expect centre of inversion in a ring compound having even number of atoms that constitute the skeleton of the ring. Since inversion centre is a point, only one atom or no atom may be at the centre. In the above two examples no atom lies at the centre and all other atoms occur in pairs. When the centre is occupied by an atom, which is unique, the other atoms must occur in pairs.

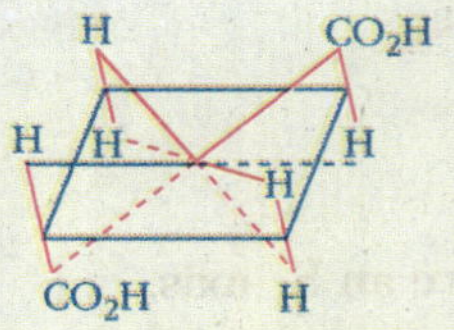

Fig. 9.60

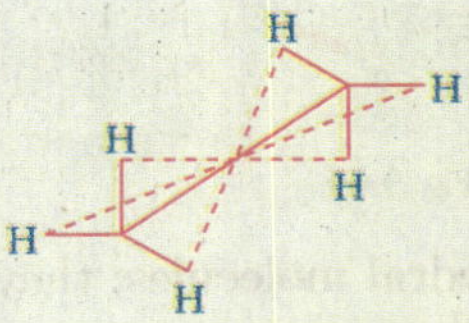

Fig. 9.61

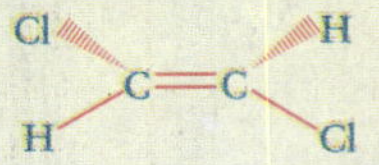

i is the mid point of C=C, no atom is at *i*

Fig. 9.62

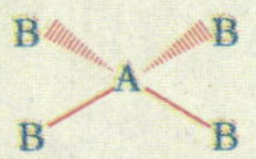

planar, A is at *i*

Fig. 9.63

The inversion operation is also denoted by i (or $\hat{i}$). When the inversion operation is carried out n times, it is denoted by i^n. When n is odd, $i^n = i$ and when n is even, $i^n =$ E.

Tetrahedral AB_4 type molecules have A at the centre and Bs occur in even number (in pairs) but they have no centre of symmetry. Thus CH_4, CCl_4 etc. have no centre of symmetry.

9.7.4 Improper rotation axis of order n or n-fold alternating axis of symmetry or Rotation-reflection axis or S_n axis

A species is said to possess an improper rotation axis of order n if rotation of the species about the axis by $\frac{2\pi}{n}$ or $\left(\frac{360°}{n}\right)$ followed by reflection through a plane perpendicular to this axis produces a structure indistinguishable from the original. The improper rotation axis is denoted by the symbol S_n and the combination of operations, rotation and reflection, is symbolised as S_n or $\hat{S}_n$. Since S_n operation combines both rotation around an axis and reflection to a plane perpendicular to the axis, it can be represented as the 'multiplication' of C_n operation and σ_h operation.

$$S_n = C_n\sigma_h = \sigma_h C_n$$

Naturally if in a species a C_n axis and a σ_h plane with respect to the C_n axis (a plane perpendicular to the C_n axis) exist independently, there should be an S_n axis in the species.

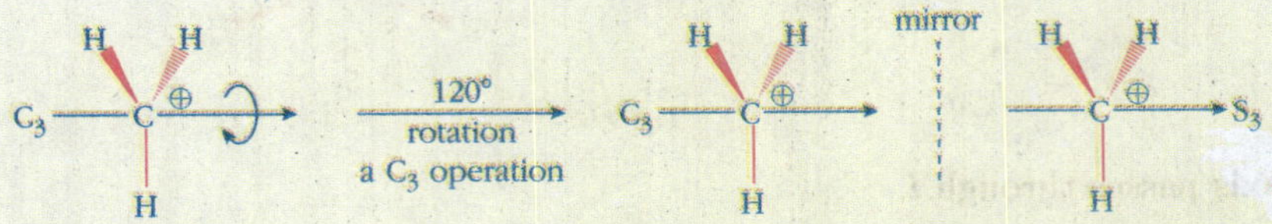

Fig. 9.64

Methyl cation or methyl radical has a planar structure (AB_3 type). Sec. 9.7.1 tells us that both have a C_3 axis which passes through the C atom and perpendicular to the plane of all the atoms. The C_3 axis is the principal axis. The plane containing all the atoms is therefore the σ_h, the plane of symmetry perpendicular to the C_3 axis. So a C_3 operation followed by a σ_h operation give a figure indistinguishable from the original. So σ_h C_3 = S_3. Hence methyl cation or methyl radical contains an S_3 axis. Furthermore, the C_3 axis is also C_1 axis and σ_h is also present in these species. Hence $C_1\sigma_h = S_1$. So, they have S_1 axis also. In fact σ plane $\equiv S_1$ axis.

Here it is important to note that an S_n axis may exist even if there exists neither a C_n (except C_1) axis nor a σ_h separately in a species.

Methane is a tetrahedral AB_4 type molecule. It has $4C_3$ axes and $3C_2$ axes but it has no C_4 axis and a σ plane perpendicular to the C_4 axis; yet it has an S_4 axis.

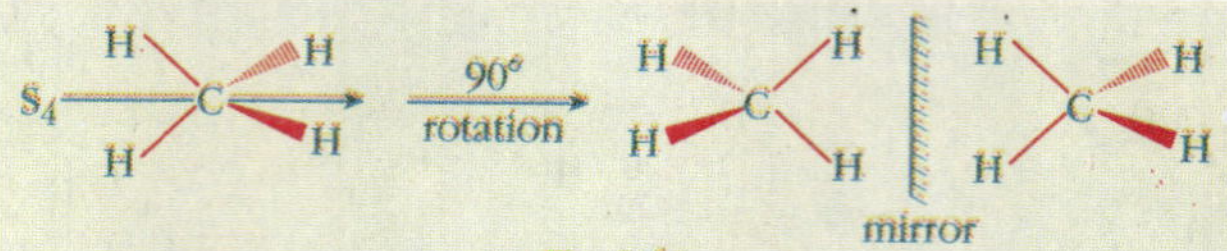

Fig. 9.65

$CHCl_3$ and CH_3Cl are AB_3C type tetrahedral molecules; they have an S_1 axis.

Similarly CH_2Cl_2 (tetrahedral AB_2C_2 type) molecule also has an S_1 axis. It has a σ plane; we may recall that $S_1 \equiv \sigma$.

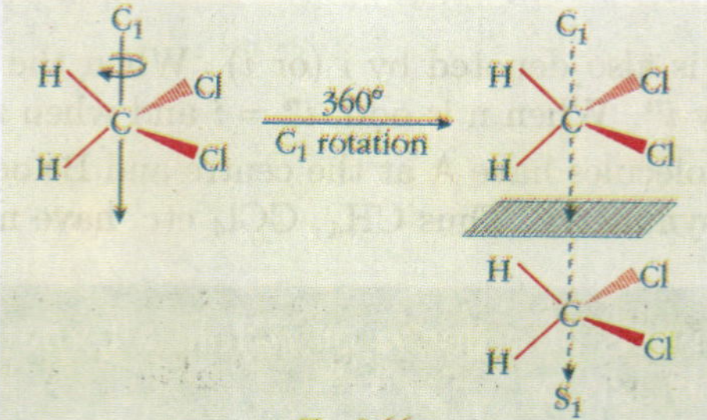

Fig. 9.66

A centre of symmetry is equivalent to an S_2 axis. The inversion operation, i, is equivalent to S_2 operation; but S_2 operation is a multiplication (combination) of C_2 operation and σ_h operation. Hence

$$i = S_2 = C_2\sigma_h = \sigma_h C_2$$

Trans-1, 2-dichloroethene has a centre of inversion and hence it has an S_2 axis.

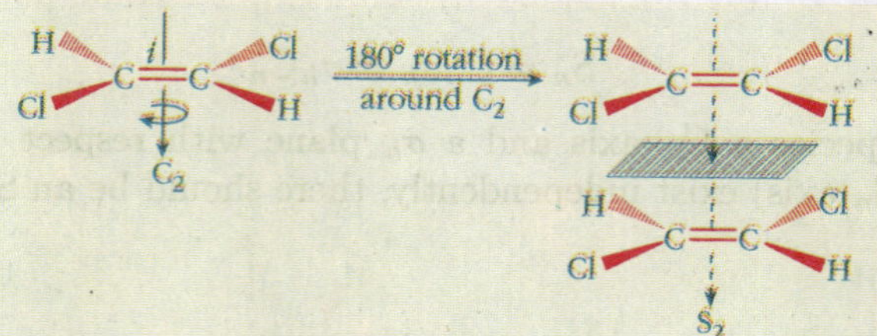

Fig. 9.67

The S_2 axis passes through i.

Benzene molecule has S_6, S_3 and S_2 axes. Its C_6 axis being C_3 and C_2 axes and its σ_h with respect to C_6, C_3 and C_2 axes, C_6 axis is the S_6, S_3 and S_2 axes (Fig. 9.30 and Fig. 9.31).

The compound of Fig. 9.68 has an S_1 axis.

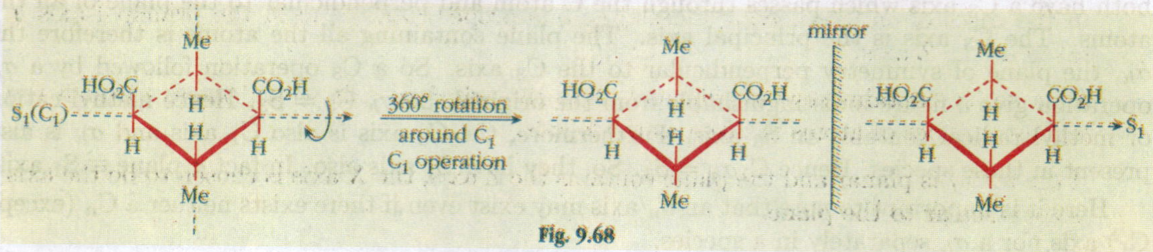

Fig. 9.68

There are a large number of chemical entities which do not have any S_n axis. A few examples are given below :

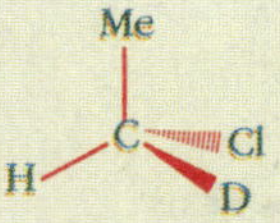

Fig. 9.69

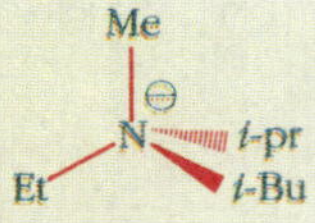

Fig. 9.70

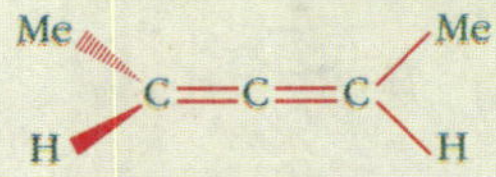

Fig. 9.71

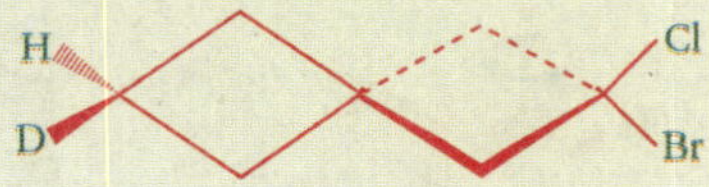

Fig. 9.72

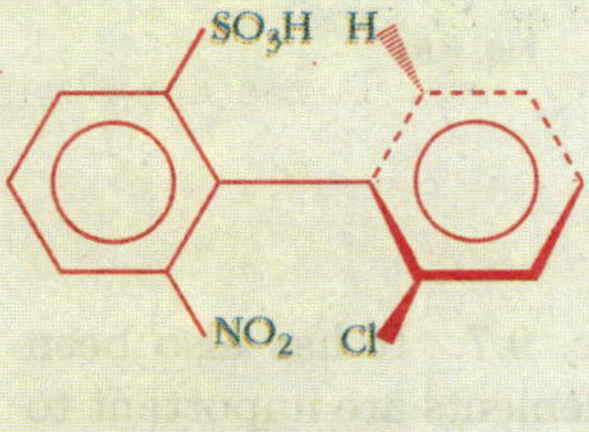

Fig. 9.73

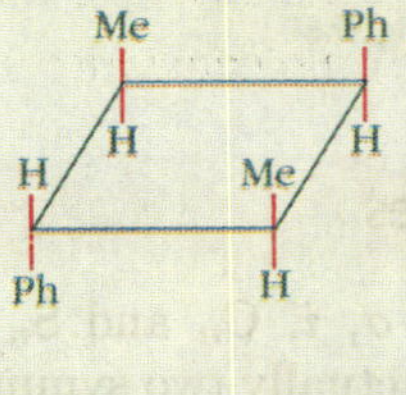

Fig. 9.74

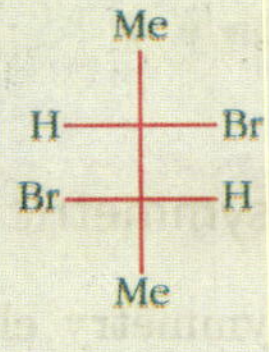

Fig. 9.75

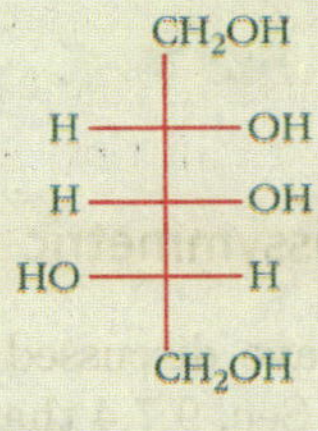

Fig. 9.76

Problem 9.5. Show, wherever possible, σ, C_n axis, i and S_n axis of the followings molecules :

i. BF_3, ii. NH_3, iii. H_2O, iv. $\ddot{\bar{C}}H_3$, v. C_2H_2, vi. H_2O_2.

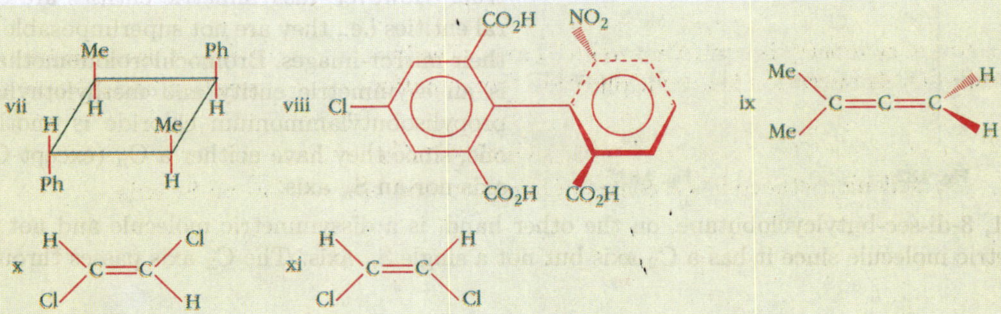

Fig. 9.77

Note

(i) To study the symmetry elements in a chemical entity, the entity is oriented in a cartesian co-ordinate system. The centre of gravity of the molecule is placed at the origin of the co-ordinate system.

(ii) If there is only one proper rotation axis in the entity, the Z axis is taken as the proper rotation axis.

(iii) In the cases of entities having more than one proper rotation axis, the principal axis is taken as the Z axis.

(iv) If an entity contains several principal axes, the principal axis which passes through the highest number of atoms is taken as the Z axis.

(v) If the entity is planar and the plane contains the Z axis, the X axis is chosen to be the axis perpendicular to the plane.

(vi) If the entity is planar and Z axis is normal to the plane, the X axis must lie on the plane and it must pass through the largest number of atoms. Examples are given below.

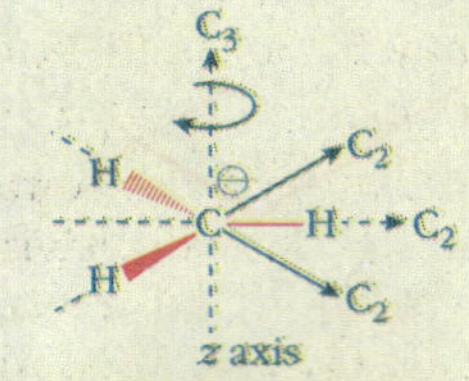

Fig. 9.78

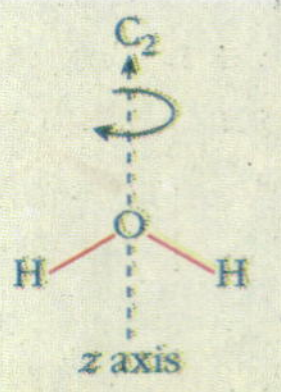

Fig. 9.79

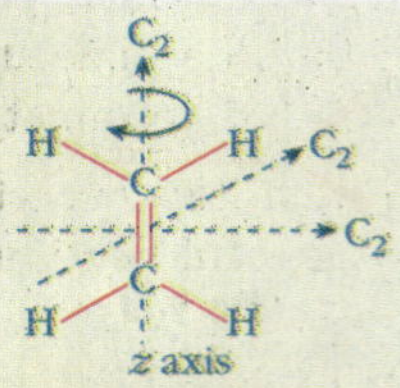

Fig. 9.80

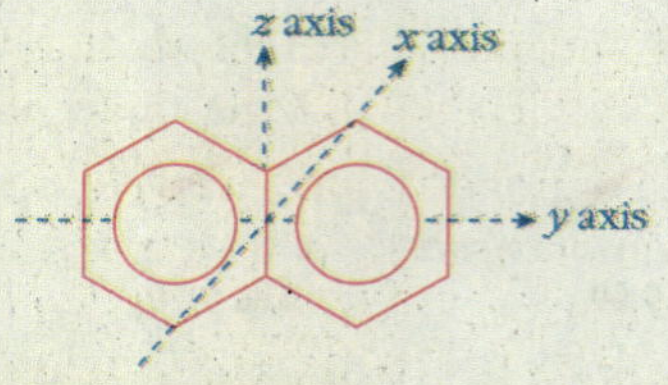

Fig. 9.81

9.8 Dissymmetric and Asymmetric Entities

We have discussed four symmetry element, σ, i, C_n and S_n in Sec. 9.7. It has also been stated in Sec. 9.7.4 that $\sigma \equiv S_1$ and $i \equiv S_2$. So, actually two symmetry elements are important to determine the symmetry of an entity — C_n and S_n. Entities having S_n, axis are called symmetric entities. Entities devoid of S_n, axis (i.e., devoid of S_n axis, σ plane and i) are called dissymmetric entities; whereas those devoid of both C_n (except C_1, which is present in all entities) and S_n axes are called asymmetric entities. So all asymmetric entities are dissymmetric but the reverse is not true. However, dissymmetric entities are chiral entities i.e., they are not superimposable on their mirror-images. Bromochloroiodomethane is an asymmetric entity and methylethylisopropylisobutylammonium chloride is another one, since they have neither a C_n (except C_1) axis nor an S_n axis.

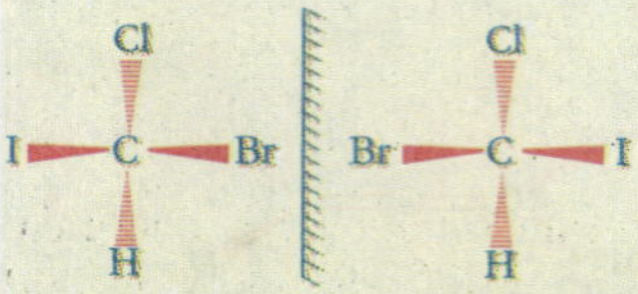

Fig. 9.82

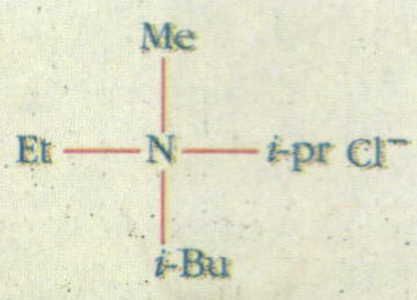

Fig. 9.83

Cis-1, 3-di-sec-butylcyclobutane, on the other hand, is a dissymmetric molecule and not an asymmetric molecule since it has a C_2 axis but not a single S_n axis. The C_2 axis passes through

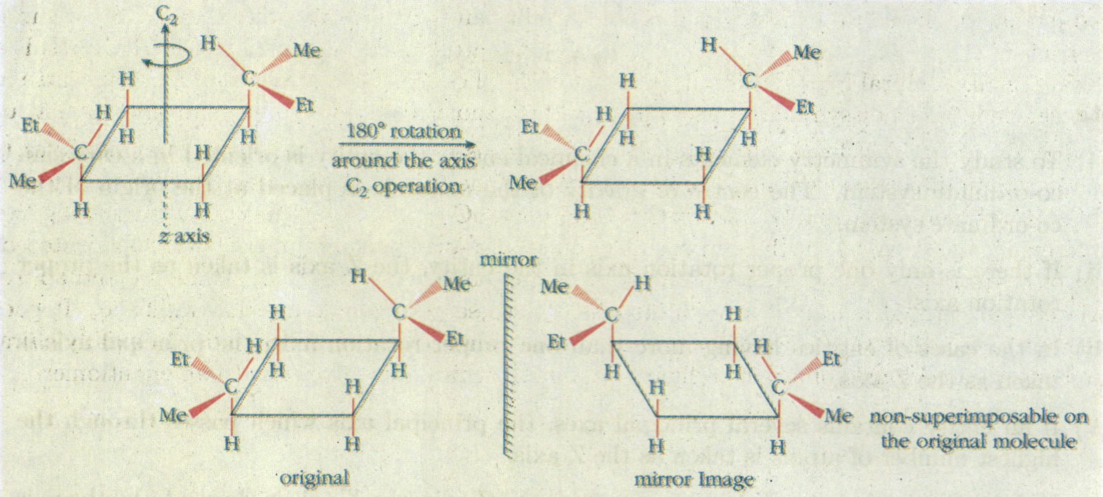

Fig. 9.84

the centre of the cyclobutane ring and perpendicular to the plane of the ring. Its mirror-image is not superimposable on the molecule (Fig. 9.84). Optically active tartaric acids are dissymmetric molecules since they have a C_2 axis which passes through the mid-point of C_2—C_3 bond and perpendicular to this bond, but the molecules do not contain any S_n axis. (+)- and (−)-$CH_3CHClCHClCH_3$ molecules are also dissymmetric for the same reason.

Dissymmetric and hence asymmetric entities i.e., chiral entities have special behavior toward plane polarised light which will be discussed in Section 9.10 under the heading 'optical activity'.

Dissymmetry arises in a chemical entity (or in a body) owing to the presence of any one of the following elements of chirality, called stereogenic units, in the entity:

(i) Chiral centre, (ii) Chiral axis, (iii) Chiral plane and (iv) Helicity.

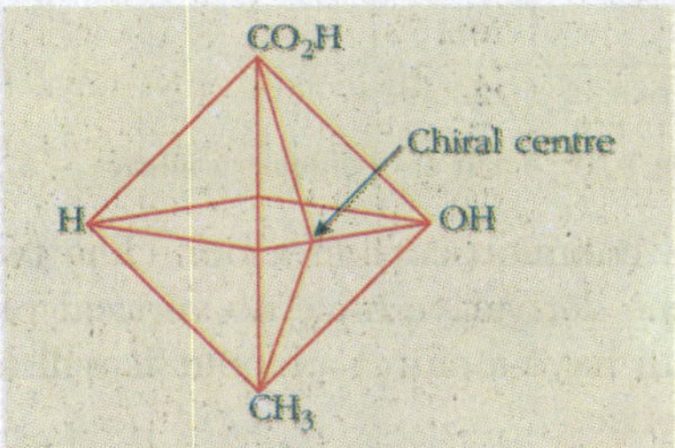

Fig. 9.85. Chiral centre

(i) Chiral centre: If a three-dimensional space is occupied dissymmetrically by a chemical entity (or a body) about the zero-dimensional element of symmetry i.e., point, the point is said to be chiral centre. A chiral carbon atom occupies the centre (a point) of tetrahedron and its four different ligands (groups) occupy the four vertices of the tetrahedron. The centre being dissymmetrically surrounded by the ligands, the space occupied by the tetrahedron dissymmetrically about the centre and the chiral carbon atom is a chiral centre. The transposition of any two ligands of the chiral carbon reverses the chirality and gives a new stereoisomer. So, a chiral centre is a stereogenic centre or stereocentre. The C-2 atom of lactic acid ($CH_3CHOHCO_2H$) is a chiral centre. A chiral N or P or Si atom is also a chiral centre. The presence of a chiral centre in a molecule leads to dissymmetry or chirality and the molecule is said to be centrodissymmetric.

(ii) Chiral axis: When a three-dimensional space is occupied by an entity dissymmetrically about the unidimensional element of symmetry i.e., line (an axis), the axis is called chiral axis. abC=C=Cab (where a ≠ b), abC=C=C ad and abC=C=C de types of allenes having two different ligands on each of the end carbons, bulky *o*, *o*′-disubstituted or *o*, *o*, *o*′-trisubstituted or *o*, *o*, *o*′, *o*′ -tetrasubstituted biphenyls without having any vertical plane of symmetry and appropriately substituted spiranes have chiral axis. So these compounds are dissymmetric. It goes without saying that these molecules do not also contain σ plane, *i* and S_n axis. In axially chiral molecules, an exchange of a pair of ligands about the chiral axis gives rise to an enantiomer.

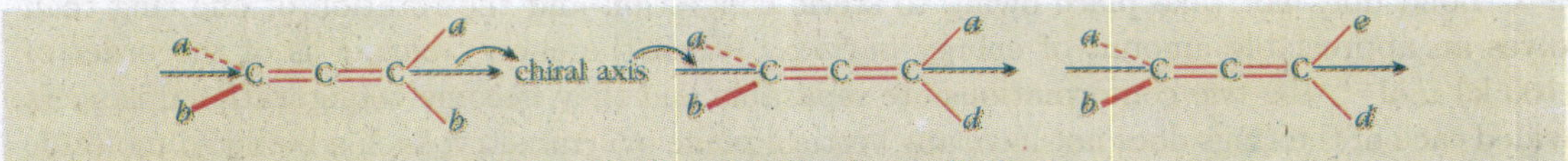

Fig. 9.86. Allene derivatives having chiral axis

The geometry of allene has already been discussed in the problem 2.13a. In allene two carbon-carbon double bonds lie on two mutually perpendicular planes. The axis passing through the C atoms is a C_2 axis, the principal axis, besides this there are two more C_2 axes perpendicular to this axis. Principal C_2 axis is also the S_4 axis and allene itself is achiral. However the substituted allenes having two different ligands on each of the two end carbons have no σ plane, i and S_n axis but they have chiral axis which passes through the three double bonded carbons as shown in Fig. 9.86. abC=C=Cab type allenes (where a $\neq$ b) have a C_2 axis (see Fig. 9.100 for spirane derivative and draw a C_2 axis for the allene). These allenes are dissymmetric molecules on the other hand abC=C=Ced type allenes are asymmetric molecules since they have no σ, i, S_n and C_n (except C_1) axes. All these molecules have non-superimposable mirror-images.

The object molecule and its mirror image constitute a pair of enantiomers (Fig. 9.87).

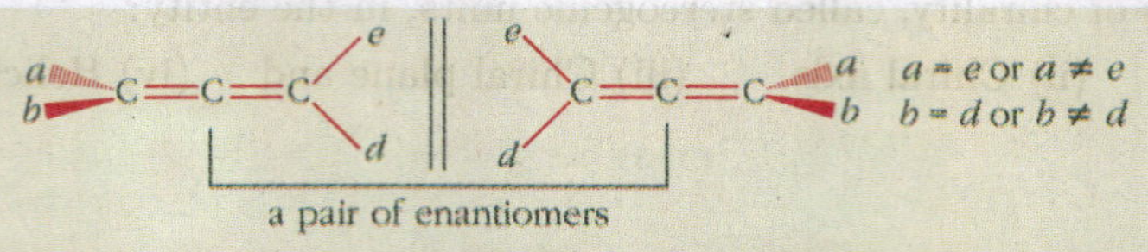

Fig. 9.87. A pair of enantiomeric allenes

Spiranes have two rings with a common carbon atom. The two rings lie on two orthogonal planes. If the two rings of a spirane derivative have no vertical plane of symmetry as shown in Fig. 9.88, the spirane derivative will have a chiral axis but it will have no σ plane, i and S_n axis and the molecule will be dissymmetric.

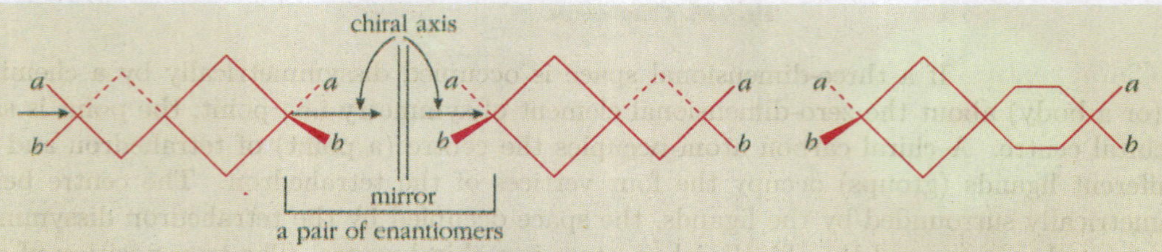

Fig. 9.88. Chiral axis of spiranes

Appropriately substituted alkylidenecycloalkanes also have chiral axis.

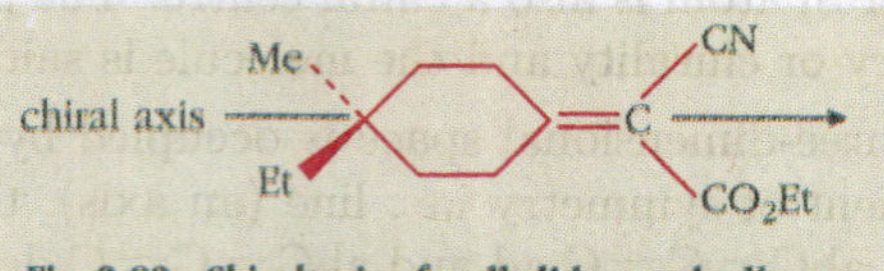

Fig. 9.89. Chiral axis of a alkylidenecycloalkane

The two benzene rings of biphenyl are not exactly coplanar. The two planes of the rings make an angle 45°. One ring can rotate about the pivotal C—C bond almost freely and the rotation gives rise to conformations. However, when ortho positions are occupied by bulky substituents the two rings lie on two mutually perpendicular planes and the rotation of one ring about the C—C bond does not take place owing to steric congestion and the rotation of one ring then involves an appreciable amount of energy barrier. When the energy barrier is of the order of 80–100 kJ mol^{-1}, the two conformations are separable and they become configurational isomers provided each of the rings does not have any vertical plane of symmetry; such a biphenyl molecule as stated has a chiral axis which passes though four para-carbons and owing to the presence of axis of chirality it becomes dissymmetric. It is needless to mention that the molecule will have a non-superimposable mirror-image and no σ plane, i and S_n axis (Fig. 9.90).

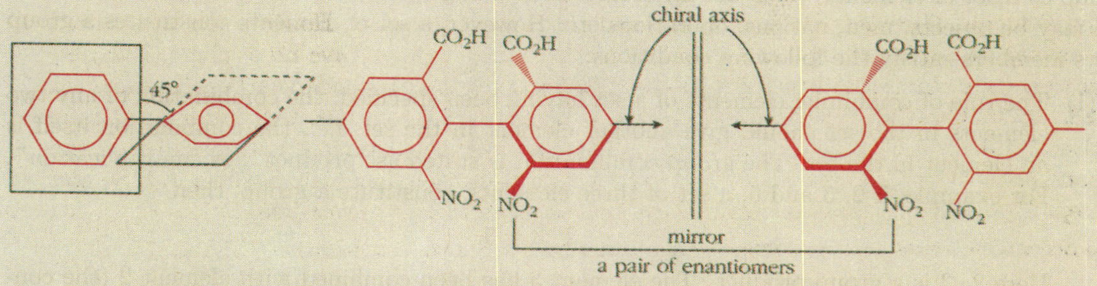

Fig. 9.90. Chiral axis of a biphenyl derivative

However, the decreasing order of steric hindrance produced by different groups is I > Br > Me > Cl > NO_2 > CO_2 > OMe > F. In fact, when ortho-positions are occupied by F or OMe groups, the biphenyl molecule is not dissymmetric where as the presence of only two very bulky substituents at o, o'-positions, for example SO_3H, in biphenyl makes the molecule dissymmetric.

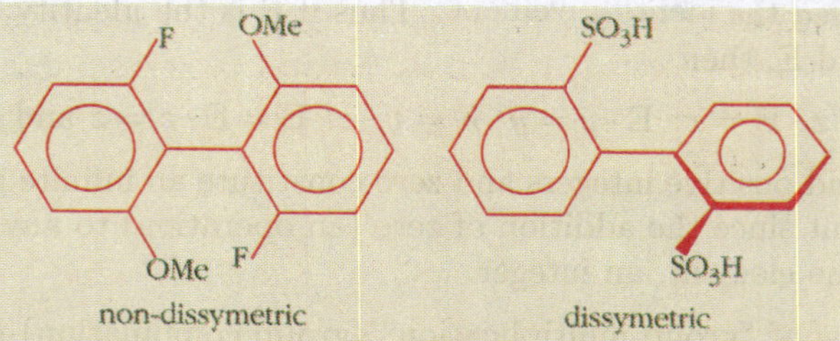

Fig. 9.91

(iii) Chiral plane: When an entity occupies a three-dimensional space dissymmetrically about the two- dimensional element of symmetry, a plane, the plane is said to be a chiral plane. The molecule does not contain σ, i and S_n axis. Trans-cyclooctene is an example.

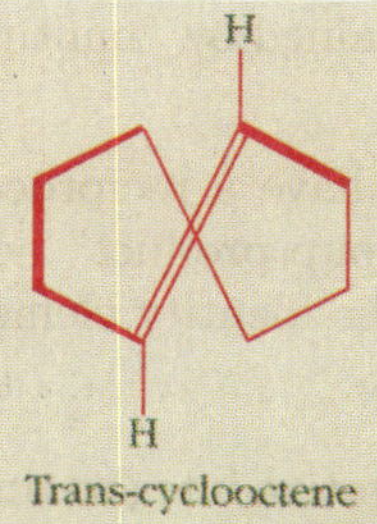

Fig. 9.92. A molecule with a plane of chirality (not shown)

(iv) Helicity: Helixes are inherrently chiral since the axis of a helix is a chiral axis. A helix is non-superimposable on its mirror-image—so helixes are dissymmetric.

9.9 Symmetry-based Classification of Chemical Entities : Point Groups

The purpose of this section is to classify non-dissymmetric, dissymmetric and asymmetric chemical entities, molecules and polyatomic ions, on the basis of symmetry operations that can be carried out on them. Since the symmetry-based classification involves mathematical groups, at first we shall develop an elementary concept of mathematical groups. Mathematicians define a group as a set of elements which are interrelated according to certain rules. The elements of the set may be objects, men, nations, operations etc. However, a set of elements constitutes a group if its members satisfy the following conditions:

(i) The rule of combining elements of a set having been specified, the combination of any two elements in the set should give another element in the set, i.e., the combination itself is an element in the set. The group combination is stated as "product", or "multiplication". For example, if 2, 3 and 5, a set of three elements, constitute a group, then

$$2 \cdot 3 = 5.$$

Here $2 \cdot 3$ is a group product. The element 3 has been combined with element 2 (the convention is to combine the right hand element with the left hand element) to give the third element 5; the sign (dot) indicates some rule (operation) according to which the combination has been done—in this case, summation is the operation, the rule of combination.

(ii) There should be at least one element in the set which must combine with every other element in the set and leave them unchanged; this element is designated by the letter E (or I) and it is called the identity element. Thus if E is the identity element of a group of elements x, y, z and E, then

$$\mathrm{E} \cdot x = x \cdot \mathrm{E} = x;\ \mathrm{E} \cdot y = y \cdot \mathrm{E} = y;\ z \cdot \mathrm{E} = \mathrm{E} \cdot z = z \text{ and } \mathrm{E} \cdot \mathrm{E} = E.$$

All the negative and positive integers and zero constitute an infinite group in which zero is the identity element since the addition of zero (an operation) to any element of the group does not change the element, an integer.

(iii) The associative law of "group multiplication" (group combination) must hold for a set of elements (say, x, y and z) to constitute a group. Thus

$$x \cdot (y \cdot z) = (x \cdot y) \cdot z.$$

This law states that if the element which is a "product" i.e. combination of two group elements z and y is "multiplied by" (i.e.), combined with another element x, the resulting element is the same as would be obtained by "multiplying" the element z by the "product" of elements y and x.

(iv) Every element (x) in the set must have a reciprocal i.e., inverse element (x^{-1}) which is also an element of the set. The "group product" i.e., the combination of an element, and its inverse element is nothing but the identity element of the group.

Thus, $x \cdot x^{-1} = x^{-1} \cdot x = E$.

In the infinite group of positive and negative integers and zero, any positive integer is a inverse element of the same negative integer and vice versa and the combination or "product" of the two same integers with opposite signs gives the identity element, zero (0).

Thus, $+10 + (-10) = 0$; $+n + (-n) = 0$; where $n = an$ integer.

9.9.1 Point groups

Point groups are mathematical groups. Either the symmetry elements or the symmetry operations are the elements of the point group. However, a point group may be defined as follows:

If the complete set of symmetry operations that can be performed on a chemical entity or if the complete set of symmetry elements that are present in the entity, constitutes a mathematical group, the group is called symmetry group, or symmetry point group or simply point group of the entity; since the symmetry operations are carried out on the entity keeping its centre of gravity fixed at a point, usually at the origin of the co-ordinate system, the symmetry group is called point group. In the cases of crystals other symmetry groups are used. The point group of a molecule or a polyatomic ion is a short-hand notation for specifying its symmetry class. If we know the point group of an entity, we can write down all the symmetry elements present in the molecule and naturally all the symmetry operations that can be performed on the molecule can also be written down. For example, the point group of dichloromethane (CH_2Cl_2) is C_{2v}; this means it belongs to the symmetry class C_{2v} and this molecule has one C_2 axis (Fig. 9.33), σ_v and $\sigma_{v'}$, (Fig. 9.40) planes and the identity element E(E = C_2^2, E = σ_v^2, E = $\sigma_{v'}^2$). Thus the symmetry operations that can be carried out on dichloromethane are E, C_2, σ_v, and σ_v' and these symmetry operations constitute the mathematical point group C_{2v}. Whether these four symmetry operations constitute the complete set of elements of the point group C_{2v} or not that can be verified by forming a "Multiplication Table" (actually a combination table consisting of symmetry operations), and their combinations as shown below:

Table 9.4 Multiplication table for C_{2v}, point group

Operation A → Operation B ↓ Column	Column I E	Column II C_2	Column III σ_v	Column IV $\sigma_{v'}$	Row
E	E	C_2	σ_v	$\sigma_{v'}$	Row I
C_2	C_2	$C_2^2 = E$	$\sigma_{v'}$	σ_v	Row II
σ_v	σ_v	$\sigma_{v'}$	$\sigma_v^2 = E$	C_2	Row III
$\sigma_{v'}$	$\sigma_{v'}$	σ_v	C_2	$\sigma_{v'}^2 = E$	Row IV

To construct this multiplication table each of the operations (elements of the set) is placed in the first row and again in the first column. The other rows and columns are obtained by "multiplying" (row elements with column elements and the convention in this regard is that the intersection of a row with a column gives the product (combination) AB of the row element A and the column element B. The product AB implies that operation B is performed first, and then operation A is performed on the result. The group product AB is not a new element in the set; it is another element already present in the set to which the group product is equivalent. Here it is important to note that the group product AB does not necessarily equal to BA although it may. When AB=BA we say that A and B follow the cumulative law. However, when the nth element of the first row is combined with ("multiplied by") the rth element of the first column, we get an element (of the set) which appears at the crossing of the $(r+1)$th row and $(n+1)$th column. For example, if we multiply (combine) the row element σ_v (the third element of the first row, $n = 3$) by the column element σ_v (the 3rd element of the first column $r = 3$. implying operation σ_v followed by σ_v, we get, $\sigma_v \cdot \sigma_v$ = E, and element which is already present in the set and it appears at the intersection of the 4th $(r + 1 = 3 + 1)$ row and the 4th $(n + 1 = 3 + 1)$ column. However, if a new element appears in the multiplication table by the combination of any two elements in the set of symmetry operations, we shall reach to the conclusion that the set is incomplete and we have missed one or more symmetry operations that can be carried out on the entity. Since no new element appears in the multiplication table for C_{2v}, point group, the set of symmetry operations, E, C_2, σ_v and $\sigma_{v'}$ constitutes a complete set. Now, let us see whether the set of symmetry operations, E, C_2, σ_v and $\sigma_{v'}$, constitutes a mathematical group or not. If each of the members of the set satisfies the four conditions for forming a mathematical group, then E_2, C_2, σ_v and $\sigma_{v'}$, will constitute a point group.

(i) The first condition for forming a mathematical group is to specify the rule of combining of any two elements in the set with another element in the same set. In this case, the rules are the symmetry operations. When the C_2 operation (the second element of the first row) is multiplied by σ_v (the third element of the first column), we get $\sigma_{v'}$, another element of the set. This is true for all other elements of the multiplication table. Hence the members of the set (E, C_2, σ_v, $\sigma_{v'}$), satisfy the first condition.

(ii) The second condition is that there must be an identity element in the set. The set of elements in the case of dichloromethane contains the identity element E, where $E \equiv C_2^2$ or σ_v^2 or $\sigma_{v'}^2$.

(iii) The third condition is that elements of the set must satisfy the associative law of group multiplication. In the case of dichloromethane, for example,

$$\sigma_v \cdot (\sigma_{v'} \cdot C_2) = \sigma_v \cdot \sigma_v = \sigma_v^2 = E$$

again, $(\sigma_v \cdot \sigma_{v'}) \cdot C_2 = C_2 \cdot C_2 = C_2^2 = E$

$\therefore \quad \sigma_v \cdot (\sigma_{v'} \cdot C_2) = (\sigma_v \cdot \sigma_{v'}) \cdot C_2$

Similarly,

$$E \cdot (C_2 \cdot \sigma_v) = E \cdot \sigma_{v'} = \sigma_{v'} \quad \text{and} \quad (E \cdot C_2) \cdot \sigma_v = C_2 \cdot \sigma_v = \sigma_{v'}$$

$\therefore \quad E \cdot (C_2 \cdot \sigma_v) = (E \cdot C_2) \cdot \sigma_v$

and so on. Therefore the set of elements (E, C_2, σ_v and $\sigma_{v'}$) follows the associative law of group multiplication.

(iv) The forth condition is that the inverse (reciprocal) of an element in the, set must also be an element in the set. In the case of dichloromethane C_2 is the inverse operation (C_2^{-1}) of C_2; σ_v and $\sigma_{v'}$ are the inverse operations of σ_v and $\sigma_{v'}$, respectively. For other cases, σ_h is the inverse element of σ_h, σ_d is the inverse element of σ_d; i is the inverse element of i, C_3^2 is the inverse element of C_3. S_4^3 is the inverse element of S_4^1 etc.

So the complete set of operations (E, C_2, σ_v and $\sigma_{v'}$) that can be performed on dichloromethane obeys the four conditions required to constitute a mathematical group. Hence E, C_2, σ_v and $\sigma_{v'}$ elements constitute a point group of dichloromethane and it is designated as C_{2v}. Formaldehyde, water etc. molecules have C_{2h}, point group. Trans-1, 2-dichloromethene has C_{2h} group. The multiplication table for the C_{2h}, group is given below (Table 9.5).

Table 9.5 Multiplication table for the C_{2h} point group

Operation A→ Operation B ↓	E	C_2	σ_h	i
E	E	C_2	σ_h	i
C_2	C_2	E	i	σ_h
σ_h	σ_h	i	E	C_2
i	i	σ_h	C_2	E

Thus the complete set of symmetry elements E, C_2, σ_h and i constitutes the point group C_{2h}.

Problem 9.6. Chloroform belongs to the C_{3v}, point group; chloroform possesses E, C_3, σ_1, σ_2 and σ_3 symmetry elements. Construct the multiplication table for chloroform.

9.9.2 Classification and notations of point groups

Point groups are usually subdivided into four types:

(a) Cubic or isotropic (having the same properties irrespective of direction) point groups : K_h, [the symmetric point group of an atom] I_h, I, O_h, O, T_d, T_h, T.

(b) Dihedral point groups : D_n, D_{nh} and D_{nd} (including $D_{\infty h}; n = 1, 2 \ldots \infty$)

(c) Axial point groups : C_n, C_{nh} and C_{nv} (including $C_{\infty v}, n = 1, 2 \ldots \infty$)

(d) Improper axial point groups : $S_n (n = 1, 2 \ldots \infty)$

S_1 is denoted by the symbol C_s and S_2 by the symbol C_i. Cubic point groups possess the largest number of symmetry elements $C_{\infty v}$ and $D_{\infty h}$ are called continuous symmetry point groups. However, here the notations of Schoenflies system are used. Herman-Manguin system has wide applications in the cases of crystal symmetry in which individual units move from one location to another and thus the symmetry groups are not point groups. In the above notations of point groups different letters and subscript letters stand for:

Table 9.6

Letters	Stand for	Subscript letter	Stand for
C	Simple axis of symmetry	s	Only σ plane
D	n simple axes of symmetry perpendicular to a principal axis	i	Only inversion centre
S	Rotation-reflexion axis	n	Only the order of rotation of the principal axis
I	Icoshahedral symmetry	nv	Symmetry planes σ_v, that contain the principal axis.
O	Octahedral symmetry	nh	Symmetry planes σ_h perpendicular to the principal axis.
T	Tetrahedral symmetry	nd	Symmetry planes that contain the principal axis and bisect the angles between two C_2 axes.

9.9.3 General discussions on point groups

We shall start our discussion in the order given below:

(i) Continuous symmetry point groups

(ii) Cubic point groups

(iii) Improper axial groups

(iv) Axial point groups other than $C_{\infty v}$

(v) Dihedral point groups other than $D_{\infty h}$

(i) *Continuous point groups:* Linear molecules belong to this type of symmetry point groups. The line joining the nuclei is the simple axis of symmetry of an infinite order since there are innumerable angles of rotation which carry the molecules into their equivalent structures. The linear molecules simultaneously contain an infinite number of vertical mirror planes (σ_v) passing through the constituent atoms. Acetylene, hydrogen cyanide, carbon dioxide, nitrogen, oxygen etc. molecules belong to this category. When a linear molecule does not have any horizontal mirror plane (σ_h; or no C_2 axis perpendicular to the C_∞ axis), the molecule belongs to the $C_{\infty v}$ group; the symmetry class of HCN, HCl, CO etc. is denoted by $C_{\infty v}$. The $C_{\infty v}$, point group does not have inversion centre (i). When a linear molecule also has a horizontal plane of symmetry (∞_h), the point group of the molecule is denoted by $D_{\infty h}$, since there are mutually perpendicular C_2 axes which are also perpendicular to the C_∞ axis of the molecule; these molecules also have inversion centre (i). Acetylene, carbon dioxide, oxygen, nitrogen etc. molecules belong to $D_{\infty h}$ point group.

(ii) *Cubic point groups:* Cubic point groups are so called because each group has the symmetry elements characteristic of a cube. The characteristic feature of cubic point groups is the presence of several C_n axes of order higher than 2. The cubic groups are further classified as:

(a) Icosahedral symmetry point groups. I_h and I

(b) Octahedral symmetry point groups. O_h and O

(c) Tetrahedral symmetry point groups. T_d, T_h and T.

(a) Icosahedral point groups (I_h & I): The characteristic elements of this symmetry class are the presence of six C_5 axes. A molecule having several C_5 axes and planes of symmetry belongs to I_h point group. The I_h point group involves 120 symmetry operations. On the other hand a molecule with several C_5 axes and no σ plane or inversion centre (i) is said to possess I point group. This point group involves 60 symmetry operations. Some boron compounds, $B_{12}N_{12}^{2-}$,$B_{12}Cl_{12}^{2-}$ etc., belong to Icosahedral symmetry.

(b) Octahedral symmetry point groups (O_h & O): A regular octahedron has $3C_4$ axes, $4C_3$ axes, 3 σ_h and $6\sigma_d$ planes. It has inversion centre (i). The presence of octahedral symmetry is indicated by the presence of C_4 axis and simultaneously by the absence of C_5 axis. A molecule is said to possess O_h point group if it has several C_4 axes and i (or σ planes). The O_h point group involves 480 symmetry operations. When σ planes (or i) are absent in a molecule having several C_4 axes and no C_5 axis, the molecule belongs to the O point group. It has 24 symmetry operations. UF_6, SF_6 etc. belong to the Octahedral symmetry point group.

(c) Tetrahedral symmetry point groups (T_d, T_h & T): The regular tetrahedron has $4C_3$ axes $3C_2$ axes (which are also S_4 axes) and $6\sigma_d$ planes. When a molecule possesses $4C_3$ axes but no C_4 and C_5 axes, it belongs to the tetrahedral symmetry point group. A molecule possessing $3C_4$ axes, $3C_2$ axes and $6\sigma_d$ planes is classified under T_d point group. This involves 24 operations. When a tetrahedral symmetry point group does not have i or σ_h, it becomes T; while a tetrahedral point group with a centre of inversion (i) is designated as T_h point group. The T point group involves 12 operations. Carbon tetrachloride and methane belong to the T_d point group. Molecules belonging to I, O and T point groups are expected to be dissymmetric since they do not have σ planes, i and S_n axes.

(iii) *Improper axial groups (S_n; the order is even) :* The S_{2n} groups are generated by the S_n operation. When n is odd $S_{n/2}^n = E$. When n is even i.e., 2, 4 etc. new groups may generate by the S_n operation; so the point group is designated as S_n. These groups possess one improper rotation-reflection axis S_n. An S_2 group is nothing but the combination of E and i operations, it is often denoted by C_i. Molecules belonging to this point group are very rare; 1, 4-dibromo-2,5-dichlorocyclohexane having all the halo groups in the equatorial positions possesses the S_2 point group.

(iv) *Axial point groups (C_n, C_{nh} and C_n; $n \neq \infty$):* These point groups possess only one n-fold simple axis of symmetry. Axial point groups are so called because they are generated by the repetition of C_n operation. n repetition of C_n operation gives back the original configuration, i.e., $C_n^n = E$. So the order of the group is n and thus it has n operations. C_n groups possess neither σ planes, nor S_n axis, nor inversion centre. Naturally molecules possessing the C_n point group are dissymmetric and optically active. The simplest C_n point group is C_1. A molecule having only the trivial C_1 axis and no other C_n axis (S_n axis, i and σ plane are also absent) is said to belong to the C_1 point group. Since there is only C_1 axis, such a molecule is asymmetric. Lactic acid belongs to the C_1 point group. abC=C=ab type allenes belong to the C_2 point group since they have only C_2 axis as the symmetry element, on the other hand abC=C=Ced type of allenes donot have any C_2 axis and hence they belong to the C_1 point group.

C_{nh} groups: If a molecule has only one C_n axis and one σ_h (a plane perpendicular to C_n), molecule belongs to the C_{nh} point groups. Trans-1,2-dichloroethene has one C_2 axis which is perpendicular to the molecular plane and passes through the mid-point of C=C. This molecule also has σ_h and i, so it possesses C_{2h} point group. It has an S_2 axis. Hence the molecule is optically inactive. However, the simplest C_{nh} group is C_{1h} which is also denoted as C_s since the molecule has S_1 axis. HOCl molecule has C_{1h}, i.e., C_s point group. However C_{nh} point groups have $2n$ elements of symmetry or symmetry operations.

C_{nv} groups: A molecule with only one C_n axis ($n > 1$) and n σ_v planes is classified under the C_{nv} point group. Since the molecules belonging to this point group have σ planes, they are achiral and optically inactive. Chloroform, methyl chloride etc. have the C_{3v} point group. The simplest point group of this class is C_{3v}; m-bromochlorobenzene belongs to the C_{1v}, point group. Cis-dichloroethene, dichloromethane, water etc. belong to the C_{2v} point group. C_{nv} groups have 2_n operations or elements of symmetry.

(v) *Dihedral point groups (D_n, D_{nh} and D_{nd}; $n \neq \infty$)*

D_n point groups: The D_n point group has C_2 axis which are perpendicular to the principal C_n axes; this group does not contain any σ plane, i and S_n axis. There are $2n$ elements in this group. Molecules possessing the D_n point groups are dissymmetric as they do not have any S_n axis; so, they are optically active. Compounds having D_n group are very rare.

D_{nh} point groups: A D_n group with a σ_h plane is designated as D_{nh} point group. So the D_{nh} point group has the principal C_n axis, nC_2 axes and one σ_h plane and σ_v planes. There are $4n$ elements in the D_{nh} group. Ethylene (Fig. 9.80) and naphthalene (Fig. 9.81) belong to the D_{2h} point group. Benzene possesses D_{6h} point group. Its principal C_n axis is C_6 and it has a σ_h plane and several C_2 axes which are naturally perpendicular and also perpendicular to C_6 axis (Fig. 9.30 & 31) since the D_{nh} point group has σ_h, the molecules belong to this group are symmetric and optically inactive.

D_{nd} point groups: A D_n group with $n\sigma_d$ planes is designated as the D_{nd} point group. The D_{nd} point group has a C_n axis as principal axis, mutually perpendicular nC_2 axes and $n\sigma_d$ planes but no σ_h plane. This point group has $4n$ symmetry elements or symmetry operations. Since the D_{nd} group contains $n\sigma_d$ planes, the molecules possessing this point group are achiral and optically inactive. Allene itself belongs to the D_{2d} point group. The staggered conformation of ethane belongs to the D_{3d} point group. In this case C—C bond axis is the C_3 axes. Perpendicular to C_3 axis, there are mutually perpendicular $3C_2$ axes and $3\sigma_d$ planes.

9.9.4 Relation between point groups and asymmetry and dissymmetry of molecules

From the discussion made in Sec. 9.9.3 the following points are to be noted in this regard:

(i) Molecules belonging to the C_1 point group are asymmetric since C_1 point group does not have any other symmetry element except C_1 axis.

(ii) Molecules possessing C_n and D_n point groups do not have σ_v or σ_h or σ_d planes and they do not also contain S_n axis but they have C_n axis ($4 > n > 1$) along with C_1 axis. So molecules belonging to C_n or D_n, groups are dissymmetric.

(iii) T, O and I point groups do not also contain σ planes and S_n axis, or i; hence molecules containing T, O and I groups are expected to be dissymmetric but not asymmetric since all of these point groups have C_n axis higher than C_1. However, dissymmetric molecules with T or O or I point groups are not usually encountered.

9.9.5 Point groups and dipole moments

Molecules possessing point groups involving inversion centre (i), cannot have dipole moment since a vector cannot lie on a point. Molecules, therefore, belonging to C_i, I_h, O_h & T_h have zero dipole moments. Molecules with more than one non-coincident simple axis of rotation cannot have a dipole moment, since the vector cannot coincide with two distinct axes. Hence I, O, T_d, D_n, D_{nh}, D_{nd}, C_{nh} point groups when present in molecules, they will be of zero dipole moment. Molecules belonging to the following point groups C_1, C_s, C_n and C_{nv} have dipole moments.

9.9.6 Assigning of point groups to chemical entities

The following procedure and Table 9.6 will help a student to assign point group to a chemical entity.

Step I: Find out whether the given molecule is linear or not; if it is linear, the molecule will belong either to $C_{\infty v}$, or $D_{\infty h}$. If there is σ_h plane perpendicular to C_∞ axis the group is $D_{\infty h}$; otherwise $C_{\infty v}$.

Step II: If the molecule is non-linear, then try to find out a C_5 axis, If it is present, the molecule belongs to either I or I_h. If there is inversion centre, the molecule is of I_h, point group; and if it is otherwise, I is the point group of the molecule.

Step III: If the non-linear molecule contains C_4 axis, the molecule belongs to the O_h or O point group. The presence of i in the molecule indicates its point group as O_h. If the molecule does not contain i, the point group is O.

Step IV: If the non-linear molecule has several C_3 axes but no C_4 or C_5 axis, it belongs to the tetrahedral symmetry group. If there is no inversion centre or σ_h, but $6\sigma_d$ elements in the molecule, the latter belongs to the T_d point group. If the molecule has i or σ_h along with several C_3 axes, it is a member of the T_h point group. If there is none of i and σ_h and only C_3 axes are present, the molecule belongs to the T point group.

Step V: If the non-linear molecule do not have a C_n axis other than C_1, the molecule belongs to

(i) C_1 if there is no σ_h, or i present in it

(ii) C_s if there is a σ_h present in it

(iii) C_i if there is no σ but an i present in it.

Step VI: If there is only C_n axis present in the molecule and no other elements of symmetry, the molecule belongs to C_n point group. If there is a σ_h plane along with C_n axis the point group of the molecule is C_{nh}. On the other hand a molecule with only one C_n axis (except C_1) and σ_v planes but devoid of σ_h planes is classified under C_{nv} point group. Molecules with only C_n and one S_n axis belong to the S_n point group.

Step VII: If a non-linear molecule with one principal axis C_n and mutually perpendicular nC_2 axes which are also perpendicular to the principal axis, the molecule belongs to either D_n or D_{nh}, or D_{nd} group. If such a molecule also possesses a σ_h plane, the point group of the molecules will be D_{nh}; when a molecule with D_n group has $n\sigma_d$ planes and no σ_h, the molecule is said to possess the D_{nd} point group.

Symmetry number of an entity is a number which tells us that how many equivalent configurations could be obtained by rotating the entity around one or more axes; those which are counted for one operation should not be counted for other operations. For example $AB_3(BF_3)$ type planar trigonal molecules have one C_3 axis and $3C_2$ axes. When the molecule is rotated around C_3, we get three equivalent configurations by $\hat{C}_3$, $\hat{C}_3^2$ and $\hat{C}_3^3(\equiv E)$ operations. When C_2 operations are carried out around each C_2 axis (passing through one F and B), we get two equivalent configurations by $\hat{C}_2$ and $\hat{C}_2^2 \equiv E$ and E has already been obtained by $\hat{C}_3^3$, we cannot count the configuration obtained by $\hat{C}_2^2$. So rotation of AB_3 around $3C_2$ axes develop three equivalent configurations. Thus AB_3 has symmetry numbers 6, 3 for C_3 and 3 for $3C_2$ (one for each C_2). Symmetry numbers are related to point groups as given below.

Point groups	Symmetry no.
C_∞, C_1, C_s, C_i	1
C_n, C_{nh}, C_{nv}	n
D_n, D_{nd}, D_{nh}	$2n$
$D_{\infty h}$	$2n$
T_d	12
O_h	24

Problem 9.7. To which point group the following molecules belong?

(i) 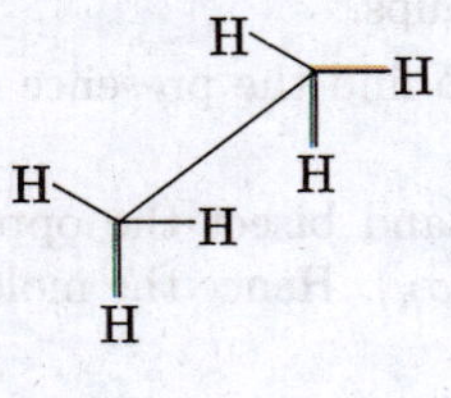 (ii) 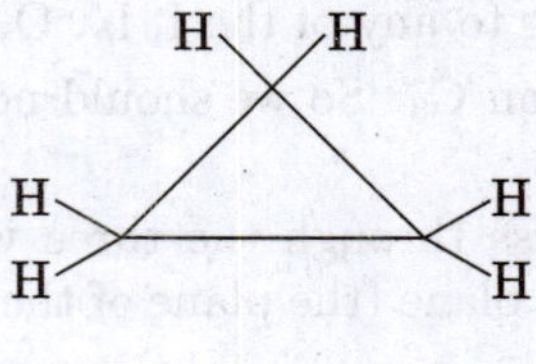 (iii) CH_2BrCl (iv) 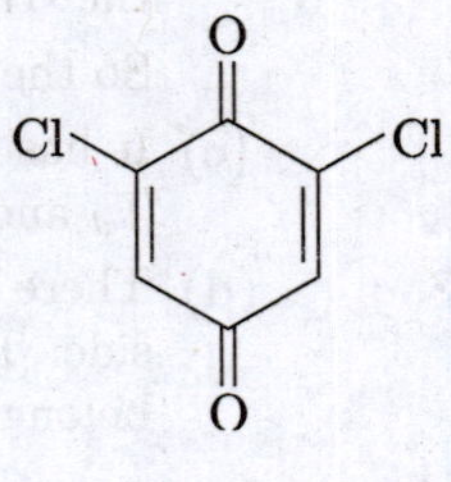

(v) 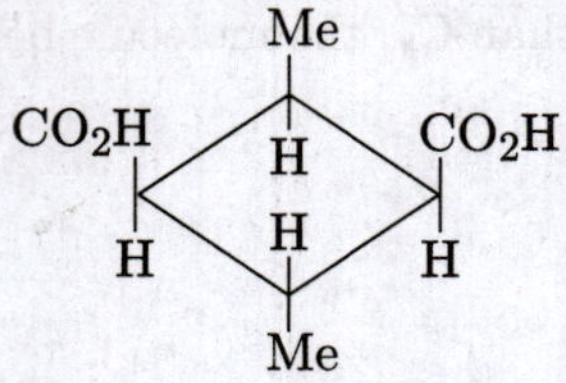
(vi) 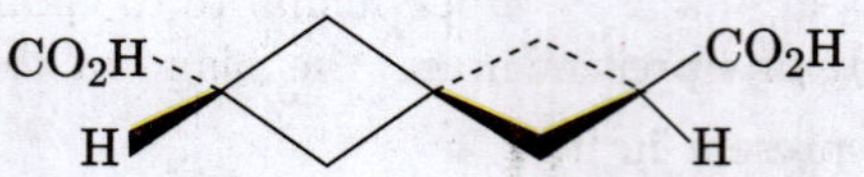
(vii) 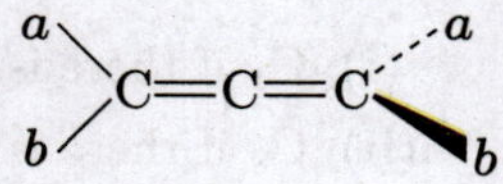

(viii) 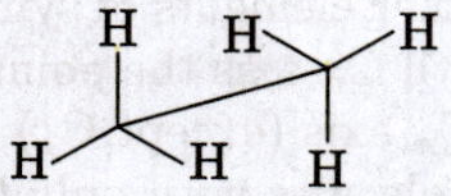
(ix) H—C≡C—Cl
(x) Cl—C≡C—Cl

Solution:

(i) (a) It is not a linear molecule. So it does not belong to $C_{\infty v}$ or $D_{\infty h}$.

(b) The molecule does not have C_5 or C_4 axis; so the molecule does not come under I or O point group.

(c) It has only one C_3 axis which coincides with the C—C σ-bond. It has $3C_2$ axes perpendicular to C_3 axis and they are also mutually perpendicular.

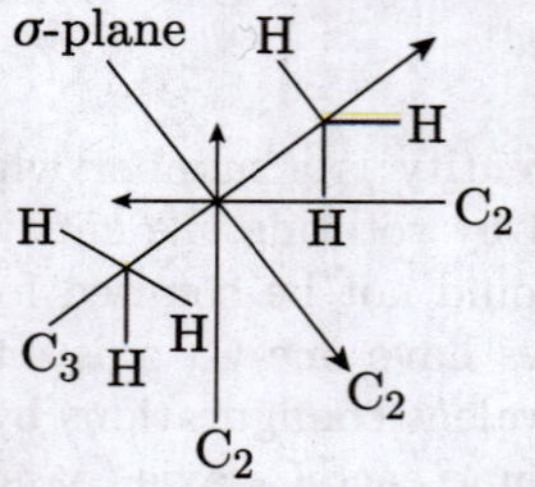

So the molecule belongs to either D_3 or D_{3h} or D_{3v}, group. Now, the molecule has a σ_h plane. So it belongs to the D_{3h} point group.

(ii) (a) Non-linear: It cannot have $C_{\infty v}$ or $D_{\infty h}$ group.

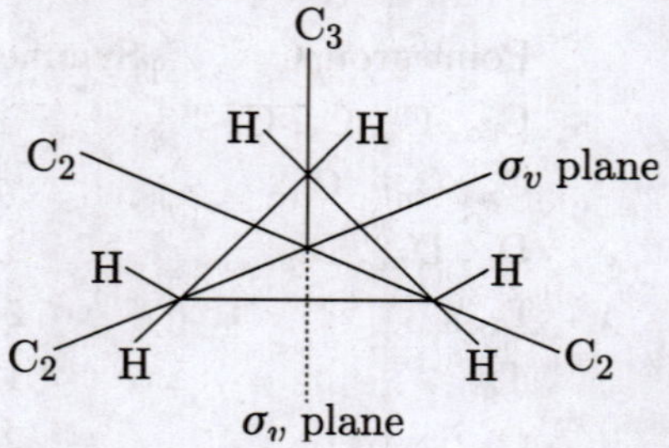

(b) The principal axis is C_3 (the highest order axis) which passes through the centre of the triangle and perpendicular to the plane of the ring constituent carbons.
So the molecule cannot belong to any of the I, I_h, O, O_h groups.

(c) It has no other axis higher than C_3. So we should not try to find the presence of T, T_d and T_h point groups.

(d) There are $3C_2$ axes which pass through the three vertices and bisect the opposite side. The molecule has one σ_h plane (the plane of the ring is σ_h). Hence the molecule belongs to D_{3h}

(iii) **(a)** It is a non-linear molecule.

(b) It has only C_1 axis which is perpendicular to the plane of the paper. There is a σ_h plane which lies on the plane of the paper. So it belongs to C_{1h}, i.e. $\mathbf{C_s}$, point group.

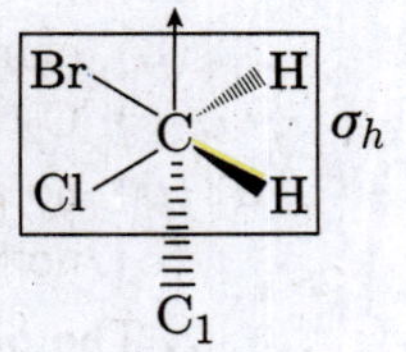

(iv) **(a)** Nonlinear: $C_{\infty v}$ and $D_{\infty h}$ are absent.

(b) It has only one C_2 axis which passes through two C=O moieties. There is no C_5, C_4 and C_3 axes. So the molecule does not belong to I, O, T groups.

(c) It has no other C_2 axis. So the question of D_n, D_{nh} and D_{nv} does not arise. It may belong to C_n, C_{nh}, and C_{nv}

(d) It has two σ planes; both contain the C_2 axis. One σ_v lies on the plane of paper and the other perpendicular to it, so the point group of the molecule is C_{2v}.

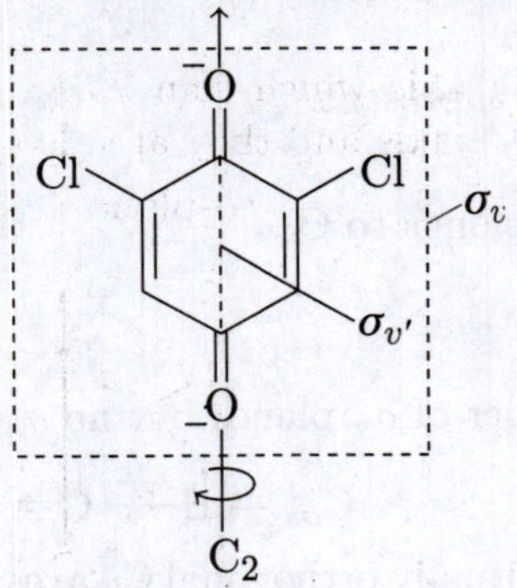

(v) The point group is $C_{1h} \equiv C_s$

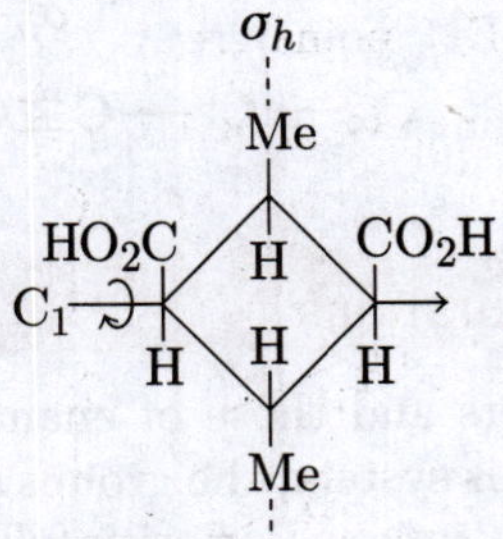

(vi) **(a)** No σ plane

(b) Chiral axis present

(c) C_2 axis present

(d) no i

(e) no S_n axis.

$\therefore$ the point group is C_2. The molecule is chiral.

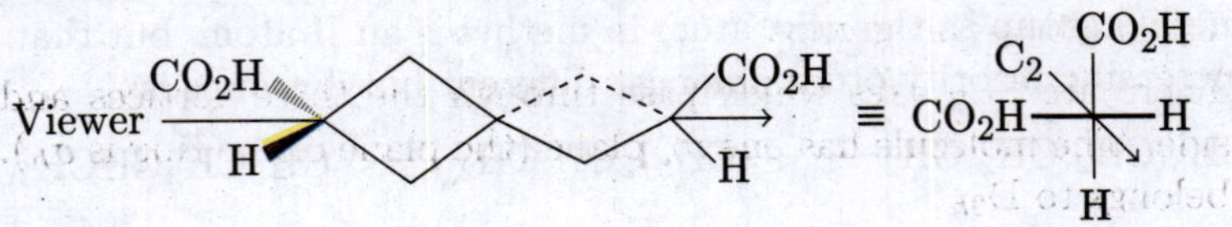

(vii) (a) No σ plane
(b) chiral axis present
(c) C_2 axis present
(d) no i
(e) no S_n axis

∴ The point group is C_2.

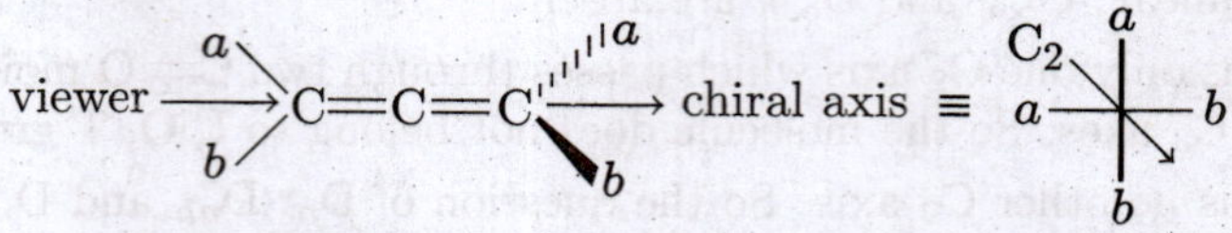

(viii) (a) Non-linear: $C_{\infty v}$ and $D_{\infty h}$ are absent.
(b) C_2 axis present: The C_2 axis passes through the mid-point of C—C bond and perpendicular to the paper.
(c) No σ plane
(d) Inversion center, i, present.
(e) S_2 axis present.

Hence the molecule belongs to C_i.

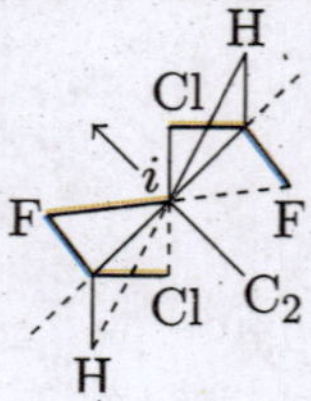

(ix) (a) Linear molecule
(b) An infinite member of σ_v planes but no σ_h. Hence its point group is $C_{\infty v}$.

$$C_\infty \rightarrow H—C \equiv C—Cl \rightarrow$$

(x) Linear molecule : 2 mutually orthogonal C_2 axes which are perpendicular to C_∞ axis which passes though Cl, C, C, Cl. It has a σ_h plane. So it belongs to $D_{\alpha h}$.

$$\sigma_h$$
$$C_\infty \rightarrow Cl—C \equiv C—Cl \rightarrow$$

9.10 Specification of Configurations

Configurations of diastereomers and those of enantiomers can easily be specified by Cahn-Ingold-Prelog (CIP) system. In this system, the groups and atoms attached to the double-bonded atoms and to the chiral atoms are assigned priority following certain rules as given below.

I. The higher the atomic number of the atom attached to the involved atom, the higher is its priority; e.g., N has priority over C, O over N, Br over Cl, I over Br etc.

II. For isotopes, the atom with higher mass has the priority, e.g., C^{14} over C^{12}, deuterium over hydrogen.

III. If the first-bonded atom is identical in at least two groups, the priority is determined by considering the next atom or atoms present in those groups, e.g. ethyl group has priority over methyl group as the next atom in methyl is an H atom, but that in ethyl is a C atom. The increasing priority order amongst different butyl groups is:

$$CH_3.CH_2.CH_2.CH_2— < (CH_3)_2CH.CH_2— < CH_3CH_2\underset{|}{C}HCH_3 < (CH_3)_3C—$$

IV. In assigning priority to multiple-bonded atoms the following changes are made:

a. 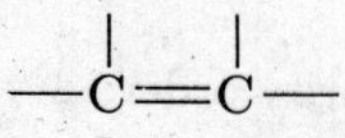 is replaced by ≡ 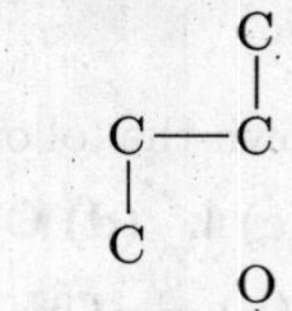

b. 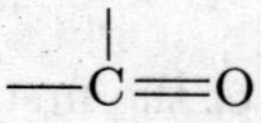 is replaced by ≡ 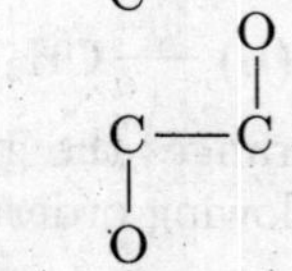

c. 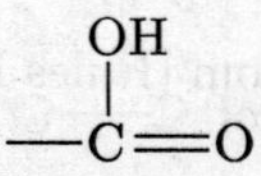 is replaced by ≡ 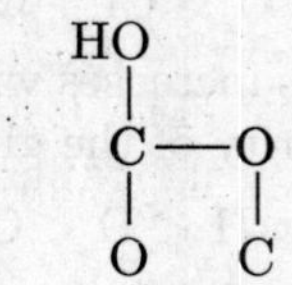

d. 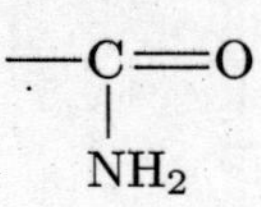 is replaced by ≡ 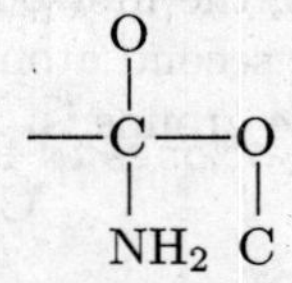

e. 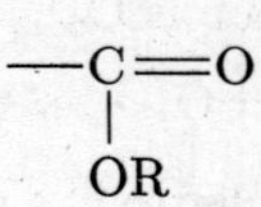 is replaced by ≡ 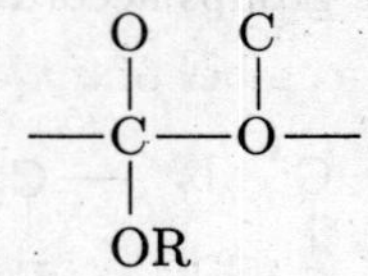

f.  is replaced by ≡ 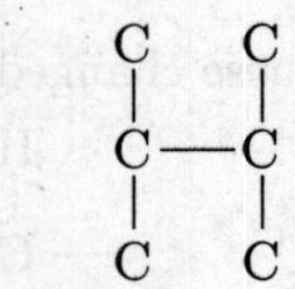

g. 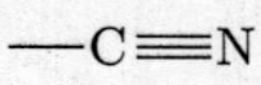 is replaced by ≡ 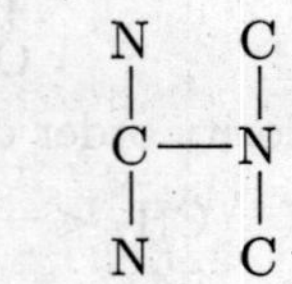

h. 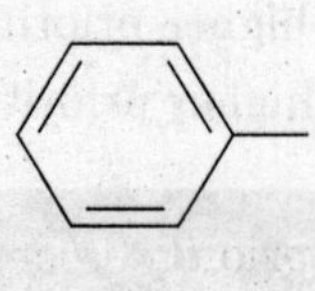 is replaced by ≡ 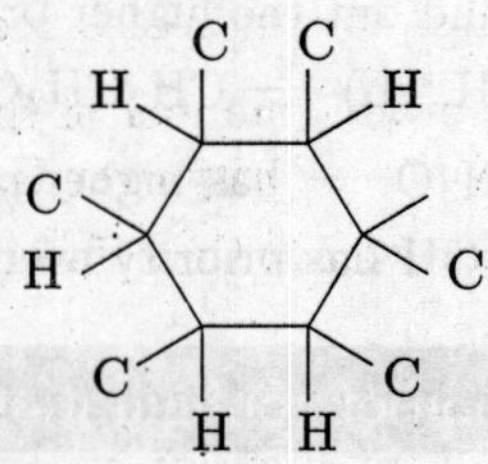

Thus the increasing priority order in each series is from left to right:

(i) —CHO, —$COCH_3$, —$CONH_2$, —COOH, —$COOCH_3$

(ii) —CH=CH_2, —C≡C—H, —C_6H_5, —C≡N

(iii) —$CH(CH_3)_2$, —CH=CH_2, —$C(CH_3)_3$, —C≡CH

(iv) $—CH_2CH{=}CH_2$, $—CH{=}CH_2$, $—CHCH_2CH_3$ (with CH_3 on CH), $—CH{=}CH—CH_3$

Problem 9.8. Write down the following groups in an increasing priority order:

(i) **(a)** CH_3, **(b)** Cl, **(c)** I, **(d)** OH, **(e)** NH_2, **(f)** D, **(g)** H.

(ii) **(a)** $—CH{=}CH$, **(b)** $—CH_2F$, **(c)** $—CHO$, **(d)** $—COCH_3$.

Solution: (i) If one arranges the groups in a decreasing order of mass of the first atom of the groups, one gets the following order:

$$I > Cl > OH > NH_2 > CH_3 > D > H$$

As the priority of groups increases with the increasing mass of the first atom (Rules I and II), the increasing order of priority of the given substituents is:

$$H < D < CH_3 < NH_2 < OH < Cl < I$$

(ii) In each of the cases, the first-bonded atom is a C atom. Hence, in determining priority order, we need consider the second atom and even the third atom. The increasing priority order of the second atoms in these groups is:

$$C < O < F \text{(Rule I)}$$

If we now rewrite the groups according to rule IV, we get:

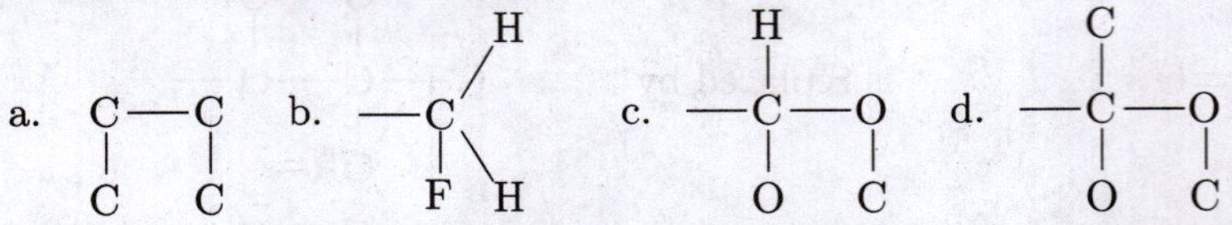

Applying rule (III) on these changed groups, we see that the increasing priority order is:

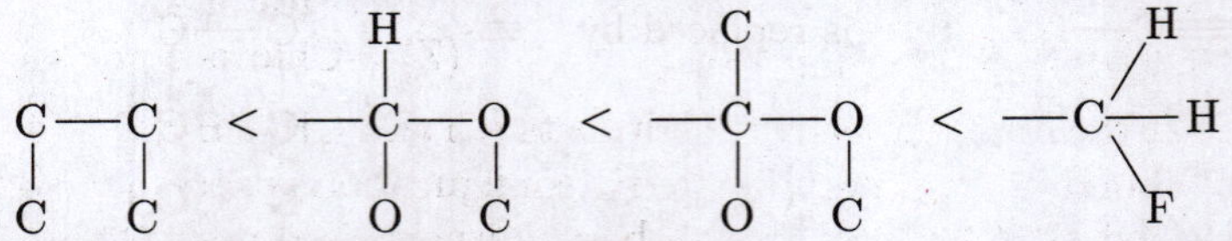

Hence the increasing priority order of the given groups is:

$$—CH{=}CH_2 < —CHO < —COCH_3 < —CH_2F$$

Problem 9.9. Find out the higher priority group from each of the given pairs:

(a) CH_3O and OH, **(b)** $—CH_2CH_2OH$ and $CH_3.CH_2.CH_2—$

Solution: (a) $CH_3O—$ has higher priority over $—OH$ as C has higher priority over H.

(b) $—CH_2.CH_2OH$ has priority over $—CH_2.CH_2.CH_3$ as O has higher priority over C.

9.10.1 Specification of configuration of diastereomers involving rigid part

To specify configurations of the class by Cahn-Ingoid-Prolog method the following procedures are adopted:

1. The structural formula for the molecule is drawn on the plane of the paper clearly showing the double bond or the ring with their substituents attached to the involved atoms of the rigid part.

2. Priority of the substituents on each of the involved atom is then determined (by the priority rule) and marked accordingly by assigning number 2 to the higher priority substituent and number 1 to the lower priority substituent.
3. If the higher priority substituents are found on the same side of the double bond or the ring, the configuration of the molecule is said to be Z (is the first letter of the German word, Zusamen meaning together); on the other hand if the higher priority substituents are placed on the opposite sides of double bond or the ring, the configuration of the molecule is described as E (E is the first letter of the German word, entgegen meaning opposite).

For example, 2-chloro-2-pentene can exist as a pair of diastereomers as written below:

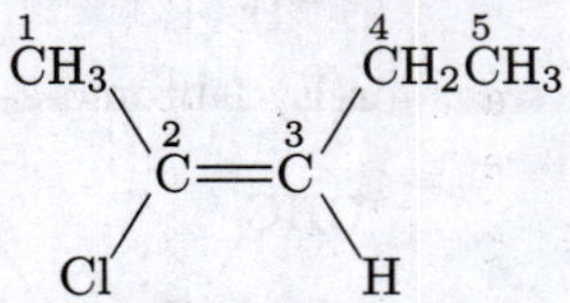

form I

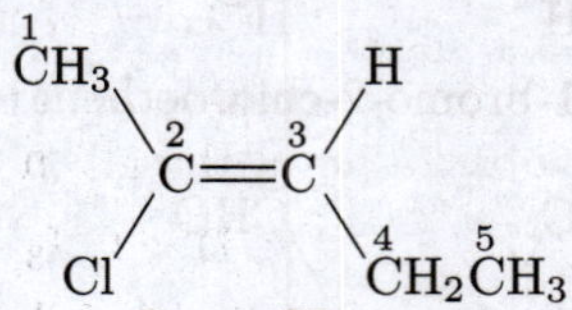

form II

To specify the configurations of forms I and II, let us determine the priority of the substituents on each double-bonded C. C_2 atom possesses two substituents; CH_3 and Cl; the Cl has priority over CH_3 (priority rule I). The two substituents on C_3 atom are $CH_3.CH_2$ and H; $CH_3.CH_2$ has priority over H (priority rule I). The substituents may now be marked according to their priorities as shown below.

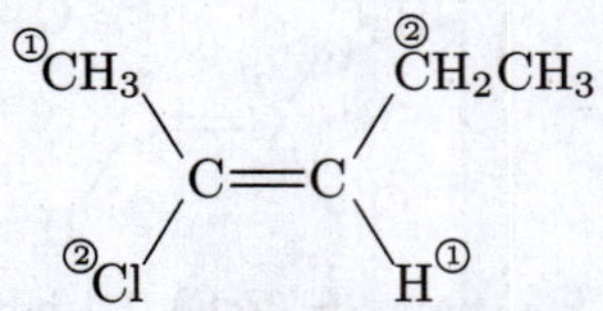

form I

(E)-2-Chloro-2-pentene

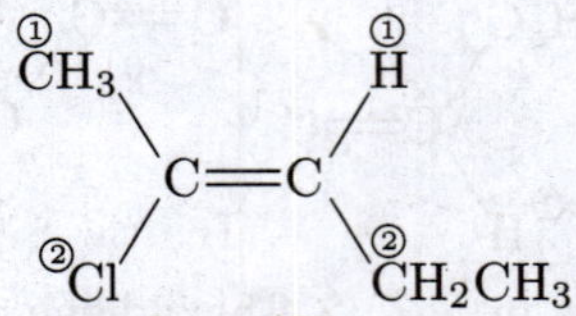

form II

(Z)-2-Chloro-2-pentene

Thus we see that the higher priority substituents (Cl and CH_2CH_3) are present on the opposite sides of the C=C bond in form I and hence its configuration is specified as E; the configuration of form II is specified as Z because two such substituents are present on the same side of the double bond.

Now if one wants to indicate the configuration of a geometrical isomer in naming such a molecule, E or Z is to be prefixed within braces to the IUPAC name of the compound placing a hyphen in between the descriptors, E or Z, and the name. Thus the name of the form I is (E)-2-chloro-2-pentene and that of the form II is (Z)-2-chloro-2-pentene.

The configurations of simple molecules about a rigid part having two identical substituents, one on each involved atom, may also be designated as cis, indicating like substituents on the same side of the rigid part, and trans indicating those on the opposite sides of that part. Thus forms III and IV of 2-butene may be designated as cis and trans respectively as shown below :

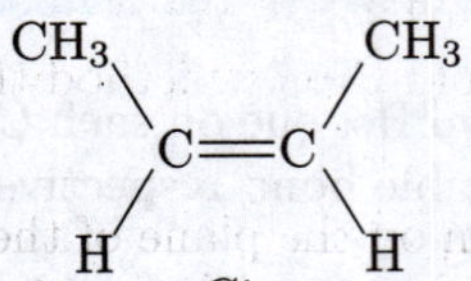

Cis

form III

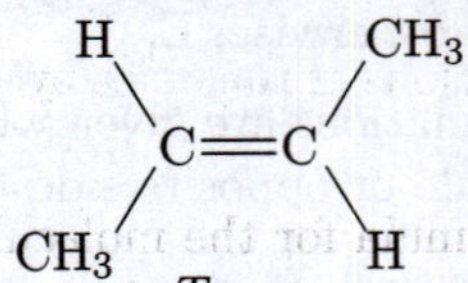

Trans

form IV

As geometrical isomers were specified by cis and trans in the early days, these are also called cis-trans isomers. Form III is thus named as cis-2-butene and form IV as trans-2-butene.

Problem 9.10. Name the diastereomers of the following compounds :

a. $CHCl{=}CHBr$; **b.** $CHO.C(CH_3){=}C(CH_3)CHO$; **c.** $CH_3.CH{=}CH(CHO){=}CH.CH_3$

Solution:

a.

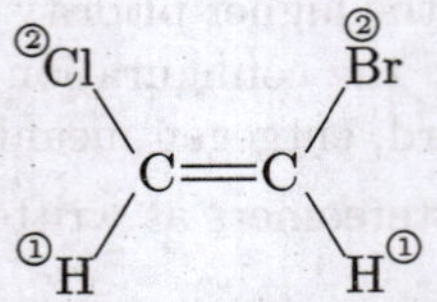

cis- or(Z)-1-bromo-2-chloroethene

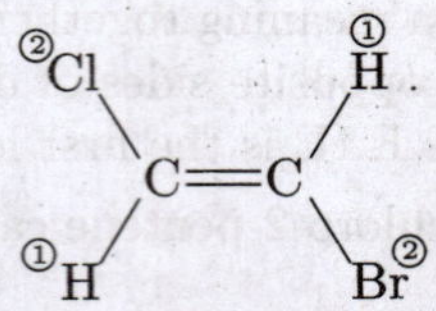

trans -or(E)-1-bromo-2-chloroethene

b.

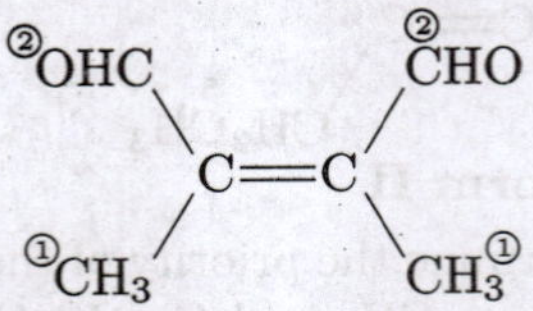

cis -or(Z)-2,-3-dimethylbutene-dial

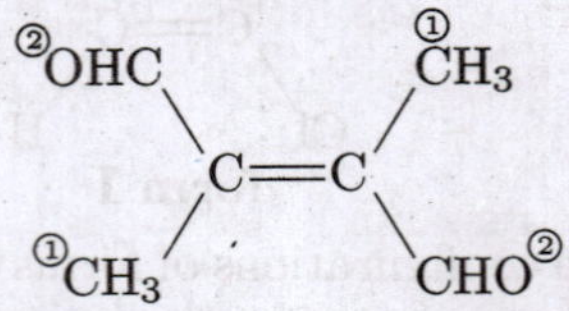

trans -or(E)-2,3-dimethylbutene-dial

c.

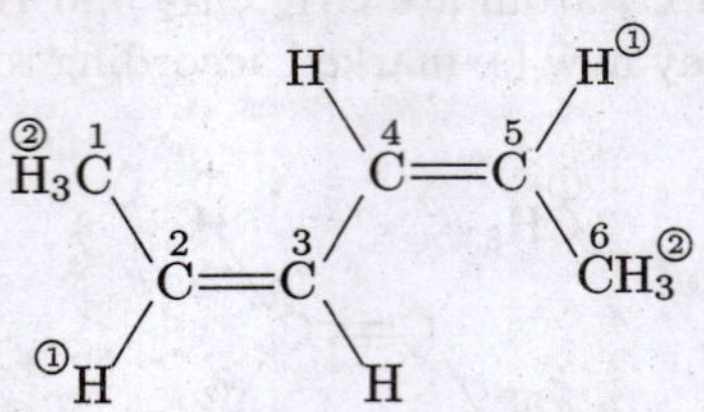

cis,cis- or(Z,Z)-2,4-hexadiene

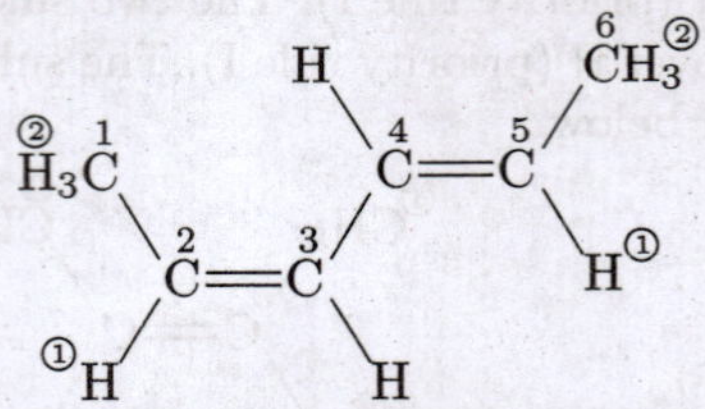

cis,trans- or (Z,E)-2,4-hexadiene

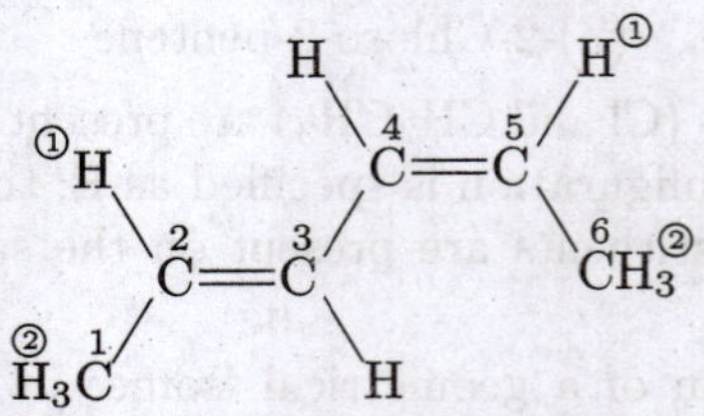

trans,cis- or(E,Z)-2,4-hexadiene

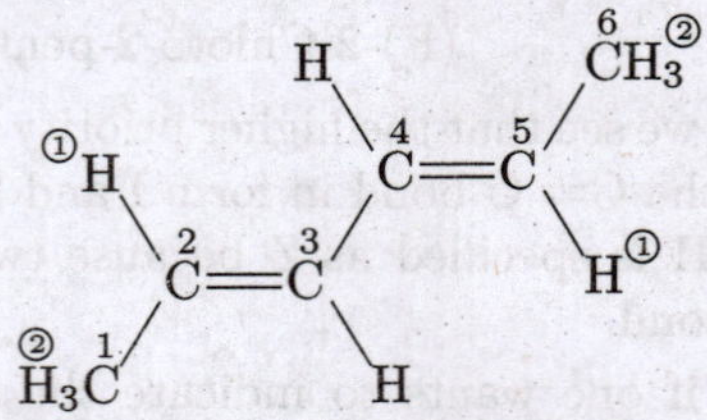

trans,trans- or (E,E)-2,4-hexadiene

To name the configuration in c, the following procedure is adopted:

(i) cis and trans, and Z and E are listed in the same way as the C atoms are numbered.

(ii) When configuration about one C=C bond is determined, $-CH{=}CH.CH_3$ is considered as one substituent and H as another, and certainly $-CH{=}CH.CH_3$ has priority over H. Thus when CH_3 and $-CH{=}CH.CH_3$ are on the same side of the double bond, it is designated as Z, if otherwise, E.

(iii) cis and trans designations are given when the two Hs, one on each C atom of C=C bond, are on the same side or opposite sides of the double bond respectively.

9.10.2 Specification of configurations about a chiral atom

To specify the configurations of enantiomers, i.e. spatial arrangements of groups and atoms about a chiral atom the following procedure is adopted:

(i) The configuration of the molecule is first drawn on the plane of the paper by Fischer projection formula.

(ii) The atoms and groups attached to the chiral atom are then assigned an order of priority according to the sequence rule (a few more sequence rules are listed below). The priorities of atoms and groups are given by numbers within parentheses, using (1) for the lowest priority and (4) for the highest priority.

(iii) The lowest priority group or atom must project away from the observer and be in the back of the paper, i.e., the lowest priority group or atom must have the lower or higher vertical position in the projection formula; if it is not so, the lowest priority group or atom should be brought to that position by making two interchanges of positions of groups.

(iv) Now, for the remaining three atoms or groups, if the sequence of decreasing priority, (4) to (3) to (2), is anticlockwise, the configuration about that chiral atom is specified as S (it is the first letter of the Latin word sinister meaning left); if it is clockwise, the configuration of this is designated as R (it is the first letter of the Latin word rectus meaning right).

The procedure is illustrated for 1-chloroethanol using Fisher projection formula.

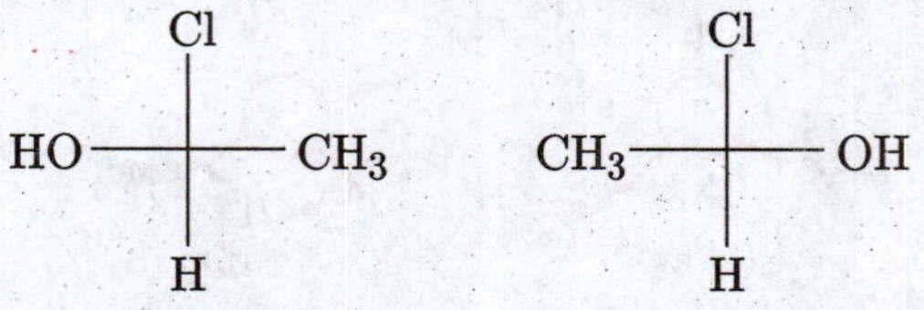

from I from II

(i) In this compound the priorities of groups and atom are: Cl(4), OH(3), CH_3(2) and H(1).

(ii) The lowest priority atom, H(1) is placed on the lower vertical position of the both forms written above. Now let us rewrite the forms indication priorities of the atoms and groups:

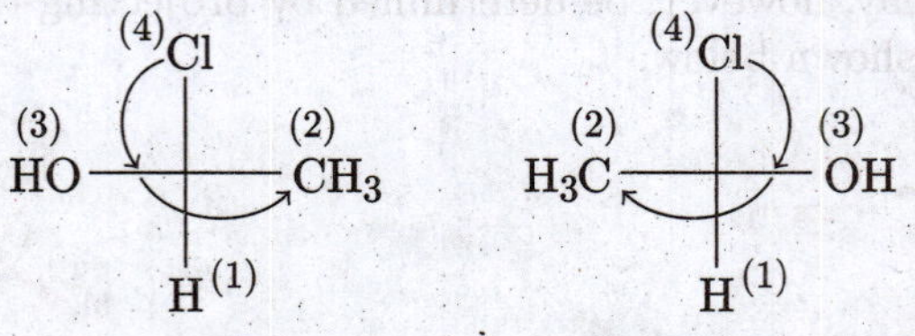

Form I Form II

(iii) The sequence of decreasing priority of the other atom (Cl) and groups (OH, CH_3) is anticlockwise in the form I and hence it is designated as S; in the form II, the said sequence is clockwise and thus its configuration is specified as R.

However, if the lowest priority group does not have a position on the vertical line, we can also assign the configuration without interchanging its position. In this case we shall have to move our eye from the top prior group to the 3rd top prior group via the 2nd prior one, in doing so if we find the configuration as S, actually it will be R and vice versa. In III on the next page an example is given

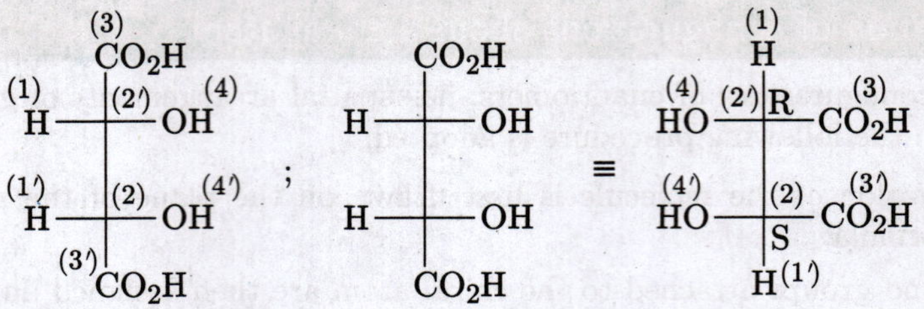

III

C_2 atom seems to have S configuration when we move our eye from (4) to (3) to (2); actually it has R configuration. Similarly C_3 seems to have R configuration but it has actually S configuration.

The R and S designations may also be done by observing models of molecules as shown below:

(a) The atoms and groups on the chiral C are to be given priority as per sequence rule.

(b) The model should be viewed in such a way that the group of the lowest priority in the sequence is pointed away from the viewer.

(c) Now if the other groups, in order of priority, are oriented anticlockwise, the chiral C is to be designated as S, if it is clockwise, R,

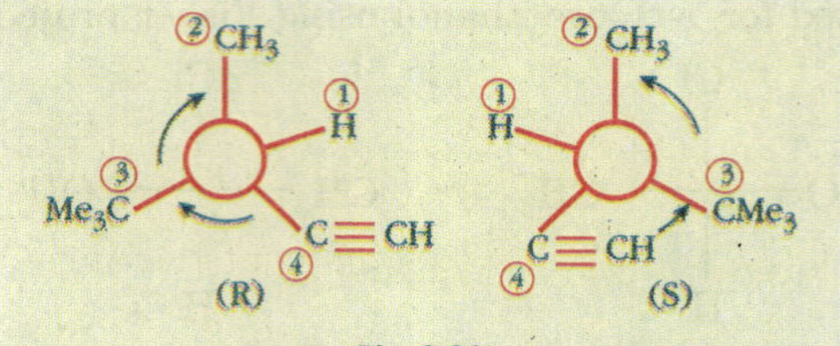

Fig. 9.93

In the left-hand-side figure H is away from the viewer. Here —C≡CH has priority over CMe_3; the decreasing priority order is —C≡CH, CMe_3, Me, H. So the configuration of the chiral C in the figure is R and that in the other figure is S.

The configurations may, however, be determined by projecting the model on the plane of the paper in the manner as shown below.

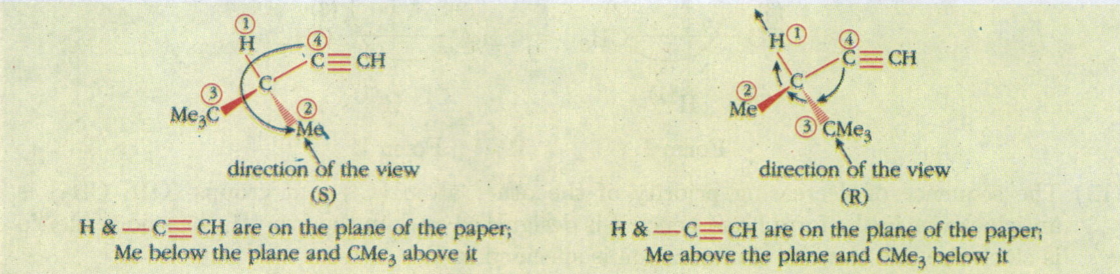

Fig. 9.94

To name an enantiomer, its configurational designation is indicated by prefixing it to the name in parentheses and separating from the latter by a hyphen. Thus form I is (S)-1- chloroethanol and form II is (R)-1-chloroethanol.

9.10.3 A few more priority sequence rules

For specifying configurational designations of species a few more sequence, i.e., priority rules are given below:

(a) When groups attached to an involved atom differ only in their geometric configuration, cis or Z has priority over the trans or E, e.g.,

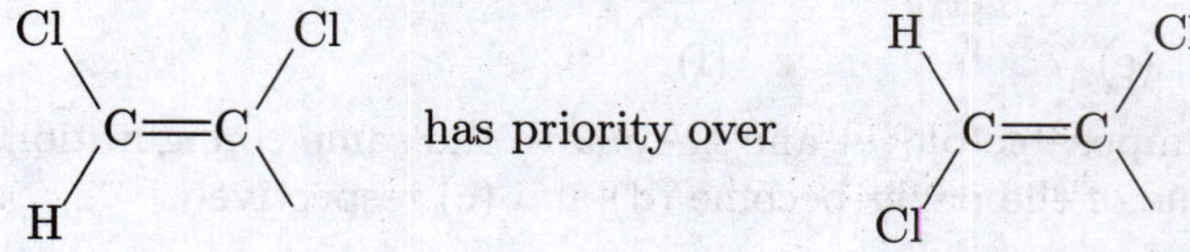

Cis or Z *Trans* or E

(b) When groups differ only in configuration about a chiral atom, R has priority over S, e.g.,

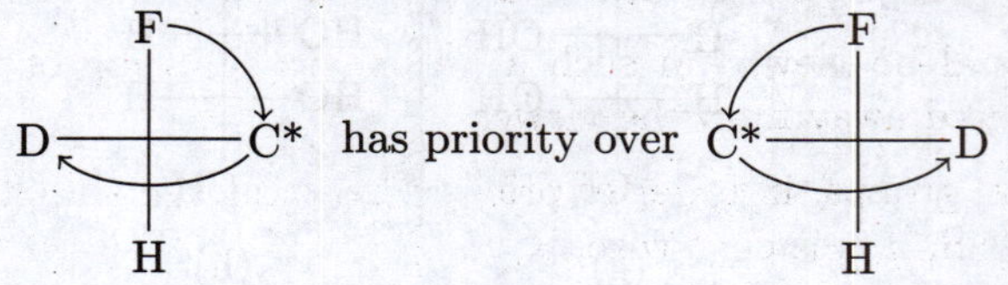

(R) (S)

here C* is another chiral C.

Let us consider the configurational isomers of 2, 3, 4-trihydroxypentane, $CH_3\overset{*}{C}HOH\overset{*}{C}HOH\overset{*}{C}HOHCH_3$. It has the following configurations:

(i)

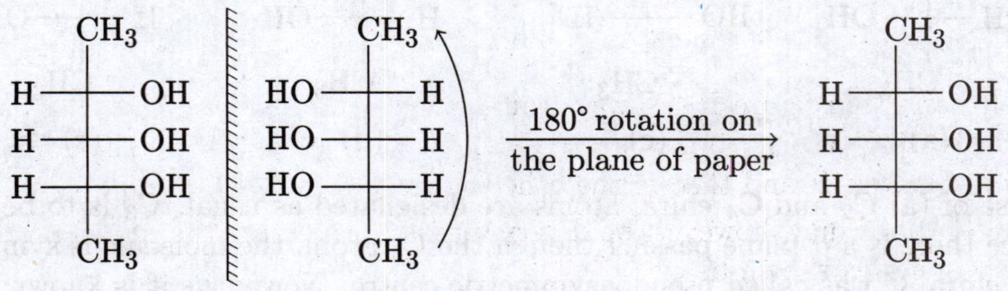

(a) (b) (a)

So, (b) is superimposable on (a) and (a) and (b) have the same configuration (Homomers).

(ii)

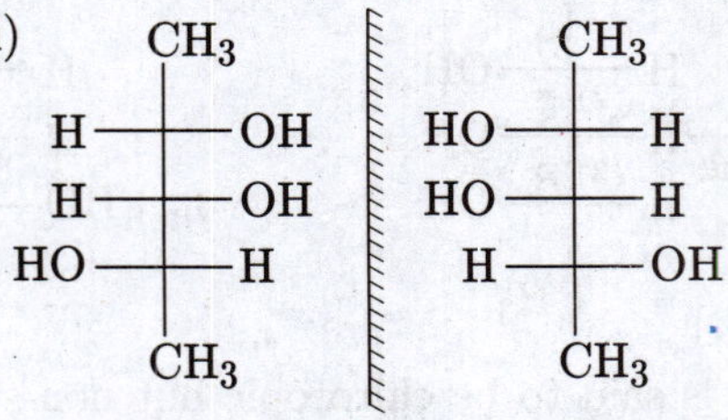

(c) (d)

(c) and (d) are non-superimposable. Hence they represent a pair of enantiomers.

(iii)

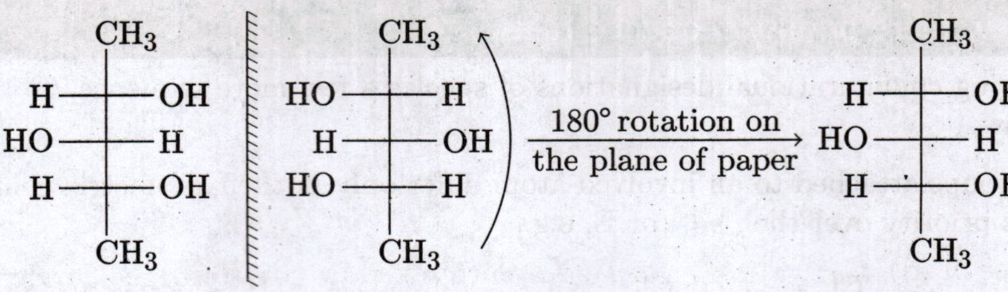

(e) (f) (e)

So, (f) is superimposable on (e) and they have the same configuration. (g) and (h) on 180° rotation on the plane of the paper become (d) and (c) respectively.

(iv)

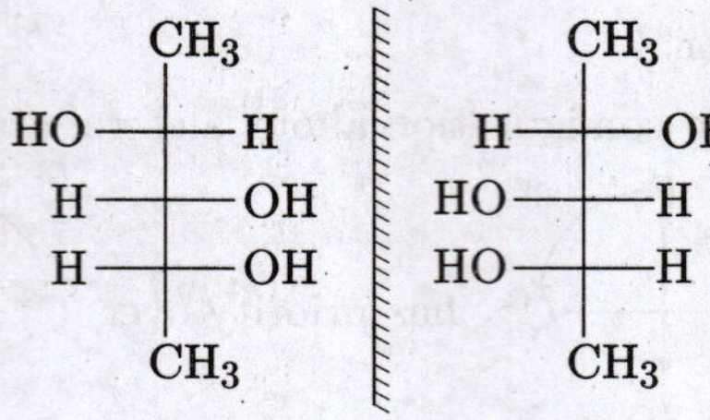

(g) (h)

So, there are only four configurational isomers of 2, 3, 4-trihydroxypentane.

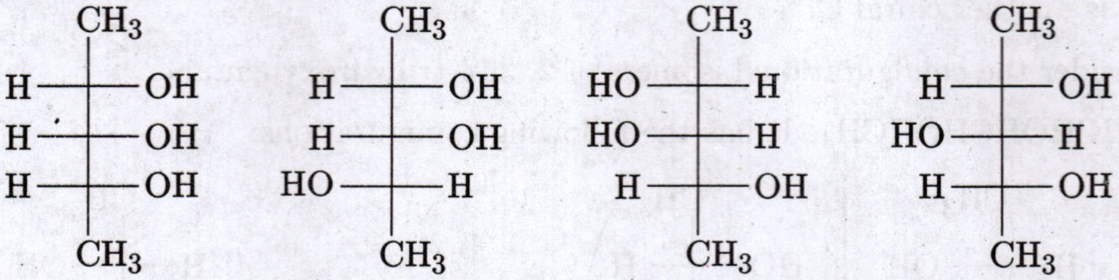

(a) (c) (d) (e)

In the case of (a) C_2 and C_4 chiral atoms are designated as usual. C_3 is to be designated as sinister. Since there is a σ plane passing though the C_3 atom, the molecule is symmetric. Previously such a chiral C was called pseudoasymmetric centre. Nowadays it is known as achirotopic but stereogenic C centre and designated as s^*.

(a) (c)

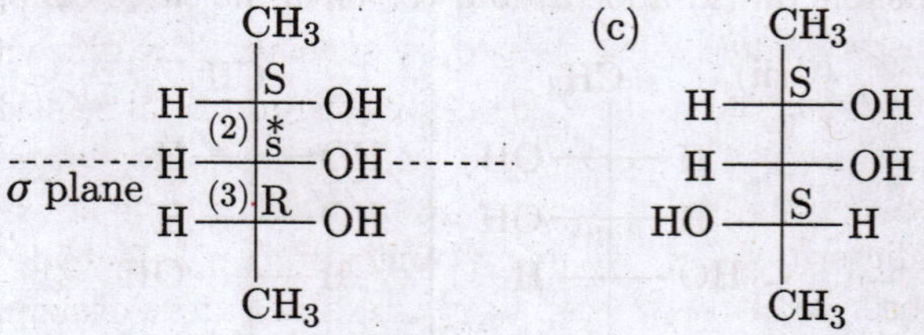

In the case of (c), C_3 atom is said to be chirotopic but non-stereogenic centre. Since it has two identical ligands with S, S designated chiral C.s, it cannot be designated as R or S.

(d) 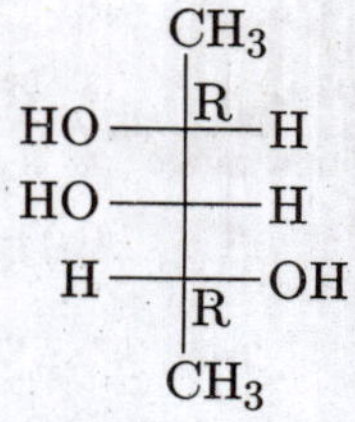

(e) 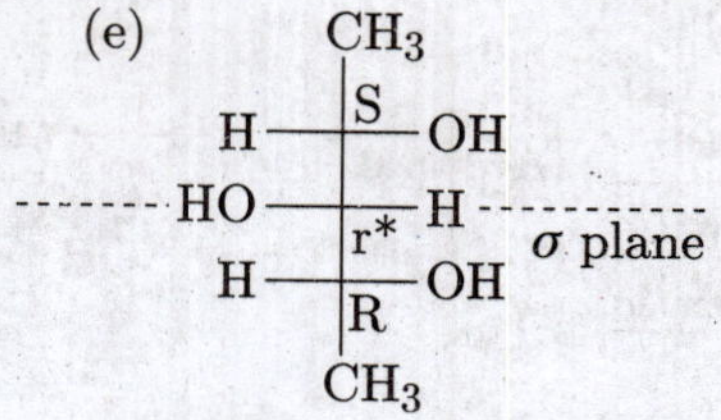

In the case of (d), C_2—R; C_4—R; C_3: chirotopic but non-stereogenic centre. It cannot be designated for the reason noted in the case of (c).

In the case of (e), C_2—S; C_4—R; C_3—r^* (achirotopic but stereogenic centre).

C_3 atom of (e) is stereogenic since interchange of two groups on it gives rise to a new configuration. It is achirotopic because the molecule is achiral about C_3.

C_3 atom of (d) is non-stereogenic since interchange of two groups or ligands does not give rise to a new configuration. However, C_3 atom in (d) is chirotopic as the molecule is chiral about C_3.

Problem 9.11. (a) Designate the configurational isomers of 2, 3, 4-tribromopantanedioic acid as R, S and also designate (b), (c), (d), (e), (f) and (g) as R and S.

(b) 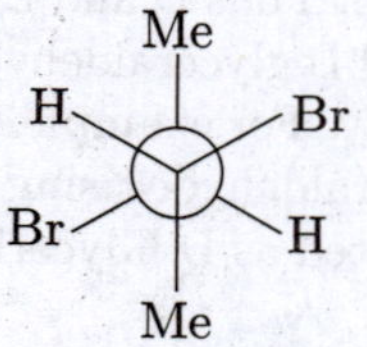

(c) 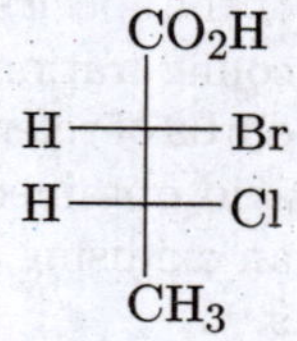

(d) 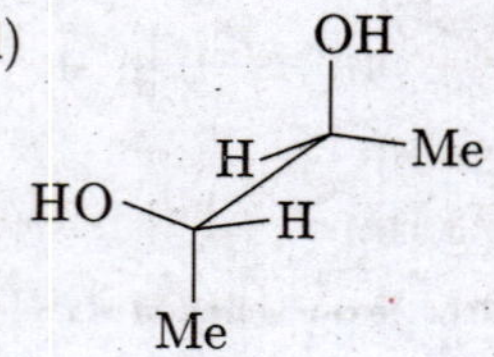

(e) 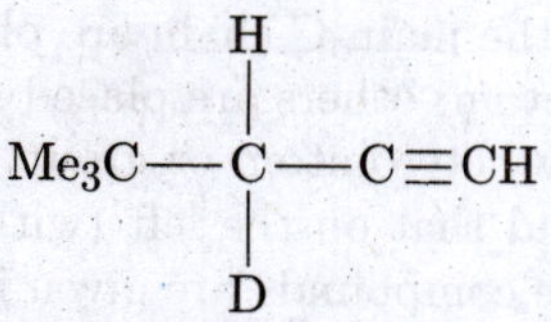

(f) 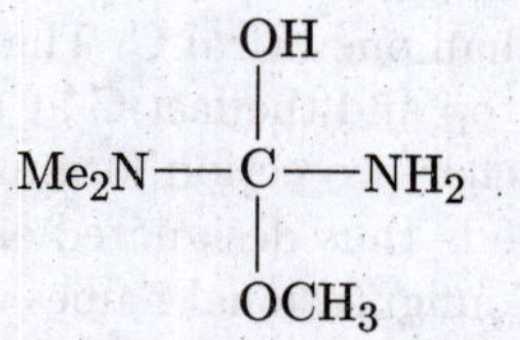

(g) 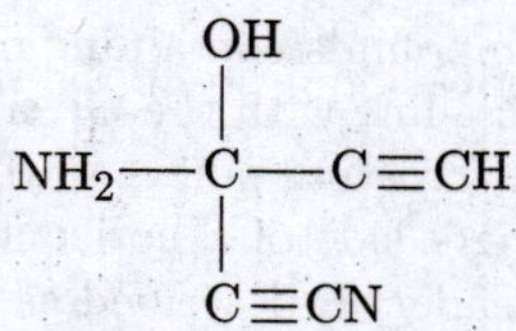

9.10.4 Absolute and relative configurational notations of a chiral C

The descriptor R and S specify the **absolute configuration** meaning actual orientation of groups in space about a chiral atom, because this method of specifying configuration does not require any other species as reference, i.e., their use does not depend on a relationship to any other species. Absolute configurations are usually determined by special X-ray diffraction technique.

Sometimes chemists require the knowledge of configurational relationship between two compounds which R or S designation does not indicate. For this, knowledge of **relative configuration**, i.e., the configurational designation of a chiral atom in relation to another species taken as standard, is essential. Configurations of two enantiomers of glyceraldehyde are taken as standard of reference for this purpose. Arbitrarily form I of glyceraldehyde is designated as D-glyceraldehyde, i.e. its chiral C is of D configuration and form II is specified as L-glyceraldehyde. However D- and L-glyceraldehydes may be specified as R and S respectively as shown in Fig. 9.95,

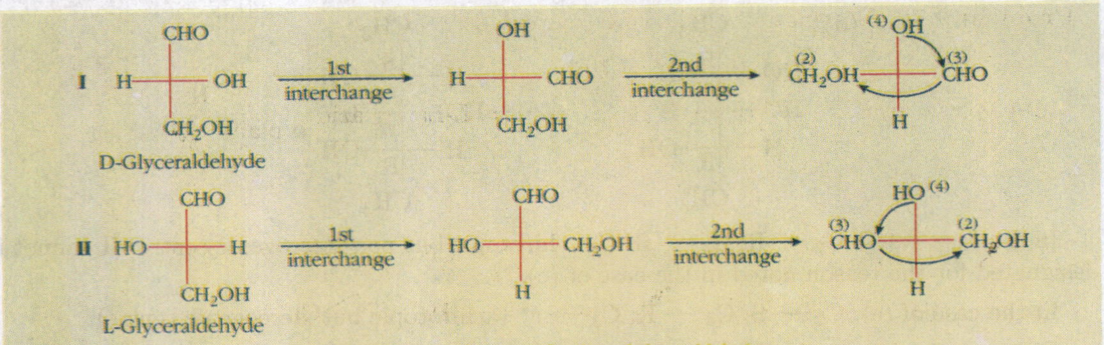

Fig. 9.95. Absolute specification of glyceraldehydes

Thus D-glyceraldehyde is (R)-glyceraldehyde and L-glyceraldehyde is (S)-glyceraldehyde.

If a species with one chiral C is obtained from or converted into D-glyceraldehyde by a reaction or reactions without breaking any bond on the chiral atom, it is said to be of D-configuration; similarly, if L-glyceraldehyde is converted into or obtained from a species without undergoing any bond breaking process on the chiral C, the configuration of the species is specified as L. Thus D and L are relative configurations taking D- and L-glyceraldehydes as standards of reference respectively. For example, the glyceric acid obtained from D-Glyceraldehyde using bromine as an oxidising agent is designated as D-Glyceric acid (Fig. 9.96).

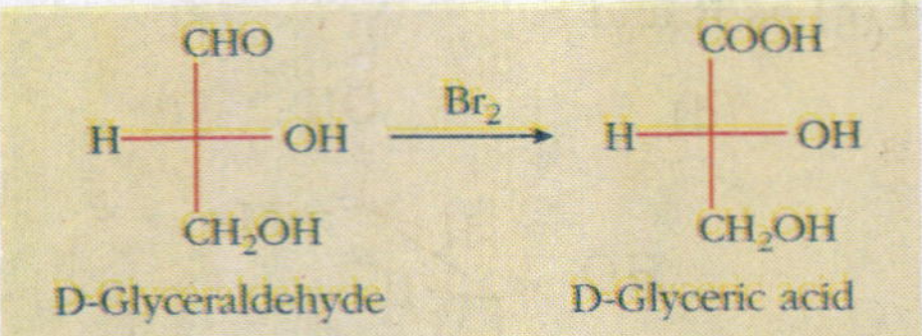

Fig. 9.96. Relative configuration of (−)-glyceric acid

The configurational descriptors, D and L, may be used in the configurational nomenclature of compounds containing more than one chiral C. The Cs of the main C-chain are placed on the vertical line with the C_1, at the top and the last C at the bottom; others are placed sequentially at the crossing points in the Fischer projection formula. Any hetero atom or group that lies on the right side of a horizontal line is thus designated as D- and that on the left (with respect to the reader) is described as L-. Configurational names of some compounds are given in Fig. 9.97.

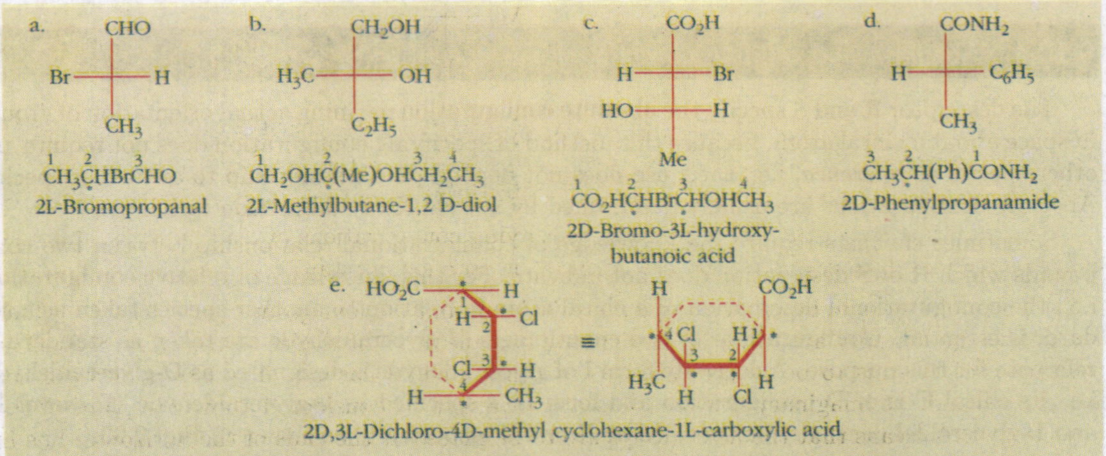

Fig. 9.97. Configurational nomenclature of compounds containing chiral C.s using D- and L- as notations.

D- and L-designations are conventional and sometimes create ambiguity.

Problem 9.12. Designate the following compounds as R or S.

a. D-Glyceric acid b. 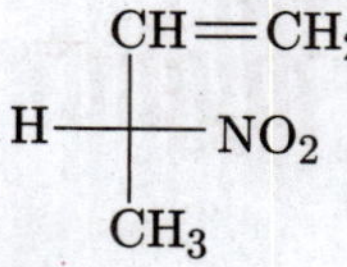 c. L-Lactic acid d. 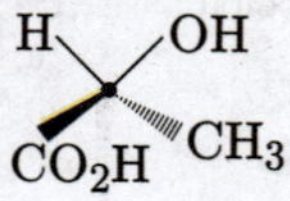

e. 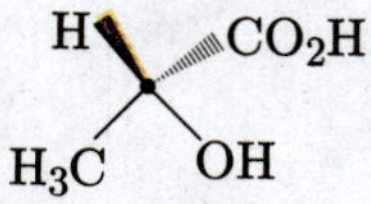 f. 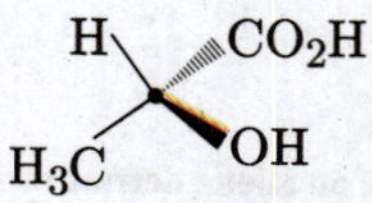 g. 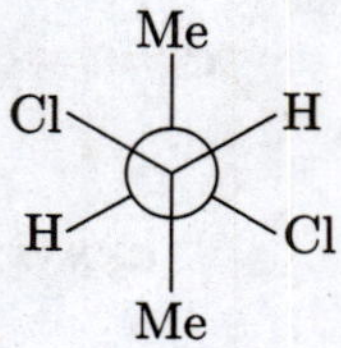

Solution:

a. 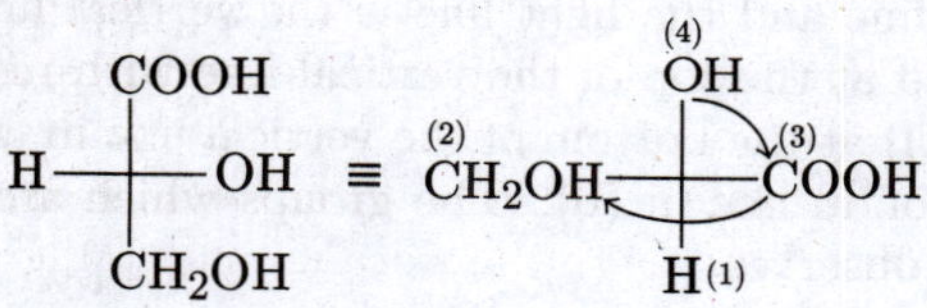

R

D-Glyceric acid

b. 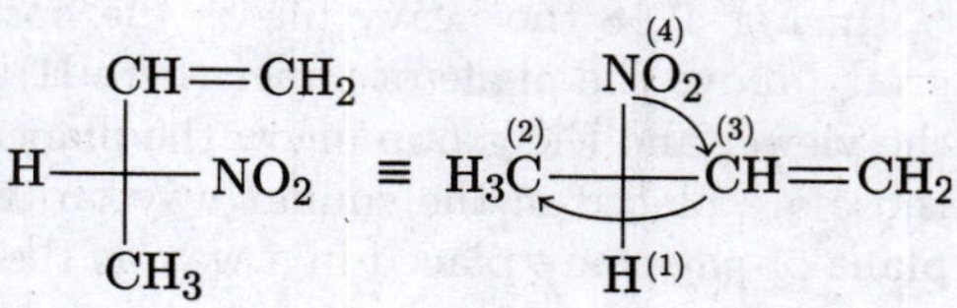

R

Here priority order is
4 > 3 > 2 > 1

c. 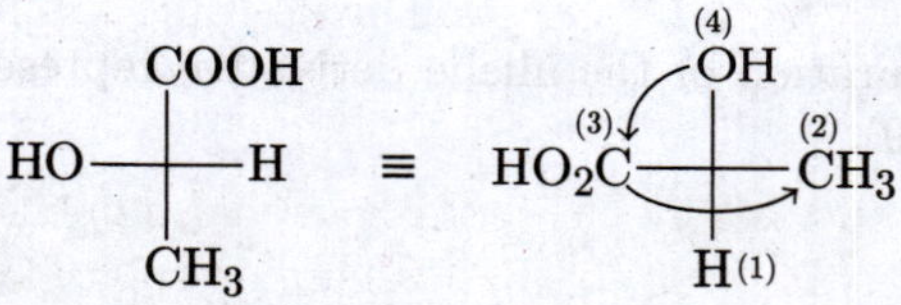

S

L-Lactic acid

d. 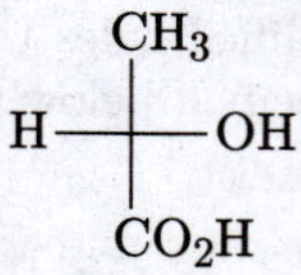

S

e. 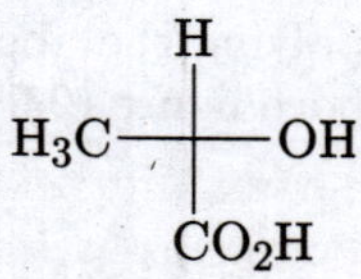

R

f. 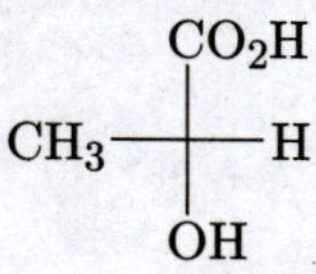

R

g. 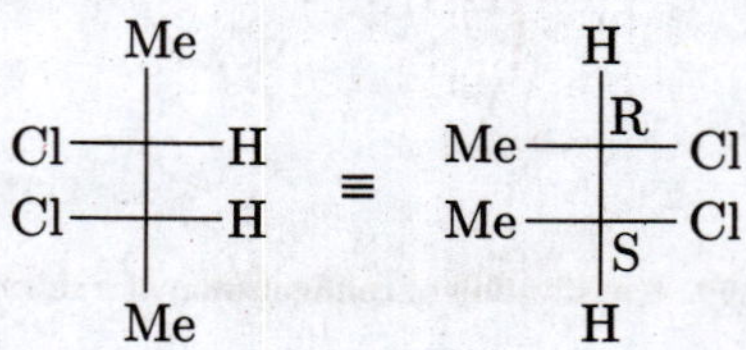

9.10.5 Specification of R, S configurations of chiral allenes, spiranes, biphenyls and alkylidenecycloalkanes

The standard rules in the CIP system for specifying configurations of chiral allenes, spiranes, biphenyls and alkylidenecycloalkanes are given below in a simple way.

(i) The molecule is to be viewed along the chiral axis and the four groups (substituents) which give rise to the chirality are to be projected onto a plane at right angles to the chiral axis. The resultant projection is to be regarded as representing a chiral carbon. The bonds between the near groups and the imaginary chiral carbon are always to be considered above the plane of the projection (this means that the near groups are to be placed at the ends of the horizontal line in projection as shown below in Fig. 9.98.

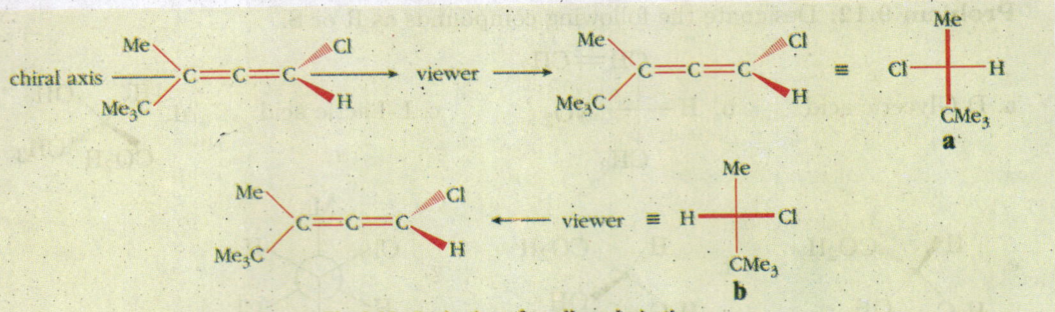

Fig. 9.98. Projection of an allene derivative

Note

In Fig. 9.98 the heavy line is the horizontal line and the light line is the vertical line. The group above the plane of paper (here H) is placed at the top of the vertical line with respect to the viewer and the group below the plane (here Cl) at the bottom of the vertical line in (a) while those are placed in the similar way on the horizontal line in (b). The groups which are on the plane of paper are placed in a way as the viewer observes.

(ii) The near groups with respect to the viewer are to be given priority over the far groups.

(iii) The order of priority between near groups and that between far groups are to be determined by the priority sequence rules as usual.

Applying the above simplified rules the configuration of the allene derivative represented in the above figure (9.98) is specified below (Fig. 9.99).

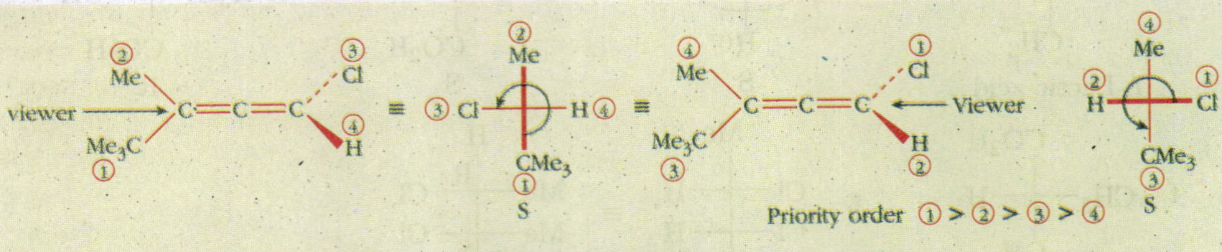

Fig. 9.99. Specification of configuration of a chiral allene

The procedure followed to specify the configuration of a chiral allene is also followed to specify the configurations of chiral spirans, biphenyls and alkylidenecycloalkanes as shown in Fig. 9.100.

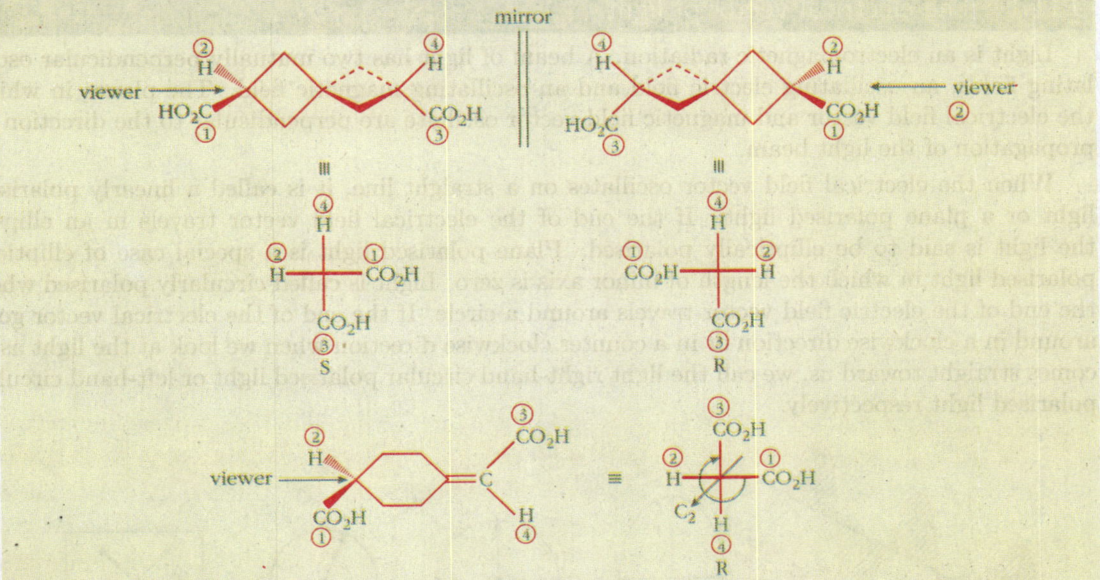

Fig. 9.100

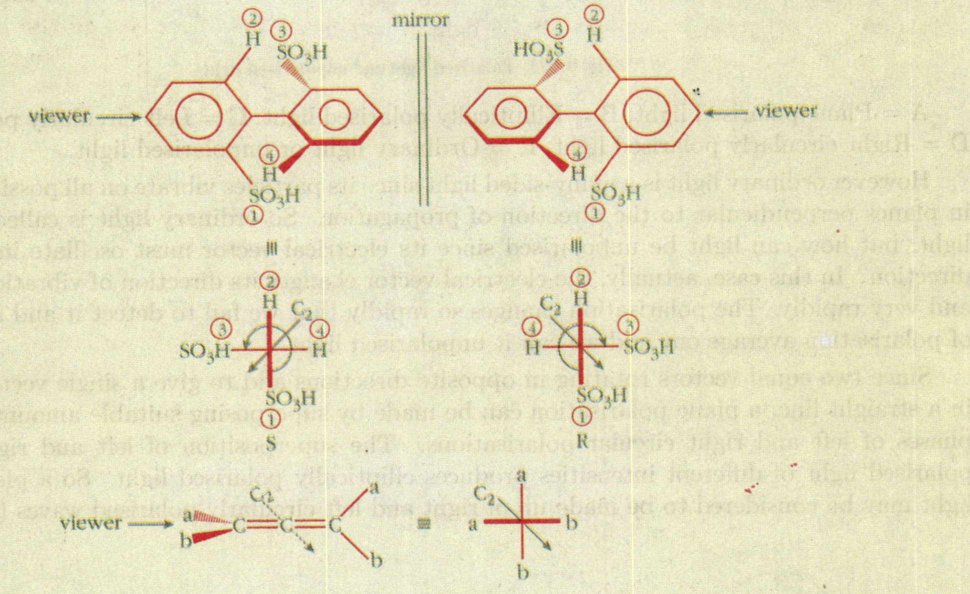

Fig. 9.101

9.11 Optical Activity

To understand the meaning of optical activity we shall have to discuss the electrical vector of light in brief,

9.11.1 The electrical vector of light

Light is an electromagnetic radiation. A beam of light has two mutually perpendicular oscillating fields, an oscillating electric field and an oscillating magnetic field. The planes in which the electrical field vector and magnetic field vector oscillate are perpendicular to the direction of propagation of the light beam.

When the electrical field vector oscillates on a straight line, it is called a linearly polarised light or a plane polarised light. If the end of the electrical field vector travels in an ellipse, the light is said to be elliptically polarised. Plane polarised light is a special case of elliptical polarised light in which the length of minor axis is zero. Light is called circularly polarised when the end of the electric field vector travels around a circle. If the end of the electrical vector goes around in a clockwise direction or in a counter clockwise direction when we look at the light as it comes straight toward us, we call the light right-hand circular polarised light or left-hand circular polarised light respectively.

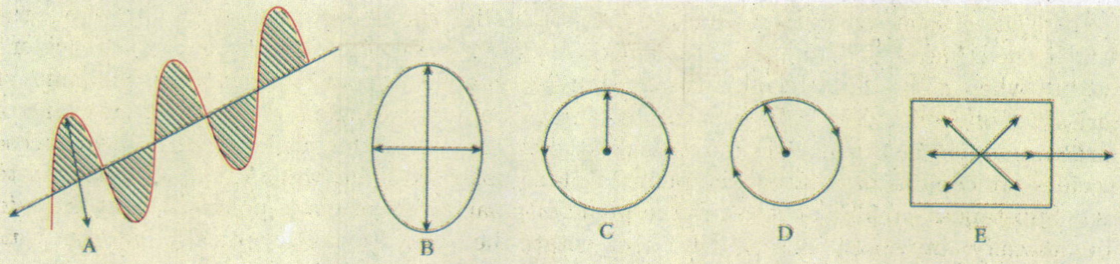

Fig. 9.102. Polarised light and unpolarised light

A = Plane polarised light, B = Elliptically polarised light, C = Left circularly polarised light, D = Right circularly polarised light, E = Ordinary light or unpolarised light.

However ordinary light is a many-sided light since its particles vibrate on all possible directions in planes perpendicular to the direction of propagation. So ordinary light is called unpolarised light, but how can light be unpolarised since its electrical vector must oscillate in a particular direction? In this case, actually, the electrical vector changes its direction of vibration constantly and very rapidly. The polarisation changes so rapidly that we fail to detect it and all the effects of polarisation average out and we call it unpolarised light.

Since two equal vectors rotating in opposite directions add to give a single vector oscillating in a straight line, a plane polarisation can be made by superposing suitable amounts at suitable phases of left and right circular polarisations. The superposition of left and right circularly polarised light of different intensities produces elliptically polarised light. So a plane polarised light may be considered to be made up of right and left circularly polarised waves (Fig. 9.103).

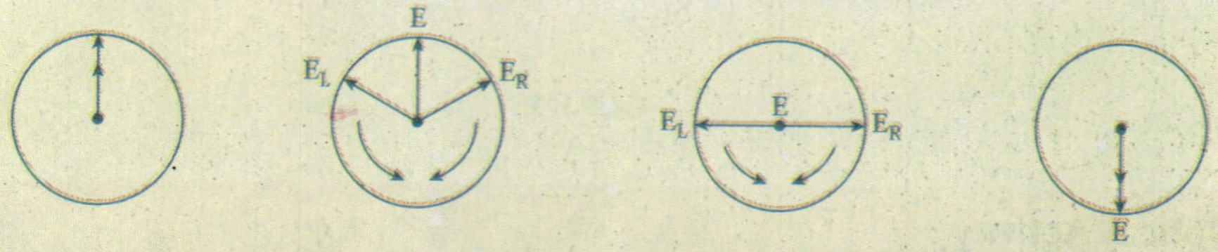

Fig. 9.103. Combination of electric field vectors of right circularly and left circularly polarised lights to give electrical field vector of plane polarised light. E_L = Electrical vector of left circularly polarised light, E_R = Electrical vector of right circularly polarised light, E = Electrical vector of plane polarised light i.e., plane of polarisation.

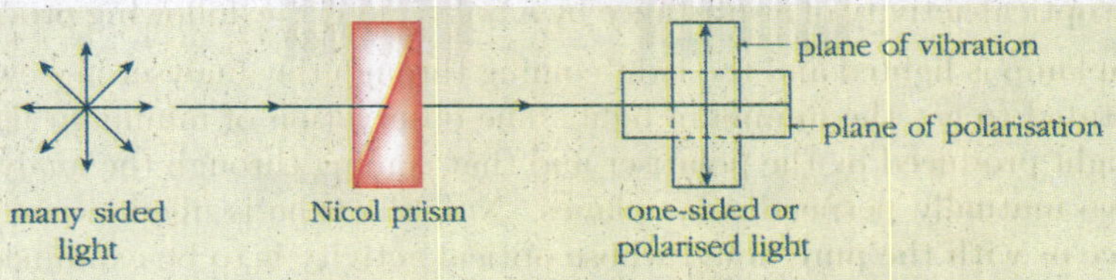

Fig. 9.104. Plane of vibration and plane of polarisation of a plane polarised light

9.11.2 Meaning of optical activity and its measurement

Unpolarised light can be converted to a plane polarised light by passing through a specially constructed prism, called Nicol prism. The plane through the ray and perpendicular to the direction of vibration of the plane polarised light is called the plane of polarisation of the plane polarised light (Fig. 9.104).

When plane polarised light produced from sodium light is passed through a solution of urea in water, the plane of polarisation of the emergent light remains the same as that of the incident light; but when such a light beam is passed through an aqueous solution of glucose, the plane of polarisation of the incident light gets rotated through a certain angle and therefore the plane of polarisation of the emergent light becomes different from that of the incident light. Thus glucose molecules are capable of rotating the plane of plane polarised light, while urea molecules fail to do so. Substances capable of rotating the plane of polarisation of plane polarised light are said to be optically active and those which fail to rotate the same are called optically inactive; thus urea is an optically inactive substance but glucose is an optically active substance.

An optically active substance may rotate the plane of polarisation of a plane polarised light either to the left (anticlockwise) or to the right (clockwise). The substance which rotates the plane of polarisation of a plane polarised light to the left is said to be **levorotatory** (Latin word laevus meaning left) and that which rotates the same to the right is called **dextrorotatory** (Latin word dexter meaning right) as the observer looks the light along the path of the beam toward the source.

The optical activity, i.e., the power of rotating the plane of plane polarised light, of a substance is detected and estimated by an instrument known as polarimeter. It consists of a sodium lamp, two Nicol lenses, and, in between the lenses, a tube for holding the substance whose optical activity is to be examined. The sodium lamp is used as a light source; the Nicol lens in front of the lamp, which produces the polarised light, is called polariser, and the lens which is placed after the tube serves as an analyser through which light ray reaches the eye of the observer. The movable analyser bears a circular scale as shown in figure 9.105.

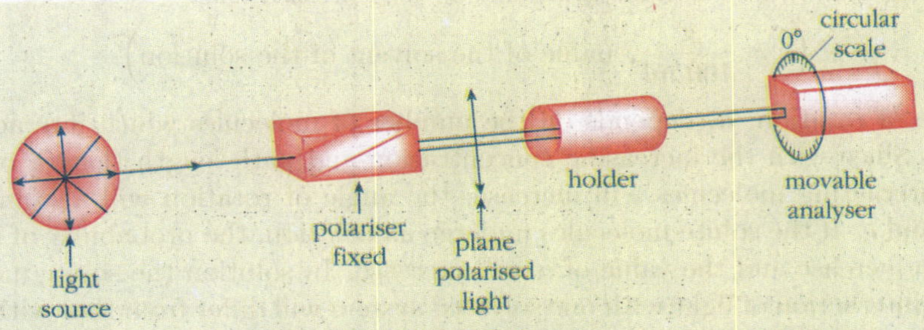

Fig. 9.105. Polarimeter

To measure optical activity of a substance by a polarimeter the following procedure is adopted :

The sodium lamp is lighted and the light coming through the analyser is viewed. The analyser lens is then rotated to get the minimum light. The observation of minimum light indicates that the polarised light produced by the polariser and that coming through the analyser to the eye are vibrating in two mutually perpendicular planes. Now the tube is filled either with the solution of the substance or with the pure liquid whose optical activity is to be examined. If the observer now finds the light of minimum intensity coming through the analyser, the compound will be optically inactive; on the other hand if the light coming to the eye is not of minimum intensity, the compound will be optically active. This means that the plane of polarisation of the incident polarised light and that of emergent light which comes to the eye through the analyser are not now vibrating in two mutually perpendicular planes, i.e., the substance taken in the tube has rotated the plane of polarisation through a certain angle. Now to find light of minimum intensity, the observer requires rotating the analyser lens; to get this if it is rotated to the left, the compound is levorotatory; if to the right, the compound is dextrototatory. This is because the direction of rotation of the lens indicates the direction of rotation of the plane of polarisation of the polarised light by the substance. The dextrotation is designated as (+) - or *d* and the levorotation is indicated by (−) - or *l*.

Polarimeter not only indicates the direction of rotation of the plane of polarised light but also gives a measure of its rotation and its unit is degree (°). Optical activity of a substance depends on (i) the wave length of the polarised light, (ii) the thickness of the sample to be examined, (iii) the solvent used, and (iv) the density or concentration of the substance; it is reported either as specific rotation or as molecular rotation. These two terms may be defined as follows :

Specific rotation is the number of degree of rotation of the plane of polarisation of a plane polarised light of specific wave length caused by a layer of 1 dm thickness of pure liquid substance or of its solution in a particular solvent containing 1 g of the substance per ml at a definite temperature in °C.

The specific rotation of a pure liquid is represented as :

$$[\alpha]_D^T = \frac{\alpha}{l \times d} \text{ (neat, } d \text{ in g/l)}$$

where α = observed rotation in degree with sign

T = temperature in °C

D = D-line of sodium light, wave length 589.6 nm.

l = length of the tube in decimeter

d = density of the liquid.

The specific rotation of a substance in a solution is expressed as

$$[\alpha]_D^T = \frac{\alpha}{l \times c}\left(c = \frac{\text{g}}{100 \text{ ml}}, \text{ name of the solvent of the solution}\right)$$

The angle of rotation, α, depends on the number of molecules which interacts with the polarised light. Since with the increasing concentration and path length (length of the tube), the number of interacting molecules will increase; the angle of rotation will also increase with the increasing l and c. If the solute molecules undergo association, the probability of interaction with light will also increase and the value of α will increase. In solution the solute molecules will get solvated, the interaction of light with one solvated system will differ from that with another. Thus the angle of rotation depends also on the nature of the solvent. Since the motion of molecules changes with temperature, the interaction of the plane polarised light with the molecules will also change and as a consequence the value of α will also change with temperature. The dependence of α on wave length of light will be discussed later on.

Here c is the concentration of the solution.

Problem 9.13. What do you mean by the following two expressions?

(a) $[\alpha]_D^{20} = -76.3° \left(c = \dfrac{2g}{100\ \text{ml}}, \text{ ethanol}\right)$

(b) $[\alpha]_D^{20} = +40°$ (neat, $l = 1$ dm.)

Solution: (a) This means that 1 dm. layer of an ethanolic solution of an optically active substance of concentration 0.02 g/ml at 20 °C rotates the plane of polarised sodium light (D) through an angle 76.3° to the left; the substance is thus levorotatory.

(b) This means that 1 dm layer of a pure liquid substance at 20 °C rotates the plane of polarised sodium light (D) through an angle 40° to the right, i.e., the substance is dextrorotatory.

Molecular rotation is the number of degrees of rotation of the plane of polarisation of a polarised light of specific wave length caused by a pure liquid of 1 dm thickness or by a solution of the substance of 1 dm layer in a given solvent containing 1 centi mole of the substance per ml at a given temp. and is expressed as:

$$[M]_D^T \text{ or } [\phi]_D^T = \frac{\alpha}{1(\text{dm}) \times c(\text{moles}/100\text{ml})} = \frac{[\alpha]_D^T \times M}{100}$$

$$\text{Thus Molecular rotation} = \frac{\text{Specific rotation} \times \text{mol. wt.}}{100}$$

Just to avoid excessive high value of molecular rotation 1 centi mole/ml of concentration is taken instead of mole/ml.

Specific rotation and molecular rotation of an optically active substance are in principle constant quantities and do not depend on l and c. However the observed value of α is actually indistinguishable from the values given by the equation $\alpha \pm 180n°$; where $n = 0, 1, 2, \ldots$ etc.

For example, let α be +40°; then we cannot say from a single experiment that it is +40° and not any one of the followings:

(i) $+40° + 180° = 220°$; (ii) $+40° - 180° = -140°$; (iii) $+40° - 2 \times 180° = -320°$ etc. We know that α is inversely proportional to the length of the tube and concentration of the solution. Now if we either change the concentration of the solution or the path length, we shall get another set of values for α which will help us determine the exact value with sign of the specific or molecular rotation. For example if we change c to $c/5$ or l to $l/5$, then the values for α will be:

$$+40°/5 = +8°, +220°/5 = +44°, -140°/5 = -28° \text{ [which appears to be } (180° - 28°) = +152°$$

and $\quad -320°/5 = -64°$ [which appears to be $(180° - 64°) = +116°$]

The value of specific rotation of the experimental substance will be the product of 5 and the value obtained in the second experiment.

Problem 9.14. The specific rotation of a compound of molecular weight 70 is +30°; what is its molecular rotation?

Solution: As molecular rotation of a compound is its specific rotation multiplied by the hundredth part of the molecular weight, the molecular rotation of the given compound $= + \left[\frac{30\times70}{100}\right]^{\circ}$ or $+ 21^{\circ}$.

9.11.3 The cause of optical activity

Dissymmetric compounds show optical activity but symmetric compounds do not. To account for the difference in property between dissymmetric and symmetric compounds we require to discuss physical theories which is beyond the scope of the book. However, an explanation may be put forward as given in the next page.

The very nature of the plane polarised light tells us that it can be resolved into two components, right circularly and left circularly polarised lights. The circularly polarised lights have three significant properties:

(i) They are dissymmetric in nature. The left and right circularly polarised lights trace out dissymmetric helical paths which are non-superimposable mirror-images of each other.

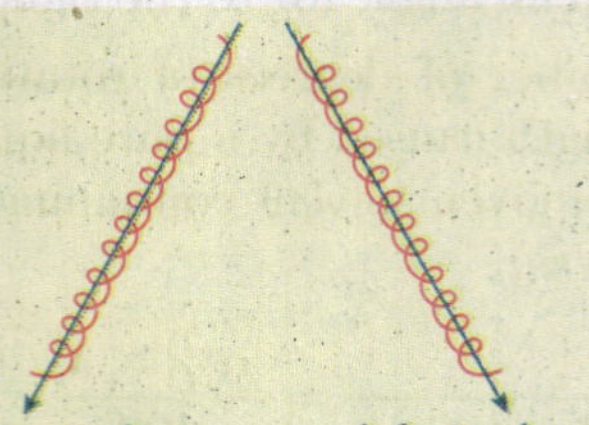

Fig. 9.106. Helical paths traced by circularly polarised lights

(ii) When these circularly polarised lights interact with dissymmetric entities, two diastereomeric (non-mirror-image) interactions develop.

(iii) Circularly polarised lights are anisotropic in nature; this means they have different properties in different directions.

Dissymmetric molecules have special features too:

(i) The molecule have dissymmetric polarisabilities.

(ii) They are circularly birefringent; this means they have different refractive indices for the left and right circularly polarised lights.

(iii) The two circularly lights have different molar absorptivity (Sec. 10.5.2.)

The refractive index of a material is related to its polarisability. Therefore circular birefringence is also related to the dissymmetry of polarisability of the dissymmetric entities. Polarisability is a consequence of electron oscillation caused by an external field. In a molecule, an electron does not freely oscillate equally in all directions; the polarisability of an electron is different in different directions i.e., anisotropic in nature. When a plane polarised light strikes a dissymmetric molecule, the electrons in the molecule oscillate out of plane of the polarisation because of the anisotropic polarisability. The direction of the electrical vector of the plane polarised light changes because of its interaction with the oscillating electrons and the plane of the plane polarised light rotates slightly. In a large collection of dissymmetric molecules which remain in solution or in the liquid state, very many similar interactions occur and a measurable amount of optical rotation takes place. If there were equal numbers of mirror-image molecules (not a single mirorr-image molecule is present in a collection of dissymmetric molecules), they would have been rotated the plane in the opposite direction and as a consequence, the optical rotation would have been zero (which is indeed the case with recemic modification).

The explanation can also be given in another way:

The diastereomeric interactions of the two dissymmetric circularly polarised lights with dissymmetric compounds are reflected in the difference in the refractive indices (n_L and n_R) as well as in the molar absorptivity (ϵ_L and ϵ_R) of the left and right circularly polarised lights respectively. So, under the condition,

$$n_L \neq n_R \quad \text{and} \quad \epsilon_L \neq \epsilon_R$$

Again, we know that the velocity of light, v, in a medium, is related to the refractive index of the medium by the equation $v = \frac{c}{n}$

Therefore, the velocity of the left circularly polarised light in the dissymmetric medium is not equal to that of the right circularly polarised light since,

$$v_L = \frac{c}{n_L} \quad \text{and} \quad v_R = \frac{c}{n_R} \quad \therefore \quad v_L \neq v_R$$

Now if $n_L < n_R$, $v_L > v_R$ and in this case the right circularly polarised light slips behind the left in phase and naturally, the direction of linear polarisation changes as shown below:

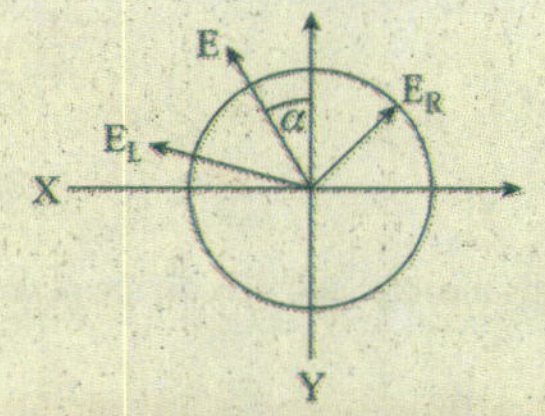

Fig. 9.107. Rotation of the plane of plane polarised light; projection of the plane is a line

As a result we shall get a direction of the electrical vector of the plane polarised light which will be something different form the Y plane and it will make an angle α with the resultant vectorial plane (E). The angle is negative when $v_L > v_R$ and it is positive when $v_L < v_R$. In the former case the compound under experiment is levorotatory while dextrorotatory in the latter case. The angle α in degrees per centimeter can be calculated by the equation.

$$\alpha = \frac{180}{\lambda}(n_L - n_R);$$

where symbols have usual meanings.

Since molar absorptivity of the left and right circularly polarised lights are different when they interact with dissymmetric entities, the two vectors E_L and E_R are actually not equal in length and the resultant vector, E, no longer oscillates in a plane but traces out a flattened helix, which is projected as an ellipse. The small difference in ϵ_L and ϵ_R values is expressed by the molecular elipticity,

$$[\theta]_\lambda^T = 3300(\epsilon_L - \epsilon_R)$$

The plane polarised light, therefore, becomes elliptically polarised owing to the difference in molar absorptivity of dissymmetric entities towards the left and right circularly polarised lights and the phenomenon is known as circular dichroism. Thus circular dichroism is intimately connected with the rotation of the plane of plane polarised light.

$[\theta]$ and $[\phi]$ depend similarly on temperature, solvent, solute and wave length of light. By convention, if $\epsilon_L > \epsilon_R$, $[\theta]$ is positive; if it is otherwise, $[\theta]$ is negative.

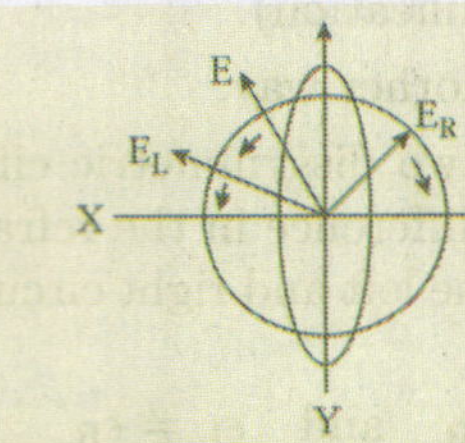

Fig. 9.108. The change of the plane polarised light to an elliptically polarised light : circular dichroism

However, in the cases of achiral molecules, interaction between the electrical vector of light and oscillating electrons also takes place and the plane of the plane polarised light changes its direction; but in a bulk for any orientation of an achiral molecule one can expect, and actually it happens, another mirror-image molecule i.e., of opposite orientation. The effects of the two molecules on the plane of polarisation will be equal in magnitude but opposite in directions and, therefore, the plane of the plane polarised light will remain unchanged when the light comes out of the material medium.

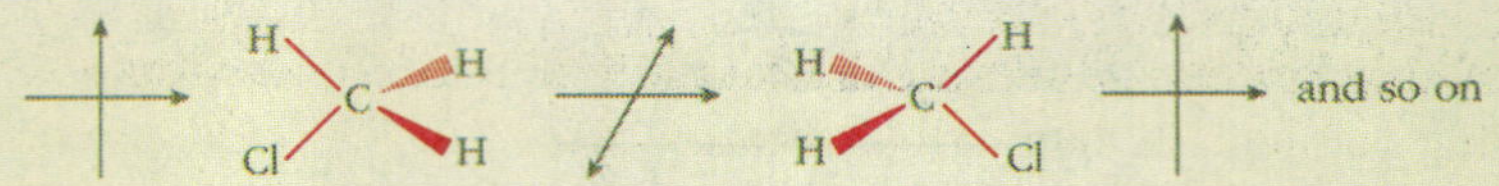

Fig. 9.109. How an achiral compound acts on the plane polarised light

9.12 Chiral Molecules and Isomer Numbers

Number of optical isomers that can be determined theoretically from the three-dimensional structure of a chiral molecule is very important in the study of stereochemistry; so this is discussed under different heads as follows :

9.12.1 Chiral molecules with one chiral C

The simplest type of chiral organic molecules contains one chiral C atom, i.e., an sp^3-hybridised C atom having four different groups or atoms arranged tetrahedrally. A molecule with one chiral C atom has two opposite configurations D and L or R and S. Both being chiral, they are optically active. If one is dextrorotatory, the other is levorotatory and hence they are optical antipodes. These two forms have non-superimposable mirror-image relation between them and are thus called a pair of enantiomers (Fig. 9.110).

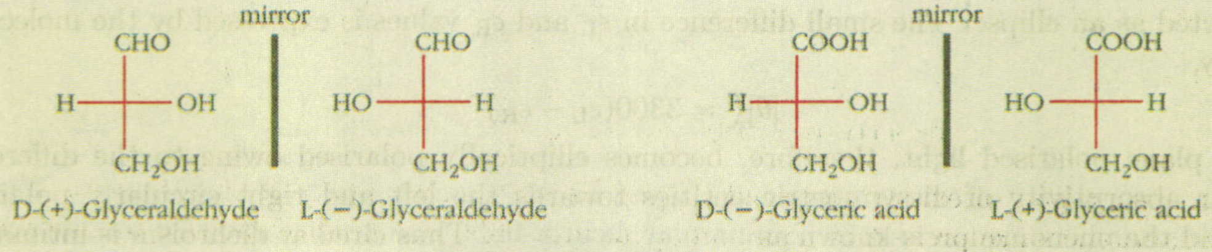

Fig. 9.110. Pairs of enantiomers

From the above examples it is clear that the sign of rotation is not at all related to the notation of configuration of a compound; D and L designations are conventions, while + and − are the signs that indicate the very property of a compound by virtue of which it rotates the plane of a plane polarised light in a specific direction and can be determined only by experiment in a polarimeter. Of course, one can relate the configuration to the sign of rotation in the case of enantiomers : enantiomers have opposite configurations and opposite signs of rotation. Thus D-glyceraldehyde has (+) sign of rotation whereas L-glyceraldehyde has (−) sign of rotation.

When equimolecular amounts of two enantiomers are present in a bulk, it becomes optically inactive; this is because the dextro rotation made be the (+) - isomer of the bulk is compensated by the levorotation of the (−) -isomer as their rotatory powers are equal in magnitude but opposite in directions. Optically inactive substances that are composed of equimolecular amounts of two enantiomers are said to be externally compensated modifications or racemic modifications and designated as (±) - or *dl*-pair.

Thus a compound with one chiral C has three optical isomers: (+) -isomer, (−) -isomer and racemic modification in the solid state, the former two are optically active, while the latter is optically inactive.

Lactic acid molecule contains one chiral C; hence it has three optical isomers : (+)-lactic acid (m.p. 26°C), (−)-lactic acid (m.p. 26°C) and (±) lactic acid (m.p. 18°C).

9.12.2 Compounds containing two or more unlike chiral carbon atoms

Naturally occurring compounds like carbohydrates, terpenes, alkaloids, steroids, peptides, etc. contain more than one chiral C atom. To understand their stereochemistry, one must have a knowledge about chiral molecules with two or more chiral C atoms.

When there are two unlike chiral C atoms, meaning different atoms and groups attached to both in a molecule, one should find two different configurations about each of them; obviously $2 \times 2 = 4$ or 2^2 different configurations of the molecule will be possible. Similarly a molecule containing three unlike chiral C atoms will exhibit $2 \times 2 \times 2$ or $8(2^3)$ different configurations. In this way, we can say that a molecule having n different chiral C atoms must have 2^n different configurations. Each of these configurations will have a mirror-image configuration and thus 2^{n-1} pairs of enantiomers will be found for such a formula having 2^n configurational isomers. Thus, four different configurations can be drawn out of the formula $CH_3CHOH.CHOH.COOH$ as given below (Fig. 9.111).

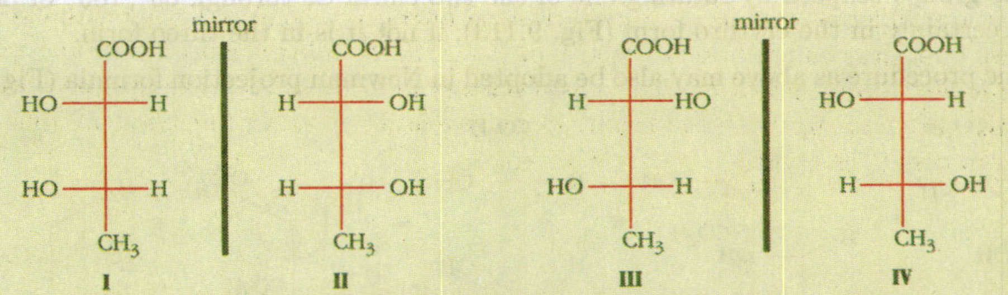

Fig. 9.111. Four configurations of $CH_3 \cdot (CHOH)_2 \cdot COOH$

I and II form a pair of enantiomers and III and IV from another pair of enantiomers. When a pair of enantiomers are present in equal amounts in a bulk, one kind of racemic modification will result in the solid state. Thus a compound composed of two unlike chiral C atoms must have six optical isomers : two pairs of enantiomers and two racemic modifications, each consisting of one pair of enantiomers. Therefore, a molecular formula consisting of n different chiral C atoms must show 2^n configurational isomers and 2^{n-1} racemic forms; hence, the total number of optical isomer will be $2^n + 2^{n-1}$.

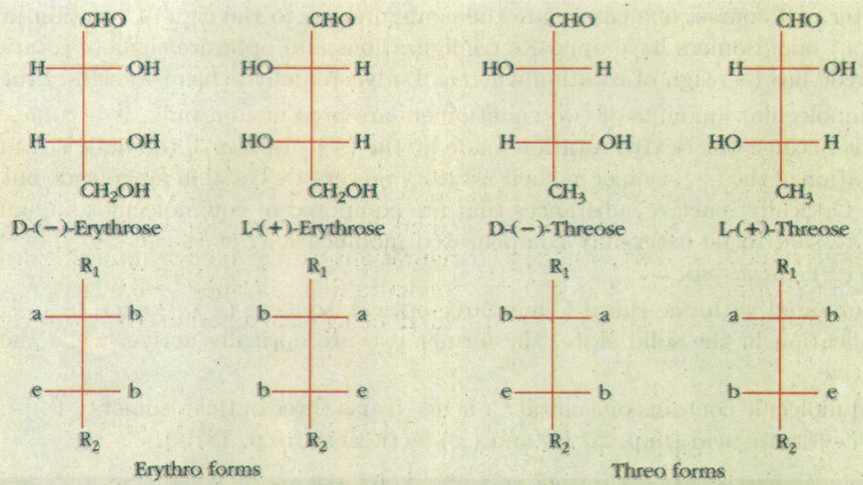

Fig. 9.112. Erythro and threo forms

Note : Molecules with two chiral Cs are frequently handled in stereochemistry, and for them a separate nomenclature system with separate forms of notations has been developed. The system of nomenclature is derived from the names of aldotetroses and applied to R_1-Cab-Cbe-R_2 type of molecules. When two like groups, b, in a Fisher projection formula are drawn on the same side of the vertical line, as there are hydroxyl groups in erythrose, the isomer is called 'erythro' form; if these are placed on the opposite sides as the OH groups in threose, the isomer is said to be 'threo' form (Fig. 9.112).

As for projections, Fischer projection formula is inadequate because in this system only eclipsed forms are represented, but reactions take place mostly when molecules are present in staggered forms. So we are much interested in staggered forms, and sawhorse, Newman and flying wedge projection formulae may be used for the same. In a sawhorse projection formula if one gets like groups eclipsed by rotating one of the two chiral Cs through 60°, 180° or 300°, the molecule is certainly in the erythro form (Fig. 9.113); if not it is in the threo form.

The same procedure as above may also be adopted in Newman projection formula (Fig. 9.114).

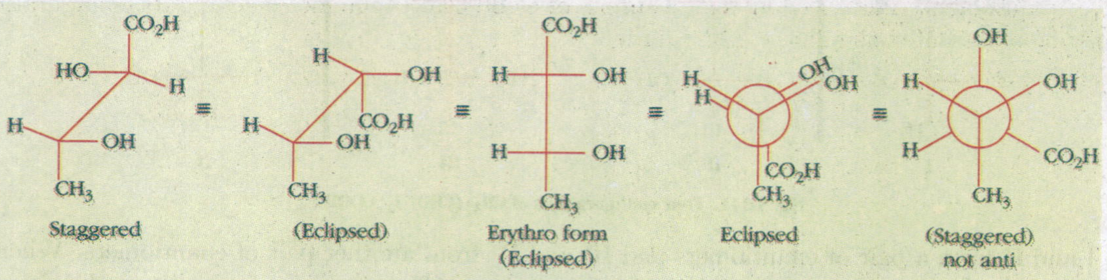

Fig. 9.113. Newman and sawhorse projections of erythro forms

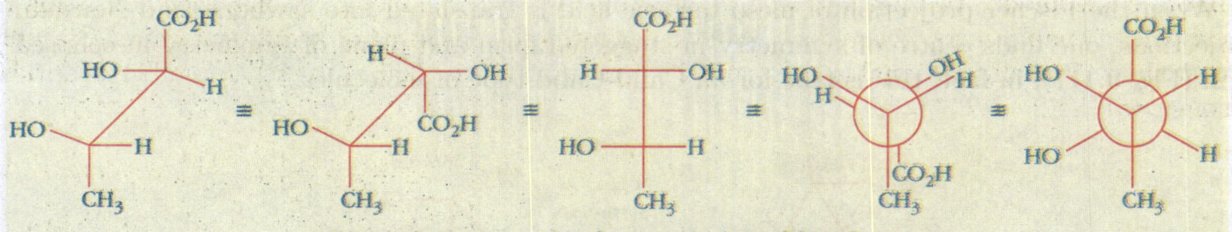

Fig. 9.114. Newman and sawhorse projections of threo forms

9.12.3 Compounds with two like chiral C atoms

In the case of compounds of the type Cabd-Cabd, in which two chiral Cs are alike, the number of optical isomers and that of its different configurations cannot be determined by the formulae $(2^n + 2^{n-1})$ and 2^n respectively. The different configurations of Cabd·Cabd when projected, form the following configurations (Fig. 9.115) : I and II being non-superimposable mirror-images of

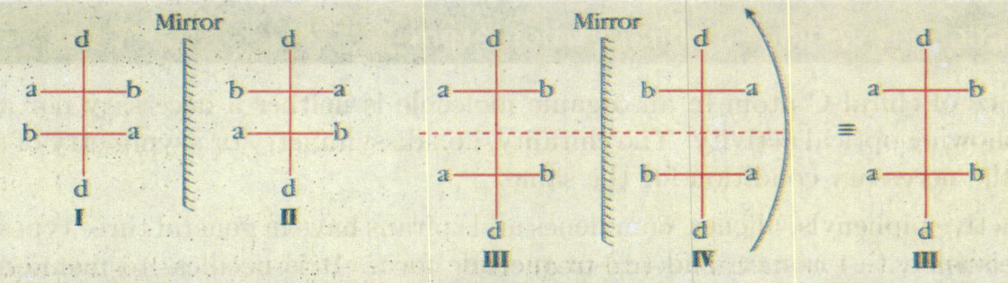

Fig. 9.115. Differnt configurations of Cabd-Cabd type of molecules. 180° rotation on the plane of the paper

each other, they are optically active and represent a pair of enantiomers. III & IV have a plane of symmetry and hence they are optically inactive by internal compensation; this is because the dextrorotation of the plane of plane polarised light made by one chiral C is compensated by the levorotation made by the other chiral C. Thus if an optical isomer containing more than one like chiral Cs is found optically inactive owing to the presence of symmetry elements (plane or centre of symmetry), the molecule is said to be meso form or internally compensated form. In fact, achiral configurational isomers of molecules having chiral Cs are called meso forms. However III and IV are super-imposable on each other; if we rotate III or IV 180° on the plane of paper (which is permissible because thereby, no configurational change takes place), we shall get III converted to IV and vice versa. Thus III and IV are identical. III and IV being identical, the actual number of different configurations is three; the number of optical isomers are four: a pair of enantiomers, the meso form and the racemic modification containing equimolecular amounts of the two enantiomers. Tartaric acid is an example of Cabd·Cabd type molecule and its configuration may be represented as follows (Fig. 9.116).

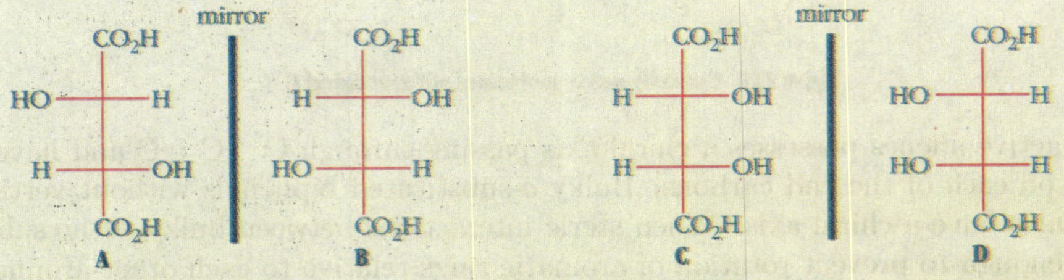

Fig. 9.116. Configurational isomers of tartaric acid; C = D

When the Fischer projection of meso tartaric acid is translated into sawhorse and Newman projections, one finds centre of symmetry in staggered form and plane of symmetry in eclipsed form (Fig. 9.117); in fact, this is true for all Cabd-Cabd type of molecules.

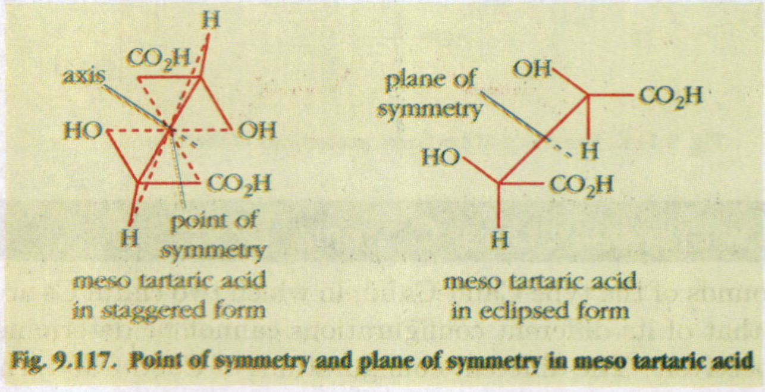

Fig. 9.117. Point of symmetry and plane of symmetry in meso tartaric acid

9.12.4 Compounds without a chiral carbon atom

The presence of chiral C atom in an organic molecule is neither a necessary nor a sufficient condition for showing optical activity. The chirality, i.e., dissymmetry or asymmetry of a molecule as a whole is the necessary condition for the same.

Optically active biphenyls, allenes, cumulenes and spirans have in general three types of optical isomers : (+)-isomer, (−)-isomer, and (±)-or racemic form. It is needless to mention that (+) -and (−)- isomers are nothing but a pair of enantiomers and thus they have non-superimposable mirror-image relations between them (Fig. 9.118).

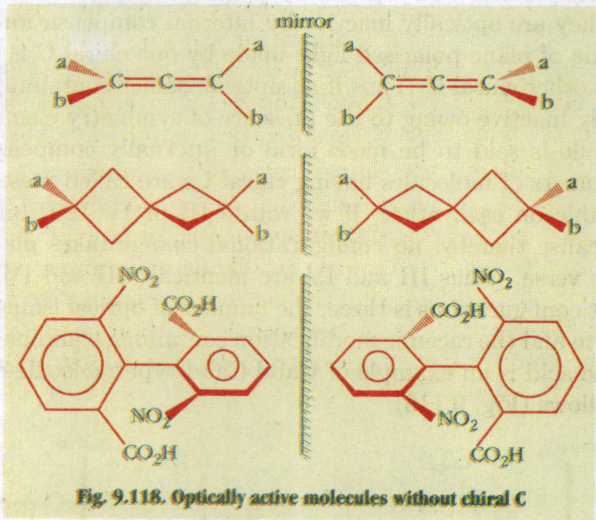

Fig. 9.118. Optically active molecules without chiral C

Optically active allenes possesses a chiral axis passing through C=C=C and have two different groups on each of the end carbons. Bulky *o*-substituted biphenyls without vertical plane of symmetry also have a chiral axis. When steric interactions between bulky ortho substituents are sufficient enough to prevent rotation of aromatic rings relative to each other around C—C bond, two preferred conformations are obtained in which the two rings are orthogonal and they

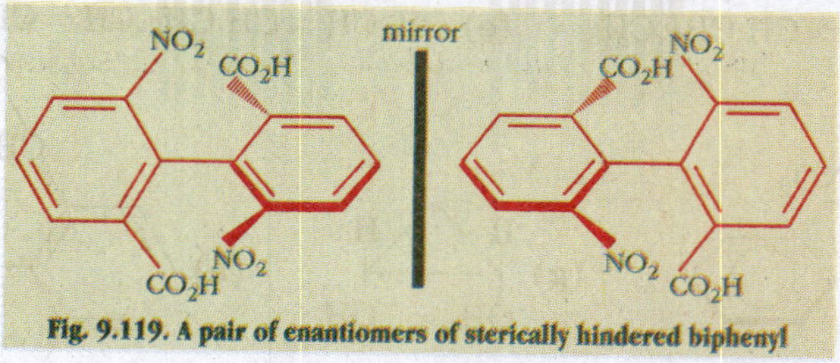

Fig. 9.119. A pair of enantiomers of sterically hindered biphenyl

constitute a pair of enantiomers and can be separated by energy barrier. Thus conformational isomers become configurational isomers and this kind of isomerism is known as **atropisomerism** and the enantiomers are called atropisomers. The racemic modification of atropisomers may be resolved at room temperature when the energy barrier is of the order 80 kJ/mole or more.

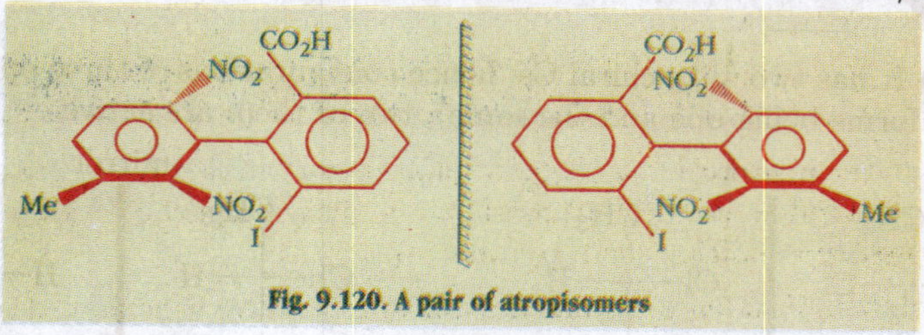

Fig. 9.120. A pair of atropisomers

9.12.5 Necessary and sufficient conditions for showing optical activity

The necessary condition for showing optical activity is the chirality or dissymmetry of a molecule as a whole, but the sufficient conditions for showing optical activity are : **(i)** The molecule should have measurable amount of optical activity; ethyl-*n*-butyl-*n*-propyl-*n*-hexylmethane does not show optical activity though the molecule is dissymmetric—it is called de facto Optically inactive. It has immeasurably small optical activity at all accessible wave lengths.

(ii) The medium must maintain the dissymmetry of the compound during the period of measurement of specific rotation. (dextro)-menthyl (levo)-menthyl-2, 6, 2′, 6′-tetranitro-4, 4′-diphenate (X) does not have an S_n axis, still it shows optical inactivity. The lack of activity is attributed to the fact that its mirror-image can readily be made superimposable on the object molecule by rotation around the bonds linking the carboxylate groups to the rings. Perhaps this happens during the time of measurement of the specific rotation of the compound and thus a dissymmetric molecule becomes symmetric and thereby fails to exhibit optical activity.

X =

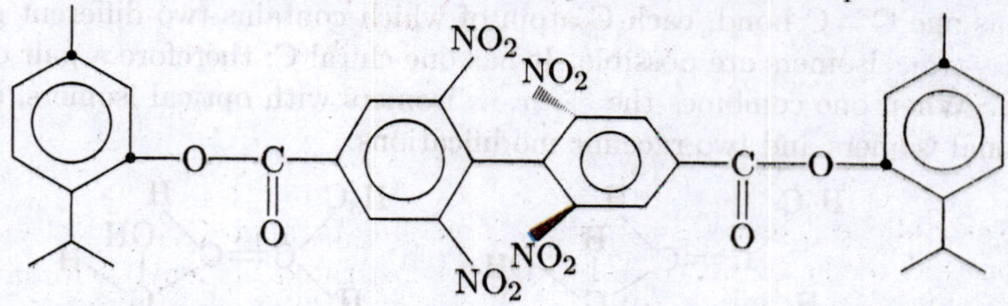

Problem 9.15. Draw configurations of all enantiomers and diastereomers for each of the following and indicate which is which:

(a) $CH_3.CHCl.CHCl.CH_3$ **(b)** $CH_3CHBr.CHCl.CH_3$ **(c)** $CH_3CH{=}CH.CHOH.CH_3$

(d) $CH_3.CH{=}C{=}CH.CH_2.CH_3$ (e) $CH_3CH{=}CH.CH_2.CH{=}CHCH_3$

(f) 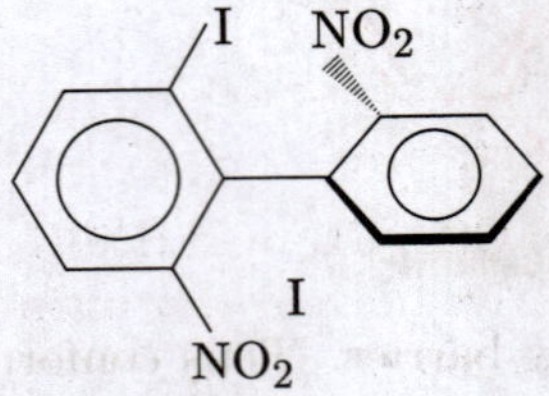

(g) 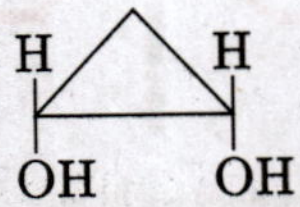

(h) 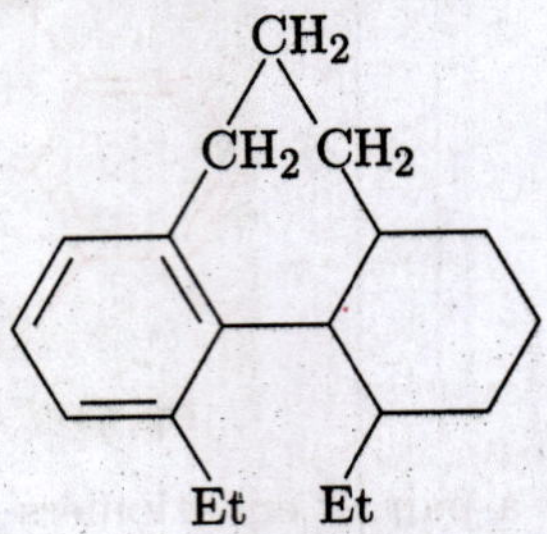

(i) 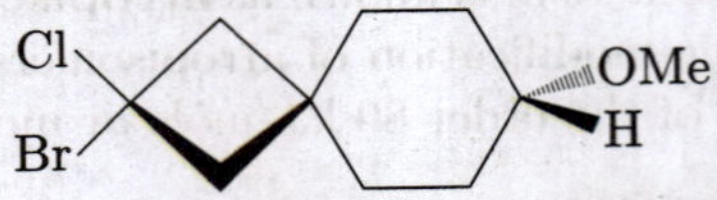

Solution: (a) It has two alike chiral Cs; hence configurations of three optical isomers can be drawn (two meso forms being one and the same); two of them are active.

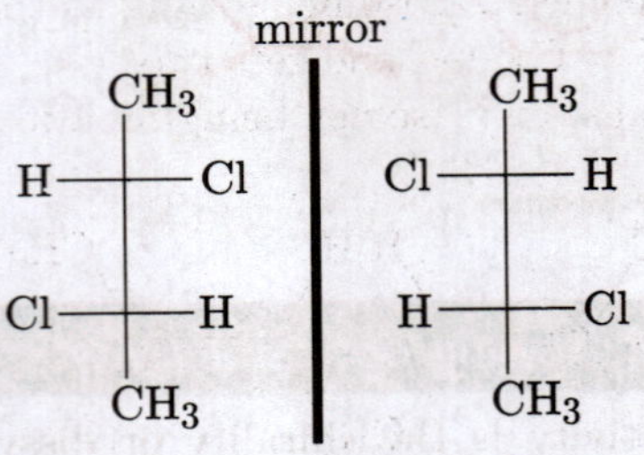

a pair of enantiomers

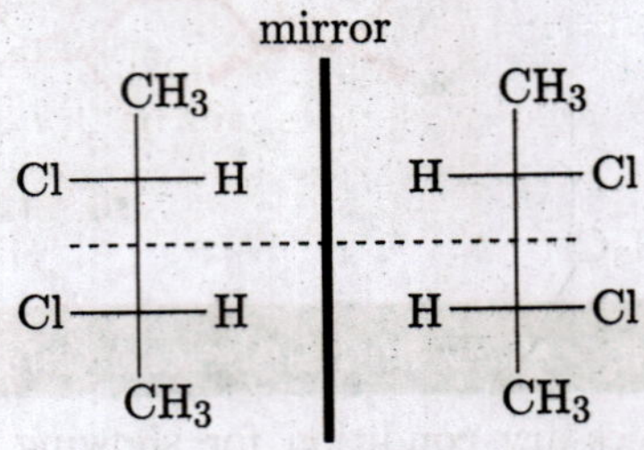

meso form (identical)

(b) It has two unlike chiral Cs; hence 2^2 or 4 optical isomers are possible—all are active.

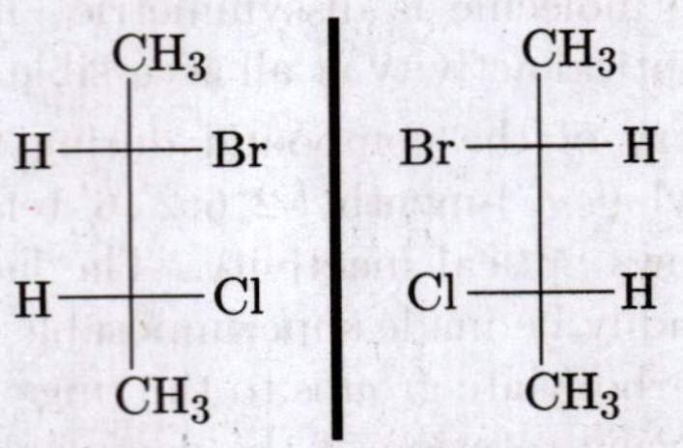

one pair of enantiomers

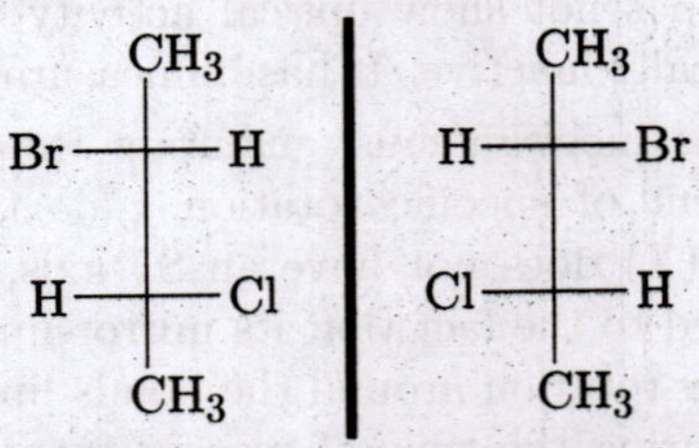

another pair of enantiomers

(c) It has one C=C bond, each C atom of which contains two different groups and atoms; hence two cis-trans isomers are possible. It has one chiral C; therefore a pair of enantiomers will be obtained. When one combines the cis-trans isomers with optical isomers, one gets $2 \times 2 = 4$ configurational isomers and two racemic modifications.

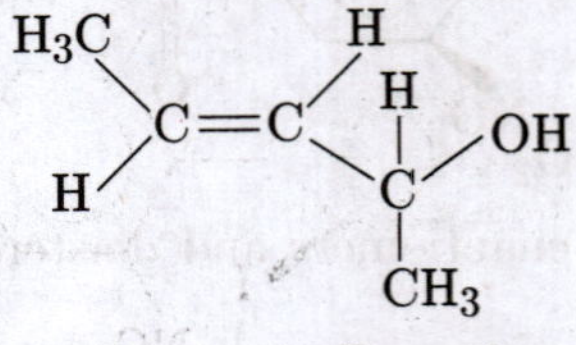

trans, optically active

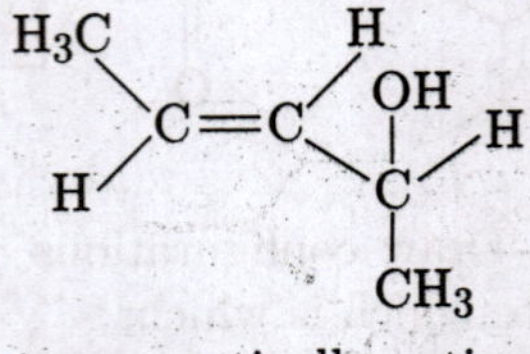

trans, optically active

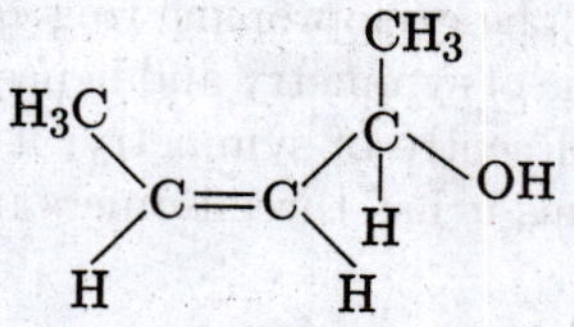

cis, optically active

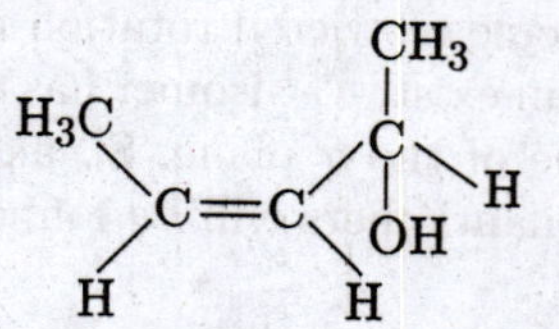

cis, optically active

(d) It is an allene derivative with two different groups on each end C(H, CH_3 and H, CH_2CH_3); so it is a chiral molecule without chiral C. It is therefore optically active (devoid of σ plain, point of symmetry and S_n axis of symmetry). Thus a pair of enantiomers can be drawn for the given allene.

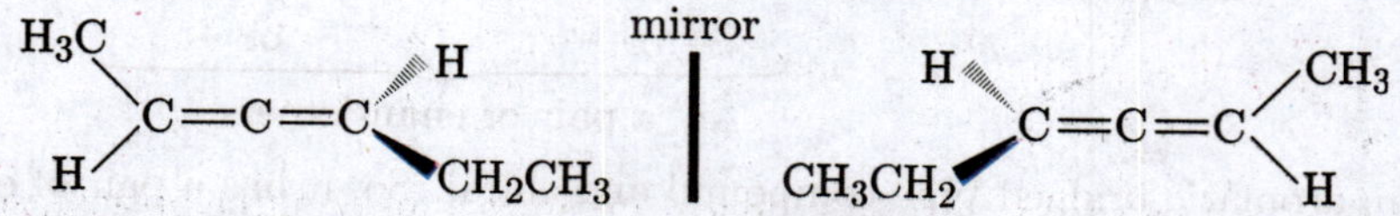

(e) It is an achiral molecule as it has a plane of symmetry and it is the plane of the double bonds. Therefore, it has no enantiomers; but it has geometrical isomers owing to the restricted rotation about each of the C═C bonds. About each C═C bond one gets two cis-trans isomers; hence total number of isomers will be 2 × 2 = 4. (trans, cis) -isomer being identical with (cis, trans) -isomer, actually three diastereomers will exist.

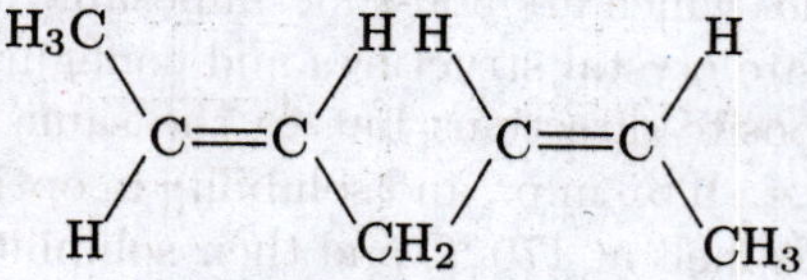

trans, cis

I

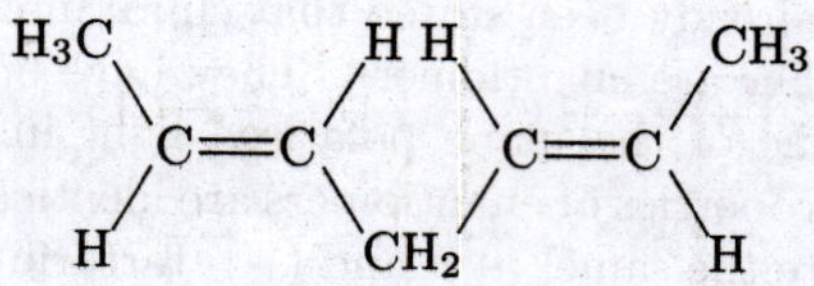

trans, trans

II

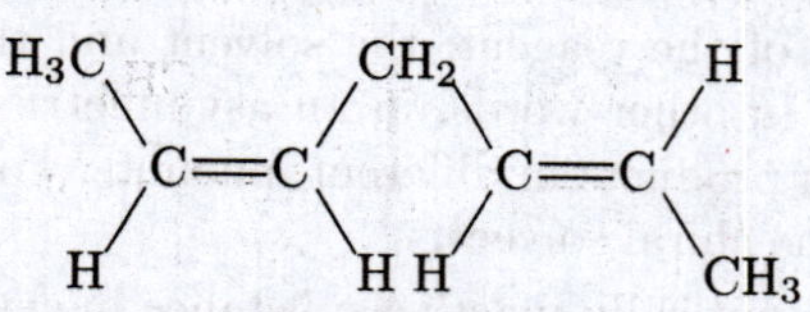

cis, trans

III

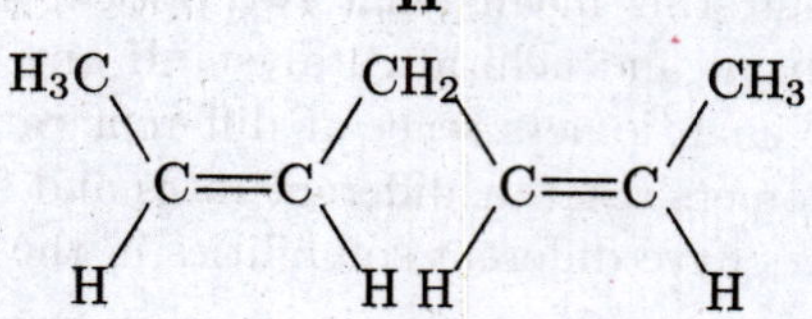

cis, cis

IV

(f) It is a bulky ortho substituted biphenyl derivative without any vertical plane of symmetry in each ring. Such a molecule has two rings in two planes at right angles to each other; Steric effect of the bulky *o*- and *o′*-substituents restricts the rotation of one of the ring C—C single bond. Thus the molecule does not have σ plane, S_n axis and point of symmetry but it has a chiral axis which passes through C_4, C_1, C_1, and C_4, atoms, i.e., it becomes a chiral molecule. Therefore it has a pair of enantiomers.

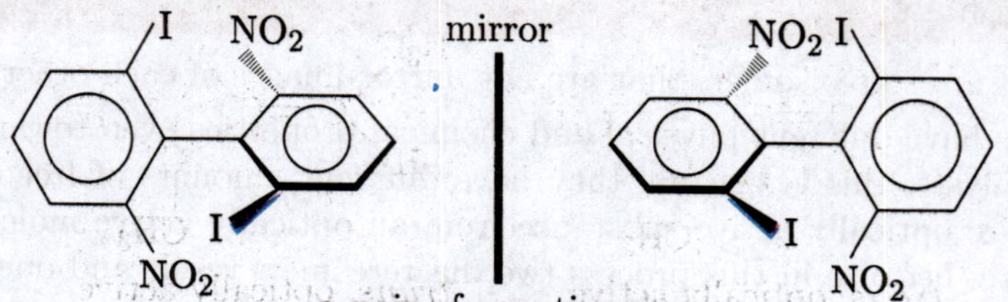

a pair of enantiomers

(g) In this molecule restricted rotation about the cyclopropane ring can take place; thus cis- and trans-isomers can exist. Cis-isomer has a plane of symmetry and hence it is optically inactive. Trans-form has none of the σ plane, S_n axis and centre of symmetry; it is, therefore, optically active and its two enantiomers will be found. Thus it has three isomers : one cis and two trans, (+)- and (−)-forms.

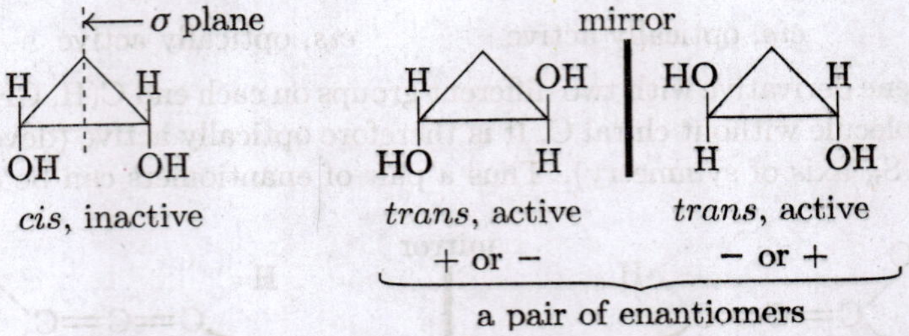

(h) It is an aromatic bridged type compound and chiral. So, it has a pair of enantiomers.

(i) It is a spirane derivative without vertical plane of symmetry in each ring. The two rings are at right angles to each other. Hence a pair of enantiomer, (+)- and (−)-forms, will exist.

9.13 Enantiomers

It has already been stated that chiral molecules which are non-superimposable mirror-images of each other are enantiomers. They have opposite crystal structures and configurations. They rotate plane of the plane polarised light in opposite directions but to the same extent. The physical properties of enantiomers are identical; e.g., b.p., m.p., and solubility in optically inactive solvents are the same; (+)- and (−)- tartaric acids melt at 170 °C and their solubility in water at 20 °C is 1390 g/ litre. The chemical properties of the enantiomers are identical in a symmetrical environment; this means that two optical antipodes behave similarly towards achiral reagent, achiral solvent and achiral catalyst. If any one of the reagent, the solvent and the catalyst is chiral, the enantiomers react at different rates. In other words, in an asymmetric environment the enantiomers react at different rates and form products in different amounts. For this reason, enantiomers have different solubilities in the same chiral solvent.

If one tries to prepare an enantiomer from an optically inactive substance in symmetric environment, usually a racemic modification results; but if one tries to synthesise an enantiomer by the use of an optically active reagent, optically active products are obtained; this may contain : (i) one enantiomer in excess over the other or (ii) a pure enantiomer. Optically active substances may also be obtained from a racemic modification using optically active reagent, catalyst and solvent. In nature one often finds pure optical isomers instead of racemic forms; this is because optically active enzymes are involved in such bio-synthetic processes.

9.14 Diastereomers

Diastereomers are stereoisomers that are not mirror-images of each other.

Diastereomers have different physical and chemical properties even towards achiral reagents, solvents and catalysts; this is because they have different amounts of free energy. Hence it is possible to prepare optically active substance from an optically active molecule creating a new asymmetric centre; because in this process two diastereomers result and one of which will be in excess over the other.

We know that (+)-tartaric acid is diastereomeric with meso tartaric acid; these two have different melting points, e.g. 170 °C and 206 °C respectively.

Problem 9.16. (+)-Glyceraldehyde is allowed to react with **HCN**; is the product an equimolecular mixture of two epimers?

Solution :

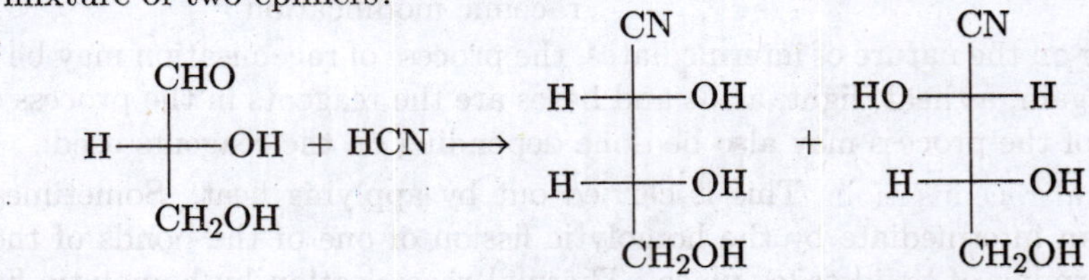

(+)-Glyceraldehyde (A) (B)

Epimers are diastereomers containing more than one chiral C but differing in configuration at a single chiral C. Compound A and B are epimers as they differ from each other in configuration at C_2 atom only. In this reaction, the chiral C present in (+)-glycealdehyde remains intact and a new chiral C is generated by the addition reaction of HCN, whereby two diastereomers are produced. As the products of the reaction are diastereomers, the transition states which lead to the products are also diastereomeric. Therefore transition states have different free energies and two products must be formed at different rates. Hence it cannot be an equimolecular mixture of two epimers.

9.15 Racemic Modification

A substance composed of equimolecular amounts of a pair of enantiomers is a *racemic modification*

A racemic modification may be prepared by the following methods:

(a) By mixing equimolecular amounts of two optical antipodes.

(b) By generating a chiral centre in a symmetric molecule in symmetric environment; when propionic acid is heated with red phosphorus and bromine, (±) – α-bromopropionic acid results:

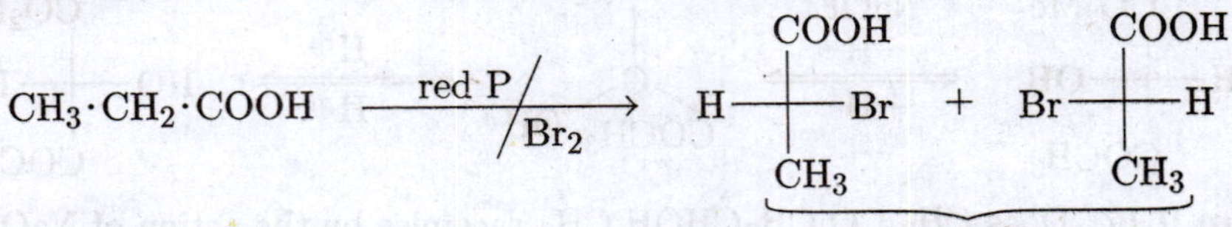

racemic form

(c) By racemisation: The process of producing a racemic modification starting with one of the pure enantiomers is termed racemisation. Thus if one starts with $2n$ moles of (+)- or (–)-lactic acid and ultimately gets n moles of each of the enanticmers together, then the process is to be called racemisation. In racemisation, one-half of the total molecules of the enantiomer does not remain as such and the others get converted into the other enantiomer; all the molecules, actually, change either to a reaction intermediate (like, free radical, carbenium ion, carbanion; See Ch. 11) or to optically inactive molecules and then by a reaction on the intermediate species recemised product results. Thus in racemisation, bond braking and bond reforming processes are involved, the process of racemisation may be pictorially represented in Fig. 9.121.

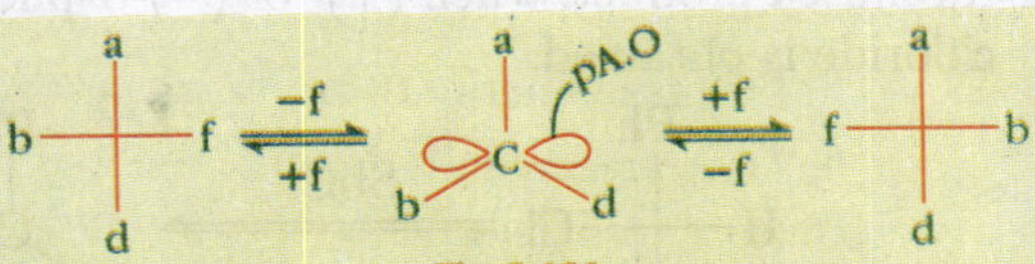

Fig. 9.121

$$2n(-)\text{-Lactic acid} \longrightarrow \underbrace{n(-)\text{-Lactic acid} + n(+)\text{-Lactic acid}}_{\text{racemic modification}}$$

$$2n(+)\text{-Lactic acid} \longrightarrow \underbrace{n(+)\text{-Lactic acid} + n(-)\text{-Lactic acid}}_{\text{racemic modification}}$$

Depending on the nature of intermediates, the process of racemisation may be discussed under four heads. Again, as heat, light, acids and bases are the reagents in the process of racemisation, classification of the process may also be done depending on the reagents used:

(i) Thermal racemisation: This is carried out by applying heat. Sometimes, a free radical is formed as an intermediate by the homolytic fission of one of the bonds of the chiral C atom and then reforming of bond takes place. Thermal racemisation by homolytic fission of bond is rare; perhaps racemisation of (+)- or (−)-α-phenethyl chloride takes place by this process when distilled at atmospheric pressure.

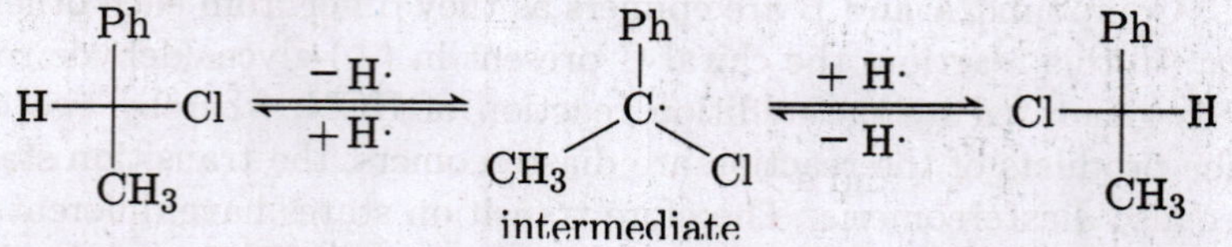

H free radical, formed by the homolytic fission of C—H bond, attacks the intermediate from two diametrically opposite sides giving rise to racemic modification; this is because the probability of attacking the intermediate from either side is 50%.

(ii) Base- catalysed racemisation: When a compound containing acidic hydrogen attached to the chiral C is heated with a base (like methoxide, ethoxide etc.), a carbanion is formed as an intermediate by the removal of the hydrogen as proton; the proton then attacks the carbanion from either side which leads to racemisation. The ease of racemisation increases with the increasing acidity of the H atom. Thus, (+)- or (−)- $CH_3CO.CHOH.CO_2Me$ is racemised by heating the compound with NaOEt; the adjacent carbonyl group and ester group make the H on the chiral C highly acidic.

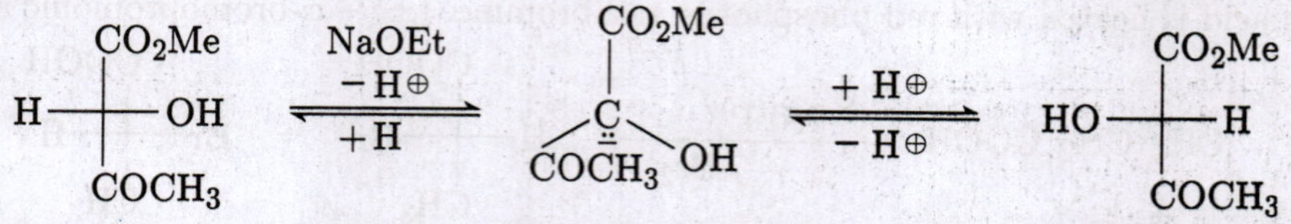

Problem 9.17. Does $CH_3.CO.CH_2CHOH.CH_3$ racemise by the action of NaOEt?

Solution: No, here the acidic H is present in CH_2 group and not in CHOH; because the former is attached to the carbonyl group. Hence racemisation will not occur.

(iii) Acid-catalysed racemisation: A carbenium ion is generated by withdrawing an anion from the chiral C atom; being a Lewis base, an anion can be withdrawn temporarily by Lewis acid like $SbCl_5$, $AlCl_3$, BF_3 etc. The sp^2-hybridised carbenium ion so formed has a planar structure and the anion (from the salt) then attacks the carbenium ion from either side which gives rise to the racemised product; when (+)- or (−)-α-phenethyl chloride is heated with $SbCl_5$, (±)-α-phenethyl chloride is obtained.

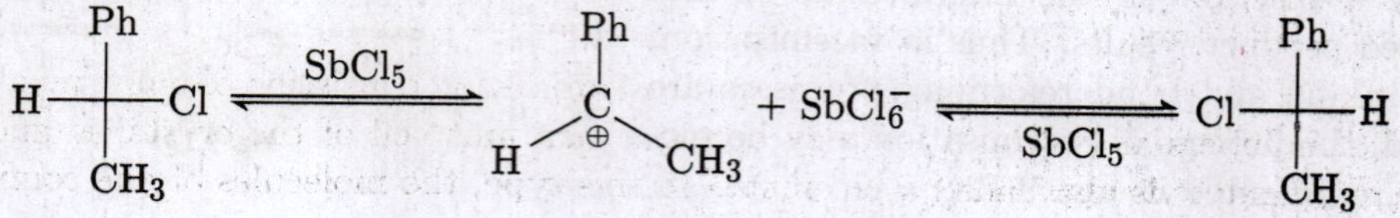

Problem 9.18. (+)-PhCH(Me)Et racemises in the presence of $AlCl_3$; what type of intermediate do you expect?

Solution: The reaction takes place via carbenium ion formation; $AlCl_3$ abstracts the H atom of the chiral C as hydride ion; this leads to the formation of $\bar{A}lHCl_3$ and the carbenium ion. So the carbenium ion is the expected intermediate.

(iv) Racemisation via reversible formation of optically inactive intermediates: When (+)- or (−)- form of 2-butanol is treated with sodium sec-butoxides, racemisation takes place; here butanone is the inactive stable intermediate.

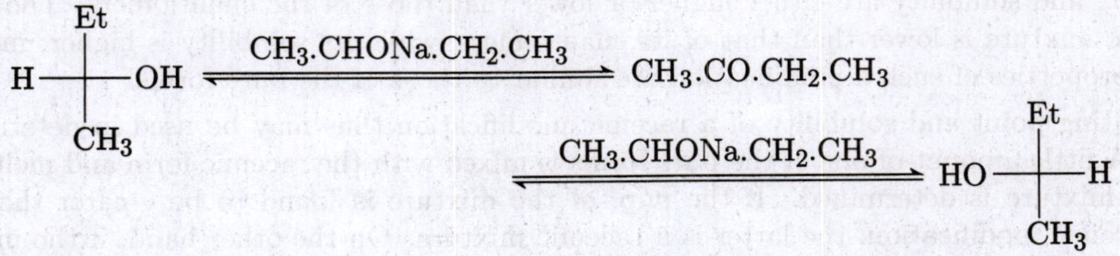

In the thermal, base-catalysed and acid-catalysed racemisation the intermediates formed have planar trigonal orientation about the chiral carbon. The chiral carbon is sp^2 hybridised and a p orbital, remains perpendicular to the plane of the intermediate. We get a half-filled pA.O. in the case of thermal racemisation. In the of base-catalysed and acid-catalysed racemisation the p orbital is filled and vacant respectively. the p orbital then overlaps the orbitals of breakaway radical (H·) or $SbCl_6^-$ by its either lobes to develop a pair of enantiomers.

A racemic modification is generally an ideal mixture of equal numbers of dextro and levorotatory molecules in the gaseous or liquid states or in solution, provided no hydrogen bonding is there. Since the physical and chemical properties of a pair of enantiomers are identical in symmetric environments, the properties of a racemic modification in gaseous or liquid state or in solution are also identical with those of the enantiomers. Thus, the b.p., refractive index, density in the liquid state of a racemic modification are the same as those of the pure enantiomers in liquid state.

However, a racemic modification in solid state is not always an ideal mixture of equal numbers of enantiomeric molecules. In the crystal lattice of a racemic modification, the enantiomeric molecules may arrange in three different ways and thus three types of racemic modifications happen to occur in the solid state :

(a) Racemic mixture or (±)-conglomerate: It is a mechanical mixture of two types of crystals; one is composed of (+)-forms and the other is composed of (−)-forms. This is because a dextrorotatory molecule has greater affinity toward another dextrorotatory molecule than toward a levo one, and a levorotatory molecule possesses the same toward its own class in such a modification. (±)-Sodium ammonium tartrate crystallised from water below 27 °C is a racemic mixture.

(b) Racemic solid solution: A racemic modification consisting of crystals that are formed by the combination of both of the enantiomers at random is known as racemic solid solution; this nearly behaves as an ideal mixture. Perhaps, constituent molecules of the crystals have a little difference in affinity between molecules of the same or opposite configurations; (±)-camphor oxime crystallised above 103 °C is a racemic solid solution.

(c) Racemic compound or racemate: A racemic modification consisting of one type of crystals having equal numbers of (+)- and (−)-molecules in each unit cell of the crystal is known as a racemic compound; it is also called a recemate. In this type, the molecules of one configuration have greater affinity for those of the opposite configuration than for their own kind. Racemates behave as a true compound. (±)-Sodium ammonium tartrate crystallised from water at a temperature above 27 °C is a racemic compound. It contains 2 moles of water per mole of the compound; while its racemic mixture contains 4 moles of water per mole of the racemic modification.

The properties of a racemic modification in solid state thus depend on the type in which it belongs. If it is a racemic solid solution, its properties will be identical with those of the pure enantiomers; its m.p. and solubility are also almost same or same as those of the pure enantiomers. The properties of a racemic compound in solid state are different from those of the pure forms; its m.p., and solubility are either higher or lower than those of the enantiomers. The m.p. of a racemic mixture is lower than that of its enantiomers, while its solubility is higher; most of the other properties of such a modification are similar to those of the pure forms.

Melting point and solubility of a racemic modification thus may be used in determining its type. A little amount of one of the pure forms is mixed with the racemic form and melting point of the mixture is determined. If the m.p. of the mixture is found to be greater than that of the racemic modification, the latter is a racemic mixture. On the other hand, if the m.p. of the mixture is less than that of the racemic form, it is a racemic compound. The racemic modification will be a solid solution if the m.p. of the mixture does not differ appreciably from that of the racemic modification.

9.16 Resolution of Racemic Modifications

Many of the chemical compounds occurring both in plants and in animals happen to be there as pure enantiomers. After the isolation of a pure enantiomer from its natural source, one must establish its structure. Again without total synthesis of the compound, the structure cannot be established. Unfortunately, the synthesis of a pure enantiomer usually leads to racemic modification. Hence a total synthesis of a naturally occurring enantiomer must involve the separation of enantiomers in the pure state from their racemic modification. The process of separation of pure enantiomers from their racemic modification is called **resolution** and it has a great practical importance in the study of the naturally occurring optically active compounds. There are a number of methods of resolution and usually quantitative recovery of the enantiomers is not possible by any one of the methods. However, some of these methods are discussed in this section for their theoretical and practical importance. It is needless to mention that the separation of the components of a substance depends on the differential properties of the components; the same is also true in the case of resolution of the racemic forms. Since most of the physical and chemical properties of a pair of enantiomers are identical, the process of resolution is a difficult task.

9.16.1 Mechanical separation

Since (+)- and (−)- forms have opposite crystalline shapes, this property can be used in separating two forms from a racemic mixture; because this type of racemic modification contains two distinct crystals of opposite shapes. In fact, Pasteur used this process to separate (+)- tartaric acid and (−)- tartaric acid from (±)-conglomerate of tartaric acid obtained by crystallisation of the same from water at a temperature below 27 °C.

To resolve the enantiomers by this process, crystals of opposite facets are to be seen distinctly under a microscope; then with the help of a tweezer the two types of crystals are to be picked up one by one and to be collected in separate containers. However, this process has some demerits:

(a) The method of mechanical separation has no value in practice. This is because it is not applicable to racemic compounds and solid solutions.

(b) Distinctive crystal facets are seldom found in racemic mixtures; this makes the separation troublesome and incomplete.

(c) The process is tedious.

Therefore the process of mechanical separation has historical interest only.

9.16.2 Preferential crystallisation by inoculation

This procedure is mainly applied to racemic mixture. The distinctive property which is utilised in this case is the greater affinity of each enantiomer for the molecules of its own kind than for those of the other enantiomer. Thus, if a saturated solution of a racemic mixture is inoculated with a pure crystal of one enantiomer, the molecules of this enantiomer from the solution deposit on the crystal and it grows in size; sometimes, instead of a crystal of one enantiomer, a crystal of an isomorphous substance may be used for inoculation.

Racemic compounds of greater solubility in a solvent than that of the pure enantiomers can also be separated from its saturated solution by this process.

The process is seldom used for the purpose of resolution, but it has great practical importance in purification of enantiomers. In fact, phenylmethylcarbinyl hydrogen phthalate after resolution to the extent of 95% is purified by preferential crystallisation by inoculation using a small crystal of its racemic modification in CS_2 medium.

9.16.3 Chromatographic separation

Enantiomers react with an optically active reagent at different rates and produce diastereomers of different stabilities. This property is utilised in chromotagraphic separation of enantiomers.

The process consists in allowing a solution of the racemic form in a chosen solvent to pass slowly down an optically active adsorbent like starch. Then the enantiomers are adsorbed by the active adsorbent at different rates to form diastereomers of different stabilities. Now the column is eluted (washed) with the solvent. The enantiomer forming less stable diastereomer comes down the column faster than the other and the elution is collected in a separate vessel. In this way a partial separation of enantiomers may be possible. (±)-Mandelic acid is resolved almost completely using starch as an adsorbent. Paper chromatography can also be used for the purpose. Here cellulose acts as an optically active substance.

The process is very useful in deciding whether a given substance is a *dl*-pair or a compound containing symmetric molecules. If the elution is found to be optically active, the substance under investigation is definitely a *dl*-pair ; if not, it consists of symmetric molecules.

9.16.4 Biochemical separation

When living organisms are added to or developed in a racemic modification, they eat up one of the enantiomers preferentially leaving the other behind. Thus resolutions of racemic form can be done using this specific property of living organisms. Living organisms contain optically active bio-catalysts, called enzymes; these being optically active, react at different rates with a pair of enantiomers. Thus a particular enzyme of an organism helps it to digest one of the enantiomers. As chemical reaction occurs in this process with the help of bio-catalysts, the process is known as biochemical separation. Penicillium glaucum, a mould, eats up preferentially (+)- ammonium tartrate when it is added to (±)- ammonium tartrate.

This method has limited use because:

(a) it is difficult to find the proper organism,

(b) one gets one.of the enantiomers at the expense of the other.

(c) dilute solutions must be used; if otherwise, organisms may die and hence small amount of resolved product is obtained.

However, since biological processes are highly specific, the method leads to a high extent of resolution provided proper organism is applied for the same. Biochemical separation was discovered by Pasteur.

9.16.5 Separation by conversion into diastereomers

Enantiomers have identical physical properties while diastereomers have different physical properties. This fact is utilised in resolving racemic modifications by this process. The racemic form to be resolved is allowed to react with an optically active substance, two diastereomeric derivatives are then formed. These have different solubilities, boiling points etc. So, by fractional crystallisation, the solid diastereomers can be separated. Each of the enantiomers may then be regenerated and isolated from the diastereomers. Nowadays gas-chromatography is also used to separate the diastereomers. Fractional distillation has minimum scope. The overall procedure may be shown schematically as given below :

(±)-Form —2(+)-reagent→ (+)·(+)·derivative + (−)·(+)·derivative

A pair of diastereomers

↓ separated

(+)·(+)·derivative ↓ regenerated → (+)-form + (+)-reagent ↓ isolated → (+)-Form

(−)·(+)·derivative ↓ regenerated → (−)-form + (+)-reagent ↓ isolated → (−)-Form

The process is mainly applied to acids and bases; these can also be applied to alcohols, aldehydes, ketones, hydrocarbons, aminoacids, etc. The reagents used and the procedure adopted to resolve the racemic forms of the classes mentioned above are discussed below :

(a) Acids : Racemic modifications of acids are converted into two diastereomeric salts by the reaction of optically active bases. The bases used are : alkaloids e.g., strychnine, ephedrine, quinine, quiniline, morphine, brucine, etc.; menthylamine and α-phenethylamine are also used. The diastereomers are then separated by fractional crystallisation. Each of the diastereomers is then hydrolysed with dil. mineral acids (e.g. HCl) to regenerate the enantiomer. The reagent and the enantiomer are then isolated :

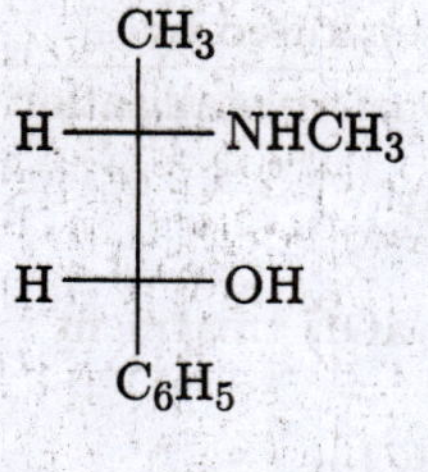

(−)-Ephedrine

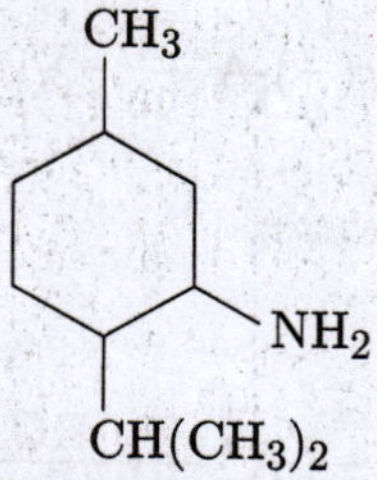

Menthylamine

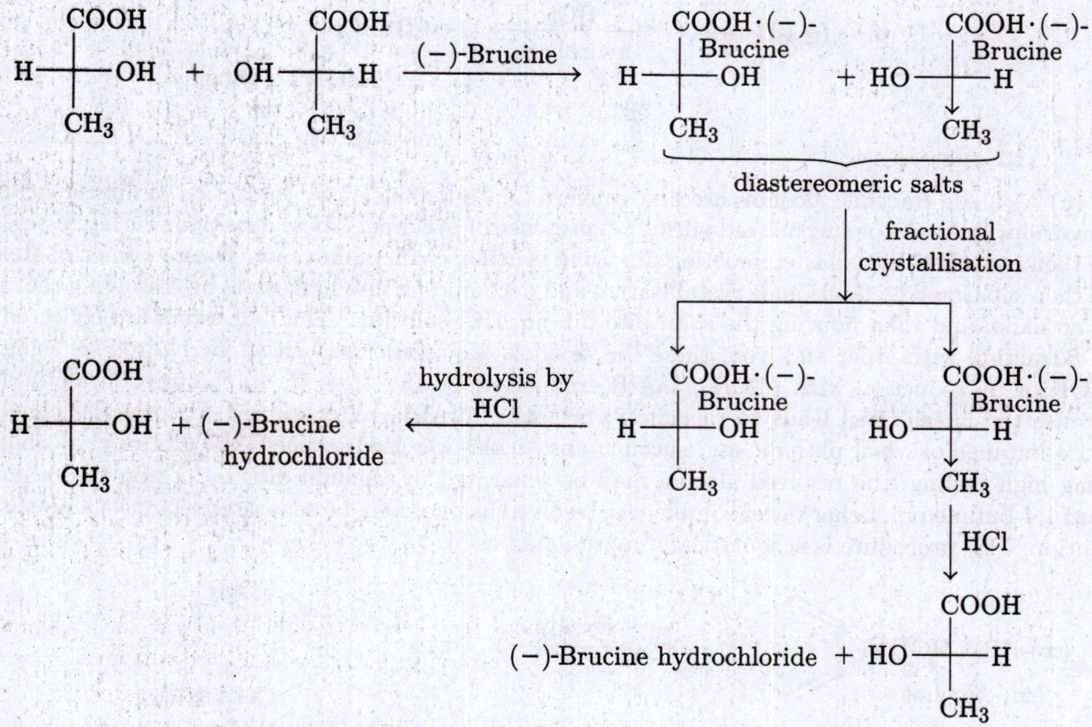

(b) Bases: Optically active bases are converted into two diastereomeric salts by the reaction of optically active acids, e.g., tartaric acid., malic acid, comphoric acid, derivatives of camphoric

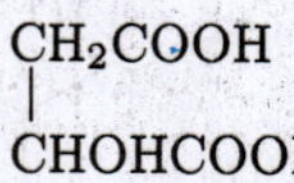

Malic acid

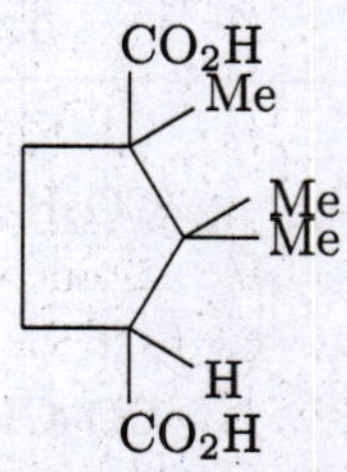

Camphoric acid

acid, etc. The salts are separated by fractional crystallisation. The enantiomers are then obtained from the product of hydrolysis of the salts. If (±)-B is the racemic form of a base to be resolved and (−)-A is an optically active acid, then the procedure may be expressed as :

$$(\pm)\text{-B} + (-)\text{-A} \longrightarrow (+)\text{-B}\cdot(-)\text{-A} + (-)\text{-B}\cdot(-)\text{-A} \xrightarrow[\text{fractional crystallisation}]{\text{separated by}} \text{(i)} + \text{(ii)}$$

$$\text{(i)}\quad (+)\text{-B}\cdot(-)\text{-A} \xrightarrow[\text{hydrolysis}]{\text{HCl}} (+)\text{-B}\cdot\text{HCl} + (-)\text{-A}$$

$$(+)\text{-B}\cdot\text{HCl} \xrightarrow{\text{NaOH treatment}} (+)\text{-B}$$

$$\text{(ii)}\quad (-)\text{-B}\cdot(+)\text{-A} \xrightarrow[\text{hydrolysis}]{\text{HCl}} (-)\text{-B}\cdot\text{HCl} + (-)\text{-A}$$

$$(-)\text{-B}\cdot\text{HCl} \xrightarrow{\text{NaOH treatment}} (-)\text{-B}$$

(c) Alcohol: Racemic alcohols are first converted into half esters by the action of dibasic acid anhydrides (phthalic or succinic anhydride) in presence of pyridine. These enantiomeric half esters are then converted into diastereomeric salts using brucine, cinchonidine, etc. Diastereomeric salts are then separated by fractional crystallisation and decomposed into half esters by dissolving them in methanol and then pouring the soln. into dil. aq. HCl solution. The half esters are collected by extraction with ether and from these the alcohols are regenerated either by hydrolysis using NaOH or by reduction with $LiAlH_4$; $LiAlH_4$ reduction is safer than the saponification method because the latter often leads to racemic form again. $LiAlH_4$ reduction gives phthalyl alcohol or 1,4-butanediol when phthalic and succinic anhydrides are used respectively. Phthalyl alcohol being high boiling, the resolved alcohol may be separated by vacuum distillation; on the other hand 1,4-butanediol, being water soluble, the resolved alcohol may be separated from the aqueous solution. The procedure is schematically represented as

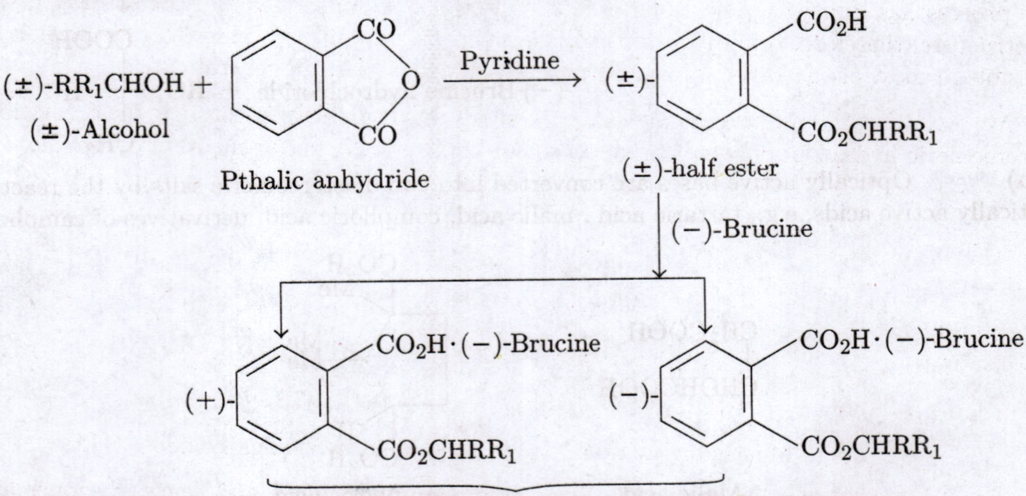

diastereomeric salts, separated by fractional crystallisation

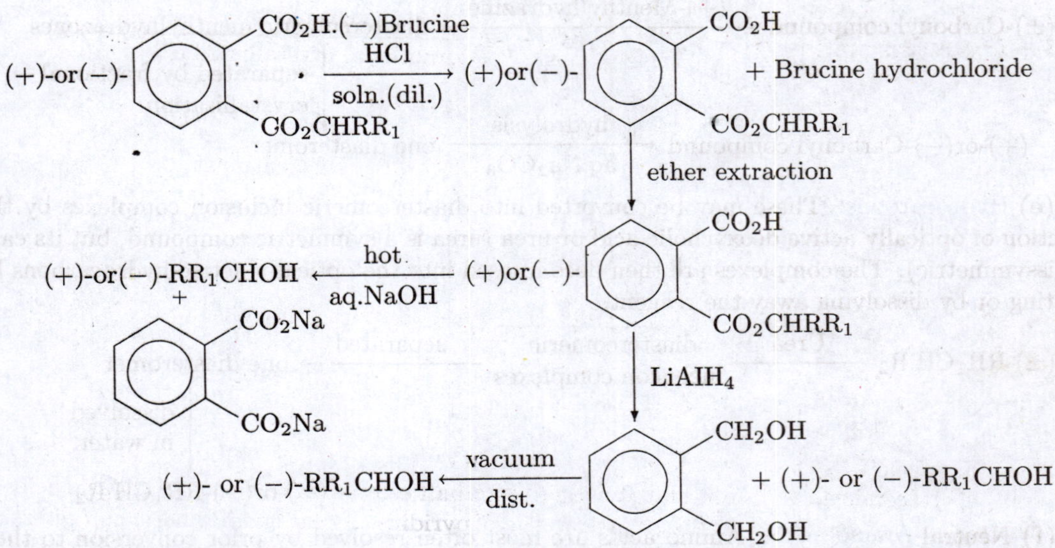

Alcohols may also be resolved through the formation of esters and glycosides; but these processes have limited applications.

Problem 9.19. Suggest a procedure for resolving racemic alcohols through the formation of ester. Explain why this process has limited application.

Solution: The (±)- alcohol is at first to be converted into an ester by the action of an optically active acid in the presence of conc. H_2SO_4. The diastereomeric esters then are to be separated by fractional crystallisation. The (+)- and (−)- forms of alcohols may then be regenerated from the two esters by reducing them with $LiAlH_4$.

The process has limited applications because esters are seldom found in satisfactorily crystalline structure; therefore, the separation of diastereomeric esters by fractional crystallisation is not possible in most of the cases. Thus, the process cannot be used in general.

(d) Aldehydes and ketones: Aldehydes and ketones may be resolved through the formation of diastereomeric hydrazones or semicarbazones. For this optically active menthyl hydrazine,

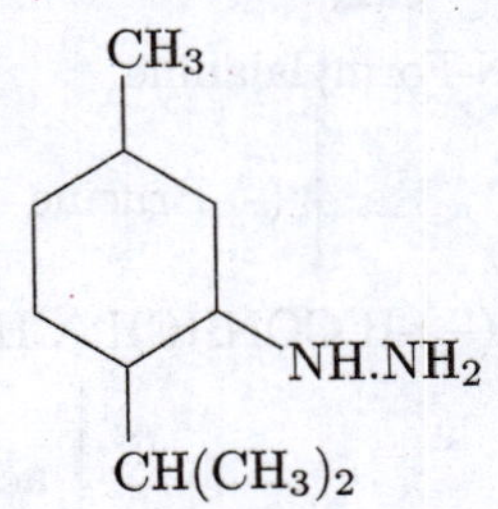

Menthylhydrazine

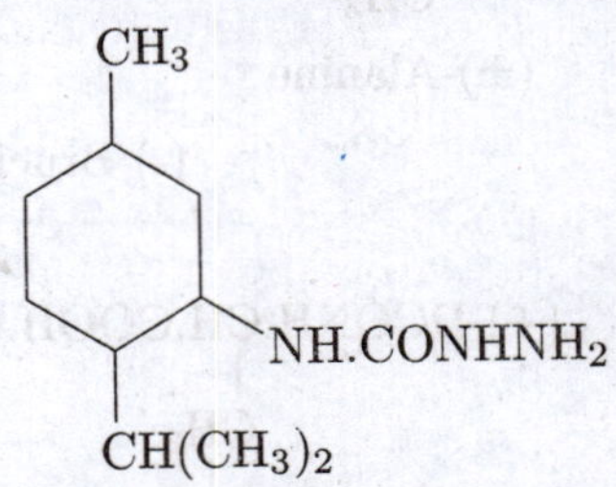

Menthyl semicarbazide

menthyl semicarbazide etc. are used. The diastereomeric hydrazones and semicarbazones are separated as usual by fractional crystallisation. Then optically active forms of the carbonyl compounds are obtained by the hydrolysis of hydrazones and semicarbazones with aqueous solution of Na_2CO_3

(±)-Carbonyl compound $\xrightarrow{\text{(+)-Menthylhydrazine}}$ diastereomeric menthylhydrazones

↓ separated by fractional crystallisation

(+)-or(−)-Carbonyl compound $\xleftarrow[\text{aq.}Na_2CO_3]{\text{hydrolysis}}$ one diasteromer

(e) Hydrocarbons: These may be converted into diastereomeric inclusion complexes by the reaction of optically active deoxycholic acid or urea (urea is a symmetric compound, but its cage is dissymmetric). The complexes are then decomposed into the optically active hydrocarbons by melting or by dissolving away the reagent.

(±)-$RR_1CH.R_2$ $\xrightarrow{\text{Urea}}$ diastereomeric inclusion complexes $\xrightarrow{\text{separated}}$ one diasteromer

↓ dissolved in water

(+)-or(−)-$RR_1CH.R_2$

(f) Neutral Amino acids: Amino acids are most often resolved by prior conversion to their acyl or benzoyl derivatives, which are formed by the reaction of the amino acid with acyl chloride or benzoyl chloride respectively. The acyl or benzoyl derivatives are then converted into diastereomeric salts by treating them with optically active bases. The salts are then separated by fractional crytallisation. Separated salts are subjected to acid hydrolysis when both the acyl (or benzoyl) group and the optically active base get removed from the salt leaving the optically active amino acid behind. In this step, sometimes, racemisation occurs because the removal of benzoyl or acyl groups requires a moderately strong condition in which bond breaking and bond reforming processes happen to occur on the chiral C. One can formylate the amino acid instead of benzoylation or acylation (other than formylation) by means of formic acid and acetic anhydride; because the removal of formyl group can be done under a mild condition in which racemisation does not take place. Thus alanine can be resolved by this process.

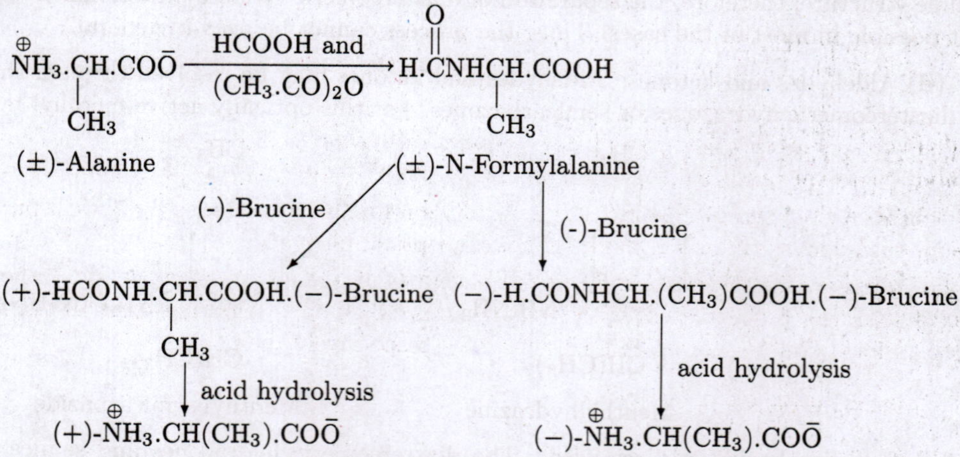

Problem 9.20. Why is the amino group of an amino acid protected by formylation prior to resolution as an acid? Can an amino acid be resolved as a base or an acid directly? Give reasons for your answer.

Solution: Amino acids form inner salts or dipolar ions in solution; e.g. alanine remains as $\overset{\oplus}{N}H_3CH(CH_3)COO^-$ in aqueous solution. The acidic carboxyl group and the basic amino group of the same molecule neutralise each other producing the inner salt or dipolar ion or zwitter ion. For this reason, they cannot be resolved as a base or as an acid. When the amino acid is formylated its base character is effectively eliminated owing to the $-I$ effect of the formyl group. Base character being lost, the amino acid derivative no longer behaves as an inner salt and acts as an acid.

Basic amino acids containing a zwitter ion (dipolar ion) as well as a free amino function can be resolved by converting them into diastereomeric salts with an optically active acid. Thus (±)-tartaric acid is used to resolve (+)-histidine, a basic amino acid.

Similarly, acidic amino acids which contain a free carboxyl group as well as a zwitter ionic carboxyl function may be resolved as an acid using an optically active strong base.

Resolution by formation of diastereomers is apparently easy; but in fact, involves many difficulties as given below:

1. pure optically active reagents are costly, rare and very difficult to prepare.
2. diastereomers may not have wide range of difference in solubility and thus complete separation of diastereomers may not be possible.
3. decomposition of diastereomers into enantiomers sometimes leads to racemisation.
4. pure optically active reagents being costly and rare, their recovery is essential; but recovery of the reagents often fails.

Note

(a) Optical purity: By optical purity we mean the excess of one enantiomer in a partially recemised or a partially resolved substance which is expressed as a percentage of the total. It is the product of the ratio of the specific rotation of the substance under test to the specific rotation of a pure enantiomer (in both the cases signs are omitted) and one hundred. Mathematically,

$$\text{Optical purity} = \frac{[\alpha]_D^T \text{ of the substance under test}}{[\alpha]_D^T \text{ of an enantiomer}} \times 100$$

Let a compound (+) -A has the specific rotation $= +50°$. Let a partially resolved sample of (±)-A is +40° Then the optical purity of the partially resolved substance,

$$\text{Optical purity} = \frac{40°}{50°} \times 100 = 80.$$

So (+)-A = 80%; (±)-A = 20%. Hence total (+)-A = (80 + 10) or 90% and (−)-A = 10%.

In a racemic form, none of the enantiomers remains in excess; so its optical purity is zero. A pure enantiomer, therefore, has cent percent optical purity.

(b) Enantiomeric excess (*ee*): The excess of one enantiomer over the other in a sample expressed as a percentage of the total sample is called enantiomeric excess. If the mole fractions of (+)-form and (−)-form are [(+)] and [(−)] respectively then

$$ee = \frac{[(+)] - [(-)]}{[(+)] + [(-)]} \times 100$$

$$\% \text{ of } [(+)] \text{ or } [(-)] = \frac{ee + 100}{2}, \text{ for the major component}$$

$$\text{and} \quad \% \ [(+)] \text{ or } [(-)] = \frac{100 - ee}{2}, \text{ for the minor component}$$

9.17 Asymmetric Synthesis

The creation of a new asymmetric atom of a dissymmetric molecule under conditions where the two stereoisomers are formed in unequal amounts is known as asymmetric synthesis; the starting substance may be optically either inactive or active.

It has already been stated that synthesis of an optically active substance in a symmetric environment always leads to a racemic modification; but if dissymmetric reagents, solvents etc. are used for the same, two stereoisomers of opposite configurations are formed in unequal amounts, i.e., either (+)- or (−)- form predominates in the product. Thus an optically active product is obtained without recourse to resolution.

Asymmetric synthesis may be classified under two heads :

(a) partial asymmetric synthesis and (b) absolute asymmetric synthesis.

(a) Partial asymmetric synthesis: **This is a process in which two stereoisomers of opposite configurations are formed under asymmetric condition in unequal amounts and thus an optically active product is obtained without recourse to resolution.** As reactions occur at different rates in an asymmetric environment, the diastereomeric transition states are formed in unequal amounts; again, as the diastereomers have different stabilities, the product resulted from the transition states contains two stereomers in different amounts.

Some examples of the process are given below :

(i)

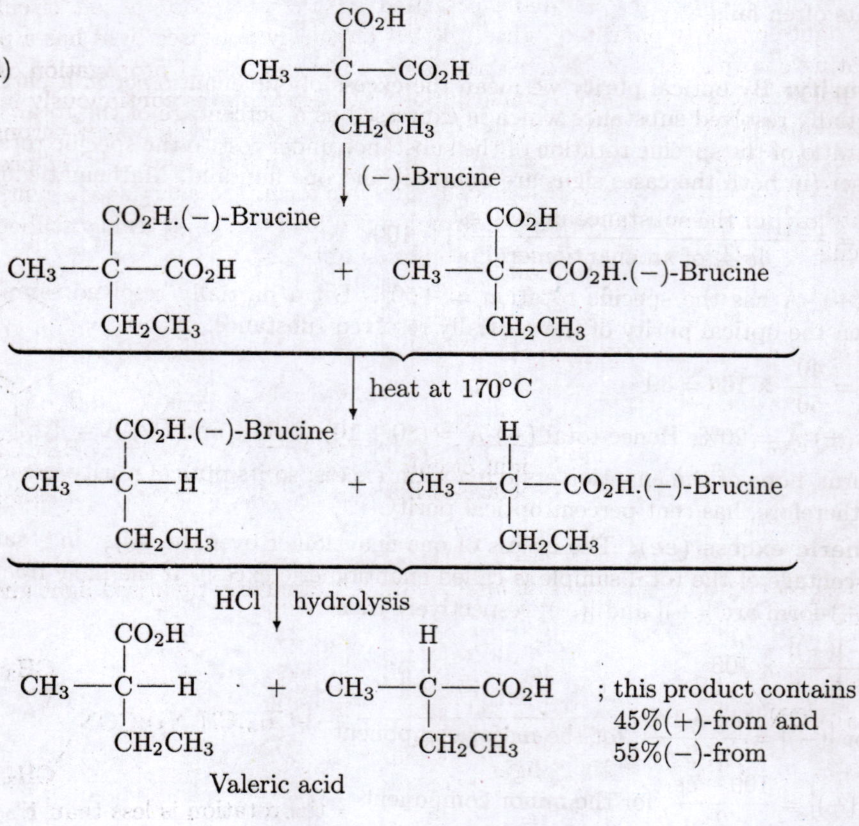

; this product contains 45%(+)-from and 55%(−)-from

Valeric acid

(ii)

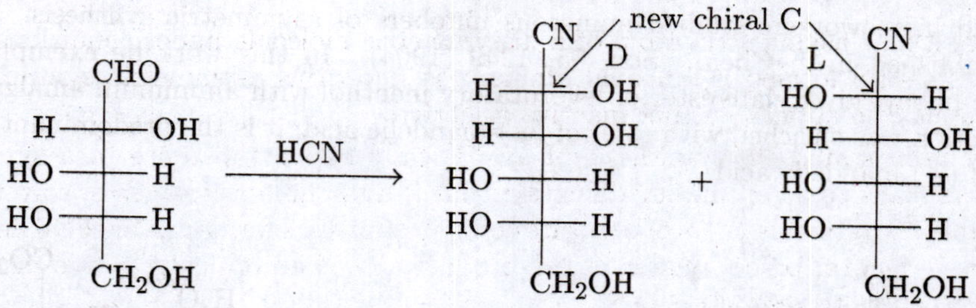

-(+)-Arabinose — Mannononitrile (major product) — Gluconitrile (minor product)

(iii)

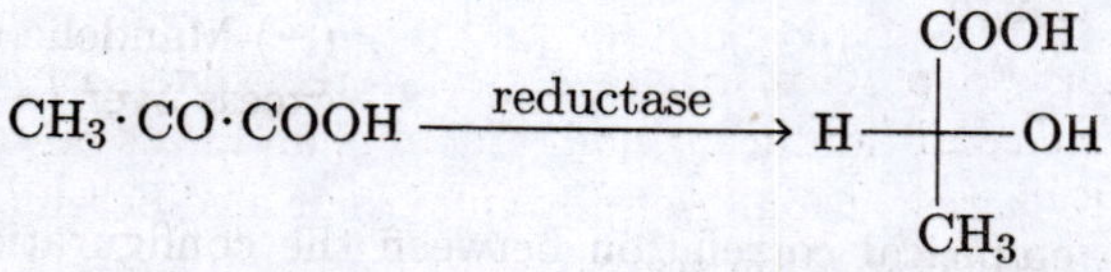

Pyruvic acid — (−)-Lactic acid

Reductase is an optically active enzyme. Being optically active, enzymes react at different rates with a substrate.

(b) Absolute asymmetric synthesis: A plane polarised light is composed of left circularly polarised light and right circularly polarised light. The left circularly polarised light has a plane of polarisation rotating continuously in the left direction around the axis of propagation of the wave; the plane of polarisation of the right circularly polarised light rotates continuously in the right direction. It is found that when left or right circularly polarised light is passed through a racemic modification, only one form of the enantiomers is obtained in pure state. As the process gives rise to a 100% pure enantiomer from an optically inactive form, the asymmetric synthesis is said to be absolute asymmetric synthesis. When bromine is added to 2, 4, 6-trinitrostilbene in a beam of light circularly polarised right, a dextro product is obtained.

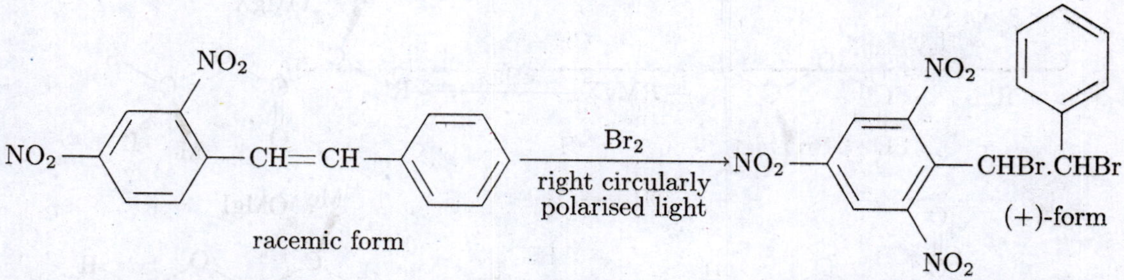

Irradiation of (±)-α-azidopropionic dimethyl amide with left circularly polarised light gives a product which is slightly levorotatory.

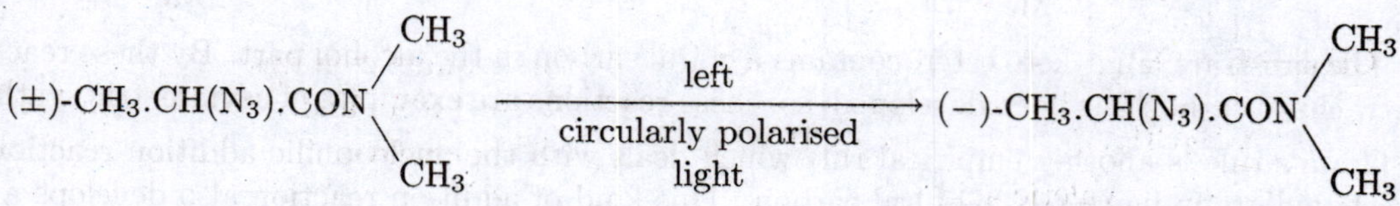

the rotation is less than 1°

Mckenzie and his coworkers studied numerous numbers of asymmetric synthesis. His first example was published in J. Chem. Soc, 85, 1249 (1904). In this work he exemplified the reduction of the phenyl glyoxylate ester of levorotatory menthol with aluminium amalgam. The product of reduction was enriched with ester of (−)-mandelic acid; it is the predominant product over the ester of (+)-mandelic acid.

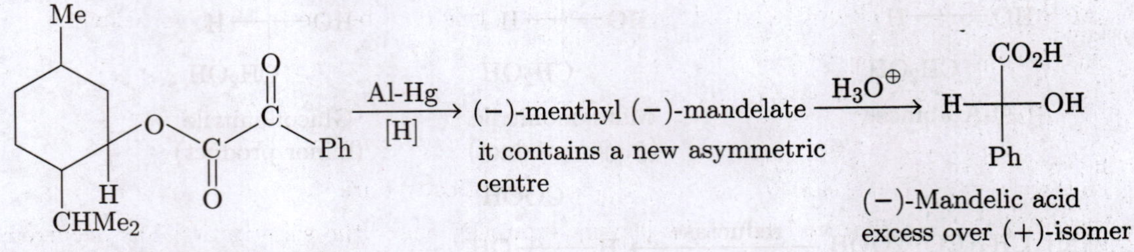

(−)-Menthyl phenylglyoxylate

In this connection Prelog stated an empirical correlation between the configuration of the predominant isomer [here ester of (−)-mandelic acid] formed and the corresponding configuration of the optically active alcohol [here (−)-menthol]. The statement made by Prelog may be extended for similar reactions to correlate the configurations of reactants and products and the rule is called Prelog's rule. The substrates are α-keto esters of optically active alcohols whose chiral carbons are bonded to oxygen atom of the ester group.

To understand the course of the reaction and the correlation made by Prelog we shall have to follow the convention for the representation of the substrate on the plane of the paper. The convention is : (i) ketonic carbonyl group and the ester carbonyl group should remain antiparallel, (ii) the three ligands bonded to the chiral carbon of the alcohol part are to be designated as small (S), medium (M) and large (L) according to their size. Prelog's rule states that the nuclophilic part of the reagent will attack the carbonyl carbon from the side of M to give the predominant product. An example is given below :

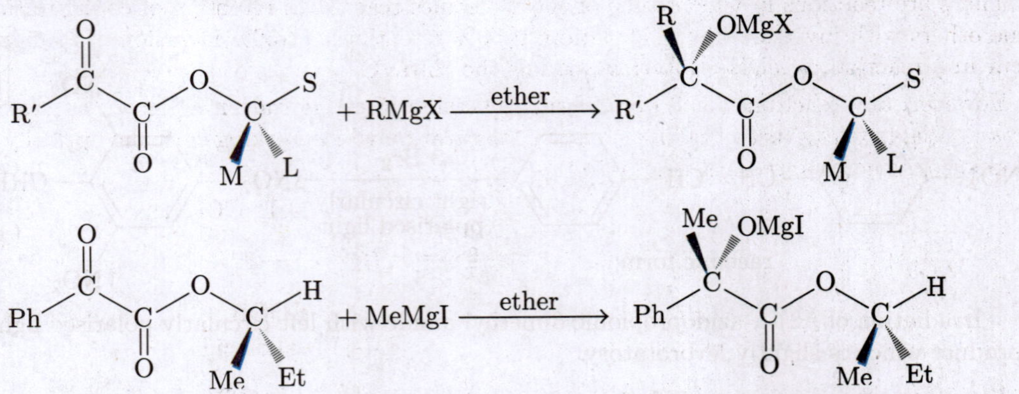

The substrate, an α-keto ester, contains a chiral carbon in the alcohol part. By these reactions a new chiral carbon has been developed; so, these reactions are examples of asymmetric synthesis.

Cram's rule is another empirical rule which deals with the nucleophilic addition reactions to a carbonyl group having an α-chiral carbon. This kind of addition reaction also develops a new chiral centre and comes under asymmetric synthesis. Cram's rule has been discussed in Sec. 16.3 in detail.

9.18 Substitution Reaction at a Chiral C

When a substitution reaction or a rearrangement reaction occurs on a chiral C, the configuration of the chiral C in the product molecules may or may not be the same as that of the substrate. There are some reactions in which the new substituent takes the position of the old in all the product molecules and the reaction is said to occur with retention of configuration; in this case, the product will be 100% optically pure. Schemetically this may be shown as given below :

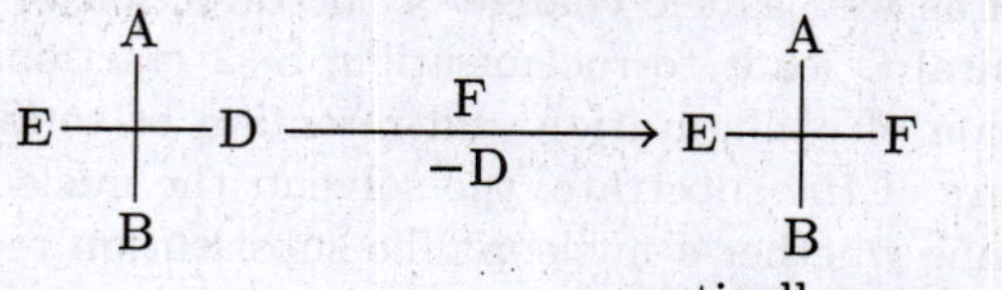

optically pure

Here F has taken the position of the leaving group D; hence the substitution has occurred with retention of configuration.

In some other reactions one may find that the incoming group has taken any position other than that of the leaving group in all the molecules of the product and the reaction is then said to occur with inversion of configuration. If all the molecules of the product have configurations opposite to those of the substrate molecules, the product will be optically pure. This is represented as shown below:

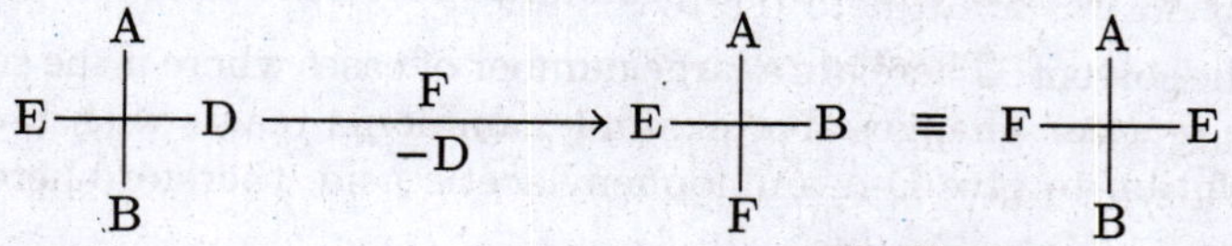

optically pure

Here the new substituent F has taken the position other than that of the old group D, i.e., the reaction has occurred with inversion of configuration; to indicate inversion, the curled arrow (⟶) or an arrow with a circle (⟶) is used.

There are reactions in which some of the molecules react with retention of configuration and some others with inversion of configuration; if 50% retention and 50% inversion of configuration occur in a reaction, it leads to racemisation at the chiral C.

However, any reaction which occurs with inversion of configuration is known as Walden inversion. Walden discovered the inversion of configuration while working with the optically active chlorosuccinic acids and malic acids:

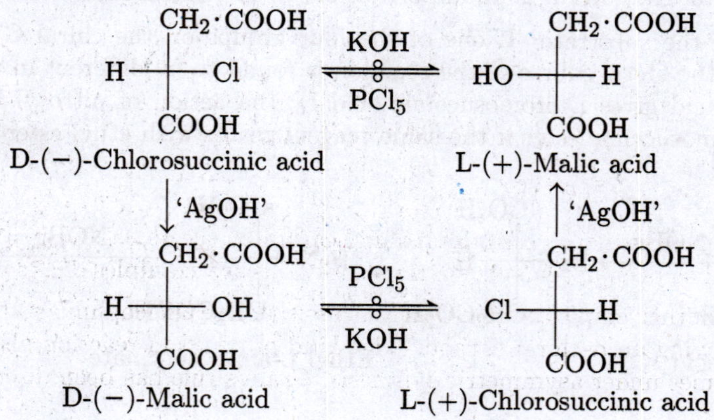

D-(−)-Chlorosuccinic acid; L-(+)-Malic acid; D-(−)-Malic acid; L-(+)-Chlorosuccinic acid

Walden argued: Since the conversion of (−)-malic acid into (+)-malic acid via (+)-chlorosuccinic acid involves conversion of one enantiomer into the other, inversion of configuration must have taken place in any one of the two steps—either in the conversion of (−)-malic acid into (+)-chlorosuccinic acid or in the conversion of the latter into (+)-malic acid. In this connection, it is to be remembered that the change of the direction of rotatory power is not related to the change of configuration, but conversion of one enantiomer into the other always indicates inversion of configuration as well as the change of the direction of rotatory power. It is found that S_NI reaction at a chiral C leads to racemisation; S_N2 reaction on a chiral C, on the other hand, occurs with inversion of configuration and retention of configuration is a feature of S_Ni reaction. Again, the nature of the substrate, the solvent, the nucleophile and the concentration of the nucleophile determine whether a nucleophillic substitution reaction will be S_N1 or S_N2 or S_Ni. Thus whether a substitution reaction will occur with or without inversion of configuration depends on the conditions of the reaction which are discussed below:

(a) Nature of the reagent: If we closely look at the configurations of substrates involved in the Walden's original experiment and the reagents used therein, we shall see that some reagents react with inversion of configuration of the substrate, while others react with retention of configuration. The reaction of D- or L-chlorosuccinic acid with KOH occurs with inversion of configuration, while the reaction of the same compound with moist silver oxide takes place with retention of configuration. Thus we see reagents have a profound effect on Walden inversion.

(b) Nature of the solvent: There are a large number of cases where if the solvent is changed, the steric course of the reaction changes. For example, ammonia reacts with L-α-bromophenylacetic acid in aqueous solution to give D-α-aminophenylacetic acid; therefore here the reaction occurs

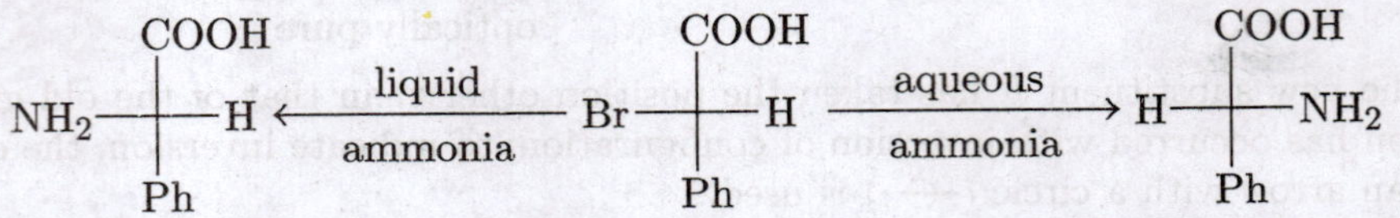

L-α-Aminophenylacetic acid — L-α-Bromophenylacetic acid — D-α-Aminophenylacetic acid

with inversion of configuration. On the other hand, one gets L-α-aminophenylacetic acid by the action of liquid ammonia on the L-α-bromophenylacetic acid; hence here the reaction occurs with the retention of configuration.

So, the nature of the solvent is an important factor in Walden inversion.

(c) Nature of the substrate: If one of the four groups on the chiral C is different in two chiral molecules, the steric course of the reaction is found to be different in many a case. Thus L-aminosuccinic acid gives L-bromosuccinic acid by the action of nitrosyl bromide, while the product is D-Bromosuccinic ester if the same reagent reacts with ethyl ester of L-aminosuccinic acid.

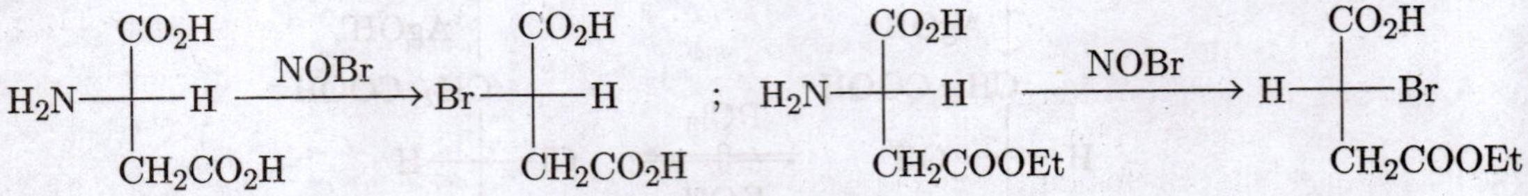

L-aminosuccinic acid — Ethyl L-aminosuccinate — D-bromosuccinate

9.19 Diastereomerism Involving Double Bonds And Rings

The phenomenon of existence of more than one compound having identical constitutional formula but differing in spatial arrangements of groups or atoms owing to the restricted rotation about a double bond (C=C, C=N etc.) or about a ring is known as **cis-trans** isomerism and the different compounds are called diastereomers (previously called geometric isomers). Thus, 1, 2-dibromoethene exists in nature in two forms (I & II) as given here, one being different form the other in the spatial arrangement of two bromine and two H atoms (Fig. 9.122).

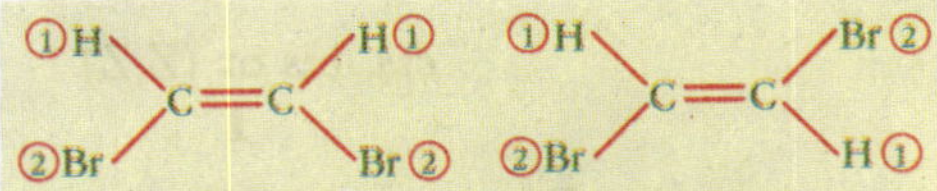

cis or (Z)– I *trans* or (E)– II

Fig. 9.122. Diastereomers of 1,2-dibromoethene

The two Cs being sp^2 hybridised, both of the forms are planar. The two Br atoms in the form I are on the same side of the double bond and the two H atoms are on the other side of it. This form, therefore, is designated as cis (as the same group are on the same side); this may also be named as (Z)-1, 2-dibromoethene as the two higher priority groups are on the same side. In form II, the higher priority groups are arranged in the opposite sides of the double bond and hence it is designated as (E)-1, 2-dibromoethene; this may also be called trans-form, the like groups being in the opposite sides of the double bond.

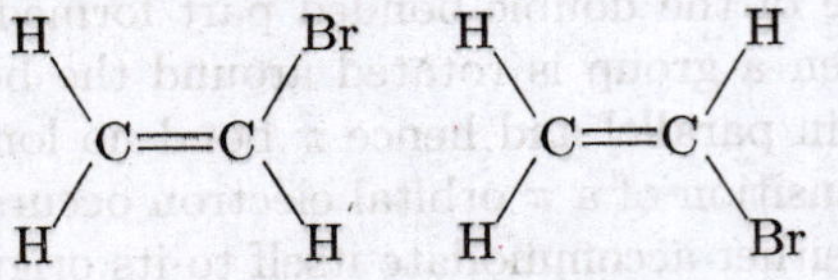

If any one of the double-bonded C atoms contains two identical groups or atoms, then geometrical isomerism does not arise because in that case two different spatial arrangements are not possible; thus, 1-bromoethene has no geometrical isomer since there are two H atoms on C_2 atom.

If there are n number of conjugated or isolated double bonds, there will be 2^n geometrical isomers provided all the substituents are different; if some of the substituents are identical, isomer number decreases. For example, 1-bromo-5-chloro-1, 4-pentadiene has four isomers while 1, 5-dibromo-1, 4-pentadiene has three such isomers (here cis, trans and trans, cis are identical).

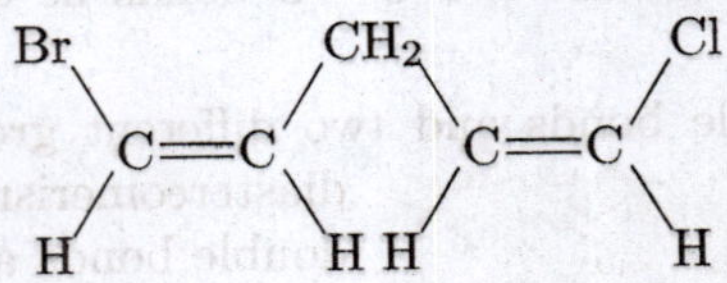

cis, cis or (Z,Z)

I

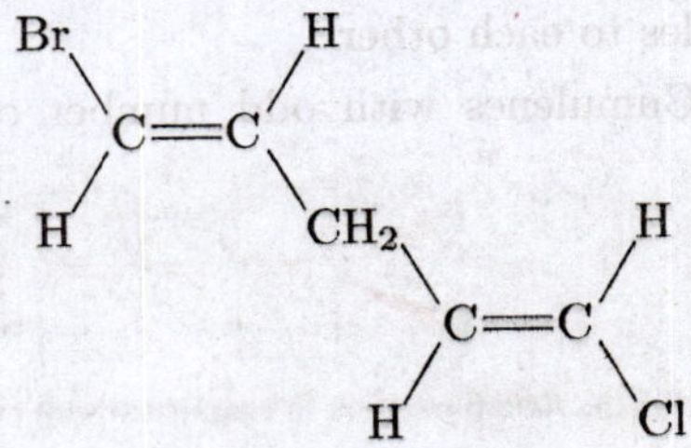

trans, trans or (E,E)

II

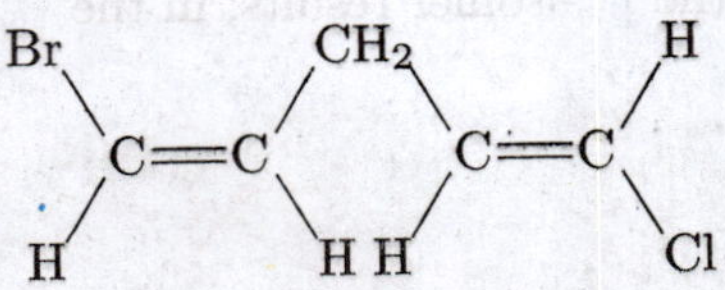

cis, trans or (Z,E)

III

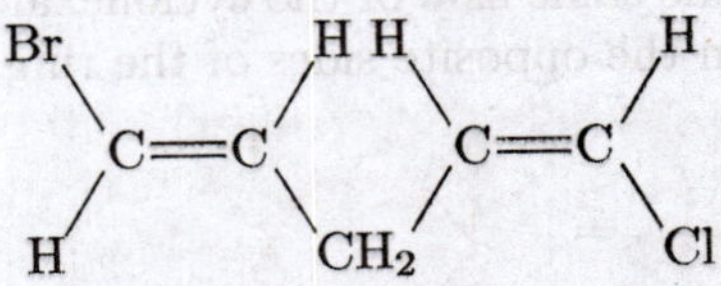

trans, cis or (E,Z)

IV

Geometric isomers of 1-bromo-5-chloro- 1,4-pentadiene

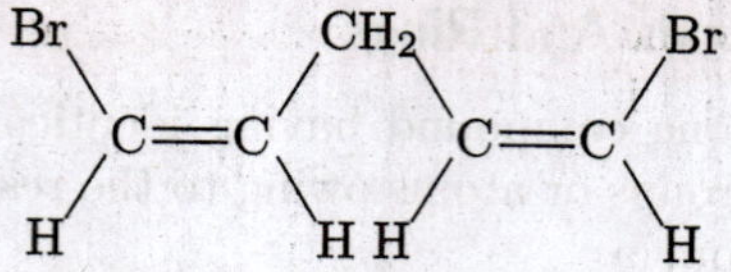

cis, cis or (Z,Z)
I

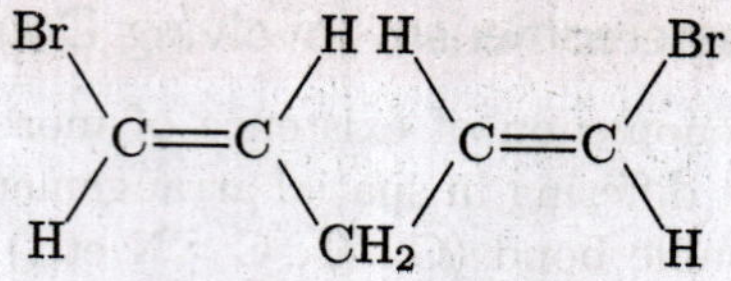

trans, trans or (E,E)
II

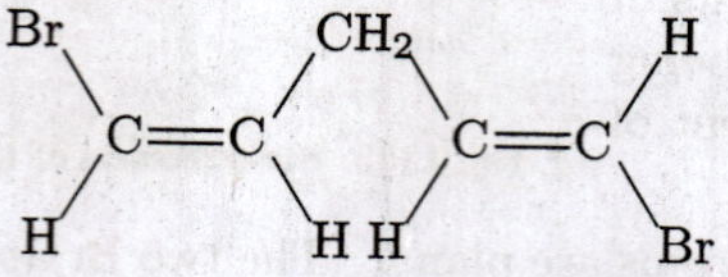

cis, trans or (Z,E)
III

≡

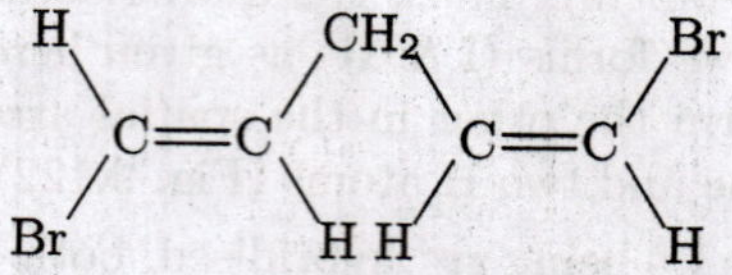

trans, cis or (E,Z)
IV

Geometric isomers of 1,5-dibromo-1,4-pentadiene

Now the question is: why do geometric isomers exist? The answer is: A double bond consists of a σ bond and a π-bond. σ bond has symmetrical cylindrical charge density and rotation of a group about the σ bond axis does not make any change in the charge density. A π bond orbital has charge density below the plane and above the plane of the double-bonded part formed by the side-on overlap of two parallel *p* orbitals. Thus when a group is rotated around the bond axis, π bond breaks, the parallel *p* orbitals do not remain parallel and hence π bond no longer exists. Actually when a cis or trans isomer is heated, transition of a π orbital electron occurs to the antibonding π orbital. When it comes down, it may either accommodate itself to its original orbital giving rise to the original configuration or it may accommodate itself to a π orbital formed after rotation of a part of the molecule around C—C single bond which was previously a double bond and this gives rise to the diastereomer of the original one. Therefore, thermally a mixture of cis and trans isomers can be obtained from a cis or trans isomer.

Allenes do not exhibit diastereomerism as the two C=C bonds lie on two planes at right angles to each other.

Cumulenes with odd number of double bonds and two different groups on end Cs show diastereomerism since alternate double bonds are on the same plane, those with even number of double bonds show enantiomerism (Fig. 9.123).

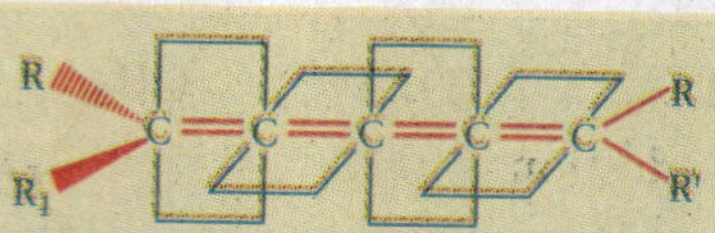

Fig. 9.123. Enantiomerism in cumulenes with even number of double bonds

Owing to the restricted rotation about a ring, diastereomerism may also develop. Thus, 1, 4-cyclohexanedicarboxylic acid exists in cis- and trans-forms. When the two carboxyl groups are present on the same side of the cyclohexane ring, the cis-isomer results; in the trans-isomer they are found on the opposite sides of the ring.

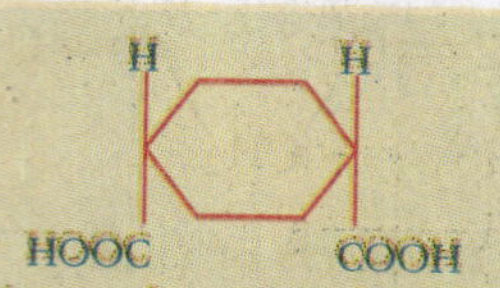

cis (being on the same side of the ring)

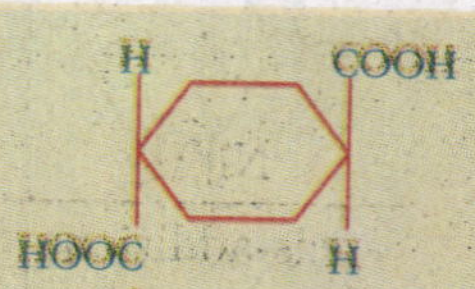

trans (being on the opposite side of the ring)

Fig. 9.124. Diastereomerism in ring compounds

Compounds containing carbon-nitrogen double bonds exhibit diastereomerism owing to the restricted rotation about the double bond; this we find in aldoximes and ketoximes (Fig. 9.125).

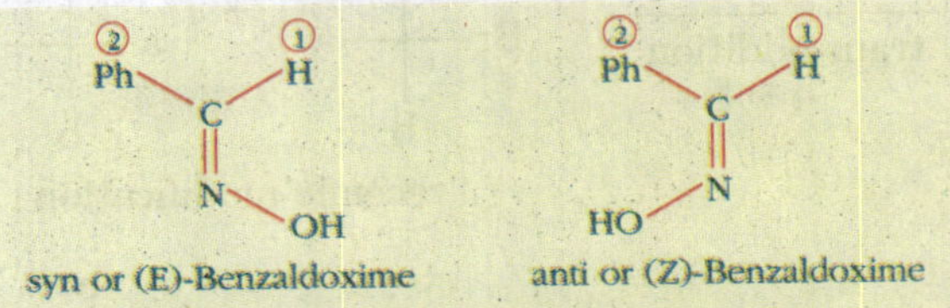

Fig. 9.125. Diastereomerism involving C=N

Configurations of geometrical isomers of oximes are denoted either as syn and anti (instead of cis or trans) or as (Z) and (E).

Diastereomers have different physical properties. As trans-isomers are more symmetrical than cis-isomers, trans-isomers have usually higher m.p., and higher solubilities in inert solvents than cis- isomers. The cis-isomers usually has a higher density, refractive index, dipole moment, heat of combustion than its trans-isomer; the higher heat of combustion of cis-isomer indicates its less thermal stability than that of the trans-isomer. These are the general properties and one may find exceptions.

As geometric isomers are not mirror-images of each other, these are called diastereomers and they have different properties.

As both of the isomers have the same functions, most of their chemical properties are identical. However, these differ in stereospecific reactions.

The two parts of an addendum may add to the C=C double bond either from the same side or from opposite sides; the former process is called cis-addition and the latter is termed as trans-addition. For example, catalytic hydrogenation of a double bond is a cis-process, whereas addition of bromine to a double bond is a trans-process. It has been found that if a cis-compound of the type abC=Cab (where $a \neq b \neq$ H) is catalytically hydrogenated, the product is a meso-compound. On the other hand its trans-isomer gives rise to racemic modification on catalytic hydrogenation. Again cis-isomer of abC=Cab ($a \neq b \neq$ Br) forms a racemic modification on bromination. A meso product results when the trans-isomer of the same structure is brominated. The steric courses of the reactions are shown below.

In fact, cis-addition to a cis-isomer and trans-addition to a trans-isomer give rise to meso-forms, while cis-addition to a trans-isomer and trans-addition to a cis-isomer form racemic modifications.

Owing to the greater thermal stability of trans-isomer than that of the cis-form, the latter can be converted into the former by heating below its melting point. The conversion of trans-form into the cis-form is not so easy; usually irradation of ultraviolet ray on a trans-form in the presence of bromine converts it into the cis-form partially and an equilibrium mixture is obtained (Sec. 13.12.1).

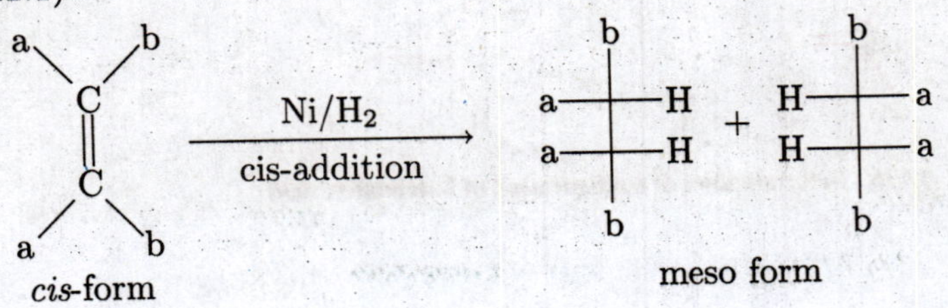

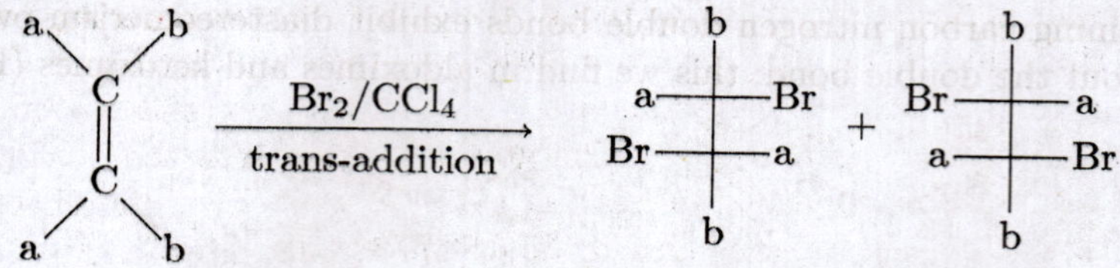

cis-form racemic modification

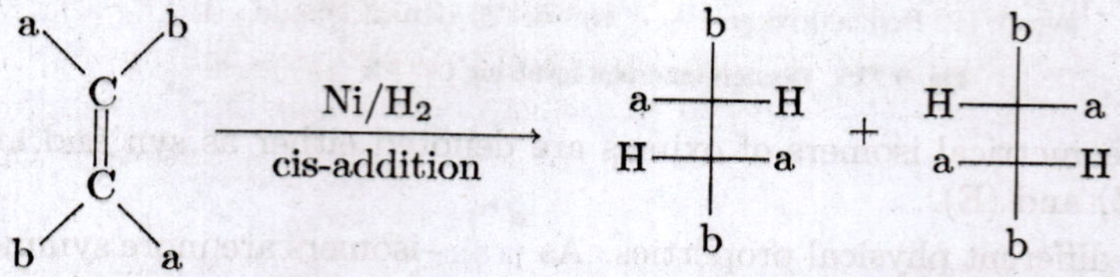

trans-form racemic modification

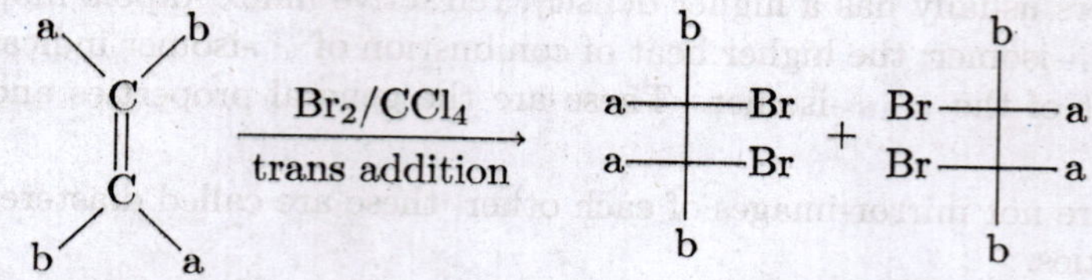

trans-form meso form

The determination of configuration of diastereomers (i.e., which one of a pair of isomers is cis and which one is trans) is done by both physical and chemical methods.

The most authentic physical methods are X-ray and electron diffraction studies. Nuclear magnetic resonance, infra-red and ultra-violet spectral analyses and comparison of melting points, solubilities, dipole moments etc. are also used for the purpose. The cis-isomer of abC═Cab type molecule has dipole moment but trans-isomer of the same type has no dipole moment.

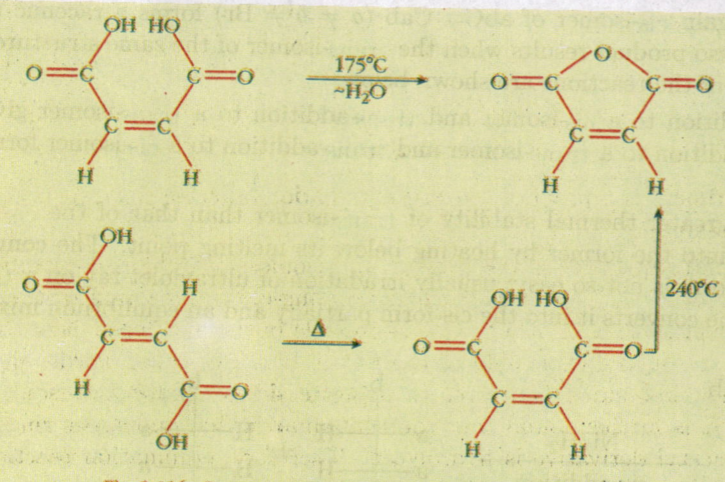

Fig. 9.126. Determination of configuration of butenedioic acid

The chemical method of ascertaining configuration of a pair of cis-trans isomers depends on the functional groups present in the structure. The most dependable chemical method is that of closing and opening of a ring; thus, if two groups are cis-oriented, they are capable of combining with each other to form a ring. If their orientations are trans, either they will not combine or will combine under drastic conditions (when trans-form gets converted into the cis-form). For example, cis-butenedioic acid forms anhydride on heating (175°C) while trans-butenedioic acid does not form the anhydride at 175°C. It forms anhydride at 240°C (Fig. 9.126).

The configurations of diastereomers of the structure abC═Cab may be determined by performing cis- and trans-addition reactions and then examining the product. If the addition is cis and the product is meso, the configuration of the starting compound is definitely cis; but if the product is a racemic modification, the compound under examination is of trans-configuration. Similarly, a meso product in a trans-addition indicates the trans-configuration, while the formation of racemic form indicates the cis configuration in trans-addition.

To determine the configurations of aldoximes, one first converts the given compound into its acetyl derivative and then hydrolyses the acetylated product with aqueous Na_2CO_3 solution. If the product of hydrolysis is the original compound, the configuration of the isomer is syn; if the same is a nitrile (formed by an elimination reaction), the configuration of the compound under examination is anti. This is because elimination of a molecule takes place when its constituent parts are in anti-positions with respect to each other (under basic conditions).

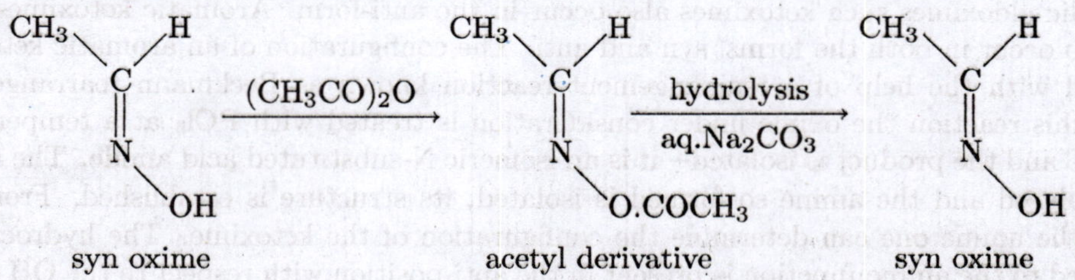

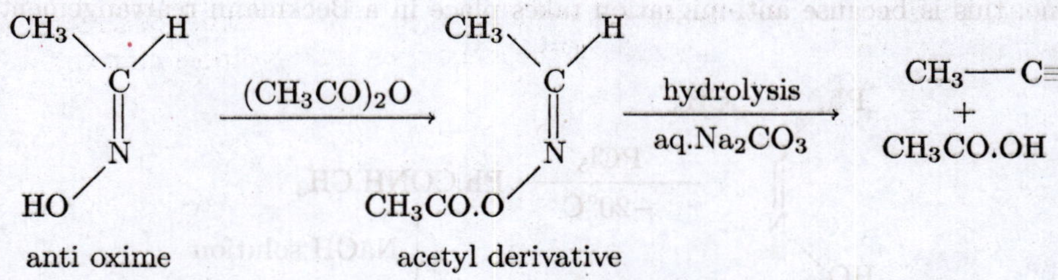

The following experiment is a proof of the anti elimination reaction that the acetyl derivative of an anti aldoxime undergoes:

When fused with solid NaOH one of the two syn anti-isomers of 2-chloro-5-nitrobenzaldoxime undergoes ring-closure reaction. That isomer undergoes ring-closure reaction which contains the Cl and OH groups very close to each other. This is possible in the anti form: here the OH group and the benzene ring are on the same side of the C═N bond; in other words, the OH group and the H atom on double-bonded C are on its opposite sides. Hence anti-isomer undergoes ring-closure reaction. It is further found that the compound which undergoes ring closure also gives nitrile when its acetyl derivative is hydrolysed. Therefore, elimination reaction occurs in the acetyl derivative of the anti-isomer.

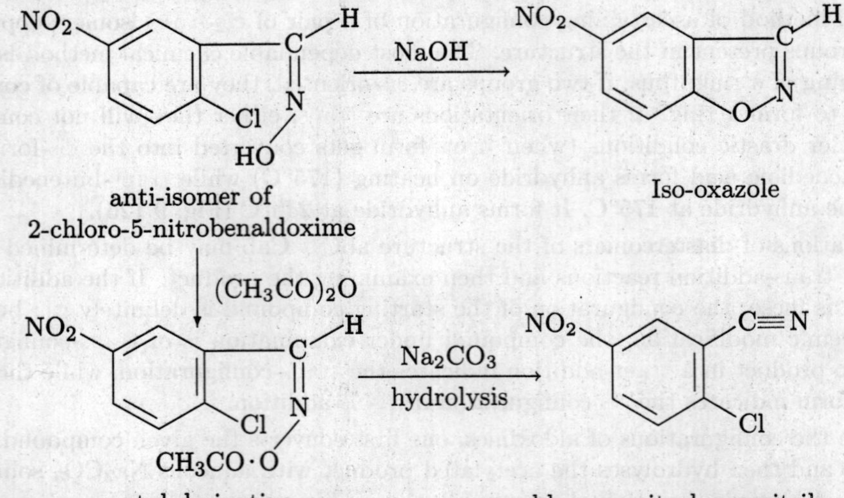

anti-isomer of 2-chloro-5-nitrobenaldoxime

Iso-oxazole

acetyl derivative

o-chloro-*m*-nitrobenzonitrile

Note: Aliphatic aldoximes occur in one form only and it is the anti-form. The question of assigning configurational designations arises in the case of the aromatic aldoximes only.

Like acyclic aldoximes such ketoximes also occur in the anti-form. Aromatic ketoximes have been found to occur in both the forms, syn and anti. The configuration of an aromatic ketoxime is determined with the help of a rearrangement reaction known as Beckmann rearrangement reaction; in this reaction the oxime under consideration is treated with PCl_5 at a temperature below -20 °C and the product is isolated—it is an isomeric N-substituted acid amide. The amide is then hydrolysed and the amine so formed is isolated; its structure is established. From the structure of the amine one can determine the configuration of the ketoxime. The hydrocarbon group attached to the amino function is present in the anti-position with respect to the OH group in the ketoxime; this is because anti-migration takes place in a Beckmann rearrangement (Sec. 14.5).

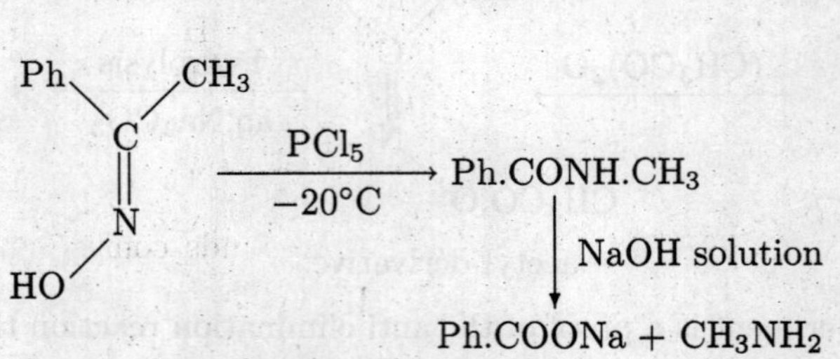

Problem 9.21 A ketoxime of the formula $C_9H_{11}ON$ gave aniline on Beckmann rearrangement and subsequent hydrolysis. Establish the configuration of the oxime.

Solution: Isolation of aniline ($C_6H_5NH_2$) indicates that the oxime contains a phenyl group and it is in the anti-position with respect to the —OH group; because in a Beckmann rearrangement, the anti-group migrates. Besides the C_6H_5 moitey, there is another hydrocarbon group consisting of two C atoms; the last C atom is of C═N group. The unknown hydrocarbon group must have five H atoms (11H – 6H; 5H of the C_6H_5 group and 1H of the OH group); therefore, the hydrocarbon group is CH_3CH_2. The ethyl group is in syn-position with respect to the OH group, the phenyl group being in anti-position.

Hence the configuration of the oxime is

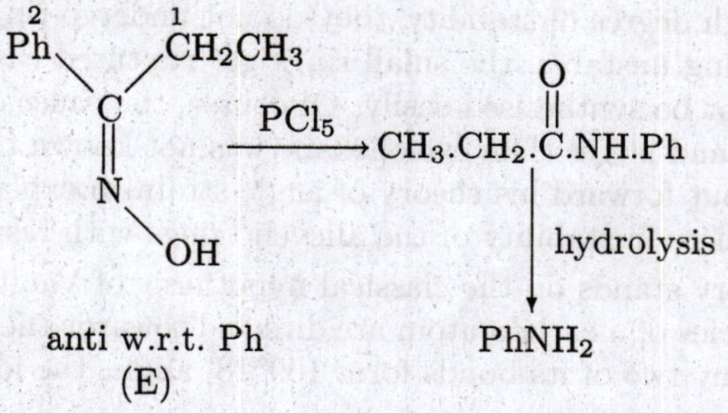

9.20 Isomerism of Cyclohexane and Its Derivatives

The very name of cyclohexane indicates that it has a cyclic structure. It consists of six sp^3-hybridised C atoms which constitute the ring; each of the C atoms is linked to two other C atoms and two H atoms by σ bonds. Naturally C—C—C, H—C—C and H—C—H bond angles are roughly 109°28′. Thus the molecule is non-planar, i.e., puckered, so that all the C atoms do not belong to one and the same plane.

The ring of an alicyclic compound being a rigid part, restricted rotation about a cyclohexane ring gives rise to diastereomers provided two or more of its H atoms are replaced by groups. Hence cis-trans isomerism of substituted cyclohexanes will be considered in this section.

Cyclohexane itself is an achiral molecule, because it has σ planes and centre of symmetry; but suitably substituted cyclohexane derivatives may be chiral and thus can exhibit enantiomerism, which too will be discussed here. There are $1C_3$ axis, $3C_2$ axes, $1S_6$ axis, $3\sigma_v$ planes, in cyclohexane.

Being comprised of six C—C σ bonds about which rotation is not free, cyclohexane and its derivatives remain in different conformations. We are specially interested in their conformational analysis. Diastereomerism and enantiomerism will be discussed here along with the conformations of cyclohexane derivatives. Before this we shall discuss Baeyer strain theory and Sachse-Mohr theory about the cycloalkane rings. Cyclohexane belongs to D_{3d} point group.

9.20.1 Baeyer strain theory

Before 1881 it was observed that alicyclic compounds consisting of five or six C atoms in the ring were frequently found among natural products; these compounds had also been easily synthesised in the laboratory. In 1881 Markownikoff and in 1882 Freund synthesised three and four membered ring compounds respectively but with difficulty. Perkin junior prepared both types of compounds in 1883. These compounds were found highly reactive. For example, cyclopropane reacts readily with bromine in carbon tetrachloride even in the dark to form 1, 3-dibromopropane. This reaction is not at all a substitution reaction as would be expected of a saturated hydrocarbon; instead, it is a ring-opening reaction which leads to addition of bromine to the terminal Cs. Cyclobutane too was found to undergo ring opening reaction like cyclopropane, but under more drastic condition; whereas cyclopentane and cyclohexane were found resistant toward ring opening reaction. At that time great difficulty had also been encountered in systhesising rings smaller than cyclopentane or larger than cyclohexane. In fact, rings containing more than seven Cs were not known. Thethen scientists related the ease of opening small rings and the difficulty in forming rings other than cyclopentane or cyclohexane to their stability. Since cyclopentane and cyclohexane have a high degree of stability, they do not undergo ring-rupturing reaction and can be formed readily. Being unstable, the small rings get ruptured easily; for the same reason small and large rings cannot be synthesised easily. Of course, the cause of instability of the rings smaller than cyclopentane and larger than cyclohexane was not known to them. At this juncture Adolf Von Baeyer (1885) put forward his theory of angle strain to explain the high reactivity of small rings and the differences in stability of the alicyclic rings with respect to size.

The Baeyer strain theory stands on the classical hypothesis of Van't Hoff and Le Bell which tells us that the four valencies of a carbon atom are directed towards the four corners of a regular tetrahedron and thereby any two of its bonds form 109°28′ angle, the ideal tetrahedral angle, at the centre of the tetrahedron.

The assumptions Baeyer made in his theory are:

(i) All the alicyclic rings are planar, i.e., these rings are regular polygons, e.g., pentagon, hexagon, heptagon, etc.

(ii) Any deviation of C—C—C bond angles of the ring carbons from the ideal tetrahedral angle develops strain, called bond angle strain, within the ring which, in its turn, brings instability to the ring.

(iii) The greater the angle strain in a ring, the less is its stability.

(iv) Bond angle strain is expressed in terms of the angle deviation, *d*, according to the following relationship:

$d = \frac{1}{2}$(109°28′—value of the C—C—C angle of the planar ring)

Since the strain is spread over two bonds, the factor, $\frac{1}{2}$ is put in the above expression.

On the basis of the Baeyer's assumptions one could easily calculate the angle strains of alicyclic rings and therefrom the differences in stability of the rings with respect to their size might at ease be drawn as shown below :

According to the older theory carbon-carbon double bond is nothing but a ring system consisting of two C atoms joined to each other by two single bonds only. The angle between these two Cs is thus 0°. Therefore, for a diatomic ring of carbons,

$$d = \frac{1}{2}(109°28' - 0°) = +54°44'.$$

The + value of the strain indicates that the bonds are forced together.

Angle strain in C=C being very high (+54°44′), it should be very unstable and highly reactive. Actually C=C is a highly reactive part of a molecule.

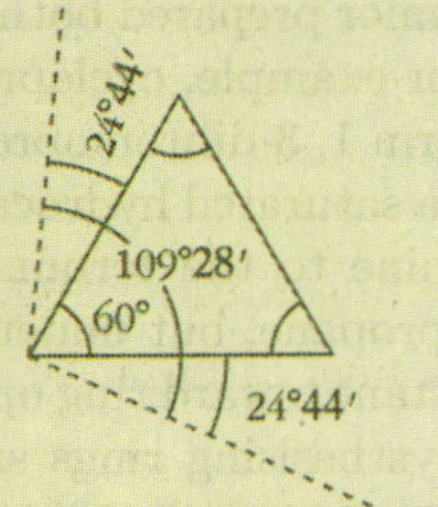

Fig. 9.127. Angle deviation in cyclopropane

If the three C atoms of cyclopropane lie on a plane it must have an equilateral triangular structure. Thus each of the C—C—C bond angles should be 60° instead of 109°28′, the ideal tetrahedral angle (Fig. 9.127). Therefore, for cyclopropane,

$$d = \frac{1}{2}(109°28' - 60°) = +24°44'$$

Because of this positive strain, cyclopropane should have low stability and high reactivity. This conclusion is in conformity with the experimental results.

If one assumes planar structure for the cyclobutane ring, then it should have a square planar structure in which C—C—C angle would be 90°. Thus for cyclobutane,

$$d = \frac{1}{2}(109°28' - 90°) = +9°44'$$

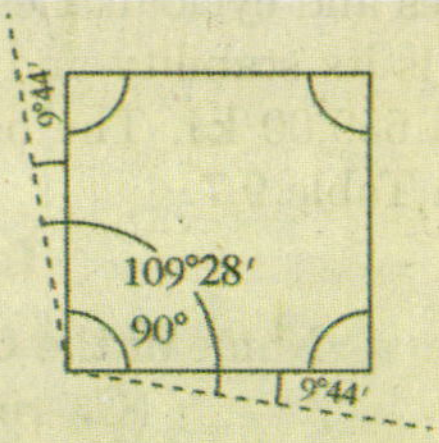

Fig. 9.128. Angle deviation in cyclobutane

Here also the angle strain is positive and of appreciable amount. Naturally cyclobutane should be unstable, but its stability must be higher than that of cyclopropane. Experiments show that cyclobutane undergoes ring opening reaction but more drastic conditions are required in comparison to cyclopropane.

For planar cyclopentane ring (C—C—C bond angle = 108°) having regular pentagonal structure,

$$d = \frac{1}{2}(109°28' - 108°) = +0°44'$$

Thus cyclopentane should be very stable since there is minimum of angle strain, +0°44′ only. Its low reactivity and ease of formation support this proposition.

For planar cyclohexane ring (C—C—C bond angle = 120°), which should assume regular hexagonal structure,

$$d = \frac{1}{2}(109°28' - 120°) = -5°44'$$

Therefore, cyclohexane ring has low negative angle strain, i.e., its bonds are forced apart to the extent of 5°44′. Hence it should be fairly stable, but its stability should be less than that of cyclopentane.

Similarly, for planar cycloheptane (a regular heptagon, ∠C—C—C=128°34′)

$$d = \frac{1}{2}(109°28' - 128°34') = -9°33'$$

Hence cycloheptane ring should be, unstable since it possesses a reasonable amount of negative strain. It is needless to mention that difficulty had been encountered to synthesise the compound.

As cycloalkane rings consisting of more than seven C atoms were unknown, Baeyer assumed that beyond cycloheptane with increasing ring size negative angle strain should go on increasing and as a consequence the stability of the rings should decrease in a regular manner. Thus according to his proposition cyclopentane should be the most stable cycloalkane because it possesses the minimum of angle strain. The position of cyclohexane is next to cyclopentane in the stability order. From cycloheptane onward with increasing ring size, the stability decreases because of the increasing negative angle strain. Small rings undergo ring-opening reaction because of their high positive angle strain. Therefore, all alicyclic rings other than cyclopentane or cyclohexane should be strained rings in so far as Bayer strain theory is concerned.

However, the theory was at once questioned by the contemporary scientists for the reasons noted below:

(a) No doubt an alkene is highly reactive but it can be prepared easily in excellent yield.

(b) Cyclohexane should not be strainless since it has appreciable amount of negative strain (−5°44′); but it is found to be more stable than cyclopentane.

(c) Cycloheptane should be more stable than cyclobutane since the former has less angle strain than the latter.

Nowadays the heat of combustion values are considered to be a dependable measure of stability of alkanes and cycloalkanes. The greater the heat of combustion relative to a similar compound, the less is its stability and *vice versa.* The heat of combustion value of a n-alkanes per mole of $>CH_2$ is 659.00 kJ. The heat of combustion values per mole of CH_2 (H_c) for cycloalkanes are noted in Table 9.7.

Table 9.7 Heat of combustion of cycloalkanes

Name of the Cycloalkane (C_n rings)	H_c in kJ mol^{-1}	Total strain $n(H_c—659)$ in kJ mol^{-1}
Ethene (C_2)	711.5	105.00
Cyclopropane (C_3)	696.9	113.70
Cyclobutane (C_4)	685.8	107.20
Cyclopentane (C_5)	664.0	25.00
Cyclohexane (C_6)	659.0	0.00
Cycloheptane (C_7)	662.4	23.80

From Table 9.7 it is found that (i) cyclohexane is the most stable cycloalkane since its total strain is nil, (ii) cycloheptane is more stable than cyclopentane. If we tabulate the total strain of the higher cycloalkanes, it will be revealed that the higher rings are more stable than cyclopentane except cyclononane. Thus, the strain theory is found to be defective and hence it cannot be accepted.

In spite of the defects of the theory it correctly explains the high reactivity of cyclopropane ring. This is because of the fact that cyclopropane actually possesses a planar triangular structure. The C.s that constitute the ring have no other alternative than to lie on a plane. All the C—C bonds being equal in length, the 3C.s constitute an equilateral and hence an equiangular triangle in which C—C—C bond angles are 60°. All the C.s being in sp^3 hybridised state, the angles should have been 109°28′ and hence there is really high internal angle strain. This makes the molecule unstable and highly reactive.

The reactivity of the cyclopropane ring may also be explained by the modern theory of covalent bond formation. Cyclopropane ring being planar, the C—C bonds are formed by the oblique overlap, of sp^3 A.O.s to form bent bond or banana bond (Sec. 4.8.2). The overlap being neither head-to-head (as in σ-bond) nor side-to-side (π-bond), the bent bond has a strength intermediate between the σ and π-bonds. The charge density along the bond axis is low and hence the C—C bonds in cyclopropane are weak, i.e., reactive. The C—C bonds in cyclobutane are less bent than those in cyclopropane. This makes cyclobutane more stable than cyclopropane (Fig. 9.129).

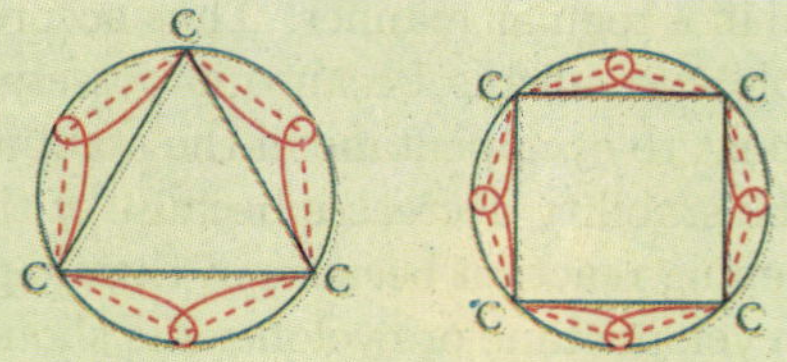

Fig. 9.129. Bent bonds in cyclopropane and in cyclobutane

However, Baeyer strain theory fails miserably to explain the stability of C_5 and higher rings. This is because his assumptions were wrong:

(i) C_5 and higher rings are not planar, instead these have non-planar structures; even cyclobutane ring need not be planar.

(ii) Angle strain of large cycloalkane rings is not at all responsible for the difficulties that arise in synthesising them. The formation of cyclic structure from an open chain compound involves intramolecular ring-closure reaction. If one wants to synthesise a large ring, a long chain starting material is to be chosen. In that case the probability of collision between two ends of the same chain for the intramolecular ring-closing reaction is far less than the probability of collision between two ends of two different chains which leads to intermolecular reaction. With increasing dilution of the solution the probability of intramolecular ring-closing reaction increases and that of intermolecular reaction decreases. In fact, in very dilute solution large rings have been prepared, though slowly; this technique is called high dilution principle.

9.20.2 Sachse-Mohr theory : the theory of strainless rings

From the consideration of heat of combustion values of alicyclic rings it is obvious that rings larger than C_4 are almost strainless and quite stable though according to Baeyer strain theory rings larger than cyclohexane are strained rings. Sachse in 1890 suggested that such rings can be free of angle strain if all the ring carbons are not forced into one plane. In other words he proposed non-planar puckered structures for cycloalkanes beyond cyclopentane in which all the C atoms could display their ideal tetrahedral bond angle. Thus one might not consider the angle strain of the rings which was held responsible for their instability. He pointed out that two non-planar, i.e., puckered models of cyclohexane, called chair and boat forms, could be constructed in which tetrahedral angles were retained by the ring carbons and therefore those models would be free of angle strain (Fig. 9.130 & 9.131).

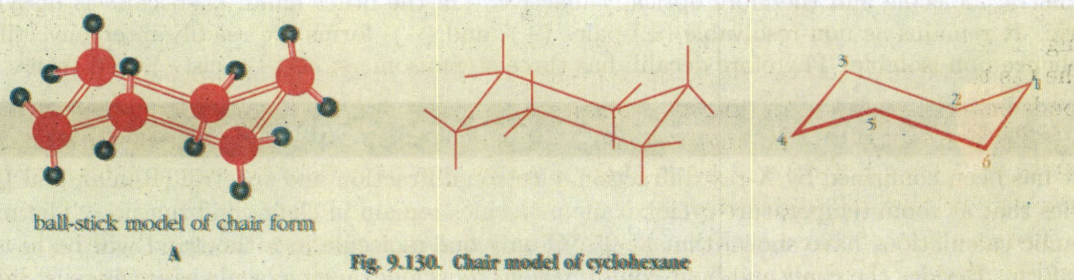

Fig. 9.130. Chair model of cyclohexane

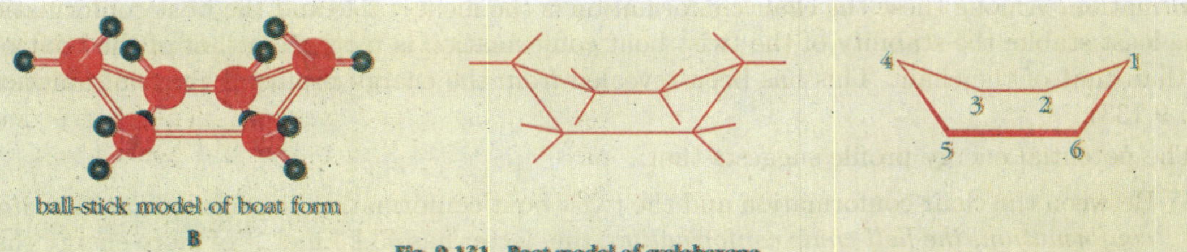

Fig. 9.131. Boat model of cyclohexane

The model A is said to be the chair form since it looks like a chair. Let us imagine that C_2—C_3 and C_5—C_6 form the seat of the chair, C_2—C_1—C_6 forms the back rest while C_3—C_4—C_5 forms the foot rest.

The model B is popularly known as boat form. C_1 and C_4 form the keel of the boat, and C_2—C_3 and C_5—C_6 form the sides of the boat.

Sachse suggested that these forms might be isolated but none was able to isolate them. Partly for this erroneous prediction Sachse's ideas about the three dimensional geometry of cyclohexane were not then accepted. In 1918, E. Mohr applied Sachse's theory to decalin (decahydronaphthalene, $C_{10}H_{18}$) and predicted that the cis- and trans-forms of decalin might exist. W. Huckel in 1925 experimentally confirmed the prediction of E. Mohr about cis- and trans-decalins (Fig. 9.132).

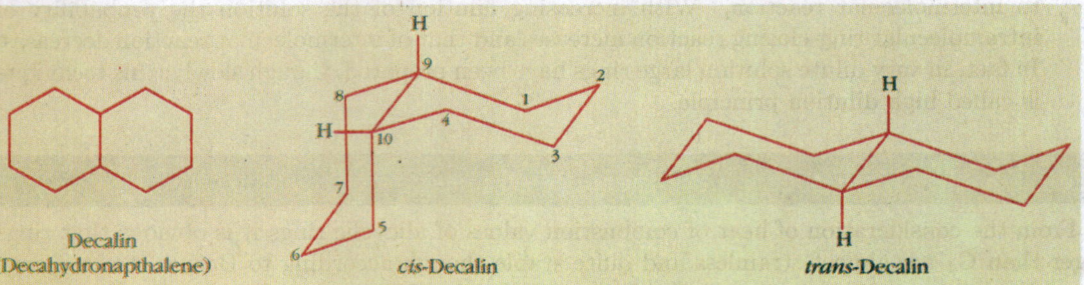

Fig. 9.132. *cis-* **and** *trans-***Decalins**

Sachse's theory being thus revived by E. Mohr, the theory of strainless puckered rings are known as Sachse-Mohr theory. Following the theory we can conclude that from C_5 all higher rings, being puckered, are strainless rings.

Note: cis-Decalin contains about 11.07 kJ mol^{-1} of extra energy than trans-decalin; this means that trans-decalin is more stable than cis-decalin. trans-Decalin possesses both centre of symmetry (midway between C_9—C_{10} bond) and twofold alternating axis of symmetry; thus it is a symmetric molecule and therefore optically inactive. On the other hand, cis-decalin is dissymmetric. It remains as non-resolvable (±)-pair. (+)- and (−)- forms are readily interconvertible and hence non-isolable. Therefore decalin has three stereoisomers; cis-(+), cis(−), and trans.

9.20.3 Conformations and conformational analysis of cyclohexane

It has been confirmed by X-ray diffraction, electron diffraction and spectral (Raman and IR) studies that at room temperature cyclohexane molecules remain in chair conformation. Thermodynamic calculations have shown that at 25 °C only one molecule in a thousand will be in the boat form. Besides the chair and boat conformations a cyclohexane molecule assumes twist boat conformation. Among these the chair conformation is the most stable and the boat conformation is the least stable; the stability of the twist boat conformation is more than that of the boat but less than that of the chair. This has been revealed from the energy profile of the conformations (Fig. 9.133).

The potential energy profile suggests that:

(i) Between the chair conformation and the twist boat conformation there is a *transition state conformation, the half chair conformation*; the, latter has 43 kJ mol^{-1} of more energy than the chair form.

(ii) The difference in energy between the chair form and the twist form is 23.00 kJ mol^{-1}; while that between the chair and the boat is 26.7 kJ mole^{-1}.

(iii) Between the two twist boat forms lies one boat form and between the two boat forms one twist boat form appears.

(iv) Only 43 kJ mol^{-1} of energy is sufficient to interconvert the conformations.

(v) Since independent existence of isomers requires an energy difference of at least 80 kJ mol^{-1}, these conformations are not isolable, rather these are interconvertible.

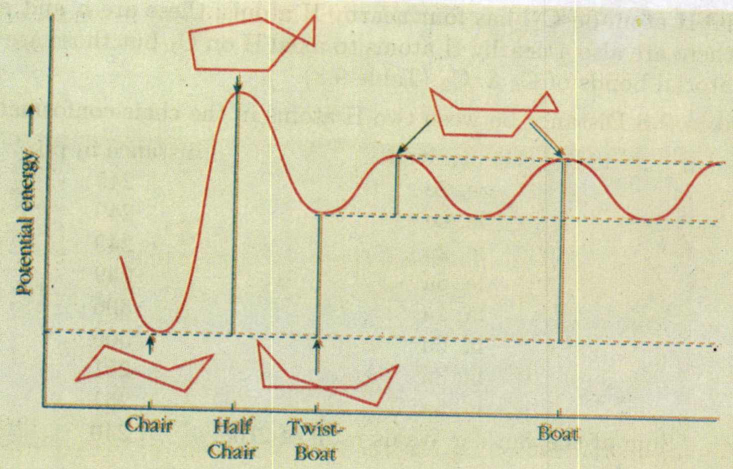

Fig. 9.133. The potential energy profile for conformations for cyclohexane

However, to account for the stability order of the conformations of cyclohexane we shall have to discuss these in detail.

Chair conformation of cyclohexane

To avoid strain, rotation of several C—C bonds of a cyclohexane molecule occurs at a time and thereby the molecule assumes a conformation in which alternate C atoms occupy two separate planes to produce a rigid puckered structure, which is popularly called chair conformation because its shape resembles that of a chair (Fig. 9.134). A close look at the drawing of the chair conformation reveals that there are two possible kinds of C—H bonds:

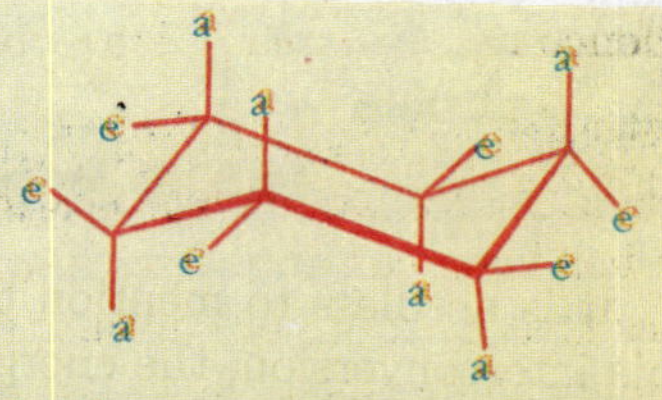

Fig. 9.134. Chair conformation of cyclohexane

(i) Axial bonds—those pointing up and down (denoted by **a**).

(ii) Equatorial bonds—those pointing sideways (denoted by **e**).

If the chair conformation is supposed to be enclosed in an imaginary sphere (comparable to the earth), then one can conceive a middle belt of the sphere, the equator, and a polar axis which are mutually perpendicular to each other (Fig. 9.135). A set of six C—H bonds are then found to be parallel to the polar axis and so these can be named as axial bonds; three of these axial Hs on alternate Cs point up and other three Hs point down. There is another set of six C—H bonds each of which makes an angle 109°28′ with the axis, i.e. these are almost parallel to the equator (they make 19°28′ angle with it) and are called equatorial bonds. Each C in the chair conformation has one equatorial C—H bond and one axial C—H bond. One of these two bonds points above the imaginary 'plane' of the molecule and the other below it. Thus three axial Hs on three alternate Cs and three equatorial Hs on the other three Cs point above the plane while the remaining three alternate axial Hs and three equatorial Hs point below the plane.

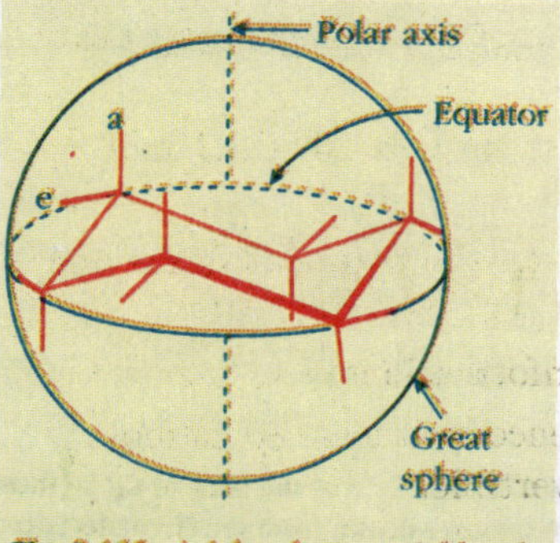

Fig. 9.135. Axial and equatorial bonds

The equatorial H atom on C_1, has four nearby H atoms; these are ***a*** and ***e*** on C_2 and those on C_6, whereas there are also 4 nearby H atoms to axial H on C_1 but those are axial bonds of C_3 and C_5 and equatorial bonds of C_6 & C_2 (Table 9.8).

Table 9.8 Distance between two H atoms in the chair conformation.

Hs involved	distance in pm
le, 2e	249
le, 6e	249
le, 2a	249
le, 6a	249
la, 2a	306
la, 6a	306
la, 3a	251
la, 5a	251
Sum of the van der Waals radii of 2Hs	240

However, the rigid chair conformation of the cyclohexane changes into another chair conformation (Fig. 9.136) and when this occurs all the six equatorial C—H bonds of the previous chair form become axial and axial C—H bonds become equatorial. This interconversion requires 43 kJ mol^{-1} of energy though the two forms are identical.

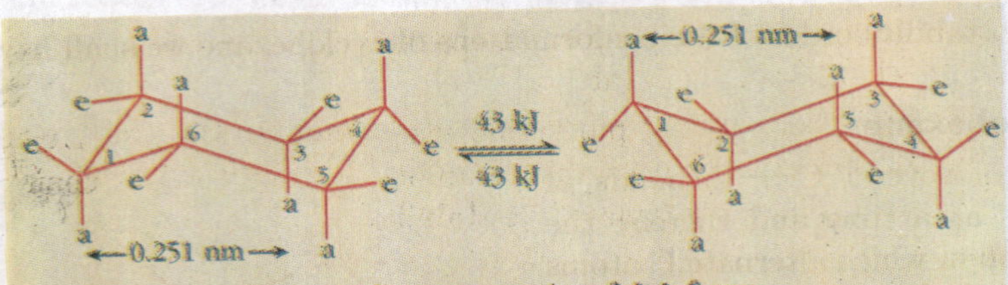

Fig. 9.136. Interconversion of chair forms

It is needless to mention that the half chair transition state conformation being involved in this interconversion, the energy requirement is so high.

Now the question is : Why is the Chair conformation free of strain? To answer the question we shall have to study the chair model and its Newman projection formula as given below (Fig. 9.137).

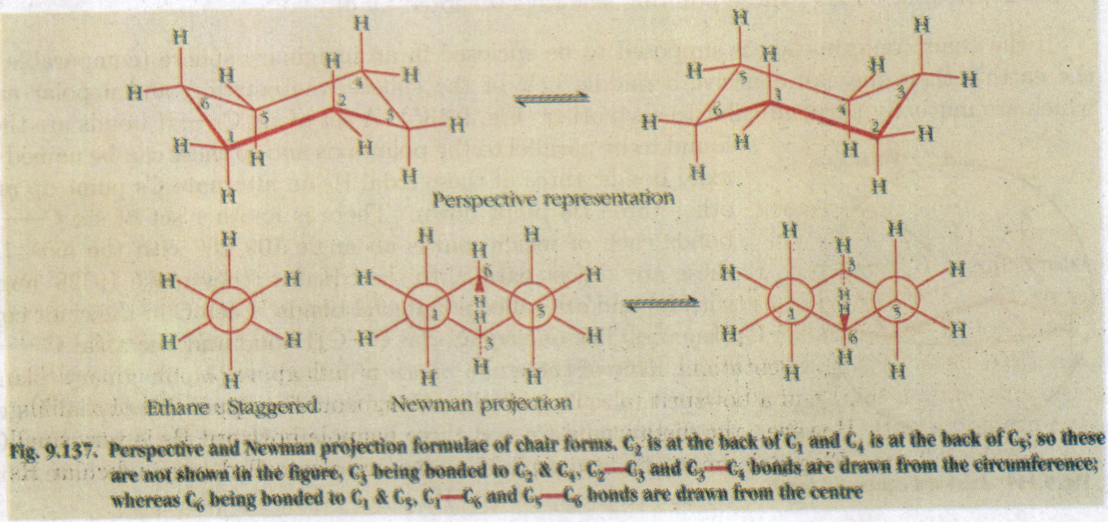

Fig. 9.137. Perspective and Newman projection formulae of chair forms. C_2 is at the back of C_1 and C_4 is at the back of C_5; so these are not shown in the figure, C_3 being bonded to C_2 & C_4, C_2—C_3 and C_3—C_4 bonds are drawn from the circumference; whereas C_6 being bonded to C_1 & C_5, C_1—C_6 and C_5—C_6 bonds are drawn from the centre

From the Newman projection formulae it is clear that all the C—H bonds in the chair conformation of cyclohexane are staggered (comparable to the staggered conformation of ethane) and hence there is no bond opposition or torsional strain; but there are six units of *n*-butane gauche interactions equivalent to 3.3×6 or 19.8 kJ mol^{-1}. The *n*-butane gauche interactions are between two $>CH_2$ groups at two ends of the following *n*-butane units: $C_1—C_2—C_3—C_4$, $C_2—C_3—C_4—C_5$, $C_3—C_4—C_5—C_6$, $C_4—C_5—C_6—C_1$, $C_5—C_6—C_1—C_2$ and $C_6—C_1—C_2—C_3$. It is needless to mention that van der Waals radii of $—CH_3$ and $>CH_2$ are one and the same and the value is 200 pm.

Again, owing to the puckered shapes, each of the Cs exhibits ideal tetrahedral bond arrangements and therefore this conformation is also free of angle strain.

Furthermore, the distance between an H on one C and that on another C of the chair form being greater than the sum of the van der Waals radii of two Hs, (240 pm; Table 9.8), the van der Waals (steric) strain is also absent in the chair conformation.

Being composed of C and H atoms, a cyclohexane molecule does not have any bond dipole practically. The question of intramolecular dipole-dipole interaction, therefore, does not arise. So the chair form is free of dipole-dipole interaction too.

All sorts of strain, except *n*-butane gauche interactions, being absent in the chair form, it is strainless, which, in its turn, makes the chair form very stable.

Note: In general most of the six membered rings, pyran, piperidine, etc. remain in the preferred chair form because of its high stability. The atomic crystal of diamond is composed of chair cyclohexane rings, and that is why diamond possesses remarkable stability.

Boat conformation of cyclohexane

Owing to the rotation of several C—C bonds at a time, a cyclohexane molecule may assume a flexible puckered ring structure in which two C atoms of two opposite corners of the non-planar hexagon are bent up leaving the remaining four Cs in a plane to give a boat-like shape to the ring which is called boat conformation (Fig. 9.138).

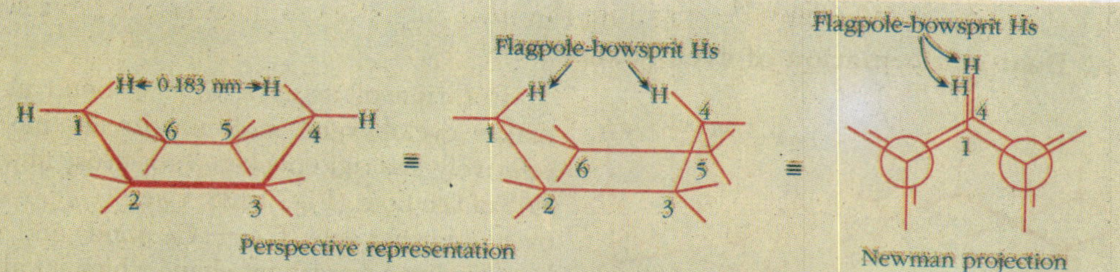

Fig. 9.138. Boat conformation of cyclohexane

Newman projection formula of the boat conformation indicates clearly that there are four pairs of eclipsed H atoms on $C_2—C_3$, $C_5—C_6$ parts, Two C—H bonds, one on C_1, and the other on C_4, are very close to each other and are called bowsprit-flagpole bonds; two pairs of eclipsed C—C bonds ($C_1—C_2/C_3—C_4$ and $C_4—C_5/C_1—C_6$) and these C—H bonds are so called because their orientations in space resemble those of a flagpole (a pole on which a flag is displayed in a ship) and a bowsprit (a stout pole projecting from the upper end of a sailing vessel for holding sail). However, the distance between the two flagpole-bowsprit Hs is very small and it is only 183 pm. Flag pole-bowsprit ($fp - bs$) H atoms are also called perpendicular Hs (perp).

Let us now discuss why boat is the least stable, i.e., the most flexible conformation among boat, chair and twist boat conformations of cyclohexane ring.

Boat form is free of angle strain because of its puckered ring structure; the C—C—C and H—C—C bond angles are approximately 109°28′ (as usual for sp^3 C atom).

It is also free of dipole-dipole interaction since there is no polar bond within the cyclohexane molecule.

On the other hand, it has four pairs of eclipsed C—H bonds on C_2—C_3 and C_5—C_6 parts and two pairs of eclipsed C—C bonds; thus the boat conformation experience bond eclipsing strain involving two, C_1—C_2—C_3—C_4 & C_1—C_6—C_5—C_4, eclipsed *n*-butane units and four units of *n*-butane gauche strain involving two terminal >CH_2 groups of C_2—C_3—C_4—C_5, C_3—C_4—C_5—C_6, C_5—C_6—C_1—C_2 and C_6—C_1—C_2—C_3.

This brings instability to the boat cyclohexane ring (see Newman projection formula, Fig. 9.138).

Furthermore, flagpole-bowsprit H atoms being at a distance (183 pm) less than the sum of their van der Waals radii (240 pm), there is a strong van der Waals strain in the boat conformation. This again brings instability to the boat cyclohexane ring.

Taking all these above said points under consideration, we can say that boat form is sufficiently strained while chair form is strainless. That is why the boat contains 28.5 kJ mol^{-1} of more energy than the chair as shown below (here 21 kJ is meant for eclipsed Me/Me interaction). However,

$$E_{boat} - E_{chair} = 2 \times E_{eclipsed\ n\text{-butane}} + 4 \times E_{n\text{-butane gauche}} + E_{fb\text{-bs}} - 6 \times E_{n\text{-butane gauche}}$$

$$\text{or, } 26.7 \text{ kJ mol}^{-1} = 2\times 21 \text{ kJ mol}^{-1} + 4\times 3.3 \text{ kJ mol}^{-1} - 6\times 3.3 \text{ kJ mol}^{-1} + E_{fb\text{-bs}} \text{ interaction}$$

$$\text{or, } E_{fb\text{-bs}} = -42 \text{ kJ mol}^{-1} + 6.6 \text{ kJ mol}^{-1} + 26.7 \text{ kJ mol}^{-1}.$$

$$= 5.2 \text{ kJ mol}^{-1}.$$

Twist Boat conformation of cyclohexane

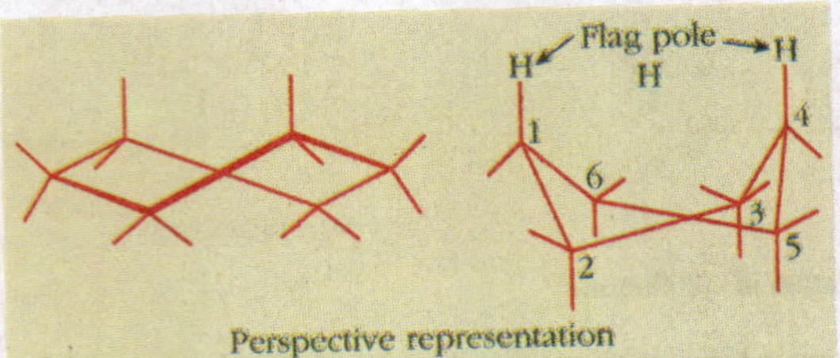

Fig. 9.139. Twist boat conformation

Rotation of many a C—C bond at a time, makes cyclohexane molecule to assume another relatively less rigid puckered structure in which one side of the boat (Fig. 9.138; C_2—C_3 bond) moves over the other side (C_5—C_6 bond) and thus giving the ring a twist boat shape which is called twist boat conformation (Fig. 9.139).

Four pairs of eclipsed C—H bonds in the boat become partly eclipsed in the twist boat, at the same time the two flagpole-bowsprit H atoms in the boat form become farther apart from each other in the twist boat. Therefore the twist boat has eclipsing strain. but it is less than that in the boat form; flagpole-bowsprit interaction (van der Waals strain) too is less in the twist boat form than that in the boat form. Moreover the twist boat does not have any angle strain and dipole- dipole interaction for reasons as stated for chair and boat conformations. Therefore the twist boat is less strained than the boat, but it is more strained than the chair since the latter form is strainless. This makes the twist form less stable than the chair form but more stable than the boat. Its energy is 3.37 kJ mol^{-1} less than that of the boat but 23.00 kJ mol^{-1} more than that of the chair.

Problem 9.22. What conformation of cyclopentane do you expect—planar pentagonal or puckered? Give reasons for your answer.

Solution: If cyclopentane takes up a planar structure, each of its C atoms will then display 108° bond angle; thus it will experience minimum of angle strain $\left[d - \frac{1}{2}(109°28' - 108°) = +0.44'\right]$. On the other hand there will be 5 pairs of eclipsed C—H bonds in the planar form; a substantial amount of bond opposition strain will, therefore, bring instability to the planar ring.

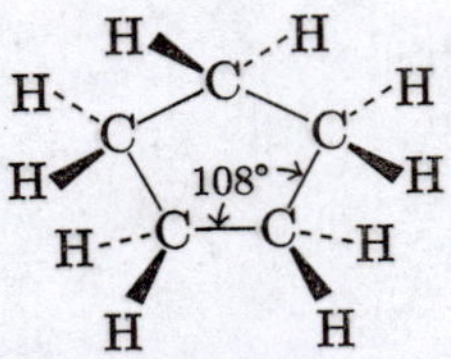

planar ring

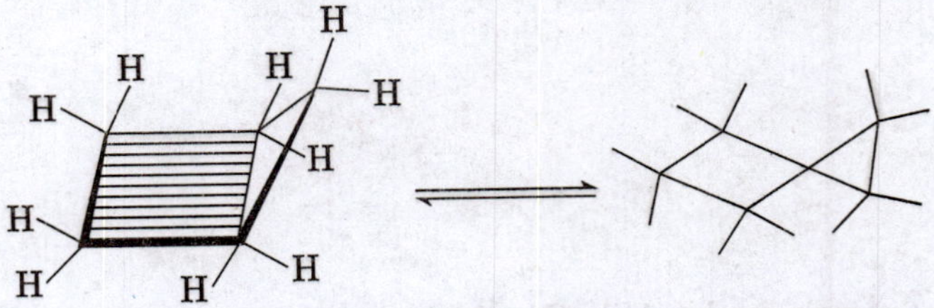

envelope & half-chair conformations puckered rings

When cyclopentane ring assumes a puckered structure, it does no longer experience any bond opposition strain but deviation of bond angles occurs in the ring. Torsional strain of five pairs of C—H bonds of the planar form is expected to be greater than the angle strain of the puckered form since deviation of the bond angle in the latter is not large. Thus cyclopentane will definitely prefer puckered structure.

The energy difference between planar and puckered forms is 17 to 21 kJ mol^{-1}.

In fact, there are two noteworthy puckered conformations of cyclopentane ring amongst others, the envelope and the half-chair.

In the envelope conformation 4C atoms lie on one plane while the fifth C puckers out of the plane. The puckering is not fixed but rotates around the ring which has been termed by Pitzer a pseudorotation. In the half chair form 3C atoms lie on one plane and the other two C atoms lie on another plane. The interconversion between the envelope form and the half-chair form does not involve substantial amount of potential energy and hence an equilibrium may exist between them. Of course, substituted cyclopentanes assume envelope conformation for the greater stability.

Problem 9.23. Account for the non-coplanarity of ring C atoms of cyclobutane.

Solution: In the planar structure of cyclobutane there are angle strain as well as four pairs of C—H bond eclipsing strain. Puckering of the ring reduces considerably the eclipsing strain and presumably for this reason the C atoms of the cyclobutane ring are not coplanar. Actually cyclobutane remains in two equivalent puckered conformations in an equilibrium mixture. In each form the C atoms remain on two planes with an angle of about 20° between them.

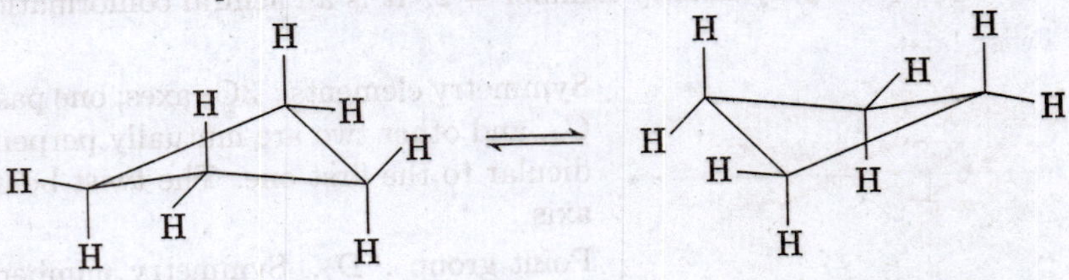

9.20.4 Symmetry elements and point groups of conformations of cyclohexane

(i) Chair conformation

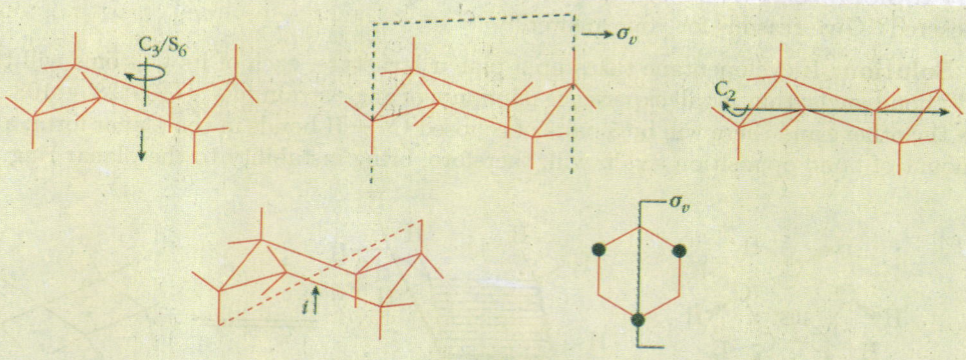

Fig. 9.140

(a) principal axis is C_3 which passes through the centre of the chair conformation and parallel to the axial bonds.

(b) $3C_2$ axes; each bisects a pair of opposite sides (C_1—C_2 & C_4—C_5; C_2—C_3 & C_5—C_6 and C_3—C_4 & C_6—C_1).

(c) $3\sigma_v s$: each bisects C_1 >CH_2 and C_4 >CH_2 units and holds the principal axis.

(d) i : i is the centre of chair conformation.

(e) S_6 : C_3 is S_6.

Point group: D_{3d}; symmetry number: 6. It is an achiral conformation.

(ii) Boat conformation

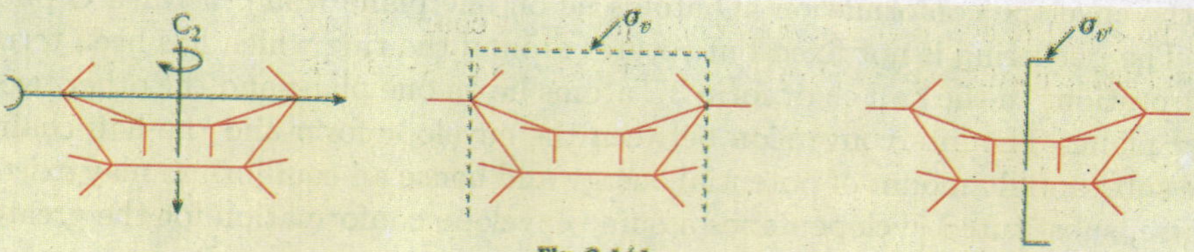

Fig. 9.141

Symmetry elements : a. C_2 axis the principal axis which passes through the centre of the boat and perpendicular to the plane of C_2—C_3 & C_5—C_6 bonds.

b. $2\sigma_v s$: One bisects C_2—C_3 & C_5—C_6 bonds and holds the principal axis; the other bisects C_1 >CH_2 and C_4 >CH_2 groups and holds the principal axis C_2. It has S_1 axis.

Point group : C_2; symmetry number = 2. It is an achiral conformation.

(iii) Twist boat

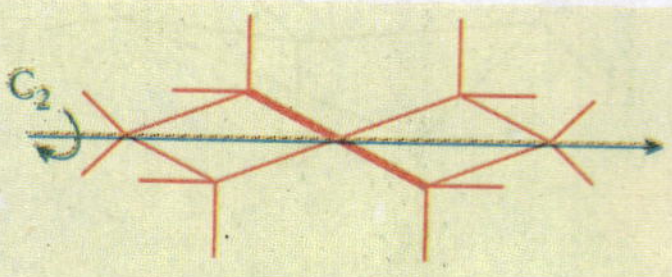

Fig. 9.142

Symmetry elements : $3C_2$ axes; one passing through C_1 and C_4, and other two are mutually perpendicular and perpendicular to the first one. The twist boat has no σ, i and S_n axis.

Point group : D_2. Symmetry number = 4. It is a chiral conformation.

9.20.5 Stereoisomerism of monosubstituted cyclohexanes

A. Conformational analysis

Being the most stable of all conformations, chair is the preferred conformation of a monosubstituted cyclohexane in general. When an H atom of cyclohexane is replaced by a group or atom (CH_3, OH, Cl, etc.), an equilibrium mixture of two chair forms is obtained. One chair form holds the substituent in the equatorial position and the other in the axial position.

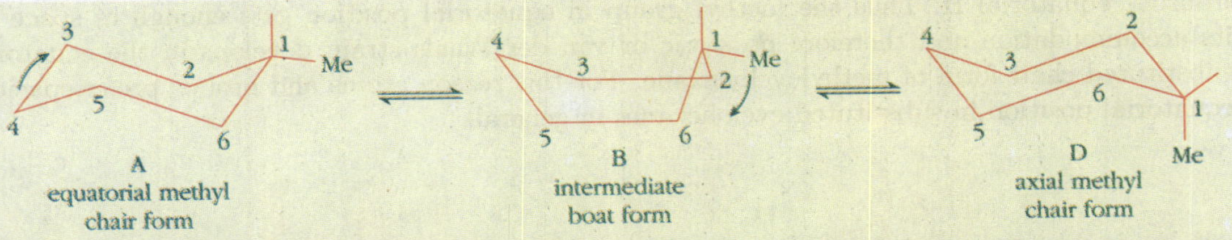

Fig. 9.143. Conformation of methyl cyclohexane. When C_4 of A is pushed up, it becomes B, the boat form; when C_1 of the latter is moved down, D results.

It has been found in general that the equatorial substituted chair form is more stable than the axial substituted chair form. For methylcyclohexane the axial substituted chair conformation has 7.56 kJ mol^{-1} of more energy than the equatorial substituted chair conformation (Fig. 9.143). Quantitative analysis of their equilibrium mixture reveals that it contains 95% equatorial chair form and 5% axial chair form. This too indicates the greater stability of the former than that of the latter. Now what are the causes behind the stability order?

Since both the forms are chairs, the question of angle and eclipsing strains do not arise; at the same time methyl being the only substituent, dipole-dipole interaction too is absent in both of the forms. Obviously bulk of the methyl group is the cause for the difference in the energy content of the two forms. Either the bulky methyl group develops greater van der Waals (steric) strain in the axial chair form than in the equatorial chair form or it creates gauche interaction (similar to that encountered in gauche conformations of n-butane) in the axial chair.

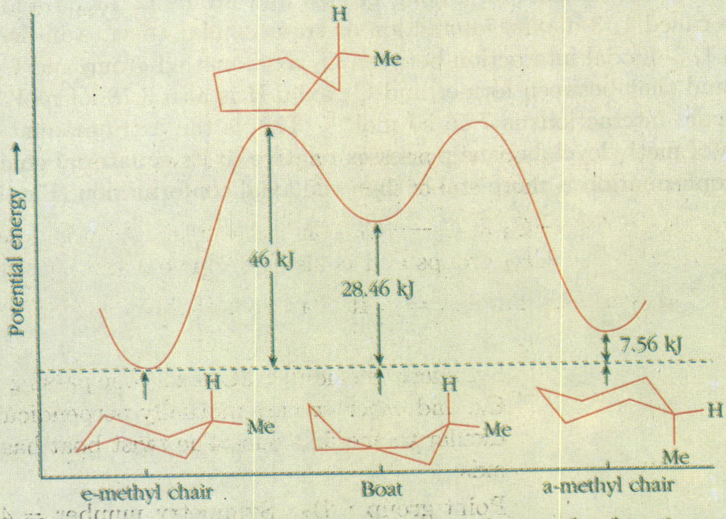

Fig. 9.144. Energy profile for methylcyclohexane as a function of conformation

If steric strain brings instability to the axial chair form, then a group or atom other than H must have more room for its accommodation in an equatorial position than in an axial position in the chair form.

The distance between C_1 equatorial H and one of its neighbouring H is only 249 pm, a value slightly greater than the sum of the van der Waals radii of 2H atoms (Table 9.8). If C_1, equatorial H is replaced by an atom or a group, one might think that there must be crowding of atoms. Of course this does not occur because all the neighbouring H atoms are extended away from C_1 equatorial H. Thus the methyl group in equatorial position gets enough of space for its accommodation and therefore no steric or van der Waals strain develops in the equatorial substituted chair form of methylcyclohexane. For this reason atoms and groups prefer spacious equatorial position in substituted cyclohexanes in general.

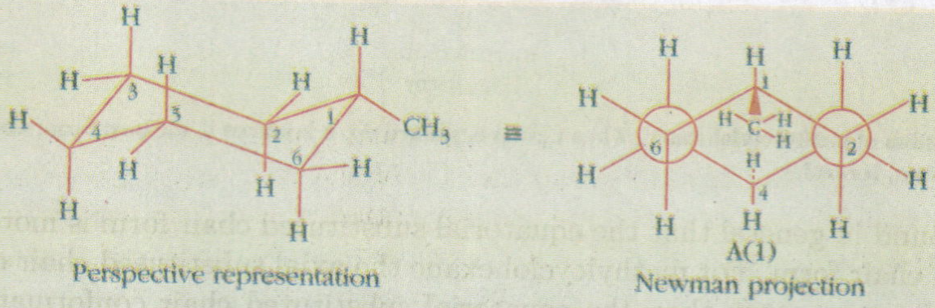

Fig. 9.145. Equatorial substituted methylcyclohexane in the chair form

On the other hand, though the distance between C_1 axial H and one of its nearby Hs (C_3 axial and C_5 axial; Table 9.8) is 251 pm yet $1a$, $3a$ and $5a$ bonds are parallel to each other and extended in the same direction. On the other hand, the sum of van der Waals radii of one H atom and a methyl group is 320 pm. For this reason when C_1 axial H is replaced by a methyl group, minimum of space is available for the latter and thus crowding occurs. This develops steric strain within the axial substituted chair form and because of this strain the form is unstable. This is also true for any other bulky group. Crowding is of course maximum when all the three axial bonds on C_1, C_3 and C_5 hold bulky groups. However, the steric strain that develops in an axial substituted chair form owing to the interaction among groups and atoms at $1a, 3a$ and $1a, 5a(2a, 4a,$ and $2a, 6a)$ bonds is called **1, 3-diaxial interaction or transannular effect**. van der Waals strain that develops due to 1, 3-diaxial interaction between C_1 axial methyl group and C_3 axial H in D(1) is 3.78 kJ mol^{-1} and that between former and C_5 axial H is also 3.78 kJ mol^{-1}; thus the sum of the two 1, 3-diaxial interactions is 7.56 kJ mol^{-1}. This is the extra amount of energy that the axial chair form of methylcyclohexane possesses relative to its equatorial chair form. Therefore, the equatorial conformation is more stable than the axial conformation (Fig. 9.146).

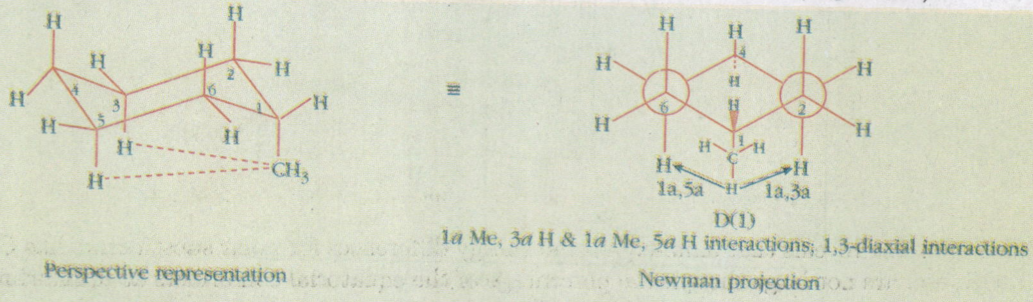

Fig. 9.146. 1,3-diaxial interaction in the axial substituted chair form of methylcyclohexane

Let us now consider the gauche interaction, if any, in both of the chair forms of methylcyclohexane. For this ball-and-stick models of the two forms are constructed and Newman projections of there models are drawn below:

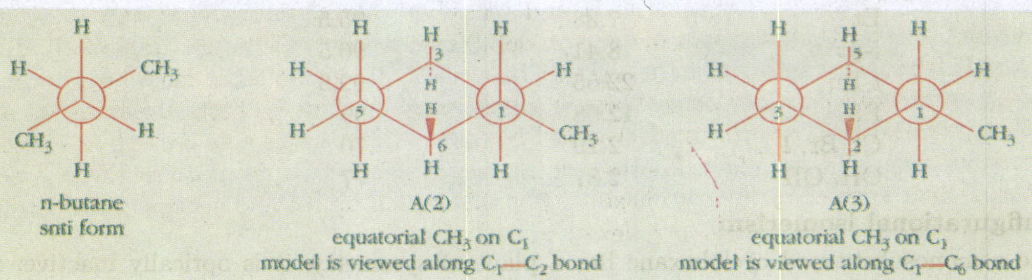

Fig. 9.147. Newman projection of the equatorial chair conformation of methylcyclohexane

Newman projection of the equatorial chair conformation of the methylcyclohexane indicates (Fig. 9.147) that the equatorial methyl group and C_3 of the ring [A(2)] have the same relative position as the two methyl groups in the anti conformation of *n*-butane which is free of gauche interaction. In A (3) *e*- methyl group on C_1 and C_5 of the ring also have anti positions. Therefore equatorial chair conformation does not experience any extra gauche interaction.

Newman projection of the axial chair conformation of the methylcyclohexane (Fig. 9.148) shows that the axial methyl and C_3 of the ring [D(2)], and the former and C_5 of the ring [D(3)] occupy the same relative locations as the two methyl groups in the gauche form of *n*-butane. Thus the axial chair conformation of methylcyclohexane experiences two gauche interactions ($1a - Me/C_3$ and $1a - Me/C_5$). 3.78 kJ mol^{-1} being the amount of energy assigned for a gauche interaction in *n*-butane, the axial chair conformation of methylcyclohexane should have 2×3.78 or 7.56 kJ mol^{-1} of excess energy than its equatorial conformation. These *n*-butane gauche interactions are usually written as axial Me on C_1 and C_3 >CH_2 or C_5 >CH_2 in the *n*-butane units, Me—C_1—C_2—C_3 and Me—C_1—C_6—C_5. Therefore, the latter is more stable than the former conformation. This is also true for conformations of other monosubstituted cyclohexane derivatives. Table 9.9 shows axial-equatorial energy differences for monosubstituted cyclohexanes and the % of the equatorial conformation present at equilibrium at 25 °C.

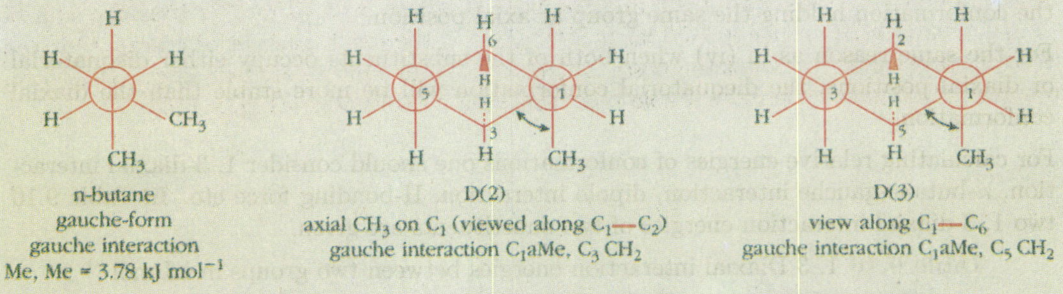

Fig. 9.148. Newman projection of the axial chair conformation of methylcyclohexane

Table 9.9 further reveals that axial-equatorial energy differences for polar substituents like Cl, Br, OH, OR, etc. are not large though the percentage of the equatorial conformers at equilibrium is high.

Table 9.9 Axial-equatorial energy differences and % of equatorial conformation at equilibrium

Substituents	ΔG in kJ mol^{-1}	~% of *e*-conformation
Me	6.97	95
Et	7.38	95.5
i-pr	8.41	96.5
t-Bu	22.55	99.9
Ph	12.08	99
Cl, Br, I	2.19	70
OH, OR	2.87	77

B. Configurational isomerism

Since a monosubstituted cyclohexane has a plane of symmetry, it is optically inactive; no enantiomer exists. These compounds do not have any cis-trans isomers also.

9.20.6 Stereoisomerism of disubstituted cyclohexanes

To discuss the stereoisomerism of disubstituted cyclohexanes following points are to be considered:

(i) Configurational isomers, geometrical and optical, may be designated by drawing planar structures.

(ii) Consideration of plane of symmetry and centre of symmetry may enable one to determine whether an isomer is optically active or not.

(iii) Chair is the preferred conformation of each of the disubstituted cyclohexanes in general since torsional strain and flagpole-bowsprit interactions are absent in this form. *n*-Butane-gauche interaction may act in a chair form.

(iv) When a disubstituted cyclohexane has one substituent in equatorial position and the other in axial position, then the chair form with the more bulky group in equatorial position will generally be more stable conformation than the chair form with the same group in the axial position. This is because of the less steric crowding in equatorial position than that in axial position. In other words, since an equatorial position has no 1, 3-diaxial interaction, the conformation holding the more bulky group at equatorial position is more stable than the conformation holding the same group at axial position.

(v) For the same reason as in (iv) when both of the substituents occupy either diequatorial or diaxial positions, the diequatorial conformation will be more stable than the diaxial conformation.

(vi) For calculating relative energies of conformations one should consider 1, 3-diaxial interaction, *n*-butane-gauche interaction, dipole interaction, H-bonding force etc. In Table 9.10 two 1, 3-diaxial interaction energies of two substituents are given.

Table 9.10 1, 3 Diaxial interaction energies between two groups in kJ mol^{-1}

OH/OH : 7.98, OAc/OAc : 8.4, Me/Me : 15.54, Me/OH : 9.64

(vii) Relative stabilities of the isomers may be determined by conformational analysis. This requires drawing of chair conformations and corresponding Newman projections in general. It has already been stated that six C—H bonds, three axial on the alternate C atoms and three equatorial on the other three Cs, point above the imaginary plane of the cyclohexane ring; other six similar bonds point below the plane. Thus if one draws $1a$ above the plane, then $2e, 4e, 6e$ and $3a, 5a$ will also extend above the plane; whereas $1e, 3e, 5e$ and $2a, 4a, 6a$ will extend below the plane. However relative positions of the two substituents in chair forms of geometric isomers are given below in Table 9.11.

Table 9.11 Relative positions of substituents in chair forms of geometric isomers

Relative positions	Designation of geometric isomers
$1a, 2a$ or $1e, 2e$	trans
$1a, 2e$ or $1e, 2a$	cis
$1a, 3e$ or $1e, 3a$	trans
$1a, 3a$ or $1e, 3e$	cis
$1a, 4a$ or $1e, 4e$	trans
$1a, 4e$ or $1e, 4a$	cis

Let us discuss a specific disubstituted derivative of cyclohexane, 1, 3-dimethylcyclohexane, to understand the stereoisomerism of disubstituted cyclohexane derivatives.

We can draw the following planar structures for 1, 3-dimethylcyclohexane.

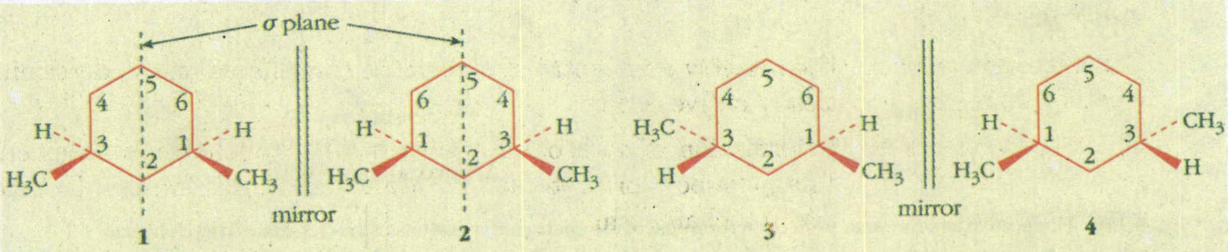

Fig. 9.149. Planar structures of 1,3-dimethylcyclohexane. A broken line indicates a bond pointed below the plane of the paper whereas a wedge above it.

In 1 and 2 (Fig. 9.149) two CH_3 groups are above the plane and two Hs are below it. So each of these two structures represents cis-1, 3-dimethylcyclohexane. In 3 and 4 one methyl group is above the plane and the other below it. So each of these structures represents the trans-form of the compound.

Structure 2 is a superimposable mirror-image of the structure 1 and each of them possesses a (σ plane; hence these two structures are identical and thus the cis-form is optically inactive. Therefore cis-1, 3-dimethylcyclohexane may be called a meso form.

Of course 3 and 4 have non-superimposable mirror-image relation to each other and each of them neither possesses a σ plane nor a centre of symmetry; thus these two structures represent a pair of enantiomers and each form is optically active. These two forms can therefore exist as discrete unit. trans-1, 3-Dimethylcyclohexane is thus a resolvable *dl*-pair and as such it is optically inactive.

However, the cis-form of 1, 3-dimethylcyclohexane is not a mirror-image of either structure 3 or structure 4. Again one cannot convert the cis-form to the trans (and the trans to the cis) without breaking and remaking of the bonds. Thus 1, 3-dimethylcyclohexane exists as a pair of configuration diastereomers, the cis- and trans-isomers. The trans-isomer, in turn, exists as a pair of configurational enantiomers.

Since in general cyclohexane derivatives prefer chair conformations, let us now discuss the chair conformations of 1, 3-dimethylcyclohexane.

As is clear from Table 9.11, the cis-form exists in two chair conformations:

A. Diequatorial chair conformation—in which two methyl groups occupy equatorial positions $(1e, 3e)$.

B. Diaxial chair conformation—in which two methyl groups occupy axial positions $(1a, 3a)$. These are shown in Figure 9.150.

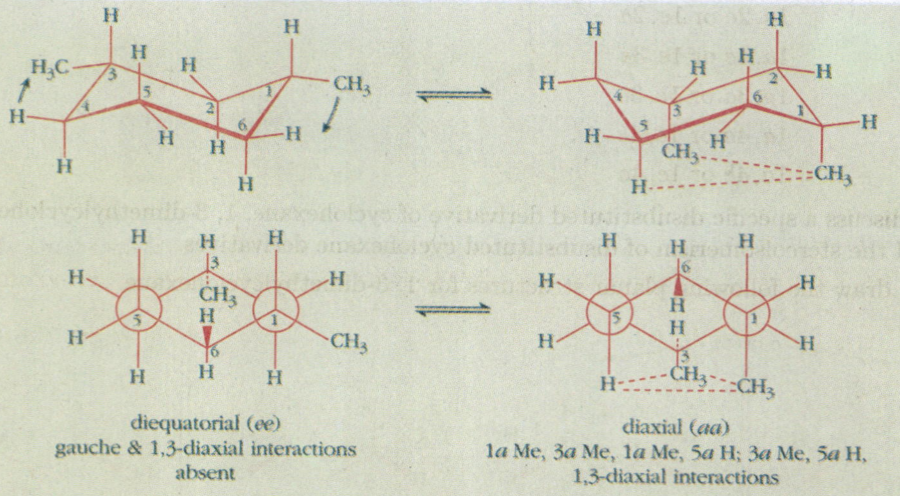

Fig. 9.150. Perspective and Newman projection formulae of the chair conformations of *cis*-1,3-dimethylcyclohexane

Perspective representations of *ee* and *aa* forms show that these two chair forms do not have mirror-image relation between them but one can be converted into the other by rotation of bonds only; therefore the cis-form of 1, 3-dimethylcyclohexane exists as a pair of conformational diastereomers. Since there is a σ plane through C_2 and C_5 (Fig. 9.150), both of them are symmetric and therefore they represent optically inactive meso forms.

In the diequatorial form there is neither any 1, 3-diaxial interaction nor any butane-gauche interaction; but there are three 1, 3-diaxial interactions in the diaxial form $(1a$ Me, $3a$ Me; $1a$ Me, $5a$ H; $3a$ Me, $5a$ H). Thus these interactions confer van der Waals strain on the diaxial conformations while the diequatorial form is free of strain. Therefore the diequatorial form is more stable than its diaxial diastereomer.

Now how much more stable is the diequatorial form? In the diaxial form the 1, 3-diaxial interactions are : $1a$ Me, $3a$ Me $= 15.54$ kJ mol^{-1}, $1a$ Me, $5a$ H $= 3.78$ kJ mol^{-1} and $3a$ Me, $5a$ H $= 3.78$ kJ mol^{-1}. Therefore the total steric strain that 1, 3-diaxial interactions confer on the cis-diaxial form is $(15.54 + 2 \times 3.78)$ or 23.1 kJ mol^{-1}. Steric strain being absent in the diequatorial form, it is more stable than the diaxial form by 23.1 kJ mol^{-1}.

As is revealed from Table 9.11, the trans-1, 3-dimethylcyclohexane also exists in two chair conformations: axial equatorial $(1a$ Me, $3e$ Me) and equatorial axial $(1e$ Me, $3a$ Me) as shown below:

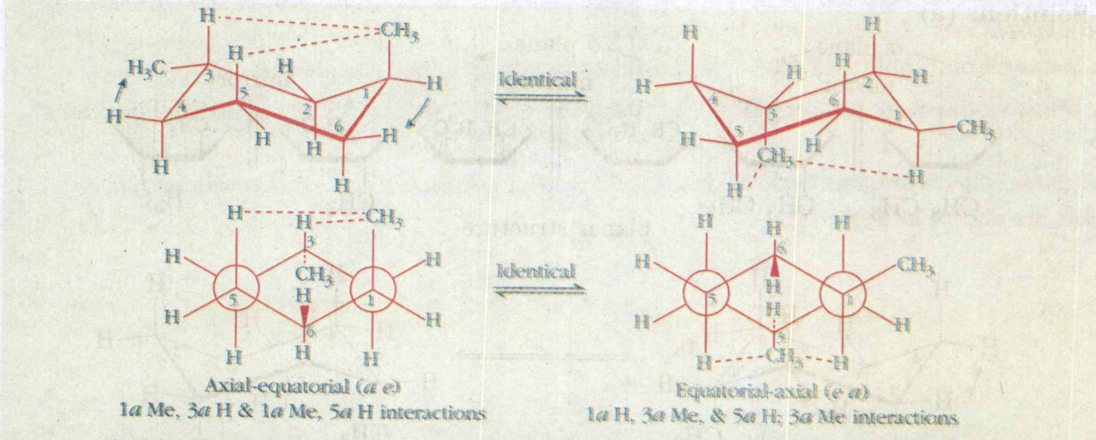

Fig. 9.151. Perspective and Newman projections of chair conformations of *trans*-1,3-dimethylcyclohexane

ae- and *ea*-chair conformations here are superimposable, so these two forms are identical; but the mirror-image of the *ae* or that of the *ea* is not superimposable with the original molecule as shown in Fig. 9.152.

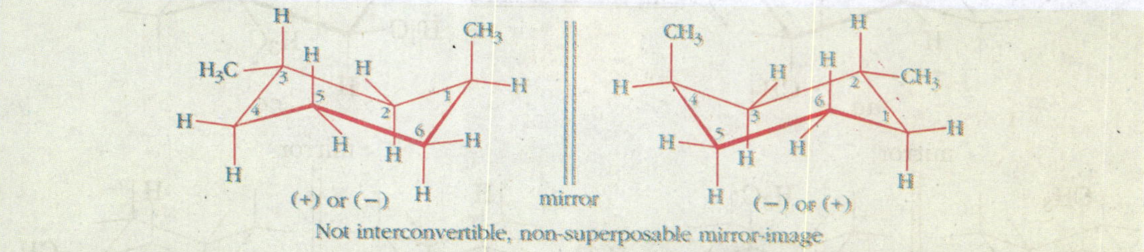

Fig. 9.152. Relation between the axial-equatorial chair conformations of 1,3-dimethylcyclohexane and its mirror image

Therefore *ea*- or *ae*-chair form is dissymmetric. Hence it is optically active. Thus each enantiomer of the trans-1, 3-dimethylcyclohexane exists in one conformation and that is *ae* or *ea*. Since this conformer exists as a *dl*-pair (*ae* or *ea* and its mirror-image), the trans-form of 1, 3-dimethylcyclohexane is optically inactive but it is resolvable.

Let us now determine the amount of steric strain that 1, 3-diaxial and *n*-butane gauche interactions develop in the trans-isomer of 1, 3-dimethylcyclohexane. Newman projection reveals that there is no *n*-butane-gauche interaction between two methyl groups, but there are 1*a* Me, 3*a* H and 1*a* Me, 5*a* H diaxial interactions; this amounts to $3.78 \times 2 = 7.56$ kJ mol^{-1} of energy. On the other hand, there is no such interactions in the more stable diequatorial chair conformation of cis-1, 3-dimethylcyclohexane. Thus the cis-isomer is more stable than the trans-isomer by $(7.56 - 0.00)$ or 7.56 kJ mol^{-1}.

Problem 9.24. (a) Draw planar structures and chair conformations for 1, 2-dimethylcyclohexane.

(b) State the number of its configurational isomers and discuss their optical activity.

(c) How many chair conformations of the compound will exist? State how they are related to each other.

(d) Indicate the most stable and the least stable chair conformations of the compound.

(e) Calculate the energy difference between cis-and trans- isomers of the compound.

Solution: (a)

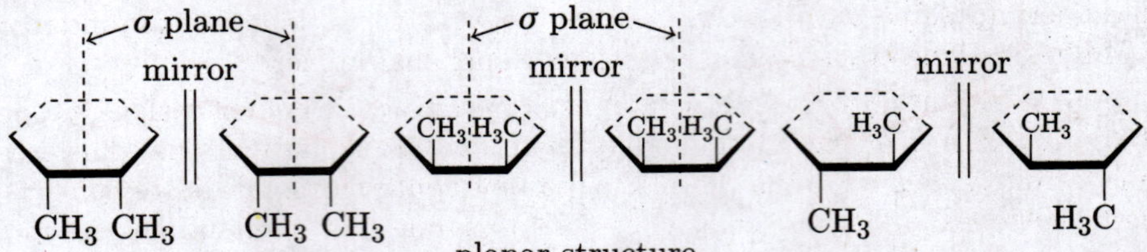

planar structure

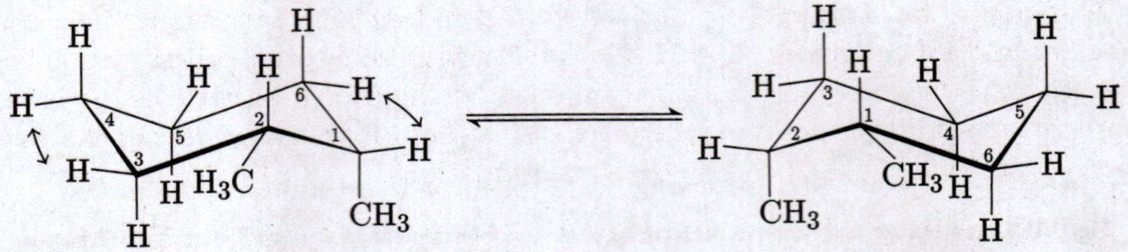

ae *ea*

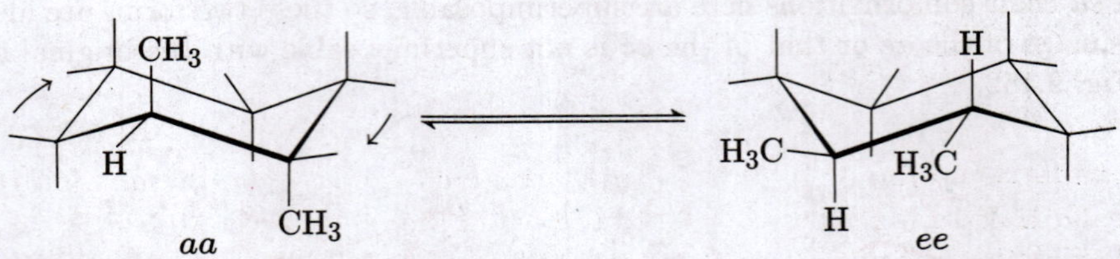

aa *ee*

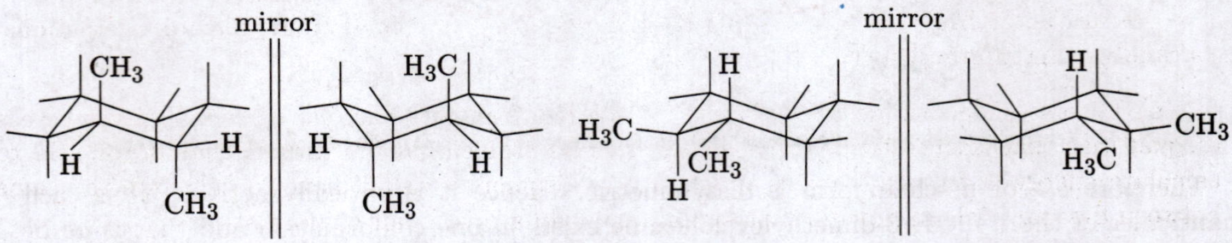

aa — mirror image of *aa* — *ee* — mirror image of *ee*

(+) or (−) (−) or (+) (+) or (−) (−) or (+)

(b) From the planar structures of 1, 2-dimethylcyclohexane we can say that its cis-form has a superimposable mirror-image; whereas its trans-forms are two in number, one is a non-superimposable mirror-image of the other. Thus there are two entiomeric trans-forms, cis- and two trans-forms cannot be interconverted without breaking and reforming of bonds. So these three are configurational isomers. The cis-form does not bear mirror-image relation with the trans-form; thus cis- and trans-isomers are a pair of **configurational diastereomers**.

The cis-form possesses a σ plane, whereas there is neither σ plane nor centre of symmetry in the trans-forms. So each of them is optically active. Since the trans-form exists as a *dl*- pair, it is also optically inactive. Being a symmetric molecule, the cis-form is also optically inactive, it may be called a meso form. However, the **chair conformation** of cis-1, 2-dimethylcyclohexane does not have any S_n axis. Hence it is expected that the cis-form should exist as a *dl*-pair but every attempt to separate them have failed; perhaps the energy barrier for their interconversion is very low and hence they are inseparable.

(c) The cis-1, 2-dimethylcyclohexane exists in two chair forms, *ae* and *ea*. These two forms bear non- superimposable mirror-image relation between them, but they are interconvertible by rotation of bonds. Hence these two are conformational enantiomers.

The trans- 1, 2-dimethylcyclohexane also exists in two chair conformations, *aa* and *ee*. They do not have any mirror-image relation; thus these two are conformational diastereomers. The mirror-images of *aa* and *ee* forms are also a pair of conformational diastereomers. Therefore, 1, 2-dimethyleyclohexane has six chair conformations, one pair of enantiomers and two pairs of diastereomers.

(d) Out of six, two chair conformations are *ae* and *ea*; and other four chair conformations are *aa*, *ee*, and their mirror images. A chair-form with equatorial substituted bulky groups is more stable than the corresponding axial substituted chair form because of the lesser crowding. Therefore, diequatorial chair form is the most stable conformation and diaxial chair conformation is the least stable among the possible chair conformations.

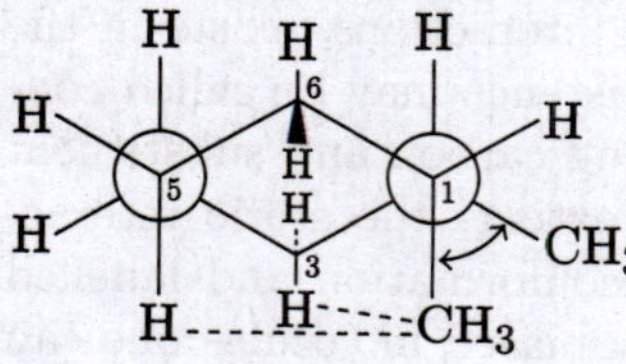

cis-form (*a e*)
1*a* Me, 3*a* H & 1*a* Me, 5*a* H diaxial interactions & 1*a* Me, 2*e* Me butane gauche interactions

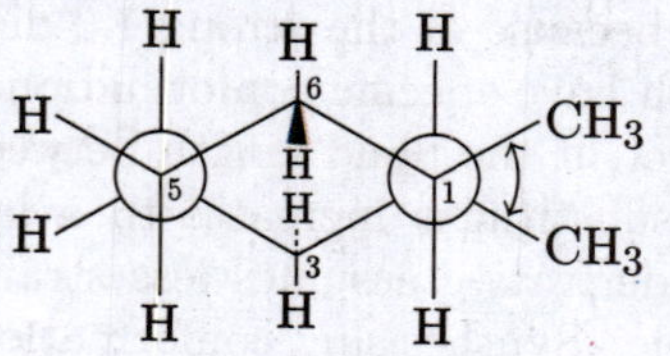

trans-form (*e e*)
Only 1*e* Me, 2*e* Me butane gauche interactions

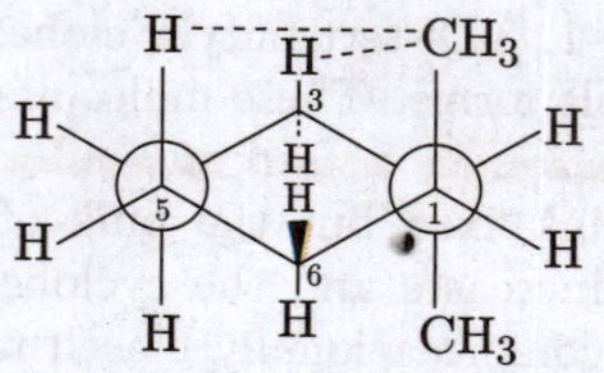

trans-form (*a a*)
1*a* Me, 3*a* H; 3*a* Me, 5*a* H; 2*a* Me, 6*a* H; 2*a* Me, 4*a* H (four) diaxial interactions

(e) The two chair conformations of the cis-isomer, *ae* and *ea*, have the same stability, whereas diequatorial chair form of the trans-isomer is more stable than its diaxial chair form. So to calculate energy difference between cis- and trans-isomers we shall have to calculate the same between diequatorial chair conformation and axial-equatorial chair conformation.

In the diequatorial chair conformation, there is no 1, 3-diaxial interaction but there is one butane gauche interaction between two methyl groups; thus the total steric strain confers on the trans-form (*ee*) is 3.78 kJ mol^{-1}. In the axial-equatorial chair conformation there are two 1, 3-diaxial interactions : 1*a* Me, 3*a* H ; 1*a* Me, 5*a*, H. There is also one butane gauche interaction between two methyl groups since these are in the gauche position. The two 1, 3-diaxial interactions amount to 2×3.78 kJ mol^{-1} of energy while one butane gauche interaction is equivalent to 3.78 kJ mol^{-1} of energy. Hence these three interactions confer 3×3.78 or 11.34 kJ mol^{-1} of van der Waals strain on the cis-form.

Thus the trans-form is more stable than the cis-form by (11.34 – 3.78) or 7.56 kJ mol^{-1}

9.20.7 Conformationally biased cyclohexane derivatives and anchoring substituents

It has already been stated that owing to 1, 3-diaxial interactions bulky groups accommodate themselves in the equatorial positions. For example, *t*-butylcyclohexane almost exclusively exists in the equatorial conformation.

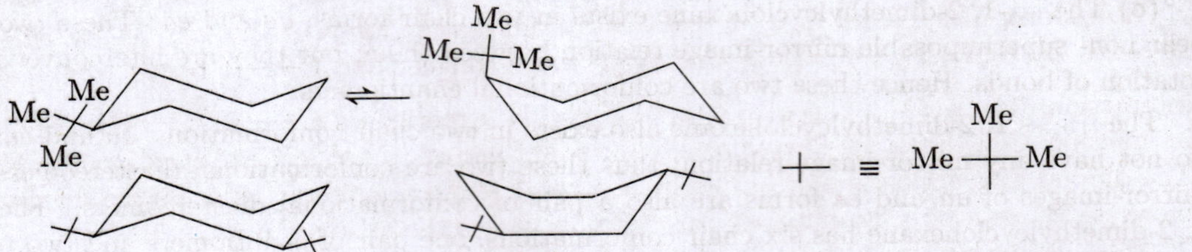

cis   *trans*

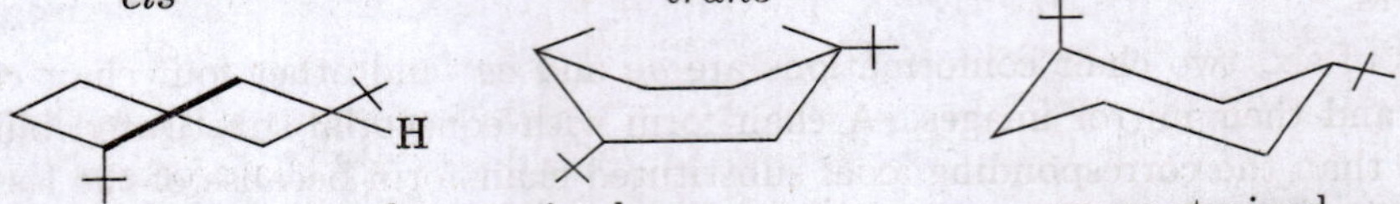

less strained   more strained

trans-1,3-di-tert-butylcyclohexane

trans-1, 3-Di-tert-butylcyclohexane, because of the strong 1,3-diaxial interactions, exists in the flexible forms. These molecules which have specific conformational existence may be called **conformationally biased molecules**. Infact, if the bond length between ring carbon and substituent remains fixed but the bulk of the substituent increases to a large extent, the steric interaction increases and the cyclohexane derivative assumes less strained conformation and labelled as a conformationally biased molecule. Synthesising conformationally biased molecules one can study the differences in reactivity of axial versus equatorial substituents. The bulky substituent which makes the molecule conformationally fixed or at least biased is called **'anchoring' substituent**, *t*-butyl group is an anchoring substituent. For example cis- and trans- isomers of 4-*t*-butylcyclohexanecarboxylic acid have the following conformations :

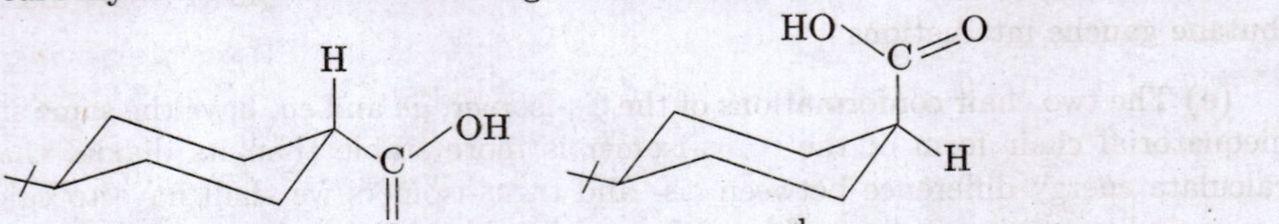

a   b

trans   *cis*

Anchoring of conformation is very important to study reactivities of substituents. If we do not fix the conformation, conformational interconversion will take place; as a consequence the differences in reactivity of axial and equatorial substituents cannot be studied. In this way, cis-isomer of 4-*t*-butylcyclohexanecarboxylic acid has been found to be less acidic than its trans-isomer in the aqueous medium. The hydrated carboxylate ion of the cis-isomer bears higher 1, 3-diaxial strain than the acid itself and this makes the former less stable than the hydrated carboxylate ion of the trans-isomer which does not bear appreciable strains. Thus we can say, in general, that the axial substituted caboxyl group is less acidic than the equatorial substituted carboxyl group in the aqueous medium.

Problem 9.25. In CCl_4 at 30°C the ratio $a : b = 2 : 1$—explain.

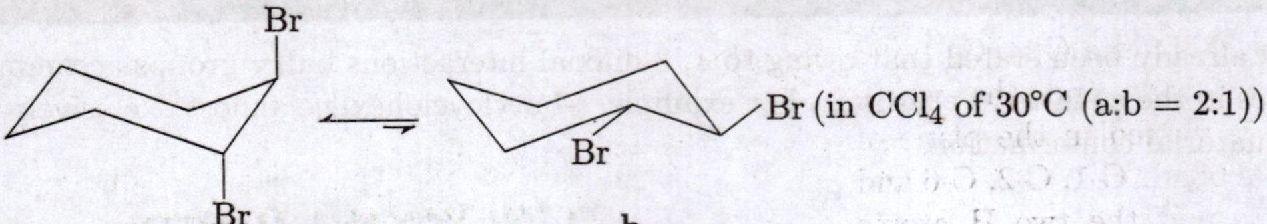

a   b

Solution: The diaxial form, a, has l, 3-diaxial interactions (1*a* Br/3*a*H, 1*a* Br/5*a*H and 2*a* Br/4*a*H, 2*a* Br/6*a*H). C—Br bonds being much longer than C—H bonds, the 1, 3-diaxial interactions are weak. The diequatorial form, b, experiences $\overset{\delta+}{C}—\overset{\delta-}{Br}/\overset{\delta+}{C}—\overset{\delta-}{Br}$ dipole-dipole interaction; CCl_4 has a very low dielectric constant and it cannot insulate the dipolar interaction. The dipolar interaction is here stronger than the 1, 3-diaxial interactions. So the diaxial form is more stable than the diequatorial form and this makes the ratio a : b = 2 : 1.

Problem 9.26. Draw the stable conformations of 1, 3-di-tert-butylcyclohexane.

Solution: A *t*-butyl group being very bulky, it experiences intolerable 1, 3-diaxial interaction between axial *t*-Bu group and axial H atoms. So, in a chair conformation it takes up the equatorial position and its cis-isomer exists in a diequatorial chair form. In a chair form of the trans-1, 3-di-tert-butylcyclohexane one of the two *t*-butyl groups will have to take up an axial position and in that case the conformation will be highly unstable. So the stable conformation of the trans-isomer is a boat form or, twist boat form.

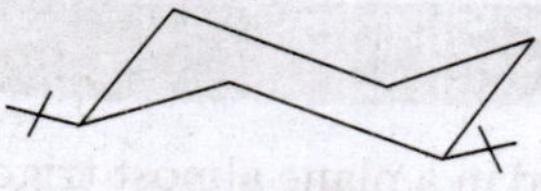

cis

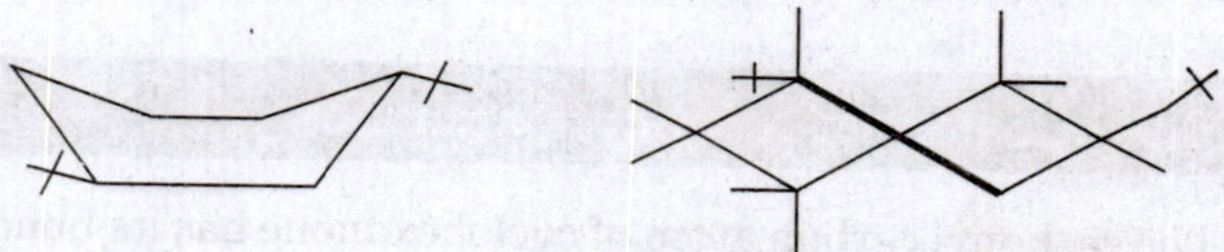

trans

Problem 9.27. Explain why 1, 2, 2, 6, 6-pentamethyl-4-hydroxy-4-phenylpiperidine is boat shaped.

Solution : The chair form of the given compound is highly strained and it has no scope to form intramolecular H-bonding which brings extra stability to a system as shown below. However, in the boat form, though there is enough strain yet internal H-bonding brings stability to the system.

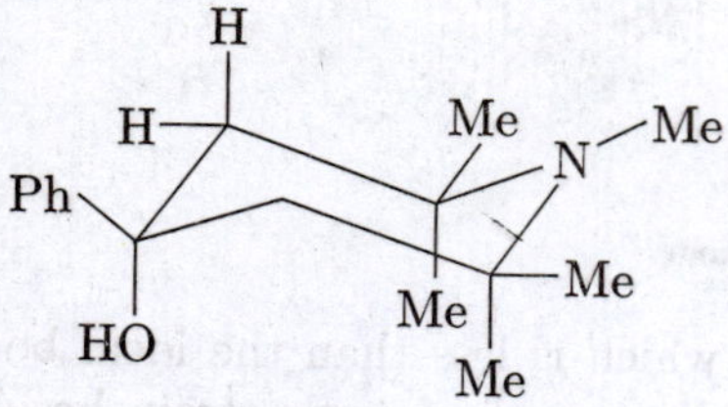

unstable

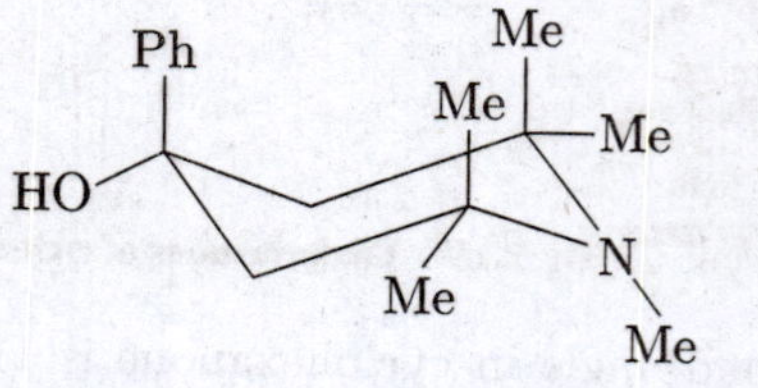

unstable

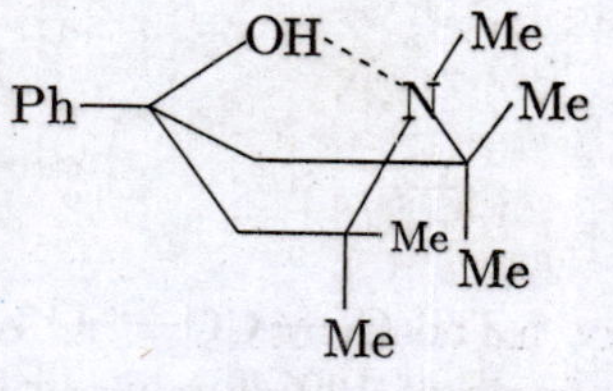

stable

9.20.8 Conformation of cyclohexene

Electron defraction studies of cyclohexene in the vapour phase and X-ray crystallographic data of cyclohexene derivatives show beyond doubt that cyclohexene has a half-chair conformation as given below:

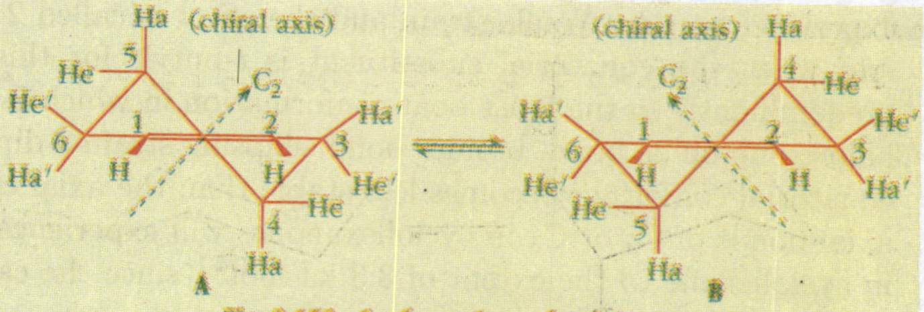

Fig. 9.153. Conformations of cyclohexene.

In Fig. 9.153 the $C_1{=}C_2$ bond is seen in front and the whole molecule is viewed in the plane of the double bond. C-1, C-2, C-6 and C-3 atoms and the two H atoms which are bonded to C-1 and C-2 atoms are coplanar; in A C-4 and C-5 atoms are below and above the plane of the double bond respectively while in B they are respectively above and below this plane. The C—H bonds marked e and e' are equatorial and pseudoequatorial respectively and those marked a and a' are respectively axial and pseudoaxial. The values of dihedral angles He′—C_6—C_1—H and He′—C_3—C_2—H are one and the same (calculated value 44°) while those for Ha′—C_6—C_1—H and Ha′—C_3—C_2—H is 76°. Thus the C—H bonds on C_3 and C_6 are not truly axial and equatorial; if they were so, then the dihedral angles would have been 60°. The C—H bonds on C_3 and C_6, are therefore designated as pseudoaxial (a') and pseudoequatorial (e'). On the other hand C_4 and C_5 atoms have almost normal axial (a) and equatorial (e) C—H bonds. However, both A and B have a chiral C_2 axis which bisects the double bond but they have no σ, i or S_n axis. So A and B are dissymmetric molecules. They are non-superimposable mirror-images of each other but A and B are interconvertible and the experimental energy barrier is only 22.2 kJ mol^{-1} and thus they are inseparable. So, A and B constitute a pair of conformational enantiomers.

9.20.9 Conformation of cyclohexanone

The carbonyl carbon atom of cyclohexanone has its bonds oriented in a plane almost trigonally. So, the hybridised state of the carbonyl carbon is supposed to be sp^2. The carbonyl C, C_2, C_6 and the O atoms are therefore coplanar. C_2—H_e, and C_6—H_e bonds are partially eclipsed (Fig. 9.154) with the C=O bond; where C_2—Ha and C_6—Ha are staggered with the C=O bond.

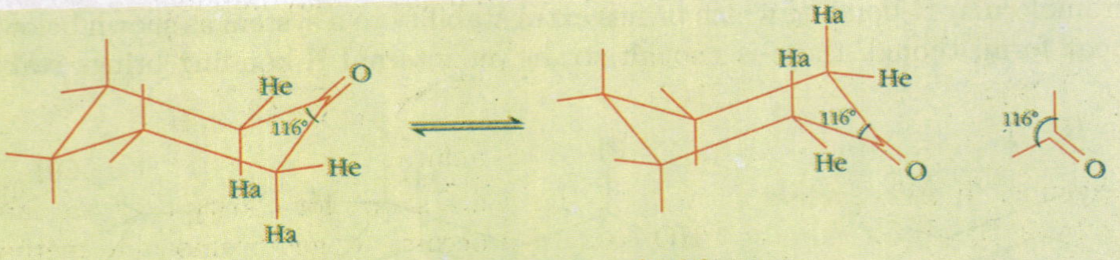

Fig. 9.154. Conformations of cyclohexanone

The C—CO—C bond angle in cyclohexanone is 116° which is less than the ideal bond angle of 120° for an sp^2C. So, the conformations have a small amount of angle strain besides the partial bond eclipsing strain. The combined effect of these two strains makes cyclohexanone less stable than cyclohexane in their most stable chair conformations. However, 2-alkylketones (except 2-methylketone) with the alkyl group at the equatorial position are less stable than the corresponding alkylcyclohexanes with the equatorial alkyl group because in the equatorial conformers of cyclohexanone, the alkyl group and the carbonyl O are partially eclipsed and experience van der Waals strain and the effect is called 2-alkylketone effect. The effect is manyfold when the equatorial substituent is t-butyl; for this reason 2-t-butylcyclohexanone exists predominantly in the twist boat conformation in which 2-alkyl effect is avoided. When there is a halo group on 2e bond, besides bond eclipsing strain a dipole-dipole interaction develops and the equatorial conformer becomes less stable than the axial conformer. However, any substituent on axial bonds of C_3 or C_5 in cyclohexanones will experience less 1, 3-diaxial interaction than those in cyclohexane to the extent of 3.3 kJ mol^{-1} since the carbonyl carbon atom does not hold any axial bond.

9.21 Homotopicity, Heterotopicity and Prochirality

Topicity is a term introduced in the literature of stereochemistry to understand the stereochemical relationships between different parts of the same entity i.e., molecule (in particular). The word 'topic' comes from the word 'topos' meaning 'place'. 'Homo' means 'same' and 'hetero' means 'different'. Therefore, homotopicity indicates the identical stereochemical relationship between two ligands of a molecule; whereas heterotopicity indicates different stereochemical relationship between two such ligands. Homotopic and heterotopic are words related to 'identical places' and 'different places' respectively. 'Pro' is a prefix which means 'in the favour of'. 'Chirality' means 'handedness'. So, 'prochirality' means 'in the favour of handedness i.e. in the favour of assuming dissymmetric center'. 'Topicity' is tested by the actual or imaginary replacement of a ligand of the molecule by another which is different from the ligands already present at the concerned part of the molecule (replacement of ligand does not always mean removal of the entire ligand and putting in a new one in its place; a part of the ligand may be substituted by another so that the new one is something different from the original, e.g., —CH_3 may be converted to —CH_2Br). However, we shall discuss in this section homotopic ligands, heterotopic ligands, homotopic faces and heterotopic faces and a prochiral centre.

9.21.1 Homotopic ligands and faces

Homotopic ligands are those which are stereochemically identical insofar as their places in the molecule are concerned and which do not give rise to stereochemically different products when such ligands are separately replaced by a group (different from ligands present at the concerned part). The two H atoms and two Cl atoms of dichloromethane may be labelled as two pairs of

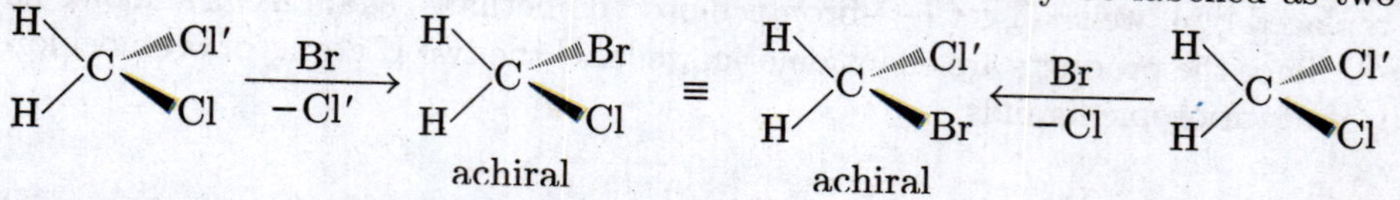

homotopic ligands. If one replaces Cl and Cl′ (both are same; for clarity they are labelled as Cl and Cl′) by a Br group through two separate processes to get bromochloromethane, two stereochemically different bromochloromethanes will not be obtained; they will be one and the same entity. Hence the two Cl atoms in CH_2Cl_2 are homotopic ligands or groups.

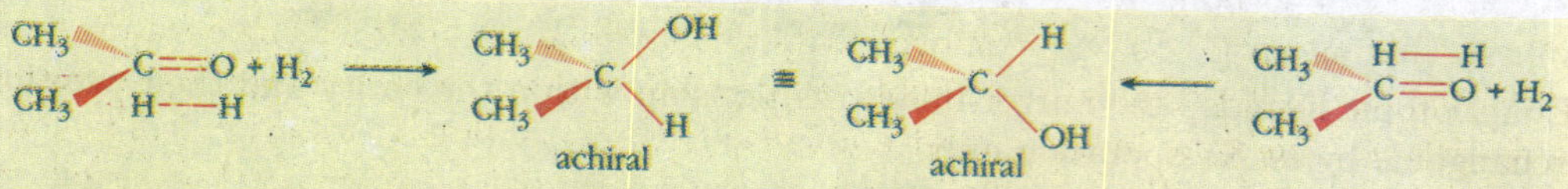

Fig. 9.155. An example of homotopic faces

In the cases of flat entities like ethylene, acetone, acetaldehyde, the faces of multiple bond are involved when a chemical reaction takes place. The two faces are said to be homotopic if a chemical reaction on two faces does not give rise to stereochemically different products, i.e., isomers. When acetone is catalytically hydrogenated, we get one and the same alcohol, isopropyl alcohol; therefore the two faces of acetone are homotopic.

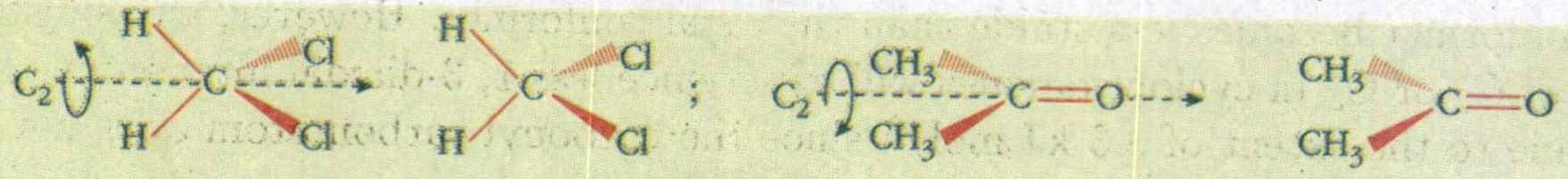

Fig. 9.156. A C_2 operation on CH_2Cl_2 and $(CH_3)_2C{=}O$

Homotopic ligands get interchanged insofar as their places are concerned when the molecule is rotated through a C_n (where $\propto > n > 1$) axis; this means that we shall get an equivalent structure indistinguishable from the original one after the C_n operation.

9.21.2 Heterotopic ligands and faces

Those ligands of an entity are said to be heterotopic ligands which on replacement by a different group, different from the already present ligands at the concerned part of the entity, gives rise to stereoisomeric products, a pair of enantiomers or diastereomers; accordingly heterotopic ligands are further classified as enantiotopic ligands and diastereotopic ligands.

Enantiotopic ligands: The word enantiotopic means "in mirror-image places". Enantiotopic ligands of an entity are those which are stereochemically different and replacement of one or the

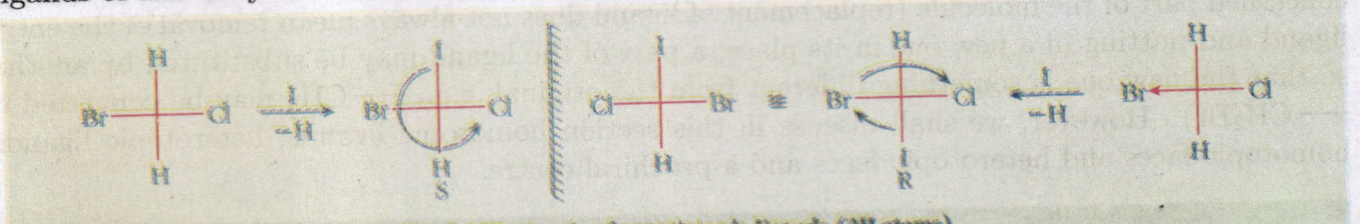

Fig. 9.157. A pair of enantiotopic ligands (2H atoms)

other of them by a different group, different from the already present ligands at the concerned part, gives rise to one or the other of a pair of enantiomers as the product. The two H atoms of bromochloromethane constitute a pair of enantiotopic ligands. When one of them is replaced by I, we get (+)-bromochloroiodomethane, a chiral molecule (say of R configuration); if the other H atom is replaced by I, we shall get (−)-bromochloroiodomethane, again a chiral molecule of S configuration. Since the products are enantiomeric, we label the two H atoms of bromochloromethane as a pair of enantiotopic ligands.

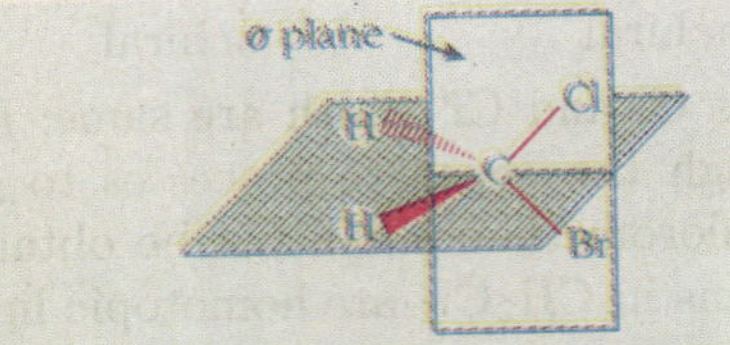

Fig. 9.158. CH_2ClBr has a σ plane and S_1 axis perpendicular to the σ plane

Enantiotopic ligands bear mirror-image relationship within the entity and their positions are interchangable by an S_n operation only.

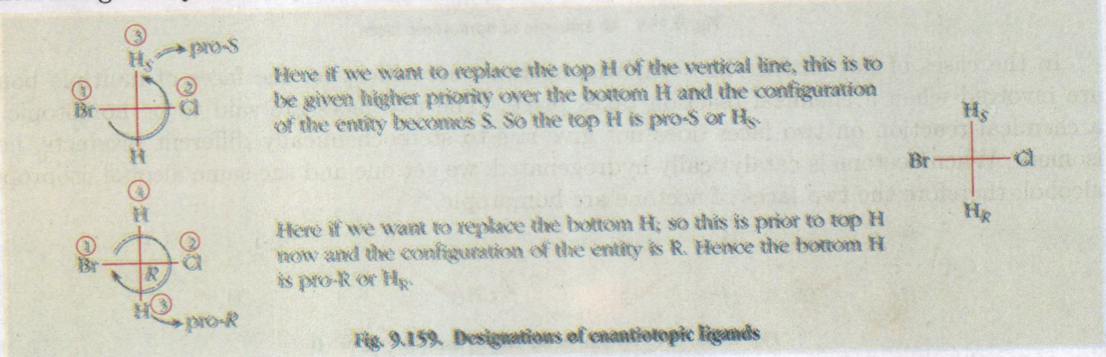

Fig. 9.159. Designations of enantiotopic ligands

Enantiotopic ligands are designated as pro-R or pro-S in accordance with the priority sequence rules. The enantiotopic ligand which is to be replaced by a group, actually or imaginary, is given the higher priority over the other and then configuration of the entity is determined. If the configuration is R, the ligand is labelled as pro-R; if the configuration is S, the ligand is designated as pro-S. However, the products may have any configuration on substitution (R or S) and the configuration of the products are in no way related to pro-R or pro-S (Fig. 9.159).

Diastereotopic ligands: Diastereotopic ligands of an entity are those when one or the other of them is replaced by a different group, the entity gives one or the other isomer of a set of diastereomers. Diastereotopic ligands exist in environments which are neither identical with nor mirror-images of each other. 1-chloroethene has two diastereotopic H atoms (H′ and H″) as ligands as shown in the next page.

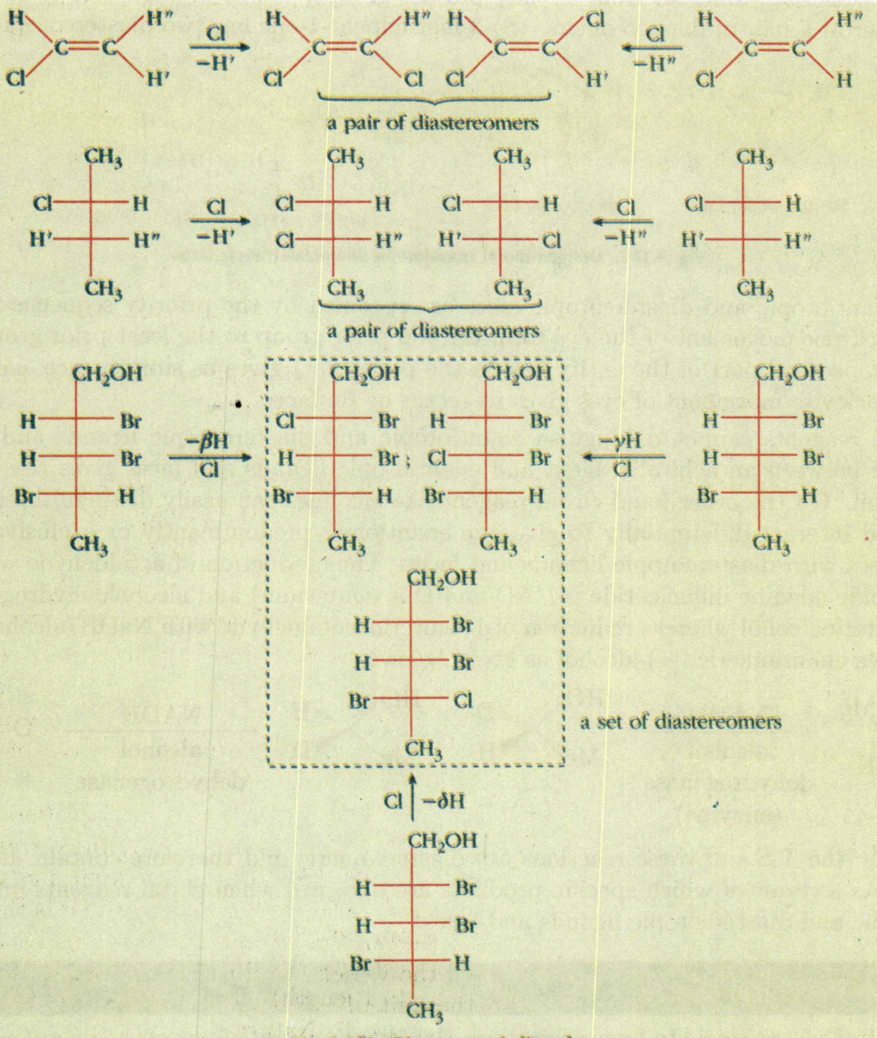

Fig. 9.160. Diastereotopic ligands

Diastereotopic ligands are not interchangeable by any symmetry operation.

Enantiotopic faces: An entity is said to have a pair of enantiotopic faces if attachment of a group to one face or the other gives one isomer or the other of a pair of enantiomers. Acetaldehyde has two enantiotopic faces. When it is deuterated, the attachment of D_2 to opposite faces gives two enantiomeric products.

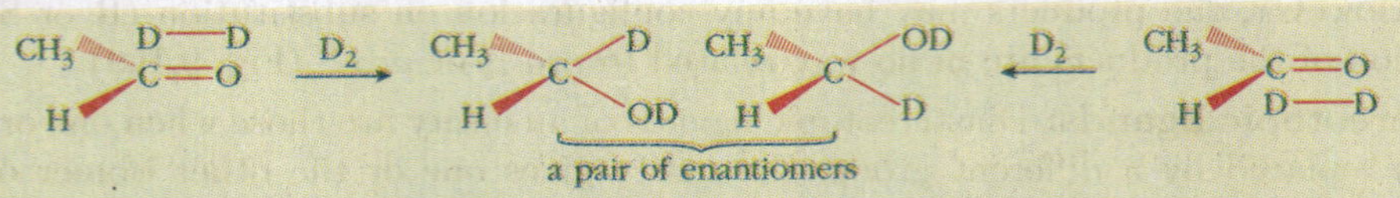

Fig. 9.161. a pair of enantiotopic faces in acetaldehyde

Diastereotopic faces: An entity is said to have a pair of diastereotopic faces if on attachment of groups, which are different from already present groups, to one or the other gives one or the other isomer of a pair of diastereomers. (S)-3-methylpent-1-ene has two diastereotopic faces.

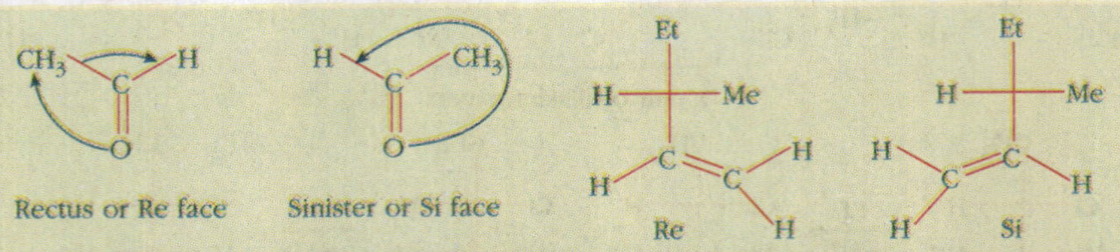

Fig. 9.162. Designation of enantiotopic and diasterotopic faces

The enantiotopic and diastereotopic faces are specified by the priority sequence rules. The counterclockwise movement of the eye from the top prior group to the least prior group (considering the concerned part of the entity lies on the paper flat) gives us sinister face, called Si face and the clockwise movement of eyes gives us rectus or Re face.

Achiral reagents cannot distinguish enantiotopic and diastereotopic ligands and faces; the interaction between an achiral reagent and enantiotopic ligands and faces gives rise to racemic modification. On the other hand chiral reagent like enzymes can easily distinguish enantiotopic ligands and interact differentially to give one enantiomer predominantly or exclusively; similar are the cases with diastereotopic ligands and faces. Thus reduction of actaldehyde with NADD (nicotinamide adenine dinucleotide is NAD and D is deuterium) and alcoholdehydrogenase gives a (+) -deuterioalcohol whereas reduction of 1-deuterioacetaldehyde with NaDH/alcohol dehydrogenase gives enantiomeric (−)-alcohol as given below:

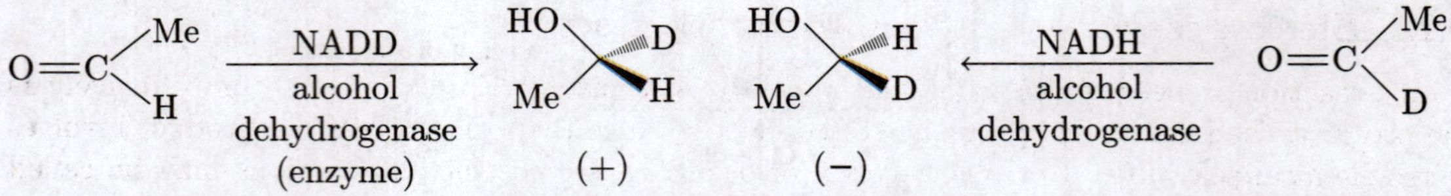

Actually the T.S.s of these reactions are diastereomeric and therefore contain different free energies. As a result of which specific products are obtained when chiral reagents interact with enantiotopic and diastereotopic ligands and faces.

9.21.3 A Prochiral centre

An achiral centre is said to be prochiral one if the replacement of one of the pair of enantiotopic ligands by a group, which is different from the ligands already present on the achiral centre, creates a chiral centre. The C atom of bromochloromethane is achiral; however, it holds two H atoms which are enantiotopic ligands. When one of them is replaced by I, we get a chiral molecule, the C atom becomes the chiral centre. Hence the C atom of bromochloromethane is a prochiral centre. The double-bonded C atoms of 2-butenes, E and Z, are also prochiral since the addition of appropriate reagents to the C=C (such as Br_2) gives rise to two chiral centers.

There are certain molecules with specific C atoms having bonded to two different achiral ligands and two other identical chiral ligands, which may or may not have the same absolute configuration. The example of such a molecule is 2, 3, 4-trihydroxypentanedioic acid, $CO_2HCHOHCHOHCHOHCO_2H$. The C_3 atom holds two achiral ligands H and OH and two other identical chiral ligands CO_2HCHOH—. We can get four configurational isomers having the constitutional formula given above.

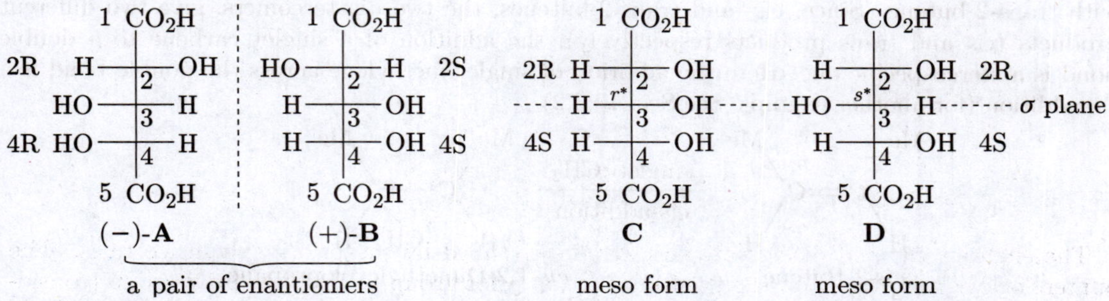

Structure A and B are chiral and optically active; they constitute a pair of enantiomers. On the other hand C and D are achiral having a σ plane. The absolute configurations of the C_3 atom of A, B, C and D are noted in the figure drawn above. In A and B, the C_3 has two identical chiral ligands having the same absolute configuration and we cannot specify its configuration. Here C_3 atom is said to be chirotopic (since the molecules are chiral) but non-stereogenic (since we cannot specify its absolute configuration). In C and D, the C_3 has two identical chiral ligands having different absolute configurations. Here the C_3 is said to be achirotopic since chirality is absent in the molecules but stereogenic since we can specify its configuration. Achirotopic stereogenic Cs are specified by the sequence rules; r^* is used for R and s^* for S (Sec Secs. 9.8.1 & 9.10.3).

9.22 Stereoselectivity and Stereospecificity of Reactions

If a reaction proceeds to form only one of a set of stereoisomers predominantly or exclusively as the product, the reaction is said to be stereoselective since the formation of the product involves stereochemical selection. The phenomenon of occurring stereoselective reactions may be called stereoselectivity of a reaction. Here 'one of a set of stereoisomers' means, it may be either one enantiomer or a pair of enantiomers or a diastereomer of several possible diastereomers. The reactions which take place with random formation of stereoisomeric products, may be called a non-stereoselective reaction. Let us consider an example for clarity of the term 'stereoselective reactions'.

Dibromosuccinic acid has two identical chiral Cs and hence it has three configurational isomers, (+) and (−) and meso. So it may exist in four isomeric forms (+)-, (−)-, (±)- and meso. The addition of bromine to maleic acid (cis-butenedioic acid) gives (±)-dibromosuccinic acid since trans addition occurs to the C=C part. As only (±) i.e., racemic form of dibromosuccinic acid is formed and not (+)- or (−)- or meso isomer alone, the reaction is said to be stereoselective. Stereoselective reactions are therefore product selective reactions.

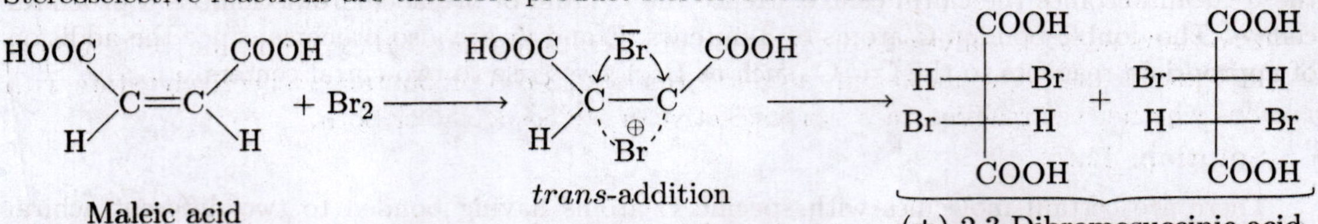

Maleic acid — *trans*-addition — (±)-2,3-Dibromosuccinic acid

When each of the stereoisomers of a set of substrate in a reaction gives a different stereoisomeric product, it is known as stereospecific reaction; e.g., methylene (carbene) adds across a double bond to form cyclopropane. When singlet methylene adds to cis-2-butene, the product is cis-1, 2-dimethylcyclopropane; whereas the same reaction produces trans-1, 2-dimethylcyclopropane with trans-2-butene. Since, cis- and trans-2-butenes, the two diastereomers, give two different products (cis and trans products respectively), the addition of a singlet carbene to a double bond is a stereospecific one. Here the addition of singlet methylene across the double bond is a cis-addition (for another example See Sec. 13.3.2).

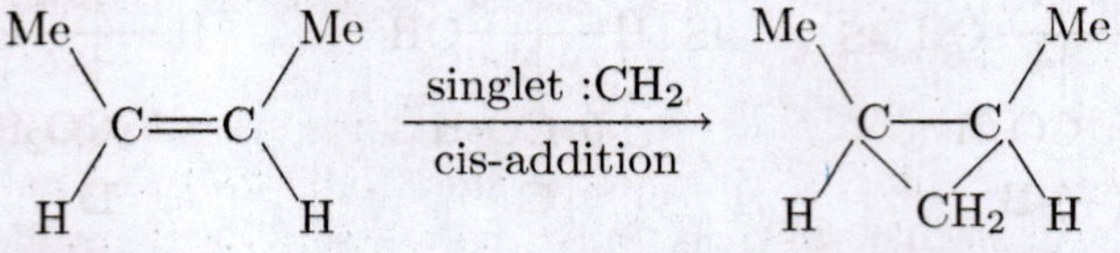

cis-2-Butene — *cis*-1,2-Dimethylcyclopropane

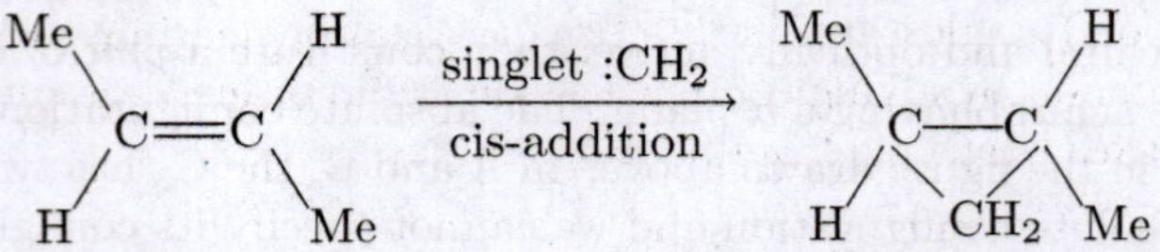

trans-2-Butene — *trans*-1,2-Dimethylcyclopropane

However, in the biochemical reactions different parts of a molecule react differently and those reactions are highly or completely stereospecific. In fact, stereoselectivity of a reaction may be either partial or moderate or high or cent percent. Nowadays both stereoselective and stereospecific reactions are further classified as enantioselective, enantiospecific, diastereoselective and diastereospecific. When the selection between two enantiomers occurs in a stereoselective reaction, it is said to be enantioselective; similarly, if the selection occurs between two diastereomeric products, it is called diasteroselective. S_N2 reactions (Chapter 12) are enantioselective whereas addition of Br_2 to alkenes (Chapter 13) are diastereoselective. Biochemical reactions are highly enantiospecific and diastereospecific. Under a chiral environment enzymes, the chiral reagents, react enantiospecifically and diastereospecifically. The resolution of racemic modifications by biochemical separation (9.15.2) is possible because of the enantiospecificity of the reactions.

It is evident from the definitions of stereoselectivity and stereospecificity that all stereospecific reactions are stereoselective but the converse is not true. The formation of cis-1, 2-dimethylcyclopropane by the reaction of singlet methylene on cis-2-butene is also stereoselective since one of the two geometric isomers, cis- and trans-1, 2-dimethylcyclopropanes, is formed by this reaction.

Problem 9.28. Maleic anhydride gives maleic acid and not fumaric acid on hydrolysis. Is the reaction stereoselective? Give reasons for your answer.

Solution: Yes, Maleic and fumeric acids are the two diastereomers. Since maleic anhydride produces only one stereoisomer i.e., maleic acid and not fumaric acid, it is a complete stereoselective reaction.

Problem 9.29. Catalytic hydrogenation of cis-2, 3-dibromo-2-butene and its trans- isomer in the presence of Ni give meso and racemic forms of 2, 3-dibromobutane respectively; state with reasons whether the reactions are stereoselective or stereospecific or both.

Solution: Both.

9.23 Strains In Organic Entities And Strainless Rings

Owing to the special molecular geometry and three dimensional structural feature of organic entities (molecules, ions and radicals), cyclic or acyclic, have to exist under strain (pressure) which they want to give up as far as practicable. Strains in organic entities are of various kinds and some of them are discussed below:

(i) **van der Waals strain:** When two non-bonded atoms or groups of an organic entity are so oriented in space that the distance between them is less than the sum of their van der Waals radii, a steric repulsion occurs between the two groups and the molecule experiences a strain, called van der Waals strain. For the van der Waals strain, the gauche conformation of n-butane is less stable than the anti conformation.

(ii) **B-strain:** B-strain is back-strain. When four bulky groups are attached to an sp^3C, they are forced together by the enforced tetrahedral angle (109°28′) and as a consequence a strain develops in the molecule which is called back-strain or B-strain. The central C atom of a molecule with back-strain gets a relief of the strain by assuming sp^2-hybridised state in which the bond angles are 120°. Thus tri-isopropylcarbinyl chloride ionises to tri-isopropylcarbenium ion by giving up the chloride ion in order to assume sp^2-hybridised state and thereby makes itself free from back-strain.

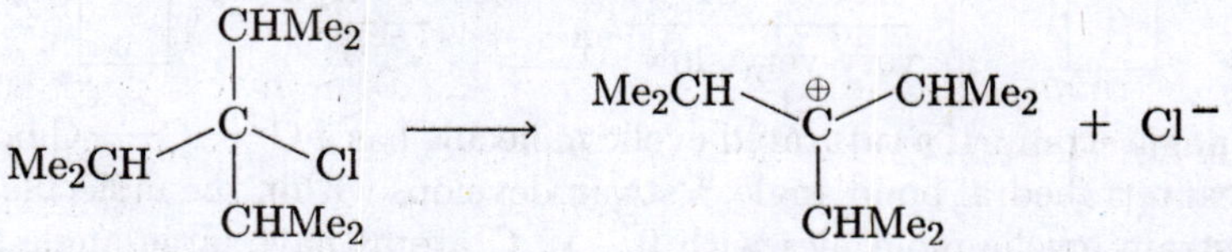

Naturally, a B-strained halide is expected to undergo S_N1, E1 and rearrangement reactions since they form carbenium ions through ionisation.

(iii) **F-strain:** When a σ bond begins to form by a coupling reaction between two ions or radicals, each having appropriate bulky groups on involved atoms, a strain develops at the time of the formation of the bond which is called face-strain or F-strain. Thus the coupling between two tri-tert-butylcarbon radical involves F-strain.

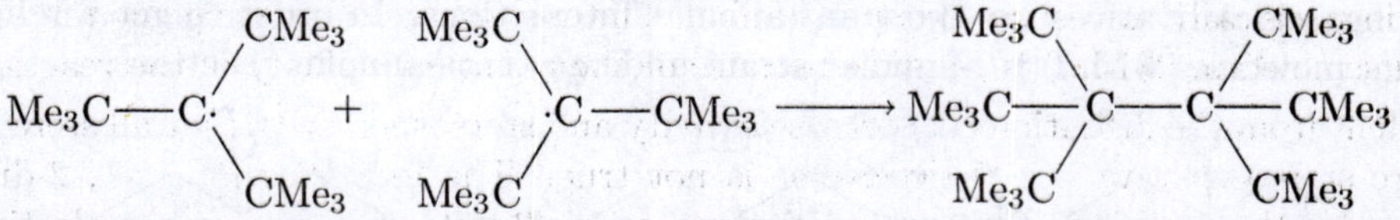

(iv) **Eclipsing strain or Pitzer strain:** The strain resulting from eclipsing of bonds in the eclipsed conformations is called eclipsing or Pitzer strain. Alicyclic rings having 5 and 7 to 10 carbon atoms bear the eclipsing or Pitzer strain even in their more stable conformations. For eclipsing strain and flag pole-bow sprit interaction six-membered alicyclic rings remain in chair conformations; in acyclic molecules chemical reactions usually occur on staggered conformations because such molecules want to get a relief of the Pitzer strain and usually remain in the staggered conformation in most of the time.

(v) **I-strain:** I-strain is internal strain; because of the special molecular geometry of cyclopropane ring, the C— C—C bond angles in cyclopropane and its derivatives are 60° though the ring C atoms are in sp^3-hybridised state (the expected bond angles are 109°28′). This deviation of ideal bond angle develops a strain within the cyclopropane molecule which is known as **small angle strain** or **internal strain** or **I-strain**. A molecule with I-strain gives up the strain through ring opening. Cyclobutane and its derivatives also experience I-strain. A substrate with I-strain cannot undergo S_N1, S_N2, E1 etc. reactions readily. In S_N1 and E1 reactions, the reaction intermediates are carbenium ions which have 120° bond angles. If a C atom with I-strain gets converted to a cyclic carbenium ion, it will have to assume sp^2-hybridised state with 120° bond angle; in this case the C atom will be more strained than in the sp^3-hybridised state and naturally the T.S. will be of high energy, i.e., the energy of activation will be very high. So, they undergo S_N1 and E1 reactions very slowly. In an S_N2 reaction on such a C atom, the T.S. contains three formal bonds in one plane while the forming and breaking bonds remain in a plane perpendicular to the plane of the former bonds; the C atom of a molecule with I-stain cannot assume such an orientation as required in the T.S. of an S_N2 reaction and as a consequence S_N2 reactions occur on I-strained molecules very slowly.

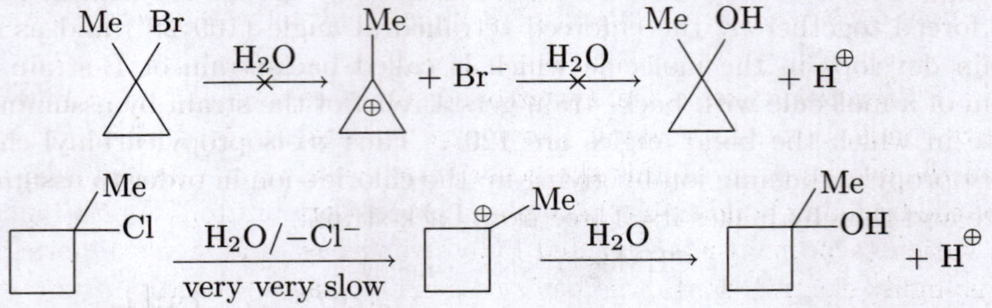

(vi) **Large angle strain:** If a saturated cyclic molecule has a C—C—C bond angle larger than the ideal tetrahedral bond angle, a strain develops within the molecule which is called large angle strain; cyclic molecules with 9 – 12 C atoms have large angle strain. It is perhaps needless to mention that small and large angle strains are **Baeyer strain**.

(vii) **Trans-annular strain:** Trans means across and annular means ring-shaped; any strain that develops owing to the presence of quasi-axial H atoms across the ring of a cyclic compound is called trans-annular strain. This is because those H atoms do not have enough internal space to fit in without coming into interaction. Rings with 7 to 12 carbon atoms experience trans-annular strain. 1, 3-diaxial interaction in the chair form of axial substituted cyclohexane derivatives are the trans-annular interactions. In order to get a relief of the strain, molecules with trans-annular strain undergo trans-annular reactions.

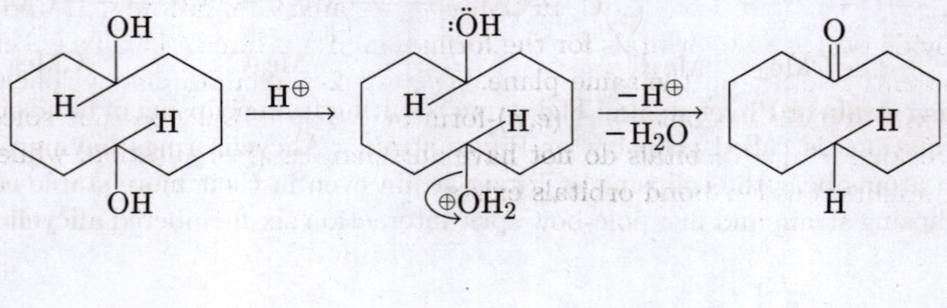

Thus we see that:

(i) small rings have I-strain-and Pitzer strain.

(ii) rings larger than cyclohexane and smaller than C_{13}-rings have large ring strains, Pitzer strain and trans-annular strain.

(iii) rings with six carbon atoms and with thirteen or more carbon atoms are almost free of Pitzer, I-, large angle, etc. strains and these may be called strainless rings. This is also supported by the heat of combustion values.

9.24 Stereoelectronic Factors and Steric Factors in Reactions

When a chemical reaction occurs, one or more bonds are in the process of breaking and one or more bonds are in the process of forming in the T.S. of the rate-determining step; the bonds are breaking to provide orbitals (filled, vacant or half filled) for the bonds which are forming. The bond formation cannot take place if the orbitals involved in the bond making process are not properly oriented; for example, to form a σ bond the overlapping orbitals must have an end-on orientation while such orbitals should have a plane parallel orientation to form a π bond. The factor that deals with the orientations of the orbitals of breaking bonds and those of the forming bonds in the T.S. is called **stereoelectronic factor**; it is actually the conformational requirements of the groups involved in the reaction.

Let us now consider a few examples to understand what is meant by the stereoelectronic factor.

When a E2 reaction takes place, two σ bonds are breaking and one π bond is forming in the T.S. Two σ bonds, through breaking, provide two orbitals to form the π orbital. Since to form a π-orbital the overlapping orbitals should orient themselves parallel to each other, the σ bond orbitals of the breaking bonds must have plane parallel orientation. Again, since in a E2 reaction usually a cation (a porton) is abstracted by a strong base and since the other leaving group in this case is highly electron dense, the basic reagent abstracts the cation from the opposite side of

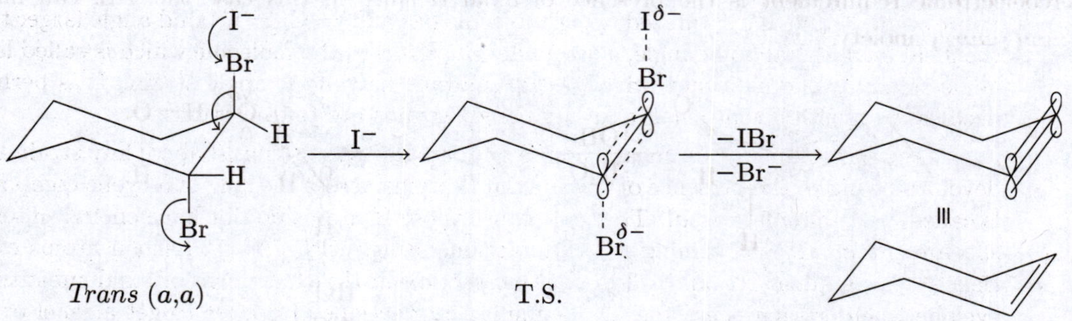

the highly electron-dense leaving group and thereby the two breaking σ bond orbitals must have anti (or trans-) periplanar orientation. In the cis-periplanar orentation of H and leaving group does not provide two parallel orbitals for the formation of a π bond. This has been proved to be so. The base unit also lies on the same plane. Trans-1, 2-diaxialdibromocyclohexane undergoes E2 reaction with iodide ion, but trans (e, e)-form or cis (ae)-form does not since in the latter forms the breaking σ bond orbitals do not have anti periplanar orientation; while in the former form proper orientation of σ bond orbitals exists.

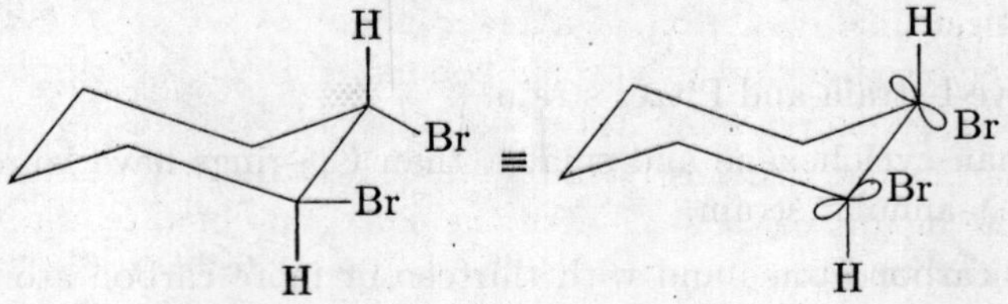

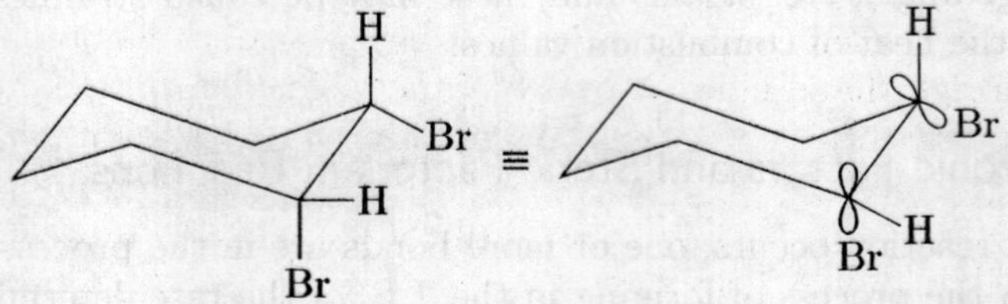

In an S_N2 reaction, there is a single concerted step in which both bond breaking and bond making processes take place. Its T.S. involves the breaking of C—G (G : leaving nucleophile) σ bond orbital and forming of Nu—C (Nu : incoming nucleophile) σ bond orbital. To ensure end-on overlap between the lobe of the σ bond orbital and that of the incoming nucleophile, the incoming nucleophile with its filled orbital attacks the σ bond orbital from the rear and thus leads to inversion of configuration. If the incoming nucleophile overlaps from the front, it will come on to a node of the anti-bonding LUMO.

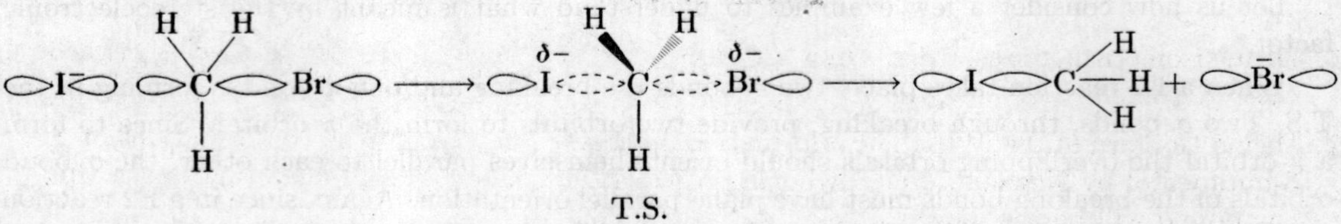

For the enolisation of cyclic ketones in which the carbonyl group is a part of the ring, the stereoelectronic requirement is the presence of axial H since in this case the T.S. can have $C\cdots C\cdots O$ moiety.

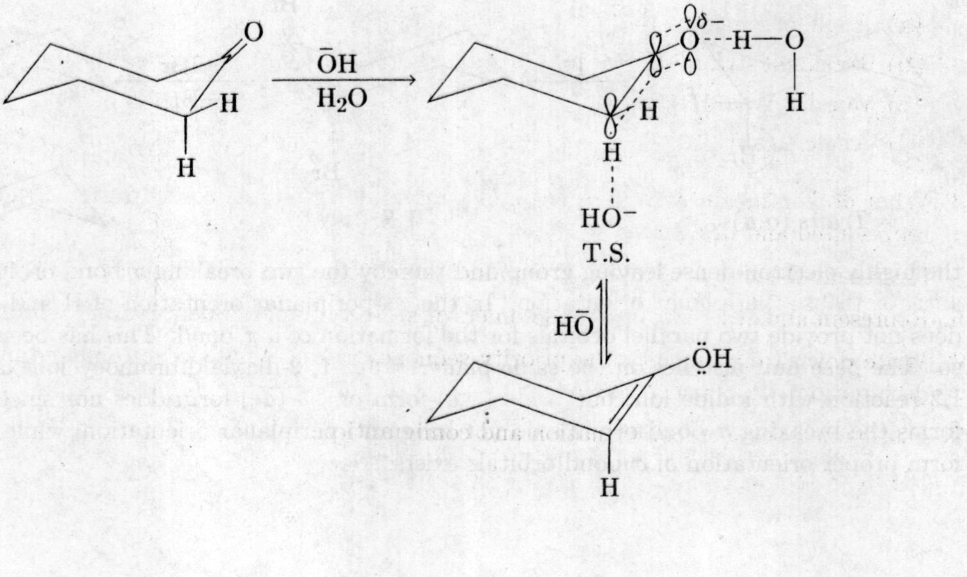

However, the stereoelectronic condition of a reaction is not all for the reaction to occur. The conformation with favourable stereoelectronic condition may give a poor yield if it bears steric strain. Steric strain develops owing to the unfavourable orientation of the non-reacting groups in the substrate. Thus to occur a reaction stereoelectronic as well as steric conditions should be favourable. The factor which deals with the conformational requirements of the non-reacting groups of the substrate may be called steric factor of the reaction; the following two conformations have favourable stereoelectronic conditions for E2 reactions but the steric factor for the same is favourable in the left hand figure only, so trans-product is obtained predominantly. Ph/Ph interaction in the right hand figure shows why the yield from that conformation is poor.

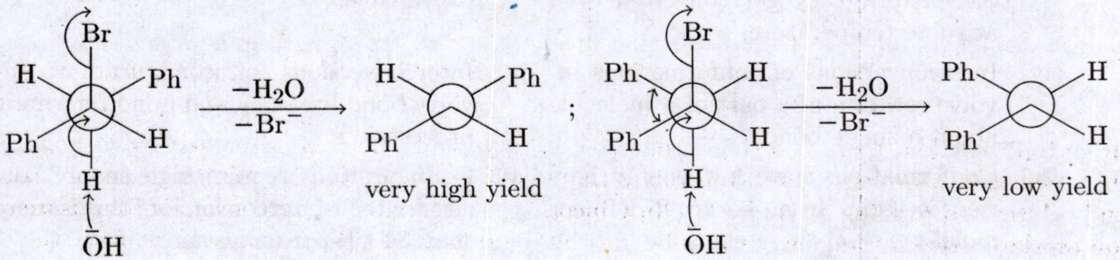

Exercise

1. Define isomerism and isomers. On what basis isomers are classified? Give examples of each class of isomers.

2. Define Conformation. What is conformational analysis? Represent the conformations of 2, 3-butanediol by Newman and Sawhorse projection formulae. Which conformation of this compound will be the most stable?

Ans. Each of the gauche conformations is the most stable because internal H-bond stabilises it.

3. Explain with example:

(a) Angle of torsion,

(b) Torsional strain,

(c) van der Waals strain,

(d) Angle strain

4. What do you mean by configuration of a molecule? How the configuration of a chiral C be represented and designated?

Represent D-lactic acid and L-glyceraldehyde by Newman and Fischer projection formulae.

5. Represent and designate cis-trans isomers of 1, 2-dibromocyclopentane and 1-bromopropene.

6. Write down from memory the priority sequence rules. What do you understand from the designations, R, S and Z, E?

7. Distinguish between conformation and configuration.

[**Ans.** of Q.7.

Conformation	Configuration
i) Conformations of a molecule are the different spatial arrangements of its atoms and groups that arise owing to the rotation of a part of the molecule about a single bond.	i) Configuration of a molecule is the spatial arrangement of atoms and groups about a rigid part of the molecule. i.e. about a double bond or ring and about a dissymmetric part of the molecule.
ii) Conformations are non-isolable.	ii) Configurations are isolable.
iii) Conformations are easily interconvertible; they get converted even at room temperature.	iii) Configurations are not easily interconvertible.
iv) Interconversions of conformations involve rotation of a part of a molecule about a single bond.	iv) Interconversions of configurations involve bond breaking and bond reforming processes.
v) Conformations have low energy barrier, it varies from 4.2 to 46 kJ. per mole.	v) Configurations require high energy barrier for their interconversion; this is more than 84 kJ. per mole.

Note. The points of differences written above are to be developed giving appropriate examples.

8. What are meant by absolute configuration and relative configuration of a molecule? Specify absolute and relative configuration of (+)-and (−)-tartaric acid.
9. Write notes on:
   (a) Specific rotation,
   (b) Molecular rotation,
   (c) Elements of symmetry,
   (d) Chiral atom and chiral molecule,
   (e) Enantiomers and diastereomers,
   (f) Dissymmetric molecule and asymmetric molecule,
   (g) Racemisation,
   (h) Resolution,
   (i) Asymmetric synthesis,
   (j) Walden inversion,
   (k) Isomerism of aldoximes and ketoximes,
   (l) Tautomerism,
   (m) Epimer.
10. The presence of a chiral C in a molecule is not a necessary condition for showing optical activity—explain.
11. Give an account of the various types of carbon compounds which exhibit optical isomerism.
12. (a) What is the necessary condition for the existence of enantiomers?
    (b) What are the necessary and sufficient conditions for the measurement of optical activity?

(c) Are all substances with chiral atoms optically active and resolvable?

(d) Are enantiomers possible in molecules that do not have chiral carbon atoms?

13. (a) Which of the following compounds are resolvable?

(i) cis-Cyclohexane-1, 2-dicarboxylic acid

(ii) trans-Cyclohexane-1, 2-dicarboxylic acid

(iii) 6, 6′-Dinitrodiphenic acid

(b) Predict the nature and number of isomers of the following compounds and give your reasons:

(i) $CH_3.CH{=}CH.CHOH.CH_3$

(ii) HOOC.CHBr.CHBr.COOH

(iii) $CH_3.CH{=}C{=}CH.COOH$

(iv) 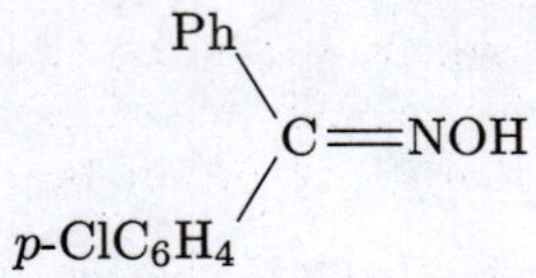

14. (a) You are given a racemic alcohol, say, 2-butanol, and a pure optically active organic base. How would you proceed to resolve the alcohol? Explain the principle.

(b) Write down the structures of two organic molecules of the same molecular formula, C_5H_8, one of which shows enantiomerism and the other diastereomerism (geometrical isomerism). Draw structure of each isomer and give the reason for isomerism.

15. (a) Write down the correct Fischer projection formulae for the following compounds: D-2-Butylmethyl ether; D-α-hydroxysuccinic acid; L-2-phenyl-2-butanol; L-2-methyl-1-butanol.

(b) Draw the projection formulae of all the stereoisomers of $CO_2H.CHOH.CHOH.CHOH.CO_2H$ and explain the optical activity of each of them.

(c) How can you resolve the following racemic compounds?

(i) $CH_3CHBrCOOH$

(ii) $CH_3CH_2CH(CH_3)CH_2NH_2$.

16. Discuss chair, boat and twist boat conformations of cyclohexane molecule. What do you mean by axial and equatorial bonds? Why is its equatorial substituted chair conformation more stable than the axial substituted chair conformation.

17. Discuss 1, 3-diaxial interaction and n-butane-gauche interaction in connection with the chair conformation of methylcyclohexane.

18. Draw planar structures, chair conformations and Newman projections for 1, 4-dimethylcyclohexanes. How many chair conformations are possible? Discuss their optical activity. Apply conformational analysis to determine the most stable conformation of 1, 4-dimethylcyclohexane.

Chapter 10

UV-Visible, IR and ^{1}HNMR Spectroscopy

Objectives

- To know the merits of spectroscopic analysis of molecular structures [10.1] p. 415
- To know the particle motion and the wave motion [10.2–10.3] p. 415 & 416
- To know the nature of electromagnetic radiation [10.4] p. 417
- To understand certain terms and laws regarding absorption of the electromagnetic radiations [10.5] p. 420
- To know the effects of electromagnetic radiations on molecules [10.6] p. 422
- To get ideas about spectra, spectrophotometers and presentation of spectra [10.7–10.11] p. 423–426
- To know the UV and visible spectral range, visible and complementary colours and certain terms [10.12] p.426
- To understand the theory of UV and IR spectroscopy and different types of UV spectral bands [10.13–10.14 and 10.20] p. 428–433 & 453
- To understand the UV and IR spectra-structure relationship of organic molecules [10.17 and 10.20.1] p. 434 & 463

- To get an idea about the applicability of UV and IR spectroscopy in organic chemistry [10.18 and 10.20.2] p. 439 & 467
- To know the ORD and CD curves [10.21] p. 471
- To understand ^{1}HNMR (PMR) spectroscopy [10.22] p. 472

10.1 Introduction

The study of the organic compounds depends on the knowledge of their molecular structures. The determination of the structure of an organic compound is, therefore, an essential task for an organic chemist. The classical methods applied for this not only are cumbersome, laborious and time-consuming but also require a large quantity of the compounds. These in general consist of the following sequential steps :

(a) (i) Establishment of the molecular formula of the compound by its elemental analysis and molecular weight determination.

(ii) Extensive study, of the chemical reactions ranging from usual tests for functional groups to degradation reactions of the compounds into simple and detectable substances.

(b) Information obtained from chemical investigation are then fitted together to predict one or more structures for the compound under consideration.

(c) Ultimately, confirmation of one of the predicted structures is done by synthesising a compound, identical with the original one, from structurally known starting materials and through well-established routes involving species of known structures.

Nowadays various types of instrumental analyses of a minute amount of an organic compound by physical methods most often give necessary and sufficient information to establish its molecular structure undoubtedly. Molecular spectroscopy deserves special mention amongst these. In order to understand the subject matter of the molecular spectroscopy and its applications, one requires a knowledge of some fundamental terms, laws and phenomena. At first these are discussed below in brief.

10.2 Particle Motion

A particle may exhibit three general kinds of motions

(a) translational. (b) rotational and (c) vibrational.

Molecules, atoms, electrons, etc. being particles, they also have motions.

A molecule may change its position; i.e., it can move from one position to another, and this is known as translational motion (Fig. 10.1).

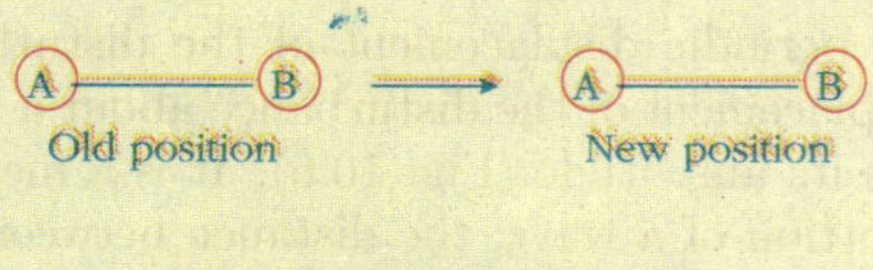

Fig. 10.1. Translational motion

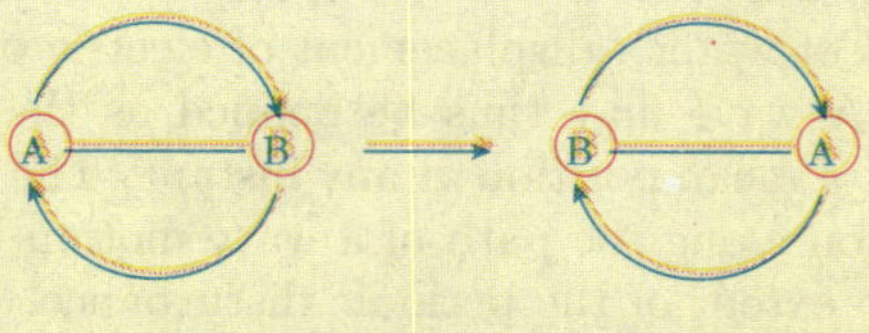

Fig. 10.2. Rotational motion

A molecule may rotate about an axis and the motion is then called rotational motion (Fig. 10.2). An atom or a group of atoms of a molecule may rotate about a bond axis (Fig. 10.3).

The spin motion is a special kind of rotational motion of a particle in which the particle revolves about its own axis. A revolving top has a spin motion since it revolves about its own axis, the nail of the top. Electrons and nucleons have spin motion. Every such particle spins either clockwise or anticlockwise about its own axis (Fig. 10.4).

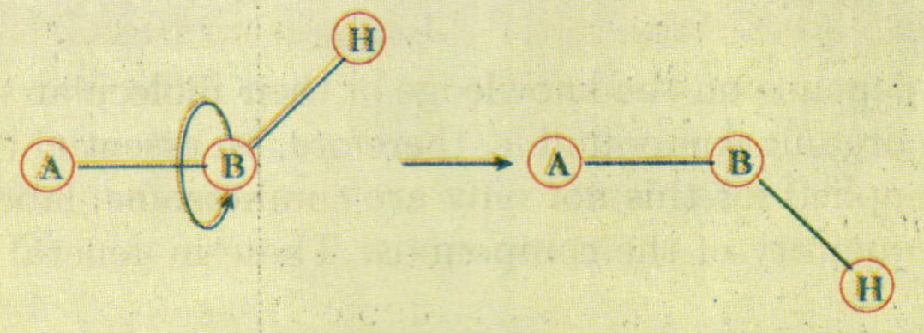

Fig. 10.3. Rotational motion of an atom of a molecule

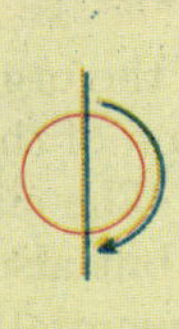

Fig. 10.4. Spin motion

The periodic to and fro motion of a particle is called its oscillatory or vibrational motion. If a particle, during its motion, retraces its path after a constant interval of time, the motion of the particle is said to be periodic, and the fixed interval of time, after which the retracement takes place, is called the period of the moving particle. The rotational motion of the earth round the sun is periodic, the period here covers 365 days (approximately). Covalently bonded atoms in molecule possess vibrational motion. A bond between two atoms in a molecule vibrates (oscillates) through a point at which electron-electron and nucleus-nucleus repulsions are balanced by the electron-nucleus attraction (Fig. 10.5, stretching of bonds.)

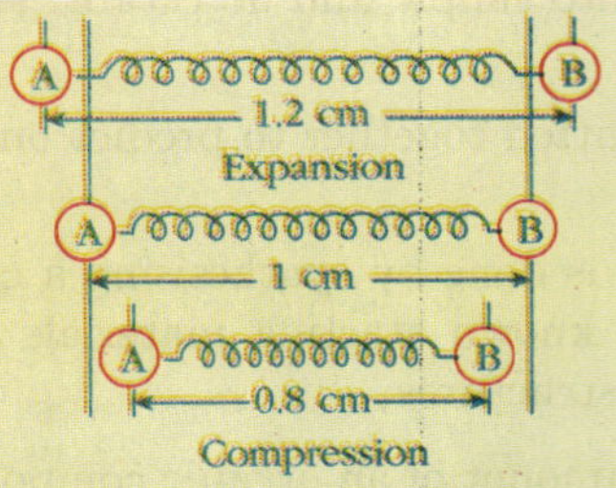

Fig. 10.5. Vibration of atoms within a molecule

However, molecules are in a continuous motion even at absolute zero; at this temperature they possess vibrational motion only.

10.3 Wave Motion

A periodic disturbance in a material medium or space, which may involve either periodic changes in some physical quantities such as temperature, pressure, electromagnetic field strength, etc., of the elastic displacement of material particles of the medium, is called a wave. When a stone is dropped into a pond of calm water, we observe waves. A wave propagates with a definite velocity.

The propagation of a wave carrying energy is called **wave motion**. Along the path of a wave motion at any point a periodic displacement of the disturbance about a mean position may take place. This may consist of a displacement of material medium (water molecules in the case of waves of water), displacement of electric or magnetic vectors (Sec. 10.4, electromagnetic waves) etc. A wave may thus be defined as the locus of a periodic displacement of the disturbance about a mean position at any instant. The highest displacement of the disturbance about a mean position along the path of a wave motion is known as its amplitude (Fig. 10.6). It is a measure of the extent of the periodic disturbance. In every portion of a wave, the distance between two successive crests or troughs is one and the same, and this distance is known as wavelength. The wavelength is usually represented by the Greek letter, λ, the unit of which is the unit of length. The frequency (f or ν) of a wave motion is the number of oscillation of the wave of energy in unit time, usually second. It is obtained when the velocity (V, in the unit of length per sec.) of the wave is divided by the wavelength (in the unit of length). The reciprocal of the wavelength is termed wave number and symbolised by the letter $\bar{\nu}$. Naturally, the unit of the wave number is the inverse of the unit of length, say cm^{-1}. The energy of a wave is determined by the amplitude of the wave, and it is proportional to the square of the amplitude.

$$\text{Energy} \propto (\text{amplitude})^2$$

The basic equation of a wave motion is $V = \nu\lambda$. (10.1)

where V, ν and λ have usual significance.

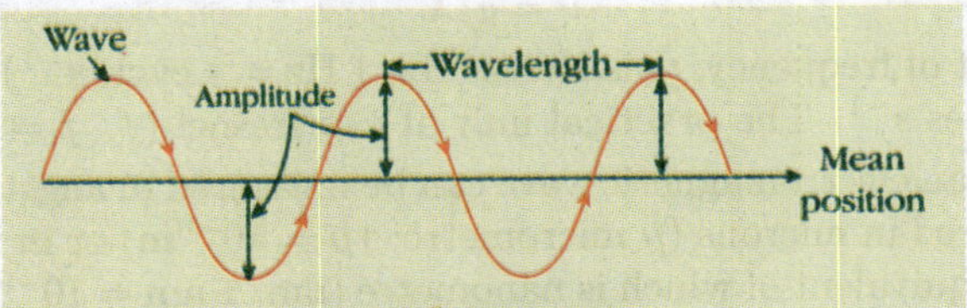

Fig. 10.6. Wave motion

A wave motion may be characterised by the amplitude, the frequency, the wavelength, the wave number, etc. of the wave.

10.4 Electromagnetic Radiation

Emission of energy (light, heat, etc.) or particles (α, β, etc.) is called radiation. The radiation, resulted from the acceleration of an electric charge, consists of waves of energy. Each of the waves is composed of a magnetic component and an electric component. These components oscillate in mutually perpendicular planes and the waves propagate at the same velocity through space (vacuum or material medium) in a direction perpendicular to the planes of the two fields (Fig. 10.7).

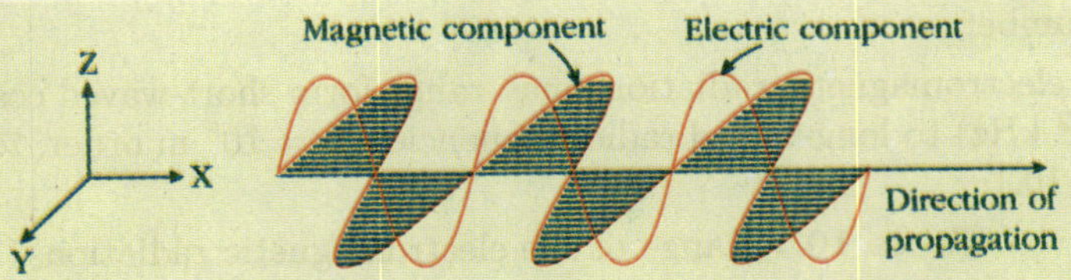

Fig. 10.7. Propagation of the electromagnetic wave of radiation

The electromagnetic radiation travels through vacuum with a uniform velocity of 2.9979×10^8 m sec^{-1} (denoted by c, the velocity of light). In a material medium, the velocity of propagation of the radiation is given by $2.9979 \times 10^8/n$ m sec^{-1}; where n is the refractive index of the medium. It is naturally smaller than the velocity of the radiation in vacuum.

The emission of waves of energy in the electromagnetic radiation occurs in discrete units or packets called photons or quanta (plural form of quantum). The magnitude of the energy (E) of a photon is given by the product of Planck's constant (h) and the frequency of radiation (ν) appropriate for the description of its propagation. Thus, $E = h\nu$ (10.2)

The units of energy are:

(i) 1 electron Volt (1 eV) = 23.053 cal/mole,

(ii) 1 cal/mole = 4.34×10^{-5} ergs

(iii) 1 joule (1 J) = 10^{-7} ergs

(iv) 1 calorie = 4.1868 joule.

The SI unit of energy is one Joule (J). The value of Planck's constant in SI unit is 6.62559×10^{-34} Js.

Since the velocity of an electromagnetic radiation is that of light, the equation for a wave of electromagnetic radiation can be written as:

$$c = \nu\lambda \tag{10.3}$$

$$\nu = c/\lambda. \tag{10.4}$$

The derived SI unit of frequency, ν, is Hertz (Hz; 1 Hz = 1 cycle s^{-1}). It is needless to mention that 1 kHz = 10^3 cycles s^{-1}. The practical unit of ν is fresnel (f, $f = 10^{12}$ Hz).

The wavelength of an electromagnetic wave can be measured in angstrom unit ($1Å = 10^{-10}$ m). It is commonly measured in microns (μ micrometre; $1\mu = 10^{-6}$m) or in millimicrons (mμ; 1 mμ = 10^{-9} m) the popular equivalent of which is nanometre (nm, 1 nm = 10^{-9} m) Thus, 1 nm = 1mμ = 10^{-9}m = $10^{-3}\mu$ = 10Å.

The SI unit of wavelength is m.

The wave number of a radiation of frequency, ν, is $\overline{\nu}$. Thus

$$\overline{\nu} = \frac{1}{\lambda} \tag{10.5}$$

The SI unit of the wave number is, therefore, m^{-1}; cm^{-1} is also used as its unit. 1 cm^{-1} is called 1 kayser (K).

So the electromagnetic radiation may be described in terms of:

(a) the energy of its photon

(b) the frequency of its waves,

(c) the wavelength and

(d) the wave number.

However, the electromagnetic radiations may range form short-waved cosmic rays ($\lambda = 10^{-17}$ m order; $\overline{\nu} = 10^{22}$ kHz) to long-waved radio frequencies ($\lambda = 10^5$ m order; $\overline{\nu}$ = 1 kHz); the range is given in Table 10.1:

Table 10.1 Range of the electromagnetic radiations

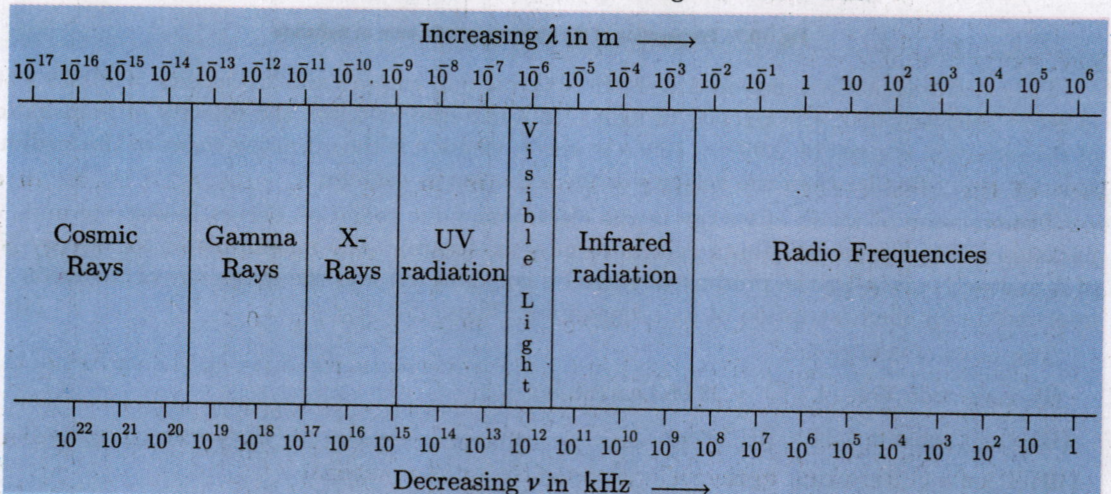

Visible (white) light represents only a very small portion of the total electromagnetic radiations. This consists of the seven colours (violet to red) of the rainbow. These colours, their frequencies, wavelengths and energies are given in Table 10.2.

Table 10.2 Visible portion of the electromagnetic radiations

Colour	Wavelength (λ in 10^{-9} m)	Frequency (ν in 10^{14} Hz)	Photon's energy (energy in 10^{-20}J)
Violet	400.0—424.0	7.495—7.115	49.66—47.14
Blue	424.0—491.2	7.115—6.104	47.14—40.44
Green	491.2—575.0	6.104—5.215	40.44—34.55
Yellow	575.0—585.0	5.215—5.125	34.55—33.956
Orange	585.0—647.0	5.125—4.634	33.956—30.703
Red	647.0—700.0	4.634—4.284	30.703—25.374

A beam of electromagnetic radiation may contain waves of:

(a) only one discrete wavelength which is said to be monochromatic.

(b) several wavelengths and then the beam is said to be polychromatic or heterochromatic.

Problem 10.1. Calculate the frequency of red light if its wavelength is 700 nm.

Solution: 700 nm = 700×10^{-9} m

$$\nu = c/\lambda; \text{ where } c = 3 \times 10^8 \text{ ms}^{-1} \text{ and } \lambda = 700 \times 10^{-9} \text{ m}$$

Therefore,
$$\nu = \frac{3 \times 10^8 \text{m s}^{-1}}{700 \times 10^{-9}\text{m}} = 4.285 \times 10^{14} \text{ s}^{-1} = 4.285 \times 10^{14} \text{ Hz.}$$

Problem 10.2. Calculate the energy of a photon of the violet light. Given, $c = 3 \times 10^8$ m s^{-1}, $\lambda = 400$ nm and $h = 6.624 \times 10^{-34}$ Js.

Solution: Frequency, ν, of the violet light $= c/\lambda$

$$= \frac{3 \times 10^8 \text{m s}^{-1}}{400 \times 10^{-9}\text{m}} = 7.5 \times 10^{14}\text{s}^{-1}$$

$$= 7.5 \times 10^{14} \text{ Hz.}$$

Now, energy of a photon of the violet light, $E = h\nu$

$$= 6.624 \times 10^{-34}\text{Js} \times 7.5 \times 10^{14} \text{ s}^{-1}$$

$$= 4.968 \times 10^{-19}\text{J.}$$

Fate of the electromagnetic radiations after impinging on a substance taken in a container

When a beam of electromagnetic radiation strikes upon a material taken in a container, the following may take place:

(a) The direction of the propagation of the beam may change (Fig. 10.8) due to

(i) Reflection : throwing back of the beam of electromagnetic radiation by the surface of the container following the laws of reflection of light.

(ii) Refraction : propagation of the beam in a different direction in passing from the medium of the incident beam to the medium consisting of the material.

(iii) Diffraction : bending of radiant waves around obstacles in their path.

(iv) Scattering : deflection of any radiation as a result of its interaction with the minute particles of matter.

(b) The radiant waves may pass through the matter with a minimum of absorption and thus the emergent beam may be of almost identical energy with that of the incident beam, i.e., the energy loss will be of minimal quantity.

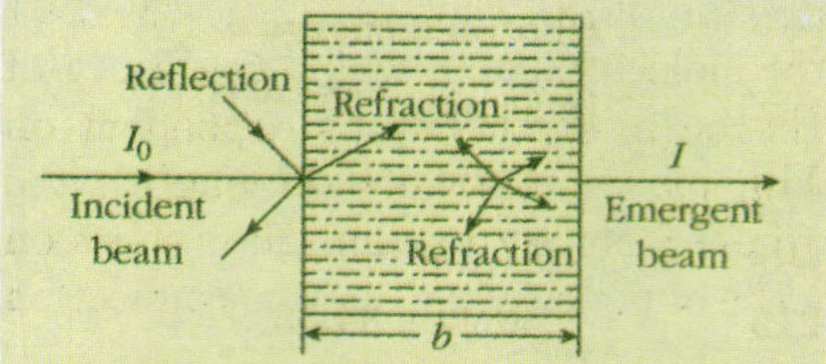

Fig. 10.8. Incident radiation striking on a substance

(c) The energy of the incident beam may be absorbed either entirely or partly by the container and the material within it. The amount of energy in the emergent beam in that case will be much less than that of the incident beam of the radiant energy.

10.5 Certain Terms and Laws Regarding Absorption of Electromagnetic Radiation

The amount of radiant energy absorbed by a chemical substance depends, amongst other factors, on the wavelength of the electromagnetic radiation passing through the substance and on the nature of the constituent chemical units. Thus the process of absorption of radiant energy is a characteristic phenomenon directly related to the specific molecular structure.

However, the amount of radiant energy absorbed by a given amount of substance may be expressed by many a term involving the strength, more correctly the intensity, and the radiant power of the the beam passing through it. Let us now discuss the terms like radiant power, intensity of radiation, etc., and the laws involved.

10.5.1 Certain terms : radiant power, intensity, transmittance and absorbance

The intensity of a beam of electromagnetic radiation is characterised by its radiant power which is proportional to the number of photons propagating in the beam per second. If the intensity and the radiant power of the incident beam of a radiant energy are I_0 and P_0, respectively and those of the emergent beam transmitted by the substance are I and P, respectively, then the ratio of the radiant power of the emergent beam to the radiant power of the incident beam is known as transmittance, which is symbolically represented as T. Thus,

$$\text{Transmittance, } T = \frac{P}{P_0} = \frac{I}{I_0} \; [P \propto I]. \tag{10.6}$$

If 1-cm layer of a solution which contains 1 percent by weight of solute absorbing radiant energy, its transmittance is represented by $T_{1-\text{cm}}^{1\%}$.

On the other hand, the logarithm to the base 10 of the ratio of the radiant power of the incident beam to the radiant power of the emergent beam is called absorbance (A).

$$\text{Thus,} \quad A = \log_{10}(P_0/P). \tag{10.7}$$

$$\text{Again,} \quad T = \frac{P}{P_0} \quad \text{or,} \quad \frac{P_0}{P} = \frac{1}{T}.$$

$$\text{Therefore,} \quad A = \log_{10}\left(\frac{1}{T}\right). \tag{10.8}$$

Hence, the logarithm to the base 10 of the reciprocal of transmittance is the absorbance, (formerly called optical density). However, $A_{1-\text{cm}}^{2\%}$ represents the absorbance of a 1-cm layer of solution, which contains 2% by weight of the absorbing solute.

10.5.2 Fundamental Laws : Lambert's and Beer's Laws

The ratio I_0/I or P_0/P is dependent on the number of molecules with which the beam of radiation encounters. Number of molecules with which the beam encounters is dependent on the length of the solution through which the radiant beam passes (optical path-length), the temperature and the concentration of the solution. Thus these factors have a marked effect on the ratios said above, i.e., on the transmittance and absorbance of the solution. in this connection there are two laws:

1. Lambert's Law

It states that when a beam of monochromatic light (plane-parallel) passes through a homogeneous absorbing medium at right angles to the plane-parallel surfaces of the medium, the intensity of the light diminishes in geometric or exponential progression as the thickness of the medium increases arithmetically. Mathematically, the law may be simply expressed as:

$$\log \frac{I_0}{I} = a'b \tag{10.9}$$

where b = optical path-length in cm and a' is a constant.

2. Beer's law

It states that the intensity of a beam of plane-parallel monochromatic light decreases in a similar manner as the concentration of the light-absorbing constituent increases in the medium through which the light passes.

Thus, $$\log \frac{I_0}{I} = a''c, \tag{10.10}$$

where c = concentration (g/l); a'' = a constant.

Beer's law is applicable to very dilute solutions only.

Now, if we combine the two laws stated above, we get:

$$\log \frac{I_0}{I} = abc \tag{10.11}$$

where a is a constant, called absorptivity index, or specific extinction or extinction coefficient. It depends on the wavelength of the radiation and the nature of the absorbing substance.

The product of the absorptivity and the molecular weight of the substance is termed as 'molar absorptivity'. It is symbolised by ϵ (epsilon; the fifth letter of the Greek alphabet).

Thus, $$\log \frac{I_0}{I} = \epsilon bc, \tag{10.12}$$

where $\epsilon = a \times$ molecular weight of the substance.

This combined Lambert-Beer law may be used for the absorption of electromagnetic radiations in general and I_0/I term may be replaced by P_0/P term.

Hence, Eqn. (10.11) and Eqn. (10.12) may be respectively written as

$$\log \frac{P_0}{P} = abc \tag{10.13}$$

and $$\log \frac{P_0}{P} = \epsilon bc. \tag{10.14}$$

Since $\log P_0/P = A$, the absorbance, we can write,

$$A = abc \quad \text{or,} \quad A = \epsilon bc \tag{10.15}$$

Problem 10.3. When a beam of radiant energy is passed through a given solution, the radiant power gets reduced to 50%—calculate the absorbance.

Solution: According to the problem, if the radiant power of the emergent beam is P, then that of the incident beam is 2P.

We know that, $A = \log \dfrac{P_0}{P}$; here $P_0 = 2P$ and $P = P$.

Therefore, $A = \log \dfrac{2P}{P} = \log 2 = 0.30203.$

10.6 Molecular Absorption of Electromagnetic Radiation

In the gaseous state of a molecule its energy may be expressed as the summation of its translational energy and internal energy. By the internal energy of a molecule we mean electronic energy, rotational energy and vibrational energy, and energy of nuclear and electronic spin states. However, being a function of temperature, the translational energy of a molecule is not expressed in terms of energy quantum, i.e., it is not quantised. A molecule, therefore, may gain or lose any finite quantity of this form of energy; on the other hand, other forms of energy are always gained or lost in discrete units, and therefore, are quantised.

When a molecule remains at the lowest energy state under a given set of conditions it is said to be in the ground state of the molecule.

When a molecule absorbs radiant energy, its kinetic energy and, consequently, its internal energy increases, this brings it to a higher energy level and then the molecule is said to be in the excited state.

So by the absorption of electromagnetic radiation molecular excitation takes place. The excited molecule is profoundly different from the ground state molecule in physical and in many of its chemical properties, such as in shape, bond length, etc.

According to the quantum theory, a molecule will absorb radiation of a particular energy ($h\nu$) which corresponds to the energy difference, ΔE, between two quantised (discrete) energy levels (the ground state energy level and the excited state energy level) in the molecule, so that the quantum of energy absorbed,

$$\Delta E = h\nu. \tag{10.16}$$

Depending upon the structure of the molecule and ν of the radiation absorbed by it, different categories of molecular excitation may occur:

(a) Rotational, (b) Vibrational, (c) Electronic and (d) Nuclear (See Sec. 10.22).

10.6.1 Rotational excitation

This involves the rotations of the atoms and groups about their bond axes within the molecule.

Approximately 4.1868 kJ mol^{-1} of energy, which comes from photons of far infra-red radiation, is required for the rotational excitation. This brings about a molecular transition between two quantised rotational energy levels, no matter in which electronic and vibrational states it may have.

10.6.2 Vibrational excitation

This involves vibration of the constituent atoms of the molecule together with their rotational motion. About 41.868 kJ mol^{-1} of energy is sufficient for the vibrational excitation of a molecule, i.e., molecular transition between two quantised vibrational energy levels (of the ground state and the excited state of the molecule). The absorption of near infra-red radiation supplies the required energy.

10.6.3 Electronic excitation

This involves the promotion of an electron from a lower energy molecular orbital, to a higher energy molecular orbital. Usually 41.686 to 1046.700 kJ mol^{-1} of energy is required for electronic excitation of a molecule. Owing to this large amount of energy change, rotational and vibrational excitations also occur. Since photons of ultra-violet and visible radiations possess the energy of this order, electronic excitation of molecules occurs in ultra-violet and visible regions.

Problem 10.4. Calculate the energy of a radiation in k cal/mole, the wavelength of which is 10 Å, $c = 3 \times 10^8$ ms^{-1} and $h = 6.624 \times 10^{-34}$ Js.

Solution: 10 Å= 10^{-9}m, the value of λ

$$\nu = \frac{c}{\lambda} = 3.00 \times 10^8 \text{ ms}^{-1}/10^{-9} \text{ m.}$$
$$= 3 \times 10^{17} \text{ s}^{-1}.$$

Now, the energy of 1 mole of photon having $\lambda = 10^{-9}$ m is

$E = N_0 h\nu$ mol^{-1}, where N_0 is Avogadro's number

$$= 6.02 \times 10^{23} \times 6.624 \times 10^{-34} \text{ Js} \times 3 \times 10^{17}\text{s}^{-1} \text{ mol}^{-1}$$
$$= 6.02 \times 6.624 \times 3 \times 10^{6}\text{J mol}^{-1}$$
$$= \frac{6.02 \times 6.624 \times 3 \times 10^6}{4.1868} \text{ cal/mole [since 1 cal} = 4.1868 \text{ J]}$$
$$= 28.48 \times 10^6 \text{ cal/mole}$$
$$= 28.48 \times 10^3 \text{ k cal/mole.}$$

10.7 Spectrum

Spectrum is a Latin word meaning an appearance. What will appear? The appearance of an ordered arrangement of the things under consideration in accordance with a specific property may be called a spectrum. The spectrum of visible or white light is an orderly arrangement of the constituent colours ranging from violet to red according to increasing wavelength.

The electromagnetic spectrum is an orderly arrangement of the radiant energy according to wavelength ranging from short-waved cosmic rays to long-waved radio frequencies. Naturally, the spectrum of white light is only a part of the electromagnetic spectrum.

A record of the amount of energy emitted by a sample as a function of its wavelength or frequency or wave number is called an emission spectrum, while a record of the quantity of electromagnetic radiation absorbed by a sample as a function of its wave number or wavelength or frequency is said to be an absorption spectrum.

The act of molecular absorption of energy in the ultra-violet or visible or infra-red region at a definite wavelength is a function of a molecular structural feature having the capability of absorbing specific photon. Consequently it indicates the presence of a particular structure. Thus the absorption of a radiant energy by a molecule, when recorded as spectrum, gives valuable information about its structure.

There are two particularly useful regions of the electromagnetic absorption spectrum, amongst many to detect the structural features of organic molecules.

1. Ultra-violet and visible regions

These involve absorption of energy in ultra-violet and visible regions and cause electronic excitation of the molecule. The absorption spectrum of such an excitation when recorded is known as ultra-violet and visible spectra (UV and visible spectra).

2. Infra-red-region

This involves absorption of infra-red radiation which causes vibrational and rotational excitations of the molecules. The spectrum thus formed is called infra-red (IR) spectrum.

When each of the absorption spectra is observed directly through an instrument, called spectroscope, absorption bands are seen; these bands are composed of lines. In a particular excitation process, a molecule absorbs only one discrete quantity of energy, thus it absorbs radiation of only one wavelength or wave number or frequency. If all the molecules of a given bulk of material were in the same rotational and vibrational energy levels, the result of absorption of radiation would have been a series of absorption lines. Since a group of molecules exists in a number of different rotational and vibrational energy levels, each differing from another by a very small amount of energy, the group of molecules absorbs energy over a small range and gives rise to a very closely spaced group of absorption lines which appear as a band.

10.8 Molecular Spectroscopy

The detection of the structural feature of a chemical species (atoms, molecules or ions) by the study of its internal energy change after analysing its emission spectrum or absorption spectrum is known as spectroscopy The spectroscopic study of the molecules is called molecular spectroscopy.

10.9 Instruments used in Spectroscopy

Any instrument used to produce spectrum may be called a spectroscope. It consists of the following parts:

1. Radiant energy source: To produce visible spectra, tungsten lamp is used as a radiant energy source; while hydrogen or deuterium discharge lamp is used as a radiant energy source for producing ultra-violet spectrum.
2. Associated optics: It consists of lens, mirror, slits, etc. These together constitute an arrangement called associated optics which produces narrow beam of radiant energy.
3. Resolving device : Prism, grating, etc. are used to resolve the radiant beam into its component wavelengths so that only one discreate wavelength at a time can be focussed on the sample area. Usually prisms made of quartz and glass are used in UV and visible spectroscopic instruments.
4. Sample and solvent holders : These are also called absorption cells. These may be either test tubes of uniform wall-thickness and of diameter or rectangular cuvettes of high quality material with uniformly clear and polished plane-parallel faces. Sample holder materials are usually quartz and glass or quartz for UV and visible spectroscopic instruments respectively.

A very simple spectroscope uses a prism to separate the visible light into its constituent colours. An eye piece may then catch each of the colours separately. Human eye acts here as receptor.

The spectroscope attached with a scale for reading the wavelength of radiant energy is called spectrometer.

A spectrograph is an instrument by which photographic record of the spectrum is done.

A spectrophotometer uses a photoelectric cell, usually barrier-layer cell, photomultiplier tubes for comparing quantitatively two radiations wavelength by wavelength.

10.10 Use of Spectrophotometer

The most generally used instrument for spectroscopy is a double beam spectrophotometer. An electromagnetic radiation source generates a radiation beam consisting of a continuum of wavelength which becomes a narrow one after passing through an optical system called associated optics. Then the narrow beam of radiation passes through a rotating prism so that radiation of one wavelength at a time can be focussed on the sample under consideration. The beam finally gets split into two beams of equal intensities. One of the beams then passes through the sample (usually in solution) and the other passes through the reference (usually the pure solvent). Detectors convert each of the two emergent radiant energies into electrical energy. The difference in electrical energies between the sample and the reference beams coming out through them gives the difference in absorbance or transmittance (Sec. 10.5.1) between them. This is recorded automatically on a moving graph paper. In this way one may get graphical plot of wavelengths of radiant energy absorbed by the substance versus its absorbance or transmittance.

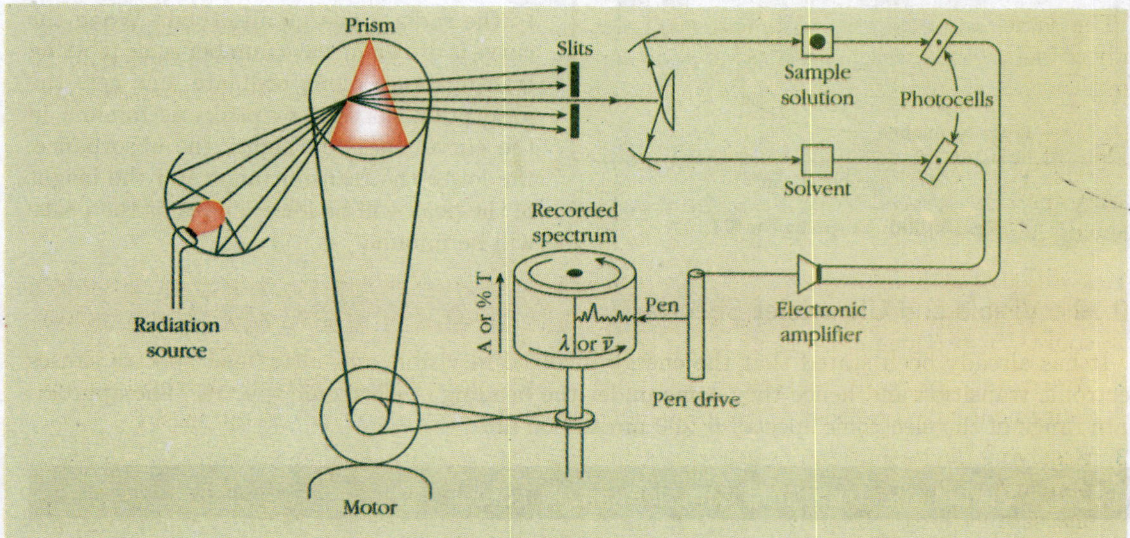

Fig. 10.9. A double beam spectrophotometer

10.11 Presentation of Spectrum

Absorption spectra are displayed in many a way.

(a) increasing values of percentage transmittance as ordinate may be plotted against increasing values of wavelength (in nm or mμ) or decreasing values of frequency (usually in the form of wave number ($\bar{\nu}$ in cm^{-1}).

(b) Increasing values of absorbance may be plotted as ordinate against increasing values of wavelength (in nm or mμ) or decreasing values of $\overline{\nu}$ (in cm^{-1}).

(c) Increasing values of molar absorptivity (ϵ) against increasing λ values or decreasing $\overline{\nu}$ values taking the former as ordinate and the latter as abscissa.

(d) Increasing values of log A (logarithm of absorbance) or log a (logarithm of absorptivity) or log ϵ (logarithm of molar absorptivity may be plotted against increasing values of λ (in nm or mμ) or $\overline{\nu}$ (in cm^{-1}). This is done when A or a or ϵ values are large.

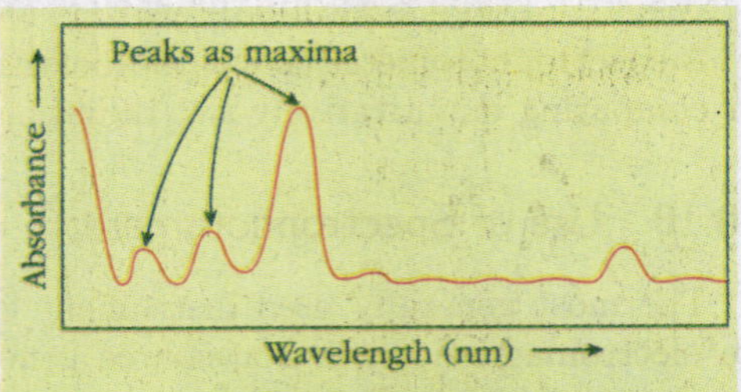

Fig. 10.10. UV spectra λ vs A

Recording of UV or visible Spectra

Usually UV and visible spectra are plotted taking absorbance or absorptivity or molar absorptivity against the wavelength of the radiant energy absorbed. The peaks of the curves of UV and visible spectra, therefore, indicates the values of the radiation absorbed maximum and it is denoted by λ_{max}; λ, of course, is inversely proportional to energy. Here absorption peaks are maxima in the curve (Fig. 10.10). Many a spectrocopist now reports UV bands in $\overline{\nu}$ (cm^{-1}). This unit is directly proportional to the radiant energy absorbed. When the curve is drawn in wave number scale plotting transmittance along ordinate, one gets the absorption maxima or peaks as minima in the curve since the higher the absorbance, the lower the transmittance and the height of the peak will be lower and thus the peaks will be minima.

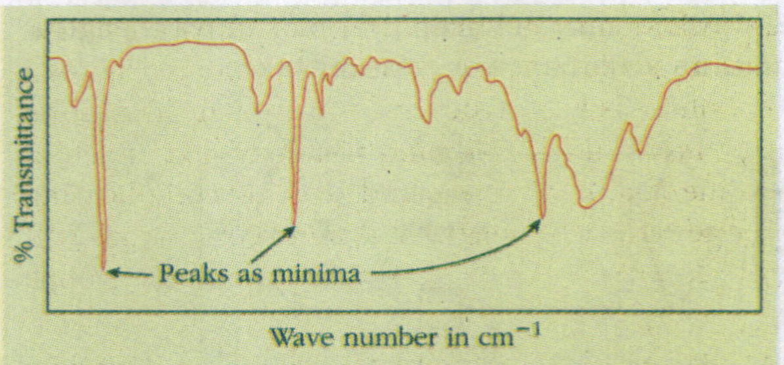

Fig. 10.10(a). UV spectra $\overline{\nu}$ vs % T

10.12 Visible and Ultraviolet Spectra

It has already been stated that the energy involved in visible and ultraviolet spectra causes electronic transition and hence they come under the heading of electronic spectra. The approximate range of the electronic spectra is 200 nm to 800 nm.

10.12.1 Visible spectrum and complementary colour

When a beam of electromagnetic wave (λ = 400 nm–800 nm) falls on a substance, the light may be completely absorbed or completely reflected and consequently the substance appears black or white respectively. If a certain portion of the visible light (400 nm–800 nm) is absorbed by the substance and the rest reflected, it has a colour of the reflected light. If a substance reflects only a single band and absorbs the rest of the visible light, it will have the colour of that reflected light. On the other hand, if only a single band is absorbed by the substance and the others are reflected, it has the complementary colour of the absorbed band which is the combination of reflected colours. For example, if a substance appears purple, it must have absorbed either green light (purple being the complementary colour of the green light; Table 10.3) or all the visible light except green which is reflected by the substance under consideration. Table 10.3 represents the relation between absorption of light and the complementary visible colour.

Table 10.3 Relation between absorption of visible light and complementary colour

Wavelength region in nm	Colour absorbed	Complementary colours (visible)
400–435	Violet	Yellowish green
435–480	Blue	Yellow
480–490	Greenish blue	Orange
490–500	Bluish green	Red
500–560	Green	Purple
560–580	Yellowish green	Violet
580–595	Yellow	Blue
595–650	Orange	Greenish blue
650–780	Red	Bluish green

10.12.2 Certain terms involved in the electronic spectra

There are many coloured compounds at our disposal. The colour appears in them owing to the presence of certain functional groups, usually with multiple bonds. These groups are called chromophores. Chromophore is a Greek word meaning colour bearer (Chroma: colour; Photos : bearer). In the language of spectroscopy, a **chromophore** is any structural feature which exhibits absorption of characteristic electromagnetic energy in the ultraviolet and visible region. Each of the members of a series of compounds having the same functional group and no other complicating factors will generally absorb energy at very nearly the same wavelength and those compounds will have almost identical molar absorptivity. $>C{=}C<$, $-C{\equiv}C-$, $>C{=}O$ functional groups are the examples of chromophores.

There are certain groups which do not themselves impart colour to an otherwise colourless compound but do increase the colouring power of chromophores—these are known as auxochromes (Gr. meaning increase colour; auxo : increase). In spectroscopy, **auxochromes** are groups which do not absorb light above 200 nm but when attached to chromophores, generally they cause a shift in absorption to longer wavelength. Some of the halogens, $-OH$, $-NH_2$, $-OR$, $-CH_3$, etc. groups are the examples of auxochromes. Nitrobenzene itself is a colourless compound but the presence of one $-NH_2$ group in nitroanilines make them yellow in colour.

The combination of an auxochrome (or a chromophore) and another chromophore is a new chromophore. The latter absorbs light at a longer wavelength than the former chromophore. Thus the presence of an auxochrome or a chromophore attached to a particular chromophore causes to shift the absorption to a longer wavelength than what would have been if one chromophore were present alone and this shift is said to be **bathochromic shift**. Since in the visible region the bathochromic shift causes the absorption to change in the order, violet—blue—greenish blue—bluish green—green—yellowish green—yellow—orange—red, this kind of shift is also known as **red shift**. As one sees the complementary colour of the band absorbed, the bathochromic shift in the visible region causes the colour of the absorbing material to shift in the opposite order—yellow—orange—red—purple—violet—blue—greenish blue—bluish green—black. Thus bathochromic shift in the visible region brings about the deepening of colour and this effect is said to be **bathochromic effect**. In general, the greater the conjugation, the greater is the deepening of colour.

The combination of a chromophore and another group sometimes shifts the absorption to a shorter wavelength than the absorption by the chromophore itself and this kind of shift is said to be **hypsochromic shift (blue shift)**, this shifting effect is known as **hypsochromic effect**. It is opposite to the bathochromic shift and brings about lightening of colour, if it occurs in the visible region. **Hyperchromic effect** and **hypochromic effect** are those which increases or decreases the intensity of absorption respectively (ϵ).

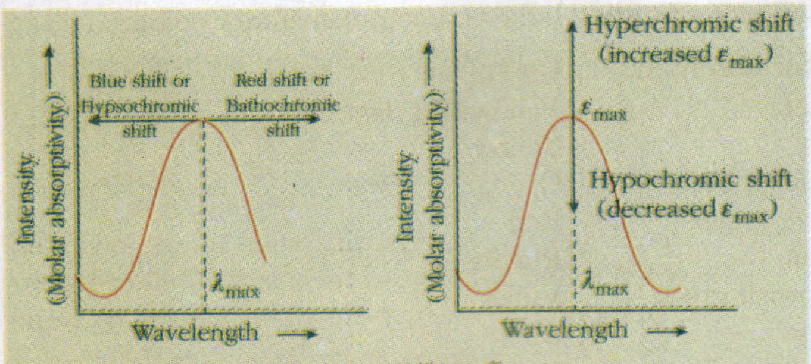

Fig. 10.11. Shifting effects

The four shifting effects discussed above are presented in Fig. 10.11.

Problem 10.5. There are three polyenes, A, B, and C, written in the increasing order of conjugation. Their colours are bluish green, blue and purple. What are their individual colours?

Solution: We know (from Table 10.3) that the complementary colour pairs are : bluish green-red, blue-yellow and purple-green. So, bluish green polyene absorbs red light, blue one absorbs yellow light and purple coloured polyene absorbs green light. Therefore, the increasing order of wavelength of light absorbed by the three given polyenes is : green, yellow, red. Again, we know that the greater the conjugation, the longer is the wavelength of light absorbed. According to the question, compound C is the most conjugated and compound A is the least conjugated. So, compound C must have absorbed light of the longest wavelength, i.e., red; and A has definitely absorbed light of the shortest wavelength, green. The complementary (visible) colours of green, yellow and red are purple, blue and bluish green, respectively. Therefore, the colour of A is purple, that of B is blue and C is a bluish green coloured polyene.

10.13 Theory of The Electronic (UV) Spectroscopy

This deals with the mode of electronic excitation of molecules in the ultraviolet-visible region, and can be best understood in terms of molecular orbital energy diagram. The latter requires detailed mathematical treatment which is beyond the scope of this section. Here we shall present qualitatively the part of the theory essential to point out correlations between spectra and structures of simple organic molecules.

The absorption of light energy in the UV and visible regions causes electronic excitation in a molecule and promotes one or more of its electrons from occupied bonding or non-bonding molecular orbitals of lower energy in the ground state to unoccupied antibonding molecular orbitals of higher energy in the excited state. Of course, as a first approximation we assume that only one of the electrons is promoted to the higher energy level, and that all other electrons remain unaffected. The excited state thus produced has a short life-time, which is usually of the order 10^{-6} to 10^{-9} seconds. The most probable electronic transition consists of the promotion of the one electron from the highest occupied bonding M.O. or non-bonding M.O. to the unfilled anti-bonding M.O. of the lowest energy. Since the electron density between the two nuclei in an anti-bonding M.O. is much lower than that in the bonding M.O. (Fig. 10.12), the former is less stable than the latter; and so electronic transition occurs from the bonding or non-bonding M.O. to anti-bonding M.O. when a molecule absorbs energy. This is thus HOMO$\longrightarrow$LUMO transition of *e*s.

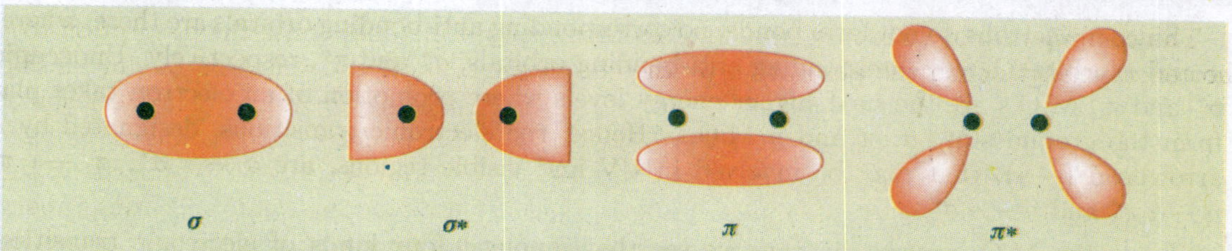

Fig. 10.12. The electronic charge distribution in $\sigma, \sigma^*, \pi, \pi^*$, M.O.s

Naturally, the wavelength of the light absorbed depends on the difference in the energies between the ground state and excited state energy levels and not just on their individual energies. The relation between the wavelength of light and amount of the energy absorbed is given by Eqn. (10.16).

$$\lambda = hc/\Delta E. \tag{10.17}$$

This relation shows that the greater the difference between the energies of the two electronic states (ΔE), the shorter is the wavelength of the light absorbed and vice versa. ΔE is also the difference in energy between LUMO and HOMO of the species under consideration.

M.O.s. involved in electronic transition

Organic molecules, in their ground state, have electrons in three main levels; σ, π and n M.O.s. are those levels written in the order of increasing energy. Electrons of a molecule that are not directly involved in bonding are called non-bonding *e*s and are designated as n electrons; these are unshared valence electrons on nitrogen, oxygen, sulphur or halogen atoms and are located in their A.O.s. Three types of *e*s are shown in the electron-dot structural formula of formaldehyde written below:

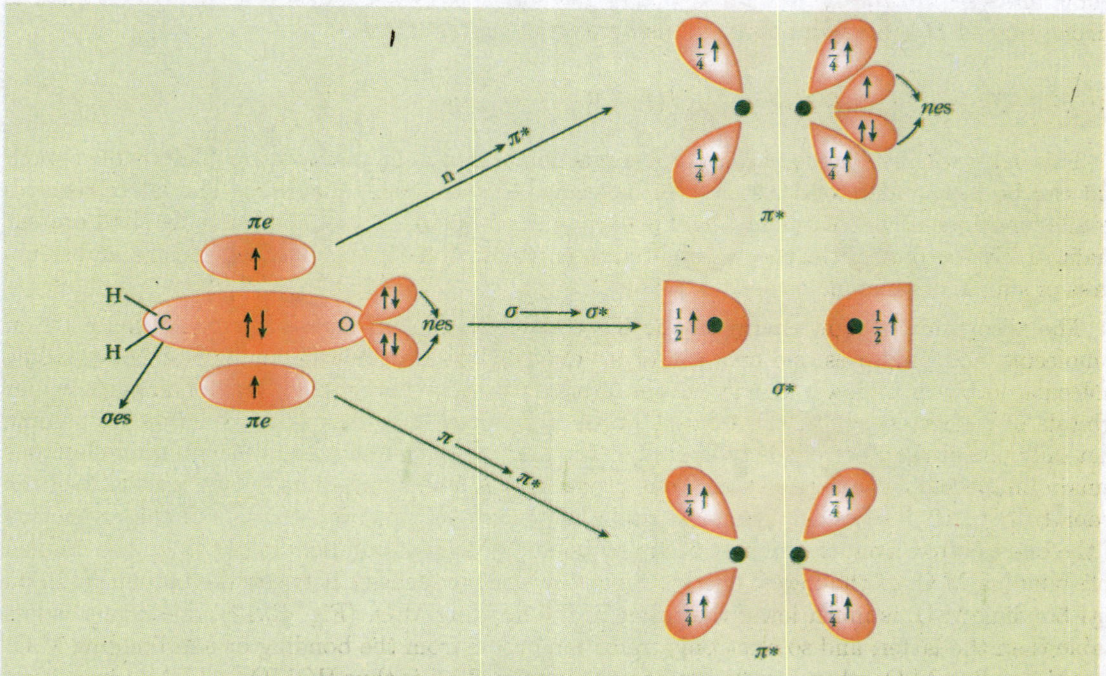

Fig. 10.12(a). σ, π and *nes* in the structural formula of formaldehyde

Since n electrons do not form bonds, no corresponding anti-bonding orbitals are there; whereas σ and π orbitals have corresponding anti-bonding orbitals, σ^* and π^*, respectively. Unoccupied σ^* and π^* M.O.s are the next higher energy levels where promotion of an electron takes place from the ground state σ, π and n M.O.s. Hence, the electronic transitions, designated by an arrowhead ($\longrightarrow$), that may be involved in UV and visible regions, are $\sigma \longrightarrow \sigma^*$, $\pi \longrightarrow \pi^*$, $n \longrightarrow \pi^*$ and $n \longrightarrow \sigma^*$.

From the M.O. energy diagram we see that amongst four kinds of electronic transitions, $\sigma \longrightarrow \sigma^*$ transition involves the highest amount of energy and hence this is associated with the shortest wavelength of light absorption; whereas the least amount of energy and consequently the longest wavelength of light absorption are involved in $n \longrightarrow \pi^*$ transition.

Molecules containing sigma bonds only and no atom with unshared valence electrons undergo $\sigma \longrightarrow \sigma^*$ transitions at wavelength shorter than 165 nm by absorbing light energy greater than 712 kJ (170 k cal) mol^{-1}. Thus it is a high energy process and occurs at far UV region. Since oxygen of the air absorbs light energy below 200 nm, this region is not readily accessible by ordinary spectrophotometer. Vacuum spectrophotometer is used to record far UV spectrum. $\sigma \longrightarrow \sigma^*$ transition is found in alkanes because only σ and no other M.O.s are involved in those compounds. Ethane and propane show λ_{max} at about 140 nm and 135 nm, respectively.

$n \longrightarrow \sigma^*$ transitions are of lower energy than $\sigma \longrightarrow \sigma^*$ transitions. Molecules containing non-bonding electrons on oxygen, nitrogen, sulphur or halogen atoms and no π M.O. are capable of showing mainly $n \longrightarrow \sigma^*$ transitions which can be recorded at ordinary UV region. This type of transition is associated with energy less than 628 kJ (150 k cal) mol^{-1} and occurs at wavelength around or greater than 185 nm. For example λ_{max}^{vapour} of trimethylamine is 227 nm.

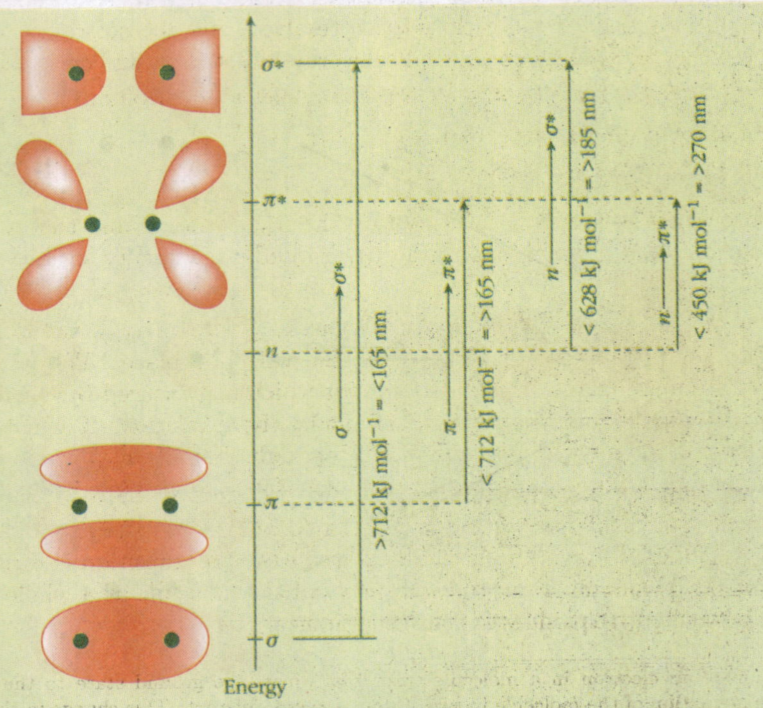

Fig. 10.13. M.O. energy diagram (not shown in scale) showing energy-wavelength relation for an electronic transition

$\pi \longrightarrow \pi^*$ transition is found in molecules containing π M.O.s. Single $\pi \longrightarrow \pi^*$ transition involves energy less than 712 kJ (170 k cal) mol^{-1}; the corresponding λ value is greater than 165 nm. In alkenes $\pi \longrightarrow \pi^*$ is the most important amongst other transitions. It shows absorption around 170–190 nm. Aldehydes and ketones also show $\pi \longrightarrow \pi^*$ transition at about 180 nm.

If π and n M.O.s are present in a molecule, it shows $n \longrightarrow \pi^*$ transition along with $\pi \longrightarrow \pi^*$ transition. This requires about 450 kJ (105 k cal) mol^{-1} of energy and absorption occurs around 270 nm. In simple aldehydes and ketones this takes place at about 280 nm.

It has already been stated (Sec. 10.6) that vibrational, rotational and electronic energies are quantised. Since UV spectrum involves quantised electronic energy, one should expect discrete line spectrum for a single electronic transition. In practice, however, we do not get a discrete line spectrum, and not even a discrete absorption band; instead we get an absorption curve. Now the question is : why is it so? In spectrophotometry we use a reasonable amount of compound containing thousands of its molecules. The electron involved in the electronic transition has one and the same electronic energy in the ground state of all the molecules; but every electronic state is associated with a series of vibrational energy levels, each of which in turn is associated with a series of rotational energy levels and thus vibrational and rotational energy levels of the involved electron differ slightly from molecule to molecule. Again, the electronic excitation energy produces not only electronic excitation but also vibrational and rotational changes. Thus the involved electron has the same electronic energy in all the excited molecules, but its vibrational and rotational energies have different values in different molecules. This gives a series of electronic transitions of slightly different energies instead of a single transition involving equal electronic, vibrational and rotational energies. Each of the transitions of different energy produces a single line or narrow peak of slightly different wavelength. The series of electronic transitions thus produce a series of lines which are too close to be resolved, and the net result is the appearance of an absorption envelope or curve instead of a discrete absorption line or band.

However, the principal characteristics of an Ultraviolet absorption curve are

1. Position of absorption radiation and
2. Intensity of absorption radiation.

The position of the absorption is indicated by the λ_{max} value whose energy is equal to that required for the promotion of an electron from the ground state energy level to the excited state energy level.

The intensity of the absorption, indicated by the ϵ_{max} (or $\log \epsilon_{max}$) value in UV spectrum, is proportional to the **probability** that the transition will take place. The latter in its turn is proportional to the square of the transition moment which is measured by the magnitude of the change in molecular dipole[1] moment brought about by the transition. If a molecule has a large transition moment, its intensity of absorption will be high. Usually ϵ_{max} value greater than 10^4 indicates high intensity of absorption, while that below 10^3 indicates low intensity of absorption.

However, a transition of low probability is called a **'forbidden'** or **infrequent transition**, whereas others are known as **allowed transitions**. The allowedness of a transition is directly linked with the geometries of the M.O.s involved, the symmetry of the molecule as a whole and the electric dipole of the incident light that induces the transition.

[1]When transition of an electron in a molecule takes place from the ground state to the excited state, the electronic charge distribution of the molecule in two states becomes different. This change in the electronic charge distribution of the molecule brings about the change in its molecular dipole value.

In general, $n \longrightarrow \pi^*$ transition occurs with low intensity ($\epsilon_{max} \leq 100$). n and π^* orbitals being perpendicular to each other, the jump of one electron from n M.O. to σ^* or π^* M.O. is geometrically poor one and probability is thus low. So an $n \longrightarrow \pi^*$ transition is said to be 'forbidden' and is associated with low ϵ_{max} value. 10 to 100 is the range of ϵ_{max} value for $n \longrightarrow \pi^*$ transition with simple carbonyl compounds.

$\pi \longrightarrow \pi^*$ transitions, on the other hand, are of high intensity (ϵ_{max} around 10^4). Since both of the orbitals, π and π^*, lie in the same plane, the transition is geometrically favourable. $n \longrightarrow \sigma^*$ transitions are, however, of moderate intensity.

10.14 Types of Absorption Band in UV Spectroscopy

Four kinds of absorption bands are recognised in the UV spectra of organic molecules.

1. K-bands: (K is the first letter of the German word Konjugierte) : This arises from π—π conjugated structure owing to $\pi \longrightarrow \pi^*$ transition and characterised by high molar absorptivity ($\epsilon_{max} > 10^4$). Butadiene, mesityl oxide and chromophoric substituted derivatives of aromatic molecules such as styrene, benzaldehyde, etc. show K-bonds.
2. R-bands: (R is the first letter of the German word Radikalartig) : This results from $n \longrightarrow \pi^*$ transition. R-bands are charactcrised by low molar absorptivity ($\epsilon_{max} \leq 10^2$). When H atoms of unsaturated groups are substituted by —OH, —NH_2, —SH, etc., $n \longrightarrow \pi^*$ transitions undergo bathochromic shifts.
3. B-bands: (B stands for benzenoid) : This is characteristic of aromatic and heteroaromatic molecules and may contain multiple peaks or fine vibrational sub-bands. Benzene shows a broad absorption band in the ordinary ultraviolet region between 230 to 270 nm. When chromophoric groups are present in an aromatic ring, B-bands are absorbed at longer wavelength than the more intense $\pi \longrightarrow \pi^*$ transition. Styrene shows $\pi \longrightarrow \pi^*$ transition at $\lambda_{max} = 244$ nm; the corresponding ϵ_{max} is 12×10^3. Styrene has a B-band at $\lambda_{max} = 282$ nm (ϵ_{max} value is 450).
4. E-bands: (E stands for Ethylinic) : E-bands arise from oscillations of electrons in aromatic ring systems and thus are characteristic of aromatic structures. Benzene shows two E-bands near 180 nm (E_1-band) and 200 nm (E_2-band).

Problem 10.6. What type of electronic transition do you expect from each of the following compounds?

(a) Ethane, **(b)** Cyclopropane, **(c)** Methanol, **(d)** Acetaldehyde.

Solution: (a) Ethane, CH_3—CH_3, contains only σ bonding M.O. Therefore, one should only expect $\sigma \longrightarrow \sigma^*$ transition.

(b) Cyclopropane also contains only σ bonding M.O. Hence the possible electronic transition that may take place is $\sigma \longrightarrow \sigma^*$ transition.

(c) Methanol, CH_3OH, contains σ bonding M.O. and n electrons on oxygen atom. So, in this case, $\sigma \longrightarrow \sigma^*$ transition as well as $n \longrightarrow \sigma^*$ transition will take place.

(d) Acetaldehyde, $CH_3CH{=}O$, possesses σ bonding M.O., π bonding M.O. and n electrons on O atom. Therefore, possible electronic transitions are : $\sigma \longrightarrow \sigma^*$, $\pi \longrightarrow \pi^*$, $n \longrightarrow \sigma^*$ and $n \longrightarrow \pi^*$.

Problem 10.7. The UV spectrum of acetone in hexane (solvent) shows two peaks: (i) $\lambda_{max} = 279$ nm, $\epsilon_{max} = 15$ and (ii) $\lambda_{max} = 189$ nm, $\epsilon_{max} = 900$.

(a) Identify the electronic transition associated with each λ_{max}. **(b)** Which one is more intense? **(c)** Name the bands of the transitions.

Solution: (a) Usual electronic transition that occurs in ketones are $\pi \longrightarrow \pi^*$ and $n \longrightarrow \pi^*$. $n \longrightarrow \pi^*$ transition requires less energy than $\pi \longrightarrow \pi^*$ transition. We know that smaller energy is associated with longer wavelength of absorption. Hence $\lambda_{max} = 279$ nm, $\epsilon_{max} = 15$ is due to $n \longrightarrow \pi^*$ transition and $\lambda_{max} = 179$ nm, $\epsilon_{max} = 900$ is for $\pi \longrightarrow \pi^*$ transition.

(b) ϵ_{max} determines the intensity of absorption. Higher ϵ_{max} means more intense is the absorption. Therefore, the $\pi \longrightarrow \pi^*$ transition ($\epsilon_{max} = 900$) is the more intense than the $n \longrightarrow \pi^*$ transition ($\epsilon_{max} = 15$).

(c) $\pi \longrightarrow \pi^*$ transitions give rise to K-bands when there is conjugation in the system. Acetone does not possess conjugated structure. So the peak that arises from $\pi \longrightarrow \pi^*$ transition should not be designated as K-band. $n \longrightarrow \pi^*$ transition produces R-band. Hence $\lambda_{max} = 279$ nm is for the R-band.

10.15 Effect of Temperature and Solvent on Fineness of The Structure of Absorption Band

Vibrational and rotational energy states depend on temperature. With the lowering of temperature, lowering of vibrational and rotational energy states of the molecules occur. So, if absorption of light occurs at a lower temperature, smaller distribution of excited states results. This produces finer structure in the absorption band than what would have been at a higher temperature (Fig. 10.14).

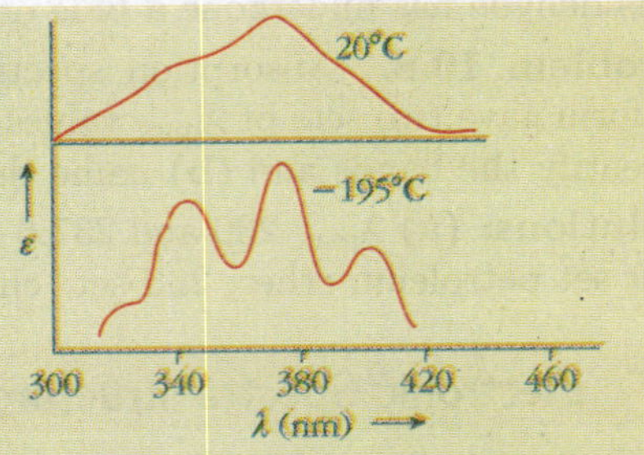

Fig. 10.14. Absorption spectrum of dodecapentaenic acid in ether-alcohol medium at 20°C and −195°C (not in scale)

The fineness of absorption band structure also depends on the nature of the solvent used. The higher the dielectric constant of a solvent, the stronger is the solvent-solute interaction. Rotational and vibrational energy states of molecules increase with increasing solvent-solute interaction, and thus absorption spectrum becomes less fine (Fig. 10.15).

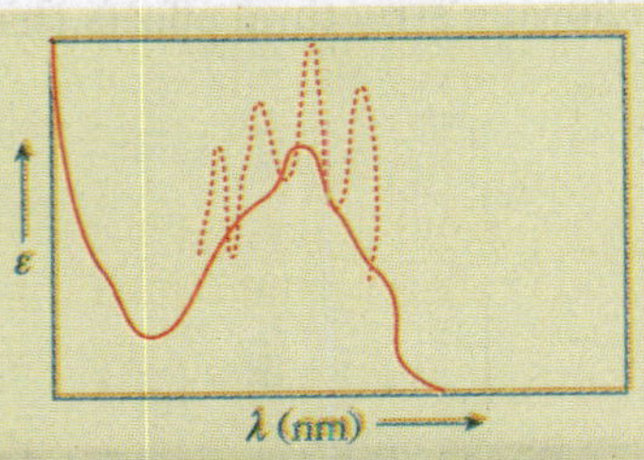

Fig. 10.15. Absorption spectrum (not in scale) of benzene in water (solid line) and in petroleum ether (broken line)

10.16 Solvent Effects on the Position and Intensity of an Absorption Band

Solvent polarity has a marked effect on the position and intensity of absorption bands of α, β-unsaturated carbonyl compounds. This is not so important in the cases of conjugated dienes and aromatic hydrocarbons.

However, π, π^* and n M.O.s get stabilised in polar solvents. The greater the polarity of the solvent, the greater is the stabilisation of orbitals. In hydroxylic solvents (like water, ethanol) an n M.O. gets maximum stability. A π^* M.O. being more polar than a π M.O., the former is more stabilised in polar solvents than the latter. Thus the increasing stability order of the M.O.s said above in the polar hydroxylic solvent is : π, π^*, n. Naturally the electronic energy of each of those M.O.s in a more polar solvent becomes less than what it would have been in a less polar solvent. Consequently, electronic excitation energy, ΔE, of $\pi \longrightarrow \pi^*$ transition should be lower in a more polar solvent than that in a less polar solvent. Lower ΔE means greater λ_{max}. Hence in a more polar solvent bathochromic shift (red shift) will take place. Since an n M.O. gets maximum stability in polar solvents, an $n \longrightarrow \pi^*$ transition in a more polar solvent involves greater ΔE, and hence λ_{max} decreases. So, in a more polar solvent an $n \longrightarrow \pi^*$ transition experiences blue shift or hypsochromic shift. For example, crotonaldehyde ($CH_3CH{=}CHCHO$) experiences an $\pi \longrightarrow \pi^*$ and $n \longrightarrow \pi^*$ transitions.

λ_{max} associated with $\pi \longrightarrow \pi^*$ transition is 214 nm in petroleum ether, but this is 220 nm in ethanol. Ethanol being a more polar solvent than petroleum ether, bathochromic shift has occurred. On the other hand, λ_{max} values corresponding to $n \longrightarrow \pi^*$ transitions of crotonaldehyde in petroleum ether and in ethanol are 329 nm and 322 nm, respectively. So the R-band of crotonaidehyde has undergone a blue or hypsochromic shift.

Problem 10.8. Absorption spectra of mesityl oxide, an enone, in petroleum ether and chloroform gave two sets of λ_{max} values : (i) 230 nm and 321 nm and (ii) 237.6 nm and 314 nm. **(a)** Identify the bands and **(b)** name the solvent associated with each set of λ_{max} values.

Solutions: (a) λ_{max} 230 and 237.6 nm— K-bond. λ_{max} : 321 nm and 314 nm— R-bond, **(b)** 1st set petroleum ether; 2nd set: chloroform.

10.17 Electronic Spectra-structure Relationships

Organic chemists study electronic spectra of compounds of known structure in detail and therefrom draw certain generalised relationships between spectra and molecular structures in an empirical manner. Then they examine the generalisation of a good many number of compounds of similar structure and try to ascertain some theoretical basis for the relationships. However, on the basis of empirically derived generalisations, structures of new compounds are predicted. In this section we shall discuss the following structural effects that have remarkable influence on the electronic spectra:

I. Resonance effect.

II. Hyperconjugative effect.

III. H-bonding effect.

IV. Steric effect.

10.17.1 Resonance effect on electronic spectra

Resonance stabilises both the ground state and the excited state of a species; the stabilisation effect may be greater in one than in the other. If resonance stabilises the excited state more in comparison to the ground state, the electronic excitation energy (ΔE) will be lower than what it would have been if the resonance had stabilised the ground state more relative to the excited state; and consequently, the λ_{max} value of the former case will be greater than that of the latter case. Thus, it is not the magnitude of the resonance energy, but whether resonance stabilises primarily the excited state or the ground state that decides the effect of resonance on electronic spectra. For example, naphthalene and azulene are two isomeric compounds; resonance stabilises the ground state of naphthalene relative to its excited state, while it stabilises the excited state of azulene in comparison to its ground state. As a consequence the electronic excitation energy of the latter is lower than that of the former, i.e., ΔE of azulene $< \Delta E$ of naphthalene. Since the wavelength of absorption depends on the ΔE value, azulene absorbs light at a longer wavelength than naphthalene does. Accordingly, azulene is deep blue in colour while naphthalene is colourless.

Conjugated systems have increased resonance stabilisation in the excited states. Greater number of charge separated structures that can be written for the excited state perhaps bring about the resonance stabilisation of the state. Thus a conjugate system has lower ΔE value and hence larger λ_{max} value than a similar non-conjugated system. In fact, as the conjugation of chromophores increases, usually deepening of colour also increases. Thus $C_6H_5—CH=CH—C_6H_5$ (λ_{max} 319) is colourless but $C_6H_5(CH=CH)_3C_6H_5$ (λ_{max} 377) and $C_6H_5(CH=CH)_5.C_6H_5$ (λ_{max} 424) are pale yellow and orange, respectively (Ref. : R. Kuhn, J. Chem. Soc., 605; 1938). Molar absorptivity also increases in a stepwise manner with increasing chromophoric conjugation. Thus, the longer the length of the conjugated system, the longer is the wavelength of absorption and the larger is molar absorptivity.

Equivalent resonance involving charged structures stabilises the excited state relative to the ground state since excited states are more polar than ground state, as a result of which absorption occurs at a longer wavelength in the case of a polar species with respect to its non-polar counterpart.

For this reason phenolphthalein is colourless in the solid state while it is deep red in an alkaline medium. Under strong alkaline medium it is colourless because cross-conjugation brings instability to the system.

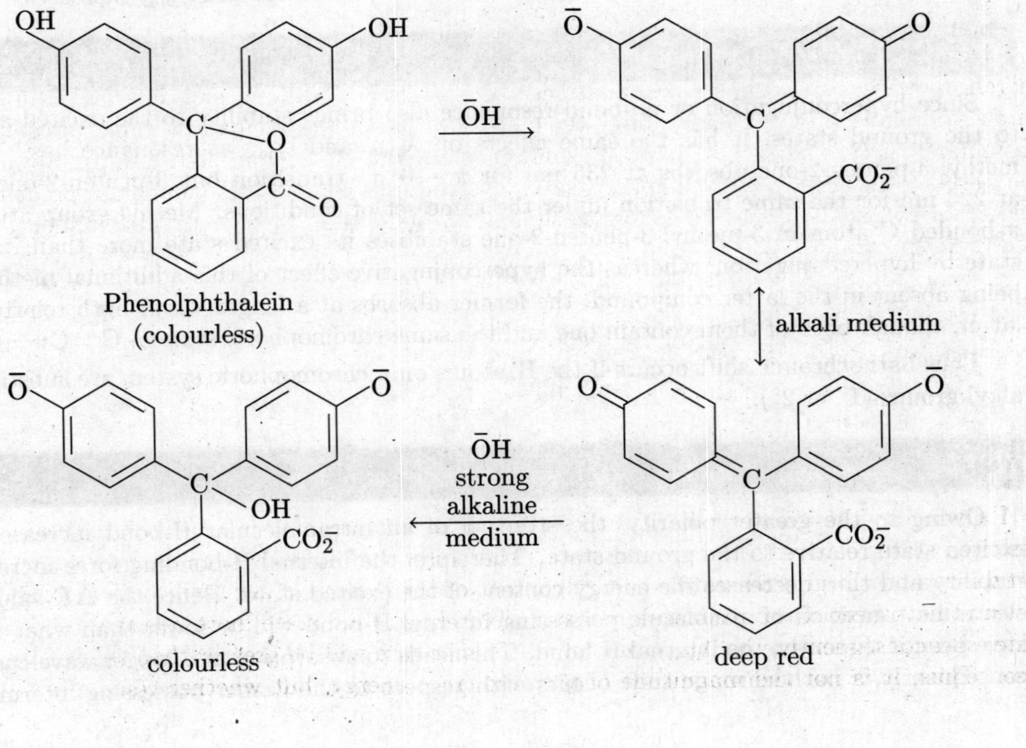

Thus, more resonance stabilisation of the excited state leads to the bathochromic shift, and the hypsochromic shift occurs if the ground state is more resonance stabilised. In the case of uncharged species, resonance brings more stability to the ground state than to the excited state.

Problem 10.9. Benzene is colourless but its isomer fulvene is yellow—explain.

Solution:

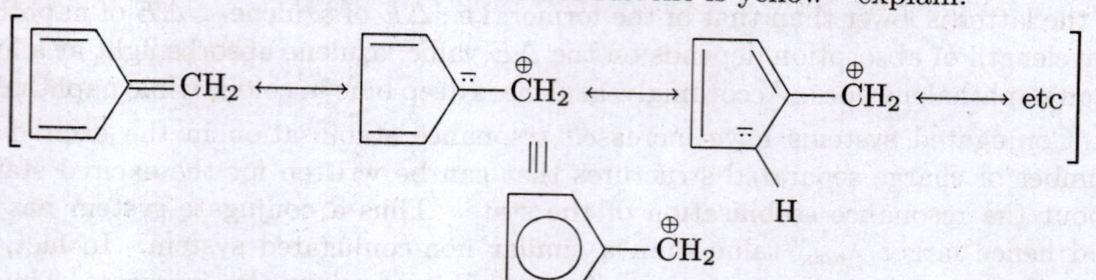

Fulvene has a charge-separated structure. Benzene has an aromatic uncharged structure, resonance stabilises the ground state of benzene relative to its excited state while reverse is the order of resonance stabilisaiion in the case of fulvene. Thus the electronic excitation energy required for fulvene is lower than that for benzene, and consequently, absorption at a longer wavelength makes fulvene yellow and that at a shorter wavelength makes benzene colourless.

Problem 10.10. λ_{max} for $CH_3.(CH{=}CH)_3CH_3$ is 263 nm but that for $CH_3(CH{=}CH)_6CH_3$ is 352 nm. Give reason for the bathochromic shift of the latter compound with respect to the former compound.

Solution: Since $CH_3.(CH{=}CH)_6.CH_3$ has a longer length of conjugation than $CH_3.(CH{=}CH)_3.CH_3$, the former absorbs at a longer wavelength than the latter. So we notice the bathochromic shift. In fact with increasing conjugation the energy difference between HOMO of the ground state and LUMO of the excited state decreases i.e., ΔE decreases and λ_{max} increases.

10.17.2 Hyperconjugative effect

Since hyperconjugation or no-bond resonance also brings stability to the excited as well as to the ground states, it has the same effects on λ_{max} and ϵ_{max} as resonance has. Thus, 3-methyl-3-penten-2-one absorbs at 236 nm for $\pi \longrightarrow \pi^*$ transition but 3-penten-2-one absorbs at 224 nm for the same transition under the same set of conditions. Methyl group attached to π-bonded C atom in 3-methyl-3-penten-2-one stabilises its excited state more than its ground state by hyperconjugation; whereas the hyperconjugative effect of this additional methyl group being absent in the latter compound, the former absorbs at a longer wavelength relative to the latter, though both of them contain one and the same chromophoric system, $C{=}C{-}C{=}O$.

Thus bathochromic shift occurs if the H atoms on a chromophoric system are substituted by alkyl groups (1° or 2°).

10.17.3 Hydrogen-bonding effect on electronic spectra

Owing to the greater polarity, the strength of an intramolecular H-bond increases in the excited state relative to the ground state. Therefore, the internal H-bonding force increases the stability and thus decreases the energy content of the excited state. Hence the ΔE value of the electronic transition of a molecule possessing internal H-bond will be lower than what it would have been if there were no internal H-bond. This leads to absorption at a longer wavelength, and consequently, a bathochromic shift occurs with respect to the isomer possessing intermolecular

H-bond. For example, *o*-nitrophenol with internal H-bond is yellow in colour while its *p*-isomer is a pale yellow coloured substance having intermolecular H-bond.

Problem 10.11. *o*-Nitroacetanilide is yellow but its para isomer is pale yellow. Why is the colour of *o*-isomer deeper than that of the *p*-isomer?

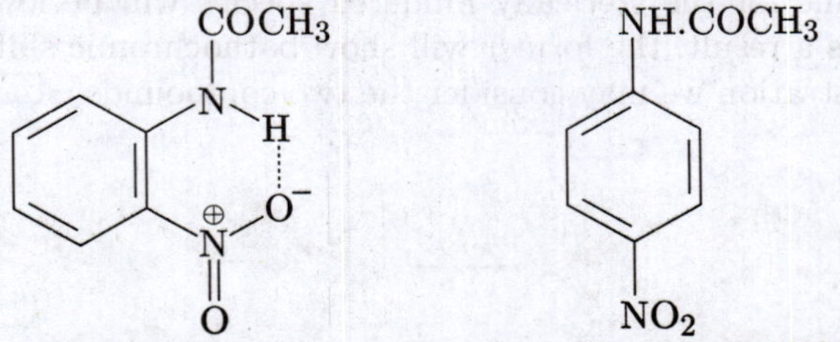

o-Nitroacetanilide *p*-Nitroacetanilide

Solution: The *o*-isomer possesses internal H-bond but the *p*-isomer has intermolecular H-bond; since the internal H-bond stabilises the excited state of a molecule to a greater extent than the intermolecular H-bond, ΔE of the electronic transition of the *o*-isomer is smaller than that of the *p*-isomer. Smaller ΔE value means absorption at a longer wavelength and as a result deepening of colour occurs. So *o*-isomer is deeper in colour than the *p*-isomer.

10.17.4 Steric effect on electronic spectra (Stereochemistry and electronic spectra)

As stated earlier the position and intensity of absorption depend on the length of a conjugated system; the longer the length of a system, the longer is the wavelength of absorption and the larger is the molar absorptivity value. If the steric effect hinders the coplanarity of a system, delocalisation of π electrons will be hindered; and as a consequence, the length of conjugation will decrease.

If steric hindrance affects primarily the excited state relative to the ground state, the electronic excitation energy of the sterically hindered species will be larger than that of the sterically unhindered isomer; thus the former will absorb at a shorter wavelength than the latter, and this will lead to a hypsochromic shift in the former case with respect to the latter. Molar absorptivity, will also change in the same order. For example, trans-stilbene absorbs at a longer wavelength and shows larger ϵ_{max} than cis-stilbene. This is because the former has a coplanar structure and

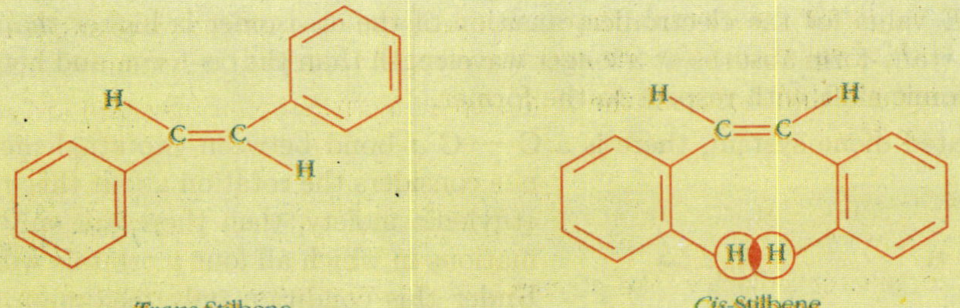

Trans-Stilbene *Cis*-Stilbene

Fig. 10.16. Steric hindrance to delocalisation of πes in cis-stilbene in the excited state

delocalisation of πes occur completely across the molecule while in cis-stilbene the delocalisation is hindered because it loses coplanarity owing to the overlap of two ortho-H atoms as shown in Fig. 10.16. In this case steric hindrance being primary in the excited state relative to the ground state, the ΔE value for the electronic transition of cis-stilbene is greater than that of trans-stilbene. Therefore, λ_{max} (for cis) < λ_{max} (for trans) and ϵ_{max} (for cis) < ϵ_{max} (for trans). Thus, the trans-isomer shows a bathochromic shift with respect to the cis-isomer.

On the other hand, if the steric effect is primary in the ground state in comparison to the excited state, the ΔE value for the sterically hindered species will be lower than that for the unhindered species; and as a result, the former will show bathochromic shift with respect to the latter compound. For illustration we may consider the two compounds, A and B as shown below (Fig. 10.17).

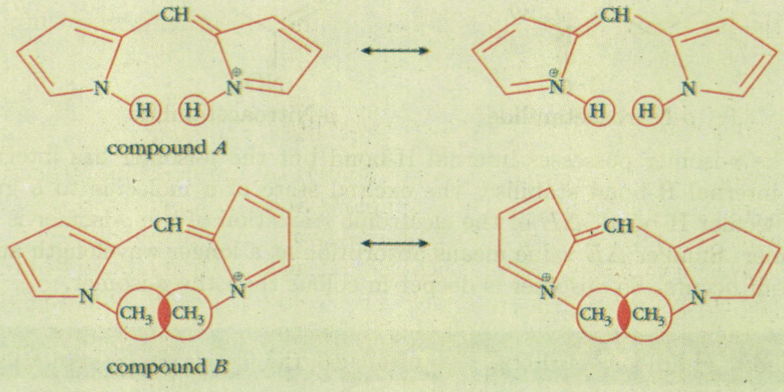

Fig. 10.17. Steric hindrance to delocalisation of πes in the ground state of compound

Since compound B is primarily affected in the ground state by the steric hindrance, ΔE for its electronic transition is lower than that of compound A, where there is no steric hindrance. It is needless to mention that the overlap of two methyl groups here produces the steric hindrance to resonance across the molecule. Thus λ_{max} for compound A is 470 nm, whereas that for compound B is 510 nm.

Problem 10.12. cis-Cinnamic acid absorbs at 268 nm (ϵ_{max} : 10,700) while trans-cinnamic acid absorbs at 272 nm (ϵ_{max} : 15,900)—explain the hypsochromic shift in the UV spectrum of the cis-isomer with respect to the trans-isomer.

Solution: Owing to the overlap of ortho-H atom of the phenyl group and the carboxyl group in the cis-form, the delocalisation of π *e*s is hindered, and the hindrance is primary in the excited state. Thus ΔE value for the electronic transition of the cis-isomer is higher than that of the trans-form. So, trans-form absorbs at a longer wavelength than the cis-form, and hence the latter shows hypsochromic shift with respect to the former.

In a conjugated diene system, there is a C—C σ bond between two ethylenic linkages. If one considers the rotation about this σ bond of one ethylenic moiety, then there are only two conformations in which all four p orbitals will be parallel. Under this condition only maximum overlap of p orbitals is possible and consequently the diene will be stable. These two conformations are known as *s*-cis (cisoid) and *s*-trans (transoid) conformations (Fig. 10.18); here *s* stands for the geometric isomerism with respect to the C—C single bond.

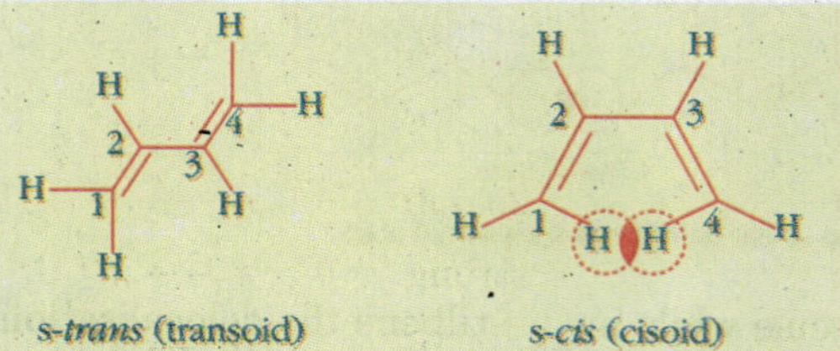

Fig. 10.18. Cisoid and transoid conformations of butadiene

In the *s*-cis conformation the two H atoms on C_2 and C_3 are in eclipsed condition and two other H atoms, one on C_1 and the other on C_4, are under high steric strain (repulsion). These two effects make the *s*-cis form less stable than the *s*-trans form in which no such effects are found. Thus, butadiene prefers the *s*-trans form. However, monocyclic diene systems are forced into cisoid structures (Fig. 10.19).

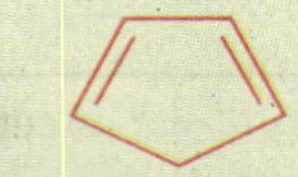

Fig. 10.19. Monocyclic diene systems have cisoid conformations

It is interesting to note that cisoid monocyclic conjugate dienes absorb at longer wavelengths than transoid acyclic conjugated dienes. However, ϵ_{max} values for the former compounds are less than those for the latter compounds. This is because the transition moment of a cisoid form is less than that of the transoid form.

10.18 Applicability of Electronic Spectroscopy in Organic Chemistry

Since the electronic spectral range is 200 nm–780 nm, the applicability of electronic spectra is limited to those organic molecules which are capable of absorption within this wavelength region or slightly below it (185 nm). As compounds with only σ bond undergo $\sigma \longrightarrow \sigma^*$ transition and absorb light of wavelength below 165 nm, organic chemists are not directly interested in those spectra. Thus the spectra of molecules involving $\pi \longrightarrow \pi^*$ $n \longrightarrow \pi^*$ and $n \longrightarrow \sigma^*$ transitions are important to them. However, wavelengths of light absorbed by molecules, each containing a single chromophoric group and involving one or more of the electronic transitions said above, are noted in Table 10.4.

A species with two or more insulated (not directly linked) same chromophoric groups absorbs radiation at nearly the same wavelength as a species containing only a single specific chromophoric group. Of course, the intensity of the absorption is then proportional to the number of that type of chromophore present in the species. Here it is important to mention that the interposition of a methylene group or 1, 3-orientation of two groups in an aromatic system is enough to insulate two chromophores almost cent per cent from each other. Hence only directly linked chromophores have electronic interactions which can influence the position of absorption.

Table 10.4 Absorption data for individual chromophores (values are approximate)

Chromophoric group	Example	λ_{max} (nm)	ϵ_{max} $\times 10^{-2} m^{-2} mol^{-1}$	Transition
$>C{=}C<$	Ethylene	165	15,000	$\pi \longrightarrow \pi^*$
$-C{\equiv}C-$	Acetylene	190	8,000	$\pi \longrightarrow \pi^*$
		175	6,000	$\pi \longrightarrow \pi^*$
$>C{=}O$	Acetone	188	900	$\pi \longrightarrow \pi^*$
		279	15	$n \longrightarrow \pi^*$
		160	20,000	$n \longrightarrow \sigma^*$
$\overset{\vert}{HC}{=}O$	Acetaldehyde	210	strong	$\pi \longrightarrow \pi^*$
		290	16	$n \longrightarrow \pi^*$

Contd.

Chromophoric group	Example	λ_{max} (nm)	ϵ_{max} $\times 10^{-2} m^{-2} mol^{-1}$	Transition
$-C(=O)-OH$	Acetic acid	204	60	$n \longrightarrow \pi^*$
$-O-$	Diethyl ether	210		$n \longrightarrow \sigma^*$
$-N=O$	Nitrosobutane	302	100	$n \longrightarrow \pi^*$
$-N=N-$	Azomethane	348	4	$n \longrightarrow \pi^*$
$-C\equiv N$	Acetonitrile	160	5	$\pi \longrightarrow \pi^*$
$-NO_2$	Nitromethane	200	5,000	$\pi \longrightarrow \pi^*$
		274	15	$n \longrightarrow \pi^*$
$-(C=C)_2-$	1, 3-butadiene	217	21,000	$\pi \longrightarrow \pi^*$
$>C=C{\cdot}C\equiv C-$	Vinylacetylene	219	76,000	$\pi \longrightarrow \pi^*$
$>C=C{\cdot}C=O$	Crotonaldehyde	218	18,000	$\pi \longrightarrow \pi^*$
		315	30	$n \longrightarrow \pi^*$
$>C=C-NO_2$	1-Nitropropene	229	9,500	$\pi \longrightarrow \pi^*$
$-ONO_2$	Ethyl nitrate	212	12	$n \longrightarrow \pi^*$
$-ONO$	Amyl nitrite	218	1,120	$\pi \longrightarrow \pi^*$

Electronic interaction between chromophoric groups can give rise to either π-π conjugation or n-π conjugation. Systems involving σ-π conjugation, i.e., hyperconjugation also experience electronic interaction. So, electronic spectra are specially applicable to those compounds which involve conjugation.

Though cumulated systems have directly linked multiple bonds, yet their spectra are mere superpositions of the spectra of individual chromophores. This is because of hindered resonance in those systems owing to the presence of adjacent multiple bonds on mutually perpendicular planes.

Therefore, the applicability of electronic spectra is limited to the conjugated systems (acyclic, alicyclic and aromatic). Identification of conjugated systems is the main use of the electronic spectra. However, these are also used for:

(i) The determination of the structure of unknown compounds.

(ii) The identification of unknown compounds and intermediates in reactions.

(iii) The determination of percentage of a compound in a mixture.

(iv) The determination of configuration of geometric isomers.

The presence of an absorption band in the spectrum of an unknown compound where a particular chromophoric group is known to absorb is generally a dependable evidence that the unknown compound contains the very chromophoric group; of course this is not a conclusive one since other groups may also produce the identical band. Moreover a great deal of negative information about the structure of an unknown compound may be obtained from its electronic spectra, e.g.,

(a) The absence of a particular band indicates the absence of the chromophore which is expected to be present.

(b) If the compound is highly transparent throughout the region (200 nm–780 nm), it does not contain conjugated (unsaturated or benzenoid) system; and carbonyl (aldehydic or ketonic), nitro, bromo, iodo groups are also absent in the compound.

(c) If the structure of the unknown compound is suspected, the electronic spectra of its pure sample may be compared with that of the known sample under the identical set of conditions; if they do not coincide, the suspected structure for the unknown is definitely proved incorrect.

However, to determine the structure of an unknown compound or to identify it certain empirical rules have been formulated. These have been discussed in the next section.

10.19 Application of UV Spectra and Empirical Rules For Calculating The Absorption Position

The empirical rules are additive and very useful for solving structural problems, specially those of conjugated dienes and unsaturated carbonyl compounds. These rules were formulated by Woodward and finally developed by Fieser.

The normal procedure is to start with a base wavelength for the principal chromophore present in the molecule under consideration and then different wavelength values corresponding to the secondary chromophores or auxochromes, which are found attached to the principal chromophoric group, are added to it. Application of the electronic spectra of some relevant types of compounds is discussed below:

10.19.1 Compounds containing ethylenic ($>C=C<$) chromophore

The first absorption, due to $\pi \longrightarrow \pi^*$ transition of one electron, in the isolated ethylenic chromophore occurs in the far ultraviolet region. Ethylene absorbs at 165 nm (intense absorption, Table 10.4). A second band appears about 200 nm; this is also due to $\pi \longrightarrow \pi^*$ transition but for two electrons. Since ethylenic double bonds are nonpolar in nature, the absorption position is independent of solvent.

However, with increasing number of alkyl groups on the double bonded Cs, bathochromic shift increases relative to the unsubstituted compound; hyperconjugative effect brings about the red shift as stated in section 10.17.2.

Bathochromic shift is also noticed when auxochromes with n electrons are linked with the π-bonded C atoms. In this respect N and S atoms are the most effective and they shift the absorption at near ultraviolet region.

Cycloalkenes show absorption like alkenes. Ring size has no effect on the absorption position.

Problem 10.13. Ethylenic chromophore usually absorbs at 160 nm ($\pi \longrightarrow \pi^*$ transition involving one e) and at 200 nm ($\pi \longrightarrow \pi^*$ transition involving two es), but $CH_3CH{=}CH.SH$ absorbs at 228 nm. How do you account for the bathochromic shift?

Solution: Methyl group and the thiol group are the two substituents on the ethylenic chromophore. The hyperconjugative effect of the former and the effect of n electrons and d orbital of the S atom of the latter group stabilise the excited state of the given compound relative to its ground state. For this reason it absorbs less energy than ethylene for the $\pi \longrightarrow \pi^*$ transition. Hence the absorption takes place at a longer wavelength with respect to ethylene and so a bathochromic shift is noticed.

10.19.2 Conjugated dienes, trienes and polyenes

The conjugated diene part of an alkadiene is usually planar (steric effect hinders coplanarity in cis-forms) and this permits delocalisation of π electrons by effective π-π overlap to form π-π conjugated system. Thus there are four π orbitals (two from each ethylenic linkage, π and π^*), $\pi_1, \pi_2, \pi_3^*, \pi_4^*$. Now in this system one electron from the π_2 orbital, the higher occupied bonding M.O. (HOMO), transits to the lower unoccupied antibonding M.O. (LUMO), π_3^*, in $\pi \longrightarrow \pi^*$ transition. This requires less energy than $\pi_1 \longrightarrow \pi_3^*$ transition or $\pi_2 \longrightarrow \pi_4^*$ transition. Naturally a conjugated alkadiene absorbs at a longer wavelength than an alkene; usually it does so at 215–300 nm (Fig. 10.20). Other $\pi \longrightarrow \pi^*$ transitions are also possible provided those are not forbidden transitions. For the same reason a non-conjugated alkadiene system absorbs at a shorter wavelength than a conjugated one. Since steric effect hinders coplanarity and consequently inhibits delocalisation of π es, sterically hindered cis-form absorbs at a shorter wavelength with smaller intensity relative to its trans-isomer (See. 10.17.4; steric effect). Because of the greater transition moment of the trans-form in the excited state, it absorbs with larger intensity than the corresponding cis-form.

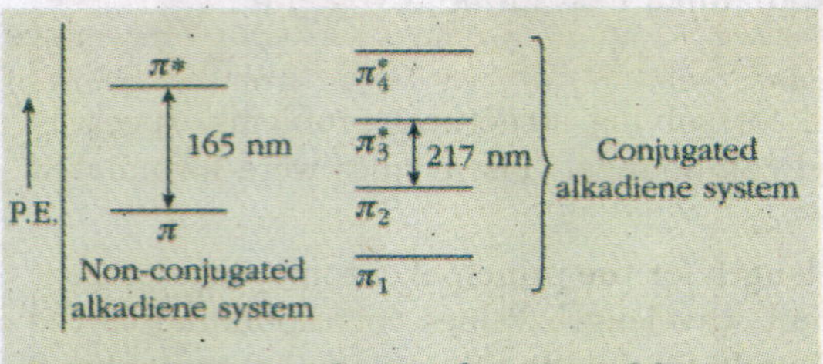

Fig. 10.20. M.O. energy diagram of a conjugated diene system

It is needless to mention that alkyl substitution and substitution of groups containing n electrons (—SH, NH_2, etc.) on a conjugated diene system bring about bathochromic shifts.

However, as the conjugation increases in open chain compounds, λ_{max} and ϵ_{max} values also increase (Fig. 10.21 and Table 10.5). The dependence of ϵ_{max} on the number of conjugated ethylenic chromophores (n) of open chain triene, tetraene and polyene systems is approximately given by the relation,

$$\epsilon_{max} = 17\,000(n-1). \tag{10.18}$$

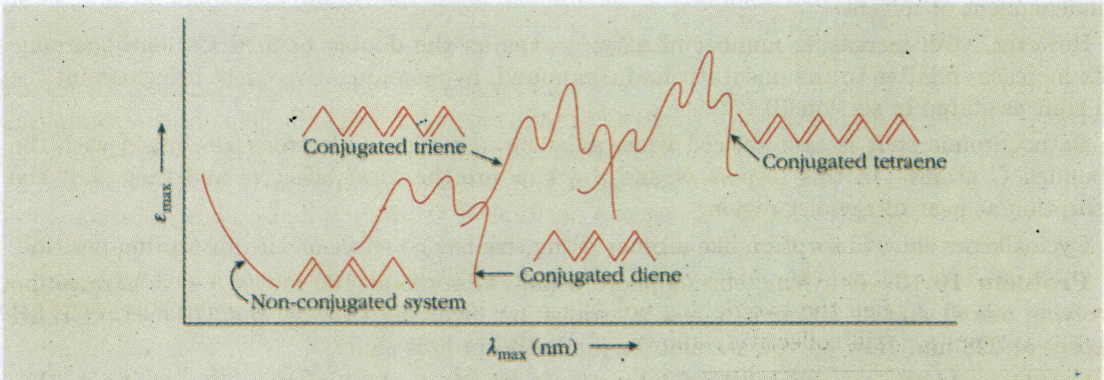

Fig. 10.21. UV spectra of diene, triene and tetraene (not in scale)

Table 10.5 Electronic absorption bands for conjugated diene, triene and tetraene

Chromophore	λ_{max}	ϵ_{max}
—(C=C)$_2$—	210–230 nm	21,000
—(C=C)$_3$—	260 nm	35,000
—(C=C)$_4$—	300 nm	52,000
1, 3-cyclopentadiene	239 nm	7,400
1. 3-cyclohexadiene	300 nm	8,000

Table 10.5 shows that monocyclic conjugated diene systems absorb at longer wavelength with decreased intensity than acyclic systems, though the monocyclic systems exist in *s*-cis (cisoid) conformation while the latter remains in *s*-trans (transoid) conformation.

Problem 10.14. 1-Pentene absorbs at 176 nm. The absorption data for three isomeric dienes A, B and C of molecular formula C_5H_8 is : λ_{max} for A = 178 nm; λ_{max} for B = 211 nm and that for C = 215 nm. Write down their structures with reasons.

Solution: We know that if a compound possesses two identical insulated ethylenic chromophores, the molecule absorbs at the same position as the single ethylenic chromophore. Since compound *A* absorbs almost at the same position as 1-pentene, it may be an isolated diene of the structure,

$$CH_2{=}CH{-}CH_2{-}CH{=}CH_2 \text{ (Compound A)}$$

1, 4-pentadiene

Since conjugated dienes absorb at longer wavelengths than an ethylenic chromophore, B and C are expected to be conjugated dienes.

Again, we know that a cis-isomer absorbs at a shorter wavelength relative to its tr -isomer. Since compound B absorbs at a shorter wavelength than compound C, compound C is trans-1, 3-pentadione and compound B is cis-1, 3-pentadiene.

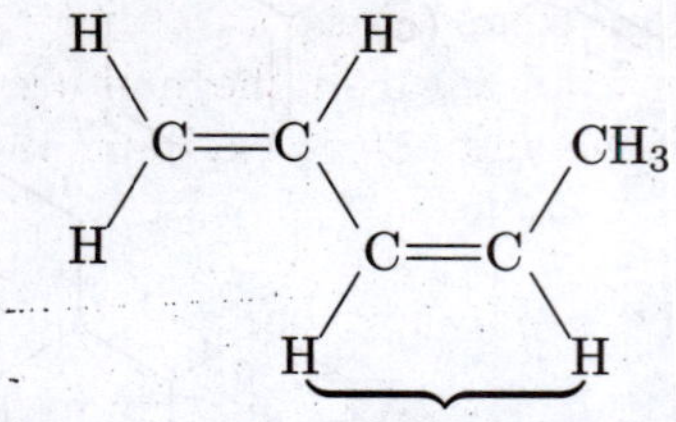

Compound B: *cis*-1,3-Pentadiene

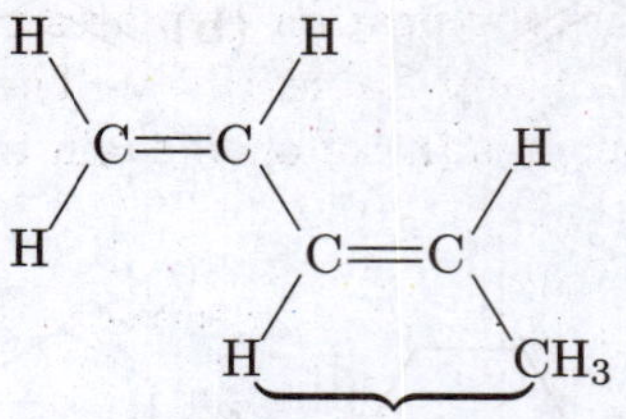

Compound C: *trans*-1,3-Pentadiene

However, the position of the most intense band for substituted conjugated dienes, trienes and polyenes (acyclic or cyclic) can be predicted from their structure with the help of certain empirical rules. R. B. Woodward, a Nobel Laureate, first enunciated the rules in 1941 for predicting the bathochromic shift due to alkyl substitution in 1, 3-butadiene. Afterwards L. M. Fieser and M. Fieser extended the rules for predicting the position of the most intense absorption for cyclic dienes. Cyclic dienes may be homoannular or heteroannular. When both of the double bonds are present in the same ring, the system is known as **homoannular**; while a system is said to be **heteroannular** when the two double bonds are not present in the same ring. However, these empirical rules are still collectively known as Woodward rules.

[1] When one C atom of the double-banded Cs is in the ring with the double bond extending outside of the ring, it is said to be an **exocyclic double bond**.

As stated earlier, the prediction starts by taking a base value for the parent system. To this are added various increments depending on the nature of the substituents and ring residues—a part of the carbocyclic ring directly linked to the diene moiety is called a ring residue. In the solution of problem 10.15 a ring residue is marked by letter '*r*'.

The base values for conjugated diene systems, and increments for substituents and ring residue are given in Table 10.6.

Table 10.6 Rules for predicting λ_{max} for conjugated dienes

Base value for acyclic dienes	214 nm
Base value for heteroannular dienes	214 nm
Base value for homoannular dienes	253 nm
Increments for each substituents	
Alkyl,—	5 nm
Ring residue	5 nm
Alkoxyl, —OR	6 nm
Alkyl mercapto, —SR	30 nm
Chloro, bromo	5 nm
N, N-dialkylamino, —NR_2	60 nm
[1]Exocyclic double bond	5 nm
Additional conjugation, —CH=CH—	30 nm
Cross conjugation	0 nm
Solvent shift	0 nm

To illustrate Woodward rules following problems are discussed.

Problem 10.15 Find out the probable absorption position of the most intense band of the following compounds using Woodward empirical rules.

(a) 

(b) 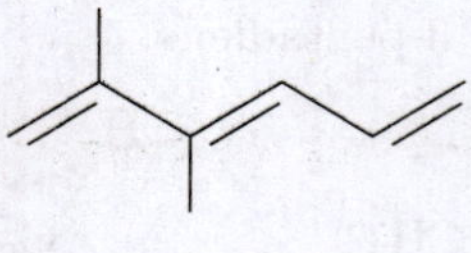

(c) 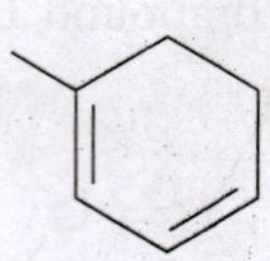

(d) 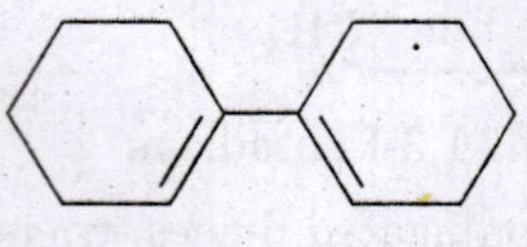

(e) 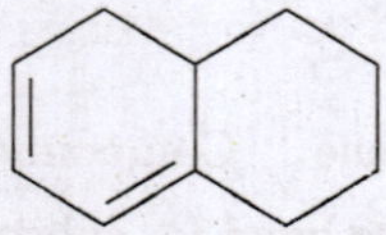

(f) 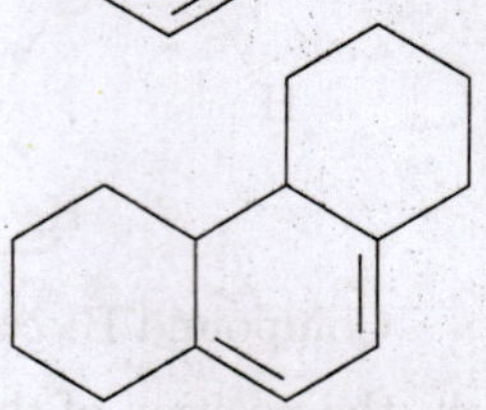

(g) 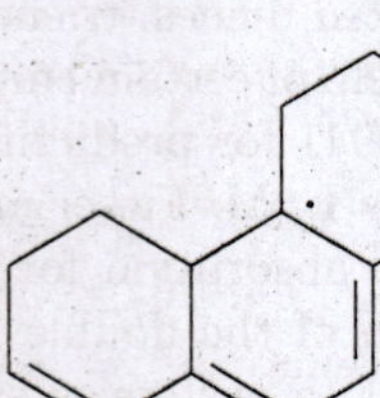

(h) 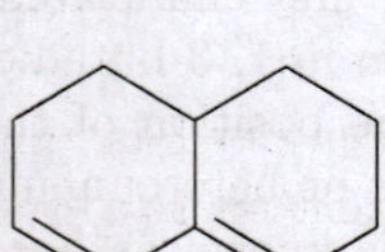

(i) 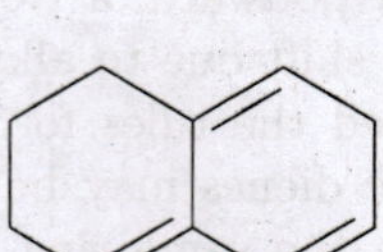

(j) 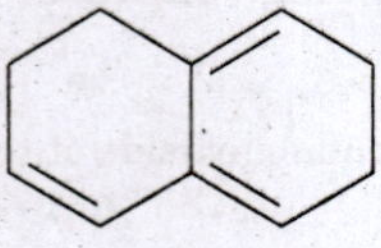

Solution: (a) It is an isolated diene. The absorption position of its most intense band will be that of alkene. An alkene absorbs below 200 nm. Here it will also absorb below 200 nm.

(b) It is an acyclic conjugated triene with two alkyl substitute on the two double-bonded carbon atoms.

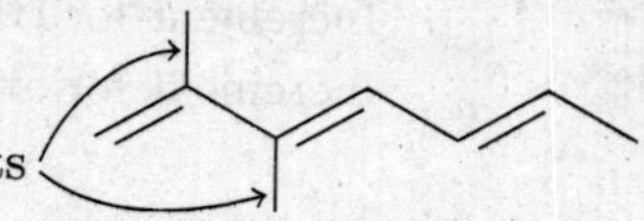

Following Table 10.6 we can calculate λ_{max} value for the given compound as written below:

Base value	...	=	214 nm
Increment for two R groups	...	=	10 nm (2 × 5)
Increment for additional conjugation	...	=	30 nm
	Total	:	254 nm

(c) It is a homoannular cyclic conjugated diene with one alkyl substituent only. Its λ_{max} value is 258 as shown below :

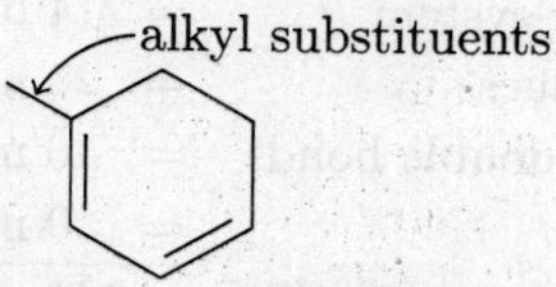

Base value	...	=	253 nm
Increment for one R	...	=	5 nm
	Total	:	258 nm

(d) It is a simple bicyclic heteroannular conjugated diene having four ring residues; but it has no exocyclic double bond :

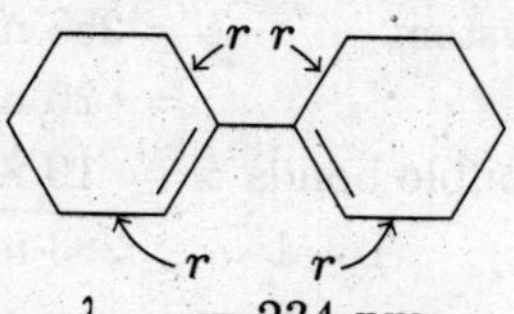

$\lambda_{max} = 234$ nm

Base value	...	=	214 nm
Increment for 4 ring residues	...	=	20 nm
	Total	:	234 nm

(e) It is a homoannular bicyclic fused ring system with three ring residues and one exocyclic double bond (*) :

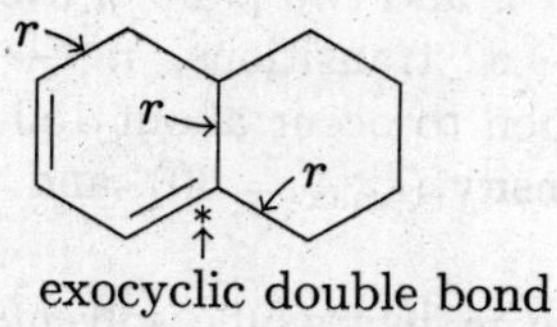

$\lambda_{max} = 273$ nm

Base value	...	=	253 nm
Increment for 3 ring residues	...	=	15 nm
Increment for exocyclic double bond	...	=	5 nm
	Total	:	273 nm

(f) It is a homoannular tricyclic fused ring system with four ring residues(r) and two exocylic double bonds (*) as shown below:

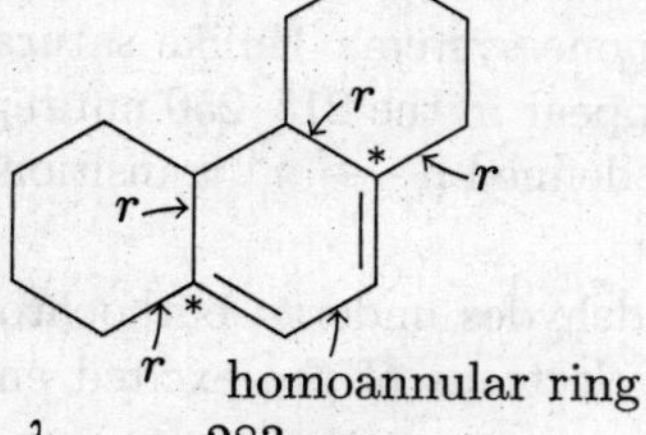

$\lambda_{max} = 283$ nm

Base value	...	=	253 nm
Increment for four ring residues	...	=	20 nm
Increment for two exocyclic double bonds		=	10 nm
	Total	:	283 nm

(g)

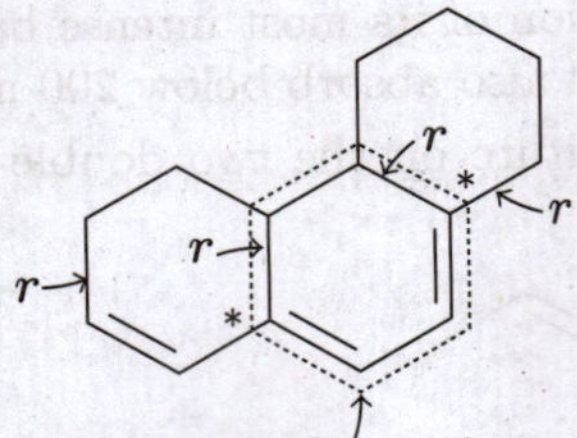

homoannular moiety

Base value for the homoannular diene	...	= 253 nm
Increment for four ring residues	...	= 20 nm
Increment for two exocyclic double bonds		= 10 nm
Increment for an additional conjugation		= 30 nm
	Total	: 313 nm

(h)

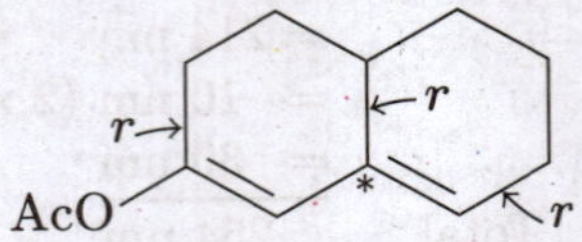

λ_{max} = 234 nm

Base value for the heteroannular system	= 214 nm
Increment for the three ring residues	= 15 nm
Increment for the one exocyclic double bond	= 5 nm
Total	: 234 nm

(i)

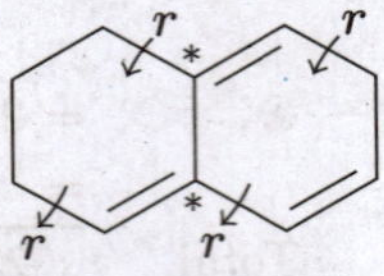

λ_{max} = 244 nm

Base value for the heteroannular system	= 214 nm
Increment for the four ring residues	= 20 nm
Increment for the two exocyclic double bonds	= 10 nm
Crossed conjugation	= 0 nm
Total	: 244 nm

(j)

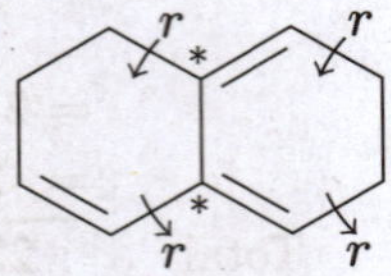

λ_{max} = 283 nm

Base value for the homoannular system	= 253 nm
Increment for the four ring residues	= 20 nm
Increment for the two exocyclic double bonds	= 10 nm
Total	: 283 nm

10.19.3 α, β-Unsaturated ketones and aldehydes

Saturated carbonyl compounds, ketones and aldehydes, contain σ, π and two pairs of n electrons. Naturally these may undergo $n \longrightarrow \sigma^*$, $n \longrightarrow \pi^*$ and $\pi \longrightarrow \pi^*$ transitions. $n \longrightarrow \sigma^*$ transition is of strong intensity. This and $\pi \longrightarrow \pi^*$ transitions happen to occur about 150 nm and 190 nm, respectively; while $n \longrightarrow \pi^*$ transition is of weak intensity ($\epsilon_{max} = 30$), and this takes place in the 270–300 nm region.

Since H-bonding force stabilises n electrons in the ground state in hydroxylic solvents, a carbonyl compound shows blue shift in such a solvent with respect to a non-hydroxylic solvent. Thus R-band of acetone appears at 279 nm in n-hexane but this is found at 264.5 nm when water is used as solvent.

α, β-unsaturated carbonyl compounds possess enone systems. When a carbonyl group is in conjugation with an ethylenic group, the system is known as enone system. Unlike saturated carbonyl compounds the $\pi \longrightarrow \pi^*$ transitions of enone systems appear in the 215–250 nm region and these K-bands are of strong intensity. The weak and poorly defined $n \longrightarrow \pi^*$ transitions of enone systems occur in 310–330 nm.

The $\pi \longrightarrow \pi^*$ transitions of α, β-unsaturated ketones and aldehydes undergo bathochromic shifts with increasing solvent polarity. This is because the π^* electrons of the excited enone systems get stabilised relative to ground states in polar solvents.

However, there are Woodward rules for calculating the absorption position of maximum intensity of enones which are summarised in Table 10.7.

Table 10.7 Rules for predicting λ_{max} for $\pi \longrightarrow \pi^*$ transition of enone and dienone systems

		λ_{max} (nm)
Base value for acyclic α,β-unsaturated ketones	—C=C—C=O	215
Base value for acyclic α,β-unsaturated aldehydes	—C=C—C(H)=O	207
Base value for α,β-unsaturated acids and esters	—C=C—CO_2H(R)	195
Base value for six membered cyclic enones		215
Base value for five membered cyclic enones		202
Increments for each substituent		
Double bond extending conjugation :	—C=C—C=C—C=O (δ γ β α)	+30
Homodiene system:	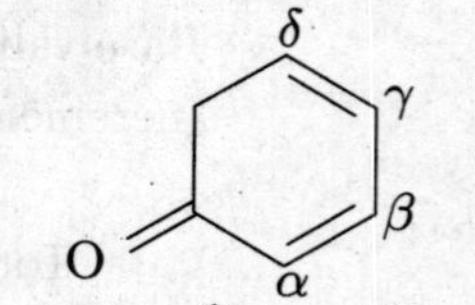	+39
Hetero annular :	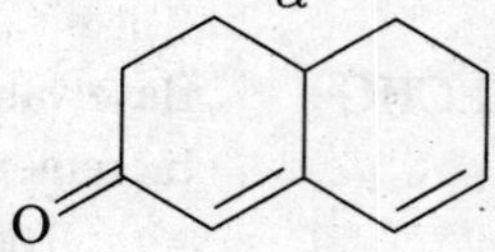	0
Alkyl group, ring residue:	α	+10
	β	+12
	γ and higher	+18
—OH	α	+35
	β,γ	+30
	δ	+50
—OR	α	+35
	β	+30
	γ	+17
	δ	+31
—OCOR	$\alpha,\beta,\gamma,\delta$	+6
—SR	β	+85
—Cl	α,γ,λ	+15
	β	+12
—Br	$\alpha,\gamma\,\delta$	+25
	β	+30
$—NH_2$, —NHR	α,γ,δ	0
	β	+95
For each exocyclic double bond		+5

Solvent shifts : Standard of reference : Methanol and ethanol

Cyclohexane, *n*-hexane	−11
Ether	−7
Dioxane	−5
Chloroform	−1
Methanol, ethanol	0
Water	+8

It has been found in most of the cases that experimental λ_{max} values for enones are same or nearly the same as those calculated on the basis of Table 10.7.

Examples:

(1) 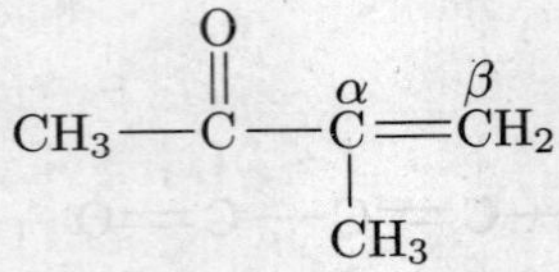

Base value: 215 nm (for acyclic α,β-unsaturated ketone)
Increment: 10 nm (for α-CH_3)
Total: 225 nm

(2) 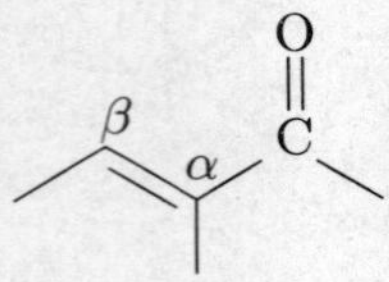

Base value: 215 nm
Increments: 10 nm (for α-substitution)
12 nm (for β-substitution)
Total: 237 nm

(3) $CH_3.\overset{\beta}{C}H{=}\overset{\alpha}{C}H.CHO$

Base value: 207 nm (for acyclic α,β-unsaturated aldehyde)
Increment: 12 nm
Total: 219 nm

(4) 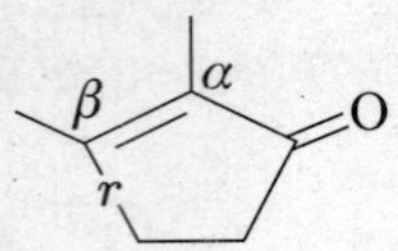

Base value: 202 nm (for 5-membered cyclic enone)
Increments: 10 nm (for α-substitution)
12 nm (for β-substitution)
12 nm (for β-ring residue)
Total: 236 nm

(5) $-\overset{\delta}{C}H{=}\overset{\gamma}{C}H-\overset{\beta}{C}H{=}\overset{\alpha}{C}H.CHO$

Base value: 207 nm (for α,β-unsaturated acyclic aldehyde)
Increments: 30 nm (for double bond extending conjugation)
18 nm (for δ-substitution)
Total: 255 nm

(6a) 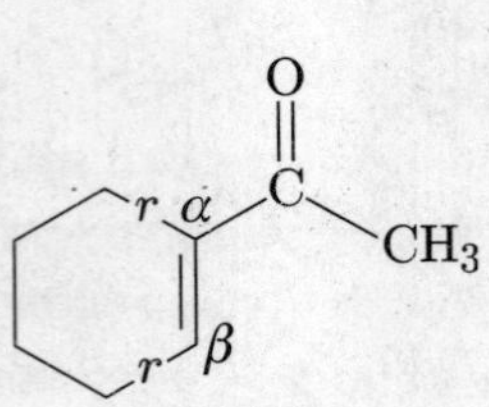

in etOH

Base value: 215 nm
Increments: 10 nm (α-ring residue)
12 nm (β-ring residue)
Total: 237 nm

(6b)

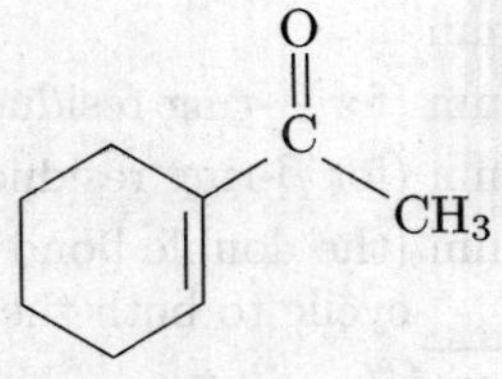

in ether

237 nm—7 nm (for solvent shift) = 230 nm

(7)

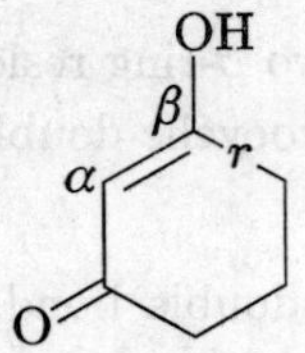

Base value:	215 nm (for 6-memberd system)
Increments:	30 nm (for β-OH)
	12 nm (for β-ring residue)
Total:	257 nm

(8)

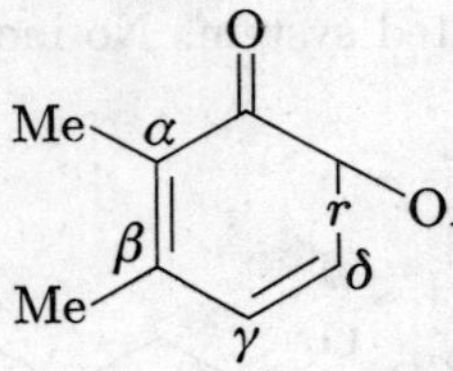

Base value:	215 nm. No increment for the—OAc group since it is not present on double-bonded C.s.
extended conjugation	30 nm
homodiene system	39 nm
α-methyl	10 nm
β-methyl	12 nm
δ-ring residue	18 nm
Total:	324 nm

(9)

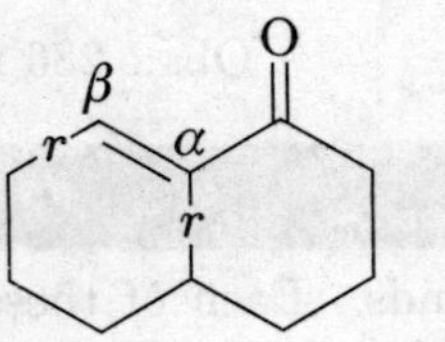

Base value:	215 nm
Increments:	10 nm (for α-ring residue)
	12 nm (for β-ring residue)
	5 nm (for exocyclic double bond)
Total:	242 nm

(10)

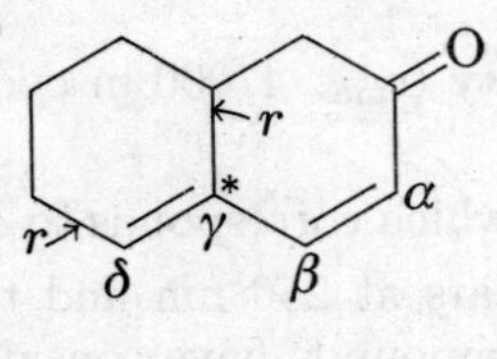

Base value:	215 nm
Increments:	18 nm (for γ-ring residue)
	18 nm (for δ-ring residue)
	30 nm (for extended conjugation)
	5 nm (for exocyclic double bond marked by *)
Total:	286 nm

(11)

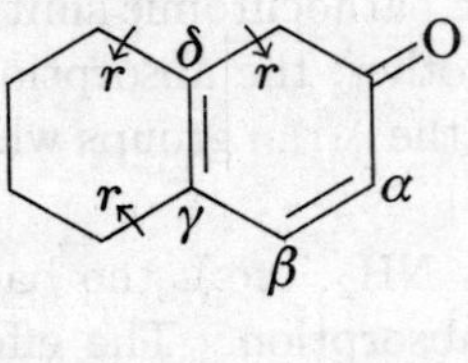

Base value:	215 nm
Increments:	30 nm (for extended conjugation)
	39 nm (for homodiene system)
	18 nm (for γ-ring residue)
	36 nm (for 2 δ-ring residues)
Total:	338 nm

(12)

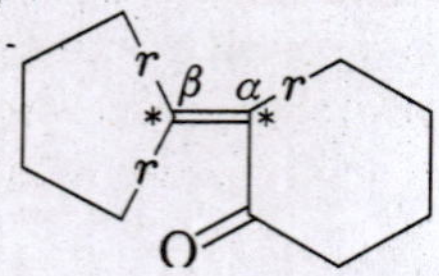

Base value:	215 nm
Increments:	10 nm (for α-ring residue)
	24 nm (for β-ring residues)
	10 nm (the double bond is exocyclic to both the rings)
Total:	259 nm

(13)

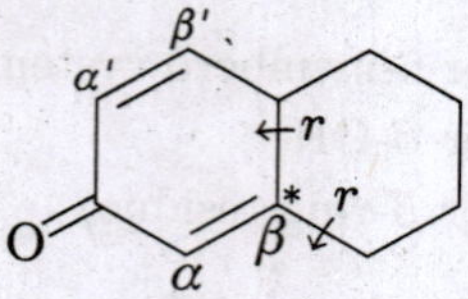

Base value:	215 nm
Increments:	24 nm (for two β-ring residues)
	5 nm (for exocyclic double bond)
Total:	244 nm

α', β'-unsaturated part has only one ring residue and no exocyclic double bond, whereas α, β-unsaturated system has two ring residues as well as an exocyclic double bond. Usually absorption takes place at a longer wavelength in a crossed conjugated system. No increment is allowed for crossed conjugation.

Problem 10.16. Explain the observed λ_{max} values for $\pi \longrightarrow \pi^*$.

(a)

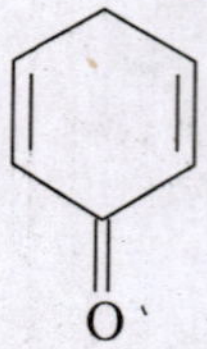

Obs : 227 nm

(b)

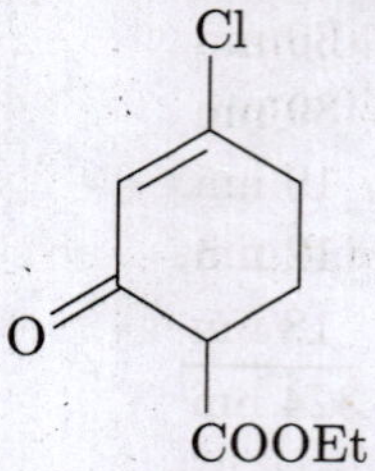

Obs : 239 nm

(c)

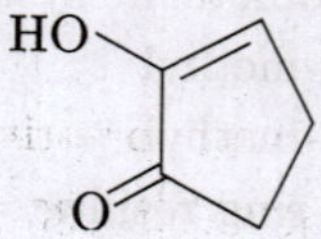

Obs : 249 nm

(d)

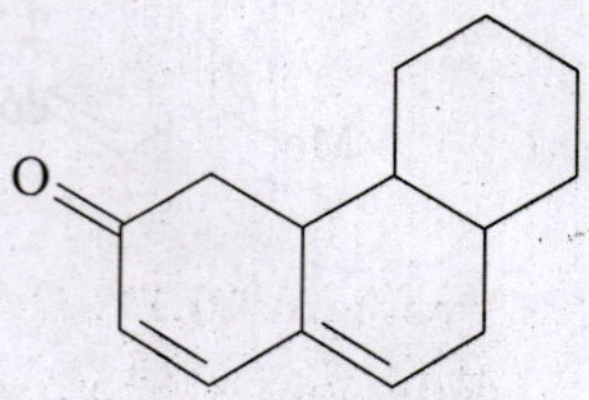

Obs : 286 nm

10.19.4 Aromatic or Benzenoid systems

The electronic spectrum of benzene displays three distinct bands. Each of these involves $\pi \longrightarrow \pi^*$ excitation, of course, the nature of the $\pi \longrightarrow \pi^*$ transition for one band differs from the others.

The first band appears at 184 nm and it is of strong intensity (ϵ_{max} 47000 in cyclohexane); this is designated as E_1 or λ_1.

The second band, known as E_2 or λ_2 band, has λ_{max} 202 nm which corresponds to ϵ_{max} 7,000.

The third band is of weak intensity (ϵ_{max} 200) which appears at 256 nm and this is said to be B or λ_3 band; B bands of benzene and its homologous compounds have considerable fine structures and are associated with vibration effects on the $\pi \longrightarrow \pi^*$ transitions.

Owing to the hyperconjugative effect of alkyl groups, alkylbenzenes produce red shift of the B band. If two alkyl groups are para to each other, the most effective bathochromic shift takes place with the largest intensity. When these groups are ortho to each other, the absorption occurs at the shortest wavelength, this is because of the steric hindrance of the ortho groups which reduces the hyperconjugative effect considerably.

If there are nucleus-substituted auxochromes with n es (OH, NH_2, etc.), the bathochromic shift of the B band, relative to benzene, occurs with intense absorption. The effect of such auxochromes in the electronic spectra of benzene derivatives is reported in Table 10.8. However, the position of maximum absorption of aromatic aldehydes, Ketones, acids and esters and their nucleus-substituted derivatives can be calculated with the help of some empirical rules as noted in Table 10.8. It may be noted here that the prediction of λ_{max} values for benzene derivatives are very difficult and actual values differ from the calculated values to a large extent in many a case.

Table 10.8 Calculation of λ_{max} for $\pi \longrightarrow \pi^*$ transition of benzene derivatives

Base value for PhCHO	250 nm	
Base value for PhCOR	246 nm	
Base value for PhCOOH	230 nm	
Base value for PhCOOR	230 nm	
Increments for substituents	*o*, *m*	*p*
Alkyl or ring residue	3 nm	7 nm
—OH, —OR	7 nm	25 nm
O^-	15 nm	80 nm
—Cl	0 nm	10 nm
—Br	2 nm	15 nm
$—NH_2$	13 nm	45 nm
—NHAc	20 nm	45 nm
$—NHR_2$	20 nm	85 nm
—NHR	···	73 nm

Problem 10.17. Calculate the λ_{max} for principal band of 3, 4-dichlorobenzoic acid.

Solution:

Base value for PhCOOH	230 nm
For *m*-chloro	0 nm
For *p*-chloro	10 nm
Total:	240 nm

Problem 10.18. An aromatic ketone of the molecular formula $C_9H_{10}O_2$ gives positive iodoform test; it does not give any test for phenolic OH. Observed λ_{max} value of the compound is 274 nm. Identify the compound.

Solution: Since the given ketone gives the positive iodoform test, there must have an acetyl group in the compound. The molecular formula of $C_6H_5—COCH_3$, aectophenone, is C_8H_8O. Therefore, the compound should have $—OCH_3$ part because the compound does not give any test for phenol. Thus the given compound is any one of the following:

o-methoxyacetophenone	Base value for $PhCOCH_3$	246 nm
	for *o*-OMe	7 nm
	Total:	253 nm
m-methoxyacetophenone	Base value for $PhCOCH_3$	246 nm
	for *m*-OMe	7 nm
	Total:	253 nm
p-methoxyacetophenone	Base value for $PhCOCH_3$	246 nm
	for *p*-OMe	25 nm
	Total:	271 nm

Since the calculated λ_{max} for *p*-methoxyacetophenone is very close to the value given in the problem, the compound is *p*-methoxyacetophenone.

Problem 10.19. Why do we use methyl orange and methyl red as acid-base indicators?

Solution: Methyl Orange exists in the solid state as orange coloured crystals of the structure (I). Methyl Red is of the structure (II).

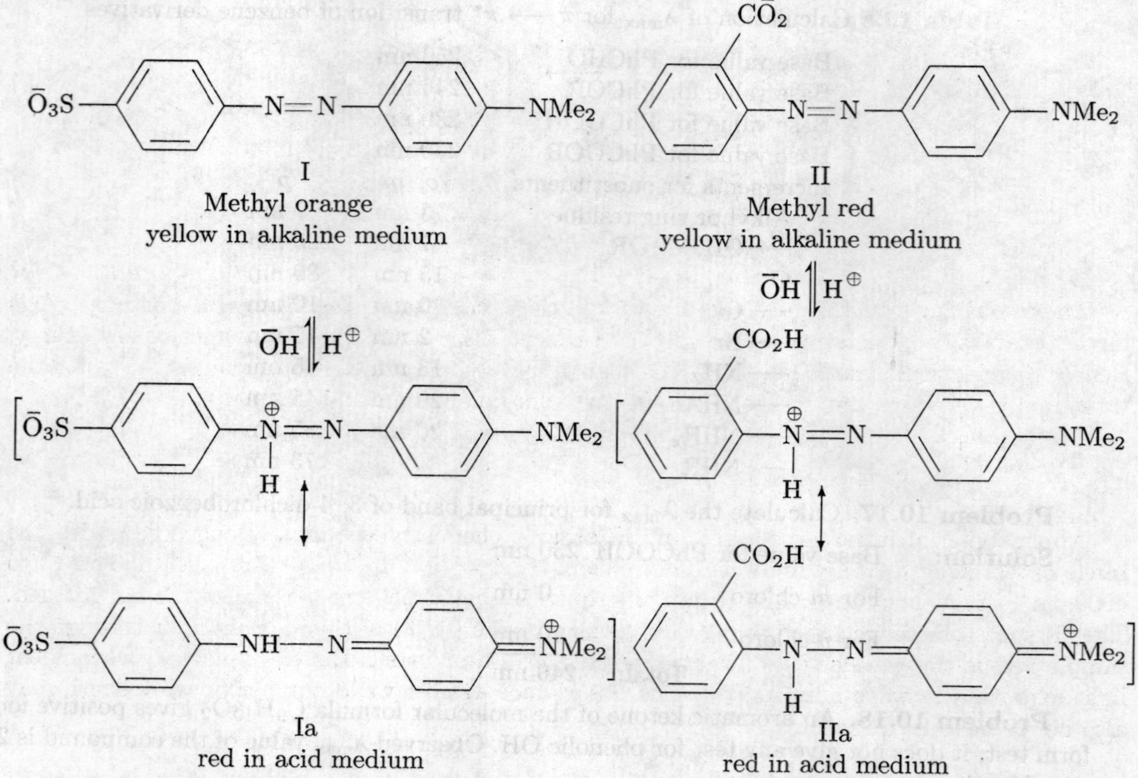

I
Methyl orange
yellow in alkaline medium

II
Methyl red
yellow in alkaline medium

Ia
red in acid medium

IIa
red in acid medium

Both the compounds have extended conjugation. Hence HOMO to LUMO transition requires small energy and, therefore, λ_{max} is high. So they have colours.

Methyl orange in solutions that are more basic than pH 4.4 exists almost entirely as the negative ion which is yellow in colour. In a solution of pH less than 3.1 it combines with a proton and forms a dipolar ion (Ia). It is a charged species; resonance brings extra stability to the excited state of the dipolar ion. HOMO to LUMO electronic transition requires less energy than that for the negative ion (I). Hence, bathochromic shift occurs in the pH 3.1 and deepening of colour occurs.

Methyl red in a solution of pH $\leq$ 4.1 remains as a cation ion (IIa). Being a resonance stabilised charged species, its LUMO at excited state is more stable than its HOMO. So the electronic transition from HOMO to LUMO requires small energy and, therefore, it absorbs at a longer wavelength, i.e., λ_{max} is high. So in the acid medium having 4.1 or less pH value we see red colour. A solution more alkaline than pH 6.1 does not contain protonated cation, it remains as an uncharged molecule. The excited state of the molecule is not so stable as that of the cation. So blue shift occurs and we see yellow colour of the solution having pH 6.2. In both cases, SO_3^- and CO_2^- do not take part in resonance.

Problem 10.20. Give an explanation for the dark blue colour of indigotin.

Solution: The trans-form of indigotin is given below which is a resonance hybrid of the following two resonating structures:

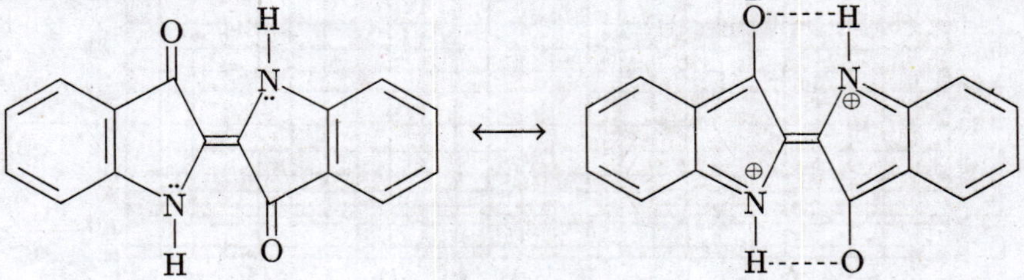

III
trans-form

IIIa

The resonating structure IIIa gets stability in the excited state; the internal H-bonding brings further stability to the excited state. So the LUMO of the excited state becomes of low energy. Hence, the electronic transition from HOMO to LUMO is very low. Consequently, λ_{max} for the transition is very high and indigotin shows dark blue colour.

10.20 Infrared Spectrcoscopy

Applications of infrared spectroscopy in organic chemistry are mostly concerned with the range of wavelength 2.5 to 15.4 micrometre (μm, 1μm $= 10^{-6}$ m $= 10^{-4}$ cm), i.e., 650 cm^{-1} to 4000 cm^{-1} frequency range. The region of frequencies below 650 cm^{-1} is called the far infrared. The IR spectrometer records the per cent transmittance (% T) of the incident light through the sample versus the wavelength of light expressed in μm or versus the frequency of the incident light expressed in wave number ($\overline{\nu}$). The absorbance against wave number or wavelength may also be recorded in IR spectra.

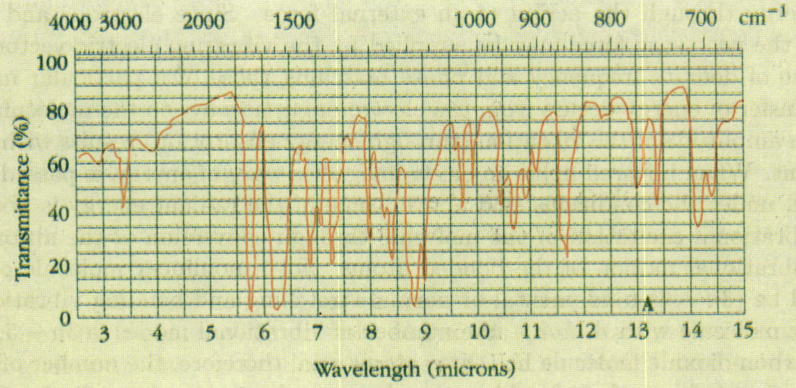

Fig. 10.22

Covalently bonded atoms in a molecule possess vibrational motion at ordinary temperature. The molecular vibrations are of two types: (i) two bonded atoms vibrate along the bond axis through a point at which the electron-electron and the nucleus-nucleus repulsions are balanced by

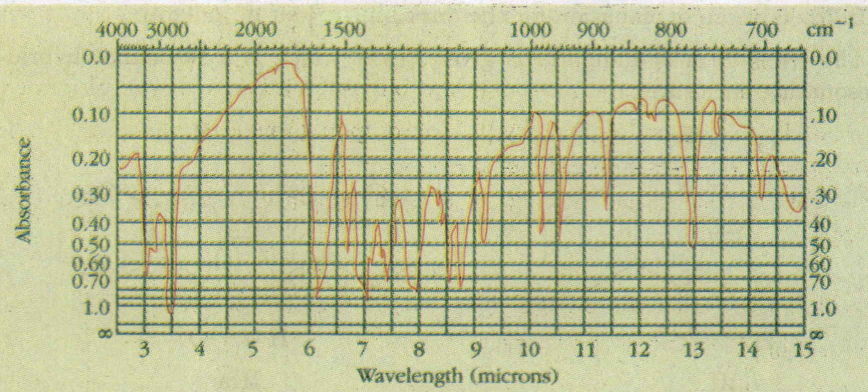

Fig. 10.23

the electron-nucleus attraction and thus lengthening and compression of the bond occur, and the mode of vibration is called the stretching vibration; (ii) bonded atoms or groups of atoms move but away from the bond axis, and the movements are called the bending vibrations. Stretching vibrations (str) are of two types—symmetric and antisymmetric.

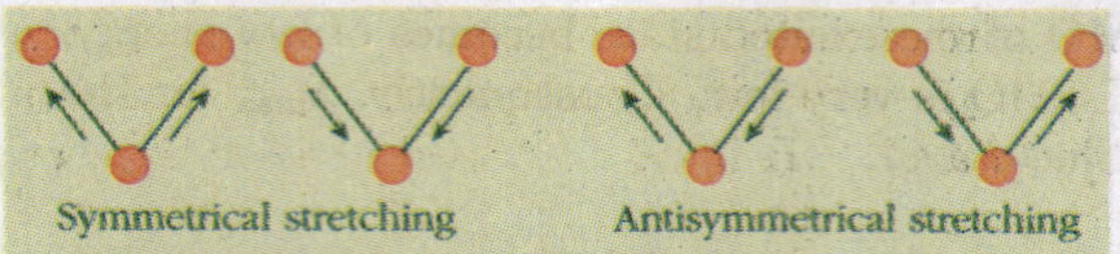

Fig. 10.24

Bending vibration, called deformations (def), are of many types; some of them are in-plane deformations and some of them are out-of-plane deformations.

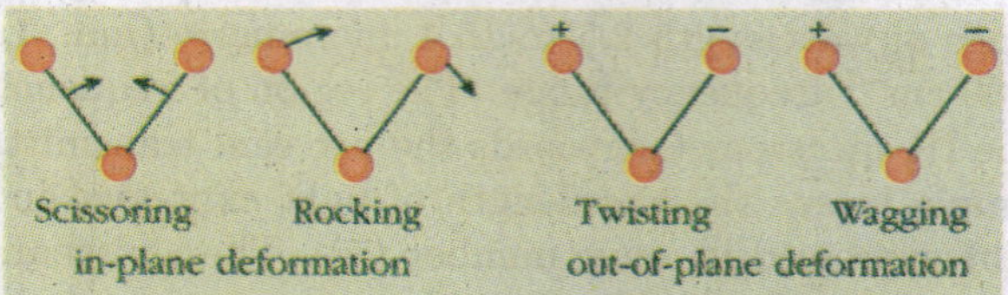

Fig. 10.25

These vibrations have natural periods which depend on the mass of the bonded atoms and the strength of the bond in question. The amplitude of the vibrations may be increased without altering their frequencies by supplying energy through the action of an external force. Since electrons and nuclei bear electric charges, the necessary force may be supplied by the vibrating electric vector of an electromagnetic wave of definite frequency and phase matching those of a particular molecular vibration. The transfer of energy occurs from the electromagnetic wave to the molecular vibration when a change in amplitude of the vibration (through energy absorption) results with a change of molecular moment. When infrared light, an electromagnetic wave of energy, is passed through an organic molecule, under the conditions said above, some of the radiant energy is absorbed; this results in the vibrational excitation of the molecule through conversion of the absorbed energy into increased vibrational motion of the bonded atoms. For a nonlinear molecule containing n atoms, there will be $(3n - 6)$ fundamental or normal stretching and bending vibrational modes; while for a linear molecule with n atoms, the number of vibrational modes is $3n - 5$. For example, the linear carbon dioxide molecule has three atoms and, therefore, the number of vibrational modes are $3 \times 3 - 5 = 4$; for methyl chloride, a non-linear molecule, there are $3 \times 5 - 6 = 9$ modes of vibration. Theoretically, one may expect to observe all these fundamental vibrations and their overtones (integral multiples of the frequencies of the fundamental vibrations) in an IR spectrum of a given molecule; but in practice, all of them are not observed because:

(i) each of them does not accompany with a change in the dipole moment of the molecule,

(ii) all of them do not lie in the 650-4000 cm^{-1} range.

(iii) some of them are of very weak mode and

(iv) two or more such vibrational modes overlap to develop a band instead of a fine peak.

However, a bond between two atoms may be looked upon as a spring joined by two balls which can stretch and bend.

There is the Hooke's law for calculating the stretching vibration of a spring and ball system which states that:

$$\nu = \frac{1}{2\pi}\left(\frac{k}{m_1 m_2/(m_1+m_2)}\right)^{\frac{1}{2}},$$

where ν = vibrational frequency of the string.

k = force constant of the spring.

m_1 & m_2 = masses of the two balls.

This equation may be used to calculate the stretching (str.) vibration of the bonds in which k will denote the force constant of the bond, a measure of the bond strength; m_1 and m_2 are the two masses of the bonded atoms and $m_1 m_2/(m_1+m_2)$ is the reduced mass of the system. As an example, let us calculate the stretching frequency of a C—H bond expressed in wave number, given that:

mass of carbon $(m_1) = 20 \times 10^{-24}$g

mass of hydrogen $(m_2) = 1.6 \times 10^{-24}$g

the force, constant (k) of C—H bond $= 5.0 \times 10^5 gs^{-2}$

the velocity of light (c) $= 3.0 \times 10^8\ ms^{-1}$

$$\bar{\nu} = \frac{\nu}{c} = \frac{1}{2 \times \frac{22}{7} \times 3 \times 10^8 ms^{-1}}\left[\frac{5.0 \times 10^5 gs^{-2}}{20 \times 10^{-24}g \times 1.6 \cdot 10^{-24}g/20 \times 10^{24}g + 1.6 \times 10^{-24}g}\right]^{\frac{1}{2}}$$

$$= 3100\ cm^{-1}.$$

The analogy between spring-ball system and bond-atom system tells us that:

(i) Stretching energy of a bond is greater than its bending energy just as the stretching energy of a spring is greater than its bending energy. This means that the frequency of a stretching absorption of a bond will be higher than that of the bending absorption of the bond.

(ii) Since $\bar{\nu}$ is directly proportional to the force constant of the bond (which is a measure of the bond strength), as the bond strength increases, the frequency of absorption energy increases. This means : the stronger the bond, the higher is the vibrational frequency of the bond. Thus, C=O str. will have higher frequency than C—O str.

(iii) Since, $\bar{\nu}$ is inversely proportional to the reduced mass, as the masses of the bonded atoms increase, the vibrational frequency of the bond decreases. For example, C—H str. will be of higher frequency than C—D str. It is needless to mention that from the consideration of mass only, the decreasing order of the stretching frequencies is :

$$\text{C—H} > \text{N—H} > \text{O—H} > \text{F—H}.$$

However, the decrease or increase in stretching frequencies depends on several other factors related to the steric and electronic conditions of the molecule; we shall discuss these under spectra-structure relation. In Table 1.9 given on the next page stretching and bending frequencies of some common bonds are listed.

Table 10.9 Characteristic IR bands of functional groups

	Group and vibrational mode	Most useful bands λ in μm	$\bar{\nu}$ in cm^{-1}	Intensity s = strong, w = weak m = medium, v = variable
1.	C—H (str.) in RH	3.38—3.51	2962—2853	m—s
	=C—H (str.) in olefins	3.23—3.32	3095—3010	m
	≡C—H (str.) in acetylenes	3.03	3300	s
	Aromatic nuclear C—H (str.)	3.30	3030	v
	OC—H in aldehyde	3.45—3.55	2900—2820	w
		3.60—3.70	2775—2700	w
2.	C—H (def) in R—H	7.46	1340	w
	=C—H (def) in olefines	7.04—7.09	1420—1410	s
	≡C—H (def) in acetylenes	15.9	630	s
	Aromatic C—H (def.)	13.3—11.3	750—880	v, s—w
3.	Symmetric and asymmetric	7.22	1385	s
	C—CH_3 (str.) in $(CH_3)_2C<$	7.32	1366	s
	C—CH_3(str.) in $(CH_3)_2C$—	7.30—7.33	1370—1365	s
		7.17—7.22	1395—1385	m
		near 7.33	~ 1365	s
4.	C—C multiple bond (str.)			
	C=C in non-conjugated alkenes	5.96—6.17	1680—1620	s
	C=C in vinyl alkenes	6.08	1645	m
	C=C in cis-alkenes	6.03	1658	m
	C=C in trans-alkenes	5.92	1675	m
	C=C in gem-alkenes	6.08	1653	m
	C=C in trisubstituted alkenes	5.99	1669	m
	C=C in tetrasubstituted alkenes	5.99	1669	w
	C=C in dienes	6.06	1650	w
		4.75	1600	
	C=C in alkenes	5.1	1960	m
		9.4	1060	m
	C=C in aromatic system	6.25	1600	v
		6.33	1580	v
		6.67	1500	m
		6.9	1450	m
	C≡C in R—C≡CH	4.67—4.76	2140—2100	m
	C≡C in R—C≡C—R	4.42—4.57	2260—2190	v, w
5.	C=O (str.)			
	in saturated acyclic systems	5.80—5.87	1725—1705	s
	in saturated cyclic systems:			
	six & higher membered ring	5.80—5.87	1725—1705	s
	five-membered ring	5.71—5.75	1750—1740	s
	four-membered ring	5.63	1775	s
	α,β-unsaturated acyclic	5.94—6.01	1685—1665	s

Contd.

Group and vibrational mode	Most useful bands λ in μm	$\overline{\nu}$ in cm^{-1}	Intensity s = strong, w = weak m = medium, v = variable
α, β-unsaturated cyclic systems:			
six and higher membered ring	5.94—6.01	1685—1665	s
five-membered ring	5.80—5.85	1725—1708	s
in aryl ketones	5.88—5.95	1700—1680	s
in diaryl ketones	5.99—6.02	1670—1660	s
in α-diketones	5.78—5.85	1730—1710	s
in β-diketones	6.10—6.50	1640—1540	s
in 1,4-quinones	5.92—6.02	1690—1660	s
in aldehydes (saturated and aliphatic)	5.75—5.81	1740—1720	s
in α, β-unsaturated aliphatic aldehydes	5.87—5.95	1705—1680	s
in aryl aldehydes	5.83—5.90	1715—1695	s
in saturated acyclic esters and in δ-lactone (and larger):	5.71—5.76	1750—1735	s
in γ-lactones	5.62—5.68	1780—1760	s
in vinyl esters	5.56—5.65	1800—1770	s
in α, β-unsaturated and aryl esters	5.78—5.82	1730—1717	s
in α, β-unsaturated δ-lactones	5.68—5.75	1760—1740	s
in α, β-unsaturated γ-lactones	5.56	1880	s
in γ-ketoesters	5.70—5.75	1775—1740	s
in β-ketoesters (enolic)	6		
in carboxylic acids (saturated aliphatic)	5.80—5.88	1725—1700	s
in α, β-unsaturated aliphatic acid	5.83—5.92	1715—1690	s
in aryl acids	5.88—5.95	1700—1680	s
in carboxylate ions str.	6.21—6.45	1610—1550	s
	7.15—7.69	1400—1300	s
in saturated acyclic acid anhydride	5.41—5.56	1850—1800	s
	5.59—5.75	1790—1740	s
in a, β-unsaturated acyclic	5.47—5.62	1830—1780	s
and aryl acid anhydrides	5.65—5.82	1770—1720	s
saturated 5-membered ring	5.35—5.49	1870—1820	s
andydrides	5.56—5.71	1800—1750	s
α, β-unsaturated cyclic			
acid andydride : 5-membered ring	5.41—5.56	1850—1800	s
in RCOF	5.41	1850	s
in RCOCl	5.57	1795	s
in RCOBr	5.53	1810	s
in α, β-unsaturated acyl and in benzoyl halides	5.61—5.72	1780—1750	s
	5.72—5.82	1750—1720	s
in amides:			
1° solid and conc. soln.	6.06	1650	s
1° dil. soln.	5.92	1690	s

Contd.

Group and vibrational mode	Most useful bands λ in μm	$\overline{\nu}$ in cm^{-1}	Intensity s = strong, w = weak m = medium, v = variable
2° solid and conc. soln.	5.95—6.14	1680—1630	*s*
2° dil. soln.	5.88—5.99	1700—1670	*s*
3° solid and conc. and dil. soln.	5.99—6.14	1670—1630	*s*
cyclic δ-lactams, dil. soln.	5.95	1680	*s*
cyclic γ-lactams, dil. soln.	5.88	1700	
6. C—O (str.)	7.22—9.52	1085—1050	*s*
in 1° alcohols	8.90—9.20	1124—1087	*s*
in 2° alcohols	8.30—8.90	1205—1124	*s*
in 3° alcohols	7.93—8.48	1260—1180	*s*
in phenols	9.55	1042	*s*
in anhydrides (acyclic)	10.75	930	*s*
in cyclic anhydrides	8.0	1250	*s*
in esters	2.7—9.5	1300—1050	*s*
in ethers	8.7—9.35	1150—1070	*s*
7. C—N (str.)			
in aliphatic amines	8.2—9.8	1220—1020	*w*
in aromatic 1° amines	7.1	1410	*w*
in aromatic 2° amines	7.46—8.00	1340—1250	*s*
	7.41—7.81	1350—1280	*s*
in aromatic 3° amines	6.36—7.64	1360—1310	*s*
in nitroarenes	6.37—6.67	1570—1500	*s*
	7.30—7.70	1370—1300	*s*
in nitroalkanes	6.37—6.45	1570—1550	*s*
	7.25—7.30	1380—1370	
in nitrosocompounds	6.25—6.67	1600—1500	*s*
8. —C=N str.			
In alkylimines and oximes	5.92—6.10	1690—1620	*v*
in α, β-unsaturated alkylimines and oximes	6.02—6.14	1660—1630	*v*
9. —C≡N str.			
in alkyl nitriles	4.42—4.46	2260—2240	*m*
in α, β-unsaturated alkylnitriles	4.47—4.51	2235—2215	*m*
in arylnitriles	4.46—4.50	2240—2220	*m*
in isonitriles	4.50—4.83	2220—2070	*m*
10. C—X str.			
in fluoro compounds	7.1—10.0	1400—1000	*s*
in chlorocompounds	12.5—16.6	800—600	*s*
in bromocompounds	16.6—20.0	600—500	*s*
in iodocompounds	~ 20	~ 500	*s*
11. C—S str. in thiols, thiophenols	14.5—17	700—600	*w*

Contd.

Group and vibrational mode	Most useful bands λ in μm	$\bar{\nu}$ in cm^{-1}	Intensity s = strong, w = weak m = medium, v = variable
12. C═S str. in thiols, thiophenols	8.33—9.52	1200—1050	*s*
13. O—H str.			
O—H without H-bonding	2.74—2.790	3650—3590	*ν*
intermolecular H-bonded O—H			
in a bridge compound	2.82—2.90	3550—3450	*ν*
in polymeric association	2.94—3.13	3400—3200	*s*
intramolecular H-bonded O—H,			
chelate compounds	3.1—4.0	3200—2500	*w*
in carboxylic acids (bonded,			
dimeric state)	3.4—4.0	2940—2500	*w—m*
14. O—H def.			
in 1°, 2°, 3° alcohols and in phenols	7.1—7.9	1410—1260	*s*
15. N—H str.			
in 1° free amines	2.86	3500	*m*
in 2° free amines	2.94	3400	*m*
in imines ═N—H	2.86—3.2	3500—3310	*m*
in amine salts	2.94—3.03	3400—3300	*m*
	3.2—3.3	3130—3030	*m*
in 1° H-bonded amides	2.99	3350	*m*
	3.15	3180	*m*
in 2° free amides	2.90	3448	*m*
in 2° H-bonded amides	3.15	3180	*m*
16. N—H def.			
in 1° amines	6.06—6.29	1650—1590	*s—n*
in 2° amines	6.06—6.45	1650—1550	*w*
in amine salts	6.25—6.35	1600—1575	*s*
	6.67	1500	*s*
in 1° amides, dil. soln.	6.17—6.29	1620—1590	*s*
in 2° amides, dil. soln.	6.45—6.62	1550—1510	*s*
17. S—H str.			
in thiols, thiophenols	3.85—3.92	2600—2550	*w*
18. N—O str.			
in nitrites	6.15—6.21	1625—1610	*s*
	5.95—6.06	1680—1650	*s*
in nitrates	6.06—6.25	1650—1600	
	7.70—8.00	1300—1250	*s*
19. S—O str.	11.1—14.2	900—700	*s*

Contd.

Group and vibrational mode	Most useful bands λ in μm	$\bar{\nu}$ in cm^{-1}	Intensity s = strong, w = weak m = medium, v = variable
20. S=O str.			
in sulphoxides	9.35—9.71	1070—1030	s
in sulphones	8.62—8.77	1160—1140	s
in sulphonic acids	8.27—8.70	1210—1150	s
	9.43—9.71	1060—1030	s
in suphonyl chlorides	15.4	650	0
	8.44—8.89	1185—1165	s
	7.30—7.46	1370—1340	s
in sulphonamides	8.48—8.77	1180—1140	s
	7.41—7.69	1350—1300	s

Let us now discuss in brief some of the characteristic bands of a few functional groups in Table 10.9.

1. **Alkenes:** For alkenes the following bands deserve to be mentioned.

(i) (a) =C—H[C(sp^2)—H(s)] stretching vibration at about 3000 cm^{-1} (small but sharp band). This appears at 3095–3075 cm^{-1} in the cases of vinyl and H—CH=CR_2; for cis- and trans-alkenes =C—H str. appears at 3040–3010 cm^{-1}.

(b) =C—H[C(sp^2)—H(s)] out-of-plane bending and wagging vibrations occur at about 1000–800 cm^{-1}. However, for the cis-isomer this appears at around 700 cm^{-1} ; while for the trans-isomers at about 970 cm^{-1}.

C=C—H[C(sp^2)—H(s)] in-plane vibration appears at 1420–1290 cm^{-1}.

(ii) (a) C=C stretching vibration appears in the region 1680–1620 cm^{-1} in the cases of simple alkenes; its intensity is much less than the intensity of >C=O str. which appears in the region 1725–1680 cm^{-1}. This is because of the less polar character of C=C than >C=O. In general, the higher the substitution on the double-bended carbons, the higher is the frequency of absorption and the lower is the intensity. Symmetrically tetra-substituted alkenes, $R_2C{=}CR_2$ either show no C=C stretching band or a weak band.

(b) When polar groups are attached to double bonded Cs, the C=C stretching frequency gets lowered to 1610–1590 cm^{-1}.

(c) The vibrational stretching frequencies of exocyclic double bonds depend on the size of the ring. In the cases of

(i) three-membered rings : $\bar{\nu}$C=C str. appears at 1736 cm^{-1}

(ii) four-membered rings : $\bar{\nu}$C=C str. appears at 1678 cm^{-1}

(iii) five-membered rings : $\bar{\nu}$C=C str. appears at 1657 cm^{-1}

and **(iv)** six-membered rings : $\bar{\nu}$C=C str. appears at 1651 cm^{-1}

The above order of wave number values show that, as the angle strain involving C—C(=C)—C angle decreases, i.e., the 's' character of the double-bonded central C atom decreases, the $\bar{\nu}$C—C str. decreases.

(d) Symmetrically substituted trans alkenes, for example, 1, 2-dichloroethene, fumaric acid etc., do not show any band for C=C str., even if the substituted groups are highly polar. In these cases, there is no transition dipole moment (i.e., the oscillating dipole moment does not change) because of the symmetry of the molecules; for the same reason O=C=O, S=C=S, $H_2C{=}C{=}CH_2$ do not show symmetric C=Z (Z = O, S or C) stretching. However, they do exhibit antisymmetric stretchings.

(i) $\bar{\nu}\vec{C{=}C}\vec{=}C<$: 1950 cm^{-1}; (ii) $\bar{\nu}\vec{C{=}C}\vec{=}O$: 2350 cm^{-1};

(iii) $\bar{\nu}\vec{C{=}C}\vec{=}O$: 2150 cm^{-1} (iv) $\bar{\nu}\vec{N{=}C}\vec{=}O$: 2250 cm^{-1}.

(e) When a C=C is conjugated with a carbonyl group, $\bar{\nu}$C=C appears at a lower wave number.

2. **Alkynes: (a)** Terminal alkynes show the following peaks.

(i) ≡C—H[C(*sp*)—H(s)] stretches at about 3300 cm^{-1} (strong and sharp absorption). Hydrogen bonded N—H and O—H bands also appear at this region but those are broad in contrast.

(ii) C≡C stretches at about 2200 cm^{-1}, (weak).

(iii) ≡C—H bending vibrations appear at 680–610 cm^{-1} and its overtone is found in the region 1300–1200 cm^{-1}.

(b) Non-terminal alkynes naturally do not show ≡C—H str. and bend bands. However, non-terminal unsymmetrical alkynes show C≡C str.; while non-terminal symmetrical alkynes like Br—C≡C—Br, CO_2H—C≡C—CO_2H, do not show C≡C str. for the reason as noted in the symmetrical trans-alkenes (1d above).

3. **Alcohols:** See spectra-structure relation (10.20.1).

4. **Ethers and cyclic ethers:** A very strong band appears at 1160–1150 cm^{-1} for asymmetric C—O stretching in the cases of dialkyl ethers. However, in the cases of aryl alkyl ethers and vinyl alkyl ethers the band appears at 1270–1230 cm^{-1}. In these cases, the C—O bond gets enough of C=O character owing to the resonance effect which brings the $\bar{\nu}$C—O str. at a higher frequency than 1160 cm^{-1}.

5. **Amines: (i)** 1° amines: **(a)** $\bar{\nu}_{N—H}$ for symmetric stretching appears at about 3400 cm^{-1}, whereas the asymmetric N—H stretching occurs at about 3500 cm^{-1}. In the cases of 2° amines N—H stretching is found at about 3300 cm^{-1} and that for a 3° amine the question does not arise.

So, $\bar{\nu}_{N—H}$ str. helps us to differentiate three types of amines. The bands appear at lower wave numbers in the condensed systems owing to the intermolecular H-bonding; but the bands are less broader than those for the hydroxyl compounds owing to the weaker nature of H-bonding in the cases of amines than in the cases of hydroxyl compounds.

(ii) The scissoring vibration of H—N—H in the cases of 1° amines is recorded in the region 1650–1590 cm^{-1}; twisting and wagging deformations appear in the region of 850–750 cm^{-1}. Secondary amines and tertiary amines do not show these bands.

(iii) $\bar{\nu}_{N—H}$ str. appears at a slightly higher wave number in the cases of aromatic amines than in the cases of aliphatic amines since in the former compounds, the N atom assumes a hybridised state between sp^2 and sp^3 and the N—H bond in the aromatic compounds are stronger than those in the aliphatic compounds.

(b) $\bar{\nu}_{C—N}$ str. appears at 1190–1020 cm^{-1} in the cases of aliphatic compounds whereas that appears at 1300–1250 cm^{-1} in the cases of aromatic amines. The increase in wave number for C—N str. is attributed to the resonance effect.

(c) $\bar{\nu}_{N—H}$ str. and $\bar{\nu}_{N—H}$ def. occurs at about 3200–2800 cm^{-1} and 1600–1400 cm^{-1}, respectively in amine salts and amino acids.

6. *Unsaturated nitrogen compounds*: —C≡N :, —$\overset{\oplus}{N}$≡$\bar{C}$:, —$\ddot{N}$=C=$\ddot{C}$: etc.

(i) $\bar{\nu}_{C \equiv N}$ str. appears at about 2250 cm^{-1}. Any effect which decreases the dipole moment of the C≡N bond also decreases the intensity of the $\bar{\nu}_{C \equiv N}$ str. It is needless to mention that when the group is conjugated with an aromatic ring or other unsaturated groups, the $\bar{\nu}_{C \equiv N}$ str. decreases. However, no change in wave number occurs in the cases of conjugated isocyanates for which $\bar{\nu}$ str. = 2275–2240 cm^{-1}.

(ii) In the cases of oximes and imines $\bar{\nu}_{>C=N}$ occur at about 1690–1590 cm^{-1}; oximes also show a N—O str. at 3200 cm^{-1}.

(iii) The azo compounds exhibit $\bar{\nu}_{C=N—}$ str. at 1600 cm^{-1} which shifts to a lower wave number by conjugation. This band has importance in aromatic compounds since the band is usually masked by the aromatic vibrations.

7. *Nitro compounds*: $\bar{\nu}_{\overset{\oplus}{N}\cdots\cdots O^-}$ str. for symmetric and asymmetric vibrations appear at about 1560–1500 cm^{-1} and 1350–1300 cm^{-1}, respectively. However, the involvement of conjugation with other groups decreases the values.

8. *Amides*: (i) $\bar{\nu}_{C=O}$ str. occurs at about 1690 cm^{-1} in dilute solutions. In the solid phase the wave number decreases to some extent (amide I band) perhaps owing to the intermolecular H-bonding.

(ii) $\bar{\nu}_{N—H}$ str. appears in the regions where those of 1° and 2° amines appear. The values decrease on passing from a dilute solution to the solid phase.

(iii) $\bar{\nu}_{N—H}$ str. bending vibrations in the cases of 1° and 2° amides appear near 1650 cm^{-1} and 1550 cm^{-1}, respectively in the solid phase (this is called amide II band). These bands help us to distinguish between primary and secondary amides.

9. *Ketones*: See the spectra-structure relationship (10.20.1).

(i) Normal ketonic carbonyl stretching vibrations appear at 1725 cm^{-1} in the cases of acyclic compounds. However, as the branching at α-Cs increases C—C(=O)—C angle increases and the carbonyl C assumes more '*p*' character. As a consequence, the electronegativity of the carbonyl C decreases and naturally the wave number for C=O str. decreases. For example, $\bar{\nu}_{>C=O}$ str. in di-tert-butyl ketone takes place at 1697 cm^{-1}. On the other hand, when C—C(=O)—C angle decreases owing to angle-strain, the carbonyl carbon assumes more '*s*'-character and its electronegativity increases; as a consequence $\bar{\nu}_{>C=O}$ band becomes appreciably strong and $\bar{\nu}_{>C=O}$ str. increases. Thus the $\bar{\nu}_{>C=O}$ str. in the cases of cyclobutanone, cyclopentanone and cyclohexanone are 1775 cm^{-1}, 1750 cm^{-1} and 1725 cm^{-1}, respectively.

(ii) In the conjugated systems, however, the carbonyl stretching frequency decreases to the extent of 40 cm^{-1}, α,β-unsaturated carbonyl compounds exists in two conformations, s-cis and s-trans and two bands appear at around 1680 cm^{-1} and 1720–1715 cm^{-1} for s-trans and s-cis conformations, respectively. In the latter case C=C π-cloud resists the carbonyl groups to assume $\overset{\oplus}{C}$—$\bar{O}$ form for minimising the repulsion between C=C and C=O groups.

10. ***Aldehydes***: **(i)** $\overline{\nu}_{>C=O}$ str. appears at a slightly higher frequency than ketones in saturated acyclic compounds. It shows peak in the region 1740–1730 cm^{-1}. However, aldehydes can be distinguished from ketones by means of aldehydic C—H stretch; this gives two weak bands near 2820 cm^{-1} and 2720 cm^{-1} owing to Fermi resonance between the fundamental aldehydic C—H stretching vibration and the first overtone of the aldehydic C—H def. band which appears at 975—780 cm^{-1}. The band near 2820 cm^{-1} is frequently not observed in the cases of alphatic saturated aldehydes because of masking by the strong saturated C—H absorption in the same region.

(ii) ***Carboxylic acids***: **(a)** Carboxylic acids remain in the dimeric state in the concentrated solution. Owing to the presence of dimer, a broad band appears in the region 3300 cm^{-1} to 2500 cm^{-1}. This broad band is attributed to the bi-molecular H-bonded O—H stretching vibration.

(b) In very dilute solutions the presence of monomeric carboxylic acid shows $\overline{\nu}_{>C=O}$ str. at 1725–1700 cm^{-1}. However, internal and bi-molecular $\overline{\nu}_{>C=O}$ str. is much less than the usual value. For example, in salicylic acid there is internal H-bonding involving O—H and C=O and $\overline{\nu}_{>C=O}$ str. occurs at 1655 cm^{-1}.

(iii) $\overline{\nu}_{>C=O}$ str. also decreases by conjugation to the extent of 20–30 cm^{-1}.

11. ***Lactones and esters*** : See the spectra-structure relationship (Sec. 10.20.1).

12. ***Carboxylic acid anhydrides*** : They show two $\overline{\nu}_{>C=O}$ str. one for antisymmetric and the other for symmetric >C=O stretching modes. Saturated anhydrides absorb around 1818 cm^{-1} (antisymmetric) and 1750 cm^{-1} (symmetric). Conjugation decreases the values. Five-membered rings (succinic anhydride) absorb at a higher frequency owing to angle-strain.

10.20.1 Spectra-structure relationship

The force constant, i.e., the strength of a particular type of bond is not actually a constant; electronic effects of the atoms and groups within the molecule, steric strain and field effects have profound influence on the strength of a bond, i.e., on its force constant. We can qualitatively determine the effects of electronic and steric conditions of the molecule on the bond. For this the following effects are to be considered:

(i) Resonance effect.
(ii) I effect.
(iii) H-bonding effect.
(iv) Steric effect.
(v) Electronegativity.
(vi) Field effect.

(i) Resonance and IR spectra

It has already been stated that owing to p-π and π-π conjugation single bonds get partial double bond character and double bonds get partial single bond character. It is also known to us that single bonds are weaker than the corresponding double bonds. So, when a single bond acquires a partial double bond character through resonance, the vibrational frequency of the bond is expected to be greater than the normal value because of the shortening of bond length. On the other hand, when a double bond possesses a partial double bond character, its vibrational frequency is expected to be lower than the normal value because of the weakening of the bond strength through resonance. Let us understand the resonance effect considering a few examples.

(a) C═O stretching in saturated acyclic ketones and α, β-unsaturated acyclic ketones

In an α, β-unsaturated acyclic ketone π-π conjugation takes place; owing to this conjugation, C═O group has a partial double bond character. Thus the C═O bond becomes longer than that in a saturated ketone. As a consequence, the C═O bond strength decreases and its vibrational frequency will be less than that of the C═O group in the saturated acyclic ketone.

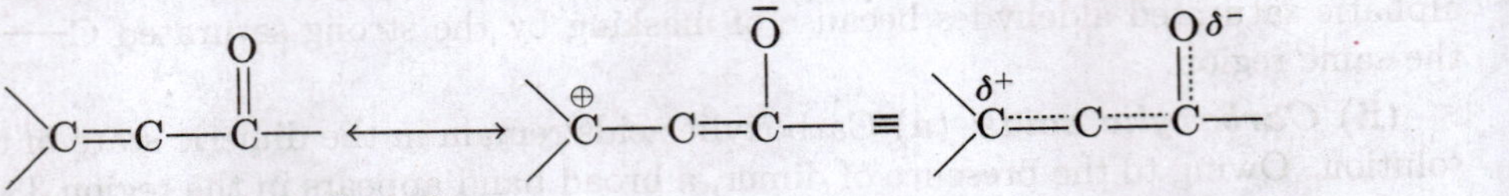

here C═O is C┈O

Table 10.9 shows that the C═O str. in a saturated acyclic ketone is 1725–1705 cm^{-1}; whereas that in an α, β-unsaturated acyclic ketone is 1685–1665 cm^{-1}. Thus the π-π conjugation here decreases the vibrational frequency by 40 cm^{-1}.

(b) C═O stretching in acetophenone and in *p*-aminoacetophenone

The C═O str. in acetophenone is 1686 cm^{-1}, whereas that in *p*-aminoacetophenone is 1677 cm^{-1}. The resonance interaction is supposed to be the cause of lowering of vibrational frequency.

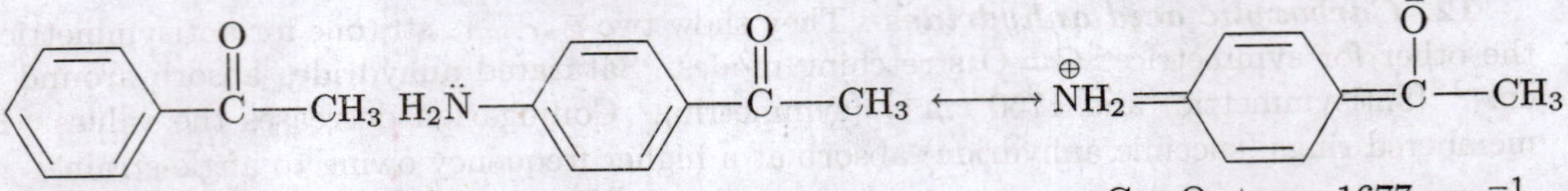

C═O str. = 1686 cm^{-1} C═O str. = 1677 cm^{-1}

(c) C═O stretching in an amide, in an ester and in lactones

An NH_2 group has −I effect as well as +R effect but the latter predominates. Owing to the +R effect, the C═O group in an amide gets a partial double bond character and possesses lower vibrational frequency (C──O str. is 1650 cm^{-1} in a 1° solid amide or in its concentrated solution in acyclic saturated acid) than that in the corresponding ester (C═O str. is 1750–1735 cm^{1}). In the case of an ester the −I effect of the alkoxy group predominates over its +R effect (see later). Lactones are inner esters. We expect ester carbonyl str. in lactones. In these compounds angle-strain factor has a profound effect

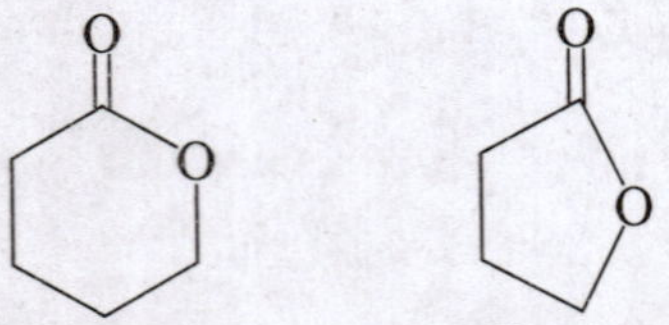

δ-Lactone γ-Lactone

In δ-lactone no angle strain is there while angle strain is present in γ-lactone. As a consequence $\bar{\nu}_{>C=O}$ str. in γ-lactone is greater than that in δ-lactone.

(d) C──O stretching in alkyl and aryl ethers

The bond order in an aryl ether gets increased by *p*-π conjugation; while in saturated acyclic ethers no such change takes place owing to the absence of π bond orbital. Thus in aryl ethers C──O bond has some double bond character; because of this the C──O bond strength increases and the vibrational frequency of the bond is found to be greater than that of the alkyl analogue.

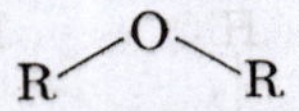

C—O str. = 1150–1070 cm^{-1}

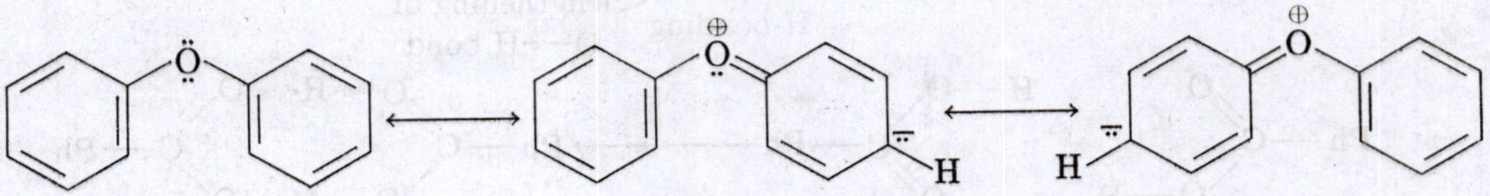

Ar–O–Ar C—O str = 1260–1130 cm^{-1}

(ii) I effects and IR Spectra

I effects have profound influence on vibrational frequencies of groups with multiple bonds. For example, in acyl halides the –I effect of the halogen atom makes the corbonyl C electron-deficient and because of the electron-deficiency of the carbonyl C, the C=O moiety cannot assume a partial single bond character. Thus the C=O bond in acyl halides is stronger than that in ketones and naturally the C=O stretching frequency in acyl halides is greater than that in ketones.

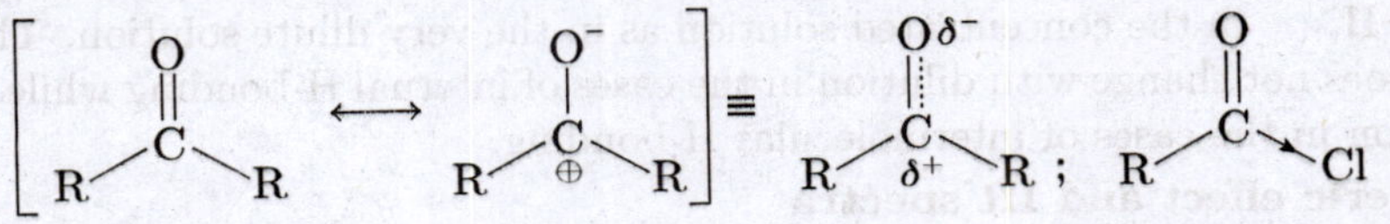

$\bar{\nu}$ = 1725–1705 cm^{-1} $\bar{\nu}$ = 1795 cm^{-1}

The influence of –I effect may also be demonstrated with α-halo and α, α'-dihaloketones; in these cases the wave number of C=O str. raises by 20 cm^{-1} and 40 cm^{-1} (approximately) respectively owing to the shortening of C=O bond length.

(iii) H-bonding and IR Spectra

Groups containing active Hs, such as O—H, N—H, etc., form intermolecular H-bond and intramolecular H-bond. If a group X—H forms H-bond with a group Z—H or X—H, both X—H and Z—H bonds become weaker than those bonds without H-bonding, i.e., free X—H and Z—H bonds. It is already known to us that a weaker bond has a lower vibrational frequency. For this reason an 1% solution of 1-butanol in CCl_4 has 3500 cm^{-1} as O—H str. value, whereas at a very low concentration (in CCl_4) a sharp band is found at 3650 cm^{-1} along with a broad band at 3350 cm^{-1}. In a concentrated solution, a number of alcohol molecules associate through intermolecular H-bonding; this decreases the O—H bond strength and consequently O—H str. frequency decreases. At a very low concentration or in vapour state free O—H group, i.e., a O—H group without H-bond exists ; this being stronger than H-bonded O—H group, it has a O—H str. at 3650 cm^{-1}. This has also been reflected in the cases of carboxylic acids; they exist in dimeric state in which O—H bonds get deformed owing to the formation of H-bonding and as a consequence the O—H str. occurs at 2700–2500 cm^{-1}.

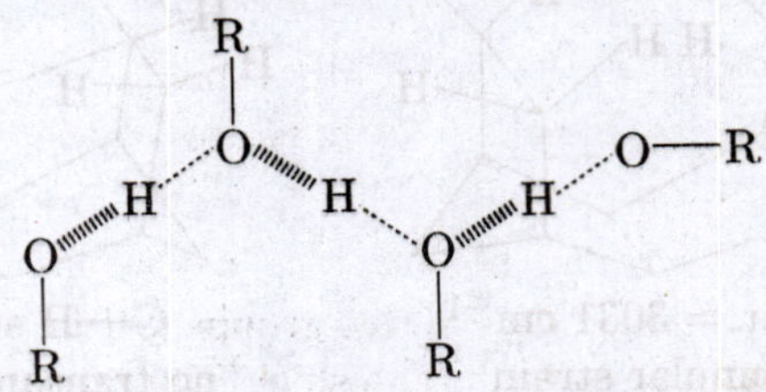

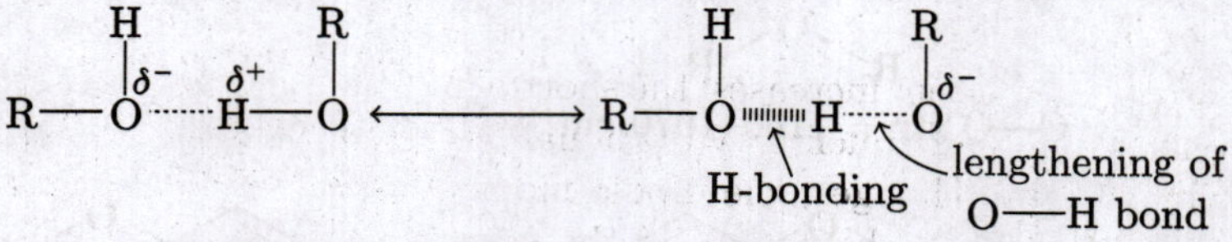

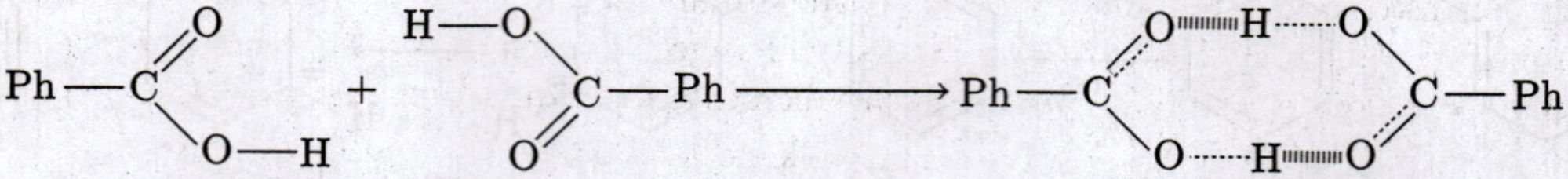

H||||||O ≡ H-bonding

O······H ≡ O—H bond stretching

Benzoic acid in dimeric state

The low O—H str. has also been seen in the cases of chelated enols; the O—H str. frequency is found to be around 2800 cm^{-1}.

Here it is important to note that compounds with intramolecular H-bonded groups do not lose their internal H-bonding on dilution with inert solvents like CCl_4, CS_2, etc., but intermolecular H-bonds break on dilution with them. So, a compound with internal H-bonding shows almost the same X—H str. in the concentrated solution as in the very dilute solution. This means X—H str. value does not change with dilution in the cases of internal H-bonding while the same changes with dilution in the cases of intermolecular H-bonding.

(iv) Steric effect and IR spectra

Steric effects such as steric inhibition of resonance, transannular effect, etc. influence stretching frequencies of bonds. For example, the C=O str. value in 2, 7-dimethylene-4, 5-benzotropone is 1596 cm^{-1}, whereas tetramethylene-4, 5-benzotropone has a C=O str. at 1724 cm^{-1}. In the latter compound, the C=O group is forced out of the plane of the molecule and being non-coplanar, steric inhibition of resonance occurs and the compound shows normal value for C=O str., whereas in the former case C—O double bond lengthens owing to conjugation of the C—O π bond with the rest of the molecule and as a consequence C=O str. comes down.

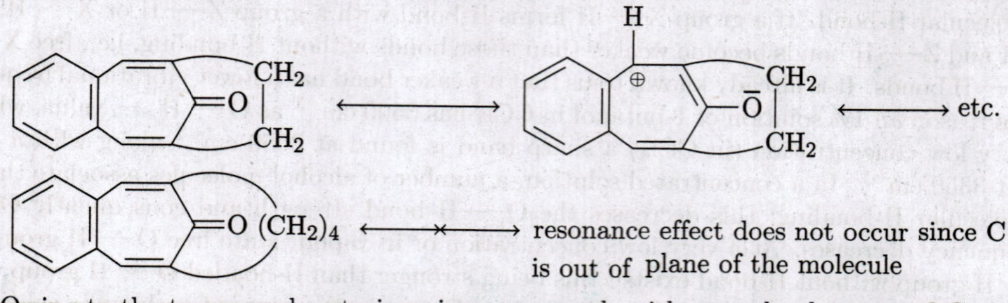

resonance effect does not occur since C=O is out of plane of the molecule

Owing to the transannular strains, ring compounds with very closely opposed C—H bonds show unusually high C—H str. values.

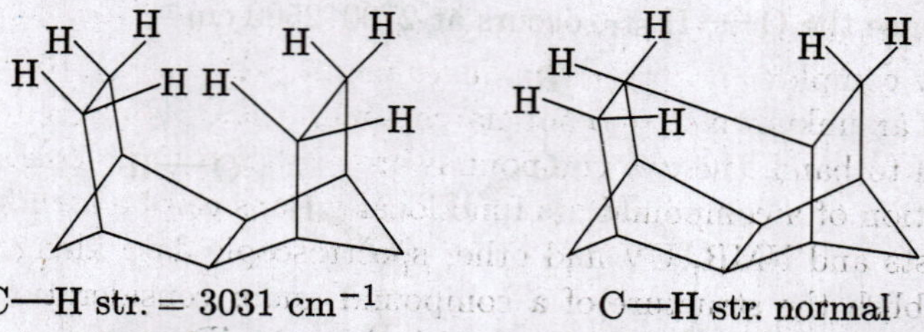

C—H str. = 3031 cm^{-1} transannular strain | C—H str. normal no transannular strain

(v) Electronegativity and IR spectra

As the electronegativity of a bonded atom increases, the shortening of bond length and increase in force constant of the bond occur; consequently vibrational frequency increases. Again, as the electronegativity of a bonded atom changes, the dipole moment of the bond changes. The absorption intensity being proportional to the square of the change in dipole moment, as the electronegativity changes, absorption intensity changes. For example, the decreasing *s* character, i.e., the decreasing electronegativity of C in alkane, alkene and alkyne is:

$$\text{alkyne} > \text{alkene} > \text{alkane};$$

the C—H str. values of these compounds are also in the same order (see Table 10.9)

$$\equiv\text{C—H str.} > =\text{CH—H str.} > -\text{CH}_2-\text{H str.}$$

In cyclopropane C—C—C bond angle is 60° and there is a strong angle strain; because of this strain the C—H bond shortens and the *s* character of the bond increases and naturally C—H str. occurs at a higher frequency (3040-3070 cm^{-1}) than the normal C—H str. frequency (2962–2853 cm^{-1}).

In cyclobutanone C—CO—C bond angle being much less than 120°, C=O has more *s* character, which leads to the shortening of C=O bond length and as a consequence the C=O str. occurs at 1785 cm^{-1}. On the other hand, in di-tert-butyl ketone, C—CO—C bond angle is greater than 120° because of the big volume of the tert-butyl groups. The C=O acquires more *p* character and the carbon-oxygen bond acquires a partial double bond character and thus it has a very low C=O str. ($\overline{\nu} = 1697$ cm^{-1}).

(vi) Field effects and IR spectra

Electrostatic or steric through-space interaction between two groups within a molecule often influences the vibrational frequencies of both the groups and the effect of the interaction may be called field effect. It has been observed that the C=O str. is higher when the Cl atom of α-chloroketone derivatives of steroids occupies the equatorial position than when it is in the axial position. It is said that in the equatorial position, the electron cloud of the Cl atom and that of the O atom experience electronic repulsion and as a result of which carbon oxygen bond acquires a less partial double bond character. If it does so, it will have to face greater repulsion.

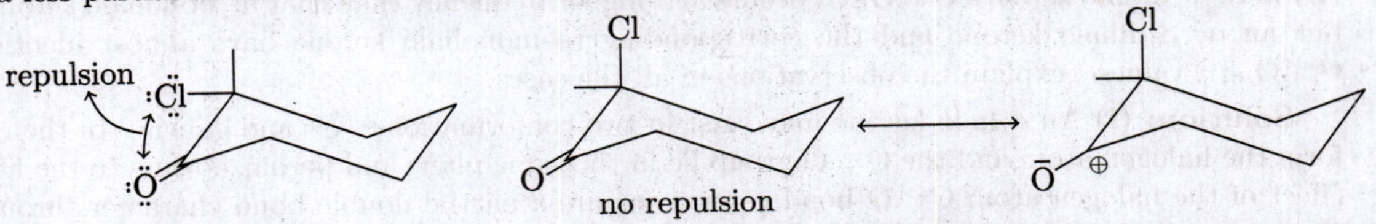

10.20.2 Application of IR Spectroscopy

IR spectra of a compound usually contains a large number of peaks, bands, shoulders, etc. It is very difficult to predict vibrational modes from an IR spectra. However, 1400 cm^{-1} to 900 cm^{-1} region is very complex and called the finger print region because if the finger print region of the IR spectra of an unknown compound matches with that of a structurally known compound peak to peak, band to band, the two compounds are expected to be one and the same. Besides complete identification of a compound, its functional groups can be identified by IR spectra. Of course chemical tests and NMR, UV and other spectroscopic data along with IR spectral data enable one to establish the structure of a compound under consideration. Intermolecular and intramolecular H-bondings can also be distinguished by the IR spectra.

Problem 10.21. The IR spectra of ethyl benzoate differs from that of methyl phenylacetate in many ways. Guess at least one point of difference and give reasons for the same.

Solution: In methyl phenylacetate the ester group has no scope for conjugation, but in ethyl benzoate the ester group undergoes conjugation with the π orbitals of the benzene ring, as a consequence the C═O group acquires a single bond character and thus the C═O bond strength in ethyl benzoate becomes less than that in methyl phenylacetate. Since the weaker the bond, the lower is the stretching absorption. C═O str. in ethyl benzoate will occur at a lower wave number than that in methyl phenylacetate, though they are isomeric compounds.

Problem 10.22. A close look at the IR spectra of ethyl acetoacetate reveals that there are strong IR absorptions at 3050 cm^{-1}, 1748 cm^{-1}, 1724 cm^{-1} and at 1650 cm^{-1}. Explain the peaks.

Solution: Ethyl acetoacetate exhibits keto-enol tautomerism. The enol form remains as a chelate as shown below:

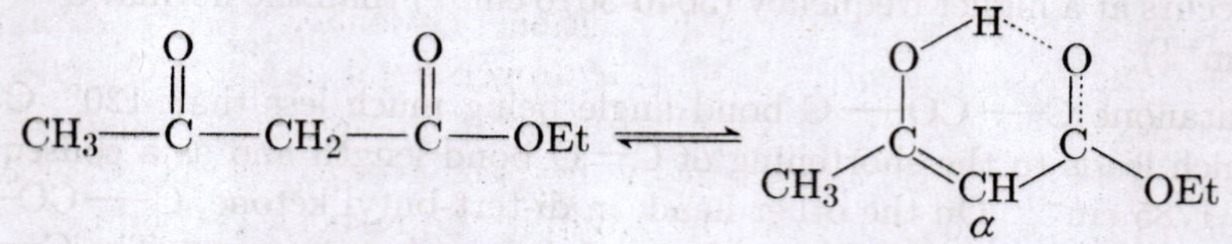

92% keto-form 8% enol-form

The absorption at 1724 cm^{-1} is the normal absorption for ketonic C═O str. The C═O str. of ester group occurs at 1748 cm^{-1} owing to the –I effect of the OEt group. The latter bond becomes weak in the enol form owing to the α, β-unsaturation as well as owing to the internal H-bonding; because of the lowering of bond strength, C═O str. in the enol form drops to 1650 cm^{-1} from 1748 cm^{-1}. The free O—H str. in an alcohol occurs at 3650 cm^{-1}; the enol form cannot exhibit this vibrational absorption simply because the chelation makes the O—H bond weaker than a free O—H bond and as a consequence the IR absorption occurs at a lower wave number, 3050 cm^{-1}.

Problem 10.23. **(i)** A monohalo ketone exhibits two closely spaced bands for C═O str. **(ii)** In α, α'-dihalo ketones C═O str. occurs at a higher frequency than that in monohalo ketones, but an α, α-dihalo ketone and the corresponding (α-monohalo ketone have almost identical C═O str. values—explain the observations in all the cases.

Solution: **(i)** An α-halo ketone may exist in two conformations—cis and gauche. In the cis-form the halogen atom and the C═O group lie in the same plane and perhaps owing to the field effect of the halogen atom C═O bond cannot acquire a partial double bond character through resonance. So in the cis-form C═O str. occurs at a higher frequency than the normal C═O str. On the other hand, in a gauche-form field effect cannot play its role but –I effect of the Cl atom may increase the partial double bond character of the C═O group and a slight increase in the C═O stretching frequency may be observed. Thus α-monohalo ketones exhibit two closely spaced bands, one for the cis-form and the other for the gauche form.

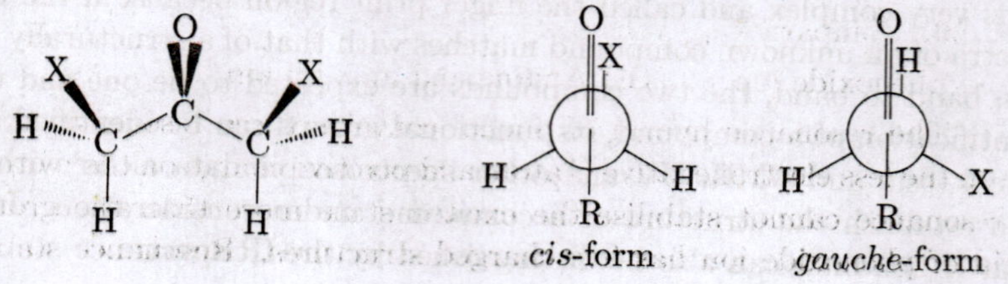

cis-form *gauche*-form

(ii) When α-and α'-halogen atoms are in the cis positions with respect to the C═O group, the field effect due to two halogen atoms stops C═O group to acquire partial double bond character and the effect is surely greater than the same effect in the cis-form of an α-monohalo ketone. So the C═O group in α, α'-dihalo compounds absorbs at a higher frequency than that in the monohalo compound. In α, α-dihalo compound two halogen atoms at a time cannot remain in the cis-conformation. So the field effect due to one halogen atom only can play its role and as a consequence the α-monohalo ketone and the corresponding α, α-dihalo ketone have almost identical C═O str. values.

Problem 10.24. The O—H str. value does not change with dilution in the case of ethylene glycol in carbon tetrachloride, but that increases in the case of ethyl alcohol on dilution—Explain.

Solution: Ethylene glycol has internal H-bond but EtOH forms intermolecular H-bond. When a solution of ethylene glycol is diluted with CCl_4, internal H-bond does not break and so the O—H str. does not change with dilution. In the case of EtOH, the intermolecular H-bond of the associated complex breaks with dilution. H-bonded O—H group has a greater oxygen-hydrogen bond length than the oxygen-hydrogen bond in a free O—H group. Associated EtOH molecules become free on dilution with carbon tetrachloride. So in the dilute solution of EtOH in CCl_4, the O—H str. occurs in the free OH group. The bond length of the latter being shorter than that of the H-bonded OH group in the associated complex, free O—H str. occurs at a higher frequency than the H-bonded O—H str. in the concentrated solution.

Problem 10.25. Compare the C═C str. in the following pair:

(i) $CH_3CH_2CH═CH_2$ and **(ii)** 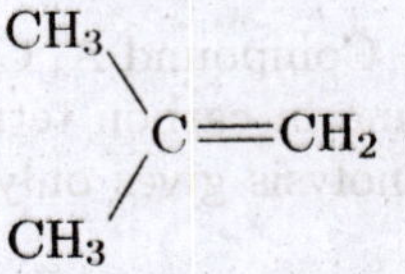

Solution: +I effect of gem dimethyl group as well as the hyperconjugative effect of two methyl groups make the C═C bond in (ii) acquire more single bond character than that in (i), where I-effect of one Et group and the hyperconjugation involving two H atoms take place.

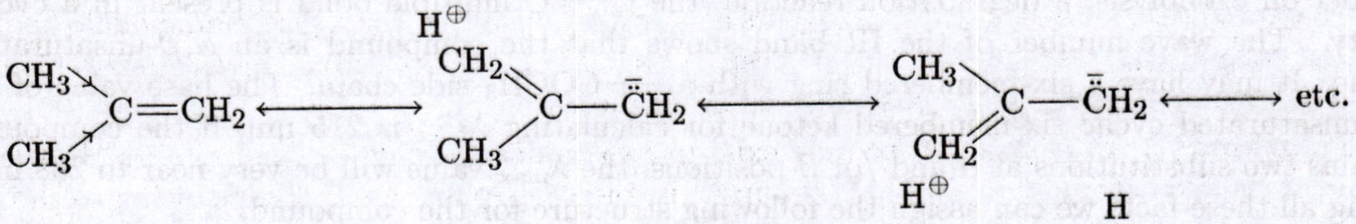

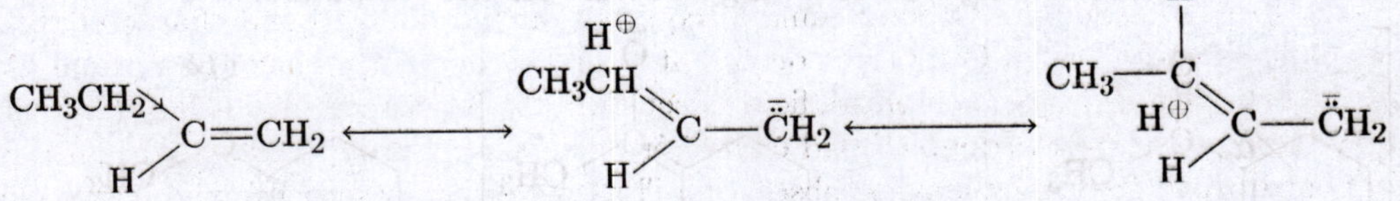

So, the C═C str. value in (ii) must be less than that in (i).

Problem 10.26. Compare λ_{max} values of :

(i) Phenol and phenoxide ion, **(ii)** Aniline and anilium ion.

Solution: (i) The resonance hybrid of phenol contains three resonating structures having negative charge on the less electronegative C atom and positive charge on the more electronegative O atom; thus resonance cannot stabilise the excited state more than the ground state. The resonance hybrid of phenoxide ion has five charged structures. Resonance stabilises a charged structure in the excited state more than the ground state structure. For this reason the energy difference between the ground state and the excited state of phenoxide ion will be less than the energy difference between the ground state and the excited state of phenol. The lower the absorption energy, the higher is the wavelength of the absorbed energy. Hence the λ_{max} of phenoxide ion will be greater than that of phenol.

(ii) The N atom of anilinium ion being tetracovalent, it has no lone pair of *es* on N atom; whereas the N atom of aniline has l.p *es*. They undergo p-π conjugation; because of the p-π conjugation, the excited state of aniline is more stable than the excited state of anilinium ion. So, the energy of UV absorption in the case of aniline is less than that in the case of anilinium ion. Therefore, the λ_{max} value of aniline is greater than that of anilinium ion. The latter shows spectra of benzene.

Problem 10.27. C≡N str. value of acrylonitrile is less than that of acetonitrile.—Explain.

Solution: p-π conjugation in acrylonitrile develops a partial double bond character in C≡N group; but CN group of acetonitrile does not experience such an effect. Therefore, the force constant of C—N triple bond in acrylonitrile is less than that in acetonitrile. So, the C≡N str. value of acrylonitrile is less than that of acetonitrile.

$$\underset{\text{Acrylonitrile}}{CH_2{=}CH{-}C{\equiv}N} \longleftrightarrow \overset{\oplus}{C}H_2{-}CH{=}C{=}\overset{-}{\ddot{N}}: \equiv \overset{\delta^+}{CH_2}\cdots CH\cdots C\cdots \overset{\delta^-}{N}$$

$$\underset{\text{Acetonitrile}}{CH_3{\rightarrow}C{\equiv}N:}$$

Problem 10.28. Compound A, $C_8H_{12}O$, shows IR band at 1686 cm^{-1} and λ_{max} at 238 nm. It decolourises bromine in carbon tetrachloride and gives iodoform test. However, one mole of compound A on ozonolysis gives only one mole of a single product. What is the structure of compound A?

Solution: Since compound A decolourises bromine in carbon tetrachloride, it has C—C multiple bond. Since the compound gives positive iodoform test, it contains a —$COCH_3$ or —$CHOHCH_3$ group. Again, since one mole of the compound gives only one mole of a single product on ozonolysis, a degradation reaction, the C—C multiple bond is present in a cyclic moiety. The wave number of the IR band shows that the compound is an α,β-unsaturated ketone. It may have a six-membered ring with a —$COCH_3$ side chain. The base value of an α,β-unsaturated cyclic six-membered ketone for calculating λ_{max} is 215 nm; if the compound contains two substitutions at α and /or β-positions, the λ_{max} value will be very near to 238 nm. Taking all these facts we can assign the following structure for the compound.

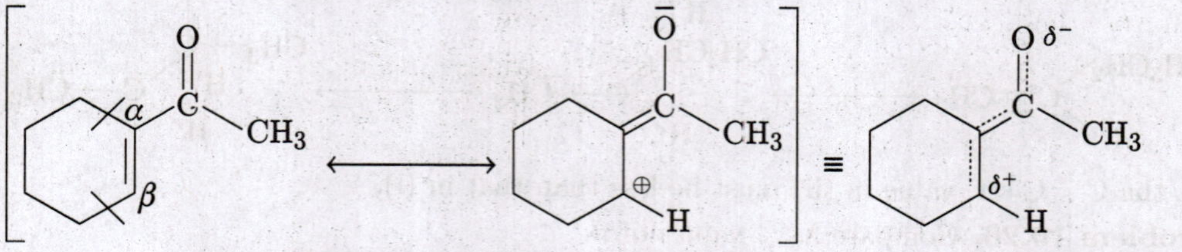

The normal C=O str. value for a ketone is 1725 to 1705 cm^{-1}; here the value is 1686 cm^{-1}. The low value of the C=O str. may be due to the π-π conjugation which lengthens C=O bond length. The above figure shows that there are two ring residues one at α-C and the other at β-C for them 10 nm and 12 nm respectively are to be added to 215 nm, the base value, to get the calculated λ_{max} and here it is 237 nm.

10.21 ORD and CD Curves

ORD and CD are the abbreviations of Optical Rotatory Dispersion and Circular Dichroism respectively. The word 'dispersion' means the scattering of values of a variable from the average. The specific rotation $[\alpha]_\lambda^T$ and the molecular rotation, $[M]_\lambda^T$ vary with the wavelength of light used though conventionally we measure the quantities using sodium light (D-line, 589 nm). However, the dependence of specific rotation or molecular rotation on wavelength is called optical rotatory dispersion (ORD).

When specific or molecular rotation is measured in a spectro-polarimeter, then this is not measured at a single chosen wavelength but over a wide spectral range. The results obtained are expressed in ORD curves which are drawn by plotting wavelengths along X-axis and specific or molecular rotations along Y-axis. ORD curves may be either plain or anomalous.

Plain ORD curves: ORD curves that ascend or descend from long to short wavelengths are called plain ORD curves. If ORD curves ascend with decreasing wavelength, they are called positive curves, while if they descend from long to short wavelength, we get negative ORD curves.

Cotton effect curves: Anomalous ORD curves are called cotton effect curves. The cotton effect is the name of the combined phenomena of circular birefringence and circular dichroism. A cotton effect curve has a peak and a trough. A cotton effect is said to be positive or negative, depending on whether a peak or trough is met first in going from long to short wavelength.

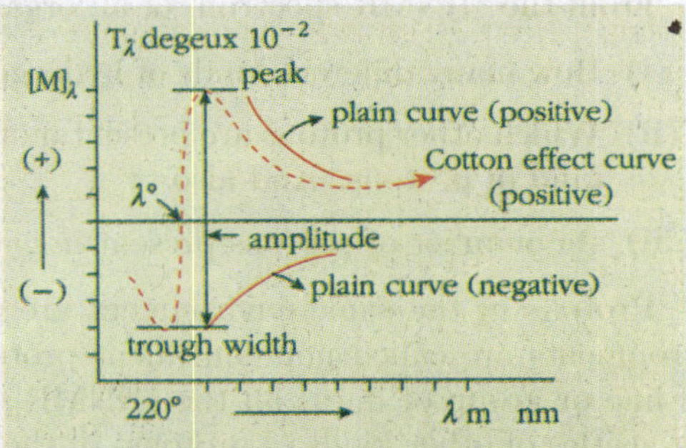

Fig. 10.26. ORD curves : Plain and Cotton

When an optically active compound contains a chromophore which absorbs in the region under study, the ORD curve of the compound is a cotton effect curve. The crossing point of the curve with the zero-line is designated as λ_0, which is the λ_{max} of the chromophore present in the compound. The difference between a peak and a trough is called the amplitude of the curve; it is a measure of the magnitude of the optical rotatory power of the compound. ORD curves give us a lot of information :

(i) An optically active substance may have positive specific rotation at one wavelength and negative specific rotation in another wavelength.

(ii) We can understand the nature of the chromophore and its position relative to the chiral center. λ_{max} can be determined from the curve.

(iii) The relative configurations of the dissymmetric centres of two compounds can be understood from the ORD curves. The two enantiomeric compounds have mirror-image ORD curves with opposite cotton effects.

(iv) We can also get an idea about the conformation of the compound provided an absolute configuration of the compound is known.

The variation of wavelength with circular dichroism is represented in a CD curve. When molecular ellipticity along Y-axis is plotted against wavelength along X-axis, we get a CD curve (Sec. 9.11.3). Nowadays we can measure CD at any wavelength more than 165 nm. The amplitude of the curve, the perpendicular distance from the peak to the zero-line, is a measure of λ_{max} of the optically active substance having a chromophoric group that absorbs under the region of study.

We get the same information from both the curves, ORD and CD as shown below.

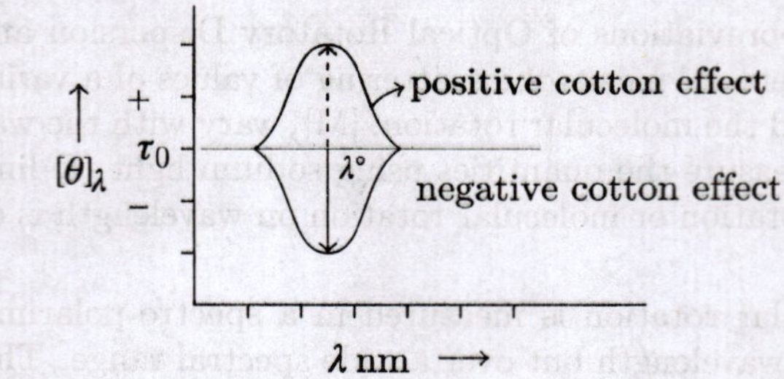

10.22 Proton Magnetic Resonance (PMR or ^{1}HNMR) Spectroscopy

From the ^{1}HNMR spectrum of an organic compound we know the following:

(i) How many different kinds of hydrogen (proton) environments are present in the molecules.

(ii) Which other protons are present on the neighbouring C aoms with respect to the particular kind of protons. And also

(iii) the number of protons present in each of these environments.

Protons in the same environment are called equivalent protons while those in different environments are called non-equivalent protons. For a kind of equivalent proton we get a signal (a line or group of lines) on the ^{1}HNMR spectrum. The relative intensity (with respect to the intensities of other kinds of protons) of the signal makes us know the number of protons that are present in the environment.

In *p*-xylene there are two kinds of non-equivalent protons:

(i) 4 H atoms (marked a) bonded to C atoms of the benzene ring,

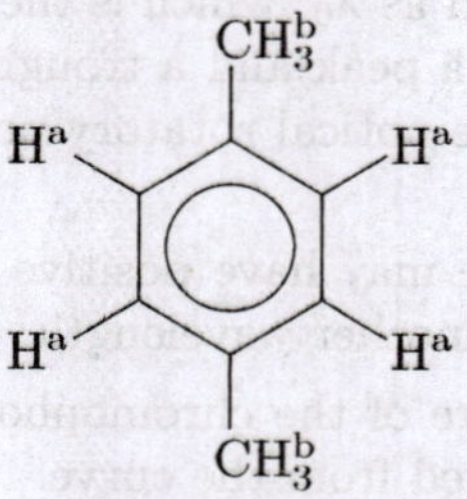

p-xylene

(ii) 6 H atoms (marked b) present in two methyl groups.

So, we shall get two signals on the ^{1}HNMR spectrum of *p*-xylene and the intensity ratio will be 4 : 6. Fig. 10.28 is the PMR spectrum of *p*-xylene.

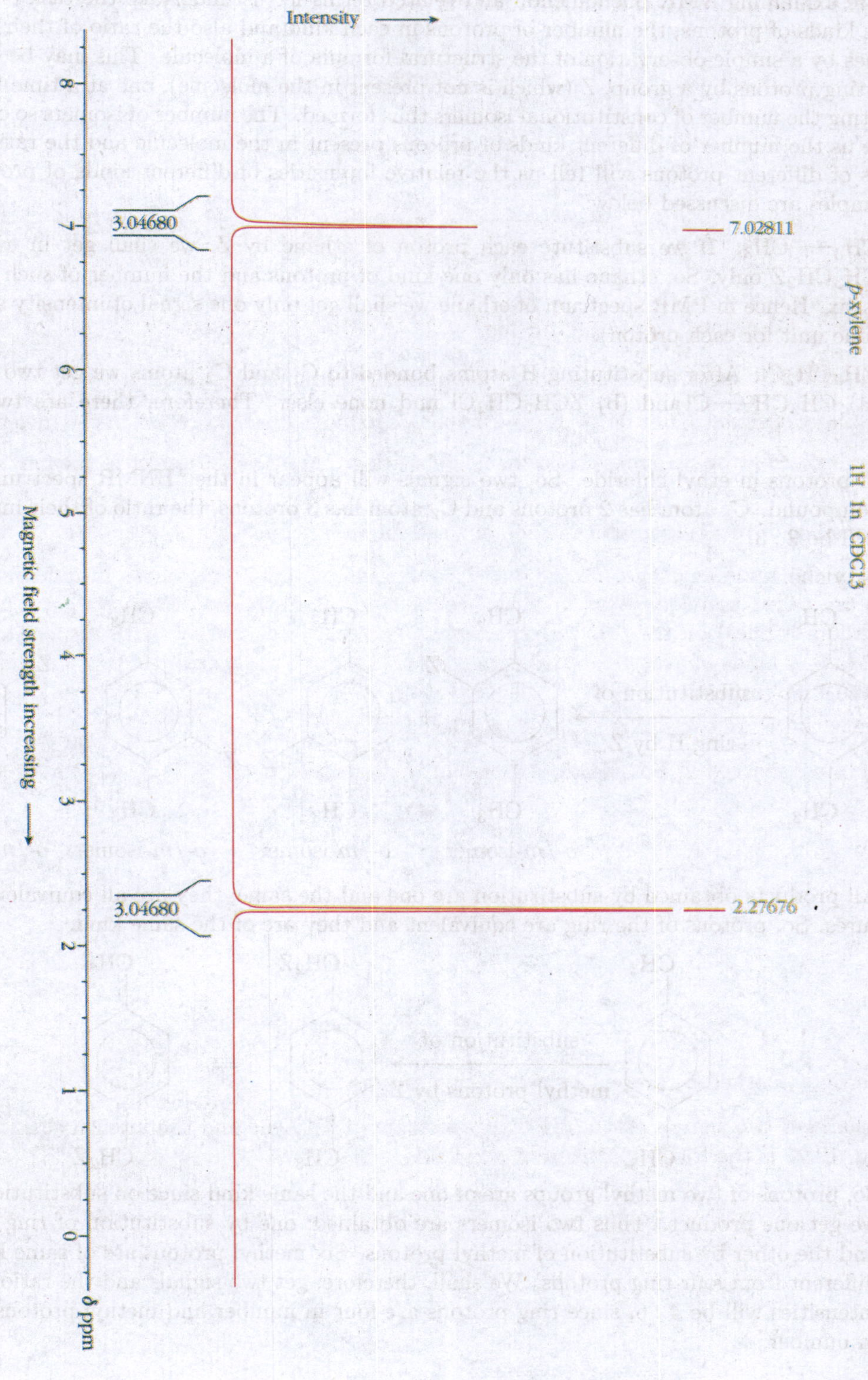

Fig. 10.28. ^{1}**HNMR spectrum of *p*-xylene (by coursey Dr. G. Mazumder (IICB))**

Before explaining NMR phenomenon and related terms let us undertake the task of finding different kinds of protons, the number of protons in each kind and also the ratio of their relative intensities by a simple observation of the structural formula of a molecule. This may be done by substituting protons by a group, Z (which is not present in the molecule), one at a time followed by counting the number of constitutional isomers thus formed. The number of isomers so obtained will give us the number of different kinds of protons present in the molecule and the ratio of the numbers of different protons will tell us the relative intensities of different kinds of protons. A few examples are discussed below:

(i) $CH_3—CH_3$: If we substitute each proton of ethane by Z, we shall get in every act CH_3CH_2Z only. So, ethane has only one kind of protons and the number of such protons is six. Hence in PMR spectrum of ethane we shall get only one signal of intensity six units (one unit for each proton).

(ii) CH_3CH_2Cl: After substituting H atoms bonded to C_1 and C_2 atoms we get two isomers (a) $CH_3\underset{\substack{|\\ Z}}{CH}—Cl$ and (b) ZCH_2CH_2Cl and none else. Therefore, there are two kinds of protons in ethyl chloride. So, two signals will appear in the ^{1}HNMR spectrum of the compound. C_1 atom has 2 protons and C_2 atom has 3 protons, the ratio of their intensities will be 2 : 3.

(iii) *p*-xylene:

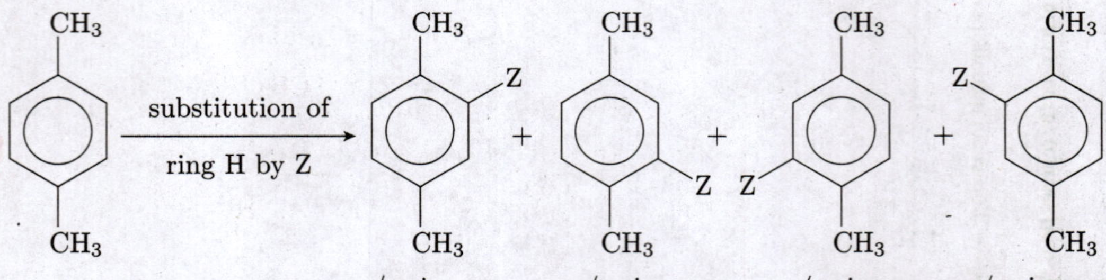

o-/*m*-isomer *o*-/*m*-isomer *o*-/*m*-isomer *o*-/*m*-isomer

All products obtained by substitution are one and the same; they are all equivalent structures. So, protons of the ring are equivalent and they are of the same kind.

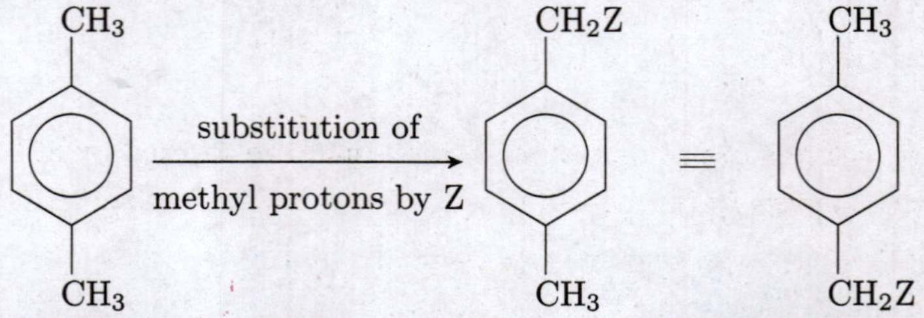

So, protons of two methyl groups are of one and the same kind since on substitution by Z, we get one product. Thus two isomers are obtained; one by substitution of ring protons and the other by substitution of methyl protons. Six methyl protons are of same kind but different from four ring protons. We shall, therefore, get two signals and the ratio of their intensities will be 4 : 6, since ring protons are four in number and methyl protons are six in number.

(iv) Propene

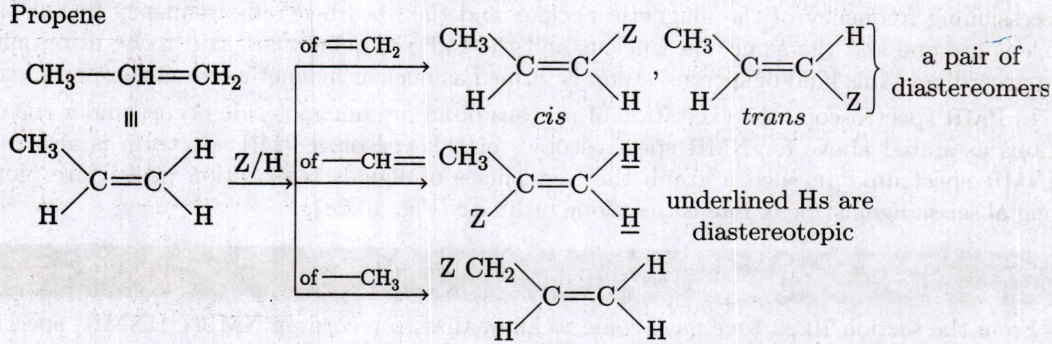

So, four isomers are obtained by substitution; hence there are four kinds of protons: two diastereotopic in $=CH_2$, one in —HC= and three in $—CH_3$. The compound will, therefore, give four ¹HNMR signals and their intensity ratio will be 1 : 1 : 1 : 3.

Problem 10.30. How many ¹HNMR signals do you expect and what will be their ratio of intensities for each of the following compounds?

i. $\overset{c}{CH_3}\overset{b}{CH_2}\overset{a}{OH}$ **ii.** $\overset{b}{CH_2Cl}\ \overset{a}{CHCl_2}$ **iii.** 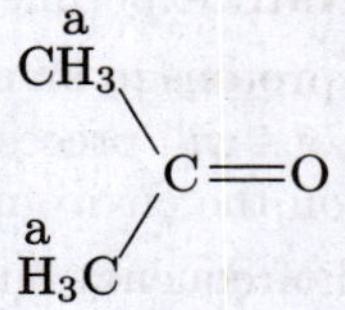 **iv.** 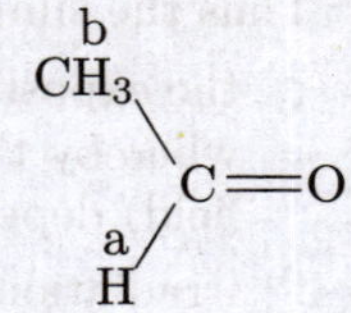

v. 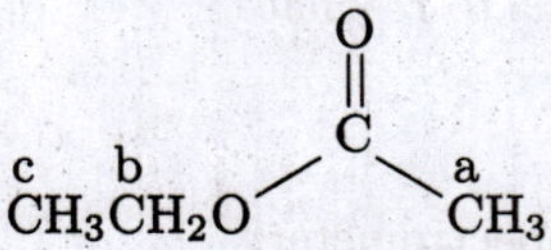 **vi.** 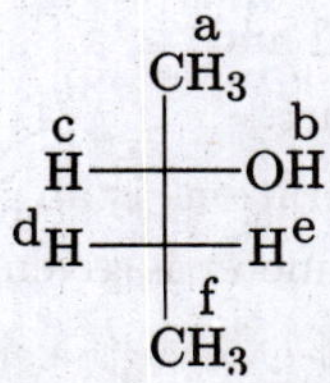 **viii.** 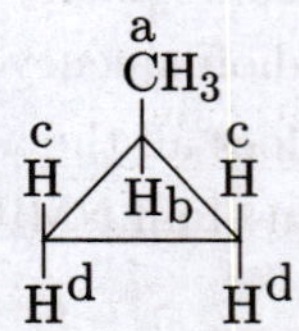

ix. 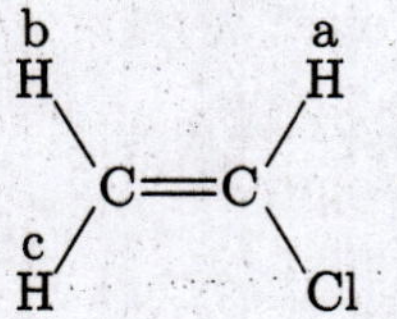 **x.** 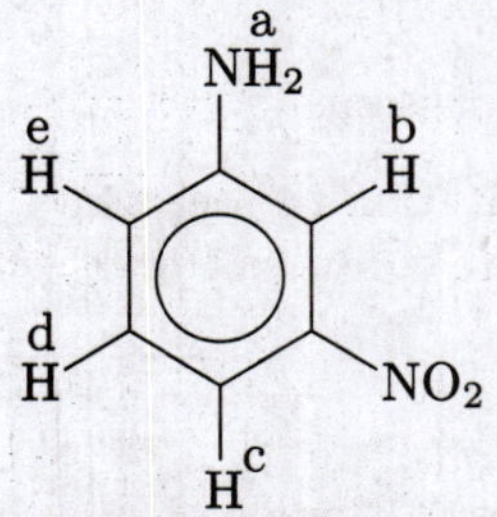

Hint: Different kinds of protons in each compound are marked by a, b, c, d, e and f. Try to get the isomers from each compound by the method of substitution.

10.22.1 Meaning of the term Nuclear Magnetic Resonance (NMR) spectroscopy

In NMR spectroscopy excitation of atomic nuclei occurs instead of electronic excitation (UV and visible spectra) and vibrational excitation (IR spectra). When a spinning nucleus in an appropriate magnetic field absorbs radio frequency which is exactly equal to the precessional frequency of the spinning nucleus, the excitation of the spinning nucleus takes place by absorbing energy and transits to a higher energy level from a lower energy level. We record the absorbed energy in the form of an NMR spectrum as it has been shown in Fig. 10.28 for *p*-xylene. Since the precessioning frequency of the magnetic nucleus and the absorbed radiofrequency by the same nucleus are one and the same, the nucleus and the radio frequency are said to be in resonance and accordingly this kind of spectral study is termed as nuclear magnetic resonance spectroscopy.

In PMR spectroscopy the excitation of protons of an organic molecule occurs under the conditions as stated above for NMR spectroscopy; for this reason a PMR spectrum is also called ^{1}HNMR spectrum. In such a graph the frequencies of signals (absorption peaks) are plotted along abscissa against peak intensities along ordinate (Fig. 10.28).

10.22.2 NMR experiment and spectrometer

From the section 10.22.1 we have come to know that to record an NMR (^{1}HNMR) spectrum of a compound we require:

(i) spinning nuclei in the molecule (protons in the case of ^{1}HNMR).
(ii) a magnetic field to be applied perpendicular to the molecule.
(iii) a radiofrequency source for absorption by the spinning nuclei (proton) in the molecule.

Thus the simple ^{1}HNMR experiment involves:

(i) the exposition of protons in an organic molecule to an appropriate external magnetic field; whereby the protons will precess in the same or different frequencies (proportional to the field) depending on the environment of the protons.
(ii) irradiation of radiofrequency energy on the precessing protons to resonate.
(iii) detection of the radiofrequency absorbed and
(iv) recording of the radiofrequency on a chart.

The instrument that does all the above operations is an NMR spectrometer.

The schematic diagram of an NMR spectrometer is given below:

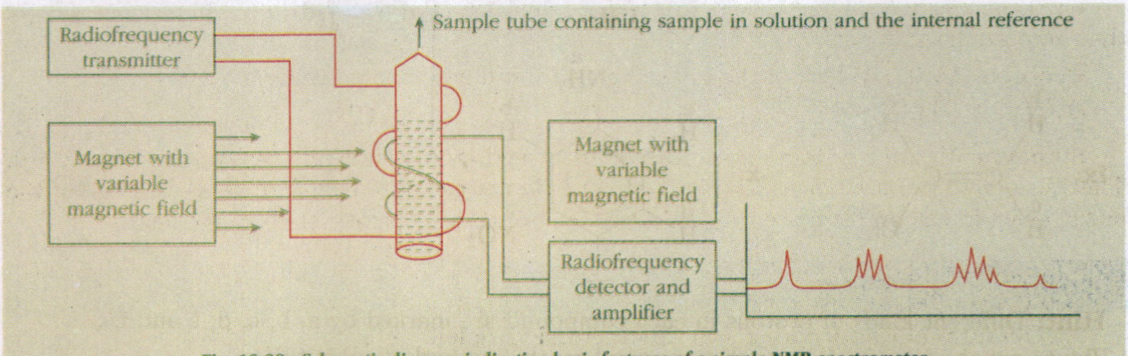

Fig. 10.29. Schematic diagram indicating basic features of a simple NMR spectrometer

NMR spectrometers are calibraled in frequency unit (Hz). The machine of different radiofrequencies are available, 60 MHz, 100 MHz, etc. Each kind of protons has a characteristic precessional frequency which resonates with radiofrequency of the source which is kept transmitting steady 60 MHz or 100 MHz or anything else. So different kinds of protons resonate with different radiofrequencies. This variation gives rise to different absorption position generally within the range of about 750 Hz in a magnetic field of 60 MHz. The range is within 3750 Hz in a 300 MHz spectrometer. For this reason spectrometers have variable electromagnetic coils. The difference in absorption radiofrequency is very small and high resolution instruments are able to detect the difference in one part per 100 million, i.e., 1 ppm in 10^8 parts (in Hz or cps). As the radiofrequency (ν) source is kept steady, the applied magnetic field (H_0) is increased ($\nu \propto H_0$). As the strength of the magnetic field increases, the precessional frequency of each proton increases until its precessional frequency rasonates with the radiofrequency of the operating machine; the absorbed frequencies are not only detected but also amplified. The amplified frequency draws out NMR spectrum on a chart by plotting the positions of the absorption peak signals along abscissa against intensity of the peaks along ordinate. The intensities of each peak are measured by electronic integrators which draw continuous lines from left to right and from signal to signal indicating steps. The ratio of the step-heights is the ratio of peak areas of different kinds of proton. In fig. 10.28 peak areas are measured in a different way. The value for the ring protons is a bit less giving 13.1% error. 2% error is accepted.

10.22.3 A magnetic property of nuclei

Every atomic nucleus is associated with charge. In some of nuclei this charge spins on the nuclear axis; consequently the circulating charge develops a magnetic dipole along the axis and they act as a tiny bar magnet with a magnetic moment μ. These are magnetic nuclei. The angular momentum of the spinning nucleus is described in terms of spin quantum number I, which has any value of $0, \frac{1}{2}, 1, \frac{3}{2}, \frac{5}{2}$, etc. The numerical value of the spin quantum number of a nucleus is related to its mass number and atomic number as follows:

Mass Number	Atomic Number	I (Spin quantum member)	Example
Odd	Odd or even	$\frac{1}{2}, \frac{3}{2}, \frac{5}{2}$, etc.	${}_1H^1\ (\frac{1}{2})$, ${}_8O^{17}\ (\frac{5}{2})$, ${}_7N^{15}\ (\frac{1}{2})$
Even	Even	0	${}_6C^{12}(0)$, ${}_8O^{16}(0)$, ${}_{16}S^{34}(0)$
Even	Odd	1, 2, 3, etc.	${}_1H^2(1)$, ${}_7N^{14}(1)$, ${}_5B^{10}(3)$

I = 0 denotes no spin and the nucleus is non-magnetic. The nuclei with a spin-number greater than zero are magnetic nuclei.

Problem 10.31. Which of the following nuclei will generate a magnetic field?

(i) ${}_6C^{13}$ **(ii)** ${}_{16}S^{32}$ **(iii)** ${}_{15}P^{31}$ **(iv)** ${}_9F^{19}$ **(v)** ${}_6C^{12}$ **(vi)** ${}_1H^2$

Solution: (ii) and (v) have both mass number and atomic number even, so they have I = 0. Therefore these two are non-magnetic nuclei and do not generate magnetic field. The rest of the nuclei given in the problem are magnetic and each of them generates a magnetic field.

10.22.4 The effect of applied magnetic field (external) on a magnetic nucleus

(i) Orientation and transition of magnetic nuclei

If a magnetic nucleus is placed in a uniform magnetic field (H_0), the nucleus will respond to the influence of the external applied magnetic field and the magnetic nucleus (dipole) assumes only a discrete set of orientations and the system is said to be quantised. The number of orientations will be (2I + 1) with respect to the direction of the applied field. I of ${}_1H^1$ being $\frac{1}{2}$, protons assume $(2 \times \frac{1}{2} + 1)$ or 2 orientations : (i) either

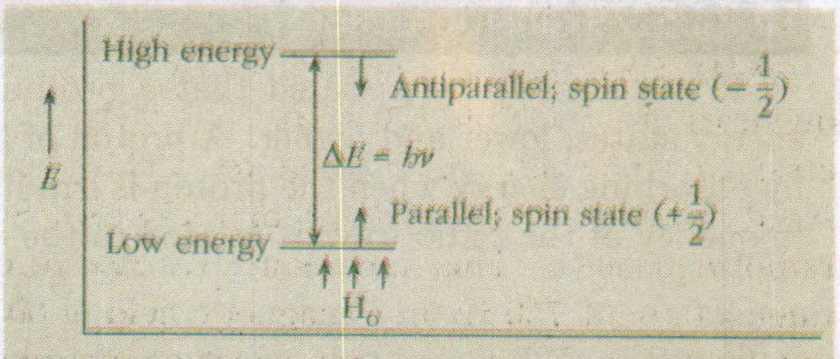

Fig. 10.30. ΔE transition from low energy state to high energy state

against the direction of the external field, called antiparallel to the direction of the applied field (H_0); this orientation is of higher energy state and (ii) or aligned with, i.e., parallel to the direction of the applied field (H_0); this orientation is of lower energy state.

Problem 10.32. How many orientations are possible when the following nuclei are placed in a magnetic field? **(i)** $_6C^{13}$ **(ii)** $_9F^{19}$

Solution: Both have $I = \frac{1}{2}$; hence there will be $\left(2 \times \frac{1}{2} + 1\right)$ or 2 orientations for each nucleus.

(ii) Precessional motion of the magnetic nuclei

The spinning magnetic nucleus (proton) not only orient itself aligned against or with the direction of the applied field but also takes up a precessional motion in a way that its rotating magnetic axis draws out a circle perpendicular to the direction of the applied field. The precessional motion of the magnetic nucleus may be compared with the gyroscopic motion of a top which precesses about the vertical axis of the earth's gravitational force when the top spins with an initial axis of rotation different from the gravitational field acting downward.

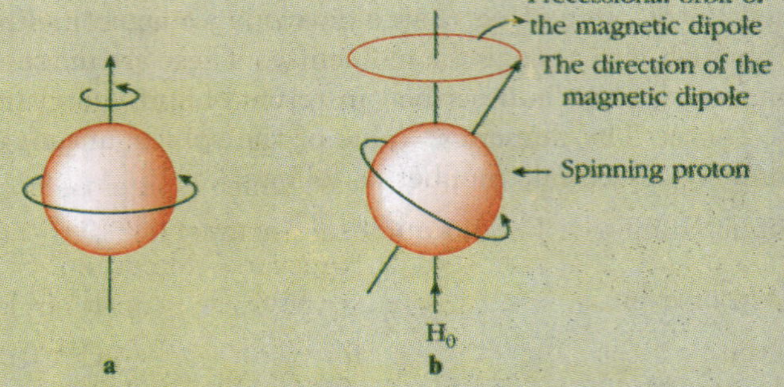

Fig. 10.31. a. Generation of magnetic moment owing to spinning charge in proton
b. Precessional motion of a proton in an applied magnetic field H_0

However, it is important to note that the spinning frequency of a magnetic nucleus does not change but precessional speed changes. The precessional frequency ν is directly proportional to the applied external field,

$$\nu \propto H_0$$

as the field strength increases, the precessional frequency of nucleus increases.

For example, a proton in an external magnetic field of 1.4 Tesla (T) [1] which is equal to 14 000 Gauss (G) precesses around sixty million cycles per second (cps); so that $\nu \approx 60$ Mc or 60 MHz. The same proton in the applied field 2.3T will have the precessing frequency $\nu \approx 100$ Mc or 100 MHz.

10.22.5 Transition energy

It has already been stated that a proton in a homogenous magnetic field can take up two energy states, lower and higher. A proton of lower energy state can go to the higher energy state by absorbing energy when the proton is irradiated with radio wave of appropriate frequency. The frequency of the radio wave is related to the energy change by the usual equation,

$$\Delta E = h\nu$$

[1] Tesla is an SI term and Gauss is a CGS unit. $1T = 10^4G$. Hz is a modern unit of frequency whereas cps is an old one.

We know that the different kinds of proton have different precessional frequencies. A particular proton absorbs that radio frequency which is exactly equal to the precessional frequency of the very proton. As soon as the absorbed radio frequency resonates with the precessional frequency of the proton, the proton 'flips' from the lower energy state to higher energy state and we get a signal in ^{1}HNMR spectrum (Fig. 10.30).

10.22.6 Saturation of the system

At the outset of ^{1}HNMR experiment, the lower energy state is more populated than the excited state (higher energy state) to the extent of 0.001%. It is this negligible but finite excess of protons in the lower energy level that helps recording net absorption of radiofrequency energy. As the process of absorption of radiofrequency energy goes on, the higher energy state becomes more and more populated than at the outset of the experiment and a time comes when there are either a few molecules or no molecules with precessing protons remain in the lower energy state for flipping up. As a consequence either a weak signal or no signal is observed in ^{1}HNMR spectrum. This is called the saturation of the system. The system, however, comes down from saturation by emitting energy by the higher energy molecules in two ways: (a) by spin-spin relaxation and (b) by spin-lattice relaxation. Relaxation here means giving up of energy.

(a) Spin-spin relaxation: The higher energy nucleus could come down to the lower energy state by transferring energy to a neighbouring nucleus, which is in close proximity to the former, whereby mutual exchange of spin energy takes place. Thus the higher energy nucleus gets relaxed and the process of relaxation is called spin-spin or transverse relaxation. Since the higher energy proton gives up energy to the other, net energy of the system does not change. This precess shortens the life time of the spinning nucleus in the higher energy state; as a consequence the spin-spin relaxation practically does not contribute to the maintenance of the required excess of nuclei in the lower energy state. If it contributes, we get broad line in the spectrum.

(b) Spin-lattice relaxation: By lattice we mean the total surroundings of a precessing nucleus under experiment. The surrounding consists of molecules of the sample, solvent and any other things. All these molecules are in constant motions of several types and they, too, have magnetic properties. So, the lattice contains a wide and varied number of magnetic fields. A properly oriented small magnetic field in the lattice is able to induce a transition in a particular precessing nucleus from its higher energy state to the lower one. The energy being given up to the lattice by the spinning nucleus, the process is called spin-lattice relaxation or longitudinal relaxation. The nucleus, in this way, comes down to the lower energy state and population of the molecules in lower energy state becomes excess over the higher energy state which is a necessary condition for observing NMR signals. Here also the total energy of the system remains unaltered. By spin-lattice relaxation the precessing nucleus gets a longer life than it would have in spin-spin relaxation and we expect a sharper line by the process of spin-lattice relaxation than that by spin-spin relaxation.

10.22.7 The effect of external magnetic field on the environment of protons

H atoms in a molecule are not bare protons though they are called so in PMR. They are bonded to other atoms by σ bond. In a magnetic field not only the protons spin but also σ and π electrons circulate and the circulating electrons develop induced magnetic fields. The induced field may oppose or reinforce the external magnetic field. When the induced field opposes the applied field, it is called antiparallel to or aligned against the applied field; reinforcing of field by the induced field means the induced field is aligned with or parallel to the applied field. If the induced field is aligned against the applied field, the proton under consideration needs a higher field strength for flipping up than what it would have required if there were no opposing field; in this case the proton is said to be shielded. An extra magnetic field is required to overcome the effect of the opposing induced magnetic field. On the other hand, when the induced magnetic field is aligned with, i.e., reinforcing the applied magnetic field, then proton flips up at a lower strength because of the reinforcing effect of the induced magnetic field, this effect is called the deshielding effect. Let us now consider three cases (Fig. 10.32) pictorially.

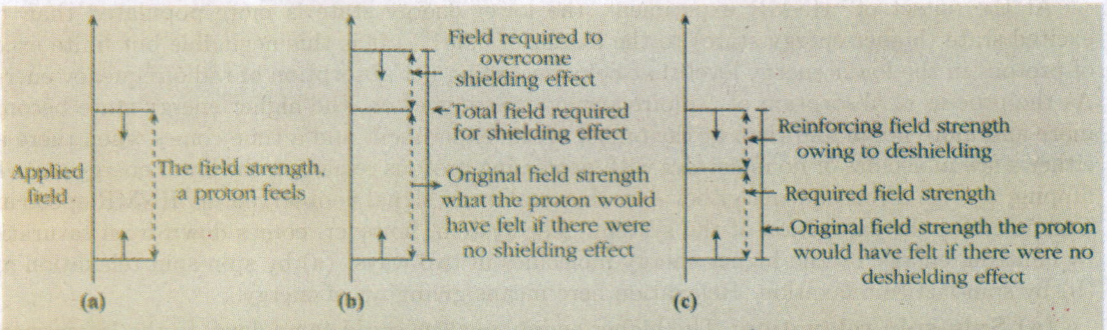

Fig. 10.32. (a) A proton without shielding or deshielding effect (b) the proton sensing shielding effect (c) the proton sensing deshielding effect

The circulation of σ-electrons develops an induced magnetic field which is antiparallel with the applied field. (Fig. 10.33)

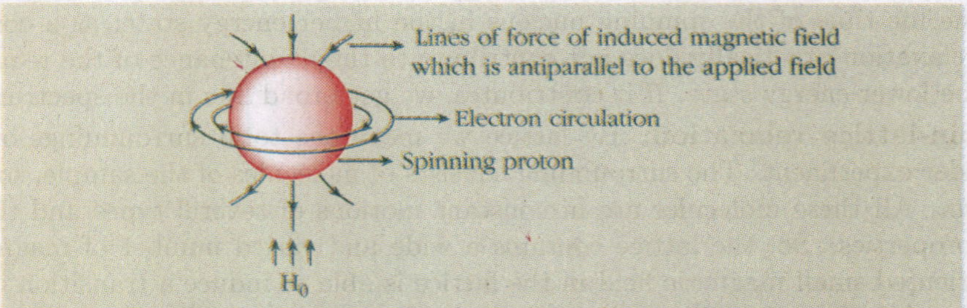

Fig. 10.33. Circulation of σ-electrons that develops induced magnetic field

(a) Shielding and deshielding effects owing to circulation of σ electrons.

(i) Shielding effect

A C—H σ bond may be electron rich or deficient owing to the presence of electron-pushing or electron-withdrawing groups respectively on the C atom. Groups with +I effect make the C—H σ bond electron dense. The induced magnetic field generated by the circulation of σ *es* becomes more and more strong as the +I effect increases. Thus the proton of the C—H gets more and more shielded. To overcome the shielding effect the applied field strength is to be increased. In tetramethylsilane (Me_4Si) Si atom has +I effect, as a consequence methyl protons flip at such a high magnetic field that usually most of the protons of the organic compounds flip below this field strength.

(ii) Deshielding effect

On the other hand, electron withdrawing groups (−I or −*R* groups) bonded to C atom make the C—H σ bond electron deficient and deshield the proton because the induced magnetic field developed by the circulation of σ electrons is weak. Consequently, the applied field strength required to flip the proton is low. H atoms of —OH, —NH_2 —SH, —CO_2H groups and of C—H moiety which holds these groups (including F, Cl, Br and I) absorb at a lower applied magnetic field.

(b) Anisotropic shielding and deshielding effects owing to the circulation of π electrons

The effects owing to circulation of π electrons are anisotropic, i.e., different in different directions, in nature. There are regions where the induced magnetic field produced by the circulation of π es opposes (diamagnetic effect) the applied field, while in other regions the same reinforces (paramagnetic effect) the applied field. The protons which come under the diamagnetic region get shielded and those under the paramagnetic region get deshielded. The former protons absorb at a higher applied magnetic field, whereas the latter protons do so at a lower field. These effects are shown below for alkenes, alkynes, aldehydes and aromatic compounds.

(i) Alkenes

Fig. 10.34 tells us that protons bonded to π-bonded C atoms feel a paramagnetic deshielding effect since the lines of the force of induced magnetic field are aligned with the applied field in

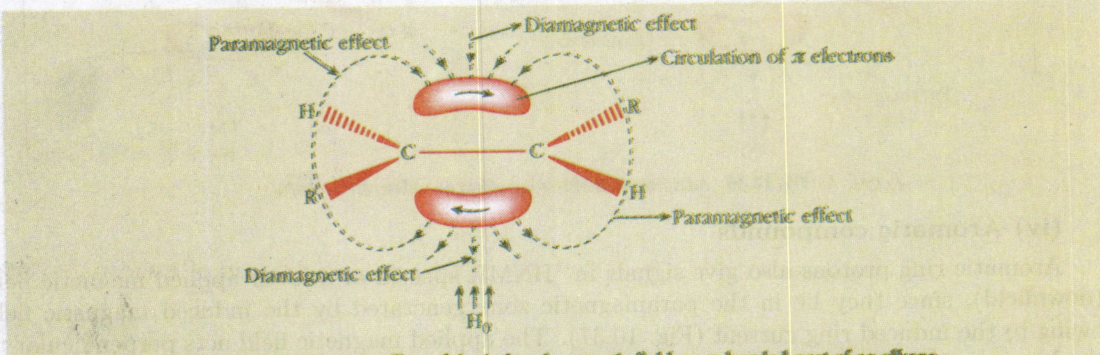

Fig. 10.34. *Anisotropic effect of the induced magnetic field on π-bonded part of an alkene*

regions where protons are located. So, they absorb radiofrequencies at a lower magnetic field than protons of alkanes. On the other hand, any proton held below or above the plane of the π bond will sense a shielding effect since lines of force of the induced magnetic field are diamagnetic in these areas. Protons of one of the two geminal methyl groups of α-pinene are present within the diamagnetic or shielding zone. Those protons flip at a higher magnetic field than the protons of the other methyl group.

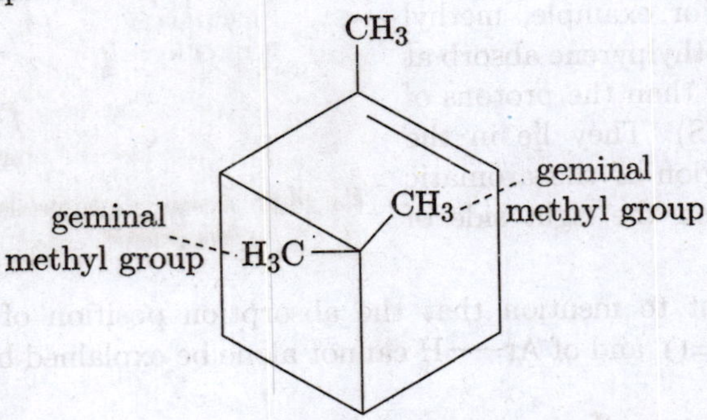

α-Pinene

(ii) Alkynes

The protons bonded to —C≡C— sense a diamagnetic effect (shielding effect) of the induced magnetic field generated by the circulation of π es (Fig. 10.35). As a consequence, an alkyne proton gives an ^{1}HNMR signal at a higher applied magnetic field than a proton bonded to C=C of an alkene.

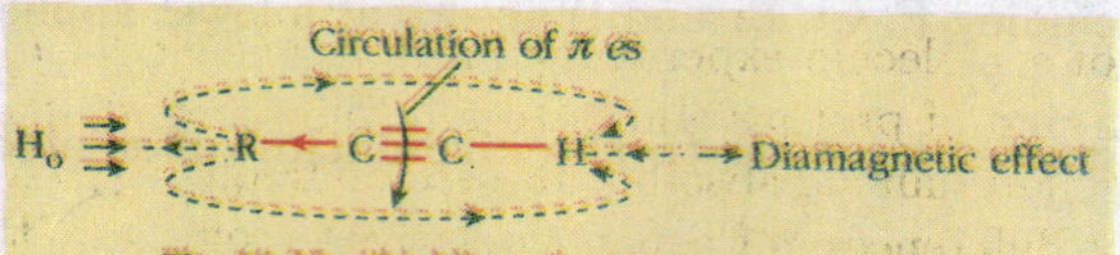

Fig. 10.35. Shielding of a proton on —C≡C—

(iii) Aldehydes

Aldehydic protons and protons of formic esters lie in the deshielding region (paramagnetic region) when the applied magnetic field acts perpendicular to the plane of the trigonal carbonyl C (Fig. 10.36). So, the aldehydic proton absorbs at a lower field (downfield).

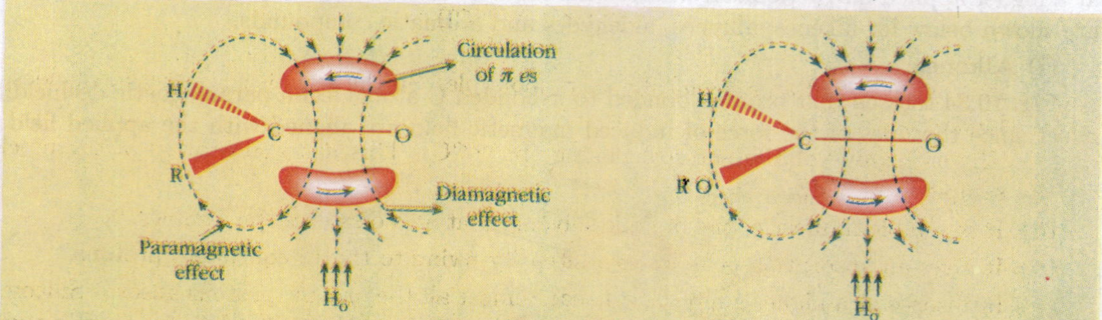

Fig. 10.36. Anisotropic deshielding effect on aldehydic protons

(iv) Aromatic compounds

Aromatic ring protons also give signals in ^{1}HNMR spectra at a lower applied magnetic field (downfield), since they lie in the paramagnetic zone generated by the induced magnetic field owing to the induced ring current (Fig. 10.37). The applied magnetic field acts perpendicular to the benzene ring. So the aromatic ring protons get deshielded.

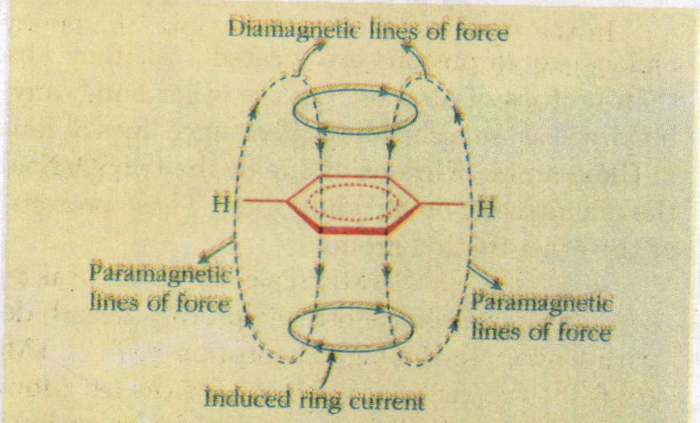

Fig. 10.37. Anisotropic deshielding effect that plays on the aromatic ring protons

It is needless to mention that any proton of the side chain that comes in the diamagnetic region owing to steric reasons will experience a strong upfield effect, i.e., will absorb at a higher field than the aromatic ring protons. For example, methyl protons of dihydrodimethylpyrene absorb at a higher magnetic field than the protons of tetramethylsilane (TMS). They lie in the strong diamagnetic region of the aromatic ring and show signal on the right side of TMS signal.

Here it is important to mention that the absorption position of protons of C C—H, C≡C—H, H—C(=O)— and of Ar—H cannot alone be explained by the electronegativity of bonded carbons; anisotropic effects are the major cause.

10.22.8 Scale for determining absorption positions (signals) in an ^{1}HNMR spectrum and the chemical shift

We have come to know that the protons of a molecule experience shielding and deshielding effects depending on the environment of each kind of protons and naturally they have different radiofrequency absorption positions. Instead of determining absolute absorption frequency, which is difficult to determine and rarely required, the differences in frequency relative to some standard reference group of protons are measured. Universally accepted standard reference is the protons of TMS (tetramethylsilane). Definitely the choice is not arbitrary. The compound has many an advantage.

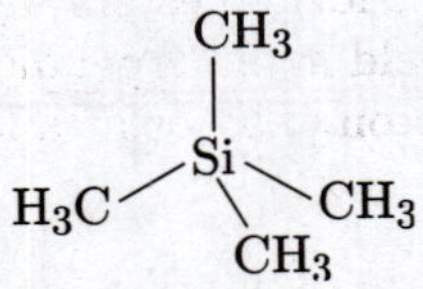

(i) It is symmetrical and chemically inert molecule. So the sample for experiment does not interact.

(ii) This is highly volatile and low boiling (bp 27°C). This helps separation of the precious sample from TMS.

(iii) It is soluble in most of the organic solvents that also dissolves the samples.

(iv) It gives an absorption peak of high intensity owing to the 12 equivalent protons.

(v) It absorbs at a higher applied field than almost all the organic protons absorb. Silicon has +I effect, it strongly shields the methyl protons because the molecule is symmetric and the shielding effect is the cause for absorption at a very high field.

(vi) It is generally used as an internal standard of reference; this means the sample, TMS and solvent together form a homogeneous solution.

(vii) It is insoluble in water or deuterium oxide. When aqueous solution of the sample is used, TMS is used as external reference in a concentric capillary.

In the aqueous solution of sample, the methyl protons of sodium 2,2-dimethyl-2-silapentane-5-sulphonate (DSS) are sometimes used.

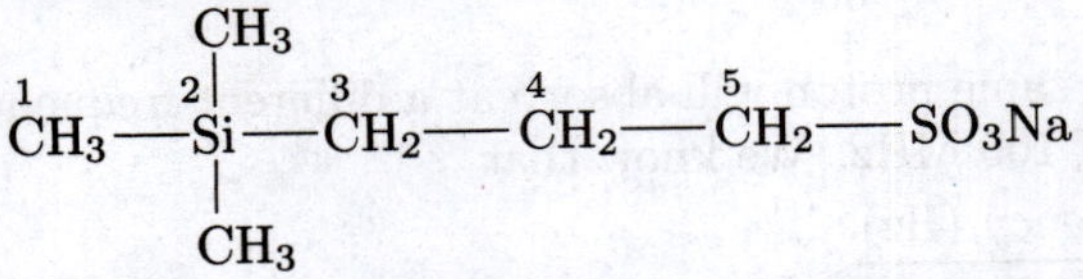

Sodium 2,2-Dimethyl-2-silapentane sulphonic acid (DSS)

In setting up ^{1}HNMR scale, the TMS peak is kept at 0 Hz at the right hand edge of the chart so that almost all the organic protons absorb downfield and signals appear at the left hand side with respect to the high absorption peak of TMS. The magnetic field increases toward the right (upfield) and that decreases toward the left (downfield). Absorption of radiofrequency at a higher magnetic field indicates more shielding while that at a lower field indicates more deshielding. However, **the difference in absorption position of a particular proton from the absorption position of TMS (reference) proton is called the chemical shift of the particular proton**. Different kinds of protons have different absorption positions and so they have different chemical shifts.

Since spectra are calibrated in frequency unit, Hz (cps), the chemical shift is also expressed in frequency unit (Hz or cps) and the radiofrequency of the transmitter, i.e., the operating frequency of the instrument is specified, because the chemical shift is dependent on the operating frequency

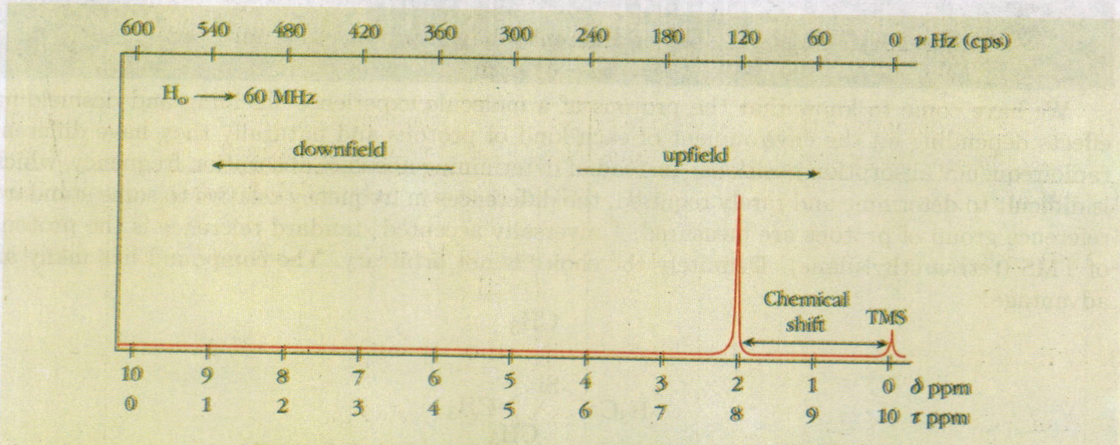

Fig. 10.38. Scales for measuring proton chemical shifts in HNMR spectra

of the spectrometer. Let us consider a kind of protons in a molecule that absorbs at 120 Hz in an NMR spectrometer of 60 MHz frequency. Since TMS protons absorb at zero Hz, the chemical shift of the proton is 120 Hz; chemical shifts may also be reported in dimensionless units which are independent of the applied field strength or the operating frequency of the machine. There are two such units in use. (i) δ unit, parts per million (ppm) and (ii) τ (tau) unit, parts per million (ppm).

$$\delta = \frac{\text{chemical shift (Hz)}}{\text{Transmitter frequency (Hz)}} \times 10^6$$

$$\text{or,}\quad \text{chemical shift (Hz)} = \frac{\delta \times \text{transmitter frequency Hz}}{10^6}$$

For example, the proton that absorbs radiofrequency 120 Hz in an NMR spectrometer of 60 MHz frequency will have $\delta = 2$ as calculated below:

$$\delta = \frac{120\text{Hz}}{60 \times 10^6\text{Hz}} \times 10^6 = \frac{120}{60} \text{ or } 2.$$

δ being independent of operating frequency, the same proton will absorb at a different frequency in a spectrometer having operating frequency of 100 MHz. We know that

$$\text{chemical shift (Hz)} = \frac{\delta \times \text{ transmitter frequency (Hz)}}{10^6}$$

$$= \frac{2 \times 100 \times 10^6 \text{ Hz}}{10^6} = 200 \text{ Hz}.$$

So, the proton that absorbs at 120 Hz in 60 MHz spectrometer has absorbed at 200 Hz in 100 MHz instrument.

In the δ-scale the absorption by the TMS is assigned 0 ppm; while in the τ- scale the same is assigned 10 ppm. Thus, $\tau + \delta = 10$ or $\tau = 10 - \delta$. Increasing δ-value indicates increasing deshielding, whereas increasing τ- value tells increasing shielding. The higher the δ-value, the lower is the field strength; the lower the δ-value, higher is the strength of the applied field. In the cases of τ- values, the increasing and decreasing effects are reverse. However, some of the proton chemical shefts are listed in Table 10.10

Table 10.10 Chemical shift of different kinds of protons in δ unit, ppm

$\tau = (10 - \delta)$ ppm

Sl. No.	Environment of the proton underlined	Range of δ, ppm	Sl. No.	Environment of the proton underlined	Range of δ, ppm
1.	1°H of R $C\underline{H}_3$	0.9	13.	α-H of R—O—$C\underline{H}_2$R (ether)	3.0–3.4
2.	2°H of $R_2C\underline{H}$	1.3	14.	α-H of $\underline{H}$—C(—)(—)—CO_2H (carboxylic acid)	2–2.6
3.	3°H of R_3C—$\underline{H}$	1.5	15.	α-H of $\underline{H}$—C(—)(—)—C(=O)—R (ketone)	2–2.7
4.	H of >C=C(—$\underline{H}$)—$C\underline{H}_3$	1.7	16.	α-H of alkyl O in R—C(=O)—O—C(—)(—)—$\underline{H}$ (ester)	3.7–4.1
5.	H of $RC\underline{H}_2I$	2.0–4.0	17.	α-H of $\underline{H}$—C(—)(—)—C(=O)—O—R (ester)	2–2.2
6.	H of $RC\underline{H}_2Br$	2.5–4.0	18.	H of (H)(R)C=O (aldehydic)	9.0–10.0
7.	H of $RC\underline{H}_2Cl$	3–4.0	19.	H of $RN\underline{H}_2$ (amine)	1.0–5.0
8.	H of RCH_2—F	4.0–4.5	20.	H of R—O—$\underline{H}$ (alcoholic H)	1.0–5.5
9.	H of >C=C<($\underline{H}$)	4.6–5.09	21.	H of R—C(=O)—O—$\underline{H}$ (carboxylic H)	10.5–12
10.	H of —C≡C—$\underline{H}$	2.0–3.0	22.	H of —C=C—O—$\underline{H}$ (enolic H)	15–17
11.	H of Ar—$\underline{H}$	7–8.5	23.	H of Ar—O—$\underline{H}$ (phenolic H)	12.0–4.0
12.	H of C_6H_5—C$\underline{H}$ (benzylic)	2.2–3.0	24.	H of cyclopropane—$\underline{H}$ (cyclopropane)	0.2
			25.	H of TMS (reference)	0.0

The factors are : (i) Electronegativity and I effects (ii) Anisotropic effects (iii) van der Walls deshielding effect and (iv) H-bonding effect.

(i) Electronegativity and I effect

Higher electronegativity of the bonded atom to proton and groups having –I effect deshield a proton. As the electronegativity of the groups increases, the deshielding effect increases and the proton chemical shift occurs more and more downfield with the higher δ-value. If we give a close look on Table 10.10, we shall see that amongst halides proton chemical shift order in δ-scale (Sl. nos. 5–8) is F > Cl > Br > I; this is also the order of decreasing electronegativity of the halogen atoms. −I effect deshields more an α-H than a β-H and so on. In CH_3CH_2Cl a methylene H absorbs at a lowerfield (higher δ-value) than the methyl H. Proton adjacent to $N^{\oplus}$ is strongly deshielded and they have high δ-values owing to the greater electronegativity of cationic N than that of a neutral N.

+I effects shield a proton and make is signal appear at a low δ-value. Because of the +I effect of Si, the protons of TMS are highly sheilded and they absorb at $\delta = 0$. A proton very close to anionic C, as in the methyl group of $-\overset{\ominus}{\underset{|}{C}}-CH_3$, gets strongly shielded and their δ-value is very small.

(ii) Anisotropic effects

These have already been discussed in Section 10.22.7(b) under the heading shielding and deshielding effects owing to the circulation of π-electrons. The order of deshielding effect is aldehydic proton > aromatic proton > alkene proton; whereas acetylinic protons get shielded and appear at an upfield.

Therefore, the order of δ-values is aldehydic proton > aromatic proton > alkene proton > alkyne proton.

(iii) van der Waals deshielding effect

Protons in some molecules have crowded environment owing to the presence of bulky groups in their neighbourhood. The electronic charge cloud of those bulky groups repel the electrons surrounding the crowded protons. As a consequence a slight deshielding effect of the protons is generated and they appear at a bit high δ value, usually of the order 1 ppm. This happens to occur in highly substituted alkaloids and steroids.

(iv) H-bonding effect

A hydrogen-bonded proton ($CH_3-O-\underline{H}\cdots\underset{\underset{H}{|}}{O}-CH_3$) has one covalently bonded strong electronegative atom on one side and on its other side a weakly bonded strong electronegative atom. Both electronegative atoms pull the electrons from the surroundings of proton. Naturally an H-bonded proton gets more deshielded than a non-H-bonded proton and that absorbs at a downfield showing a higher δ-value. δ-value of a proton having intermolecular H-bonding gets lowered with dilution since the degree of H-bonding changes on dilution; while the δ-value of an intramolecular H-bonded proton does not show any change with dilution.

10.22.10 Solvents used in ^{1}HNMR

If the sample is a non-viscous liquid, the neat liquid is used for its NMR experiment. For solids and viscous liquids the choice of solvent arises. Around 10% solubility of the sample in a solvent is required for the experiment. This is an important criterion for choosing the solvent. Normally aprotic polar solvents having no H atom will be our first choice since they will not give any peak in PMR spectra for their presence. Common solvents are: (i) Carbon tetrachloride (CCl_4), (ii) Carbon disulphide (CS_2), (iii) Deuteriochloroform ($CDCl_3$), (iv) Hexadeuteriobenzene (C_6D_6), (v) deuterium oxide (D_2O), (vi) Hexadeuterioacetone [$(CD_3)_2C{=}O$], (vii) Hexachloroacetone [$(CCl_3)_2C{=}O$], (viii) Hexadeuteriodimethyl sulphoxide [$(CD_3)_3S{=}O$].

Problem 10.33. A negligibly impure deuteriochloroform has been used as a solvent in an ^{1}HNMR experiment because of the high cost of pure $CDCl_3$. What problem has arisen in the spectra? How that has been rectified?

Solution: An impure $CDCl_3$ contains a small amount of $CHCl_3$. The proton of $CHCl_3$ has given a signal in the experiment and that is the problem. The δ-value of $CHCl_3$ proton is known. Any signal that appears at that δ-value may be considered as that of $CHCl_3$ and to be ignored as a signal of the sample. Of course, superimposition of signals may occur; in that case another ^{1}HNMR experiment has been carried out with another solvent which does not give signal at the absorption position of chloroform. If we get a signal where the signal of $CHCl_3$ appears, then the sample must have a proton at that δ-value. In this way the problem has been solved.

10.22.10 Intensity of absorption and peak area

The same kind of protons in a molecule absorb radiofrequency at a definite δ-value. As the number of protons of the kind increases, the intensity of absorption also increases and the height of the peak rises. The area underneath a peak is called peak area. If the NMR spectrum is redrawn on a graph-paper and squares underneath a peak is counted, we can get the peak area. The peak areas are proportional to the number of protons. However, the NMR spectrometer automatically integrates the peak areas as given below:

If there is no signal, it draws a horizontal line from left to right. When a signal is reached, the line ascends and when another signal comes in the way, the line again ascends. Thus steps are formed. The relative height from a step to another gives the peak area. There are three signals (Fig. 10.39) at δ-values 8, 6 and 3 and corresponding step heights are 1,2 and 3. In this spectrum absorption at

(i) $\delta = 8$ indicates the presence of one kind of proton whose number is 1.
(ii) $\delta = 6$ indicates the presence of another kind of proton which are 2 in number.
(iii) $\delta = 3$ tells us that there are 3 equivalent protons of the 3rd kind.

Fig. 10.39 has been drawn to make the readers understand the peak areas and the method of drawing the integrals.

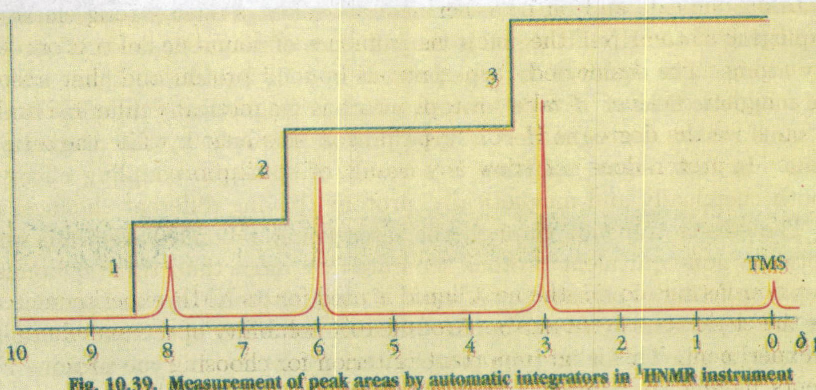

Fig. 10.39. Measurement of peak areas by automatic integrators in ^{1}HNMR instrument

We may get peak areas as fractions. They are to be converted to integers. Say, for *p*-xylene, we get a signal of step height 1 and another of step height 1.5. The ratio of step heights is 1:1.5, i.e., 2:3 or 4:6. Ratio of integrals obtained from the spectrometer (Fig. 10.28) in the case of *p*-xylene is seen as 3.468:6 which is very close to 4:6 (perhaps an integration error is there).

10.22.11 Multiple peak signals

The ^{1}HNMR spectrum of *p*-xylene (Fig. 10.28) shows two signals, each consisting of a single peak (called singlet). There are compounds whose ^{1}HNMR spectra exhibit not only singlets but also multiplets consisting of multiple peaks like doublet (a group of two peaks), triplet (a group of three peaks) and so on. The multiplets usually appear on high resolution spectra. Let us illustrate the multiplets of 1,1,2-trichloroethane. It has

$$\text{Cl}-\overset{\text{b}}{\text{C}\underline{\text{H}}_2}-\overset{\text{a}}{\text{C}\underline{\text{H}}\text{Cl}_2}$$

two kinds of proton : 'a' type and 'b' type. One 'a' type proton is bonded to C_1 having 2 Cl atoms and two '*b*' type protons bonded to C_2 atom having one Cl atom. We expect two signals.

(i) one for '*a*' type proton which will have higher δ-value (downfield) because the two-fold $-$I effect of two Cl atoms; and

(ii) the other for 'b' type protons which will have lower δ-value (upfield) because of only one-fold $-$I effect of the single Cl atom. Ratio of their areas is 1 : 2. Actually we get an one-proton triplet (downfield) and a two-proton doublet (upfield) in its ^{1}HNMR spectrum.

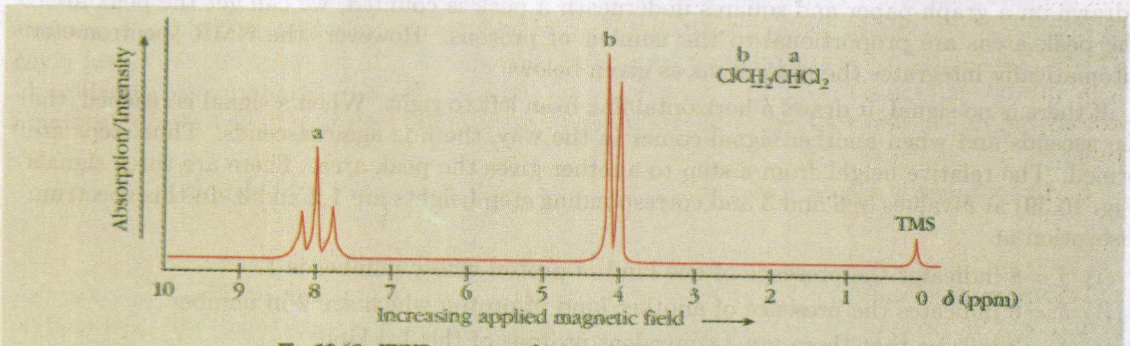

Fig. 10.40. HNMR spectrum of 1,1,2-Trichloroethane (not to scale)

10.22.12 Spin-spin coupling and spin-spin splitting

The cause of splitting of an ^{1}HNMR signal is the influence of magnetic fields of non-equivalent protons on nearby atoms. The signal of 'b' type protons in 1, 1, 2-trichloroethane splits because of the influence of magnetic field of 'a' type protons which is magnetically different from 'b' type protons. For the same reason the signal for 'a' type proton gets split by the magnetic influence of 'b' type protons. In fact, signal splitting is a result of spin-spin coupling effect involving non-equivalent, both chemically and magnetically, protons (having different chemical shifts) on adjacent atoms. The effects transmit through the intervening σ-bonded electrons and do not generally arise between non-equivalent protons separated by more than three σ bonds. We do not observe spin-spin splitting of signals for 'X' and 'Y' type protons that are present in methyl acetate since they are separated by five σ bonds; whereas the NMR spectrum of ethyl chloride shows two split signals, a quartet for 'a' type protons and triplet for 'b' type protons

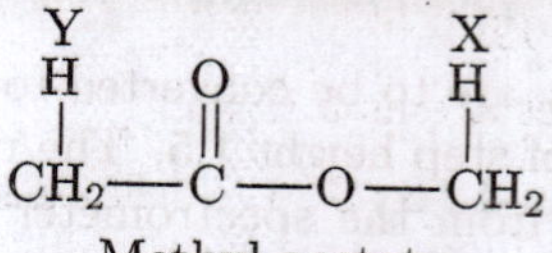

Methyl acetate

$$\overset{\text{b}}{\text{CH}_3}\ \overset{\text{a}}{\text{CH}_2\text{Cl}}$$

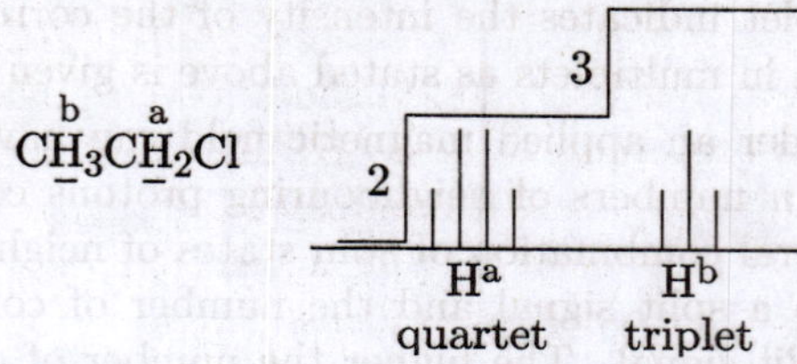

Chemically equivalent protons and enantiotopic protons (in optically inactive solvents) do not give split signals since chemically equivalent protons are also magnetically equivalent. For protons of ethane, we get one singlet though it contains 3 protons on each carbon atom.

Number of peaks in a multiplet and $(n+1)$rule

The number of peaks in an ^{1}HNMR signal depends on the number of equivalent protons on adjacent atoms. If the number of neighbouring equivalent protons is n, the multiplet will have $(n+1)$ peaks. This is $(n+1)$ rule. The split signals may further split if the proton that gives the signal has another set of equivalent protons but different from the protons for which primary splitting has occurred. Further splitting also follows the $(n+1)$ rule. In ethyl chloride 'a' and 'b' protons are magnetically non-equivalent but two 'a' protons are equivalent and three 'b' protons are also equivalent. 'a' protons have 3 equivalent neighbouring 'b' protons. So, the signal given by 'a' proton splits into $(3+1)$ or 4 peaks which constitute a quartet. Similarly, the signal for 'b' protons, which has two equivalent neighbouring protons, splits into $(2+1)$ or 3 peaks forming a triplet.

1-Nitropropane is a good example for showing further splitting of split signals

$$\overset{c}{CH_3}\cdot\overset{b}{CH_2}\cdot\overset{a}{CH_2}\text{—}NO_2 \qquad \text{1-Nitropropane}$$

1-Nitropropane molecule has three kinds of chemically non-equivalent protons marked as 'a', 'b' and 'c'. 'b' protons have two kinds of chemically non-equivalent protons— 'a' type and 'c' type. The signal of 'b' protons will split into quartet by the spin-spin coupling of 'c' protons. Each peak of the quartet will be further split by 'a' protons into a triplet since there are two 'a' protons. 4 peaks of the quartet after splitting will develop 4×3 or 12 peaks as shown below. The meaning of J_{bc} and J_{ba} will be dealt with later on under coupling constant (J).

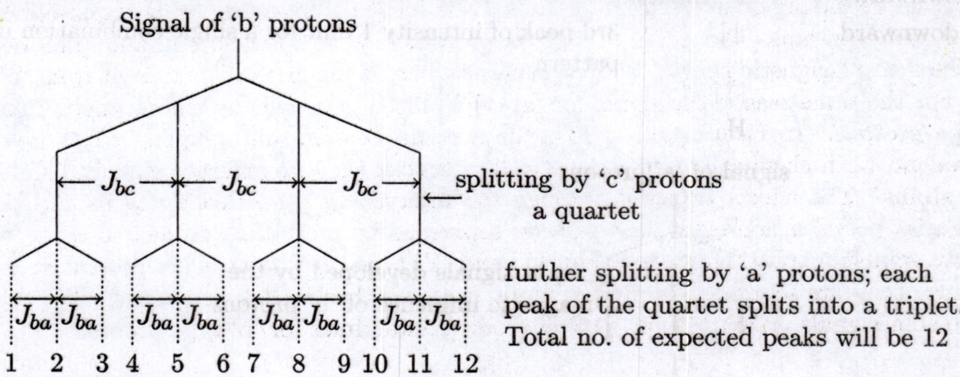

Ratios of relative peak areas of a multiplet are as follows:

For a doublet : 1 : 1; for a quartet : 1 : 3 : 3 : 1 For a sextet : 1 : 5 : 10 : 10 : 5 : 1

For a triplet : 1 : 2 : 1; For a quintet : 1 : 4 : 6 : 4 : 1

Each peak area of a multiplet indicates the intensity of the corresponding split peak. The reason for arriving at the ratios in multiplets as stated above is given below:

A proton in a molecule under an applied magnetic field can have either (↑) spin state or, (↓) spin state. If the spins of n numbers of neighbouring protons couple with the spin of the proton whose signal splits, several combination of spin states of neighbouring protons will arise. Each combination will develop a split signal and the number of combinations in the pattern dictates the intensity of the split signal. The higher the number of combinations in a pattern, the higher will be its intensity, i.e., peak area. Let us again consider the ^{1}HNMR spectrum of 1, 1, 2-trichloroethane as an example.

$$\overset{}{\text{Cl}}\ \overset{b}{\text{C}\underline{\text{H}}_2}\ \overset{a}{\text{C}\underline{\text{H}}}\ \text{Cl}_2$$

The signal given by '*b*' protons split owing to the magnetic influence of '*a*' proton. The spin states of '*a*' proton are ↑ and ↓. Two orientation patterns of '*a*' protons tell us that the signal of '*b*' protons will split into a doublet. Each pattern has one orientation; so, the intensity ratio will be 1 : 1.

signal of 'b' protons

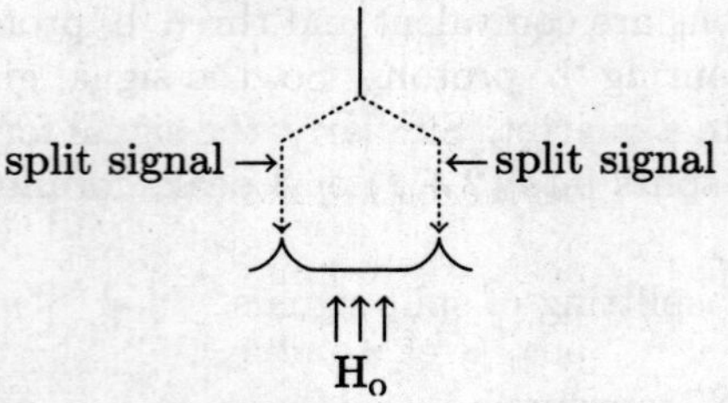

↑ ↓ orientations of 'a' proton

↑↑↑
H$_o$

Similarly the signal given by 'a' proton will split owing to three combined orientations of two 'b' protons.

both upward	↑↑	first peak of intensity 1 unit for a single combination in the pattern
one upward one downward	↑↓ ↓↑	2nd peak of intensity 2 units for two combinations
both downward	↓↓	3rd peak of intensity 1 unit for a single combination in the pattern

↑↑↑
H$_0$

signal of 'a' protons

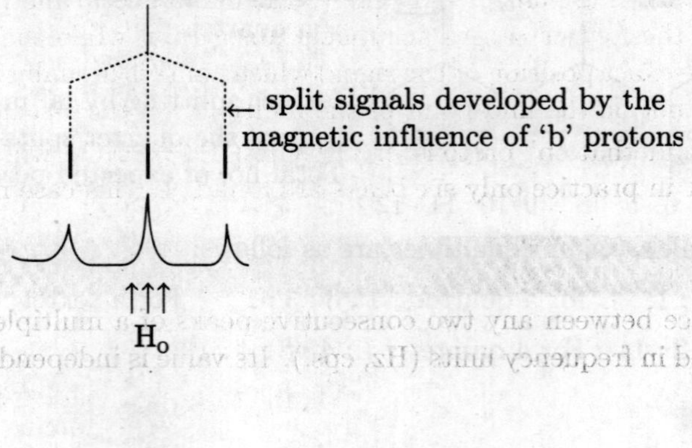

In ethyl bromide $\overset{b}{C}H_3\overset{a}{C}H_2Br$, the 'a' proton's signal will split by the magnetic influence of 3 'b' protons. The spin states, i.e., the magnetic orientations, of 3 'b' protons have the following combination.

all upward	↑↑↑	one combination; peak area 1 unit
two upward and one downward	↑↑↓ ↑↓↑ ↓↑↑	3 combinations; peak area 3 units
one upward and two downward	↑↓↓ ↓↑↓ ↓↓↑	3 combinations; peak area 3 units
all downward	↓↓↓	one combination; peak area 1 unit.

↑↑↑
H_0

So, 'a' proton signal will form a quartet since there are four combination patterns. The ratio of their peak areas will be 1 : 3 : 3 : 1 corresponding to the number of combinations in each pattern.

In this way spin state of four neighbouring protons will develop 5 combination patterns:

↑↑↑↑	peak area 1 unit
↑↑↑↓ ↑↑↓↑ ↑↓↑↑ ↓↑↑↑	peak area 4 units
↑↑↓↓ ↑↓↑↓ ↑↓↓↑ ↓↑↑↓ ↓↑↓↑ ↓↓↑↑	peak area 6 units
↑↓↓↓ ↓↑↓↓ ↓↓↑↓ ↓↓↓↑	peak area 4 units
↓↓↓↓	peak area 1 unit

↑↑↑
H_0

10.22.13 Shape and nature of multiplets

The ratio of peak areas of multiplet as calculated theoretically tells us that the multiplet will be symmetrical. In practice, however, we get multiplets which are not symmetrical and they have leaning appearance either toward upfield or toward downfield. It has been found, in general, that protons having a multiplet inclined in the downfield direction, it must have coupled with protons having a multiplet inclined in the upfield direction and vice versa. Thus this general pattern of inclination of multiplets indicates which type of protons couple with which type in a particular spectrum. The slanting multiplets is due to the diamagnetic and paramagnetic effects of neighbouring protons; the former causes an upfield absorption while the latter effect causes downfield absorption. The exact position of the signal which splits is actually the centre of gravity of the multiplet and may not be the mid-point of the multiplet on the δ-scale.

It has already been said that 'b' protons of $CH_3^cCH_2^bCH_2^a—NO_2$ will have a split signal consisting of 12 peaks but in practice only six peaks are found. In this case many peaks overlap.

10.22.14 Coupling constant (J)

The spacing or distance between any two consecutive peaks of a multiplet is called coupling constant, J. It is measured in frequency units (Hz, cps.). Its value is independent of the operating frequency of the spectrometer. The value of coupling constant of a multiplet measured on 30 MHz machine will be the same as that measured on 60 MHz or 100 MHz instrument. If the signal given by the proton 'a' gets split by the neighbouring proton 'b', the coupling constant, J, is designated as J_{ab}.

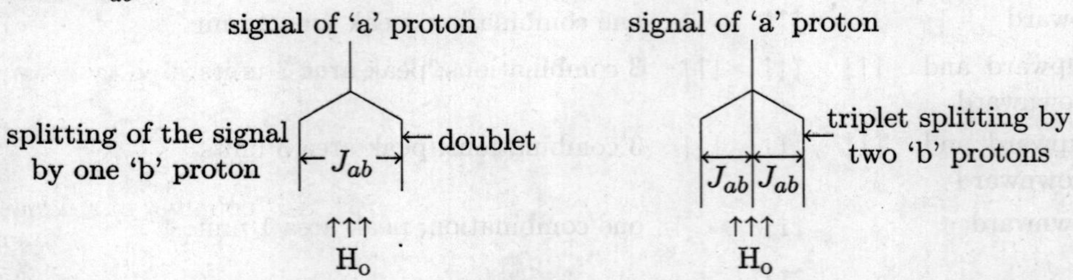

However, coupling constants have different values for coupling of two types of protons since their environments are different. J-values thus characterise the nature of protons coupling. So they give clue in determining structural features of a compound. Some values are given in Table 10.11.

Table 10.11 Range of values of coupling constants

Types of protons coupling		J_{ab} in Hz	Remarks
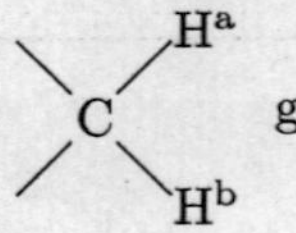	geminal	10–18	It depends on the electronegativity of the bonded groups
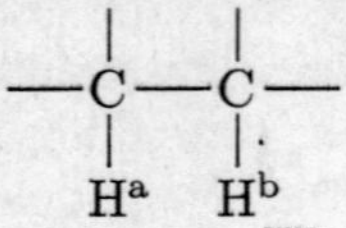	Vicinal	~ 7	The value depends on the dihedral angle
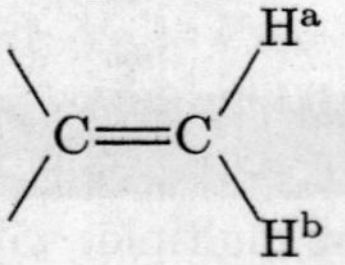	geminal	0–3	Around 2Hz in simple alkenes
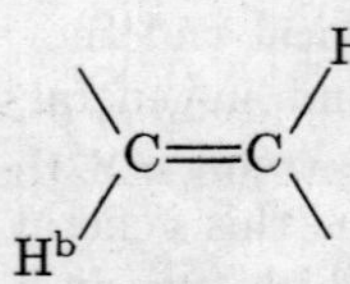	*trans*	13–18	*trans* coupling is stronger than *cis* coupling
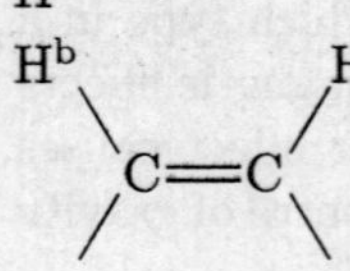	*cis*	7–12	*cis* coupling is weaker than *trans* coupling
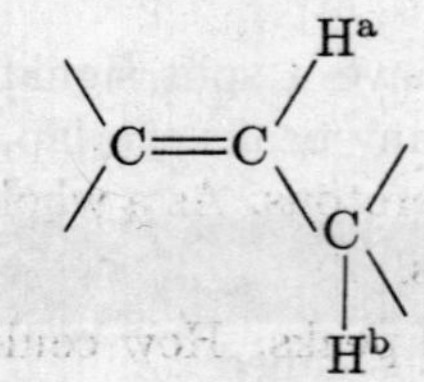		4–10	Coupling between vinylic and allylic H separated by 3 σ bonds

Contd.

Types of protons coupling	J_{ab} in Hz	Remarks
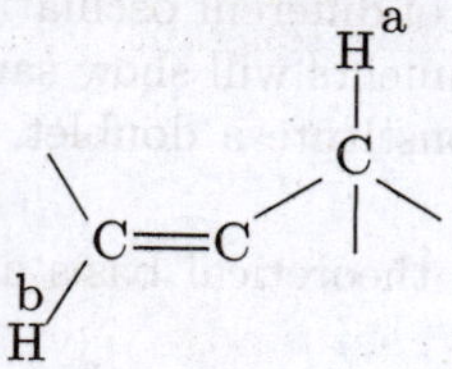	0–2	Allylic H separated by 4 bonds
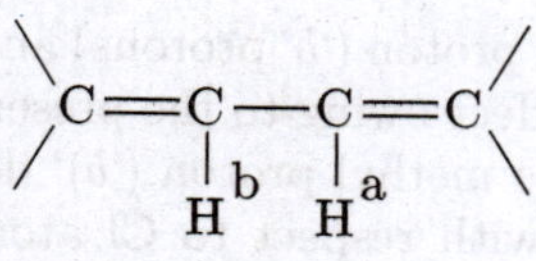	10–13	β-H, β-H coupling of a diene moiety
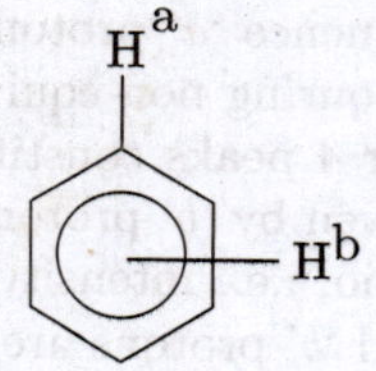	7–10	*ortho* protons couple
	2–3	*meta* protons couple
	0–1	*para* protons couple

10.22.15 ^{1}HNMR spectra of aromatic compounds

Aromatic protons absorb at δ 6–8.5. The anisotropic effect of induced magnetic field has paramagnetic effect on ring protons and they absorb at downfields. In the cases of monosubstituted benzenes *o*-, *m*- and *p*-protons get shielded or deshielded in different magnitude depending on the nature of substitutent. Any substituted group that has electron-pushing effect, either by +I or by +R makes *o*-positions most electron dense while *m*-positions least. The order of electron density at *o*-, *m*- and *p*-position is

o-> *p*-> *m*- $\longrightarrow$ Decreasing electron density

o-> *p*-> *m*- $\longrightarrow$ Decreasing shielding effect

So, the order of the shielding effect owing to the presence of electron-pushing group is *o*-> -*p* >-*m*. Naturally an *o*-proton absorbs at a slightly higher δ-value than a *p*-proton and the *p*-proton absorbs at a slightly higher δ-value than the *m*-proton.

An electron-withdrawing substituent decreases the electron density in the order *o*->-*p* >-*m* and deshields them in the same order.

— upfield $\longrightarrow$ *m*-, *p*- and *o*-	for +I and +R substituents (alkyl, —OH, —NH_2, etc.)
$\longleftarrow$ downfield — *o*-, *p*-, *m*-	for –I and –R substituents (—NO_2, —CHO, —CCl_3 etc.)

Besides the above effects spin-spin splitting occurs involving o-,m- and p- protons. As a whole ^{1}HNMR spectra of aromatic compounds are complex and give multiple peaks.

Problem 10.34. An ^{1}HNMR spectrum of a compound shows two close peaks. How could you identify whether they constitute a doublet or two singlets for two different kinds of protons?

Solution: We know that the frequency of absorption is dependent on the operating frequency of the spectrometer; but coupling constants are independent of operating frequency of the instrument. So we are to take two spectra of the compound in two spectrometers of different oscillating frequencies, say one of 60 MHz and the other of 100 MHz. The two instruments will show same spacing of peaks (J of the same Hz), if the twin peaks under question constitute a doublet. If the spacing changes, the twin peaks will be two singlets.

Problem 10.35. Draw the ^{1}HNMR spectrum of CH_3CHCl_2 on the theoretical basis and indicate **(i)** relative field strengths and **(ii)** relative peak areas.

Solution: In $\overset{b}{C}H_3\overset{a}{C}HCl_2$ there are two kinds of protons, one methyl proton ('b' protons) and a proton of —$CHCl_2$ ('a' proton). 'a' proton experiences deshielding effect owing to the presence of two electronegative Cl atoms on the C atom bearing the proton; the methyl proton ('b)' does not experience −I-effect to that extent because it is on the β—C with respect to Cl atoms. So 'b' protons will be much less deshielded than 'a' protons. As a consequence 'a' protons will absorb at a lower field strength than 'b' protons. 'a' proton has 3 neighbouring non-equivalent 'b' protons. So the signal given by the 'a' proton will split into $(3 + 1)$ or 4 peaks constituting a quartet; 'b' protons have one non-equivalent 'a' proton. So the signal given by 'b' protons will split into $(1 + 1)$ or two peaks constituting a doublet. The peak area ratio, i.e., intensity ratio for 'a' and 'b' protons is $a : b :: 1 : 3$, since 'a' proton is one in number and 'b' protons are three in number as shown below.

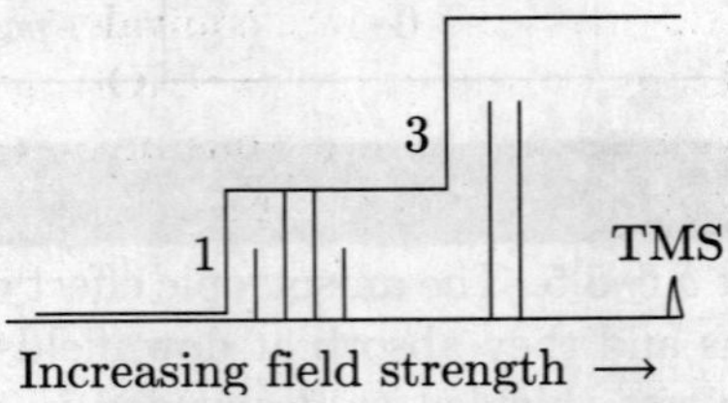

Problem 10.36 A compound of molecular formula $C_8H_8O_2$ shows HNMR peaks at δ-values 7.21, 3.53 and 11.95. Suggest a structure for the compound.

Solution: δ-value 7.21 indicates that the compound is a substituted benzene since the proton of benzene ring absorbs at δ-value ~7 ppm. 11.95 δ value indicates the presence of carboxyl group (which absorbs within the range 10.5–12 (See Table 10.10). So the compound may have any of the following structures.

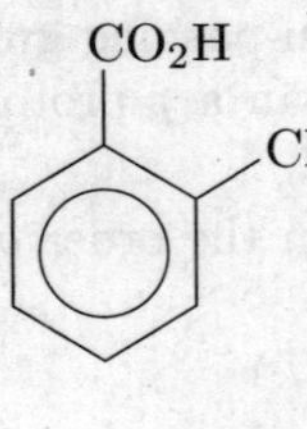

1
,
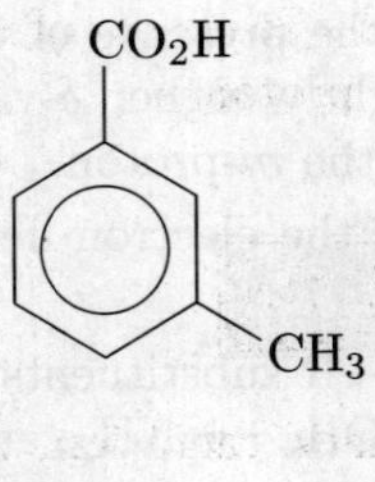

2
,

3
or
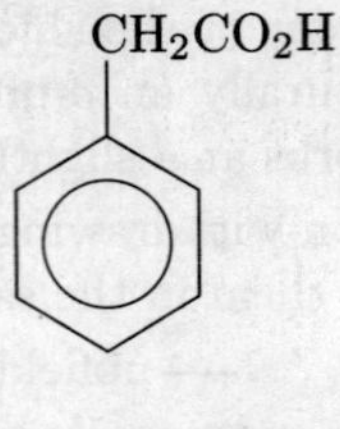

4

All the four compounds will show 3 peaks and all contain -CO_2H and ring protons. Methyl protons, however, absorb at a higher field than $\delta = 3.53$. So, structures 1, 2 and 3 are discarded. The methylene proton of structure 4 may absorb at $\delta = 3.53$ owing to the deshielding effect of —CO_2H [methyl and methylene protons are benzylic protons which absorb at (2–3)δ]. The compound is, therefore, phenylacetic acid (4).

Problem 10.37. A compound C_2H_2XY has two doublets and the coupling constant is 9Hz. Suggest a structure for the compound.

Solution: Three possible structures could be drawn :

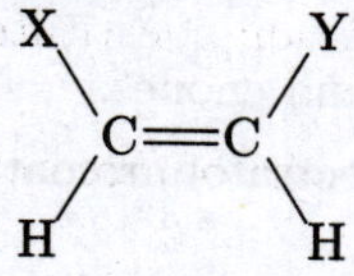

cis

J for *cis* Hs coupling = 7 – 12 Hz

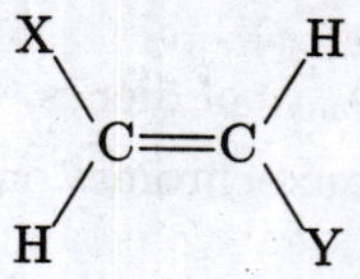

J for *trans* Hs coupling 13 – 18 Hz

and

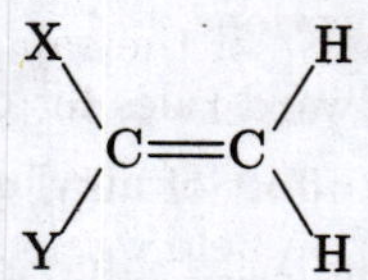

J for geminal Hs coupling = 13 – 18 Hz

The value of J indicates that the compound is a cis-isomer.

Problem 10.38. How could you distinguish between *o*-dinitrobenzene and *p*-dinitrobenzene by ^{1}HNMR spectroscopy?

Solution:

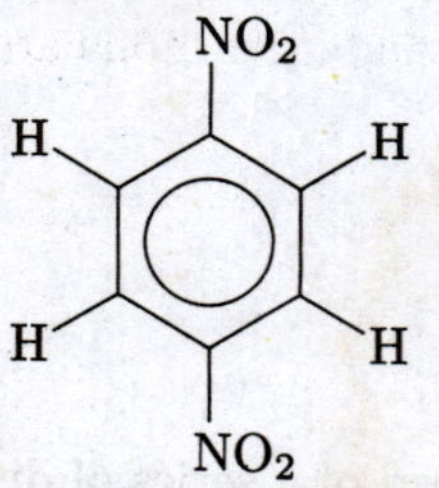

p-dinitrobenzene

All the ring protons are equivalent; each is ortho to one —NO_2 group and meta to the other —NO_2 group. So, we shall get a singlet at a downfield than the protons in C_6H_6 (benzene) because of –I and –R effects of —NO_2 groups.

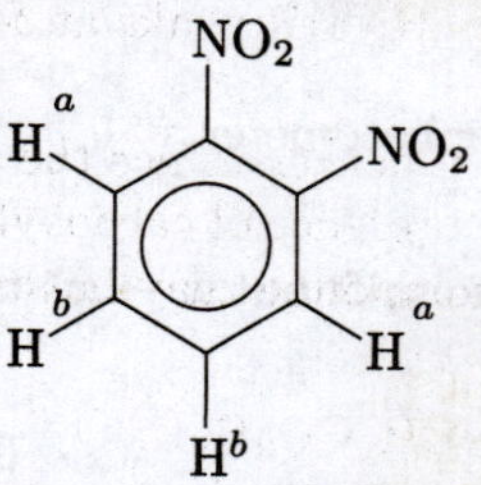

o-dinitrobenzene

This has two kinds of protons marked as H^a and H^b. H^a is ortho to one —NO_2 group and meta to other. H^b is para to one —NO_2 group and meta to other. So we shall get two signals, each will develop a doublet because of spin-spin interaction between H^a and H^b. J_{ab} will be within the range of 7–10 Hz. Each kind of proton will absorb at a downfield than the protons of benzene. H^a will absorb at a downfield than H^b because –I and –R effects of —NO_2 groups deshield ortho protons more than para protons.

Exercise

1. Write notes on: Wave motion, electromagnetic radiation, radiant power and intensity of electromagnetic radiation, transmittance, absorbance, Lambert-Beer's Law, molar absorptivity, molecular excitation, chromophore, auxochrome, bathochromic (red) shift, hypsochromic (blue) shift, hyperchromic shift, hypochromic shift, forbidden and allowed transitions, λ_{max}, ϵ_{max}.
2. Make a table on different units of wavelength, wave number, frequency and energy.
3. Discuss the theory of electronic spectroscopy.

4. Discuss different types of electronic transitions and classify different types of bands.
5. Write down the effect of temperature and solvent on UV Spectra.
6. Relate resonance effect, hyperconjugative effect, H-bonding effect and steric effect to position of absorptions.
7. State Woodward rules for calculating λ_{max} of dienes, trienes and enones.
8. Discuss the effect of alkyl groups and auxochromes on UV spectra of aromatic systems.
9. Calculate the λ_{max} value in each case :

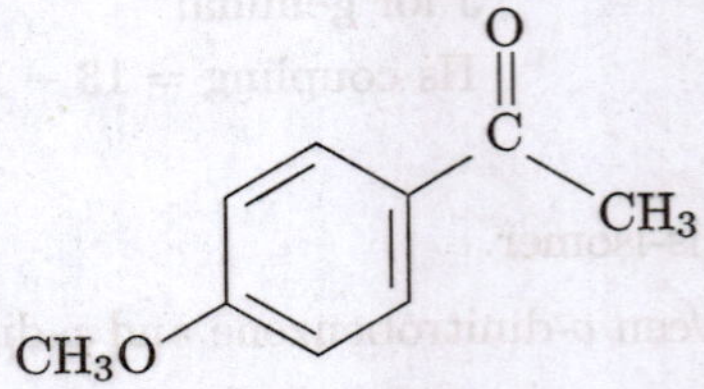

Ans. (a) 276 nm;

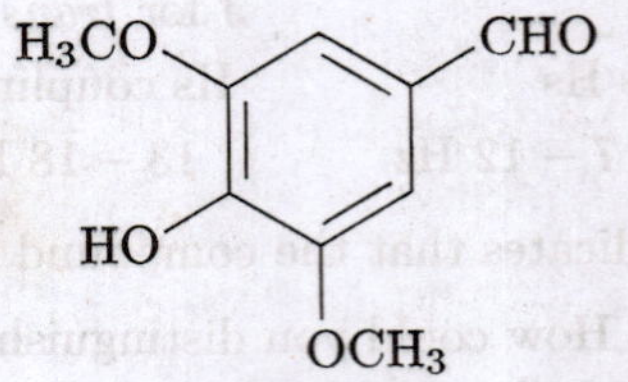

Ans. (b) 289 nm;

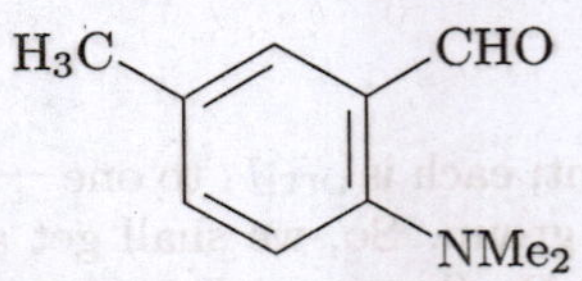

Ans. (c) 273 nm;

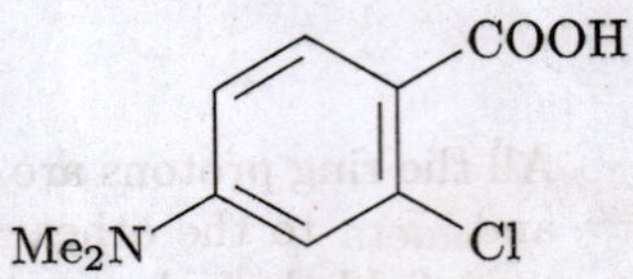

Ans. (d) 315 nm.

10. Explain why:
   (a) Spectrophotograph of an electronic excitation is a curve instead of a series of discrete lines.
   (b) $n \rightarrow \pi^*$ transition is a forbidden transition.
   (c) $\sigma \rightarrow \sigma^*$ transitions cannot be accessible by ordinary spectrophotometer.
   (d) Ethylene shows two λ_{max} values.
   (e) In a more polar solvent $n \rightarrow \pi^*$ transitions experience a blue shift but $\pi \rightarrow \pi^*$ transitions show red shift.
   (f) With increasing conjugation deepening of colour occurs.
   (g) cis-isomer absorbs at a shorter wavelength than the trans-isomer.
11. What do you mean by the following terms
   (a) NMR phenomenon,
   (b) precessional frequency,
   (c) chemical shift,
   (d) coupling constant,
   (e) shielding,
   (f) deshielding,
   (g) upfield,
   (h) downfield.

12. (a) Why are δ- and τ-scales are used instead of frequency-scale in ^{1}HNMR spectroscopy?
    (b) What will be the ν-values in 60 MHz and 100 MHz instruments, if the absorption by a proton appears at $\tau = 6$?
13. Write down a structure for a compound $C_3H_5O_2Br$ which shows a signal at τ between -2.0 to -0.5, a doublet around $\tau = 8.5$ and a quartet around $\tau = 6$.
14. Describe the ^{1}HNMR spectrum of
    (i) isopropylbenzene
    (ii) 2-chloropropane.
15. Methanol shows two singlets in carbon tetrachloride solvent in its ^{1}HNMR spectrum; but methanol gives a doublet and a quartet in $(CD_3)_2S{=}O$. Explain.
16. Whtat is 'end absorption'?

Ans. For measuring absorptions in any spectroscopy a scale is constructed. When the scale gets exhausted during the measurement, it is called, 'end absorption'.

For example, in UV and visible spectroscopy the absorption after $\sigma \rightarrow \sigma^*$ transition is 'end absorption'.

Chapter 11

Fundamentals of Organic Reaction Mechanism

Objectives

◈ To understand:

* What is meant by the mechanism of an organic reaction [11.1] p. 501
* Classification of organic reactions [11.2] p. 502
* Bond-breaking and bond-making processes [11.3] p. 503
* Concerted and stepwise reactions [11.3.4] p. 505
* Reactivity and orientation [11.6] p. 539

◈ To have a clear idea about reaction intermediates [11.4] p. 506

* Carbon radicals [11.4.1] p. 506
* Carbenium ions [11.4.2] p. 510
* Carbanions [11.4.3] p. 514
* Carbenes [11.4.4] p. 517
* Nitrenes [11.4.5] p. 522
* Benzynes [11.4.6] p. 524

- ✱ π-Complexes [11.4.7] p. 525
- ✱ σ-Complexes [11.4.8] p. 527

- To understand:
  - ✱ Types of reagents [11.5] p. 531
    - Electrophiles [11.5.1a] p. 531
    - Nucleophiles [11.5.1b] p. 532
    - Hard and soft electrophiles and nucleophiles [1.5.1c] p. 534
    - Ambident nucleophiles [11.5.1d] p. 535
    - Ambident electrophilic centres p. 537
  - ✱ The position of transition state [11.7] p. 540
    - The selectivity principle [11.7.1] p. 541
    - The microscopic reversibility principle [11.7.2] p. 542
    - NGP and anchimeric effect [11.7.3] p. 542
- To know the methods of determining reaction mechanisms [11.8] p. 546
  - Nature of the products and their proportions [11.8.1] p. 550
  - Uses of isotopes [11.8.2] p. 548
  - Kinetic study and kinetic isotope effect [11.8.3] p. 549
  - The study of reaction intermediates [11.8.4] p. 553
  - Cross-over experiments [11.8.7] p. 554

11.1 Introduction

A chemical reaction is a process which changes the structure of at least one of the starting materials and hence gives rise to new species. Usually this involves collision between two chemical units of the starting materials in which one or more of their bonds break and new bond or bonds form and thus new species develop.

In an organic reaction, the main organic entity which is converted into a new one by the processes of breaking and making of bonds is known as a reactant or substrate and the new one formed is named as a product; the chemical entity which causes the change is called a reagent.

The bond breaking and bond making processes of an organic entity may take place in three ways :

(a) These may occur as separate steps.

(b) The two processes may overlap—before breaking of a bond completely, a new bond may begin to form.

(c) These may occur at a time—this is known as a concerted step.

After the process of bond breaking, intermediate entities are often formed momentarily during the course of chemical reactions; these are very important to understand chemical reactions and are said to be reaction intermediates.

The nature of reaction intermediate and that of the product in a chemical reaction depend on:

I. Constitutional features of the entities involved, i.e., mode of arrangement of their constituent atoms and electron availability within the entities.

II. Stereochemical features including stereoelectronic features, i.e., orientation of constituent atoms and that of the orbitals of the entities in space.

III. Nature of the reagents that are taking part in the reaction.

IV. Nature of the solvent that is employed for the reaction.

Of course thorough understanding of a chemical reaction requires the knowledge of energy changes involved in the system as a whole. This includes:

(i) Thermodynamics of the reaction—to know the ratio of reactant and product in a particular step when the reaction goes to the equilibrium.

(ii) Kinetics of the reaction—to know the rate at which the reaction occurs.

Most of the organic reactions consist of more than one step and a number of products may result in a single reaction. Why one of these possible products actually perdominates and why other products form in practice are determined by the detailed description of the overall course of the reaction including:

(a) Constitutional and stereochemical features of the entities.

(b) Bond-breaking and bond-making processes.

(c) Thermodynamics and kinetics of the reaction.

This overall description of the reaction course is termed as reaction mechanism.

The mechanism of a reaction must tally with all the facts available about the reaction. If new facts about the very reaction are discovered, the mechanism proposed may change. In fact, the mechanism for a reaction is never known to be complete; and for this reason, we call the mechanisms for most of the reactions as plausible mechanisms.

However, the substrates and the products of the organic reactions are mostly molecules and their structural features have already been discussed in different chapters. In this chapter we shall devote our attention to the processes involved in between the reactant state and the product state. This requires the discussion of the following subjects:

A. Elementary classification of reactions.

B. Breaking and making of bonds.

C. Types of reaction intermediates.

D. Types of reagents.

E. Thermodynamics of the reactions (Chapter five).

F. Kinetics of the reactions (Chapter six).

G. Stereochemical features (Chapter nine).

H. Relationship between these things and structures.

I. Methods of determining mechanisms.

11.2 Elementary Classification of Reactions

Organic reactions may be broadly classified under four heads:

1. Substitution or displacement reaction—In which an atom or a group of atoms are replaced by another atom or group; for instance, hydrolysis of methyl chloride to methyl alcohol by aqueous caustic soda solution. Here the Cl atom is replaced by the OH group.

$$CH_3Cl \xrightarrow{\text{aq. NaOH}} CH_3OH$$

2. Addition reaction—In which two molecules unite to produce a single molecule; this is common to compounds containing multiple bonds. Thus bromine in carbon tetrachloride undergoes addition reaction with ethylene to give rise to ethylene bromide.

$$CH_2{=}CH_2 \xrightarrow{Br_2/CCl_4} CH_2Br.CH_2Br$$

3. Elimination reaction—In which two atoms or groups are removed from a single molecule; the process is the reverse of addition reaction. If elimination of two atoms or groups happens to occur from one and the same atom of the substrate, it is known as α-elimination reaction. The process of elimination may occur from two adjacent atoms of the molecule and this is known as β-elimination; in this case formation of multiple bond is the result. Elimination may also take place from atoms separated by one, two or more atoms; these are then called γ, δ, etc. elimination reactions. In such cases usually rings are formed. When ethyl alcohol is heated with conc. H_2SO_4 at 170°C we get ethylene and water which gets absorbed by conc. H_2SO_4—this is an example of β-elimination reaction.

$$\text{Conc. } H_2SO_4 + CH_3CH_2OH \xrightleftharpoons{170^\circ C} CH_2 = CH_2 + [H_2SO_4 + H_2O]$$

4. Rearrangement reaction—In which one or more isomeric compounds are formed, e.g., when phenetole is heated at 150°C in the presence of anhydrous $AlCl_3$, *o*-ethylphenol and *p*-ethylphenol are produced.

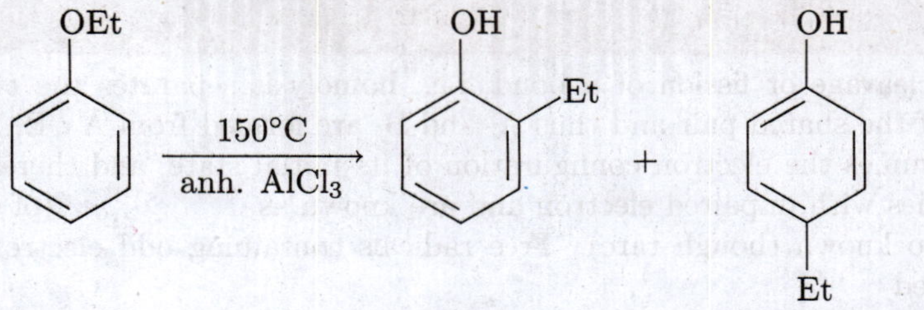

Phenetole o-Ehylphenol p-Ethylphenol

Besides these four general types, one may find the names of several other reactions; e.g.,

(a) Oxidation reduction or redox reaction—In which change in oxidation number of an atom or atoms occur; thus conversion of methyl alcohol into formaldehyde involves increase in oxidation number of the C atom from —2 to 0. Aldehydic C has +I oxidation state in acebaldehyde; in ethyl alcohol, that C has −I oxidation state.

$$\overset{-II}{CH_3OH} \xrightarrow{\text{oxidation}} \overset{0}{H}CHO;\ CH_3\overset{+I}{C}HO \xrightarrow{\text{reduction}} CH_3\overset{-I}{C}H_2OH$$

(b) Condensation reaction—In which two or more molecules unite with or without loss of small units; thus condensation of two molecules of acetaldehyde gives rise to one molecule of aldol.

$$CH_3CHO + CH_3CHO \xrightarrow{OH^-} CH_3CHOH{\cdot}CH_2CHO$$

(c) Polymerisation reaction—In which a large number of species unite to form a macro species. A large number of ethene molecules unite to form polyethene. These may be considered as reactions involving one or more of the substitution, addition or elimination types of reactions.

Reaction intermediates undergo certain reactions which cannot be explained by the general types of reactions; e.g.,

(i) Fragmentation or decomposition reaction—in which a species cleaves into two other species.
(ii) Coupling reaction—in which two species unite to form a single species.
(iii) Atom abstraction reaction—in which one atom, H atom in most of the cases, is taken out from a species.
(iv) Insertion reaction—in which one species is inserted between two σ-bonded atoms in another species.

11.3 Breaking and Making of Bonds

As stated earlier, a covalent bond between two atoms in a molecule, A—B, may be looked upon as a shared pair of electrons and thus may be represented as A : B.

When the bond between A and B is broken, the molecule breaks into two species. The way in which a bond is broken determines the nature of the species formed. Both theory and practice tell us that a covalent bond can be broken in two ways:

1. Homolytically ; A—B, i.e., $A : B \xrightarrow[\text{fission}]{\text{homolytic}} A\cdot + B\cdot$

2. Heterolytically ; A—B, i.e., $A : B \xrightarrow[\text{fission}]{\text{heterolytic}} A^{\oplus} + \overset{-}{B:}$ or $\overset{-}{A:} + B^{\oplus}$

11.3.1 Homolytic fission

Homolytic cleavage or fission of a bond, i.e., homolysis separates the two atoms with one electron each of the shared pair and thus A· and B· are formed from A : B. Each of the species formed now assumes the electron configuration of its initial state, and therefore, A· and B· are uncharged species with unpaired electron and are known as free radicals (of course charged free radicals are also known though rare). Free radicals containing odd electrons on C atoms are collectively called carbon radicals.

Homolysis of a bond is carried out either in the gas-phase or in the non-polar solvent by means of heat, light, radioactive elements, organic peroxides, azo compounds, redox reagents, etc. These are called initiators. Reactions involving carbon radicals are called non-polar reactions.

It is obvious that in the homolytic fission of bonds, the reagent primarily attacks those bonds which are most easily breakable, i.e., possessing low bond dissociation energy. Homolytic cleavage of a bond is the most characteristic reaction of the paraffins. The C—H bonds of alkanes are first attacked and then by the process of chain reactions C—C bonds also break. The energy required to break up C—H bonds in some of the hydrocarbons are given below in Table 11.1.

Table 11.1 Homolytic bond dissociations energy of C—H bonds

Compound	Energy in kJ mol^{-1}	Compound	Energy in kJ mol^{-1}
$H_3C—H$	431.2	$CH_2{=}CH.CH_2—H$	322.4
$CH_3.CH_2—H$	410.3	$C_6H_5—H$	427.1
$(CH_3)_2CH—H$	393.7	$C_6H_5.CH_2—H$	341.1
$(CH_3)_3C—H$	380.9	$Cl_3C—H$	372.7
$F_3C—H$	427.1	$HC{\equiv}C—H$	506.6

11.3.2 Heterolytic fission

When the fission of a bond gives rise to two ions, a cation and an anion, the process is named as heterolytic fission or cleavage, or heterolysis. Which one of the bonded atoms will be cation or anion that depends on whether A or B retains the bonded electron pair.

If in a cation, formed by the heterolysis of a bond, a C atom bears the positive charge, it is known as a **carbenium ion**; a carbenium ion is an electron-deficient centre. On the other hand, if there is a negative charge on a C atom in an anion, it is said to be a carbanion; it is an electron rich centre. Reactions involving carbenium ions or carbanions are termed as polar or ionic reactions. **Carbenium ions** are also called **carbocations.** Carbonium ions are $\overset{\oplus}{CH_5}$ and its derivatives.

Heterolysis of bonds are generally carried out in polar solvents by means of acids, bases or ions. Bond breaking process requires input of energy.

Nature of the reagent, the solvent and the atoms that are present in the substrate collectively determine the bond to be broken in the substrate.

11.3.3 Bond-making process

The bond-formation process is the reverse of bond-breaking process. Attack by the radicals or ions to other species gives rise to new bonds. Coupling reaction, abstraction of ions, atoms or groups, etc. form new bonds. Bond formation processes liberate energy.

11.3.4 Concerted and stepwise reactions

The chemical reactions in which more than one bond is broken or formed are usually classified under two heads—(i) Concerted step reactions and (ii) Stepwise reactions.

(i) Concerted step reactions : In which all the bond forming and bond breaking processes occur in concert, i.e., simultaneously so that at one step the conversion of reactants to products takes place without the formation of any intermediate, a stable molecule or a highly unstable transient species (called a reaction intermediate). The concerted step reactions are also called multicentered reactions since several bonds take part at the same time at more than one centre. When ethyl acetate is pyrolysed, acetic acid and ethylene are formed; this is in example of concerted step reaction as shown below. Here one C—O π bond, one C—H σ bond and one O—C σ bond break and one C—O π bond, one C—C π-bond and one O—H σ bond form in concert.

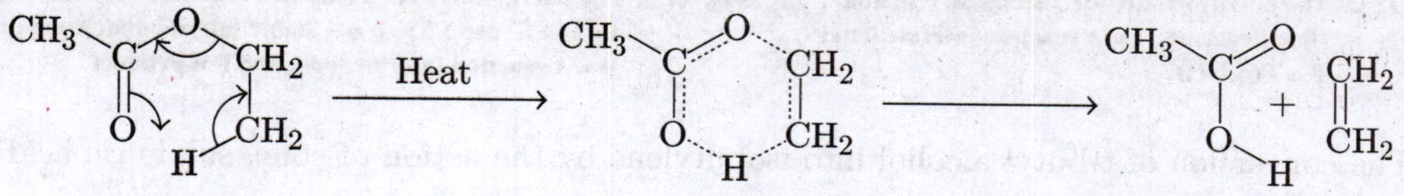

a multi-centred T.S

In a concerted step reaction all bond breaking and bond making processes are inter-dependent so that both processes are going on at the same time and the energy required for the bond breaking is somewhat compensated by the energy evolved through the bond making. However, the two processes may be synchronous, i.e., can occur exactly at the same rate or it may not be. There are concerted step reactions in which one process starts before the other but does not end before the other process begins. The two processes will overlap at one time and the bond breaking or bond making process will couple with and controlled by the other process. Though there will be a time lag between their start, the reactions will still be considered as the concerted step reactions. A concerted step reaction involves only one energy barrier and a single transition state. The energy profile for such a reaction is drawn below :

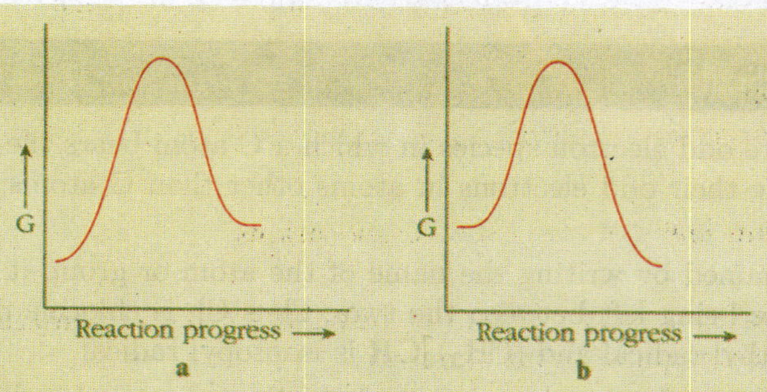

Fig. 11.1. a. The energy profile for an endothermic concerted step reaction
b. The energy profile for an exothermic concerted step reaction

(ii) Stepwise reactions : In which the bond making and the bond breaking processes occur consecutively and two or more intermediates are involved. The intermediates may either be stable molecules capable of isolation, or highly unstable species of transient existence which may be called the reaction intermediates. If the intermediate is a stable molecule, the reaction may be considered as a multistep reaction consisting of two or more concerted or discrete steps. The energy curve for a stepwise reaction is Fig. 11.2 having a reaction intermediate (I), and Fig. 11.3 represents a multistep reaction involving a stable intermediate compound (B).

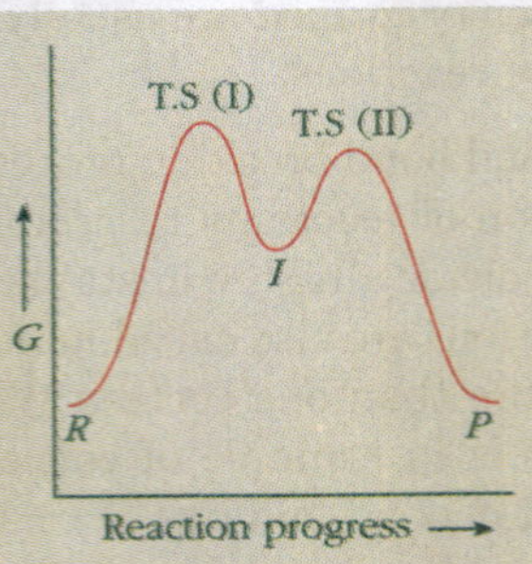

Fig. 11.2. The energy profile for a stepwise reaction. R = Reactants, I = A reaction intermediate, P = Products

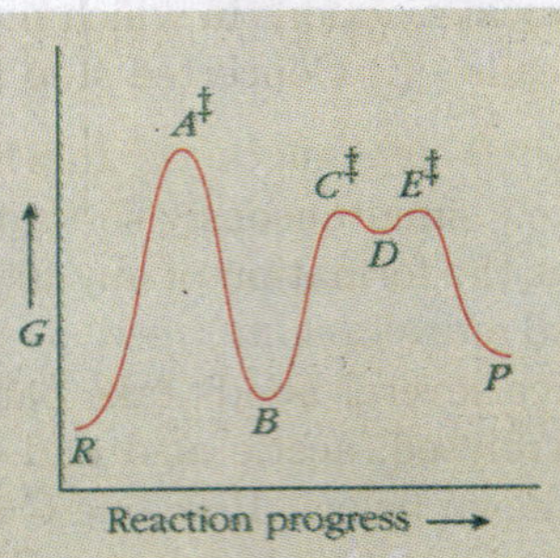

Fig. 11.3. The energy curve for a stepwise reaction. R : Reactant, $A^{\ddagger}$, $C^{\ddagger}$ and $E^{\ddagger}$ are T.S.s, B = A stable intermediate compound, D = A reaction intermediate, and P = Product

The conversion of t-butyl alcohol into isobutylene by the action of conc. sulphuric acid is an example of stepwise reactions.

11.4 Reaction Intermediates

Reaction intermediates are generated by the breaking of bonds. These are short-lived and highly reactive species. Of course, very short-lived species cannot get the time to react. Reaction intermediates are said to be stable in the sense that they get time to undergo the reaction.

There are eight important types of reaction intermediates—(a) carbon radicals, (b) carbenium ions, (c) carbanions, (d) carbenes, (e) nitrenes, (f) benzynes, (g) pi-complexes, (h) sigma-complexes.

We shall discuss here the structure, reactivity, formation and properties of reaction intermediates.

11.4.1 Carbon radicals

Carbon radicals are odd electron species in which a C atom bears the odd electron. Organic free radicals may have their odd electrons in atoms other than C atoms; for example, benzoxy radical (PhCOO·).

A free radical is named by writing the name of the atom or group it contains and then the word 'radical', a space being left between the two. Thus Cl· is chlorine radical, $\cdot CH_3$ is methyl radical, $CH_3\dot{C}H_2$ is ethyl radical and $(CH_3)_2\dot{C}H$ is isopropyl radical.

A carbon radical may also be named as a derivative of methyl radical, e.g., $(C_6H_5)_3C\cdot$ is triphenylmethyl radical.

Homolytic fission of a bond gives rise to free radicals. This cleavage is initiated either by physical aids or by chemical reagents.

(a) Physical aids : Heat, light, radioactive radiation. These are high energy processes.

(b) Chemical reagents : (i) Organic peroxides, namely, benzoyl peroxide; (ii) Azo compounds, namely, azobis-isobutyronitrile; (iii) Reductants and oxidants, e.g., metals.

Carbon radicals may also be formed by the reaction of other free radicals. For example :

1. **By the fragmentation or decomposition of a free radical**—Benzoxy free radical decomposes and thus phenyl radical is generated.

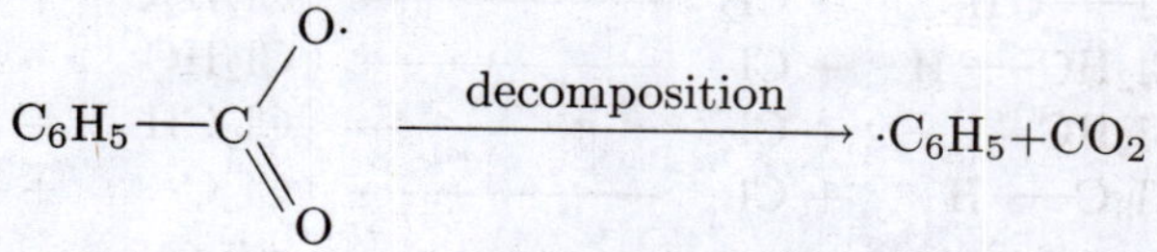

2. **By the abstraction of an atom from another species**—Chlorine free radical abstracts an H atom from methane and thereby methyl free radical is formed.

$$H_3C—H + Cl\cdot \xrightarrow{\text{abstraction}} H_3C\cdot + HCl$$

3. **By the addition reaction of a free radical to a multiple bond**—When bromine free radical is added to ethylene, a carbon radical forms.

$$CH_2{=}CH_2 + Br\cdot \xrightarrow{\text{addition}} \cdot CH_2—CH_2Br$$

4. **By the rearrangement of another carbon radical**—Neopentyl radical rearranges to tert-pentyl free radical.

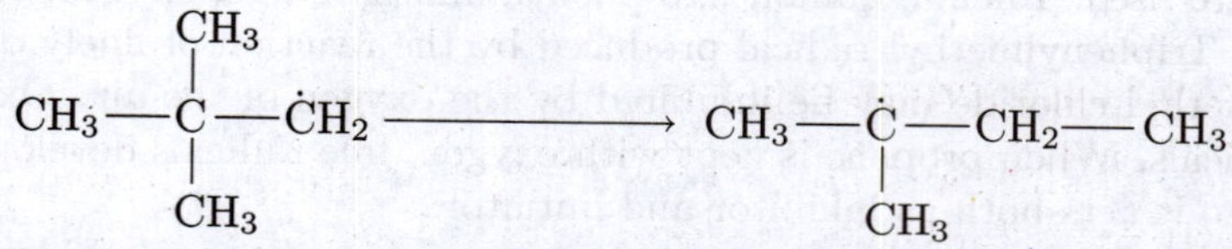

Owing to the presence of odd electron, a carbon radical, like other free radicals, is paramagnetic in nature; for the same reason these are highly reactive. They undergo coupling reactions, disproportionation reactions and other redox reactions. Atom abstraction and addition reactions are also common to them. They also undergo rearrangement reactions which have already been discussed along with the formation of free radicals. Here we shall discuss coupling reaction and disproportionation reaction.

Two carbon radicals couple to form a molecule—thus two methyl radicals unite to form ethane molecule :

$$H_3C\cdot + \cdot CH_3 \xrightarrow{\text{dimerisation}} H_3C—CH_3$$

Ethyl radicals undergo disproportionation reaction in which one radical gets oxidised to ethylene and the other gets reduced to ethane.

$$CH_3\dot{C}H_2 + \cdot CH_2CH_3 \xrightarrow{\text{disproportionation}} CH_2{=}CH_2 + CH_3—CH_3$$

A free radical reaction takes place in three steps; the reaction between methane and Cl_2 in diffused sunlight may be represented as follows :

I. Initiation—In this step a free radical is first produced, generally by thermolysis (action of heat) or photolysis (action of light) or by redox reaction.

$$Cl—Cl \xrightarrow{\text{photolysis}} Cl\cdot + Cl\cdot$$

II. Propagation—In this step further free radicals are produced by the action of the first formed free radical. In this step a chain of reactions happen to occur if it is energetically favourable.

$$CH_4 + Cl\cdot \longrightarrow H_3C\cdot + HCl$$
$$H_3C\cdot + Cl_2 \longrightarrow H_3CCl + Cl\cdot$$
$$ClH_2C—H + Cl\cdot \longrightarrow \cdot CH_2Cl + HCl$$
$$Cl—\dot{C}H_2 + Cl_2 \longrightarrow Cl_2H_2C + Cl\cdot$$
$$Cl_2HC—H + Cl\cdot \longrightarrow Cl_2HC\cdot + HCl$$
$$Cl_2HC\cdot + Cl_2 \longrightarrow Cl_3CH + \cdot Cl$$
$$Cl_3C—H + Cl\cdot \longrightarrow Cl_3C\cdot + HCl$$

III. Termination step—Here further formation of free radical is stopped either by coupling or by disproportionation reactions :

$$\cdot CH_3 + \cdot CH_3 \longrightarrow CH_3—CH_3$$
$$\cdot CH_3 + Cl\cdot \longrightarrow CH_3—Cl$$
$$\cdot CH_2Cl + Cl\cdot \longrightarrow CH_2Cl_2$$
$$Cl_2HC\cdot + Cl\cdot \longrightarrow Cl_3HC$$
$$Cl_3C\cdot + Cl\cdot \longrightarrow CCl_4$$

Therefore, a free radical reaction gives rise to a mixture of products. In practice, sometimes we want to inhibit the propagation step; for this, **inhibitors**, i.e., substances which terminate the reactions are used. Phenol, quinones, diphenylamines, iodine, oxygen, etc. are generally used as inhibitors. Triphenylmethyl radical produced by the reaction of finely divided metallic silver on triphenylmethyl chloride may be inhibited by the oxygen of the air. Oxygen is also an initiator of free radicals. When propene is kept with oxygen, free radicals develop. Molecular oxygen is a diradical; so it acts both as inhibitor and initiator.

The structure of the carbon radicals are very difficult to predict. They have planar to pyramidal structure depending on the groups and atoms attached to the C atom bearing the odd electron. It is, therefore, assumed that such a C atom is either sp^2 or sp^3 hybridised or the percentage of *s* character of the C atom varies from 33.3% to 25%. For example, methyl radical is essentially planar; its C atom is considered to be sp^2 hybridized with the odd electron in the *p* A.O. as shown in the diagram (11.4). On the other hand, trifluoromethyl radical is pyramidal, the involved C atom is considered to be sp^3 hybridised (Fig. 11.4). In the methyl radical bond pair—bond pair interaction of electrons is stronger than the bond pair—lone electron interaction; whereas in the trifluoromethyl radical, the lone electron—bond pair interaction is stronger than bond pair—bond pair interaction because of the high electronegativity of F and the length of C—F bond (which is longer than C—H bond).

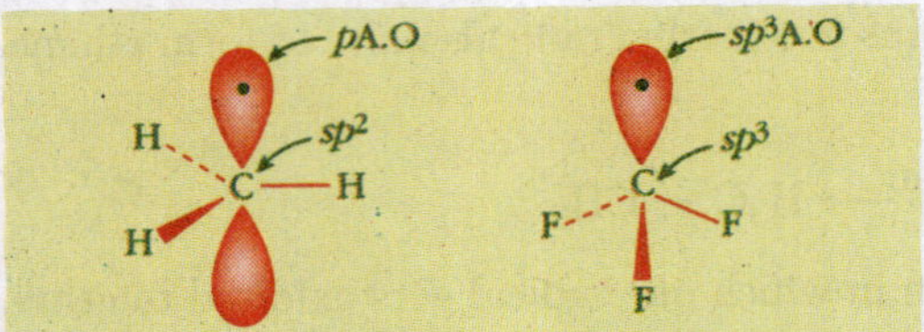

Planar trigonal shape of methyl radical

Pyramidal shape of trifluoromethyl radical

Fig. 11.4. Geometry of carbon radicals

Allylic and benzylic free radicals are also planar; the involved C atoms are supposed to be sp^2 hybridised.

The stability of a carbon radical depends on the degree of delocalisation of the odd electron. As the delocalisation increases, the stability of the radical increases. The delocalisation of an odd electron may take place in two ways : (a) By hyperconjugation and (b) by conjugation.

The stability order of the alkyl radicals, as given below, is explained by the hyperconjugative effect (see problem 4.20).

$$Me_3\dot{C} > Me_2\dot{C}H > Me\dot{C}H_2 > \dot{C}H_3$$

As we move from the tert-butyl radical to methyl radical, we get decreasing number of no-bond resonating structures in their resonance hybrids and thus their stability falls down the series.

The stability of allylic and benzylic radicals are explained by the $p - \pi$ conjugation as given below :

$$[\dot{C}H_2{=}CH{-}\dot{C}H_2 \longleftrightarrow \dot{C}H_2{-}CH{=}CH_2] \equiv [CH_2{\cdots}CH{\cdots}CH_2]^{\cdot}$$

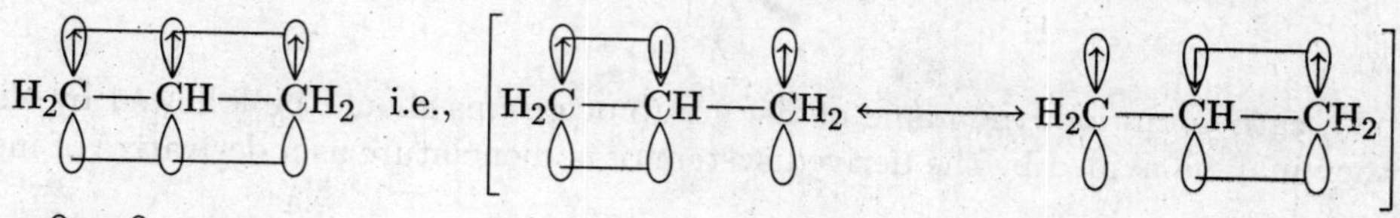

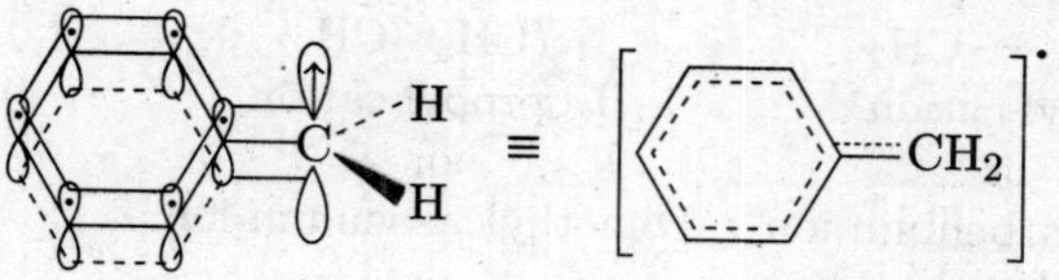

We know that bond formation energy and bond dissociation energy are equal in magnitude but opposite in sign. Table 11.1 also represents the bond formation energy of those radicals. High energy content means instability; hence radicals with higher energy are less stable than those with lower energy.

A free radical may be detected by the study of magnetic susceptibility and electron-magnetic-resonance.

Problem 11.1. Write down the stability order of the following radicals : Benzyl, diphenylmethyl and triphenylmethyl.

Solution : The structures of the given radicals are as follows :

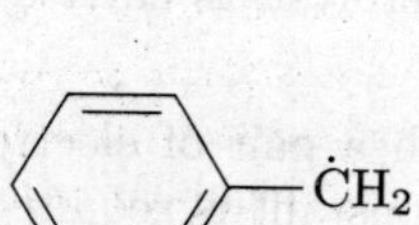

Benzyl radical

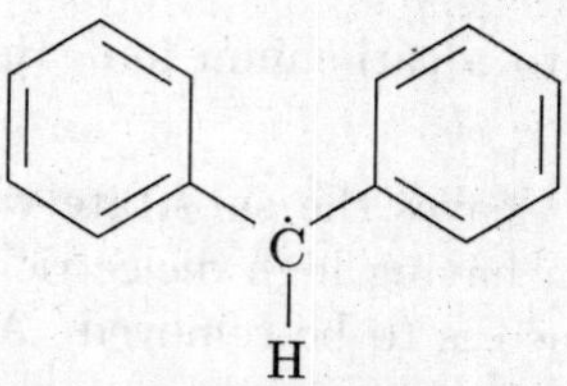

Diphenylmethyl radical

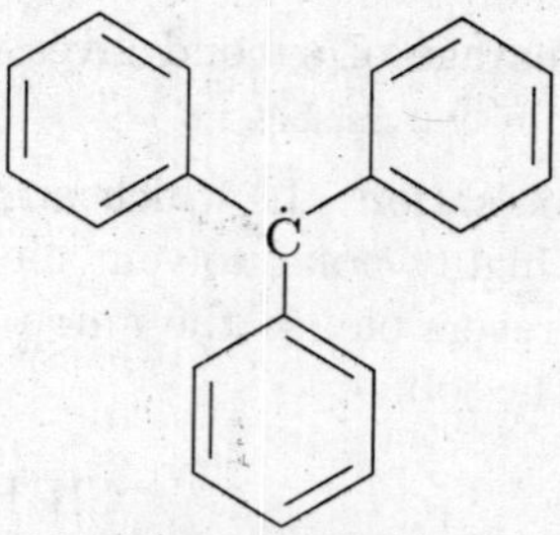

Triphenylmethyl radical

The odd electron of each of the radicals can get delocalised by $p - \pi$ conjugation because the C atoms bearing the odd electrons are in sp^2 hybridised state and Cs of the benzene rings are also in the same state. Again, as the number of benzene ring increases from benzyl to triphenylmethyl radical, the probability of delocalisation of the electron over a greater range increases. We know that greater delocalisation of electrons means greater stability of the species. Hence stability order of the given radicals is :

$$Ph\dot{C}H_2 < Ph_2\dot{C}H < Ph_3\dot{C}$$

It is found that with increasing number of Ph group, the interaction between two ortho-H atoms of two rings increases. This disturbs the coplanarity of the rings though the three C-Ph bonds in the triphenylmethyl radical are very much planar; it has a propeller-like shape.

Carbenium ions or carbocations

Carbenium ions are positively charged species in which a C atom bears a positive charge and six bonded electrons, i.e., three bonds. Hence a carbenium ion may be represented as

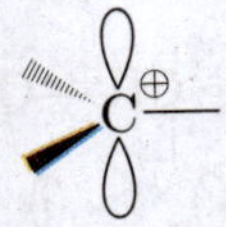

To name a carbenium ion, the name of the group or groups should be followed by the word cation; or they may be named by the derived system of nomenclature as a derivative of methane.

$\overset{\oplus}{C}H_3$	$CH_3—\overset{\oplus}{C}H_2$	$(CH_3)_2\overset{\oplus}{C}H$
Methyl cation (carbenium ion) or carbocation	Ethyl cation or Methylcarbenium ion or methyl carbocation	Isopropyl cation or Dimethylcarbenium ion or dimethyl carbocation

Problem 11.2. Write down the names of the following species :

(i) $Me_3\overset{\oplus}{C}$ (ii) 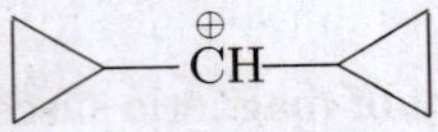 (iii) 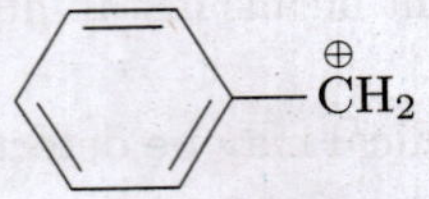

Solution :

(i) tert-Butyl cation or trimethylcarbenium ion (trimethyl carbocation).

(ii) Dicyclopropylmethyl cation or dicyclopropylcarbenium ion.

(iii) Benzyl cation or phenylcarbenium ion.

Heterolytic cleavage of a bond gives rise to a carbenium ion. Heterolysis is carried out by several methods :

1. **By direct ionisation**—In which a group leaves the substrate with a pair of electrons. In this case a highly polar solvent medium having high dielectric constant is required. This reaction is reversible, so the anion formed is to be removed. Ag^+, Zn^{++}, Hg^{++}, etc. can readily do the job.

$$CH_3Br \xrightleftharpoons{Ag^+} AgBr + \overset{\oplus}{C}H_3$$

2. **By the addition of acids to multiple bonds**—Acids used may be either protic acid or lewis acid.

(a) $CH_2{=}CH_2 \xrightleftharpoons{H^{\oplus}} CH_3—\overset{\oplus}{C}H_2$ (b) $CH_3—\underset{H}{\overset{|}{C}}{=}O \xrightleftharpoons{H^{\oplus}} CH_3—\underset{H}{\overset{\oplus}{C}}—OH$

(c) $CH_3C{\equiv}N\!: \xrightleftharpoons{AlCl_3} CH_3\overset{\oplus}{C}{=}N—\bar{A}lCl_3$

Addition of a cation to a multiple bond also generates a carbenium ion.

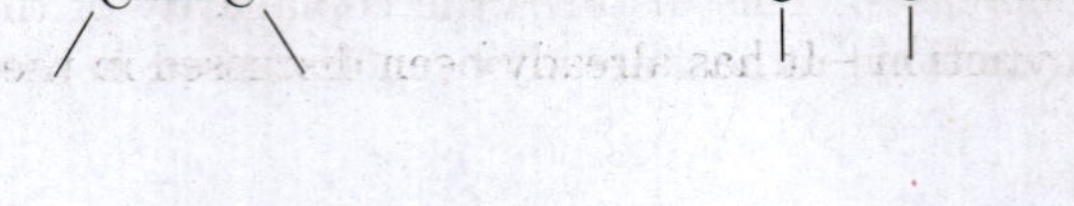

3. **By the abstraction of an atom or group using protic or lewis acids**—In which a lone pair of electrons of the substrate coordinates either with the proton or with the lewis acid and then decomposes to a carbenium ion.

(a) 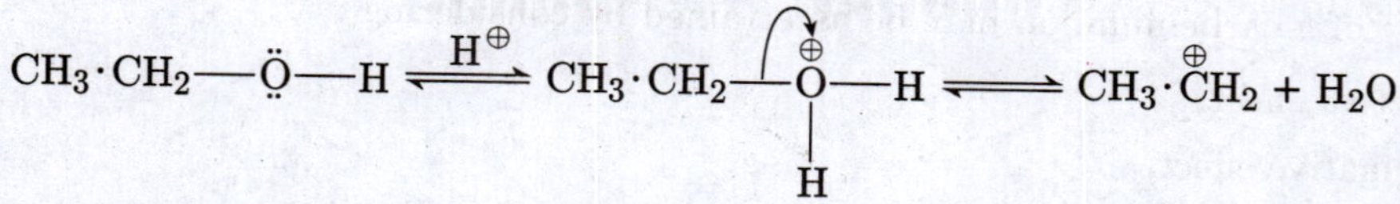

(b) $CH_3-\ddot{Cl}: \xrightleftharpoons{SbCl_5} CH_3-\overset{\oplus}{Cl}-\overline{Sb}Cl_5 \rightleftharpoons \overset{\oplus}{C}H_3 + SbCl_6^-$

4. **By the decomposition reaction**—In which a species decomposes to produce a carbenium ion. Thus diazonium ions decompose and carbenium ions form.

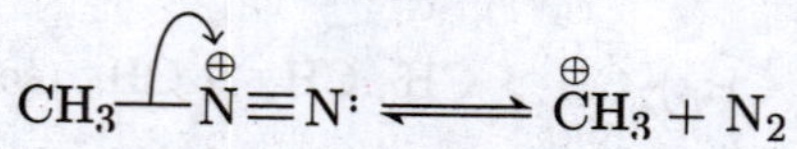

5. **By the rearrangement of a carbenium ion**—In which a carbenium ion rearranges to produce a more stable carbenium ion. Neopentyl cation ion rearranges readily.

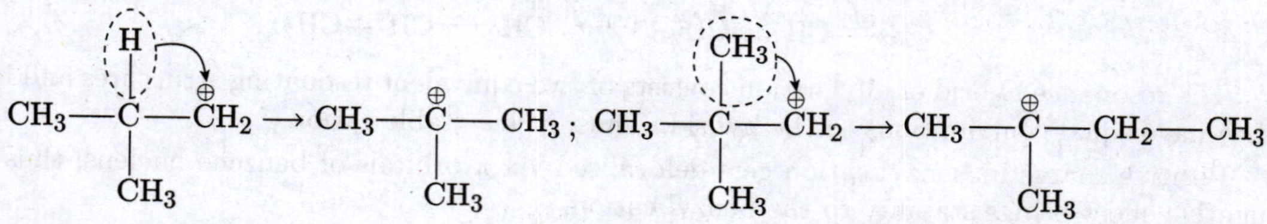

Neopentyl cation
highlly strained

Carbenium ions are short-lived species. They can accept electrons and hence may be looked upon as lewis acids. They undergo four types of reactions :

1. **Combination with an anion**—In which one often gets a stable product. Thus, when HCl is added to ethylene, first ethyl cation is formed which combines with Cl^- ion to form ethyl chloride.

$$CH_2{=}CH_2 \xrightleftharpoons{H^{\oplus}} CH_3\cdot\overset{\oplus}{C}H_2 \xrightarrow{Cl^-} CH_3CH_2Cl$$

2. **Elimination of a proton**—In which a stable unsaturated compound is formed; for instance, isopropyl cation gives up a proton to form propylene.

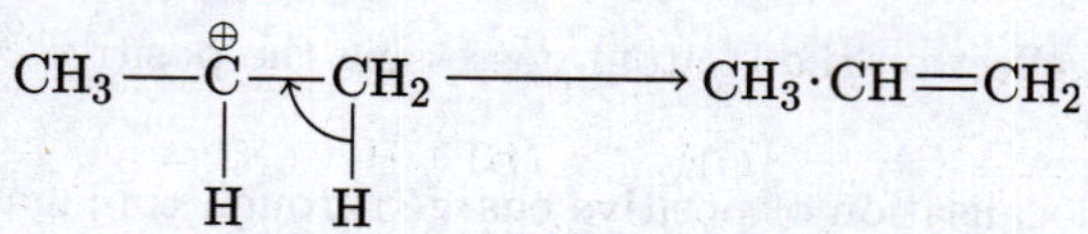

3. **Addition to a multiple bond**—In this case another cation is formed which then produces a stable compound by another reaction.

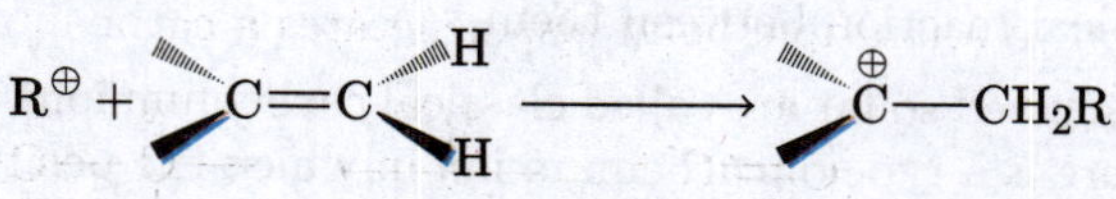

4. **Rearrangement reaction**—It has already been discussed in the formation of carbenium ions.

The three bond axes of a cationic carbon have been found to have a planar trigonal orientation like those of boron in BF_3 molecule. So the C atom carrying the positive charge is considered to be sp^2 hybridised in a carbenium ion; one of its p A.O.s. remains vacant.

Stability of a carbenium ion may be ascertained by considering :

(a) Inductive effect,

(b) Conjugative effect,

(c) Hyperconjugative effect,

(d) Solvation effect.

The stability order of alkylcarbenium ions are well explained by considering inductive and hyperconjugative effects and the order is found to be :

$$(CH_3)_3\overset{\oplus}{C} > (CH_3)_2\overset{\oplus}{C}H > CH_3.\overset{\oplus}{C}H_2 > \overset{\oplus}{C}H_3 \text{ (see Problem 4.20)}$$

The stabilities of allyl and benzyl cations are found to be high; conjugative effect can explain their stability :

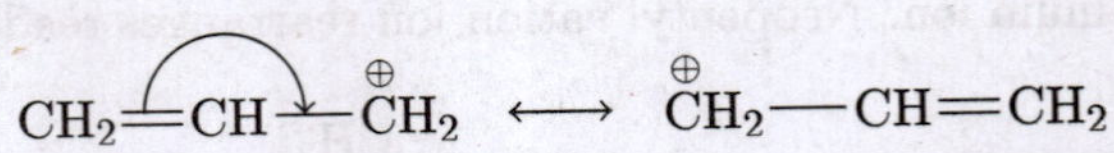

The resonance hybrid of allyl cation consists of two equivalent resonating structures and hence they have equal contributions to the hybrid. Thus, it is a stable cation.

Positive charge in benzyl cation gets delocalised via π orbitals of benzene nucleus; thus conjugative effect brings stability to the benzyl cation.

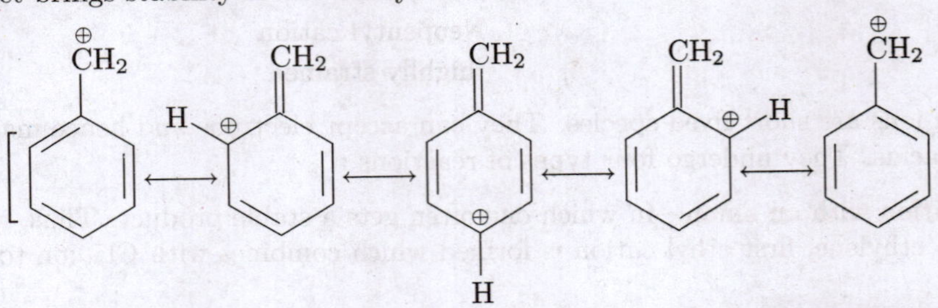

In a nutshell, the following rules are to be remembered for comparing the stability of carbenium ions (carbocations).

1. The more the +I effect on the C atom bearing the positive charge, the more stable is the carbenium ion.
2. The more the −I effect on the C atom possessing the positive charge, the less stable is the carbenium ion.
3. The more the delocalisation of positive charge through conjugation, the more stable is the ion.
4. The more polar the solvent, the more stable is the carbenium ion through solvation, provided there is no chemical reaction between them.

Carbenium ions discussed so far are called classical carbenium ions which contain two-electron two-centre bonds. There is a type of carbenium ion in which the positive charge does not remain on a single C atom but spreads over at least three atoms and those three atoms form a cyclic cation and are thus called bridged carbenium ions or non-classical carbenium ion or, carbonium ion. β-phenethyl cation is an example of non-classical carbenium ion :

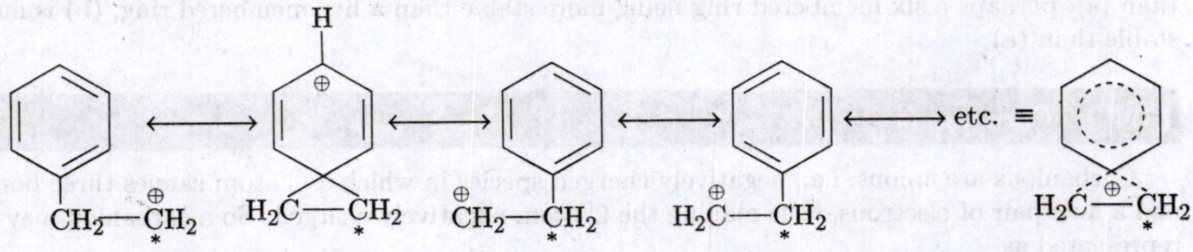

Bridged cations with delocalised bonding σ electrons are called non-classical carbenium ions or carbocations which contain two-electron three-centre bonds. These are now called carbonium ions.

Non-classical norbornyl cation forms as an intermediate when exo- and endo-norbornyl brosylates (brosyl group is p-$BrC_6H_4SO_2$) are separately subjected to acetolysis in the acetic acid medium containing potassium acetate. Norbornyl cation then gives racemic norbornyl acetate though the two substrates, the exo- and endo-forms, are diastereomeric. However, the former one

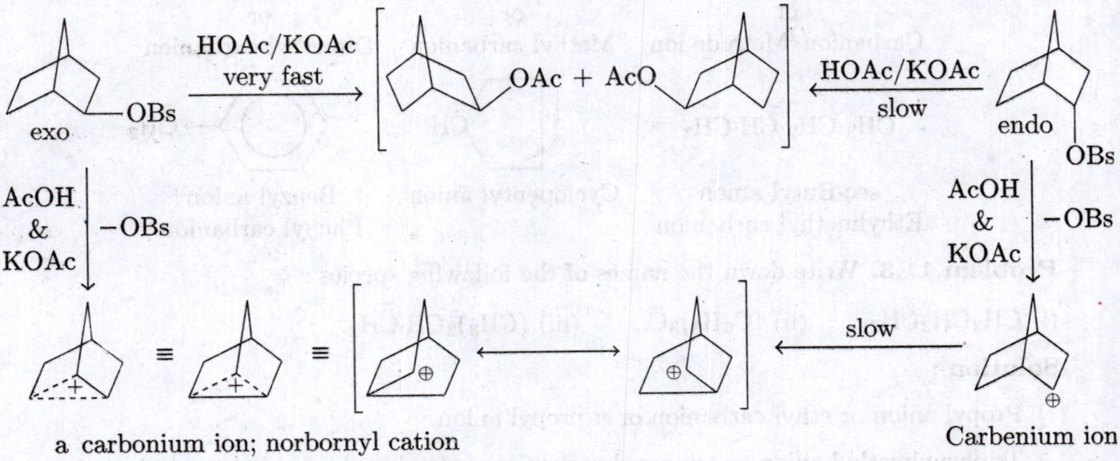

reacts much faster than the latter, and direct non-classical carbocation formation by the neighbouring group participation (Sec. 11.7.3) is said to be the cause for the very fast rate of the reaction. The endo-isomer, under the same set of reaction conditions, undergoes the reaction but slowly. It is supposed that a classical carbenium ion forms first which then changes slowly to the non-classical carbenium ion before the reaction with acetic acid and potassium acetate.

Bridge-head carbenium ions or carbocations are different from bridged or non-classical carbenium ions. C atoms at the junctions of a bridged ring system are called bridge-head carbons. When a bridge-head carbon bears a positive charge, it is called a bridge-head carbocation as shown below :

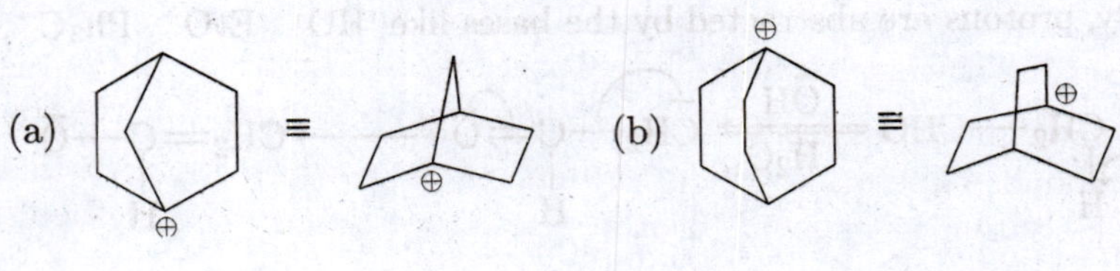

Bridge-head carbenium ions are highly unstable since the cationic carbon cannot assume its usual planar trigonal orientation of its bonds. However, (b) is less strained and hence less unstable than (a); perhaps a six membered ring being more stable than a five membered ring, (b) is more stable than (a).

11.4.3 Carbanions

Carbanions are anions, i.e., negatively charged species in which a C atom carries three bonds and a lone pair of electrons, thus making the C atom negatively charged. So a carbanion may be represented as

$$-\overset{|}{\underset{|}{C}}:^{-}$$

In naming a carbanion, the name of the group or groups should be followed by the word anion; or these may be named as derivatives of methane :

$H_3\bar{C}$:	$CH_3\cdot\bar{\bar{C}}H_2$	$CH_3\cdot\bar{\bar{C}}H\cdot CH_3$
Methyl anion	Ethyl anion	Isopropyl anion
or	or	or
Carbanion/Methide ion	Methyl carbanion	Dimethyl carbanion

$CH_3\cdot CH_2\cdot\bar{\bar{C}}H\cdot CH_3$

sec-Butyl anion
Ethylmethyl carbanion

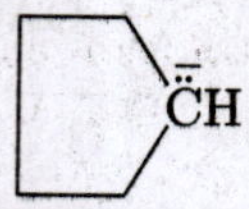

Cyclopentyl anion

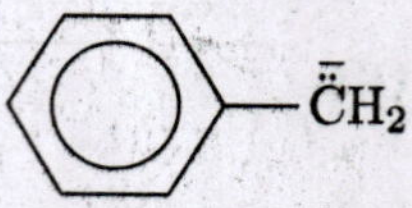

Benzyl anion
Phenyl carbanion

Problem 11.3. Write down the names of the following species :

(i) $CH_3CH_2\bar{\bar{C}}H_2$, (ii) $(C_6H_5)_3\bar{C}$:, (iii) $(CH_3)_2CH\cdot\bar{\bar{C}}H_2$.

Solution :

(i) Propyl anion or ethyl carbanion or *n*-propyl anion.

(ii) Triphenylmethyl anion or triphenyl carbanion or triphenylmethide ion,

(iii) Isobutyl anion or isopropyl carbanion.

Like carbenium ions, carbanions can be generated by the heterolytic cleavage of bonds; the heterolysis may be carried out in several ways which are parallel with the methods of generating carbenium ions.

1. **By the abstraction of an atom :** Using a base in which a cation is removed.

$$R—X \rightleftharpoons \bar{\bar{R}} + X^{\oplus}$$

Generally, protons are abstracted by the bases like, HO^-, EtO^-, Ph_3C^-, etc.

(a)

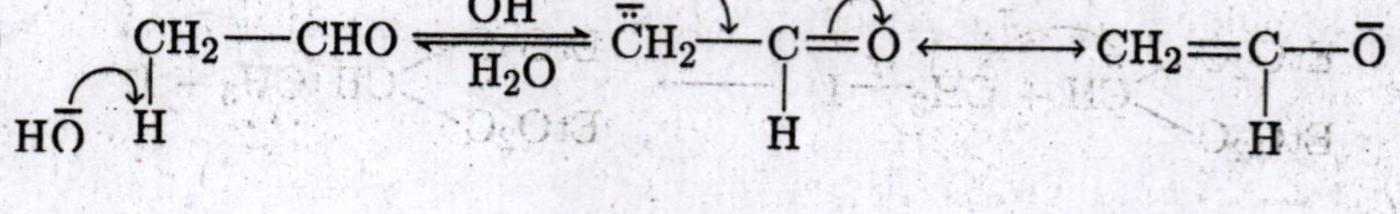

(b) 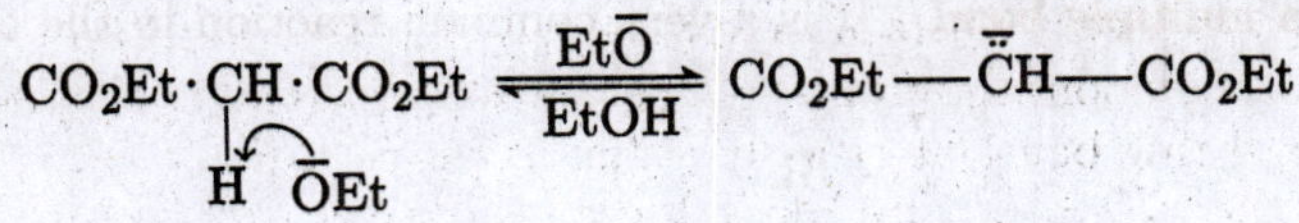

(c) 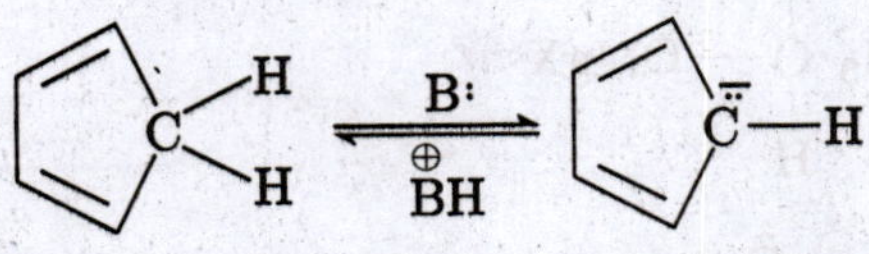

Note : Atoms other than hydrogen are also removed.

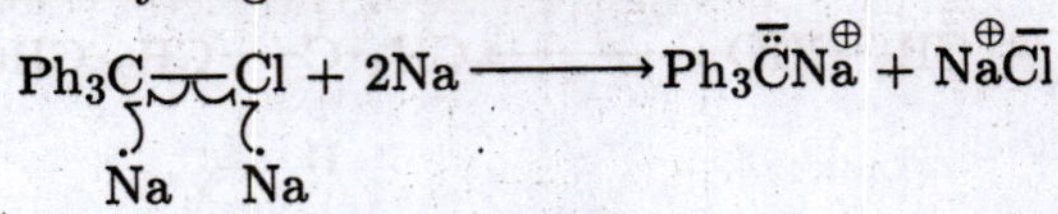

Perhaps this is an example of homolytic cleavage.

2. By the decomposition of anions : Usually carboxylate ions decompose to carbanions.

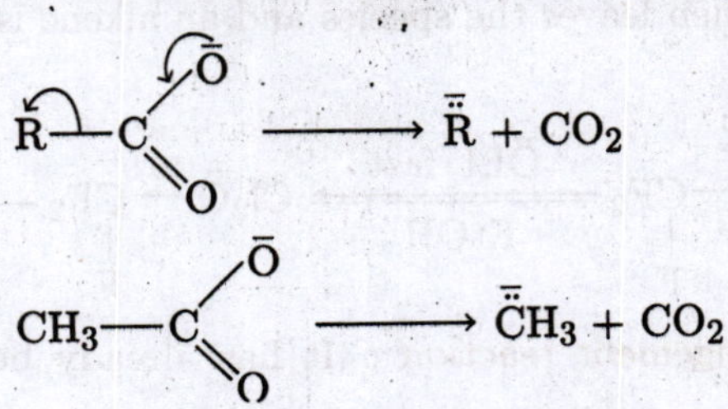

3. By the addition of anions to a multiple bond : When highly electron-withdrawing groups are present on multiple-bonded C atoms, the latter atoms can add up anions and thereby carbanions can form :

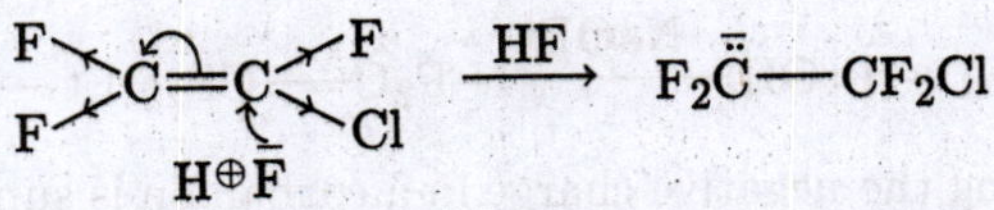

4. By the rearrangement of carbanions : A less stable carbanion rearranges to a more stable carbanion.

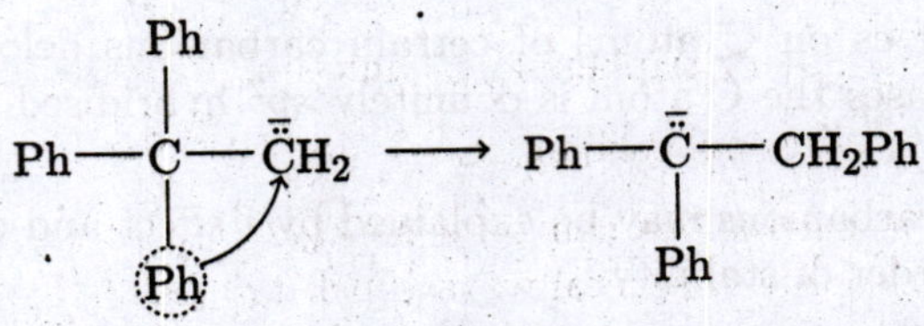

A carbanion is capable of donating electrons to other species. Therefore, it behaves as a Lewis base. These are short-lived species. As an intermediate, a carbanion may undergo all the four types of general reactions. Amongst these displacement and carbanion addition reactions are very common; whereas carbanion rearrangement reactions are rare.

1. Displacement reaction : It displaces halogen atom from an alkyl halide and forms an alkylated product.

2. **Carbanion addition to a multiple bond :** It is a very common reaction in the chemistry of carbonyl compounds.

(a) 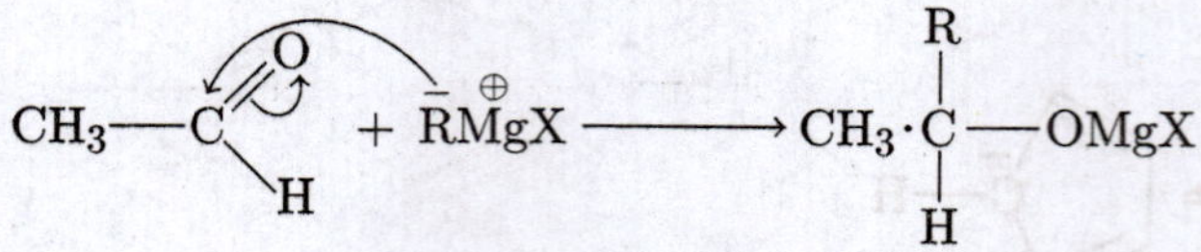

(b) 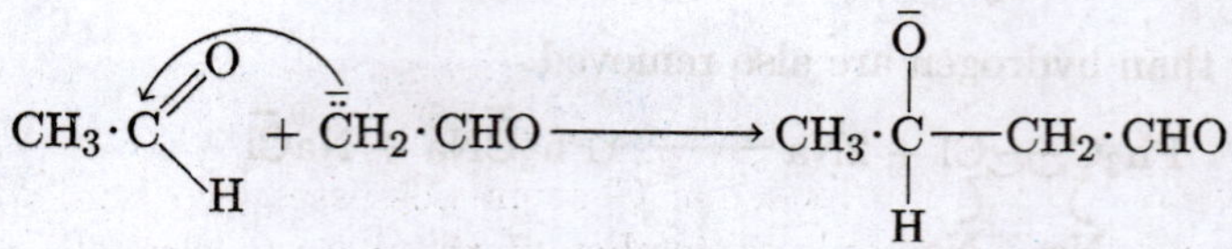

Aldol condensation reactions involve carbanion additions.

3. **Elimination reaction :** When Cl_2CHCF_3 is treated with alcoholic alkali a carbanion results. The fluoride ion then leaves the species and an alkene is formed.

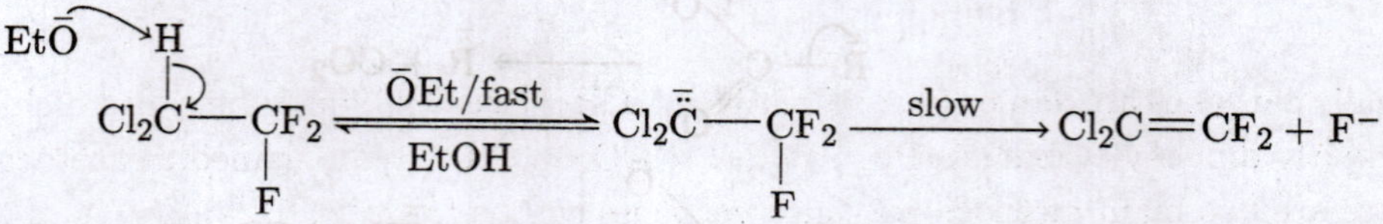

4. **Carbanion rearrangement reaction :** It has already been discussed in the formation of carbanion.

5. **Combination with cation :** The most common reaction of the carbanions is the combination with cations, e.g., protons combine readily with them.

$$F_2C{=}CF_2 + EtOH \xrightarrow{NaOEt} F_2\bar{\bar{C}}{-}CF_2OEt \xrightarrow{H^{\oplus}} F_2CH{-}CF_2OEt$$

The C atom bearing the negative charge in a carbanion is supposed to be sp^3 hybridised since it has a pyramidal structure like ammonia. In fact, it is isoelectronic with amines. The two unshared electrons reside in one of the four sp^3 A.O.s.

The lone pair of *e*s on C atoms of certain carbanions delocalise involving resonance and therefore, in those cases the C atom is definitely sp^2 hybridised and the lone pair occupies a *p*. A.O..

The stability of carbanions may be explained by I effect and conjugative effect. Alkyl anions have the following order of stability :

$$\bar{\bar{C}}H_3 > CH_3\bar{\bar{C}}H_2 > (CH_3)_2\,\bar{\bar{C}}H > (CH_3)_3\,\bar{\bar{C}}$$

This order can well be explained by inductive effect. +I effect increases the electron density on the negatively charged C atom of an anion; thus as the number of methyl group increases on the anionic C the negative charge also increases on it and consequently the stability of the species decreases. Therefore, the stability decreases down the series.

The high stability of allyl and benzyl anions are explained by the conjugative effects. We know that the greater the delocalisation of negative charge, the greater is the stability.

$$CH_2{=}CH{-}\bar{\bar{C}}H_2 \longleftrightarrow \bar{\bar{C}}H_2{-}CH{=}CH_2$$

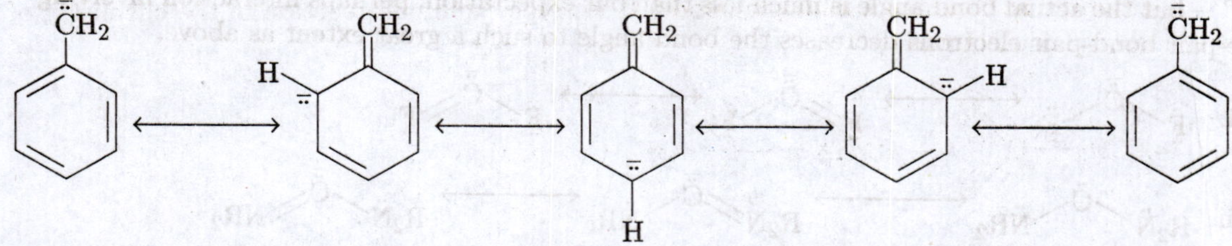

11.4.4 Carbenes or Methylenes

Carbenes are very short lived species in which one C atom possesses two bonds and two electrons, either paired or unpaired. The simplest member of the class is methylene, a non-isolable species of the formula H_2C:; besides this, the most common carbene is $:CCl_2$. The formulae of a few more carbenes are given below:

$$R_2C{=}C{=}C:,\qquad C_6H_5{-}\ddot{C}{-}C_6H_5,\qquad HC{-}C({=}O){-}R$$

Usually carbenes are named as a derivative of $:CH_2$; thus $:CCl_2$ may be called dichlorocarbene and similarly diphenylcarbene is $Ph_2\ddot{C}$; $CH_3{-}CO{-}\ddot{C}H$ may be named as acetocarbene.

There are two main methods by which carbenes can be formed :

(a) By elimination reaction : Formation of dichlorocarbene by the action of alcoholic KOH on chloroform is an example of this method. This may also be obtained from trichloroacetate by the process of thermolysis.

$$CHCl_3 \xrightarrow[-\,HCl]{\text{alc.KOH}} Cl_2C: \xleftarrow[-\,CO_2\ \&\ -\,Cl^-]{\text{heat}} CCl_3COO^-$$

(b) By decomposition reaction : When diazomethane or ketene are pyrolysed or irradiated with ultraviolet light, they give rise to carbenes :

$$CH_2{=}C{=}O \xrightarrow[\text{or, } h\nu]{\text{heat}} :CH_2 \xleftarrow[\text{or, } h\nu]{\text{heat}} CH_2{=}\overset{\oplus}{N}{=}\ddot{\bar{N}}:$$

The two non-bonded electrons of a carbene may be either unpaired or paired. When these are unpaired, the carbene is said to be in a triplet state (spectrally). If they are paired, then the carbene is called singlet ('triplet state' indicates two unpaired spins, whereas 'singlet state' refers to paired spins) carbene.

Both singlet and triplet carbenes have bent shape; bond angles are around 103° and 136° respectively. There are a few singlet carbenes which have resonance stabilisation involving an α-atom with unshared pair of *es* as shown in the next page.

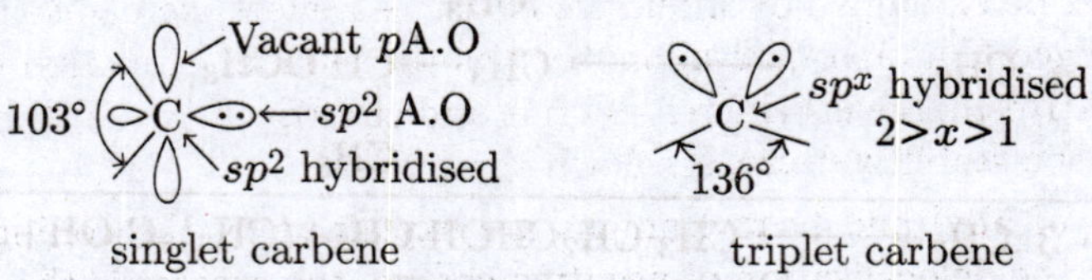

singlet carbene triplet carbene

At least in the above cases the C atom holding the lone pair of *es* is expected to be in the sp^2 hybridised state and it has a vacant p A.O. for back bonding. So, the expected bond angle is

120° —but the actual bond angle is much less than our expectation; perhaps interaction involving lone-pair bond-pair electrons decreases the bond angle to such a great extent as above.

i. 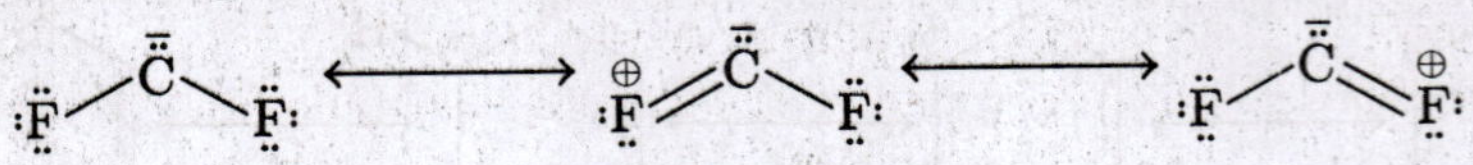

ii. 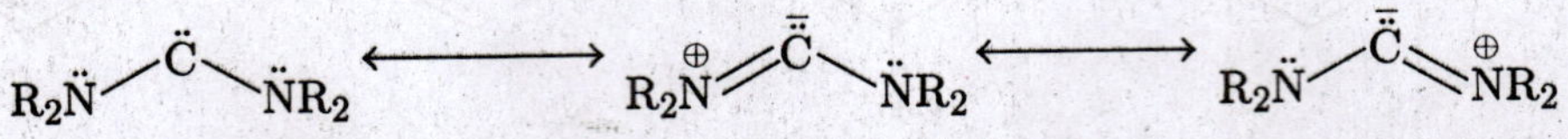

Usually a triplet carbene is more stable than the corresponding singlet carbene because in the latter two electrons get accommodation in a single orbital though there is a vacant one of almost same energy. However, singlet carbenes having resonance stabilisation are found to be more stable than the corresponding triplet state. Singlet difluorocarbene has been found to be more stable than its triplet state owing to resonance as shown above.

A singlet carbene is usually generated in the liquid phase in situ, i.e. within the reaction medium where carbene is to be reacted with other reagents, by α-elemination reactions and by thermolysis or photolysis as stated above.

By the same methods, thermolysis and photolysis, triplet carbenes are formed in the gas phase in the presence of an inert gas like nitrogen, argon, etc. which acts as a sensitiser. Through collision between the singlet carbene, which forms first, and the inert gas molecule, triplet carbenes, the ground state carbenes are generated.

Carbenes have a lifetime less than one second. They are, therefore, unstable and highly reactive. A singlet carbene may act as a Lewis base as it can donate its non-bonded electron pair. A singlet carbene also acts as a Lewis acid by accepting two electrons in its vacant orbital; thus monochlorocarbene reacts with butyllithium and donates its electron pair to lithium.

$$CH_3CH_2\,CH_2CH_2Li + :\underset{\text{(singlet)}}{CHCl} \longrightarrow CH_3CH_2CH_2CH_2{-}\underset{\underset{\displaystyle Li}{|}}{CHCl}$$

However, singlet carbene undergoes the following types of reactions :

(a) Insertion, (b) Addition, (c) Rearrangement, (d) Abstraction and (e) A reaction leading to dimerisation.

A triplet carbene undergoes addition reactions. In some cases not exactly the carbenes but some carbene-like, i.e., carbenoid species undergo the addition reaction.

(a) Insertion : Singlet carbenes undergo unusual insertion reactions but triplet carbenes do not; the reaction occurs in C—H, bonds but does not at all occur in a C—C bond.

$$H{-}CH_2{\cdot}CHOH{\cdot}CH_3 + :CH_2 \longrightarrow CH_3{\cdot}CH_2{\cdot}CHOHCH_3$$

$$CH_3{-}\underset{\underset{\displaystyle H}{|}}{C}(OH){\cdot}CH_3 + :CH_2 \longrightarrow CH_3{-}\underset{\underset{\displaystyle CH_3}{|}}{C}(OH){\cdot}CH_3$$

$$CH_3{\cdot}\underset{\underset{\displaystyle CH_3}{|}}{CH}{-}O{-}H + :CH_2 \longrightarrow CH_3{-}\underset{\underset{\displaystyle CH_3}{|}}{CH}{\cdot}OCH_3$$

$$3\ CH_3{\cdot}CHOH{\cdot}CH_3 + 3\ :CH_2 \longrightarrow CH_3{\cdot}CH_2CHOH{\cdot}CH_3 + (CH_3)_3C{\cdot}OH + (CH_3)_2\ CH{\cdot}OCH_3$$

Perhaps owing to the stabilisation of the dihalocarbenes in the singlet state they do not undergo insertion reaction, which is a characteristic reaction of unsubstituted singlet carbenes.

(b) Addition : Both the singlet and the triplet carbenes undergo addition reactions to a multiple bond. The addition of a singlet carbene to a multiple bond is a stereoselective and a stereospecific reaction. The addition is syn. When cis-but-2-ene is reacted with chloroform in the presence of potassium tert- butoxide in the liquid phase, cis-1,1-dichloro-2, 3-dimethylcyclopropane is obtained; under the same set of conditions, trans-2-butene gives trans-1,1-dichloro-2, 3-dimethylcyclopropane.

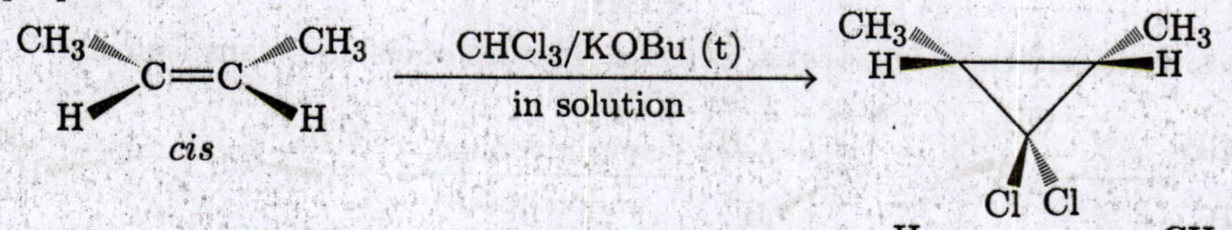

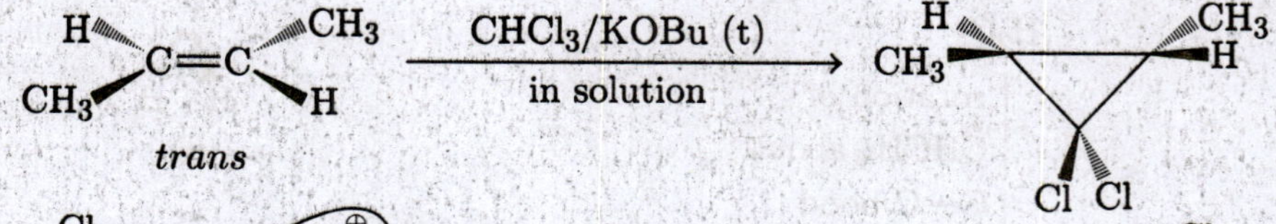

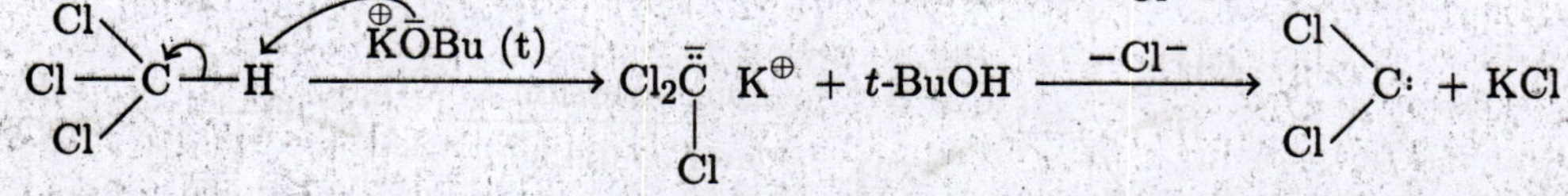

In this step through an α-elimination reaction on chloroform a singlet dichlorocarbene is formed. The electron-deficient (i.e., electrophilic) $:CCl_2$ accepts π electron of the alkene in its empty orbital and donates its lone pair of es simultaneously to the more cationic C atom of the two double bonded carbons and those become partially positively charged in the T.S.

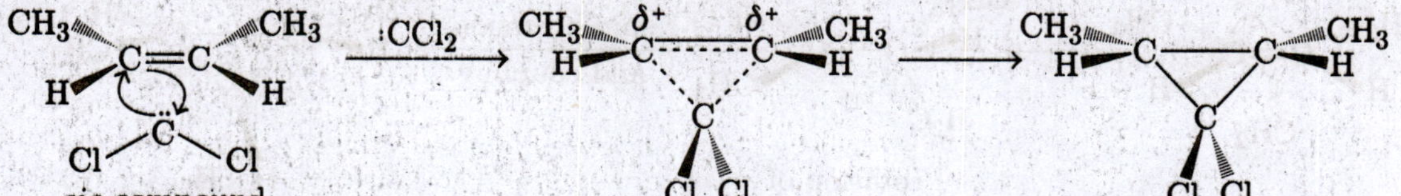

cis compound
cis addition
T.S.
cis product

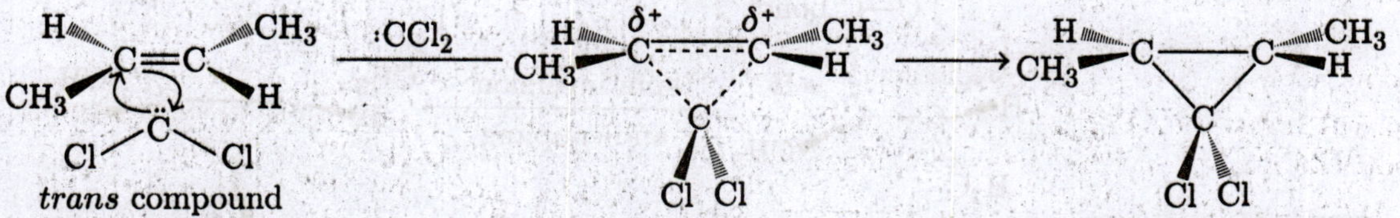

trans compound
cis addition
T.S.
trans product

A triplet carbene undergoes radical addition reaction with a C—C multiple bond since it is a diradical. It is generated in the gas phase in situ by the process of photolysis on a mixture of diazomethane and alkenes; the latter compounds are the substrates for the addition reactions.

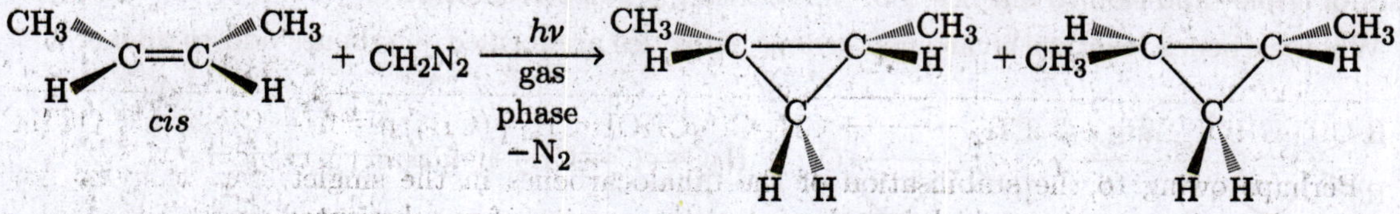

cis *trans*

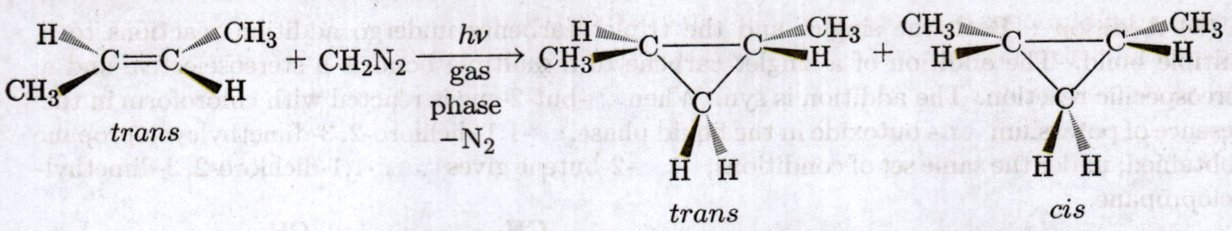

In both the cases of cis and trans substrates we get a mixture of cis and trans isomers.

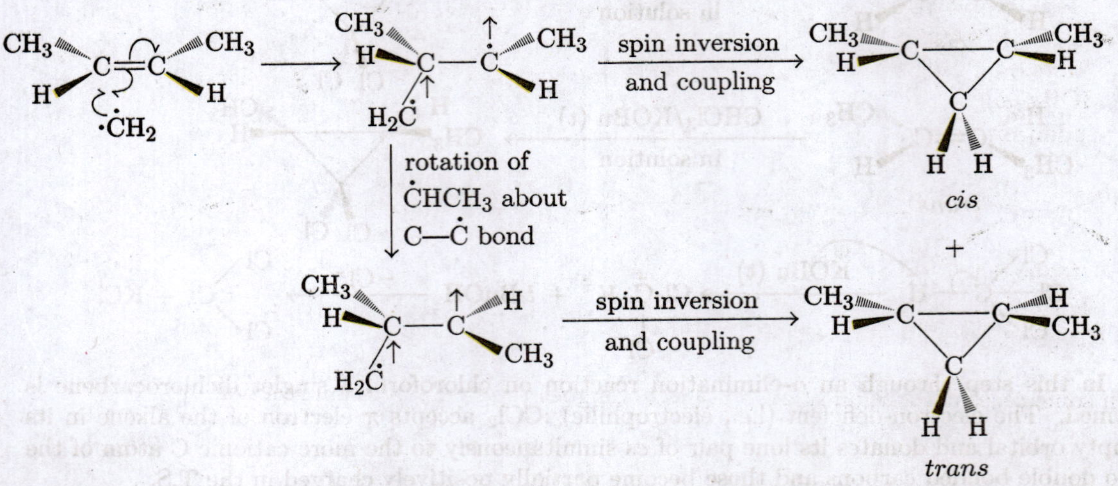

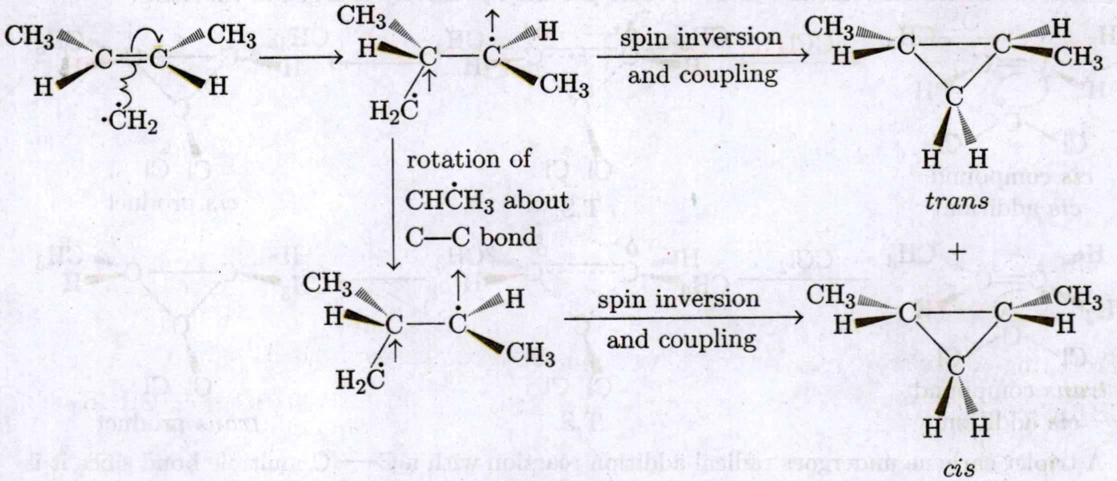

None of the two products, cis or trans, has been formed as the predominant product in all the cases of triplet addition to alkenes. So, the reaction is non-stereoselective.

When carbenes react with carbonyl groups, epoxides are formed. Carbenes add to allene.

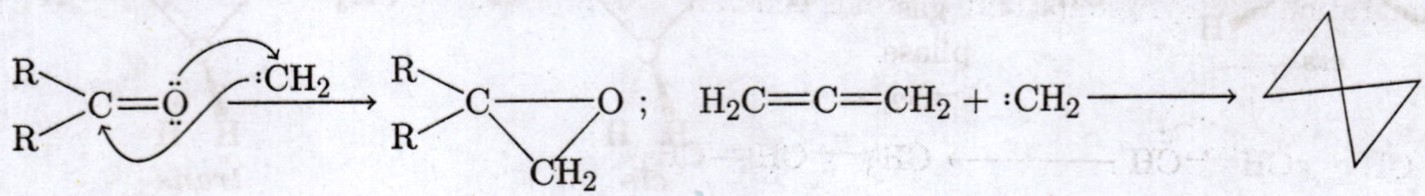

However, in Simmons-Smith reaction not a free carbene but a carbenoid species, an organozinc compound, adds to an alkene and a cyclopropane derivative forms. The reaction is stereoselective as well as stereospecific.

$$CH_2I_2 + Zn(Cu) \xrightarrow{\quad} ICH_2\,ZnI$$

Zinc-Copper couple

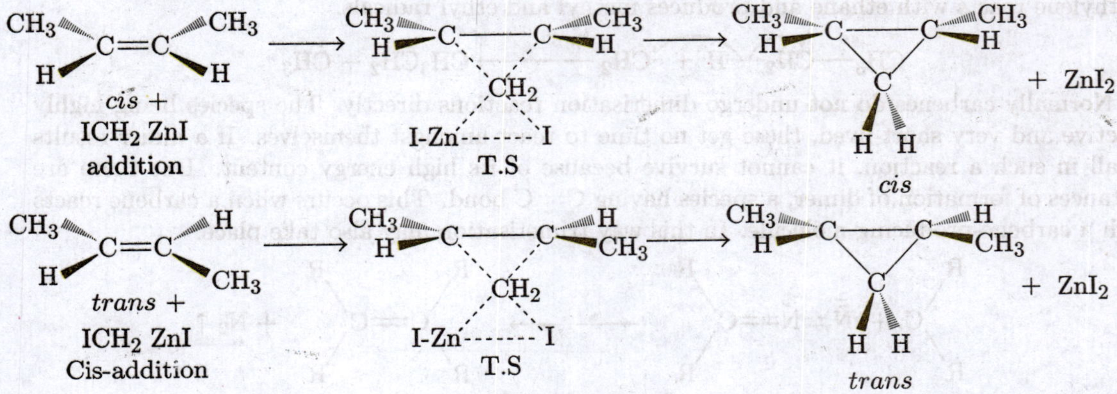

In some cases carbene addition product undergoes the ring expansion reaction.

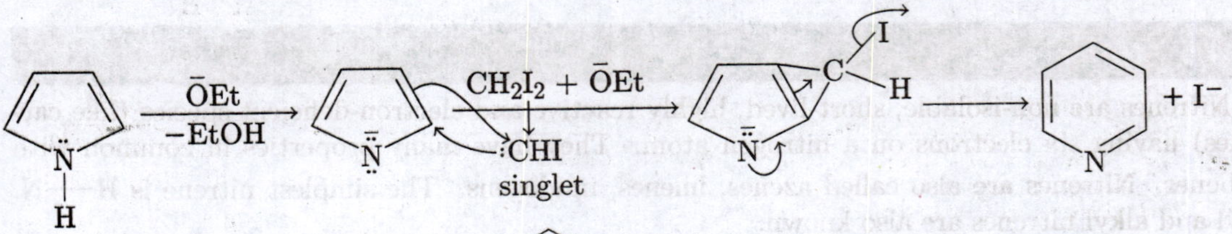

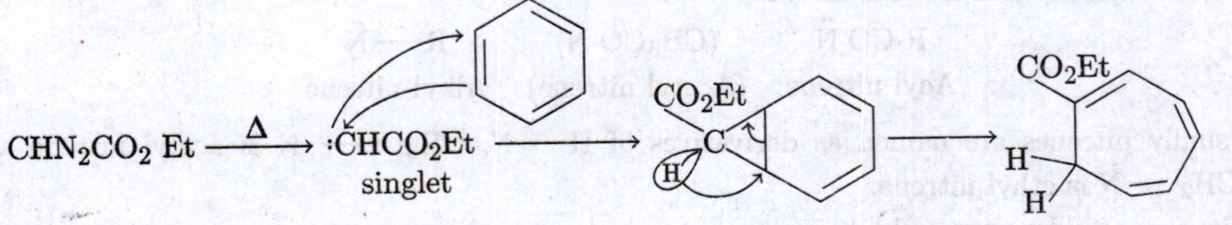

From ring addition products several other compounds can be synthesised. e.g.,

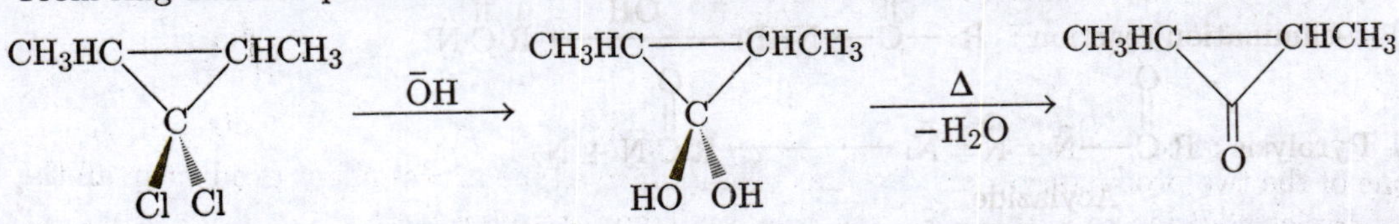

Carbenes undergo rearrangement reactions and stable molecules are obtained. In these reactions migration of alkyl groups and H atoms generally takes place.

(i) 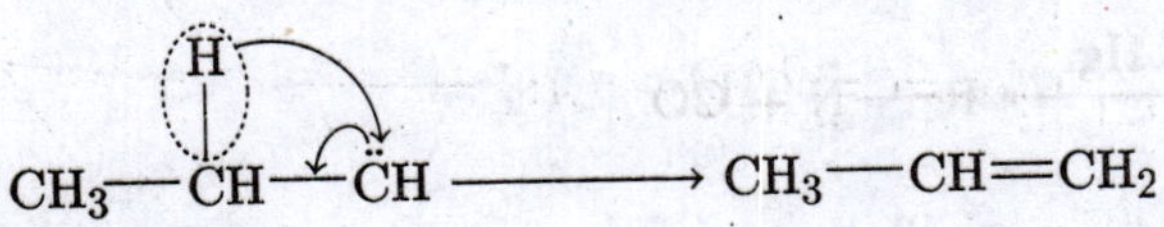

(ii) 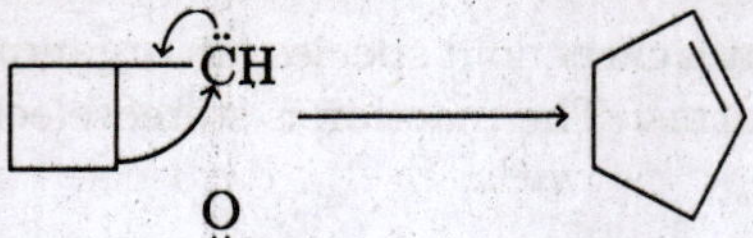

(iii) 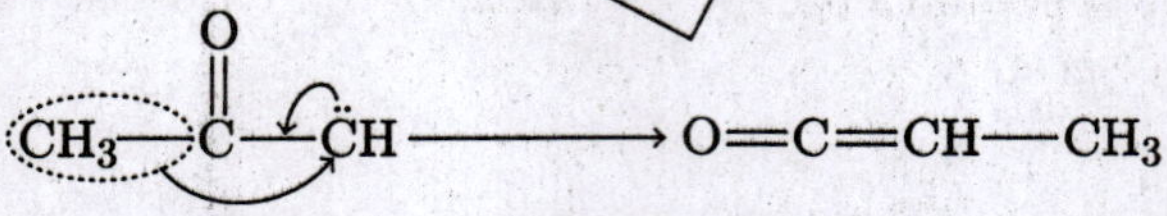

Triplet carbenes also take part in abstraction reactions and thus form free radicals, e.g., methylene reacts with ethane and produces methyl and ethyl radicals.

$$CH_3{-}CH_2{-}H + \dot{C}H_2 \longrightarrow CH_3\dot{C}H_2 + \dot{C}H_3$$

Normally carbenes do not undergo dimerisation reactions directly. The species being highly reactive and very short-lived, these get no time to react amongst themselves. If a dimer results at all in such a reaction, it cannot survive because of its high energy content. But there are instances of formation of dimer, a species having C=C bond. This occurs when a carbene reacts with a carbene-producing molecule. In this way trimerisation may also take place.

$$R_2C: + :\bar{N}{=}\overset{\oplus}{N}{=}CR_2 \longrightarrow R_2C{=}CR_2 + N_2\uparrow$$

Singlet carbenes are diamagnetic in nature but triplet carbenes are paramagnetic to the extent of two odd electrons.

11.4.5 Nitrenes

Nitrenes are non-isolable, short-lived, highly reactive and electron-deficient species (like carbenes) having six electrons on a nitrogen atom. They have many properties in common with carbenes. Nitrenes are also called azenes, imenes, imidogens. The simplest nitrene is $H{-}\ddot{N}$. Acyl and alkyl nitrenes are also known.

$R{\cdot}CO{\cdot}\ddot{N}$	$(CH_3CO{\cdot}\ddot{N})$	$R{-}\ddot{N}$
Acyl nitrene	(Acetyl nitrene)	Alkyl nitrene

Usually nitrenes are named as derivatives of $H{-}\ddot{N}$; $CH_3CO{-}\ddot{N}$ is acetyl nitrene and $CH_3CH_2{-}\ddot{N}$ is ethyl nitrene.

Nitrenes may be prepared by :

(i) α-elimination reaction : $R{-}\overset{O}{\overset{\|}{C}}{-}NHBr \xrightarrow{\bar{O}H} R{\cdot}\overset{O}{\overset{\|}{C}}{\cdot}\ddot{N}$

(ii) Pyrolysis : $R{\cdot}\overset{O}{\overset{\|}{C}}{-}\ddot{N}{=}\overset{\oplus}{N}{=}\bar{N}: \longrightarrow R{\cdot}\overset{O}{\overset{\|}{C}}{\cdot}\ddot{N} + N_2$

Acylazide

(iii) Photoysis : $H{-}\bar{N}{-}\overset{\oplus}{N}{\equiv}N: \xrightarrow[UV]{Hg.\ h\nu} H{-}\ddot{N} + N_2$

Hydrazoic acid

$R{-}\ddot{N}{=}C{=}O \xrightarrow{h\nu,\ Hg} R{-}\ddot{N} + CO$

Alkyl isocyanate

(iv) **Oxidation-reduction method :** When compounds containing —NH group are oxidised with lead tetraacetate, activated mercuric oxide, manganese oxide etc., nitrenes are obtained. Aromatic nitro and nitroso compounds on heating with $(EtO)_3$ P (triethyl phosphate, a reducing agent) give nitrene.

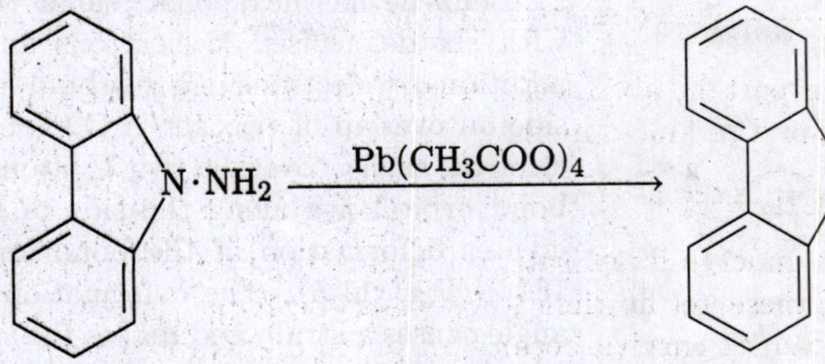

$$Ar\cdot NO_2 \xrightarrow{(EtO)_3P} Ar\cdot NO \xrightarrow{(EtO)_3P} Ar\cdot \ddot{N}$$

There are two types of nitrenes : triplet nitrene and singlet nitrene; triplet is the ground state of nitrene since two degenerated orbitals accommodate one electron each. A singlet nitrene has both the electrons in an orbital.

Nitrenes behave as electrophiles and they undergo usually addition, insertion, rearrangement and H-abstraction reactions.

(a) **Addition reaction :** Nitrenes add to the C=C double bond

(i) 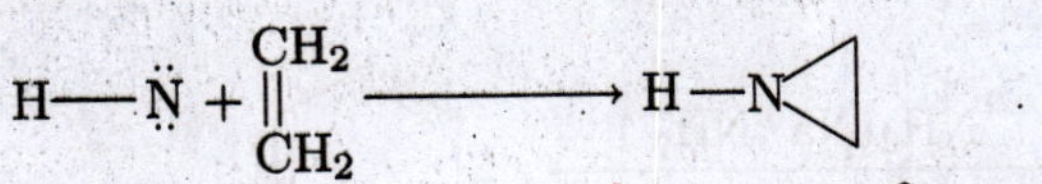

(ii) 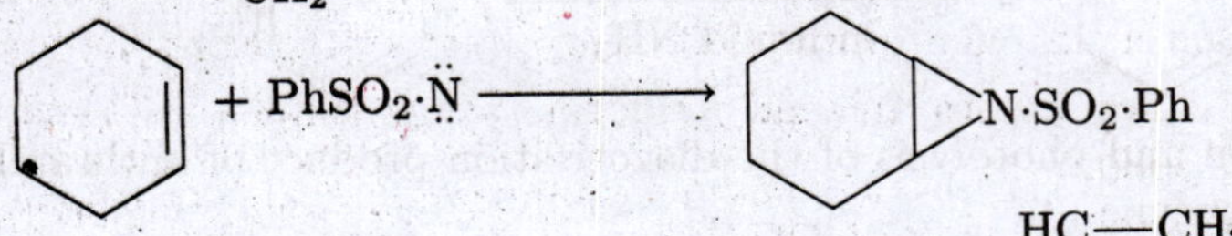

(iii) 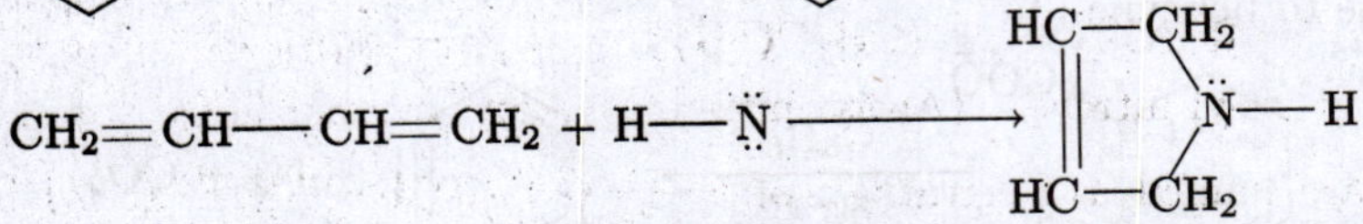

(b) **Insertion reaction :** Nitrenes are found to undergo insertion reaction in C—H bonds :

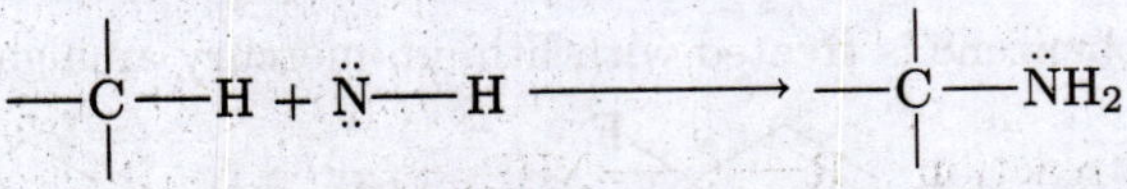

(c) **Rearrangement reaction :** The most easily formed nitrenes usually rearrange :

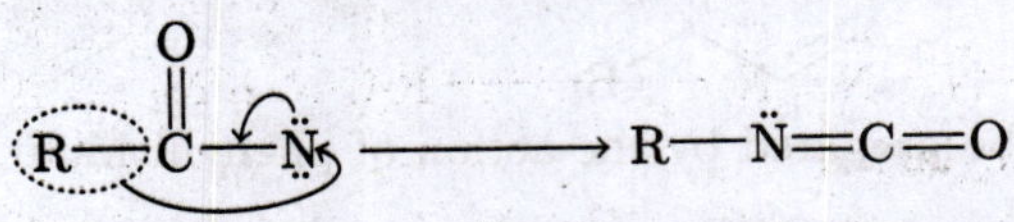

Nitrene Alkyl isocyanate

Note : This is a proposed step in Hofmann, Lossen and Curtius rearrangements; perhaps the formation of nitrene from RCONHBr and its rearrangement to alkyl isocyanate are concurrent steps and no free nitrene forms, but it has been argued that nitrene exists for a short time in the acid-catalysed Lossen rearrangement.

11.4.6 Benzynes

1,2-dehydrobenzene, C_6H_4, and its derivatives may be called benzynes or arynes; the simplest member is benzyne (Fig.11.5). It is a non-isolable highly reactive species.

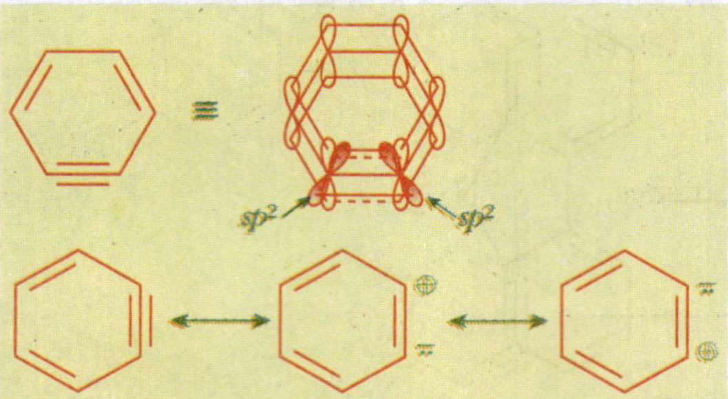

Fig. 11.5

Benzyne has hexagonal planar ring structure with six delocalised π electrons and two other additional π electrons in a π orbital formed by the side-on overlap of two sp^2 A.O.s (containing one electron each) covering two C atoms only; this bond orbital lies along the side of the ring and creates deformation of the bond angle (120° to 180°) (Fig. 11.5). The deformation of the bond angle causes a strain and makes the species highly reactive. Its structure may also be looked upon as a resonance hybrid of some resonating structures (Fig.11.5); dipolar structures explain the electrophilic character of benzyne and the six-membered ring structure containing a triple bond explains the instability and hence the high reactivity of the species. Thus the structure tells us that these two additional electrons do not interact with the π cloud involving Huckel number of π electrons and hence they do not affect the aromaticity of the benzyne molecule.

There are four important methods for preparing benzynes :

(i) When halobenzenes or their derivatives are treated with metal amide ion in liquid ammonia, benzynes are formed.

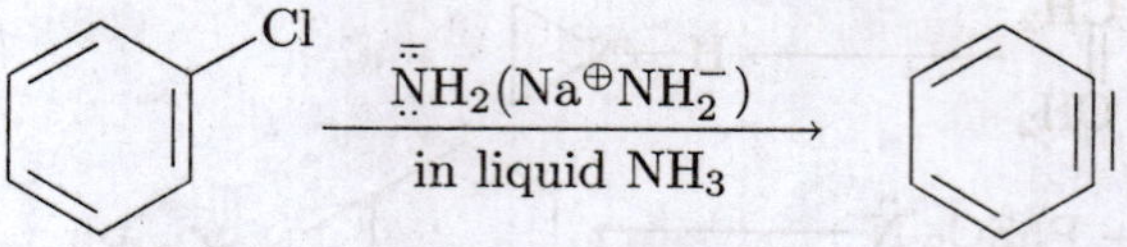

(ii) Thermal decomposition and photolysis of the diazotisation product of anthranilic acid in benzene give rise to benzyne :

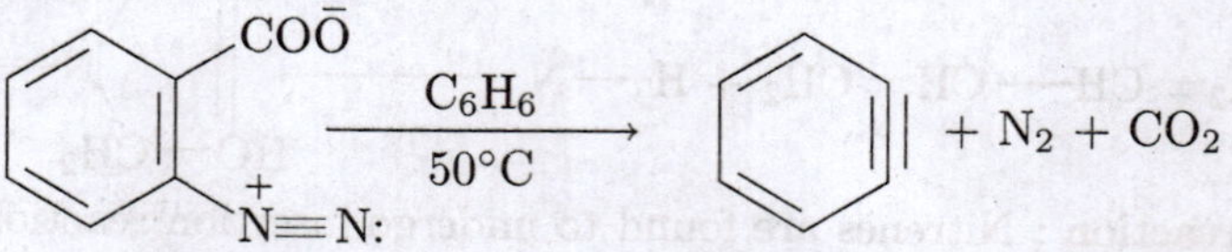

(iii) When *o*-dihalobenzene is treated with lithium-mercury amalgam, benzyne results.

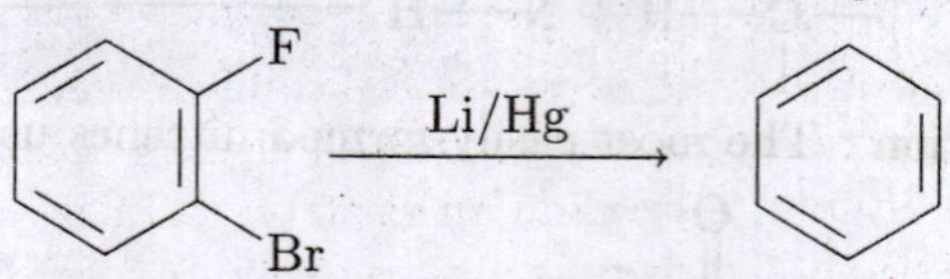

(iv) Benzynes can also be prepared by the action of phenyllithium.

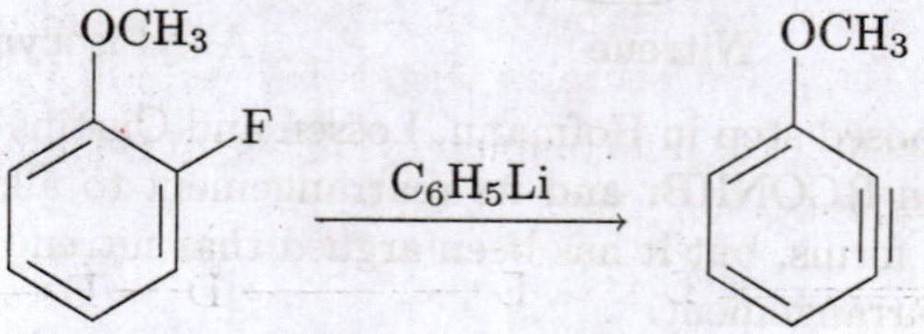

Following properties of benzynes are discussed below :

(i) Benzynes being electrophilic, they undergo nucleophilic addition reactions.

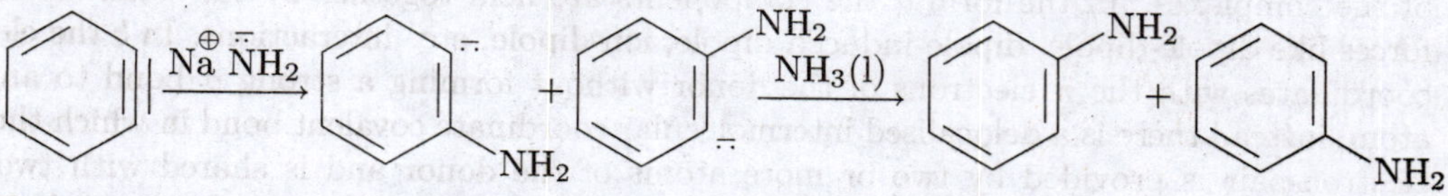

(ii) Benzynes undergo 1,4-addition to conjugated dienes.

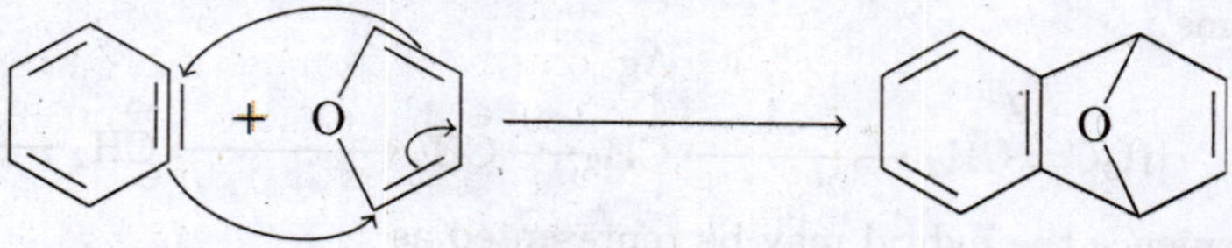

The transient existence of benzynes as intermediates is detected spectroscopically and by trapping them, i.e., making them react with conjugated dienes and isolating the products.

Nowadays a number of syntheses are done via benzyne; e.g., one can prepare phenyl benzoate from anthranilic acid via benzyne formation :

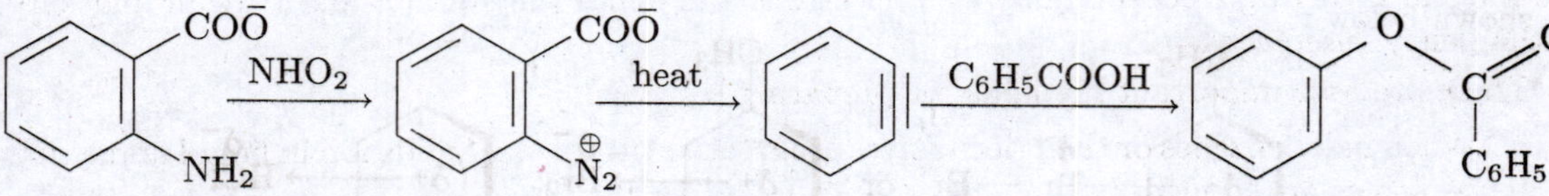

Biphenyles can be prepared by the action of phenyllithium.

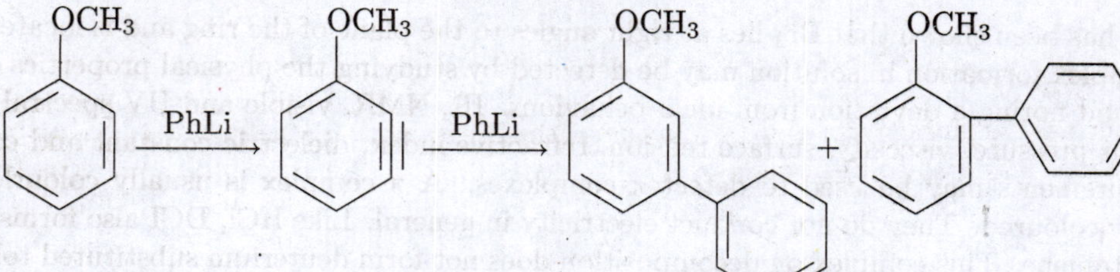

For detailed reactions of benzynes consult aromatic substitution reaction nucleophilic via arynes (Chapter 12).

11.4.7 π-complexes

A weakly bonded charge-transfer complex which exists in solution only and is formed by the association of an electrophilic species (E) and an electron-donating species (D) which utilises its π orbital electrons for association is known as a π-complex. Arenes, alkenes, alkynes, cycloalkenes etc., are the electron-donors and metal ions, molecular halogens, hydrogen halides etc. are the common electrophilic reagents. For example, at $-78°C$ toluene forms a 1 : 1 π-complex with HCl alone. Similarly ethylene forms a π-complex with silver cation ($Ag^{\oplus}$).

The structure of a π-complex is not completely known. It is supposed to be a resonance hybrid of the following resonating forms :

$$D + E \rightleftharpoons \underset{a}{\overset{\delta+}{D}\cdots\overset{\delta-}{E}} \longleftrightarrow \underset{b}{\overset{+}{D}-\overset{-}{E}} \longleftrightarrow \underset{c}{\overset{-}{D}-\overset{+}{E}}$$

Form *a* contributes predominantly in the ground state of the complex, while forms *b* and *c* contribute primarily in the excited state; contribution of *c* is much less than that of *b* in most of the complexes. In the form *a* the components are held together by the weak intermolecular forces like dipole-dipole, dipole-induced dipole, ion-dipole, etc. interactions. In *b* the electrophile coordinates with the π electrons of the donor without forming a strong σ bond to any specific atom, instead there is a delocalised intermolecular coordinate covalent bond in which the bonding electron-pair is provided by two or more atoms of the donor and is shared with two or more atoms in the electrophile. For example, ethylene-silver ion π-complex is a resonance hybrid of the following forms :

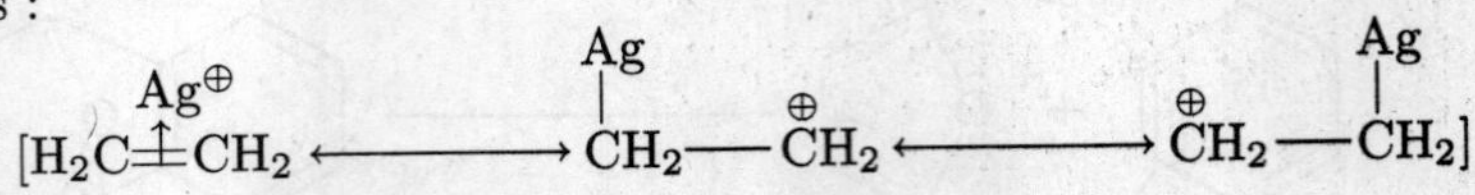

For convenience the hybrid may be represented as

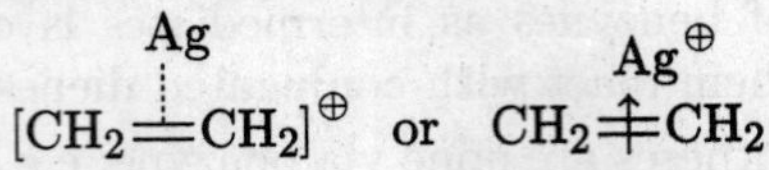

Similarly, toluene-bromine and benzene-hydrogen chloride π-complex may be represented as shown below :

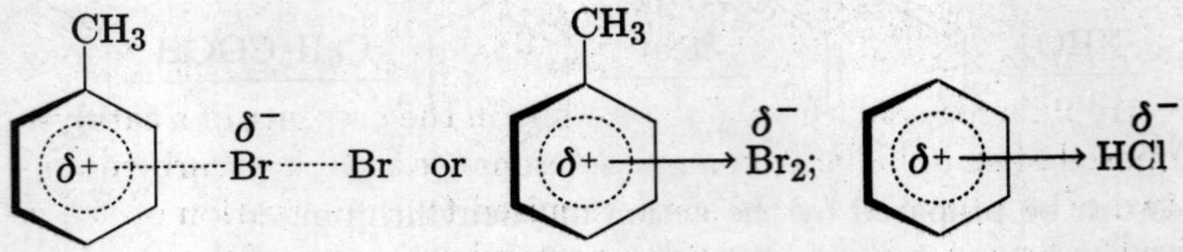

It has been shown that Br_2 lies at right angles to the plane of the ring and is located centrally. π-complex formation in solution may be detected by studying the physical properties of the solution and noting a deviation from ideal behaviour. IR, NMR, visible and UV spectral study, and vapour pressure, viscosity, surface tension, refractive index, dielectric constant and conductivity measurements may be used to detect π-complexes. A π-complex is usually colourless or light yellow coloured. They do not conduct electricity in general. Like HCl, DCI also forms π-complex with toluene. This complex on decomposition does not form deuterium substituted toluene. This indicates that a π-complex does not undergo isotope exchange reactions.

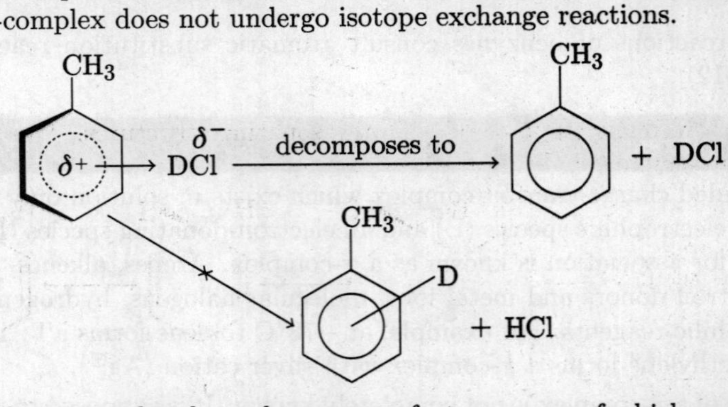

The stability of a π-complex depends on many factors, some of which are discussed below.

The greater the polarisability of the donor component, the more stable is the π-complex. The strength of the π-complex increases with decreasing ionisation potential of the donor and increasing electron affinity of the electrophile. Electron-releasing groups which increase electron density in the donor species and electron-withdrawing groups which decrease electron density in the electrophile increase the strength of the π-complex. In a π-complex the distance between the two planes of the components is usually 0.3 nm to 0.35 nm. If either component is non-planar, the distance between the planes increases, which decreases the atractive force and thereby decreases the stability of the complex.

N.B. : Charge-transfer complexes may exist in solid state or in solution. In such a complex the donor component may utilise its πes or nes to form the complex. Thus a π-complex is a charge-transfer complex but a charge-transfer complex is not always a π-complex.

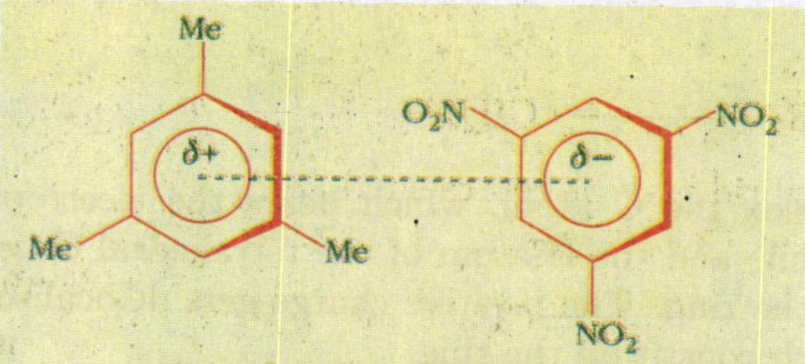

Fig. 11.6. A charge-transfer complex

11.4.8 σ-complexes

When an electrophile reacts with an arene usually in the presence of a catalyst, a salt is formed. This salt is composed of an anion and a complex resonance stabilised carbenium ion (arenium ion) in which only two of the total πes, which are required for preservation of the aromaticity of the arene, are utilised to form a σ-bond between a particular C atom of the ring and the electrophile; such a salt is called a σ-complex salt and the carbenium ion is called a σ-complex or Wehland intermediate. For example, when benzene is treated with HCl in the presence of anhydrous $AlCl_3$, a Lewis acid, a salt of benzenium ion and $\bar{A}lCl_4$, the σ-complex salt, is formed. H—C—E moiety of a σ-complex lies on a plane perpendicular to the plane of other ring carbons as given below.

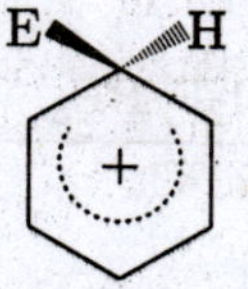

A σ-complex is a resonance hybrid of several resonating structures. The benzenium ion may be represented as a resonance hybrid.

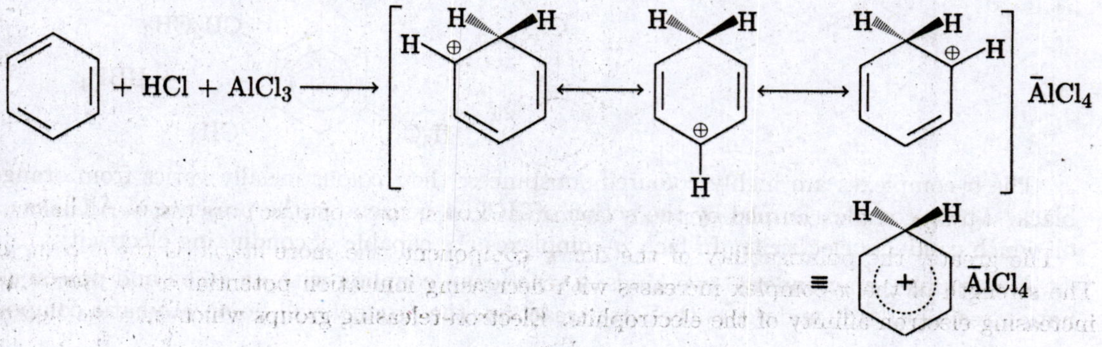

Similarly, in the presence of conc. H_2SO_4 benzene forms a σ-complex with nitronium ion.

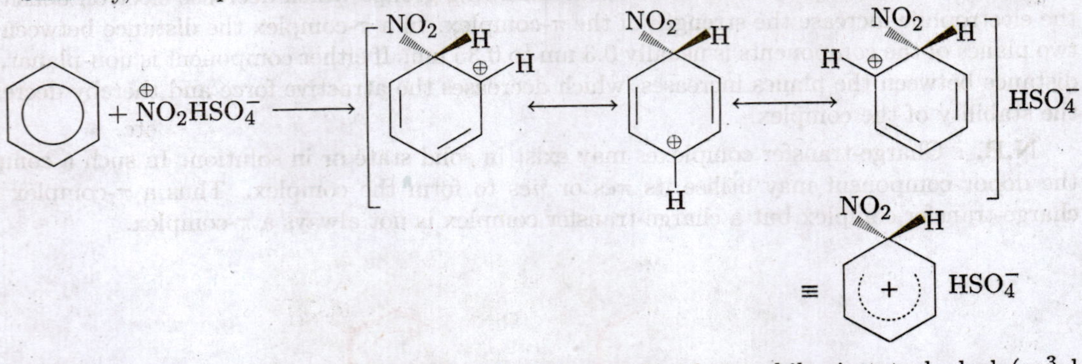

Thus in the σ-complex the C atom, which bears the electrophile, is tetrahedral (sp^3 hybridised). The electrophile and the H atom of the tetrahedral C lie in a plane which is perpendicular to the plane of the ring. The positive charge gets delocalised over the remaining atoms, through π orbitals, which constitute the ring.

It is assumed that a σ-complex may form via a π-complex and an equilibrium exists between the reactants, the π-complex and the σ-complex.

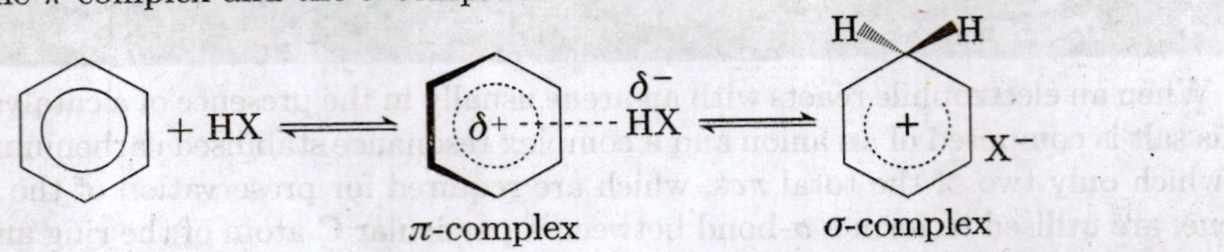

Nowadays several sigma-complex salts have been isolated. For example, the σ-complex salt formed by the reaction of ethylfluoride on 1,3,5-trimethylbenzene in the presence of catalyst BF_3 at $-80°C$ has been isolated as an orange solid which melts at $-15°C$.

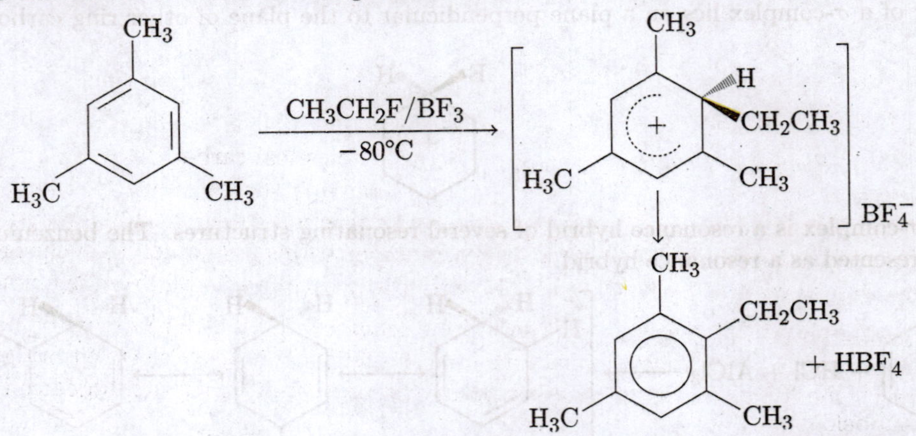

The σ-complexes are highly coloured substances; their colour usually varies from orange to black. The σ-complex formed by the action of HCl on benzene in the presence of $AlCl_3$ is a red oil which conducts electricity. In fact, σ-complexes are capable of conducting electricity.

A σ-complex may undergo two kinds of reactions—coupling with an anion and proton elimination reaction. If it undergoes the latter reaction, the product will be an aromatic compound;

while the coupling product of a sigma-complex is not an aromatic compound. An aromatic compound being more stable than a non-aromatic compound having double bonds, σ-complexes give up proton to gain stability. For example, A gives nitrobenzene by giving up a proton.

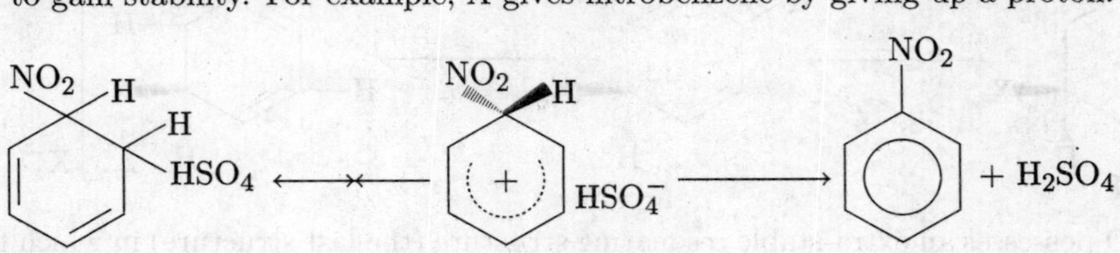

A

σ-complexes undergo isotope exchange reactions. When benzene is treated with DCl—AlCl$_3$, there is an exchange of deuterium for hydrogen—C_6D_6 has been prepared by this method.

The stability of a σ-complex mainly depends on three factors : I effect, R effect and hyperconjugative effect. An arenium ion being a positively charged species, +I effect, +R effect and hyperconjugative effect of alkyl groups increase its stability; while –I effect and –R effect decrease it. If a group with +I or +R effect is situated at the *o*- or *p*-position instead of the *m*-position with respect to the tetrahedral C atom, the σ-complex gets extra stability.

For example, σ-complex B is more stable than σ-complex C.

B

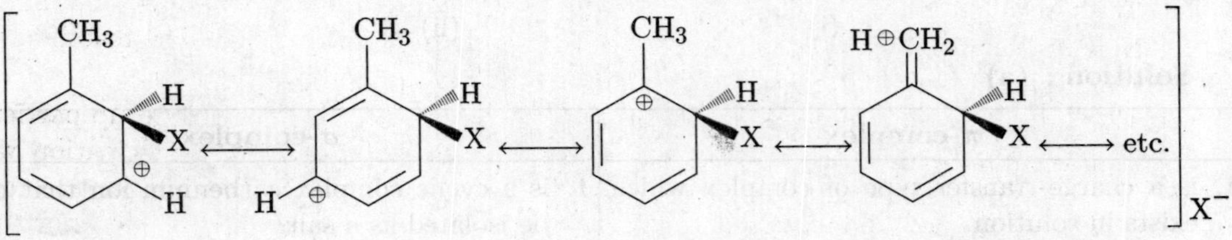

C

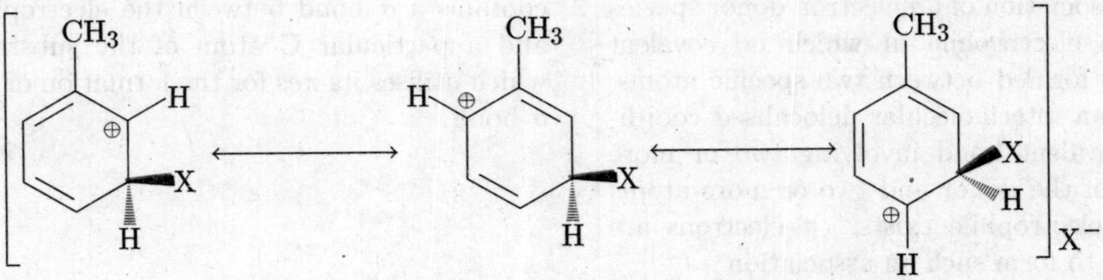

Thus in B the positive charge gets further delocalised by the hyperconjugative effect of the methyl group; whereas in C there is no such effect. We know that the greater the delocalisation of charge, the greater is the stabilisation. So B is more stable than C.

Similarly, D is more stable than E :

D

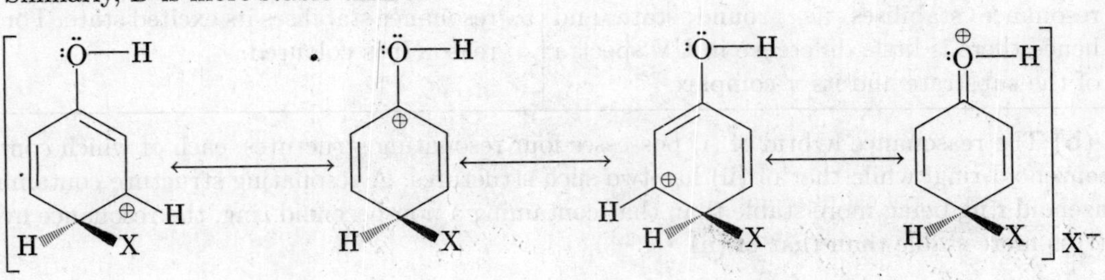

E 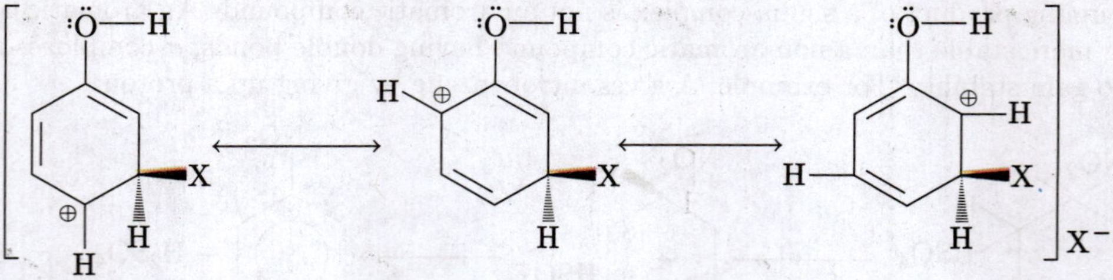

Here D possesses an extra-stable resonating structure (the last structure) in which the O atom and all the C atoms are in octet state while E does not have such an extra-stable structure. This stabilises D more than E.

Problem 11.4. **(a)** Distinguish between a π-complex and a σ-complex; **(b)** which one is more stable —(i) or (ii)?

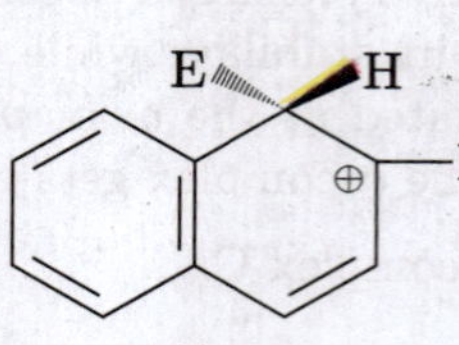

(i)

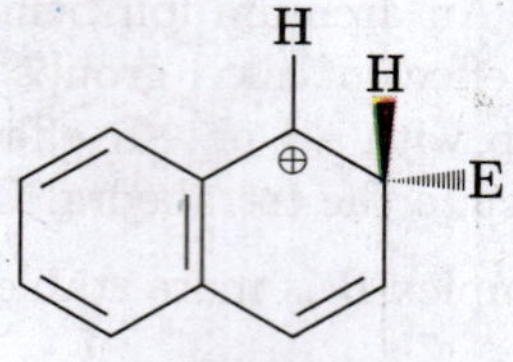

(ii)

Solution : (a)

π-complex	σ-complex
1. is a charge-transfer type of complex which exists in solution.	1. is a cyclic complex carbenium ion that can be isolated as a salt.
2. is an association of an electron donor species and an electrophile in which no covalent bond is formed between two specific atoms, rather an intermolecular delocalised coordinate covalent bond involving two or more atoms of the donor and two or more atoms of the electrophile exists. π electrons are utilised to form such an association.	2. contains a σ bond between the electrophile and a particular C atom of the substrate which utilises its πes for the formation of the σ bond.
3. is usually colourless or light yellow coloured.	3. is orange or deeper in colour.
4. does not conduct electricity.	4. conducts electricity.
5. does not undergo the isotope exchange reaction.	5. undergoes the isotope exchange reaction.
6. resonance stabilises its ground state and hence there is little difference in UV spectra of the substrate and its π-complex.	6. resonance stabilises its excited state. For this reason it is coloured.

(b) The reasonance hybrid of (i) possesses four resonating structures, each of which contains a benzenoid ring; while that of (ii) has two such structures. A resonating structure containing a benzenoid ring being more stable than that containing a non-bezenoid ring, the resonance hybrid of (i) is more stable than that of (ii).

(i) 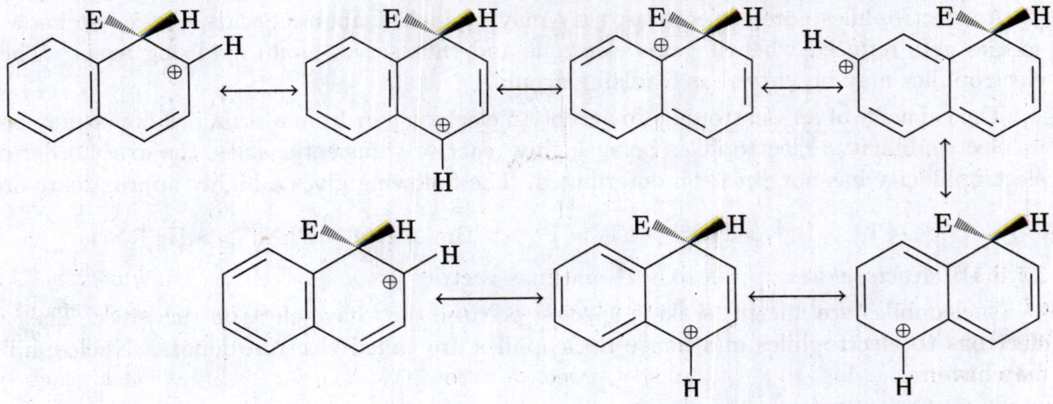

(ii) 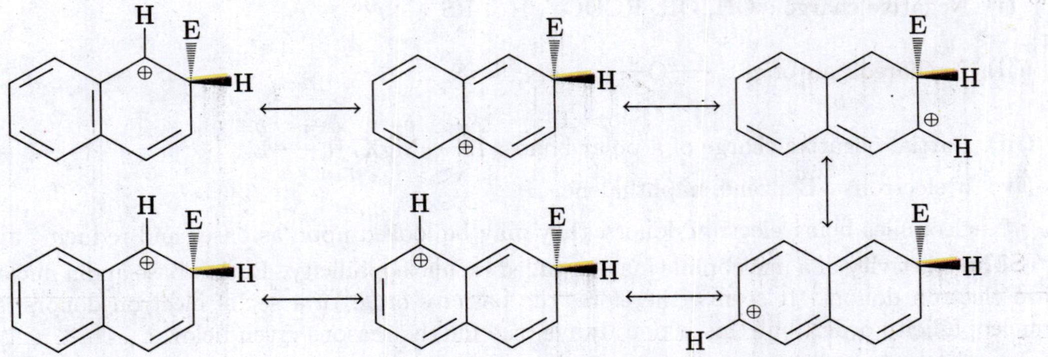

11.5 Types of Reagents

Reagents in organic reactions are generally classified under three heads; (a) electrophiles and (b) nucleophiles and (c) radicals. There are ambident nucleophiles and electrophiles also.

11.5.1 Electrophiles and Nucleophiles

11.5.1a. *Electrophiles*

Electrophiles are electron seeking species. They seek electrons because they are electron-deficient species. Therefore, they accept electrons from other species; for this reason electrophiles are also called electron acceptors. A particular site of an electrophile is generally found to be electron deficient and that site is called electrophilic site. Electrophilic sites are of three categories:

(i) They may have positive charge : $\overset{\oplus}{NO_2}$, $\overset{\oplus}{H}$, $\overset{\oplus}{NO}$, $R_3\overset{\oplus}{C}$,

(ii) They may have less than 8 *e*s in the outermost shell of the electrophilic site such as : B of BF_3, Al of $AlCl_3$,

(iii) They may be the positive pole of a polar bond :

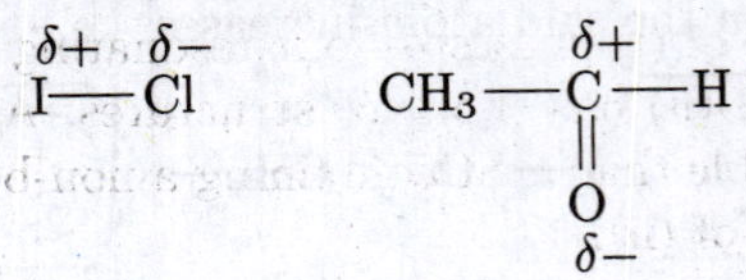

As electrophiles accept electrons, they may be looked upon as acids. Again we know that a species gets reduced when it gains electrons and thus serves as an oxidising agent. Therefore, electrophiles may be viewed as oxidising agents.

The capacity of an electrophile to accept an electron pair from a standard donor may be called its electrophilicity. Electrophiles being highly reactive transient species, the exact order of their electrophilicity has not yet been determined. The following gives a highly approximate order :

$$Cl^{\oplus} > Br^{\oplus} > NO_2^{\oplus} > Cl_2 = I^{\oplus} >> Br_2 > \overset{\oplus}{N}O > PhN_2^{\oplus} >> Hg^{\oplus} > I_2$$

11.5.1b. *Nucleophiles*

Nucleophiles are nucleus seeking species, because they have electron-rich sites. They donate electrons to electrophiles and hence nucleophiles are called electron donors. Nucleophilic sites may have :

(i) Negative charge : $\bar{O}H$, $\bar{O}R$, $RCOO^-$, H^-, RS^-

(ii) Unshared pair of *es* : $—\ddot{O}—$, $—\ddot{N}:$ (with two further bonds), $=\ddot{S}:$

(iii) Partial negative charge of a polar bond : $\overset{\delta-}{R}—\overset{\delta+}{MgX}$, $\overset{\delta-}{R}—\overset{\delta+}{Li}$

(iv) π electrons : Benzene, naphthalene.

Nucleophiles being electron donors, they may be looked upon as bases and reducing agents.

The reactivity of a nucleophile may be called its nucleophilicity. Both a base and a nucleophile are electron donors. It seems, therefore, the basicity order of a set of electron donors is their nucleophilicity order; but this is not always true for the reasons given below :

(i) A base coordinates mainly with a proton whereas a nucleophile coordinates with an atom other than proton and very often with carbon.

(ii) An acid-base reaction is an equilibrium controlled, i.e., a thermodynamically controlled reaction whereas an electrophile-nucleophile interaction is a rate controlled, i.e. kinetically controlled reaction.

However, nucleophilicity depends on many factors as given below :

(i) Basicity, (ii) Polarisability, (iii) Solvation, (iv) Nucleofugacity, (v) α effect.

To understand the nucleophilicity order among a set of nucleophiles the following rules, which are related to the factors said above, are discussed :

1. Basicity and Nucleophylicity :

a. Under a given set of reaction conditions a nucleophile with an anionic attacking atom is more nucleophilic than its conjugate acid which is also a nucleophile. Thus $\bar{O}H$ is more nucleophilic than H_2O and $\bar{N}H_2$ than NH_3. Here the basicity order and the nucleophilicity order are one and the same.

$$H\ddot{\underset{..}{O}}:^- > H_2O;\ H_2\ddot{N}:^- > \ddot{N}H_3$$

b. In a set of nucleophiles having the same attacking atom, the nucleophilicity increases as the negative charge density on the said atom increases, i.e., the basicity increases:

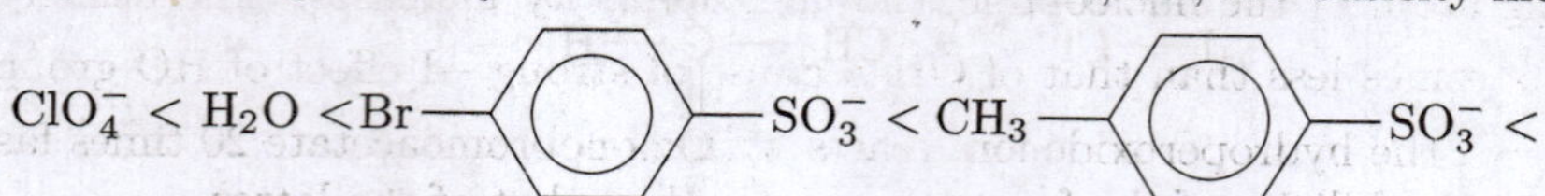

$CH_3CO_2^- < CO_3^{--} < \bar{O}Ph < \bar{O}H < \bar{O}R$

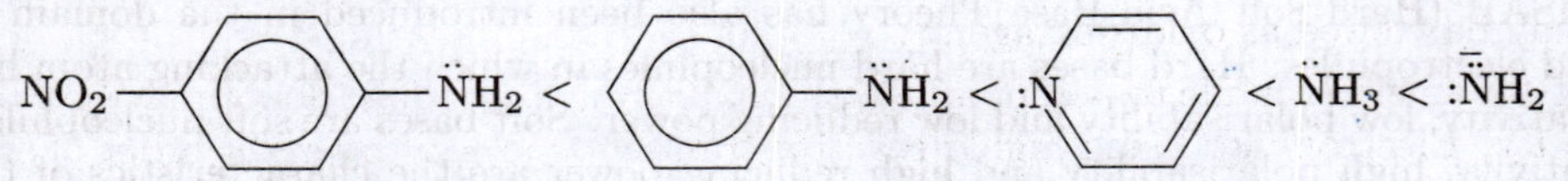

c. Since electronegativity of atoms increases along a period from left to right, the basicity and the nucleophilicity decreases similarly.

(i) $\bar{C}H_3 > \bar{N}H_2 > \bar{O}H > F^-$

(ii) $R_3C^- > R_2N^- > RO^- > F^-$

2. Polarisability : The greater the distortion of the charge cloud of the valence electrons around the attacking atom of the nucleophile by the electrophilic centre, i.e., the greater the polarisability of the attacking atom, the greater is the nucleophilicity. Again the greater the volume of the attacking atom, the higher is its polarisability. For example, the nucleophilicity of PhS^- is higher than that of PhO^-; but this is not always true. From the point of polarisability the nucleophilicity order among halide ions parallels their order of volume only in a protic solvent, i.e., $I^- > Br^- > Cl^- > F^-$. Actually nucleophilicity depends on solvation also.

3. Solvation : In a protic solvent smaller groups with higher charge density become more solvated than larger groups with less charge density. The smaller groups being strongly solvated, they are less free for attack than the larger groups which are more free being less solvated. The more freer the nucleophile, the more higher is its nucleophilicity. In a protic solvent, the degree of solvation decreases down a group. So F^- is the most solvated nucleophile and I^- is the least in a protic solvent. Therefore, from F^- to I^-, in a protic solvent, solvent effect decreases, i.e., freeness increases and polarisability increases; and the nucleophilicity order is $I^- > Br^- > Cl^- > F^-$. In an aprotic solvent none of them gets solvated (see Ch. 7). Hence all are free to move. So, the charge density order is $F^- > Cl^- > Br^- > I^-$ and this is also the nucleophilicity order.

4. Nucleofugacity : If in a reaction the leaving group of the substrate is electron dense, it is called nucleofuge. The ability of nucleofuge to leave the substrate in a reaction is called its nucleofugacity. As the nucleofugacity increases, the C atom bearing the nucleofuge becomes more electrophilic and its ability to polarise the incoming nucleophile also increases. The more polarisable nucleophile, therefore, gets polarised more and its nucleophilicity increases. In general a highly polarisable nucleofuge is most readily displaced by a highly polarisable nucleophile. For example, the rate of the reaction between pyridine and ethyl iodide is 5.75 times faster than that between pyridine and ethyl bromide; the nucleofugacity of I^- is greater than that of Br^- from the consideration of volume and polarisability. Similarly, reaction A is 31 times faster than the reaction B as given below :

(A) $I^- + Et—I^* \longrightarrow I—Et + I^{*-}$;

(B) $I^- + Et—Br \longrightarrow I—Et + Br^-$

5. α-Effect : It has been found that if there is a lone pair of electrons on the α-atom with respect to the attacking atom of the nucleophile, the nucleophilicity increases. The basicity of hydroperoxide ion is 10^4 times less than that of $\bar{O}H$ because of strong −I effect of HO group in $H\ddot{O}—O^-$; but $H\ddot{O}—\bar{O}$, the hydroperoxide ion, reacts with monobromoacetate 20 times faster than $\bar{O}H$ because the nucleophilicity of the former is greater than that of the latter.

11.5.1c. *Hard and soft nucleophiles and electrophiles*

Pearson's HSAB (Hard-Soft Acid-Base)Theory has also been introduced in the domain of nucleophiles and electrophiles. Hard bases are hard nucleophiles in which the attacking atom has high electronegativity, low polarisability and low reducing power. Soft bases are soft nucleophiles; low electronegativity, high polarisability and high reducing power are the characteristics of the attacking atom of the soft nucleophiles. So I^- is a soft nucleophile and $\bar{O}H$ is a hard one. Proton and all proton donors are hard electrophiles while a saturated carbon is a soft electrophile. A few electrophiles and nucleophiles are presented in Table 11.2 as hard and soft.

Table 11.2 Some hard and soft electrophiles and nucleophiles

Electrophiles (Acids)	Nucleophiles (Bases)
Hard	**Hard**
$H^{\oplus}$, $Li^{\oplus}$, $Na^{\oplus}$, $K^{\oplus}$, $Si^{4\oplus}$	H_2O, $\bar{O}H$, F^-
BF_3, $B(OR)_3$, $AlCl_3$, AlH_3	$CH_3CO_2^-$, PO_4^{3-}, SO_4^{2-}
$RSO_2^{\oplus}$, $ROSO_2^{\oplus}$, SO_3	Cl^-, CO_3^{2-}, ClO_4^-, NO_3^-
$RC^{\oplus}{=}O$, CO_2	ROH, RO^-, R_2O
HX (hydrogen bonding molecules)	NH_3, RNH_2, N_2H_4
Borderline	Borderline
$Cu^{2\oplus}$, $Zn^{2\oplus}$, $Pb^{2\oplus}$, $B(CH_3)_3$, SO_2	$PhNH_2$, Pyridine, N_3^-
$NO^{\oplus}$, $M_3C^{\oplus}$	
$C_6H_5^{\oplus}$	Br^-, SO_3^{2-}, N_2
Soft	**Soft**
Saturated carbon	R_2S, RSH, RS^-
$Cu^{\oplus}$, $Ag^{\oplus}$, $Hg^{\oplus}$	I^-, SCN^-
$I^{\oplus}$, $Br^{\oplus}$, I_2, Br_2	CN^-, RNC, CO
Trinitrobenzene, Chloranil	C_2H_4, Benzene
Quinones, etc.	H^-, R^-
Tetracyanoethylene	
Metal atom	
Carbenes	
O, Cl, Br, I, N, RO, RO_2.	

However, there is an empirical rule which governs acid-base and nucleophile-electrophile interactions. The rule states that hard-likes-hard and soft-likes-soft. This means hard-hard as well as soft-soft interactions are strong, i.e., these lead to strong bonds but hard-soft interactions are weak. For example, $H^{\oplus}$, a hard acid, will react faster with $\bar{O}H$ (a hard base) than with I^- which is a soft base. Along a period from left to right the hardness of nucleophiles increases and the softness decreases; on the other hand down a group the softness of nucleophiles increases and the hardness decreases. Thus the order of decreasing softness, i.e., increasing hardness amongst the following site of nucleophiles is:

(i) $H^- < :\bar{C}N < \bar{O}H < F^-$; (ii) $I^- < Br^- < Cl^- < F^-$; (iii) $C^- < N^- < O^- < F^-$.

Furthermore, a nucleophile with a negative charge on the attacking atom is harder than its conjugate acid. Thus

$$\bar{\ddot{N}}H_2 > \ddot{N}H_3,\ \bar{O}H > H_2O \quad \text{and} \quad HS^- > H_2S.$$

Again, a positively charged carbon is a harder centre (electrophile) than a partially positively charged carbon, while the latter is a harder electrophile than a neutral or less partially positively charged carbon. Thus

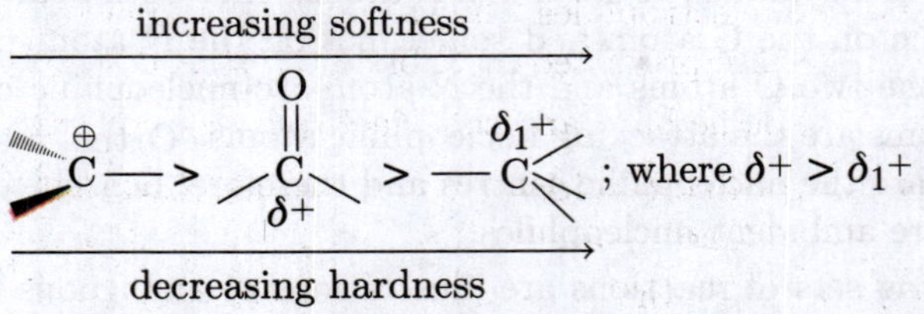

Let us now discuss hardness and softness of electrophiles and nucleophiles and their interactions in the light of Frontier Orbital Theory. The occupied M.O. of the highest energy of a chemical species is called HOMO (Highest Occupied Molecular Orbital) and the unoccupied orbital of the lowest energy is called LUMO (Lowest Unoccupied Molecular Orbital). When a nucleophile-electrophile interaction occurs, the most effective interaction is that between the HOMO of the nucleophile and the LUMO of the electrophile. So, HOMOs and LUMOs are called the frontier orbitals. When two chemical species collide, the orbitals of one reactant overlap those of the other and the energy gained or lost (ΔE) by the interaction may be expressed as:

$$\Delta E = E_{core} + E_{electrostatic} + E_{overlap}, \tag{11.1}$$

where $E_{core} = +$ ve, this arises owing to repulsion between occupied orbitals of the two species.

$E_{electrostatic} = -$ ve, this arises owing to the attraction of any + ve or − ve charge on one by any − ve or + ve charge respectively on the other.

$E_{overlap} = -$ ve, this arises owing to the overlap of occupied orbitals, especially the HOMO, of one with unoccupied orbitals, especially the LUMO of the other.

Here it is needless to mention that the negative ΔE value leads to the effective interaction. An electrophile-nucleophile interaction is also governed by the above expression.

However, a hard nucleophile has a low energy HOMO and usually a negative charge on the attacking atom; whereas a soft nucleophile has a high energy HOMO but it may or may not have a negative charge on the attacking atom. A hard electrophile possesses a high energy LUMO and usually a positive charge. On the other hand, a soft clectrophile has a low energy LUMO and it may or may not have a positive charge.

A hard-hard interaction is strong and fast because it contributes largely to the $E_{electrostatic}$ term of the expression whereby ΔE becomes negative. A soft-soft interaction is strong and fast because of the large contribution to the $E_{overlap}$ term which again makes the ΔE value negative. $E_{overlap}$ term will be large when a symmetry compatible HOMO and LUMO, which are close in energy of the nucleophile and the electrophile respectively, interact.

11.5.1d. *Ambident nucleophiles*

'Ambi' is a prefix occuring in loan words from Latin where it meant 'ambiguous' (uncertain) and 'dent' means tooth. Therefore, an ambident nucleophile is a species which has two (or more) different or similar types of nucleophilic centres through which it can attack electrophilic centres. Cyanide, nitrite and enolate ions are the examples of ambident nucleophiles.

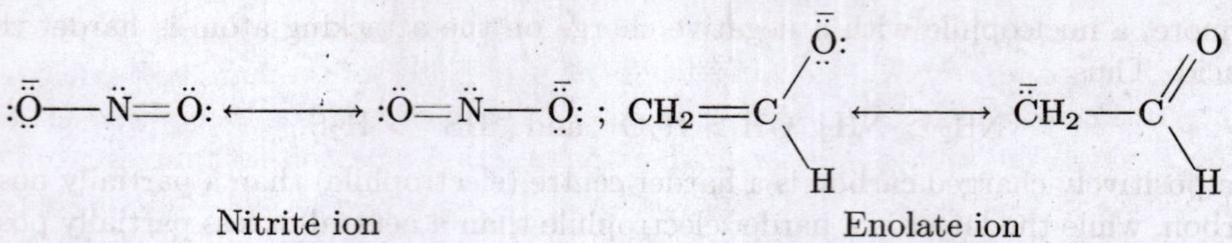

Nitrite ion Enolate ion

Nucleophilic centres in a cyanide ion are C and N atoms since both bear a lone pair of es; $:\bar{C}N:$ is sometimes nucleophilic on the C atom and sometimes on the N atom. The same is the case with nitrite ion where the two O atoms and the N atom are nucleophilic centres. In an enolate ion anionic O and C atoms are the attacking nucleophilic atoms. Ortho, meta and para positions of a benzene derivative are the nucleophilic centres and therefore, benzene derivatives (and other aromatic compounds) are ambident nucleophiles.

However, the following sets of reactions are illustrations of ambiguous behavior of ambident nucleophiles :

(a) **(i)** $CH_3Br + KCN \xrightarrow[\text{CN}^- \text{ through C atom}]{\text{attack by}} CH_3{-}C{\equiv}N: + CH_3{-}\overset{\oplus}{N}{\equiv}\bar{C}:$

Ethanenitrile (major pdt.) Ethaneisonitrile (trace)

(ii) $CH_3Br + AgCN \xrightarrow[\text{CN}^- \text{ through N atom}]{\text{attack by}} CH_3{-}\overset{\oplus}{N}{\equiv}\bar{C}: + CH_3{-}C{\equiv}N:$

major trace

(b) **(i)**

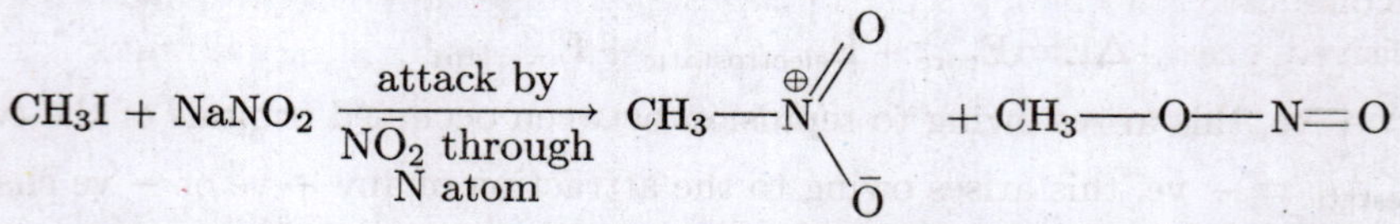

Nitromethane (major pdt.) methyl nitrite (minor pdt.)

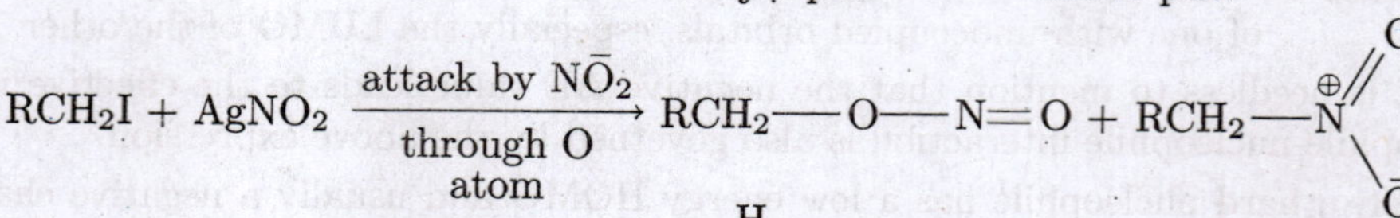

(c) **(i)**

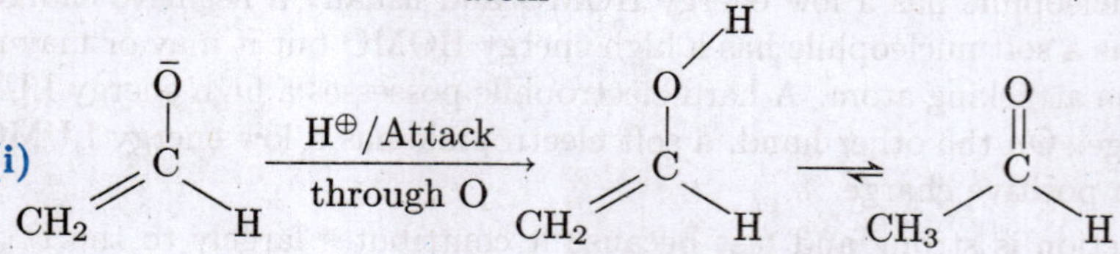

a b (b forms via a)

Enolate Ethenol Acetaldehyde

(ii)

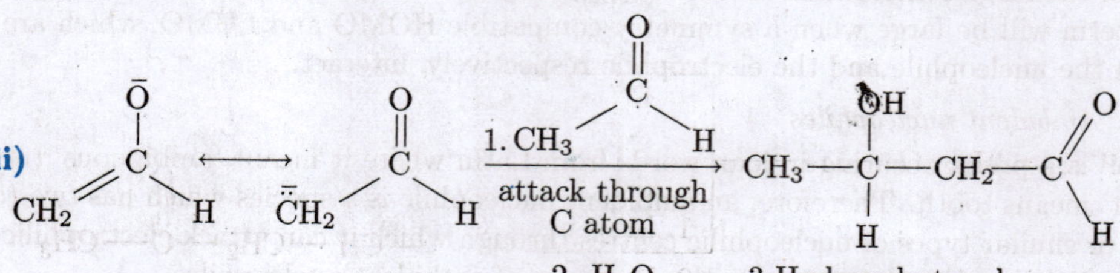

3-Hydroxybutanal

Now the question is : are really the behaviours of the ambident nucleophiles ambiguous? The answer is: no. Let us now examine the answer from the standpoint of the hard-soft nucleophile-electrophile interactions as discussed in section 11.5.1c. In each case of the before-said examples (a,b and c) the reaction involves either a hard-hard interaction or a soft-soft interaction. The hard and soft centres of each nucleophile are labelled here following the fact that the hardness increases along a period from left to right.

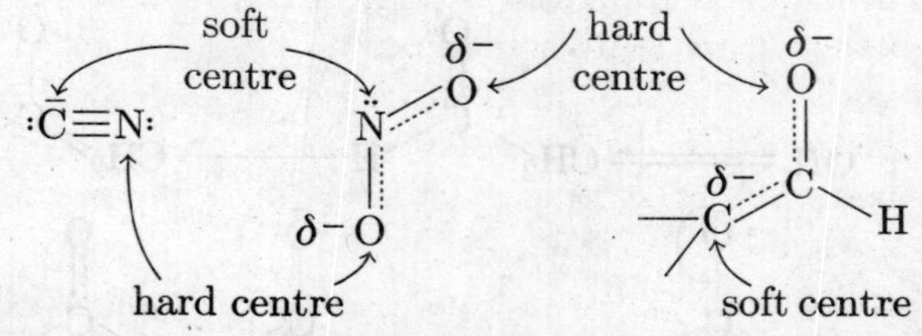

Ambident nucleophiles labelled as hard and soft centres

The C atoms of CH_3Br, CH_3I and RCH_2I are the soft centres. So, CN^- and NO_2^- attack the electrophilic Cs through soft centres, C and N respectively and we get methyl cyanide, nitromethane, etc. as major products. When potassium cyanide and sodium nitrite are the reagents, 1° and 2° alkyl halides react with nucleophiles without having carbocation generating counter part ($Ag^{\oplus}$, $Hg^{\oplus\oplus}$ etc.) in the same way as methyl halides. In the presence of $Ag^{\oplus}$ or $Hg^{\oplus\oplus}$, which are generated from the reagent AgCN, $AgNO_2$, $HgCl_2$ etc. as shown below, 1° and 2° alkyl halides become more carbenium ion-like in which cationic carbons are hard centers (being appreciably positively charged). So the nucleophiles attack the cationic Cs through hard centres (N of CN^- and O of NO_2^-) and naturally the major products are alkyl isocyanides and alkyl nitrites.

$$\underset{\text{solid}}{AgCN} \rightleftharpoons \underset{\text{in soln.}}{AgCN} \rightleftharpoons \underset{\text{in soln.}}{Ag^{\oplus} + CN^-}; \; AgNO_2 \rightleftharpoons \underset{\text{in solution}}{Ag^{\oplus} + NO_2^-}$$

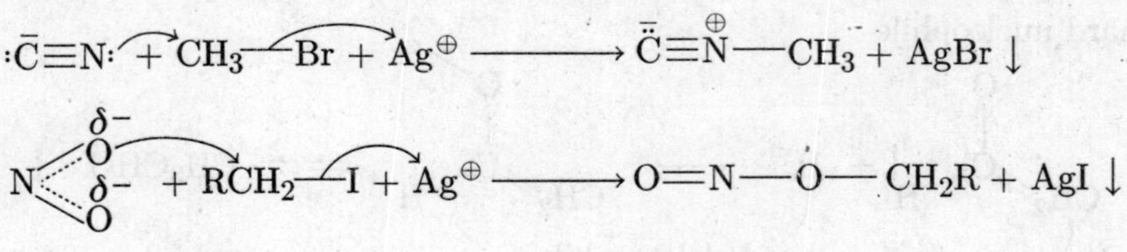

11.5.1e. *Ambident electrophiles*

α, β-unsaturated carbonyl compounds are ambident electrophiles. For example, in

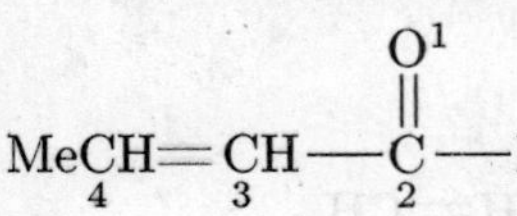 there are two electrophilic centres. They are 2 and 4 positions.

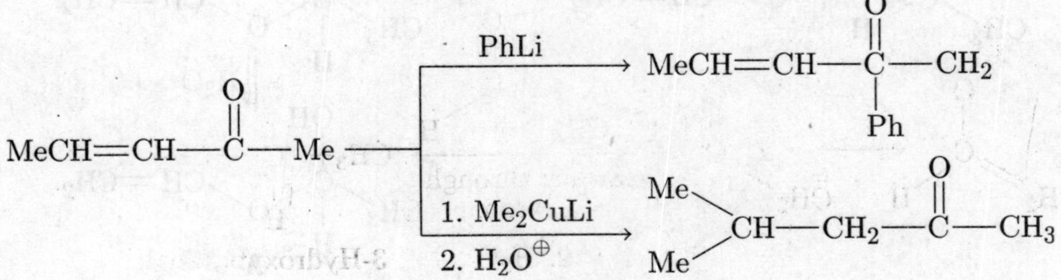

11.5.1 Radicals

See Sec. 11.4.1.

Problem 11.5. (a) Which one will be formed first the enol or the carbonyl compound when one adds acidulated water in excess in an enolate ion? Give a reason for your answer.

(b) When acetaldehyde is treated with a dilute caustic soda solution, we get 3-hydroxybutanal as shown below :

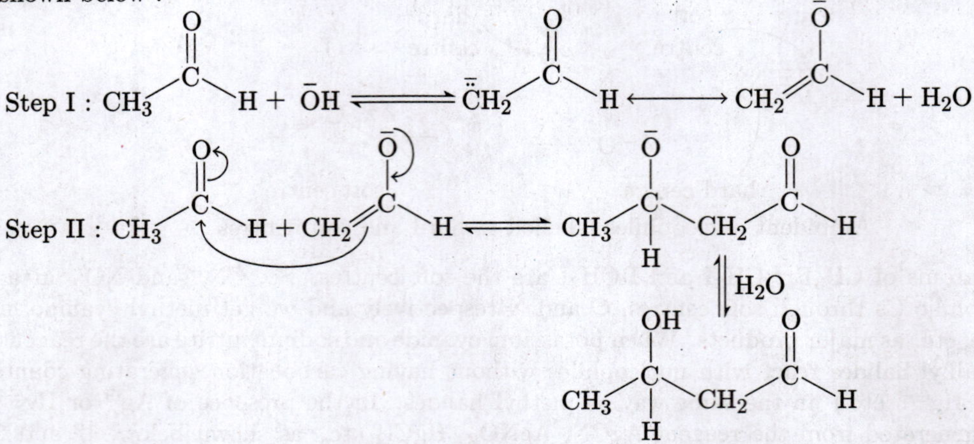

Why do we not get $CH_3CHOH—O—CH{=}CH_2$ (compound A)?

Solution : (a) $H_3O^{\oplus}(\equiv H^{\oplus})$ is a hard electrophile. In an enolate ion the anionic O is the hard centre and the C atom (anionic or neutral) is a soft centre. Since hard-hard interactions are stronger and faster than hard-soft interactions, the enolate ion will combine with the $H^{\oplus}$, a hard electrophile, through anionic O which is also a hard centre and enol will be formed first and then the enol will tautomerise to the isomeric carbonyl compound.

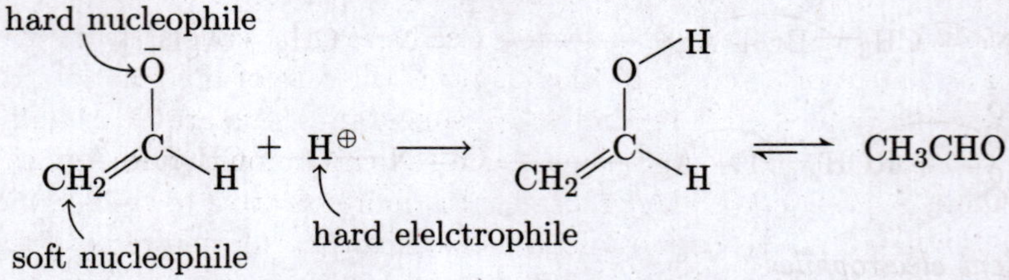

(b) Compound A will form if the enolate ion attacks the electrophilic carbonyl C in step II through the O atom instead of the C atom as shown below :

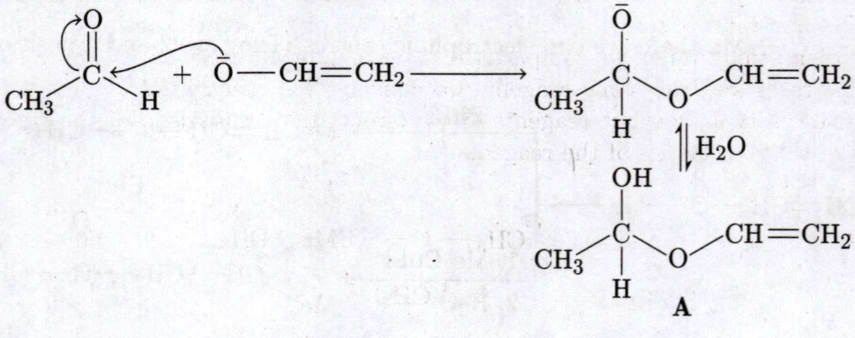

With respect to the O^-, a hard nucleophilic centre, the electrophilic C atom of acetaldehyde is a soft centre; a hard-soft interaction being weaker and slower than a soft-soft interaction, the anionic C of the enolate ion attacks the electrophilic C of the acetaldehyde and 3-hydroxybutanal is formed and no hydroxyether is formed by the hard-soft interaction.

This reaction may also be explained from the thermodynamic stand point. In this reaction an αC—H bond and the π bond of the C=O bond are broken, whereas a C—C and a O—H bonds are formed. The calculated heats of reaction (in $kJmol^{-1}$) are :

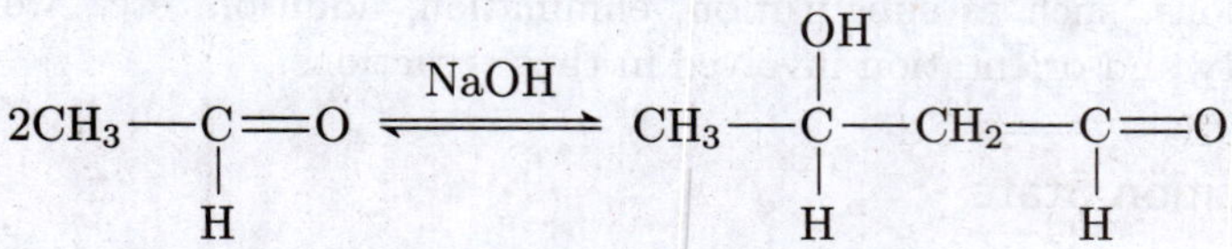

$$\Delta H = 378.9 + 413.2 - (463 + 345.8) \text{ kJ mol}^{-1}$$
$$= -16.7 \text{ kJ mol}^{-1},$$

where ${}^{\Delta H}C{=}O_{\pi}$: 378.9; ${}^{\Delta H}C{-}H$: 413.2; ${}^{\Delta H}O{-}H$: 463; ${}^{\Delta H}C{-}C$: 345.8; ${}^{\Delta H}C{-}O_{\sigma} = 358$; ${}^{\Delta H}C{-}C_{\pi} = 264.6$ (in kJ mol^{-1}).

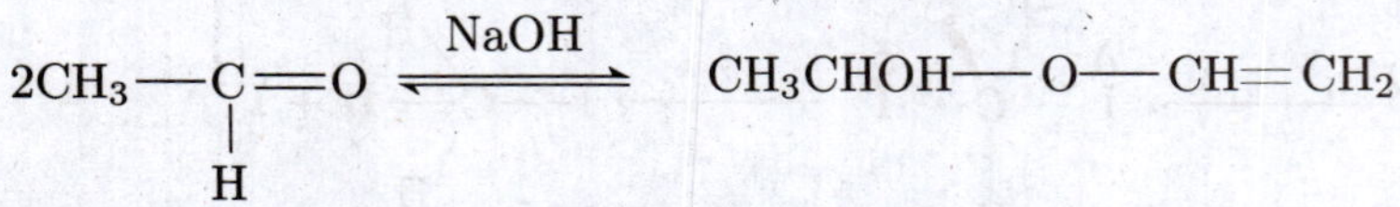

$$\Delta H = 2 \times 378.9 + 413.8 - (463 + 358 + 264.6) \text{ kJ mol}^{-1}$$
$$= 85.4 \text{ kJ mol}^{-1}.$$

So, 3-hydroxybutanal formation reaction is exothermic to the extent of 16.7 kJ but hydroxyether formation is endothermic to the extent of 85.4 kJ. Hence the hydroxyaldehyde formation is thermodynamically more favourable than the hydroxyether formation.

11.6 Reactivity and Orientation

The ease with which a substrate or a reagent reacts may be called its reactivity. The higher the reactivity of a reagent or a substrate, the higher is the rate of the reaction. A substrate may have different reactivities at its different sites. In isopentane there are $1°, 2°$ and $3°$ carbons and hydrogens and the reactivity order towards halogen radicals amongst three types of hydrogens is $3° > 2° > 1°$. On the other hand, chlorine radical is more reactive towards a particular type of hydrogen in isopentane than bromine radical. This means a chlorine radical reacts faster than a bromine radical and therefore, the reactivity of the former is greater than that of the latter. However, when we compare the rates of reactions at different sites of a substrate with the same reagent, we consider it as a problem of orientation though different sites have different reactivities. The comparison of rates of reactions of chlorine radicals on different H atoms of isopentane is the problem of orientation. When we compare the rates of similar reactions at similar sites involving different substrates and the same reagent, we consider the reactivities of different substrates. If the substrate is the same but reagents are different, the comparison of the rates means the comparison of the reactivities of the reagents.

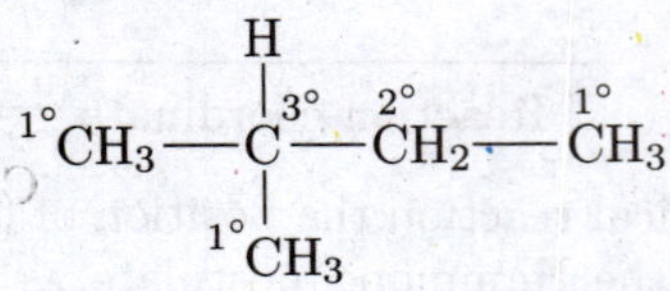

However, in a given reaction if we want to determine the reactivity order amongst the members of a set of substrates or reagents or solvents, we have to consider the rate-determining step only. The higher the $\Delta G^\ddagger$ of the R.D. step, the lower will be the reactivity. $\Delta G^\ddagger$, in its turn, depends on the energy content of the substrate, reagent, T.S. and the solvent. To compare the energy content of these species their structures, factors influencing the electron availability (I,R, hyperconjugative, etc. effects; Chapter 4), the strain factors, the stereoelectronic and steric factors (Chapter 9), the solvation effect (Chapter 7), etc. are to be considered. When we shall study different classes of reactions, such as substitution, elimination, addition, etc., we shall discuss the problems of reactivity and orientation involved in those reactions.

11.7 The Position of Transition State

A reaction or a step of a multi-step reaction may be symmetrical or unsymmetrical. The conversion of methyl iodide into methyl iodide by the action of potassium iodide is a symmetrical reaction while the conversion of methyl iodide into methyl alcohol by the alkaline hydrolysis is an example of unsymmetrical reaction.

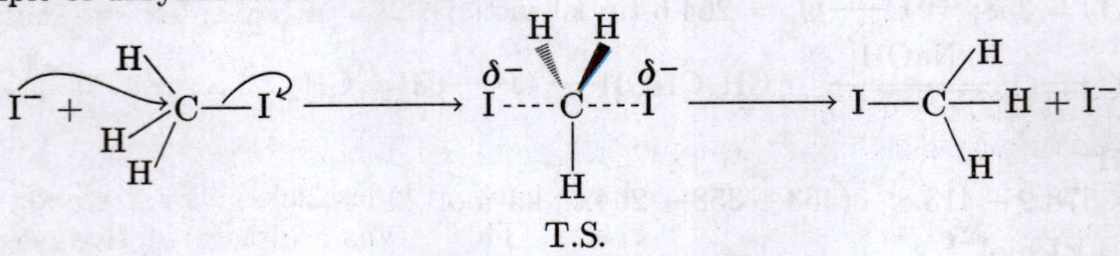

a symmetrical reaction

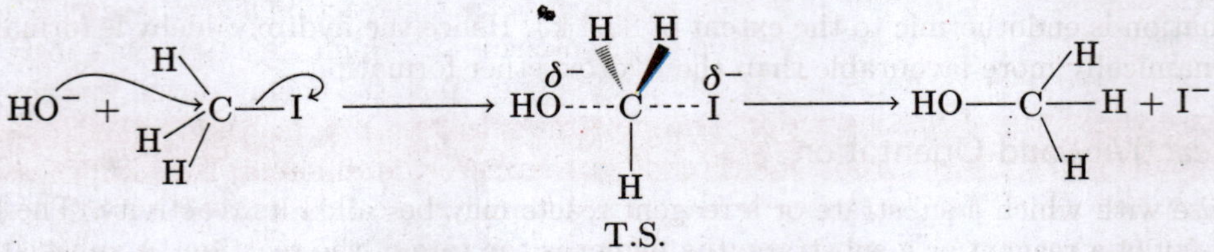

an unsymmetrical reaction

For a symmetrical reaction the determination of position of the T.S. on the reaction co-ordinate is very simple. It will be much above the reactant and the product and it will occupy the mid-point of the energy curve.

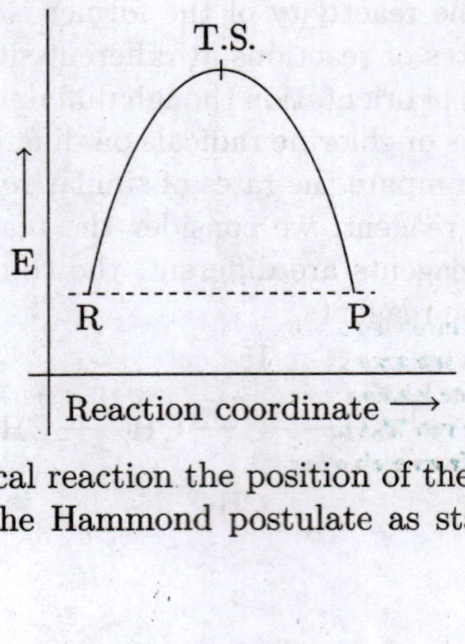

In the case of an unsymmetrical reaction the position of the T.S. on the reaction co-ordinate may usually be determined by the Hammond postulate as stated earlier in the sixth chapter.

However, the postulate tells us that the position of the T.S. of a reaction on the reaction coordinate varies with the relative energies of the reactants, T.S. and products and the position changes in a continuous manner as their energies are perturbed by the changes in structure, solvent, catalyst, etc. In fact, to locate the T.S. on the energy curve the following things are to be considered mainly :

(i) Relative energies of the reactant, T.S. and product (Sec. 6.6)
(ii) The Hammond postulate (Sec. 6.4.2)
(iii) The effect of catalyst (Sec. 6.3)
(iv) The effect of solvent (Sec. 7.7.5)
(v) The kinetic control and the thermodynamic control of a reaction (Sec. 6.7)
(vi) The selectivity reactivity principle
(vii) The microscopic reversibility principle
(viii) The neighbouring group participation.

Since (i) to (v) have already been discussed in the sixth and seventh chapters, we shall discuss here the last three topics only.

11.7.1 Selectivity reactivity principle

The principle states that : the more reactive a reagent, the less selectivity it reacts; and the less reactive a reagent, the more selectivity it reacts. This means a highly reactive reagent will react indiscriminately and randomly with different substrates in a reaction, whereas a less reactive reagent will differentiate between two or more substrates in the reaction and will react selectively. When the reagent is highly reactive, its energy content is very high. Though the substrates are different, yet the energy of the reactant side will be high since the energy of the reagent is high. In this case the different T.S.s for different substrates will be relatively early and reagent-like insofar as Hammond's postulate is concerned; furthermore the different T.S.s will not be very dissimilar in energy. When the reagent will attack different sites of the same substrate, it will also attack randomly and the T.S.s for the different sites will differ in energy but to a small extent. Therefore, in both the cases, in the cases of different substrates or sites, the rates of formation of different products will not vary appreciably. The reagent will thus show low selectivity.

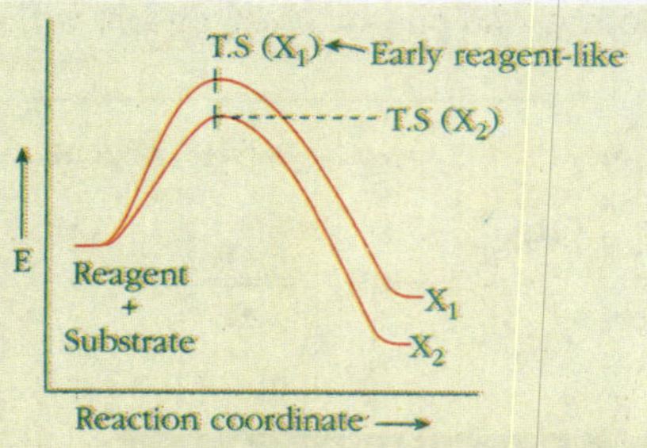

Fig. 11.7. The selectivity reactivity principle involving a highly reactive reagent and two different substrates or two different sites of the same substrate leading to two different products, X_1 and X_2. The two T.S.s are early and reagent-like and differ from each other in energy to a small extent

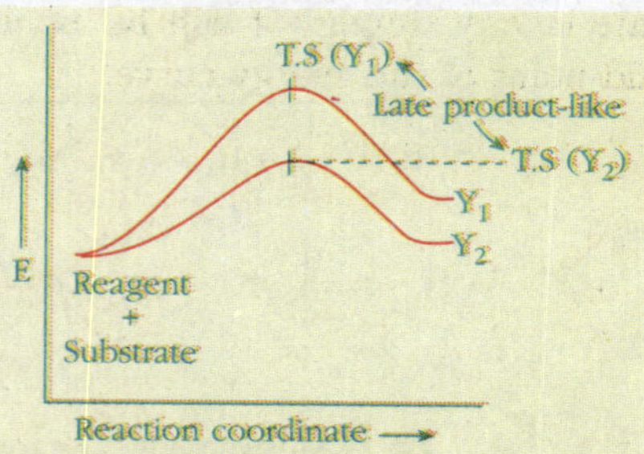

Fig. 11.8. The selectivity reactivity principle involving a less reactive reagent and two different substrates or two different sites of the same substrate leading to two different products, Y_1 and Y_2. T.S.s have appreciable difference in energy and they are late and product-like

When the reagent is less reactive, no matter whether it reacts with two different substrates or it attacks at different sites of the same substrate, the T.S.s will be late and more product-like and the reaction will be less exothermic. There will be an appreciable energy difference between the T.S.s and therefore, the product having the lower energy T.S. will be formed faster than the other (Fig. 11.8). Hence high selectivity will be shown by the reagent.

Chlorine radicals are much more reactive than bromine radicals insofar as substitution reactions on alkanes are concerned. A 3°H in an alkane reacts only 4.2 times faster than a 1°H with chlorine radicals; whereas the former type of H reacts 1700 times faster than the latter type of H with bromine radicals. Bromine radicals being less reactive, they react more selectively with 3°H than 1°H; while chlorine radical, being more reactive, they react randomly with 1°H and 3°H atoms.

11.7.2 The microscopic reversibility principle

The principle tells us that if a reaction is a reversible one and if a particular pathway is energetically the most favourable route for the forward reaction (from reactants to products), the lowest energy pathway for the reverse reaction will also be along the same route but with all bond-forming and bond-breaking processes will be reversed and the forward and reverse reactions will involve the same T.S.s, reaction intermediates and mechanism. Therefore, the T.S.s and intermediates for the forward and reverse reactions will have the same locations on the energy curve. When 3° alcohols are heated with concentrated sulphuric acid at a temperature 170°C, alkenes are formed. The reaction is reversible; when alkenes are subjected to acid-catalysed hydration, alcohols are formed. The principle of microscopic reversibility can be applied to these reactions and the route for the forward reaction will also be the route for the reverse reaction. Both the reactions involve the same carbenium ion and the T.S. and they have the same positions on the energy curve.

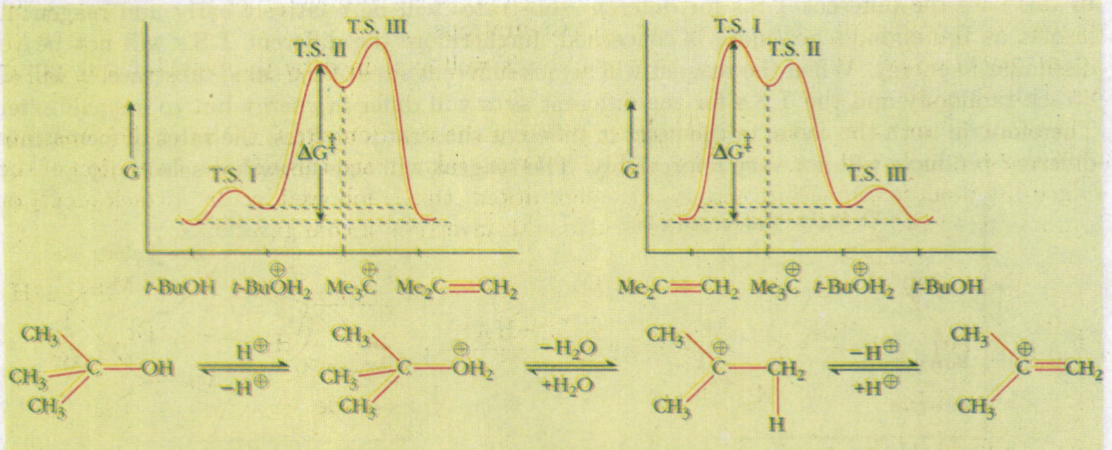

Fig. 11.9. Energy profile for a reaction which follows the microscopic reversibility principle

11.7.3 Neighbouring Group Participation (NGP) and Anchimeric Assistance

A group close to the site of the reaction in a substrate is called a neighbouring group; when a neighbouring group participates in the reaction without necessarily giving rise to an intramolecular rearrangement, the participation is known as neighbouring group participation. Usually the neighbouring group remains on the β-C, but it may remain farther away and in this case the group comes near to the reaction site owing to the zig-zag nature of the carbon-chain. However, NGP is defined as the direct interaction of the reaction site with the lone pair of *es* of an atom or with σ or π *es* which are already present in the substrate molecule but not conjugated with the reaction site. NGP develops cyclic T.S.s or intermediate ring structures—3,4,5,6 membered rings are common; 5 and 6 membered rings are the most stable.

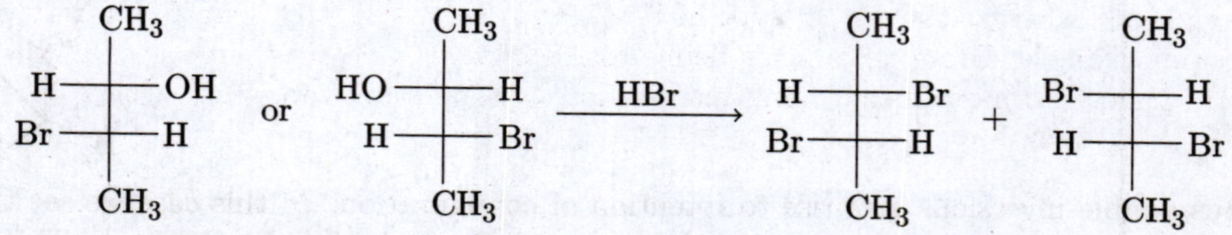

(+)-or(−)-threo forms (±)-threo forms

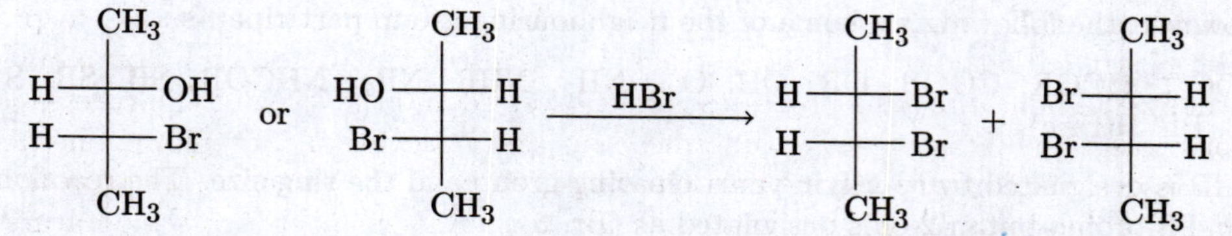

(+)-or(−)-erythro forms erythro (meso) forms

The reactions between 3-bromo-2-butanols, active threo and erythro forms, and hydrobromic acid are one set of the classical examples of neighbouring group participation. The active threo form of the alcohol gives only (±) 2,3-dibromobutane on treatment with hydrobromic acid; while the same reaction with active erythro form gives pure meso product. Thus we see that the reactions have taken place with retention of configurations.

The retention of configuration in this reaction is well explained by the neighbouring group participation as shown below :

In this case Br atom is the neighbouring group of the reaction site $C_2(OH)$ of the substrate 3-bromo-2-butanol. The removal of the OH as H_2O has taken place through the formation of the bridged bromonium ion with inversion of configuration;. this is followed by Br^- attack on C_2 or C_3 atom and cleavage of the bromonium bridge with inversion of configuration.

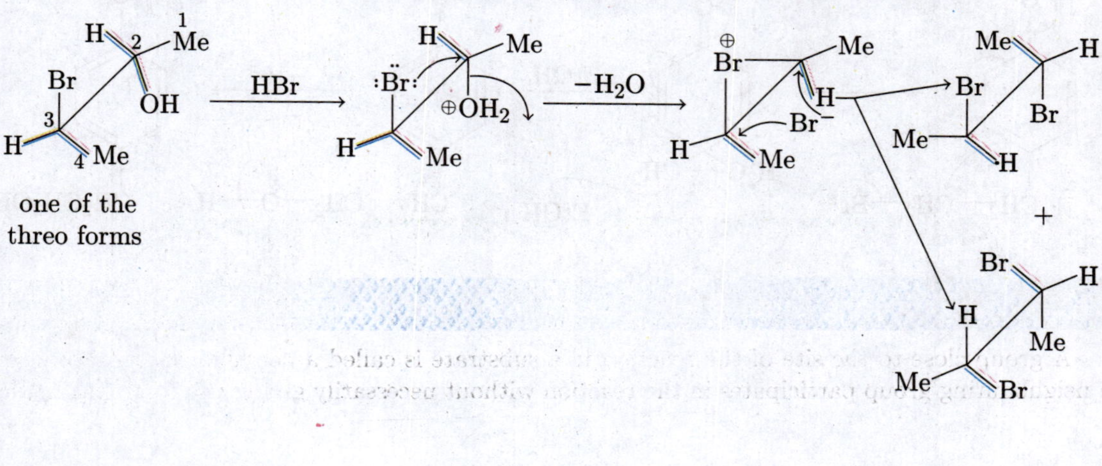

one of the threo forms

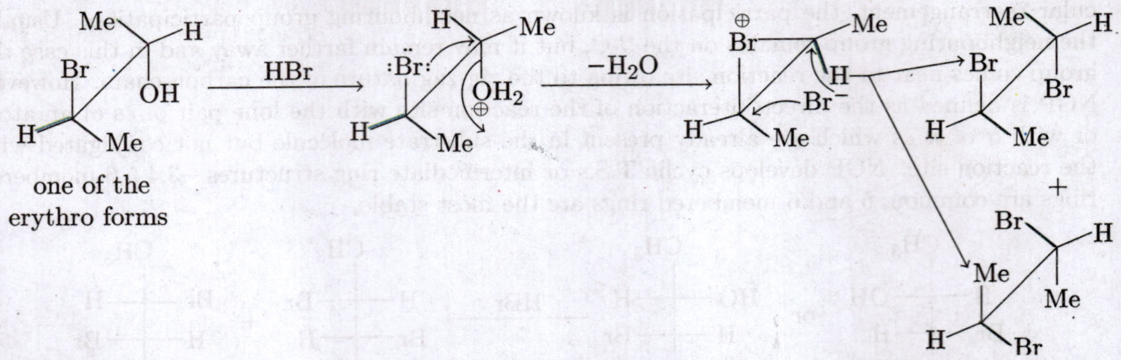

Thus double inversions give rise to retention of configuration. In this case we see that a full covalent bond formation has taken place between the Br atom and the C_2 atom. This is not at all the necessary requirement; there are examples where only partial bond forms.

However, the following are some of the neighbouring group participants :

COO^-, OCOR, COOR, OR, OH, O^-, NH_2, NHR, NR_2, NHCOR, SH, SR, S^-, Br, I, C=C, Ph, H(rare).

NGP is designated by specifying participating group and the ring size. The reaction between HBr and 3-bromo-butan-2-ol is designated as (Br, 3).

Now, the question is : what are evidences in favour of the neighbouring group participating mechanisms? The answer is : only in the case of neighbouring group participation with rearrangement one can get the direct evidence; in other cases, stereochemical criterion—retention of configuration and manyfold enhancement of rate over what might be expected normally are the evidences. The enhancement of rate of a reaction involving neighbouring group participation is called anchimeric assistance (meaning partaining to the adjacent part; Gr. anchi and meros). For example, the conjugate base of the β-(p-hydroxyphenyl)ethyl bromide undergoes ethanolysis about 106 times faster than β-(p-methoxyphenyl)ethyl bromide; the enhancement of rate is attributed to the anchimeric assistance of the p-phenoxy group.

II (p 545) is much less stable than I because of + charge on the O atom. So in this case no effective bridged species forms.

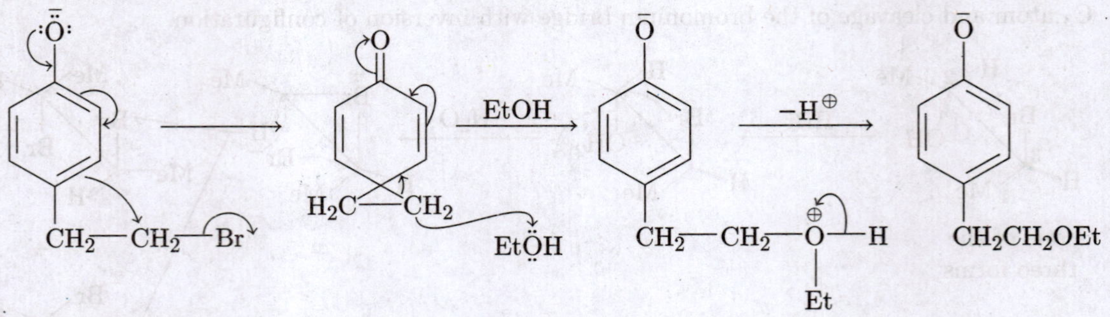

I

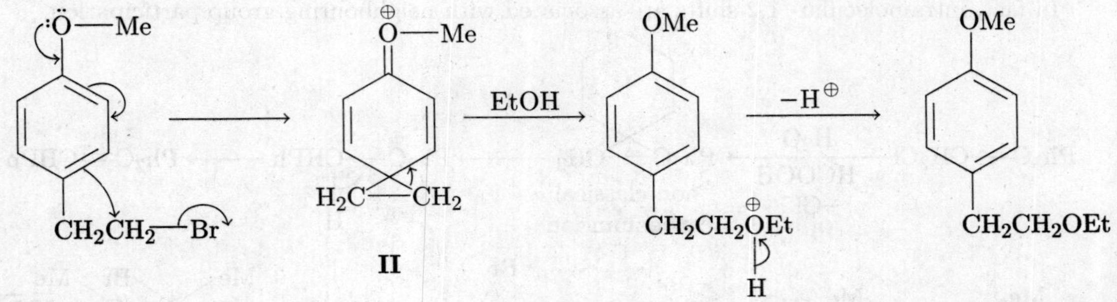

Now we must get an explanation for the enhancement of rate in an anchimeric assistance. In a chemical reaction, the reaction site and the reagent require collision with each other. The probability of such a collision in this case is less than the availability of the neighbouring group to the reaction site. Furthermore, in a bimolecular or termolecular reactions, the T.S. involves 2 or 3 species respectively and therefore, in such a reaction the entropy decreases reasonably. On the other hand, in a neighbouring group mechanism since the neighbouring group attacks the reaction site intramolecularly, the T.S. does not involve any species other than the substrate and therefore, entropy decreases but to a little extent owing to conformational reason. Here it is needless to mention that a reaction with or without NGP has the same $\Delta H^{\ddagger}$ when NG and the reagent form the same type of bond with the reaction site. Therefore, $\Delta S^{\ddagger}$ is the main criterion to ascertain $\Delta G^{\ddagger}$.

Whether the rate of a reaction increases due to anchimeric assistance or not can be ascertained by comparing the rates of the reaction of a substrate with and without the neighbouring group. If the rate of the reaction with the neighbouring group is many fold more than that of the substrate without the neighbouring group, neighbouring group participation may be considered.

However, neighbouring group participations are found in addition, elimination, rearrangement and substitution reactions as shown below :

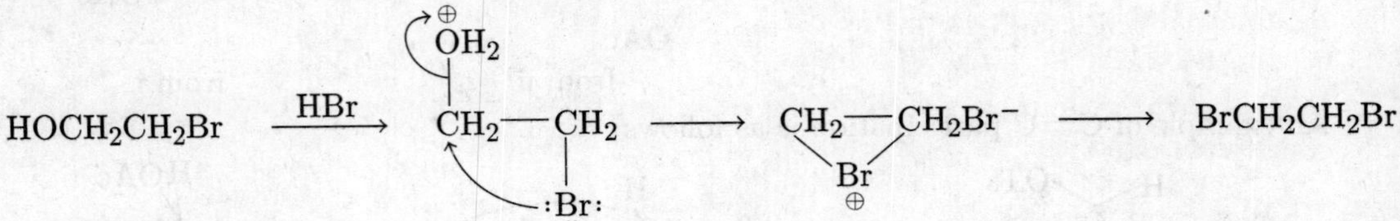

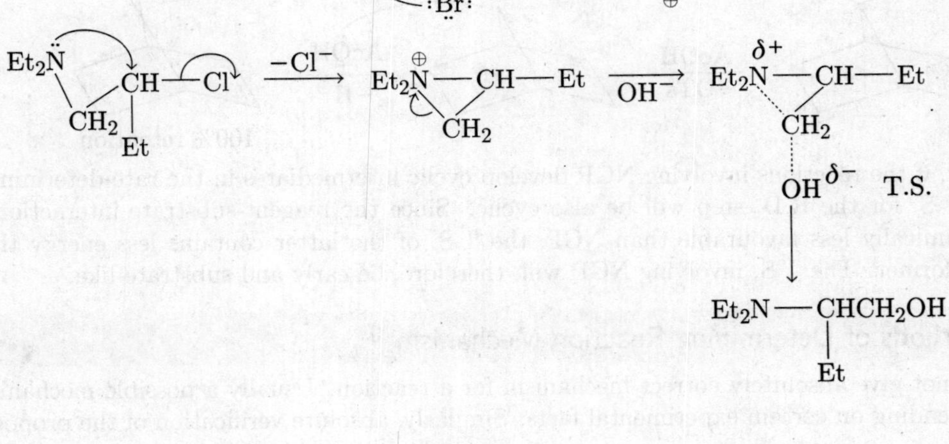

In fact, intramolecular- 1,2-shifts are associated with neighbouring group participation.

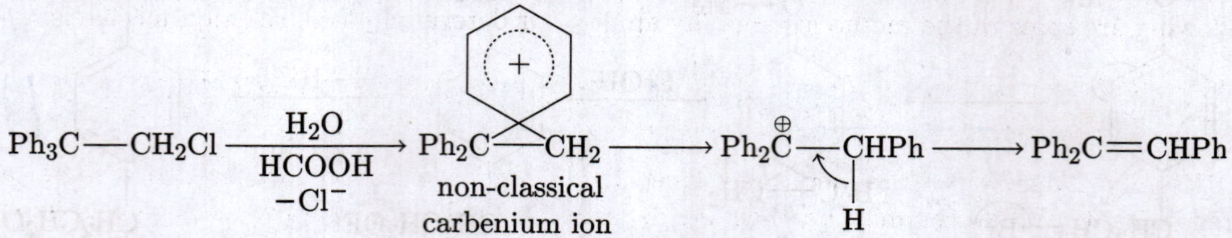

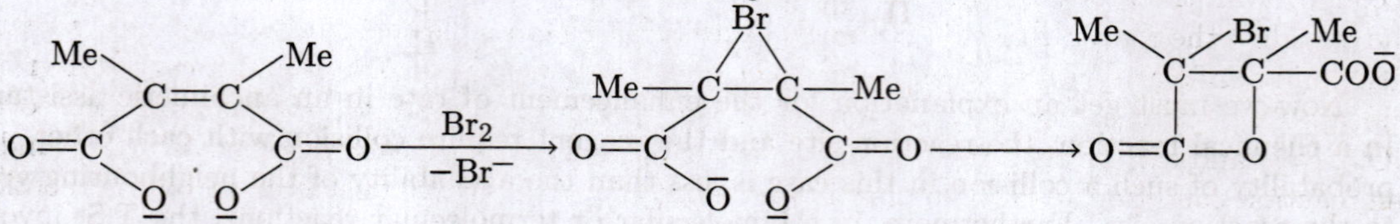

In the cyclohexane series NGP requires 1,2-diaxial (trans) orientation.

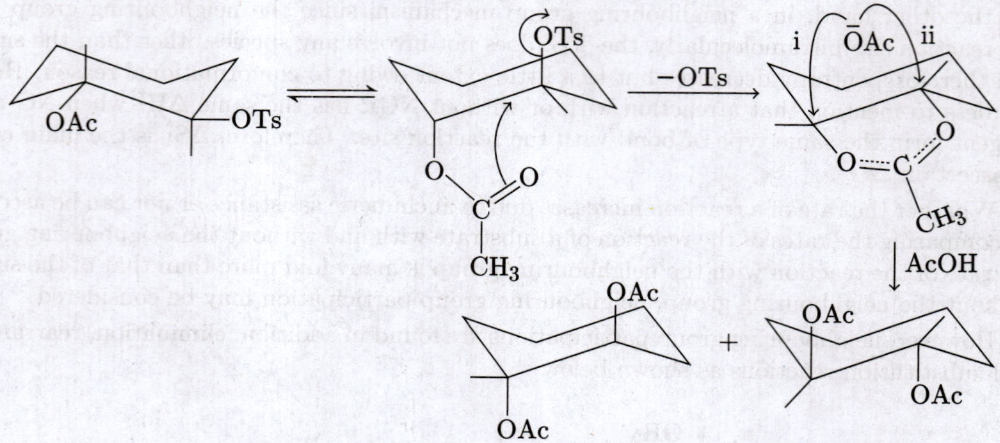

from ii from i

An example of C=C participation is as follows :

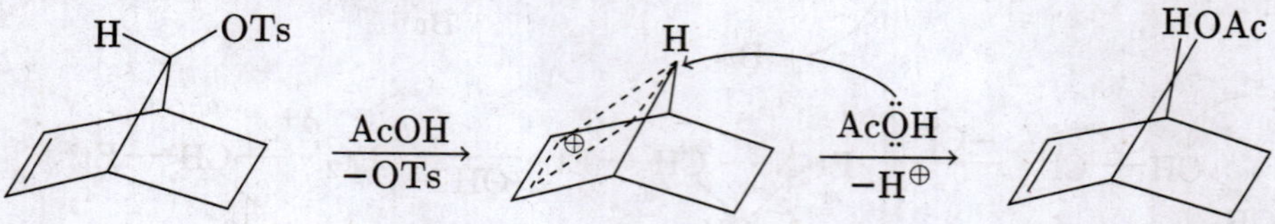

100% retention

However, if the reactions involving NGP develop cyclic intermediates in the rate-determining steps, the T.S. for the R.D. step will be also cyclic. Since the reagent-substrate interaction is thermodynamically less favourable than NGP, the T.S. of the latter contains less energy than that of the former. The T.S. involving NGP will, therefore, be early and substrate-like.

11.8 Methods of Determining Reaction Mechanism

One cannot give absolutely correct mechanism for a reaction. Usually a possible mechanism is given depending on certain experimental facts. Similarly, absolute verification of the proposed mechanism is not also possible by a single experiment or by a set of experiments. However, the following are some of the methods generally applied for determining the reaction mechanism.

11.8.1 Nature of the products and their proportions

Usually a number of products are formed in an organic reaction, one or two of them are predominant. Whatever may be their yield, large or small, any proposed mechanism should account for all the products and their proportions. If the mechanism fails to predict the products and roughly their proportions, it cannot be taken as a possible mechanism. When aromatic nitrocompounds are treated with cyanide ion, the nitro group is displaced and a carboxyl group appears ortho to the displaced nitro group. The products of the reaction was supposed to be NO_2^-, NH_3 and carboxylic acid. Since CN^- is the reagent and aryl cyanide gets hydrolysed under the reaction condition to carboxylic acid and ammonia, ArCN was considered to be the reaction

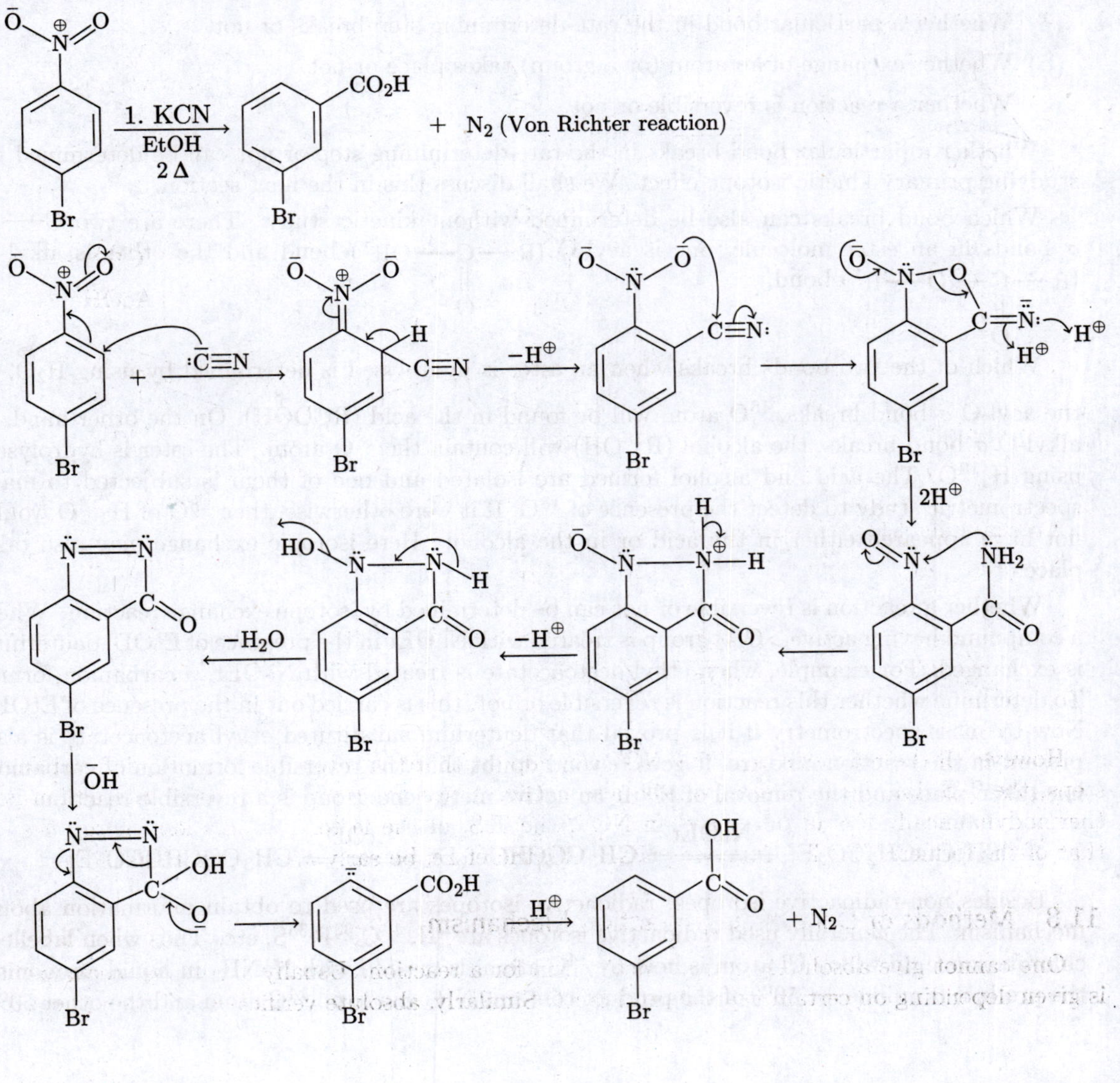

intermediate. Now it is found that besides NO_2^- and NH_3, molecular nitrogen is formed. It is also found that not the NO_2^- but N_2 is the major product of the reaction. To substantiate these facts, a quite different mechanism, has been proposed for the reaction which does not involve ArCN as an intermediate. The reaction is known as Von Richter reaction. Its mechanism is given in the page 545.

Therefore, the isolation of products, the elucidation of their structure and determination of their proportions give much information about the mechanism of a reaction. Nowadays generally the products are isolated by chromatographic processes and their structures are established by spectroscopic methods and from their syntheses.

11.8.2 Use of isotopes

By using isotopes many information about reaction mechanism can be obtained; some of them are given below :

(a) Whether a particular bond in the rate-determining step breaks or not.

(b) Whether exchange of an atom (or a group) takes place or not.

(c) Whether a reaction is reversible or not.

Whether a particular bond breaks in the rate-determining step or not can be determined by studying primary kinetic isotope effect. We shall discuss this in the next section.

Which bond breaks can also be determined without kinetic study. There are two C—O σ bonds in an ester molecule, one is acyl-O $(R-\underset{\underset{O}{\|}}{C}-OR^1)$ bond and the other is alkyl-O $(R-\underset{\underset{O}{\|}}{C}-O-R^1)$ bond.

Which of the two bonds breaks when an ester is hydrolysed is determined by using $H_2\overset{18}{O}$. If the acyl-O σ bond breaks, ^{18}O atom will be found in the acid $(RCO\overset{18}{O}H)$. On the other hand, if alkyl-O σ bond breaks, the alcohol $(R^{18}OH)$ will contain the ^{18}O atom. The ester is hydrolysed using $H_2{}^{18}O$. The acid and alcohol formed are isolated and one of them is subjected to mass spectrometric study to detect the presence of ^{18}O. If it were otherwise, then ^{18}O of $H_2{}^{18}O$ would not have appeared either in the acid or in the alcohol. Here isotopic exchange may also take place.

Whether a reaction is reversible or not can be determined by isotope exchange reaction. When a compound having active $>CH_2$ group is treated with NaOEt in the presence of EtOD, deuterium is exchanged. For example, when ethyl acetoacetate is treated with NaOEt, a carbanion forms. To determine whether this reaction is reversible or not, this is carried out in the presence of EtOD. Now by mass spectrometry if it is proved that deuterium substituted ethyl acetoacetate is also present in the reaction mixture, it goes beyond doubt that the reversible formation of carbanion has taken place and the removal of $H^\oplus$ from active methylene group is a reversible reaction.

$$CH_3COCH_2CO_2Et \underset{}{\overset{NaOEt}{\rightleftharpoons}} CH_3CO\overset{-}{C}HCO_2Et \underset{}{\overset{EtOD}{\rightleftharpoons}} CH_3COCHDCO_2Et$$

Besides non-radioactive isotopes, radioactive isotopes are used to obtain information about mechanism. The generally used radioactive isotopes are ^3H, ^{14}C, ^{32}P, ^{35}S, etc. Thus when labelled chlorobenzene, in which Cl atom is held by ^{14}C atom, is treated with $NaNH_2$ in liquid ammonia, the product is aniline—50% of the product contains NH_2 group on ^{14}C atom and the other 50%

contains that on ^{12}C atom. This information proves clearly that the reaction is not a simple case of direct substitution reaction and that there must be a reaction intermediate.

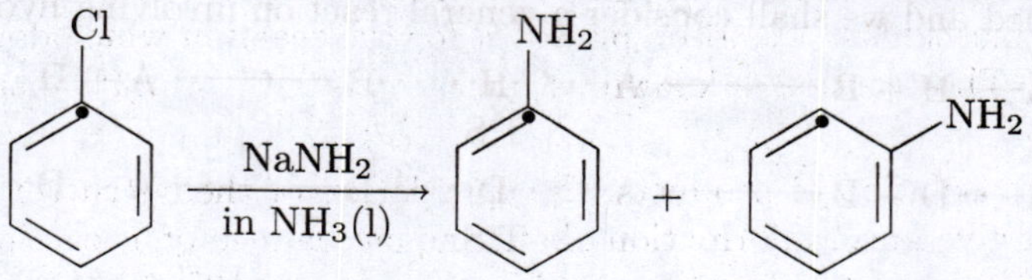

• indicates ^{14}C atom

11.8.3 Kinetic study

The second major type of information in determining the reaction pathway comes from the rate measurements. From the rate measurement data an empirical rate equation can be constructed. For simple cases unimolecular, bimolecular and termolecular reactions follow first, second and third order kinetics, respectively. However, the kinetic study involves a large number of things under consideration such as the solvent effect, the identification of the effective reacting species and its concentration, etc. Nowadays the measurement of kinetic isotope effects helps us understand which bond breaks and forms in a rate-determining step and we shall discuss this in detail in this section.

(a) Kinetic isotope effects (KIEs)

Changes in reaction rate owing to the isotopic substitution, either on the substrate or on the solvent molecules, are known as kinetic isotope effects.

KIE has been observed in the chromic acid oxidation of isopropyl alcohol to acetone.

$$CH_3CHOHCH_3 + H_2CrO_4 \longrightarrow CH_3COCH_3 + H_2O + H_2CrO_3$$

Let the rate of a reaction involving a substrate containing a light isotope of mass number X be k_x and that involving the substrate with heavy isotope of mass number Y be k_y. Then, if $k_x \neq k_y$, we say that there exists KIE. KIE is expressed as k_x/k_y. When isopropyl alcohol and α-deuterated isopropyl alcohol are subjected to chromic acid oxidation, the KIE, i.e. k_H/k_D is found to be 7.

However, there are three kinds of KIE.

i. Primary kinetic isotope effect (PKIE), ii. Secondary kinetic isotope effect (SKIE) and iii. Solvent isotope effect.

Again, when $k_x/k_y > 1$, the KIE is said to be normal and when $k_x/k_y < 1$, we call it inverse KIE.

i. Primary kinetic isotope effect (PKIE)

A PKIE is defined as the change observed in the rate of a reaction when the bond to the substituted isotopic atom of the substrate is broken or formed in the rate-determining step. In the oxidation of isopropyl alcohol to acetone by chromic acid, the α C-H bond breaks in the R.D. step. So, the reaction shows a normal PKIE and it is 7.

$$Me_2CHOH + HCrO_4^- + H^{\oplus} \xrightarrow{k_H} Me_2C{=}O + H_2O + H_2CrO_3$$

$$Me_2CDOH + HCrO_4^- + H^{\oplus} \xrightarrow{k_D} Me_2C{=}O + H_2O + HDCrO_3$$

PKIE, i.e., $k_H/k_D = 7$

H_2CrO_3 & $HDCrO_3$ disproportionate later.

KIEs involving isotopes of hydrogen, H(^{1}H), D(^{2}H) and T (tritium, ^{3}H) have been most thoroughly examined and we shall consider a general reaction involving hydrogen isotope:

$$\text{A—H} + \text{B} \longrightarrow \underset{\text{T.S}}{\text{A}\cdots\cdots\text{H}\cdots\cdots\text{B}} \longrightarrow \text{A} + \text{H} \quad \text{B}$$

$$\text{A—D} + \text{B} \longrightarrow \underset{\text{T.S}}{\text{A}\cdots\cdots\text{D}\cdots\cdots\text{B}} \longrightarrow \text{A} + \text{D—B}$$

Here A—H and A—D bonds are in the process of breaking and H—B and D—B bonds are in the process of forming in the T.S.s. It is needless to mention that k_H and k_D are dependent on the corresponding energy of activation; the lower the energy of activation, the faster is the rate of the reaction. Again, the activation energy is the energy of the T.S. relative to the average energy of the reactants. The energy of the T.S. and that of the reactants depend on their respective zero point vibrational energy. In the second chapter (p. 39-40) it has already been stated that,

$E = \left(v + \dfrac{1}{2}\right) h\nu$; where E = energy, v = vibrational quantum number (0, 1, 2, etc).

h = Planck's constant and ν = vibrational frequency of the A—H or A—D bond. In the ground state $v = 0$, and if ν is ν_0, then the zero point vibrational energy of the bond is

$E_0 = \dfrac{1}{2} h\nu_0$. Again, $\nu_0 = \dfrac{1}{2\pi}\sqrt{\dfrac{f}{\mu}}$; where f = force constant and μ = reduced mass.

$\therefore \quad \nu_0 \propto \dfrac{1}{\sqrt{\mu}}$; but $E_0 \propto \nu_0$. So, $E_0 \propto \dfrac{1}{\sqrt{\mu}}$.

Since the reduced mass of A—D is greater than that of A—H, the zero point vibrational energy of A—D will be less than that of A—H. Therefore, C—D, N—D, O—D, etc. have lower ground state energies than the corresponding C—H, N—H, O—H, etc. bonds in the same environment. However, the isotopic difference in energies of the corresponding vibrational levels ($v = 1, 2, \ldots$) decreases as the value of ν increases and at the bond dissociation limit it is zero but their potential energy curves are of the same character (Fig. 11.9).

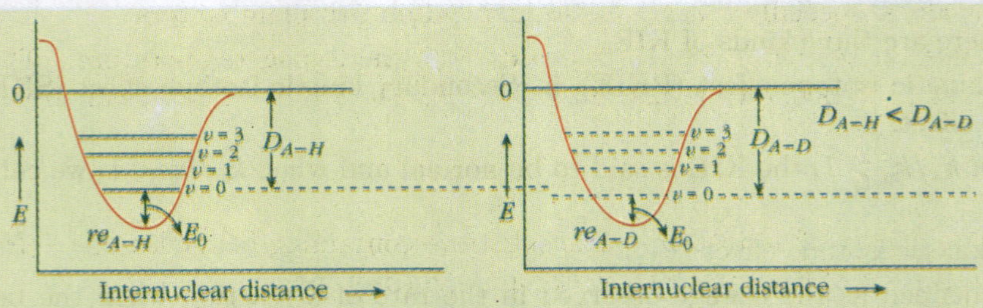

Fig. 11.9. Potential energy curve : for A—H bond and A—D bond breaking processes

The zero point vibrational energy of the T.S. A···H···B··· or A···D···B is always less than the A—H or A—D bond dissociation energy respectively since the energy of rupture of the bonds is partly compensated by the energy liberated during the formation of H—B and D—B bonds. However, for the higher reduced mass of A···D···B than that of A···H···B, the zero point energy of A···D···B is lower than that of A···H···B (in the limiting case they will be equal); but the difference in the zero point energy between the two T.S.s is somewhat less than that between the two sets of reactants (A—H + B and A—D + B) in their ground states. So, the $\Delta G^{\ddagger}$ of the reaction between A—H and B will be less than that of the reaction between A—D and B as shown in the figure given below (Fig. 11.10). So A—H bond breaking will occur faster than A—D bond breaking, i.e., $k_H > k_D$. PKIEs have also been observed in the reactions involving simple breaking or formation of one bond instead of atom transfer as shown in the example above. Retro Diels-Alder reaction is an example of such a reaction.

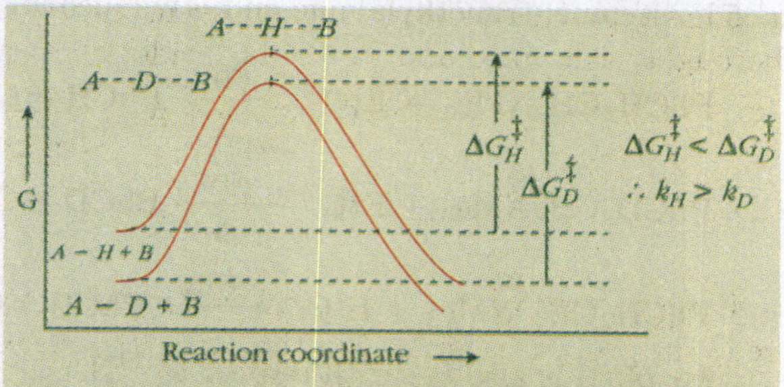

Fig. 11.10. Free energy curves for isotopic atom transfer reactions

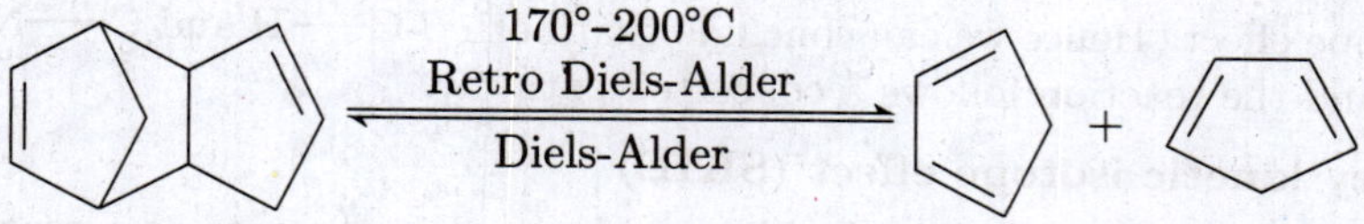

Here bond breaking and bond formation occur in a concerted step.

PKIEs are essentially negligible for isotopes of mass number greater than 20 to 30, but they may be very high for isotopes of hydrogen and k_H/k_D ranges from 1 to 8 and even 9. The maximum normal hydrogen-deuterium isotope effect at 0°C is 8.3. PKIE values for different isotopes are given in Table 11.3.

Table 11.3 Maximum normal values of PKIE.s at 25°C

Element	k_x/k_y	value	Element	k_x/k_y	value
Hydrogen	k_H/k_D	6.3	Nitrogen	k_{14}/k_{15}	1.03
Hydrogen	k_H/k_T	13.0	Oxygen	k_{16}/k_{18}	1.02
Carbon	k_{l2}/k_{13}	1.04	Sulphur	k_{32}/k_{34}	1.01
Carbon	k_{l2}/k_{14}	1.07	Chlorine	k_{35}/k_{37}	1.01

The PKIE values, specially k_H/k_D values, depend on two main factors :

(i) It depends on temperature—It decreases with increasing temperature. The maximum normal k_H/k_D at 25°C is 6.4 while that at 0°C is 8.3. The PKIE—temperature relation is given by equation 11.2.

$$k_x/k_y = e^{\Delta\Delta G^{\ddagger}/RT} \qquad (11.2)$$

$\Delta\Delta G^{\ddagger}$ = Free energy of activation of the substrate containing heavy isotope —free energy of activation of the substrate containing light isotope.

R = Universal gas constant and T = Temperature in degree kelvin.

(ii) PKIE values depend on the structure of the T.S. The maximum isotope effect is expected from a linear and symmetrical T.S. having the isotopic atom placed centrally between the two partially bonded atoms.

PKIEs play an important role in determining the reaction mechanism. If no PKIE is observed for a reaction, we can conclude that the R.D. step does not involve the breaking or making of a particular bond. A high PKIE value indicates, on the other hand, that the particular bond breaks or forms in the R.D. step. When trimethyl-β-phenylethylammonium ion is heated with sodium ethoxide, styrene, trimethylamine and ethyl alcohol are obtained as shown below :

$$\text{(i)}\ PhCH_2CH_2\overset{\oplus}{N}Me_3 + EtO^- \xrightarrow[\Delta]{k_H} PhCH{=}CH_2 + NMe_3 + EtOH$$

$$\text{(ii)}\ PhCD_2CH_2\overset{\oplus}{N}Me_3 + EtO^- \xrightarrow[\Delta]{k_D} PhCD{=}CH_2 + NMe_3 + EtOD$$

$$\text{(iii)}\ PhCH_2CH_2^{15}\overset{\oplus}{N}Me_3 + EtO^- \xrightarrow{k_{15}} PhCH{=}CH_2 + {}^{15}NMe_3 + EtO$$

k_H of (i) is also k_{14} with respect to ^{14}N.

Here k_H/k_D is 4.6 and k_{14}/k_{15} is 1.009. So the reaction has hydrogen-deuterium isotope effect and ^{14}N-^{15}N isotope effect. Hence we can conclude that both β-C—H and C—N bonds break in the R.D. step and the reaction follows a concerted pathway.

(b) Secondary kinetic isotope effect (SKIE)

A secondary kinetic isotope effect is the change observed in the rate of a reaction when an atom of the substrate is replaced by its isotope at a bond which is not broken or formed during the reaction.

This also gives information regarding the mechanism of the reaction. SKIEs can be subdivided under many heads. If the isotopes are bonded respectively to α-C, β-C, γ-C, etc. with respect to the atom undergoing bond breaking, SKIEs are designated as α, β, γ, etc, SKIEs.

When a SKIE is observed by substituting an isotope for an atom (α to the leaving groups), not undergoing bond cleavage, it is called α-SKIE. It varies from 0.94 to 1.16, i:e., it has inverse and normal KIE. S_N1 reactions (Sec. 12.4.5) have an α-deuterium normal SKIE.

A β-SKIE has been observed in the solvolysis of isopropyl bromide.

$$(CH_3)_2CHBr + H_2O \xrightarrow{k_H} (CH_3)2CHOH + HBr$$

$$(CD_3)_2CHBr + H_2O \xrightarrow{k_D} (CD_3)_2CHOH + HBr$$

In this reaction; $k_H/k_D = 1.34$

The Diels-Aider reaction also show SKIE.

Hyperconjugative effect is supposed to be the cause for the β-SKIE. Reactions which exhibit β-SKIE perhaps involve a carbocationic T.S. in the R.D. step and the T.S. gets stability through hyperconjugation involving β-H. Since the difference in the vibrational energy between the C—H bond and the C—D bond in the T.S. is somewhat less than it is in the ground state, the involvement of β-C—D bond in no-bond resonance brings less stability to the carbocationic T.S. than the involvement of β-C—H bond in no-bond resonance and hence substitution of D for H in the β-C slows down the reaction.

However, SKIE values are much smaller than PKIE values.

(c) Solvent isotope effect : When a KIE is observed on mere changing the solvent to one which differs only isotopically, we say the KIE as solvent isotope effect.

An example of solvent isotope effects is the acid-catalysed hydrolysis of acetals (and ketals) in H_2O (D_2O) medium.

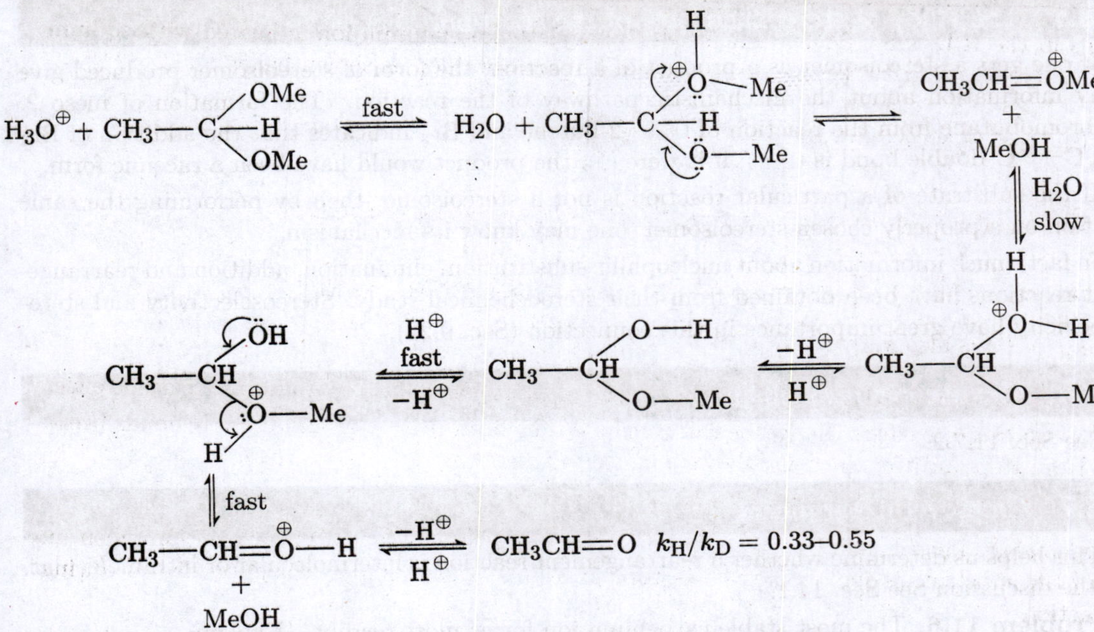

The solvent isotope effect here (k_H/k_D) varies from 0.33—0.55.

The solvent isotope effect also plays an important role in the elucidation of the reaction mechanism. From the solvent isotope effect, the slowest step in the hydrolysis of acetals has been determined. H_2O addition step is the R.D. step.

11.8.4 The study of reaction intermediates

Very often a reaction intermediate is proposed to draw the reaction pathway of an organic reaction. Either the isolation of the intermediate or the detection of its presence may be a strong evidence for such a proposed mechanism.

For example, the mechanism of the reaction between ethyl picrate and methoxide ion has been established by the isolation of Meisenheimer salt which is a complex intermediate in the reaction.

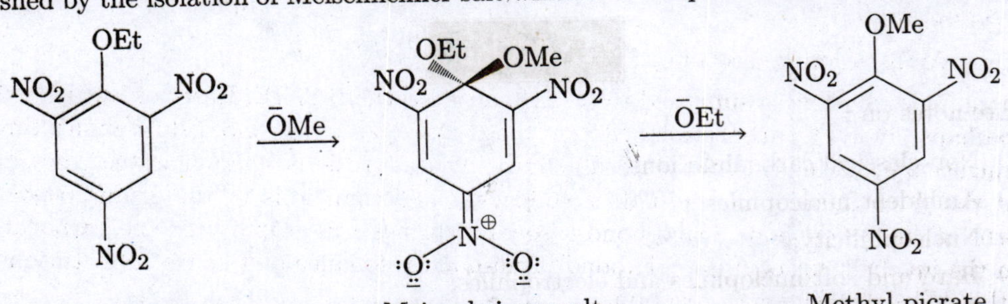

Ethyl picrate Meisenheimer salt Methyl picrate

The detection of reaction intermediates may be done by infrared, nuclear magnetic resonance, electron paramagnetic resonance or by other spectral studies. They may be detected also by trapping. The detection of nitronium ion ($\overset{\oplus}{N}O_2$) by Raman spectra is regarded as a strong evidence that the nitration of benzene involves $\overset{\oplus}{N}O_2$.

11.8.5 The study of stereochemistry of the reaction

If one gets a stereoisomer as a product of a reaction, the form of stereoisomer produced give many information about the mechanistic pathway of the reaction. The formation of meso 2, 3-dibromobutane from the reaction of trans-2-butene and Br_2 indicates that the addition of Br_2 to a C—C double bond is trans; if it were cis, the product would have been a racemic form.

If the substrate of a particular reaction is not a stereoisomer then by performing the same reaction on a properly chosen stereoisomer, one may know its mechanism.

In fact, much information about nucleophilic substitution, elimination, addition and rearrangement reactions have been obtained from their stereochemical study. Stereoselectivity and stereospecificity have great importance in this connection (Sec. 9.22).

11.8.6 Microscopic reversibility

See Sec. 11.7.2.

11.8.7 Crossover experiments

This helps us determine whether a rearrangement reaction is intermolecular or intramolecular. For the discussion See Sec. 14.1.

Problem 11.6. The most stable carbenium ion forms most readily.—Explain.

Solution : The formation of carbenium ions from stable compounds is an endothermic reaction. According to Hammond's postulate the T.S. will be carbenium ion-like. The most stable carbenium ion, therefore, will have the most stable T.S. This means its $\Delta G^\ddagger$ will be the lowest and hence it will be the easiest to overcome the energy barrier. In this case the factors, I, R, no-bond resonance, steric, effects, etc., stabilise both the T.S. and the intermediate. So the most stable carbenium ion will form most readily.

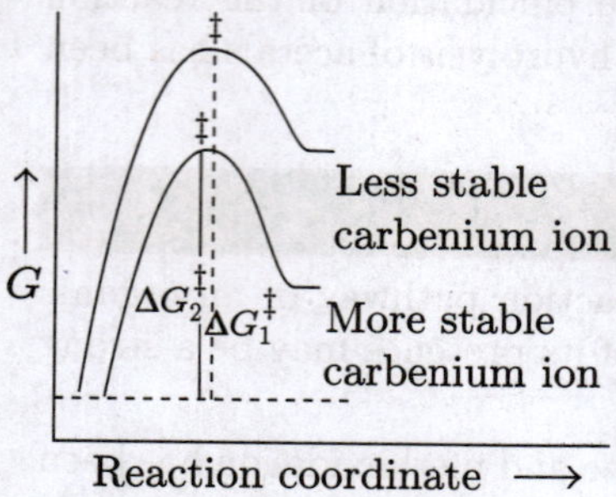

Exercise

1. Write notes on :

   (a) Non-classical carbenium ions;
   (b) Ambident nucleophiles;
   (c) Nucleophilicity;
   (d) Hard and soft nuclophiles and electrophiles;
   (e) Reactivity and orientation;
   (f) Concerted step reactions and stepwise reactions;
   (g) Microscopic reversibility principle;
   (h) Reactivity-selectivity principle;
   (i) Neighbouring group participation and anchimeriç effect.

2. Write down the products from the following multi-centred T.S.s :

(i) 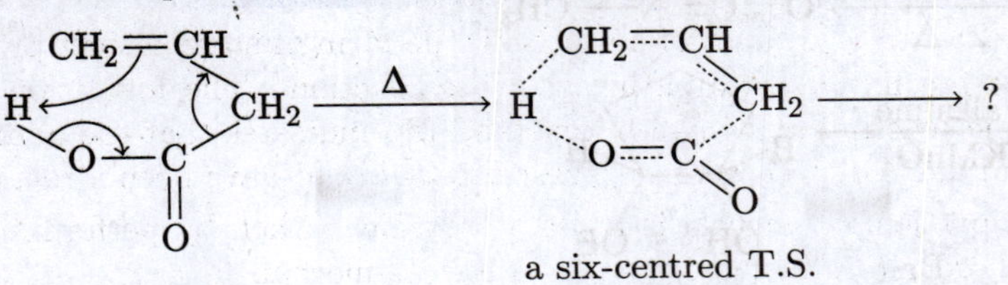

a six-centred T.S.

(ii) 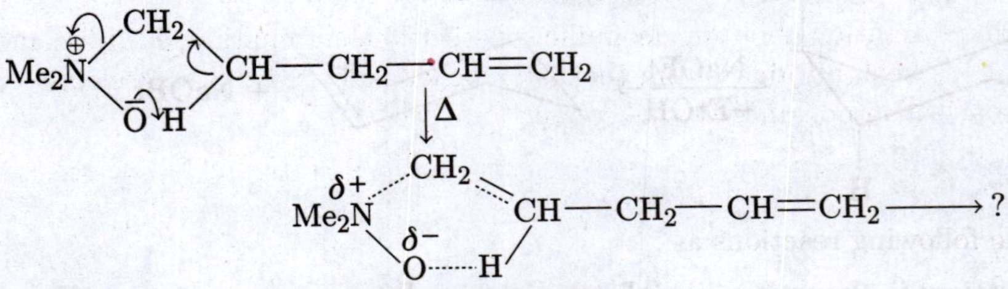

a five-centred T.S.

(iii) 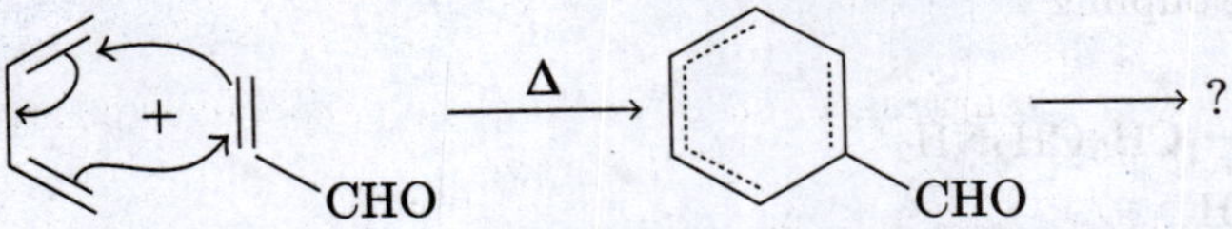

a six-centred T.S.

(iv) 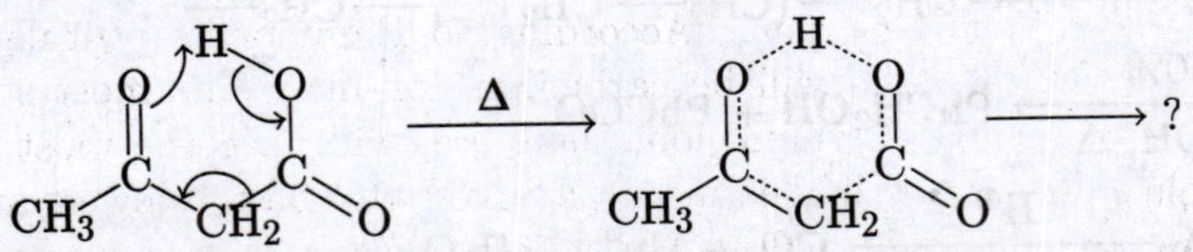

a six-centred T.S.

(v) 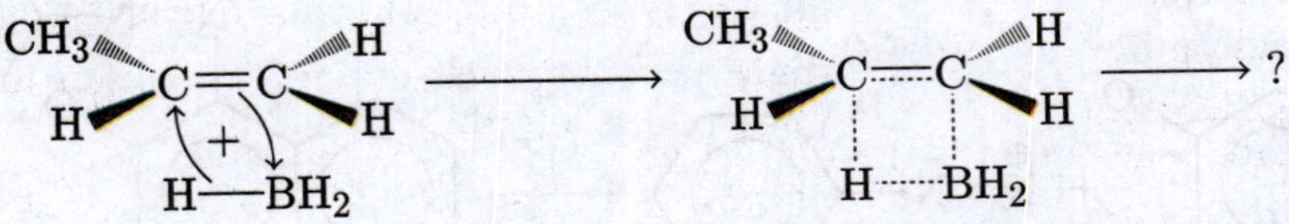

a four-centred T.S.

(vi) 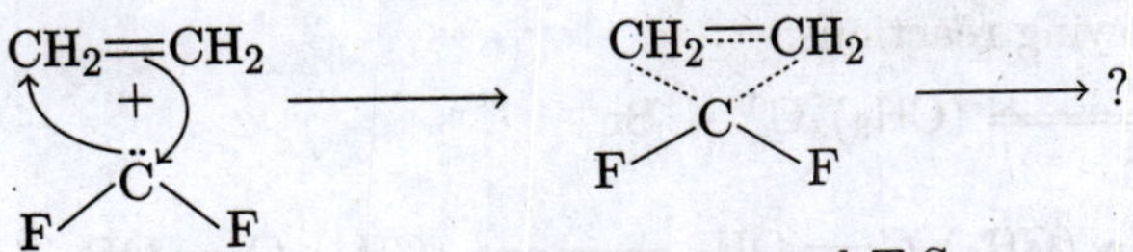

singlet

a three-centred T.S.

3. Classify the following reactions as :

i. Substitution; ii. Elimination; iii. Addition; iv. Rearrangement.

(a) $CHCl_3 \xrightarrow{\bar{O}Et} :CCl_2 + EtOH + Cl^-$.

(b) $CH_3Cl \xrightarrow{\bar{O}H} CH_3OH$

(c) $CH_3COCl \xrightarrow{NaN_3} CH_3CON_3 + NaCl$

(d) $CH_3CO\ddot{N} \xrightarrow[2.\ \Delta]{1.\ CHCl_3} O{=}C{=}N{-}CH_3$

(e) 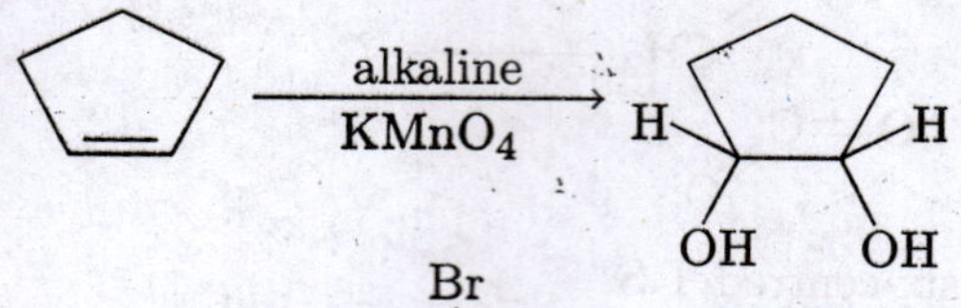

(f) 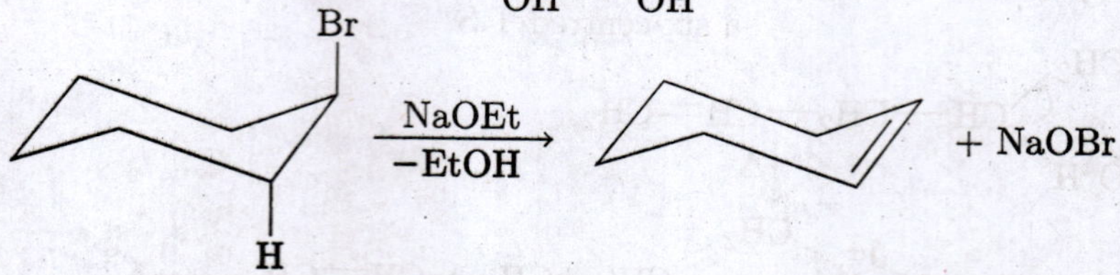

4. Classify the following reactions as :

i. Condensation; ii. Insertion; iii. Abstraction; iv. Disproportionation; v. Fragmentation; vi. Redox reaction; vii. Coupling :

(a) $CH_3NH_2 \xrightarrow{:CH_2} CH_3CH_2NH_2$

(b) $CH_3{-}CHO \xrightarrow{\bar{O}H} \bar{C}H_2{-}CHO + H_2O$

(c) $nCH_2{=}CH_2 \longrightarrow {-}CH_2{-}(CH_2{-}CH_2)_{n-1}{-}CH_2{-}$

(d) $PhCHO \xrightarrow[NaOH,\ \Delta]{50\%} PhCH_2OH + PhCOO^-$

(e) $CO_2H{-}CO_2H \xrightarrow[KMnO_4]{H^{\oplus}} CO_2 + Mn^{++} + H_2O$

(f) 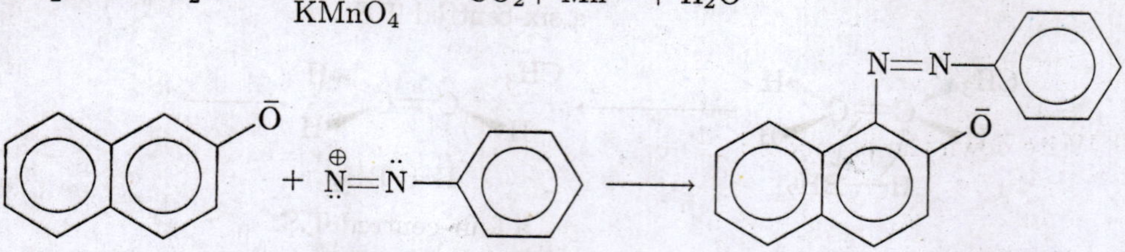

5. (i) Draw T.S.s for the following reactions.

(a) $(CH_3)_3C{-}Br \rightleftharpoons (CH_3)_3C^{\oplus} + Br^-$

(b) $(CH_3)_3C^{\oplus} \overset{H_2O}{\rightleftharpoons} (CH_3)_3C{-}\overset{\oplus}{O}H_2 \rightleftharpoons (CH_3)_3C{-}OH + H^{\oplus}$

(c) $CH_3O^- + CH_3Br \longrightarrow CH_3OCH_3 + Br^-$ (a single-step reaction)

(d) 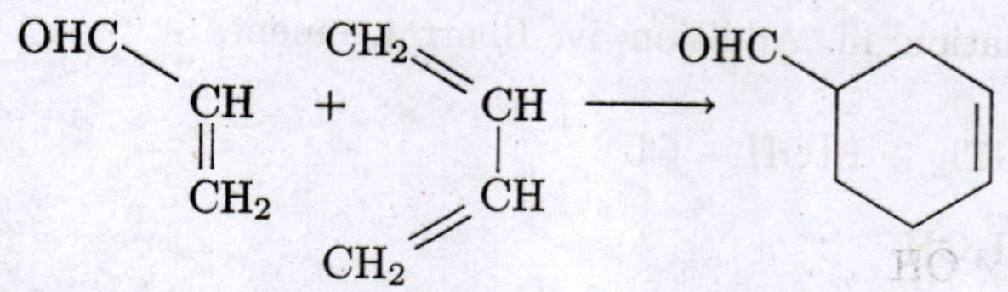 (a six-centred reaction)

(ii) Draw energy profiles for the above reactions.

6. Write down the stability order of the following sets :

(i) $CH_3\overset{\oplus}{C}H_2$, $CH_2{=}CH{-}\overset{\oplus}{C}H_2$, 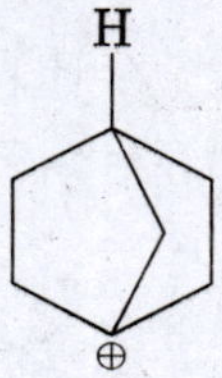

(ii) $\bar{C}H_2{-}C(=O){-}H$, $CH_3{-}\ddot{O}{-}\bar{C}H_2$

(iii) $:CCl_2$, $:CHCl$, $:CH_2$ (all in the singlet state)

(iv) 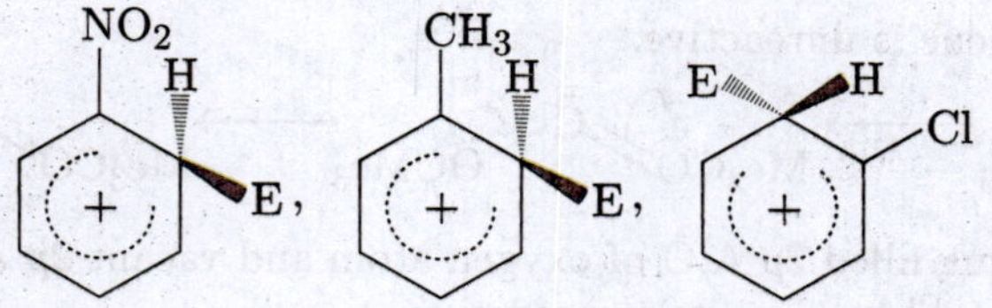

(v) t-Bu cation in water and in gas-phase

(vi) 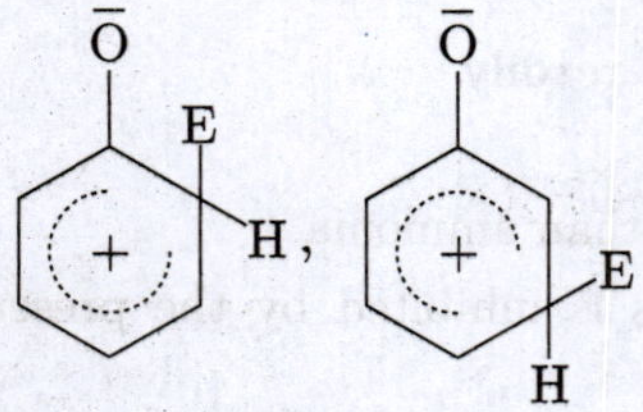

7. Write down the products :

(i) $CH_3\underset{}{\overset{CH_3}{\overset{|}{C}}}H{-}Br \xrightarrow{Zn^{\oplus\oplus}}$

(ii) PhCH=CHPh (trans) + $\cdot\dot{C}H_2 \longrightarrow$

(iii) $CH_2{=}C{=}CH_2 \xrightarrow{H^{\oplus}} \xrightarrow{H_2O}$

(iv) 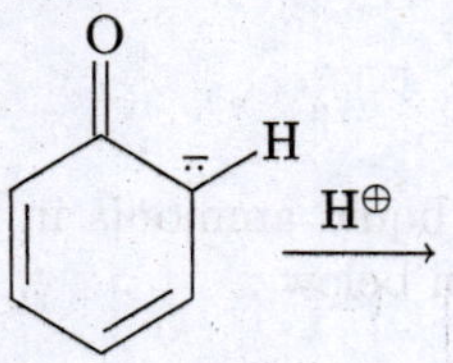

(v) $CH_3{-}\underset{CH_3}{\overset{CH_3}{\overset{|}{\underset{|}{C}}}}{-}\overset{\oplus}{C}H_2 \xrightarrow{\bar{O}H}$

(vi) 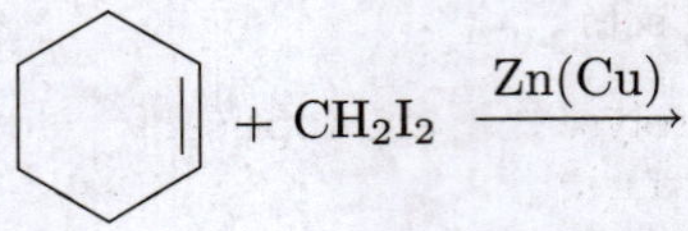

(vii) 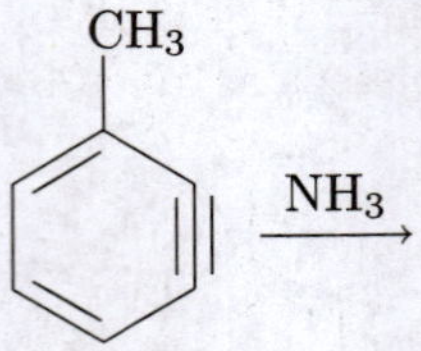

(viii) 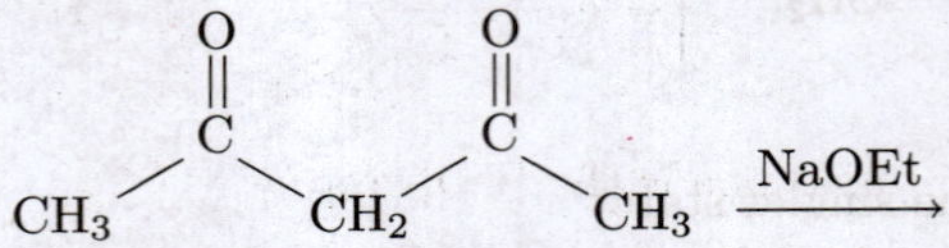

8. Explain why di-t-butoxycarbene is unreactive.

Ans. 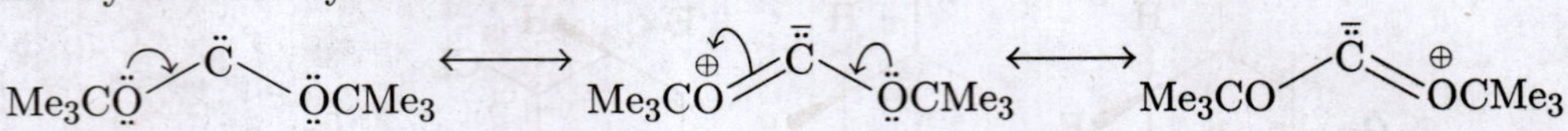

Owing to conjugation involving filled $2p$ A.O of oxygen atom and vacant $2p$ A.O of carbon atom, the carbene is highly stable and so it is unreactive.

9. Explain :

(i) The least stable carbon radical forms least readily.

(ii) Toluene is an ambident nucleophile.

(iii) Methylhydrazine is a stronger nucleophile than ammonia.

(iv) The vapour phase chlorination of alkanes is inhibited by the presence of oxygen (which is a diradical).

(v) When diazomethane is decomposed in the gas-phase by light, ethylene and cyclopropane are formed.

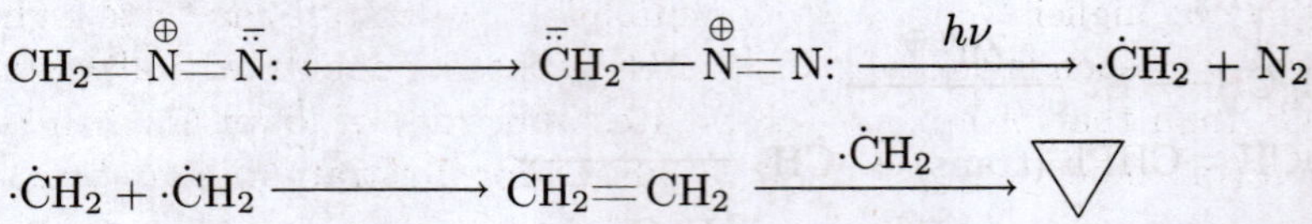

(vi) The lower the C—H bond dissociation energy, the higher is the stability of the carbon radical formed [see Table 11.1].

10. What do the following facts indicate :

(i) When chlorobenzene is treated with sodamide in liquid ammonia in the presence of furan, a 1,4-addition product is obtained as shown below :

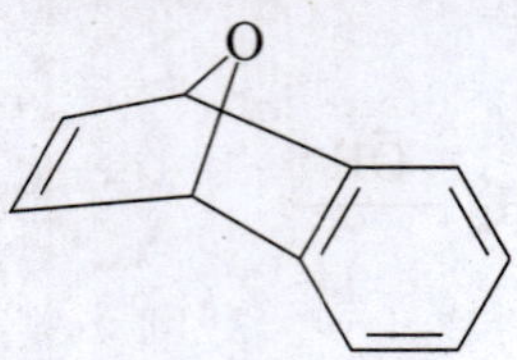

(ii) When $CH_3-\overset{O}{\overset{\|}{C}}-OCHDMe$ is hydrolysed, an optically active alcohol (MeCHDOH) is obtained in cent percent purity.

(iii) The decarboxylation of $PhCOCH_2CO_2^-$ shows $k_{12}/k_{13}=1.04$.

(iv) The mechanism for nitration of benzene involves two steps as given below :

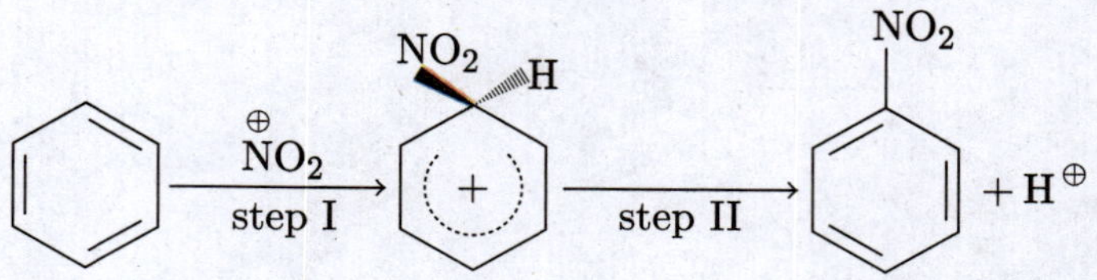

In this reaction $k_H/k_D=1.00$; which one is the R.D. step?

(v) trans-But-2-ene + $Ph_2\bar{\bar{C}}-\overset{\oplus}{N}\equiv N \xrightarrow{\Delta}$

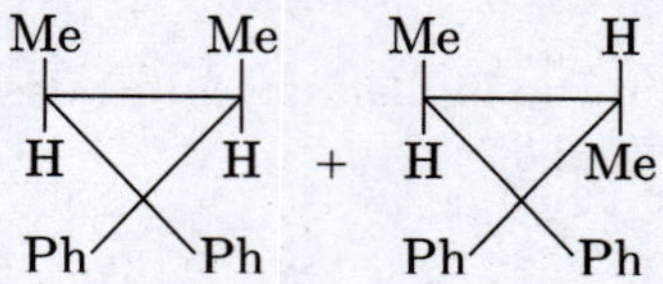

What kind of carbene generates in the reaction?

(vi) Explain the stereochemistry for the reaction :

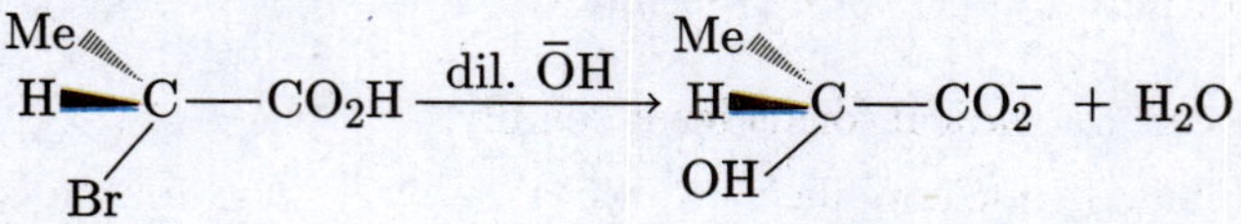

Retention of configuration

11. Arrange in the order of increasing nucleophilicity : H_2O, $CH_3CO_2^-$, CH_3OH, CH_3O^-, $\bar{O}H$

12. What is Curtin-Hammett principle?

Ans. This states that ground state population of the stable conformation or tautomer in rapid equilibrium does not always give the predominant product. Less stable conformation (tautomer) having higher G and less population in ground state may give higher yield. The yield of a product depends on $\Delta G^\ddagger$. If the $\Delta G^\ddagger$ of the less populated and unstable substrate has less $\Delta G^\ddagger$ than that of the more populated substrate of lower G, then less populated conformation (tautomer)will give major product. For illustration see p. 576.

What kind of [illegible] is there in the reaction?

Chapter 12

Nucleophilic Substituition Reactions

Objectives

- To understand the mechanism of:
  - S_N1 reactions on sp^3C [12.3] p. 563
  - S_N2 reactions on sp^3C [12.4] p. 573
  - S_Ni reactions on sp^3C [12.5] p. 585
  - S_N reactions on sp^2C [12.6] p. 587
  - Aromatic S_N reactions [12.7] p. 591
  - Aromatic S_N1 reactions [12.8] p. 593
  - Aromatic S_N2 reactions [12.9] p. 594
  - Aromatic S_N reactions involving aryne [12.10] p. 602

12.1 Introduction

Since there are three types of reagents—nucleophiles, electrophiles and radicals, the substitution reactions may be classified under three heads :

(a) substitution Nucleophilic, denoted by S_N; where a nucleophile is displaced by another nucleophile.

$$CH_3\text{—}Cl \xrightarrow[-Cl^-]{\bar{O}H} CH_3OH$$

(b) Substitution Electrophilic, denoted by S_E; where an electrophile is displaced by another.

$$C_6H_6 \xrightarrow[-H^{\oplus}]{\overset{\oplus}{N}O_2} C_6H_5NO_2$$

(c) Substitution radical, where a radical is displaced by another.

$$CH_4 \xrightarrow{2Cl\cdot} CH_3Cl + HCl$$

12.2 Substitution Nucleophilic

A nucleophilic substitution reaction may be schematically represented as :

$$R\text{—}G + :\bar{N}u \longrightarrow R\text{—}Nu + \bar{G}:$$

where G is a leaving nucleophile and $\bar{N}u$: is an incoming nucleophile.

An incoming nucleophile may either be negatively charged or it may be an uncharged species with unshared pair of electrons :

$$NO_2^-, N_3^-, RCOO^-, \bar{C}N, I^-, Br^-, Cl^-, \bar{O}H, \bar{O}Me, H_2O, ROH, \ddot{N}H_3, R\ddot{N}H_2, \text{etc.}$$

The reactivity of a nucleophile is called its nucleophilicity. The nucleophilicity increases with the increasing basicity and polarisibility. Nucleophilicities can be compared as given in Sec. 11.5.1b.

However, in a nut shell we can say :

1. Nucleophilicity of an atom increases with the increasing negative charge on it. Thus $\bar{O}H$ is more nucleophilic than H_2O.
2. Nucleophilicity increases with the increasing atomic number down a group in the periodic table. Thus the nucleophilic reactivity of $I^- > Br^- > Cl^- > F^-$ in a protic polar solvent.
3. It decreases with the increasing atomic number along a period in the periodic table. Thus the nucleophilicity of $C^- > N^- > O^- > F^-$ and similarly, that of $NH_3 > H_2O > HF$.

Leaving nucleophiles are usually, I^-, Br^-, Cl^-, H_2O, $N_2^{\oplus}$, $-NR_3$, $SO_4^{=}$, RSO_3^-, $RCOO^-$, NO_3^- etc. The most polarisable and the least basic groups are generally good leaving groups.

Depending on the molecularity of the S_N reactions they are further classified under two subheads in the case of sp^3 C : S_N1 and S_N2.

12.3 S_N1 Reaction : Substitution Nucleophilic Unimolecular

A nucleophilic substitution reaction on an sp^3 C in which only one species is involved in the transition state of the rate-determining step is known as nucleophilic substitution reaction unimolecular and designated as S_N1. This type of reaction takes place in two steps.

Step I : At this step slow ionisation of the substrate occurs; a carbenium ion and the leaving nucleophile get separated.

$$\text{R—G} \xrightleftharpoons{\text{slow}} \text{R}^{\oplus} + \bar{\text{G}}\text{:}$$

Step II : The carbenium ion so formed invites the incoming nucleophile and subsequently a bond between them forms rapidly.

$$\text{R}^{\oplus} + \text{N}\bar{\text{u}}\text{:} \xrightarrow{\text{rapid}} \text{R—Nu}$$

Step I being slow, it is the rate-determining step.

The transition state of this step is a partially bonded species. It has a broken bond between a C atom and the leaving nucleophile : R···G. Thus the overall reaction may be represented as :

$$\text{R—G} \xrightleftharpoons{\text{slow}} \underset{\text{T.S}}{\left[\overset{\delta+}{\text{R}}\cdots\overset{\delta-}{\text{G}}\right]} \xrightleftharpoons{\text{slow}} \text{R}^{\oplus} + \bar{\text{G}}\text{:} \xrightarrow[\text{rapid}]{\text{N}\bar{\text{u}}} \text{R—Nu} + \bar{\text{G}}\text{:}$$

The first step being endothermic, the T.S. resembles more to the carbenium ion than to the reactants. The T.S. for the second step also resembles the carbenium ion because this step is exothermic.

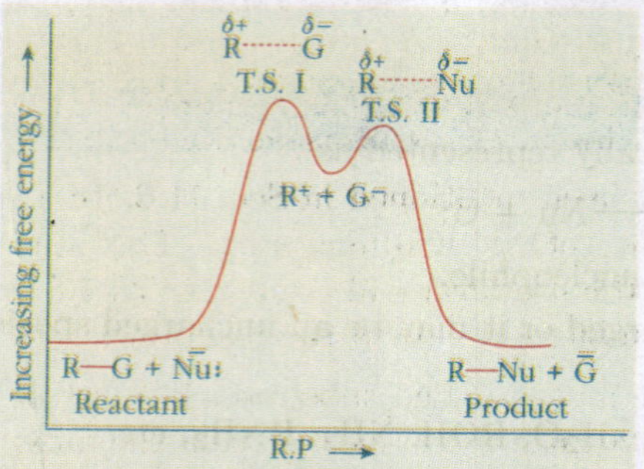

Fig. 12.1. Energy profile of an S_N1 reaction

The energy profile of the reaction may be drawn as :

Since the rate of a reaction is the rate of the slowest step, and since in an S_N1 reaction this step involves the substrate only,

$$\text{Rate} = k\,[\text{R—G}] \tag{12.1}$$

Naturally an S_N1 reaction should follow the first order kinetics. Experimental results, in general, support this. Of course if the solvent acts as a nucleophile, the rate law will not be so simple.

The involved C atom of the substrate is in sp^3-hybridised state. A carbenium ion is formed at the rate-determining step. It is in sp^2-hybridised state and has a planar trigonal structure with an empty p orbital lying perpendicular to the plane of the species; one lobe of the p orbital lies below the plane and the other above it (Fig. 12.2).

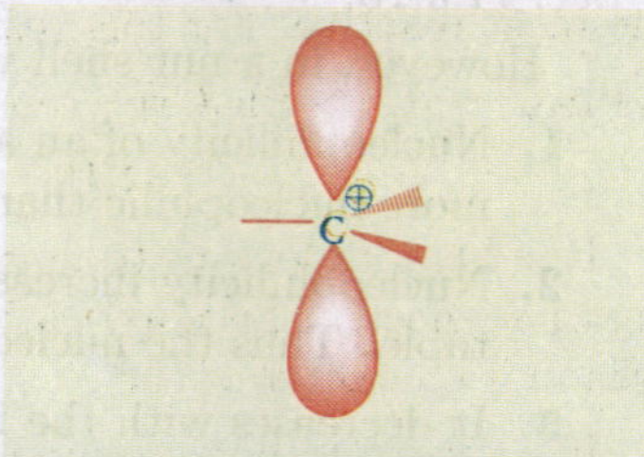

Fig. 12.2. Intermediate carbenium ion

Now the filled orbital of the nucleophile may overlap either lobe of the empty p orbital of the carbenium ion to form R—Nu bond. This means that the overlapping of orbitals may take place from either side of the substrate, i.e., from the side of the leaving group or from its opposite side, Thus it leads to 50% retention of configuration and 50% inversion of configuration. In many a case the stereochemical study of chiral alkyl halides undergoing S_N1 reaction gives partial racemised products instead of complete racemised products. The products usually consist of 5-20% inverted species and 80-95% racemised species. This may occur if the carbenium ion is not sufficiently stable and if the leaving nucleophile is not far enough removed from the carbenium ion. In that case, the leaving nucleophile blocks the incoming nucleophile on that side and thus a product with partial inversion of configuration is formed.

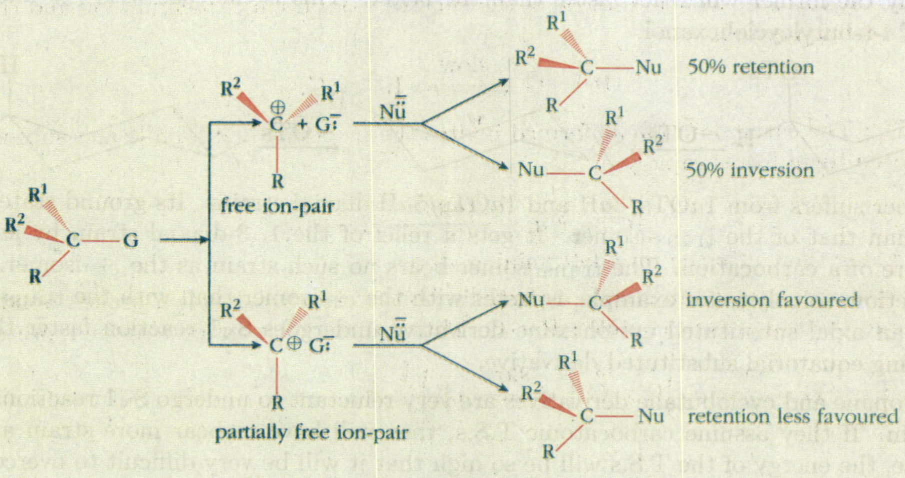

Fig. 12.3. Stereochemistry of the S_N1 reaction

12.3.1 Factors influencing S_N1 reaction rate

The factors which influence a reaction rate have already been mentioned in Sec. 11.6. In an S_N1 reaction the carbocation formation step (Step I) is the rate-determining step. This step involves a substrate, a carbocationic T.S., a carbocation (as the product) and a solvent. We have to discuss all these in connection with the $\Delta G^\ddagger$ which determines the rate of the reaction.

(i) The effect of substrate structure and leaving group on rate : The substrate, in a set of substrates, which forms the most stable carbenium ion (carbocation) will also develop the most stable carbocationic T.S. (Hammond's postulate; problem 11.6), and the $\Delta G^\ddagger$ for the substrate will be the lowest in the set. Naturally the reaction will take minimum of time to overcome the energy of the T.S. Hence the reaction rate will be highest, i.e., the substrate in the set will be the most reactive. For this reason the reactivity order amongst the following compounds is :

$$Ph_3C—G \gg Ph_2CH—G \gg PhCH_2—G \approx CH_2{=}CH—CH_2—G > 3° \gg 2° > 1° >$$
$$CH_3—G \gg CH_2{=}CH—G > CH{\equiv}C—G$$

It is here important to note that +R effect stabilises a cation much more than the hyperconjugative effect. The vinyl cations and acetylenyl cations do not have conjugative stability; the greater *s* character of their cationic Cs make them more electronegative and less stable.

If a substrate bears more strain than another, the ground state energy of the former will be more than that of the latter and the former will try more than the latter to get a relief of the strain.

If, by assuming a carbocationic T.S., the more strained substrate gets the relief, the energy of the T.S. will be appreciably lower than our expectation and therefore the $\Delta G^\ddagger$ for the reaction with the more strained species will be less than the $\Delta G^\ddagger$ for the reaction with the less strained species as shown in Fig. 12.4.

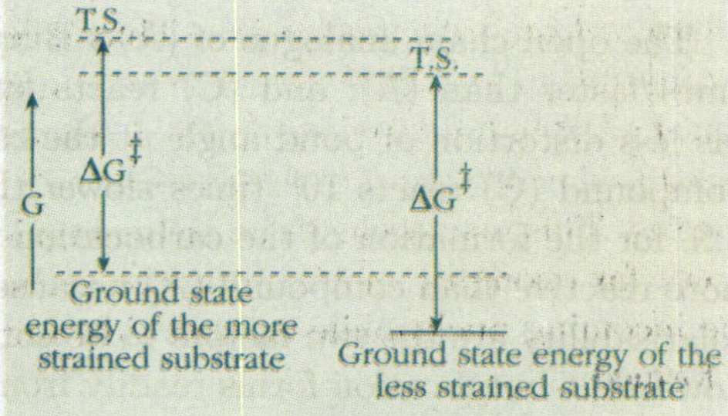

Fig. 12.4

Naturally the former will react faster than the latter. This is the case with cis- and trans-tosylates of 4-t-butylcyclohexanol.

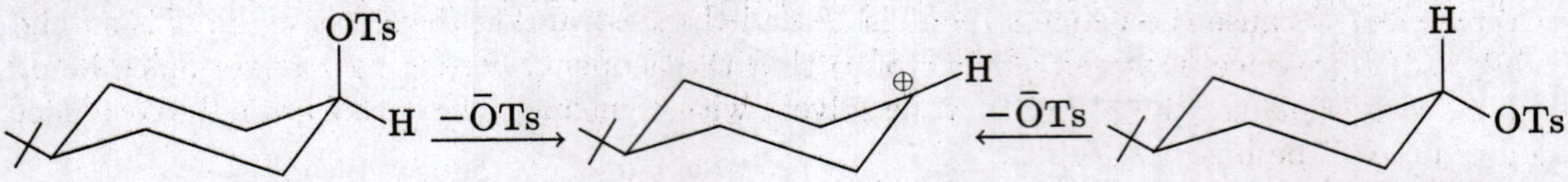

Cis-isomer suffers from 1aOTs/3aH and 1aOTs/5aH diaxial strains. Its ground state energy is higher than that of the trans-isomer. It gets a relief of the 1, 3-diaxial strain by assuming the structure of a carbocation. The trans-isomer bears no such strain as the cis-isomer. Hence an S_N1 reaction, solvolysis for example, is faster with the cis-isomer than with the trans-isomer. In general an axial substituted cyclohexane derivative undergoes S_N1 reaction faster than the corresponding equatorial substituted derivative.

Cyclopropane and cyclobutane derivatives are very reluctant to undergo S_N1 reactions. They bear I-strain. If they assume carbocatonic T.S.s, they will have to bear more strain and as a consequence, the energy of the T.S.s will be so high that it will be very difficult to overcome the energy barriers.

It has also been stated in Sec. 11.4.2. that bridgehead carbocations are highly unstable. Therefore, the T.S.s for the formation of such carbocations have very high energy and it is difficult for a bridgehead substituted compound to reach the energy of the T.S. So bridgehead substituted compounds undergo S_N1 reactions but very very slowly. When 1-chloro-7, 7-dimetylbicyclo [2.2.1] heptane (A), 1-bromobicyclo [2.2.1] heptane (B) and 1-bromobicyclo [2.2.2] octane (C) are solvolysed in the presence of silver nitrate, we see the following results :

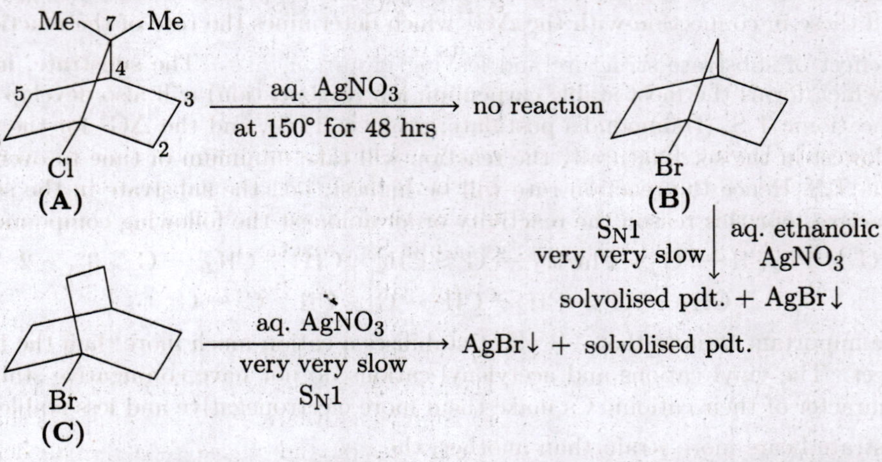

The open chain analogue of (A) t-Butyl choride has been found to undergo the reaction 10^9 times faster than (A); and (C) reacts faster than (B). Bicyclo [2.2.2] octyl cation [from (C)] has less distortion of bond angle at the cationic bridgehead carbon than that formed from (B). Compound (C) reacts 10^6 times slower than t-butyl bromide because of the high strain in the T.S. for the formation of the carbocation from compound (C). However, compound (B) is much more reactive than compound (A) because Br^- is a better leaving group than Cl^-. Basicity and polarisability are the two factors to be considered to determine the reactivity of a leaving group. However, t-butyl cation forms readily from t-butyl bromide not only because the cation is stable but also because t-butyl bromide gets a relief of the B-strain it has owing to 109° bond angle—

through ionisation the carbon atom holding Br atom becomes planar trigonal in orientation with 120° bond angle. Neopentyl bromide, on the other hand, does not undergo S_N1 reactions without rearrangement because the neopentylic C is 1° and the B-strain on the neopentylic C is so high (owing to the presence of large t-butyl group) that the neopentyl cation rearranges much before coupling with a solvent; the coupling of the solvent with 1° neopentylic carbocation if takes place, the product will be highly strained.

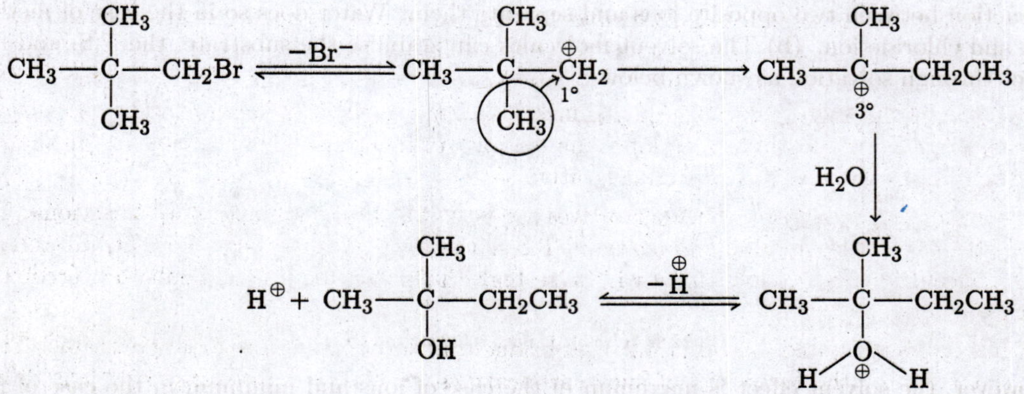

(ii) The effect of leaving group character on rate : The best leaving groups (G) are those which are least basic, i.e., their conjugate acids are most acidic. Thus the order of the leaving group reactivity of the following species is :

$$H_2O > ROH > AcO^- > \bar{O}H$$

The more polarisable a group the more is its reactivity as leaving group. Thus the reactivity order amongst the halides is :

$$I^- > Br^- > Cl^- > F^-$$

Leaving groups having non-bonded *es* (*n es*) and long C—G bond have good polarisability. The better the leaving group ability, the higher is the ionic character of the T.S. The factors which stabilise the incipient ions will also stabilise the T.S. and will decrease its energy; therefore $\Delta G^{\ddagger}$ will also decrease and the rate will increase in S_N1 reactions.

(iii) The effect of solvents on rate : A solvent with high dielectric constant and ion-solvating ability enhances the rate of S_N1 reactions. Polar hydroxylic solvents like water, methyl alcohol, formic acid have high dielectric constants as well as ion-solvating ability; so they are the good solvents for S_N1 reactions and they increase the rate. A carbocation and an anion form in the R.D. step of an S_N1 reaction. The step is, therefore, highly endothermic. For example, in the gas-phase the ionic dissociation of methyl chloride to methyl cation and chloride ion is endothermic to the extent of 950.4 kJ mol^{-1} (so the gas-phase S_N1 reaction involving methyl chloride is unexpected). However, the T.S. for this step, according to Hammond's postulate, will be a carbocation-like. Anything which can stabilise the incipient carbocation and anion will also stabilise the T.S., and $\Delta G^{\ddagger}$ for the reaction will decrease. Hydroxylic polar solvents solvate both cations and anions and hence they are capable of stabilising the carbocationic T.S. In fact, these solvents stabilise the substrates of S_N1 reactions as well as the T.S.s; but they bring more stability to the T.S. than to the substrate. For this reason the heat of ionisation of methyl chloride in water as solvent is +263.76 kJ mol^{-1} in contrast to +950.4 kJ mol^{-1}, the gas-phase ionic dissociation energy of methyl chloride. This indicates that the stabilising effect of water on methyl cation and chloride ion is about 686.64 kJ mol^{-1}. However, it has already been stated in chapter seven that solvent effects are primarily of two kinds :

(a) The solvent molecules tend to orient themselves in such a way as to decrease the electrostatic forces between solute particles (ions or molecules), i.e., the solvent molecules act as insulators; the greater the dielectric constant of a solvent, the higher is the insulating capacity of the solvent. Water has a dielectric constant, $\varepsilon = 78.5$. So, it can decrease the electrostatic forces of attraction between two opposite ions and separate them. Water does so in the case of methyl cation and chloride ion. (b) The solvent molecules can stabilise the substrate, the T.S. and the product through solvation as shown below :

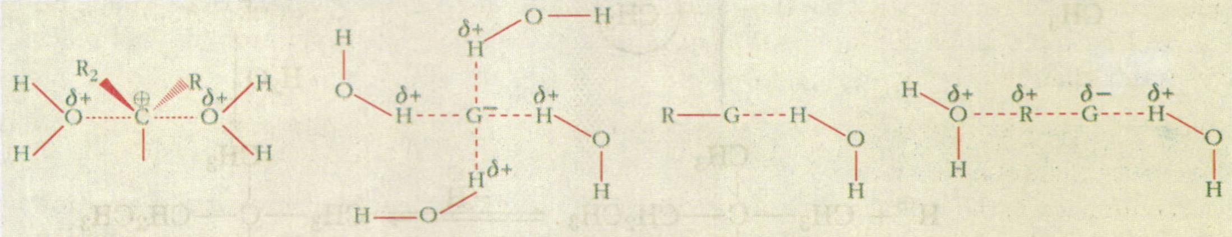

Fig. 12.5. Solvation of ions, substrate and T.S.

However, the solvent effect is maximum in the case of ions and minimum in the case of the substrate. The T.S. gets stability much better than the substrate in an S_N1 reaction through solvation and $\Delta G^{\ddagger}$ decreases appreciably as shown in Fig. 12.6.

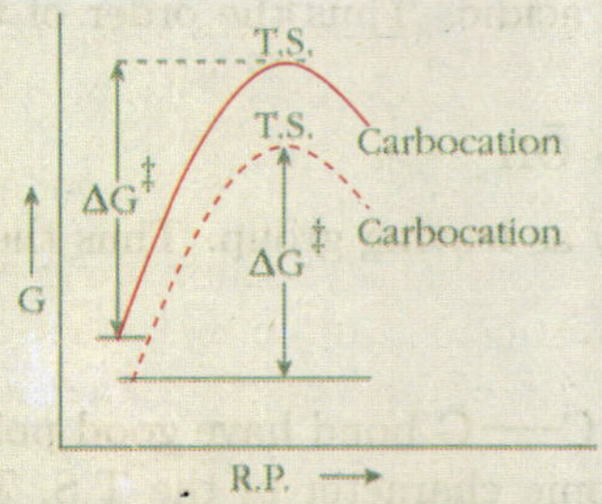

Fig. 12.6. The solid curve represents the R.D. step for an S_N1 reaction in a non-hydroxylic solvent. The broken curve represents the same in a hydroxylic polar solvent

Thus the higher the dielectric constant and ion-solvating ability, the higher will be the S_N1 reaction rate. t-Butyl chloride in methyl alcohol at 25°C reacts eight times faster than in ethyl alcohol because of the higher dielectric constant and greater solvating ability of methyl alcohol than ethyl alcohol; for the same reason in the aqueous medium the rate of the reaction of t-butyl chloride will be still higher.

(iii) The effect of addition of non-common ion salt : When a non-common ion salt is added in the reaction vessel the rate of S_N1 reaction increases—this is because the non-common ion salt increases the dielectric constant of the medium and this is known as salt-effect

(iv) The effect of nucleophiles on rate : Since the rate-determining step of an S_N1 reaction does not involve the incoming nucleophile, neither its nucleophilicity, nor its concentration has any effect on the rate of S_N1 reactions. So an S_N1 reaction can proceed with weak nucleophiles of low concentration.

(v) The common ion effect : When an S_N1 reaction is carried out in the presence of large excess of common ion, the rate falls. The rate of S_N1 reaction involving Ph_2CHCl and F^- in $SO_2(l)$ falls when Cl^- ion is added to the reaction mixture from outside.

Problem 12.1. Which one will react faster in the S_N1 reactions?

(a) $ClCH_2—OCH_2CH_3$ or **(b)** $CH_3O—CH_2CH_2Cl$

(c) $CH_2=CH—CH_2Cl$ or **(d)** $Ph_3C—Cl$

(e) t-Butyl chloride or **(f)** t-Butyl bromide

(g) $(CH_3)_2CH$—Br in water or **(h)** $(CH_3)_2CH$—Br in acetone

Solution : (a) will develop a resonance stabilised carbocation but **(b)** will develop no-bond resonance stabilised carbocation. The former effect is stronger than the latter. The T.S. of **(a)** will be more stable than that of **(b)** $\Delta G_a^{\ddagger} < \Delta G_b^{\ddagger}$. Hence **(a)** will react faster than **(b)**.

(d) will react faster than **(c)** because of the higher stability of the T.S. **(f)** has Br as a leaving group and **(e)** has Cl. Br is a better leaving group than Cl. The better the leaving group, the higher is the stability of the T.S. and the higher is the reaction rate; in **(g)** water is a hydroxylic polar solvent but in **(h)** acetone is a non-hydroxylic solvent; the former can stabilise both cation and anion but the latter can stabilise the cation only and that to a lesser extent than water. So the T.S. will be more stabilised in water than in acetone. Hence, isopropyl bromide will react faster in water than in acetone.

12.3.2 Competitive reactions

A carbenium ion is the intermediate in S_N1 reactions. It may give up a $H^{\oplus}$ from the β-position with respect to the leaving group. This leads to the formation of a double bond and the reaction is then called a β-elimination reaction, designated as E1 reaction. A solvent molecule usually acts as a base and abstracts the β-H.

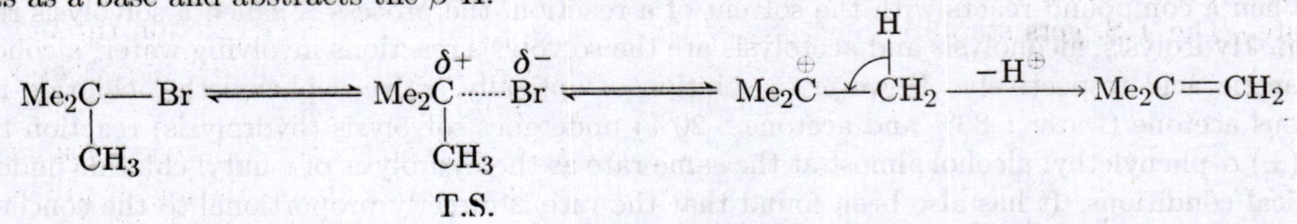

On the other hand, a carbenium ion may rearrange to give a stable carbenium ion (Sec. 11.4.2). Thus elimination and rearrangement reactions compete with substitution in S_N1 reactions. Since the rates of those reactions are determined by the slow ionisation of the substrate, they all proceed with the same rate. Strong nucleophiles being strong bases, they can abstract a $H^{\oplus}$ and thus favours elimination reaction (E2). It is, therefore, better to use weak nucleophiles in S_N1 reaction.

12.3.3 S_N1' reactions

Sometimes an S_N1 reaction occurs with allylic rearrangement—this may be called S_N1' reaction.

$$CH_3{\cdot}CH{=}CH{\cdot}CH_2Cl \xrightarrow{-Cl^-} \left[CH_3{\cdot}CH{=}CH{-}\overset{\oplus}{C}H_2 \longleftrightarrow CH_3{\cdot}\overset{\oplus}{C}H{-}CH{=}CH_2\right]$$

$$(1)\ CH_3{\cdot}CH{=}CH{-}CH_2OH + (2)\ CH_3{\cdot}CHOH{\cdot}CH{=}CH_2 \xleftarrow{\bar{O}H} CH_3{\cdot}\overset{\delta+}{C}H \cdots CH \cdots \overset{\delta+}{C}H_2$$

Product (1) and (2) are formed by S_N1 and S_N1' reactions respectively.

12.3.4 S_N1 reaction conditions

If one finds the presence of the following in a nucleophilic substitution reaction, the reaction can be designated as S_N1 reaction :

(i) A substrate which is able to develop a stable carbocation.

(ii) Ag$^{\oplus}$ or Zn$^{\oplus\oplus}$ or Hg$^{\oplus\oplus}$ or etc., ion, which helps to develop a carbocation; the latter may or may not be stable.

$$RCH_2\text{-}Br + AgNO_3 \longrightarrow R\overset{\oplus}{C}H_2 + AgBr + NO_3^-$$

(iii) A very weak nucleophile of low concentration or a very weakly nucleophilic solvent which can bring stability to the T.S. and to the incipient carbocation.

12.3.5 Evidence for a limiting S_N1 reaction

(i) Stereochemical evidence : It has already been stated in discussing the stereochemistry of S_N1 reactions that some reactions give cent per cent racemised products and some others partially racemised products. The former reactions are said to be pure or limiting S_N1 reactions. So, the formation of complete racemised product from an S_N reaction is a clear evidence for a limiting S_N1 reaction.

(ii) Kinetic evidence : S_N1 reactions usually follow first order kinetics with respect to substrates.

(iii) Secondary kinetic isotope effect : see Sec. 11.8.3 and Sec. 12.4.5(ii).

12.3.6 Solvolysis reactions

When a compound reacts with the solvent of a reaction, the process is called a solvolysis reaction. Hydrolysis, alcoholysis and acetolysis are the solvolysis reactions involving water, alcohol and acetic acid respectively. However, a solution of optically active α-phenylethyl chloride in aqueous acetone (water : 80% and acetone : 20%) undergoes solvolysis (hydrolysis) reaction to give ($\pm$) α-phenylethyl alcohol almost at the same rate as the hydrolysis of t-butyl chloride under identical conditions. It has also been found that the rate is directly proportional to the concentration of α-phenylethyl chloride; the rate is not also affected by the change in concentration of the solvent. Thus, the rate = k [$PhCHClCH_3$].

Since a racemic product is obtained from an optically active substrate, (+) or (−), the above reaction has been designated as S_N1 solvolysis reaction and the mechanistic pathway is given below.

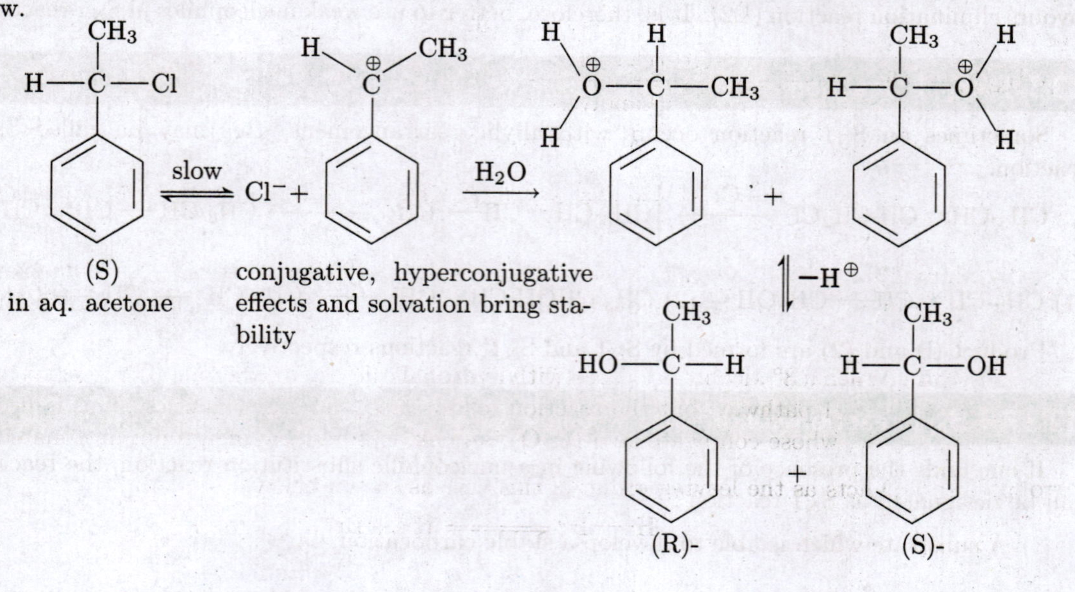

$$\text{Overall reaction : Ph CHClCH}_3 + 2\text{H}_2\text{O} \xrightarrow{\text{aq. Acetone}} \text{Ph CHOHCH}_3 + \text{H}_3\text{O}^{\oplus} + \text{Cl}^-$$

(+) or (−) [98% (±)+ 2% inverted pdt.]

It is needless to mention that the greater the stability of the carbocation, the higher is the rate of S_N1 solvolysis reaction. Solvolytic nucleophilic substitution reactions are usually unimolecular. As the nucleophilic solvent is present in large excess over the substrate concentration, the concentration of solvent essentially does not change during the reaction and by convention the solvent concentration is not included in the rate equation. Therefore, first order kinetics is observed in S_N1 solvolysis reactions. Thus the kinetic order of a solvolysis reaction cannot be considered as a criterion of the mechanism. Since carbenium ions are the intermediates in a solvolysis S_N1 reaction, E1, β- elimination unimolecular, also takes place along with the substitution reaction.

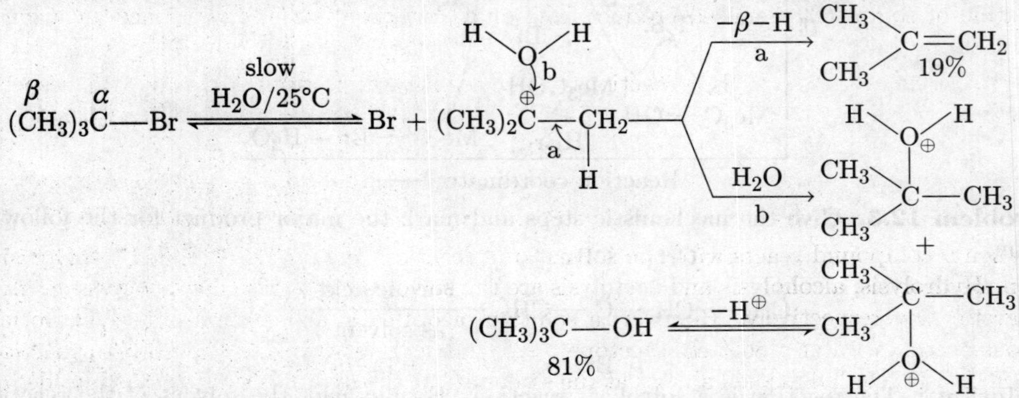

Usually in high dielectric solvents which are highly ionising media, the substitution product is major; but in low ionising solvent (of low dielectric constant) and high temperature, E1 products can compete with S_N1 products and even E1 products are major products. The following examples will tell the truth.

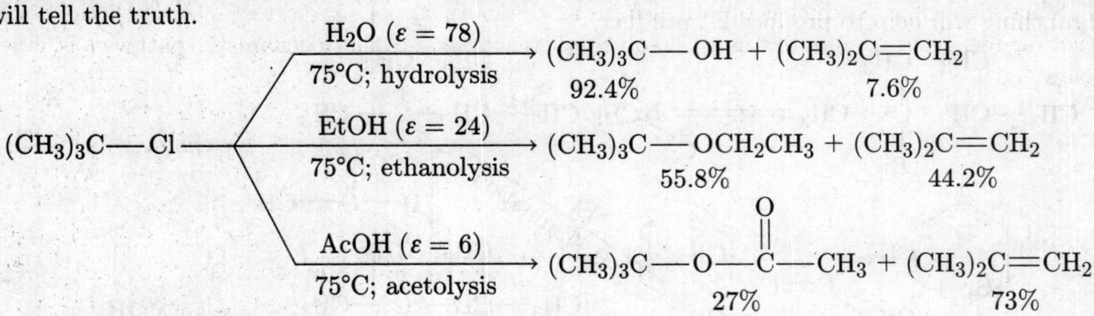

Problem 12.2. "All S_N1 reactions do not follow the first order kinetics."—Give an example where the S_N1 reaction does not follow the first order kinetics. Draw the energy profile for the reaction.

Solution : When a 3° alcohol is treated with hydrohalic acids, for example HBr, t-butyl halide is formed by the S_N1 pathway; but the reaction follows a second order kinetics. This is because $\bar{O}H$, a strong base, whose conjugate acid (H_2O) is weak, is a bad leaving group; protonated OH group, i.e., $H_2\overset{\oplus}{O}$ acts as the leaving group in this case as shown below.

$$\text{H—Br} \rightleftharpoons \text{H}^{\oplus} + \text{Br}^-$$

$$\mathrm{Me_3C{-}OH} \xrightleftharpoons{\mathrm{H^{\oplus};\ fast}} \mathrm{Me_3C{-}\overset{\oplus}{O}H_2} \xrightleftharpoons{\mathrm{R.D.Step,\ slow}} \mathrm{H_2O + Me_3\overset{\oplus}{C}} \xrightleftharpoons{\mathrm{fast;\ Br^-}} \mathrm{Me_3C{-}Br + Br{-}CMe_3}$$

So, the rate = $k_3[\mathrm{Me_3COH}][\overset{\oplus}{\mathrm{H}}]$.

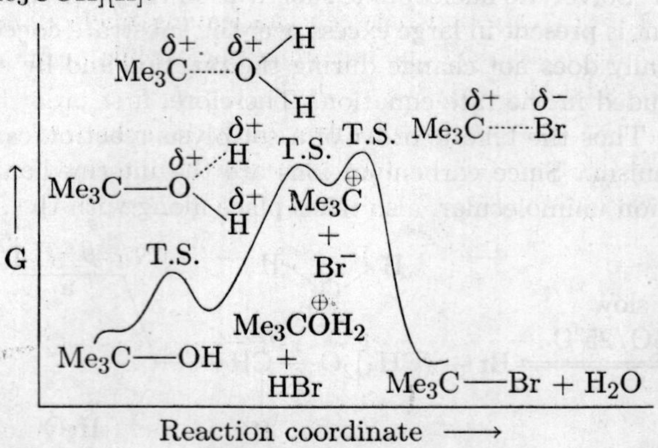

Problem 12.3. Give the mechanistic steps and mark the major product for the following reaction.

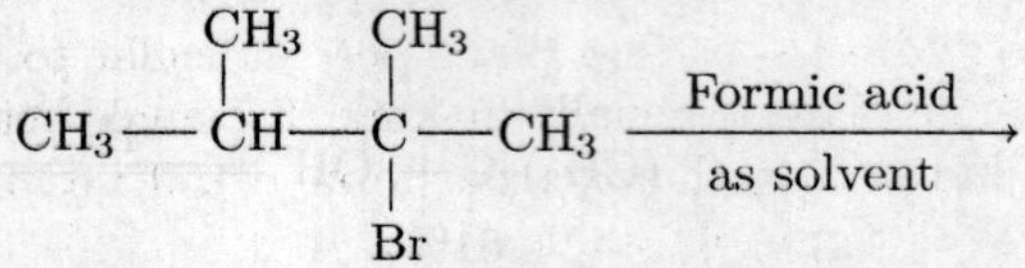

Solution : The reaction is a solvolysis reaction. Formic acid, the solvent of the reaction, has high dielectric constant ($\varepsilon = 59$) and therefore, it has high ionising power. In a medium of high ionising power the S_N1 product predominates over the E1 product. Furthermore the nucleophilicity of formic acid is very low; so S_N2 and E2 products will not form. However, α and β branching will help to produce E1 product.

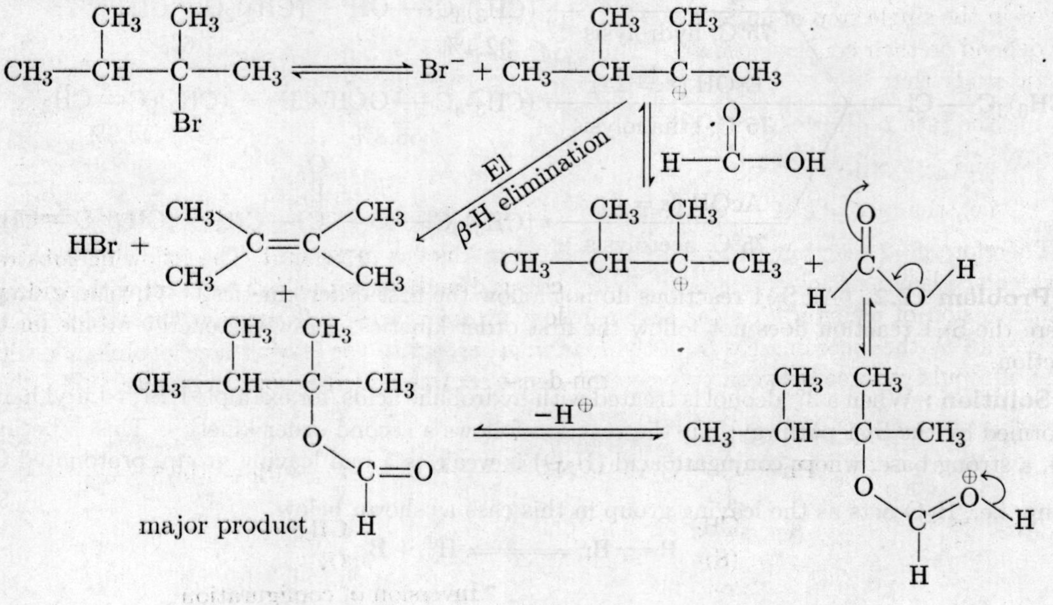

12.4 S_N2 Reaction : Substitution Nucleophilic Bimolecular

If the transition state of a one-step nucleophilic substitution reaction on sp^3C involves two species, it is said to be substitution nucleophilic bimolecular and is symbolically designated as S_N2.

In an S_N2 reaction breaking of bond between the leaving nucleophile and the C atom and making of bond between the latter and the incoming nucleophile occur simultaneously, i.e., an S_N2 reaction has a concerted step.

The transition state of an S_N2 reaction involves both the substrate and the incoming nucleophile. The latter attacks the carbon site of the substrate from a position diametrically opposite to that of the leaving nucleophile.

$$R \cdots G + \bar{N\ddot{u}} \longrightarrow \overset{\delta-}{Nu} \cdots R \cdots \overset{\delta-}{G} \longrightarrow Nu—R + G^{\bar{:}}$$

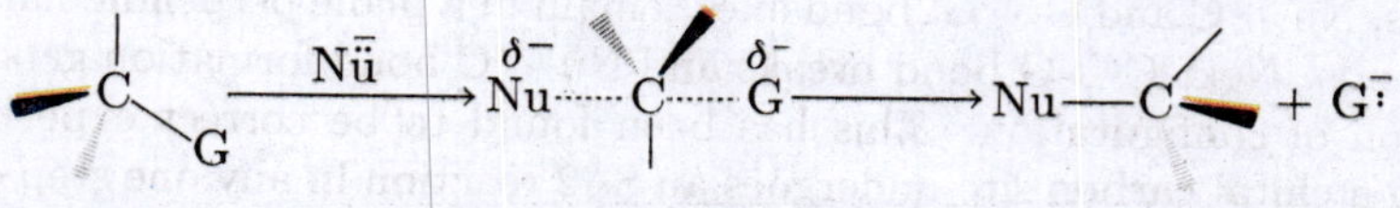

Thus the carbon site of the T.S. has five bonds, three of them are complete and two of them, C—G and Nu—C bonds, are partial and represented by broken lines. The three groups on the carbon site lie in one plane along with the involved C and the two nucleophiles, incoming and leaving, along with the involved C lie in a plane perpendicular to the former plane. Thus, the three groups are oriented in a plane trigonally making 120° angle between two such groups in the T.S.; whereas in the ground state each angle was around 109°; each of the two nucleophiles makes an angle 90° with any of the three groups. Therefore, the geometry of the T.S. is trigonal bipyramidal.

The energy profile of an exothermic S_N2 reaction is drawn as given in Fig. 12.7 :

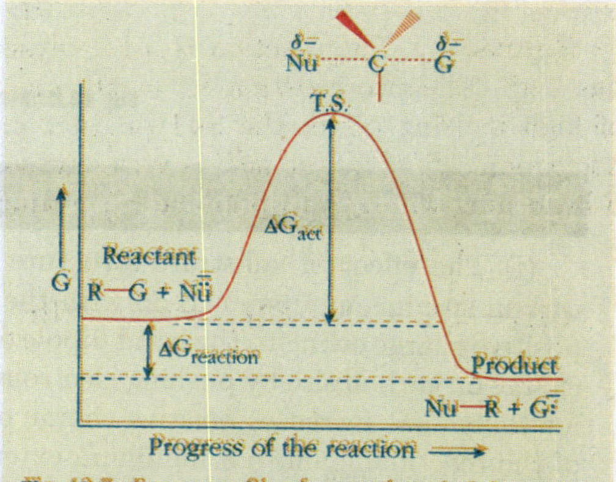

Fig. 12.7. Energy profile of an exothermic S_N2 reaction

Since the substrate and the nucleophile are involved in the single step of an S_N2 reaction, its rate will depend on their concentrations. It is found from kinetic study that

the rate $\propto$ [R—G]

and $\propto$ [$\bar{N\ddot{u}}$]

i.e., the rate = k[R—G] [$\bar{N\ddot{u}}$] (12.2)

Therefore, S_N2 reactions follow the second order kinetics in general. The following substrate in which the leaving nucleophile is positively charged undergoes the S_N2 reaction with azide ion with inversion of configuration (as shown below within braces) tells us that the attack on the carbon site by the incoming $\bar{N}$u from the rear with respect to the leaving neucleophile is not the cause of repulsive force between two electron-dense centres, but it is something else.

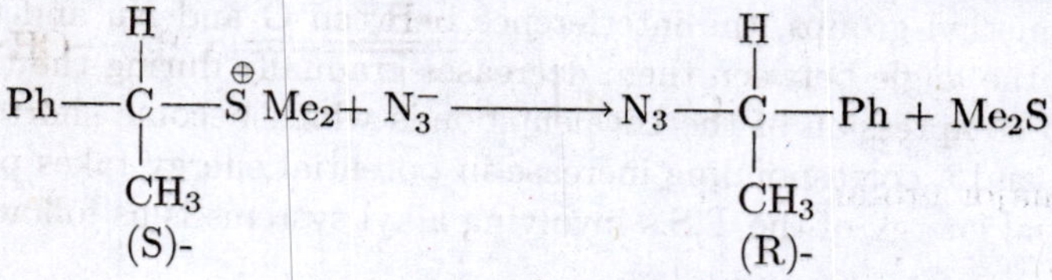

(S)- (R)-

Inversion of configuration

Frontier Molecular Orbital theory gives a good explanation for the back-side attack. The incoming nucleophile utilises its HOMO to form the bond; it overlaps the substrate carbon site LUMO which will be an antibonding σ orbital between the leaving nucleophile and the carbon site. An attack from the side of leaving nucleophile will be minimal since the HOMO will be brought into the node of the antibonding LUMO of antibonding σ orbital. The rear lobe does not contain any node.

If groups and atoms of small bulk are present on the carbon site, an enough of space is available for the attack by the nucleophile. As the nucleophile approaches the carbon site from the rear, the leaving group is pushed away. The bigger lobe of the orbital that holds the leaving nucleophile shrinks and the smaller lobe of that gradually expands to overlap the filled orbital of the incoming nucleophile. In this way, the transition state is reached where the carbon site has two-half bonds, Nu$\cdots$C and C$\cdots$G, bond axes remain in a plane perpendicular to the plane of the other three bonds. Next C$\cdots$G bond breaks and Nu$\cdots$C bond formation gets completed leading to the inversion of configuration. This has been found to be correct experimentally. When a substrate with a chiral carbon site undergoes an S_N2 reaction in any one step of a reaction chain, D-substrate gets converted into L-product and vice versa.

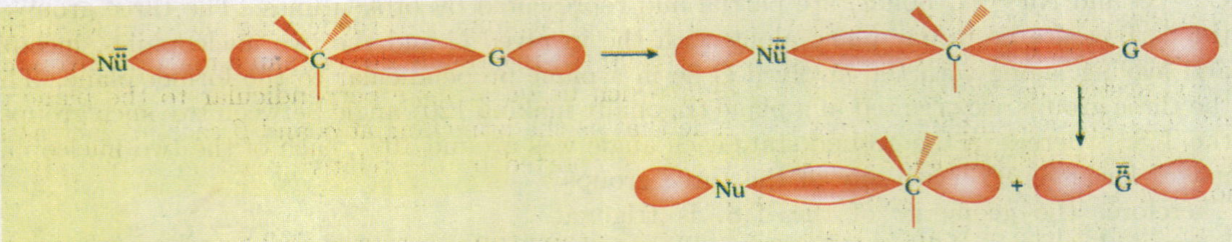

Fig. 12.8. Stereochemistry of the S_N2 reaction

12.4.1 Factors influencing S_N2 reaction rate

(i) The effect of substrate structure on rate : The electron donating polar effects on the carbon site has minimum to do with the S_N2 reaction rate. The C atom usually bears a partial positive charge owing to the bond dipole of the C—G bond. Anything that decreases the positive charge on carbon site by pushing electrons, decreases the reaction rate to a minimum extent; while anything that increases positive charge on the same by withdrawing electron, increases the rate of the reaction again to a minimum extent. On the other hand, steric factor has profound effect on the S_N2 reaction rate. Availability of space for attack on the carbon site by the nucleophile decreases with the increasing steric hindrance and consequently the rate of the reaction diminishes largely.

During the transition of the ground state substrate to the T.S. of an S_N2 reaction involving methyl systems H—C—H angles increase from 109° to about 120°; at the same time H—C—G angles and H—C—Nu angles decrease to near 90°. When Hs are replaced by much larger methyl groups, the interference between G and Nu and the methyl group goes on increasing as the angle between them decreases gradually during the transition of the substrate. This leads to compression of the covalent bonds which become shorter in the T.S. than in the ground state and a corresponding increase in potential energy takes place. Therefore, the order of the potential energy of the T.S.s involving alkyl systems is as follows :

$$\xrightarrow{\text{increasing } E^{\ddagger}}$$

$$CH_3—G < CH_3CH_2—G < CH_3CH_2CH_2G < (CH_3)_2\,CH—G < (CH_3)_3C—G < (CH_3)_3C—CH_2—G$$

$$\xrightarrow[\text{decreasing rate}]{}$$

Hence, the usual rate order amongst the alkyl systems in S_N2 reactions is the reverse of the above order. However, the steric strain factor may be understood from the geometry of the T.S.

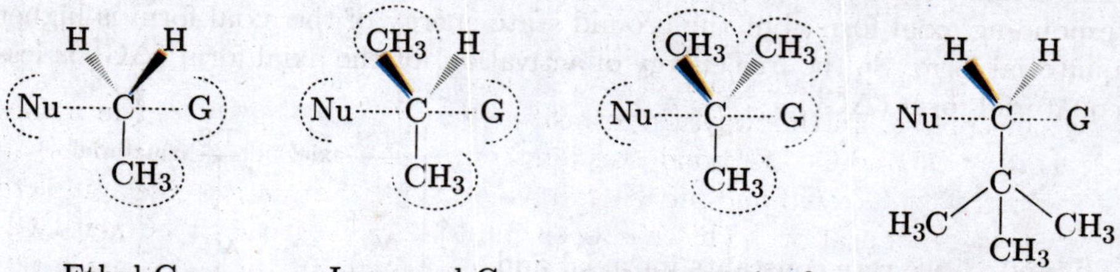

Ethyl-G less strained | Isopropyl-G strained | more strained t-Butyl-G | most strained neo-Pentyl-G

In the case of the primary neopentyl system, the incoming and the leaving nucleophiles in the T.S. experience a substantial steric hindrance and the reaction is slower than a reaction involving t-butyl system having three α-substituents which lie on a plane perpendicular to the plane of Nu· · · C· · · G. In summary we may conclude that as the branching at α and β carbons increases, the S_N2 reaction rate decreases. This is also supported by the relative reaction rate study as shown in Table 12.1.

Table 12.1 Effect of branching at the α and β carbons on the rate of S_N2 reactions relative to $CH_3—G$ (rate is 30)

α-branching	Relative rate	β-branching	Relative rate
$CH_3CH_2—G$	1	$CH_3\,CH_2\,CH_2—G$	0.4
$(CH_3)_2\,CH—G$	0.02	$(CH_3)_2\,CH—CH_2—G$	0.03
$(CH_3)_3\,C—G$	~ 0.0	$(CH_3)_3\,C—CH_2—G$	0.000

Vinyl, acetylenic and aryl systems are very unreactive towards S_N2 reactions; they have carbon sites of high electronegativity and the leaving group wants to leave the species with bonded *e*s. However, the allyl and benzyl systems undergo S_N2 reactions very readily because of resonance stabilisation of the T.S. Allyl and benzyl systems react 30 and 1[illegible] times faster than ethyl systems usually.

However, the presence of α and β-halogens decreases the rate of S_N2 reactions. For example, the rate of S_N2 reactions on chloromethanes with all nucleophiles decreases in the order : $CH_3Cl > CH_2Cl_2 > CHCl_3 > CCl_4$. As the number of Cl atoms increases in the substrate, an electronic effect (along with the steric strain factor) increases the energy of the T.S. and thus decreases the rate of the reaction. When one H atom of CH_3Cl is replaced by another chlorine atom, two major effects take place : (i) the carbon atom becomes a better acceptor of the nucleophile and (ii) C—Cl bond becomes more stable in the T.S.; the former effect stabilises the T.S. but the latter effect destabilises the T.S. more. So the reaction rate decreases.

Cyclopropyl derivatives are relatively unreactive towards S_N2 reactions. In these compounds the C—C—C bond angle is close to 60° (instead of 109°) and the molecules bear high I-strain.

In the T.S. of an S_N2 reaction the angle will be about 120° and the T.S. will have to bear an additional angle strain. Cyclobutyl systems are relatively more reactive than cyclopropyl systems since the additional angle strain is less in the T.S. Cyclopentyl systems undergo S_N2 reactions at comparable rates as open chain compounds and yields are also good. However, cyclohexyl systems undergo S_N2 reactions rather slowly. 1,3-diaxial interactions appear to be significant in the system. Axial substituted cyclohexanes react more rapidly than the corresponding equatorial substituted cyclohexanes. Since the axial approach of the incoming nucleophile is more sterically hindered than the equatorial approach, the T.S. for the equatorial form has higher energy than the corresponding axial form; but the ground state energy of the axial form is higher than that of the equatorial form. So the free energy of activation for the axial form ($\Delta G_a^\ddagger$ is less than that for the equatorial form ($\Delta G_e^\ddagger$).

k_a and k_e are rate constants for axial and equatorial forms respectively

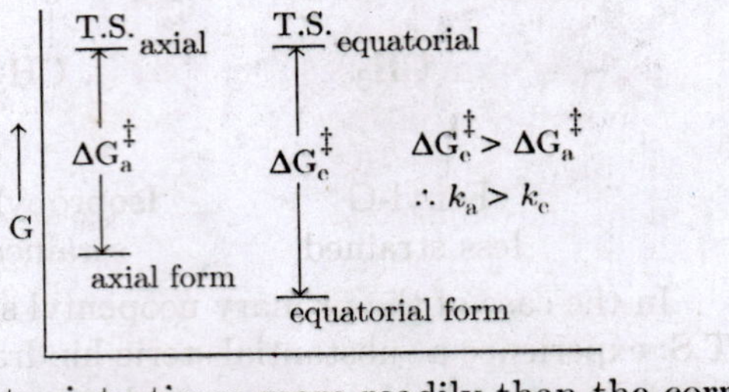

For example, cis-4-t-butylcyclohexyl bromide reacts sixty times more readily than the corresponding trans-isomer with sodium thiophenoxide at 87°C in ethyl alcohol; cis-4-t-butylcyclohexyl tosylate also reacts thirty times faster than its trans-isomer with sodium thiophenoxide. These reactions show that the interactions involving approaching nucleophile is more important than the similar interaction involving leaving nucleophile.

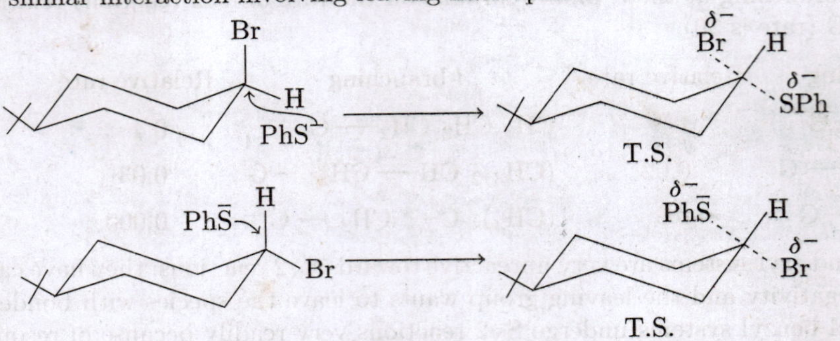

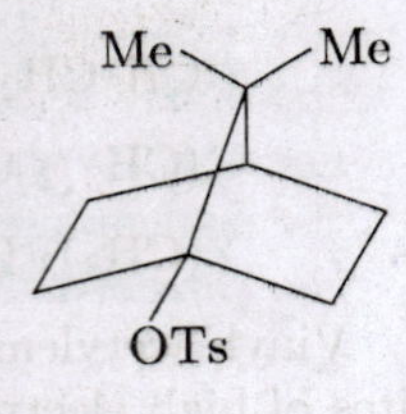

The bridgehead carbons of rigid bicyclic systems do not fulfil the stereoelectronic requirement for the S_N2 reaction, i.e., the T.S. involving bridgehead carbon cannot assume trigonal bipyramidal geometry and hence they are inert to S_N2 reactions. For example., 7,7-dimethyl [2.2.1] bicycloheptyl-l-*p*-tosylate does not show inversion of configuration with a nucleophile under S_N2 reaction conditions.

(ii) The effect of solvents on the reaction rates : Depending on the nature of incoming nucleophiles and that of leaving group, S_N2 reactions may be classified under four heads. The solvent effect on rates is not the same for all the types of S_N2 reactions. So let us first classify the S_N2 reactions as given below (Nu is the incoming nucleophile and G is the leaving nucleophile) :

(a) Nu is neutral but G leaves as G^-; here the T.S. is more polar than the reactants.

$$H_2O + RI \longrightarrow H_2\overset{\delta+}{O}\cdots R \cdots \overset{\delta-}{I} \longrightarrow H_2\overset{\oplus}{O}R + I^-$$

(b) Nu is neutral and G leaves as a neutral species; here the T.S. is less polar than the reactants.

$$H_3N: + R—\overset{\oplus}{S}R_2 \longrightarrow \overset{\delta+}{H_3N}\cdots R\cdots \overset{\delta+}{SR_2} \longrightarrow H_3\overset{\oplus}{N}R + SR_2$$

(c) Nu is negatively charged but G leaves as a neutral unit; the T.S., in this case, is less polar than the reactants.

$$Cl^- + R—\underset{\displaystyle H}{\overset{\oplus}{N}}R_2 \longrightarrow \underset{\text{T.S.}}{\overset{\delta-}{Cl}\cdots\cdots R\cdots\cdots \underset{\displaystyle H}{\overset{\delta+}{N}}R_2} \longrightarrow ClR + R_2NH$$

(d) Nu is negatively charged and G leaves as G^-; here the T.S. is less polar than the reactants.

$$I^- + R—Cl \longrightarrow \overset{\delta-}{I}\cdots R\cdots \overset{\delta-}{Cl} \longrightarrow IR + Cl^-$$

When an S_N2 reaction belongs to class 'a', the replacement of a less polar protic solvent by a more polar protic solvent increases the rate. This is because the more polar solvent stabilises the more polar T.S. more than the less polar reactants. However, if the nucleophiles are not small and charge-localised, the effects are not appreciable. When the solvent is changed from a protic to a dipolar aprotic one, the rate of the S_N2 reactions do not change appreciably except in those reactions of type 'a' involving the small, charge-localised nucleophiles such as $\bar{O}H$, F^-, Cl^- in which cases the rate increases dramatically. Polar protic solvents solvate the small charge-localised nucleophiles so strongly that they cannot attack the substrate carbon readily; but aprotic polar solvents (DMSO, DMF, etc.) cannot solvate the nucleophiles (See Sec. 11.5.b) and as a consequence the nucleophiles attack the carbon site very readily and the rates of the reactions increase dramatically. For example, the reaction of acetate ion on methyl iodide is ten million times faster in DMF than in MeOH. When the S_N2 reactions belong to types b, c and d, the solvent polarity practically has no effect on the reaction rate—the rate decreases slightly with the increasing polarity of the solvent.

(iii) The effect of nucleophiles on rate : Since the single-step S_N2 reaction involves the substrate and the nucleophile, the rate of the reaction depends largely on the concentration of the nucleophile and on its nucleophilicity. The rate decreases with the decreasing concentration of the nucleophile and that increases with the increasing concentration. Strong nucleophiles increase the rate of the S_N2 reaction while weak nucleophiles decrease it. Thus the hydrolysis of alkyl halides with water as a nucleophile have slower rates than those with an alkali. However, the nucleophilicity order is substrate dependent. When the substrate C is a soft centre, the reactivity order is $\overline{C} > N^- > O^-$; but when the substrate C is a comparatively hard centre, the reactivity order becomes $\overline{O} > N^- > C^-$. For example, when methyl iodide is the substrate, we find the former reactivity order ($\overline{C} > N^- > O^-$) but towards *n*-propyl tosylate the latter reactivity order, $\overline{O} > \overline{N} > \overline{C}$ is observed. Furthermore, the solvent effect on nucleophiles is also important in this regard. In fact, the greater the nucleophilicity under a given set of conditions, the higher is the rate (the order of nucleophilicity has already been discussed in Section 11.5.1b) of the reaction.

(iv) The effect of leaving groups on rate : The rate of an S_N2 reaction is dependent on the nature of the leaving group. Groups of lower basicity and higher polarisability increase the rate of the reaction. Thus the rate order of S_N2 reactions for halide ions as leaving group is similar to their polarisability order, i.e.,

$$I^- > Br^- > Cl^- > F^-$$

On the other hand, depending on the basicity the reactivity order is :

$$H_2O > ROH > \bar{O}H$$

Conjugate bases of weak acids are very poor leaving groups since they are strong bases; CN^-, the conjugate base of weak acid HCN, does not act as a leaving group. For the same reason CH_3COO^- (The cB of acetic acid) and N_3^- are bad leaving groups in the cases of S_N2 reactions on sp^3 Cs. However, it is needless to mention that conjugate bases of strong acids are good leaving groups (this has already been discussed in Sec. 12.3.1.).

(v) Neighbouring group participation : A second function usually participates using its unshared electron and internal cyclisation often takes place with elimination of the leaving nucleophile. The cyclised product then undergoes substitution reaction with the incoming nucleophile. The electron cloud on the neighbouring group blocks the incoming nucleophile to attack the carbon site from the back side. Thus the final substitution occurs from the side of the leaving nucleophile and naturally retention of configuration occurs. In fact, double inversion of configurations leads to retention. However, instead of a cyclic intermediate, a cyclic T.S. may form in an S_N reaction with NGP.

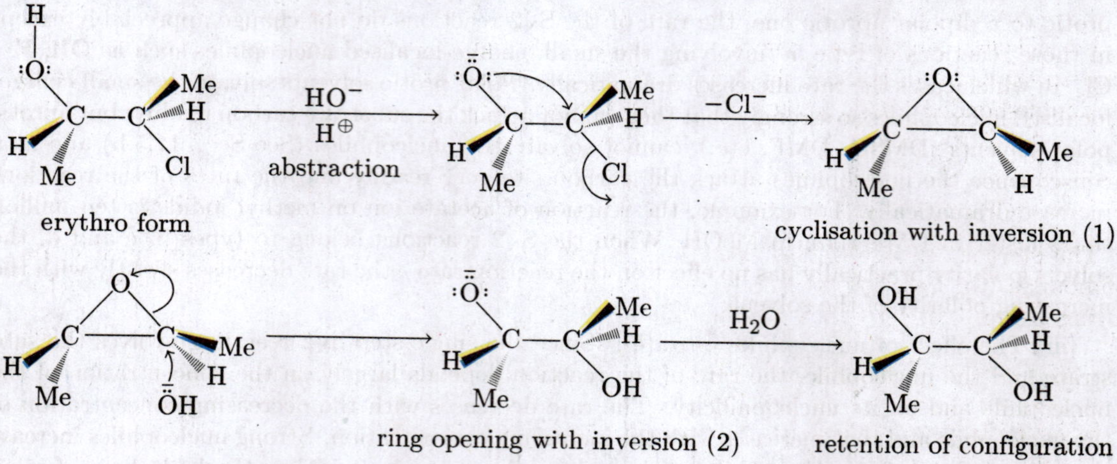

In the cases of diastereomeric halohydrins as shown above, the reaction is stereospecific in so far as the formation of epoxides are concerned; erythro and threo forms produce trans- and cis-epoxides respectively. It is also clear from the above that during the formation of the epoxides both the erythro and the threo forms adopt conformations in which the hydroxyl group and the chlorine atom remain anti to each other.

12.4.3 Competitive reaction

The elimination reaction competes with the S_N2 reaction. A strong base usually abstracts a proton from the β-position with respect to the leaving group and simultaneously leaving group itself leaves the species to form an alkene by E2 pathway. For example, 2-bromo-3-methylbutane in acetone reacts with Cl^- ion to give both S_N2 and E2 products as shown below :

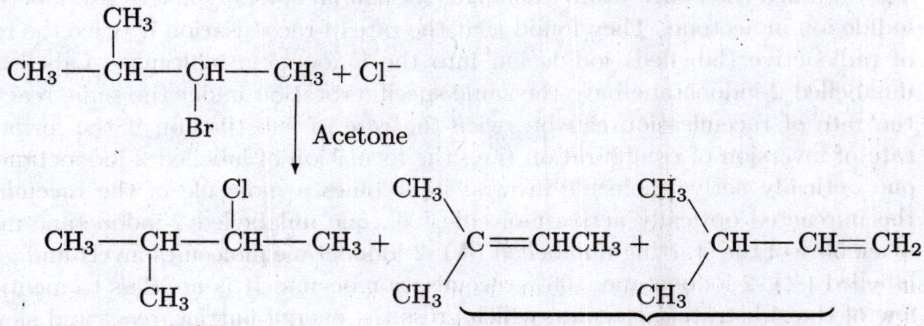

50% S_N2 product 50% E2 product

If an S_N2 reaction is carried out at a lower temperature, the yield of E2 product may be negligible under same set of conditions.

12.4.4 S_N2' reaction

Bimolecular substitution reactions of allylic leaving groups involving migration of the double bond to the C atom holding the leaving group and substitution of the incoming nucleophile to the double-bonded 1°C atom is known as S_N2' reaction.

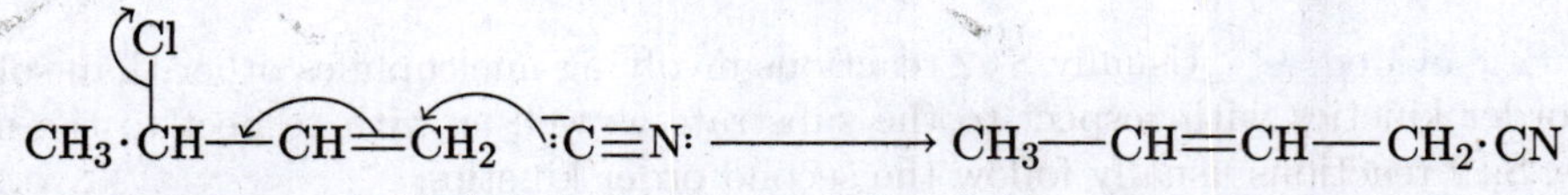

The stereochemistry of S_N2' reactions is not clear. However, in the cases of open chain compounds, the reaction is usually stereospecific; the nucleophile generally enters syn to the leaving group.

12.4.5 S_N2 reaction conditions

S_N2 reactions usually are favoured when

(i) the carbon site of the substrate is 1° and 2°.

(ii) the nucleophile, i.e., the base is comparatively less hard and small in volume, i.e., weakly basic and highly polarisable, e.g., Cl^-, Br^-, I^-, RS^-, CH_3COO^-, etc.

(iii) the reaction temperature is low.

(iv) there is no $Ag^{\oplus}$, $Hg^{\oplus\oplus}$ etc. ions which make a reaction more S_N1-like.

However if (i) the carbon site is 2°, (ii) α- and β-branches are on the carbon site, (iii) the nucleophile is a comparatively hard base, (iv) the reaction temperature is high, and (v) strong nucleophile in high concentration is present, usually E2 reactions are favoured (See the 15th chapter). A 3° carbon site in the presence of a base undergoes E2 reaction.

12.4.6 Evidences in favour of S_N2 reactions

(i) Stereochemical evidence : The cent per cent inversion of configuration of the substrate carbon in a nucleophilic substitution reaction is the evidence in favour of the S_N2 pathway. Huge and his co-worker first demonstrated that S_N2 reactions occur with inversion of configuration.

They studied the iodine atom exchange reaction on optically active 2-iodooctane with redioactive iodide ion in acetone. They found that the rate of racemisation is twice the rate of incorporation of radioactive (labelled) iodide ion into the iodooctane molecule. Labelled 2-iodooctane and unlabelled 2-iodooctane have the same specific rotation under the same reaction conditions. So, the rate of racemisation can be twice the rate of substitution if the former rate is twice the rate of inversion of configuration (i.e., the formation of labelled 2-iodooctane molecules). When one optically active molecule inverts, it becomes a molecule of the racemic pair, the other is the unreacted optically active molecule, i.e., one unlabelled 2-iodooctane molecule. Therefore, when 50% of the starting unlabelled (+) -2-iodooctane molecules invert and get converted to 50% labelled (−) -2-iodooctane, 100% racemisation occurs. It is needless to mention that at a time a few of the substrate molecules, which cross the energy barrier, react and slowly the equilibrium is established between the forward and reverse reactions involving substrate 2-iodooctane and labelled 2-iodooctane respectively. This reaction being neither endothermic, nor exothermic, the energy barriers of the two pathways are one and the same, the equilibrium constant is unity.

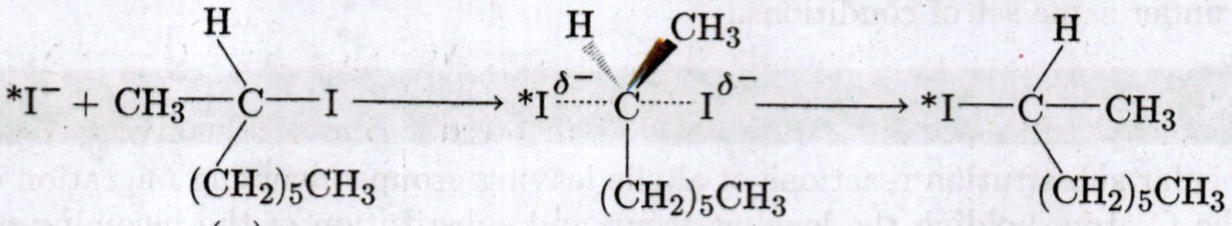

(+)-say T.S. (−)-

(ii) Kinetic evidences : Usually S_N2 reactions involving nucleophiles other than solvents show the first order kinetics with respect to the substrate as well as with respect to the nucleophile. Therefore, S_N2 reactions usually follow the second order kinetics.

(iii) Secondary kinetic isotope effect (α-deuterium isotope effect) : Usually S_N2 reactions show α-deuterium isotope effect of approximately unity; whereas limiting S_N1 reactions show this effect ranging from 1.13 to 1.25 depending on the nature of the leaving group. In an S_N2 reaction the substrate changes its geometry from tetrahedral to trigonal bipyramidal on going from the starting molecule to the transition state. This makes C—H (C—D) bond bending easier but the presence of both the incoming nucleophile and the leaving nucleophile at angles 90° with the C—H (C—D) bond make the bending of the bond difficult; this two opposing effects cancel each other. Thus the α-deuterium isotope effect is approximately of unity. In the limiting S_N1 reaction there is no nucleophile in the rate-determining step but the substrate molecule changes its geometry from tetrahedral to planar trigonal on going to T.S. So, β-SKIE is observed in the cases of limiting S_N1 reactions.

12.4.7 The S_N1—S_N2 mechanistic spectrum

Theoretically we may consider the following rate equation (12.3) for the nucleophilic substitution reaction :

$$R—G + N\bar{u} \longrightarrow R—Nu + G^-$$

The rate $= k_1[R—G] + k_2[R—G]\,[N\bar{u}]$ (12.3)

When $k_1 \gg k_2$, the reaction becomes a limiting S_N1 reaction.

When $k_2 \gg k_1$, the reaction is an S_N2 reaction.

When k_1 and k_2 have appreciable values, the reaction may be a mixed S_N1 and S_N2. However, there are several theories about the mechanistic spectrum of S_N1—S_N2 pathways :

(i) According to Swain there is only one mechanism. In all the cases there is a pulling of G by the solvent molecule and pushing of it by the $N\bar{u}$. When there is no push, the reaction is pure S_N1; when there is no pull, the reaction is pure S_N2.

(ii) According to some others : (a) R—G may undergo direct S_N2 reaction, (b) it may ionise without nucleophilic assistance and may follow the limiting S_N1 pathway and (c) it may ionise with nucleophilic assistance.

Problem 12.4. Compare S_N1 and S_N2 mechanisms.

Solution :

Points	S_N1	S_N2
1. Steps	Two Steps $R—G \underset{}{\overset{-\bar{G}}{\rightleftharpoons}} R^{\oplus} \xrightarrow{N\bar{u}} R—Nu$	One step $N\bar{u} + R—G \longrightarrow Nu—R + \bar{G}:$
2. Molecularity	Unimolecular, R—G involved.	Bimolecular, R—G and $N\bar{u}$: involved.
3. Kinetics and rate	1st order; Rate = k[R—G].	2nd order; Rate = k[R—G][$\ddot{\bar{N}}$u]
4. Stereochemistry	Racemisation and partial inversion.	Complete inversion.
5. Effect of substrate structure on rate.	Stabilisation of carbenium ion and crowding of groups increase the rate.	Steric hindrance decreases the rate of the S_N2 reaction.
6. Effect of solvents on rate.	Good ionising solvents increase the rate.	Hydroxylic solvents decrease the rate. Polarity of solvents has negligible importance.
7. Nucleophilicity and concentration of nucleophiles.	Weak nucleophiles of low concentration too can make the reaction to occur.	Strong nucleophiles of high concentration increase the rate.
8. Leaving group effect.	Weakly basic and highly polarisable groups increase the rate.	Weakly basic and highly polarisable groups increase the rate.
9. Transition State	$\overset{\delta+}{R}\cdots\cdots\overset{\delta-}{G}$	$\overset{\delta-}{Nu}\cdots\cdots R\cdots\cdots\overset{\delta-}{G}$
10. Reaction Intermediate	Carbenium ion	None
11. Competitive reaction	Elimination (E1) and Rearrangement	Elimination (E2)
12. Catalyst	$Ag^{\oplus}$, $AlCl_3$, $ZnCl_2$, etc.	None

Problem 12.5. Write down the mechanistic steps for the following reactions :

(a) $Me_3COH + HCl \longrightarrow ?$ (b) $CH_3OH + HI \longrightarrow ?$

Solution : (a) It is an S_N1 reaction. The reaction intermediate is a carbenium ion, $(CH_3)_3C^{\oplus}$, stabilised by +I effect and hyperconjugative effect.

Step I : $Me_3COH + HCl \xrightleftharpoons{\text{fast}} Me_3C—\overset{\oplus}{O}H_2 + Cl^-$

Step II : $Me_3C—\overset{\oplus}{O}H_2 \xrightleftharpoons[\text{rate-determining step}]{\text{slow}} Me_3C^{\oplus} + H_2O$

Step III : $Me_3C^{\oplus} + Cl^- \xrightarrow{\text{fast}} Me_3C—Cl$

(b) It is an S_N2 reaction. There is no steric hindrance on the carbon site. I^-, a good nucleophile, attacks the carbon site from the rear.

Step I : $CH_3OH + HI \xrightleftharpoons{\text{fast}} CH_3—\overset{\oplus}{O}H_2 + I^-$

Step II : 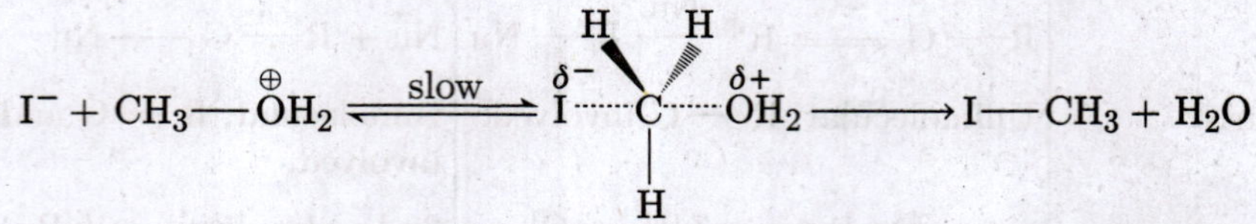

Problem 12.6. ROH reacts with NaBr in the presence of H_2SO_4 to give RBr. Why is H_2SO_4 required ? What is the mechanism of the reaction?

Solution : HO^- being a stronger nucleophile than Br^-, the latter cannot displace the former. H_2SO_4 protonates ROH to form $R\overset{\oplus}{O}H_2$. H_2O being a weaker nucleophile than Br^-, it is easily displaced by the latter.

$$R—OH + H_2SO_4 \rightleftharpoons R—\overset{\oplus}{O}H_2 + HSO_4^-$$

$$R—\overset{\oplus}{O}H_2 + Br^- \rightleftharpoons R—Br + H_2O$$

If R is a tertiary alkyl group, it will follow the S_N1 pathway; if it is a primary one, the S_N2 reaction will occur.

Problem 12.7. CH_3CH_2Cl hydrolyses slowly in the aqueous medium but the reaction is rapid in the presence of catalytic amount of KI.—Explain why it is so.

Solution : H_2O is a weak nucleophile and Cl^- is not a good leaving group. So the hydrolysis of CH_3CH_2Cl in aqeueous medium is slow.

I^- from KI is a powerful nucleophile, it displaces Cl^- to form $CH_3{\cdot}CH_2{\cdot}I$. Now I^- being a good leaving group, H_2O can displace it rapidly.

$$CH_3CH_2Cl \xrightarrow[\text{fast}]{I^-} CH_3CH_2I \xrightarrow[\text{fast}]{H_2O} MeCH_2\overset{\oplus}{O}H_2 \xrightarrow{-H^{\oplus}} MeCH_2OH$$

Here I^- acts as a nucleophilic catalyst.

Problem 12.8. The reactivity order in S_N2 reactions of halide ions in LiX salts in acetone is $I^- > Br^- > Cl^-$ but that in $Bu_4N^{\oplus}X^-$ in the same medium is $Cl^- > Br^- > I^-$ —explain.

Solution : The freer the nucleophile, the greater is the reactivity. In LiX halide ions are less free than those in $Bu_4N^{\oplus}X^-$. Again as the size of the nucleophile increases, the freeness of the halide ion increases in a strongly associated salt. Therefore, in LiX the decreasing order of freeness of halide ions is $I^- > Br^- > Cl^-$ and this is the reactivity order towards S_N2 reactions. In $Bu_4N^{\oplus}X^-$ halide ions being much less associated, the reactivity order is the basicity order, i.e., $Cl^- > Br^- > I^-$.

Problem 12.9. "The kinetic order of a solvolysis reaction cannot be used as a criterion of its mechanism."—Comment on the statement.

Solution : Solvolysis reactions usually follow first order kinetics. An S_N2 solvolytic reaction also follows first orders kinetics since the solvent, the nucleophile, remains present in such a large excess that its concentration practically does not change during the reaction. Usually S_N1 reactions follow first order kinetics. Therefore, from the kinetic order we cannot designate a particular solvolytic reaction as S_N1 or S_N2.

Problem 12.10. 4-Bromobutylamine cyclises in a very dilute solution : (a) What is the mechanism for the reaction? (b) What is the order of the reaction?

Solution :

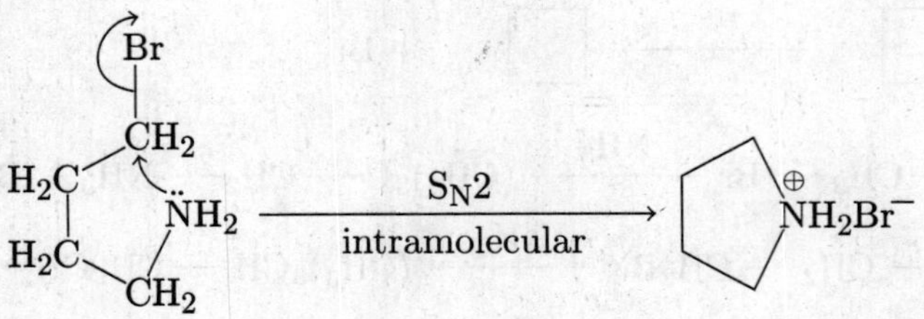

Here NH_2 group is the attacking nucleophile and Br is the leaving nucleophile. Since the solution is very dilute and no other nucleophile is present in the system, an internal S_N2 reaction takes place. Naturally the order of the reaction will be one.

$$\text{The rate} = k\,[\text{Substrate}].$$

Problem 12.11. The absolute configuration of the product of an S_N2 reaction involving an optically active substrate is opposite to that of the substrate—comment and give example in favour of your answer.

Solution : In an S_N2 reaction inversion of configuration occurs about the involved chiral carbon. Its relative configuration changes but absolute configuration may or may not change since the absolute configuration depends only on the groups and atoms present on the carbon site.

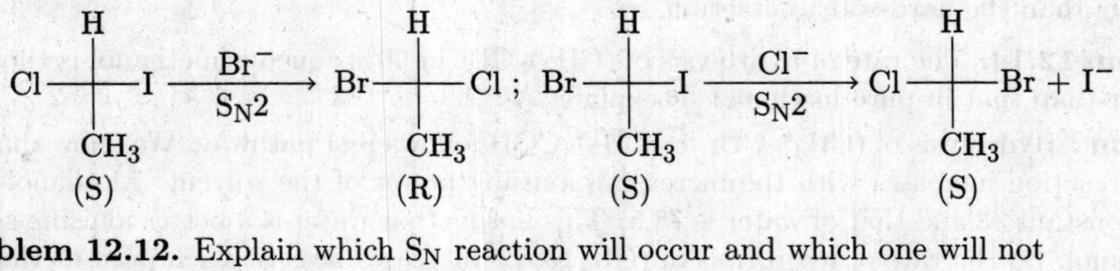

(S) (R) (S) (S)

Problem 12.12. Explain which S_N reaction will occur and which one will not

(i) $CH_3CH_2CN + I^- \longrightarrow CH_3\,CH_2I + CN^-$ **(ii)** $CH_3I + N_3^- \longrightarrow CH_3N_3 + I^-$

Solution : (i) will not occur while **(ii)** will occur.

CN^- does never act as a leaving group whereas I^- is a good leaving group.

Problem 12.13. Explain which reaction will take place faster than the other in the following pairs (assume S_N2 pathway in all the cases):

(a) **(i)** $CH_3CH_2I + CN^- \xrightarrow[\text{solvent}]{\text{EtOH}} CH_3CH_2CN + I^-$

(ii) $CH_3CH_2I + CN^- \xrightarrow[\text{solvent}]{\text{DMF}} CH_3CH_2CN + I^-$

(b) **(i)** $CH_3CH_2CH_2Br + I^- \xrightarrow{\text{Acetone}} CH_3CH_2CH_2I + Br^-$

(ii) $CH_3CH_2CH_2Br + Cl^- \xrightarrow{\text{Acetone}} CH_3CH_2CH_2Cl + Br^-$

(c) **(i)** $PhS^- +$ 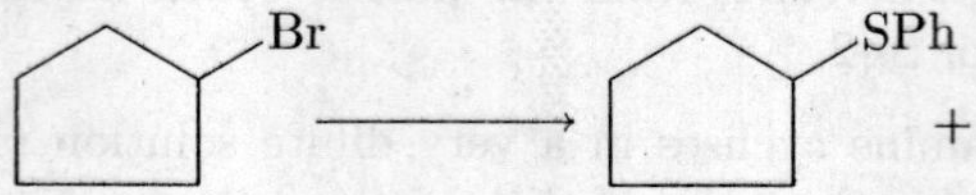 $+ Br^-$

(ii) $PhS^- +$ 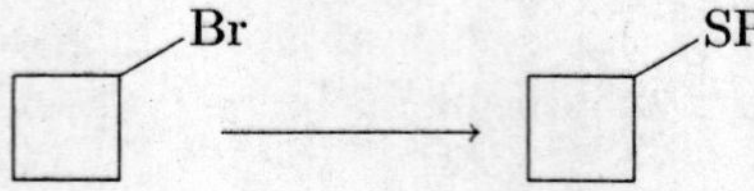 $+ Br^-$

(d) **(i)** $(CH_3)_3C—CH_2—Br \xrightarrow{NH_3} (CH_3)_3C—CH_2—\overset{\oplus}{N}H_3 + Br^-$

(ii) $(CH_3)_2CH—CH_2—CH_2Br \xrightarrow{NH_3} (CH_3)_3CH—CH_2CH_2—\overset{\oplus}{N}H_3 + Br^-$

(e) **(i)** $CH_3CH_2Br + CN^- \longrightarrow CH_3CH_2—CN + Br^-$

(ii) $CH_3CH_2Br + CN^- \longrightarrow CH_3CH_2—NC + Br^-$

Solution : (a) (ii); since DMF does not solvate CN^- but ethanol does.

(b) (ii); since in a nonhydroxylic solvent Cl^- is a stronger nucleophile than I^-.

(c) (ii); since the T.S. of S_N2 reactions involving cyclopentyl systems are more stable than that involving cyclobutyl systems.

(d) (ii); since a neopentyl system will have a very high energy T.S. because of the steric congestion owing to the presence of bulky tert-butyl group.

(e) (i) CN^- is an ambident nucleophile; its C atom is the soft centre while its N atom is the hard centre. The C atom of CH_2Br is a soft centre. So the soft-soft interaction will take place more readily than the hard-soft interaction.

Problem 12.14. The rate of hydrolysis of $(CH_3)_3CBr$ in 50% aqueous methanol is thousand times faster than that in pure methanol — explain.

Solution : Hydrolysis of $(CH_3)_3CBr$ to $(CH_3)_3COH$ follows S_N1 pathway. We know that rate of an S_N1 reaction increases with the increasing ionising power of the solvent. Methanol has a dielectric constant 33 and that of water is 78.5. This means that water is a better ionising solvent than methanol. So the rate of hydrolysis of $(CH_3)_3CBr$ increases largely when pure methanol is replaced by 50% aqueous methanol since more polar water molecules bring extra stability to the T.S. of the carbenium ion formation step, the R.D. step, and therefore, $\Delta G^\ddagger$ becomes less than before.

Problem 12.15. Relative rates of hydrolysis of CH_3Br, CH_3CH_2Br, $(CH_3)_2CHBr$ and $(CH_3)_3CBr$ in a given solvent at a given temperature are 2140, 171, 4.99 and 1010 respectively. Explain the differences in their rates.

Solution : Hydrolysis of CH_3Br, CH_3CH_2Br, $(CH_3)_2CHBr$ follow S_N2 pathway. The rate of an S_N2 reaction decreases as the crowding increases on the carbon site. CH_3Br to $(CH_3)_2CHBr$, bare H atoms are replaced by bulky CH_3 groups. This makes H_2O difficult to attack the carbon site from the back side and naturally the rate falls.

On the other hand, $(CH_3)_3CBr$ hydrolyses by S_N1 mechanism. In a hydroxylic solvent like water with high dielectric constant (78.5), $(CH_3)_3C^{\oplus}$ ion gets stability by solvation and H_2O molecules attack the intermediate rapidly. Because of the high stability of the intermediate and the T.S. of its formation, $\Delta G^{\ddagger}$ decreases sharply, and the sharp rise in the rate occurs in the case of $(CH_3)_3CBr$.

12.5 S_Ni Reaction

If in a nucleophilic substitution reaction a part of the leaving group attacks the carbon site of the substrate and detaches itself from the rest of the leaving group, it is known as substitution nucleophilic internal or S_Ni since the attacking nucleophile has already been present on the substrate and its attack has occurred internally.

The best example of an S_Ni reaction is that of an alcohol with thionyl chloride in ether medium.

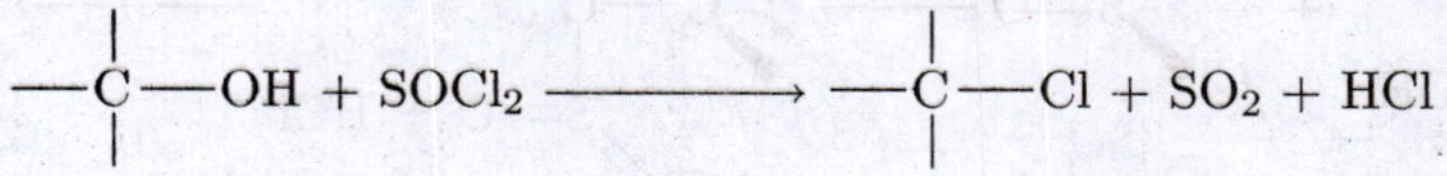

This reaction takes place in three steps :

Step I. Formation of alkyl chlorosulphite as an intermediate compound

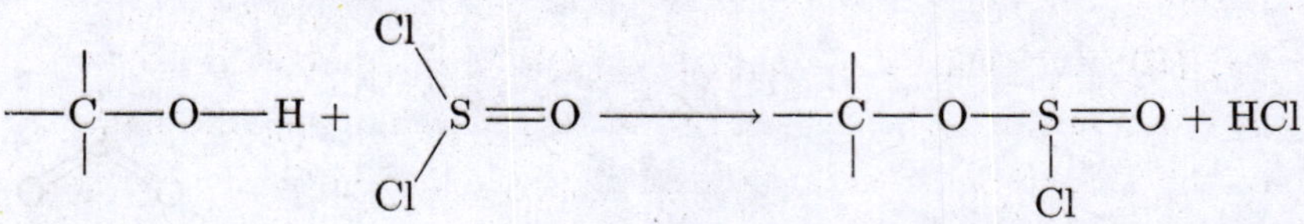

Alcohol Thionyl chloride Alkyl chlorosulphite intermediate

Step II. Alkyl chlorosulphite then dissociates slowly into an intimate ion pair, a carbenium ion and $\bar{O}SOCl$:

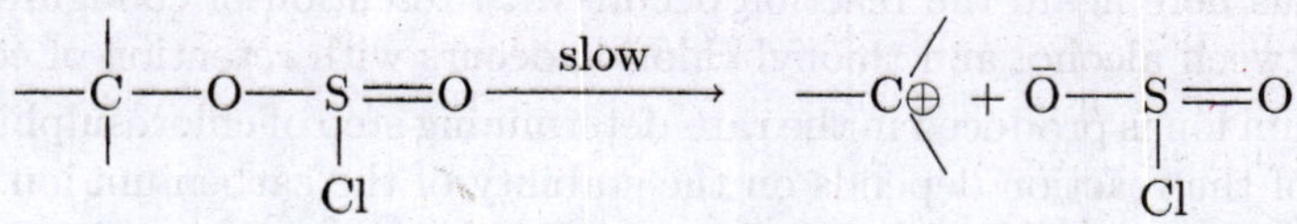

The formation of the ion pair has been proved by the fact that 3-methyl-2-butanol, $(CH_3)_2CH{\cdot}CHOHCH_3$, on reaction with thionyl chloride also gives $(CH_3)_2CClCH_2CH_3$ which is possible only if an intermediate carbenium ion forms and subsequently it rearranges :

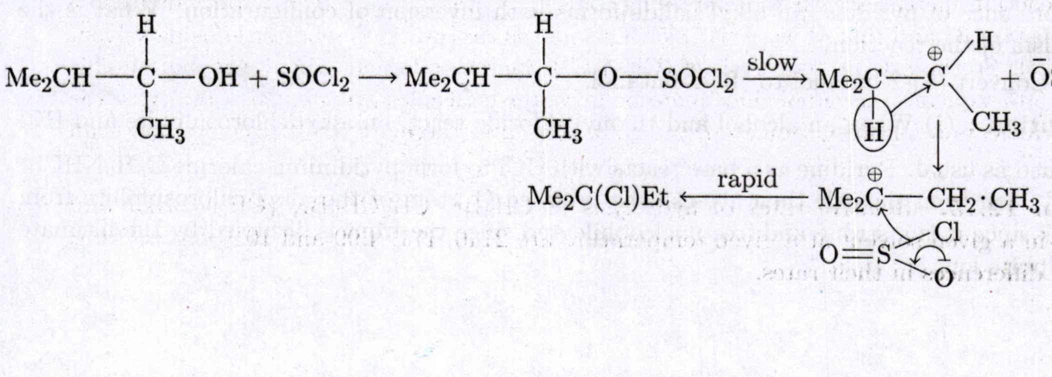

Step III. Now the nucleophile Cl^-,a part of the leaving group ($\bar{O}$—SOCl) attacks the carbenium ion from the side of the leaving group since it is unable to get the rear.

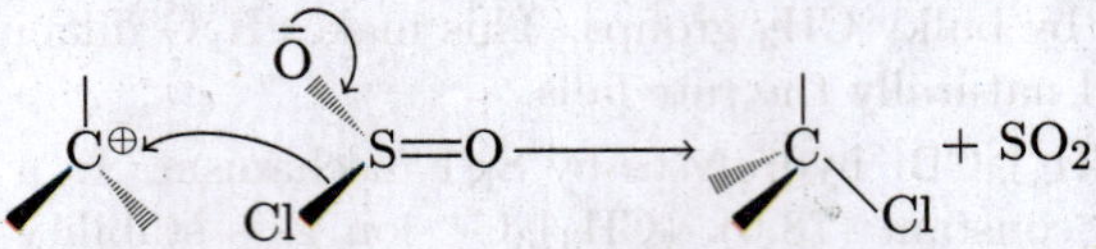

The rate of the overall reaction is found to follow the second order kinetics :

The rate = k [alcohol] [thionyl chloride]

On the other hand, when the S_Ni reaction takes place by the decomposition of the alkyl chlorosulphite, the rate follows the first order rate equation :

The rate = k [alkyl chlorosulphite]

The C—O bond does not break when alkyl chlorosulphite, a true intermediate, forms by the reaction of alcohol and thionyl chloride, i.e., the reaction takes place with retention of configuration. When the chlorosulphite dissociates into an intimate ion-pair, no solvent molecules remain between the two ions as shown below :

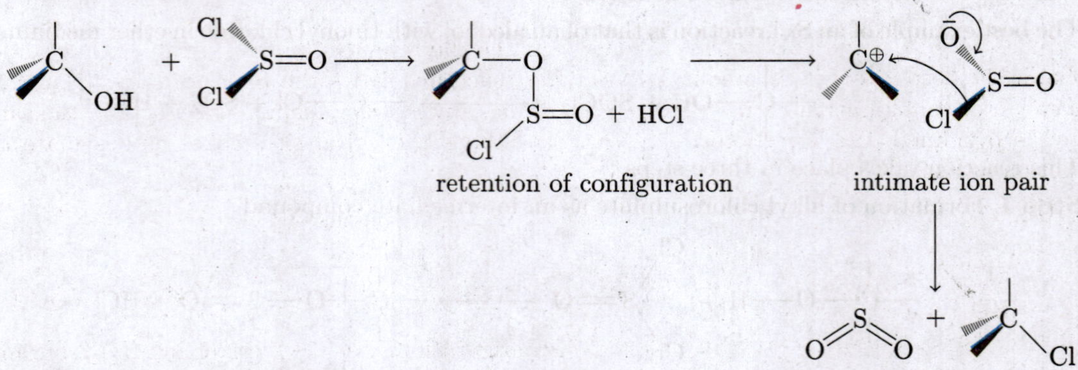

Naturally Cl^- from $\bar{O}SOCl$ ion cannot attack the carbenium ion from the rear. It is, therefore, very likely to attack the carbenium ion from that side of the $C^{\oplus}$ from which $\bar{O}SOCl$ left the chlorosulphite. Thus here again the reaction occurs with retention of configuration. Hence, the overall reaction between alcohol and thionyl chloride occurs with retention of configuration.

Since a carbenium ion is produced in the rate-determining step of chlorosulphite decomposition reaction, the rate of the reaction depends on the stability of the carbenium ion. The greater the stability of the carbenium ion, the greater is the rate. For the same reason the increasing ionisation power of the solvent increases the rate also.

Problem 12.16. **(i)** When the reaction of an alcohol and thionyl chloride is carried out in the presence of pyridine, an alkyl halide forms with inversion of configuration. What is the mechanism of the reaction?

(ii) Convert (S)-2-butanol to (R)-2-butanol.

Solution : **(i)** When an alcohol and thionyl chloride react, an alkyl chlorosulphite and HCl are formed as usual. Pyridine as a base reacts with HCl to form pyridinium chloride $C_5H_5\overset{\oplus}{N}HCl^-$ salt. Cl^- ion from the salt then attacks the involved C atom of the alkyl chlorosulphite from the back since it is a strong and free nucleophile and since the front is occupied by the intimate chlorsulphite ion.

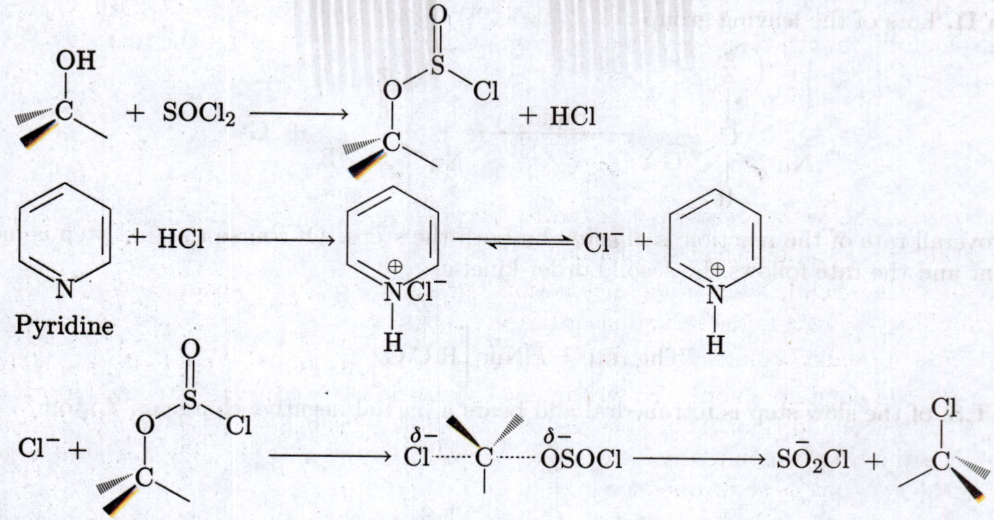

Since the T.S. of the slow step involves Cl^- and the alkyl chlorosulphite, the rate of the reaction follows the second order rate equation and the reaction follows the S_N2 pathway. Naturally inversion of configuration occurs. Not only pyridine, any tertiary amine (R_3N) acts in the same way as pyridine. The inversion of configuration about the chiral site makes us designate the reaction as S_N2.

(ii)

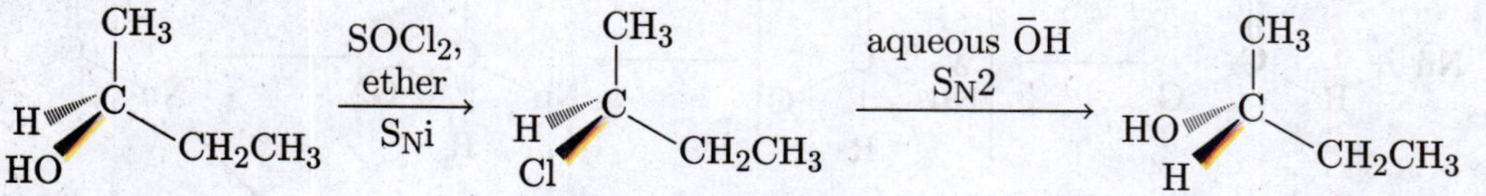

(S)-2-Butanol　　retention of configuration　　inversion (R)-2-butanol

12.6 S_N Reaction on sp^2-Hybridised Aliphatic C Atom

Compounds of the type $R.\overset{\overset{Z}{\|}}{C}.G.$ undergo nucleophilic substitution reactions, where Z may be O, N, S, etc. and G may be —Cl, —OH, —OR, —OCOR, —NH_2, etc. groups.

These reactions are found to follow the second order mechanism usually. Essentially ionic compounds produce carbenium ions and thus display S_N1 mechanism which is no doubt very rare.

Second order nucleophilic substitution reactions may occur without catalyst, but acid-catalysed reactions take place readily. The reaction proceeds by two steps :

Step I. Nucleophilic attack on the sp^2-hybridised C and the formation of a tetrahedral intermediate, an actual compound :

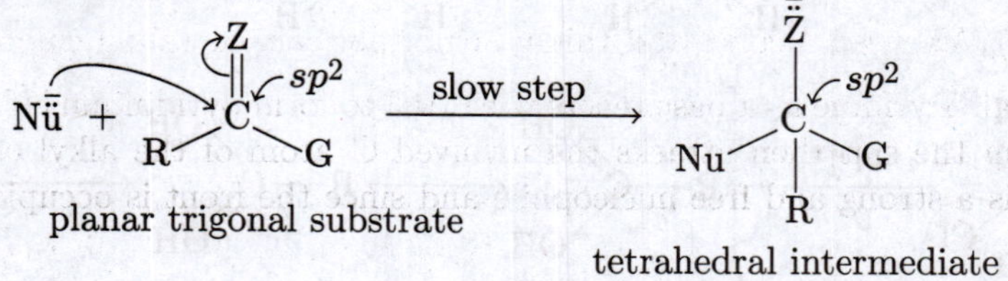

planar trigonal substrate　　tetrahedral intermediate

Step II. Loss of the leaving group :

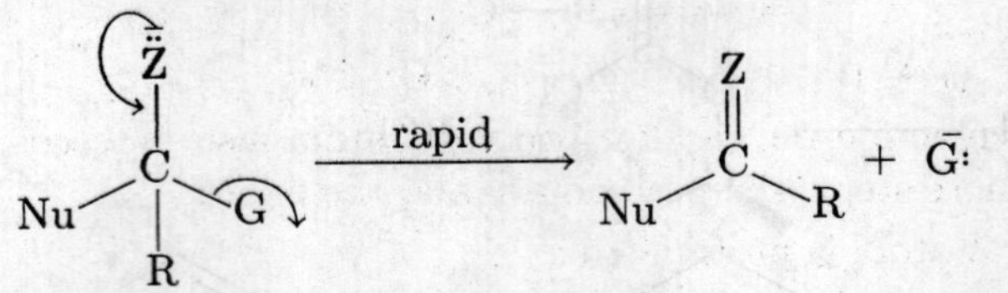

The overall rate of the reaction is affected by both the steps. Of course the first step is more important and the rate follows the second order kinetics.

$$\text{The rate} = k[\ddot{\bar{N}u}:]\left[\text{R.}\overset{\overset{\text{Z}}{\|}}{\text{C}}\text{.G.}\right]$$

The T.S. of the slow step is tetrahedral and bears a partial negative charge on Z atom.

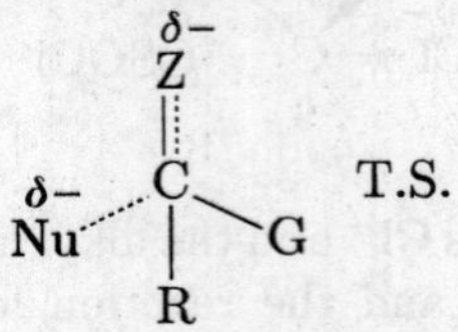

The overall reaction may thus be written as :

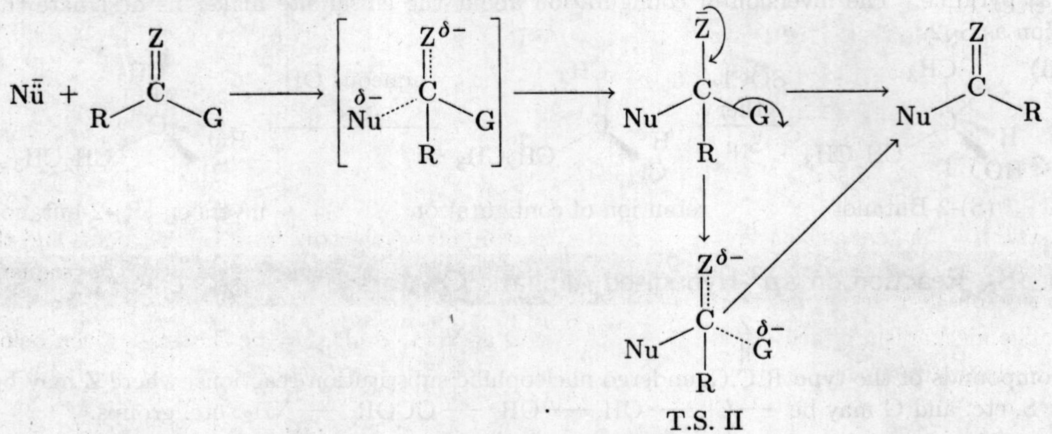

The hydrolysis of acyl chloride may be cited as an example :

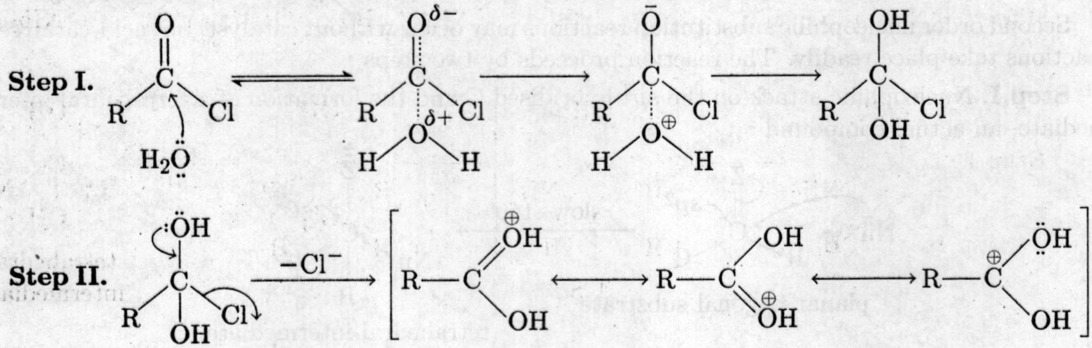

Step III. 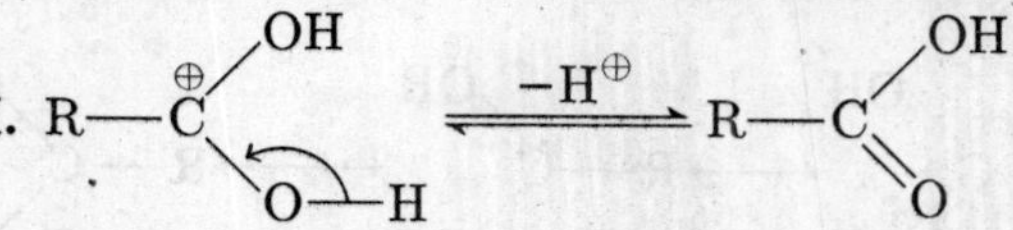

Acid-catalysed S_N reaction on an aliphatic sp^2C atom also proceeds by two steps. Here of course one finds two more steps, the preliminary and the final :

Preliminary step : Z atom is protonated :

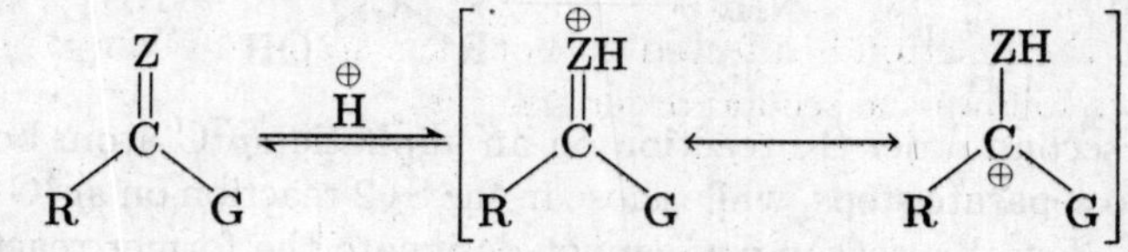

Step I. 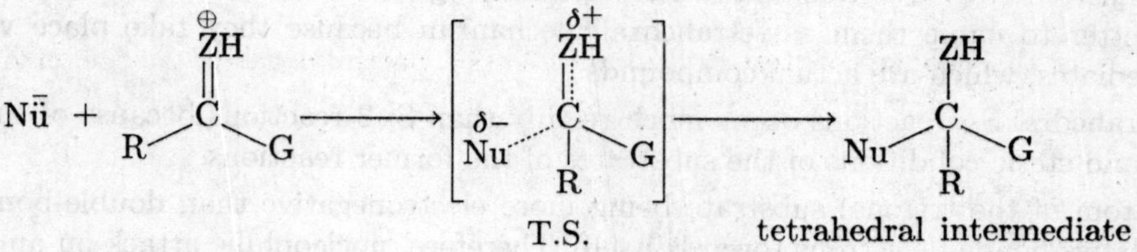

Step II. 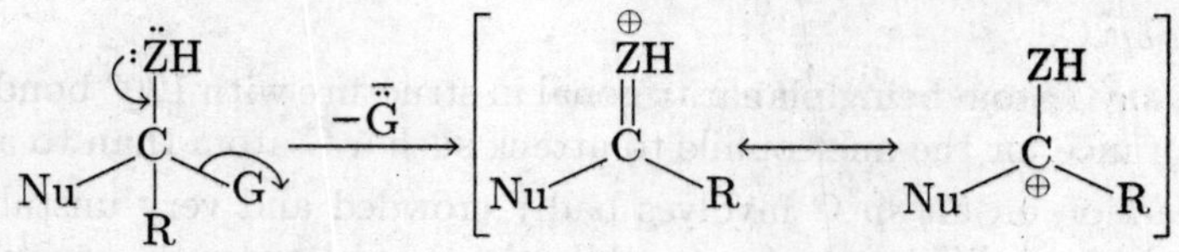

Step III. 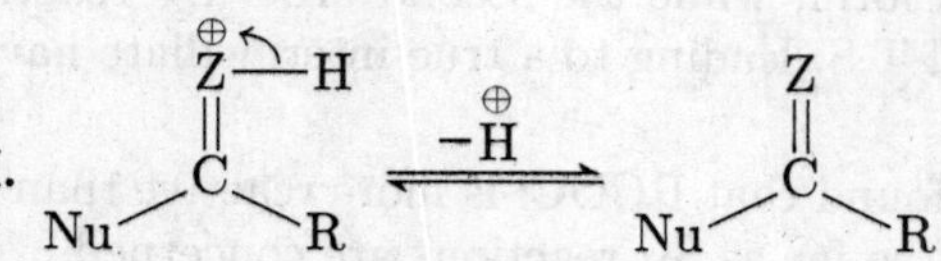

The $H^{\oplus}$ ion acts as a catalyst; protonated Z atom draws electron from C—Z bond and thus decreases the electron density on sp^2C. This facilitates the nucleophile to attack and consequently the rate of the reaction increases.

The mechanism of acid-catalysed hydrolysis of an acid amide may be shown as given below :

Preliminary step :

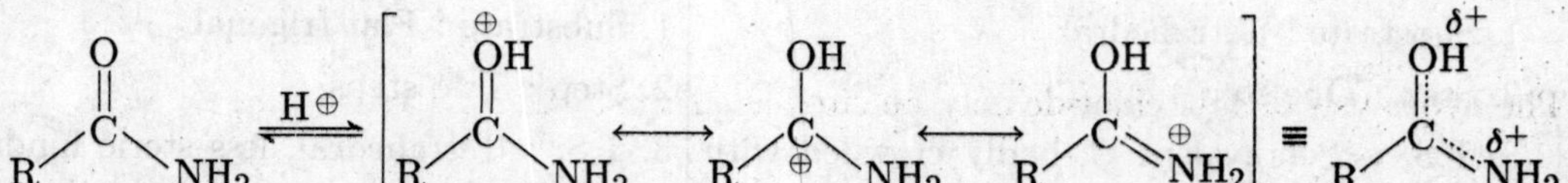

Step I. 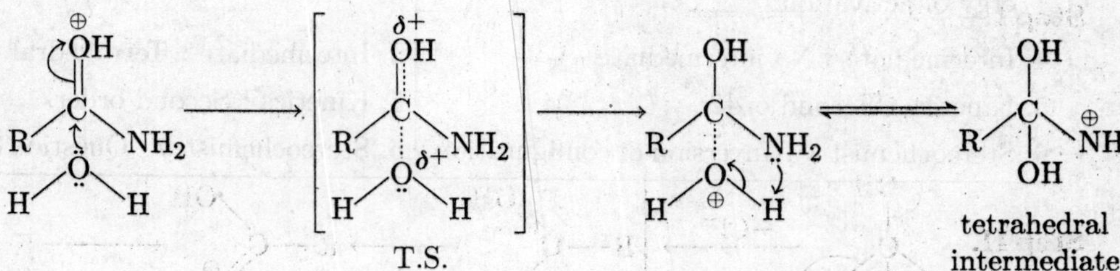

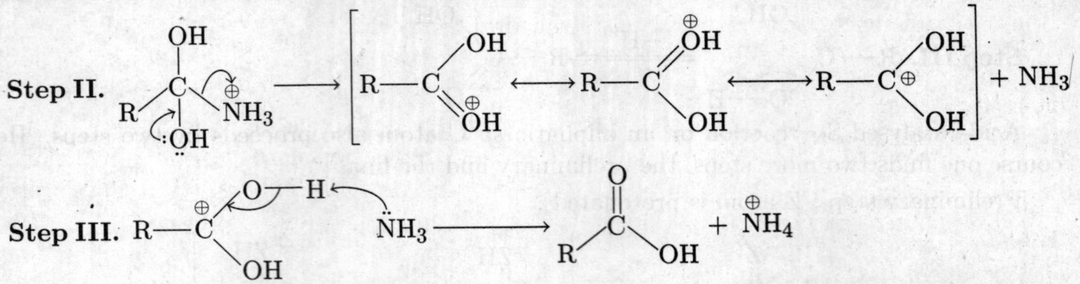

Therefore, in a second order S_N reaction on an aliphatic sp^2C atom bond making and bond breaking are the two separate steps, while those in the S_N2 reaction on sp^3C atom happen to occur in a concerted step. For this reason one cannot designate the former reactions as S_N2 reaction. It is better to name them as tetrahedral mechanism because they take place via tetrahedral intermediates, which are actual compounds.

Tetrahedral S_N reactions occur much readily than S_N2 reactions because of the special electronic and steric conditions of the substrates of the former reactions.

Z atom of the trigonal substrate being more electronegative than double-bonded C, the Z-atom draws bonded electrons towards itself. Therefore, nucleophilic attack on an sp^2C becomes easier than that on an sp^3C.

On the other hand, sp^2C atom being planar trigonal in structure with 120° bond angle roughly, there is less steric hindrance for the nucleophile to attack such a C atom than to attack an sp^3C.

Lastly, the S_N2 reaction on an sp^3C involves badly crowded and very unstable pentavalent transition state which is very difficult to form; while the second order S_N reaction on an sp^2C involves relatively unhindered tetrahedral T.S. leading to a true intermediate having tetrahedral arrangement.

For the reasons discussed above, it is found that RCOCl is more reactive than RCl, $RCONH_2$ than RNH_2, RCOOR than R—O—R′ so far as S_N reactions are concerned.

Problem 12.17. Compare S_N2 mechanism with tetrahedral S_N mechanism.

Solution :

S_N2	Tetrahedral
1. Substrate : Tetrahedral	1. Substrate : Flat trigonal
2. Step : One step	2. Step : Two steps
3. T.S. : Pentavalent C, badly crowded with more steric hindrance; requires higher energy of activation.	3. T.S. : Tetrahedral, less steric hindrance; requires lower energy of activation.
4. Intermediate : No intermediate	4. Intermediate : Tetrahedral intermediate.
5. Kinetics : Second order.	5. Kinetics : Second order.
6. Stereochemistry : Inversion of configuration.	6. Stereochemistry : Question does not arise.

Pictorial differences between S_N2 and tetrahedral S_N mechanisms.

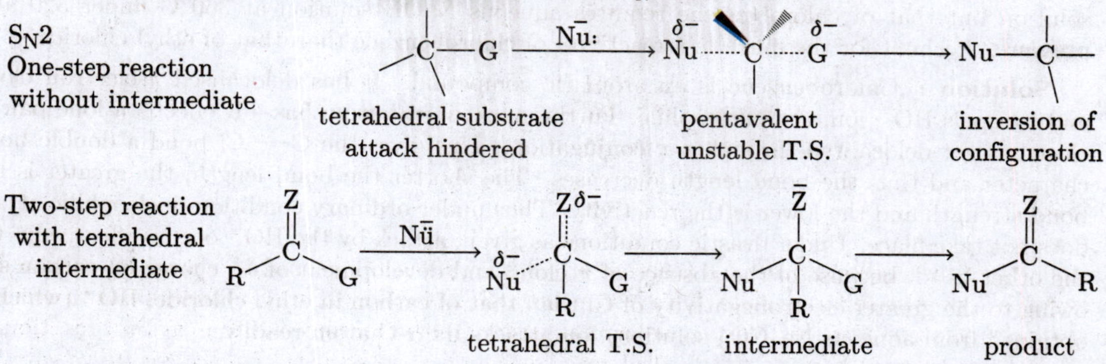

Problem 12.18. Write down the mechanistic steps of alkaline hydrolysis of an acid amide.

Solution : Tetrahedral S_N mechanism.

Step I. Nucleophilic attack by strong $\bar{O}H$ on the sp^2C.

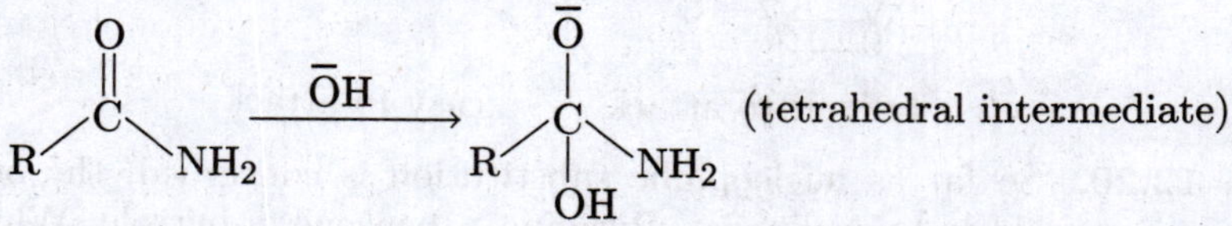

Step II. Decomposition of the intermediate to a carboxylate ion and ammonia.

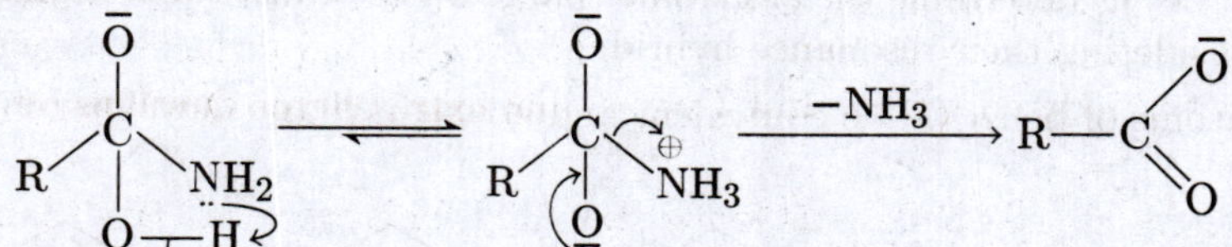

12.7 Aromatic Nucleophilic Substitution

Every aromatic system has annular electronic cloud below and above its plane because of the presence of delocalised π-electron cloud. When a nucleophile approaches the nucleus of an aromatic system, a repulsion takes place between them. Because of this repulsion and because of the less ability of π orbitals of the complex system developed by the interaction between the aromatic system and the nucleophile to delocalise two more electrons of the latter, the nucleophilic attack is very difficult on aromatic species. Of course under drastic conditions, i.e., under high temperature and pressure, strong nucleophiles may be substituted. On the other hand, substituted aromatic compounds having properly placed strong electron-withdrawing groups like F, NO_2, Cl, OH, etc. or aromatic nucleus consisting of electronegative hetero atom (O, N, S, etc.) undergo nucleophilic substitution with less difficulty as the electron-withdrawing group or the electronegative atom respectively draws electron towards itself making the aromatic nucleus less electron dense.

However, three main types of aromatic S_N reactions are found to occur.

(a) Unimolecular via a carbenium ion intermediate (S_N1 aromatic).

(b) Bimolecular via intermediate complex anion (S_N2 aromatic).

(c) Elimination—addition via aryne intermediate.

Problem 12.19. Hydrolysis of ethyl chloride can be carried out with aqueous Na_2CO_3 solution but that of chlorobenzene requires aqueous NaOH solution at 360°C under 320 atm pressure. Account for the much less reactivity of chlorobenzene than that of ethyl chloride.

Solution : Chlorobenzene is an aromatic compound. It has delocalised π-electron cloud and it repels HO^- ion, the nucleophile. Furthermore, its Cl atom has +R effect—a lone pair of electrons get delocalised through p-π conjugation. This gives the C—Cl bond a double bond character and thus the bond length decreases. The shorter the bond length, the greater is the bond strength and the lower is the reactivity. Thus under ordinary condition nucleophilic attack does not take place. Under drastic conditions as given, attack by the HO^- occurs effectively. On the other hand, because of the absence of π cloud and development of δ^+ charge on carbon site owing to the greater electronegativity of Cl than that of carbon in ethyl chloride, HO^- (which is obtained from aqueous Na_2CO_3 solution) ion attacks its α-C atom readily.

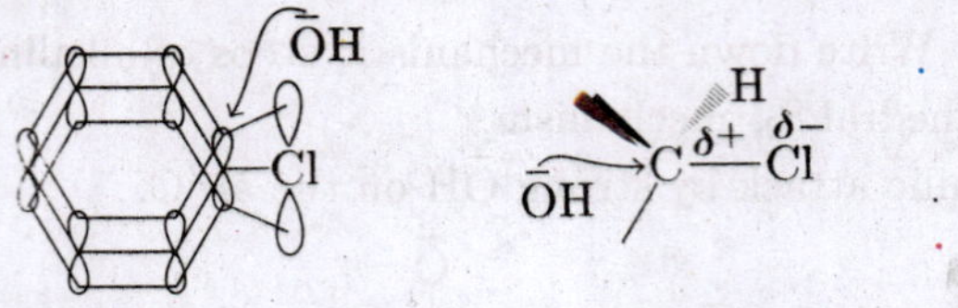

difficult to attack easy to attack

Problem 12.20. So far as nucleophilic substitution is concerned, the order of reactivity amongst benzene, pyrrole and pyridine is : Pyridine > benzene > pyrrole. Why is it so?

Solution : Let us determine the electronic charge on the constituent atoms of the three given compounds considering their resonance hybrid.

\ll the C atoms of benze e equivalent and no extra charge develops on anv one of the six

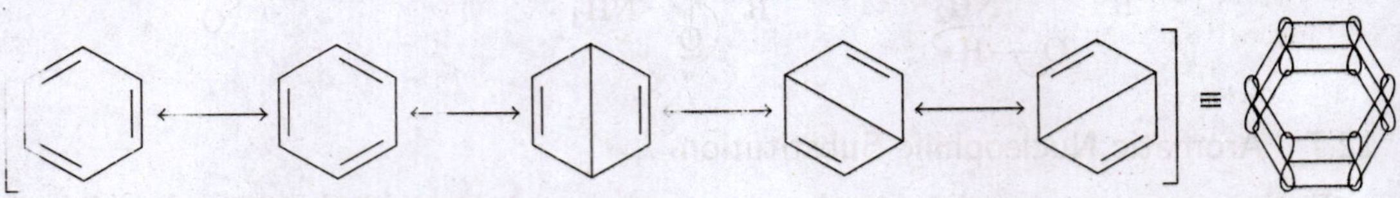

p_z orbital of the N atom contains 2 es in pyrrole. It, therefore, cannot accommodate further electron. To preserve aromaticity, its electrons move away from it. Thus negative charge appears on each of the Cs in pyrrole which prevents nucleophilic attack.

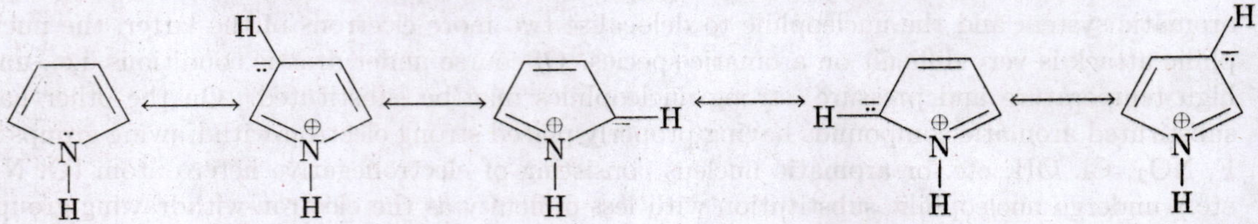

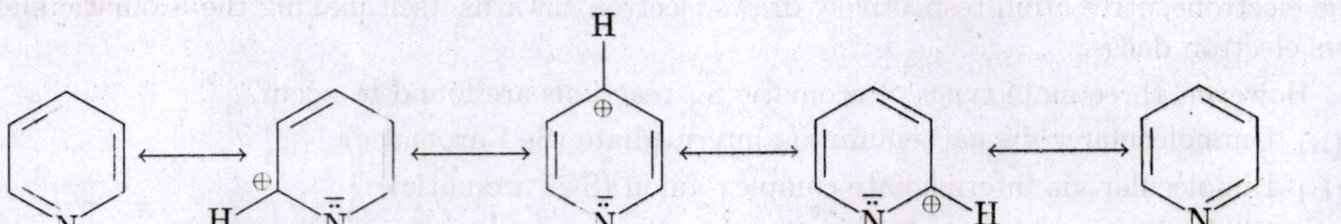

Because of the greater electronegativity of N atom than that of C atom, the former draws electron towards itself without disturbing aromaticity in pyridine. The N atom has one e in its p_z orbital. It accommodates another electron from neighbouring C atom. Thus the N atom becomes negatively charged and C_2, C_4 and C_6 atoms become positively charged. Therefore, these C atoms are less electron dense than C_3 & C_5 atoms. Since nucleophilic substitution occurs at less electron dense site, it will occur most readily in pyridine at C_2, C_4 and C_6 atoms. In pyrrole it will occur least readily and the position of benzene in this respect is intermediate between pyridine and pyrrole. So, in the nucleophilic substitution, the reactivity order is : pyridine > benzene > pyrrole.

Problem 12.21. Hydrolysis of *o*-nitrochlorobenzene occurs much readily than that of chlorobenzene by alkali.—Explain.

Solution : Nitro group being electron withdrawing, its $-I$ effect decreases the electron density on the C atom of the C—Cl bond in *o*-nitrochlorobenzene. This effect being absent in chlorobenzene, HO^- ion attacks the C atom of the C—Cl bond in the former compound much readily than that in chlorobenzene and hence the hydrolysis occurs in *o*-nitrochlorobenzene much readily than in chlorobenzene.

12.8 Unimolecular Aromatic Nucleophilic Substitution

Aromatic S_N1 reactions are very rare. The most common example is the displacement of N_2 in the reactions of diazonium salts in a strong polar medium.

The reaction proceeds in two steps :

Step I. Reversible formation of aryl cation

$$\text{Ar}-\overset{\oplus}{\text{N}}\equiv\text{N} \underset{}{\overset{\text{slow}}{\rightleftharpoons}} \text{Ar}^{\oplus} + \text{N}_2$$

The reversible formation of aryl cation is proved by using $\text{Ar}-{}^{15}\overset{\oplus}{\text{N}}\equiv\text{N}$ ion. The decomposition product on examination is found to contain not only $\text{Ar}-{}^{15}\overset{\oplus}{\text{N}}\equiv\text{N}$ but also $\text{Ar}-\overset{\oplus}{\text{N}}\equiv{}^{15}\text{N}$. This can be possible only if the recombination of $\text{Ar}^{\oplus}$ and N_2 takes place.

The aryl cation so formed is very unstable since the positive charge cannot get delocalised. This is because the breaking of C—N σ bond leaves a vacant sp^2 A.O. with the cationic C atom; this being perpendicular to π orbitals, delocalisation of the charge is not possible.

Step II. Nucleophilic attack on the cationic C atom :

$$\text{Ar}^{\oplus} + \text{N}\bar{\text{u}} \xrightarrow{\text{rapid}} \text{Ar}-\text{Nu}.$$

The first step being the slow step, it is the rate-determining step.

The T.S. is $\text{Ar}^{\delta+}\cdots\cdots\overset{\delta+}{\text{N}}\equiv\text{N}$.

The reaction is found to follow the first order kinetics, and the rate $= k\,[\text{Ar}-\overset{\oplus}{\text{N}}\equiv\text{N}]$.

Addition of nucleophiles to the reaction medium only changes the nature of the product but the rate of the reaction remains unaltered.

An aryl cation being very unstable, it combines with any available nucleophile. A question arises here—if aryl cation is very unstable, then why does the reaction occur at all? Because of the high stability of molecular nitrogen (N≡N bond dissociation energy is 946 kJ/mole), N_2 acts as a good leaving group and this makes the generation of aryl cation easy.

Because of the unselective reactivity of aryl cation the reaction is very useful for synthesis.

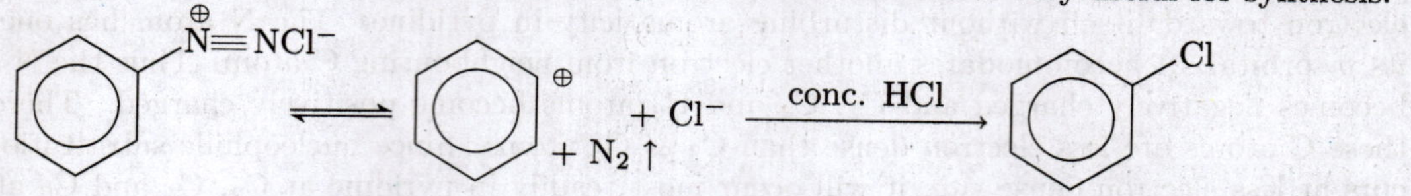

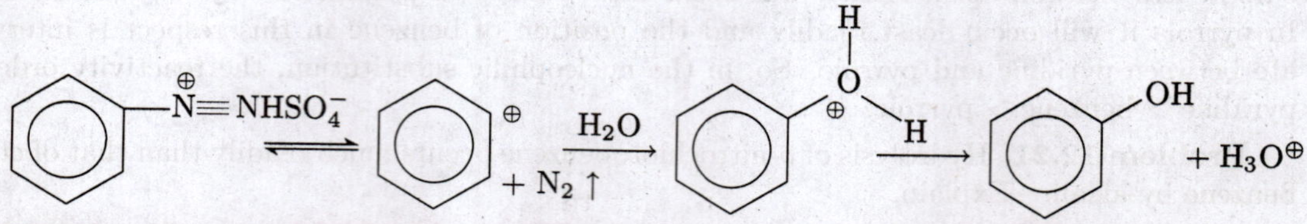

12.9 Bimolecular Aromatic S_N Reactions

This is perhaps the most common type of aromatic S_N reaction. X^-(X=F, Cl, Br, I), SO_3^{--}, R_2N^-, H^- are the leaving groups in order of importance. RO^-, HO^- and $\bar{N}H_2$ are the incoming nucleophiles usually.

The mechanism consists of two steps :

Step I. Nucleophilic attack on a carbon site and formation of a resonance-stabilised intermediate complex carbanion via a T.S. which involves the substrate and the incoming nucleophile. This is a slow step and hence the rate-determining step.

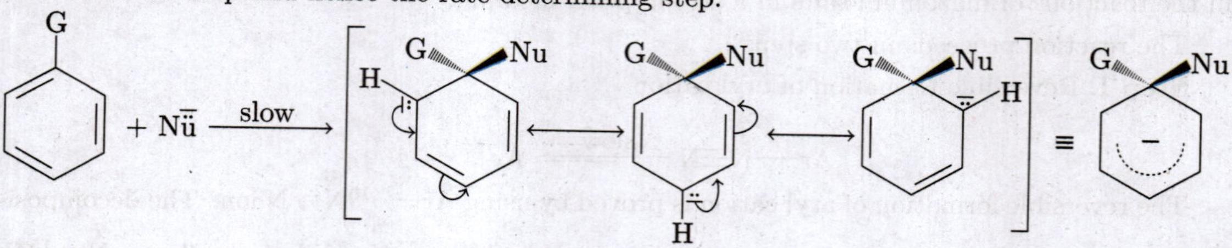

T.S. I is 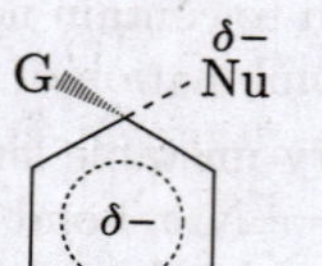

When the aromatic nucleus is benzene and its derivative, the intermediate is called cyclopentadienyl carbanion or cyclohexadienyl anion.

Step II. Loss of the leaving group from the intermediate anion to yield the product :

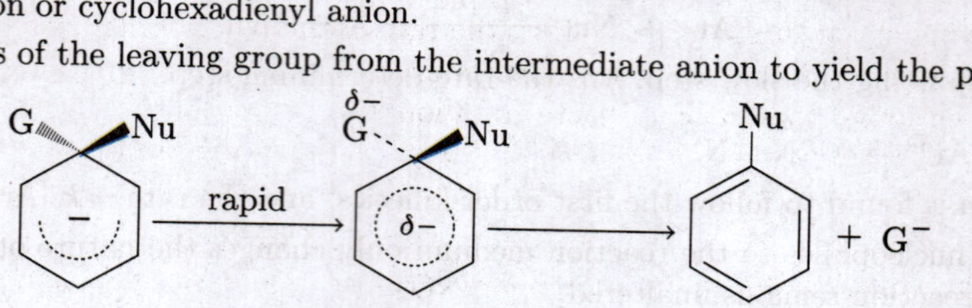

T.S. II

Since Step I is the slow rate-determining step and its T.S. involves both the nucleophile and the substrate, it is a bimolecular reaction. The kinetics of the reaction follows the second order rate equation.

The rate = k[Ar G] [Nü$^-$]

The energy profile of the reaction (See Fig. 12.9) :

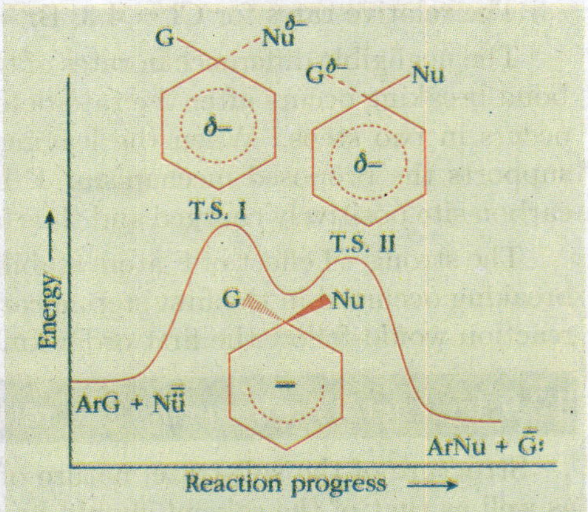

Fig. 12.9. Energy profile of an aromatic S_N2 reaction

In this case C—Nu bond formation takes place in the rate-determining step when the bond between the leaving group and the C atom remains in tact; the bond breaking occurs in Step II. Hence, this mechanism cannot be called S_N2 since bond making and bond breaking do not take place simultaneously. Over and above this, here the nucleophile attacks the carbon site from the side of the leaving group whereas in an S_N2 reaction nucleophilic attack occurs from the opposite side of the leaving group leading to inversion. The mechanism may be called intermediate complex mechanism, since a complex intermediate, which we do not get in an S_N2 reaction on an sp^3C atom, is formed. This may also be named as S_N2 aromatic.

However, there is a good deal of evidence in support of this two-step mechanism as discussed above. We shall consider two of them here.

1. **Isolation and detection of the intermediate complex anion :** Meisenheimer in 1902 isolated the intermediate complex carbanion of the reaction between ethyl picrate and MeO$^-$ ion. Similar type of intermediates are found stable and a number of such compounds have been isolated since 1961.

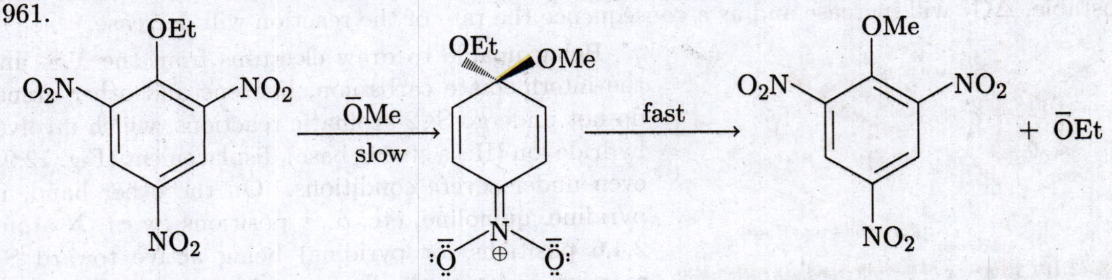

Ethyl picrate intermediate meisenheimer salt Methyl picrate

Intermediate complex anions have also been detected by N.M.R. spectroscopy and X-ray diffraction study.

2. **Studies of the effect of the leaving group on the reaction rate :** Since bond breaking does not take place in the rate-determining step (this occurs in the second step), whatever may be the nature of the leaving group, it should not affect the rate of the reaction. In fact, it is found to be so. Taking different derivatives of 2,4-dinitrobenzene as substrates and piperidine as nucleophile, the rates have been found to differ by a factor of about 5.

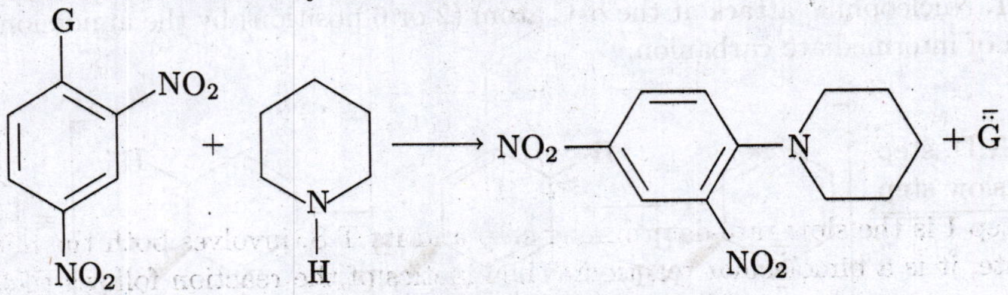

G = Cl, Br, I, SOPh, SO_2Ph, etc.

The relative rates for Cl = 4.3; Br = 4.3; I = 1; F = 3300.

The negligible difference in rates of the above reactions (except for F) proves undoubtedly that bond breaking occurs after the rate-determining step and this in its turn proves that the reaction occurs in two steps. When the leaving group is F, the rate increases tremendously. This also supports the proposed mechanism. F being a strong electron withdrawing group, it makes the carbon site positively charged and thus facilitates the nucleophile to attack the same much readily.

The strong −I effect of F atom stabilises the anionic T.S. of the rate-determining step. If bond breaking occurred in the first step, a carbenium ion would have been the result and naturally the reaction would follow the first order kinetics.

12.9.1 Factors influencing the rate of aromatic bimolecular S_N reactions

Structure of the substrate, nature of the leaving group and that of the incoming nucleophile as well as that of the solvent should be taken into consideration as usual.

(1) Effect of the substrate structure on rate : The first step of bimolecular aromatic S_N reaction is the formation of intermediate complex carbanion. Anything which stabilises the carbanion will increase the rate of the reaction. A group that draws electrons toward itself, either by −I effect or by resonance effect, will stabilise the T.S. and the intermediate carbanion and therefore, will activate the ring towards nucleophilic substitution. $\Delta G^{\ddagger}$ will decrease and consequently the rate of the reaction will increase. On the other hand, a group or atom that releases electrons will intensify the negative charge of the T.S. and that of the intermediate; thus it will make them unstable, $\Delta G^{\ddagger}$ will increase and as a consequence the rate of the reaction will decrease.

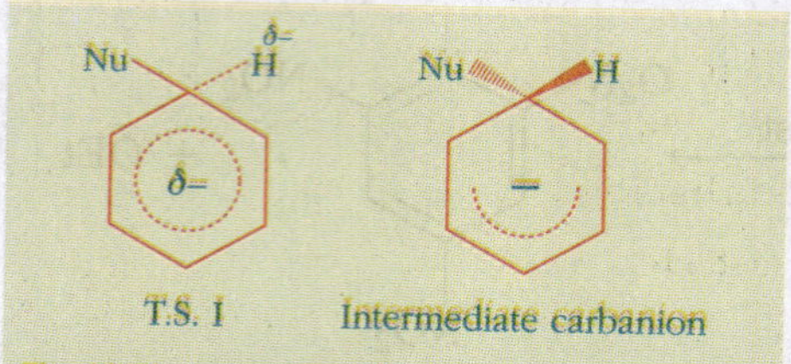

Fig. 12.10. Structure of the intermediate carbanion and that of the T.S.

Being unable to draw electrons from the T.S. and the intermediate carbanion, benzene and other arenes do not undergo S_N2 aromatic reactions, which involves hydride ion ($\bar{H}$:, a strong base) displacement (Fig. 12.9), even under severe conditions. On the other hand, in pyridine, quinoline, etc. α, γ positions (w.r.t. N atom; 2,4,6 positions for pyridine) being active toward S_N reaction, owing to the electron-withdrawing effect of the electronegative N atom, one may expect bimolecular S_N reaction aromatic involving $\bar{H}$: displacement by a strong base. Chichibabin (Tschitschibabin) reaction is an example of the category.

It is a reaction by which pyridine and other heterocyclic nitrogen compounds are aminated mainly at α-positions with respect to the N atom (2 and 6 positions for pyridine) by alkali metal amide ($NaNH_2$, KNH_2) in liquid ammonia at an elevated temperature (100-200°C) where $\bar{H}$: gets displaced by :NH_2 ion as shown below mechanistically.

Step I. Nucleophilic attack at the α-C atom (2 or 6 position) by the amide ion leading to the formation of intermediate carbanion.

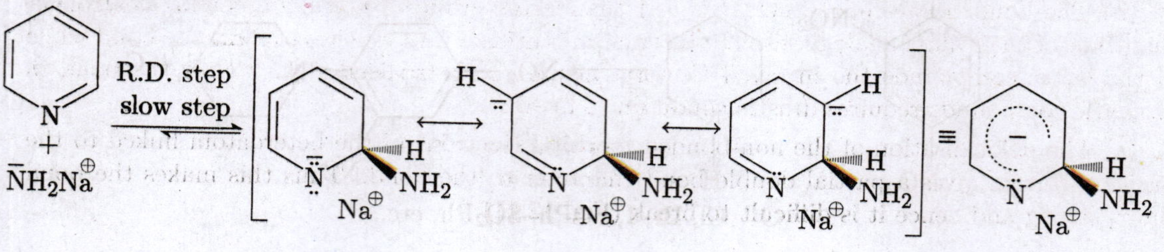

Step II. Elimination of $\bar{H}$: from the carbanion to form 2-aminopyridine.

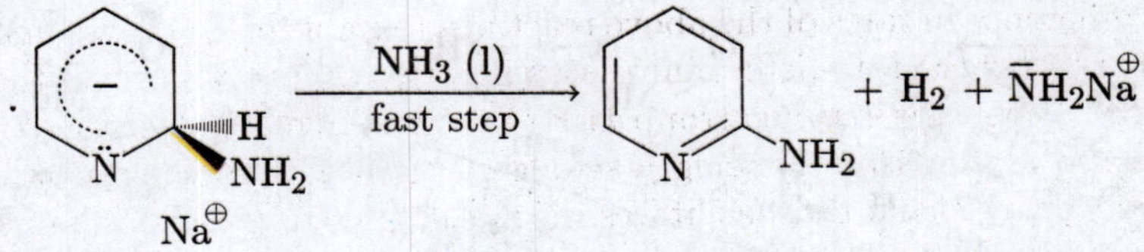

Step III. Abstraction of $H^{\oplus}$ from the amino group by $\bar{N}H_2$ to form sodio salt.

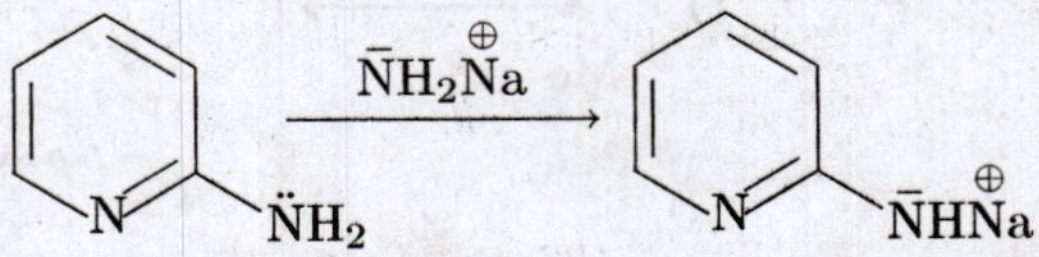

However, under drastic conditions arene derivatives carrying at least one electron withdrawing group (—NO_2, —CN, —COOH, etc.) undergo S_N2 aromatic reaction involving $\bar{H}$: displacement from α and δ positions with respect to the group (*o*- and *p*- positions for benzene). Thus when nitrobenzene is fused with KOH in the presence of air, a mixture of *o*-Nitrophenol (major product) and *p*-nitrophenol (small amount) is obtained. Here oxygen of the air oxidises $\bar{H}$: to H_2O and thus helps to remove it.

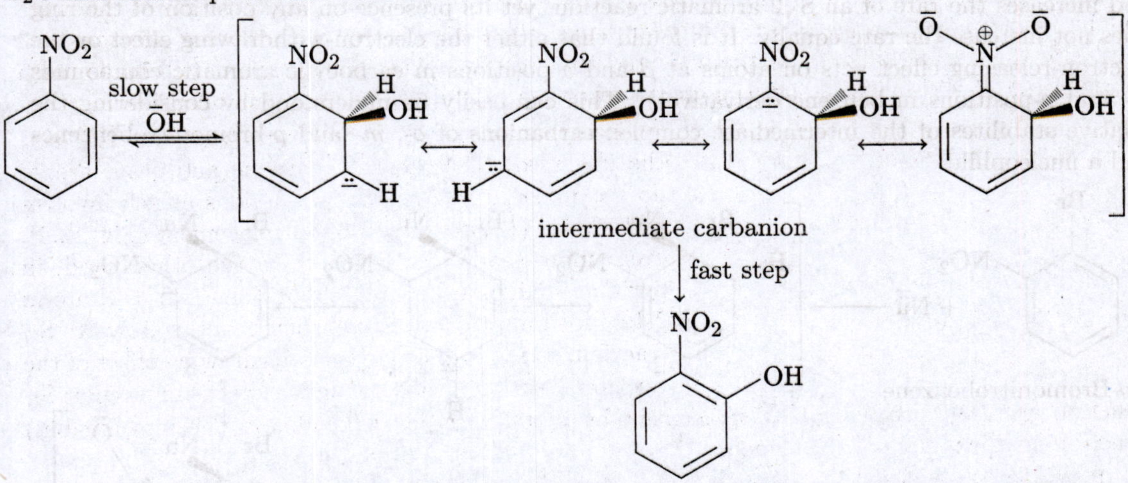

Of course bare aryl halides (they may follow benzyne mechanism), aryl ethers, arenesulphonic acids, etc. undergo S_N aromatic bimolecular reactions under severe conditions. Nucleophiles in these cases are $\bar{O}H$, NH_3, etc. and the leaving groups are X, OR, SO_3H, etc. Drastic conditions are essential for the reasons noted below :

(a) The π cloud of the aromatic system repels the incoming nucleophile.

(b) The bond between the nuclear C and the leaving group of these compounds is stronger than that of their alkyl analogues since the nuclear C utilises its sp^2 A.O. to form the bond while in the latter compounds the involved C atoms are in sp^3 hybridised state. Thus the bond in aromatic compounds requires drastic condition to break.

(c) The delocalisation of the non-bonded p orbital electrons of the heteroatom linked to the nuclear C atom gives a partial double bond character to the bond. Thus this makes the bond short, strong and hence it is difficult to break (Fig. 12.11).

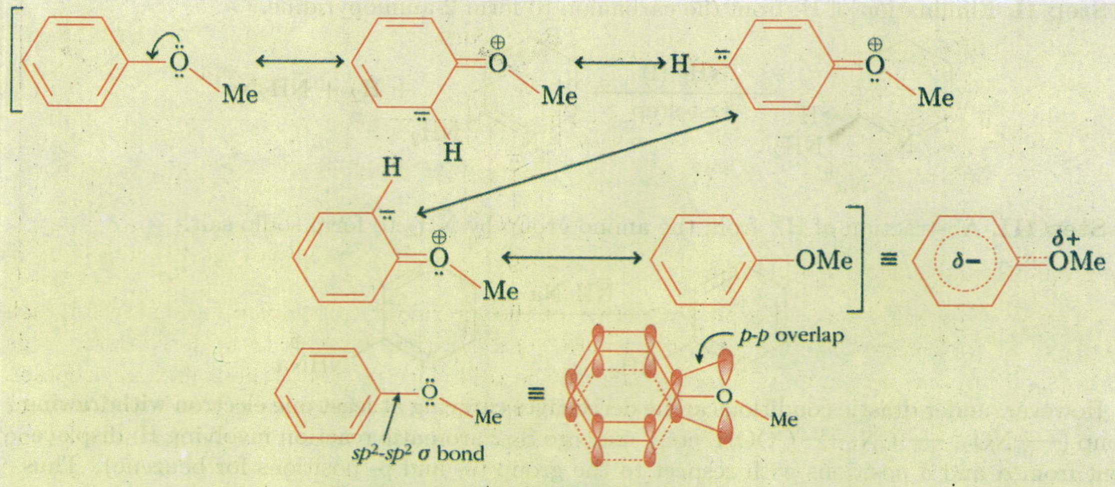

Fig. 12.11. Partial double bond character of C—OMe

Although the presence of an electron-withdrawing group on the aromatic ring activates it and increases the rate of an S_N2 aromatic reaction, yet its presence on any position of the ring does not increase the rate equally. It is found that either the electron-withdrawing effect or the electron-releasing effect acts on atoms at β and δ positions in carbocylic aromatic compounds (*o*- and *p*-positions in benzene derivatives). This can easily be understood by considering the relative stabilites of the intermediate complex carbanions of *o*-, *m*- and *p*-bromonitrobenzenes and a nucleophile.

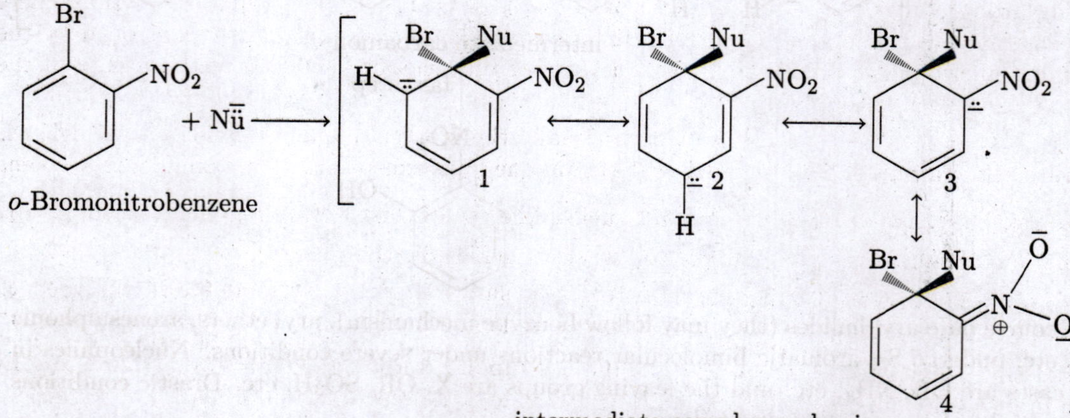

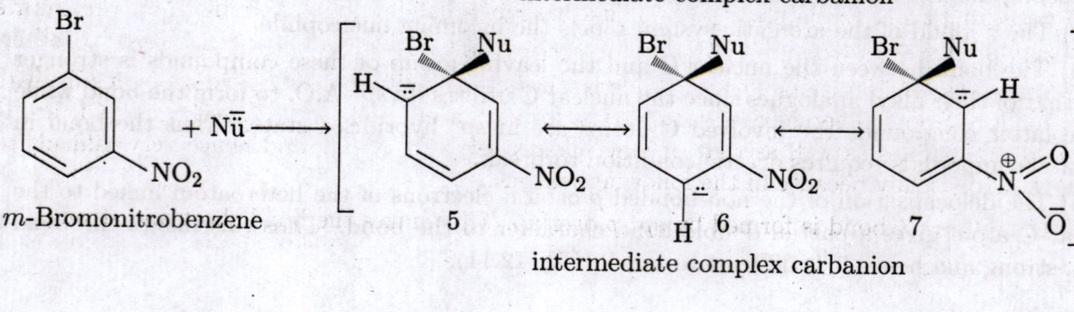

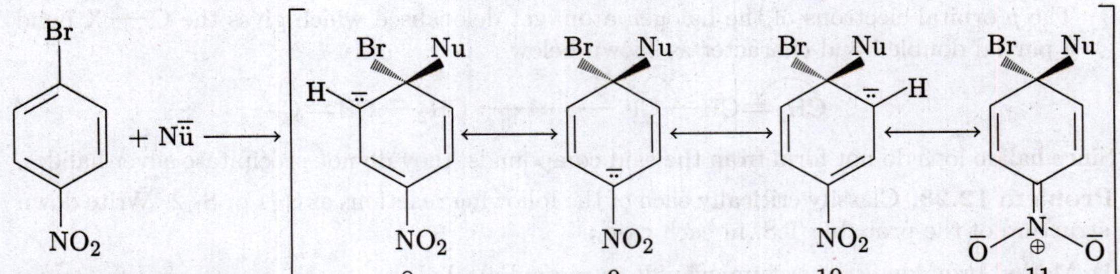

p-Bromonitrobenzene 8 9 10 11

Thus we get one extra resonating structure of the intermediate complex carbanion for each of the ortho- (structure 4) and para- (structure 11) derivatives relative to *m*-derivatives. This extra structure has a remarkable contribution to the resonance hybrid of the complex carbanion since resonance occurs, in those cases, over a wider range involving the nitro group with respect to that of the intermediate complex carbanion of the *m*-derivative having no participation of the nitro group.

Therefore, it is needless to mention that with the increasing substitution of electron-withdrawing groups on the β and δ positions of a substrate, the rate of aromatic S_N reaction will increase. For this reason 2, 4, 6-trinitrochlorobenzene hydrolyses even in dilute aqueous Na_2CO_3 solution, while hydrolysis of chlorobenzene requires severe conditions like 320 atm pressure, 15% NaOH solution, etc. Hydrolysis of chlorobenzene, however, may follow benzyne mechanism (See Sec. 12.10 and 10(c) of the exercise).

(2) Effect of the leaving group : Halide, —NO_2, —N_3, —$\overset{\oplus}{N}R_3$, —OR, —SR, —SO_2R, —OAr, etc. are the usual leaving groups in S_N2 aromatic reactions. The usual order of the leaving group ability is F > NO_2 > Cl, Br, I > N_3 > OAr, OR, SR, SO_2R, —NH_2. There are exceptions of this order and many a time leaving group ability is greatly controlled by the incoming nucleophile. However, usually the rate of the reaction increases as the ability of the leaving group increases.

(3) Effect of the incoming nucleophile : Usually the greater the nucleophilicity, the higher is the rate of the aromatic S_N2 reaction. Here again one finds exceptions. For example, HO^- being a stronger base than $\bar{O}Ar$, it is a stronger nucleophile than $\bar{O}Ar$ in all S_N reactions except in S_N2 aromatic reactions—where HO^- acts as a weaker nucleophile than $\bar{O}Ar$.

(4) Effect of the solvent : With the increasing polarity of aprotic solvents the rate of the S_N2 aromatic reaction increases. Thus chlorobenzene undergoes S_N2 aromatic reaction with Me_3CO^- in dimethyl sulphoxide (DMSO, $Me_2\overset{\oplus}{S}—O^-$; a dipolar aprotic solvent) solvent more readily than in methanol. In dimethyl formamide ($HCONMe_2$) the reaction occurs readily.

Problem 12.22. Explain why aryl halides and vinyl halides do not give insoluble precipitate of silver halides when these are treated with alcoholic silver nitrate solution even for days together.

Solution : Precipitation of silver halide by the action of silver nitrate solution on a substrate is an ionic reaction. Therefore, it requires the formation of halide ion. The C—X (Cl, Br, I) bond of aryl halides as well as that of vinyl halides are very strong and hence very difficult to break heterolytically because of the following reasons :

(i) The C—X bond is formed by sp^2-p overlap. This makes the bond short and strong.

(ii) The *p* orbital electrons of the halogen atom get delocalised which gives the C—X bond a partial double bond character as shown below :

$$CH_2{=}CH{-}\ddot{\underset{..}{C}}l{:} \longrightarrow CH_2{-}CH{=}\overset{\oplus}{\underset{..}{C}}l{:}$$

Since halide ions do not form from the said compounds, they do not precipitate silver halides.

Problem 12.23. Classify critically each of the following reactions as S_N1 or S_N2. Write down the structure of the probable T.S. in each case :

(i) Methyl bromide + potassium cyanide in methyl alcohol.
(ii) Allyl bromide + sodium acetate in the aqueous medium.
(iii) Benzyl chloride + aqueous sodium hydroxide solution.
(iv) 1-chloro-2-butene + alcoholic potassium cyanide
(v) Neopentyl chloride + aqueous sodium hydroxide.

Solution : (i) $Me{-}Br + KCN \longrightarrow Me{-}CN + KBr.$

It is an S_N2 reaction on sp^3C atom of Me. Br^- is the leaving group and $:\bar{C}{\equiv}N:$ is the attacking nucleophile. This is because :

(a) owing to the presence of small H atoms on the substrate C atom, cyanide ion can attack it from the rear with respect to the Br atom.

(b) solvent is not highly polar.

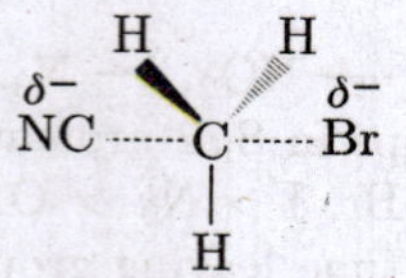

probable T.S.

(ii) The reaction partly follows S_N1 mechanism and partly S_N2 mechanism though the involved C atom is 1°.

This is because :

allyl bromide ionises in the aqueous medium slowly to form resonance stabilised allyl cation :

$$\overset{\gamma}{C}H_2{=}\overset{\beta}{C}H{\cdot}\overset{\alpha}{C}H_2Br \xrightarrow[\text{slow}]{H_2O} H{-}O{-}\overset{\delta+}{H}\cdots\overset{\delta-}{Br} + [CH_2{=}CH{-}\overset{\oplus}{C}H_2 \longleftrightarrow \overset{\oplus}{C}H_2{-}CH{=}CH_2]$$

The CH_3COO^- ion from sodium acetate may then attack the allyl cation either at the α-C atom or at the γ-C atom. The former one thus follows S_N1 pathway but the latter one follows S_N1' pathway.

T.S.s of S_N1 and S_N1' reactions are one and the same :

$$\overset{\delta-}{C}H_2\cdots\cdots CH\cdots\cdots\overset{\delta+}{C}H_2\cdots\cdots\overset{\delta-}{Br}$$

Allyl acetate may also be formed by the S_N2 and S_N2' pathways.

(iii) 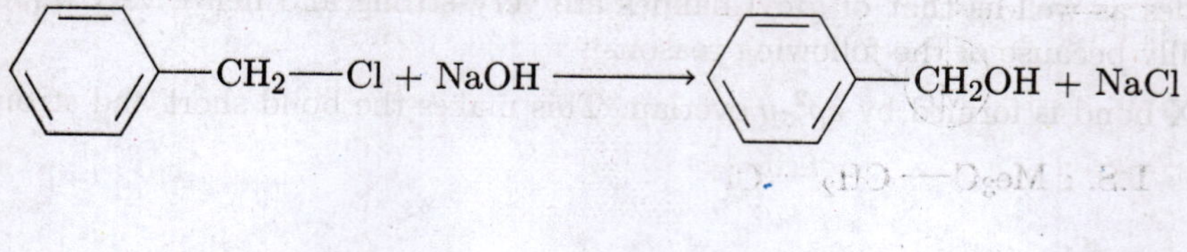

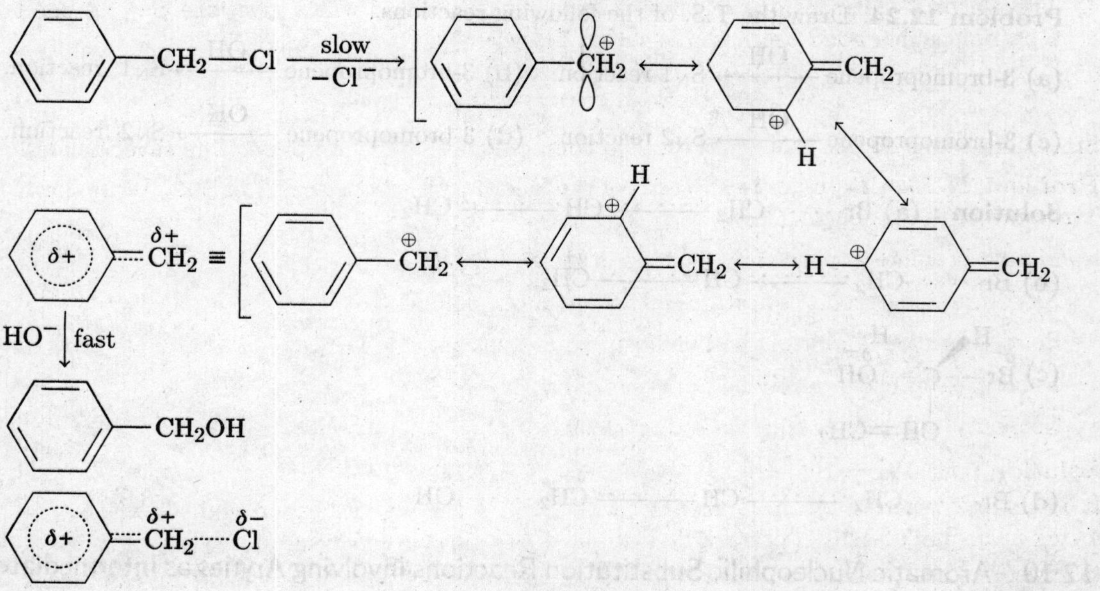

T.S.

This reaction also follows the S_N2 pathway, as shown below :

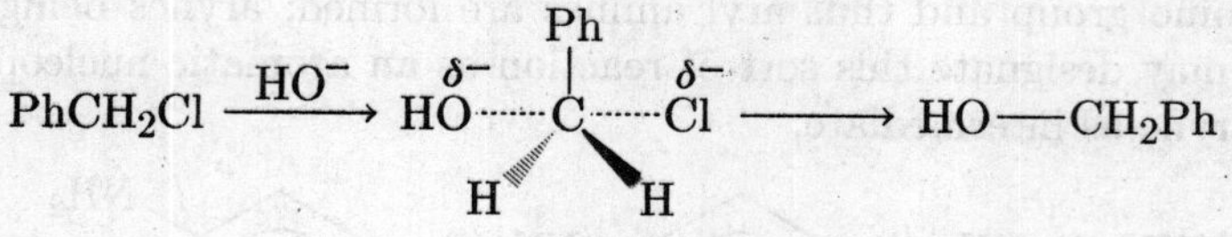

T.S.

(iv) $CH_3 \cdot CH{=}CH \cdot CH_2Cl + KCN \longrightarrow CH_3 \cdot CH{=}CH \cdot CH_2 \cdot CN + CH_3{-}CH(CN) \cdot CH{=}CH_2$

S_N1 product; S_N1' product

T.S. : $\overset{\delta+}{CH_3 \cdot CH} \cdots\cdots CH \cdots\cdots \overset{\delta+}{CH_2} \cdots\cdots \overset{\delta-}{Cl}$

(v) $Me_3C{-}CH_2{-}Cl + NaOH \longrightarrow Me_2C(OH){-}CH_2 \cdot CH_3$ (rearranged product)

Since there is a bulky tert-butyl group on the substrate C atom, back side attack by the HO^- ion is not possible.

This may be called S_N1 reaction with rearrangement.

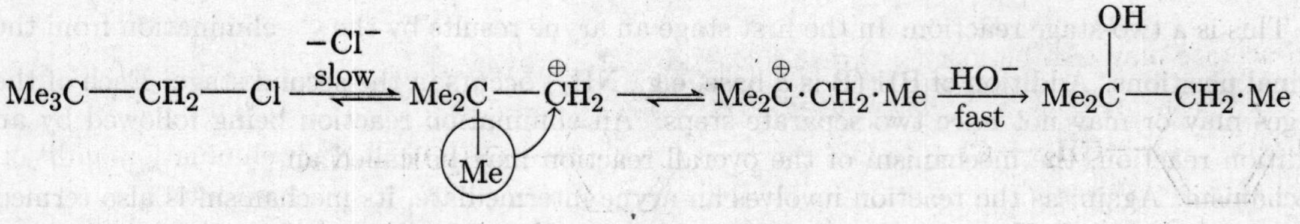

T.S. : $Me_3C{-}CH_2 \cdots\cdots Cl$

Problem 12.24. Draw the T.S. of the following reactions.

(a) 3-bromopropene $\xrightarrow{\bar{O}H}$ S_N1 reaction. (b) 3-bromopropene $\xrightarrow{\bar{O}H}$ S_N1' reaction.

(c) 3-bromopropene $\xrightarrow{\bar{O}H}$ S_N2 reaction. (d) 3-bromopropene $\xrightarrow{\bar{O}H}$ S_N2' reaction.

Solution : (a) $\overset{\delta-}{Br}\cdots\cdots\overset{\delta+}{CH_2}\overset{\cdots\cdots}{=\!=}CH\overset{\cdots\cdots}{=\!=}\overset{\delta+}{CH_2}$

(b) $\overset{\delta-}{Br}\cdots\cdots\overset{\delta+}{CH_2}\overset{\cdots\cdots}{=\!=}CH\overset{\cdots\cdots}{=\!=}\overset{\delta+}{CH_2}$

(c) 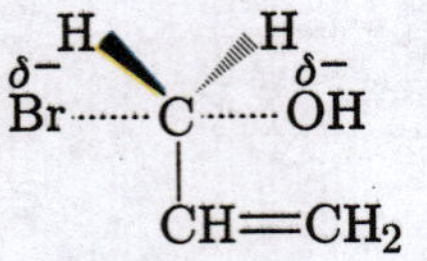

(d) $\overset{\delta-}{Br}\cdots\cdots\overset{\delta+}{CH_2}\overset{\cdots\cdots}{=\!=}CH\overset{\cdots\cdots}{=\!=}\overset{\delta+}{CH_2}\cdots\cdots\overset{\delta-}{OH}$

12.10 Aromatic Nucleophilic Substitution Reactions Involving Arynes as Intermediates

When aryl halides (and their alkyl and alkoxy derivatives) with at least one ortho-hydrogen are treated with a strong base like $NaNH_2$ or KNH_2 in liquid ammonia, the halogen group gets displaced by the amino group and thus aryl amines are formed; arynes being intermediates in these reactions, we may designate this sort of reaction as an aromatic nucleophilic substitution reaction involving aryne as intermediate.

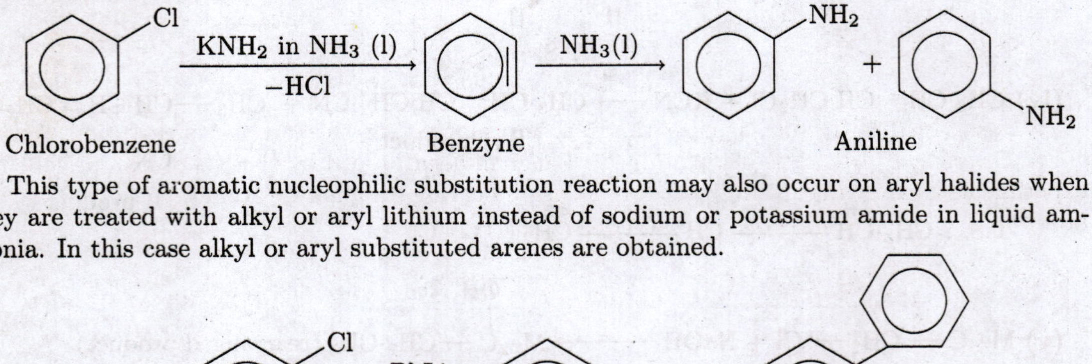

Chlorobenzene Benzyne Aniline

This type of aromatic nucleophilic substitution reaction may also occur on aryl halides when they are treated with alkyl or aryl lithium instead of sodium or potassium amide in liquid ammonia. In this case alkyl or aryl substituted arenes are obtained.

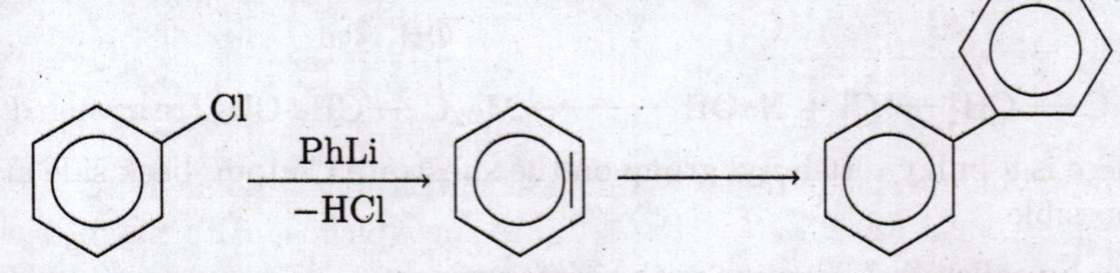

Chlorobenzene Benzyne Biphenyl

This is a two-stage reaction. In the first stage an aryne results by the cis-elimination from the vicinal positions. Addition of BH ($\ddot{\bar{B}}$ is a base, e,g., $\ddot{\bar{N}}H_2$) occurs in the second stage. Each of the stages may or may not have two separate steps. An elimination reaction being followed by an addition reaction, the mechanism of the overall reaction may be called an elimination-addition mechanism. Again, as the reaction involves an aryne intermediate, its mechanism is also termed benzyne mechanism. However, the mechanism may be schematically represented as :

1st stage : cis-Elimination of HX from the vicinal position and formation of an aryne.

1st step : Abstraction of proton from the aryl halide by the base from the ortho position with respect to the halogen atom.

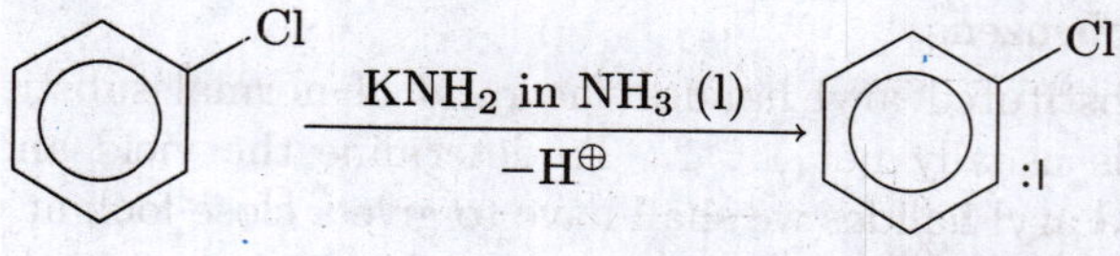

carbanion

2nd step : Elimination of the halide ion and formation of the aryne.

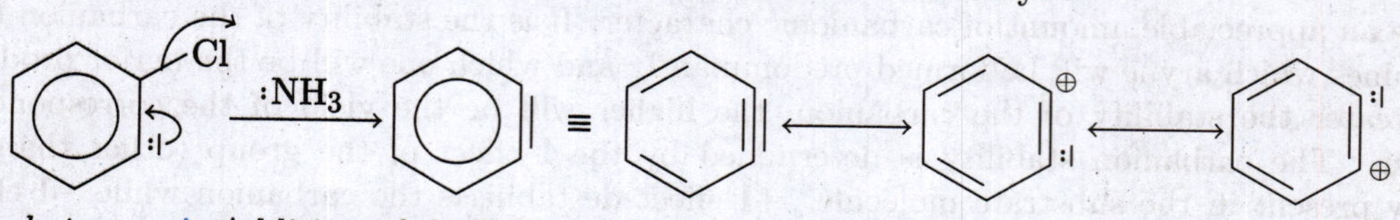

2nd stage : cis-Addition of the BH to the aryne intermediate.

3rd step : Nucleophilic attack at any one of the triple-bonded Cs by the base.

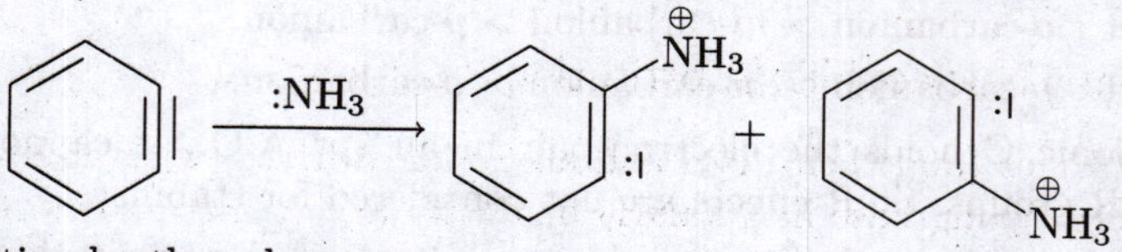

4th step : Protonation by the solvent.

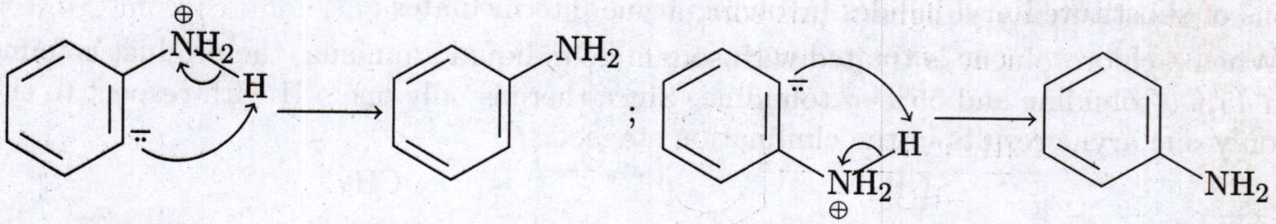

It is interesting to note that in this reaction along with the normal substitution product cine substitution product is also obtained. This has been proved by the use of isotope labelling experiments as shown below; the asterisked carbon bearing the chlorine atom is an isotopic carbon atom.

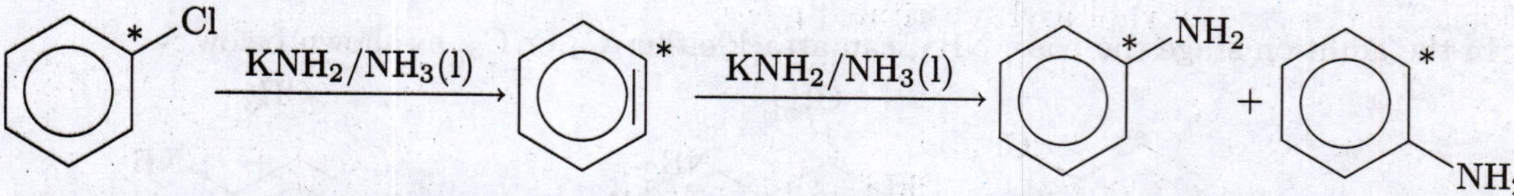

Since the aryne intermediate possesses two identical triple-bonded Cs, the nucleophilic attack by the base may occur at either C atom of the aryne which leads to the formation of 50% normal substitution product and 50% cine substitution product. Of course in the isotope labelling experiment cine substitution product will be slightly in excess over the normal substitution product because of the primary kinetic isotope effect. When a substitution occurs on an aromatic ring by displacing ortho hydrogen w.r.t. the leaving group the substitution is called cine substitution.

However, the deprotonation stage is found to be the slower step and hence this is the rate-determining step. An experiment to study primary kinetic isotope effect is a definite proof regarding this. Under the same set of conditions chlorobenzene reacts 10^6 times faster than the *o*-deuterochlorobenzene does. It has already been stated in Ch. 11 that a C—D bond being stronger than a C—H bond, the latter bond breaks much faster than the former. If the elimination step involving ortho-C—H (or C—D) bond breaking were not the rate-determining step, the rate of the reaction of *o*-deuterochlorobenzene would not have been 10^6 times slower than that of the chlorobenzene.

In the cases of substituted aryl halides the ratio of normal substitution product and cine substitution product is usually not 1 : 1. To determine the yield and the orientation in the reactions of substituted aryl halides we shall have to give a close look at the reaction mechanism.

When each of the elimination and addition stages occurs in two separate steps, a carbanion is an intermediate; even if each stage occurs in a concerted step, the transition state is supposed to assume an appreciable amount of carbanionic character. It is the stability of the carbanion that determines which aryne will be formed predominantly and which one will be the major product. The greater the stability of the carbanion, the higher will be the yield of the corresponding product. The carbanion stability is determined by the I effect of the group (other than X) already present in the substrate molecule. +I effect destabilises the carbanion while –I effect stabilises the same. Since I effects decrease with the increasing chain length, the stability order of the intermediate carbanions is :

(a) for –I effect : *o*-carbanion > *m*-carbanion > *p*-carbanion.

(b) for +I effect : *p*-carbanion > *m*-carbanion > *o*-carbanion.

Since the anionic C holds the electron-pair in an sp^2 A.O., it cannot undergo resonance interaction with R groups. So R effects are not considered for stability.

Let us now discuss some specific examples to understand the orientation and the yield in the reactions of substituted aryl halides involving aryne intermediates :

1. When *o*-chlorotoluene is treated with sodamide in liquid ammonia, the product is found to contain 45% *o*-toluidine and 55% *m*-toluidine. Since there is only one *o*-H with respect to the Cl atom, only one aryne results in the elimination stage.

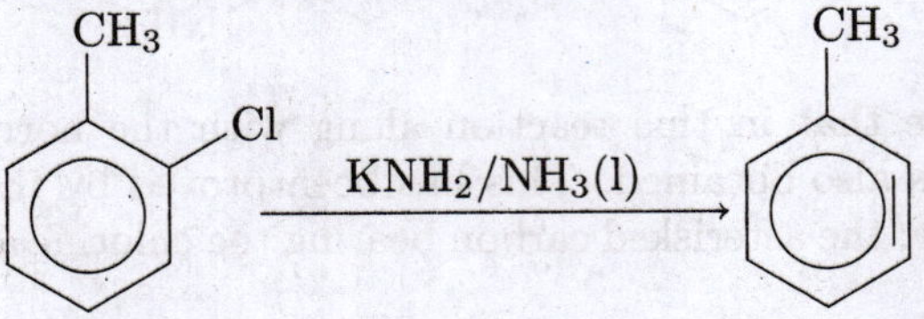

In the addition stage the base, $\ddot{N}H_3$, can attack either C_2 or C_3, as shown below :

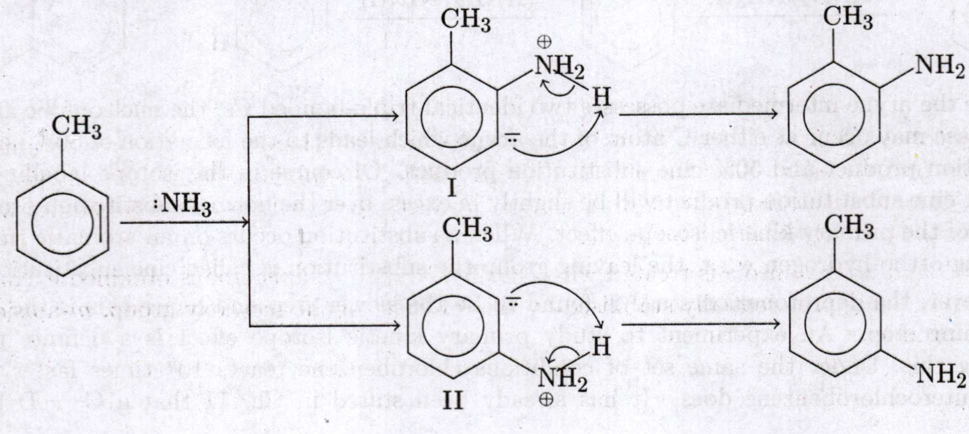

+I effect of the methyl group destabilises carbanion II more than carbanion I. So it is expected that the yield of *o*-toluidine will be more than that of *m*-toluidine. Perhaps steric hindrance at *o*-position decreases the yield.

2. When *m*-chlorotoluene is treated with $KNH_2/NH_3(l)$, *m*-toluidine is formed in large excess (56%) over *o*-and *p*-toluidines which are formed in equal amounts (22% each). Here and in all *m*-halocompounds two arynes may form since there are two *o*-Hs with respect to the halo atom.

Since +I effect destabilises *o*-carbanion more than *p*-carbanion, it is expected that aryne IV will be the major intermediate and III will be the minor one. As discussed in the case of example one, aryne III will give *m*-toluidine in excess over *o*-toluidine. Aryne IV also can form two products, *m*-toluidine and *p*-toluidine; the former is formed via carbanion IVe and the latter via IVd.

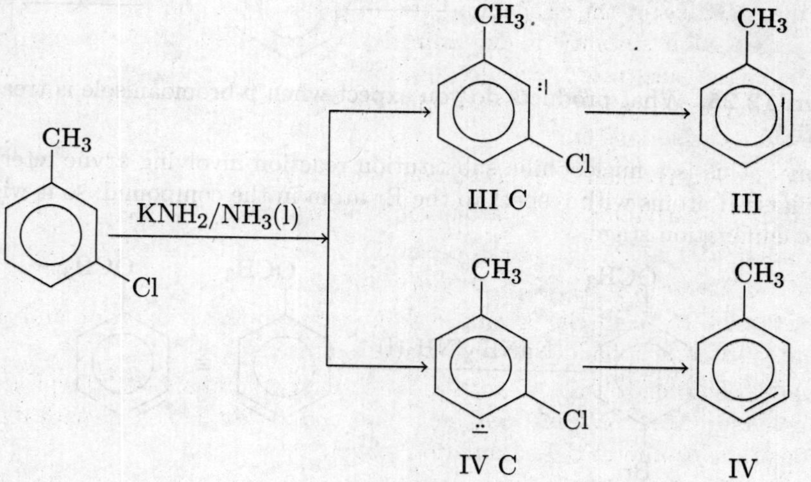

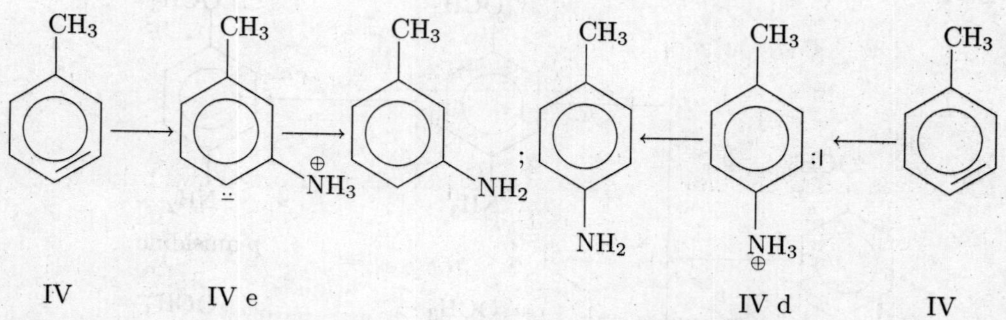

+I effect of the methyl group being less active at *p*-position than at *m*-position, carbanion IVe is formed preferentially. This leads to the higher yield of *m*-toluidine than that of *p*-toluidine. Thus *m*-chlorotoluene gives *m*-toluidine (major product) along with *o*- and *p*-toluidines when it is treated with KNH_2 in liquid ammonia.

3. When *o*-bromoanisole is treated with $NaNH_2/NH_3(l)$, *m*-anisidine is obtained exclusively. *o*-carbanion being preferentially stabilised by the −I effect of the methoxy group, *m*-anisidine is the sole product.

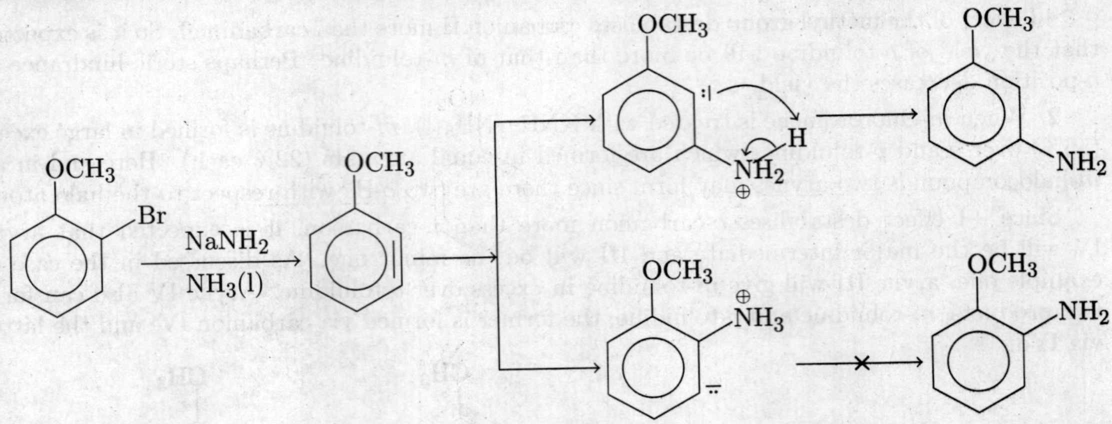

Problem 12.25. What products do you expect when *p*-bromoanisole is treated with KNH_2 in liquid NH_3?

Solution : This is a nucleophilic substitution reaction involving aryne intermediate. There are two similar *o*-H atoms with respect to the Br atom in the compound; so it will form only one aryne in the elimination stage,

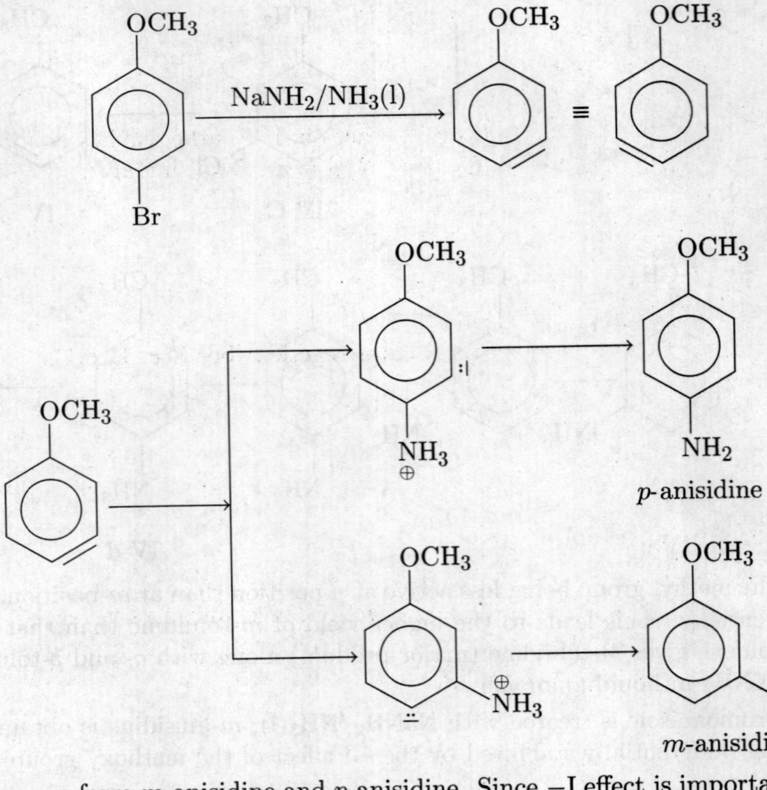

This aryne can form *m*-anisidine and *p*-anisidine. Since –I effect is important up to C_2 atom, *m*- and *p*-derivatives will be formed almost in equal amounts. *p*-Anisidine may be obtained slightly in excess over the *m*-isomer.

Problem 12.26. Write down the main product(s) and the mechanism for the following reactions :

(a) 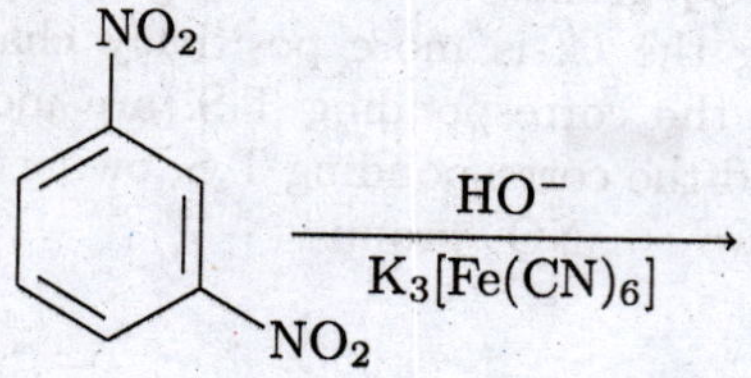

(b) 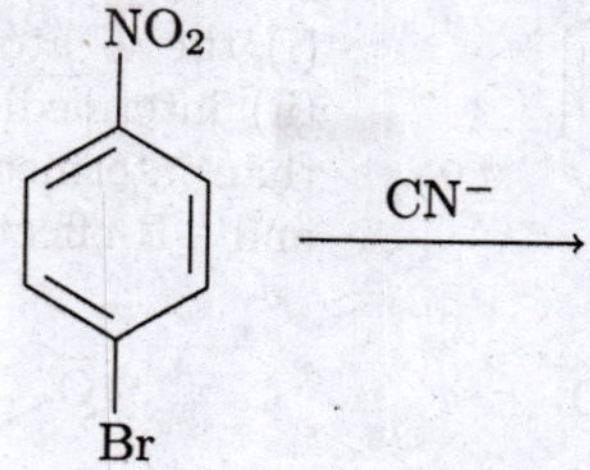

(c) 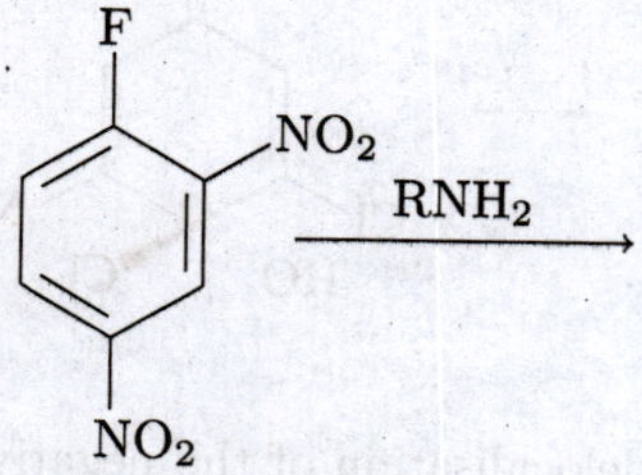

(d) 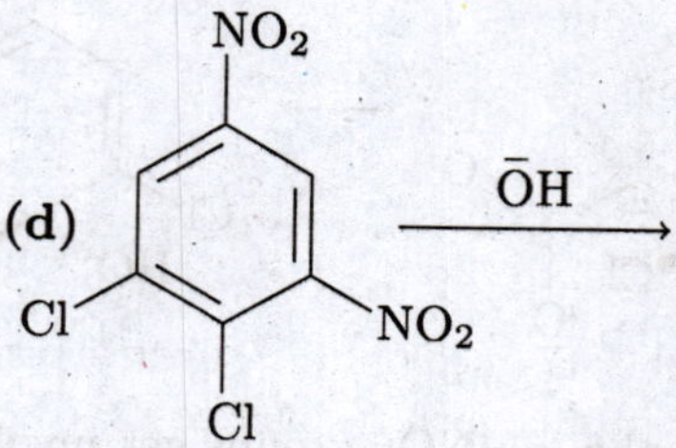

Solution :

(a) 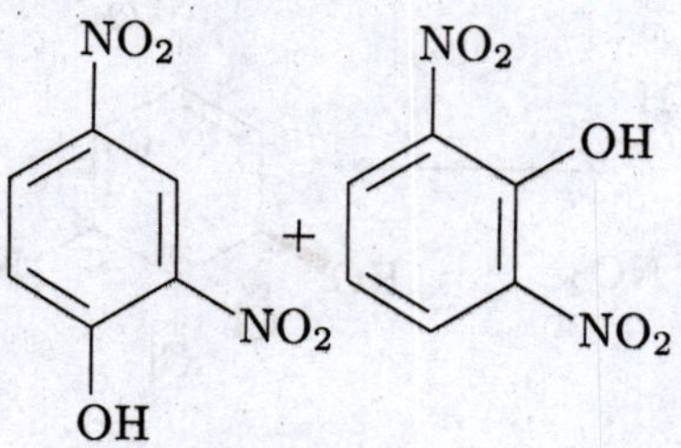

Ar.S_N2 reaction; H^- is the leaving group and $K_3[Fe(CN)_6]$ is the oxidising agent

(b) 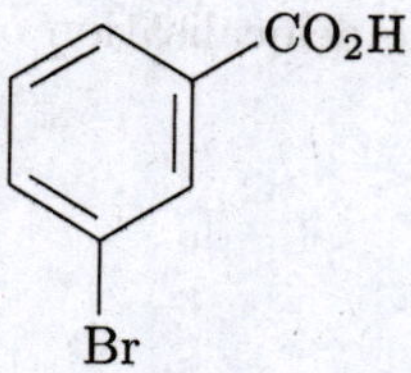

Von Richter reaction (See Sec. 11.8.1)

(c) 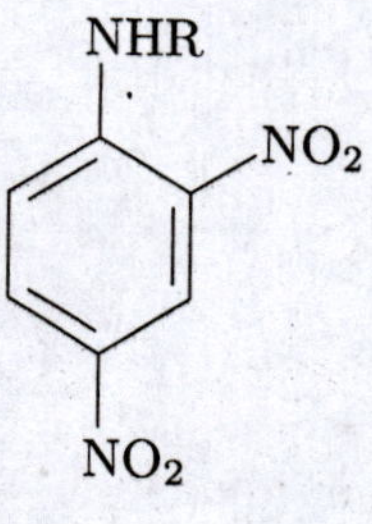

Ar.S_N2 reaction; mechanism as usual, F^- is the leaving group.

(d) 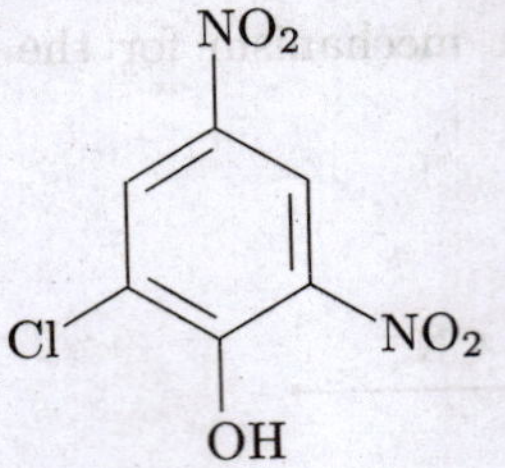

Ar.S_N2 reaction; mechanism as given in the text. The Cl atom *o*- and *p*- to the —NO_2 groups is the leaving group because (i) the C atom bearing the Cl is more positively charged and (ii) intermediate I and the corresponding T.S. are more stable than intermediate II and the corresponding T.S. owing to the –I and –R-effects of the two —NO_2 groups.

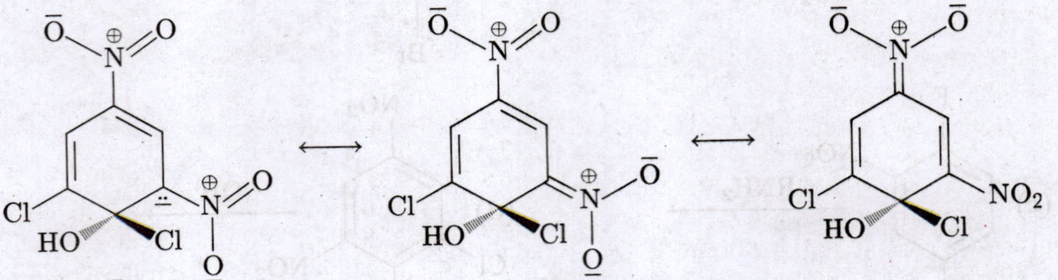

I

(I) Here both the —NO_2 groups get involved in the delocalisation of the negative charge. *o*-NO_2 group is somewhat out of plane.

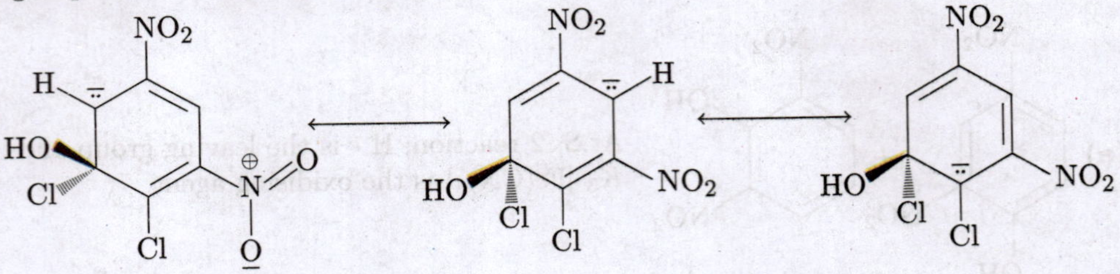

II

(II) In this case there is no involvement of —NO_2 group in the delocalisation of negative charge.

Exercise

1. Which one will react more rapidly by an S_N2 pathway?

(a) CH_3CH_2Br or $CH_3CH(Br)CH_3$

(b) CH_3CH_2Cl or CH_3CH_2I

(c) $CH_3CH(CH_3)CH_2CH_2Br$ or $CH_3CH_2CH(CH_3)CH_2Br$

(d) $CH_2{=}CHBr$ or CH_3CH_2Br

2. Which S_N1 reaction of the following pairs will take place more rapidly?

(a) $Me_3C{-}I \xrightarrow{H_2O} Me_3COH + HI$ or $Me_3C{-}Cl \xrightarrow{H_2O} Me_3COH + HCl$

(b) $Me_3CBr \xrightarrow{C_2H_5OH} Me_3COCH_2CH_3 + HBr$

or

$Me_3C{-}Br \xrightarrow{H_2O} Me_3C{-}OH + HBr$

(c) $Me_2C{=}CH{-}Cl \xrightarrow{H_2O}$ or $Me{-}CH{=}CH{-}CH_2Cl \xrightarrow{H_2O}$

(d) $Me_2CH{-}Br \xrightarrow{MeOH} Me_2CH{-}OMe + HBr$

or

$CH_3CH_2CH_2Br \xrightarrow{MeOH} CH_3CH_2CH_2\ OMe + HBr$

3. Which S_N2 reaction of the each pair will occur more readily?

(a) $CH_3Br + \bar{S}H \longrightarrow CH_3SH + B\bar{r}$ or $CH_3Br + H_2S \longrightarrow CH_3SH + HBr$

(b) $CH_3CH_2Cl + B\bar{r} \longrightarrow CH_3CH_2Br + Cl^-$ or $CH_3CH_2Cl + I^- \longrightarrow CH_3CH_2I + Cl^-$

(c) 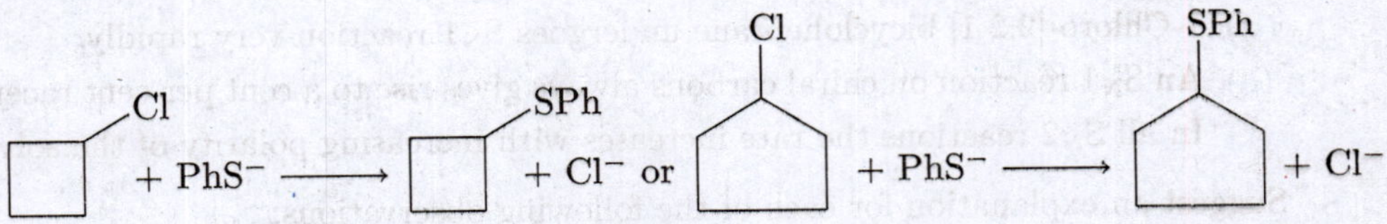

(d) $CH_3CH_2Br + I^-\ (1.0M) \longrightarrow CH_3CH_2I + Br^-$

or

$CH_3CH_2Br + I^-\ (0.01M) \longrightarrow CH_3CH_2I + Br^-$

(e) $CH_3I + Me_3N: \longrightarrow Me_4\overset{\oplus}{N}I^-$ or $CH_3I + Me_3P \longrightarrow Me_4\overset{\oplus}{P}I^-$

(f) $MeBr + \bar{O}H \longrightarrow MeOH + Br^-$ or $MeBr + MeO^- \longrightarrow Me{-}O{-}Me + Br^-$

(g) $CH_3CH_2CH_2Br + SCN^- \longrightarrow CH_3CH_2CH_2SCN + Br^-$

or

$CH_3CH_2CH_2Br + SCN^- \longrightarrow CH_3CH_2CH_2NCS + Br^-$

4. Which one will be the predominant product? Mention the pathway :

(a) $CH_3\underset{\displaystyle Br}{\underset{|}{C}}HCH_3 \xrightarrow[EtOH\ 55°C]{\bar{O}Et} CH_3{-}\underset{\displaystyle OEt}{\underset{|}{C}H}{-}CH_3$ or $CH_3{-}CH{=}CH_2$

(b) $CH_3CH_2Br \xrightarrow[EtOH\ 55°C]{\bar{O}Et} CH_3CH_2OCH_2CH_3$ or $CH_2{=}CH_2$

(c) $CH_3—C(CH_3)_2—Br \xrightarrow[25°C]{EtOH} CH_3—C(CH_3)_2—OEt$ or $(CH_3)_2C{=}CH_2$

(d) $CH_3—CH(Br)—CH_3 + CH_3COO^- \longrightarrow CH_3—CH(OCOCH_3)—CH_3$ or $CH_3—CH{=}CH_2$

(e) $CH_3—CH(Br)—CH_3 + CH_3CO_2H \longrightarrow CH_3—CH(OCOCH_3)—CH_3$ or $CH_3—CH{=}CH_2$

(f) $CH_3—CH(Br)—CH_3 + RS^- \longrightarrow CH_3—CH(SR)—CH_3$ or $CH_3—CH{=}CH_2$

5. Comment on the following statements :
   (a) All S_N2 reactions follow the second order kinetics.
   (b) The order of all S_N1 reactions is one.
   (c) cis- and trans-4-tert-Butylchlorocyclohexanes undergo the S_N2 reaction at the same rate.
   (d) 1-Chloro-[2.2.1] bicycloheptane undergoes S_N1 reaction very rapidly.
   (e) An S_N1 reaction on chiral carbons always gives rise to a cent per cent racemic product.
   (f) In all S_N2 reactions the rate increases with increasing polarity of the solvent.
6. Suggest an explanation for each of the following observations.

(i) $CH_3—C(CH_3)_2—CH_2Br \xrightarrow[EtOH,\ \Delta]{H_2O} CH_3—C(CH_3)(OH)—CH_2CH_3 + CH_3—C(CH_3)(OEt)—CH_2CH_3$

(ii) 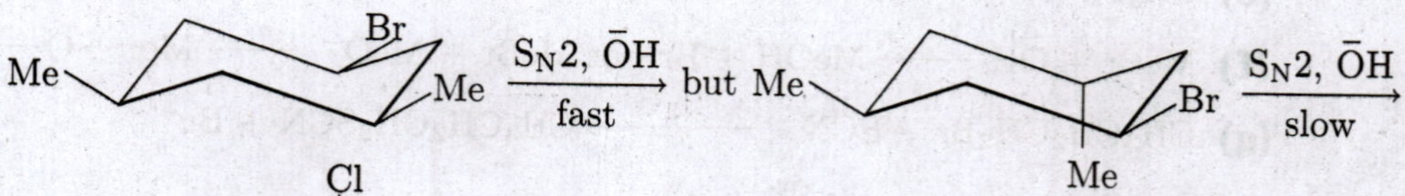

(iii) 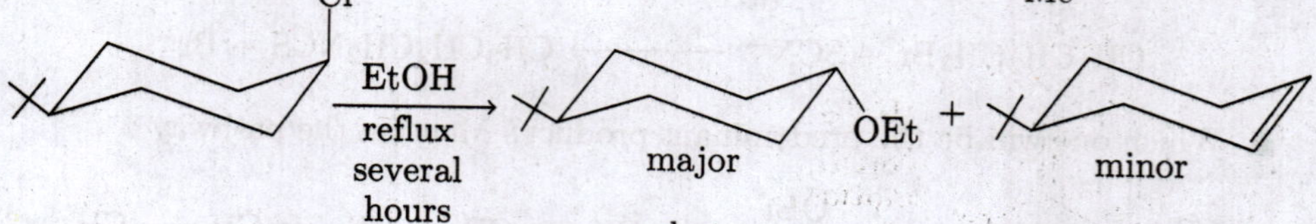

but

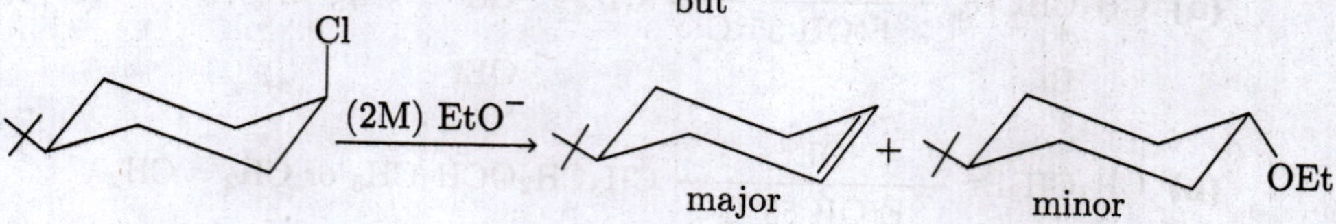

(iv) In a dilute alkaline medium the erythro form of 3-bromo-2-butanol gives the erythro form of 2,3-butanediol.

(v) The hydrolysis of methylbromide takes place much faster in the presence of I^- than in its absence.

7. Predict the configuration of the product(s).

(a) 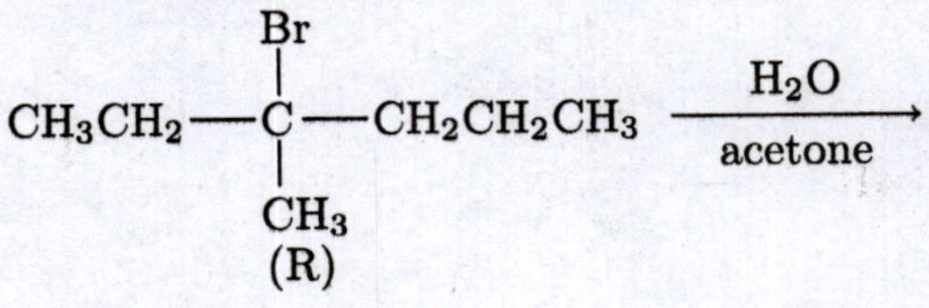

(b) 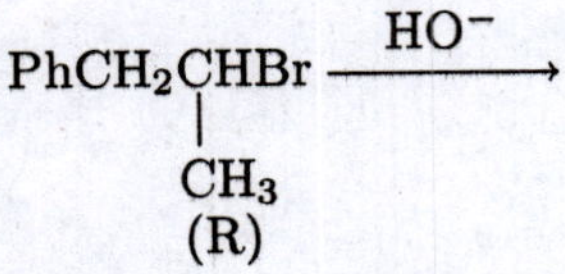

(c) 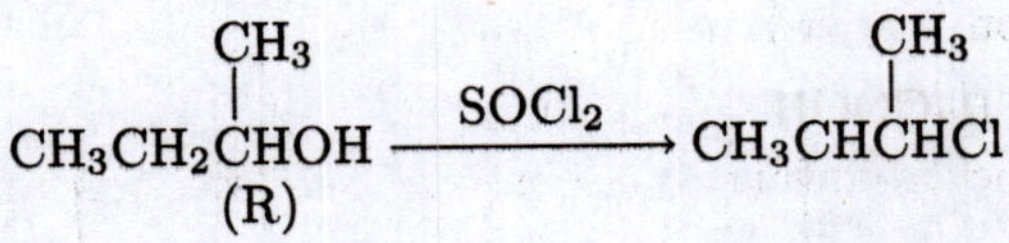

8. Predict the product (s) and give the probable mechanism for the following reactions :

(a) 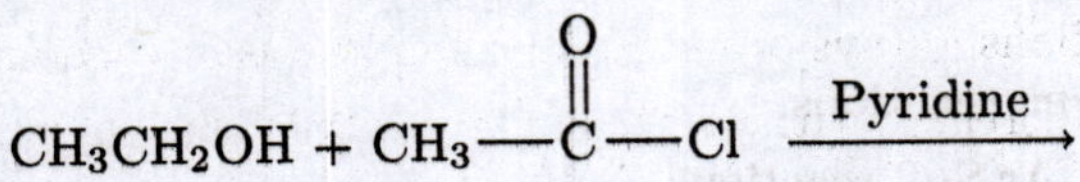

(b) 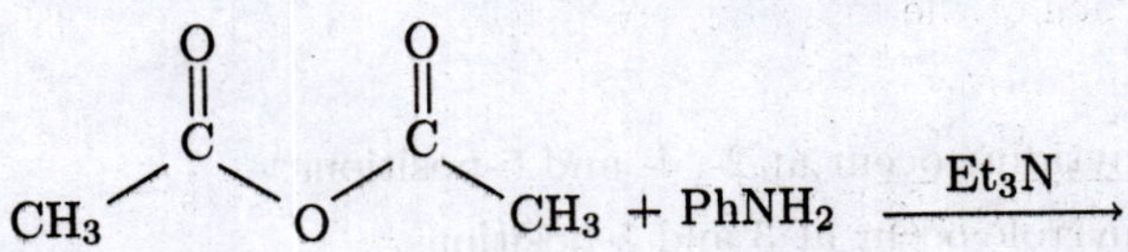

(c) 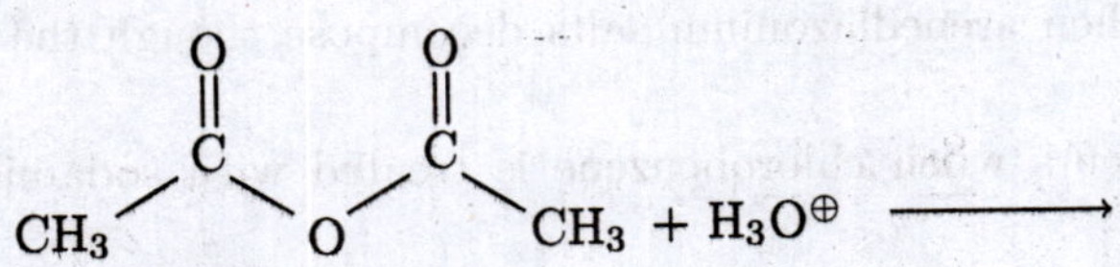

(d) 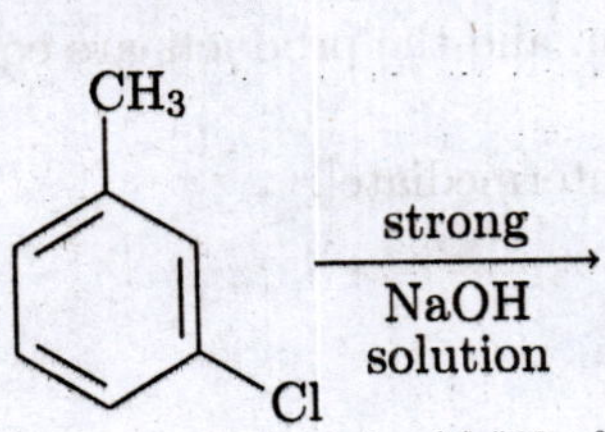

(e) 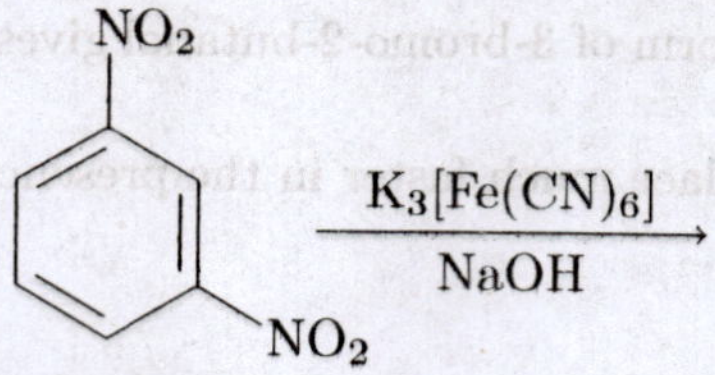

(f) 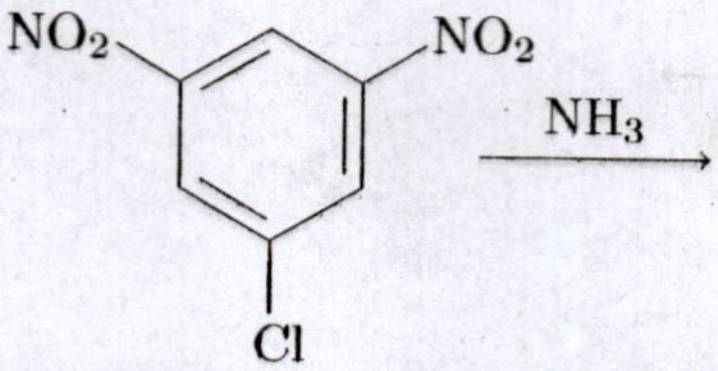

(g) $MeCH(Cl)-CH{=}CH_2 \xrightarrow{RS^-}$

(h) $Me_2C{=}CH-CH_2Cl \xrightarrow{CH_3COOH}$

9. Distinguish between :

   (a) Ar.S_N2 reactions and benzyne reactions.

   (b) S_N2 reactions on sp^2 C and Ar.S_N2 reactions.

10. Explain with reasons ;

    (a) Ar.S_N2 reactions on pyridine occur at 2-, 4- and 6-positions.

    (b) Ar.S_N2 reactions on pyrrole occur at 3 and 5-positions.

    (c) Phenyl cations form when arenediazonium salts decompose though the formers are unstable.

    (d) A cine substitution occurs when chlorobenzene is treated with sodamide in liquid ammonia.

    (e) *p*-chlorotoluene is boiled with strong NaOH soln. and the products are equal amounts of *p*-cresol and *m*-cresol.

    [**Hint :** The reaction takes place via benzyne intermediate].

Chapter 13

Addition to C—C Multiple Bond

Objectives

- To know the types of addition reaction [13.1] p. 615
- To discuss :
  - Electrophilic addition to C=C in general (Ad_E2 [13.2] p. 615
  - Electrophilic addition of halogens [13.3] p. 621
  - Electrophilic addition of hydrogen halides [13.4] p. 625
  - Hydration of alkenes [13.5] p. 628
  - Hydration of alkynes [13.6] p. 631
  - Hydroxylation of alkenes [13.7] p. 633
  - Hydroxylation by OsO_4 [13.7.1] p. 633
  - Hydroxylation by permanganate [13.7.2] p. 635
  - Hydroxylation via epoxidation [13.7.3] p. 637
  - Ozonolysis [13.8] p. 641
  - Catalytic hydrogenation [13.9] p. 645
  - Hydroboration [13.10] p. 648
  - Oxymercuration-deoxymercuration [13.11] p. 654

* Radical addition reaction (Ad_H2) [13.12] p. 656
* Radical addition of halogens [13.12.1] p. 657
* Radical addition of hydrogen halides [13.12.2] p. 659
* Nucleophilic addition to C=C (Ad_N2) [13.13] p. 662
* Cyanoethylation [13.13.1] p. 663
* Michael condensation [13.13.2] p. 664
* Addition reaction to a conjugated diene [13.14] p. 674
* Cycloaddition reaction [13.15] p. 678
* Diels-Alder reaction [13.15.1] p. 684

13.1 Introduction

A C—C multiple bond is composed of a σ-bond and one or two π bonds. Being weakly held by the two atomic nuclei, a pair of πes are readily available for reaction. Usually addition is the characteristic reaction of a π bond. Ions and free radicals may add across a multiple bond; accordingly the additions are said to be (i) ionic or polar addition and (ii) radical addition. The former type of additions take place in a polar solvent and the latter type in a non-polar solvent. In the case of an ionic reaction an electrophile or a nucleophile (or a negative or positive pole of a dipole) may initiate the process of addition. When an electrophile (or a positive pole of a dipole) initiates the process, the reaction is termed as electrophilic addition bimolecular (Ad_E2); whereas a reaction initiated by a nucleophile or a negative pole of a dipole is said to be a nucleophilic addition reaction bimolecular (Ad_N2). An addition reaction to a C—C multiple bond may take place in two separate steps or in a single step. However, we shall discuss in this chapter electrophilic, nucleophilic and radical additions (Ad_H2) across C—C double and triple bonds. Addition reactions are usually bimolecular, In the mechanistic designations for addition reactions 'Ad' stands for addition; E, Nu and H stand for electrophile, nucleophile and radical respectively and 2 indicates bimolecularity.

13.2 Electrophilic addition to C=C (Ad_E2)

Two double-bonded C atoms and four other atoms bonded to the Cs of C=C moiety lie in a plane. The π orbital with its weakly held electron-pair remains perpendicular to the plane (Fig. 13.1); one lobe of the π orbital lies above the plane and the other below it. Thus the two double-bonded C atoms become sandwiched between negative charge clouds; these, in their turn, shield the C atoms from attack by nucleophiles and make them react readily with electrophiles.

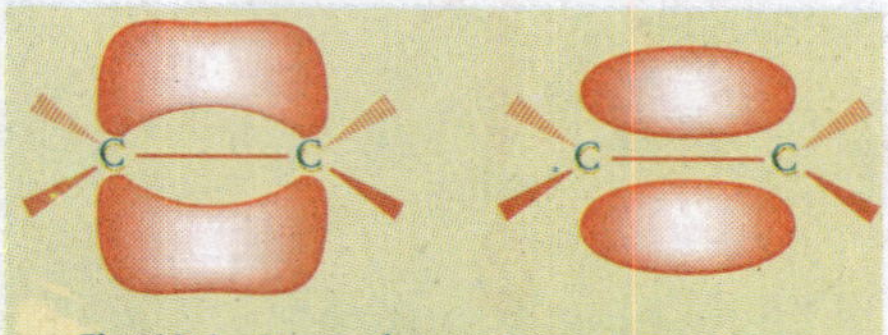

Fig. 13.1. Stereoelectronic structure of C=C

In fact, electrophilic addition to the C—C double bond is the most common type of reactions which this function undergoes. Electrophilic addition to a C=C usually occurs in two steps :

Step I : The incoming electrophile with its vacant orbital attacks the π cloud in a perpendicular direction from above or below the plane to form a π-complex (Sec. 11.4.7) which may or may not be an intermediate. The π-complex then may change to a carbenium ion through the formation of a σ bond between an atom of the electrophile and any one of the double bonded Cs as shown in Fig. 13.2.

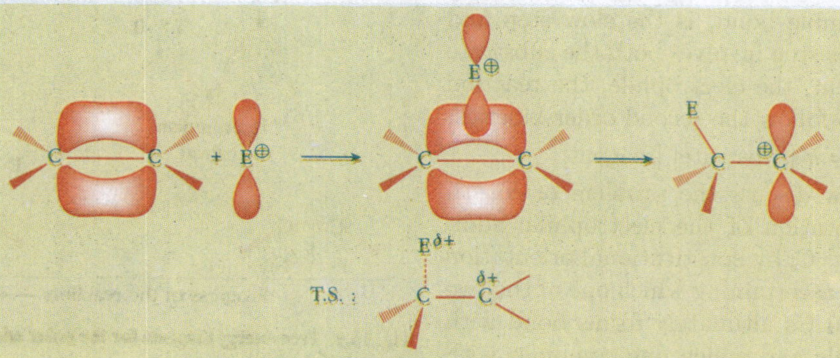

Fig. 13.2. Initial electrophilic attack at a C=C moiety

If the electrophile possesses a lone pair of non-bonded *e*s, then a three membered cyclic cation results in which the electrophile forms two σ bonds with two C atoms of the C$=$C and bears a positive charge. The cyclic cation may either be an intermediate (e.g., in the case of polar addition of halogens) or a true cyclic product (e.g., addition products of carbene, nitrene, etc.). However, the cyclic intermediate cation has, to some extent, a carbenium ion-like character.

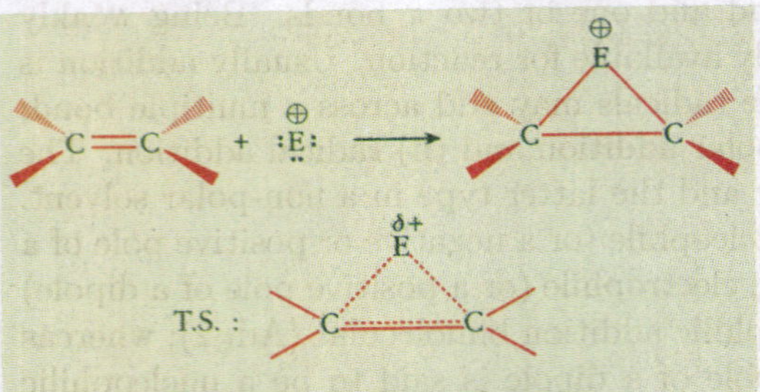

Fig. 13.3. Bridged cation formation

This step is the slow step and hence this is the rate-determining step.

Step II : In this step the intermediate, the carbenium ion or the three membered cyclic cation, combines with a nucleophile to produce the addition product. It is the fast step.

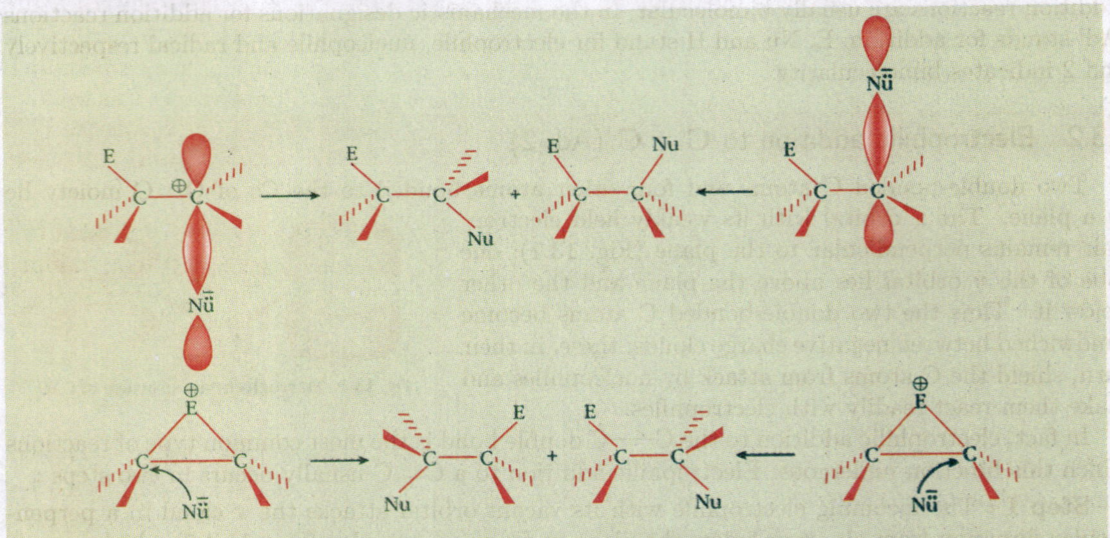
Fig. 13.4. Nucleophilic attack on the intermediate cyclic cation

The free energy diagram for the polar addition to a C$=$C may be represented by Fig. 13.5.

Since the step I, the electrophilic addition to C$=$C double bond, is the slow step and the T.S. of this step involves both the substrate and the reagent, the electrophile, the reaction is expected to follow the second order kinetics.

The rate $= k$ [substrate] [reagent]

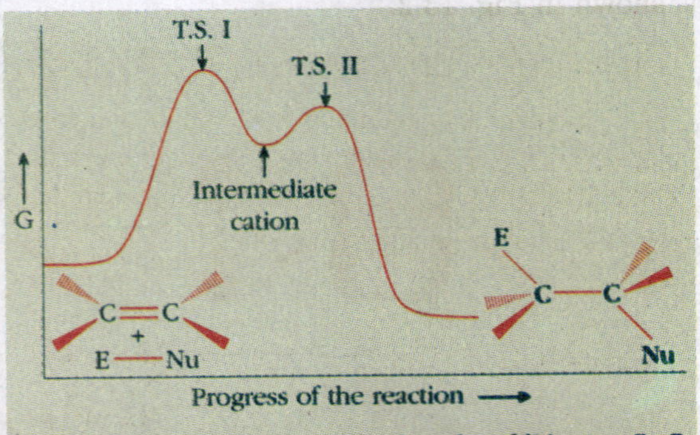

Fig. 13.5. Free energy diagram for the polar addition to a C$=$C

Let us now discuss the problem of constitutional orientation of the electrophilic addition to the C$=$C; by constitutional orientation here we mean ascertaining which one of the two double-bonded Cs ultimately forms bond with the electrophile and which one combines with the nucleophile. The problem of constitutional orientation does not arise in the cases of addition of symmetrical reagents (addendum) to both symmetrical and unsymmetrical substrates and also the addition of unsymmetrical reagents to the symmetrical substrates; but this is very important for the addition of an unsymmetrical reagent to an unsymmetrical C=C (that contains different groups on two double-bonded carbons; e.g., $RCH{=}CHR^1$). In most of the cases the problem may be solved by applying the Markownikoff's rule.

For electrophilic additions to unsymmetrical C—C multiple bonds the Markownikoff's rule states that the positive part of the unsymmetrical addendum attaches itself to that multiple-bonded C atom which has more H atoms This means that its negative part goes to that multiple-bonded C atom which bears less H atoms. According to this rule, and in fact, when HBr is added to propene in the absence of air, 2-bromopropane is the product and not 1-bromopropane. Here

$$CH_3.CH_2CH_2Br \xleftarrow{\times} \overset{3}{C}H_3{-}\overset{2}{C}H{=}\overset{1}{C}H_2 + HBr \longrightarrow CH_3.CHBr.CH_3$$

C_1 holds two H atoms and C_2 has one H atom; the positive part and the negative part of HBr are $H^{\oplus}$ and Br^- respectively. The addition has taken place according to the rule; $H^{\oplus}$ has attached itself to C_1 and Br^- to C_2. The most acceptable explanation for the Markownikoff's rule is : $E^{\oplus}$ adds to that C atom which gives the more stable carbenium ion. The stability of the carbenium ion is determined by considering +I effect, conjugative effect and hyperconjugative effect as usual. It has already been discussed in chapter 4 that the order of stability of carbenium ions is :

$$Ph\,\overset{\oplus}{C}H_2 \approx CH_2{=}CH{-}\overset{\oplus}{C}H_2 > (CH_3)_3\overset{\oplus}{C} > (CH_3)_2\overset{\oplus}{C}H > CH_3\overset{\oplus}{C}H_2 > \overset{\oplus}{C}H_3$$

Thus when $H^{\oplus}$ adds to C_2 of propene, it develops *n*-propyl cation (1°); whereas if it adds to C_1, it gives rise to isopropyl cation (2°). Being more stable, the isopropyl cation, a 2° carbenium ion, results by the attack of $H^{\oplus}$ on C_1 (see problem 4.20 and Sec. 11.6.3).

$$CH_3CH{=}CH_2 \xrightarrow{H^{\oplus}} \begin{cases} \xrightarrow{\times} CH_3CH_2\overset{\oplus}{C}H_2 \\ \longrightarrow CH_3\overset{\oplus}{C}HCH_3 \end{cases}$$

However, the Markownikoff's rule does not hold good for all electrophilic addition reactions to multiple bonds. For example, the addition of HI to allyl chloride follows the rule and forms 1-chloro-2-iodo-propane; but the addition of HOCl to the same substrate gives 2, 3-dichloropropan-1-ol instead of 1,3-dichloropropan-2-ol —this is contrary to the Markownikoff's rule.

$$CH_2{=}CH{-}CH_2Cl \xrightarrow{HI} CH_3.CHI.CH_2Cl \text{ (Markownikoffs rule)}$$

$$CH_2{=}CH{-}CH_2Cl \xrightarrow{HOCl} CH_2.OH.CHCl.CH_2Cl \text{ (against Markownikoff's rule)}$$

No plausible explanation has yet been given for such anomalous results. However, the formation of CH_3CHICH_2Cl may be explained by the hyperconjugative effect while the formation of $CH_2OHCHClCH_2Cl$ may be explained by the electromeric effect which is governed by the −I effect of Cl atom as shown below :

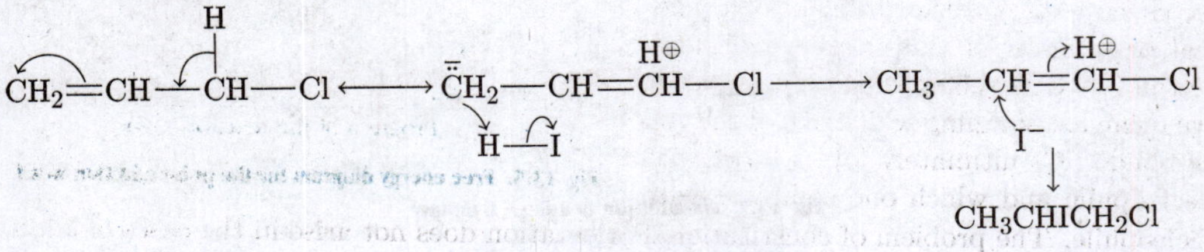

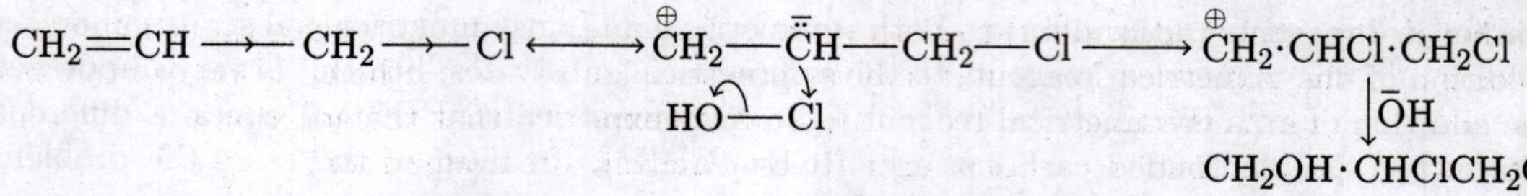

Note : The addition of **HBr** in the presence of organic peroxides or oxygen to unsymmetrical multiple bonds also occurs contrary to the Markownikoff's rule. This reaction is not an electrophilic addition but it is an example of radical addition. This will be discussed later on.

Now to understand the stereochemical orientation, i.e., the stereochemistry of the reaction, we require to discuss two points :

(a) Whether both the electrophile and the nucleophile will join themselves to the two double-bonded Cs from the same side of the double bond or from the opposite sides.

(b) The geometrical orientation of the two parts of the addendum, $E^{\oplus}$ and $N\bar{u}$, to each other and the rest of the product molecule, i.e., the configuration of the addition product.

If the electrophile forms a cyclic cation as an intermediate in an electrophilic addition, the nucleophile has no other alternative but to attack the intermediate from the opposite side and naturally the addition is trans or anti.

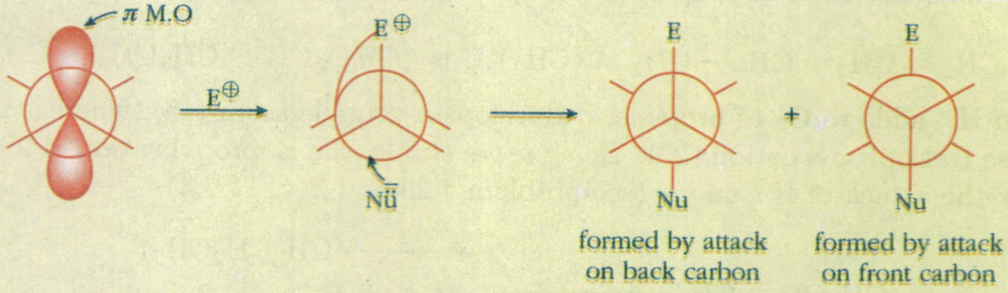

Fig. 13.6. ***trans*-addition to a C ═ C**

On the other hand, if the intermediate is a classical carbenium ion, it is difficult to predict the stereochemistry of the addition. When the carbenium ion has a relatively long life, this may undergo rotation about a $C^{\oplus}$—C σ bond and as a consequence the reaction will be non-stereospecific. Relatively short-lived carbenium ions undergo stereospecific reactions. For example, when the reagent is a dipole, after the addition of the electrophilic part the nucleophile may form an intimate ion-pair with the carbenium ion and in that case the addition will be cis or syn.

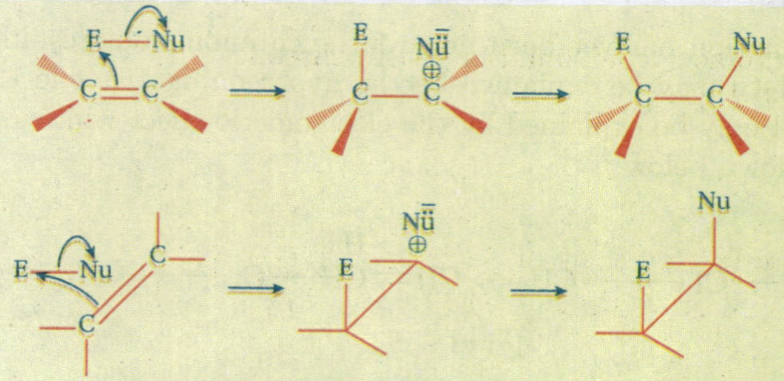

Fig. 13.7. ***cis*-addition to a C ═ C moiety**

Thus, if the configurations of the substrate and that of the product are known, the mode of the addition and the probable mechanism may be guessed. For example, the addition of bromine to trans-2-butene forms the meso product; it is therefore expected that the addition is trans which, in its turn, indicates that the reaction may involve a cyclic bromonium ion.

If the $E^{\oplus}$ and $\bar{Nu}$ enter from the same side of the double bond (or triple bond), the addition will be cis and stereospecific; whereas if they enter from the opposite sides, the addition will be stereospecific but trans (Chapter 9). Sometimes the addition is found to be non-stereospecific.

To ascertain the nature of addition, cis or trans, of a particular reagent (ENu), it is very often allowed to add to a substrate of the type XYC=CXY (where $E \neq X \neq Y$ & $Nu \neq X \neq Y$, but E may or may not be identical to Nu).

Since both the double-bonded Cs are equivalent and since each of them has 50% probability of being attacked by $E^{\oplus}$, the following results are obtained from cis and trans isomers.

(a) When E and Nu are different,

(i) cis-addition to a cis compound gives (±)-erythro form :

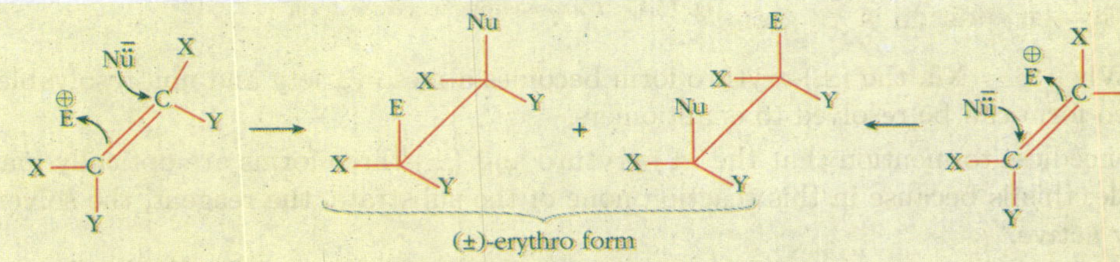

Fig. 13.8. ***cis*-addition to a *cis*-form**

(ii) cis-addition to a trans compound forms(±)-threo form :

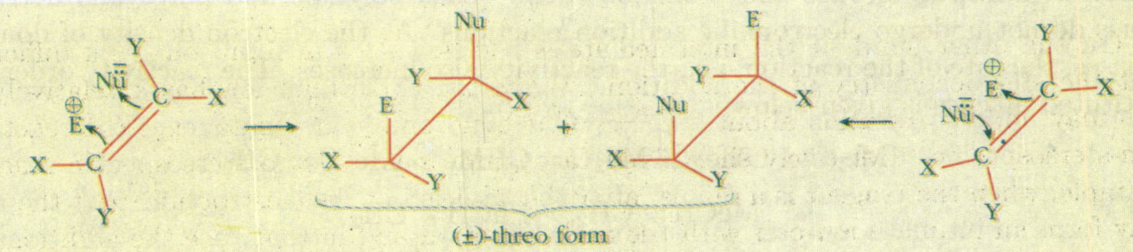

Fig. 13.9. ***cis*-addition to a *trans*-form**

(iii) trans-addition to a cis compound gives (±)-threo form :

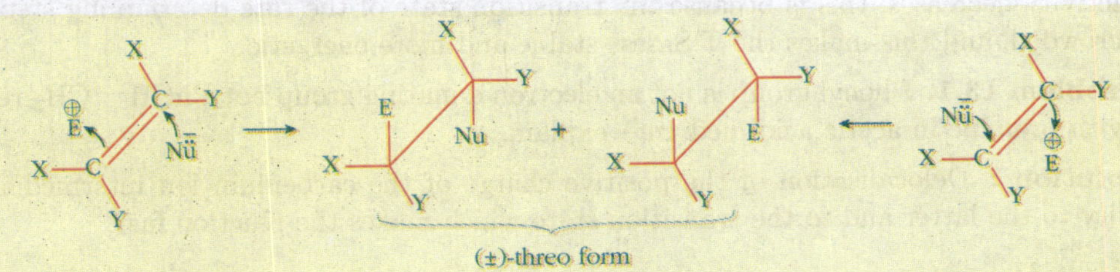

Fig. 13.10. ***trans*-addition to a *cis*-form**

(iv) trans-addition to a trans compound produces (±)-erythro form.

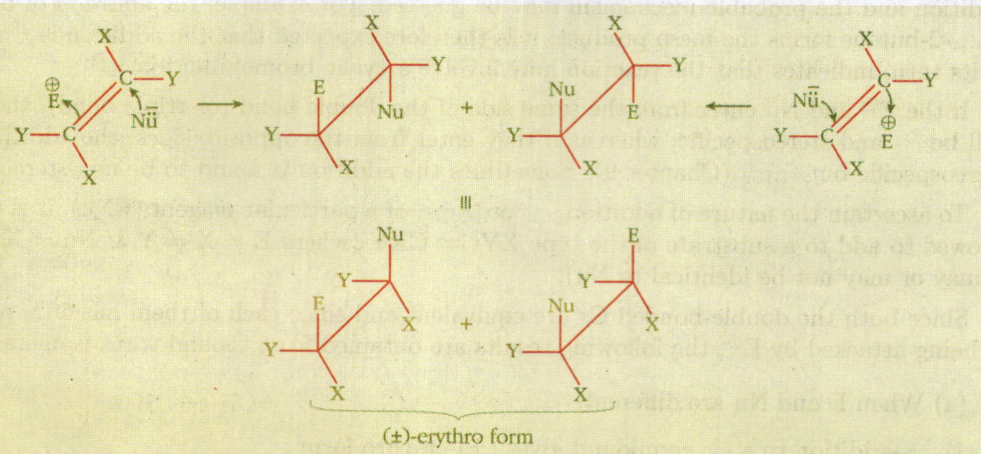

Fig. 13.11. *trans*-addition to a *trans*-form

(b) When E = Nü, the (±)-erythro form becomes a meso variety and non-resolvable; but the (±)-threo form can be resolved to enantiomers.

It is needless to mention that the (±)-erythro and (±)-threo forms are optically inactive but resolvable; this is because in this reaction none of the substrate, the reagent, the solvent, etc. is optically active.

The electrophilic attack on the double bond being the rate-determining step of an electrophilic addition reaction, the reactivity of the substrate increases with the increasing number of electron-donating substituents on the double-bonded Cs, while it decreases owing to the presence of the electron-withdrawing groups and atoms on them. Thus polyhalo and polycyano derivatives of alkenes do not undergo electrophilic addition reactions. As the electron density of double bond increases, the rate of the reaction, i.e., the reactivity also increases. The reactivity order for some substituted alkenes is given below :

$$Me_2C{=}CMe_2 > Me_2C{=}CHMe > Me_2C{=}CH_2 >$$
$$MeCH{=}CH_2 > PhCH{=}CH_2 >$$
$$CH_2{=}CH_2 > CH_2{=}CH{-}Br.$$

Here it is important to mention that with the increasing number of bulky electron-donating substituents on the double-bonded Cs, though the reactivity increases yet the rate of increase of reactivity decreases; this is because the transition state of the rate-determining step becomes over-crowded, and this makes the T.S. less stable and more energetic.

Problem 13.1. Phenyl group is not an electron donating group but $PhCH{=}CH_2$ reacts very fast with bromine in acetic acid medium—explain.

Solution : Delocalisation of the positive charge of the carbenium ion intermediate brings stability to the latter and to the transition state which makes the reaction fast.

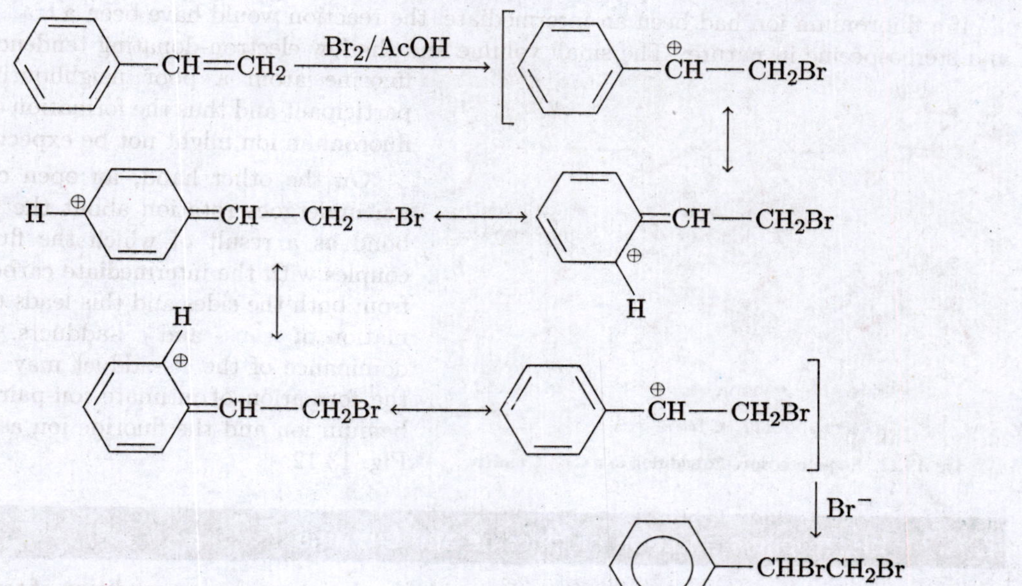

13.3 Electrophilic Addition of Halogens

Fluorine, chlorine, bromine and iodine undergo electrophilic addition reaction with C—C double bond in the absence of any catalyst. Fluorine and chlorine add to the C—C double bonds more rapidly than bromine does. Iodine adds to C—C double bonds reversibly and the equilibrium lies to the left, i.e., the reactant side. So it is usually considered that iodine does not enter into addition reactions of this type. However, the additions of individual halogens are discussed below :

13.3.1 Electrophilic addition of fluorine

The electrophilic addition reaction between fluorine and a C—C double bond is carried out at a very low temperature (−78°C) and xenonfluoride may be used as a reagent under controlled condition.

This is a non-stereospecific reaction in which cis-addition predominates. This suggests that an acyclic carbenium ion may be the intermediate and the reaction perhaps takes place in two steps :

Step I :

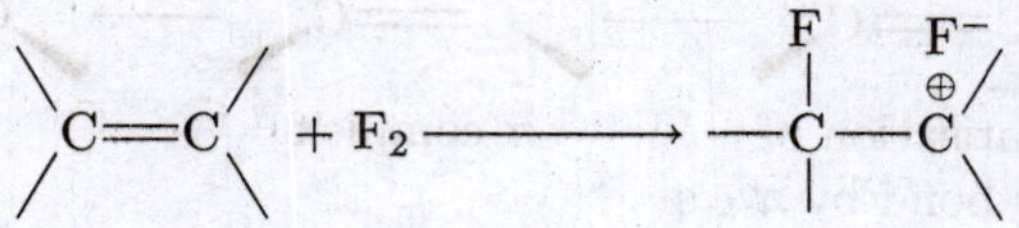

Step II :

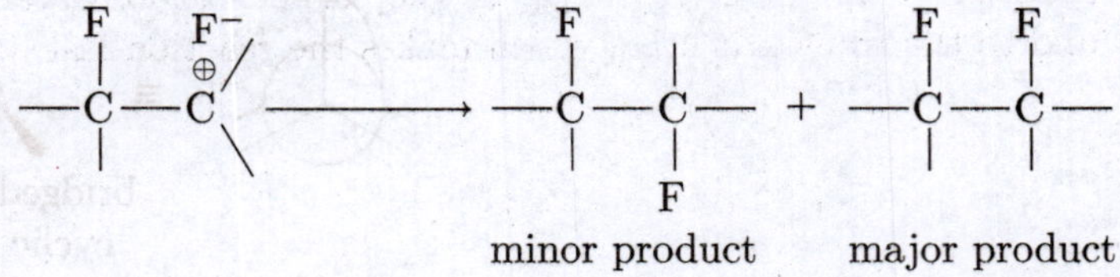

If a fluoronium ion had been an intermediate, the reaction would have been a trans-addition and stereospecific in nature. The small volume and the low electron-donating tendency make a fluorine atom a poor neighbouring group participant and thus the formation of bridged fluoronium ion might not be expected.

On the other hand, an open carbenium ion undergoes rotation about the C—C σ bond as a result of which the fluoride ion couples with the intermediate carbenium ion from both the sides and this leads to the formation of trans- and cis-adducts. The predominance of the cis-adduct may be due to the formation of intimate ion-pair, the carbenium ion and the fluoride ion as shown in Fig. 13.12.

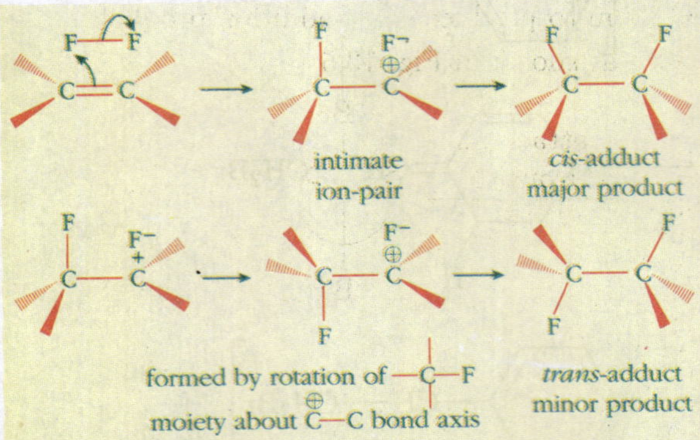

Fig. 13.12. Non-stereospecific addition to a C═C moiety

13.3.2 Electrophilic addition of chlorine and bromine

Chlorine and bromine add to the C—C double bond under ordinary condition of temperature and pressure in a solvent medium such as water, alcohol, acetic acid or carbon tetrachloride. This minimises the tendency toward substitution reaction. In carbon tetrachloride medium the reaction is slow and may be accelerated by using a catalyst like $AlCl_3$. On the other hand, in the hydroxylic solvents (H_2O, ROH, AcOH) the reaction is rapid and mixed products are obtained which consist of monohalo and dihalo derivatives.

The most familiar of the addition reactions is the addition reaction of bromine in CCl_4 or CH_3COOH to C—C multiple bonds. In this reaction decolourisation of bromine occurs without evolution of hydrogen bromide and thus constitutes a classical test for unsaturation. For this reason many scientists have studied the reaction extensively. On the basis of their findings a two-step mechanism involving a three-membered cyclic bromonium ion as an intermediate has been proposed which has been widely accepted.

Step I : Electrophilic attack by the reagent (bromine or chlorine) and the formation of cyclic bromonium (or chloronium) ion via a π-complex.

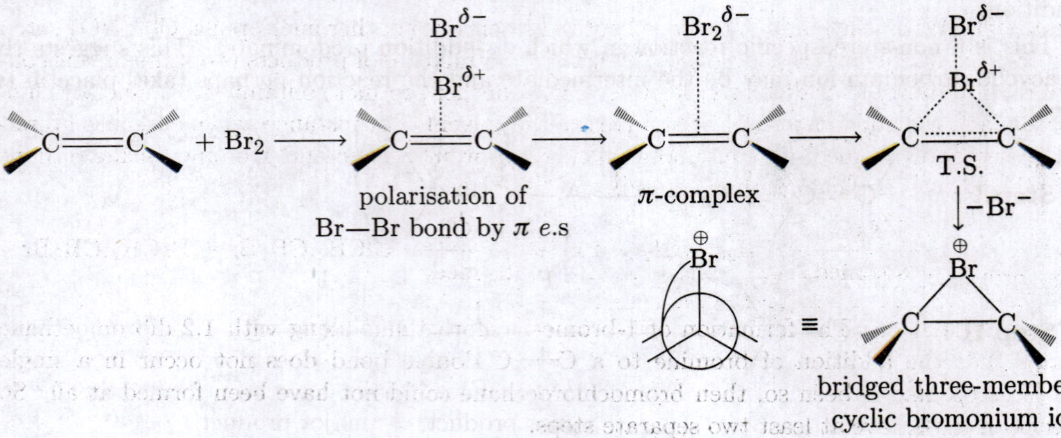

This is the slow step and hence the rate-determining step.

Step II : Nucleophilic attack by the Br^- (Cl^-) ion from the sterically unhindered position, i.e., from the opposite side of the ring on either C atom to produce trans-addition product.

The energy profile for the reaction may be drawn as shown in Fig. 13.5.

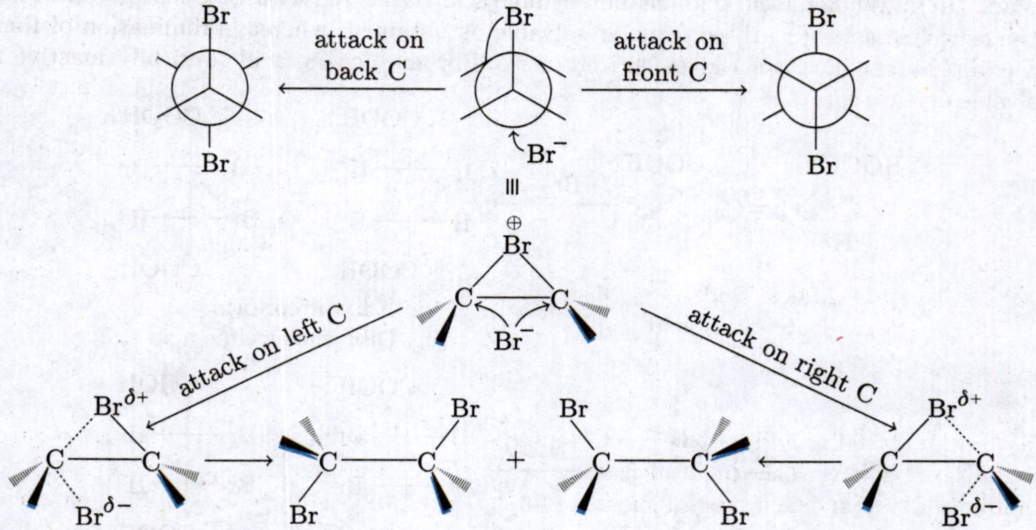

The reaction follows the second order kinetics.

The rate = k [alkene] [Br_2 or Cl_2]

The following experimental findings may be considered in support of the proposed mechanism.

Fact (i) : The rate of the reaction increases

(a) with the increasing number of electron-donating groups on multiple bonded Cs,

(b) with the increasing polarity of solvents,

(c) in the presence of catalysts, e.g., $AlCl_3$.

Conclusion : Since the above three polarising effects increase the rate of the reaction, it proceeds through a polar mechanism.

Fact (ii) : When the reaction is carried out in presence of another nucleophile, Cl^-, NO_3^-, etc., or in a hydroxylic solvent (H_2O, MeOH, AcOH, etc.), a mixture of products is obtained; this consists of dihalo (dibromo or dichloro) derivative and another product containing bromine (or chlorine) atom and the added nucleophile or a part of the solvent. For instance, when ethene is brominated in the aqueous medium containing NaCl 1,2-dibromoethane and 1-bromo-2-chloroethane are obtained; but 1,2-dichloroethane does not at all form.

$$\underset{\text{does not form}}{ClCH_2CH_2Cl} \xleftarrow[\text{aq. med.}]{\text{NaCl}} CH_2{=}CH_2 + Br_2 \xrightarrow[\text{aq. med.}]{\text{NaCl}} ClCH_2.CH_2Br + BrCH_2.CH_2Br$$

Conclusion : (a) The formation of 1-bromo-2-chloroethane along with 1,2-dibromoethane indicates that the addition of bromine to a C═C double bond does not occur in a single concerted step; had it been so, then bromochloroethane could not have been formed at all. So the reaction must have at least two separate steps.

(b) Since no 1,2-dichloroethane forms in the reaction, Cl^-, i.e., a nucleophile does not attack the C—C double bond at the first step. This means that an electrophile, $Br^{\oplus}$, must be the attacking reagent at the first step of the reaction; then the addition of a nucleophile, Br^- or Cl^-, occurs in the second step.

Fact (iii) : When maleic acid (cis-butenedioic acid) is treated with Br_2, cent percent racemic dibromosuccinic acid [(±)-threo form—resolvable] is obtained; whereas bromination of fumaric acid (trans-butenedioic acid) gives entirely meso dibromosuccinic acid (optically inactive non-resolvable erythro form).

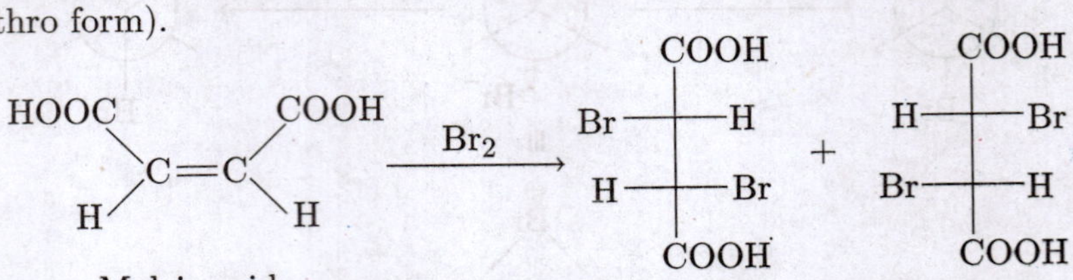

Maleic acid
cis-Butendioic acid

(±)-threo form
(±)-Dibromosuccinic acid

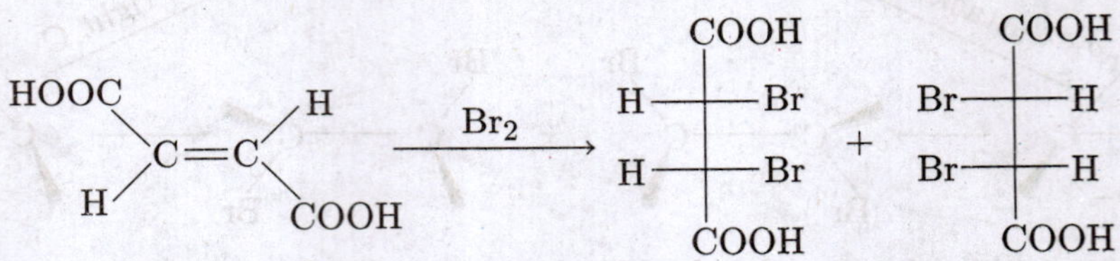

Fumaric acid
trans-Butendioic acid

Erythro form
meso Dibromosuccinic acid

Conclusion : The reaction is cent percent stereospecific.

Since trans-addition of a symmetrical reagent to a cis-isomer forms (±)-threo derivative and that to a trans-isomer produces (±)-erythro form (here meso isomer, since two double-bonded Cs are equivalent), the addition of Br_2 to a C—C double bond must be a cent percent **trans**-additon. Cent percent trans-addition rules out the possibility of the formation of an intermediate carbenium ion because both the isomers, cis and trans, would then form one and the same carbenium ion which, in its turn, would give rise to identical product distribution from both the isomers. Furthermore, to account for the cent percent streospecificity and the product distribution, a bridged bromonium ion may be considered; not that bromonium ion alone can explain the nature of the products, but it can explain the fact effectively well as shown in Fig. 13.13.

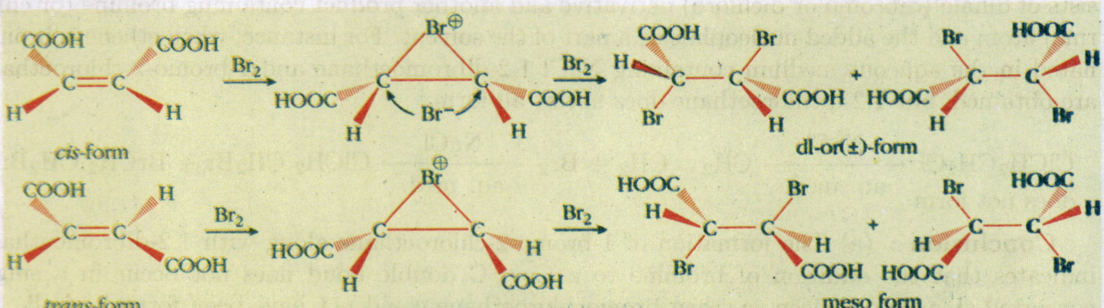

Fig. 13.13. Stereospecific addition of $Br_2(Cl_2)$

Recently one bridged bromonium ion has been detected by N.M.R. spectroscopy.

Fact (iv) : Addition of chlorine as well as of bromine to C—C double bonds in dilute solutions has been found to follow the second order kinetics—the first order, with respect to the substrate and the halogen.

Conclusion : the transition state of the rate-limiting step must involve both the substrate and the halogen. This is also consistent with the proposed mechanism.

13.3.3 Addition of iodine

It has already been stated that the addition reaction between an alkene and iodine is a reversible one and the equilibrium lies over the reactant side. It is supposed that the reaction occurs in two steps via bridged iodonium ion since iodine is a good neighbouring group participant. So far as stereochemistry of the reaction is concerned, no specific work has yet been done. The addition of iodine fluoride and iodine isocyanate proves that the additions are trans.

13.3.4 Addition of halogens to alkynes

The addition of bromine to an alkyne is found to be stereoselective—trans-product is the predominant one. This leads to assume that addition takes place via a bridged halonium ion in two steps.

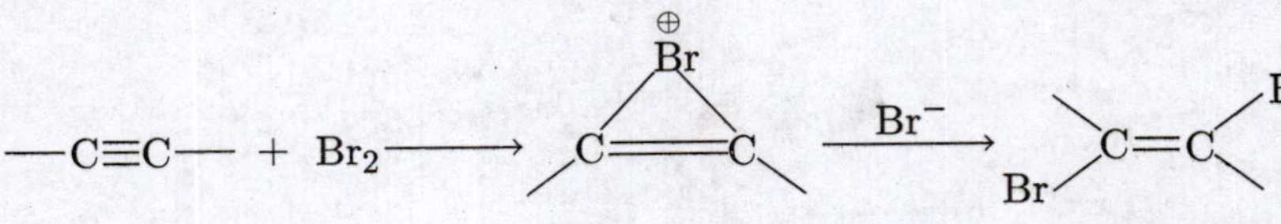

trans (70%)

13.4 Electrophilic Addition of Hydrogen Halides

Hydrogen halides add to alkenes by polar mechanism usually to form 2° and 3° alkyl halides.

$$CH_3{-}CH{=}CH_2 + HX \longrightarrow CH_3CHXCH_3$$

$$CH_3{-}\underset{\displaystyle CH_3}{\underset{|}{C}}{=}CH_2 + HX \longrightarrow (CH_3)_3CX$$

Unsaturated natural products and alkenes from petroleum are converted to alkyl halides by this reaction and hence the reaction has some preparative value. The reaction is usually carried out by passing dry hydrogen halide either into the alkene directly or into a solution of the alkene in a moderately polar solvent like acetic acid which dissolves the polar hydrogen halides. Aqueous solutions of hydrogen halides are not generally employed for the reaction because this may lead to hydration of alkene.

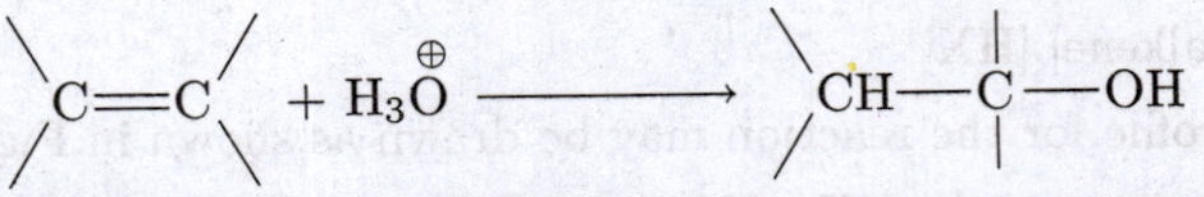

In the electrophilic addition reaction of HX to a multiple bond, $H^{\oplus}$ is the electrophile and X^- is the nucleophile. When dry HX is passed into the alkene or into the solution of the alkene in non-polar solvent, the proton is supplied by the HX; whereas if hydroxylic polar solvents are used, conjugate acid of the solvent, e.g., $H_3O^{\oplus}$ in the aqueous solution, supplies the proton.

The reaction takes place in two discrete steps :

Step I : Electrophilic attack by the H$^{\oplus}$ to the πes to form either a more stable carbenium ion or a π-complex (protonium ion complex).

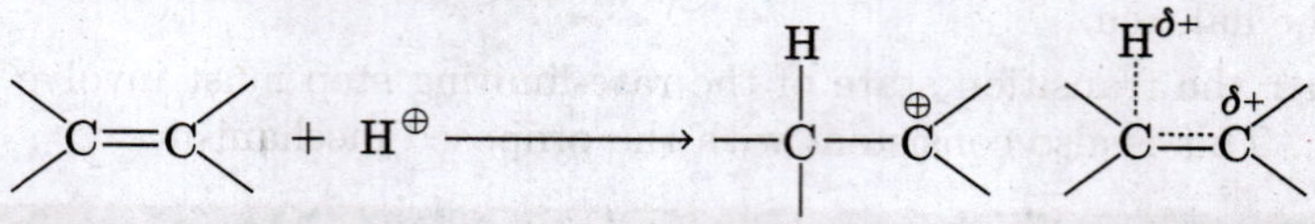

T.S.

Since there is no non-bonded electron-pair on a proton, the formation of bridged protonium ion is not possible. The electrophilic attack may or may not occur via a π-complex formation as shown in figure 13.14.

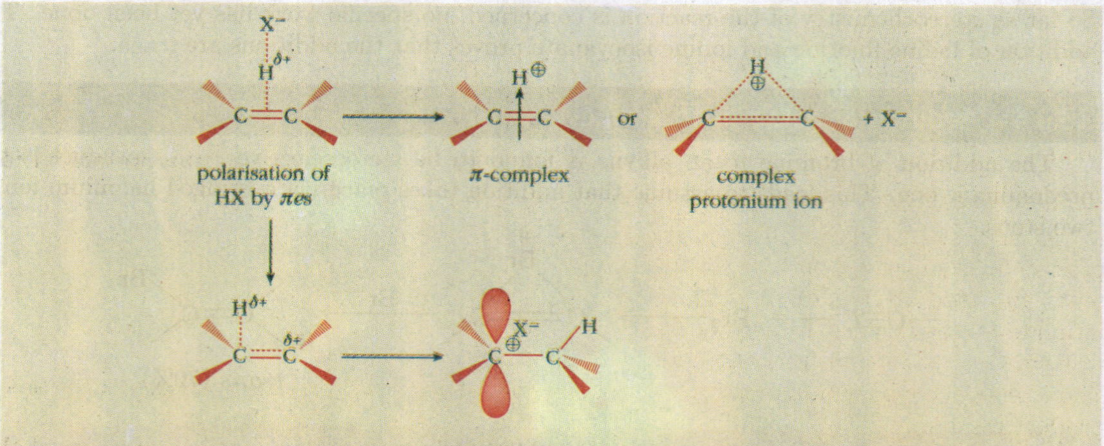

Fig. 13.14. Electrophilic addition of hydrogen halides

Between the two double-bonded C.s, that C atom will acquire the positive charge which will produce the more stable carbenium ion; the stability of the carbenium ion will be determined by the I effect, the conjugative effect and the hyperconjugative effect. The carbenium ion formation step is the slow step—the rate-determining step.

Step II : The rapid attack at the carbenium ion by the nucleophile X$^-$ to give the addition product.

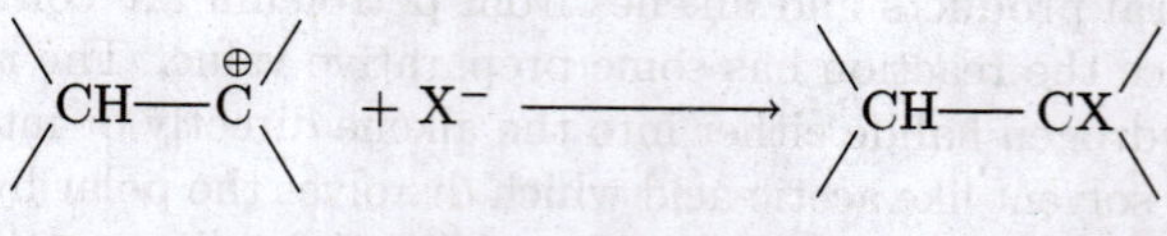

Since the first step, the rate-determining step, involves the substrate and the reagent, the reaction is expected to follow the second order kinetics.

The rate= k [alkene] [HX]

The energy profile for the reaction may be drawn as shown in Fig. 13.5.

The stereochemistry of the HX addition to a C=C double bond is not well established. The reaction is less stereoselective than the addition of halogens. There are instances where trans-addition occurs; at the same time predominant cis-products are also obtained in good many number of cases.

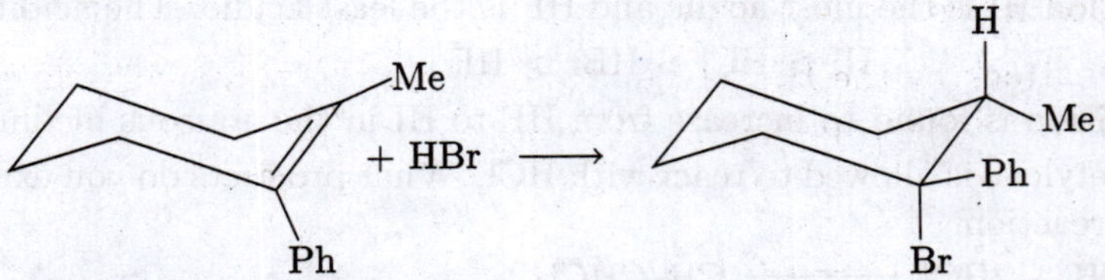

(*trans*-addition)

1-Methyl-2-phenylcyclohexene

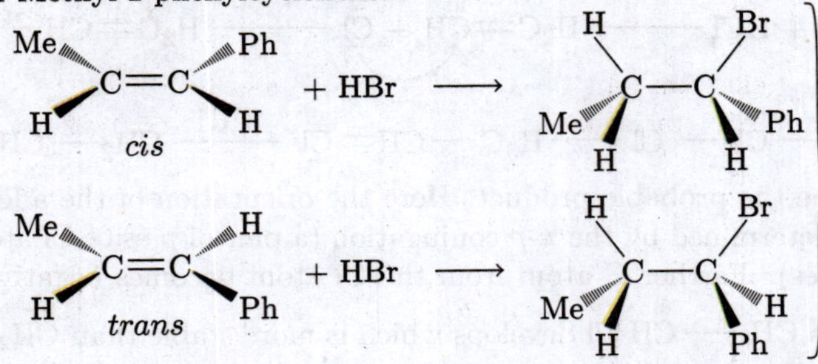

cis-products predominate

For the trans-addition, it is assumed that the nucleophile, X^-, attacks the protonium ion complex from behind at the more stable carbenium ion site.

To account for the predominant cis-product it is supposed that the carbenium ion and the halide ion form an intimate ion-pair which then collapses to form C—X σ bond from the same side of the double bond and thereby cis-product results.

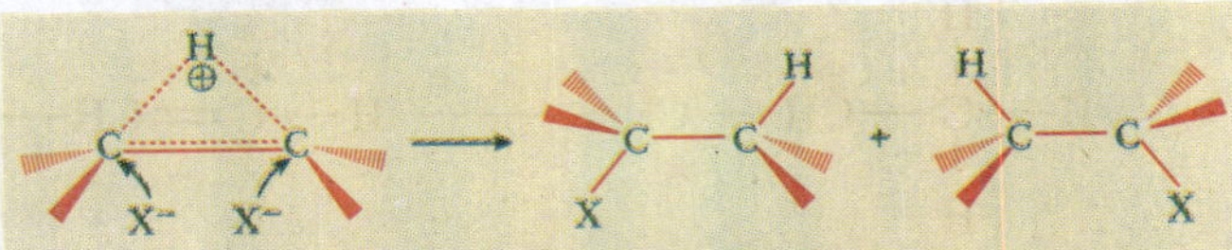

Fig. 13.15. Stereochemistry for the *trans*-addition of HX to a C═C moiety

It is further assumed that when the carbenium ion is considerably stable, a portion of the carbenium ion rotates about C—C⊕ σ bond and then forms the trans-product.

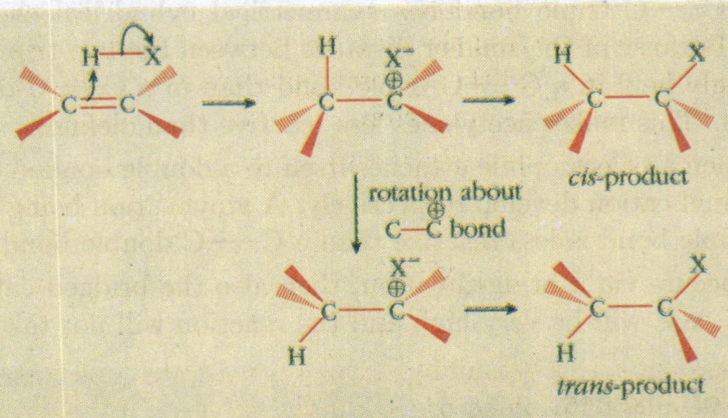

Fig. 13.16. Streochemistry for the predominant *cis*-addition of HX to a C═C moiety

The rate of the reaction depends on :

(a) the substrate structure and (b) the electrophilicity or acidity of HX.

The electron-donating groups on the double-bonded Cs increase the rate. The rate of the reaction decreases in the order :

$$Me_2C{=}CMe_2 > Me_2C{=}CHMe > MeCH{=}CHMe > MeCH{=}CH_2 > CH_2{=}CH_2$$

In the aqueous solution HI is the most acidic and HF is the least acidic. The acidity order is :

$$HI > HCl > HBr > HF$$

The rate of the addition is found to increase from HF to HI in the aqueous medium.

Problem 13.2. Acetylene is allowed to react with HCl. What products do you expect? Draw the mechanism for the reaction.

Solution : $HC{\equiv}CH + 2HCl \longleftrightarrow CH_3CHCl_2$

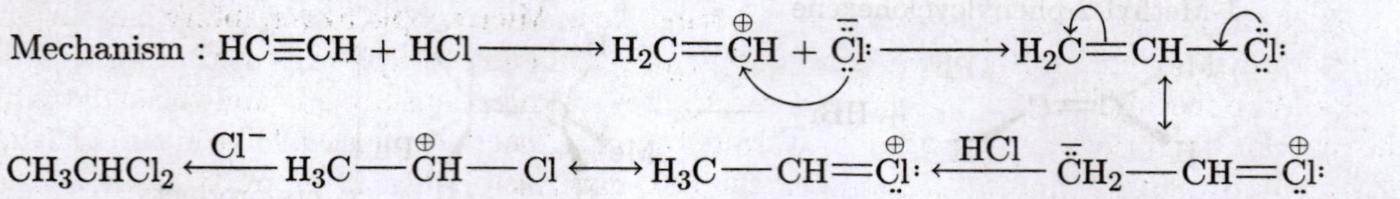

1,1-Dichloroethane will be the probable product. Here the orientation of the addition of $H^{\oplus}$ and Cl^- to vinylchloride is determined by the π-p conjugation (a pair of p es of Cl atom undergo conjugation with C—C π es). Farther C atom from the Cl atom becomes negatively charged and invites the proton. Thus $CH_3{-}\overset{\oplus}{C}H{\cdot}Cl$ develops which is more stable than $\overset{\oplus}{C}H_2CH_2{-}Cl$. The higher stability of $CH_3{\cdot}\overset{\oplus}{C}HCl$ is due to the +I effect, the hyperconjugative effect and the conjugative effect as shown below :

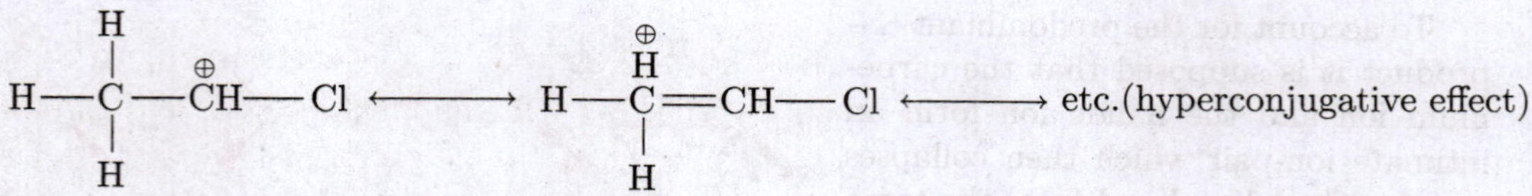

$$CH_3{-}\overset{\oplus}{C}H{-}\ddot{C}l{:} \longleftrightarrow CH_3{\cdot}CH{=}\overset{\oplus}{C}l{:} \quad \text{(conjugative effect)}$$

Problem 13.3. Acetylenes are less reactive than olefines so far as electrophilic additions are concerned—explain.

Solution : A C—C triple bond has symmetrical cylindrical charge density around the C—C bond axis. Because of the smaller distance between the two triple bonded Cs, the charge cloud is more strongly held in a C—C triple bond than in a C—C double bond which has a greater bond length. This makes acetylenes less reactive than olefines.

Furthermore, when an electrophile attaches itself to a double bonded C and to a triple bonded C, an alkyl and a vinyl cation develop respectively. A vinyl cation being less stable than an alkyl cation, a C—C triple bond is less reactive than a C—C double bond.

If the reaction occurs via a bridged cation, then also the bridged cation and its T.S. will be highly strained and $\Delta G^{\ddagger}$ will be very high and the reaction will not take place.

13.5 Hydration of Alkenes

The addition of water molecule to alkenes is called hydration of alkenes. Alkenes on hydration produce alcohols. The reaction is extensively used to manufacture alcohols from cracked petroleum alkene fractions by vapour phase hydration with steam over acid catalysts. Therefore the process has high industrial value.

$$CH_3CH{=}CH_2 \underset{\text{conc. } H_2SO_4}{\overset{\text{dil. } H_2SO_4}{\rightleftharpoons}} CH_3CHOHCH_3$$

This is a reversible reaction. In the presence of concentrated acids, alcohols get dehydrated to alkenes.

The reaction is carried out at a low temperature with dilute sulphuric acid which provides high concentration of water. The high concentration of water shifts the equilibrium to the right. On the other hand, the backward reaction is favoured by shifting the equilibrium to the left; this requires high temperature and regular distilling out of the alkene formed by the process.

Since the reaction is a reversible one, the principle of **Microscopic Reversibility** may be applied to determine the mechanism of the hydration of alkenes. The principle tells us that for a reversible reaction the mechanisms for the forward and the reverse reactions are one and the same provided the reaction conditions are identical. Naturally both the reactions must involve identical T.S.s and intermediates. Huges and Ingold studied the acid-catalysed dehydration of 3° alcohols in detail and on the basis of their experimental results they proposed a two-step mechanism involving a carbenium ion intermediate for the hydration of alkenes as shown below;

Preliminary step : $H_2SO_4 + H_2O \rightleftharpoons H_3O^{\oplus} + HSO_4^-$

Step I : Electrophilic attack by the proton to form the more stable carbenium ion. This may or may not occur via a π-complex or protonium ion. This is the rate-determining step.

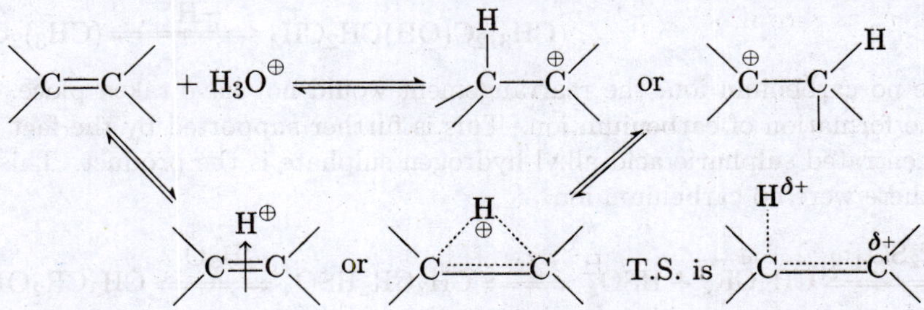

Step II : Nucleophilic attack by H_2O on the positively charged carbon atom.

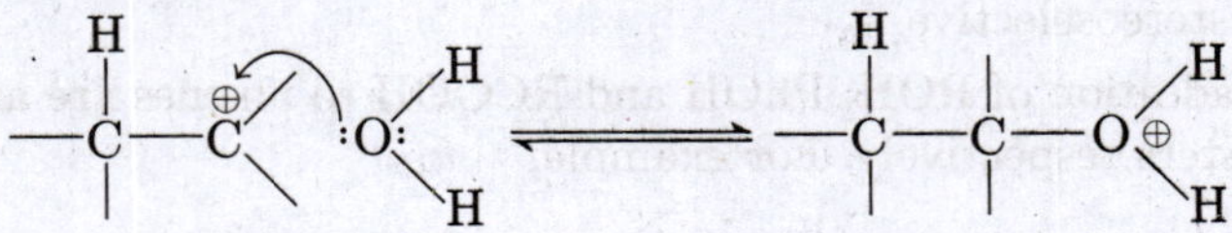

HSO_4^- being a weaker nucleophile than water, the former does not attack the cationic carbon. Furthermore, water molecules envelop the HSO_4^- ion by solvation and thus it fails to come in contact with the carbenium ion as shown in Fig. 13.17.

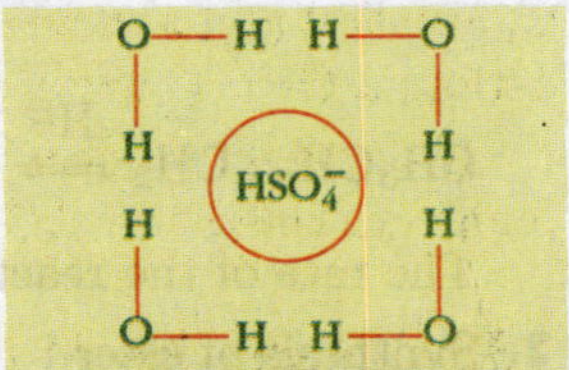

Fig. 13.17. Solvation of HSO_4^-

Final Step : Loss of proton from the $H_2O^{\oplus}$ moiety.

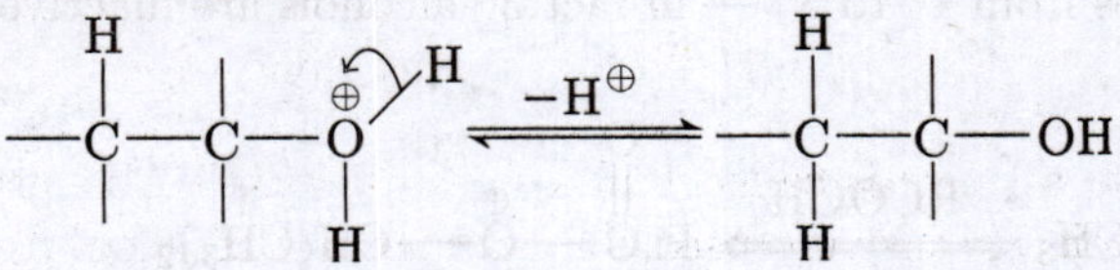

The reaction is supposed to follow the second order kinetics.

The rate= k [alkene] [$H^{\oplus}$]

The energy profile for the reaction may be drawn as shown in Figure 13.5.

The reaction occurs in accordance with the Markownikoff's rule.

The proposed mechanism is supported by the following facts:

(i) Hydration does not occur in alkaline medium—this is a strong evidence in favour of the first step which involves the electrophilic attack by a proton.

(ii) Water alone does not attack the C—C double bond. Water is a weak electrophile; it cannot transfer a proton. So the proton must come from the acid.

(iii) There are alkenes which on hydration form alcohols through the rearrangement of intermediate carbenium ions, e.g.,

$$(CH_3)_2CH{\cdot}CH{=}CH_2 \xrightleftharpoons{H_3O^{\oplus}} (CH_3)_2C(OH)CH_2CH_3$$

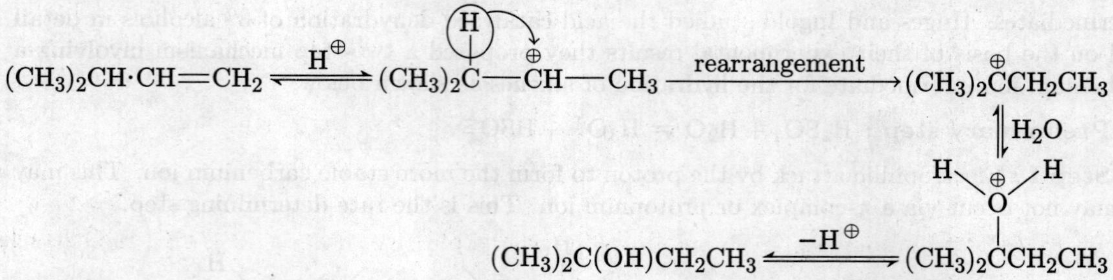

If there were no carbenium ion, the rearrangement would not have taken place. Therefore, this supports the formation of carbenium ion. This is further supported by the fact that in the presence of concentrated sulphuric acid alkyl hydrogen sulphate is the product. This would not have formed if there were no carbenium ion.

$$CH_2{=}CR_2 \xrightleftharpoons{H_2SO_4} CH_3\overset{\oplus}{C}R_2 + HSO_4^- \rightleftharpoons CH_3CR_2HSO_4 \xrightleftharpoons{H_3\overset{\oplus}{O}} CH_3CR_2OH + H_2SO_4$$

This fact also indicates that H_2O is involved in the second step of the mechanism.

The reaction is not stereoselective.

The acid-catalysed addition of ROH, PhOH and RCOOH to alkenes are analogous reactions; they form ethers and esters respectively. For example,

1. Synthesis of ether :

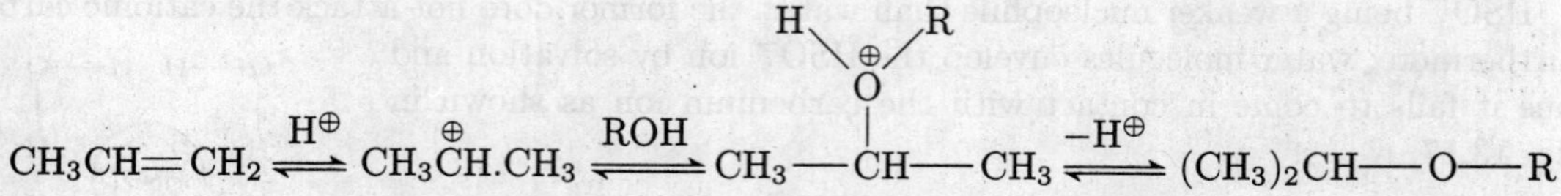

The rate of the reaction decreases from 1° to 3° — in fact 3° alcohols are inactive.

2. Synthesis of ester :

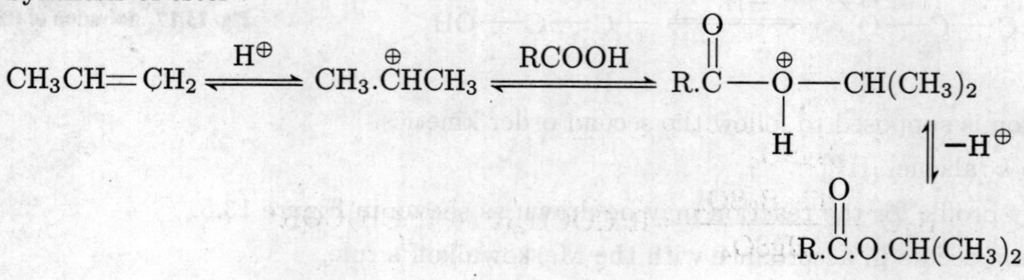

Problem 13.4. Give mechanistic steps for the following reactions :

(a) $CH_3CH{=}CH_2 + HOCl \longrightarrow$

(b) $PhCH{=}CH_2 + ICl \longrightarrow$

(a) Preliminary step : $HOCl + H^{\oplus} \rightleftharpoons H_2\overset{\oplus}{O}{-}Cl \rightleftharpoons H_2O + \overset{\oplus}{Cl}$

Step I : R.D. step : $CH_3CH{=}CH_2 + Cl^{\oplus} \longrightarrow$ 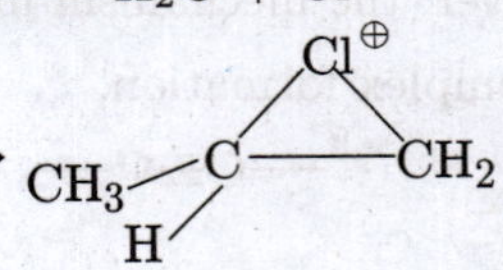

Step II : Fast step : 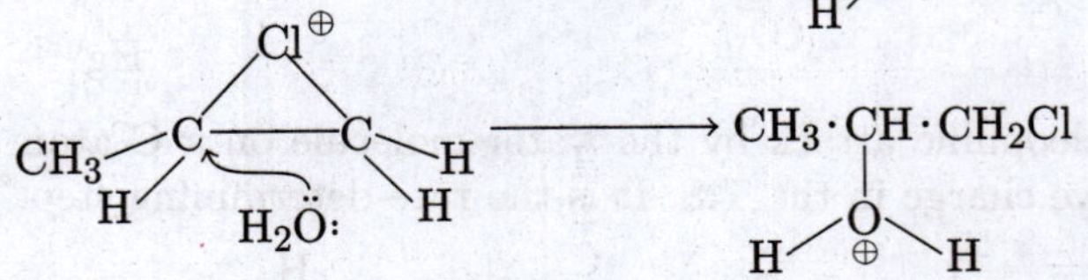

Final step : 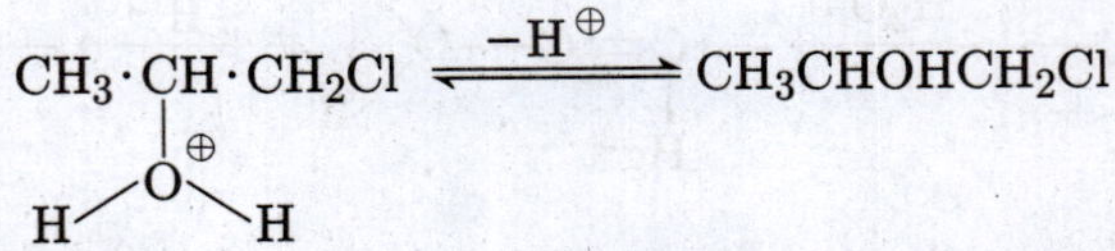

Since a 2°C atom better accommodates the partial positive charge than a 1°C, H_2O attacks the second C atom.

(b)

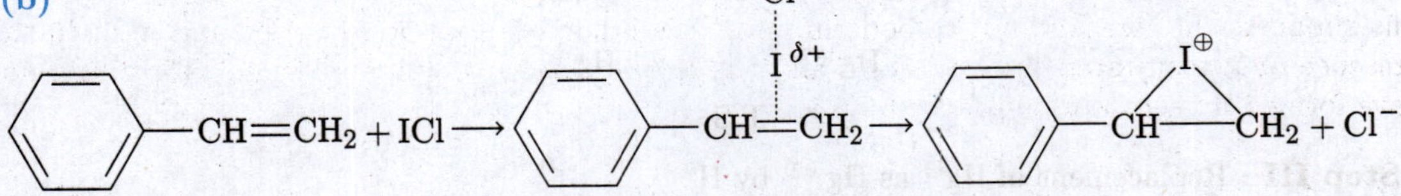

I is more electropositive than Cl

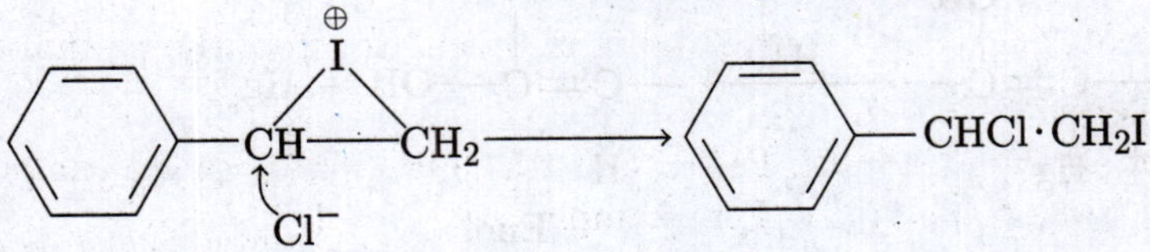

For explanation consult Problem 13.5.

13.6 Hydration Of Alkynes

When an alkyne is treated with dil. H_2SO_4 in the presence of catalytic amount of mercuric sulphate at 60°C to 80°C, hydration of the alkyne occurs and a carbonyl compound is formed. Hydration of acetylene produces acetaldehyde; monosubstituted acetylenes give methylketones and disubstituted acetylenes form a mixture of ketones by this process.

$$CH{\equiv}CH + H_2O \xrightarrow[HgSO_4]{\text{dil. } H_2SO_4} CH_3CHO$$

$$RC{\equiv}CH + H_2O \xrightarrow[HgSO_4]{\text{dil. } H_2SO_4} R.CO.CH_3$$

$$R^1C{\equiv}CR^2 + H_2O \xrightarrow[HgSO_4]{\text{dil. } H_2SO_4} R^1COCH_2R^2 + R^1CH_2COR^2$$

Since a large number of compounds can be synthesised from acetaldehyde, the reaction has a great preparative value.

The mechanism of the reaction is uncertain. It is supposed that the hydration occurs via a π- or mercury (II)-acetylene complex. $Hg^{\oplus\oplus}$ being a large cation, it can form a bridged cationic complex. However, the mechanism may be shown as follows:

Step I : Complex formation.

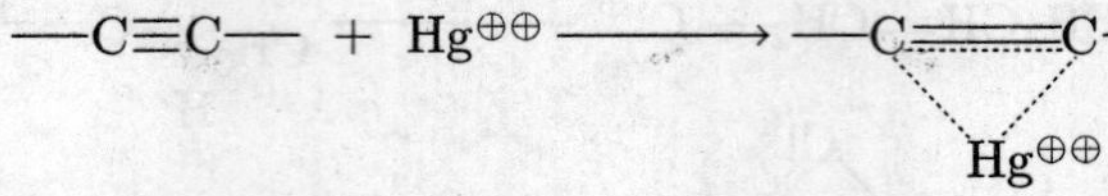

Step II : Nucleophilic attack by the water molecule on a C atom which can better accommodate the positive charge in the T.S. It is the rate-determining step.

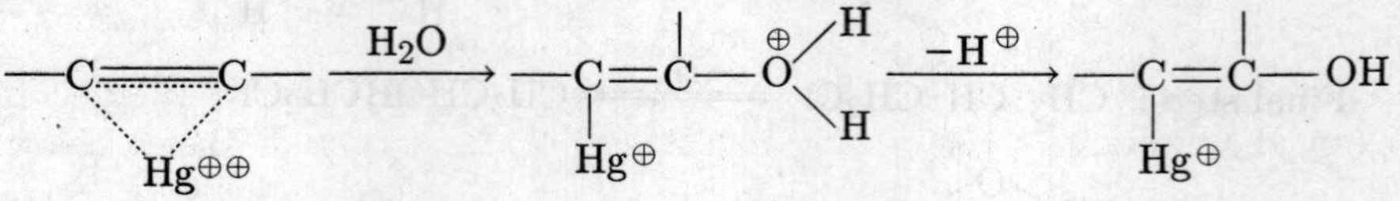

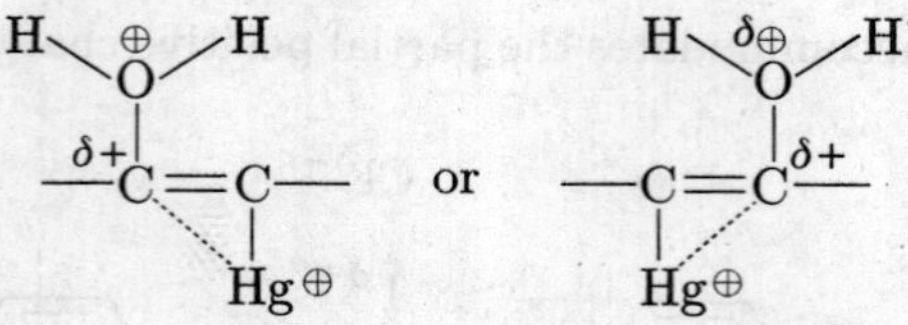

T.S.

Step III : Replacement of $Hg^{\oplus}$ as $Hg^{\oplus\oplus}$ by $H^{\oplus}$

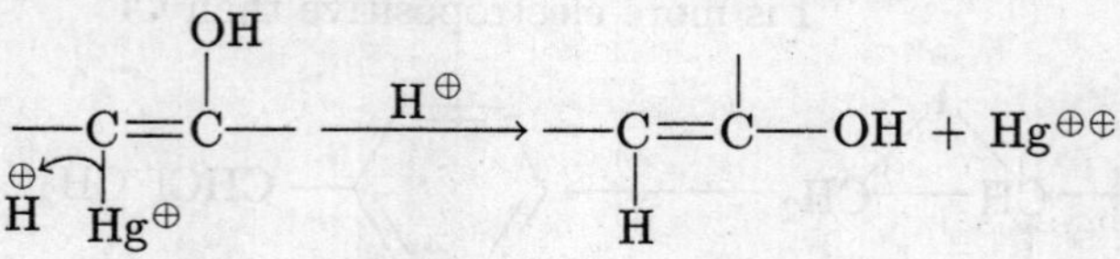

Enol

Step IV : Tautomerisation of the enol to a carbonyl compound.

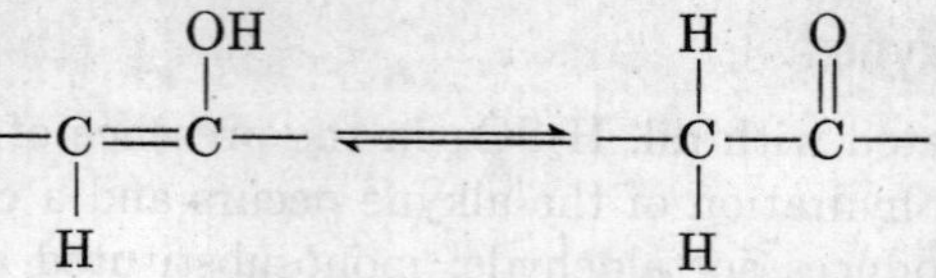

Problem 13.5. What product do you expect from the reaction between phenylacetylene and dil. sulphuric acid in the presence of mercuric ion?

Give the mechanism of the reaction with explanation.

Solution : Methyl phenyl ketone is the expected product.

Since the partial positive charge developed on the C_2 atom of the T.S. gets delocalised by the conjugative effect of the phenyl group, the nucleophilic attack on that C atom occurs following

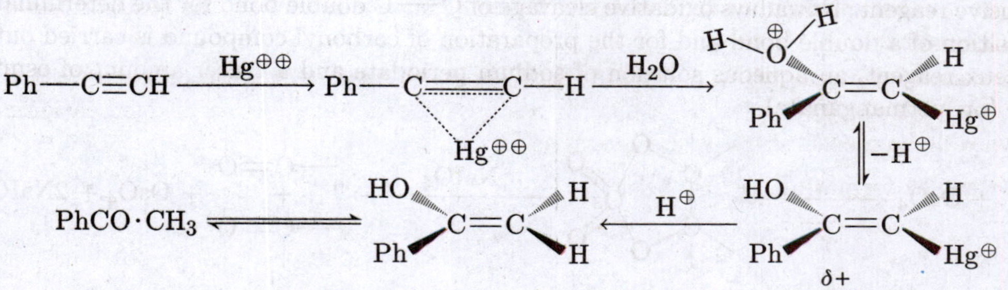

T.S :

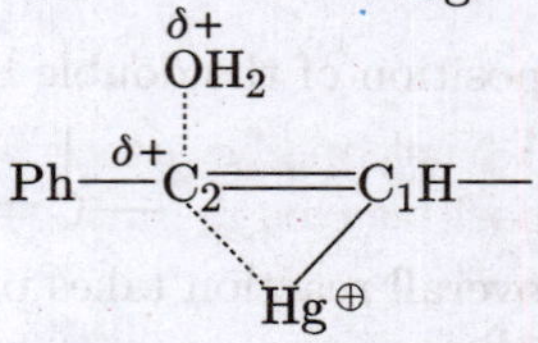

S_N2 pathway and consequently the final product of hydration is methyl phenyl ketone. If the nucleophilic attack by H_2O were on the C_1 atom, the T.S. would have been as shown below :

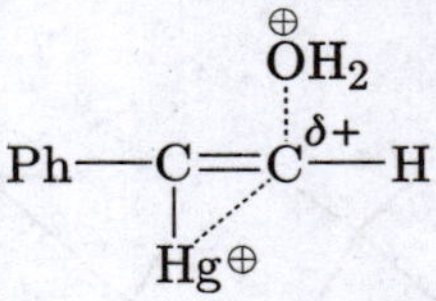

The partial positive charge on the C_1 atom of this T.S. has no scope for delocalisation since +I effect, conjugative effect and hyperconjugative effect are absent.

13.7 Hydroxylation of Alkenes

The conversion of alkenes to 1,2-diols by oxidation and subsequent hydrolysis is known as hydroxylation of alkenes. Usual oxidising agents are osmium tetroxide, ruthenium tetroxide, potassium permanganate, potassium dichromate, peroxycarboxylic acid, etc. Hydrolysing agents are different for different oxidising agents. Besides peroxycarboxylic acid all the agents mentioned above produce cis-addition products while peroxyacids give rise to trans-addition products. However, the detailed discussion on individual processes is given below.

13.7.1 Hydroxylation by osmium tetroxide

At room temperature osmium tetroxide adds very readily to a C—C double bond to form a cis-cyclic ester of osmic acid, the osmic ester, which on refluxing with an aqueous ethanolic sodium bisulphite solution hydrolyses to a cis-1,2-diol.

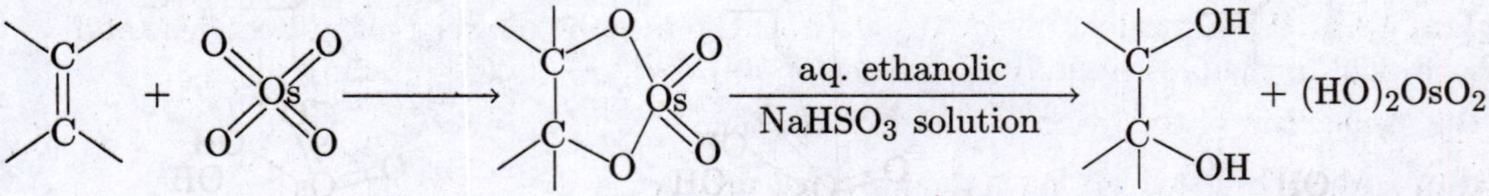

Osmium tetroxide is a highly expensive and toxic chemical. Its use as a reagent for the preparation of cis-1,2-diol is not usually preferred. However, this may be used in catalytic amount, to avoid toxic action as far as possible, in association with hydrogen peroxide which oxidises the oximic acid formed in the reaction to osmium tetroxide. Thus hydrogen peroxide helps to recover the expensive reagent. Nowadays oxidative cleavage of C—C double bond for the determination of the position of a double bond and for the preparation of carbonyl compound is carried out by the Lemieux reagent, an aqueous solution of sodium periodate and a tracer amount of osmium tetroxide (or, permanganate).

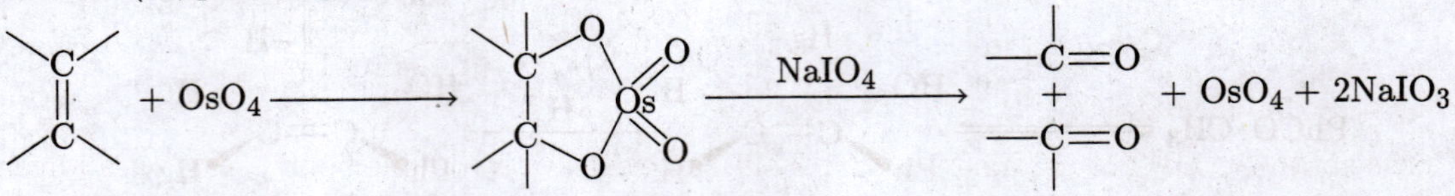

The position of the double bond is determined as follows :

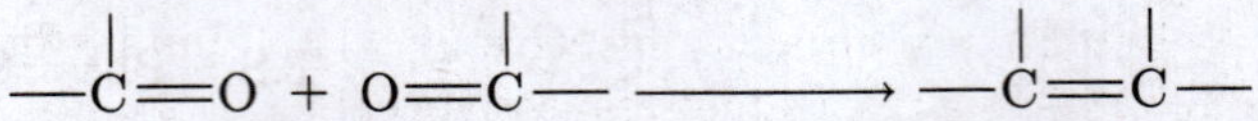

The overall reaction takes place in two steps via a cyclic intermediate, an ester of osmic acid and the alkene.

Step I : This consists of syn or cis-addition of osmium (VIII) tetroxide to the alkene to form a cis-cyclic osmic (VI) ester. This is a slow step.

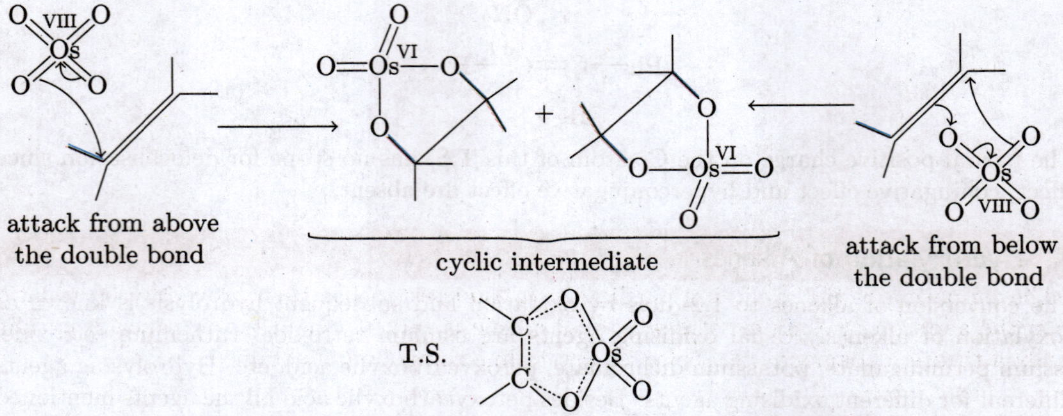

Step II : Fast hydrolysis of the cyclic intermediate to form a cis-1,2-diol.

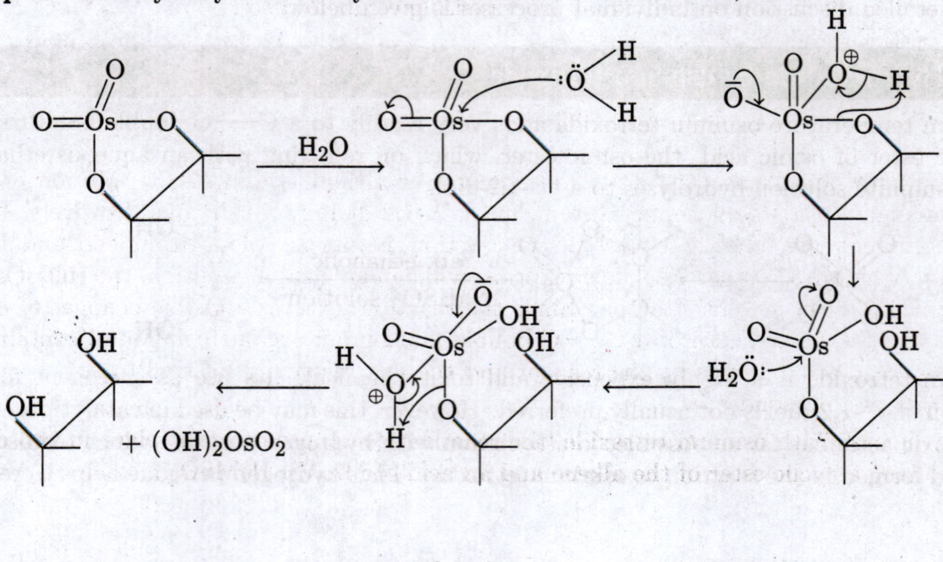

Similarly :

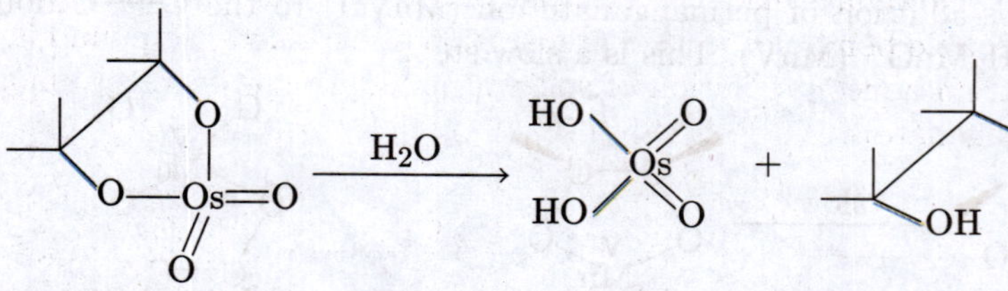

Since both the OH groups form bond from the same side of the double bond,cis or syn-addition products are obtained and hence the reaction is stereoselective. It has also been found that the reaction is stereospecific. The stereospecificity of the reaction is proved by the fact that the cis-2-butene and trans-2- butene give optically inactive meso 2, 3-butanediol and optically inactive racemic 2,3-butanediol (which is resolvable) respectively (Sec 13.2).

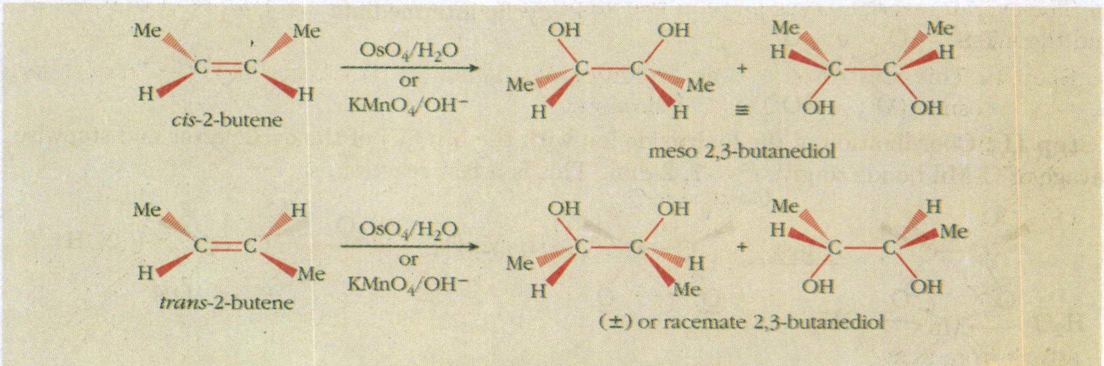

Fig. 13.18. Stereochemistry of hydroxylation

The formation of the cyclic intermediate, the osmic ester, has been proved by its isolation. When the addition of osmium tetroxide to an alkene is carried out in the presence of pyridine, usually a theoretical amount of a coloured crystalline compound is obtained.

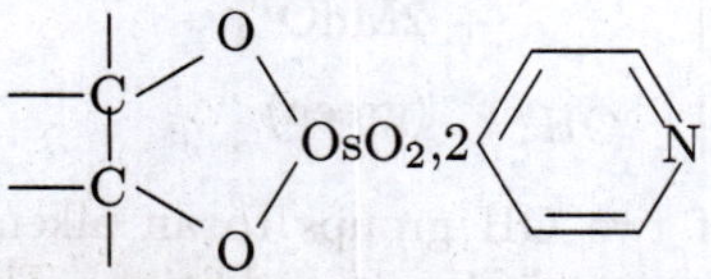
crystalline coloured compound

13.7.2 Hydroxylation by permanganate

Treatment of an alkene with a neutral aqueous or alkaline (*p*H : 12-13) solution of potassium permanganate at room temperature forms a cis-1,2-diol practically quantitatively. When a neutral permanganate solution is used for the reaction, its purple colour disappears and a brown precipitate of hydrated manganese dioxide forms; while in an alkaline solution, the colour changes from purple to green by this reaction. Since the reaction involves a visible change (of colour), it is used to test the presence of a C—C double bond in an organic compound containing no other easily oxidisable groups by the reagent and the test is usually referred to as Baeyer's test for a C—C double bond.

As is the case with osmium tetroxide, the reaction is supposed to take place in two discrete steps involving a cyclic ester of the alkene and an acid of Mn(V), $H_2MnO_4^-$.

Step I : Syn or cis-addition of permanganate ion (MnVII) to the C—C double bond to form a cyclic ester of $H_2MnO_4^-$ (MnV). This is a slow step.

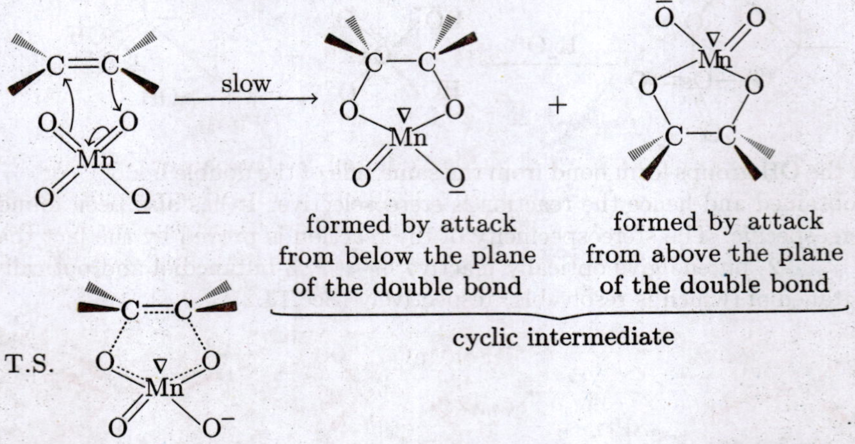

Step II : Coordination of the hydroxide ion with the Mn (V) of the cyclic ester and stepwise cleavage of O-Mn bonds to give cis, -1, 2-diol. This is a fast reaction.

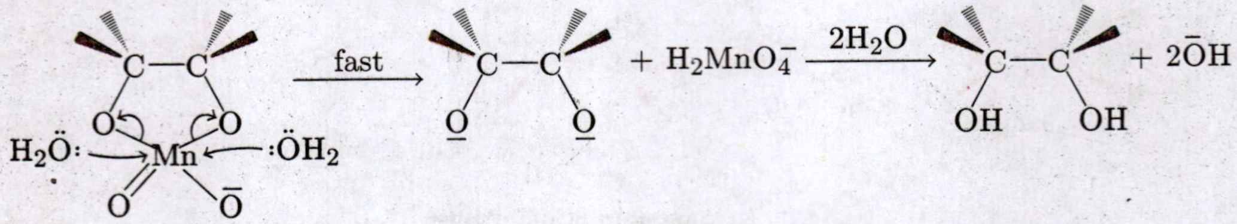

$$H_2MnO_4^- + 2\bar{O}H + MnO_4^- \longrightarrow 2MnO''_4 + 2H_2O$$

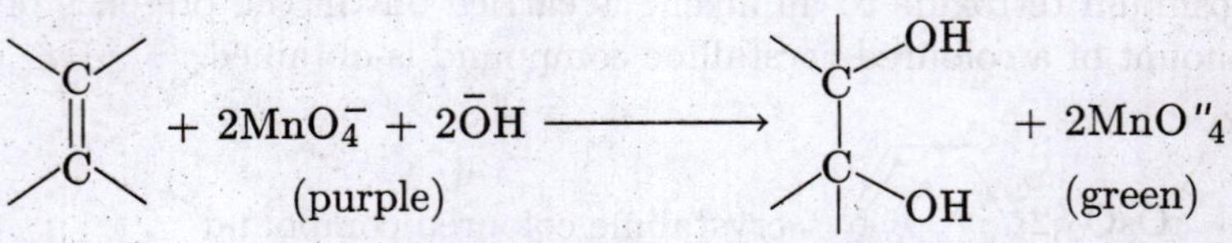

Thus the overall reaction involves the addition of two OH groups to an alkene from the same side of the double bond, i.e., the addition is a stereoselective cis-addition. This is also a stereospecific reaction; cis-2-butene forms meso 2,3-butanediol and trans-2-butene gives dl-2,3-butanediol (see Fig. 13.18).

The formation of the cyclic ester has been assumed in analogy with the reaction between an alkene and osmium tetroxide but this has not yet been isolated. Of course there is a strong evidence in favour of this assumption. When an alkene is treated with $Mn^{18}O_4^-$ instead of ordinary MnO_4^-, the isolated diol contains two ^{18}OH groups which proves that the two oxygen atoms of the diol have come from the $Mn^{18}O_4^-$ and neither from the reagent ($\bar{O}H$) nor from the solvent, H_2O.

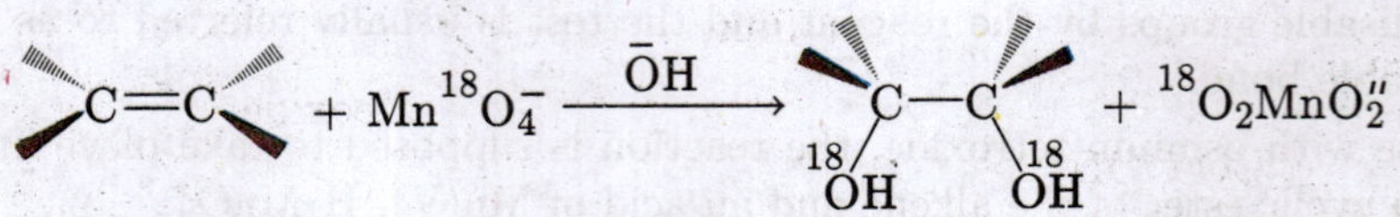

The syn-addition is proved by the fact that cyclopentene forms only the cis-diol on reaction with alkaline $KMnO_4$.

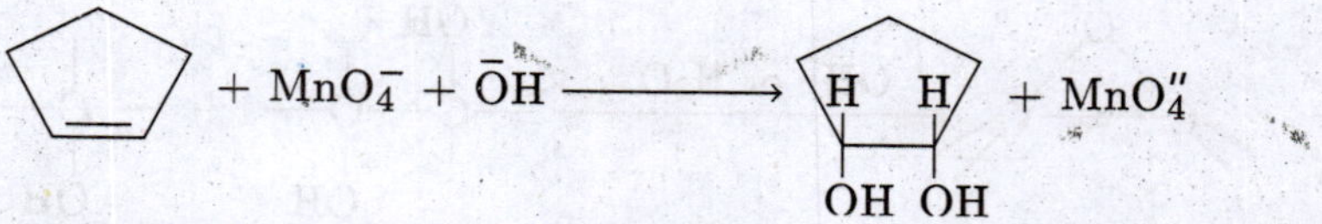

The disadvantage of the reaction is that the oxidation does not always stop at the diol stage—further cleavage of the C—C σ bond usually occurs to form carbonyl compounds. In fact, in determining the position of a double bond in an alkene and in preparing carbonyl compounds sodium periodate along with a trace of potassium permanganate is used as Lemieux reagent.

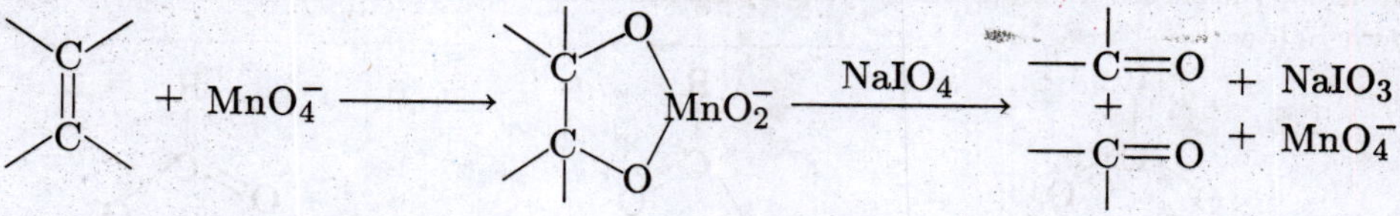

Problem 13.6. In the alkaline medium chromate reacts with alkenes to give cis-diols draw mechanistic steps.

Solution :

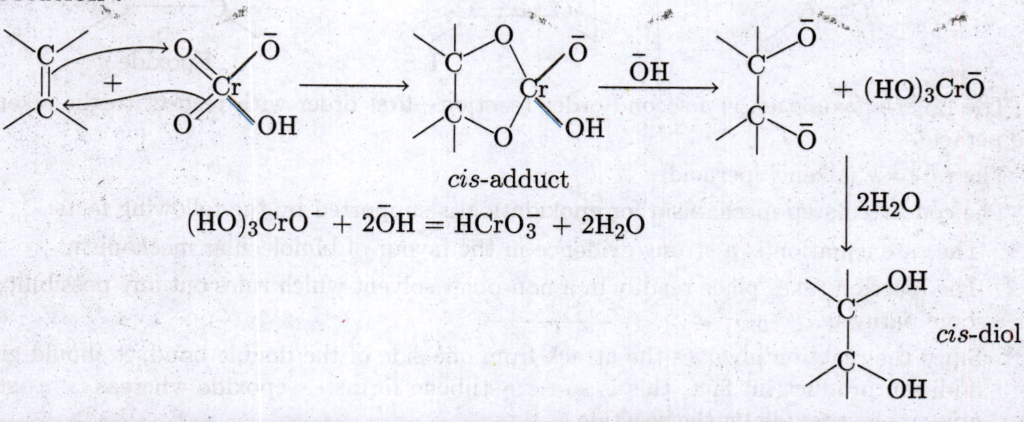

13.7.3 Hydroxylation via epoxidation

A good many organic peracids such as peroxybenzoic acid, peroxytrifluoroacetic acid, etc. stereospecifically react with alkenes in non-polar solvents like dichloromethane (CH_2Cl_2) under refluxing condition to give epoxides (three-membered cyclic compounds in which the ring consists of one oxygen and two carbon atoms) and the process is known as polar epoxidation. Nowadays mcpb, meta-chloroperbenzoic acids (mcpb) is used as better reagent for epoxidation.

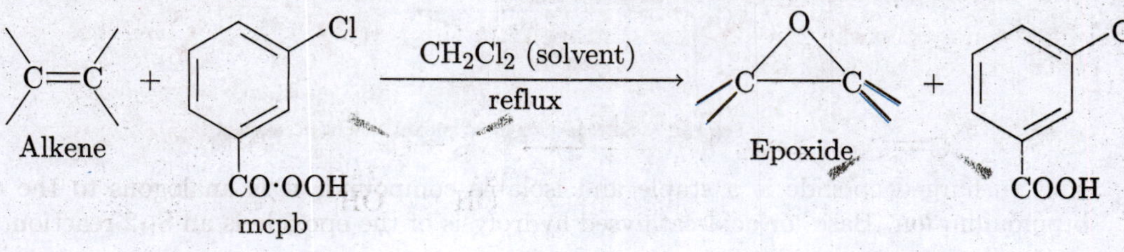

The acid- or base-catalysed hydrolysis of the epoxide so formed leads to an anti- or trans-1, 2-diol. The hydrolysis of an epoxide is also a stereospecific reaction.

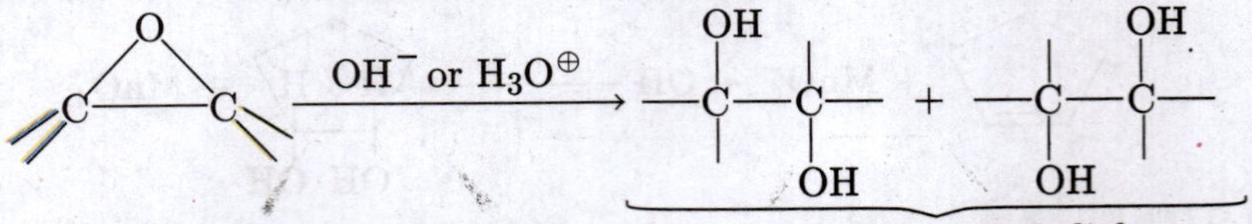

Thus the overall reaction may be called hydroxylation via epoxidation. Naturally the process is stereospecific.

The most plausible mechanism for the epoxidation is a bimolecular concerted step mechanism as shown below :

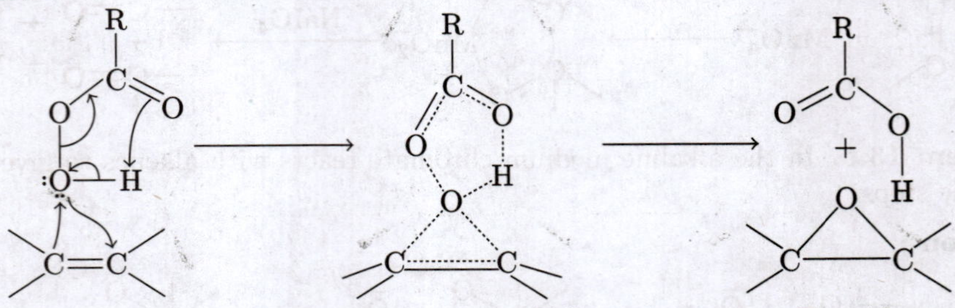

The polar epoxidation is a second order reaction—first order with respect to the alkene and the peracid.

The rate= k [alkene] [peracid]

The concerted step mechanism for epoxidation is supported by the following facts:

1. The rate equation is a strong evidence in the favour of bimolecular mechanism.
2. The reaction takes place readily in a non-polar solvent which rules out any possibility of an ionic pathway.
3. Since the reaction involves the attack from one side of the double bond, it should give cis-addition product; in fact, this is so—cis-stilbene forms cis-epoxide whereas trans-stilbene gives trans-epoxide by the reaction.

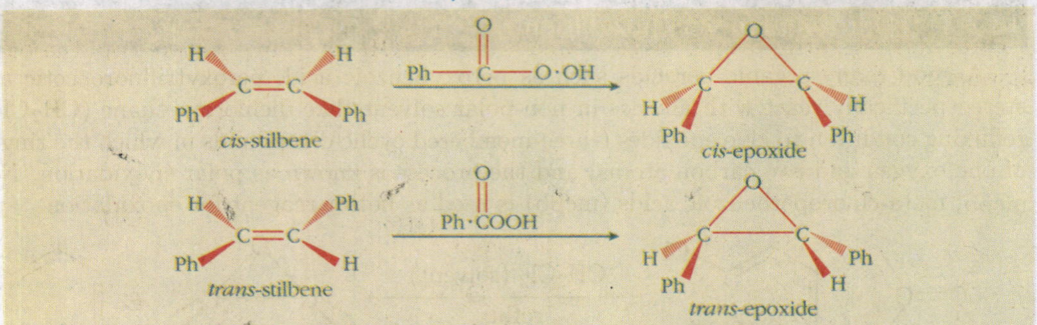

Fig. 13.19. Stereochemistry of epoxidation by peroxyacid

The uncharged epoxide is a stable and isolable compound; it is analogous to the cyclic bromonium ion. Base- or acid-catalysed hydrolysis of the epoxide is an S_N2 reaction.

Acid-catalysed hydrolysis :

Initial step : Protonation to the oxygen atom—it is a fast step.

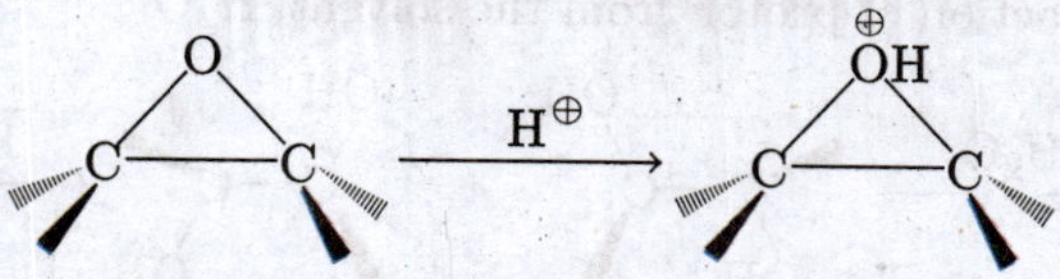

Rate-determining step : Nucleophilic attack by H_2O on a ring carbon from the back side of the epoxy linkage.

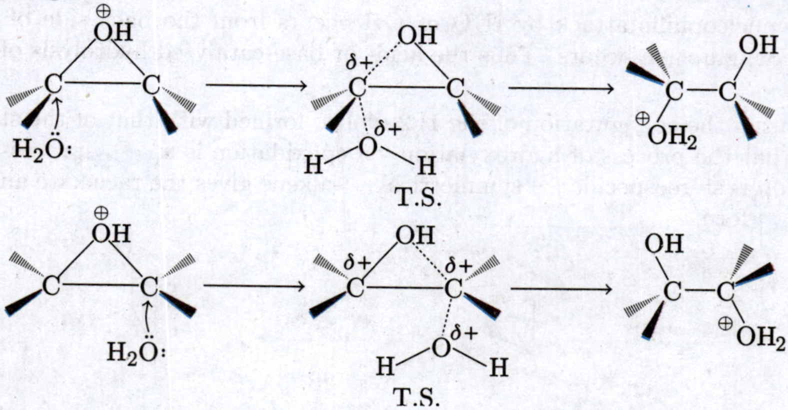

The C—$\overset{\oplus}{O}H$ bond breaking and the C—$\overset{\oplus}{O}H_2$ bond making take place in a concerted step by the S_N2 pathway. If the epoxide is not a symmetrical one, the nucleophilic attack occurs on that C atom which better accommodates the partial positive charge in the transition state.

Final Step : Loss of proton.

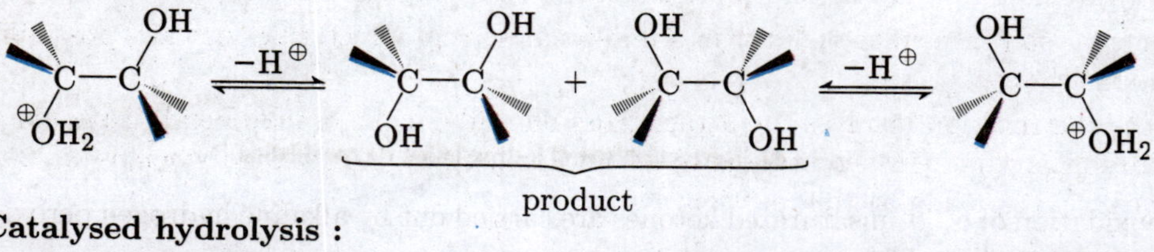

Base-Catalysed hydrolysis :

Rate-determining step : Nucleophilic attack by $\bar{O}H$ on a ring carbon from the back side following S_N2 pathway :

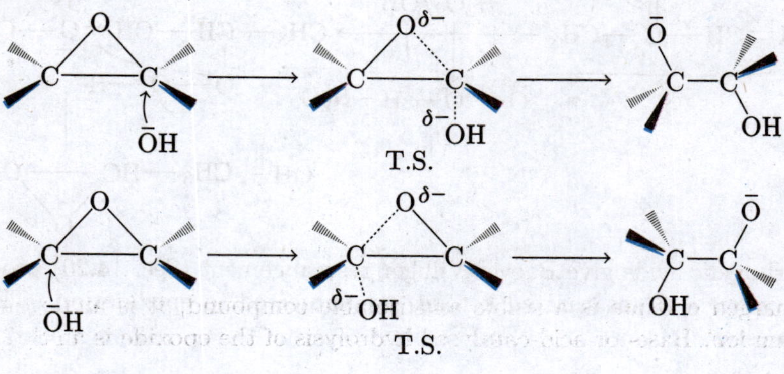

$\bar{O}H$ ion attacks the less sterically hindered epoxy C more readily than the more sterically hindered C.

Final Step : Abstraction of proton from the solvent, H_2O :

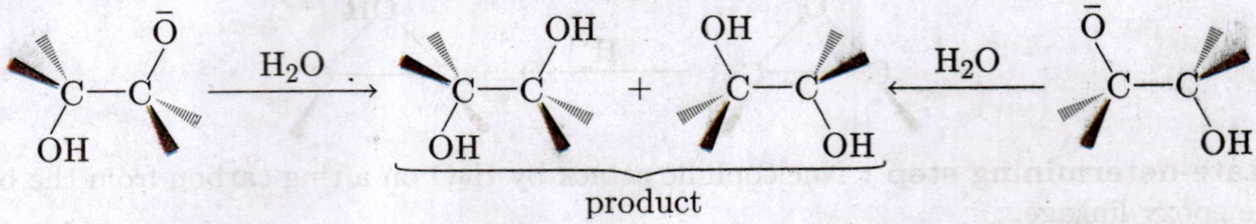

Since the nucleophilic attack by H_2O or $\bar{O}H$ occurs from the back side of the epoxy ring, inversion of configuration occurs. Thus the acid- or base-catalysed hydrolysis of an epoxide is a trans-process.

By comparing the configuration of the 1, 2-diol so formed with that of the starting alkene, it will be seen that the process of hydroxylation via epoxidation is a trans-process. Therefore, the overall reaction is stereospecific : a symmetrical cis-alkene gives the racemate and a trans-alkene yields the meso form.

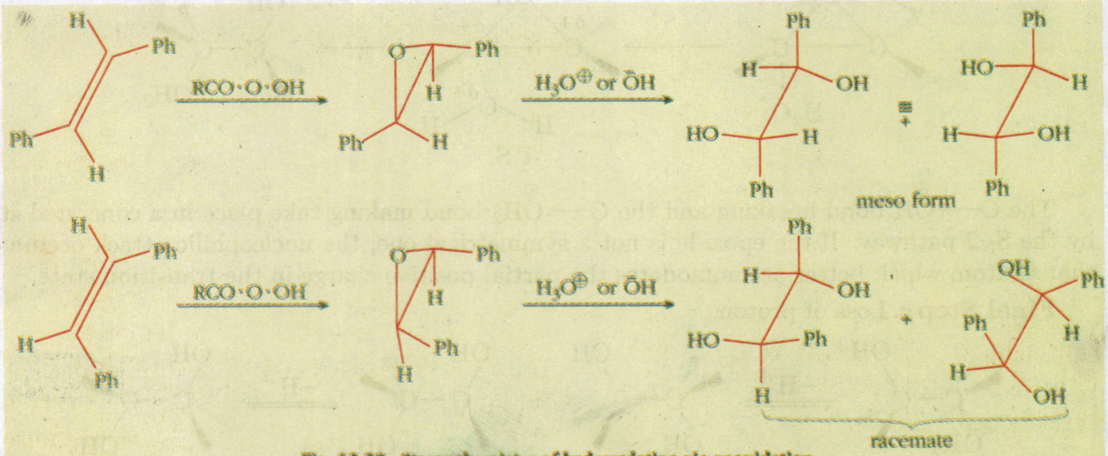

Fig. 13.20. Stereochemistry of hydroxylation via epoxidation

Epoxidation of α, β-unsaturated ketones are carried out by alkaline hydrogen peroxide instead of peroxycarboxylic acid.

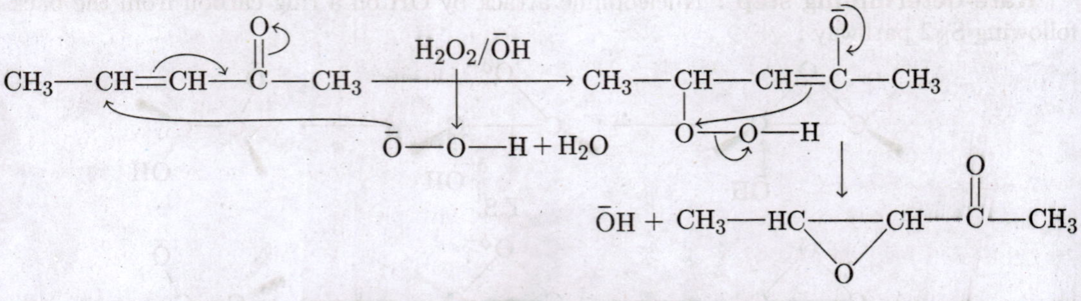

Peroxycarboxylic acids give Baeyer-Villiger rearrangement (Sec. 14.20) products along with the epoxides when α β-unsaturated ketones are subjected to epoxidation. The rearrangement product is an ester.

13.8 Ozonolysis

Ozonolysis means cleavage by ozone. When ozone is added to a C—C multiple bond and subsequently the addition product is decomposed, the C—C multiple bond gets ruptured; because of this rupturing of C—C multiple bond, the process is called ozonolysis. Thus one molecule of a compound on ozonolysis gives two or more molecules of products, each of which contains fewer number of C atoms than the starting substrate and hence it is classified as a degradation reaction. The process of ozonolysis involves three stages :

(i) Addition of ozone to a C—C multiple bond and formation of ozonide, the addition product :

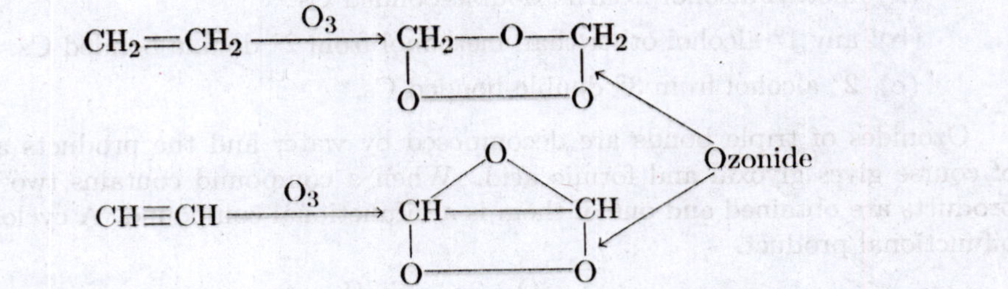

(ii) Reductive or oxidative decomposition of the ozonide :

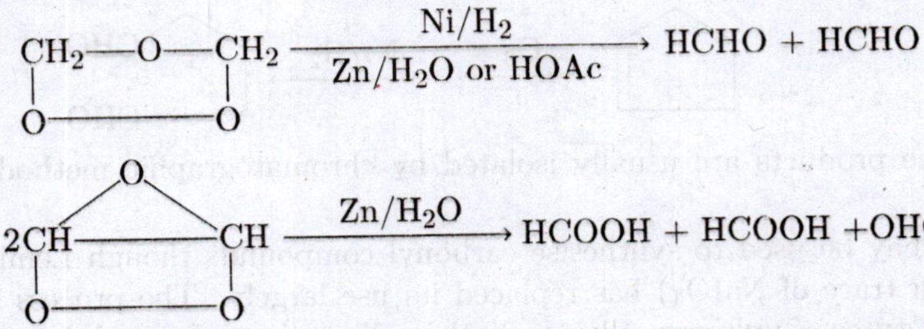

(iii) Identification of the decomposition products :

$$\text{Ethylene} + O_3 \longrightarrow \text{ozonide} \longrightarrow \text{Formaldehyde} + \text{Formaldehyde}$$

In fact, the complete process of preparing ozonide by adding ozone to a C—C multiple bond, and subsequently decomposing the ozonide and identifying the products formed is called **ozonolysis**.

The ozonide is prepared by passing a stream of ozonised air or oxygen through an unsaturated compound in the liquid state or through its non-aqueous solution in an inert solvent such as chloroform, carbon tetrachloride, glacial acetic acid, light petrol, etc. (with which the ozone does not react) and then evaporating out the solvent. The ozonide then remains as a viscous oil.

The ozonide is then directly subjected to decomposition without isolation and purification because of its unstable explosive nature. The decomposition may be carried out in several ways.

1. Oxidative decomposition : Oxygen, silver oxide, hydrogen peroxide or peracids may be used for this purpose; in this case :

(a) 1° double-bonded Cs give formic acid.

(b) 2° double-bonded Cs give carboxylic acids other than formic acid.

(c) 3° double-bonded Cs give ketones.

2. Reductive decomposition : Zn/H_2O, Zn/CH_3COOH, Pt/H_2, Ni/H_2, $LiAlH_4$, or $NaBH_4$ may be used for this.

(i) When the first four reducing agents are used,

(a) 1° double-bonded Cs give formaldehyde.

(b) 2° double-bonded Cs give aldehydes other than formaldehyde.

(c) 3° double-bonded Cs give ketones.

(ii) When $LiAlH_4$ or $NaBH_4$ is used, the products are :

(a) methyl alcohol from 1° double-bonded Cs.

(b) any 1° alcohol other than methanol from 2° double bonded Cs.

(c) 2° alcohol from 3° double-bonded Cs.

Ozonides of triple bonds are decomposed by water and the products are acids. Acetylene of course gives glyoxal and formic acid. When a compound contains two double bonds, three products are obtained and one of them is a bifunctional compound. A cycloalkene also gives one bifunctional product.

$$CH_2{=}CH.CH{=}CH_2 \xrightarrow{O_3} \xrightarrow{Ni/H_2} HCHO + CHO.CHO + HCHO$$

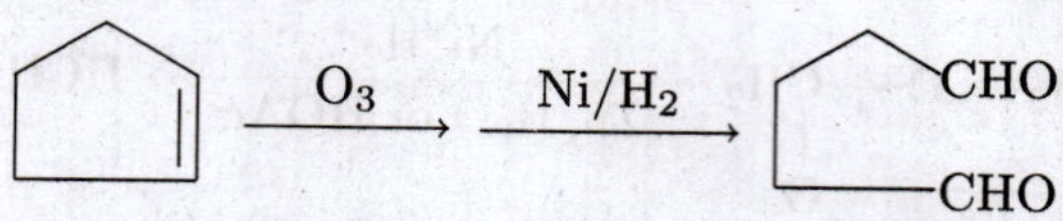

However, the products are usually isolated by chromatographic methods and identified as usual.

Ozonolysis may be used to synthesise carbonyl compounds though Lemieux reagent (OsO_4 or $KMnO_4$ with trace of $NaIO_4$) has replaced its use largely. The process was widely used in elucidating structures of unknown alkenes, both cyclic and acyclic, and determining the position of unsaturation. But with the advent of instrumental analysis, e.g., NMR, IR, etc., the process is no longer used. However, for elucidating structures the following procedure is followed :

The ozonides are usually decomposed by Zn/acid or Ni/H_2. The decomposition products are isolated and identified; Then the structural formulae of the compounds are written down side by side in such a way that C=O functions remain face to face. Now by placing a double bond between the two carbonyl carbon atoms and dropping the oxygen atoms, the structure of the original compound is obtained. Following any system of nomenclature locant of the double bond, i.e, the position of the double bond is determined; for example, a hydrocarbon on ozonolysis gave one mole of glyoxal, one mole of formaldehyde and one mole of acetone. We know that a bifunctional product is obtained from a compound containing two double bonds. Therefore, the compound is a diene and its structure is written below :

$$H_2C{=}O + O{=}CH.CH{=}O + O{=}\underset{\displaystyle CH_3}{\underset{|}{C}}{-}CH_3 \longrightarrow H_2C{=}CH.CH{=}\underset{\displaystyle CH_3}{\underset{|}{C}}{-}CH_3$$

Hence locants of the two C—C double bonds are 1 and 3.

The mechanism for the formation of ozonide has not yet been established with certainty. However, the most plausible mechanism involves three steps.

Step I : 1, 3-dipolar cis-addition of ozone to the double bond and formation of a very unstable compound called initial ozonide or molozonide.

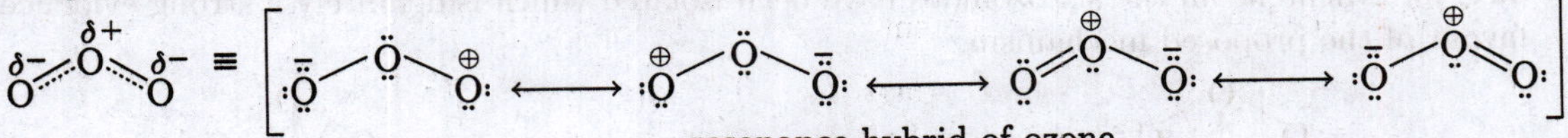

resonance hybrid of ozone

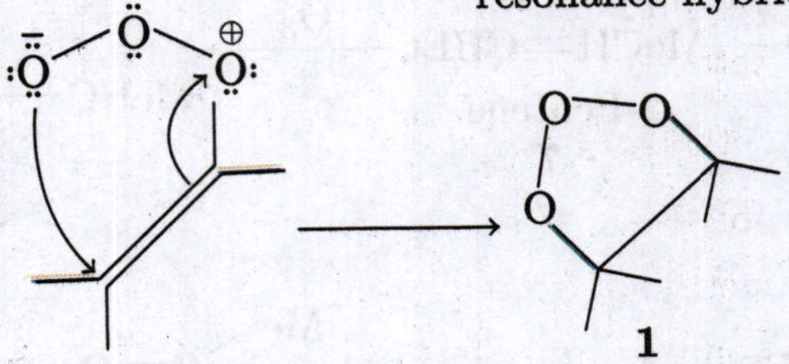

initial ozonide or molozonide

1

Step II : Decomposition of the molozonide (1) to a carbonyl compound (2) and a zwitterion (3).

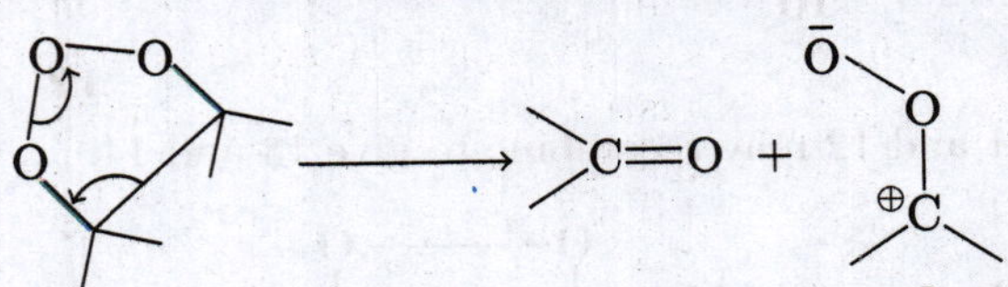

1 2 3

Step III : Recombination of the carbonyl compound (2) and the zwitterion (3) to form the normal ozonide.

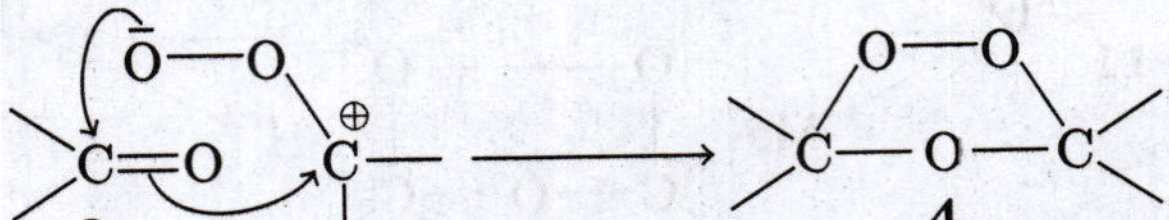

2 3 4

normal ozonide

The following facts support the formation and the structure of the molozonide (1).

(a) When ozone is added to 2,2,5,5-tetramethylhex-3-ene, an additive compound (5) is formed which possesses a structure analogous to that of (1); this has been confirmed by NMR spectroscopy.

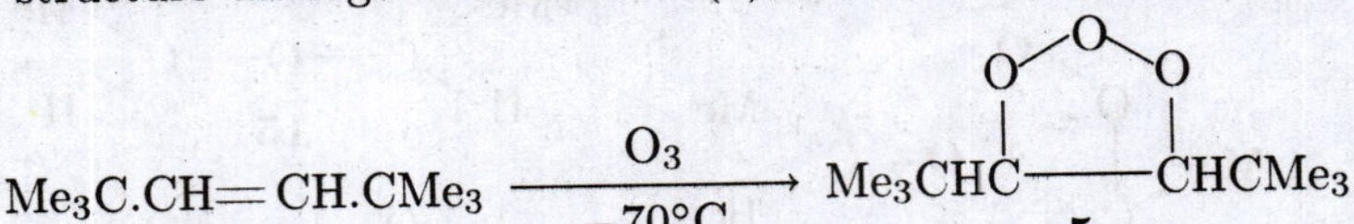

5

The decomposition of the molozonide (1) to the carbonyl compound (2) and the zwitterion (3) is supported by the following facts :

(i) Along with the normal ozonide (4) a dimer of the zwitterion (3), a bis-peroxide (6), has been isolated for $Me_2C{=}CMe_2$. A bis-peroxide can form if the decomposition of the molozonide (1) develops the zwitterion (3).

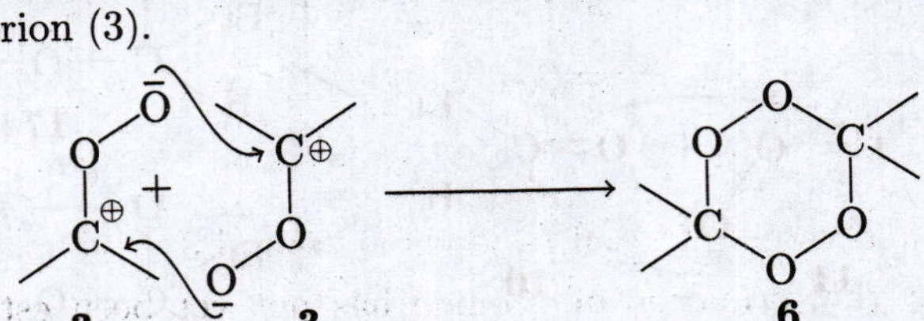

3 3 6

bis-peroxide

(ii) When an unsymmetrical alkene is treated with ozone, six ozonides for the cis- and trans-isomers may form if the proposed decomposition step and the recombination step are correct. In fact, for 2-pentene all the six ozonides have been isolated which is definitely a strong evidence in favour of the proposed mechanism.

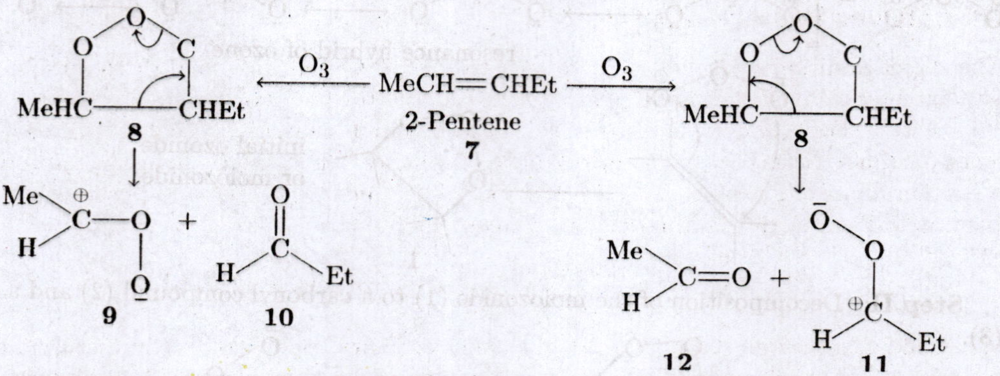

Now 9 and 10 or 11 and 12 may recombine to give 13 and 14.

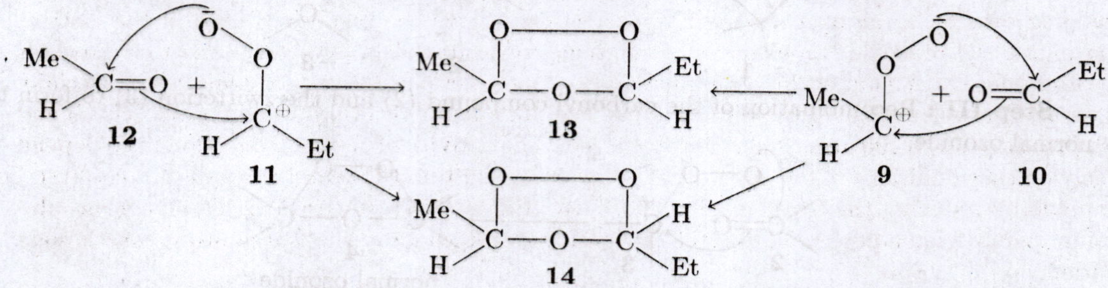

Again, 9 and 12 may be combined to give 15 and 16; similarly, 10 and 11 may be combined to give 17 and 18.

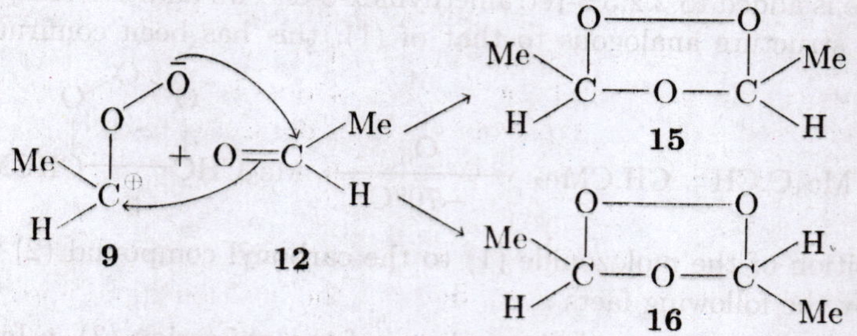

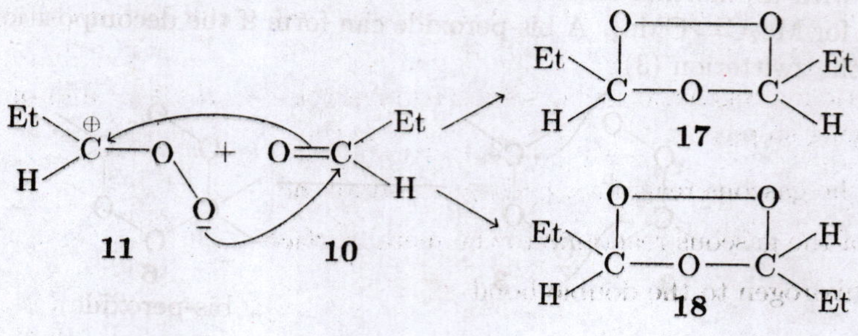

Thus the six ozonides are 13, 14, 15, 16, 17 and 18 which have been obtained from 2-pentene.

Though the mechanism explains most of the facts of ozonolysis yet it fails to explain correctly some of the facts involving stereochemistry of the reaction.

13.9 Catalytic Hydrogenation

The direct addition of molecular hydrogen to a C—C multiple bond in the presence of a transition metal catalyst such as Ni, Pt, Pd, Ru, Rh, etc. (or their compounds) is called catalytic hydrogenation. For example, 1 mole of ethene adds up 1 mole of molecular hydrogen in the presence of Raney Ni as catalyst to give 1 mole of ethane. The catalytic hydrogenation reaction is an exothermic reaction—usually 111—137.5 kJ of heat energy is liberated per mole of C—C double bond. The reaction is almost quantitative.

$$CH_2{=}CH_2 \xrightarrow[\text{Pt or Pd/H}_2]{\text{Ni/H}_2\text{ or}} CH_3{-}CH_3 + 126.6 \text{ kJ.}$$

When the unsaturated compound is a gas under ordinary condition of temperature and pressure, a mixture of the gas with hydrogen may be passed over the finely divided catalyst. Liquid and solid hydrocarbons may be hydrogenated by shaking their solutions in inert solvents with powdered catalyst in an atmosphere of hydrogen gas. Since the catalyst is used in the distinct solid phase, the process is an example of heterogeneous catalysis. However, in most of the cases the reaction is carried out at 25°C and just above atmospheric pressure. For some cases, specially for tri- and tetra-substituted C—C double bonds, high temperature (not more than 275°C) and high pressure (up to 100 atmosphere) are required. The activity of the catalyst is found to depend largely on the conditions under which they are prepared. Pt and Pd catalysts (Adam's catalysts) are made by reducing their oxides with hydrogen. Raney Ni is prepared by dissolving out aluminium partly from a powdered mass of Ni-Al alloy by its reaction with hot aqueous caustic soda solution.

The process of hydrogenation has analytical, industrial and preparative importance. Since the reaction adds up hydrogen to a multiple bond almost quantitatively and since the volume of hydrogen consumed can be measured easily without error, hydrogenation becomes an important analytical tool. For example, the amount of an unsaturated compound of known structure in a mixture with saturated compounds, and the number of double or triple bonds per mole of compounds can be determined from the experimental data obtained by the process of catalytic hydrogenation. Hydrogenation of fats and oils is an example of its industrial and preparative applications.

The reaction is supposed to involve chemisorption—the process of exothermic absorption of gases by solids in which a chemical link is formed between the absorbing and absorbed materials. Here the absorbing material is a solid transition metal in finely divided state which provides enormous surface area, and hydrogen gas and an unsaturated compound in the gaseous state are the absorbed materials.

When a catalytic hydrogenation of an alkene involves gaseous reactant and product, it occurs through the following steps :

(a) diffusion of the gaseous reactants, hydrogen and alkene

(b) adsorption of the gaseous reactants to the metal surface

(c) addition of hydrogen to the double bond

(d) desorption of the gaseous product, an alkene, and

(e) diffusion of the product, the alkane.

The steps may be represented pictorially as shown below :

(a) Diffusion of gases :

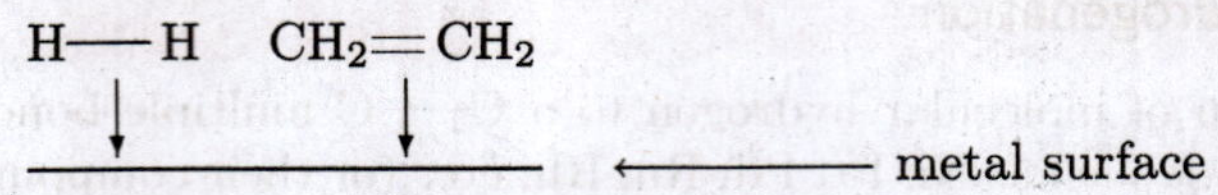

(b) Adsorption of gases on the metal surface :

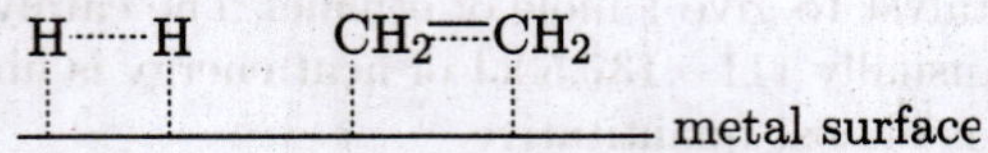

(c) Stepwise addition of hydrogen to the double bond :

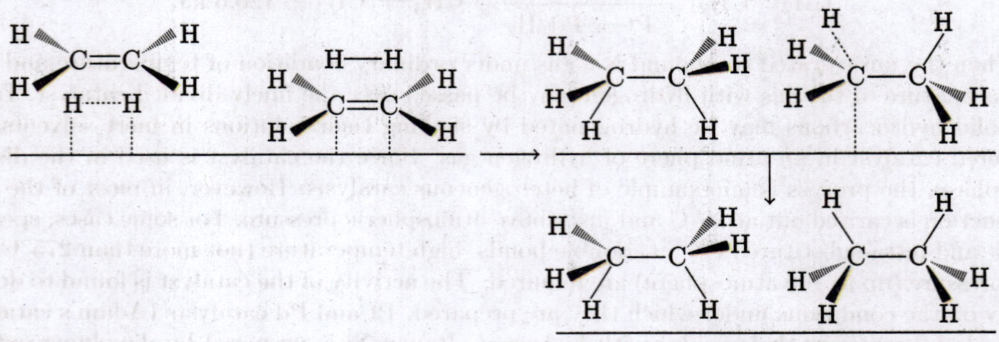

(d) Desorption of the gaseous product :

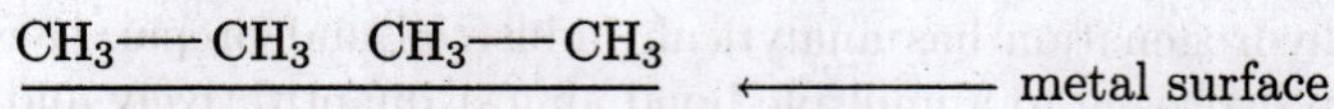

(e) Diffusion of the product :

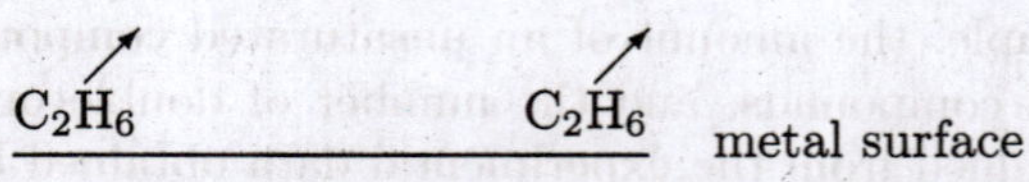

The details of the mechanism have not yet been known. It is believed that prior to addition of hydrogen, the H—H σ bond and C—C π bond break. The total reaction is, however, very complex. It has been found that exchange of hydrogen takes place between the alkene and molecular hydrogen.

In most of the cases syn- or cis-addition of hydrogen occurs from the less sterically hindered side of the double bond. Of course cent percent cis-addition does not occur—it is about 80%. This supports that the two H atoms do not add to the double bond at a time; after the addition of one H atom to a double-bonded C atom, the C atom with its groups rotates about C—C σ bond and then the other H atom adds, and this gives rise to an anti- or trans-product. Since the addition of two H atoms is rapid, the addition is stereospecific and almost cis.

Enthalpy consideration of the addition of hydrogen to ethylene tells us that it should be an exothermic reaction and will occur spontaneously in the absence of a catalyst. However, this does not happen. This is perhaps owing to the very large energy of activation for the reaction.

It is the metal surface that speeds up the reaction. A metal atom at the surface has some extra reactivity in comparison with an atom well within the metal—it lowers the energy of activation by activating the reactant making chemical link with the latter. The lowering of energy of activation makes the reaction to occur even at room temperature.

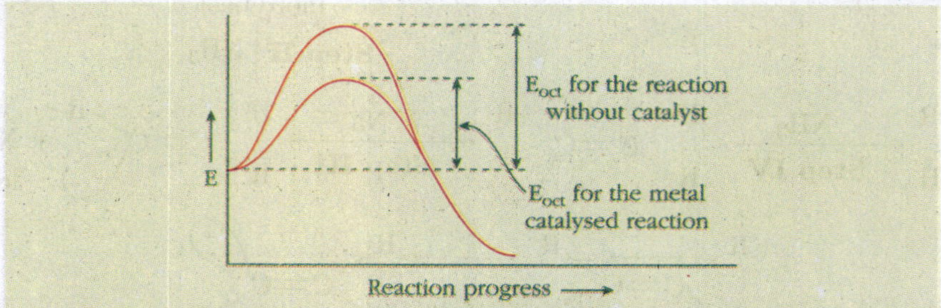

Fig. 13.21. Energy profile for catalytic and non-catalytic hydrogenation

If metal catalysts lower the energy of activation for the hydrogenation of alkenes, they will also lower the energy of activation for the dehydrogenation of alkene, the reverse reaction. If this is true, then the catalysts used for hydrogenation could also be used for dehydrogenation of alkane, of course, under suitable set of conditions. In fact, this is indeed so; one and the same catalyst brings about hydrogenation of an alkene as well as dehydrogenation of the alkane.

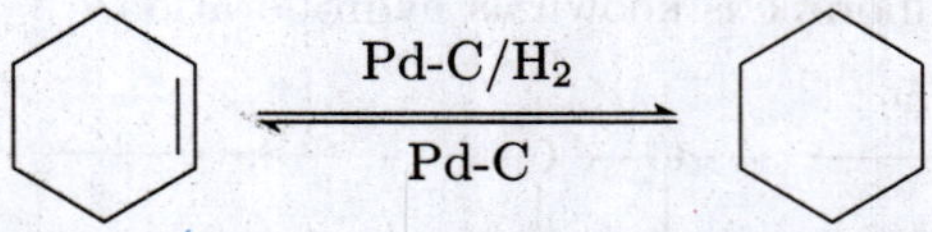

However, an alkyne may be converted to almost cent percent cis-olefins by means of Lindlaar catalyst, H_2/Pd partially poisoned with lead acetate or quinoline (which leads to partial hydrogenation).

$$MeO_2C(CH_2)_3—C\equiv C—(CH_2)_3CO_2Me + H_2$$

$\downarrow$ (1 atm); 5% Pd on $BaSO_4$ quinoline; MeOH 25°C

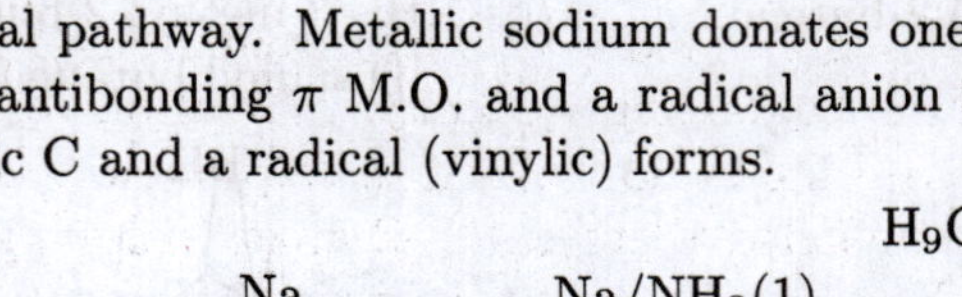

Alkynes without H—C (sp) bond can be partially reduced to trans alkene by using Na or Li in liquid ammonia as shown below :

The reaction involves radical pathway. Metallic sodium donates one electron at a time and the first electron goes into an antibonding π M.O. and a radical anion forms. Liquid ammonia gives up a proton to the anionic C and a radical (vinylic) forms.

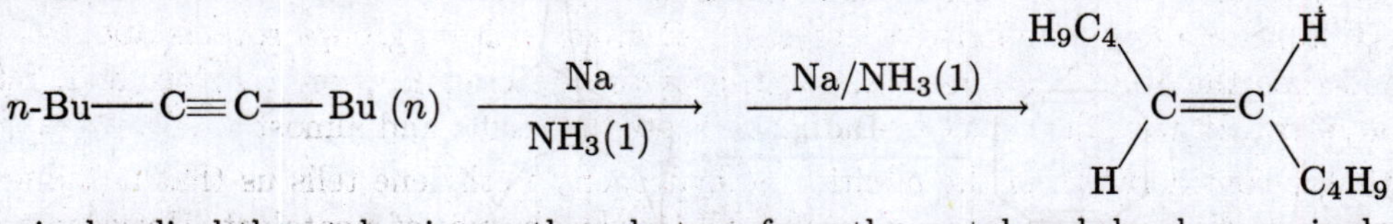

The vinyl radical then takes in another electron from the metal and develops a vinyl anion. The anion abstracts a proton from ammonia molecule.

A trans-vinyl radical being more stable than its cis isomer, because of steric congestion (non-bonded interaction between R groups in cis-position), the former forms preferably.

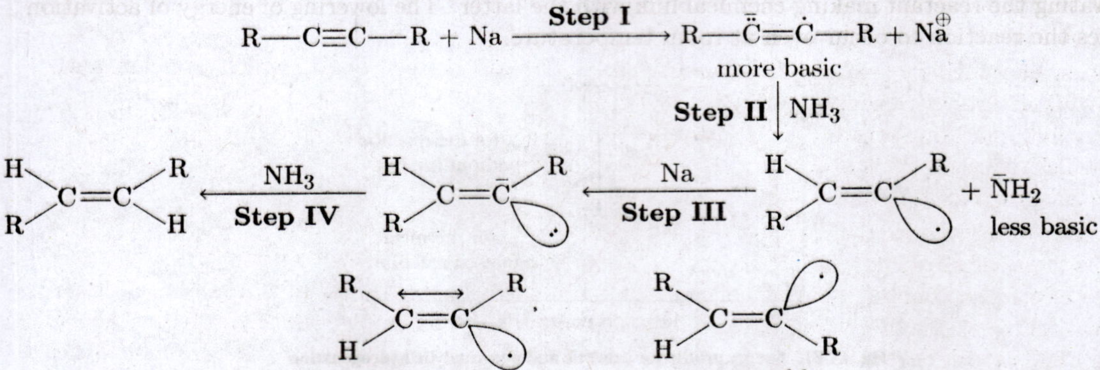

less stable *cis*-vinyl radical

more stable *trans*-vinyl radical

13.10 Hydroboration

The addition of borane across a C=C double bond by the action of diborane (B_2H_6) on an olefinic compound to form alkylborane is known as hydroboration.

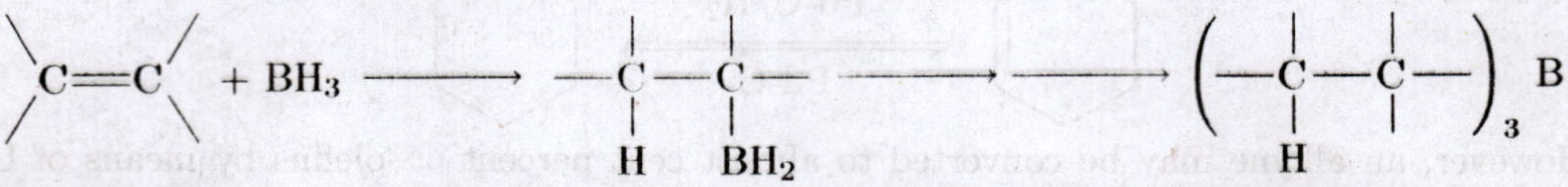

When propylene is treated with diborane, tri-*n*-propylborane forms. Here the boron atom attaches itself to that double-bonded C atom which contains the greater number of H atoms, because B is more electropositive than H.

$$3\,CH_3.CH{=}CH_2 \xrightarrow[\text{ether}]{B_2H_6} (CH_3.CH_2.CH_2)_3B$$

Propylene Tri-*n*-propylborane

similarly,

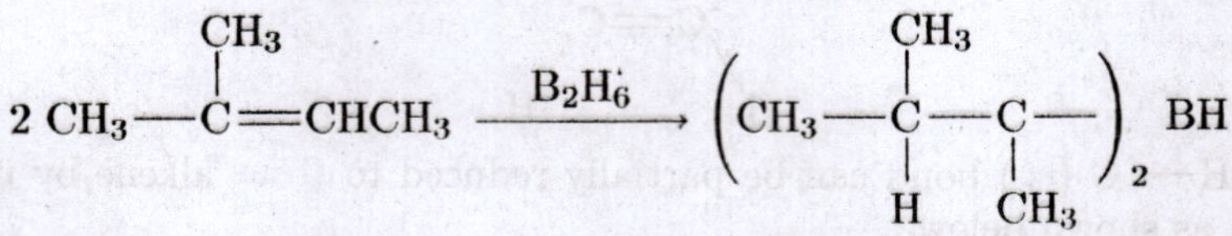

2-Methyl-2-butene Bis-3-methyl-2-butylborane [Disiamyl borane (Si α)$_2$ BH]

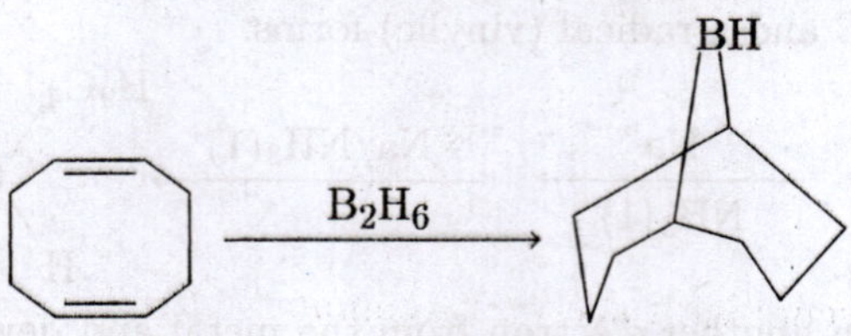

Cyclooocate-1,5-diene 9-BBN : 9-Borabicyclo [3.3.1]nonane

When a solution of an alkene in ether, commonly in tetrahydrofuran or diglyme (diethylene glycol methyl ether; $CH_3O{\cdot}CH_2CH_2O{\cdot}CH_2CH_2{\cdot}O{\cdot}CH_3$). is treated with sodium borohydride ($NaBH_4$) and borontrifluoride (BF_3) whereby diborane forms in situ and adds to the C—C double bond as if it were borane, BH_3.

The stereospecific cis-addition of borane to a C—C double bond, which appears anti-Markownikoff wrongly, occurs in a single step involving a four-centre transition state which consists of the alkene and borane as shown in Fig. 13.22.

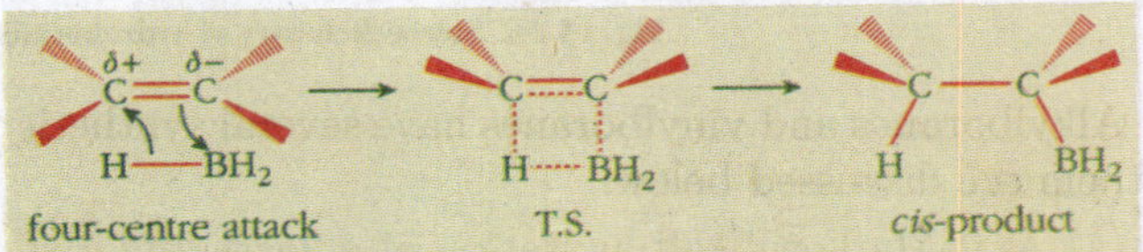

Fig. 13.22. Stereochemistry of hydroboration reaction

In borane, boron atom possesses six *e*s in its outer shell while each H atom has the electronic configuration of helium. This makes the boron atom in borane an electron-deficient centre and more electropositive than the Hs. For this reason the π *e*s of the C—C double bond attack the more electropositive boron atom while an H atom of the borane attaches itself to the more electropositive carbon. Thus, actually addition of borane across a double bond is a typical example of Markownikoff addition. Here inductive effect determines the direction of the electromeric effect; alkyl groups increase electron density on the less alkylated double-bonded C atom.

Since the attack of B and H atoms proceeds from the less sterically hindered side of the double bond, the reaction is stereospecific and the result is a cis-product. A symmetrical cis-alkene gives an erythro product while its trans-isomer gives a dl-pair on hydroboration. For example,

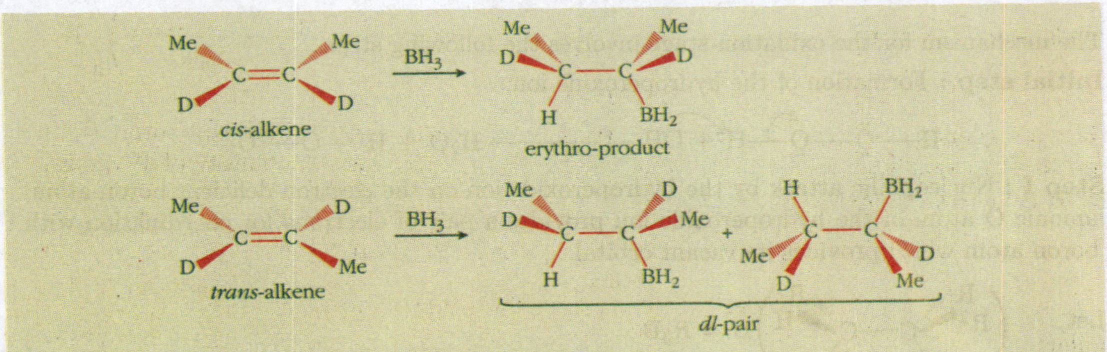

Fig. 13.23. Stereospecificity of hydroboration reaction

However, only one molecule of the alkene does not react with a molecule of borane; two more molecules of the alkene can react :

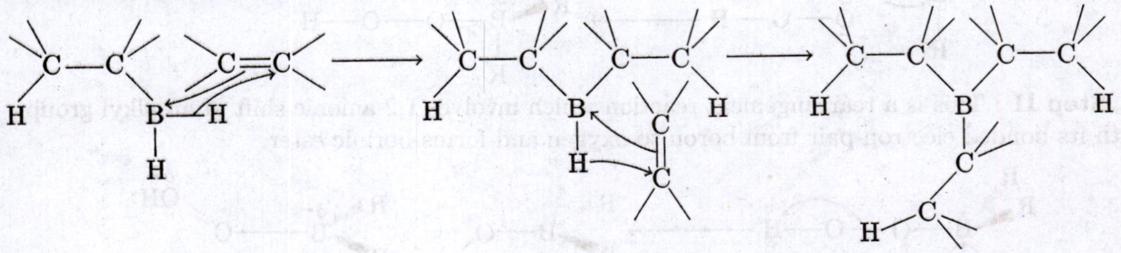

The addition of borane also occurs with acetylenic compounds which react faster than the olefinic compounds. A wide and varied type of reactions may occur with a C—C triple bond;

however, the reaction can be stopped at the vinylborane stage by allowing it to occur with stoichiometric amount of reactants. The reaction yields a cis-product.

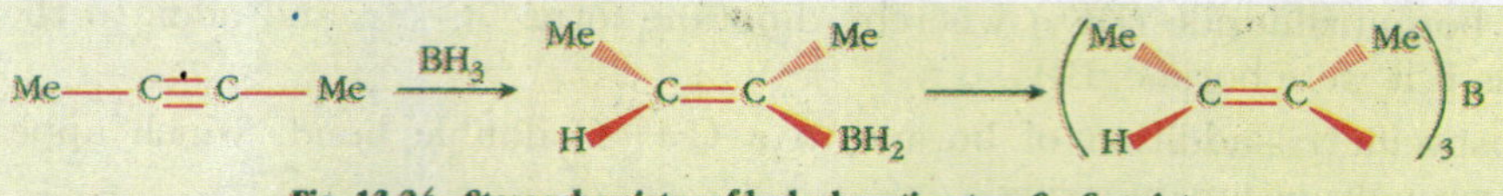

Fig. 13.24. Stereochemistry of hydroboration to a C≡C moiety

Alkylboranes and vinylboranes have several synthetic uses as intermediate compounds. A few of them are discussed below:

(a) Anti-Markownikoff hydration of alkenes and alkynes by hydroboration-oxidation : When alkenes and alkynes are treated with diborane and then subjected to oxidative hydrolysis using alkaline hydrogen peroxide solution, anti-Markownikoff hydration occurs and cis-alcohols are obtained.

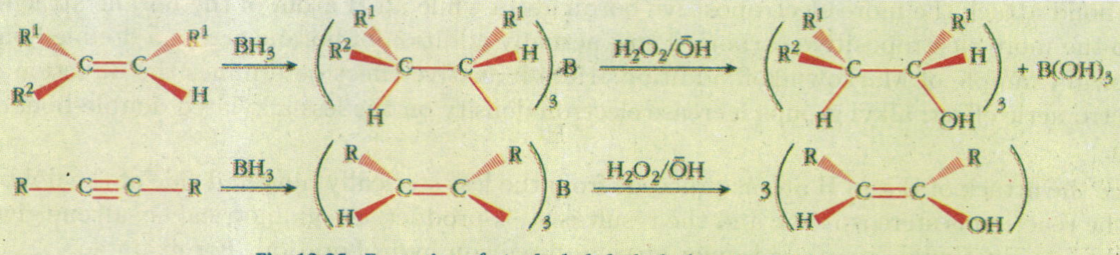

Fig. 13.25. Formation of *cis*-alcohols by hydroboration-oxidation

The mechanism for the oxidation stage involves the following steps.

Initial step : Formation of the hydroperoxide ion.

$$H-\ddot{O}-\ddot{O}-H + \bar{O}H \longrightarrow H_2O + H-\ddot{O}-\bar{\ddot{O}}:$$

Step I : Nucleophilic attack by the hydroperoxide ion on the electron-deficient boron atom; the anionic O atom of the hydroperoxide ion provides a pair of electrons for coordination with the boron atom which provides its vacant orbital.

Let, 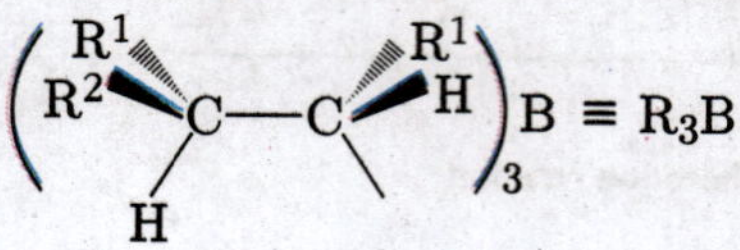

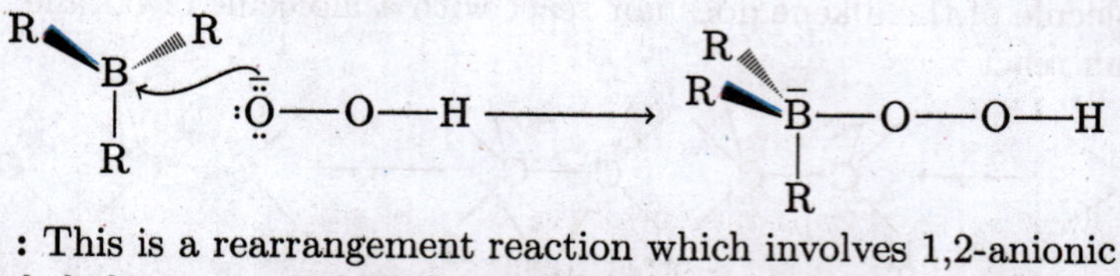

Step II : This is a rearrangement reaction which involves 1,2-anionic shift of an alkyl group with its bonded electron-pair from boron to oxygen and forms borinic ester.

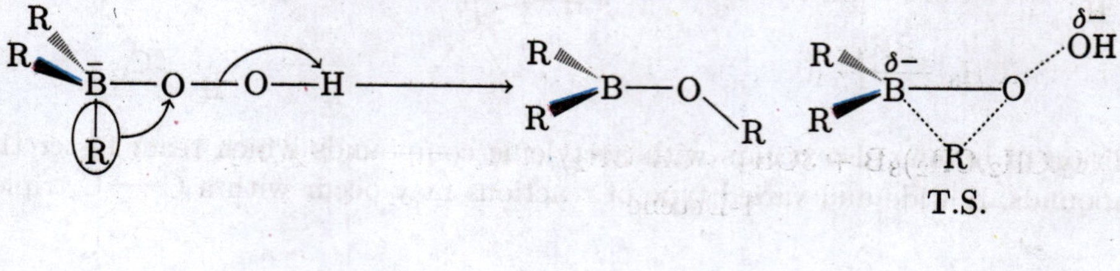

Step III : Hydrolysis of the borinic ester to the corresponding alcohol.

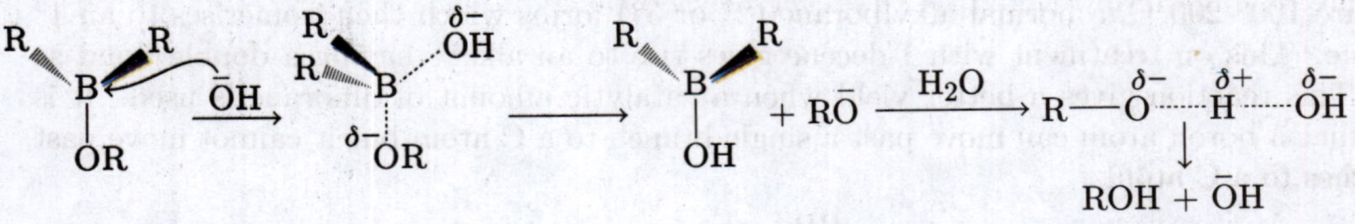

$$R_2BOH \xrightarrow{\bar{O}OH} \xrightarrow{\bar{O}H} \xrightarrow{H_2O} RB(OH)_2 + ROH \xrightarrow{\bar{O}OH} \xrightarrow{\bar{O}H} \xrightarrow{H_2O} B(OH)_3 + ROH$$

The overall reaction is :

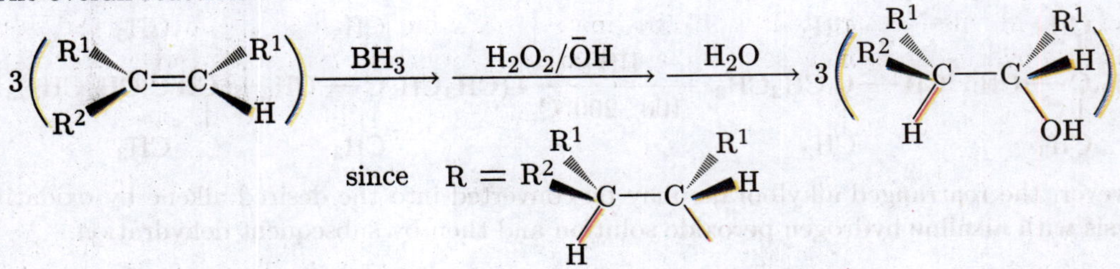

Similarly,

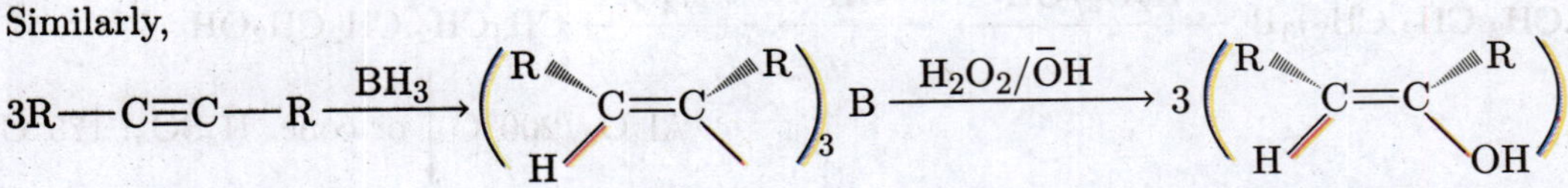

Since the boron atom attaches itself to the less substituted double-bonded or triple-bonded C atom, the method provides a means for hydrating a double or triple bond anti-Markownikoffically.

$$CH_3CH{=}CH_2 \xrightarrow{BH_3} (CH_3CH_2CH_2)_3B \xrightarrow{H_2O_2/\bar{O}H} \xrightarrow{\bar{O}H} \xrightarrow{H_2O} CH_3CH_2CH_2OH$$

(b) Non-catalytic hydrogenation of alkenes and alkynes : When alkylboranes and vinylboranes are treated with protic acids, alkanes and cis-alkenes are produced respectively. Acetic acid may be used for the purpose.

$$Me_2C{=}CHMe \xrightarrow{BH_3} (Me_2CH{-}CHMe)_3\,B \xrightarrow{CH_3COOH} 3\ Me_2CH{-}CH_2Me$$

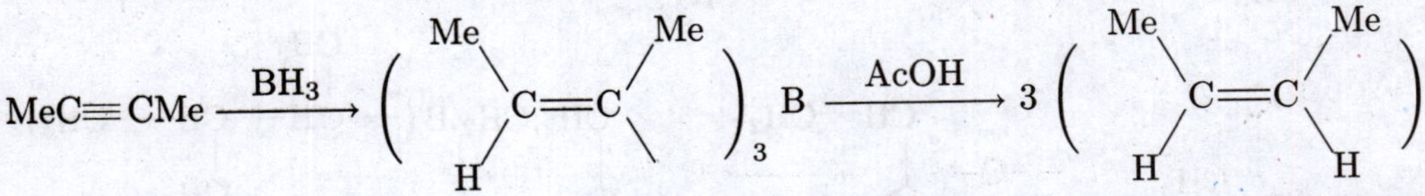

(c) Conversion of a non-terminal double bond into a terminal double bond : This involves: (i) hydroboration of the alkene, (ii) rearrangement of the 2° or 3° alkylborane into 1° alkylborane and (iii) exchange of the double bond with another alkene or oxidative hydrolysis to an alcohol followed by dehydration. Thus 2-butene can be converted to 1-butene as shown below:

$$CH_3CH{=}CHCH_3 \xrightarrow{BH_3} (CH_3CH_2.\underset{|}{C}HCH_3)_3\,B \xrightarrow{100°\text{-}200°C} (CH_3.CH_2.CH_2.CH_2)_3B$$

$$(CH_3.CH_2.CH_2.CH_2)_3B + 3\underset{\text{1-Decene}}{CH_2{=}CH(CH_2)_7.CH_3} \longrightarrow CH_2CH_2CH{=}CH_2 + (C_{10}H_{21})_3B$$

When the hydroboration of an alkene having an internal double bond is carried out at a temperature 100°-200°C, a normal alkylborane (2° or 3°) forms which then isomerises to an 1° alkylborane. This on treatment with 1-decene gives rise to an alkene having a double bond at the end. This reaction gives a better yield when a catalytic amount of diborane is used. It is reported that a boron atom can move past a single branch to a C atom but it cannot move past two branches to a C atom.

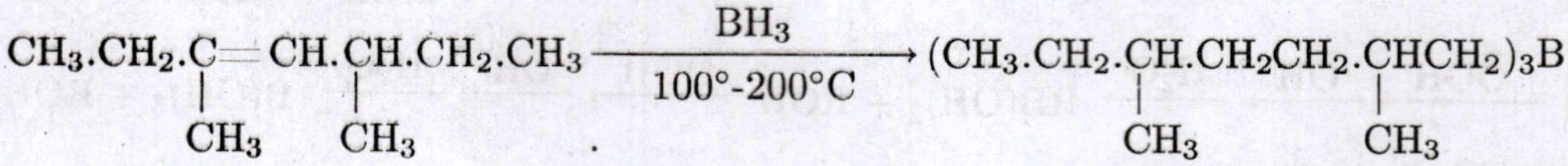

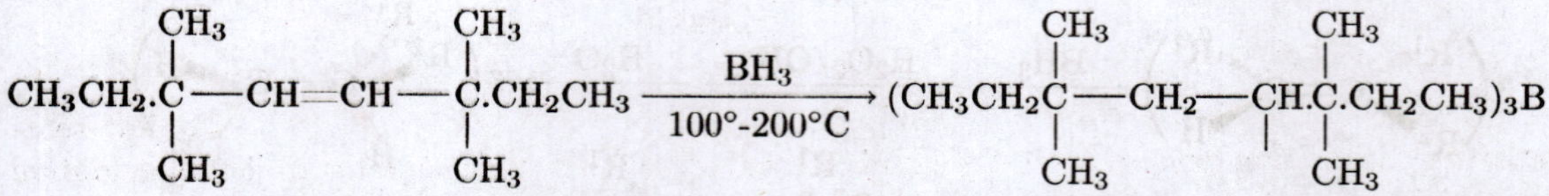

However, the rearranged alkylborane may be converted into the desired alkene by oxidative hydrolysis with alkaline hydrogen peroxide solution and then by subsequent dehydration.

$$(CH_3.CH_2.CH_2.CH_2)_3B \xrightarrow{H_2O_2/\bar{O}H} \xrightarrow{\bar{O}H} \xrightarrow{H_2O} CH_3CH_2.CH_2CH_2OH \xrightarrow[\text{or conc. } H_2SO_4;\ 175°C]{Al_2O_3/200°C} CH_3.CH_2.CH{=}CH_2$$

(d) Selective reduction and hydration by bis-3-methyl-2-butylborane [$(Si\ \alpha)_2$ BH : disiamylboron hydride] : When 2-methyl-2-butene is treated with diborane, bis-3-methyl-2-butylborane results. This can hydroboronate selectively a less hindered double bond. From the product of hydroboration desired alcohol and alkane can be prepared as usual.

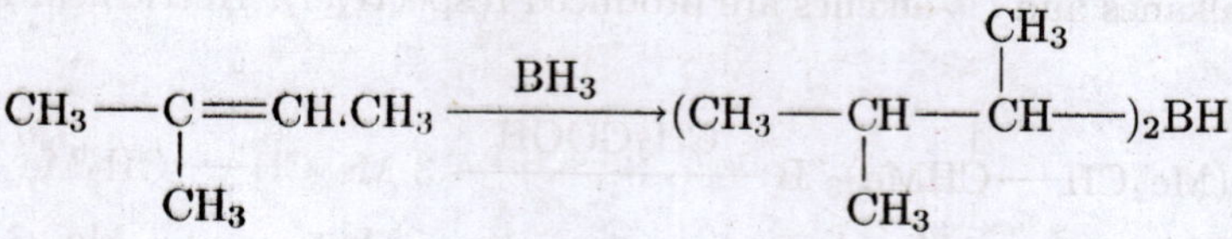

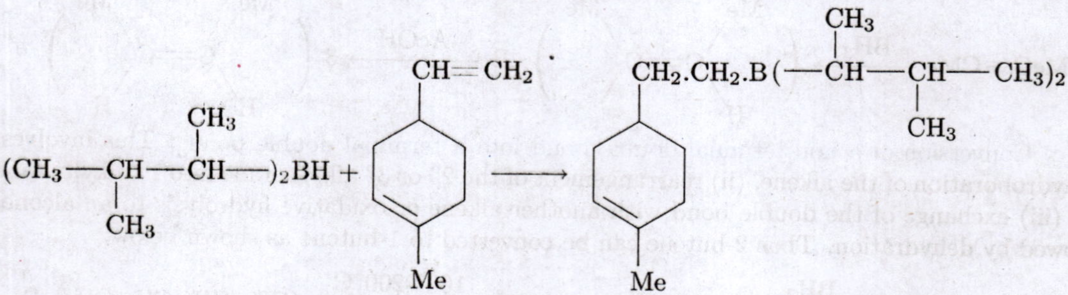

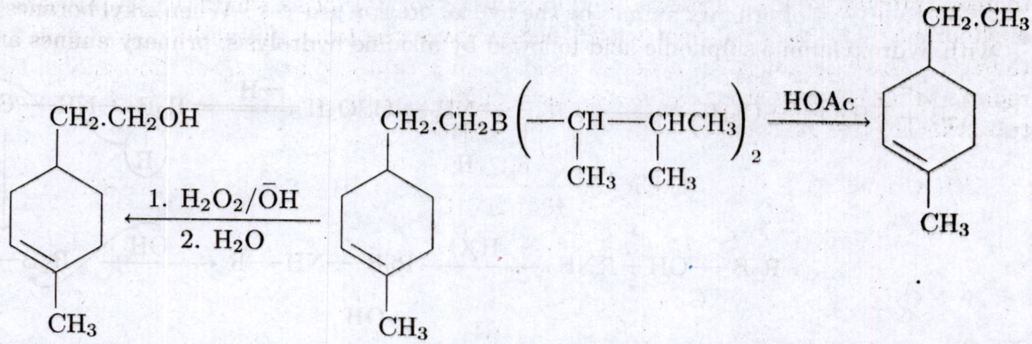

(e) Synthesis of tertiary alcohol by the hydroboration-cabonylation process : **At low temperature carbon monoxide reacts with tri-alkylboranes yielding intermediates which on oxidative hydrolysis with alkaline hydrogen peroxide give tertiary alcohols as products.**

$$R_3B + CO \longrightarrow R_3C—B=O \xrightarrow{H_2O_2/\bar{O}H} R_3COH.$$

The plausible mechanism perhaps involves :

Step I : Carbonylation; nucleophilic attack by the C atom of CO on the electron deficient B atom.

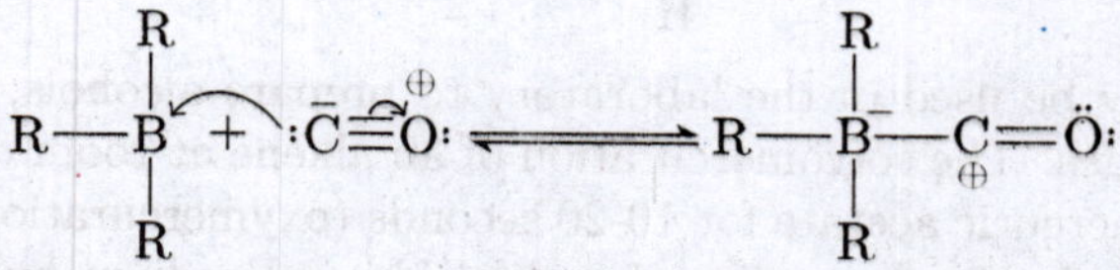

Step II : Anionic 1,2-shift of an alkyl group to electron-deficient C atom

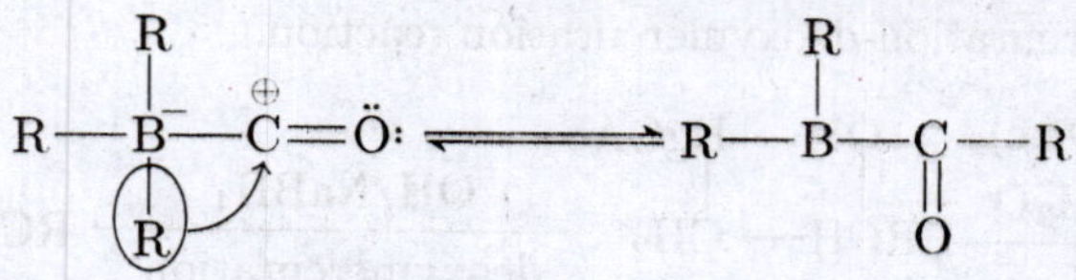

Step III : Cyclisation and further rearrangement to $R_3C—B=O$.

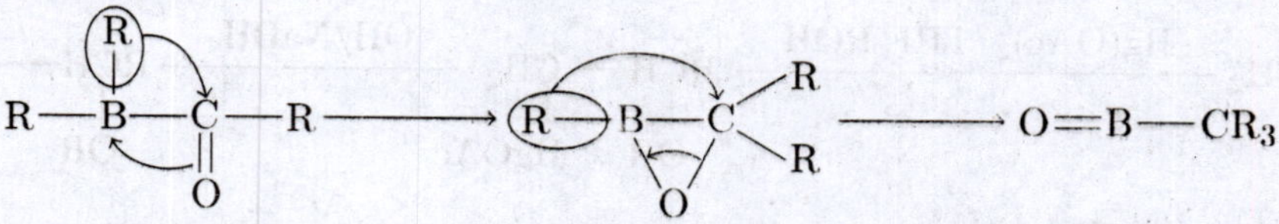

Step IV : Oxidative hydrolysis of $O=B.CR_3$.

$$O=B.CR_3 \xrightarrow{H_2O_2/\bar{O}H} R_3COH.$$

The reaction can be stopped at R_2BCOR stage by carrying out the reaction in the presence of $LiBH_4$. Then R_2BCOR may be converted to an 1° alcohol having one C atom more than the starting alkene.

$$R_2BCOR \xrightarrow[\bar{O}H]{H_2O} R_2BOH + RCH_2OH.$$

(f) Synthesis of primary amines by the hydroboration process : When alkyl boranes are treated with hydroxylamine sulphonic acid followed by alkaline hydrolysis, primary amines are formed.

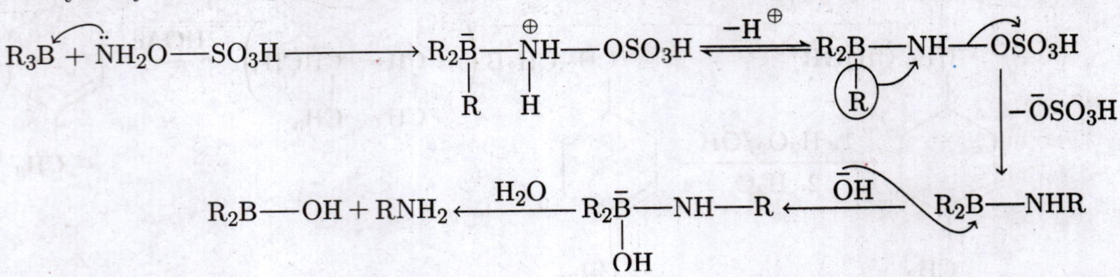

13.11 Oxymercuration-Deoxymercuration

When alkenes are treated with mercuric acetate in acetic acid, we get 1-acetoxy-2-mercuriacetate derivatives.

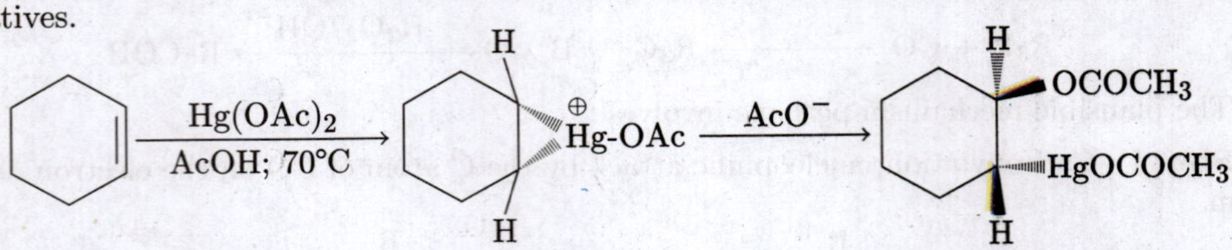

This reaction may be used in the laboratory to prepare alcohols, ethers, etc. The reactions take place in two stages. The solvomercuration of an alkene at room temperature or below using THF (solvent) and mercuric acetate for 10-20 seconds (oxymercuration stage) and then reducing the mercuriacetate derivative by sodium borohydride and sodium hydroxide for an hour or less (deoxymercuration stage) are the steps to produce an alcohol or an ether and the overall procedure may be called oxymercuration-deoxymercuration reaction.

$$RCH{=}CH_2 \xrightarrow[\text{20°C, 10-20 secs}]{\text{1.Hg (OAc)}_2\text{, THF-H}_2\text{O}} RCH(OH)-CH_2(HgOAc) \xrightarrow[\text{deoxymercuration, 1 hour}]{\bar{O}H/NaBH_4} RCHOHCH_3 + Hg + AcO^-$$

$$RCH{=}CH_2 \xrightarrow{Hg(OAc)_2/THF\text{-}ROH} RCH(OR)-CH_2(HgOAc) \xrightarrow{\bar{O}H/NaBH_4} RCH(OR)-CH_3$$

The reaction follows Markownikoff's rule; $\overset{\oplus}{Hg}OAc$ attaches itself to the C atom containing the higher number of H atoms.

$$\underset{\text{Hex-1-ene}}{CH_3CH_2CH_2CH_2CH{=}CH_2} \xrightarrow[\text{2. }\bar{O}H/NaBH_4;\ 25°C]{\text{1. }Hg(OAc)_2\ /THF\text{-}H_2O} \underset{\text{96\%, Hexan-2-ol}}{CH_3CH_2CH_2CH_2CHOHCH_3} + Hg$$

$$CH_3CH{=}CH_2 \xrightarrow[\text{2. }\bar{O}H/NaBH_4;\ 25°C]{\text{1. }Hg(OAc)_2\ /THF\text{-}MeOH} CH_3CH(OMe)-CH_3$$

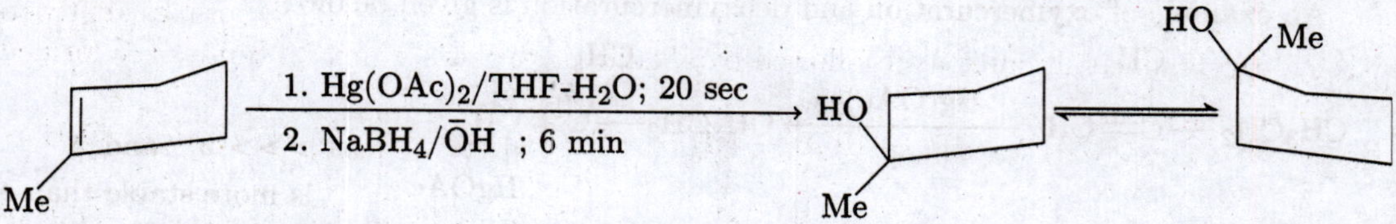

The mechanism of the reaction is as follows :

$$Hg(OAc)_2 \rightleftharpoons \overset{\oplus}{Hg}OAc + \bar{O}Ac$$

Step I : 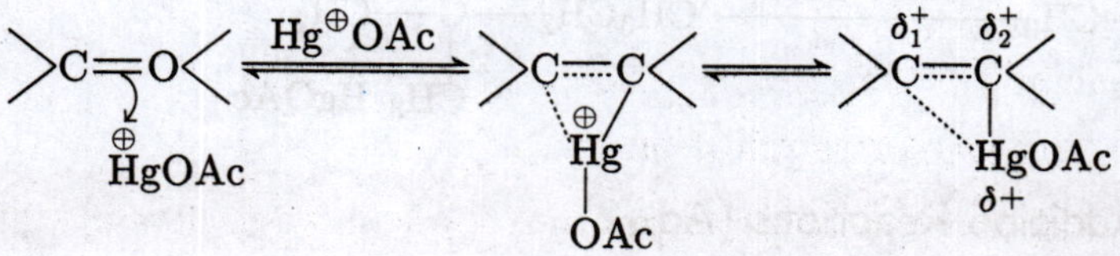

$\delta+ >>> \delta_1^+$ or δ_2^+

Step II : 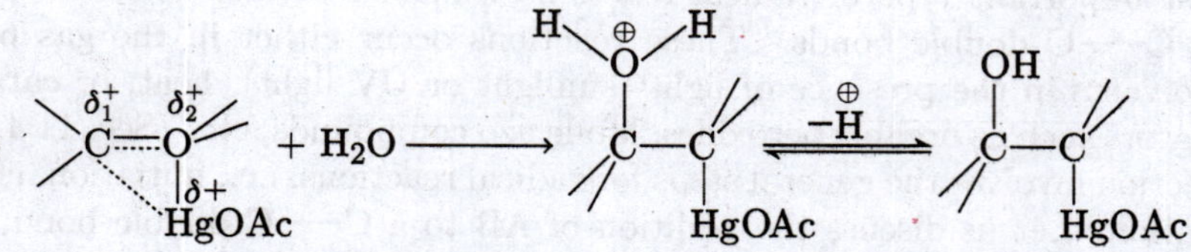

A (hydroxyalkyl) mercury compound or an (alkoxyalkyl) mercury compound forms in the oximercuration step. It is a rapid step. The positive charge resides mostly on Hg in step I and mercuriacetate cation may attack the double bonded Cs either from the hindered or from the unhindered side of the double bond. However, the nucleophilic solvent, H_2O or ROH, attacks that C which better accommodates the δ_1^+ or δ_2^+ charge; the attack occurs from the less hindered side of the double bond. In fact the more alkyl group substituted C atom holds, more positive charge than the less substituted C atom and hence the oxymercuration step follows the Markownikoff's addition :

Deoxymercuration involves alkaline reduction of the (hydroxy) or (alkoxy)alkylmercury compound by $\bar{O}H/NaBH_4$.

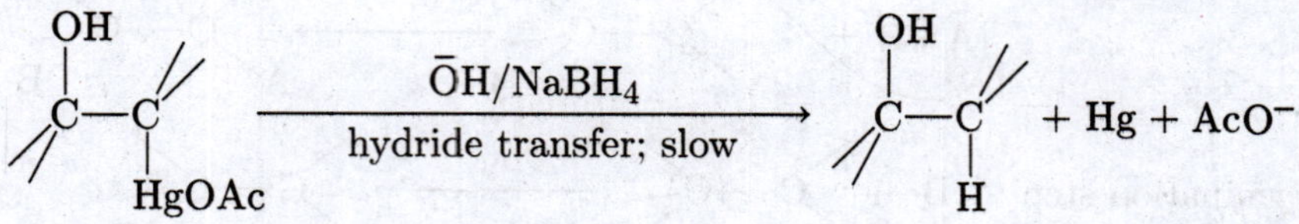

NGP has also been observed in the oxymercuration step.

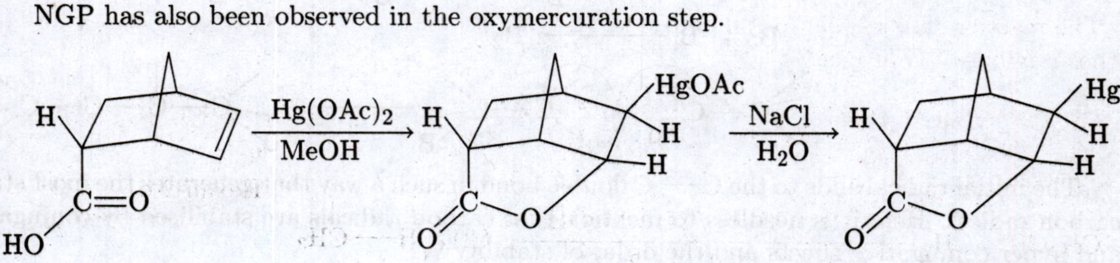

94% yield

An example of oxymercuration and deoxymercuration is given below :

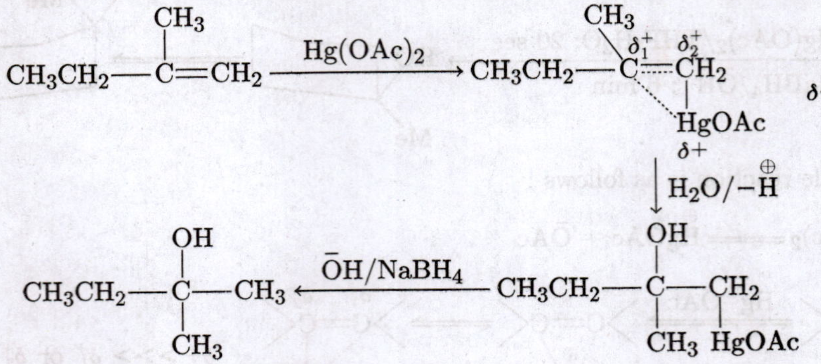

13.12 Radical Addition Reactions (Ad_H2)

The most important type of radical reactions which deserve special mention is the radical addition to C—C double bonds. These reactions occur either in the gas phase or in inert non-polar solvents in the presence of light (sunlight or UV light), heat, or catalytic amount of radical initiators such as organic peroxides, labile azo compounds, etc. (Sec. 11.4.1). The mode of addition reaction involves the general steps for radical reactions, i.e., initiation, chain propagation and termination. Let us discuss the addition of AB to a C—C double bond, where A and B may or may not be identical.

Chain initiation step : $A:B \xrightarrow[\text{light}]{\text{heat or}} A\cdot + \cdot B$

or, $\text{Initiator} \xrightarrow[\text{light}]{\text{heat or}} \text{Radical (Rad}\cdot)$

$\text{Rad}\cdot + A:B \longrightarrow \text{Rad}:A + \cdot B$

Chain propagation step :

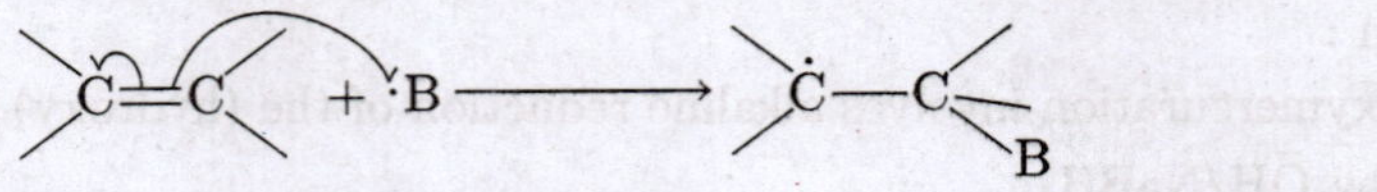

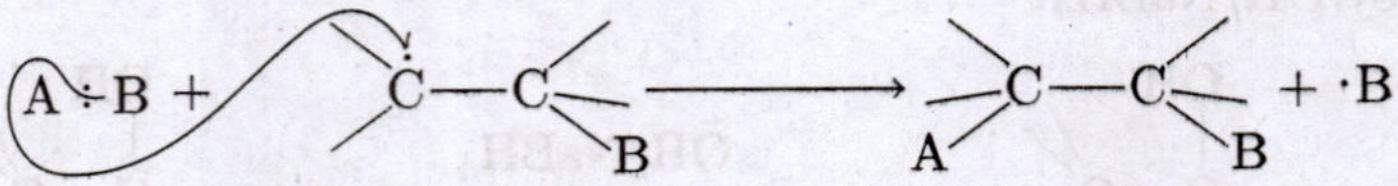

Chain termination step :

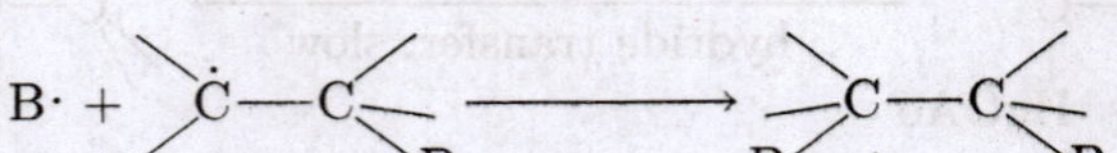

$B\cdot + \cdot B \longrightarrow B_2$

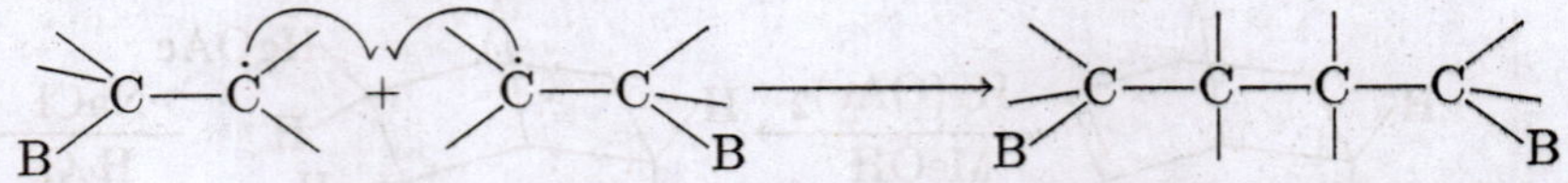

The initial radical adds to the C—C double bond in such a way that generates the most stable carbon radical. Here it is needless to mention that carbon radicals are stabilised by conjugative and hyper conjugative effects and the order of stability is :

$$\text{Benzyl} \approx \text{allyl} > 3^\circ \text{ alkyl} > 2^\circ \text{ alkyl} > 1^\circ \text{ alkyl.}$$

Those radical addition reactions may occur which involve exothermic chain propagation steps. When both the steps of the chain propagation are exothermic, the reaction occurs most readily. In Table 13.1 the heats of radical additions are given. In this table negative signs are used for exothermic reactions and positive signs are meant for endothermic reactions.

Table 13.1 The heats of radical addition reactions in kJ mol^{-1}

A—B A=B or A ≠ B	ΔH of $B\cdot + CH_2{=}CH_2 \rightarrow BCH_2{—}\dot{C}H_2$	ΔH of $BCH_2{—}\dot{C}H_2 +$ A : B $\rightarrow BCH_2{—}CH_2A + B\cdot$	A—B A=B or A ≠ B	ΔH of $B\cdot + CH_2{=}CH_2 \rightarrow BCH_2{—}\dot{C}H_2$	ΔH of $BH_2{—}\dot{C}H_2$ + A : B $\rightarrow BCH_2{—}CH_2A + B\cdot$
H—H	−167.5	+25.1	HO—Cl	−134	−71.2
H—F	−188.7	+154.9	Cl—Cl	−79.5	−79.5
H—Cl	−79.5	+20.9	Br—Br	−20.9	−71.2
H—Br	−20.9	−46.0	I—I	+37.7	−54.5
H—I	+37.7	−113.0	H—OH		+154.9

Table 13.1 tells us that the radical addition of HBr, HOCl, Cl—Cl and Br—Br are only energetically favourable since these reagents involve exothermic chain propagation steps.

Most of the radical addition reactions are non-stereospecific. A very few stereospecific reactions have been reported which take place at a temperature much below 0°C. However, there are examples of stereoselective reactions even at higher temperature. In this section we shall discuss radical additions of halogens and hydrogen halides to C—C double bonds.

13.12.1 Radical addition of halogens

The most common radical addition involving halogen is the addition of bromine in carbon tetrachloride to a C—C double bond. In this reaction the brown colour of bromine gets discharged without evolution of any gas. This reaction is regarded as a quantitative test to indicate the presence of unsaturation in a compound.

$$>C{=}C< + Br_2 \text{ (in } CCl_4) \longrightarrow Br{—}\underset{|}{\overset{|}{C}}{—}\underset{|}{\overset{|}{C}}{—}Br$$

The method for adding halogens to a C—C double bond is the general procedure as discussed in Sec. 13.3.2.

The reaction steps are :

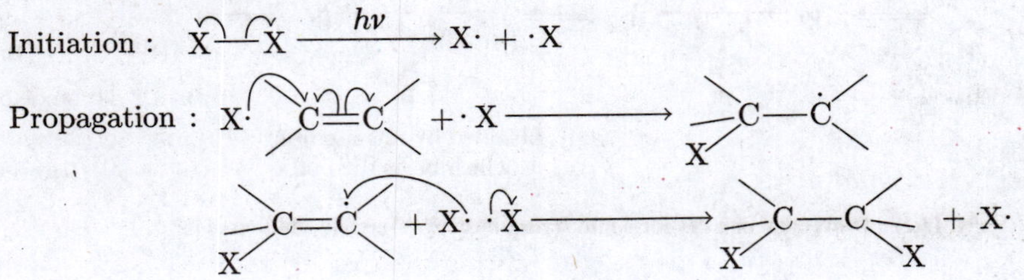

Termination : $X\cdot + \cdot X \longrightarrow X_2$

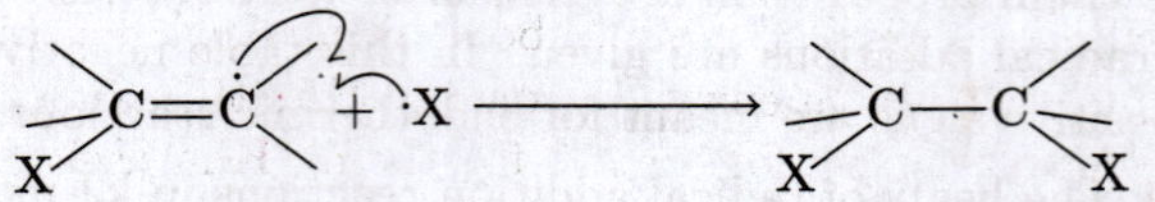

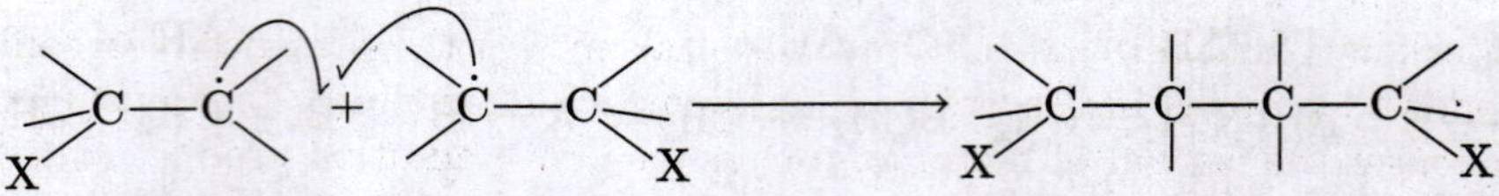

However, fluorine adds to an alkene vigorously even in the absence of light and initiators and gives a variety of products including substitution products and hence the reaction has no practical importance. For example, the radical addition of fluorine to benzene gives a mixture of products containing octa-, nona- and deca-fluorobenzenes. Chlorine adds readily to an alkene and gives addition products below 200°C; the reaction is not readily reversible and occurs with long chains. The radical addition of bromine occurs readily but reversibly with short reaction chains. Whereas iodine addition is readily reversible and very difficult to perform since the first propagation step is endothermic and since vicinal diiodides are very unstable.

The radical addition of halogens has some preparative value. In fact, it is a general method for preparing vicinal dihalides :

$$Cl_2 + CH_3—CH=CH_2 \xrightarrow{h\nu} CH_3—CHCl—CH_2Cl$$

When a mixture of chlorine and benzene is irradiated with UV light (λ = 360 – 400 nm), a mixture of isomeric benzene hexachlorides is obtained. The mixture contains, amongst others, a strong insecticide—γ-benzenehexachloride (13 to 18%; popularly known as gammexane).

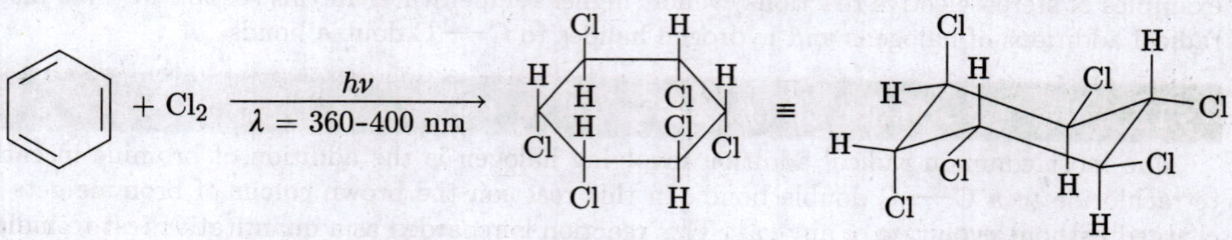

γ-Benzenehexachloride or Gammexane

The reversible addition of bromine and iodine to a C—C double bond may be used to convert a less stable geometric isomer to a more stable one. Thus cis-1, 2-dibromoethane may be converted to the trans-isomer.

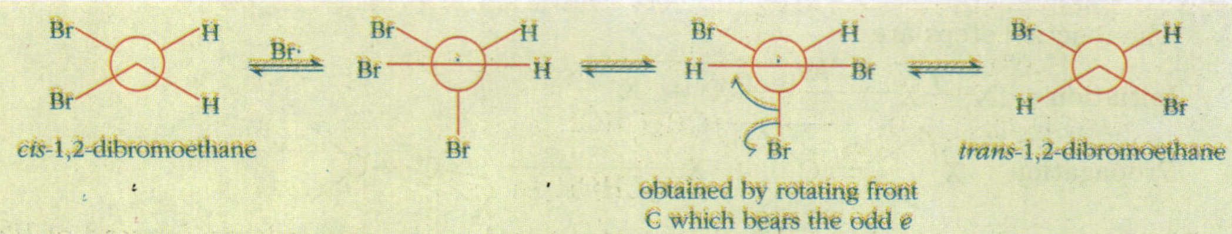

Fig. 13.26. Conversion of a *cis*-form into *trans*-form by reversible addition of Br·

13.12.2 Radical addition of hydrogen halides

The radical addition of HBr to a C—C double bond occurs in the presence of UV light or a small amount of initiators such as dibenzoyl peroxide. HF, HCl and HI do not undergo the reaction because it is not energetically favourable for them. Kharasch in 1932 studied extensively the radical addition of HBr across a C—C double bond and found that it occurs anti-Markownikoffically with unsymmetrical alkenes in the presence of organic peroxides. Thus HBr adds to propene in the presence of dibenzoylperoxide at a temperature 60°C-80°C to give 1-bromopropane instead of 2-bromopropane (the ionic addition product of HBr to propene).

$$HBr + CH_3\cdot CH{=}CH_2 \xrightarrow[\text{60°C-80°C}]{\text{dibenzolyl peroxode}} \begin{cases} \longrightarrow CH_3\cdot CH_2\cdot CH_2\cdot Br \\ \not\longrightarrow CH_3\cdot CHBr\cdot CH_3 \end{cases}$$

According to Kharasch the change in orientation is due to the presence of organic peroxide and so the change in orientation is said to be peroxide effect and the principle of addition is called **Kharasch principle.**

The mechanistic steps for the reaction are :

Step I : Thermolysis of dibenzoyl peroxide to benzoylperoxy radical

$$\text{PhCO.O.O.CO.Ph} \xrightarrow{\text{60°C-80°C}} 2\text{PhC(=O)O}\cdot$$

Step II : Generation of bromine radical from HBr.

$$\text{PhC(=O)O}\cdot + \text{H}\cdot\cdot\text{Br} \longrightarrow \text{PhCOOH} + \cdot\text{Br}$$

Step III : Electrophilic addition of bromine radical to a double-bonded C atom which generates the more stable carbon radical.

$$CH_3{-}CH{=}CH_2 + \cdot Br \longrightarrow CH_3\dot{C}HCH_2Br$$

Step IV : Abstraction of a hydrogen radical from another molecule of HBr to give 1-bromopropane and further bromine radical.

$$CH_3\dot{C}HCH_2Br + H\cdot\cdot Br \longrightarrow CH_3CH_2CH_2Br + \cdot Br$$

Termination

Step :

$$Br\cdot + \cdot Br \longrightarrow Br_2$$

$$CH_3\dot{C}HCH_2Br + \cdot Br \longrightarrow CH_3CHBrCH_2Br$$

$$2CH_3\dot{C}HCH_2Br \longrightarrow \begin{matrix} CH_3CHCH_2Br \\ | \\ CH_3CHCH_2Br \end{matrix}$$

Thus we see that bromine and hydrogen atoms add to the double bond (instead of $H^{\oplus}$ and Br^-) and a carbon radical is an intermediate (instead of a carbenium ion) in the addition of HBr in the presence of a peroxide to a C=C.

The above drawn mechanism is supported by the following facts :

1. A small amount of peroxide is sufficient to add HBr across a double bond anti-Markownikoffically. This is in good agreement with the proposed chain reaction mechanism—once a free radical is formed it can run the chain until an inhibitor, such as O_2, hydroquinone or heat stops the chain.
2. Since a catalytic amount of an inhibitor stops the reaction, the reaction most probably has taken place through a chain reaction.
3. When a mixture of HBr and propene is irradiated with UV light, 1-bromopropane results. We know that UV light generates free radical. Therefore, this reaction involves radical addition. Since in the presence of a peroxide, HBr adds to propene to give 1-bromopropane, this reaction may also involve radical addition.
4. Recently the presence as well as the structure of free radicals have been established by electron- spin-resonance spectroscopy.

However, when the reaction is carried out by UV irradiation, it is the Br· radical and not the H· radical which starts the propagation step. Obviously the Br· radical attaches itself to that double-bonded C atom which generates the more stable carbon radical. Since the stability order of the alkyl radicals is 3° > 2° > 1°, the Br· radical attaches itself to the least substituted double-bonded C atom and thereby generates 2° or 3° carbon radical and consequently anti-Markownikoff's addition occurs through the abstraction of H· radical from HBr by the incipient carbon radical in the next propagation step.

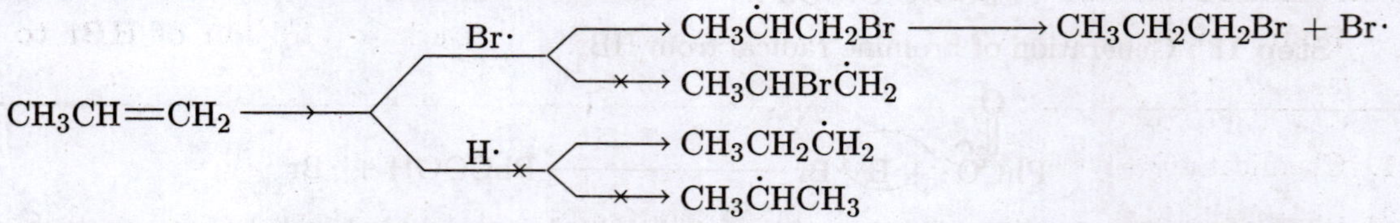

It has already been stated that only HBr and none of the hydrogen halides can add to a C—C double bond by radical mechanism. If we give a close look on Table 13.1, it will reveal that the first propagation step for H—F is exothermic ($\Delta H = -188.7$ kJ mol^{-1}) and favours the reaction, while the second propagation step is highly endothermic ($\Delta H = +154.9$ kJ mol^{-1}) owing to the high strength of the H—F bond. It is thus difficult to break and therefore, the reaction is energetically unfavourable. For HCl, too, the first propagation step is exothermic ($\Delta H = -79.9$ kJ mol^{-1}), but the second one is again endothermic ($\Delta H = +20.9$ kJ mol^{-1}) but to a lesser extent and hence the reaction is energetically less favourable. In the case of HI the second propagation step is exothermic and favourable ($\Delta H = -113.0$ kJ mol^{-1}), but the I—C bond formation step is endothermic ($\Delta H = +37$ kJ mol^{-1})—this indicates that the energy gained in the I—C bond formation is not enough for homolytic fission of the C—C π bond and hence the overall reaction is not favourable energetically. Only in the case of HBr, both the propagation steps are exothermic and the reaction takes place readily.

At room temperature the radical addition of HBr to an acyclic alkene, cis or trans, is found to give 78% trans-addition product and 22% cis-addition product. Therefore, the reaction is trans-stereoselective. Perhaps after the formation of bromoalkyl radical there is time for the radical to assume a different conformation owing to the rotation of a part about C—C σ bond.

On the other hand, when liquid H—Br is added to an acyclic alkene at −80°C, the reaction is found to be almost stereospecific; in this case stereoselective trans-addition occurs whereby trans-alkanes give 92% meso product but cis-isomers give almost cent percent dl-pair. To account for the stereospecificity of the reaction a bridged bromonium radical is considered as an intermediate in analogy to the bridged bromonium ion in the ionic addition of HBr to alkenes.

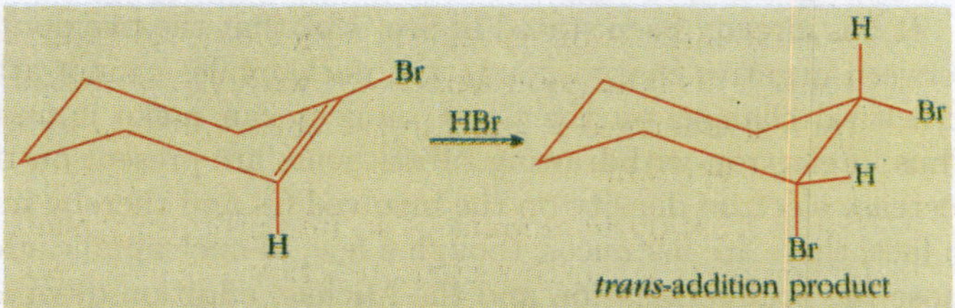

Fig. 13.27. Stereospecific radical addition of HBr to cyclohexene

The radical addition of HBr to 1-bromocyclohexene is also found to be stereospecific and involves stereoselective trans-addition.

Here perhaps the second chain propagation step occurs so rapidly that the chair form which takes part in the first propagation step gets no time to flip to the other chair form.

Since alkenes are easily obtained from cracking of natural gas and petroleum and since the radical addition of hydrogen bromide yields anti-Markownikoff addition product, 1° alkyl bromide and 1° alcohols (by the hydrolysis of alkyl bromides) can be synthesized directly.

However, it is very difficult to conduct a pure radical addition; moreover, it does not give a pure product. It is also difficult to isolate the required one single product from the mixture of products obtained. So nowadays 1° alcohols are prepared by hydroboration-oxidation process.

Now we are in a position to compare the two addition processes, the ionic addition of HBr to alkenes and the radical addition of HBr to alkenes. This has been done in the following chart:

Points of comparison	Ionic addition of HBr to an alkene	Radical addition of HBr to an alkene
1. Chemical units which add	$\overset{\oplus}{H}$ and Br^-	Br· and H·
2. Intermediate	Protonium ion complex or Carbenium ion	Carbon radical
3. Which chemical unit adds first	$\overset{\oplus}{H}$	Br·
4. Orientation of addition	According to Markownikoff's empirical rule.	Against Markownikoff's rule.
5. Solvent	Polar solvent, like HOAc.	Non-polar solvent, e.g., CCl_4. The reaction may occur in the gas phase.
6. Initiator		Peroxides and/or heat or light
7. Stereoselectivity	Less stereoselective.	Less stereoselective.
8. Purity of the product	Almost pure.	Not at all pure.
9. Importance	It has some preparative value; alkyl halides may be prepared from natural products and petroleum	It has no preparative value.

13.13 Nucleophilic Addition To C=C (Ad_N2)

It has already been stated in See. 13.2 that the two double-bonded C atoms being sandwiched between negative charge clouds, the nucleophiles cannot attack them easily. However, favourable structural changes on the C=C moiety can make it undergo nucleophilic addition reactions. Thus, if electron-withdrawing substituents are present on the multiple bonded C atom, they will decrease electron density on the involved Cs and thereby nucleophilic addition will be facilitated. In fact, there are instances, though a few, of nucleophilic addition across a double bond. Amongst those the cyanoethylation and the Michael addition deserve special mention.

However, a nucleophilic addition takes place in two separate steps:

Step I : The incoming nucleophile attacks the more electropositive double-bonded C atom and forms an intermediate carbanion.

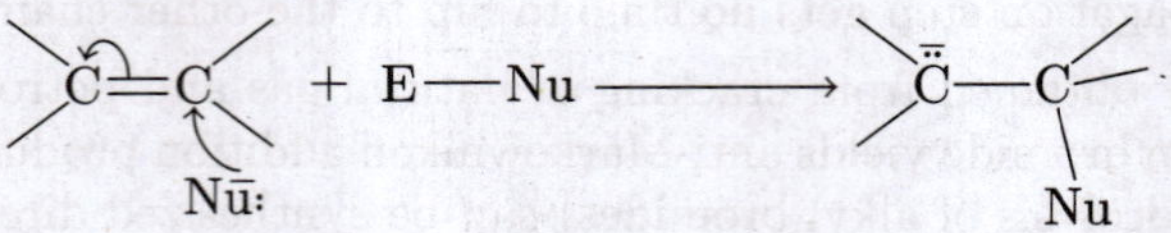

intermediate carbanion

Step II : The intermediate carbanion then couples with a positively charged or electron deficient species which can provide a vacant orbital for a pair of *e*.s on the anionic C atom.

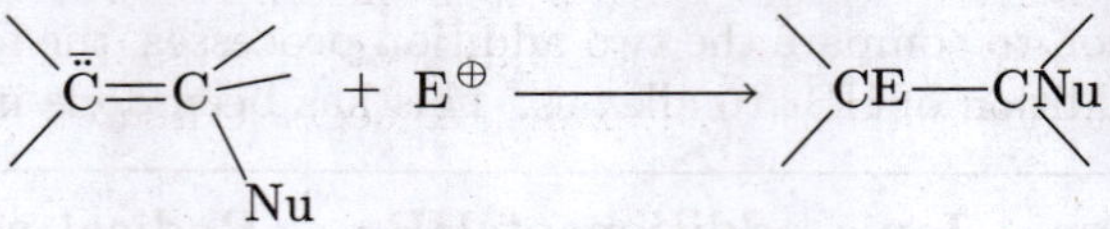

Thus the overall reaction is :

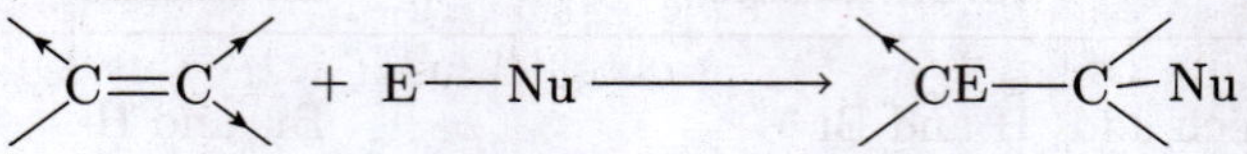

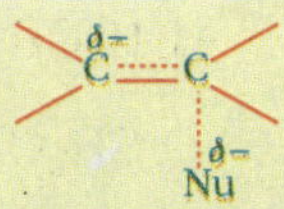

Fig. 13.28. The T.S. for neucleophilic addition

The first step being slow, it is the rate-determining step. The T.S. of the rate-determining step will be Fig. 13.28.

Since the T.S. of the rate-determining step involves two species, the reaction is expected to follow the second order kinetics. Thus, the rate = k [substrate] [$\bar{\ddot{\text{Nu}}}$].

The free energy diagram for the nucleophilic addition to a C=C may be represented by Fig. 13.29.

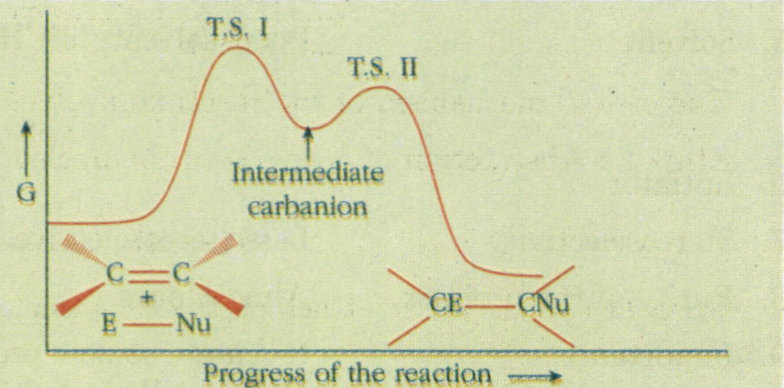

Fig. 13.29. Energy profile for the nucleophilic addition to a C=C moiety

Simple alkenes or alkenes with electron-releasing substituents do not undergo nucleophilic addition. Polyhalo, polycyano alkenes or alkenes with strong electron-withdrawing substituents such as —CHO, —COR, —NO_2, etc. undergo nucleophilic addition reaction.

Very little is known about the constitutional orientation in the nucleophilic addition except for Michael type of addition reaction. In the latter case negative part of the addendum, the nucleophile, always adds to the C atom which does not carry the —CHO, or —COR, etc. group.

Since very little work has been done on the nucleophilic addition reaction, its stereochemistry is not known to us. However, in the Michael type of condensation with cyclic substrates, exclusive cis-addition in some cases and in some others trans-addition have been reported and thus the reactions are stereoselective.

13.13.1 The cyanoethylation

The addition of cyanoethyl group, —CH_2CH_2CN, to a nucleophile is said to be cyanoethylation process. The reagent for the reaction is acrylonitrile, $CH_2{=}CHCN$. The substrates are usually compounds containing active hydrogen, e.g., alcohol, phenol, amine, etc.

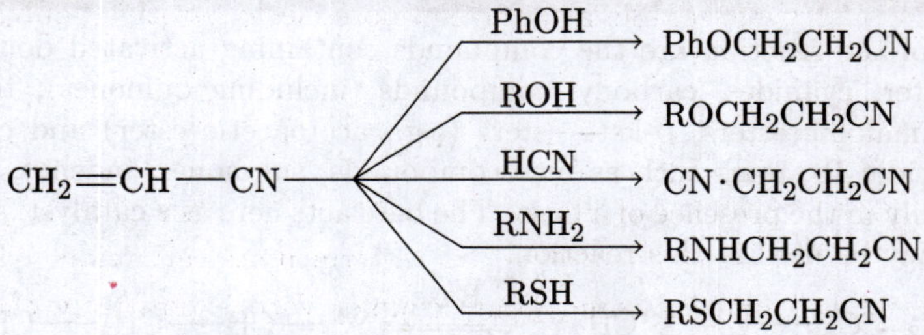

The reaction is often carrried out in a polar solvent in the presence of a base which generates a stronger nucleophile by abstracting an active H as proton. A good catalyst for these reactions is triton-B, trimethylbenzylammonium hydroxide. Aqueous ethanolic alkali solutions may also be used as bases.

The reaction has immense synthetic utility. It is a typical means by which three C atoms can be incorporated in a compound at a time. Furthermore, the cyano group can be reduced, oxidised and converted to other functional groups. Thus it helps to prepare a large number of compounds.

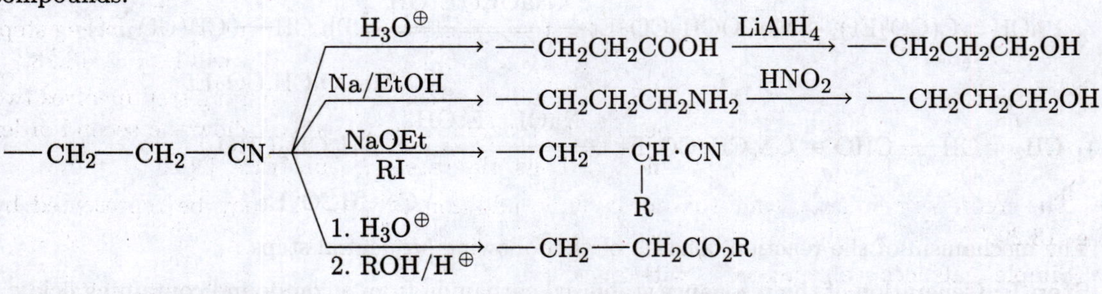

The overall mechanism of the reaction involves the following steps :

Step I : Abstraction of active H of the nucleophile, H—Nu, by the base.

$$H{-}Nu \xrightleftharpoons{\bar{B}:} HB + N\ddot{\bar{u}}$$

Step II : Nucleophilic attack on the C_3 atom and formation of a resonance stabilised carbanion.

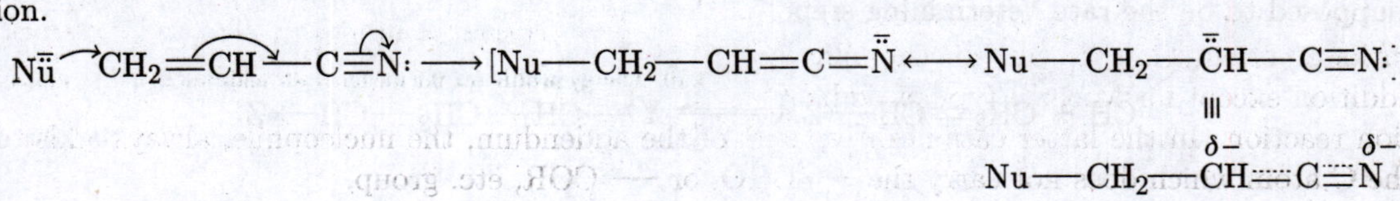

Step III : Abstraction of proton from the conjugate acid or from the solvent.

$$\text{Nu—CH}_2\text{—CH}=\text{C}=\ddot{\text{N}}\text{:} \underset{\ddot{\text{B}}\text{:}}{\overset{\text{HB}}{\rightleftharpoons}} \text{Nu—CH}_2\text{—CH}=\text{C}=\ddot{\text{N}}\text{H}$$

Step IV : Tautomerisation of the enamine to the nitrile.

$$\text{Nu—CH}_2.\text{CH}=\text{C}=\ddot{\text{N}}\text{H} \rightleftharpoons \text{NuCH}_2\text{CH}_2.\text{CN}.$$

Thus the overall addition seems 3, 4 but it occurs via 1, 4- or, conjugate addition.

13.13.2 The Michael addition (condensation) reaction

The substrates of the reaction are the compounds containing activated double bonds, e.g., α,β-unsaturated ester, cyanides, carbonyl compounds (including quinones); the reagents are cyanoacetic esters, malonic esters, β-keto esters (e.g., acetoacetic ester) and compounds containing relatively acidic H atoms such as nitrocompounds, sulphones, indenes, fluorenes which form carbanion readily in the presence of a base. The base acts here as a catalyst; alkali, alkoxides, secondary amines may be used in this reaction.

$$\text{CH}_2=\text{CH—Z} + \text{Y}^1\text{CH}_2\text{Y}^2 \overset{\text{Base}}{\rightleftharpoons} \text{Y}^1\text{—}\underset{\text{Y}^2}{\underset{|}{\text{CH}}}\text{—CH}_2\text{—CH}_2\text{—Z}$$

where Z, Y^1 and y^2 are electron-withdrawing groups.

For example :

1. $$\text{CH}_2=\text{CH—CN} + \text{CH}_2(\text{CO}_2\text{Et})_2 \overset{\text{NaOMe/MeOH}}{\rightleftharpoons} \text{CO}_2\text{Et.}\underset{\text{CO}_2\text{Et}}{\underset{|}{\text{CH}}}\text{.CH}_2.\text{CH}_2.\text{CN}$$

2. $$\text{PhCH}=\text{C}(\text{CO}_2\text{Et})_2 + \text{MeCOCH}_2\text{CO}_2\text{Et} \overset{\text{NaOEt/EtOH}}{\rightleftharpoons} \text{Ph.}\underset{\text{MeCO.CH.CO}_2\text{Et}}{\underset{|}{\text{CH}}}\text{—CH}(\text{CO}_2\text{Et})_2$$

3. $$\text{CH}_2=\text{CH—CHO} + \text{CN.CH}_2\text{CO}_2\text{Et} \overset{\text{NaOEt/EtOH}}{\rightleftharpoons} \underset{\text{CNCHCO}_2\text{Et}}{\underset{|}{\text{CH}_2}}\text{CH}_2\text{CHO}$$

The mechanism of the reaction consists of the following sequential steps :

Step I : Generation of the resonance stabilised carbanion from a compound containing acidic H atom by the base-catalysed abstraction of proton :

$$\text{Y}^1\text{CH}_2\text{Y}^2 \overset{\text{base}}{\rightleftharpoons} \text{Y}^1\overset{-}{\ddot{\text{C}}}\text{HY}^2 + \text{BH}^{\oplus}$$

Step II : Forced addition of the carbanion, a nucleophile, to the double bond and formation of the resonance stabilised carbanion. This step involves the carbon-carbon bond formation and is supposed to be the rate determining step.

$$(\text{Y}^1)(\text{Y}^2)\overset{-}{\ddot{\text{C}}}\text{H} + \text{CH}_2=\text{CH—Z} \rightleftharpoons \text{Y}^1\text{—}\underset{\text{Y}^2}{\underset{|}{\text{CH}}}\text{—CH}_2\text{—}\overset{-}{\ddot{\text{C}}}\text{H—Z}$$

Step III : The carbanion then combines with a positively charged species, usually a proton, to produce 3, 4-addition product. In practice actually 1, 4-addition product is formed first which then tautomerises to the 3, 4-addition product.

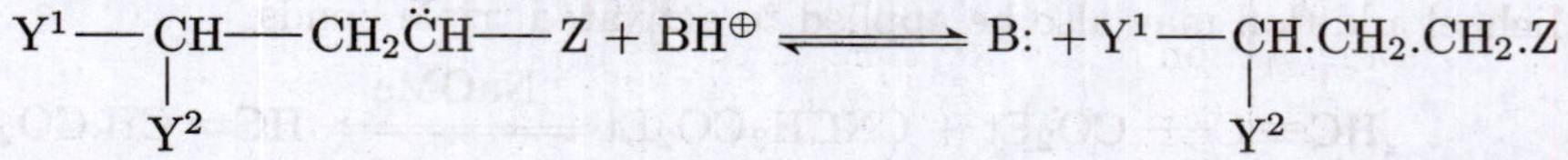

Let us consider the mechanism of a particular reaction, the reaction between methyl vinyl ketone and nitroethane in the presence of sodium methoxide; here nitroethane is the reagent and methyl vinyl ketone is the substrate.

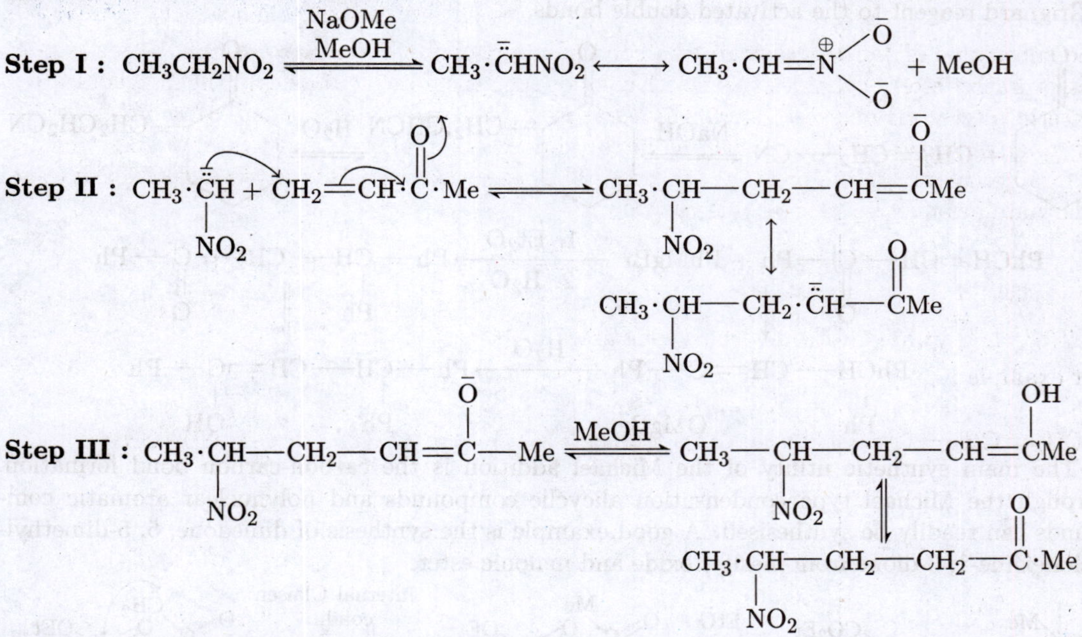

The Michael additions are reversible reactions. The Michael 3, 4-addition products may often be decomposed by heating with or without alkali to produce compounds different from the starting substrate and the phenomenon is known as the Michael retrogression or reversal. Thus the Michael addition product of ethyl benzylidenemalonate and ethyl acetoacetate on decomposition with NaOEt gives rise to the reagent (ethyl acetoacetate), the substrate (ethyl benzylidenemalonate), and ethyl malonate and benzylideneacetoacetate (retrogression products); a portion of the addition product also remains in equilibrium with the decomposition products.

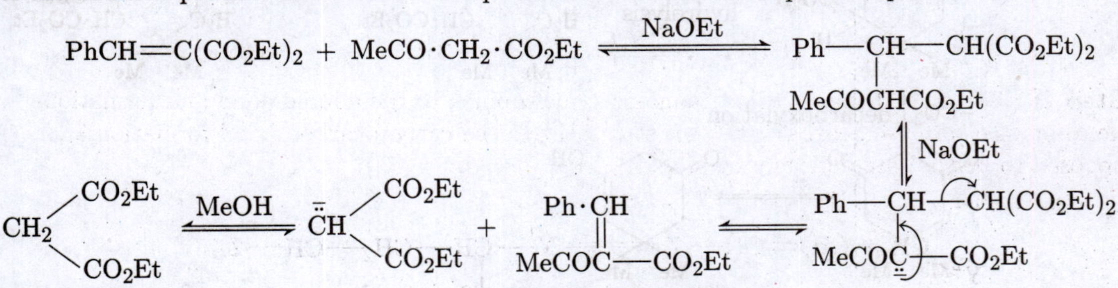

Ethyl malonate Benzylideneacetoacetate

In this reaction ethyl malonate and Bezylidene acetoacetate are the retrogression products.

To avoid the formation of retrogression products usually mild bases are employed to carry out Michael condensation.

The Michael addition may also be applied to activated triple bonds.

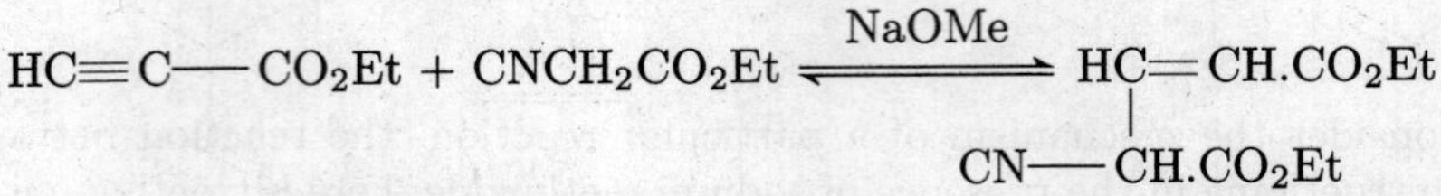

Reactions closely related to the Michael addition reaction are cyanoethylation and the addition of Grignard reagent to the activated double bonds.

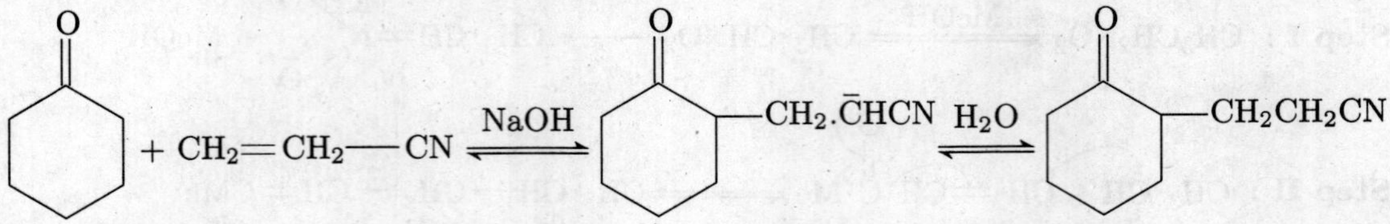

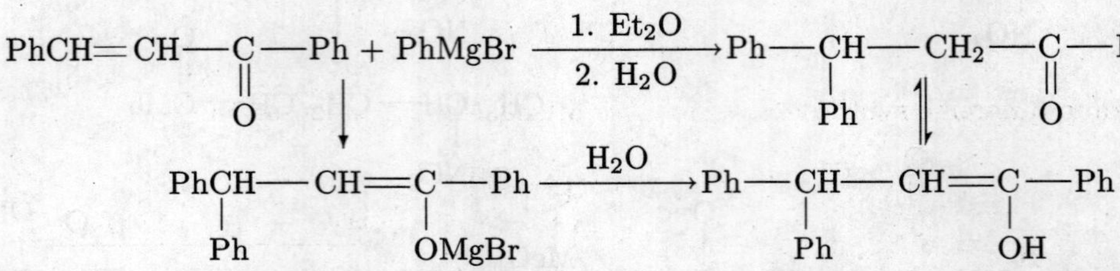

The main synthetic utility of the Michael addition is the carbon-carbon bond formation. Through the Michael type condensation alicyclic compounds and polynuclear aromatic compounds can readily be synthesised. A good example is the synthesis of dimedone, 5, 5-dimethylcyclohexane-1, 3-dione from mesityl oxide and malonic ester.

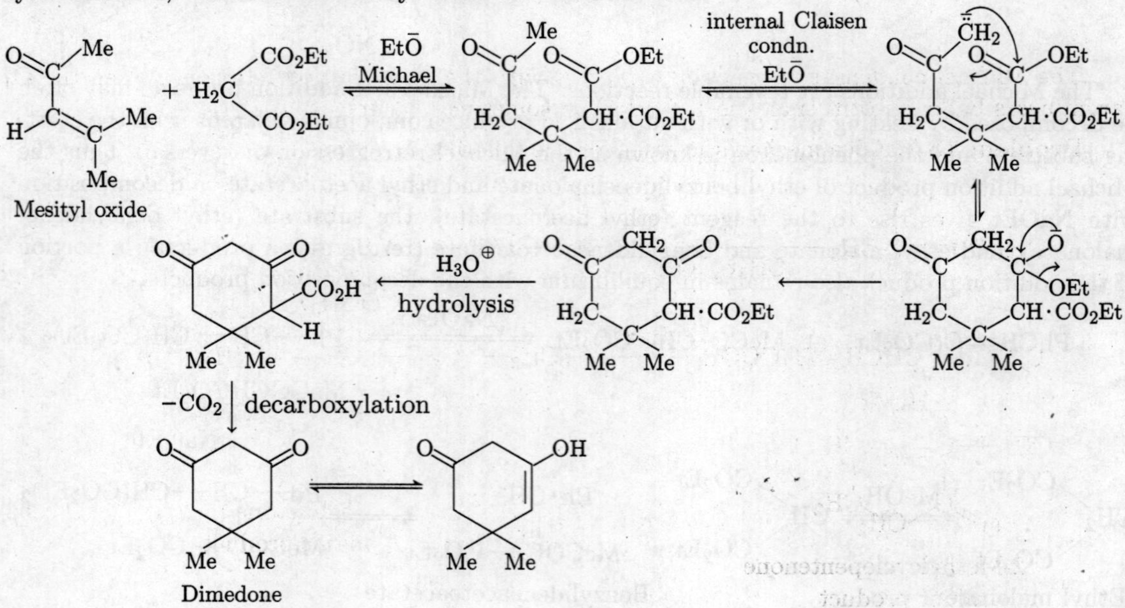

Michael addition reaction followed by ring formation or annelation is Robinson annelation reaction as we have seen in the dimedone synthesis. The other example is given below :

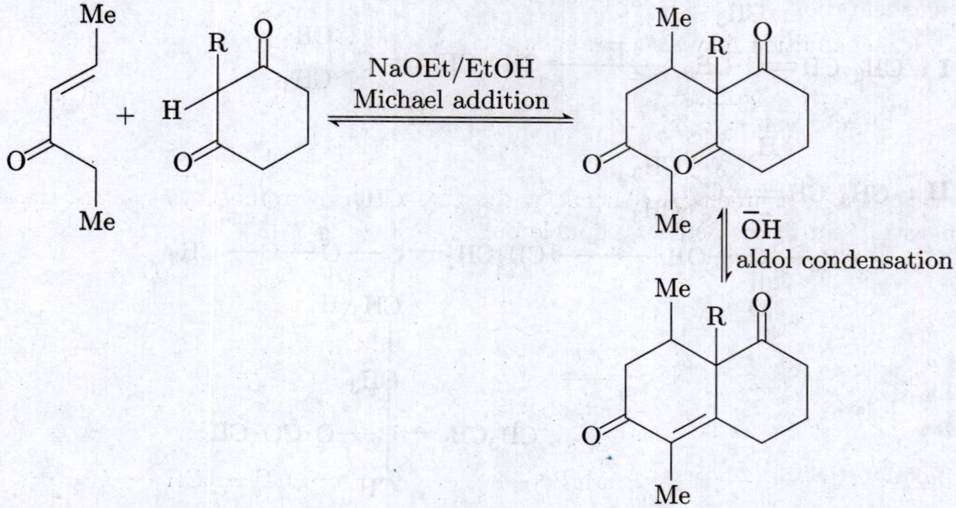

The Skraup quinoline synthesis also involves Michael-type condensation.

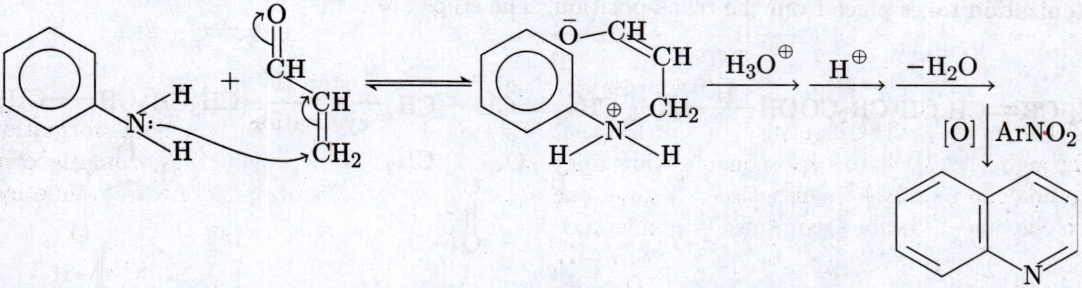

The Michael condensation reaction competes with the 1, 2-addition reaction. When the C_4 atom bears bulky substituents, the 1, 2-addition sometimes occurs.

Problem 13.7. Give mechanistic steps for the following reactions :

(i) $CH_3.CH{=}C(CH_3).CH_3 \xrightarrow[CH_3COOH]{H^{\oplus}} CH_3{-}CH_2{-}C(CH_3)(OCOCH_3).CH_3$

(ii)

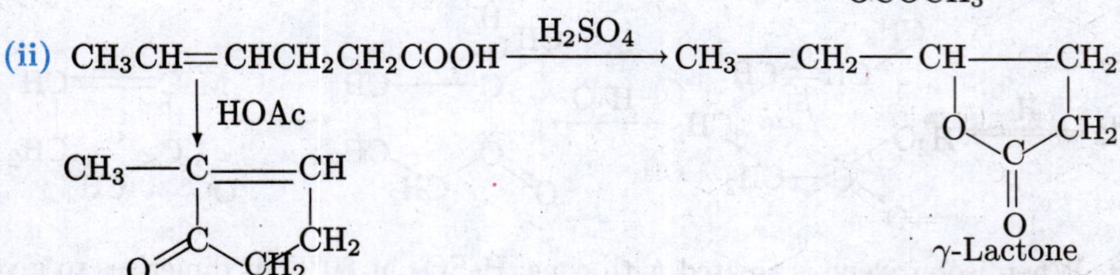

γ-Lactone
major product

2-Methylcyclopentenone
minor product

Solution : (i) This is an acid-catalysed addition reaction of a carboxylic acid across a C—C double bond. The steps for the reaction are :

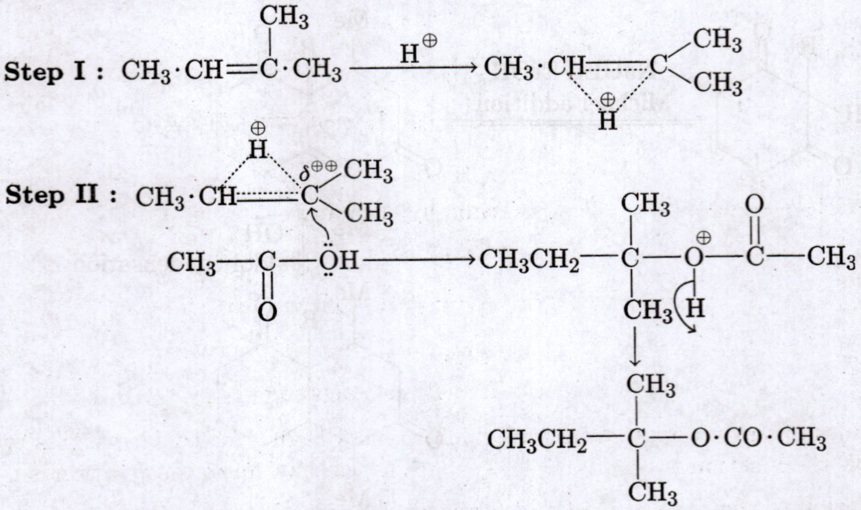

(ii) When an enoic acid is treated with a strong acid, H_2SO_4, internal esterification, i.e., lactonization takes place from the trans-position. The steps are:

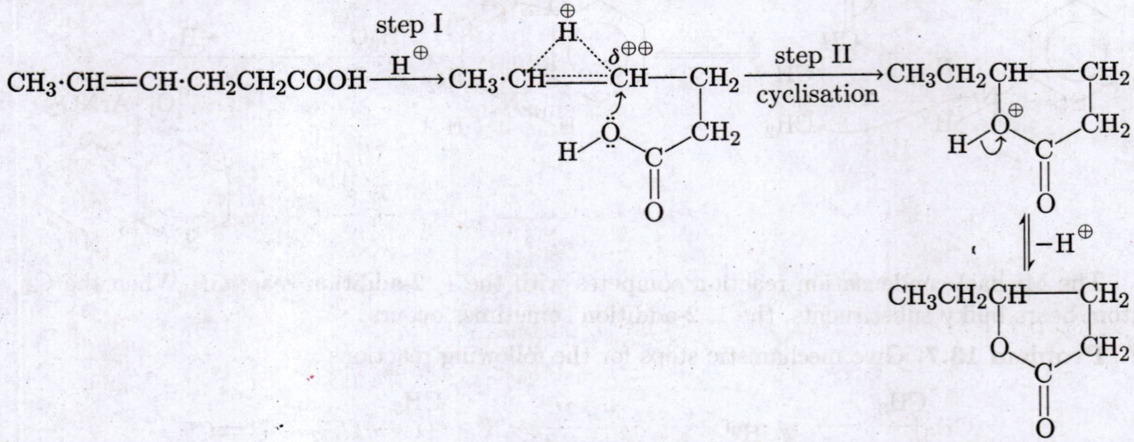

When a Lewis acid or a weak protic acid is used, instead of lactonization, a cyclic ketone results.

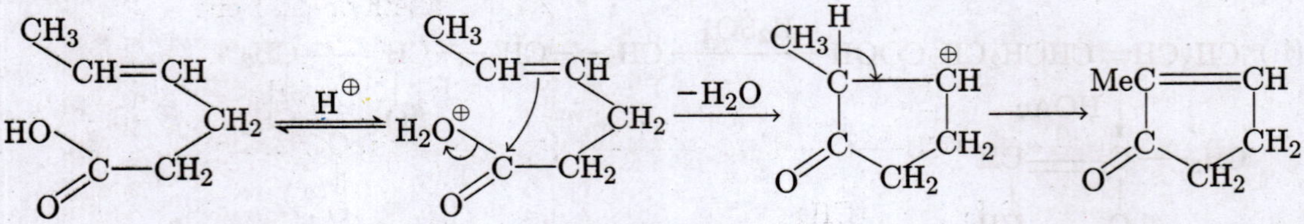

Problem 13.8. When isobutylene is treated with conc. H_2SO_4 at 80°C, it dimerises to give two products—name the products and give the mechanism for the reaction. What happens when the products are catalytically reduced?

Solution : This reaction involves acid-catalysed addition of one molecule of alkene to the other molecule.

The products are :

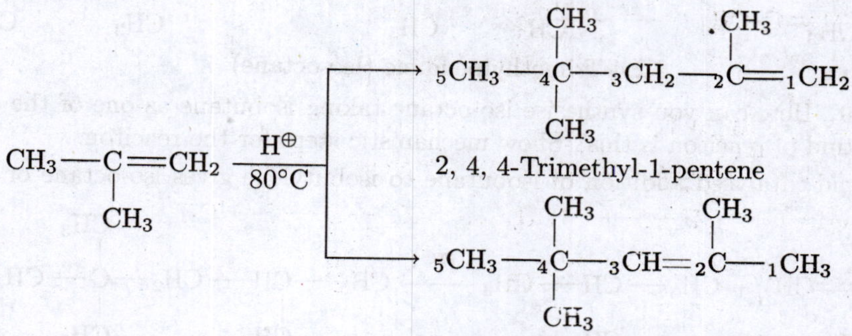

2, 4, 4-Trimethyl-1-pentene

2, 4, 4-Trimethyl-2-pentene

Since two molecules of the substrate couple to give products having molecular weights twice that of the substrate, the reaction is called dimerisation. The mechanism of the reaction is :

Step I : Protonation to the double bond and formation of trimethylcarbenium ion

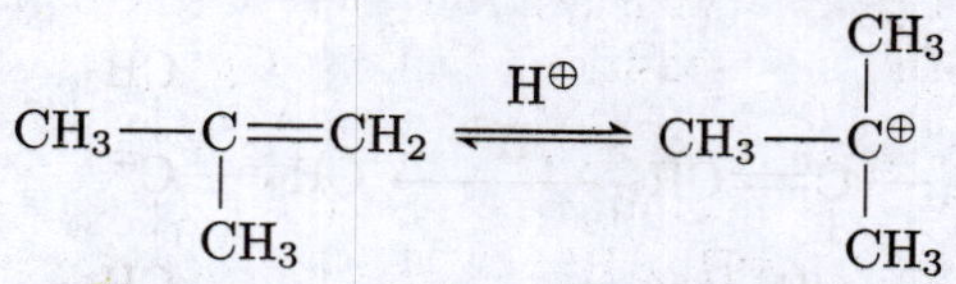

Step II : Electrophilic addition of the carbenium ion to the double bond.

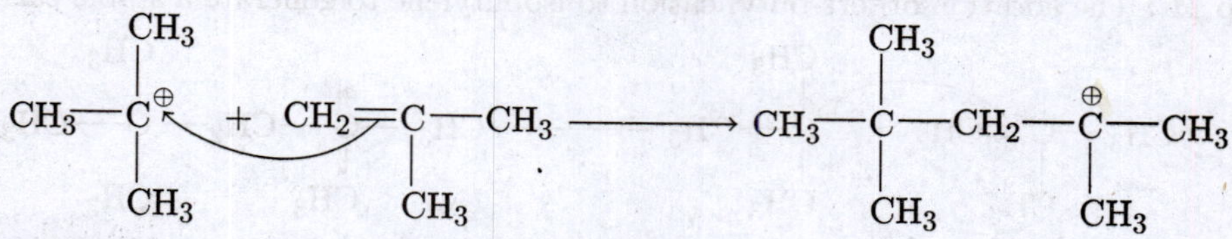

Step III : Deprotonation and formation of alkenes.

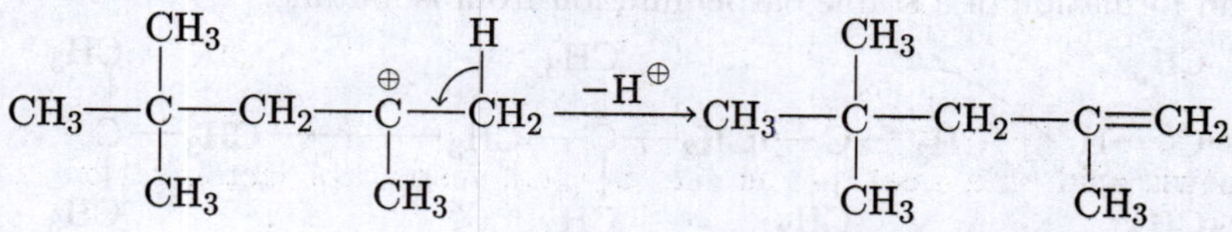

2,4,4-Trimethylpent-1-ene

+

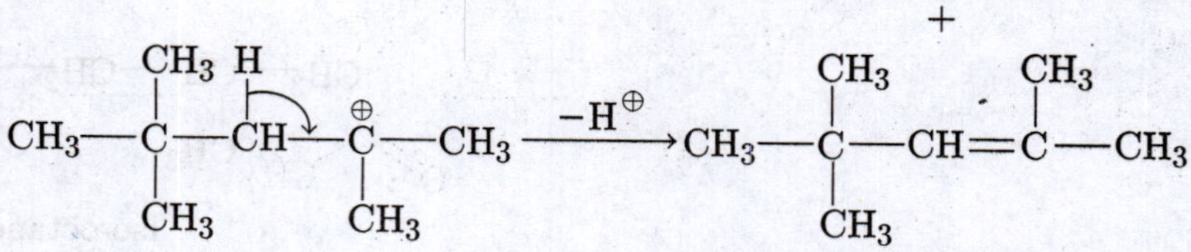

2,4,4-Trimethylpent-2-ene

When these two isomeric alkenes are catalytically hydrogenated, they form 2, 2, 4-trimethylpentane.

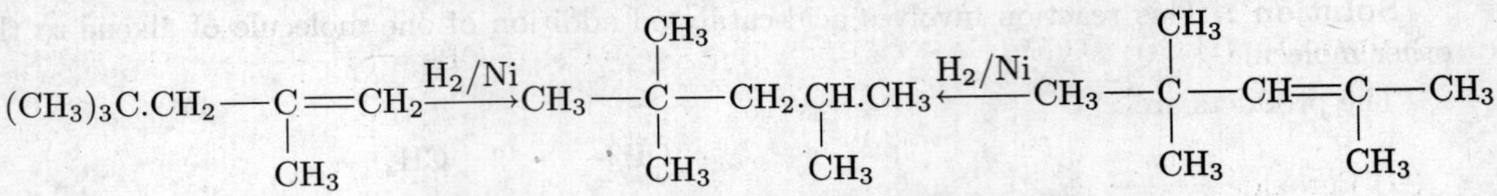

2,2,4-Trimethylpendane (Iso-octane)

Problem 13.9. How can you synthesise iso-octane taking isobutane as one of the starting materials. What kind of reaciton is this? Show mechanistic steps for the reaciton.

Solution : Acid-catalysed addition of isobutane to isobutylene gives iso-octane or 2, 2, 4-trimethylpentane.

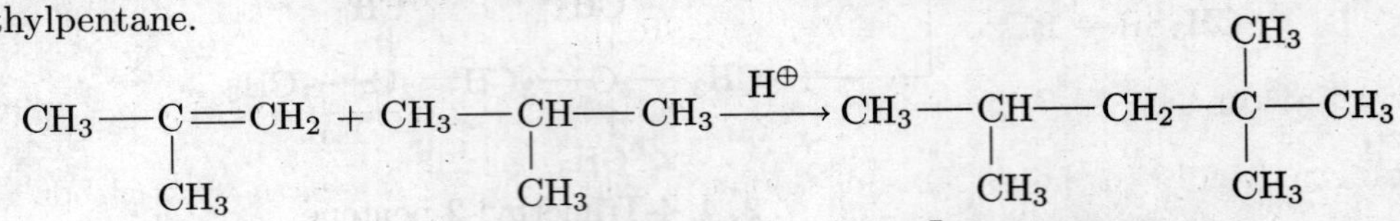

Iso-octane

This reaction may be regarded as the addition of an alkane to an alkene.

The stepwise mechanism of the reaction is as follows :

Step I : Electrophilic attack by the proton on that double-bonded C atom which generates the more stable carbenium ion.

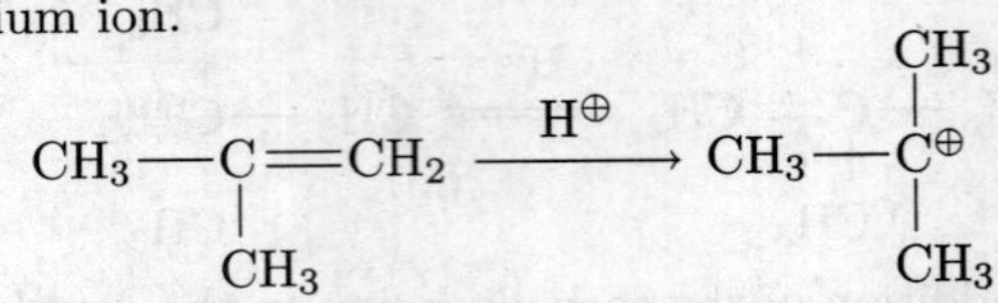

Step II : The addition of tert-butyl cation to isobutylene to generate a stable carbenium ion.

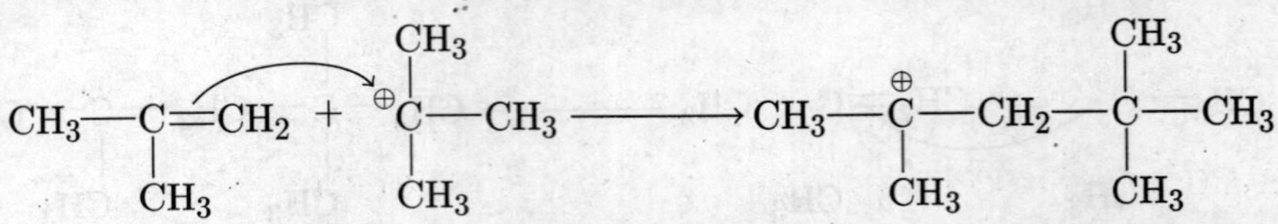

Step III : Intermolecular abstraction of a hydride ion from isobutane by the dimeric carbenium ion and formation of a stable carbenium ion from isobutane.

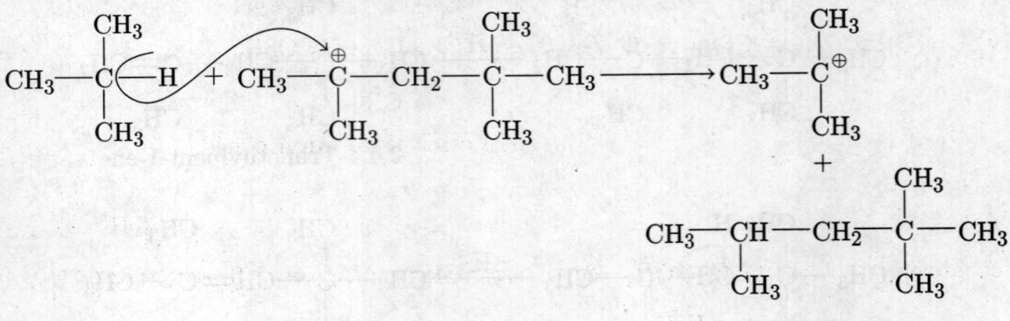

Iso-octane

Problem 13.10. Draw mechanistic steps for the following reactions :

(i) $CH_2{=}CH_2 + H_2S \xrightarrow{H^\oplus} CH_3CH_2SH$ **(ii)** $CH_2{=}CH_2 + RSH \xrightarrow{\text{peroxide}} CH_3.CH_2SR$

Solution : (i) $CH_2{=}CH_2 \xrightarrow{H^{\oplus}} CH_2 \overset{\oplus}{\underset{H}{\cdots}} CH_2 \xrightarrow{H_2S} CH_3{-}CH_2{-}\overset{\oplus}{S}H(H) \xrightarrow{-H^{\oplus}} CH_3CH_2SH$

(ii) Peroxide $\xrightarrow{h\nu}$ Rad·
Rad· + RSH ⟶ Rad·H + RS· } Initiation

$CH_2{=}CH_2 + RS\cdot \longrightarrow \dot{C}H_2{-}CH_2SR$
$\dot{C}H_2{-}CH_2SR + RSH \longrightarrow CH_3{\cdot}CH_2SR$ } Propagation

Termination : $RS\cdot + \cdot SR \longrightarrow RS{-}SR$; $2\dot{C}H_2{-}CH_2SR \longrightarrow RSCH_2CH_2CH_2CH_2SR$

Problem 13.11. In the presence of sulphuric acid, propene is treated with carbon monoxide (carbonylation) at 300°C under 500 atm pressure and subsequently the product is treated separately with **(i)** H_2O, **(ii)** RNH_2, **(iii)** ROH, **(iv)** RSH.

Write down the products and the mechanism of each reaction.

How can you make the reaction occur at room temperature under atmospheric pressure?

Solution : (i) $CH_3CH{=}CH_2 + CO + H_2O \xrightarrow[300°C/500\ atm]{H^{\oplus}} CH_3.\underset{\displaystyle CH_3}{CH}.COOH$

Iso-butyric acid

(ii) $CH_3CH{=}CH_2 + CO + RNH_2 \longrightarrow CH_3.\underset{\displaystyle CH_3}{CH}{-}CONHR$

N-Alkyl iso-butyramide

(iii) $CH_3CH{=}CH_2 + CO + ROH \longrightarrow CH_3.\underset{\displaystyle CH_3}{CH}.COOR$

Alkyl iso-butyrate

(iv) $CH_3CH{=}CH_2 + CO + RSH \longrightarrow CH_3.\underset{\displaystyle CH_3}{CH}{-}COSR$

Thiol ester of iso-butyric acid

Mechanism :

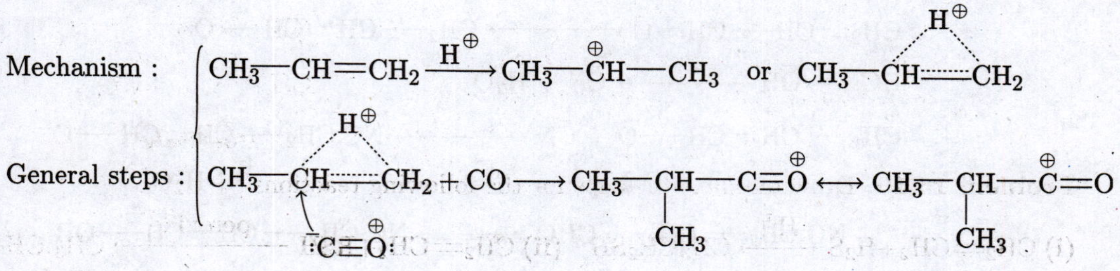

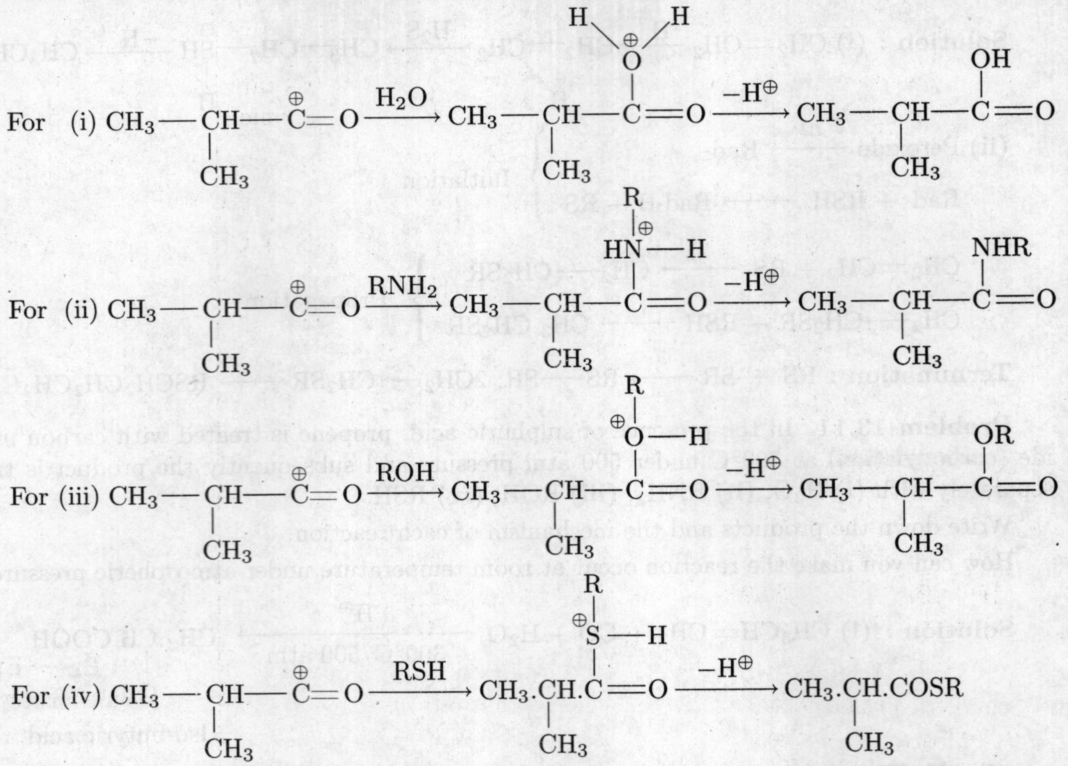

If formic acid is used as a source of carbon monoxide as well as that of water, the reaction takes place at room temperature under atmospheric pressure.

$$HCOOH \longrightarrow H_2O + CO$$

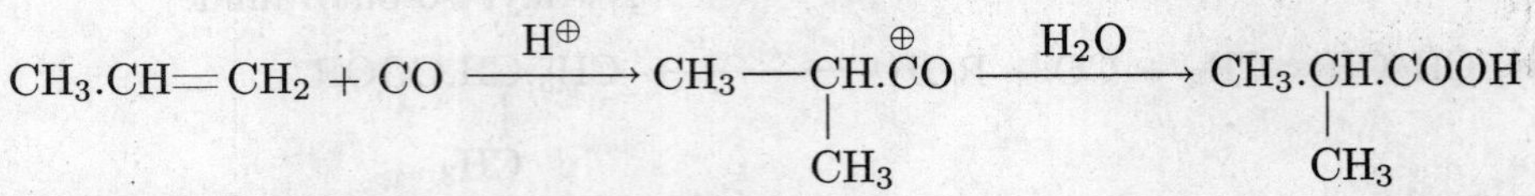

Problem 13.12. HCN does not add normally to an alkene. What structural changes can make HCN to add? State the condition for such an addition and classify the reaction.

Solution : Electrophilic addition of HCN to an alkene is not known. In the presence of a base HCN can be converted into cyanide ion, $:\bar{C}{\equiv}N:$, which can add to a C—C double bond as a nucleophile provided the π *es* are drawn away from the carbon site. This can be done by substituting the Hs of the double bonded Cs with electron withdrawing groups like F, Cl, —CHO, —COR, etc.

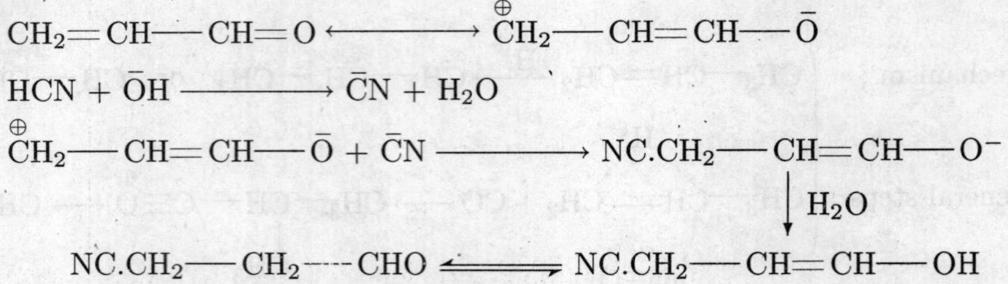

So under a basic condition, HCN can be added to a C—C double bond. The reaction then will be a nucleophilic addition reaction. HCN is a very weak acid. It does not form enough $H^{\oplus}$ to form the protonium ion complex with an alkene. So the electrophilic addition does not occur.

Problem 13.13. A carbanion never attacks the C_3 atom of an α,β-unsaturated carbonyl compound in a Michael condensation.—Explain.

Solution : If the nucleophilic addition occurs at the C_3 atom, the C_4 atom will have to bear the negative charge and this carbanion will be very unstable since no resonating structure will bring stability to the species. On the other hand, when the C_4 atom forms bond with the carbanion, the C_3 atom becomes negatively charged which gets stability through resonance.

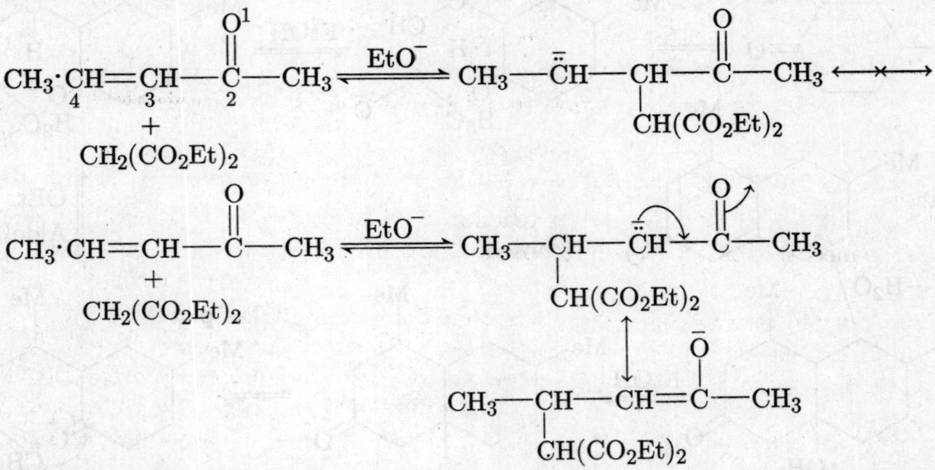

Problem 13.14. Suggest a mechanism to account for the following transformations:

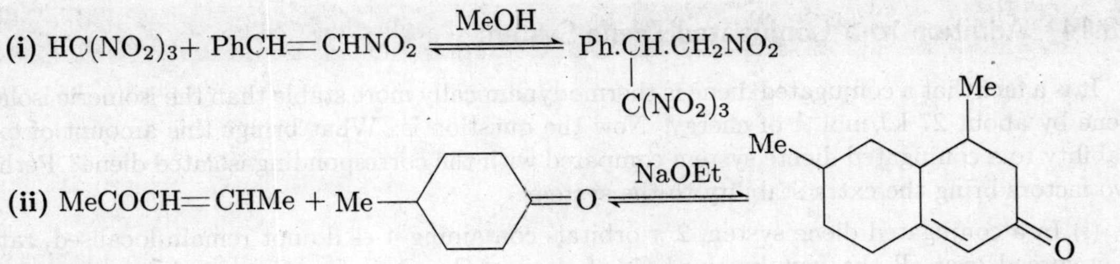

Solution :

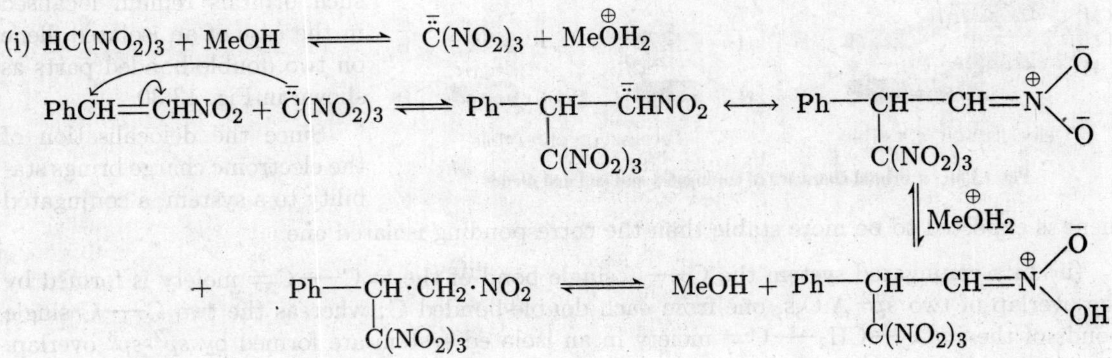

(ii) Robinson annelation

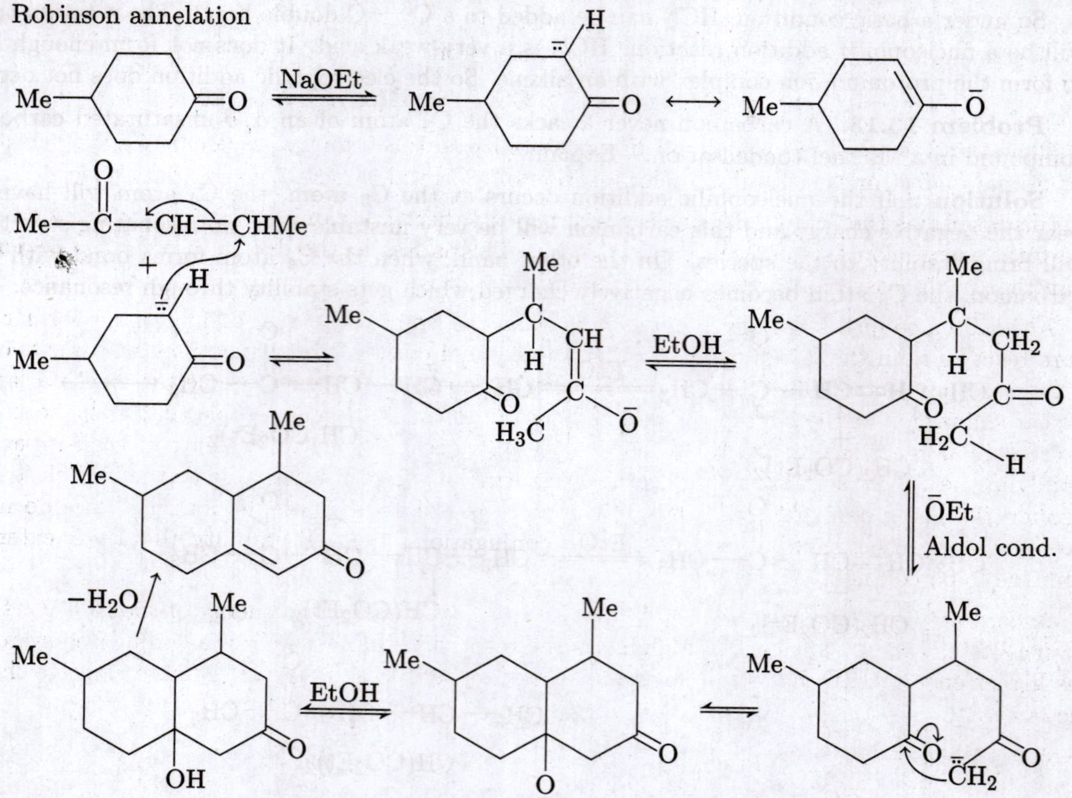

13.14 Addition to a Conjugated Diene System

It is a fact that a conjugated diene is thermodynamically more stable than the isomeric isolated diene by about 27 kJ mol^{-1} of energy. Now the question is: What brings this amount of extra stability to a conjugated diente system compared with the corresponding isolated diene? Perhaps two factors bring the extra stability to the system:

(i) In a conjugated diene system 2 π orbitals containing 4 *es* do not remain localised, rather they spread over all the four involved Cs of the —C=C—C=C— moiety; whereas two such orbitals remain localised in the case of an isolated diene on two double-bonded parts as shown in Fig. 13.30.

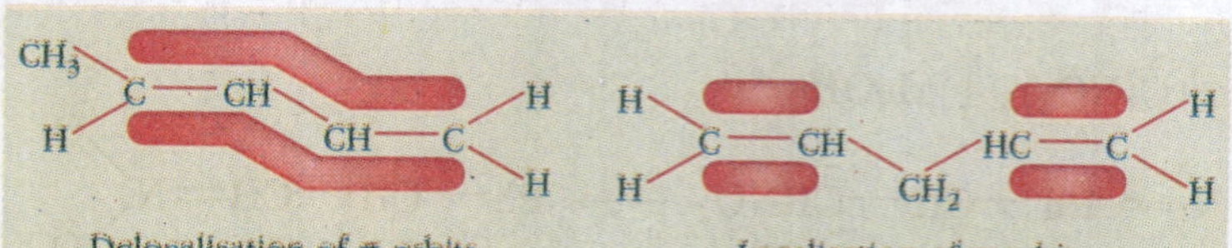

Fig. 13.30. π orbital character of conjugated and isolated dienes

Since the delocalisation of the electronic charge brings stability to a system, a conjugated diene is expected to be more stable than the corresponding isolated one.

(ii) In a conjugated system the C—C single bond of the =C—C= moiety is formed by the overlap of two sp^2 A.O.s, one from each double-bonded C, whereas the two C—C single bonds of the =C—CH_2—C= moiety in an isolated system are formed by sp^2-sp^3 overlap.

The σ bond formed by sp^2-sp^2 overlap being shorter than that formed by the sp^3-sp^2, the former one is stronger than the latter one. This also reveals from the bond dissociation values :

$$C(sp^2)\text{—}C(sp^2) = 383.3 \text{ kJ mol}^{-1};\ C(sp^2)\text{—}C(sp^3) = 357.6 \text{ kJ mol}^{-1}.$$

Thus the stronger C—C σ bond brings stability to the conjugated diene— in fact, this is the primary effect on the stability; delocalisation of π orbital is merely a secondary effect.

(iii) Insofar as the F.M.O. Theory is concerned the thermodynamic stability is determined by the total energy content of the occupied M.O.s. This is less in the case of a conjugated diene than that in the case of an isomeric isolated diene.

Although a conjugated diene is more stable than the isomeric isolated diene, yet the former is more reactive than the latter so far as ionic, radical and cyclo addition reactions are concerned. In the cases of electrophilic addition of radical addition to a conjugated diene the intermediate, the carbenium ion or the carbon radical, and the T.S formed are more stable than those formed in the similar reactions of the isolated diene. This is because the intermediate of the former is an allylic one which is stabilised by the delocalisation of the positive charge or the odd electron through the $p-\pi$ conjugation. While in the cases of isolated dienes, the intermediates do not have any scope of being stabilised by the $p-\pi$ conjugation—they get stability, to a lesser extent, from the hyperconjugative effect.

According to the Frontier Molecular Orbital theory the higher the energy of the HOMO, the lower is its kinetic stability and the higher is the reactivity. The HOMO of a conjugated system has higher energy than that of an isomeric isolated system. So the former is more reactive than the latter.

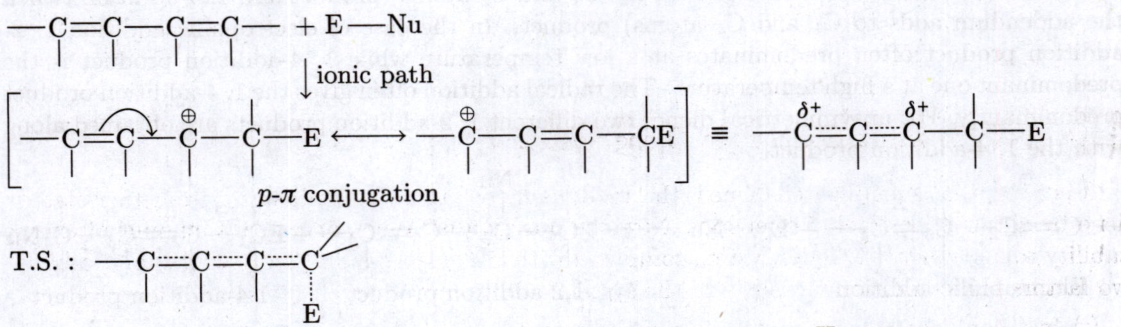

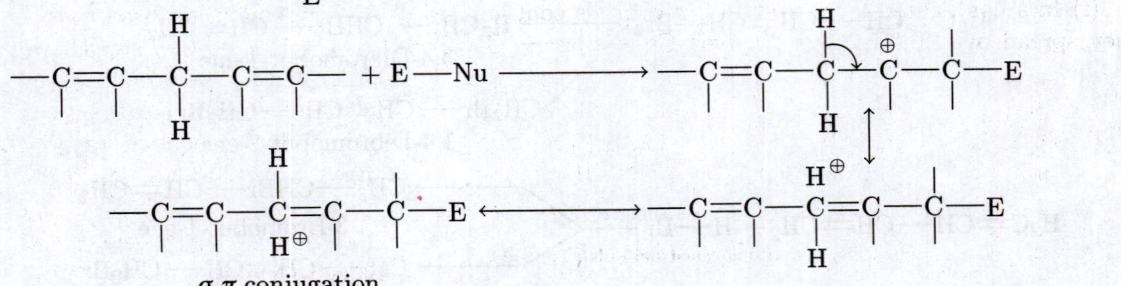

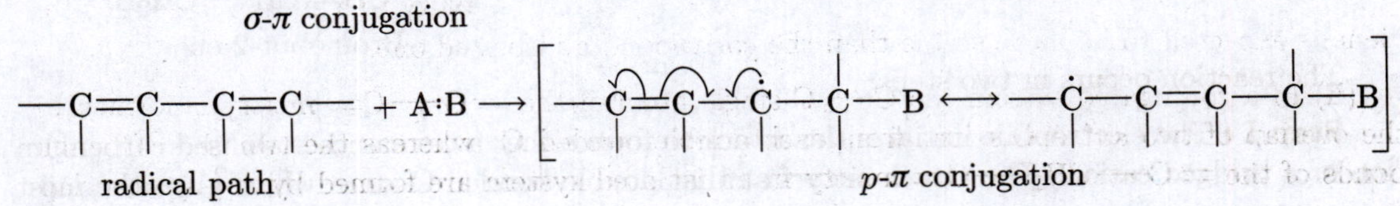

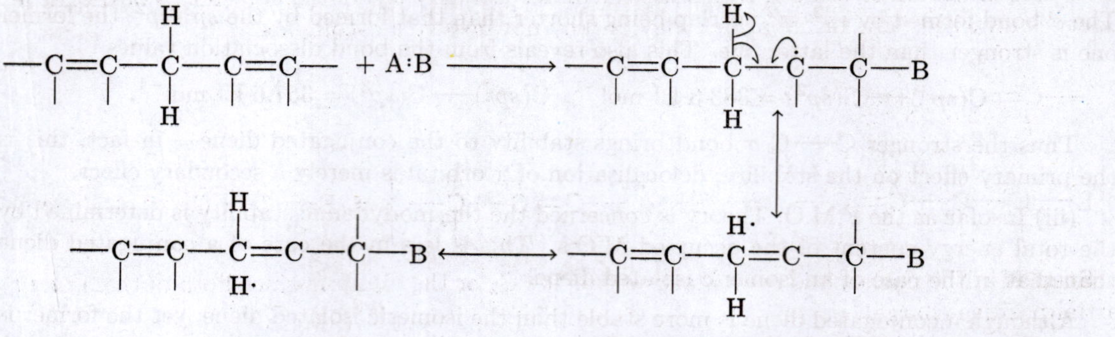

σ-p (half field) conjugation

However, besides ionic and radical addition reactions, conjugated dienes undergo cycloaddition reaction, one kind of percyclic reaction. These reactions are so named because they are reversible intermolecular cyclisation reactions which neither follow the ionic pathway nor follow the radical pathway and involve a cyclic T.S. The reactions only occur thermally or photochemically. We shall discuss this later on.

13.14.1 Electrophilic and radical additions

When a conjugated diene is subjected either to an electrophilic addition reaction or to a radical addition reaction, a mixture of products is often obtained. The mixture consists of 1, 2-addition (when the addendum adds to the C_1 and C_2 atoms) product and 1, 4-addition (when the addendum adds to C_1 and C_4 atoms) product. In the case of electrophilic addition 1, 2-addition product often predominates at a low temperature while 1, 4-addition product is the predominant one at a high temperature. The radical addition often gives the 1, 4-addition product predominantly. For unsymmetrical dienes two different 1, 2-addition products are obtained along with the 1, 4-addition product.

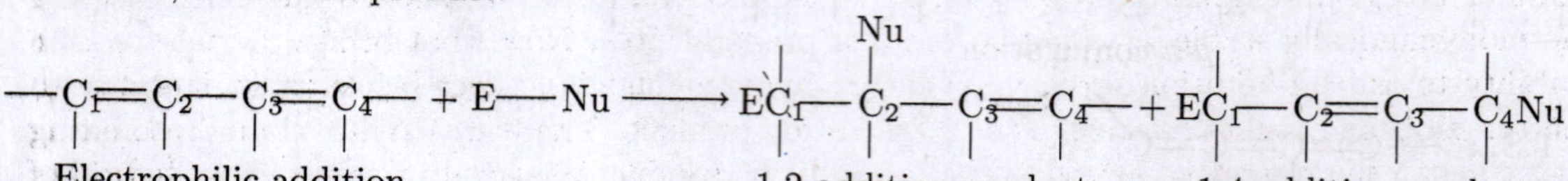

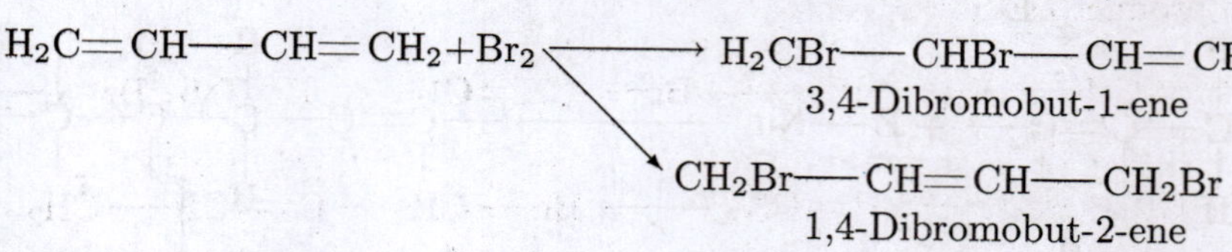

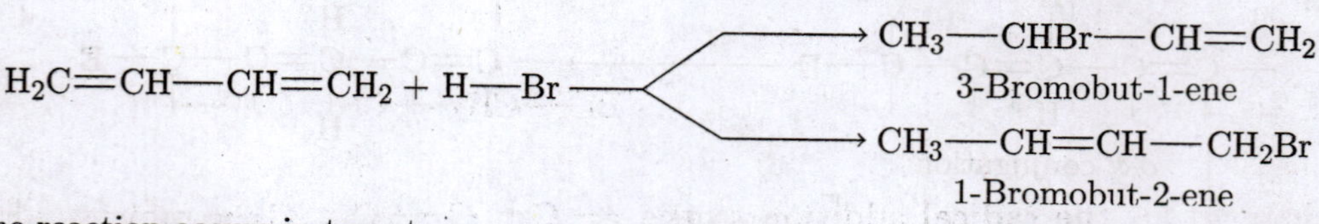

The reaction occurs in two steps:

Step I : The electrophile initiates the attack to form either a resonance stabilised carbenium ion or a bridged cation. The attack occurs at that double-bonded C atom which gives the most stable carbenium ion or bridged cation intermediate. Actually one of the two end Cs of the diene moiety is attacked. The initial attack may or may not involve a π-complex.

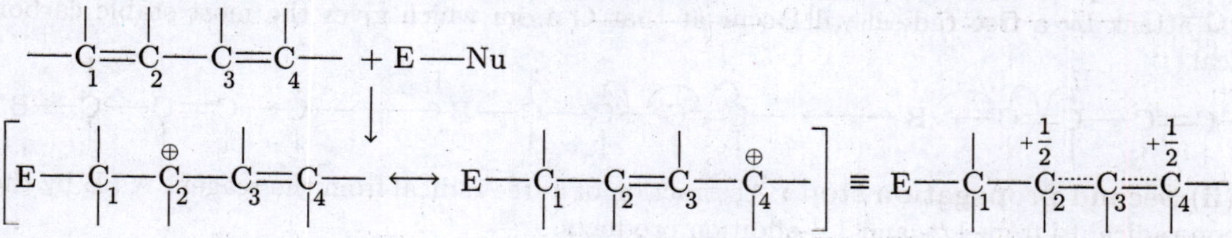

Step II : The nucleophile now attacks either the C_2 or the C_4 atom since both of them carry partial positive charge.

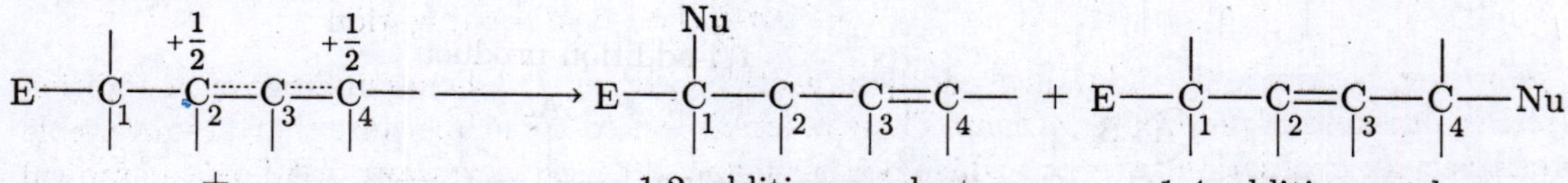

1,2-addition product 1,4-addition product

When the intermediate is a bridged cation, the 1, 2-trans-addition occurs as usual owing to the back-side attack by the nucleophile in the second step.

The first step is the rate-determining step and it is expected that the reaction will follow the second order kinetics.

Now the question is: Why does the 1,2-addition product predominate at a low temperature? This is because the resonating structure which bears the positive charge on C_2 has a greater contribution in the resonance hybrid of the carbenium ion than the other resonating structures. Perhaps $p - \pi$ conjugation stabilises the former resonating structure. The greater the stability of the carbenium ion, the greater is the reactivity. Furthermore, it is said that 1,2-addition is kinetically controlled and hence at a low temperature the 1,2-addition product predominates, while at a higher temperature the reaction is thermodynamically controlled and naturally the thermodynamically stable 1,4-addition product predominates. Now what brings thermodynamic stability to the 1,4-addition product? Perhaps hyperconjugative effect brings extra stability to the 1,4-addition product relative to the 1,2-addition product. There are extra no-bond resonating structures in the resonance hybrid of the 1,4-addition product relative to the resonance hybrid of the 1,2- addition product as shown below :

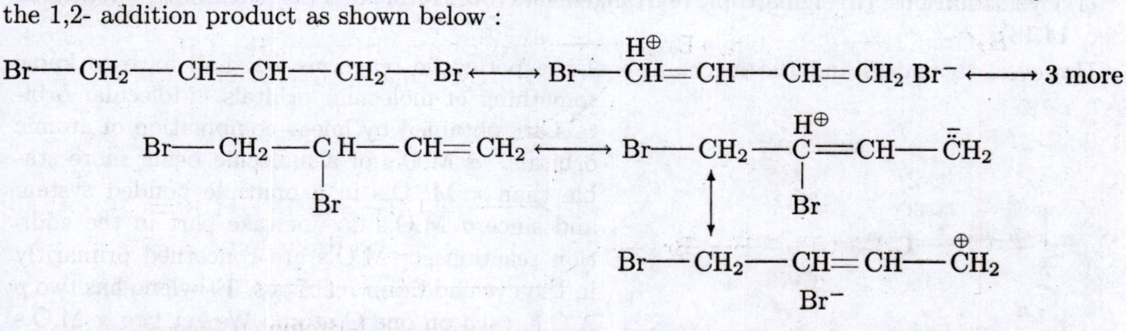

Mechanism for the radical addition to the —C=C—C=C— moiety

The reaction involves all the general steps for radical reactions. The initiation and the termination steps are analogous to those for the radical addition to alkenes.

Two propagation steps are:

(i) First propagation step : Generation of the resonance stabilised carbon radical. The initial attack by a free radical will occur at that C atom which gives the most stable carbon radical.

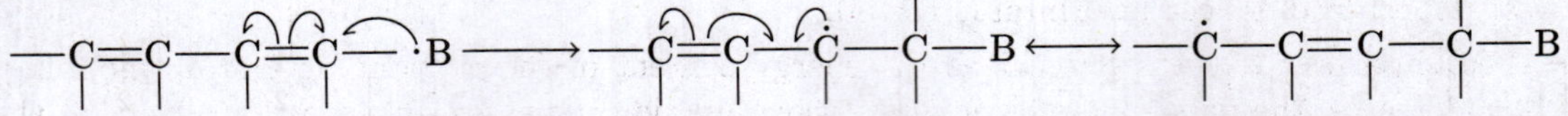

(ii) Second propagation step : Abstraction of a free radical from the reagent A : B by the carbon radical to from 1,2- and 1,4-addition products.

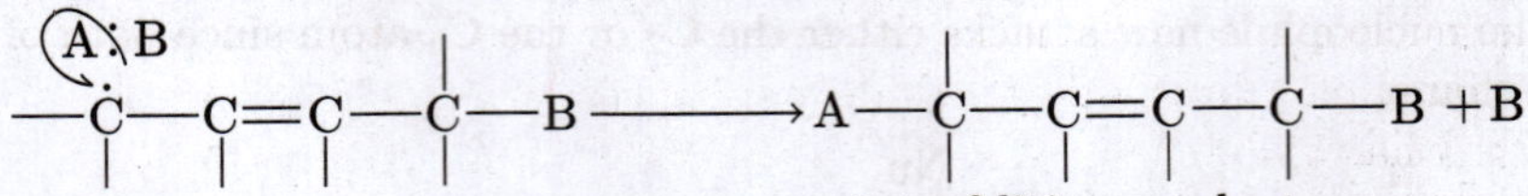

1,4-addition product

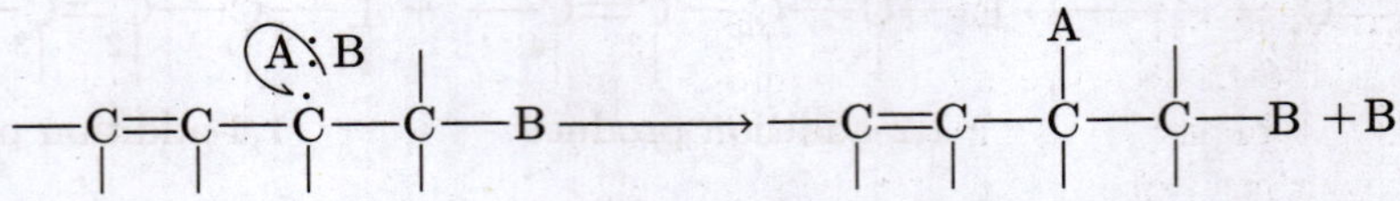

1,2-addition product

13.15 Concerted Step Cycloaddition Reactions

Cycloaddition reactions are a kind of pericyclic reactions. There are some reversible reactions which follow neither the ionic nor the radical pathway,. Therefore, the reactions do not involve:

(i) electrophiles and nucleophiles and radicals as reagents,
(ii) ions and radicals as intermediates,
(iii) catalysts, initiators or polar solvents.

Instead, these reversible reactions are caused by the action of heat or light, as a result of which two or more bonds form and break in a single concerted step through a cyclic transition state. This kind of reactions are collectively known as pericyclic reactions. 'Peri' is a Greek word meaning 'around'; since the T.S.s of such reactions possess cyclic structures and since the reactions follow an altogether different pathway, these reactions are called pericyclic.

There are three kinds of pericyclic reactions:

(i) Cycloadditions, (ii) sigmatropic rearrangements (Sec. 14.17) and (iii) electrocyclic reactions (Sec. 14.18).

However, to understand the mechanism of the pericyclic reactions we shall have to know something of molecular orbitals. Molecular orbitals are obtained by linear combination of atomic orbitals. σ M.O.s of a molecule being more stable than π M. O.s in a multiple bonded system and since σ M.O.s do not take part in the addition reactions, π M.O.s are concerned primarily in the cycloaddition reactions. Ethylene has two p A.O.s. each on one C atom. We get two π M.O.s by linear combination of these two p A.O.s, one is bonding (π) with phase symmetry and the other is antibonding (π^*) without phase symmetry. The

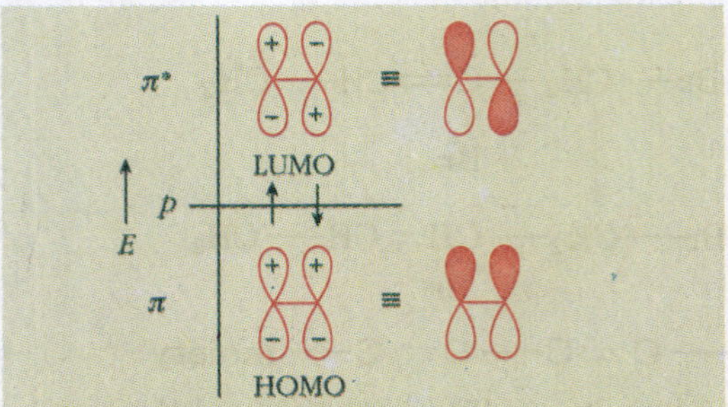

Fig.13.31. π Molecular orbital energy diagram for ethylene

phase of an atomic orbital is designated by plus (+) and minus (−) signs (Fig. 13.31). A bonding M.O. results when two orbitals of the same phase overlap. The bonding M.O. of ethylene is called the highest occupied molecular orbital (HOMO). When two orbitals of opposite phases overlap antibonding orbital results. The antibonding π orbital of ethylene (π^*) is called the lowest unoccupied molecular orbital (LUMO).

It is important to mention that as the energy content of the π (bonding and antibonding) M.O.s increases, the number of nodes in the wave function increases. The wave function for the π orbital of ethylene has no node. In fact, the nth π orbital, bonding or anti bonding, has $(n-1)$ nodes. So the wave function of π^* orbital of ethylene has one node since π^* is the 2nd π orbital. Thus the wave function of the 4th π orbital of 1,3-butadiene (π_4) has 3 nodes.

The π orbital of ethylene has a σ plane. It is said to be σ symmetric (S) with respect to the σ plane. The σ plane bisects the molecule and perpendicular to the plane of the molecule. It has no C_2 axis; so the π orbital of ethylene is antisymmetric (A) with respect to C_2 axis. π^* orbital is antisymmetric (A) with respect to σ plane since it does not contain a σ plane but it is symmetric (S) with respect to C_2-axis which lies on the plane of the molecule and perpendicular to the mid-point of the C—C bond.

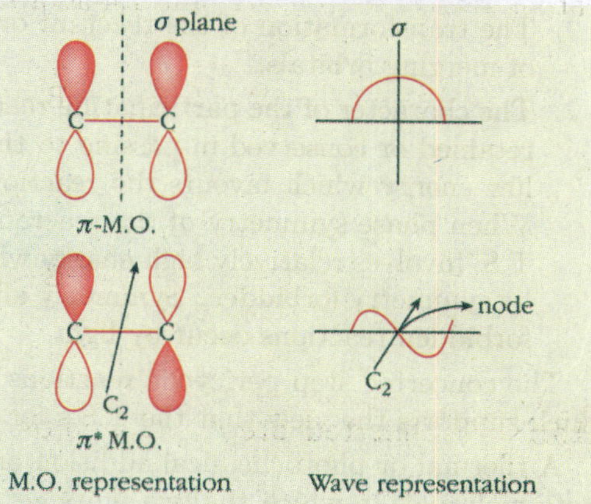

Fig.13.32. Symmetry elements in π and π^* M.O.s of ethylene

However, a linear combination of four p A.O.s of butadiene will give rise to four M.O.s, two bonding and two antibonding (Fig. 13.33). These are designated as π_1, π_2, π_3^* and π_4^*. Depending

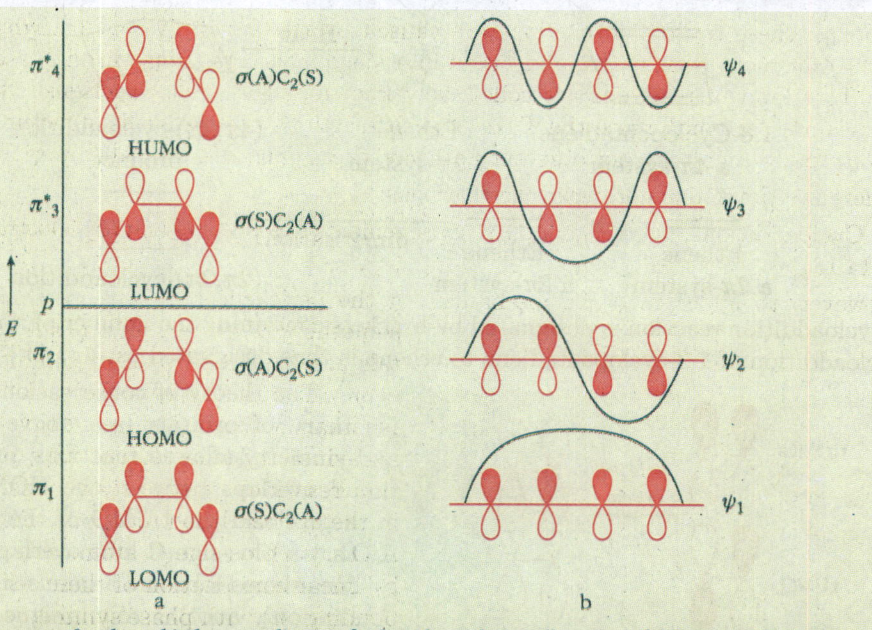

Fig.13.33. a. π-molecular orbital energy diagram for butadiene b. Standing wave of a linear 1,3-butadiene

on phase symmetry and energy content π_1 is the LOMO (the lowest occupied molecular orbital), π_2 is the HOMO, π_3^* is the LUMO and π_4^* is the HUMO (the highest unoccupied molecular orbital). It is needless to mention that a bonding orbital results by overlapping of two orbitals of the same phase.

Now we shall discuss the salient points of pericyclic reactions which are put forward by Woodward and Hoffmann :

1. Simultaneous overlapping of the participating bond orbitals of the reactant molecules occurs in such a way that they may pass smoothly over to the bond orbitals of the product molecule.
2. The transformation of the reactant orbitals into the product orbitals involves a cyclic T.S. of merging orbitals.
3. The character of the participating reactant orbitals, i.e., their phase symmetry is somewhat retained or conserved in passing to the product orbitals and thus the T.S. has a relatively low energy which favours the reaction, and the reaction is said to be symmetry allowed. When phase symmetry of the reactant orbitals is not retained in the product orbitals, the T.S. involves relatively high energy which disfavours the reaction and the reaction is said to be symmetry forbidden. Symmetry allowed reactions usually occur by heat while symmetry forbidden reactions occur by light.

The concerted step pericyclic reactions are characterised by a negative entropy of activation which supports the view that the T.S.s for the reactions will be cyclic.

A thermal or photochemical addition or coupling reaction between double bonds in two separate molecules in which the two units get joined by two σ bonds created from the two original π bonds to form a cyclic product via a cyclic T.S. is called cycloaddition reaction of the common type. This does not require a polar solvent, a catalyst or a radical initiator and does not follow ionic or radical pathway.

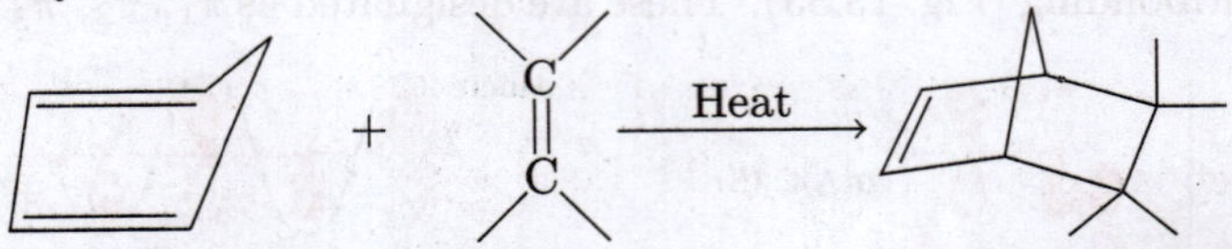

1,3-Cyclopentadiene a 4π-system | Ethene a 2π-system | $(4\pi,2\pi)$-cycloaddition product

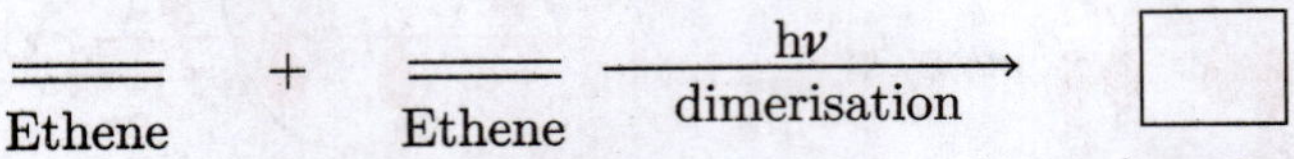

Ethene a 2π-system | Ethene a 2π-system | $(2\pi,2\pi)$-cycloaddition

A cycloaddition reaction is designated by brackets containing the number of π es in each unit; the cycloaddition of 1,3-cyclopentadiene to ethene is thus designated as $(4\pi,\ 2\pi)$-cycloaddition.

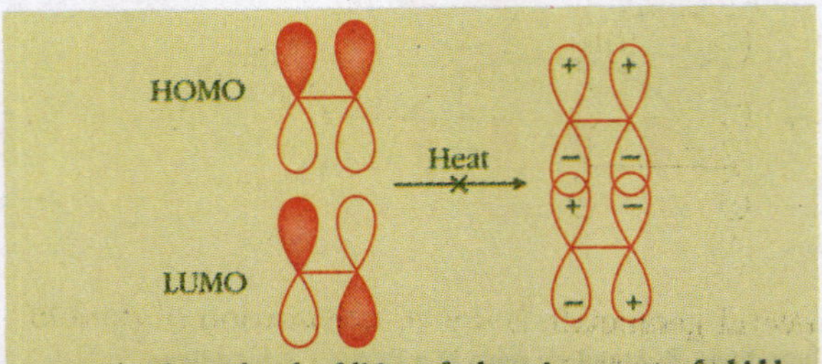

Fig. 13.34. Thermal cycloaddition of ethene is symmetry forbidden

The theory of conservation of phase symmetry of orbitals (see above) tells us that symmetry-allowed reactions occur only when all overlaps between the HOMO of one reactant and the LUMO of the other are such that a plus-signed lobe overlaps only another plus-signed lobe and a minus-signed lobe overlaps only another minus-signed lobe. Let us consider the case of concerted step dimerisation of ethene to cyclobutane. Here the LUMO of one molecule should overlap the HOMO of another molecule. Here one (+)-lobe overlaps another (−)-lobe (Fig. 13.34) which is not allowed; so thermal cycloaddition will not occur here.

Of course, this can occur in two steps by radical pathway.

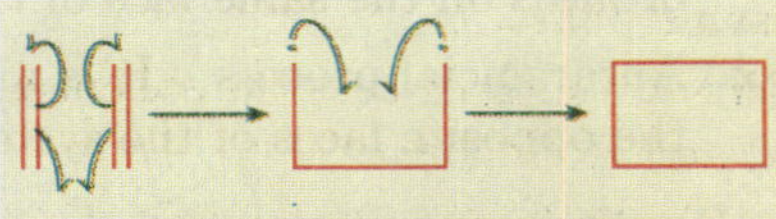

Fig. 13.35. Radical cycloaddition

On the other hand, the reaction may be made to occur photochemically. By the absorption of light energy one molecule will be in the excited state; one electron from its HOMO transits to its LUMO in the excited state. Thus the LUMO becomes the HOMO of the excited state and the HOMO of the molecule becomes so oriented that it overlaps the LUMO of another molecule to form cyclobutane. The HOMO of the excited state is called SOMO (Single-electron Occupied MO). These HOMO, LUMO and newly formed orbitals of the excited product molecule are symmetric(S) with respect to C_2 axis but antisymmetri (A) with respect to σ plane. In this way the symmetry of orbitals is somewhat conserved.

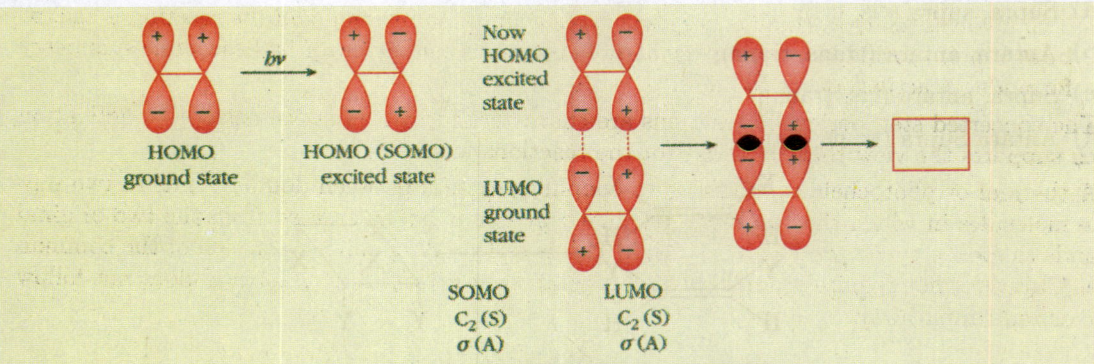

Fig. 13.36. Photochemical cycloaddition of ethene

The concerted step mechanism for cycloaddition between butadiene and ethene is as follows.

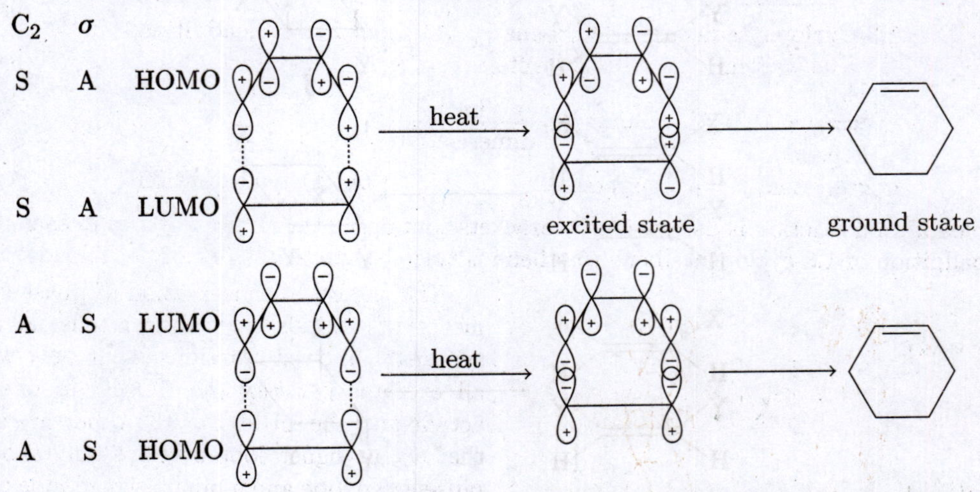

A concerted step cycloaddition gives rise to several geometric isomers; orientation of double bonds and the molecules during addition are responsible for this.

So far orientation of double bonds during addition are concerned Woodward and Hoffmann formulations are important. In this connection they formulated two processes:

1. Suprafacial process : It is a process in which the bonds that are formed or cleaved have orbitals on the same face of the system that undergoes the reaction.
2. Antarafacial process : It is a process in which the bonds that are formed or broken lie on the opposite faces of the system that undergoes the reaction.

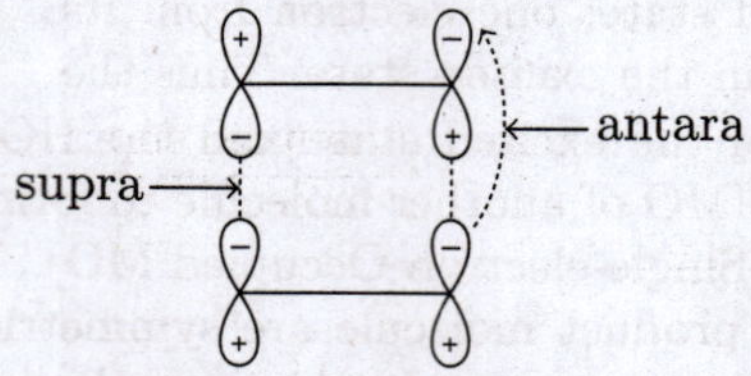

By the combination of these two processes four different modes of addition are obtained.

(a) Supra, supra (cis, cis);

(b) Antara, antara (trans, trans);

(c) Supra, antara (cis, trans);

(d) Antara Supra (trans, cis);

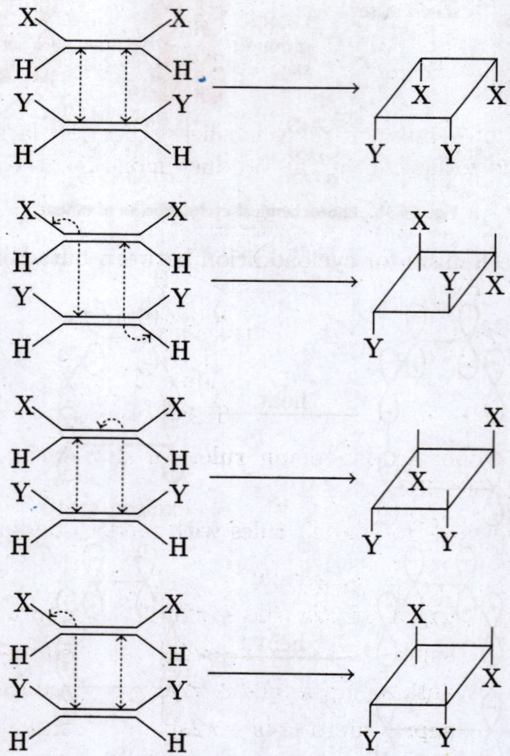

Orientation of reactant molecules may be of different types and thus different geometric isomers may result.

(a) Head to head orientation :

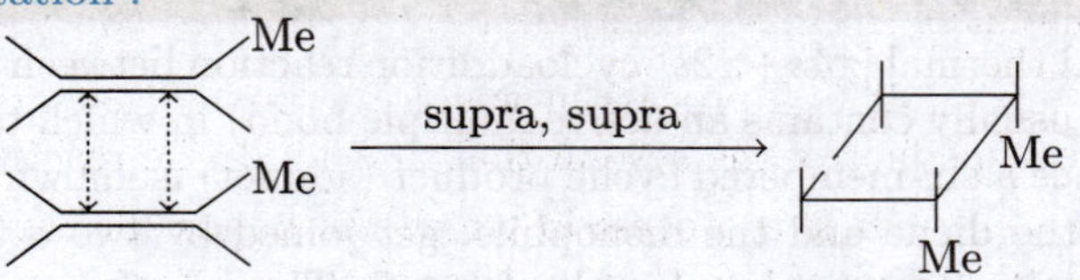

(b) Head to tail orientation :

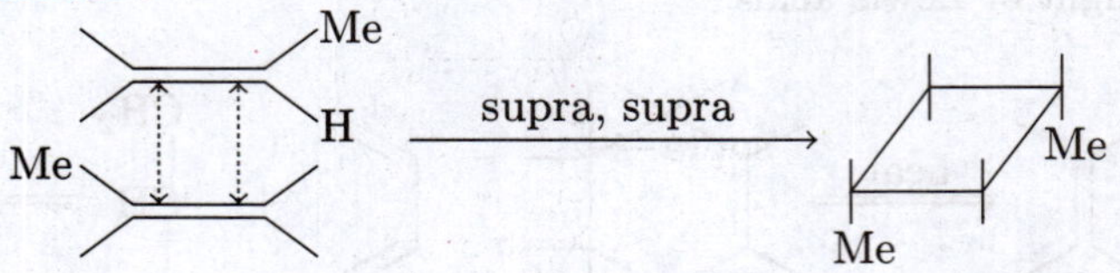

(c) Endo-cycloaddition :

If the larger side of a reactant molecule lies under the larger side of another reactant molecule during the formation of the product orbitals, the cycloaddition is called endo-addition.

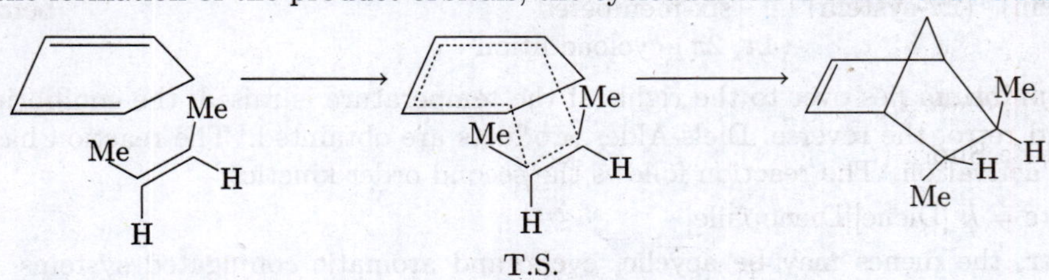

(d) Exo-cycloaddition :

If the smaller side of a reactant molecule lies under the larger side of another reactant molecule during the formation of the product molecule, the cycloaddition is called exo-addition.

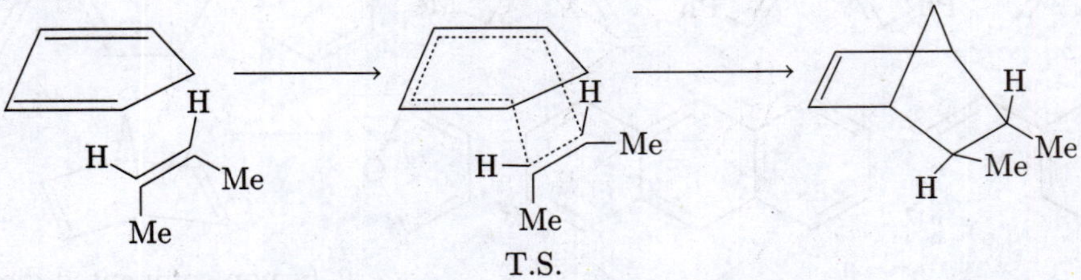

However, there are Woodward-Hoffmann rules on suprafacial and antarafacial processes. These are tabulated in Table 13.2.

Table 13.2 Woodward-Hoffmann's rules with modern designations within braces

Cycloaddition	Thermal	Photochemical
$(2\pi,\ 2\pi)$	supra, antara $[\pi 2s + \pi 2a]$	supra, supra $[\pi 2s + \pi 2s]$
$(4\pi,\ 4\pi)$	supra, antara $[\pi 4s + \pi 4a]$	supra,supra $[\pi 4s + \pi 4s]$
$(6\pi,\ 2\pi)$	antara, supra $[\pi 6a + \pi 2s]$	antara antara $[\pi 6a + \pi 2a]$
$(4\pi,\ 2\pi)$	supra, supra $[\pi 4s + \pi 2s]$	supra, antara $[\pi 4s + \pi 2a]$
$(6\pi,\ 4\pi)$	antara, antara $[\pi 6a + \pi 4a]$	antara, supra $[\pi 6a + \pi 4s]$
$(8\pi,\ 2\pi)$	antara, antara $[\pi 8a + \pi 2a]$	antara, supra $[\pi 8a + \pi 2s]$

43 BC

13.15.1 The Diels-Alder reaction

The reversible and thermal [$\pi 4s+\pi 2s$]-cycloaddition reaction between a diene and a dienophile (diene loving species usually contains an active multiple bond) in which the dienophile adds, 1, 4, to the diene to produce a six-membered cyclic product (adduct) is known as Diels-Alder reaction. Here the two units, the diene and the dienophile, get joined by two σ bonds created from the two π bonds in two units; a new π bond is also formed. The reaction may also be caused by the catalytic action of light or Lewis acids.

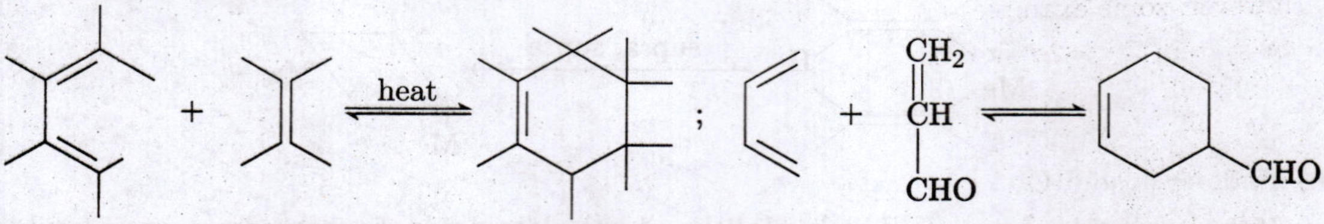

Diene (4π-system) Dieneophile (2π-system) cyclic product six-membered (4π, 2π)-cycloaddition ; Butadiene Propenal Tetrahydro-benzaldehyde

The equilibrium lies over to the right. If the temperature is raised, the equilibrium shifts to the left and retro, the reverse, Diels-Alder products are obtainted. The reaction has a negative entropy of activation. The reaction follows the second order kinetics.

The rate = k [Diene][Dienophile]

However, the dienes may be acyclic, cyclic and aromatic conjugated systems. Even non-conjugated dienes of appropriate geometry may undergo the Diels-Alder reaction. Some of the dienes are noted below :

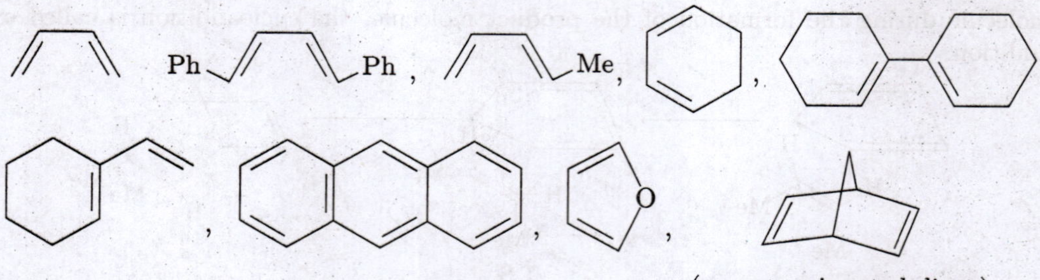

(a non-conjugated diene)

Several types of dienophiles may be used in the Diels-Alder reaction. For example,

(a) Activated monoenes : —C=C—CHO, —C=C—COR, —C=C—COOH, COOH—C=C—COOH, —C=C—CN, —C=C—NO_2, —C=CCH_2OH, —C=C—CH_2Cl, —C=C—CH_2NH_2, —C=C—CO_2R, 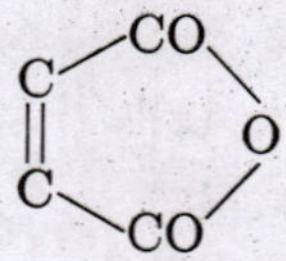

(b) Dienes : —C=C—C=C—, —C=C=C—,

(c) Activated monoynes : MeO_2C—C≡C-—CO_2Me, 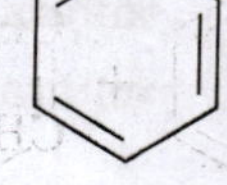

(d) Carbon-hetero multiple bonds : N≡C—, —N=C—, —C=O.

(e) Multiple bonds other than, C—C : CO_2Me—N=N—CO_2Me, O=N—, O=O, etc.

Tetracyanoethene is known to be the best dienophile till to date.

The procedure for the Diels-Alder reaction is very simple—only the mixture of the diene and the dienophile needs heating with or without solvent; the common solvent is benzene.

However, some examples are given below :

4π-system 2π-system

1. 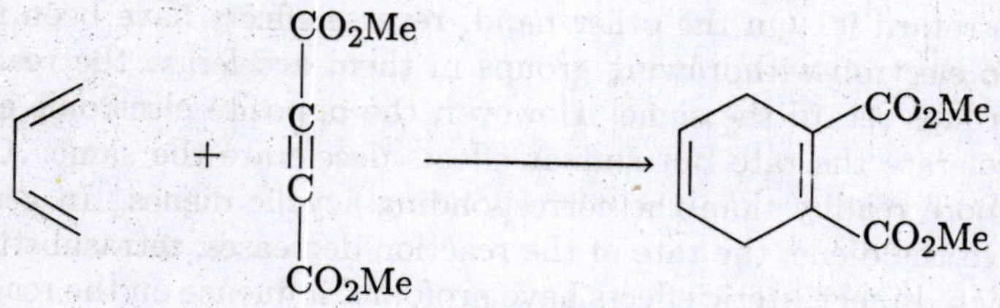

2. 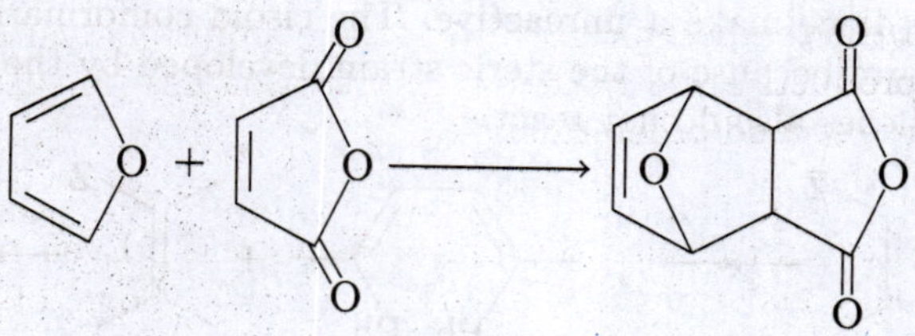

3. 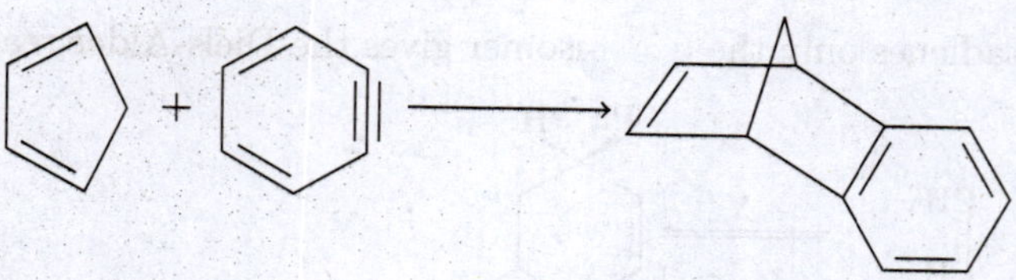

4. 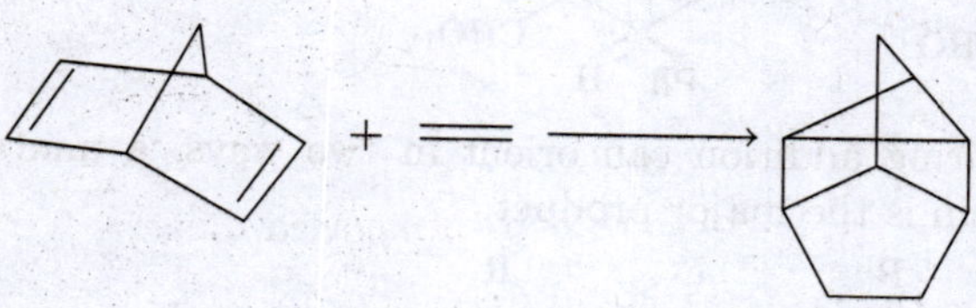

5. 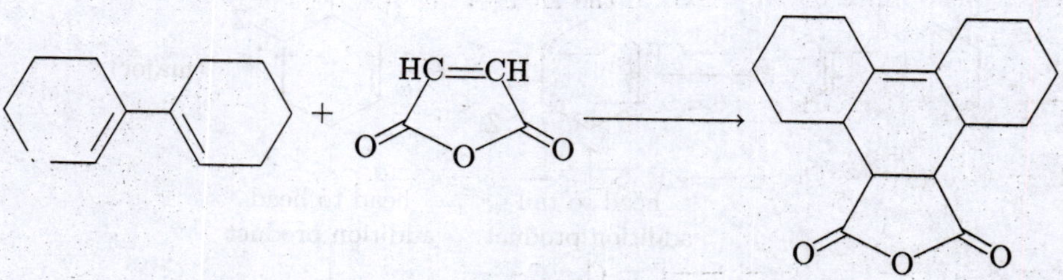

6. 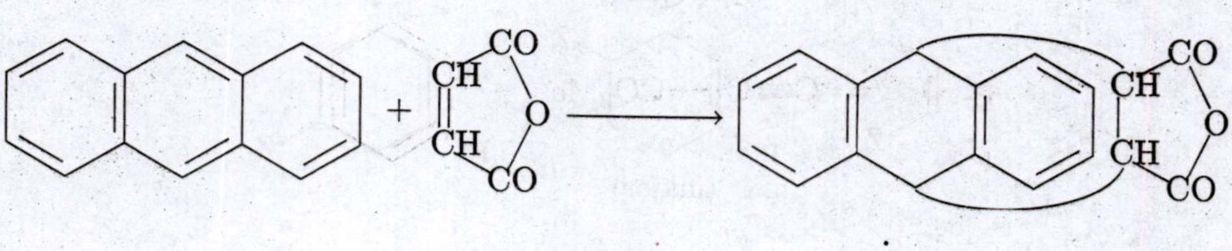

7. 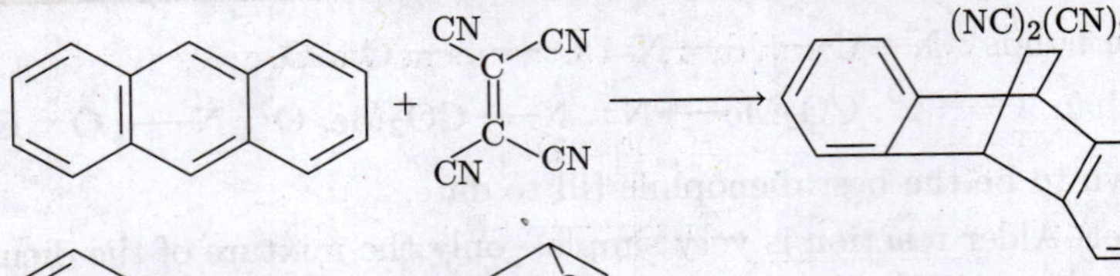

8. 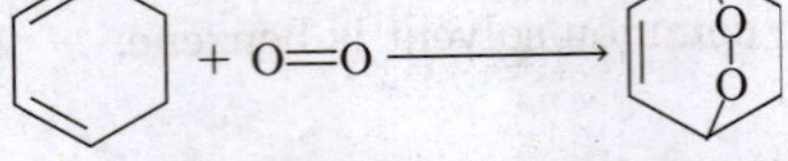

It has been found that the electron-donating groups in the dienes accelerate the reaction while the electron-withdrawing groups retard it. On the other hand, reverse effects have been found in the cases of dieneophiles—the electron-withdrawing groups in them accelerate the reaction, whereas the electron-donating groups retard the same. However, the opposite electronic effects, on the diene and dienophile accelerate the rate but similar effects decelerate the same. Cyclic dienes in the cisoid-form react more readily than the corresponding acyclic dienes. In general, with the increasing substitution in the diene, the rate of the reaction decreases; tetrasubstituted dienes in 1, 4-positions do not react. In fact, steric effects have profound influence on the reaction. Large groups on 2- and 3-positions of the diene make it unreactive. The cisoid conformation of such a diene gets frozen to the transoid-form because of the steric strain developed by the large groups. Bulky cis, cis-1, 4-disubstituted dienes also do not react.

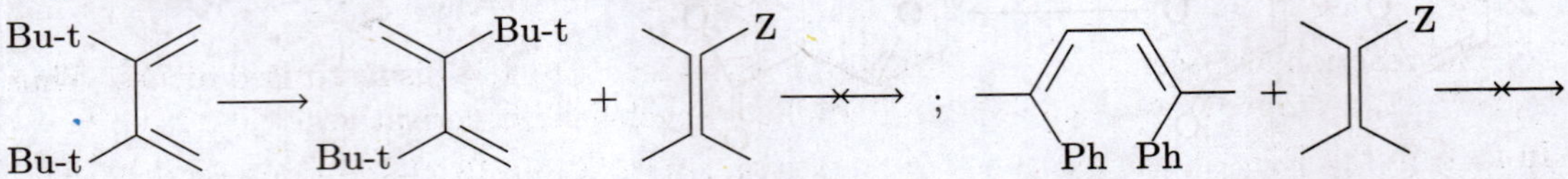

Among the four 1, 4-diphenylbutadienes only the trans-isomer gives the Diels-Alder reaction.

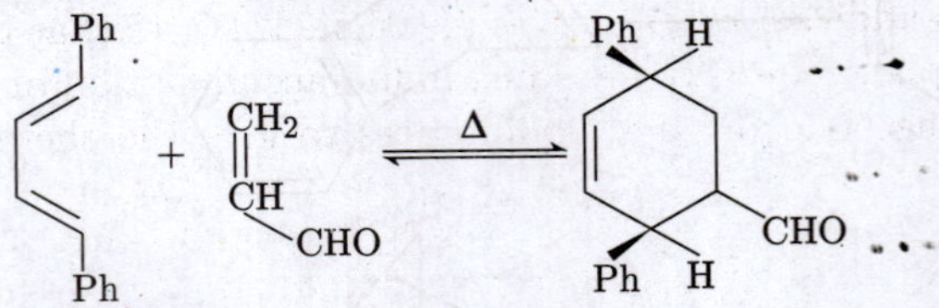

Since the diene and the dienophile during addition can orient in two ways, a mixture of products are obtained; of course one of them is the major product.

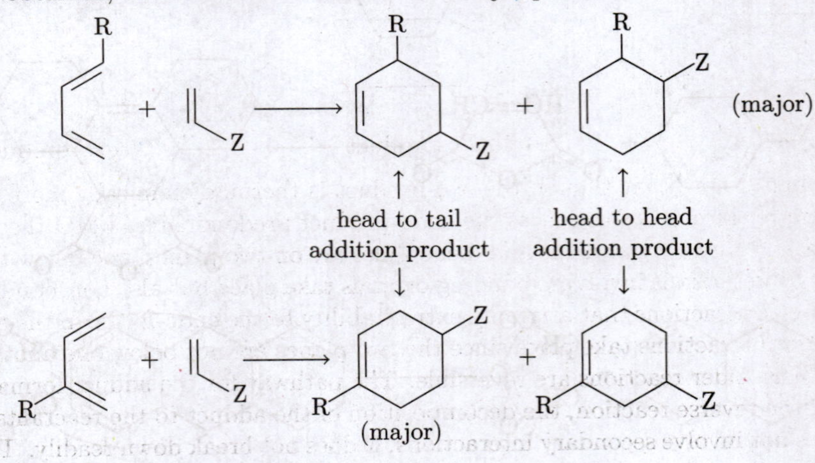

There are examples in which the dienophile adds 1, 2 to the diene which gives rise to a four membered cyclic structure.

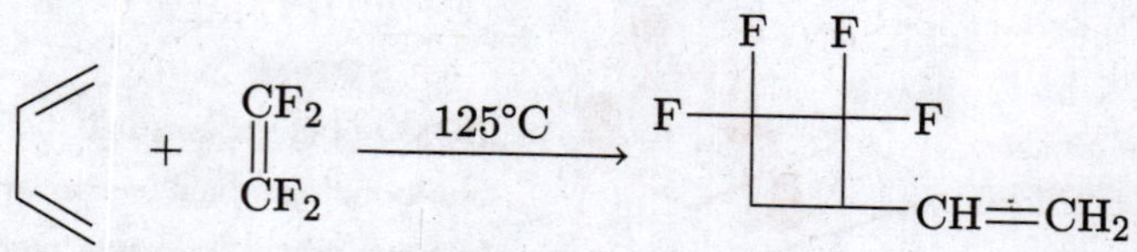

The discussion on the stereochemistry of the Diels-Alder reaction involves several aspects.

(a) Only conjugated dienes in the cisoid conformation undego the reaction. Any factor that makes the diene assume the transoid conformation will hinder the reaction.

(b) The reaction is stereospecific with respect to the dienophile and the addition is always cis—the cis groups remain cis and the trans group remains trans in the product.

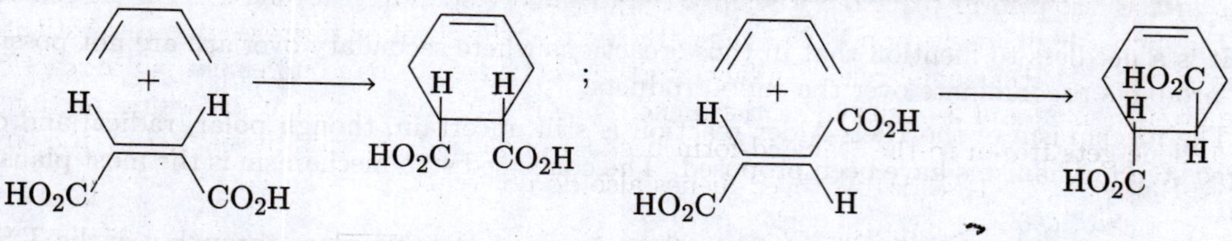

(c) The reaction also seems to be stereospecific with respect to 1, 4-disubstituted dienes. Thus trans, trans-1, 4-diphenylbutadiene yields cis-diphenyl-cyclohexene derivatives.

In fact, in the Diels-Alder reaction cis, cis (supra, supra; Table 13.2)-addition takes place.

(d) In the cases of cyclic dienes kinetically controlled endo-products are obtained. In this connection **Alder's endo-addition rule** deserves mention. It states that the reactants orient themselves in those parallel planes in which the maximum accumulation of double bonds take place. However, on keeping, the endo product equilibrates to give the thermodynamically controlled exo-product.

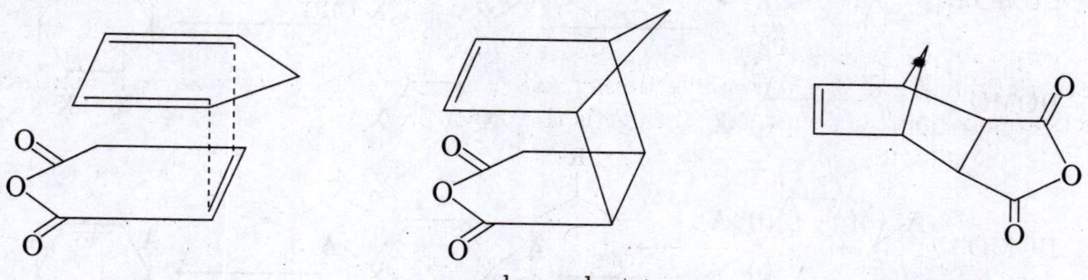

endo-product exo-product

It is interesting to note that though the exo-product is thermodynamically more stable than the endo-product yet in most of the cases the endo-product predominates. FMO theory gives an explanation in this regard. When diene and dienophiles are on two planes, one below the another, not only primary interactions involving bonding orbitals take place but also non-bonding orbitals undergo secondary interactions that attribute extra stability to the endo-form. In the exo-product no such secondary interactions take place since the two planes are not below one another exactly. We know that Diels-Alder reactions are reversible. The pathway for the adduct formation is also the pathway for the reverse reaction, the decomposition of the adduct to the reactants. Since the exo-product does not involve secondary interactions, it does not break down readily. The primary and secondary interactions are shown in Fig. 13.37 for the reaction between cyclopentadiene and maleic anhydride.

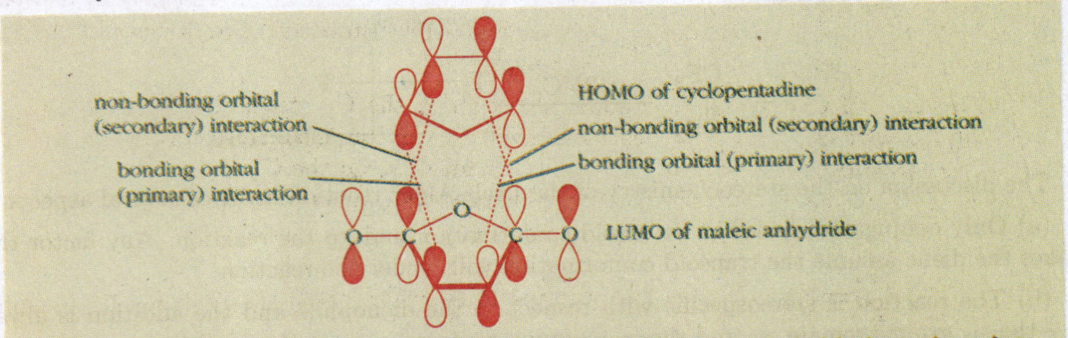

Fig. 13.37. Primary and secondary intereactions of orbitals in the diels-Alder reaction which leads to endo-product

It is a needless to mention that in those reactions where secondary overlaps are not possible, exo-products predominate over the endo-products.

The mechanism of the Diels-Alder reaction is still uncertain, though polar, radical and concerted step mechanisms have been proposed. The concerted step mechanism is the most plausible one.

The concerted step mechanism tells us that the reaction takes place through a cyclic T.S. in which the two σ bonds are formed simultaneously or almost simultaneously. As discussed in the cycloaddition reaction (See Sec. 13.14), the simultaneous overlapping of orbitals may take place in two ways in which one reactant will be in the HOMO state and the other one in the LUMO state and vice versa. In both the cases the adduct is formed in the excited state and then the adduct comes down to the ground state and preservation of orbital symmetry is maintained.

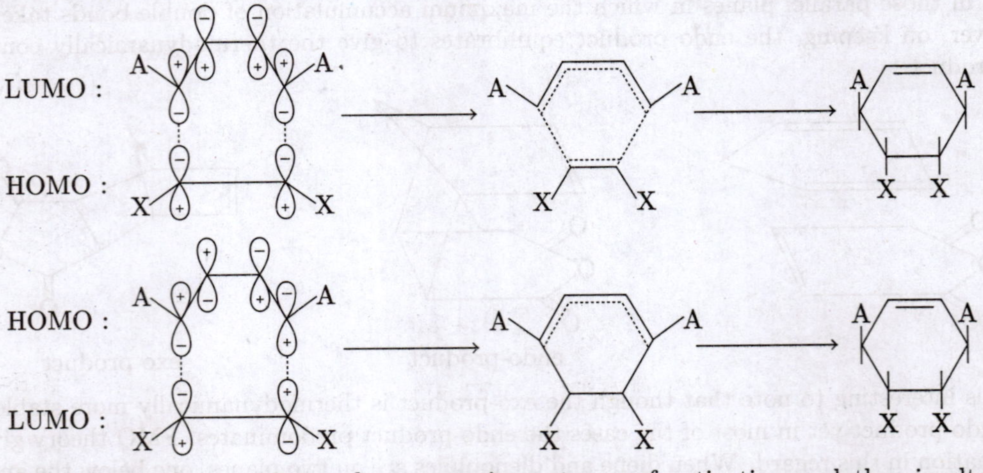

The stereospecific cis-addition, large negative entropy of activation and the simultaneous cleavage of two σ bonds during decomposition of the Diels-Alder product support the concerted step mechanism. Here it is important to note that the interaction between HOMO of the diene and LUMO of the dienophile involves less energy difference between them than if the HOMO of the dienophile interacts with the LUMO of the diene. Thus the former interaction is more energy compatible than the latter and this makes Diels-Alder reactions regeoselective.

Since the reaction procedure is very simple and since wide and varied types of compounds take part in forming new C—C σ bonds, it has a good preparative value. For example, Woodward synthesised cis-decalin by the Diels-Alder reaction which served as an intermediate in the synthesis of reserpin. It contains six-chiral Cs out of which four are present in the cis-decalin moiety.

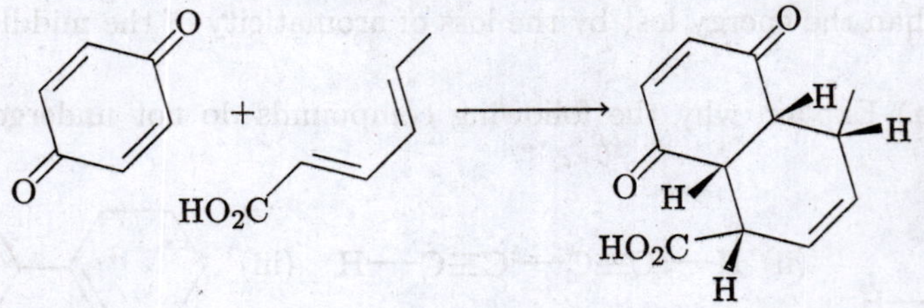

a derivative of *cis*-decalin

Cantharidin, a natural product, has been synthesised by using the Diels-Alder reaction twice in a chain of fifteen steps.

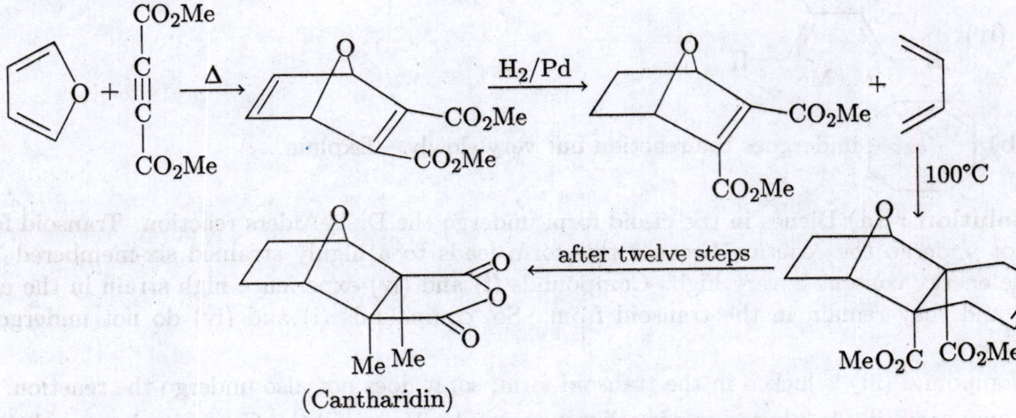

(Cantharidin)

Polynuclear aromatic compounds may be prepared by the Diels-Alder reaction.

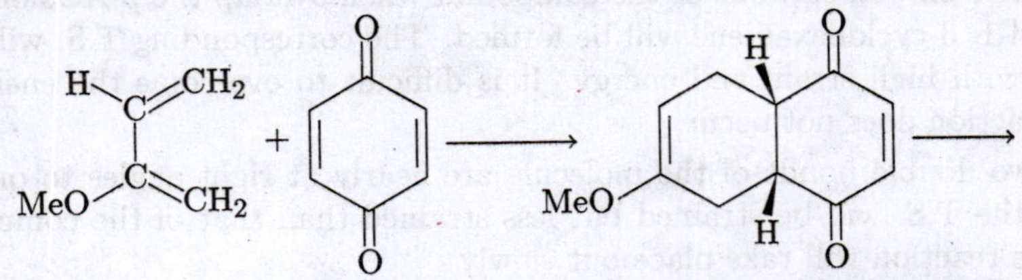

Since the Diels-Alder reaction is a reversible one, this reaction may be used to protect the double bond of a compound.

Non-isolable arynes have been trapped by the Diels-Alder reaction.

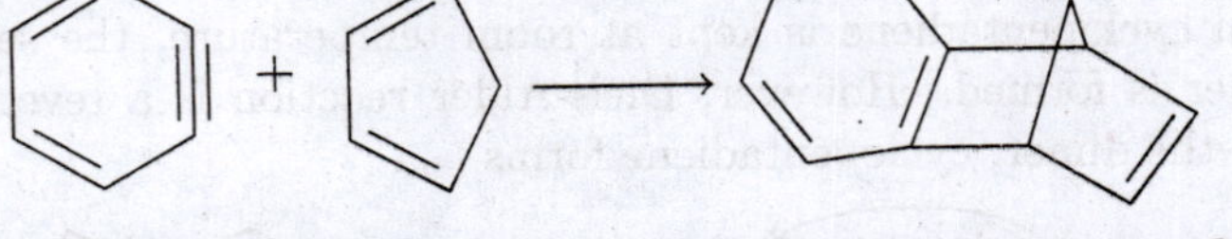

Problem 13.15. **(a)** Furan acts as a diene in the Diels-Alder reaction but pyrrole normally does not.—Explain.

(b) Why does anthracene undergo the Diels-Alder reaction but naphthalene does not?

Solution : (a) Furan being less aromatic than pyrrole it takes part in the Diels-Alder reaction.

(b) The energy gained by the cycloaddition reaction being less than the energy lost by the naphthalene molecule owing to the loss of aromaticity by one of the two benzene rings, naphthalene does not take part in the Diels-Alder reaction. In the case of anthracene, the energy gained by the reaction is more than the energy lost by the loss of aromaticity of the middle ring and so the reaction occurs.

Problem 13.16. **(a)** Explain why the following compounds do not undergo Diels-Alder reaction :

(i) 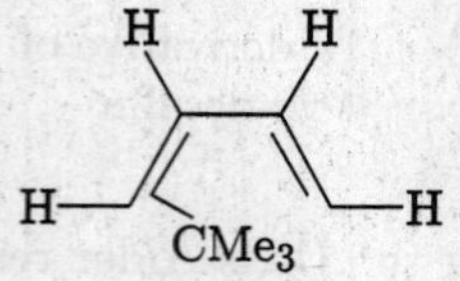

(ii) H—C≡C—C≡C—H (iii) 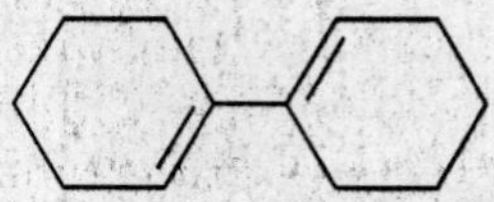

(iv) 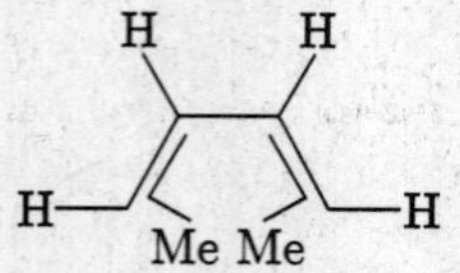

(b) 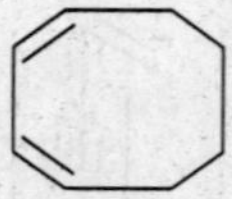 undergoes the reaction but very slowly.—Explain.

Solution : (a) Dienes in the cisoid form undergo the Diels-Alders reaction. Transoid forms do not undergo the reaction because this form leads to a highly strained six-membered T.S., whose energy content is very high. Compounds (i) and (iv) experience high strain in the cisoid form and they remain in the transoid from. So, compounds (i) and (iv) do not undergo the reaction.

Compound (iii) is locked in the transoid form; so, it does not also undergo the reaction.

Compound (ii), 1, 3-butadiyne is a linear molecule. Hence all the Cs are in the *sp*-hybridised state. The *p* AOs on end carbons of the compound when overlap the *p* AOs of the dienophile, a highly strained 1, 3-cyclohexadiene will be formed. The corresponding T.S. will have trienyl-like linear moiety with high strain and energy. It is difficult to overcome the energy barrier of the T.S. So the reaction does not occur.

(b) The two double bonds of the molecule are nearly at right angles to one another. So in this case also the T.S. will be strained but less strained than that of the compounds (i) to (iv). Hence here the reaction will take place but slowly.

Problem 13.17. Explain why **(a)** cyclopentadiene remains as dimer but the former can be obtained from the latter on distilling at a temperature at 170°C-200°C. **(b)** Tetraphenylcyclopentadiene is monomeric at room temperature.

Solution : (a) Cyclopentadiene is highly reactive as a diene and it is moderately so as a dienophile. When cyclopentadiene is kept at room temperature, the self Diels-Alder reaction occurs and a dimer is formed. However, Diels-Alder reaction is a reversible reaction; for this reason on heating the dimer, cyclopentadiene forms.

(b) The molecule has four bulky substituents. Its dimeric form experiences high steric repulsion. So, its dimer cannot exist at room temperature.

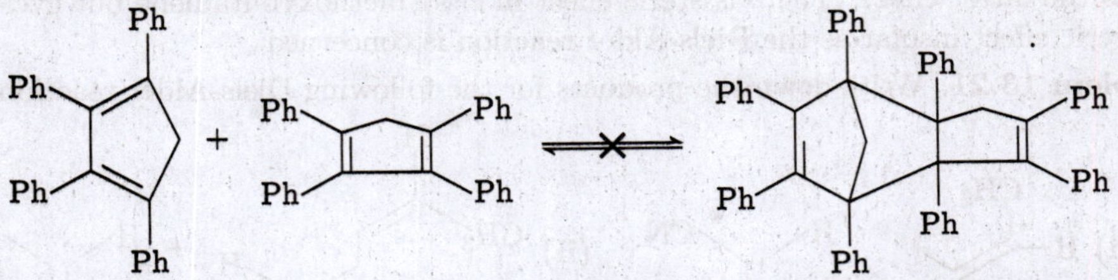

Problem 13.18. How can you synthesise all cis-cyclopentanetetracarboxyiic acid?

Solution : The product can be obtained through a reaction which is highly stereospecific. Diels-Alder reaction may be used to synthesise the compound.

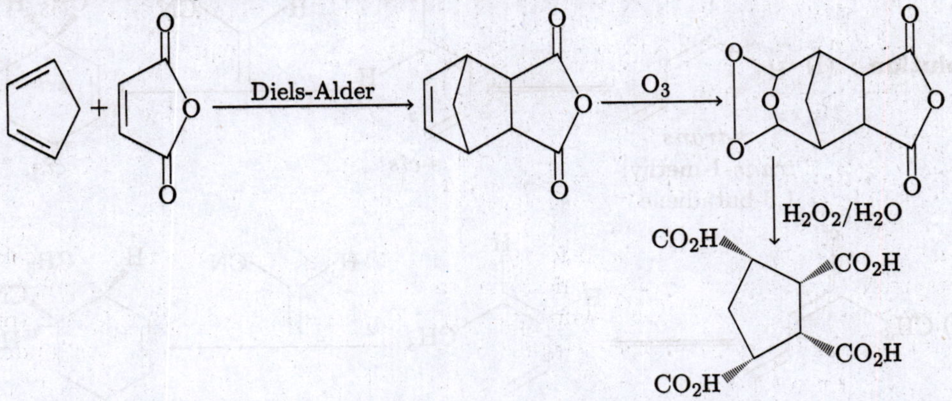

Problem 13.19. The reactivity order of the following dienophiles towards cyclopentadiene in Diels-Alder reaction is as follows :

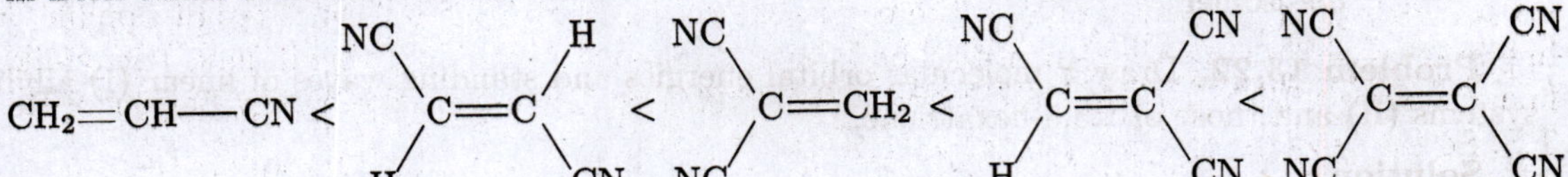

1 < 100 < 45 000 < 500 000 < 43000 000

Explain the order.

Solution : The higher the electron-withdrawing effects in a dienophile, the higher is its reactivity. Ethylene has no electron-withdrawing substitutent. Cyano group withdraws electron by –I and –R effects; as the number of cyano group increases in ethylene the reactivity of cyanoethylenes increases.

Problem 13.20. Explain the reactivity order of the following 1,3-butadienes towards tetracyanoethylene as dienophile in the Diels-Alder reaction :

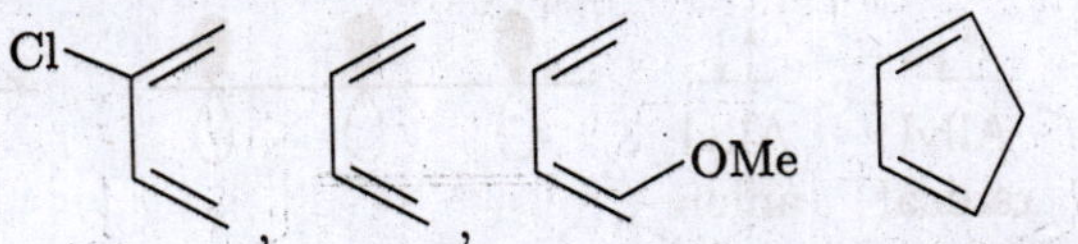

1, 500, 500 000, 410 000 000

Solution : Electron-withdrawing groups in a diene decrease its reactivity while electron-donating groups increase the same in a Diels-Alder reaction. Cl has the former effect but MeO group has the latter effect. There is steric effect in cis-1-methoxybutadiene but cyclopantadiene has no steric effect insofar as the Diels-Alder reaction is concerned.

Problem 13.21. Write down the products for the following Diels-Alder reactions :

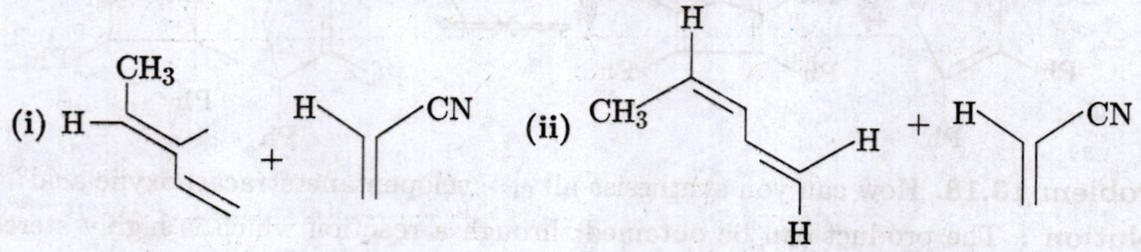

Solution : (i)

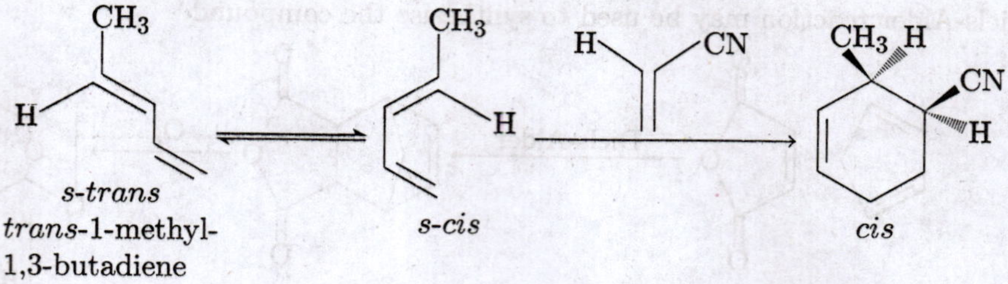

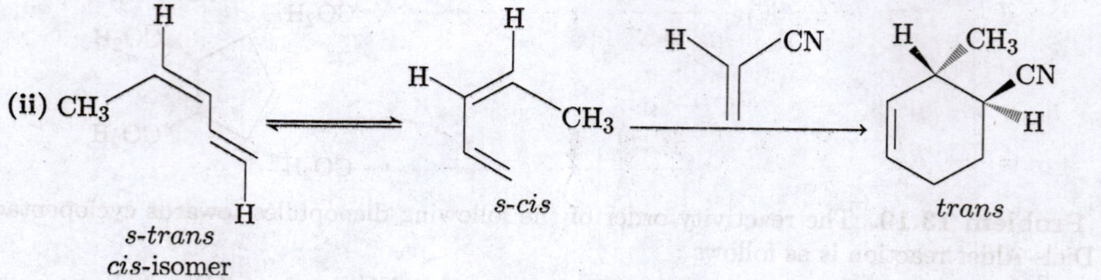

Problem 13.22. Draw π molecular orbital energies and standing waves of linear (i) allylic systems (ii) and those of 1,3,5-hexatriene.

Solution :

(i)

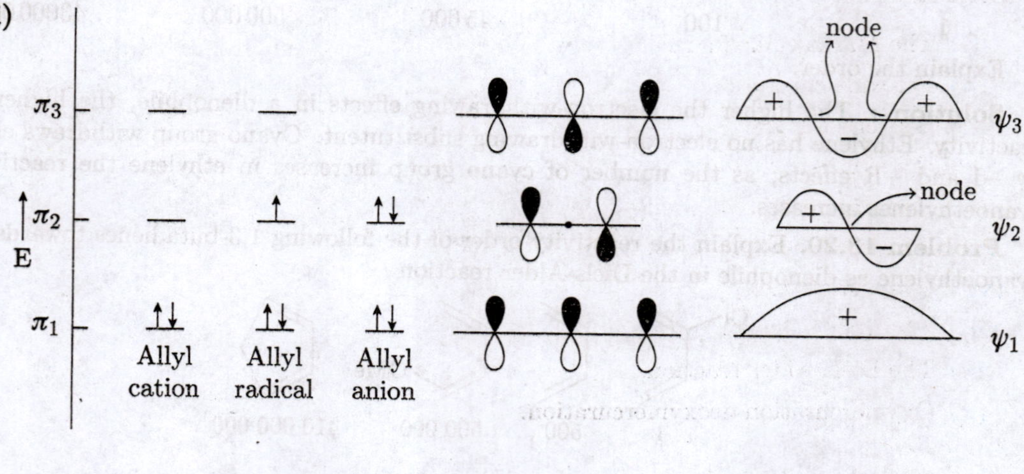

(ii)

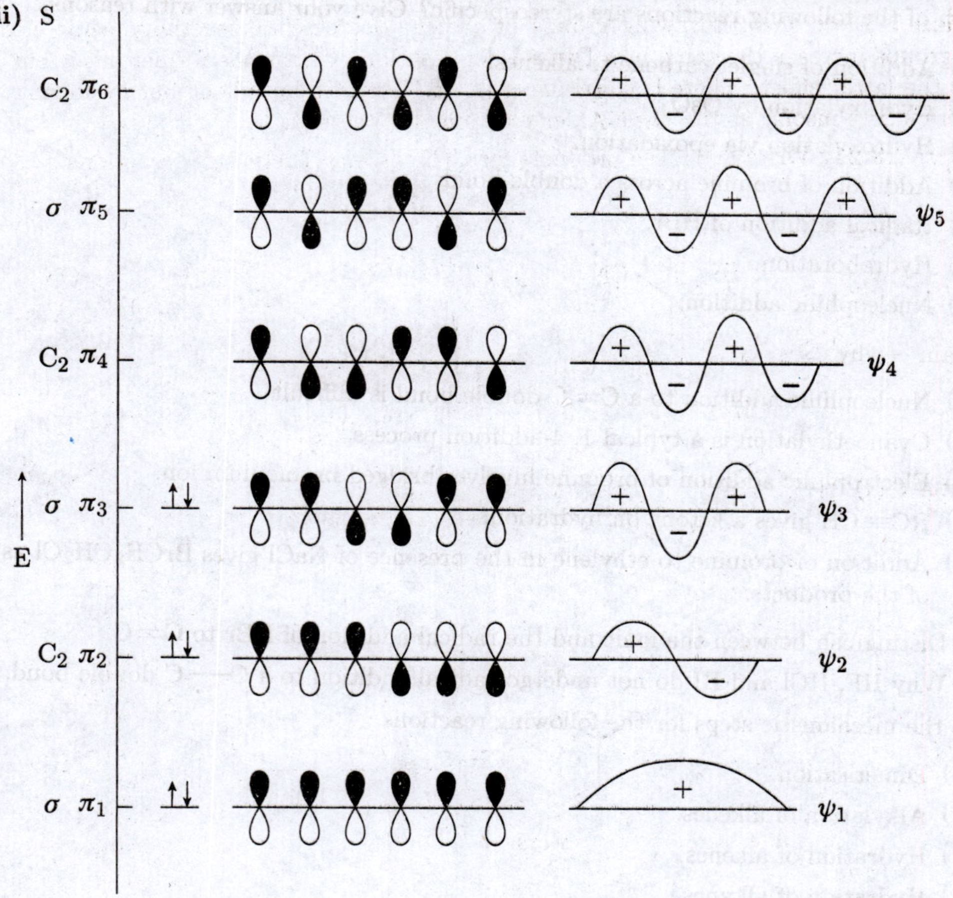

Exercise

1. Write short notes on :

   (i) The Markownikoff's rule,
   (ii) Peroxide effect,
   (iii) Baeyer's test,
   (iv) Epoxidation,
   (v) Hydroboration-oxidation,
   (vi) Hydroboration-carbonylation,
   (vii) Cyanoethylation,
   (viii) The Michael condensation,
   (ix) The Diels-Alder reaction,
   (x) Oxymercuration-deoxymercuration.

2. Which of the following reactions are stereospecific? Give your answer with reasons.

(i) Addition of singlet carbene to alkenes.
(ii) Hydroxylation by OsO_4.
(iii) Hydroxylation via epoxidation.
(iv) Addition of bromine across a double bond.
(v) Radical addition of HBr.
(vi) Hydroboration.
(vii) Nucleophilic addition.

3. Explain —why

(i) Nucleophilic addition to a C=C double bond is difficult.
(ii) Cyanoethylation is a typical 1, 4-addition process.
(iii) Electrophilic addition of bromine involves bridged bromonium ion.
(iv) RC≡CH gives a ketone on hydration.
(v) Addition of bromine to ethylene in the presence of NaCl gives $BrCH_2CH_2Cl$ as one of the products.

4. (a) Distinguish between the ionic and the radical addition of HBr to C=C.
(b) Why HF, HCl and HI do not undergo radical addition to a C—C double bond.

5. Give the mechanistic steps for the following reactions :

(i) Dimerisation.
(ii) Alkylation of alkenes.
(iii) Hydration of alkenes.
(iv) Hydration of alkynes.

6. How can you convert the following ?

(a) $CH_3.CH_2.CH{=}CH.CH_2\ CH_3 \longrightarrow CH_2{=}CH_2.CH_2CH_2CH_2CH_3$
(b) $H_2S \longrightarrow SH.CH_2.CH_2CH_2OH.$

7. Complete the following reactions.

(a) $CH_3.CH{=}CH.CO_2Et + CH_2(CN)_2$
(b) $CH_2{=}CH{-}CHO + CH_2(CO_2Me)_2$

Solution :

(a) $CH_3.CH{=}CH.CO_2Et + CH_2(CN)_2 \xrightleftharpoons{\text{EtOH \& NaOEt}} CH_3{-}\underset{\underset{CH(CN)_2}{|}}{CH}{-}CH_2CO_2Et$

(b) $CH_2{=}CH{-}CHO + CH_2(CO_2Me)_2 \xrightleftharpoons{\text{EtOH \& NaOEt}} \underset{\underset{CH(CO_2Me)_2}{|}}{CH_2}{-}CH_2.CHO$

8. What happens when an α, β-unsaturated ketone is treated with alkaline H_2O_2?

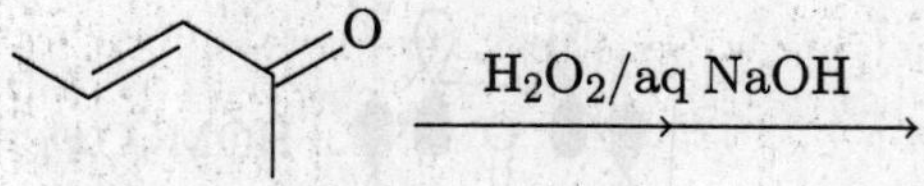

Give the mechanism and explain why a peroxycarboxylic acid is not used in this reaction.

9. Using frontier orbital overlap, explain why $\pi 4s + \pi 2s$ reaction between butadiene and ethylene is thermally allowed but not photochemically.

Solution : According to FMO approach, HOMO of one π-system and LUMO of the other are involved in $[\pi 4s + \pi 2s]$ cycloaddition. In such a thermal addition LUMO of butadiene (4π) and HOMO of ethylene (2π) can undergo supra, supra interactions involving same phase overlap as shown below :

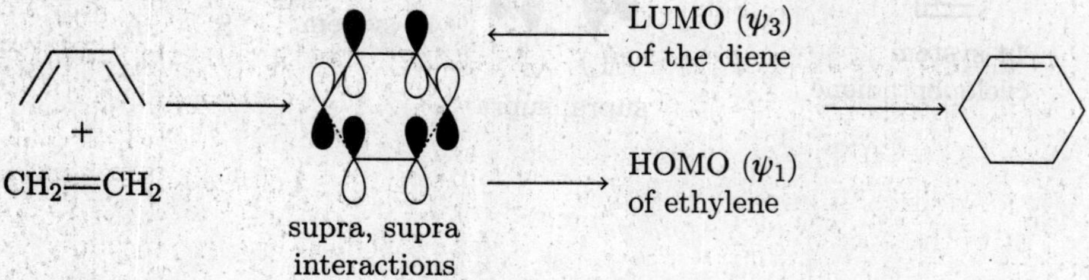

In a photochemical reaction, excitation of the HOMO of one system or the other takes place. In either case supra, antara interactions take place instead of supra, supra as asked in the question. So photochemical $(4\pi s + 2\pi s)$ addition is not allowed.

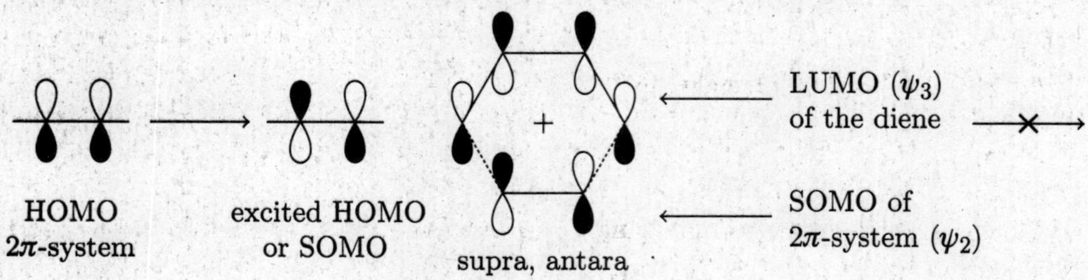

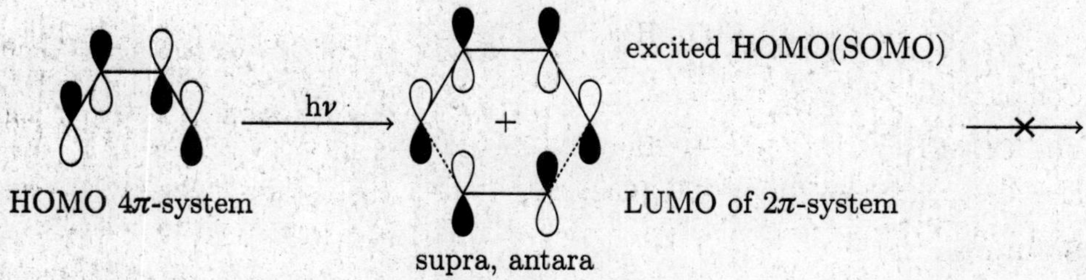

10. Is the following reaction symmetry allowed?

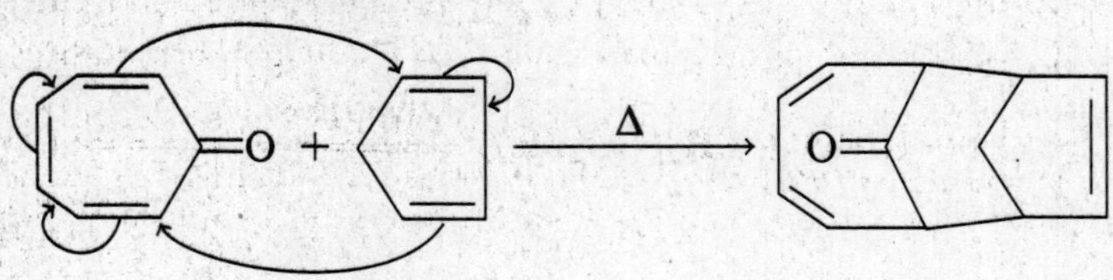

Solution : Yes, it is a $[6\pi s + 4\pi s]$ cycloaddition as shown below :

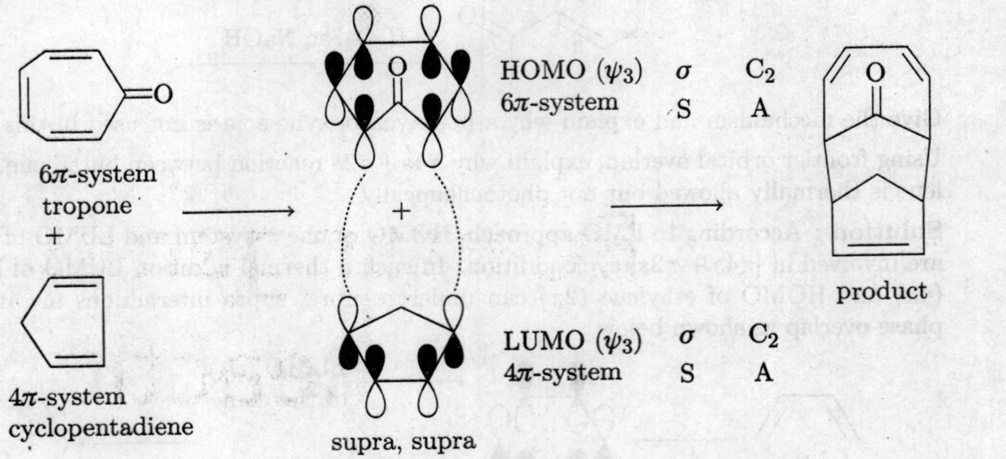

Chapter 14

Rearrangement Reactions

Objectives

- Types of rearrangement reactions [14.1] p. 699
- The pinacol-pinacolone rearrangement [14.2] p. 701
- The benzilic acid rearrangement [14.3] p. 709
- The Demyanov rearrangement [14.4] p. 711
- The Beckmann rearrangement [14.5] p. 715
- The Hofmann rearrangement (14.6] p. 720
- The Curtius rearrangement [14.7] p. 725
- The Schmidt rearrangement of acid azides [14.8] p. 727
- The Lossen rearrangement [14.9] p. 728
- The Wolff rearrangement and the Arndt-Eistert syntheses [14.10] p. 730
- The benzidine and semidine rearrangements [14.11] p. 733
- The Hofmann-Martius rearrangement [14.12] p. 742
- The Orton or the chloramine rearrangement [14.13] p. 744
- The Fischer-Heff rearrangement [14.14] p. 746

- The phenylhydroxylamine–*p*-aminophenol rearrangement [14.15] p. 745
- The allylic rearrangement [14.16] p. 750
- The sigmatropic rearrangement [14.17] p. 753
- The Claisen rearrangement [14.17.1] and the Cope rearrangement [14.17.2] and sigmatropic [1, j] hydrogen shift [14.17.3] p. 753–763
- The electrocyclic rearrangement [14.18] p. 764
- The Fries rearrangement [14.19] p. 769
- The Baeyer-Villiger oxidation and Dakin reaction [14.20] p. 773
- The Favorskii rearrangement [14.21] p. 777

14.1 Introduction

When a chemical unit, an atom or an ion or a group of atoms, migrates from one atom to another in the same species or in another species of the same kind and thereby develops a new species, the reaction is called a rearrangement reaction. The atom from which the group migrates is known as migration origin and the atom to which it migrates is termed migration terminus. For example, in the conversion of G—A—B to A—B—G, G is the migrating group, A and B are the migration origin and the migration terminus respectively.

$$\text{Migrating group} \rightarrow \text{G—}\underset{\substack{\uparrow \\ \text{Migration origin}}}{\text{A}}\text{—B} \xrightarrow{\text{rearrangement}} \text{A—}\underset{\substack{\uparrow \\ \text{Migration terminus}}}{\text{B}}\text{—G}$$

However, the migrating group may either be an anion or a cation or a free radical, and accordingly the rearrangement reactions are broadly classified as :

1. Anionotropic rearrangement—where an anion migrates.
2. Cationotropic rearrangement—where a cation migrates.
3. Free radical rearrangement—where a free radical migrates.

When the migration of a group from an atom to another atom occurs within a molecule, it is said to be intramolecular or molecular rearrangement. In this case the migrating group never detaches itself completely from the rearranging species and there lies a continuing tie between the migrating group and the remaining skeleton of the substrate; on the other hand, when the migration of a group from an atom of a molecule to an atom of another molecule of the same or a similar compound takes place, the migrating group definitely detaches itself from the substrate and the reaction is termed as intermolecular rearrangement.

Whether a rearrangement is intramolecular or intermolecular that may be determined by cross-over experiments : when a mixture of two structurally related substrates, G—A—B and X—A—Y (where G is closely related to X and B to Y), is subjected to a rearrangement reaction, it is known as cross-over experiment. If the reaction is an intermolecular one, the product will be a mixture of four different species : A—B—G and A—Y—X, A—B—X, A—Y—G.

$$\text{G—A—B + X—A—Y} \xrightarrow[\text{rearrangement}]{\text{intramolecular}} \underset{\text{intramolecular products}}{\text{A—B—G + A—Y—X}}$$

$$\text{G—A—B + X—A—Y} \xrightarrow[\text{rearrangement}]{\text{intermolecular}} \underset{\text{non-cross-over products}}{\text{A—B—G + A—Y—X}}$$

$$\underset{\text{cross-over products}}{\text{A—B—X + A—Y—G}}$$

In an intermolecular reaction A—B—G and A—Y—X are the non-cross-over products and A—B—X and A—Y—G are called the cross-over products; whereas in an intramolecular reaction of the above cross-over experiment no cross-over products will be obtained—only A—B—G and A—Y—X, the normal intramolecular products, will form.

However, the migration may take place from an atom to the adjacent atom and this sort of migration is called the Whitemore 1,2-shift. The rearrangement of G—A—B to A—B—G is an example of 1,2-shift; 1,3-shift, 1,4-shift, 1,5-shift, etc. which are also known. The conversion of G—A—B—D to A—B—D—G is a 1,3-shift.

Stereochemistry of the rearrangement reactions and the migratory preferences of the groups depend on the nature of the substrate; so these will be dealt with under individual reaction.

14.1.1 Anionotropic rearrangements

When an anion migrates in a rearrangement, it is called anionotropic rearrangement. An anion can migrate to an electron-deficient centre. So, this sort of rearrangement involves the generation of an electron-deficient migration terminus containing six electrons in its valence shell at the first step. In the second step the migration of a group with its bonded electrons takes place and the migration origin becomes a new electron-deficient centre. These two steps may be discrete or concerted. Usually an initial step precedes these two steps in which the migration terminus is made active to undergo the first step. However, in the third step the eletron-deficient migration origin usually combines with a nucleophile. Sometimes a non-classical electron-deficient centre, a bridged intermediate, forms during the transition of the active substrate to the rearranged species consisting of the electron-deficient migration origin. Thus an anionotropic rearrangement may involve the following steps :

(i) Initial step : Activation of the migration terminus.

(ii) First step : Generation of the electron-deficient migration terminus.

(iii) Second step : Migration of the group; this may develop a non-classical bridged intermediate and may occur along with the first step simultaneously.

(iv) Third step : Nucleophilic coupling involving the migration origin and a nucleophile.

(v) Final step : Formation of the stable product.

However, the migrating groups may be

(a) an alkyl, an aryl, a ring residue or, a side chain,

(b) an H atom, a halogen atom, a hydroxyl or an amino group.

(c) a double or a triple bond.

The migration origin and the migration terminus may be C, O, N, etc.

Anionotropic rearrangements are usually carried out in an acid or neutral or alkaline medium. However, most of the rearrangements are anionotropic and are often called electron-deficient skeletal rearrangements.

14.1.2 Cationotropic rearrangements

An electron-deficient group migrates in this type of rearrangement. Naturally the initial step involves activation of the migration terminus so that this can develop to an electron-rich centre in the first step. Then usually a non-classical bridged intermediate forms in the second step. In the third step migration is completed and then the electron-rich migration origin becomes neutral by any means. Cationotropic rearrangements are also called electron-rich skeletal rearrangements.

14.1.3 Free radical rearrangements

A free radical rearrangement involves the migration of a free radical. Here also several steps are involved. The initiation of a free radical constitutes the initial step. The first step is the generation of a free radical migration terminus through propagation. This free radical may get stability through the formation of a bridged intermediate. In the next step actual migration occurs and the migration origin becomes a new free radical. In the last step the new free radical stabilises itself by a termination reaction.

In a free radical rearrangement aryl, vinyl, chloro, bromo, etc., groups often migrate; a hydrogen atom and an alkyl group usually do not migrate in the 1,2-shift. The direction of migration is normally toward the more stable free radical though there are exceptions. 1,3-, 1,4-, 1,5- and 1,6-shifts of hydrogen have been observed; among these 1,5-shift is the most common.

Free radical rearrangements are usually carried out in the gas phase.

14.1.4 Pericyclic rearrangements

Besides ionic and radical rearrangements there are certain rearrangements involving concerted step which occurs with the conservation of orbital symmetry via cyclic T.S.s by the action of heat or light with or without inert solvents. These are called pericyclic rearrangement reactions.

There are two kinds of pericyclic rearrangements :

1. Sigmatropic rearrangements,
2. Electrocyclic rearrangements.

However, here we have no scope to discuss all sorts of rearrangements. We shall discuss here some of the rearrangements which have been studied in detail.

14.2 The Pinacol-Pinacolone Rearrangement

Pinacols are ditertiary 1,2-diols. The simplest member of the class is $Me_2C(OH)\cdot C(OH)Me_2$. When this is treated with dilute or moderately concentrated H_2SO_4, a rearrangement reaction takes place which leads to the formation of $Me_3C\cdot CO$. Me (pinacolone), and the rearrangement is known as the pinacol-pinacolone or pinacol rearrangement.

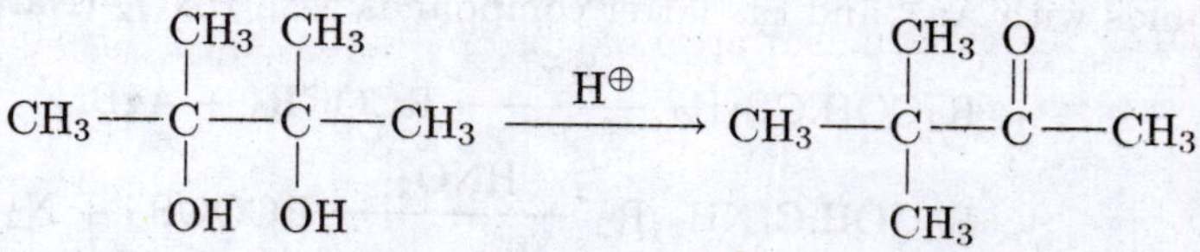

Pinacol Pinacolone

Nowadays the acid-catalysed rearrangement reactions of 1,2-diols (vic-diols) to oxo compounds, aldehydes or ketones, are called the pinacol-pinacolone rearrangements. For example :

(i) 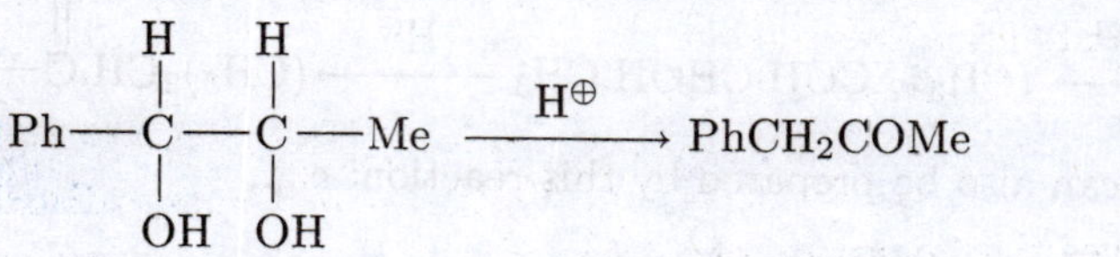

(ii) 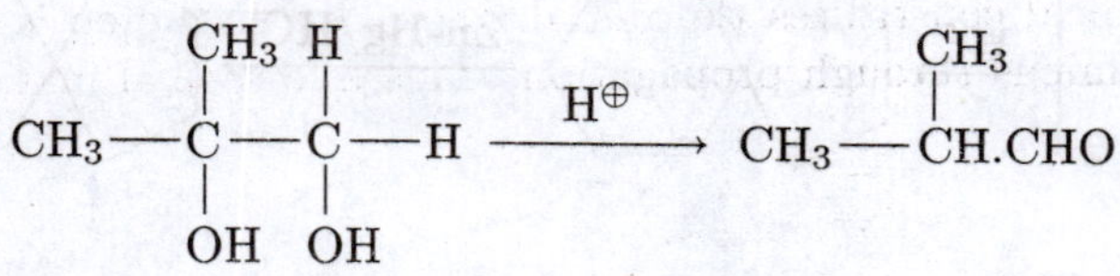

(iii) 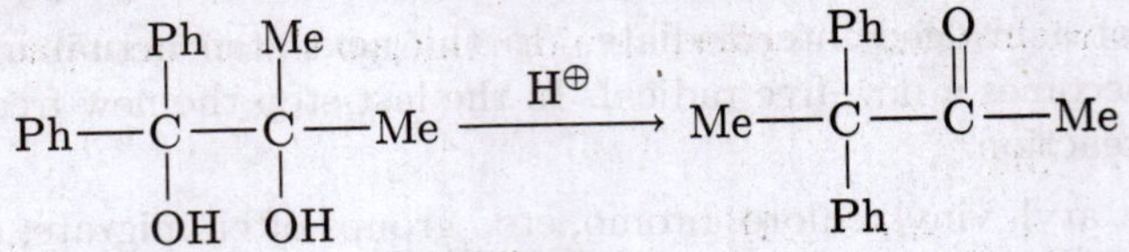

(iv) 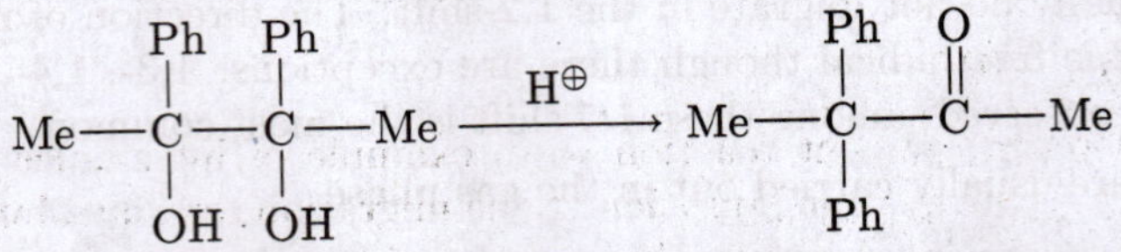

The above examples show that the migration origin and migration terminus are the two adjacent C atoms, and the migrating group may be an aryl or an alkyl group, or an H atom etc. Migration of a bond may also occur in this rearrangement whereby ring expansion and ring contraction reactions may take place.

Ring expansion 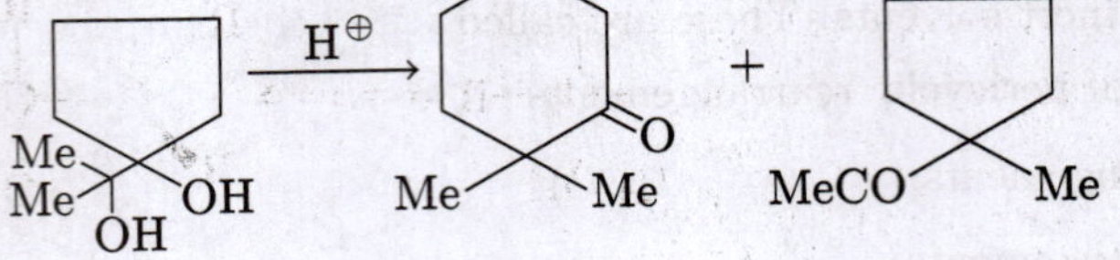

Ring contraction 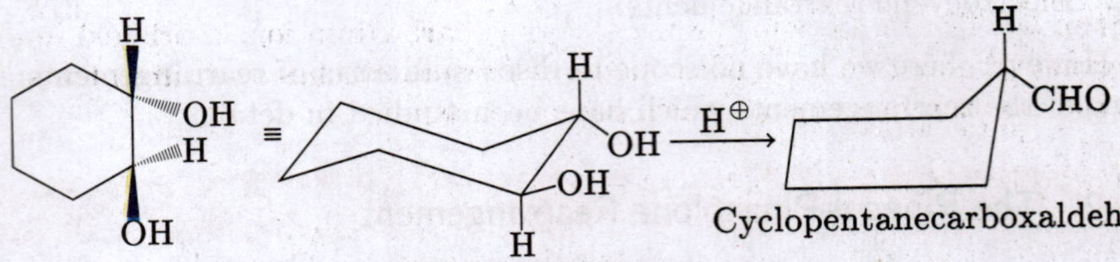

Cyclopentanecarboxaldehyde

trans-1,2-Cyclohexanediol *e,e* form;
the most stable conformation

Rearrangement reactions of 1,2-halohydrines and 1,2-amino-alcohols to pinacolones are analogous reactions and called the pinacolic rearrangements; these are carried out by treating the former compounds with $Ag^{\oplus}$ and the latter compounds with HNO_2 ($NaNO_2$/HCl).

$$R_2COH.CBrR_2 \xrightarrow{Ag^{\oplus}} RCOCR_3 + AgBr\downarrow$$

$$R_2COH.C(NH_2)R_2 \xrightarrow{HNO_2} RCOCR_3 + N_2\uparrow$$

Since highly branched oxo compounds are very difficult to prepare by the other reactions, this rearrangement has interesting applications in synthesis.

For example, methyl isopropyl ketone is easily prepared from 2,3-dichloro-2-methylbutane by this rearrangement.

$$(CH_3)_2C\ Cl.CHCl.CH_3 \xrightarrow[\text{heat}]{H_3O^{\oplus}} (CH_3)_2.COH.CHOH.CH_3 \xrightarrow{H^{\oplus}} (CH_3)_2CH.\overset{O}{\overset{\|}{C}}-CH_3$$

Spiranes and their derivatives can also be prepared by this reaction; e.g.,

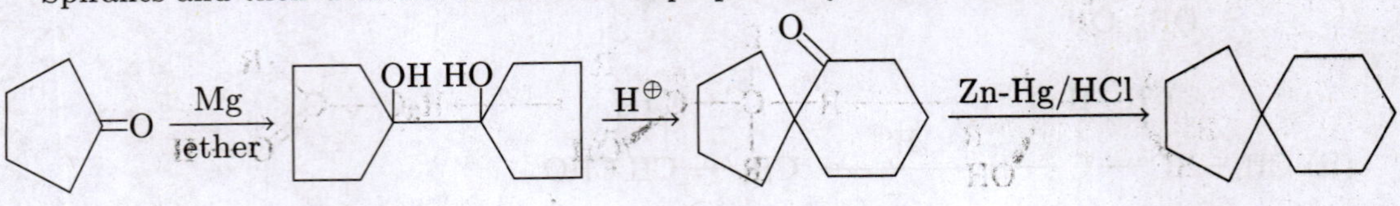

1,2-diols containing substituents with strong −I effect do not undergo the pinacol-pinacolone rearrangement reaction.

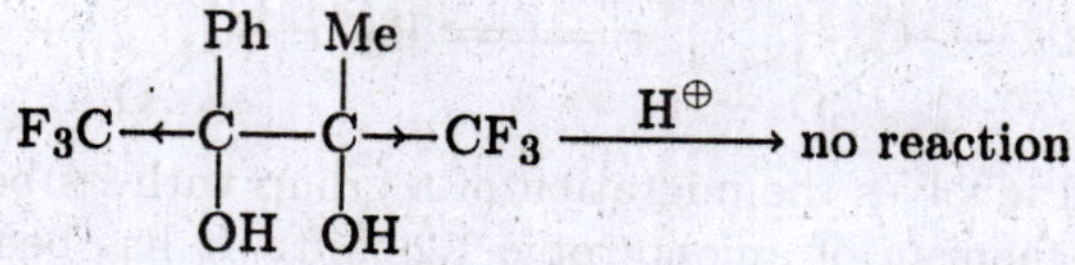

This electron-deficient skeletal rearrangement reaction is an example of intramolecular anionotropic 1,2-shift which involves inversion of configuration at the migration terminus but retention of configuration of the migrating group. However, this consists of four steps, two or more of which may be concerted.

Step 1. Reversible protonation to a hydroxyl group and the elimination of water molecule; an electron-deficient carbenium ion is thus formed in this step :

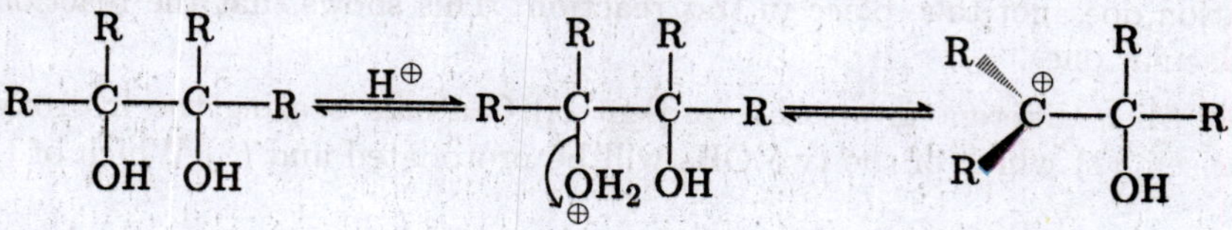

Step 2. The formation of a non-classical carbenium ion, a bridged intermediate. Here it is important to note that the migrating group never detaches itself from the substrate skeleton and thus the reaction becomes an intramolecular one.

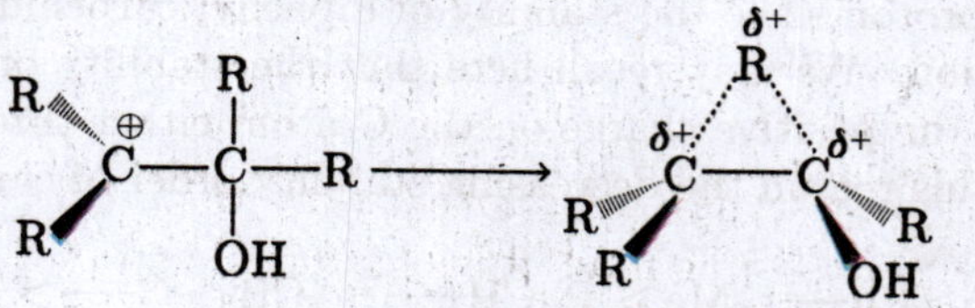

Bridged intermediate

When an aryl group migrates, the bridged intermediate is an aryl cation (e.g., phenyl cation) which may be an actual compound since it is reasonably stable.

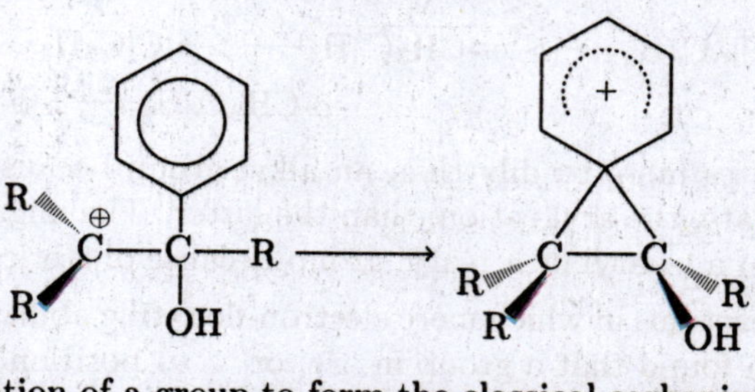

Step 3. Actual migration of a group to form the classical carbenium ion; thus the migration origin becomes an electron-deficient atom, a resonance stabilised carbenium ion.

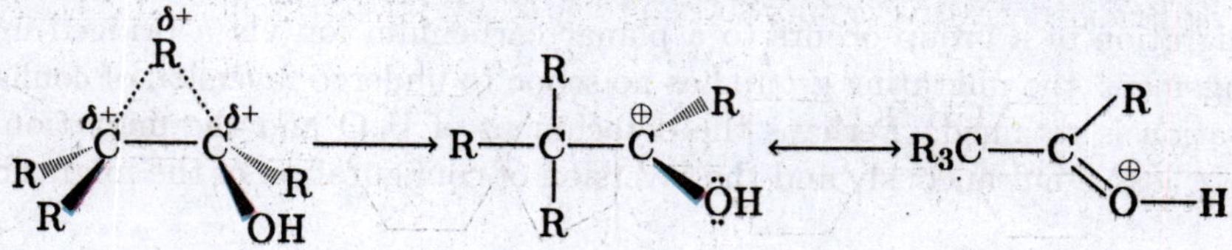

Step 4. The loss of proton and the formation of oxo compound :

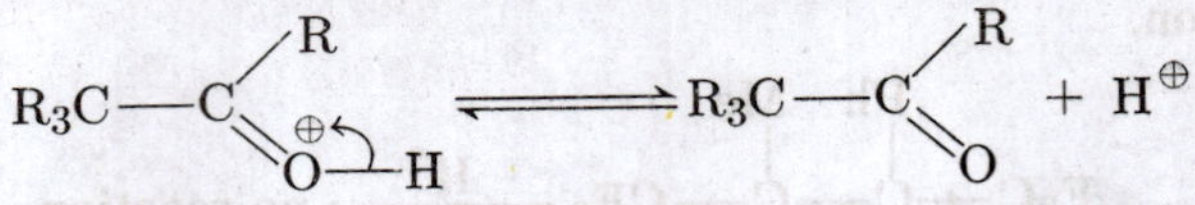

Since the mechanism involves the migration of a group with its bonded *e*s to an adjacent C atom, it follows the mechanism of anionotropic 1,2-shift. It has been known from the kinetic study that the step in which elimination of water molecule occurs from the protonated 1,2-diol is the slow step and hence it is the rate-determining step.

The intramolecular nature of the reaction is supported by the cross-over experiment in which a mixture of $Ph_2COH{\cdot}COHMe_2$ and $Ph_2COH{\cdot}COHEt_2$ has been treated with acid. In this reaction only intramolecular products: $Ph_2CMe{\cdot}COMe$ and $Ph_2CEt{\cdot}COEt$, have been obtained; $Ph_2CEt{\cdot}COMe$ and $Ph_2CMeCOEt$, the possible cross-products, have not yet been isolated. Thus cross-migration does not take place in this reaction. This shows that the reaction is essentially an intramolecular one.

However, when different groups are present on C atoms bearing the hydroxyl groups, two questions arise : (1) which of the two OHs will be protonated and (2) Which of the groups will migrate?

The answer of the first question lies with the stability order of the carbenium ions. Usually that OH receives the proton which produces the more stable carbenium ion by the elimination of water molecule. Thus in $Ph_2COH{\cdot}COH\ Me_2$ the OH group on the C atom holding the phenyl groups will receive the proton since the stability of diphenyl carbenium ion is greater than that of dimethyl carbenium ion. We may recall here that the stability of a carbenium ion depends on the delocalisation of the positive charge on the C atom either through resonance or through hyperconjugation. For this reason the decreasing stability order of carbenium ions is :

$$Ph_2\overset{\oplus}{C}{-} > Ph{-}\overset{\oplus}{\underset{|}{C}}Me > Ph\overset{\oplus}{C}H{-} > (CH_3)_2\overset{\oplus}{C}{-} > CH_3\overset{\oplus}{C}H{-}$$

There is no clear-cut answer in so far as migratory preference is concerned.

In terms of electron-donating ability the usual migratory preference order is :

$$p\text{-}CH_3OC_6H_4{-} > p\text{-}CH_3.C_6H_4{-} > m\text{-}CH_3C_6H_4{-} > p\text{-}ClC_6H_4{-} > m\text{-}CH_3OC_6H_4{-} > o\text{-}CH_3C_6H_4{-} > o\text{-}ClCH_2C_6H_4{-} > \text{alkyl, H}$$

An aryl group migrates more readily than an alkyl group because the former forms a more stable bridged intermediate, the aryl cation, than the latter. The migratory aptitude of an *o*-aryl group is less than that of a *m*-aryl or a *p*-aryl group because of steric hindrance.

However, there are reactions in which mere electron-donating ability does not decide which one will migrate. It has been found that a group in anti or trans position with respect to the leaving group, H_2O, in the more stable conformation of the protonated substrate migrates preferentially (for the evidence see the stereochemistry of the reaction).

Since migration of a group occurs to a planar carbenium ion via a bridged intermediate in this rearrangement, the migrating group has no scope to undergo inversion of configuration; thus its configuration is retained. Perhaps the detachment of H_2O and the migration of the group occur more or less simultaneously and the inversion of configuration of the migration terminus is observed.

Thus when optically active 1,1-diphenyl-2-amino- 1 -propanol is treated with $NaNO_2/HCl$, the pinacolic deamination reaction occurs with inversion of configuration at the asymmetric migration terminus.

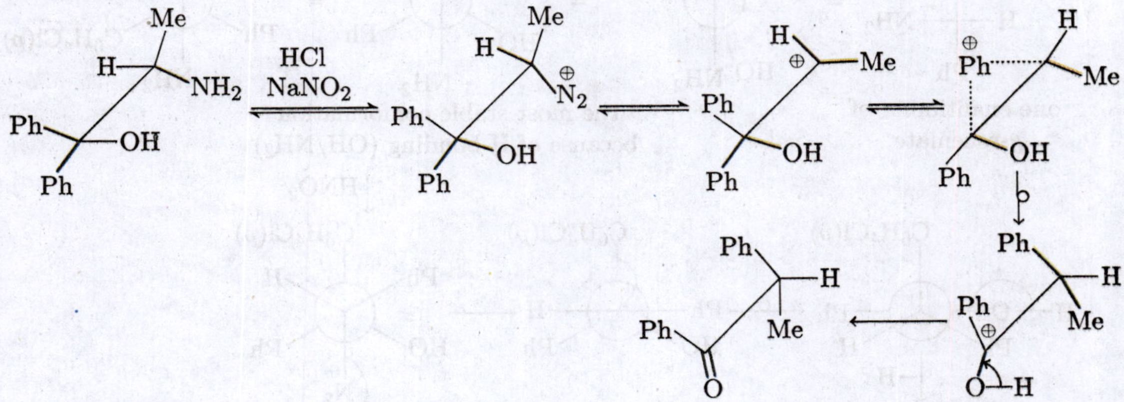

Pollak and Curtin carried out the deamination reaction on α- and β-racemates of 1,2-diphenyl-1 -(*p*-chlorophenyl)-2-aminoethanol and found that in the case of α-racemate the migrating group is phenyl whereas *p*-chlorophenyl group migrates in the case of β-racemate. From the consideration of migratory aptitude in terms of electron-donating ability, phenyl group should be the only migrating group in both the cases. Here actually the group which is present in the anti-position (with respect to the amino group) in the most stable conformation migrates.

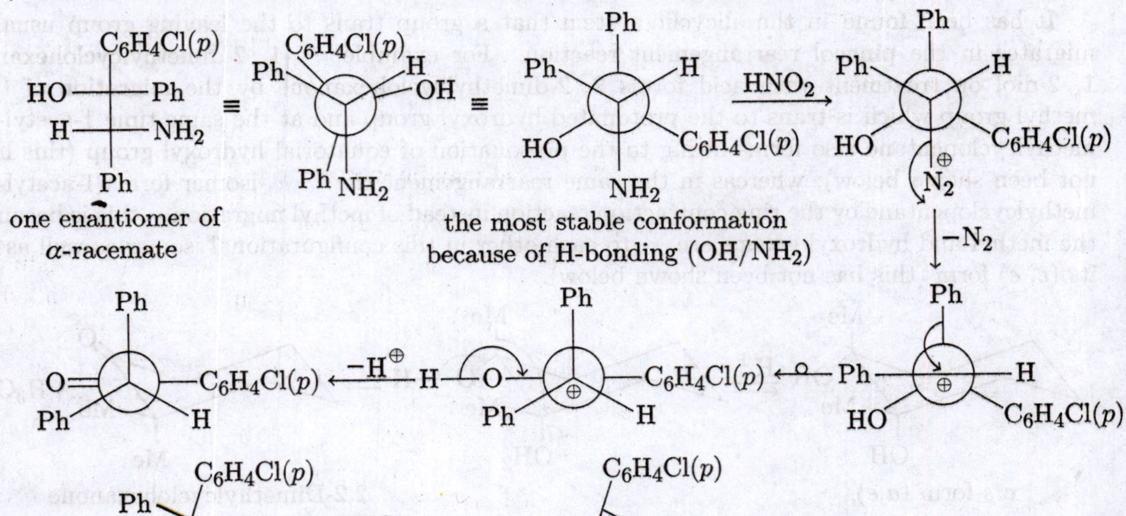

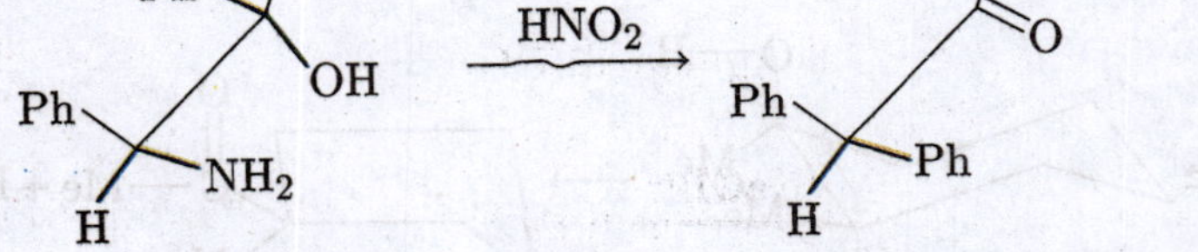

Thus, where there are two bulky substituents of comparable migratory aptitude (in so far as electron-donating ability is concerned), the actual migrating group is one which occupies the anti-position (with respect to the leaving group) in the most stable conformation.

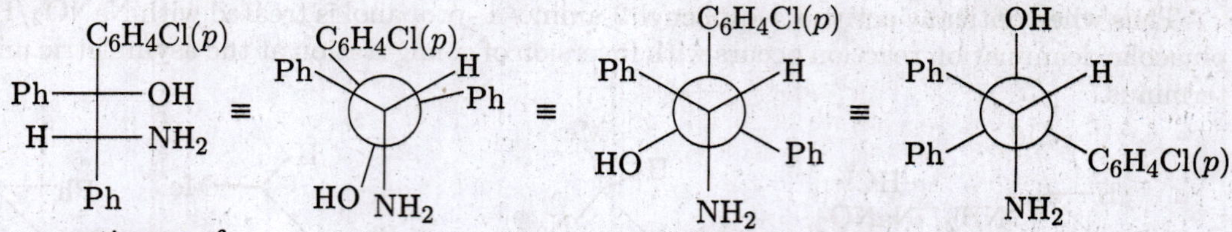

one enantiomer of β-racemate

the most stable conformation because of H-bonding (OH/NH_2)

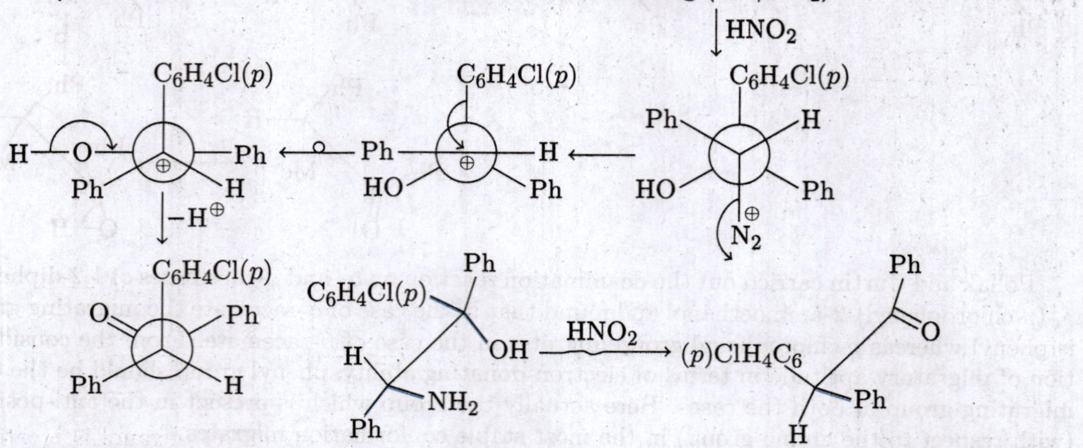

It has been found in the alicyclic system that a group trans to the leaving group usually migrates in the pinacol rearrangement reaction. For example, cis-1, 2-dimethylcyclohexane-1, 2-diol on treatment with acid forms 2, 2-dimethylcyclohexanone by the migration of the methyl group which is trans to the protonated hydroxyl group and at the same time 1-acetyl-1-methylcyclopentane also forms owing to the protonation of equatorial hydroxyl group (this has not been shown below); whereas in the same rearrangement the trans-isomer forms 1-acetyl-1-methylcyclopentane by the ring contraction reaction instead of methyl migration— this is because the methyl and hydroxyl groups are cis to each other in this configurational isomer as well as in its (*e*, *e*) form (this has not been shown below).

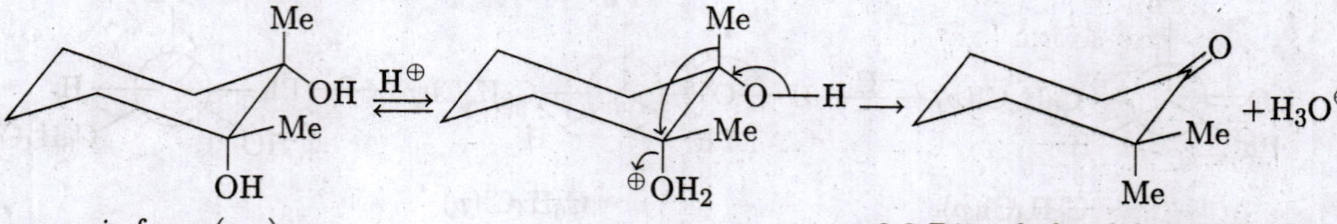

cis-form (*a*,*e*)

2,2-Dimethylcyclohexanone

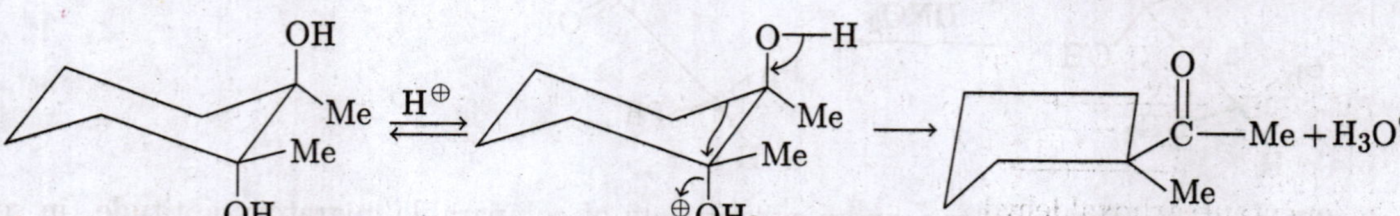

trans-form

(OH/OH : *a*,*a*; Me/Me : *e*,*e*)

1-Acetyl-1-methyl-cyclopentane

Problem 14.1. How can you synthesise :

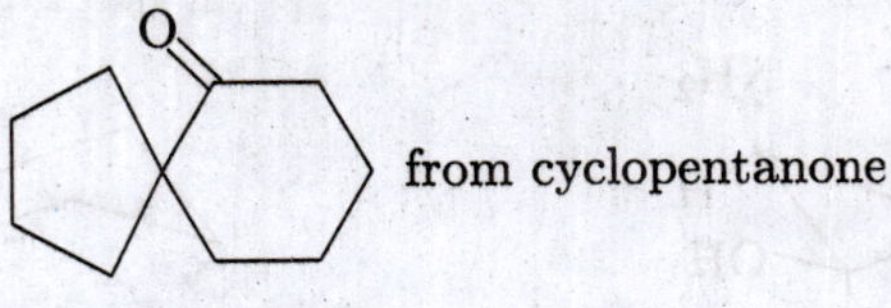

Give mechanistic steps for the synthesis.

Solution :

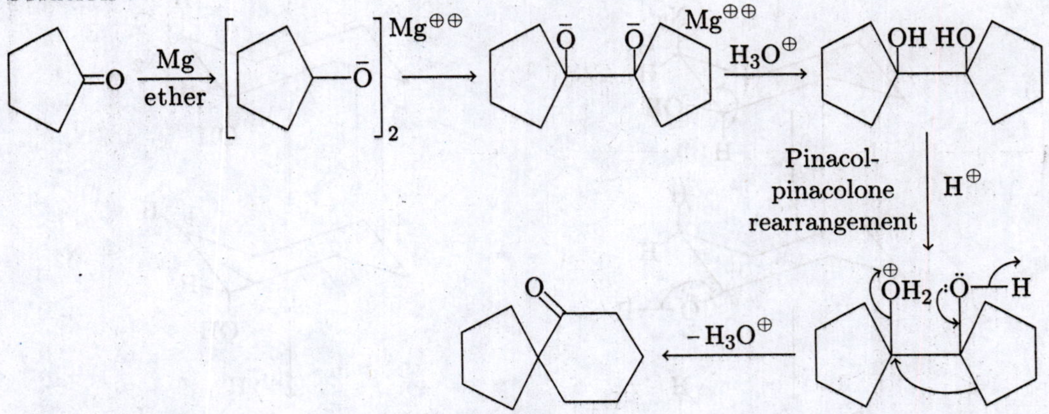

Problem 14.2. What product do you expect when trans-2-aminocyclohexanol is treated with aqueous $NaNO_2$ and dil. HCl?

Solution : Trans-2-aminocyclohexanol will exist in diequatorial chair conformation since this is the most stable conformation among others. Since this *ee*-form does not contain an H atom trans to diazonium group, ring contraction will occur to form cyclopentanecarboxaldehyde.

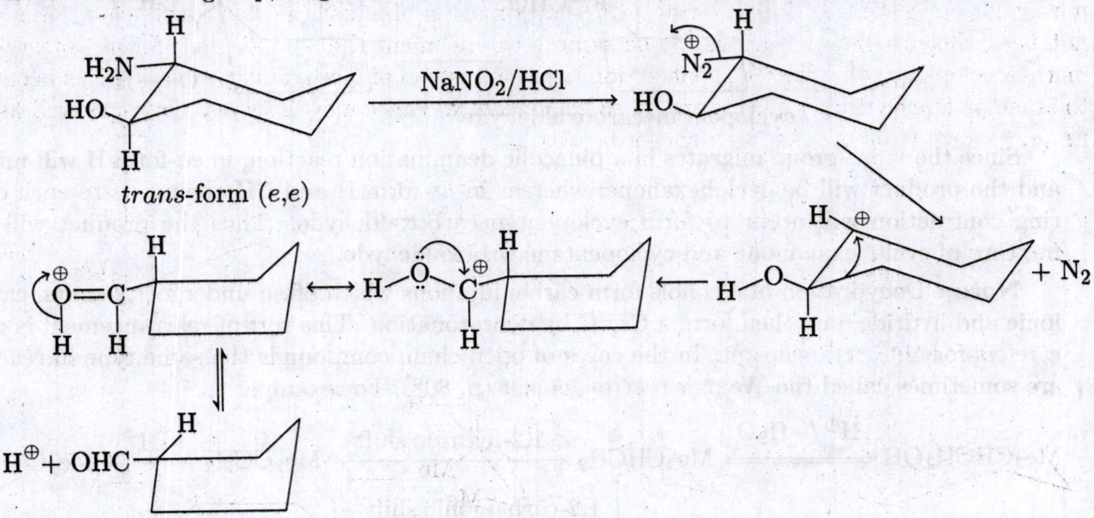

Cyclopentanecarboxaldehyde

Problem 14.3. What happens when cis-2-aminocyclohexanol is subjected to pinacolic rearrangement reaction? Write down the mechanistic steps.

Solution : cis-2-aminocyclohexanol may exist in two chair conformations, *ae* and *ea*, as drawn below :

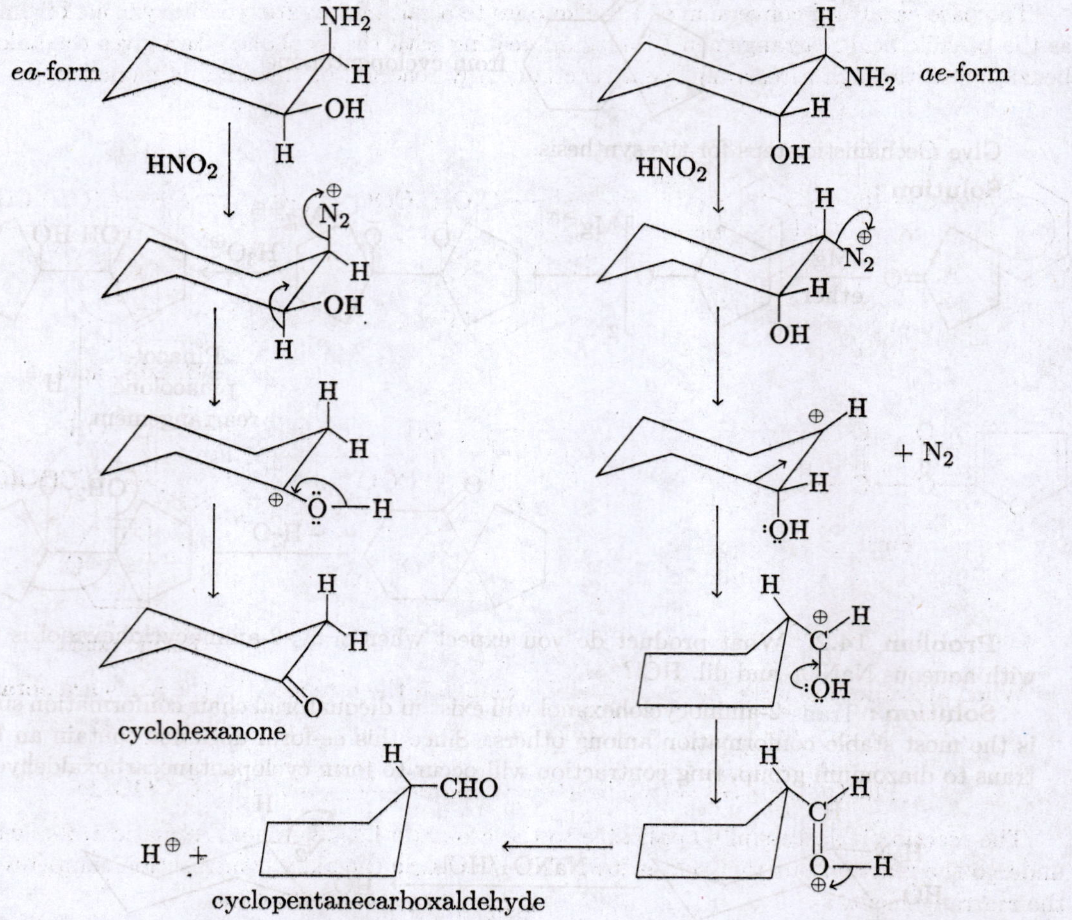

Since the trans-group migrates in a pinacolic deamination reaction, in *ea*-form H will migrate and the product will be cyclohexanone; whereas in *ae*-form H and NH_2 being cis to each other, ring contraction will occur to form cyclopentanecarboxaldehyde. Thus the product will be a mixture of cyclohexananone and cyclopentanecarboxaldehyde.

Note : Dehydration of alcohols form carbenium ions which often undergo 1,2-shifts, carbanionic and hydride, and then form a C═C by deprotonation. This sort of rearrangement is called a retropinacol rearrangement. In the cases of open chain compounds the same type of reactions are sometimes called the Wagner rearrangement (p. 808). For example,

$$Me_2CHCH_2OH \xrightleftharpoons{H^{\oplus}/-H_2O} Me_2CH\overset{\oplus}{C}H_2 \xrightarrow{\text{1,2-hydride shift}} Me_2\overset{\oplus}{C}CH_3 \xrightarrow{-H^{\oplus}} Me_2C{=}CH_2$$

$$\downarrow \text{1,2-carbanionic shift}$$

$$\underset{\text{minor}}{CH_2{=}CH{-}CH_2Me} \xleftarrow{-H^{\oplus}} Me\overset{\oplus}{C}H{-}CH_2Me \xrightarrow{-H^{\oplus}} \underset{\text{major}}{MeCH{=}CHMe}$$

14.3 The Benzilic Acid Rearrangement

The base-catalysed conversion of 1,2-diketones to a salt of 2-hydroxycarboxylic acid is known as the benzilic acid rearrangement. Benzil on heating with the alcoholic alkali gives the anion of benzilic acid through a rearrangement reaction. Any reaction of this kind is named as benzilic acid rearrangement.

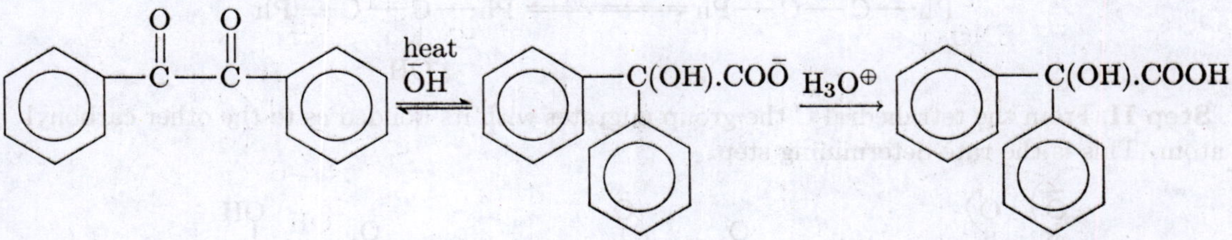

Benzilic acid

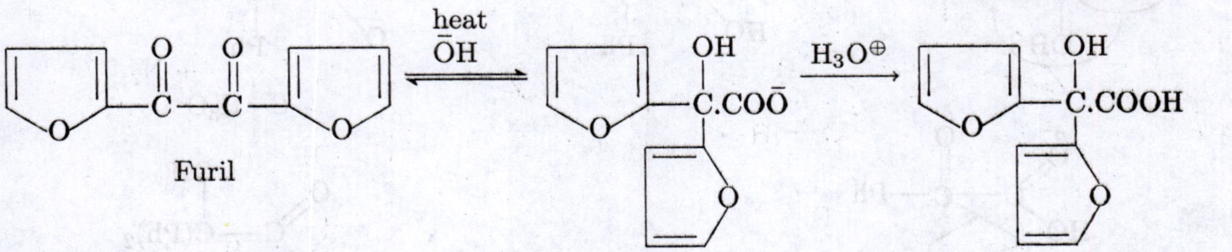

Furil

Furilic acid

When alkoxides are used, instead of alkalis, as bases, the esters of the the acids are obtained.

$$\text{PhCOCOPh} \xrightleftharpoons{\text{MeO}^-} (\text{Ph})_2\overset{\overset{\bar{\text{O}}}{|}}{\text{C}}\text{—COOMe} \xrightarrow{\text{H}_3\text{O}^\oplus} (\text{Ph})_2\overset{\overset{\text{OH}}{|}}{\text{C}}\text{.COOMe}$$

The reaction is successfully applicable to the aromatic 1,2-diketones. Alphatic 1, 2-diketones undergo the reaction but the yield is low. This is because the aldol condensation competes with the rearrangement.

$$\text{HOOC.CH}_2\text{.}\overset{\overset{\text{O}}{\|}}{\text{C}}\text{—}\overset{\overset{\text{O}}{\|}}{\text{C}}\text{—CH}_2\text{COOH} \xrightleftharpoons[\text{KOH/H}_2\text{O}]{\text{reflux}} \bar{\text{O}}\text{OC.CH}_2\text{—}\underset{\underset{\text{CO}\bar{\text{O}}}{|}}{\overset{\overset{\text{OH}}{|}}{\text{C}}}\text{.CH}_2\text{CO}\bar{\text{O}}$$

$$\downarrow \text{H}_3\text{O}^\oplus$$

$$\text{HOOC.CH}_2\text{—}\underset{\underset{\text{COOH}}{|}}{\overset{\overset{\text{OH}}{|}}{\text{C}}}\text{.CH}_2\text{COOH}$$

The reaction is usually found to follow the second order kinetics.

The rate $= k$ [substrate][base]

This intramolecular reaction involves 1,2-carbanionic shift. The migration origin and the migration terminus are the C atoms and the migrating group migrates with a pair of bonded *es* to an electron-deficient carbonyl carbon atom. The reaction is a reversible one and consists of the following two steps :

Step I. Rapid and reversible nucleophilic attack by the base on that carbonyl carbon atom which accommodates positive charge less effectively.

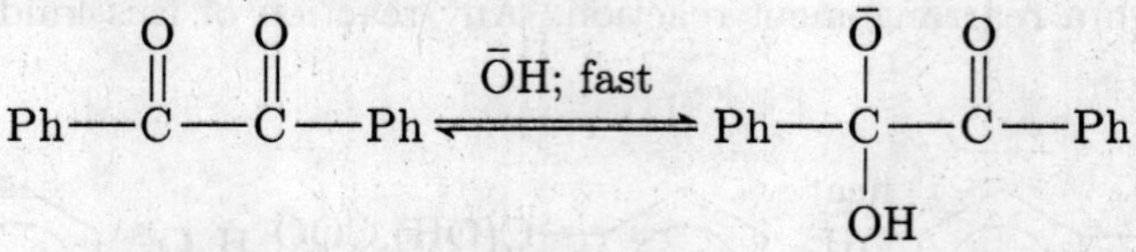

Step II. From the tetrahedral C the group migrates with its bonded *es* to the other carbonyl C atom. This is the rate-determining step.

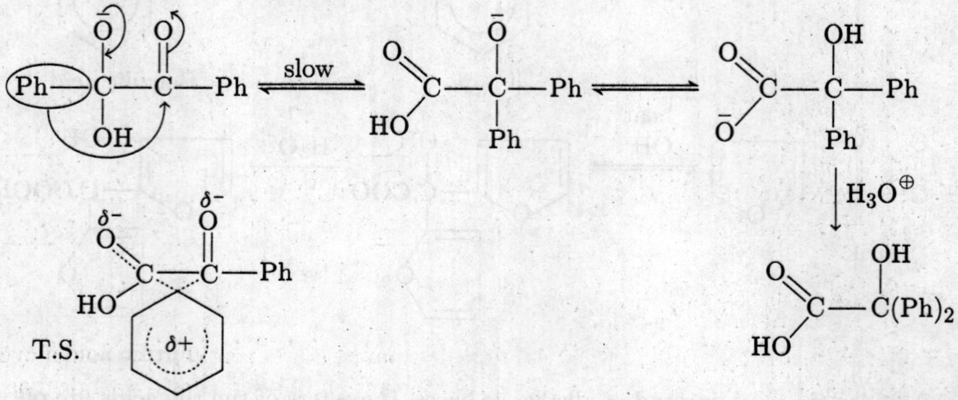

The reversibility of the reaction has been established by the isotope exchange experiment.

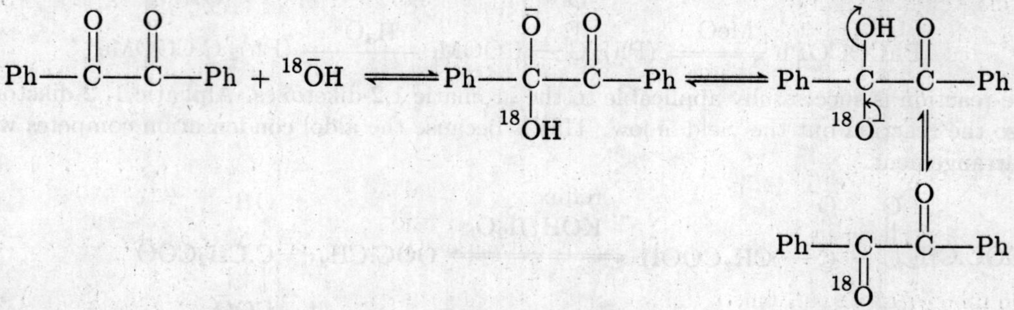

No cross-over products in cross-over experiments have been obtained; this supports the intramolecularity of the reaction.

In the cases of *p*-methylbenzil and *p*-methoxybenzil, the phenyl group migrates in the reaction.

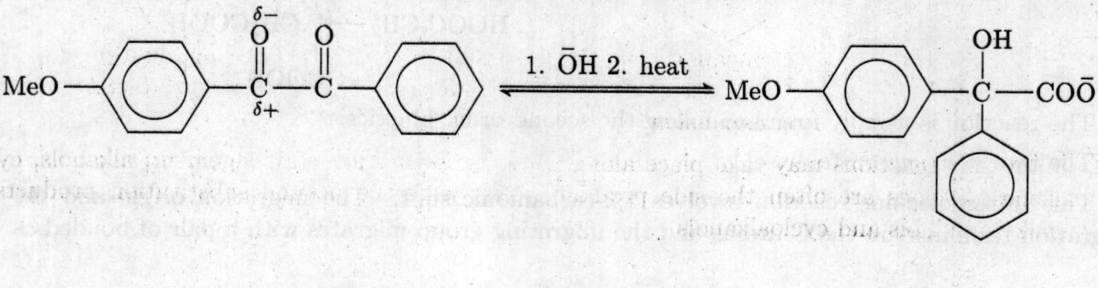

This is perhaps *p*-anisyl and *p*-tolyl groups can better accommodate the positive charge by +R or hyperconjugative effect on the adjacent carbonyl C atom than the phenyl group does and consequently the base attacks that carbonyl carbon which has more positive charge.

Problem 14.4. Give mechanistic steps for the following transformation :

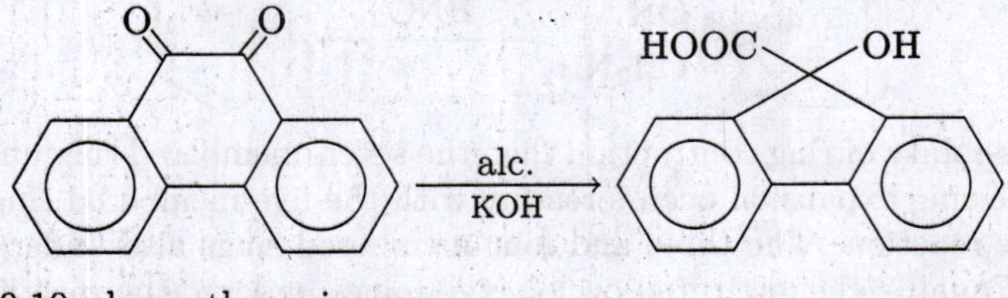

9,10-phenanthraquinone

Solution :

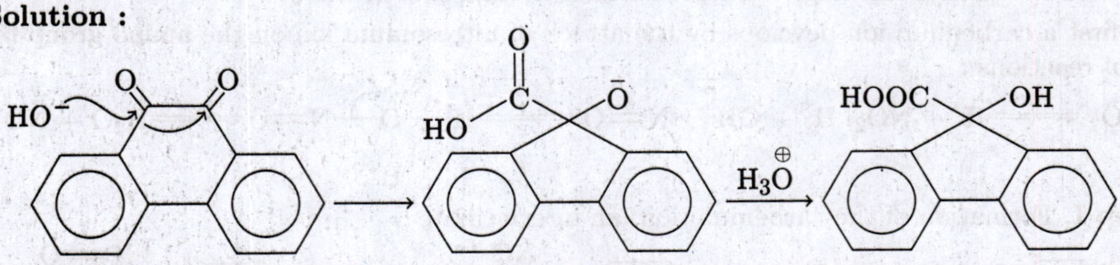

14.4 The Demyanov Rearrangement

The isomerisation of alicyclic rings involving ring expansion and ring contraction of cycloalkylmethylamines and cycloalkylamines respectively by the action of nitrous acid is known as the Demyanov rearrangement in the name of discoverer of the reaction, N.Ya. Demyanov. For example, cycloheptylamine on nitrous acid treatment gives cyclohexylcarbinol, a ring contraction product, along with other products.

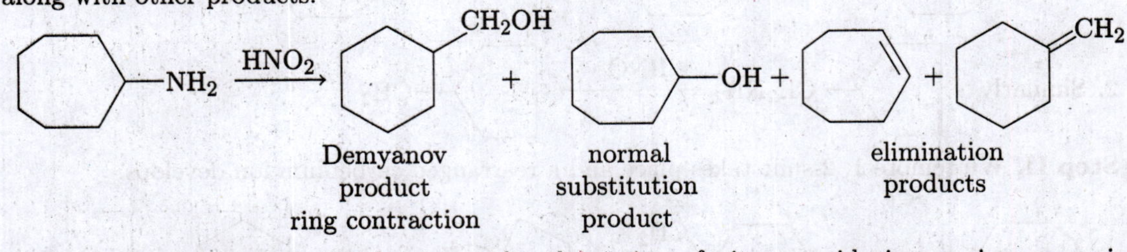

Similarly, cyclopentylmethylamine under the action of nitrous acid gives a ring expansion product amongst others.

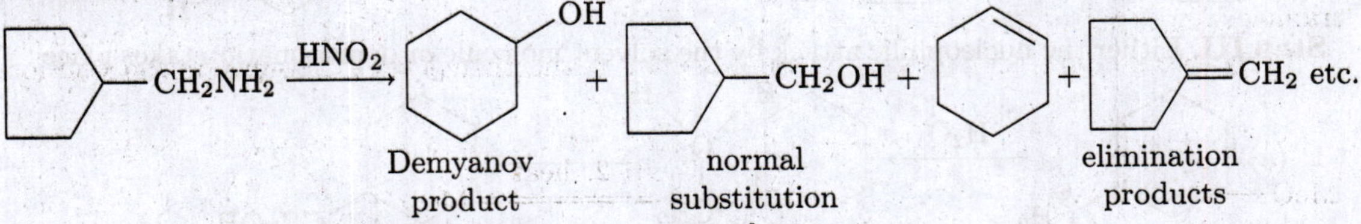

Many side reactions may take place along with the Demyanov rearrangement; alkanols, cycloalkenes, alkenes are often the side products along with the normal substitution products, cycloalkylcarbinols and cycloalkanols.

The rearrangement may be extended to alicyclic hydroxy-amines which undergo a rearrangement analogous to the pinacolic rearrangement and is known as Tiffenev-Demyanov ring expansion.

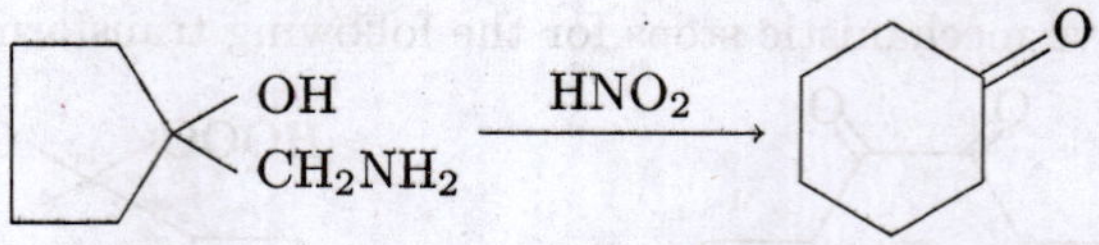

If there is any possibility of ring contraction then the seven-membered ring undergoes ring contraction readily while ring expansion occurs readily with the five-membered ring if the substrate structure favours the reaction. The three and four membered rings also undergo the rearrangement with an ease. Equatorial substituted cyclohexylamines undergo the reaction, whereas axial substituted cyclohexylamines give the products of normal substitution.

At first a carbenium ion develops by the attack of nitrosonium ion on the amino group by a series of reactions :

$$HNO_2 \rightleftharpoons H^{\oplus} + \overset{-}{NO_2};\ H^{\oplus} + OH{-}N{=}O \rightleftharpoons H{-}\overset{\oplus}{\underset{|}{\underset{H}{O}}}{-}N{=}O \rightleftharpoons H_2O + \overset{\oplus}{N}O$$

Step I. Formation of the carbenium ion, an intermediate :

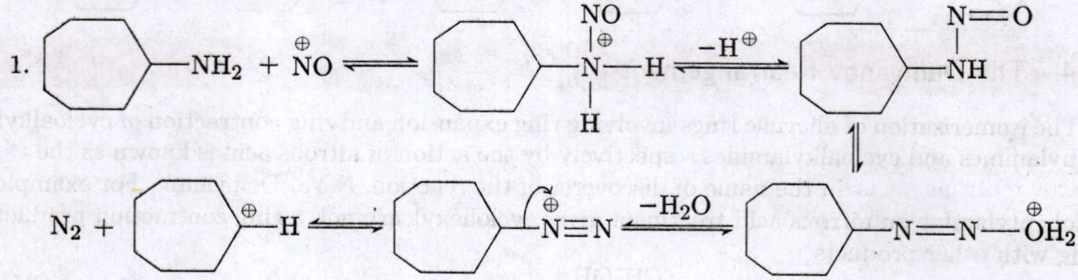

2. Similarly,

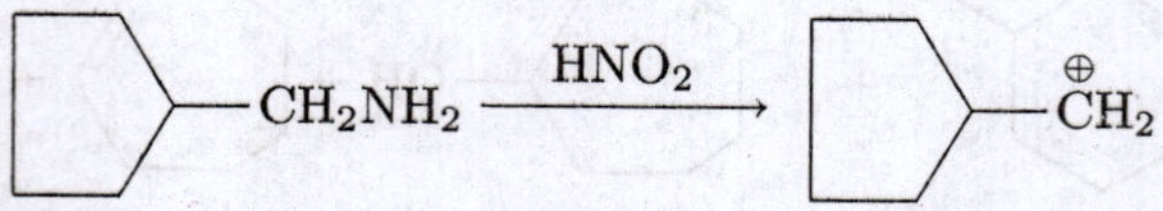

Step II. Whitemore 1, 2-shift takes place and a rearranged carbenium ion develops.

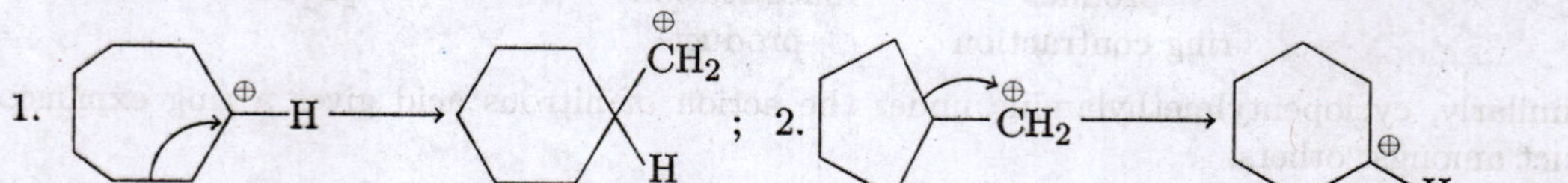

Step III. Either the nucleophilic attack by the solvent molecule or deprotonation takes place :

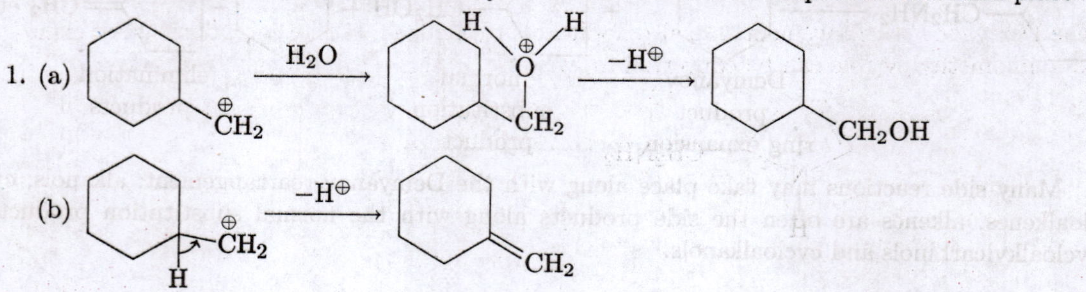

2. (a) 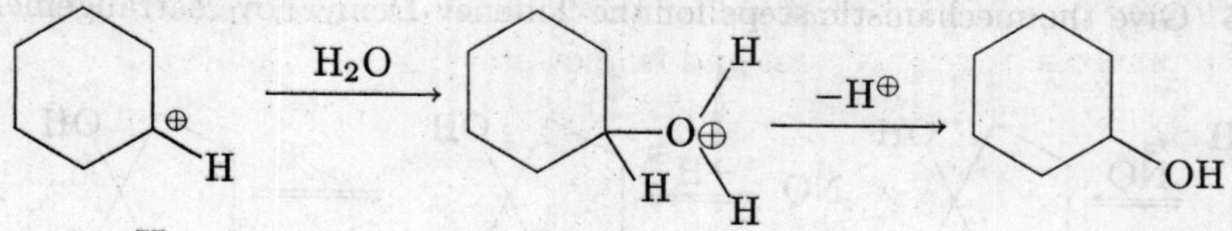

(b) 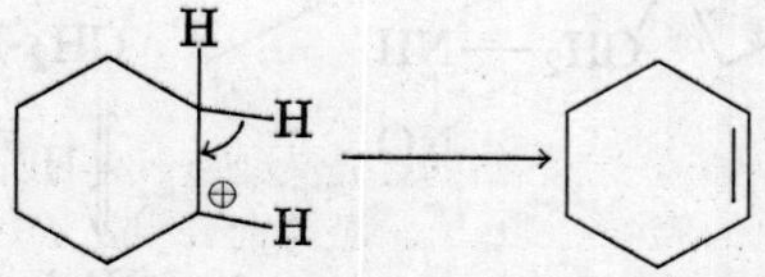

Since the carbenium ion is the intermediate, it can produce several other products:

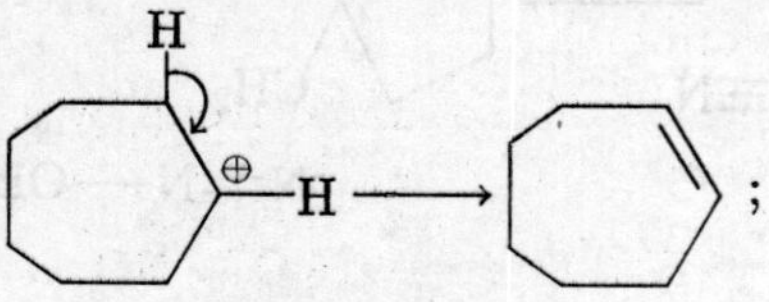

deprotonation

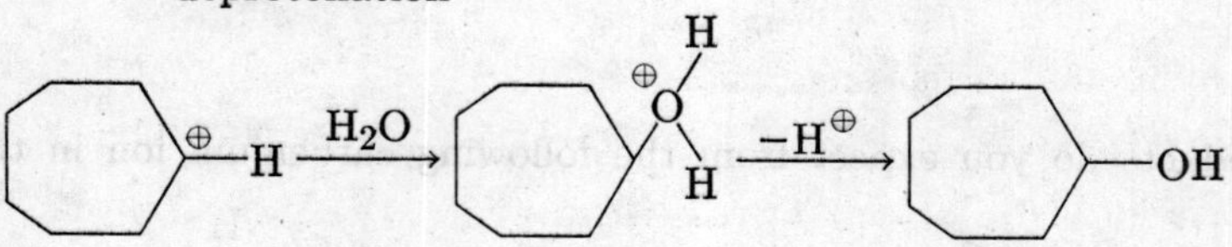

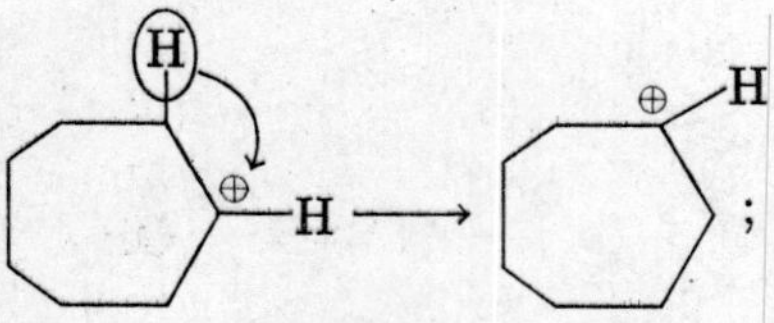

hydride shift

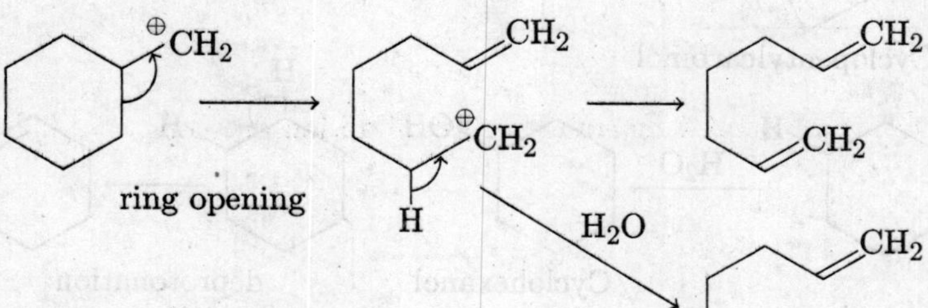

ring opening

The Demyanov rearrangements are also applicable to the heterocyclic compounds. For example, 2-aminomethylpyrrole can be converted to pyridine by this rearrangement reaction :

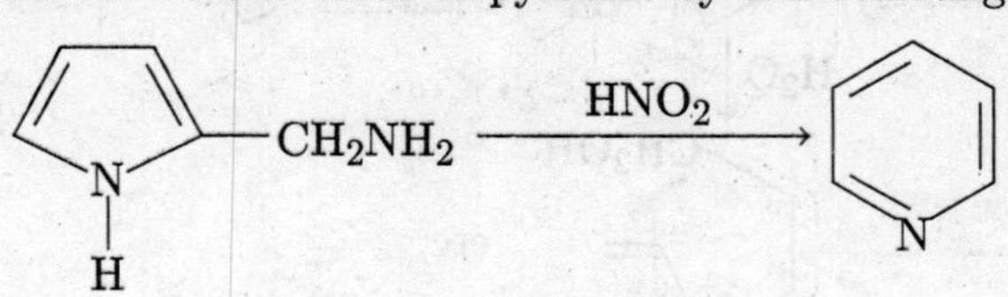

Problem 14.5. Give the mechanistic steps for the Tiffenev-Demyanov rearrangement.

Solution :

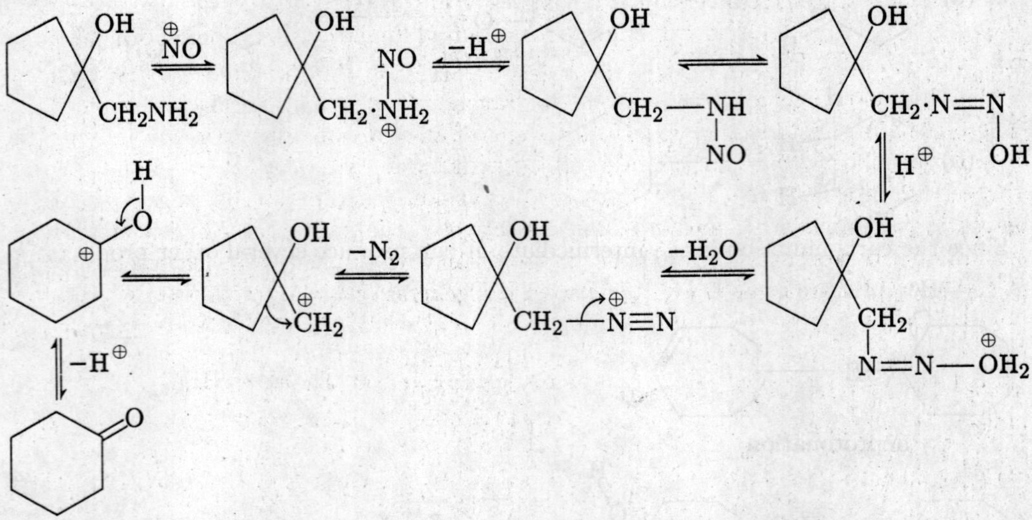

Problem 14.6. What products do you expect from the following carbenium ion in the aqueous medium?

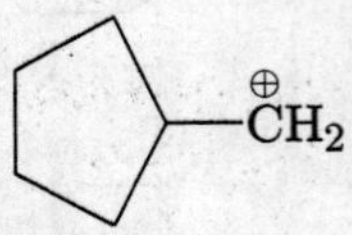

Solution :

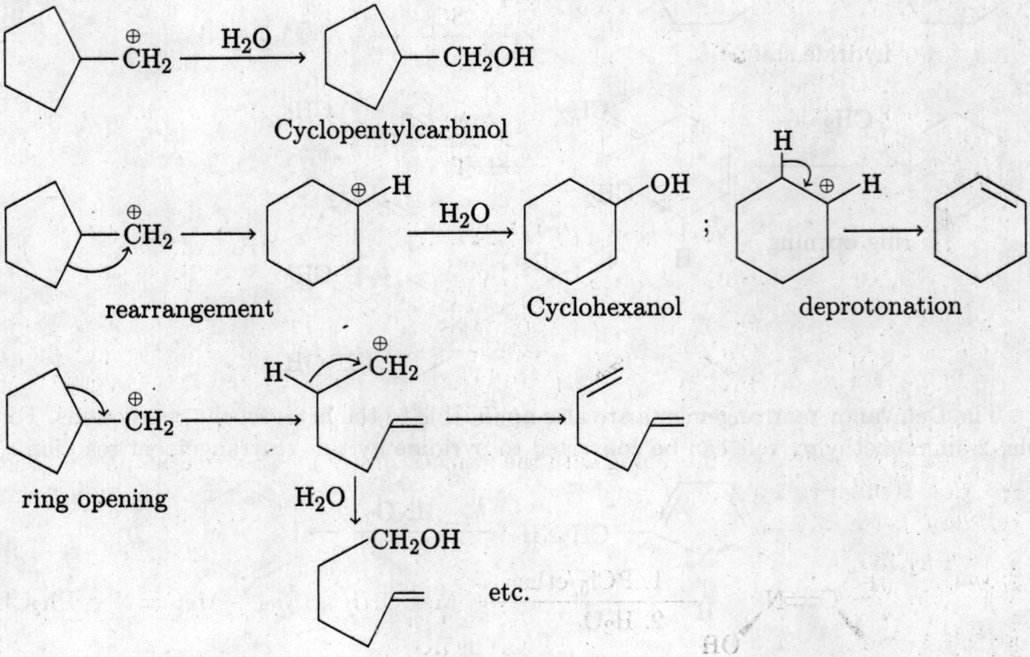

14.5 The Beckmann Rearrangement

The acid-catalysed conversion of ketoximes to N-substituted amides is usually called the Beckmann rearrangement. Acid catalysts used are of wide and varied types, both protic and Lewis; e.g., H_2SO_4, HCl, PCl_5, $PhSO_2Cl$, $SOCl_2$, RCOCl, SO_3, BF_3, polyphosphoric acid, etc.

In this rearrangement reaction the migrating group is in general alkyl or aryl or alicyclyl. There are a few instances in which an H atom of an aldoxime migrates, specially in the presence of polyphosphoric acid. In the cases of cycloalkanone oximes a C—C bond on the carbonyl carbon migrates which leads to the ring expansion reactions. The migration origin and the migration terminus are C and N atoms respectively. The migration is highly stereoselective—the migrating group is that which has the anti-position with respect to the leaving group in the oxime. A few examples are cited below along with the general scheme :

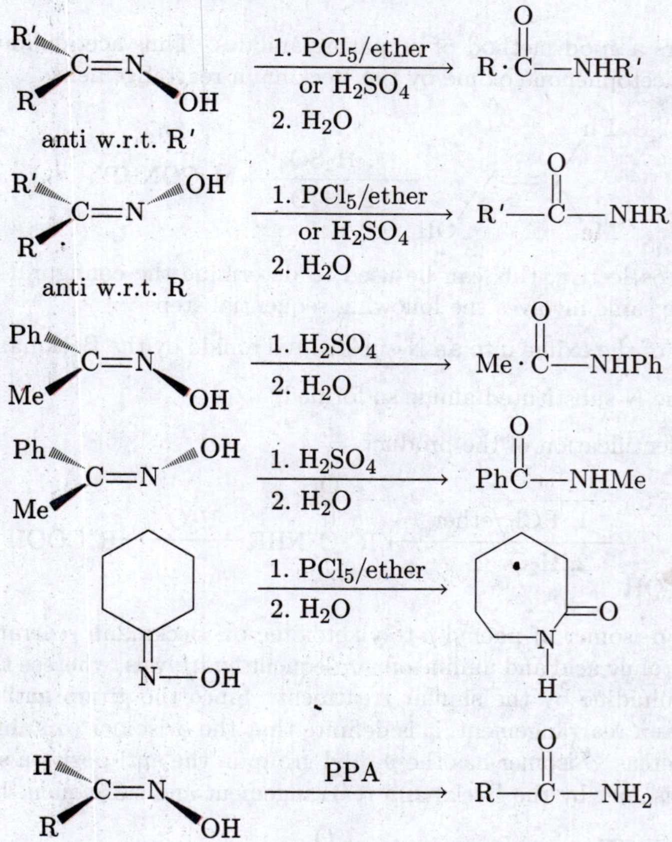

where R = alkyl group

Sometimes a nitrile results along with the isomeric amide in a Beckmann reaction and this side reaction is often called an abnormal Backmann rearrangement; this is actually a fragmentation reaction.

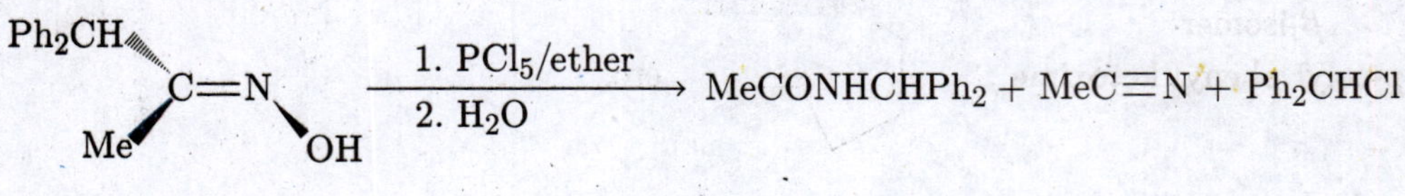

This reaction is extended to O-esters (both inorganic and organic) of the ketoximes.

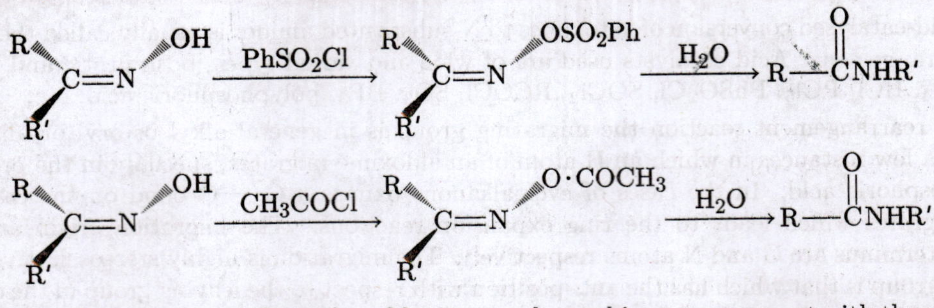

It has already been stated that a few aldoximes undergo this rearrangement with the migration of the H atom.

This reaction offers a good method of preparing anilides. Thus acetanilide can be prepared from the anti-methylacetophenone oxime by the Beckmann rearrangement.

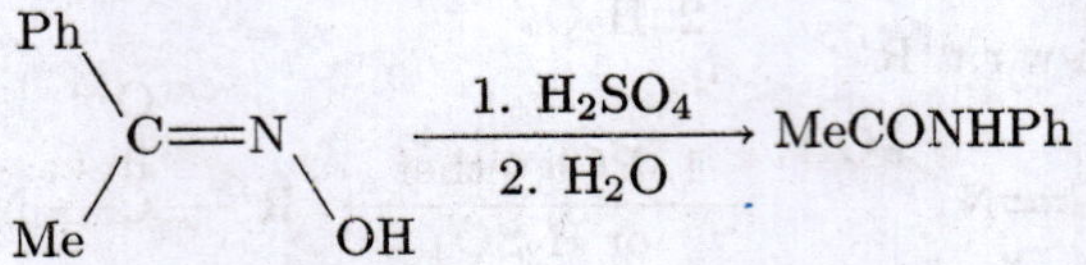

Being highly stereoselective, this can be used to determine the configurations of ketoximes. The procedure for the same involves the following sequential steps :

(a) Transformation of the oxime into an N-substituted amide by the Beckmann rearrangement.

(b) Hydrolysis of the N-substituted amide so formed.

(c) Isolation and identification of the product.

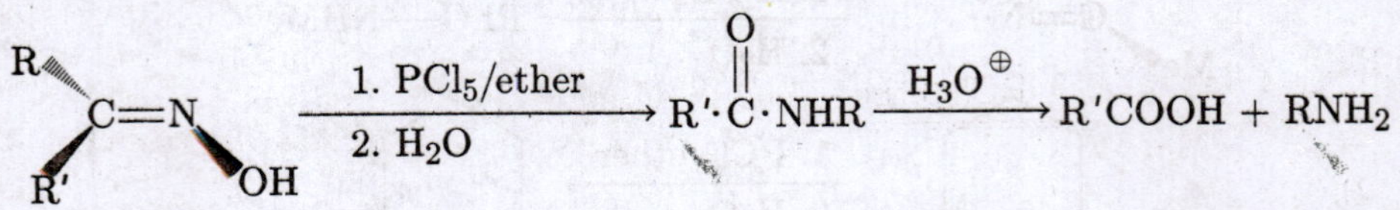

For example, the α-isomer of phenyl-p-tolylketoxime on Beckmann rearrangement yields an anilide which gives p-toluic acid and aniline on subsequent hydrolysis, whereas the β-isomer yields benzoic acid and p-toluidine by the similar treatment. Since the group anti to the OH group migrates in a Beckmann rearrangement, it is definite that the α-isomer contains phenyl group in the anti-position; whereas, β-isomer has the p-tolyl group in the anti position since p-toluidine is obtained from the β-isomer by the Beckmann rearrangement and subsequent hydrolysis :

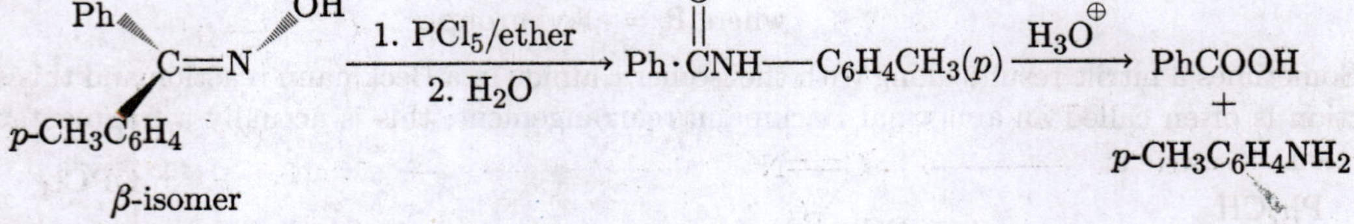

β-isomer

anti-p-tolyl phenyl ketoxime

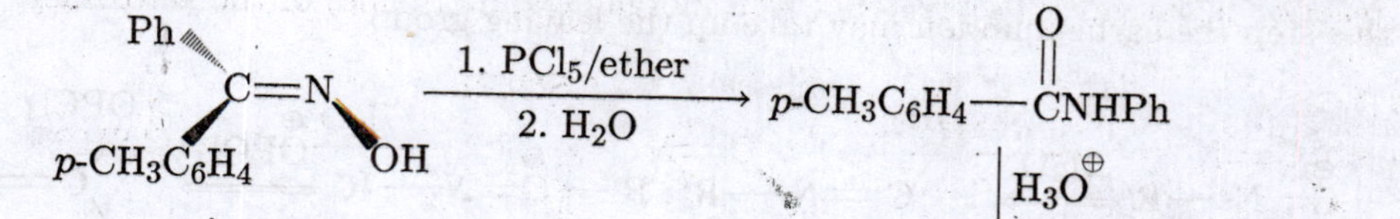

α-isomer

anti-phenyl-*p*-tolyl ketoxime

So the α-isomer is anti-phenyl-*p*-tolyl ketoxime and β-isomer is syn-phenyl-*p*-tolyl ketoxime.

This reaction has a very important synthetic application. Nowadays Nylon-6, a worthful textile polymer, is synthesised in large scale by this rearrangement.

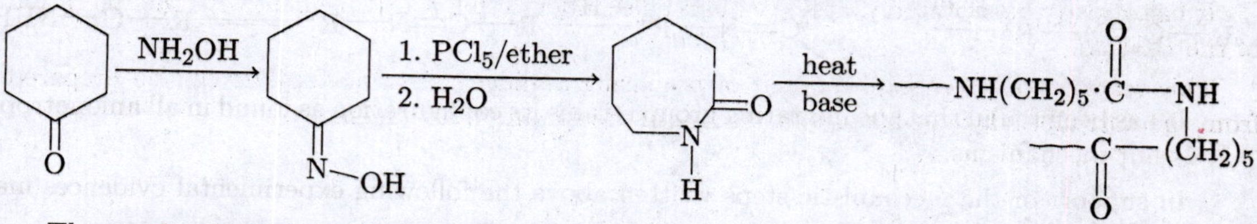

The mechanism of the rearrangement is a subject of much discussion. However, the most acceptable pathway tells us that it follows an intramolecular anionotropic 1, 2-shift mechanism involving the following sequential steps :

Step 1. Formation of a better leaving group :

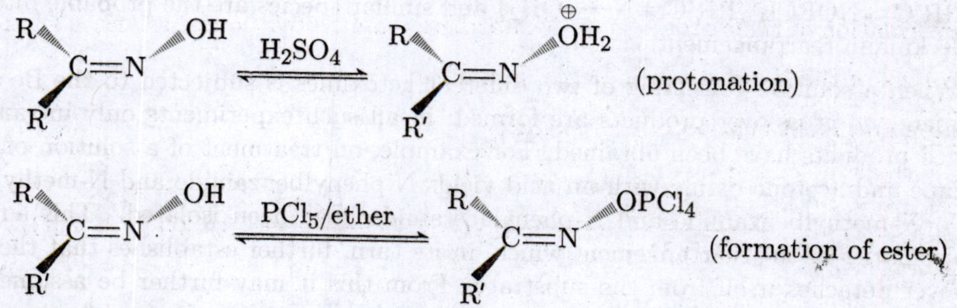

Step 2. Migration of the anti group (w.r.t. to the leaving group) and the loss of the leaving group take place simultaneously. It may be called ionisation step; it is the rate-determining step.

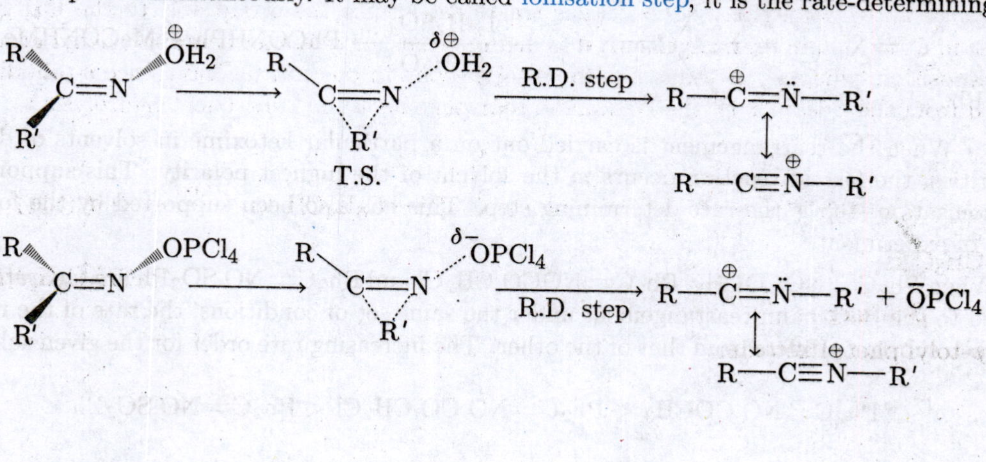

In this step the carbenium ion may take up the leaving group.

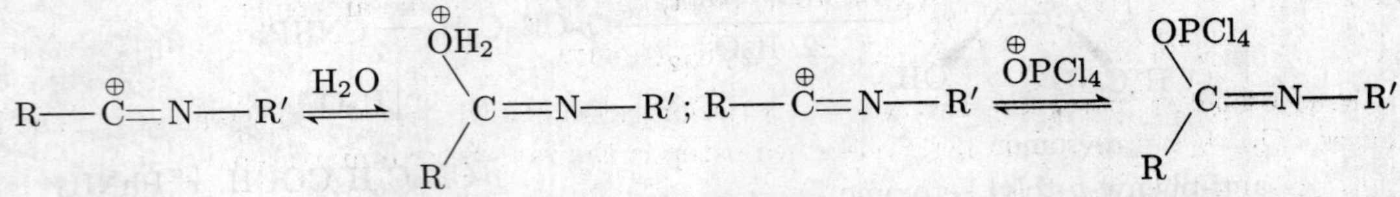

Step 3. Nucleophilic attack by water molecule to the carbenium ion to form an enol of the amide which rapidly isomerises to yield an N-substituted amide.

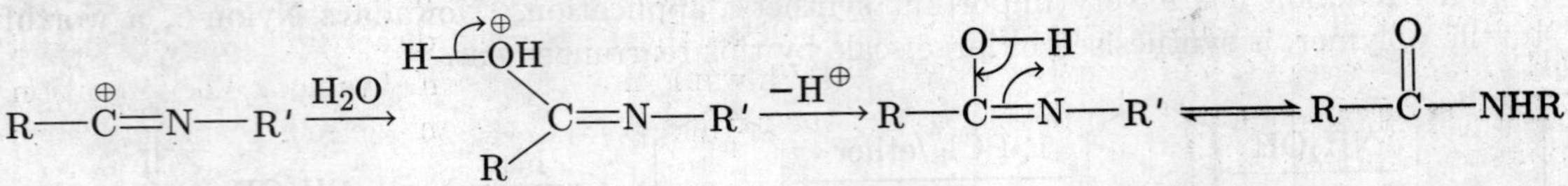

It has been found that the migrating group retains its configuration as found in all anionotropic 1, 2-shift mechanisms.

In support of the mechanistic steps written above the following experimental evidences may be cited :

(a) A number of intermediates of the formula RR′C=NOX (X=PCl_4, $COCH_3$, SO_2Ph) have been prepared and found to yield expected amides in the neutral medium in the absence of catalysts through the rearrangement reaction.

So, RR′C=$NOPCl_4$ (RR′C=N—$\overset{\oplus}{O}H_2$) and similar species are the probable intermediates of the Beckmann rearrangement.

(b) When a solution consisting of two different ketoximes is subjected to the Beckmann rearrangement, no cross-over products are formed; in all such experiments only intramolecularly rearranged products have been obtained. For example, on treatment of a solution of benzophenone oxime and acetone oxime with an acid yields N-phenylbenzamide and N-methylacetamide only. No N-methylbenzamide and N-phenylacetamide have been isolated. This supports the intramolecularity of the rearrangement which, in its turn, further establishes that the migrating group never detaches itself from the substrate. From this it may further be assumed that the migration of the group and the loss of the leaving group occur in a synchronous step.

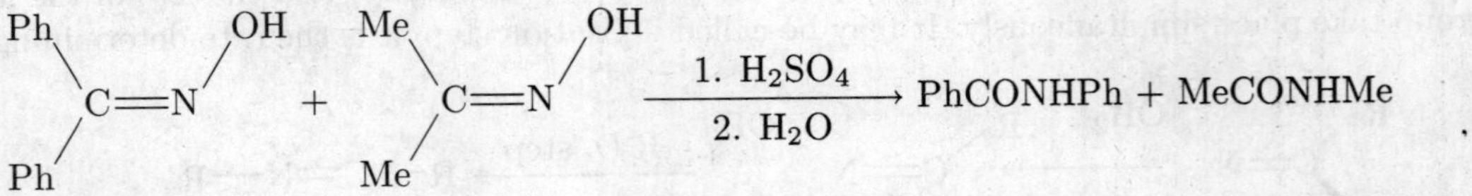

(c) When the rearrangement is carried out on a particular ketoxime in solvents of different polarities, the fastest reaction occurs in the solvent of the highest polarity. This supports that the ionisation step is the rate-determining step. This has also been supported by the following type of experiment.

When Ph_2C=$NOCOCH_3$, Ph_2C=NOCO.CH_2Cl and Ph_2C=NO.SO_2Ph are separately subjected to the Beckmann rearrangement under the same set of conditions, the rate of the reaction for each substrate differs from that of the other. The increasing rate order for the given substrates is :

$$Ph_2C{=}NO.COCH_3 < Ph_2C{=}NO.CO.CH_2Cl < Ph_2C{=}NO.SO_2Ph$$

To account for this rate order the stability order of the leaving group anions (CH_3COO^-, $ClCH_2COO^-$, $PhSO_3^-$) is to be considered; the stability order of the anions is :

$$CH_3COO^- < ClCH_2COO^- < PhSO_3^-$$

Thus with the increasing leaving group (anion) stability, the rate of the rearrangement increases. This is possible only if the ionisation step is the rate-determining step.

(d) That the direct exchange of the leaving group and the migrating group do not occur between N and C atoms have been established by carrying out the rearrangement on benzophenone in the aqueous medium containing $H_2{}^{18}O$. N-phenylbenzamide (benzanilide) isolated therefrom contains PhCONHPh along with $PhC^{18}ONHPh$. Had there been any direct exchange of Ph and OH groups between C and N atoms, $PhC^{18}ONHPh$ would not have been obtained. The formation of $PhC^{18}ONHPh$ supports beyond doubt that the oxygen atom of the anilide so formed must have come from the medium.

(e) 2-Bromo-5-nitroacetophenone oxime exists in two forms. One of them undergoes ring-closure reaction in cold on NaOH treatment but the other under drastic conditions. The form which undergoes ring-closure reaction readily gives N-methyl derivative of 2-bromo-5 nitrobenzamide by the Beckmann rearrangement. The ketoxime which contains the methyl group in the anti-position with respect to the OH group can undergo the ring-closure reaction readily since Cl and OH groups are reasonably close to each other in the form, whereas in the syn-form (w.r.t. the methyl group) their positions being not close , ring closure occurs but under drastic condition.

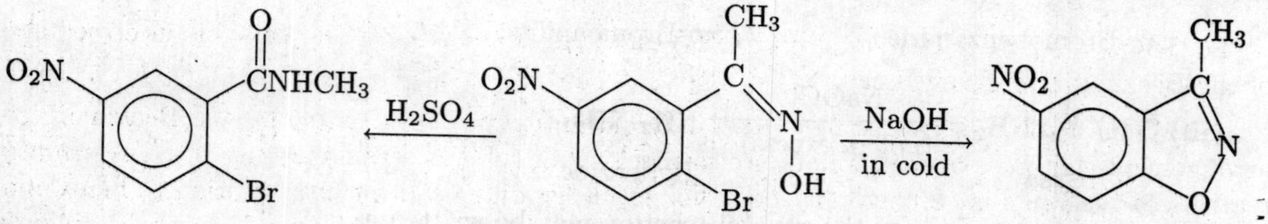

anti-form (w.r.t.Me)

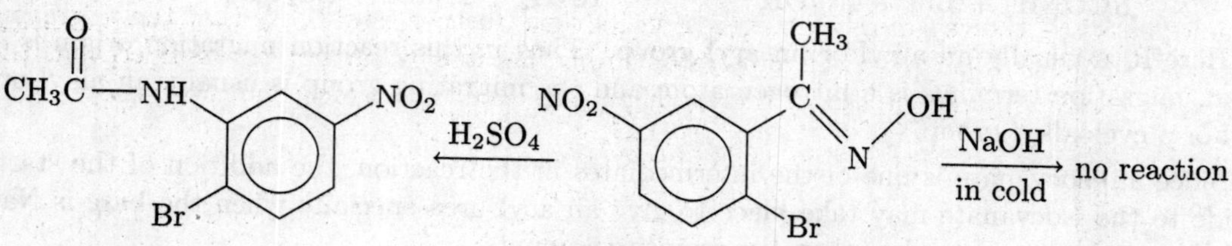

syn-form (w.r.t.Me)

Thus it is concluded that the anti group with respect to the leaving group migrates in a Beckmann rearrangement.

(f) The following reaction sequence supports that the migration of group occurs with retention of configuration in the Beckmann rearrangement.

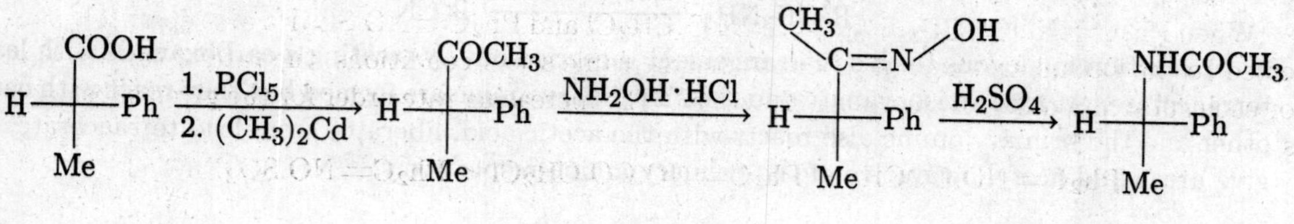

14.6 The Hofmann Rearrangement (or Degradation of Amide)

The conversion of a carboxamide to a primary amine with a C atom less by the action of alkali and halogen (chlorine or bromine) or hypohalites (NaOCl or NaOBr) is known as the Hofmann degradation of amide. This involves a rearrangement reaction of an N-haloamide into an isocyanate which hydrolyses rapidly, under the conditions of the reaction, into a primary amine and carbonate ion; because of the intermediate rearrangement, the reaction is also termed the Hofmann haloamide rearrangement. A few examples of the reaction are given below along with the general reaction.

$$RCONH_2 \xrightarrow{NaOBr} RCONHBr \xrightarrow{\bar{O}H} R.CO\bar{N}Br \longrightarrow O{=}C{=}N{-}R \xrightarrow[OH]{H_2O} RNH_2$$

(i) $CH_3.CH_2.CH(CH_3)CONH_2 \xrightarrow{NaOBr} CH_3.CH_2.CH(CH_3)NH_2$

2-Methylbutanamide → 1-Methylpropanamine

(ii) 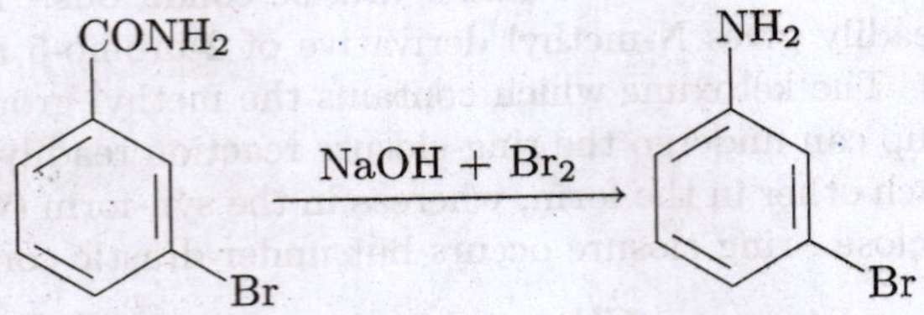

m-Bromobenzamide → *m*-Bromoaniline

(iii) $NH_2.CO.NH_2 \xrightarrow[\text{(not in excess)}]{NaOCl} NH_2.NH_2$

Urea → Hydrazine

The overall equation for the general reaction may be written as :

$$RCONH_2 + Br_2 + 4NaOH \longrightarrow RNH_2 + 2NaBr + Na_2CO_3 + 2H_2O$$

Here R is usually an alkyl or an aryl group. Thus in this reaction migration origin is a C atom, migration terminus is a nitrogen atom and the migrating group is usually an alkyl or an aryl or a cycloalkyl group.

Since an isocyanate is one of the intermediates in the reaction, the addition of the starting amide to the isocyanate may take place to give an acyl urea specially when the base is NaOH and the reagents used are less than the requisite amounts.

$$RCONH_2 \xrightarrow{NaOBr} O{=}C{=}N{-}R \xrightarrow{RCONH_2} RCONH.CO.NH.R$$

If the primary amine formed has a 1° alkyl group ($R'CH_2{-}NH_2$), it may sometimes get dehydrogenated by NaOBr to give a nitrile.

$$R'CH_2NH_2 \xrightarrow{NaOBr} R'CN$$

A reaction analogous to this rearrangement is carried out by treating a carboxamide with lead tetraacetate. At first an isocyanate and then a primary amine are formed but they react with each other and the primary amine also reacts with the acetic acid, liberated from lead tetraacetate, to give ureas and acetyl derivative of the primary amine respectively.

A primary amine consisting of more than seven Cs cannot be directly prepared by the Hofmann rearrangement; a nitrile is usually formed which may then be reduced to a primary amine.

$$CH_3(CH_2)_7CONH_2 \xrightarrow{KOH + Br_2} CH_3(CH_2)_6CN \xrightarrow{[H]} CH_3(CH_2)_6CH_2NH_2$$

However, when a methanolic solution of such a long chain amide containing sodium methoxide as a base is treated rapidly with bromine, N-substituted urethan is formed which on alkaline hydrolysis produces the desired primary amine.

$$CH_3(CH_2)_n.CH_2CO.NH_2 \xrightarrow[Br_2]{MeONa} CH_3(CH_2)_nCH_2NHCO_2CH_3 \xrightarrow{NaOH} CH_3(CH_2)_nCH_2NH_2$$

Here, $n > 5$.

This reaction has a wide scope and varied applications; a few are discussed below :

(i) It is a good method for preparing pure primary amines; e.g., *m*-haloaniline or *m*-dihalobenzenes can be prepared in good yield by this reaction; these are difficult to prepare by other means.

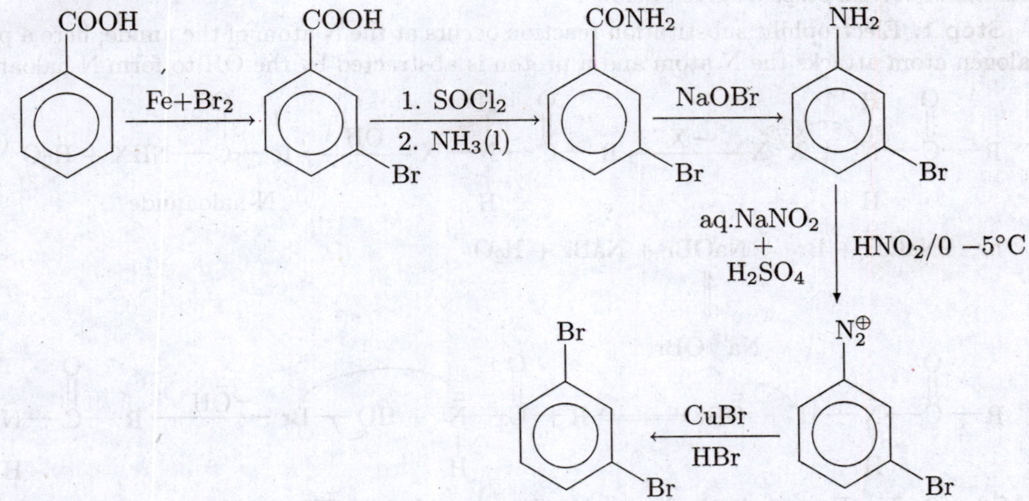

(ii) A compound can be descended to its lower homologue by this reaction; e.g.,

$$CH_3CH_2CH_2OH \xrightarrow{[O]} CH_3CH_2COOH \xrightarrow[2.\ NH_3(l)]{1.\ SOCl_2} CH_3CH_2CONH_2 \xrightarrow[Br_2]{NaOH} CH_3CH_2NH_2$$

$$CH_3CH_2NH_2 \xrightarrow{HNO_2} CH_3.CH_2.OH \xrightarrow{[O]} CH_3COOH$$

(iii) Hydrazine and anthranilic acid, the two important chemicals, are now manufactured applying the Hofmann rearrangement.

(a) $NH_2CONH_2 \xrightarrow[\text{requisite amount}]{NaOCl} NH_2.NH_2$

(b) 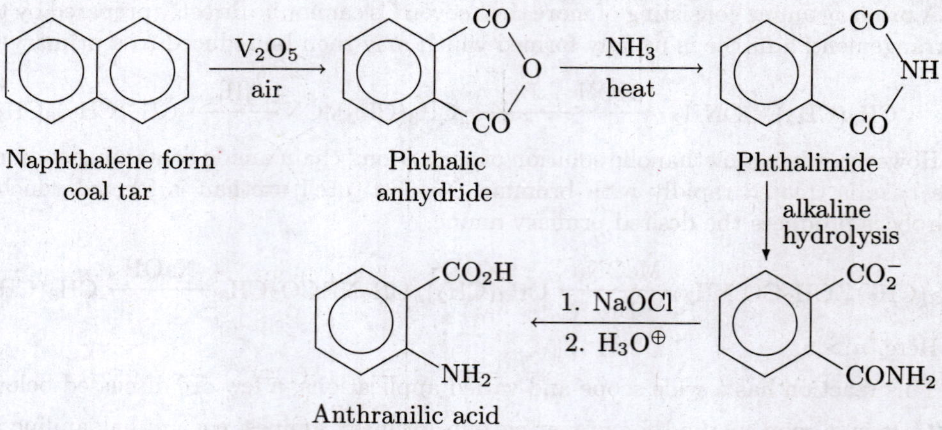

This is an intramolecular anionotropic 1, 2-shift rearrangement reaction involving migration of an aryl or an alkyl group from a C atom to an N atom with retention of configuration. It consists of several steps as given below :

Step 1. Electrophilic substitution reaction occurs at the N atom of the amide; here a positive halogen atom attacks the N atom and a proton is abstracted by the $\bar{O}H$ to form N-haloamide.

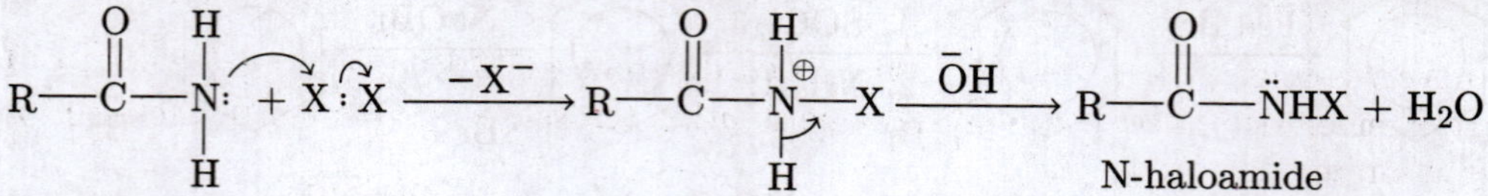

or, $2NaOH + Br_2 = NaOBr + NaBr + H_2O$

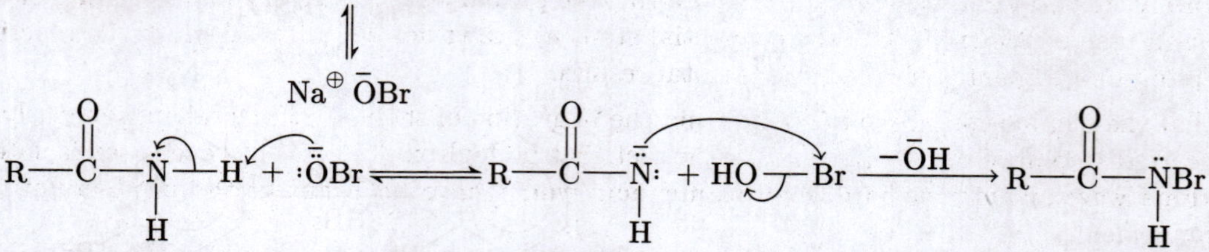

Step 2. The H atom on the N atom is abstracted by the $\bar{O}H$; the electron-withdrawing effect of the carbonyl and bromo groups makes the H highly acidic.

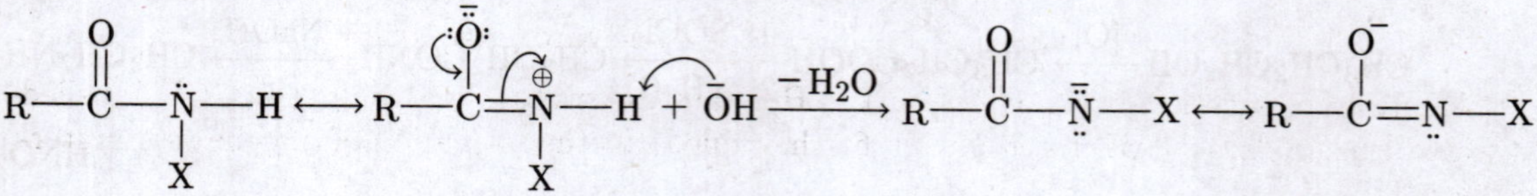

Step 3. Elimination of the halide ion and migration of the R occur in this concerted step to form isocyanate. Probably this is the R.D. Step.

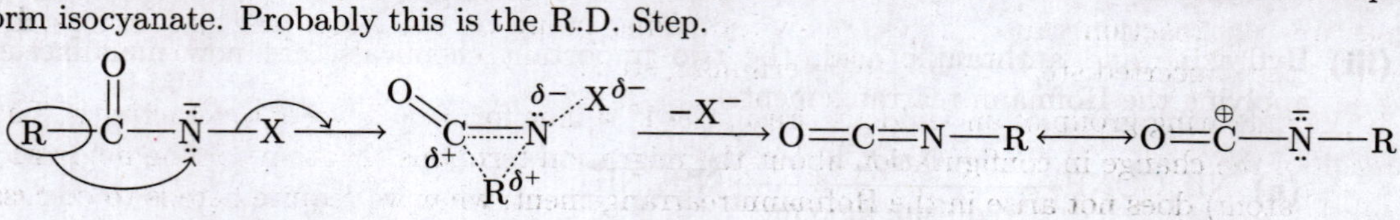

bridged T.S.

However, when an aryl group migrates, the bridged T.S. might be a bridged intermediate.

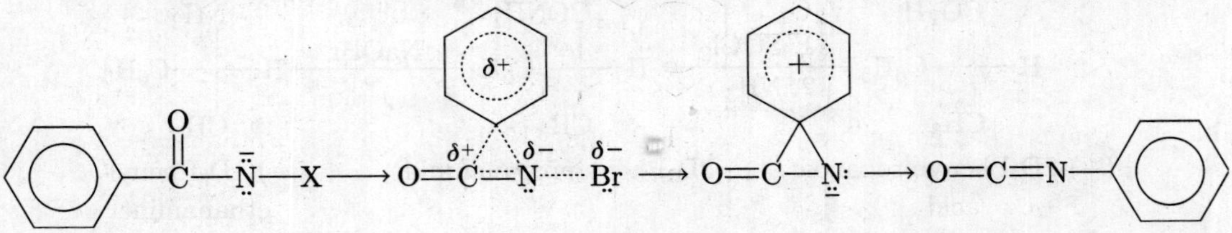

Step 4. In this step isocyanate hydrolyses to 1° amine and carbonate ion by the aqueous alkali.

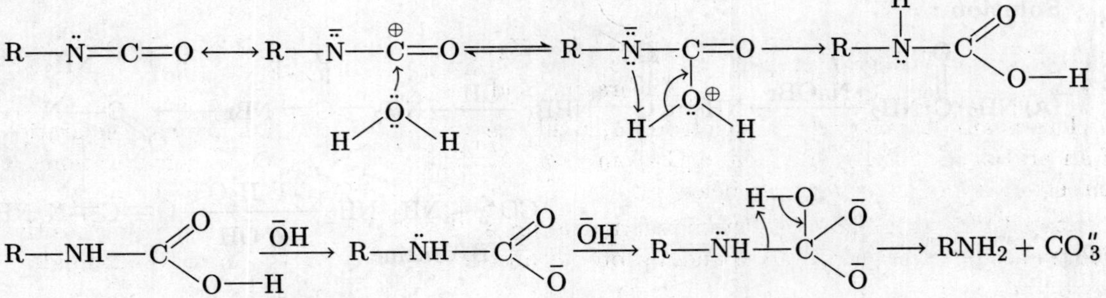

Since no cross-migration is observed when two different amides are subjected to the Hofmann rearrangement simultaneously in the same solution (cross-over experiment), the reaction is undoubtedly an intramolecular one.

N-haloamides, their salts and isocyantes have been isolated. When the isolated products are separately treated with alkali, each of them forms the primary amine, the ultimate product of the Hofmann rearrangement. Thus the sequential steps as drawn for the mechanism of the reaction are quite in agreement with the experimental results.

That the elimination of the halide ion and the migration of the R take place simultaneously in a concerted step have been supported by the fact that no hydroxamic acid has ever been isolated. If nitrene were an intermediate, hydroxamic acid would have been one of the products of the rearrangement :

$$\underset{\text{Nitrene}}{R-\overset{O}{\overset{\|}{C}}-\ddot{N}:} + H_2O \longrightarrow \underset{\text{Hydroxamic acid}}{R-\overset{O}{\overset{\|}{C}}-NHOH}$$

The study of kinetic isotope effect on the Hofmann rearrangement of phenyl-1-C^{14} labelled N-bromobenzamide by Wright and Fry (1968) has ruled out the formation of nitrene as an intermediate.

Since the presence of an electron-releasing group at the para position of an aromatic amide accelerates the reaction while an electron-withdrawing group on the same position retards the reaction, the concerted step is the rate-determining step.

If the migrating group of an amide is chiral, the 1° amine formed shows optical activity. The question of the change in configuration about the migration terminus (N atom) or the migration origin (C atom) does not arise in the Hofmann rearrangement; what we require here is to discuss the configuration of the migrating group. The migrating group retains its configuration in this rearrangement. This has been established by the following reaction sequence :

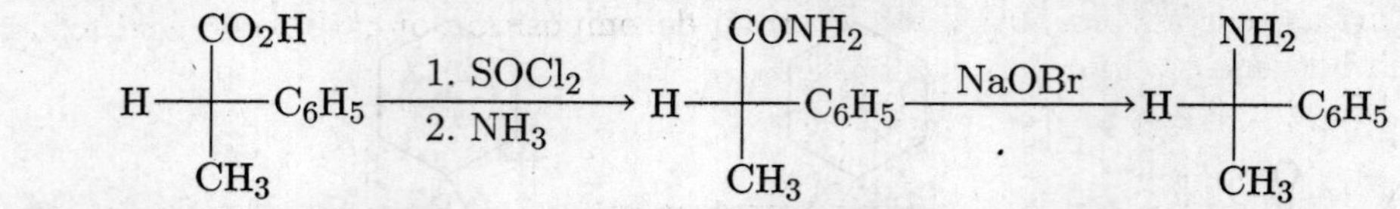

(+)-2D-Phenylpropanoic acid — 2D-Phenylpropananmide — (−)-1D-Phenyl-ethanamine

Problem 14.7. Give the mechanistic steps for the synthesis of **(a)** hydrazine and **(b)** anthranilic acid from urea and phthalimide respectively.

Solution :

(a)

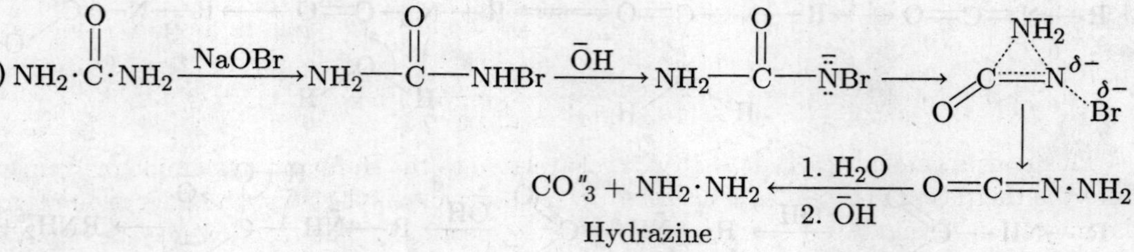

(b)

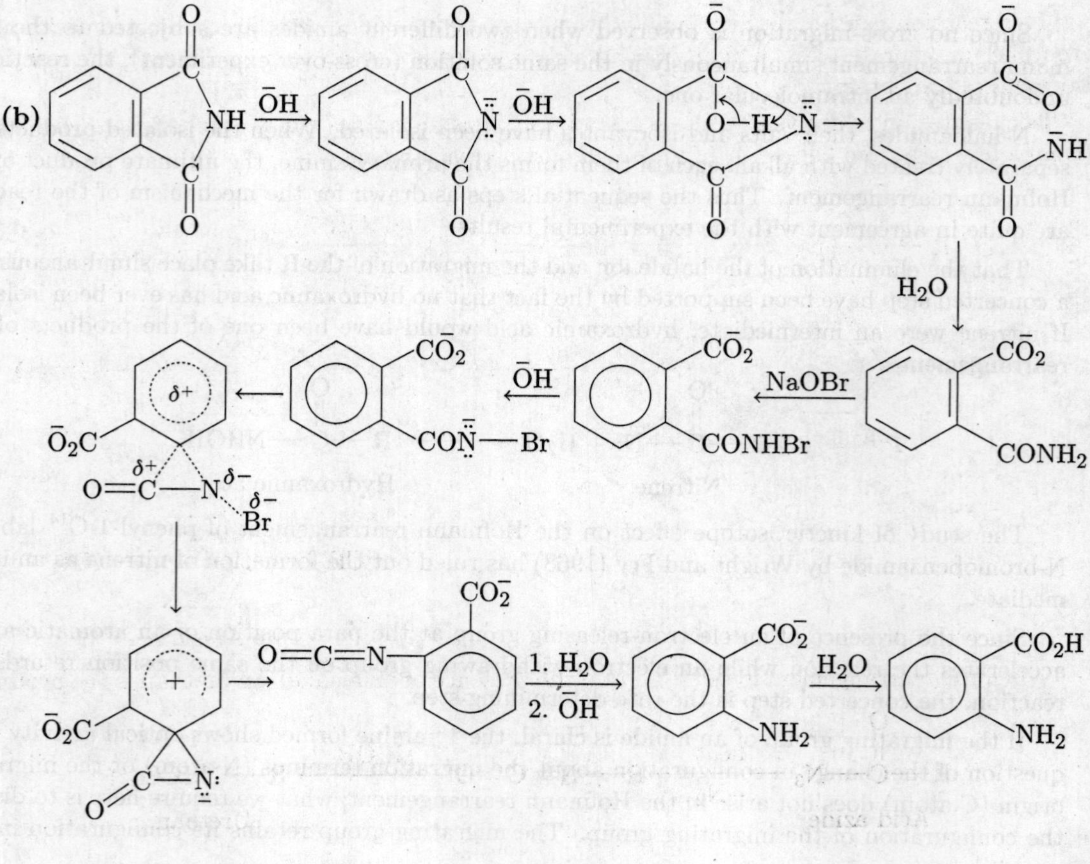

14.7 The Curtius Rearrangement

The Curtius rearrangement involves thermal decomposition of carboxylic acid azides to isocyanates in benzene or chloroform solution.

$$R{-}\overset{\overset{\displaystyle O}{\|}}{C}{-}N_3 \xrightarrow[\text{benzene medium}]{\text{heat}} N_2\uparrow + R{-}N{=}C{=}O$$

Thus acetyl azide and benzoyl azide may be converted to methyl isocyanate and phenyl isocyanate respectively.

$$CH_3{-}\overset{\overset{\displaystyle O}{\|}}{C}{-}N_3 \xrightarrow[\text{benzene medium}]{\text{heat}} CH_3{-}N{=}C{=}O$$

$$Ph{-}\overset{\overset{\displaystyle O}{\|}}{C}{-}N_3 \xrightarrow[\text{benzene medium}]{\text{heat}} Ph{-}N{=}C{=}O$$

This rearrangement reaction is very closely related to the Hofmann haloamide rearrangement. Here the migrating group is usually alkyl or aryl or cycloalkyl group; the migration origin and the migration terminus are C atom and N atom respectively. The reaction may be catalysed by acids, Lewis or protic.

This reaction can be applied to all sorts of carboxylic azides; acyclic, cyclic, aromatic, heterocyclic, unsaturated and polyfunctional. Since water is not present in the reaction medium, isocyanates are obtained in good yield.

Isocyanates on hydrolysis produce 1° amines. Thus 1° amines can be prepared in pure state (completely free from 2° and 3° amines) via curtius rearrangement.

$$R{-}\overset{\overset{\displaystyle O}{\|}}{C}{-}N_3 \xrightarrow[-N_2]{\text{heat}} R{-}N{=}C{=}O \xrightarrow[H_2O]{\bar{O}H} RNH_2 + CO_3''$$

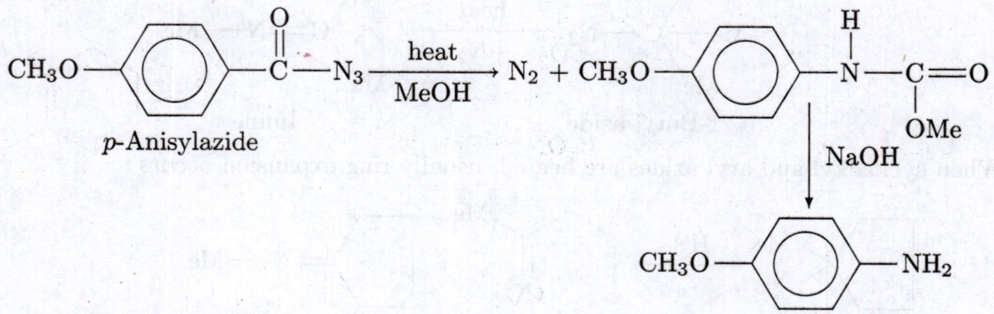

p-Anisylazide

p-Anisidine

When the rearrangement is carried out in the presence of an alcohol, an urethan is the product

$$\underset{\text{Acid azide}}{R{-}\overset{\overset{\displaystyle O}{\|}}{C}{-}N_3} \xrightarrow[-N_2]{\text{heat}} R{-}N{=}C{=}O \xrightarrow{R'OH} \underset{\text{Urethan}}{R{-}NHCO_2R}$$

This reaction may be utilised to step down a series. For example, butanoic acid may be stepped down to propanoic acid through the Curtius rearrangement as shown below :

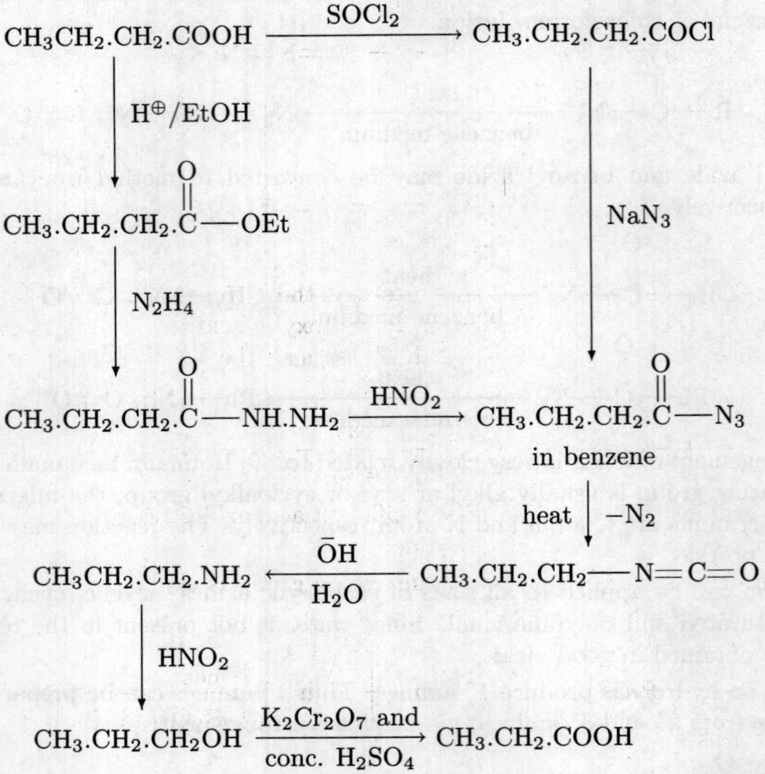

This rearrangement reaction may be extended to prepare alkyl imines from alkyl azides.

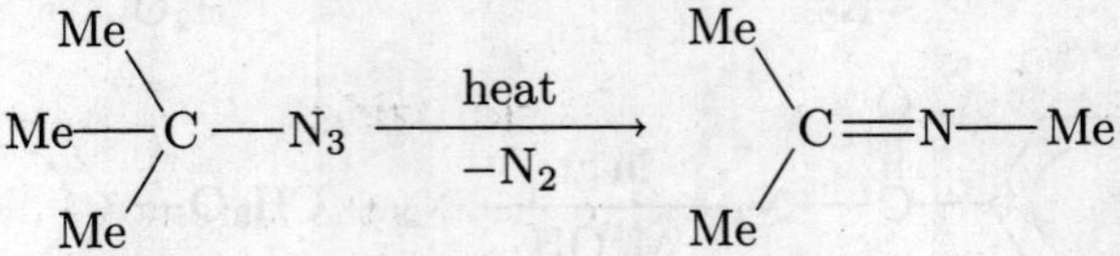

t-Butyl azide Imine

When cycloalkyl and aryl azides are heated, usually ring expansion occurs :

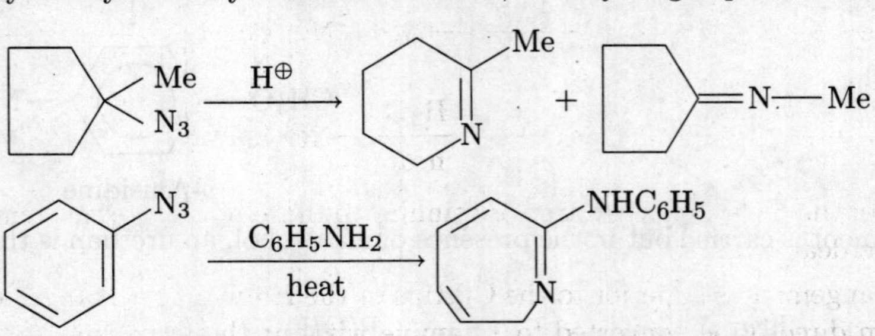

The mechanism of the rearrangement is very similar to that of the Hofmann haloamide rearrangement. This also involves an intramolecular anionotropic 1, 2-shift in which the elimination of molecular nitrogen and migration of the group with its bonding electrons occur perhaps in a concerted step; the migrating group retains its configuration.

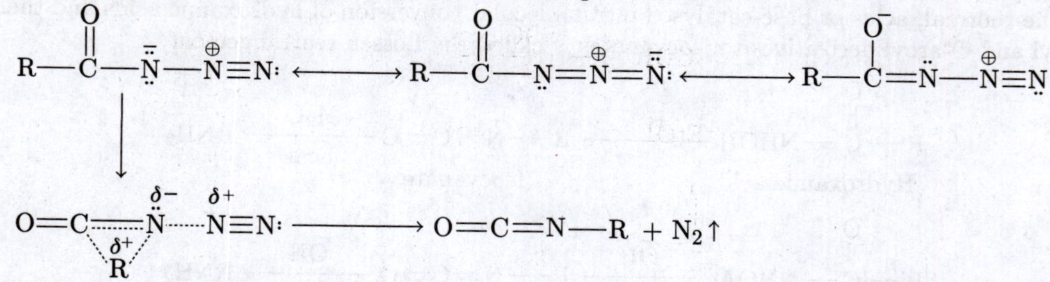

The isotope labelling experiment supports that the nitrogen bonded to C atom does not get eliminated in this thermal decomposition. A carboxylic acid azide containing terminal N^{15} on pyrolysis liberates elemental nitrogen which contains all the N^{15} atoms.

$$R-\overset{O}{\overset{\|}{C}}-\ddot{N}-\overset{\oplus}{N}\equiv N^{15} \xrightarrow[\text{benzene}]{\text{heat}} R-N=C=O + :N\equiv N^{15}$$

The cross-over experiment involving two different acid azides gives rise to only two isocyanates which are expected from the individual azide.

The retention of configuration by the migrating group is established by the following set of reactions :

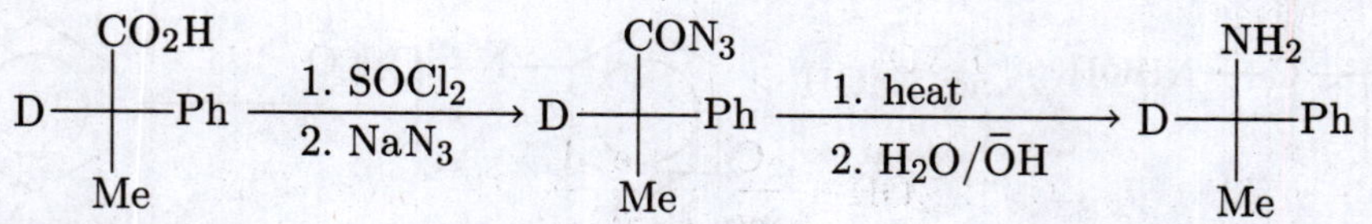

(+)-2D-Phenylpropanoic acid (−)-1D-Phenyl-ethanamine

14.8 The Schmidt Rearrangement of Acid Azides

The most common type of the Schmidt reaction is the intramolecular conversion of acid azides to isocyanates by the catalytic action of sulphuric acid; Lewis acids may also be used.

$$R-\overset{O}{\overset{\|}{C}}-OH \xrightarrow{HN_3} R-\overset{O}{\overset{\|}{C}}-N_3 \xrightarrow{H_2SO_4} R-N=C=O \xrightarrow[\text{2. heat}]{\text{1. } H_2O} RNH_2$$

$$RCOOH + HN_3 \xrightarrow{H^{\oplus}} R-N=C=O \xrightarrow[\text{2. heat}]{\text{1. } H_2O} RNH_2$$

The other aspects of the rearrangement are very similar to those of the Curtius rearrangement and hence are not here dealt with.

However, the rearrangement is superior to the Curtius or the Hofmann rearrangement because in practice an acid can directly be converted to 1° amine without the formation of the azide or amide, respectively provided this gives better yield than the others.

14.9 The Lossen Rearrangement

The thermal, acid- or base-catalysed intramolecular conversion of hydroxamic acids and their O-acyl and O-aroyl derivatives to isocyanates is called the Lossen rearrangement.

$$\underset{\text{Hydroxamic acid}}{R{-}\overset{\overset{\displaystyle O}{\|}}{C}{-}NHOH} \xrightarrow{H^{\oplus}} \underset{\text{Isocyanate}}{R{-}N{=}C{=}O} \xrightarrow[\text{heat}]{H_2O} RNH_2$$

$$\underset{\text{O-acetyl hydroxamic acid}}{R{-}\overset{\overset{\displaystyle O}{\|}}{C}{-}NHOAc} \xrightarrow{\bar{O}H} \underset{\text{Isocyanate}}{R{-}N{=}C{=}O} \xrightarrow[\text{heat}]{\bar{O}H} RNH_2$$

The isocyanates, so formed, get hydrolysed instantly in aqueous medium to the carbamic acids which do not exist in the free state and readily decompose to 1° amines. However, isocyanates may be isolated by carrying out the reaction of the hydroxamic acid with thionyl chloride.

$$R{-}\overset{\overset{\displaystyle O}{\|}}{C}{-}NHOH + SOCl_2 \longrightarrow R{-}N{=}C{=}O + SO_2\uparrow + HCl\uparrow$$

Thus benzohydroxamic acid on heating with aqueous caustic soda solution gives aniline via phenyl isocyanate, the intermediate of the reaction.

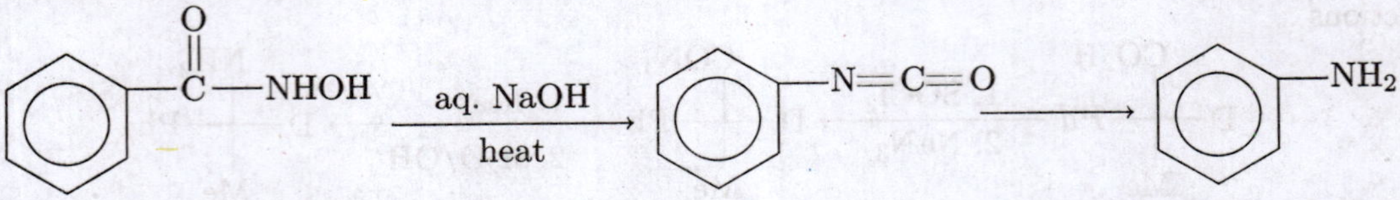

The reaction is analogous to the Hofmann haloamide rearrangement. Here again the migrating group is usually alkyl or aryl or cycloalkyl; the migration origin and the migration terminus are the C atom and the N atom respectively.

This reaction has much less application than the Hofmann haloamide, the Curtius and the Schmidt rearrangements. However, carboxylic acids may the converted to 1° amines through this rearrangement.

$$RCOOH \xrightarrow{SOCl_2} RCOCl \xrightarrow{NH_2OH} R{-}CONHOH \xrightarrow{\bar{O}H} R{-}N{=}C{=}O$$

$$R{-}N{=}C{=}O \xrightarrow{\bar{O}H/\text{heat}} \underset{1^\circ\ \text{Amine}}{RNH_2} + CO_3'' \xrightarrow{HNO_2} ROH \xrightarrow[H_2SO_4]{K_2Cr_2O_7} \text{A carboxylic acid}$$

Thus the reaction may be used to step down a series.

$$CH_3CH_2COOH \xrightarrow{SOCl_2} C_2H_5COCl \xrightarrow{NH_2OH} C_2H_5CONHOH \xrightarrow{\bar{O}H} C_2H_5{-}C{=}N{=}O$$

$$C_2H_5{-}C{=}N{=}O \xrightarrow{\bar{O}H/\text{heat}} C_2H_5NH_2 \xrightarrow{HNO_2} C_2H_5OH \xrightarrow[\text{[O]}]{K_2Cr_2O_7/H_2SO_4} CH_3COOH$$

The mechanism of the rearrangement is very alike to that of the Hofmann haloamide rearrangement. This also involves an intramolecular anionotropic 1, 2-shift in which the elimination of the O-acyl group and the migration of the alkyl or aryl group occur perhaps in a concerted step. Of course, the acid-catalysed rearrangement occurs in three steps; a nitrene is formed as an intermediate and then the migration occurs to give the isocyanate. However, the migrating group retains its configuration in all the examples of the Lossen rearrangement

Mechanism :

(a) The base-catalysed rearrangement of hydroxamic acids

Step I. Abstraction of the proton from the N atom of the hydroxamic acid :

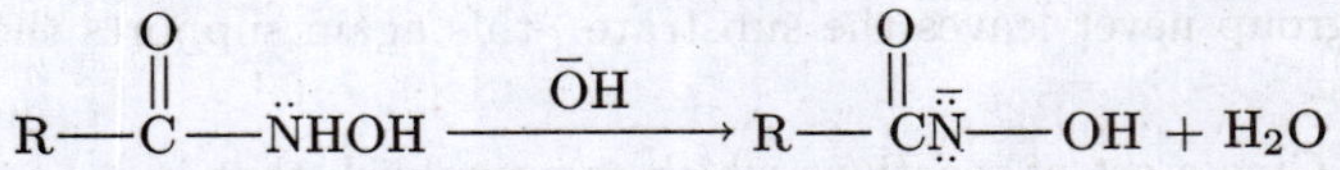

Step II. The detachment of the OH group and the migration of the R with its bonded electrons :

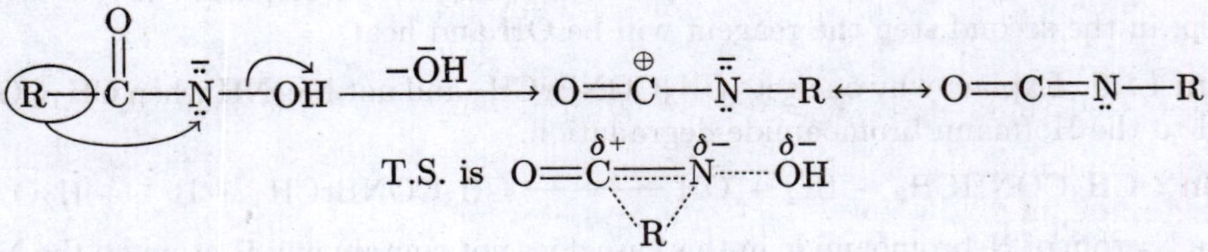

(b) The base-catalysed rearrangement of hydroxamic esters

Step I. Abstraction of the proton from the N atom :

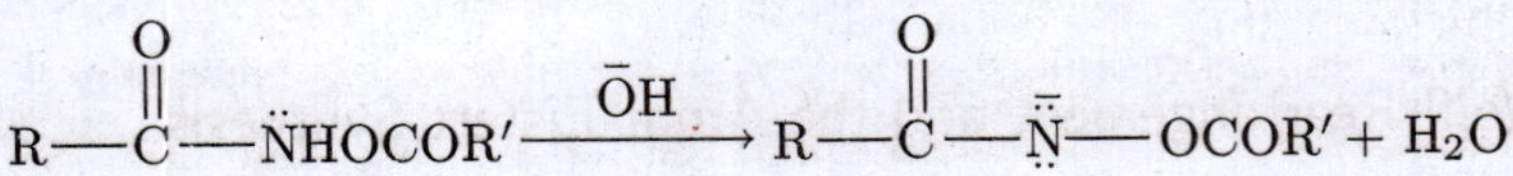

Step II. The detachment of the OCOR′ group and the migration of the R—.

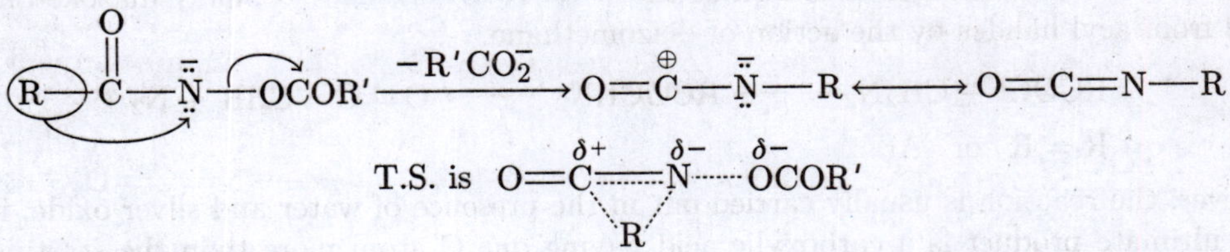

(c) The acid-catalysed rearrangement of hydroxamic acids

Step I. Protonation of the hydroxamic acid and elimination of water molecule :

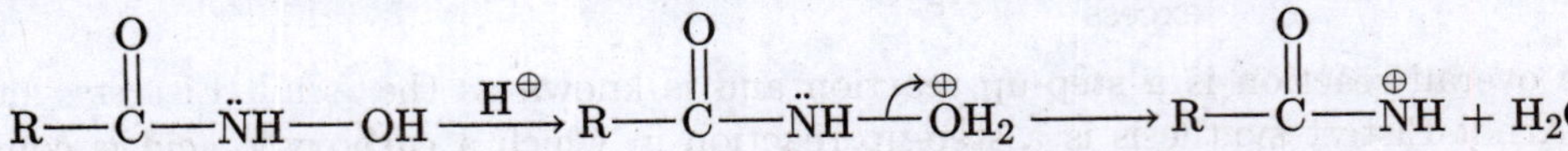

Step II. Deprotonation and formation of the nitrene :

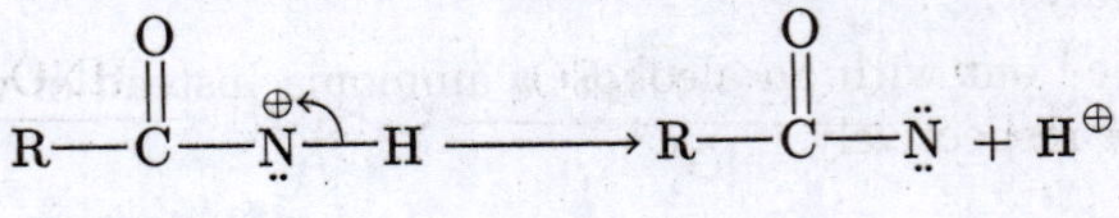

Step III. Migration of the R with its bonded *es* to the electron-deficient N atom :

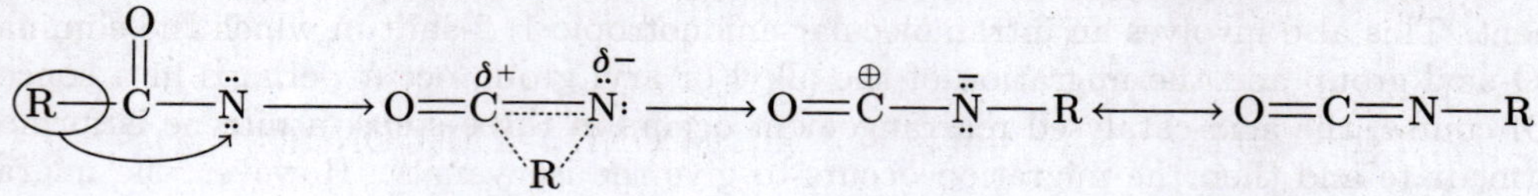

T.S.

The cross-over experiment with two different hydroxamic acids does not give any cross-over product; this supports the intramolecularity of the rearrangement. It has been found that an electron-donating group in the appropriate position of a migrating aryl group accelerates the reaction appreciably; this supports the formation of a non-classical carbenium ion; which indicates that the migrating group never leaves the substrate—this again supports the intramolecularity of the rearrangement.

Problem 14.8. Give a set of reactions which can establish that in a Lossen rearrangement the migrating group retains its configuration after migration.

Solution : See the last portion of the Curtius rearrangement. Replace NaN_3 by NH_2OH in the first step; in the second step the reagent will be $\bar{O}H$ and heat.

Problem 14.9. Explain why one gets $CH_3CONBrCH_3$ and not Me_2NH when $CH_3CONH.CH_3$ is subjected to the Hofmann bromoamide degradation.

Solution : $CH_3CONHCH_3 + Br_2 + \bar{O}H \longrightarrow CH_3CONBrCH_3 + Br^- + H_2O$

Since the N atom of N-bromoamide in this case does not contain any H atom on the N atom, it cannot undergo deprotonation (to form an active moiety which can undergo the rearrangement). So the requirement for the Hofmann rearrangement is not fulfilled and hence no further reaction will take place; the ultimate product will be $CH_3CONBrCH_3$.

14.10 The Wolff Rearrangement and the Arndt-Eistert Synthesis

The thermal or photochemical or catalytic conversion of a diazoketone to a ketene with the liberation of molecular nitrogen is known as the Wolff rearrangement. Usually diazoketones are prepared from acyl halides by the action of diazomethane.

$$RCOCl + CH_2N_2 \longrightarrow RCOCHN_2 \longrightarrow O{=}C{=}CHR + N_2\uparrow$$

$$R = R \text{ or } Ar$$

However, the reaction is usually carried out in the presence of water and silver oxide; in this case the ultimate product is a carboxylic acid having one C atom more than the starting acid halide.

$$RCOCl + \underset{\text{excess}}{CH_2N_2} \xrightarrow[Ag_2O]{H_2O} RCH_2COOH + N_2\uparrow + HCl$$

So, the overall reaction is a step-up reaction and is known as the Arndt-Eistert synthesis. Thus the Arndt-Eistert synthesis is a step-up reaction in which a carboxylic acid is converted into its higher homologue by the reaction of its acid halide with diazomethane in the presence of water and silver oxide catalyst.

If the reaction is carried out with an alcohol or ammonia instead of water, the product will be a higher ester or amide respectively.

$$RCOCl + CH_2N_2 \xrightarrow[Ag_2O]{R'OH} RCH_2CO_2R' + N_2 \uparrow$$

$$RCOCl + CH_2N_2 \xrightarrow[Ag_2O]{NH_3(R'NH_2)} RCH_2CONH_2(RCH_2CONHR') + N_2 \uparrow$$

When the reaction is carried out in the presence of water and formic acid hydroxymethyl ketones are obtained.

$$RCOCl + CH_2N_2 \xrightarrow[HCOOH]{H_2O} RCOCH_2OH$$

Thus the Arndt-Eistert synthesis is a good means for lengthening carbon chain and the reaction has a wide scope since alkyl, aryl and cycloalkyl diazoketones with other functions can undergo the reaction.

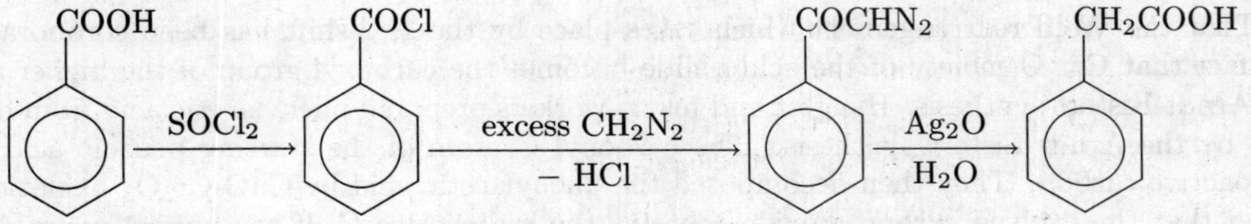

The possible mechanism of the Arndt-Eistert synthesis is as follows :

Stage I : The conversion of the acid halide into diazoketone via a tetrahedral intermediate :

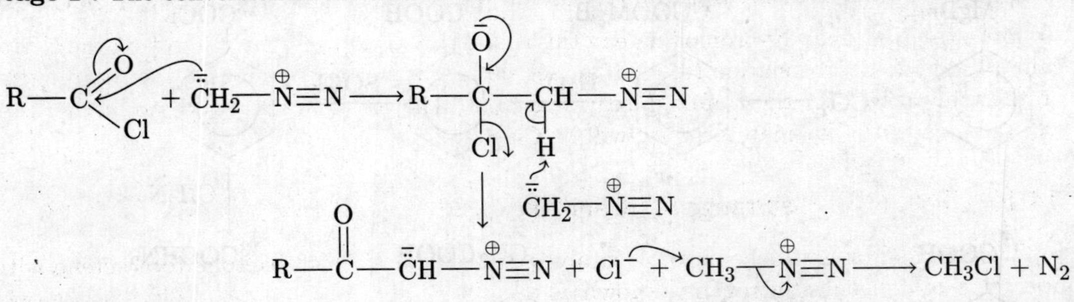

Stage II : Wolff rearrangement : In this stage molecular nitrogen gets eliminated and the R group migrates to the anionic C more or less in a concerted step; this involves 1, 2-shift and forms an unstable compound called ketene.

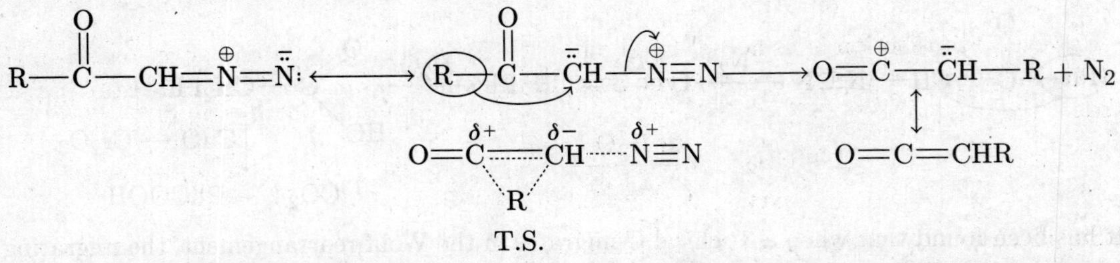

Stage III : The addition of water to the ketene and the formation of the carboxylic acid.

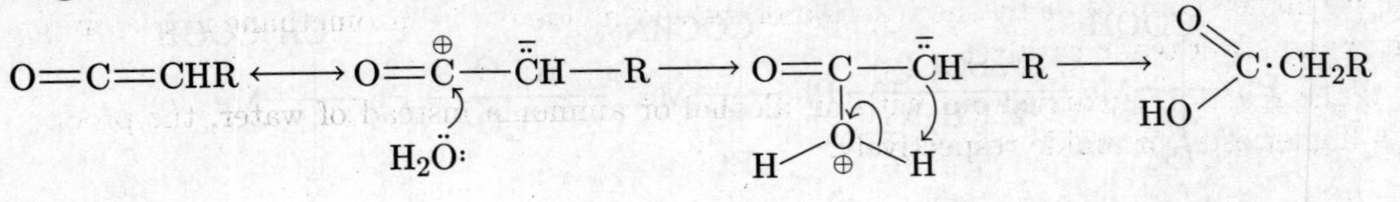

The excess use of diazomethane molecules per molecule of RCOCl is essential for the Wolff rearrangement and Arndt-Eistert synthesis. If we do not use an excess of diazomethane, $RCOCH_2Cl$ will be formed instead of diazoketone as shown below :

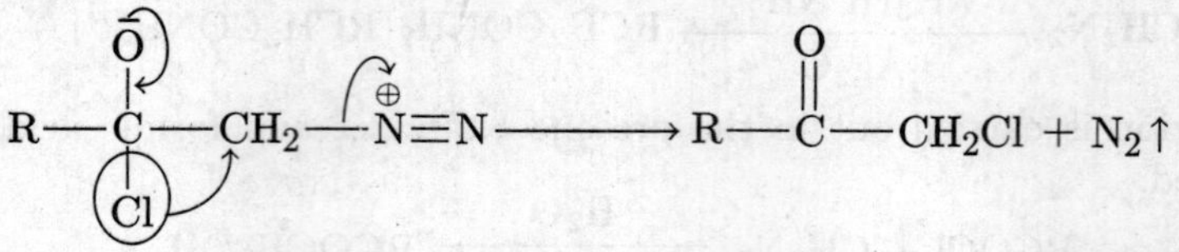

The first molecule undergoes bimolecular S_N reaction on the sp^2C of RCOCl and the second molecule abstracts the α-H; Cl^- ion gets eliminated to form the diazoketone. In the absense of excess diazomethane, molecular nitrogen evolves out and the Cl atom with its bonded *e*s. migrates to the N-bonded C atom to give $RCOCH_2Cl$.

That the Wolff rearrangement which takes place by the 1, 2-shift has been coroborated by the fact that C=O moiety of the acid halide becomes the carboxyl group of the higher acid in the Arndt-Eistert synthesis. Huggett and his co-workers prepared phenylacetic acid from benzoic acid by the Arndt-Eistert synthesis. The carbonyl C atom of the starting benzoic acid was a radioactive carbon. They then decomposed the phenylacetic acid by CuO-Cr_2O_3 oxidation and found that the evolved carbon dioxide contains the radioactive C. If the phenyl group did not migrate from the carbonyl C atom to the α-C atom, this would not have been the result.

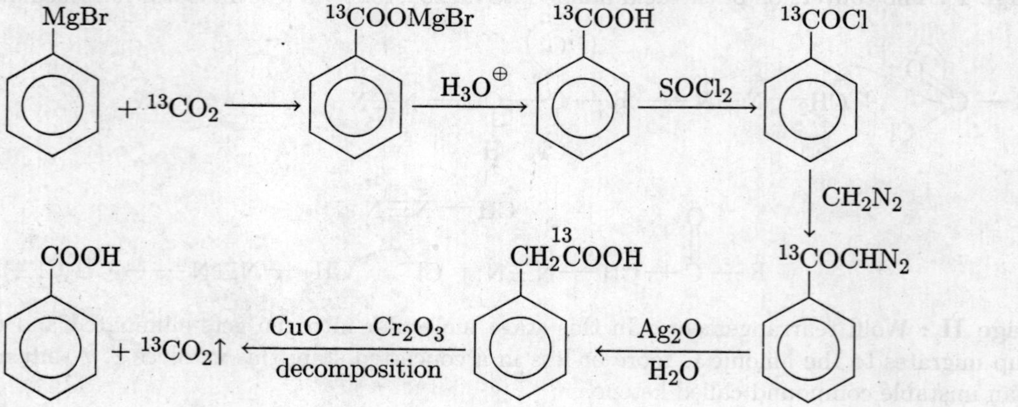

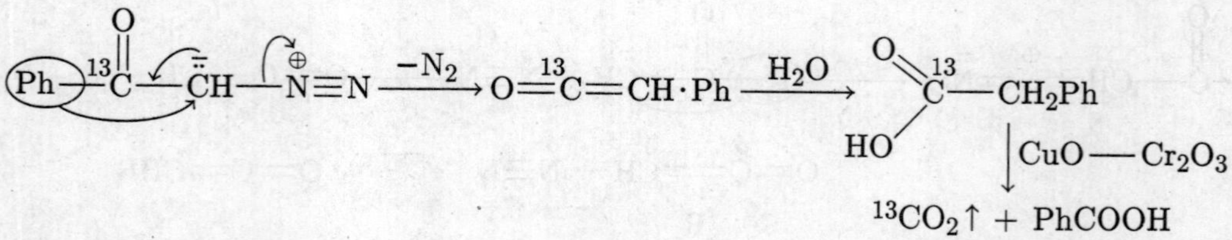

It has been found that when a 3° chiral C migrates in the Wolff rearrangement, the migrating group retains its configuration.

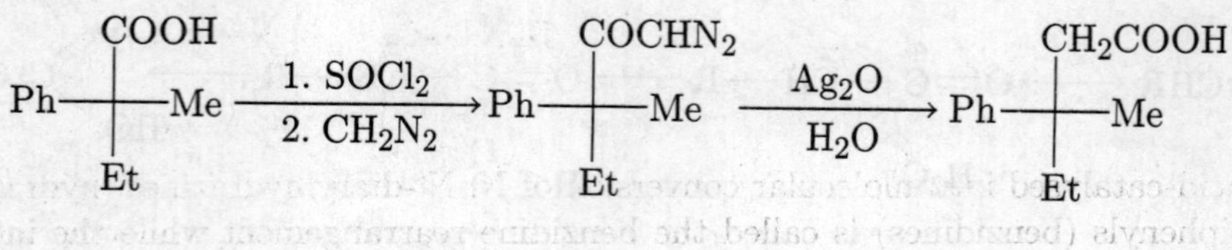

Note : The photochemical or thermal Wolff rearrangement is supposed to occur via carbene, an intermediate, though no carbene has yet been isolated or detected.

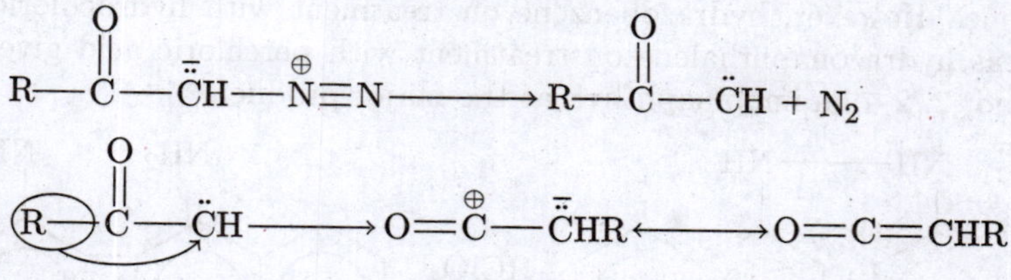

Problem 14.10. Write down the mechanistic steps for the synthesis of ethyl propanoate from ethanoyl chloride and diazomethane.

14.12 The Benzidine and Semidine Rearrangements

The common name of 4, 4′-diaminobiphenyl is benzidine whereas 2, 2′-diaminobiphenyl is known as *o*-benzidine; 2-phenylaminoaniline and 4-phenylaminoaniline are called *o*-semidine and *p*-semidine respectively. These compounds and 2, 4′-diaminobiphenyl, called diphenyline, may be obtained by the acid-catalysed rearrangement of hydrazobenzene.

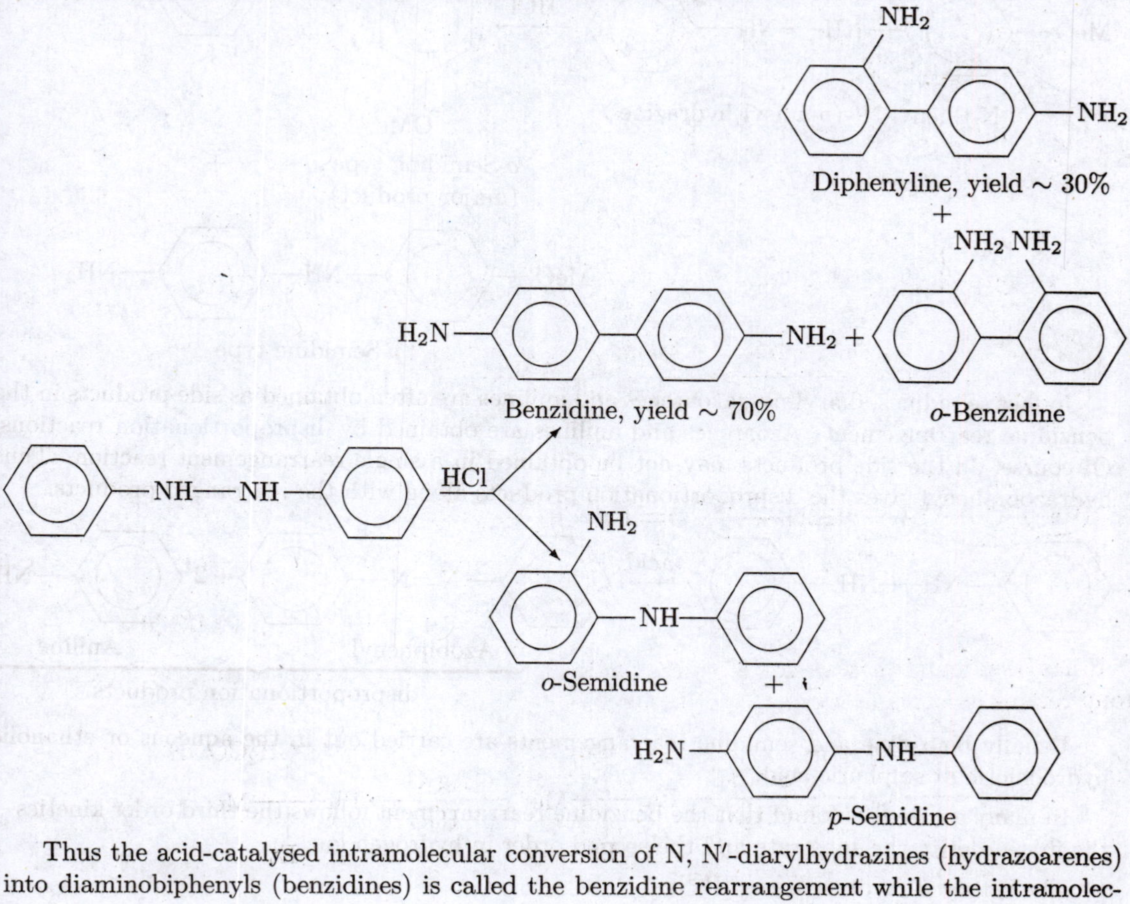

Thus the acid-catalysed intramolecular conversion of N, N′-diarylhydrazines (hydrazoarenes) into diaminobiphenyls (benzidines) is called the benzidine rearrangement while the intramolecular conversion of hydrazoarenes into 2- and 4-aminoarylanilines (semidines) is called the semidine rearrangement. The names have been derived from the names of the rearranged products of hydrazobenzene. However, hydrazobenzene on treatment with hydrochloric acid gives 70% benzidine whereas hydrazonaphthalene on treatment with perchloric acid gives an *o*-benzidine type of compound, 2, 2′-diaminobinaphthyl, as the major product.

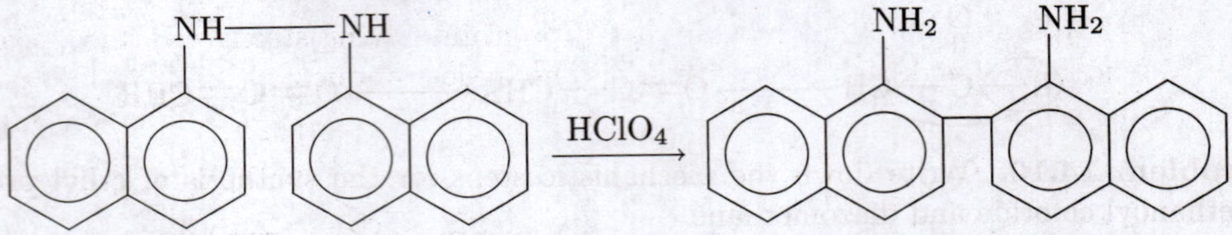

2,2′-Diaminobinaphthyl

On the other hand, N-phenyl-N′-(*p*-methoxyphenyl) hydrazine on acid treatment forms *o*-semidine (major product) and *p*-semidine types of compounds.

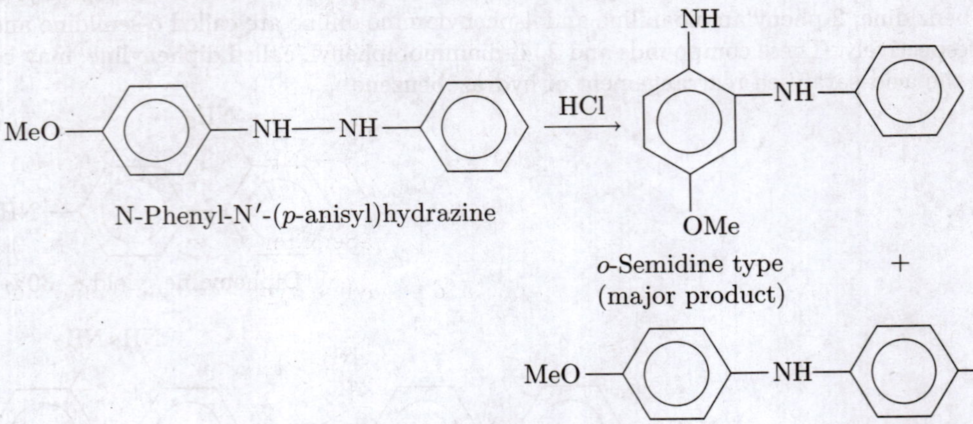

N-Phenyl-N′-(*p*-anisyl)hydrazine

o-Semidine type (major product) +

p-Semidine type

In fact semidines, diarylines, azoarenes and anilines are often obtained as side products in the benzidine rearrangement. Azoarenes and anilines are obtained by disproportionation reactions. Of course, all the side products may not be obtained in a single rearrangement reaction. Thus hydrazobiphenyl gives the disproportionation products along with the rearranged products.

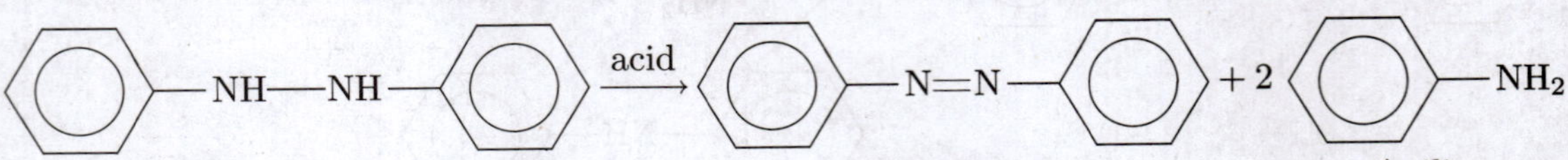

Azobiphenyl Aniline

disproportionation products

Usually benzidine and semidine rearrangements are carried out in the aqueous or ethanolic hydrochloric or sulphuric acid.

In many a case it is found that the benzidine rearrangement follows the third order kinetics—the first order in the substrate and the second order in hydrogen ion.

The rate $= k$ [hydrazoarene] $[H^{\oplus}]^2$

There are also reactions where the second order kinetics are followed; in those cases,

The rate = k [hydrazoarene] [$H^{\oplus}$]

Hydrazoarenes containing substituents at the *o*- and /or *m*-positions give benzidines as major products on rearrangement, while those containing one or two substituents at the *p*-positions give side products, mainly *o*-semidines as the major products. Thus,

(i) If the *p*-position of one nucleus is filled with a group having strong +R effect like OMe, OEt, etc., the *o*-semidine type compound is the major product.

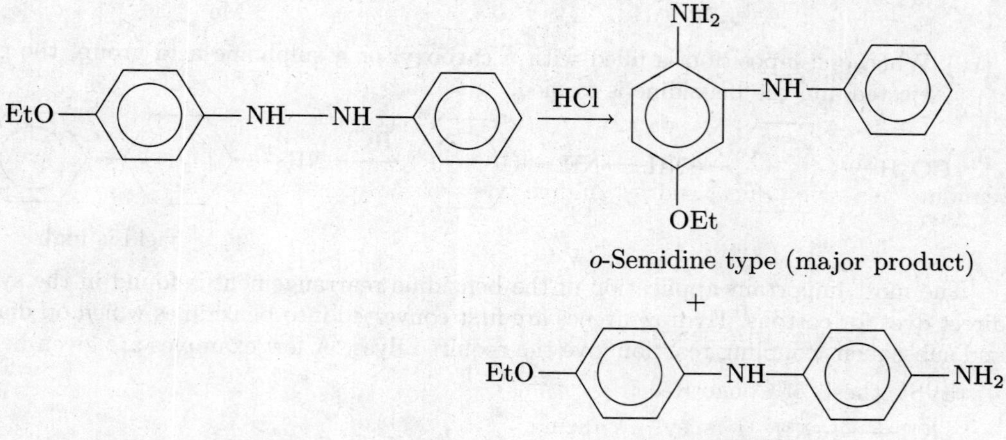

o-Semidine type (major product) + *p*-Semidine type

(ii) If a strong −I group occupies one *p*-position, a diphenyline type compound and an *o*-Semidine type compound are formed.

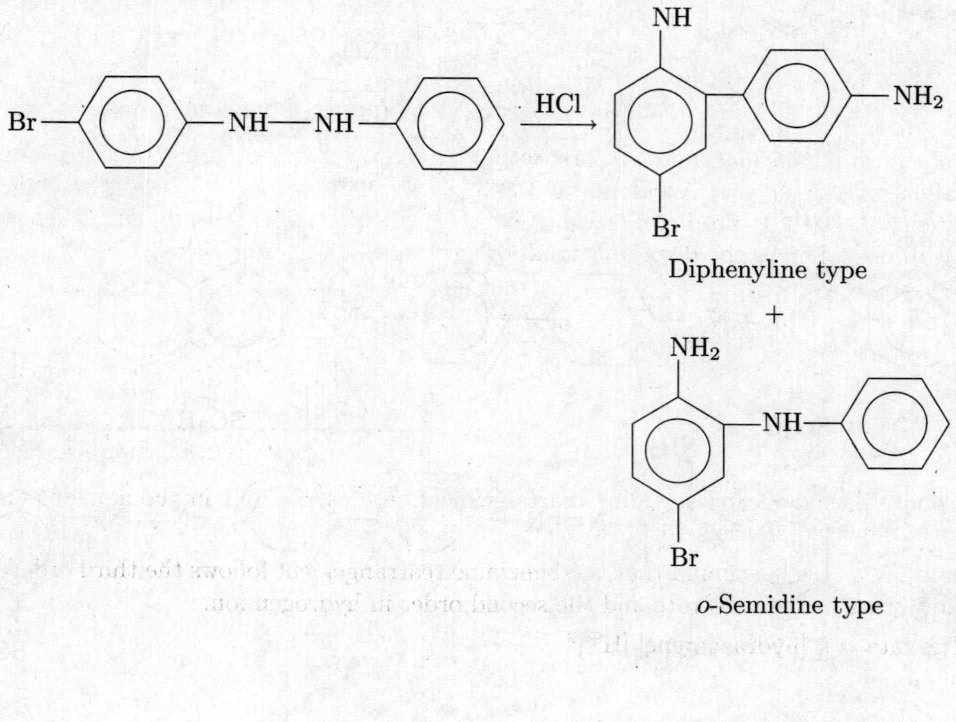

Diphenyline type + *o*-Semidine type

(iii) When both the *p*-positions are blocked, the major product is an *o*-semidine type of compound.

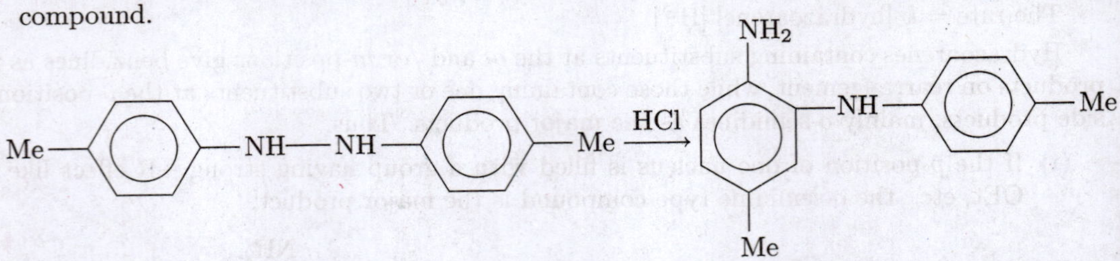

(iv) When one *p*-position is filled with a carboxyl or a sulphonic acid group, the group gets ejected and the benzidine is formed.

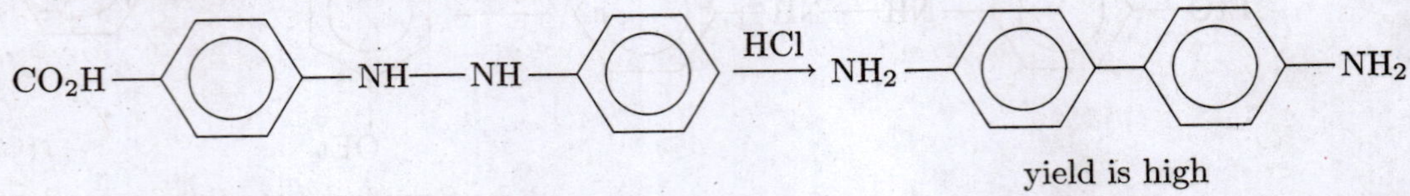

The most important application of the benzidine rearrangement is found in the syntheses of direct dyes for cottons. Hydrazoarenes are first converted into benzidines which on diazotisation and subsequent coupling reaction give the required dyes. A few examples are given below :

(a) Synthesis of Congo Red :

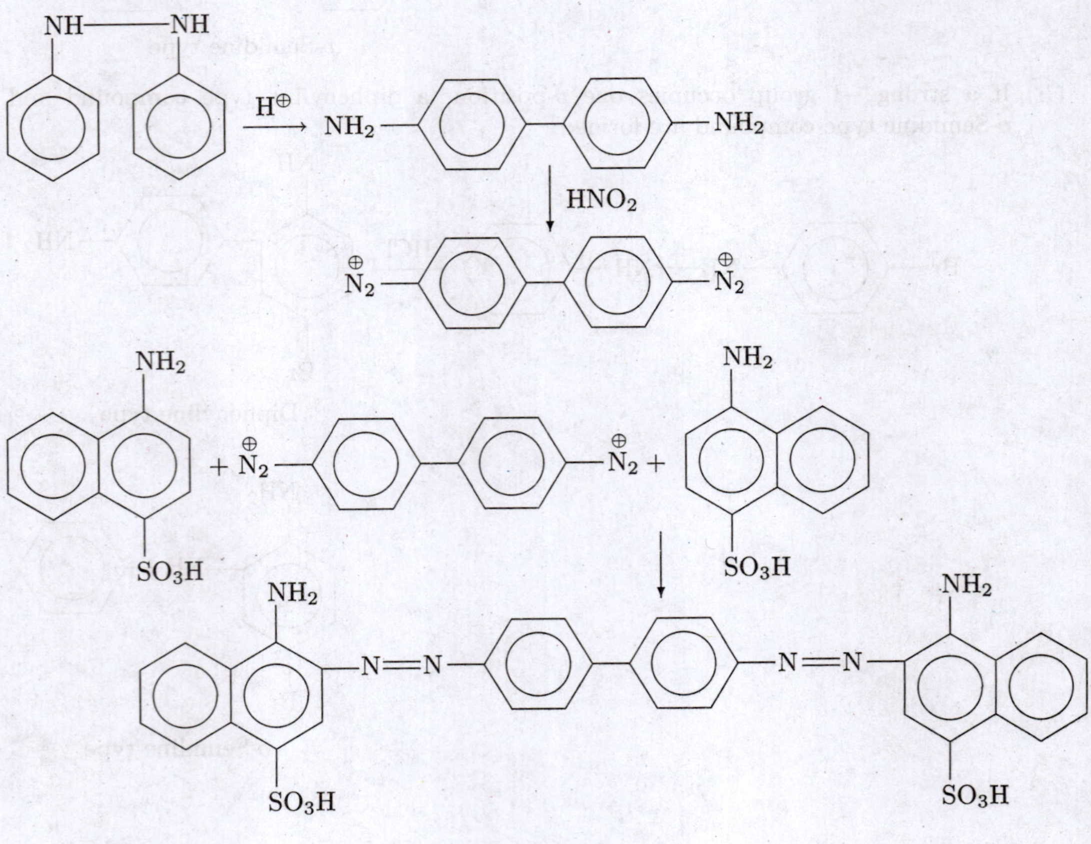

(b) Synthesis of Direct Red 2 :

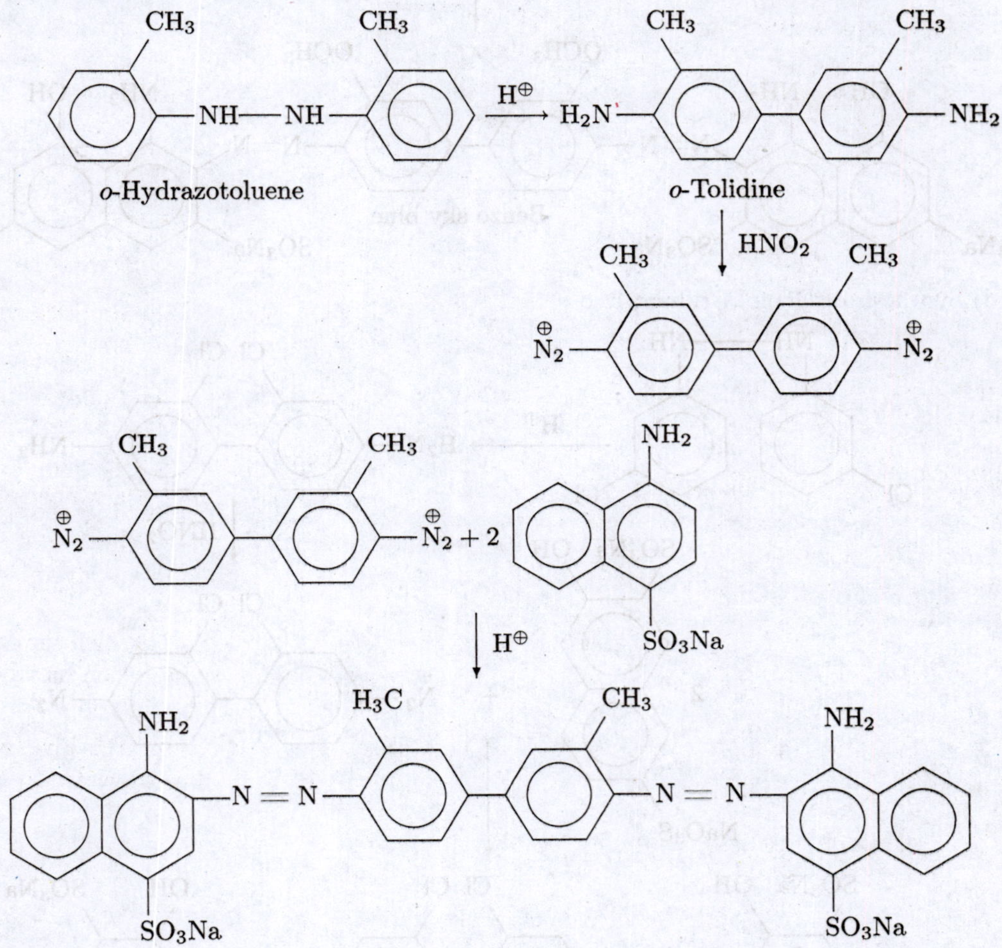

(c) Synthesis of Benzo Sky Blue :

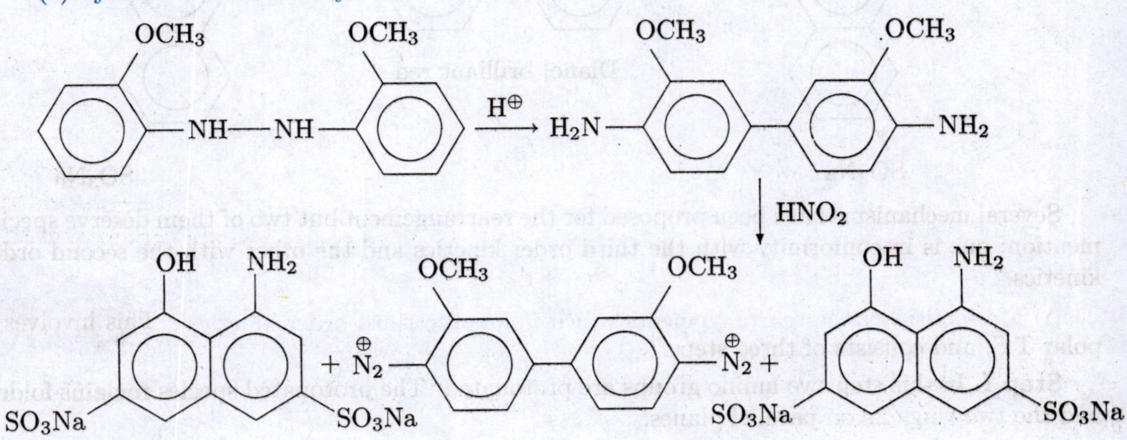

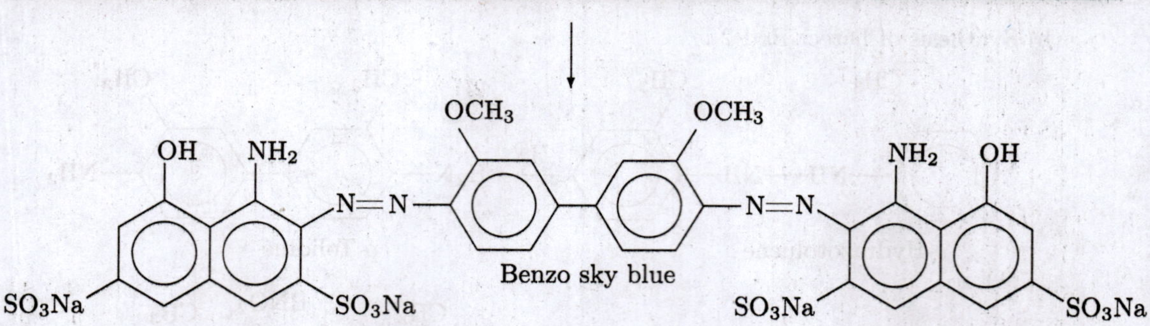

Benzo sky blue

(d) Synthesis of Dianol Brilliant Red :

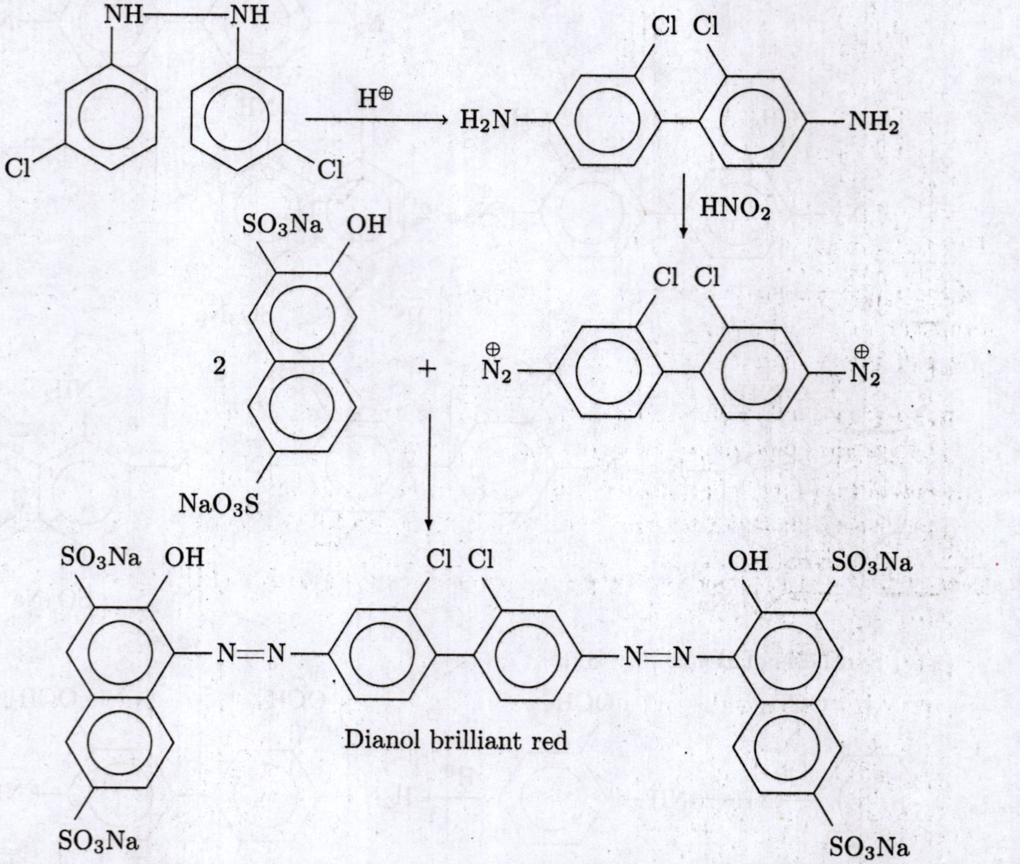

Dianol brilliant red

Several mechanisms have been proposed for the rearrangement but two of them deserve special mention; one is in conformity with the third order kinetics and the other with the second order kinetics.

(i) Mechanism of the rearrangements which follow the third order kinetics : This involves a polar T.S. and consists of three steps.

Step I. In this step two amino groups are protonated. The protonated species remains folded and the two rings lie on parallel planes.

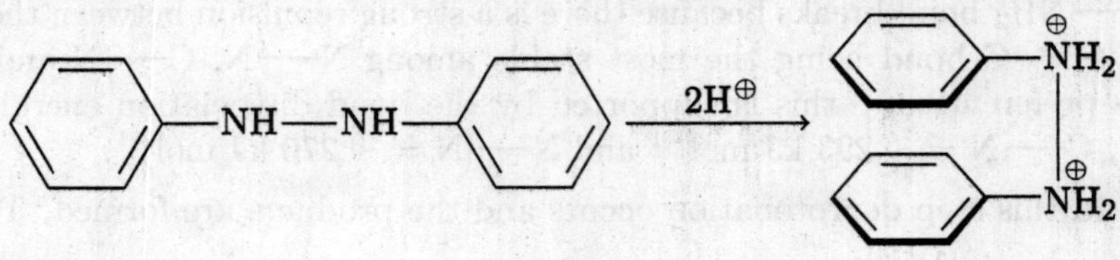

This is a fast step.

Step II. The $H_2\overset{\oplus}{N}$——$\overset{\oplus}{N}H_2$ bond breaks and C——C or C——N bond forms in a concerted step through a polar T.S. This is a slow step and hence this is the rate-determining step. For different products different possible T.S.s(I–V) are drawn below.

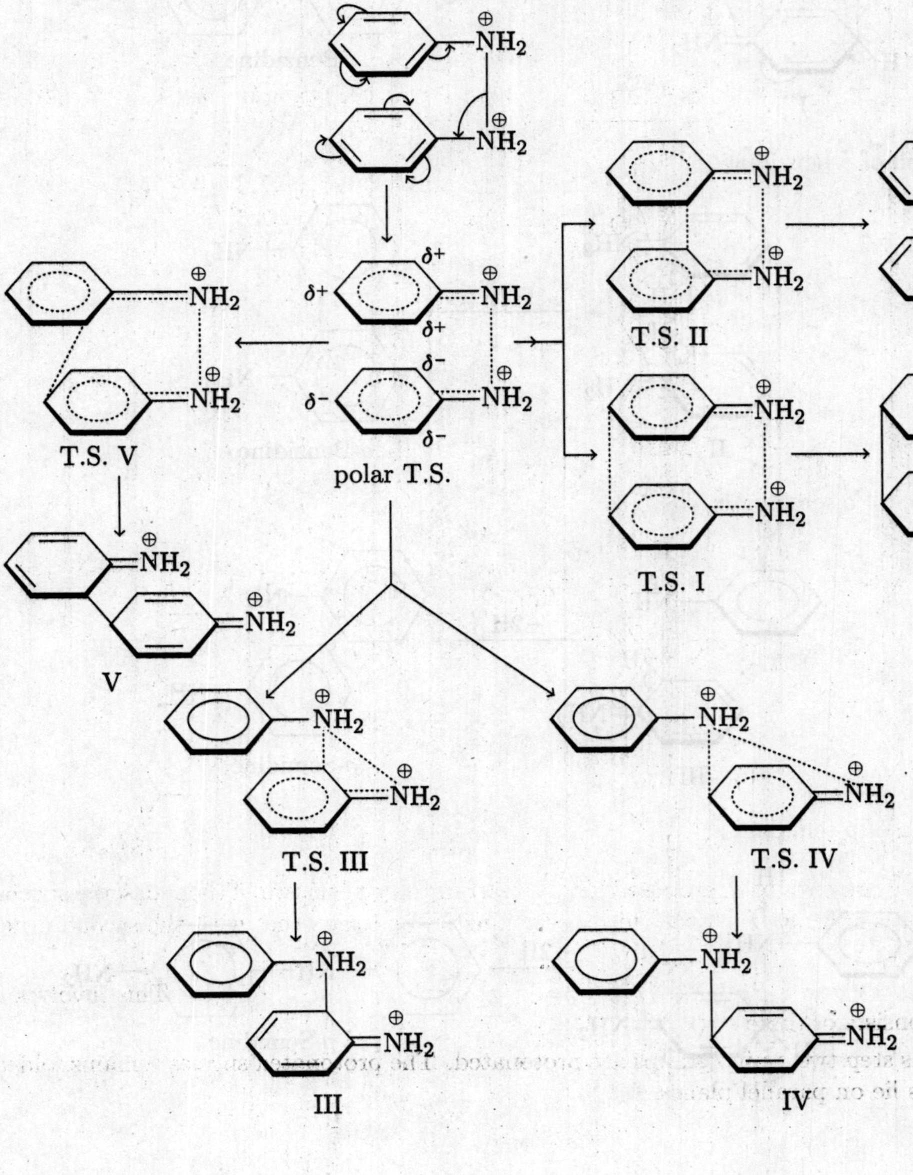

The $H_2\overset{\oplus}{N}$—$\overset{\oplus}{N}H_2$ bond breaks because there is a strong repulsion between the positive charge centers. The C—C bond being the most stable among N—N, C—N and C—C bonds, C—C forms preferentially—this is supported by the bond dissociation energies; C—C = ~ 348 kJ mol^{-1}, C—N = ~ 293 kJ mol^{-1} and N—N = ~ 270 kJ mol^{-1}.

Step III. In this step deprotonation occurs and the products are formed. This is also a fast step.

(a) Formation of benzidine :

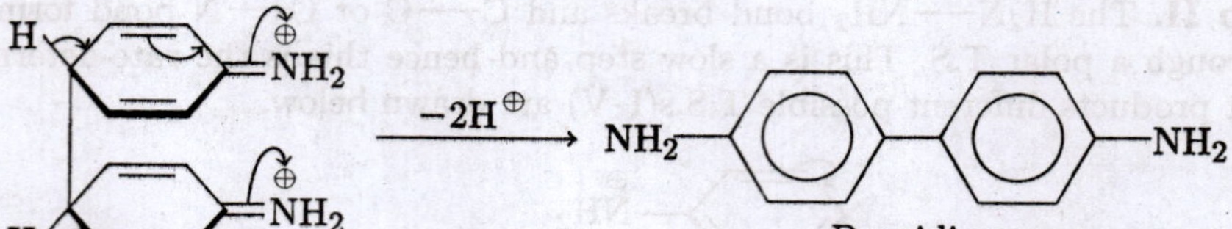

I Benzidine

(b) Formation of *o*-benzidine :

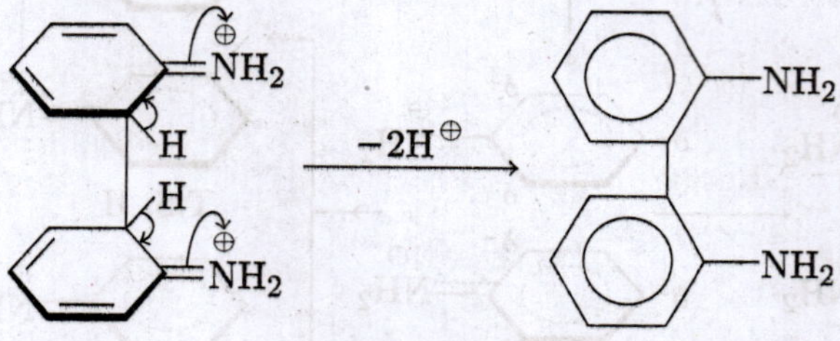

II *o*-Benzidine

(c) Formation of *o*-semidine :

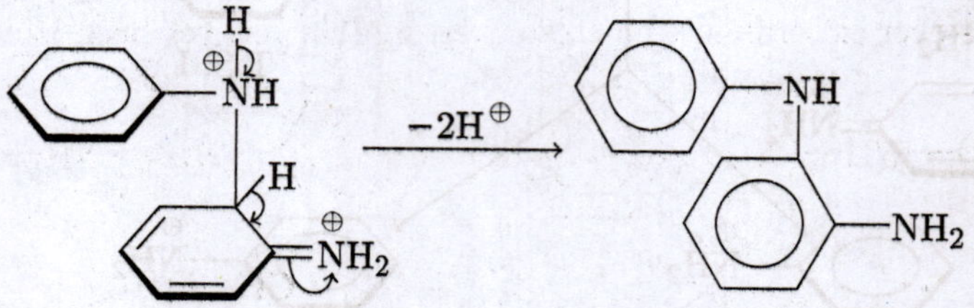

III *o*-Semidine

(d) Formation of *p*-semidine :

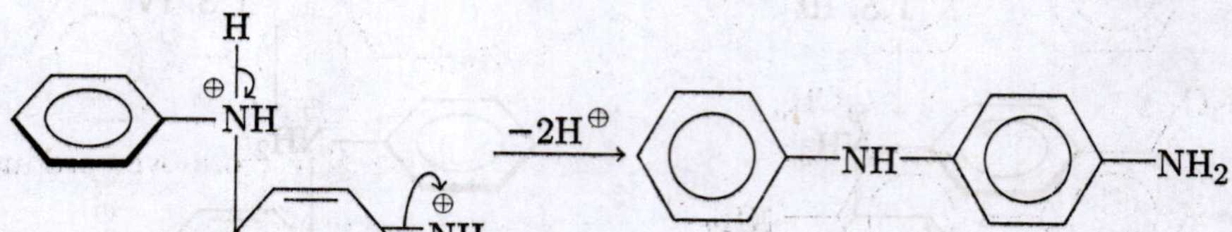

IV *p*-Semidine

(e) Formation of diphenyline :

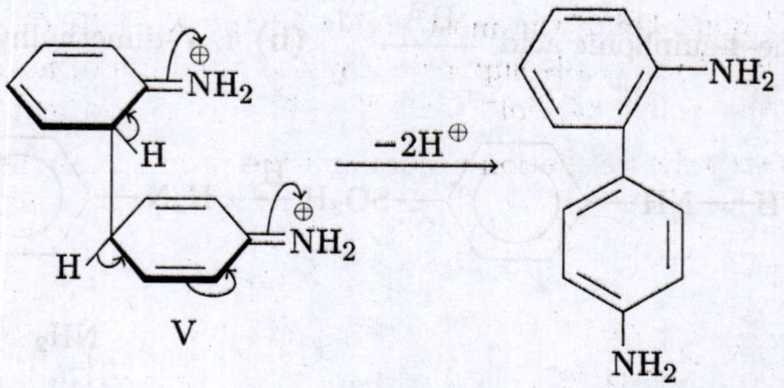

Diphenylene

(ii) Mechanism of the rearrangements which follow the second order kinetics : **This involves a π-complex as an intermediate.**

The π-complex consists of two benzene type rings and a proton. The proton is hydrogen bonded to the π electrons of each ring. Perhaps the formation of π-complex is the rate-determining step. This then gets protonated at a nitrogen atom and by the rotation of one of the ring several possible products are formed.

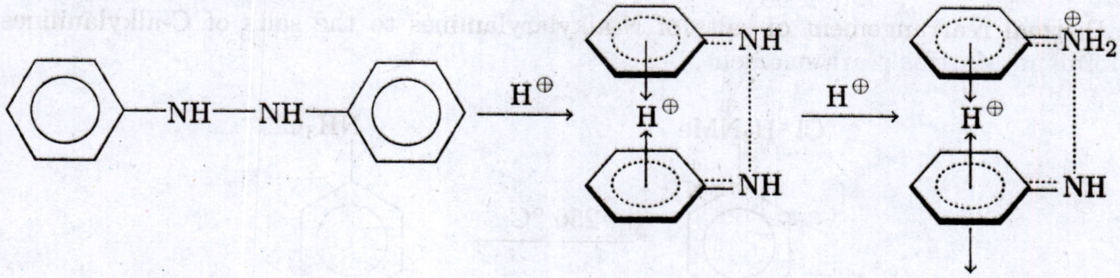

products

However, the benzidine rearrangement is an intramolecular reaction and this has been established by the cross-over experiment; no cross-over product has yet been isolated.

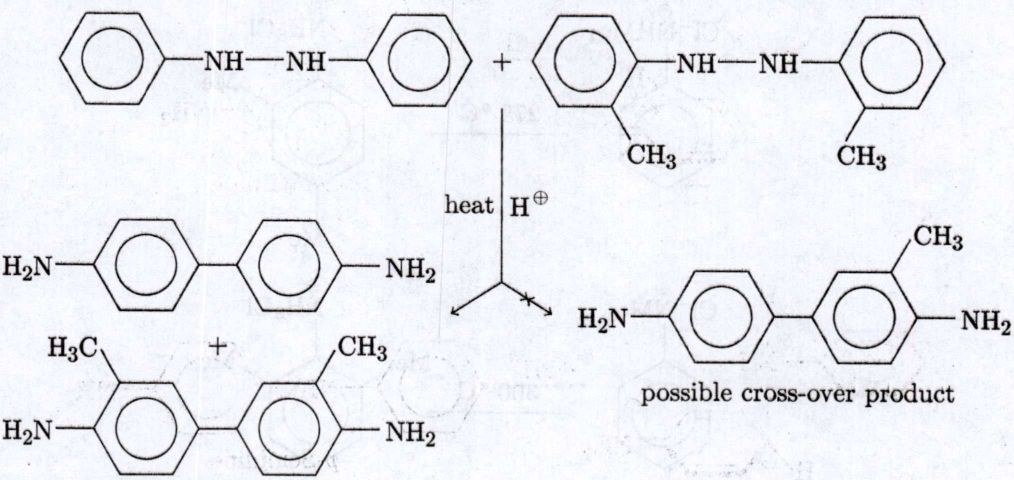

possible cross-over product

intramolecular products

Problem 14.11. Name the major products for the following rearrangements :

(a) Hydrazobenzene-4-sulphonic acid $\xrightarrow{H^{\oplus}}$ **(b)** 4, 4′-dimethylhydrazobenzene $\xrightarrow{H^{\oplus}}$

Solution :

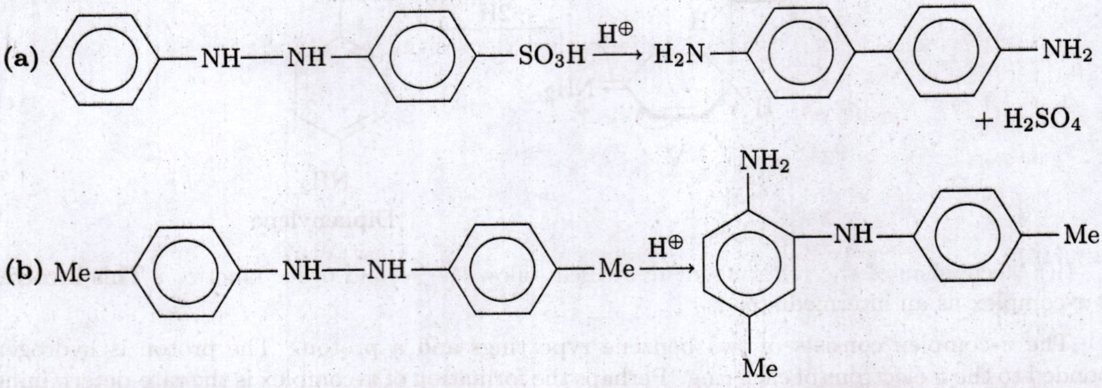

14.13 The Hofmann-Martius Rearrangement

The thermal rearrangement of salts of N-alkylarylamines to the salts of C-alkylanilines is called Hofmann-Martius rearrangement.

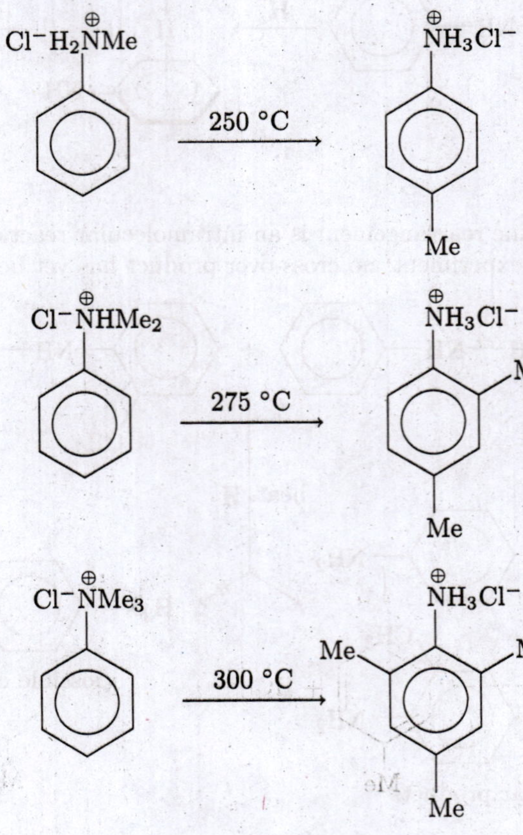

The reaction is usually carried out by heating the salt of N-alkylarylamine under high pressure. Usually *p*-products are predominant. When the *p*-position is filled, *o*-rearrangement occurs. The reaction follows an intermolecular mechanism and consists of the following steps :

Step I. Decomposition of the salt to aniline and alkyl chloride.

Step II. Electrophilic attack at the *p*-C by the positive pole of the $\overset{\delta\oplus}{R}—\overset{\delta-}{Cl}$ dipole and the formation of a σ-complex.

Step III. Aromatisation of the σ-complex :

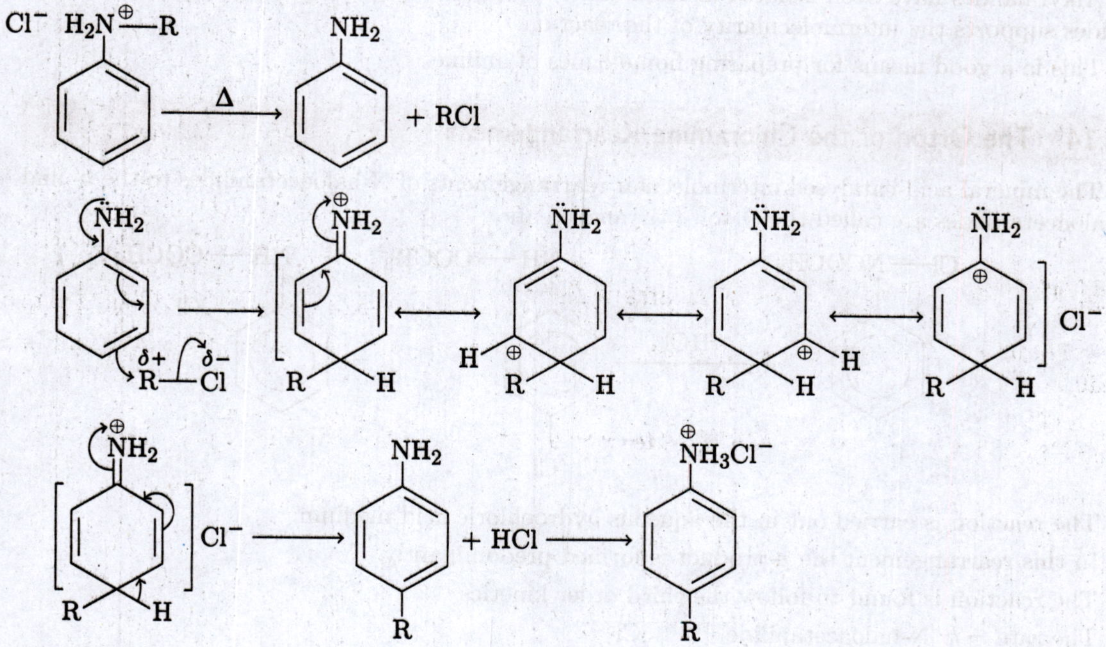

Similarly,

(i) 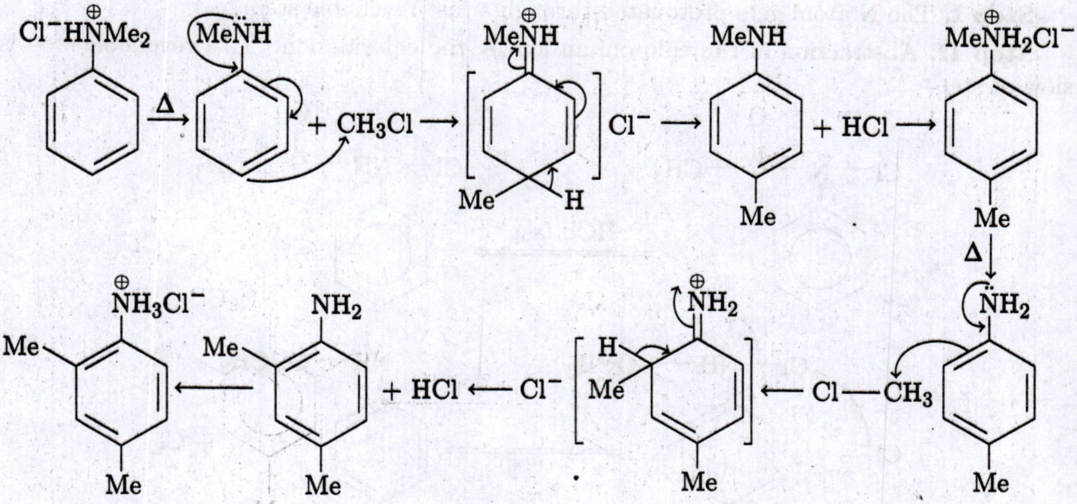

(ii) 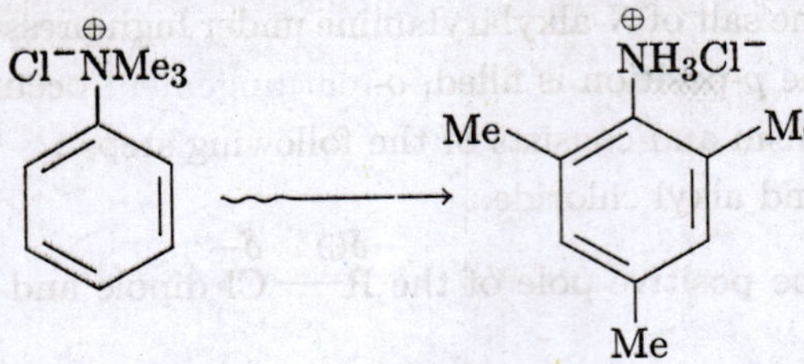

Alkyl halides have been isolated in Hofmann-Martius rearrangements. The isolation of alkyl halides supports the intermolecularity of the reaction.

This ia a good means for preparing homologues of aniline.

14.14 The Orton or the Chloramine Rearrangement

The mineral acid-catalysed intermolecular rearrangements of N-haloacetanilides to the *o*- and *p*-haloacetanilides are called the Orton rearrangements.

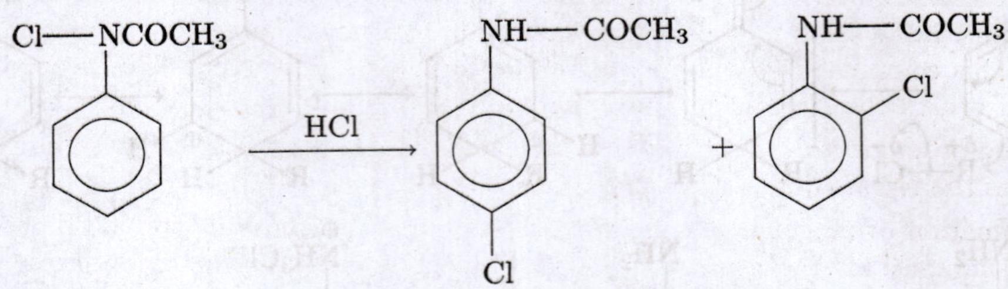

The reaction is carried out in the aqueous hydrochloric acid medium.

In this rearrangement the *p*-product is formed predominantly.

The reaction is found to follow the third order kinetics

The rate = k [N-haloacetanilide] $[H^{\oplus}]$ $[Cl^-]$

The mechanistic steps of the rearrangement are :

Step I. The N atom gets protonated through a fast reversible step.

Step II. Abstraction of the chloronium ion by the chloride ion. This reversible step is the slowest step.

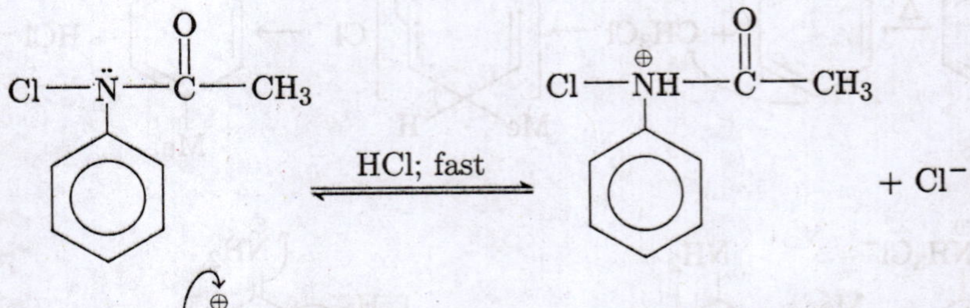

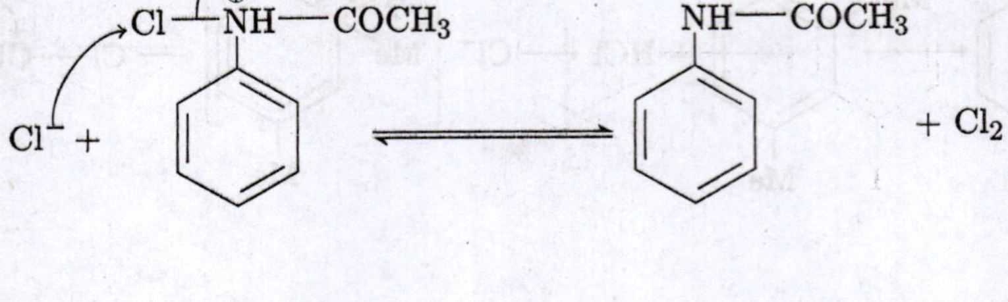

Step III. Aromatic eletrophilic substitution bimolecular—the chloronium ion gets substituted either at *p*- or at *o*-position via a sigma-complex. The step is more rapid than the step II.

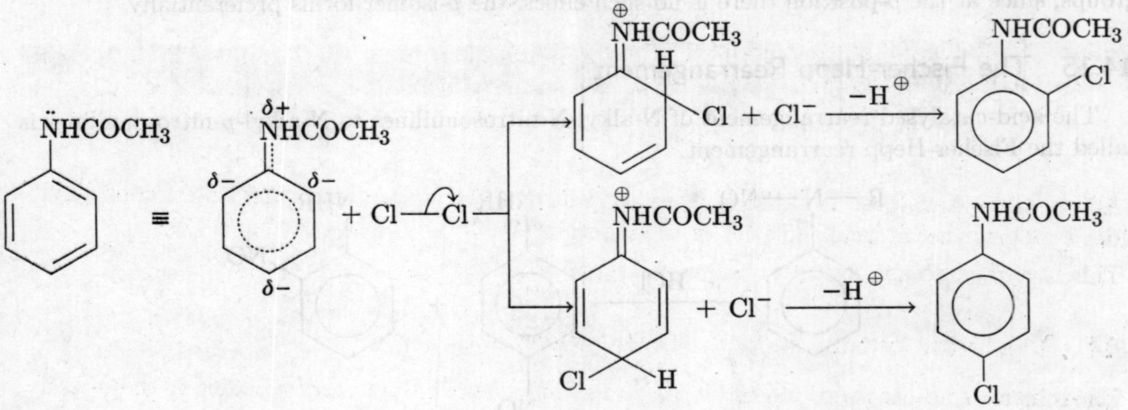

The intermolecularity of the reaction has been established by the isolation of the molecular chlorine. This, in its turn, tells us that a chloronium ion must detach itself from the rest of the molecule. This combines with the Cl^- of HCl to form Cl_2. This is possible because prior to the detachment of $Cl^{\oplus}$ the N atom gets protonated.

The third order kinetics of the reaction indicates that the step II is the rate-limiting step since this involves the N-haloanilide, Cl^- and $H^{\oplus}$ ions.

The formation of monohaloacetanilide also supports the above drawn mechanistic pathway. Since only one mole of Cl_2 forms per mole of N-haloanilide and since only one mole of $Cl^{\oplus}$ ion is available per mole of Cl_2, the monohaloacetanilide only can form.

Problem 14.12. A mixture of *o*- and *p*-iodoacetanilide are formed when N-chloroacetanilide is treated with hydroiodic acid. Give the mechanism of the reaction. Why does *p*-isomer form predominantly?

Solution :

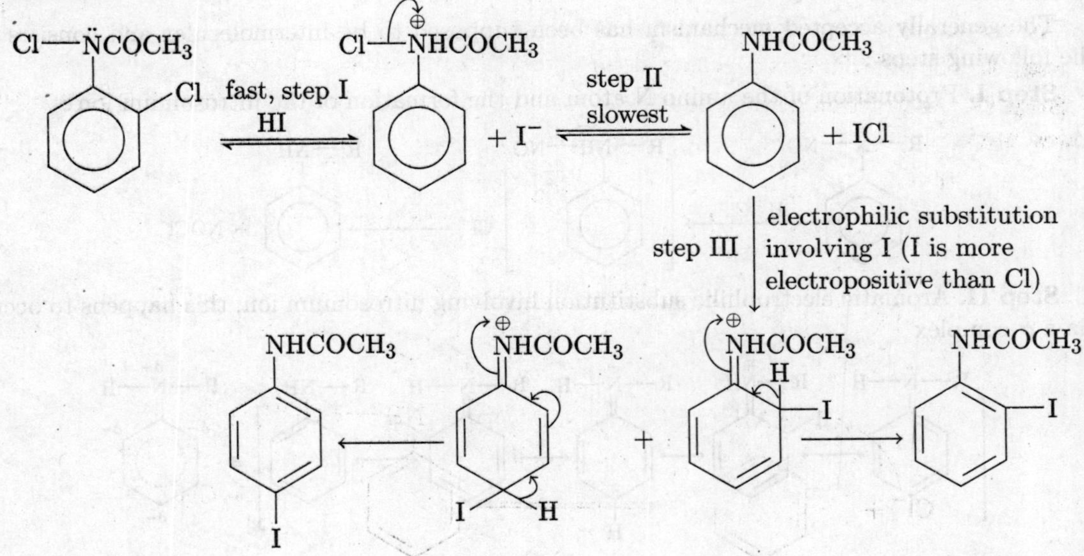

Since the I atom and the $NHCOCH_3$ are bulky groups, the I atom does not get enough room for its accommodation at the *o*-position—a spatial interaction takes place between the two bulky groups; since at the *p*-position there is no such effect, the *p*-isomer forms preferentially.

14.15 The Fischer-Hepp Rearrangement

The acid-catalysed rearrangement of N-alkyl-N-nitrosoanilines to N-alkyl-*p*-nitrosoanilines is called the Fischer-Hepp rearrangement.

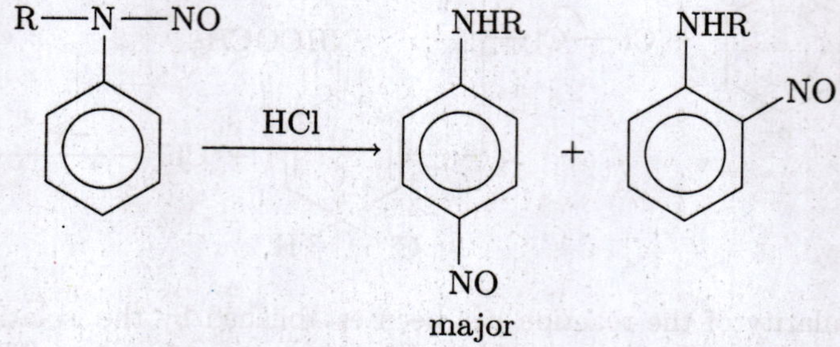

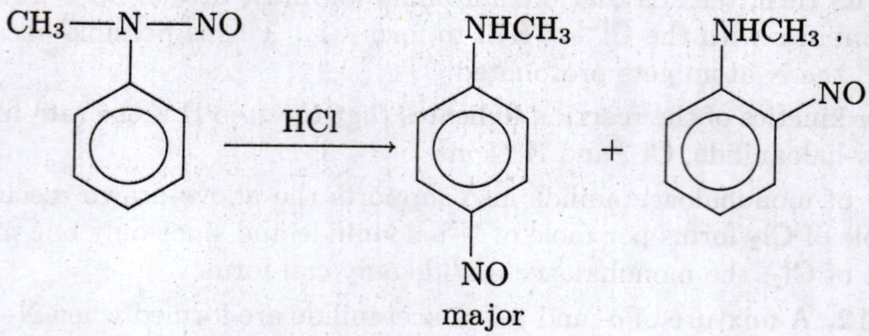

In this reaction the *p*-product is formed predominantly.

The reaction has a good preparative value since N-alkyl-*p*-nitrosoanilines cannot be prepared by the direct nitrosation process.

The generally accepted mechanism has been supposed to be intermolecular and consists of the following steps :

Step I. Protonation of the amino N atom and the formation of the nitrosonium ion :

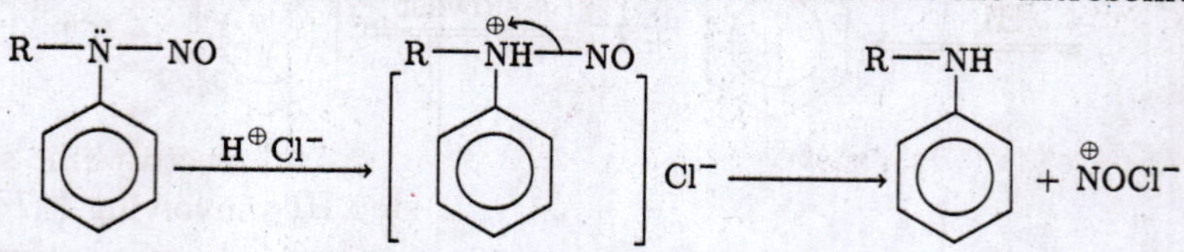

Step II. Aromatic electrophilic substitution involving nitrosonium ion; this happens to occur via a σ-complex

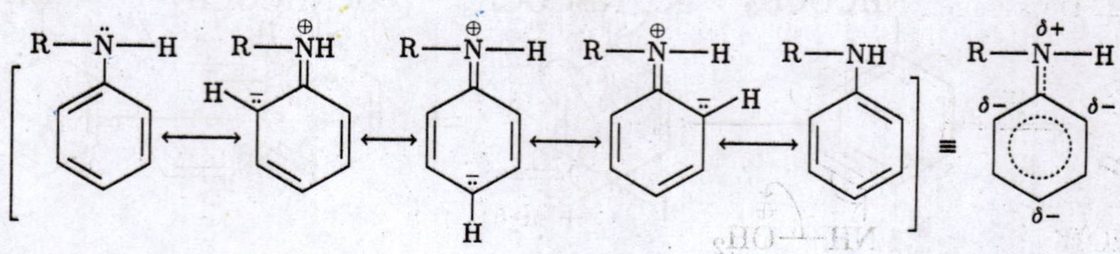

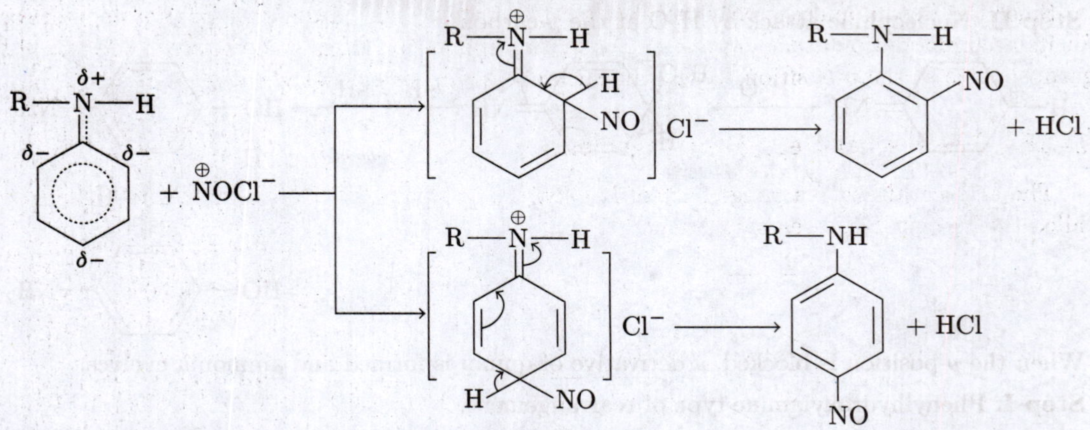

The intermolecular mechanism is supported by the formation of the predominant *p*-product. However, in 1969 Williams et al. proposed that the reaction is intramolecular. They carried out the reaction in the presence of $Na^{15}NO_2$ and H_2SO_4 and showed that no ^{15}NO was incorporated in the product which suggests that the reaction is intramolecular. The mechanistic steps which are in conformity with the above experiment may be drawn as given below:

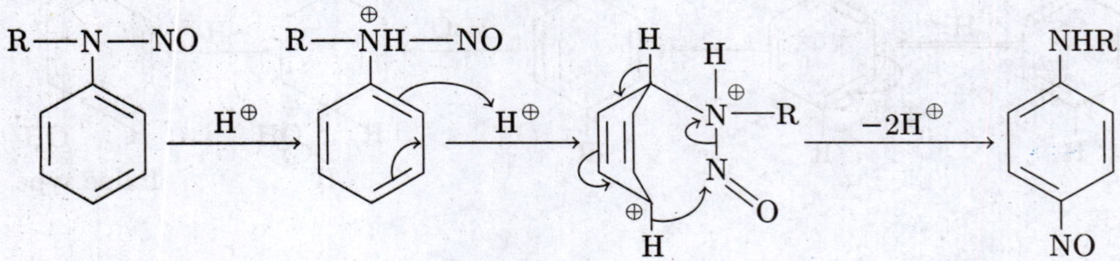

14.16 Phenylhydroxylamine—*p*-Aminophenol Rearrangement

In the acidic solution phenylhydroxylamine rearranges to *p*-aminophenol.

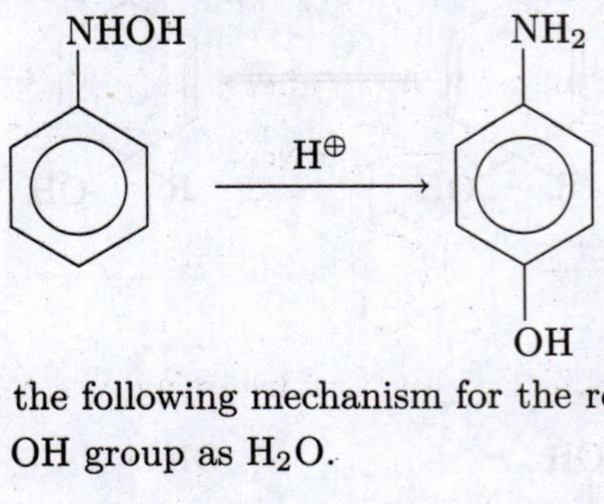

Huges and Ingold proposed the following mechanism for the rearrangement :

Step I. Detachment of the OH group as H_2O.

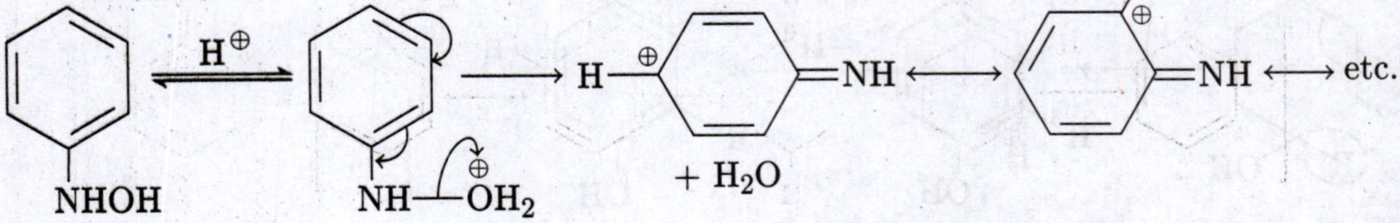

Step II. Nucleophilic attack by H_2O at the *p*-carbon.

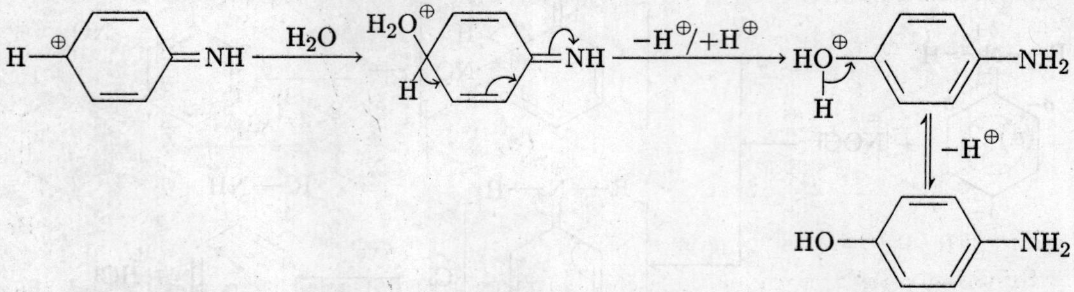

When the *p*-position is blocked, a derivative of quinol is formed and ammonia evolves.

Step I. Phenylhydroxylamine type of rearrangement.

Step II. Hydrolysis of the imine type product to give a dienone type compound.

Step III. Dienone type compound then undergoes the dienone—phenol rearrangement.

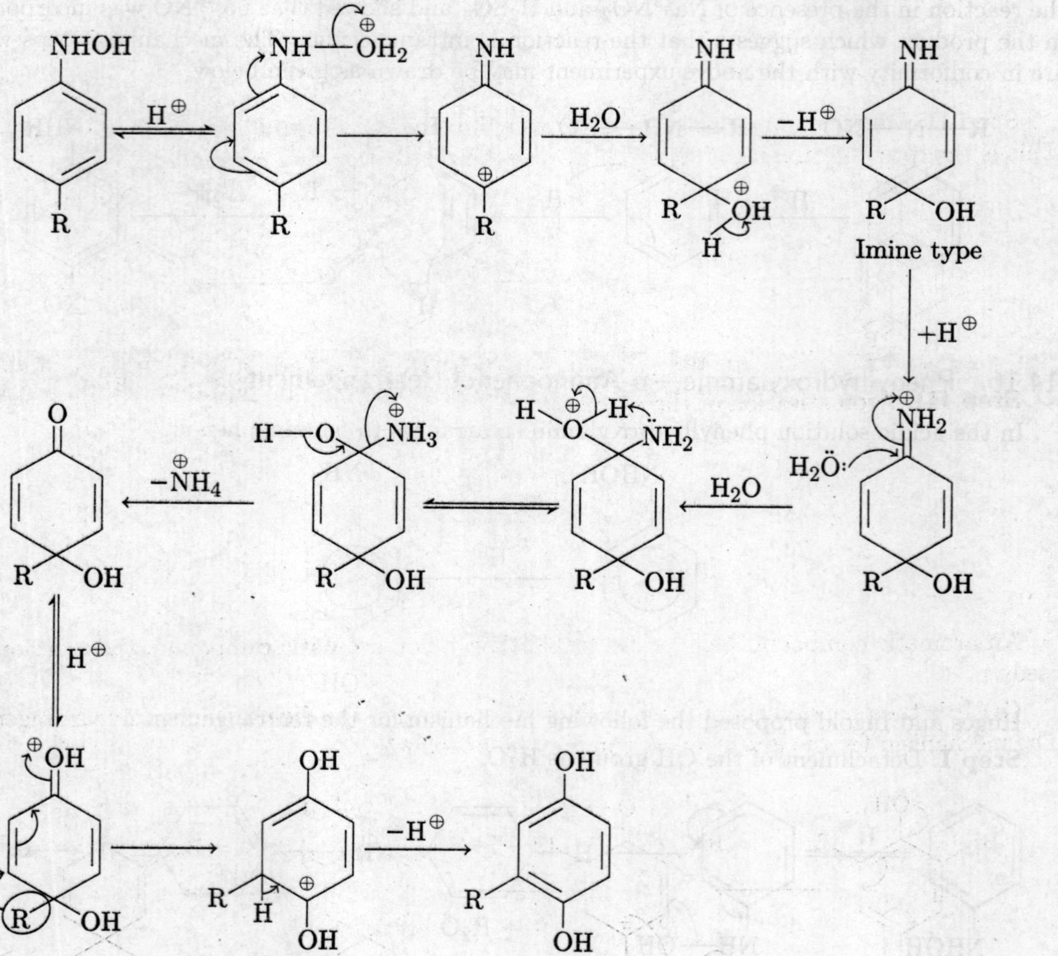

Problem 14.13. Give the mechanistic steps for the following conversions.

(a) 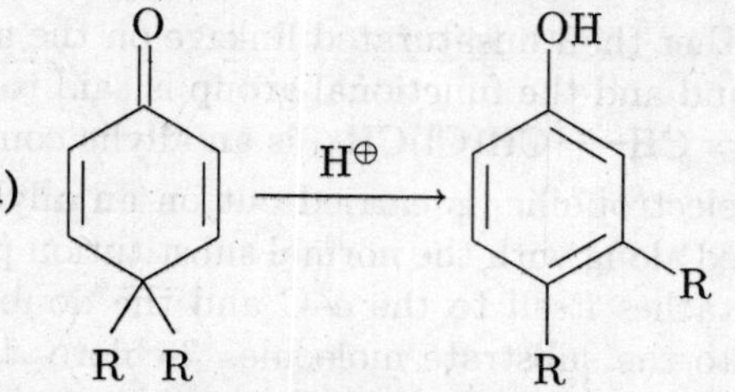 ; (b) 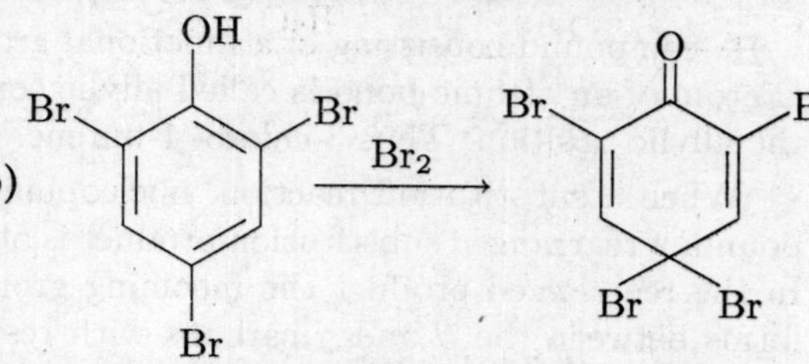

Solution : (a) It is called the dienone—phenol rearrangement, since this involves the conversion of a dienone into a phenol. However, it consists of the following steps :

Step I. Protonation of the oxygen atom.

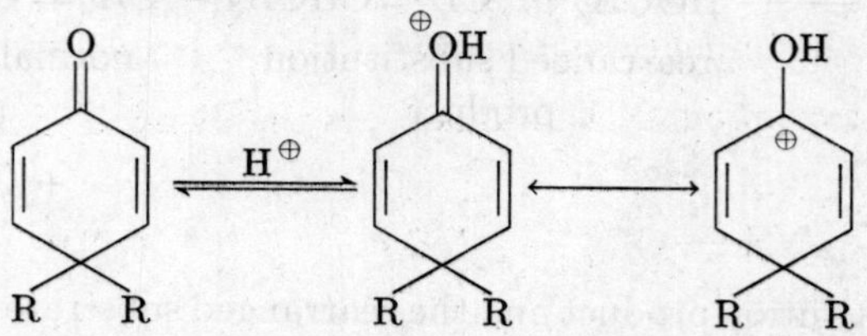

Step II. Migration of an alkyl group with its bonded *e*s to the adjacent C atom—a 1, 2-shift. This is the rate-determining step which involves the formation of a σ-complex.

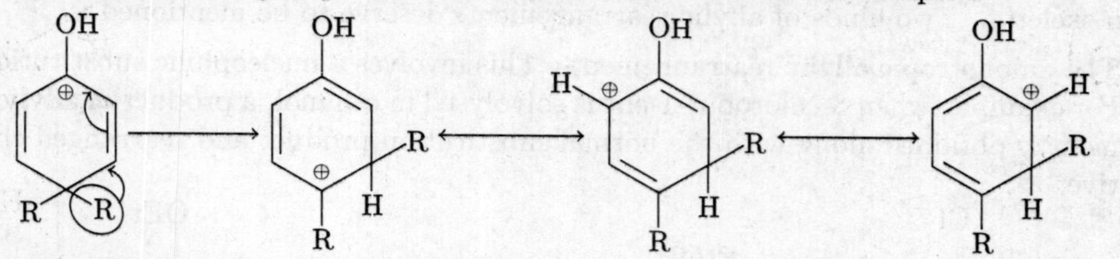

Step III. Aromatisation of the σ-complex :

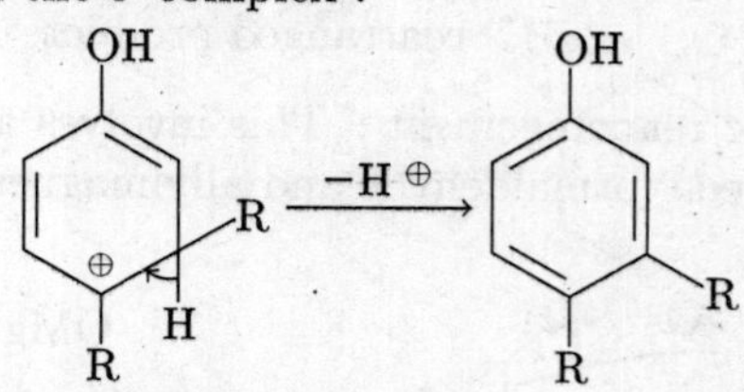

An aromatic compound being more stable than a non-aromatic compound, the reaction proceeds.

(b) It is the phenol-dienone type reaction, the reverse of the dienone—phenol rearrangement. The mechanism for the given conversion may be as follows :

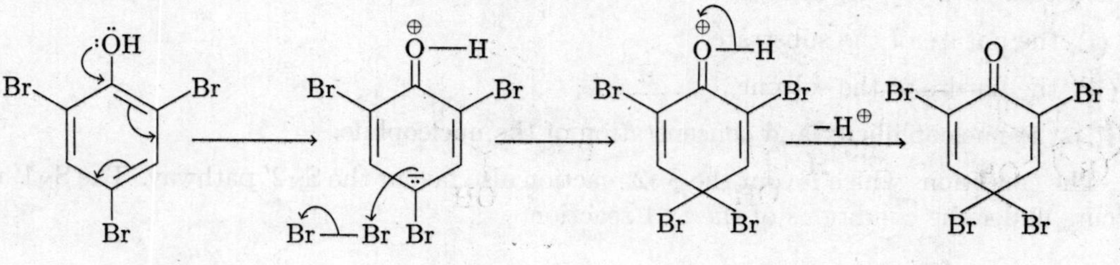

14.17 Allylic Rearrangement

A compound consisting of a functional group other than unsaturated linkage on the adjacent C atom of an olefinic bond is called allylic compound and the functional group is said to occupy the allylic position. Thus 3-chloro- 1-butene, $CH_2{=}CH{-}CH(Cl)CH_3$, is an allylic compound.

When a substitution reaction, nucleophilic or electrophilic, is carried out on an allylic compound, a rearranged substitution product is obtained along with the normal substitution product. In the rearranged product the incoming group attaches itself to the α-C and the double bond forms between the β and γ carbons with respect to the substrate molecule. 3-chloro- 1-butene on treatment with an aqueous alkali forms 2-buten- 1-ol, a rearranged substitution product, and 3-buten-2-ol along with 1-chloro-but-2-ene.

$$CH_2{=}CH.CH.Cl.CH_3 \xrightarrow{OH^-} \underset{\text{rearranged substitution product}}{HOCH_2{-}CH{=}CHCH_3} + \underset{\text{normal substitution product}}{CH_2{=}CH.CHOHCH_3}$$

$$+ \underset{\text{rearranged substrate}}{Cl{-}CH_2{-}CH{=}CHCH_3}$$

The rearranged substitution product and the rearranged substrate are obtained by a rearrangement known as the allylic rearrangement. Thus an allylic rearrangement of an allylic compound involves the interchanging of places between the double bond and the adjacent substituent on the carbon skeleton. Two kinds of allylic rearrangements deserve to be mentioned :

1. **The anionotropic allylic rearrangement :** This involves a nucleophilic substitution reaction. For example, when 3-chlorobut-1-ene is solvolysed in ethanol, a product of allylic rearrangement is obtained along with the normal substitution product and rearranged chloroderivative.

$$CH_3.\overset{\overset{\displaystyle Cl}{|}}{C}H.CH{=}CH_2 \xrightarrow{EtOH} \underset{\text{rearranged product}}{CH_3CH{=}CH.CH_2OEt} + \underset{\text{normal product}}{CH_3.\overset{\overset{\displaystyle OEt}{|}}{C}H.CH{=}CH_2}$$

2. **The cationotropic allylic rearrangement :** This involves an electrophilic substitution reaction. The reaction between formaldehyde and allylmagnesium chloride involves this sort of allylic rearrangement.

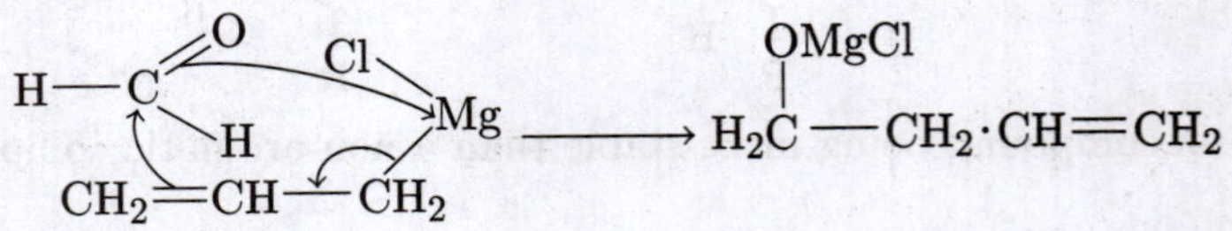

The anionotropic allylic rearrangement can take place both by the S_N2' and S_N1' pathways. The actual pathway depends on :

(i) the nature of the substrate,

(ii) the nature of the solvent,

(iii) the nucleophilicity and concentration of the nucleophile.

The conditions which favour the S_N2 reaction also favour the S_N2' pathway. The S_N1' reaction occurs under the conditions of the S_N1 reaction.

The S_N2' pathway

This involves a concerted bond breaking and bond making step :

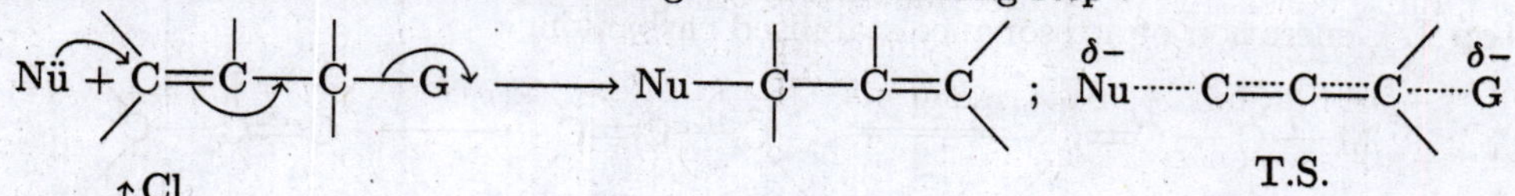

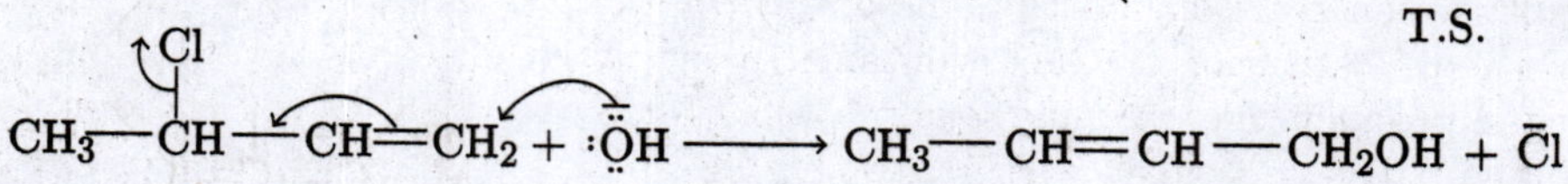

The S_N1' pathway

This occurs in two steps via a resonance-stabilised carbenium ion intermediate.

Step I. Formation of the resonance-stabilised carbenium ion :

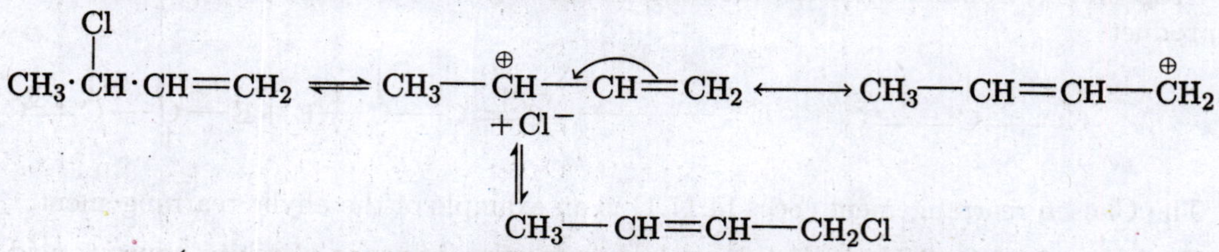

(a rearranged substrate formed by the allylic rearrangement)

This is a slow step and this is, therefore, the rate-determining step. The T.S. is :

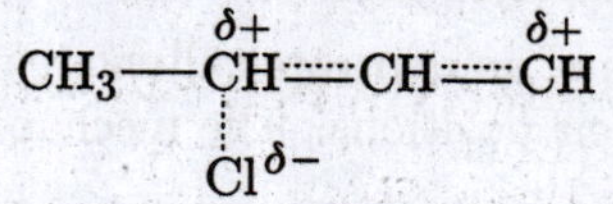

Step II. Nucleophilic attack at the positively charged C atom :

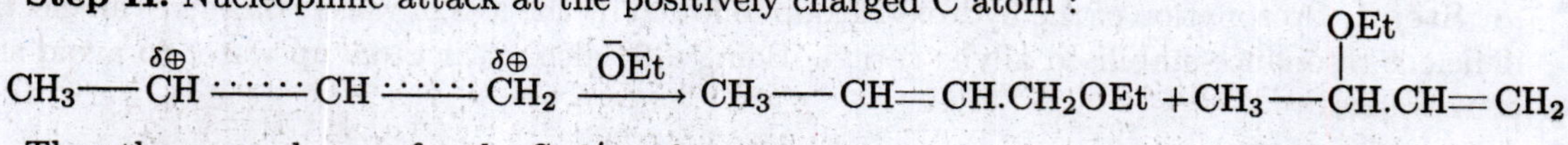

Thus the general steps for the S_N1' pathway are

(i) 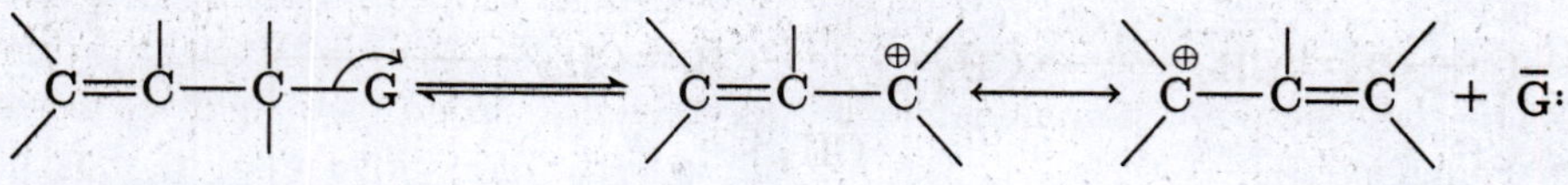

This is a slow rate-determining step.

(ii) 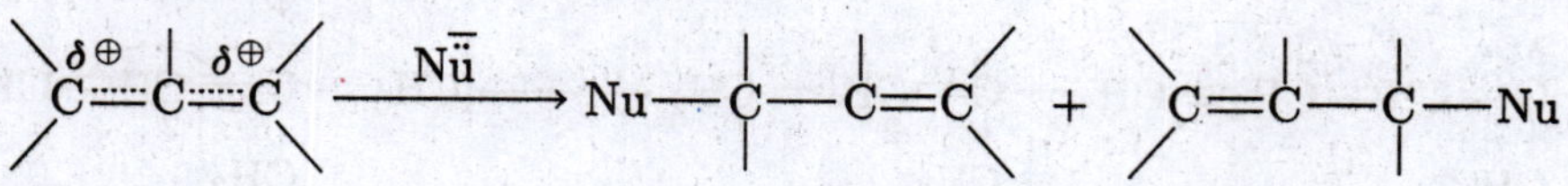

The cationotropic allylic rearrangement involves the S_E2' or S_E1' pathway :

The S_E2' pathway

The electrophilic substitution occurs in a single concerted step :

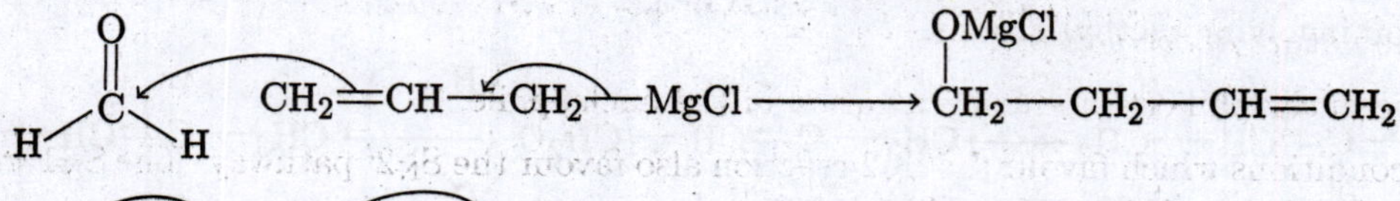

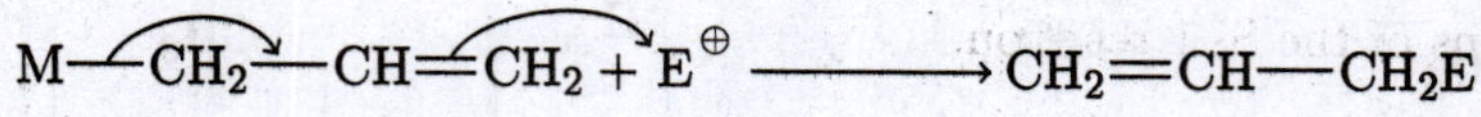

The S_E1' pathway

This occurs in two steps :

Step I. Generation of a resonance stabilised carbanion.

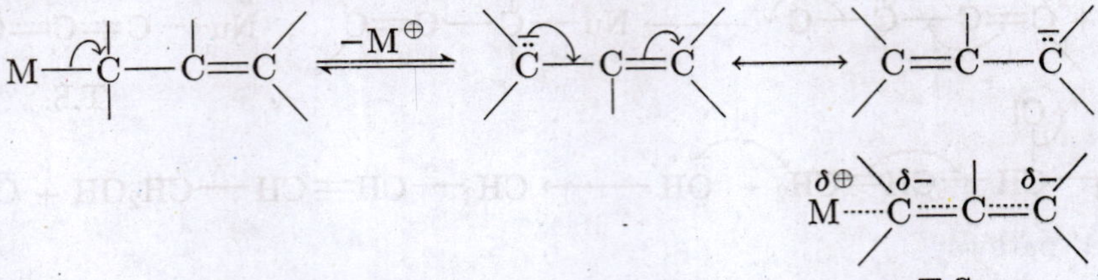

T.S.

Step II. Electrophilic attack at the negatively charged C atom and the formation of a mixture of products,

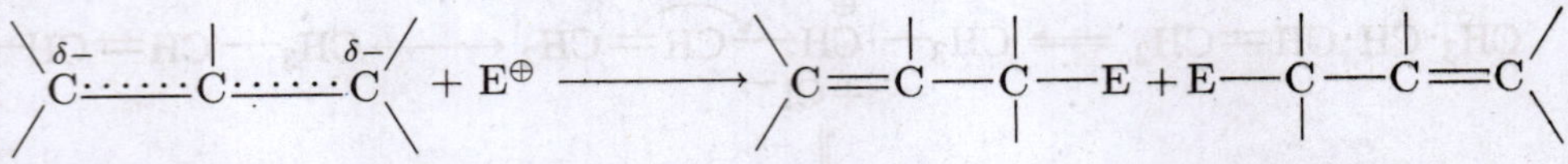

The Claisen rearrangement (Sec. 14.17.1) is an example of the allylic rearrangement.

Problem 14.14. 2-Methylbut-3-ene-1-ol under the influence of dilute aqueous acid forms 3-methylbut-2-ene-1-ol as the major product. Give the mechanism of the reaction and explain why 3-methylbut-2-ene-1-ol is the major product.

Solution : This is an example of anionotropic allylic rearrangement in which the double bond and the OH group interchange their positions. The mechanism for the reaction may be drawn as shown below :

Step I. Protonation of the hydroxyl group followed by the loss of water, this gives an electron-deficient resonance-stabilised allylic cation. Being a 3° alcohol, it gives up water to avoid steric congestion.

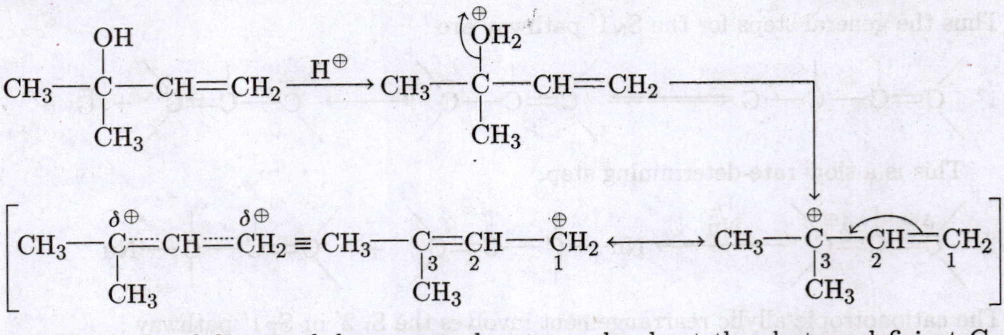

Step II. Nucleophilic attack by a neutral water molecule at the electron-deficient C atom. Since the C_3 atom is more sterically hindered than the C_1 atom, water attaches itself to the C_1 atom; here C_1 and C_3 cations are almost equally stable owing to p-π conjugation and hyperconjugation involving methyl groups.

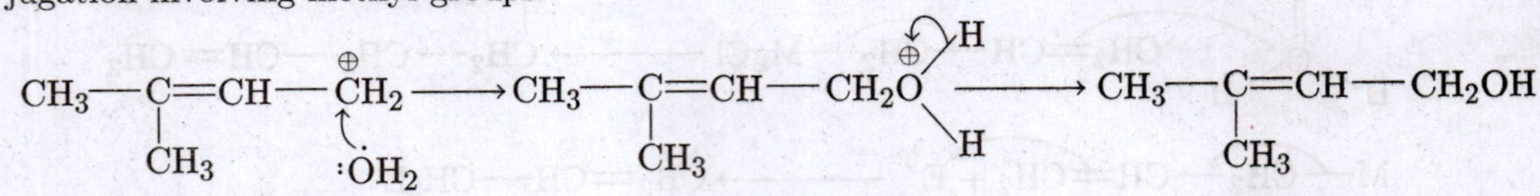

14.17 The Sigmatropic Rearrangement

The thermal or photochemical concerted step rearrangement involving a cyclic T.S. in which a σ bond, flanked by one or two π orbitals, shifts intramolecularly from one atom to another with the conservation of orbital symmetry is known as a sigmatropic rearrangement. These rearrangements are named as $[i,\ j]$ shifts, where i and j are the two ordinal numbers to which each end of the shifting sigma bond goes. Thus the thermal rearrangement of N-allylphenylamine to *o*-allylaniline is an example of sigmatropic rearrangement. Here N—CH_2 σ bond shifts and it is named as a [3, 3] shift.

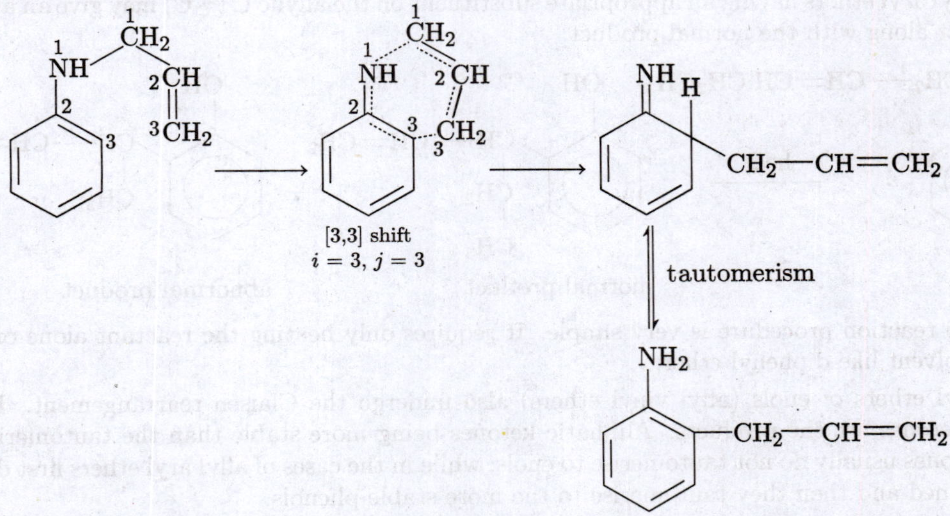

The best known example of the sigmatropic rearrangement is the Claisen rearrangement. This is discussed below in detail.

14.17.1 The Claisen rearrangement

The Claisen rearrangement is a thermal sigmatropic rearrangement involving [3, 3] shift in which allyl aryl ethers get converted into allylphenols intramolecularly via cyclic T.S.s.

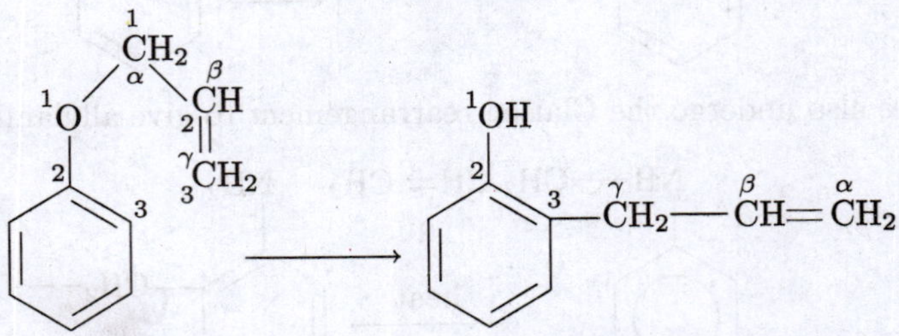

Here the O—C_α σ bond undergoes [3, 3] shift and becomes *o*-C—C_γ σ bond.

When one or both the *o*-positions are free, *o*-allylphenols are obtained, but if the *o*-positions are blocked, *p*-allylphenols are formed.

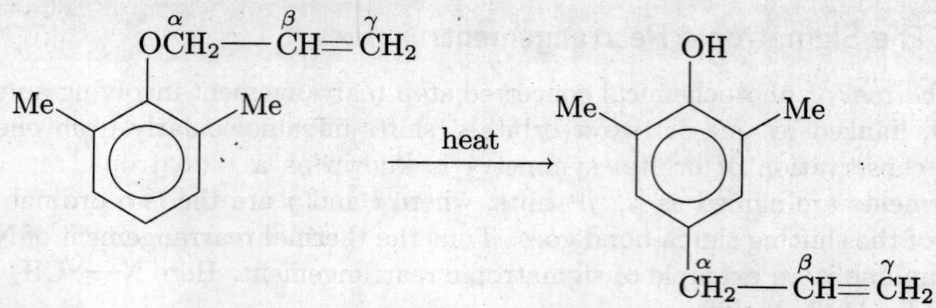

Allyl aryl ethers having an appropriate substituent on the allylic C (γ-C) may give an abnormal product along with the normal product.

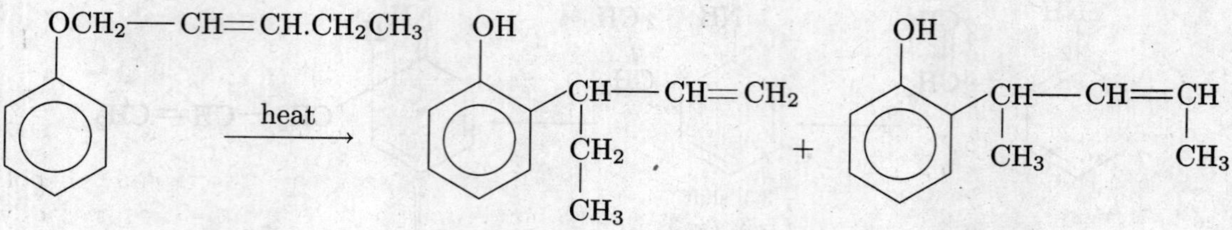

normal product abnormal product

The reaction procedure is very simple. It requires only heating the reactant alone or in any inert solvent like diphenyl ether.

Allyl ethers of enols (allyl vinyl ethers) also undergo the Claisen rearrangement. In these cases ketones are the products. Aliphatic ketones being more stable than the tautomeric enols, the ketones usually do not tautomerise to enols; while in the cases of allyl aryl ethers first dienones are formed and then they tautomerise to the more stable phenols.

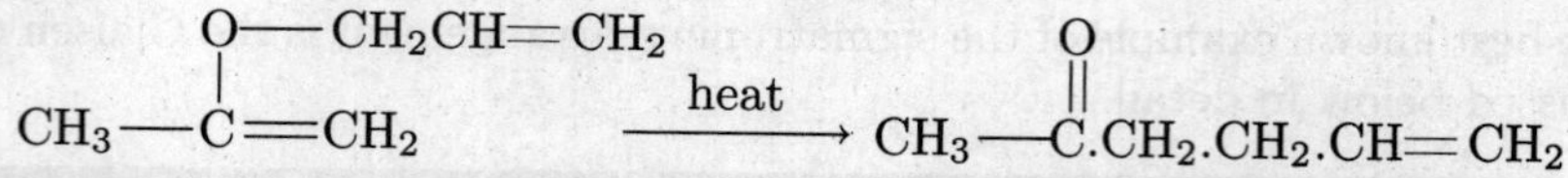

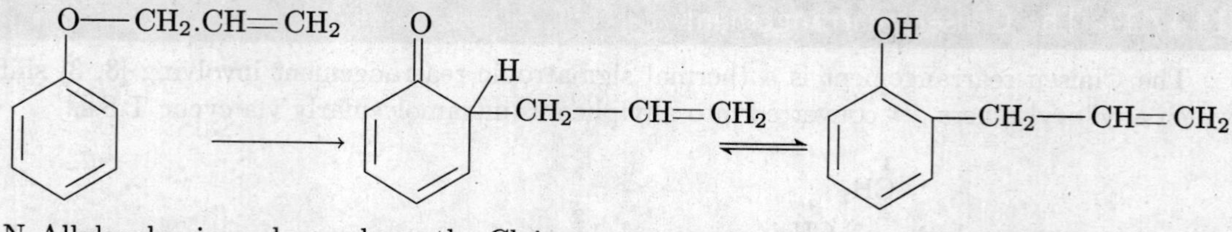

N-Allylarylamines also undergo the Claisen rearrangement to give allylanilines.

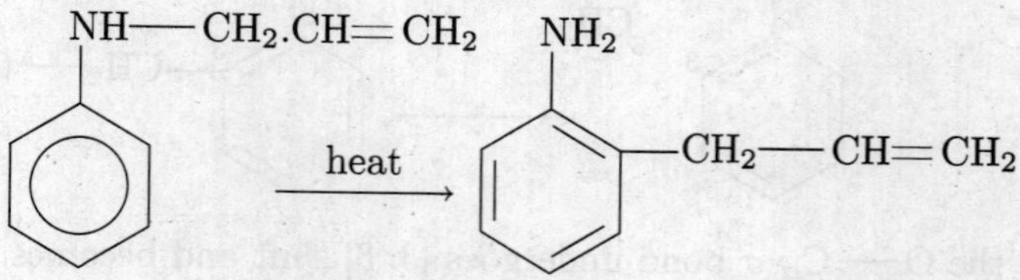

It has been found that the electron-donating substituents in the aryl moiety increases the rate of the Claisen rearrangement while the electron-withdrawing substituents decrease its rate; of course, the difference between the rates is not high enough. For example, the *p*-nitroaryl allyl ethers react only ten to twenty times slower than the *p*-aminoaryl allyl ethers.

However, the reaction follows the first order kinetics.

The rate= k [allyl aryl ether]

The Claisen rearrangement involves negative entropy of activation which indicates that the T.S. of the reaction will be more 'ordered' than the reactant.

The low electronic effect on the rate, the first order kinetics of the reaction and the negative entropy of activation tell us that the reaction follows the concerted step mechanism via a cyclic transition state involving chair conformation.

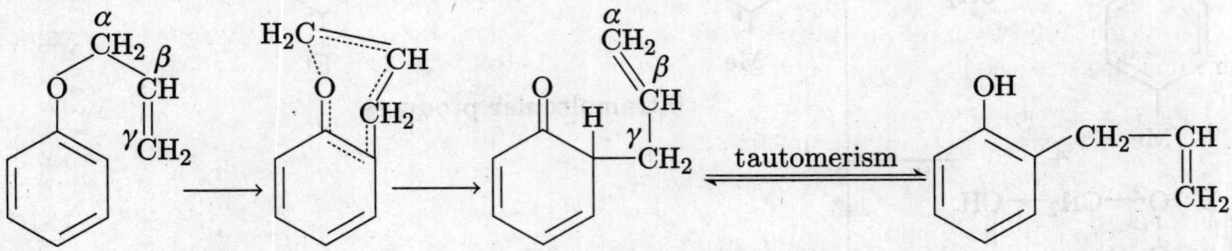

At first a dienone is formed as an intermediate which tautomerises readily to the allylphenol. The latter being aromatic, it is more stable than the non-aromatic dienone and hence it readily gets converted into the allylphenol. Here it is important to note that the γ-C atom of the allylic group becomes the α-C atom of the o-allyphenol.

When the two o-positions are filled, there will be no o-H atoms. In that case the allyl group will first migrate to the o-position and then the second migration of the group will take place to the p-position via a deformed cyclic T.S. Here the α-C in the original ether migrates to the p-C by the double migration. Since the T.S. for the p-product is a deformed one, it contains more energy than the T.S. for the o-product and hence it is less stable. For this reason when one o-position is free, the o-product is obtained instead of the p-product

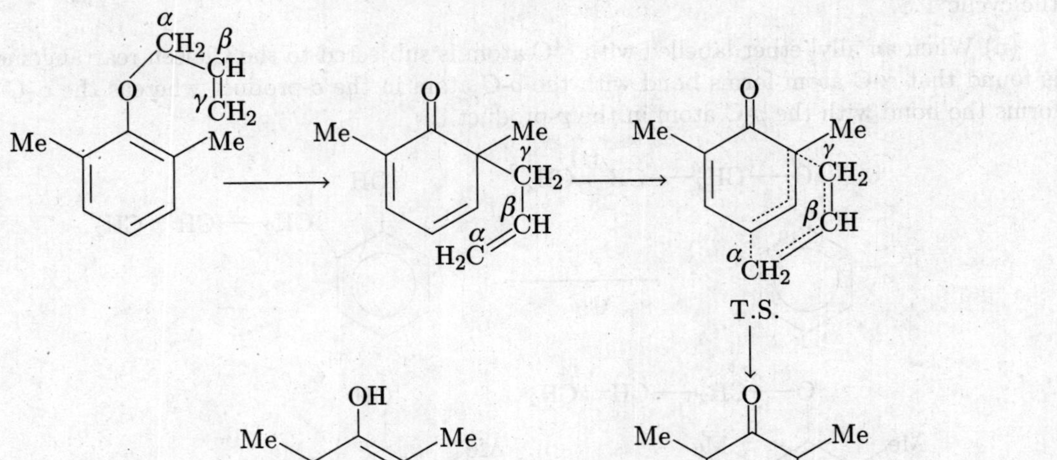

It has been found that the o- or p-H migrates to the O atom in the dienone-phenol rearrangement step to form the o- or p-product respectively.

The following facts support the mechanism as shown above :

(i) The cross-over experiment with two different allyl aryl ethers does not give the cross-over products.

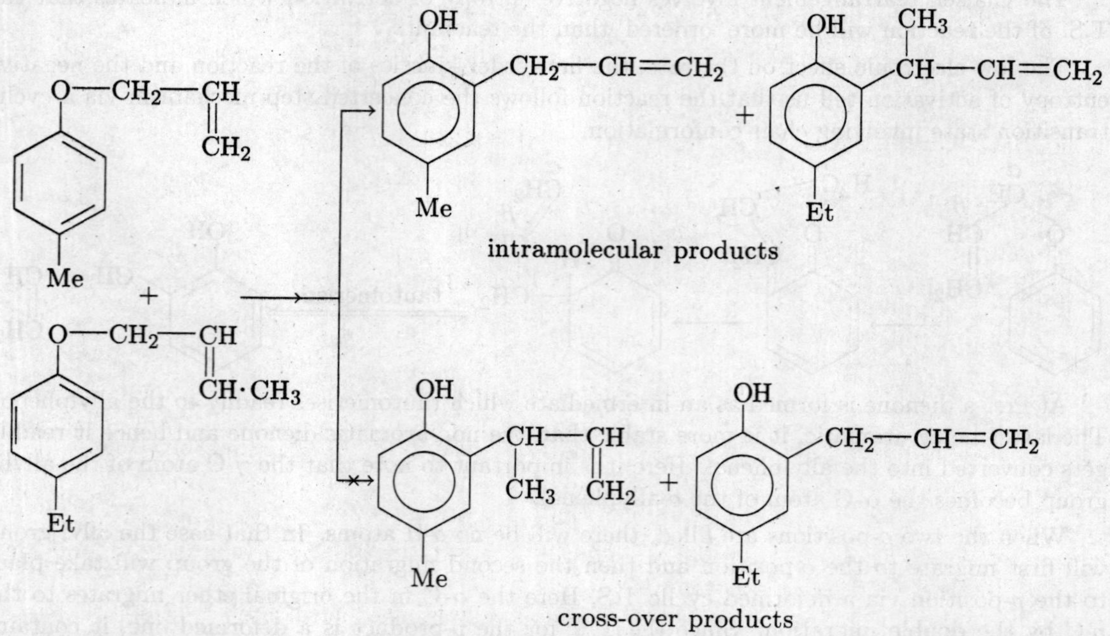

This means that the allyl group does not detach itself from the substrate; this is a strong evidence for the intramolecularity of the reaction. The negative entropy of activation supports the cyclic T.S.

(ii) When an allyl ether labelled with ^{14}C atom is subjected to the Claisen rearrangement, it is found that γ-C atom forms bond with the *o*-C atom in the *o*-product whereas the α-C atom forms the bond with the *p*-C atom in the *p*-product.

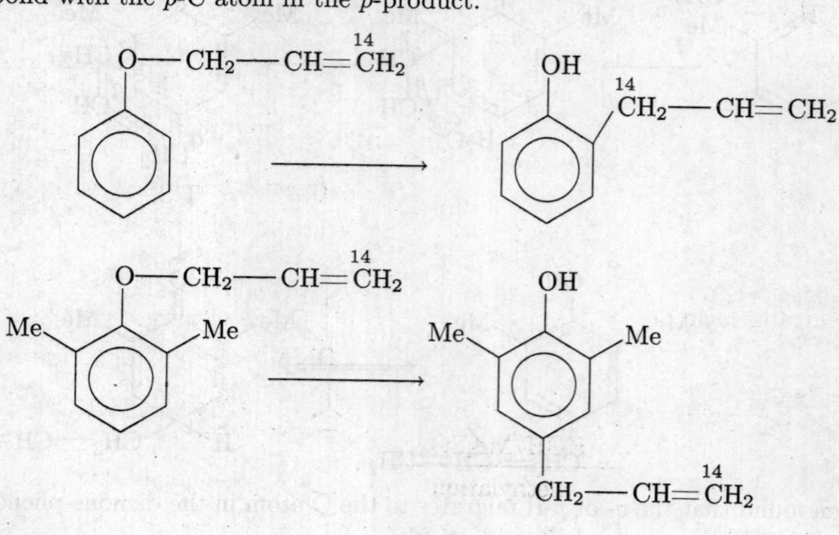

The isotope-labelled compound has been prepared as shown below :

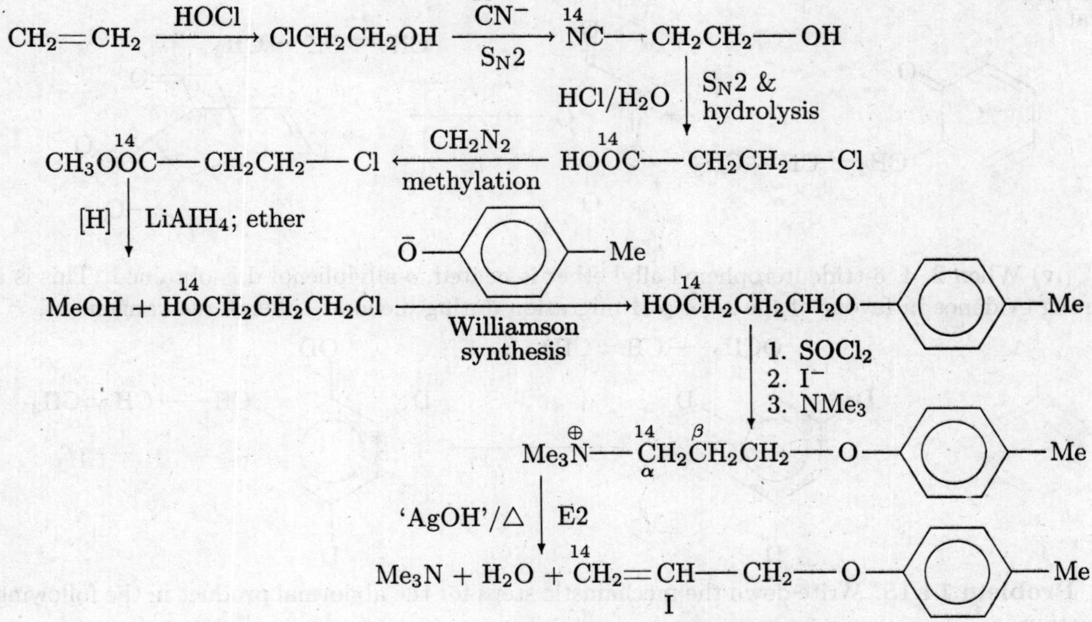

isotopically labelled substrate

When the substrate I is subjected to the Claisen rearrangement and the product is degraded, we get formaldehyde which does not show radioactivity; but the other product of degradation (II) retains total activity. If the α-C, instead of γ^{14}C were migrated, there would have been active formaldehyde as one of the degradation products; thus the experiment shows beyond doubt that it is the γ-C that migrates in the Claisen rearrangement.

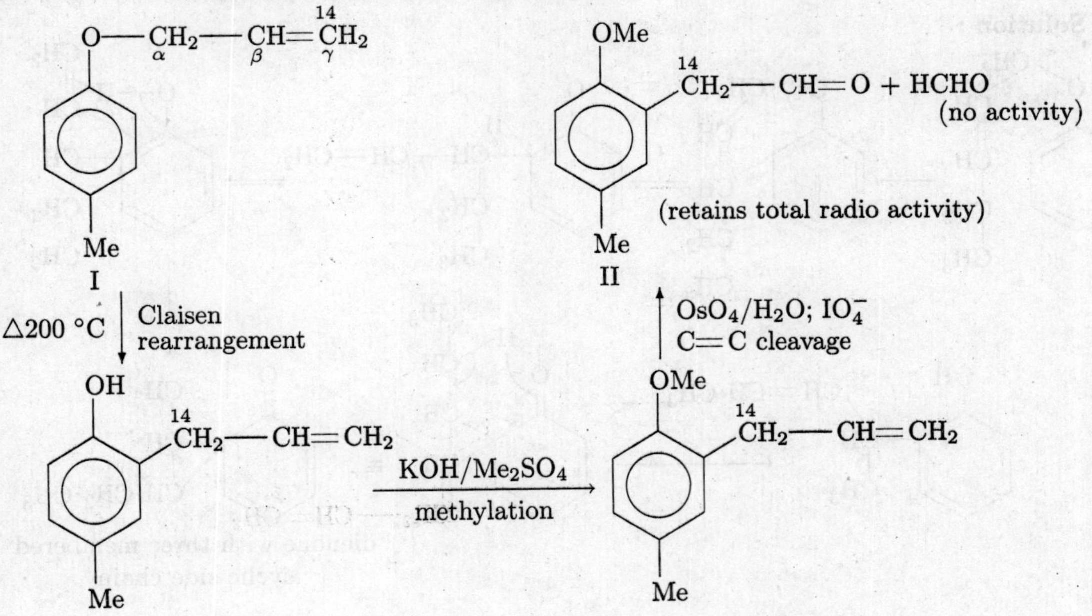

(iii) When the reaction is carried out in the presence of maleic anhydride, a small amount of Diels-Alder adduct is obtained. This supports the formation of dienone derivative as an intermediate.

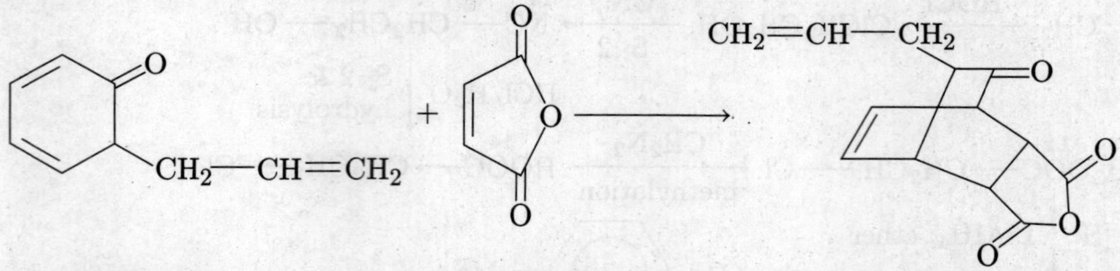

(iv) When 2, 4, 6-trideuterophenyl allyl ether is heated, *o*-allylphenol-d is obtained. This is a strong evidence in favour of the *o*- or *p*-H migration during dienone—phenol rearrangement.

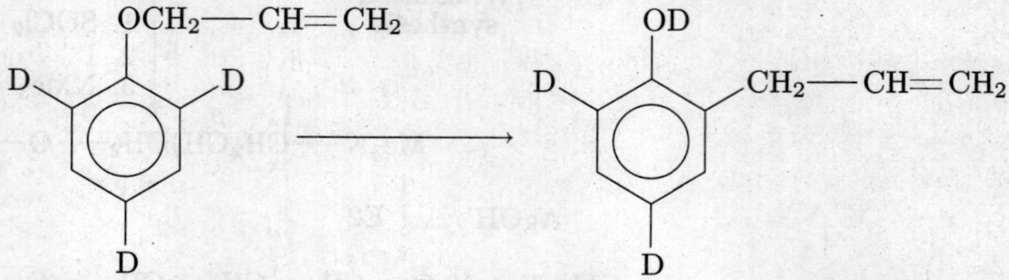

Problem 14.15. Write down the mechanistic steps for the abnormal product in the following reaction.

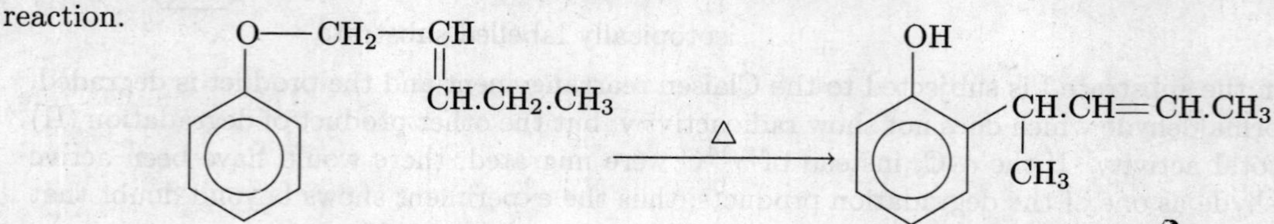

abnormal product

Solution :

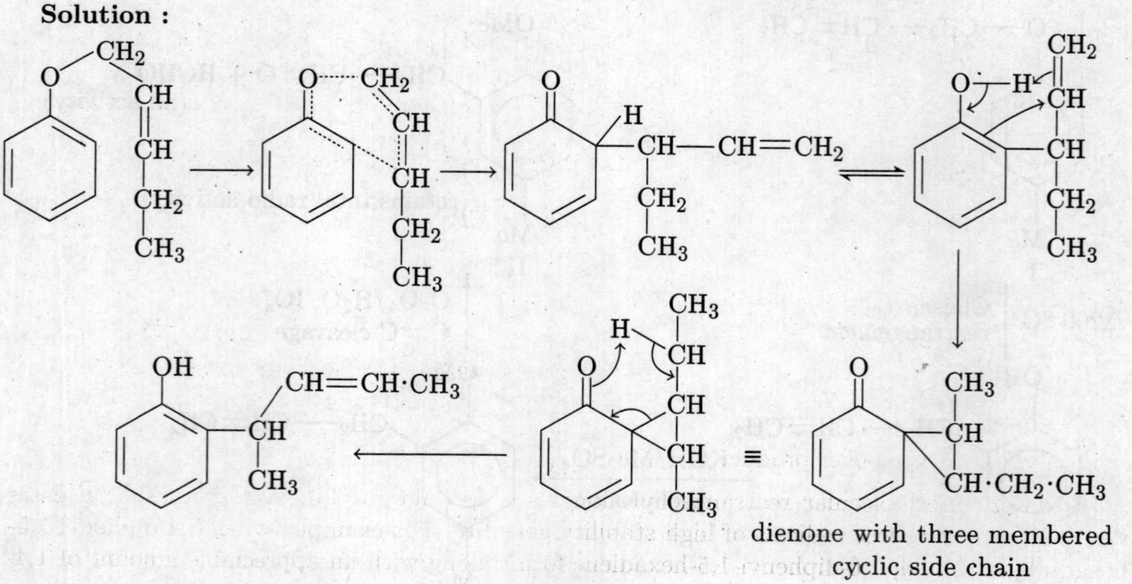

dienone with three membered cyclic side chain

[3, 3]-sigmatropic (Claisen) rearrangement of allyl vinyl ethers also take place through a cyclic six-membered T.S. having chair conformation. The substituent on α-C accommodates itself in the equatorial position and the newly formed double bond between α-C and β-C has E-configuration with respect to γ-C bonded moiety as shown below.

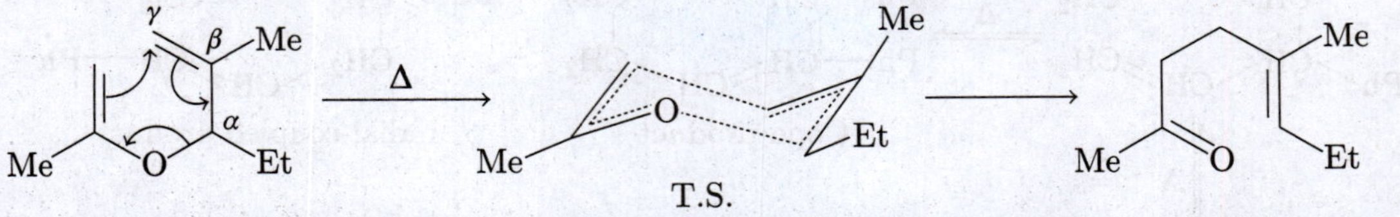

14.17.2 The Cope rearrangement

The cope rearrangement is the most common $[i,\ j]$ sigmatropic rearrangement in which both the ends of migrating σ bond move to 3 positions involving a π framework and, therefore, the reaction is designated as [3,3] sigmatropic rearrangement. When a 1,5-hexadiene is heated in solution or in the vapour phase, an isomeric 1, 5-hexadiene forms by the 3,3-migration of the sigma bond as shown below.

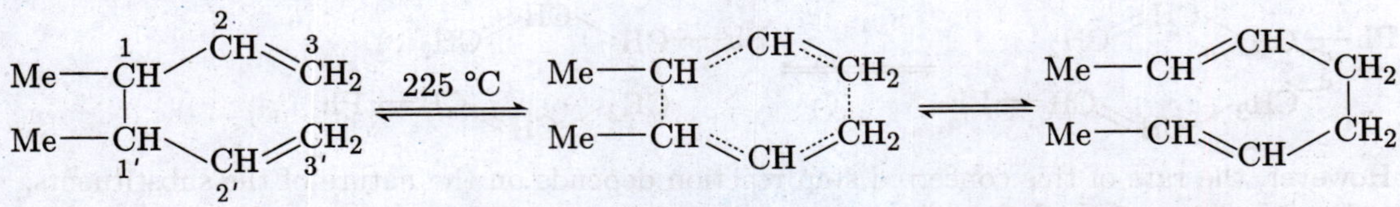

The temperature of the reaction depends on the nature of groups on the C atom that bear the breaking σ bond. It is usually around 200 °C, whereas the activation energy is 105–125 kJ mol^{-1}.

Cross-over experiments show that the rearrangement is really an intramolecular one.

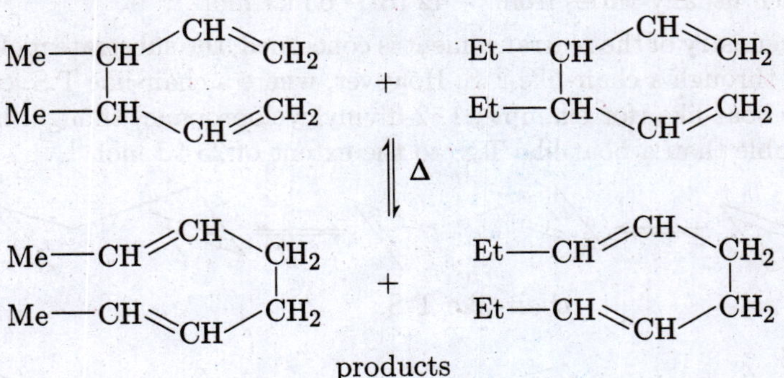

products

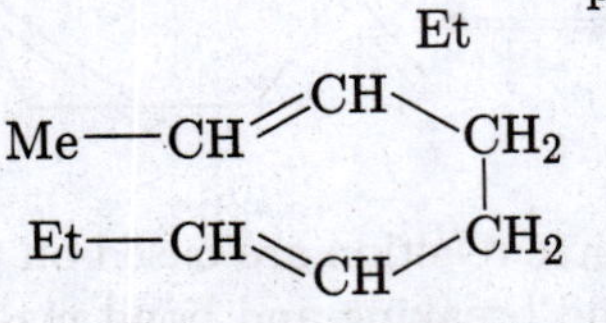

(this cross-over product does not form.)

cross-over product

However, intermolecular rearranged products are also obtained in some very special cases where intermediate free radicals of high stability generate. For example, when 3,4-diphenyl 1,5-hexadiene is heated, 1,6-diphenyl-1,5-hexadiene forms along with an appreciable amount of 1,4-

diphenyl-1,5-hexadiene. The latter product forms through the formation of free radicals followed by their coupling. In this case the radical pathway competes with the concerted step Cope rearrangement.

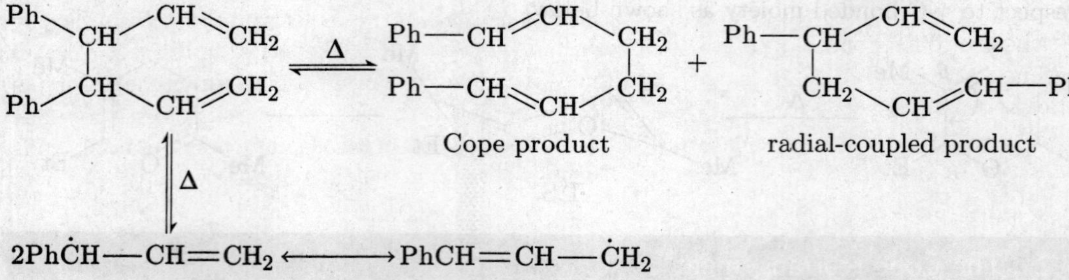

Cope product radial-coupled product

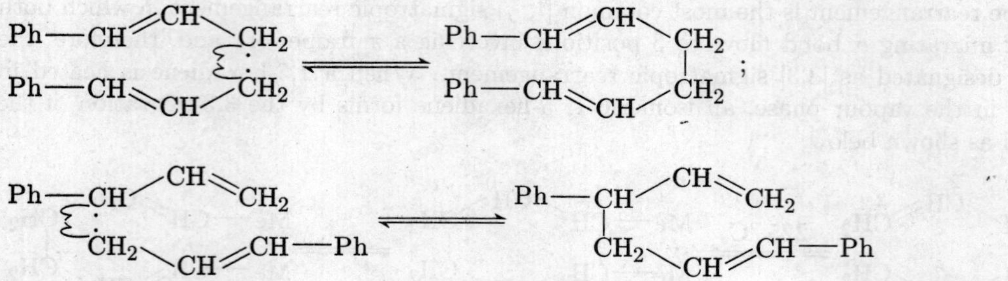

However, the rate of this concerted step reaction depends on the nature of the substituents; phenyl and ester groups and also other groups that conjugate with the π-system in the cyclic T.S. are found to accelerate the reaction.

Since this highly stereoselective reaction follows a concerted step pathway, it has a negative entropy of activation which usually varies from -42 to -63 kJ mol^{-1}.

Insofar as the stereochemistry of the rearrangement is concerned, the substrate molecule passes on to a product molecule through a chair-like T.S. However, where a chair-like T.S. cannot form, the T.S. is found to be a boat-like (for example, 1, 2-divinylcyclopropane) though, in general, a chair-like T.S. is more stable than a boat-like T.S. to the extent of 25 kJ mol^{-1}.

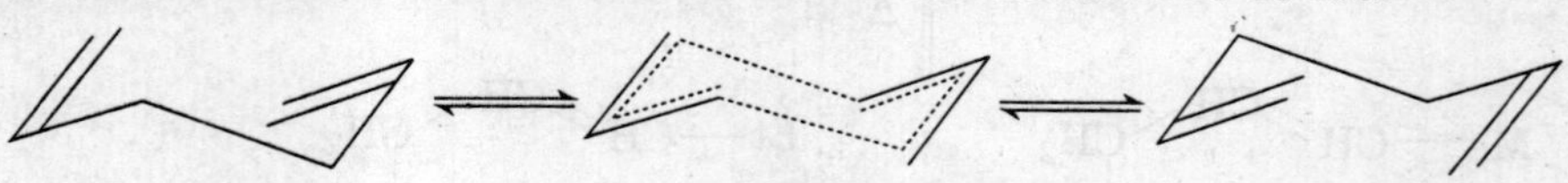

chair-like T.S.

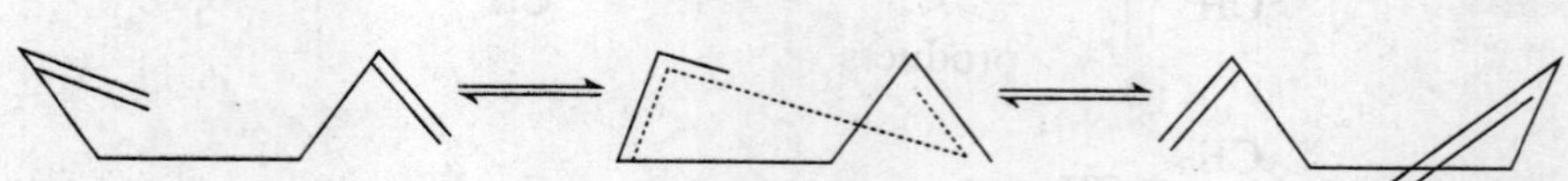

boat-like T.S.

It is needless to mention that the theory of the sigmatropic reaction are based on the assumption that the bond type overlap must exist both at the bond breaking and bond making termini simultaneously in the cyclic T.S; the above T.S.s are drawn according to this basis. However, the cope rearrangement is a $[_{\sigma}2_s + {}_{\pi}2_s + {}_{\pi}2_s]$ cycloaddition reaction (Sec. 13.15) involving a six-electron $[(4q+2);\ q = 1, 2, 3\ldots]$ system. In this highly stereoselective reaction supra-supra interaction takes place in the ground state as stated in the Woodward-Hoffmann selection rules and the mechanism can be explained by HOMO-LUMO interaction (Frontier Molecular Orbital Theory). The involved σ bond along with a π bond may be looked upon as a 4-electron component and the other π bond as a 2-electron component; thus $[_{\sigma}2_s + _{\pi}2_s + _{\pi}2_s]$ cycloaddition becomes [4+2] cycloaddition. The nodal properties of orbitals of the 4-electron component closely resemble those of butadiene. The HOMO of 4-electron component is like the ground state HOMO of butadiene. The LUMO of 2-electron component and the HOMO of 4-electron component (as shown below) are the two Frontier Orbitals that interact with each other in such a way that there always exists bonding type of overlap both at the bond breaking and bond making ends simultaneously in the chair-like T.S. in most of the cases (boat-like T.S. is rare).

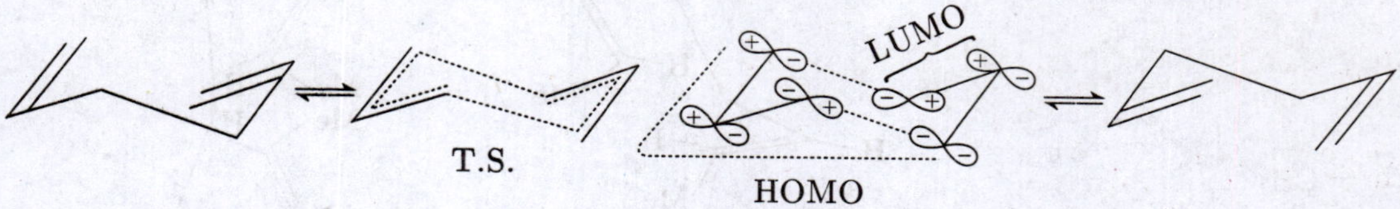

chair-like T.S.

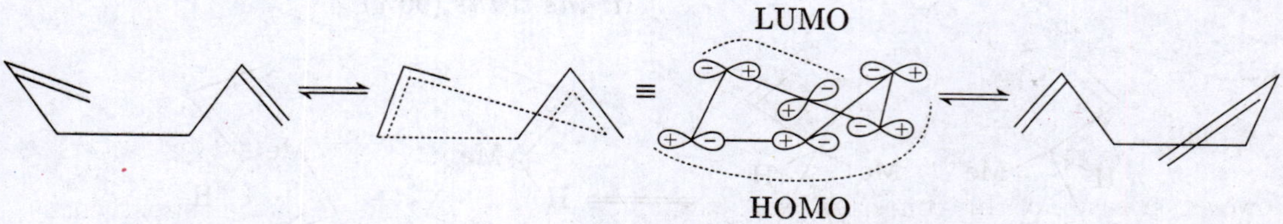

boat-like T.S.

The evidences in the favour of chair-like T.S. are as follows :

Meso 3,4-dimethyl-1, 5-hexadiene on heating isomerises to a mixture of cis, trans-2,6-octadiene (99.7%) and trans, trans–2, 6–octadiene (0.3%). The chair-like T.S. only accounts for the highly preferred cis-trans isomer. Again when (±)-3,4-dimethyl-1,5-hexadiene is heated, a mixture of 90% trans-trans, 10% cis-cis and less than 1% cis-trans 2,6-octadienes forms. This result also indicates that the T.S. of the reaction is chair-like.

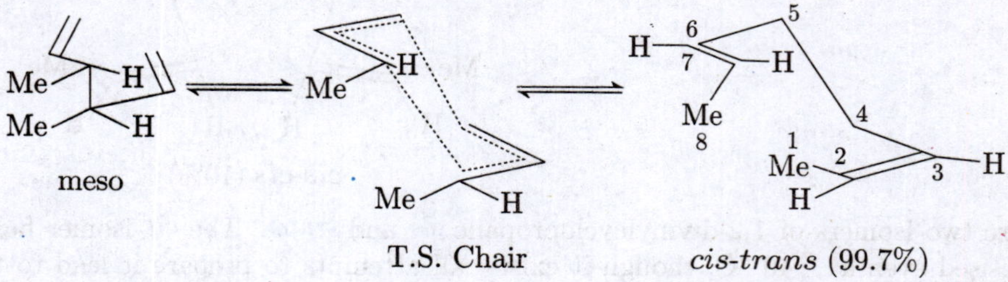

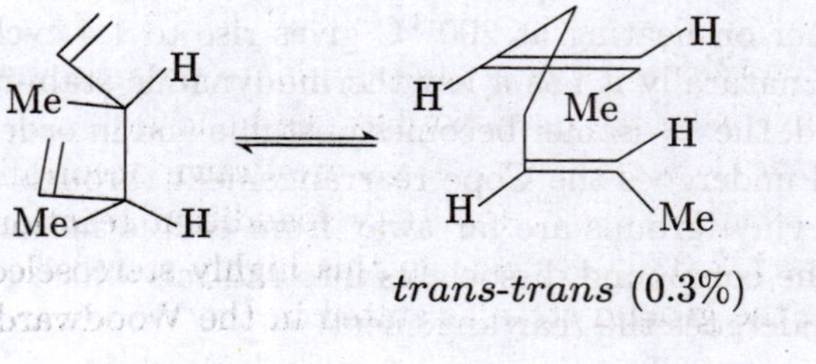

trans-trans (0.3%)

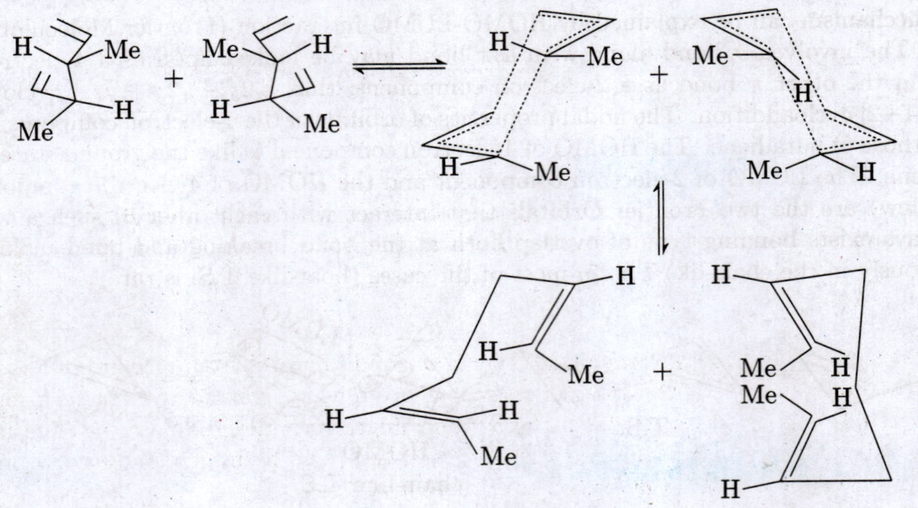

trans-trans (90%)

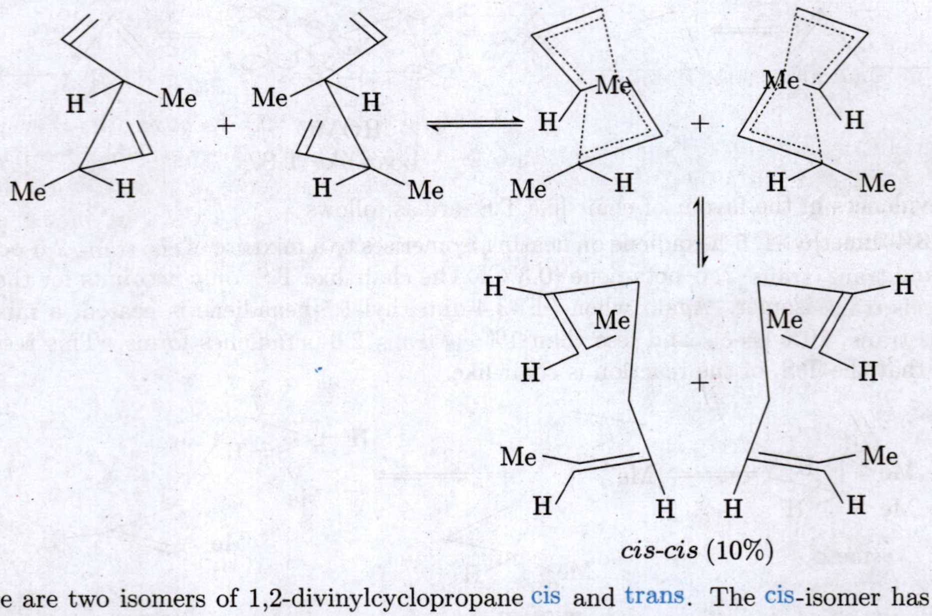

cis-cis (10%)

There are two isomers of 1,2-divinylcyclopropane cis and trans. The cis-isomer has not yet been synthesised even at −40 °C, though it exists; all attempts to prepare it lead to the Cope rearrangement product 1, 4-cycloheptadiene. The trans-isomer, on the other hand, has been synthesised; this isomer on heating at 200 °C gives rise to 1,4-cycloheptadiene. The cis-isomer bears high strain and naturally it has a low thermodynamic stability. When the temperature of the system is increased, the cis-isomer becomes unstable and in order to get stability, by releasing strain, the compound undergoes the Cope rearrangement through a cyclic boat-like T.S. In the trans-isomer the two vinyl groups are far away from each other and cannot form a cyclic T.S. However, at 200 °C the compound dissociates into radicals which couple to form the cis-isomer; the cis-isomer then undergoes the rearrangement.

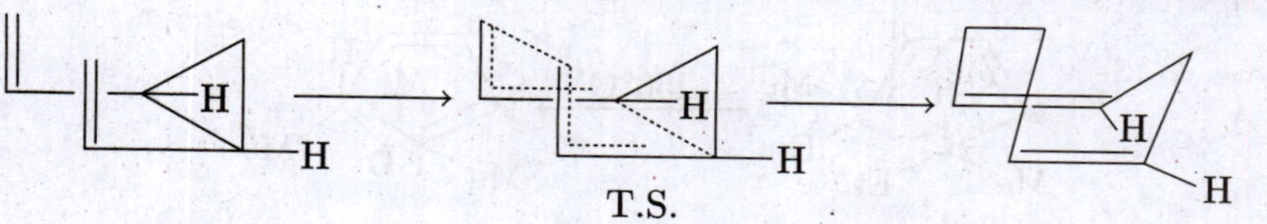

T.S.
boat-like

So, the above reaction is an example of the Cope rearrangement involving a boat-like T.S. This reaction also illustrates the fact that the rearrangement is essentially an intramolecular one.

14.17.3 Sigmatropic [1, j] hydrogen shift

In these sigmatropic rearrangements a C—H σ bond migrates from one end of a conjugated polyene system to the other. The H atom and the C atom which constitute the migrating σ bond are numbered 1. The C atom to which the σ bond migrates is designated as j. The migrating σ bond is indicated by a thick line and the shift is called [1, j] hydrogen shift. For example the equation given below is a [1, 5] hydrogen shift, here $j = 5$

$$\underset{1}{\overset{\text{H}}{\text{R}_2\text{C}}}-\underset{2}{\text{CH}}=\underset{3}{\text{CH}}-\underset{4}{\text{CH}}=\underset{5}{\text{CR}'_2} \xrightarrow{\Delta} \text{R}_2\text{C}=\text{CH}-\text{CH}=\text{CH}-\overset{\text{H}}{\text{CR}'_2}$$

Here we shall discuss [1, 5] and [1, 3] hydrogen shifts only.

(a) Sigmatropic [1, 5] hydrogen shift : According to the FMO approach the thermal [1, 5] hydrogen shift may be looked upon as a [$\sigma 2s + \pi 4s$] cycloaddition between the C—H σ bond and the remaining diene system of the molecule. In a cycloaddition we have considered HOMO of one molecule and LUMO of the other. In this case we are to consider LUMO of the diene system and HOMO of C—H σ bond. In this thermal process supra, supra interactions are symmetry allowed.

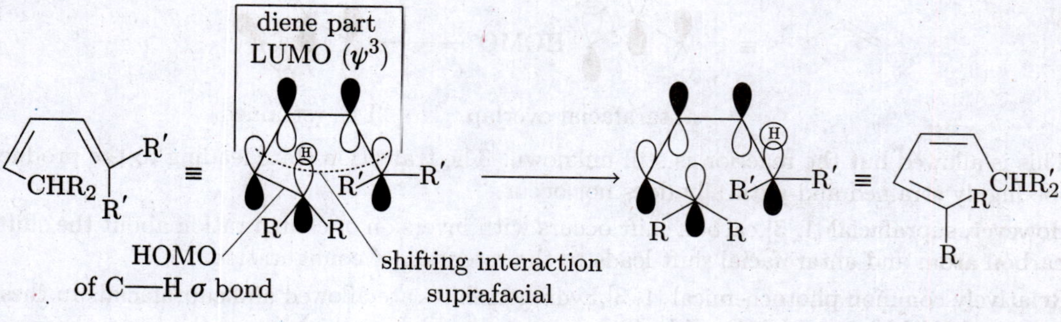

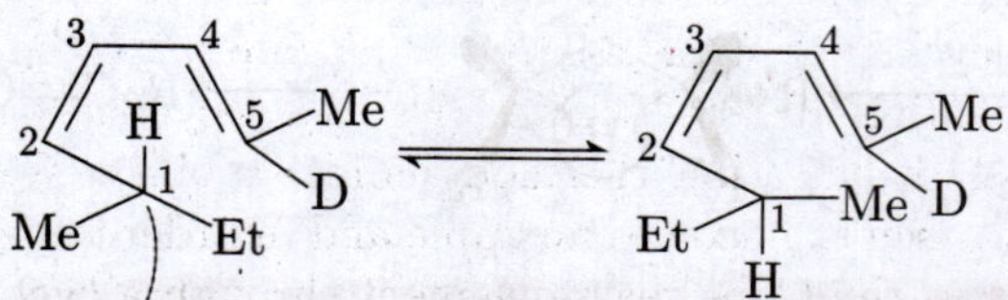

chiral centre
s-cis
1

conformational change
around chiral centre; *s*-cis
2

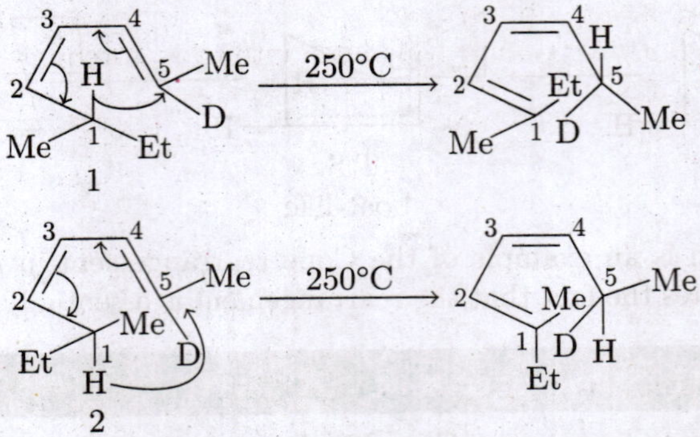

1 and 2 are conformations of the diene derivative in *s*-cis form; C_1 atom is a chiral centre. On heating them at 250°C the chirality shifts form C_1 to C_5 as [1, 5] hydrogen shift takes place. The hydrogen shift is suprafacial.

However, the most common [1, 5] hydrogen shift is found in the cases of mono-substituted cyclopentadiene system as given below.

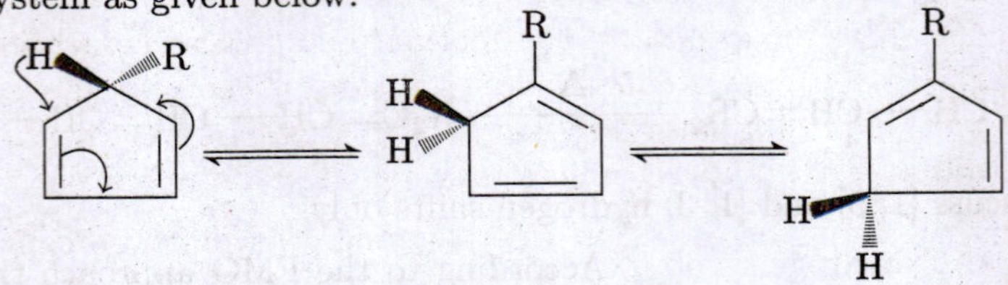

This occurs at room temperature. It has been argued that close proximity of the shifting H atom and the C atom (where migration takes place) is the reason for easy shifting of H atom.

Let us take the example of propene in which [1, 3] hydrogen shift thermally is possible or not.

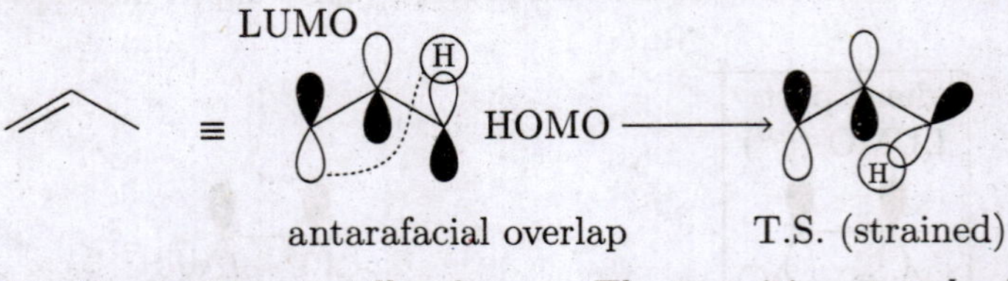

This is allowed but the reaction is still unknown. The transition state leading to the product will be highly strained and naturally does not occur.

However, suprafacial [1, 3] carbon shift occurs with inversion of configuration about the shifting carbon atom and antarafacial shift leads to the retention of configuration.

Relatively common photochemical [1, 3] hydrogen shifts are allowed and suprafacial. In these cases excited HOMO i.e., LUMO of the T.S. has terminal lobes which are of the same phase

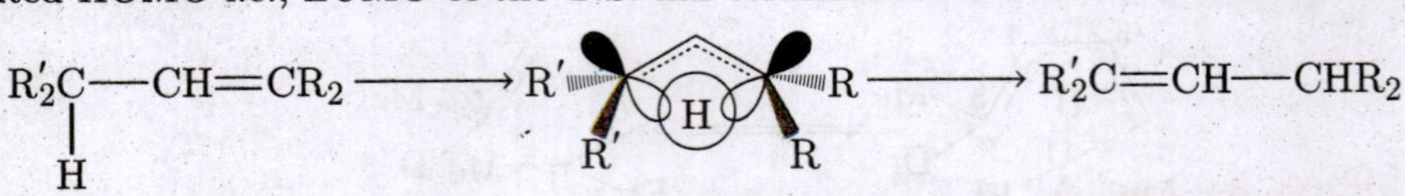

14.18 The Electrocyclic Rearrangements

An intramlecular thermal or photochemical reaction of a conjugated polyene system involving a cyclic transition state in which two π bonds are converted into a σ bond and a π bond reversibly with conservation of orbital symmetry is called electrocyclic rearrangement reaction. The reversible and thermal conversion of butadiene into cyclobutene is an example of this type of reactions.

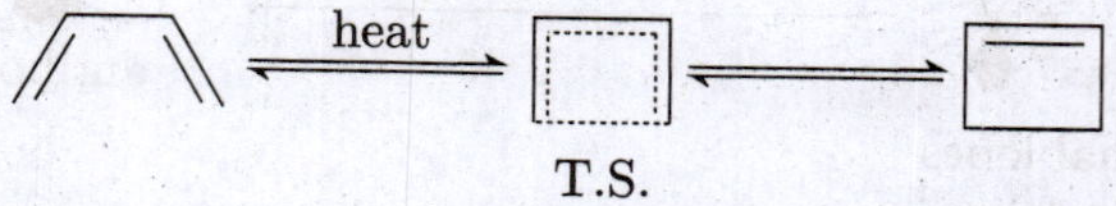

Conjugated di-, tri- and polyenes undergo this sort of reactions. FMO approach tells us that in the cases of thermal reaction HOMO of the polyene system is to be taken under consideration. On the other hand, the excited HOMO i.e., LUMO is to be considered in the cases of photochemical conversions. However, the mechanistic pathway and the stereochemistry of the electrocyclic rearrangements are well understood by the Woodward-Hoffmann's selection Rules (Table 14.1). Empirically the principle may be stated as follows.

Two terminal lobes of the conjugated polyene system rotate through an angle of 90° and then they combine to form a C—C σ bond. The rotation of the lobes should take place in such a way that the two like-sign lobes can overlap end-to-end to form a bonding σ orbital. For the thermal electrocyclic reactions involving $4n\pi$es (where $n = 1, 2, 3$, etc.), the terminal lobes undergo rotation in the same direction and the motion is known as **conrotatory**; while for the photochemical reaction in such a system, the lobes rotate in the opposite directions and the rotation of the lobes is called **disrotatory**. Systems involving $(4n + 2)\pi$es thermal conversions involve disrotatory motion, while photochemical conversions involve conrotatory motion.

Table 14.1 Woodward-Hoffmann's Selection rules (symmetry elements of involved M.O.s and mode of rotation)

No. of involved πes ($n = 1, 2, 3, \ldots$)	Reaction condition	Mode of rotation of orbitals for bonding	Involved M.O. (symmetry element)
$4n$	Thermal	Conrotatory	HOMO (C_2)
$4n$	Photochemical	Disrotatory	LUMO (σ)
$4n+2$	Thermal	Disrotatory	HOMO (σ)
$4n+2$	Photochemical	Conrotatory	LUMO (C_2)

Let us first illustrate conrotatory and disrotatory motions and the results we get therefrom.

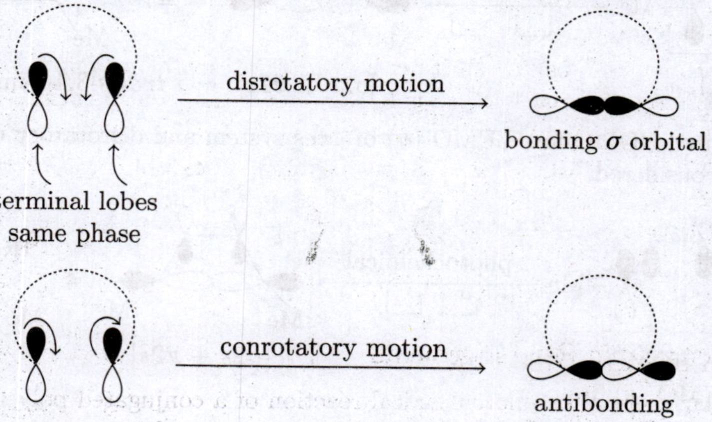

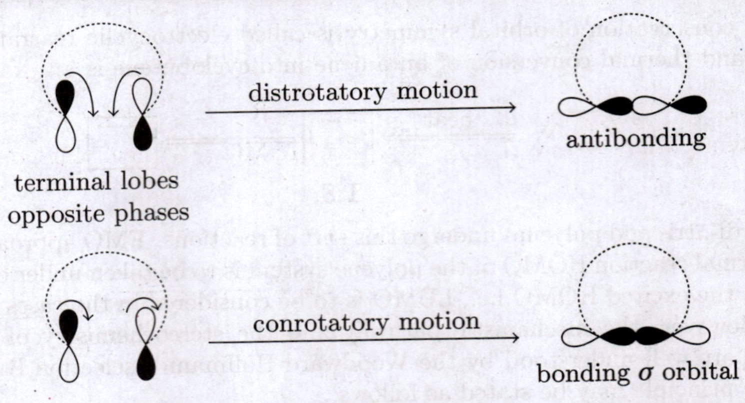

Thus we see that if the terminal lobes are of same phase, their disrotatory motion gives rise to a bonding σ orbital. In this case a conrotatory motion develops an antibonding situation. On the other hand, a conrotatory motion of terminal lobes of opposite phases gives rise to a bonding σ orbital while their disrotatory motion creates an antibonding situation. However, some examples are given below.

1. trans, trans-2,4-Hexadiene 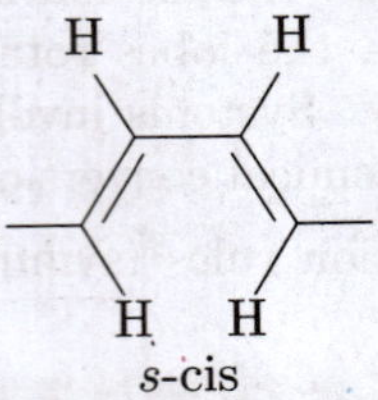 It contains $4\pi es$ $(n = 1)$

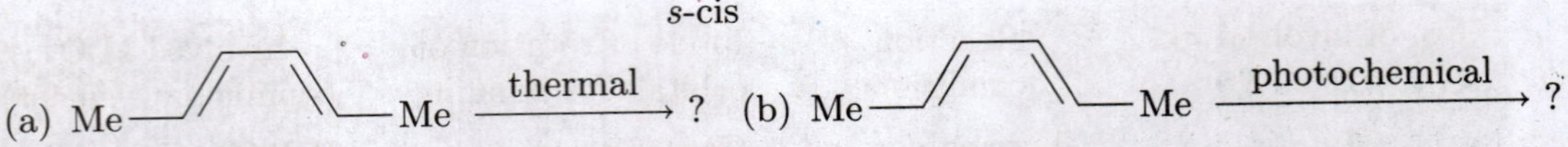

Let us apply Woodward-Hoffmann selection rule as given in Table 14.1 to get the products.

(a) We shall have to consider HOMO (C_2) and conrotatory motion of the terminal lobes in the thermal change.

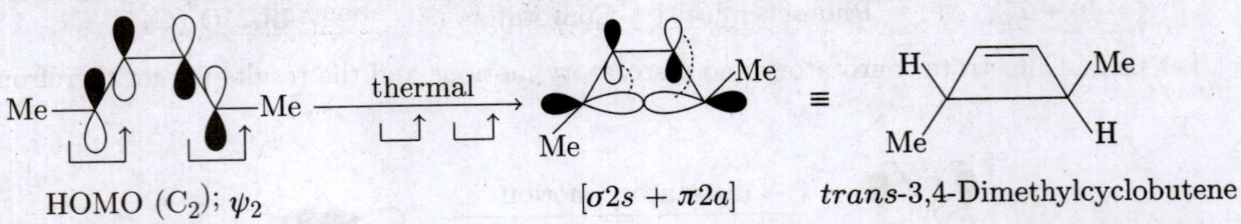

(b) Here excited HOMO i.e., LUMO (σ) of $4\pi es$ system and disrotatory motions of terminal lobes are to be considered.

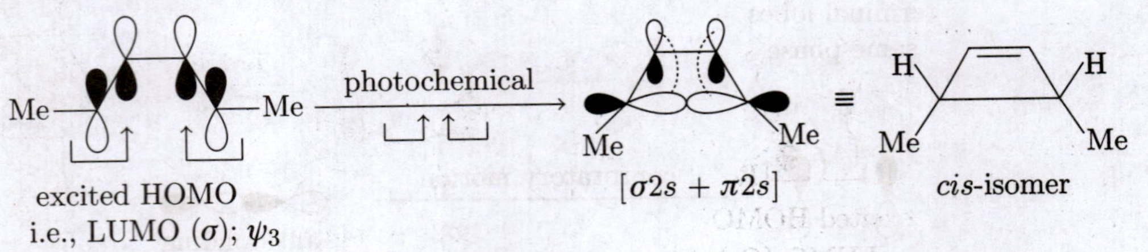

2. (a) 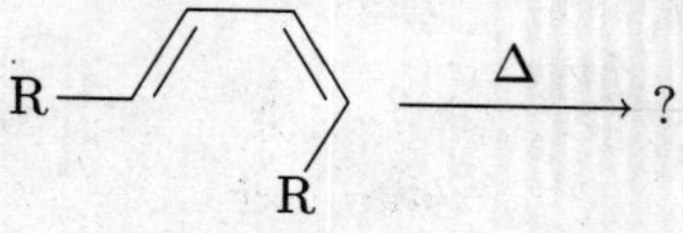

(b) 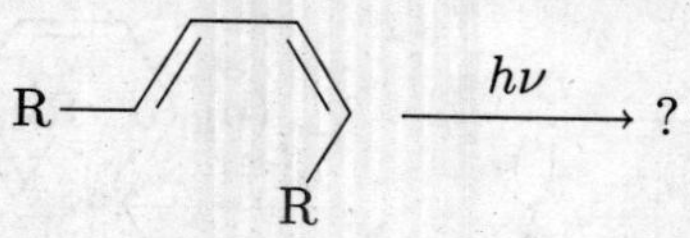

(a) $4n$ π-system. This thermal change occurs on HOMO (C_2) involving conrotatory motion of the terminal lobes.

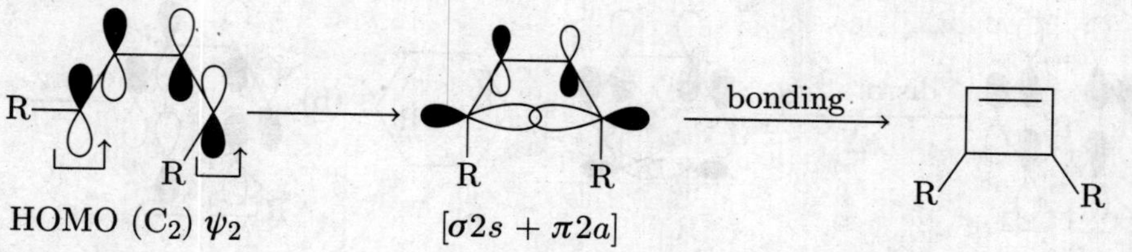

(b) $4n$ π-system. The photochemical change on excited HOMO (ψ_3) i.e., LUMO (σ).

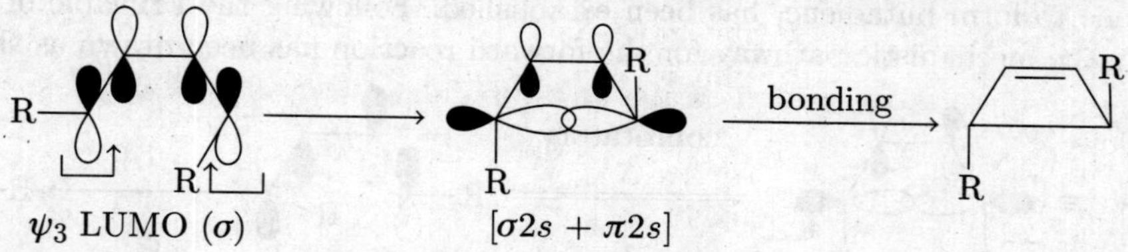

3. (a) 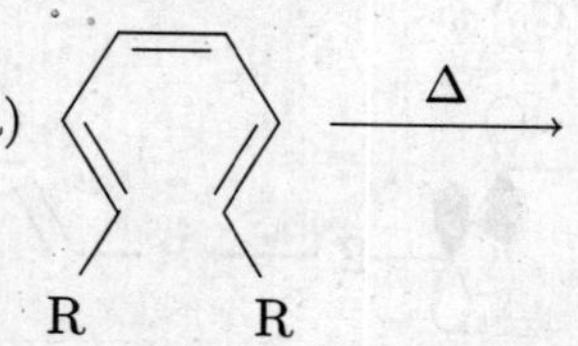

(b) 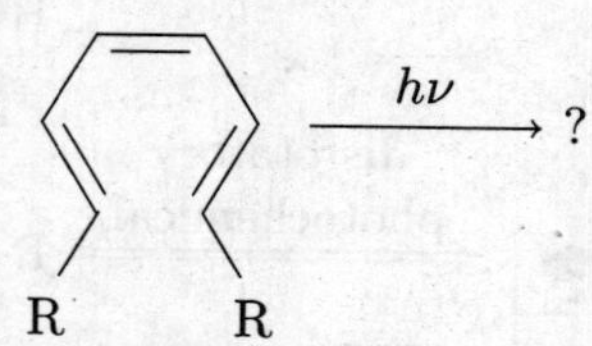

(a) this is a $(4n+2)$ π-system. According to Table 14.1, the thermal change involves HOMO (σ) of the π-system and disrotatory motion of terminal lobes.

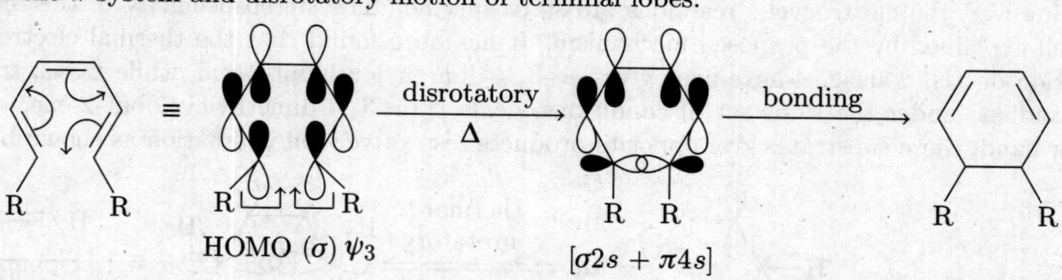

(b) According to Table 14.1, the photochemical change involves LUMO (C_2) of the π-system and conrotatory motion of the terminal lobes.

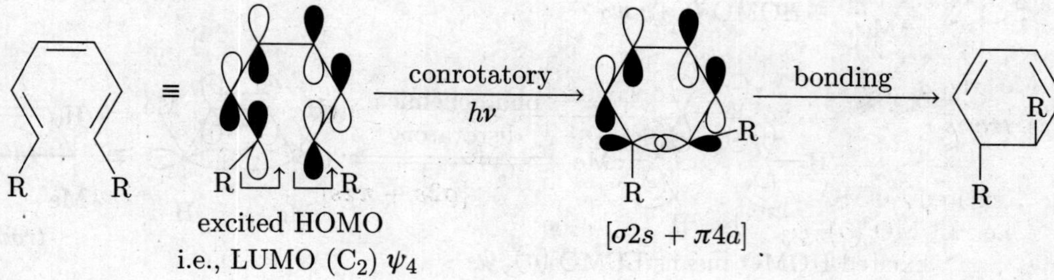

4. (a) 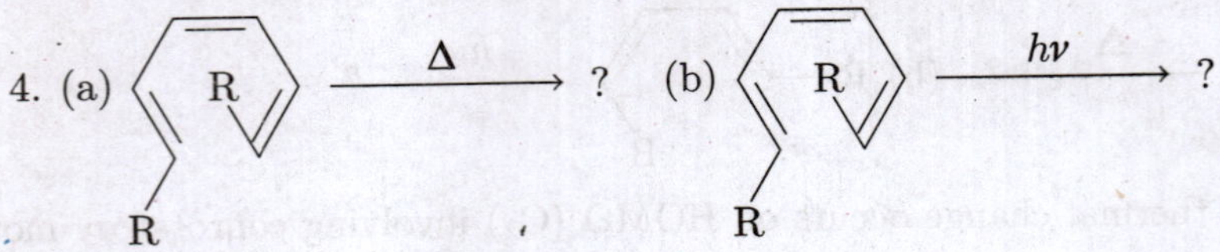

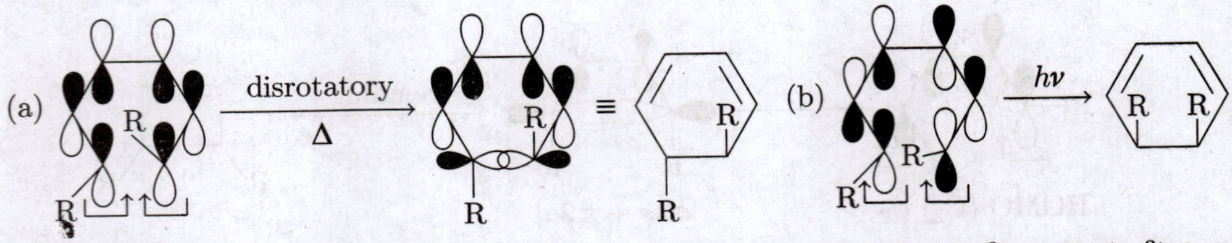

Actually, the mechanism of the reverse reaction, the cleavage of C (sp^3)—C (sp^3) bond of cyclobutene to form butadiene, has been established. Following the Principle of Microscopic Reversibility the mechanistic pathway for the forward reaction has been drawn as shown above.

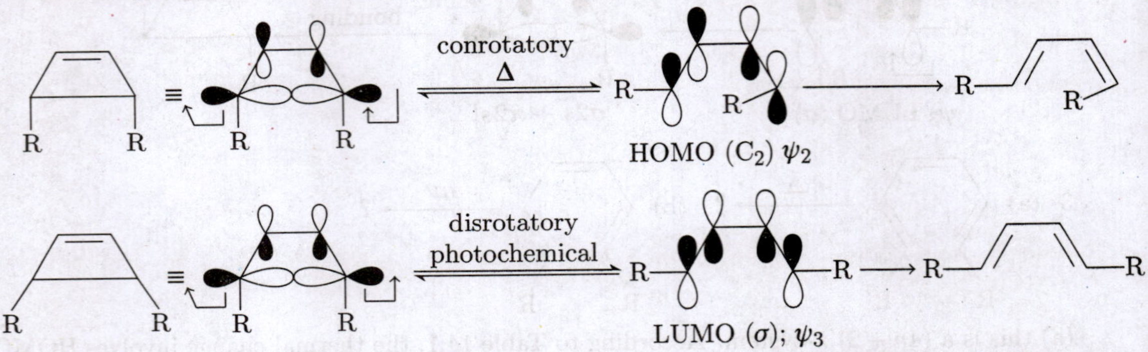

However, the electrocyclic reactions are stereospecific. The stereospecificity of the reaction is well explained by the proposed mechanism. It has been found that the thermal electrocyclic reaction on cis, trans-2, 4-hexadiene gives cis-3, 4-dimethylcyclobut-2-ene, while trans, trans-2, 4-hexadiene under the same set of conditions yields trans-3, 4-dimethylcyclobut-2-ene; on the other hand, these substrates give opposite products respectively on irradiation as shown below :

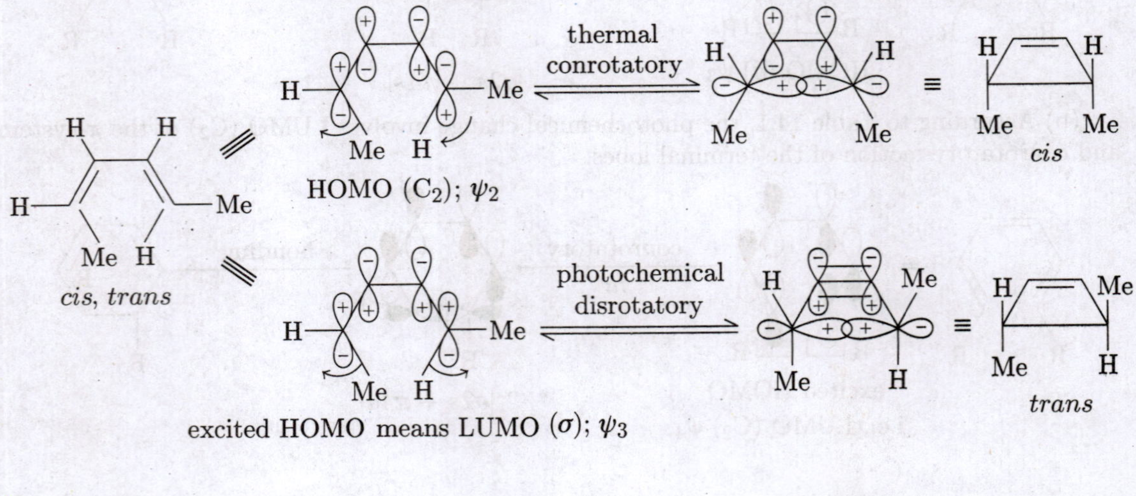

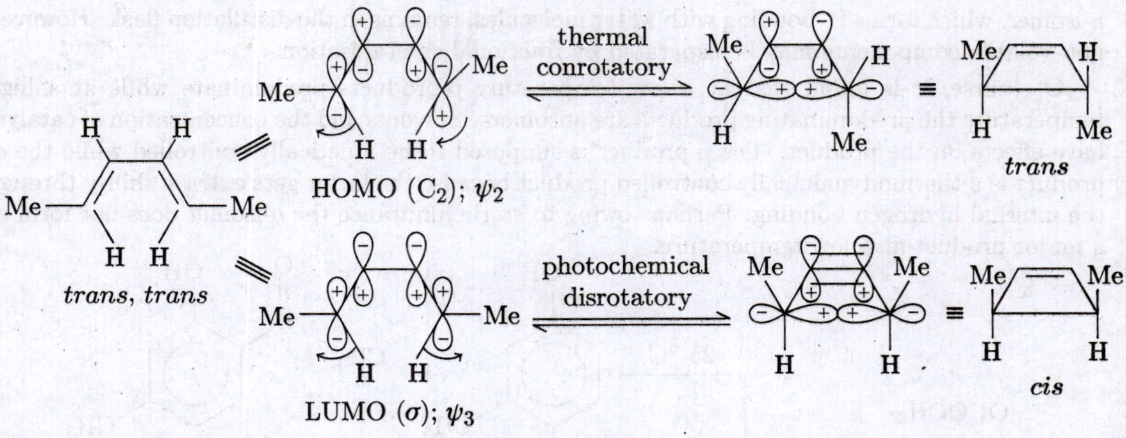

14.19 The Fries Rearrangement

When a phenolic ester is heated in an inert solvent in the presence of a Lewis acid as catalyst and the product is subsequently hydrolysed, a mixture of *o*- and *p*-acylphenols is obtained and the reaction is known as the Fries rearrangement.

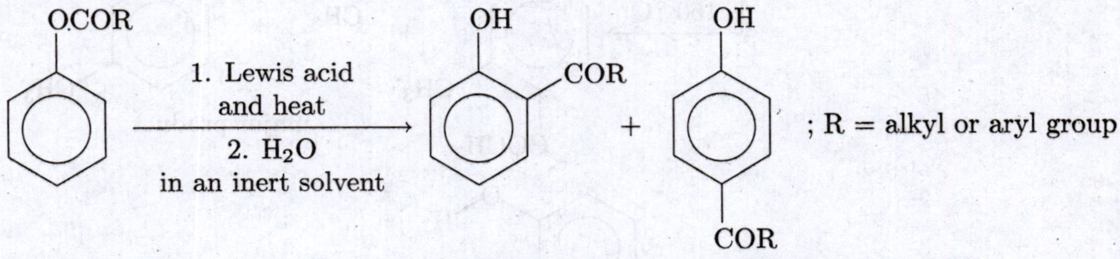

o-Acylphenol *p*-Acylphenol

The Fries rearrangement also occurs without a catalyst in the presence of ultraviolet light and the reaction is known as photo Fries rearrangement. Thus when a solution of phenyl acetate in ethyl alcohol is irradiated with ultraviolet light, *o*- and *p*-acetylphenols are formed along with a small amount of phenol.

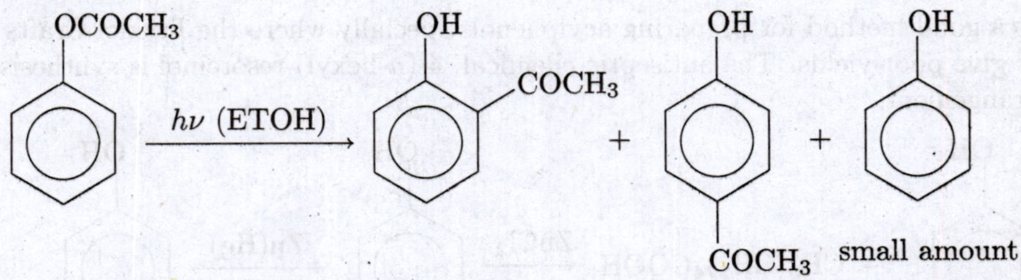

small amount

For Fries rearrangements reaction the common catalyst is anhydrous aluminium chloride and the common solvents are carbon disulphide, nitrobenzene, etc. The *o*- and *p*-isomers may be separated by the process of steam distillation. Since the *o*-isomer possesses internal hydrogen bonding but the *p*-isomer does not, the former has a lower b.p., a higher volatility and a higher vapour pressure than the latter. Naturally, the *o*-isomer distills out with steam while the *p*-isomer, which forms H-bonding with water molecules, remains in the distillation flask. However, non-volatile components may be separated by fractional crystallisation.

Of course, it is found that at a low temperature *p*-products predominate while at a high temperature the predominating products are *o*-isomers. Solvents and the concentration of catalyst have effects on the product. The *p*-product is supposed to be kinetically controlled while the *o*-product is a thermodynamically controlled product because the latter gets extra stability through the internal hydrogen bonding. Perhaps owing to steric hindrance the *o*-isomer does not form as a major product at a low temperature.

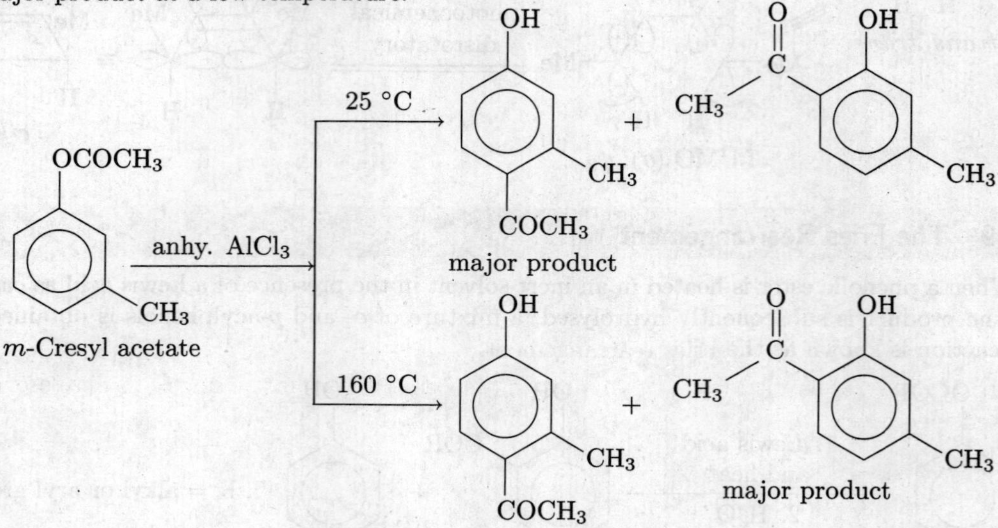

m-Cresyl acetate; major product (25 °C); major product (160 °C)

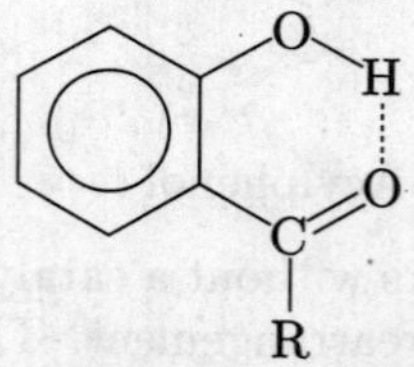

When a *m*-directing group is present in the aromatic ring, it interferes with the reaction.

In some cases *m*-products are obtained in the Fries rearrangement.

This is a good method for preparing acylphenols specially where the Friedel Crafts acylation of phenols give poor yields. The antiseptic chemical, 4-(*n*-hexyl)-resorcinol is synthesised by the Fries rearrangement.

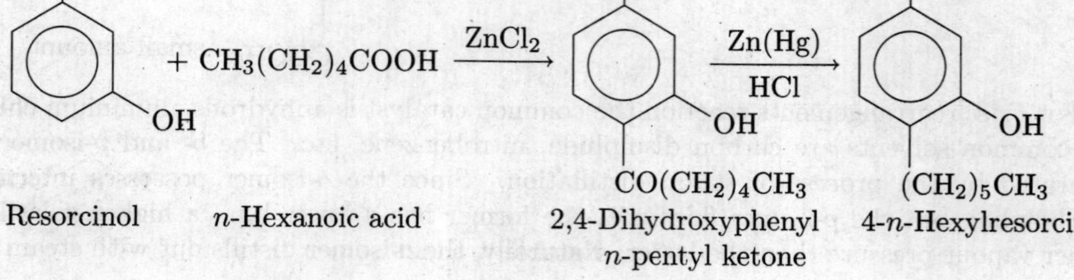

Resorcinol | *n*-Hexanoic acid | 2,4-Dihydroxyphenyl *n*-pentyl ketone | 4-*n*-Hexylresorcinol

The reaction is supposed to follow both the intermolecular and intramolecular mechanisms simultaneously. However, the initial step in both the cases is the complexation of the phenolic ester with the Lewis acid.

Initial Step : Complex formation :

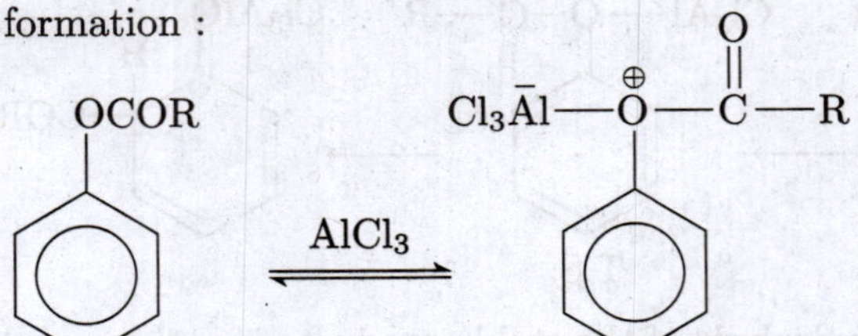

(a) Intermolecular mechanism :

Step I. Reversible formation of the acyl cation.

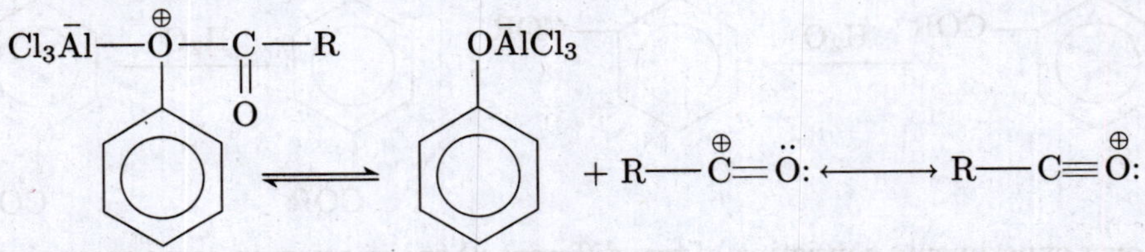

Step II. Electrophilic attack at the *o*- and *p*-positions of the aromatic ring to form a resonance stabilised σ-complex.

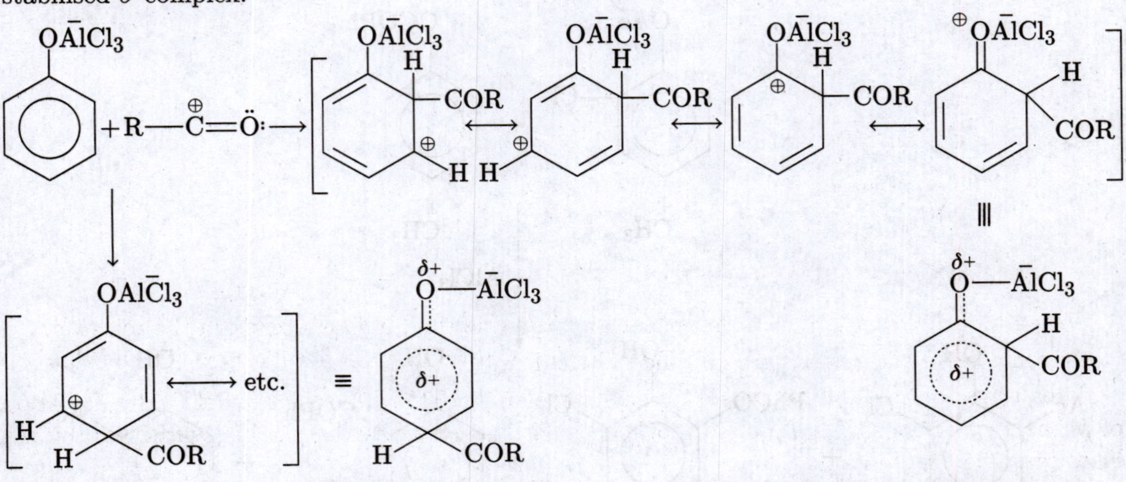

Step III. Elimination of HCl from the σ-complex to form a more stable aromatic product.

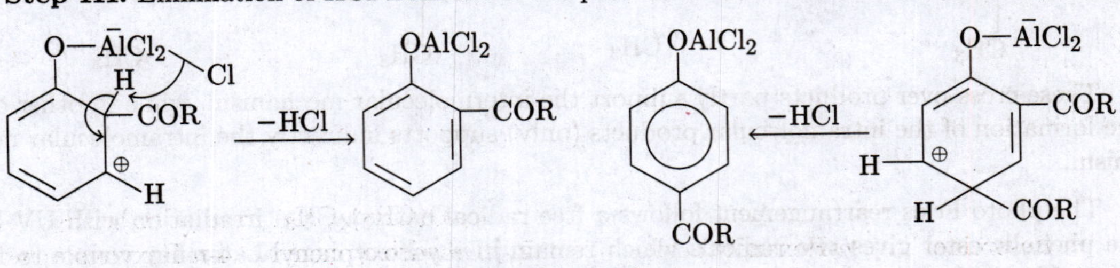

a stable product

(b) Intramolecular mechanism :

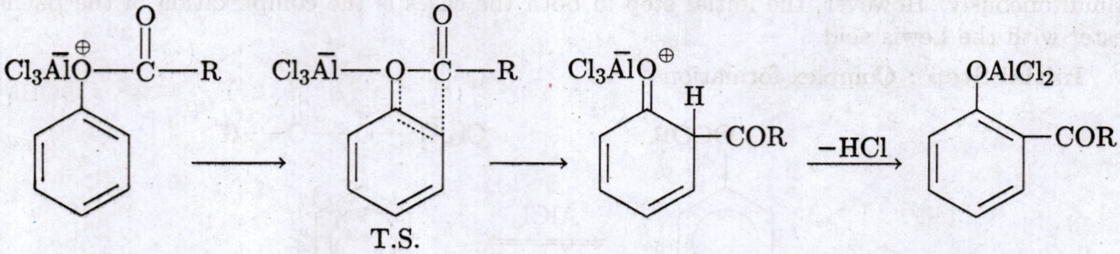

Final Step : Hydrolysis of the stable product gives the acylphenol.

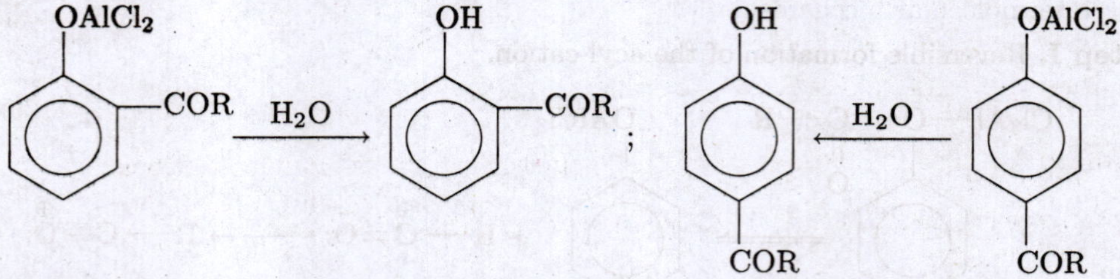

Cross-over experiments consisting of two different phenolic esters in some cases give cross-over products and in some other cases only the normal intramolecular products are obtained :

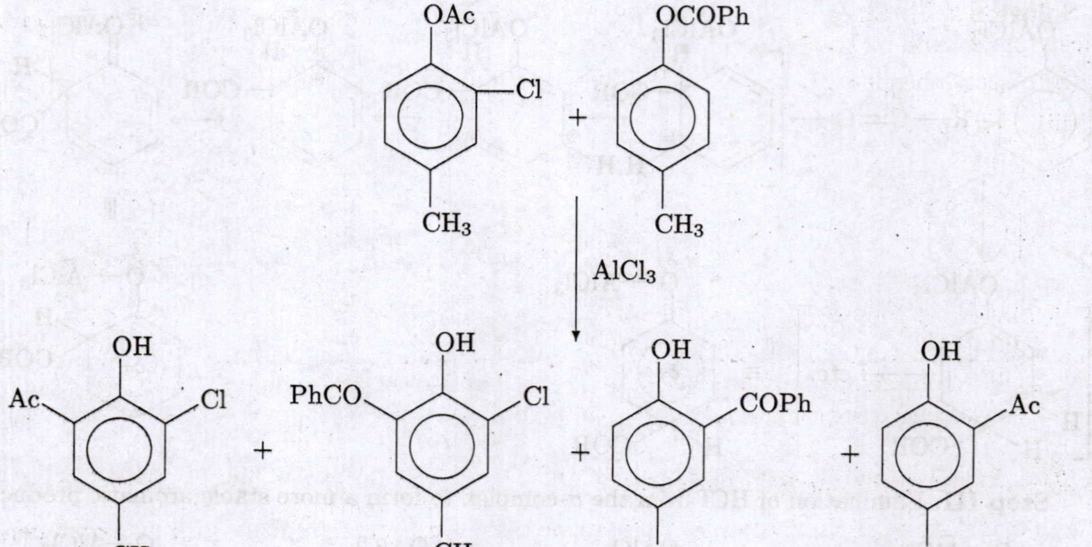

These cross-over products partly support the intermolecular mechanism, while in other cases the formation of the intramolecular products (only) supports indirectly the intramolecular mechanism.

The photo Fries rearrangement follows a free radical pathway. On irradiation with UV light the phenolic ester gives two radicals which remain in a solvent cage. Later hey couple to form the stable products.

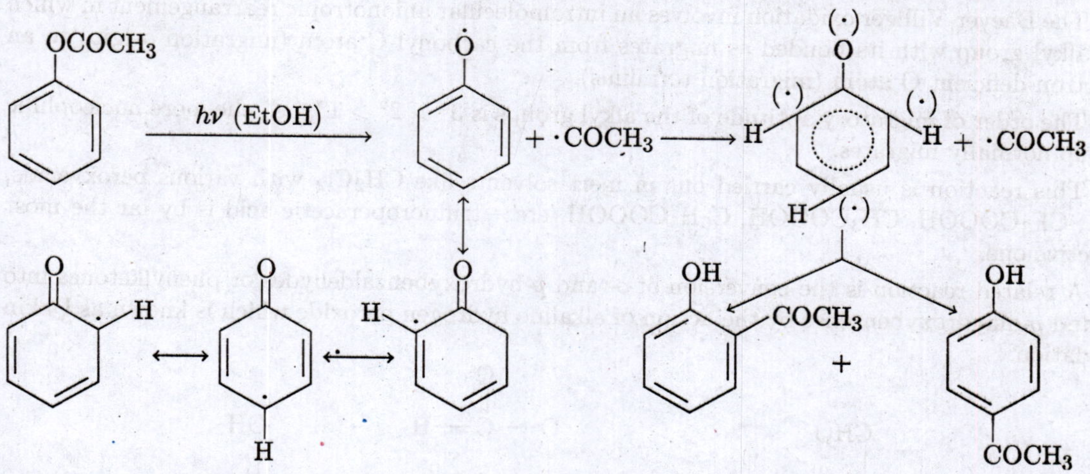

Problem 14.16. What products do you expect from the following reactions :

(i) $PhOCOCH_3 \xrightarrow[60\ °C]{AlCl_3}$? **(ii)** $PhOCOCH_3 \xrightarrow[165\ °C]{AlCl_3}$?

(iii)

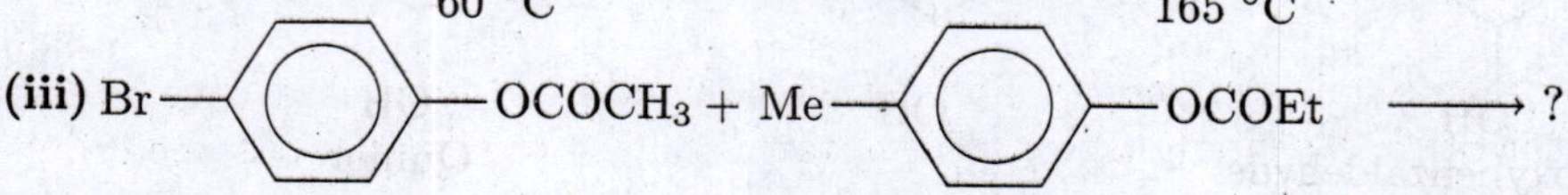

Solution : (i) Since at a low temperature the *p*-product forms in a Fries rearrangement, *p*-acetylphenol will be the predominant product.

(ii) *o*-acetylphenol.

(iii)

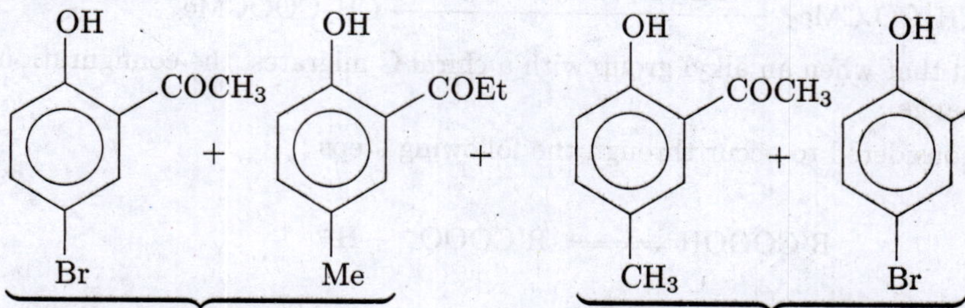

intramolecular products (non-cross-over products) cross-over products

14.20 The Baeyer-Villiger Oxidation

The oxidation of aliphatic ketones to esters or their hydrolysed products with hydrogen peroxide or organic peroxyacids is called the Baeyer-Villiger oxidation. For example, when acetone is treated with peroxyacetic acid, methyl acetate forms.

$$CH_3-\overset{\overset{O}{\|}}{C}-CH_3 \xrightarrow{CH_3COOOH} CH_3-\overset{\overset{O}{\|}}{C}-OCH_3 + CH_3COO^-$$

$$R-\overset{\overset{O}{\|}}{C}-R \xrightarrow{R'COOOH} R-\overset{\overset{O}{\|}}{C}-OR + R'CO\bar{O}$$

The Baeyer-Villiger oxidation involves an intramolecular anionotropic rearrangement in which an alkyl group with its bonded *es* migrates from the carbonyl C atom (migration origin) to an electron-deficient O atom (migration terminus).

The order of migratory aptitude of the alkyl groups is 3° > 2° > 1°, i.e., the more nucleophilic group normally migrates.

This reaction is usually carried out in inert solvents like CH_2Cl_2 with various peroxyacids, e.g., CF_3COOOH, CH_3COOOH, C_6H_5COOOH, etc.—trifluoroperacetic acid is by far the most reactive one.

A related reaction is the conversion of *o*- and *p*-hydroxybenzaldehydes or phenylketones into *o*- and *p*-dihydroxybenzenes by the action of alkaline hydrogen peroxide which is known as Dakin oxidation.

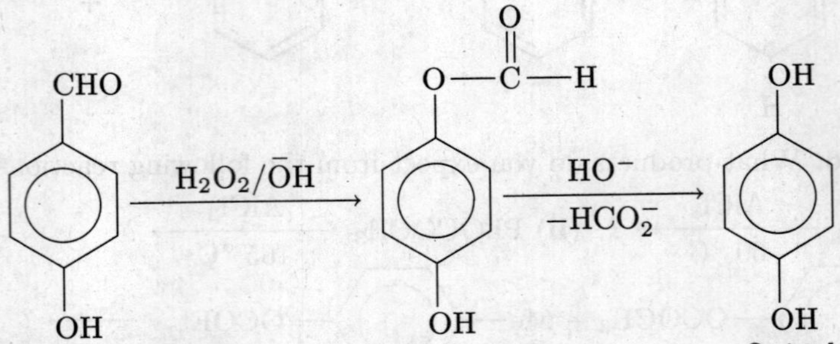

p-Hydroxybenzaldehyde Quinol

The Baeyer-Villiger reaction gives a good yield of esters or acids. So this may be used to prepare esters and carboxylic acids.

$$CH_3CO.CMe_3 \xrightarrow{CH_2Cl_2,CH_3COOOH} CH_3COOCMe_3$$

It has been found that when an alkyl group with a chiral C migrates, the configuration of the chiral C does not change.

The reaction is considered to occur through the following steps :

Initial Step :

$$R'COOOH \rightleftharpoons R'COOO^- + H^{\oplus}$$

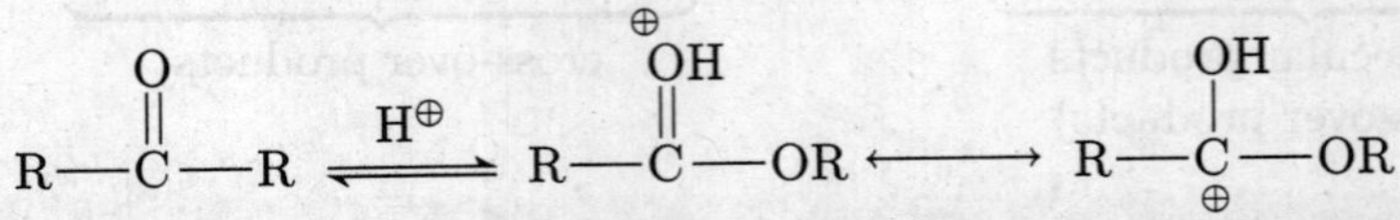

Thus, in this step the O atom of the ketone is protonated and a resonance-stablised carbenium ion forms.

Step I. Nucleophilic attack at the carbonyl carbon by the peroxycarboxylate ion.

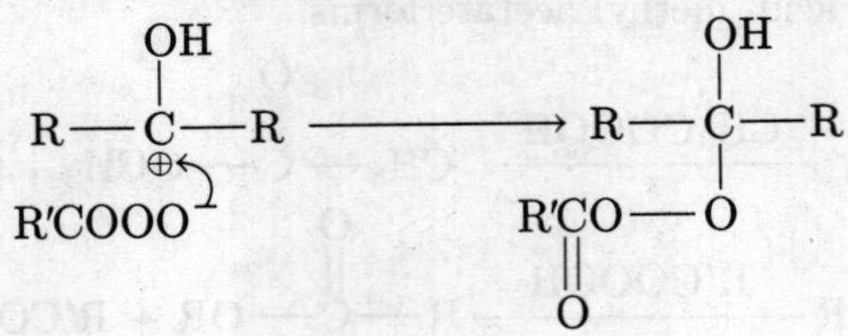

Step II. Migration of the alkyl group and loss of $RCOO^-$, a good leaving group, in a concerted step.

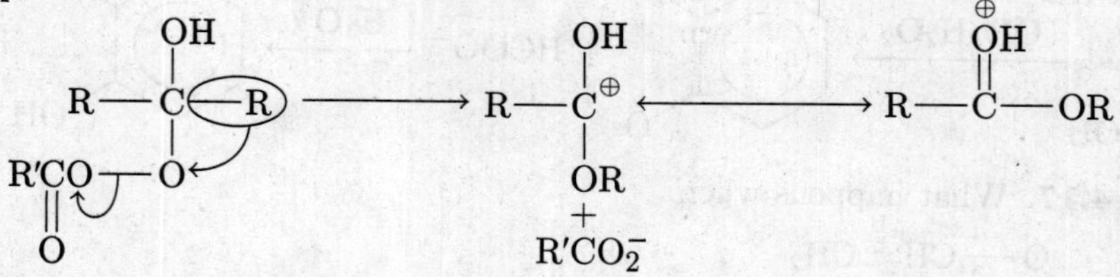

This is the slower step and hence the rate-limiting step.

Final Step : Loss of proton and the formation of ester.

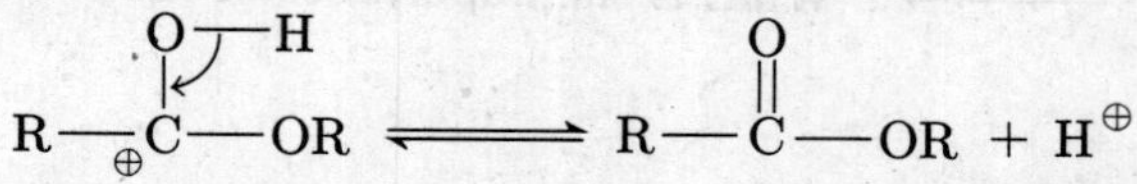

When a labelled ketone with ^{18}O is subjected to the Baeyer-Villiger rearrangement, the resulting ester contains ^{18}O. This supports the proposed mechanism.

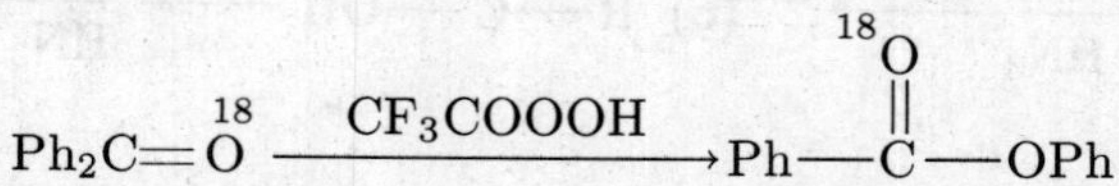

That the concerted step is the rate-limiting step is supported by the following facts :

(i) An electron-withdrawing group in the R′COOO group increases the rate of the reaction.

(ii) An electron-releasing group in the migrating alkyl group also increases the rate.

The intramolecularity of the reaction has been supported by the cross-over experiment—no cross-over product has been obtained.

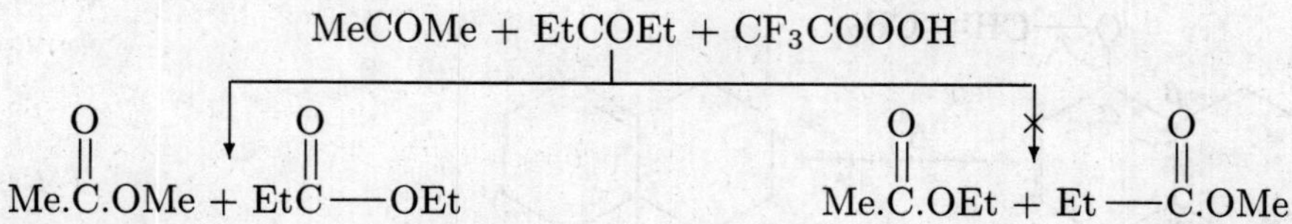

products of intramolecular rearrangement — no crossover products formed

However, the mechanism for the Dakin reaction may be as follows :

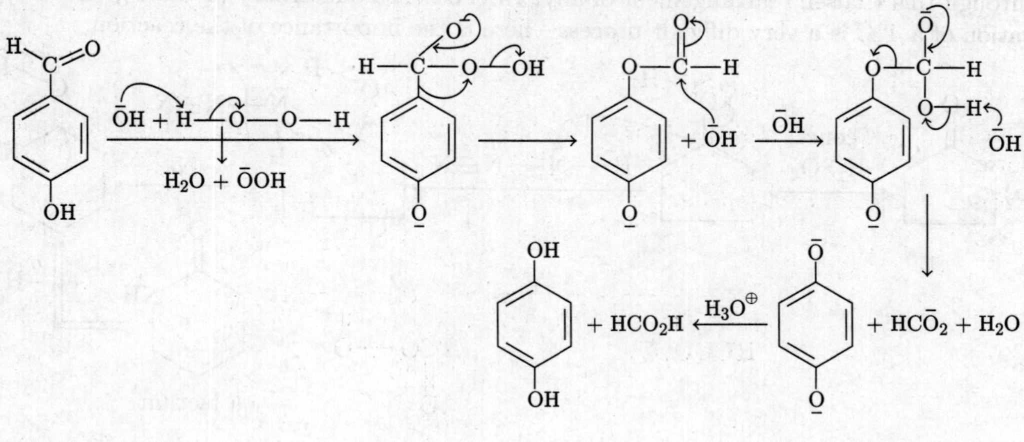

Similarly,

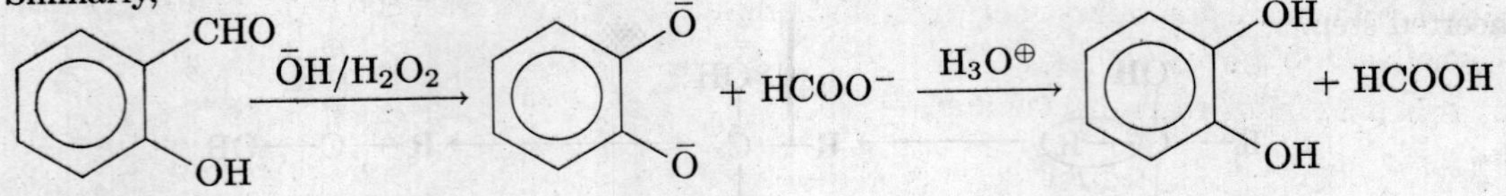

Problem 14.17. What happens when

(a) 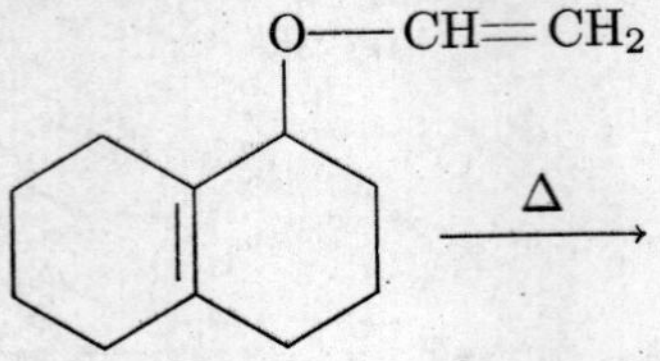 ? What is the importance of the reaction ?

(b) 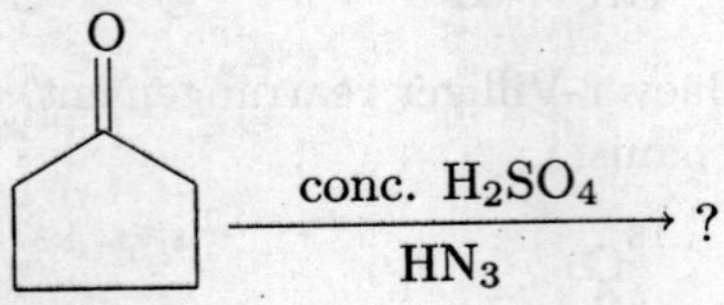 ? **(c)** 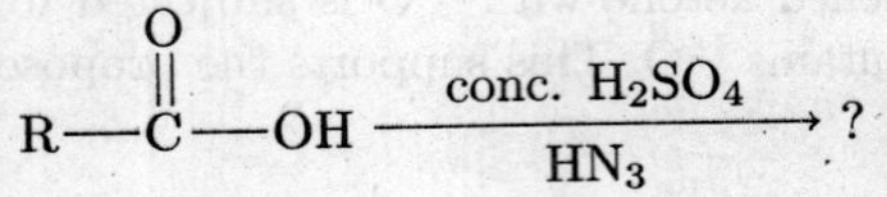 ?

(d) 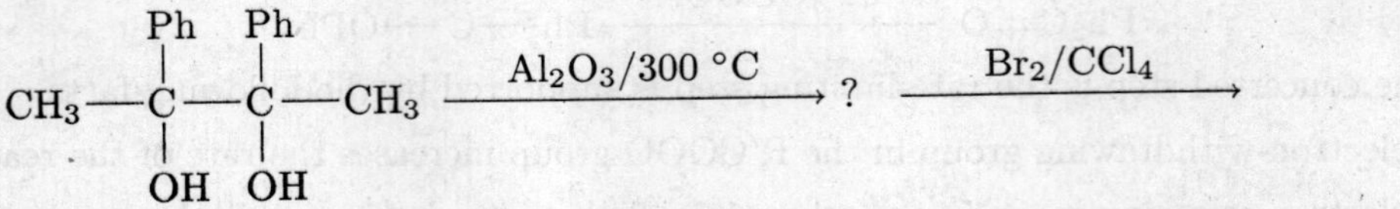

Write down the mechanism for the reactions given above.

Solution :

(a) 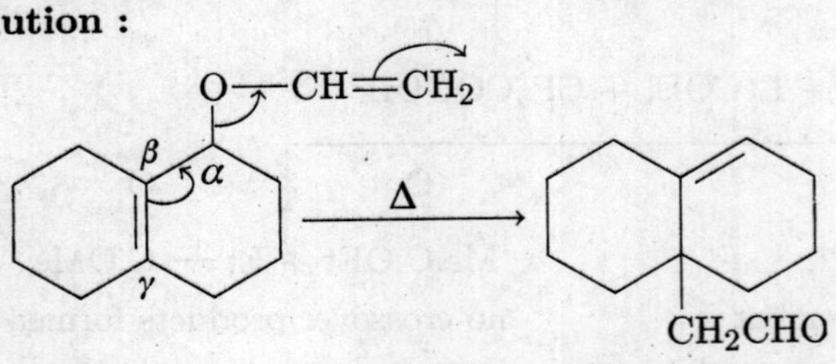

Through this Claisen rearrangement of allyl vinyl ether, a quaternary (4°)C is generated; the generation of a 4°C is a very difficult process—here is the importance of the reaction.

(b) 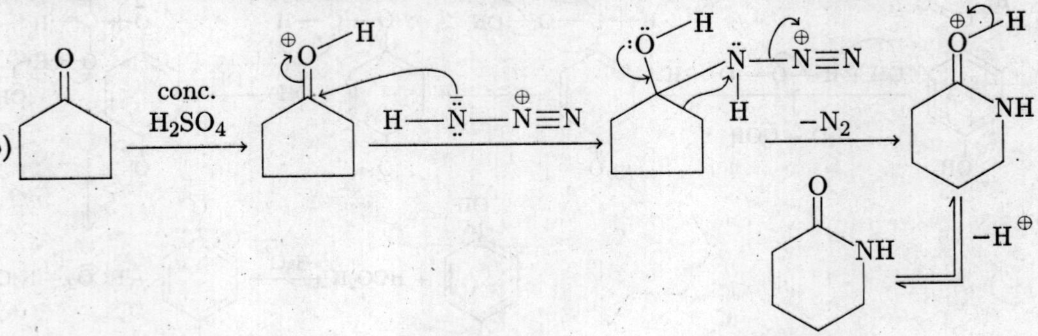

a lactum

This is a Schmidt reaction on a ketone—an intramolecular carbanionic 1, 2-shift rearrangement in which C and N are the migration origin and migration terminus respectively. The migration step is the R. D. Step.

(c) It is a Schmidt rearrangement :

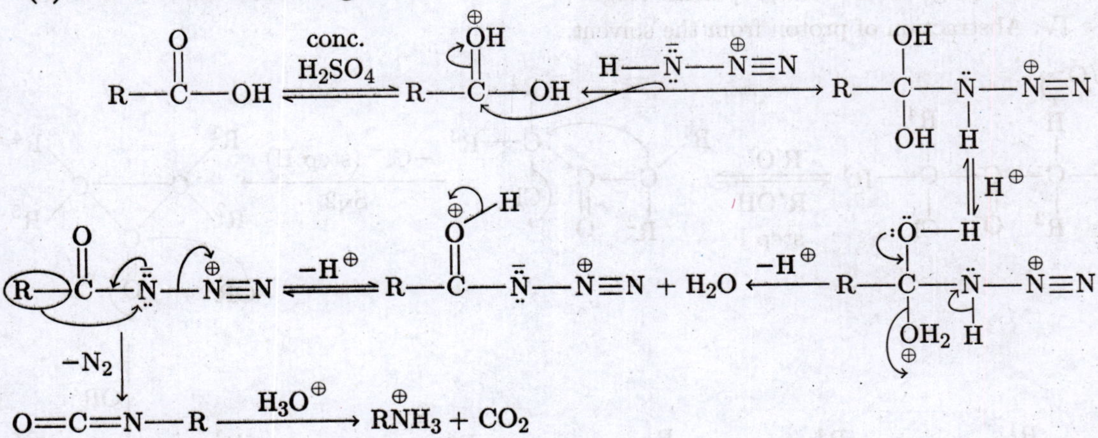

(d)

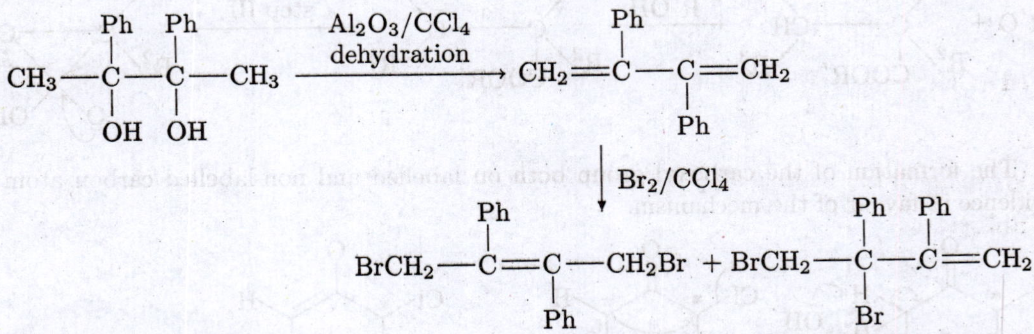

14.21 The Favorskii Rearrangement

The reaction of α-chloro- or bromoketones with a base (alkoxide, hydroxide, amines, etc.) give rearranged acids or esters and the rearrangement is called the Favorskii rearrangement; Cycloalkanones with α-halo group give ring contracted products while rearranged acid is the product when hydroxides or amines are used as bases.

$$PhCH_2COCH_2Cl \xrightarrow{H\bar{O}} PhCH_2CH_2COOH; \; PhCH(Cl)COCH_3 \xrightarrow{MeO^-} PhCH_2CH_2COOMe$$

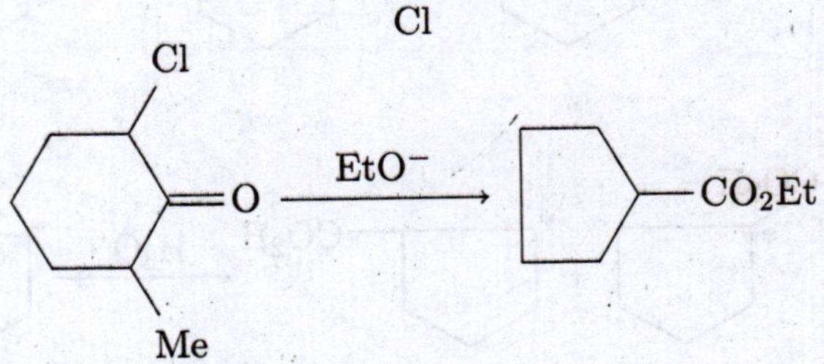

The mechanism involves the following steps :

I. Abstraction of α-H.

II. S_N2 reaction and formation of a cyclopropanone ring.

III. Cleavage of the cyclopropanone ring.

IV. Abstraction of proton from the solvent.

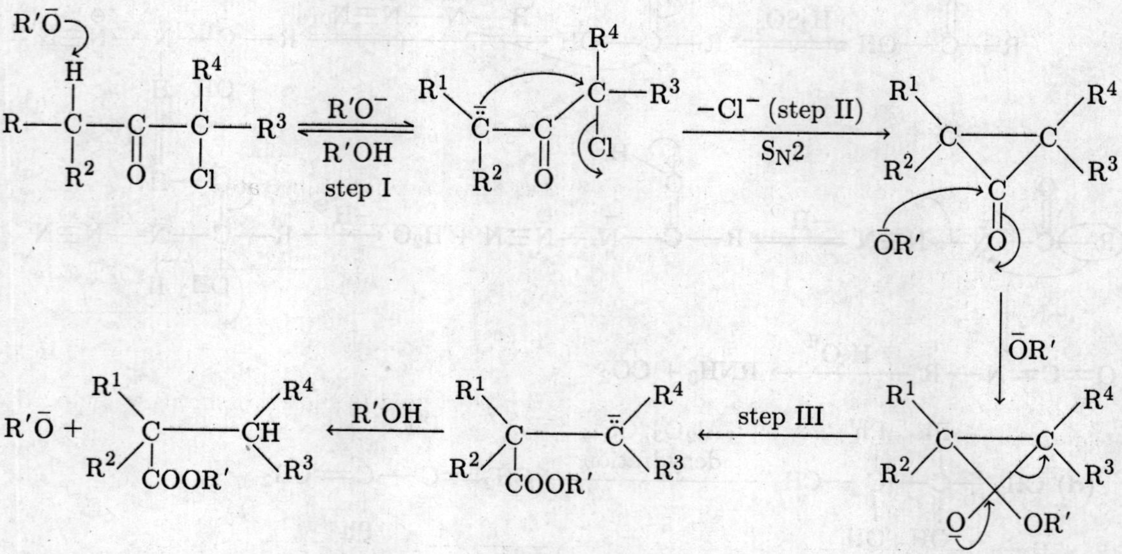

The formation of the carboxyl group both on labelled and non-labelled carbon atom is an evidence in favour of the mechanism.

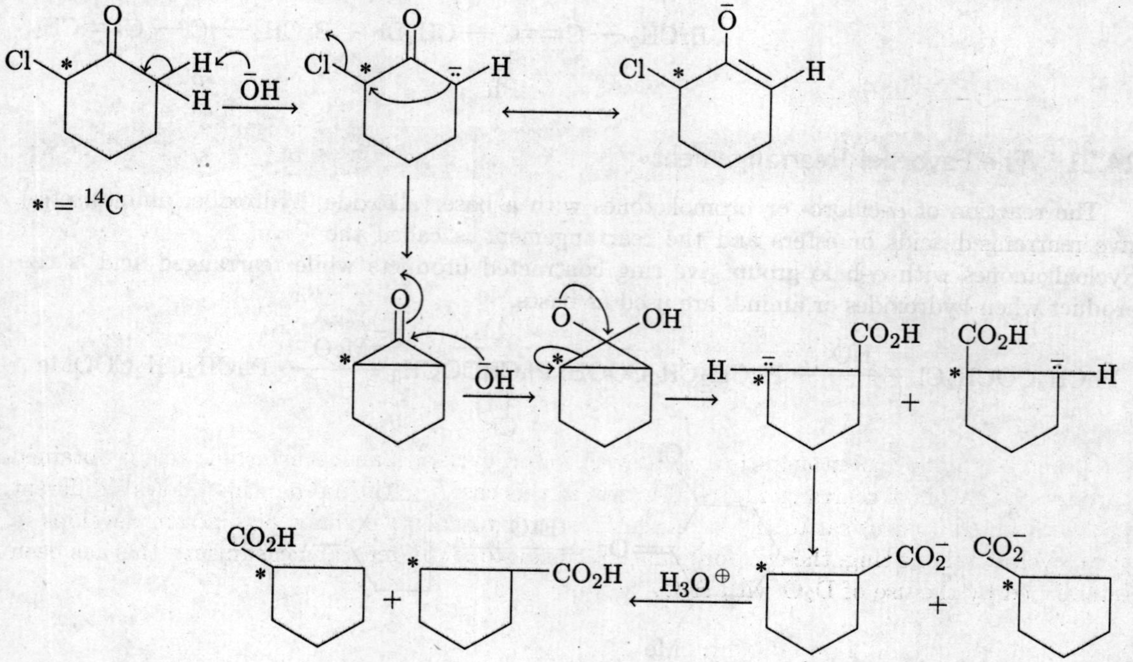

$* = {}^{14}C$

The following are further evidences in favour of the mechanism :

(i) Cyclopropanone intermediate has been trapped as an adduct with furan.

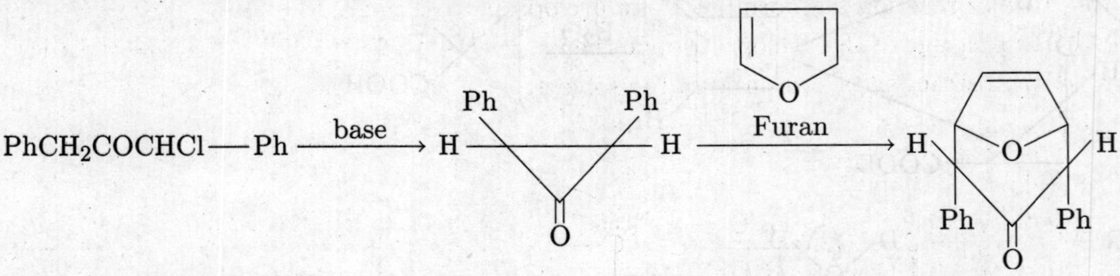

(ii) Stereospecific reactions take place in the cases of appropriate substrates.

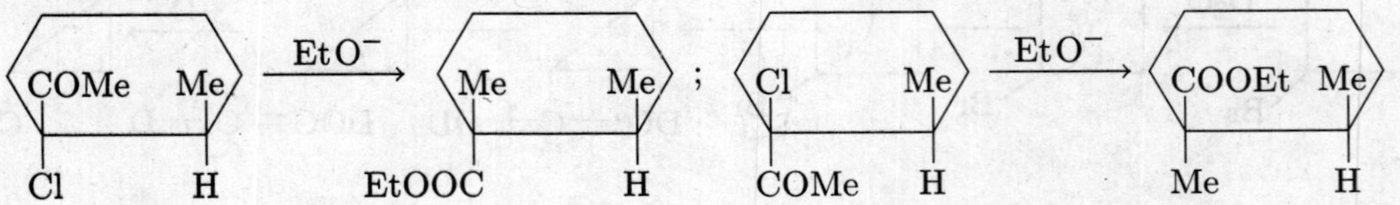

When α, α- or α, α'-dihaloketones with an α-H is subjected to the Favorskii reaction, α, β-unsaturated esters are formed.

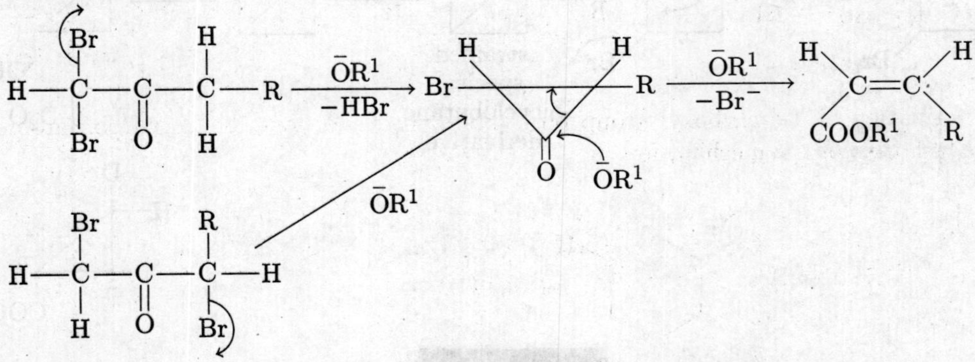

However, α-haloketones without α-H atom also undergo the Favorskii reaction. Of course, the mechanism is quite different in this case :

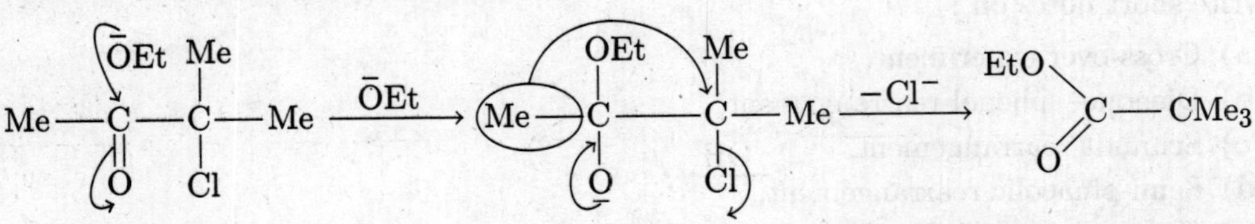

When 2-bromocyclobutanone is treated with water, cyclopropanonecarboxylic acid is obtained through the Favorskii rearrangement. However, in this case also the mechanism follows a different pathway which is identical to the semi-benzilic rearrangement; perhaps high strain develops in the bicyclobutanone ring, the cyclopropane intermediate of the general mechanism—this has been established by the use of D_2O with H_2O.

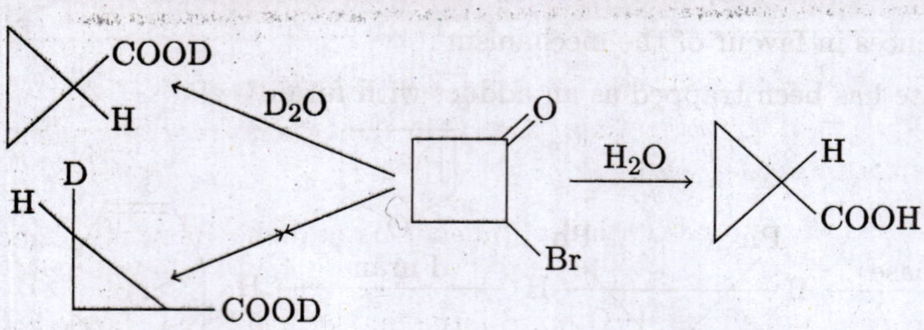

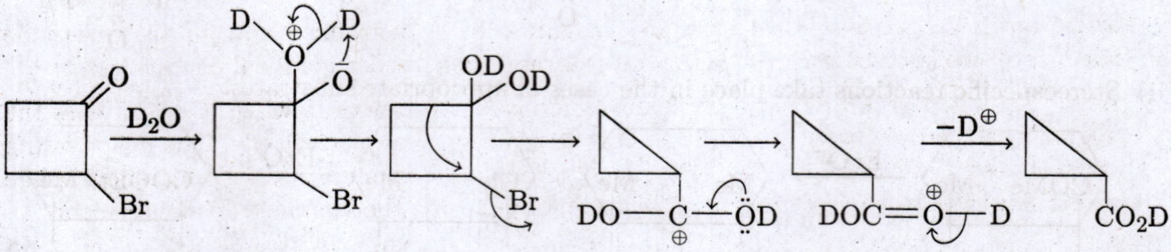

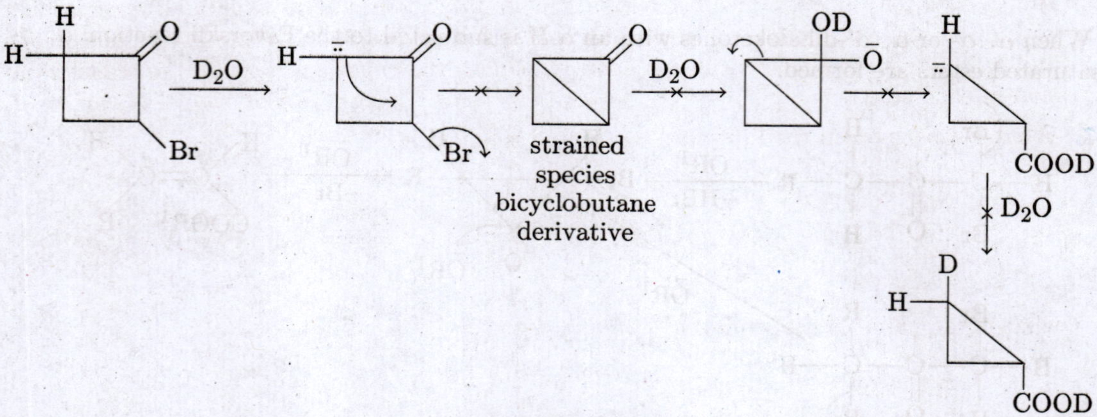

Exercise

1. Write short notes on :
   (a) Cross-over experiment,
   (b) Dienone—phenol rearrangement,
   (c) Schmidt rearrangement,
   (d) Semi-pinacolic rearrangement,
   (e) Conrotatory and disrotatory motions.
2. Give evidences in favour of
   (a) the anti-migration in the Beckmann rearrangement.
   (b) the intramolecularity of the pinacol-pinacolone rearrangement.
   (c) the reversibility of the benzilic acid rearrangement.
   (d) the carbenium ion intermediate in the Demyanov rearrangement.

(e) the concerted step in the Hofmann rearrangement is the rate-limiting step.

(f) the retention of configuration of the migrating chiral C in the Curtius rearrangement.

(g) the Wolff rearrangement involves 1, 2-shift.

3. Write down the migratory preferences for the pinacol-pinacolone and the Baeyer-Villiger rearrangements.

4. Why do the N—H bond breaks and the C—C bond forms in the benzidine rearrangement?

5. Write down the relative rates of amine formation from benzamide and *p*-nitrobenzamide when they are subjected to the Hofmann rearrangement under identical conditions.

Ans. The nitro group in the *p*-nitrobenzamide is electron-withdrawing; it deactivates the migrating *p*-nitrophenyl group and thus the migration takes place slowly. The deactivating effect is absent in the migration of the phenyl group in benzamide. Hence *p*-nitrobenzamide reacts slower than benzamide when they are subjected to the Hofmann rearrangement.

6. An organic compound A, $C_9H_{18}O_2$, is readily oxidised by HIO_4 to two compounds $C_6H_{10}O$ and C_3H_6O both of which furnish crystalline semicarbazones. The compound A on boiling with dil. H_2SO_4 furnishes two isomeric ketones identified as 2, 2, 3-trimethylcyclohexanone and 2, 2, 6-trimethylcyclohexanone. Identify the compound A and explain the transformation in each case.

7. What happens when the following compound is acidified?

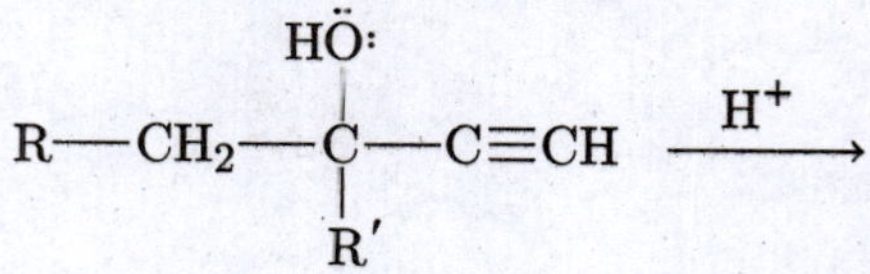

what is the name of the reaction?

Ans.

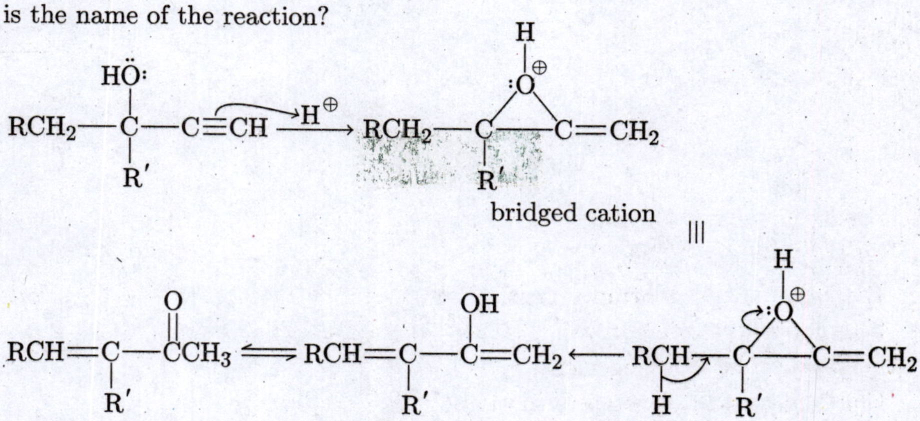

The name of the reaction is **Rupe rearrangement**.

α-Acetylinic alcohols having α-H with respect to alcoholic OH undergo Rupe rearrangement via bridgedcations.

Chapter 15

1, 2-Elimination Reactions

Objectives

- To understand :
  - The meaning of the elimination reactions [15.1] p. 785
  - The meaning of the 1, 2-elimination reactions [15.2] p. 785
- To have a clear idea about :
  - E1 reactions in general [15.2.1] p. 788
  - E2 reactions in general [15.2.2] p. 791
  - E1cB reactions in general [15.2.3] p. 797
  - The acid-catalysed dehydration of alcohols [15.3] p. 807
  - The Hofmann's exhaustive methylation [15.4] p. 809
  - Dehalogenation of α, β-dihalides [15.5] p. 812
  - Internal elimination (Ei) reactions [15.6] p. 813
  - Pyrolysis of esters [15.6.1] p. 813
  - The Chugaev reaction [15.6.2] p. 814
  - The Cope reaction [15.6.3] p. 815

15.1 Introduction

A reaction involving the loss of two atoms or groups from a molecule, without their being substituted by other atoms or groups, is known as elimination reaction. Thus it is the reverse process of addition reaction. The loss of one H atom and the Br atom from 2-bromopropane by the action of alcoholic alkali is an example of elimination reaction.

$$CH_3CHBrCH_3 + KOH(alc.) \longrightarrow CH_3{-}CH{=}CH_2 + KBr + H_2O$$

Through the elimination reactions π bonds can be generated, such as C=C, C≡C, C=N, C≡N, C=O, C=S, N=N, etc.; cyclic compounds can also be prepared from acyclic compounds by the elimination reactions.

However, on the basis of the relative positions of the two eliminating groups or atoms, the elimination reactions may be classified as : (a) 1, 1- or α-elimination reaction, (b) 1, 2- or, β-elimination reaction, (c) 1, 3- or γ-elimination reaction, etc.

(a) 1, 1- or α-elimination reaction : When the two groups are lost from one and the same atom of a substrate in an elimination reaction, the reaction is said to be a 1, 1- or an α-elimination reaction. A 1, 1-elimination reaction usually gives rise to a carbene. Thus, dichlorocarbene is obtained when chloroform is heated with alcoholic KOH solution.

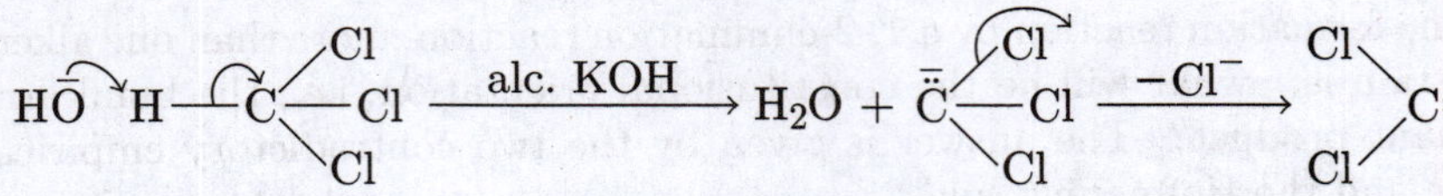

(b) 1, 2- or β-Elimination reaction : When two atoms or groups are lost from the two adjacent atoms in an elimination reaction, it is said to be 1, 2- or β-elimination reaction.

In a 1, 2-elimination reaction a π bond is formed. The formation of ethylene from ethyl alcohol by the action of concentrated sulphuric acid at 170 °C is an illustration of 1, 2-elimination reaction.

$$CH_3CH_2OH + H_2SO_4(conc.) \longrightarrow CH_2{=}CH_2 + [H_2SO_4 + H_2O]$$

(c) 1, 3- or γ-elimination reaction : When in an elimination reaction the two groups of atoms are lost from two different atoms which remain separated by another atom in a chain, the reaction is known as γ- or, 1, 3-elimination reaction. Usually in a γ-elimination reaction a three-membered ring forms.

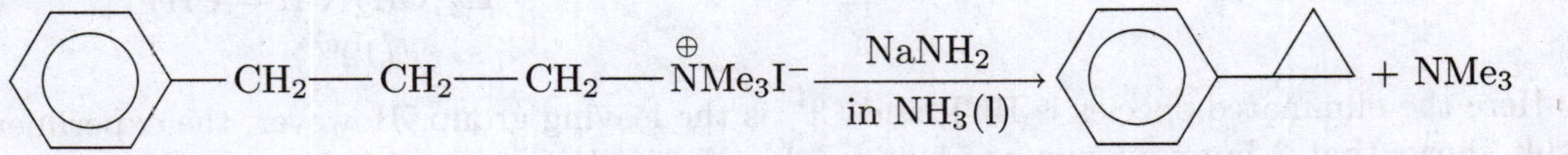

However, the most common type of elimination reaction is the 1, 2-elimination reaction. We shall discuss this here in detail.

15.2 1, 2-Elimination Reactions

Through the 1, 2-elimination reactions C=C, C≡C, C=N, C≡N, C=S, C=O, N=N can be formed. Among these the formation of C=C and C≡C are the most useful processes.

The common substrates are alkyl and cycloalkyl halides, vic-dihalides, alcohols, ethers, esters, tetra alkyl ammonium halides and hydroxides, tetra alkyl phosphonium halides and hydroxides, sulphonium halides and hydroxides, alkyldiazonium halides, aldoximes and ketoximes and their acyl derivatives, etc.

In most of the cases one of the two leaving groups is proton and the other, called a leaving group, may be one of the following :

$$H_2O,\ ROH,\ RCOO^-,\ NR_3,\ PR_3,\ SR_2,\ N_2,\ \text{etc.}$$

Ionic 1, 2-elimination reactions are carried out by heating the substrate with or without catalyst in a weak nucleophilic solvent whereas pyrolytic 1, 2-eliminations are carried by the application of heat only. In the former cases acids or bases are used as catalysts.

The driving force of a 1, 2-elimination reaction is supposed to be the back strain on the carbon site where elimination will take place. The decreasing order of B-strain in the compounds, which undergo the elimination reaction, is :

$$3° > 2° > 1°.$$

Hence the reactivity order is also expected to be :

$$3° > 2° > 1°.$$

In the cases where proton loss is involved in the rate-determining step, one should expect that with the increasing acidity of the H atom, which will leave the substrate as proton, the reactivity increases. On the other hand, if the rate-determining step involves the loss of the leaving group only, the reactivity will increase with the increasing ability of the group to leave the substrate.

In an alkene formation reaction by a 1, 2-elimination reaction, more than one alkene may form. Now the question is : what will be the constitutional orientation, i.e., the bond arrangement of the predominant product? The answer is given by the two contradictory empirical rules—the Saytzeff's rule and the Hofmann's rule.

The Saytzeff's rule deals with the constitutional orientation of the predominant alkene which is formed by the elimination reaction. This states that in a 1, 2-elimination reaction the most substituted alkene will be the predominant product. This means that the two carbons which bear the newly formed π bond should contain the least number of H atoms. For example, when 2-chlorobutane is heated with alcoholic KOH, a mixture of 1-butene and 2-butene is obtained; according to the Saytzeff's rule, 2-butene will be formed in excess over 1-butene, i.e., 2-butene will be the predominant product.

$$CH_3\cdot CH_2CHClCH_3 \xrightarrow{\text{alc. KOH}} \begin{cases} CH_3\cdot CH{=}CH\cdot CH_3 & \text{major product (81\%)} \\ CH_3\cdot CH_2\cdot CH{=}CH_2 & (19\%) \end{cases}$$

Here the eliminated species is HCl, and Cl^- is the leaving group. However, the experimental result shows that 2-butene forms predominantly. Now, what makes 2-butene to form predominantly? 2-Butene is the thermodynamically stable product. The hyperconjugative effect brings extra stability to 2-butene over 1-butene and because of this extra stability, the yield of 2-butene predominates. 2-Butene has six no-bond resonating structures in its resonance hybrid involving σ-π conjugation while the resonance hybrid of 1-butene contains only two such structures.

$$[CH_3CH{=}CHCH_3 \longleftrightarrow \overset{H^{\oplus}}{CH_2}{=}CH{-}\bar{C}HCH_3 \longleftrightarrow CH_2{=}CH{-}\bar{C}HCH_3 \ \ (H^{\oplus})$$

$$\updownarrow$$

$$\text{etc.} \longleftrightarrow CH_3{-}\bar{C}H{-}CH{=}\overset{H^{\oplus}}{CH_2} \longleftrightarrow H^{\oplus}CH_2{=}CH{-}\bar{C}HCH_3]$$

no-bond resonating structures of 2-butene

$$[CH_3—CH_2.CH{=}CH_2 \longleftrightarrow CH_3—\overset{H^{\oplus}}{CH}{=}CH—\bar{C}H_2 \longleftrightarrow CH_3—\underset{H^{\oplus}}{CH}{=}CH—\bar{C}H_2]$$

no-bond resonating structures of 1-butene

Usually 1, 2-elimination reactions involving alkyl halides, alcohols, etc. follow the Saytzeff's rule.

The Hofmann's rule also deals with the constitutional orientation of the alkenes formed by 1, 2-elimination reactions. This states that the least substituted alkene is the predominant product in a 1, 2-elimination reaction involving alkene formation. For example, when sec-butyltrimethylammonium hydroxide is heated, 1-butene will be formed predominantly insofar as the Hofmann's rule is concerned.

$$\overset{\gamma}{CH_3}\cdot\overset{\beta}{CH_2}\underset{\alpha}{\overset{\overset{\beta}{CH_3}|}{CH}}\cdot\overset{\oplus}{N}Me_3\bar{O}H \xrightarrow{-H_2O} CH_3CH_2\cdot CH{=}CH_2 + NMe_3 \text{ major product (95\%)}$$

$$\xrightarrow{-H_2O} CH_3CH{=}CHCH_3 + NMe_3 \text{ (5\%)}$$

In fact, 1-butene forms as a major product when sec-butyltrimethylammonium hydroxide is heated. In this reaction NMe_3 is the leaving group and the HO^- abstracts a proton from the β-C atom to form a molecule of water. In this case there are two β-Cs in the substrate. If the —CH_3 group gives up a proton, 1-butene will be the product, whereas when —CH_2— group gives up the proton, 2-butene will be formed. Here 1-butene forms preferentially. Now, the question is : why does 1-butene form predominantly? Two explanations have been given to account for the yield.

(i) The more acidic β-H gets removed faster than the less acidic β-H. In the above said example the β-H of the —CH_3 group is more acidic than the β-H of the —CH_2— group since the +I effect of the γ-methyl moiety makes the β-H of the —CH_2— group less acidic than the β-H of the CH_3 group where the +I effect is absent. Thus β-H of the —CH_3 group is abstracted by the $\bar{O}H$ ion and 1-butene forms predominantly.

(ii) The less the crowding on the β-C, the more effectively the $\bar{O}H$ ion attacks the β-H and the more readily the product is formed. Thus the steric effect makes the crowded β-C unfavourable for giving up the H atom. In the case of sec-butyltrimethylammonium hydroxide, the β-methylene group has two H atoms, and one methyl group, whereas β-methyl group has three H atoms only. Therefore, the β-methyl group is less crowded than the β-methylene group and hence the β-H from the former is more readily abstracted by the $\bar{O}H$ ion than that from the latter and thus 1-butene is formed predominantly.

Usually 'onium' (ammonium, phosphonium, sulphonium, etc.) compounds give the Hofmann products by 1, 2-elimination reactions.

On the basis of the steric course of the 1, 2-elimination reactions, these can be classified as cis or syn and trans or anti elimination reactions. When the two leaving groups get eliminated from the same side of a single or double bond, the process is called syn- or cis-elimination; whereas if they leave the substrate from the opposite sides of a single or double bond, the process is termed as anti- or trans-elimination. Two examples of the processes are given below :

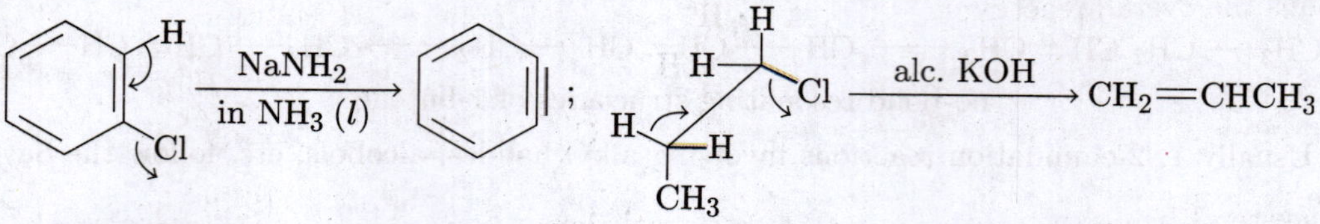

cis-elimination *trans*-elimination

Nucleophilic substitution reactions usually compete with the 1,2-elimination reactions. At a high temperature elimination reaction products are obtained while at a relatively low temperature substitution reaction products are formed.

$$CH_3CH_2CH_2Cl \xrightarrow{\bar{O}H} CH_3CH{=}CH_2 \text{ (elimination product)}$$
$$\phantom{CH_3CH_2CH_2Cl \xrightarrow{\bar{O}H}} CH_3CH_2CH_2OH \text{ (substitution product)}$$

However, depending on the mechanistic pathways of ionic 1, 2-eliminations, these are classified further under three heads :

(i) **E1 reactions**—1, 2-elimination reactions in which the T.S. of the rate-determining step involves only one species is called E1 reaction. The E1 reaction usually follows the first order kinetics.

(ii) **E2 reactions**—1, 2-elimination reactions in which the T.S. of the single concerted step involves two species is called E2 reaction. The E2 reaction generally follows the second order kinetics.

(iii) **E1cB reactions**—This involves an intermediate carbanion, the conjugate base (cB) of the substrate, which gives up a leaving group through a T.S. involving only one chemical unit.

We shall discuss here in detail these three types of reactions.

15.2.1 E1 reactions

E1 reactions take place in two discrete steps via a carbenium ion intermediate.

Step I : This involves the formation of carbenium ion.

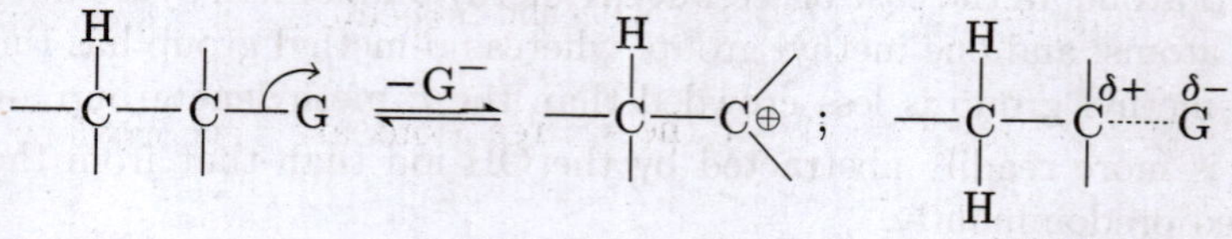

This is the slow step and hence the rate-determining step.

Step II : This involves rapid loss of a proton from the carbon atom adjacent to the C atom bearing the positive charge. This step is faster than the step I.

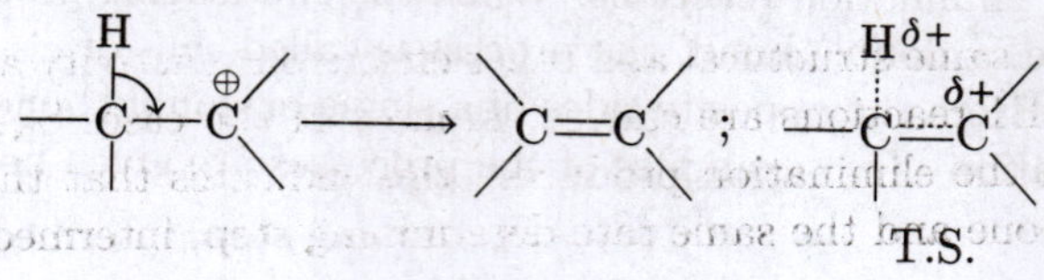

Thus the overall reaction is:

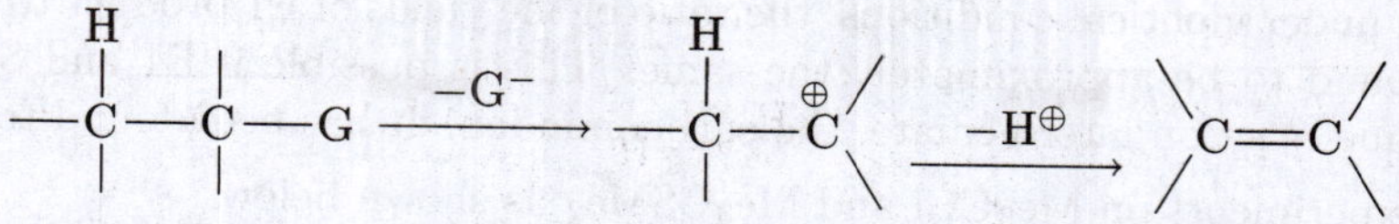

The step I being the slower step, this is the rate-determining step; since the T.S. of the rate-determining step involves only one species, the E1 reaction is expected to follow the first order kinetics and actually it does so.

The rate = k [substrate].

As E1 reactions occur via a carbenium ion intermediate, the energy profile for such a reaction (exothermic) may be drawn as shown in Figure 15.1.

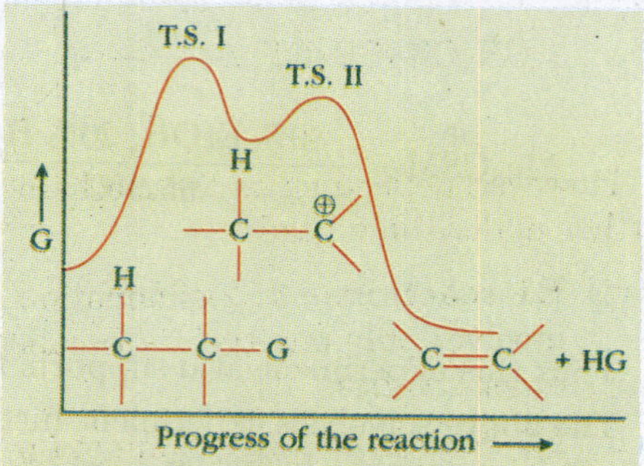

Fig. 15.1. The energy profile for the E1 reaction (exothermic)

E1 reactions usually proceed in the absence of bases; the proton is abstracted by the weakly basic solvent molecules.

So far as the constitutional orientation is concerned E1 reactions give the Saytzeff's products, i.e., relatively the more stable alkenes form through E1 reactions. This is because in this reaction the leaving group leaves the species before the choice is made as to which direction the double bond will be formed. Of course, there are exceptions—in some cases the Hofmann's products are also obtained; for example :

(i) If the leaving group forms an ion-pair with the carbenium ion and does not leave the latter before the departure of the β-H, the Hofmann's product results.

(ii) If one of the β-Cs with respect to the leaving group (G) is crowded, then the solvent molecule cannot take up a proton from the sterically strained C atom; and in such a case the proton is lost from the less sterically strained β-C which gives rise to the predominant Hofmann's product. Thus 2-chloro-2, 4, 4-trimethylpentane gives 81% Hofmann's product and 19% Saytzeff's product in the E1 reaction.

Since the carbenium ions are the intermediates in the E1 reactions and since carbenium ion can adopt the most stable conformation, the E1 reactions are usually non-stereospecific. However, when the leaving group forms an intimate ion-pair with the carbenium ion, a stereospecific product may result.

The mechanistic steps as drawn above for the E1 reactions are supported by the following evidences :

(i) The reactions, which are designated as E1, are found to follow the first order kinetics.

(ii) The products of some El reactions are such that they could not be formed if there were no carbenium ion rearrangement before the loss of proton—this is a strong evidence in favour of the intermediate carbenium ion formation.

(iii) E1 reactions have the same structural and other effects on reactivity as the S_N1 reactions have. In fact, when E1 reactions are carried, in most of the cases S_N1 reaction products also form along with the elimination products—this indicates that the S_N1 reaction and the El reaction have one and the same rate-determining step, intermediate, etc.

(iv) If two substrates, which differ from one another only in the leaving group, are subjected to E1 reaction under identical conditions, the ratio of the yields of E1 product to S_N1 product is always found to be approximately the same. This is possible if E1 and S_N1 reactions have the same reaction intermediate, the carbenium ion. It has been found to be so in the solvolysis experiments on Me_3CCl and $Me_3\overset{\oplus}{C}SMe_2$ as shown below :

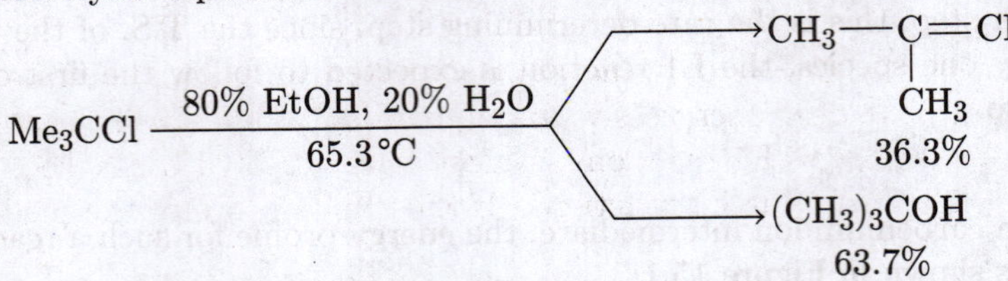

$$\frac{\text{E1 prodt.}}{S_N1 \text{ prodt.}} = \frac{36.3}{63.7} = 0.57$$

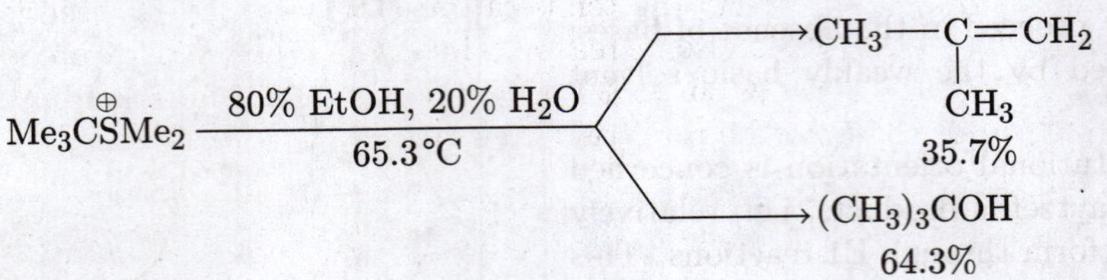

$$\frac{\text{E1 prodt.}}{S_N1 \text{ prodt.}} = \frac{35.7}{64.3} = 0.555$$

Thus this experiment also supports the formation of carbenium ion in E1 reactions.

The above solvolysis experiment further shows that S_N1 reactions compete with and is analogous to E1 reactions. In fact, in most of the cases these two types of reactions occur simultaneously and S_N1 products are often found to be the predominant products, specially when the reactions are carried out at low temperatures.

It has already been said that the factors that influence the rate of S_N1 reactions also influence the rate of E1 reactions. This is because in both the reactions the intermediate carbenium ion formation step is the rate-determining step. However, the factors are the substrate structure, the nature of the leaving group, the nature of the solvent, etc.

(a) Substrate structure and the reactivity : A carbenium ion being an intermediate in the E1 reaction, the rate of the reaction depends on the stability of the carbenium ion. Since +I-effect, hyperconjugative effect and conjugative effect stabilise carbenium ions, any structure which forms stable carbenium ion will be reactive. Again, we know that a sterically strained molecule tries to give up the strain and for that it is even ready to assume ionic structure. Taking all these the order of reactivity of the compounds is expected to be :

$3° > 2° > 1°$ which is also the stability order of the carbenium ions.

Thus alkyl or aryl substitutions on α and β positions with respect to the leaving group increase the rate of E1 reactions. As the strain increases, the yield of E1 products increases. This is perhaps through elimination reaction, the substrate becomes free from steric strain while in the S_N1 reactions the strain is reintroduced in the product. However, bridgehead halides do not undergo the reaction since the intermediate carbocation cannot assume planar trigonal orientation and the T.S. will be highly strained as well as energetic.

(b) Basicity and concentration of the base and the reactivity : Since E1 reactions do not usually require an added base (the solvent molecules serve the purpose), the strength and the concentration of the base have nothing to do with the rate of E1 reactions. This is also the case with S_N1 reactions. Usually the S_N1 product predominates over the E1 product because of the weak base character of the solvent.

(c) Nature of the leaving group and the reactivity : Reactivity of the substrate depends largely on the nature of the leaving group. The best leaving groups are those which are least basic and more polarisable. Thus the order of leaving group reactivity of the following two sets of species is :

$$H_2O > ROH > CH_3COO^- > HO^-$$

$$I^- > Br^- > Cl^- > F^-$$

(d) Nature of the solvents and the reactivity : Since E1 reactions involve ionic intermediates, the carbenium ions, the rate of E1 reaction increases with the increasing polarity of the solvents. However, S_N1 reactions are favoured over E1 reactions in most solvents. Hydroxylic solvents of low polarity and poor nucleophilicity favour E1 reactions. If the temperature of the medium is increased, E1 reactions are also favoured.

(e) Effect of temperature on the rate : As the temperature increases, the E1 reaction rate increases and consequently the $E1/S_N1$ ratio also increases. Thus, when the temperature is raised from 25 °C to 55 °C in the solvolysis of Me_3Cl in 80% EtOH, the $E1/S_N1$ ratio rises from 0.20 to 0.57; if the temperature is raised to 75°C, the ratio rises to 0.80.

15.2.2 E2 reactions

E2 reactions take place in a single concerted step in which two sigma bonds break and a π bond forms simultaneously with the simultaneous departure of two groups, one of which is usually a proton. An E2 reaction requires an added base which pulls out the proton.

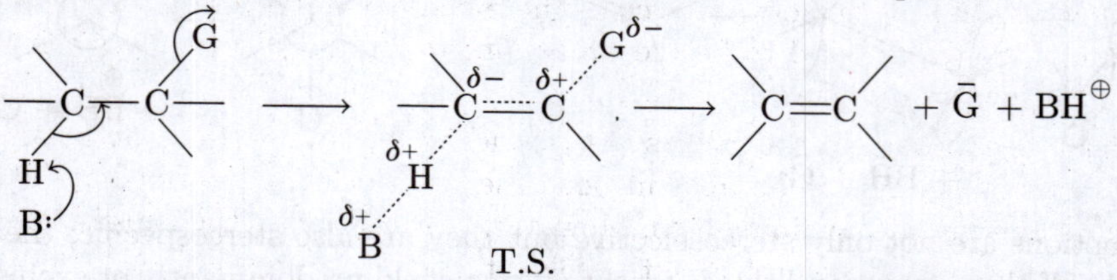

Since both the base and the substrate are involved in the T.S. of the single concerted rate-determining step of the E2 reaction, it is kinetically second order —first order with respect to the substrate and first order also with respect to the base; thus the rate = k [substrate] [base].

The energy profile for the E2 reaction (exothermic) is given in Figure 15.2.

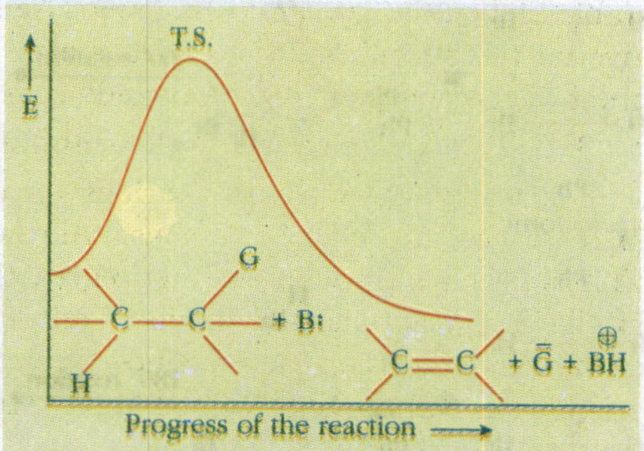

Fig. 15.2. The energy profile for the E2 reaction (exothermic)

In the E2 reaction the leaving group leaves the substrate with a pair of *e*s whereby a vacant *p* orbital remains with the α-C of the substrate; on the otherhand, an electrophile, usually a proton, leaves the substrate as a cation, whereby a filled *p* orbital lies with the β-C of the substrate. If the vacant and the filled *p* orbitals so left are parallel to each other, then only the desired sidewise overlap between the orbitals can occur to give rise to a π bond orbital. For the parallel orientations of the *p* orbitals, the leaving group, the electrophile, the α-C and the β-C, require to lie on the same plane. It will be energetically favourable for the attacking base to be on the common plane with the electrophile. Thus, for the E2 reaction periplanar (around planar) orientation of the said groups, atoms and the base is essential. Again, the molecule suffering the elimination reaction requires the two leaving groups in anti-periplanar orientation. Anti-periplanar orientation means the two groups are on the same plane but in the anti-position with respect to each other ; while when the two coplanar groups are in the syn-position, their orientation is said to be syn-periplanar. When the two leaving groups are in the anti-periplanar arrangement, the substrate is in the staggered conformation instead of the eclipsed conformation. The energy content of the former is certainly less than that of the latter. In this orientation the π bond forming orbitals are very much parallel to each other—this is the stereoelectronic condition for the reaction.

In the syn-orientation of the orbitals, on the other hand, will make an angle other than 0°, between their axes. If the π bond forms it will not follow the priciple of maximum overlap and the bond will be very weak.

In fact, in most of the cases acyclic compounds undergo 1, 2-elimination reactions by the E2 pathway assuming anti-periplanar arrangement. It is needless to mention that in this arrangement the dihedral angle between the two leaving groups is 180°, while that in the syn-periplanar arrangement is 0°. However, the discussion made above tells us clearly that the E2 reactions are stereoselective.

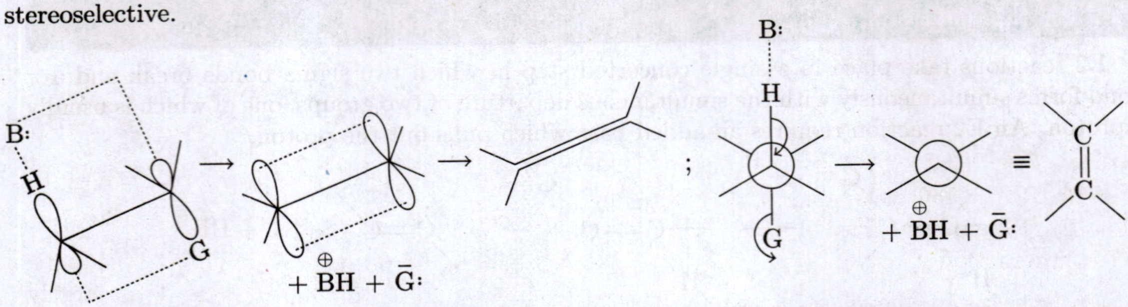

The E2 reactions are not only stereoselective but they are also stereospecific; the acyclic di-astereomers, the *dl*-threo forms and the *dl*-erythro forms, yield predominantly or exclusively **trans** and **cis** olefins respectively. For example, meso (erythro)-form of 1, 2-dibromo-1, 2-diphenylethane

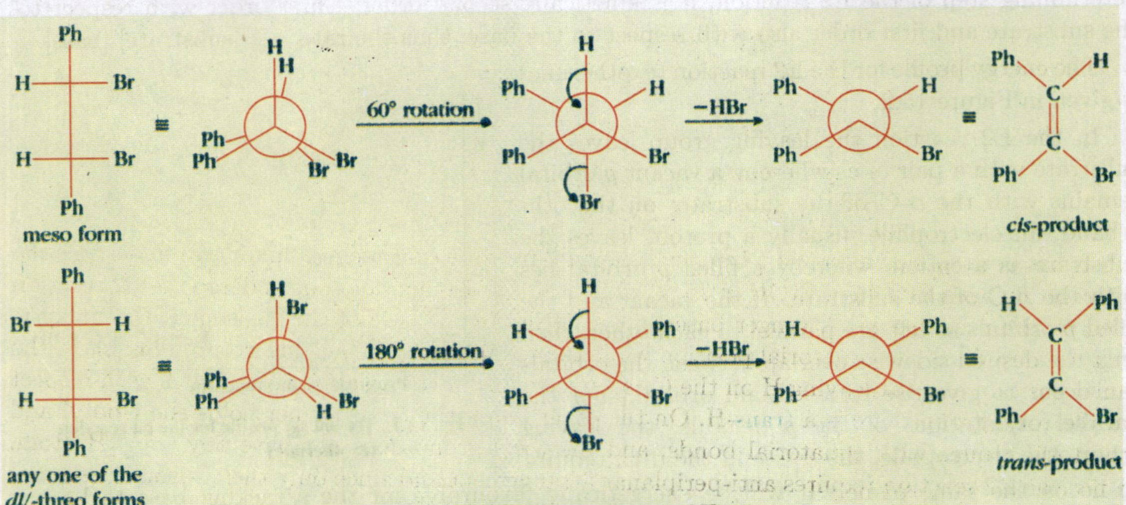

Fig. 19.3. Stereospecificity of the E2 reaction

on dehydrohalgenation yields cis-product, while *dl*-thero forms give trans-product through the E2 reaction.

However, it has been found that the threo forms undergo 1, 2-elimination reaction by the E2 pathway more readily than the erythro forms under the same set of conditions. It is the conformational stability of the substrate that makes a threo form react more readily than an erythro form. To facilitate anti-elimination each of the thero and erythro forms takes up that conformation in which the two leaving groups are in anti-positions with respect to each other. In the case of an erythro form the gauche conformation holds the two leaving groups in the anti-position, while in the case of a threo form the anti conformation holds the two leaving groups in the anti-position. Anti conformations are usually more stable than gauche conformations because of the presence of gauche interaction in the latter. Thus in the case of an E2 reaction, the T.S. of the conformation of the threo form contains less energy than that of the conformation of the erythro form. The lower energy content of the T.S. of threo form than that of the erythro form makes the former react more readily than the latter and this sort of gauche interaction is popularly called eclipsing effect (Sec. 9.24; steric factor).

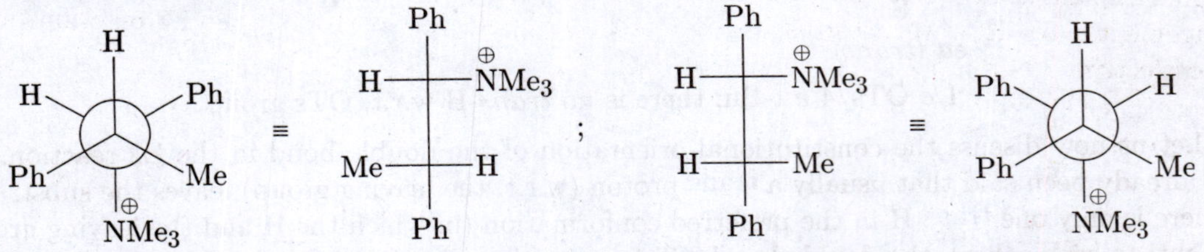

ϕ(Ph/Ph) = 180° no Ph/Ph gauche interaction

one of the two threo forms

one of the two erythro forms

ϕ(Ph/Ph) = 60° Ph/Ph gauche interaction

In the cases of alicyclic compounds whether the elimination will be anti or syn is largely determined by the relative rigidity of the ring. Thus for a series of 1, 2-eliminations by the E2 pathway the following degrees of stereoselectivity have been observed (Table 15. 1).

Table 15.1 Ring size and the stereoselectivity in the E2 reaction

Ring size	% of anti-elimination
Cyclobutane derivative	10%
Cyclopentane derivative	54%
Cyclohexane derivative	96%
Cycloheptane derivative	63%

It has been found that cis-4-t-butylcyclohexyl *p*-tosylate (*p*-toluene sulphonate) undergoes the E2 reaction with EtONa in EtOH at 75 °C while its trans-isomer does not undergo the E2 reaction even in the presence of a stronger base than EtONa. In the case of the cis-isomer, the bulky t-butyl group takes the equatorial position while the tosyl group, the leaving group, remains in the axial position and the leaving H on the C_2 atom remains in the axial position. Thus with respect to the tosyl group there is a trans-H. On the other hand, the trans-isomer holds the t-butyl and the tosyl groups with equatorial bonds, and the tosyl group does not have any trans H atom. Since the E2 reaction requires anti-periplanar arrangement and since only the cis-isomer assumes this arrangement, it undergoes the E2 reaction while the trans-isomer does not.

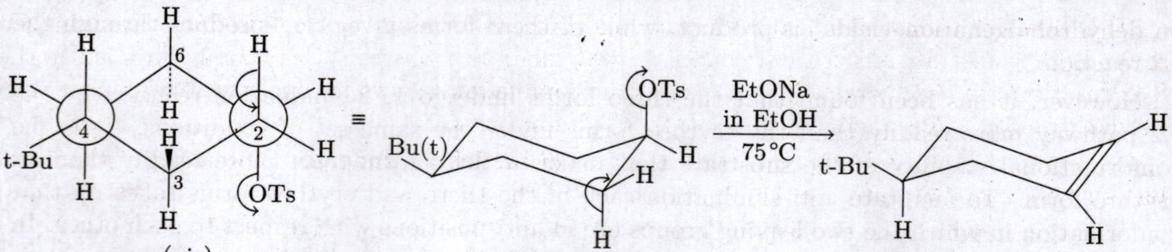

ea (*cis*)

1 *a* OTs/4 *e* t-Bu; there is *trans*-H w.r.t. OTs group

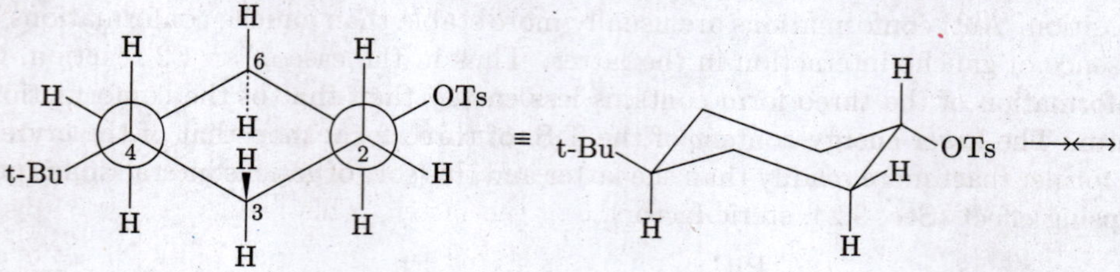

ea (*trans*)

1 *e* OTs/4 *e* t-Bu; there is no *trans*-H w.r.t. OTs group

Let us now discuss the constitutional orientation of the double bond in the E2 reaction. It has already been said that usually a trans proton (w.r.t. the leaving group) leaves the substrate; if there is only one trans H in the preferred conformation (in which the H and the leaving group are anti to each other), the double bond will form to that direction. On the other hand, if there are more than one trans H atoms, the question is : which one of them will leave the substrate and in which direction the double bond will be formed? The direction of the double bond formation largely depends on the nature of the leaving group in the E2 reaction. If the leaving group bears a positive charge and leaves the substrate as a neutral unit, its strong –I effect which will operate on the β-C, through the α-C, will make the β-Hs acidic. In this case the most acidic β-H will be abstracted by the base more readily than any less acidic β-H and as a result of which the Hofmann's products will be formed. For example, substrates with $\overset{\oplus}{N}R_3$, $\overset{\oplus}{S}R_2$, etc., leaving groups give the Hofmann's product in the E2 reactions. However, to compare the acidity of the β-Hs we shall have to consider inductive and conjugative effects as usual. Furthermore, the base usually attacks the least sterically hindered β-H and thus the least substituted alkene is formed. On the other hand, if the leaving group is a neutral one and leaves the species as an anion, the Saytzeff's product is obtained predominantly. Thus in this case, not the acidity of the β-H but the stability of the alkene determines the orientation of the double bond.

It has also been found that as the bulk of the leaving group of the substrate or that of the attacking base increases, the yield of the Hofmann elimination product increases. Thus when $CH_3CH_2CHXCH_3$ type of substrates are subjected to the E2 reactions under the same set of conditions, the following results are obtained :

Increasing size of the leaving group →

Leaving group (X)	Br	I	OTs	$\overset{\oplus}{S}Me_2$	SO_2Me	$\overset{\oplus}{N}Me_3$
Yield of the Hofmann's product	31%	30%	48%	87%	89%	98%

Increasing yield →

It is needless to mention that S_N2 reactions compete with and analogous to E2 reactions. In so far as the substrate structure is concerned, the difference between the two pathways is : in the S_N2 reaction the base attacks the α-C as a nucleophile and shows its nucleophilicity whereas in the E2 reaction it abstracts a β-H and exhibits its basicity.

Reaction conditions : Substrates with α and β branching undergo, E2 reactions with hard bases at high temperatures.

The given mechanistic step for the E2 reaction has been supported, though not always, by the following facts.

(i) The 1, 2-elimination reactions which are designated as E2 reactions have been found to follow the second order kinetics.

(ii) Primary kinetic isotope effect is a strong evidence in favour of the trans or anti-elimination. Thus when the trans H is replaced by a deuterium atom, the rate of the E2 reaction becomes less than that of the original substrate; the C—D bond being stronger than the C—H bond, the former breaks more slowly than the latter. This result also indicates that the C—H bond breaks in the rate-determining step.

(iii) Since no rearrangement products have yet been obtained along with the E2 reaction products, it indicates that the E2 reaction does not involve any reaction intermediate and that the reaction occurs in a sinchronous step.

(iv) The threo forms give trans-olefin and the erythro forms give cis-olefin usually by the E2 pathway. This fact supports the trans-elimination and the stereospecificity of the reaction.

Now we shall discuss the effects of the substrate structure, the strength and concentration of the added base, and the effect of the leaving group on the rate of the E2 reactions.

(a) The substrate structure and rate : It has been found that with the increasing α- and β-branching the rate of the E2 reaction increases. This is because as the number of alkyl or aryl groups increases on the C atoms of the developing double bond, the stability of the T.S. increases. Table 15.2 shows this clearly.

Table 15.2 Substrate structure and the rate of the E2 reactions

Substrate	temperature	% yield	Rate
CH_3CH_2Br	55 °C	0.9	1.0×10^5
$(CH_3)_2CH\ Br$	25 °C	80.3	0.237×10^5
$(CH_3)_3C\ Br$	25 °C	97	4.7×10^5
$CH_3CH_2CH_2Br$	55 °C	8.9	5.3×10^5
$(CH_3)_2CH.CH_2Br$	55 °C	59.5	8.5×10^5

Table 15.2 shows that the rate of the E2 reaction on isopropyl bromide is less than that on ethyl bromide; perhaps the lower temperature applied for the former reaction is responsible for its lower rate. Table 15.2 also reveals that the order of the rates is $3° > 2° > 1°$. Since the T.S. of an S_N2 reaction involves high steric strain, the α- and β-branches slow down the S_N2 reaction rate while they speed up the E2 reaction rate. Thus, with increasing branching at α- and β-Cs, the E2/S_N2 ratio increases.

It has also been found that the electron-withdrawing groups on the β-C increase the rate of the E2 reaction. This is because the $-I$ effect of such a group increases the acidity of the β-H and stabilises the carbanion character of the T.S. For example, the presence of an aryl group or a carbonyl group on the β-C increases the rate; 3-bromobutanone undergoes the E2 reaction very much slowly than 4-bromobutanone. In the latter case the electron-withdrawing carbonyl group makes the β-H more acidic than that of the former and consequently the T.S. of the former is less stable than that of the latter. As a consequence $\Delta G^{\ddagger}$ for the former is more than that of the latter.

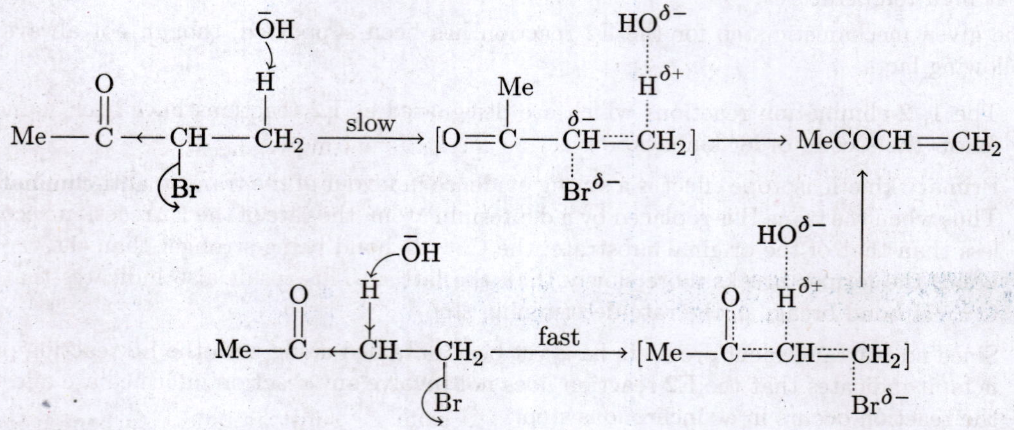

Here, not only the –I-effect of the carbonyl group but also the conjugative effect make the T.S. of the 4-bromobutanone more stable than the T.S. of the 3-bromobutanone where there is no conjugative effect.

(b) The nature of the leaving group and rate : In general, the better the leaving group, the higher is the rate of the E2 reaction. With the increasing size, the leaving group ability increases. Thus when β-phenethyl halides are treated with sodium ethoxide in ethyl alcohol, the rate of the E2 reaction increases with the increasing size of the halogen atom.

Substrate	$PhCH_2CH_2F$	$PhCH_2CH_2Cl$	$PhCH_2CH_2Br$	$PhCH_2CH_2I$
The rate	1	0.007×10^3	4.2×10^3	27×10^3

It has also been found that with the increasing size of the halogen atom E2/S_N2 ratio increases but to a minimum extent.

(c) The strength and the concentration of the added base and rate : With the increasing basicity and the concentration of the added base, the rates of the E2 reactions have been found to increase. The order of basicity among sodium hydroxide, sodium ethoxide and sodium amide is :

$$\bar{N}H_2 > \bar{O}Et > \bar{O}H$$

The rate order of E2 reactions on a given substrate with the above said bases under identical conditions is also found to be :

$$\bar{N}H_2 > \bar{O}Et > \bar{O}H$$

Thus the order of basicity is the rate order.

When the base is weak but strongly nucleophilic toward carbon, E2/S_N2 ratio is low but in the presence of a strong base the E2/S_N2 ratio increases.

Since the T.S. involves the base, its concentration is expected to influence the rate. With the increasing concentration of the base, the rate of the E2 reaction will increase.

(d) The nature of the solvent and rate : The yield of the E2 reaction product increases with the decrease in solvent polarity. Thus the E2/S_N2 ratio in the reaction of Me_2CHBr at 55 °C with NaOH has been found to increase as the solvent polarity decreases :

Solvent	E2/S_N2 ratio
60% EtOH & 40% H_2O	1.17
80% EtOH & 20% H_2O	1.44
100% EtOH	2.45

(e) The effect of temperature and rate : It has already been stated that as the temperature of an E2 reaction increases, the rate of the reaction increases; naturally the E2/S_N2 ratio will also increase with the increasing temperature. This has been found to be so. Thus, E2/S_N2 ratio in the reaction of Me_2CHBr with NaOH in 60% ethyl alcohol is found to rise from 1.14 to 2.42 as the temperature is raised from 45 °C to 100 °C.

15.2.3 E1cB reactions

This is a two-step base-catalysed 1, 2-elimination reaction which involves a carbanion, the conjugate base of the substrate, as the reaction intermediate.

Step I : The base abstracts the most acidic β-H from the substrate and a carbanion forms. This is a reversible and fast step.

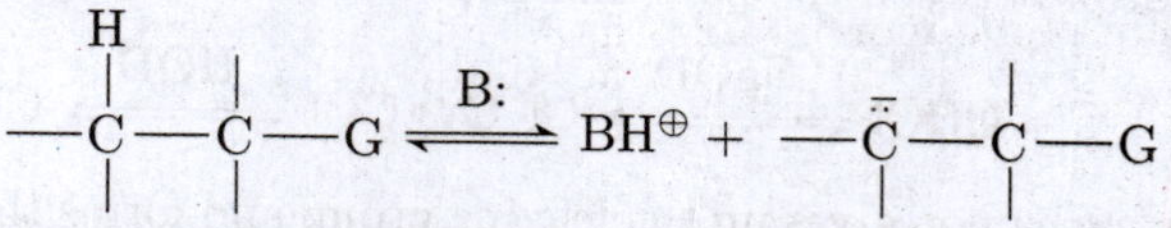

Step II : This leaving group leaves the carbanion, and a C—C double bond forms :

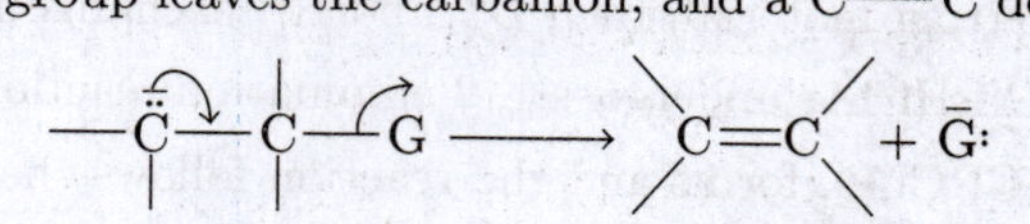

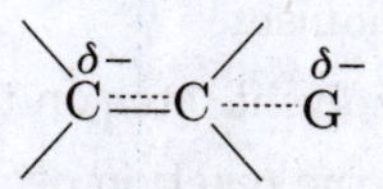

T.S. of an E1cB reaction

Since the T.S. for this step involves the conjugate base (cB) of the substrate only.

The rate of this step = k [cB].

This sort of 1, 2-elimination reaction is popularly called E1cB–elimination unimolecular with respect to the conjugate base of the substrate.

However, in some reactions the first step is the R.D. step and in some other reactions the second step is the R.D. step.

The overall reaction is :

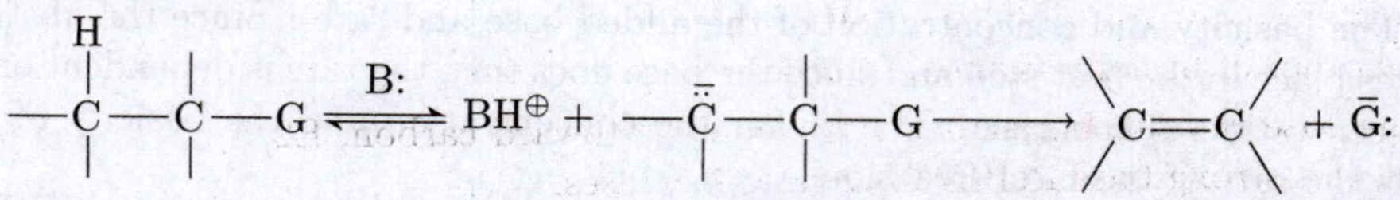

Since the generation of the carbanion involves both the base and the substrate and since the rate depends on the concentration of the carbanion, the overall reaction follows the second order kinetics; the rate = k [B] [substrate].

As the E1cB reaction occurs via a carbanion intermediate, the energy profile for the exothermic reaction may be drawn as given in Figure 15.4.

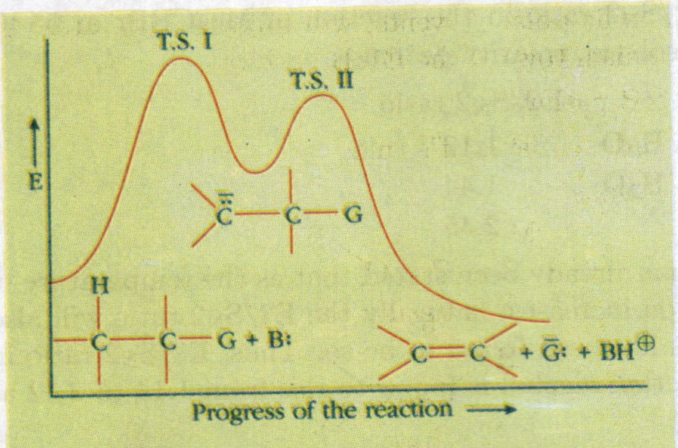

Fig. 15.4. The energy profile for the E1cB (exothermic) reaction

E1cB reactions usually proceed in the presence of concentrated strong bases in polar media at elevated temperatures. However, pure E1cB reactions are very rare.

Since the most acidic β-H is abstracted by the base before the leaving group departs the substrate, the Hofmann's product is expected to be formed predominantly.

Since carbanions are the intermediates in the E1cB reactions and since a carbanion can adopt the most stable conformation, the E1cB reactions are non-stereospecific.

The mechanistic steps as drawn above for the E1cB reaction are supported by the following two evidences :

(i) When $Cl_2C{=}CHCl$ is heated with NaOD, the recovered starting material is found to contain deuterium. This supports the reversible formation of a carbanion and the exchange of deuterium and proton between the base and the substrate.

$$Cl_2C{=}CHCl \underset{}{\overset{NaOD}{\rightleftharpoons}} Cl_2C{=}\bar{\ddot{C}}Cl \overset{HOD}{\rightleftharpoons} Cl_2C{=}CDCl$$

However, if the carbanion gives up the leaving group and forms the product much more readily than it reverts to the starting material, then $D^{\oplus}$ and $H^{\oplus}$ exchange may not take place.

(ii) When ArCOCHClCHClAr undergoes 1, 2-elimination reaction to give ArCOCCl=CHAr, a carbanion, $ArCO\bar{\ddot{C}}Cl.CHClAr$, forms and the reaction follows the second order kinetics which can be predicted on the basis of the E1cB mechanism.

Let us now discuss the effects of structure, solvent, etc. on the rate of the E1cB reactions.

(a) The substrate structure and rate : Since the reaction occurs via a carbanion intermediate, the rate of the reaction depends on the stability of the carbanion. The presence of electron-withdrawing groups on the β-C and any group which is able to undergo π-p(filled) conjugation will increase the rate of the reaction. Thus when any of the —NO_2, $PhSO_2$, C=O, aryl, etc. groups is present on the β-C, the rate of the 1, 2-elimination reaction by the E1cB pathway increases.

Furthermore, since an H atom on an sp^3 C is less acidic than an H atom on an sp^2 C, the formation of a C-C triple bond from a C-C double bond occurs more readily than the formation of a C-C double bond from a saturated compound.

(b) The basicity and concentration of the added base and rate : Since the abstraction of the β-H takes place in the first step and since the base does this, the rate is dependent on the strength and concentration of the base. The higher the concentration and the basicity of the base, the higher is the rate of the E1cB reaction.

(c) The nature of the leaving group and rate : Since the rate depends on the acidity of the β-H, the electron-withdrawing leaving groups, which are either positively charged or strongly electronegative, increase the rate of the E1cB reactions.

(d) The nature of the solvent and rate : Since carbanions are the intermediates in the E1cB reactions, the polar solvents, specially non-hydroxylic polar solvents, will increase the rate. A β-elimination reaction in DMSO moves the mechanism toward the E1cB pathway.

Problem 15.1. (a) Compare E1, E2 and E1cB pathways.

(b) Distinguish between the Hofmann's rule and the Saytzeff's rule.

(c) Account for the yields of the following reactions :

(i) 2-Bromopentane

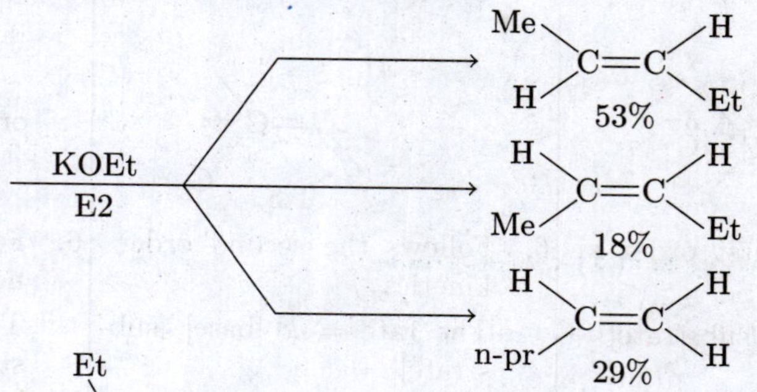

(ii)

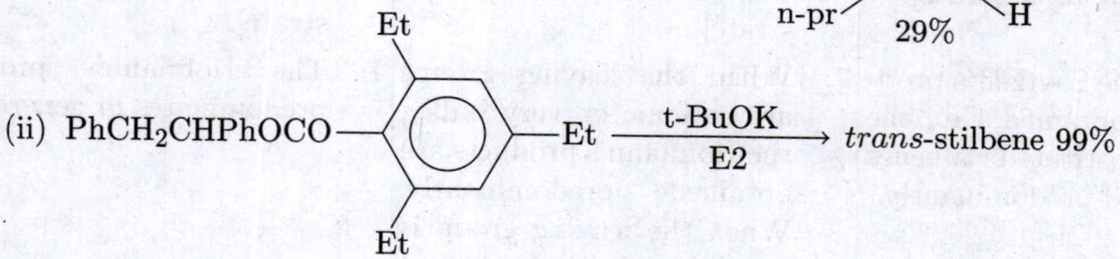

Solution (a) :

E1	E2	E1cB
1. A two-step process 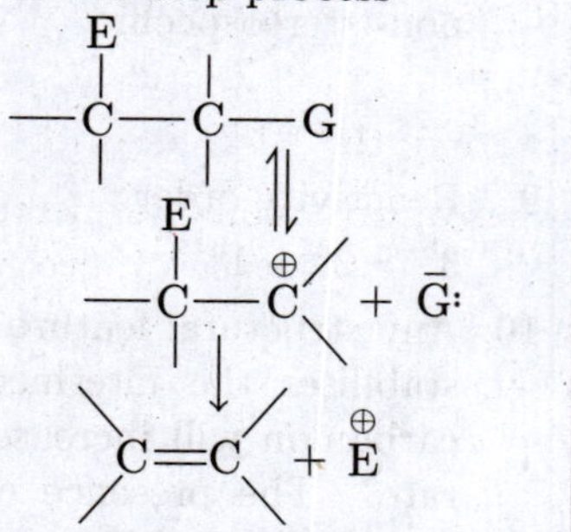	1. A one-step process 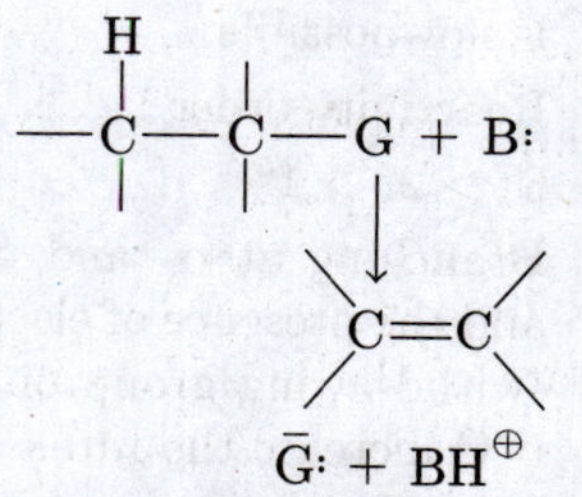	1. A two-step process 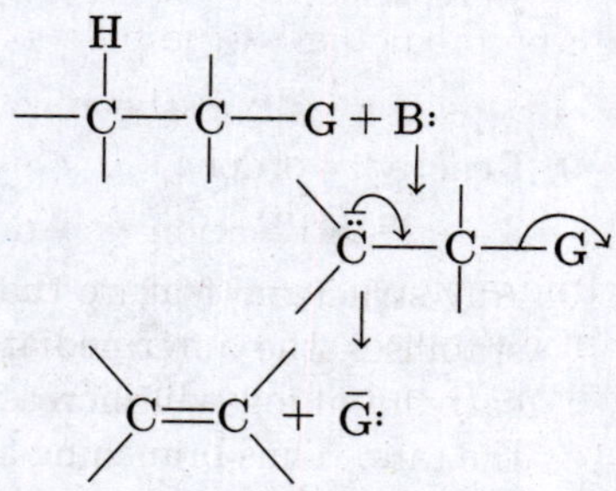
2. A carbenium ion forms as an intermediate.	2. No intermediate forms.	2. A carbanion forms as an intermediate.
3. In the second step the electrophile leaves the carbenium ion.	3. It does not have a second step	3. The leaving group leaves the carbanion either as an anion or as a neutral group.
4. Usually the solvent acts as the base and takes up the electrophile and no added base is required for this.	4. A strong base abstracts a proton or an electrophile.	4. A strong base abstracts a proton or an electrophile.

Contd.

E1	E2	E1cB
5. The first step is the rate-determining step. The T.S. involves only one species. 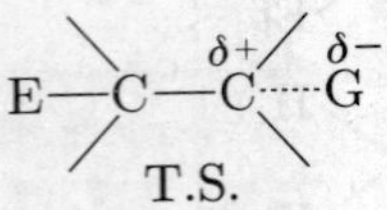 T.S.	5. The concerted step is the rate-determining step. The T.S. involves two species, the base and the substrate. 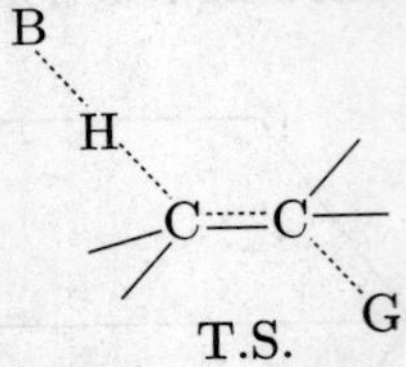 T.S.	5. Either the first step or the second step is the rate-determining step. Either 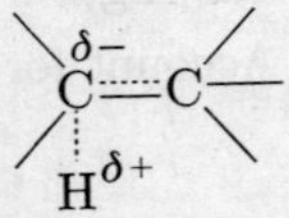 or, 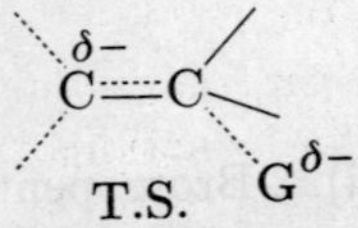 T.S.
6. Follows the first order kinetics The rate = k [substrate]	6. Follows the second order kinetics The rate = k [base] [substrate]	6. Follows the second order kinetics The rate = k [base] [substrate].
7. Usually the Saytzeff's products are obtained, i.e., the most substituted alkenes are formed predominantly.	7. When the leaving groups are cationic or very bulky, the Hofmann's products are obtained predominantly. When the leaving group is neutral and the base used is not bulky, the Saytzeff's product predominates.	7. The Hofmann's product predominates in general.
8. Non-stereospecific and non-stereoselective.	8. Stereoselective and stereospecific. Anti-elimination takes place; syn when anti is not possible.	8. Non-stereoselective and non- stereospecific.
9. Reactivity order : 3° > 2° > 1°	9. Reactivity order : 3° > 2° > 1°	9. Reactivity order : 3° > 2° > 1°
10. Any structural feature that stabilises the intermediate carbenium ion will increase the rate. Thus branching at α- and β-Cs increases the rate.	10. Branching at α- and β-Cs and the presence of electron -withdrawing group on the α-C increase the rate.	10. Any structural feature that stabilises the intermediate carbanion will increase the rate. The presence of an α-aryl group shifts a β-elimination reaction towards E1cB pathway.
11. Polar solvents increase the rate.	11. In a polar solvent the rate decreases.	11. Non-hydroxylic polar solvents increase the rate.
12. The concentration and the basicity of the solvent have no influence on the rate.	12. With increasing concentration and basicity of the added base, the rate increases.	12. With increasing concentration and basicity of the added base, the rate increases.

Contd.

E1	E2	E1cB
13. As the temperature increases, the rate increases.	13. As the temperature increases, the rate increases.	13. As the temperature increases, the rate increases.
14. S_N1 reactions compete with and analogous to E1 reactions.	14. S_N2 reactions compete with and analogous to E2 reactions.	14. S_N reactions do not compete with and analogous to E1cB reactions
15. Common.	15. The most common.	15. Very rare.

Solution (b) :

The Saytzeff's rule	The Hofmann's rule
1. Deals with the orientation of C-C double and triple bonds in β-elimination reactions.	1. Deals with the orientation of C-C double and triple bonds in β-elimination reactions.
2. Tells that the most substituted alkene will be formed predominantly in a β-elimination reaction.	2. Tells that the least substituted alkene will be formed predominantly in a β-elimination.
3. Thermodynamic stability of the product determines the direction of the formation of the double bond.	3. The acidity of the β-H and the steric strain on the β-C determine the direction of the formation of the double bond.
4. Usually E1 reactions follow the Saytzeff's rule.	4. Usually E2 and E1cB reactions follow the Hofmann's rule.
5. E2 reactions involving neutral and small leaving groups usually follow the Saytzeff's rule.	5. E2 reactions involving bulky positively charged leaving groups follow the Hofmann's rule, e.g., $\overset{\oplus}{N}R_3$, $\overset{\oplus}{S}Me_2$.
6. As the bulk of the leaving group increases, the yield of the Saytzeff's product decreases.	6. As the bulk of the leaving group increases, the yield of the Hofmann's product increases.
7. With the increasing bulk of the bases the yield of the Saytzeff's product decreases.	7. With the increasing bulk of the bases the yield of the Hofmann's product increases.

Solution (c) : (i) Since the reaction follows the E2 pathway, the trans-elimination will take place. The most stable conformation of 2-bromopentane in which one H atom is at anti-position with respect to the leaving group Br, a neutral group will give the major Saytzeff's product. Any gauche interaction, i.e., eclipsing effect will decrease the yield.

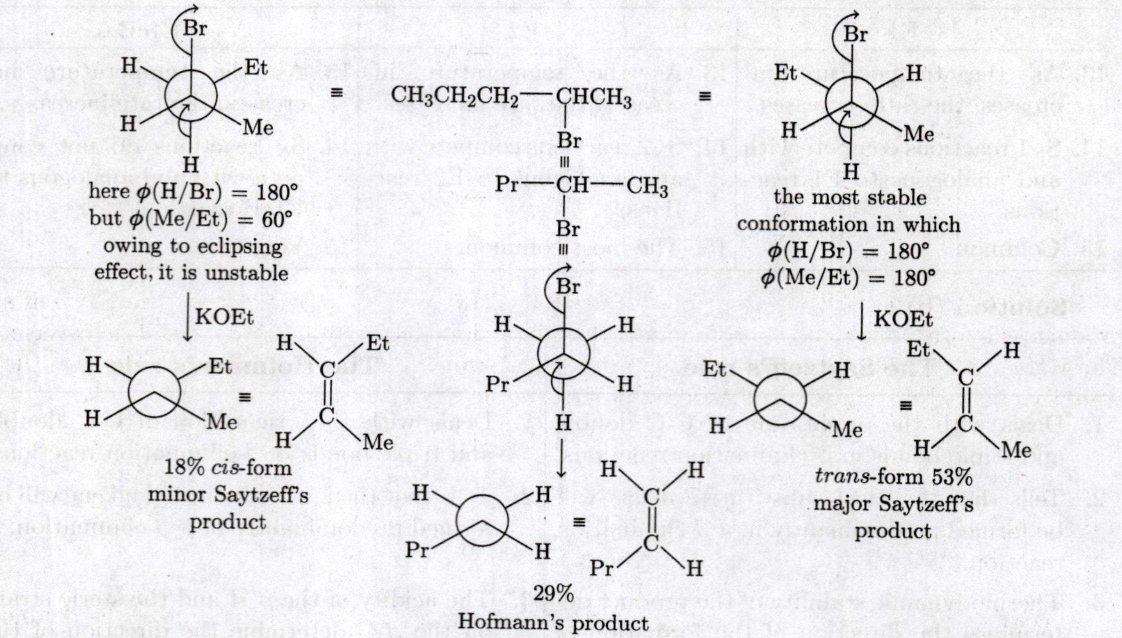

(ii) In this case the leaving group is 2, 4, 6-triethylbenzoate. Here the direction of the double bond is fixed since there is only one β-C; here the alkene will be PhCH$=$CHPh. Now which one will be formed predominantly—cis- or trans-form? Since it is an E2 reaction the trans H will leave the substrate as proton from the more stable conformation.

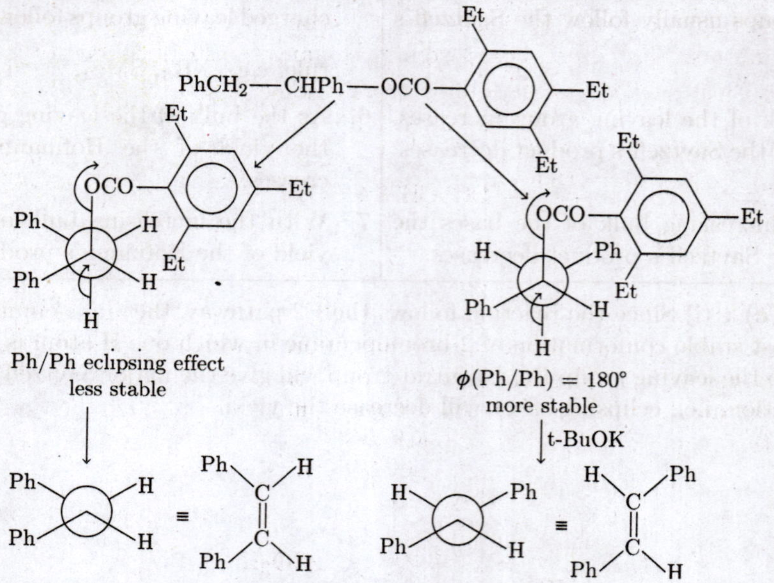

Owing to the Ph/Ph eclipsing effect [ϕ(Ph/Ph) = 60°], the left side conformation is much less stable than the right side conformation in which ϕ(Ph/Ph) = 180°. So 99% of the product is trans and 1% is cis.

Problem 15.2. Account for the following yield :

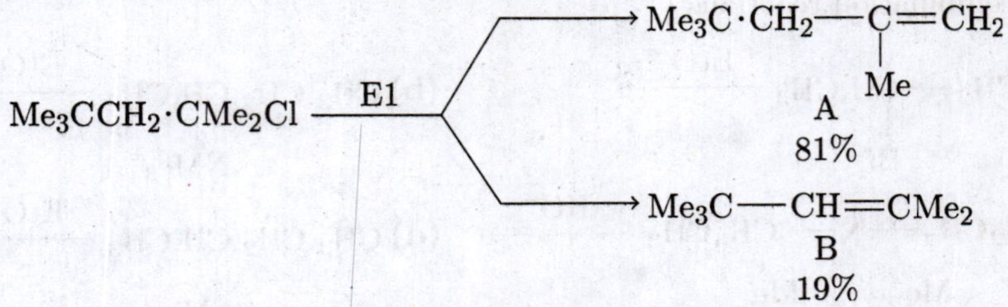

Solution : Since the reaction follows the E1 pathway, it is expected that the Saytzeff's product will predominate, i.e., compound B will predominate but here the result is reverse. The Hofmann's product, compound A, is formed in large excess over compound B. Perhaps the stability of B is less than that of A because of the steric effect.

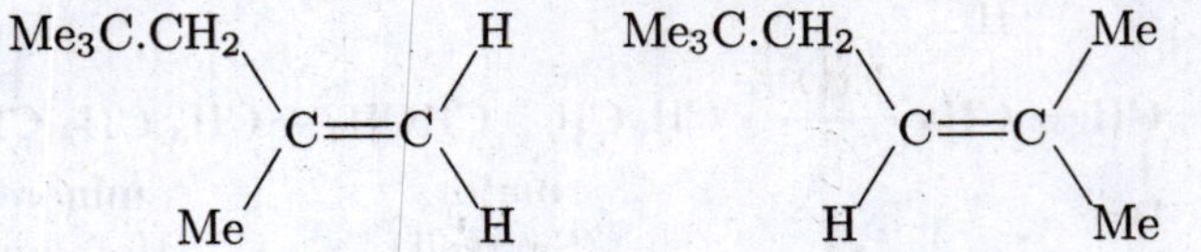

A B

In compound A, there are one H atom and a neopentyl group on the same side of the double bond— because of the small volume of the H atom, no steric strain develops. On the other hand, in compound B, a neopentyl group and a methyl group are present on the same side of the double bond; because of their bulk, heavy steric strain develops and makes compound B very unstable. Owing to the instability of compound B, compound A forms in large excess.

Problem 15.3. Dehydrobromination of bromofumaric acid to yield butynedioic acid proceeds much more rapidly than the corresponding elimination from its diastereomer, bromomaleic acid, in the alkaline medium—suggest the mechanistic pathway and the stereochemistry for the reaction.

Solution : Let us draw the structures for the two isomeric substrates :

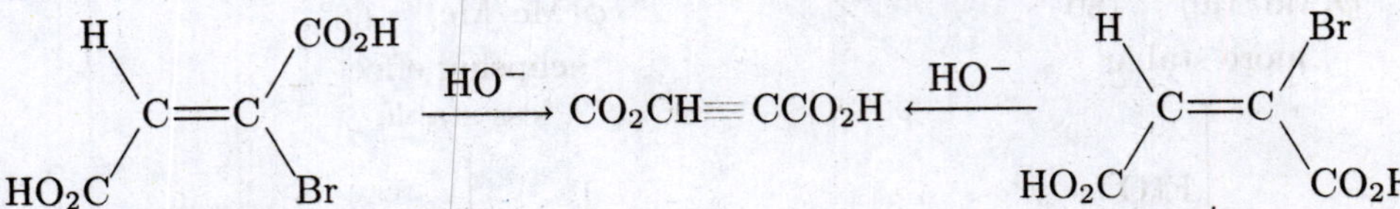

trans Bromofumaric acid — *cis* Bromomaleic acid

The two compounds differ only in configuration. In bromofumaric acid the leaving groups, Br and H, are trans to each other while in bromomaleic acid, they are cis to each other. Perhaps trans-elimination is favourable in this 1, 2-elimination reaction. Since trans-elimination is favoured in an E2 reaction, this reaction follows the E2 pathway.

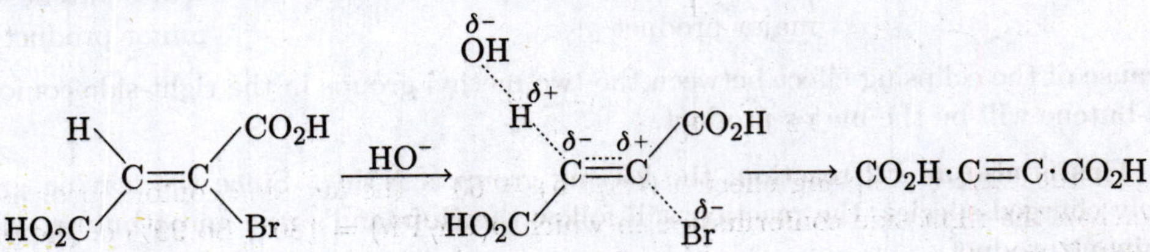

T.S.

Problem 15.4. Write down the names and the structures of the predominant products for the following elimination reactions :

(a) $CH_3.CH_2—CH(Br).CH_3 \xrightarrow{EtO^-}$ (b) $CH_3.CH_2.CH(\overset{\oplus}{N}Me_3).CH_3 \xrightarrow{EtO^-}$

(c) $PhCH_2CH_2—\overset{\oplus}{N}(Me)(Me)—CH_3CH_3 \xrightarrow{HO^-}$ (d) $CH_3.CH_2.CH(\overset{\oplus}{S}Me_2).CH_3 \xrightarrow{EtO^-}$

Give reasons for your answer.

Solution : (a) Since a base, EtO^-, is used, the reaction is expected to follow the E2 pathway. Here Br, a neutral group, is a leaving group. So the Saytzeff's product will form predominantly.

$$CH_3—\underset{H}{CH}—\underset{Br}{CH}—\underset{H}{CH_2} \xrightarrow{EtO^-} \underset{\substack{\text{major}\\ \text{(Saytzeff)}}}{CH_3CH{=}CHCH_3} + \underset{\substack{\text{minor}\\ \text{(Hofmann)}}}{CH_3.CH_2.CH{=}CH_2}$$

Now, the Saytzeff's product may be a mixture of cis- and trans-butenes.
Since trans-elimination takes place in the E2 reaction, the molecule may assume the following conformations :

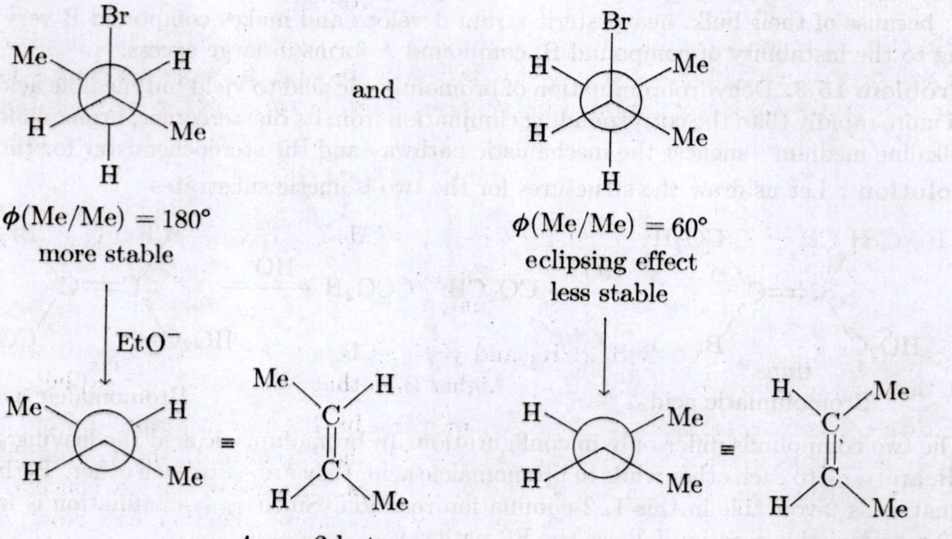

Because of the eclipsing effect between the two methyl groups in the right-side conformation, trans-2-butene will be the major product.

(b) This is also an E2 reaction; the leaving group is $\overset{\oplus}{N}Me_3$. Since the leaving group is a positively charged species, the reaction will follow the Hofmann's rule and 1-butene will be the predominant product.

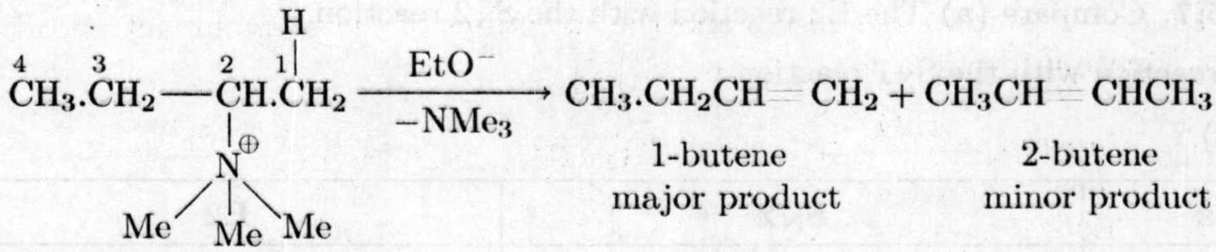

The base EtO^-, will abstract more acidic and the less sterically hindered H atom. The H atom on C_1 is more acidic and less sterically hindered than the H atom on C_3. This is because (i) the CH_3 group on C_3 has a +I effect which decreases the acidity of the H atom on C_3; (ii) this group creates steric hindrance to the base for abstracting the H atom and (iii) 3 β-H atoms on methyl group could remain in the anti-position with respect to the NMe_3 group, while only 2 β-H atoms are available for the stereoelectronic condition of the reaction and therefore, statistical probability also helps form Hofmann's product.

(c) This is an E2 reaction. The more acidic H is on $PhCH_2$— moiety. −I effect of the Ph group makes the H more acidic than that of CH_3 group. Furthermore, $PhCH{=}CH_2$, the elimination product, is more stable than $CH_2{=}CH_2$, the other product of elemination. Because of these two reasons, $PhCH{=}CH_2$, styrene, will be the major product.

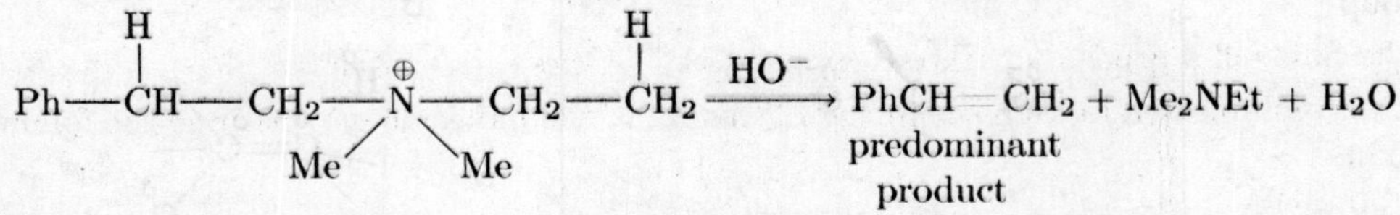

(d) In this E2 reaction, a positively charged group is a leaving group. So, the Hofmann's product will be the predominant product—the base will abstract the more acidic H which is on C_1 atom. The statistical factor as stated in (b), should also be taken under consideration.

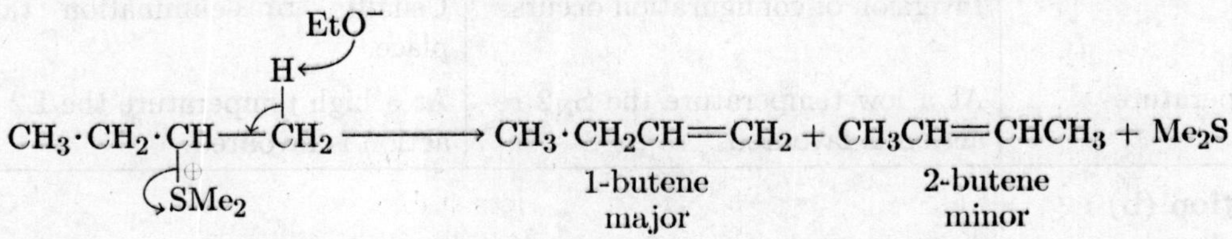

Problem 15.5. When $CH_3CHBrCH_3$ and $(CH_3)_3CBr$ are separately treated with 0.05N NaOEt, the yield of propylene is found to be higher than that of isobutylene—explain.

Solution : The concentration of the base is low here. In this concentration isopropylbromide undergoes β-elimination reaction by the E2 pathway. On the other hand, t-butyl bromide ionises more readily to t-butyl cation than isopropyl bromide does. Because of the low concentration of the base, it undergoes the S_N1 reaction more rapidly than the E1 reaction, as a result of which the yield of the E1 product of Me_3CBr becomes less than that of the E2 product from Me_2CHBr.

Problem 15.6. In one vessel t-butyl chloride is treated with 80% ethyl alcohol at 25 °C; in another vessel the same substrate is heated with 80% ethanol at 65°C—which one of the two will give you the higher yield of isobutylene and why ?

Solution : In both the cases S_N1 and E1 reactions will take place. At a low temperature, $E1/S_N1$ ratio is very low. As the temperature is raised, $E1/S_N1$ ratio rises. So at 65 °C the yield of isobutylene will be higher than that at 25 °C.

Problem 15.7. Compare **(a)** The E2 reaction with the S_N2 reaction.

(b) The E1 reaction with the S_N1 reaction.

Solution (a) :

Points	S_N2	E2
1. Step	One 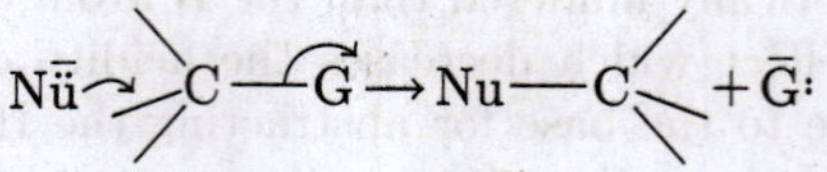	One 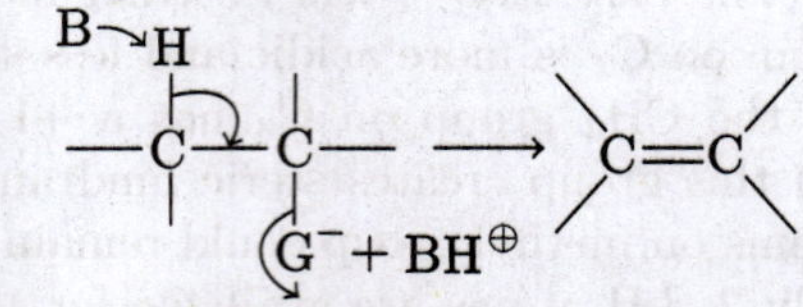
2. Reagent	Strong nucleophile	Strong base
3. Solvent	Non-hydroxylic solvents of high polarity increase the rate.	Non-hydroxylic solvents of low polarity increase the rate.
4. Structure and reactivity: nature of the alkyl group	$1° > 2° > 3°$	$3° > 2° > 1°$
5. T.S.	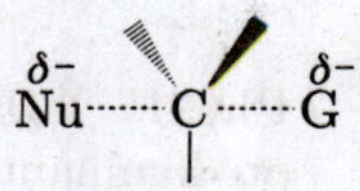	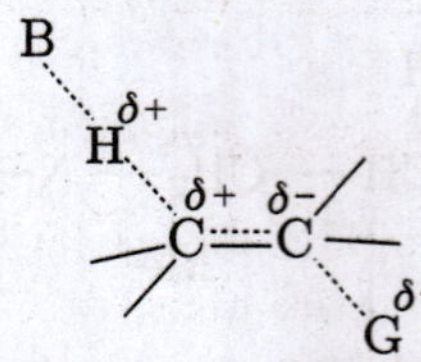
6. Kinetics	The rate = k [substrate] [Nu].	The rate = k [substrate] [base].
7. Stereospecificity	Stereospecific Inversion of configuration occurs.	Stereospecific Usually trans-elimination takes place.
8. Temperature	At a low temperature the S_N2 reaction is favoured.	At a high temperature the E2 reaction is favoured.

Solution (b) :

Points	E1	S_N1
1. Step	Two steps 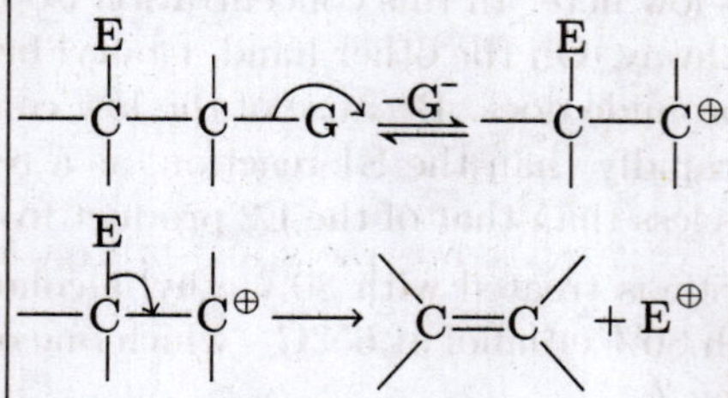	Two steps 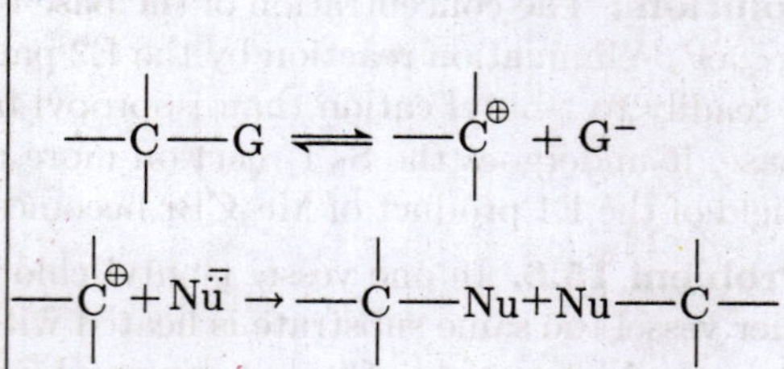
2. Reagent	Solvent plays the role of the reagent.	Solvents and weakly basic reagents of low concentration.

Contd.

Points	E1	S_N1
3. Solvent	Polar hydroxylic solvents of low dielectric constant.	Polar hydroxylic solvents of high dielectric constant.
4. Structure and reactivity: Nature of the alkyl group.	3° > 2° > 1°	3° > 2° > 1°
5. Kinetics	The first step is the rate-determining step. It follows the first order kinetics.	The first step is the rate-determining step. It follows the first order kinetics.
6. T.S.	$-\overset{\displaystyle E}{\overset{\vert}{C}}-\overset{\vert}{\underset{\vert}{C}}{}^{\delta+}\cdots G^{\delta-}$	$-\overset{\vert}{\underset{\vert}{C}}{}^{\delta+}\cdots G^{\delta-}$
7. Stereochemistry	Non-stereospecific.	Racemisation and partial inversion occur.
8. Temperature	High temperature favours.	Low temperature favours.

15.3 Acid-Catalysed Dehydration of Alcohols

When an alcohol is heated at a specific temperature with a non-volatile acid such as concentrated sulphuric, phosphoric or *p*-toluenesulphonic acid, an alkene forms. The alkene is either distilled out from the reaction mixture or separated from that as an insoluble layer. The acids catayse the reaction. If the alcohol is a volatile one, the reaction may be carried out by passing the vapour of the alcohol through a tube packed with pumice stone impregnated with the acid and heated to the requisite temperature. The usual temperature ranges from 150 °C to 200 °C.

The protic acid-catalysed dehydration of alcohols follow the E1 pathway. It consists of the following steps :

Initial step : Dissociation of the acid to proton and the conjugate base :

$$H_2SO_4 \rightleftharpoons H^{\oplus} + HSO_4^-$$

Step I : This involves the protonation of the alcoholic hydroxyl group.

$$(CH_3)_3C-OH \xrightleftharpoons{H^{\oplus}} (CH_3)_3C-\overset{\oplus}{O}H_2$$

Step II : $\overset{\oplus}{O}H_2$ being a good leaving group, at this step it leaves the species and a carbenium ion develops.

$$(CH_3)_3C-\overset{\oplus}{O}H_2 \rightleftharpoons (CH_3)_3\overset{\oplus}{C} + H_2O$$

Step III : The conjugate base of the acid catalyst abstracts a β-proton and usually the Saytzeff's product form

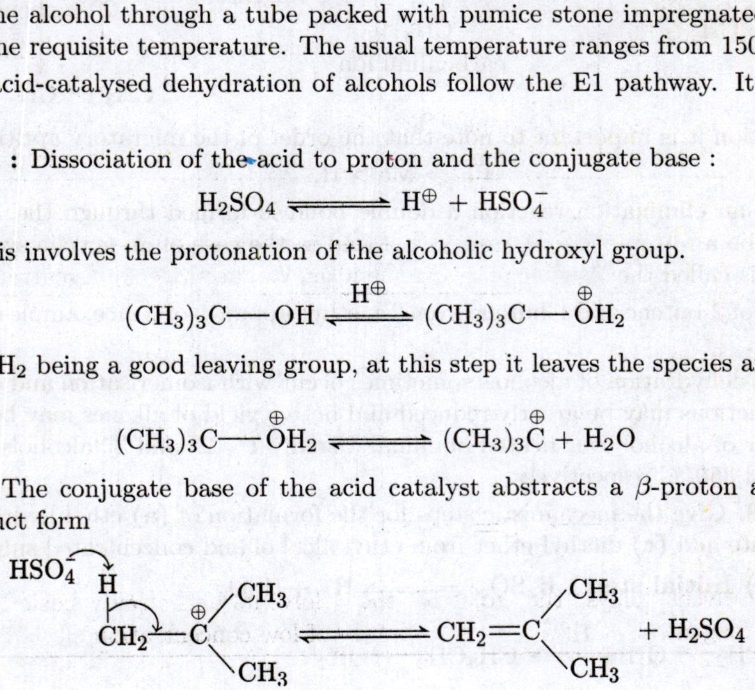

The reaction follows the first order kinetics : the rate = k [substrate]

However, Ethyl alcohol undergoes the E2 reaction.

The reactivity order of the alcohols is : 3° > 2° > 1°. The dehydration of 3° and 2° alcohols often occurs with boiling aqueous acid solutions.

Since carbenium ions are the intermediates in the dehydration of alcohols, the thermodynamically stable products, the Saytzeff's products, are formed predominantly. For the same reason, the reaction will be non-stereoselective.

$$CH_3CHOHCH_2CH_3 + H^{\oplus} \longrightarrow \begin{cases} CH_3CH{=}CHCH_3 \quad \text{major product} \\ CH_3CH_2CH{=}CH_2 \quad \text{minor product} \end{cases}$$

Carbenium ions being the intermediates in this reaction, the nucleophilic substitution and the rearrangement reactions compete with the reaction. For example, when 2-methylpropan-1-ol is heated with sulphuric acid, 2-butene forms as a major product along with 1-butene and 2-methylpropene. This shows that a carbenium ion is an intermediate and this rearranges to a more stable carbenium ion before the loss of proton takes place as shown below :

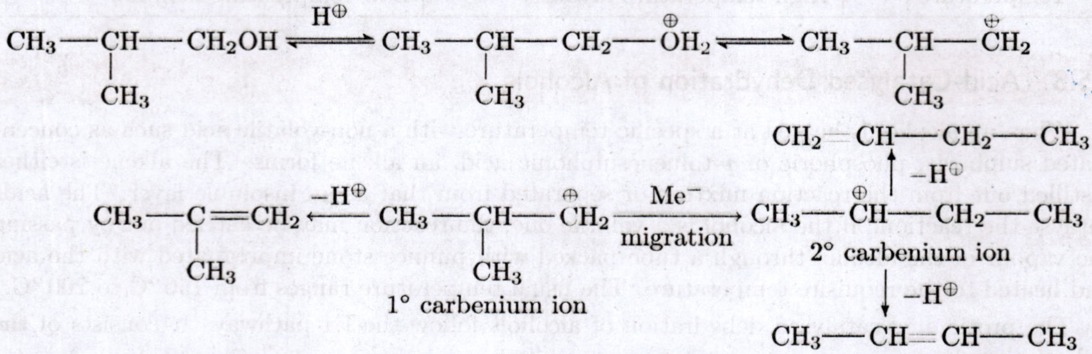

In this connection it is important to note that the order of the migratory aptitude is :

$$Ph \gg Me > H.$$

However, if in an elimination reaction a double bond is formed through the 1, 2-shift, the process is said to be a retropinacol rearrangement and in the case of an acyclic system this sort of rearrangement is called the Wagner rearrangement or, Wagner-Meerwein rearrangement.

The formation of 2-butene and 1-butene from 2-methylpropan-1-ol is an example of a retropinacol rearrangement.

Acid-catalysed dehydration of alcohols sometimes occur with isomerisation and etherification. These two side reactions may be greatly reduced and better yield of alkenes may be obtained by passing the vapour of alcohol over heated alumina. Usually 1°, 2° and 3° alcohols dehydrate at 150 °C, 250 °C and 350 °C respectively.

Problem 15.8. Give the mechanistic steps for the formation of **(a)** ethyl hydrogensulphate, **(b)** diethyl sulphate and **(c)** diethyl ether from ethyl alcohol and concentrated sulphuric acid.

Solution : (a) Initial step : $H_2SO_4 \rightleftharpoons H^{\oplus} + HSO_4^-$

Step I : $CH_3CH_2{-}OH \overset{H^{\oplus}}{\rightleftharpoons} CH_3CH_2{-}\overset{\oplus}{O}H_2$

Step II : S_N2 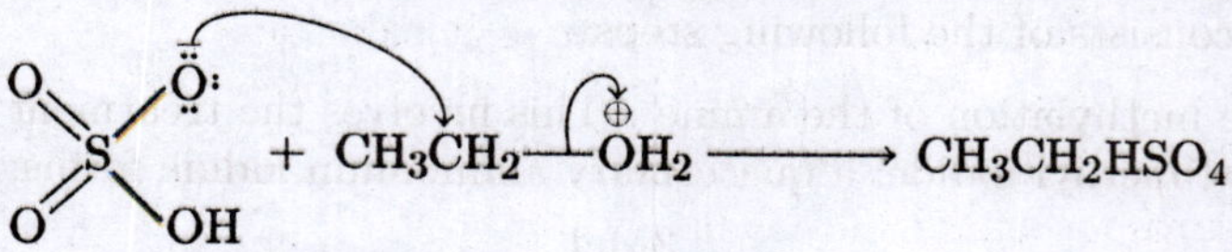

(b) At first ethyl hydrogensulphate forms as shown above and then diethyl sulphate forms through the following steps.

Step III : 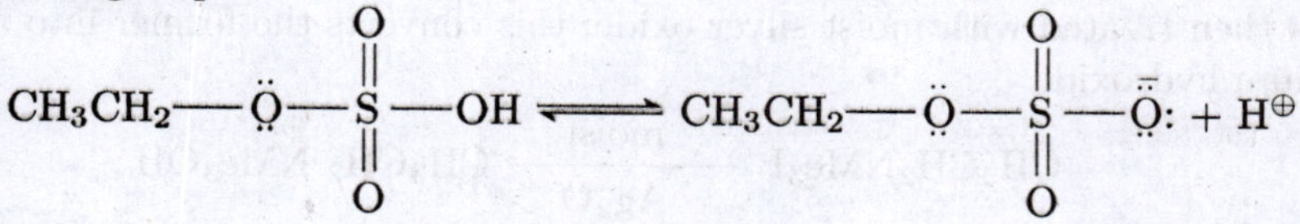

Step IV : $CH_3CH_2—OH \underset{}{\overset{H^{\oplus}}{\rightleftharpoons}} CH_3CH_2—\overset{\oplus}{O}H_2$

Step V : S_N2

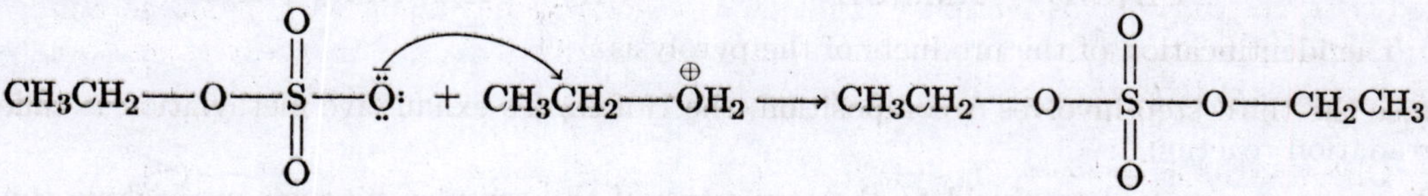

(c) Initial Step : $H_2SO_4 \rightleftharpoons H^{\oplus} + HSO_4^-$

Step I : $CH_3CH_2—OH + H^{\oplus} \rightleftharpoons CH_3CH_2—\overset{\oplus}{O}H_2$

Step II : S_N2 : 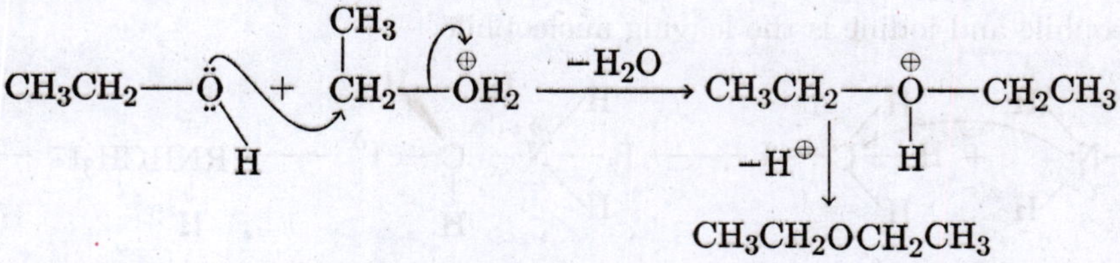

15.4 The Hofmann's Exhaustive Methylation

The complete process of converting an amine, 1°, 2° or 3°, to an unsaturated compound by the pyrolytic decomposition of the quaternary ammonium hydroxide (obtained by treating the amine with the excess of methyl iodide and then hydrolysing the resulting quaternary ammonium iodide with moist silver oxide) and identifying the products of pyrolysis of corresponding quaternary ammonium hydroxide so formed is called the Hofmann's exhaustive methylation or the Hofmann's degradation. The general reaction sequence may be given as shown below :

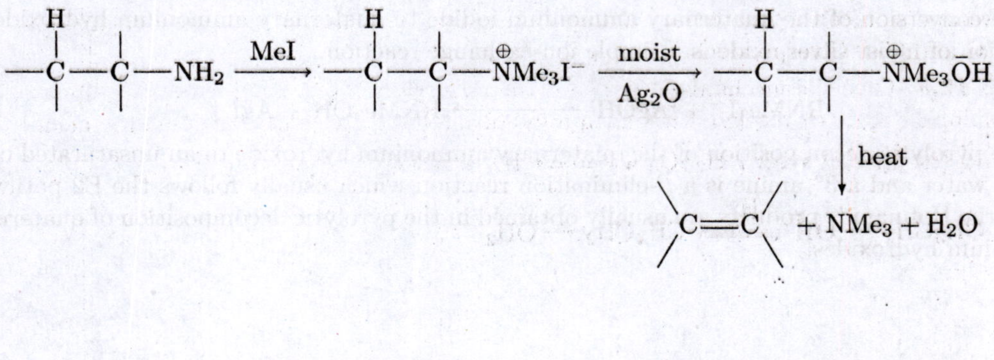

Thus the process consists of the following steps :

(i) The exhaustive methylation of the amine : This involves the treatment of the amine with a large excess of methyl iodide; a quatermary ammonium iodide forms.

$$CH_3CH_2NH_2 \xrightarrow{3MeI} CH_3CH_2\overset{\oplus}{N}Me_3I^-$$

(ii) The formation of the quaternary ammonium hydroxide : The quarternary ammonium iodide is then treated with moist silver oxide; this converts the former into a quarternary ammonium hydroxide.

$$CH_3CH_2\overset{\oplus}{N}Me_3I^- \xrightarrow[Ag_2O]{moist} CH_3CH_2\ \overset{\oplus}{N}Me_3\overset{-}{O}H$$

(iii) The pyrolysis of the quaternary ammonium hydroxide : The quarternary ammonium hydroxide is heated at 100 °C in an aqueous or alcoholic solution.

$$CH_3CH_2{-}\overset{\oplus}{N}Me_3\overset{-}{O}H \xrightarrow{heat} CH_3{=}CH_2 + NMe_3 + H_2O$$

(iv) The identification of the products of the pyrolysis :

Since the third step involves decomposition, the Hofmann's exhaustive methylation is called a degradation reaction.

The process was used to elucidate the structures of the complex nitrogenous organic compounds such as alkaloids. We shall discuss this later on.

Mechanism : The conversion of an amine to a quaternary ammonium iodide by treating the former with an excess of methyl iodide is an S_N2 reaction; here the amine is the incoming nucleophile and iodine is the leaving nucleophile.

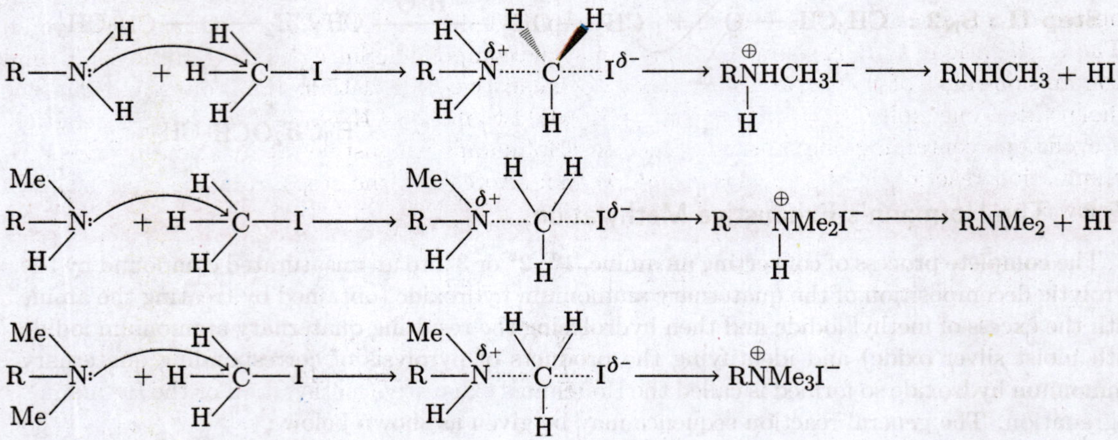

The conversion of the quaternary ammonium iodide to quaternary ammonium hydroxide by the action of moist silver oxide is a simple ion-exchange reaction.

$$R\overset{\oplus}{N}Me_3I^- + \text{'AgOH'} \longrightarrow R\overset{\oplus}{N}Me_3\overset{-}{O}H + AgI\downarrow$$

The pyrolytic decomposition of the quaternary ammonium hydroxide to an unsaturated compound, water and a 3° amine is a β-elimination reaction which usually follows the E2 pathway; and so the Hofmann's products are usually obtained in the pyrolytic decomposition of quaternary ammonium hydroxides.

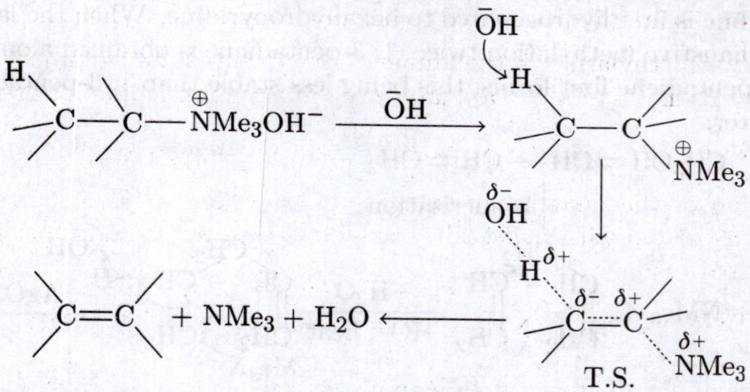

In this case the most acidic H from the less sterically hindered β-carbon gets eliminated. For this elimination the most stable conformation of the substrate requires to assume the anti- or trans-periplanar arrangement. The hydroxide ion also lies on the trans-periplanar plane of the substrate because this helps the hydroxide ion to abstract the proton easily.

As stated earlier the Hofmann's exhaustive methylation was used to elucidate the structures of alkaloids. This process may also be used to prepare cycloalkenes. However, the elucidation of structure often involves the repetition of the four steps of the process. Let us discuss this with a suitable example : A basic nitrogenous organic compound of molecular formula C_4H_5N gave C_4H_9N on complete hydrogenation. When the hydrogenated product, C_4H_9N, is subjected to the Hofmann's exhaustive methylation twice, 1, 3-butadiene, trimethylamine and two moles of water are obtained, what is the structure of the organic base?

Since the compound absorbs 2 moles of H_2 ($C_4H_9N - C_4H_5N = 2H_2$), the compound is expected to contain either 2 double bonds or 1 triple bond, and C_4H_9N is a saturated compound. The C : H ratio in C_4H_9N shows that this is a cyclic compound. Since one mole of the compound contains one mole of N atom and since repeated exhaustive methylation of the compound gave one mole NMe_3, one mole $CH_2{=}CH{-}CH{=}CH_2$ and two moles of H_2O, the compound is definitely a cyclic one containing one N atom. Since the Hofmann's exhaustive methylation involves a β-elimination reaction in the pyrolysis step, we can stretch out the steps from the end as shown below :

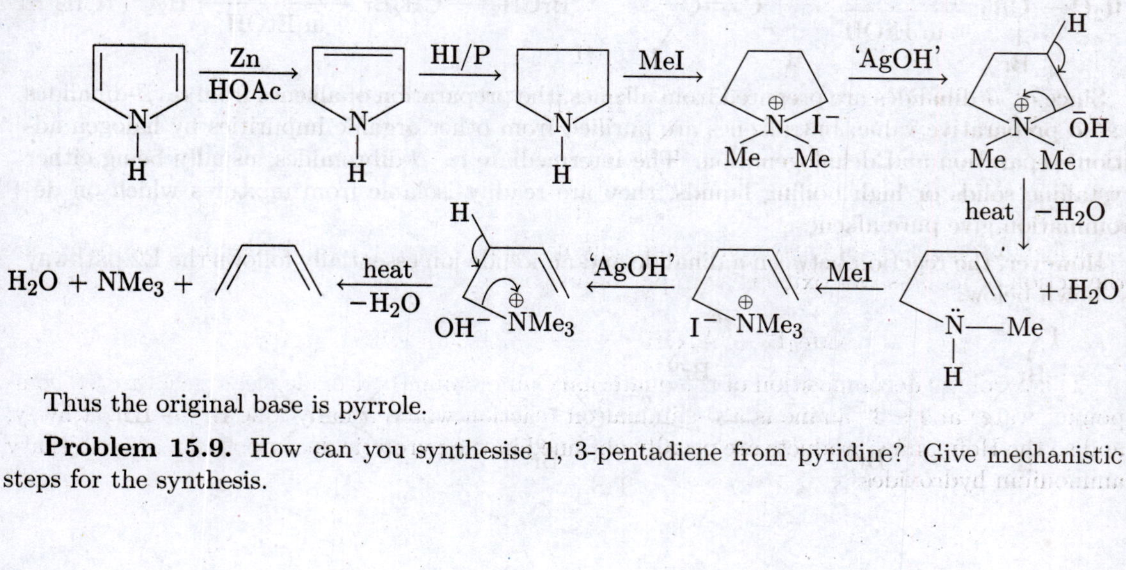

Thus the original base is pyrrole.

Problem 15.9. How can you synthesise 1, 3-pentadiene from pyridine? Give mechanistic steps for the synthesis.

Solution : Pyridine is first hydrogenated to hexahydropyridine. When the latter is subjected to the Hofmann's exhaustive methylation twice, 1, 3-pentadiene is obtained along with trimethylamine. In fact, 1,4-pentadiene first forms; this being less stable than 1, 3-pentadiene, the former isomerises to the latter.

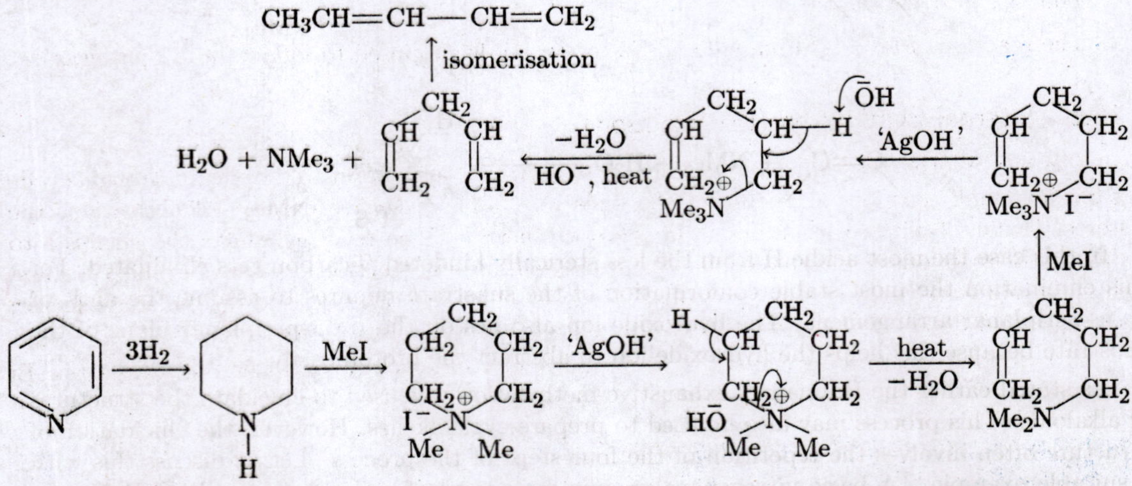

Problem 15.10. What products do you expect when tetramethylammonium hydroxide is heated **(i)** with an aqueous caustic soda solution and **(ii)** in the absence of a solvent?

Solution : (i) When tetramethylammonium hydroxide is heated in water, methyl alcohol is obtained; **(ii)** In the absence of a solvent dimethyl ether is the product.

15.5 Dehalogenation of α, β-Dihalides

Dehalogenation of α, β-dihalides, mainly dibromides may be carried out with Zn or Mg in ethanol or sodium iodide in ethanol or with phenyllithium, etc.

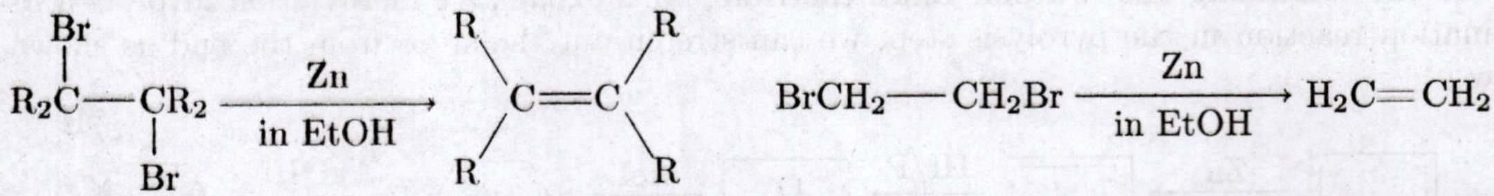

Since α, β-dihalides are prepared from alkenes, the preparation of alkenes from α, β-dihalides has no preparative value; but alkenes are purified from other organic impurities by halogen addition, separation and dehalogenation. The intermediate α, β-dibromides, usually being either crystalline solids or high boiling liquids, they are readily isolable from mixtures which on debromination give pure alkenes.

However, the reaction between a dihalide and an iodide ion essentially follows the E2 pathway as shown below.

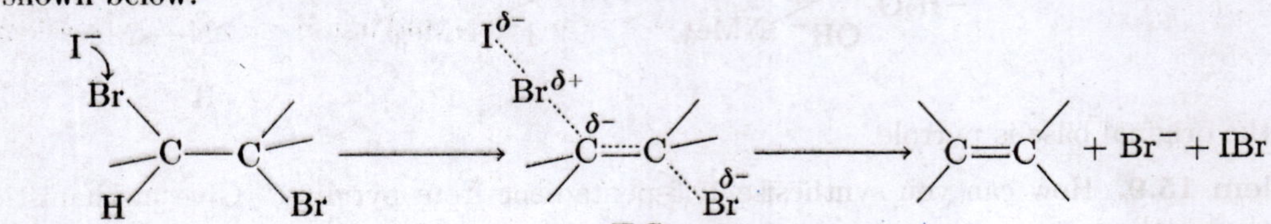

T.S.

The two Br atoms, the two involved Cs and the I^- ion are in a trans-periplanar arrangement in the T.S. and through a β-elimination reaction the alkene is obtained. This is stereospecific trans-elimination reaction. The reaction follows the second order kinetics.

The rate = k [substrate) $[I^-]$

The reaction of an α, β-dihalide and Zn in alcohol is supposed to follow the E2 pathway.

15.6 Internal Elimination (Ei) Reactions

There is another type of elimination reactions which are supposed to occur through cyclic T.S.s involving only one molecule of the substrate. The negative entropy of activation, the stereospecific cis-elimination and the first order kinetics of the reactions make the chemists to classify the reactions as intramolecular or internal elimination reactions involving cyclic T.S.s and thus they are designated as Ei reactions. The rate of Ei reactions depend on the steric factor; and the yield depends on steric as well as on statistical probability factor.

Ei reactions are, therefore, pericyclic elimination reactions. We shall here discuss three Ei reactions.

15.6.1 Pyrolysis of esters

When an acetate ester of an alcohol is heated at 500 °C, an alkene and acetic acid are obtained. This is a reversible reaction.

$$R_2CH.CH_2O.CO.Me \xrightleftharpoons{\text{pyrolysis } 500\,^\circ C} R_2C{=}CH_2 + CH_3COOH$$

$$MeCH_2.\underset{\substack{|\\ Me}}{CH}.O.CO.Me \xrightleftharpoons{500\,^\circ C} MeCH{=}CHMe + MeCO_2H + \underset{\text{major}}{MeCH_2CH{=}CH_2}$$

Alkenes may be prepared by this reaction.

The pyrolysis of acetate esters is designated as an Ei reaction. It is a single-step reaction which is supposed to involve a six-membered cyclic T.S.

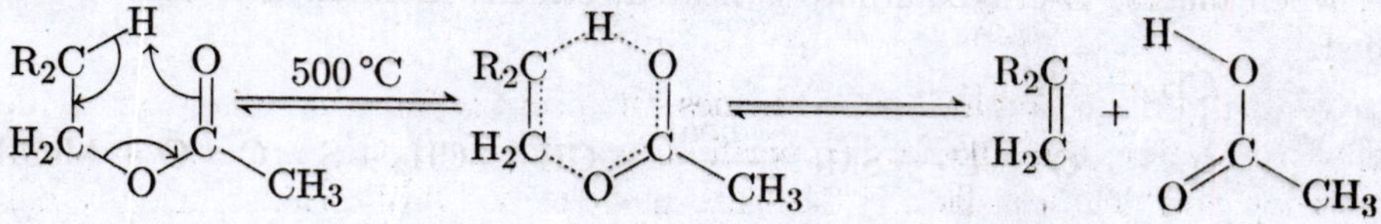

The reaction follows the first order kinetics :

The rate = k [substrate].

This is a stereospecific cis-elimination reaction. Erythro and threo forms of 2-deuterio-1, 2-diphenylethyl ethanoate on pyrolysis give mainly deuterio- trans- and non-deutereio- trans-alkenes respectively.

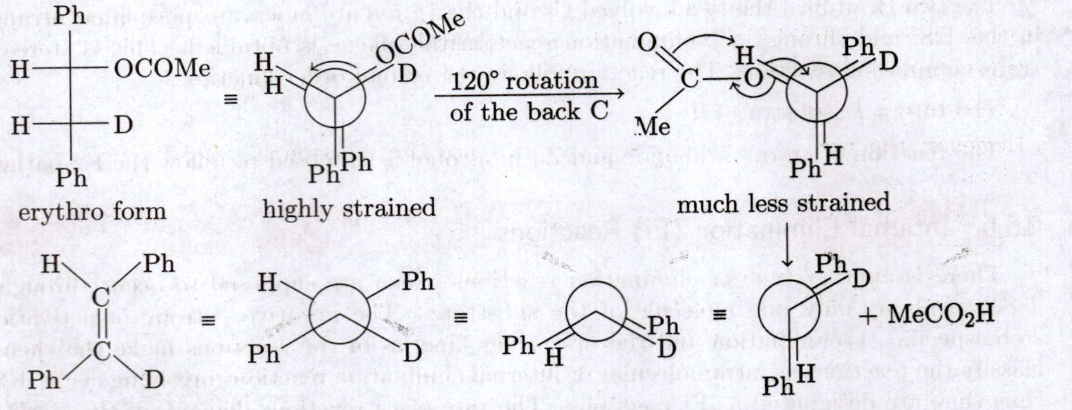

The cis-elimination, the first order kenetics and the negative entropy of activation support the formation of the cyclic T.S. and the Ei pathway.

15.6.2 The Chugaev reaction

This involves pyrolytic decomposition of xanthate esters.

O-Alkyldithiocarbonic acids, $R{-}O{-}C({=}S){-}SH$, are called alkylxanthic acids. Metallic salts of O-alkylxanthic acids are called xanthates; e.g., $R{-}O{-}C({=}S){-}\bar{S}Na^{\oplus}$ is sodium O-alkylxanthate. $R{-}O{-}C({=}S){-}SR'$ is a xanthate ester. Xanthate esters are prepared as follows:

$$CS_2 + ROH + NaOH \longrightarrow H_2O + R{-}O{-}C({=}S){-}\bar{S}Na^{\oplus} \xrightarrow{R'I} R{-}O{-}C({=}S){-}SR'$$

When a xanthate ester with a β-H in the O-alkyl part is pyrolysed, an alkene, carbonyl sulphide and a thiol form and the reaction is known as the Chugaev (Tschugaev) reaction. For example, when methyl O-ethylxanthate is heated, ethene, methanethiol and carbonyl sulphide are formed.

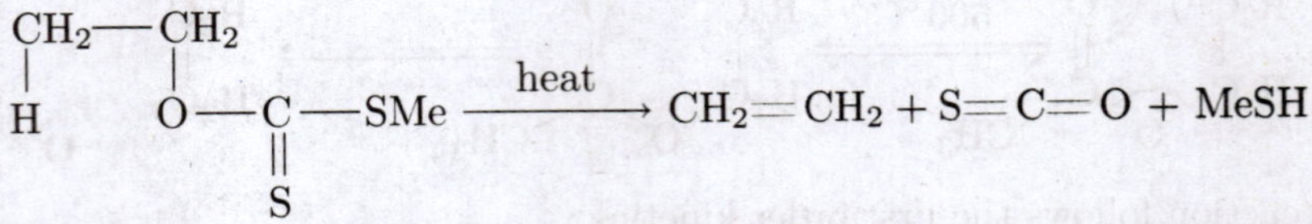

The yield has been found to increase when the xanthate is heated with chloroacetic acid and then the product is refluxed in slightly acidic solution. The rate of the reaction and the yield are found to increase when it is carried out in the presence of a Lewis catalyst, e.g., BF_3.

The reaction is supposed to be an intramolecular elimination reaction (Ei) involving a six-membered cyclic T.S.

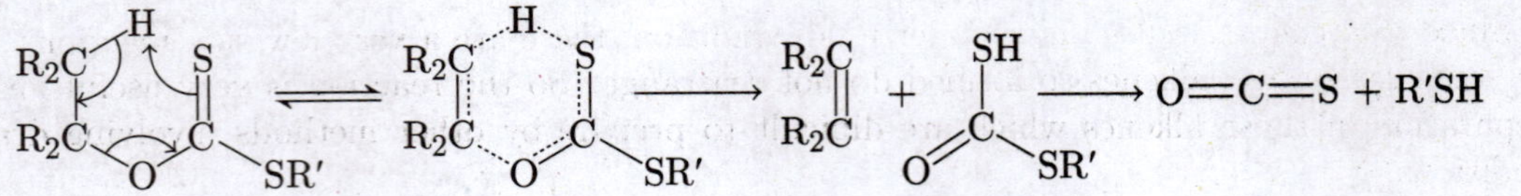

In the T.S, C=S attacks the β-H of the O-alkyl group : the reaction is not cent percent stereospecific—cis-elimination takes place predominantly.

It is a first order reaction.

The rate $= k$ [substrate].

However, the predominant cis-stereospecificity of the reaction has been supported by the reactions of diastereomeric xanthates.

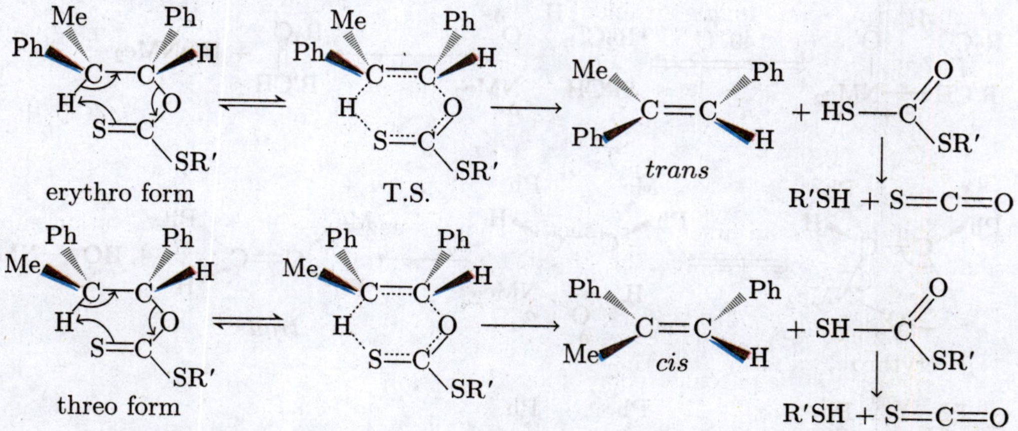

It is needless to mention that the negative entropy of activation, the first order kinetics of the reaction and the predominant cis-elimination support the formation of the cyclic T.S. and the pericyclic Ei pathway of the reaction. Here threo form reacts slower than the erythro form.

15.6.3 The Cope reaction

The thermal decomposition of a 3° amine oxide containing at least one β-H to an alkene and N, N-disubstituted hydroxylamine is known as the Cope reaction.

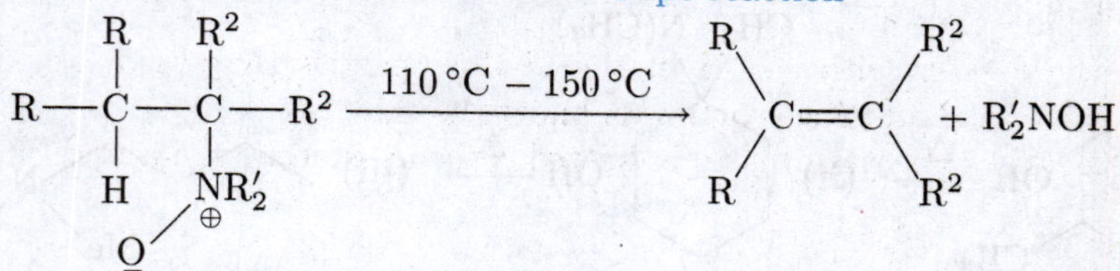

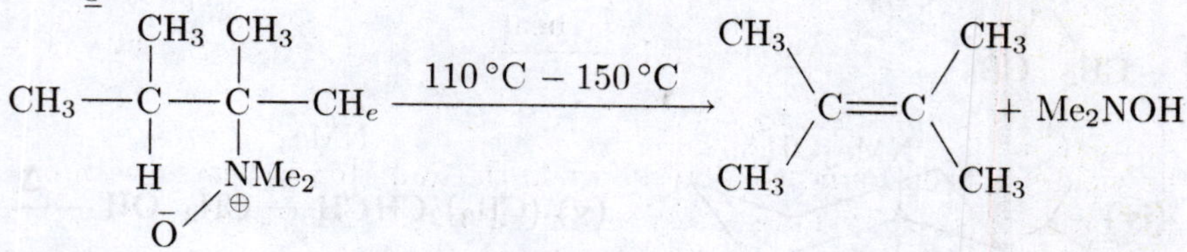

The reaction is not performed directly on the amine oxide but a mixture of the amine and an oxidising agent, e.g., H_2O_2, is heated at a temperature 110 °C – 150 °C. The reaction can also be carried out at room temperature in dry DMSO or THF (tetrahydrofuran).

Since the reaction takes place under mild condition, there are a very few side reactions. For the same reason the alkenes so formed do not rearrange. So the reaction is very useful for the preparation of those alkenes which are difficult to prepare by other methods involving drastic conditions.

The Cope reaction follows the Ei pathway involving a five-membered cyclic T.S. The reaction follows the first order kinetics, possesses negative entropy of activation and gives cis-elimination product stereospecifically. All these, in their turn, support the Ei pathway of the reaction.

When erythro form of 1-(N, N-dimethylamino)-1, 2-diphenylpropane oxide is pyrolysed, almost 93% trans-alkene is obtained, whereas the same treatment on the threo form gives almost 97% cis-alkene. This supports the predominance of the cis-elimination over the trans.

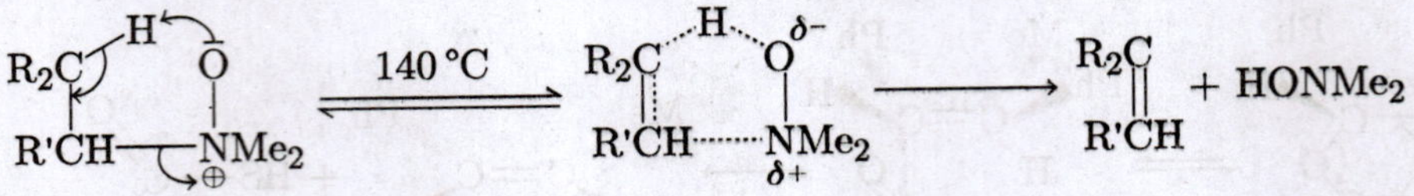

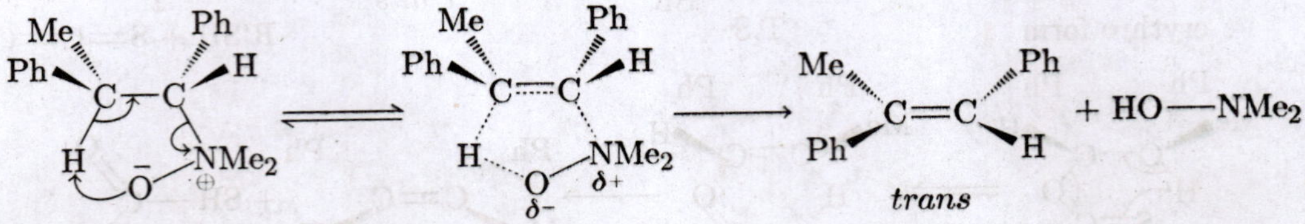

erythro

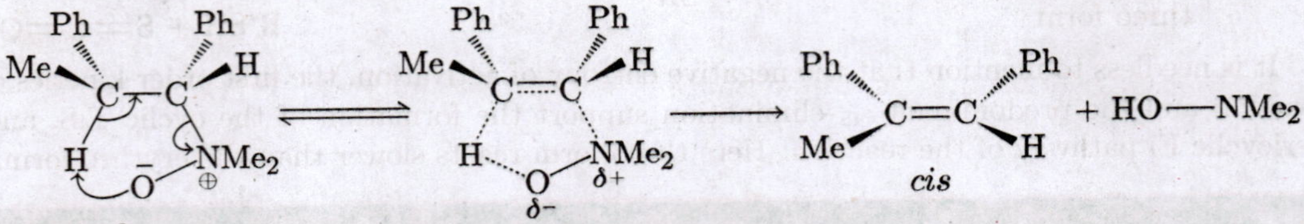

threo

Problem 15.11. Draw the constitutional formulae of the products for the following reactions :

(i) 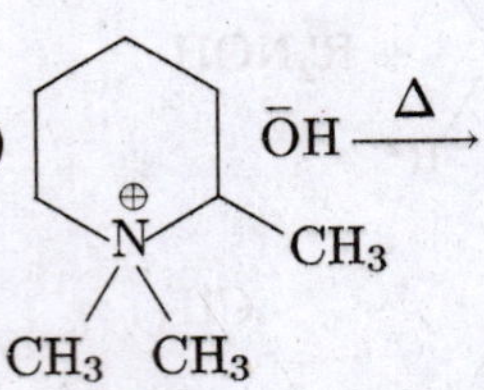 (ii) 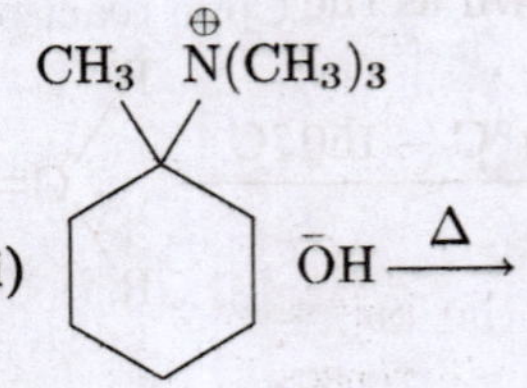 (iii) 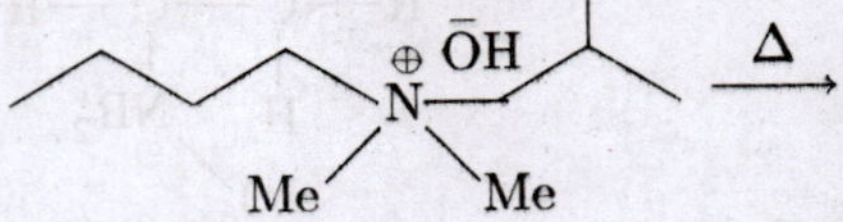

(iv) 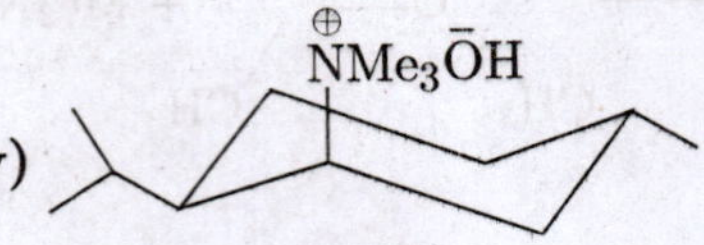 (v) $(CH_3)_2CH\underset{}{\overset{\oplus}{\overset{NMe_3}{CH}}}—CH_3\ \bar{O}H \xrightarrow{\Delta}$

Solution : (i) 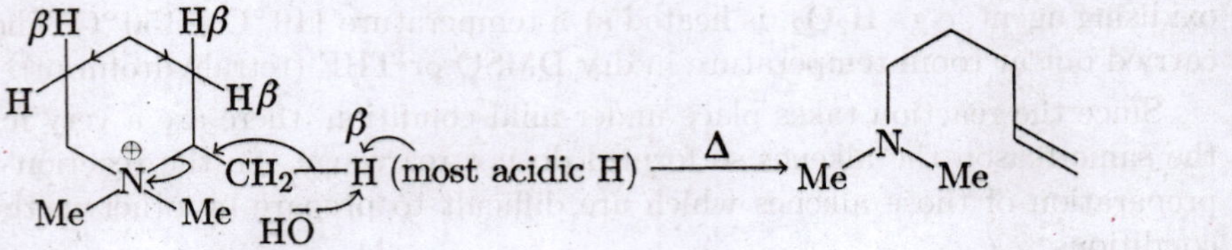

(ii) 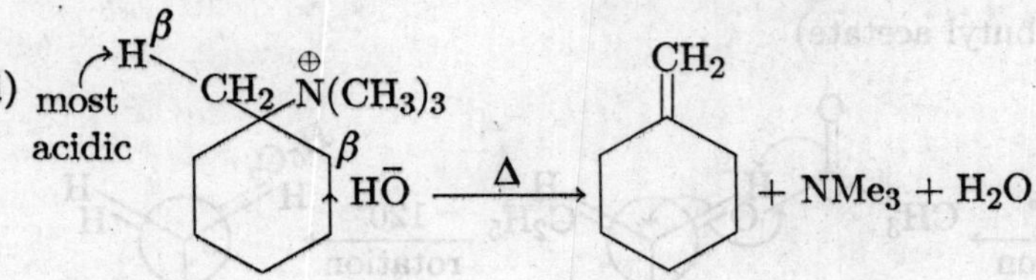

(iii) 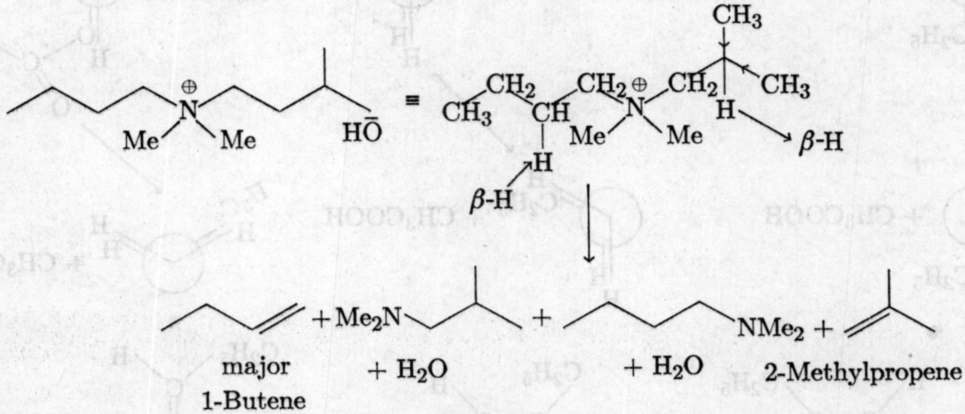

(iv) The compound has an anchoring substituent, the t-butyl group, and hence it has a fixed conformation. There are two β-Hs—one 3° and the other 2°. For steric reasons the 3° H is abstracted predominantly.

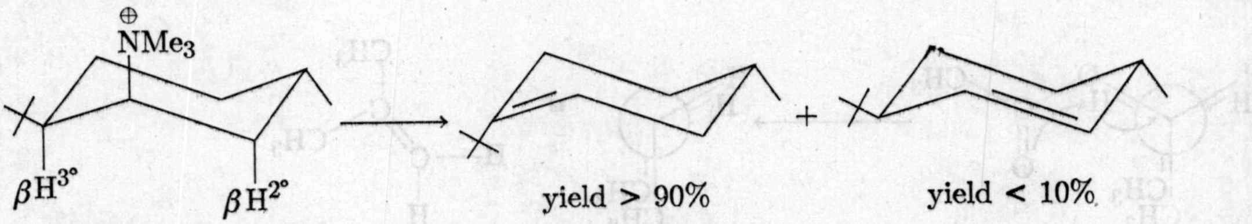

(v) $Me_2CHCH{=}CH_2$ (major) + $Me_2C{=}CH{-}CH_3 + 2NMe_3 + 2H_2O$

Problem 15.12. What product(s) do you expect when sec-butyl acetate is heated. Give the mechanism and name the major product.

Solution : We shall get a mixture of two elimination products along with acetic acid. The two elimination products are 1-butene and 2-butene. 1-butene will be the major product since in this Ei reaction the C_1 atom has 3 H-atoms for eclipsed conformation while the C_3 atom has 2 such H atoms. The statistical factor is the cause for the major yield of 1-butene as shown below :

$CH_3CH_2CH(CH_3)OCOCH_3$ (sec-butyl acetate)

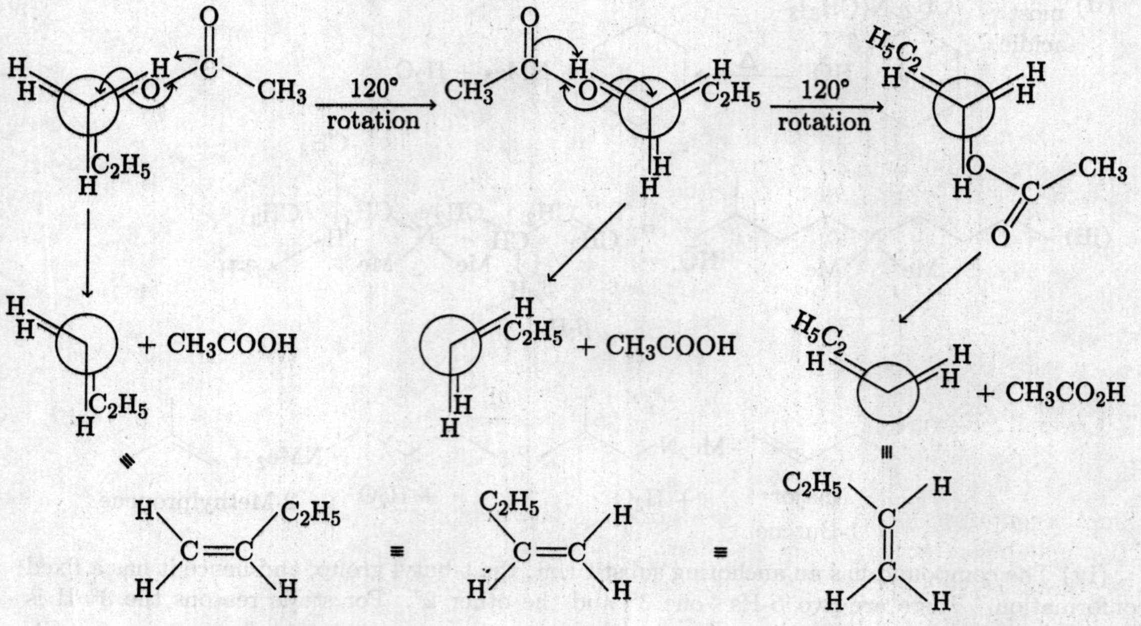

1-Butene major

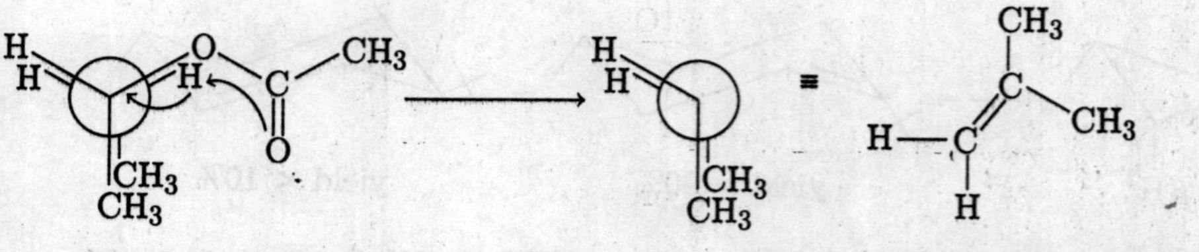

highly strained

2-Methylpropene (trace)

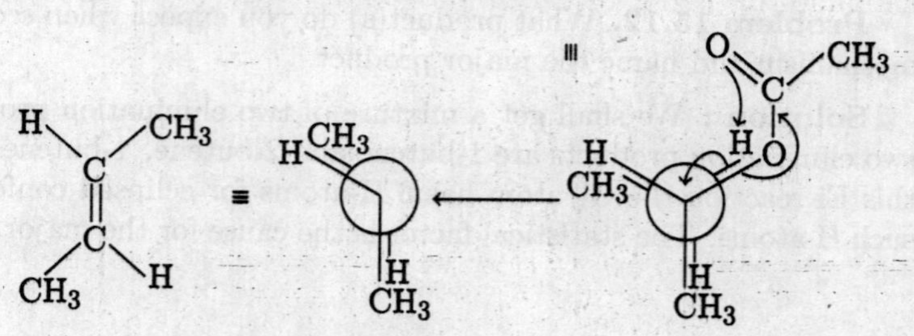

2-Butene

Problem 15.13. 3-Hydroxybutanal undergoes an elimination reaction and gives rise to 2 molecules of acetaldehyde in the presence of a base. Draw the mechanistic steps for the reaction and designate the reaction pathway.

Solution :

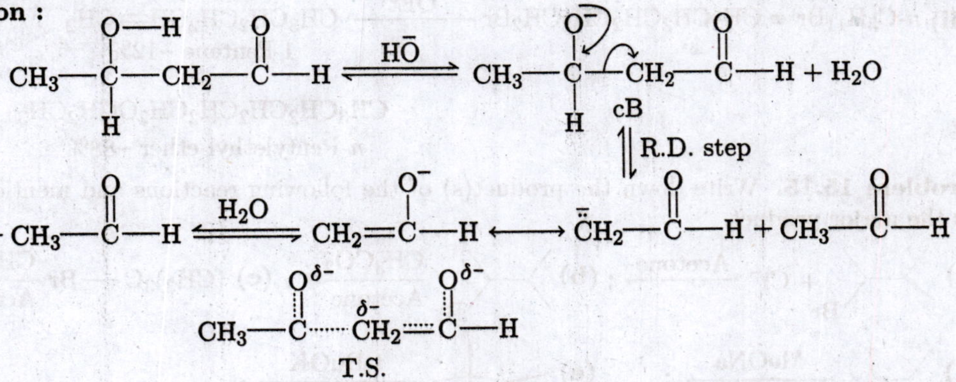

The overall reaction :

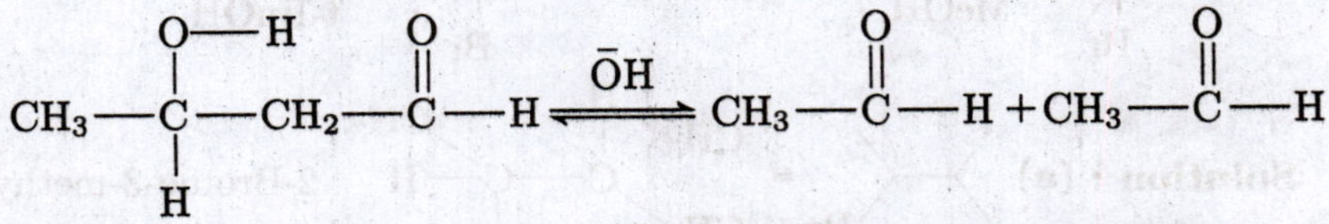

Since a conjugate base (cB) forms as a reaction intermediate, this retroaldol (reverse aldol) reaction may be designated as an E1cB reaction.

Problem 15.14. What products do you expect when the following substrates are treated with a hard base (say EtONa)?

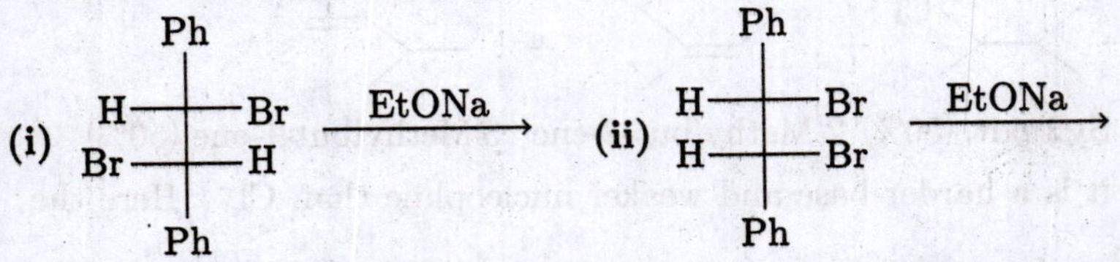

(iii) $n\text{-}C_5H_{11}Br \xrightarrow[\text{EtOH, } 55\,°C]{\text{EtONa}}$

(1S, 2S)-1,2-Dibromodiphenylethane (1S, 2R)-1,2-Dibromo-1,2-diphenylethane

Solution : All the substrates will undergo E2 reactions since a hard base is employed. Compound (iii) will give an S_N2 product as a major product since a 1° bromide is the substrate.

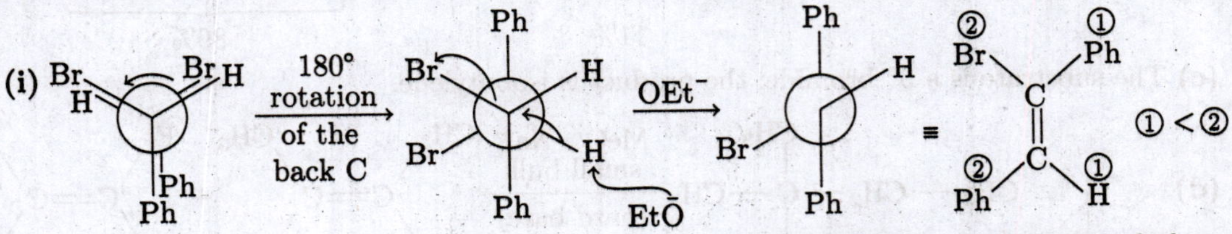

(Z)-1-Bromo-1,2-diphenylethene

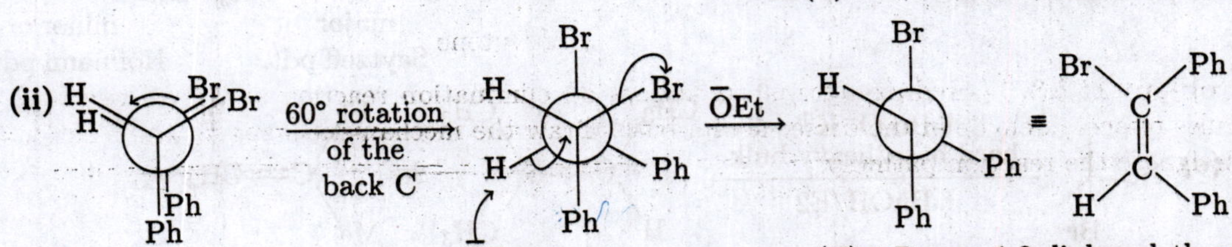

(E)-1-Bromo-1,2-diphenylethene

(iii) $n\text{-}C_5H_{11}Br \equiv CH_3CH_2CH_2CH_2CH_2Br \xrightarrow{\bar{O}Et} CH_3CH_2CH_2CH{=}CH_2$ (1-Pentene ~12%) + $CH_3CH_2CH_2CH_2CH_2OCH_2CH_3$ (*n*-Pentylethyl ether ~88%)

Problem 15.15. Write down the product(s) of the following reactions and mention which one is the major product.

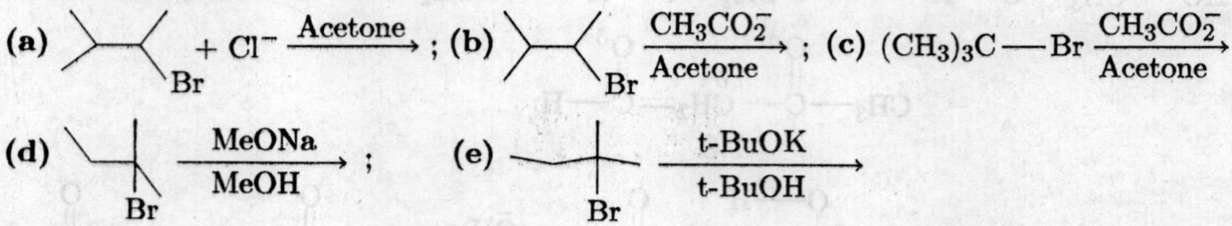

Solution : (a)

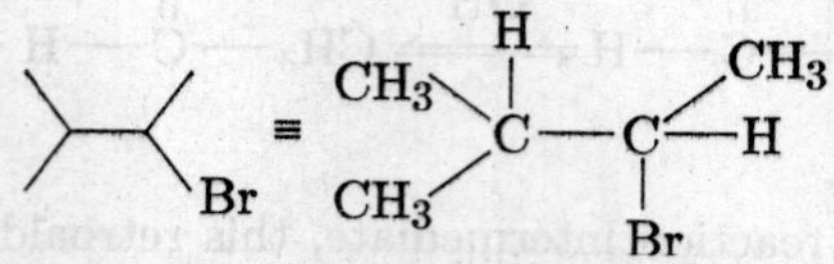

2-Bromo-3-methylbutane

(i) The substrate is a 2° bromide, with α and β branching; (ii) The dielectric constant of acetone is not very low ($\varepsilon = 21$) and (iii) the base is not very hard (Cl^-); so we shall expect both S_N2 and E2 products.

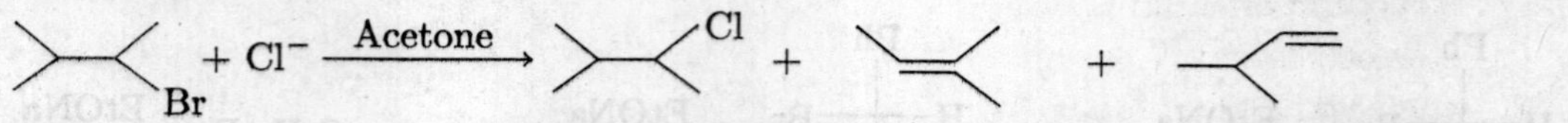

S_N2 pdt. (50%) 2-Methylbut-2-ene 3-Methylbut-1-ene (50%)

(b) The base is acetate ion; it is a harder base and weaker nucleophile than Cl^-. Here the E2 product will be predominant.

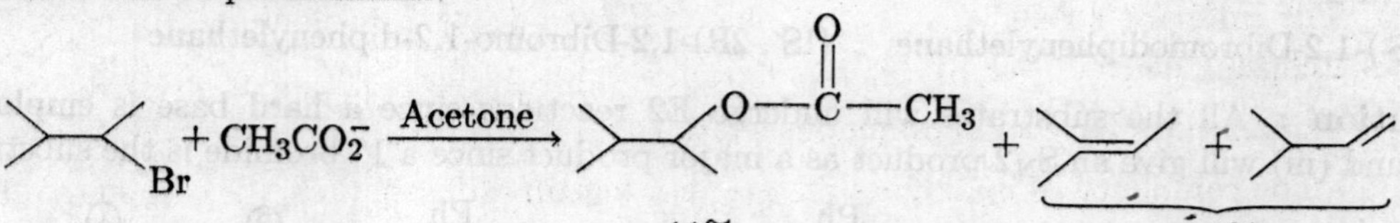

11% 89%

(c) The substrate is a 3° bromide; the product is isobutylene.

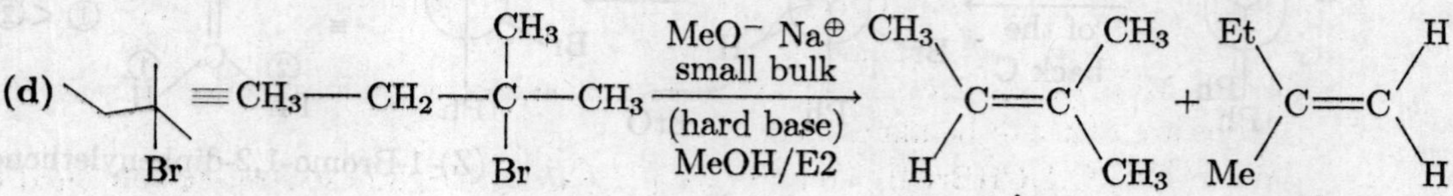

major Saytzeff pdt. minor Hofmann pdt.

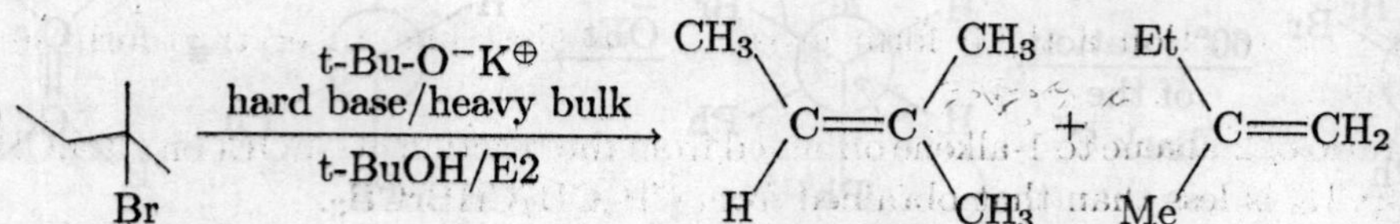

minor major

Problem 15.16. What could you predict from the following fact about the mechanism of the reaction?

$$\left.\begin{array}{l} CH_3CHBrCH_3 \xrightarrow{EtO^-} CH_3CH{=}CH_2 \\[1em] CD_3CHBrCD_3 \xrightarrow{EtO^-} CD_3CH{=}CD_2 \end{array}\right\} k^H/k^D = 6.7$$

Solution : As there is a normal PKIE, the C-H bond breaking occurs in the R.D. step. So the reaction follows either an E2 pathway or an E1cB pathway. Since the substrate does not contain any electron-withdrawing group, a carbanion cannot form. So, the reaction cannot follow E1cB pathway. Therefore, the reaction follows E2 pathway.

Exercise

1. Discuss :

(i) E1 reactions,

(ii) E2 reactions and

(iii) E1cB reactions.

2. Write an essay on dehydrohalogenation reactions touching the following points :

(a) nature of the reactions,

(b) reactivity,

(c) structural orientation,

(d) stereochemistry,

(e) competitive reactions. **[Hint : Sec. 15.2]**

3. Write notes on :

(a) The Saytzeff's rule,

(b) The Hofmann's rule,

(c) Eclipsing effect.

4. Give explanations for the following observations :

(i) The compound B undergoes E2 reaction ten times faster than the compound A :

$$\underset{\textbf{A}}{CH_3-CH_2-\underset{\overset{\oplus}{N}(CH_3)_3}{\underset{|}{CH}}-C(CH_3)_3} \qquad \underset{\textbf{B}}{CH_3-\underset{\overset{\oplus}{N}(CH_3)_3}{\underset{|}{CH}}-C(CH_3)_3}$$

(ii) When Me_3CCl and $Me_3C\overset{\oplus}{S}Me_2$ are separately subjected to E1 reaction, the $E1/S_N1$ ratios are found to be almost same.

(iii) When $CH_3.CH_2CHBrCH_3$ and $CH_3CH_2CH(\overset{\oplus}{N}Me_3)CH_3$ are separately subjected to E2 reaction, the Hofmann's product in the former case is 31%, while that in the latter case is 98%.

(iv) In the E2 reaction a threo form gives trans-olefin while an erythro form gives a cis-olefin.

(v) The ratio of 2-alkene to 1-alkene obtained from the reaction of NaOEt on $CH_3.CH_2CH_2$ $CHBrCH_3$ is less than that obtained from $CH_3CH_2CHBrCH_3$.

5. Discuss :

(a) The Cope reaction,

(b) The Chugaev reaction and

(c) Pyrolysis of esters.

Why are these reactions called Ei reactions? What are the general evidences in favour of Ei reactions?

6. Why do we get 1, 3-pentadiene instead of 1, 4-pentadine when hexahydropyridine is subjected to the Hofmann's exhaustive methylation?

7. When 1-butanol is treated with concentrated sulphuric acid, 2-butene forms as the major product—explain.

8. 3-hydroxybutanal can be dehydrated by alkali as well as by acid—draw the mechanistic steps.

9. Write down the products and the mechanistic pathways for the following reactions :

(i) $Br—CH_2—CH_2—Br + Mg \longrightarrow$

(ii) 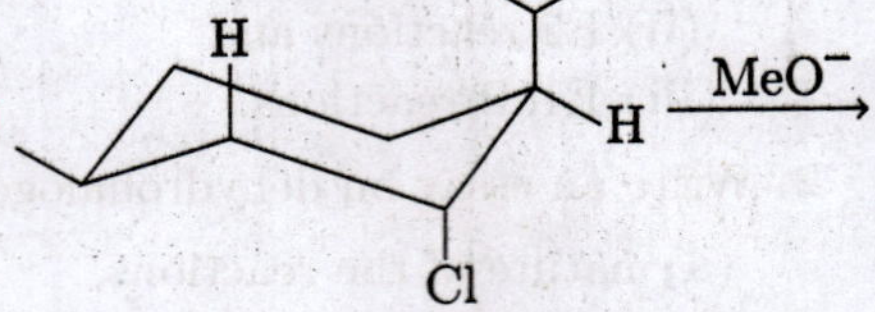

(iii) 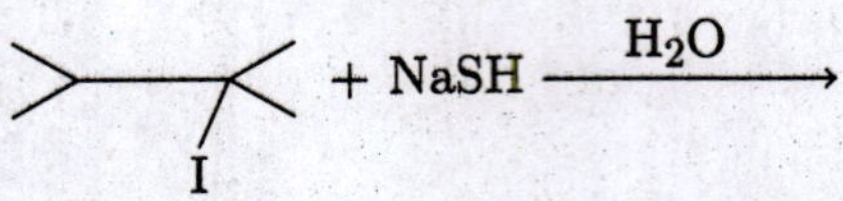

(iv) 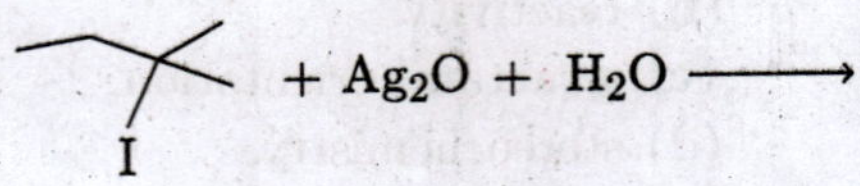

(v) $CH_3—C(OH)(H)—CN \xrightarrow{HO^-}$

10. Identity A, B, C

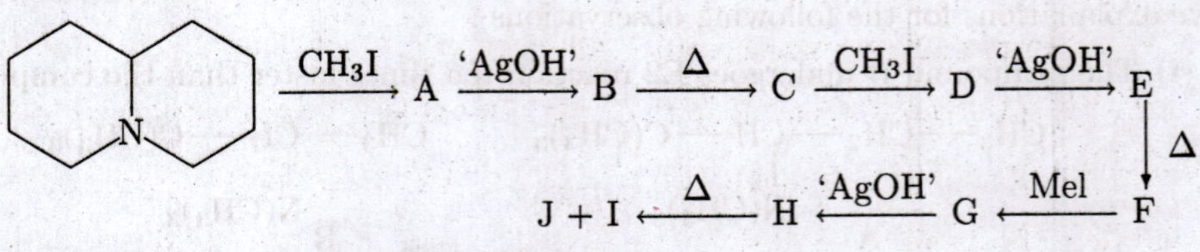

11. Distinguish between S_N1 and E2 pathways.

12. (2R, 3S)-2, 3-dibromobutane gives trans-but-2-ene, but (2S, 3S)-2, 3-dibromobutane gives cis-but-2-ene on dehalogenation with Zn—explain.

Chapter 16

Reactions Involving >C=O Group

Objectives

- To know the types of reaction of a compound containing >C=O group [16.1] p. 825
- To understand :
  - The structural effect on rates of nucleophilic additions [16.2] p. 826
  - The stereochemistry of addition to >C=O group [16.3] p. 827
- To discuss :
  - Simple addition reactions [16.4] p. 828
    - HCN [16.4.1] p. 828
    - $R{-}C{\equiv}\bar{C}$ [16.4.2] p. 831
    - $NaHSO_3$ [16.4.3] p. 832
    - ROH [16.4.4] p. 833
    - RMgX [16.4.5] p. 838
    - The Reformatsky reaction [16.4.6] p. 847
    - H_2O [16.4.7] p. 850
  - Addition-elimination reactions [16.5] p. 851

* Reactions involving hydride ion transfer [16.6] p. 854
    * Reaction of $LiAlH_4$ [16.6.1] p. 854
    * The Meerwein-Pondorf-Verley reduction and the Oppenauer oxidation [16.6.2] p. 857
    * The Cannizzaro reaction [16.6.3] p. 860
* The Wolff-Kishner and Huang Minlon reduction [16.7] p. 865
* The Wittig reaction [16.8] p. 867
* Reactions involving anionic radicals [16.9] p. 871
    * The Bouveault-Blanc reduction [16.9.1] p. 871
    * Bimolecular reduction of carbonyl compounds [16.9.2] p. 872
    * Acyloin condensation or synthesis [16.9.3] p. 873
* Direct or 1, 2-addition versus conjugate or Michael type or 1, 4-addition [16.10] p. 876
* Halogenation and the haloform reaction [16.11] p. 880
* The Hell-Volhard-Zelinsky reaction [16.12] p. 889
* Condensation reactions [16.13] p. 890
    * Aldol condensation [16.13.1] p. 891
    * The Claisen condensation [16.13.2] p. 894
    * The Stobbe condensation [16.13.3] p. 897
    * The Darzen glycidic ester condensation [16.13.4] p. 899
    * Benzoin condensation [16.13.5] p. 900
    * The Perkin condensation [16.13.6] p. 903
    * The Knoevenagel condensation [16.13.7] p. 908
    * The Mannich reaction and the Strecker synthesis [16.13.8] p. 911

16.1 Introduction

The nature of the >C=O moiety has already been discussed in Section 2.16.6. Since it contains a σ and a π bonds and since the electronegativity of the O atom is greater than that of the C atom, the C atom is partially positively charged while the O atom is partially negatively charged. However, the >C=O moiety may be represented as a resonance hybrid of the following two structures :

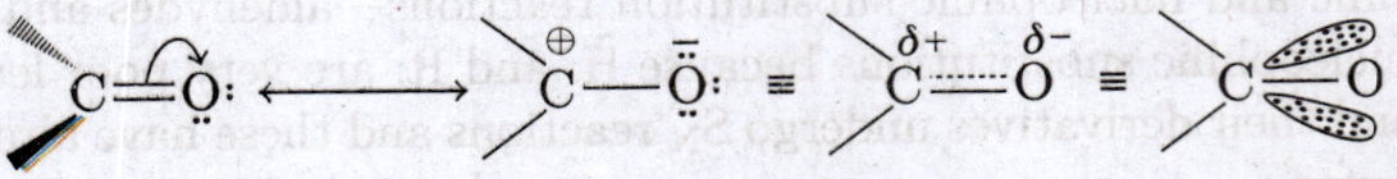

Owing to the presence of π bond, the >C=O moiety undergoes addition reaction. Here the addition may either be nucleophilic or electrophilic. In the former case the point of the nucleophilic attack is the partially positively charged C atom; while the electrophilic attack occurs on the partially negatively charged O atom. The electrophilic addition is rare; usually strong acids undergo the electrophilic addition. The common additions to >C=O are nucleophilic in nature. This sort of reactions may be acid- or base-catalysed. When hydra acids are used as catalysts, the carbonyl O atom forms H-bond with the catalyst. If the reaction is carried out in a hydroxylic solvent, then also the carbonyl O forms H-bond with the solvent molecule as shown below :

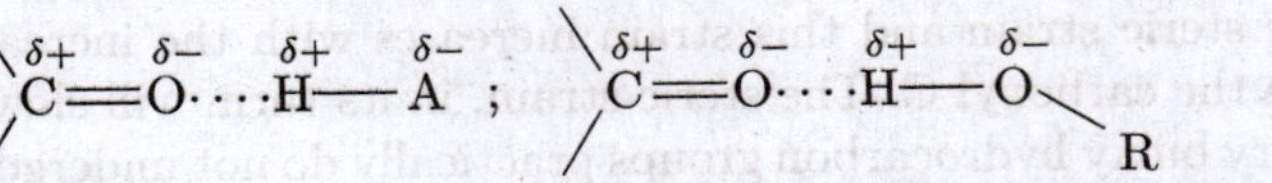

In both the cases the effect is the same; the carbonyl O atom further pulls the C—O bonded *e*s towards itself which makes the carbonyl C more electron-deficient, i.e., more positively charged than what it was before H-bonding and thus facilitates the nucleophilic attack on the C atom of the >C=O group.

The positive character of the carbonyl C atom further increases when a pair of the non-bonded *e*s of the carbonyl O atom coordinate with a proton or a Lewis acid as shown below :

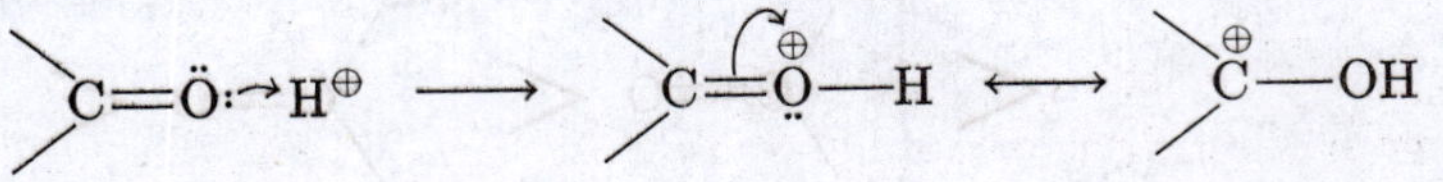

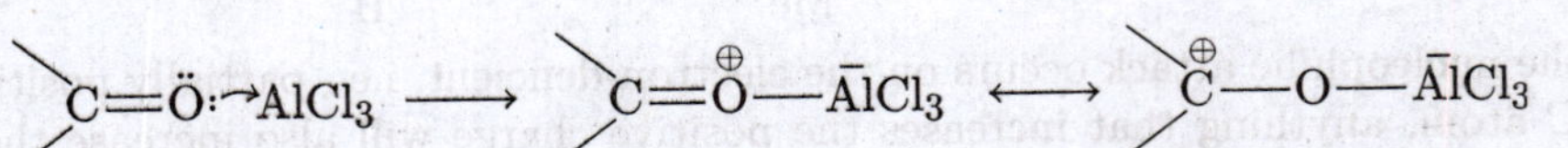

This will facilitate the nucleophilic attack on the carbonyl C to a greater extent than the H-bonding effect.

When a base is used as a catalyst along with a poorly acidic addendum, the basic catalyst develops a strong nucleophile by taking in a proton from the addendum. Then the strong nucleophile couples with the weakly electrophilic carbonyl C easily.

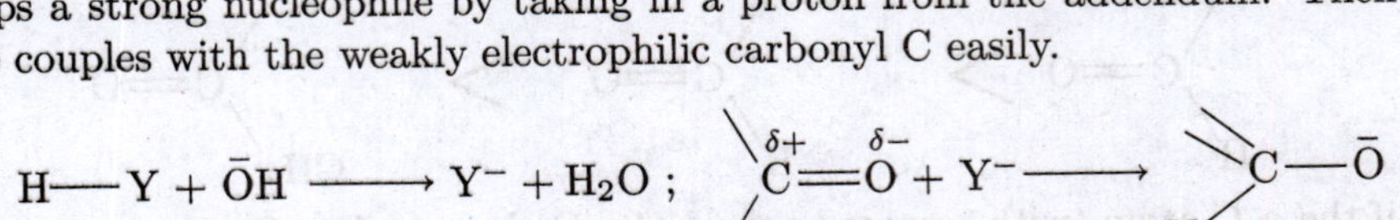

Since >C=O has −I effect, the α-H atom, with respect to >C=O moiety, has an acidic character. This acidic H can be reversibly abstracted by a strong base ($\bar{O}H$, Et$\bar{O}$, etc.) and thereby a resonance-stabilised carbanion can be formed.

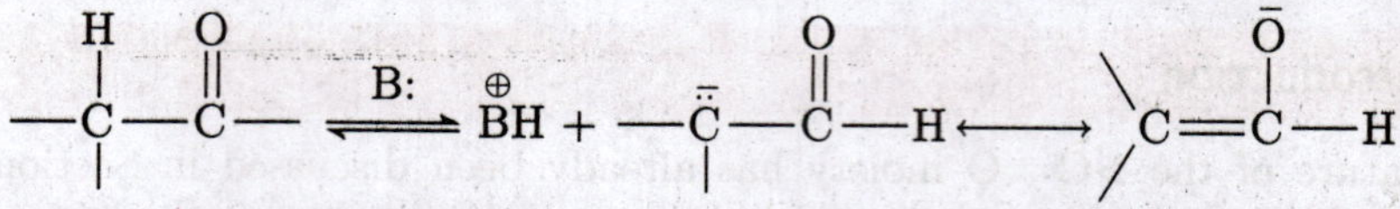

The resonance-stabilised carbanion may undergo a wide and varied types of condensation reactions.

Besides the addition and condensation reactions, compounds containing >C=O moiety may undergo electrophilic and nucleophilic substitution reactions—aldehydes and ketones, of course, do not undergo nucleophilic substitutions because $\bar{H}$: and $\bar{R}$: are very poor leaving groups; while carboxylic acids and their derivatives undergo S_N reactions and these have already been discussed in the twelfth chapter.

16.2 Nucleophilic Addition and the Effect of Structure on Rate

Both the steric and electronic effects have profound influence on the rate of nucleophilic additions.

The carbonyl C atom is in sp^2 hybridised state and the bond angles are 120° approximately; while in the adduct this C becomes sp^3 hybridised and the bond angles on it are around 109°28′. The T.S. of such an addition reaction being more similar to the adduct than to the substrate, it suffers from the steric strain and this strain increases with the increasing bulk of the groups already present on the carbonyl C. The steric strain, in its turn, will slow down the reaction and ketones bearing very bulky hydrocarbon groups practically do not undergo addition reactions with bulky nucleophiles. Thus the rate of addition will go on increasing with the following compounds.

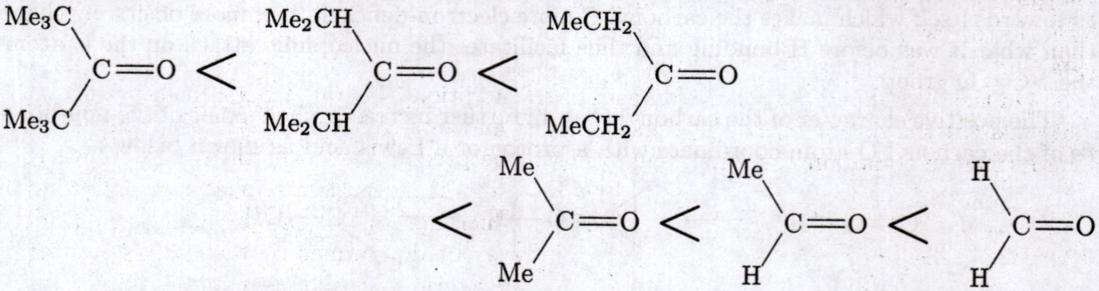

Since the nucleophilic attack occurs on the electron-deficient, i.e., partially positively charged carbonyl C atom, anything that increases the positive charge will also increase the rate; while anything that decreases the same will decrease the rate of the reaction. Since alkyl groups have +I effect, their presence slows down the reaction; whereas the presence of electron-withdrawing groups on the carbonyl C increases the rate by making it more positively charged. Thus the rate order will be :

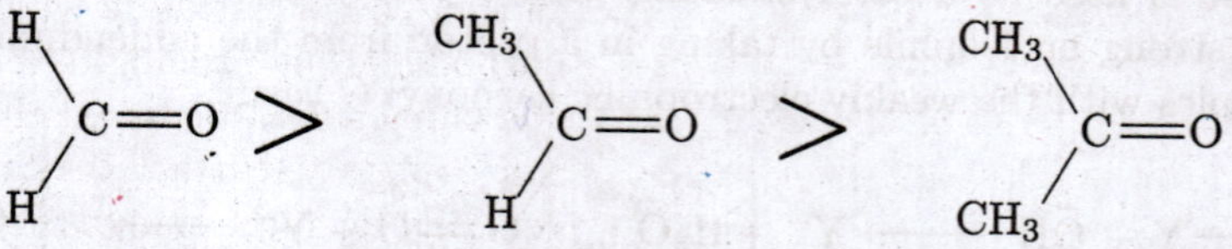

Again, if the α-C atom (with respect to the >C=O) bears a C=C group or a benzene ring, the conjugative effect will bring stability to the system; after the addition of the nucleophile to the carbonyl group the product will no longer be a conjugated one. Since through addition the conjugation is lost, these sorts of compounds will show slow addition rate. However, the presence of an electron-withdrawing group on the C=C part or at *o*- or *p*-position of such a benzene ring will relatively increase the rate.

16.3 Stereochemistry of the Nucleophilic Addition to the >C=O Group

The two parts of an addendum, YZ, may add to the C=O group either from the same side of the double bond (cis-addition) or from the opposite sides (trans-addition); but the mode of addition will not be possible to determine since OZ group will rotate around C=O bond axis and will adopt the most stable conformation instantaneously.

Fig. 16.1. *cis*- and *trans*-addition

−I and −R effects of nitro group in *p*-nitrobenzaldehyde increases the electrophilicity of the aldehydic C and thus increases the value of formation constant. +R groups, *p*-dimethylamino and *p*-methoxy, decrease the clectrophilicity of the aldehydic C and thereby decrease the value of formation constant. In the cases of ethyl methyl ketone, acetophenone and benzophenone, the steric congestion around the carbonyl C decreases the reactivity of the carbonyl compounds and we get a small formation constant for the former and for the latter two, $K_{formation}$ is immeasurably small.

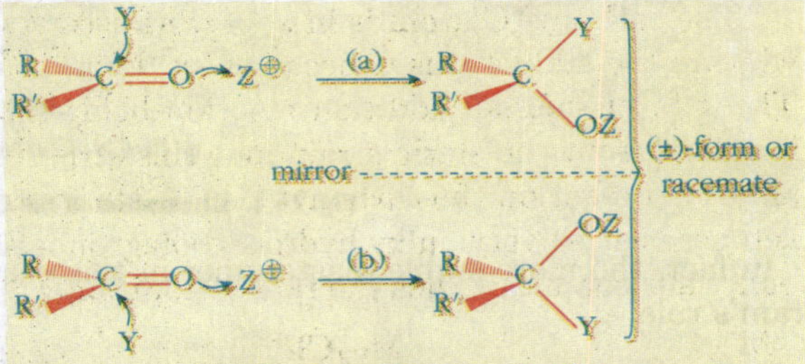

Fig. 16.2. Stereochemistry of addition

On the other hand, if the environment of the addition is chiral, T.S.s formed by the pathways (a) and (b) will not be formed in equal amounts and as a consequence the products will not also be formed in equal amounts, i.e., one form will be predominant over the other. When the α-C atom is asymmetric in a carbonyl compound, which one will be the predominant product that may be determined by the **Cram's empirical rule** of "steric control of asymmetric induction". This rule only considers the steric effect and tells nothing about the electronic effects. For this reason the predictions made on the basis of the Cram's rule do not always found correct. However, the rule states that a carbonyl compound will react in that conformation in which the O atom of the carbonyl group orients itself anti to the largest of the three substituents on the α-C atom and the nucleophile will attack the carbonyl C atom preferentially from the side of the plane containing the smallest group (Fig. 16.3).

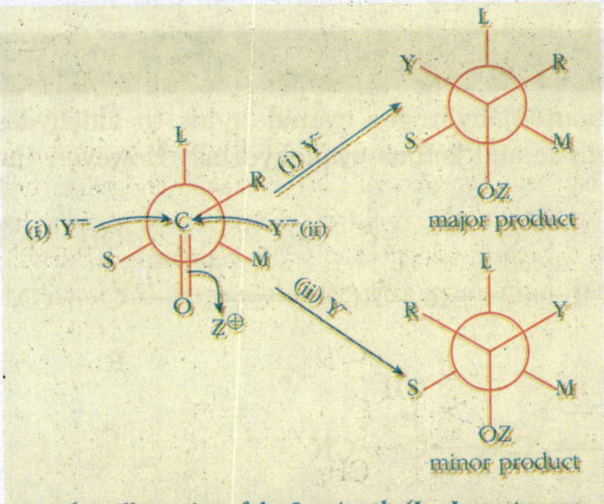

Fig. 16.3. Illustration of the Cram's rule (L = largest group, S = smallest group, M = medium group)

For example, when HCN is added to 2-methylbutanal, an optically active compound, one of the two diastereomeric forms of the cyanohydrin will be predominant. According to the Cram's rule, that conformation of the aldehyde will react in which the O atom of C═O will orient itself anti to the ethyl group (this being the largest among Et, Me and H on the α-C) and $\bar{C}N$ will attack the carbonyl carbon preferentially from the side of the plane containing the H atom, the smallest among Et, Me and H.

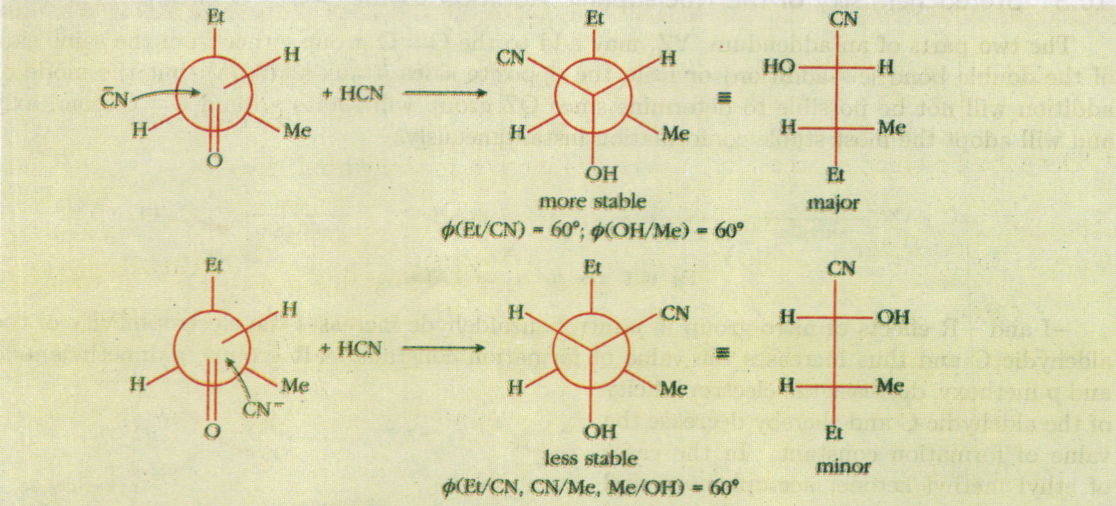

Fig. 16.4. Illustration of the Cram's rule

In fact, the more stable diasteromer in this case predominates which is also predicted by the Cram's rule.

16.4 Simple Nucleophilic Addition Reactions

Several simple nucleophilic addition to aldehydes and ketones occur in which the carbonyl C atom is the point of attack. We shall discuss here the nucleophilic addition of HCN, $NaHSO_3$, acetylide ion, alcohols, the Grignard reagent, the organozinc compound and water.

16.4.1 Addition of HCN

In the slightly acidic medium (at a pH of about 4) hydrogen cyanide adds to aldehydes and ketones to yield α-hydroxynitriles, called aldehyde and ketone cyanohydrins. However, the reaction is carried out in a slightly basic medium.

$$R{-}\overset{O}{\overset{\|}{C}}{-}H + HCN \rightleftharpoons R{-}\underset{H}{\underset{|}{\overset{OH}{\overset{|}{C}}}}{-}CN \qquad R{-}\overset{O}{\overset{\|}{C}}{-}R + HCN \rightleftharpoons R{-}\underset{R}{\underset{|}{\overset{OH}{\overset{|}{C}}}}{-}CN$$

$$CH_3{-}\overset{O}{\overset{\|}{C}}{-}H + HCN \rightleftharpoons CH_3{-}\underset{H}{\underset{|}{\overset{OH}{\overset{|}{C}}}}{-}CN$$

(Acetaldehyde cyanohydrin)

The reaction gives appreciable yield with aldehydes and aliphatic and alicyclic ketones; but it gives poor yield with ArCOR and does not occur with ArCOAr.

However, the reaction has some preparative value. α-Hydroxy acids can be prepared from aldehydes and ketones via cyanohydrin.

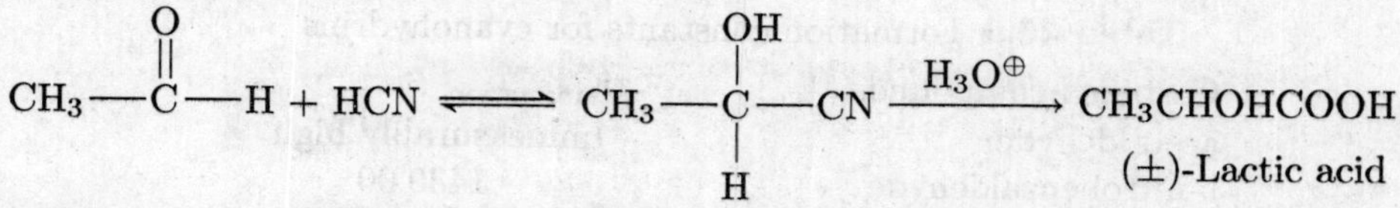

The Kiliani method of ascending sugar series involves cyanohydrin.

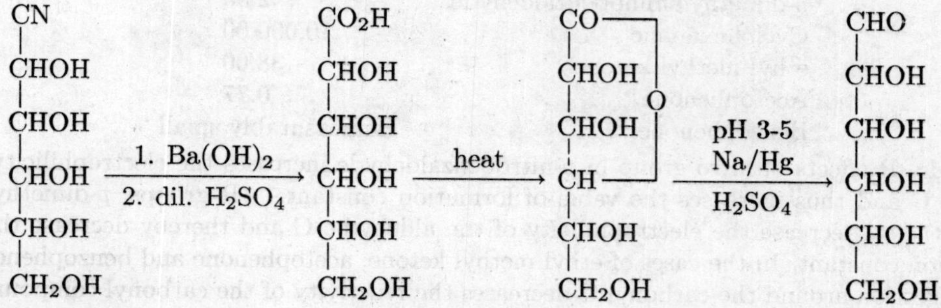

The addition of HCN to aldehydes and ketones is a reversible reaction in which $\bar{C}N$ is the nucleophile. The reaction tends to lie over to the left but it can be made to lie over to the right by the use of a proton donor. However, the reaction takes place in two steps.

Step I : Nucleophilic attack by the cyanide ion on the electron-deficient carbonyl C atom

$$HCN \rightleftharpoons H^{\oplus} + :\bar{C}{\equiv}N$$

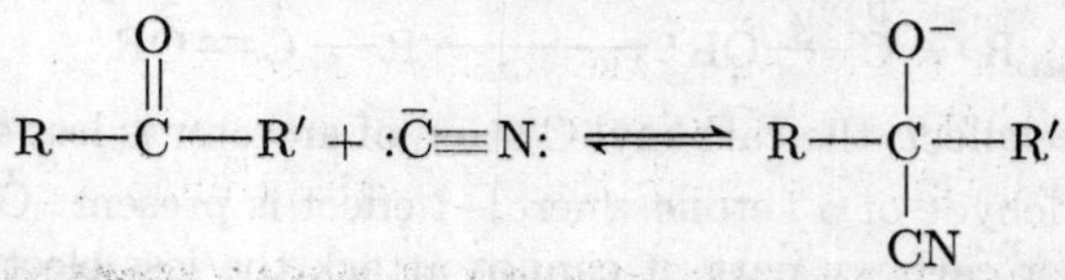

Step II : Protonation of the anionic O atom by the HCN or by the solvent (dilute acid).

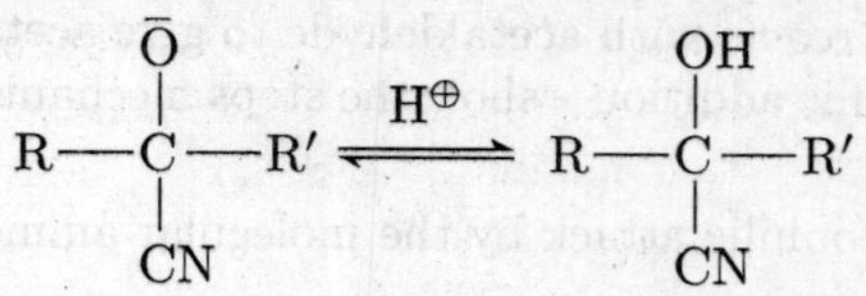

The base-catalysed reaction follows the second order kinetics.

The rate = k [substrate] [$\bar{C}N$]

In this case the addition of the $\bar{C}N$ is the slow step while protonation is a fast step. It is a reversible reaction; the back reaction follows the E1cB path.

There is a good evidence in favour of the mechanism. The reaction rate retards in the presence of high concentration of the acid while the presence of base accelerates the rate. The findings are in agreement with the mechanism given above. The high acid concentration suppresses the dissociation of hydrogen cyanide since it is a weak acid and consequently low $\bar{C}N$ concentration retards the reaction. In the presence of a base, hydrogen cyanide gives cyanide ion and thus accelerates the rate :

$$HCN + \bar{O}H = H_2O + :\bar{C}N$$

As usual the extent of cyanohydrin formation depends on the steric and stereoelectronic factors. The cyanohydrin formation constants can well explain the fact.

Table 16.1 Formation constants for cyanohydrins

Carbonyl compounds	$K_{formation}$
acetaldehyde	Immeasurably high
p-nitrobenzaldehyde	1430.00
p-methoxybenzaldehyde	210.00
p-dimethylaminobenzaldehyde	2.50
Cyclohexanone	10,000.00
ethyl methyl ketone	38.00
Acetophenone	0.77
Benzophenone	Immeasurably small

–I and –R effects of nitro group in *p*-nitrobenzaldehyde increases the electrophilicity of the aldehydic C and thus increases the value of formation constant. +R groups, *p*-dimethylamino and *p*-methoxy, decrease the electrophilicity of the aldehydic C and thereby decrease the value of formation constant. In the cases of ethyl methyl ketone, acetophenone and benzophenone, the steric congestion around the carbonyl C decreases the reactivity of the carbonyl compounds and we get a small formation constant for the former and for the latter two, $K_{formation}$ is immeasurably small.

Problem 16.1. Why does HCN not add to the >C=O part of an ester?

Solution : An ester is a resonance hybrid of the following two structures :

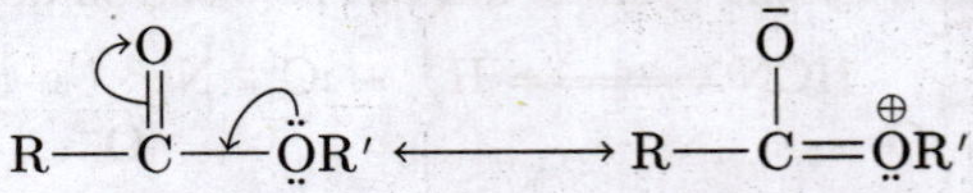

Owing to the resonance effect, the carbonyl C atom of an ester is less electron-deficient than a carbonyl C atom of an aldehyde or a ketone where +I effect is present. $\bar{C}N$ being not sufficiently necleophilic to add to ester carboxy part, it cannot attack the less electron-deficient C atom of the ester.

Problem 16.2. Ammonia reacts with acetaldehyde to give acetaldehyde ammonia; it follows the mechanism of the nucleophilic addition—show the steps mechanistically. What is the ultimate product of the reaction?

Solution : Step I : Nucleophilic attack by the molecular ammonia on the electron-deficient carbonyl C atom :

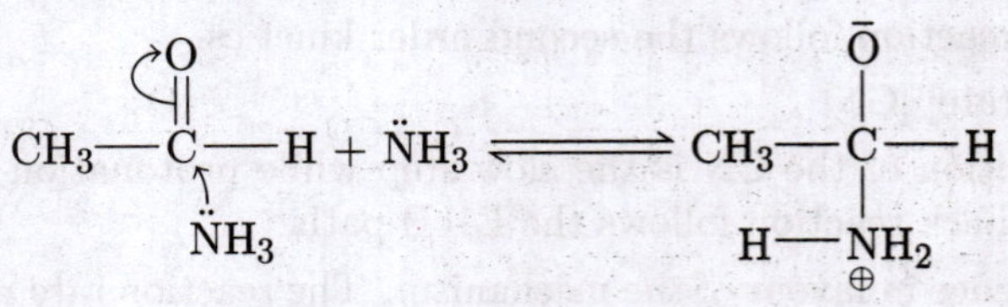

Step II : Proton transfer from the cationic N to anionic O :

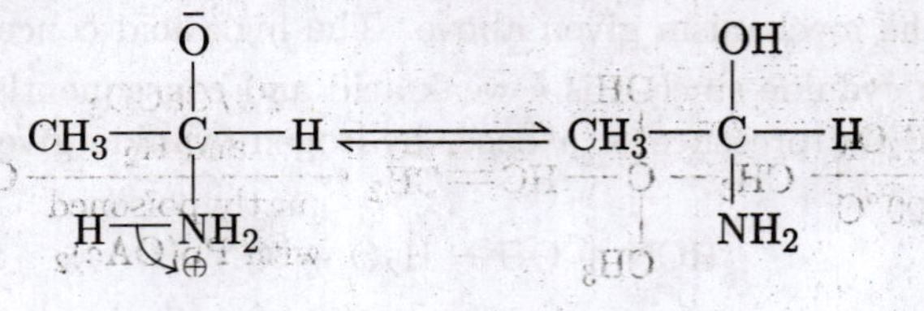

Step III : Loss of water and formation of aldimine :

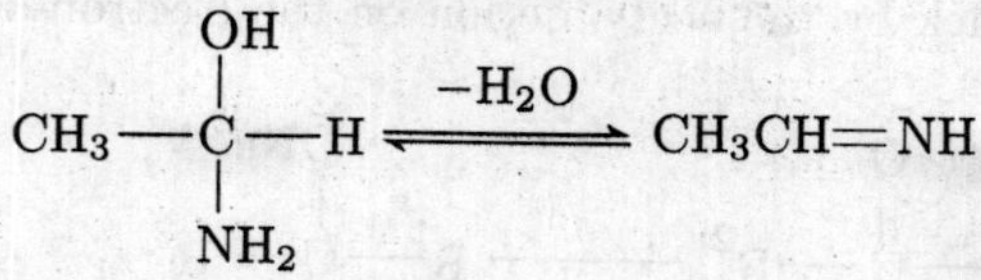

Step IV : Trimerisation of the aldimine :

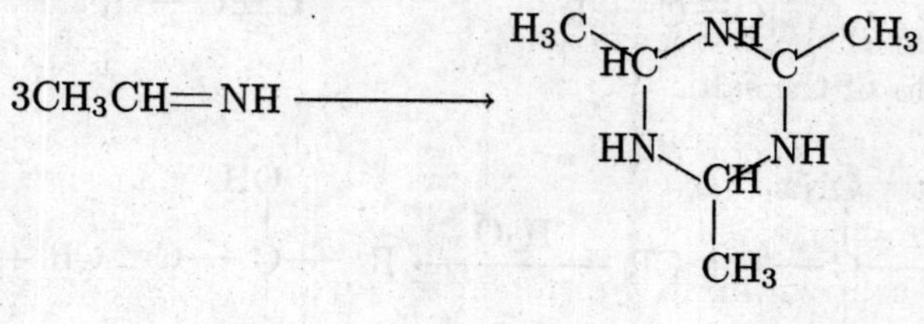

Acetylide ions are strong nucleophiles and readily attack the electron-deficient carbonyl C atom to give metal ynolate salts which on hydrolysis form ynols.

$$RC{\equiv}CH \xrightarrow[\text{in } NH_3(l)]{NaNH_2} R{-}C{\equiv}\bar{C}Na^{\oplus}$$

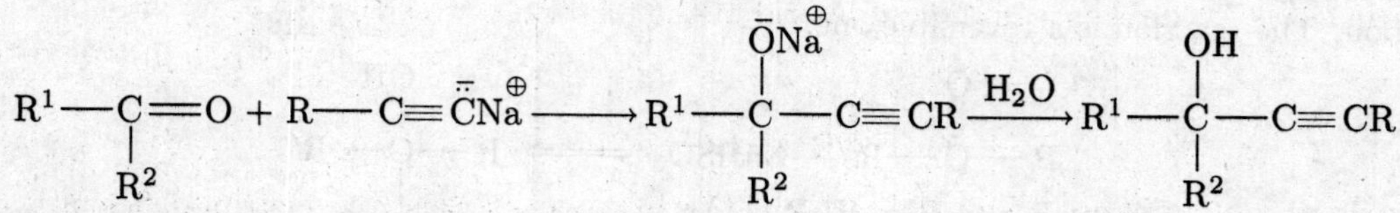

R^1, R^2 and R may be H, R or Ar, etc.

Ynols may be converted to other compounds of synthetic value. For example, the addition product of acetone and sodium acetylide may be converted to isoprene, which polymerises to a substance resembling natural rubber. Ynols may also be converted to α-hydroxy ketones and dienes.

$$HC{\equiv}CH \xrightarrow[\text{in liquid } NH_3]{NaNH_2} CH{\equiv}\bar{C}Na^{\oplus}$$

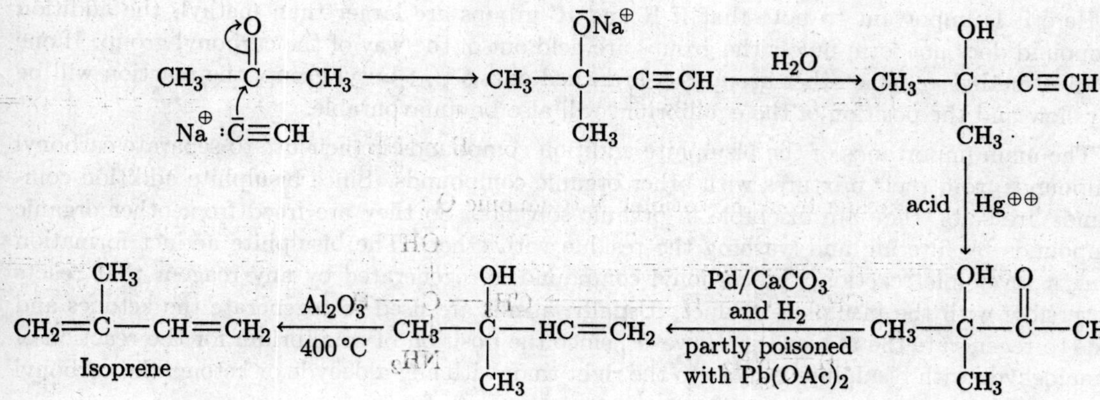

The mechanism of ynol formation involves two steps :

Step I : Nucleophilic attack by the acetylide ion on the electron-deficient carbonyl carbon atom to yield ynolate salt :

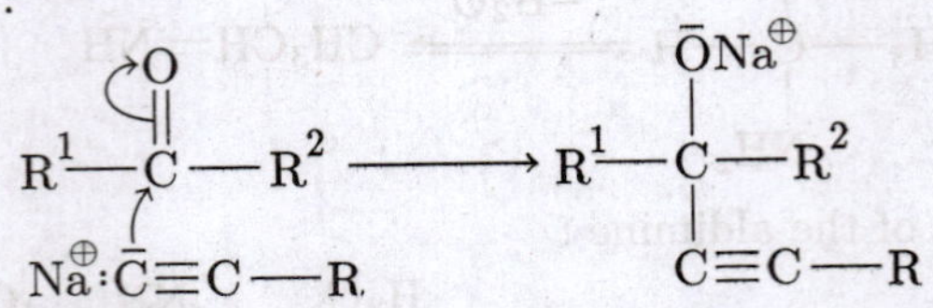

Step II : Hydrolysis of the salt :

$$R^1—\underset{R^2}{\overset{\bar{O}Na^{\oplus}}{C}}—C\equiv CR \xrightarrow{H_2O} R^1—\underset{R^2}{\overset{OH}{C}}—C\equiv CR + NaOH$$

16.4.3 Addition of sodium bisulphite

When aldehydes and ketones are shaken with an aqueous solution of sodium bisulphite (saturated), slightly soluble bisulphite addition compounds, called aldehyde bisulphite and ketone bisulphite, are formed; here hydrogen adds to the oxygen and sodium sulphonate to the carbonyl carbon. The reaction is a reversible one.

$$R—\overset{O}{\overset{\|}{C}}—R' + NaHSO_3 \rightleftharpoons R—\underset{SO_3Na}{\overset{OH}{C}}—R'$$

R and R'=H, R or Ar

$$CH_3—\overset{O}{\overset{\|}{C}}—H + NaHSO_3 \rightleftharpoons CH_3—\underset{H}{\overset{OH}{C}}—SO_3Na$$

Acetaldehyde sodium bisulphite

Here it is important to note that if R and R' groups are larger than methyl, the addition compound does not form unless the groups are held out of the way of the carbonyl group. If one group is methyl and the other group is a branched one, e.g., t-butyl group, the reaction will be very slow and the position of the equilibrium will also be unfavourable.

The main importance of the bisulphite addition compounds is their use to separate carbonyl compounds from their mixtures with other organic compounds. Since bisulphite addition compounds are salts, they are insoluble in organic solvents. So they are freed from other organic compounds by filtering and washing the residue with ether. The bisulphite adduct formation being a reversible reaction, the carbonyl compound is regenerated by any reagent that reacts irreversibly with the bisulphite adduct. Usually, alkalis are used to regenerate the ketones and acids to regenerate the aldehydes. However, since the position of equilibrium for the reaction of formaldehyde with bisulphite is more to the right than with any aldehyde or ketone, the carbonyl compound can be regenerated by heating a bisulphite adduct with a slight excess of an aqueous solution of formaldehyde through an exchange reaction as shown below :

$$\underset{\text{RCH—SO}_3\text{Na}}{\overset{\text{OH}}{|}} + \text{HCHO} \rightleftharpoons \text{RCHO} + \underset{\text{HCH.SO}_3\text{Na}}{\overset{\text{OH}}{|}}$$

When a carbonyl compound is present as impurity in other compounds, it is precipitated as a bisulphite adduct and then the bisulphite is removed by filtration.

The mechanism of the reaction is given below :

$$NaHSO_3 \longrightarrow Na^{\oplus} + HSO_3^- \qquad H_2O + HSO_3^- \rightleftharpoons H_3O^{\oplus} + SO_3^{--}$$

Step I : Nucleophilic attack by the sulphite ion through sulphur (S being more nucleophilic than O because of its greater volume) on the electron-deficient carbonyl C atom :

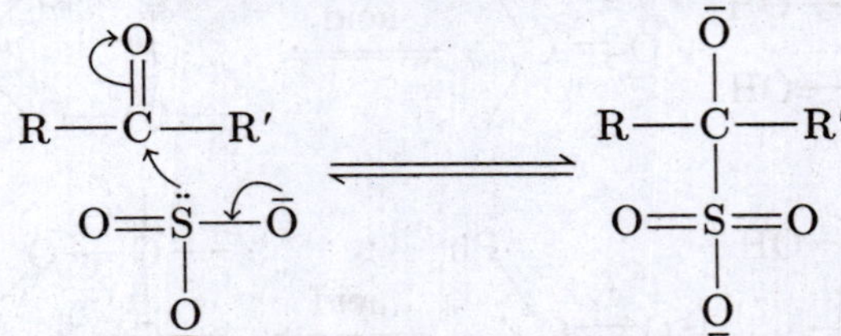

Step II : Protonation of the carbonyl anionic O by bisulphite ion :

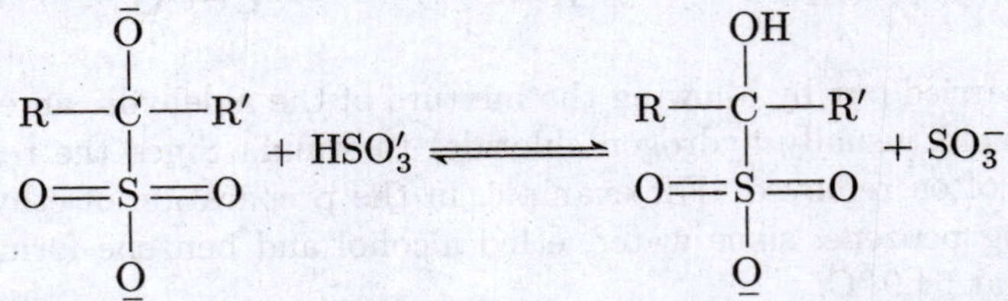

Problem 16.3. Though an aqueous solution of glucose contains free aldehydic glucose (0.02%), yet it does not form bisulphite adduct.—Explain.

Solution : Addition of bisulphite to a carbonyl compound is a reversible reaction. The reaction being reversible, it cannot disturb the equilibrium appreciably among α-glucose ($\sim$ 36%), open chain glucose (0.02%) and β-glucose ($\sim$ 64%) in the aqueous solution of glucose. That is why glucose does not form the bisulphite adduct.

16.4.4 Addition of alcohols

Alcohols add to aldehydes and ketones in the presence of specific acid-catalysts to yield acetals via hemiacetals. Acetals are 1, 1-diethers and hemiacetals are α-hydroxyethers, i.e., half acetals.

$$\text{R—}\overset{\text{O}}{\overset{\|}{\text{C}}}\text{—R}' + \text{R}^2\text{OH} \underset{}{\overset{\text{dry HCl}}{\rightleftharpoons}} \underset{\text{hemiacetal}}{\text{R—}\underset{\text{OR}^2}{\overset{\text{OH}}{\text{C}}}\text{—R}'} \overset{\text{R}^2\text{OH, dry HCl}}{\rightleftharpoons} \underset{\text{acetal}}{\text{R—}\underset{\text{OR}^2}{\overset{\text{OR}^2}{\text{C}}}\text{—R}'}; \ \Delta H = -20 \text{ to } 40 \text{ kJ mol}^{-1}$$

R & R′═H or alkyl group

$$\text{CH}_3\text{—}\overset{\text{O}}{\overset{\|}{\text{C}}}\text{—H} + \text{EtOH} \overset{\text{dry HCl}}{\rightleftharpoons} \text{CH}_3\text{—}\underset{\text{OEt}}{\overset{\text{OH}}{\text{C}}}\text{—H} \overset{\text{EtOH, dry HCl}}{\rightleftharpoons} \text{CH}_3\text{—}\underset{\text{OEt}}{\overset{\text{OEt}}{\text{C}}}\text{—H} + \text{H}_2\text{O}$$

Usually aldehydes undergo the acetal formation reaction readily since the equilibrium lies to the right at room temperature. The position of equilibrium of the reaction between ketones and monohydric alcohols at room temperature is so far to the left that ketals do not form appreciably. At a low temperature, however, the reaction between methyl alcohol and unhindered ketones is somewhat favourable. For example, acetone reacts with methyl alcohol at −27 °C in the presence of a strong acid ion-exchange resin as catalyst.

$$\text{MeCOMe} + 2\text{MeOH} \xrightleftharpoons{-27\,^\circ\text{C; acid resin}} \text{Me}_2\text{C(OMe)}_2 + \text{H}_2\text{O}$$

Of course both aldehydes and ketones react readily with 1, 2- and 1, 3-diols to form cyclic acetals. Aromatic aldehydes specially react with 1, 3-diols.

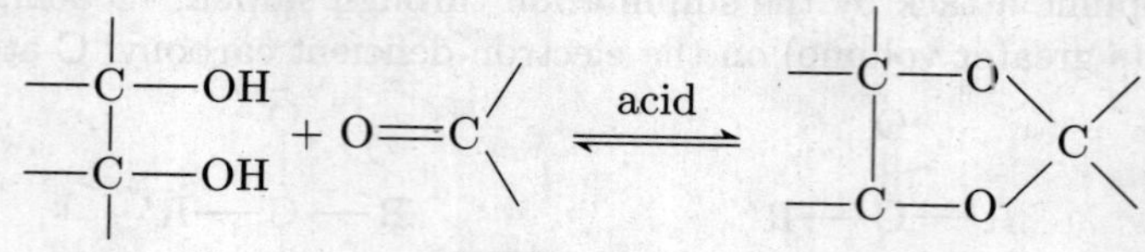

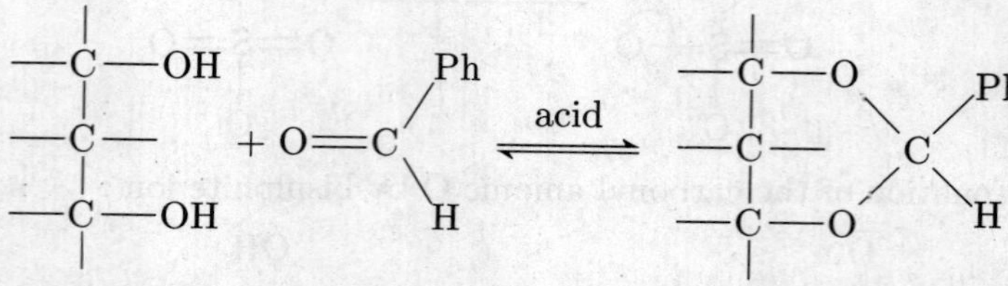

The reaction is carried out by allowing the mixture of the aldehyde, an excess of the alcohol and the anhydrous acid (usually hydrogen chloride) to stand. Since the reaction is reversible, the water formed is often removed. For example, in the preparation of ethyl acetals the water is removed by adding benzene, since water, ethyl alcohol and benzene form a constant boiling mixture which boils at 64.9 °C.

However, the acetal formation reaction is reversible and the hydrolysis of the acetal back to the starting carbonyl compound may easily be effected with dilute acid. Since an acetal does not contain any H atom bonded to O atoms, it does not hydrolyse in the presence of a base. Thus acetals are stable towards bases.

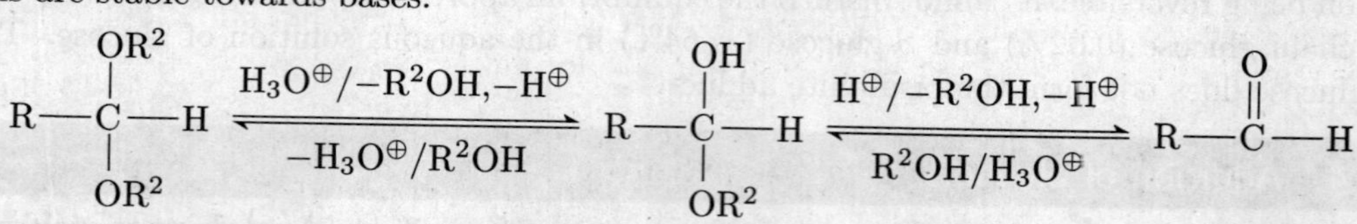

A similar reaction is the addition of thiols to carbonyl compounds in the presence of hydrogen chloride to yield hemithioacetals and thioacetals.

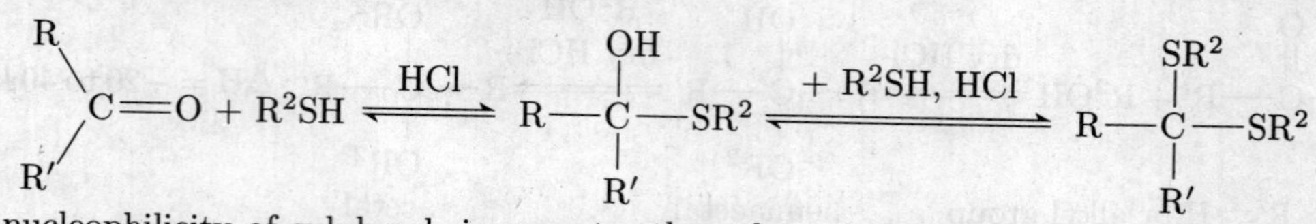

The nucleophilicity of sulphur being greater than that of oxygen, thiols undergo the acetal formation reaction more readily than alcohols. The thioacetals are relatively stable to dilute acids and bases. However, they may be decomposed readily by mercuric chloride and cadmiun carbonate mixture.

1, 3-dithianes are used to protect aldehydic group and to synthesise ketones from aldehydes.

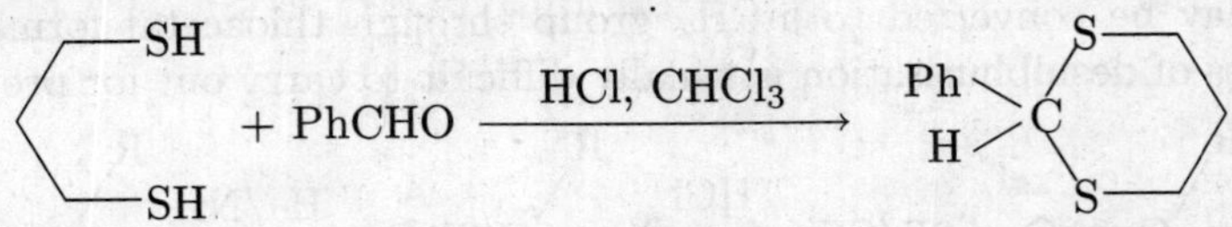

Thioacetal so formed when treated with a strong base a resonance-stabilised carbanion results involving *dπ-pπ* back bonding.

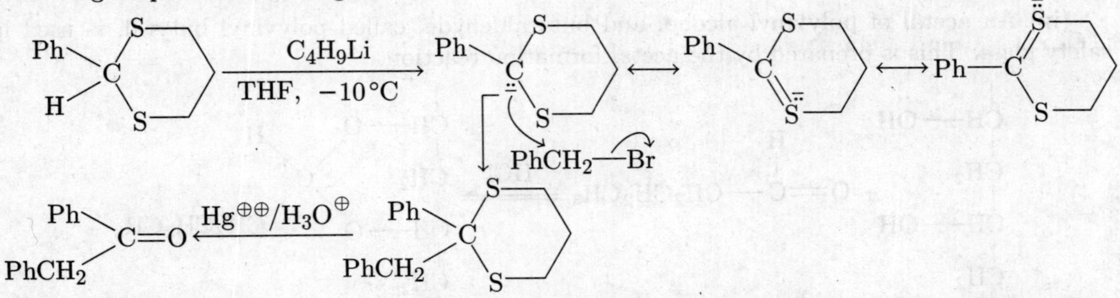

The carbanion then readily undergoes S_N reaction and the product on treatment with $Hg^{\oplus\oplus}$/$H_3O^{\oplus}$ gives a ketone. In this way via thioacetal an aldehyde not only can be converted to a ketone but formaldehyde to any other aldehyde also. We know that carbonyl carbon is electrophilic ($\delta+$) and oxygen is nucleophilic ($\delta-$); by the use of 1, 3-dithiane the carbonyl carbon has been converted to a negative pole—thus a reversal of polarity (called **Umpolung**) has been done. However, an H on carbonyl carbon of an acetal cannot be abstracted by a base because a carbanion, if formed, would not get stability by resonance since an oxygen atom does not have a *d* orbital for back bonding.

The acetal formation reaction has some preparative and analytical values.

(i) Acetals being stable towards bases and they being readily hydrolysed to the parent carbonyl compounds by dilute acids, the acetal formation reaction may safely be used for protecting carbonyl groups in the presence of bases.

For example, acetyl derivative of 3-hydroxypropanal may be converted to 3-hydroxypropanal by base-catalysed hydrolysis after protecting the aldehyde group. The protection of the aldehyde group is required to avoid the aldol condensation reaction under basic conditions.

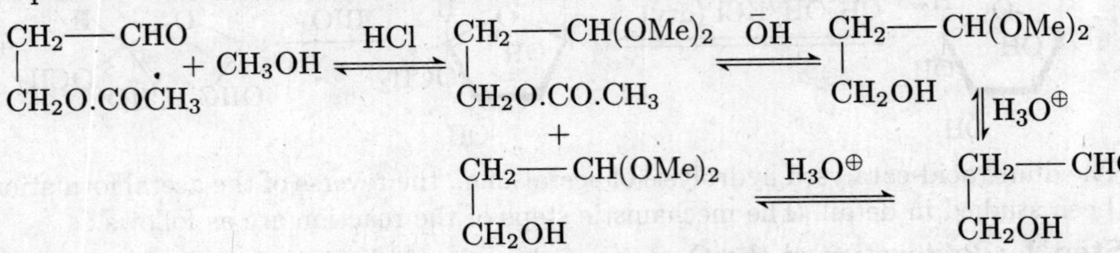

Ketonic carbonyl groups are usually protected by forming cyclic acetals with 1, 2 and 1, 3-diols.

However, when the carbonyl group requires protection for carrying reaction in the acid medium, thioacetal formation reaction is used, since thioacetals are stable towards dilute acids.

(ii) Since acetals consume water to give back the carbonyl compound, the water, formed in other reactions, is removed by acetals as reagents, e.g., $Me_2C(OMe)_2$ is used to remove free water in the esterification reaction.

(iii) Since thioacetals can be desulphurised by hydrogen in the presence of Raney nickel, >C=O group may be converted to $>CH_2$ group through thioacetal formation reaction. Of course, the process of desulphurisation is usually difficult to carry out for preparative processes.

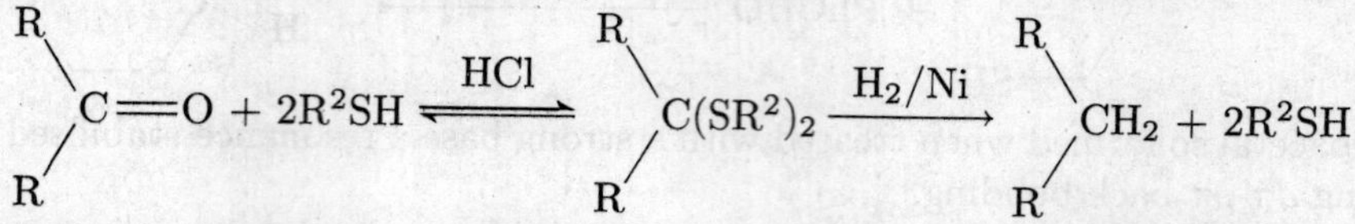

(iv) An acetal of polyvinyl alcohol and butaraldehyde, called polyvinyl butyral, is used in safety glass. This is prepared by the acetal formation reaction.

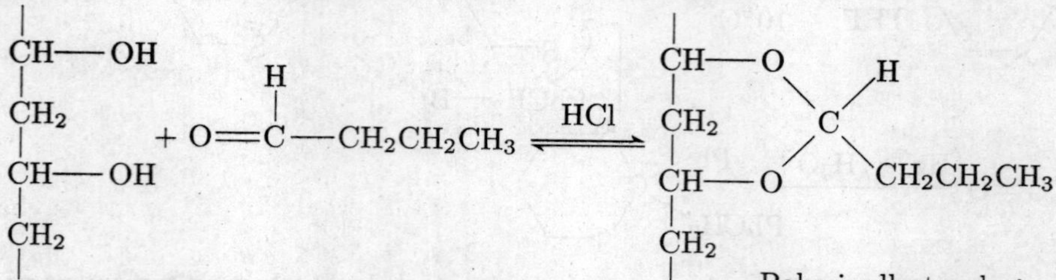

Polyvinylbutyral

(v) Reducing monosaccharides are cyclic hemiacetals. They are converted to acetals by allowing them to react with methyl alcohol in the presence of dry HCl. The acetals then are subjected to HIO_4 oxidation. If formic acid is obtained as an oxidative degradation product the sugar ring is considered to contain at least three alcoholic OH groups in three side-by-side C atoms and the ring is a six membered one. On the other hand, the absence of formic acid in the degradation product indicates that the ring is a five membered one. Thus HIO_4 oxidation of acetals of reducing sugars helps us to determine the size of the ring of the monosaccharides.

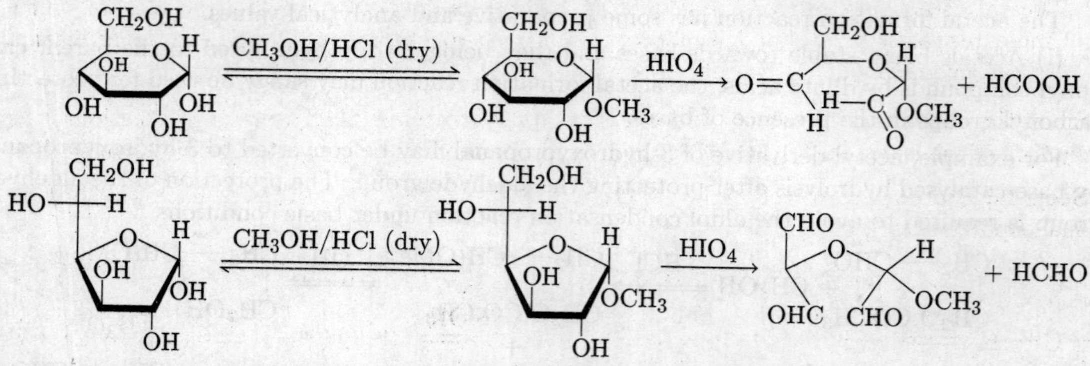

The dilute acid-catalysed hydrolysis of acetals, i.e., the reverse of the acetal formation reaction has been studied in detail. The mechanistic steps of the reaction are as follows :

Step I : Protonation of the O atom of the ether linkage and reversible formation of the resonance-stabilised carbenium ion :

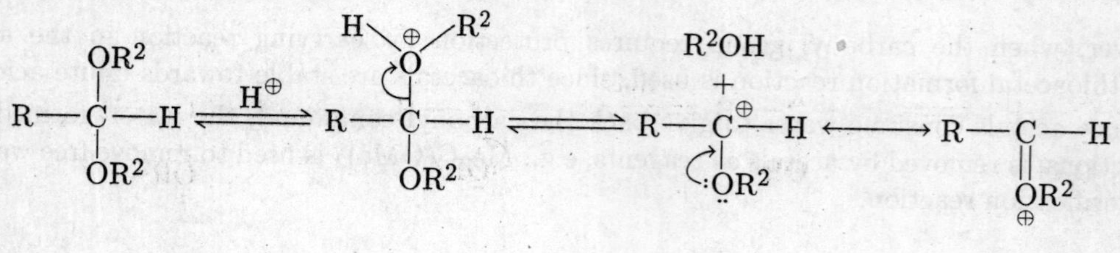

Step II : Nucleophilic attack of water on the resonance-stabilised cationic carbon to form hemiacetal :

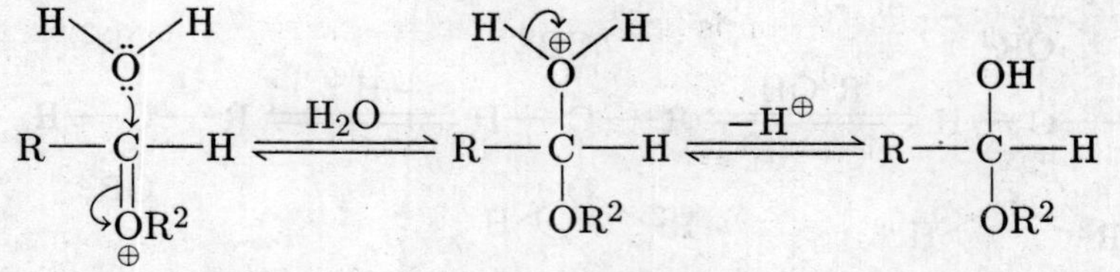

Step III : Protonation of the O atom of OR^2 group and subsequent loss of R^2OH :

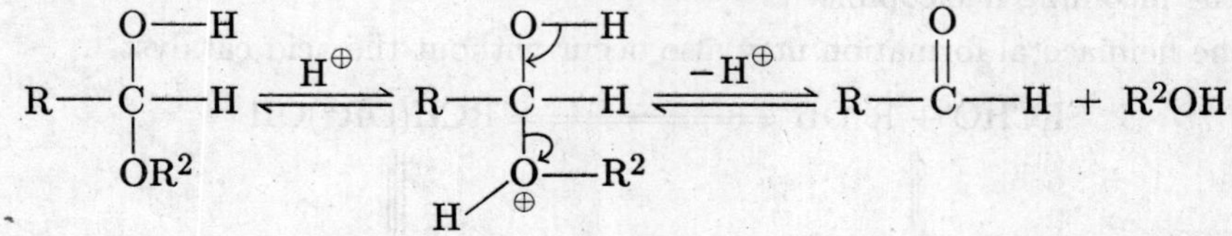

The first and the second steps constitute an S_N1 reaction where R^2OH is the leaving nucleophile and H_2O is the attacking one.

The following facts support the mechanism of the hydrolysis of acetals :

(i) The reaction proceeds with the specific acid-catalyst, $H_3O^{\oplus}$.

(ii) The reaction has been found to be faster with D_2O than with H_2O. So, Step II may be R.D. step (solvent KIE effect).

(iii) The R^2—O bond does not break even when R^2 is a tert-butyl group—this has been established by ^{18}O labelling. Therefore, C—OR^2 bond breaks.

(iv) Acetal of an optically active alcohol on hydrolysis does not give racemic alcohol. This indicates that $^{\oplus}R^2$ cation does not form. If it were so, there would have been the formation of racemic alcohol, $(\pm)$-R^2OH.

Since the hydrolysis of acetals and the formation of acetals are the reverse reactions, the mechanism of the latter reaction may be given following the Principle of Microscopic Reversibility.

Step I : Protonation of the carbonyl O atom and the formation of the hemiacetal through the nucleophilic attack of R^2OH :

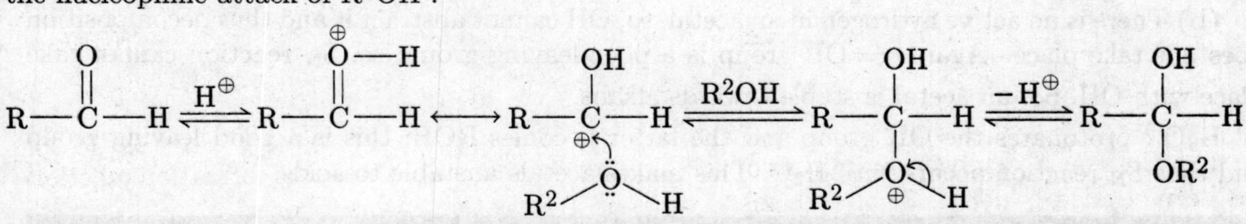

Step II : Protonation of the —OH group of the hemiacetal and the formation of the resonance-stabilised carbenium ion :

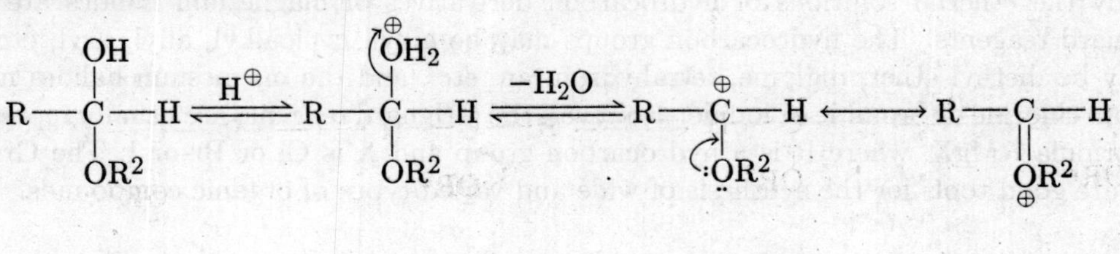

Step III : Nucleophilic attack by the second R^2OH molecule on the cationic carbon and the subsequent loss of proton.

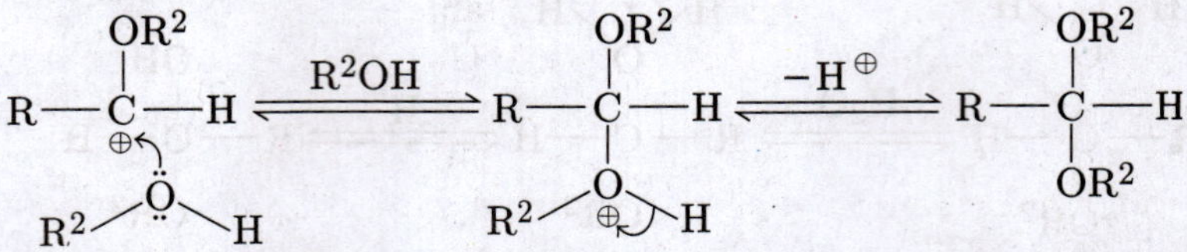

The second and the third steps constitute an S_N1 reaction where H_2O is the leaving nucleophile and R^2OH is the incoming nucleophile.

However, the hemiacetal formation may also occur without the acid catalyst.

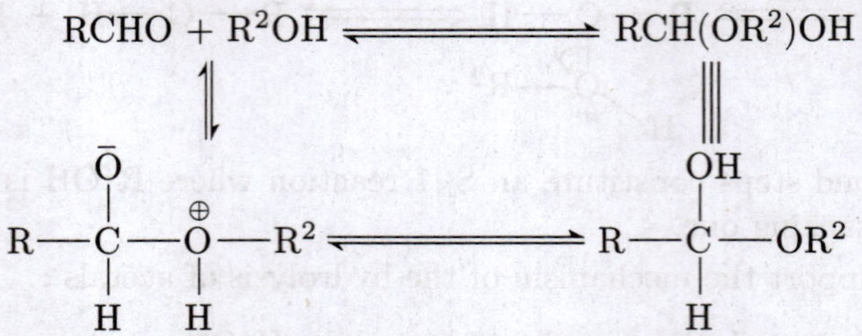

Problem 16.4. (a) Convert acraldehyde to glyceraldehyde. **(b)** Why are acetals stable towards alkalis but not to acids?

Solution : (a)

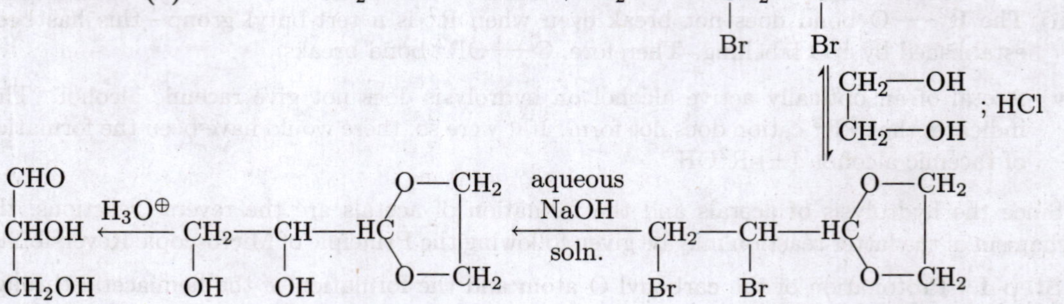

(b) There is no active hydrogen in an acetal, so, $\bar{O}H$ cannot abstract it and thus decomposition does not take place. Again —OR group is a poor leaving group; so, S_N reaction cannot take place with $\bar{O}H$. So, an acetal is stable towards alkalis.

$H_3O^{\oplus}$ protonates the OR group and the latter becomes $R\overset{\oplus}{O}H$; this is a good leaving group and thus S_N reaction occurs with H_2O. This makes acetals unstable to acids.

16.4.5 Addition of the Grignard reagent and alkyl metals

Usually the etherial solutions of hydrocarbon derivatives of magnesium halides are called the Grignard reagents. The hydrocarbon groups may be alkyl, cycloalkyl, allyl, aryl, etc. The ether may be diethyl ether, diglyme, tetrahydrofuran, etc., and the magnesium halides may be magnesium chloride or bromide or iodide. However, the Grignard reagents are usually represented by the formula RMgX, where R is a hydrocarbon group and X is Cl or Br or I. The Grignard reagents are good tools for the synthesis of wide and varied types of organic compounds.

A Grignard reagent is prepared by shaking an etherial solution of halogen derivative of a hydrocarbon with pure and dry pieces of magnesium in ether in the presence of a piece of iodine which initiates the reaction. The apparatus, the reactants and the solvent should be free from moisture, alcohol and anything which reacts with the Grignard reagent to be formed. For this reason all these are washed, purified and dried perfectly. However, at the start, the ether becomes cloudy and begins to boil gently and then the ether refluxes. When the reaction is complete, a clear solution of the Grignard reagent in ether is formed and this is then treated with any chosen reagent in situ to carry out the synthesis of a required compound. After the reaction is over, in many a case, the product is treated with acidulated water or aqueous ammonium chloride solution.

The structure of the Grignard reagent is still not known. The crystalline Grignard reagents, which have been isolated from ethers, are dietherates (like hydrates) of the formula $RMgX(OEt_2)_2$ and thus remain as monomer. X-xay analysis has shown that the monomer has a tetrahedral arrangement in which magnesium atom takes the centre of the tetrahedron, and the R, the X and the two Et_2O groups remain in the four corners of the tetrahedron as in the Fig. 16.5.

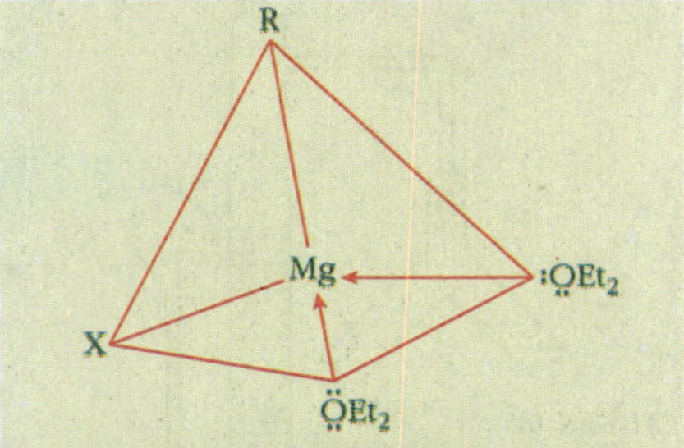

Fig. 16.5. The monomer of the Grignard reagent

The R—Mg bond is a polar covalent bond while Mg—X bond is supposed to be ionic; Et_2O forms two coordinate covalent bonds with Mg through the lone pair of *es* of the O atoms. However, in the etherial solution, the Grignard reagent remains in association and the extent of association depends on the concentration. The association is extensive above about 0.3 M. In the dimeric state, the halogen atom of one RMgX unit coordinates with the Mg atom of the other unit and Mg atom of the former coordinates with two ether molecules while that of the latter coordinates with only one ether molecule so that both of them contain eight electrons in their outer-most shell as shown in Figure 16.6.

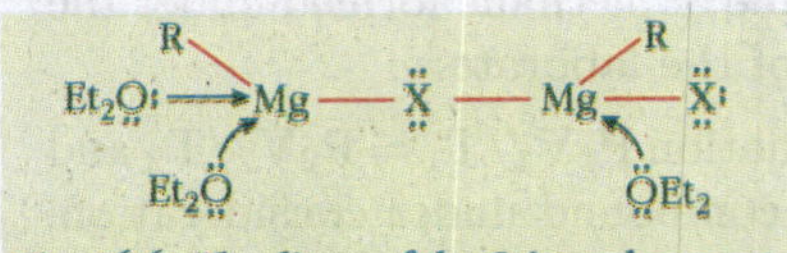

Fig. 16.6. The dimer of the Grignard reagent

However, for our purpose we shall consider the Grignard reagent as R—MgX in which R—Mg bond is a polar one having a partial negative pole on R and a partial positive pole on Mg; thus it is :

$$\overset{\delta-}{R}—\overset{\delta+}{MgX}.$$

The Grignard reagent is versatile in the sense that it reacts with a wide and varied types of compounds to give a overwhelming number of synthetic routes. It reacts with oxygen and adds to carbon-hetero multiple bonds, e.g., $>C{=}O$, $>C{=}S$, $—N{=}O$, $>S{=}O$, $C{=}N$, etc. It also undergoes the nucleophilic substitution reaction on sp^2C; for instance with CO_2Et, COCl etc. It racts with active hydrogens, i.e., acidic hydrogens—by acidic hydrogens we mean the H atoms of acetylene, active methylene and those of NH_3, H_2O phenolic OH, enolic OH, alcoholic OH, amino group, SH group, etc., which can be abstracted as proton. $R^{\delta-}$ pole of the reagent abstracts the acidic hydrogen of a compound to yield a hydrocarbon :

$$H—O—H + \overset{\delta-}{R}—\overset{\delta+}{MgX} \longrightarrow RH + HOMgX$$

$$H_2NH + \overset{\delta-}{R}—\overset{\delta+}{MgX} \longrightarrow RH + NH_2MgX$$

$$R^1\ddot{O}H + R{-}MgX \longrightarrow RH + R^1OMgX$$

$$Ph{-}O{-}H + RMgX \longrightarrow RH + PhOMgX$$

$$R^1NH_2 + RMgX \longrightarrow RH + R^1NHMgX$$

$$R^1{-}C{\equiv}C{-}H + RMgX \longrightarrow R^1{-}C{\equiv}CMgX + RH$$

These reactions may be considered as acid-base reactions; the weaker acid, RH, being liberated by the stronger acid. This type of reaction is valuable for the estimation of active hydrogen and consequently for the determination of groups containing active hydrogen. For instance, hydroxyl groups, and the process is known as the Zerewitinoff's method of active hydrogen determination. A simple form of the apparatus used in the method is shown in Fig. 16.7. A solution of methylmagnesium iodide in a high boiling ether (say, iso-amyl ether) is taken in one leg of the bifurcated tube. Any accurately weighed amount of the compound is taken in ether solvent in the other leg. By opening the stopcock, the apparatus is brought to temperature equilibrium. The cock is then closed and the height of the mercury column is noted. The two solutions are then mixed and the apparatus is brought to temperature equilibrium and again the height of the mercury column is noted. The difference between the two readings gives the volume of methane formed under the temperature and pressure of the laboratory.

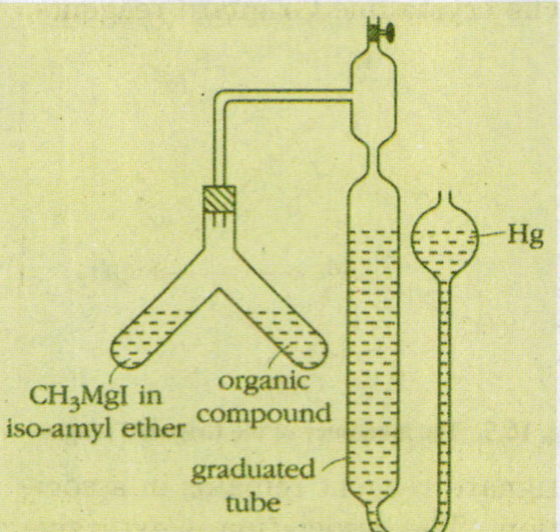

Fig. 16.7. Zerewitinoff's apparatus for the determination of active hydrogen

Reducing the volume of methane at N.T.P. according to the equation $P_1V_1/T_1 = P_2V_2/T_2$ and calculating the mole of methane formed per mole of the organic compound under consideration, it is possible to calculate the number of mole of active hydrogen, i.e., the number of mole of the group (bearing active H) per mole of the compound; here it is important to note that n mole of active H give n mole of methane at N.T.P. from one mole of the compound.

However, the Grignard reagents add to the carbonyl group of aldehydes and ketones to yield addition products which on hydrolysis with acidulated water give alcohols. In this reaction formaldehyde gives 1° alcohol, other aldehydes give 2° alcohols and ketones give 3° alcohols. The hydrocarbon group of a Grignard reagent attaches itself to the carbonyl C atom and the magnesium halide part goes to the carbonyl O atom.

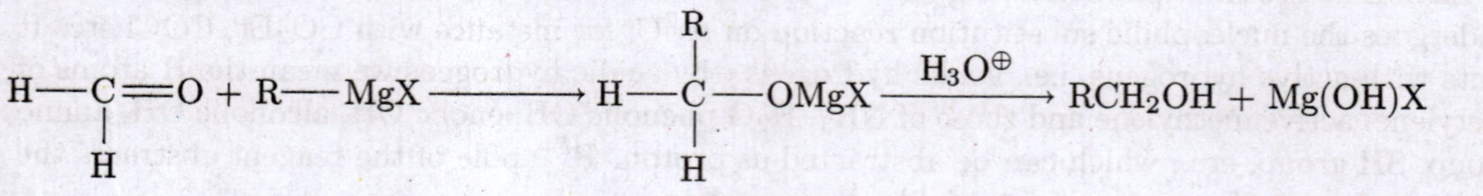

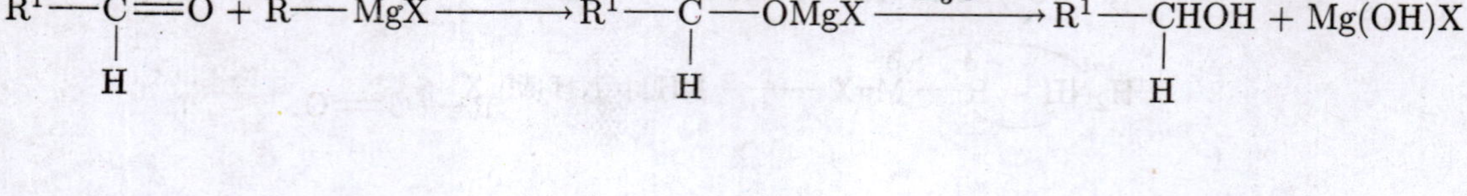

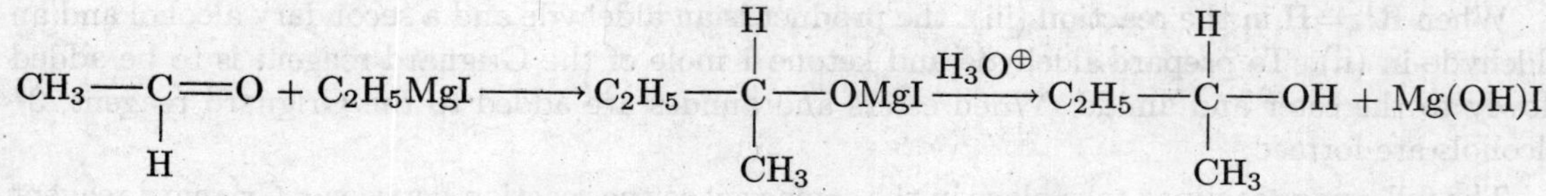

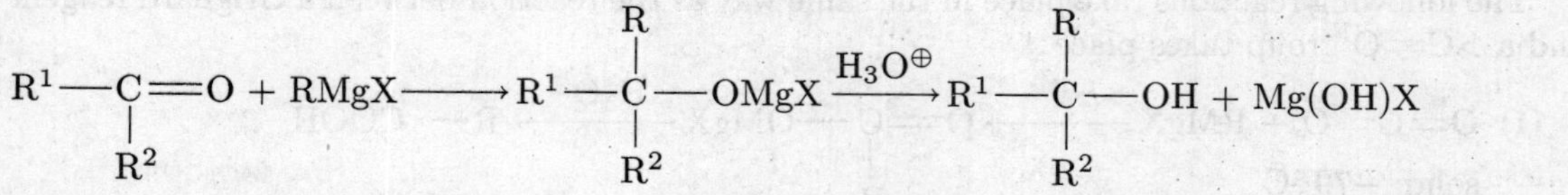

Acid halides, esters, amides, cyanides undergo nucleophilic substitution reaction with the Grignard reagent to form ketones; the ketones then may undergo the addition reactions to yield tertiary alcohols.

(i) 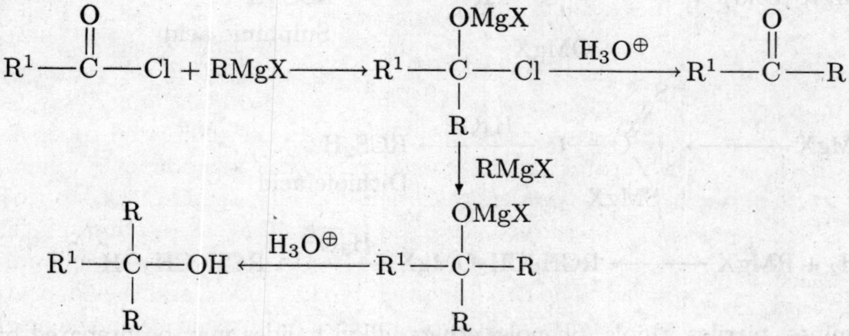

(ii) 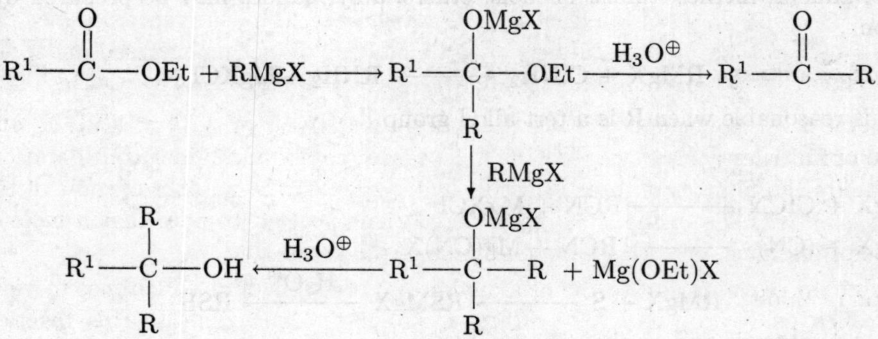

(iii) 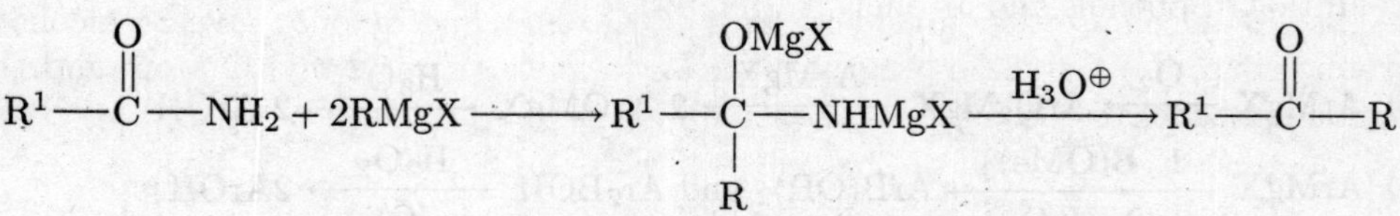

(iv) 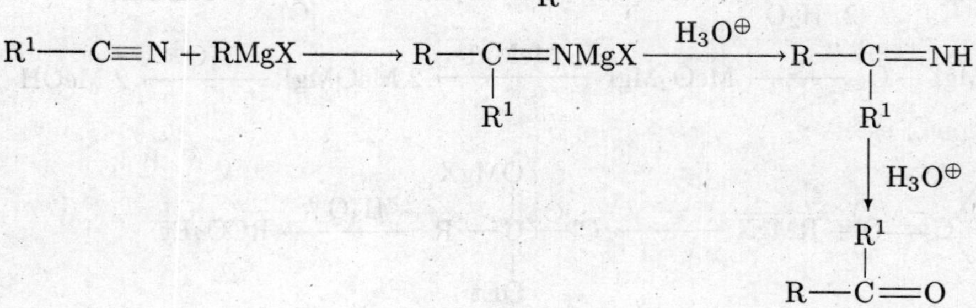

When $R^1 = H$ in the reaction (iii), the product is an aldehyde and a secondary alcohol and an aldehyde in (ii). To prepare aldehyde and ketone 1 mole of the Grignard reagent is to be added slowly to the ester and amide. When esters and amides are added to the Grignard reagent, 3° alcohols are formed.

The following reactions take place in the same way as the reaction between a Grignard reagent and a >C=O group takes place.

(i) $O{=}C{=}O + RMgX \longrightarrow O{=}C(R){-}OMgX \xrightarrow{H_3O^{\oplus}} R{-}COOH$

solid; −79 °C

(ii) 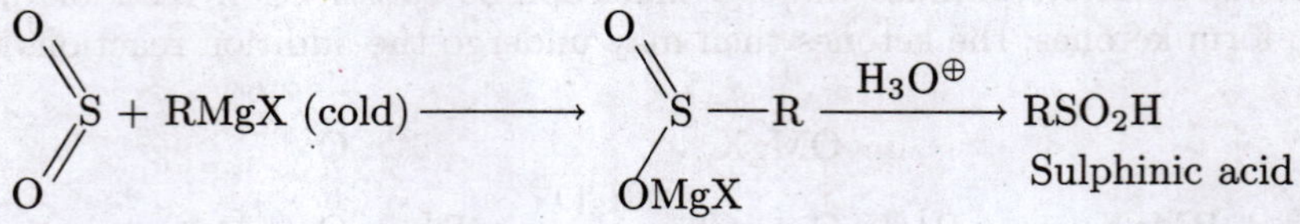

(iii) 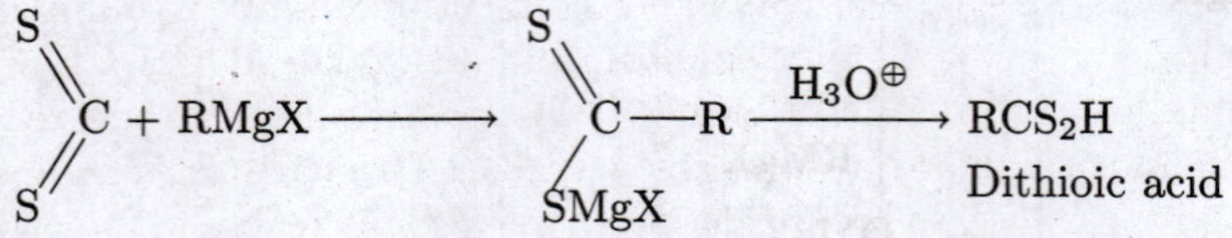

(iv) $H_2C{-}CH_2$ (epoxide, bridged by O) $+ RMgX \longrightarrow RCH_2CH_2OMgX \xrightarrow{H_3O^{\oplus}} RCH_2CH_2OH$

Besides these, amines, nitriles, thiols, phenols, ethers, alkyl halides may be prepared by the Grignard reaction.

(i) Preparation of amines : $RMgX + ClNH_2 \longrightarrow RNH_2 + MgXCl$

The yield is reasonable when R is a tert-alkyl group.

(ii) Preparation of nitriles :

(a) $RMgX + ClCN \longrightarrow RCN + MgXCl$

(b) $RMgX + (CN)_2 \longrightarrow RCN + Mg(CN)X$

(iii) Preparation of thiols : $RMgX + S \longrightarrow RSMgX \xrightarrow{H_3O^{\oplus}} RSH$

(iv) Preparation of phenols and alcohols :

(a) $ArMgX \xrightarrow{O_2} ArO_2MgX \xrightarrow{ArMgX} 2\ ArOMgX \xrightarrow{H_3O^{\oplus}} 2\ ArOH$

(b) $ArMgX \xrightarrow[2.\ H_2O]{1.\ B(OMe)_3} ArB(OH)_2$ and $Ar_2BOH \xrightarrow[{[O]}]{H_2O_2} 2ArOH$

(c) $MeMgI + O_2 \longrightarrow MeO_2MgI \xrightarrow{MeMgI} 2\ MeOMgI \xrightarrow{H_3O^{\oplus}} 2\ MeOH$

(v) Preparation of esters :

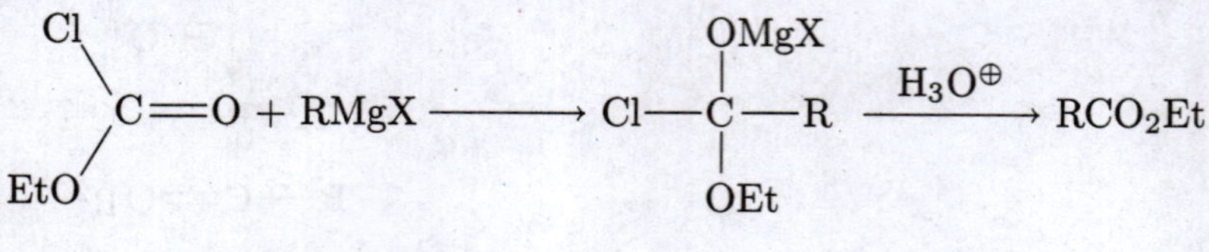

(vi) Preparation of alkyl halide :

$$RMgCl + X_2(X{=}Br \text{ or } I) \longrightarrow RX + MgClX$$

(vii) Preparation of ether :

$$R^1OCH_2Cl + RMgCl \longrightarrow R^1OCH_2R + MgCl_2$$

(viii) Preparation of hydrocarbons : This has already been discussed along with the determination of active hydrogen. In this case the hydrocarbon group present in the Grignard reagent will give the corresponding hydrocarbon.

(ix) Preparation of organocadmiums :

$$2RMgX + CdCl_2 \longrightarrow R_2Cd + 2MgXCl$$

However, the reactivity order of the substrates is:

Active H compounds > CHO ≫ CO > COCl > CO_2R > CH_2X compounds.

Since the Grignard reagents are very sensitive, a specific reaction on a compound containing more than one functional group is very difficult to perform. For example, $HOCH_2CH_2CHO$ does not undergo the Grignard addition to the CHO group; the active H on OH group reacts much faster than the CHO group. Furthermore, to prepare a specific Grignard reagent the halogen compound is to be chosen with utmost care. Since a Grignard reagent reacts with active hydrogens, carbonyl compounds, esters, nitriles, nitro compounds, etc., halogen compounds containing these functions cannot be converted to the Grignard reagent. Besides these, branching of the carbon chain near the functional group of the substrate and the bulky hydrocarbon group in the Grignard reagent exert steric effect and as a consequence a functional group capable of undergoing the Grignard reaction fails to react. For example, steric hindrance in the substrate makes hexamethylacetone, $Me_3CCOCMe_3$, inactive toward the reaction with MeMgI or MeMgBr. On the other hand, isopropyl methyl ketone does not react with t-butyl-magnesium chloride because of the bulk of t-butyl group while methylmagnesium iodide reacts readily with isopropyl methyl ketone. There are, again, some instances where the reactions take place but abnormal products are obtained instead of expected products; these are enols and reduced products. These are formed owing to the (i) bulk of the reagent and (ii) steric hindrance in the substrate. For enolisation, a sterically hindred carbonyl compound requires to possess an α-H; while for reduction, the Grignard reagent requires a β-hydrogen. For example, when isopropylmagnesium bromide is added to diisopropyl ketone, no triisopropylcarbinol (a 3° alcohol) forms, instead diisopropylcarbinol (a 2° alcohol) forms through reduction. The other products of the reaction are starting ketone (from enol), propane and propene.

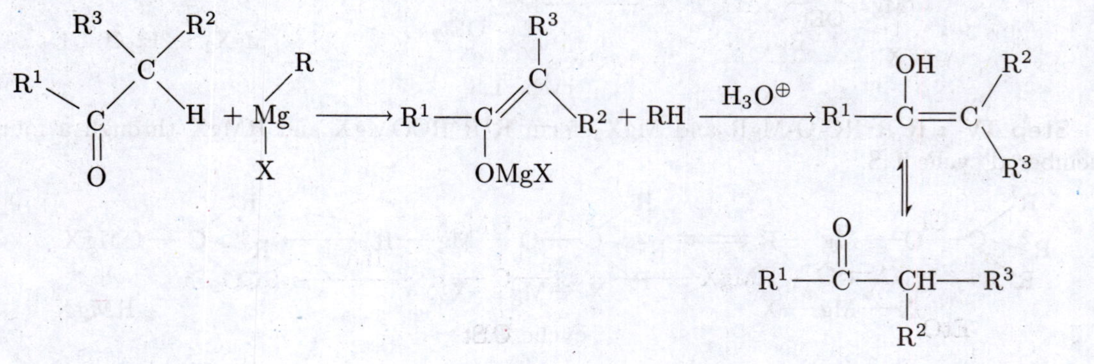

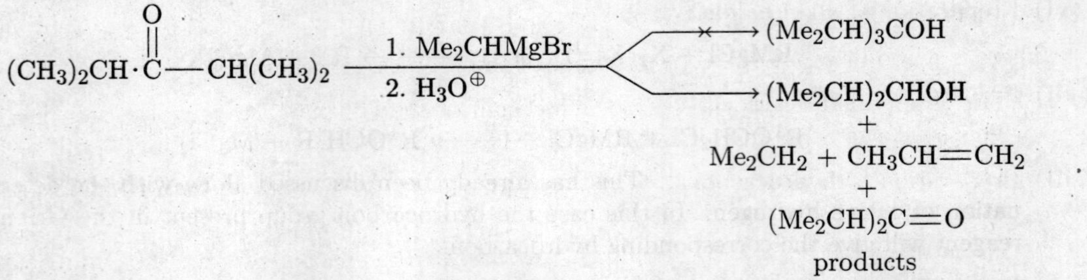

The mechanism of the addition of the Grignard reagent to an aldehyde or a ketone has been studied in detail. Still the mechanism is uncertain. Many mechanisms have been put forward. However, the mechanism proposed by Ashby and co-workers (1967) is a plausible one. This is given below. His proposition is based on the monomeric structure of the Grignard reagent in the dietherate form.

Step I : The carbonyl O coordinates with the magnesium of the Grignard reagent which is present in the dietherate form.

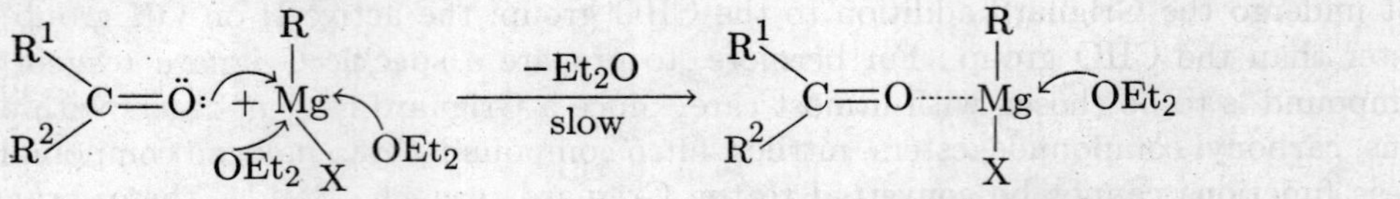

coordinated complex

Step II : The halogen atom of the coordinated complex then coordinates with the Mg of a second molecule of the Grignard reagent in the dietherate form :

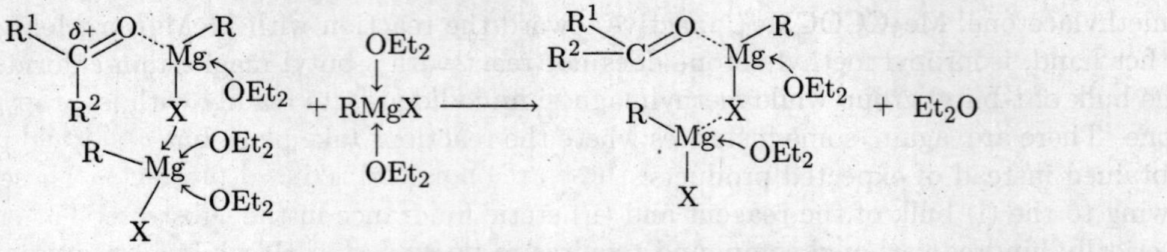

Step III : Formation of $R^1R^2RC\text{-}OMgR$ and MgX_2 via a cyclic six-membered T.S.

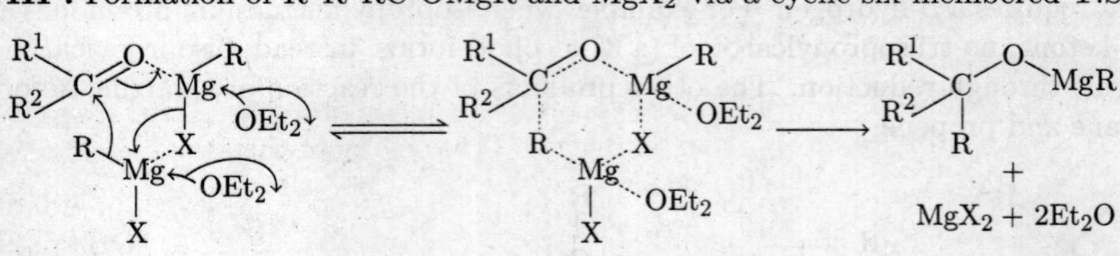

cyclic T.S.

Step IV : $R^1R^2RC\text{-}O\text{-}MgR$ and MgX_2 form $R^1R^2RCOMgX$ and RMgX through a four-membered cyclic T.S. :

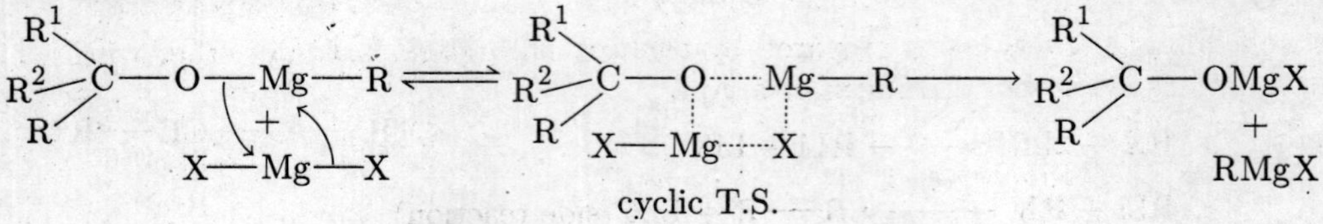

cyclic T.S.

Final step : Hydrolysis of $R^1R^2RCOMgX$ to R^1R^2RCOH.

There is a good deal of evidence in favour of the proposed mechanism:

(i) The rate of the reaction decreases when half of the Grignard reagent has been used up. This supports the participation of two RMgX units in the reaction.

(ii) The rate of the reaction increases when $MgBr_2$, a better Lewis acid than RMgX, is added. This indicates that step I is a slow step.

(iii) When R_3COMgR and $MgBr_2$ taken together in the ether medium, $R_3COMgBr$ forms. This supports step III and step IV.

(iv) The abnormal Grignard reactions, i.e., the enolisation and the reduction reactions, can be explained by considering the coordinated complex between one molecule of the carbonyl compound and one molecule of the Grignard reagent.

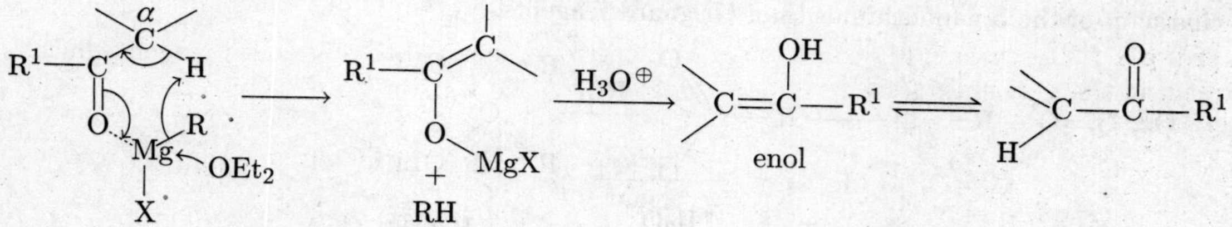

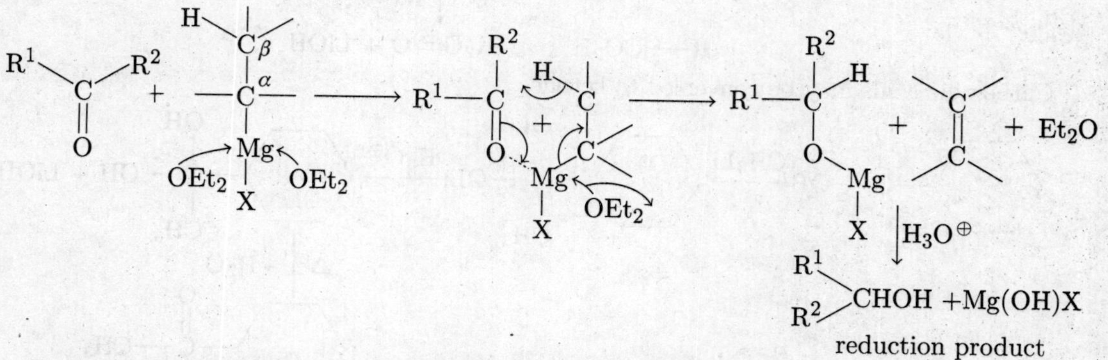

Problem 16.5. Synthesise $CH_3CH_2CH_2OH$ from CH_3MgI.

Solution : The given alcohol is a primary alcohol and it containts two more C atoms than the starting Grignard reagent. The reaction between ethylene oxide and a Grignard reagent followed by hydrolysis gives a primary alcohol containing two C atoms more than the Grignard reagent.

$$\underset{\text{(O bridged)}}{H_2C—CH_2} + CH_3MgI \longrightarrow CH_3—CH_2—CH_2OMgI \xrightarrow{H_3O^{\oplus}} CH_3CH_2CH_2OH$$

However, Grignard reagents are less reactive than organolithiums. The latter compounds undergo some reactions which the former compounds do not; here is the importance of organolithiums. The reactions of organolithiums are very alike to those of Grignard reagents.

Organolithiums are very often prepared by treating an organic halide in ether with lithium metal in an atmosphere of dry nitrogen or argon.

$$RX + 2Li \longrightarrow RLi + LiX$$

$$RLi + RX \longrightarrow R—R + LiX \text{ (side reaction)}$$

To minimiise the side reaction, the product mixture is cooled to $-10\,°C$ in the cases of alkyllithiums which are highly reactive; aryllithiums being less reactive towards neucleophiles, they are synthesised at the b.p. of the solvent. Aryl and vinyl halides being less reactive towards metallic lithium, butyllithiums are used to prepare aryl and viniyllithiniums.

$$Bu{-}X + 2Li \longrightarrow LiX + BuLi \xrightarrow{Ar-X} ArLi + BuX$$

Organolithiums can react more or less readily where Grignard reagents fail to react owing to steric hindrance. For example, isopropylmagnesium bromide does not add to diisopropyl ketone but iso-propyllithnum adds effectively to form triisopropylcarbinol.

$$(CH_3)_2CHLi + [(CH_3)_2CH]_2C{=}O \longrightarrow [(CH_3)_2CH]_3C{-}OLi \xrightarrow{H_3O^{\oplus}} [(CH_3)_2\ CH]_3COH$$

Carbon dioxide reacts with organolithiums to give ketones but with Grignard reagents to give carboxylic acids; the difference in the reaction mode is owing to the stronger nucleophilic character of the organolithiums than Grignard reagents.

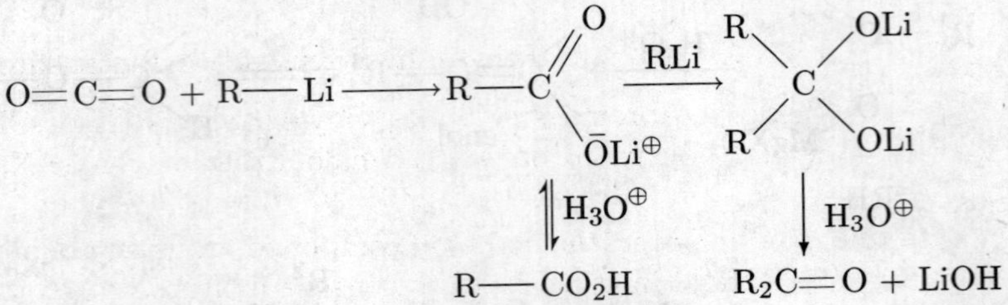

Carboxylic acids may be converted to ketones.

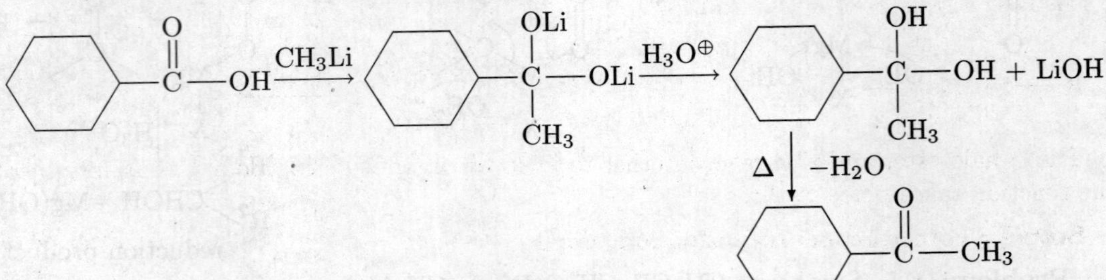

Organolithiums react with alkenes but Grignard reagents do not.

$$R{-}Li + CH_2{=}CH_2 \xrightarrow{\text{high pressure}} R{-}CH_2{-}CH_2{-}Li \xrightarrow{CH_2=CH_2} R(CH_2)_4Li \rightarrow$$

However, in the cases of conjugated polyenes the reaction takes place under the atmospheric pressure. These reactions proceed indefinitely if the substrate site has no steric hindrance. In the cases of sterically hindered compounds, the reactions stop at the first stage.

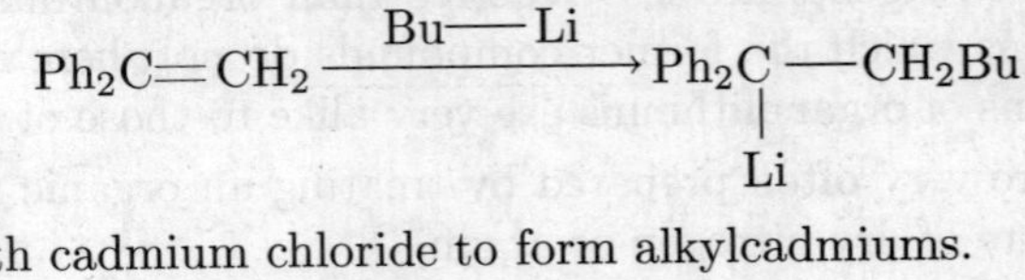

Alkyllithiums react with cadmium chloride to form alkylcadmiums.

$$2RLi + CdCl_2 = R_2Cd + 2LiCl$$

[See Secs. 16.10 and 19.6.4(v)c]

16.4.6 The Reformatsky reaction

The Grignard-like addition of an α-halo-ester to an aldehyde or a ketone in the presence of metallic zinc to yield a β-hydroxy-ester is known as the Reformatsky reaction.

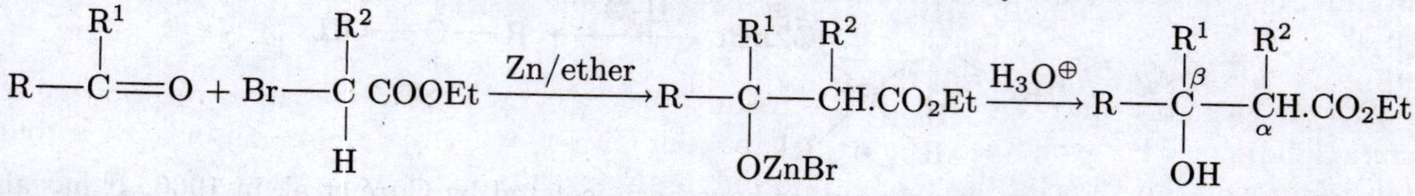

R=H or alkyl or aryl, R^1= alkyl or aryl, R^2= alkyl or aryl or H β-hydroxy-ester

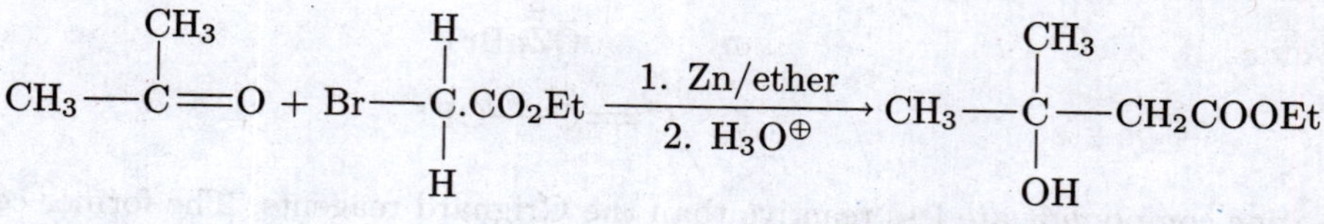

Ethyl 3-hydroxy-3-methylbutanoate

The α-halo ester in most of the cases is α-bromoester or its vinylog, e.g., RCHBr.CH=CHCO$_2$Et. However, the reaction may also be carried out with α-halonitriles or α-halo-N, N-disubstituted amides.

The procedure is very simple. The α-bromoester, the carbonyl compound and zinc are mixed together at one time and the mixture is refluxed in ether or ether-benzene mixture.

The reaction between a nitrile and an α-halo-ester in the presence of zinc is a similar reaction. In this case a β-keto-ester is obtained.

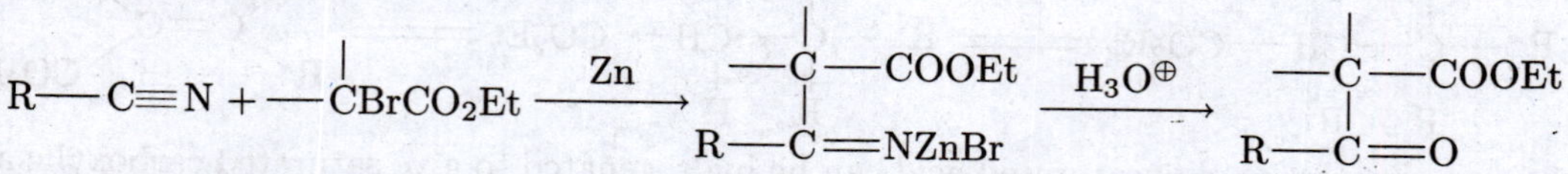

The α-halo-ester and zinc react in ether to yield an organozinc compound as an intermediate. The reaction takes place in three steps.

Step I : Formation of organozinc compound :

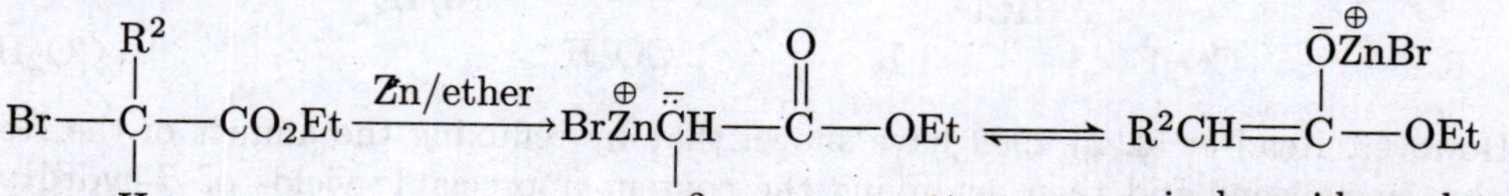

zincbromide enolate

Step II : Addition of the zincbromide enolate to the carbonyl group :

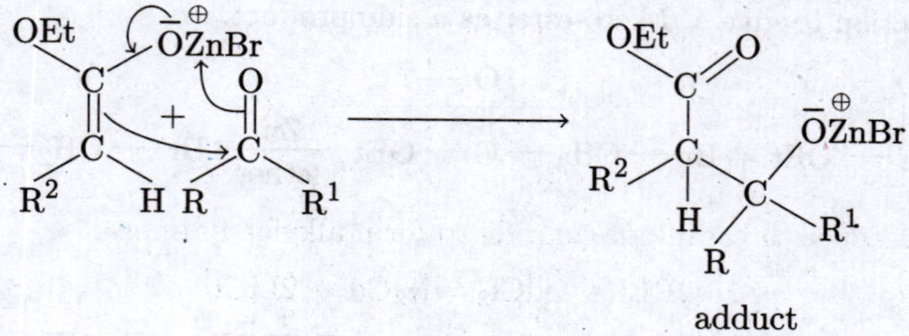

adduct

Step III : Hydrolysis of the adduct by dilute acid :

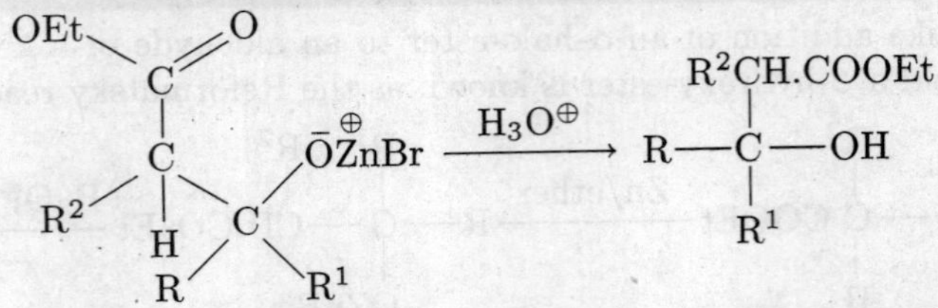

The intermediate organozinc bromides have been isolated by Curé et al. in 1966. It has also been established by Zimmermann and Traxler and others that the structure of the organozinc compound is :

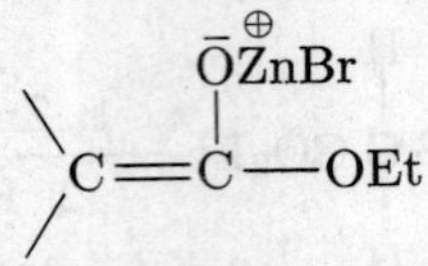

Organozinc compounds are less reactive than the Grignard reagents. The former compounds react very slowly with esters but very readily with aldehydes and ketones. By taking the advantage of this preferential reactivity of organozinc halides β-hydroxy-esters are prepared by the Reformatsky reaction. β-hydroxy-acids can also be prepared from the β-hydroxyesters either directly or indirectly. However, there is a problem in the direct conversion of esters to the corresponding acid. When the ester is subjected to hydrolysis by a base or by an acid, an α, β-unsaturated acid is formed by dehydration, specially in the cases of β-aryl-β-hydroxy-esters.

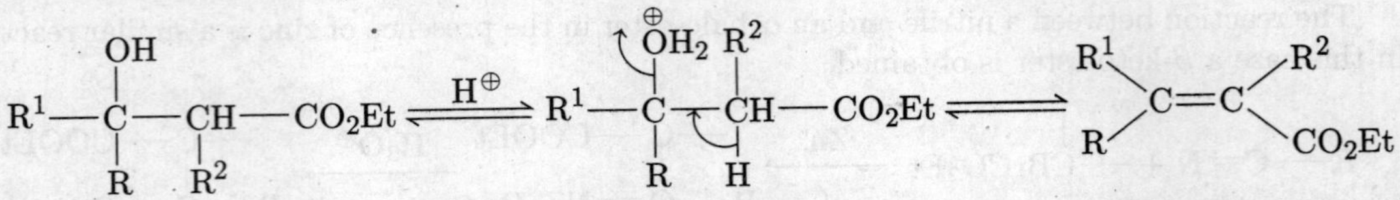

Of course, the α, β-unsaturated acid can be hydrogenated to give saturated carboxylic acids.

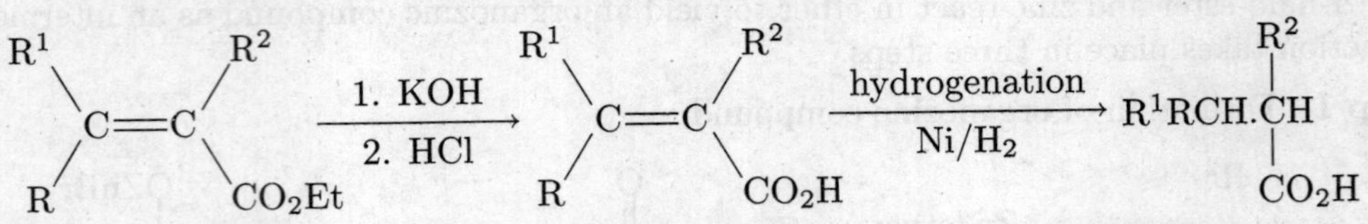

However, Read et al. in 1969 have shown that by refluxing the adduct of the Reformatsky reaction in benzene and then acidifying the content appreciable yields of β-hydroxy-acids are obtained.

There is a drawback of the Reformatsky reaction—the α-Bromo-ester may undergo self-condensation reaction to give a β-keto-ester as a side product.

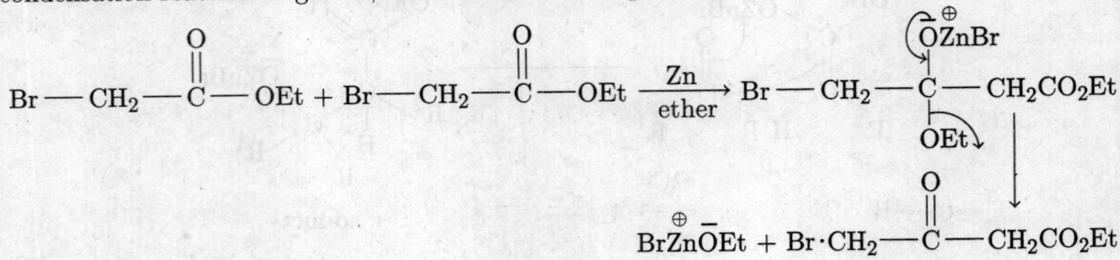

The reaction may also be extended to the vinylogous compounds of α-bromo-esters. For example, ethyl 4-bromo-2-butenoate undergoes the reaction.

$$Me_2C{=}O + BrCH_2{-}CH{=}CH{-}CO_2Et \xrightarrow[2.\ H_2O]{1.\ Zn} Me_2C(OH){-}CH_2{-}CH{=}CH{-}CO_2Et$$

However, the zinc compounds of α-bromoesters are one kind of organozinc compounds. The other organozinc compounds are alkylzincs. Alkylzincs are prepared from alkyl iodides and a zinc-copper couple.

$$2RI + 2Zn(Cu) \longrightarrow R_2Zn + ZnI_2$$

The properties of alkylzincs resemble those of alkylcadmiums. Both kinds of compounds react with acid halides to give ketones, since they react much faster with acid chlorides than with ketones. However, alkylzincs ignite spontaneously in air and this makes the compounds difficult to handle. Alkylcadmiums have no difficulty to handle and they are used to prepare the ketones from acid chlorides.

$$2RCOCl + R'_2Zn \longrightarrow 2\ R'RC{=}O + ZnCl_2$$

Good method :

$$2RCOCl + R'_2Cd \longrightarrow 2\ R'RC{=}O + CdCl_2$$

Besides Reformatsky reaction, a useful application of organozinc compounds is in Simmon-Smith reactions which involve a carbenoid zinchalide as an addendum (See. 11.4.4, p. 521).

However, the reactivity order amongst organolithiums, Grignard reagents and organozincs is organozincs < Grignard reagents < organolithiums.

Problem 16.6. How can you synthesise cinnamic acid by the Reformatsky reaction?

Solution : Cinnamic acid is 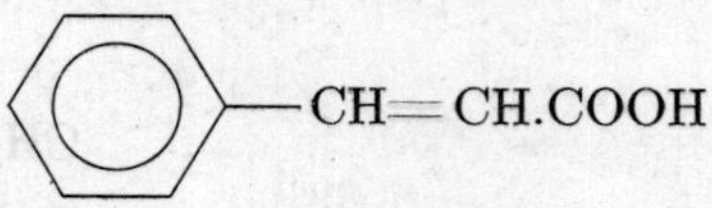

It contains PhCH = and CHCOOH parts. PhCH = part can be obtained from benzaldehyde and CH.COOH part can be obtained from $BrCH_2CO_2Et$ as shown below :

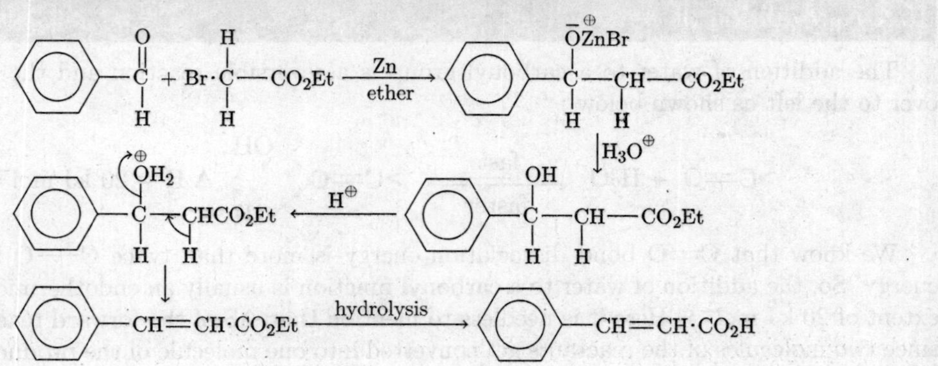

Problem 16.7. Show mechanistic steps for the following reactions :

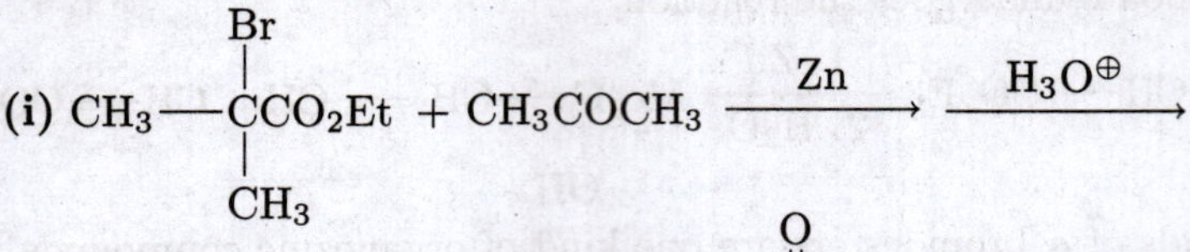

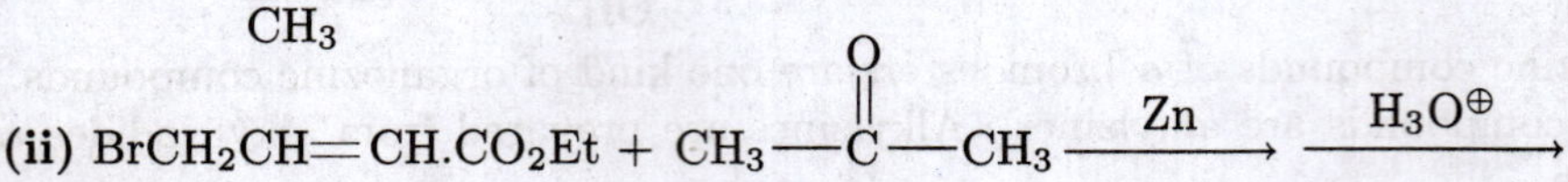

Solution :

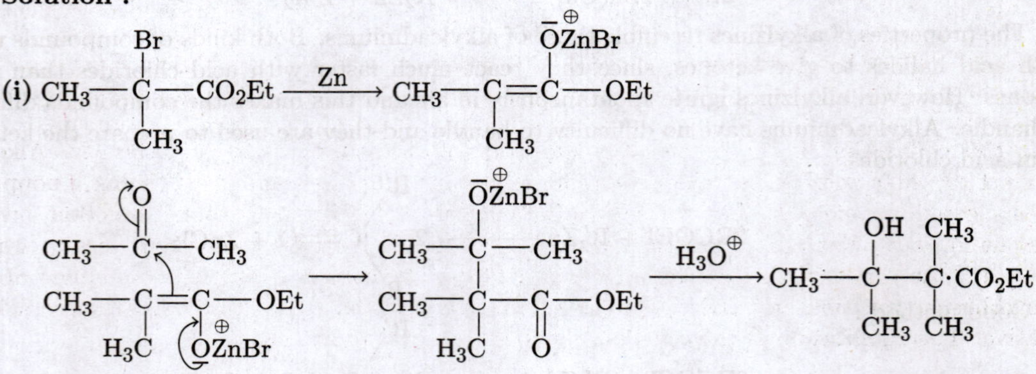

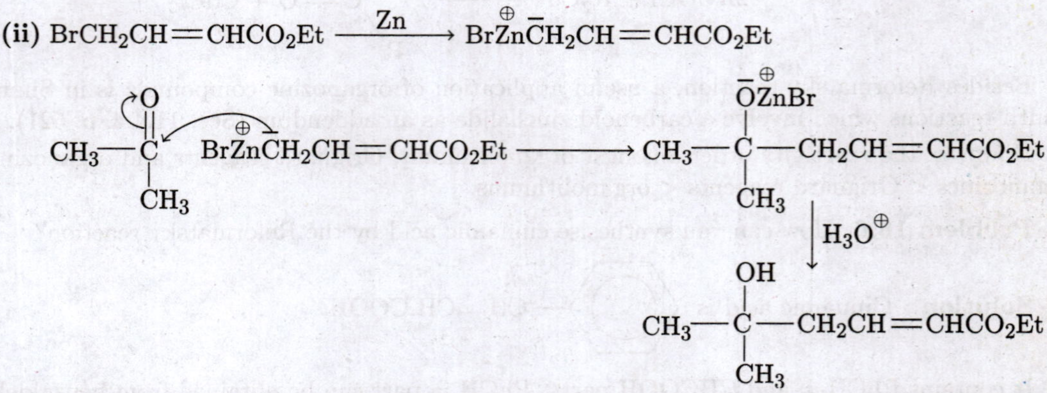

16.4.7 Addition of water

The addition of water to a carbonyl group is a reversible reaction and the equilibrium lies over to the left as shown below :

$$>C{=}O + H_2O \underset{\text{faster}}{\overset{\text{fast}}{\rightleftharpoons}} >C{=}C\begin{matrix}OH\\OH\end{matrix} \qquad \Delta H = 20 \text{ kJ mol}^{-1}$$

We know that C=O bond dissociation energy is more than twice C—O bond formation energy. So, the addition of water to a carbonyl function is usually an endothermic reaction to the extent of 20 kJ mol^{-1}. Here it is needless to mention that ΔS of the forward reaction is negative since two molecules of the reactants get converted into one molecule of the product. The reaction of formation of hydrated acetaldehyde has been found to be very slow at pH 7 but it occurs much more rapidly at pH 4 and pH 11. This means, the acid- or base-catalysed hydration is rapid.

(i)

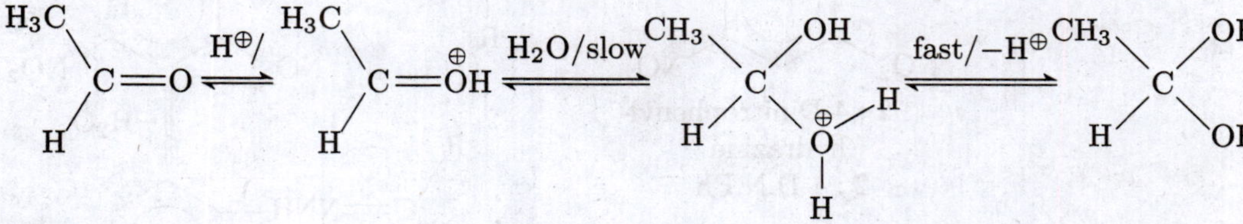

(ii)

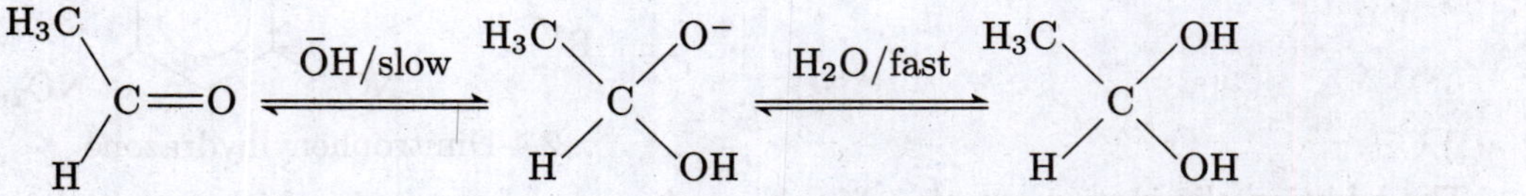

In this connection let us compare the hydration of alkenes with that of carbonyls. Alkenes are inert to water at pH 7 but aldehydes and ketones are not. Perhaps C=C being a nonpolar and nucleophilic moiety, it does not add H_2O (another nucleophile). Carbonyl function, on the other hand, is a polar group and the carbonyl C being electrophilic, a nucleophile, like water, can attack the carbonyl C. However, the acid-catalysed hydration of an olefinic double bond is an exothermic reaction and $\Delta H = -56$ kJ mole^{-1}, though the ΔS of the reaction here also is negative. The equilibrium of the reaction, however, lies over to the right.

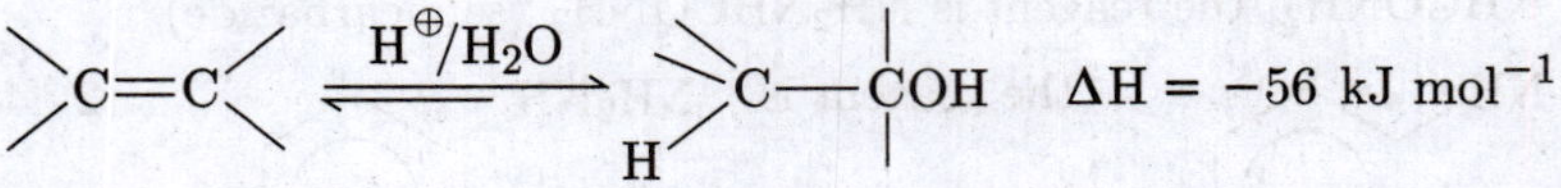

16.5 Addition-Elimination (Ad—E) Reactions

Certain derivatives of ammonia undergo addition reactions with aldehydes and ketones to yield α-hydroxy derivatives. These, in their turn, undergo water elimination reaction to give compounds containing C-N double bond.

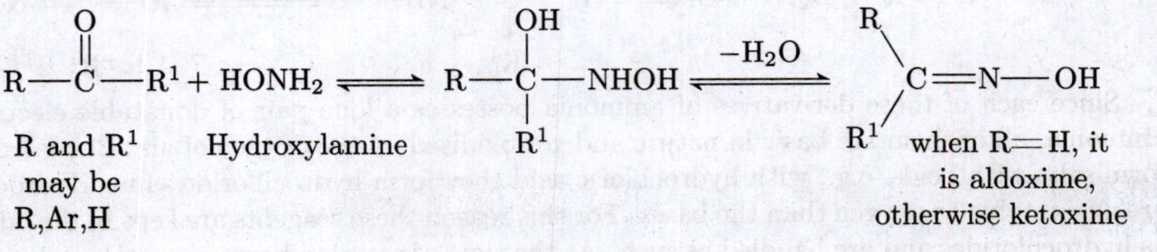

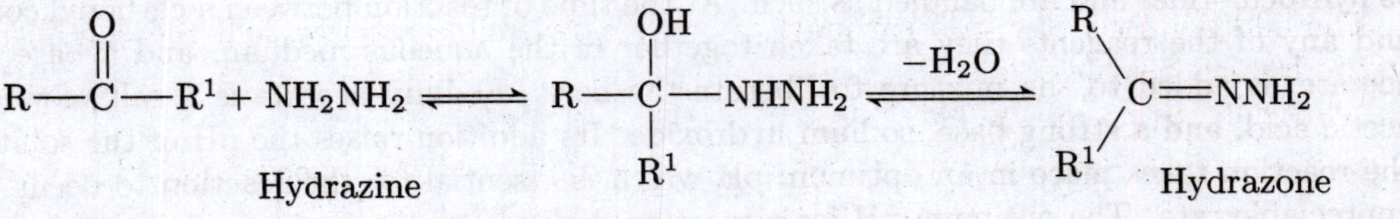

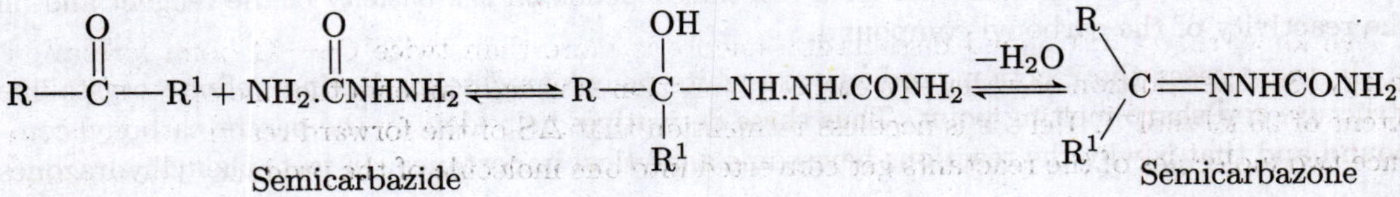

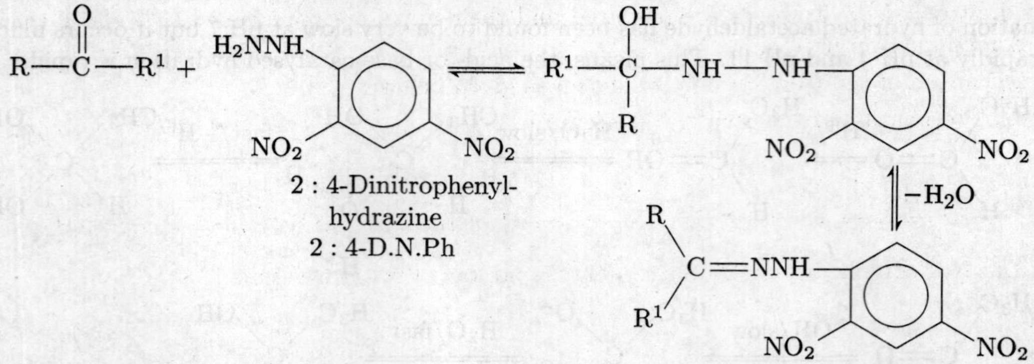

2 : 4-Dinitrophenylhydrazine
2 : 4-D.N.Ph

2,4-Dinitrophenylhydrazone

The addition-elimination reactions involve derivatives of ammonia which may be represented in general as :

GNH_2

when G is – OH, the reagent is NH_2OH (hydroxylamine)

G is – NH_2 the reagent is NH_2NH_2 (hydrazine)

G is – NHPh, the reagent is NH_2NHPh (phenylhydrazine)

G is – $NHCONH_2$, the reagent is $NH_2NHCONH_2$ (semicarbazide)

G is – NH– (2,4-dinitrophenyl), the reagent is NH_2NH– (2,4-dinitrophenyl) (2 : 4-D.N.Ph)

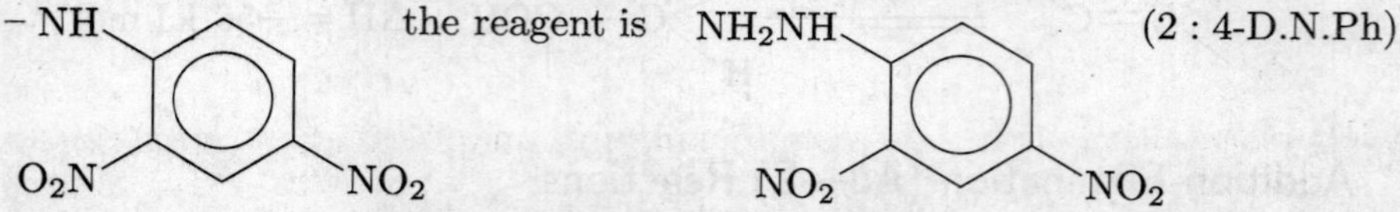

Thus, the general equation of this class of reaction may be represented as :

$$R-\overset{\overset{\displaystyle O}{\|}}{C}-R^1 + H_2N-G \rightleftharpoons R-\underset{\underset{\displaystyle R^1}{|}}{\overset{\overset{\displaystyle OH}{|}}{C}}-NH-G \xrightleftharpoons{-H_2O} R-\underset{\underset{\displaystyle R^1}{|}}{C}=N-G$$

Since each of these derivatives of ammonia possesses a lone pair of donatable electrons like ammonia, all of them are basic in nature and got oxidised by the oxygen of air. Being basic, they form salts with acids, e.g., with hydrochloric acid they form hydrochloride salts. Hydrochlorides are more stable to oxygen than the bases. For this reason these reagents are kept in the laboratory as hydrochlorides and are handled as such. At the time of reaction between a carbonyl compound and any of the reagents they are taken together in the aqueous medium, and a base, sodium acetate, is added to the mixture to liberate the base. Sodium acetate is a salt of weak acid, acetic acid, and a strong base, sodium hydroxide. Its addition raises the pH of the solution and the reaction takes place in an optimum pH which is essential for the reaction to occur with an appreciable rate. The optimum pH for a reaction depends on the basicity of the reagent and on the reactivity of the carbonyl compound.

In these cases the reaction products are white or coloured solids having definite crystalline structure and sharp melting point. Thus these derivatives help identify the parent carbonyl compound and that is why the reactions have some analytical importance. In fact, phenylhydrazones and 2 : 4-dinitrophenylhydrazones are prepared and their m.p.s are determined to identify the carbonyl compounds very often.

The mechanism of the reaction may be shown as given below :

Step I : Protonation of the carbonyl O atom :

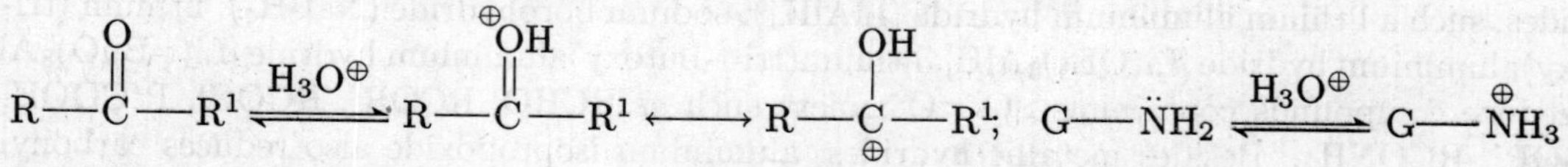

Step II : Nucleophilic attack by the reagent; rapid loss of water molecule :

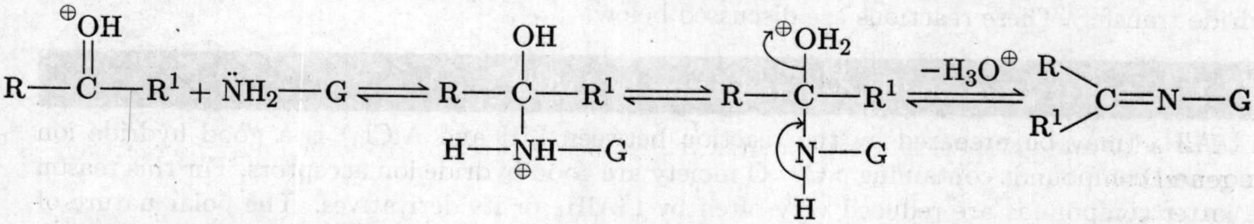

In the acid medium, attack by the nucleophile, the reagent, is the slow and rate-determining step. In the neutral or slightly alkaline medium the reaction mechanism is as follows :

Step I : Nucleophilic attack by the reagent :

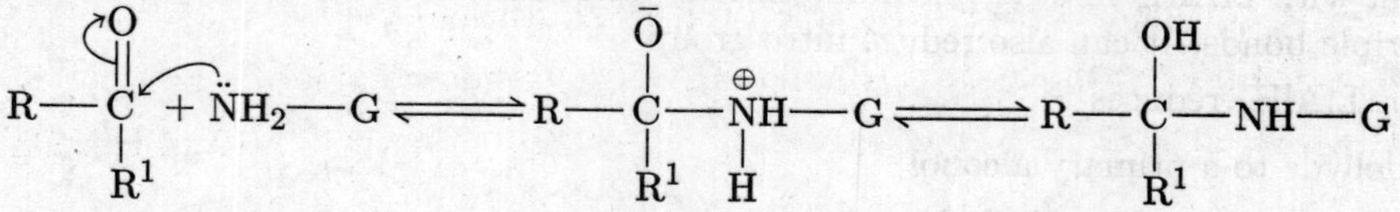

Step II : Protonation of the OH group by dil. acid and elimination of $H_3O^{\oplus}$:

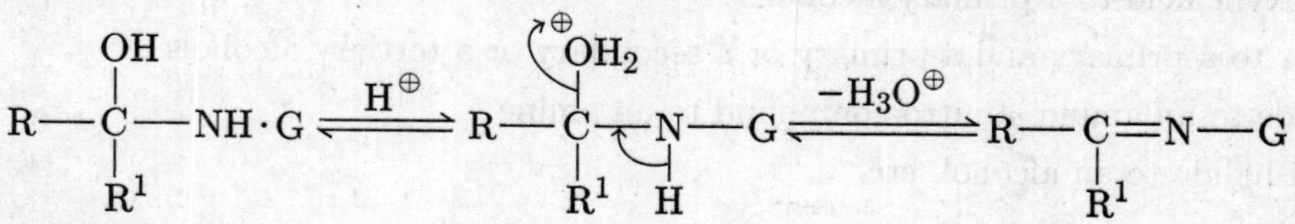

In the neutral or feebly alkaline medium, the dehydration step is the slow rate-determining step.

Now the question is : why does each of these reactions require an optimum pH? The answer is (i) at a low pH, the O atom of the carbonyl group will be readily protonated and consequently the carbonyl C will be strongly electron-deficient and the nucleophilic attack will be easy; but NH_2—G will also be protonated and will form $\overset{\oplus}{N}H_3$—G—this will then have no nucleophilicity and will fail to attack the carbonyl C; (ii) at a high pH, the carbonyl C will not be enough electrophilic for attack by the nucleophiles. So, an optimum pH is required to carry out this sort of reactions. For example, to prepare oxime of acetone the optimum pH is 4.5. At an optimum pH the free base remains in equilibrium with its salt and the carbonyl O gets protonated. The free base reacts with the strongly electrophilic carbonyl C.

However, when oximes and hydrazones are refluxed with dilute HCl, the carbonyl compound is regenerated. When a phenylhydrazone is treated with a highly reactive carbonyl compound, the phenylhydrazone of the reactive carbonyl compound is obtained and the parent carbonyl compound is set free. This shows the reversibility of the reaction.

16.6 Reactions Involving Hydride Ion Transfer

There are reactions in which one or more hydride ions ($:H^-$) get transferred from one species to the other directly. A hydride ion being a good electron donor, the reactions involving hydride ion transfer are red-ox reactions and the hydride donors are good reducing agents. Metallic hydrides, such a lithium aluminium hydride ($LiAlH_4$), sodium borohydride ($NaBH_4$), lithium (triethoxy)aluminium hydride [$Li(OEt)_3AlH$], lithium(tri-t-butoxy)aluminium hydride [$Li(t\text{-}BuO)_3AlH$], etc. reduce compounds containing >C=O moiety such as RCHO, RCOR′, RCOCl, R′COOH, RCOOR′, $RCONH_2$. Besides metallic hydrides, aluminium isopropoxide also reduces carbonyl compounds by hydride transfer. The Cannizzaro reaction, a self red-ox reaction, also involves hydride transfer. These reactions are discussed below.

16.6.1 The Nystrom and Brown reduction by $LiAlH_4$

$LiAlH_4$ (may be prepared by the reaction between LiH and $AlCl_3$) is a good hydride ion donor and compounds containing >C=O moiety are good hydride ion acceptors. For this reason the latter compounds are reduced very often by $LiAlH_4$ or its derivatives. The polar nature of >C=O moiety makes the C atom electrophilic toward the nucleophilic hydride ion and enables the compounds containing >C=O moiety to be reduced by $LiAlH_4$ (or similar compounds). Thus aldehydes, ketones, carboxylic acids, esters, amides, acid halides, nitriles, etc. got reduced on treatment with $LiAlH_4$ in ether medium. $LiAlH_4$, of course, does not usually reduce C—C double or triple bonds. It can also reduce nitro group.

However, $LiAlH_4$ reduces

(a) an aldehyde to a primary alcohol
(b) a ketone to a secondary alcohol
(c) a carboxylic acid to a primary alcohol
(d) an ester to a primary and a primary or a secondary or a tertiary alcohols
(e) an amide to an amine; a nitro-compound to an amine
(f) an acid halide to an alcohol, etc.

For example,

$$4\ CH_3CHO \xrightarrow[\text{ether}]{LiAlH_4} (CH_3CH_2O)_4\bar{Al}Li^+ \xrightarrow{4H^\oplus} 4CH_3CH_2OH + Li^+ + Al^{+++}$$

$$4\ CH_3COCH_3 \xrightarrow[\text{ether}]{LiAlH_4} (Me_2CHO)_4\bar{Al}Li^+ \xrightarrow{4H^\oplus} 4Me_2CHOH + Li^+ + Al^{+++}$$

$$CH_3COOH \xrightarrow[\text{2. } H^\oplus]{\text{1. } LiAlH_4/\text{ether}} CH_3CHO \xrightarrow[\text{2. } H^\oplus]{\text{1. } LiAlH_4/\text{ether}} CH_3CH_2OH$$

$$4\ PhCOOCH_3 \xrightarrow[\text{ether}]{LiAlH_4} 4PhCHO + 4CH_3\bar{O} \xrightarrow{H^\oplus} 4CH_3OH$$

$$4PhCHO \xrightarrow{LiAlH_4/\text{ether}} (PhCH_2O)_4\bar{Al}Li^\oplus \xrightarrow{H^\oplus} PhCH_2OH$$

$LiAlH_4$ being soluble in ether, the reaction is usually carried out in ether (ether, THF, etc.). At first a lithium aluminium alkoxide type complex is formed which on acidification liberates the reduction product.

Since $LiAlH_4$ does not usually reduce C-C multiple bonds, unsaturated alcohols are prepared from unsaturated carbonyl compounds, acids and esters.

$$CH_3CH{=}CH{-}CHO \xrightarrow[THF]{LiAlH_4} \xrightarrow{H^{\oplus}} CH_3CH{=}CH{-}CH_2OH$$

However, a C-C multiple bond in conjugation with an aromatic system on one side and a carbonyl group on the other side gets reduced by $LiAlH_4$,

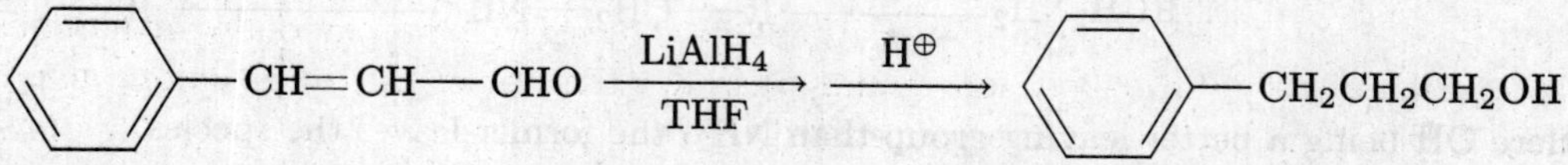

Since $LiAlH_4$ neutralises a carboxylic acid, the reduction of a carboxylic acid requires a large excess of the reagent. To avoid the misuse of the reagent, the sodio salt of the carboxylic acid is usually treated with the reagent.

The mechanism of the $LiAlH_4$ reduction of a compound containing >C=O moiety involves a four-centred T.S. through which the hydride ion is transferred from the reagent to the electrophilic carbon atom. $Li^{\oplus}$ complexes with the carbonyl O and increases the electrophilicity of carbonyl carbon atom.

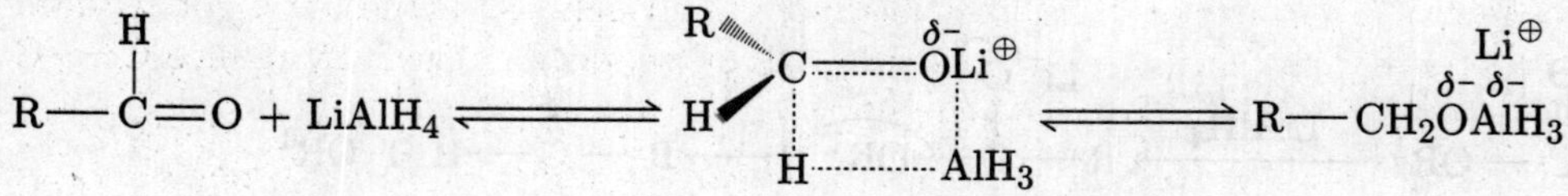

$$R{-}CH_2{-}\overset{\delta-\ \delta-}{O}\overset{Li^{\oplus}}{AlH_3} \xrightarrow{3R-\overset{H}{\overset{|}{C}}=O} (RCH_2O)_4\bar{Al}Li^{\oplus} \xrightarrow{4H^{\oplus}} RCH_2OH + Li^{+} + Al^{+++}$$

However, lithium (tri-t-butoxy)aluminium hydride may be used successfully to reduce a specific carbonyl group in a dicarbonyl compound. The less sterically hindered carbonyl group gets reduced preferentially than the more sterically hindered one.

$$Me_3C.CO.CH(Me).CH_2CHO \xrightarrow[2.\ H^{\oplus}]{1.\ Li(t\text{-}BuO)_3AlH} Me_3C.COCH(Me)CH_2CH_2OH$$

Problem 16.8. Why does $LiAlH_4$ usually fail to reduce a C-C multiple bond? Why does such a bond in conjugation with an aromatic system on one side and a carbonyl group on the other side get reduced by the same reagent?

Solution : A C-C multiple bond is an electron-dense moiety where electrophilic attack occurs. A hydride ion is an electron-dense species and thus behaves as a strong nucleophile. So, a hydride ion experiences a repulsion when it comes close to a C-C multiple bond and thus fails to reduce it usually. On the other hand, the electron-withdrawing effect of an aromatic system on one side and that of the carbonyl group on the other side make the C-C multiple bond less electron-dense and enable the hydride ion to attack the C-C moiety. For this reason such a C-C multiple bond gets reduced by $LiAlH_4$.

Problem 16.9. Give mechanistic steps for the $LiAlH_4$ reduction of the following compounds :

(i) $RCONH_2$ (ii) RCOOH (iii) $RCOOR^1$ (iv) RCOCl

Solution : (i) 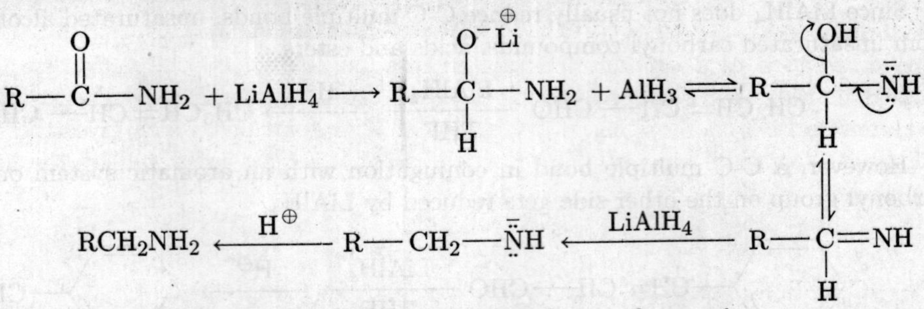

Here OH being a better leaving group than NH_2, the former leaves the species.

(ii) 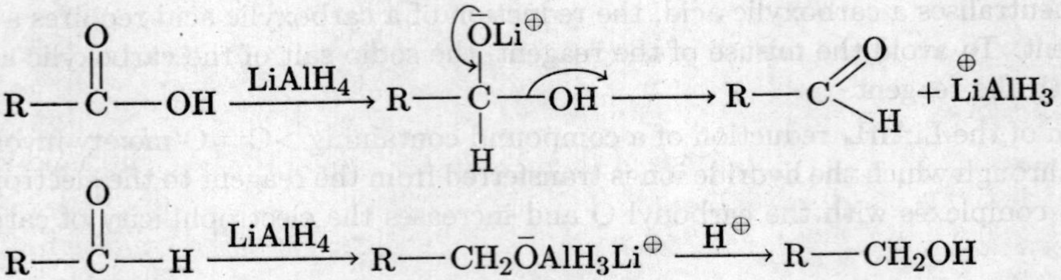

(iii) 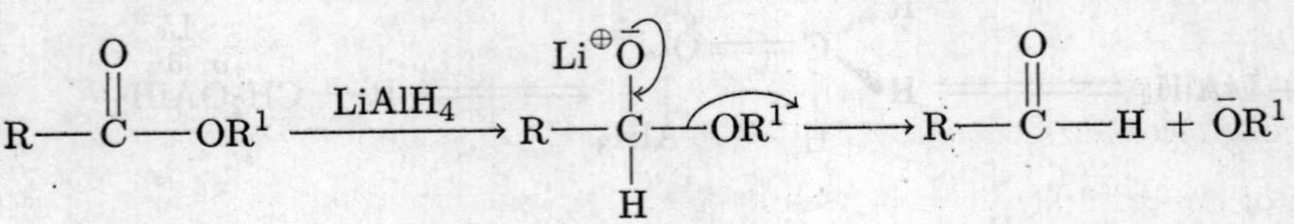

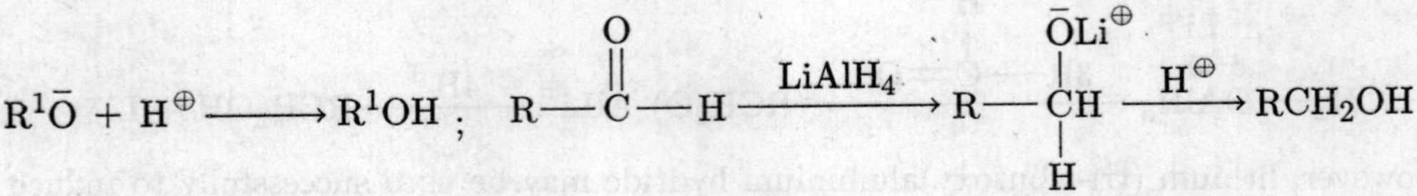

(iv) 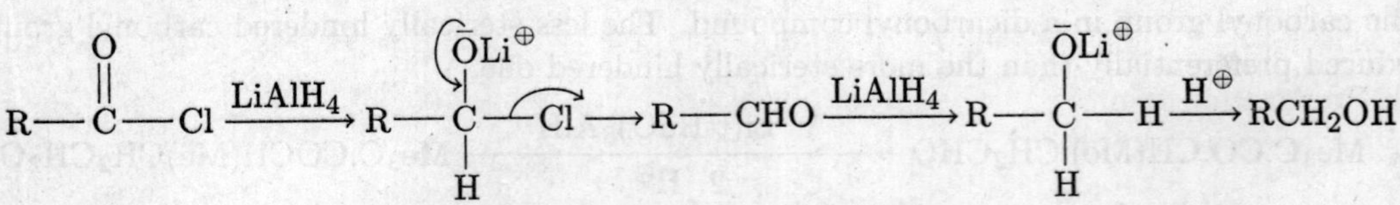

Problem 16.10. $Li(OEt)_3AlH$ reduces RCOCl to RCHO while $LiAlH_4$ reduces the same compound to RCH_2OH—give reason for the different products in the two cases.

Solution : $Li(OEt)_3AlH$ being a weaker reducing agent than $LiAlH_4$, the former reduces RCOCl to the aldehyde stage only whereas the latter reduces RCOCl to the alcohol through the aldehyde.

Note : $NaBH_4$ is another metallic hydride which can reduce aldehydes, ketones, esters, etc. to alcohols. However, it does not reduce CN, NO_2 COOH, C=C and C≡C groups. So nitro-alcohols, alcohols containing cyano group can be prepared from the corresponding aldehydes and ketones by $NaBH_4$ reduction ($LiAlH_4$ cannot be used since it reduces nitro and cyano groups).

$NaBH_4$ does not react with protic solvents like H_2O, ROH, etc. So, it is easy to use $NaBH_4$ in H_2O, ROH, etc. Since $LiAlH_4$ reacts with protic solvents, ether is used as a solvent for $LiAlH_4$ reduction.

16.6.2 The Meerwein-Pondorf-Verley reduction and the Oppenauer oxidation

The conversion of a carbonyl compound into an alcohol by the action of aluminium isopropoxide in isopropyl alcohol medium is called the Meerwein-Pondorf-Verley Reduction. At first, aluminium alkoxide of the newly formed alcohol is formed which gives the alcohol by the reaction with isopropyl alcohol. However, the alcohol is now generated by the action of dilute sulphuric acid on the aluminium alkoxide.

$$3R{-}\overset{\overset{\displaystyle O}{\|}}{C}{-}R' + (Me_2CHO)_3Al \rightleftharpoons 3Me_2CO + (RR'CHO)_3Al$$

$$(RR'CHO)_3Al + 3CH_3CHOHCH_3 \rightleftharpoons 3RR'CHOH + [(CH_3)_2CHO]_3Al$$

$$2(RR'CHO)_3Al + 3H_2SO_4(\text{dilute}) \longrightarrow 6RR'CHOH + Al_2(SO_4)_3$$

(R & R' = alkyl or aryl or cycloalkyl group or H atom)

This is a reversible reaction. The equilibrium of the reaction is shifted to the right hand side almost completely by distilling out the acetone formed in the reaction.

However, the Meerwein-Pondorf-Verley Reduction takes place under very mild condition. The reaction is highly specific for aldehydes and ketones; other groups which are susceptible to reduction, such as $C{=}C$, $C{\equiv}C$, NO_2, COOR, etc., do not get reduced by this reagent. This specific ··ure of the reagent and the mild reaction conditions give a wide scope for preparing unsatura.. alcohols, nitro-alcohols, aldols, etc.; for example :

(a) Preparation of unsaturated alcohols from unsaturated carbonyl compounds :

$$\underset{\substack{\text{2-Butenal}\\ \text{(Crotonaldehyde)}}}{CH_3CH{=}CHCHO} \xrightleftharpoons[]{\substack{(Me_2CHO)_3Al\\ \text{in } Me_2CHOH}} \underset{\substack{\text{2-Butenol or, But-2-ene-1-ol}\\ \text{(Crotyl alcohol)}}}{CH_3CH{=}CH.CH_2OH}$$

(b) Preparation of nitro-alcohols from nitro-carbonyl compounds :

$$NO_2{-}CH_2COCH_3 \xrightleftharpoons{(Me_2CHO)_3Al/Me_2CHOH} NO_2CH_2CHOHCH_3$$

(c) Selective reduction of a carbonyl group of a dicarbonyl compound via acetal formation :

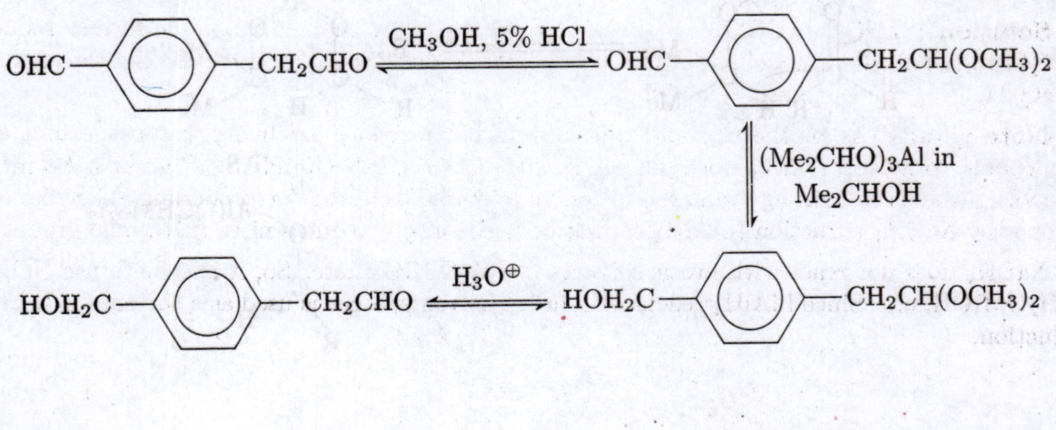

Of course, carbonyl compounds which possess high percentage of enol, such as, β-ketoesters, β-diketones, etc. do not undergo the Meerwein-Pondorf-Verley reduction. Perhaps aluminium isopropoxide gets converted into isopropyl alcohol by the action of the enol.

$$\underset{\text{keto-form}}{CH_3.CO.CH_2.CO.CH_3} \rightleftharpoons \underset{\text{enol-form}}{CH_3—C(OH)=CH.COCH_3} + (Me_2CHO)_3Al$$

$$\rightleftharpoons (CH_3—\underset{|}{\overset{}{C}}(O)=CHCOCH_3)_3Al + Me_2CHOH$$

The mechanism of the Meerwein-Pondorf-Verley reduction is considered to involve the direct hydride transfer from the α-C of aluminium isopropoxide to the carbonyl C of the carbonyl compound reversibly through a cyclic six-centered T.S.

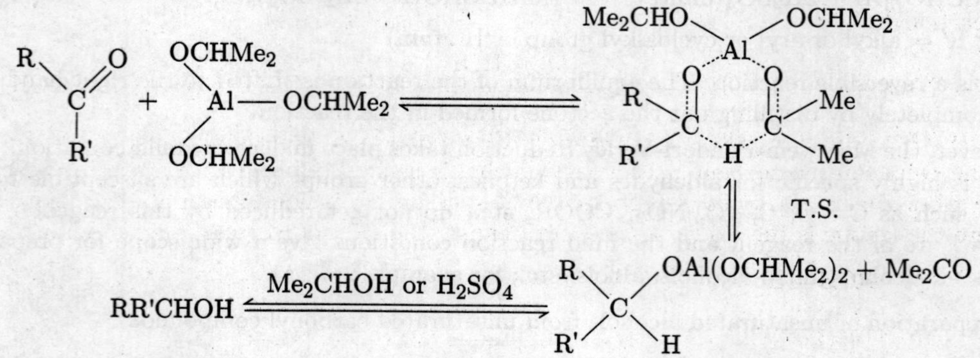

However one may consider a three-stepped mechanism for the reaction.

(i) Step I : Coordination of the ketone with the aluminium isopropoxide :

$$RR'C=\ddot{O}: + Al(OCHMe_2)_3 \rightleftharpoons RR'C=\overset{\oplus}{O}—\bar{Al}(OCHMe_2)_3$$

(ii) Step II : The hydride transfer from the α-H of the aluminium isopropoxide to the carbonyl C atom of the carbonyl compound via a cyclic T.S.

(iii) Step III : Separation of the new complex :

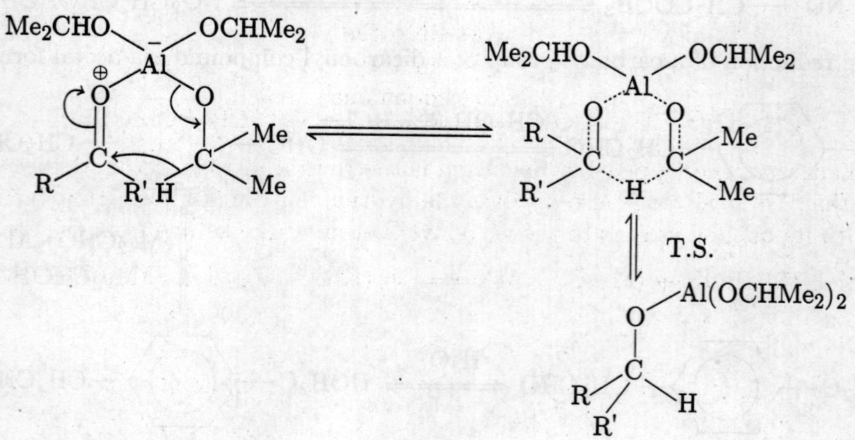

Final Step : Generation of the alcohol from the new complex.

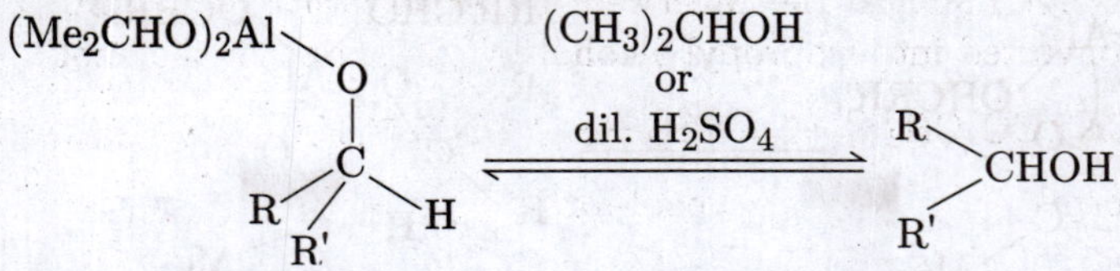

There are reactions in which the rate is second order with respect to the aluminium isopropoxide. In this case it has been considered that one molecule of the alkoxide coordinates with the ketone and the other transfers the :H^- ion.

When the reaction is carried out with α-deutero-aluminium isopropoxide, the product is found to contain deuterium at the carbonyl C atom of the starting carbonyl compound. This is an evidence in favour of the direct hydride transfer from the α-C atom of the aluminium isopropoxide to the carbonyl C.

$$(Me_2CDO)_3Al + CH_3CH_2COCH_3 \xrightleftharpoons{Me_2CHOH} CH_3CH_2CDOHCH_3 + CH_3COCH_3$$

The reverse reaction of the Meerwein-Pondorf-Verley reduction is the Oppenauer Oxidation.

In the Oppenauer oxidation a secondary alcohol is usually refluxed with a specific base, aluminium t-butoxide, in a liquid ketone medium—such as, propanone, butanone, cyclohexanone, etc. Here the ketone is the oxidising agent, it oxidises the alcohol to a carbonyl compound.

$$RCHOHR' + (Me_3CO)_3Al \rightleftharpoons (RR'CHO)_3Al + Me_3COH$$

$$(RR'CHO)_3Al + Me_2CO \rightleftharpoons RR'CO + (Me_2CHO)_3Al$$

The reaction is highly selective. Usually ketones are prepared through this reaction but it may be used for the preparation of aldehydes. When the aldehydes are prepared from 1° alcohols, *p*-benzoquinone is used as hydrogen acceptor instead of aliphatic ketones. the process is very useful for oxidising unsaturated secondary alcohols to ketones.

$$CH_3.CH{=}CH.CHOH.CH_3 \xrightleftharpoons[]{\text{Al t-Butoxide in } p\text{-Benzoquinone}} CH_3CH{=}CH.COCH_3$$

$$CH_3.CH{=}CH.CH_2OH \xrightleftharpoons[]{\text{Al t-Butoxide in } p\text{-Benzoquinone}} CH_3.CH{=}CH.CHO$$

The mechanism of the Oppenauer oxidation is the reverse of that of the Meerwein-Pondorf-Verley reduction. Naturally this also involves the hydride ion transfer through a cyclic T.S.; the α-H atom with its bonded *e*s goes to the carbonyl C atom of the ketone from the alcohol

$$3RR'CHOH + (Me_3CO)_3Al \rightleftharpoons (RR'CHO)_3Al + 3Me_3COH$$

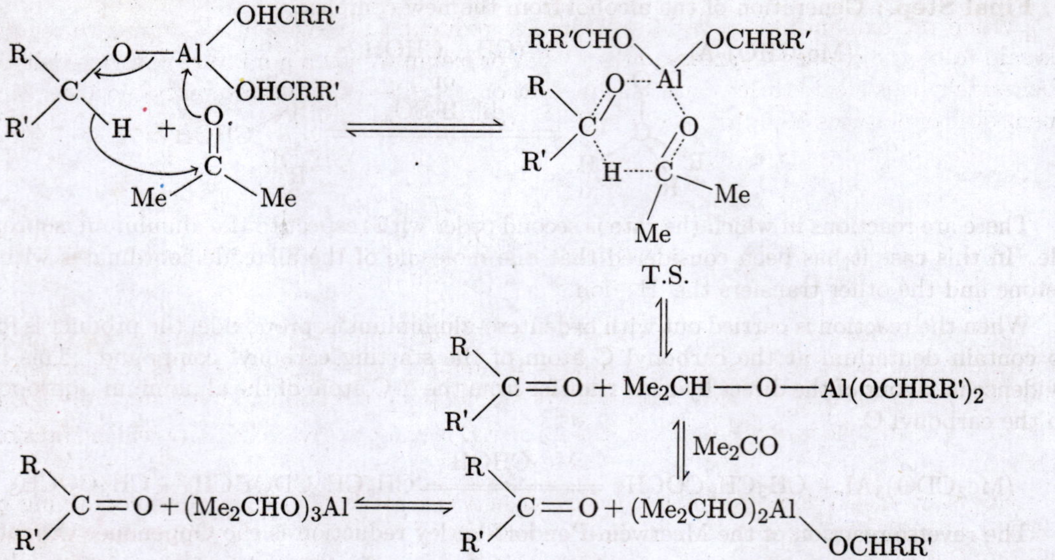

16.6.3 The Cannizzaro Reaction

When aldehydes, which do not possess α-H atom, are treated with strong bases (usually 50% aqueous or ethanolic NaOH), they undergo self oxidation-reduction reaction in which one molecule of the aldehyde acts as an oxidising agent and gets reduced to the corresponding alcohol while another molecule gets oxidised to the corresponding carboxylate ion and the reaction is known as the Cannizzaro reaction.

$$R_3CCHO + R_3CCHO \xrightarrow{\bar{O}H} R_3CCH_2OH + R_3CCO\bar{O}$$

Since aromatic aldehydes do not contain α-H atoms, the Cannizzaro reaction is a characteristic reaction of the aromatic aldehydes.

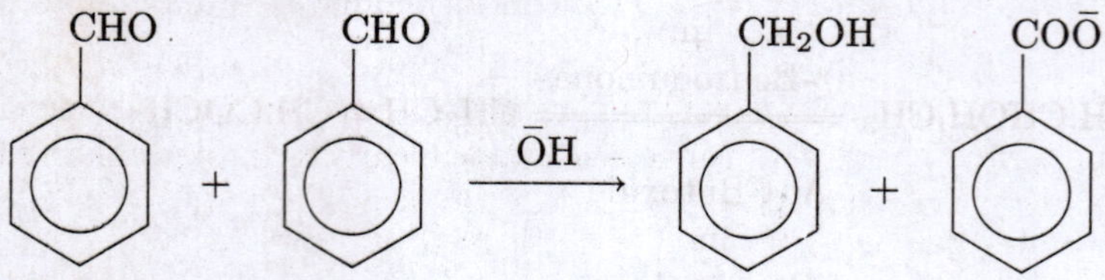

Benzaldehyde

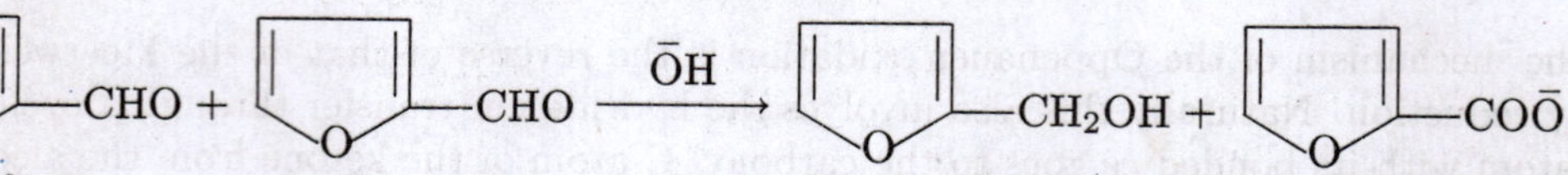

Furfural　　　　Furfuryl alcohol　　　　Furoate

Aliphatic aldehydes that have no α-H atom also undergo the Cannizzaro reaction.

$$2HCHO \xrightarrow{\bar{O}H} HCO\bar{O} + HCH_2OH$$

$$2Me_3CCHO \xrightarrow{\bar{O}H} Me_3CCH_2OH + Me_3CCO\bar{O}$$

When the oxidant aldehyde differs from the reductant aldehyde in a Cannizzaro reaction, it is said to be the **crossed Cannizaaro** reaction. For example, when a mixture of formaldehyde and benzaldehyde is treated with caustic soda solution, a crossed Cannizzaro reaction takes place and benzyl alcohol forms along with sodium formate.

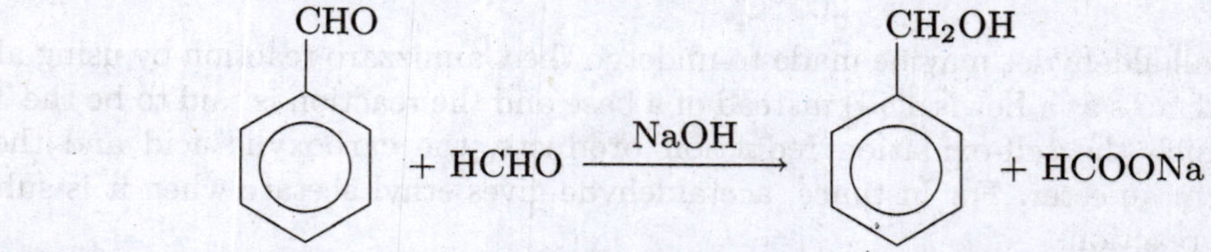

In this connection it is important to note that in a crossed Cannizzaro reaction involving formaldehyde, the latter gets oxidised to formate and the other gets reduced to the corresponding alcohol; this is because the C of formaldehyde is more electrophilic and less sterically congested than that of the other and $\bar{O}H$ atacks the former C readily to give $HOCH_2\bar{O}$ which transfers H^- directly.

However, the **Tollen's reaction**, a reaction between a carbonyl compound containing one or more α-H atoms and formaldehyde in the presence of $Ca(OH)_2$ (or a similar base), is also an example of the crossed Cannizzaro reaction. Here a 1, 3-glycol and formate are formed if the carbonyl compound contains one α-H. On the other hand, if the aldehyde contains more than one α-H, they may be replaced one at a time.

$$R_2CH.C(=O)—R' + 2HCHO \xrightarrow{Ca(OH)_2} R_2C(CH_2OH)—CHOHR' + HCO\bar{O}$$

$$CH_3CHO + 4HCHO \xrightarrow{Ca(OH)_2} HOCH_2—C(CH_2OH)_2—CH_2OH + HCO\bar{O}$$

In the Tollen's reaction the product alcohol contains $(n+1)$ —CH_2OH group if the parent aldehyde contains n α-H atoms and naturally $(n+1)$ mole of formaldehyde are required per mole of the carbonyl compound.

α-keto aldehydes give internal or intramolecular Cannizzaro reaction on treatment with a strong base. In this case the formyl group gets converted into the carboxylate group and the keto group gets changed to 2° alcoholic moiety.

$$R—C(=O)—CHO \xrightarrow{\bar{O}H} RCHOHCO\bar{O}$$

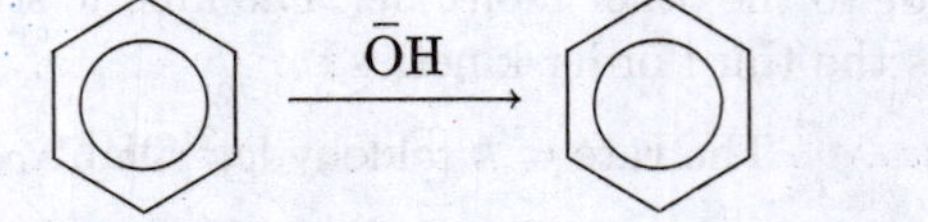

Phenylglyoxal

$$CHO—CHO \xrightarrow{\bar{O}H} HOCH_2—CO\bar{O}$$

Glyoxal

However, there are α-monoalkylated aldehydes containing α-H atoms which on heating with aqueous sodium hydroxide solution at 170 °C – 200°C give the Cannizzaro reaction products.

$$2Me_2CHCHO \xrightarrow[200\,°C]{\text{NaOH soln.}} Me_2CHCO_2^- + Me_2CHCH_2OH$$

Of course, all aldehydes may be made to undergo the Cannizzaro reaction by using aluminium ethoxide (which acts as a Lewis acid) instead of a base and the reaction is said to be the **Tischenko reaction** in which the self-oxidation reduction products, the carboxylic acid and the alcohol, combine to form an ester. For instance, acetaldehyde gives ethyl acetate when it is subjected to the Tischenko reaction.

$$CH_3CHO + CH_3CHO \xrightarrow{Al(OEt)_3} CH_3COOCH_2CH_3$$

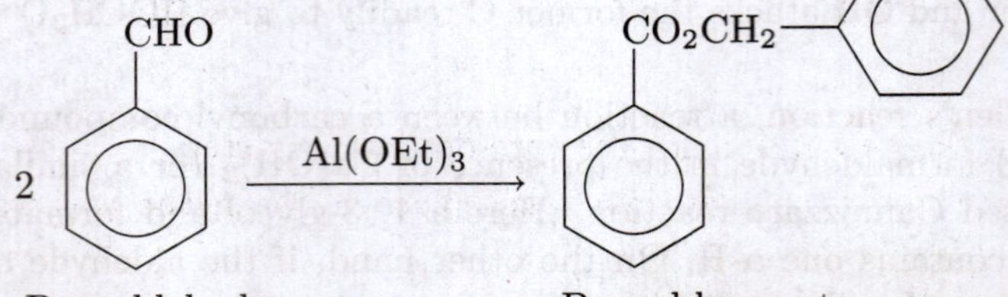

Benzaldehyde   Benzyl benzoate

The crossed Cannizzaro reaction involving formaldehyde is a good synthetic tool since formaldehyde has a high tendency to undergo oxidation. For example, the crossed Cannizzaro reaction between one molecule acetaldehyde and four molecules of formaldehyde in the presence of calcium hydroxide gives rise to pentaerythritol (a white soluble powder, m.p. 260 °C) which is used in the manufacture of plastics, plasticisers and explosives.

$$CH_3CHO + 4HCHO \xrightarrow{Ca(OH)_2} (CH_2OH)_4C + HCO\bar{O}$$

p-Methoxybenzyl alcohol may be prepared from *p*-Methoxybenzaldehyde by the crossed Cannizzaro reaction :

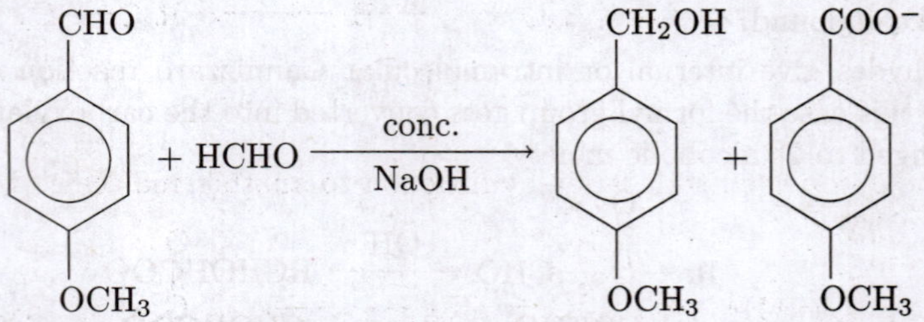

The most probable mechanism of the reaction tells us that it involves direct hydride shift from one molecule of the aldehyde to the other molecule. The kinetic study of the reaction shows that the reaction usually follows the third order kinetics :

$$\text{The rate} = k\,[\text{aldehyde}]^2[\bar{O}H].$$

There are also instances where the reaction follows the fourth order kinetics and

$$\text{the rate} = k\,[\text{aldehyde}]^2[\bar{O}H]^2.$$

However, the steps for the reactions which follow the third order kinetics are given below :

Step I : Nucleophilic attack on the electron-deficient carbonyl C atom by the $\bar{O}H$ ion :

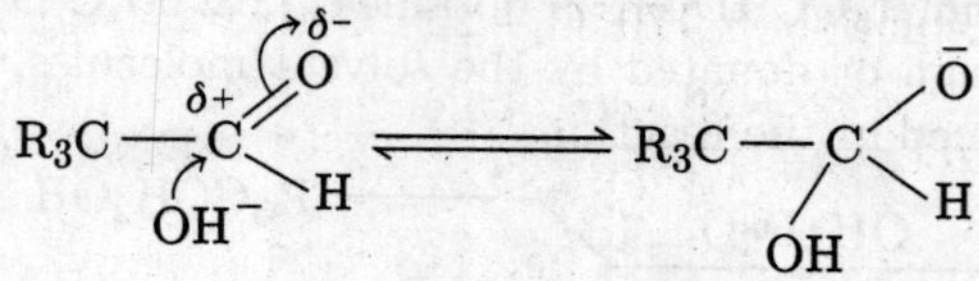

A

This step is rapid.

Step II : The species A then directly transfers a hydride ion to the electron-deficient carbonyl C atom of another aldehyde molecule.

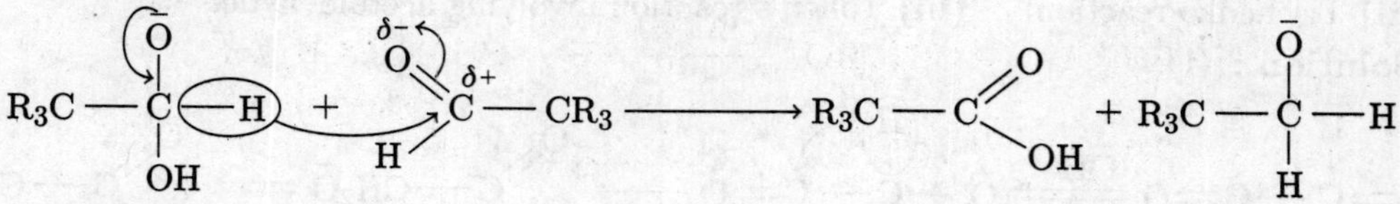

A B

This step is the slowest step and hence the rate-determining step.

Step III : Rapid proton exchange then takes place between the carboxylic acid and the alkoxide ion, B.

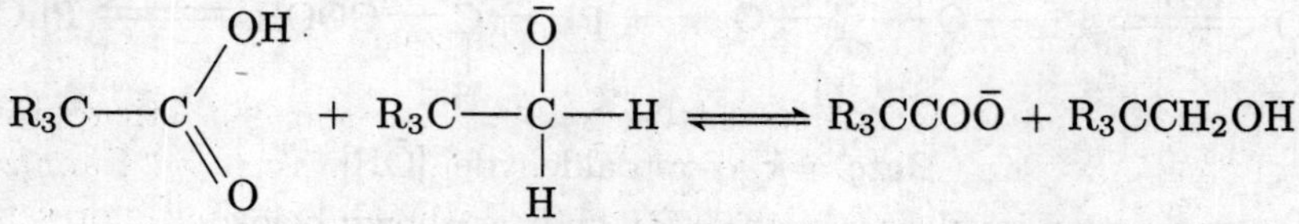

B

The mechanistic steps for the reactions which follow the fourth order kinetics are :

Step I : Nucleophilic attack on the electron-deficient carbonyl C atom and abstraction of a proton and thus forms a di-ion.

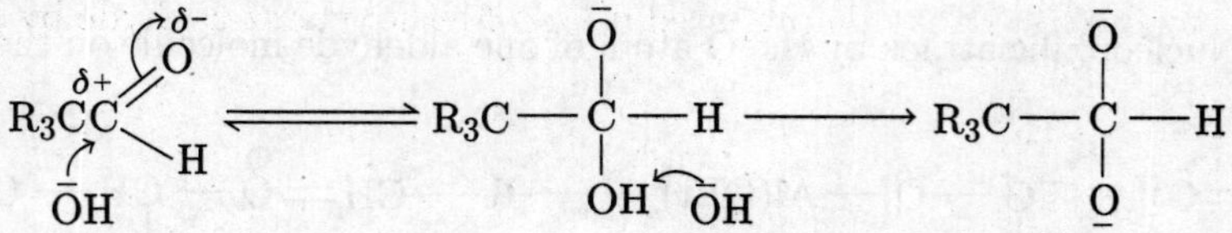

di-ion

Step II : The di-ion then transfers a hydride ion to another molecule of the aldehyde.

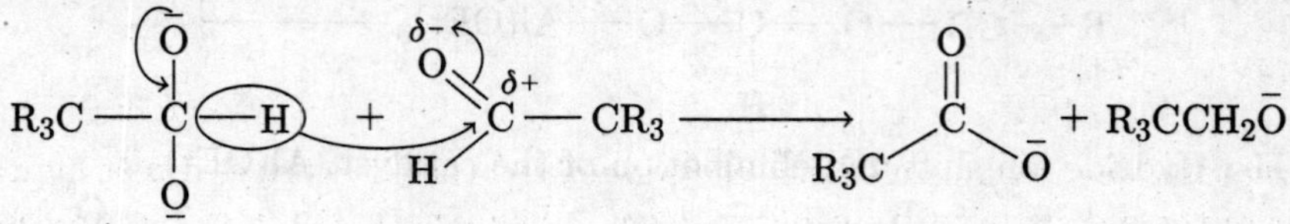

di-ion

The di-ion contains two anionic O.s. They develop high back strain within the di-ion. The di-ion, in its turn, gives up the strain by transferring the hydride ion. This is again a slow step and, therefore, it is a rate-determining step.

Step III : The alkoxide ion formed in the second step takes up a proton from the solvent molecule.

$$R_3CCH_2\bar{O} + H_2O \rightleftharpoons R_3C\,CH_2OH + \bar{O}H$$

A strong evidence in favour of the direct hydride shift is : when the reaction is carried out in the presence of D_2O, no $R_3CCHDOH$ is obtained, i.e., no C-D bond forms. If the hydride ion were transferred through or donated by the solvent molecules, deuterium would have been present in the alcohol formed in the reaction.

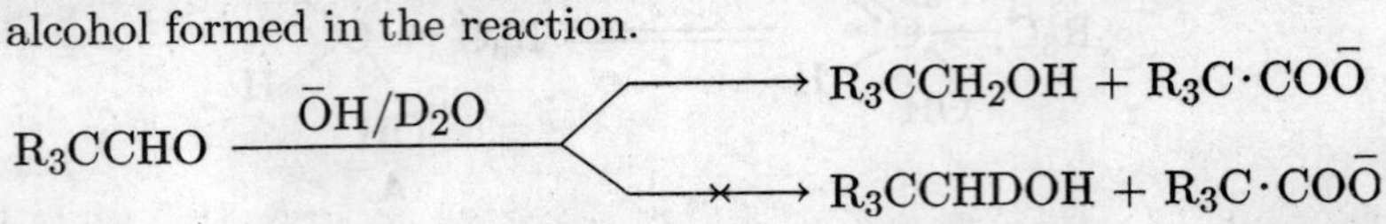

Problem 16.11. Write down the mechanism for the following reactions.
(i) Internal Cannizzaro reaction involving glyoxal and phenylglyoxal.
(ii) Tischenko reaction. **(iii)** Tollen's reaction involving acetaldehyde.

Solution : (i)

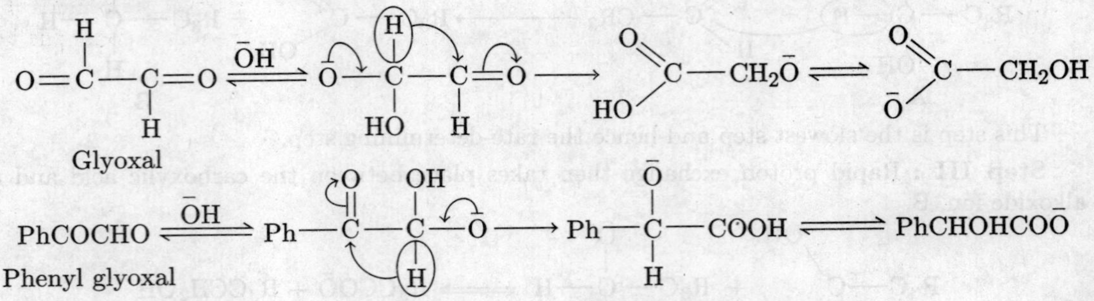

Glyoxal

Phenyl glyoxal

$$\text{Rate} = k\ [\alpha\text{-ketoaldehyde}]\ [\bar{O}H]$$

(ii) Tischenko reaction involves a hydride ion shift as shown below :

Step I : Coordination of the aldehyde with the aluminium ethoxide :

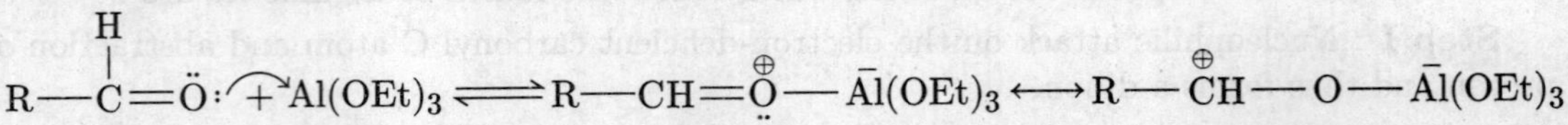

Step II : Nucleophilic attack by the O atom of one aldehyde molecule on the cationic carbon :

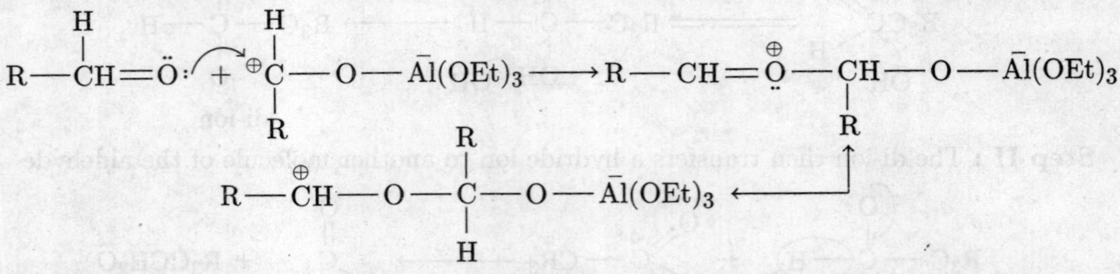

Step III : Hydride ion shift and elimination of the catalyst, $Al(OEt)_3$:

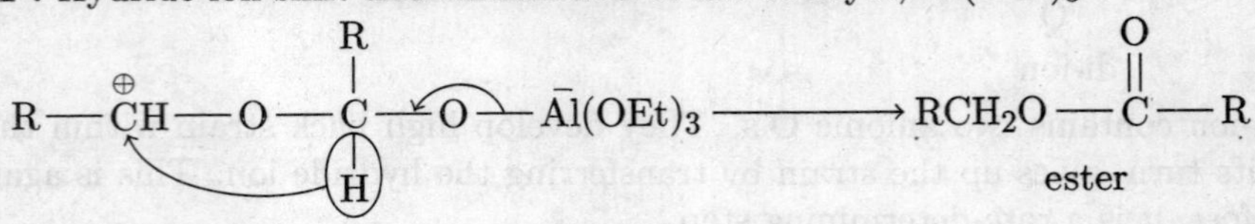

(iii) In the Tollen's reaction first crossed aldol condensation takes place; the aldol then undergoes the crossed Cannizzaro reaction. In both the steps formaldehyde is involved. When acetaldehyde undergoes the Tollen's reaction, $(CH_2OH)_3$ CCHO is formed; since $(CH_2OH)_3$ CCHO does not contain any α-H atom, it undergoes the crossed Cannizzaro reaction with formaldehyde to give formate and pentaerythritol, $(CH_2OH)_4C$. So, the mechanistic steps are those of the aldol condensation and of the Cannizzaro reaction.

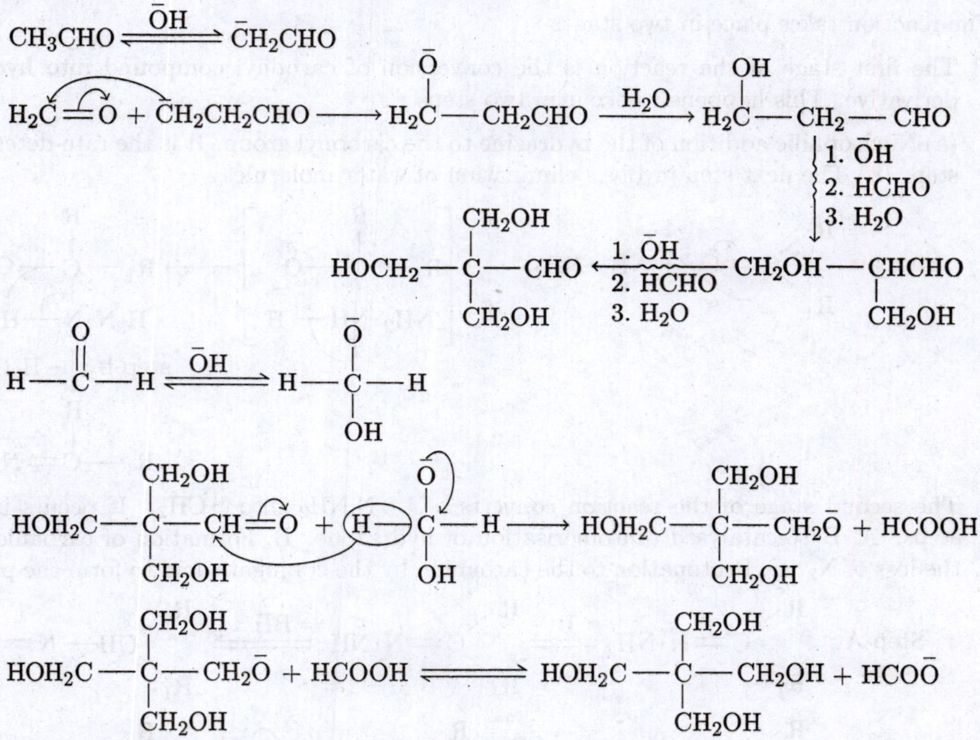

Problem 16.12. Aldehydes possessing α-H atoms usually do not undergo the Cannizzaro reaction under its reaction condition—give an explanation.

Solution : When aldehydes with α-H atoms are treated with a strong base, they form corresponding resonance-stabilised carbanions which then combine with another molecule of the aldehyde to give aldol condensation product, the aldol. The rate of the aldol condensation being much faster than the rate of the Cannizzaro reaction, aldehydes with α-H atom do not undergo the Cannizzaro reaction.

16.7 The Wolff-Kishner and Huang Minlon Reductions

The reduction of carbonyl compounds to hydrocarbons and their derivatives by heating them with hydrazine hydrate and a base is known as the Wolff-Kishner reduction. In this process the >C═O group gets converted into $>CH_2$ group via $>C{=}N.NH_2$ and nitrogen gas evolves out.

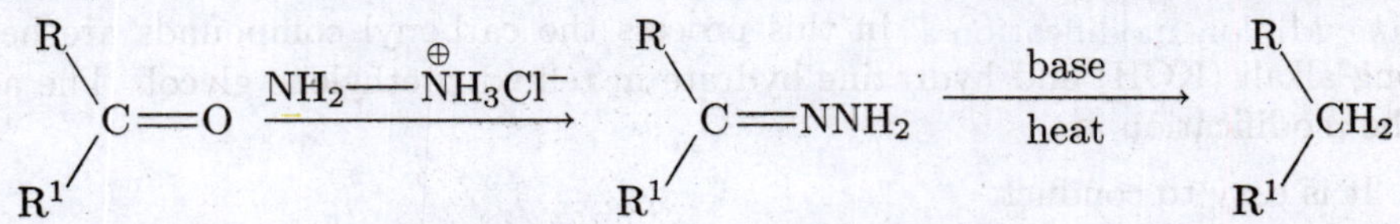

The earlier procedure of the Wolff-Kishner reduction is to heat the carbonyl compound with hydrazine hydrate and NaOEt in a sealed tube at 180 °C.

This is a specific reduction for aldehydes and ketones, i.e., other reducible groups such as C—C double bond —NO_2, etc. usually remain undisturbed under the conditions of the reaction.

The reaction takes place in two stages :

(i) The first stage of the reaction is the conversion of carbonyl compound into hydrazone derivative. This happens to occur in two steps :

(a) Nucleophilic addition of the hydrazine to the carbonyl group. It is the rate-determining step. (b) The next step involves elimination of water molecule.

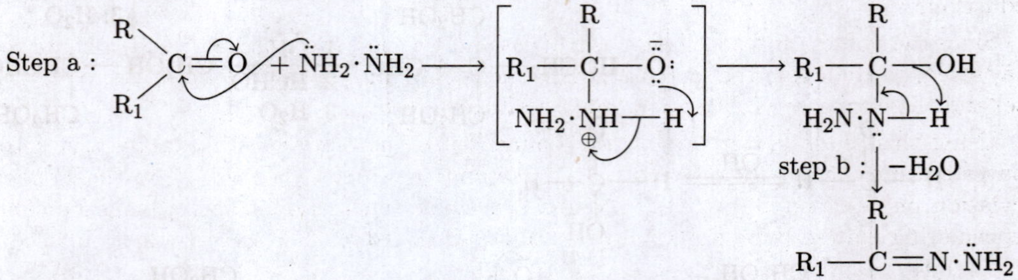

(ii) The second stage of the reaction converts >C=N.NH_2 into >CH_2. It occurs in three steps. A. Base-catalysed tautomerisation of hydrazone. B. Formation of carbanion with the loss of N_2. C. Protonation to the carbanion by the conjugate acid to form the product.

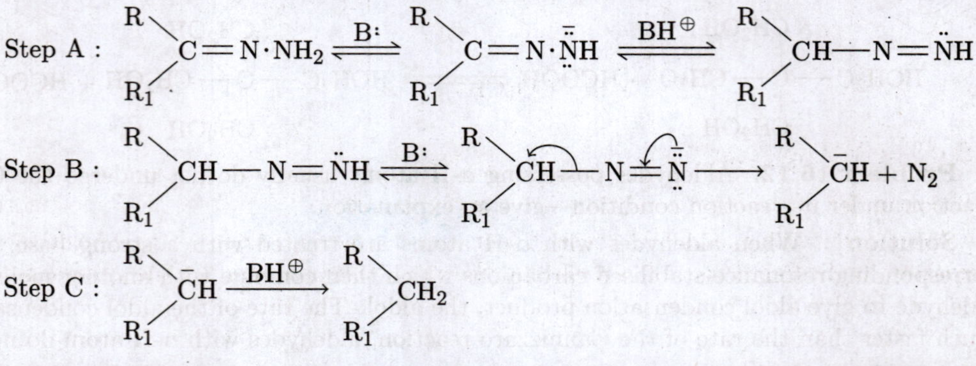

The reduction needs high temperature. It is also unsuccessful for the sterically hindered ketones. Although the process is specific for >C=O group, sometimes migration of double bond takes place with the formation of abnormal product instead of expected one when α, β-unsaturated carbonyl compounds are subjected to the Wolff-Kishner reduction.

The limitations of the process are, however, overcome by modifying the following procedures as shown below :

(i) Huang Minlon modification : In this process the carbonyl compounds are heated with strong alkali (KOH) and hydrazine hydrate in refluxing ethylene glycol. The advantages of the modification are :

(a) It is easy to conduct,

(b) Pure products can be isolated,

(c) Sterically hindered ketones can also be reduced.

(ii) Cram etal. modification : The reduction is carried in dimethyl sulphoxide medium at room temperature using potassium t-butoxide as the base catalyst.

The advantages of this modification are as follows :

(a) It can be carried out at room temperature.

(b) Sterically hindered ketones can be reduced to form expected alkene derivatives.

(c) α, β-unsaturated carbonyl compounds can be reduced to form expected alkene derivatives.

The mechanism of both the processes are supposed to be the same as that of the Wolff-Kishner reduction.

Since the reduction is specific for aldehydes and ketones, the modified Wolff-Kishner reduction may safely be applied to the carbonyl compounds containing C—C double bond, —NO_2 group, etc. and the corresponding alkenes, nitro derivatives, etc. can be prepared.

Using this reduction process, like Clemmensen reducation, generally alkyl aryl ketones are converted into Ar.CH_2.R. Since alkyl aryl ketones are directly obtained by the Friedel-Crafts acylation on benzene derivatives, one can synthesise long straight alkyl chains bonded to the benzene ring indirectly by the application of the modified Wolff-Kishner reduction.

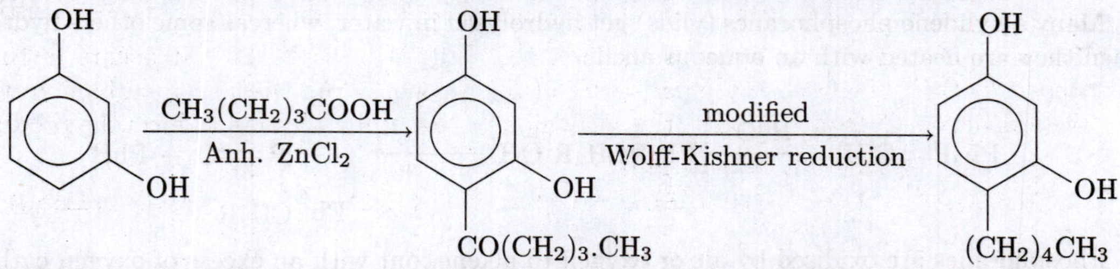

16.8 The Wittig Reaction

The reagents of the Wittig reaction are the phosphorous ylids, also called phosphoranes.

A species containing an anionic carbon (carbanion) covalently bonded to a positively charged heteroatom, N, S, P, etc., is called a ylid. A ylid may be represented by the general formula :

$$-\overset{\ominus}{\underset{|}{C}}-\overset{\oplus}{\underset{|}{X}}- \quad \text{where X=N, P or S.}$$

Phosphorus, nitrogen and sulphur ylids are known.

When 'onium' salts are treated with strong bases ylids are formed. Strong bases are usually butyl and phenyllithiums in ether or tetrahydrofuran, sodium or lithium ethoxides in ethyl acohol or dimethyl formamide, and sodium hydride in dimethyl sulphoxide.

$$Ph_3P + MeBr \longrightarrow Ph_3\overset{\oplus}{P}MeBr^- \xrightarrow[Et_2O]{PhLi} Ph_3\overset{\oplus}{P}-\overset{\ominus}{C}H_2$$

$$Me_3N + MeBr \longrightarrow Me_3\overset{\oplus}{N}MeBr^- \xrightarrow[Et_2O]{PhLi} Me_3\overset{\oplus}{N}-\overset{\ominus}{C}H_2$$

$$Me_2\overset{\oplus}{S}-CH_3I^- \xrightarrow{CH_3-\overset{\overset{O}{\|}}{S}-\overset{\ominus}{C}H_2Na^{\oplus}} Me-\underset{\underset{Me}{|}}{\overset{\oplus}{S}}-\overset{\ominus}{C}H_2$$

$$CH_2{=}CH{-}CH_2{-}Cl + Ph_3P \longrightarrow Ph_3\overset{\oplus}{P}{-}CH_2{-}CH{=}CH_2\bar{C}l \xrightarrow{\text{NaOEt} \quad \text{DMF}} Ph_3P{=}CH{-}CH{=}CH_2$$

Sulphur and phosphorus ylids are more stable than nitrogen ylids since the lone pair of *e*s on anionic C form *pπ-dπ* back bonding with the *d* orbital of sulphur and phosphorus and thus they are considered resonance hybrids of two resonating structures :

$$Ph_3\overset{\oplus}{P}{-}\overset{-}{C}H_2 \longleftrightarrow Ph_3P{=}CH_2;\ Me_2\overset{\oplus}{S}{-}\overset{-}{C}H_2 \longleftrightarrow Me_2S{=}CH_2$$

However, acyl and aryl derivatives of ylids in which the anionic C bears the acyl or aryl group get stability through delocalisation of electronic charge.

$$(CH_3)_3\overset{\oplus}{N}\overset{-}{C}H{-}\overset{O}{\overset{\|}{C}}CH_3 \longleftrightarrow (CH_3)_3\overset{\oplus}{N}{-}CH{=}\overset{\bar{O}}{\overset{|}{C}}CH_3$$

Groups with –I effect attached to the anionic C bring stability to the ylids.

Many alkylidene phosphoranes (ylids) get hydrolysed in water, whereas some others hydrolyse when they are heated with an aqueous alkali.

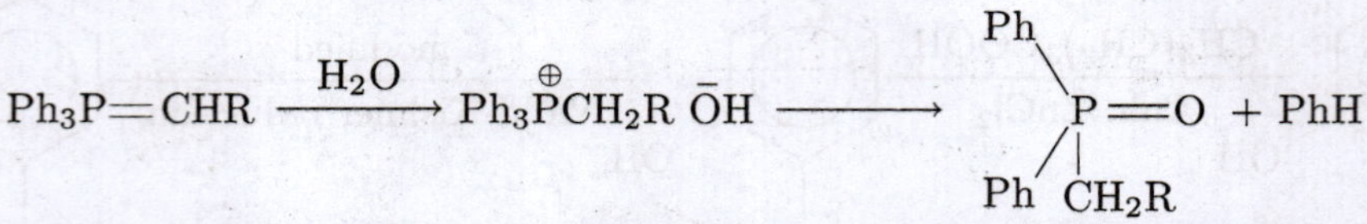

Phosphoranes are oxidised by air or oxygen to alkene, but with an excess of oxygen carbonyl compounds are obtained :

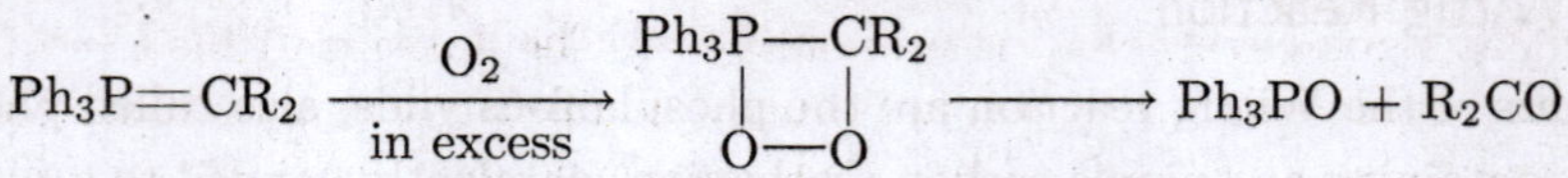

$$2Ph_3P{=}CR_2 \xrightarrow[\text{small amount}]{O_2} 2\ Ph_3PO + R_2C{=}CR_2$$

It is needless to mention that ylids are strong nucleophiles and the anionic C is the nucleophilic centre.

Besides these, phosphorus ylids of appropriate structure undergo the Wittig reaction. We shall discuss this reaction here in detail.

The reaction between an alkylidenetriphenylphosphorane, called the Wittig reagent, and an aldehyde or a ketone to yield an alkene is called the Wittig reaction. For instance, cyclohexanone reacts with methylidenetriphenylphosphorane to form cyclohexyl methylidene. The Wittig reaction may be looked upon as the replacement of carbonyl oxygen, =O, by =C< group.

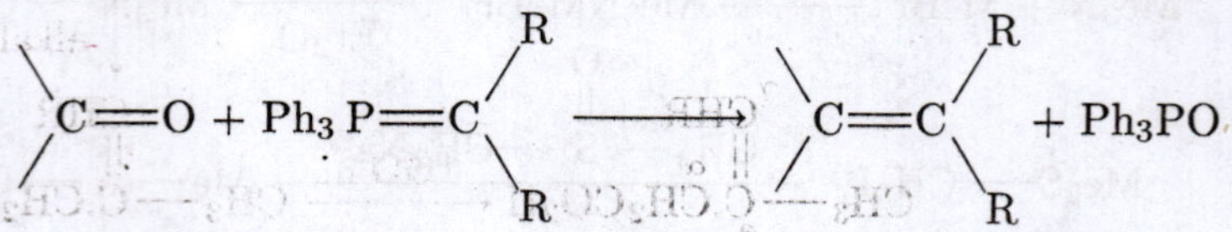

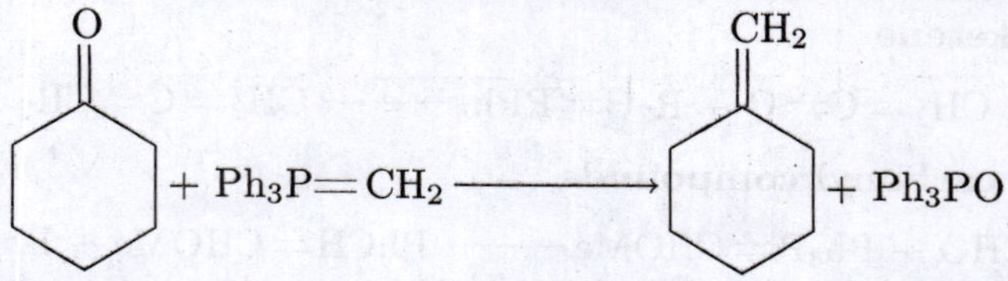

The ylids are generated in situ and are treated with carbonyl compounds in an atomosphere of nitrogen because oxygen, water and alcohol, hydrohalic acids, etc. react with the ylids.

The Wittig reaction is of general nature. Aliphatic, alicyclic and aromatic carbonyl compounds react readily. A carbonyl compound having conjugated double or triple bond reacts without interference of the π bond. Carbonyl compounds containing ester group also undergo the reaction safely since ylids react with —CO_2R group more slowly than with carbonyl group. Stable ylids containing electron-withdrawing groups, such as $R{-}C{=}O$, —CHO, —CN, —COOR, etc., at α-position react readily with aldehydes but slowly or not at all with the ketones.

Since phosphorus ylids with electron-withdrawing groups at α-position do not react with ketones, a modification has been introduced to overcome the difficulty. Instead of phosphorus ylid, a phosphonate ylid is allowed to react with ketones.

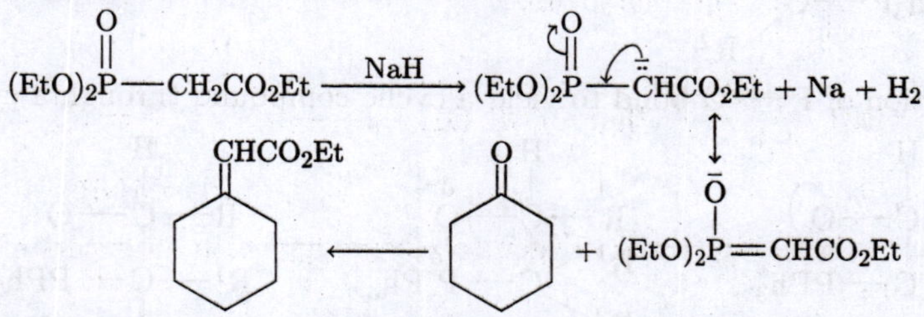

The reaction has a good synthetic value because : (i) The position of the C—C double bond in the product is unambiguous. (ii) It can introduce a double bond into positions that are difficult to form by other processes; for instance, exocyclic double bonds. (iii) The reaction helps prepare β, γ-unsaturated acids. (iv) The intramolecular Wittig reaction yields cyclic alkenes containing 5-16C atoms. (v) The reaction may be carried with ketenes, isocyanates, nitroso compounds, etc. (vi) Carbonyl compounds can be stepped up.

Formation of exocyclic double bond

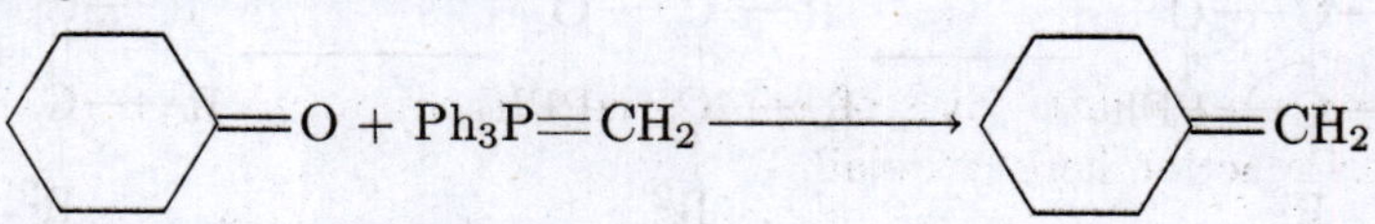

Preparation of β, γ-unsaturated acids

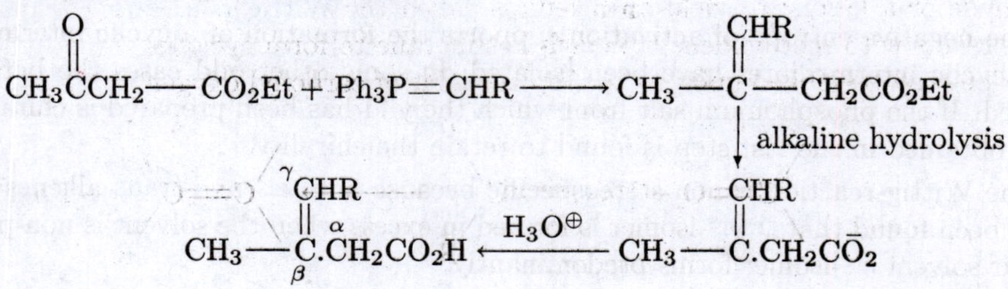

Reaction with ketene

$$CH_2{=}C{=}O + R_2C{=}PPh_3 \longrightarrow CH_2{=}C{=}CR_2$$

Stepping-up of carbonyl compounds

$$PhCHO + Ph_3P{=}CHOMe \longrightarrow PhCH{=}CHOMe + Ph_3PO$$

$$\downarrow H^{\oplus}$$

$$PhCH_2CHO \rightleftharpoons PhCH{=}CHOH$$

The most accepted mechanism of the Wittig reaction involves three steps :

Step I : Nucleophilic attack by the anionic C of the ylid on the electron-deficient carbonyl C to yield a betaine as an intermediate.

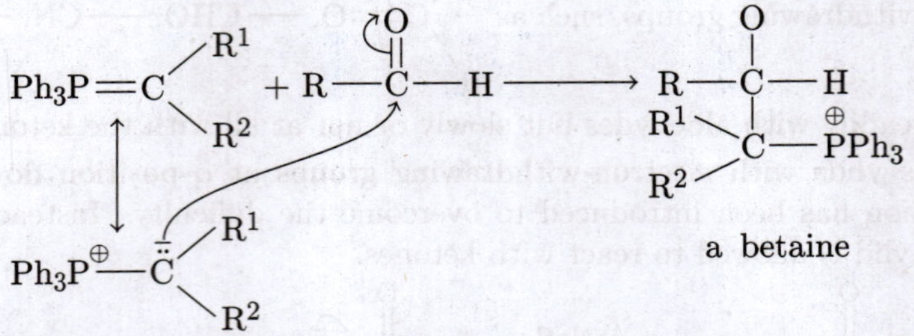

Step II : Formation of P—O bond to yield a cyclic compound through a cyclic T.S. :

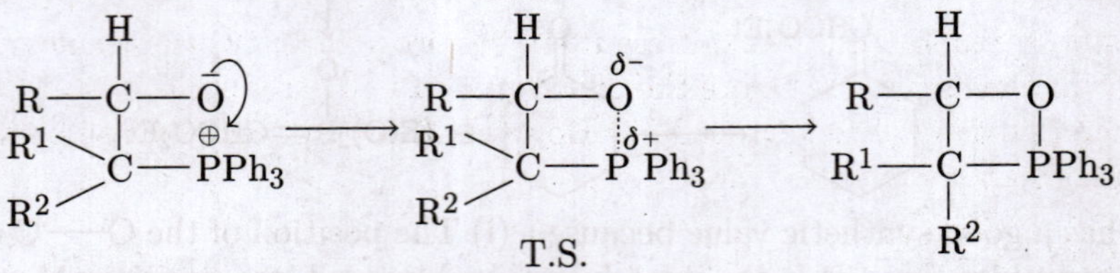

Step III : Breaking of the C—O and C—P σ bonds and the formation of C—C π bond through a four-centred cyclic T.S. The step II and the step III may be separate or concerted and the step I or the step II may be the rate-limiting step.

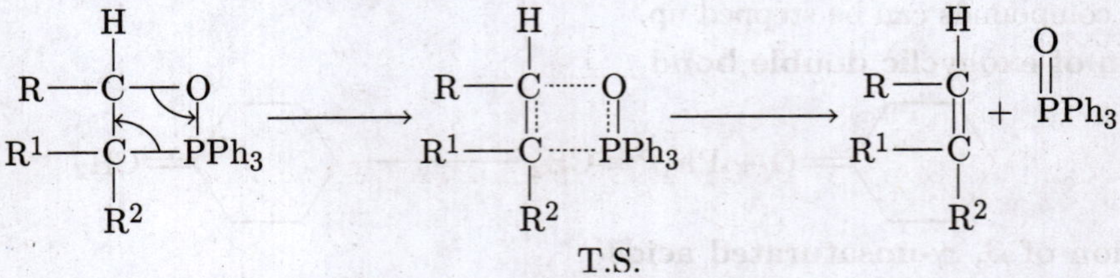

The negative entropy of activation supports the formation of a cyclic intermediate. In some cases cyclic intermediates have been isolated; in some other odd cases the betaines have been isolated. If the phosphonium salt from which the ylid has been prepared is chiral, the phosphine oxide obtained in the last step is found to retain the chirality.

The Wittig reaction is non-stereospecific because both cis and trans alkenes form. However, it has been found that trans-isomer is formed in excess when the solvent is non-polar; whereas in a polar solvent cis-isomer forms predominantly.

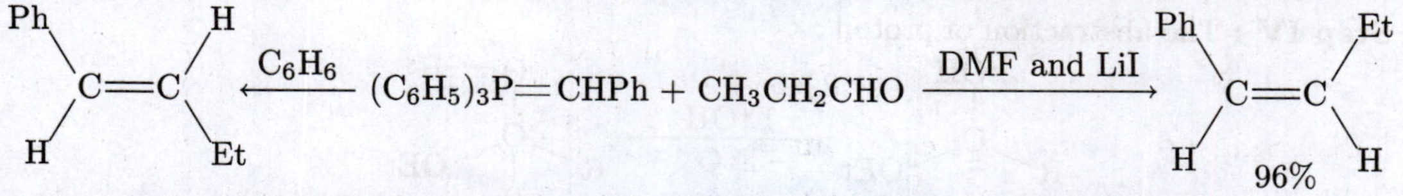

16.9 Reactions Involving Anionic Radicals

Aldehydes, ketones and esters can be reduced by metals in neutral or acid medium. These reactions are believed to proceed via anionic radicals. It is supposed that the reaction occurs at the surface of the metal and the metal atoms give up electrons which get solvated. The solvated electrons in their turn attack the electron-deficient carbonyl C atom as a nucleophile one at a time. The Bouveault-Blanc reduction, the pinacol formation reaction from ketones and the acyloin condensation reaction are considered to follow this sort of reactions.

16.9.1 The Bouveault-Blanc reduction

The reduction of aldehydes, ketones and esters by metallic sodium in EtOH or n-BuOH to the corresponding alcohol is known as the Bouveault-Blanc reduction.

$$\mathrm{RCHO} \xrightarrow{\mathrm{Na/EtOH}} \mathrm{RCH_2OH}$$

$$\mathrm{CH_3COCH_3} \xrightarrow{\mathrm{Na/EtOH}} \mathrm{CH_3CHOHCH_3}$$

$$\mathrm{CH_3CH_2COOCH_3} \xrightarrow{\mathrm{Na/EtOH}} \mathrm{CH_3CH_2CH_2OH + CH_3OH}$$

When an ester is reduced by the Bouveault-Blanc procedure, two molecules of alcohols are obtained as shown above. In fact, before the discovery of $LiAlH_4$ reduction, esters were converted to alcohols by the Bouveault-Blanc procedure. However, this reaction may also be used to prepare amines from nitriles and isonitriles.

$$\mathrm{RCN} \xrightarrow{\mathrm{Na/EtOH}} \mathrm{RCH_2NH_2} \qquad \mathrm{RNC} \xrightarrow{\mathrm{Na/EtOH}} \mathrm{RNHCH_3}$$

However, the probable mechanism of the reaction is as follows :

Step I : Nucleophilic attack by an e on the electron-deficient carbonyl C to form an anionic radical :

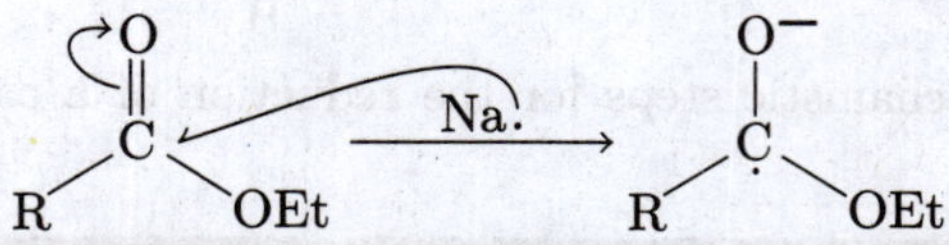

Step II : The abstraction of proton by O^- from the solvent alcohol :

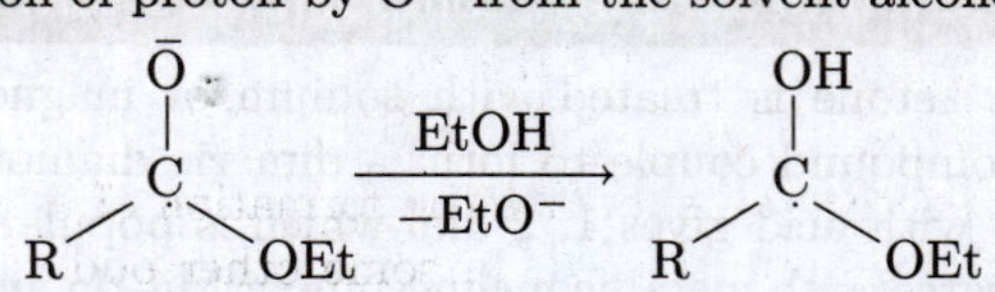

Step III : The formation of carbanion :

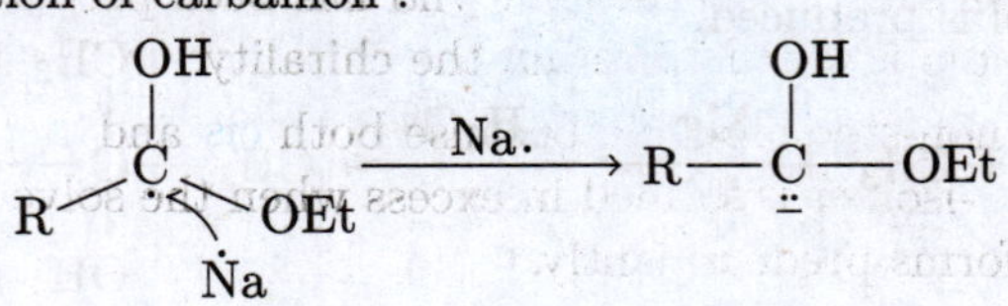

Step IV : The abstraction of proton :

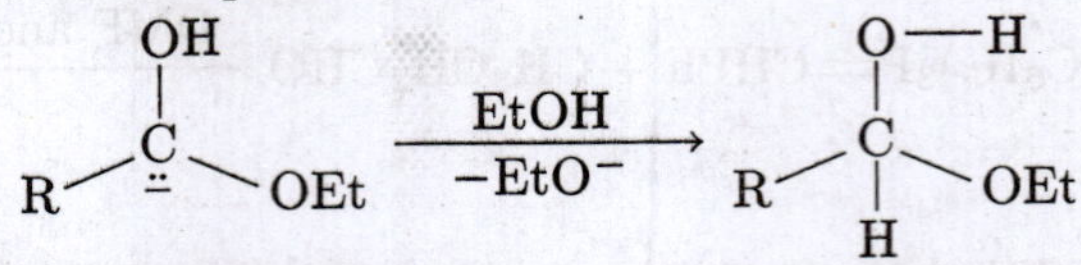

Step V : The elimination of EtOH :

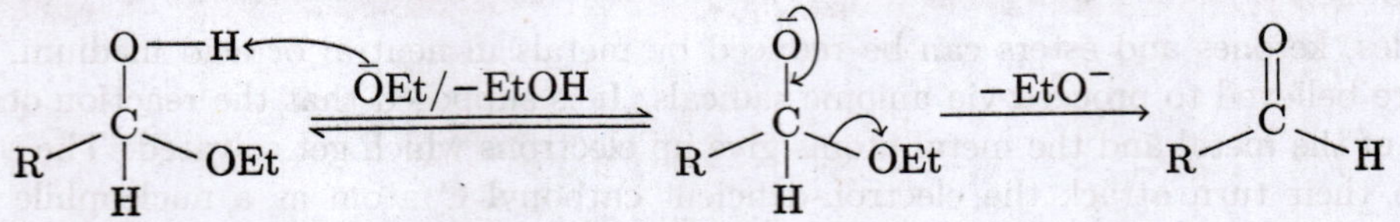

Step VI : The formation of anionic radical

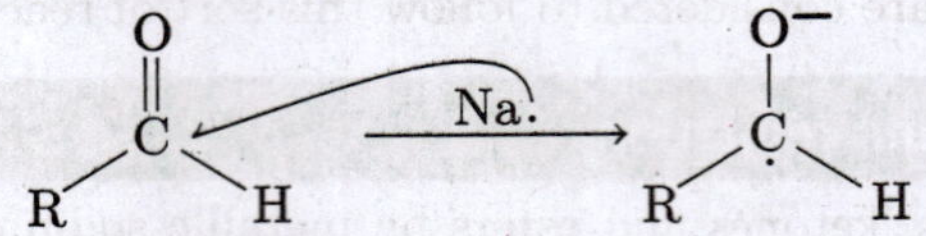

Step VII : The abstraction of proton :

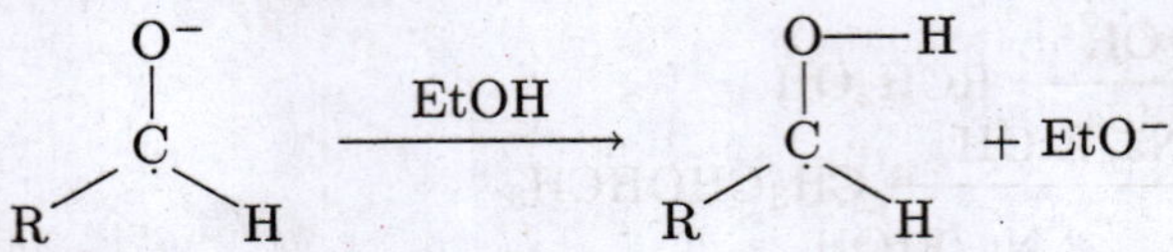

Step VIII : The formation of carbanion :

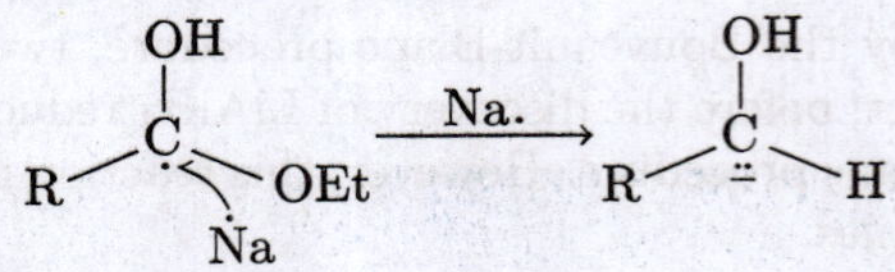

Step IX : The abstraction of proton :

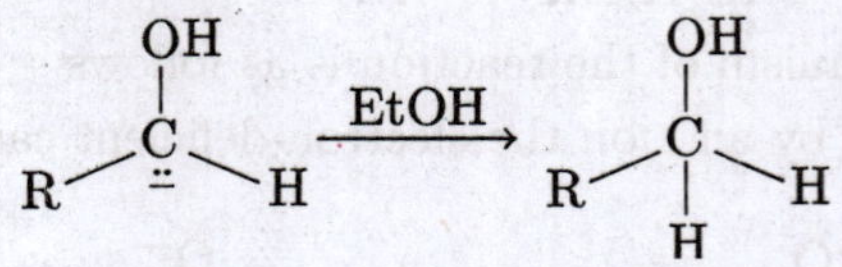

Step VI to IX are the mechanistic steps for the reduction of a carbonyl compound by the Bouveault-Blanc procedure.

16.9.2 Bimolecular reduction of carbonyl compounds to form pinacols, 1, 2-diols

When an aldehyde or a ketone is treated with sodium or magnesium or aluminium, two molecules of the carbonyl compound couple to form a dimeric dianion which exists as a metal salt. The salt on treatment with acid gives 1, 2-diol which is popularly known as pinacol. For example, when acetone is treated with metallic magnesium or Mg-Hg amalgam and then acidified, 2, 3-dimethylbutane-2, 3-diol is produced.

$$2\,CH_3-CO-CH_3 \xrightarrow{Mg} \xrightarrow{H_3O^{\oplus}} CH_3-C(CH_3)(OH)-C(CH_3)(OH)-CH_3$$

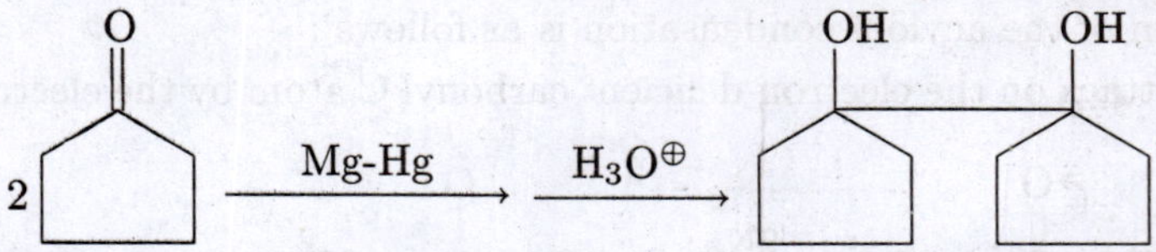

However, this reaction gives better yields with aromatic ketones than aliphatic ketones. Cyclic 1, 2-diols have been prepared from diketones by this reaction.

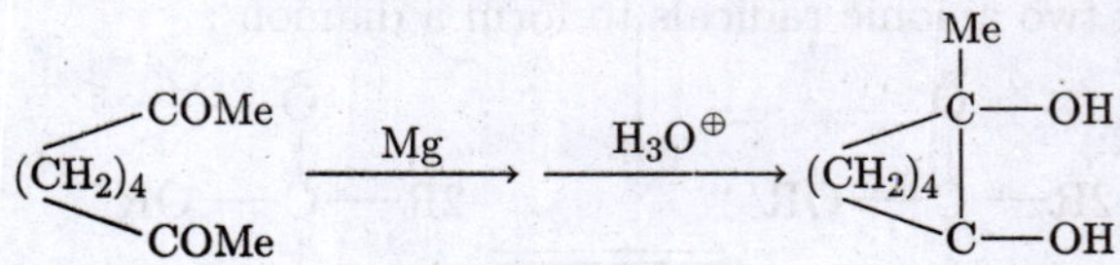

The mechanistic steps are :

Step I : Formation of anionic radical :

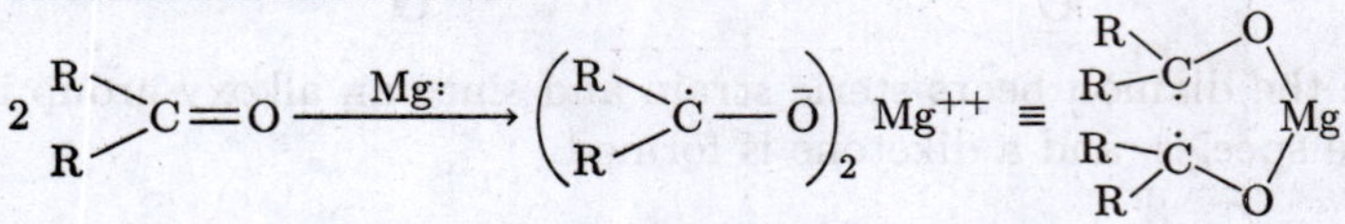

Step III : Formation of dimeric dianion which exists as a salt.

Step IV : Hydrolysis of the salt by $H_3O^{\oplus}$:

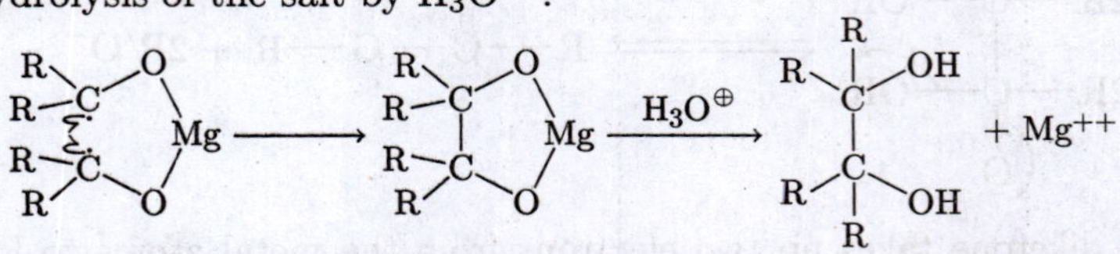

16.9.3 Acyloin condensation or synthesis

When an ester is refluxed with sodium in an inert solvent such as ether, benzene, toluene, xylene, etc. in the absence of any free alcohol and then the product of reflux is treated with an acid or alcohol, a dimeric α-hydroxyketone, called acyloin, is obtained and the reaction is called acyloin synthesis or condensation. For example, ethyl propanoate on refluxing with sodium in benzene and then on acidification gives 4-hydroxy-3-hexanone.

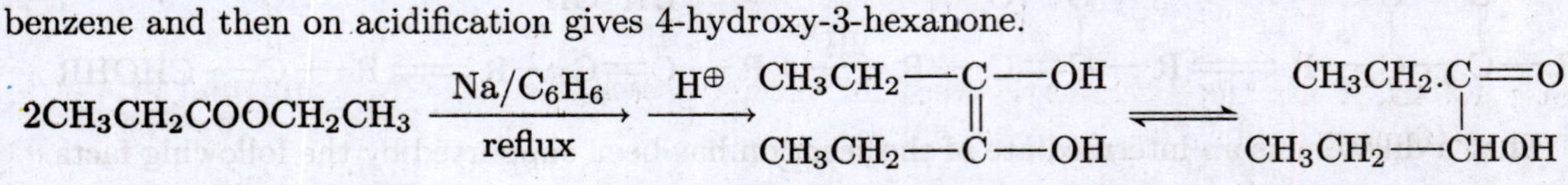

It is a good means for preparing long chain acyloins containing 18 to 20 C atoms. Ten to twenty membered cyclic acyloins have been prepared from diesters of dicarboxylic acid, e.g., $EtOCO(CH_2)_nCOOEt$ ($n = 8 - 18$), using high dilution technique. In this case the yield varies from 60—95%.

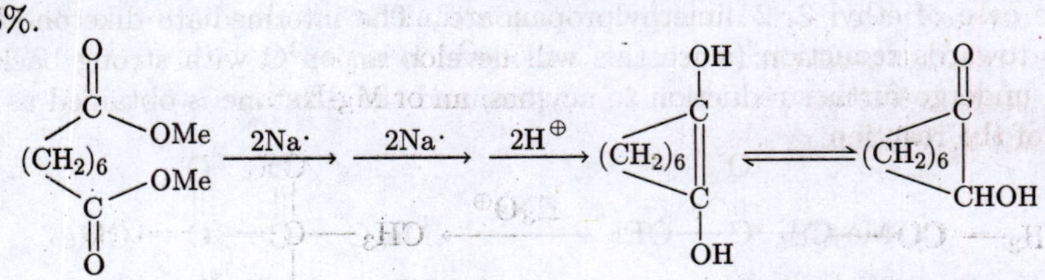

2-Hydroxycyclooctanone

The possible mechanism of the acyloin condensation is as follows :

Step I : Nucleophilic attack on the electron-deficient carbonyl C atom by the electron to form an anionic radical.

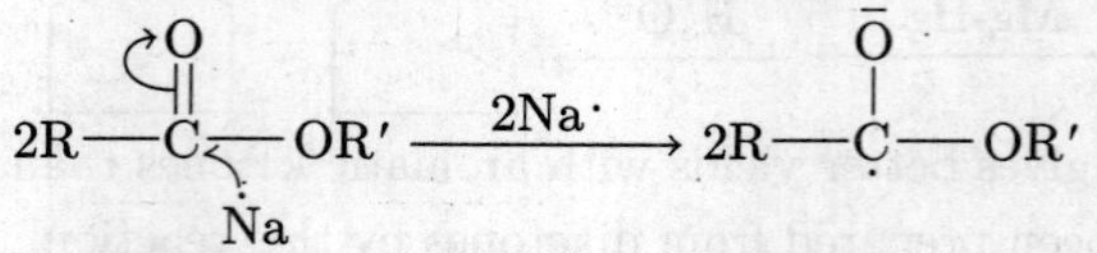

Step II : Coupling of two anionic radicals to form a dianion :

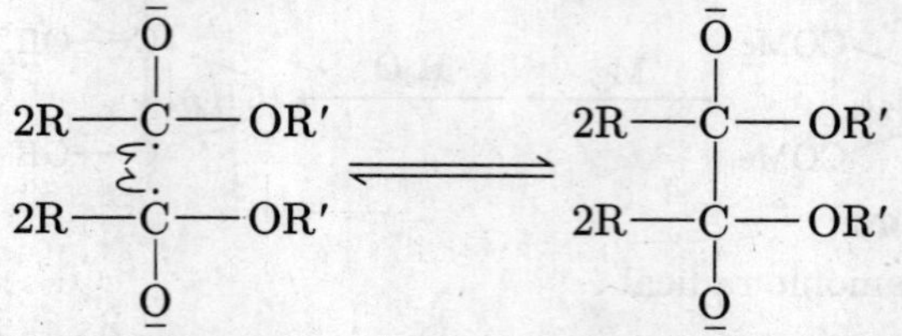

Step III : Since the dianion bears steric strain and since an alkoxy group is a good leaving group, they leave the species, and a diketone is formed.

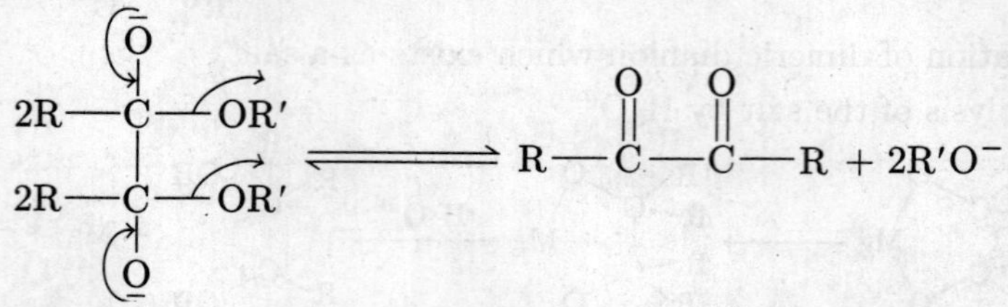

Step IV : The diketone takes up two electrons from the metal atoms to form a dianionic diradical.

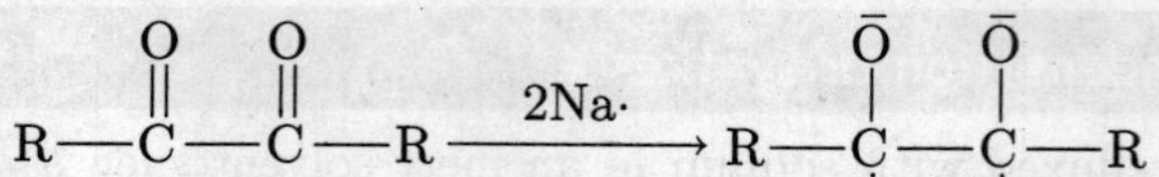

Step V : The dianionic diradical then forms an en-dioxide dianion which on acidification gives an en-diol. The en-diol then rapidly isomerises to an α-hydroxyketone, acyloin.

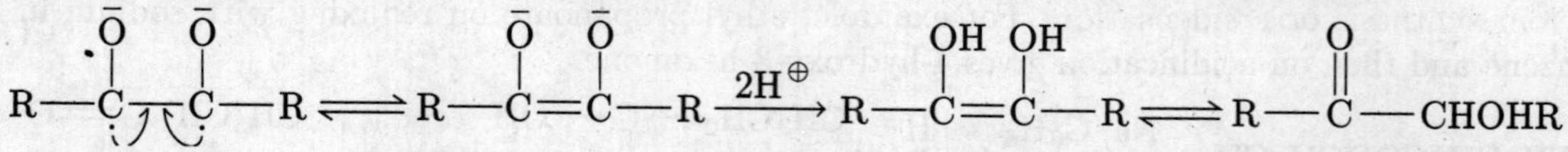

That a diketone is an intermediate of the reaction has been supported by the following facts :

(i) In a number of cases a small amount of diketone have been isolated.

(ii) If an ester on acyloin synthesis gives a diketone which is very resistant towards reduction, one cannot expect a good yield of the acyloin from the ester. This has been found to be so in the case of ethyl 2, 2-dimethylpropanoate. The intermediate diketone being very resistant towards reduction (since this will develop an sp^3C with strong back-strain) it does not undergo further reduction to acyloin, and the diketone is obtained as the major product of the reaction.

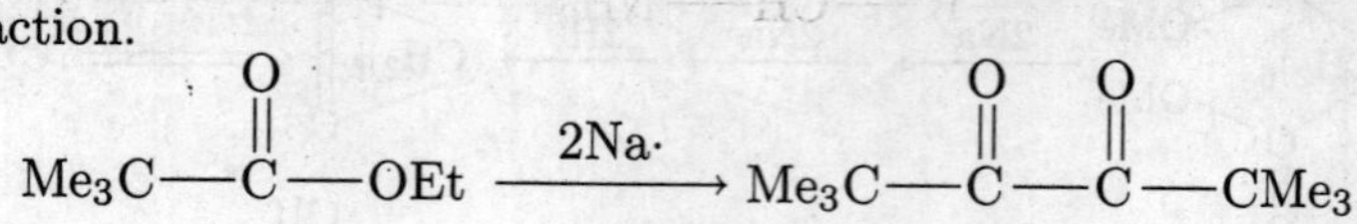

2,2,5,5-tetramethylhexan-3,4-dione

Problem 16.13. Schematically represent the mechanistic steps for the conversion :

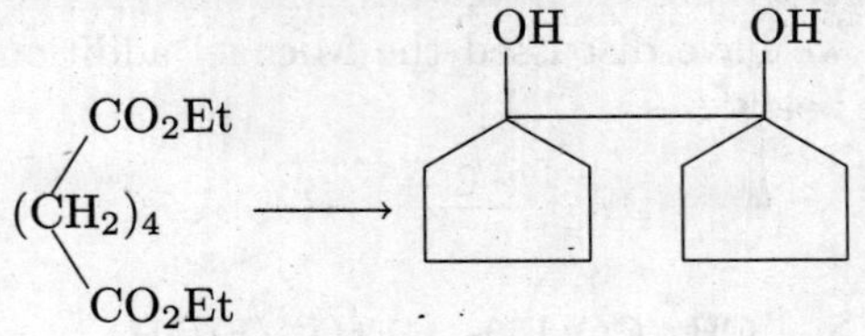

Solution :

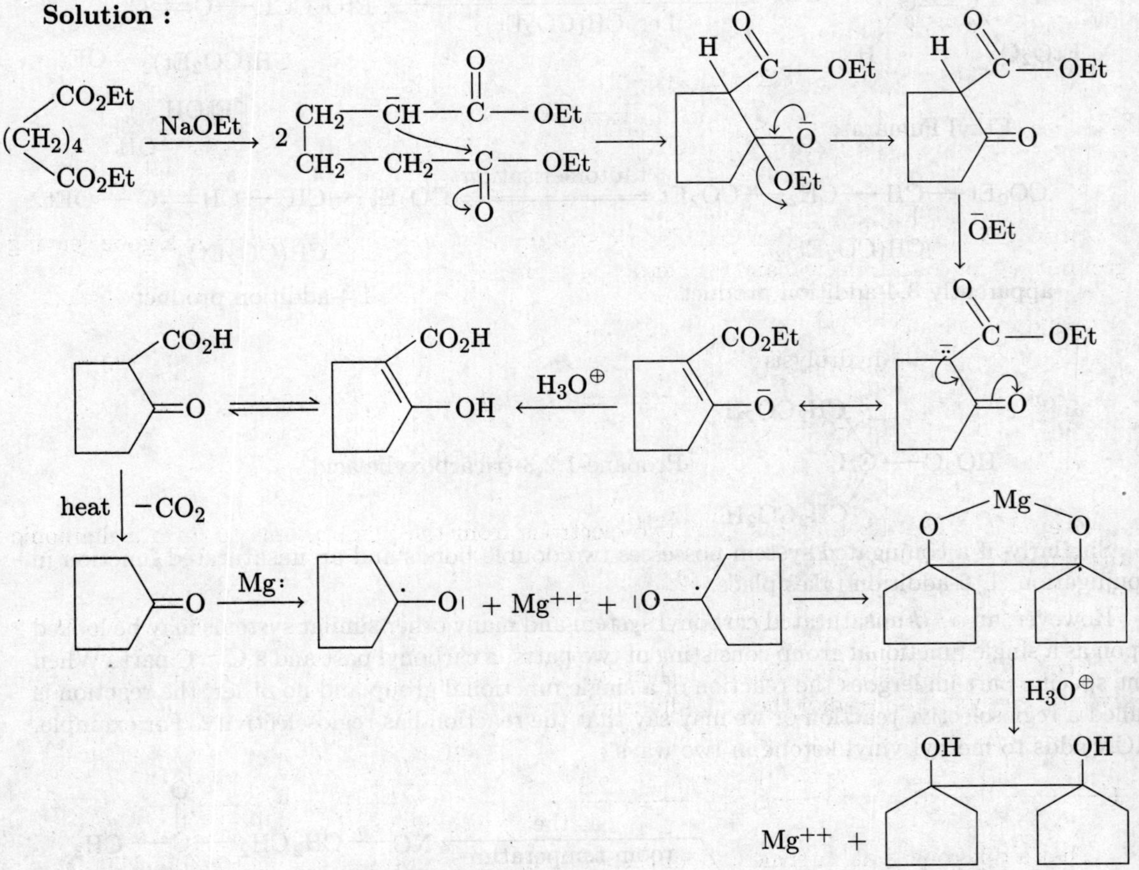

Problem 16.14. Give the mechanism of reduction of RCN by the Bouveault-Blanc procedure.

Solution :

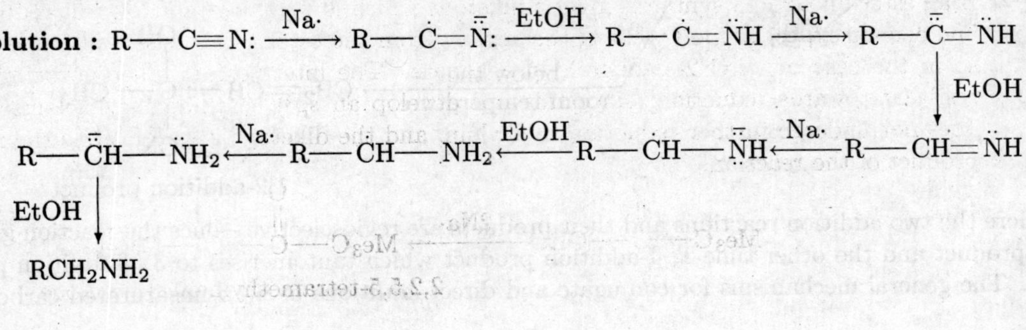

16.10 1, 2 or Direct Addition Versus 1, 4 or Conjugate or Michael Type Addition

In the thirteenth chapter we have discussed the Michael addition reaction on a conjugated bifunctional system as shown below :

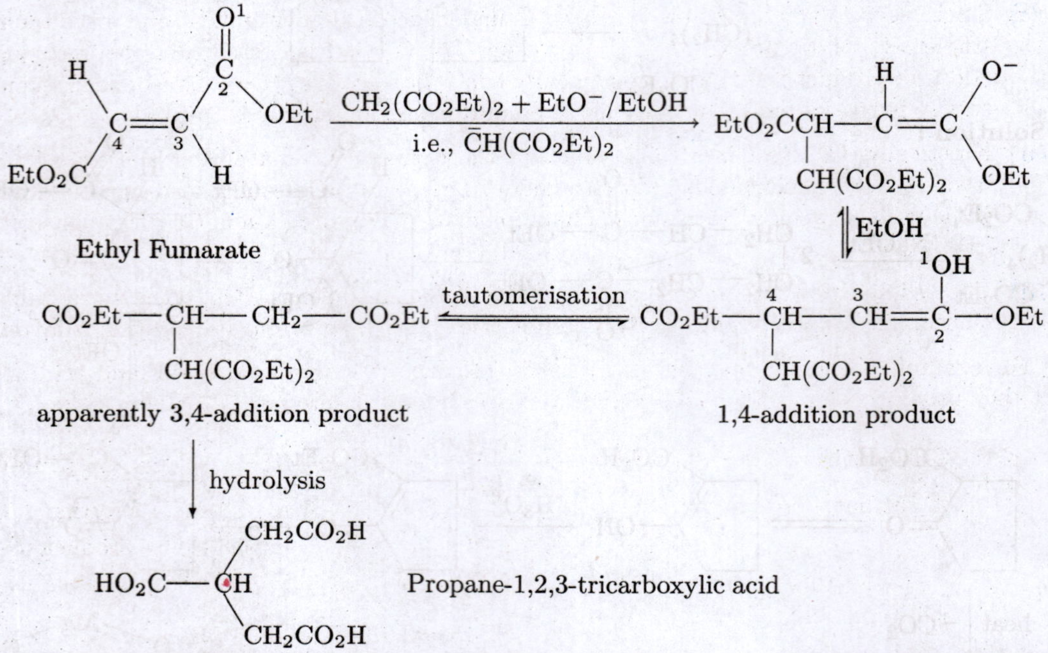

Similarly, if a conjugated system possesses two double bonds and an unsaturated function in conjugation, 1, 6-addition takes place.

However, an α, β-unsaturated carbonyl system and many other similar systems may be looked upon as a single functional group consisting of two parts, a carbonyl part and a C=C part. When one specific part undergoes the reaction of a single functional group and no other, the reaction is called a regioselective reaction or we may say that the reaction has regioselectivity. For example, HCN adds to methyl vinyl ketone in two ways :

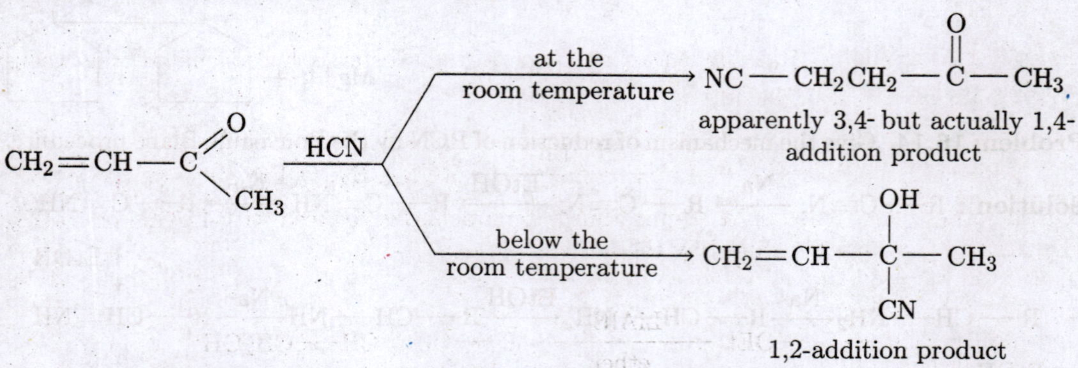

Here the two addition reactions and their products are regioselective—once the reaction gives 1, 2-product and the other time 1, 4-addition product which tautomerises to 3, 4-addition product. The general mechanisms for conjugate and direct additions in α, β-unsaturated carbonyl compounds have already been discussed. Here we are to settle where and why the regioselective addition will occur.

However, the general principles are :

(i) Since >C=O is more stable than C=C under identical environment and since the more stable >C=O is preserved and the weaker C=C bond is destroyed, 1,4- or conjugate-addition product is a thermodynamically controlled product (C—C π bond dissociation energy is 264 kJ mol^{-1} but C—O π bond dissociation energy is 390 kJ mol^{-1}).

(ii) Kinetically, C_4 atom is a softer site and the carbonyl C is a harder one. So, strongly basic or hard nucleophiles tend to undergo 1, 2-addition through hard-hard interaction, while soft or weakly basic nucleophiles tend to undergo 1, 4-addition through soft-soft interaction.

(iii) The 1, 4-addition is more easily reversed than the 1, 2-addition. Hence, the more stable the nucleophile, or the more reversible the reaction, the more conjugate addition is favoured.

For example, Grignard reagents add directly to carbonyl part in aldehydes and acid halides but they usually undergo Michael type addition to esters and ketones.

$$\underset{\text{sec-butyl crotonate}}{CH_3—CH=CH—CO_2Bu(s)} \xrightarrow[2.\ H_3O^{\oplus}]{1.\ BuMgBr} \underset{60\%}{CH_3—\underset{\displaystyle Bu}{\underset{|}{CH}}—CH_2—CO_2Bu(s)} \quad \text{+ 1,4-addition product}$$

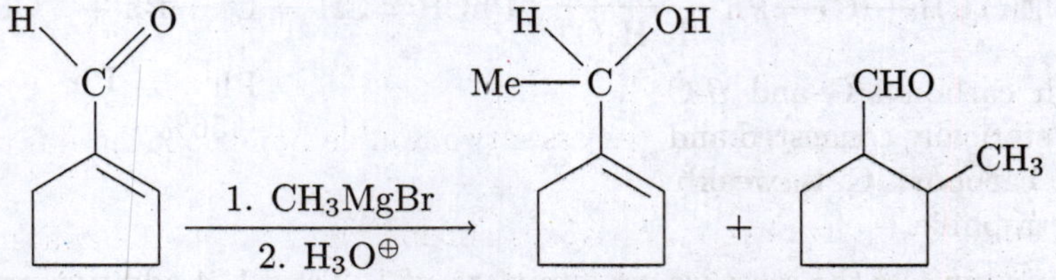

85–90% small amount

In general, RLi, $\bar{N}H_2$, RO^- and H^- (all are hard nucleophiles) are more likely to add directly; RS^-, neutral amines, RMgX and stable carbanions (less hard nucleophiles) are more likely to add Michael like.

For the reasons noted above when an α, β-unsaturated ester is hydrogenated, the weaker C=C gets reduced much readily.

$$R—CH=CH—\overset{\displaystyle O}{\overset{\|}{C}}—OEt \xrightarrow{H_2/Pd\text{-}C} RCH_2CH_2CO_2Et$$

On the other hand, when an α, β-unsathurted compound is reduced with $LiAlH_4$ or $NaBH_4$, the harder site C=O gets reduced by the hard H^- ion.

$$RCH=CH—\overset{\displaystyle O}{\overset{\|}{C}}—OEt \xrightarrow[\text{ether}]{LiAlH_4} R—CH=CH—CH_2OH$$

However, to what extent a given nucleophile undergoes the direct addition and the conjugate addition that depends on the steric and electronic factors. If the carbonyl C is more sterically congested and weakly electrophilic, the conjugate addition will occur more readily than the direct addition. On the other hand, if the β-C is more sterically congested, the direct addition will take place predominantly. A few examples are given below :

Substrate **Direct addition & Conjugate addition**

a. 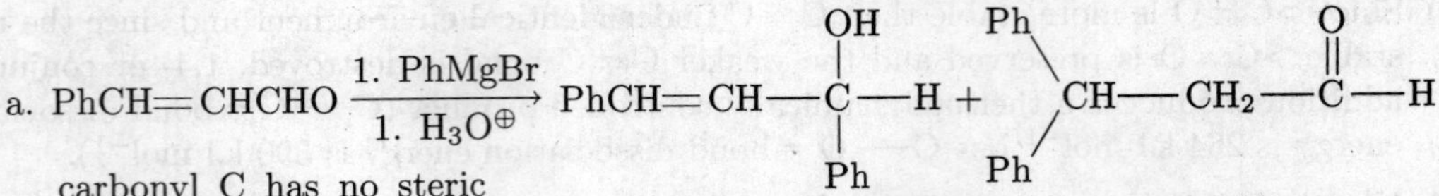

100% 0%

carbonyl C has no steric congestion but β-C is sterically congested.

b. 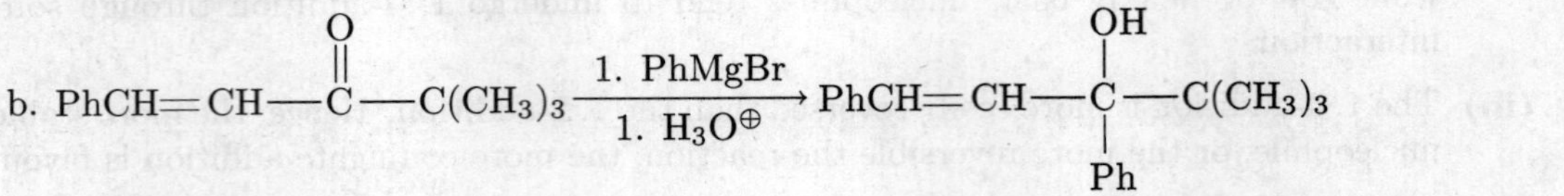

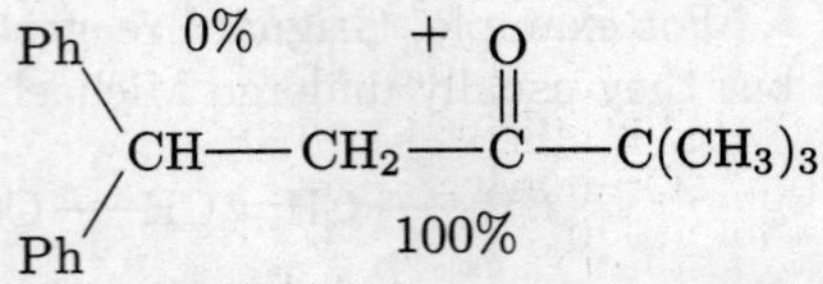

carbonyl C is sterically congested and weakly electrophilic.

c. 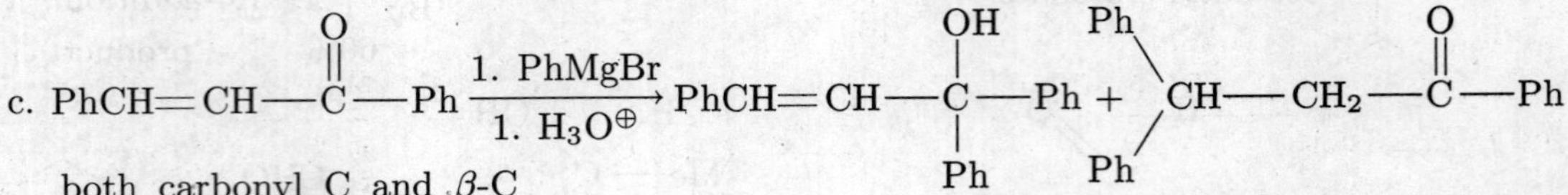

56% 44%

both carbonyl C and β-C are sterically congested and the carbonyl C is weakly electrophilic.

- In fact, in most of the cases we get a mixture of 1, 2 and 1, 4-addition products in comparable amounts. By the proper choice of the reagent some control may be exerted on the product. For example,

(i) Aliphatic Grignard reagents are more prone to 1, 4-addition than are aromatic Grignard reagents.

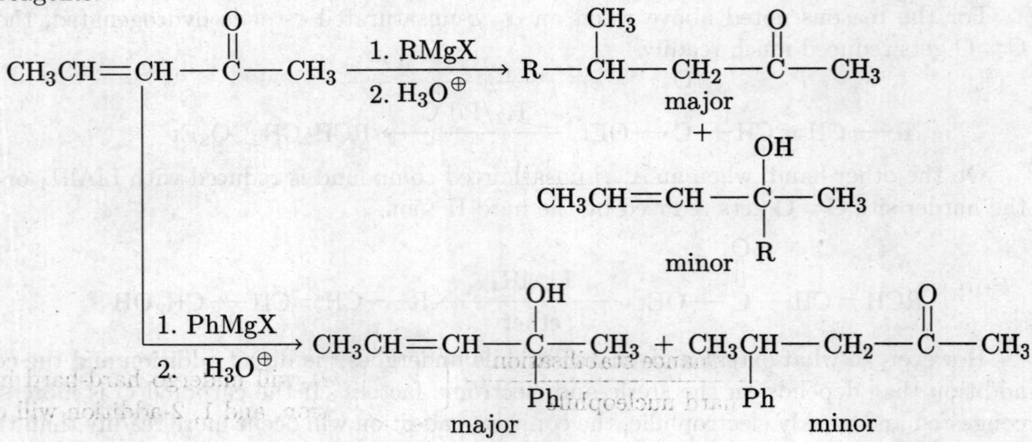

(ii) Organolithiums give better yield of the direct addition products than the corresponding Grignard reagents

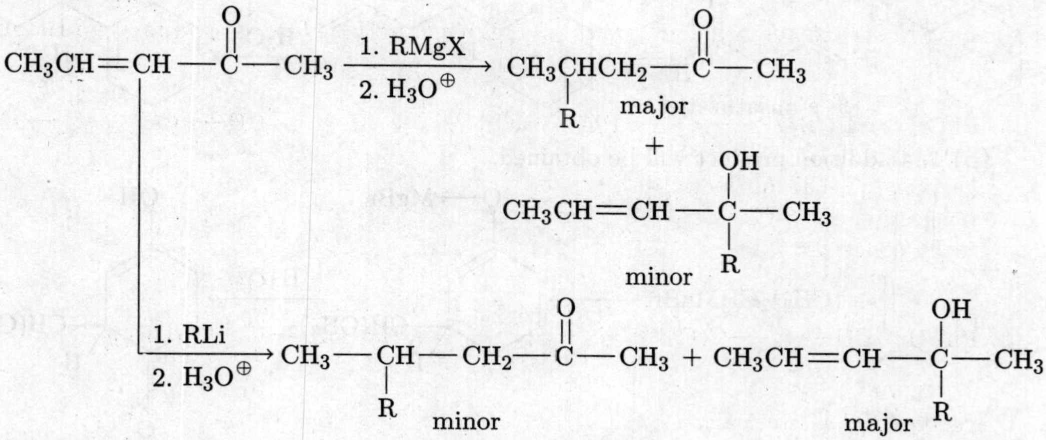

(iii) The inclusion of a copper (I or II) salt in the addition reaction of organolithiums and Grignard reagents to an α, β-unsaturated system promotes the 1, 4- or conjugate-addition.

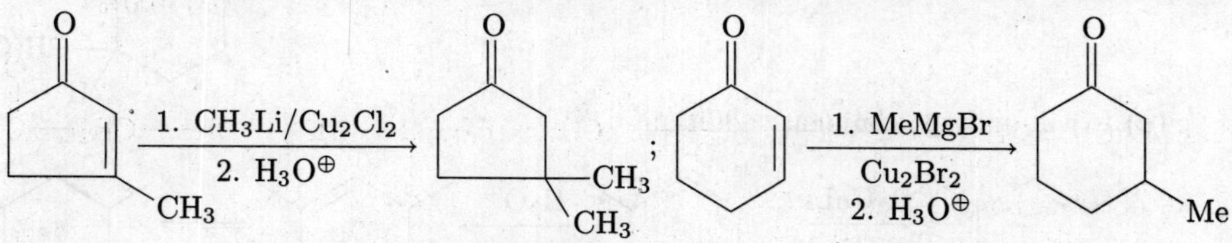

Problem 16.15. What products do you expect from the following reactions :

(a) 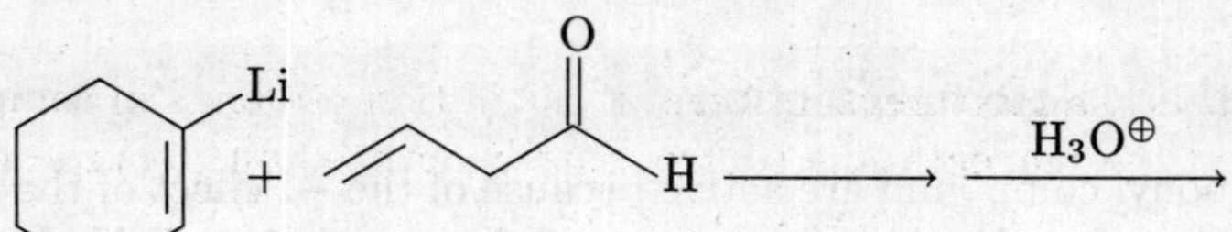

(b) 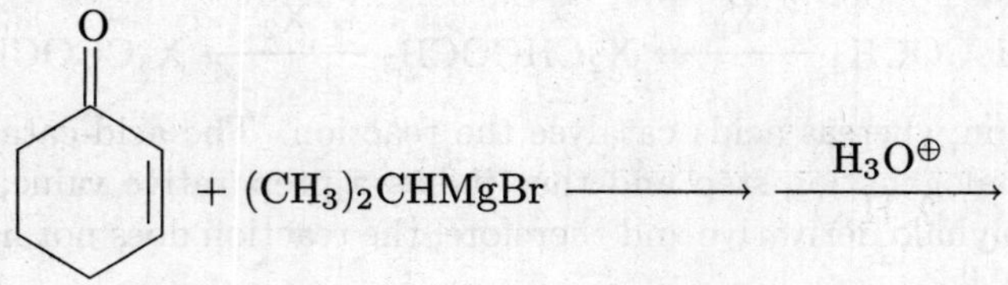

(c) 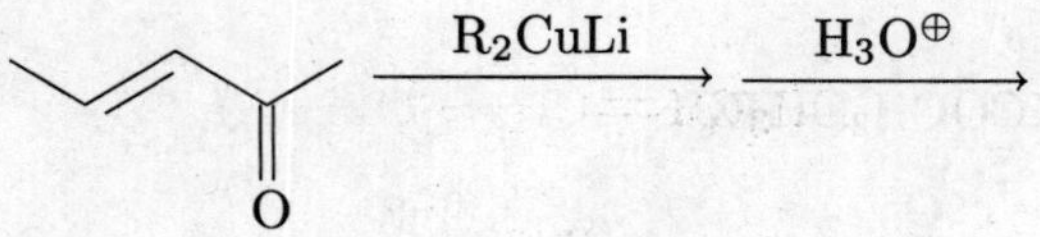

Solution :

(a) 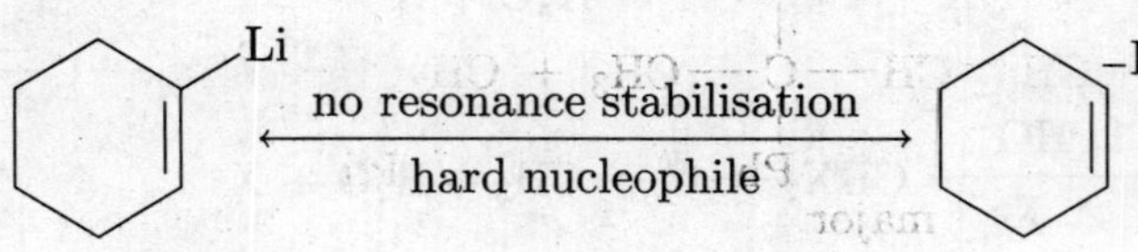

It will undergo hard-hard interaction, and 1, 2-addition will occur.

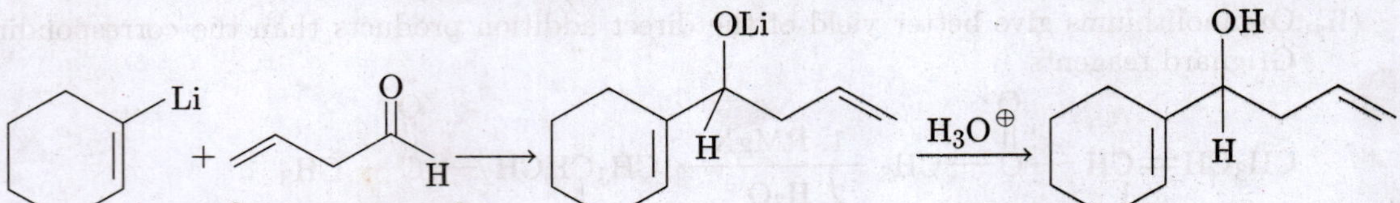

β, γ-unsaturated

(b) 1, 4-addition product will be obtained.

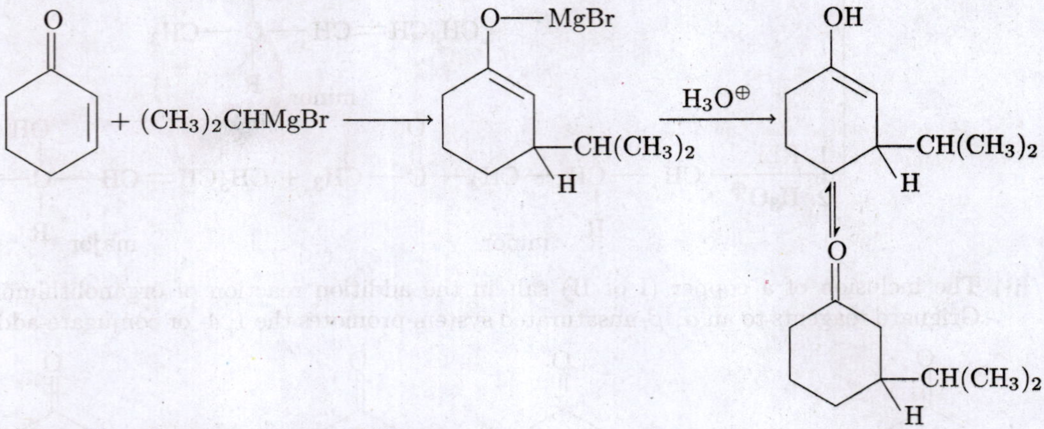

(c) It will undergo conjugate addition.

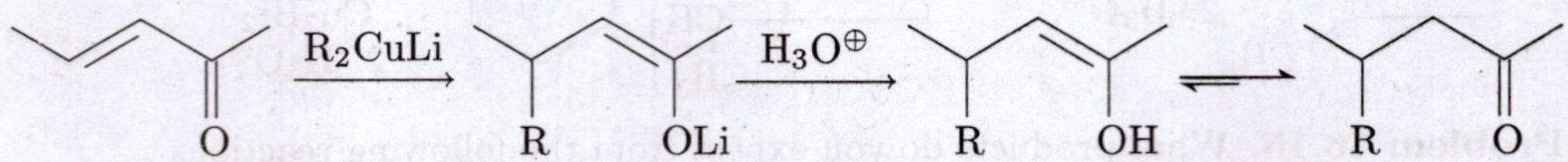

16.11 Halogenation and the Haloform Reaction

α-Hydrogen atoms of a carbonyl compound are active because of the $-$I effect of the carbonyl group. Active α-H atoms may be replaced by halogen atoms and the process is called halogenation.

$$CH_3COCH_3 \xrightarrow{X_2} XCH_2COCH_3 \xrightarrow{X_2} X_2CHCOCH_3 \xrightarrow{X_2} X_3CCOCH_3$$

Bases promote the halogenation whereas acids catalyse the reaction. The acid-catalysed reaction may be stopped at monohalogenation step and thus it has a preparative value; but the base-promoted reaction gives a polyhalo derivative and therefore, the reaction does not have little preparative value.

$$CH_3COCH_3 \xrightarrow[2.\ X_2]{1.\ H^{\oplus}} CH_2XCOCH_3 + HX$$

$$CH_3COCH_3 \xrightarrow[2.\ X_2]{1.\ HO^-} CH_2X{-}\overset{\overset{\large O}{\|}}{C}{-}CH_3 + H_2O + X^-$$

$$CH_2X{-}\overset{\overset{\large O}{\|}}{C}{-}CH_3 \xrightarrow[2.\ X_2]{1.\ HO^-} CHX_2{-}\overset{\overset{\large O}{\|}}{C}{-}CH_3 + H_2O + X^-$$

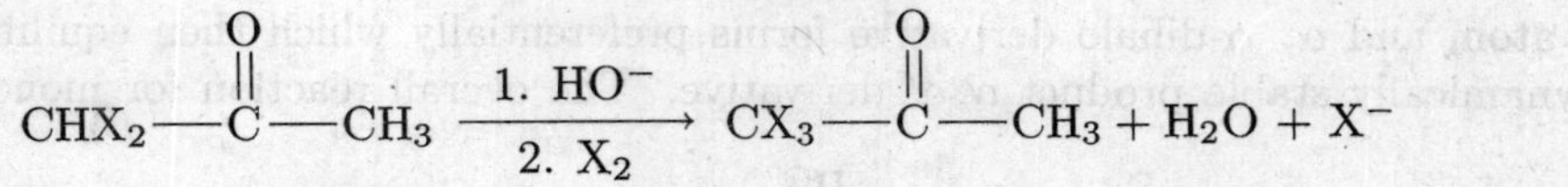

Mechanism of the acid-catalysed halogenation

The acid-catalysed halogenation reaction consists of the following steps :

Step I : Protonation of the carbonyl O atom and formation of the resonance-stabilised oxonium ion.

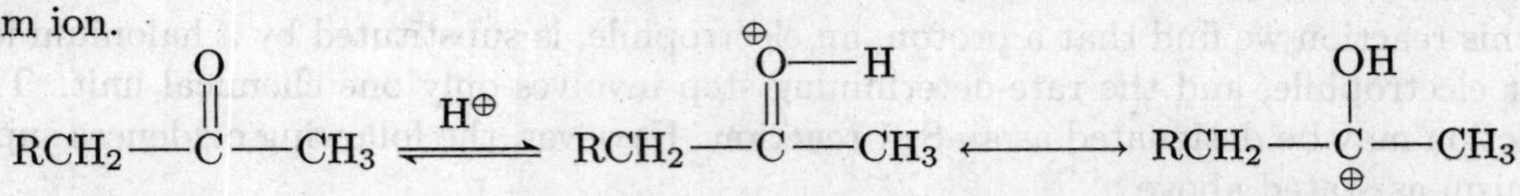

Here the oxonium ion is more stable than the carbenium ion because the cationic C is in the sextet state while the cationic O is in the octet state. However, this is a fast step.

Step II : Elimination of a proton from the α-position and formation of an enol.

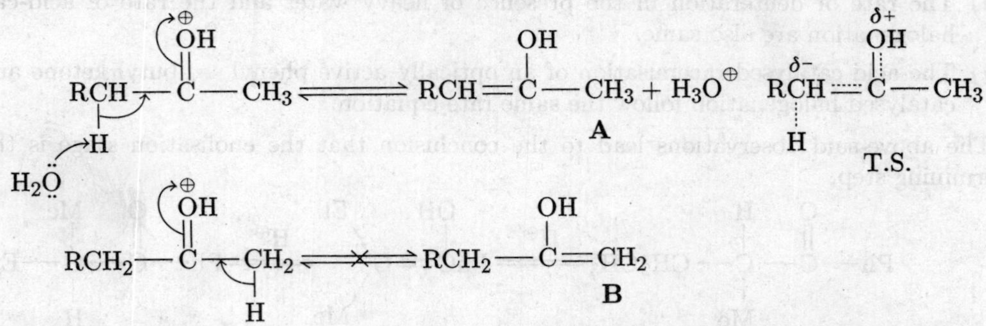

Here compound A forms preferentially since the hyperconjugative effects of the H atoms of —CH_3 group and R group stabilise compound A more than compound B which has only two H atoms in $>CH_2$ for hyperconjugation.

This is a slow step and hence the rate-determining step.

Step III : Electrophilic addition of halonium ion from the molecular halogen.

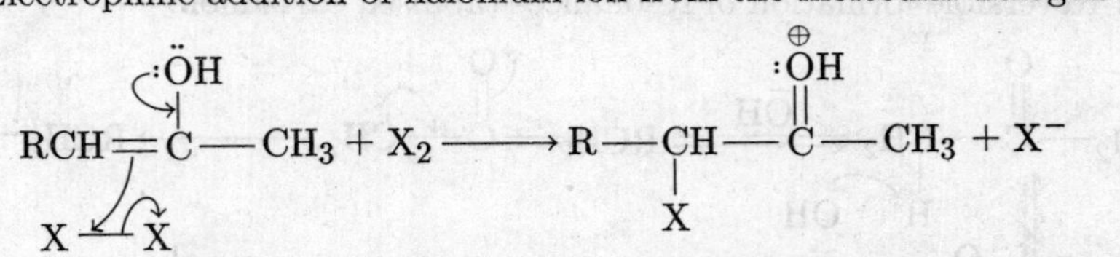

This is a fast step. As soon as the enol forms, it reacts rapidly with the halogen.

Step IV : The halide ion will then abstract the proton from the $\overset{\oplus}{O}H$ and monohalo derivative will form.

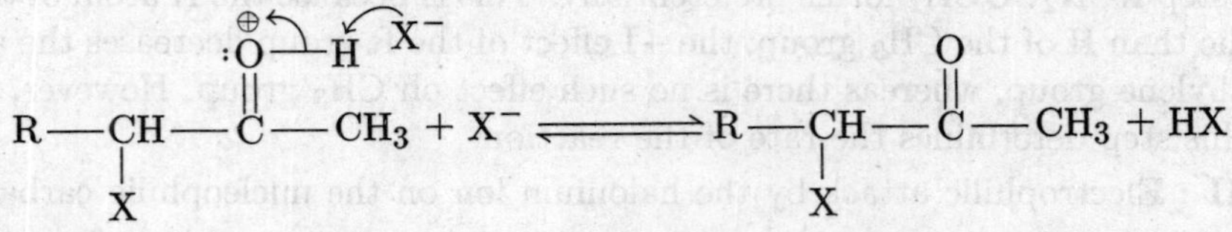

Since the X atom of the RCHX—moiety has –I effect, the O atom of the carbonyl group does not have enough electron-density for coordination with a proton and, therefore, further halogenation does not take place usually; if it at all occurs an H atom of the $>CH_2$ group gets replaced by an X atom and α, α-dihalo derivative forms preferentially which then equilibrates to give thermodynamically stable product α, α′-derivative. The overall reaction for monohalogenation is :

$$RCH_2COCH_3 + X_2 \overset{H^{\oplus}}{\rightleftharpoons} RCHXCOCH_3 + H^{\oplus} + X^-$$

Since the slowest step of the reaction is the enol formation step and since it involves the substrate and the acid, the rate $= k$[substrate] $[H^{\oplus}]$

In this reaction we find that a proton, an electrophile, is substituted by a halonium ion $(X^{\oplus})$, another electrophile, and the rate-determining step involves only one chemical unit. Therefore, the reaction may be designated as an S_E1 reaction. However, the following evidences support the mechanism as stated above :

(i) The rates of acid-catalysed chlorination, bromination and iodination of acetone are one and the same; this indicates that the rate is independent of the concentration of halogens.

(ii) The rate of deuteration in the presence of heavy water and the rate of acid-catalysed halogenation are also same.

(iii) The acid-catalysed racemisation of an optically active phenyl sec-butyl ketone and acid-catalysed halogenation follow the same rate equation.

The above-said observations lead to the conclusion that the enolisation stage is the rate-determining step.

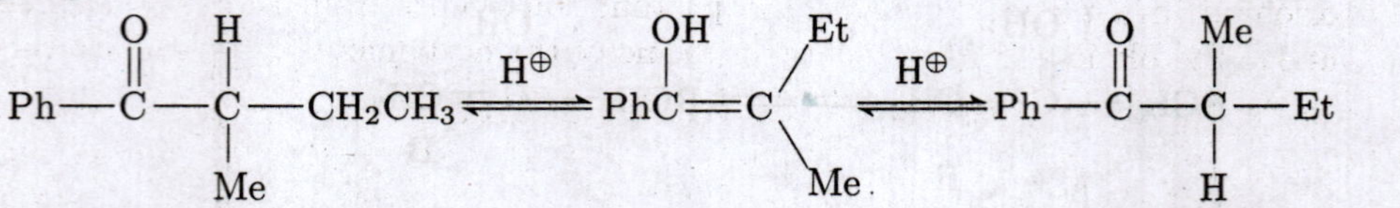

(+)- or (−)- form (−)- or (+)- form

Mechanism of the base-promoted halogenation

The base-promoted halogenation consists of the following steps and involves a resonance-stabilised carbanion as an intermediate.

Step I : Reversible formation of resonance-stabilised carbanion.

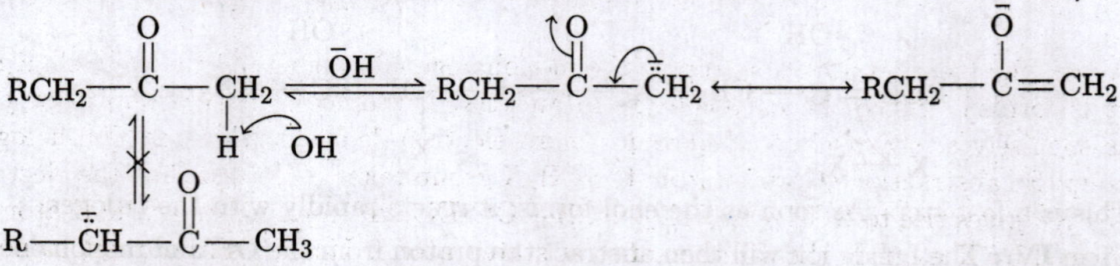

In this step $RCH_2CO\bar{C}H_2$ forms preferentially. This is because the H atom of the RCH_2 group is less acidic than H of the CH_3 group; the +I effect of the R group decreases the acidity of the H of the methylene group, whereas there is no such effect on CH_3 group. However, this is the slow step and this step determines the rate of the reaction.

Step II : Electrophilic attack by the halonium ion on the nucleophilic carbanion. This is a fast step :

The formation of monohalo derivative follows the second order kinetics : it is first order with respect to the substrate and first order with respect to the base; the rate = k [substrate] [HO^-].

This reaction may be designated as S_E1 reaction though the rate-determining step involves the concentrations of the substrate and the base (but it does not involve the halogen).

The following are the evidences in favour of the mechanism written above :

(i) The reversible formation of the carbanion is supported by the fact that when the reaction is carried out with $\bar{O}D$ in D_2O, exchange of deuterium occurs.

(ii) The kinetic study shows that the rate of the reaction is independent of the concentration of the halogen but it depends on the concentration of the substrate as well as that of the base. This supports the rate equation as predicted above.

(iii) The base-promoted halogenation of acetone shows that the rates of bromination and iodination of acetone are one and the same. This also indicates that the rate does not depend on the concentration of halogen.

(iv) The rate of base-promoted racemisation of optically active phenyl sec-butyl ketone is the same as that of the halogenation. The formation of the optically inactive phenyl sec-butyl ketone from the (+) - or (−) - phenyl sec-butyl ketone indicates that a bond of the chiral C must have broken and the formation of the carbanion fits this requirement. In the absence of halogen the carbanion takes up a proton from the conjugate acid; the carbanion being oriented in a plane trigonally, the protonation occurs from the two sides equally, from above the plane and form below the plane of the carbanion which leads to racemisation.

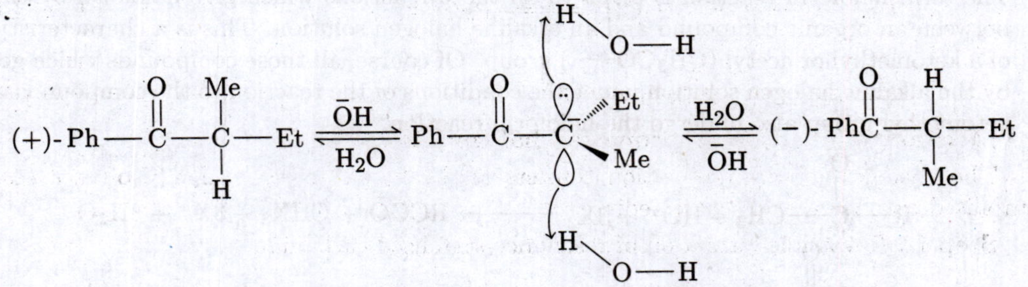

However, the reaction does not stop at the monohalogenation stage; usually a trihalogenated product is formed. Since the carbonyl group and the X atom of CH_2X group draw electron towards themselves, the α-C atom becomes more electron-deficient than the non-halogenated CH_3 group. So abstraction of proton from the CH_2X group takes place and then the electrophilic attack by $X^{\oplus}$ gives rise to RCH_2COCHX_2. RCH_2COCHX_2 still contains an acidic H atom which gets replaced by X atom in the same manner as stated above. Here it is important to note that the H atoms of RCH_2 are much less acidic than the H atoms of $>CH_2$, CH_2X and CHX_2 groups because of the +I effect of the R group and −I effect of the X atoms.

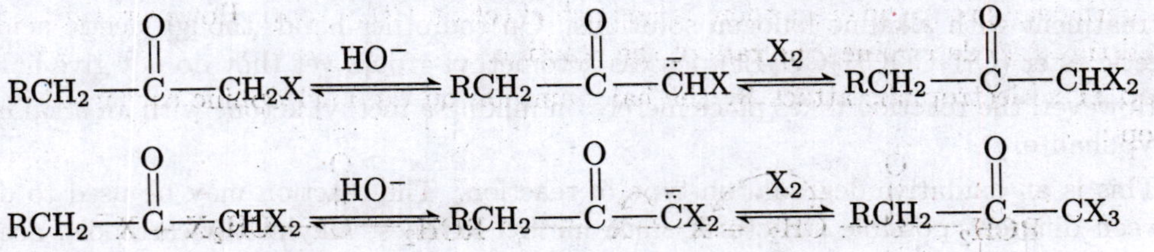

If there is an excess of base, the base now attacks the highly electron-deficient carbonyl C atom and develops the species C. The carbonyl C now bears high steric strain owing to the presence of the RCH_2, OH, CX_3 groups and the bulky anionic O atom. In order to get rid of the steric

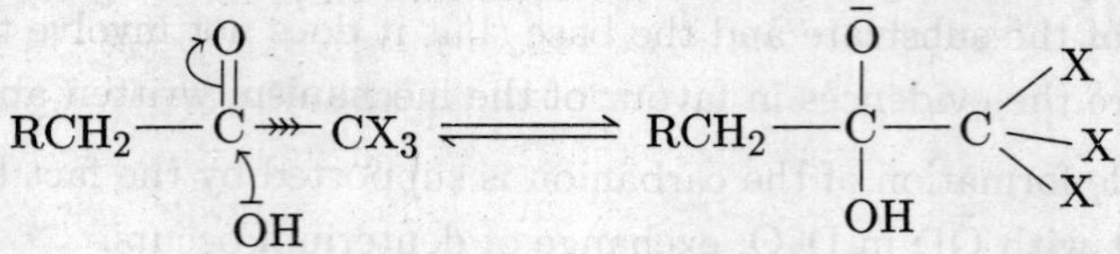

C

strain C—CX_3 bond breaks and trihalocarbanion forms. The latter is rather a stable species since the three halogen atoms draw the negative charge of carbanion towards themselves by the −I effect and by *pπ-dπ* conjugation. Thus the steric strain and the stability of trihalomethide ion help to break the C—CX_3 bond as shown below :

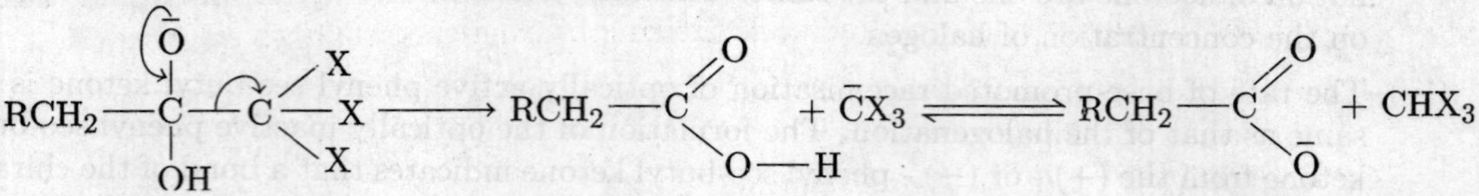

The trihalomethide ion then takes up the proton to give haloform and a carboxylate ion.

However, CHX_3, trihalomethane, on hydrolysis yields formic acid, which leads Dumas to give the general name of chloroform ($CHCl_3$), bromoform ($CHBr_3$) and iodoform (CHI_3) as haloforms. The term haloform reaction is given to all those reactions which give haloform by the reaction between an organic compound and an alkaline halogen solution. This is a characteristic reaction of a ketomethyl or acetyl (CH_3CO—) group. Of course, all those compounds which get oxidised by the alkaline halogen solution under the conditions of the reaction to the compounds containing ketomethyl group also undergo the haloform reaction.

$$R{-}\overset{O}{\overset{\|}{C}}{-}CH_3 + HO^- + 3X_2 \longrightarrow RCO\bar{O} + CHX_3 + 3X^- + 3H_2O$$

$$CH_3{-}\overset{O}{\overset{\|}{C}}{-}H + HO^- + 3I_2 \longrightarrow HCO\bar{O} + CHI_3 + 3I^- + 3H_2O$$

$$CH_3CH_2OH + 2NaOX \longrightarrow CH_3CHO + NaX + H_2O$$

$$CH_3CHO + HO^- + 3X_2 \longrightarrow HCO\bar{O} + CHX_3 + 3X^- + 3H_2O$$

$$RCH_2CHOHCH_3 + NaOX \longrightarrow RCH_2COCH_3 + NaX + H_2O$$

$$RCH_2COCH_3 + HO^- + 3X_2 \longrightarrow RCH_2CO\bar{O} + CHX_3 + 3X^- + 3H_2O$$

There are compounds which give haloform against our expectation. Isopropylamine and compounds of the type $RCOCH_2COR$, $RCHOHCH_2CHOHR$, $CH_3CH(NHCOR)COR$ give haloform on treatment with alkaline halogen solutions. On the other hand, though acetic acid and acetoacetic ester ($CH_3COCH_2CO_2Et$) possess ketomethyl group, yet they do not give haloform.

However, the reaction takes place merely on mixing a methyl ketone with an alkaline solution of hypohalite.

This is an oxidative degradation type of reaction. The reaction may be used to distinguish between different possible structures since methyl ketones, alkylmethylcarbinols, acetaldehyde (the only aldehyde) and ethyl alcohol (the only primary alcohol) give the test. For example, between $CH_3COCH_2CH_2CH_3$ and $CH_3CH_2COCH_2CH_3$ only the former gives the haloform test and not the latter since the former one only contains a ketomethyl group. So by the haloform reaction one can determine the structure of a compound. In practice iodoform test is performed instead of bromoform or chloroform test since iodoform is a yellow crystalline solid insoluble in water and can be readily identified by its melting point. The iodoform reaction is best carried out by taking the solution of the substance in dioxane and treating it first with dilute NaOH solution and then with a slight excess of iodine in potassium iodide solution, warming the mixture, and finally adding water to it. If the compound under consideration contains a —$COCH_3$ group or CH_3CHOH— group or if it is acetaldehyde or ethyl alcohol, a yellow precipitate of iodoform will be obtained.

Besides structure determination, the reaction may be used to prepare aromatic acids where other oxidising agents cannot be used. Methylnaphthalene and acetylnaphthalene are very susceptible towards oxidation. When they are treated with strong oxidising agents, naphthoic acid is not obtained instead the aromatic character of the ring bearing the group gets destroyed.

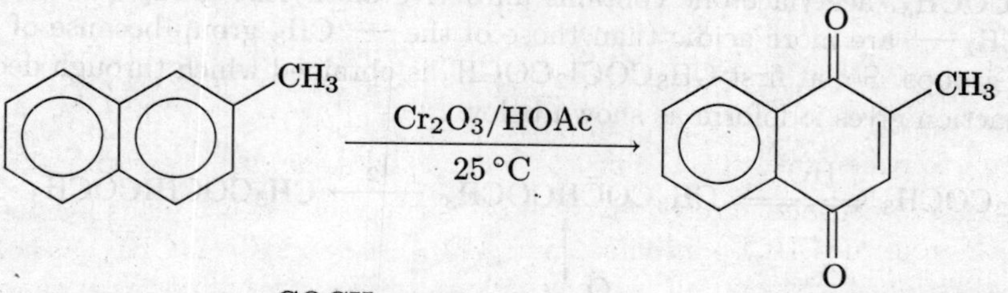

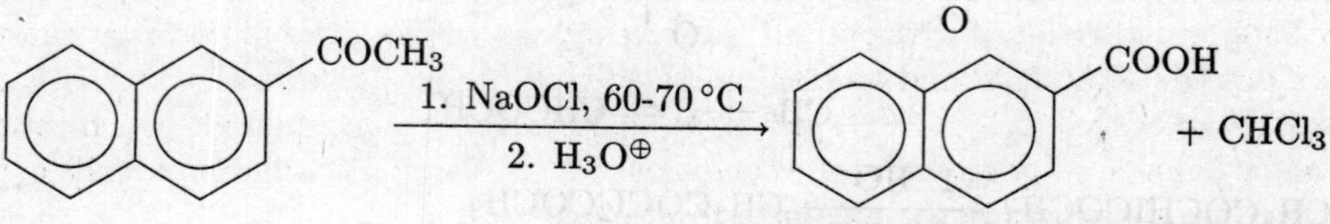

2-Naphthalenecarboxylic acid
or
β-Naphthoic acid

NaOX being a mild oxidising agent, it can oxidise the acetyl group to carboxylate group without destroying the aromatic character of the ring.

The mechanism of the reaction has already been discussed along with the mechanism of the base-promoted halogenation of carbonyl compounds. However, in a nutshell this may be represented as :

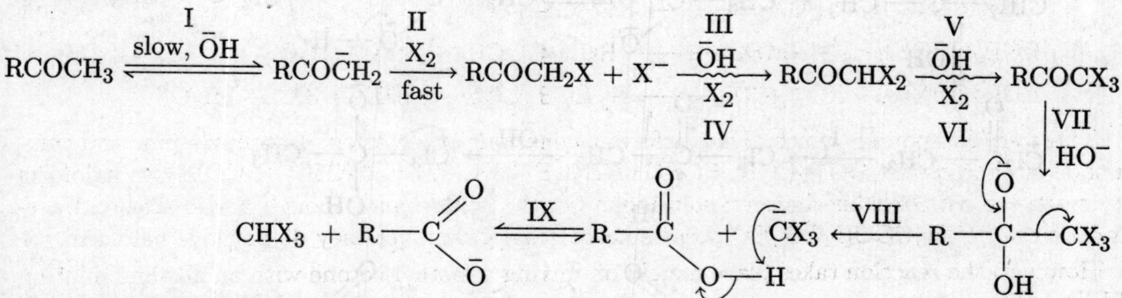

Step I : Carbanion formation.

Step II : Nucleophilic attack by the anionic C on $X^{\oplus}$ of X_2.

Steps III & V : Repetition of the step I.

Steps IV & VI : Repetition of the step II.

Step VII : Nucleophilic attack on the electron-deficient carbonyl C by HO^-.

Step VIII : Cleavage of C—CX_3 bond and formation of trihalomethide ion.

Step IX : Proton exchange between the carboxylic acid and the trihalomethide ion to form haloform.

Problem 16.16. Why do the following compounds give iodoform test?

(a) CH_3CH_2OH (b) CH_3CHO (c) $CH_3CO.CH_2CO.CH_3$ (d) $RCO.CH_2COR$
(e) $RCHOHCH_2CHOHR$ (f) $(CH_3)_2CH.NH_2$

Solution : (a) CH_3CH_2OH gets oxidised to CH_3CHO. CH_3CHO possesses a ketomethyl group; so it gives iodoform reaction.

(b) See above.

(c) $CH_3COCH_2COCH_3$, acetylacetone contains an active methylene group (—CH_2—). The Hs of the —CH_2— are more acidic than those of the —CH_3 group because of the −I effects of 2CH_3CO—groups. So, at first $CH_3COCI_2COCH_3$ is obtained which through decomposition and further reaction gives iodoform as shown below :

$$CH_3COCH_2COCH_3 \underset{}{\overset{HO^-}{\rightleftharpoons}} CH_3CO\bar{C}HCOCH_3 \xrightarrow{I_2} CH_3COCHICOCH_3$$

$$CH_3COCHICOCH_3 \xrightarrow[2.\ I_2]{1.\ HO^-} CH_3COCI_2COCH_3$$

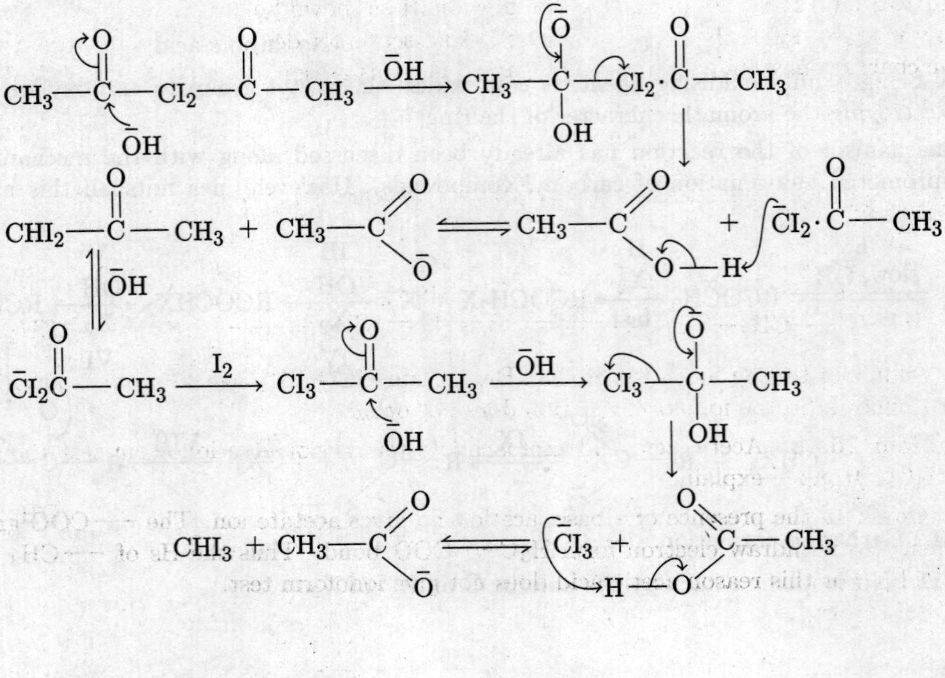

(d) 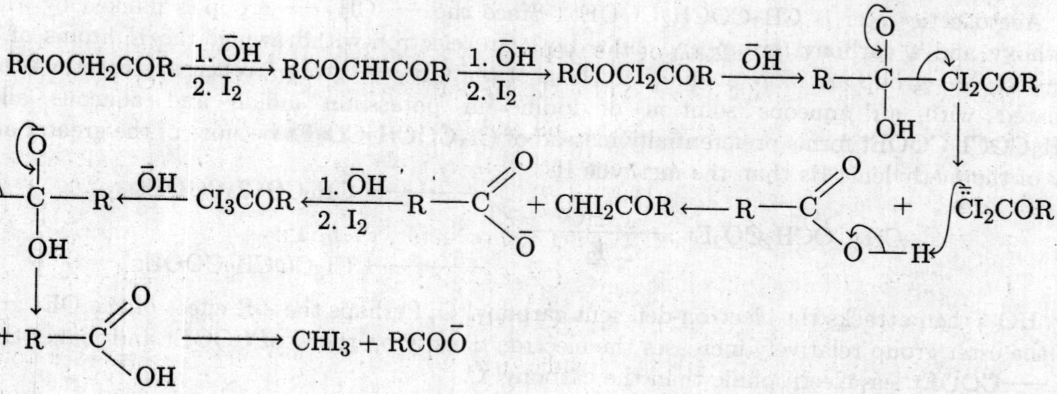

(e) $RCHOH.CH_2CHOHR \xrightarrow{2NaOI} RCO.CH_2COR \longrightarrow$ etc. (see above)

(f) 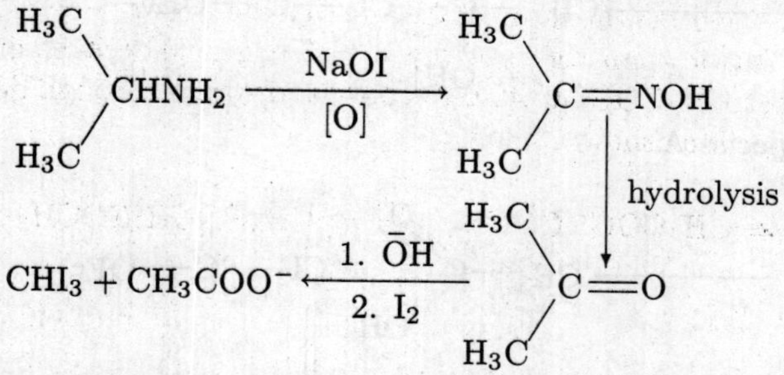

Problem 16.17. (i) Why does a base fail to remove the H atom of the aldehyde group when there is an α-H? **(ii)** Pinacolone gives bromoform test but it does not give iodoform test — explain.

Solution : (i) If a base abstracts the aldehydic H as shown below, the carbanion formed will be highly unstable since the negative charge has no scope for delocalisation; but the α-C with a negative charge gets stability through resonance. For this reason a base abstracts an α-H instead of the aldehydic H.

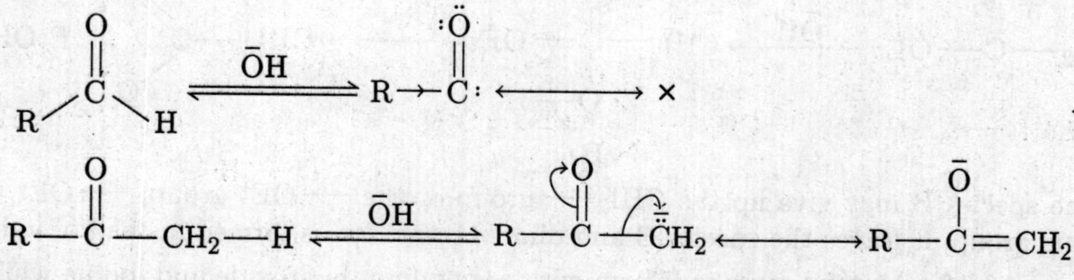

(ii) Volume of I being much larger than Br, the $Me_3CCOCHI_2$ has no room to accommodate the third bulky I. So the iodoform reaction does not occur.

Problem 16.18. Acetic acid and acetoacetic ester do not give iodoform test though they have CH_3CO group —explain.

Solution : In the presence of a base, acetic acid gives acetate ion. The $—CO\bar{O}$ group has no tendency to withdraw electron form $H_3C—CO\bar{O}$ bond. Thus the Hs of $—CH_3$ were no longer acidic. For this reason acetic acid does not give iodoform test.

Acetoacetic ester is CH_3COCH_2COOEt. Since the —CH_2— group is flanked by a carbethoxy and a carbonyl group and since both are electron-withdrawing, the H atoms of the —CH_2— group are more acidic than those of the CH_3— group. When acetoacetic ester is treated with an aqueous solution of iodine in potassium iodide and aqueous alkali, CH_3COCI_2COOEt forms preferentially instead of $CI_3COCH_2CO_2Et$ because of the greater acidity of the methylenic Hs than the methylic H.

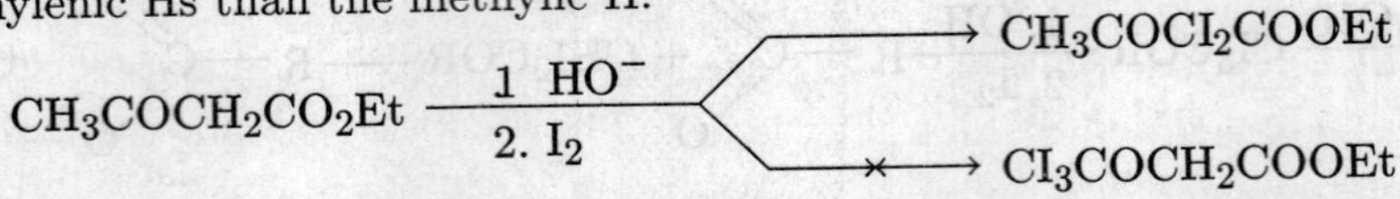

HO^- then attacks the electron-deficient carbonyl C. Perhaps the +R effect of the OEt group of the ester group relatively increases the electron density on the C of COOEt and makes the C of —COOEt less electrophilic than the carbonyl C.

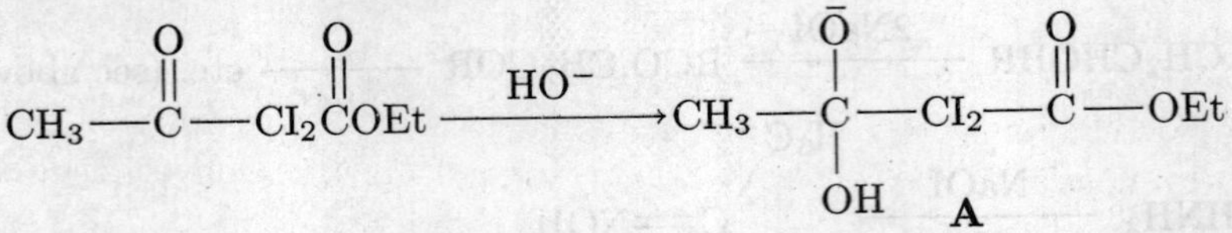

Steric strain helps decompose the species A.

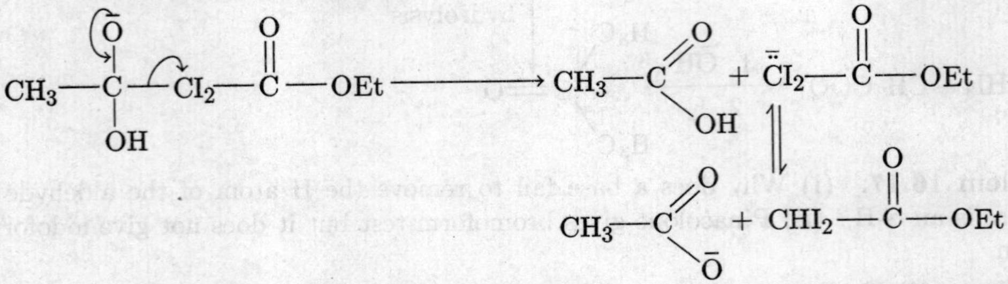

HO^- ion then attacks the C atom of the —COOEt group which is now electron-deficient owing to the presence of >CI_2 group.

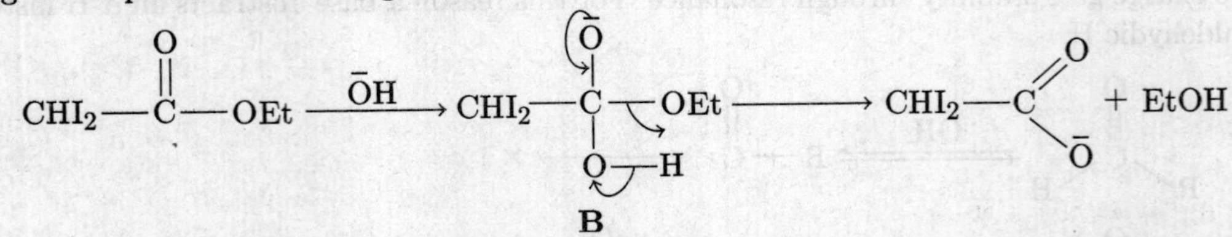

Now the species B may give up the CHI_2— group or the —OEt group; —OEt being a good leaving group, it leaves the species B and diiodoacetate ion is formed instead of iodoform.

Problem 16.19. Acetone gives iodoform with ammonium hydroxide and iodine while ethyl alcohol does not; but ethyl alcohol forms iodoform when it is allowed to react with an aqueous solution of NaOH and I_2—explain.

Solution : When ethyl alcohol gives iodoform, it first gets converted to acetaldehyde by the oxidising action of sodium hypoiodite (NaOI) which forms by the reaction of iodine with alkali.

$$2NaOH + I_2 = NaI + NaOI + H_2O$$

$$NH_4OH + I_2 \not\longrightarrow NH_4OI + NH_4I + H_2O$$

NH_4OI does not exist. So, ethyl alcohol does not get oxidised to acetaldehyde which can form iodoform. For the iodoform reaction acetone does not require any oxidising agent—it has the required ketomethyl group.

16.12 The Hell-Volhard-Zelinsky Reaction

Carboxylic acids containing α-H atoms can easily be converted to its α-haloderivate by the action of chlorine or bromine in the presence of catalytic amount of red phosphorus, and the reaction is known as the Hell-Volhard-Zelinsky reaction. The reaction does not occur with iodine and flourine. When the acid contains two α-H atoms, one or both may be replaced but it is often difficult to stop the reaction with one.

$$RCH_2COOH \xrightarrow{Cl_2/P(red)} RCHClCOOH \xrightarrow{Cl_2/P(red)} RCCl_2COOH$$

The halogenation of carboxylic acids by the Hell-Volhard-Zelinsky reaction takes place very readily and exclusively at α-position. The readiness and the exclusive halogenation at α-position give the reaction a great value in synthesis. α-haloacids can easily be converted to α-hydroxyacids, α-amino acids and α-cyano acids. The α-cyano acids in their turn may be converted to compounds having synthetic importance such as malonic acid and its derivatives.

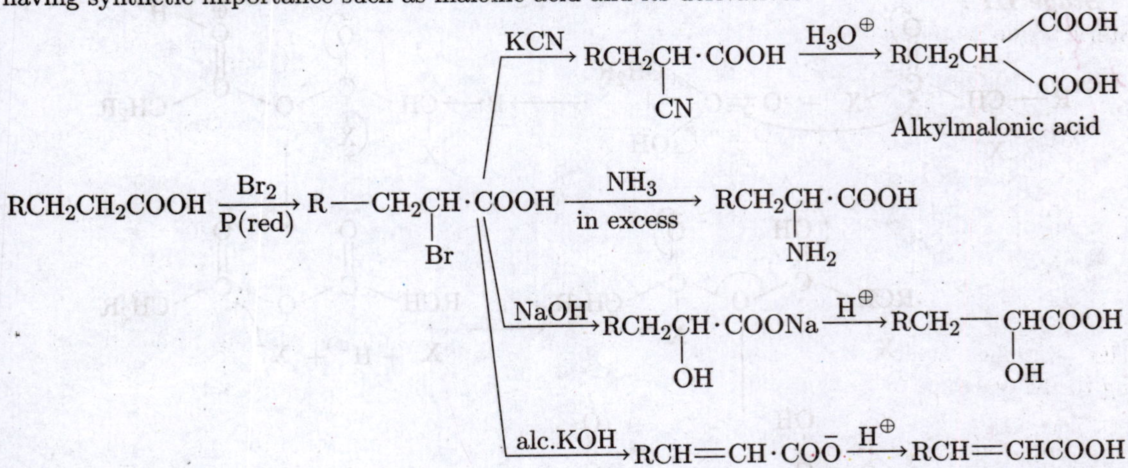

Here it is important to note that in the case of a di- or polycarboxylic acid, the α-H atoms with respect to every carboxyl group may be replaced by halogens. Since the reaction with Br_2 is specific for α-H atoms, it can be used to detect the presence of α-H in an acid.

However, the mechanism of the reaction is still uncertain.

The reaction is supposed to occur in three stages :

(i) The formation of an acid halide occurs in the first stage. This involves : (a) nucleophilic displacement of halide ion from PX_3, (b) the halide ion, a nucleophile, attacks the C atom of >C=O and forms acid halide.

(ii) The second stage involves enolisation of the acid halide. A halonium ion, $X^{\oplus}$ adds to the intermediate enol and subsequently gives up a proton to X^-.

(iii) The third stage involves exchange of acyl halogen between α-halogenated acid halide and the free acid. Thus,

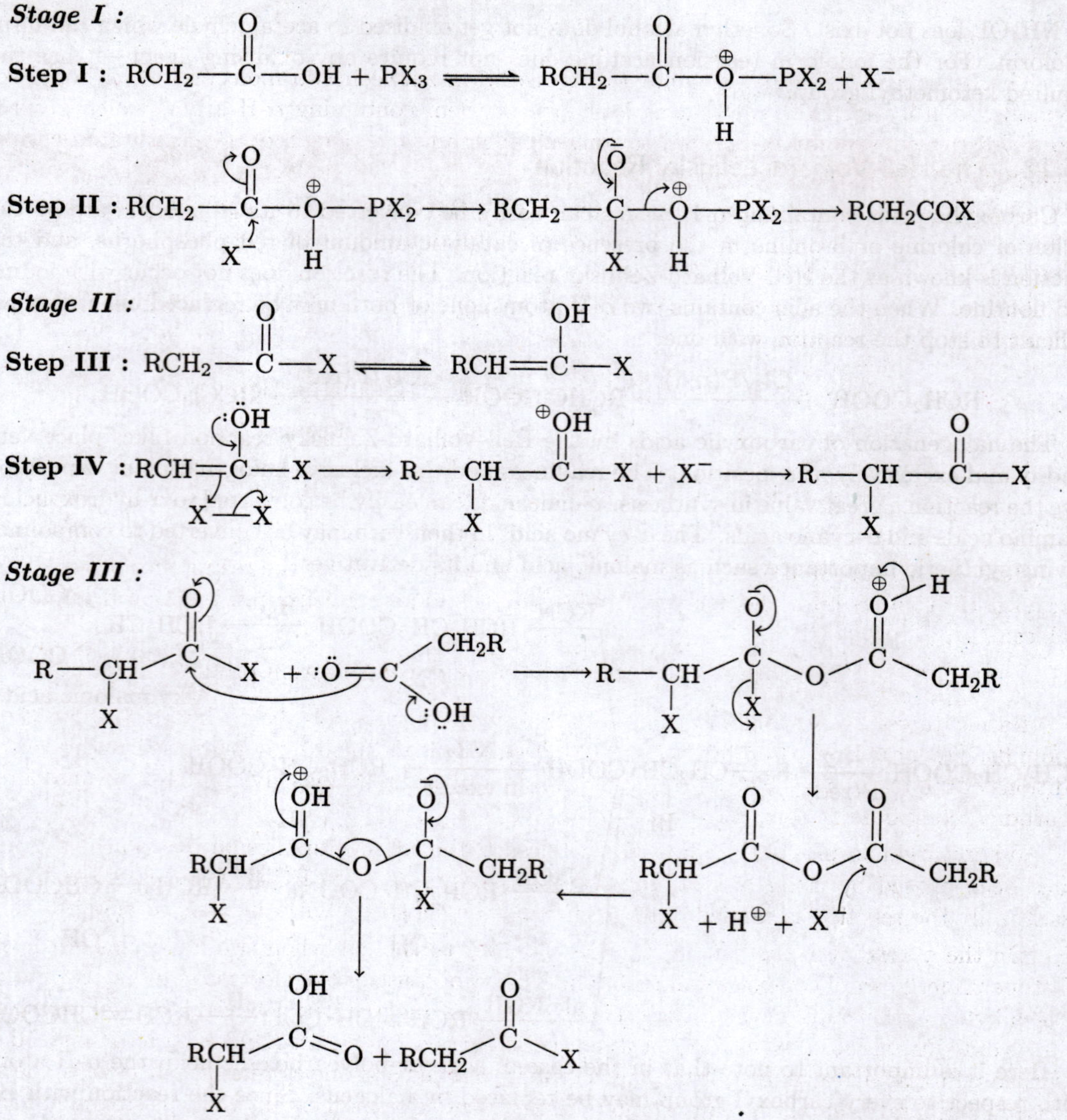

Since RCH_2COX is generated at the last reaction, the reaction can run smoothly without catalyst. The catalyst is required only to initiate the reaction. It has been found that when a carboxylic acid is chlorinated in the presence of PBr_3, α-chloro acid is obtained without any tracer amount of α-bromo acid.

16.13 Condensation Reactions

When two or more molecules, identical or different, unite together with or without elimination of simple units like H_2O, NH_3, etc., the reaction is said to be a condensation reaction. If a condensation process involves large number of molecules, it is called condensation polymerisation or polycondensation. Here we shall discuss some of the condensation reactions.

16.13.1 Aldol condensation

The base- or acid-catalysed condensation reaction between two molecules of carbonyl compounds, identical or different and at least one of them containing α-H atom, which gives rise to a β-hydroxy carbonyl compound (a dimer, called an aldol) or an α, β-unsaturated carbonyl compound is known as aldol condensation reaction.

Usually a dil. aqueous solution of NaOH, KOH, $Ba(OH_2)$, K_2CO_3 and that of HCl are used as base and acid catalysts respectively. NaOEt, potassium-t-butoxide, etc. strong bases are also used as catalysts in aldol condensation.

Most simple aldol may be prepared by condensing two molecules of acetaldehyde in the presence of dil. NaOH solution.

$$CH_3CHO + CH_3CHO \xrightleftharpoons{HO^-} \underset{\text{3-Hydroxybutanal}}{CH_3.CHOH.CH_2CHO}$$

$$CH_3CHO + CH_3CHO \xrightleftharpoons{H^{\oplus}} \underset{\text{2-Butenal}}{CH_3.CH{=}CH.CHO}$$

Here one α-H atom of one carbonyl compound is responsible for the reaction to occur. If there is no α-H atom on both the carbonyl molecules, the aldol condensation reaction does not take place.

$$2C_6H_5CHO \xrightleftharpoons{HO^- \text{ or } H^{\oplus}} \text{no aldol condensation product}$$

Base-catalysed aldol condensation reactions usually form aldols, i.e., β-hydroxycarbonyl compounds. Sometimes α, β-unsaturated derivatives are also formed by dehydration under the conditions of the reactions. On the other hand, acid-catalysed reactions give rise to α, β-unsaturated carbonyl compounds.

When two molecules of the same aldehyde react, the equilibrium lies far over to the right This has been proved by using D_2O in the reaction medium and no RCHD.CHO is found to form. Naturally the reaction is carried out by refluxing the substrate with dil. base or acid.

On the other hand, the equilibrium lies, far over to the left when two molecules of the same ketone undergo aldol condensation reaction. This is because of the electronic and steric effects of the alkyl or aryl groups on the carbonyl C atom. Consequently, very low yield of aldol is obtained. To avoid this, aldol condensation of a ketone is carried out in a soxhlet extractor instead of a simple refluxing arrangement. Here the ketone is refluxed and the condensate drips over the solid base in a separate chamber. The reaction proceeds but to a small extent which is siphoned back to the refluxing chamber when the reaction chamber is full. The b.p. of the ketone being lower than that of the dimer, the former drips back over the base and further amount of it gets converted into the dimer. Repetitions of the process gives a reasonable amount of the expected product. Thus 2 moles of acetone produces 70% diacetone alcohol by the recycling process mentioned above while the first condensate of the reaction contains roughly 2% of the product only.

$$2CH_3COCH_3 \xrightleftharpoons{Ba(OH)_2} CH_3{-}\overset{\displaystyle OH}{\underset{\displaystyle CH_3}{\overset{|}{\underset{|}{C}}}}{-}CH_2.COCH_3 \xrightarrow{\text{recycled, several times}} \text{70\% diactone alcohol}$$

first condensate ($\approx$ 2% of the product)

The generally accepted base-catalysed mechanism of aldol condensation involves three steps :

Step I : Abstraction of $H^{\oplus}$ from the α-C atom of one of the carbonyl molecules to form a resonance-stabilised carbanion :

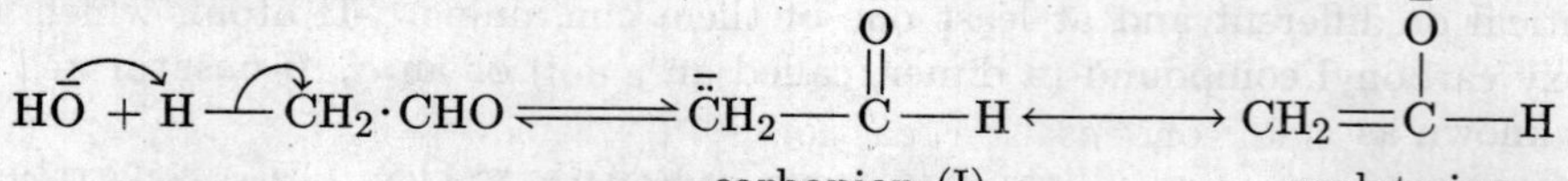

carbanion (I) enolate ion

Step II : Nucleophilic addition of the carbanion (I) to the carbonyl carbon atom of the other carbonyl molecule.

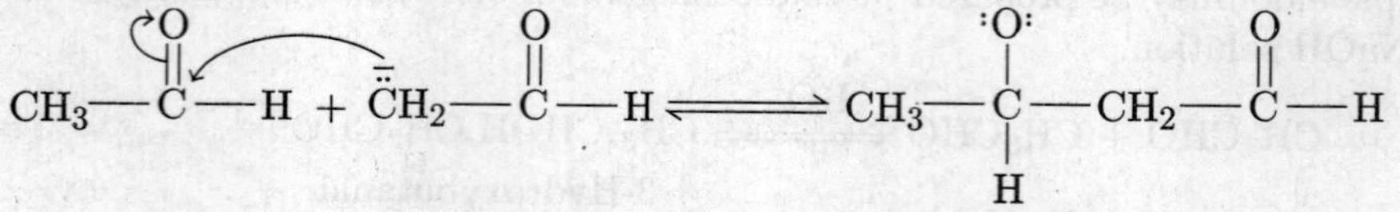

oxide ion (II)

Step III : The oxide ion (II) takes up a proton from water to form the β-hydroxycarbonyl compound.

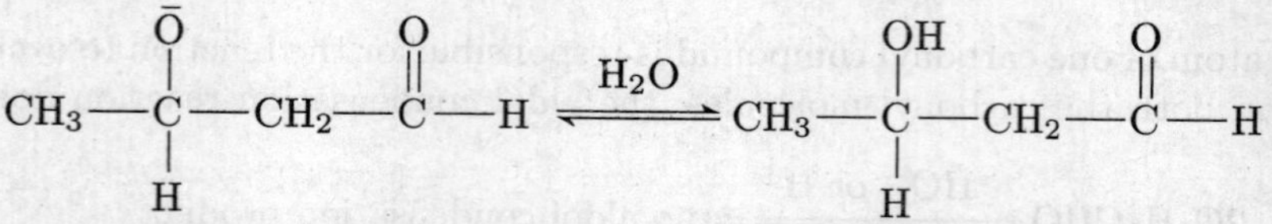

Sometimes dehydration of the aldol takes place in the presence of the base. Thus aldol formed by the condensation of benzaldehyde and acetaldehyde in the presence of 10% NaOH gets dehydrated to cinnamaldehyde :

Step IV : Dehydration of the aldol; the HO^- ion takes up the α-H, and HO^- group leaves the species :

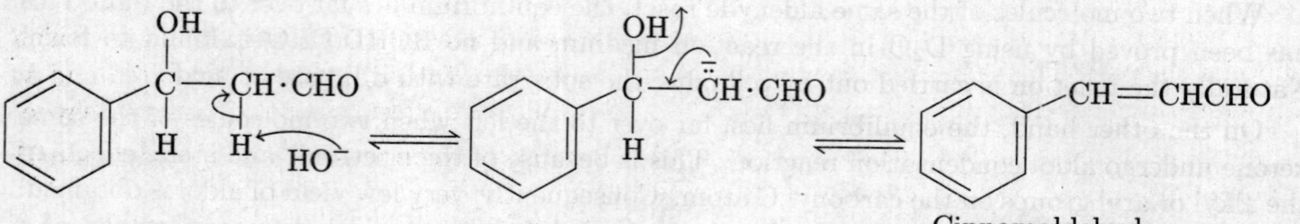

Cinnamaldehyde

The dehydration is possible because : (i) There are active α-Hs on the aldol (the presence of one α-H is required). (ii) HO^- is not a very bad leaving group. (iii) The new C-C double bond forms a conjugated system with the carbonyl group and thus the unsaturated carbonyl compound becomes stable.

The generally assumed acid-catalysed mechanism of the aldol condensation also has four steps :

Step I : One of the carbonyl molecules gets protonated to form the conjugate acid :

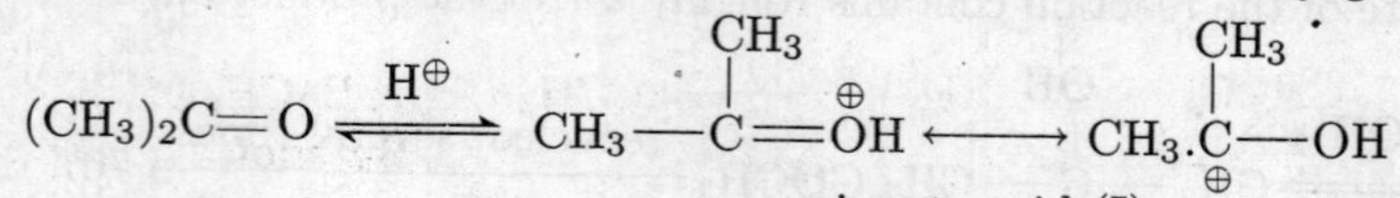

conjugate acid (I)

Step II : Electrophilic addition of the carbenium ion (I), the conjugate acid, to the enol form of the other carbonyl molecule.

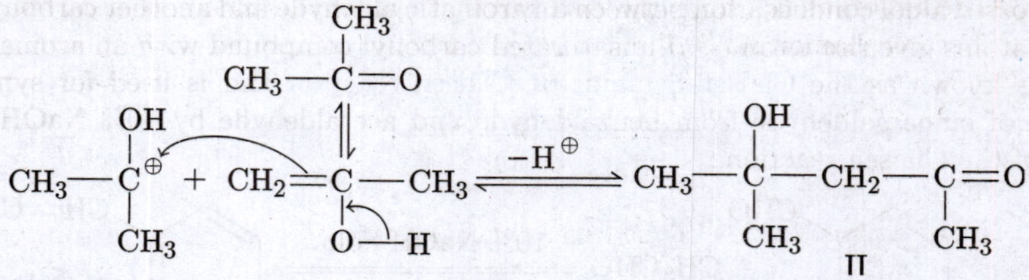

II

Step III : Protonation of the alcoholic OH of the aldol (II) and elimination of H_2O to form the carbenium ion (III).

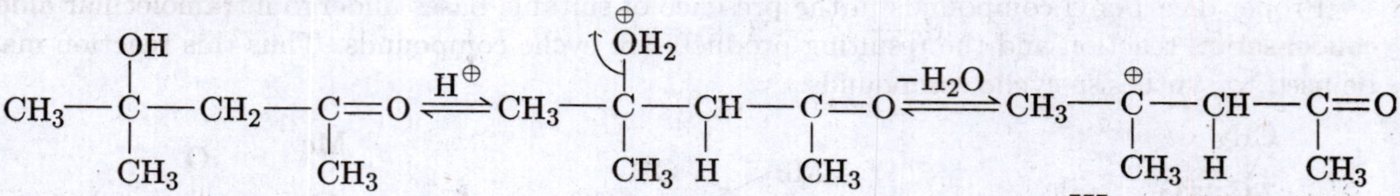

III

Step IV : Loss of $H^{\oplus}$ from the α-C atom of III and formation of the α, β-unsaturated carbonyl compound (IV).

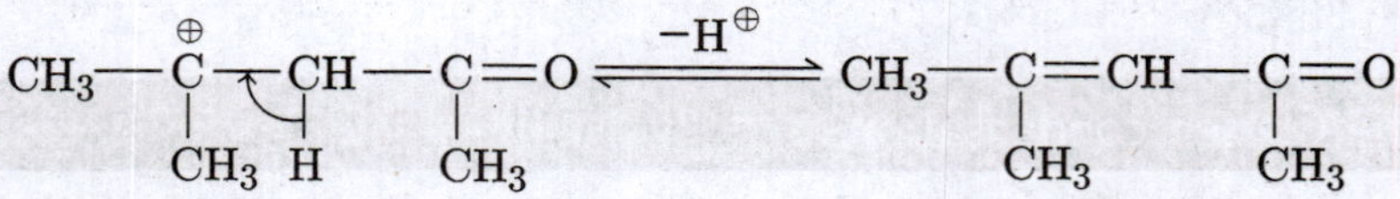

IV

Under the acid condition tertiary alcohol (II) dehydrates easily. For this reason, α, β-unsaturated carbonyl compound is the end product in the acid-catalysed reaction. Nevertheless, all the steps in both the processes are reversible.

When a ketone and an aldehyde undergo the crossed aldol condensation, it is the ketone which supplies the α-H. This is because the carbonyl C of the ketone is sterically more congested and at the same time less electrophilic (owing to the +I and conjugative effect) than that of an aldehyde.

Since an aldol is a carbonyl compound, it may undergo polycondensation if it possesses α-H atoms.

When two different carbonyl compounds undergo aldol condensation, it may be called crossed aldol condensation— four different aldols and four different α, β-unsaturated compounds are formed. Thus the reaction becomes unsuitable for synthesis.

The condensation between two molecules of the same aldehyde gives one aldol and subsequently one α, β-unsaturated carbonyl compound. Therefore, this reaction can safely be used for synthesis.

An α, β-unsaturated dimer can be reduced to a saturated alcohol by catalytic reduction.

$$\underset{\text{2-Butenal}}{CH_3.CH{=}CH.CHO} \xrightarrow{Ni/H_2} \underset{\text{1-Butanol}}{CH_3.CH_2.CH_2CH_2OH}$$

The unsaturated dimer may be converted into the corresponding unsaturated alcohol and acid by the use of specific reducing and oxidising agents respectively :

$$CH_3CH{=}CHCH_2OH \xleftarrow[\text{ether}]{LiAlH_4} CH_3CH{=}CHCHO \xrightarrow[AgNO_3]{\text{ammoniacal}} CH_3CH{=}CHCO_2H$$

The crossed aldol condensation between an aromatic aldehyde and another carbonyl compound with α-H atoms give rise to an α, β-unsaturated carbonyl compound with an aromatic ring; this reaction is known as the Claisen-Schmidt or Claisen reaction and is used for synthesis. The formation of cinnamaldehyde from benzaldehyde and acetaldehyde by 10% NaOH soln. is an example of the Claisen reaction :

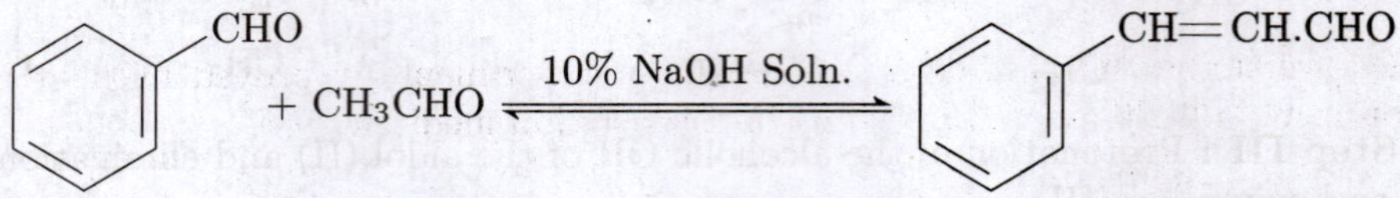

Proper dicarbonyl compounds in the presence of suitable bases undergo **intramolecular aldol condensation** reaction and the resulting products are cyclic compounds. Thus this reaction may be used to synthesise cyclic compounds.

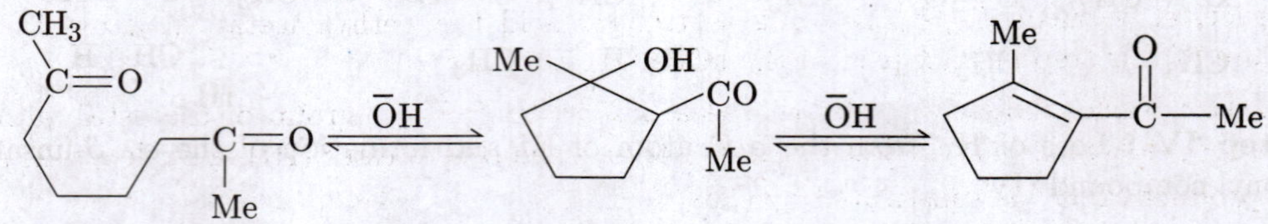

16.13.2 The Claisen condensation

An aldol type condensation reaction involving a nucleophilic attack of a carbanion (generated by abstracting an active α-H usually from ketones, esters, etc. by a strong base, e.g. sodio salt of alkoxide ion, triphenylmethyl anion) on an electron-deficient carbonyl C atom (usually of esters) followed by loss of an anion from the seat of the reaction leading to a nucleophilic substitution is known as the Claisen condensation.

Thus the Claisen condensation between a ketone and an ester gives rise to a 1, 3-diketone.

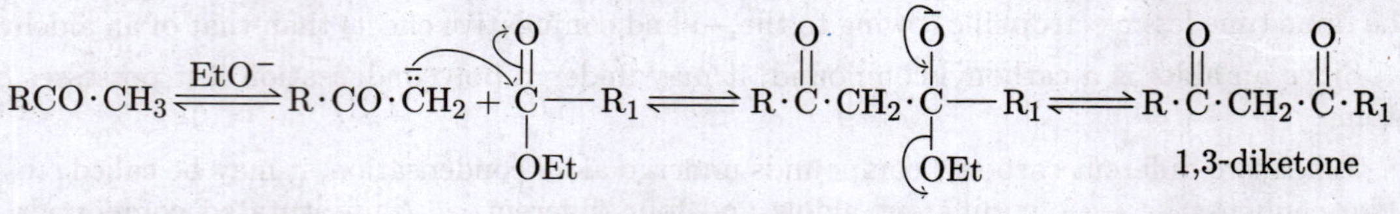

Of course, to most of the chemists, the self-condensation reaction of esters in the presence of a strong base, e.g., $Et\bar{O}Na^{\oplus}$, Ph_3CNa, to give β-ketoesters is the Claisen condensation which may be termed as the Claisen ester condensation.

$$CH_3COOEt + CH_3COOEt \xrightleftharpoons{NaOEt/EtOH} CH_3.\underset{\underset{O}{\|}}{C}CH_2.COOEt + EtO^-$$

In fact, the preparation of ethyl acetoacetate from ethyl acetate and NaOEt is a well studied example of the Claisen condensation, and we shall discuss the reaction here in detail.

To prepare ethyl acetoacetate, two moles of ethyl acetate is refluxed with one mole of Na as wire or in finely divided form and a little amount of EtOH. To isolate the final product, i.e., ethyl acetoacetate, acidification is necessary.

The generally accepted mechanism of the Claisen condensation involves three steps :

Step I : The formation of resonance-stabilised α-carbanion from one ester molecule by abstraction of $H^{\oplus}$ with EtO^-.

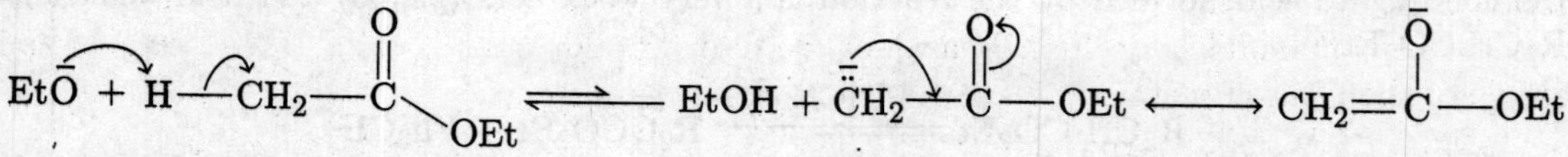

The reversible formation of the carbanion is experimentally proved by deuterium exchange reaction using a little amount of EtOD in the reaction medium.

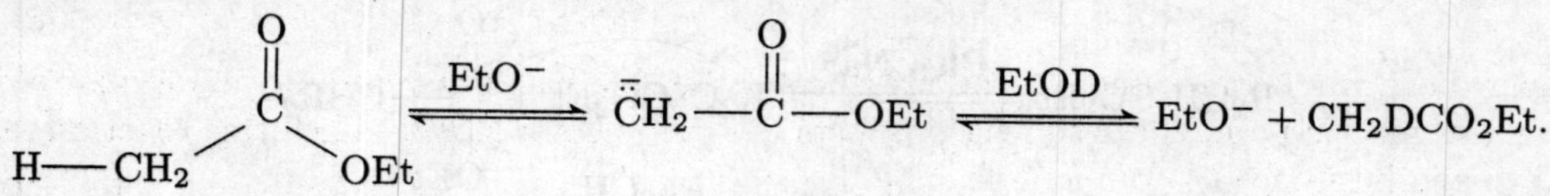

However, ethyl alcohol (pk_a15) being a stronger acid than ethyl acetate ($pk_a \sim 23$), the equilibrium of this step lies far over on the left.

Step II : Nucleophilic attack by the α-carbanion on >C=O group of the ester with the formation of α-alkoxide ion which loses EtO^- to form the β-keto ester (III), a nucleophilic substitution product (EtO^- is substituted by $\bar{C}H_2CO_2Et$).

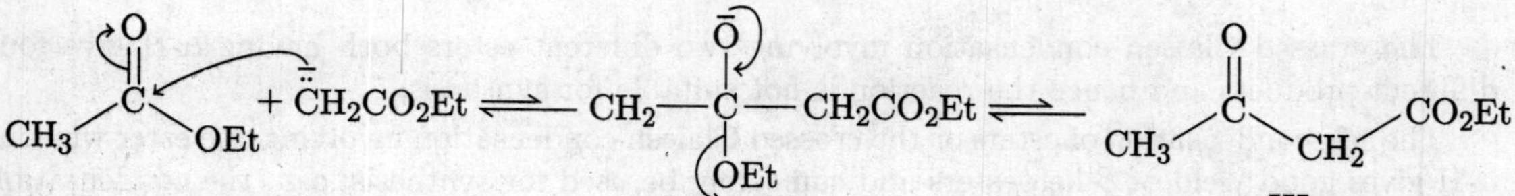

III

Step III : β-ketoester of the type (III) is a stronger acid ($pk_a \sim 11$) than EtOH. The former dissociates to form resonance-stabilised carbanion (IV), the cB of the β-ketoester. This makes the equilibrium of the overall reaction lie on the right by removing EtO^- as EtOH.

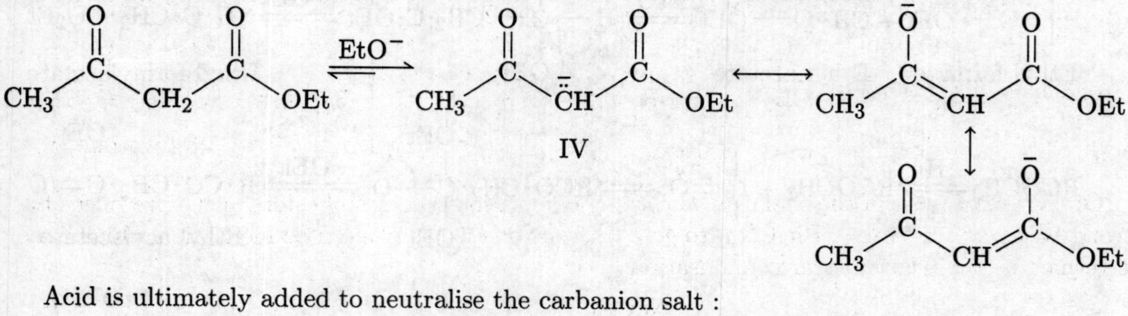

IV

Acid is ultimately added to neutralise the carbanion salt :

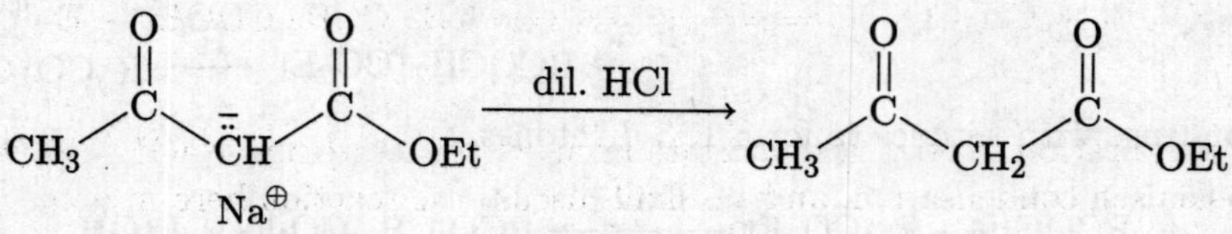

This has been proved by the fact that an ester of the type R_2CHCO_2Et (containing one α-H) does not produce the corresponding β-ketoester ($R_2CHCOCR_2CO_2Et$) in the presence of NaOEt. Since the β-ketoester so formed does not have any α-H on $>CR_2$ part, resonance-stabilised carbanion cannot form and thus the equilibrium lies far over to the left (almost entirely). If, however, a strong base like Ph_3CNa (whose cA, Ph_3CH, a very weak acid) is used instead of EtONa, the equilibrium of step I lies on the right. This is because of the fact that Ph_3CH, the counjugate acid, formed by the reaction is a very weak acid ($pk_a 33$) and it is weaker than $R_2CH.CO_2Et$.

$$R_2CH\cdot CO_2Et \xrightleftharpoons{Ph_3\bar{C}Na^{\oplus}} R_2\bar{\ddot{C}}CO_2Et + Ph_3CH$$

The equilibrium of the 1st stage being on the right, the equilibrium of the overall reaction lies on the right and $R_2CH.CO.CR_2.CO_2Et$ is formed.

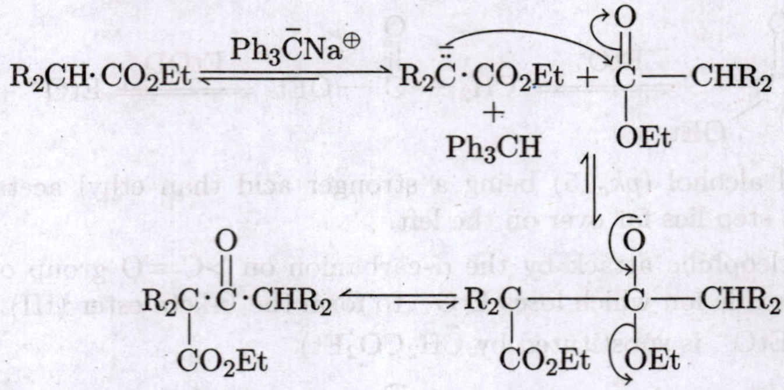

The crossed Claisen condensation involving two different esters both having α-H give four different products and hence the reaction is not suitable for synthesis.

The self-condensation of esters or the crossed Claisen condensation involving one ester without α-H gives good yield of β-ketoesters and can safely be used for synthesis; e.g., the condensation between a ketone and ethyl carbonate or oxalate gives β-ketoaester. The preferred condensing agent is NaH (H^-:).

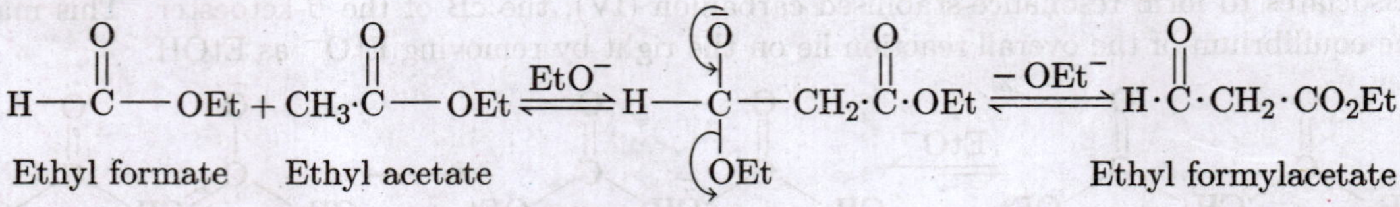

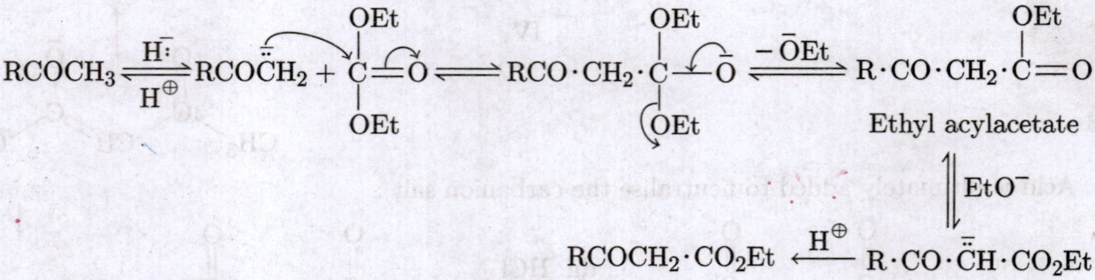

Esters condense with ketones to form 1, 3-diketones.

$$RCOCH_3 + R_1\cdot CO_2Et \xrightleftharpoons{EtO^-} RCO\cdot CH_2\cdot CO\cdot R_1 + EtOH$$

The intramolecular or internal Claisen condensation reactions involving esters of dibasic acids containing α-H atoms and 6 to 8 C atoms give rise to cyclic compounds and the reaction is known as the Dieckmann cyclisation.

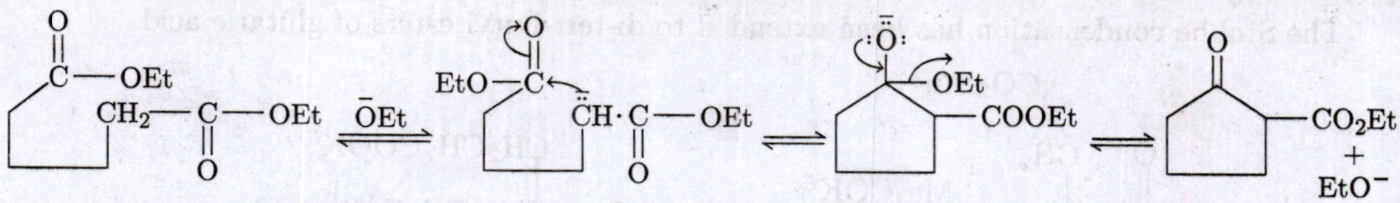

Besides these, β-ketoesters, specially ethyl acetoacetate, have wide and varied synthetic applications. Hence a number of different compounds can be prepared via Claisen condensation.

16.13.3 The Stobbe condensation

An aldehyde or a ketone undergoes condensation reaction with diethyl succinate in the presence of sodium ethoxide or sodium hydride or potassium tert-butoxide to produce the salt of an unsaturated ester-acid and, the reaction is called the Stobbe condensation.

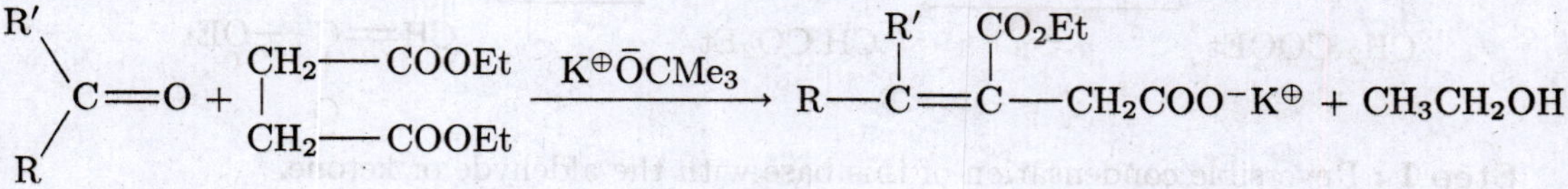

R,R′=H or alkyl or aryl

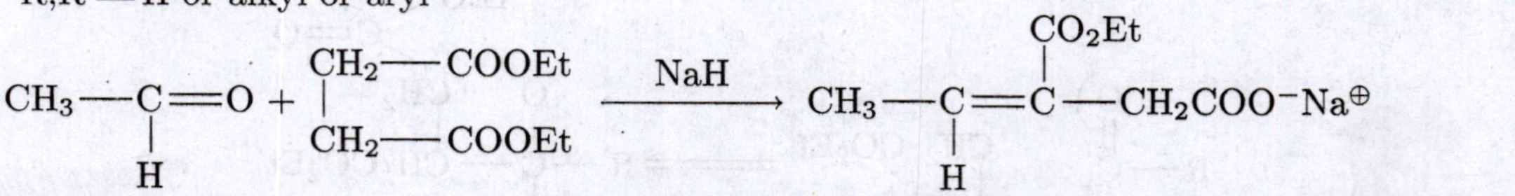

Saturated carboxylic acids may be prepared via the Stobbe condensation. When the salt of the unsaturated ester-acid is heated with hydrobromic acid in acetic acid, a β, γ-unsaturated acid forms; the latter may be hydrogenated to a saturated acid by copper-chromite catalyst, a mixture of Cr_2O_3 and CuO. For example, cyclohexanone can be converted to 3-cyclohexylpropanoic acid by the Stobbe condensation followed by hydrolysis, decarboxylation and hydrogenation steps.

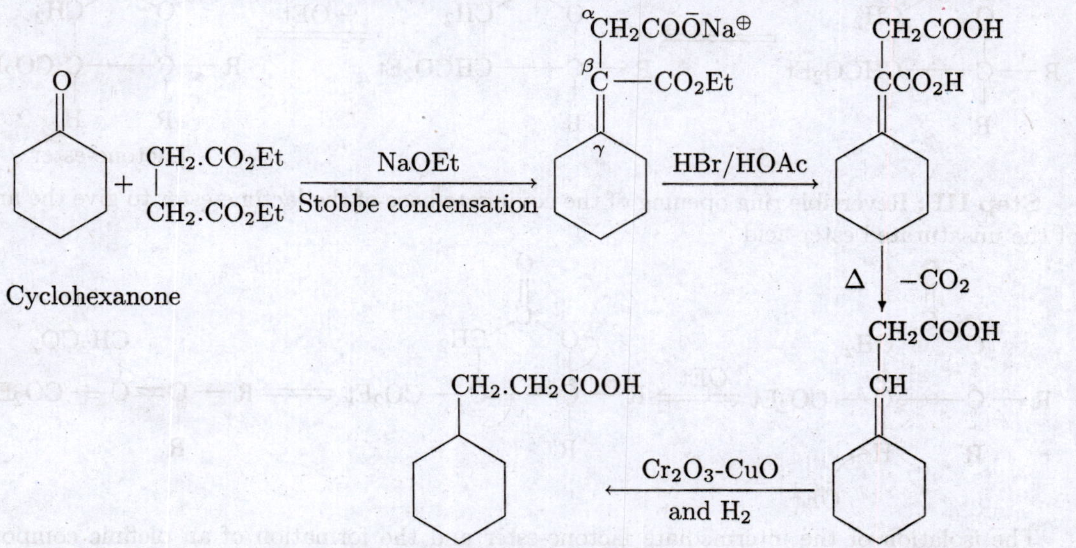

Cyclohexanone

3-Cyclohexylpropanoic acid

The Stobbe condensation has been extended to di-tert-butyl esters of glutaric acid.

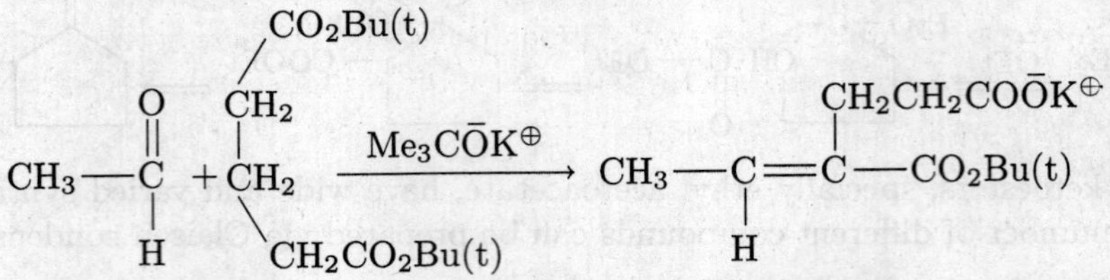

The mechanism of the reaction involves the following steps :

Initial Step : Formation of the conjugate base of ethyl succinate, a resonance-stabilised carbanion.

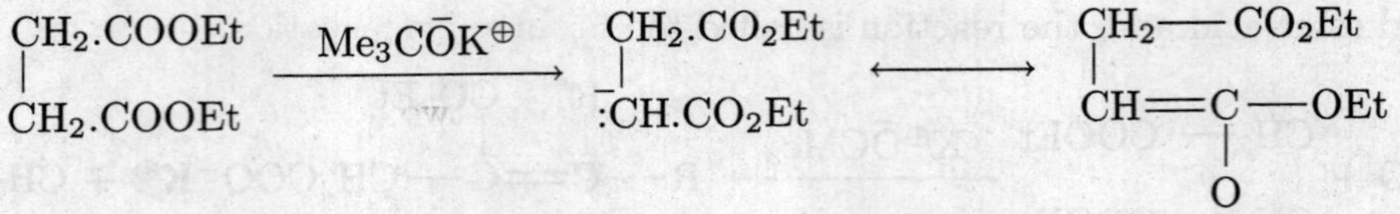

Step I : Reversible condensation of this base with the aldehyde or ketone.

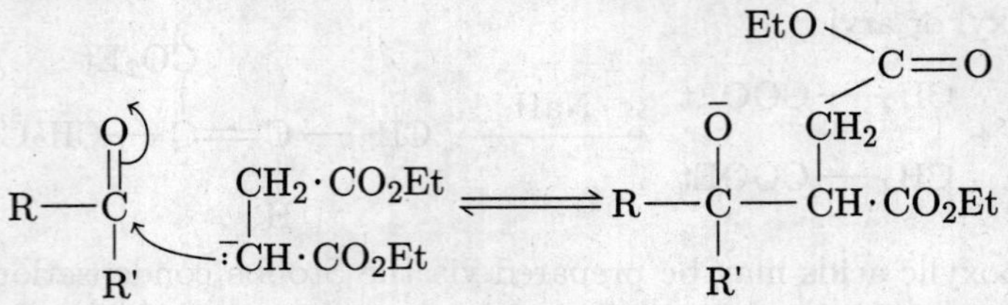

Step II : Reversible formation of a lactone-ester by the intramolecular cyclisation.

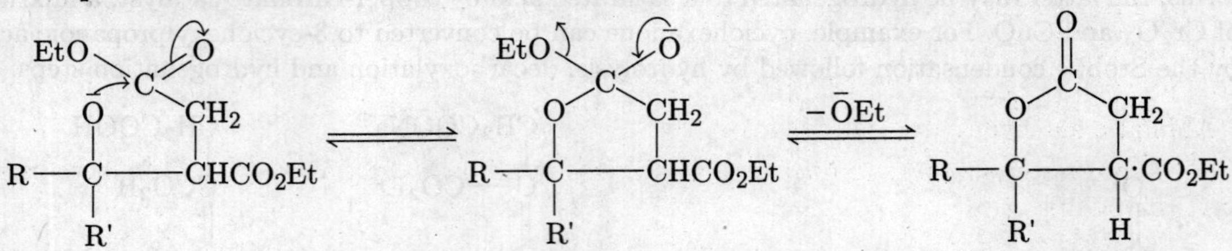

Lactone-ester

Step III : Reversible ring opening of the conjugate base of the lactone-ester to give the anion of the unsaturated ester-acid.

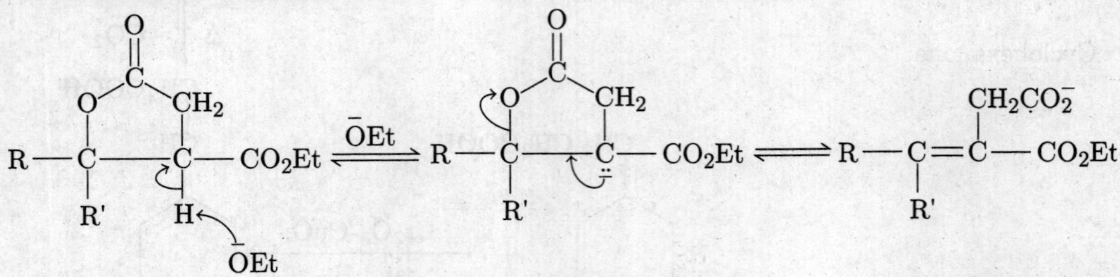

The isolation of the intermediate lactone-ester and the formation of an olefinic compound instead of an alcohol support the mechanism of the Stobbe condensation as written above.

16.13.4 The Darzen glycidic ester condensation

The Darzen glycidic ester condensation is an aldol type reaction between a carbonyl compound and an α-haloester containing α-H atoms in the presence of a strong base (NaOEt, $NaNH_2$) followed by an internal nucleophilic substitution which leads to the formation of a glycidic ester (an α, β-epoxyester).

When ethyl chloroacetate condenses with acetone in the presence of $NaNH_2$, an α, β-epoxyester is formed.

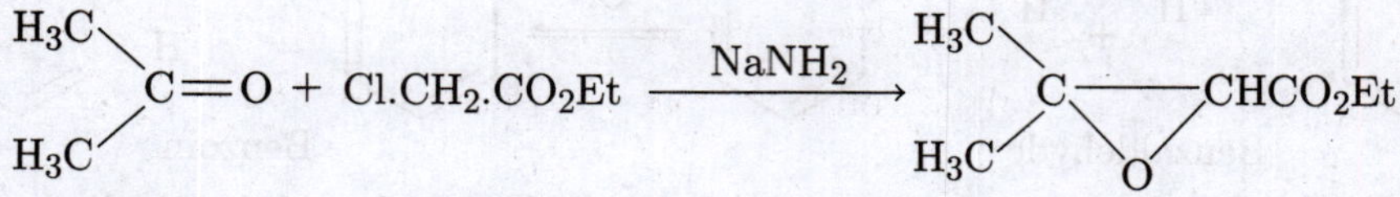

Acetone Ethyl α-Chloroacetate Glycidic ester

The generally accepted mechanism of the reaction involves two steps :

Step I : Formation of a resonance-stabilised α-carbanion from the α-haloester by the abstruction of an α-H as $H^{\oplus}$ by a strong base :

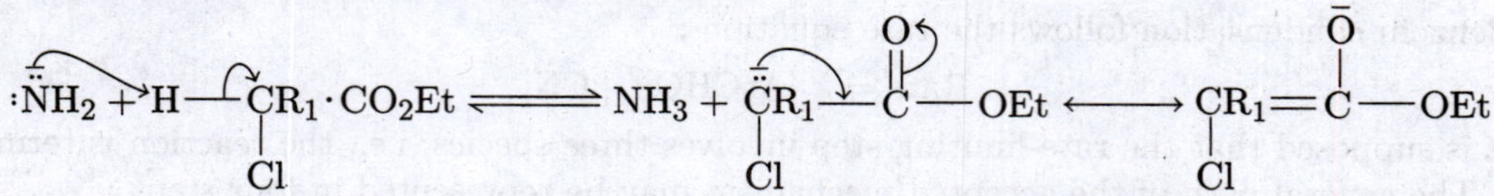

Step II : Nucleophilic attack by the carbanion on the electron-deficient carbonyl C atom, followed by internal S_N2 reaction with the loss of halide ion :

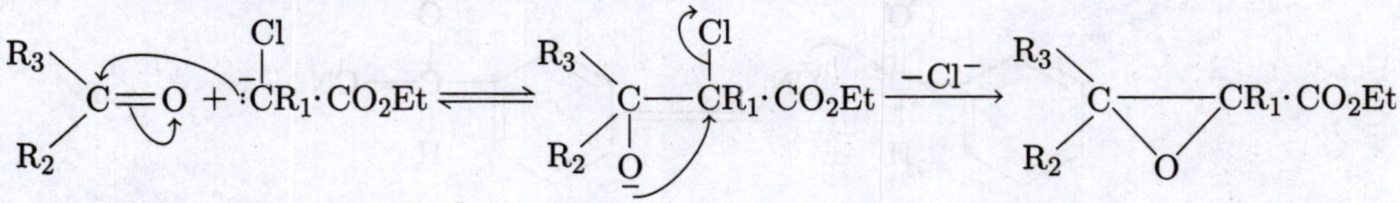

Aliphatic aldehydes give poor yields.

α, β-epoxyesters upon saponification, acidification and decarboxylation by heating give carbonyl compounds.

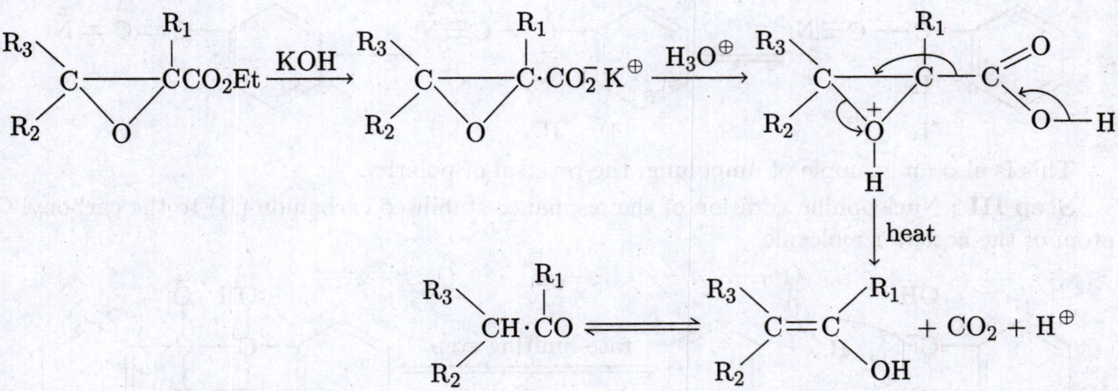

The end product depends on the nature of the α-haloester. If it is α-haloacetate, the product will be an aldehyde. If it is a substituted α-haloacetate, one gets a ketone.

16.13.5 Benzoin condensation

The self-condensation reaction of an aromatic aldehyde specifically catalysed by cyanide ion is usually referred to as benzoin condensation. An α-hydroxyketone, a dimer known as benzoin, is formed by this reaction. Thus benzaldehyde on benzoin condensation gives benzoin (1) :

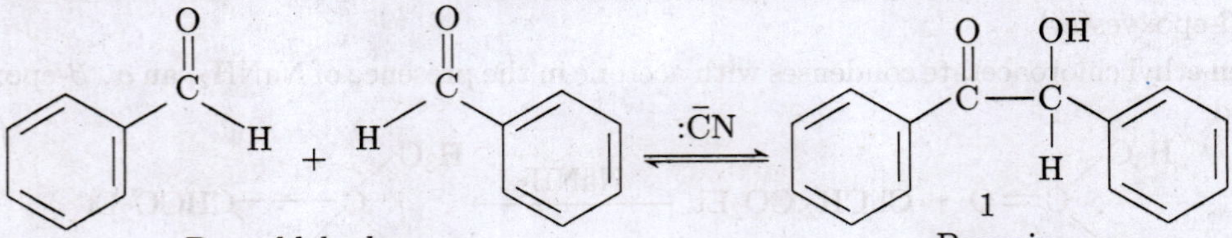

Benzaldehyde Benzoin

The condensation is carried out by boiling the alcoholic solution of the aldehyde containing a small amount of NaCN or KCN.

Here two molecules of the aldehyde react differently. One of them donates the aldehydic H atom to the other which accepts it, and consequently the former is known as a donor and the latter called an acceptor.

Benzoin condensation follows the rate equation :

$$\text{Rate} = k\ [\text{ArCHO}]^2\ [:\bar{\text{C}}\text{N}]$$

It is supposed that the rate-limiting step involves three species, i.e., the reaction is termolecular. The general path of the accepted mechanism may be represented in four steps :

Step I : Nucleophilic addition of the cyanide ion to the carbonyl carbon atom of the donor molecule :

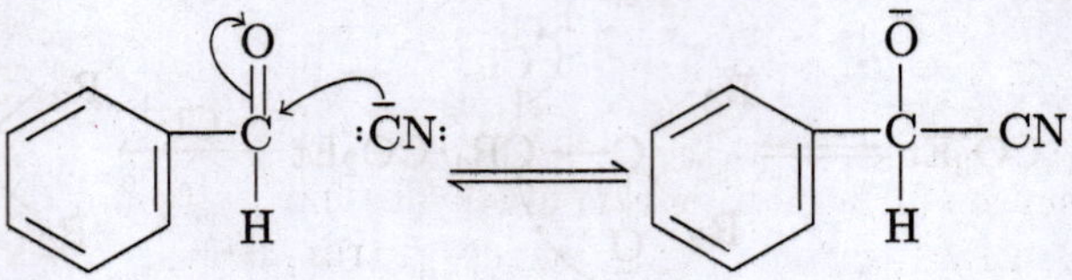

donor I

Step II : Migration of the aldehydic H of (I) to its anionic O atom to form a carbanion.

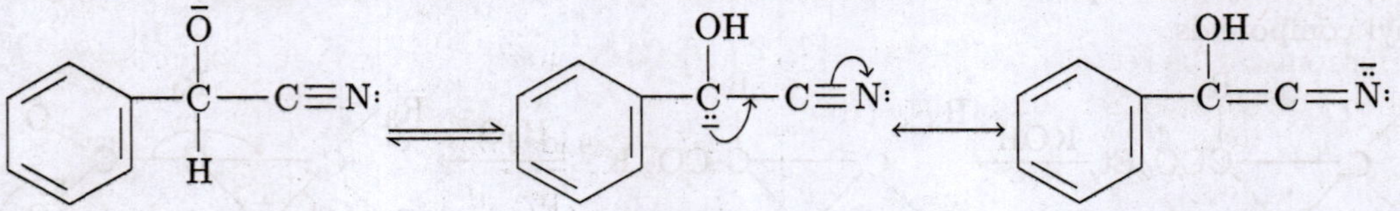

I II

This is also an example of umpolung, the reversal of polarity.

Step III : Nucleophilic addition of the resonance-stabilised carbanion (II) to the carbonyl C atom of the acceptor molecule.

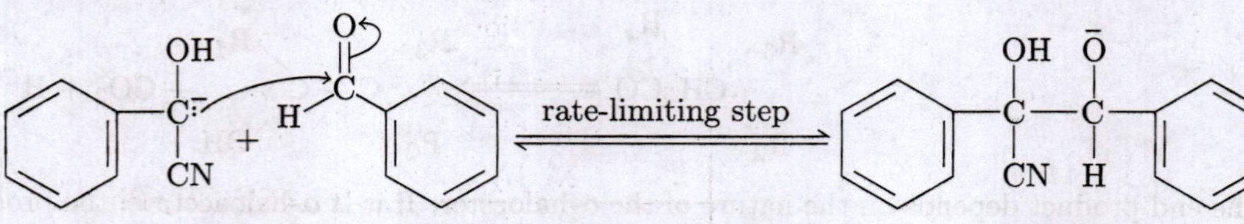

acceptor III

Step IV : Rapid proton transfer and rapid loss of :$\bar{C}N$ ion occur to form benzoin :

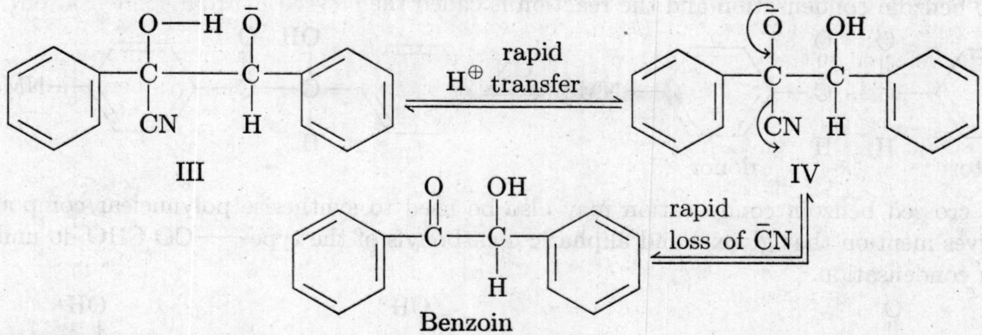

Thus $\bar{C}N$ being sufficiently nucleophilic, highly electron-withdrawing and good leaving group, it acts as a specific catalyst in the benzoin condensation.

The reaction is a complete reversible process. This has been establised by the fact that when a benzoin of an aldehyde is heated with another aldehyde in ethanolic KCN solution, a new benzoin and the aldehyde of the original benzoin are obtained.

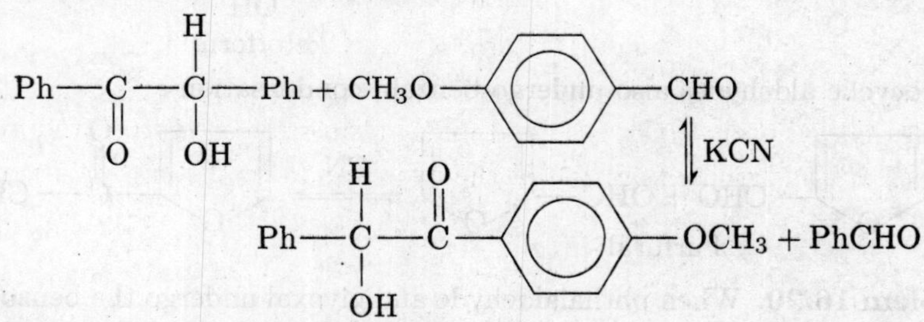

The reaction is unsuccessful with aliphatic aldehydes having α-H atom. Perhaps the enolisation of aldehyde is responsible for this. The presence of strongly electron-donating or withdrawing groups at the para positions make the reactions unsuccessful (See Q. 19 of the exercise).

The reaction fails when strongly electron-withdrawing groups are present at ortho or para to CHO group.

Benzoin can be reduced and oxidized to stable compounds and thus the condensation may be used to synthesise other compounds.

PhCO·CHOH·Ph →

- HNO_3 oxidation or, $CuSO_4$/pyridine → Ph·CO·CO·Ph (Benzil)
- Sn/Hg & HCl reduction → Ph·CO·CH_2·Ph (Deoxybenzoin)
- Na–Hg & H_2O reduction → PhCHOH·CHOH·Ph (Hydrobenzoin)
- Zn–Hg/HCl in EtOH reduction → Ph·CH=CH·Ph (Stilbene)
- Ni/H_2 reduction → Ph·CH_2·CH_2·Ph (Dibenzyl)

Instead of self-condensation, two different aldehydes, in which one can act as donor only, undergo benzoin condensation and the reaction is called the crossed benzoin condensation.

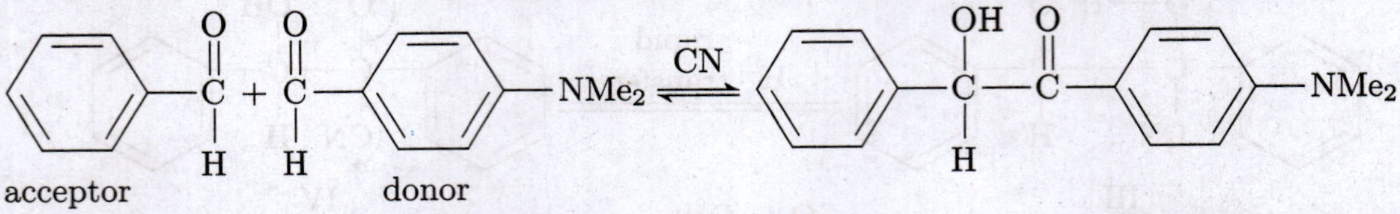

The crossed benzoin condensation may also be used to synthesise polynuclear compounds. It deserves mention that glyoxal and aliphatic dicarbonyls of the type —CO·CHO do undergo benzoin condensation.

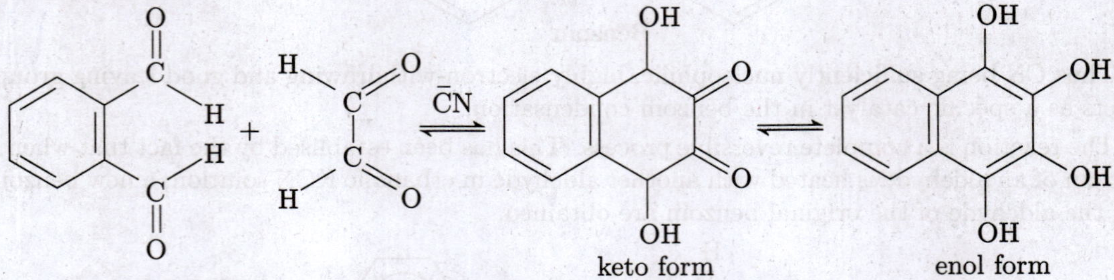

Heterocyclic aldehydes also undergo benzoin condensation :

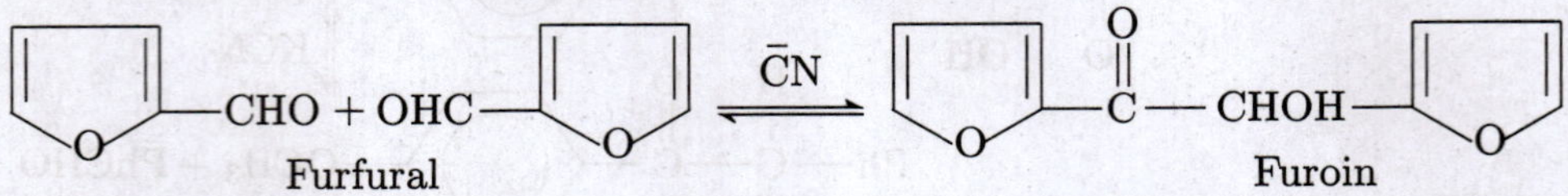

Problem 16.20. When phthalaldehyde and glyoxal undergo the benzoin condensation in the presence of $\bar{C}N$ ion, 1, 2, 3, 4-tetrahydroxynaphthalene is produced—give mechanistic steps.

Solution :

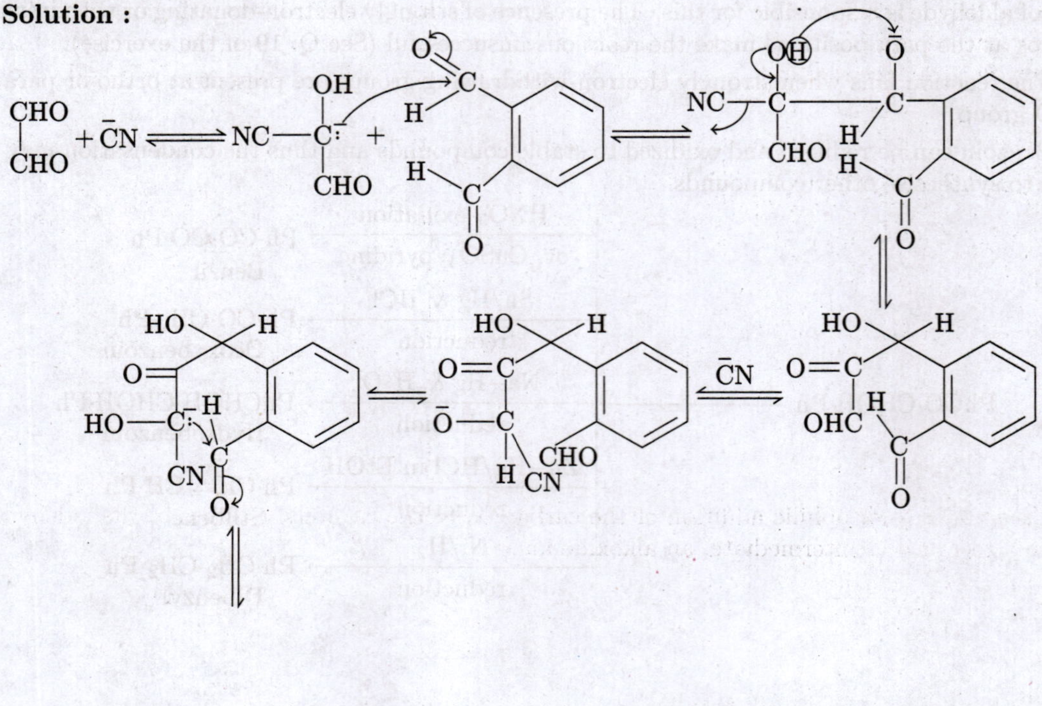

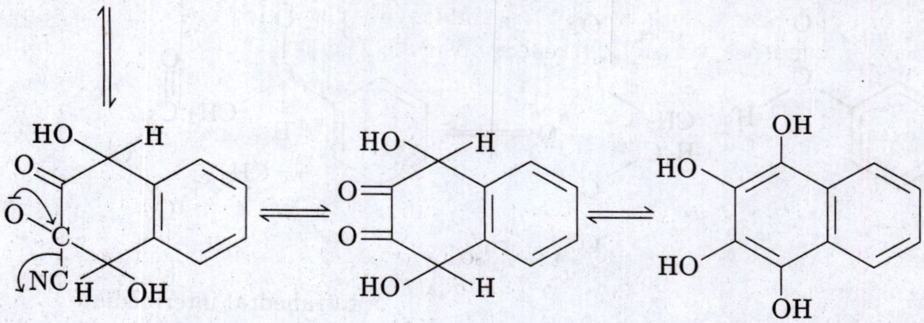

16.13.6 The Perkin reaction (condensation)

The aldol type condensation of an aromatic aldehyde with an aliphatic acid anhydride containing two α-H atoms in the presence of its carboxylate ion, which acts as a base-catalyst, to produce β-arylacrylic acid is said to be the Perkin reaction or condensation. Thus benzaldehyde condenses with acetic anhydride containing two α-H atoms in the presence of sodium acetate to give cinnamic acid.

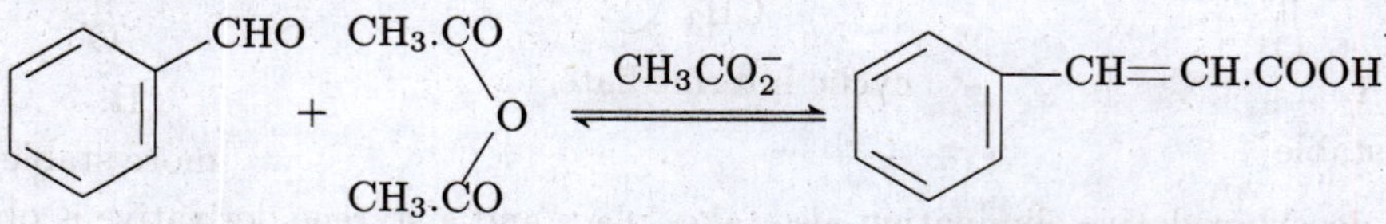

β-Phenylacrylic acid
[Cinnamic acid]

Sometimes instead of carboxylate ion, other bases like triethylamine give better result.

The reaction is carried out by heating the aldehyde with an excess of acid anhydride (which acts as the solvent) containing corresponding carboxylate salt at 180 °C. Usually dehydration takes place under the conditions of the reaction and a mixed anhydride results. The unreacted aldehyde is removed by distillation in steam and the β-arylacrylic acid is obtained by hydrolysis of the mixed anhydride with dil. HCl.

The generally accepted mechanism of the Perkin reaction may be represented by the following steps.

Step I : Abstraction of the proton by the carboxylate ion to form the resonance-stabilised carbanion.

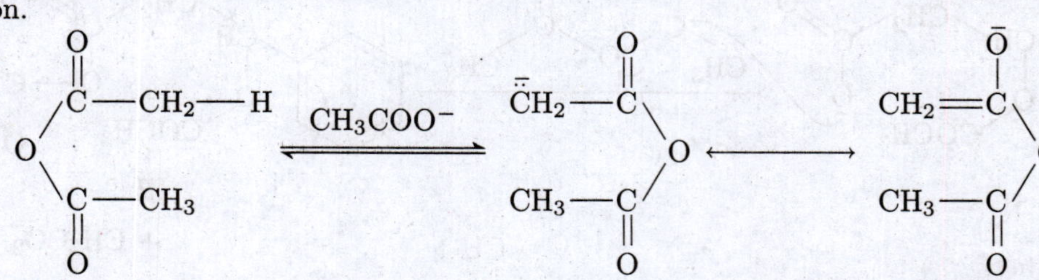

Step II : Nucleophilic addition of the carbanion to the carbonyl C atom of the aldehyde to form a tetrahedral intermediate, an alkoxide ion.

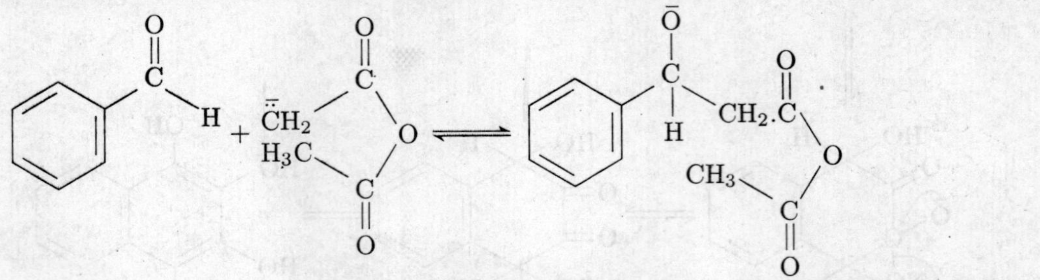

a tetrahedral intermediate

Step III : Internal transfer of the acetyl group from the carboxyl O atom to the alkoxy O atom via a cyclic intermediate.

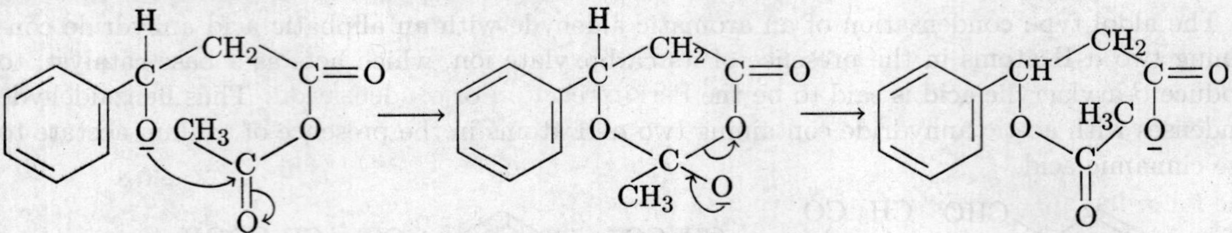

I — less stable; cyclic intermediate; II — more stable

Sometimes, decarboxylative elimination also takes place and a styrene derivative is obtained.

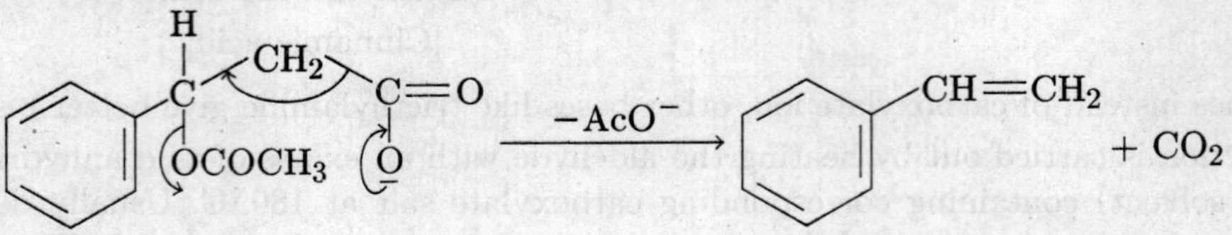

Styrene

Since COO^- group is resonance stabilised, II is more stable than I.

Step IV : II undergoes acetylation by acetic anhydride. A small amount of styrene is also formed from II by decarboxylative elimination as shown above.

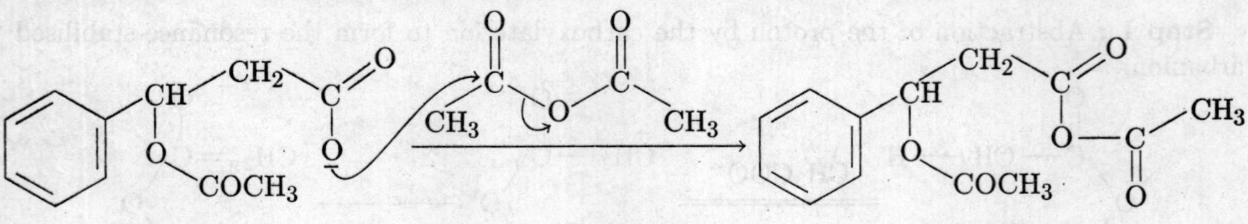

III

$+ CH_3CO_2^-$

Step V : III undergoes β-elimination reaction, $MeCO_2^-$ is the base which abstracts the H and $MeCO_2^-$ is the leaving group.

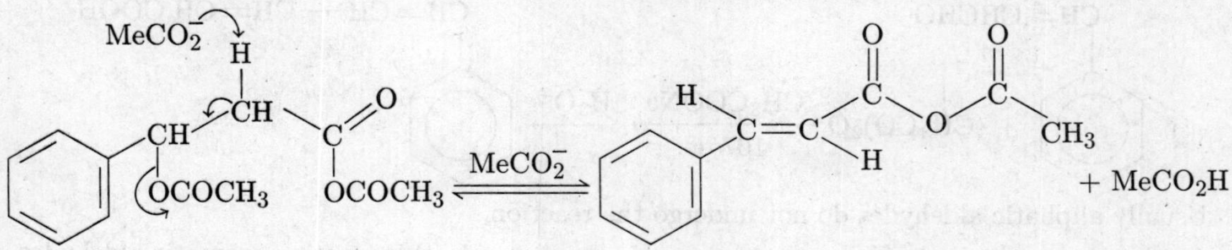

III IV

Step VI : Hydrolysis of IV by the dilute acid to give β-phenylacrylic acid.

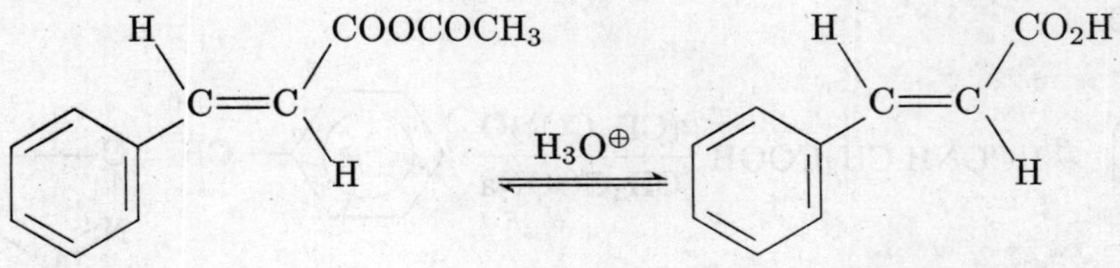

55-60% yield

The following are the evidences in favour of the mechanism :

(i) Acetic anhydride condenses with benzaldehyde in the presence of pyridine as base to yield cinnamic acid after hydrolysis; whereas sodium acetate does not condense with benzaldehyde in the presence of pyridine. This supports the formation of carbanion from the anhydride and rules out the possibility of the formation of the same from sodium acetate.

$$PhCHO + (CH_3CO)_2O \xrightarrow[2.\ H_3O^{\oplus}]{1.\ pyridine} PhCH{=}CH.COOH$$

$$PhCHO + CH_3CO\bar{O}Na^{\oplus} \xrightarrow[2.\ H_3O^{\oplus}]{1.\ pyridine} PhCH{=}CH.COOH$$

does not from

(ii) When the acid anhydride contains only one α-H atom, a β-hydroxy-acid is formed instead of α, β-unsaturated acid because of the unavailability of the active H on the α-C.

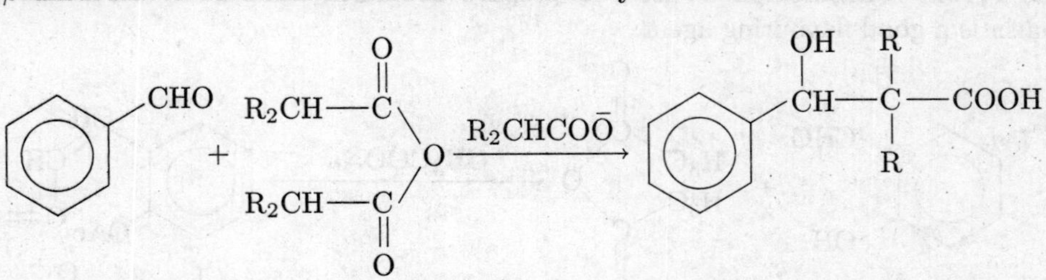

Owing to the steric effect, *o*-substituted benzaldehyde gives low yield of the product. For example, *o*-methylbenzaldehyde gives only 15% *o*-methylcinnamic acid. As regarding the effects of the substitutents on the *p*-position the rate order decreases along the series : NO_2, Cl, H, OMe, alkyl.

However, the reaction occurs with the vinylogs of aromatic aldehydes also.

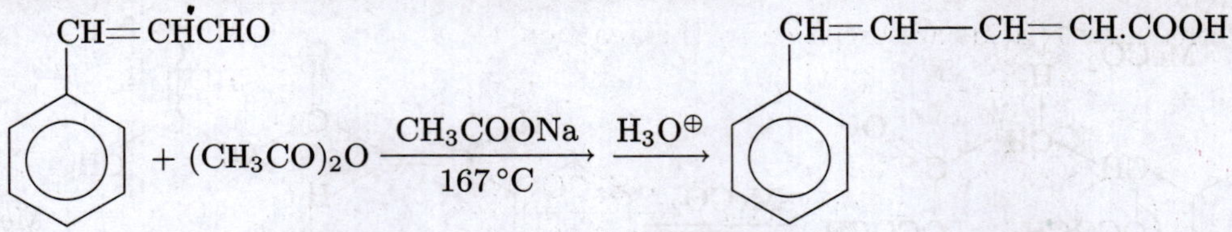

Usually aliphatic aldehydes do not undergo the reaction.

A related reaction is the Erlenmeyer azlactone synthesis. In this reaction aromatic aldehydes are condensed with N-acylglycine in the presence of sodium acetate and acetic anhydride to yield an azlactone.

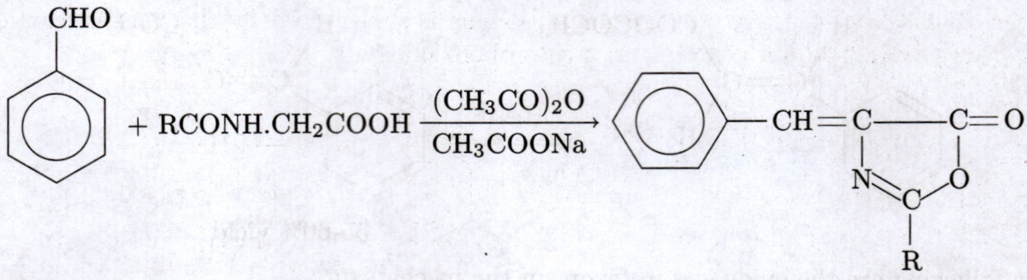

an azlactone

Sometimes, decarboxylative elimination also takes place and a styrene forms as a side product.

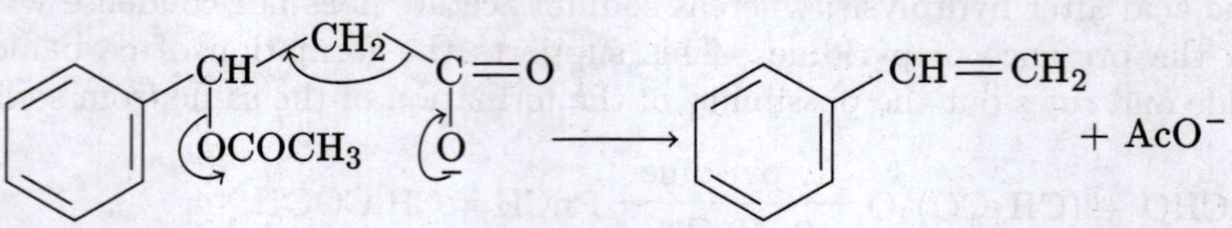

By the Perkin condensation aryl-substituted acrylic acid is prepared and through catalytic hydrogenation these can be converted to β-aryl derivatives of saturated carboxylic acids which may then be converted to different other compounds.

$$\text{Ar.CH{=}CH.CO}_2\text{H} \xrightarrow{\text{H}_2/\text{Ni}} \text{Ar.CH}_2\text{CH}_2\text{CO}_2\text{H} \xrightarrow{\text{ether, LiAlH}_4} \text{Ar CH}_2\text{CH}_2\text{CH}_2\text{OH}$$

The Perkin condensation is used to prepare coumarin industrially from salicylaldehyde. Coumarin is a good flavouring agent.

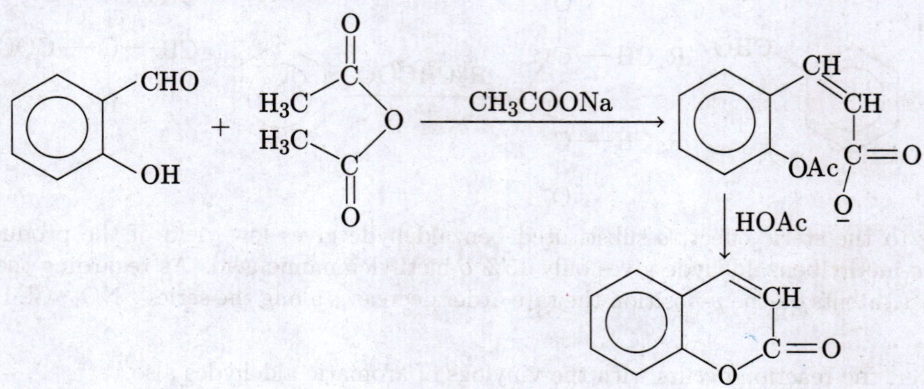

Coumarin

Phthalylacetic acids are prepared by this reaction.

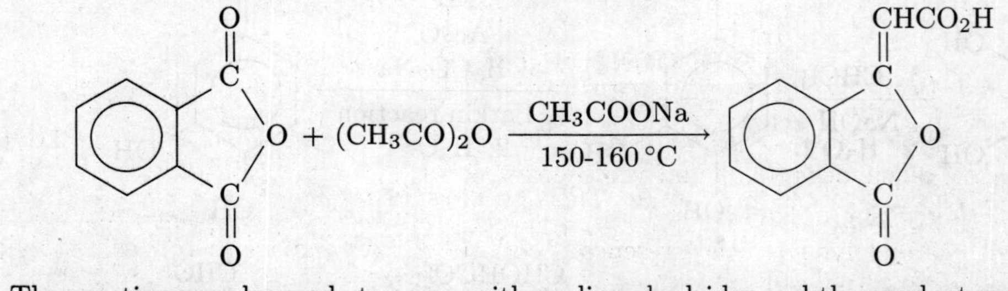

The reaction may be made to occur with cyclic anhydrides and the product may be converted to other derivatives. Thus 1, 5-dichloronaphthalene is synthesised by the Perkin reaction using succinic anhydride, sodium acetate and *o*-chlorobenzaldehyde.

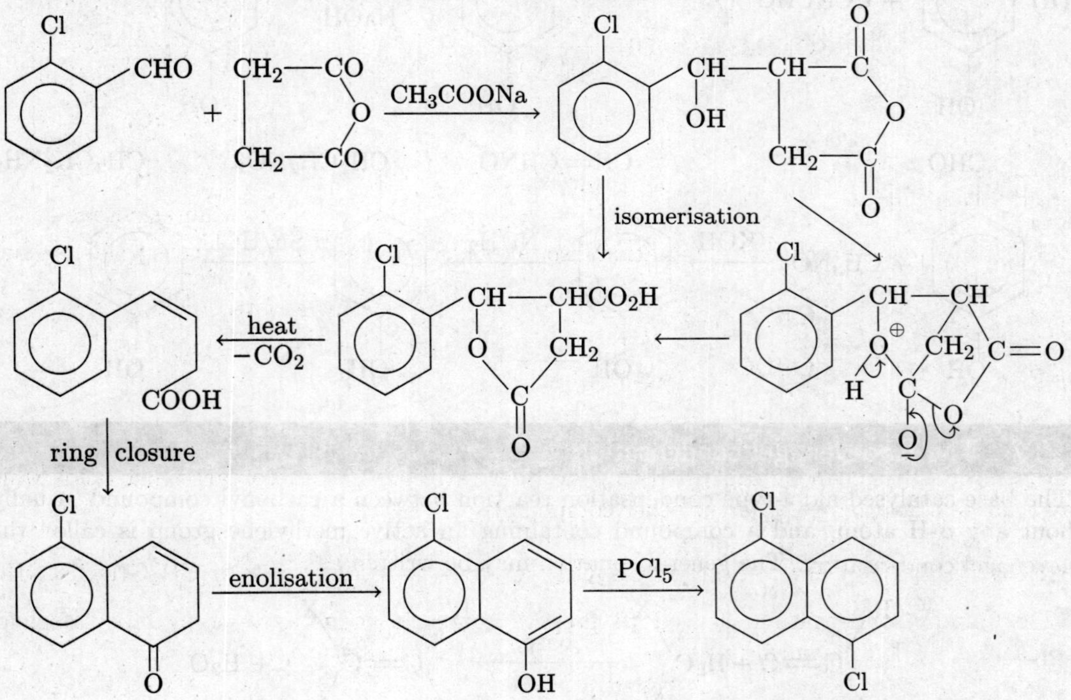

1, 5-Dichloronapthalene

Problem 16.21. Synthesise the following :

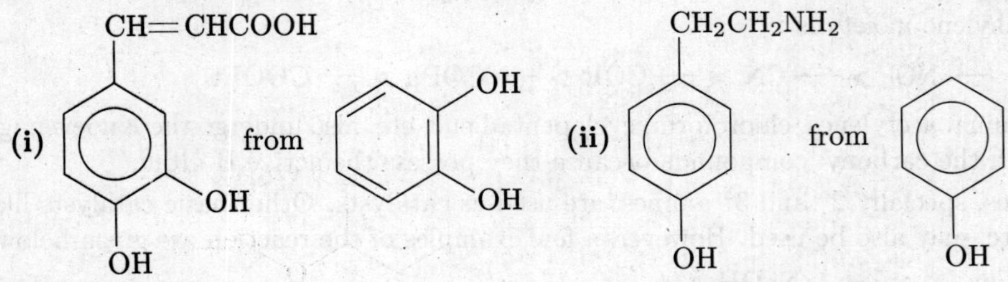

Caffeic acid Tyramine

Solution :

(i) 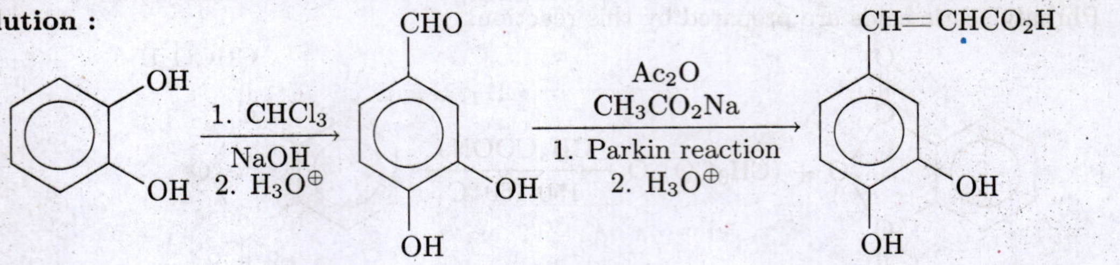

(ii) 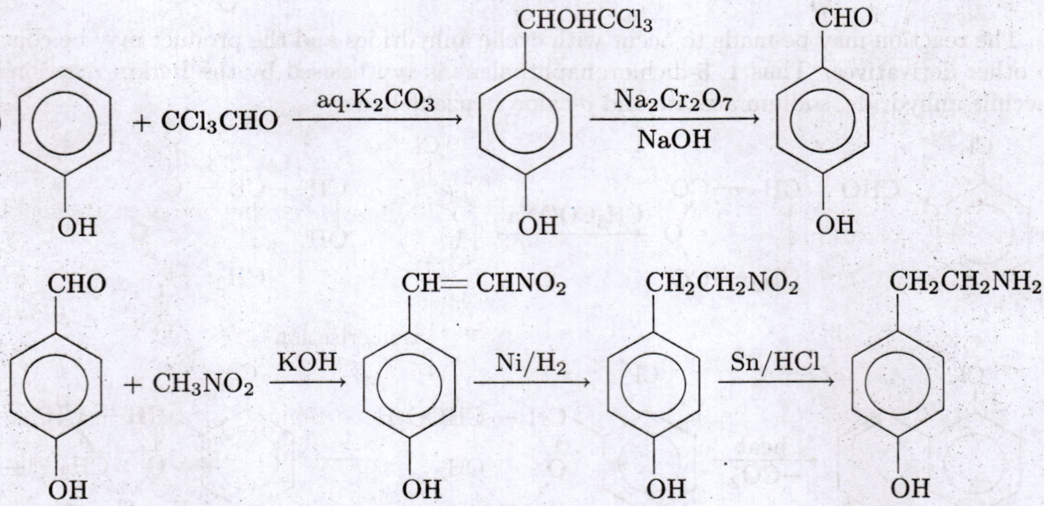

16.13.7 The Knoevenagel condensation

The base-catalysed aldol-type condensation reaction between a carbonyl compound, usually without any α-H atom, and a compound containing an active methylene group is called the Knoevenagel condensation. The general equation may be written as :

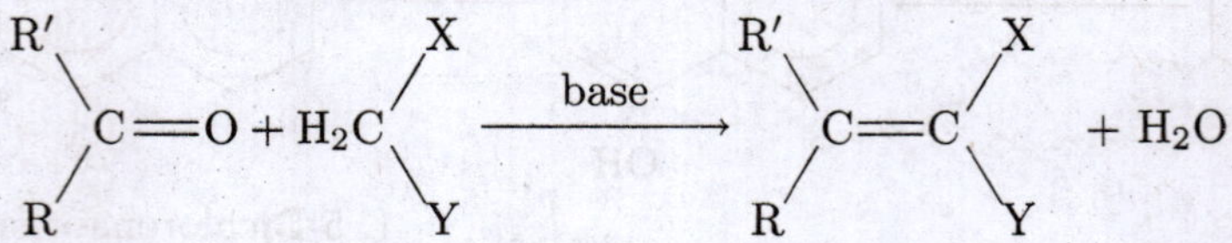

Here R′=H, R or Ar; and X and Y are the two electron-withdrawing groups like —CHO, —COR, —COOR, —CN, —NO_2, —SOR, —SO_2R, etc. The order of the activating effect on the methylene moiety is :

$$—NO_2 > —CN > —COR > —COPh > —COOEt$$

However, terminal acetylenes, chloroform, cyclopentadiene, etc. also undergo the Knoevenagel condensation with the carbonyl compounds because they possess the active H atom.

Usually amines, specially 2° and 3° amines, are used as catalysts. Other basic catalysts like, $\bar{O}R$, $\bar{N}H_2$, $\bar{O}H$, etc. may also be used. However, a few examples of the reaction are given below :

$$PhCHO + CH_3NO_2 \xrightarrow{NaOH} PhCH{=}CHNO_2$$

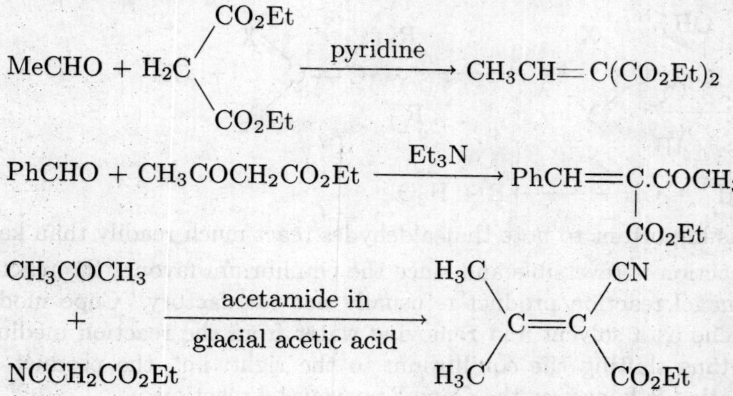

The stepwise mechanism of the reaction is given below :

Step I : Reversible abstraction of a proton by the base from the active methylene group : this develops a resonance-stabilised carbanion.

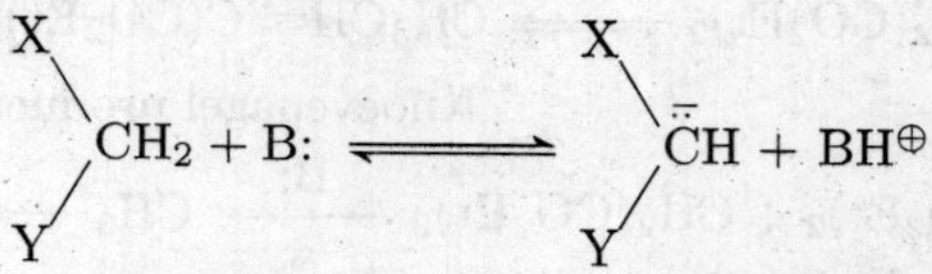

Step II : Nucleophilic attack on the electron-deficient carbonyl C atom by the carbanion.

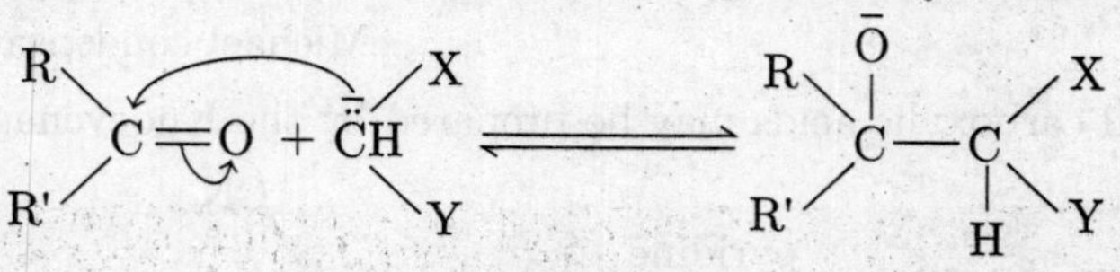

I

Step III : $BH^{\oplus}$ protonates the anionic O atom of I.

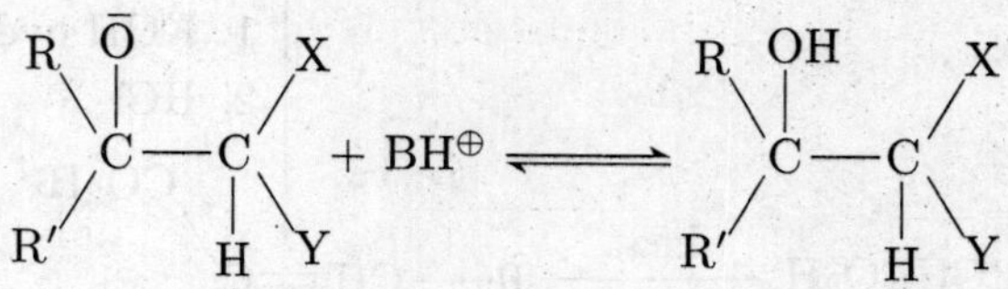

II

Step IV : X and Y being electron-withdrawing groups, H atom of the —CHXY moiety in II is active. The base, abstracts this H and forms a new carbanion III (reversibly) which gets stability through resonance.

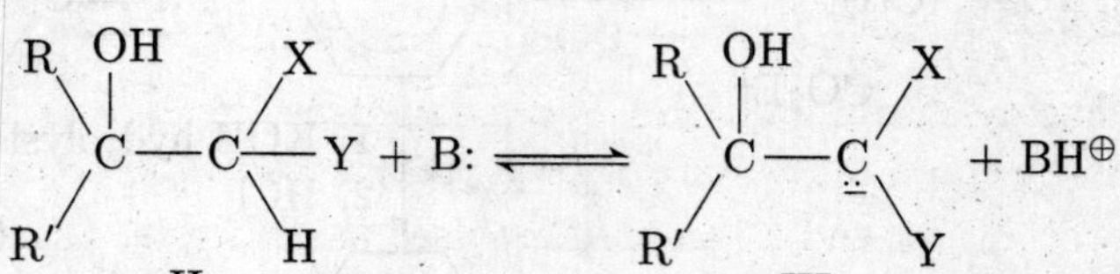

II III

Step V : The carbanion III then gives up the alcoholic OH group as a hydroxide ion. The conversion of III to IV is a β-elimination reaction and follows the ElcB pathway.

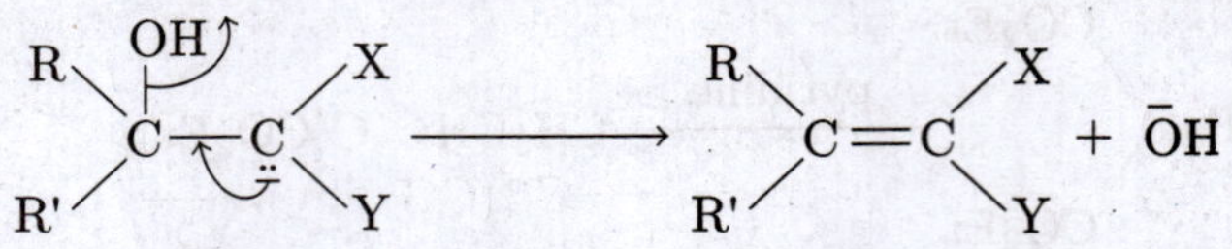

III IV

$$BH^{\oplus} + \bar{O}H \longrightarrow B + H_2O$$

In this connection it is important to note that aldehydes react much readily than ketones.

However, since the reaction is reversible and since the equilibrium favours the reactant side, the yield of the Knoevenagel reaction product is usually not satisfactory. Cope modified the procedure by using benzene as a solvent and removing water from the reaction medium as an azeotropic mixture and thus shifting the equilibrium to the right, i.e., the product side and consequently the modification is known as the Cope-Knoevenagel reaction.

In some cases the Knoevenagel condensation products undergo the Michael addition with the active methylene compound.

$$CH_3CHO + CH_2(CO_2Et)_2 \xrightleftharpoons{B:} \underset{\text{Knoevenagel product}}{CH_3CH{=}C(CO_2Et)_2}$$

$$\underset{\text{Knoevenagel product}}{CH_3CH{=}C(CO_2Et)_2} + CH_2(CO_2Et)_2 \xrightarrow{B:} \underset{\text{Michael condensation product}}{CH_3{-}CH(CH(CO_2Et)_2){-}CH(CO_2Et)_2}$$

α, β-Unsaturated carboxylic acids may be prepared by the Knoevenagel reaction.

1. 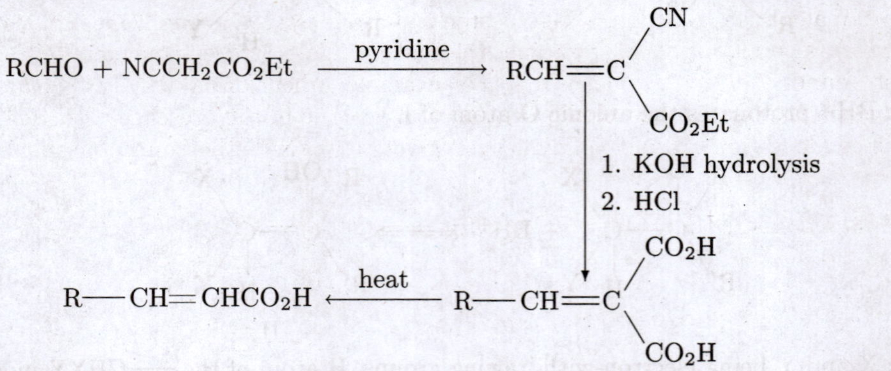

2. 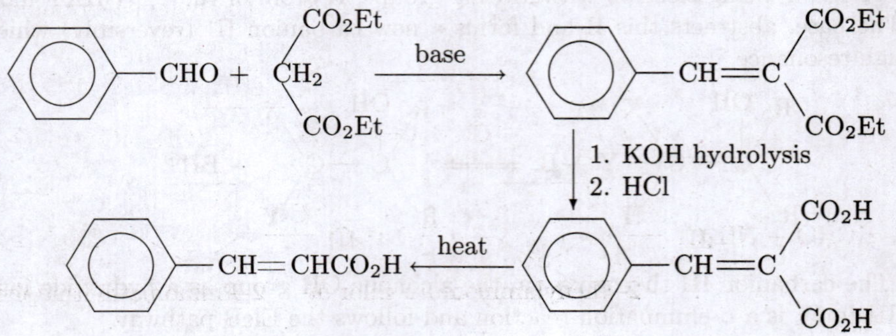

Fumaric acid, a dibasic acid, may be prepared by the Knoevenagel reaction.

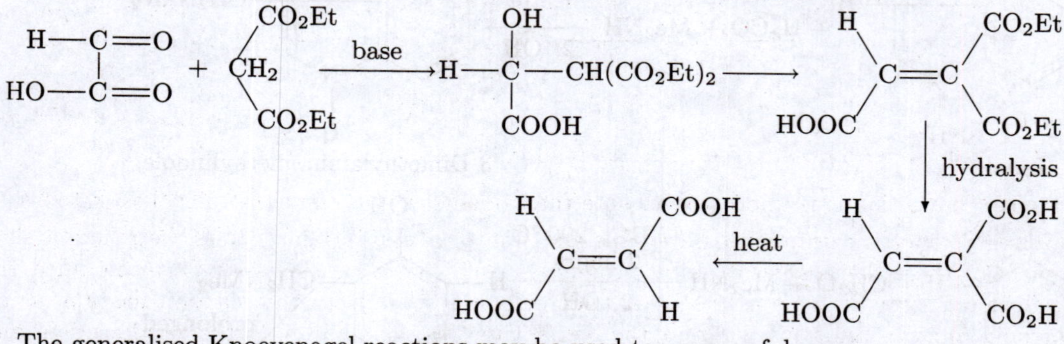

The generalised Knoevenagel reactions may be used to prepare fulvenes.

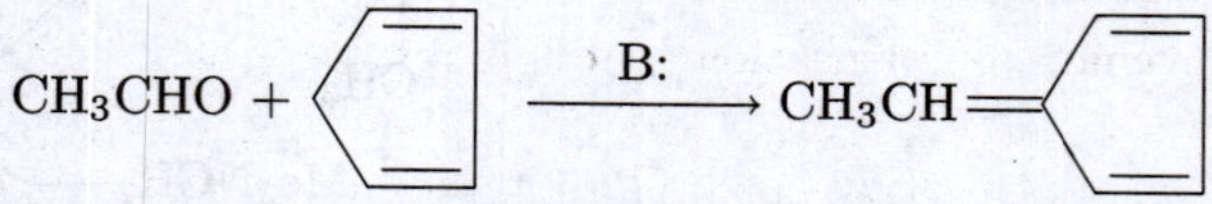

16.13.8 The Mannich reaction and the Strecker synthesis

The acid- or base-catalysed condensation reaction involving formaldehyde (or any other aldehyde), ammonia or a primary or a secondary amine and a compound containing at least one active hydrogen to yield amino or substituted amino derivative of the compound is called the Mannich reaction. In the base-catalysed reaction the product, the amino or substituted amino derivative, is called the Mannich base; in the acid-catalysed reaction the salt of the Mannich base is obtained which on hydrolysis by a base gives the Mannich base. The compound containing active hydrogen may be an aldehyde, a ketone, a β-ketoester, a β-cyanoester, a nitroalkane, a monoalkyl or aryl acetylene, an alcohol, a phenol, pyridine, thiophene, indole, hydrocyanic acid, etc. Usually ammonia or amines are used as their salts. For example, when formaldehyde is condensed with diethylamine (in the form of hydrochloride salt) and acetone in the presence of an acid and subsequently the condensation product on hydrolysis gives, 4-(N, N-diethylamino)butanone.

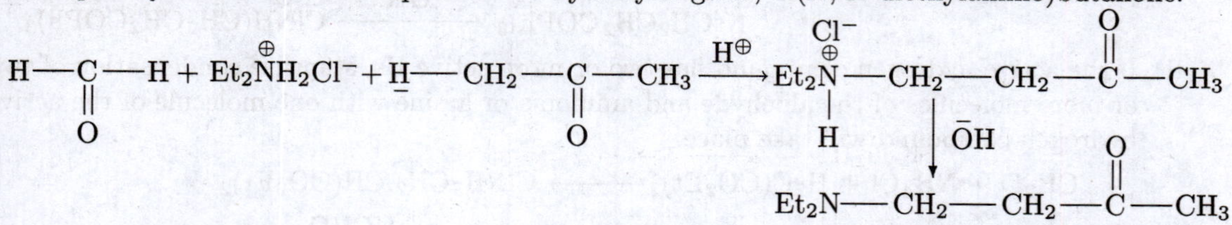

4-(N, N-Diethylamio)butanone

Similarly,

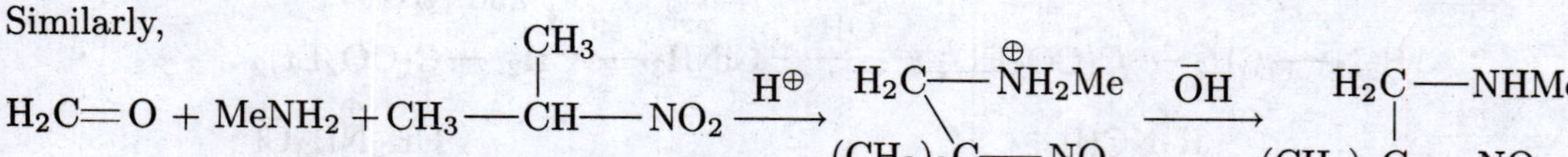

2-Nitropropane

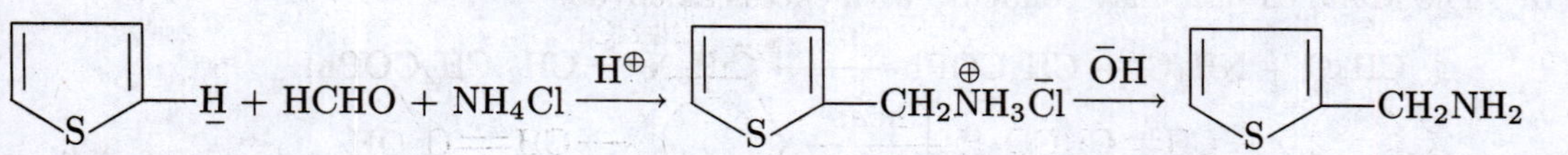

Thiophene 2-Thenylammonium chloride 2-Aminomethylthiophene

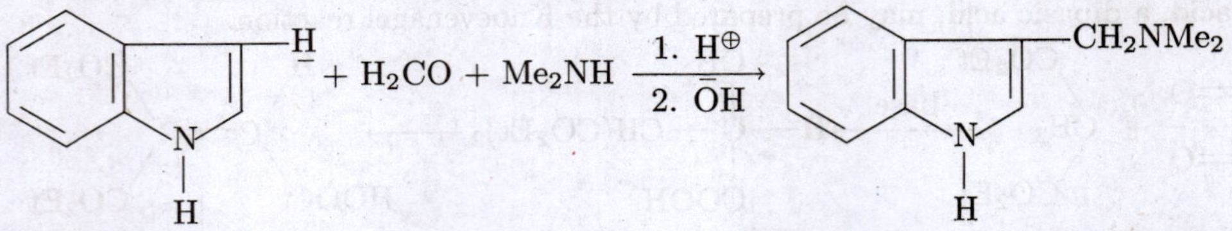

Indole 3-Dimethylaminomethylindole

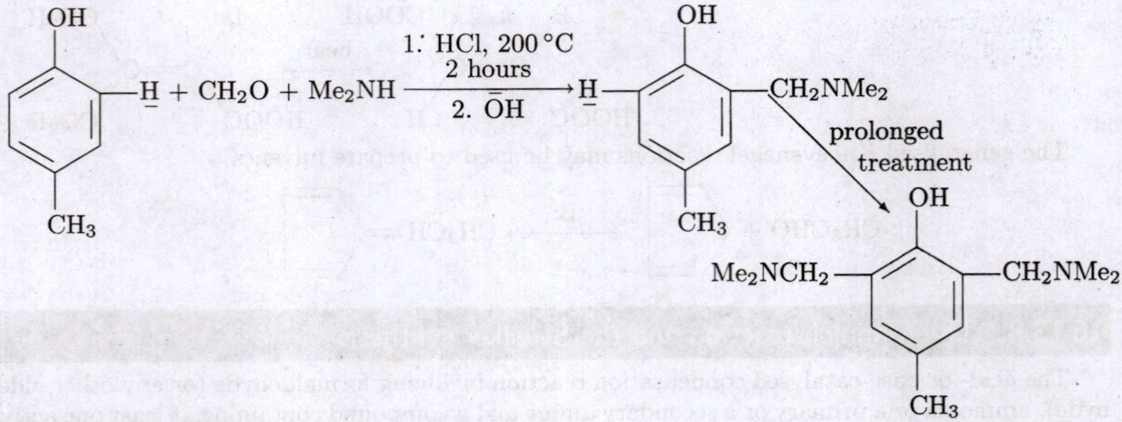

Note : Underlined H atoms are active H atoms.

However, the reaction may not stop at this stage. It may further react in three ways :

(i) If the Mannich base is a primary or a secondary amine it may further condense with one or two additional molecules of aldehyde and active hydrogen compound.

$$PhCOCH_3 + CH_2O + NH_4Cl \longrightarrow \overset{-}{Cl}H_3\overset{\oplus}{N}.CH_2CH_2COPh \xrightarrow[MeCOPh]{HCHO} \overset{-}{Cl}H_2\overset{\oplus}{N}(CH_2CH_2COPh)_2$$

$$\overset{-}{Cl}H_2\overset{\oplus}{N}(CH_2CH_2COPh)_2 \xrightarrow{\text{HCHO and MeCOPh}} \overset{-}{Cl}\overset{\oplus}{N}H(CH_2CH_2COPh)_3 \xrightarrow{\bar{O}H} N(CH_2CH_2COPh)_3$$

(ii) If the active hydrogen compound has two or more active Hs, stepwise condensation of two or more molecules of the aldehyde and ammonia or amine with one molecule of the active hydrogen compound will take place.

$$CH_2O + NH_4Cl + H_2C(CO_2Et)_2 \longrightarrow \overset{-}{Cl}\overset{\oplus}{N}H_3CH_2.CH(CO_2Et)_2$$

$$\xrightarrow{\text{HCHO and } NH_4Cl} \overset{-}{Cl}\overset{\oplus}{N}H_3—CH_2—C(CO_2Et)_2(CH_2\overset{\oplus}{N}H_3\overset{-}{Cl}) \xrightarrow{\bar{O}H} H_2N—CH_2—C(CO_2Et)_2(H_2NCH_2)$$

(iii) The Mannich base may condense with excess aldehyde.

$$CH_2O + NH_4Cl + CH_3COPh \longrightarrow \overset{-}{Cl}H_3\overset{\oplus}{N}—CH_2.CH_2COPh$$

$$\xrightarrow{\bar{O}H} H_2N—CH_2—CH_2COPh \xrightarrow{H_2C=O} H_2C=N—CH_2CH_2COPh$$

Since the Mannich bases are useful as intermediates in synthesis, the Mannich reaction has wide scope.

(i) When the Mannich base is a β-amino-carbonyl compound, an α, β-unsaturated carbonyl compound is obtained on heating the Mannich base. This may then be converted to a saturated ketone by the catalytic hydrogenation.

$$PhCOCH_2CH_2\overset{\oplus}{N}HMe_2\bar{C}l \longrightarrow PhCOCH{=}CH_2 \xrightarrow{H_2;\ Cat.} PhCOCH_2CH_3$$

$$PhCOCH_2CH_2\overset{\oplus}{N}HMe_2\bar{C}l \xrightarrow{MeCl} PhCOCH_2CH_2\overset{\oplus}{N}Me_3Cl^- \xrightarrow{\bar{O}H} PhCOCH_2CH_2\overset{\oplus}{N}Me_3OH^- \xrightarrow{heat} PhCOCH{=}CH_2$$

(ii) The amino group of a Mannich base (or from its salts) may be replaced by a nucleophile.

$$PhCOCH_2CH_2\overset{\oplus}{N}Me_3I^- \xrightarrow{\bar{C}N} PhCOCH_2CH_2CN$$

It has been found that the Mannich reaction is an important biosynthetic route to natural products, e.g., alkaloids. Some of these routes have been utilised in the laboratory—Robinson's synthesis of tropinone is one of such routes. Atropine is an ester of tropic acid and tropine, a 2° alcohol. Tropine may be oxidized to tropinone, the corresponding ketone of tropine. Robinson synthesised tropinone by the Mannich reaction involving succinaldehyde, methylamine hydrochloride and acetone dicarboxylic acid.

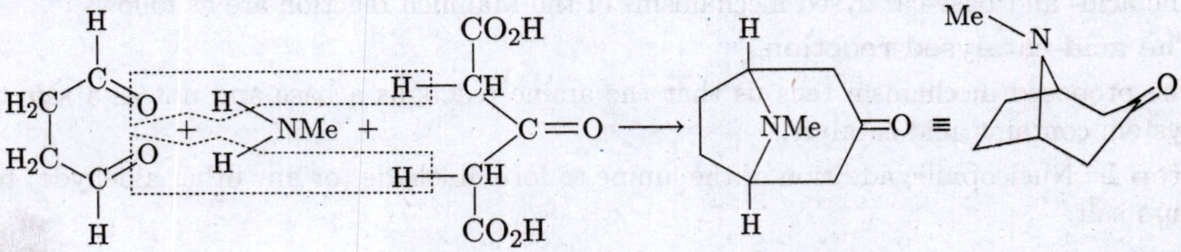

Tropinone

Robinson applied the Mannich reaction also in his annelation reaction. Michael additions followed by cyclisation is **Robinson's annelation** (minor formation).

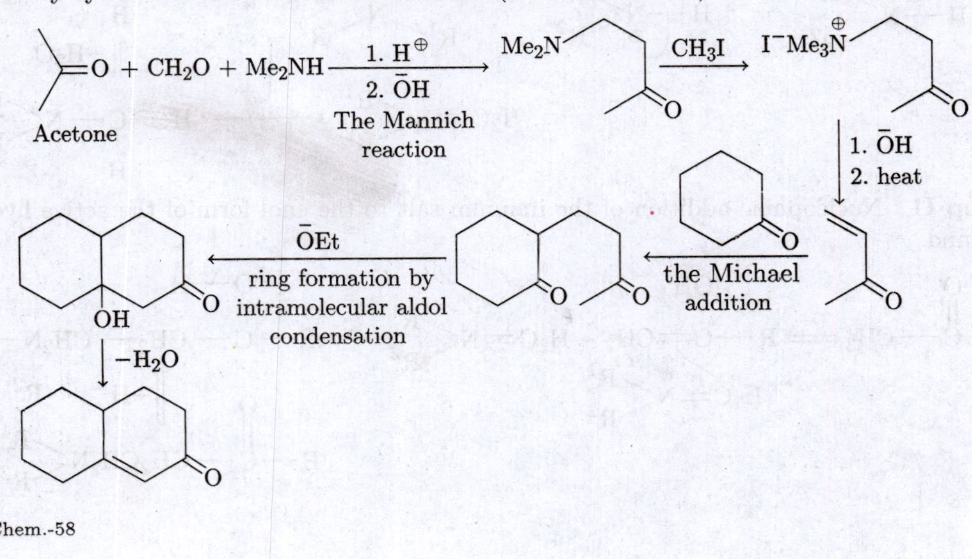

The replacement of the NMe_2 group of the Mannich base by the $\bar{C}N$ ion helps us to obtain a starting material to synthesise 3-indole derivatives, 2-pyridine derivatives, 3-thiophene derivatives etc.

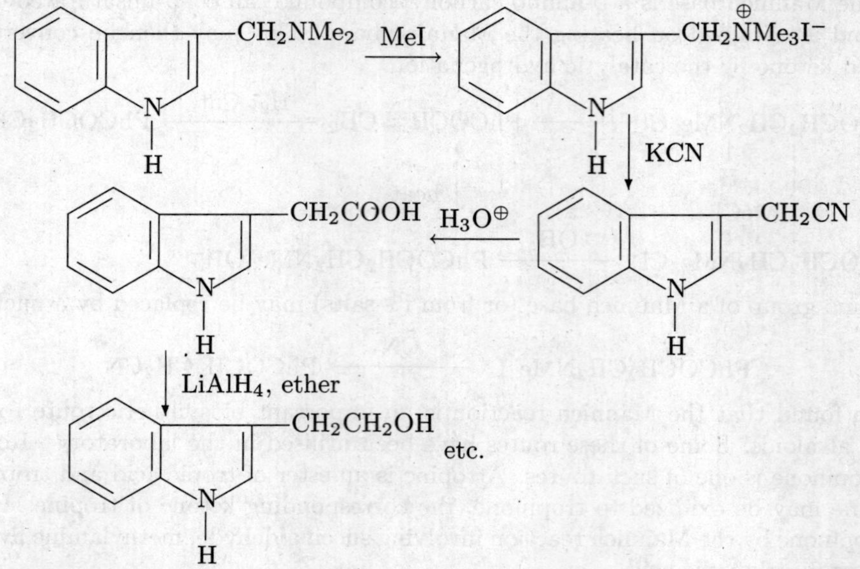

The acid- and base-catalysed mechanisms of the Mannich reaction are as follows :

The acid-catalysed reaction

The proposed mechanism tells us that the amine reacts as a base and not as a salt though the system contains acid catalyst.

Step I : Nucleophilic addition of the amine to formaldehyde (or any other aldehyde) to form iminium salt.

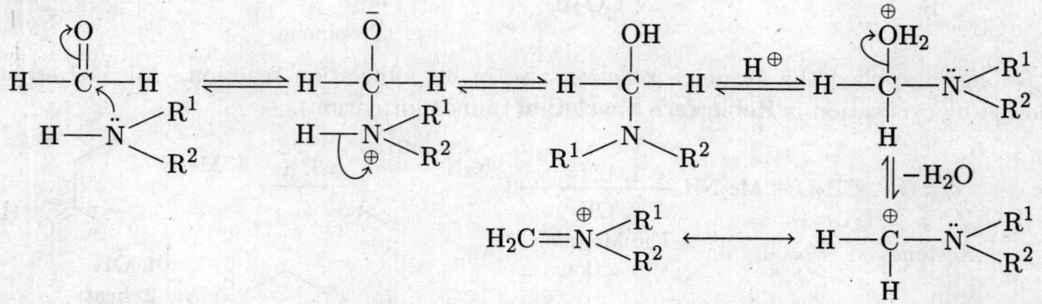

Step II : Nucleophilic addition of the iminium salt to the enol form of the active hydrogen compound.

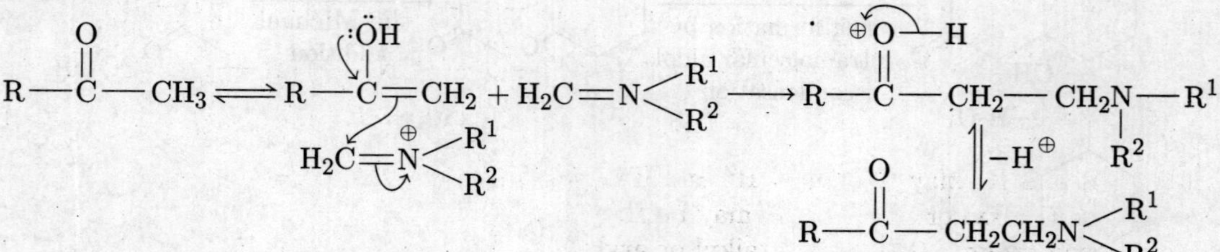

The base-catalysed reaction

There is controversy about the proposed pathway of the base-catalysed reaction. On the basis of evidences the following steps may be drawn for the proposed mechanism.

Step I : Reversible formation of the resonance-stabilised carbanion.

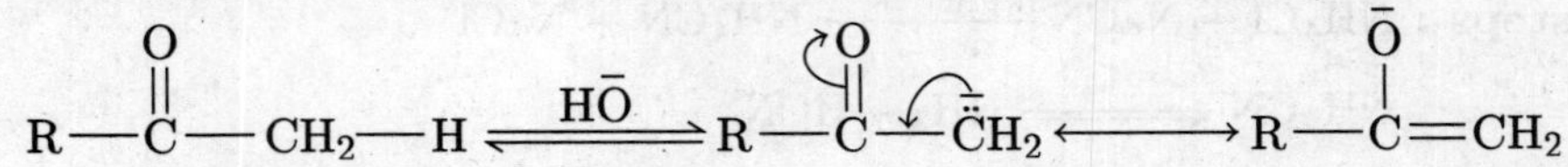

Step II : Addition of the amine to formaldehyde.

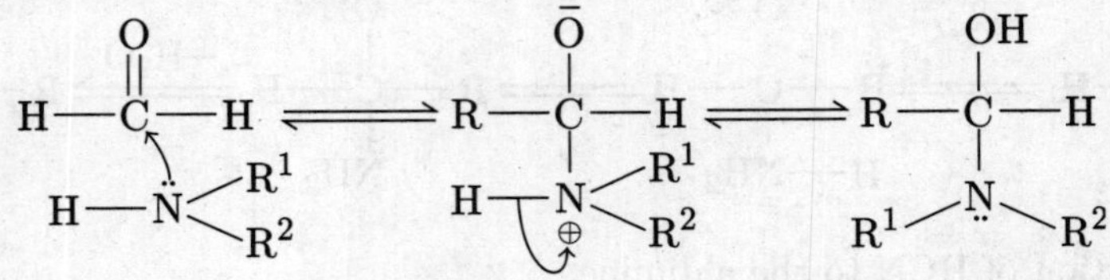

Step III : The resonance-stabilized carbanion replaces the OH group and the reaction follows the S_N2 pathway.

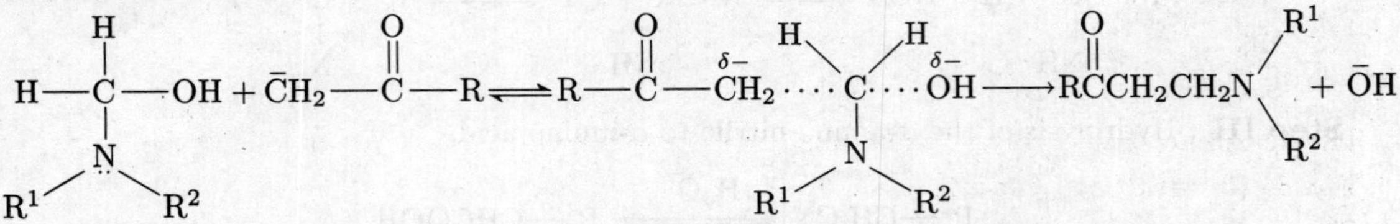

However, the rate = k [HCHO] [R_1NHR_2] [CH_3COR]

The reaction follows the third order kinetics in the acid-catalysed reaction also, the addition of the iminium salt to the enol is the rate-determining step.

The Strecker synthesis of α-amino acids is a special case of the Mannich reaction. This involves the treatment of an aldehyde or a ketone with NaCN and NH_4Cl whereby an α-aminonitrile is formed; the latter on hydrolysis gives an α-aminoacid. For example, when acetaldehyde is treated with NaCN and NH_4Cl, alanine is formed after hydrolysing the condensation product.

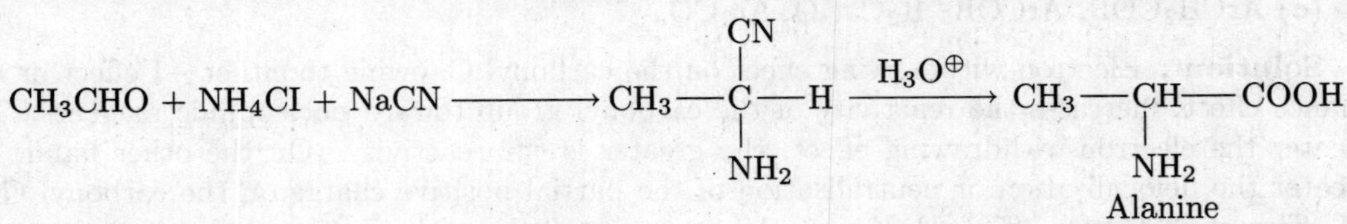

The general equation is :

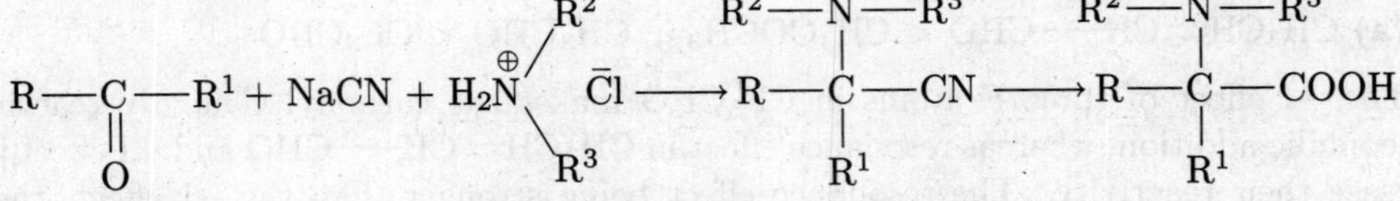

R and R^1 may be H alkyl or aryl group.

R^2 and R^3 may be H, alkyl or aryl

In the Strecker synthesis, HCN is the active hydrogen compound. NH_4CN dissociates to NH_3 and HCN; NH_4CN is formed by the double decomposition of NH_4Cl and NaCN. Instead of NH_4Cl, salts of amines may be used and in this case N-substituted amino acids form. However, the mechanistic steps of the Strecker synthesis is as follows :

Initial steps : $NH_4Cl + NaCN \longrightarrow NH_4CN + NaCl$

$$NH_4CN \rightleftharpoons NH_3 + HCN.$$

Step I : Nucleophilic addition of ammonia to the aldehyde to form aldimine :

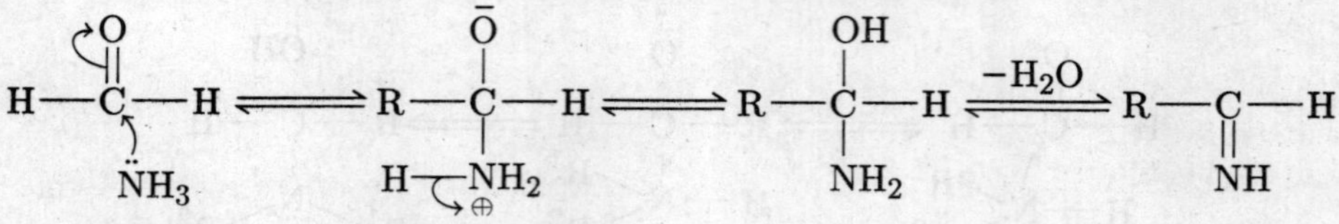

Step II : Addition of HCN to the aldimine.

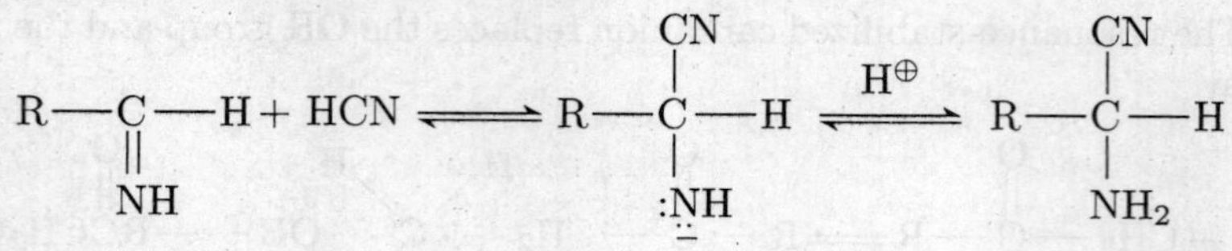

Step III : Hydrolysis of the α-amino nitrile to α-amino acid.

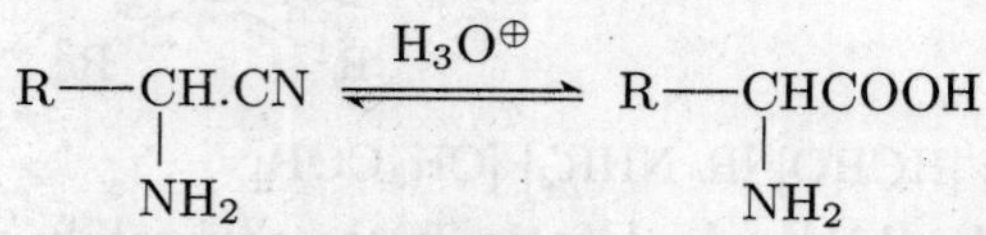

Problem 16.22. Arrange the compounds in order of increasing reactivity toward nucleophilic reagents.

(a) CH_3CHO, CH_3COCH_3; CF_3CHO, $CH_3CH{=}CH{-}CHO$

(b) $ClCH_2CHO$, $BrCH_2CHO$, $CH_2{=}CH{-}CHO$, CH_3CF_2CHO

(c) $ArCH_2COR$, $ArCOR_1$ $R_2C{=}O$, Ar_2CO.

Solution : Electron-withdrawing effect on the carbonyl C, owing to either –I effect or resonance effect, increases the reactivity of the carbonyl group toward nucleophilic reagents. The greater the electron-withdrawing effect, the greater is the reactivity. On the other hand, the greater the delocalisation or neutralisation of the partial positive charge on the carbonyl C by +I effect or resonance effect, the less is the reactivity of the carbonyl group. In the light of the discussions made above we can say that the reactivity order of the above sets are

(a) $CH_3CH{=}CH{-}CHO < CH_3COCH_3 < CH_3CHO < CF_3CHO$

The –I effect of three F atoms in CF_3CHO makes the carbonyl C highly reactive toward nucleophilic addition, whereas resonance effect in $CH_3CH{=}CH{-}CHO$ and +I effect in acetone decrease their reactivity. The resonance effect being stronger than the +I effect, the order of reactivity is as shown above.

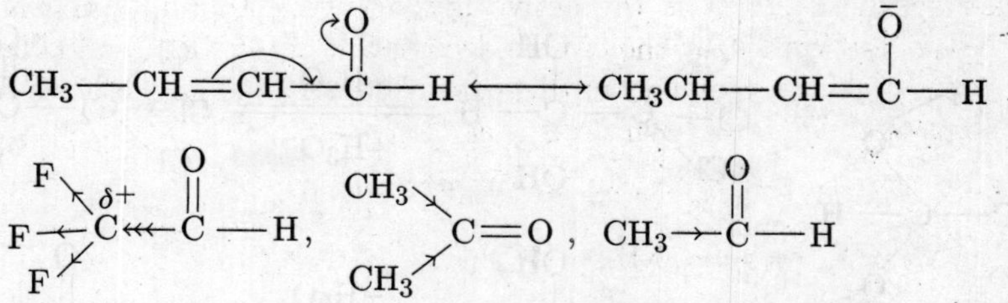

(b) $CH_2{=}CH{-}CHO < BrCH_2CHO < ClCH_2CHO < CH_3CF_2CHO$, the order of −I effect being F > Cl > Br.

(c) $Ar_2CO < ArCOR < R_2CO < ArCH_2COR$. An Ar group directly linked to a >CO group delocalises its positive charge. A R group directly linked to a >C=O decreases the positive charge by +I effect; $ArCH_2$ group has a lower +I effect than a R group owing to −I effect of Ar.

Problem 16.23. When a mixture of one mole of benzaldehyde and one mole of cyclohexanone is treated with one mole of semicarbazide, cyclohexanone semicarbazone precipitates only; but after a few hours the precipitate is the semicarbazone of benzaldehyde—explain the result.

Solution : The —CHO group of benzaldehyde is less reactive toward nucleophilic attack by semicarbazide than the >CO group of cyclohexanone since resonance effect neutralises the $\delta+$ charge on C of —CHO more than the +I effect on that of >C=O. For these reasons, cyclohexanone reacts faster than benzaldehyde and the kinetically controlled cyclohexanone semicarbazone is obtained. On standing, exchange of $=N.NH.CONH_2$ group occurs between cyclohexanone semicarbazone and benzaldehyde for two reasons :

(i) The semicarbazone formation reaction is reversible and (ii) the semicarbazone of benzaldehyde is more stable than the semicarbazone of cyclohexanone because the former gets stability through resonance; and hence it is being thermodynamically stable, it forms slowly on standing.

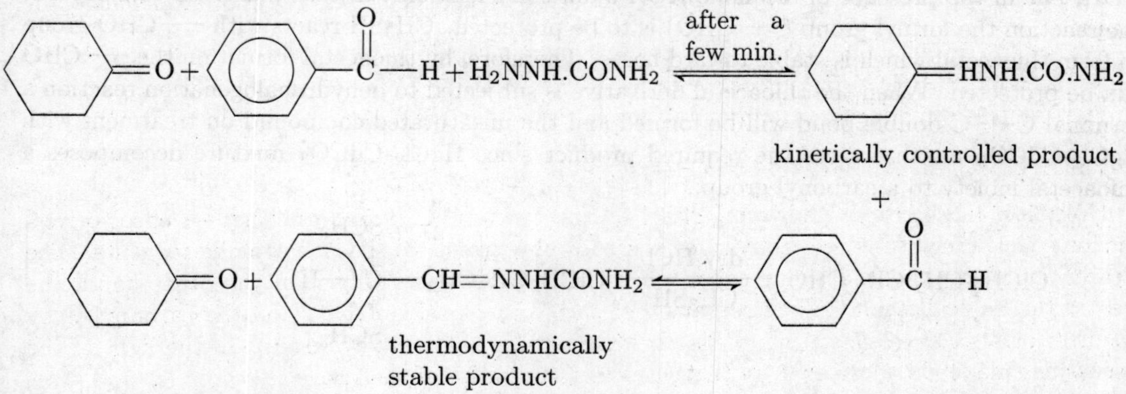

Problem 16.24. Hydrated acetaldehyde is unstable but hydrated trichloroacetaldehyde is stable—explain.

Solution : The hydrated C atom of the chloral hydrate bears a greater steric strain than the similar C atom of acetaldehyde hydrate. Therefore, it seems that chloral hydrate will be less stable than acetaldehyde hydrate. However, both the hydrates may give back the parent aldehyde since the hydration reaction is essentially a reversible reaction and the equilibrium lies to the left usually.

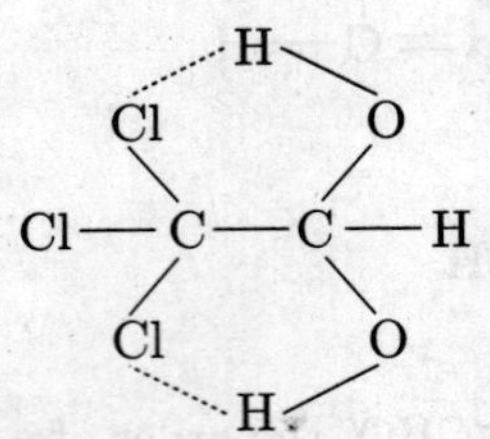

intramolecular H-bonds

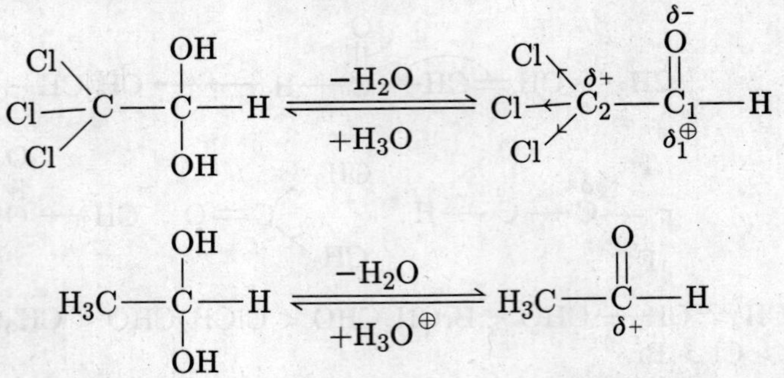

In chloral, the $-I$ effects of the three Cl atoms develop a $\delta+$ charge on C_2 and $-I$ effect of O atom develops the same on C_1. These partial positive charges repel each other and brings instability to chloral. In the chloral hydrate this effect is minimum owing to the absence of the carbonyl group. IR spectra of chloral hydrate shows that, there are internal H-bonds in chloral hydrate; the internal H-bonding brings stability to the system. So, chloral hydrate is more stable than chloral. Since hydrated acetaldehyde experiences steric strain and since acetaldehyde does not have any steric strain, acetaldehyde is more stable than acetaldehyde hydrate; and consequently acetaldehyde hydrate decomposes and gives acetaldehyde. So, chloral hydrate is more stable than acetaldehyde hydrate.

Problem 16.25. Convert the following compounds as directed :

(a) $ClCH_2CH_2CH_2CH_2CHO \longrightarrow CH_2{=}CHCH_2CH_2CHO$ using CH_3SH

(b) $BrCH_2CH_2CHO \longrightarrow HOCH_2CHOH.CHO$ using $CH_2OH.CH_2OH$

Solution : (a) The given compound can be dehydrohalogenated by the treatment of alcoholic KOH; but in the presence of alcoholic KOH aldol condensation will take place. So to carry out the reaction the formyl group (—CHO) is to be protected. CH_3SH reacts with —CHO group to form thioacetal which is stable toward bases. Therefore, by thioacetal formation the —CHO can be protected. When the thioacetal derivative is subjected to dehydrohalogenation reaction a terminal C—C double bond will be formed and the unsaturated compound on treatment with $HgCl_2$-$CdCO_3$ mixture gives the required product since $HgCl_2$-$CdCO_3$ mixture decomposes a thioacetal moiety to a carbonyl group.

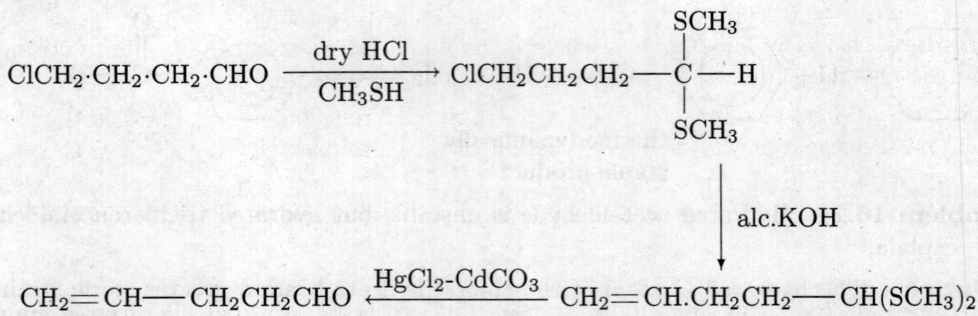

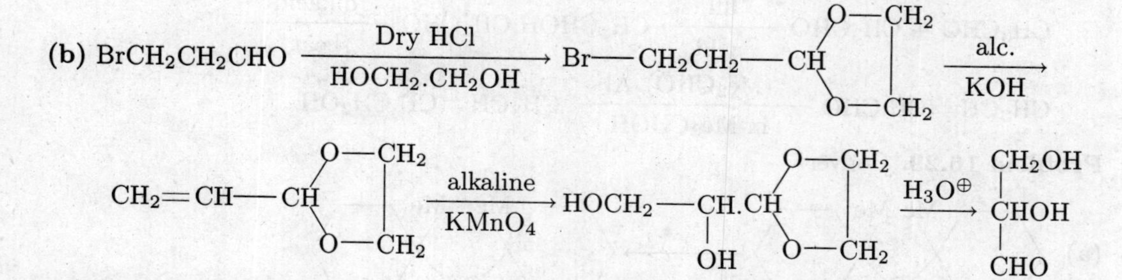

Problem 16.26. What will be the product when $HOCH_2CH_2COCH_2CH_2CO_2Et$ is treated with four moles of CH_3MgI step by step. Write down different steps and give reasons for the steps.

Solution : The given compound, $HOCH_2CH_2.CO.CH_2CH_2CO_2Et$ contains three functional groups—alcohollic OH bearing active H, ketonic carbonyl group and an ester group, —CO_2Et. The active H gets replaced by MgI and methane is evolved when the compound is treated with one mole of CH_3MgI. One mole CH_3MgI undergoes addition reaction with the carbonyl group. The ester group also undergoes addition reaction with CH_3MgI. The addition product will then react with another mole of CH_3MgI in which OEt group will be replaced by a methyl group. Now it has been found that the order of reactivity of the functional groups present in the molecule is : active H≫ >C=O >—COOEt. Considering the points said above the sequential steps of the given reaction may be written as :

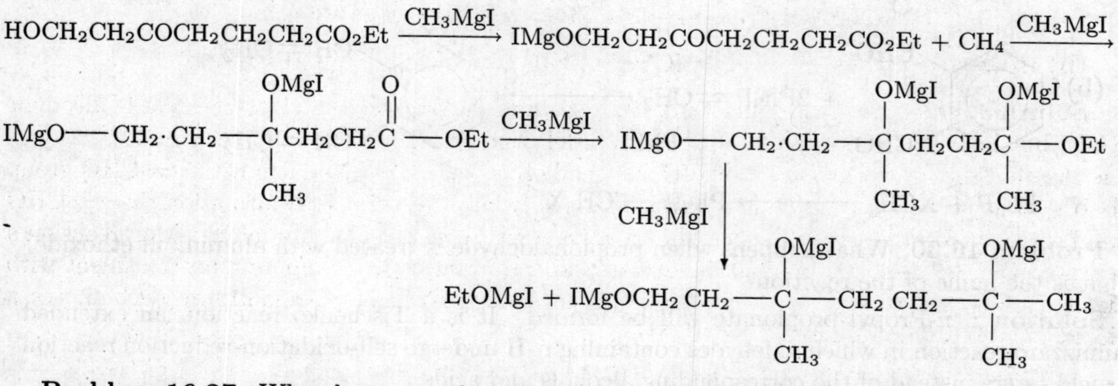

Problem 16.27. What happens when phenylmagnesium bromide is treated with methoxy-acetonitrile and the adduct is hydrolysed with acidulated water?

Solution : Phenylmagnesium bromide is a Grignard reagent; this will add to the —C≡N group. The adduct on hydrolysis gives a ketone of the formula $MeOCH_2COPh$ as shown below :

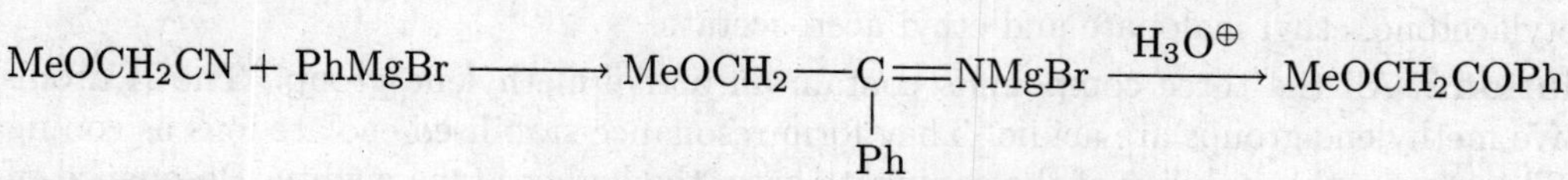

Problem 16.28. Using aluminium isopropoxide synthesise crotyl alcohol from acetaldehyde.

Solution : Dilute acid-catalysed self-condensation of acetaldehyde gives crotonaldehyde; this on treatment with aluminium isopropoxide in isopropyl alcohol reduces to crotyl alcohol selectively.

$$CH_3CHO + CH_3CHO \xrightarrow[\text{acid}]{\text{dil.}} CH_3CHOH.CH_2CHO \xrightarrow[\text{heat}]{\text{dil.acid}}$$

$$CH_3CH{=}CH.CHO \xrightarrow[\text{in } Me_2CHOH]{(Me_2CHO)_3Al} CH_3CH{=}CH.CH_2OH$$

Problem 16.29. Convert

(a) 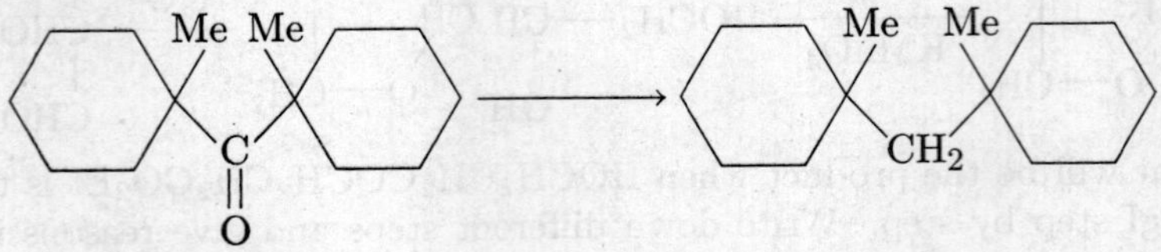

(b) 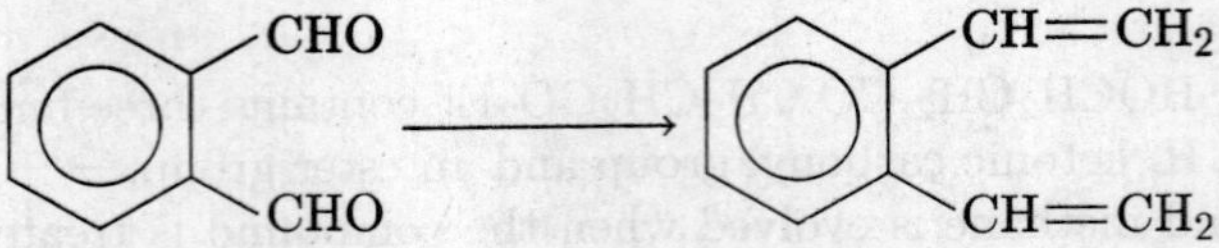

Solution :

(a) 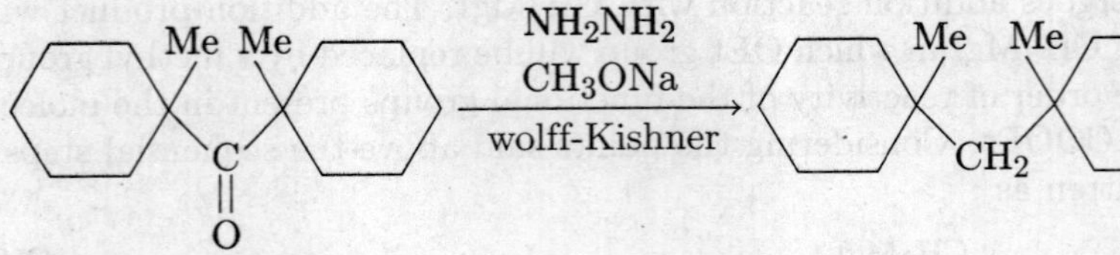

(b) 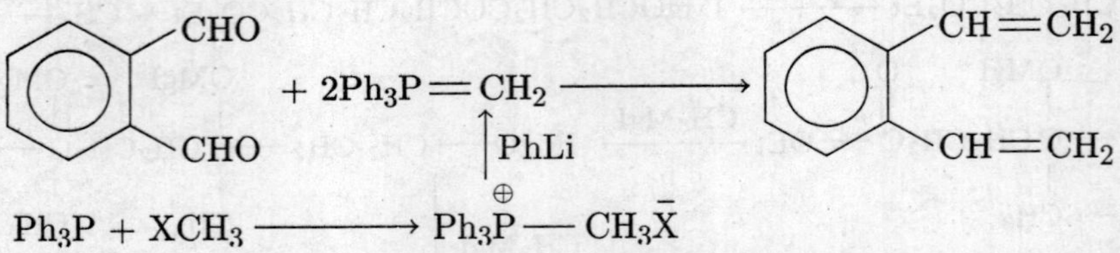

Problem 16.30. What happens when propionaldehyde is treated with aluminium ethoxide? What is the name of the reaction?

Solution : *n*-Propyl propionate will be formed. It is a Tischenko reaction, an extended Cannizzaro reaction in which aldehydes containing α-H undergo self-oxidation-reduction reaction to yield esters instead of the corresponding alcohols and acids.

$$CH_3CH_2CHO \xrightarrow{Al(OEt)_3} CH_3CH_2{-}\overset{O}{\overset{\|}{C}}{-}OCH_2CH_2CH_3$$

Problem 16.31. Write down the increasing order of the acidity of the following compounds. Acetylacetone, ethyl malonate and ethyl acetoacetate.

Solution : All the three compounds contain an active methylene group. The H atoms of the active methylene groups are acidic. They form resonance-stabilised enolate ions as conjugate bases. The greater the stability of the conjugate base, the higher is the acidity. Resonance effect brings stability to the enolate ion. Since the resonance-stabilisation is the highest in the enolate ion of acetylacetone and it is the least in the enolate ion of malonic ester, the increasing order of acidity of the given compounds is :

$$CH_2(CO_2Et)_2 < CH_3COCH_2CO_2Et < CH_3COCH_2COCH_3$$

The resonance-stabilization of the enolate ion of ethyl malonate is the lowest since the participation of the two OEt groups increases the electron density on CH part. In the enolate of the acetoacetate this effect is due to one OEt group. Since in the enolate of acetylacetone there is no such effect, it is the most stable conjugate base. Therefore, the acidity order will be that as shown in the previous page.

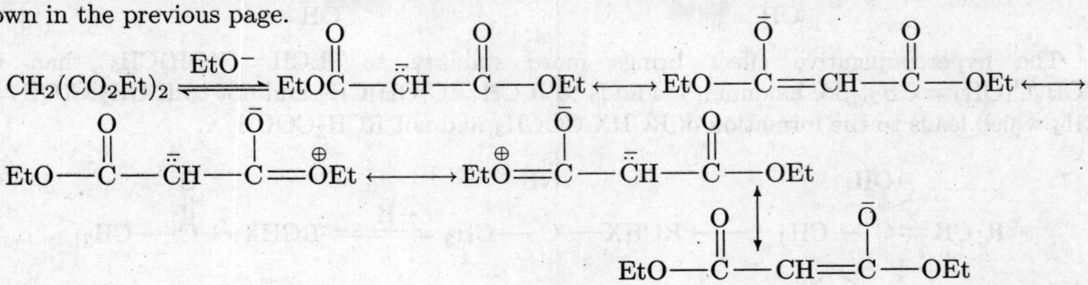

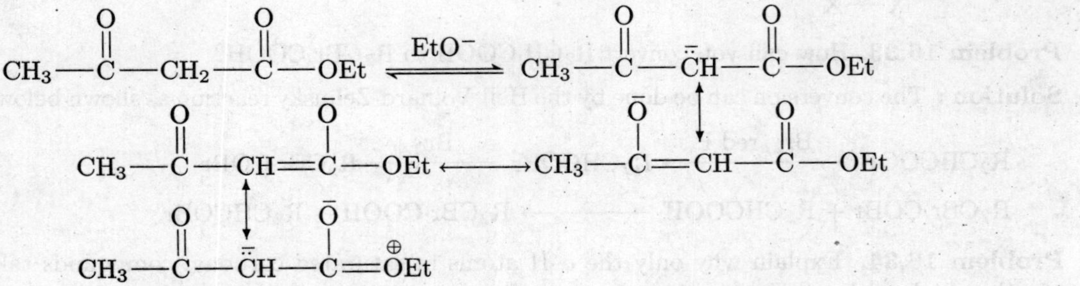

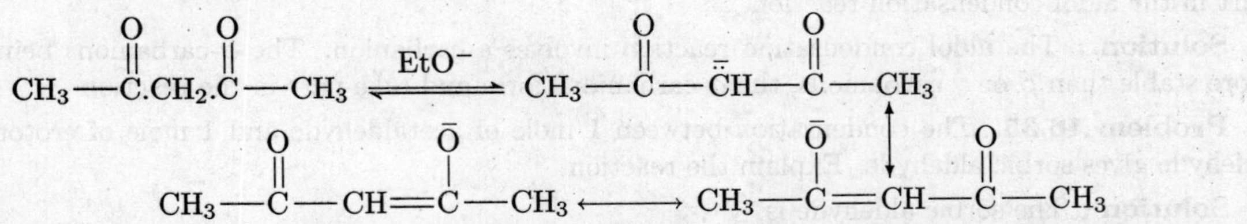

Problem 16.32. Explain why (a) the base-catalysed halogenation of acetone cannot be stopped at monohalo stage. (b) The trihaloderivative of the base-catalysed halogenation of acetone is $CH_3CO.CX_3$ and not $CH_2X.COCHX_2$. (c) The reaction follows the zero order kinetics with respect to halogen. (d) The acid-catalysed halogenation of RCH_2COCH_3 gives $RCHXCOCH_3$ and not RCH_2COCH_2X.

Solution : (a) Monohalo acetone makes the H atoms of —CH_2X more acidic than those of —CH_3 because of the –I effect of X and >C=O groups. For this reason, the halogenation occurs further and **(b)** for the same reason CH_3COCX_3 and not $CH_2XCO.CHX_2$ forms.

(c) The rate-determining step of the base-catalysed halogenation of acetone is the carbanion formation step. This step does not involve the halogen. So, the reaction follows the zero order kinetics with respect to halogen.

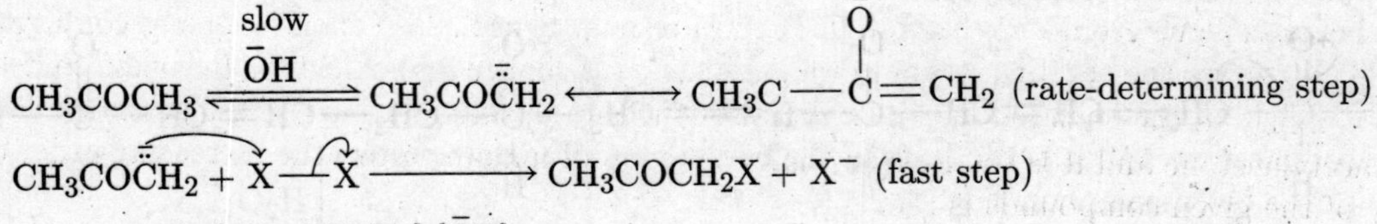

The rate = $k\,[CH_3COCH_3]\,[\bar{O}H]$

(d) The acid-catalysed halogenation takes place through enolisation. The enol forms of RCH_2COCH_3 are :

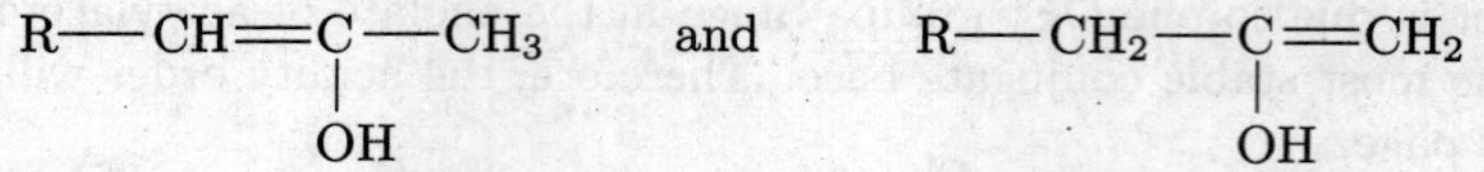

The hyperconjugative effect brings more stability to $R.CH{=}C(OH)CH_3$ than in $RCH_2C(OH){=}CH_2$. So, halonium ion adds to $R.CH{=}C(OH)CH_3$ and not to $R.CH_2.C(OH){=}CH_2$ which leads to the formation of $RCHX.COCH_3$ and not RCH_2COCH_2X.

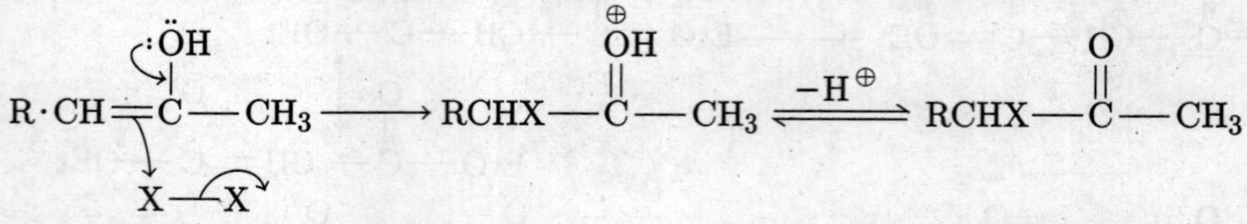

Problem 16.33. How can you convert $R_2CH.COOH$ to $R_2CBr.COOH$?

Solution : The conversion can be done by the Hell-Volhard-Zelinsky reaction as shown below :

$$R_2CHCOOH \xrightarrow{Br_2/\text{red P}} R_2CHCOBr \xrightarrow{Br_2} R_2CBr{\cdot}COBr$$

$$R_2CBr.COBr + R_2CHCOOH \longrightarrow R_2CBr{\cdot}COOH + R_2CHCOBr$$

Problem 16.34. Explain why only the α-H atoms in saturated carbonyl compounds take part in the aldol condensation reaction.

Solution : The aldol condensation reaction involves a carbanion. The α-carbanions being more stable than β or γ carbanions, the α-carbanions form and take part in the reaction.

Problem 16.35. The condensation between 1 mole of acetaldehyde and 1 mole of crotonaldehyde gives sorbic aldehyde. Explain the reaction.

Solution : The sorbic aldehyde is

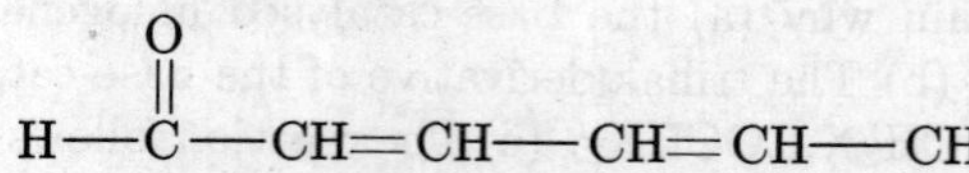

The γ-H atoms of $CH_3{-}CH{=}CH{-}CHO$, crotonaldehyde, are more acidic than the α-H atoms of acetaldehyde because of the greater resonance-stabilisation of the conjugate base of the crotonaldehyde than that of the acetaldehyde as shown below :

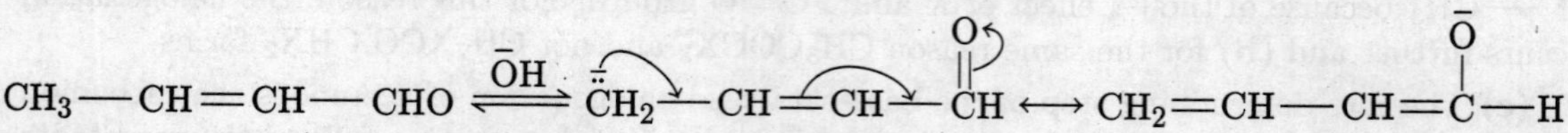

So, the carbanion of the crotonaldehyde makes nucleophilic attack on the electron-deficient carbonyl C atom of acetaldehyde.

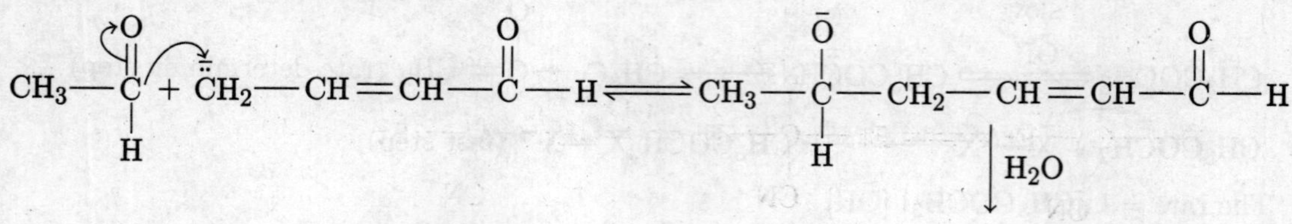

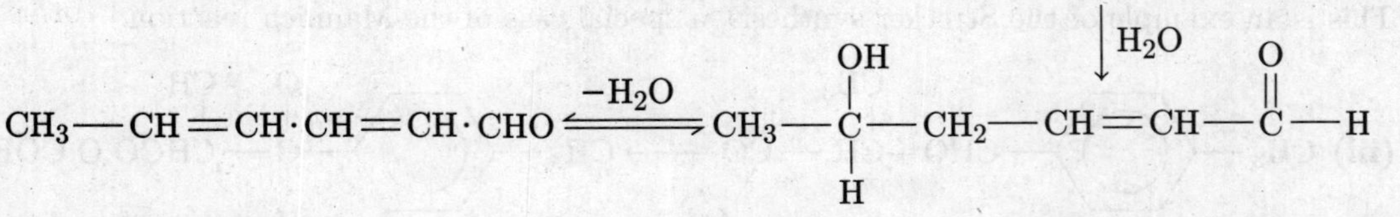

Problem 16.36. What are the products of the following condensation reactions? Name the type of reaction in each case.

(i) $PhCOMe + ClCH_2CO_2Et \xrightarrow{NaNH_2}$ **(ii)** $Et_2CO + NH_4Cl + NaCN \xrightarrow{NH_4OH}$

(iii) 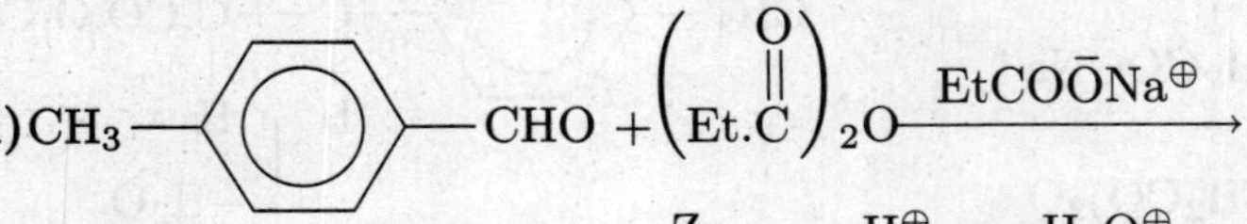

(iv) $PhCHO + BrCH_2CO_2Et \xrightarrow{Zn} \xrightarrow{H^{\oplus}} \xrightarrow{H_3O^{\oplus}}$

(v) $MeCH{=}CHCO_2Et + CH_3NO_2 \xrightarrow{EtONa} \xrightarrow{H^{\oplus}}$

(vi) $Me_2CHNO_2 + HCHO + NH_2Et \xrightarrow{HCl}$

(vii) $PhCHO + CH_2(CO_2Et)_2 \xrightarrow[MeCOO^{-}]{Me_2\overset{\oplus}{N}H_2}$

Solution :

(i) 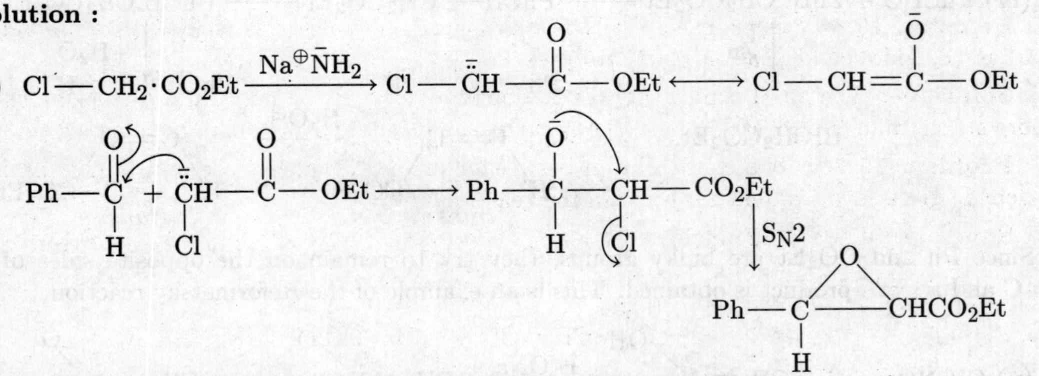

This is an example of the Darzen glycidic ester condensation.

(ii) $NH_4Cl + NaCN \longrightarrow NH_4CN + NaCl$

$NH_4CN \rightleftharpoons NH_3 + HCN$

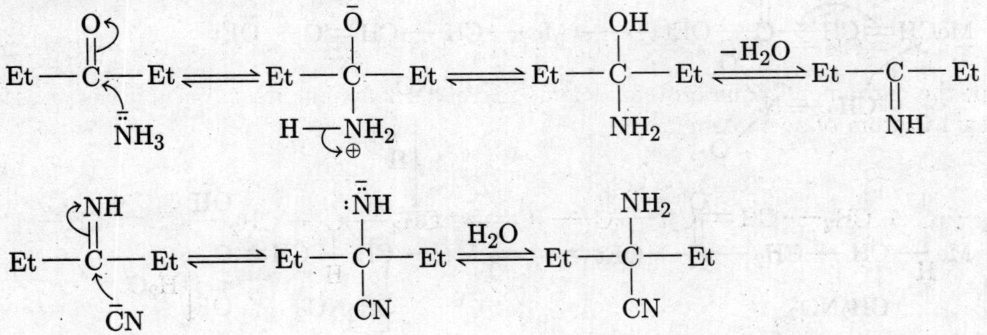

This is an example of the Strecker synthesis, a special case of the Mannich reaction.

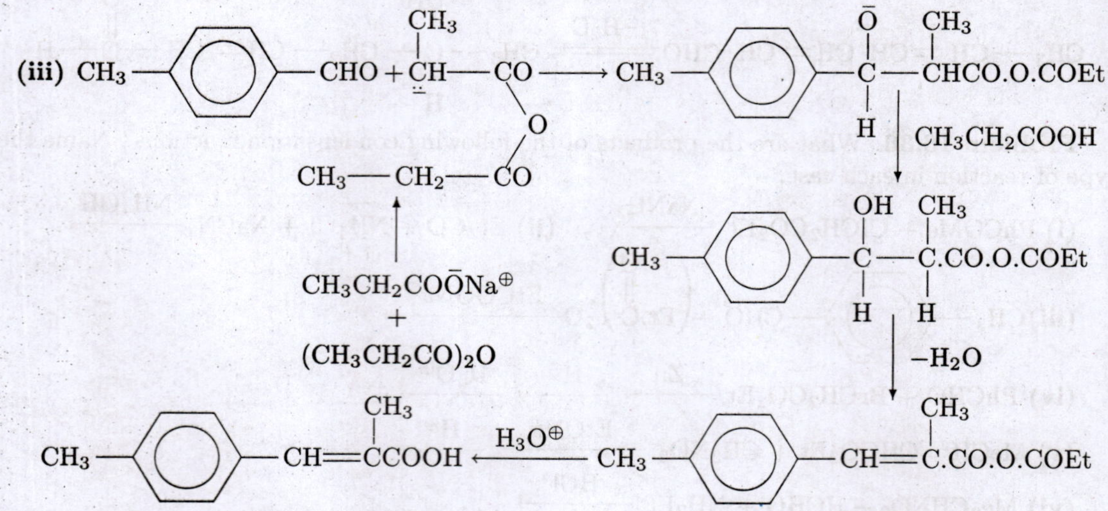

It is a Perkin reaction.

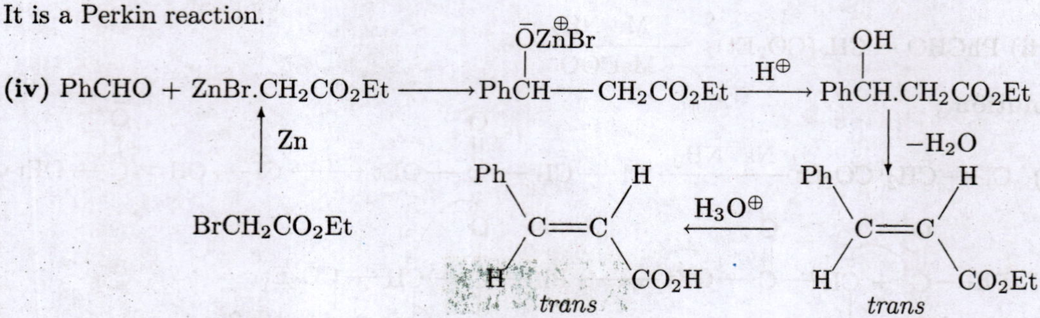

Since Ph and CO_2Et are bulky groups, they try to remain on the opposite sides of the C=C and a trans-product is obtained. This is an example of the Reformatsky reaction.

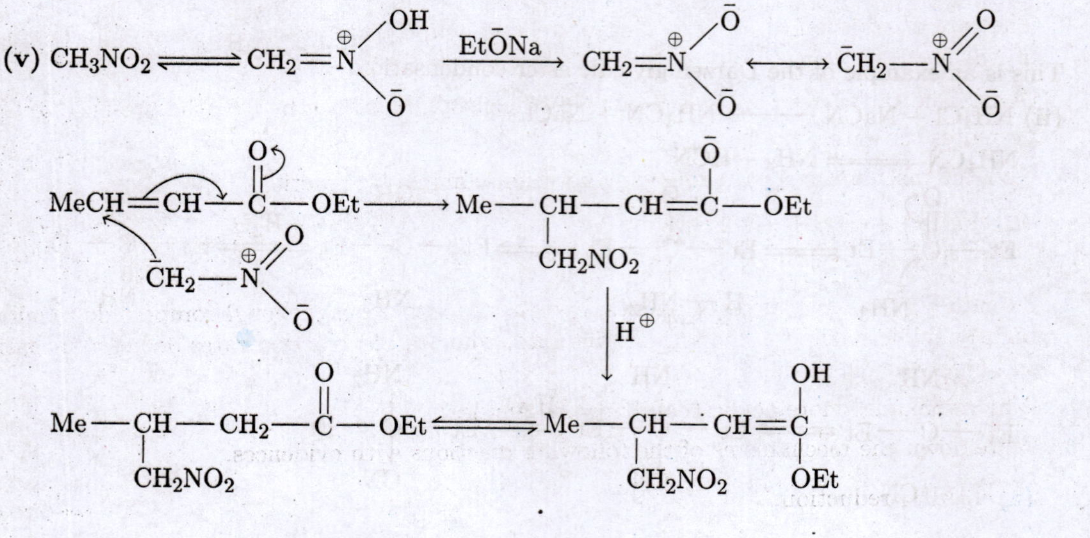

This is an example of the Michael condensation.

(vi) $Me_2CHNO_2 + HCHO + EtNH_2 \longrightarrow$

$$CH_2O + EtNH_2 \rightleftharpoons CH_2{=}\overset{\oplus}{N}HEt \longleftrightarrow \overset{\oplus}{C}H_2{-}NHEt$$

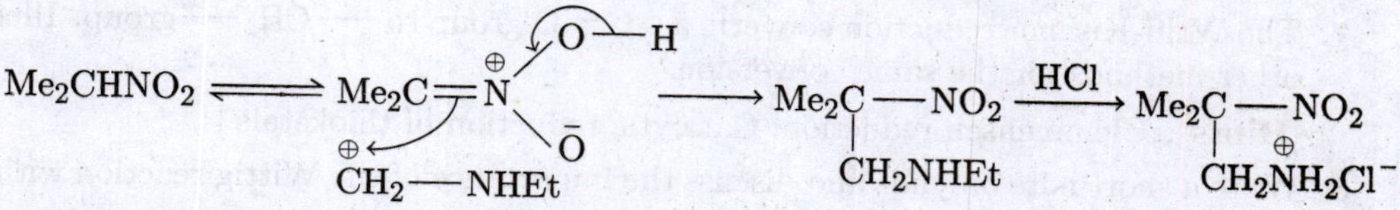

This is an example of the Mannich reaction.

(vii)

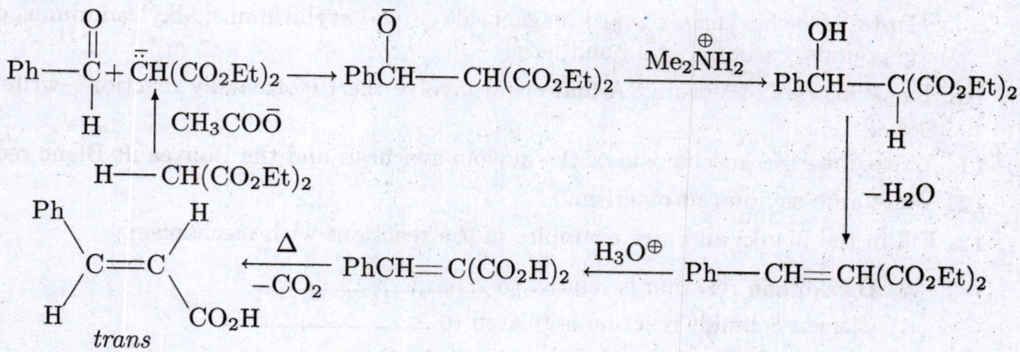

This is an example of the Knoevenagel reaction,

Exercise

1. The order of reactivity in nucleophilic addition is

$$HCHO > RCHO > R_2CO > RCO_2Et \text{ — explain.}$$

2. The rate of addition of HCN to a carbonyl compound increases in the presence of a trace of NaCN— explain.

[**Ans.** The rate-determining step being the nucleophilic attack by $\bar{C}N$, the presence of $\bar{C}N$ promotes the rate.]

3. Give the mechanisms of the following addition-elimination reactions.

   (a) $RCHO + H_2N{-}OH$

   (b) $R_2C{=}O$ + 2, 4-D.N.Ph

4. Explain why the formation of oximes, hydrazones, etc. from carbonyl compounds requires slightly acidic media for the appreciable rate, whereas the reaction rates decrease in basic and highly acidic media.

5. Why are oximes more acidic than hydroxyl amines?

6. Write down the mechanisms of the following reactions with evidences.

   (a) $LiAlH_4$ reduction,

(b) The Oppenauer oxidation,
(c) The Cannizzaro reaction,
(d) The Tischenko reaction,
(e) The Grignard addition.

7. The Wolff-Kishner reduction converts a $>C{=}O$ group to $-CH_2-$ group. Illustrate two other methods for the same conversion.

[**Hints :** Clemmensen reduction; Catalytic reduction of thioketals.]

8. Write a short note on ylids and discuss the importance of the Wittig reaction with example.

9. Write an essay on organometallic compounds,

[**Hints :** Discuss the Grignard reagent, alkyl- and aryllithiums, alkylcadmiums, organozinc compounds, carbon-metal bond, etc.]

10. The syntheses of Vitamin A and citral involve the Reformatsky reaction—write down the steps.

11. Write down the mechanism of the acyloin synthesis and the Bouveault-Blanc reduction.

12. Write an essay on tautomerism.

13. Fill in the blanks and give examples of the reactions with mechanism.

(a) Dieckmann reaction is related to ______________ .
(b) Claisen-Schmidt reaction is related to ______________ .
(c) Strecker Synthesis is related to ______________ .

14. Give mechanisms of the following reactions :

(a) The acid-catalysed condensation of acetaldehyde.
(b) Benzaldehyde is treated with KCN solution.
(c) Synthesis of coumarin through the Perkin reaction.
(d) Synthesis of 3-cyclohexylpropanoic acid from cyclohexanone by the Stobbe condensation.
(e) Synthesis of 3-dimethylaminomethylindole.
(f) The base-catalysed condensation of benzaldehyde and nitromethane.
(g) The condensation of ethyl isobutyrate in the presence of sodium triphenylmethide.
(h) The condensation between acetone and ethyl chloroacetate in the presence of $NaNH_2$.

15. Complete and name the reactions :

(a) $(C_6H_5)_3P{=}CHEt + CH_3CO.\underset{\underset{CH_3}{|}}{CH}.CO_2Et \longrightarrow$

(b) $PhCOCH_2CH_3 + CH_2O + Me\overset{\oplus}{N}H_3Cl^- \longrightarrow$

(c) $CH_3CH_2NO_2 + PhCH{=}CHCOCH_3 \longrightarrow$

(d) $(EtOOC)_2\ CH(CH_2)_2\ CH(CO_2Et)_2 + Et\bar{O} \longrightarrow$

(e) $EtOOC.CH_2COCOOEt + BrCH_2COOEt + Zn \longrightarrow$

(f) $PhCOCH_3 + EtOOCCH_2CH_2COOEt + Me_3C\bar{O}K^{\oplus} \longrightarrow$

(g) PHCHO + $NCCH_2CO_2Et$ + $\bar{O}Et$ ⟶

(h) CH_3CH_2CHO + $Br.CH_2CO_2Et$ $\xrightarrow{Me_3C\bar{O}K^{\oplus}}$

16. Synthesise the following compounds.

(a) Cyclohexanone by the Dieckmann reaction.

(b) $CN.CH.CH_2CH_2CN$ (with CO_2CH_3 on the CH) by the Michael condensation.

(c) $CH_3OOC(CH_2)_4CH{=}CHCOOCH_3$

(d)

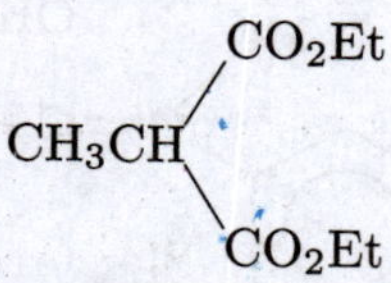

(e)

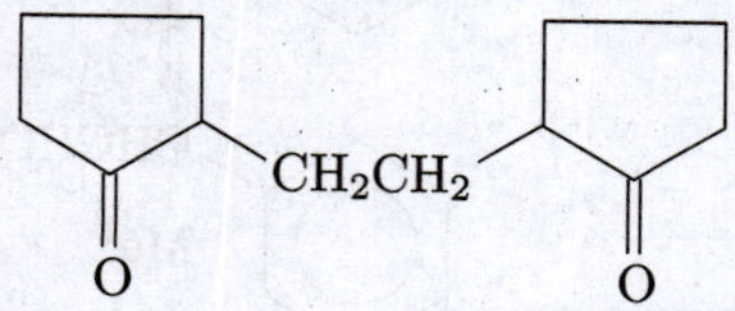

17. Convert :

(a)

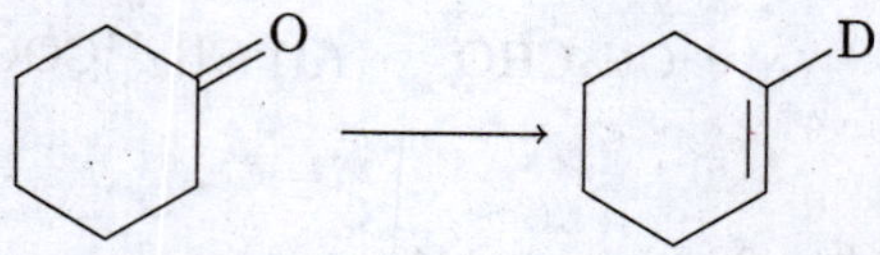

(b)

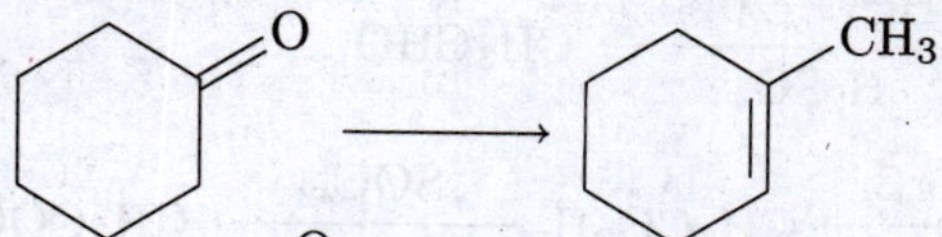

(c)

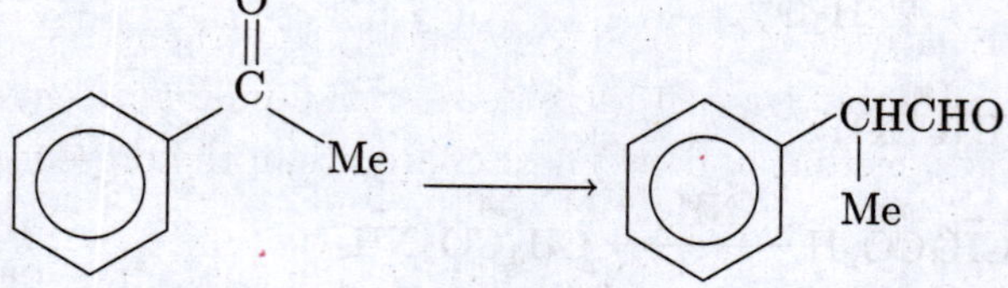

[**Answer :** a. (i) $LiAlD_4$ (ii) $H_3O^{\oplus}$ (iii) Conc. H_2SO_4 at 165 °C.

b. (i) CH_3 MgI (ii) $H_3O^{\oplus}$ (iii) Conc. H_2SO_4/Δ165 °C.

c. By Darzen glycidic ester condensation using $ClCH_2CO_2Et$.

(a)

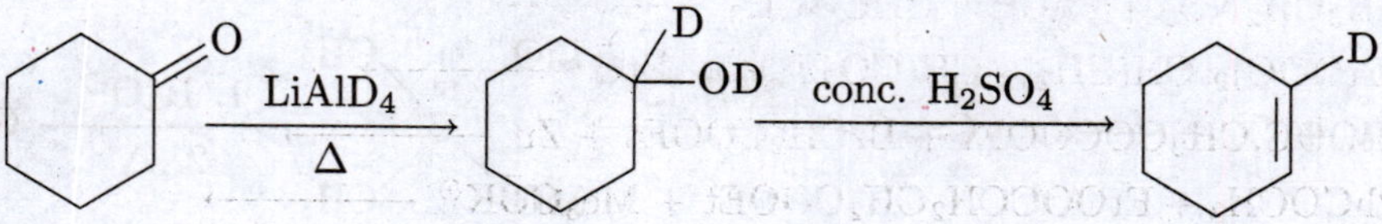

(b) 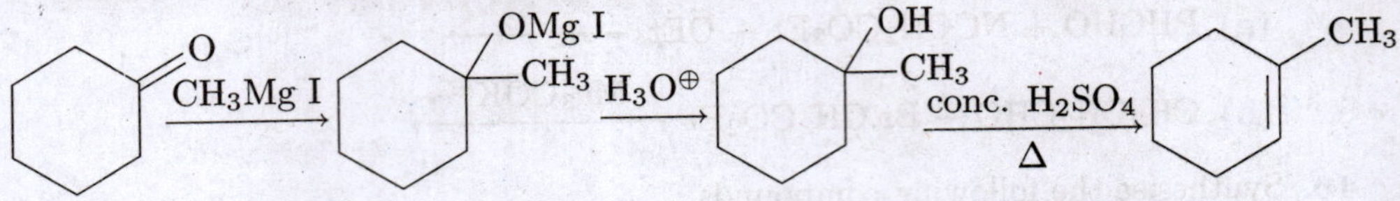

(c) 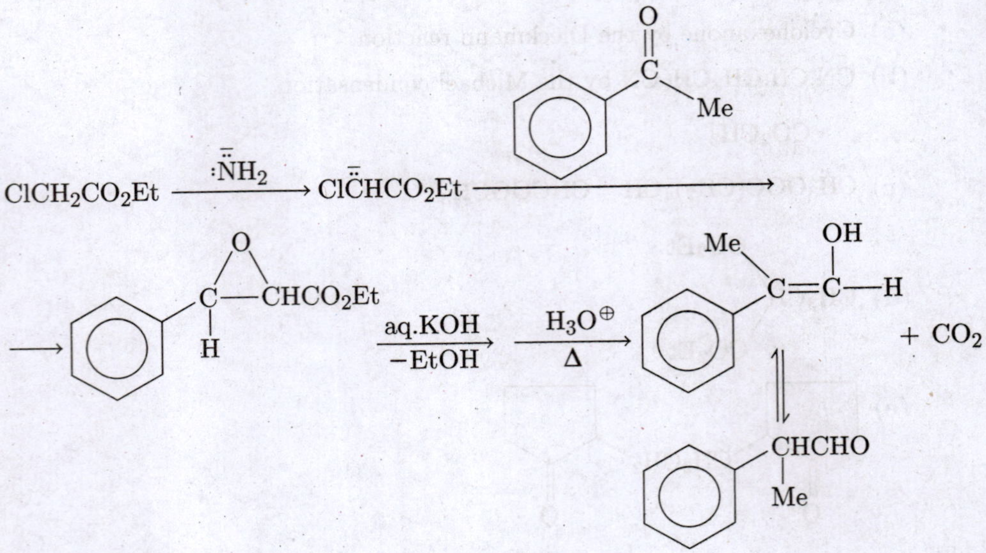

18. Synthesise

(a) $^{14}CH_3{}^{14}CHO$ (b) $CH_3{}^{14}CHO$ (c) $^{14}CH_3CHO$ (d) $CH_3{}^{14}COCH_3$

Answer :

(a) $^{14}C + H_2 \longrightarrow H^{14}C{\equiv}^{14}CH \xrightarrow[H_2SO_4]{Hg^{\oplus\oplus}/20\%} {}^{14}CH_3{}^{14}CHO$

(b) $^{14}C + O_2 \longrightarrow {}^{14}CO_2 \xrightarrow[2.\ H_3O^{\oplus}]{1.\ CH_3MgBr} CH_3{}^{14}CO_2H \xrightarrow{SOCl_2} CH_3{}^{14}COCl$

$\xrightarrow{Li(OEt)_3AlH} CH_3{}^{14}CHO$

(c) $^{14}CH_3{}^{14}CHO \xrightarrow{[O]} {}^{14}CH_3{}^{14}CO_2H \xrightarrow{NH_3} {}^{14}CH_3{}^{14}CO_2NH_4$

$\xrightarrow[-H_2O]{\Delta} {}^{14}CH_3{}^{14}CONH_2 \xrightarrow{NaOH/Br_2} {}^{14}CH_3NH_2 \xrightarrow{HNO_2} {}^{14}CH_3OH$

$\xrightarrow{PBr_3} {}^{14}CH_3Br \xrightarrow[ether]{Mg} {}^{14}CH_3MgBr \xrightarrow[2.\ H_3O^{\oplus}]{1.\ HCHO} {}^{14}CH_3CH_2OH \xrightarrow{[O]}$

(d) 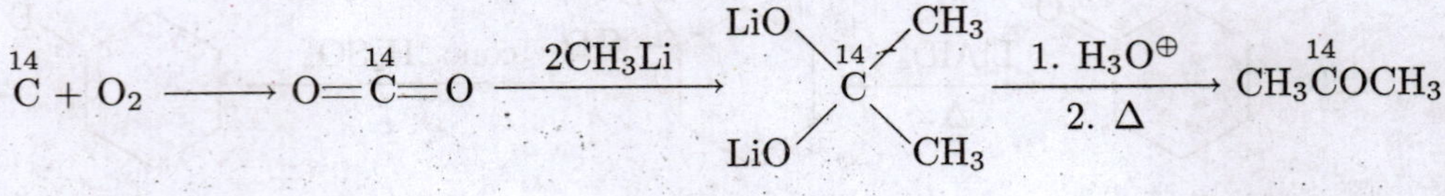

19. Explain

(a) *p*-Aminobenzaldehyde does not undergo the benzoin condensation.

[**Hints :** Aldehydic C is less electrophilic owing to resonance interaction involving amino and aldehydic groups and does not act as an acceptor molecule.]

(b) *o*-Nitrobenzaldehyde undergoes the benzoin condensation but *p*-nitrobenzaldehyde does not.

[**Hints :** $p\text{-}NO_2\phi-\ddot{\bar{C}}(OH)-CN$ has a very weakly nucleophilic anionic C because of the resonance interaction involving *p*-NO_2 group and CN group; so it cannot attack the aldehydic C atom of the other molecule. In the case of *o*-isomer, steric inhibition of resonance occurs and this makes the anionic C of $o\text{-}NO_2\phi-\ddot{\bar{C}}(OH)-CN$ a stronger nucleophile than that of the *p*-isomer—so the reaction takes place.]

20. What happens when two moles of sodium bisulphite are added to acraldehyde?

Ans.

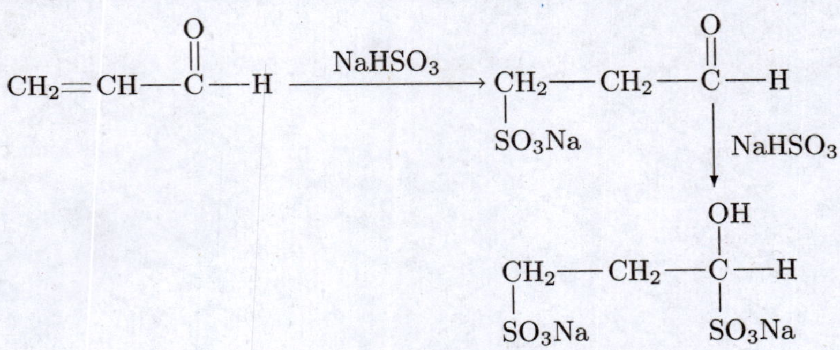

Chapter 17

Aromatic Electrophilic Substitutions

Objectives

- To understand :
  - Electrophilic substitutions [17.1] p. 933
  - Aromatic electrophilic substitutions [17.2] p. 933
- To know :
  - Rates of Ar.S_E2 reactions [17.3] p. 936
  - Orientation in monosubstituted benzenes [17.4.1] p. 938
  - Determination of orientation of disubstituted benzenes [17.4.2] p. 944
  - Orientation in disubstituted benzene [17.4.3] p. 946
  - Reactivity of and orientation in other aromatic systems [17.4.4] p. 946
  - Ortho-para ratio [17.5] p. 952
- To understand :
  - Nitration [17.6] p. 955
  - Sulphonation [17.7] p. 962
  - Halogenation [17.8] p. 968

* The Friedel-Crafts alkylation [17.9.1] p. 975
* The Friedel-Crafts acylation [17.9.2] p. 981
* Chloromethylation [17.10] p. 986
* The Gattermann-Koch aldehyde synthesis [17.11] p. 988
* The Gattermann-aldehyde synthesis [17.12] p. 989
* The Reimer-Tiemann reaction [17.13] p. 990
* The Kolbe-Schmidt reaction [17.14] p. 994
* The Vilsmeier-Haak reaction [17.15] p. 996
* The diazonium coupling reaction [17.16] p. 997
* The Fischer indole synthesis [17.17] p. 1004
* The Skraup quinoline synthesis [17.18] p. 1006

17.1 Introduction

An electrophilic substitution reaction may be represented as :

$$R{-}G + E^{\oplus} \longrightarrow R{-}E + G^{\oplus}$$

Here a leaving electrophile, G, is displaced by $E^{\oplus}$, an incoming electrophile.

An incoming electrophile may be a positively charged species like $\overset{\oplus}{N}O_2$, $\overset{\oplus}{N}O$, $Br^{\oplus}$, $Cl^{\oplus}$, $I^{\oplus}$, $R^{\oplus}$, $R\overset{\oplus}{C}O$, etc. It may also be an uncharged electron-deficient centre of a molecule or an ion. For example, the asterisked atoms of $\overset{*}{S}O_3$, $\overset{*}{S}O_3H$, $\overset{*}{B}F_3$, etc. being electron-deficient, they act as electrophiles.

The weakest Lewis acids are the best leaving electrophiles since they can best exist without a pair of electrons which are required to fill the valance shell. However, protons and metal ions are the usual leaving groups in the aromatic electrophilic substitutions. Of course, there are reactions where SO_3H, $X^{\oplus}$, etc. are found as leaving groups and the substitution reaction is called an Ipso substitution.

Aliphatic as well as aromatic compounds undergo electrophilic substitution reactions; nevertheless these are the characteristic reactions of aromatic compounds. In the cases of aliphatic compounds this type of reaction is not a common one. However, the aliphatic electrophilic substitution reactions may broadly be classified as S_E1 and S_E2. The reactions in which the T.S.s of the rate-determining steps involve one species are S_E1 reactions. On the other hand, if the T.S. of the rate-limiting step of an S_E reaction involves two species, the reaction is said to be S_E2 reaction. In an S_E2 reaction on an sp^3C the bond formation and the bond breaking occur in a single concerted step while in an S_E1 reaction there are two separate steps and the reaction takes place through a carbanion intermediate. Deuterium exchange and decarboxylation are S_E1 reactions. Organometallics undergo S_E2 reactions. Base promoted halogenation of $COCH_3$ group is an example of S_E1 reaction (p. 883).

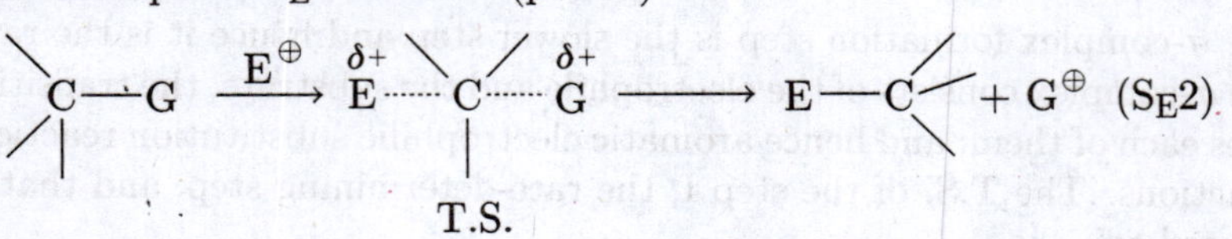

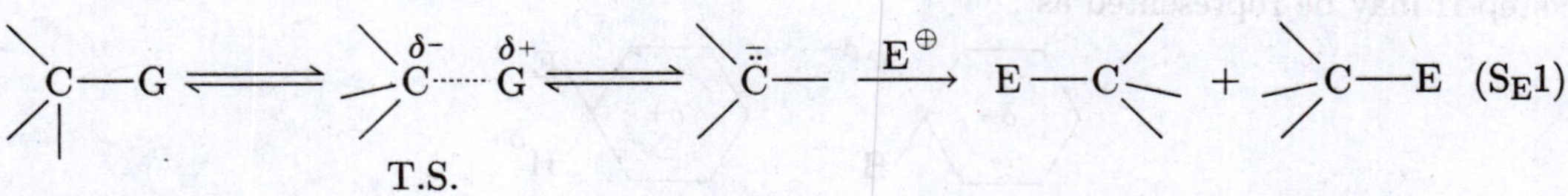

In S_E2 reactions both retention and inversion of configuration have been observed while racemisation takes place in an S_E1 reaction.

However, S_E reactions being characteristic reactions of aromatic compounds, we shall discuss here the aromatic electrophilic substitution reactions in detail.

17.2 Aromatic Electrophilic Substitution Reactions

It has been stated that an aromatic ring has negative charge cloud below and above its plane. Electrophiles being positively charged or electron-deficient species, they are very prone to attack the negative charge cloud of the delocalised πes. Therefore, it might be expected that electrophilic reactions will occur on aromatic rings much readily; but an aromatic compound wants to preserve its aromatic character and hence its stability, and therefore, it is to some extent relactant to utilise its πes. For this reason the electrophiles react with aromatic compounds but not readily, rather slowly. However, the most important type of reactions that the aromatic compounds undergo is the bimolecular electrophilic substitutions which are designated as S_E2 aromatic.

Aromatic S_E2 reactions are supposed to occur in two discrete steps :

Step I. In this step the electrophile, $E^{\oplus}$, is attacked by the aromatic substrate laterally, i.e., perpendicular to the ring and forms a sigma-complex or Wheland complex (or cyclohexadienyl cation in the case of benzene) in which a specific atom of the electrophile forms a covalent bond with a specific C atom of the ring; the involved sp^2 hybridised C atom of the ring then assumes sp^3 hybridised state. Let us represent the step taking benzene as an aromatic substrate.

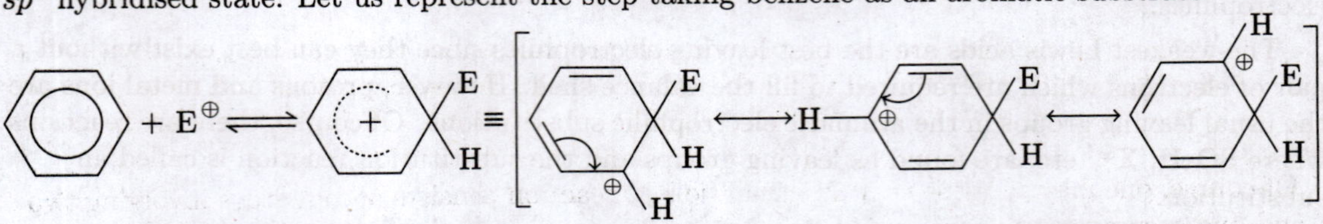

The sigma-complex so formed remains as a salt which is formed by the union of the cationic σ-complex with an anion.

Step II. In this step, the σ-complex gives up a proton and gets converted into another aromatic compound, a derivative of the aromatic substrate; the proton is removed by the anionic part of the σ-complex salt.

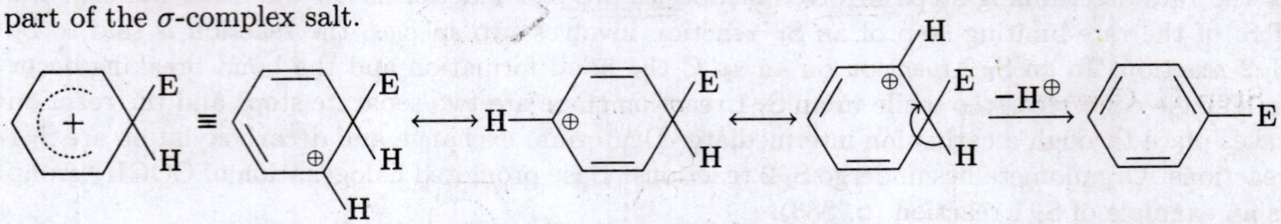

In most of the cases the σ-complex formation step is the slower step and hence it is the rate-determining step. Since the σ-complex consists of the electrophile and the substrate, the transition state of the reaction involves each of them; and hence aromatic electrophilic substitution reactions are usually bimolecular reactions. The T.S. of the step I, the rate-determining step, and that of the step II may be represented as :

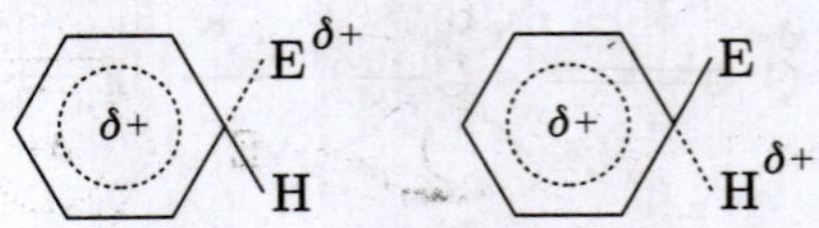

T.S. I T.S. II

Therefore the overall reaction may be written as :

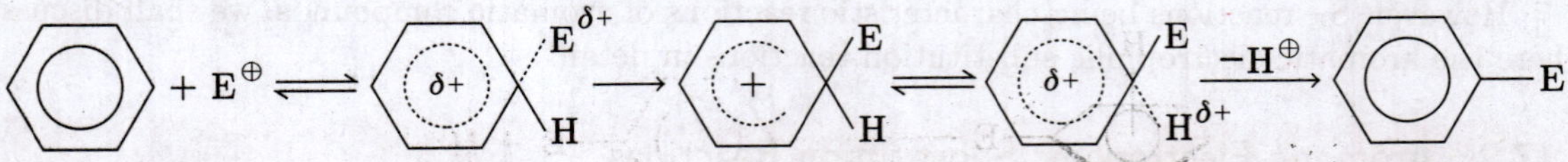

Since the T.S. of the rate-determining step of an aromatic S_E2 reaction involves two species, it is expected that the reaction will follow the second order kinetics.

The energy profile for the reaction is shown in Figure. 17.1.

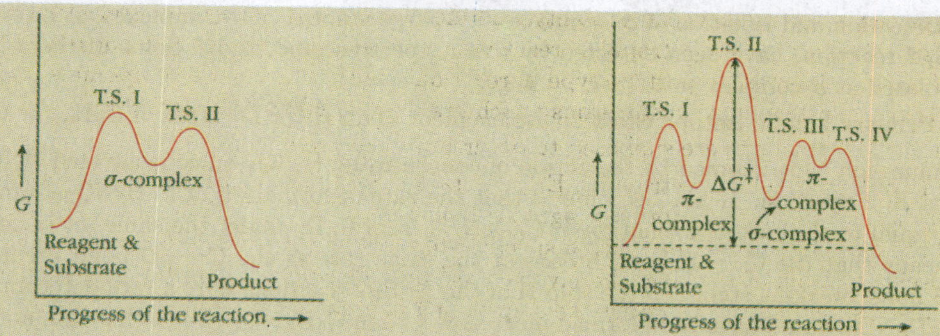

(a) The energy profile of an exothermic aromatic S_E2 reaction without the formation of a π-complex

(b) The energy profile of an exothermic aromatic S_E2 reaction involving π-complex

Fig. 17.1

Of course, one may suppose that an aromatic S_E2 reaction consists of four steps involving two π- and one σ-complexes as shown below :

Step I. Formation of a π-complex :

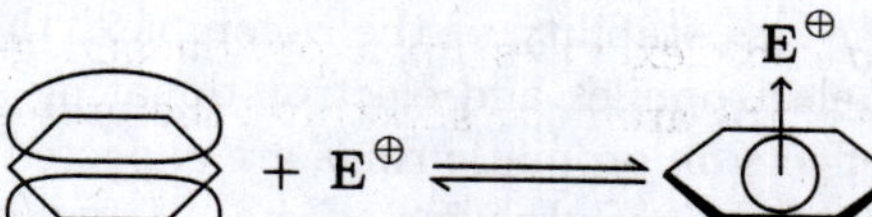

Step II. Conversion of the π-complex to a σ-complex :

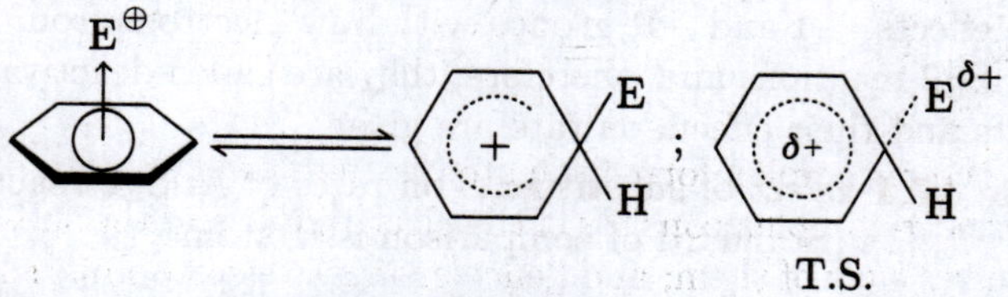

Step III. Conversion of the σ-complex into a π-complex :

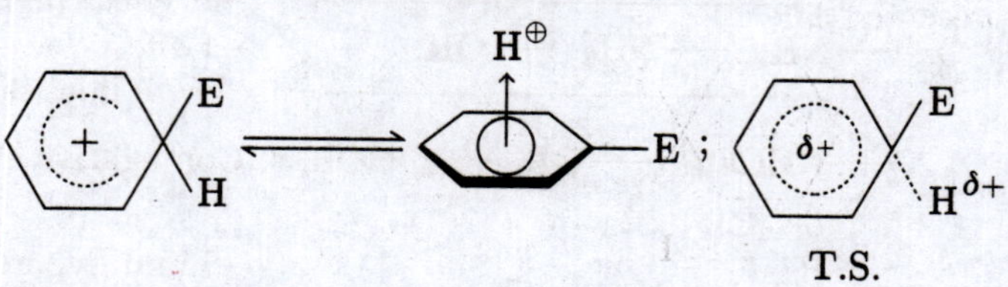

Step IV. Decomposition of the π-complex to products :

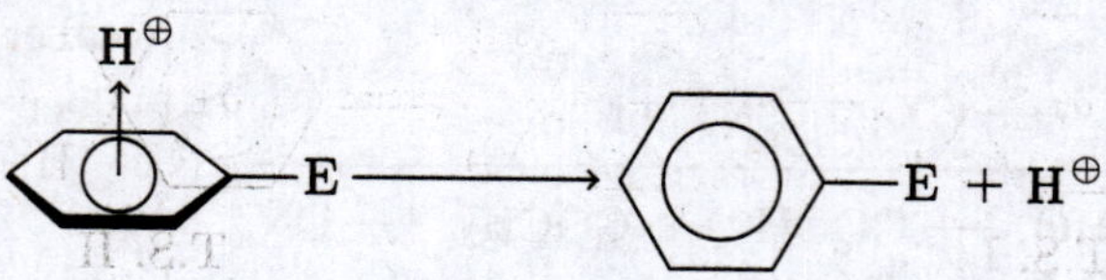

However, the most widely accepted mechanism for aromatic electrophilic bimolecular substitution reactions is the two-step mechanism.

The evidences for the two-step mechanism are:

(i) Detection and isolation of σ-complexes : A large number of σ-complexes, as intermediates in Ar.S_E2 reaction, have been detected by n.m.r. spectroscopic study. Some of them have also been isolated as σ-complex salts (Sec. 11.4.8.).

(ii) Primary kinetic isotope effect : Displacement of an H (or D) atom of C_6H_6 (or C_6D_6) by a nitronium ion $\overset{\oplus}{N}O_2$ (formed by the action of concentrated H_2SO_4 on concentrated HNO_3) is an aromatic S_E2 reaction. It has been found that the rate of formation of $C_6H_5NO_2$ from C_6H_6 is almost equal to the rate of formation of $C_6D_5NO_2$ from C_6D_6 under the same set of conditions. This means that the C—H bond breaks at the same rate as the C—D bond breaks. If the C—H bond breaking step, i.e., the step II of the Ar.S_E2 reactions were a rate-determining step, then $C_6D_5NO_2$ would have been formed more slowly than $C_6H_5NO_2$. Since this does not happen, the σ-complex formation step, i.e., the first step is the rate-determining step.

17.3 Rate of Ar.S_E2 Reactions

There is no doubt that the π-electron cloud of an aromatic system invites an electrophile to react, nevertheless the electrophilic substitution reactions do not occur readily. Owing to the delocalisation of $(4n+2)\pi$ electrons in a planar cyclic system, an aromatic substrate is more stable than the corresponding resonance-stabilised σ-complex which is formed as an intermediate in an Ar.S_E2 reaction; because of the less stability of the σ-complex, the aromatic compounds react but slowly. Of course, strong electrophiles and electron-donating substituents, which have +I, +R and/or hyperconjugative effects on aromatic rings are expected to enhance the rate. On the other hand, substituents with electron-withdrawing effects, −I and/or −R, and weak electrophiles decrease the rate of an Ar.S_E2 reaction. The groups which increase the rates of Ar.S_E2 reactions by donating electron to the ring are called activating groups. They activate the ring by +I, +R and hyperconjugative effects. −I and −R groups withdraw electrons from an aromatic ring and thereby slow down Ar.S_E2 reactions and, therefore, they are called deactivating groups. In Table 17.1, some substituents and their effects on rate are given.

Table 17.1 Effect of substituents on rates of Ar.S_E2 reactions
(Standard of comparison is H atom)

Nature of group	Decreasing order of rate	Polar effect
Activating	$—\ddot{\underset{..}{O}}:^{-}$, $—\ddot{N}R_2$, $—\ddot{N}H_2$, $—\ddot{\underset{..}{O}}H$, →	+R effect predominates over the −I effect; $—\ddot{\underset{..}{O}}:^{-}$ does not have −I effect
	$—NHCOR$, $—\ddot{\underset{..}{O}}R$, $—\ddot{\underset{..}{O}}COR$	Conjugative effect
	$—Ar$	
	$—R$	+I and hyperconjugative effects
Deactivating	$—\ddot{\underset{..}{X}}:$	−I effect predominates over the +R effect
	$—CX_3$	−I effect
	$—CHO$, $—COOH$, $—COCH_3$, $—CN$, $—SO_3H$, $—NO_2$ →	−I and −R effects
	$\overset{\oplus}{S}R_2$, $\overset{\oplus}{N}R_3$	−I effect

However, the rate of an Ar.S_E2 reaction depends on the G_{act} (free energy of activation), i.e., on the stability of the T.S. of the rate-determining step; the higher the stability of the T.S., i.e., the lower the value of the G_{act}, the faster is the reaction. In an Ar.S_E2 reaction the structure and stability of the T.S. of the rate-determining step are similar to those of the σ-complex since this step is endothermic (Hammond postulate). Therefore, the stability of the T.S. may be determined by considering the factors which stabilise or destabilise the σ-complex. Since the T.S. bears a partial positive charge on the attacking electrophile and that on the ring, +I, +R and hyperconjugative effects will bring stability to the T.S. and hence the G_{act} will be lowered by an activating, i.e., electron-donating substituent, which in its turn, will enhance the rate of the reaction. The electron-withdrawing, i.e, deactivating substituents, on the other hand, decrease the rate by destabilising the T.S., i.e., increasing the G_{act} of the T.S. For example, the decreasing order of rates of nitration on benzene, nitrobenzene and toluene is :

Toluene > benzene > nitrobenzene.

The relative rates may be determined experimentally by employing an inadequate amount of the nitrating agent (a mixture of concentrated nitric acid and concentrated sulphuric acid) to react with a mixture of benzene, nitrobenzene and toluene and then measuring the amounts of nitrobenzene, dinitrobenzene and nitrotoluenes formed in the reaction. The amount of nitrotoluenes formed is found to be the highest and that of dinitrobenzene is the lowest. Since Ar.S_E2 reactions are usually kinetically controlled, we may say that the rate of nitration of nitrobenzene is the lowest and that of toluene is the highest; while the rate of nitration of benzene is lower than that of toluene but higher than the rate of nitration of nitrobenzene. However, the rate order may be explained as follows :

+I and hyperconjugative effects of the methyl group in toluene increase the electron-density of the ring and thus minimise the positive charge that develops in the T.S. owing to the attack of the electrophile. Therefore, the electron-donating effect of the methyl group brings stability to the T.S. and lowers G_{act}. In the case of nitrobenzente, −I effect and −R effect of the nitro group help it to withdraw electrons from the ring of the T.S. and thereby make the ring more positively charged. Thus the T.S. becomes highly unstable and its G_{act} is appreciably high. The positive charge of the T.S. of the nitration of benzene is neither decreased by −I and/or −R effects nor it is increased by +I and/or +R effects. So, the order of G_{act}s of the T.S.s of the nitration reaction is :

Nitrobenzene > benzene > toluene.

Therefore, the order of the rates of nitration will be :

toluene > benzene > nitrobenzene.

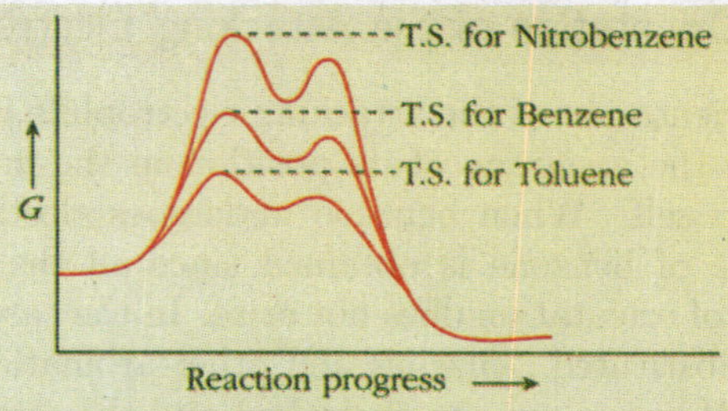

Fig. 17.2. Relative positions of the T.S.s (not in scale) in the energy profiles for the nitration of benzene, nitrobenzene and toluene

This may be represented in the energy profile as shown in Figure 17.2.

Since I effects are reasonably strong up to C_2 atom and since they virtually die down at C_4 atom, a +I group activates the *o*-positions more than the *p*-positions and a −I group deactivates the *p*-position less than the *o*-positions. Therefore, an aromatic compound with a +I substituent will undergo electrophilic substitution at the *o*-position more readily than at the *p*-position and the case is reverse when there is a −I group in the compound in place of the +I substituent. Thus when toluene is nitrated, the *o*-product predominates (yield 58%) over the *p*-product (yield 37%);

the +I effect of the Me group in toluene activates the *o*-positions more than the *p*-position and as a result of which the rate of the *o*-substitutions is greater than that of the *p*-substitution. On the other hand, when chlorobenzene is nitrated, the *p*-product is obtained as the predominant product (yield: *p*-isomer—70%, *o*-isomer—30%). The *p*-position being less deactivated than the *o*-positions by the −I effect of the chloro group, the rate of nitration at the *p*-position is more than the rate of nitration at *o*-positions.

Groups with predominant +R effect activate the *p*-position more than the *o*-positions and consequently the rate of electrophilic substitution on the *p*-position is higher than that on the *o*-positions when a +R group is the substituent. For example, —NHCOCH$_3$ group has −I and +R effects, but the +R effect is stronger than the −I effect; thus when acetanilide is nitrated, the yield of the *p*-product is almost 80%.

Problem 17.1. Write down the relative rate order of monobromination of the following compounds and assign reasons for the order.

$$PhNH_2, PhCOOH, PhH, PhBr, PhNO_2, PhCH_3.$$

Solution : Among the given compounds PhNH$_2$ and PhCH$_3$ have activating groups, —NH$_2$, and —CH$_3$ respectively. NH$_2$ has a strong +R effect and CH$_3$ has +I effect and hyperconjugative effect; +R effect being stronger than +I and hyperconjugative effects, PhNH$_2$ will undergo the reaction more readily than PhCH$_3$. Again, PhCOOH, PhBr, PhNO$_2$ have deactivating groups; the order of increasing deactivating effects is :

$$—Br < —COOH < —NO_2.$$

So, among these three the rate order will be :

$$PhBr > PhCOOH > PhNO_2.$$

PhH has neither a deactiving nor an activating group; so, it will be more reactive than the compounds with deactivating groups. Therefore, the decreasing rate order is :

$$PhNH_2 > PhCH_3 > PhH > PhBr > PhCOOH > PhNO_2.$$

17.4 Orientation of the Attacking Electrophile

By orientation of the attacking electrophile in an aromatic electrophilic substitution reaction we mean the assigning of its position in the aromatic ring in which the attacking electrophile attaches itself. When benzene undergoes electrophilic substitution, only one monosubstituted derivative of benzene is obtained since all the H atoms of benzene are equivalent. Here the question of orientation does not arise. In the cases of kinetically controlled reactions of mono and higher substituted benzenes and other aromatic rings (e.g., pyrrole, pyridine, naphthalene), all the ring H atoms are not equivalent and the attacking electrophile attaches itself to that C atom where the electron-density is relatively high. In those cases the problem of assigning position of the attacking electrophile in the ring deserves a separate discussion.

17.4.1 Orientation in monosubstituted benzenes

When a monosubstituted benzene Ph—X, undergoes an electrophilic substitution reaction, the incoming electrophile ($E^{\oplus}$) may attach itself to the ortho, or meta, or para position with respect to the substituent already present (X) in the ring. In fact, the substitutent, X, on the ring determines which position the attacking electrophile, $E^{\oplus}$, will take. When an activating group or a halogen atom is the substituent, it directs the incoming electrophile to take the ortho and the para positions and the substituent is then said to be a first class group or an ortho-para

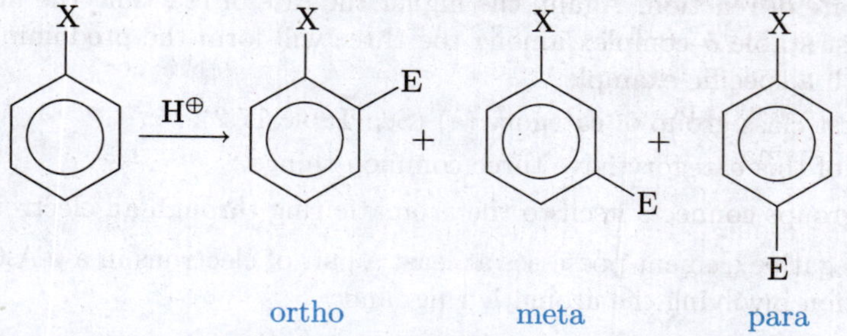

directing group. Deactivating substituents (excluding halogens), on the other hand, directs the attacking electrophile to take the meta-positions and are known as the second class groups or meta-directing groups. Polar effects (I or R) of the groups are actually responsible for their directing nature. However, a list of groups, their polar effects and directing nature are given in Table 17.2.

Table 17.2 Polar effects and directing nature of some groups

	Groups	Polar effects	Directing nature
(a)	$—O^-$	+R	*o, p*
	$—NR_2$, $—NH_2$, —OH, —NHCOR,		
	—OR, —SR, —O—, COR	−I, +R	*o, p*
	—F, —Cl, —Br, —I, —Ph	−I, +R	*o, p*
(b)	—Me, $—CH_2Me$, $—CHMe_2$	+I, hyperconjugative	*o, p*
	$—CMe_3$	+I	*o, p*
	$—CH{=}CH_2$	+I	*o, p*
(c)	$—\overset{\oplus}{N}H_3$, $—\overset{\oplus}{N}R_3$, $—\overset{\oplus}{S}R_2$, $—\overset{\oplus}{P}R_3$		
	$—CF_3$, $—CCl_3$	−I	*m*
(d)	$—NO_2$, —CN, $—SO_3H$, —COOH,		
	$—SO_3R$, $—B(OH)_2$, —CHO, —COR	−I, −R	*m*

Although the substituents have definite directing effects yet exclusively *o*-/*p*- or *m*-disubstituted products are seldom obtained from a monosubstituted benzene on further electrophilic substitution. For example, in the nitration of acetanilide ($PhNHCOCH_3$; substituent is $—NHCOCH_3$) one expects a mixture of *o*-nitroacetanilide and *p*-nitroacetanilide in so far as the directing nature of the $—NHCOCH_3$ group is concerned (See Table 17.2). In practice, however, 19% *o*-product, 79% *p*-product and 2% *m*-product are obtained. Therefore, the directing nature of a substituent tells us which product will form predominantly. However, why are some groups *o*-/*p*-directing and some others *m*-directing? There are two ways to answer the question :

(i) Explanation on the basis of the stability of the σ-complexes.

(ii) Explanation on the basis of the charge distribution in the substrate.

(i) Stability of the σ-complexes and the orientation effect : This method tells us that the resonating structures of the three possible σ-complexes for the three products, ortho, meta and para, are to be drawn and their relative stabilities are to be determined. We know from Sec. 17.3, that greater the stability of the σ-complex, the lower is the energy of activation of the T.S. and the higher is the rate of reaction. Again, the higher the rate of reaction, the higher is the yield. Therefore, the most stable σ-complex among the three will form the predominant product. Let us explain this with a specific example.

—$\ddot{N}H_2$ is a first class group of category (a) (See Table 17.2).

All the groups of this category have three common things :

(a) each of the groups connects itself to the aromatic ring through an electronegative element,

(b) this electronegative element possesses at least a pair of electrons in a p A.O. which undergo p-π conjugation involving the aromatic ring, and

(c) the electronegative element does not have any π bond.

These three common factors are responsible for the o-/p-directing nature of the first class groups of category (a). Now let us apply the method to determine the directing nature of the $\ddot{N}H_2$ group. The three possible σ-complexes are :

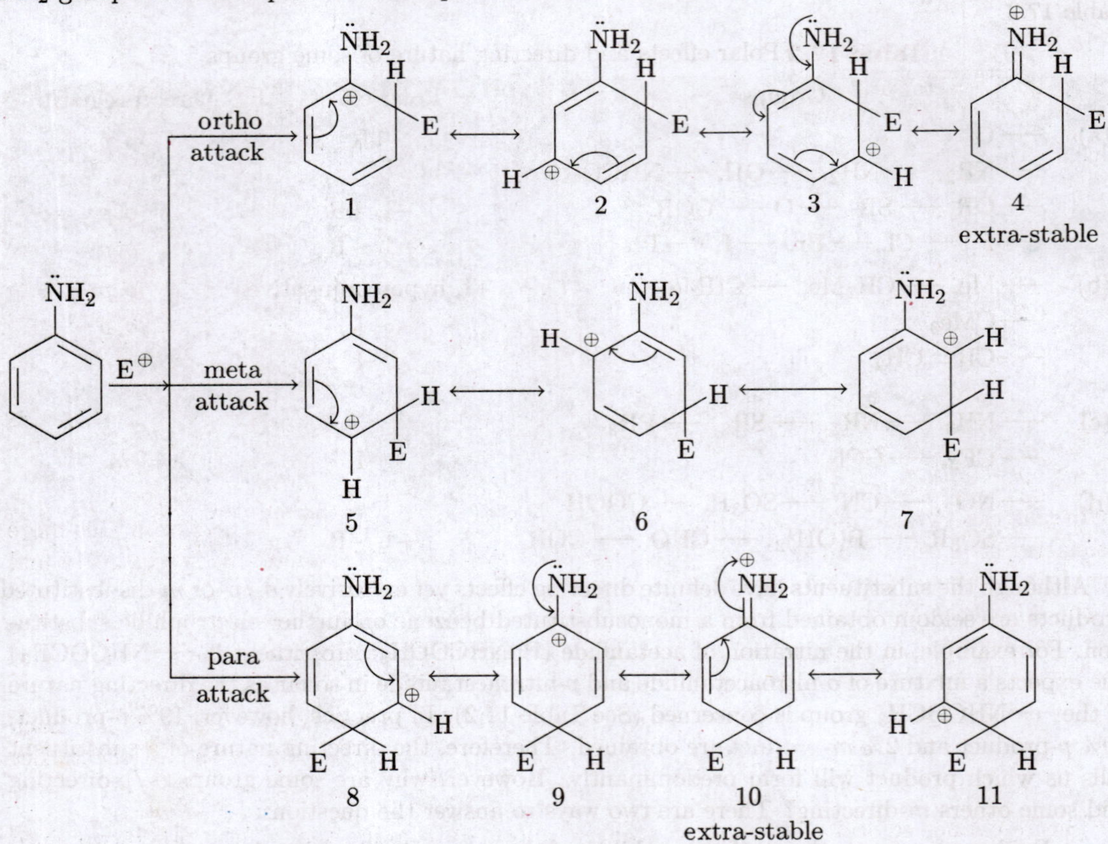

Thus, we see that each of the ortho and para complexes has extra resonating structure (4 and 10 respectively) in their resonance hybrids; these structures are extra-stable since all the C and N atoms are in the octet state. These two structures bring extra stability to the ortho and the para complexes. Therefore, two T.S.s for the o-attack and that for the p-attack will have lower energy than the T.S. for the *m*-attack. As a consequence the rates of reaction at the *o*- and *p*-positions will be greater than the rate of reaction at the *m*-position which will lead to form predominant *o*-/*p*-products. Thus, the groups of category (a) are *o*-/*p*-directing.

Let us now consider the case of a first class group of category (b) (See Table 17.2). All of them exert +I effect and hyperconjugative effect; only—CMe_3 does not exert hyperconjugative effect.

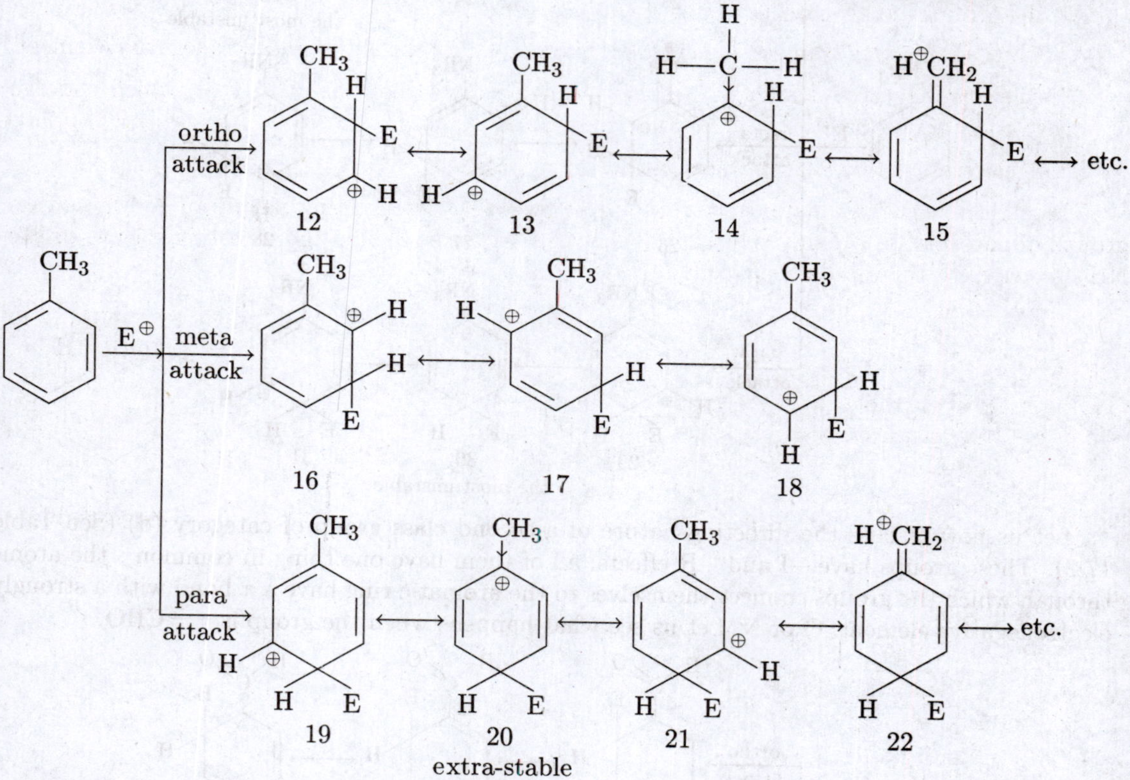

In this case we see that each of the complexes for the ortho and para attacks contains more resonating structures than the complex for the *m*-attack and at the same time structure 14 and structure 20 are extra-stable since in these two structures the positive charge is directly neutralised by the +I effect of the methyl group. Therefore, *o*- and *p*-σ-complexes are more stable than the *m*-σ-complex and naturally a+I group will direct the incoming electrophile to the ortho and para positions. Therefore, *o*-/*p*-products will also form predominantly.

Let us now consider the case of a second class group of category (c) (See Table 17.2). Each of the groups of this category withdraws electrons from the aromatic ring by the –I effect. Let us take the case of —$\overset{\oplus}{N}R_3$ group :

The resonance hybrid of the *m*-σ-complex is more stable than that of the other two since each of the resonance hybrids of *o*-and *p*-σ-complexes contains one very unstable structure (25 and 30 respectively). In these two structures the substituent —$\overset{\oplus}{N}R_3$ draws electrons from an electron-deficient positively charged C atom which make them very unstable. The *m*-σ-complex being the most stable; *m*-product will form predominantly, and the groups with –I effect will be called *m*-directing groups.

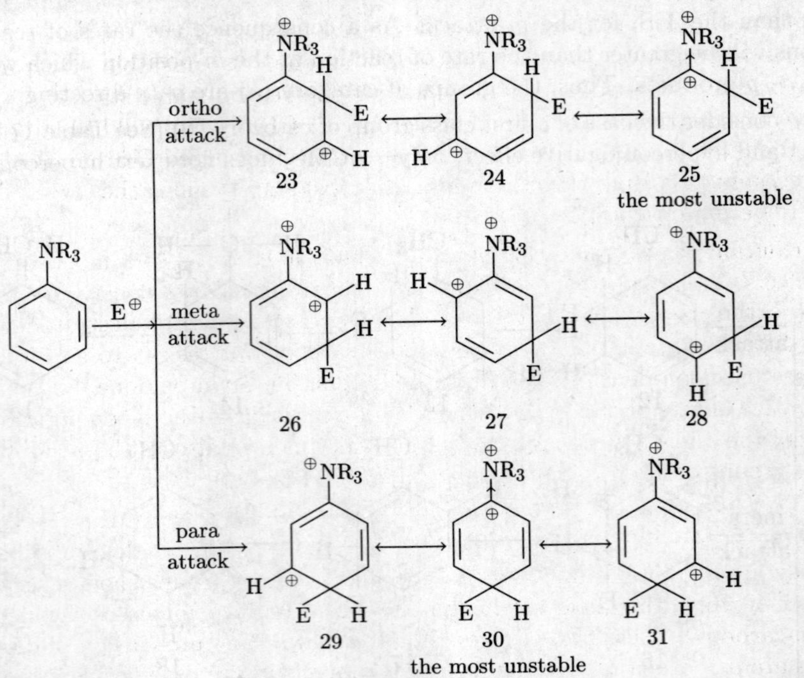

Let us now discuss the directing nature of a second class group of category (d) (See Table 17.2). These groups have $-$I and $-$R effects; all of them have one thing in common : the atoms through which the groups connect themselves to the aromatic ring have a π bond with a strongly electronegative element, O or N. Let us see what happens when the group is —CHO.

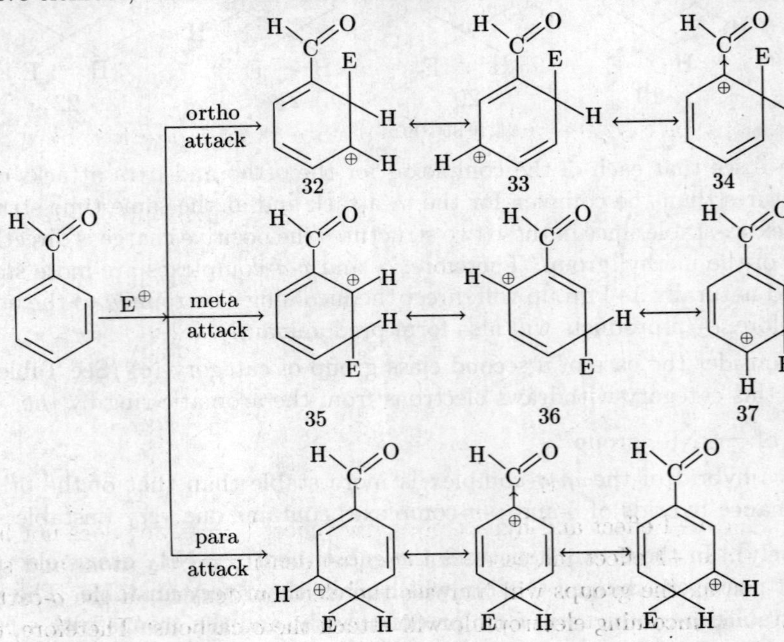

Here again we find that the m-σ-complex is more stable than the other two σ-complexes. The resonance hybrid of the o-σ-complex contains one very unstable structure (34); and the resonance hybrid of the p-σ-complex also contains one very unstable structure (39); structure 34 and structure 39 are specially unstable because the —CHO group will have to withdraw electron (by –I & –R effects) from an electron-deficient positively charged C atom. Thus the m-product will form predominantly since the m-σ-complex is most stable among the three. Therefore, groups of category (d) are m-orienting.

(ii) Distribution of electronic charge and orientation effects : In a monosubstituted benzene the polar effects of the substituent disturb the electronic charge density on the ring carbons. Some of the ring carbons become electron-dense and some others electron-deficient. Since an electrophile attacks the relatively electron-dense carbon, our task is to ascertain which carbon has the highest electron-density relative to the others. This can be done by considering I, R and hyperconjugative effects of the substituent. The orientation of the incoming electrophile will be the position of the highest electron-dense carbon. Let us now discuss the polar effects of the four categories of groups on the ring carbons,

(a) Groups of Category (a) : —$\bar{O}$, —NR_2, —NH_2, —OH, —NHCOR, —OR, —SR, —O—, —OCOR, —F, —Cl, —Br, –I, etc. (halogroups have deactivating effect yet they are o-/p-orienting for their +R effect). These groups show –I effect (except O^-) and +R effect of which the latter predominates. The +R effect increases the electron density on the o- and p-carbons. Therefore, electrophilic attack will be on o-carbon and p-carbon. Hence these groups are o/p orienting. Let us see what happens in the case of phenoxide ion.

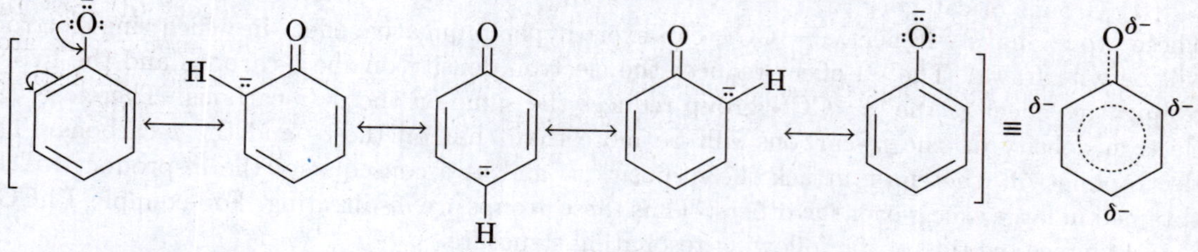

Since m-positions have no electrical charge and since o-and p-positions have partial negative charge, the —$\bar{O}$ group will be o-/p-directing.

(b) Groups of category (b) : —CH_3, —CH_2CH_3, —$CHMe_2$, —CMe_3, etc.

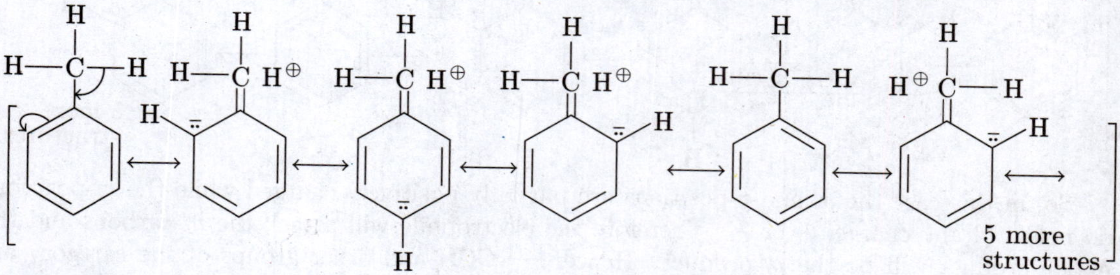

These groups have +I effect and hyperconjugative effect (—CMe_3 does not have hyperconjugative effect). Since +I effect increases the electron-density on C_2 atom and since the effect dies down at C_4 atom, the groups will increase the electron density on the o-carbons appreciably and naturally the incoming electrophile will attack the o-carbons. Therefore, the groups are o-directing. The hyperconjugative effect of the groups also increases the electron density on the

p-carbon reasonably and therefore, they are also p-orienting groups. Thus, the hyperconjugative effect of the methyl group increases.the electron density on the o-/p-positions; therefore, methyl group is an o-/p-directing group.

(c) Groups of category (c) : —NO_2, —CN, —SO_3H, —SO_3R, —COOH, —COOR, —CHO, —COR, etc. These groups have −I and −R effects; the −R effect is the predominant effect. The −R effect makes the o/p carbons less electron-dense than the m-carbons. Therefore, the incoming electrophile will attack the m-carbon and predominant m-product will be obtained; thus the groups will be m-orienting. Let us study the situation with nitrobenzene.

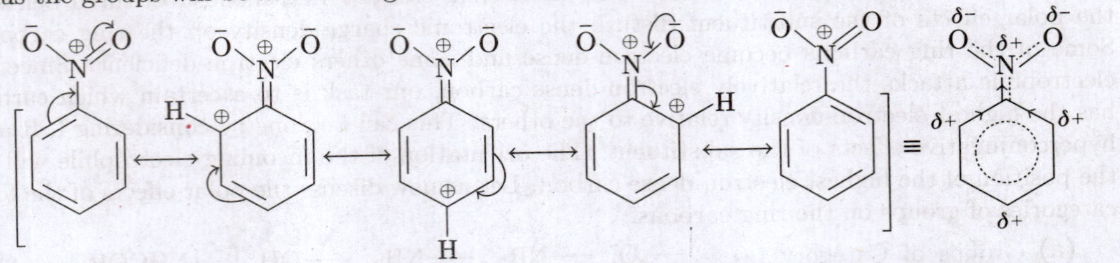

Thus, the o- and the p-positions become partially positively charged owing to the presence of —NO_2 in the benzene ring; while m-positions do not bear any charge, positive or negative. Therefore, the incoming electrophile will attack the m-carbons and the m-product will be formed predominantly. So —NO_2 group and other groups of the category (c) are m-directing.

(d) Groups of category (d) : —$\overset{\oplus}{N}H_3$, —$\overset{\oplus}{N}R_3$, —$\overset{\oplus}{P}R_3$, —$\overset{\oplus}{S}R_2$, —CCl_3, —CF_3, etc. These groups have −I effect, —CCl_3 can exert hyperconjugative effect in which ring-electrons will be withdrawn. The −I effect reduces the electron density on the o-carbons and the hyperconjugative effect of the —CCl_3 group reduces the same on the o-/p-carbons. Therefore, the electron density on the m-carbons will be more than that on the o- and the p-carbons. The electrophile will, therefore, attack the m-positions and as a consequence the m-product will be obtained in large excess over the others. Thus these groups are m-directing. For example, $PhCCl_3$ is a resonance hybride of the following resonating structure.

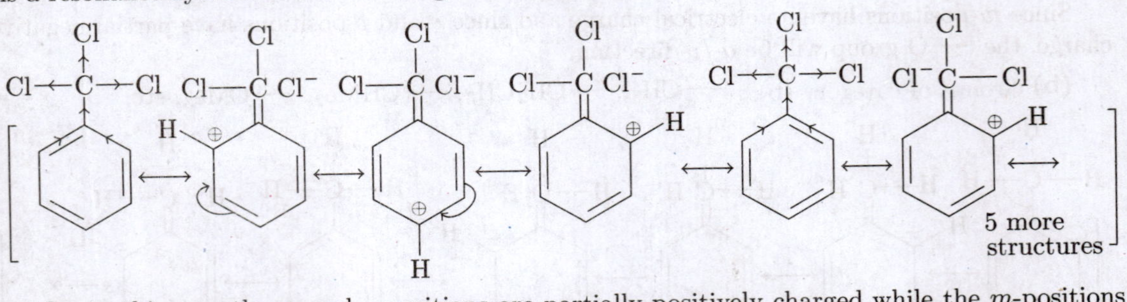

So, in this case the o- and p-positions are partially positively charged while the m-positions do not hold any charge, ⊕ or −. Therefore the electrophile will attack the m-carbons and the major product will be the m-product. Hence, —CCl_3 and other groups of the category are m-orienting.

17.4.2 Determination of orientation of groups in disubstituted benzenes

The relative positions of two groups in a disubstituted benzene may be determined in several ways. Some of them will be discussed here.

(i) Körner's absolute method : The method is based on the principle that the introduction of a third substituent by direct replacement of hydrogen of the *p*-isomer gives only one trisubstituted product, the *o*-isomer two and the *m*-isomer three trisubstituted products. By carrying out nitration or bromination on a disubstituted benzene, then separating the reaction products and subsequently establishing the number of isomers formed, one can determine the orientation of the two substituents in the disubstituted benzene under study. For example, to determine the orientation of xylenes we may introduce a bromo group as the third substituent.

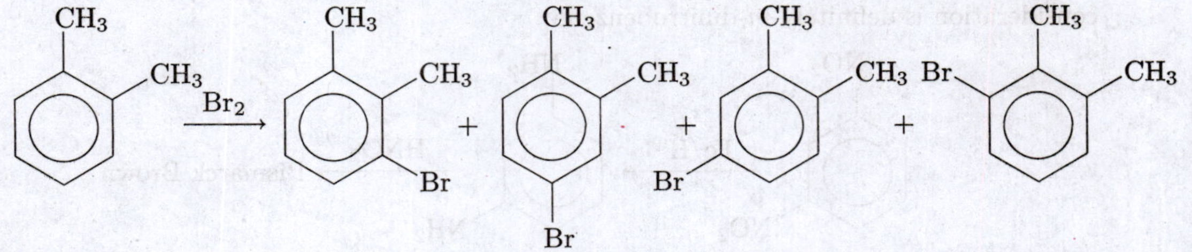

o-derivative I II III IV

I≡IV and II≡III

So, *o*-xylene will give two products :

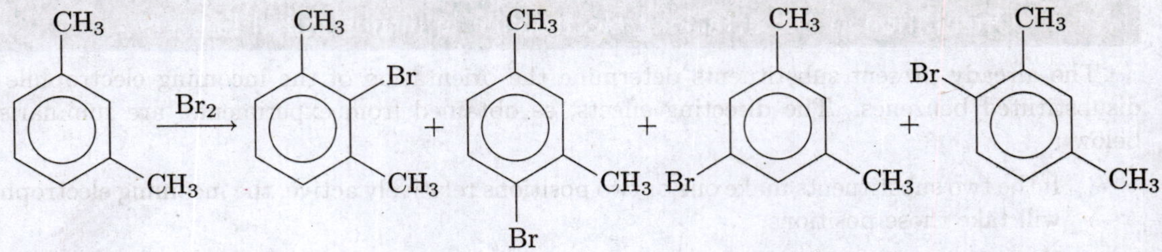

m-derivative V VI VII VIII

VI≡VIII

Hence, *m*-xylene will give three products :

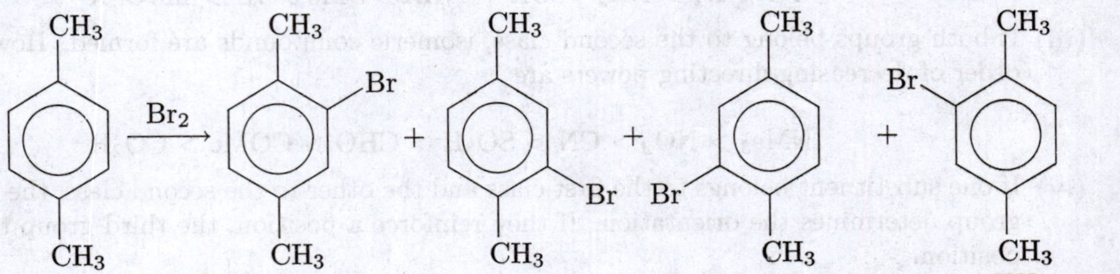

p-derivative IX X XI XII

IX≡X≡XI≡XII

p-Xylene, therefore, will give one product only.

Since products are formed in highly unequal amounts and since the separation of products is difficult, it is not a successful method though Körner applied the method successfully to establish the orientation of the isomeric dibromobenzenes. He performed nitration on them.

(ii) Relative methods : In this method the compound under study is synthesised from or converted to a compound of known orientation. The orientation of the substituents will be the same as that of the converted compound or of the starting compound from which it is synthesised.

Let us take some examples to understand the method.

(a) Among the three isomeric phenylenediamines the *m*-phenylenediamine only forms a brown dye, known as Bismarck brown, on diazotisation with nitrous acid. If a dinitrobenzene on reduction and subsequent diazotisation gives Bismarck Brown, the dinitrobenzene under consideration is definitely *m*-dinitrobenzene.

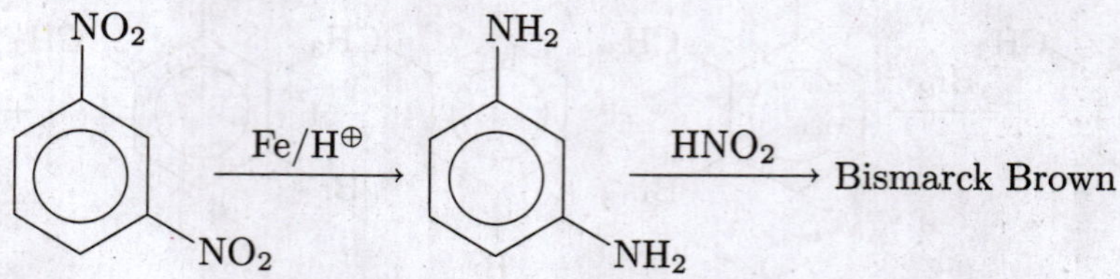

(b) *o*-Phenylenediamine is the only phenylenediamine which forms heterocyclic compounds with dicarbonyls. Therefore, if a phenylenediamine forms a heterocyclic compound with a dialdehyde or a diketone, it is beyond doubt that the compound is *o*-phenylenediamine.

17.4.3 Orientation of the attacking electrophile in disubstituted benzenes

The already present substituents determine the orientation of the incoming electrophile in disubstituted benzenes. The directing effects, as obtained from experiments, are summarised below :

(i) If the two substituents make one or two positions relatively active, the incoming electrophile will take those positions.

(ii) If the two substituents belong to the first class, the stronger *o*/*p*-director will control the orientation; nevertheless several trisubstituted isomers will form.

The order of decreasing directing powers of the first class groups is :

$$\bar{O} > NR_2 > NH_2 > OH > NHAc, OMe > Ar > Me > X$$

(iii) If both groups belong to the second class, isomeric compounds are formed. However, the order of decreasing directing powers are

$$\overset{\oplus}{N}Me_3 > NO_2 > CN > SO_3H > CHO > COMe > CO_2H$$

(iv) If one substituent belongs to the first class and the other to the second class, the first class group determines the orientation. If they reinforce a position, the third group takes that position.

(v) A very small substitution occurs between two *m*-substituents since the position is highly sterically hindered.

17.4.4 Orientation and reactivity of aromatic compounds other than benzene

In this section we shall discuss the problem of orientation and reactivity of naphthalene, anthracene, furan, pyrrole, thiophene, pyridine, indole and quinoline.

(a) Napthalene: Naphthalene contains two benzene rings fused at *o*-positions. Since it contains two benzene rings, it is expected that the resonance energy of naphthalene will be very close to $150.6 \times 2 = 301.2$ kJ mol^{-1}, but its resonance energy is actually 255.2 kJ mol^{-1}; this is much lower than our expectation. As a consequence naphthalene is less aromatic than benzene and therefore, it is more reactive than the latter. So, an electrophilic substitution on naphthalene occurs faster than that on benzene.

It has already been stated that all the H atoms in benzene are equivalent and hence benzene forms only one monosubstituted derivative. In naphthalene, however, α-H and β-H are not equivalent. The resonance hybrid of the σ-complex for the α-attack is found to be more stable than the resonance hybrid of the σ-complex for the β-attack.

Thus, we see that resonance hybrid of the σ-complex for α-attack contains 7 resonating structures while that for β-attack contains 6 such structures. Furthermore, the former contains 4 structures containing benzene ring whereas the latter contains only two structures with benzene ring. Canonical structures with benzene ring being more stable than those without benzene ring, the σ-complex for α-attack is more stable than that for β-attack. Consequently, electrophilic substitution on naphthalene occurs mostly at α-position. Of course, reversible sulphonation at high temperature occurs at β-position since this gives themodynamically stable product.

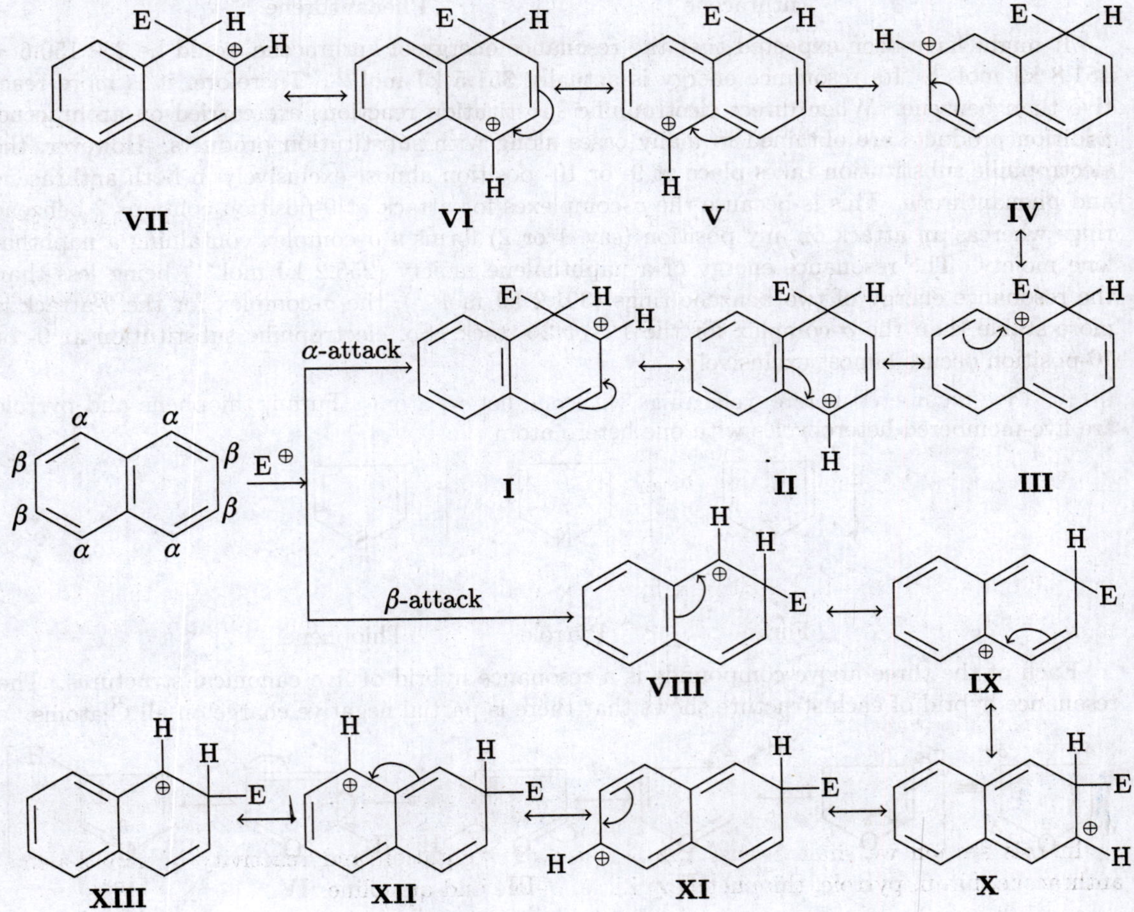

However, the predominant product for further substitution on monosubstituted naphthalenes may be predicted by the following empirical rules.

(i) An activating group at 1-position directs an incoming electrophile to 4-position preferentially and to a lesser extent to 2-position. If such a group is at position 2, it directs the electrophile to take position 1.

(ii) A deactivating group directs the attacking electrophile to take a position in the second ring. For example, when further nitration is performed on 1-nitronaphthalene, a mixture of 1, 5 and 1, 8-dinitronaphthalenes is obtained.

(b) Anthracene and phenanthrene : Anthacene has three benzene rings fused linearly.

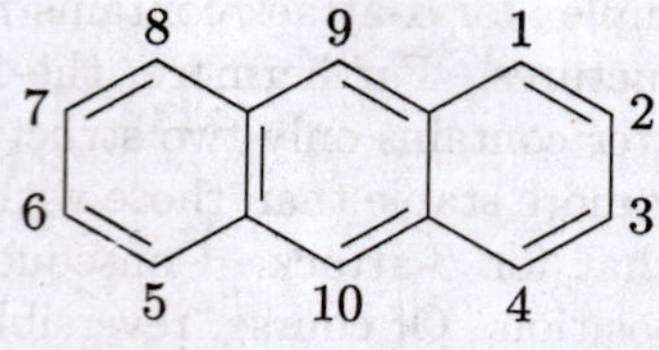
R.E. = 351.5 kJ mol^{-1}

Anthracene

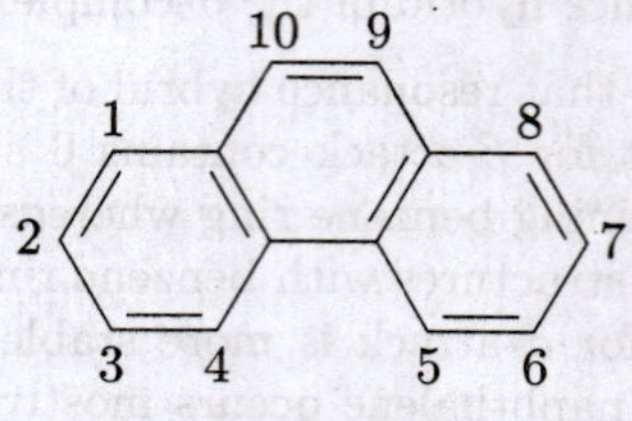
R.E. = 387.1 kJ mol^{-1}

Phenanthrene

It might have been expected that the resonance energy of anthracene would be $3 \times 150.6 = 451.8$ kJ mol^{-1}. Its resonance energy is actually 351.5 kJ mol^{-1}. Therefore, it is more reactive than benzene. When direct electrophilic substitution reactions are carried on anthracene, addition products are obtained in many cases along with substitution products. However, the electrophilic substitution takes place at 9- or 10- position almost exclusively in both anthracene and phenanthrene. This is because the σ-complexes for attack at 9-position contains 2 benzene rings whereas an attack on any position (say, 1 or 2) forms a σ-complex containing a naphthalene moiety. The resonance energy of a naphthalene moiety (255.2 kJ mol^{-1}) being less than the resonance energy of two benzene rings (301.2 kJ mol^{-1}), the σ-complex for the 9-attack is more stable than the σ-complex for the 1 - or 2-attack. So, electrophilic substitution at 9- or 10-position occurs almost exclusively.

(c) Five-membered heterocyclic rings with one hetero atom : Furan, thiophene and pyrrole are five-membered heterocycles with one hetero atom.

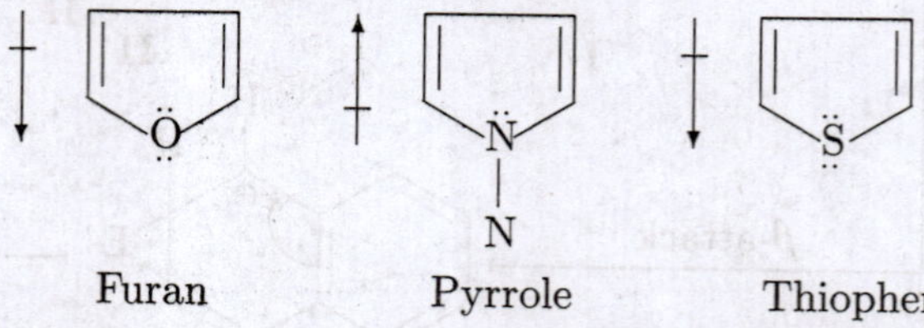

Furan Pyrrole Thiophene

Each of the three above compounds is a resonance hybrid of five canonical structures. The resonance hybrid of each structure shows that there is partial negative charge on all C atoms.

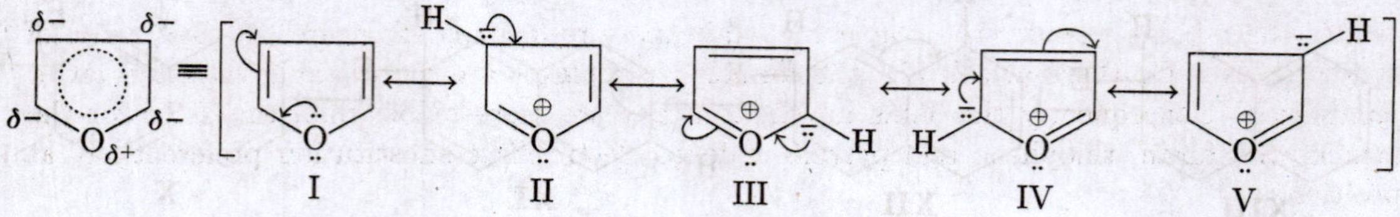

I II III IV V

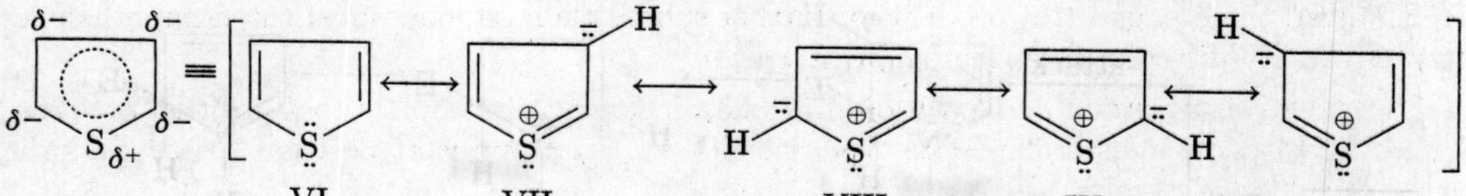

VI VII VIII IX X

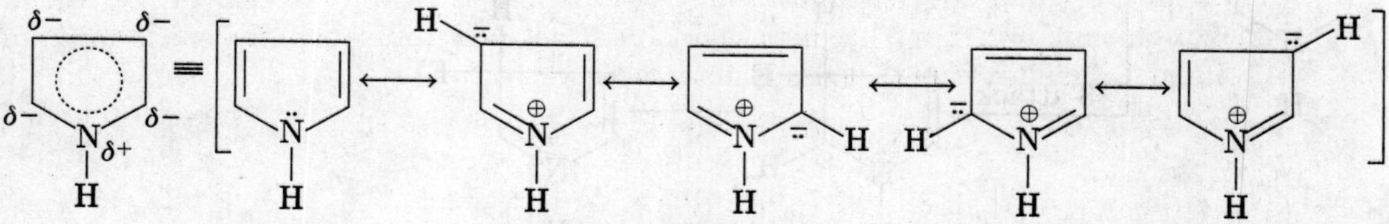

XI XII XIII XIV XV

Therefore, they will undergo electrophilic substitution reactions more readily than benzene. Since the resonating structures with unlike charges separated (II, V; VII, X and XII, XV) are less contributing than those with unlike charges not separated (III, IV; VIII, IX and XIII, XIV) to the resonance hybrid, the incoming electrophile takes 2 and 5 positions preferentially in furan, thiophen and pyrrole. Extra-stability of the σ-complexes for the 2-attack (or 5-attack) over that for the 3-attack (or 4-attack) further supports the fact.

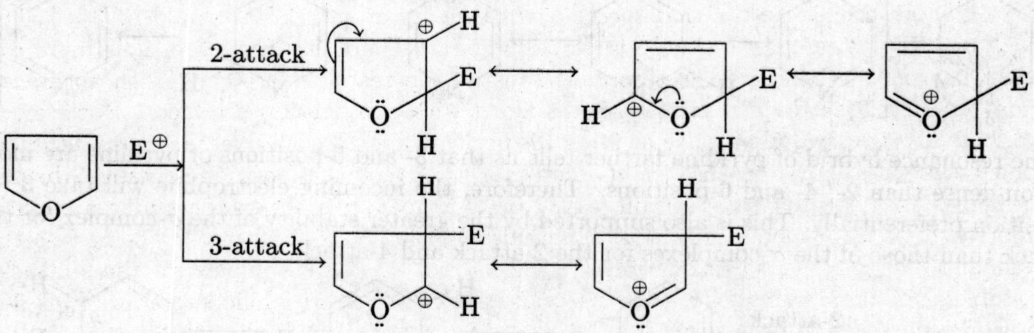

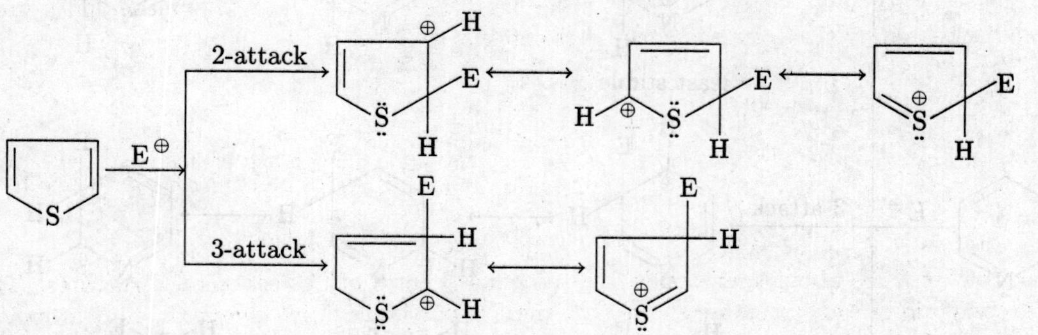

Since the σ-complexes for the 2-attack contain more resonating structure (3) than those of the σ-complexes for the 3-attack (2), the former σ-complexes are more stable than the latter σ-complexes. Consequently the T.S.s for the 2-attack are more stable than the T.S.s for the 3-attack. So, furan, thiophene and pyrrole undergo electrophilic substitution preferentially at 2-position.

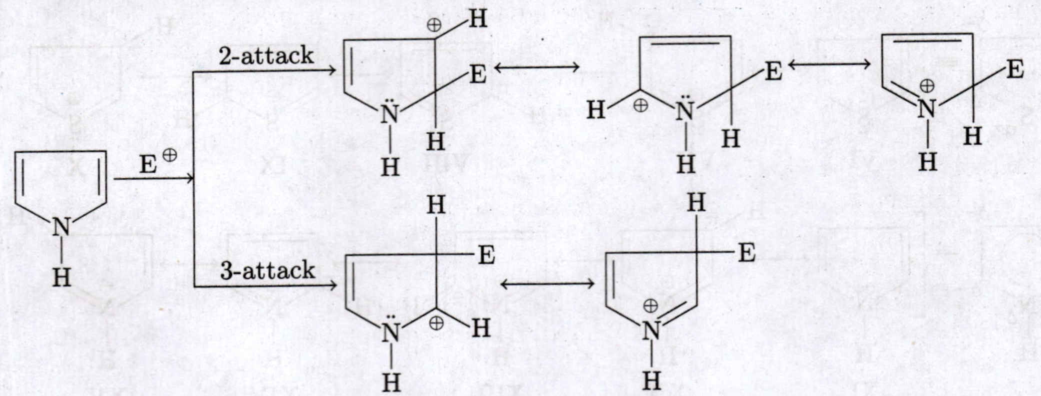

(d) Pyridine : The resonance hybrid of pyridine contains five resonating structures. We see that there are partial positive charges on the 2-, 4- & 6-positions in the resonance hybrid. Therefore, pyridine ring is a deactivated one. Direction of dipole moment also supports this. So, pyridine will undergo electrophilic substitution reactions less readily than benzene.

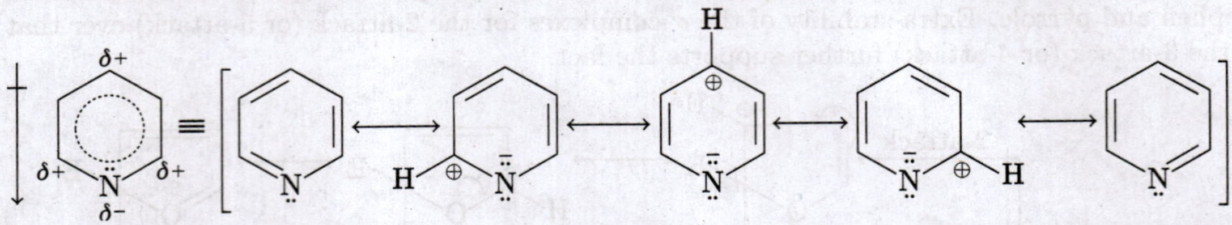

The resonance hybrid of pyridine farther tells us that 3- and 5-positions of pyridine are more electron-dense than 2-, 4- and 6-positions. Therefore, the incoming electrophile will take 3 (or 5)-position preferentially. This is also supported by the greater stability of the σ-complex for the 3-attack than those of the σ-complexes for the 2-attack and 4-attack.

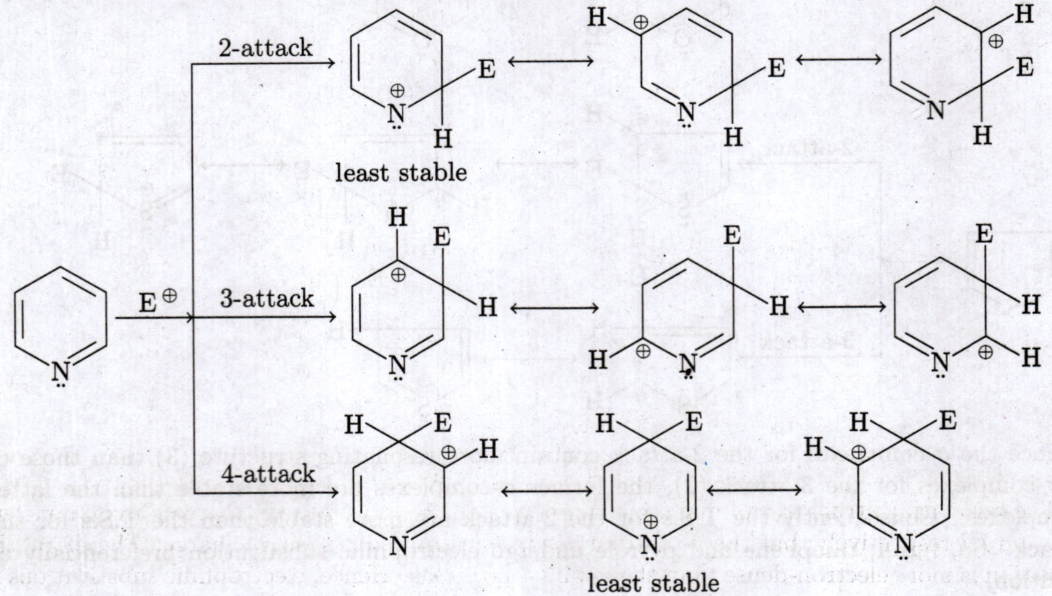

Each of the σ-complexes for the 2-attack and 4-attack contains a very unstable structure in which the more electronegative element nitrogen bears a positive charge; whereas the complex for the 3-attack does not contain such an unstable structure. Therefore, the latter σ-complex is more stable than the former two. Hence, the T.S. for the 3-attack will be more stable than the T.S.s for the 2-attack and 4-attack. So, pyridine will undergo the electrophilic substitution preferentially at 3- and 5-positions.

(e) Indole : The resonance hybrid of indole contains several charged resonating structures. There are only two stable structures (2 and 6) among these in which the 3-position acquires a negative charge. These two structures are more stable than the others, simply because each of them contains a benzene ring. So, the electrophilic substitution on indole will take place at position 3. However, if the 3-position is blocked, the substitution occurs at the 2-position. When both the 2- and 3-positions are occupied by groups other than H, the substitution takes place at position 6.

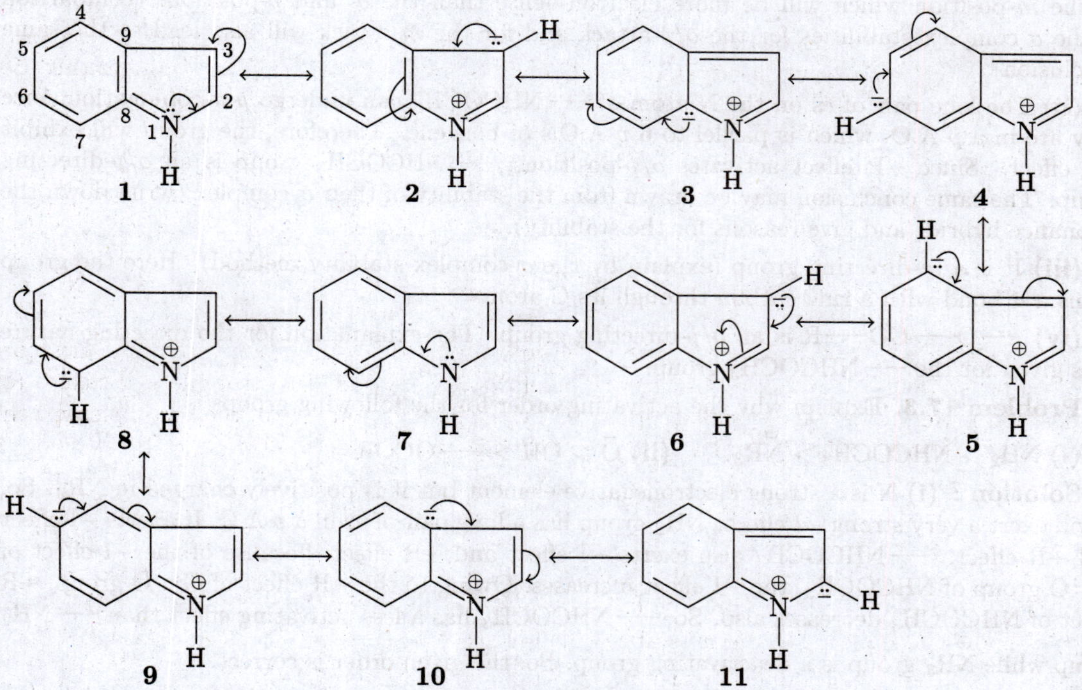

(f) Quinoline : The resonance hybrid of quinoline is written below :

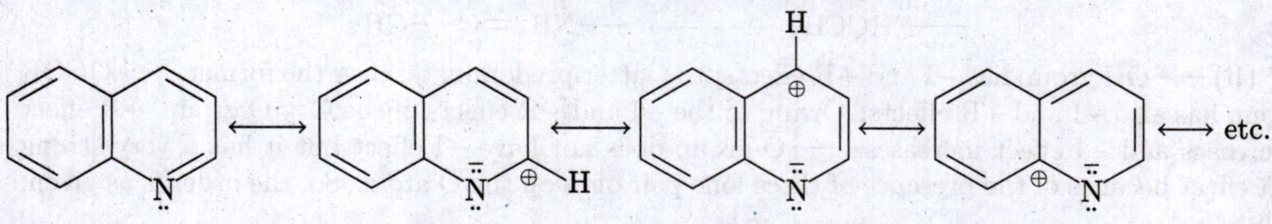

The resonating structures of quinoline show that the 2- and 4-positions are electron deficient in II and III respectively; but the 3-position has never acquired a positive charge. Therefore, the 3-position is more electron-dense than the 2- and 4-positions. Hence, electrophilic substitutions in quinoline will preferentially take place at 3-position. However, Electrophilic substitutions occur also at 5-, 8- and 2-positions. The most preferential position for electrophilic substitutions is the position 8.

Problem 17.2. Write down the directing nature of the following groups in so far as electrophilic substitution on benzene is concerned.

(i) —C≡N, **(ii)** —$\ddot{N}HCOCH_3$, **(iii)** —COOR, **(iv)** —OCOR.

Solution : (i) —C≡N group attaches itself to a benzene ring through the C atom. This, in its turn, is bonded to a strongly electronegative element, N, by a multiple bond. Owing to the greater electronegativity of N than that of C, the N atom will withdraw electron from the C≡N bond. As a consequence, the π electrons of the benzene ring will be withdrawn by the —C≡N group and thus —C≡N will act as a deactivating group through −I and −R effects. Deactivation will occur mostly at *o*- and *p*-positions. So, it will direct an incoming electrophile to the *m*-position which will be more electron-dense than the *o*- and *p*-positions (comparison of the σ-complex stabilities for the *o*/*p*-attack and for the *m*-attack will also lead to the same conclusion).

(ii) The lone pair of *e*s on the N atom of —$\ddot{N}HCOCH_3$ can undergo *p*-π-conjugation since they are in a *p* A.O. which is parllel to 6 *p* A.O.s of benzene. Therefore, the group will exhibit +R effect. Since +R effect activates *o*/*p*-positions; —$NHCOCH_3$ group is an *o*/*p*-directing group. The same conclusion may be drawn from the stability of their σ-complex (write down the resonance hybrids and give reasons for the stability).

(iii) It is a *m*-directing group (explain by the σ-complex stability method). Here the group forms a σ bond with a ring carbon through its C atom.

(iv) —O—CO—R is an *o*/*p*-directing group. The explanation for the directing nature is as given for the —$NHCOCH_3$ group.

Problem 17.3. Explain why the activating order for the following groups is :

(i) $\ddot{N}H_2 > \ddot{N}HCOCH_3 > \overset{\oplus}{N}R_3$. **(ii)** $\bar{O}$ > OH > —OCOR

Solution : (i) N is a strong electronegative element but it is positively charged in $\overset{\oplus}{N}R_3$. So, it will exert a very strong −I effect. $\ddot{N}H_2$ group has a lone pair of *e*s in a *p* A.O. It exerts −I effect and +R effect. —$NHCOCH_3$ also exerts −I effect and +R effect; because of the −I effect of C═O group of $NHCOCH_3$, the −I effect increases. Owing to the −R effect of C═O group, +R effect of $NHCOCH_3$ decreases also. So, —$NHCOCH_3$ has a less activating effect than —$\ddot{N}H_2$ group while $\overset{\oplus}{N}R_3$ group is a deactivating group. So, the given order is correct.

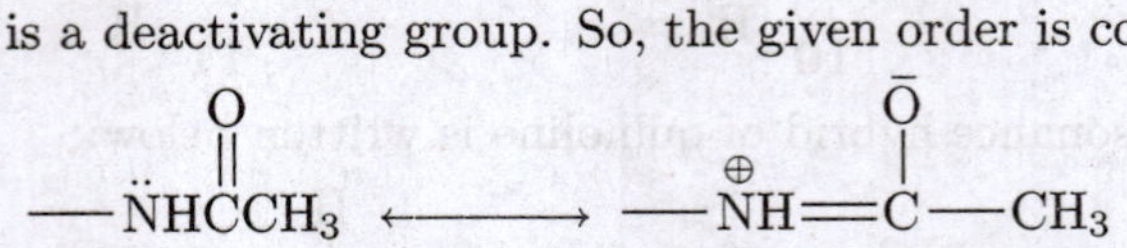

(ii) —OH group has −I and +R effects; the latter predominates over the former. —OCOR group has also −I and +R effects. Owing to the −I and −R effects of the C═O group, +R effect decreases and −I effect increases. —$\bar{O}$ group does not have −I effect but it has a very strong +R effect because of the presence of three lone pair of *e*s on the O atom. So, the order is as given.

17.5 Ortho-Para Ratio

When electrophilic substitution reactions are carried out on a benzene derivative containing an ortho-para directing group, the ratio of the yield of the ortho-product to that of the para-product is found to vary from reaction to reaction and substrate to substrate. Since there are two *o*-positions and one *p*-position relative to the substituent in a monosubstituted benzene, it may seem that the o/p ratio will be 2 : 1 in all reactions irrespective of any substituent; but this is not true. The products never form corresponding to that ratio. Several factors need consideration to explain the o/p ratios observed in aromatic substitution reactions. The factors are:

(a) Steric effect, (b) electronic effects of the substituent already present in the benzene ring, (c) interaction between the substituent and the attacking electrophile, (d) effect of temperature and (e) solvation effect.

(a) Steric effect : *Ortho*-positions being very near to the substituent, the attacking electrophile experiences a spatial congestion when it approaches the *o*-positions; but it does not experience the same when it attacks the *p*-position since this is far from the substituent. The larger the volume of the substituent or that of the attacking electrophile, the greater is the spatial congestion at the *o*-position and consequently the lower is the o/p ratio. This means with increasing spatial congestion at the *o*-position, the yield of the *o*-product decreases and that of the *p*-product increases. Let us discuss some examples to substantiate the steric effect on o/p ratio.

(i) When alkylbenzenes are nitrated under the same set of conditions, the o/p ratio is found to be the highest in the case of toluene and the lowest in the case of t-butylbenzene as shown in Table 17.3.

Table 17.3 o/p ratios in the nitration of alkylbenzenes

Substituents	% of *o*-product	% of *p*-product	o/p ratio
Me	58	37	1.59
Et	45	49	0.92
i-Pr	30	62	0.48
t-Bu	16	73	0.22

Table 17.3 reveals that as the volume of the substituent (alkyl group) increases down the series (Me to t-Bu) the yield of *o*-nitroproduct decreases and that of the *p*-nitroproduct increases and as a consequence o/p ratio decreases.

(ii) It has also been found that as the volume of the attacking electrophile increases in an electrophilic substitution reaction on a benzene derivative, the o/p ratio decreases. Alkylation of toluene by the Friedel-Crafts method and electrophilic substitution on chlorobenzene with electrophiles of different size tell us that with increasing volume of the attacking electrophile the yield of *o*-product decreases and that of the *p*-product increases (See Tables 17.4 and 17.5).

Table 17.4 Alkylation of toluene by the Friedel-Crafts reaction and the % of *o*-product

	Attacking electrophile	% of *o*-product
Increasing bulk ↓	Me	53.8
	Et	45.0
	i-Pr	37.5
	t-Bu	0

Table 17.5 Ar.S_E2 reactions on chlorobenzene and the o/p ratio

Increasing bulk ↓	Attacking electrophile	% of o-product	% of p-product	o/p ratio
	$Cl^{\oplus}$ (chlorination)	39	55	0.71
	$\overset{\oplus}{N}O_2$ (nitration)	30	70	0.43
	$Br^{\oplus}$ (bromination)	11	87	0.14
	SO_3 (sulphonation)	1	99	0.01

(b) Electronic effect : However, the steric factor alone should not be held responsible for the observed o/p ratios in different reactions. The electronic effects should also be considered among other factors. When halobenzenes are nitrated under the same set of conditions, the yield of o-product increases and that of the p-product decreases despite the increase in the size of the substituents, F to I, as shown in Table 17.6.

Table 17.6 Electronic effects and the o/p ratio in the nitration of halobenzenes

Increasing volume and decreasing –I effect ↓	Substituent	% of o-product	% of p-product	o/p ratio
	F	12	88	0.14
	Cl	30	69	0.44
	Br	37	62	0.60
	I	38	60	0.63

–I effect of the halogen groups is supposed to be responsible for the observed o/p ratios as shown in Table 17.6. We know that :

(i) The –I effect decreases on going from F to I as the electronegativity of the group decreases.

(ii) The –I effect deactivates the o-, m- and p-positions but the effect is highest on the o-positions and it is practically nil on the p-position. In fluorobenzene the o-positions being highly deactivated by the strong –I effect of the F atom, the attacking electrophile, nitronium ion, attacks the para-position preferentially and p-nitrofluorobenzene forms in large excess over the o-isomer; while in the case of iodobenzene the –I effect of the I atom being very small, it deactivates the o-positions to a minimal extent and consequently an appreciable amount of o-product forms. If there were no steric hindrance owing to the large volume of the I atom, the yield of o-product would have been still higher than the observed yield. Thus the o/p ratio increases on going from fluorobenzene to iodobenzene in the nitration of halobenzenes though the steric congestion increases on going down the series.

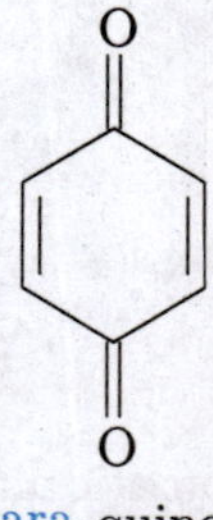

para-quinone

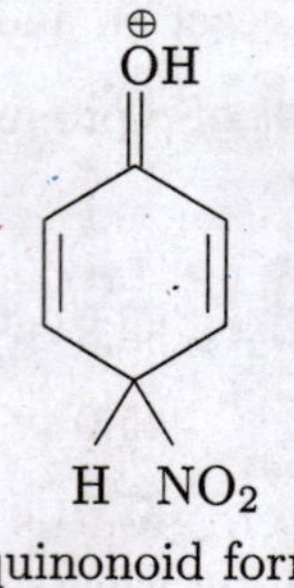

p-quinonoid form

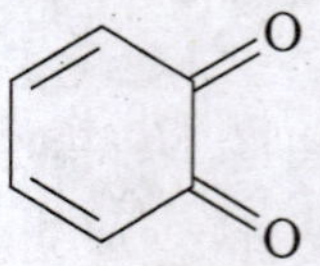

ortho-quinone

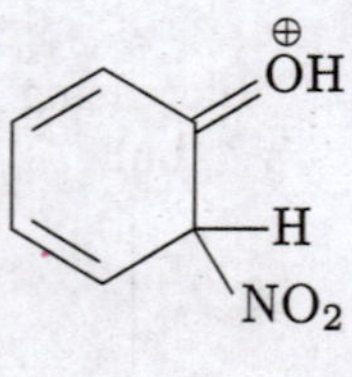

o-quinonoid form

It is said that −R effect deactivates the *p*-position more than the *o*-position and +R effect activates the *p*-position more than the *o*-position; these inferences have been drawn from the fact that para-quinonoid structures are more stable than the ortho-quinonoid structures. If this is true, the presence of a +R substituent will decrease the *o*/*p* ratio.

(c) Interaction between the substituent and the attacking eletrophile : Furthermore, when the attacking reagent undergoes some sort of interaction with the substituent before attacking the ring, usually the *o*-attack is the predominant one and the *o*/*p* ratio increases. For example, when methyl phenylethyl ether is nitrated with N_2O_5 in methyl cyanide, 69% *o*-product and 28% *p*-product form; while nitration of the same compound by the common nitrating agent (concentrated HNO_3 and concentrated H_2SO_4) gives 32% *o*-product and 59% *p*-product. It is argued that the interaction between the O atom of the ether linkage and N_2O_5 group occurs in the first stage and then a $\overset{\oplus}{N}O_2$ ion develops which attacks the ring perhaps in a concerted step; the distance between the developing nitronium ion and the *o*-position being small, the $\overset{\oplus}{N}O_2$ ion attacks the *o*-position preferentially.

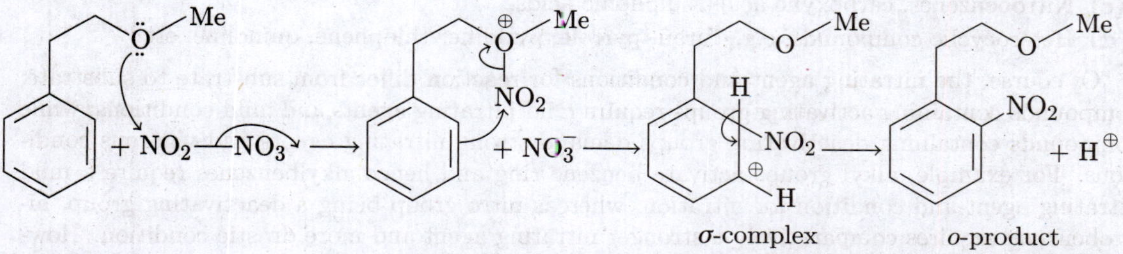

(d) Effect of temperature : Usually the aromatic electrophilic substitution reactions are kinetically controlled rather than thermodynamically. If such a reaction is thermodynamically controlled, generally the *o*/*p* ratio decreases as the temperature of the reaction is increased. For example, sulphonation of phenol gives a high yield of *o*-product at 15-20 °C while at 100 °C the yield of *p*-isomer predominates.

(e) Solvation effect : Solvents can also influence the *o*/*p* ratios of $Ar.S_E$ reactions. If a solvent through solvation brings extra stability to the T.S. for the *o*-attack relative to the T.S. for the *p*-attack, the *o*/*p* ratio will increase. On the other hand, if the solvation stabilises the T.S. for the *p*-attack more than the T.S. of the *o*-attack, the ratio will decrease. In the halogenation of toluene without a Lewis catalyst the solvents are found to play this role and the *o*/*p* ratio varies from 1.57 to 0.67 with the change of the solvent and keeping other conditions of the reaction identical.

Problem 17.4. Explain why the *o*/*p* ratio decreases in the nitration of the series given below :

Substrate :	$PhCH_3$	$PhCH_2Cl$	$PhCHCl_2$	$PhCCl_3$
o/*p* ratio :	1.57	0.75	0.53	0.23

Solution : The increasing steric congestion and decreasing +I effect (increasing −I effect from —CH_3 to —CCl_3) are the causes for the decreasing *o*/*p* ratio.

17.6 Nitration

Displacement of an H atom of an aromatic nucleus by a nitro group is called aromatic nitration process. The reaction may be carried out with several nitrating agents as given below :

(a) Concentrated nitric acid in glacial acetic acid.
(b) Nitric acid in water.
(c) Concentrated nitric acid in concentrated sulphuric acid (mixed acid).
(d) Fuming nitric acid in cocentrated sulphuric acid.
(e) Fuming nitric acid in fuming sulphuric acid.
(f) Acylnitrates. e.g., acetylnitrate, benzoylnitrate in organic solvents.
(g) Nitronium salts, e.g., nitronium fluoborate.
(h) Nitrogen pentoxide in carbon tetrachloride in the presence of phosphorous pentoxide.

Among these the common reagent is the mixed acid and others are used depending on the nature of the substrate.

However, a wide variety of aromatic compounds undergo nitration :

(a) Aromatic hydrocarbons; e.g., benzene, toluene, naphthalene.
(b) Halobenzenes, phenols, phenolic ethers, amines, anilides.
(c) Nitrobenzenes, carboxylic acids, sulphonic acids.
(d) Heterocyclic compounds, e.g., furan, pyrrole, pyridine. thiophene, quinoline, etc.

Of course, the nitrating agent and conditions for reaction differ from substrate to substrate; compounds containing activating groups require mild nitrating agents and mild conditions; while compounds containing deactivating groups demand strong nitrating agents and vigorous conditions. For example, alkyl groups activate benzene ring and hence alkylbenzenes require a mild nitrating agent and condition for nitration; whereas nitro group being a deactivating group, nitirobenzene requires comparatively a stronger nitrating agent and more drastic condition. However, a few examples of nitration. are given below :

1. 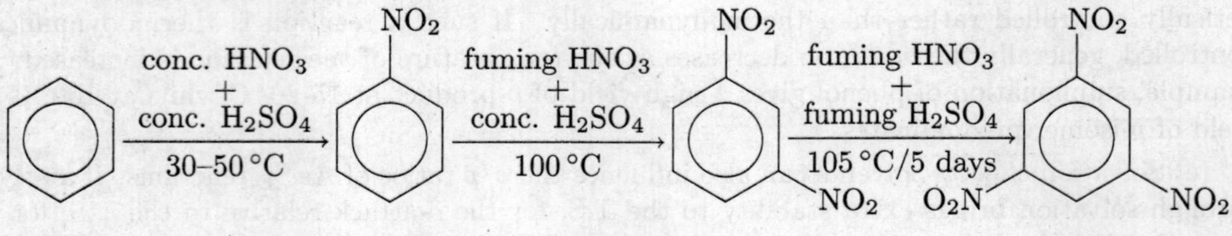

2. 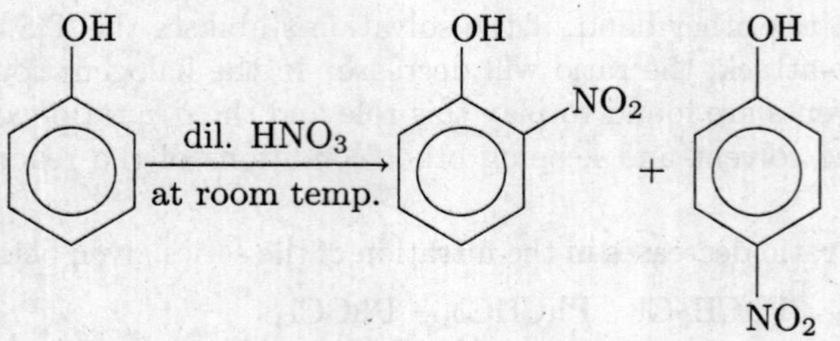

3. 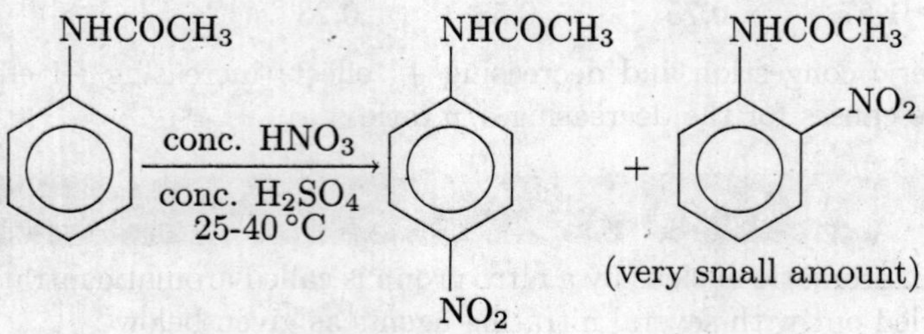

4. 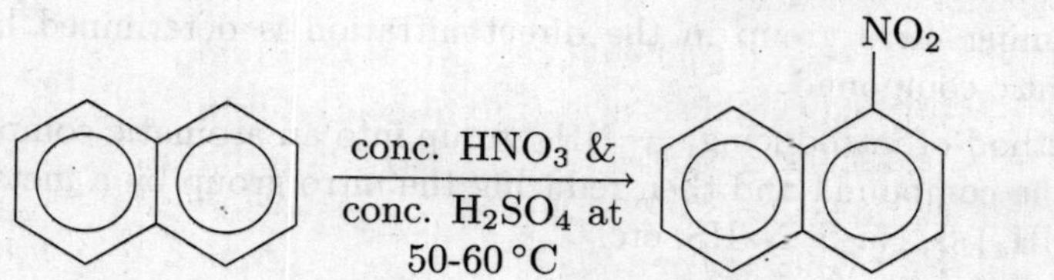

5. 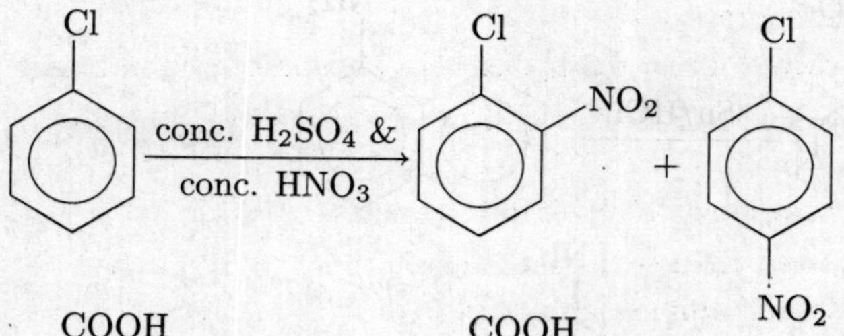

6. 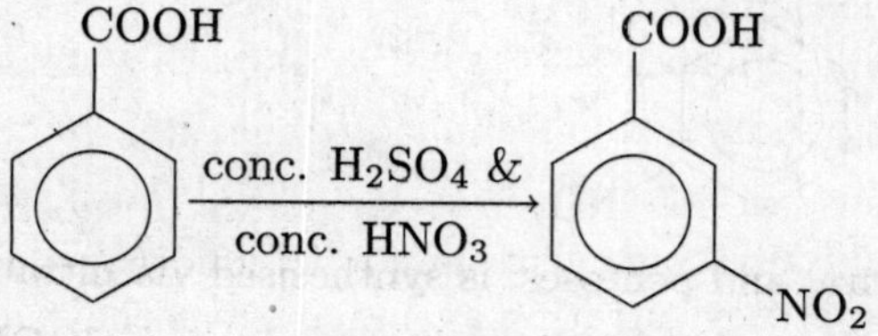

7. 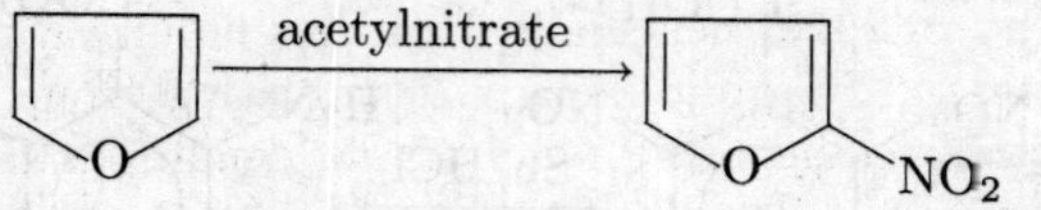

8. 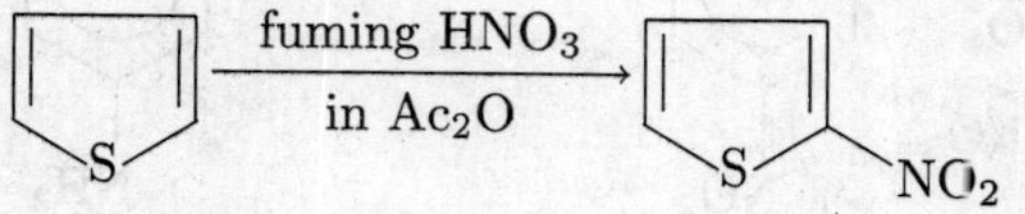

9. 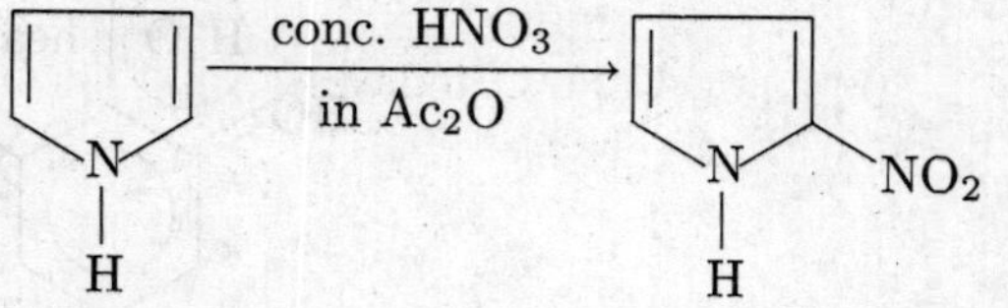

10. 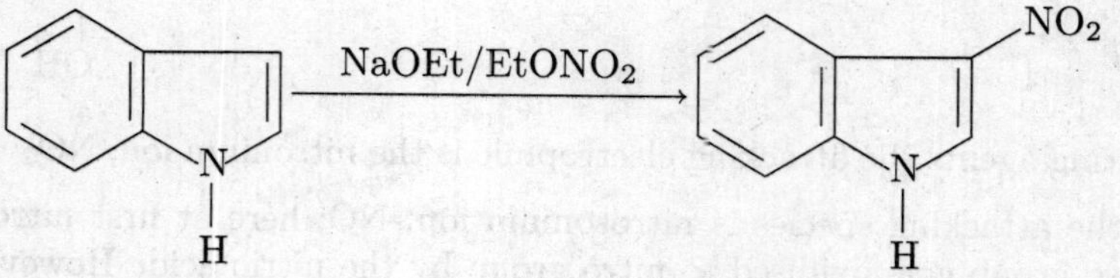

11. 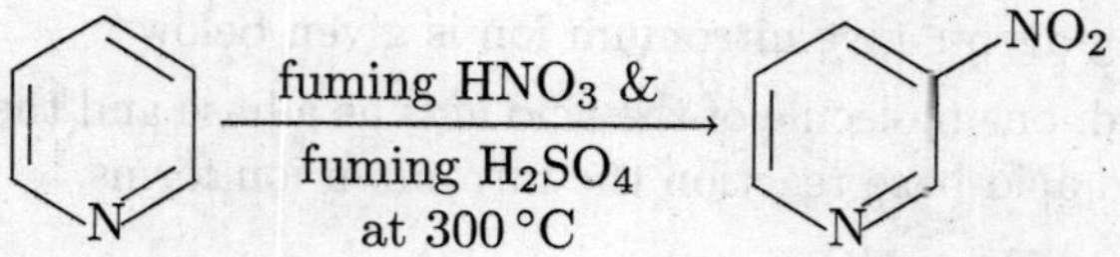

12. 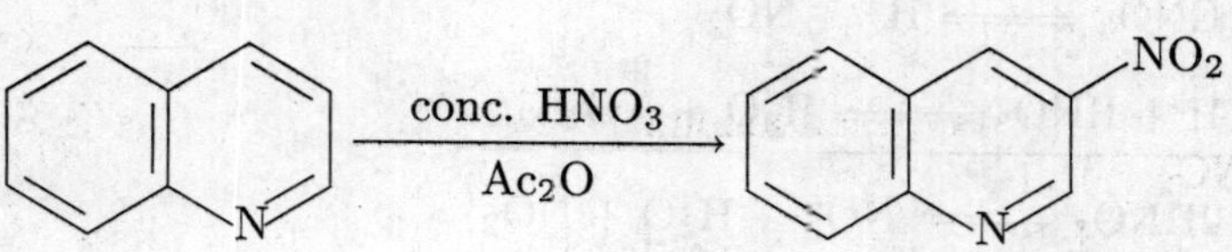

The orientation of the incoming nitro group in the direct nitration is determined by the substituent present in the aromatic compound.

There is no simple direct method of introducing —NH_2 group into an aromatic compound. This can be done by nitrating the compound and then reducing the nitro group by a metal and a mineral acid or by NH_4HS, $(NH_4)_2S$, Na_2S, NaHS, etc.

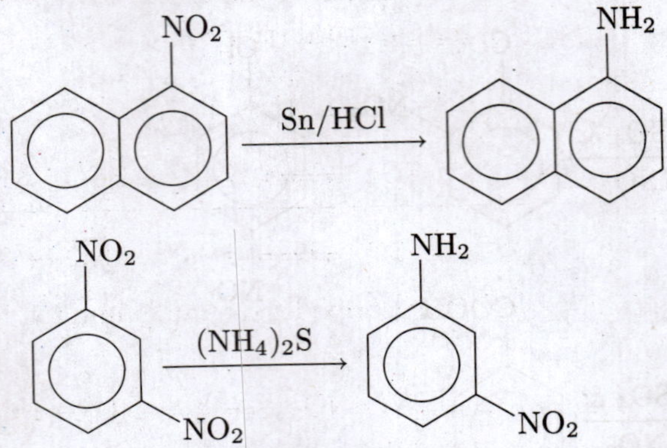

Phloroglucinol, a reagent for estimating furfural and pentoses, is synthesised via nitration,

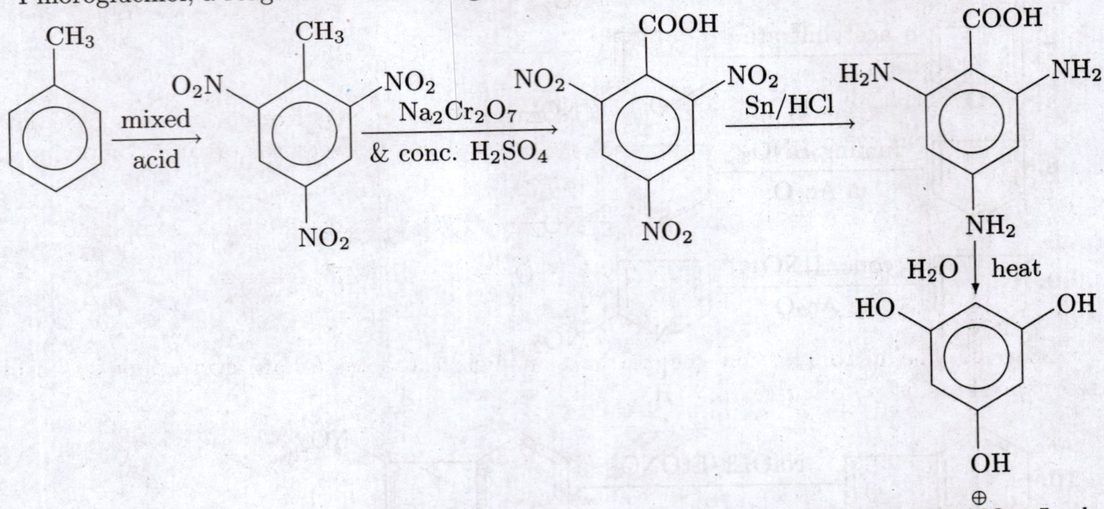

With most of the nitrating agents the attacking electrophile is the nitronium ion, $\overset{\oplus}{N}O_2$. In the case of dilute nitric acid the attacking species is nitrosonium ion, $\overset{\oplus}{N}O$; here at first nitrosation occurs and then the nitroso group gets oxidised to nitro group by the nitric acid. However, the pathways through which different nitrating agents give nitronium ion is given below :

(a) In the case of concentrated nitric acid, one molecule of the acid acts as a base and the other molecule as an acid; thus through an acid-base reaction the nitronium ion forms.

$$HNO_3 \rightleftharpoons H^{\oplus} + NO_3^-$$

$$\overset{\oplus}{H} + HNO_3 \rightleftharpoons H_2O + \overset{\oplus}{N}O_2$$

$$2HNO_3 \rightleftharpoons NO_3^- + H_2O + \overset{\oplus}{N}O_2$$

The equilibrium lies to the left but sufficient amount of nitronium ion forms for nitration to occur. When conc. HNO_3 is taken in an organic solvent, a very small amount of $\overset{\oplus}{N}O_2$ forms by the same reaction.

(b) In the mixed acid, an acid-base reaction takes place in which H_2SO_4 acts as an acid and HNO_3 as a base. The ionisation is essentially complete.

$$H_2SO_4 \rightleftharpoons H^{\oplus} + HSO_4^-$$

$$HNO_3 + H^{\oplus} \rightleftharpoons H_2O + \overset{\oplus}{N}O_2$$

$$H_2O + 2H_2SO_4 \rightleftharpoons H_3O^{\oplus} + HSO_4^-$$

$$HNO_3 + 2H_2SO_4 \rightleftharpoons H_3O^{\oplus} + \overset{\oplus}{N}O_2 + 2HSO_4^-$$

(c) When the nitrating agent is nitrogen pentoxide, the nitronium ion forms through the spontaneous dissociation of N_2O_5.

$$N_2O_5 \rightleftharpoons \overset{\oplus}{N}O_2\bar{N}O_3 \rightleftharpoons \overset{\oplus}{N}O_2 + NO_3^-$$

(d) In the case of nitronium salts, the nitronium ion is present in the system from the very beginning.

$$\overset{\oplus}{N}O_2\bar{B}F_4 \longrightarrow \overset{\oplus}{N}O_2 + \bar{B}F_4$$

(e) When the reagent is dilute HNO_3, $\overset{\oplus}{N}O$ forms instead of $\overset{\oplus}{N}O_2$ from nitrous acid; the latter forms in the system through the minor red-ox reaction between the organic compound and nitric acid.

$$HNO_3 + \text{organic compound} \longrightarrow HNO_2 + \text{oxidised organic compound}$$

$$HNO_3 \rightleftharpoons H^{\oplus} + NO_3^-;\ H{-}O{-}N{=}O + H^{\oplus} \rightleftharpoons \overset{\oplus}{N}O + H_2O$$

Common mechanistic steps

Step I. The nitronium ion then attacks a nuclear carbon as an electrophile to form a resonance-stabilised σ-complex.

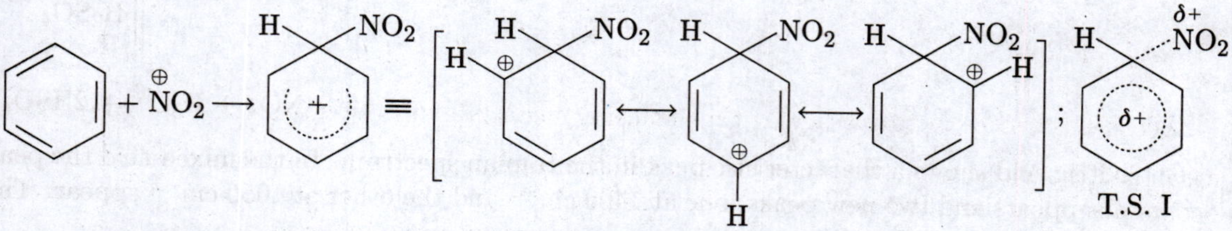

Step II. The σ-complex then gives up a proton to a base and gets aromatised to a nitroarene.

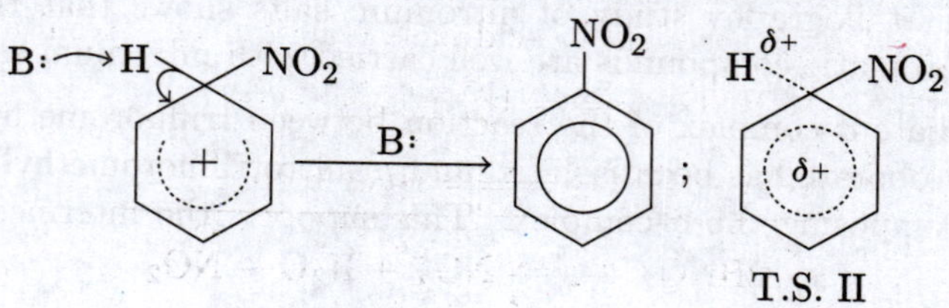

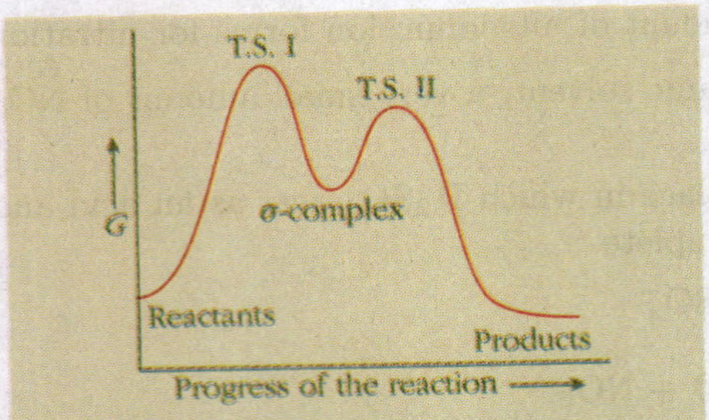

Fig. 17.3. The energy profile for nitration of benzene

The base may be HSO_4^- in the case of the mixed acid, it may be a solvent molecule in the cases of organic solvents.

The energy profile for the reaction is given in Fig. 17.3.

Since in all the cases a large excess of nitrating agent is used, the reaction is of zero order with respect to the nitronium ion. The reaction is found to be of zero order, first order or fractional order depending on the nature of the substrate.

The zero order or pseudo-zero order reaction is followed by substrates which are more reactive than benzene. In these cases the reversible formation of the nitronium ion step is slower than the electrophilic attack step. Thus the rate-determining step is the reversible nitronium ion formation step; toluene, xylene follow this pathway. When the substrates are less reactive than benzene, the nitronium ion formation step is found to be faster than the electrophilic attack step and as a consequence the reaction shows the first order kinetics with respect to the substrate. Thus in these cases the reaction is said to be the first order or the pseudo-first order :

the rate $= k$ [substrate]. Ethyl benzoate, nitrobenzene follow the first order kinetics.

When the substrate is of intermediate reactivity, a non-integral order is followed by the reactions; it is between 0 to 1. An example of this kind is the nitration of chlorobenzene. The concentration of the reagent is also responsible for the non-integral order of the reaction.

In most of the cases the nitronium ion is the attacking agent; this has been supported by a number of evidences :

(i) When conc. nitric acid is added to conc. sulphuric acid, the freezing point of H_2SO_4 is lowered about four times. The four-fold freezing point depression indicates the presence of four ions in the mixed acid; this can be accounted for by the following equilibria. The four ions are two HSO_4^-, one $\overset{\oplus}{N}O_2$ and one $H_3O^{\oplus}$.

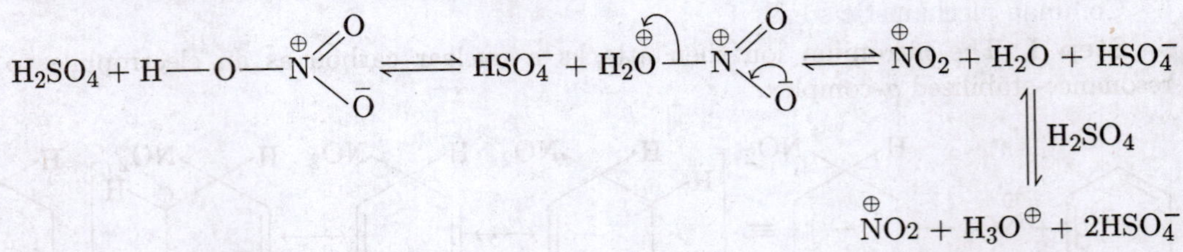

(ii) Nitric acid shows a characteristic peak in the Raman spectrum. In the mixed acid the peak disappears and two new peaks, one at 1400 cm^{-1} and the other at 1050 cm^{-1} appear. The former is due to $\overset{\oplus}{N}O_2$ and the latter due to HSO_4^-.

(iii) The X-ray crystallography study of nitronium salts shows that they contain $\overset{\oplus}{N}O_2$ ion. Nitration of aromatic compounds are well carried with nitronium salts.

(iv) The intermediate σ-complex of the reaction between trifluoromethylbenzene and nitronium tetrafluoroborate has been isolated and *m*-nitrotrifluoromethylbenzene has been prepared by decomposing the σ-complex. This supports the intermediate formation of the σ-complex.

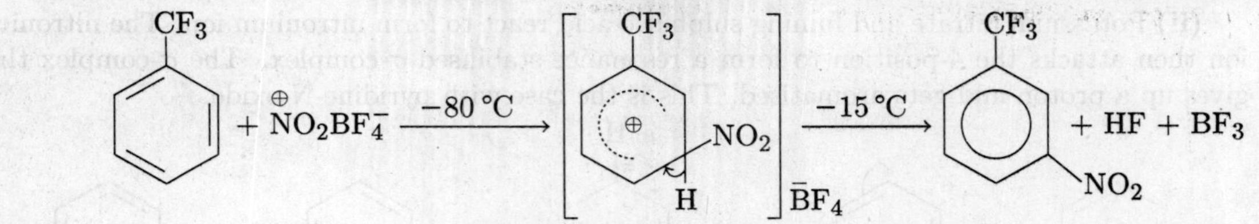

The reaction does not show any primary kinetic isotope effect; the rates of nitration of $C_6D_5NO_2$ and that of $C_6H_5NO_2$ are found to be one and the same. This shows that the C—H bond breaking step is not the rate-determining step and the two steps are not concerted.

Problem 17.5. Give mechanistic steps for the nitration of (i) phenol in dil. HNO_3 and (ii) pyridine-1-oxide by solid KNO_3 in fuming sulphuric acid.

Solution : (i) $H_2SO_4 \rightleftharpoons H^{\oplus} + HSO_4^-$

$$H^{\oplus} + H—O—N=O \rightleftharpoons H_2\overset{\oplus}{O}—N=O \rightleftharpoons H_2O + \overset{\oplus}{N}O$$

$$H_2O + H_2SO_4 \rightleftharpoons H_3O^{\oplus} + HSO_4^-$$

Step I : Electrophilic attack by the nitrosonium ion on the nuclear carbon and the formation of the σ-complex.

Step II. Deprotonation and aromatisation of the σ-complex.

Step III. Oxidation of the nitroso derivative to nitrophenol.

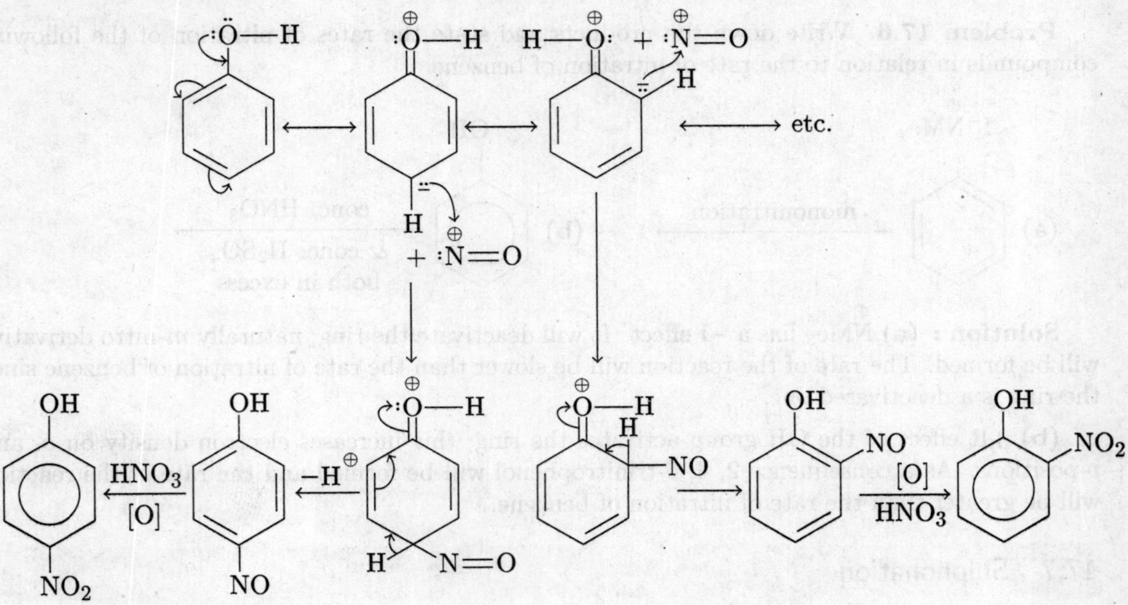

p-nitrosophenol *o*-nitrosophenol

It has been found that in the dilute nitric acid containing a very little amount of nitrous acid, the *o*- and the *p*-isomers form in the ratio 7 : 3; whereas in the presence of a large excess of nitrous acid the ratio becomes 1 : 9. However, nitric acid oxidises the nitrosophenols readily to nitrophenols.

(ii) Potassium nitrate and fuming sulphuric acid react to form nitronium ion. The nitronium ion then attacks the 4-position to form a resonance-stabilised σ-complex. The σ-complex then gives up a proton and gets aromatised. This is the case with pyridine-N-oxide.

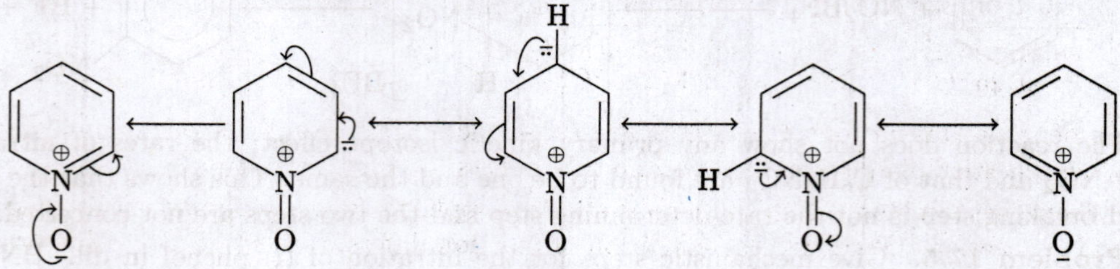

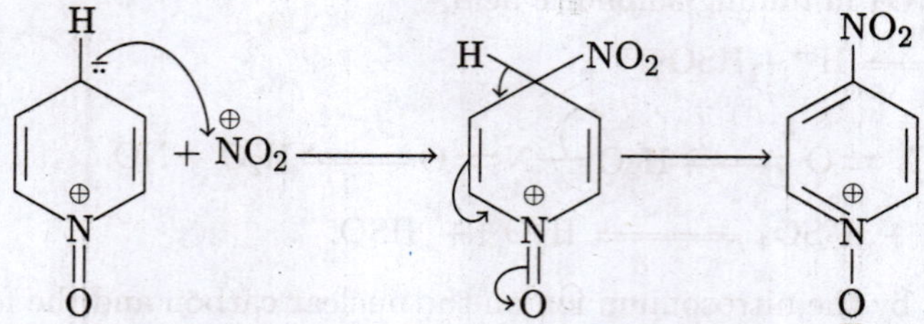

$$KNO_3 + H_2SO_4 = HNO_3 + KHSO_4$$

$$HNO_3 + 2H_2SO_4 \rightleftharpoons \overset{\oplus}{N}O_2 + H_3O^{\oplus} + 2HSO_4^-$$

Problem 17.6. Write down the products and state the rates of nitration of the following compounds in relation to the rate of nitration of benzene.

(a) 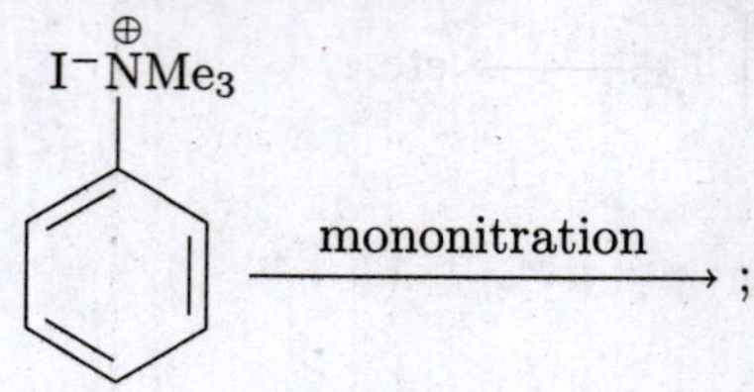

; (b) 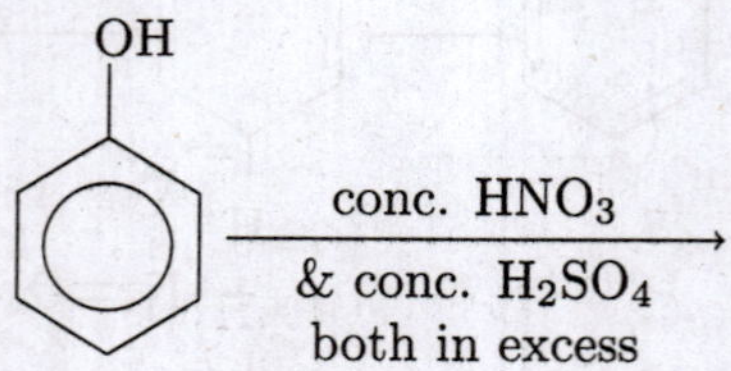

Solution : (a) $\overset{\oplus}{N}Me_3$ has a $-I$ effect. It will deactivate the ring; naturally *m*-nitro derivative will be formed. The rate of the reaction will be slower than the rate of nitration of benzene since the ring is a deactivated one.

(b) +R effect of the OH group activates the ring; this increases electron density on *o*- and *p*-positions. As a consequence 2, 4, 6-trinitrophenol will be formed and the rate of the reaction will be greater than the rate of nitration of benzene.

17.7 Sulphonation

The process of introducing SO_3H group into an aromatic nucleus by the action of concentrated sulphuric acid or fuming sulphuric acid (oleium; $H_2S_2O_7$—sulphur trioxide in sulphuric acid) or sulphur trioxide in an organic solvent (nitromethane, pyridine, etc.) or chlorosulphonic acid in carbon tetrachloride is called aromatic sulphonation.

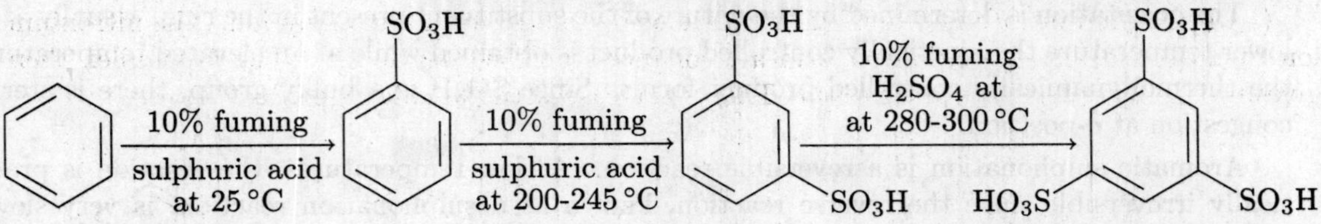

Thus benzene can be converted to benzenesulphonic acid, 1, 3-benzenedisulphonic acid and 1, 3, 5- benzenetrisulphonic acid stepwise by the process.

Alkylbenzenes, amines, phenols, phenolic ethers, thiophene, furan, pyridine, naphthalene, etc. undergo the reaction. Of course, the condition of reaction differs from substrate to substrate.

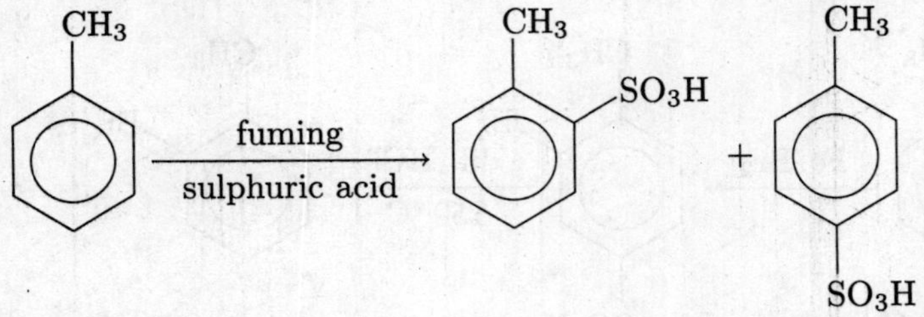

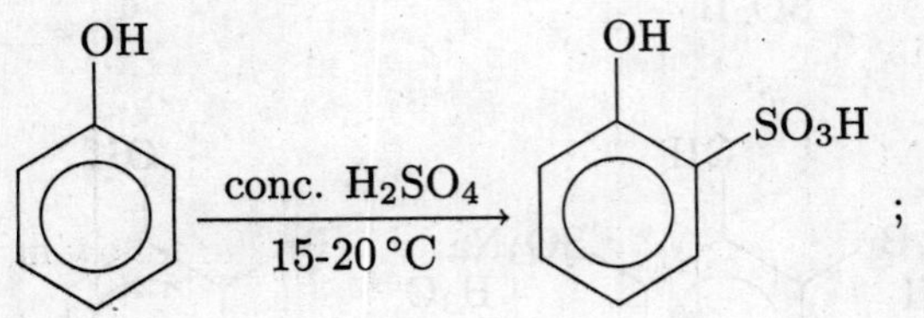

;

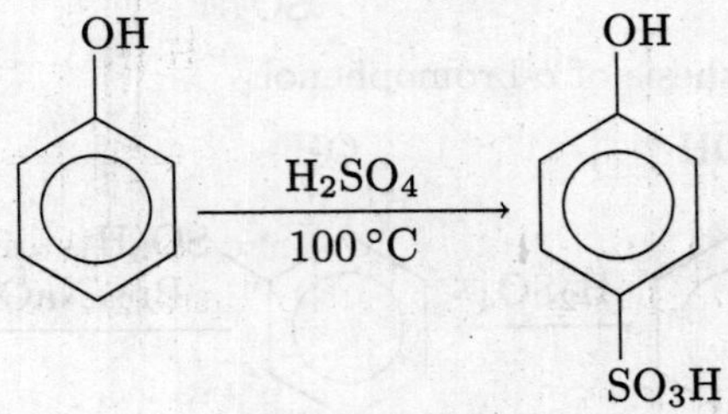

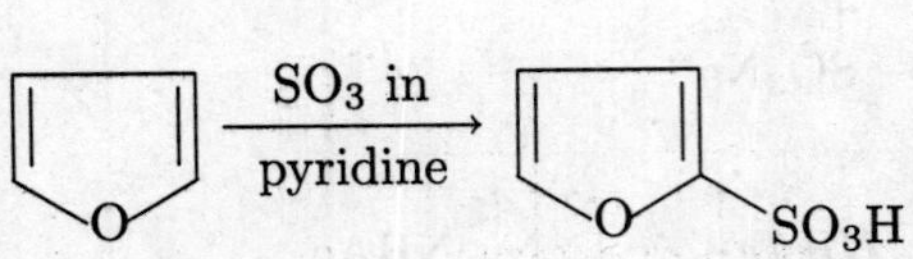

;

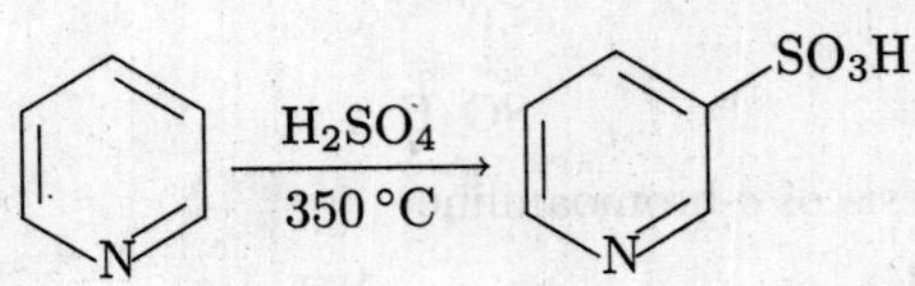

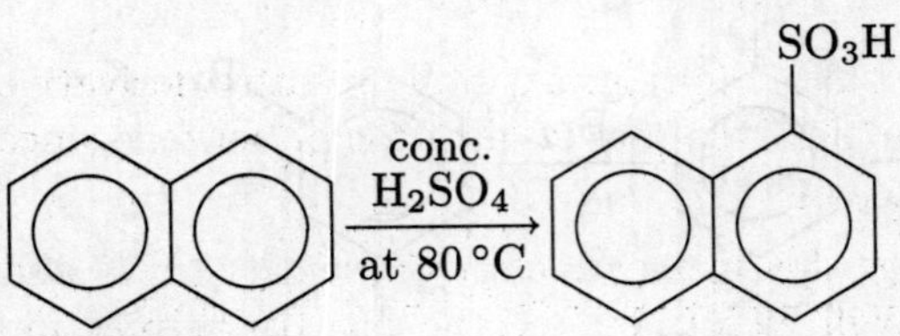

;

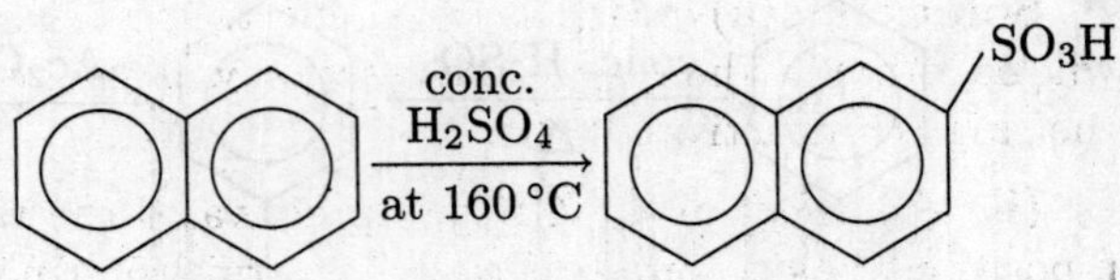

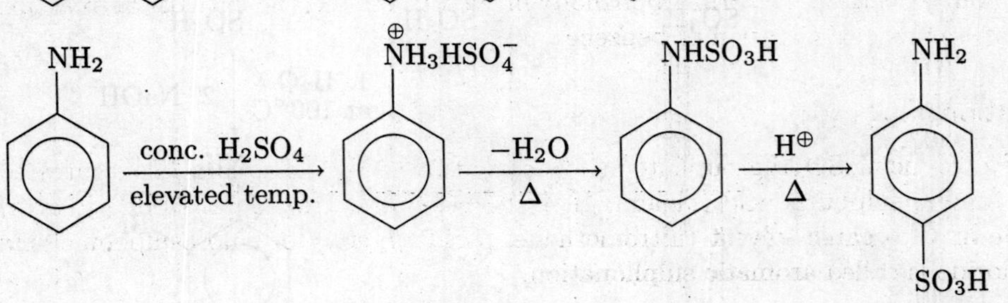

Sulphanilic acid

The orientation is determined by the nature of the substituent present in the ring. Usually at a lower temperature the kinetically controlled product is obtained while at an elevated temperature the thermodynamically controlled product forms. Since SO_3H is a bulky group, there is steric congestion at *o*-position.

Aromatic sulphonation is a reversible reaction. At low temperatures, the reaction is practically irreversible since the reverse reaction, known as desulphonation reaction, is very slow. Since sulphonation and desulphonation are very simple processes, they have some synthetic values. Several aromatic compounds and *β*-substituted naphthalenes are prepared with the help of sulphonation and desulphonation processes. Desulphonation is usually done by heating the sulphonic acid derivative with dilute HCl at an elevated temperature.

(i) Synthesis of *o*-bromotoluene.

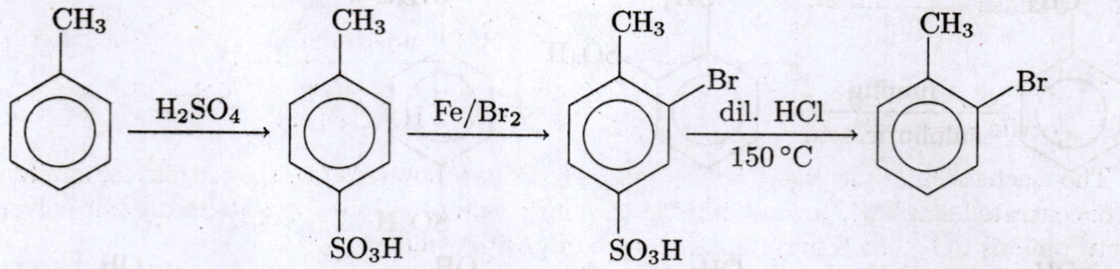

(ii) Synthesis of *o*-bromophenol.

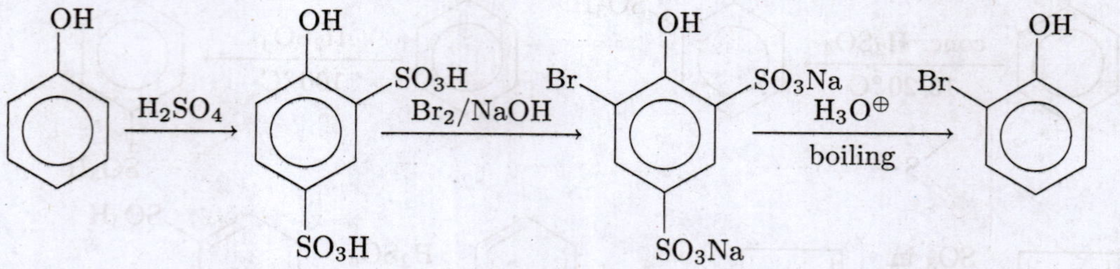

(iii) Synthesis of *o*-bromoaniline

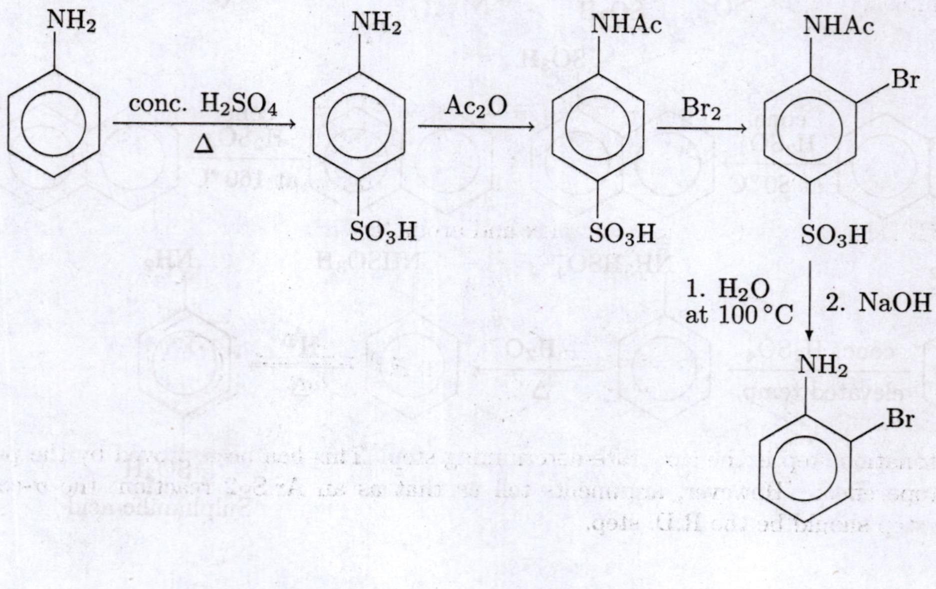

(iv) Synthesis of β-naphthol and other β-substituted naphthalenes.

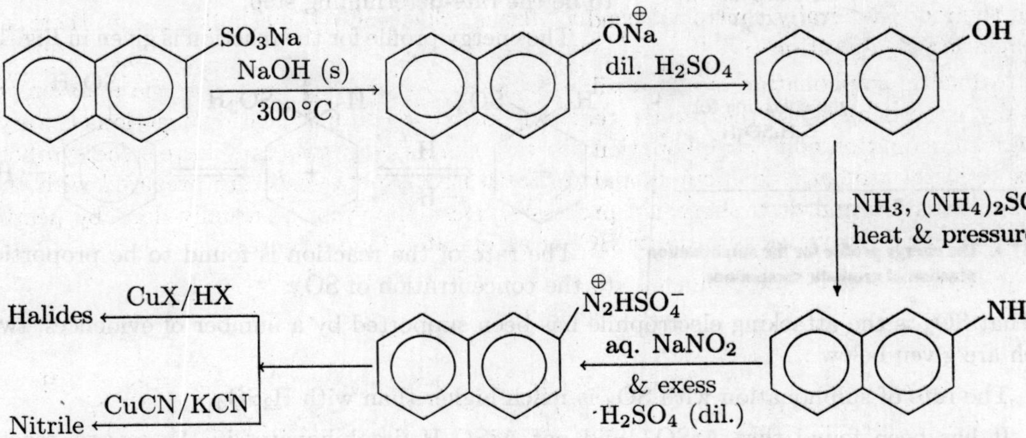

The mechanism of the reaction is supposed to be as follows. When the reagent for sulphonation is concentrated sulphuric acid or fuming sulphuric acid or sulphurtrioxide in organic solvent, the electrophile is SO_3; the S atom is the centre of positive charge.

$$2H_2SO_4 \rightleftharpoons SO_3 + H_3O^{\oplus} + HSO_4^-$$

$$H_2S_2O_7 \rightleftharpoons H_2SO_4 + SO_3$$

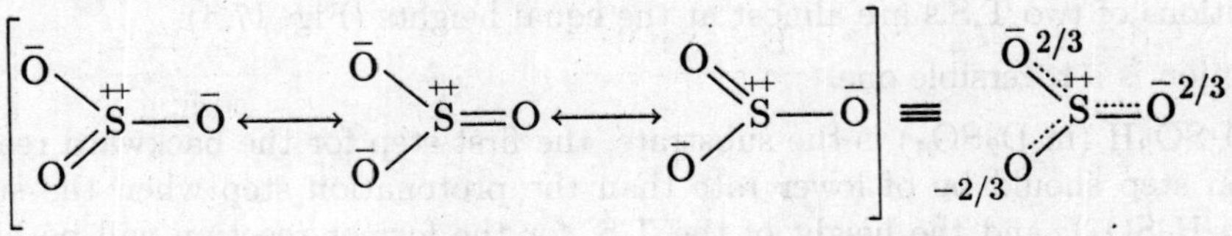

Step I. Electrophilic attack by the SO_3 on a nuclear carbon of the aromatic compound to form σ-complex.

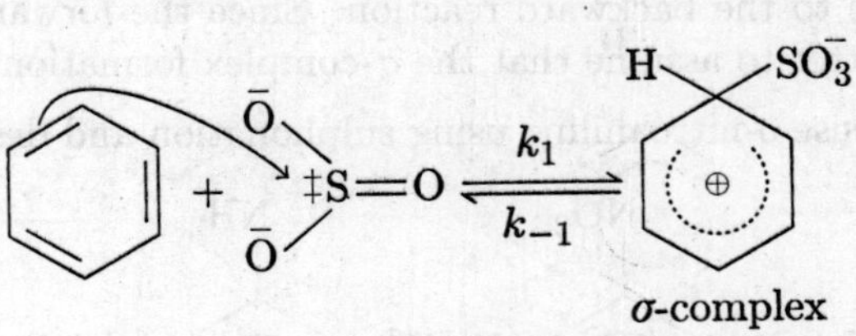

Step II. Deprotonation of the σ-complex and aromatisation.

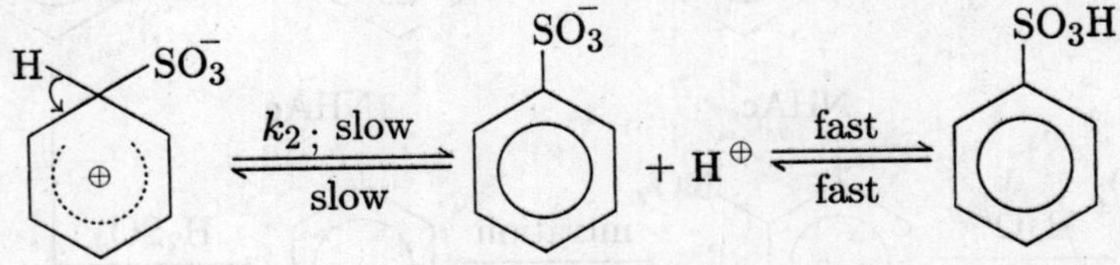

Deprotonation step is the slow rate-determining step. This has been proved by the primary kinetic istope effect. However, arguments tell us that as an Ar.S_E2 reaction, the σ-complex formation step should be the R.D. step.

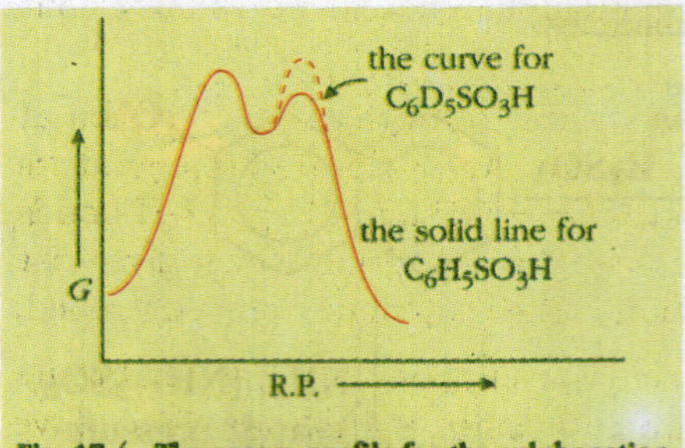

Fig. 17.4. The energy profile for the sulphonation reaction of aromatic compounds

When the reagent is oleum the second step is supposed to be the rate-determining step.

The energy profile for the reaction is given in Fig. 17.4.

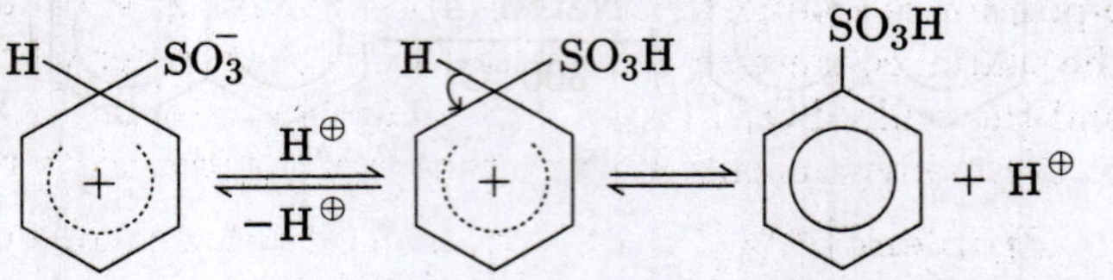

The rate of the reaction is found to be proportional to the concentration of SO_3.

That SO_3 is the attacking electrophile has been supported by a number of evidences, two of which are given below :

(a) The rate of sulphonation with SO_3 is much higher than with H_2SO_4.

(b) It has been found that $ArSO_3^-$ and not $ArSO_3H$ desulphonates in the reverse reaction. Following the Principle of Microscopic Reversibility, it can be said that, SO_3 and not $SO_3H^{\oplus}$ is the attacking electrophile.

Deprotonation step shows primary kinetic isotope effect. This indicates that C—H bond breaking is involved in the rate-determining step, i.e., $k_{-1} > k_2$.

Whether the deprotonation step is the R.D. step or not that may be assumed from the facts given below :

(a) The positions of two T.S.s are almost at the equal heights (Fig. 17.4).

(b) The reaction is a reversible one.

When $C_6D_5SO_3H$ (in D_2SO_4) is the substrate, the first step for the backward reaction, i.e., the deuteration step should be of lower rate than the protonation step when the substrate is $C_6H_5SO_3H$ (in H_2SO_4); and the height of the T.S. for the former reaction will be higher than the height of the T.S. for the latter reaction and we shall observe PKIE for the reaction. So the observed PKIE may be due to the backward reaction. Since the forward reaction is essentially an Ar.S_E2 reaction, it is better to assume that the σ-complex formation step is the R.D. step.

Problem 17.7. Synthesise *o*-nitroaniline using sulphonation and desulphonation processes.

Solution :

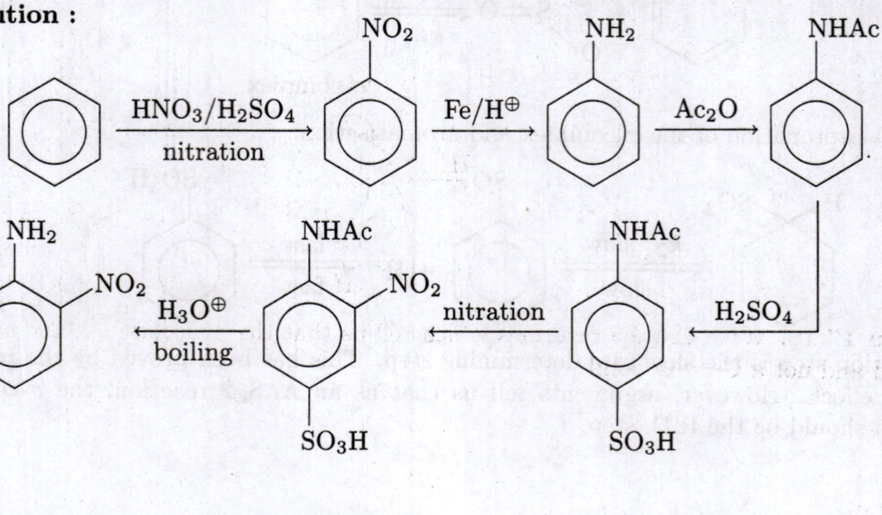

Problem 17.8. Why does phenol form *o*-hydroxybenzenesulphonic acid at 15-20 °C but *p*-hydroxybenzenesulphonic acid at 100 °C?

Solution : *o*-position has spatial congestion but *p*-position is free from this. The reaction at *o*-position is kinetically controlled; so it occurs at a low temperature. At an elevated temperature the spatial congestion increases owing to the greater vibration of C—OH and C—SO_3H bonds and the bulky SO_3H group leaves the substrate and SO_3 attacks the *p*-position which gives rise to thermodynamically stable product. Thus, the latter reaction is thermodynamically controlled.

Problem 17.9. The conversion of aniline to sulphanilic acid is a sulphonation reaction but its mechanism is something different from the mechanism of aromatic sulphonation reaction. Give a probable mechanism for the conversion.

Solution : The reaction involves the following steps :

Step I. Salt formation.

$$H_2SO_4 \rightleftharpoons H^{\oplus} + HSO_4^-$$

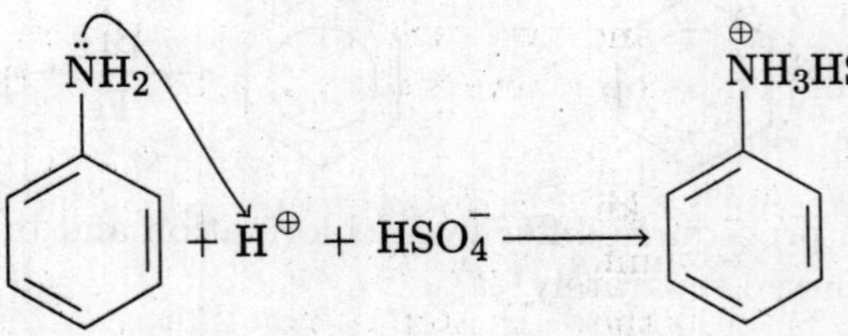

Step II. Loss of water and the formation of phenylsulphamic acid.

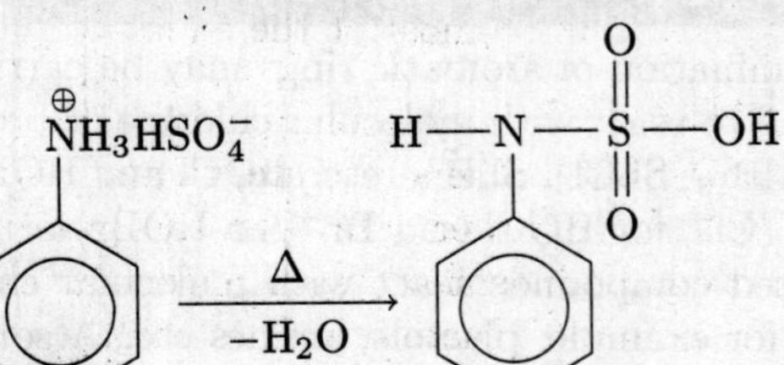

Step III. Stepwise protonation of phenylsulphamic acid and the rearrangement of the protonated phenylsulphamic acid to sulphanilic acid.

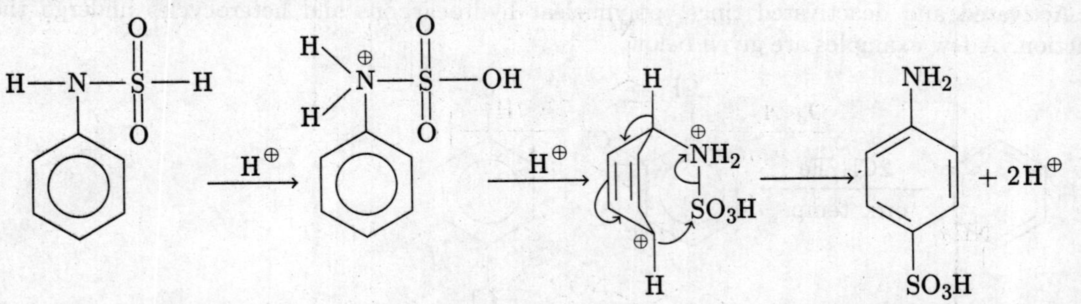

Problem 17.10. What are the reactions which tell us that the benzenesulphonic acid has a C—S bond and not a C—O bond?

Solution : The conversion of SO_3H group to SH group and vice versa tells us that C—S bond is present in benzenesulphonic acid.

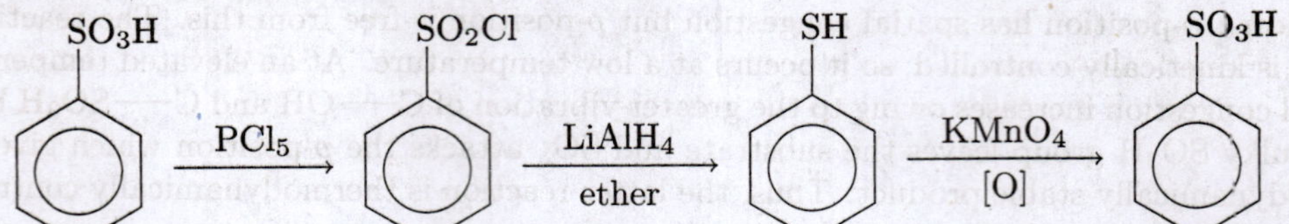

17.8 Halogenation

Replacement of an H atom from an aromatic ring by a halogen atom is called aromatic halogenation reaction.

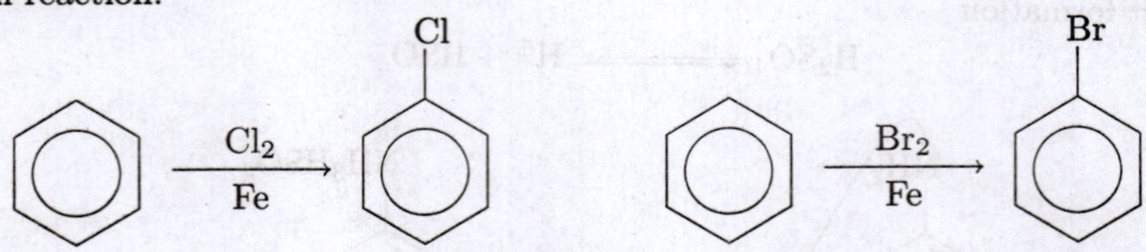

Fluorination and iodination procedures differ from chlorination and bromination procedures. So, these processes will be discussed separately.

17.8.1 Aromatic chlorination and bromination

Chlorination and bromination of aromatic rings may be carried out at ordinary temperature by allowing the compound to react with molecular chlorine or bromine in the presence of Fe, or a Lewis acid, e.g., $AlCl_3$, $AlBr_3$, $SbCl_5$, $SbBr_5$, etc. HOCl and HOBr may also be used but in these cases either a halide ion (Cl^- for HOCl and Br^- for HOBr) or an acid is to be added to speed up the reaction. Activated compounds react with molecular chlorine and bromine even in the absence of a Lewis acid, for example, phenols, amines etc. A solution of the reagent in water or in an organic solvent is used in these cases. Here it is important to mention that inter-halogens, ICl, IBr, etc. and iodine itself can act as a halogen carrier.

A wide and varied types of substrates can be halogenated by the process.

Activated and deactivated rings, polynuclear hydrocarbons and heterocycles undergo the reaction. A few examples are given below.

(a)

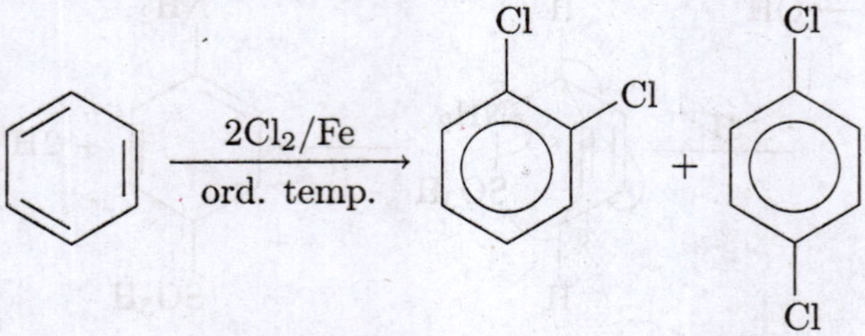

major

(b) 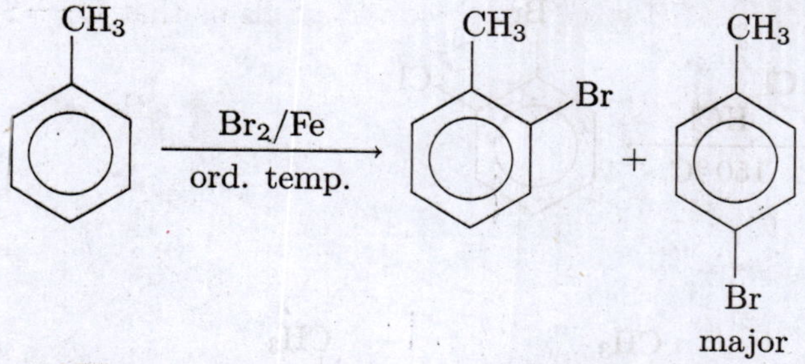
CH3
Br2/Fe
ord. temp.
CH3
Br
+
CH3
Br
major

(c) 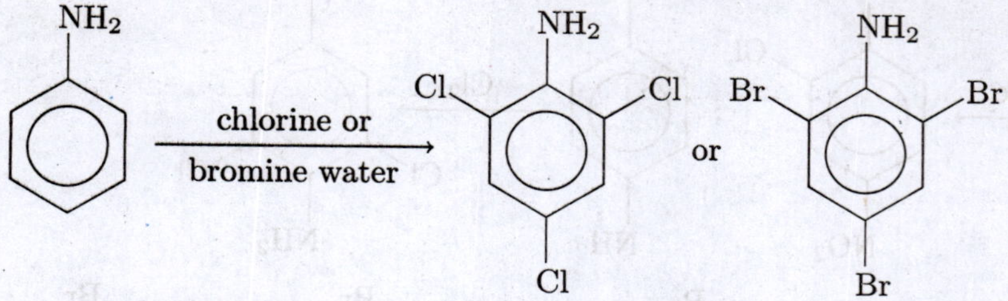
NH2
chlorine or
bromine water
NH2
Cl
Cl
Cl
or
NH2
Br
Br
Br

(d) 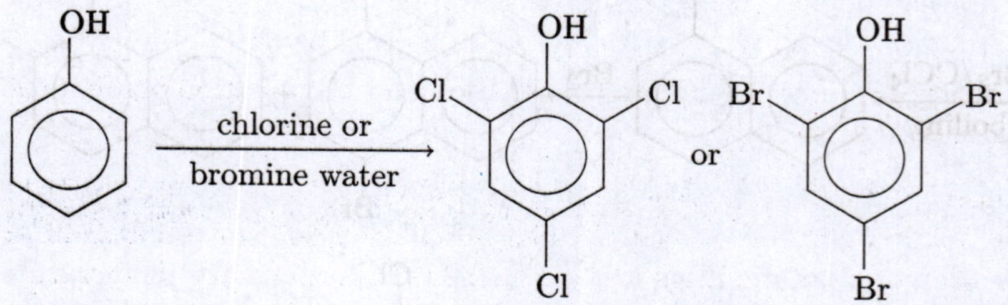
OH
chlorine or
bromine water
OH
Cl
Cl
Cl
or
OH
Br
Br
Br

(e) 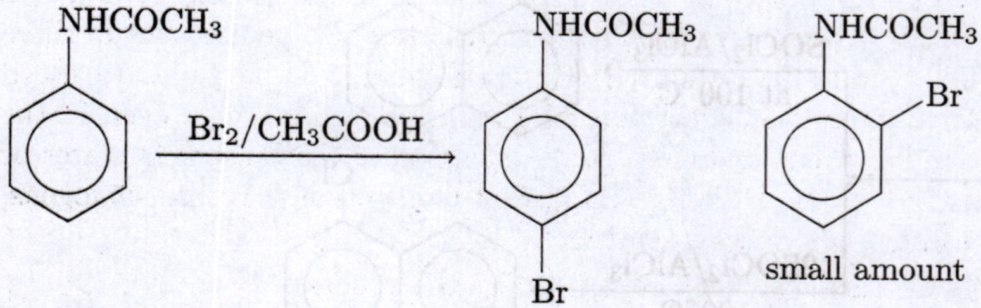
NHCOCH3
Br2/CH3COOH
NHCOCH3
Br
NHCOCH3
Br
small amount

(f) 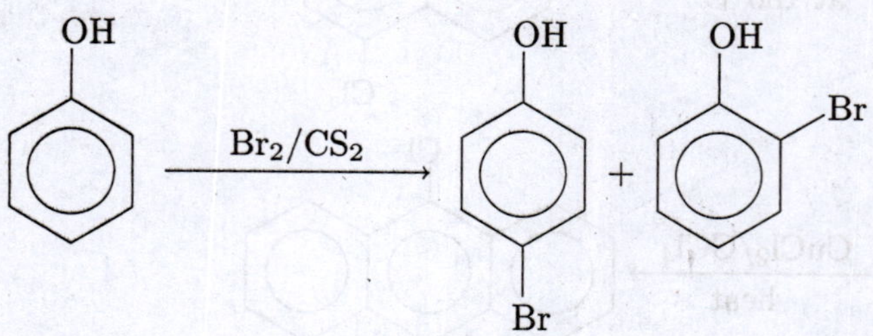
OH
Br2/CS2
OH
Br
+
OH
Br

(g) 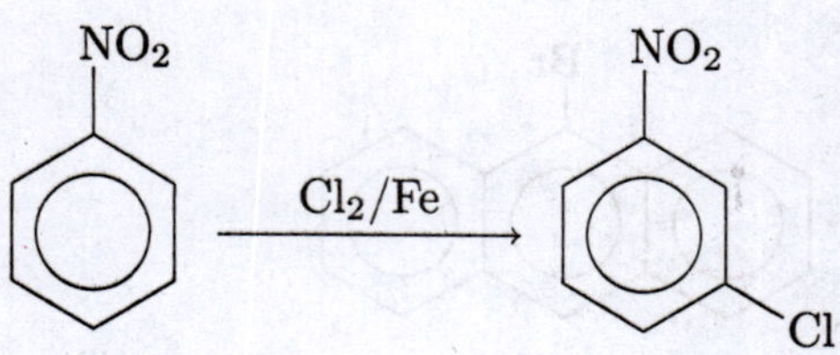
NO2
Cl2/Fe
NO2
Cl

(h)

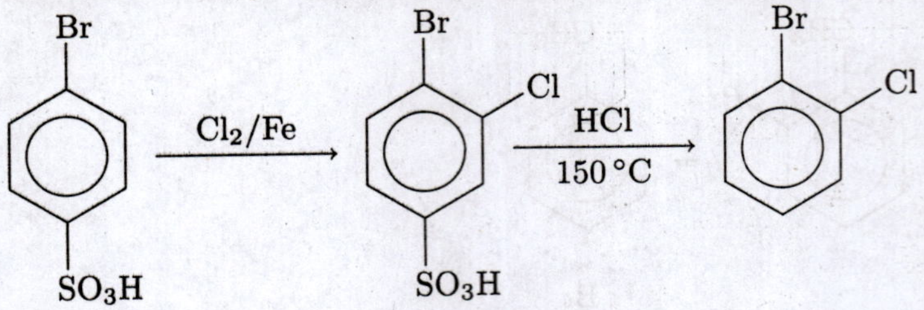
Br
SO3H
Cl2/Fe
Br
Cl
SO3H
HCl
150 °C
Br
Cl

(i)

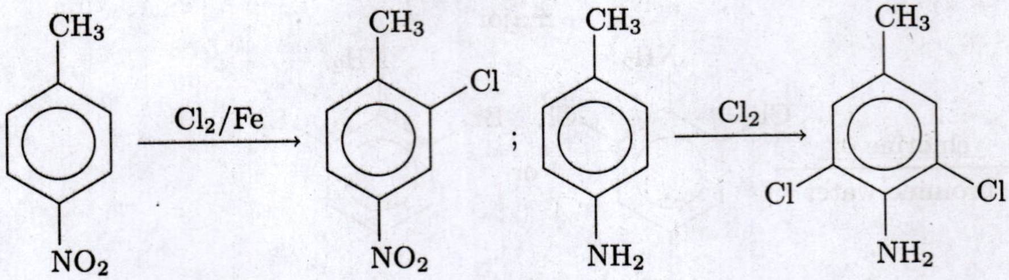
CH3
NO2
Cl2/Fe
CH3
Cl
NO2
;
CH3
NH2
Cl2
CH3
Cl
Cl
NH2

(j)

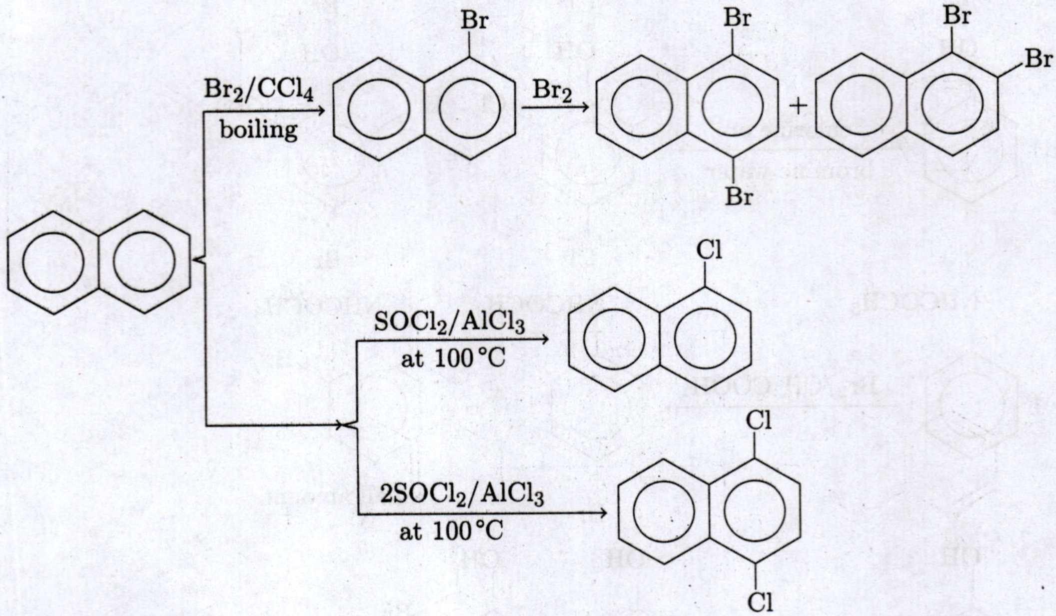
Br2/CCl4
boiling
Br
Br2
Br
Br
+
Br
Br
SOCl2/AlCl3
at 100 °C
Cl
2SOCl2/AlCl3
at 100 °C
Cl
Cl

(k)

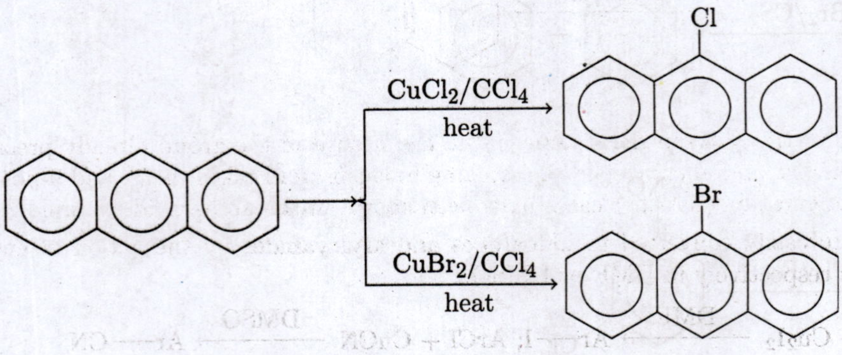
CuCl2/CCl4
heat
Cl
CuBr2/CCl4
heat
Br

(l) 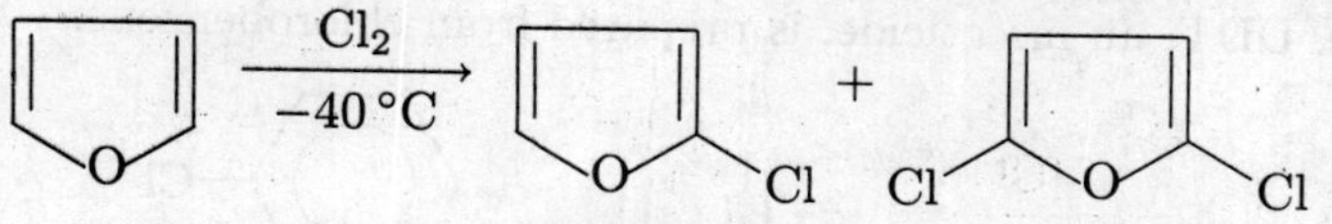

(m) 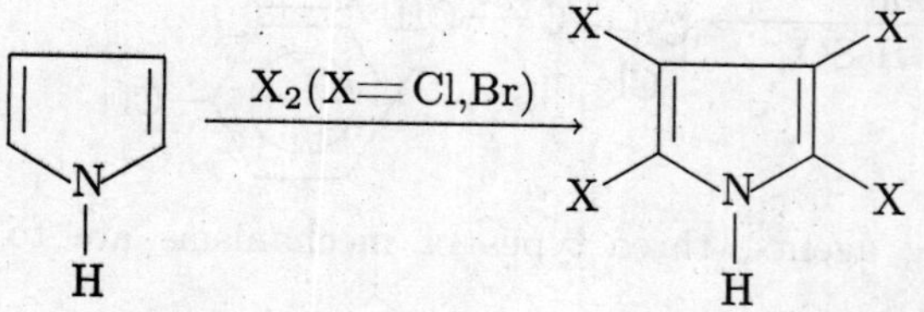

(n) 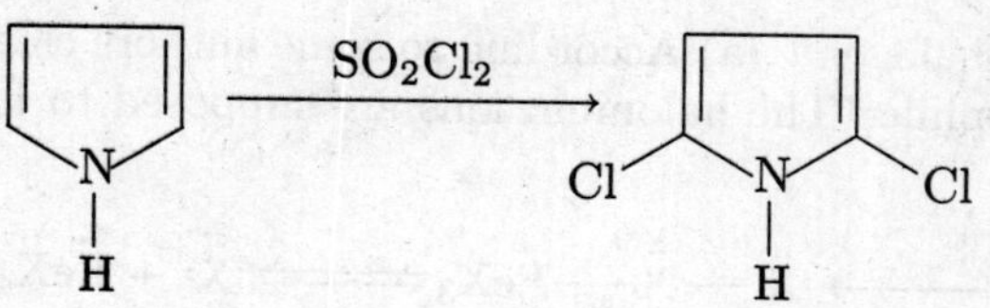

(o) 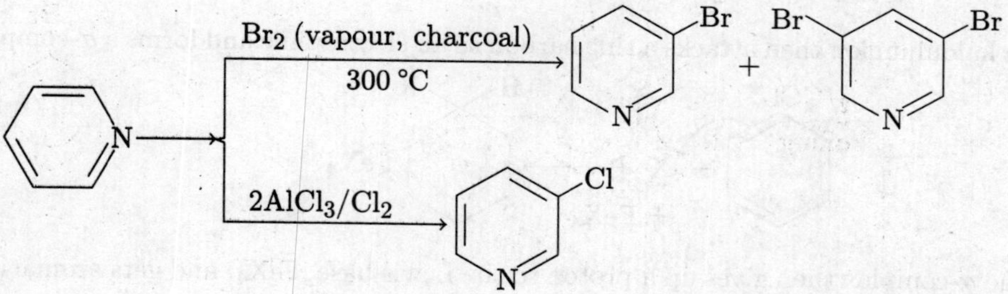

(p) 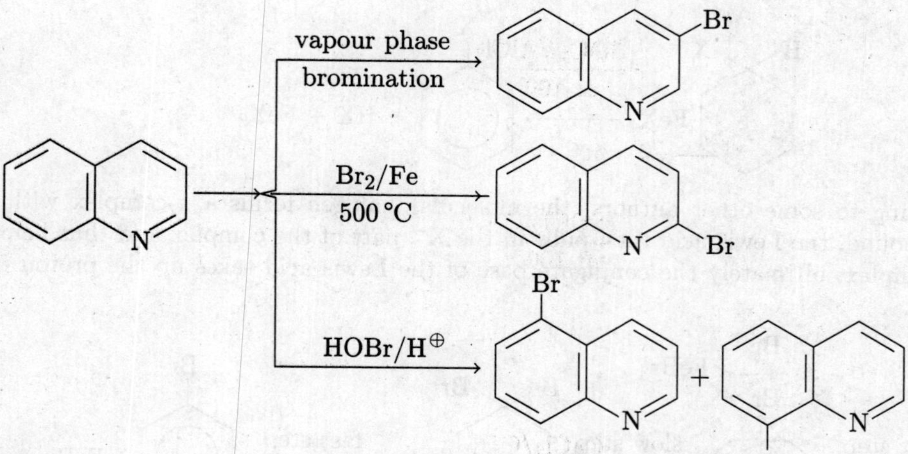

The orientation of the halogen atom depends on the nature of the group already present in the ring. In naphthalene and anthracene the incoming halogen atom takes the 1- and 9-positions respectively. The orientations in other cases have been shown in the above said examples.

Chloroarenes are easily converted to iodoarenes and arylcyanides by the action of cuprous iodide and cyanide respectively in DMF or DMSO.

$$ArCl + Cu_2I_2 \xrightarrow{DMF} Ar{-}I; \quad ArCl + CuCN \xrightarrow{DMSO} Ar{-}CN$$

Dichlorodiphenyltrichloroethane, DDT, an insecticide, is prepared from chlorobenzene.

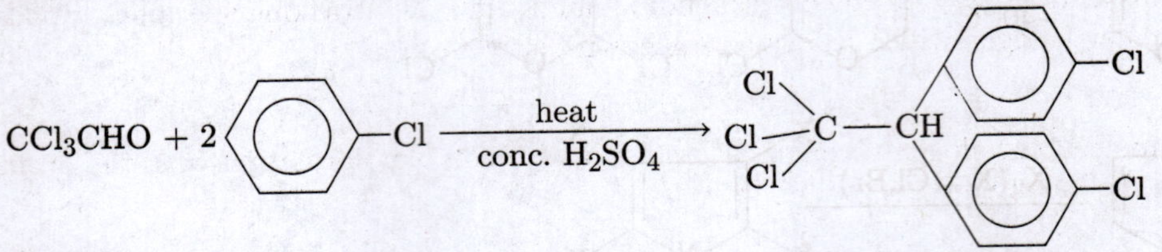

Depending on the nature of halogenating agents, three types of mechanisms are to be discussed :

(i) Mechanism of halogenation with iron and halogen : (a) According to some authors chloronium or bromonium ion is the attacking electrophile. The halonium ions are supposed to form by the following way :

$$2Fe + 3X_2 \longrightarrow 2FeX_3;\quad :\ddot{X}-\ddot{X}: + FeX_3 \longrightarrow :\ddot{X}-\overset{\oplus}{\ddot{X}}-\bar{F}eX_3 \rightleftharpoons :\overset{\oplus}{X}: + \bar{F}eX_4$$

Step I. The halonium ion then attacks a ring carbon as an electrophile and forms a σ-complex.

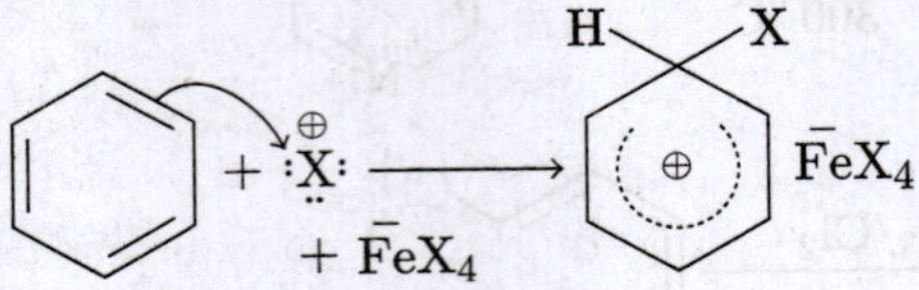

Step II. The σ-complex then gives up a proton to the Lewis base, $\bar{F}eX_4$, and gets aromatised.

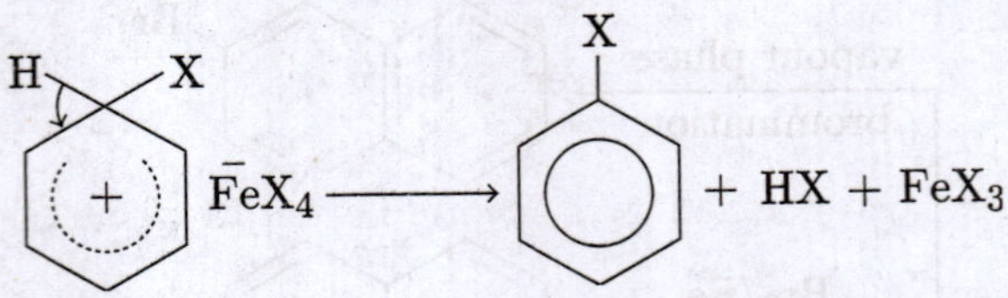

(b) According to some other authors, the molecular halogen forms a π-complex with the aromatic compound, the Lewis acid then pulls in the X^- part of the complex and thus helps to form the σ-complex; ultimately the conjugate base of the Lewis acid takes up the proton from the σ-complex.

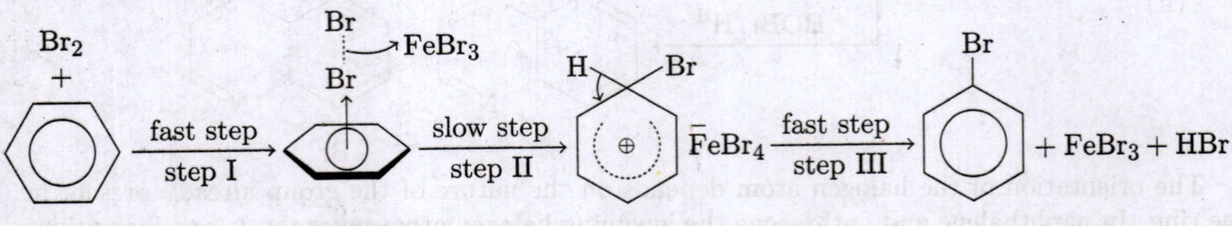

The rate = k [substrate] [X_2] [Lewis acid]

This rate order is often followed by the halogenation processes and supports the π-complex mechanism (b). Therefore, step II of (b) pathway is the rate-determining step. No primary kinetic isotope effect has yet been observed with chlorination (and rarely observed with bromination). This suggests that C—H bond-breaking step is not the rate-determining step.

(ii) Mechanism of halogenation with hypohalous acid and halide ion : When the reagent is hypohalous acid along with the corresponding halide ion, the attacking species is supposed to be the molecular halogen which is formed as shown below :

$$2H^{\oplus}\ {}^{-}O—X + X^{-} \longrightarrow H_2O + X_2.$$

This molecular halogen through π-complex and σ-complex formation with the aromatic ring produces the haloarene.

(iii) Mechanism of halogenation with hypohalous acid and a strong acid : In this case the attacking electrophile is either the protonated hypohalous acid or the halonium ion which is formed by the following pathway.

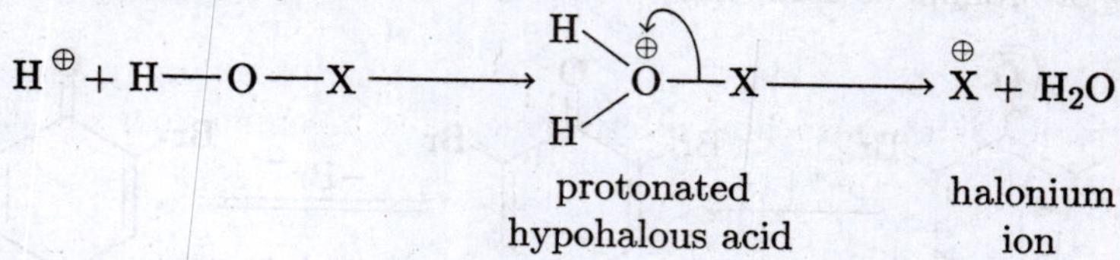

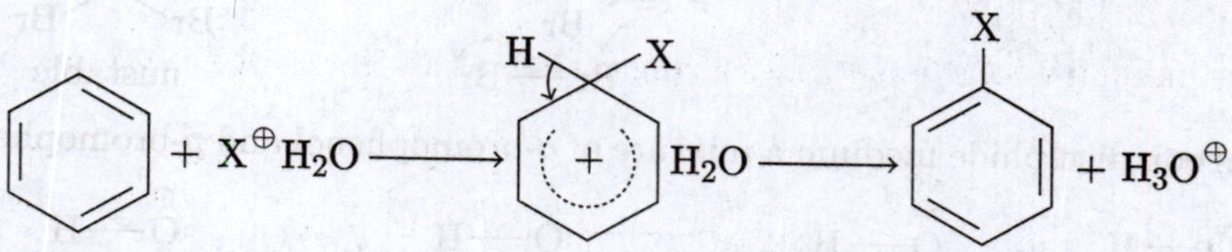

In the cases of highly reactive substrate such as phenol, anisole, aniline, etc. the reactions become very nearly of zero order in the substrate and over all second order.

The rate = k [HOX] [$H^{\oplus}$]

When the substrate is relatively unreactive but water soluble, e.g., nitrophenols, the halogenation is first order in the substrate and over all third order.

The rate = k [HOX] [$H^{\oplus}$] [substrate]

Problem 17.11. Give the mechanism of bromination of phenol in water and in carbon disulphide.

Solution :

(a)

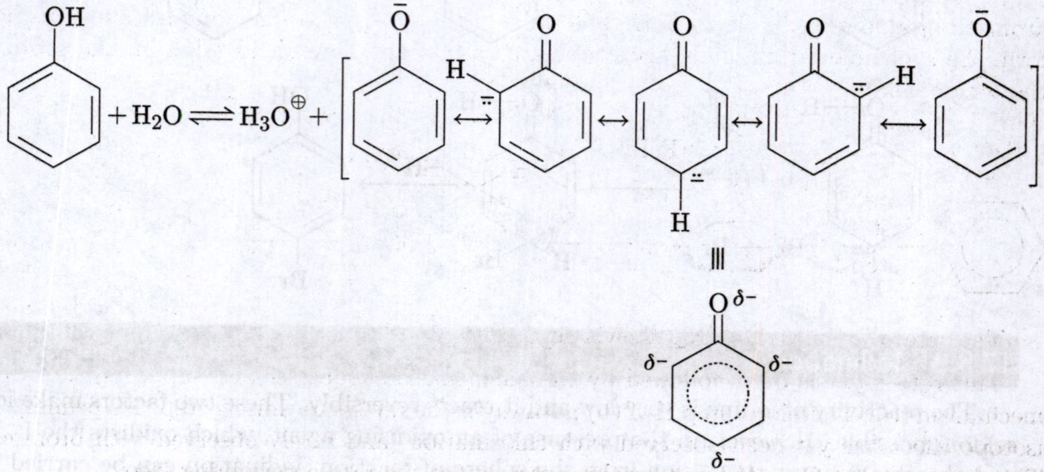

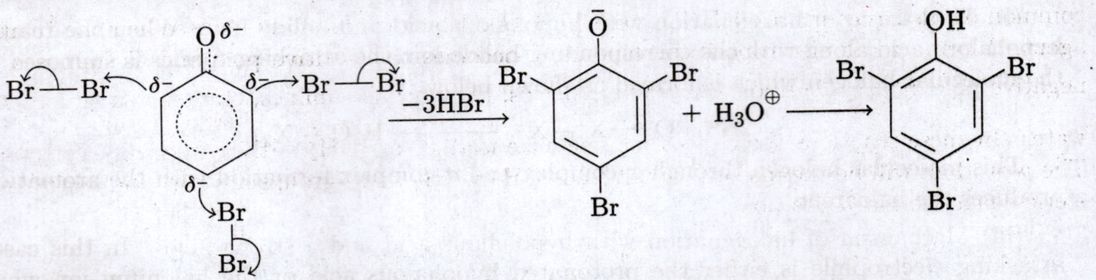

When excess of bromine is used, the product is tetrabromocyclohexadienone.

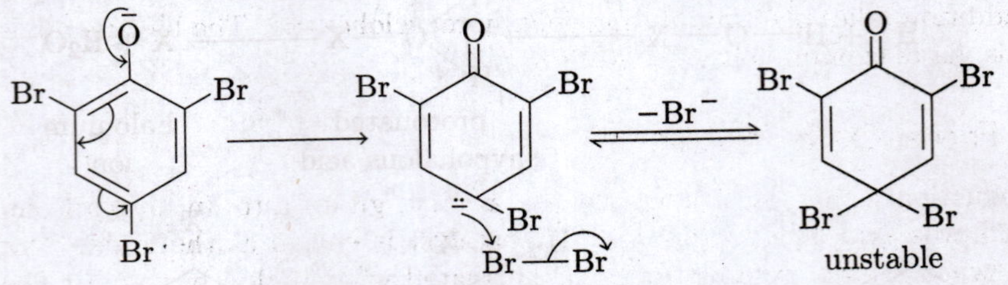

(b) In carbon disulphide medium a mixture of *o*-bromophenol and *p*-bromophenol is obtained.

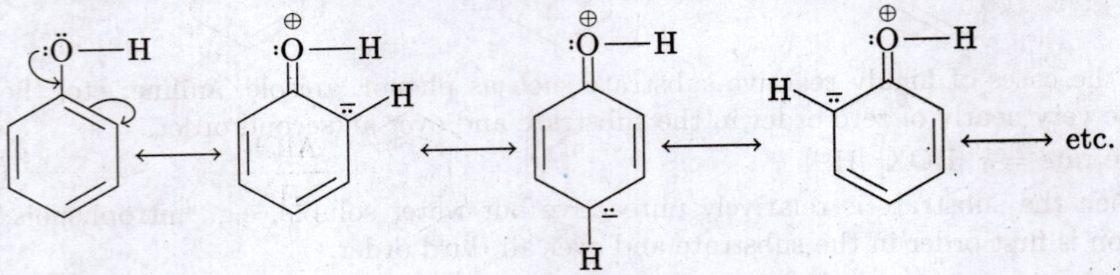

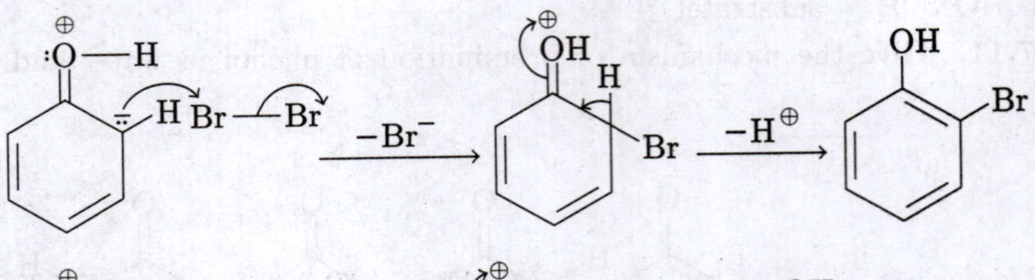

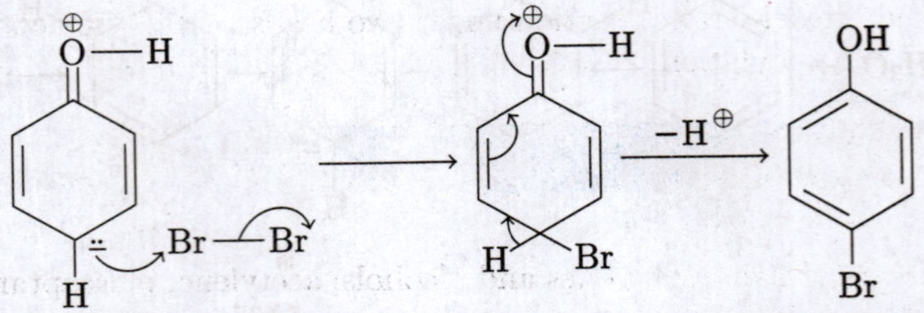

17.8.2 Iodination

The reactivity of iodine is very low and it reacts reversibly. These two factors make iodination seldom possible. However, in the presence of an oxidising agent, which oxidises the I^- ion to I_2 and thereby removes the I^- ion from the sphere of reaction, iodination can be carried out. The common oxidising agents are : HNO_3, HIO_3, H_2O_2. However, ICl may be used as an iodinating agent. I being more electropositive than Cl, I becomes the positive pole and Cl becomes the negative pole of $\overset{\delta+}{I}—\overset{\delta-}{Cl}$ dipole. Activated species, like phenols, amines, etc. react with iodine in water. In these reactions there is strong evidence that $I^{\oplus}$ or $\overset{\oplus}{I}OH_2$ is the attacking electrophile. The *p*-product will be the major product since OH or NH_2 and I groups are fairly bulky to create steric congestion at *o*-positions.

17.8.3 Fluorination

The F-F bond energy is only 154.4 kJ mol^{-1}. When fluorine is allowed to react with benzene, free radical addition takes place and forms polyfluorocyclohexane. The best method of preparing fluoroarenes is via diazonium ion (Sec. 20.1.2n).

17.9 The Friedel-Crafts Reaction

The introduction of an alkyl or an aryl or an acyl group into an aromatic nucleus in the presence of a Lewis acid or a protonic acid as catalyst is known as the Friedel-Crafts reaction. For example, when a large excess of benzene is treated with methyl bromide in the presence of anhydrous $AlBr_3$, toluene is obtained; treatment of benzene with acetyl chloride in the presence of $AlCl_3$ gives acetophenone.

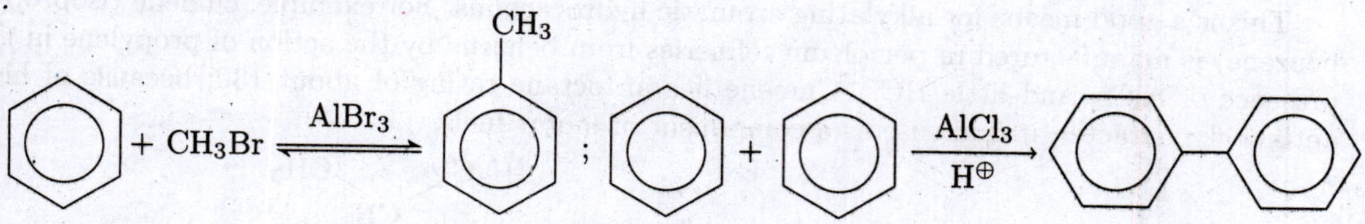

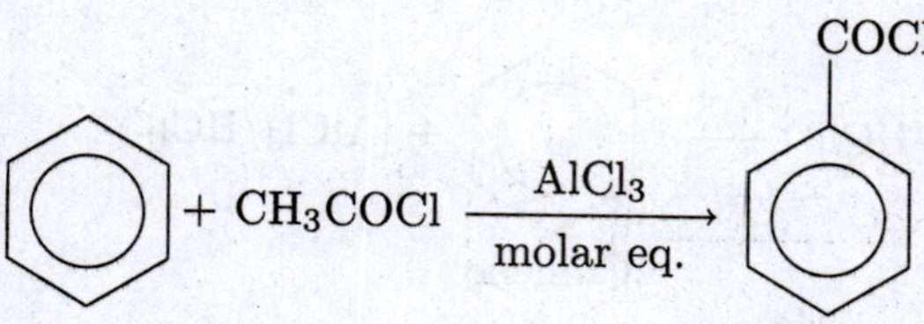

However, we shall discuss the Friedel-Crafts reaction under two heads : (i) The Friedel-Crafts alkylation and (ii) the Friedel-Crafts acylation.

17.9.1 The Friedel-Crafts alkylation

In the Friedel-Crafts alkylation reaction

(i) The alkylating agents are alkyl halides, alkenes and alcohols; acetylene, mercaptans, inorganic esters, sulphides may also be used.

However, the order of reactivity for all types of reagents is :

$$\text{allyl, benzyl} > 3° > 2° > 1°.$$

The order of reactivity among halides is :

$$\text{fluoride} > \text{chloride} > \text{bromide} > \text{iodide}.$$

(ii) The substrates are aromatic hydrocarbons, haloarenes and phenols. Arenes with deactivating groups do not undergo the Friedel-Crafts alkylation; but if they contain activating group also, the reaction occurs. For example, nitrobenzene does not undergo the Friedel-Crafts reaction but *p*-nitrotoluene can be made to undergo the reaction. Phenols give poor yields in the Friedel-Crafts reaction. Heterocycles do not usually undergo the reaction. However, furan and thiophene have been alkylated by the process.

(iii) The acid catalysts are, $AlCl_3$, $GaCl_3$, BF_3, HF, H_2SO_4 phosphoric acid, etc.

The reactivity order of Lewis acid catalysts is :

$$AlBr_3 > AlCl_3 > GaCl_3 > FeCl_3 > SbCl_5 > SnCl_4 > BCl_3, BF_3, \text{etc.}$$

Of course, the reactivity of acid catalysts changes from reagent to reagent. Metal halides are generally used in the cases of alkyl halides and alkenes. When alcohols are used as alkylating agents, HF, BF_3, H_2SO_4, are acid catalysts.

(iv) The solvents for solid substrates are usually carbon disulphide and nitrobenzene; for liquid substrates use of a solvent is not essential; the excess of the substrate may play the role of solvent. However, CH_2Cl_2, *n*-hexane, etc. may also be used successfully.

The reaction procedure is very simple; this is carried out by mixing together the substrate, the reagent and the solvent. Depending on the nature of the required product sometimes heating and sometimes cooling are done.

This is a good means for alkylating aromatic hydrocarbons. For example, cumene (isopropylbenzene) is manufactured in petroleum refineries from benzene by the action of propylene in the presence of $AlCl_3$ and little HCl. Cumene has an octane rating of about 130; because of high antiknock character, it finds use as a component of motor fuels.

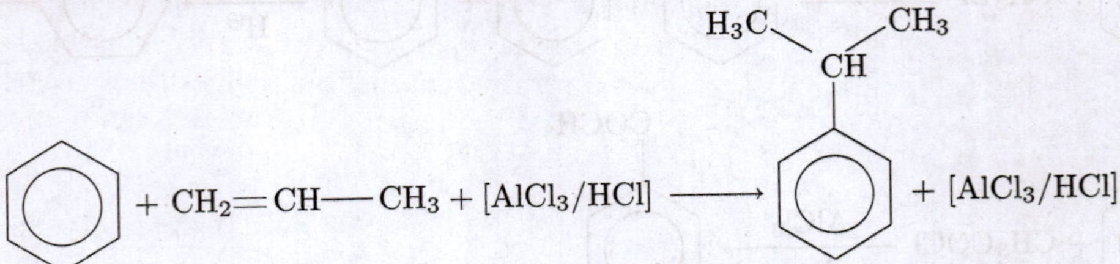

Cumene

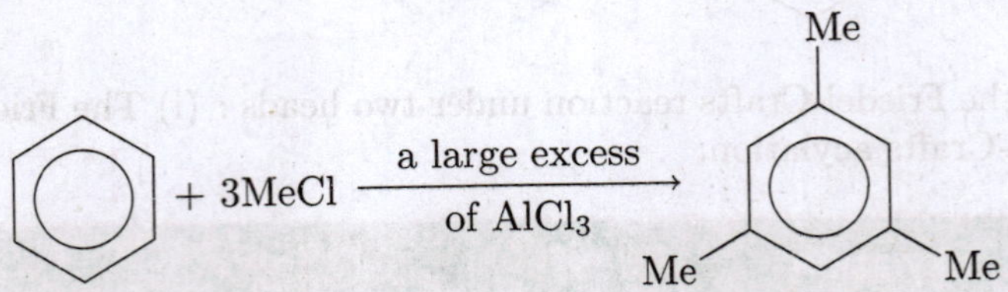

Mesitylene

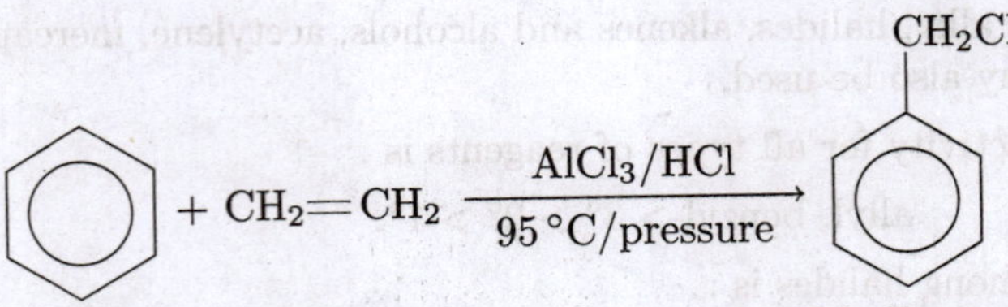

Ethylbenzene

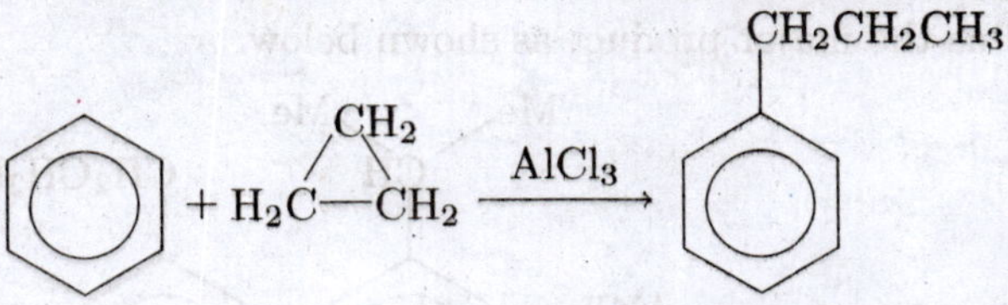

n-propylbenzene

The reaction may be used to prepare fused ring systems; six membered rings have been prepared most successfully.

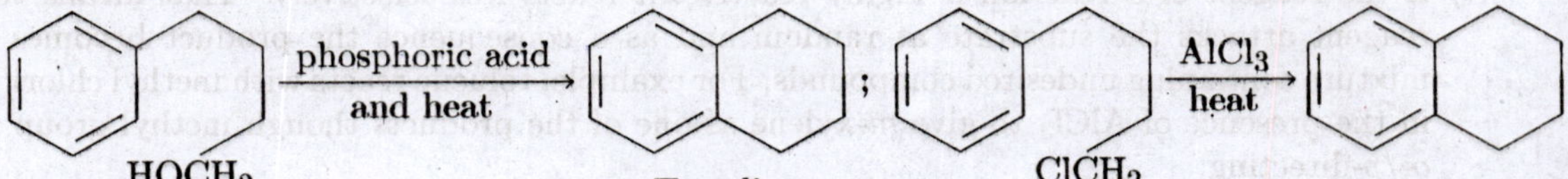

Tetralin

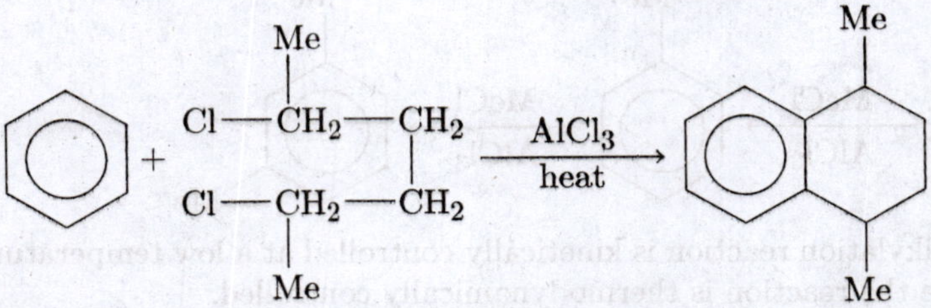

Tertiary alkyl group of an aromatic hydrocarbon may be easily removed by treating the compound with excess of benzene in the presence of $AlCl_3$-HCl; this is possible because the reaction is a reversible one. Utilising this, one can introduce a group at a less effective position; for example, when toluene is acylated, the major-product is *p*-acyltoluene this is because the *p*-position has no spatial congestion. However, by protecting the *p*-position with tert-butyl group, acylating *p*-t-butyltoluene and then removing the tert-butyl group one can prepare *o*-acyltoluene as shown below :

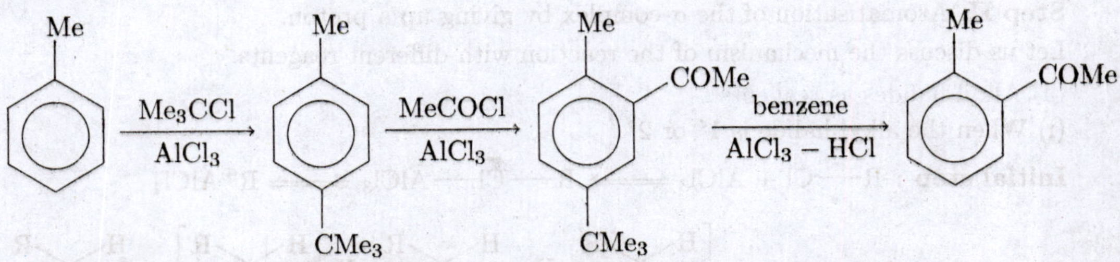

However, the Friedel-Crafts reaction has a number of limitations as shown below :

(i) Activated aromatic hydrocarbons undergo polyalkylation instead of monoalkylation. Naphthalene being more active than benzene, the former cannot be monoalkylated by the Friedel-Crafts reaction. Alkylbenzenes being more active than benzene, they undergo further alkylation by the process. This, however, can be stopped by employing excess of the substrate, a suitable solvent and high-speed stiring. Micro wave technique may be used.

(ii) In many cases isomeric alkylated products are formed instead of the desired product. For example, when *n*-propylchloride is used as reagent to prepare *n*-propylbenzene, actually isopropylbenzene is obtained as the major product as shown below.

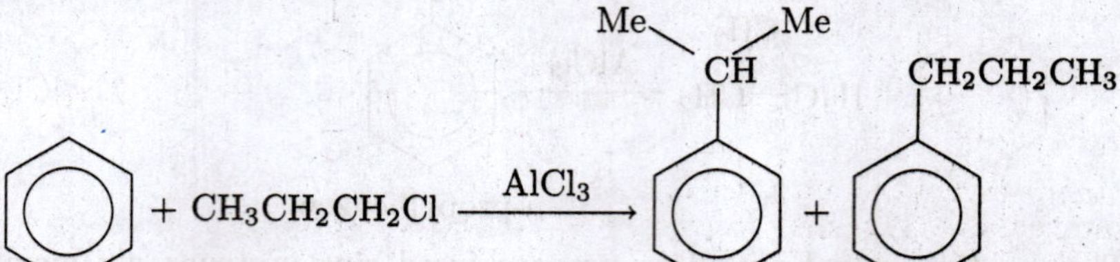

(iii) If the reagent of a reaction is highly reactive, it reacts less selectively. That means the reagent attacks the substrate at random and as a consequence the product becomes a mixture containing undesired compounds. For example, toluene reacts with methyl chloride in the presence of $AlCl_3$ to give *m*-xylene as one of the products though methyl group is *o*-/*p*-directing.

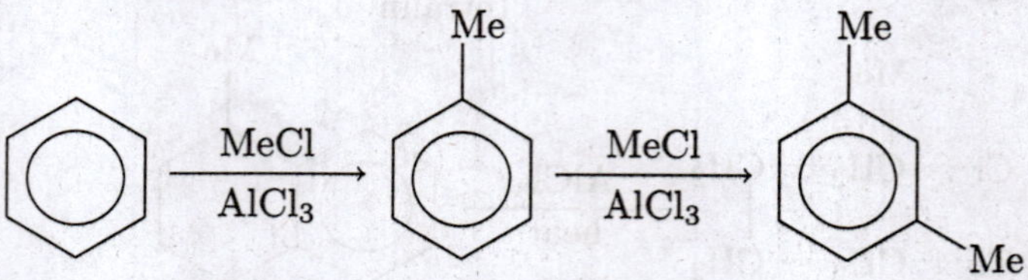

In fact, the Friedel-Crafts alkylation reaction is kinetically controlled at a low temperature while at a high temperature the reaction is thermodynamically controlled.

(iv) When a deactivating group is present in the aromatic ring, the reaction does not take place. Nitro group being a deactivating group, nitrobenzene does not undergo the reaction.

The Friedel-Crafts alkylation follows the aromatic S_E2 pathway. The general steps are :

Initial step : Formation of a an effective electrophile.

Step I. Electrophilic attack on an activated nuclear carbon of the aromatic substrate to form the σ-complex.

Step II. Aromatisation of the σ-complex by giving up a proton.

Let us discuss the mechanism of the reaction with different reagents.

(a) Alkyl halides as reagents :

(i) When the alkyl halide is 1° or 2°.

Initial step : $R{-}Cl + AlCl_3 \rightleftharpoons R{-}\overset{\oplus}{Cl}{-}\bar{A}lCl_3 \rightleftharpoons R^{\oplus}\bar{A}lCl_4$

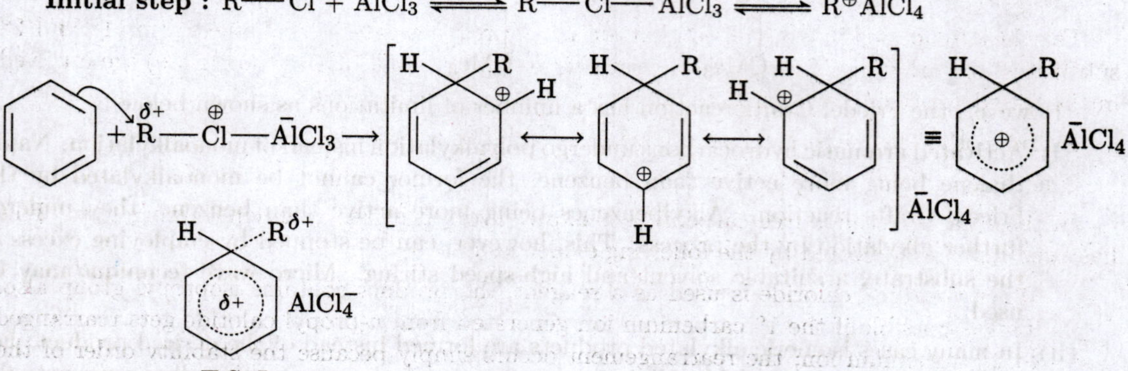

T.S. I

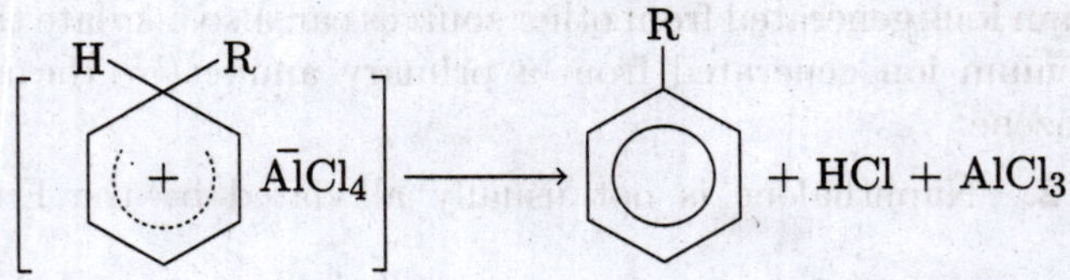

However, $R^{\oplus}\bar{A}lCl_4$ may also react as in the case of 3° alkyl halide.

(ii) When the alkyl halide is 3°.

Initial step : $RCl + AlCl_3 \longrightarrow R^{\oplus}\bar{A}lCl_4$

Step I. 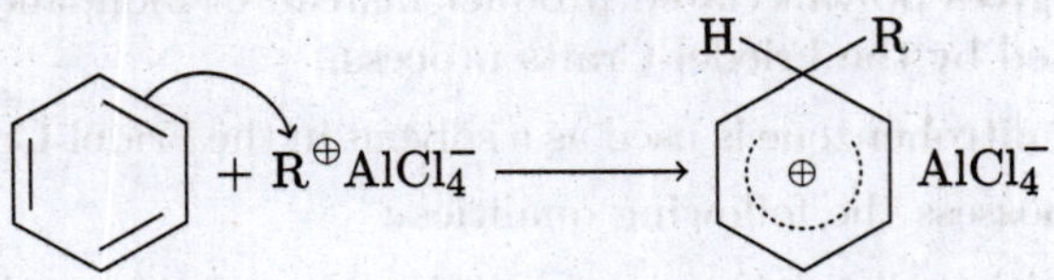

Step II. 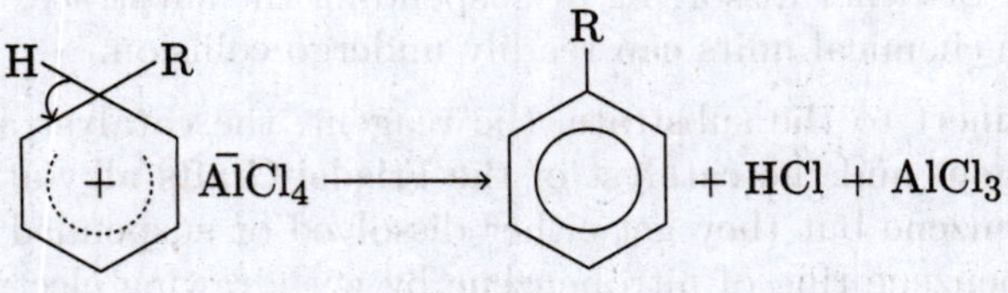

(b) Alcohols as reagents : Perhaps 1°, 2° and 3° alcohols follow the same initial step as shown below :

The other two steps are similar to those as given for 3° alkyl halides.

$$R—O—H + BF_3 \rightleftharpoons R—\overset{\oplus}{\underset{\bar{B}F_3}{O}}—H \rightleftharpoons R^{\oplus} + HO\bar{B}F_3$$

(c) Alkenes as reagents : The initial step involves the formation of a stable carbenium ion. The other steps are similar to those as given for 3° alkyl halides.

Order of the reaction

In some of the cases the reaction follows the third order kinetics.

The rate = k [substrate] [reagent] [catalyst].

For these reactions it is assumed that either the reagent-catalyst complexes (for 1° and 2° substrates) or carbenium ion-conjugate base pairs (e.g., $R^{\oplus}\ AlCl_4^-$; for 3° substrate) are involved in the rate-determining step (step I).

In some other cases the reaction follows the second order kinetics.

The rate = k [substrate] [reagent].

In these cases perhaps free carbenium ions are involved in the rate-determining step. This has been further supported by the following evidences :

(i) When n-propyl chloride is used as a reagent, the product contains isopropyl group also. This is possible if the 1° carbenium ion generated from n-propyl chloride gets rearranged to 2° carbenium ion; the rearrangement occurs simply because the stability order of the carbenium ions is 3° > 2° > 1°.

(ii) Free carbenium ions generated from other sources can also alkylate the aromatic substrates; e.g., a carbenium ion generated from a primary amine (by the action of nitrous acid) alkylates benzene.

Problem 17.12. Naphthalene is not usually alkylated by the Friedel-Crafts alkylation process—explain.

Solution : Naphthalene has two benzene rings fused at *o*-positions. Since one benzene ring has 150.7 kJ mol^{-1} of resonance energy, it is expected that the resonance energy of naphthalene is 301.4 kJ mol^{-1}. Actually it contains 255.69 kJ mol^{-1} of resonance energy which is much less than the expected value. This means it is an activated species. Being activated, the Friedel-Crafts alkylation on naphthalene gives polyalkylated product instead of monoalkylated one. So, naphthalene is not usually alkylated by the Friedel-Crafts process.

Problem 17.13. Explain why nitrobenzene is used as a solvent in the Fridel-Crafts alkylation.

Solution : A solvent should possess the following qualities :

(i) It must have the power of either dissolving or suspending the substrate, the reagent, and the catalyst so that the chemical units can readily undergo collision.

(ii) It must be chemically inert to the substrate, the reagent, the catalyst and the product. The substrate, the reagent and the catalyst of the Friedel-Crafts alkylation do not react chemically with nitrobenzene but they get either dissolved or suspended in it. The nitro group deactivates the benzene ring of nitrobenzene by withdrawing electron form it. The product of alkylation can also be readily separated from nitrobenzene. For these reasons, nitrobenzene is used as a solvent in the Friedel-Crafts reaction.

Problem 17.14. What happens when

(i) PhH is heated with CH_2Cl_2 in the presence of anhydrous $AlCl_3$.

(ii) Chlorobenzene and benzene is treated together in the presence of $AlCl_3$.

(iii) *m*-xylene is treated with tert-butyl chloride at 100 °C in the presence of $AlCl_3$.

(iv) Benzene is treated with ethylene oxide in the presence of $AlCl_3$; give the mechanism of the reaction.

Solution : (i) $PhH + CH_2Cl_2 \xrightarrow{PhH} PhCH_2Cl \xrightarrow[AlCl_3]{PhH} PhCH_2Ph.$

(ii) $PhH + PhCl \xrightarrow{AlCl_3}$ no reaction; since chlorobenzene does not form phenyl cation.

(iii) This is a Friedel-Crafts alkylation reaction; here the substrate is *m*-xylene and tert-butyl chloride is the reagent. At high temperature the thermodynamically controlled *m*-product will be formed, the tert-butyl group will take up the *m*-position with respect to the two CH_3 groups in order to avoid steric strain though the CH_3 group is *o*-/*p*-director. The product will be :

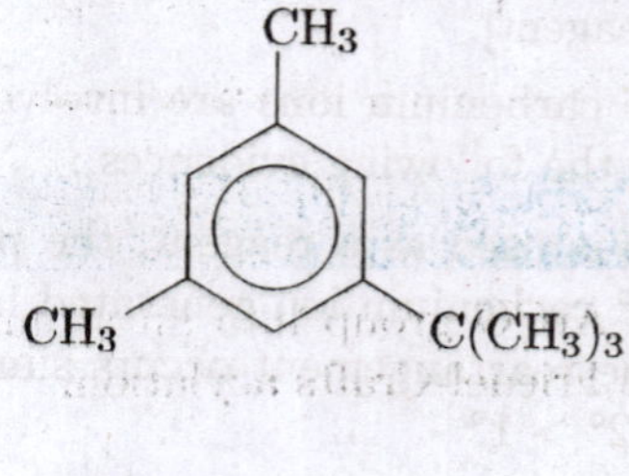

(iv) The product of the given reaction is β-phenylethyl alcohol

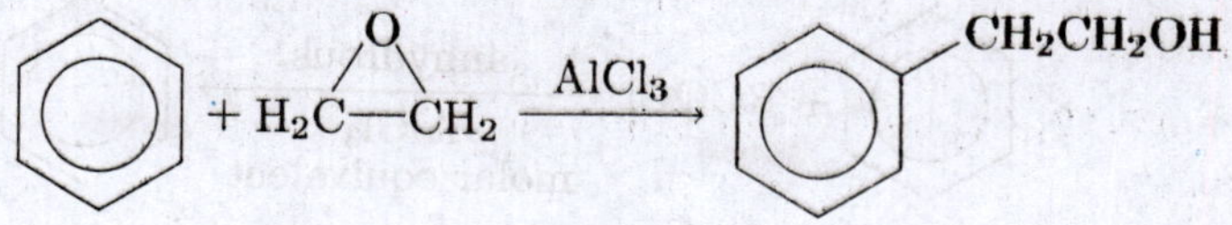

The mechanistic steps are :

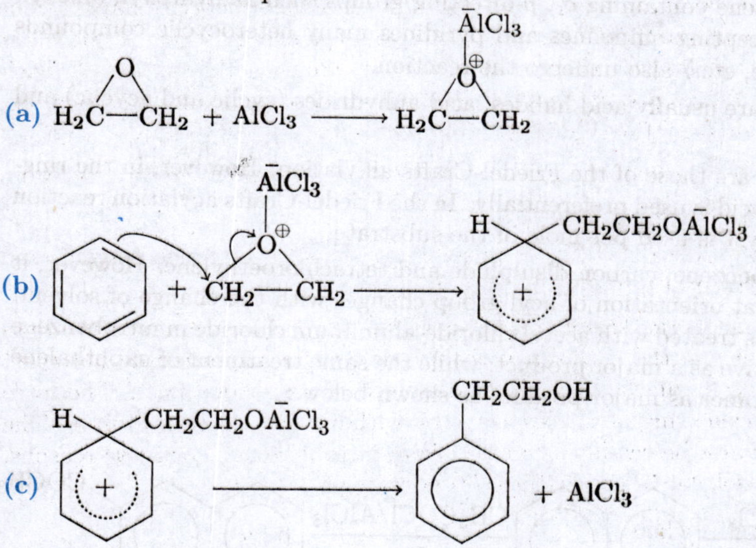

Problem 17.15. When aniline is subjected to the Friedel-Crafts alkylation in the presence of catalytic amount of $AlCl_3$, alkylation does not occur; while in the presence of large excess of $AlCl_3$, a very small amount of *m*-alkylaniline is obtained—explain.

Solution : Aniline is a base, while $AlCl_3$ is a Lewis acid. The former donates a pair of *e*s of the N atom to the Al of $AlCl_3$ to form a salt as shown below :

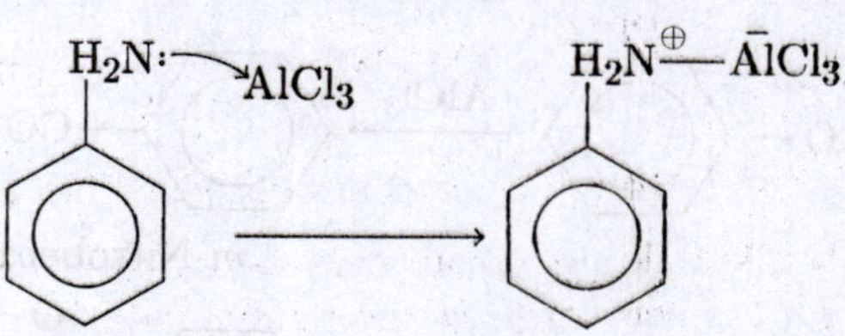

When a catalytic amount of $AlCl_3$ is used, it forms the salt and no $AlCl_3$ is left to activate the alkyl halide for the Friedel-Crafts alkylation. So, alkylation does not occur. When an excess of $AlCl_3$ is used, the alkyl halide forms the alkylating species which reacts with the salt to give a poor yield of *m*-alkylaniline since $\overset{\oplus}{N}H_2$ is a *m*-directing group.

17.9.2 The Friedel-Crafts acylation

The introduction of a RCO or ArCO group into an aromatic nucleus in the presence of a Lewis acid-catalyst is known as the Friedel-Crafts acylation.

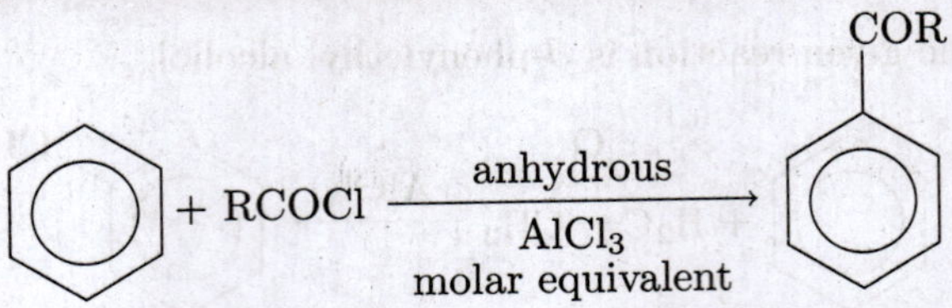

In this reaction the substrates are of various types, e.g., aromatic hydrocarbons including fused ring systems, derivatives of benzene containing *o*-, *p*-directing groups such as hydroxyl, alkoxyl, halo and acid-amido groups; excepting quinolines and pyridines many heterocyclic compounds (furan, pyran, thiophene, pyrrole, etc.) also undergo the reaction.

The reagents of the reaction are usually acid halides, acid anhydrides (cyclic and acyclic) and carboxylic acids.

The catalysts of the reaction are those of the Friedel-Crafts alkylation. However, in the ring-closure reaction polyphosphoric acid is used preferentially. In the Friedel-Crafts acylation reaction more than one mole of the catalyst is used per mole of the substrate.

The usual solvents are nitrobenzene, carbon disulphide and tetrachloroethylene. However, it has been found in some cases that orientation of acyl group changes with the change of solvent. For example, when naphthalene is treated with acetyl chloride-aluminium chloride in nitrobenzene medium, it gives β-acetyl derivative as a major product; while the same treatment of naphthalene in tetrachloroethylene gives α-isomer as major product as shown below :

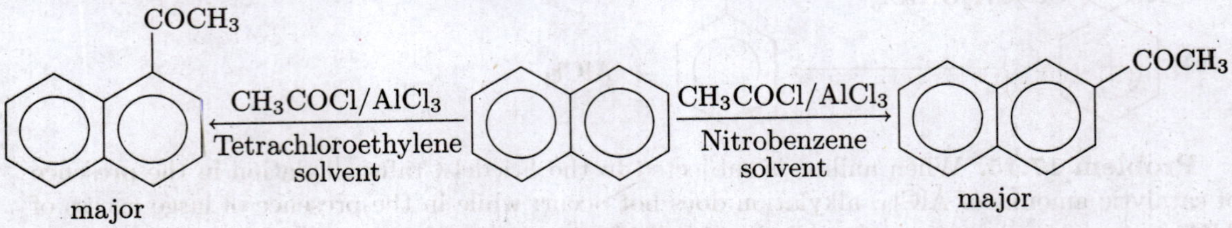

The reaction has good synthetic and preparative values.

(i) It is a good method for preparing aromatic ketones.

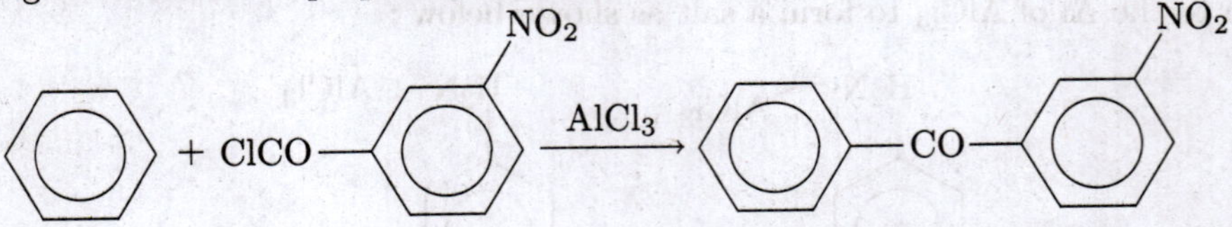

m-Nitrobenzophenone

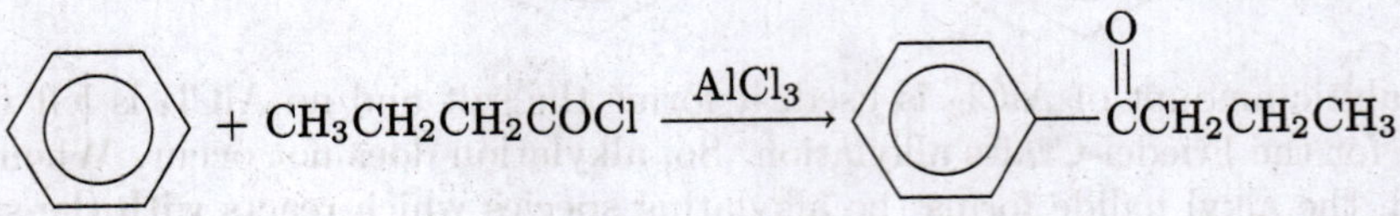

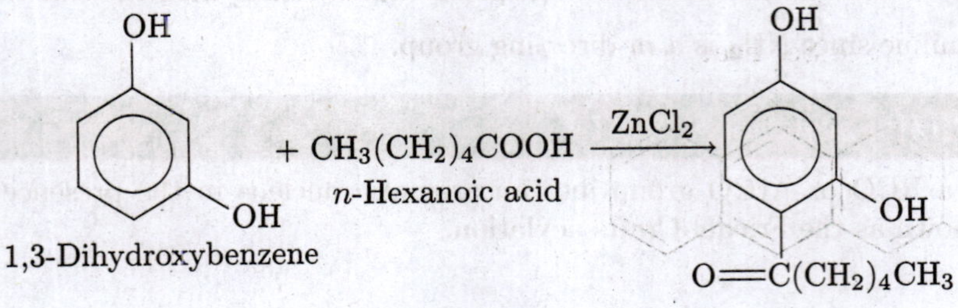

(ii) This is also a good method for synthesising polynuclear hydrocarbons having fused ring systems.

(a) Synthesis of naphthalene (Howarth synthesis)

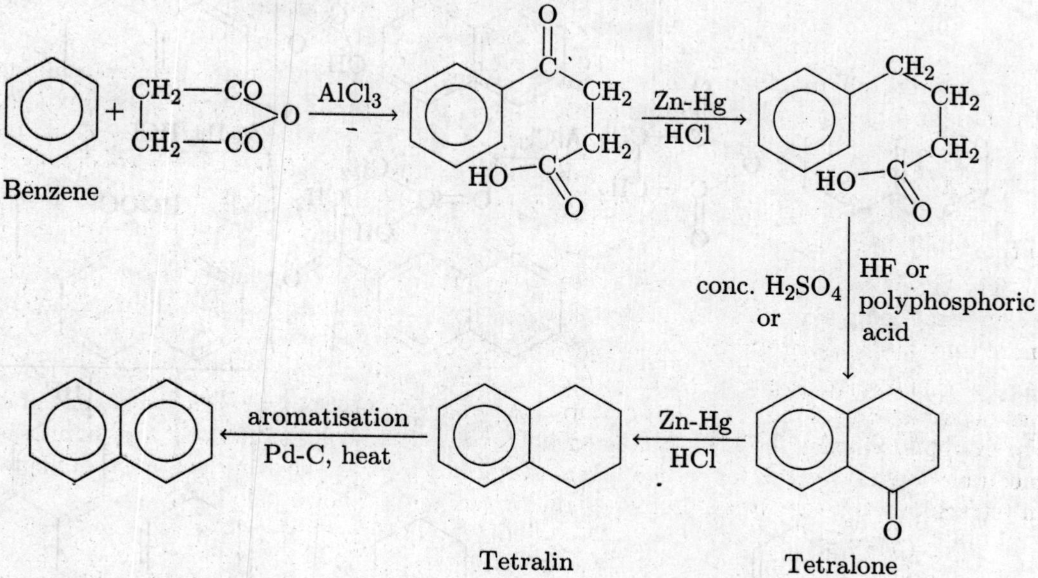

(b) Synthesis of anthracene :

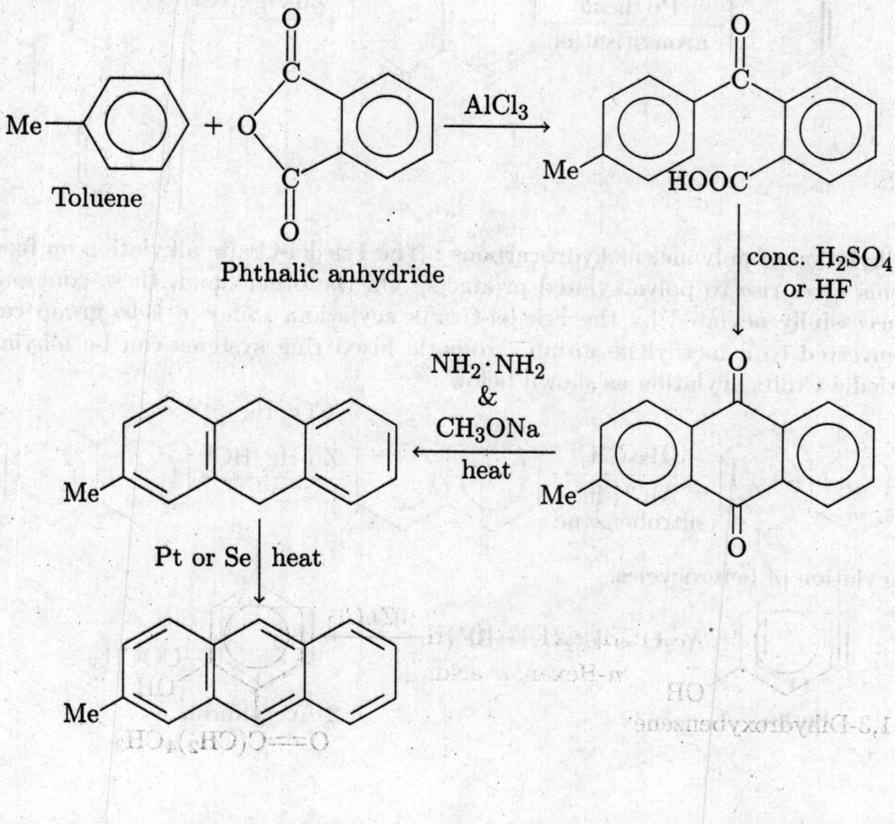

(c) Synthesis of phenanthrene :

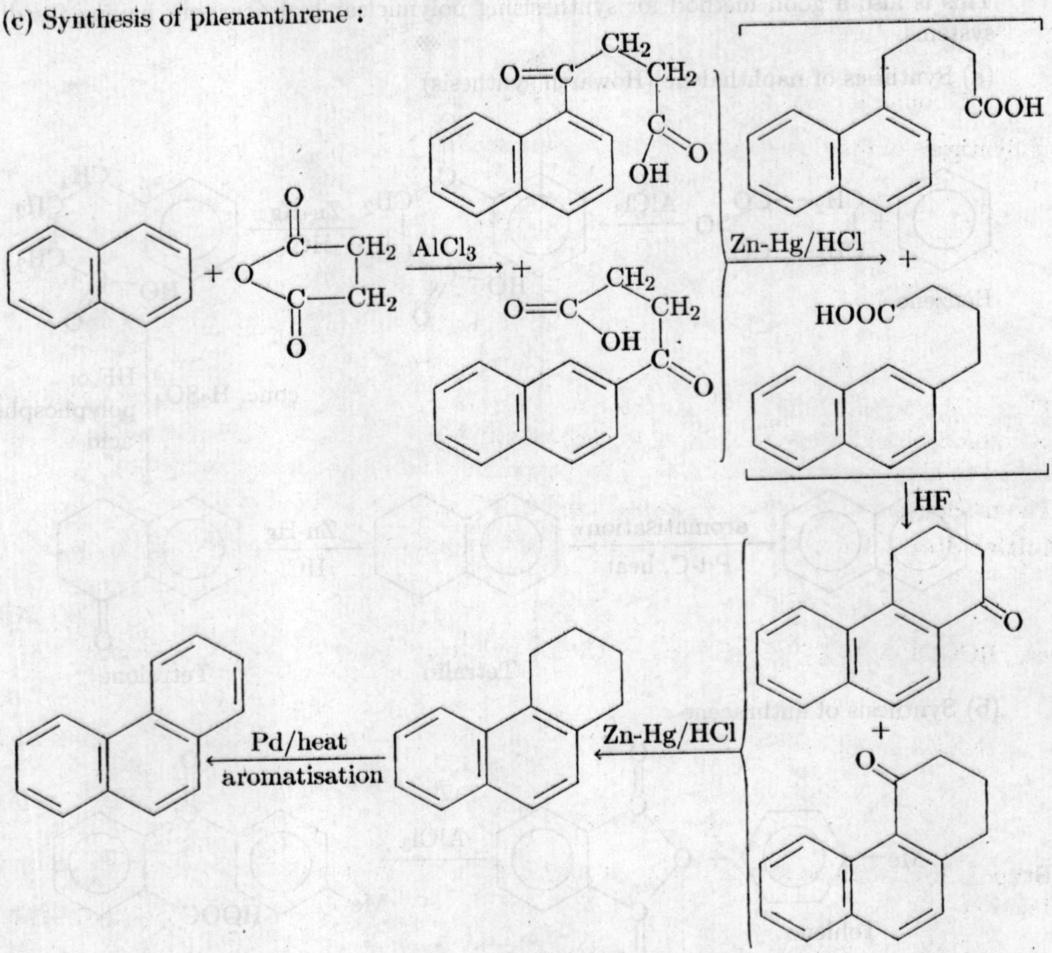

(iii) **Alkylation of polynuclear hydrocarbons :** The Friedel-Crafts alkylation on fused ring systems gives rise to polyalkylated products. On the other hand, these compounds can be successfully acylated by the Friedel-Crafts acylation. Since a keto group can easily be converted to a methylene group, aromatic fused ring systems can be alkylated via the Friedel-Crafts acylation as shown below :

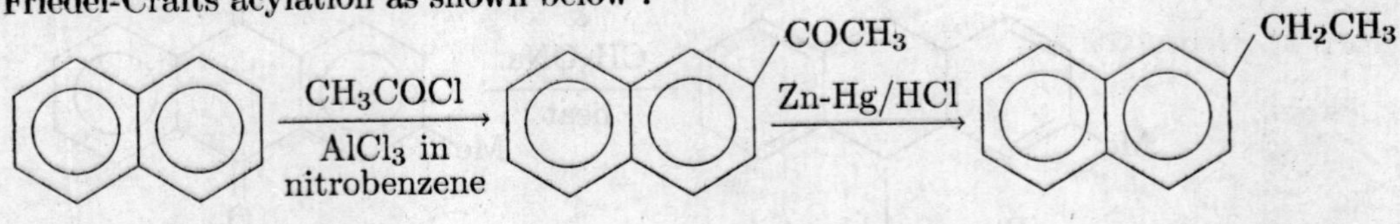

(iv) **Acylation of heterocycles.**

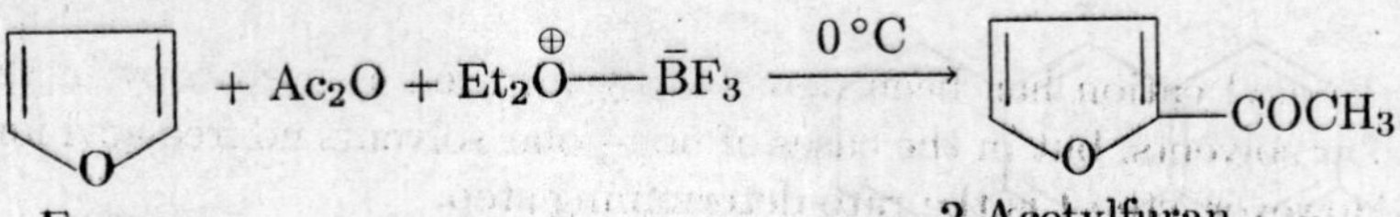

Furan 2-Acetylfuran

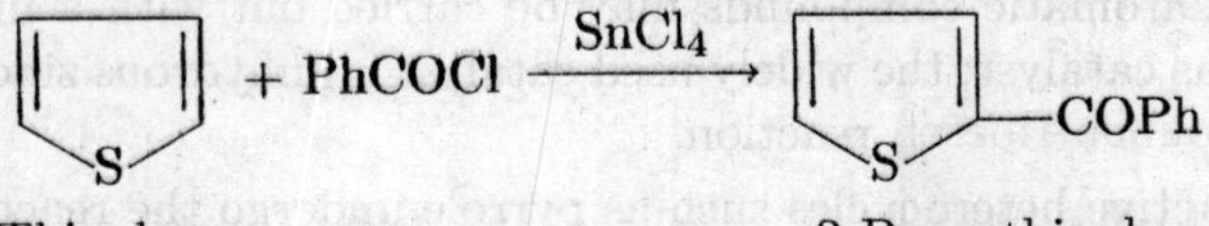

Thiophene 2-Benzothiophene

(v) Synthesis of fused ring systems with hetero atom.

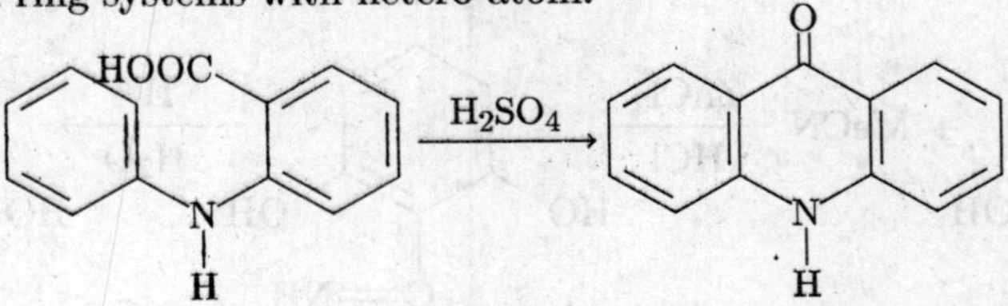

Acridone

Thus, the Friedel-Crafts acylation has a number of advantages over the Friedel-Crafts alkylation. Heterocycles and polynuclear hydrocarbons can be acylated and via acylation they can be converted to alkyl derivatives using the Clemmensen or the Wolff-Kishner reduction.

The mechanism of the reaction involves the following steps :

Initial Step : Formation of either 1 : 1 adduct or acyl cation.

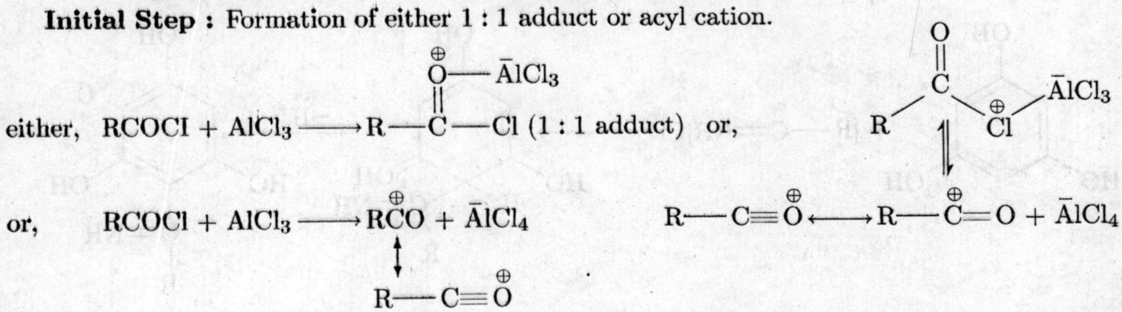

Step I. Electrophilic attack on the aromatic substrate either by the 1 : 1 adduct or by the acyl cation.

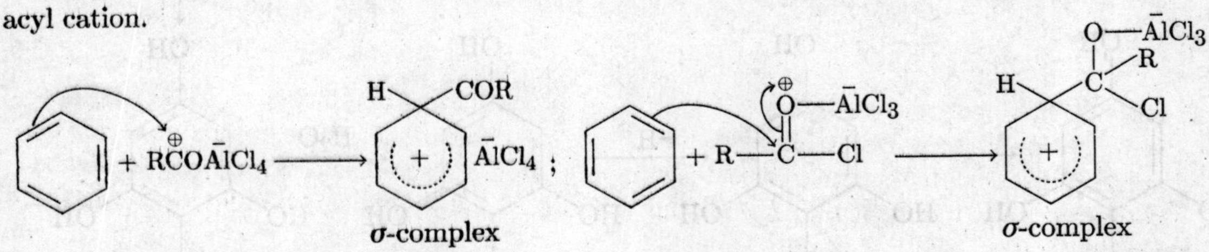

Step II. Deprotonation of the σ-complex.

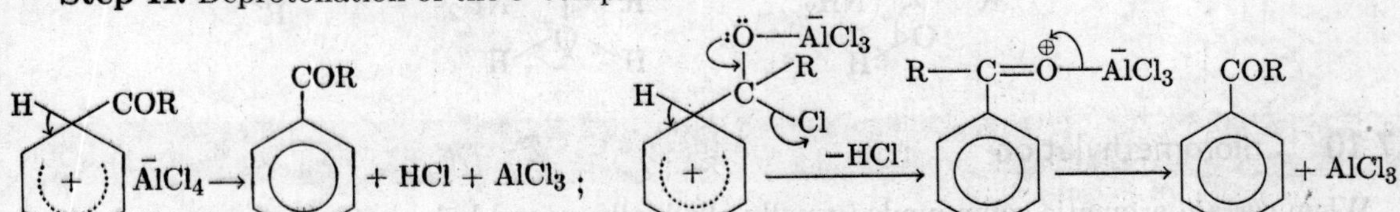

The presence of free acyl cation has been detected by infrared spectroscopy in the cases of liquid reagents and polar solvents, but in the cases of non-polar solvents no free acyl ion has been found to be present. However, step I is the rate-determining step.

The Friedel-Crafts acylation on aromatic compounds may be carried out with a nitrile and HCl in the presence of a Lewis acid as catalyst; the widely used catalyst is anhydrous zinc chloride and the reaction is known as the Houben-Hoesch reaction.

Phenols, phenolic ethers and reactive heterocycles such as pyrrole undergo the reaction.

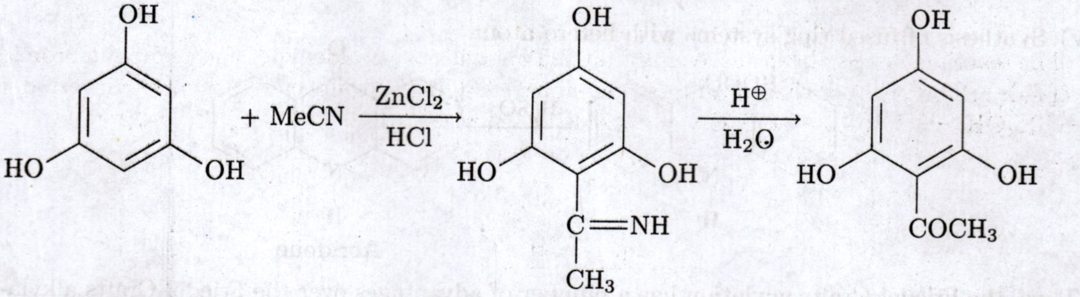

The mechanism is not certain. however, this may be represented as :

$$R—C\equiv N + HCl \rightleftharpoons R—C\equiv \overset{\oplus}{N}HCl^- \longleftrightarrow [R—\overset{\oplus}{C}=NH]Cl^-$$

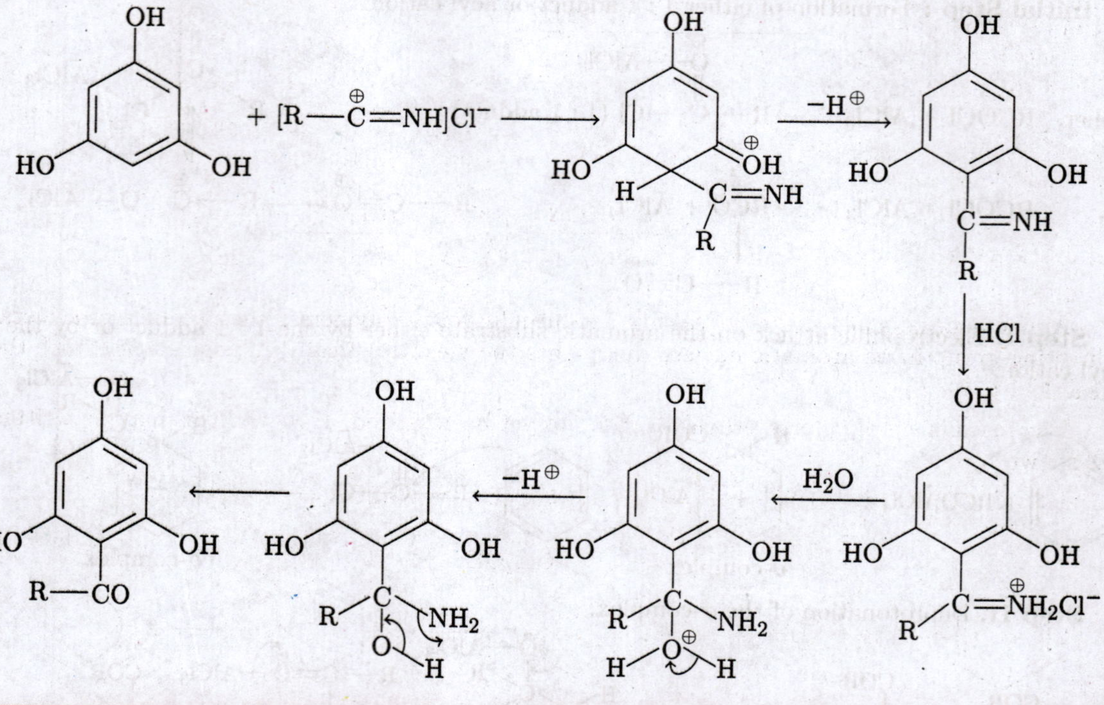

17.10 Chloromethylation

When certain aromatic compounds (usually alkyl, alkoxy and halobenzenes) are treated with formaldehyde (as formalin or paraformaldehyde) in the presence of a Friedel-Crafts catalyst, e.g, $ZnCl_2$, $SnCl_4$, $AlCl_3$, etc., one H atom of the aromatic compound gets replaced by the chloromethyl (—CH_2Cl) group and the reaction is known as chloromethylation.

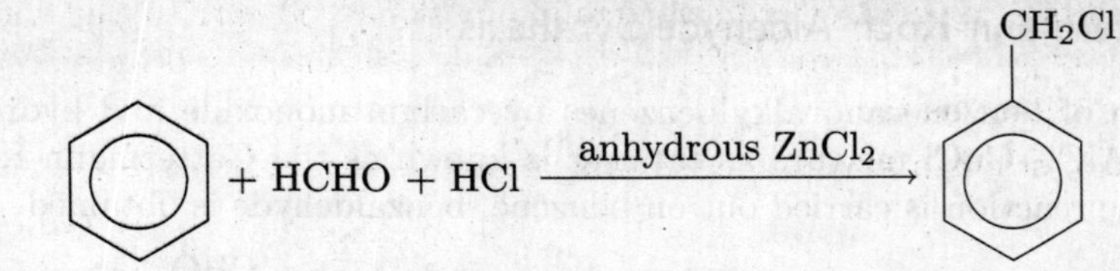

The reaction has also been successfully carried out with other aldehydes and hydrogen bromide or iodide. The reaction has a wide scope since —CH_2X group can easily be converted to —CH_2OH, —CHO, —CH_2CN, —CH_2COOH, etc. groups.

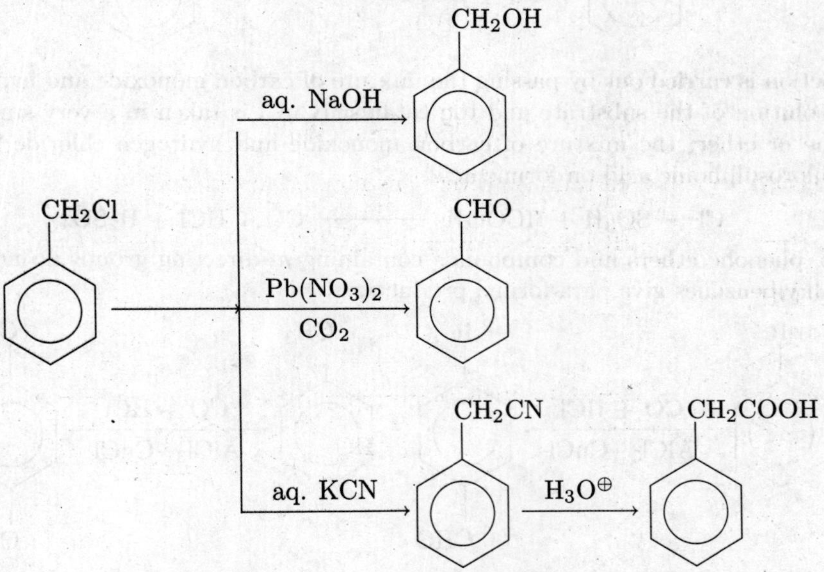

Amines and phenols are very reactive and usually form polymers. The presence of a meta-directing group in an aromatic nucleus diminishes the yield drastically and sometimes stops the reaction.

The mechanism of chloromethylation has not yet been settled. However, this may be written as shown below :

$$2CH_2{=}O + HCl \xrightarrow{ZnCl_2} [2CH_2{=}\overset{\oplus}{O}H \longleftrightarrow 2\overset{\oplus}{C}H_2{-}OH] + ZnCl_4''$$

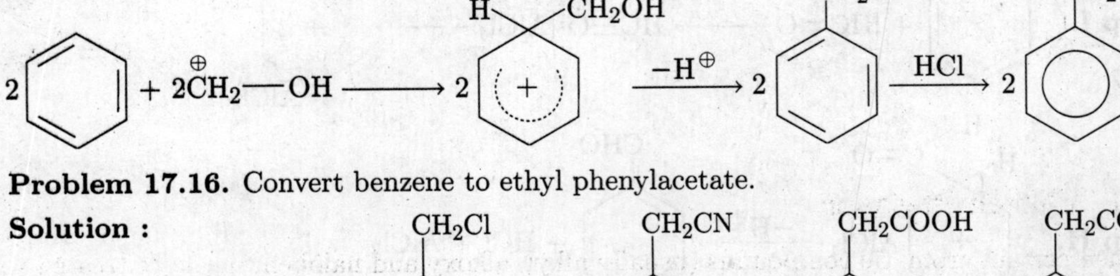

Problem 17.16. Convert benzene to ethyl phenylacetate.

Solution :

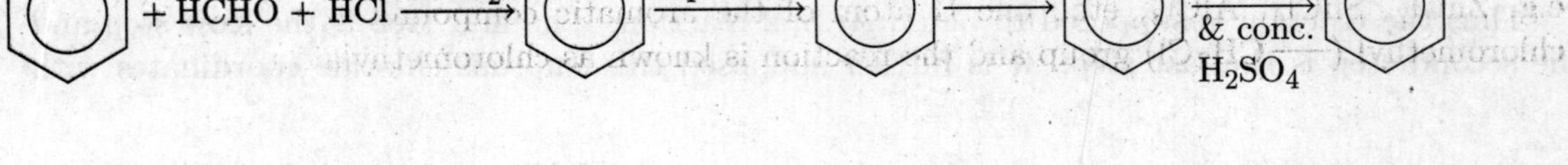

17.11 The Gattermann-Koch Aldehyde Synthesis

The formylation of benzene and alkylbenzenes by carbon monoxide and hydrogen chloride in the presence of $AlCl_3$-CuCl mixture as catalyst is known as the Gattermann-Koch aldehyde synthesis. When the reaction is carried out on benzene, benzaldehyde is obtained.

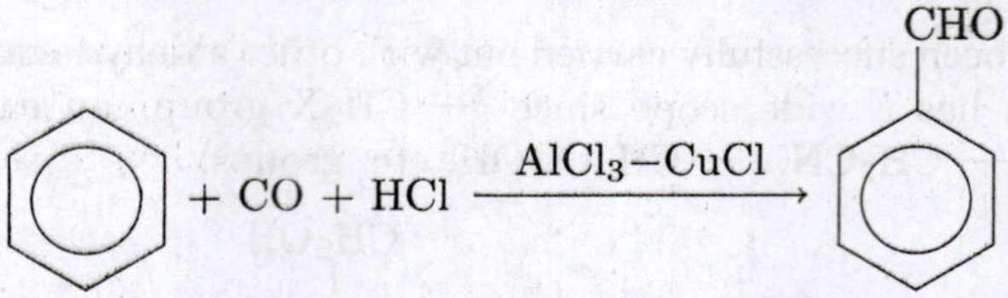

The reaction is carried out by passing the mixture of carbon monoxide and hydrogen chloride through a solution of the substrate and the catalyst (CuCl is taken in a very small amount) in nitrobenzene or ether; the mixture of carbon monoxide and hydrogen chloride is obtained by dropping chlorosulphonic acid on formic acid.

$$Cl\text{—}SO_3H + HCOOH \longrightarrow CO + HCl + H_2SO_4$$

Phenols, phenolic ethers and compounds containing *m*-directing groups do not undergo the reaction. Alkylbenzenes give para-formyl products.

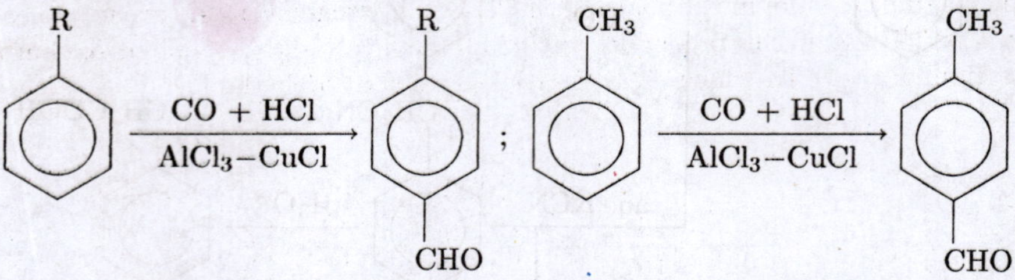

The mechanism is not certain. However, the following pathway may be drawn :

Initial Step : $H\ddot{Cl}: + AlCl_3 \rightleftharpoons \overset{\oplus}{H}\overset{-}{Al}Cl_4$

$$:\bar{C}{\equiv}\overset{\oplus}{O}: + H^{\oplus}AlCl_4^{-} \rightleftharpoons [HC{\equiv}\overset{\oplus}{O} \longleftrightarrow H\overset{\oplus}{C}{=}O]\,\overset{-}{Al}Cl_4$$

Step I.

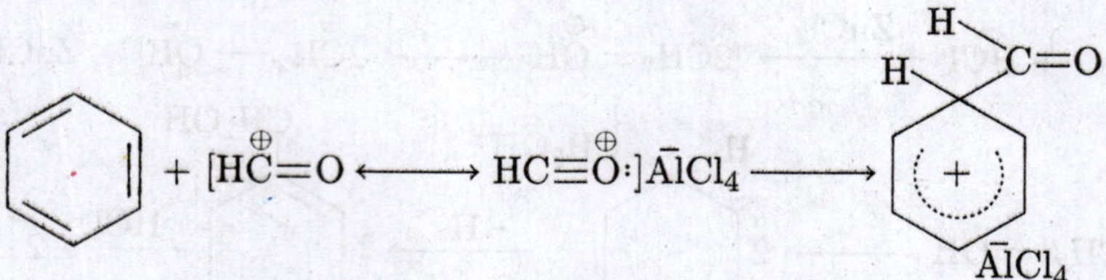

Step II.

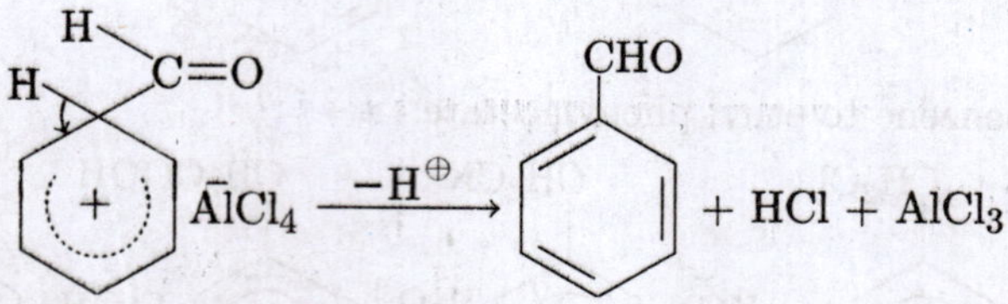

Thus the reaction is supposed to be an Ar.S_E2 reaction. The first step is the slow step and the second step is the fast step. It is further supposed that cuprous chloride coordinates with carbon monoxide molecule and thereby increases the concentration of carbon monoxide for the reaction.

17.12 The Gattermann Aldehyde Synthesis

Formylation of certain aromatic compounds with hydrogen cyanide and hydrogen chloride in the presence of aluminium chloride as catalyst is called the Gattermann aldehyde synthesis. This is regarded as the special case of the Houben-Hoesch reaction. When toluene is subjected to this reaction, *p*-methylbenzaldehyde is formed.

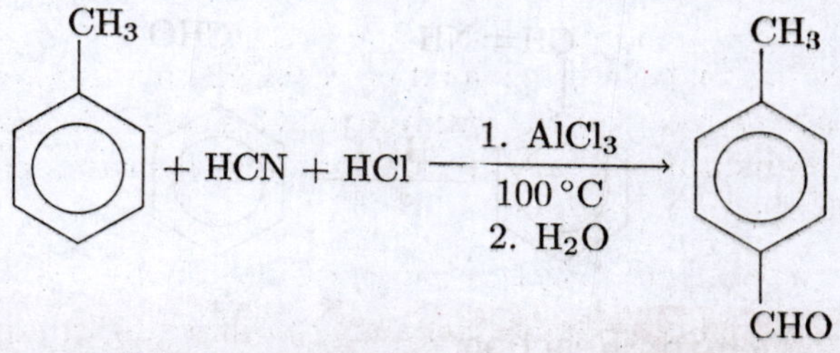

The reaction may be carried out by passing hydrogen chloride through nitrobenzene containing the substrate and zinc cyanide; in the solution $ZnCl_2$ and hydrogen cyanide are formed and $ZnCl_2$ acts as the catalyst. Through this procedure resorcinol is converted to 2, 4-dihydroxybenzaldehyde. At first an arylimine is formed which on hydrolysis gives the aldehyde.

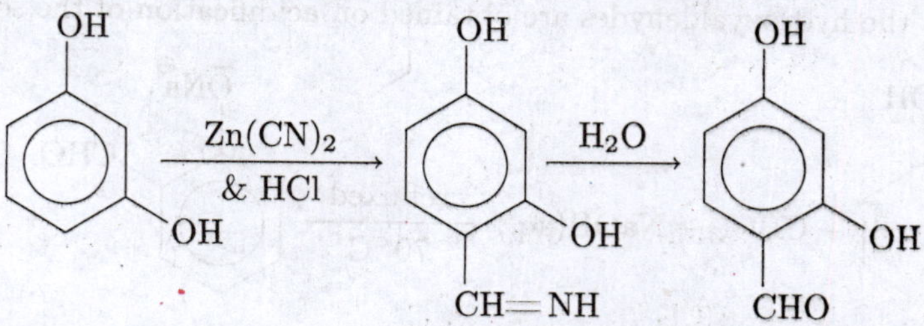

Benzene, alkylbenzenes, phenols, phenolic ethers and some heterocycles undergo the reaction. Nitrobenzene, amines and compounds containing *m*-directing groups do not take part in the reaction.

The mechanism is not yet certain. This may, however, be assumed to be as follows :

$$Zn(CN)_2 + HCl \rightleftharpoons ZnCl_2 + HCN$$

$$ZnCl_2 + HCl \rightleftharpoons \overset{\oplus}{H}\overset{-}{Z}nCl_3$$

$$H—C\equiv N + \overset{\oplus}{H}\overset{-}{Z}nCl_3 \rightleftharpoons H—C\equiv\overset{\oplus}{N}H + ZnCl_2 + Cl^-$$

$$\updownarrow$$

$$[H—\overset{\oplus}{C}=NH]Cl^-$$

Step I. Formation of σ-complex

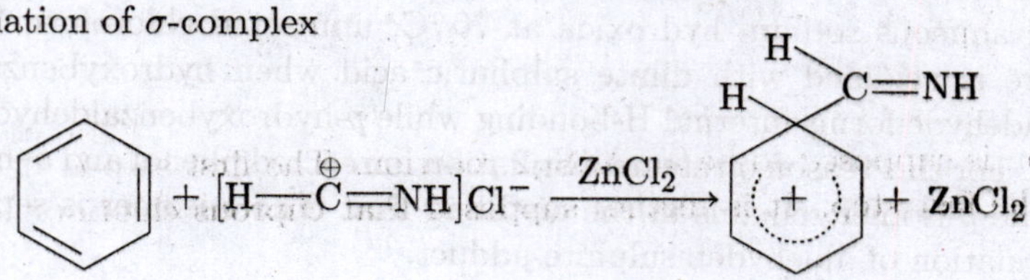

Step II. Deprotonation of the σ-complex and formation of arylimine.

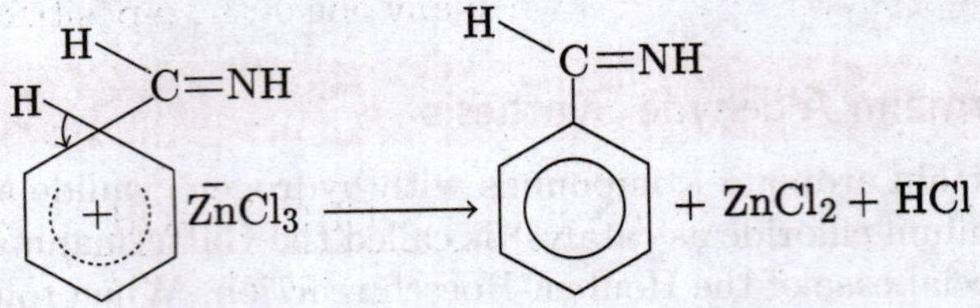

Final Step : Hydrolysis of the imine to aldehyde.

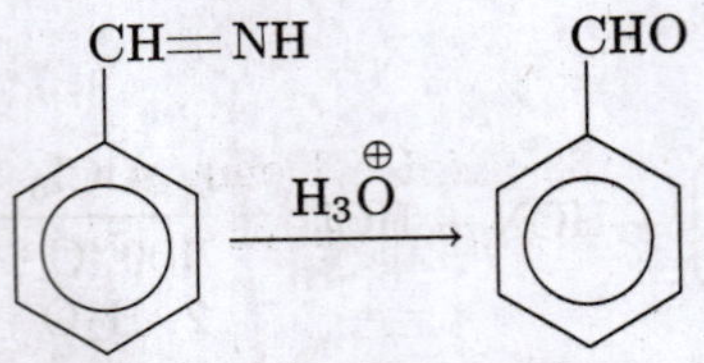

17.13 The Reimer-Tiemann Reaction

Formylation of phenols by the reaction with chloroform in strong and aqueous sodium hydroxide is known as the Reimer-Tiemann reaction. When phenol is subjected to this reaction, sodio salts of *o*-hydroxybenzaldehyde (salicylaldehyde) and *p*-hydroxybenzaldehyde are formed in the ratio of 4 : 1; the hydroxyaldehydes are obtained on acidification of the sodio salts.

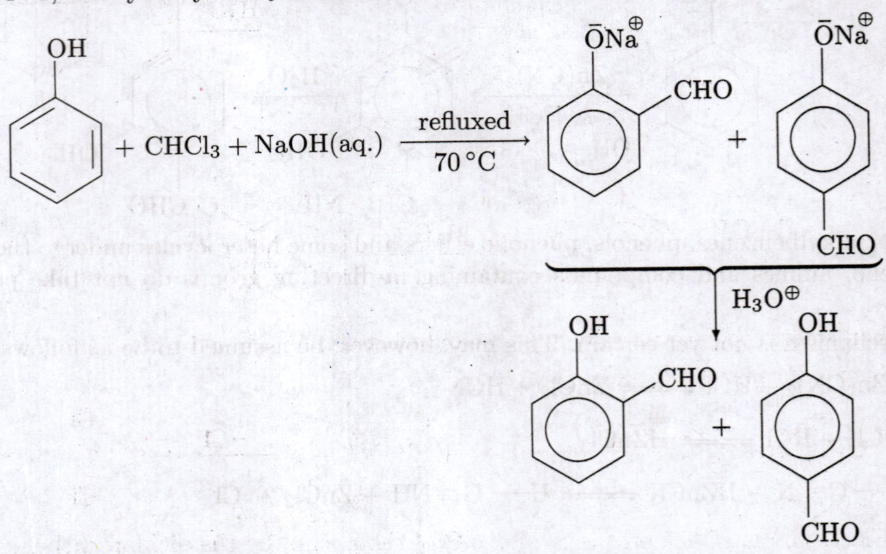

The reaction procedure is simple. This involves refluxing of a mixture of phenol, chloroform and strong and aqueous sodium hydroxide at 70 °C; unreacted chloroform is then distilled off and the mixture is acidified with dilute sulphuric acid when hydroxybenzaldehydes separate. *o*-Hdroxybenzaldehyde forms internal H-bonding while *p*-hydroxybenzaldehyde forms intermolecular H-bonding. For this reason on steam distillation unreacted phenol and *o*–hydroxybenzaldehyde distil out and the *p*-isomer remains in the distilling flux. The *o*-isomer is separated from phenol through the formation of aldehydebisulphite adduct.

In this reaction usually the *o*-product predominates; perhaps internal H-bonding gives thermodynamically stable *o*-product. Of course, when any one of the *o*-positions is blocked, *p*-product is obtained.

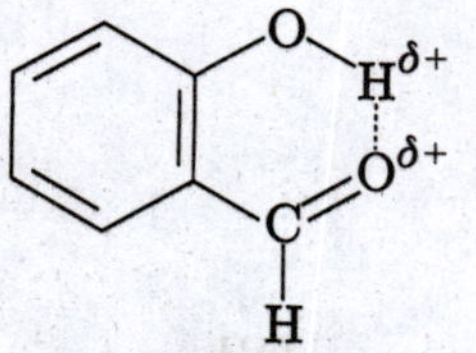

internal H-bonding

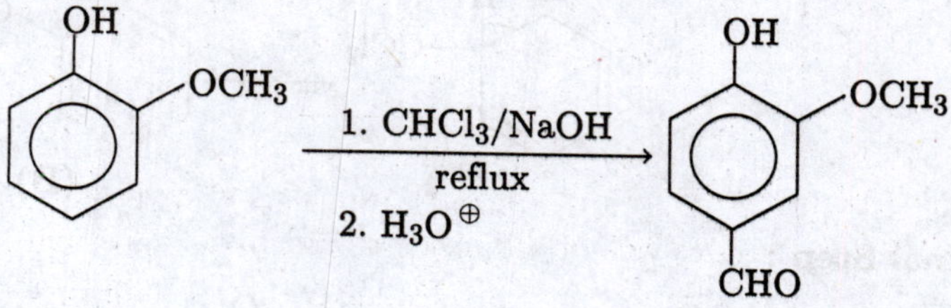

Vanilin

Pyrrole and indole also undergo the Reimer-Tiemann reaction; in these cases ring expansion products are obtained along with formylated products. Using CCl_4, instead of $CHCl_3$, salicylic acid can be perpared from phenol.

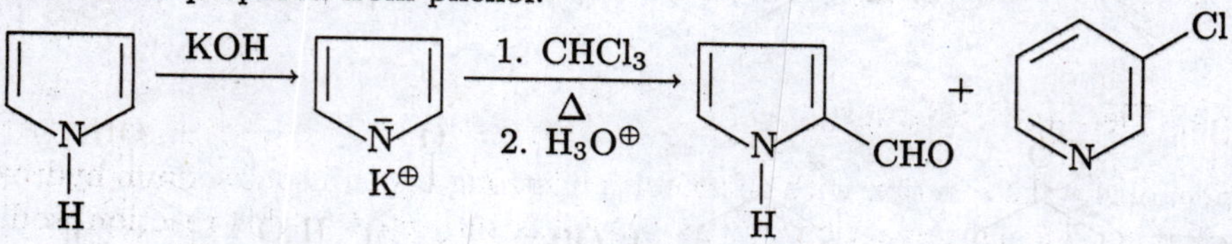

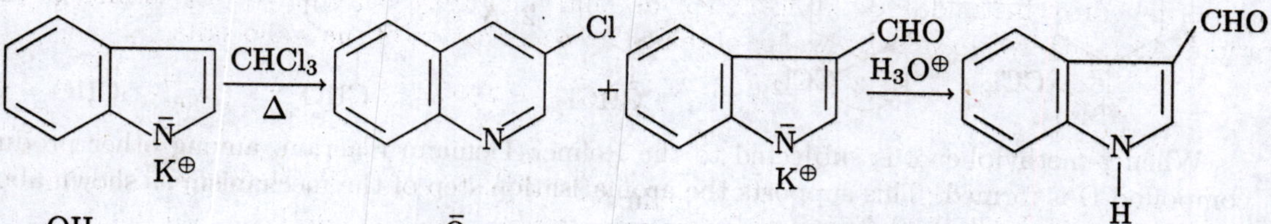

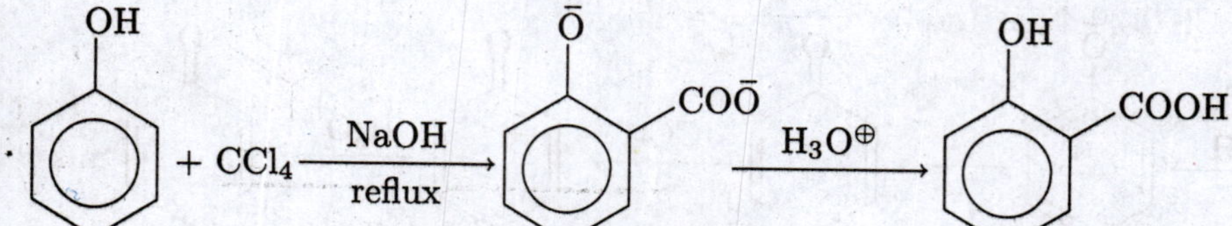

The mechanism of the reaction involves the following steps :

Initial Step : Formation of dichlorocarbene.

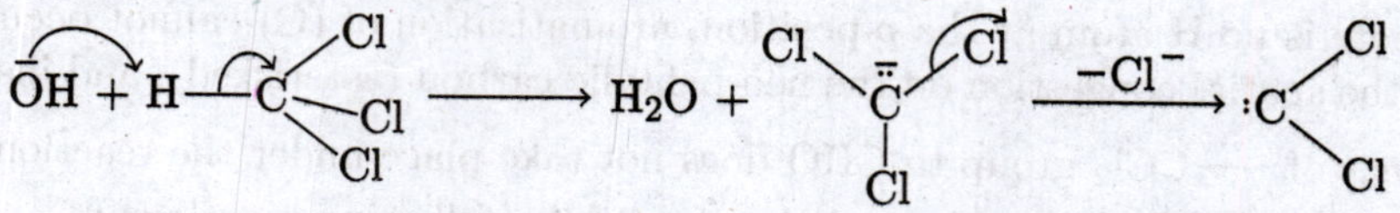

Step I. Electrophilic attack on the *o*- and *p*-C of the phenol by the dichlorocarbene :

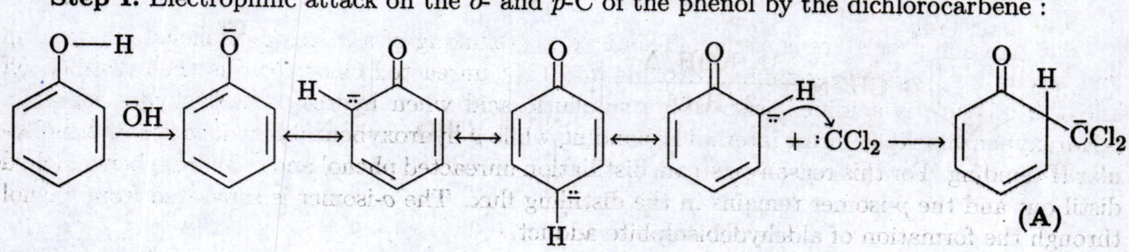

(A)

Step II. Aromatisation of A to B by $H^{\oplus}$ shift.

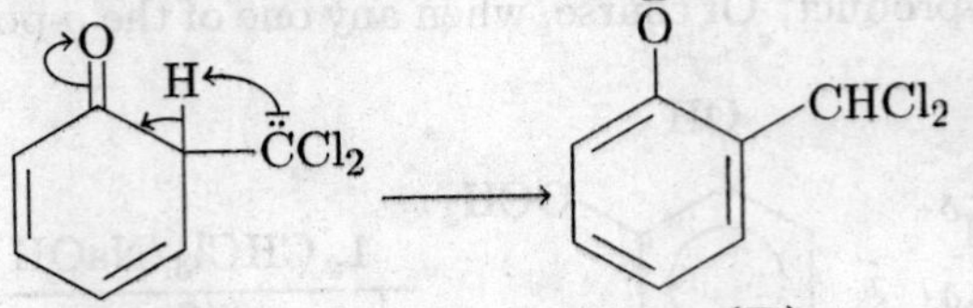

(B)

Final Step :

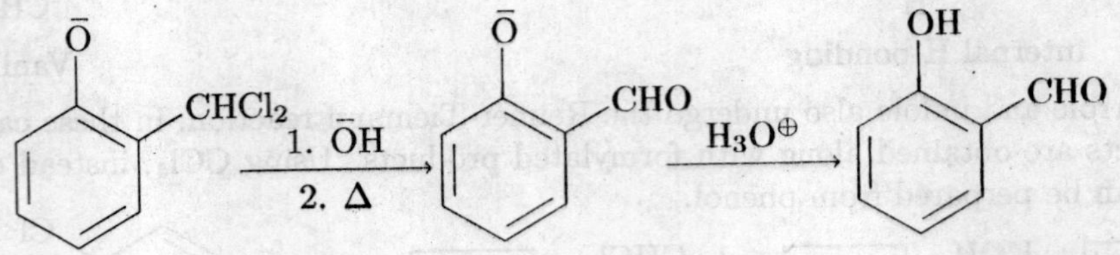

Similarly,

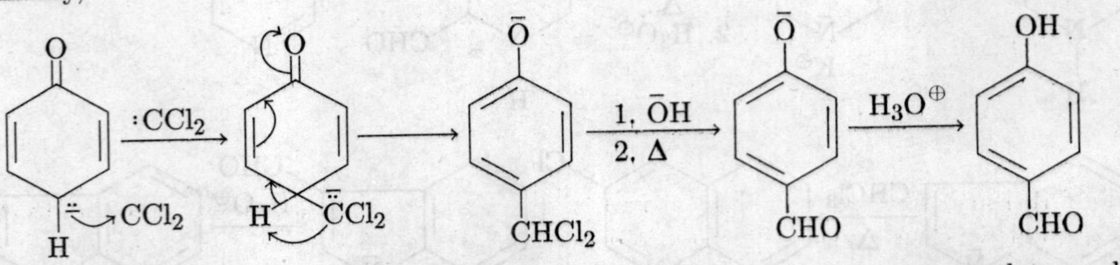

When *p*-methylphenol is subjected to the Reimer-Tiemann reaction, among other products compound D is formed. This supports the aromatisation step of the mechanism as shown above.

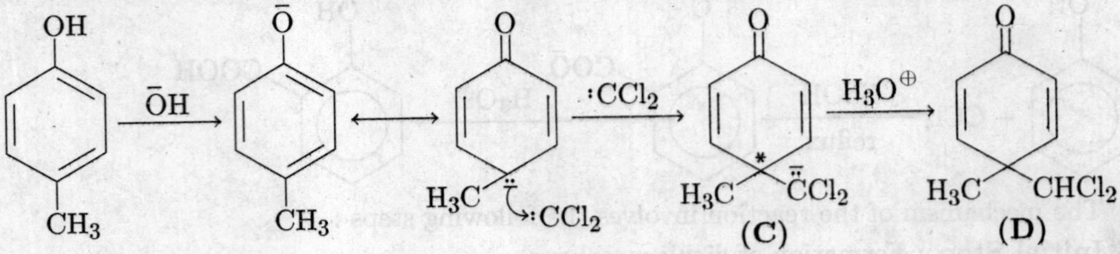

Since there is no H atom at the *p*-position, aromatisation of (C) cannot occur. Furthermore, because of the spatial congestion on the neo-pentylic carbon (asterisked), and insolubility of (C), the hydrolysis of $-\ddot{C}Cl_2$ group to CHO does not take place under the reaction conditions.

Problem 17.17. Give the mechanistic steps of the following reactions :

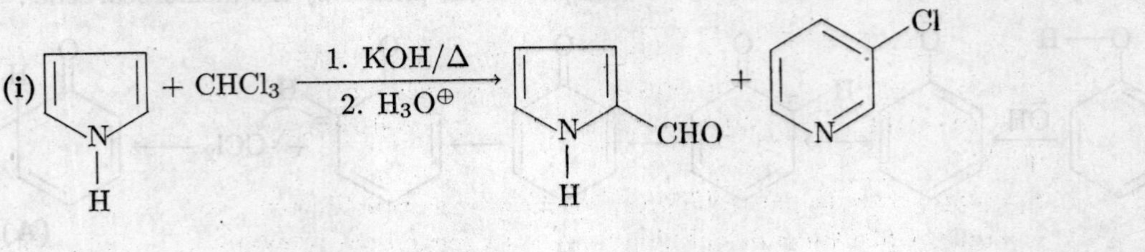

(ii) 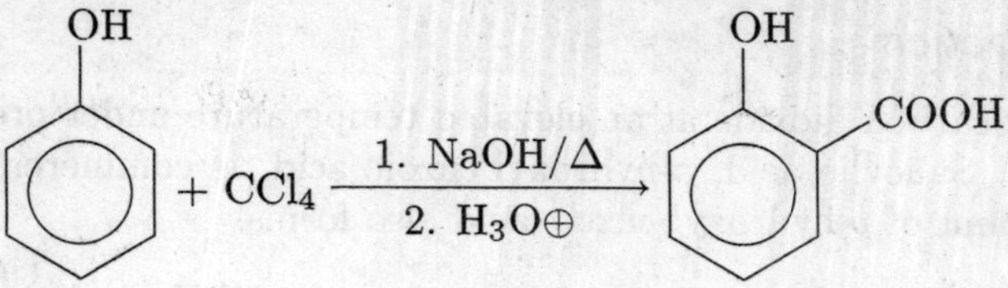

Solution : (i) $CHCl_3 \xrightarrow{\bar{O}H} :\bar{C}Cl_3 \xrightarrow{-Cl^-} :CCl_2$

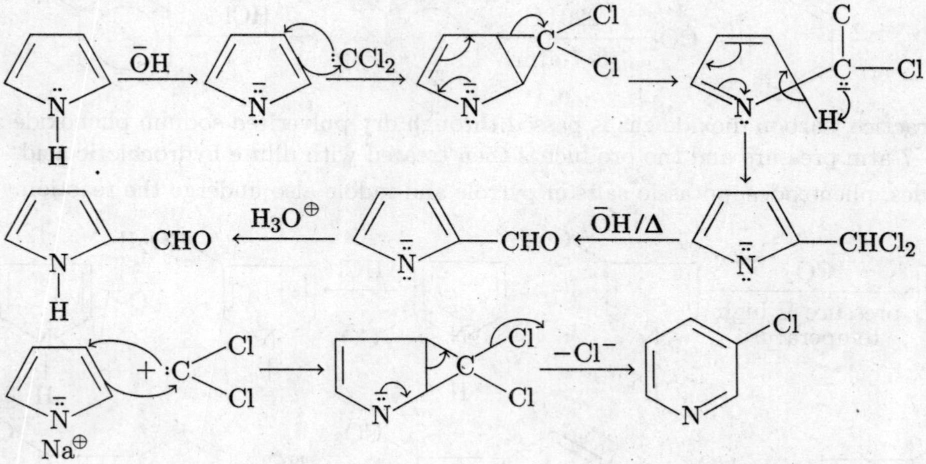

(ii) 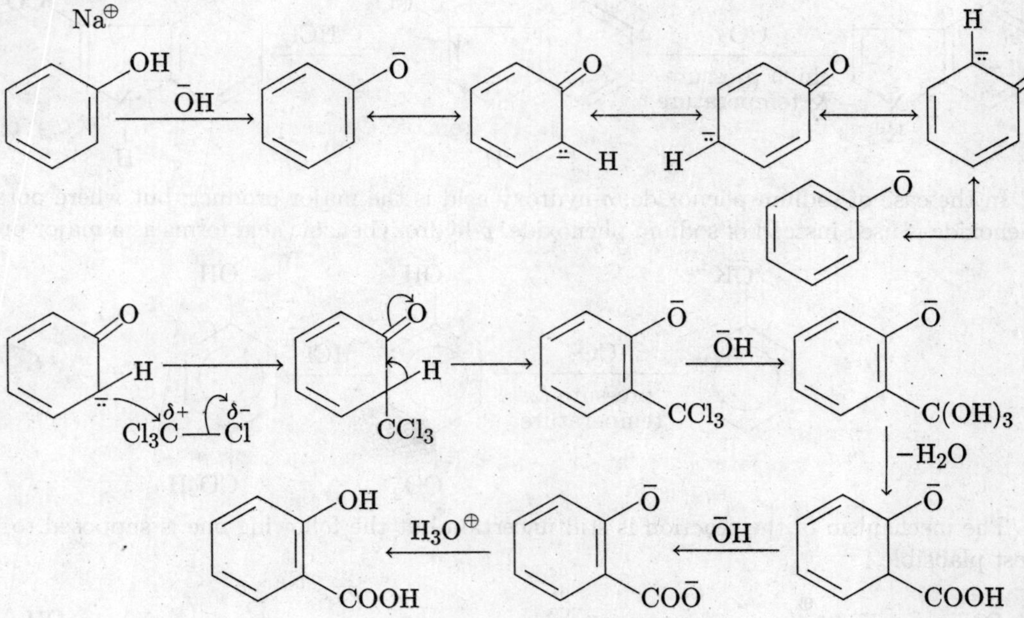

Similarly,

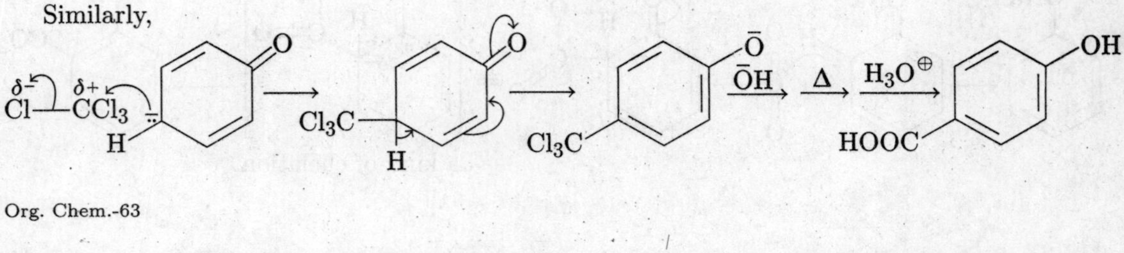

17.14 The Kolbe-Schmidt Reaction

Carboxylation of phenoxides by carbon dioxide at an elevated temperature under pressure is called the Kolbe-Schmidt reaction. Salicylic acid, *o*-hydroxybenzoic acid, is commercially prepared by this reaction; a small amount of *p*-hydroxybenzoic acid also forms.

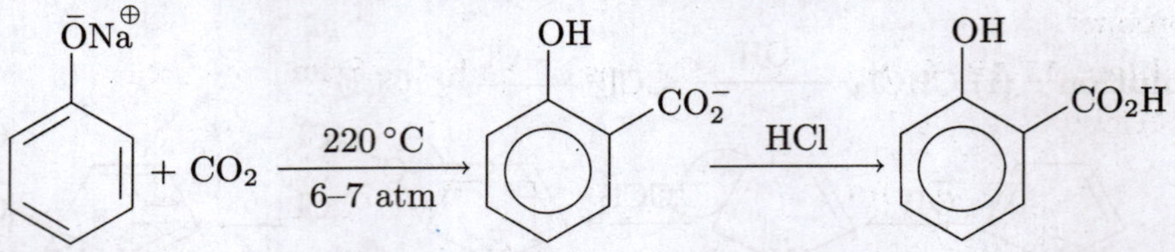

In practice, carbon dioxide gas is passed through dry pulverized sodium phenoxide at 200 °C under 6–7 atm pressure and the product is then treated with dilute hydrochloric acid.

Besides, phenoxides, potassio salts of pyrrole and indole also undergo the reaction.

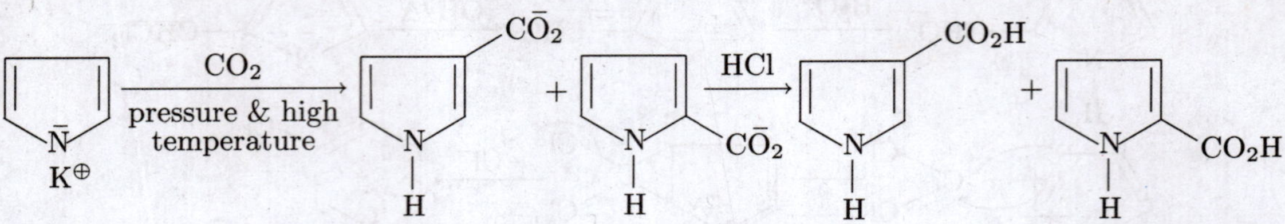

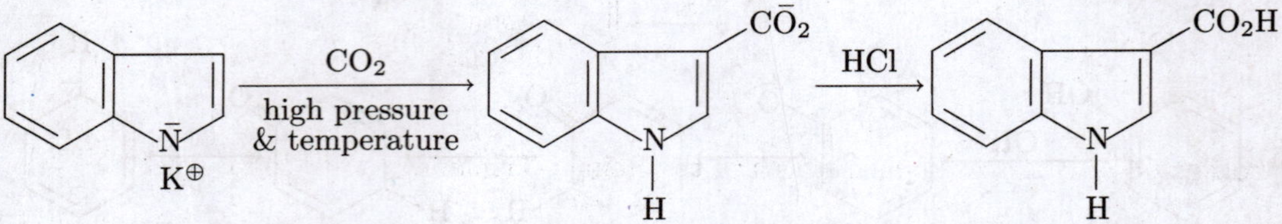

In the case of sodium phenoxide, *o*-hydroxy acid is the major product; but where potassium phenoxide is used instead of sodium phenoxide, *p*-hydroxybenzoic acid forms as a major product.

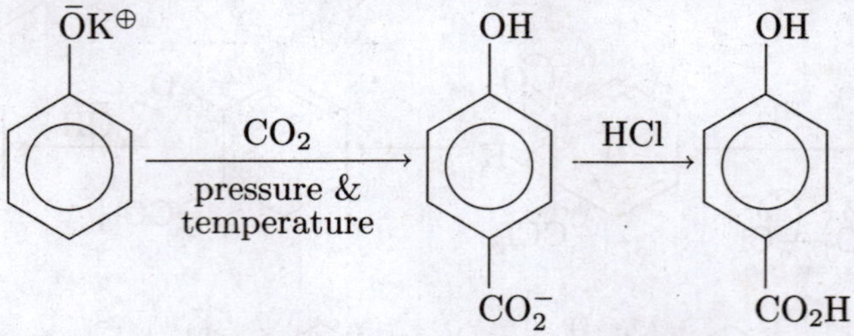

The mechanism of the reaction is still uncertain but the following one is supposed to be the most plausible :

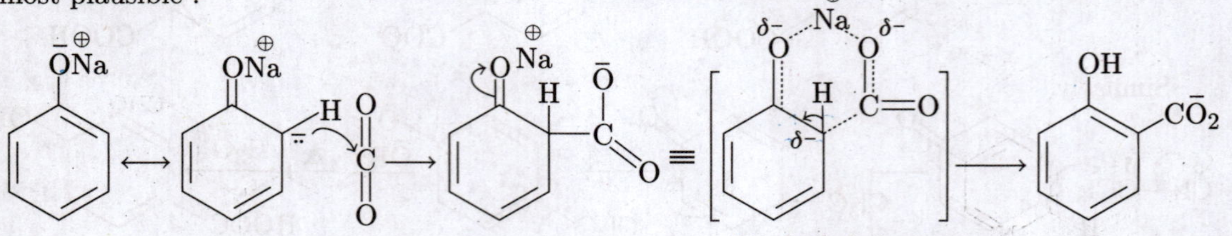

a kind of chelation

Thus the reaction is essentially an aromatic electrophilic substitution reaction. When potassium salt of phenol is used, perhaps chelation does not happen to occur because of high volume and low positive charge density on the metal ion; owing to the absence of chelation in the case of potassio salt, the *o*-product is not the major product. Since chelation brings stability to the complex of sodium phenoxide and carbon dioxide, the *o*-product is formed in large excess over the *p*-product.

Problem 17.18. Give mechanistic steps of the Kolbe-Schmidt reaction on pyrrole.

Solution :

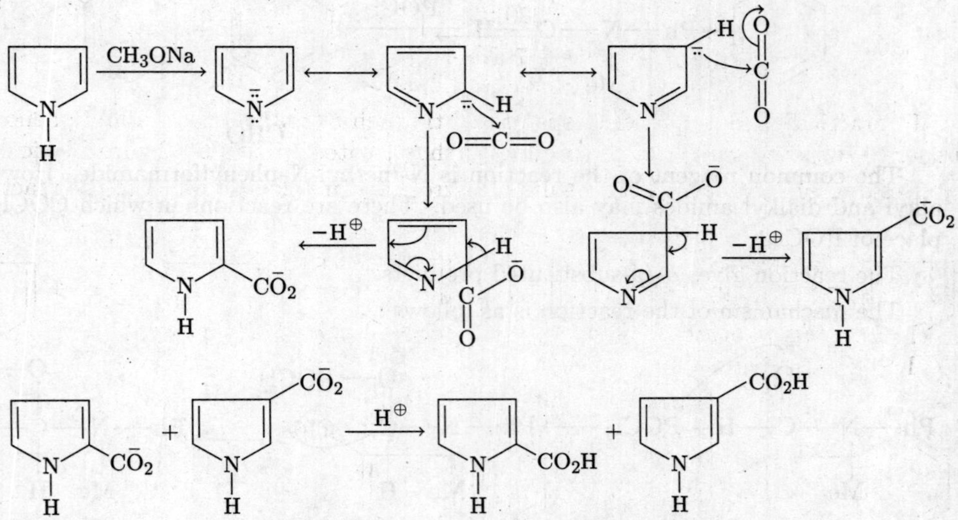

Problem 17.19. Give the mechanism of the Reimer-Tiemann reaction on indole.

Solution : $CHCl_3 \xrightarrow{\bar{O}H} \bar{C}Cl_3 \longrightarrow :CCl_2$

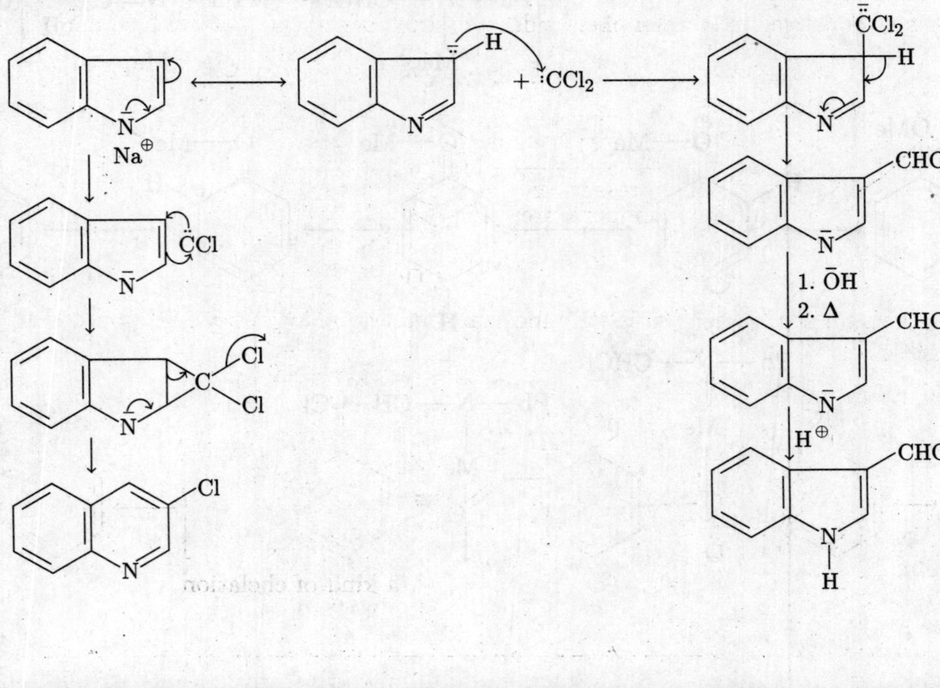

17.15 The Vilsmeier or the Vilsmeier-Haak Reaction

Formylation of phenols, phenolic ethers, aromatic amines by disubstituted formamides and phosphorus oxychloride is called the Vilsmeier or the Vilsmeier-Haak reaction. This is by far the best method of formylating activated aromatic compounds :

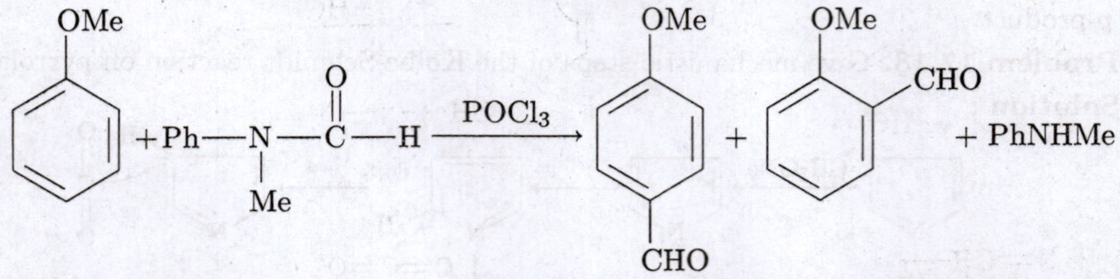

The common reagent of the reaction is N-methyl-N-phenylformamide. However, other aryl alkyl and dialkyl amides may also be used. There are reactions in which $COCl_2$ is used in the place of $POCl_3$.

The reaction gives *o*-/*p*-substituted products.

The mechanism of the reaction is as follows :

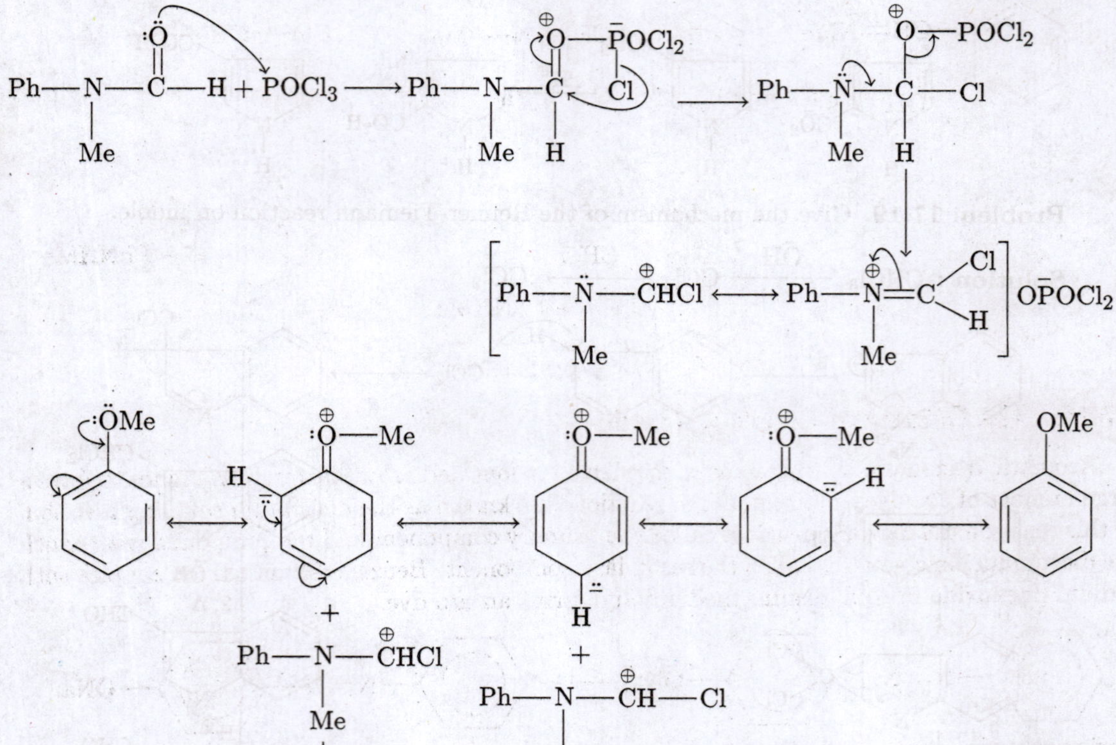

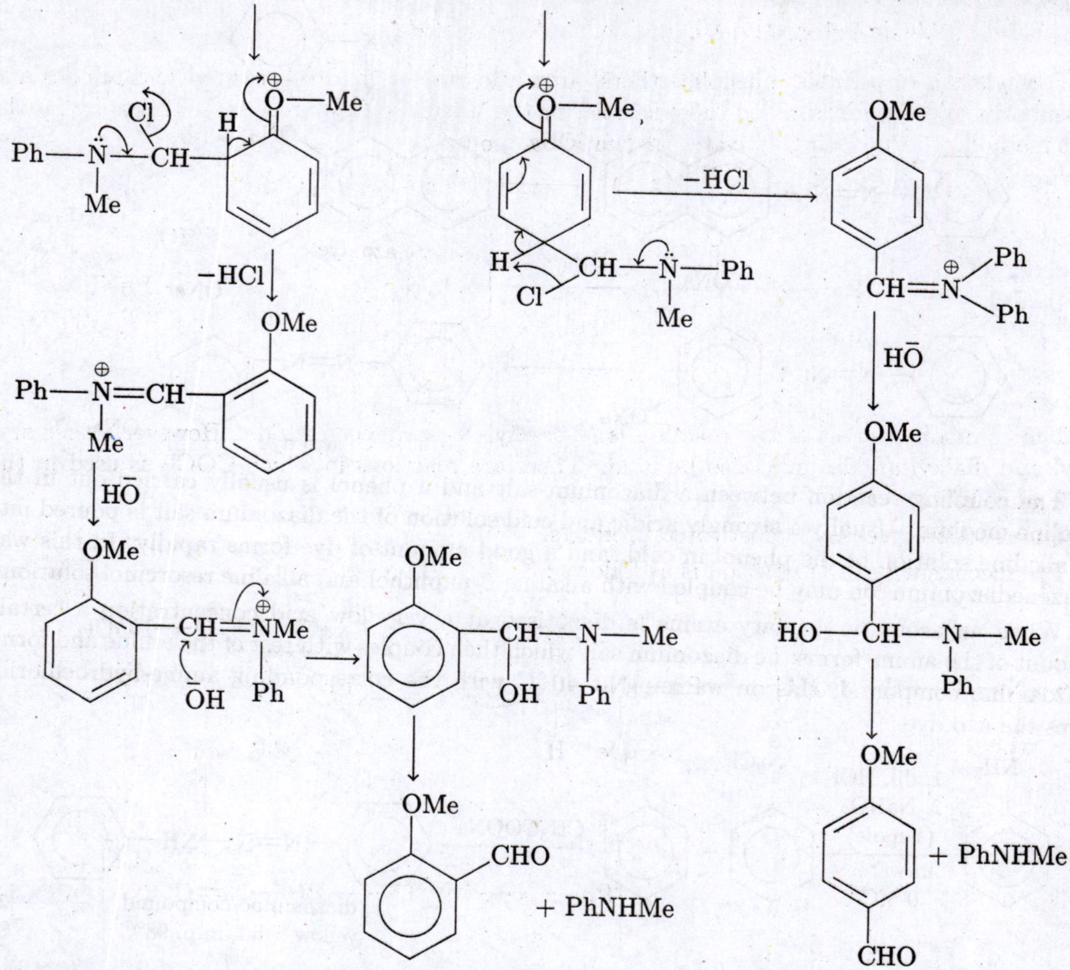

17.16 The Diazonium Coupling Reaction

Aromatic diazonium ions combine with phenoxide ions and aromatic tertiary amines to form a large number of azo dyes and this sort of reactions are known as the diazonium coupling reaction. In this reaction the diazonium salt is called the primary component and the compound with which the diazonium ion couples is called the secondary component. Benzenediazonium ion couples with sodium phenoxide in cold alkaline medium and forms an azo dye.

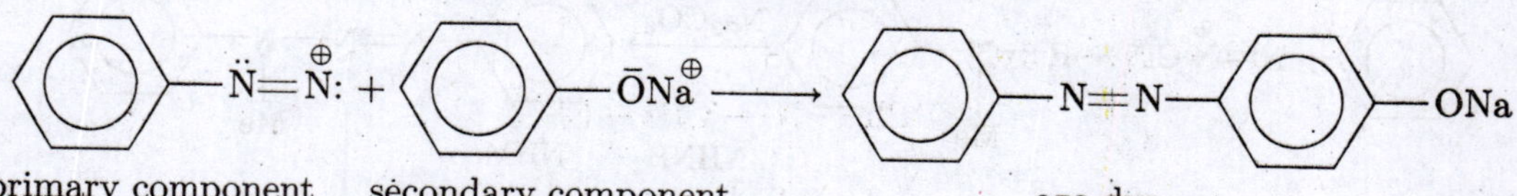

primary component secondary component azo dye

Depending on the nature of the secondary component the coupling reaction is carried out either in an alkaline or in a neutral or in a slightly acidic medium. Diazonium ions being not fairly stable, the coupling reactions are usually carried out in cold.

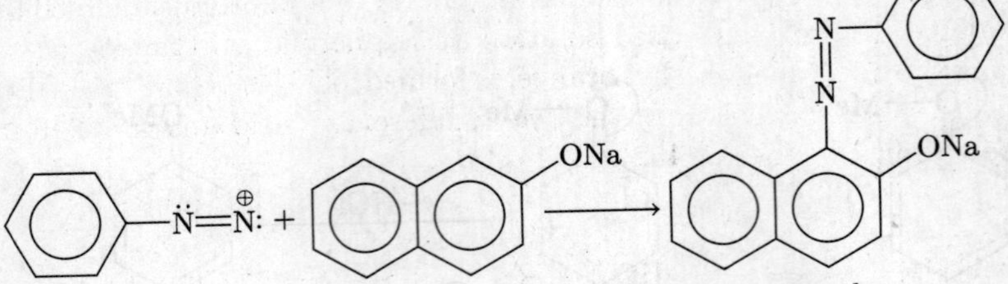

azo dye

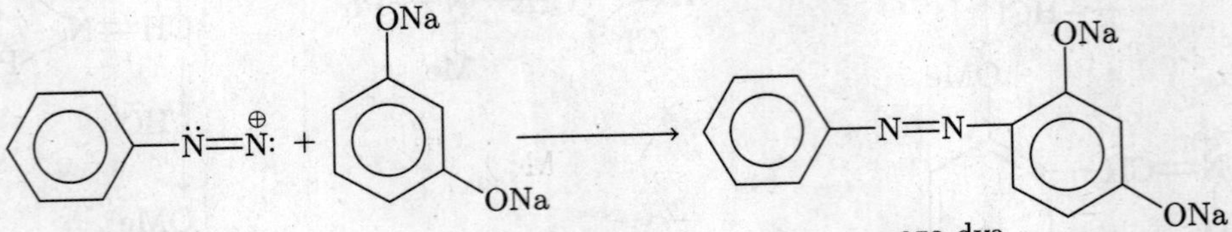

azo dye

The coupling reaction between a diazonium salt and a phenol is usually carried out in the alkaline medium. Usually a strongly acidic and cold solution of the diazonium salt is poured into an alkaline solution of the phenol in cold, and a good amount of dye forms rapidly. In this way benzenediazonium ion may be coupled with alkaline β-naphthol and alkaline resorcinol solutions.

When an aromatic primary amine is diazotised at a very low acid concentration a certain amount of the amine forms the diazonium salt which then couples with rest of the amine and forms diazoamino compound; this on warming at 40 °C with the corresponding amine-hydrochloride gives the azo dye.

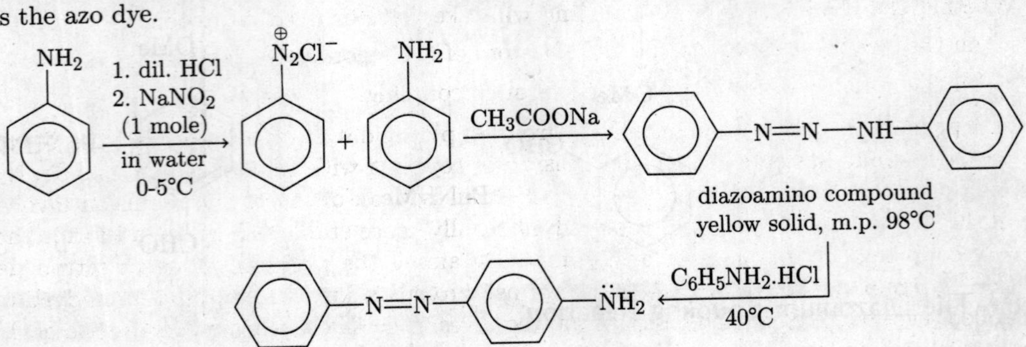

azo dye

When a secondary amine is a secondary component, the diazonium ion couples with it to form N-substituted diazoaminobenzene which on reduction with stannous chloride in hydrochloric acid gives an N-substituted amine and a hydrazine derivative as shown below :

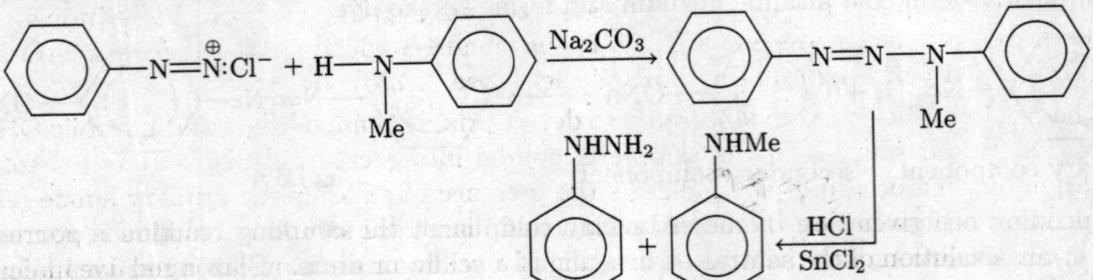

However, a tertiary amine couples with the diazonium ion in a faintly acidic medium to form an azo dye. For example, when a diazotised solution of sulphanilic acid is treated with N, N-dimethylaniline, an orange dye, called methyl orange, is formed. This is yellow in alkaline medium but red in acid solution. Since the colour of the dye is different in acid and in alkali media, this is used as an acid-base indicator (See p. 452).

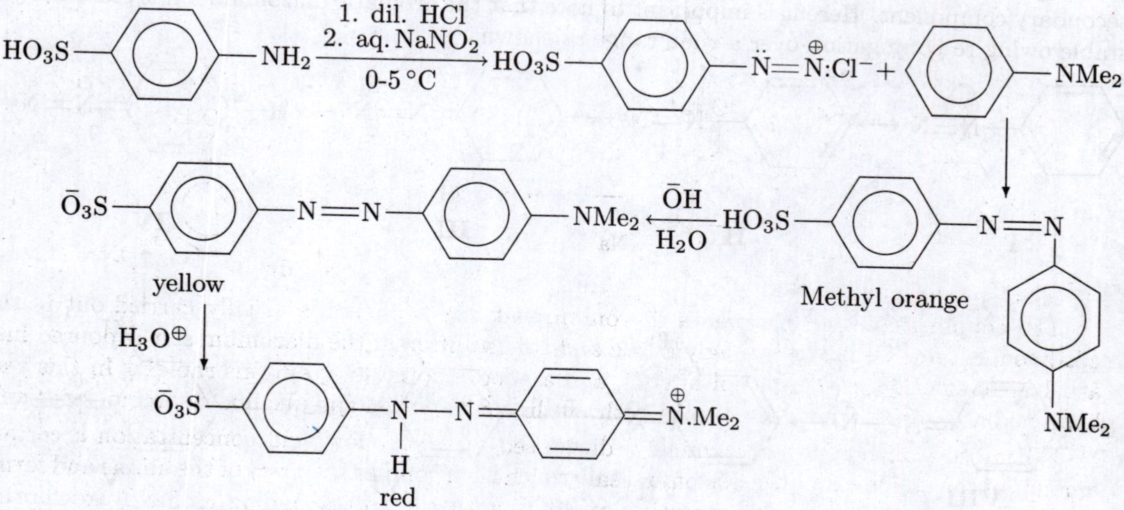

Whether the C—N or N—N coupling will take place or not that depends :

(i) on the electron density on the C or N atom of the secondary component and

(ii) on the stability of the bond formed through coupling.

In a primary aromatic amine the NH_2 group is pyramidal in orientation and the N atom of the —NH_2 group has enough electron density for coupling with the diazonium ion. So, N—N coupling occurs at first. However, the N—N bond is weak owing to the repulsion involving *n* es.; at 40 °C the bond breaks and thermodynamically more stable C—N bond forms through C—N coupling. In the case of an aromatic 3° amine the *p*-C has enough electron density for C—N coupling and a C—N bond forms through a kinetically and thermodynamically controlled reaction. When phenoxides are nucleophiles, C—N coupling occurs instead of O—N coupling; this is because C—N coupled products are kinetically as well as thermodynamically controlled.

However, the reaction has preparative and analytical values.

The diazonium coupling reaction is a good means for preparing azo dyes; almost 50% of the azo dyes used nowadays are prepared by the diazonium coupling reaction. For example, the first synthetic dye, congo red, was prepared by the diazonium coupling reaction between tetra-azotised benzidine and 4-aminonaphthalenesulphonic acid (See p. 736).

The reaction is used in the laboratory to detect primary amino group and phenolic OH group. A cold diazotised solution of the sample is poured into a cold alkaline *β*-naphthol solution—formation of a red or scarlet dye indicates the presence of an aromatic primary amine (equation has already been given); on the other hand, a cold diazotised solution of aniline is poured into a cold alkaline solution of the sample—formation of a yellow or an orange or a red dye indicates the presence of phenolic OH in the sample. For further applications, see problems 17.22 and 17.24.

The diazonium coupling reactions involving tertiary amines and phenols are aromatic electrophilic substitution reactions bimolecular. The diazonium ion acts as an electrophile and the secondary component acts as a nucleophile: The mechanistic steps of the reaction are given below :

Step I. Electrophilic attack by the aromatic diazonium ion on the nuclear carbon of the secondary component. Here it is important to note that the aromatic diazonium ion is reasonably stable owing to conjugation over a wide range as shown

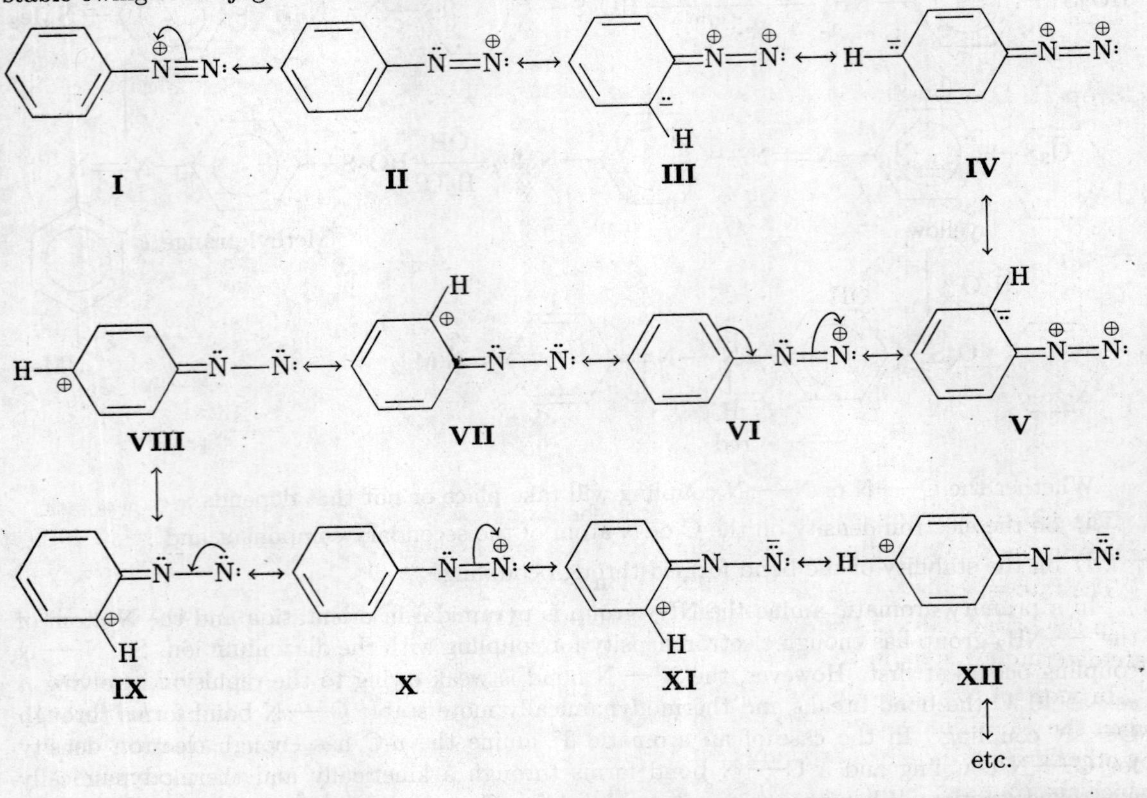

Amongst these resonating structures, structure II contributes maximum to the resonance hybrid of the aromatic diazonium ion and may be considered as taking part in the coupling reaction as an electrophile.

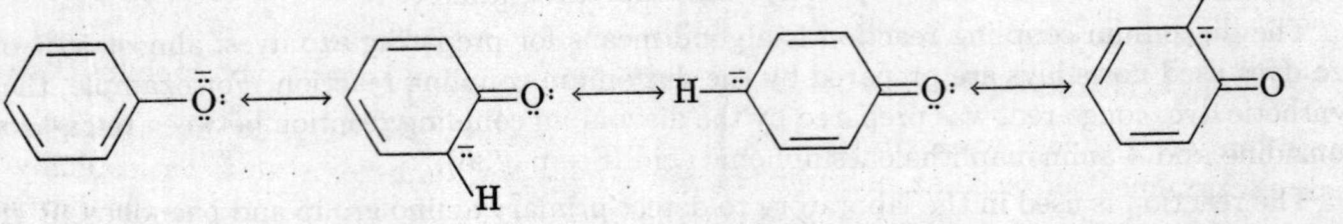

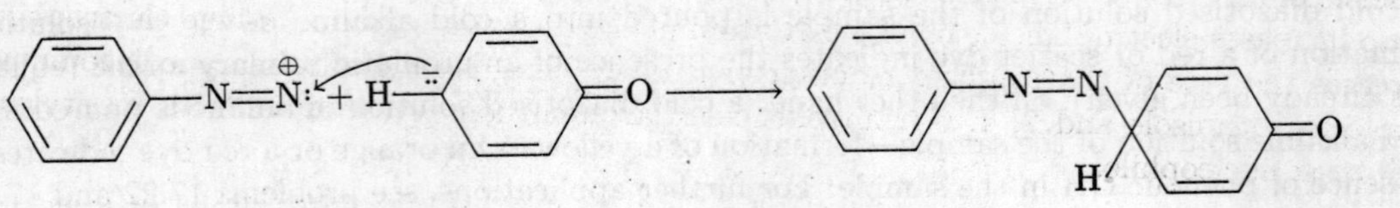

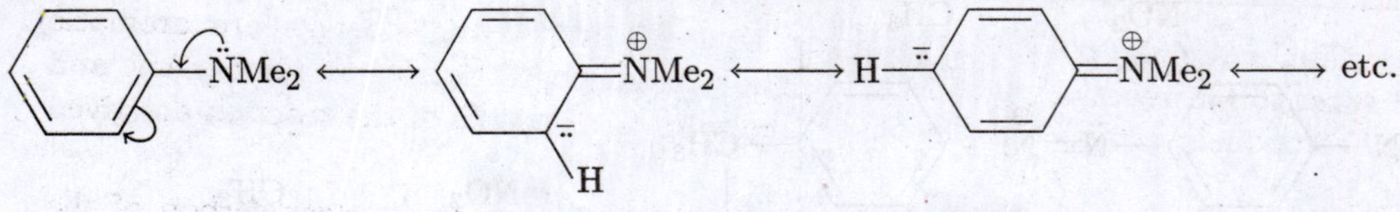

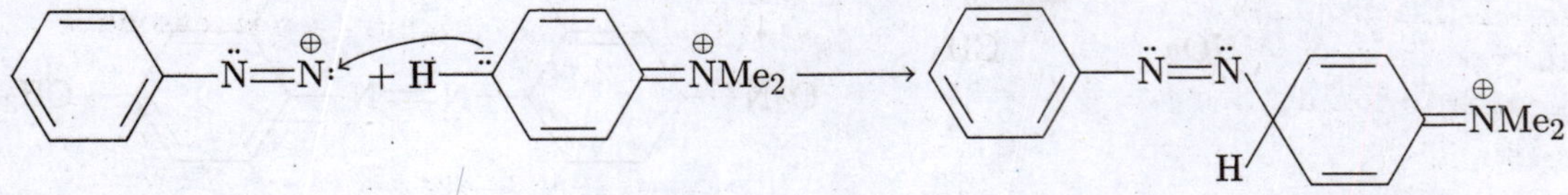

Step II. Loss of a proton form the secondary component moiety of the coupled species.

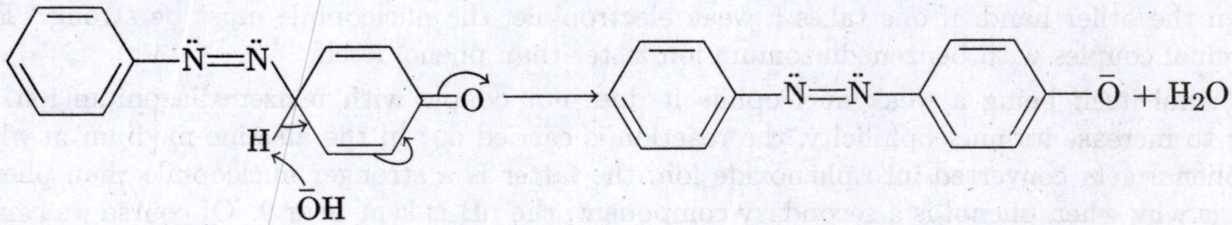

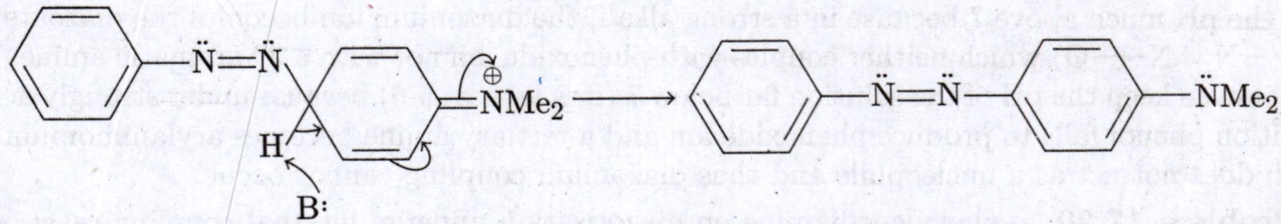

The first step is the slow step in the cases of phenoxide ion and also in the cases of 3° aromatic amines in the acidic medium; the reactions thus follow the second order kinetics.

The rate = k [diazonium salt] [phenoxide ion or 3° amine]

In the base-catalysed coupling reaction involving 3° amine the second step is found to be the rate-determining step; in this step C—H bond breaks.

In most of the cases the coupling occurs at the para-position of the secondary component, when the *p*-position is blocked, *o*-coupling occurs. When the *p*- and two *o*-positions are blocked by other groups; usually diazonium coupling does not take place. However, there are reactions in which the coupling occurs at the *p*-position and the group occupying the *p*-position gets ejected.

The diazonium coupling reaction can take place between :

(i) a strong electrophile and a strong nucleophile.

(ii) a strong electrophile and a weak nucleophile.

(iii) a weak electrophile and a strong nucleophile.

This reaction cannot take place between a weak electrophile and a weak nucleophile. Benzenediazonium ion is a very weak electrophile; it is much weaker than $\overset{\oplus}{N}O_2$, $\overset{\oplus}{Br}$, etc. Being a very weak electrophile, benzenediazonium ion cannot couple with anisole, a weak nucleophile. The presence of an electron-withdrawing group at the *o*- or *p*-position with respect to the diazonium group increases electrophilicity on the *β*-N atom of the diazonium group. As the electrophilicity increases the rate of the coupling reaction increases. Thus 2- or 4-nitrobenzenediazonium ion couples with anisole and 2, 4, 6-trinitrobenzenediazonium ion even couples with mesitylene, a very weak nucleophile.

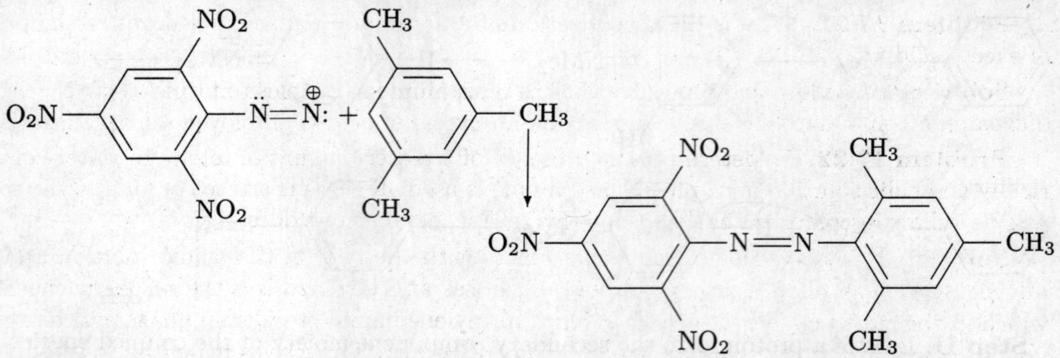

On the other hand, if one takes a weak electrophile, the nucleophile must be strong. Thus resorcinol couples with benzenediazonium ion faster than phenol itself.

Phenol itself being a weak nucleophile it does not couple with benzenediazonium ion. In order to increase its nucleophilicity, the reaction is carried out in the alkaline medium in which the phenol gets converted into phenoxide ion, the latter is a stronger nucleophile than phenol. That is why when phenol is a secondary component, the pH is kept near 9. Of course we cannot keep the pH much above 7 because in a strong alkali, the diazonium ion becomes the diazoate ion (Ar—N=N—$\bar{O}$) which neither couples with phenoxide ion nor with a 3° aromatic amine. We cannot even keep the pH of the solution far below 7 (it is kept at 5-6) because under strongly acidic condition phenol fails to produce phenoxide ion and a tertiary amine becomes arylammonium ion which does not act as a nucleophile and thus diazonium coupling cannot occur.

Problem 17.20. *o*-phenylenediamine on diazotisation undergo internal coupling reaction—give the mechanism.

Solution : $NaNO_2 + HCl \longrightarrow HNO_2 + NaCl;\ HNO_2 + HCl \longrightarrow H_2O + Cl^- + \overset{\oplus}{N}O$

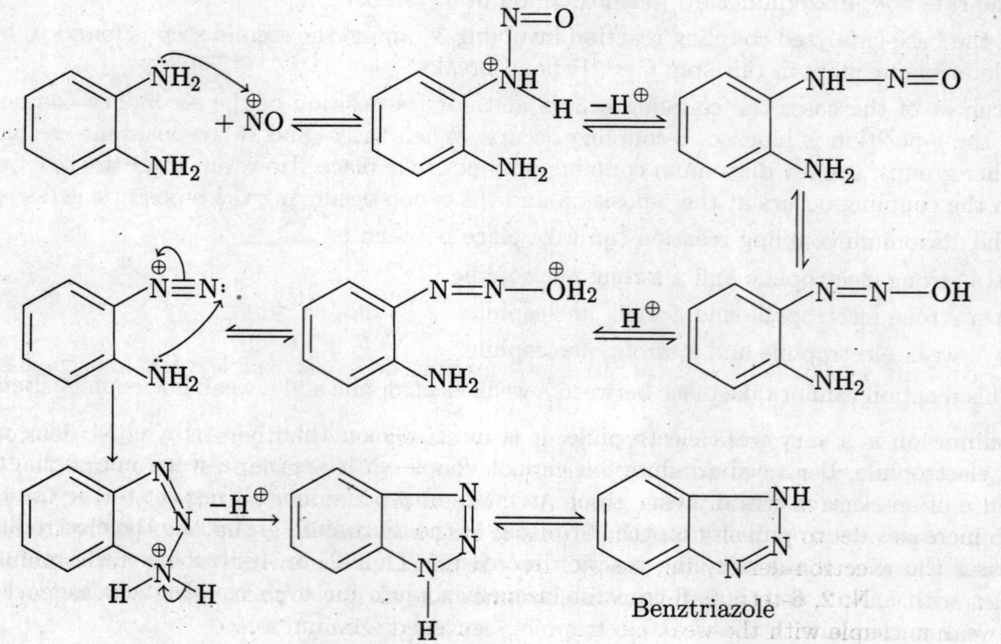

Benztriazole

Problem 17.21. Why is the diazotisation of an aromatic amine carried out in the presence of excess acid?

Solution : At a low acid concentration the diazonium ion couples to form ArN=N—NHAr, diazoamino compound. To stop this reaction an excess of acid is employed during diazotisation.

Problem 17.22. To determine the presence of a tracer amount of nitrite in water colorimetrically an acidic solution of *m*-phenylenediamine is used. If there is a trace of nitrite, the solution will be yellow in colour. What develops the yellow colour? How does it form?

Solution : Dissolved nitrite in water reacts with the acid of the acidic solution and forms nitrous acid. The nitrous acid so formed diazotises and tetraazotises the *m*-phenylenediamine; which in their turn couple with unreacted *m*-phenylenediamine to form monazo and bisazo compounds; these are brown dyes known as Bismarck Brown. These dyes make the solution yellow.

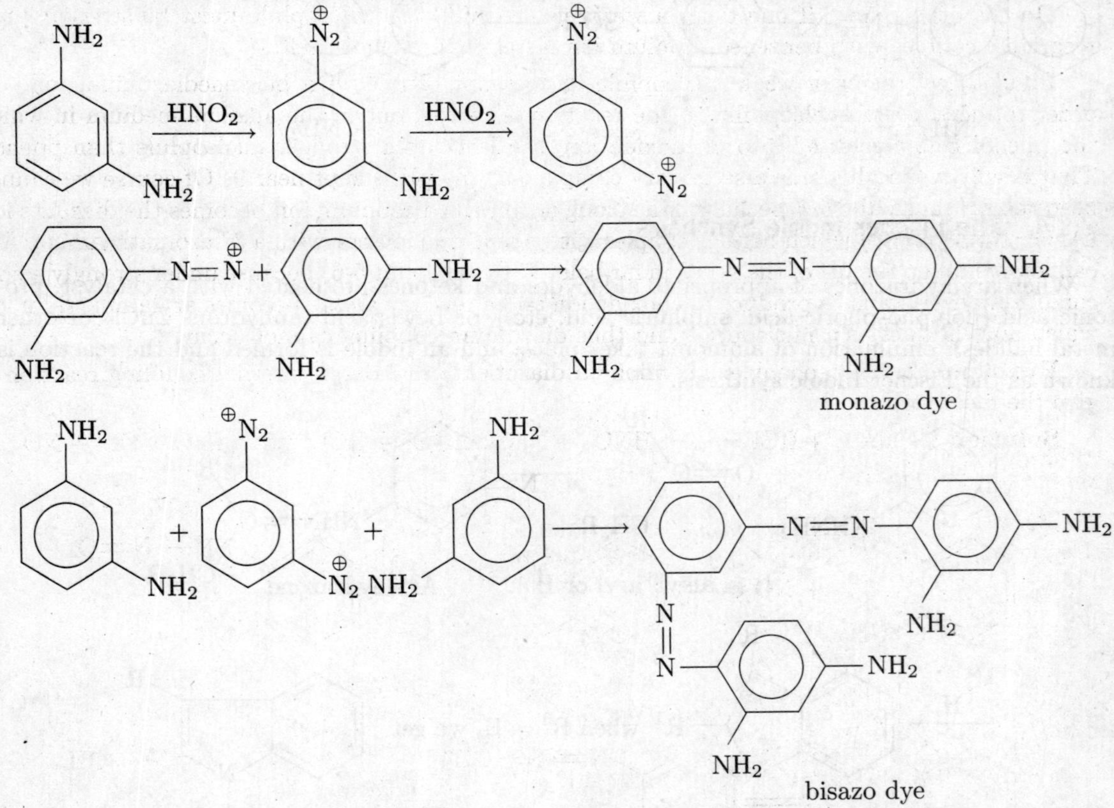

Problem 17.23. N, N, 2, 6-Tetramethylaniline does not couple with benzenediazonium ion—explain.

Solution : Four methyl groups being very closely placed, there is a good deal of steric congestion. In order to avoid steric congestion the —NMe_2 group rotates round the C—N bond axis; because of this rotation, the *p* A.O. containing the lone pair of *e*s on the N atom does no longer parallel to *p* orbitals of the aromatic ring and therefore, the *e*s of the N atom cannot increase the electron-density on the *p*-C by +R effect. Thus owing to the steric inhibition of resonance, N, N, 2, 6-tetramethylaniline becomes a weak nucleophile; being a weak nucleophile, it does not couple with the weak electrophile, benzenediazonium ion.

Problem 17.24. The presence of nitrite is detected in the laboratory by sulphanilic acid and α-naphthylamine as reagents in acetic acid. Write down the reaction.

Solution :

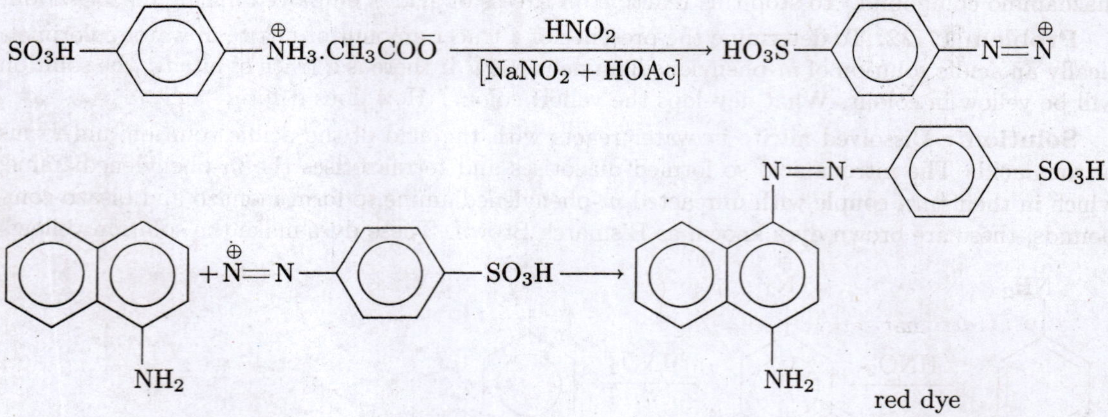

17.17 The Fischer Indole Synthesis

When arylhydrazones of appropriate aldehydes and ketones are heated with a catalyst, protonic acid (polyphosphoric acid, sulphuric acid, etc.) or Lewis acid (anhydrous $ZnCl_2$ or other metal halides), elimination of ammonia takes place, and an indole is formed and the reaction is known as the Fischer Indole synthesis.

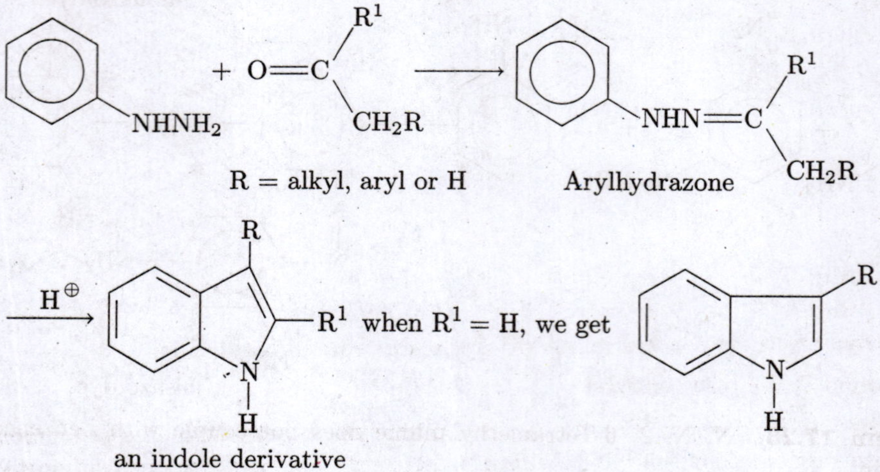

When R^1 is a methyl group, it invariably appears at the 2-position.

Arylhydrazones of acetaldehyde and CH_3COCHR_2 do not form indole derivatives.

The mechanism of the reaction is supposed to be as follows :

Step I. A Claisen type of rearrangement takes place in the tautomeric form of the arylhydrazone. This is the main step. When R and R^1 are H, compound A (see next page) becomes unstable and the reaction does not occur.

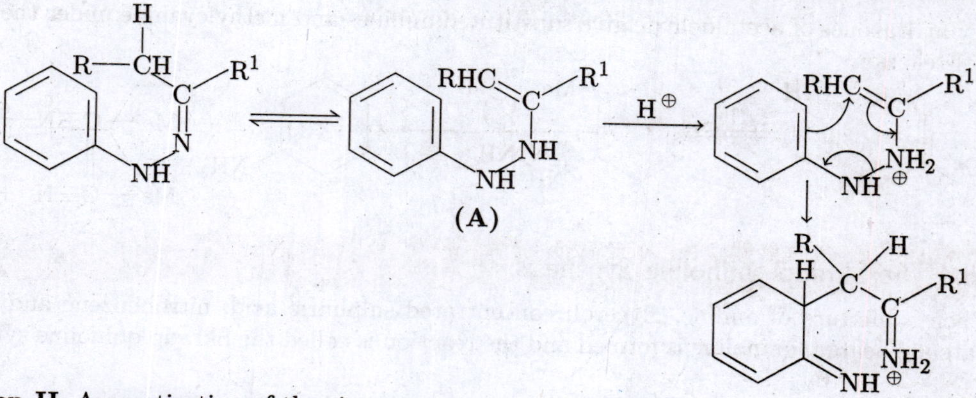

Step II. Aromatisation of the ring.

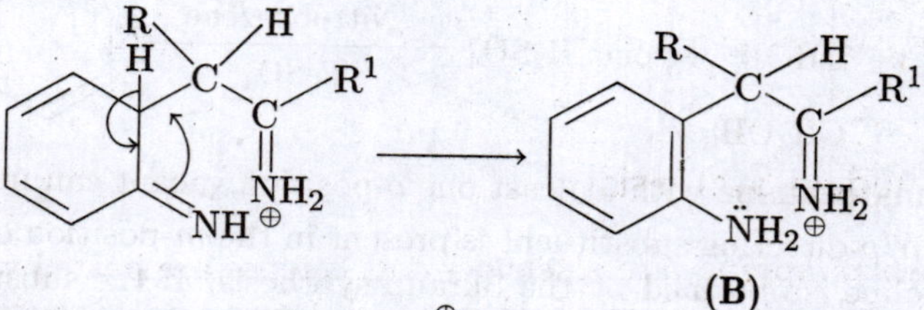

Step III. Ring closure and elimination of $\overset{\oplus}{N}H_4$ ion.

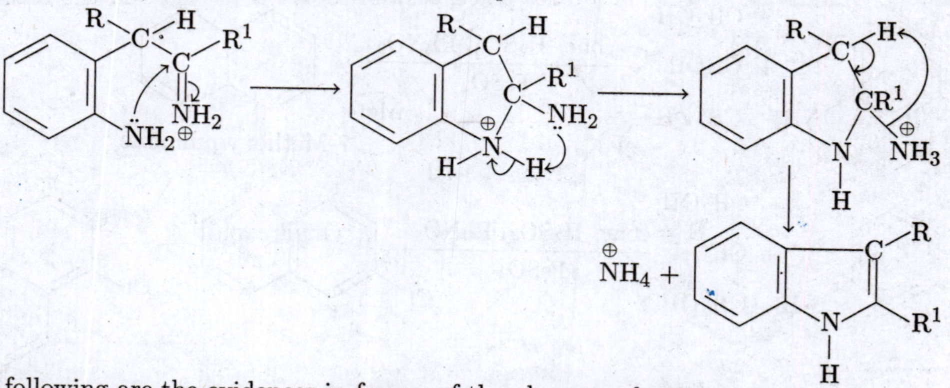

The following are the evidences in favour of the above mechanism.

(i) Compound B has been isolated.

(ii) That the $=\overset{\oplus}{N}H_2$ group on the side chain of compound B gets eliminated as ammonium ion has been supported by the ^{15}N labelling experiment. This has also been supported by the fact that the product of hydrolysis of compound B can be cyclised to the indole derivative.

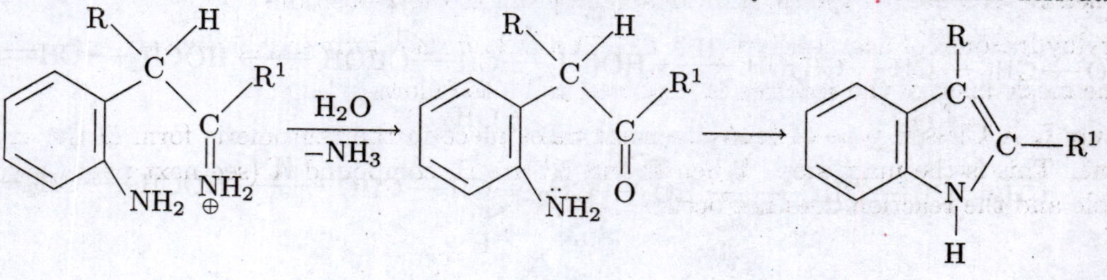

Arylhydrazones of acetaldehyde give substituted anilines and methylcyanide under the conditions of reaction

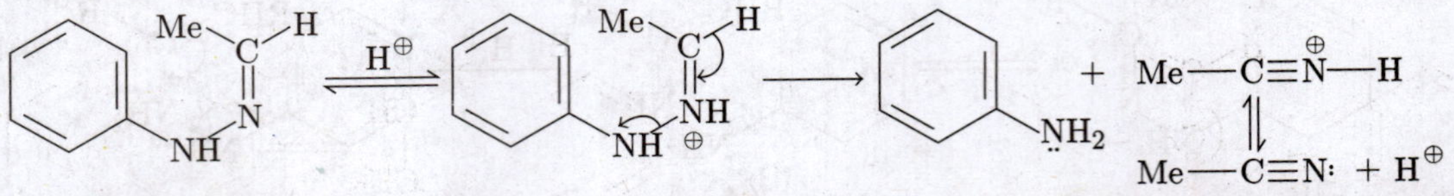

17.18 The Skraup Quinoline Synthesis

When a mixture of aniline, glycerol, concentrated sulphuric acid, nitrobenzene and ferrous sulphate are heated, quinoline is formed and the reaction is called the Skraup quinoline synthesis.

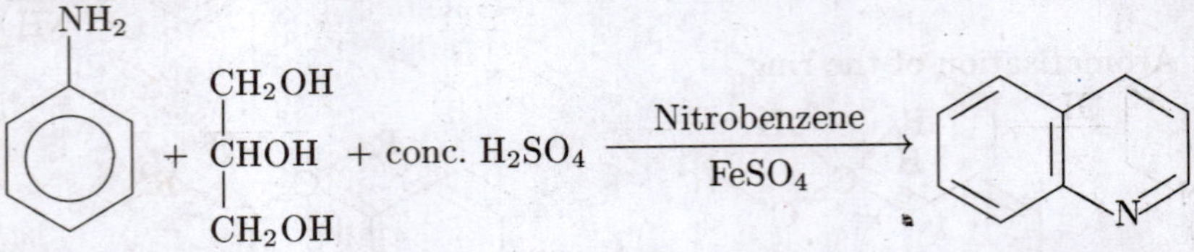

Any primary aromatic amine with at least one *o*-position vacant can undergo the reaction.

When a strongly *o*/*p*-directing substituent is present in the *m*-position of the aromatic amine, a 7-substituted quinoline is obtained by the Skraup synthesis. If the substituent is a weak *o*/*p*-director, both 5- and 7-substituted quinolines are obtained by the reaction.

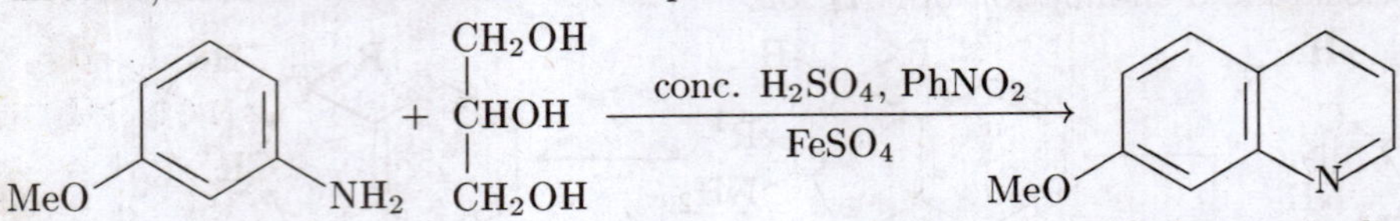

7-Methoxyquinoline

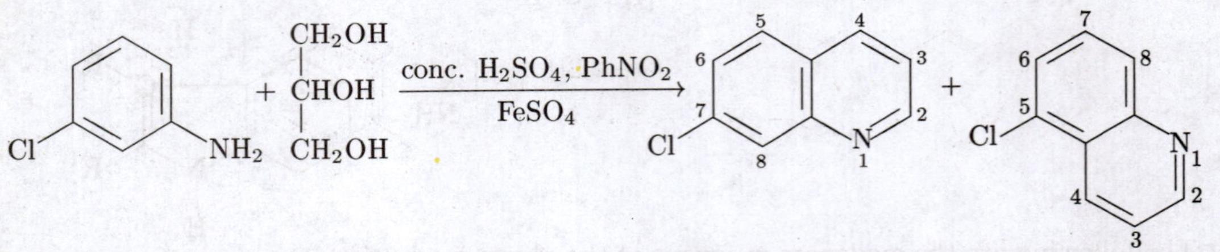

$FeSO_4$ makes the reaction less vigorous. Nitrobenzene acts as an oxidising agent in the reaction. It removes the extra hydrogen atoms.

The mechanism of the reaction consists of four steps :

Step I. Dehydration of glycerol and formation of acraldehyde.

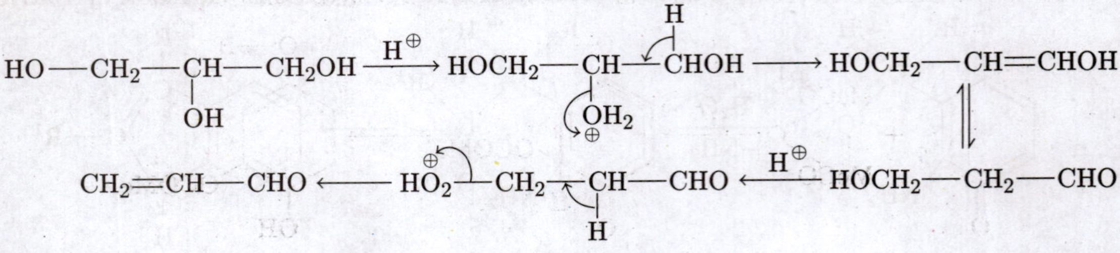

Step II. This is a Michael type addition of the primary aromatic amine to the α, β-unsaturated aldehyde, acraldehyde.

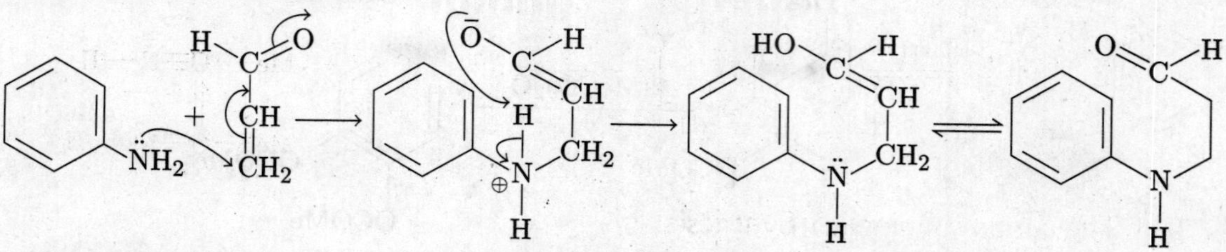

Step III. This step involves the Friedel-Crafts ring-closure reaction.

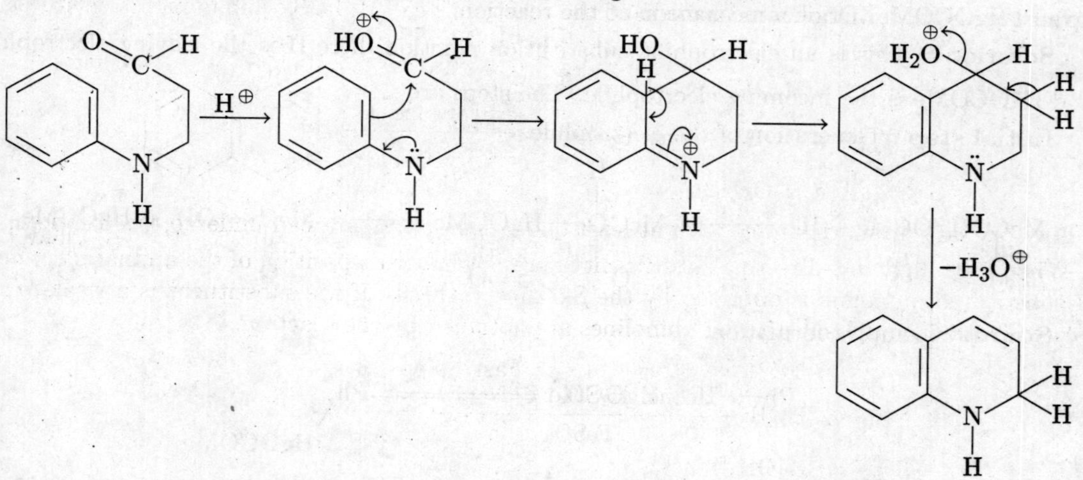

Step IV. Oxidative aromatisation; $PhNO_2$ is the oxidising agent.

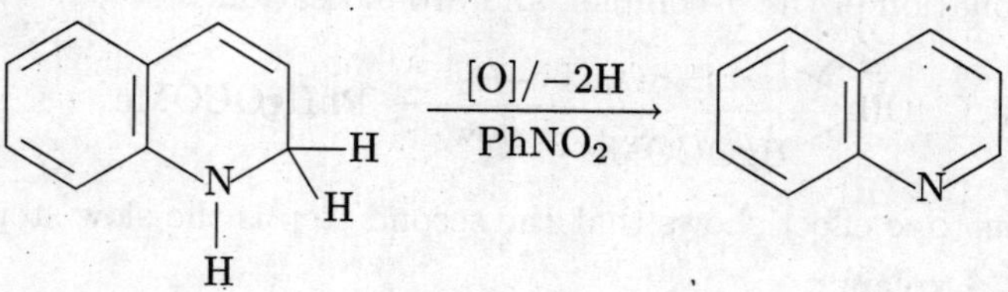

Problem 17.25. When *p*-benzoquinone is treated with acetic anhydride in the presence of sulphuric acid, hydroxyquinol triacetate is formed and the reaction is known as Thiele acetylation. Give the mechanism of the reaction.

Solution :

Step I. 1, 4-addition to the α, β-unsaturated carbonyl compound, *p*-benzoquinone.

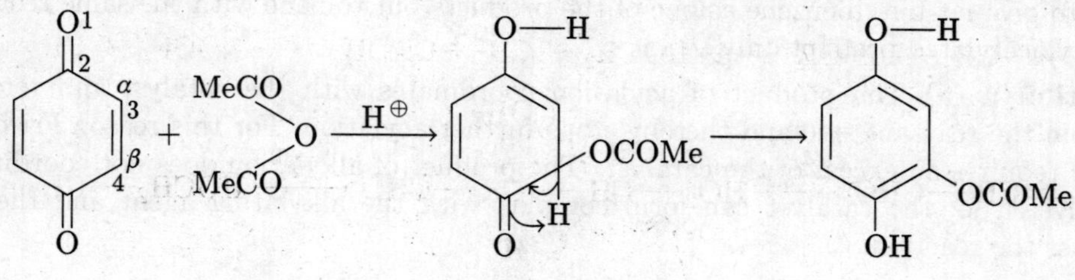

Step II. Acetylation of phenolic hydroxyl groups.

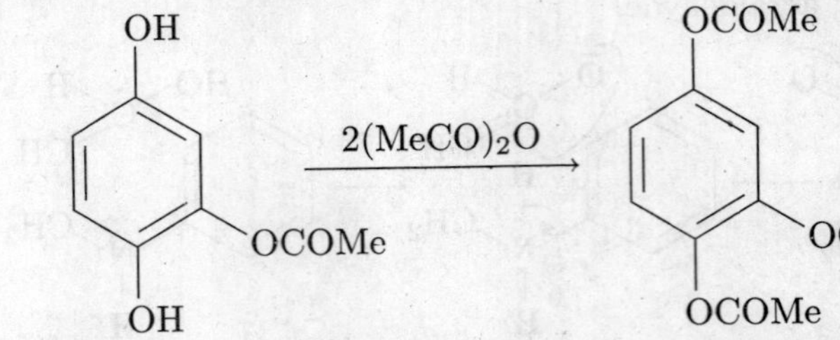

Problem 17.26. Mercuric acetate reacts with benzene in the presense of perchloric acid to give PhHgOCOMe. Draw a mechanism of the reaction.

Solution : This is an electrophilic substitution reaction; here $\overset{\oplus}{H}$ is the leaving electrophile and $\overset{\oplus}{H}$gOCOMe is the incoming electrophile. The steps are :

Initial step : Generation of the electrophile.

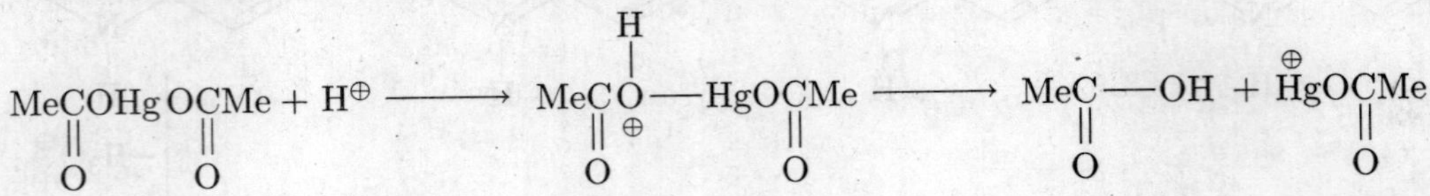

Step I. σ-complex formation.

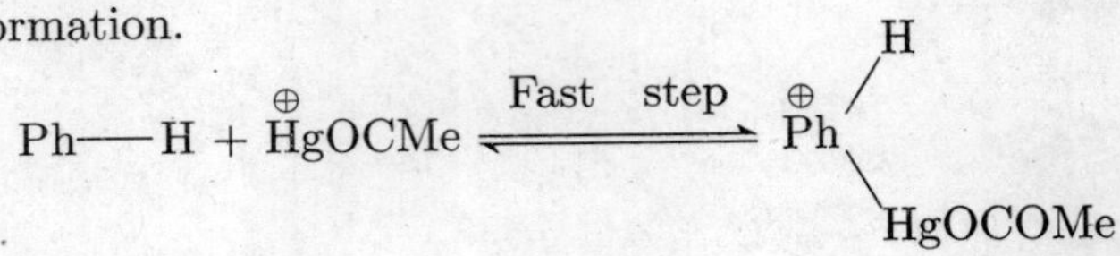

σ-complex

Step II. Deprotonation of the σ-complex and aromatisation.

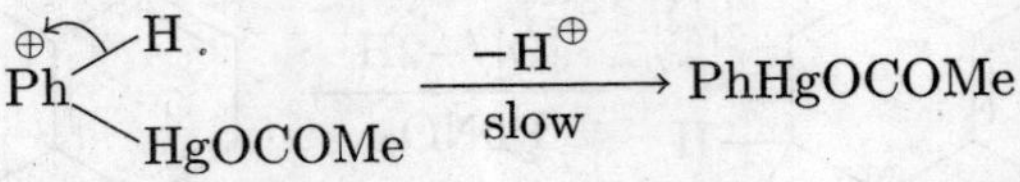

Primary kinetic isotope effect shows that the second step is the slow step.

Problem 17.27. Explain :

(a) Friedel-Crafts acylation requires an excess of the catalyst but Friedel-Crafts alkylation requires only a catalytic amount.

(b) Friedel-Crafts acylation does not give any polyacylated product.

(c) When *o*-xylene is heated at 80 °C with anhydrous $AlCl_3$ and HCl, *m*-xylene is obtained.

(d) Benzene reacts with trimethylacetyl chloride in the presence of anhydrous aluminium chloride to give *p*-t-butylbenzene as one of the products but toluene with the same reagent gives *p*-tertbutylacetylated product only.

Solution : (a) The product of acylation coordinates with the catalyst and removes the latter from the reactant side and thereby stops further acylation. For this reason Friedel-Crafts acylation requires an excess of the catalyst. The product of alkylation does not coordinate with the catalyst. So, the catalyst can form complex with the alkylating agent and the catalyst propagates the reaction.

(b) An acyl group on an aromatic ring deactivates the ring towards Ar.S$_E$2 reaction. So, the monoacylated product is less reactive than the substrate and further acylation on the monoacyl derivative does not take place. Furthermore, the monoacyl derivative coordinates with the catalyst and further deactivation of the ring occurs.

(c) Friedel-Crafts reaction is a thermodynamically controlled reaction. At 80 °C in the presence of HCl—$AlCl_3$, a methyl group dissociates and reunites at the *m*-position where steric congestion is absent.

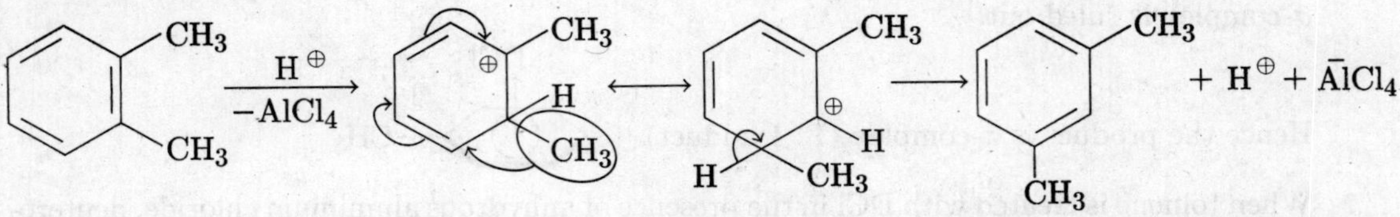

(d) Owing to the +I and hyperconjugative effects of CH_3 group, toluene is more reactive than benzene towards Ar.S$_E$2 reaction. The Friedel-Crafts acylation involving trimethylacetyl chloride and toluene takes place at a lower temperature than that involving benzene and trimethylacetyl chloride. At an elevated temperature, decarbonylation of the acyl cation occurs and t-butyl cation forms. The t-Bu cation then takes part in the reaction to give t-butylbenzene.

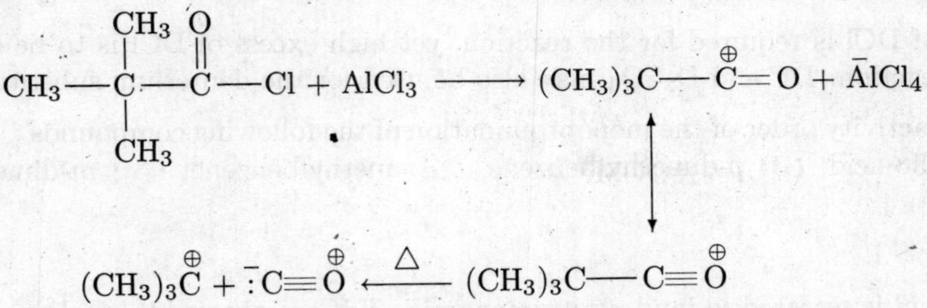

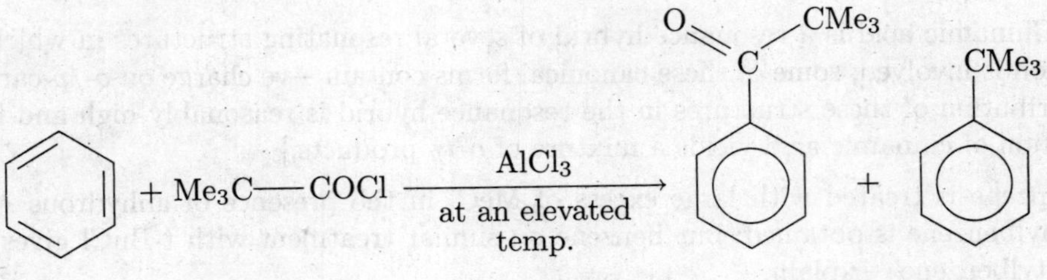

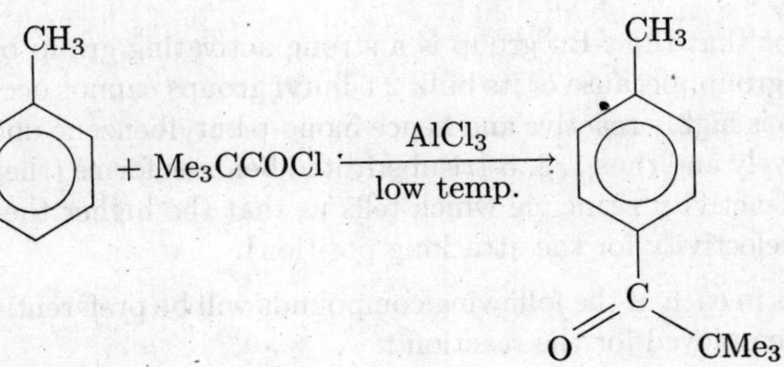

Exercise

1. When toluene is treated with DCl at $-78\,^\circ C$, no deuterium is exchanged and the solution does not conduct electricity. What is the structure of the complex?

Ans. Under the given condition either a π-complex or a σ-complex usually forms. Since no D is exchanged and since the solution does not conduct electricity, the formation of the σ-complex is ruled out.

Hence the product is π-complex (1 : 1 adduct).

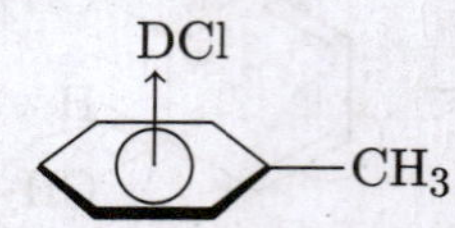

2. When toluene is treated with DCl in the presence of anhydrous aluminium chloride, deutero-toluene is formed; suggest a mechanism for the reaction.

[**Hint :** Ar.S_E2 reaction; σ-complex is an intermediate.]

3. Convert benzene to hexadeuterobenzene.

Ans. $C_6H_6 + 6DCl \xrightarrow{AlCl_3} C_6D_6 + 6HCl$

Though 6 moles of DCl is required for the reaction, yet high excess of DCl is to be taken to complete the reaction. DF and D_2SO_4 may also be used for the deuterium substitution.

4. Write down the reactivity order of the monobromination of the following compounds : (i) *p*-benzenedicarboxylic acid, (ii) *p*-dimethylbenzene, (iii) methylbenzene, (iv) *m*-dimethylbenzene.

Ans. iv > ii > iii > i.

5. When cinnamic acid is nitrated; *o*- and *p*-nitrocinnamic acids are obtained—explain.

[**Hint :** Cinnamic acid is a resonance hybrid of several resonating structures in which π-π conjugation is involved; some of these canonical forms contain $-$ve charge on *o*-/*p*-carbons. The contribution of these structures in the resonance hybrid is reasonably high and hence the nitration of cinnamic acid yields a mixture of *o*-/*p* products.]

6. When benzene is treated with large excess of MeCl in the presence of anhydrous $AlCl_3$, hexamethylbenzene is obtained; but benzene on similar treatment with t-BuCl gives 1, 3, 5-tri-t-butylbenzene—explain.

Ans. There is no doubt that the t-Bu group is a strong activating group but at the same time it is a very bullky group; because of its bulk 2 t-butyl groups cannot occupy *o*-position. Furthermore, the group is highly reactive and hence mono-t-butylbenzene undergoes further substitution less selectively and thus 1, 3, 5-trisubstituted benzene forms (there is a principle called the Selectivity Reactivity Principle which tells us that the higher the reactivity of a reagent, the less is its selectivity for the attacking position).

7. Indicate which positions in each of the following compounds will be preferentially deuterated when 1 mole of DCl is employed for the reaction :

(i) anisole, (ii) nitrobenzene, (iii) benzotrifluoride.

Ans. (i) *o*/*p*; (ii) *m* and (iii) *m*.

8. OMe directs an attacking electrophile to *o*- and *p*-positions but it deactivates the *o*-position more than the *p*-position—explain.

9. Give an explanation for the yield of the *m*-product in the following compounds when they are subjected to the Friedel-Crafts alkylation with methyl chloride and anhydrous aluminium chloride.

$$PhCH_3 : 4.4\%,\ PhCH_2Cl : 15\%,\ PhCHCl_2 : 33.8\%,\ PhCCl_3 : 64.6\%,\ Ph\overset{\oplus}{N}Me_3 : 100\%.$$

10. Account for the following yields of the *m*-product when toluene is separately treated with **(i)** $Br_2/AcOH$: 0.5%, **(ii)** $HNO_3/AcOH$: 3.5%, **(iii)** $CH_3CH_2Br/GaBr_3$: 21%.

Ans. Toluene contains CH_3 group which is *o*-/*p*-directing. Therefore, in all the cases the predominant products will be a mixture of *o*-/*p*-products and a small amount of the *m*-product is expected in each case. Now to account for the increasing yield of the *m*-product we shall have to consider the Selectivity Reactivity Principle. This states that the higher the reactivity of a reagent, the less selectively it attacks a position in the aromatic ring. Since the reactivity order of the given reagents is :

$$CH_3\overset{\oplus}{C}H_2(CH_3CH_2Br/GaBr_3) > \overset{\oplus}{N}O_2(HNO_3/AcOH) > Br_2(Br_2/AcOH),$$

the selectivity order is :

$$Br_2 > \overset{\oplus}{N}O_2 > CH_3\overset{\oplus}{C}H_2.$$

So, the yield of the *m*-product in the case of CH_3CH_2Br is the highest and that of the *m*-product in the case of Br_2 is the lowest.

11. Calculate the dipole moment of the three isomeric chlorotoluenes.

12. Explain why *p*-chloronitrobenzene has a lower dipole moment than *p*-nitrotoluene.

13. Explain why nitration of naphthalene occurs at α-position.

14. Give an evidence for the formation of nitronium ion in the nitration of aromatic compounds by mixed acids.

15. Explain why the *o*-product predominates when methyl phenylethyl ether is nitrated with N_2O_5.

16. The rate of nitration of an aromatic compound falls sharply on addition of sodium nitrate but that increases on addition of conc. H_2SO_4—explain.

17. In the nitration of halobenzenes the *o*/*p* ratio increases on going from fluorobenzene to iodobenzene—explain.

18. Write down the products of the sulphonation of *p*-bromoanisole, *p*-tert-butyltoluene, *m*-bromoaniline and *p*-nitrobenzoic acid.

19. In the aromatic sulphonation reaction primary kinetic isotope effect is observed—draw a conclusion about the mechanism of the reaction from the fact.

20. Benzene is not used as a solvent for the Friedel-Crafts alkylation of bromobenzene.

21. Explain why in the Howarth synthesis of naphthalene the C=O group is reduced to —CH_2— group before the Friedel-Crafts ring-closure reaction.

22. Apply Friedel-Crafts reaction in the synthesis of anthracene.

23. Explain why halogenation of anthracene occurs at the 9-position.

24. Nitration of chlorobenzene gives the *p*-product in excess over the *o*-product—explain.

25. What happens when :

(a) The Vilsmeier-Haak formylation is carried out on anthracene.
(b) 1-Nitronaphthalene is further subjected to nitration.
(c) Potassium phenoxide is subjected to the Kolbe-Schmidt synthesis.
(d) *p*-Methylphenol is subjected to the Reimer-Tiemann reaction.
(e) Nitrobenzene is treated with mercuric acetate and perchloric acid.

26. Explain why naphthalene forms α-naphthalenesulphonic acid at 80 °C but β-naphthalenesulphonic acid at 160 °C when naphthalene is sulphonated.

27. The chlorination of aromatic compounds with Cl_2 may be catalysed by a small amount of iodine. Write down the function of the catalyst.

28. The acylation of toluene with Me_3CCOCl in CS_2 in the presence of anhydrous $AlCl_3$ gives t-butyltoluene as a side reaction. Give a mechanism for the side reaction. The yield of the side product increases with increasing dilution—explain.

29. Synthesise salicylaldehyde from acetone, benzene and other inorganic reagents.

30. Give an evidence in favour of the fact that the diazonium coupling reaction is an electrophilic substitution reaction.

31. The diazonium coupling reaction with phenol should be carried out at pH = 9 and that with a 3° amine at pH = 5.6—explain.

32. Write short notes on :

(a) Chloromethylation,
(b) The Gattermann aldehyde synthesis,
(c) The Houben-Hoesch synthesis,
(d) The Vilsmeier-Haak formylation,
(e) The Körner's method of determining orientation,
(f) The Gattermann-Koch aldehyde synthesis,
(g) The Skraup synthesis of quinoline.